Diagnostic Ultrasound

Musculoskeletal

SECOND EDITION

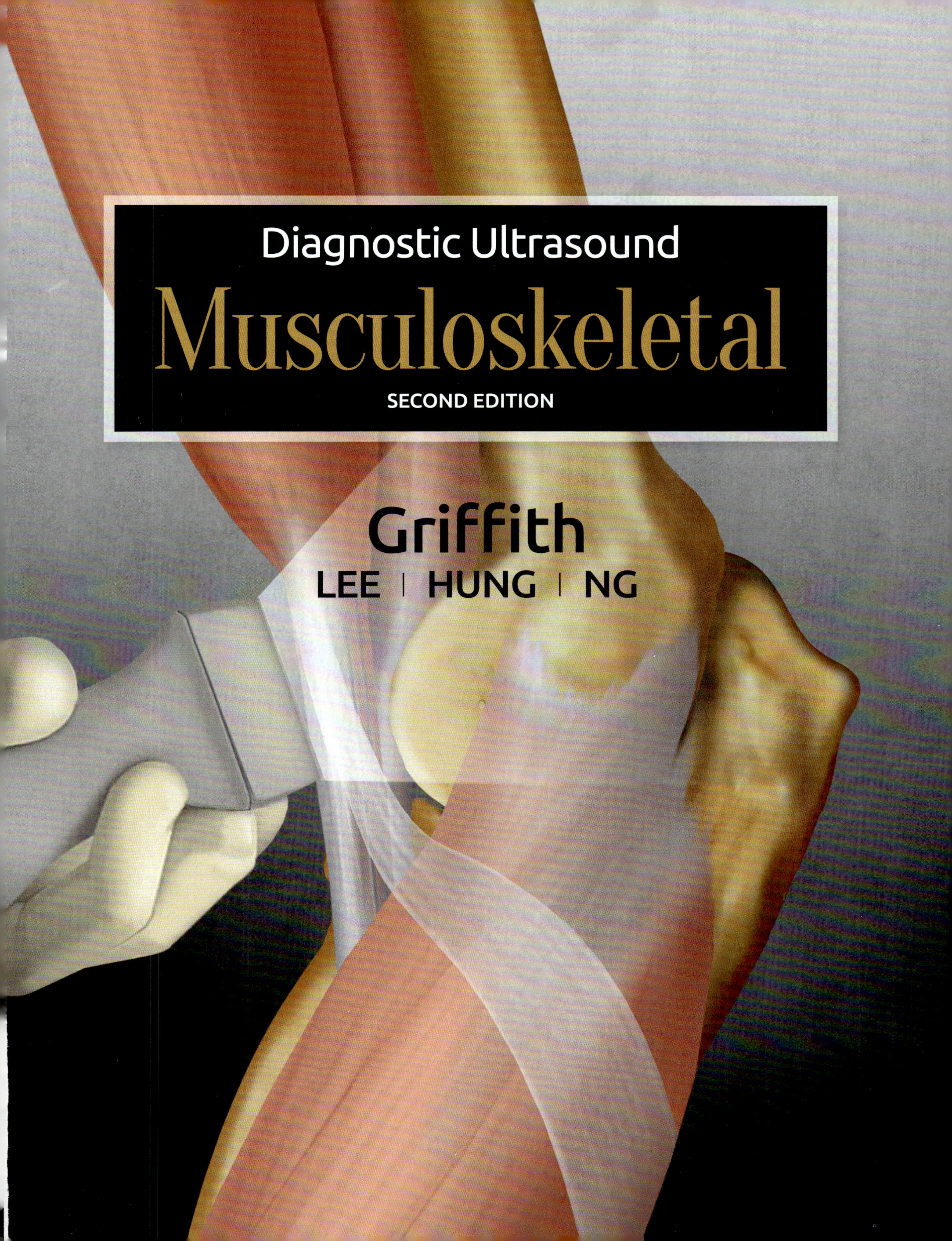

Griffith

LEE | HUNG | NG

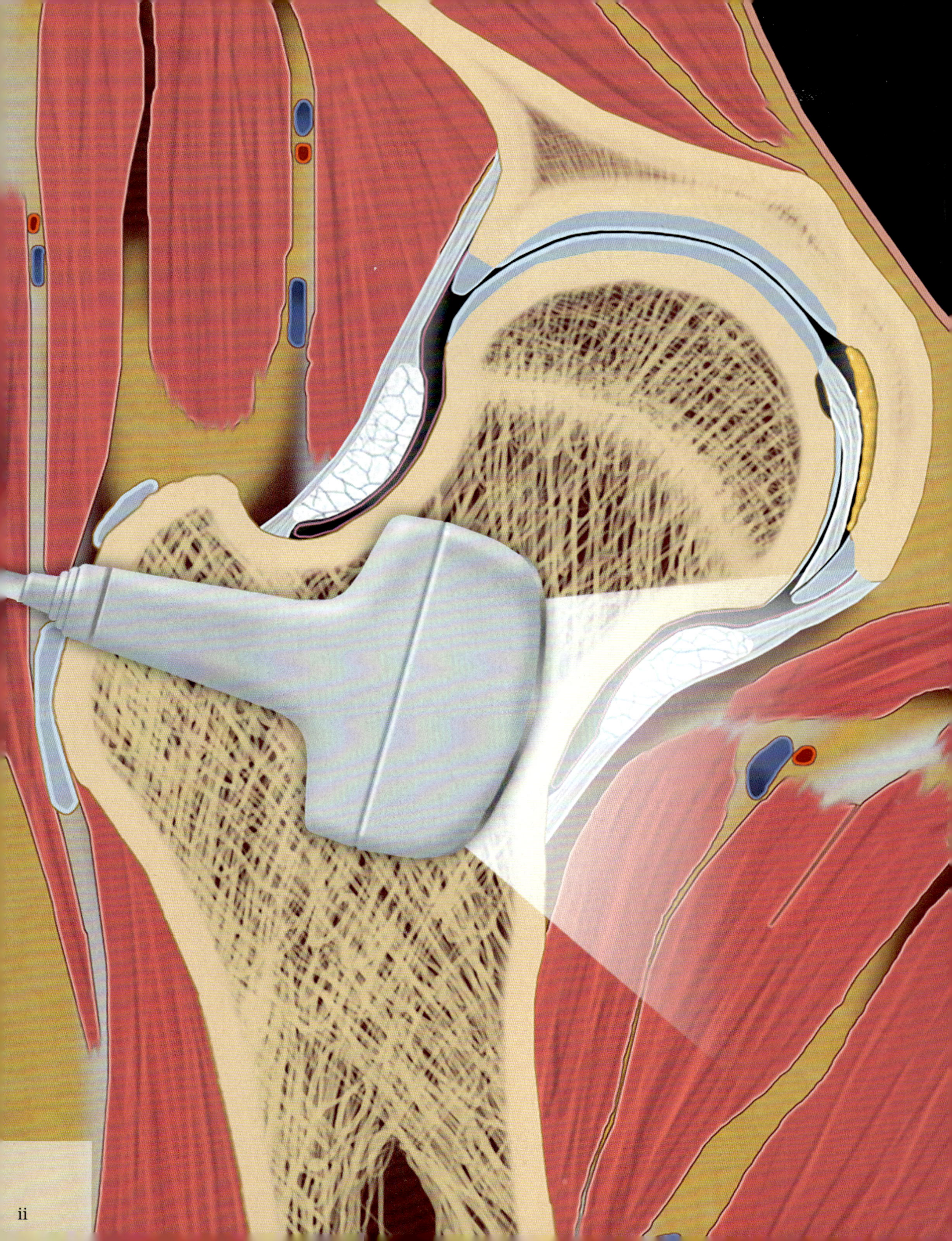

Diagnostic Ultrasound

Musculoskeletal

SECOND EDITION

James F. Griffith, MD, MRCP, FRCR
Professor
Department of Imaging and Interventional Radiology
The Chinese University of Hong Kong
Hong Kong (SAR), China

Ryan K. L. Lee, MBChB, FRCR, FHKCR, FHKAM (Radiology)
Clinical Assistant Professor (Honorary)
Department of Imaging and Interventional Radiology
The Chinese University of Hong Kong
Hong Kong (SAR), China

Esther H. Y. Hung, MBChB, FRCR, FHKCR, FHKAM (Radiology)
Clinical Associate Professor (Honorary)
Department of Imaging and Interventional Radiology
The Chinese University of Hong Kong
Hong Kong (SAR), China

Alex W. H. Ng, MBChB, FRCR, FHKCR, FHKAM (Radiology)
Clinical Associate Professor (Honorary)
Department of Imaging and Interventional Radiology
The Chinese University of Hong Kong
Hong Kong (SAR), China

ELSEVIER

1600 John F. Kennedy Blvd.
Ste 1800
Philadelphia, PA 19103-2899

DIAGNOSTIC ULTRASOUND: MUSCULOSKELETAL, SECOND EDITION ISBN: 978-0-323-57013-8

Notices

Practitioners and researchers must always rely on their own experience and knowledge in evaluating and using any information, methods, compounds or experiments described herein. Because of rapid advances in the medical sciences, in particular, independent verification of diagnoses and drug dosages should be made. To the fullest extent of the law, no responsibility is assumed by Elsevier, authors, editors or contributors for any injury and/or damage to persons or property as a matter of products liability, negligence or otherwise, or from any use or operation of any methods, products, instructions, or ideas contained in the material herein.

Library of Congress Control Number: 2018953139

Cover Designer: Tom M. Olson, BA
Printed in Canada by Friesens, Altona, Manitoba, Canada

Last digit is the print number: 9 8 7 6 5 4 3 2 1

Dedication

To Clara, Isobel and Olivia, Mum, Dad, and all my siblings.
For the very best of times, thank you.

JFG

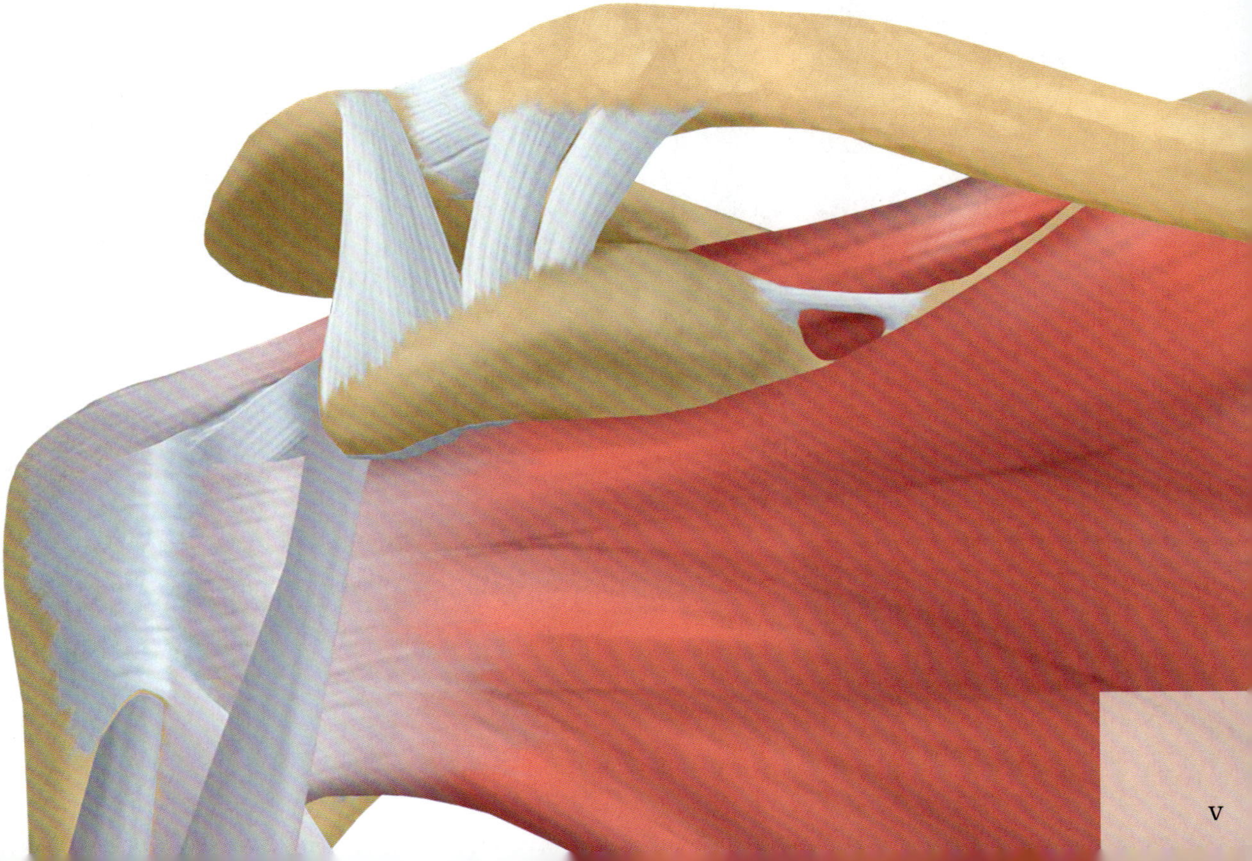

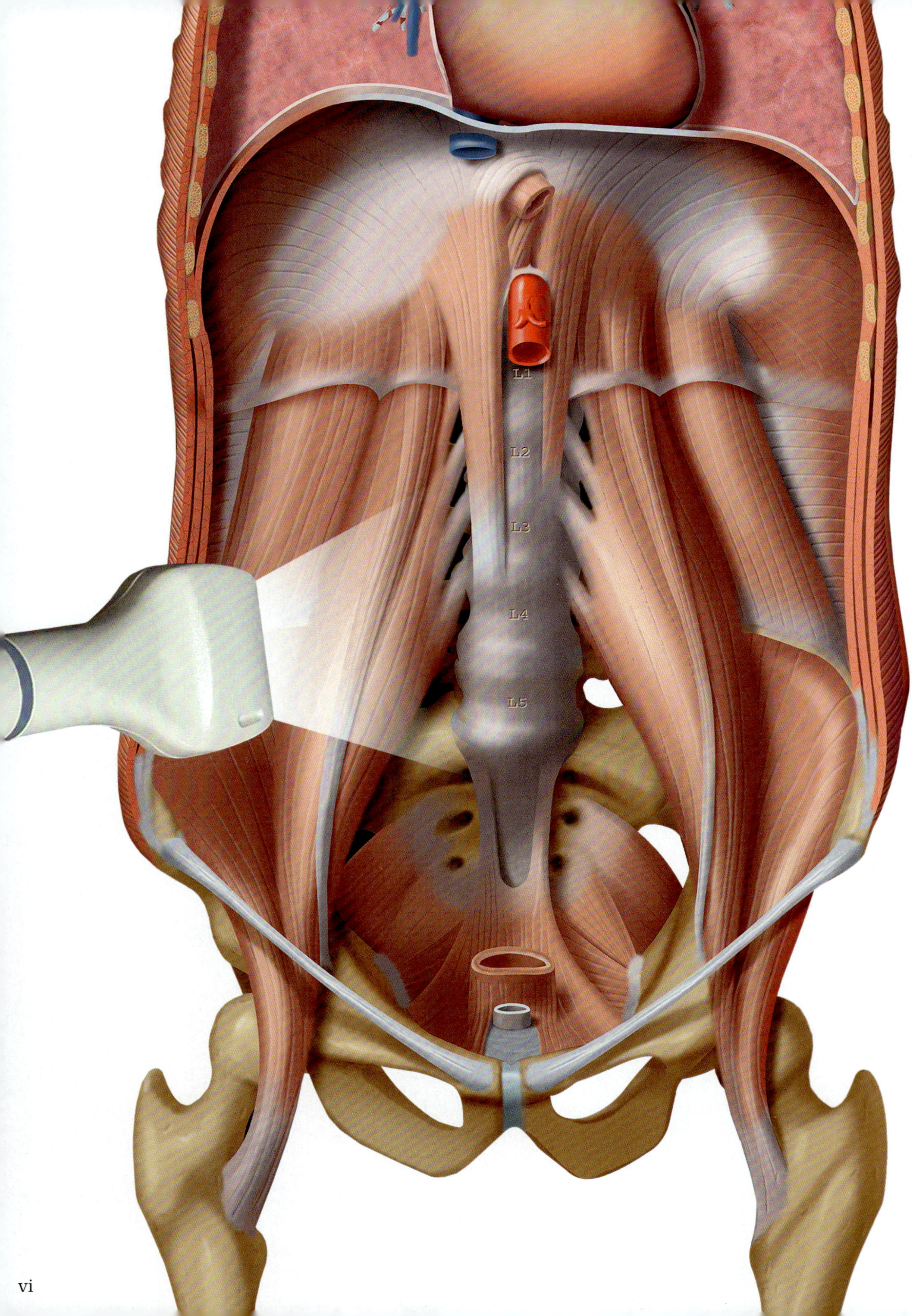
L1
L2
L3
L4
L5

Additional Contributing Authors

Jill M. Abrigo, MD

Gregory E. Antonio, MD, DRANZCR, FHKCR

Stella Sin Yee Ho, RDMS, RVT, PhD

Eric K. H. Liu, PhD, RDMS

Eugene McNally, FRCR, FRCPI

Karen Partington, MRCS, FRCR

Bhawan K. Paunipagar, MBBS, MD, DNB

K. T. Wong, MBChB, FRCR

Jade Wong-You-Cheong, MBChB, MRCP, FRCR

Paula J. Woodward, MD

Philip Yoong, FRCR

Preface

I have no doubt whatsoever that musculoskeletal ultrasound will become the most influential imaging modality in the world of musculoskeletal disease diagnosis, treatment, and monitoring. It is already well on its way to becoming that megastar. Those musculoskeletal conditions best imaged by musculoskeletal ultrasound and those best examined by other modalities, such as MR, CT, or radiography, are now quite clearly understood. The vast majority of musculoskeletal conditions outside of deeper joint structures are accessible to adequate evaluation by musculoskeletal ultrasound. Going forward, this range of clinical application is not likely to significantly expand, but the clarity with which these conditions will be seen will no doubt improve even further. Since the emergence of high-resolution transducers in the early 90s, ultrasound image resolution has improved beyond recognition. High musculoskeletal image quality is now achievable on all modern mid- to high-end ultrasound machines.

Ultrasound is different from other modalities in that you alone are the master of your destiny in realizing a high-quality ultrasound examination, acquiring readily understandable images, and arriving at the correct diagnosis. Similar to other imaging modalities, any diagnosis is formulated, not just on a single imaging characteristic, but also on myriad imaging signs evaluated within a particular clinical context. All ultrasound experts will tell you that ultrasound imaging is a lifelong learning skill. Each time you examine a patient with ultrasound, you should finish that examination a teeny weeny bit more skilled than you were before you began it. No other imaging modality seems to afford this capacity for continual improvement quite as much as ultrasound. This requires a pedantic approach armed with a thorough knowledge of ultrasound anatomy, a meticulous ultrasound technique, a clear understanding of the pertinent ultrasound findings, awareness of the most likely diagnosis, the mitigating features, and the potential differential diagnoses. Each of these aspects has been specifically addressed in this book, the 2nd edition of *Diagnostic Ultrasound: Musculoskeletal*.

Anatomy, with a particular emphasis on anatomy relevant to musculoskeletal ultrasound, is comprehensively covered in the 1st section. The next section, Technique, discusses an overall approach to musculoskeletal ultrasound and artifacts, followed by a detailed account of how to undertake a comprehensive examination of the main joints, emphasizing those areas most frequently examined by musculoskeletal ultrasound. Key elements to writing an ultrasound report are also addressed. The 3rd section, Diagnoses, looks at specific musculoskeletal conditions, the range of potential ultrasound appearances encountered, and the key facts required to make a correct diagnosis. The 4th section, Differential Diagnoses, looks at clinical ultrasound from a different perspective, outlining those diagnoses to be considered when faced with a particular ultrasound scenario, such as a hypoechoic muscle mass or a hypervascular subcutaneous mass. The 5th section addresses the fast-growing influence and range of ultrasound-guided interventional techniques in treatment and diagnosis. The overall content of the book, particularly the latter 4 sections, has been updated by at least 20% from the 1st edition, which was published in 2013.

A huge thank you goes out to Rebecca Bluth, Karen E. Concannon, and Rich Coombs at Elsevier for your help in putting this book together. The email equivalent of neural strain was understandably evident on many occasions over the past 18 months. Despite this, and having never met in person, we have nevertheless, I feel, become good friends and I remain in awe of your professional attitude, good nature, and dedication. A big, big thank you for your efforts.

A huge thank you also goes out to my close colleagues, Alex Ng, Ryan Lee, Esther Hung, and Cina Tong, as well as all the staff in the Department of Imaging and Interventional Radiology, Prince of Wales Hospital, The Chinese University of Hong Kong. Your dedication to maintaining a high standard of musculoskeletal ultrasound is admirable.

And to you, the readers, I do hope you enjoy this book and find it helpful in your daily practice. Please do persevere with musculoskeletal ultrasound. 100% guaranteed, you will not be disappointed.

James F. Griffith, MD, MRCP, FRCR
Professor
Department of Imaging and Interventional Radiology
The Chinese University of Hong Kong
Hong Kong (SAR), China

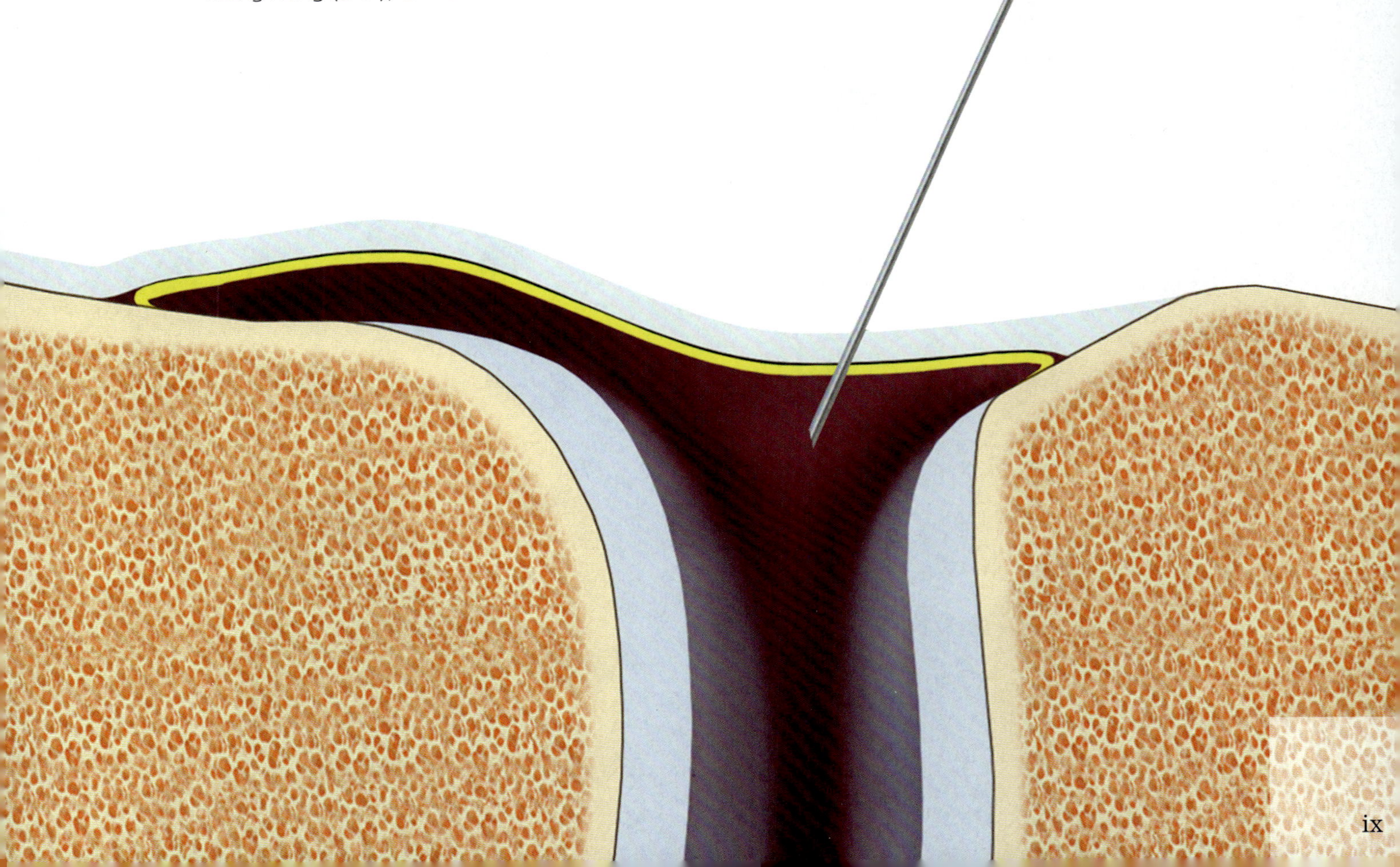

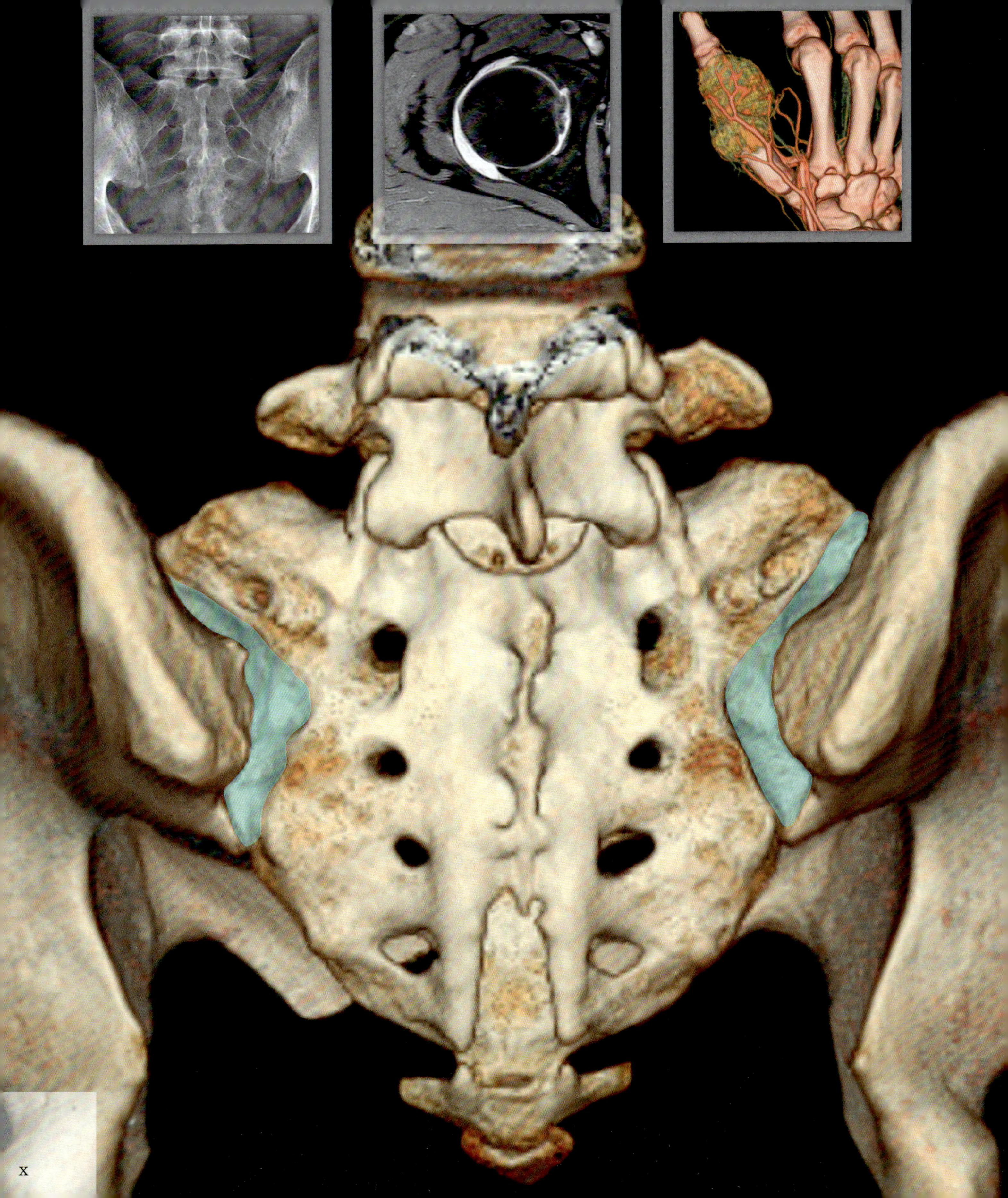

Acknowledgments

Lead Editor

Rebecca L. Bluth, BA

Text Editors

Arthur G. Gelsinger, MA
Nina I. Bennett, BA
Terry W. Ferrell, MS
Matt W. Hoecherl, BS
Megg Morin, BA
Joshua Reynolds, PhD

Image Editors

Jeffrey J. Marmorstone, BS
Lisa A. M. Steadman, BS

Illustrations

Richard Coombs, MS
Lane R. Bennion, MS
Laura C. Wissler, MA

Art Direction and Design

Tom M. Olson, BA

Production Coordinators

Emily C. Fassett, BA
Angela M. G. Terry, BA

ELSEVIER

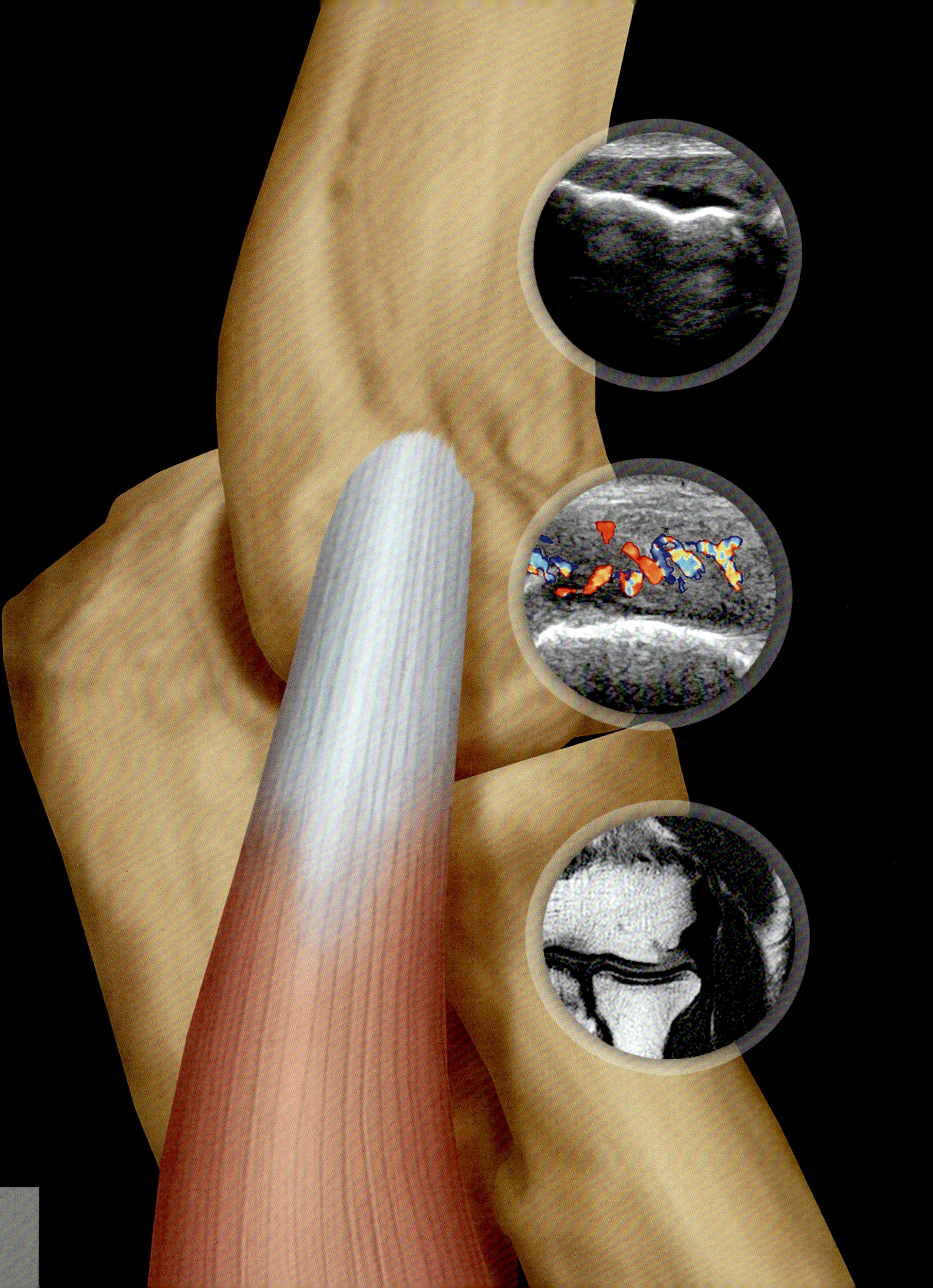

Sections

SECTION 1:
Anatomy

SECTION 2:
Technique

SECTION 3:
Diagnoses

SECTION 4:
Differential Diagnoses

SECTION 5:
Interventional Procedures

TABLE OF CONTENTS

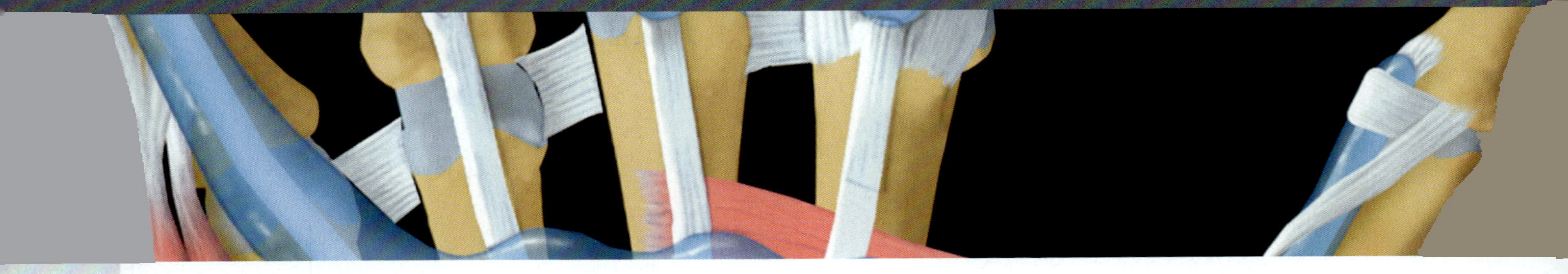

SECTION 1: ANATOMY

UPPER LIMB

4 **Sternoclavicular and Acromioclavicular Joints**
James F. Griffith, MD, MRCP, FRCR and Bhawan K. Paunipagar, MBBS, MD, DNB

10 **Shoulder**
James F. Griffith, MD, MRCP, FRCR and Bhawan K. Paunipagar, MBBS, MD, DNB

28 **Axilla**
James F. Griffith, MD, MRCP, FRCR and Bhawan K. Paunipagar, MBBS, MD, DNB

36 **Arm**
James F. Griffith, MD, MRCP, FRCR and Bhawan K. Paunipagar, MBBS, MD, DNB

44 **Arm Vessels**
James F. Griffith, MD, MRCP, FRCR and Bhawan K. Paunipagar, MBBS, MD, DNB

52 **Elbow**
James F. Griffith, MD, MRCP, FRCR and Bhawan K. Paunipagar, MBBS, MD, DNB

72 **Forearm**
James F. Griffith, MD, MRCP, FRCR and Bhawan K. Paunipagar, MBBS, MD, DNB

80 **Forearm Vessels**
James F. Griffith, MD, MRCP, FRCR and Bhawan K. Paunipagar, MBBS, MD, DNB

88 **Wrist**
James F. Griffith, MD, MRCP, FRCR and Bhawan K. Paunipagar, MBBS, MD, DNB

102 **Hand**
James F. Griffith, MD, MRCP, FRCR and Bhawan K. Paunipagar, MBBS, MD, DNB

114 **Hand Vessels**
James F. Griffith, MD, MRCP, FRCR and Bhawan K. Paunipagar, MBBS, MD, DNB

120 **Thumb**
James F. Griffith, MD, MRCP, FRCR and Bhawan K. Paunipagar, MBBS, MD, DNB

130 **Fingers**
James F. Griffith, MD, MRCP, FRCR and Bhawan K. Paunipagar, MBBS, MD, DNB

142 **Radial Nerve**
James F. Griffith, MD, MRCP, FRCR and Bhawan K. Paunipagar, MBBS, MD, DNB

150 **Median Nerve**
James F. Griffith, MD, MRCP, FRCR and Bhawan K. Paunipagar, MBBS, MD, DNB

160 **Ulnar Nerve**
James F. Griffith, MD, MRCP, FRCR and Bhawan K. Paunipagar, MBBS, MD, DNB

LOWER LIMB

170 **Hip**
Gregory E. Antonio, MD, DRANZCR, FHKCR and Eric K. H. Liu, PhD, RDMS

180 **Thigh Muscles**
Gregory E. Antonio, MD, DRANZCR, FHKCR and Eric K. H. Liu, PhD, RDMS

192 **Femoral Vessels and Nerves**
Gregory E. Antonio, MD, DRANZCR, FHKCR and Eric K. H. Liu, PhD, RDMS

206 **Knee**
Gregory E. Antonio, MD, DRANZCR, FHKCR and Eric K. H. Liu, PhD, RDMS

224 **Leg Muscles**
Gregory E. Antonio, MD, DRANZCR, FHKCR and Eric K. H. Liu, PhD, RDMS

236 **Leg Vessels**
Gregory E. Antonio, MD, DRANZCR, FHKCR, Eric K. H. Liu, PhD, RDMS, and Paula J. Woodward, MD

254 **Leg Nerves**
James F. Griffith, MD, MRCP, FRCR, Gregory E. Antonio, MD, DRANZCR, FHKCR, and Eric K. H. Liu, PhD, RDMS

258 **Ankle**
Gregory E. Antonio, MD, DRANZCR, FHKCR and Eric K. H. Liu, PhD, RDMS

276 **Tarsus**
Gregory E. Antonio, MD, DRANZCR, FHKCR and Eric K. H. Liu, PhD, RDMS

290 **Foot Vessels**
Gregory E. Antonio, MD, DRANZCR, FHKCR and Eric K. H. Liu, PhD, RDMS

296 **Metatarsals and Toes**
Gregory E. Antonio, MD, DRANZCR, FHKCR and Eric K. H. Liu, PhD, RDMS

TRUNK

302 **Brachial Plexus**
James F. Griffith, MD, MRCP, FRCR, K. T. Wong, MBChB, FRCR, and Paula J. Woodward, MD

310 **Ribs and Intercostal Space**
James F. Griffith, MD, MRCP, FRCR, Gregory E. Antonio, MD, DRANZCR, FHKCR, and Eric K. H. Liu, PhD, RDMS

314 **Abdominal Wall**
Jade Wong-You-Cheong, MBChB, MRCP, FRCR

TABLE OF CONTENTS

326 **Abdominal Wall and Paraspinal Structures**
Esther H. Y. Hung, MBChB, FRCR, FHKCR, FHKAM (Radiology), Stella Sin Yee Ho, RDMS, RVT, PhD, and Jill M. Abrigo, MD

338 **Groin**
Alex W. H. Ng, MBChB, FRCR, FHKCR, FHKAM (Radiology), Gregory E. Antonio, MD, DRANZCR, FHKCR, and Eric K. H. Liu, PhD, RDMS

348 **Gluteal Muscles**
Ryan K. L. Lee, MBChB, FRCR, FHKAM (Radiology), Gregory E. Antonio, MD, DRANZCR, FHKCR, and Eric K. H. Liu, PhD, RDMS

SECTION 2: TECHNIQUE

360 **Approach to Musculoskeletal Ultrasound**
James F. Griffith, MD, MRCP, FRCR

364 **Musculoskeletal Ultrasound Artifacts**
James F. Griffith, MD, MRCP, FRCR

372 **Shoulder Ultrasound**
James F. Griffith, MD, MRCP, FRCR

380 **Elbow Ultrasound**
James F. Griffith, MD, MRCP, FRCR

388 **Wrist Ultrasound**
James F. Griffith, MD, MRCP, FRCR

396 **Hand Ultrasound**
James F. Griffith, MD, MRCP, FRCR

400 **Groin Hernia Ultrasound**
Alex W. H. Ng, MBChB, FRCR, FHKCR, FHKAM (Radiology)

408 **Hip Ultrasound**
James F. Griffith, MD, MRCP, FRCR

416 **Knee Ultrasound**
James F. Griffith, MD, MRCP, FRCR

424 **Ankle Ultrasound**
James F. Griffith, MD, MRCP, FRCR

432 **Foot Ultrasound**
James F. Griffith, MD, MRCP, FRCR

438 **Writing an Ultrasound Report**
James F. Griffith, MD, MRCP, FRCR

SECTION 3: DIAGNOSES

TENDON DISORDERS

442 **Rotator Cuff/Biceps Tendinosis**
Esther H. Y. Hung, MBChB, FRCR, FHKCR, FHKAM (Radiology) and James F. Griffith, MD, MRCP, FRCR

454 **Rotator Cuff/Biceps Tendon Tear**
Esther H. Y. Hung, MBChB, FRCR, FHKCR, FHKAM (Radiology) and James F. Griffith, MD, MRCP, FRCR

462 **Nonrotator Cuff Tendinosis**
Alex W. H. Ng, MBChB, FRCR, FHKCR, FHKAM (Radiology) and James F. Griffith, MD, MRCP, FRCR

472 **Nonrotator Cuff Tendon Tears**
Alex W. H. Ng, MBChB, FRCR, FHKCR, FHKAM (Radiology) and James F. Griffith, MD, MRCP, FRCR

484 **Tenosynovitis**
Alex W. H. Ng, MBChB, FRCR, FHKCR, FHKAM (Radiology) and James F. Griffith, MD, MRCP, FRCR

494 **Elbow Epicondylitis**
Ryan K. L. Lee, MBChB, FRCR, FHKAM (Radiology) and James F. Griffith, MD, MRCP, FRCR

SOFT TISSUE, BONE, AND JOINT INJURY

502 **Fat Injury**
Esther H. Y. Hung, MBChB, FRCR, FHKCR, FHKAM (Radiology) and James F. Griffith, MD, MRCP, FRCR

510 **Muscle Infarction**
James F. Griffith, MD, MRCP, FRCR

514 **Muscle Injury**
James F. Griffith, MD, MRCP, FRCR

524 **Hematoma/Seroma**
Alex W. H. Ng, MBChB, FRCR, FHKCR, FHKAM (Radiology) and James F. Griffith, MD, MRCP, FRCR

532 **Ligament Injury**
Alex W. H. Ng, MBChB, FRCR, FHKCR, FHKAM (Radiology) and James F. Griffith, MD, MRCP, FRCR

540 **Bone Fracture**
James F. Griffith, MD, MRCP, FRCR

ARTHROPATHIES

550 **Osteoarthritis**
James F. Griffith, MD, MRCP, FRCR

558 **Inflammatory Arthritis**
James F. Griffith, MD, MRCP, FRCR

568 **Gout and Pseudogout**
Ryan K. L. Lee, MBChB, FRCR, FHKAM (Radiology) and James F. Griffith, MD, MRCP, FRCR

576 **Developmental Hip Dysplasia**
Alex W. H. Ng, MBChB, FRCR, FHKCR, FHKAM (Radiology) and James F. Griffith, MD, MRCP, FRCR

NEUROVASCULAR ABNORMALITIES

582 **Nerve Injury**
James F. Griffith, MD, MRCP, FRCR

590 **Nerve Sheath Tumors**
Esther H. Y. Hung, MBChB, FRCR, FHKCR, FHKAM (Radiology) and James F. Griffith, MD, MRCP, FRCR

602 **Carpal Tunnel Syndrome**
Alex W. H. Ng, MBChB, FRCR, FHKCR, FHKAM (Radiology) and James F. Griffith, MD, MRCP, FRCR

610 **Cubital Tunnel Syndrome**
Ryan K. L. Lee, MBChB, FRCR, FHKAM (Radiology) and James F. Griffith, MD, MRCP, FRCR

614 **Tarsal Tunnel Syndrome**
James F. Griffith, MD, MRCP, FRCR

620 **Vascular Dilatation or Inflammation**
Ryan K. L. Lee, MBChB, FRCR, FHKAM (Radiology) and James F. Griffith, MD, MRCP, FRCR

INFECTION

628 **Soft Tissue Infection**
Esther H. Y. Hung, MBChB, FRCR, FHKCR, FHKAM (Radiology) and James F. Griffith, MD, MRCP, FRCR

636 **Bone Infection**
Esther H. Y. Hung, MBChB, FRCR, FHKCR, FHKAM (Radiology) and James F. Griffith, MD, MRCP, FRCR

TABLE OF CONTENTS

646 **Joint Infection**
Esther H. Y. Hung, MBChB, FRCR, FHKCR, FHKAM (Radiology) and James F. Griffith, MD, MRCP, FRCR

652 **Postoperative Infection**
Esther H. Y. Hung, MBChB, FRCR, FHKCR, FHKAM (Radiology) and James F. Griffith, MD, MRCP, FRCR

ARTICULAR AND PARAARTICULAR MASSES

658 **Hemarthrosis and Lipohemarthrosis**
Ryan K. L. Lee, MBChB, FRCR, FHKAM (Radiology) and James F. Griffith, MD, MRCP, FRCR

662 **Baker Cyst**
Ryan K. L. Lee, MBChB, FRCR, FHKAM (Radiology) and James F. Griffith, MD, MRCP, FRCR

668 **Bursitis**
Alex W. H. Ng, MBChB, FRCR, FHKCR, FHKAM (Radiology) and James F. Griffith, MD, MRCP, FRCR

680 **Ganglion Cyst**
Ryan K. L. Lee, MBChB, FRCR, FHKAM (Radiology) and James F. Griffith, MD, MRCP, FRCR

690 **Parameniscal Cyst**
Ryan K. L. Lee, MBChB, FRCR, FHKAM (Radiology) and James F. Griffith, MD, MRCP, FRCR

694 **Synovial Tumor**
James F. Griffith, MD, MRCP, FRCR

SOFT TISSUE AND BONE TUMORS

706 **Plantar Fasciitis and Fibromatosis**
Ryan K. L. Lee, MBChB, FRCR, FHKAM (Radiology) and James F. Griffith, MD, MRCP, FRCR

712 **Lipoma**
Esther H. Y. Hung, MBChB, FRCR, FHKCR, FHKAM (Radiology) and James F. Griffith, MD, MRCP, FRCR

720 **Epidermoid Cyst**
Esther H. Y. Hung, MBChB, FRCR, FHKCR, FHKAM (Radiology) and James F. Griffith, MD, MRCP, FRCR

726 **Pilomatricoma**
Esther H. Y. Hung, MBChB, FRCR, FHKCR, FHKAM (Radiology) and James F. Griffith, MD, MRCP, FRCR

730 **Dermatofibrosarcoma Protuberans**
Ryan K. L. Lee, MBChB, FRCR, FHKAM (Radiology) and James F. Griffith, MD, MRCP, FRCR

734 **Vascular Leiomyoma**
Ryan K. L. Lee, MBChB, FRCR, FHKAM (Radiology) and James F. Griffith, MD, MRCP, FRCR

738 **Superficial Metastases, Lymphoma, and Melanoma**
Esther H. Y. Hung, MBChB, FRCR, FHKCR, FHKAM (Radiology) and James F. Griffith, MD, MRCP, FRCR

744 **Vascular Anomaly**
Esther H. Y. Hung, MBChB, FRCR, FHKCR, FHKAM (Radiology) and James F. Griffith, MD, MRCP, FRCR

752 **Foreign Body and Injection Granuloma**
Ryan K. L. Lee, MBChB, FRCR, FHKAM (Radiology) and James F. Griffith, MD, MRCP, FRCR

760 **Lymph Node Abnormality**
Esther H. Y. Hung, MBChB, FRCR, FHKCR, FHKAM (Radiology) and James F. Griffith, MD, MRCP, FRCR

766 **Soft Tissue Sarcoma**
Alex W. H. Ng, MBChB, FRCR, FHKCR, FHKAM (Radiology) and James F. Griffith, MD, MRCP, FRCR

774 **Bone Tumor**
James F. Griffith, MD, MRCP, FRCR

786 **Local Tumor Recurrence**
Alex W. H. Ng, MBChB, FRCR, FHKCR, FHKAM (Radiology) and James F. Griffith, MD, MRCP, FRCR

HERNIA

796 **Abdominal Wall Hernia**
Ryan K. L. Lee, MBChB, FRCR, FHKAM (Radiology) and James F. Griffith, MD, MRCP, FRCR

804 **Groin Hernia**
Ryan K. L. Lee, MBChB, FRCR, FHKAM (Radiology) and James F. Griffith, MD, MRCP, FRCR

SECTION 4: DIFFERENTIAL DIAGNOSES

GENERAL LUMPS AND BUMPS

812 **Hypoechoic Subcutaneous Mass**
Esther H. Y. Hung, MBChB, FRCR, FHKCR, FHKAM (Radiology) and James F. Griffith, MD, MRCP, FRCR

818 **Hyperechoic Subcutaneous Mass**
Esther H. Y. Hung, MBChB, FRCR, FHKCR, FHKAM (Radiology) and James F. Griffith, MD, MRCP, FRCR

824 **Hypoechoic Muscle Mass**
James F. Griffith, MD, MRCP, FRCR

830 **Hyperechoic Muscle Mass**
James F. Griffith, MD, MRCP, FRCR

836 **Cystic Soft Tissue Mass**
Ryan K. L. Lee, MBChB, FRCR, FHKAM (Radiology) and James F. Griffith, MD, MRCP, FRCR

842 **Calcified Soft Tissue Mass**
Alex W. H. Ng, MBChB, FRCR, FHKCR, FHKAM (Radiology) and James F. Griffith, MD, MRCP, FRCR

846 **Hypervascular Soft Tissue Mass**
Esther H. Y. Hung, MBChB, FRCR, FHKCR, FHKAM (Radiology) and James F. Griffith, MD, MRCP, FRCR

TENDON ABNORMALITIES

852 **Peritendinous Mass**
Alex W. H. Ng, MBChB, FRCR, FHKCR, FHKAM (Radiology) and James F. Griffith, MD, MRCP, FRCR

856 **Tendon Hypoechogenicity**
Alex W. H. Ng, MBChB, FRCR, FHKCR, FHKAM (Radiology) and James F. Griffith, MD, MRCP, FRCR

860 **Tendon Hyperechogenicity**
Alex W. H. Ng, MBChB, FRCR, FHKCR, FHKAM (Radiology) and James F. Griffith, MD, MRCP, FRCR

864 **Tendon Swelling**
Alex W. H. Ng, MBChB, FRCR, FHKCR, FHKAM (Radiology) and James F. Griffith, MD, MRCP, FRCR

NERVE, FASCIA, AND BONE

870 **Swollen Nerve**
James F. Griffith, MD, MRCP, FRCR

TABLE OF CONTENTS

876 **Fascial Lesion**
James F. Griffith, MD, MRCP, FRCR

882 **Bone Surface Lesion**
James F. Griffith, MD, MRCP, FRCR

JOINT ABNORMALITIES

888 **Paraarticular Cystic Mass**
James F. Griffith, MD, MRCP, FRCR

894 **Synovial Swelling**
James F. Griffith, MD, MRCP, FRCR

900 **Joint Effusion**
Ryan K. L. Lee, MBChB, FRCR, FHKAM (Radiology) and James F. Griffith, MD, MRCP, FRCR

CHEST AND ABDOMINAL WALL

904 **Chest Wall Lesion**
Ryan K. L. Lee, MBChB, FRCR, FHKAM (Radiology) and James F. Griffith, MD, MRCP, FRCR

910 **Abdominal Wall Mass**
Ryan K. L. Lee, MBChB, FRCR, FHKAM (Radiology) and James F. Griffith, MD, MRCP, FRCR

SECTION 5: INTERVENTIONAL PROCEDURES

BIOPSY

918 **Soft Tissue Tumor Biopsy**
Esther H. Y. Hung, MBChB, FRCR, FHKCR, FHKAM (Radiology) and James F. Griffith, MD, MRCP, FRCR

924 **Bone Tumor Biopsy**
James F. Griffith, MD, MRCP, FRCR

JOINT PROCEDURES

932 **Joint Injection: Upper Limb**
Ryan K. L. Lee, MBChB, FRCR, FHKAM (Radiology)

942 **Joint Injection: Lower Limb**
Ryan K. L. Lee, MBChB, FRCR, FHKAM (Radiology)

950 **Shoulder Procedures**
Esther H. Y. Hung, MBChB, FRCR, FHKCR, FHKAM (Radiology), Karen Partington, MRCS, FRCR, and Eugene McNally, FRCR, FRCPI

958 **Elbow Procedures**
Alex W. H. Ng, MBChB, FRCR, FHKCR, FHKAM (Radiology), Philip Yoong, FRCR, and James F. Griffith, MD, MRCP, FRCR

964 **Hand and Wrist Procedures**
Ryan K. L. Lee, MBChB, FRCR, FHKAM (Radiology), Philip Yoong, FRCR, and James F. Griffith, MD, MRCP, FRCR

970 **Hip and Pelvis Procedures**
James F. Griffith, MD, MRCP, FRCR and Philip Yoong, FRCR

978 **Knee Procedures**
Alex W. H. Ng, MBChB, FRCR, FHKCR, FHKAM (Radiology), Karen Partington, MRCS, FRCR, and Eugene McNally, FRCR, FRCPI

986 **Ankle and Foot Procedures**
Ryan K. L. Lee, MBChB, FRCR, FHKAM (Radiology), Karen Partington, MRCS, FRCR, and Eugene McNally, FRCR, FRCPI

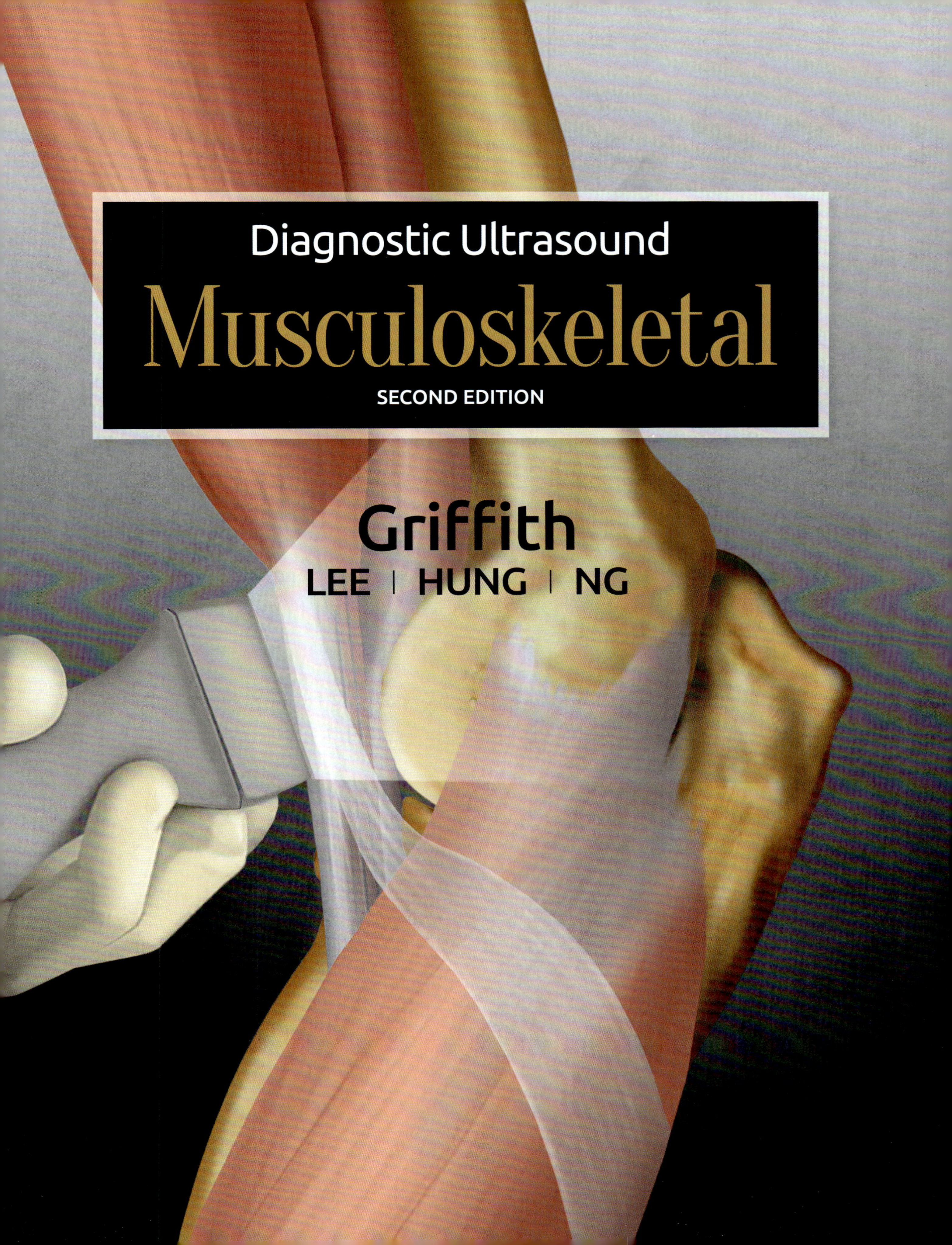

Diagnostic Ultrasound

Musculoskeletal

SECOND EDITION

Griffith

LEE | HUNG | NG

SECTION 1

Anatomy

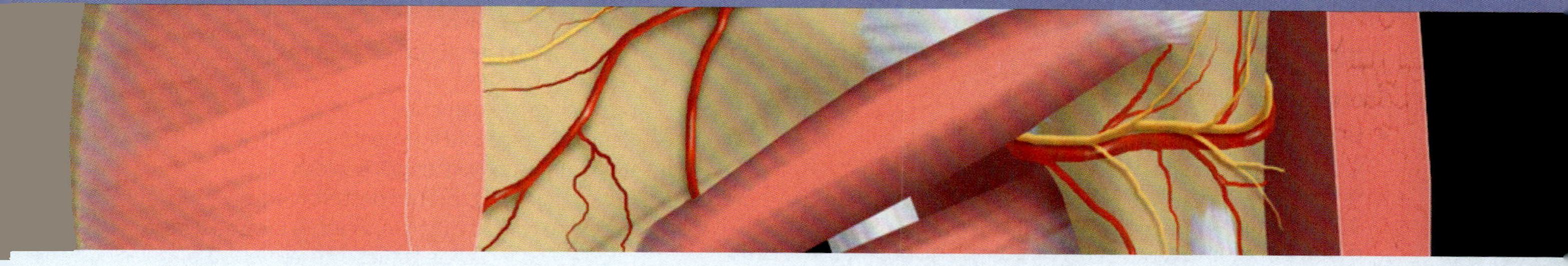

Upper Limb

Sternoclavicular and Acromioclavicular Joints 4
Shoulder 10
Axilla 28
Arm 36
Arm Vessels 44
Elbow 52
Forearm 72
Forearm Vessels 80
Wrist 88
Hand 102
Hand Vessels 114
Thumb 120
Fingers 130
Radial Nerve 142
Median Nerve 150
Ulnar Nerve 160

Lower Limb

Hip 170
Thigh Muscles 180
Femoral Vessels and Nerves 192
Knee 206
Leg Muscles 224
Leg Vessels 236
Leg Nerves 254
Ankle 258
Tarsus 276
Foot Vessels 290
Metatarsals and Toes 296

Trunk

Brachial Plexus 302
Ribs and Intercostal Space 310
Abdominal Wall 314
Abdominal Wall and Paraspinal Structures 326
Groin 338
Gluteal Muscles 348

TERMINOLOGY

Abbreviations

- Sternoclavicular (SC) joint
- Acromioclavicular (AC) joint

GROSS ANATOMY

Sternoclavicular Joint

- Between medial end of clavicle & manubrium
  - Synovial sellar-type (saddle) joint
  - Medial end of clavicle = large & bulbous
  - Much larger than manubrial concavity
  - < 1/2 of medial clavicle articulates with manubrium
    - Stability through capsuloligamentous structures
- Intraarticular disc
  - Attached to joint capsule anteriorly & posteriorly
  - Complete or incomplete ± perforations
  - Thickest posterosuperiorly (3 mm)
- Ligaments of SC joint
  - Capsular ligaments
    - Cover anterosuperior & posterior aspects of SC joint
    - Prevent upward displacement of medial clavicle, which may be caused by downward force on shoulder
    - Anterior stronger than posterior portion
  - Interclavicular ligament
    - Connects superomedial aspect of clavicle to capsular ligaments & upper manubrium
    - Covers anterosuperior & posterior aspects of joint
    - Prevents excessive upward motion of clavicle
  - Costoclavicular ligaments
    - Unite inferior surface medial end clavicle to upper surface of 1st rib
    - Anterior fibers arise from anteromedial surface of 1st rib & resist upward motion
    - Posterior fibers arise lateral to anterior fibers & resist downward motion
- Muscle attachments to medial clavicle & sternum
  - Pectoralis major from anterior aspect medial 2/3 clavicle (clavicular head)
  - Sternocleidomastoid from posterior surface medial 1/3 of clavicle (clavicular head)
  - Sternohyoid & sternothyroid muscles separate great vessels from SC joint

Acromioclavicular Joint

- Synovial joint between lateral end of clavicle & medial end of acromion
  - Articular surface of clavicle oriented posterolaterally whereas articular surface of acromion oriented anteromedially
    - Angle of inclination between opposing articular surfaces varies with clavicle overriding acromion (50%), vertical orientation between acromion & clavicle (25%), clavicle underriding acromion (5%), & mixed pattern (20%)
    - Maximum width of normal joint on US = 5 mm if < 35 years & < 4.4 mm if > 35 years
    - Maximum thickness of capsule from bony surface = 2.7 mm if < 35 years & < 3.6 mm if > 35 years
- Intraarticular disc
  - Undergoes rapid degeneration beginning in 2nd decade → marked degeneration of disc by 4th decade
- Ligaments of AC joint
  - Superior AC ligament
    - Stronger & thicker (2.0-5.5 mm) than thin or absent inferior AC ligament
    - Inserts along lateral clavicle (8 mm) & medial acromion (10 mm)
  - Coracoclavicular ligaments
    - Conoid & trapezoid ligaments
    - Vary significantly in length & width
    - Conoid ligament located posteromedially
    - Inserts to conoid tubercle, which is located where middle 1/3 of clavicle curves into lateral 1/3 of clavicle
    - Mainly prevents upward movement of clavicle
    - Trapezoid ligament located anterolaterally
    - Inserts to trapezoid ridge, which runs along inferior surface of lateral 1/3 of clavicle
    - Mainly prevents lateral compression of clavicle against acromion
  - Muscle attachments to lateral clavicle
    - Deltoid attached to anterior surface lateral 1/3 of clavicle
    - Trapezius attached to posterior surface lateral 1/3 of clavicle

ANATOMY IMAGING ISSUES

Imaging Recommendations

- High-resolution linear transducer
- Align transducer transversely along SC or AC joints
- AC joint laxity can be assessed by pulling down on arm while observing change in joint width on US
  - Compare with contralateral side
- Main clinical presentation of SC joint is painless lump
  - Mild degrees of capsular thickening is readily apparent clinically since joint just beneath skin surface
    - Clinical swelling often due to relative forward positioning of apparently swollen SC joint due to axial rotation of upper trunk
    - Occasionally due to mild capsular swelling ± mild subluxation secondary to SC osteoarthritis
  - Main clinical presentation of AC joint is pain due to osteoarthritis, AC joint impingement, inflammatory arthropathy, & subluxation/dislocation

Imaging Pitfalls

- SC or AC joints
  - Normally step-off between medial clavicle & manubrium &, to lesser degree, between lateral clavicle & acromion
  - Should not be interpreted as subluxation
  - Acromion normally elevates from rest position during arm adduction
  - AC joint index = AC joint width of uninjured side/AC joint width of injured side = 1.0 normally
  - Determine whether AC joint is
    - Not subluxed (similar to opposite side): Grade 1
    - Partially subluxed (clavicle subluxed < 50% depth of AC joint): Grade 2
    - Severely subluxed or dislocated (clavicle subluxed > 50% depth of AC joint): Grade 3

TRANSVERSE US, STERNOCLAVICULAR JOINT

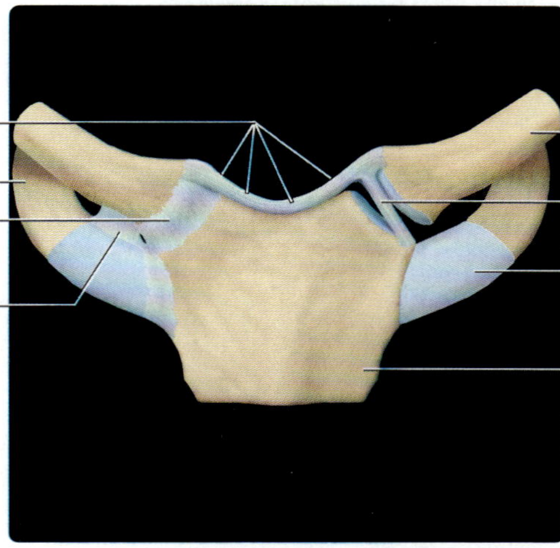

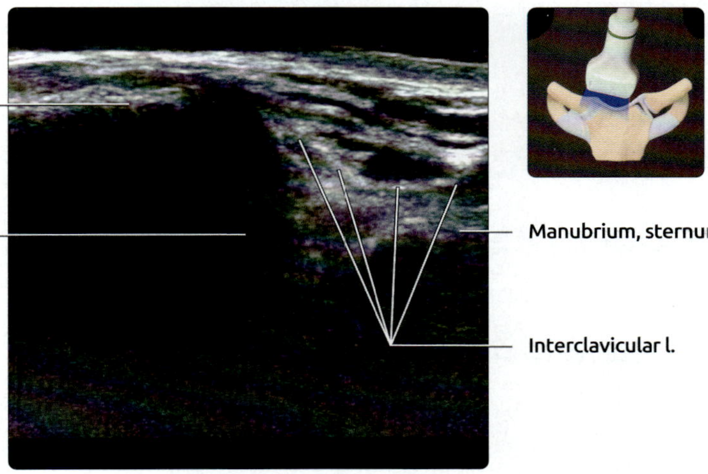

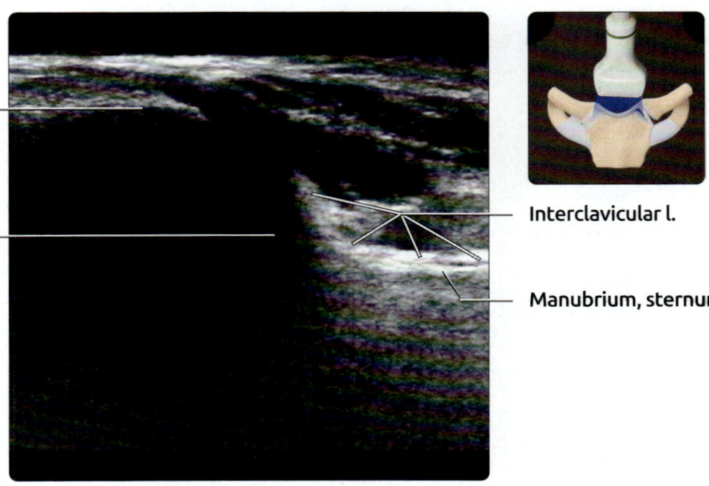

(Top) *Graphic shows the anterior aspect of the sternoclavicular joint. Note the joint capsule, articular disc, and interclavicular ligament.* **(Middle)** *Transverse grayscale US shows the anterosuperior aspect of the sternoclavicular joint. The medial clavicle is much larger than the articulating surface of the manubrium. The thin interclavicular ligament is closely applied to the superior aspect of manubrium, and its connection with the medial ends of both clavicles is depicted.* **(Bottom)** *Transverse grayscale US shows the superior aspect of the sternoclavicular joint. The costoclavicular ligament prevents upward movement of the medial clavicle when the lateral clavicle or shoulder is depressed.*

LONGITUDINAL US, STERNOCLAVICULAR JOINT

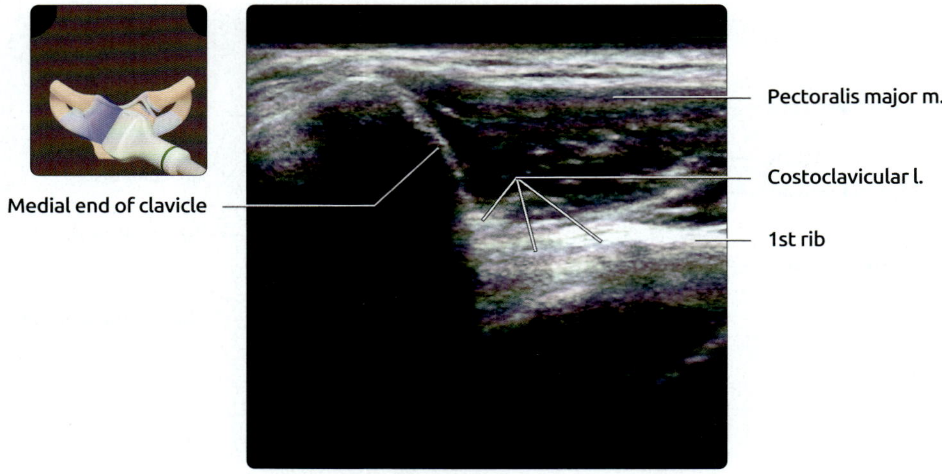

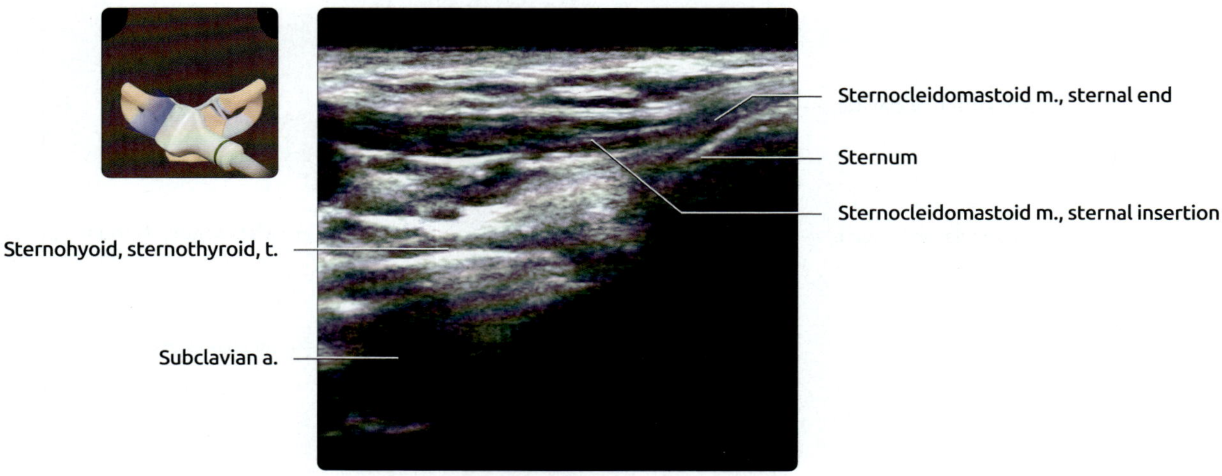

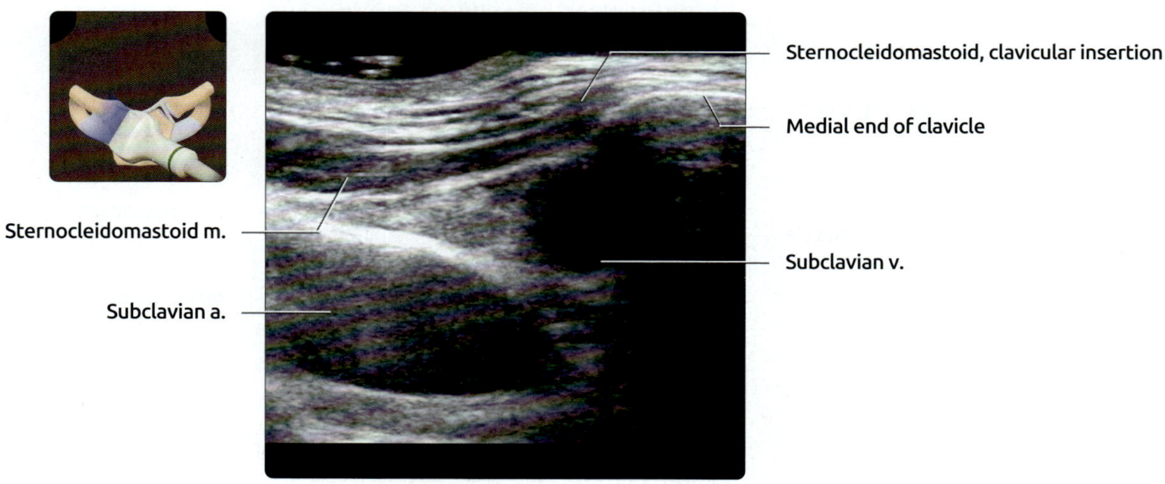

(Top) *Longitudinal grayscale US shows sternoclavicular joint. Costoclavicular ligament prevents upward movement of the medial clavicle when shoulder is depressed. Pectoralis major muscle arises from the medial 1/2 of the anterior surface of the clavicle as well as from the sternum, upper costal cartilages, and upper part of external oblique aponeurosis.* **(Middle)** *Longitudinal grayscale US shows the sternoclavicular joint region. The sternocleidomastoid is attached to the upper surface of the medial end of the clavicle as well as the upper anterior surface of the manubrium. The sternohyoid and sternothyroid are attached to the posterior aspect of the sternum as well as the clavicle and 1st costal cartilage.* **(Bottom)** *Longitudinal grayscale US shows the sternoclavicular joint. Great vessels lie posterior to the sternoclavicular joint and may get injured in posterior dislocation. All tendinous attachments should be assessed if dislocation is present, as they may also be injured.*

US, ACROMIOCLAVICULAR JOINT

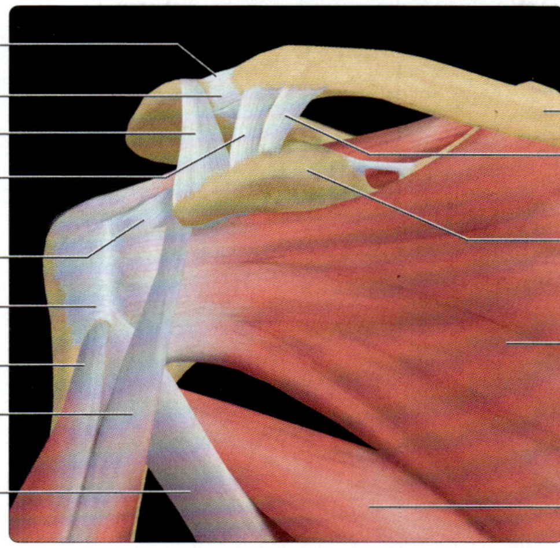

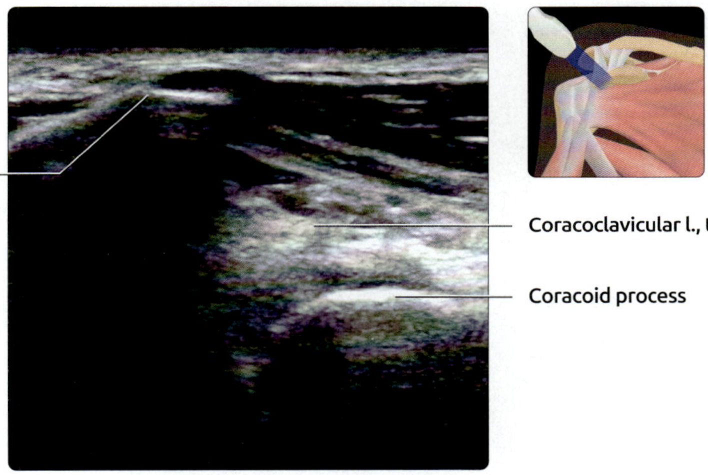

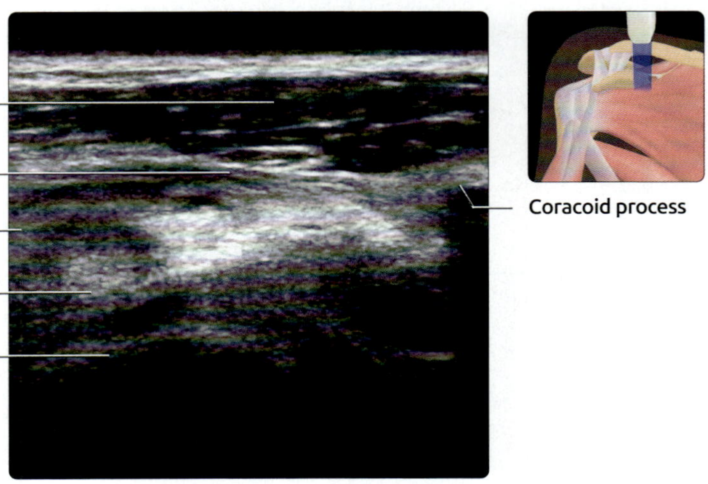

(Top) *Anterior graphic shows the shoulder in superficial dissection.* **(Middle)** *Longitudinal grayscale US shows the acromioclavicular joint region. The coracoclavicular ligament is demonstrated but is not as clearly depicted on US as it is on MR exam. These ligaments prevent upward and lateral movement of the clavicle.* **(Bottom)** *Transverse grayscale US of the acromioclavicular joint region shows the coracoacromial ligament. The supraspinatus tendon and intervening bursa can impinge against the coracoacromial ligament during arm abduction.*

TRANSVERSE US, ACROMIOCLAVICULAR JOINT

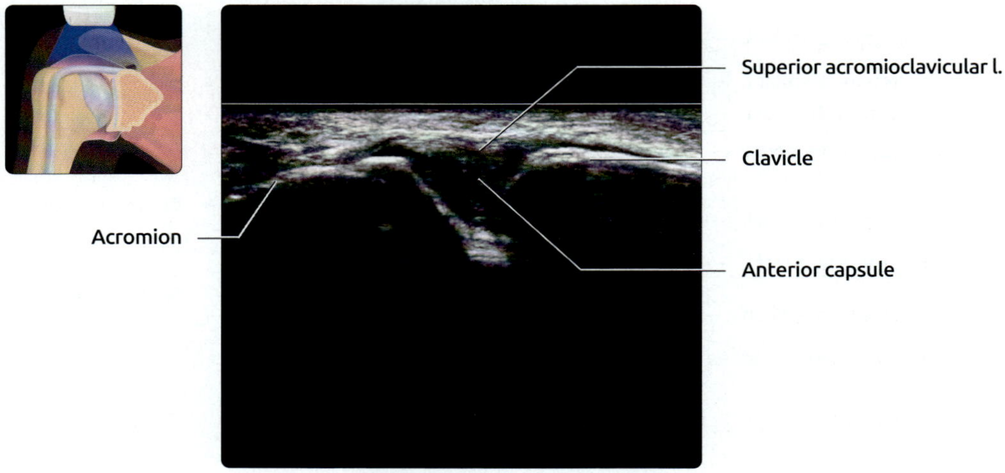

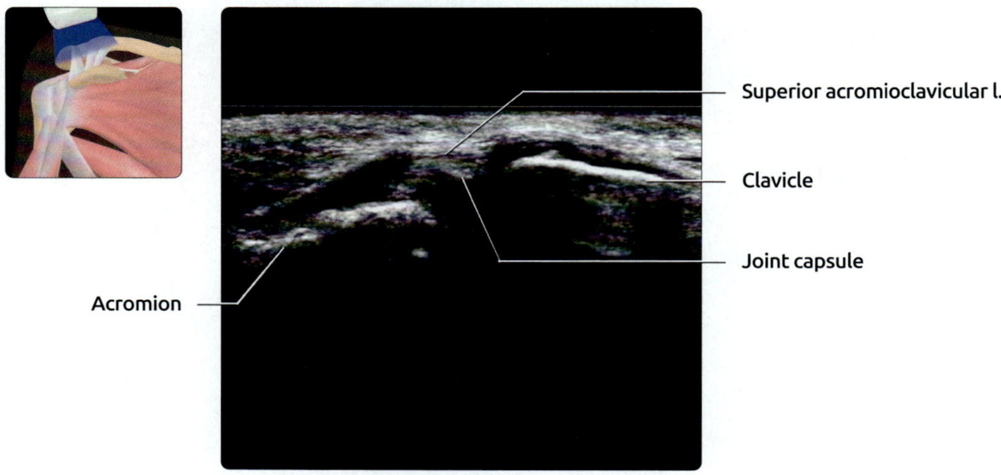

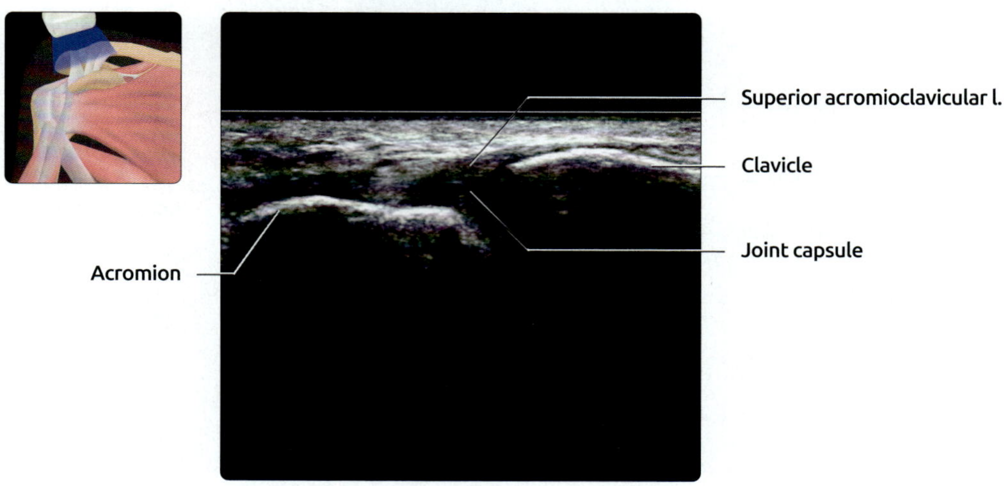

(Top) *Transverse grayscale US shows the anterior aspect of the acromioclavicular joint. The joint capsule of the acromioclavicular joint is thin with a strong supporting superior acromioclavicular ligament.* **(Middle)** *Transverse grayscale US shows the anterosuperior acromioclavicular joint. Separation of the clavicle and acromion can be readily appreciated. Note how opposing bone margins are not vertically aligned.* **(Bottom)** *Transverse grayscale US shows the superior aspect of the acromioclavicular joint. In this image, the clavicle slightly overrides the acromion. This is a normal configuration.*

TRANSVERSE US, ACROMIOCLAVICULAR JOINT

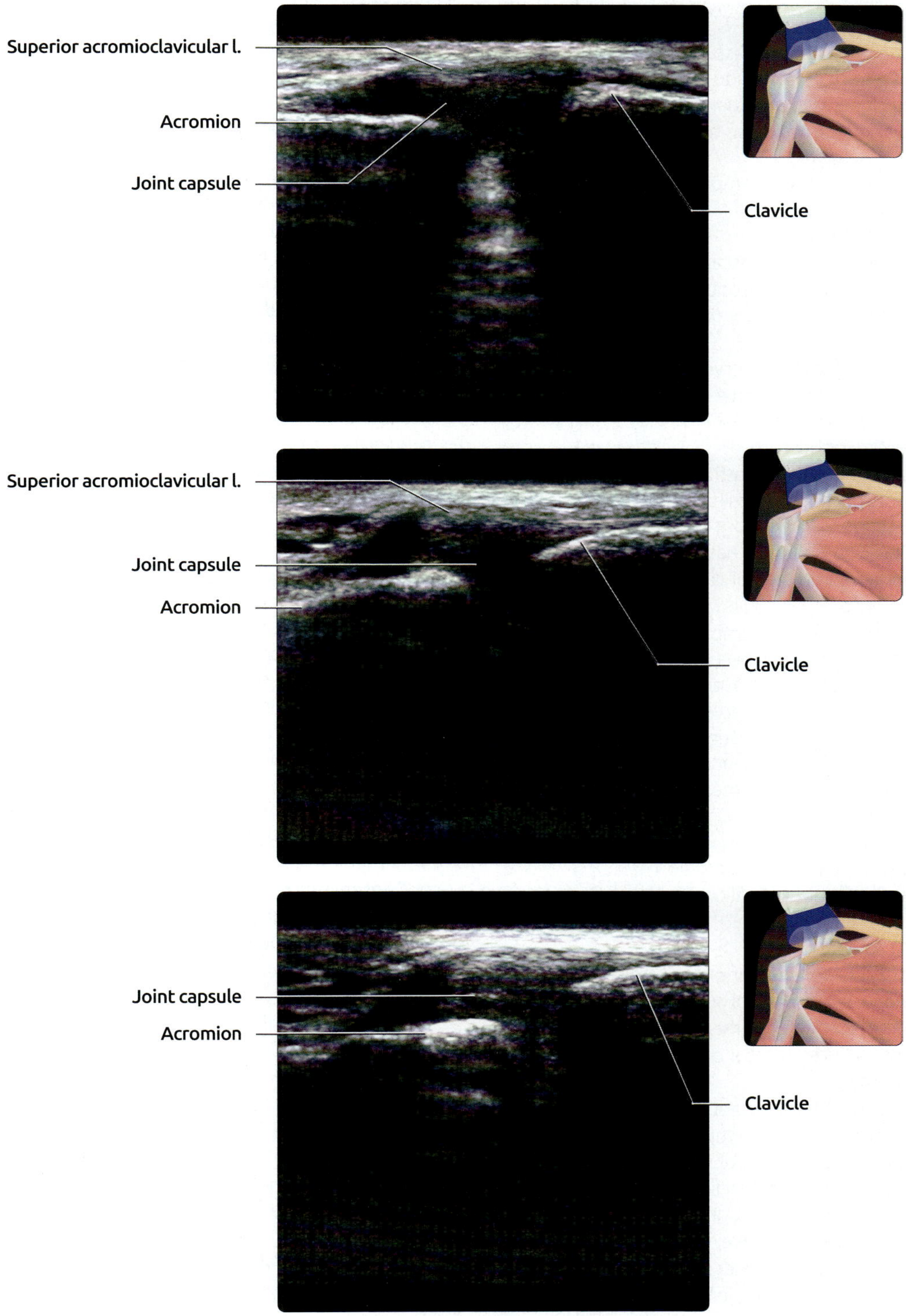

(Top) *Transverse grayscale US shows the anterosuperior aspect of the acromioclavicular joint with the arm positioned by the side of the body. Note that the clavicle slightly overrides the acromion.* **(Middle)** *Transverse grayscale US shows the acromioclavicular joint with the arm in abducted position. The acromion is now level with the lateral aspect of the clavicle. Note how the joint capsule bulges superiorly, and the opposing bones are approximated with the arm abducted.* **(Bottom)** *Transverse grayscale US shows the acromioclavicular joint with the arm in an abducted position. The acromion is now depressed relative to this lateral end of the clavicle.*

IMAGING ANATOMY

Overview

- **Rotator cuff**
  - Consists of supraspinatus, infraspinatus, teres minor, subscapularis muscles, and tendons
  - Cuff tendons blend with shoulder joint capsule
  - Supraspinatus and infraspinatus tendons are inseparable at insertion
  - Anterior 2.25 cm of tendon comprises supraspinatus tendon insertional area
- **Supraspinatus muscle**
  - Origin: Supraspinatus fossa of scapula
  - Insertion: Superior facet (horizontal orientation) and anterior portion of middle facet of greater tuberosity
    - Broad insertional area
  - Nerve supply: Suprascapular nerve
  - Blood supply: Suprascapular artery and circumflex scapular branches of subscapular artery
  - Action: Abduction of humerus
  - Muscle consists of 2 distinct portions
    - Anterior portion is larger, fusiform in shape, has dominant tendon, and is more likely to tear
    - Posterior portion is flat and has terminal tendon
  - Most commonly injured rotator cuff tendon
- **Infraspinatus muscle**
  - Origin: Infraspinatus fossa of scapula
  - Insertion: Mid to posterior aspects of middle facet of greater tuberosity; centrally positioned within tendon
  - Nerve supply: Suprascapular nerve, distal fibers
  - Blood supply: Suprascapular artery and circumflex scapular branches of subscapular artery
  - Action: External rotation of humerus and resists posterior subluxation
- **Teres minor muscle**
  - Origin: Lateral scapular border, middle 1/2
  - Insertion: Inferior facet (vertical orientation) of greater tuberosity
  - Nerve supply: Axillary nerve
  - Blood supply: Posterior circumflex humeral artery and circumflex scapular branches of subscapular artery
  - Action: External rotation of humerus
  - Least commonly injured rotator cuff tendon
- **Subscapularis muscle**
  - Origin: Subscapular fossa of scapula
  - Insertion: Lesser tuberosity and up to 40% of tendon may insert along adjacent part of surgical neck
  - Some fibers cross over to lateral lip of bicipital groove, reinforcing and blending with transverse ligament
  - Nerve supply: Subscapular nerve, upper and lower
  - Blood supply: Subscapularis artery
  - Action: Internal rotation of humerus, also adduction, extension, depression, and flexion
  - 4-6 tendon slips converge into main tendon; multipennate morphology increases strength
- **Rotator cuff tendon blood supply**
  - Derived from adjacent muscle, bone, and bursae
  - Normal hypovascular regions in tendons
    - Termed critical zone: ~ 1 cm proximal to insertion
    - Vulnerable to degeneration and calcific deposition
    - However, insertional area is more prone to tearing than critical zone
- **Biceps tendon, long head**
  - Origin: Superior glenoid labrum (biceps anchor)
    - Portions may attach to supraglenoid tubercle, anterosuperior labrum, posterosuperior labrum, and coracoid base
  - Runs along superior aspect of shoulder to bicipital groove
  - Action: Stabilizes and depresses humeral head
  - Anatomic variants: Anomalous intra- and extraarticular origins from rotator cuff and joint capsule
  - Tendon sheath communicates with glenohumeral joint and normally contains small amount of fluid
- **Subacromial-subdeltoid fat plane**
  - Subacromial and subdeltoid portions
    - ± subcoracoid extension in some patients
  - Fat plane is superficial to bursa
  - May be interrupted or absent in normal patients
  - Attached along free border of coracoacromial ligament and deep surface of deltoid muscle, and humeral neck
- **Rotator cuff interval**
  - Space between supraspinatus and subscapularis tendon through which biceps tendon passes
  - **Borders of rotator cuff interval**
    - Triangular-shaped space
    - Reflections of glenohumeral ligament and coracohumeral ligament form biceps reflection pulley
    - Biceps reflection pulley stabilizes biceps tendon within rotator cuff interval
    - Superior border: Leading edge of supraspinatus
    - Inferior border: Superior aspect of subscapularis tendon
    - Lateral border: Long head of biceps tendon and bicipital groove
    - Medial border: Base of coracoid process
  - **Contents of rotator interval**
    - Long head of biceps tendon; biceps reflection pulley
- **Coracoacromial ligament**
  - Forms coracoacromial arch along with acromion and coracoid process
    - Reinforces inferior aspect of acromioclavicular joint
  - Extends from distal coracoid to subacromial area
  - Broad insertion to undersurface acromion
    - Ligament is thicker at acromion (normal thickness < 2.5 mm) and may be associated with spurs
- **Glenoid labrum**
  - Triangular-shaped rim of fibrocartilage, which extends around periphery of glenoid

ANATOMY IMAGING ISSUES

Imaging Approaches

- Tendons best seen when on stretch
  - High-resolution linear transducer
  - Long-axis (longitudinal) & transverse view of each tendon
  - Each part of tendon needs to be examined; anisotropy prevents all parts of curved rotator cuff tendons from being seen at same time
  - Need to realign ("toggle") probe to see different parts of tendons

- Supraspinatus tendon
  - Arm extended and internally rotated behind lumbar region (Crass position)
    - If too painful, hand in behind hip ("back pocket") with elbow close to body (modified Crass position)
- Infraspinatus and teres minor tendons
  - Arm flexed and internally rotated with hand placed on contralateral shoulder
  - Teres minor tendon located posteroinferior to infraspinatus tendon
- Subscapularis tendon: Arm neutral and externally rotated
- Long head of biceps tendon
  - Arm neutral and externally rotated
  - Vary degree of external rotation for optimal view of biceps tendon
  - Check for tendon subluxation
- Subacromial-subdeltoid bursa
  - Stretching tendons may squeeze fluid from area of bursa under inspection
  - Examine in all positions and also in neutral position
  - Fluid collects preferentially just lateral to acromion and proximal humerus and near coracoacromial ligament
- Coracoid process and coracoacromial ligament
  - Neutral position
- Acromioclavicular joint
  - Neutral position
  - Can pull down on arm to assess joint laxity
- Glenohumeral joint
  - Neutral position
  - Best seen from posterior aspect of joint
  - Passive movement of arm during scanning can help in identifying posterior glenoid labrum
- Spinoglenoid notch
  - Neutral position just medial to glenohumeral joint
- Supraspinatus and infraspinatus muscles
  - Neutral position with hands resting on thigh
  - Examine thickest part of muscles from behind (in coronal and sagittal planes)
  - ↓ muscle bulk, ↑ echogenicity, and ↓ visibility of central tendon are signs of atrophy with fatty replacement
    - Compare muscle echogenicity to that of trapezius or deltoid muscle

Imaging Sweet Spots

- Look for tears particularly at anterior leading edge of supraspinatus tendon
  - Unexplained bursal fluid is good secondary sign of rotator cuff tear
- Bursal fluid is often best seen with arm in neutral position or ↓ internal rotation (hand in back pocket)

Imaging Pitfalls

- **Anisotropy**
  - Echoes are optimally reflected when transducer is parallel to tendon fibers
  - Rotator cuff tendons are prone to anisotropy due to curved course
  - If transducer is not at right angles to tendon, it will appear either isoechoic or hypoechoic to muscle
    - May simulate tendinosis or partial tear
- **Tendon edges**
  - Interfaces of tendons with adjacent structures may simulate tears
  - All pathology should be confirmed in 2 planes
- **Rotator cuff cable**
  - Thick band of fibers running perpendicular to supraspinatus tendon
  - Located on deeper aspect of tendon just proximal to insertional area
  - May reinforce critical zone supraspinatus fibers
  - Cable thicker in young subjects but more easily seen in elderly subjects due to supraspinatus tendinosis
  - Can simulate tendinosis or partial-thickness tear
- **Tendinous interspace at rotator cuff interval**
  - Interspace between leading (anterior) edge of supraspinatus and long head of biceps tendon may simulate tear
  - Overcome by recognizing ovoid or rounded shape of biceps tendon
  - Rotator cuff interval best seen with external rotation
- **Focal thinning at supraspinatus-infraspinatus junction**
  - Mild diffuse thinning of supraspinatus and infraspinatus tendon junction is normal finding
  - Should not be mistaken for tendon attenuation or partial-thickness tear
- **Musculotendinous junction**
  - Supraspinatus tendon
    - Hypoechoic muscle extending along superficial aspect of tendon may simulate subacromial-subdeltoid bursal distension
    - Interdigitating tendons of anterior and posterior portions may simulate tendinosis or tear
  - Infraspinatus tendon
    - Muscle fibers surrounding centrally positioned tendon may be confused with tear
  - Subscapularis tendon
    - 4-6 tendon slips converging into main tendon may simulate tendinosis
- **Fibrocartilaginous insertion**
  - Thin layer of fibrocartilage exists between tendon and bone at insertional area
  - Steeper tendon insertion = thicker fibrocartilaginous layer
  - This thin hypoechoic layer of fibrocartilage may simulate avulsive tear
- **Subacromial-subdeltoid fat plane**
  - Fat plane lies mainly superficial to bursa and deep to deltoid muscle
  - Normal bursa is very thin
  - Thickness of echogenic fat plane is variable among patients though usually similar from side to side
  - May be wrongly interpreted as bursal fluid
    - Look for intrabursal fluid ± hyperemia (latter is feature of inflammatory arthropathy)
- **Fluid in biceps tendon sheath**
  - Communicates with glenohumeral joint
  - Small amount of fluid is normal
    - Do not misinterpret as long head of biceps tenosynovitis
  - ↑ fluid in biceps tendon sheath usually reflects ↑ fluid in glenohumeral joint

MUSCLES AND LIGAMENTS

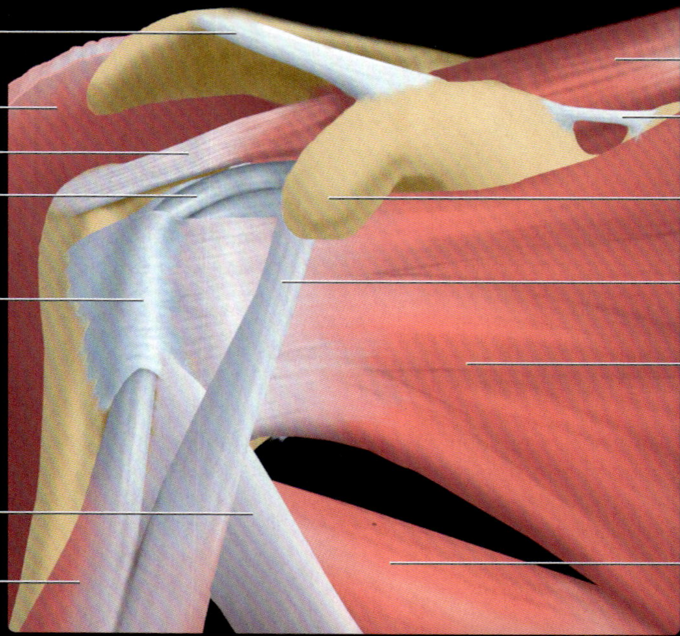

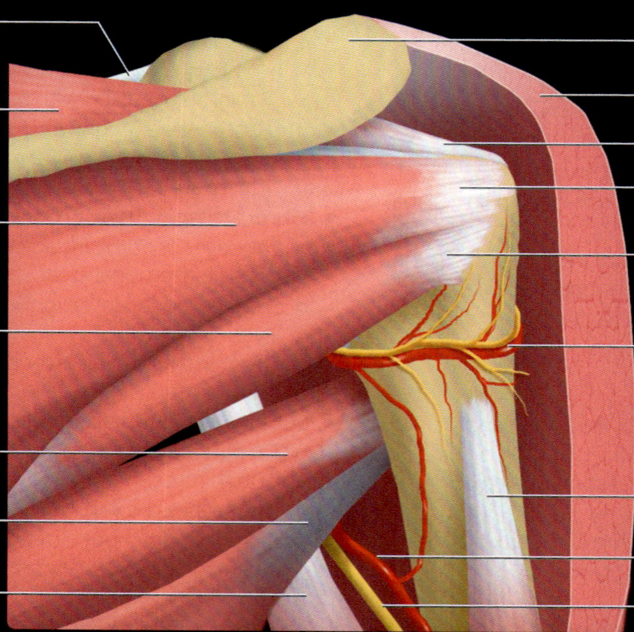

(Top) *Anterior graphic of the shoulder illustrates the rotator cuff and adjacent structures. The rotator cuff consists of supraspinatus, infraspinatus, teres minor, and subscapularis muscles and tendons. The biceps tendon courses the rotator cuff interval between the supraspinatus and subscapularis tendons, then descends along the bicipital groove, which is covered by the transverse ligament.* **(Bottom)** *Posterior graphic of the shoulder illustrates the rotator cuff and adjacent structures. The infraspinatus and teres muscles and tendons form the posterior wall of the rotator cuff. Inferior to the teres minor muscle and superior to the teres major is the axillary nerve and posterior circumflex humeral vessels running through the quadrilateral space.*

DEEP STRUCTURES

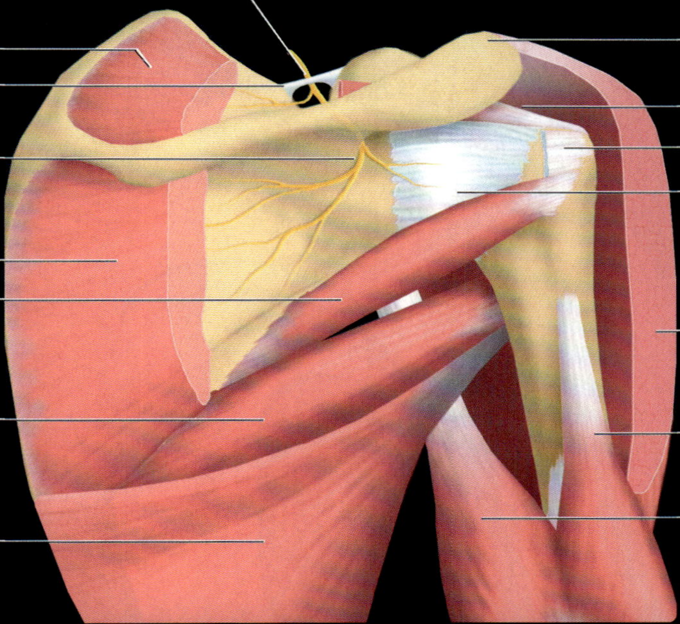

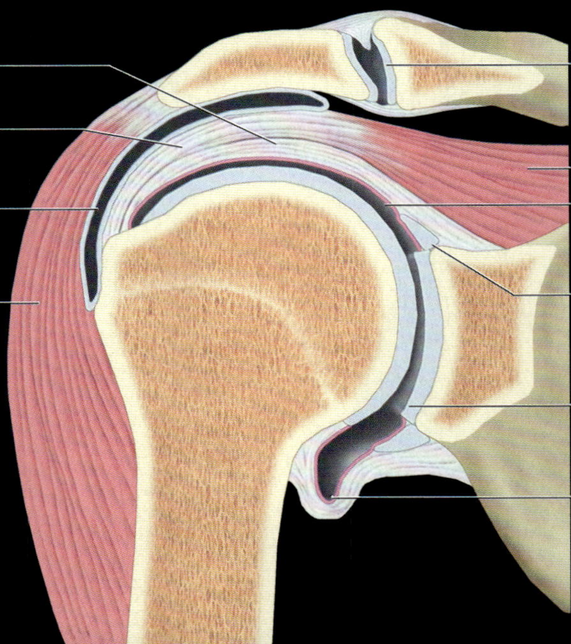

(Top) *Deep scapulohumeral dissection shows the course of the suprascapular nerve. The nerve enters the supraspinous fossa through the suprascapular notch, below the superior transverse scapular ligament. The nerve then passes beneath the supraspinatus and curves around the lateral border of the spine of the scapula to enter the infraspinous fossa.* **(Bottom)** *Graphic shows the coronal section through the midportion of the shoulder joint. Note the subacromial-subdeltoid bursa is situated in the space between the deltoid, acromion, acromioclavicular joint, distal clavicle superiorly, and the supraspinatus tendon and muscle inferiorly. There is no direct communication between the bursa and glenohumeral joint unless there is full-thickness tear of the supraspinatus tendon. The bursa is the main pain-producing structure around the shoulder. The wide expanse of the subacromial-subdeltoid bursa enables one to understand why shoulder symptoms are poorly localized. Patients often complain of generalized shoulder pain.*

SAGITTAL NORMAL LABRUM

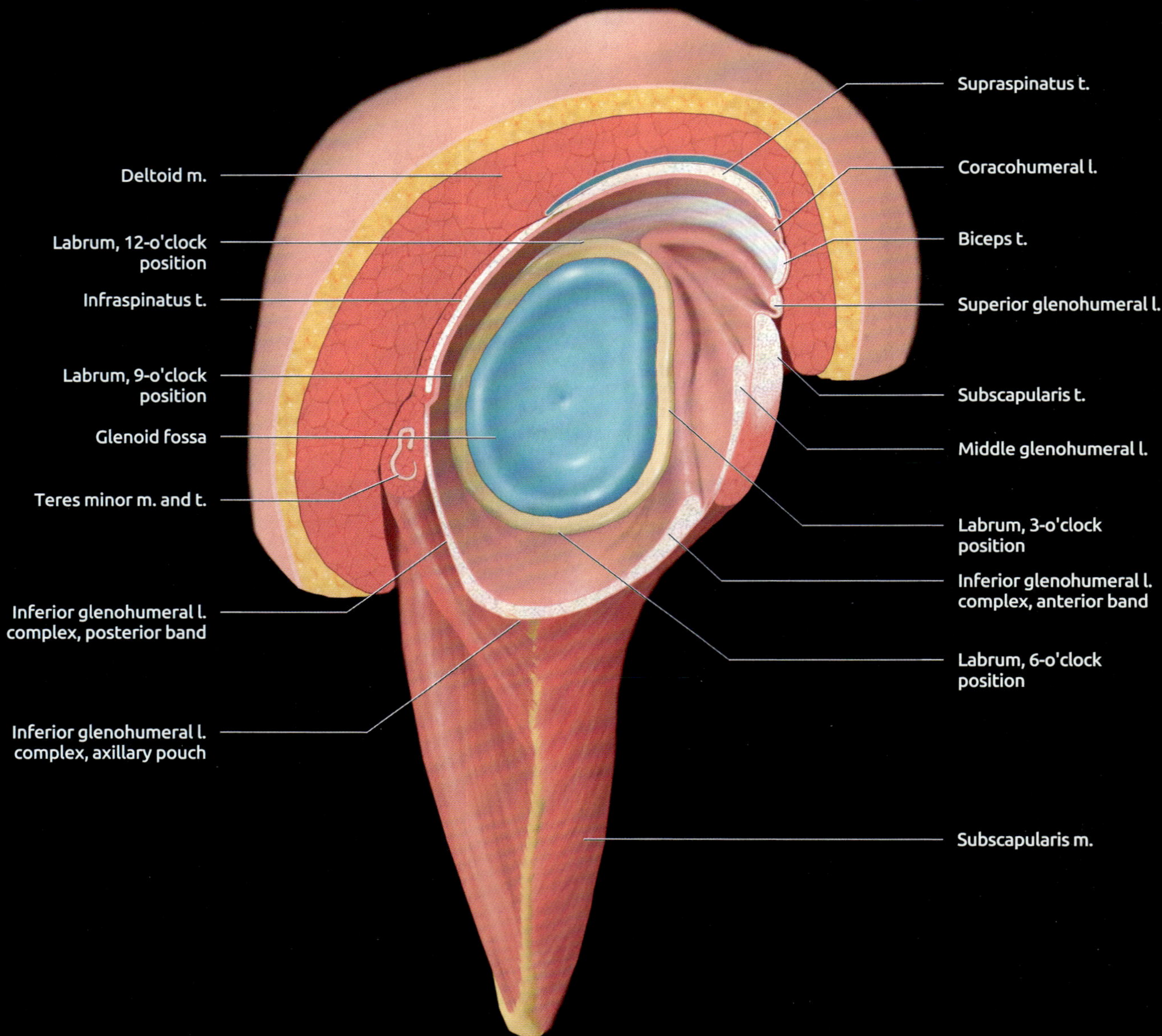

Sagittal graphic shows the glenoid fossa. The labrum lines the edge of the glenoid, increasing the circumference and depth of the shoulder joint. The labrum plays an important role in stabilization of the shoulder joint. At the inferior aspect of the glenoid, the labrum is barely discernible. The superior, middle, and inferior glenohumeral ligaments are distinctive thickenings of the anterior shoulder capsule and cannot be seen as separate structures on US examination.

LONGITUDINAL US, SUPRASPINATUS INSERTIONAL AREA

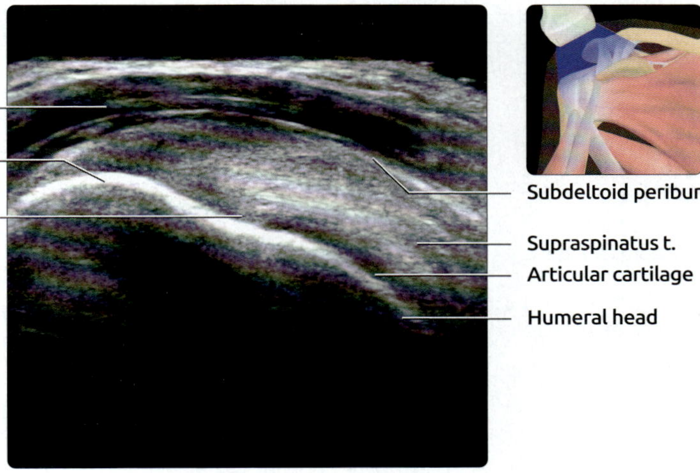

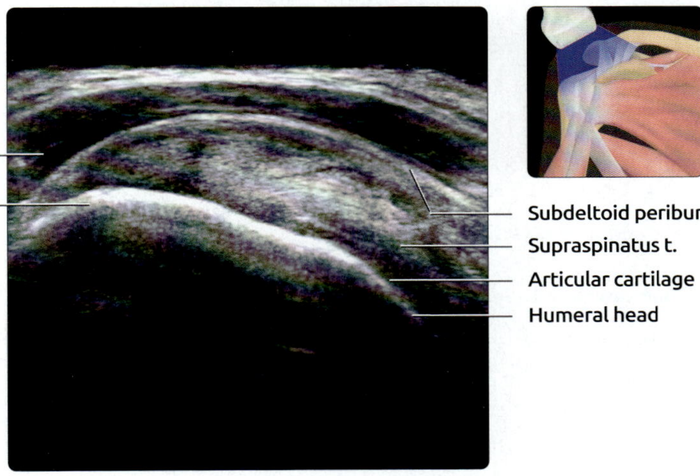

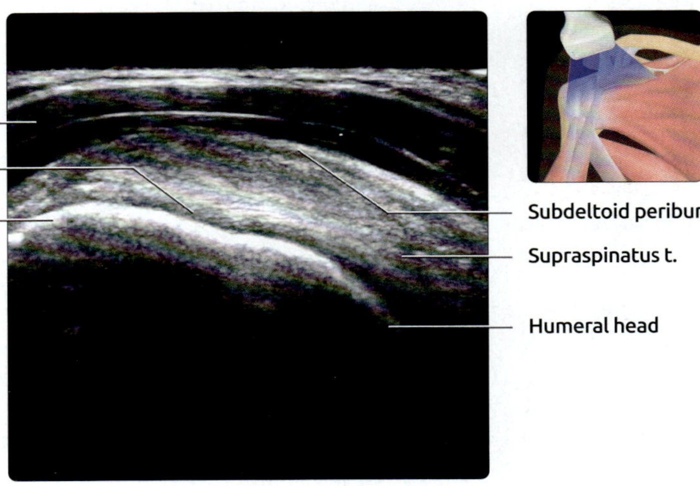

(Top) *Longitudinal grayscale US shows the anterior fibers of the supraspinatus tendon insertional area. The supraspinatus tendon inserts over a wide area (footprint) on the anterior aspect of the greater tuberosity. Many tears of the supraspinatus tendon involve avulsion of the tendon from its insertional site.* **(Middle)** *Longitudinal grayscale US shows the supraspinatus tendon midfibers. There is often a thin hypoechoic line at the insertional area. This represents Sharpey fibers and fibrocartilage.* **(Bottom)** *Longitudinal grayscale US shows the supraspinatus tendon posterior fibers. The anterior, middle, and posterior fibers should be evaluated in turn.*

TRANSVERSE US, SUPRASPINATUS INSERTIONAL AREA

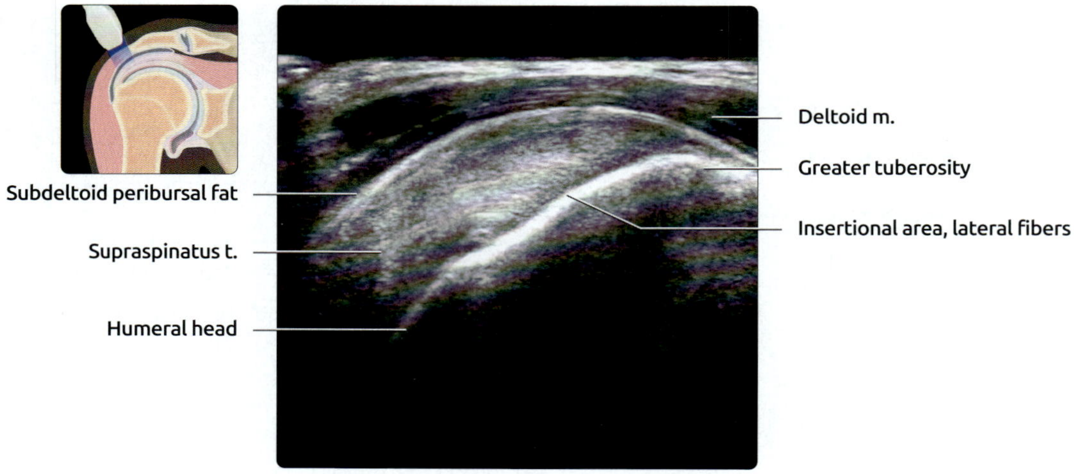

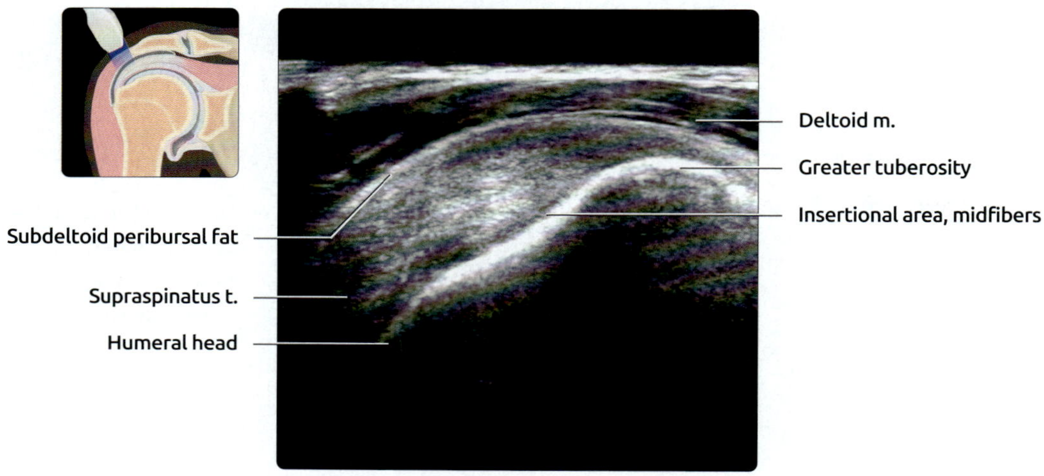

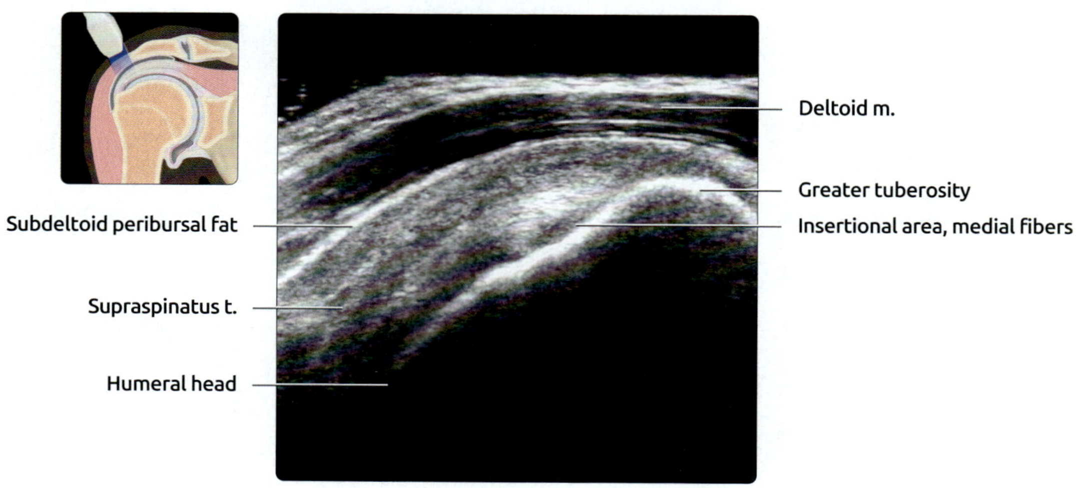

(Top) *Transverse grayscale US shows the supraspinatus tendon insertional area. Angulation and slight movement of the transducer will allow visualization of the lateral, middle, and medial fibers, respectively. The leading anterior edge of the supraspinatus is a common site of tear (rim rent tear).* **(Middle)** *Transverse grayscale US shows the supraspinatus tendon at the insertional area of midfibers. The fibrillar echotexture of the supraspinatus tendon can be seen but is prone to anisotropy.* **(Bottom)** *Transverse grayscale US shows the supraspinatus midfibers insertional area. Subdeltoid peribursal fat is variable in depth. The normal bursa cannot be depicted. It is seen only when distended with fluid or thickened due to synovitis.*

TRANSVERSE AND LONGITUDINAL US, SUPRASPINATUS

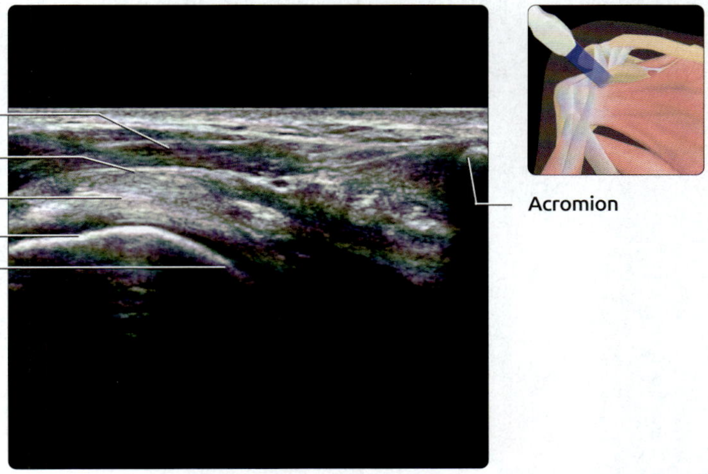

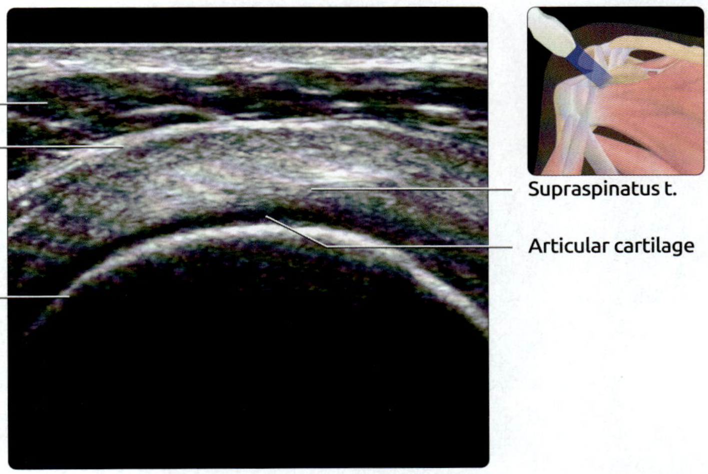

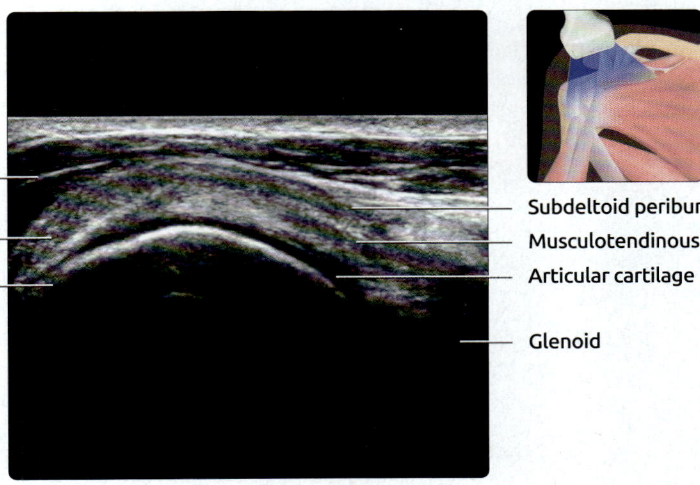

(Top) *Transverse grayscale US shows the supraspinatus tendon just medial to the insertional area. The critical zone is located just medial to the insertional area. This is a relatively hypovascular area. It is prone to calcific tendinosis and tears, though most tears tend to occur at the insertional site.* **(Middle)** *Transverse grayscale US shows the supraspinatus medial to the insertional area. The fibrillar pattern of the supraspinatus tendon is prone to anisotropy. With tendinosis, the fibrillar pattern is disrupted. The tendon becomes more hypoechoic and thickened.* **(Bottom)** *Longitudinal grayscale US shows the supraspinatus tendon at a slightly more medial aspect. Tears of the musculotendinous junction are relatively uncommon.*

GRAPHIC AND TRANSVERSE US, ROTATOR CUFF INTERVAL

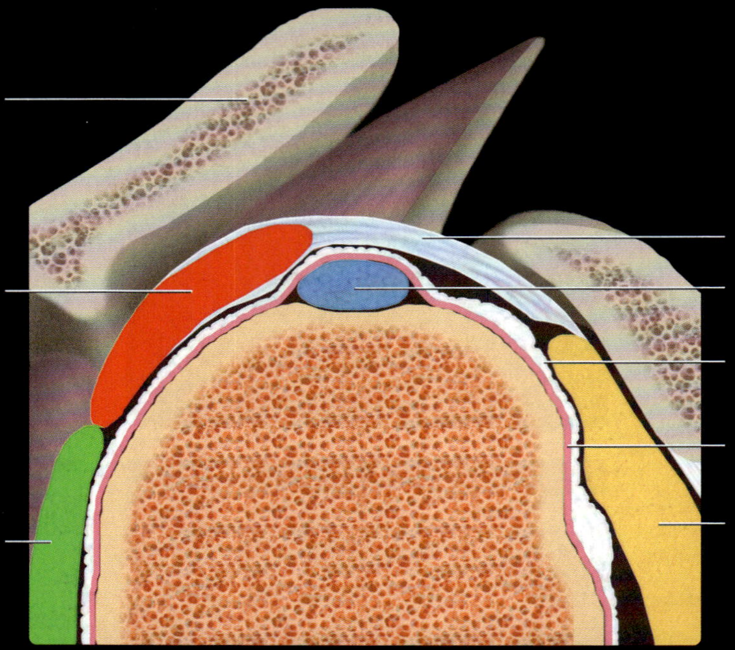

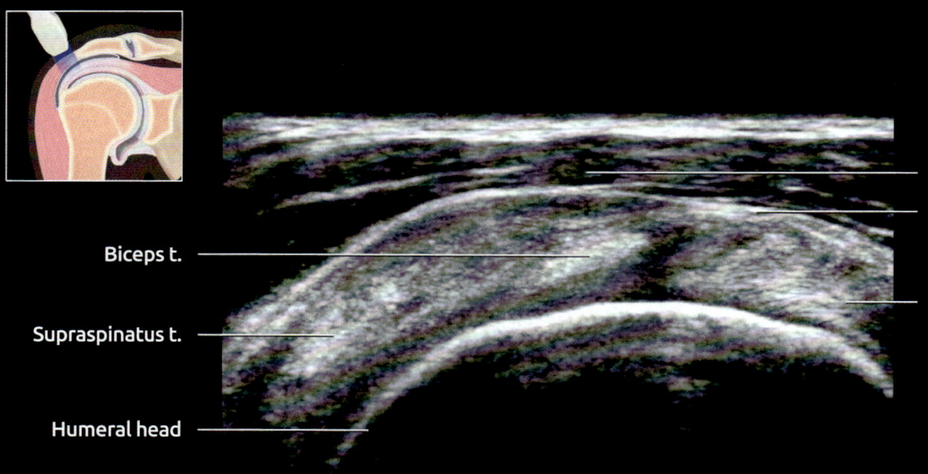

(Top) *Graphic shows the relationship of the coracohumeral ligament to the rotator cuff tendons. The coracohumeral ligament is not a true ligament but a folded portion of the glenohumeral capsule that extends from the coracoid process to the humerus. The undersurface is lined by synovium. Portions of the coracohumeral ligament pass superficial and deep to the supraspinatus tendon. The coracohumeral ligament attaches to the superior border of the subscapularis tendon as well as to the greater tuberosity.* **(Bottom)** *Transverse grayscale US shows the shoulder at the rotator cuff interval. The rotator cuff interval represents the space between the leading anterior edge of the supraspinatus tendon and the adjacent superior edge of the subscapularis. It contains the biceps tendon. This should not be mistaken for a tear of the anterior aspect of the supraspinatus tendon. Patients with adhesive capsulitis may show increased hypoechogenicity and hyperemia of the rotator cuff interval as a result of inflammatory fibrovascular soft tissue overgrowth.*

TRANSVERSE US, BICEPS TENDON

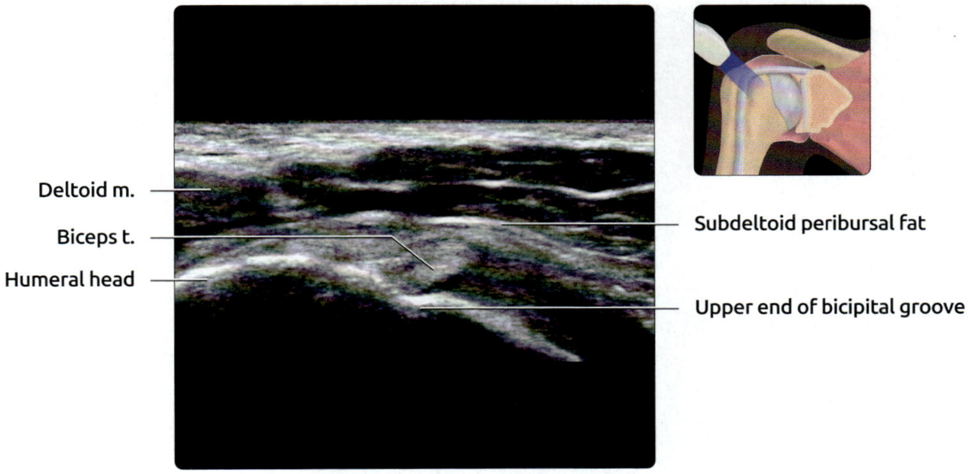

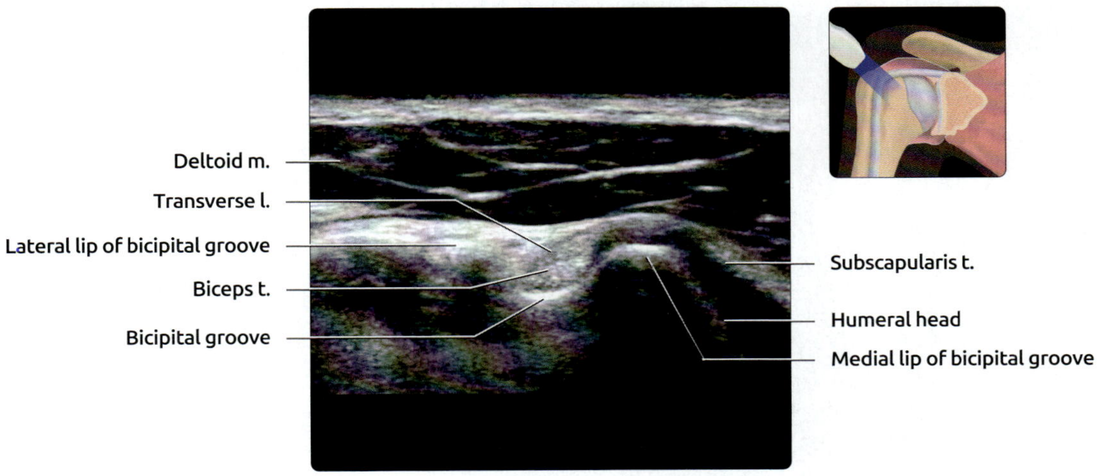

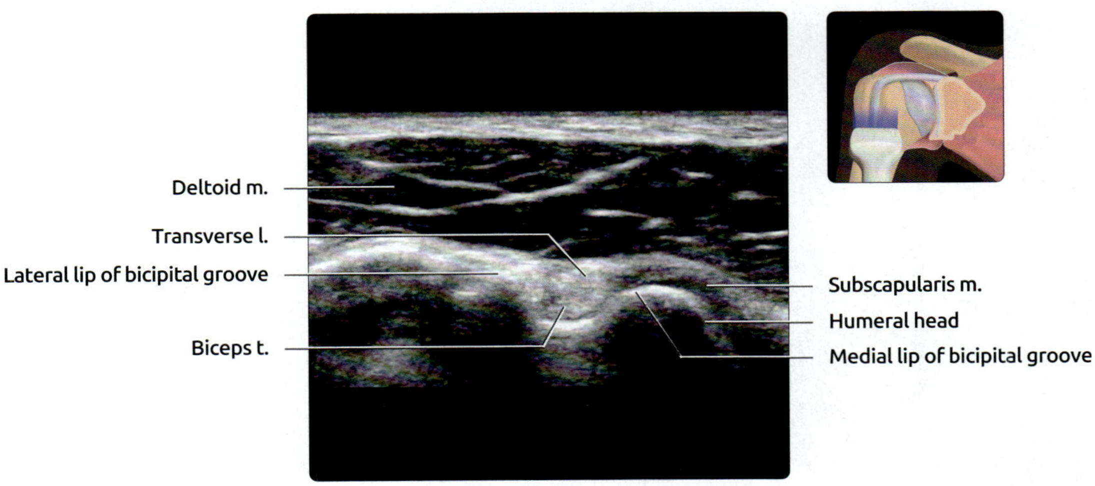

(Top) *Transverse grayscale US at the proximal aspect of the biceps tendon is shown. The bicipital groove forms a fibroosseous tunnel for the biceps tendon. The normal biceps tendon has an ovoid configuration. With tendinosis, the tendon becomes larger and more rounded in appearance.* **(Middle)** *Transverse grayscale US shows the biceps tendon. The biceps tendon is held in position by the transverse ligament. The subscapularis tendon inserts into the lesser tuberosity, which forms the medial lip of the bicipital groove. The biceps tendon may sublux medially from the groove and may be associated with subscapularis tendon injury.* **(Bottom)** *Transverse grayscale US shows relations of the biceps tendon. The transverse ligament is not a distinct entity but consists of a fibrous expansion of both the pectoralis major tendon and subscapularis tendon inserting into the lateral lip of the bicipital groove.*

LONGITUDINAL US, BICEPS TENDON

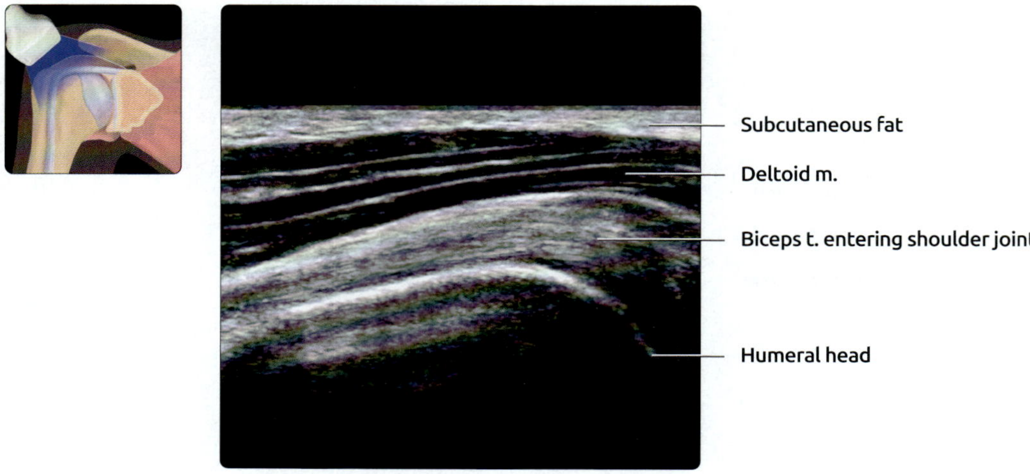

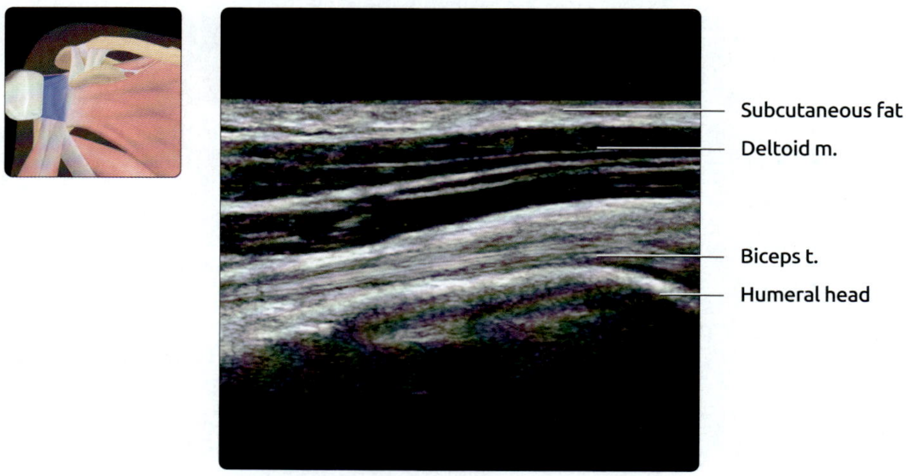

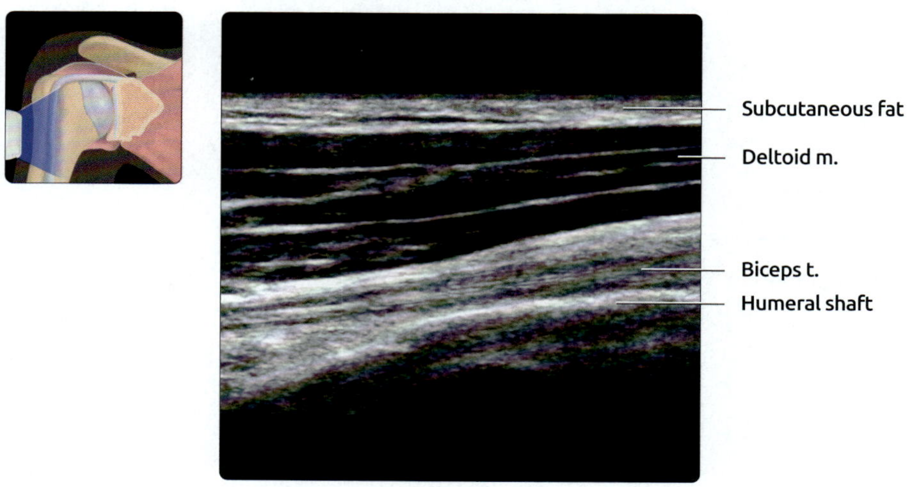

(Top) *Longitudinal grayscale US shows the uppermost section of the biceps tendon. The intraarticular portion of the biceps long head tendon cannot be fully depicted on longitudinal imaging. On longitudinal section, the biceps tendon seems to expand at its upper end as it becomes more ovoid in contour. This location is also the most common site of biceps tendinosis, which also manifests as enlargement of the biceps tendon.* **(Middle)** *Longitudinal grayscale US shows the upper section of the biceps tendon. A small amount of fluid in the biceps tendon sheath, which is continuous with the glenohumeral joint, is normal and should not be mistaken for tenosynovitis. No fluid is depicted in this image.* **(Bottom)** *Longitudinal grayscale US shows the lower aspect of the biceps tendon. Most incomplete tears of the biceps tendons are longitudinal in orientation and are best depicted on transverse imaging.*

LONGITUDINAL US, SUBSCAPULARIS

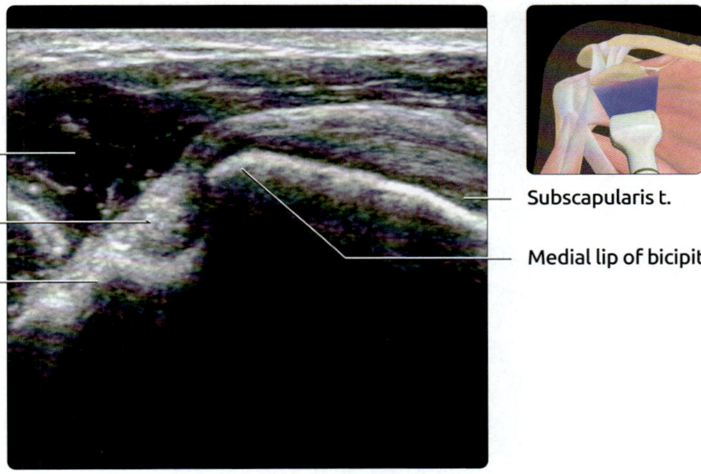

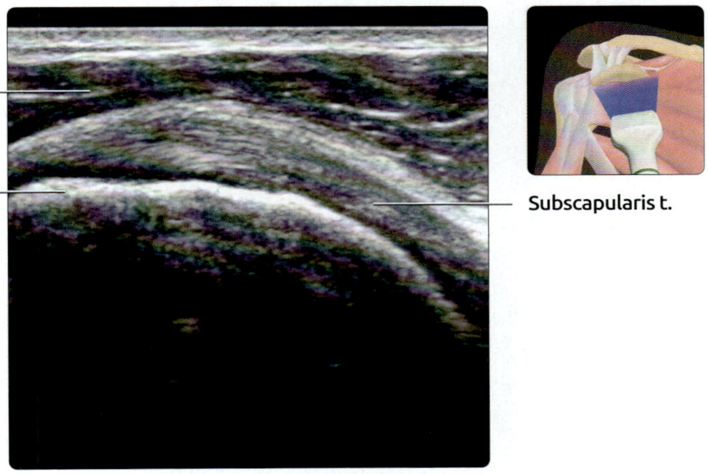

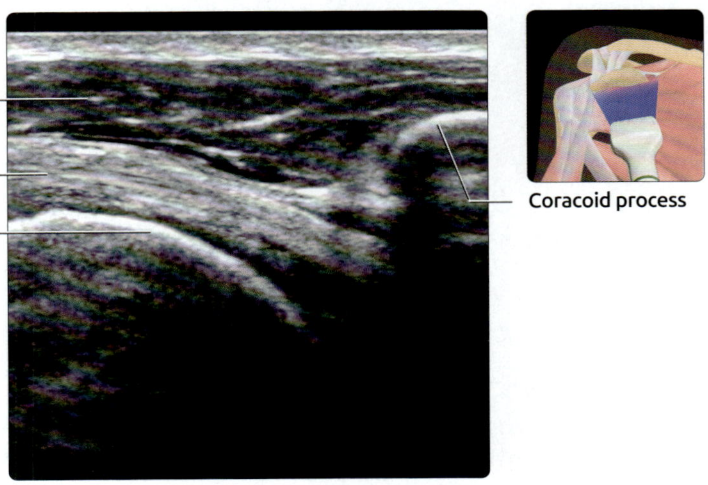

(Top) *Longitudinal grayscale US shows the subscapularis insertion. The subscapularis inserts into the medial lip of the bicipital groove but has a fibrous expansion traversing the biceps tendon through which it also gains attachment to the lateral lip of the bicipital groove.* **(Middle)** *Longitudinal grayscale US shows the subscapularis insertional area. Complete tears of the subscapularis are uncommon and usually follow a severe traumatic event. Partial tears are more common and usually involve the superior edge of the tendon.* **(Bottom)** *Longitudinal grayscale US shows the subscapularis tendon. The subscapularis moves beneath the coracoid process during internal-external rotation. Impingement may potentially occur at this location (subcoracoid impingement).*

TRANSVERSE US, SUBSCAPULARIS TENDON

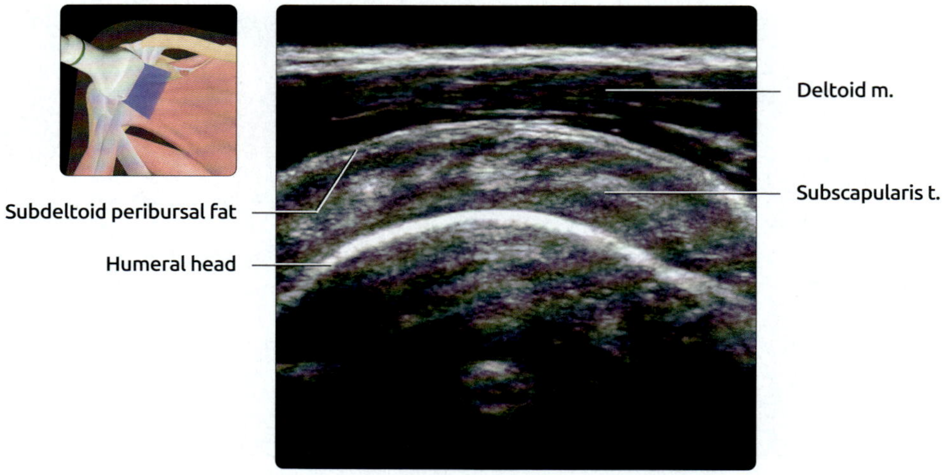

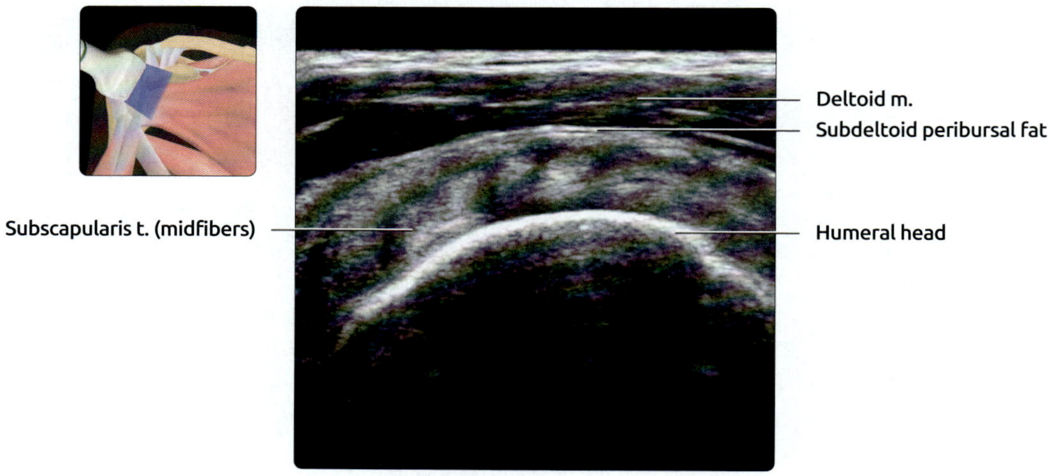

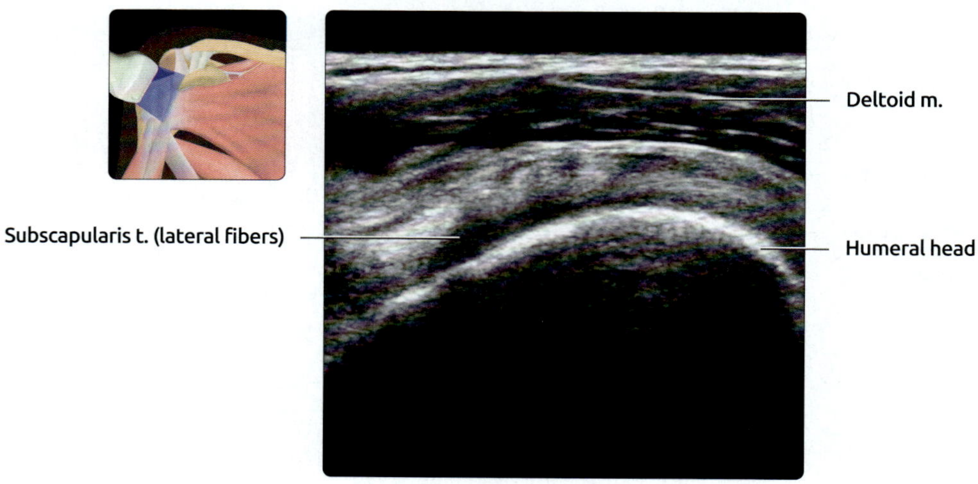

(Top) *Transverse grayscale US shows the subscapularis tendon midfibers. As they converge toward the insertion, fiber bundles of the multipennate subscapularis tendon give the tendon a mixed echogenic appearance. This is normal and should not be mistaken for tendinosis.* **(Middle)** *Transverse grayscale US shows the subscapularis at the level of midfibers.* **(Bottom)** *Transverse grayscale US shows the subscapularis tendon. Tears of the subscapularis tendon usually occur just proximal to the insertional area. These tears may involve the fascial covering of the biceps tendon, facilitating biceps tendon dislocation.*

LONGITUDINAL AND TRANSVERSE US, INFRASPINATUS

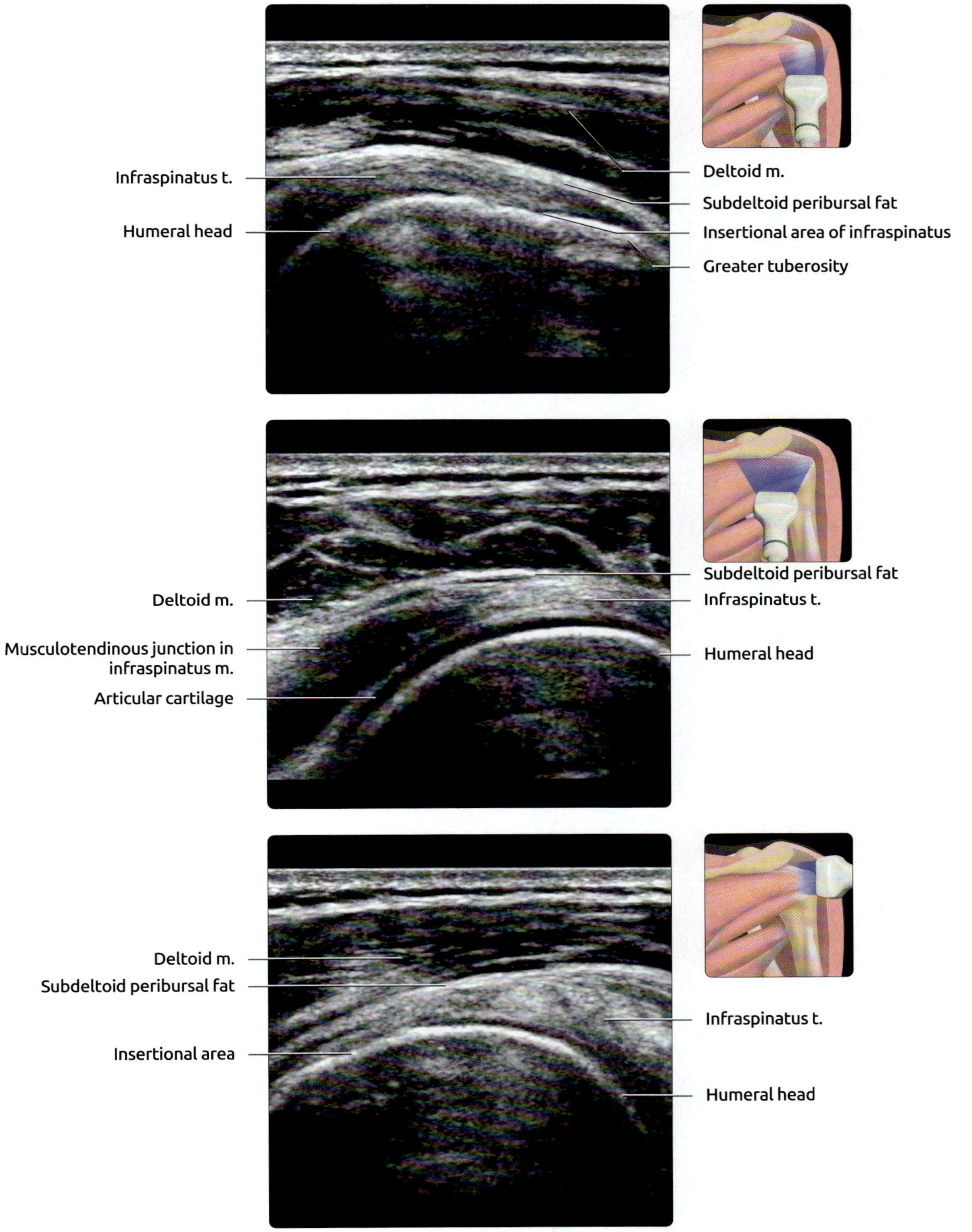

(Top) *Longitudinal grayscale US shows the infraspinatus tendon insertional area. The infraspinatus tendon is less commonly torn than the supraspinatus tendon. Most tears are avulsive-type tears involving the insertional area.* **(Middle)** *Longitudinal grayscale US shows the infraspinatus musculotendinous junction. The muscle fibers interdigitate with the tendon at the musculotendinous junction and should not be mistaken for tears/tendinosis.* **(Bottom)** *Transverse grayscale US shows the infraspinatus insertional area. All tears of the rotator cuff tendons should be confirmed in both planes (transverse and longitudinal).*

LONGITUDINAL AND TRANSVERSE US, TERES MINOR

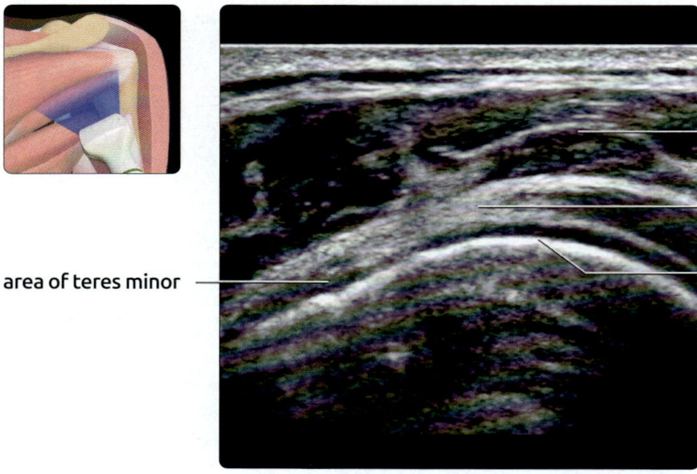

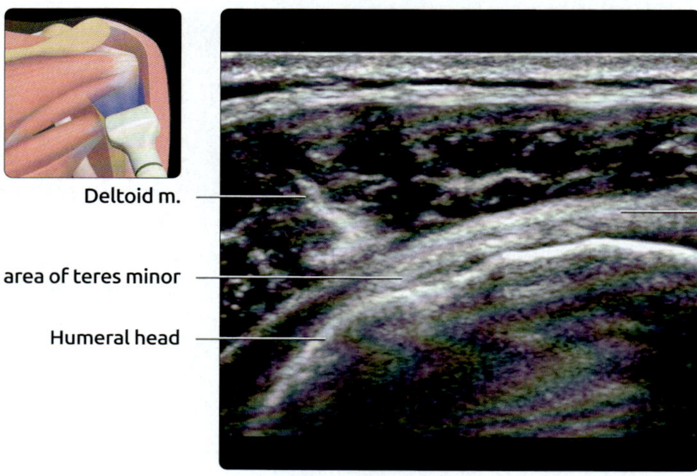

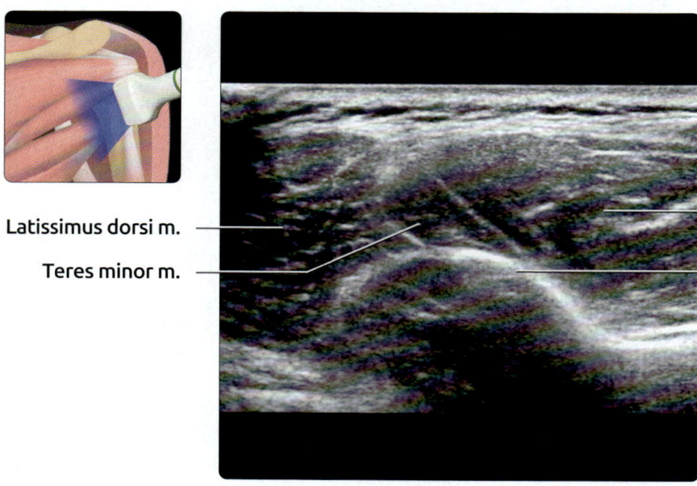

(Top) *Longitudinal grayscale US shows the teres minor muscle insertional area. The teres minor muscle is usually not torn in isolation, but it may be torn in massive rotator cuff tears. It is a small muscle and tendon seen at the posteroinferior edge of the infraspinatus tendon.* **(Middle)** *Longitudinal grayscale US shows the teres minor insertional area.* **(Bottom)** *Transverse grayscale US shows the teres minor. The teres minor muscle is best depicted on longitudinal imaging at the inferior aspect of the infraspinatus tendon. Isolated atrophy of the teres minor muscle can occur in quadrilateral space syndrome due to compression of the axillary nerve (part of the deltoid muscle may also be affected).*

SHOULDER JOINT AND POSTERIOR ASPECT

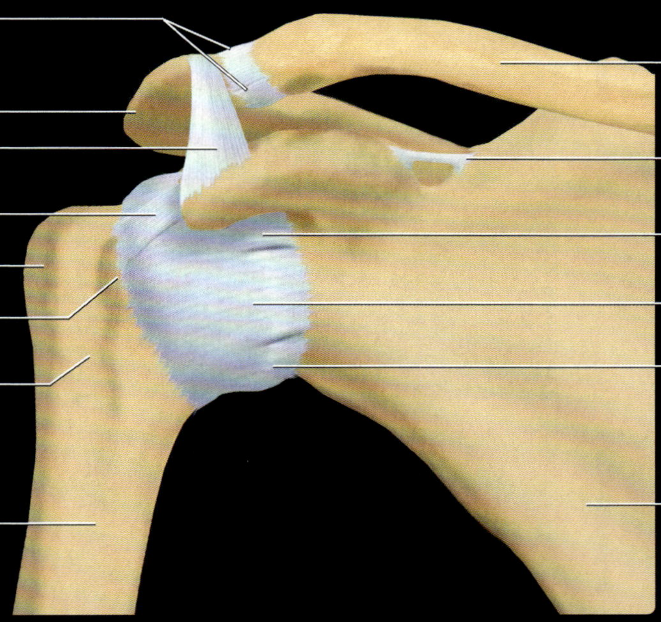

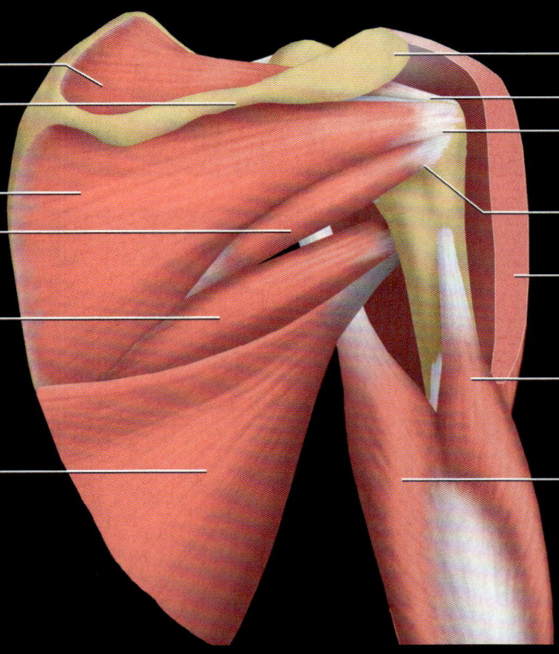

(Top) *Anterior graphic illustrates the right shoulder in deep dissection. The muscles have been removed.* **(Bottom)** *Posterior graphic illustrates the shoulder. Superficial scapulohumeral dissection demonstrates the musculature.*

LONGITUDINAL US, POSTERIOR ASPECT

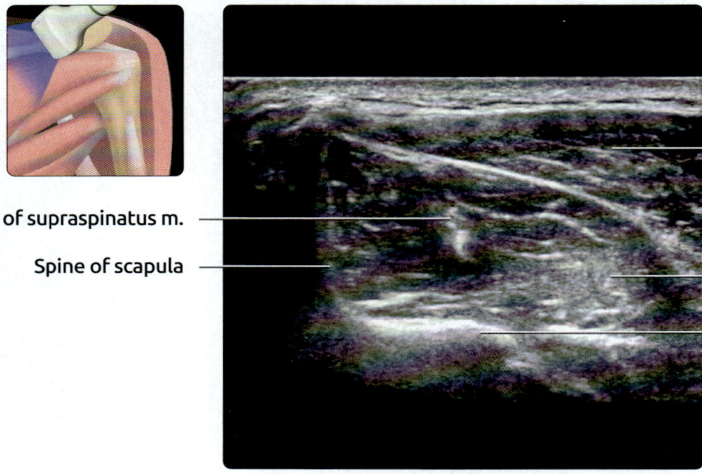

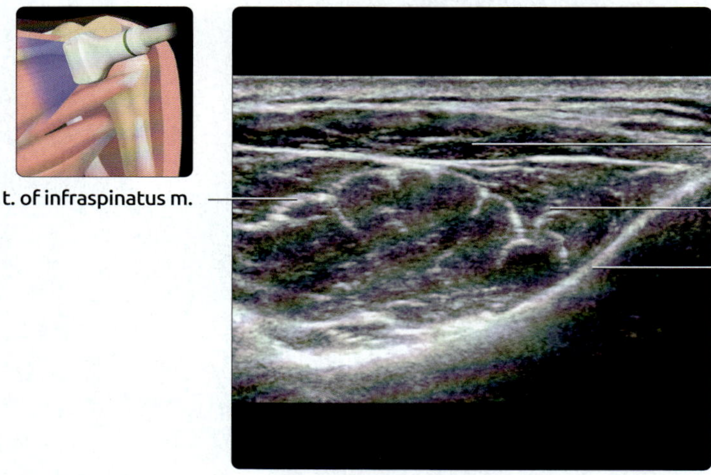

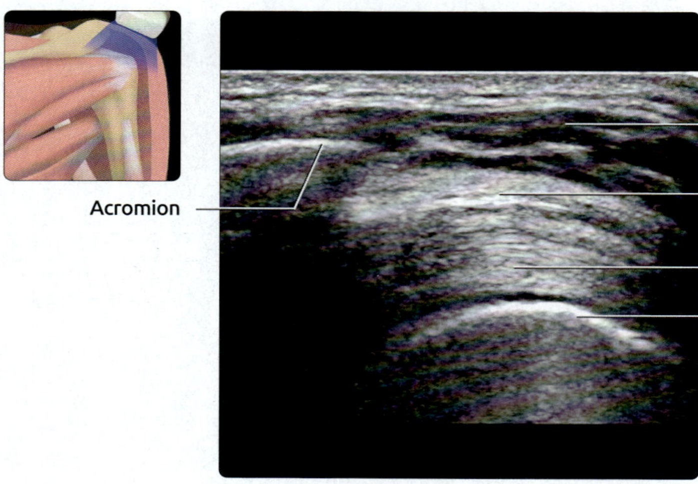

(Top) *Longitudinal grayscale US shows the posterior shoulder at the supraspinatus muscle. Muscle atrophy is common in rotator cuff pathology. US is almost as sensitive as MR in depiction of muscle atrophy. Atrophy is seen as a reduction in muscle bulk with increase in muscle echogenicity. As a result, the central tendon is less readily seen.* **(Middle)** *Longitudinal grayscale US shows the shoulder on posterior view. Infraspinatus muscle atrophy usually accompanies supraspinatus muscle atrophy to a lesser or greater degree. Similar to the supraspinatus muscle, muscle bulk is reduced and becomes more echogenic with the central tendon becoming less easy to see.* **(Bottom)** *Longitudinal grayscale US shows the supraspinatus tendon. The supraspinatus tendon slides below the acromion during abduction. This relationship may be seen from either the anterior or posterior aspect of the shoulder. Dynamic imaging may reveal impingement of the supraspinatus against the acromion during abduction.*

LONGITUDINAL AND TRANSVERSE US, POSTERIOR GLENOID LABRUM AND GLENOHUMERAL JOINT

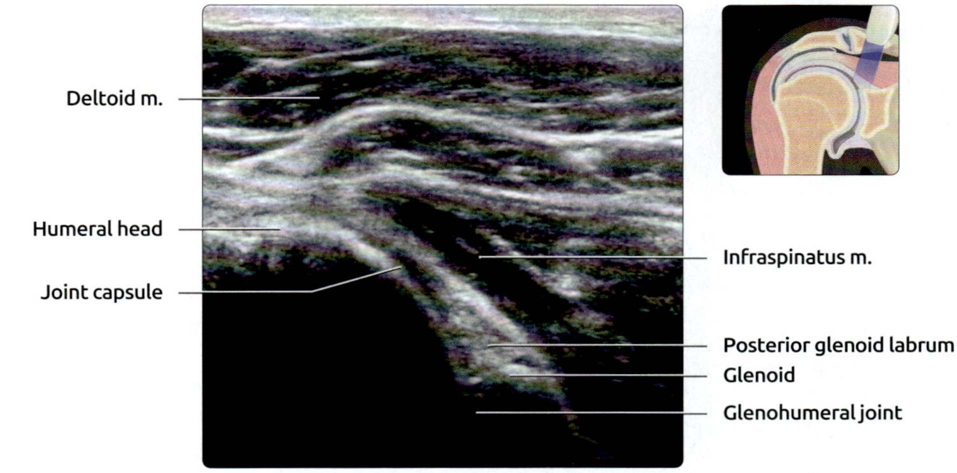

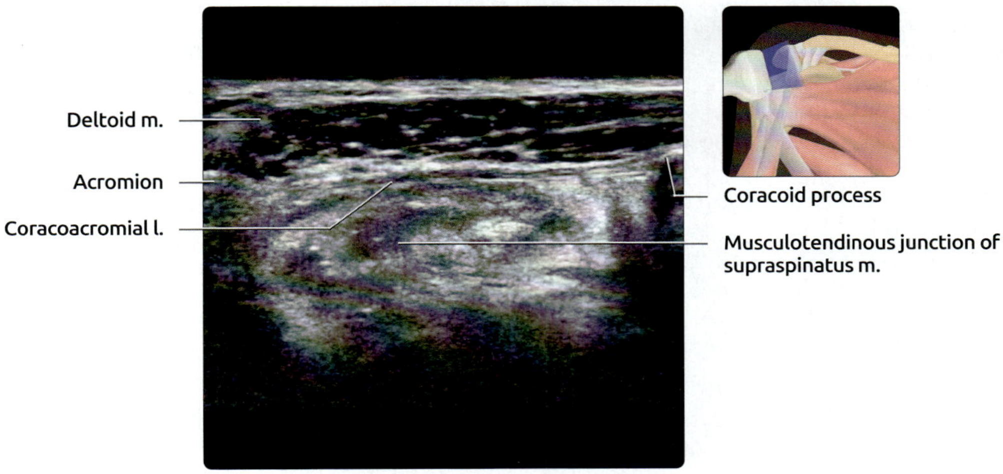

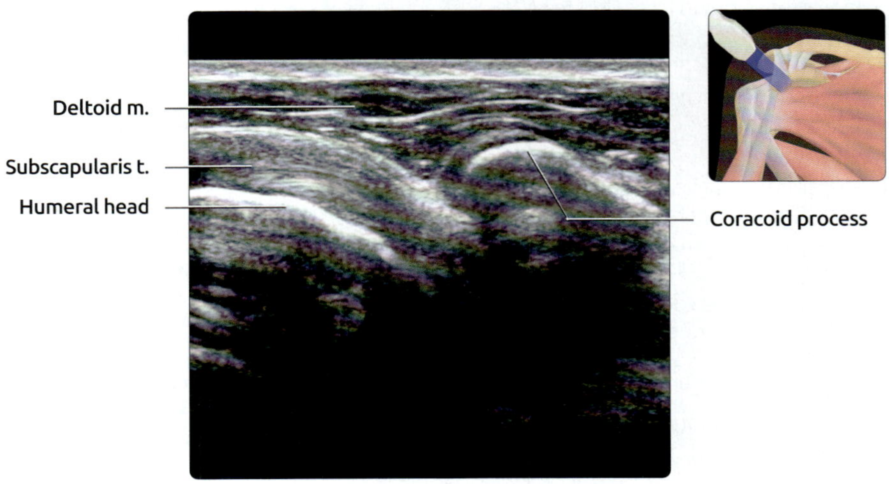

(Top) *Longitudinal grayscale US shows the posterior shoulder. The glenohumeral joint is most easily seen from the posterior aspect. This is a good site for US-guided joint injection. Joint effusions are also best seen in this area.* **(Middle)** *Longitudinal grayscale US shows the coracoacromial ligament. The coracoacromial ligament extends from the coracoid process to the anteroinferior margin of the acromion. Impingement of the supraspinatus tendon and the overlying subacromial-subdeltoid bursa may occur during shoulder abduction.* **(Bottom)** *Transverse grayscale US shows the coracoid process. The coracoid process is close to the humeral head and intervening subscapularis. Occasionally, the coracohumeral distance is reduced and subcoracoid impingement may potentially occur.*

TERMINOLOGY

Definitions

- Fat-filled space between upper arm and lateral chest wall

IMAGING ANATOMY

Extent

- Axilla is shaped like a pyramid
- Consists of apex, base, and 4 walls
    - Apex
        - Bounded by scapula, 1st rib, and mid 1/3 of clavicle
        - Passageway between arm and posterior triangle of neck
    - Anterior wall
        - Pectoralis major and minor muscles
    - Posterior wall
        - Subscapularis, teres major, and latissimus dorsi muscles
    - Medial wall
        - Serratus anterior, upper 4 ribs, and intercostal spaces
    - Lateral wall
        - Medial and lateral lips of bicipital groove into which anterior and posterior walls are inserted, coracobrachialis and short head biceps muscles
    - Base
        - Axillary fascia, subcutaneous fat, and skin
- Contents of axilla
    - Axillary artery and vein
        - 90% and 10% of subjects have accessory axillary vein and artery, respectively
    - Cords and terminal branches of brachial plexus
    - Lymph nodes and vessels
    - Fat
- Axillary artery
    - Continuation of subclavian artery
        - Termed axillary artery beyond lateral edge of 1st rib
        - Termed brachial artery beyond lower edge of teres major muscle
    - Surrounded by cords and branches of brachial plexus
    - Vein lies medial to artery
    - Arterial branches
        - Thoracodorsal, lateral thoracic, subscapular, anterior and posterior circumflex humeral arteries
            - Lateral thoracic artery seen deep to pectoralis muscles of anterior wall axilla
            - Thoracodorsal artery seen deep to latissimus dorsi and teres major muscles of posterior wall axilla
- Axillary vein
    - Continuation of brachial vein
        - Tributaries correspond to branches of axillary artery and also cephalic vein
- Brachial plexus
    - Cords and terminal branches pass through axilla
    - Lateral cord gives rise to
        - Median nerve
        - Musculocutaneous nerve: Pierces coracobrachialis to descend arm between biceps and brachialis muscles
        - Lateral pectoral nerve
    - Medial cord gives rise to
        - Ulnar nerve
        - Contributes to median nerve
        - Medial pectoral nerve
        - Medial cutaneous nerves of arm and forearm
    - Posterior cord gives rise to
        - Radial nerve
        - Axillary nerve → quadrilateral space with posterior circumflex humeral artery
        - Subscapular nerve
        - Thoracodorsal nerve
- Lymph nodes
    - 20-50 axillary lymph nodes
    - Clustered into 5 groups: Mnemonic "SCALP"
        - **S**ubscapular (posterior) group in front of subscapularis muscle
        - **C**entral group in center of axilla
        - **A**pical group at lateral border of 1st rib
        - **L**ateral group at medial side of axillary vein
        - **P**ectoral (anterior) group along lower border of pectoralis minor muscle

Anatomy Relationships

- **Quadrilateral space**
    - Superior border: Teres minor muscle
    - Inferior border: Teres major muscle
    - Lateral border: Surgical neck of humerus
    - Medial border: Long head of triceps muscle
    - Contents: Axillary nerve and posterior circumflex humeral artery
        - Axillary nerve supplies teres minor muscle, deltoid muscle, posterolateral cutaneous region of upper arm
- **Triangular space**
    - Located medial to quadrilateral space
    - Superior border: Teres minor muscle
    - Inferior border: Teres major muscle
    - Lateral border: Long head of triceps muscle
    - Contents: Circumflex scapular artery
        - Branch of subscapular artery supplying infraspinatus fossa

ANATOMY IMAGING ISSUES

Key Concepts

- Examine axilla in oblique supine position with arm abducted ("bathing beauty" position)
- Lymph nodes typically follow vessels (axillary, lateral thoracic, and thoracodorsal arteries)
    - Also frequently occur in isolation in axillary fat and down toward axillary tail of breast
- Quadrilateral space syndrome = neurovascular compression syndrome due to compression of axillary nerve or posterior humeral circumflex artery in quadrilateral space
    - Neurogenic, arterial or combined symptoms
        - Poorly localized shoulder pain ± paresthesia radiating to lateral arm
            - Aggravated by abduction and external rotation
    - Teres minor ± deltoid atrophy
    - Intermittent, ischemic-type pain
    - Due to trauma, including shoulder dislocation, paralabral cyst, muscle hypertrophy, or prolonged arm abduction during sleep

MUSCLES

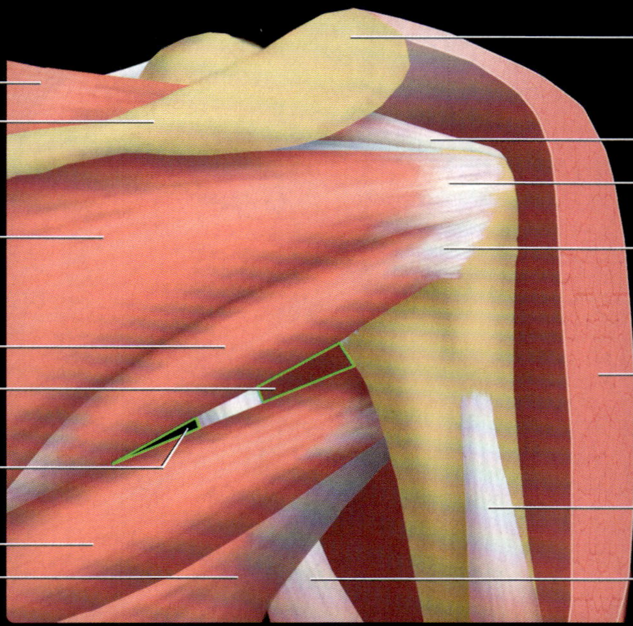

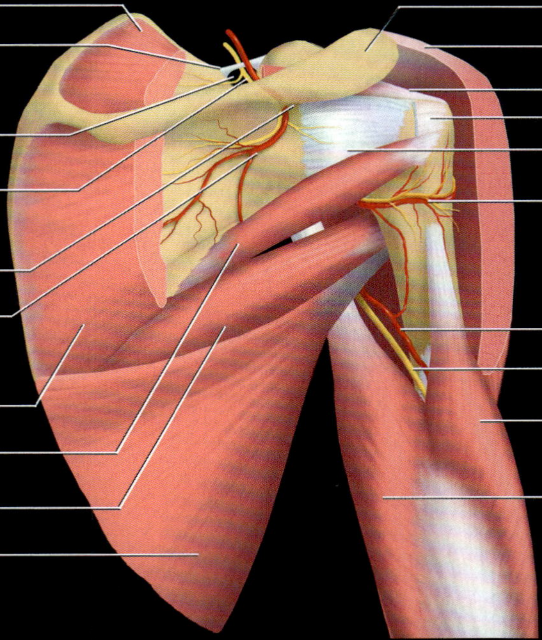

(Top) *Posterior graphic shows the shoulder. Superficial scapulohumeral dissection shows the location of the quadrilateral space and triangular space (each outlined in green). The quadrilateral space transmits the axillary nerve and the posterior circumflex humeral artery, while the much less important triangular space transmits the scapular circumflex vessels.* **(Bottom)** *Graphic of the deep scapulohumeral dissection shows the major neurovascular structures, including those in the quadrilateral space.*

DEEPER STRUCTURES

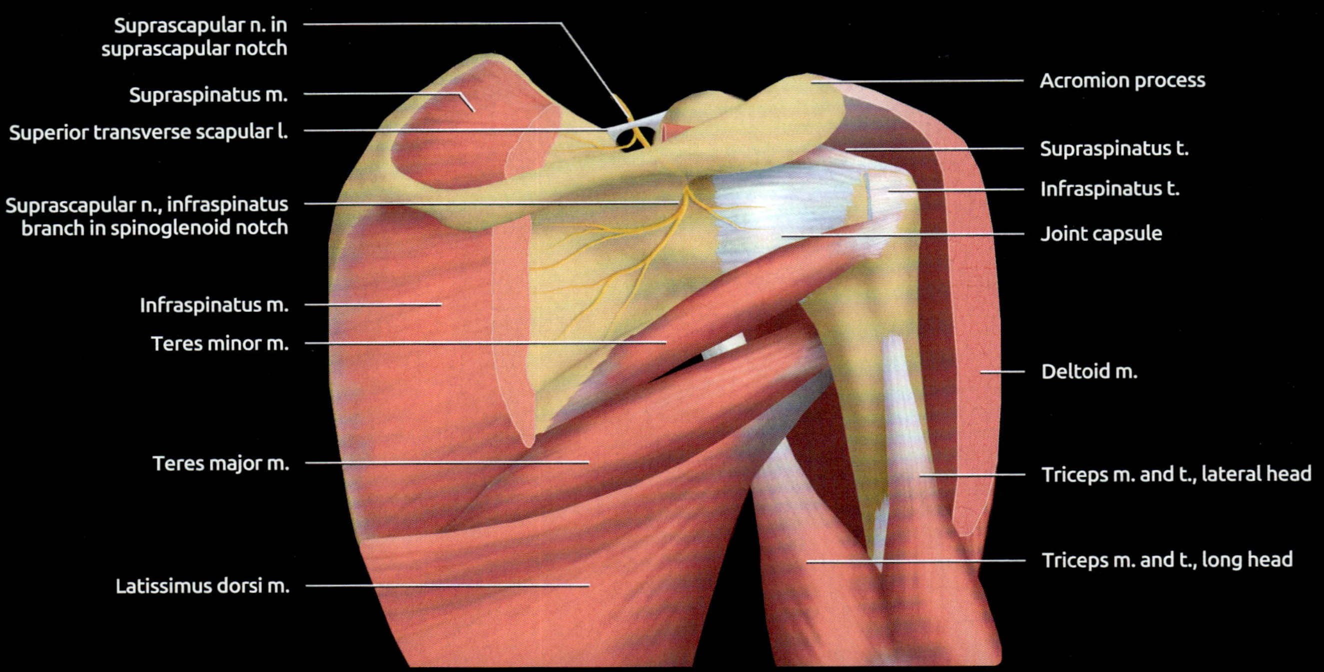

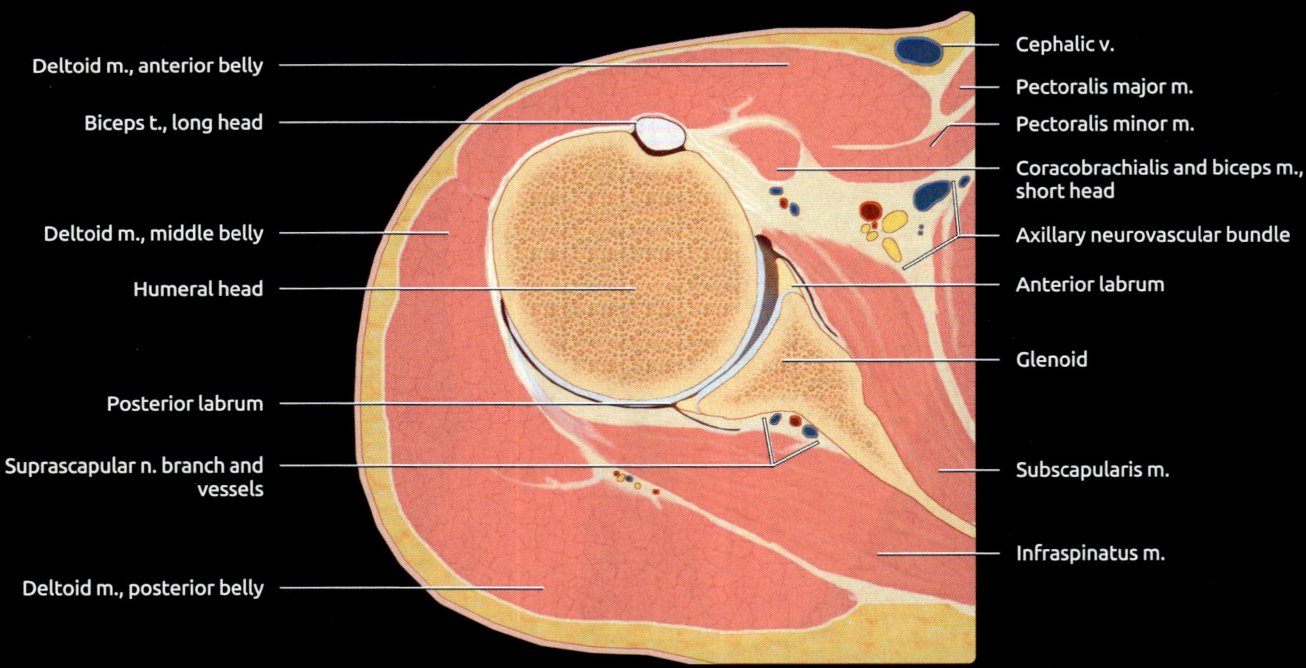

(Top) *Graphic of deep scapulohumeral dissection shows the course of the suprascapular nerve.* **(Bottom)** *Axial graphic shows the location of the suprascapular artery, nerve, and vein branches just below the level of the spinoglenoid notch. The suprascapular nerve is a mixed motor and sensory nerve. It arises from the anterior rami of the C5 and C6 roots. It passes deep to the superior transverse scapular ligament of the suprascapular notch. It then passes deep to the inferior transverse scapular ligament of the spinoglenoid notch. It supplies the supraspinatus and the infraspinatus muscles with sensory fibers to the acromioclavicular joint and the glenohumeral joint capsules.*

RELATIONSHIP OF NEUROVASCULAR STRUCTURES

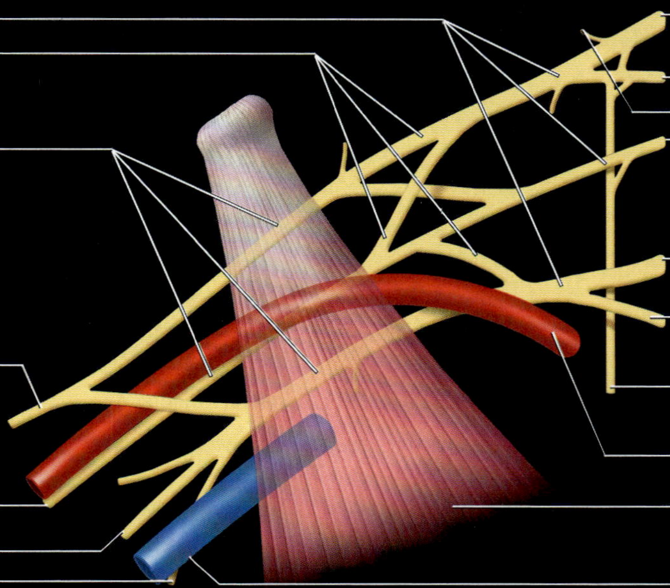

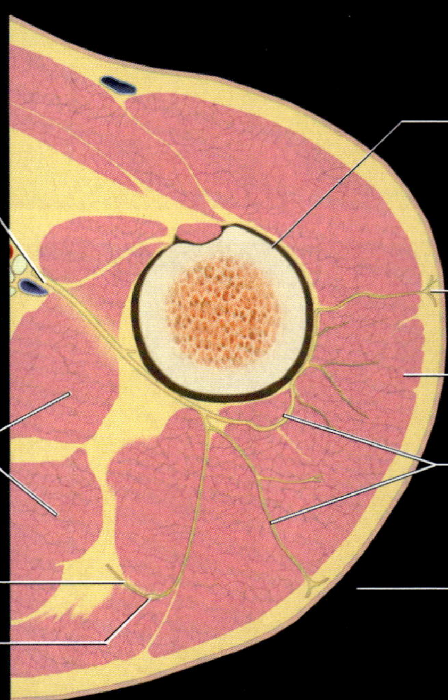

(Top) *Graphic shows relations of the brachial plexus in the axilla. The brachial plexus arises from the anterior rami of the C5, C6, C7, C8, and T1 nerves. They first unite to form the superior, middle, and inferior trunks. The trunks then divide and reunite to form the lateral, posterior, and middle cords. Beyond the lateral margin of the pectoralis minor muscle, they continue as the terminal branches of the plexus (axillary, musculocutaneous, radial, medial, and ulnar nerves).* **(Bottom)** *Graphic shows a section through the upper arm with the axillary nerve and its branches.*

TRANSVERSE US, AXILLA

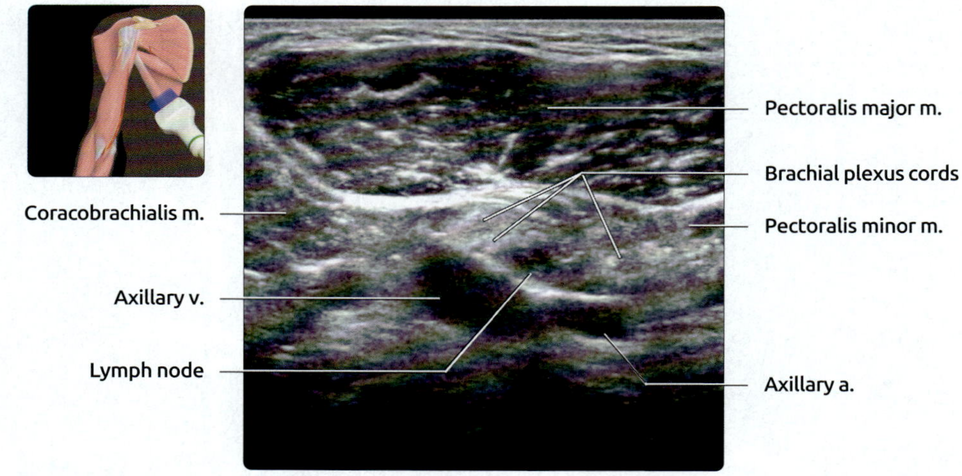

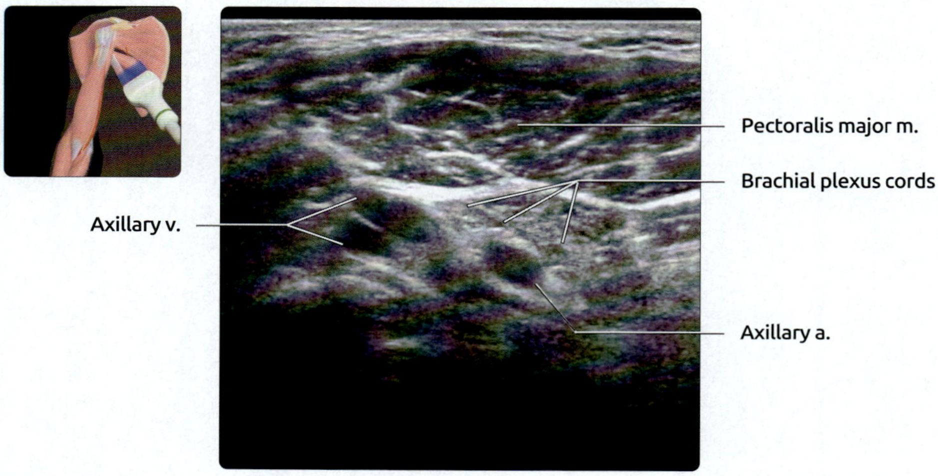

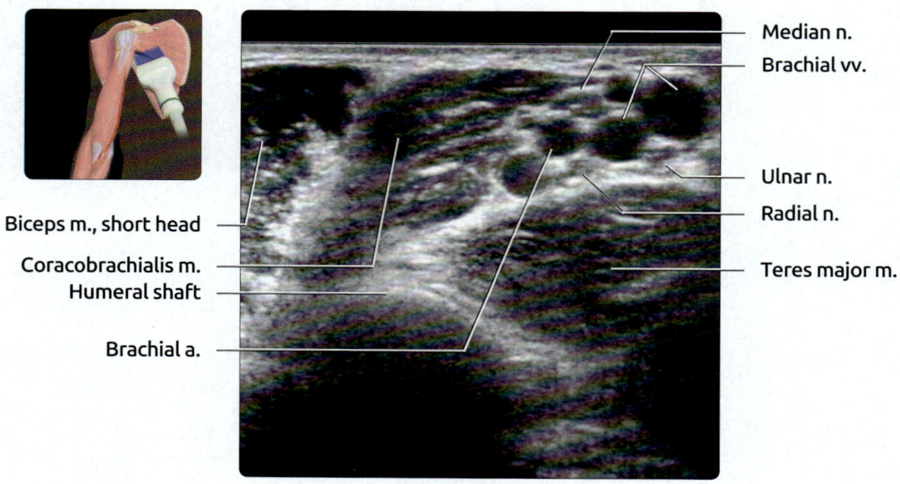

(Top) *Transverse grayscale US shows the proximal axilla with the arm abducted. The contents of the axilla are inspected with the arm fully abducted.* **(Middle)** *Transverse grayscale US shows the midportion of the axilla. The axillary artery and vein pass through the axilla surrounded by the cords of the brachial plexus. The cords (medial, lateral, and posterior) are named relative to their position alongside the axillary artery. Lymph nodes with variably fatty hila are commonly seen in the axilla around the neurovascular bundle.* **(Bottom)** *Transverse grayscale US shows the distal aspect of the axilla.*

TRANSVERSE US, ANTERIOR AXILLARY WALL

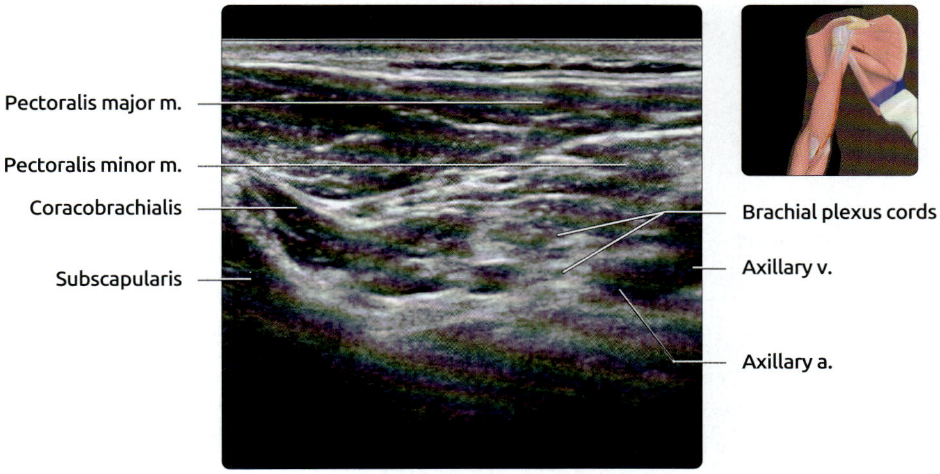

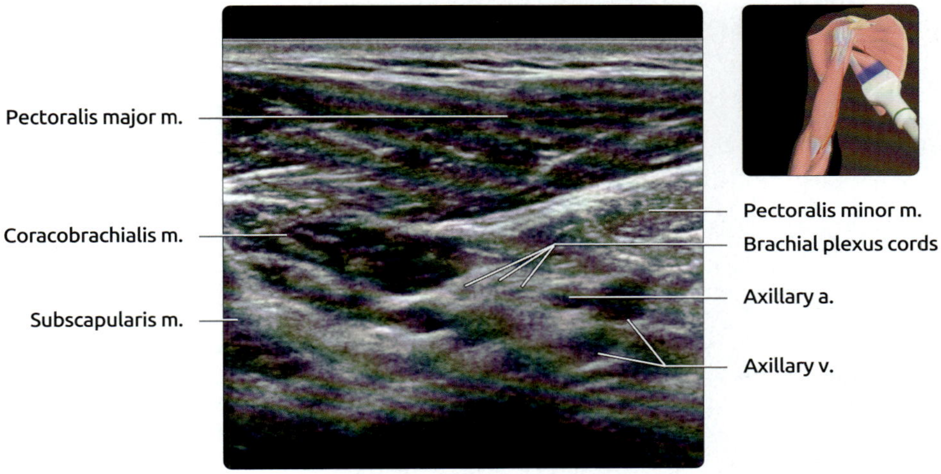

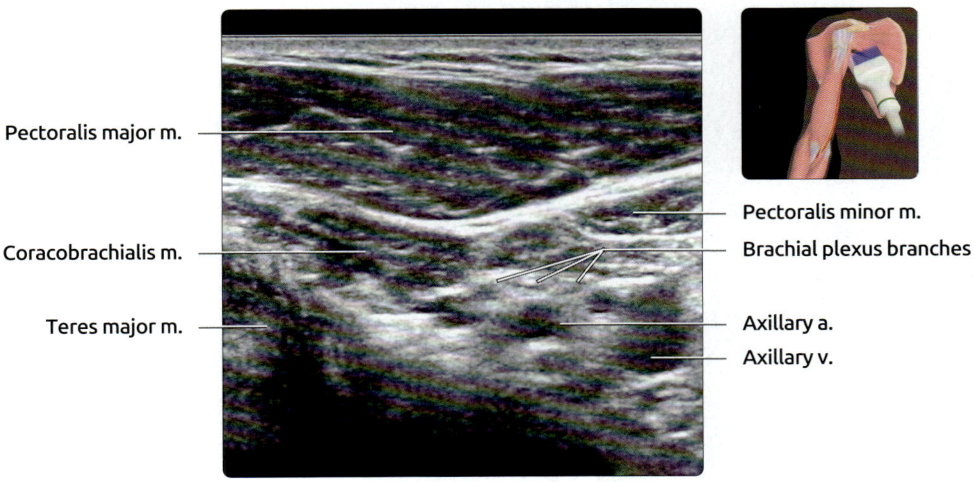

(Top) *Transverse grayscale US shows the proximal anterior wall of the axilla. The anterior wall of the axilla can be examined with the arm by the side.* **(Middle)** *Transverse grayscale US shows the midsection of the anterior wall of the axilla. The anterior wall of the axilla is formed by the pectoralis major and pectoralis minor muscles.* **(Bottom)** *Transverse grayscale US shows the distal anterior wall of axilla.*

TRANSVERSE US, QUADRILATERAL SPACE

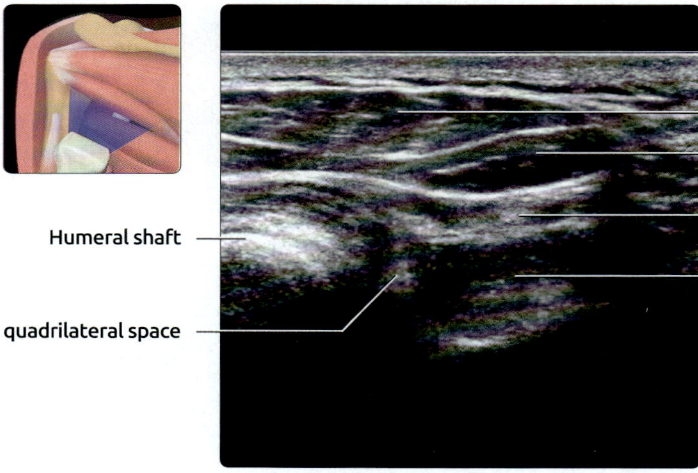

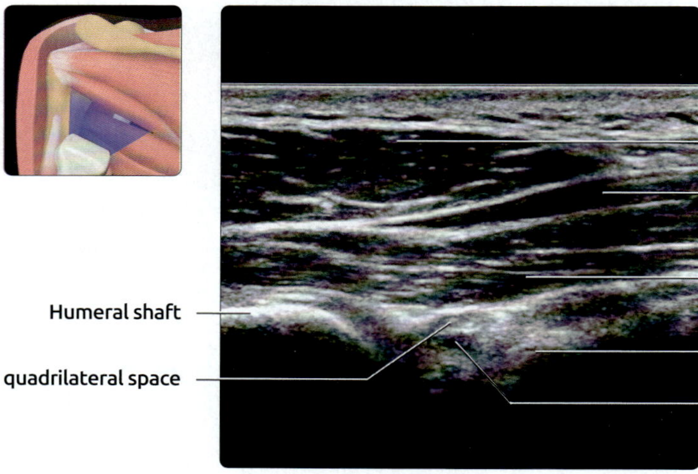

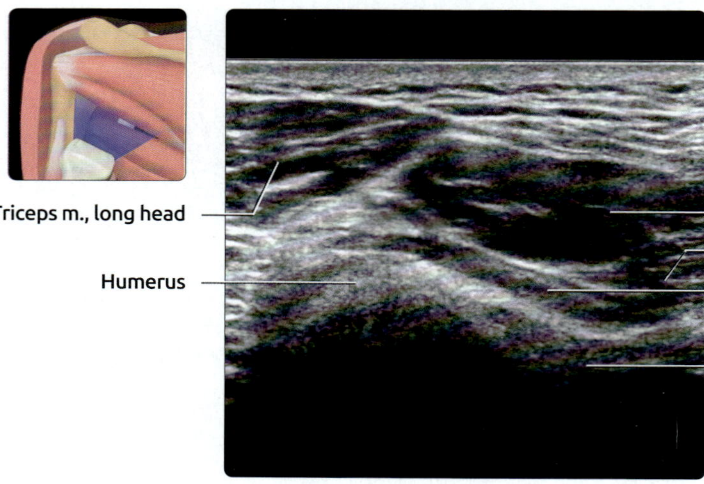

(Top) *Transverse grayscale US of the arm shows the anterior aspect of the quadrilateral space (quadrangular space) with the arm abducted. This space contains the axillary nerve and posterior circumflex humeral vessels.* **(Middle)** *Transverse grayscale US shows the quadrilateral space with the arm adducted. The contents of the space are more easily seen with the arm adducted.* **(Bottom)** *Transverse grayscale US in the more posterior position of the arm shows the posterior circumflex humeral artery posterior to the quadrilateral space.*

TRANSVERSE US, POSTERIOR AXILLARY WALL

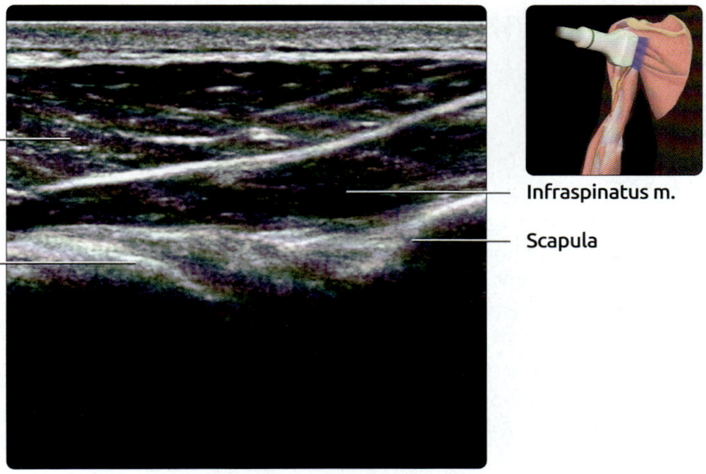

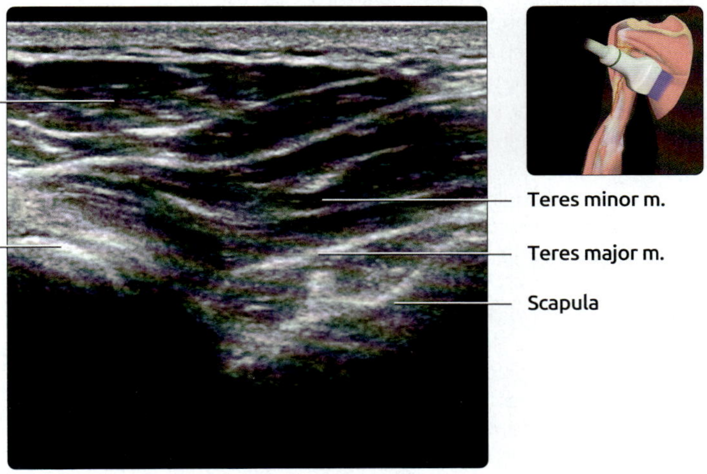

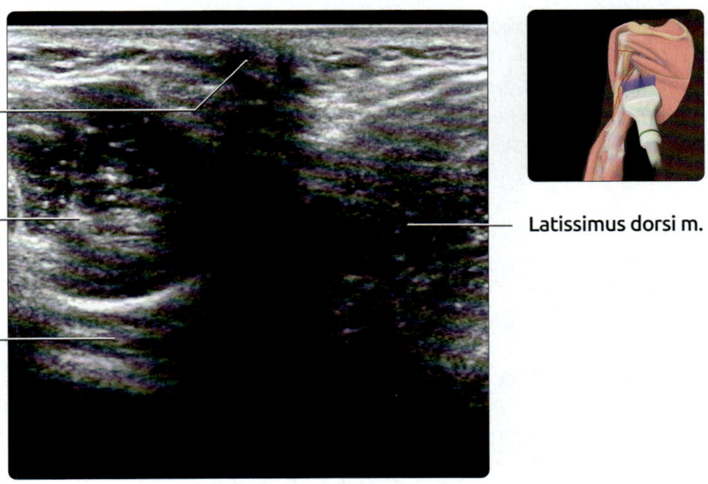

(Top) *Transverse grayscale US shows the upper aspect of the posterior axillary wall.* **(Middle)** *Transverse grayscale US shows the middle aspect of the posterior wall of the axilla. The posterior wall of the axilla is composed of the teres major, latissimus dorsi, and subscapularis muscles.* **(Bottom)** *Transverse grayscale US shows the lowermost aspect of the axilla.*

IMAGING ANATOMY

Overview

- Muscles of upper arm are divided into anterior and posterior compartments

Anatomy Relationships

- **Anterior compartment of arm**
  - Coracobrachialis muscle
    - Origin: Coracoid process tip, in common with and medial to short head biceps tendon
    - Insertion: Medial surface of humeral midshaft, between brachialis and triceps muscle origins
    - Nerve supply: Musculocutaneous nerve, perforates muscle
    - Blood supply: Brachial artery, muscular branches
    - Action: Flexes and adducts shoulder, supports humeral head in glenoid
  - Biceps muscle, short head
    - Origin: Coracoid process tip, in common with and lateral to coracobrachialis tendon
    - Insertion: Radial tuberosity after joining long head
    - Nerve supply: Musculocutaneous nerve
    - Blood supply: Brachial artery, muscular branches
    - Action: Flexes elbow and shoulder, supinates forearm
  - Biceps muscle, long head
    - Origin: Predominantly supraglenoid tubercle, also superior glenoid labrum and coracoid base
    - Insertion: Radial tuberosity after joining with short head
    - Tendon runs along bicipital groove, which is covered by transverse ligament
    - Transverse ligament spans proximal end of bicipital groove and comprises deeper layer formed from subscapularis tendon fibers and superficial fibrous layer in continuity with supraspinatus tendon and coracohumeral ligament
    - Nerve supply: Musculocutaneous nerve
    - Blood supply: Brachial artery, muscular branches
    - Action: Flexes elbow, supinates forearm
    - Variants, biceps muscle: 3rd head in 10% arising at upper medial aspect of brachialis muscle; 4th head can arise from lateral humerus, bicipital groove, or greater tuberosity
  - Brachialis muscle
    - Origin: Distal 1/2 of anterior humeral shaft and 2 intermuscular septa
    - Insertion: Tuberosity of ulna and anterior surface of coronoid process
    - Nerve supply: Musculocutaneous nerve plus branch of radial nerve
    - Blood supply: Brachial artery, muscular branches, and recurrent radial artery
    - Action: Flexes forearm
    - Covers anterior aspect of elbow joint
- **Posterior compartment of arm**
  - Triceps muscle, long head
    - Origin: Infraglenoid tubercle of scapula
    - Insertion: Proximal olecranon and deep fascia of arm after joining with lateral and medial heads
    - Nerve supply: Radial nerve
    - Blood supply: Deep brachial artery branches
    - Action: Elbow extension, adducts humerus when arm is extended
  - Triceps muscle, lateral head
    - Origin: Posterior and lateral humeral shaft, lateral intermuscular septum
    - Insertion: Proximal olecranon and deep fascia of arm after joining with long and medial heads
    - Nerve supply: Radial nerve
    - Blood supply: Deep brachial artery branches
    - Action: Elbow extension
  - Triceps muscle, medial head
    - Origin: Posterior humeral shaft from teres major insertion to near trochlea, medial intermuscular septum
    - Insertion: Proximal olecranon and deep fascia of arm after joining with lateral and long heads
    - Nerve supply: Radial and branches of ulnar nerve
    - Blood supply: Deep brachial artery branches
    - Action: Elbow extension
  - Anconeus muscle
    - Origin: Lateral epicondyle of humerus
    - Insertion: Lateral cortex of olecranon and posterior 1/4 of ulna
    - Nerve supply: Radial nerve
    - Blood supply: Deep brachial artery branch
    - Action: Assists elbow extension, abducts ulna
- **Fascia**
  - Brachial fascia
    - Continuous with fascia covering deltoid and pectoralis major
    - Varies in thickness: Thin over biceps and thick over triceps muscles
    - Lateral intermuscular septum from lower aspect of greater tuberosity to lateral epicondyle
    - Medial intermuscular septum from lower aspect of lesser tuberosity to medial epicondyle
    - Perforated by ulnar nerve, superior ulnar collateral artery, and posterior branch of inferior ulnar collateral artery
    - Provides traction on deep fascia of forearm
  - Bicipital fascia
    - a.k.a. lacertus fibrosus
    - Arises from medial side of distal biceps tendon at level of elbow joint
    - Passes superficial to brachial artery
    - Continuous with deep fascia of forearm

ANATOMY IMAGING ISSUES

Imaging Recommendations

- Examine with arm extended in supine and prone positions
- Transverse plane most helpful to delineate borders of anterior and posterior compartments and relationship to neurovascular structures

Imaging Pitfalls

- Muscle may appear echogenic, simulating fatty replacement or edema when beam not aligned parallel to muscle fibers

ARM

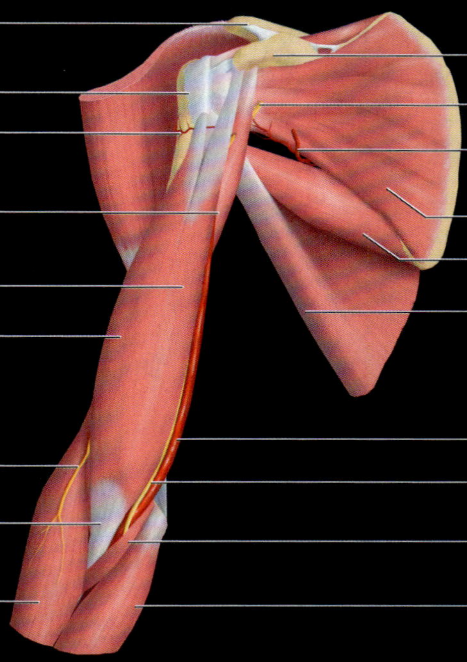

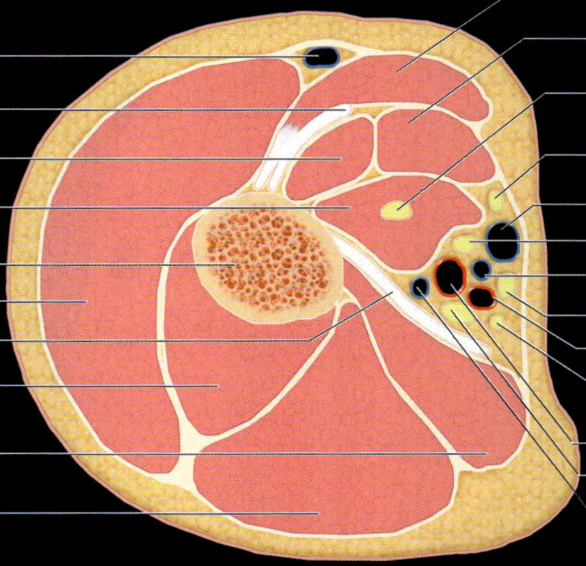

(Top) *Anterior graphic of the arm is shown. The short head of the biceps originates from the coracoid process tip, in common with the coracobrachialis tendon. The long head of the biceps tendon originates from the biceps labral complex at the supraglenoid tubercle and extends along the rotator cuff interval and bicipital groove of humerus to join with the short head and form the biceps muscle. The brachialis muscle originates from the distal 1/2 of the anterior humeral shaft. The transverse ligament spans the proximal bicipital groove from medial to lateral lip. Anatomical studies suggest that the transverse ligament is formed from fibers of the subscapularis tendon as well as a more superficial discrete fibrous band.* **(Bottom)** *Graphic shows the upper humeral level. Note that the coracobrachialis muscle is lying deep to the biceps muscle. In the posterior compartment, the lateral head and long heads of the triceps muscles can be seen at this level. All the neurovascular bundles are lying on the medial side of the upper arm.*

AXIAL, ARM

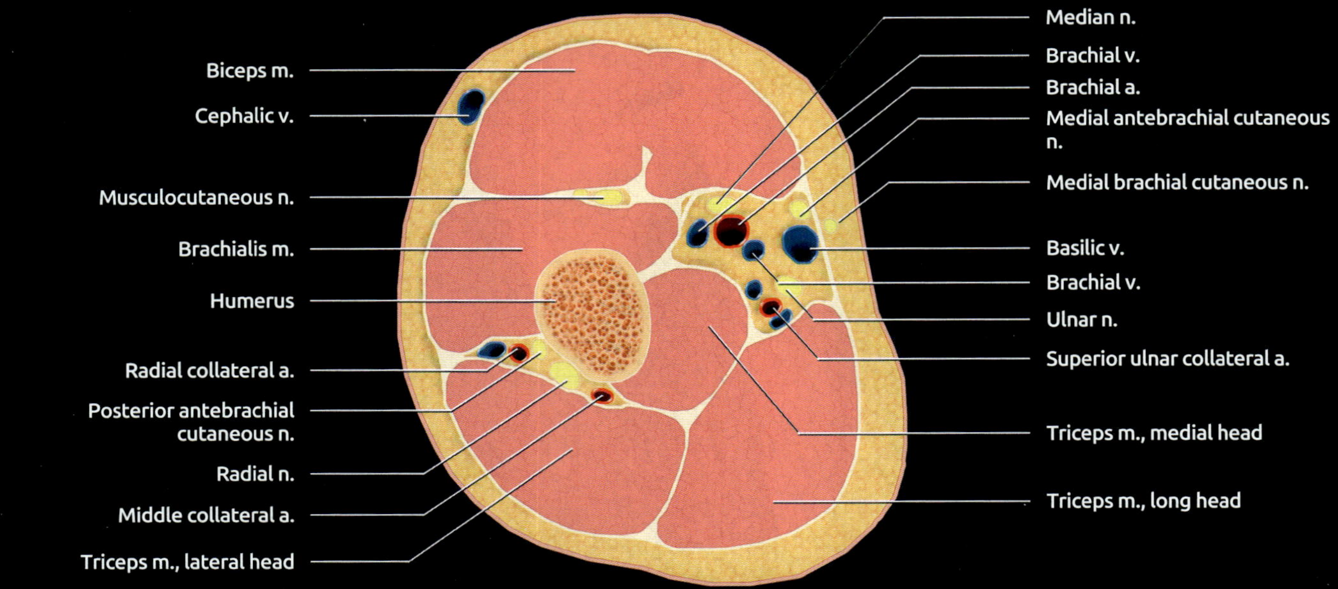

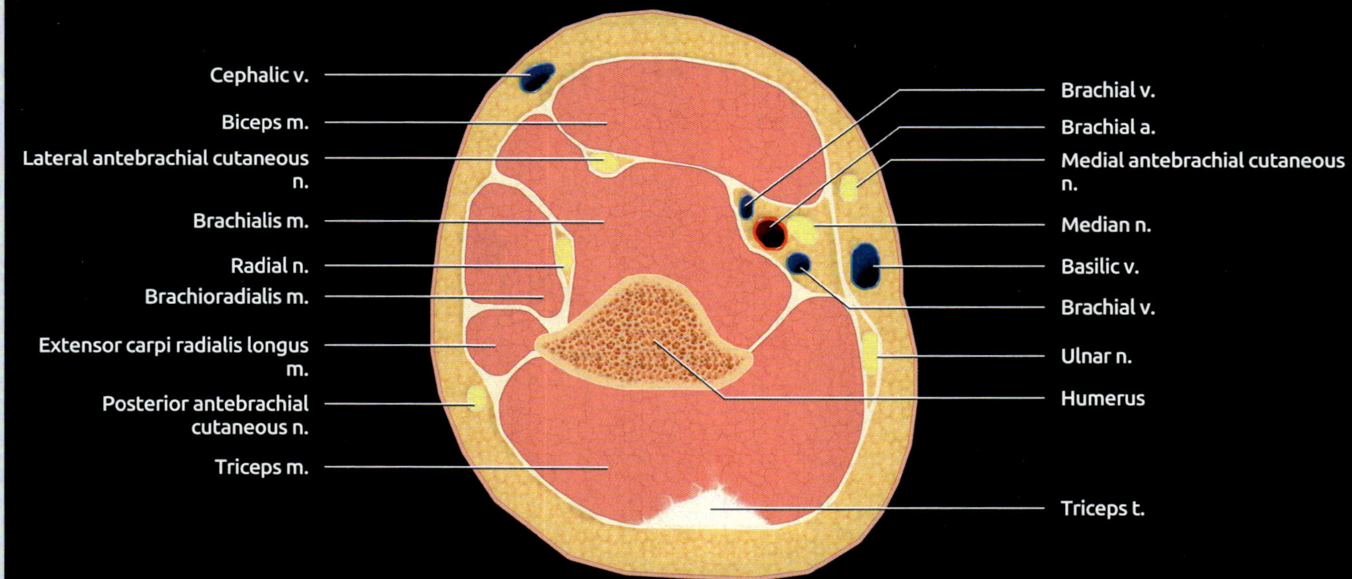

(Top) *Axial graphic shows the right arm at the midhumeral level. The posterior compartment of the arm consists of 3 heads of triceps muscle, and the medial head of the triceps can be delineated in this level. In the anterior compartment, the brachialis can be delineated at this midarm level.* **(Bottom)** *Axial section shows the distal humeral level. The deep brachial artery and radial nerve course along the posterolateral humerus reaching the anterolateral aspect of the humerus at this level. The median nerve is lying within the intermuscular septum. The biceps muscle is thinning anteriorly, and the triceps tendon appears.*

TRANSVERSE US, ANTEROMEDIAL ASPECT

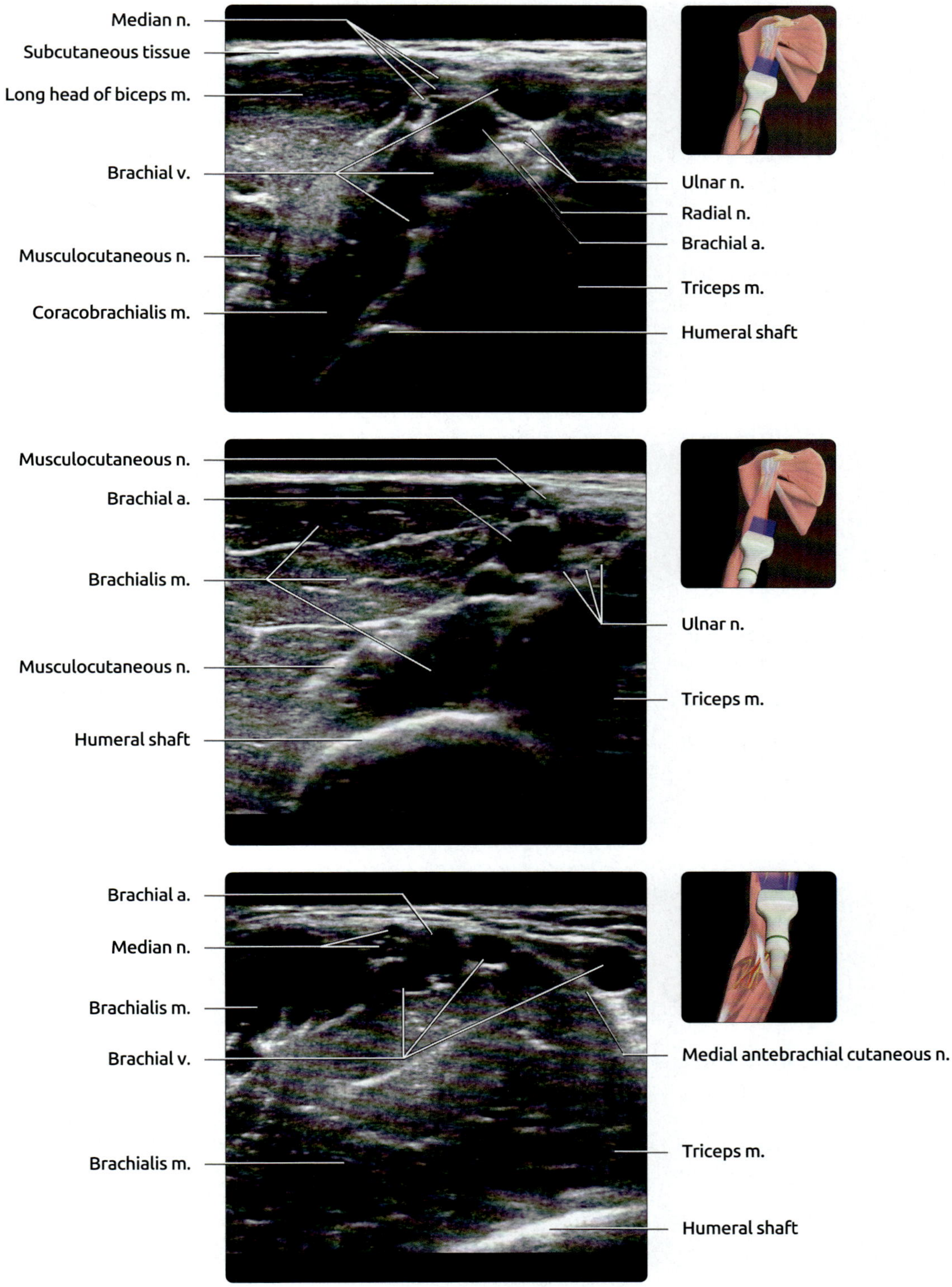

(Top) *Transverse grayscale US shows the anteromedial aspect of the upper arm. The terminal branches of the brachial plexus (namely, the median nerve, ulnar nerve, and radial nerve) all lie around the brachial vein and artery. They can be recognized by their position relative to the brachial artery: The radial nerve lying deep, median nerve laterally, and ulnar nerve medially.* **(Middle)** *Transverse grayscale US shows the anteromedial aspect of the midarm. Compression of the median or ulnar nerve is uncommon in the midarm. The neurovascular bundle helps define the separation of the anterior and posterior compartments of the arm medially.* **(Bottom)** *Transverse grayscale US shows the anteromedial aspect of the distal arm. In the distal arm, the median nerve alone is closely associated with the brachial neurovascular bundle.*

TRANSVERSE US, ANTERIOR ASPECT

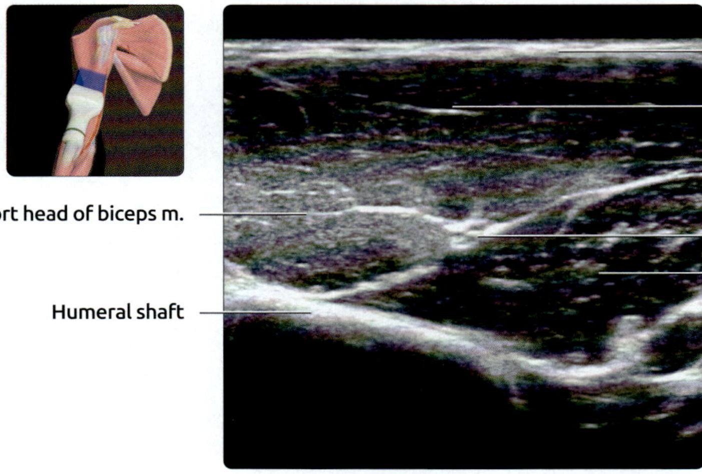

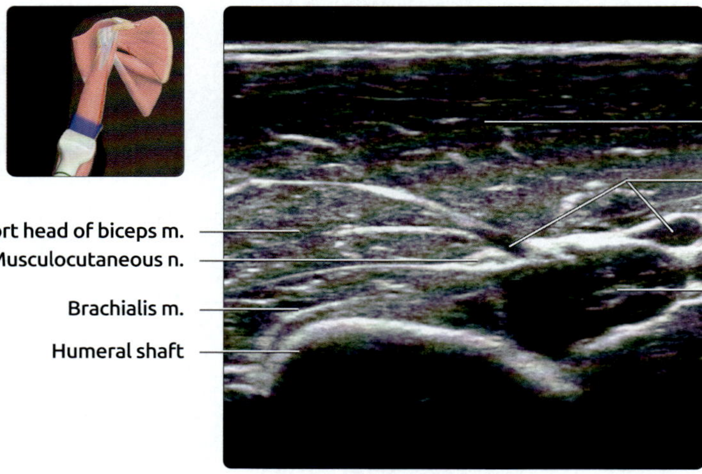

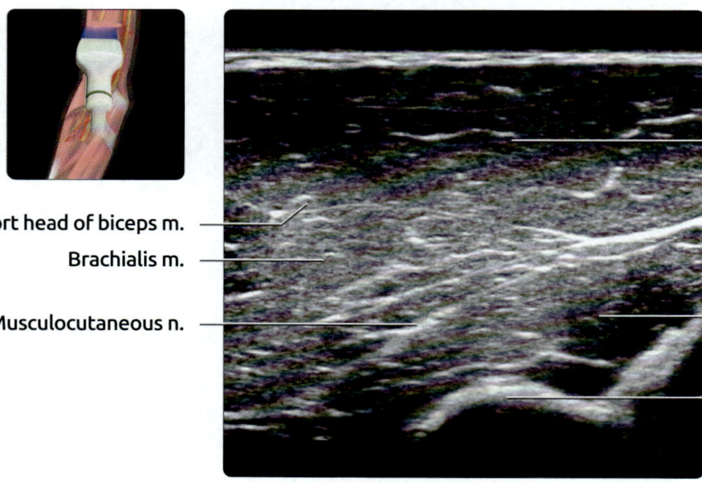

(Top) *Transverse grayscale US shows the proximal 1/3 of the arm. The anatomy of the upper arm is quite straightforward. The arm muscles are divided into 2 compartments. The anterior compartment is composed of the biceps, brachialis, and coracobrachialis muscles. The biceps muscle is midline, the brachialis muscle is midline and lateral, and the coracobrachialis muscle is midline and medial.* **(Middle)** *Transverse grayscale US shows the mid 1/3 of the anterior arm. Most of the neurovascular bundles in the arm lie anteromedially.* **(Bottom)** *Transverse grayscale US shows the distal 1/3 of the anterior arm. The musculocutaneous nerve is the main critical neurovascular structure of the anterior arm lying between the biceps, coracobrachialis, and brachialis muscles.*

TRANSVERSE US, ANTEROLATERAL ASPECT

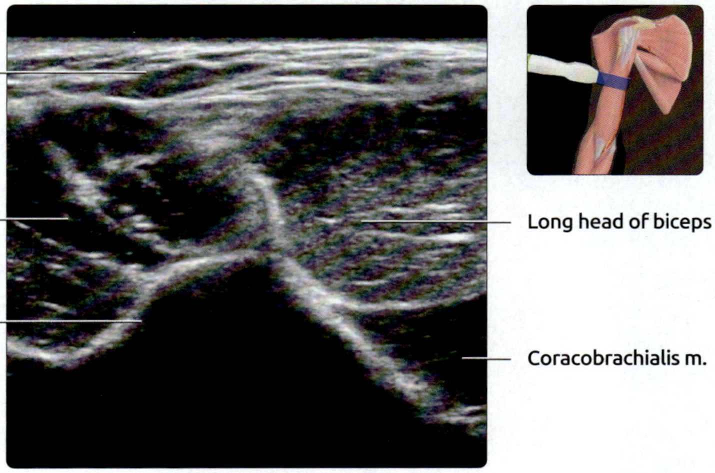

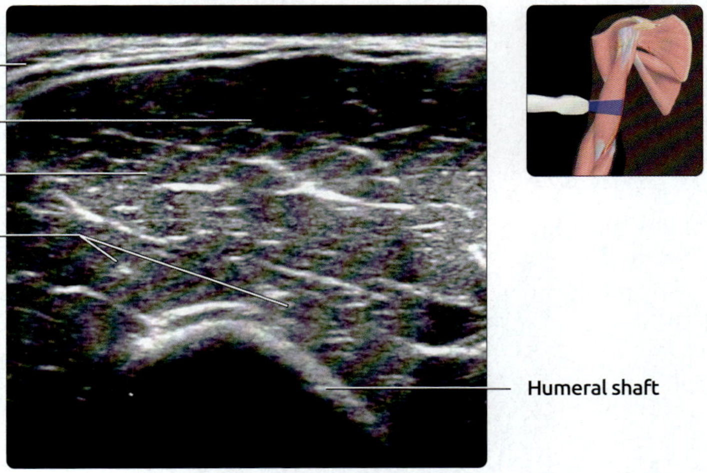

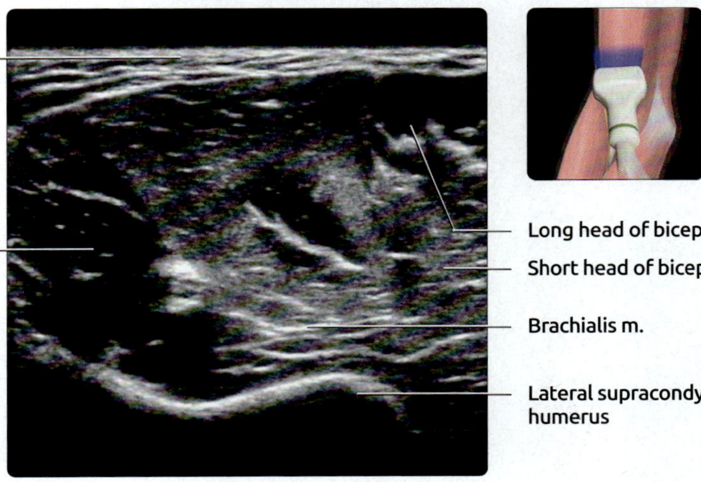

(Top) *Transverse grayscale US shows the anterolateral aspect of the proximal arm. This is an area prone to injury during contact sports. Fortunately, there is no neurovascular bundle in the area that would be exposed to injury.* **(Middle)** *Transverse grayscale US shows the anterolateral section of the midarm. Both the biceps muscle and brachialis muscle may be prone to overuse muscle injury if too vigorous exercise is undertaken. The biceps muscle may appear echogenic if the transducer is aligned at an angle to the transverse plane.* **(Bottom)** *Transverse grayscale US shows the anterolateral aspect of the distal arm. Both heads of the biceps converge to form the distal biceps tendon just proximal to the antecubital fossa.*

POSTERIOR ARM

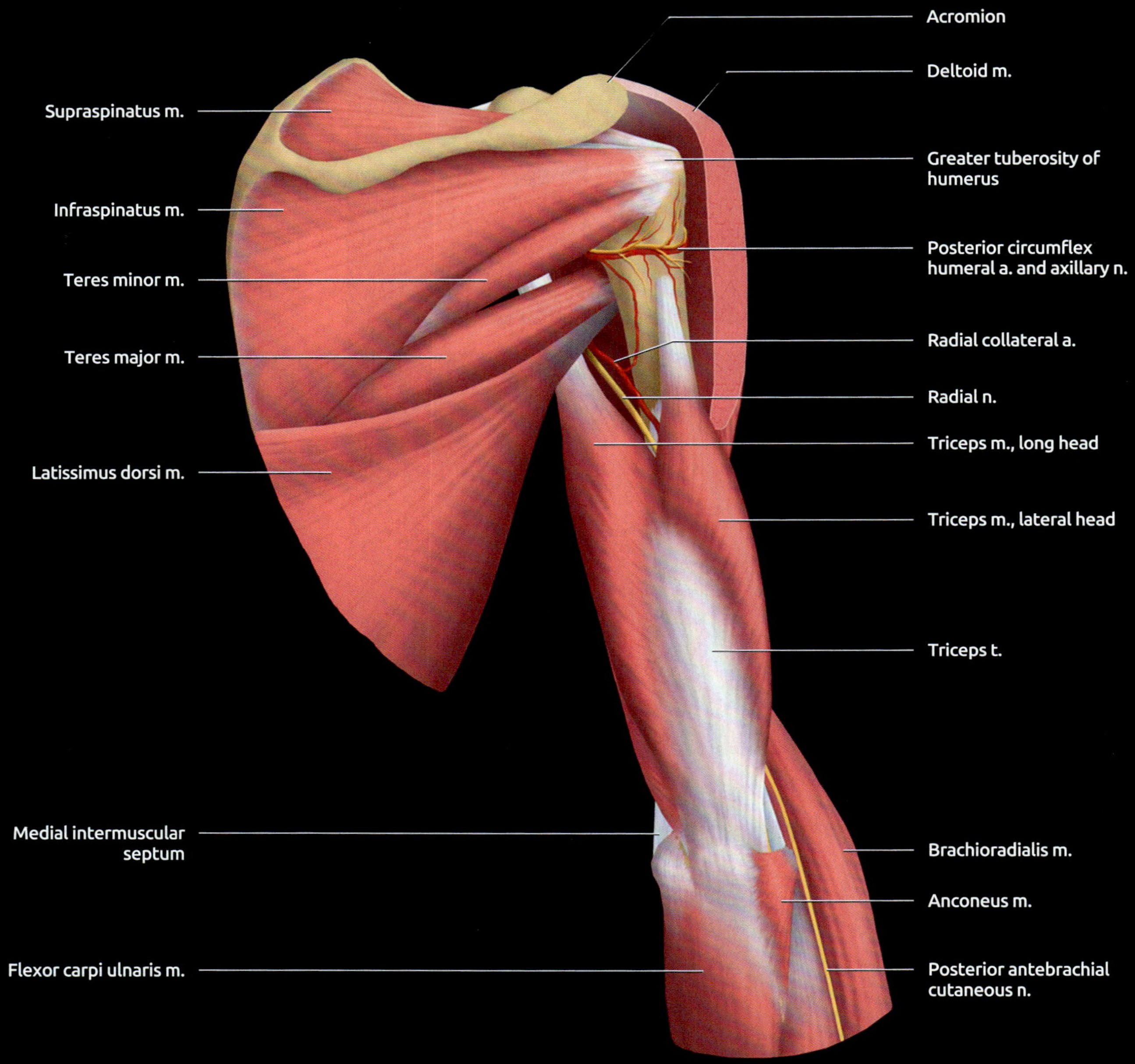

Graphic shows the superficial dissection of the posterior arm. Note that the quadrilateral space is an anatomic space bounded by the long head of the triceps, teres minor and teres major muscles, and cortex of the humerus. Axillary nerve and posterior circumflex humeral artery are within the space. Any narrowing of this space will cause quadrilateral space syndrome.

TRANSVERSE US, POSTERIOR ASPECT

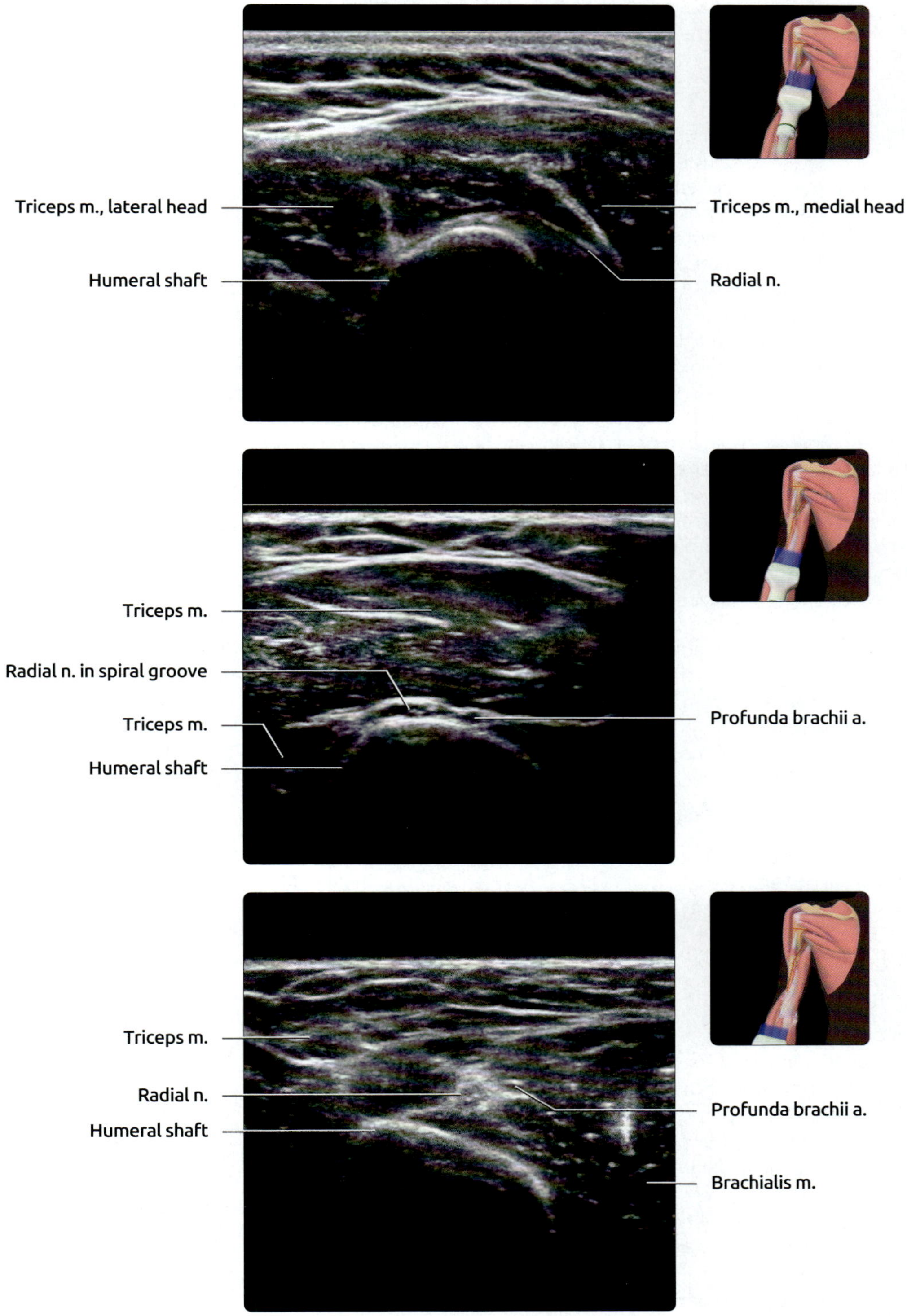

(Top) *Transverse grayscale US shows the posterior upper arm. The triceps muscle occupies nearly the whole of the posterior upper arm. The radial nerve supplies the triceps and supinator muscles.* **(Middle)** *Transverse grayscale US shows the posterior midarm. The radial nerve runs in the spiral groove on the posterior aspect of the midarm. At this level, it is prone to injury from humeral shaft fractures, fibrosis, or fibrotic bands following trauma or extrinsic compression.* **(Bottom)** *Transverse grayscale US shows the posterior aspect of the distal arm. The radial nerve pierces the lateral intermuscular septum just proximal to the lateral epicondyle. It then runs between the brachialis and brachioradialis muscles and passes anterior to the lateral epicondyle. The nerve may be compressed by the lateral intermuscular septum.*

GROSS ANATOMY

Arteries

- Brachial artery
- Continuation of axillary artery
- Extends from lower border of teres major to neck of radius
- Accompanies median nerve
- Branches of brachial artery in arm
  - Superior ulnar collateral arises medially, descends with ulnar nerve posterior to medial epicondyle, and then forms anastomosis with branches of ulnar artery
  - Inferior ulnar collateral arises distal to superior ulnar collateral artery, descends anterior to medial epicondyle, and then forms anastomosis with branches of ulnar artery
  - Profunda brachii: Main branch of brachial artery
    - Arises just distal to teres major, accompanies radial nerve, and passes backward in spiral groove on posterior surface of humerus
    - Gives off another branch, which passes upward and forms anastomosis with arteries around shoulder joint
    - Gives 2 lower medial branches, which descend to elbow joint on its medial side and take part in anterior and posterior anastomosis around elbow joint
- Brachial artery lies superficial in its course in arm
- Covered by bicipital aponeurosis at elbow, where it may be compressed against medial surface of humerus
- Brachial artery divides at neck of radius into radial artery and ulnar artery
  - High takeoff of radial artery from brachial artery is common anatomical variant
  - Radial artery passes inferiorly and laterally, lying on tendon of biceps brachii
  - Ulnar artery passes down and medially, deep to pronator teres
  - Within antecubital fossa, radial and ulnar arteries give off recurrent branches to elbow joint

Veins

- Categorized as superficial and deep venous system
  - Veins of upper limb are variable in number and position
- Brachial vein
  - Forms deep venous system
  - Singular or paired, continue into axillary vein
  - Accompanied by brachial artery and nerves of upper limb
- Cephalic vein
  - Lies anterolaterally in superficial fascia of antecubital fossa and gives off medial cubital vein, which joins basilic vein
  - Runs in deltopectoral groove before piercing clavipectoral fascia to drain into axillary vein
- Basilic vein
  - Passes upward on medial side of forearm, receiving anterior and posterior tributaries, and then passes to anteromedial aspect of elbow where it receives medial cubital vein
  - Continues on medial aspect of biceps brachii until it pierces deep fascia to run cranially, medial to brachial artery
  - Joins brachial vein and becomes axillary vein at lower border of teres major

ANATOMY IMAGING ISSUES

Imaging Recommendations

- Venous examination
  - Examination of upper limb veins is normally performed with patient supine and arm abducted to about 90°
  - Use of high-frequency linear transducer (5-10 MHz) is recommended
    - Determination of compressibility
    - Flow pattern: Normal venous waveform
  - Augmentation of flow is obtained by manual compression of forearm or upper arm
    - Alternatively, patient clenches fist to increase venous flow
    - Flow can also be augmented by deep inspiration
- Arterial examination
  - In patients with possible arterial compression syndromes, arteries are examined in various positions of shoulder abduction so that any compression may be accentuated

Imaging Pitfalls

- Clavicle can produce mirror artifact of subclavian vessels
- Brachial vein duplication is common
- Due to proximity, arm venous flow is affected by respiration
  - Deep inspiration will augment flow
- Angle of insonation should not be > 60° as this will reduce sensitivity of Doppler and color flow signals
- Color signal scale should match flow velocity
- Doppler gate should be located at center of artery or vein as this is site of peak flow velocity
- Arteriography: Gold standard for assessment of accuracy of arterial Doppler US with respect to stenoses or occlusion
  - Other methods of visualizing arterial tree are computed tomography angiography and magnetic resonance angiography

CLINICAL IMPLICATIONS

Clinical Importance

- Brachial artery may be used for access in angiography, either by high brachial approach or in antecubital fossa, where it lies medial to biceps tendon and lateral to median nerve
- Spontaneous thrombosis of axillary vein occasionally occurs following excessive movements of arm at shoulder joint
- Diagnosis of exclusion of deep vein thrombosis in upper or lower limb; spontaneous or related to indwelling catheters
- Vein mapping prior to bypass grafts
- Localization of veins for venous access and cannulation

ARTERIES

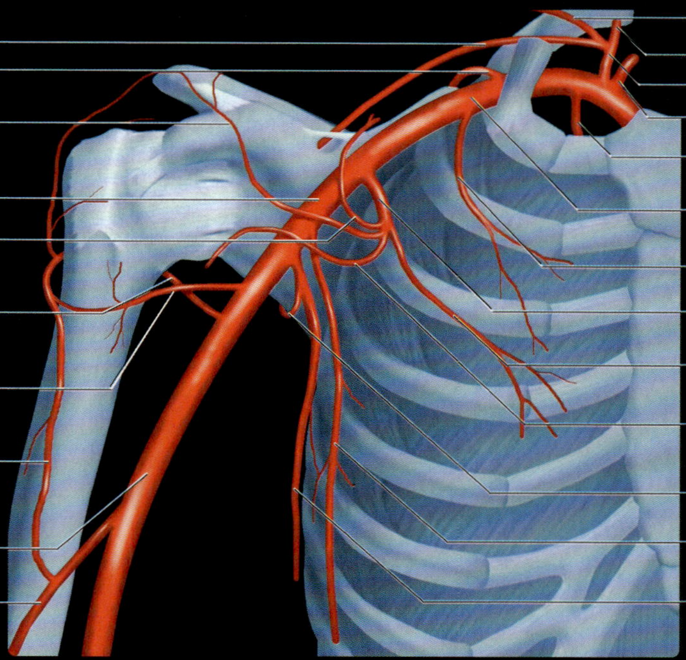

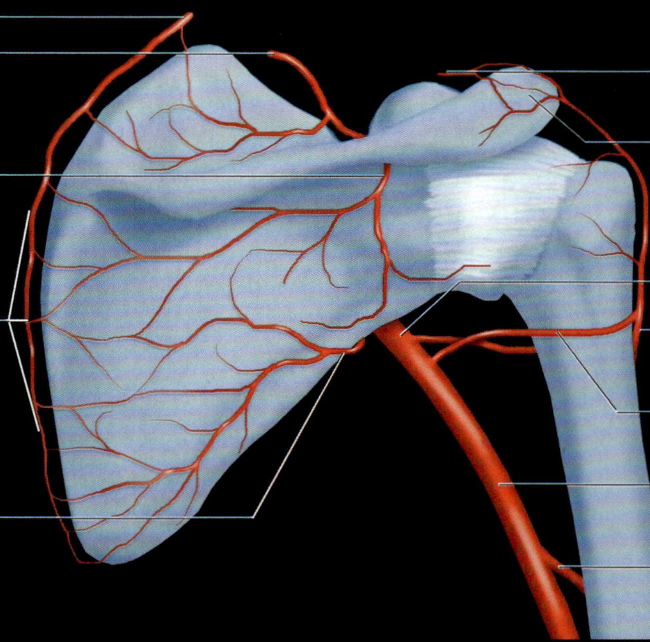

(Top) *Anterior graphic shows arterial supply to the shoulder. The shoulder is predominantly supplied by the anterior and posterior circumflex humeral, suprascapular, and circumflex scapular arteries.* **(Bottom)** *Posterior graphic shows arterial supply to the shoulder. Extensive collateral blood vessels include anastomoses with intercostal arteries.*

AXIAL, RIGHT ARM ARTERIES AND VEINS

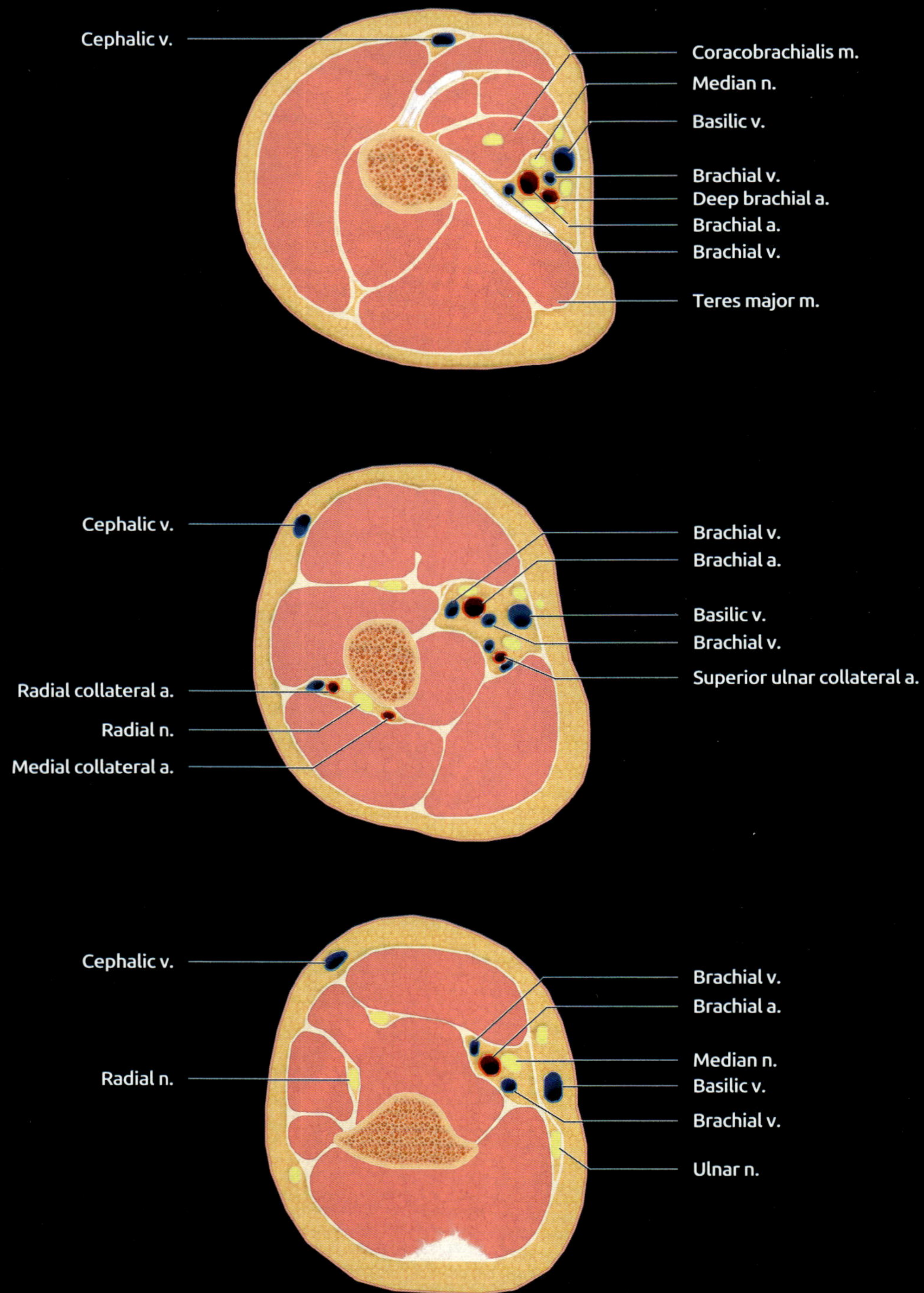

(Top) *Axial graphic shows the upper right arm. The brachial artery is a continuation of the axillary artery and accompanies the median nerve. The deep brachial artery (profunda brachii artery) is the main branch of the brachial artery and accompanies the radial nerve. The basilic vein lies medially and the cephalic vein anterolaterally.* **(Middle)** *Axial graphic shows the right arm at the midhumeral level. Note that the profunda brachii artery accompanies the radial nerve and passes backward in spiral groove of humeral shaft before dividing into medial and radial collateral arteries. Superior ulnar collateral artery arises from the brachial artery and descends with the ulnar nerve.* **(Bottom)** *Axial graphic shows the right arm at the distal humeral level. Note how the median nerve is on the medial side of the brachial artery at this level.*

TRANSVERSE US, PROXIMAL ARM

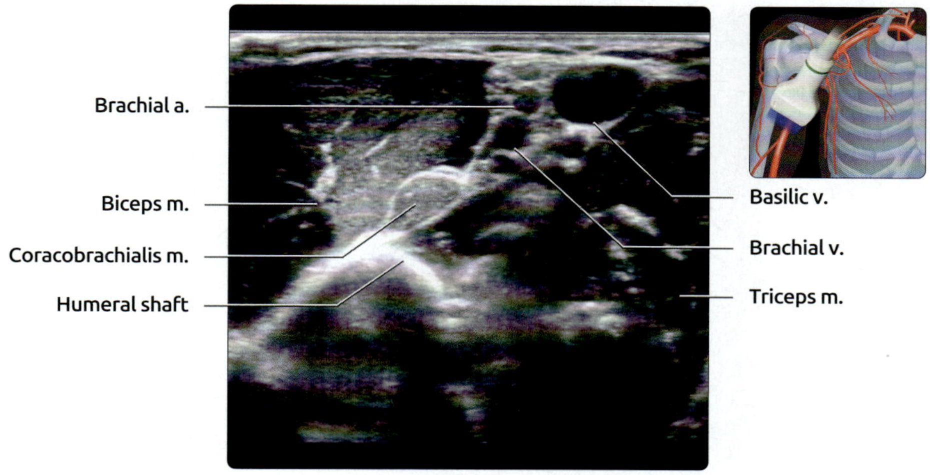

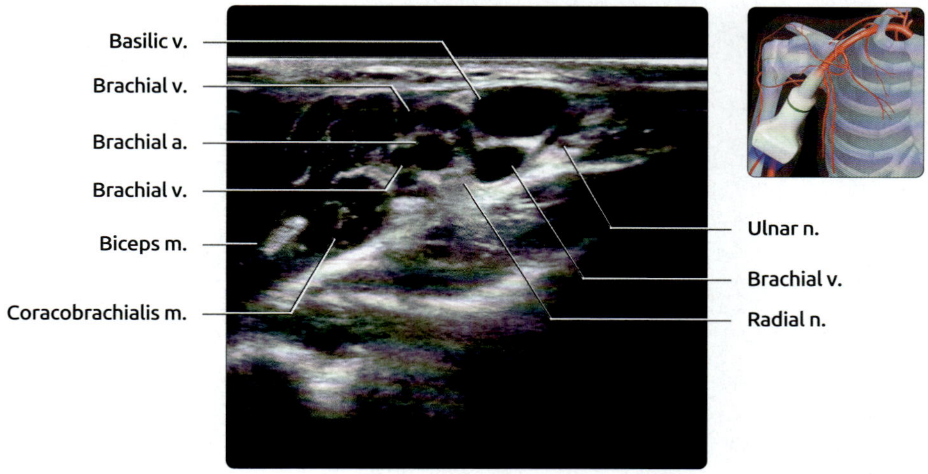

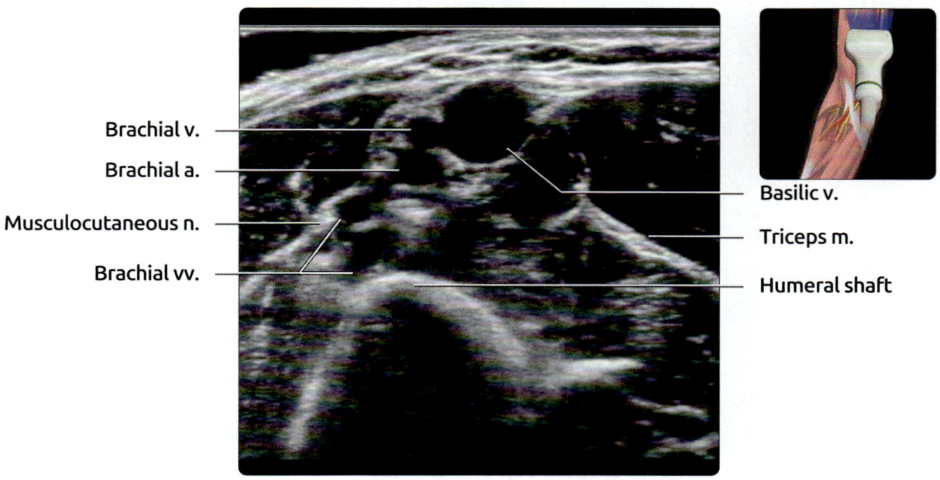

(Top) *Transverse grayscale US shows the vessels of the proximal arm. The basilic vein lies deep to the investing fascia in the arm. The veins are highly variable in location and are surrounded by nerves; hence, US guidance is helpful when cannulating the veins of the arm.* **(Middle)** *Transverse grayscale US shows the vessels of the midarm. The basilic vein is formed on the ulnar side of the wrist and is joined by the median cubital vein (from cephalic vein at the elbow). In some cases, the brachial artery may bifurcate in the midarm or higher.* **(Bottom)** *Transverse grayscale US shows the vessels of the distal arm. The brachial artery can be seen clinically as a pulsatile vessel in "locomotor brachialis," i.e., in arteriosclerosis.*

COLOR DOPPLER US, BRACHIAL ARTERY

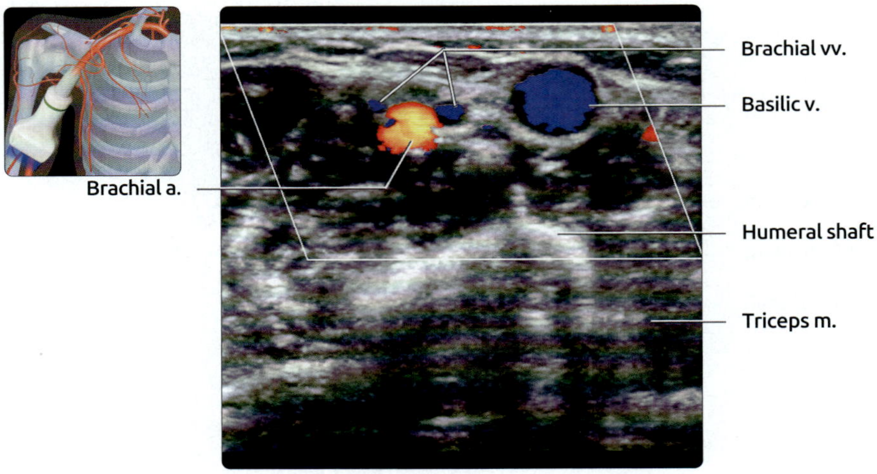

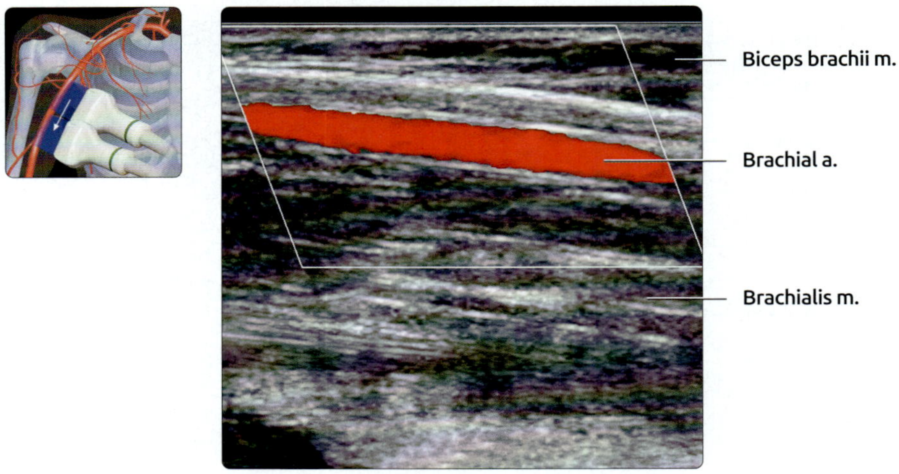

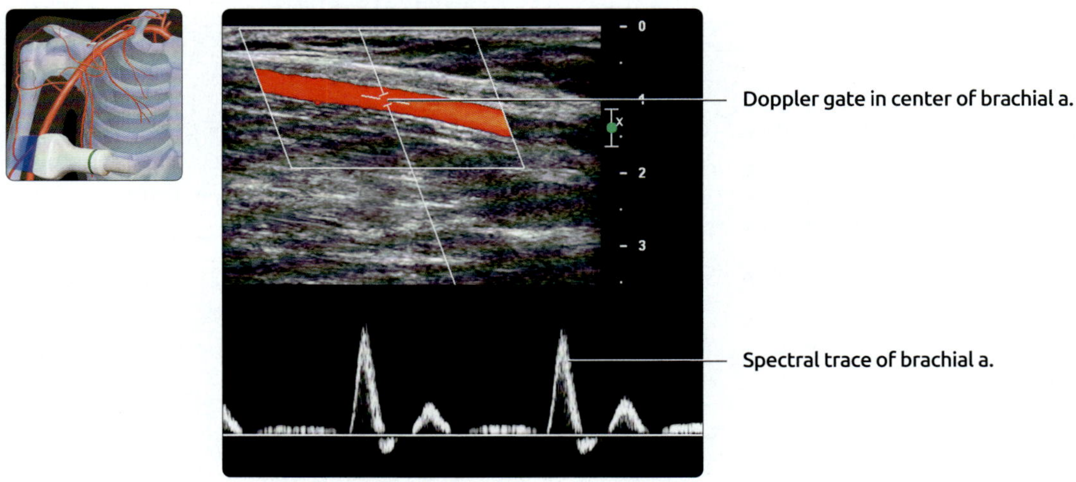

(Top) *Transverse color Doppler US shows the brachial artery. Note the accompanying paired brachial veins.* **(Middle)** *Longitudinal color Doppler US shows the brachial artery. The normal diameter of the brachial artery at the antecubital fossa ranges from 5.5-6.0 mm.* **(Bottom)** *Longitudinal color Doppler spectral analysis US shows the brachial artery. Normal triphasic flow pattern of the large arteries is seen, compared to the biphasic flow of more distal arteries, but changes in temperature or even maneuvers, such as fist clenching, may dramatically change the characteristics of upper extremity waveforms, especially in and near the hands. Peak systolic velocity of the brachial artery ranges from 50-100 cm/sec (average: 57 cm/sec); end-diastolic velocity ranges from 7-12 cm/s, and a pulsatility index around 5 is considered normal.*

LONGITUDINAL COLOR DOPPLER US, BRACHIAL VEIN

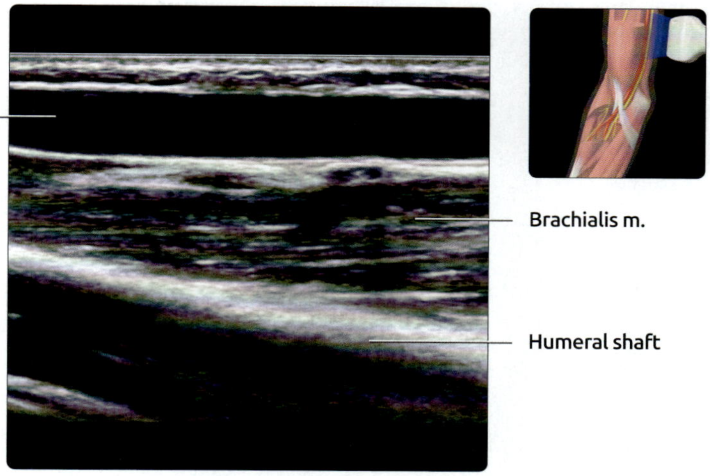

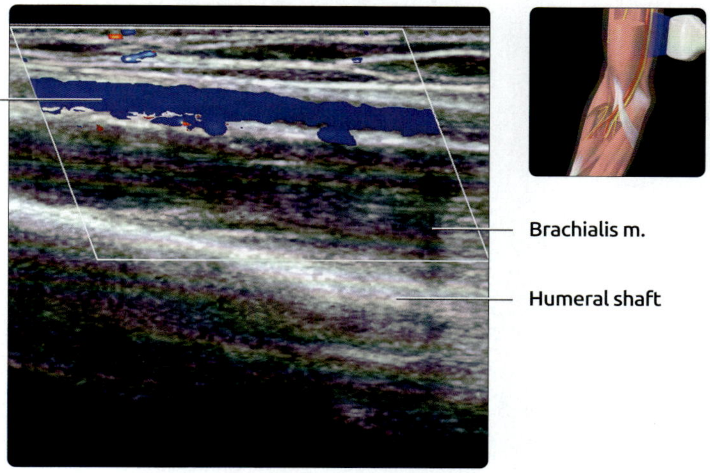

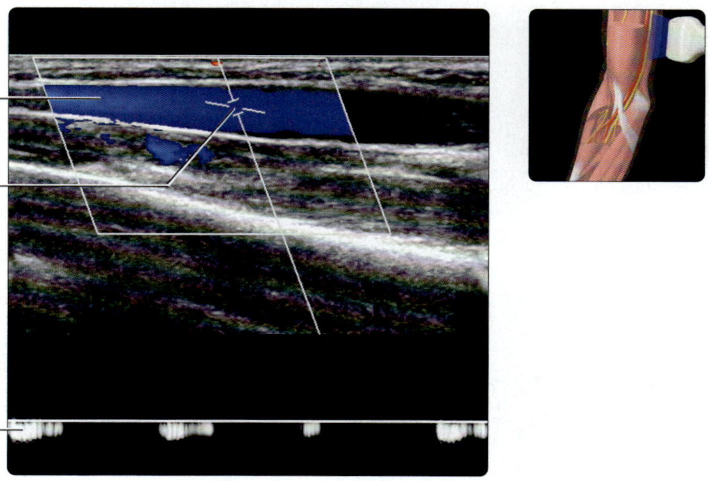

(Top) *Longitudinal grayscale US shows the brachial vein. Normal brachial vein in the upper arm measures around 2.5-2.7 mm. The venous system of the upper limb is categorized as a superficial and deep system. Brachial vein constitutes the deep venous system. Brachial veins can be paired or singular.* **(Middle)** *Longitudinal color Doppler US shows the brachial vein. The ulnar and radial veins collect blood from the deep palmar arch and run in pairs alongside the corresponding arteries. They unite slightly above the elbow joint and flow into the brachial veins, which may be either paired or singular.* **(Bottom)** *Spectral Doppler US shows the flow pattern in the brachial vein. The flow is intermittent due to respiration. Examinations using proximal compression are not clinically relevant. Distal compression, which produces augmentation sounds, helps locate the veins.*

LONGITUDINAL US, DYNAMIC

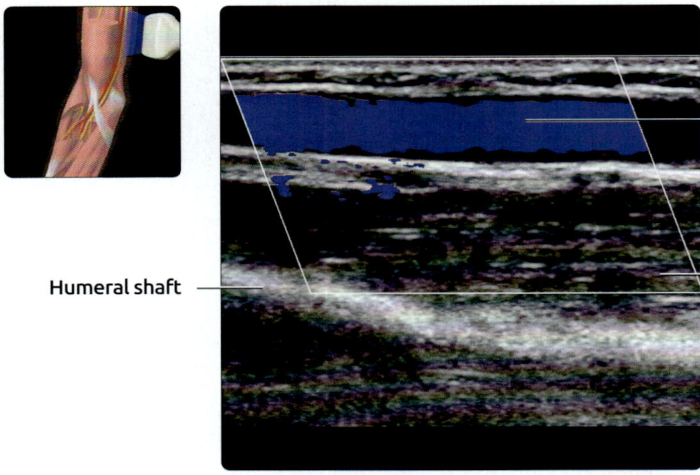

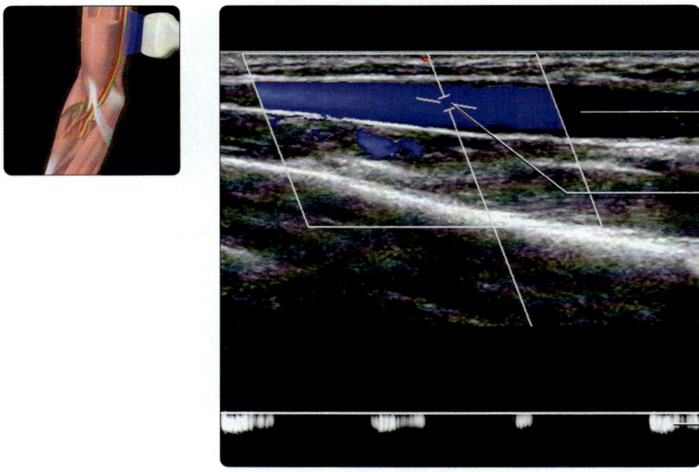

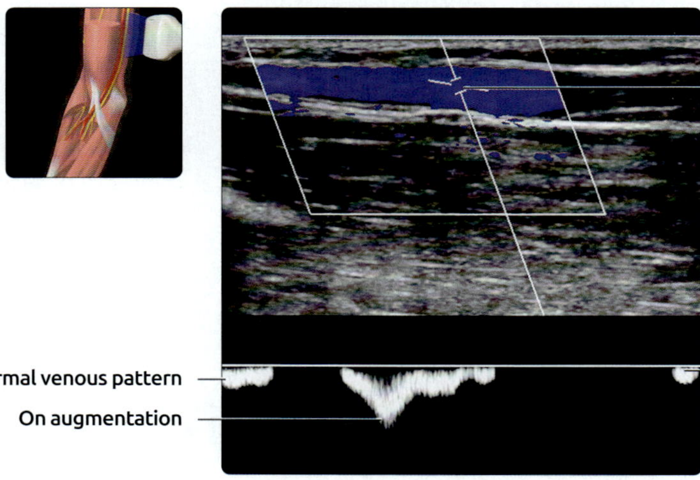

(Top) *Longitudinal color Doppler US shows the brachial vein. Using Doppler mode, phasic flow during normal respiration is observed.* **(Middle)** *Spectral Doppler US shows the flow pattern in the brachial vein. In the upper extremities, including the subclavian vein, the behavior of venous flow during respiration is the opposite of what it is in the lower body region. During inspiration, pressure in the thorax decreases, and the venous blood flows toward the heart. During expiration, the thoracic pressure increases and the flow decreases, or even ceases, during deep respiration.* **(Bottom)** *Spectral Doppler US shows the phenomenon of augmentation, demonstrating the patency of the veins. It involves squeezing of the distal arm muscles (causing motion artifacts from movement), which will increase venous flow. Deep inspiration will have the same effect.*

TRANSVERSE AND LONGITUDINAL US, CEPHALIC VEIN

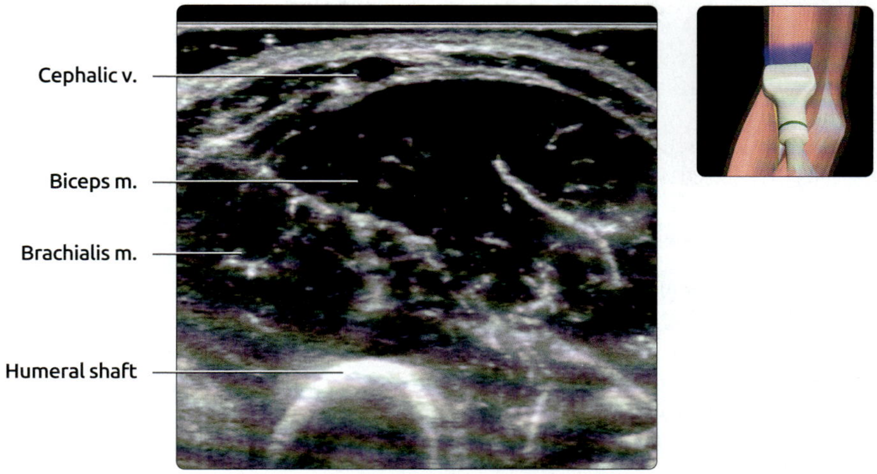

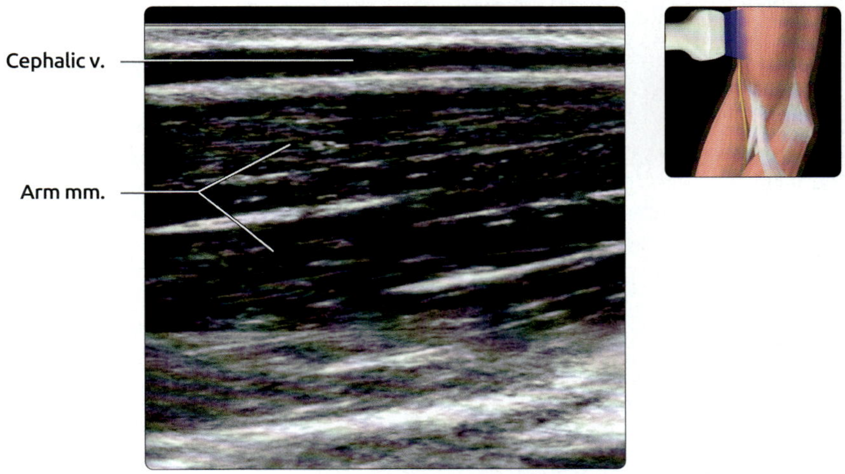

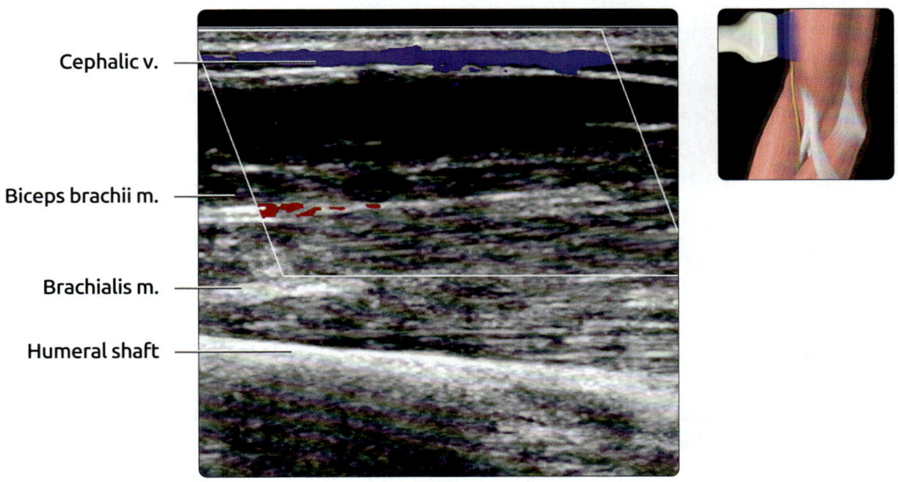

(Top) *Transverse grayscale US at the proximal arm shows the cephalic vein. The normal diameter of the cephalic vein just above the elbow measures 2.5-2.9 mm.* **(Middle)** *Longitudinal grayscale US at the proximal arm shows the cephalic vein. The most important veins of the arm are the cephalic vein and basilic veins. The superficial cephalic vein unites the deep-seated brachial and axillary veins. Assessment of valvular insufficiency has no clinical role in the upper limb vessels.* **(Bottom)** *Longitudinal color Doppler US in the distal arm shows the cephalic vein. Color flow duplex sonography allows ready assessment of the important upper arm veins (the paired brachial veins, cephalic veins, and basilic veins). Continuous color intensity makes it possible to exclude thromboses.*

GROSS ANATOMY

Joint Capsule Attachments

- Posterior
  - Proximally attaches to humerus proximal to olecranon fossa and capitellum; distally attaches to proximal ulna
- Anterior
  - Proximally attaches to humerus proximal to coronoid and radial fossae; distally attaches to coronoid process and annular ligament
- Anterior and posterior elbow fat pads are intracapsular but extrasynovial
- Capsule of elbow is weakest anteriorly; it is strengthened on both sides by ligaments
- Synovial recesses
  - Olecranon recesses
    - Largest; superior, medial, and lateral around olecranon process
  - Anterior humeral recess
    - Proximal to coronoid fossa
  - Annular recess
    - Surrounds radial neck
  - Ulnar collateral ligament recess
    - Deep to ligament
  - Radial collateral ligament recess
    - Deep to ligament
  - Capsule of elbow is weakest anteriorly
    - Capsule is strengthened on both sides by collateral ligaments

Ligaments

- **Ulnar collateral ligament**
  - Extends from distal aspect of medial epicondyle to coronoid and olecranon portions of ulna
  - Separated from overlying common flexor tendon by thin layer of fat
  - 3 bands: Anterior, posterior, and transverse
    - Anterior band: Most important; cord-like, 4-6 mm thick, taut in extension, extends from medial epicondyle to coronoid process (sublime tubercle); divided into superficial and deep layers
    - Posterior band: Less important; fan-like and thinner, taut in flexion, extends from medial epicondyle to olecranon
    - Transverse band: Functionally unimportant; forms base of triangle between anterior and posterior bands
- **Radial collateral ligament**
  - Triangular in shape
  - Apex on lateral epicondyle → blends distally with annular ligament
  - Lies deep to common extensor tendon
  - Provides origin for superficial head of supinator muscle
- **Annular ligament**
  - Attached to anterior and posterior aspects of radial notch of ulna forming collar around radial head
    - Anterior attachment becomes taut in supination
    - Posterior attachment becomes taut in extreme pronation
    - Provides origin for superficial head of supinator muscle

Tendons

- **Triceps tendon**
  - 3 heads combine to insert onto olecranon process
  - Medial fibers insert exactly on medial margin of olecranon
  - Lateral fibers may fan out as lateral cubital retinaculum to connect with ulna and antebrachial fascia
- **Biceps tendon**
  - Flat tendon forms ~ 7 cm proximal to elbow joint
  - Twists 90° as it passes deep so that anterior surface faces laterally
    - Expands close to attachment into radial tuberosity; attachment area is ~ 3 cm^2
  - Short and long heads have separate adjoined attachments
  - Also attaches to bicipital aponeurosis
    - Bicipital aponeurosis completely encircles ulnar forearm flexor muscles
    - Merges with investing fascia of forearm
    - May be important in stabilizing distal aspect of biceps tendon
- **Common extensor tendon**
  - Arises from lateral epicondyle
  - Superficial to radial collateral ligament
  - Composed of extensor-supinator group
    - Extensor carpi radialis brevis, extensor carpi radialis longus, extensor digiti minimi, extensor digitorum communis
- **Common flexor tendon**
  - Arises from medial epicondyle
  - Superficial to ulnar collateral ligament
  - Composed of flexor-pronator group
    - Flexor carpi radialis, flexor carpi ulnaris, flexor digitorum superficialis, pronator teres, palmaris longus

Bursae

- **Subtendinous olecranon bursa**
  - Between triceps tendon and olecranon
- **Subcutaneous olecranon bursa**
  - Between skin and olecranon process
- **Bicipitoradial bursa**
  - Between biceps tendon and radial tuberosity
- **Radioulnar bursa**
  - Between extensor digitorum and radiohumeral joint

ANATOMY IMAGING ISSUES

Imaging Recommendations

- High-frequency linear transducer
- Large amount of gel for superficial structures
- Examination is usually tailored to site of symptoms
- Lateral elbow is best seen with elbow fully or slightly flexed
- Medial elbow is best seen with elbow extended
- Distal biceps tendon is best seen with forearm in full supination
- Distal biceps tendon insertion is best seen from posterior part of elbow with forearm fully pronated and elbow flexed with transducer aligned transversely over radial tuberosity
- Effusion will displace fat pads away from bone

CORONAL ELBOW

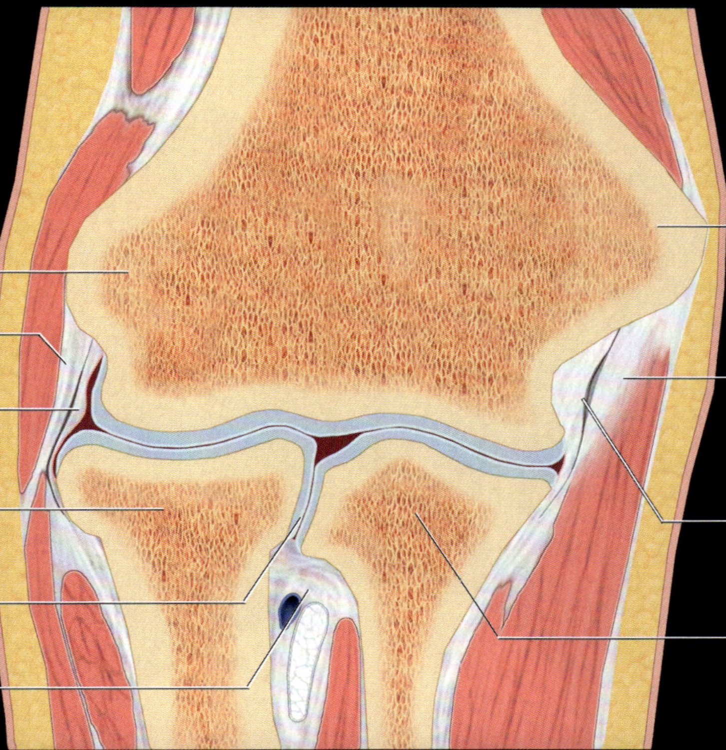

Coronal section through the level of the epicondyles shows the collateral ligaments deep to the common tendon groups. Although the radial collateral ligament is shown, the section is too anterior to show most of the ulnar collateral ligament, which originates just posterior to the radial collateral ligament.

PROXIMAL ELBOW CROSS SECTION

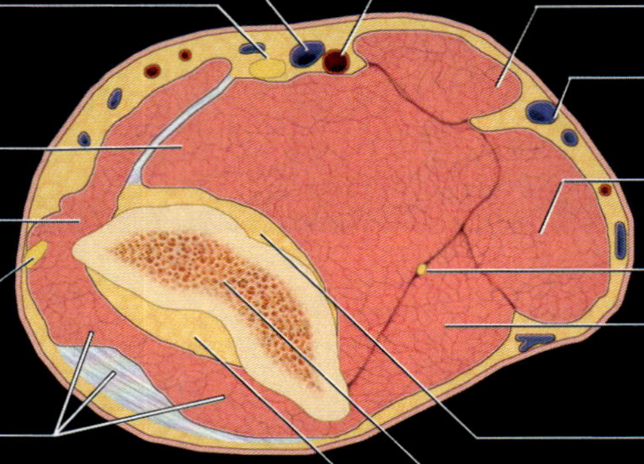

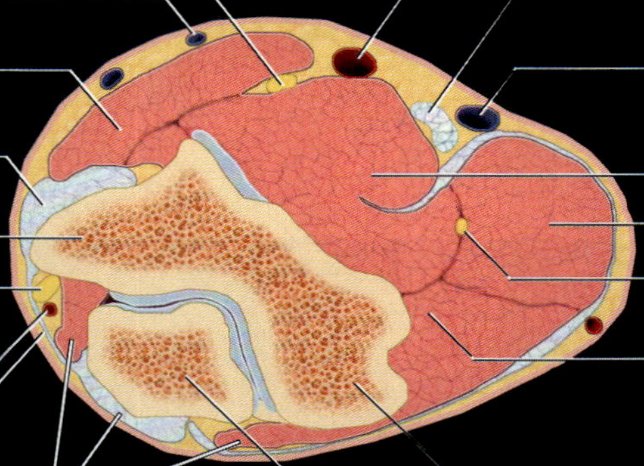

(Top) *Axial graphic shows the supracondylar region of the humerus. The anterior and posterior fat pads are seen in the coronoid and olecranon fossae, respectively. The brachialis muscle accounts for the bulk of the anterior compartment of the distal arm.* **(Bottom)** *Axial graphic shows the epicondylar region of the distal humerus. The triceps muscle thins as its tendon attaches to the olecranon. The ulnar nerve and posterior ulnar recurrent artery are held in the cubital tunnel by the cubital retinaculum (formed from the ligament of Osborne and the fascia of flexor carpi ulnaris tendon).*

MID ELBOW CROSS SECTION

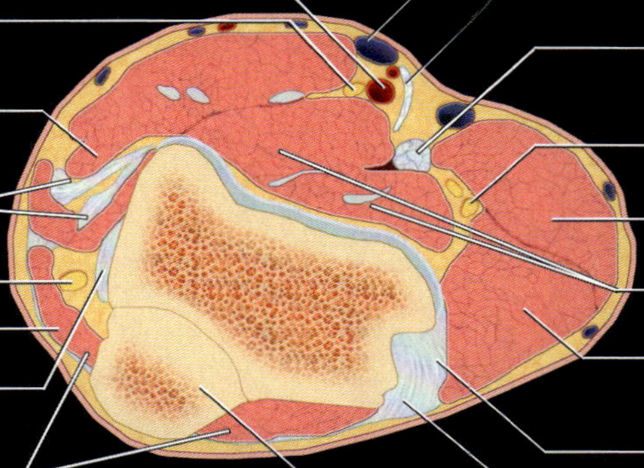

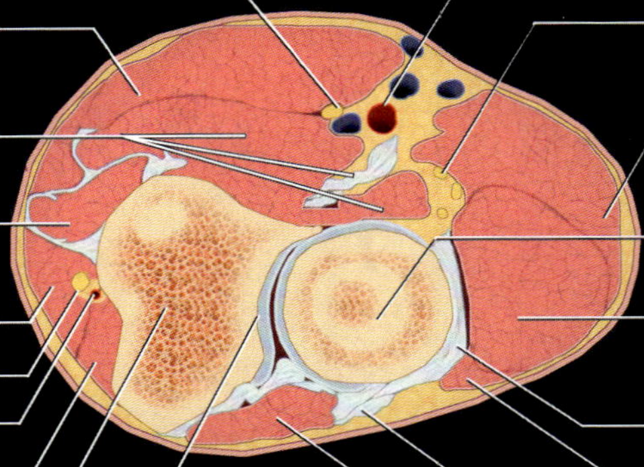

(Top) *Axial graphic immediately proximal to the elbow joint is shown. The common extensor tendon overlies the radial collateral ligament and may be difficult to distinguish at this level. The ulnar nerve has exited the cubital tunnel and is entering the flexor carpi ulnaris.* **(Bottom)** *Axial graphic at the level of the proximal radioulnar joint is shown. The articulating surfaces of the proximal radioulnar joint are well seen with the radial head being held in the radial notch of the ulna by the annular ligament. The lateral ulnar collateral ligament blends with the posterior aspect of the annular ligament.*

DISTAL ELBOW CROSS SECTION

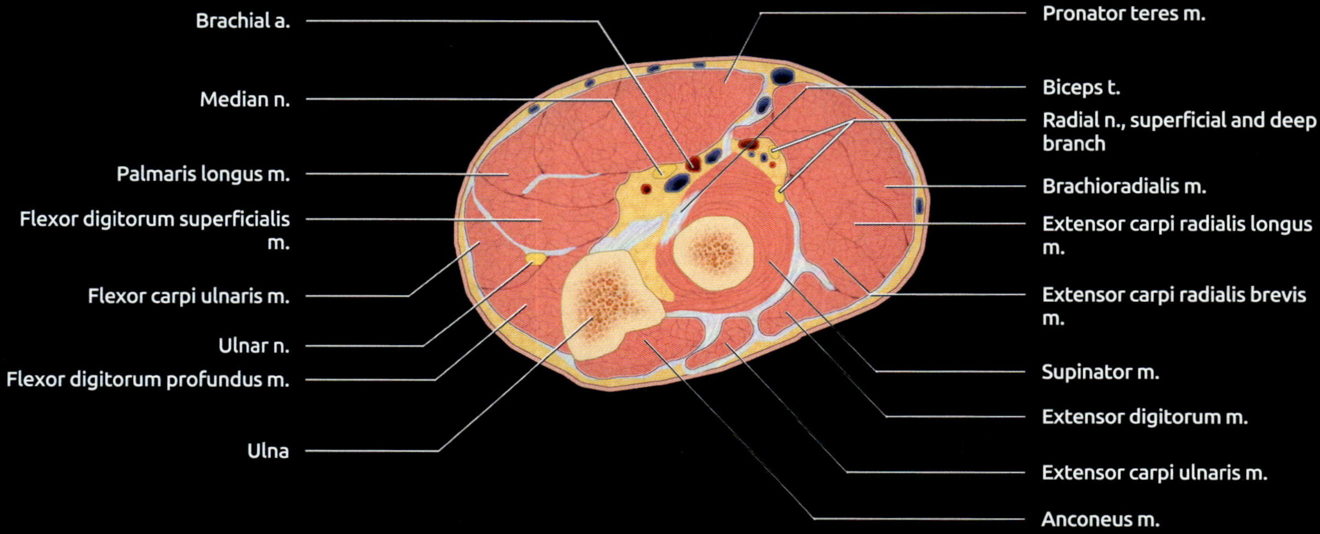

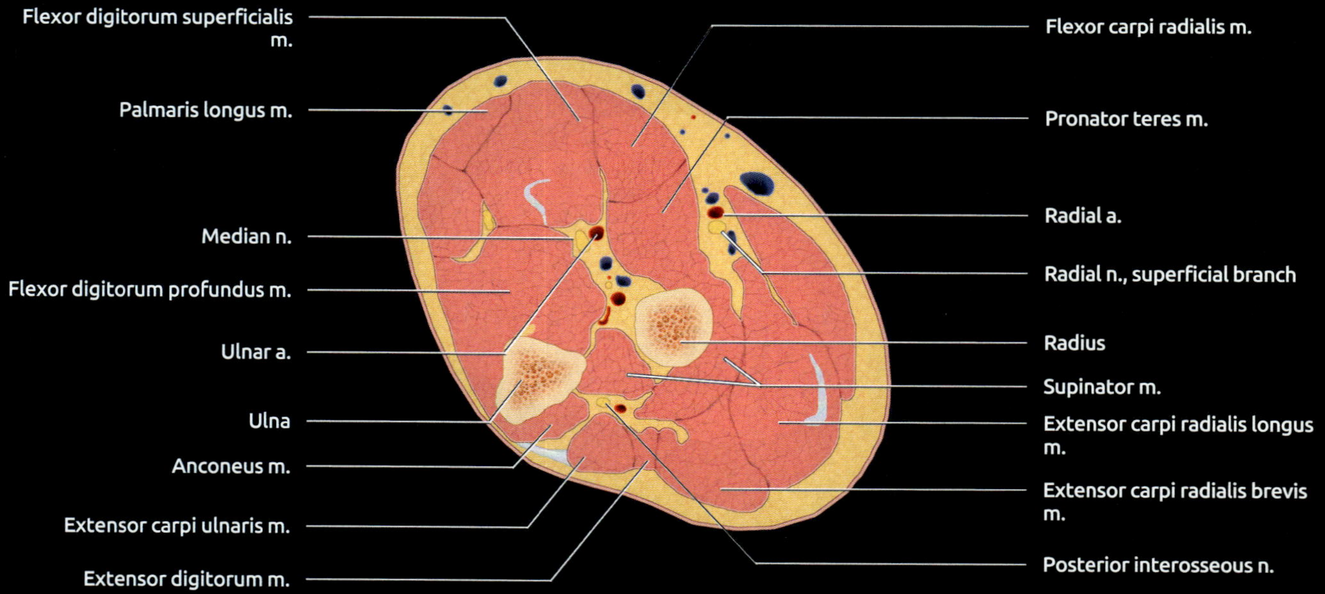

(Top) *Graphic shows the axial elbow at a level immediately above the radial tuberosity. The brachialis tendon inserts on the ulnar tuberosity, and the biceps tendon approaches its insertion on the radial tuberosity, which is more distal than the brachialis insertion.* **(Bottom)** *At the level of the proximal forearm, the muscles are starting to align themselves into the anterior (flexor) compartment and the posterior (extensor) compartment.*

LONGITUDINAL US, ANTERIOR ELBOW

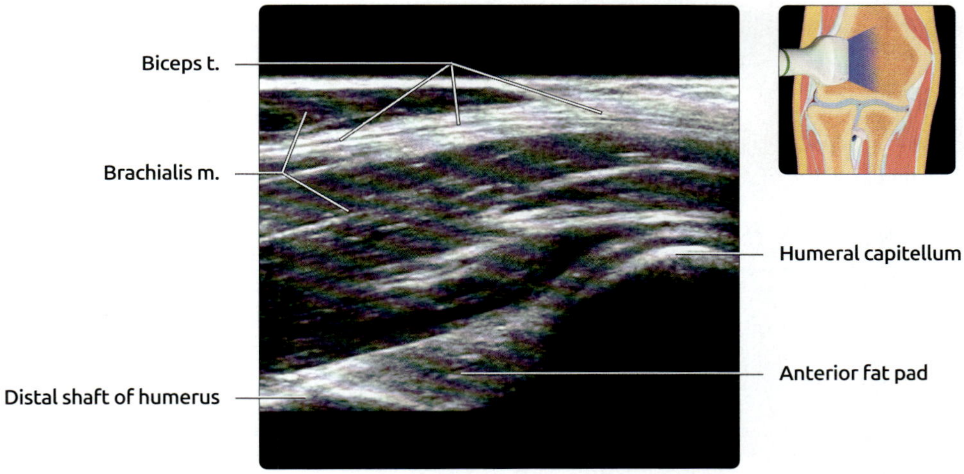

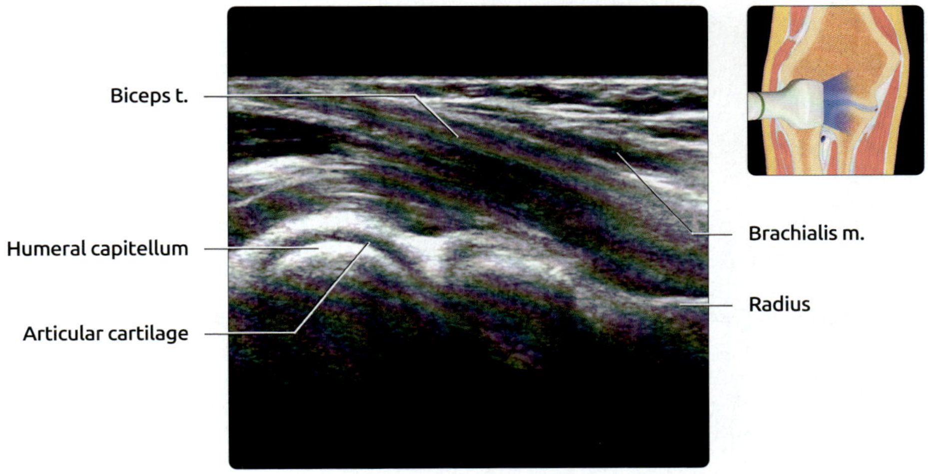

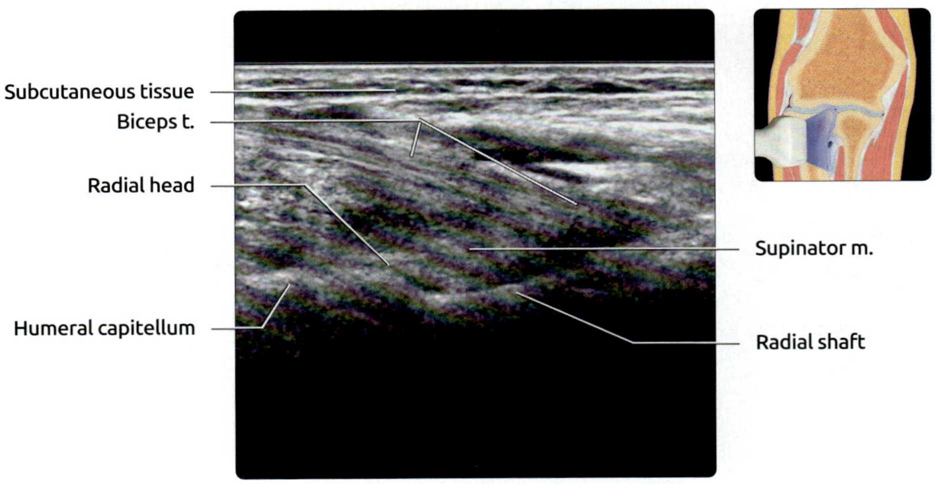

(Top) *Longitudinal grayscale US shows the biceps tendon proximal to the elbow. Most biceps tendon tears occur near the shoulder. Tears of the distal end of the biceps tendon are uncommon and mostly occur when unexpected force is applied to the bent arm.* **(Middle)** *Longitudinal grayscale US shows the biceps tendon at the elbow. With a torn biceps tendon, a bulge in the arm ("Popeye" muscle) may appear.* **(Bottom)** *Longitudinal grayscale US shows the biceps tendon just distal to the elbow. The biceps tendon flattens and rotates 90° as it approaches the insertion into the posterior aspect of the radial tuberosity. The distal biceps tendon is best visualized on longitudinal section with the forearm fully supinated. To see the insertional area of distal biceps tendon, fully pronate the forearm and flex the elbow with transducer placed posteriorly over the radial tuberosity.*

TRANSVERSE US, ANTERIOR ELBOW

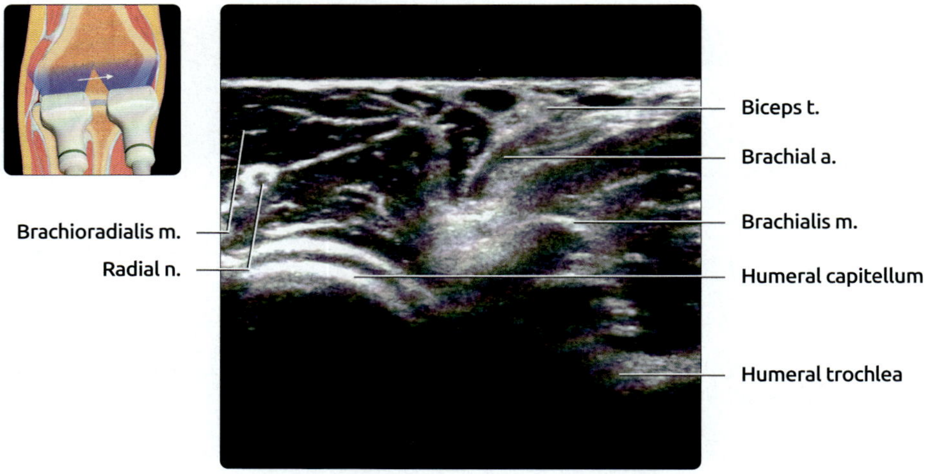

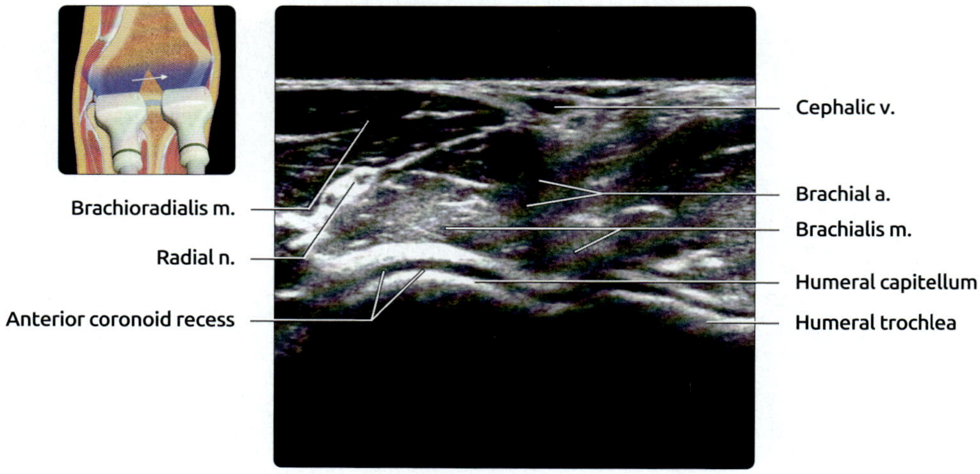

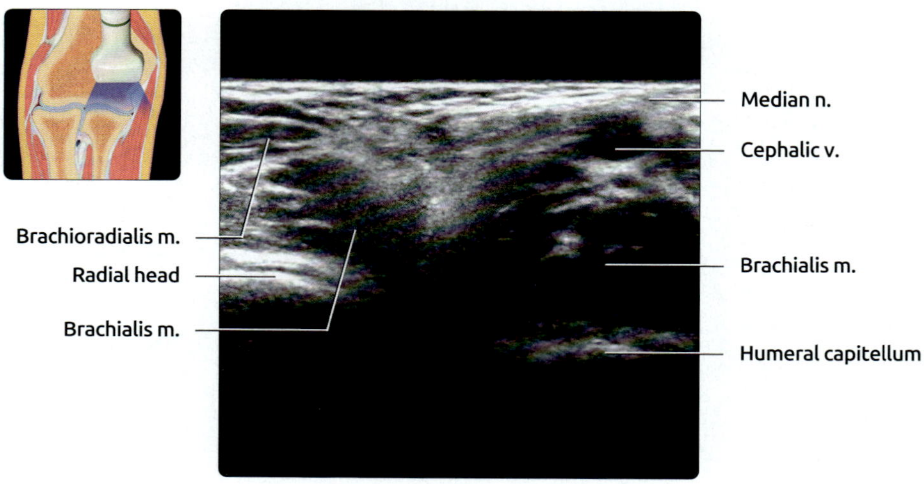

(Top) *Transverse grayscale US shows the proximal end of the antecubital fossa. Flexion and extension occur at the humeroulnar articulation, while pronation and supination involve pronation of the radius around the ulna.* **(Middle)** *Transverse grayscale US shows the midportion of the antecubital fossa. The median cubital vein unites the cephalic vein with the basilic vein. It is often used for venipuncture and venous access. It lies superficial to the bicipital aponeurosis.* **(Bottom)** *Transverse grayscale US shows the distal aspect of the antecubital fossa. The elbow joint is supplied by arteries from the profunda brachii, brachial arteries, and recurrent branches of the radial and ulnar arteries.*

TRANSVERSE US, ANTECUBITAL FOSSA

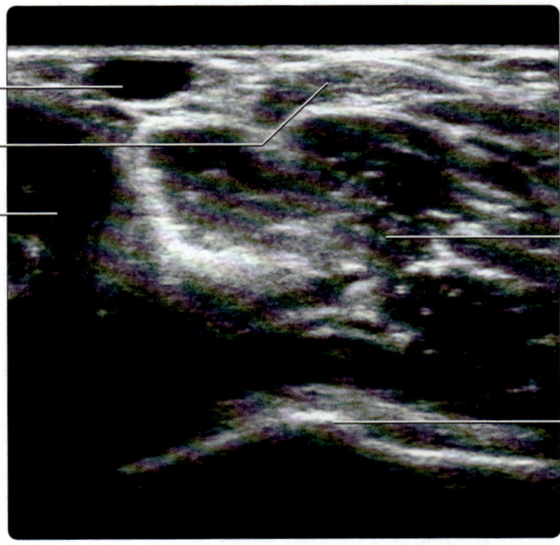

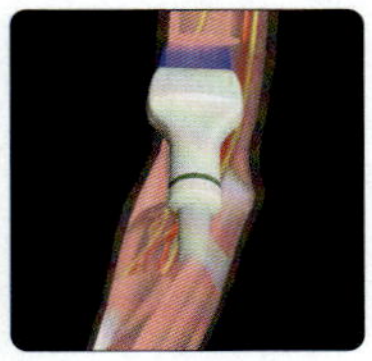

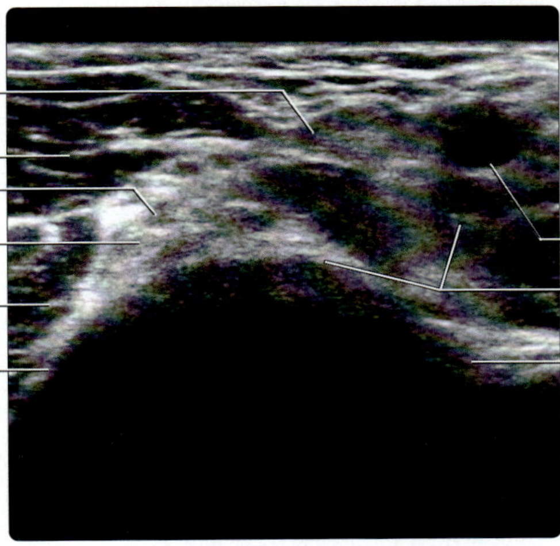

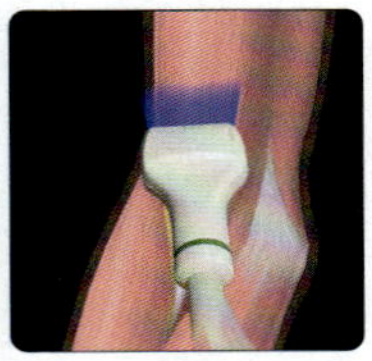

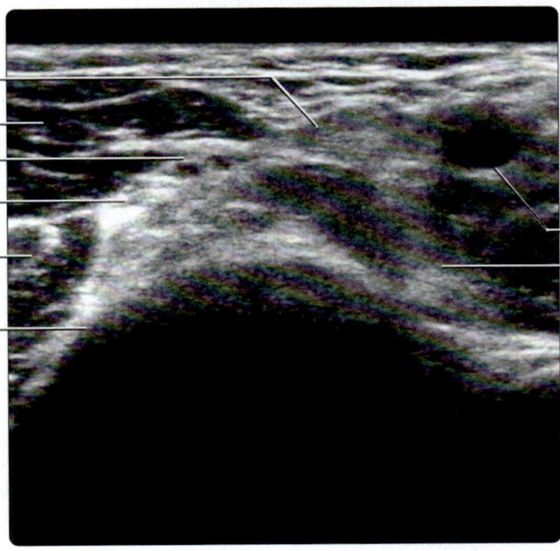

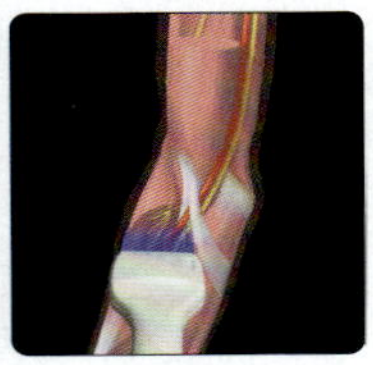

(Top) *Transverse grayscale US shows the anterior elbow and proximal antecubital fossa. The antecubital fossa or cubital fossa is the area anterior to the elbow joint.* **(Middle)** *Transverse grayscale US shows the antecubital fossa at the midportion. The antecubital fossa contains the radial nerve, biceps brachii tendon, brachial artery, and median nerve from lateral to medial orientation.* **(Bottom)** *Transverse grayscale US shows the distal antecubital fossa. The bicipital aponeurosis forms the roof of the antecubital fossa. The superficial veins of the elbow lie above the bicipital aponeurosis.*

TRANSVERSE US, BICEPS TENDON

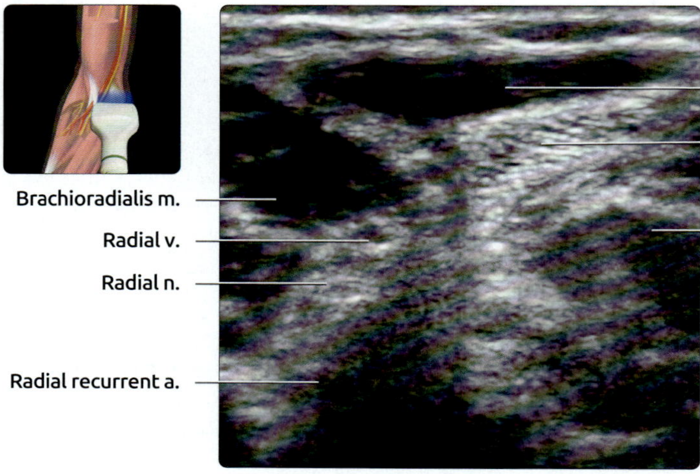

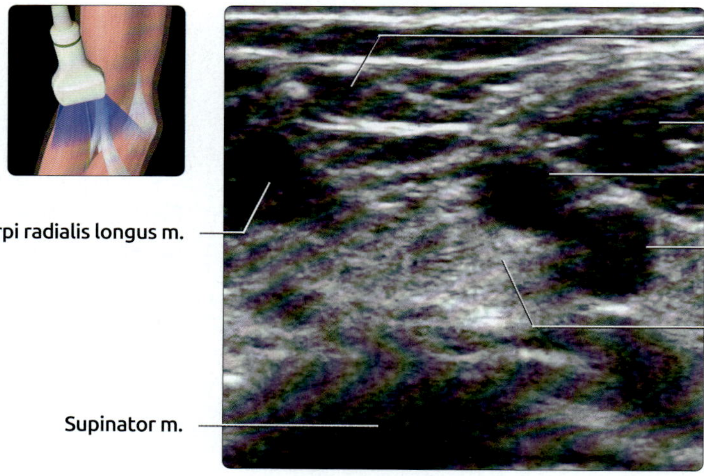

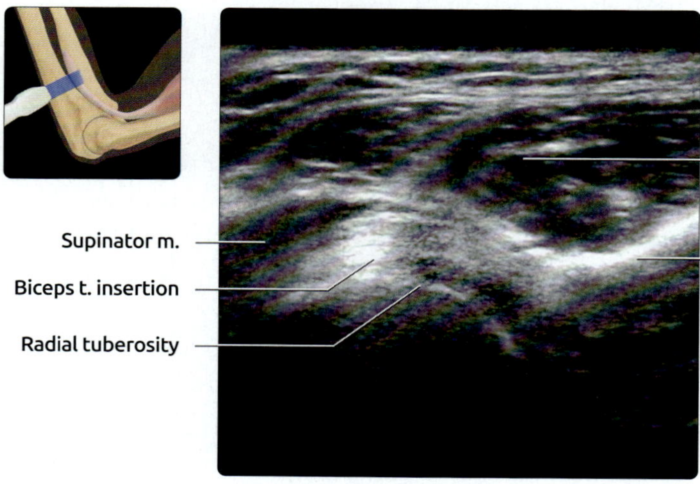

(Top) *Transverse grayscale US shows the biceps tendon at the upper antecubital fossa. The biceps tendon forms a few centimeters above the elbow joint.* **(Middle)** *Transverse grayscale US shows the biceps tendon at the midcubital fossa. The biceps tendon dips deeply and rotates as it passes to its insertion. The distal aspect of the biceps tendon and the insertion is difficult to see on US of the anterior surface of the elbow. Maximum supination will increase its visibility.* **(Bottom)** *Transverse grayscale US shows the biceps tendon insertional area. The insertional area of the biceps tendon is best seen with the elbow flexed and the wrist fully pronated. This positions the radial tuberosity posteriorly, allowing the insertional area to be viewed from the posterior aspect as shown here.*

TRANSVERSE AND LONGITUDINAL US, ANTEROLATERAL ELBOW

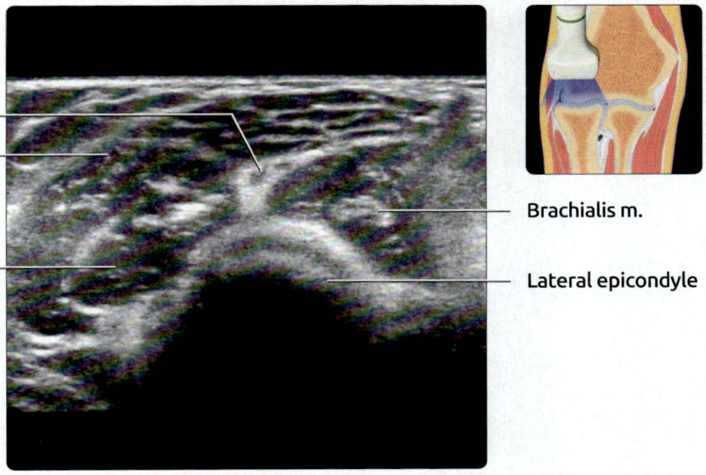

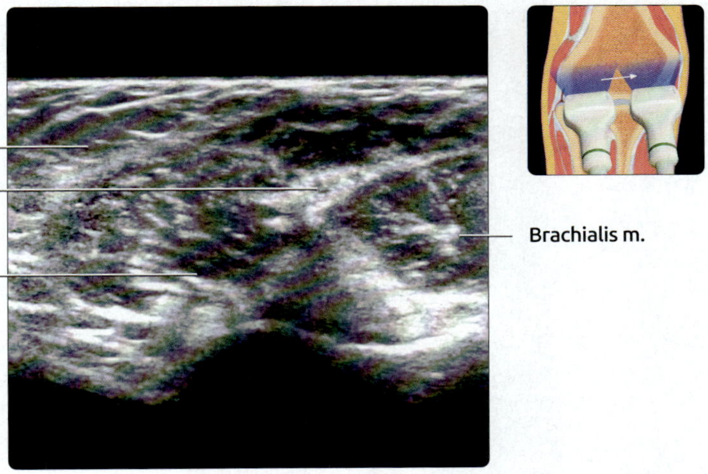

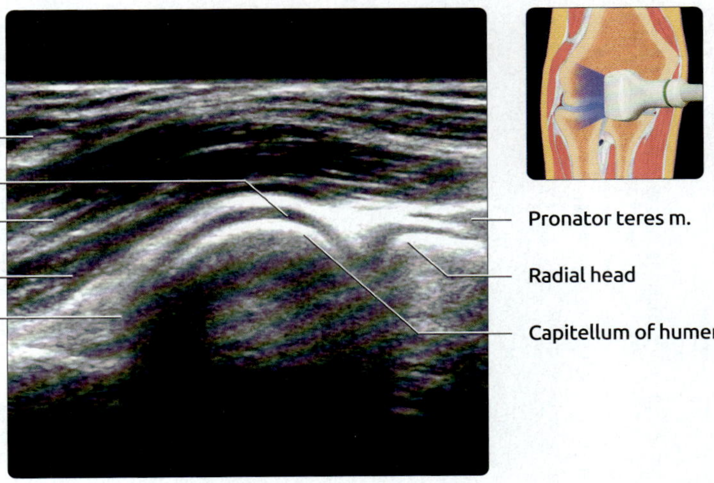

(Top) *Transverse grayscale US shows the proximal anterolateral aspect of the elbow. The radial nerve is located on the anterolateral aspect of the elbow. It can be readily identified deep to the brachioradialis muscle.* **(Middle)** *Transverse grayscale US shows the midsection of the anterolateral elbow. The radial nerve passes anterior to the epicondyle deep to the brachioradialis muscle.* **(Bottom)** *Longitudinal grayscale US shows the anterolateral aspect of the elbow. The fat pads are intracapsular but extrasynovial. The synovial space lies deep to the fat pad on the bone surface. Distension of the fat pads allows these to become visible radiographically in elbow joint effusions.*

LATERAL ELBOW

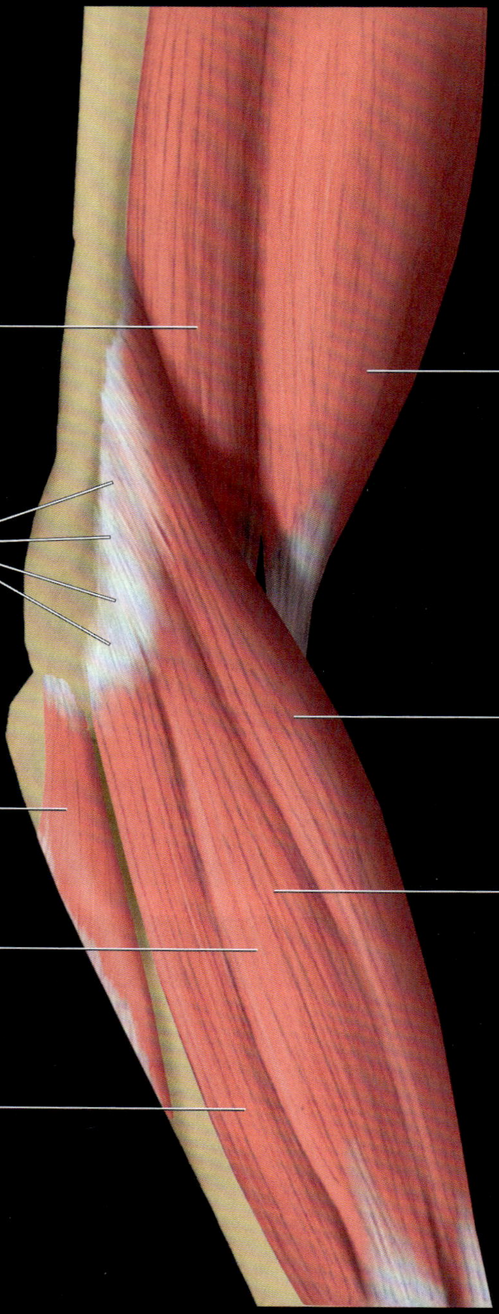

Side view of the lateral aspect of the elbow shows the extensor-supinator group and its common extensor tendon origin attaching to the lateral epicondyle and supracondylar aspect of the humerus. The common extensor tendon origin is composed of the extensor carpi radialis brevis, extensor digitorum, extensor digiti minimi, and extensor carpi ulnaris. The extensor digitorum origin and extensor carpi radialis brevis are very closely aligned in this section.

LONGITUDINAL US, COMMON EXTENSOR ORIGIN

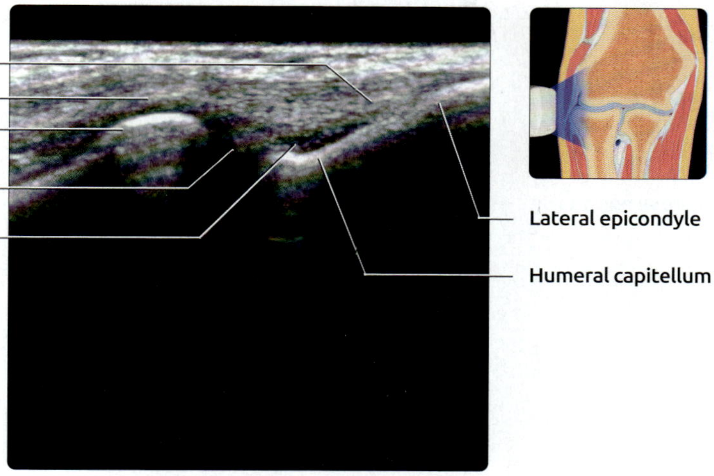

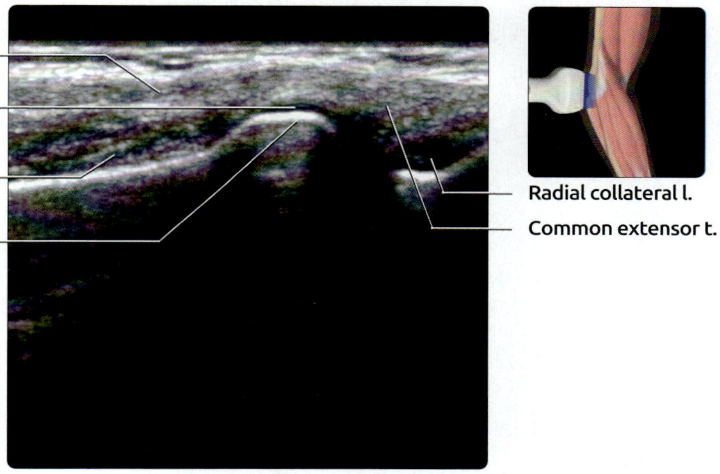

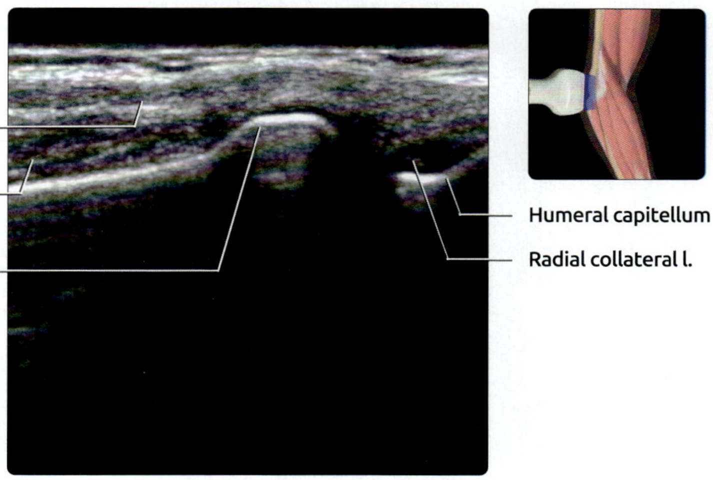

(Top) *Longitudinal grayscale US at the common extensor tendon origin is shown. Common extensor tendon origin has a broad attachment on the posteroinferior aspect of the lateral epicondyle. Tendon fibers cannot be separated into individual components at or near the insertion. Fibers from extensor digitorum make up most of the superficial portion, while fibers from extensor carpi radialis brevis make up most of the deep fibers. Contribution from extensor carpi ulnaris and extensor digiti minimi is only small.* **(Middle)** *Longitudinal grayscale US at common extensor tendon origin with the wrist in neutral position is shown. The radial collateral ligament is less distinct than the ulnar collateral ligament. It runs from the inferior aspect of the lateral epicondyle to gain attachment to the annular ligament.* **(Bottom)** *Longitudinal grayscale US at common extensor tendon origin with the wrist in extension is shown. Wrist extension slightly straightens extensor tendons, and dynamic movement during US can help separate movable tendons from fixed radial collateral ligament.*

TRANSVERSE US, COMMON EXTENSOR TENDONS

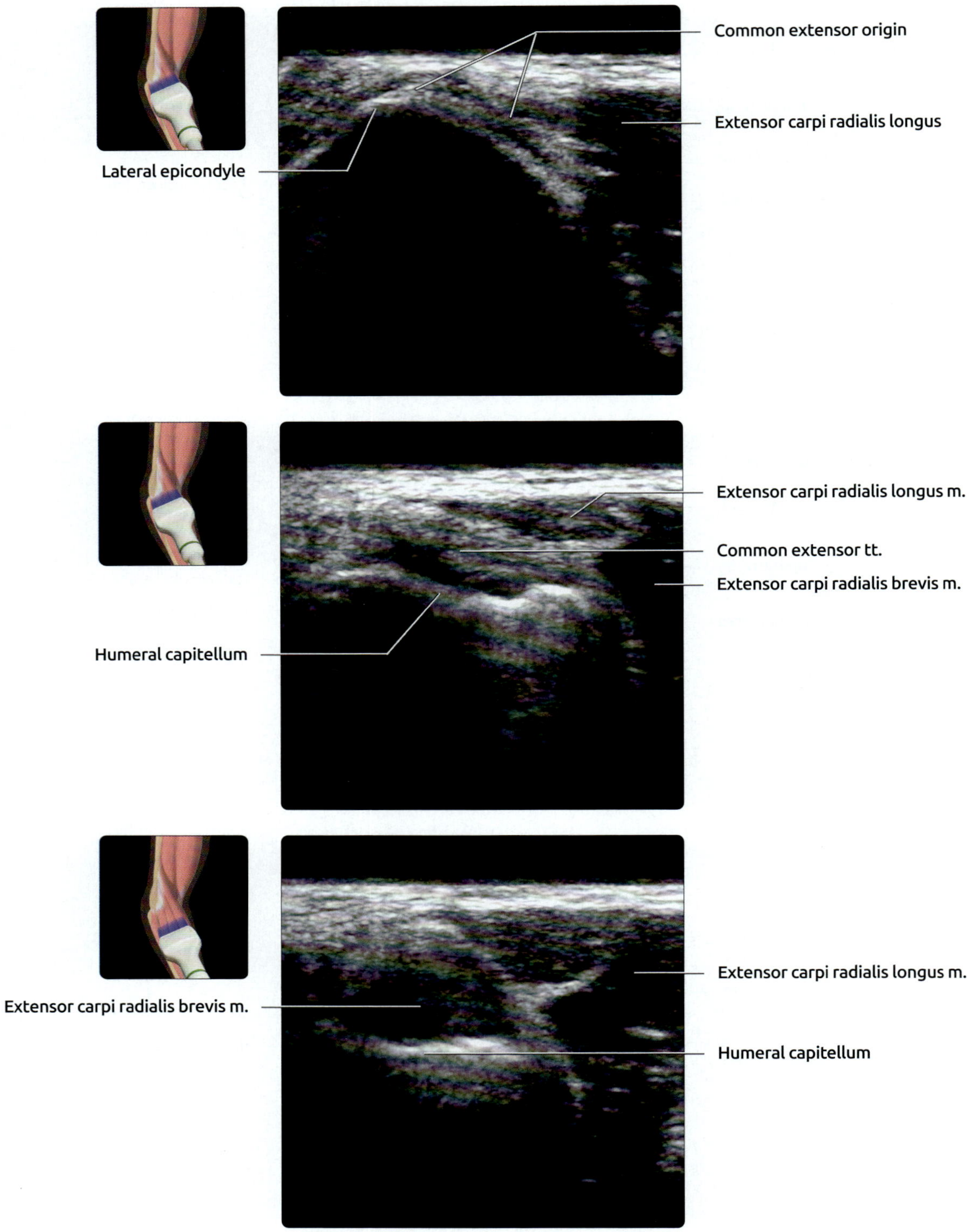

(Top) *Transverse grayscale US shows the common extensor origin. The common extensor tendon gains broad insertion onto the posterior aspect of the lateral epicondyle.* **(Middle)** *Transverse grayscale US shows the common extensor origin just distal to its attachment. The tendons begin to diverge just distal to the attachment, though they are not yet separate.* **(Bottom)** *Transverse grayscale US shows the common extensor tendons even more distal to the above section. The individual components of the common tendon origin begin to be seen near the elbow joint.*

TRANSVERSE AND LONGITUDINAL US, LATERAL ELBOW

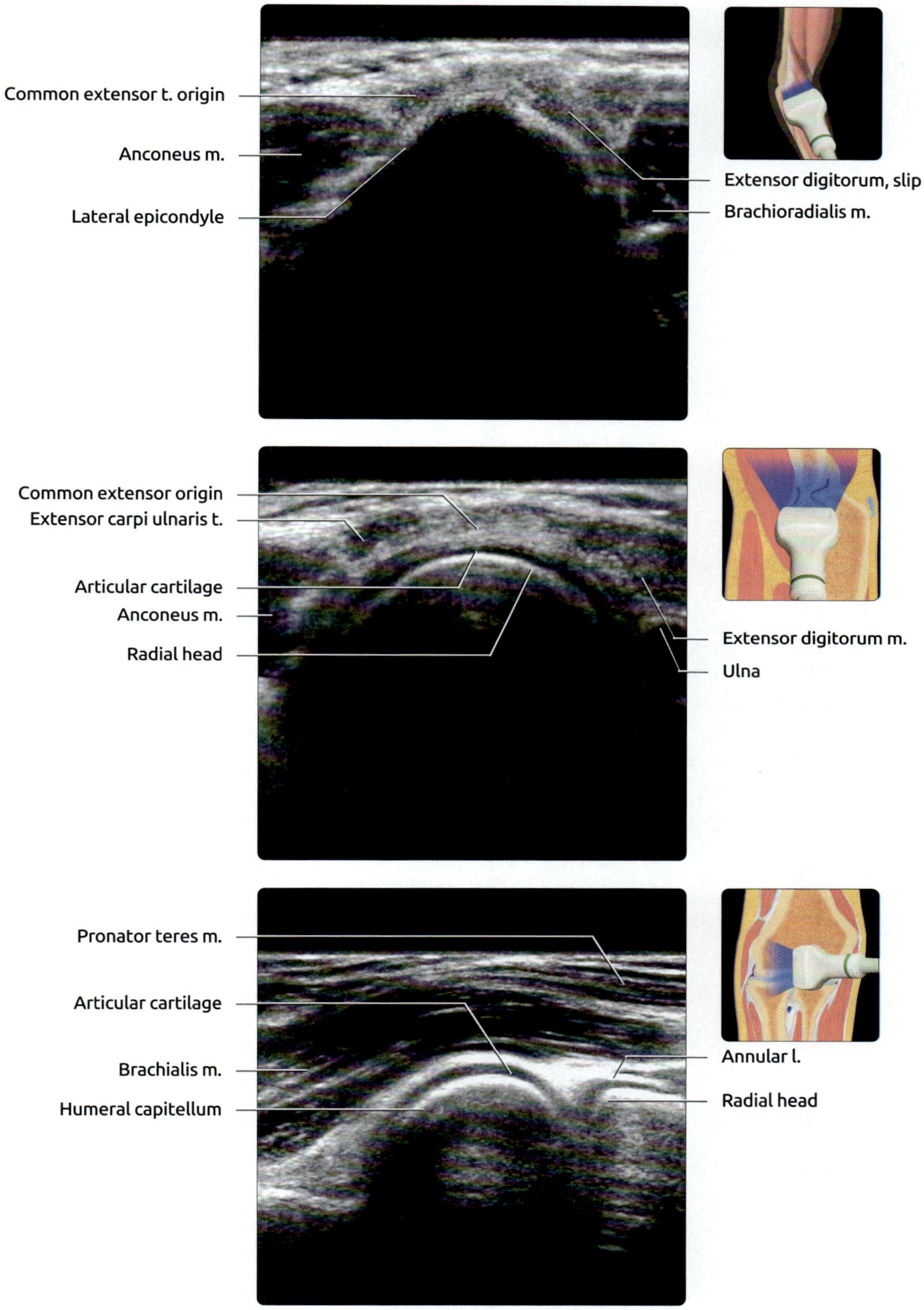

(Top) *Transverse grayscale US shows the common extensor tendon origin. The tendon caliber is often better appreciated on transverse, rather than on longitudinal, imaging.* **(Middle)** *Transverse grayscale US shows the common extensor tendons at the level of the radial head. The extensor tendons that originate from the common tendon origin can be traced distally, allowing them to be seen as separate tendons (extensor digitorum, extensor digiti minimi, extensor carpi ulnaris, and extensor carpi radialis brevis).* **(Bottom)** *Longitudinal grayscale US shows the common extensor tendon origin. The caliber of the common extensor tendon origin can be best appreciated on transverse imaging.*

TRANSVERSE AND LONGITUDINAL US, ANTEROMEDIAL ELBOW

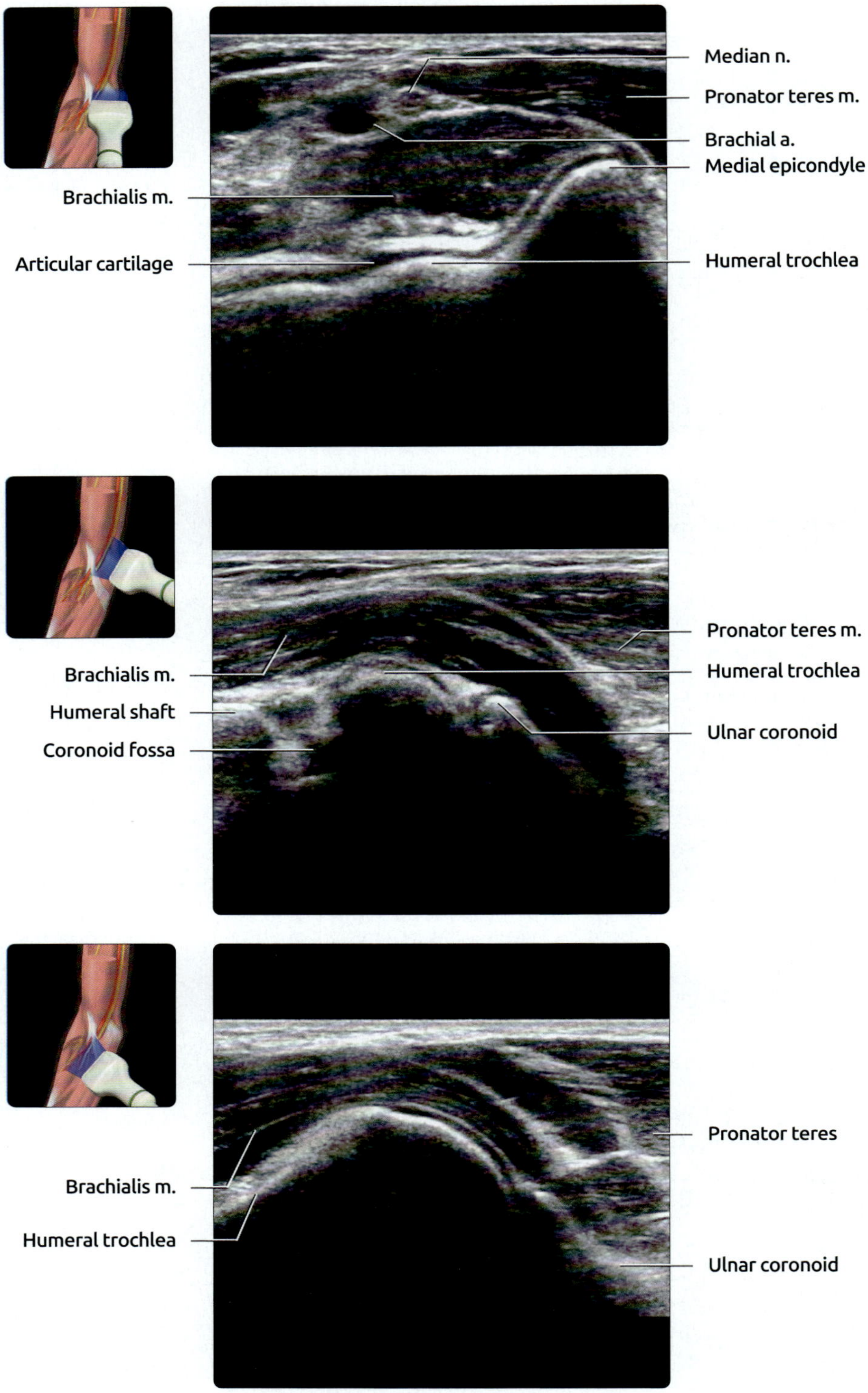

(Top) *Transverse grayscale US shows the anteromedial aspect of elbow. The pronator teres has 2 heads: 1 from the common flexor origin (medial epicondyle) and supracondylar region and 1 from the medial side of the coronoid process. The median nerve lies deep to the pronator teres and may be trapped between these 2 heads as it enters the forearm.* **(Middle)** *Longitudinal grayscale US shows the anteromedial aspect of the elbow. Centrally at the junction of the distal humerus between the coronoid fossa anteriorly and the olecranon fossa posteriorly, the humerus is very thin and potentially simulating a bone defect.* **(Bottom)** *Longitudinal grayscale US shows the anteromedial aspect of the elbow. The brachialis muscle is closely applied to the anterior aspect of the elbow joint as it descends to its insertion on the ulnar tuberosity and the coronoid process of the ulna.*

FLEXOR ASPECT

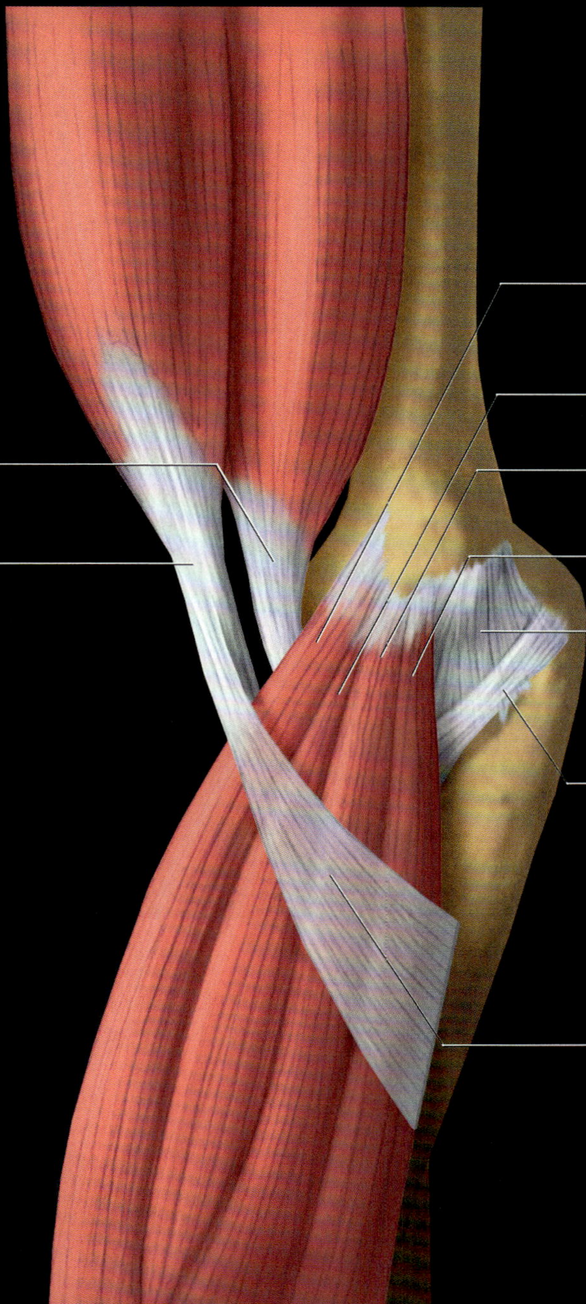

Side view of the medial aspect of the elbow shows the flexor-pronator group and its common flexor tendons attaching to the medial epicondyle. The anterior band of the ulnar collateral ligament is deep to the common flexor tendons. The biceps aponeurosis blends with the anterior aspect of the common flexor muscle mass. The flexor digitorum profundus (not shown) arises from the proximal and midulna, posterior and deep to the flexor carpi ulnaris. It does not act on the elbow and is not part of the flexor-pronator group.

LIGAMENTS AROUND ELBOW

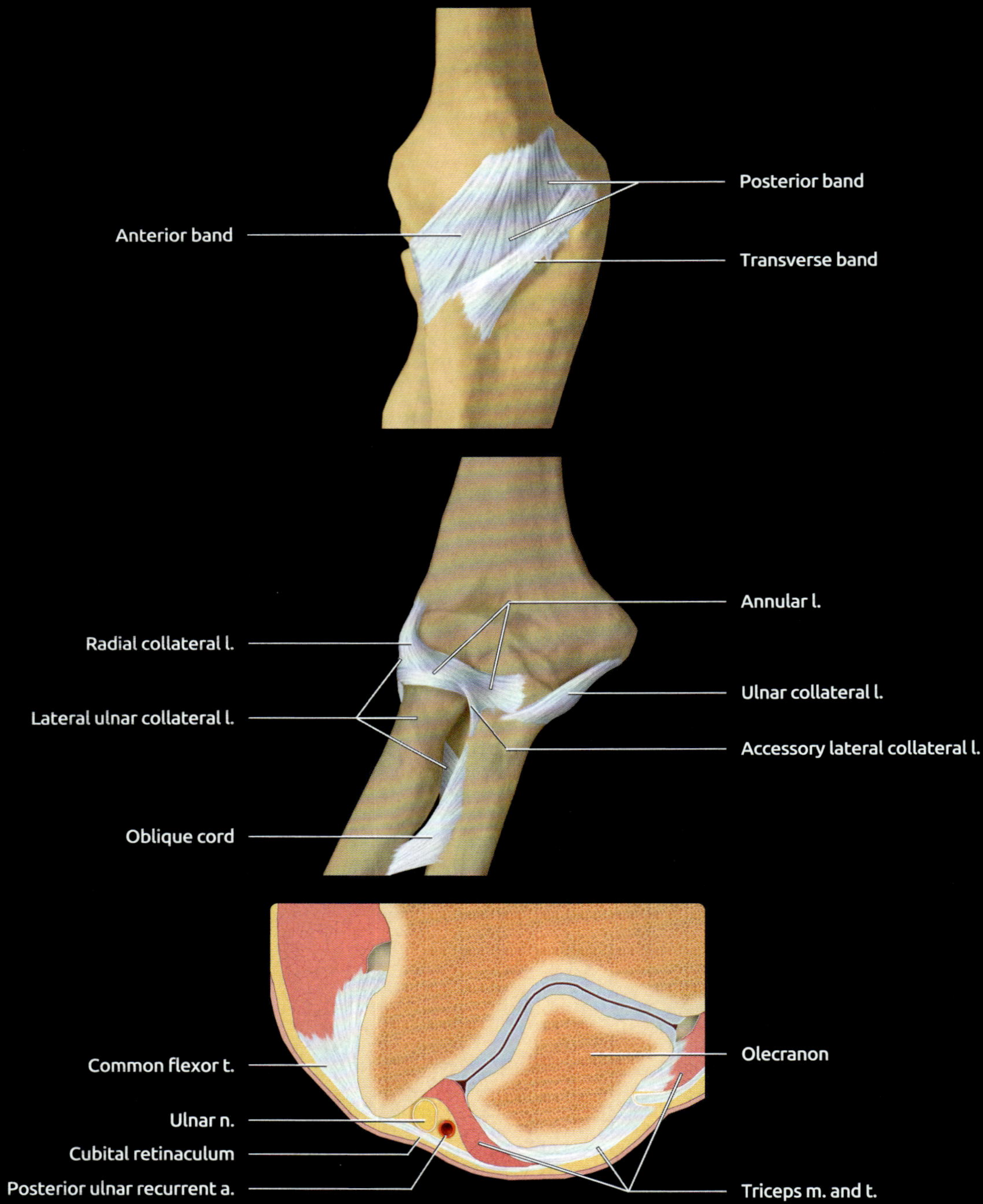

(Top) *Side view of the medial aspect of the elbow shows the 3 components of the ulnar collateral ligament: Anterior, posterior, and transverse band.* **(Middle)** *Anterior view of the elbow shows the radial collateral ligament complex, which is composed of the radial collateral ligament (provides varus stability), lateral ulnar collateral ligament (provides posterolateral stability), annular ligament (holds radial head against radial notch of the ulna), and accessory lateral collateral ligament (reinforces annular ligament). The oblique cord is part of the proximal radioulnar joint.* **(Bottom)** *Axial graphic shows the cubital tunnel. The ulnar nerve may be compressed within the cubital tunnel (cubital tunnel syndrome) by a mass, posttraumatic osseous deformity, or aneurysm of the recurrent ulnar artery. The ulnar nerve may also sublux out of the cubital tunnel leading to a low-grade traumatic neuritis and symptoms of carpal tunnel syndrome.*

LONGITUDINAL US, COMMON FLEXOR ORIGIN

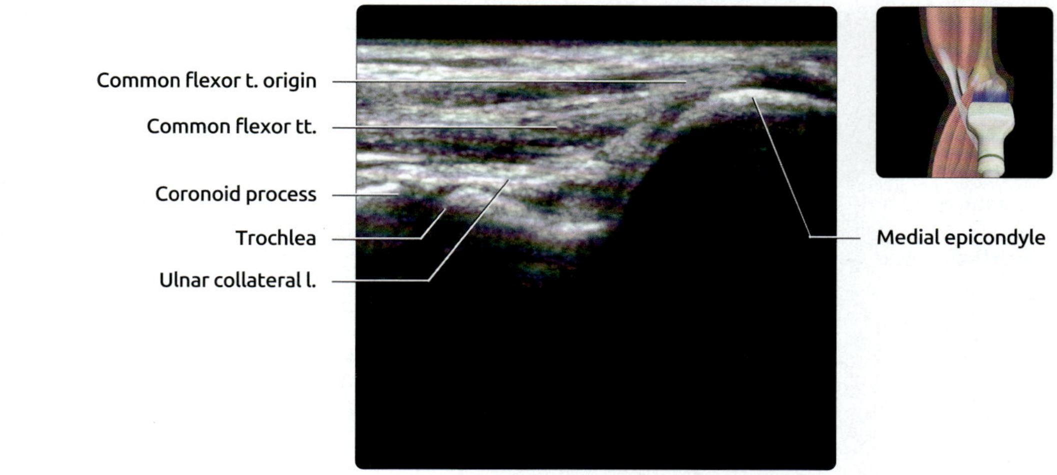

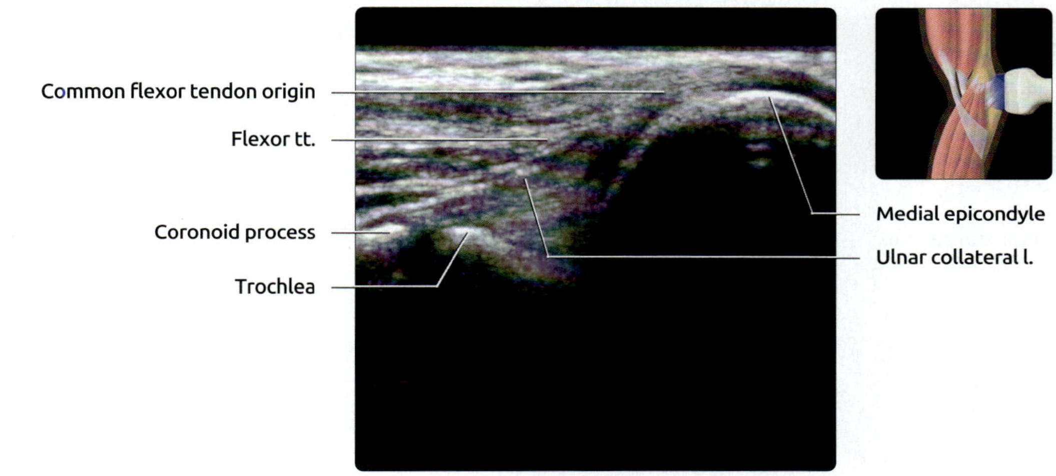

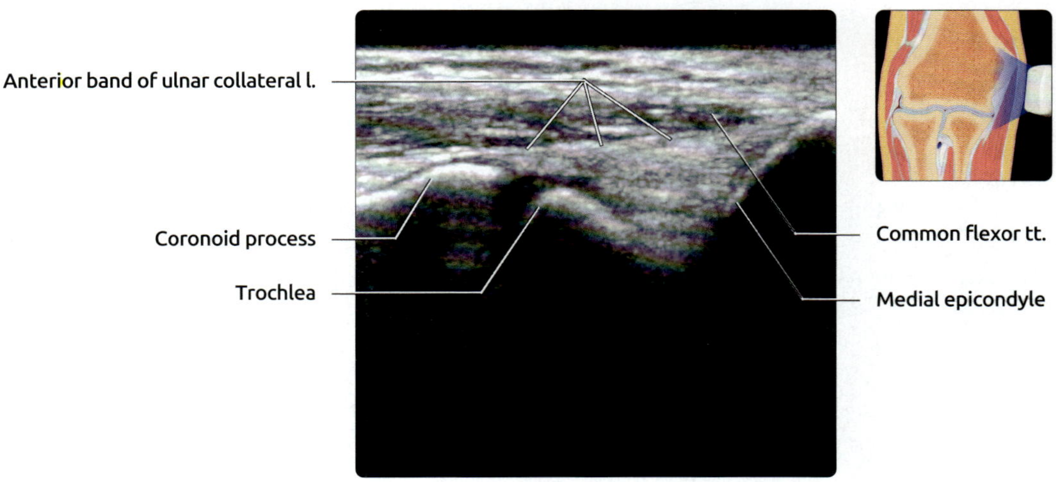

(Top) *Longitudinal grayscale US shows the common flexor tendon origin. These tendons (pronator teres, flexor carpi radialis, palmaris longus, flexor digitorum superficialis, flexor carpi ulnaris) are inseparable at or close to the origin.* **(Middle)** *Longitudinal grayscale US shows the ulnar collateral ligament. The ulnar collateral ligament runs from the medial epicondyle to the coronoid process. The ulnar collateral ligament lies just deep to the flexor tendons and may be injured along with these tendons.* **(Bottom)** *Longitudinal grayscale US shows the ulnar collateral ligament anterior band. The anterior band (medial epicondyle-sublime tubercle) is the strongest component of the ulnar collateral ligament.*

TRANSVERSE US, COMMON FLEXOR ORIGIN

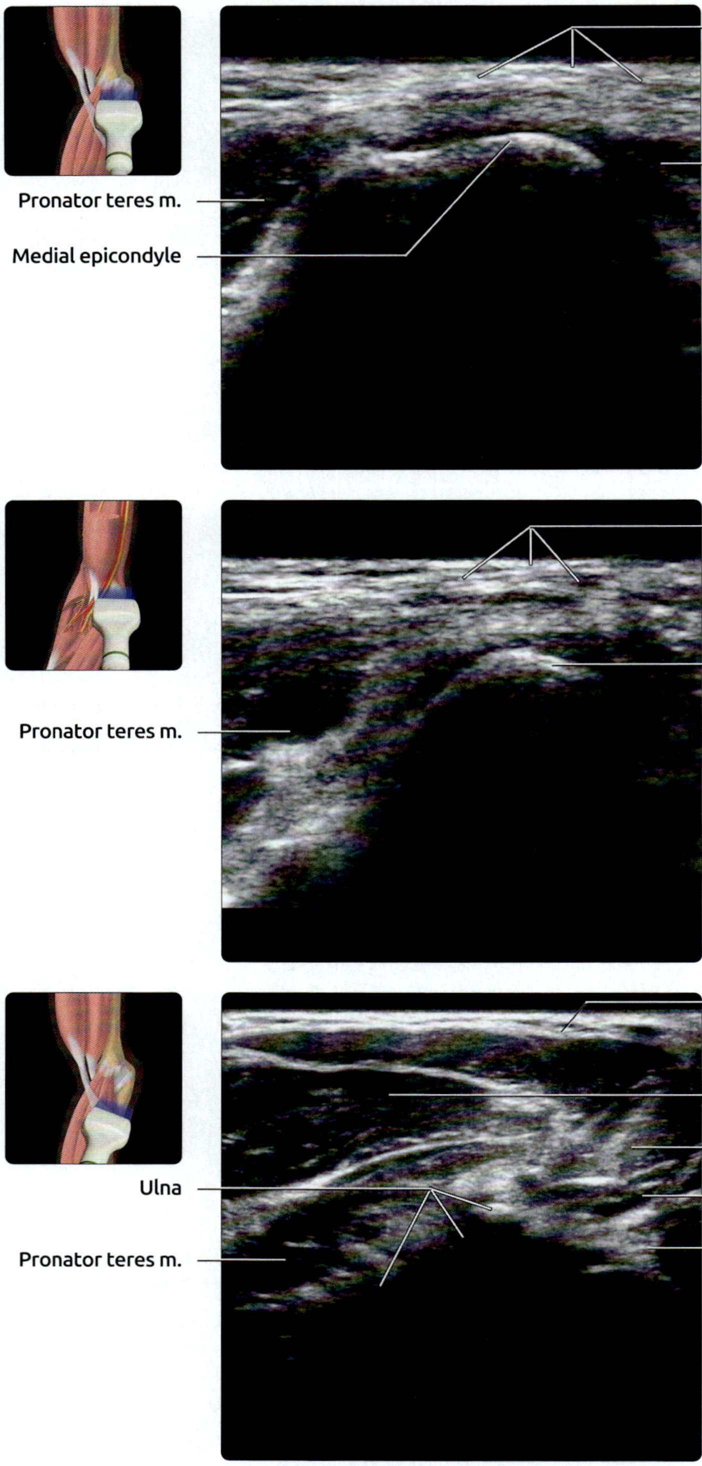

(Top) *Transverse grayscale US shows the common flexor tendon origin. Tendinosis of the common flexor tendon is termed golfer's elbow or medial epicondylitis. Yet most people who get this injury are involved in other sports, particularly tennis.* **(Middle)** *Transverse grayscale US shows the common flexor tendons just distal to the elbow. The different components of the common flexor tendons can be seen by following the tendon distally and proximally.* **(Bottom)** *Transverse grayscale US shows the forearm more distal to the common flexor origin.*

TRANSVERSE US, POSTERIOR ELBOW

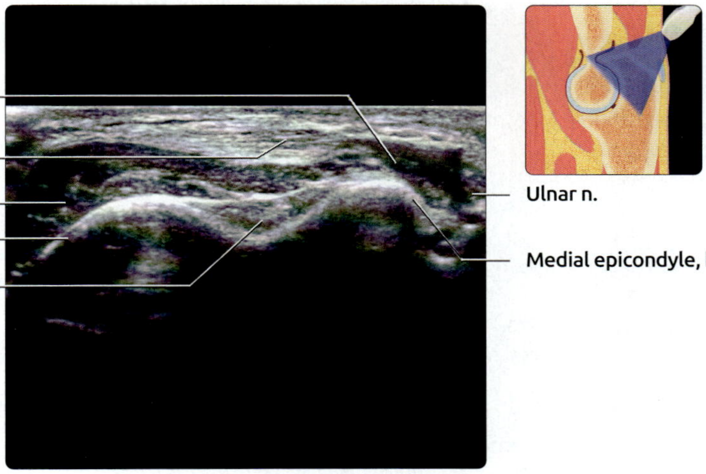

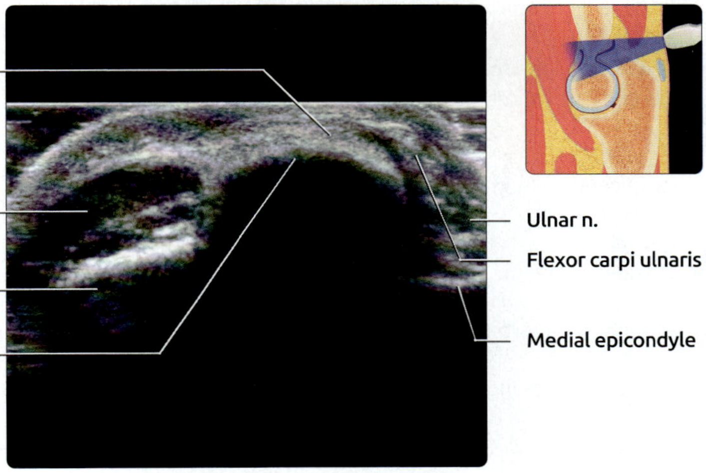

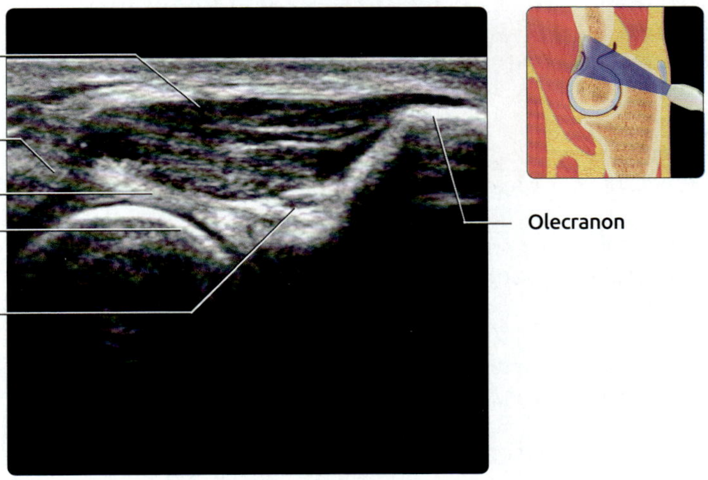

(Top) *Transverse grayscale US shows the posterior elbow. There are 3 fat pads in the elbow: 2 anterior fat pads (radial and coronoid fossa) and 1 posterior fat pad (olecranon fossa). The synovium extends from the articular surface of the humerus and lines these fossae, i.e., it lies deep to the fat pads, which are enclosed by the elbow capsule. The fat pads therefore are intraarticular but extrasynovial. Fluid distension of the synovial cavity elevates the fat pads.* **(Middle)** *Transverse grayscale US shows the elbow at the level of the olecranon. The anconeus is a small muscle on the posterior aspect of the elbow that extends from the lateral epicondyle to the olecranon.* **(Bottom)** *Transverse grayscale US shows the proximal radioulnar joint. The annular ligament is attached to the anterior and posterior margins of the radial notch of the ulna.*

IMAGING ANATOMY

Flexors

- Deep flexor group
  - Flexor digitorum profundus
    - Origin: Proximal ulna
    - Insertion: Index, middle, ring, and little finger distal phalangeal bases
  - Flexor pollicis longus
    - Origin: Radius, interosseous membrane, coronoid process
    - Insertion: Thumb distal phalangeal base
  - Pronator quadratus
    - Origin: Medial distal volar ulna
    - Insertion: Lateral distal dorsal radius
- Superficial
  - Flexor carpi radialis (FCR)
    - Origin: Medial epicondyle/common flexor tendon
    - Insertion: 2nd metacarpal (MC) base; slip to 3rd MC base
  - Palmaris longus (PL)
    - Origin: Medial epicondyle/common flexor tendon
    - Insertion: Volar distal flexor retinaculum and palmar aponeurosis
  - Flexor carpi ulnaris (FCU)
    - Origin: Medial epicondyle (humeral head) and medial proximal ulna (ulnar head)
    - Insertion: Pisiform
  - Flexor digitorum superficialis
    - Origin: Medial epicondyle and ulnar coronoid process (humeroulnar head); volar proximal radius (radial head)
    - Insertion: Middle phalanges index, middle, ring, and little fingers

Extensors

- Deep
  - Abductor pollicis longus (APL)
    - Origin: Dorsal lateral ulna, dorsal midradius
    - Insertion: 1st MC radial base
  - Extensor pollicis brevis (EPB)
    - Origin: Dorsal midradius
    - Insertion: Thumb proximal phalangeal base
  - Extensor pollicis longus (EPL)
    - Origin: Dorsal midulna
    - Insertion: Thumb distal phalanx base
    - Variant: Fused with EPB
  - Extensor indicis
    - Origin: Dorsal midulna and interosseous membrane
    - Insertion: Extensor hood of index finger
- Superficial
  - Extensor carpi radialis longus (ECRL)
    - Origin: Lateral epicondyle/common extensor tendon
    - Insertion: Dorsal radial 2nd MC
  - Extensor carpi radialis brevis (ECRB)
    - Origin: Lateral epicondyle/common extensor tendon
    - Insertion: Dorsal radial 3rd MC
  - Extensor digitorum (ED)
    - Origin: Lateral epicondyle/common extensor tendon
    - Insertion: Middle and distal phalanges of index, middle, ring, little fingers
  - Extensor digiti minimi
    - Origin: Lateral epicondyle/common extensor tendon
    - Insertion: Extensor hood of little finger proximal phalanx with slip to ring finger
  - Extensor carpi ulnaris (ECU)
    - Origin: Common extensor tendon and dorsal ulna
    - Insertion: Dorsal ulnar 5th MC base

Anomalous Muscles

- May present as soft tissue mass
  - May create neural compression
- Accessory PL
  - Superficial to flexor digitorum tendons, medial to FCR
- ED manus brevis
  - Arises from distal radius or dorsal radiocarpal ligament
  - Inserts on 2nd MC
  - May be tender or present as mass
- Extensor carpi radialis intermedius
  - Arises from humerus or as accessory slip from ECRB or ECRL
  - Inserts on 2nd &/or 3rd MC
- Extensor carpi radialis accessory
  - Arises from humerus or ECRL
  - Inserts on 1st MC bone, abductor pollicis brevis, or 1st dorsal interosseous muscle

ANATOMY IMAGING ISSUES

Imaging Pitfalls

- Many variations of flexor and extensor muscles and tendons
- EPB may be absent or fused with EPL
- PL absent in 10%; short tendon with extended muscle belly may compress median nerve
- Multiple tendon slips can mimic longitudinal tendon tears (e.g., APL)
- Small amount of fluid common in tendon sheath (e.g., ECRB, ECRL, ECU)

Imaging Issues

- APL and EPB intersect ECRB and ECRL just proximal to extensor retinaculum; may impinge at musculotendinous intersection ("intersection syndrome")
- Sites of radial nerve entrapment in forearm
  - Posterior interosseous nerve within radial tunnel under arcade of Frohse
  - Posterior interosseous nerve between deep and superficial heads of supinator muscle
  - Superficial radial nerve between brachioradialis and ECRL in mid- to distal forearm
- Sites of ulnar nerve entrapment in forearm
  - Cubital tunnel
  - Beyond cubital tunnel, it passes between humeral and ulnar origins of FCU muscle
- Sites of medial nerve entrapment in forearm
  - Between heads of pronator teres

COMMON EXTENSOR TENDON

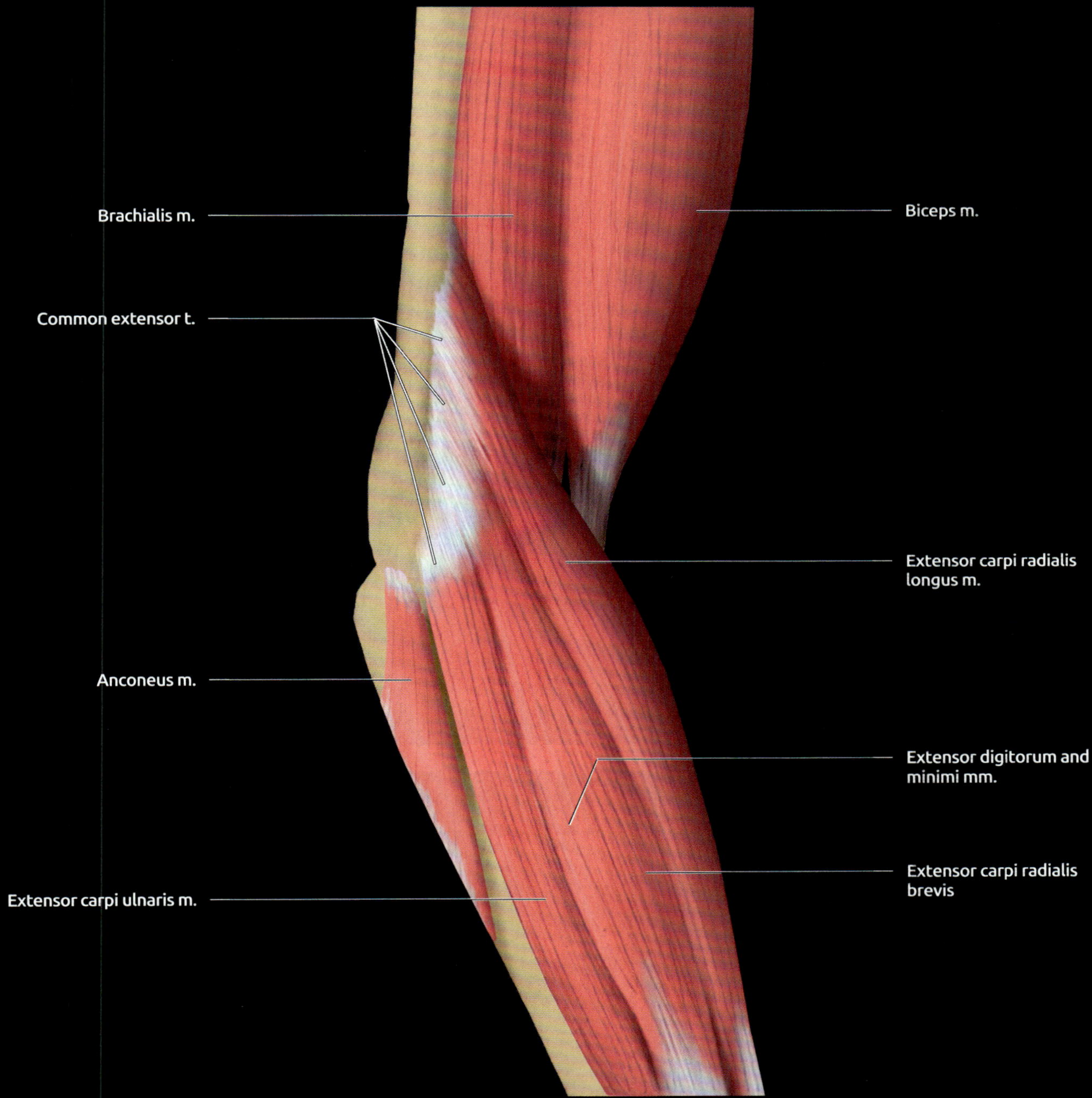

Graphic of a side view of the lateral aspect of the elbow shows the extensor-supinator group and its common extensor tendon origin attached to the lateral epicondyle and supracondylar aspect of the humerus. The common extensor tendon origin is composed of the extensor carpi radialis brevis, extensor digitorum, extensor digiti minimi, and extensor carpi ulnaris.

TRANSVERSE US, ANTERIOR FOREARM

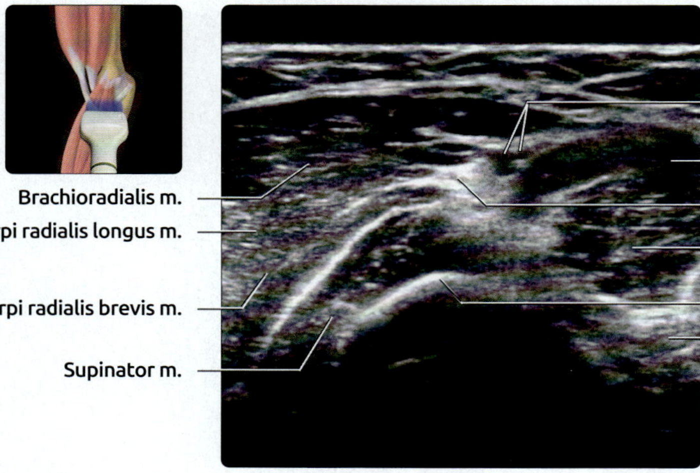

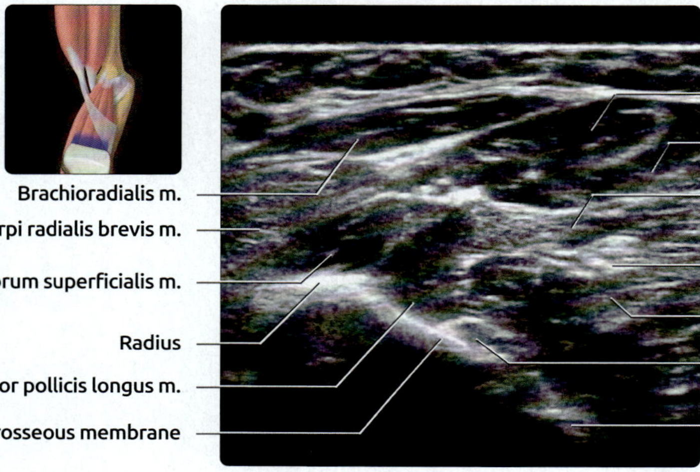

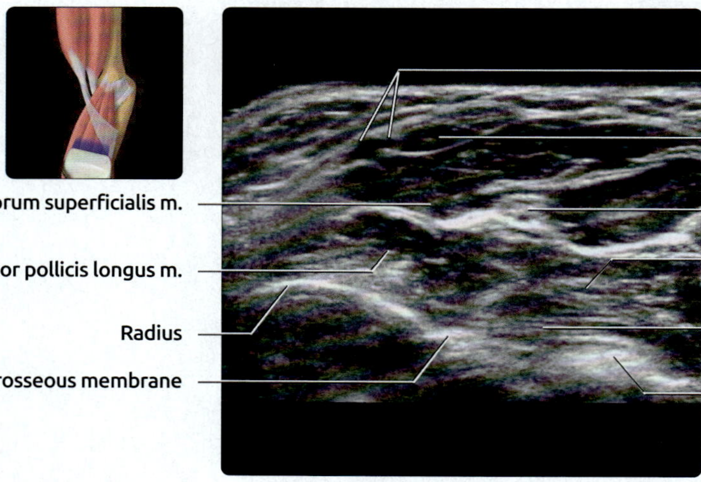

(Top) *Transverse grayscale US shows the anterior aspect of the proximal 1/3 of the forearm. The anterior aspect of the forearm consists of 5 flexor muscles: 3 superficial and 2 deep.* **(Middle)** *Transverse grayscale US shows the anterior aspect of the midforearm. The 3 superficial flexor muscles are, from ulnar to radial aspect, the flexor carpi ulnaris, flexor digitorum superficialis, and flexor carpi radialis. The 2 deep flexor muscles are flexor digitorum profundus and flexor pollicis longus.* **(Bottom)** *Transverse grayscale US shows the anterior aspect distal 1/3 of the forearm. The pronator quadratus is an easily recognized muscle that helps to pronate the forearm.*

TRANSVERSE US, ANTEROMEDIAL FOREARM

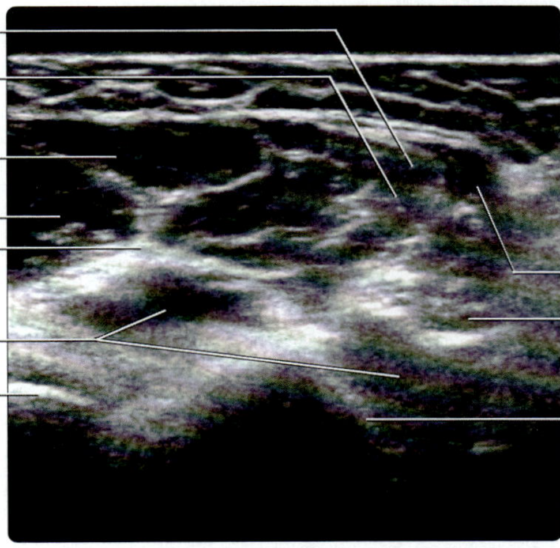

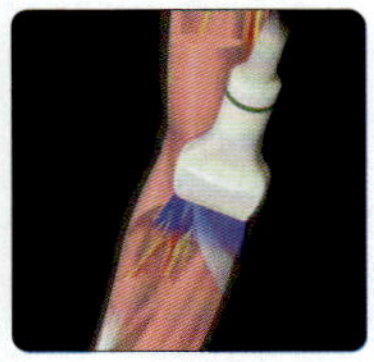

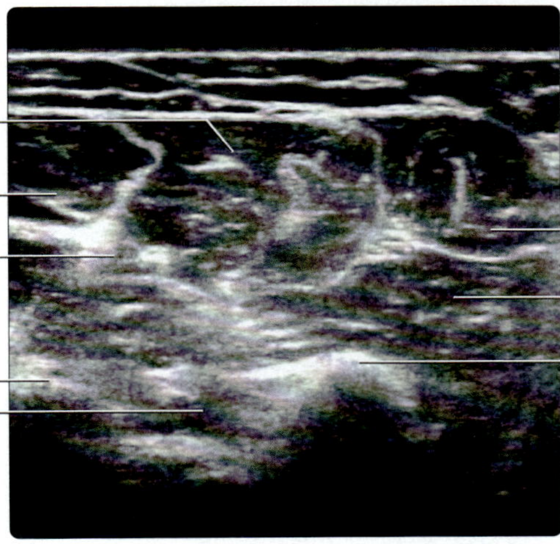

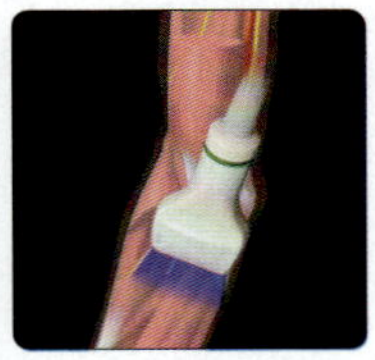

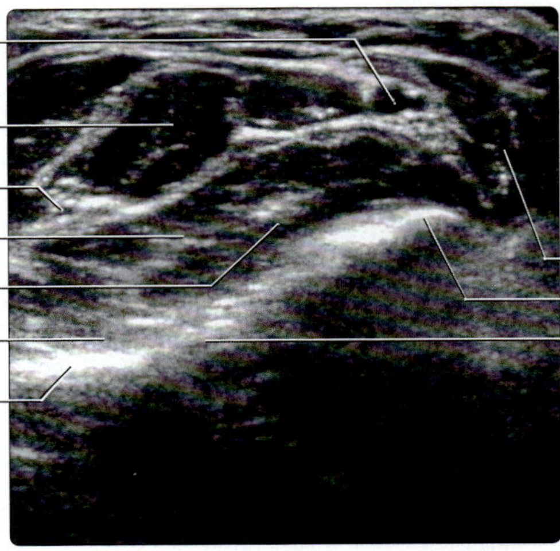

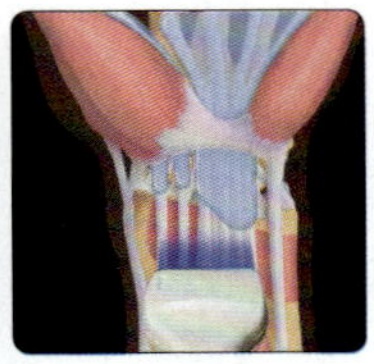

(Top) *Transverse grayscale US shows the anteromedial aspect of the proximal 1/3 of the forearm. The main flexors of the wrist are contained along the anterior and anteromedial aspect of the forearm. These are divided into 2 groups: Superficial and deep.* **(Middle)** *Transverse grayscale US shows the anteromedial aspect of the middle 1/3 of the forearm. The superficial group consists of the flexor carpi ulnaris, the flexor digitorum superficialis, and the flexor carpi radialis. The deep group consists of the flexor digitorum profundus and the flexor pollicis longus.* **(Bottom)** *Transverse grayscale US shows the anteromedial aspect of the distal 1/3 of the forearm. The ulnar nerve lies deep to the flexor carpi ulnaris, while the median nerve lies between the flexor digitorum superficialis and flexor digitorum profundus muscles.*

TRANSVERSE US, ANTEROLATERAL FOREARM

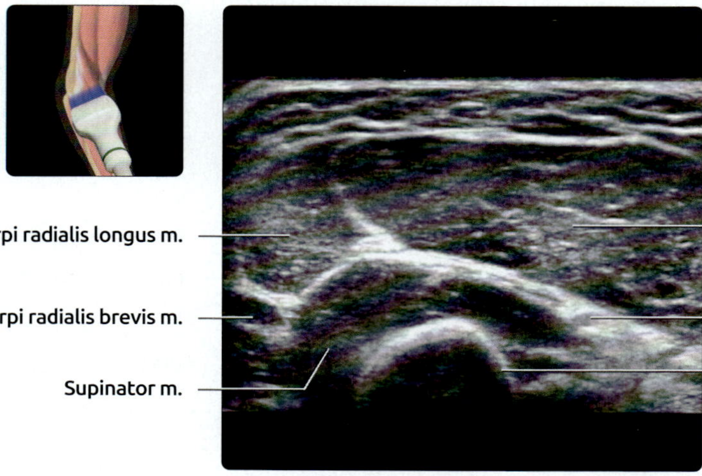

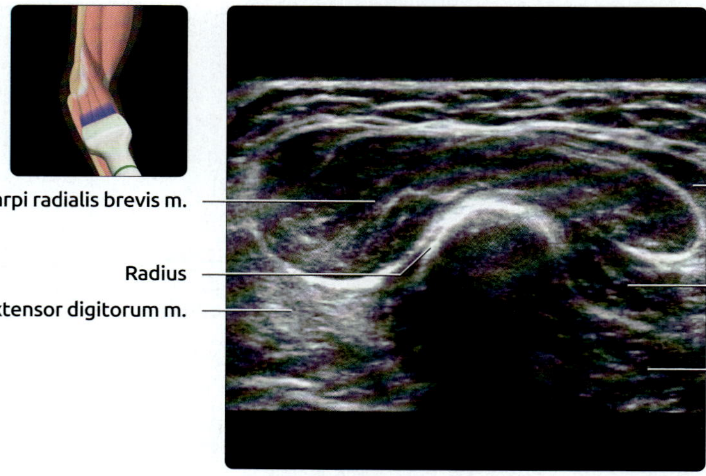

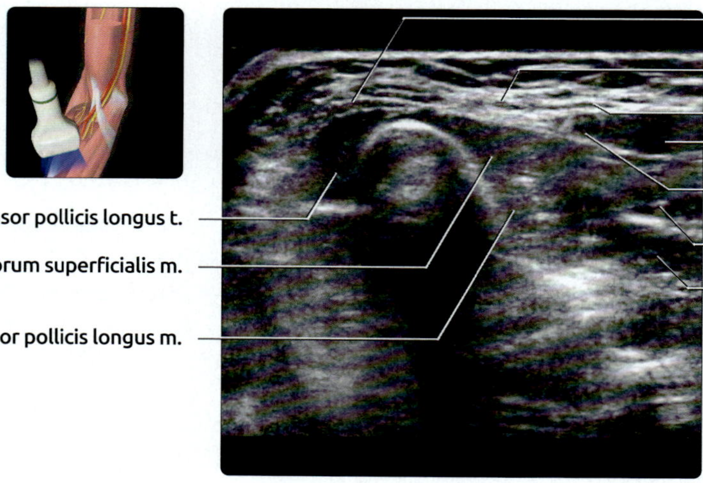

(Top) *Transverse grayscale US shows the anterolateral proximal 1/3 of the forearm. The anterolateral aspect of the forearm comprises 4 extensor muscles running along the course of the radius.* **(Middle)** *Transverse grayscale US shows the anterolateral aspect of the middle 1/3 of the forearm. The anterolateral group comprises the extensor carpi radialis brevis, extensor carpi radialis longus, brachioradialis, and the pronator teres.* **(Bottom)** *Transverse grayscale US shows the anterolateral aspect of the distal 1/3 of the forearm.*

ANTERIOR FOREARM, ARTERIES AND NERVES

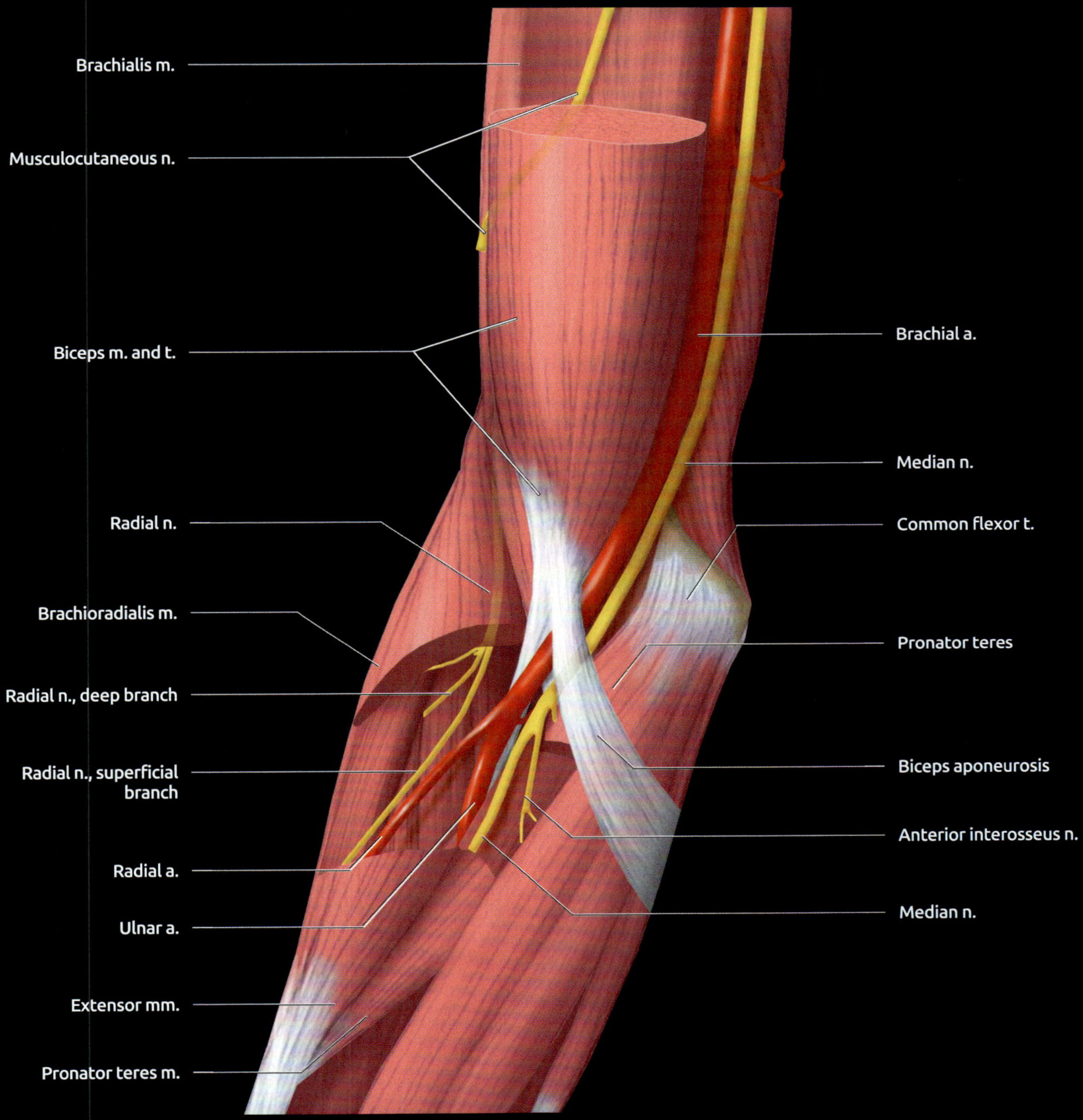

Graphic of the anterior view of the cubital fossa shows the median nerve and brachial artery passing underneath the biceps aponeurosis. The anterior interosseus nerve arises from the median nerve as the median nerve passes between the 2 heads of the pronator teres muscle. The radial nerve, deep to the brachioradialis muscle, is seen dividing into the superficial and deep branches.

COMMON FLEXOR TENDON

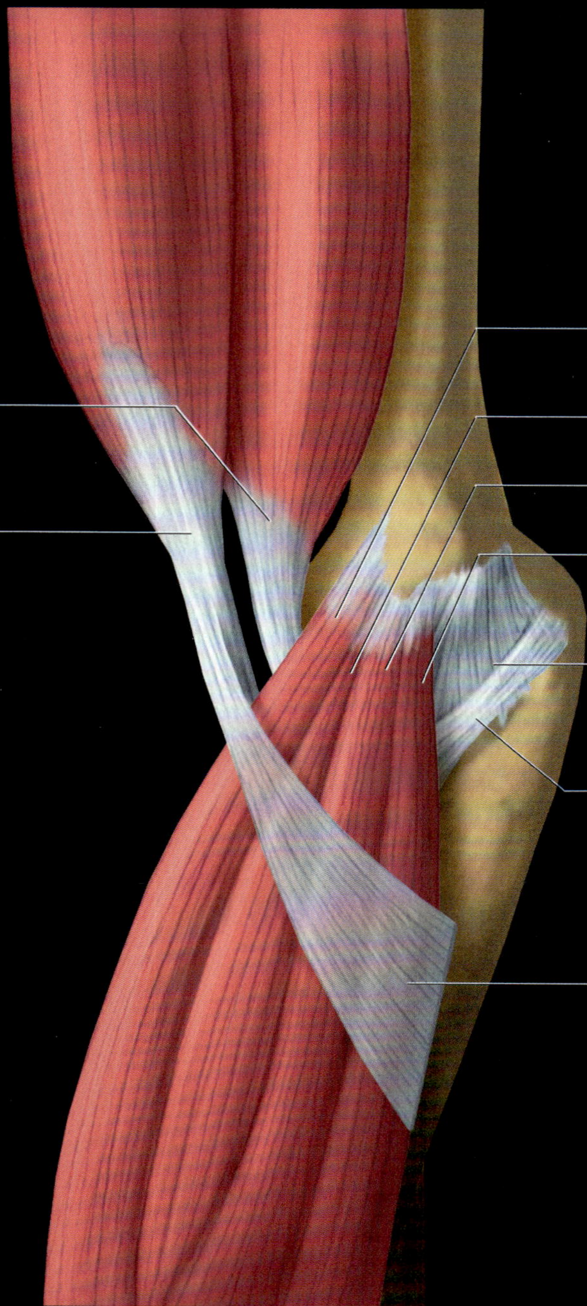

Graphic of the side view of the medial aspect of the elbow shows the flexor-pronator group and the common flexor tendon origin attaching to the medial epicondyle. The anterior band of the ulnar collateral ligament is deep to the common flexor tendon. The biceps aponeurosis blends with the anterior aspect of the common flexor muscle mass. The flexor digitorum profundus arises from the proximal ulna, posterior and deep to the flexor carpi ulnaris. It does not act on the elbow and is not part of the flexor-pronator group.

TRANSVERSE US, POSTERIOR ASPECT

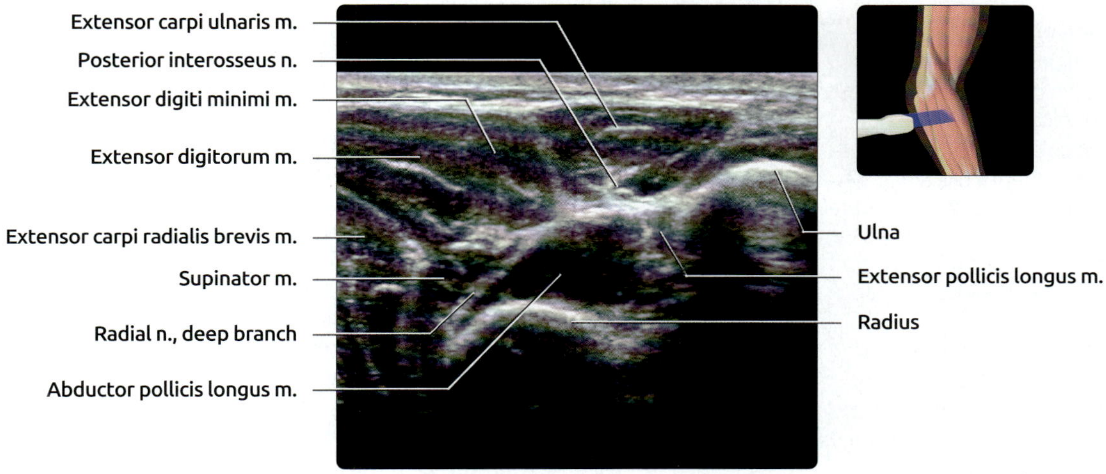

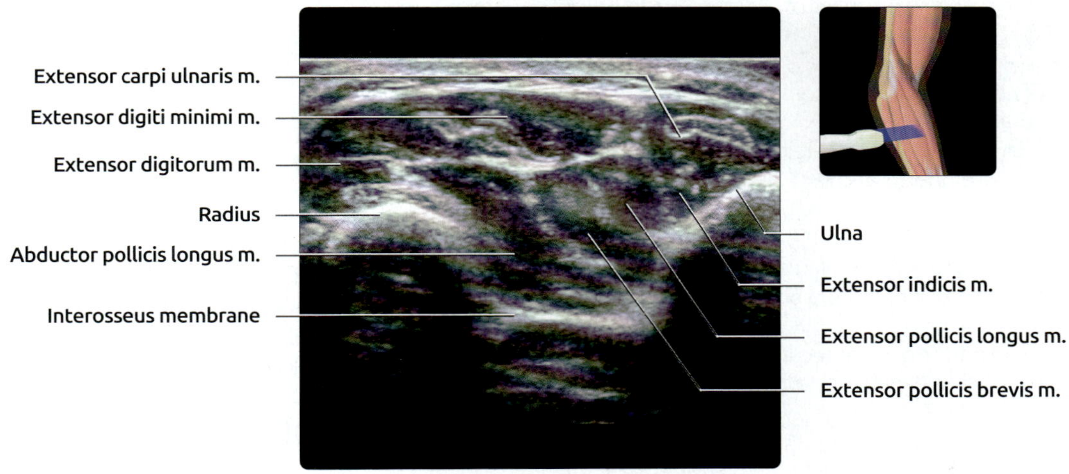

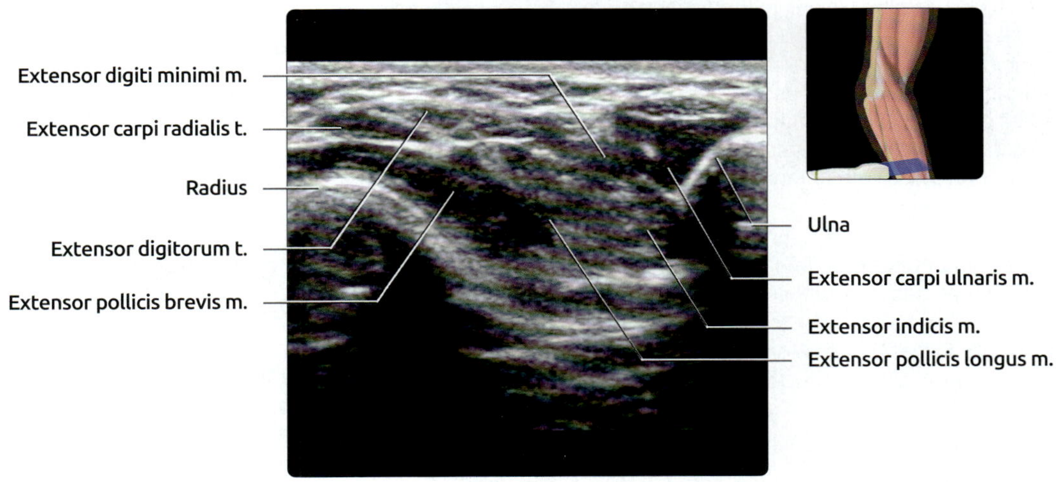

(Top) *Transverse grayscale US shows the posterior aspect of the proximal 1/3 of the forearm. The posterior aspect of the forearm contains the extensor muscles. They are contained in 2 groups: Superficial and deep.* **(Middle)** *Transverse grayscale US shows the posterior aspect of the middle 1/3 of the forearm. The superficial group comprises (ulnar to radial) the extensor carpi ulnaris, extensor digiti minimi, and extensor digitorum. The deep group comprises the extensor pollicis longus, extensor pollicis brevis, and abductor pollicis longus.* **(Bottom)** *Transverse grayscale US shows the posterior aspect of the distal 1/3 of the forearm.*

GROSS ANATOMY

Arteries

- Brachial artery
  - Continuation of axillary artery
  - Located in cubital fossa, medial to biceps tendon and deep to biceps aponeurosis
  - Accompanies median nerve
  - Gives off 2 medial branches, which descend toward elbow joint on medial side of brachial artery and contribute to anterior and posterior anastomosis around elbow
  - Brachial artery divides into radial artery and ulnar artery in cubital fossa
- Radial artery
  - Medial to distal biceps tendon
  - Covered by brachioradialis muscle
  - Distally, leaves forearm and moves laterally, crossing floor of anatomical snuffbox
  - Terminates in deep palmar arch of hand
  - Branches
    - Radial recurrent artery: Runs proximally along lateral side of elbow to form anastomosis with branches of deep brachial artery
    - Muscular branches to lateral side of forearm
    - Distal anastomotic branches: Palmar carpal arch, superficial palmar arch, and dorsal carpal arch
- Ulnar artery
  - Proximally deep to pronator teres
  - Distally lies on flexor digitorum profundus and lateral to ulnar nerve
  - Anterior and posterior ulnar recurrent arteries form anastomosis around medial side of elbow with branches of brachial artery
  - Branches
    - Common interosseus artery: Arises in distal aspect of cubital fossa, 2 cm below origin of ulnar artery from brachial artery
    - Anterior interosseus artery: Runs distally on interosseus membrane and ends in dorsal carpal arch; reaches posterior compartment by means of perforating branches; terminates at wrist joint after piercing interosseus membrane; supplies deep muscles of forearm
    - Posterior interosseus artery: Enters posterior compartment proximal to interosseus membrane, between supinator and abductor pollicis longus, and supplies posterior muscles
    - Muscular branches to medial side of forearm
    - Distal anastomotic branches: Palmar carpal arch, dorsal carpal arch

Veins

- Veins of upper limb are variable in number and position
  - In hand, there are 2 sets of veins: Deep and superficial
    - Deep veins follow arteries traveling as paired venae comitantes; radial, ulnar, and brachial veins
  - Superficial veins of hand and forearm are drained by cephalic and basilic veins
- Cephalic vein
  - Begins on lateral aspect of wrist by draining blood from venous plexus on dorsum of hand
    - Ascends along lateral side of forearm and upper arm in superficial fascia
    - As vein drains cranially, it receives tributaries from anterior and posterior aspects of forearm
    - Cephalic vein lies laterally in superficial fascia of antecubital fossa and gives off median cubital vein
    - Median vein of forearm flows into medial cubital vein
- Basilic vein
  - Begins on medial side of wrist and drains medial part of venous plexus on dorsum of hand
  - Basilic vein passes upward on medial side of forearm, receiving anterior and posterior tributaries
  - Passes to anteromedial aspect of elbow where it receives median cubital vein

ANATOMY IMAGING ISSUES

Imaging Recommendations

- Venous examination
  - Examination of upper limb veins is normally performed with patient supine and arm abducted to ~ 90°
  - Use of high-frequency linear transducer (5-10 MHz) is recommended
    - Determination of compressibility and flow pattern is normally performed
    - Phasicity with respiration noted
  - Augmentation of flow is obtained by manual compression of forearm
  - Alternatively, augmentation can be achieved by asking patient to clench fist
- Arterial examination
  - In patients with possible arterial compression syndromes, arteries are examined in various degrees of abduction so that any compression may be accentuated
  - If angle of insonation is > 60°, Doppler and color sensitivity ↓
  - Keep Doppler gate in center of lumen for peak flow velocity
  - Match color signal scale with flow velocity

Imaging Pitfalls

- Gold standard for assessment of accuracy of arterial Doppler US is usually arteriography

CLINICAL IMPLICATIONS

Clinical Importance

- Brachial artery in antecubital fossa may be used for arterial access for angiography
- High takeoff of radial artery from brachial artery is common anatomical variant
  - On arterial Doppler studies, if either radial or ulnar artery is difficult to trace distally from elbow, then artery should be sought at wrist and followed proximally
- Hemodialysis arteriovenous fistulae are generally constructed by joining cephalic veins to radial artery at wrist

FOREARM ARTERIES

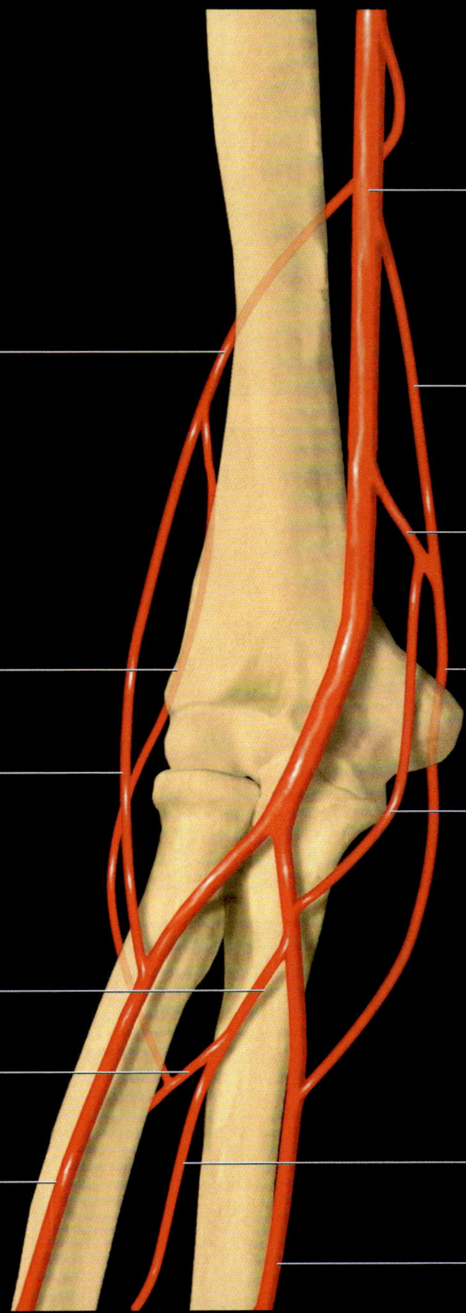

Graphic shows the brachial artery, which is the major artery of the arm. It divides into the radial and ulnar arteries in the proximal forearm.

TRANSVERSE US, PROXIMAL FOREARM

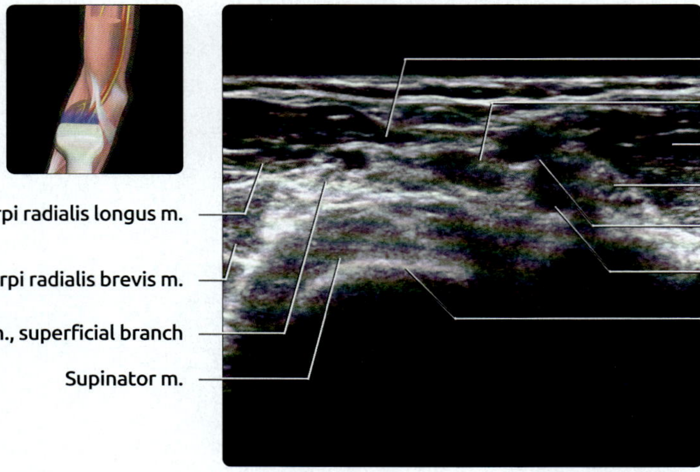

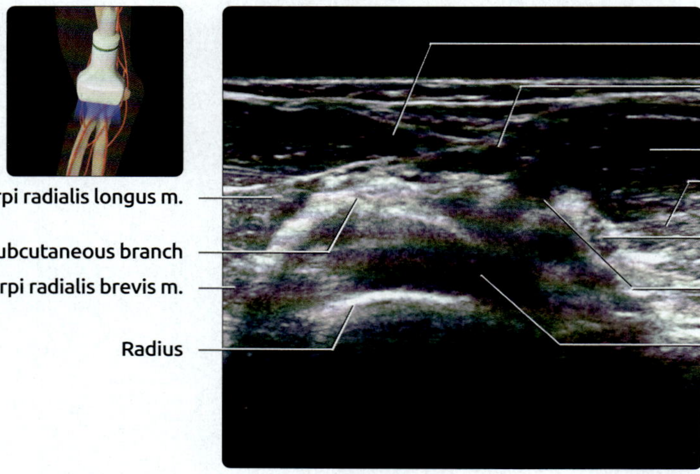

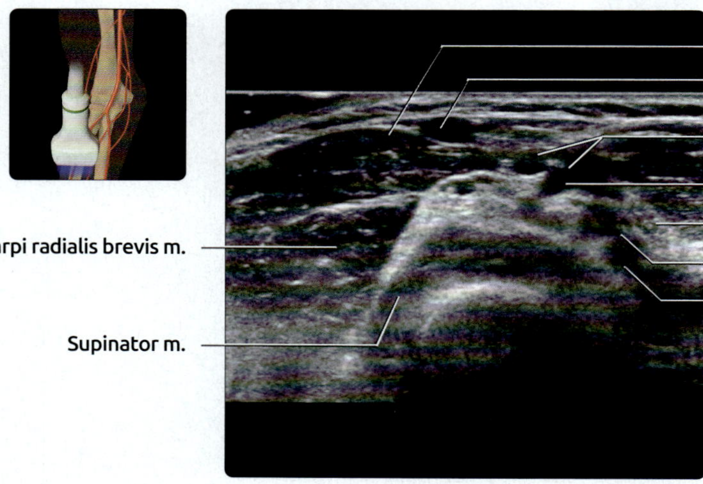

(Top) *Transverse grayscale US shows forearm vessels at the volar aspect of the proximal forearm. The brachial artery divides into the radial and ulnar arteries in the proximal arm. The forearm veins are very variable in number and appearance. Both main veins should be assessed when investigating deep vein thrombosis.* **(Middle)** *Transverse grayscale US shows radial vessels on the volar aspect of the proximal forearm just distal to bifurcation of the brachial artery.* **(Bottom)** *Transverse grayscale US shows radial vessels on the volar aspect of midforearm. The superficial branch of the radial nerve roughly follows the radial artery, while the ulnar nerve closely follows the ulnar artery.*

TRANSVERSE US, RADIAL ARTERY

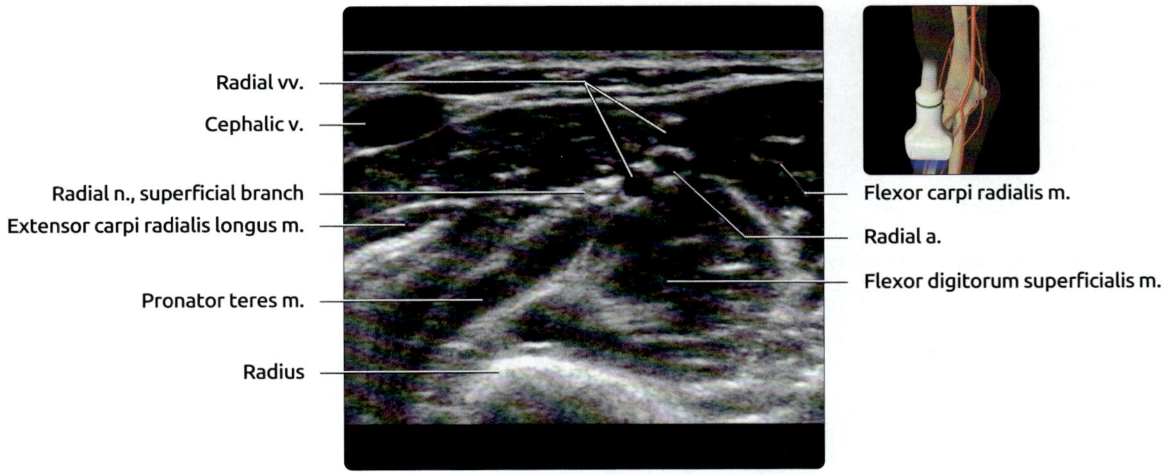

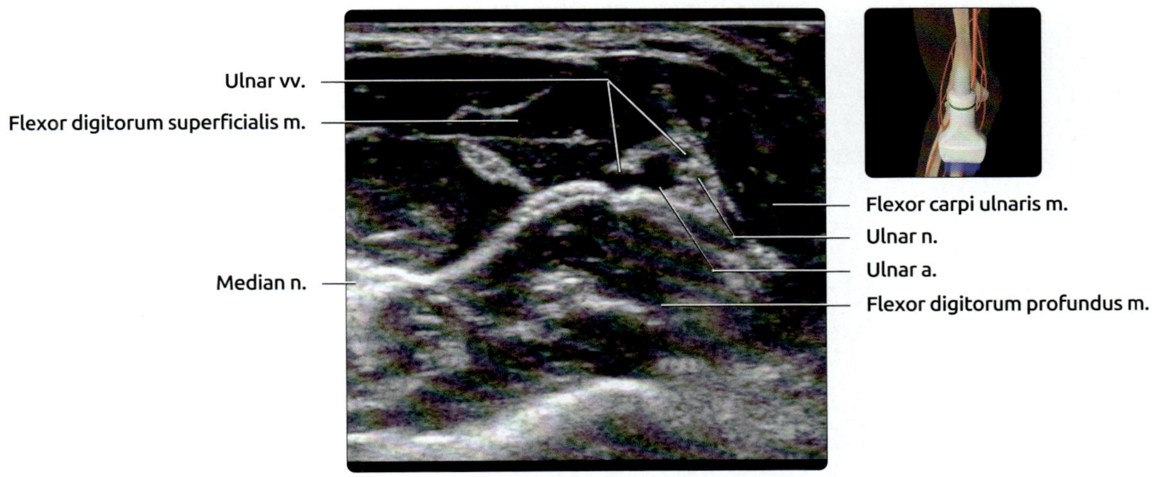

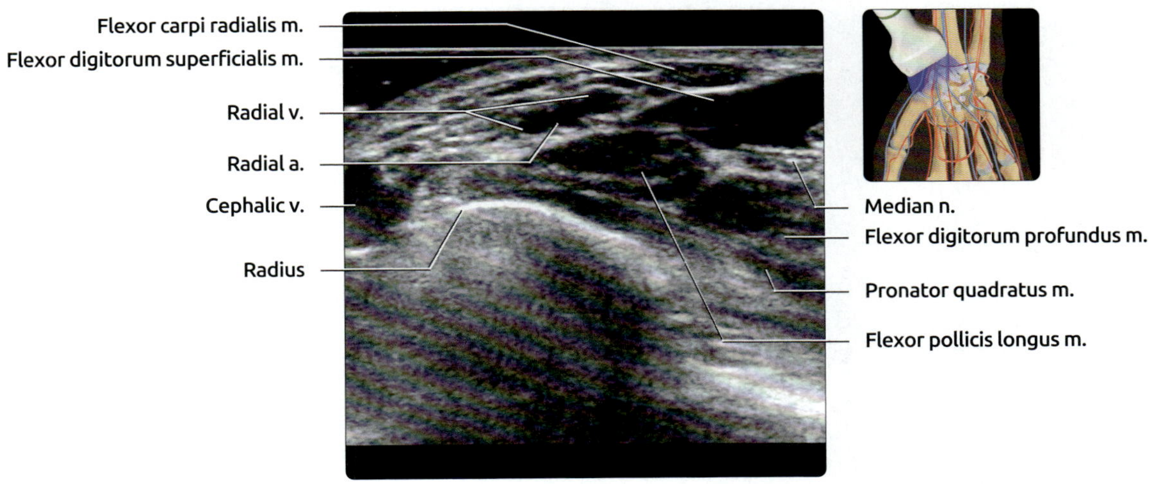

(Top) *Transverse grayscale US shows the radial artery on the volar aspect of the midforearm. The radial artery serves as a landmark to divide the anterior (flexor) and posterior (extensor) compartments of the forearm.* **(Middle)** *Transverse grayscale US shows the ulnar artery on the volar aspect of the midforearm. The ulnar artery is more variable in position than the radial artery. It may arise more proximal in the arm region. When the ulnar artery arises proximally, it often has an aberrant course and may lie superficial to the flexor muscles in the forearm.* **(Bottom)** *Transverse grayscale US shows the radial artery on the volar aspect of distal forearm. The radial artery is sometimes used instead of the long saphenous vein in coronary artery surgery. The caliber of the artery can be checked prior to surgery by US.*

TRANSVERSE AND LONGITUDINAL US, ULNAR ARTERY

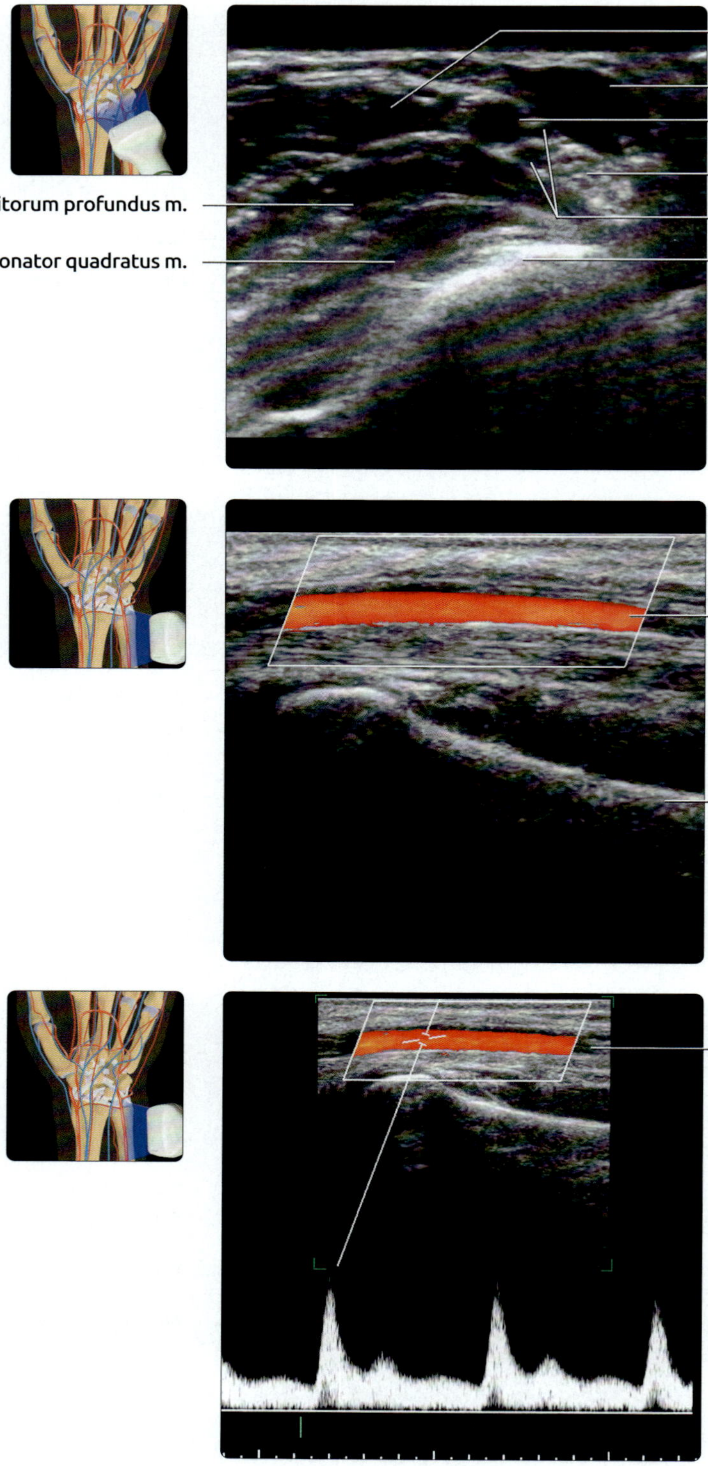

(Top) *Transverse grayscale US shows the ulnar artery on the volar aspect of the distal forearm. The Allen test is performed clinically to check dual arterial supply to the hand prior to radial artery cannulation or excision.* **(Middle)** *Longitudinal color Doppler US shows the ulnar artery on the volar aspect of distal forearm. Flow velocity and direction is denoted by the color signal. The normal diameter of the ulnar artery (1.9-2.2 mm) is measured near the wrist.* **(Bottom)** *Longitudinal spectral Doppler analysis of the ulnar artery on the volar aspect of the distal forearm is shown. The spectral gate is placed in the middle of the arterial lumen. Normal unidirectional flow is present with a systolic peak and good diastolic flow. The peak systolic velocity of the ulnar artery ranges from 40-90 cm/sec, while end-diastolic velocity ranges from 5-12 cm/sec. Pulsatility index is variable, ranging from 3.9-5.0.*

ARTERIES AT DISTAL FOREARM AND HAND

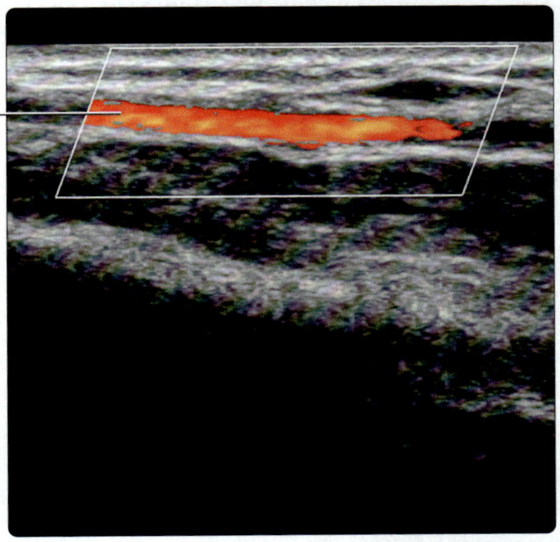

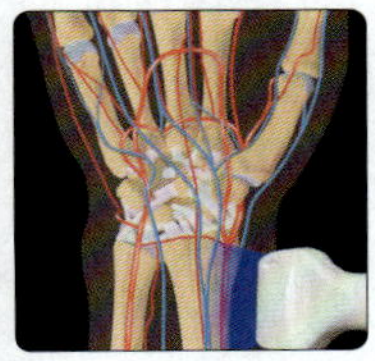

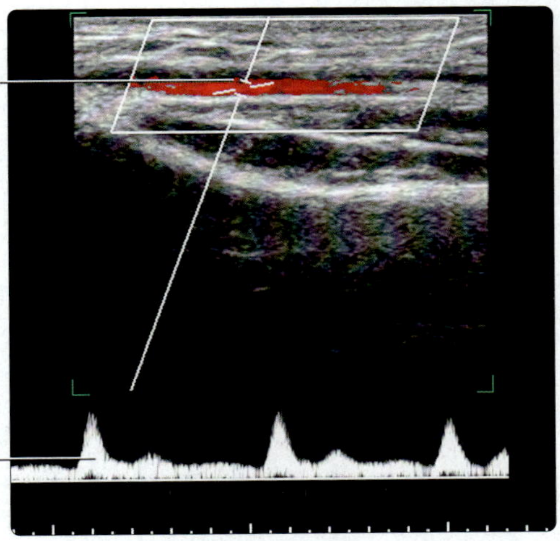

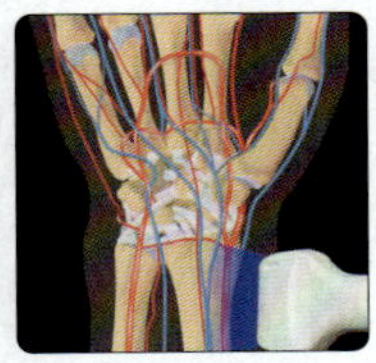

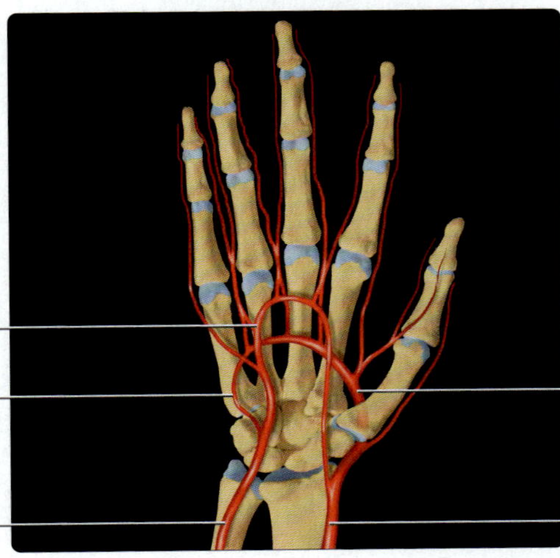

(Top) *Longitudinal color Doppler US shows the radial artery on the volar aspect of the distal forearm just proximal to the wrist. The normal diameter of the radial artery just proximal to the wrist ranges from 2.2-2.8 mm.* **(Middle)** *Spectral analysis shows the radial artery just proximal to the wrist. Spectral analysis is best performed with the artery visualized longitudinally rather then transversely. The normal angle corrected peak systolic velocity of the radial artery ranges from 40-90 cm/sec, the end-diastolic velocity ranges from 5-10 cm/sec, and pulsatility index is variable, ranging from 3.9-5.0.* **(Bottom)** *Graphic shows the distal aspect of the radial and ulnar arteries in the forearm and their terminal course in the wrist and hand to form the superficial and deep palmar arterial arches.*

VENOUS SYSTEM, ANTERIOR

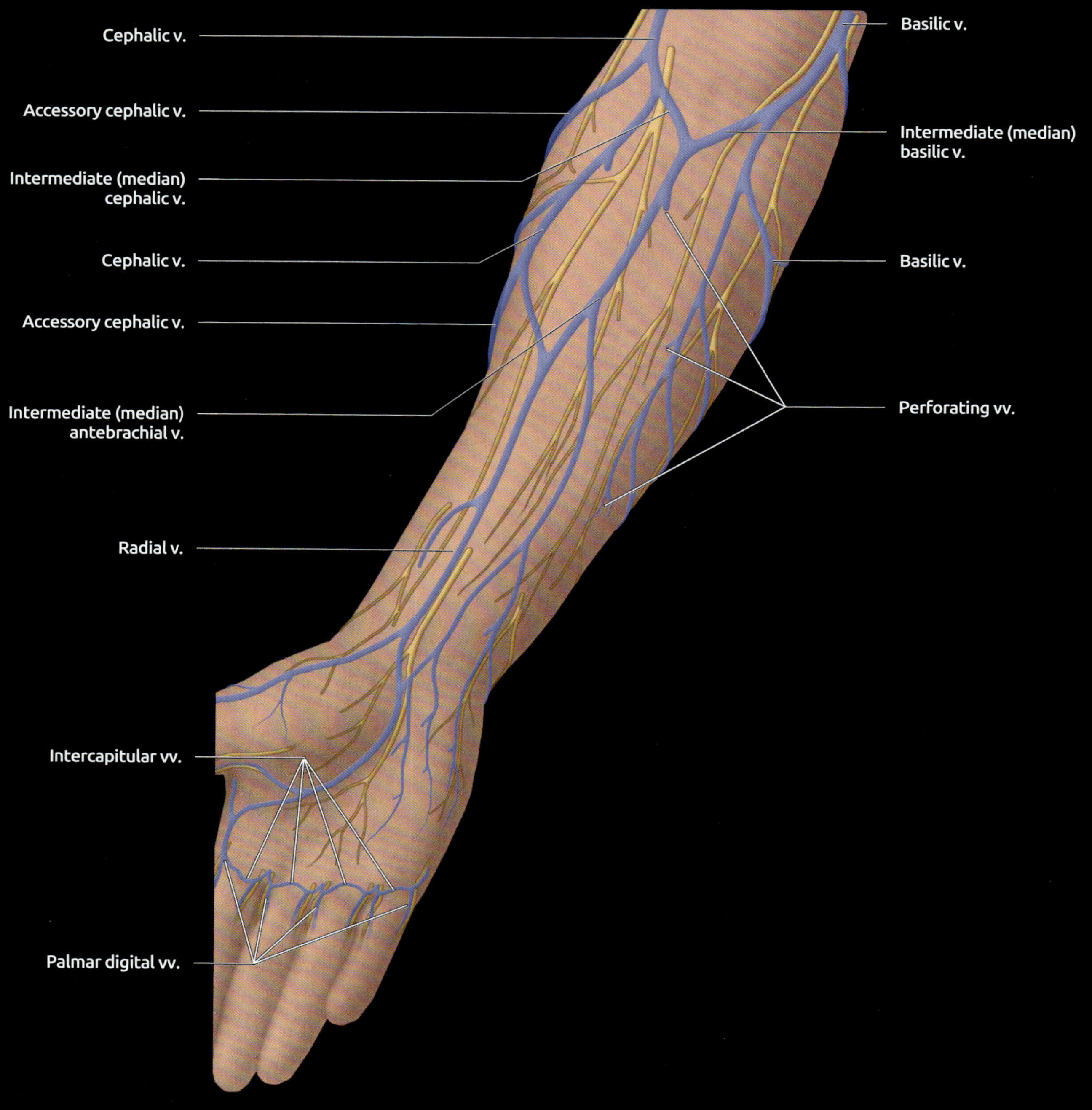

Graphic shows the superficial venous system of the forearm. The most important collecting vein of the forearm and arm is the cephalic vein. The cephalic vein proceeds from the radial margin of the dorsal venous network of the hand and follows a course diagonally above the radial margin of the forearm to the cubital fossa. From there, it continues laterally in the distal arm, moving anteromedially in the upper arm to enter the axillary vein just below the clavicle. It contains 6-10 valves. The ulnar and radial veins collect blood from the deep palmar arch and course in pairs alongside the corresponding arteries. Blood is continuously shunted from these superficial veins in the subcutaneous tissue to deep veins via the perforating veins. Although the upper extremities have many communicating veins, they have no clinical significance and are therefore not specifically named and located.

VENOUS SYSTEM, POSTERIOR

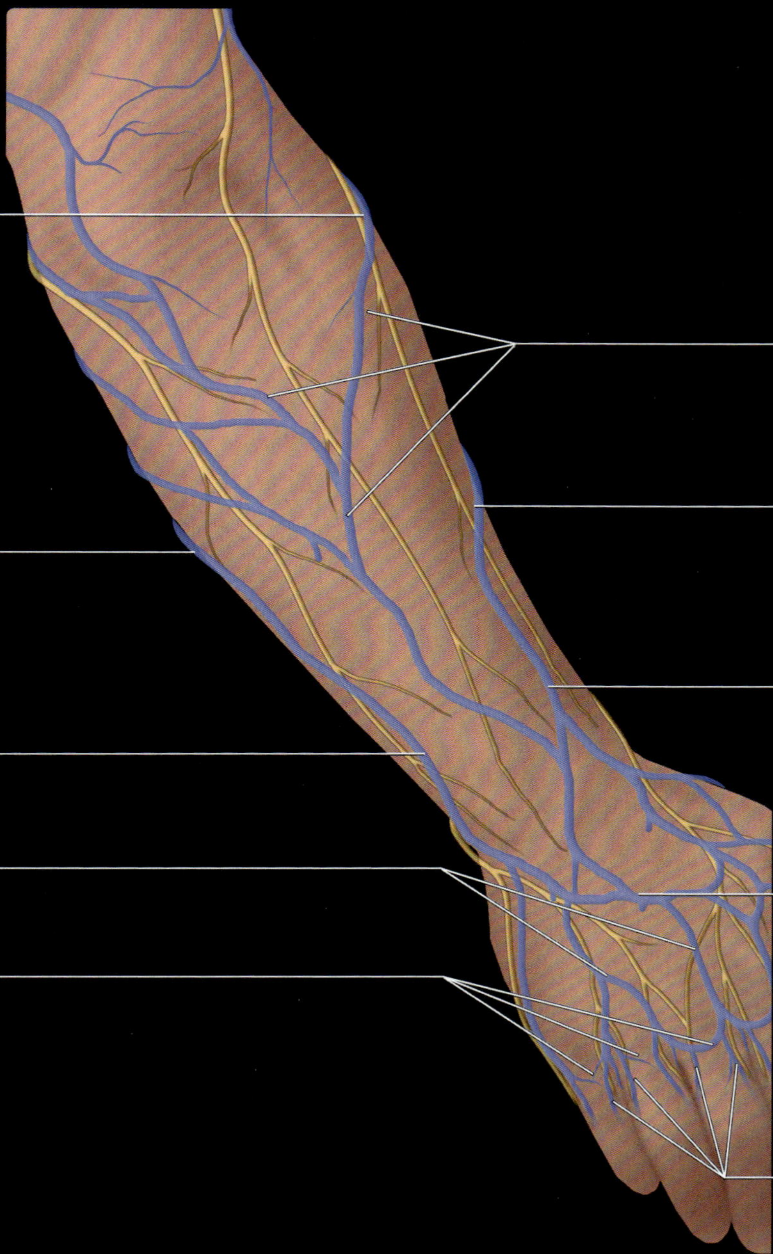

Graphic shows the posterior view of the superficial venous system of the forearm. The basilic vein starts from the ulnar margin of the dorsal venous plexus and travels onward to the bend of the elbow, where it is joined by the cephalic vein. It contains 4-8 valves. There are 2 additional small branches that pass superficially on the volar side of the forearm, the median antebrachial vein and the median cubital vein, which flow into the cephalic vein and the basilic vein somewhat above the bend of the elbow. Although the upper extremities have many communicating veins, they have no clinical significance and are therefore not specifically named and located.

TERMINOLOGY

Definitions

- Articulation complex comprising distal radioulnar, radiocarpal, ulnocarpal, pisotriquetral, midcarpal, and carpometacarpal joints

IMAGING ANATOMY

Osseous Structures

- Distal radius: Lister tubercle on dorsal surface acts as fulcrum for action of extensor pollicis longus
- Distal ulna: Ulnar variance refers to length of ulnar head relative to distal radius: Ulnar negative (25%) or ulnar positive (15%)
- Proximal carpal row: Scaphoid, lunate, triquetrum, pisiform
  - Scaphoid: Proximal and distal pole separated by waist
    - Tuberosity = volar prominence of distal pole
  - Lunate: 1/2 moon-shaped
  - Triquetrum: Triangular-shaped
  - Pisiform: Pea-shaped, sesamoid-type bone to which flexor carpi ulnaris attaches and continues distally as pisohamate and pisometacarpal ligaments
- Distal carpal row: Trapezium, trapezoid, capitate, hamate
  - Trapezium: Saddle-shaped bone linking carpus and thumb
  - Trapezoid: Wedge-shaped bone
  - Capitate: Head (proximal), neck (midportion), and body (bulky distal portion)
  - Hamate: Hook (hamulus) arises from palmar surface

Ligaments

- Extrinsic (on palmar or volar aspects of wrist) or intrinsic (between carpal bones)
- Major wrist stabilizers: Volar ligaments
- Extrinsic ligaments
  - Palmar: Radioscaphocapitate, radiolunotriquetral, radioscapholunate, ulnotriquetral, ulnolunate, scaphotriquetral
  - Dorsal: Scaphotriquetral, radiotriquetral, ulnotriquetral, radial collateral
- Intrinsic ligaments
  - Proximal interosseous: Scapholunate, lunotriquetral
  - Distal interosseous: Trapeziotrapezoid, trapeziocapitate, capitohamate

Muscles and Tendons

- **Flexors, deep**
  - Flexor digitorum profundus: Ulnar shaft → index, middle, ring, and little finger distal phalangeal bases
  - Flexor pollicis longus: Radial shaft, interosseous membrane, and coronoid process ulna → distal phalangeal base thumb
- **Flexors, superficial**
  - Flexor carpi radialis: Medial epicondyle → 2nd metacarpal base
  - Palmaris longus: Medial epicondyle → palmar aponeurosis
  - Flexor carpi ulnaris: Medial epicondyle and proximal ulna → pisiform
  - Flexor digitorum superficialis: Medial epicondyle, coronoid process ulna, and radius shaft → middle phalangeal bases of digits 2-5
- **Extensors, deep**
  - Abductor pollicis longus: Midulna → radial aspect 1st metacarpal base
  - Extensor pollicis brevis: Midradius → proximal phalangeal base of thumb
  - Extensor pollicis longus: Midulna → distal phalangeal base of thumb
  - Extensor indicis: Midulna → ulnar side extensor digitorum tendon → 2nd digit extensor hood
- **Extensors, superficial**
  - Extensor carpi radialis longus: Lateral supracondylar ridge humerus → 2nd metacarpal base
  - Extensor carpi radialis brevis: Lateral humeral epicondyle → 3rd metacarpal base
  - Extensor digitorum: Lateral humeral epicondyle → distal phalangeal bases digits of 2-5
  - Extensor digiti minimi: Lateral humeral epicondyle → extensor hood of little finger
  - Extensor carpi ulnaris: Lateral humeral epicondyle → 5th metacarpal base

Retinacula

- **Flexor retinaculum**
  - Also called transverse carpal ligament: Attached to pisiform and hook of hamate, scaphoid, and trapezium
- **Extensor retinaculum**
  - Attaches to ulnar styloid process, triquetrum, and pisiform medially; crosses obliquely to attach to Lister tubercle and radial styloid process laterally
    - Sends septa to radius, creating 6 separate compartments for extensor tendons
  - Compartment contents
    - 1st: Abductor pollicis longus and extensor pollicis brevis
    - 2nd: Extensor carpi radialis longus and brevis
    - 3rd: Extensor pollicis longus
    - 4th: Extensor digitorum and extensor indicis
    - 5th: Extensor digiti minimi
    - 6th: Extensor carpi ulnaris

Anatomic Spaces

- **Carpal tunnel**
  - Margins: Carpal bones (dorsal margin), flexor retinaculum (volar margin), pisiform and hook of hamate (ulnar margin), scaphoid and trapezium (radial margin), radiocarpal joint (proximal margin), and metacarpal base (distal margin)
    - Contents: Flexor digitorum superficialis, flexor digitorum profundus, flexor pollicis longus, median nerve
- **Guyon canal**
  - Margins: Fascial extension from flexor retinaculum, volar carpal ligament (volar margin), pisiform and flexor carpi ulnaris (ulnar margin), flexor retinaculum (radial and dorsal margins)
    - Contents: Ulnar artery and vein, ulnar nerve

CARPAL BONES AND WRIST COMPARTMENTS

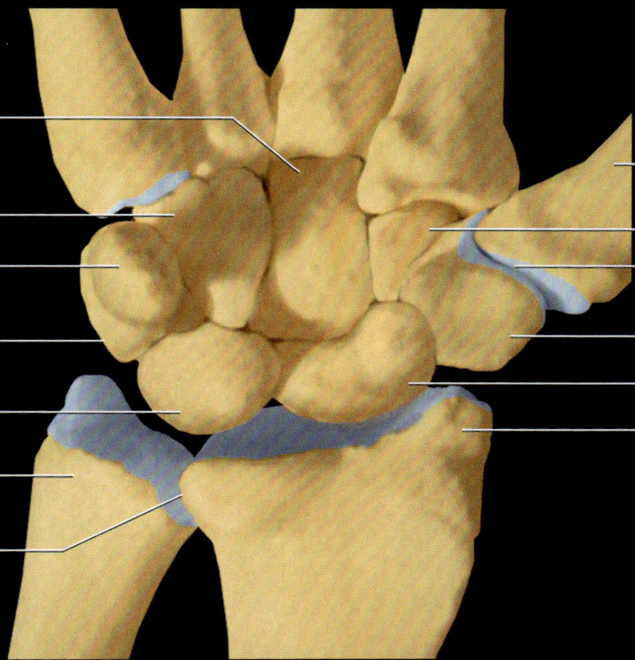

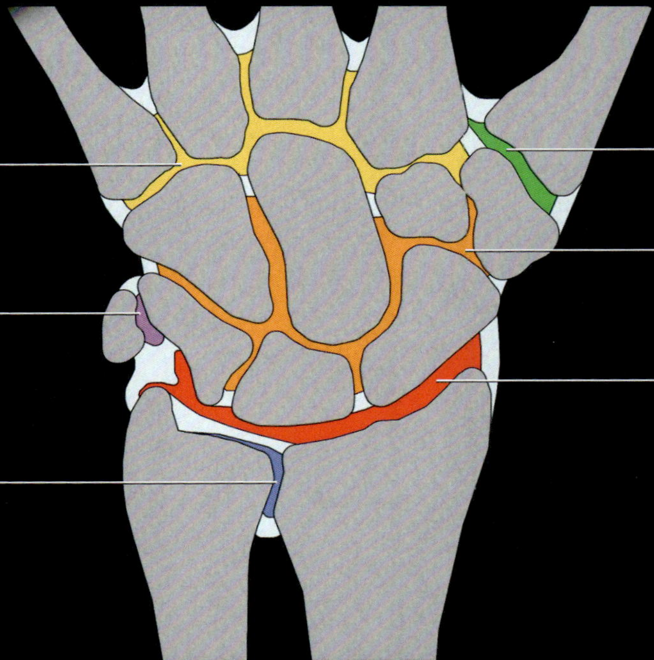

(Top) *Graphic shows the bones of the wrist joint.* **(Bottom)** *Graphic shows the 5 different wrist compartments. The distal radioulnar compartment is separated from the radiocarpal compartment by the triangular fibrocartilage complex. The pisotriquetral compartment is separated from the radiocarpal compartment in 20%. The midcarpal compartment is separated from the radiocarpal compartment by the scapholunate and lunotriquetral ligaments and typically communicates with the carpometacarpal joints. The 1st carpometacarpal compartment is separated from the common carpometacarpal compartment by the trapeziometacarpal ligament.*

DORSAL TENDONS AND TENDON SHEATHS

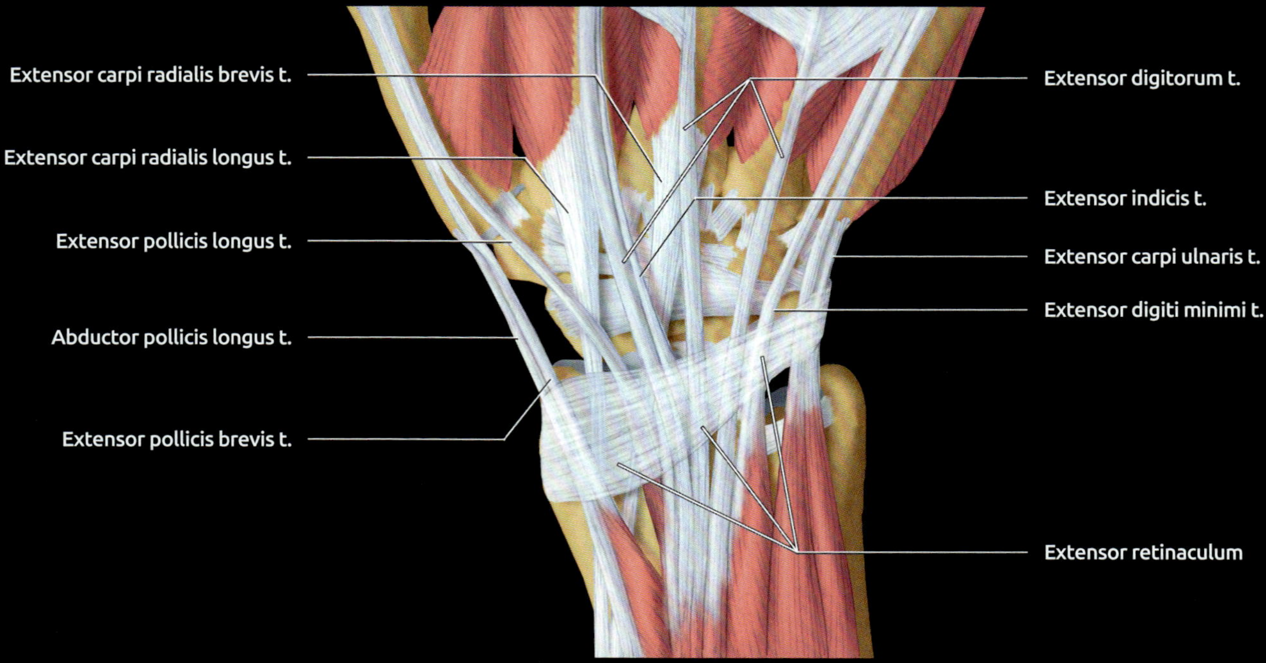

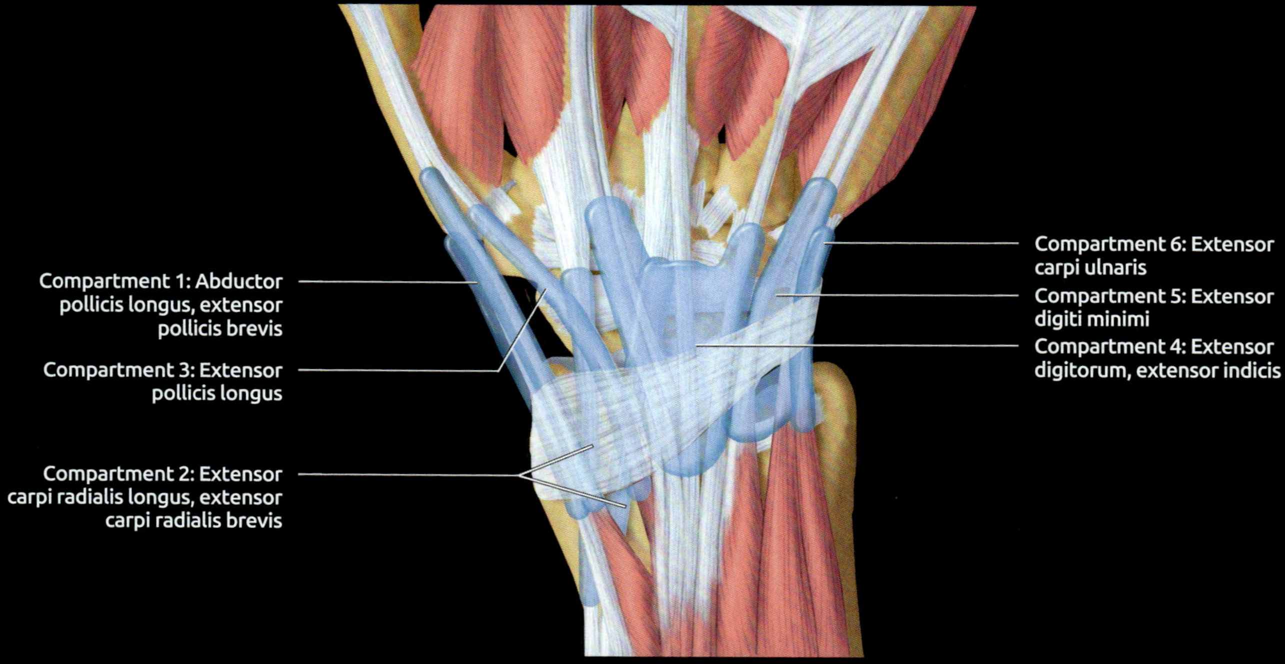

(Top) *Dorsal extensor tendons pass deep to the extensor retinaculum, separated into 6 compartments by fibrous attachments of the retinaculum to underlying bone. Compartment contents include abductor pollicis longus and extensor pollicis brevis, extensor carpi radialis longus and brevis, extensor pollicis longus, extensor digitorum and extensor indicis, extensor digiti minimi, and extensor carpi ulnaris.* **(Bottom)** *Separate tendon sheaths enclose the dorsal extensor tendons in compartments 1-6 individually.*

TENDONS: RELATIONS TO DORSAL AND VOLAR WRIST

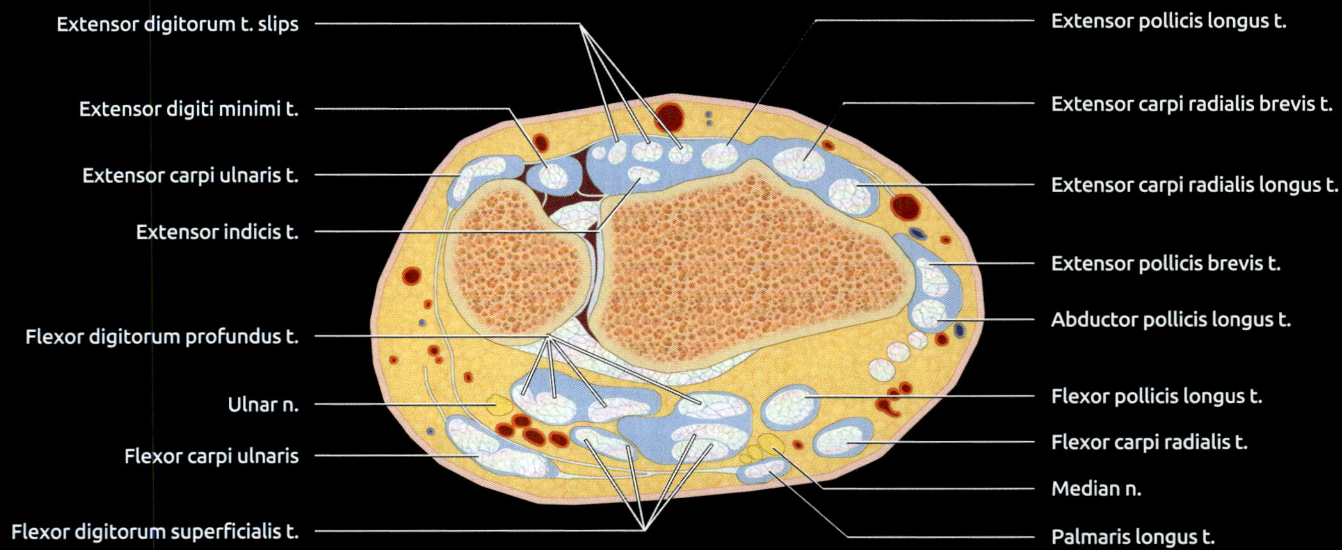

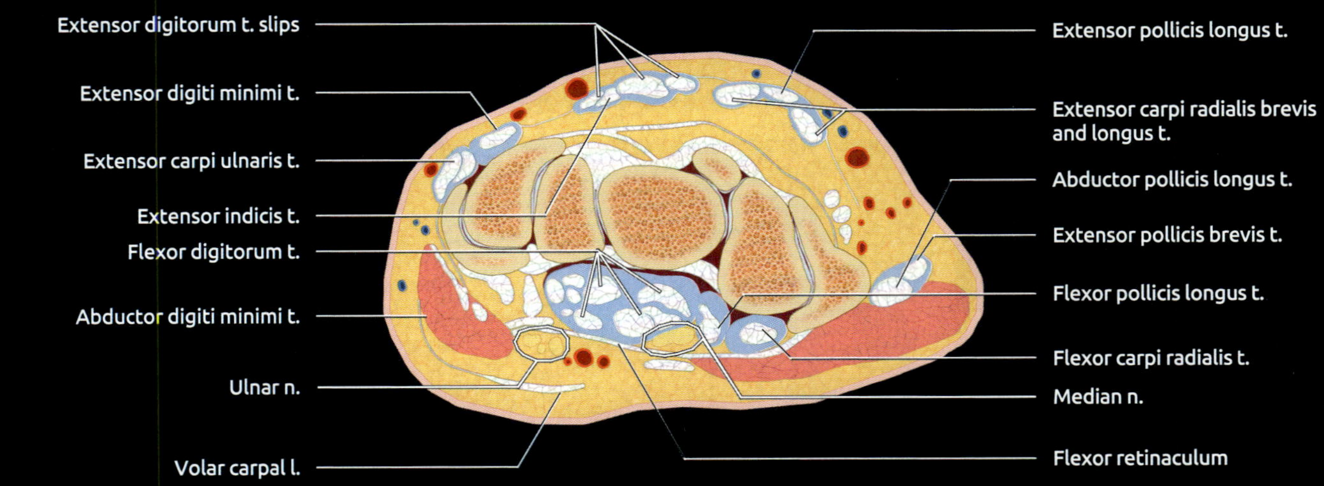

(Top) *Graphic representation shows tendons in the proximal wrist. Extensor tendons are deep to the extensor retinaculum, whereas flexor tendons are proximal to the flexor retinaculum at this level in the wrist.* **(Bottom)** *Midcarpal tunnel is shown. The median nerve is slightly flattened as it passes deep to the flexor retinaculum and remains superficial to the flexor pollicis longus. The ulnar nerve, artery, and veins lie lateral to the pisiform and may divide near the pisiform into deep and superficial branches.*

TRANSVERSE US, RADIAL ASPECT

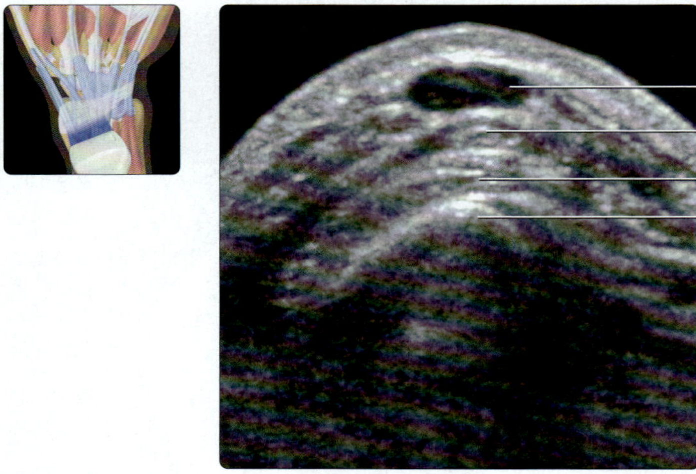

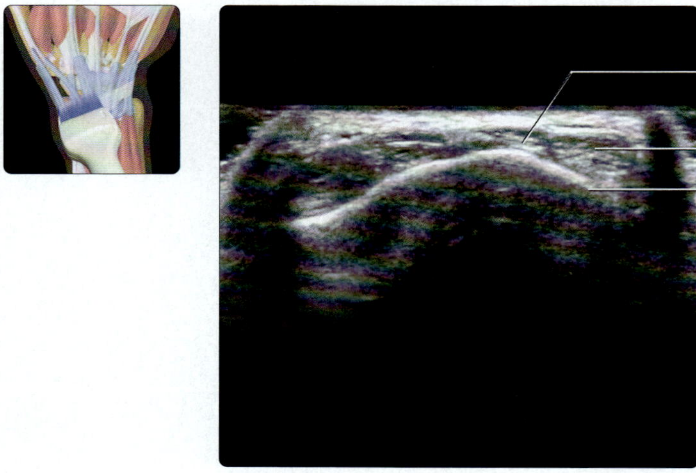

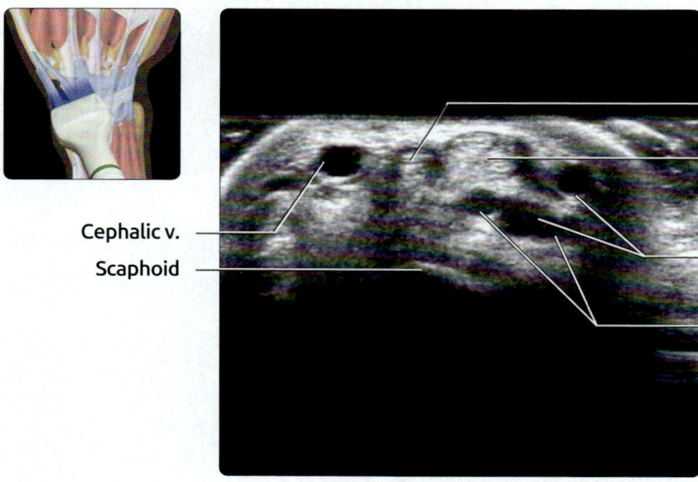

(Top) *Transverse grayscale US shows the 1st extensor compartment proximal to the distal radius. The 1st compartment contains the extensor pollicis brevis and the larger abductor pollicis longus tendons. The 1st compartment tendons run obliquely over 2nd compartment tendons a few centimeters proximal to the wrist. Pain may occur at this intersection (intersection syndrome).* **(Middle)** *Transverse grayscale US shows the 1st extensor compartment at the distal radius. Extensor pollicis brevis extends to the base of the proximal phalanx of the thumb, whereas abductor pollicis longus inserts onto the base of the 1st metacarpal bone. They form the radial aspect of the anatomical snuff box. These tendons are swollen in patients with de Quervain disease.* **(Bottom)** *Transverse grayscale US shows the 1st extensor compartment at the level of the scaphoid. The dorsal branch of the radial artery passes deep to the tendons of the 1st extensor compartment to enter the dorsum of the hand.*

TRANSVERSE US, DORSAL WRIST

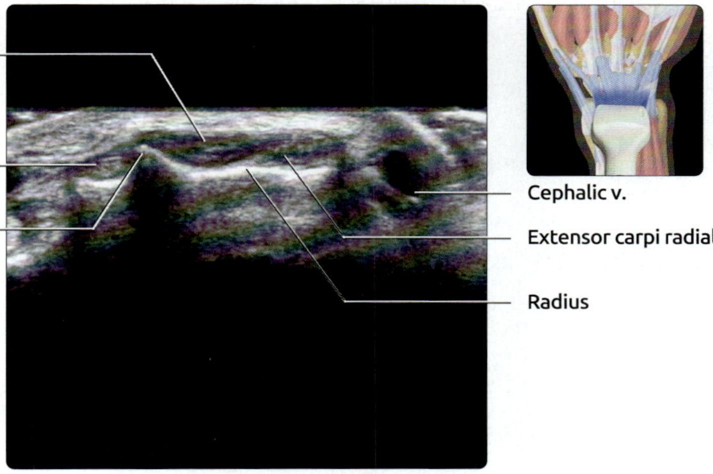

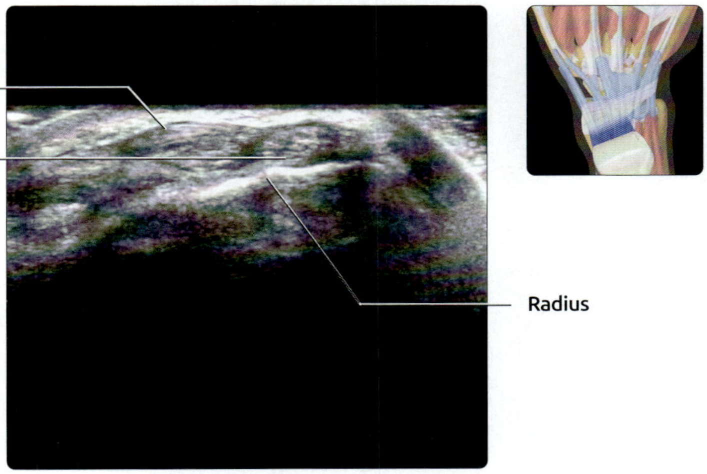

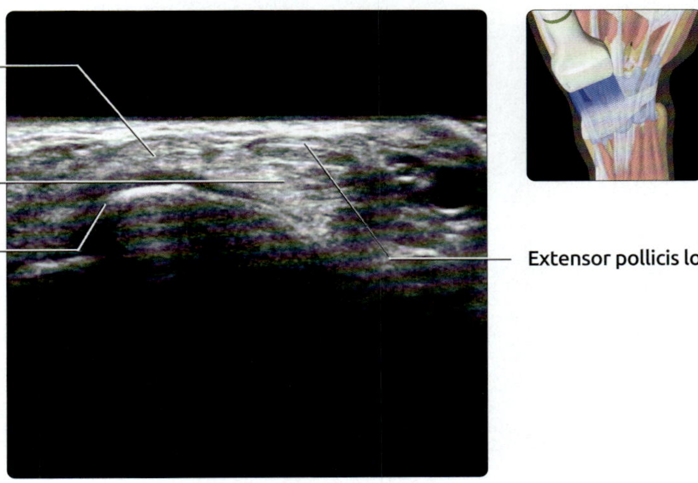

(Top) *Transverse grayscale US shows the 2nd and 3rd extensor compartments at the distal radius level. The 2nd extensor compartment comprises the extensor carpi radialis longus and extensor carpi radialis brevis. The extensor carpi radialis longus inserts into the base of the index finger metacarpal, whereas the extensor carpi radialis brevis inserts into the middle finger metacarpal. The 2nd compartment is separated from the 3rd compartment, containing the extensor pollicis longus, by the Lister tubercle.* **(Middle)** *Transverse grayscale US shows the 2nd and 3rd extensor compartments at the distal radius level. The 2nd compartment forms the dorsal aspect of the anatomical snuff box.* **(Bottom)** *Transverse grayscale US shows the 2nd extensor compartment at the scaphoid level. The extensor pollicis longus tendon hooks around the Lister tubercle and crosses superficial to the 1st compartment tendons as it runs toward its insertion on the base of the distal phalanx of the thumb. It is prone to rupture in inflammatory arthropathies and distal radial fractures.*

TRANSVERSE US, EXTENSOR DIGITORUM

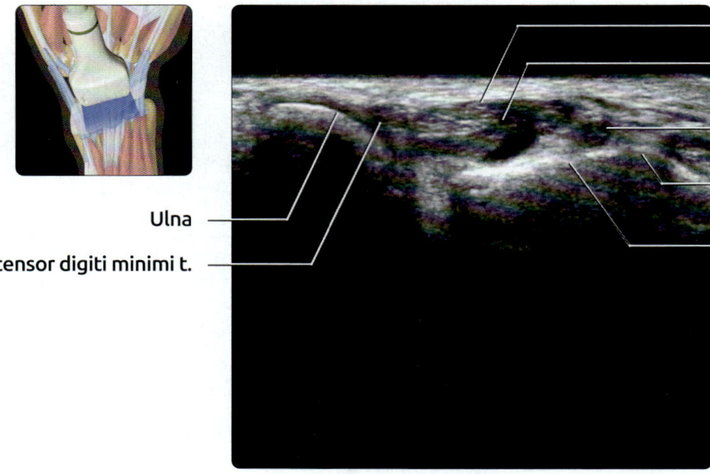

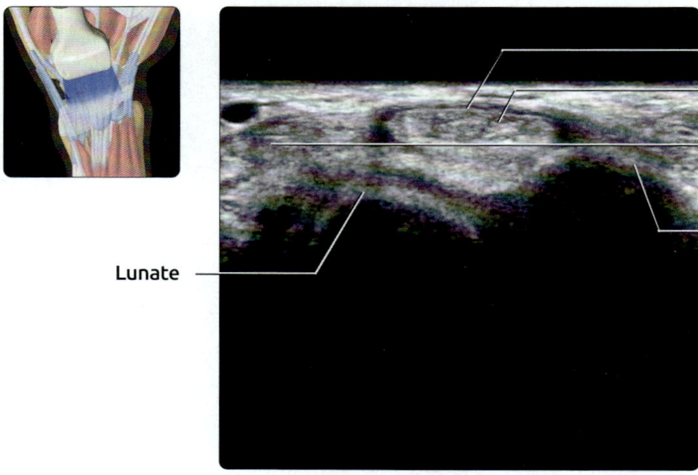

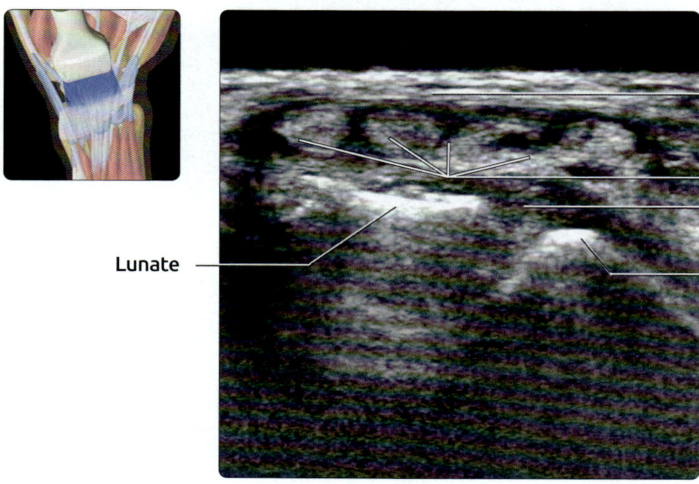

(Top) *Transverse grayscale US shows the 4th extensor compartment at the distal radius. The 4th extensor compartment contains the 4 extensor digitorum tendons and the extensor indicis tendon, which lies radial to the extensor digitorum tendons. The extensor digiti minimi has a separate compartment.* **(Middle)** *Transverse grayscale US shows the 4th extensor compartment at the level of the proximal carpal row. The extensor retinaculum is a thickened continuation of the antebrachial fascia. It is attached to the anterior aspect of the distal radius, the styloid process of the ulna, and the triquetral and pisiform bones.* **(Bottom)** *Transverse grayscale US shows the 4th extensor compartment at the level of the proximal carpal row. The extensor retinaculum holds the extensor tendons in place. Because it is closely applied to the extensor tendons, with anisotropy, it may appear hypoechoic and be confused with tenosynovitis.*

TRANSVERSE US, EXTENSOR CARPI ULNARIS

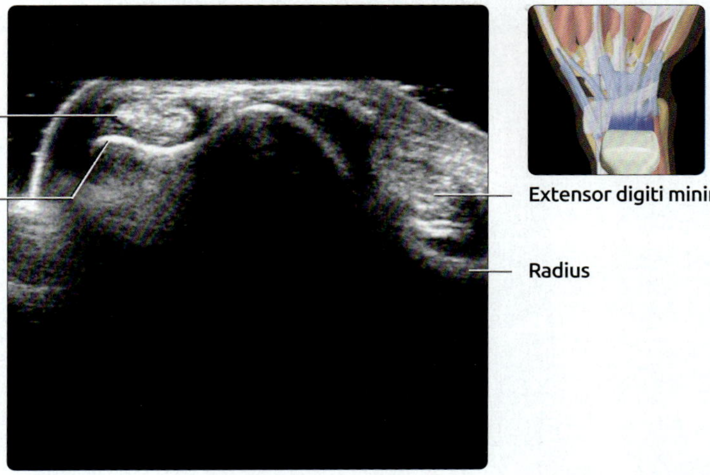

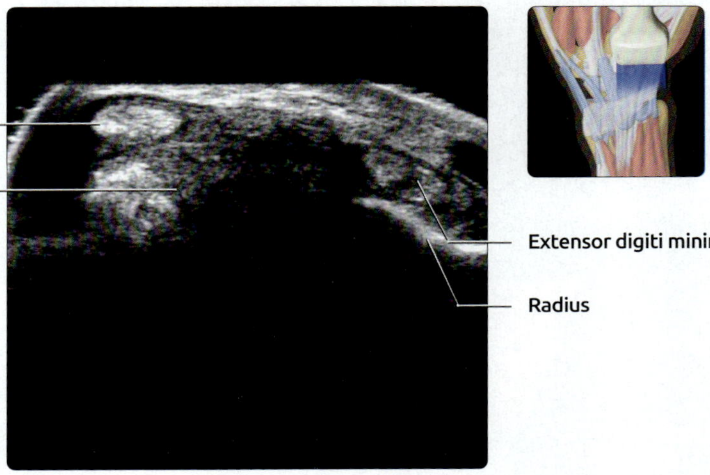

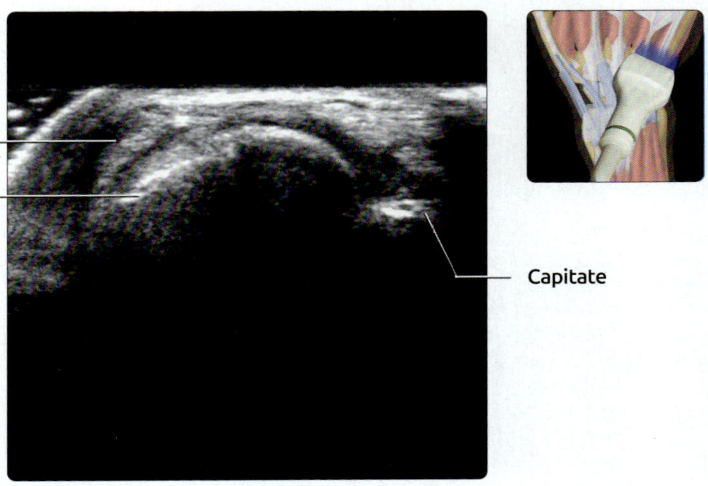

(Top) *Transverse grayscale US shows the 5th and 6th extensor compartments. The 5th compartment contains the extensor digiti minimi tendon. This tendon is joined by the extensor digitorum tendon to the little finger just proximal to the metacarpophalangeal joint. The 6th compartment contains the extensor carpi ulnaris tendon, which runs in a groove in the distal ulna. Its position in the groove will change with pronation and supination.* **(Middle)** *Transverse grayscale US shows the 5th and 6th extensor compartments. The extensor carpi ulnaris tendon often has a midline irregular hypoechoic line within its substance close to the insertion. This should not be mistaken for a longitudinal tear.* **(Bottom)** *Transverse grayscale US at insertion of the extensor carpi ulnaris tendon is shown. The extensor carpi ulnaris tendon widens as it passes the lunate bone on its way to insert into the base of the 5th metacarpal.*

VOLAR WRIST

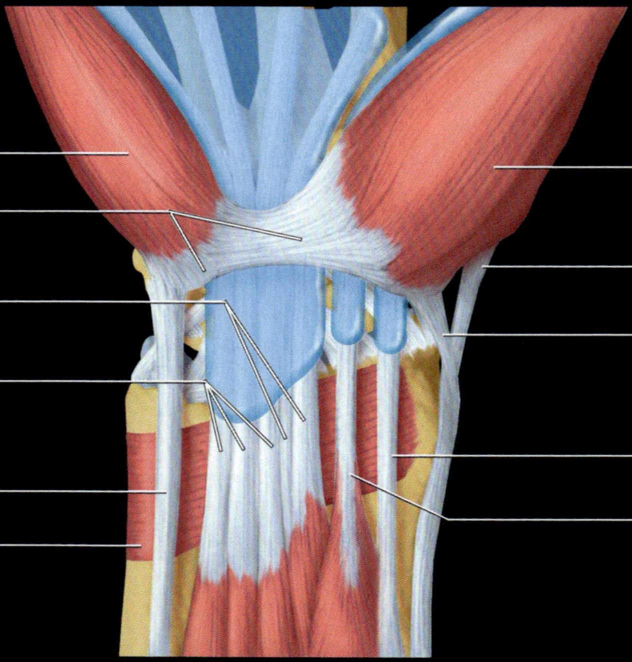

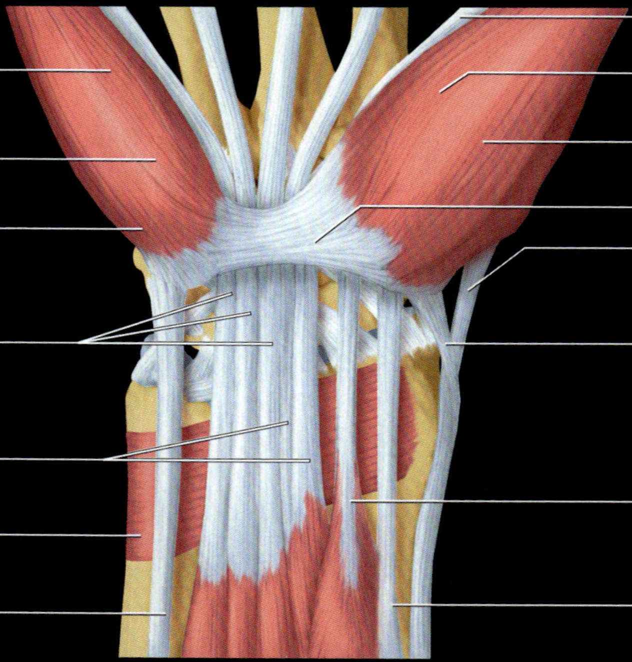

(Top) *Graphic shows tendons and retinaculum of the volar wrist. The flexor retinaculum spans the palmar arch, attaching to the tubercle of the scaphoid, pisiform, hook of hamate, and ridge of the trapezium. The thenar eminence musculature includes the abductor pollicis brevis, opponens pollicis, flexor pollicis brevis, and adductor pollicis. The hypothenar eminence musculature includes the palmaris brevis, adductor digiti minimi, flexor digiti minimi brevis, and opponens digiti minimi.* **(Bottom)** *Volar muscles and tendons are displayed with their relation to the flexor retinaculum. Note the muscles of thenar and hypothenar eminences arise from the retinaculum itself. The flexor digitorum and flexor pollicis longus tendons pass deep to the retinaculum. On the radial side, the retinaculum splits to accommodate the flexor carpi radialis tendon.*

TRANSVERSE US, VOLAR WRIST

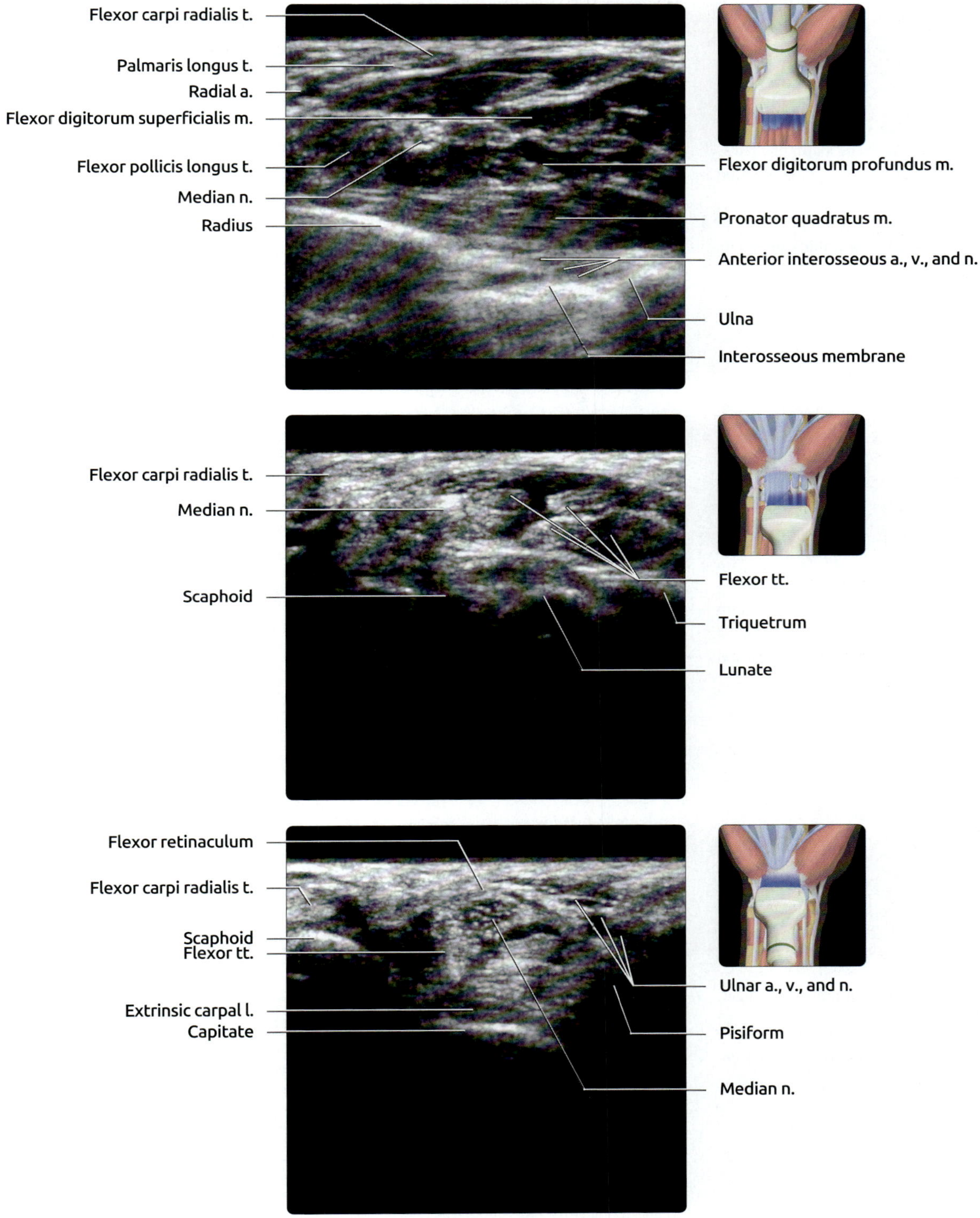

(Top) *Transverse grayscale US shows the volar aspect of the distal forearm just proximal to the wrist. In addition to the tendons that pass through the carpal tunnel, the flexor carpi ulnaris, flexor carpi radialis, and palmaris longus tendons also traverse the wrist joint.* **(Middle)** *Transverse grayscale US shows the volar aspect of the wrist just proximal to the carpal tunnel. The 4 tendons of the flexor digitorum superficialis, the 4 tendons of the flexor digitorum profundus, and the flexor pollicis longus tendons pass through the carpal tunnel. The median nerve dips deeply as it enters the carpal tunnel.* **(Bottom)** *Transverse grayscale US shows the volar aspect of the wrist at the tunnel inlet. The inlet (and outlet) of the carpal tunnel can be best recognized by identifying the proximal and distal margins of the flexor retinaculum. In evaluating carpal tunnel syndrome, the caliber of the nerve should be measured proximal to the tunnel, at the tunnel inlet, and at the tunnel outlet.*

US, VOLAR WRIST

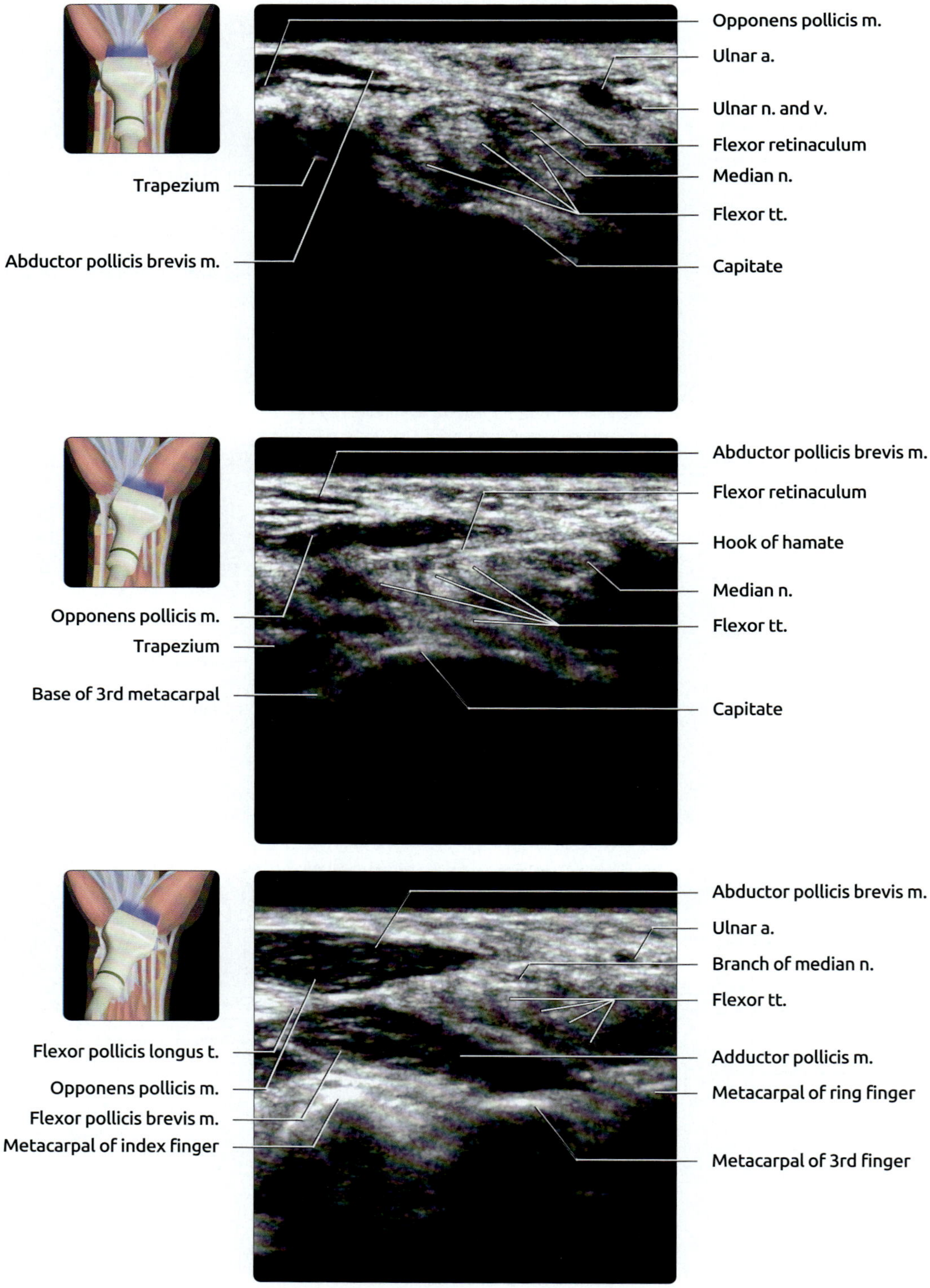

(Top) *Transverse grayscale US shows the volar aspect at midcarpal tunnel. The median nerve lies in the carpal tunnel just deep to the retinaculum. You may need to use anisotropy to clearly identify the margins of the median nerve separate from the adjacent flexor tendons.* **(Middle)** *Transverse grayscale US shows the volar aspect of the wrist at the tunnel outlet. The tunnel outlet is considered to be the narrowest part of the carpal tunnel.* **(Bottom)** *Transverse grayscale US shows the volar aspect of the wrist just beyond the tunnel outlet. The median nerve divides into its terminal branches just beyond the tunnel outlet.*

US, SCAPHOID

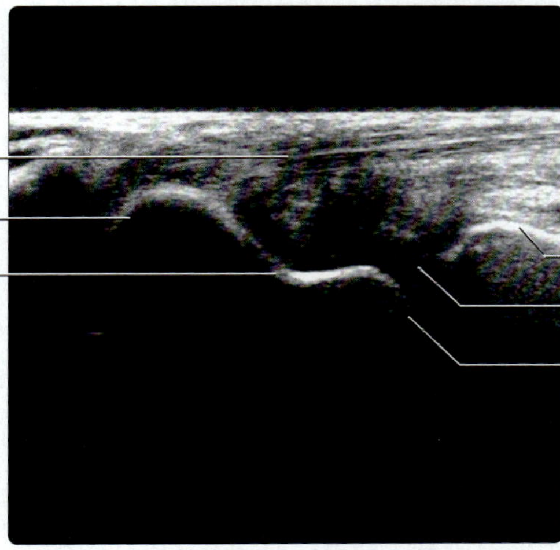

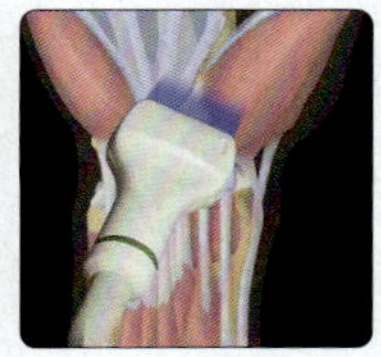

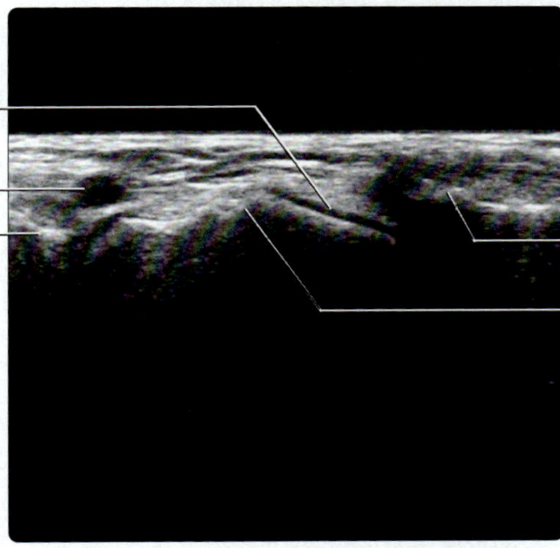

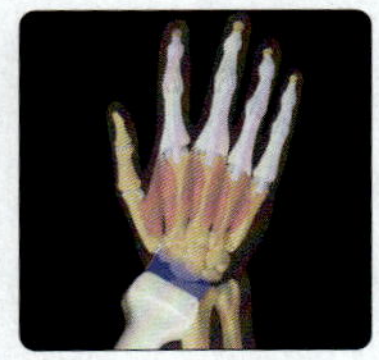

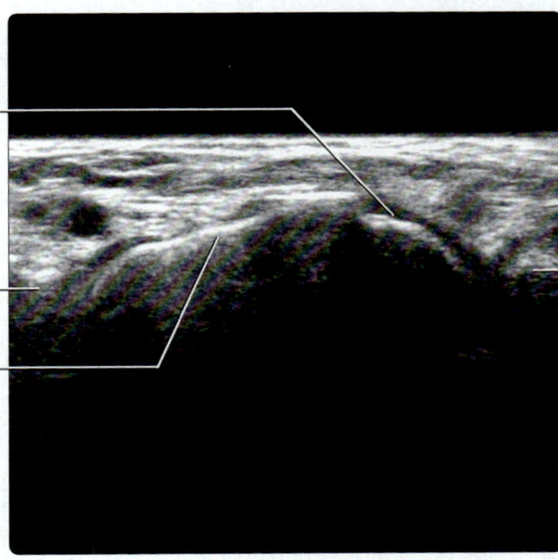

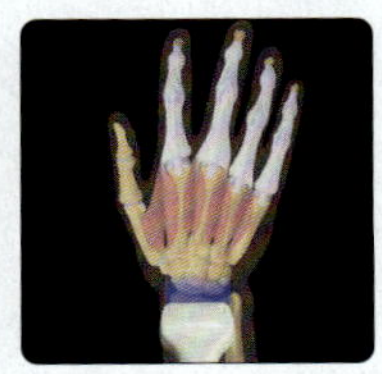

(Top) *Grayscale US longitudinal to the palmar aspect of the scaphoid bone is shown. US is a useful means of diagnosing a scaphoid fracture. Angulation of the transducer along the long axis of the scaphoid allows the palmar cortical outline to be appreciated.* **(Middle)** *Grayscale US longitudinal to the dorsal aspect of the scaphoid bone is shown. There is often mild cortical irregularity of the scaphoid surface, particularly on the dorsal side. The absence of surrounding edema, hematoma, periosteal thickening, and cortical discontinuity allows one to differentiate this normal appearance from a fracture.* **(Bottom)** *Grayscale US transverse to the dorsal aspect of the scaphoid bone is shown.*

GRAPHICS AND US, TRIANGULAR FIBROCARTILAGE

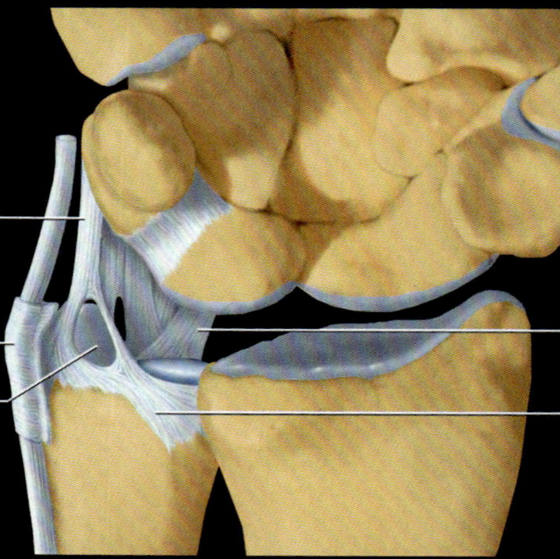

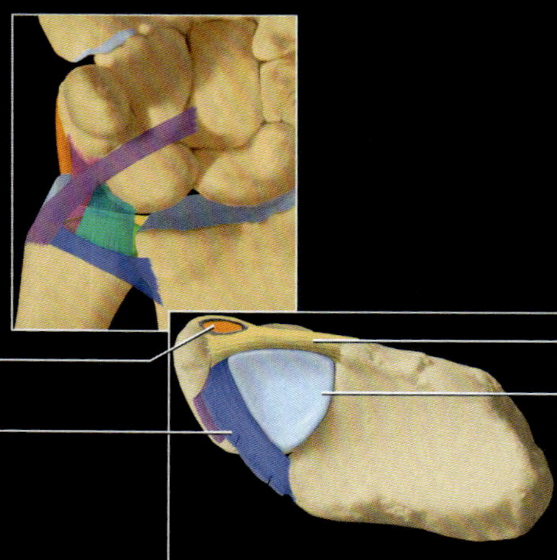

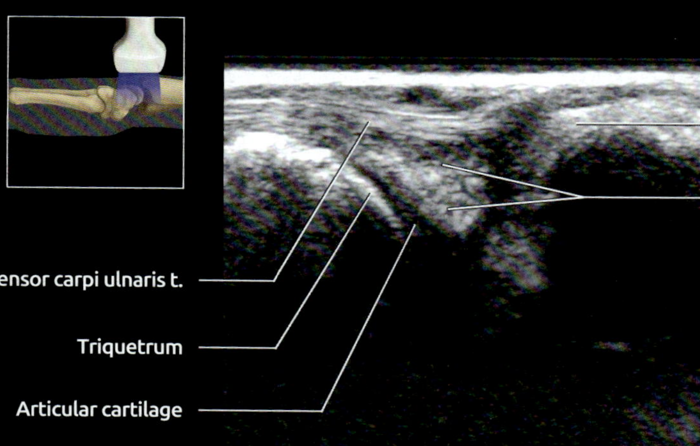

(Top) *Graphic shows supporting structures of the triangular fibrocartilage complex. The ulnocarpal ligaments and the volar radioulnar ligament are on the volar side. At the ulnar border, there is the ulnar collateral ligament. On the dorsal surface, there is the extensor carpi ulnaris tendon and its subsheath as well as the dorsal radioulnar ligament.* **(Middle)** *Graphic depicts the axial view of the articular disc of triangular fibrocartilage complex. The articular disc is inseparable from supporting dorsal and volar radioulnar ligaments. The disc is widest at its radial attachment. Central tears are more common, while peripheral tears, being better vascularized, have the capacity to heal.* **(Bottom)** *Longitudinal grayscale US shows the ulnar aspect of triangular fibrocartilage complex. The fibrocartilaginous articular disc is of different echotexture to hypoechoic hyaline cartilage. US is not as sensitive at depicting triangular fibrocartilage complex tears as MR. The extensor carpi ulnaris tendon provides an acoustic window though which to see the articular disc of the triangular fibrocartilage complex.*

VOLAR AND DORSAL LIGAMENTS

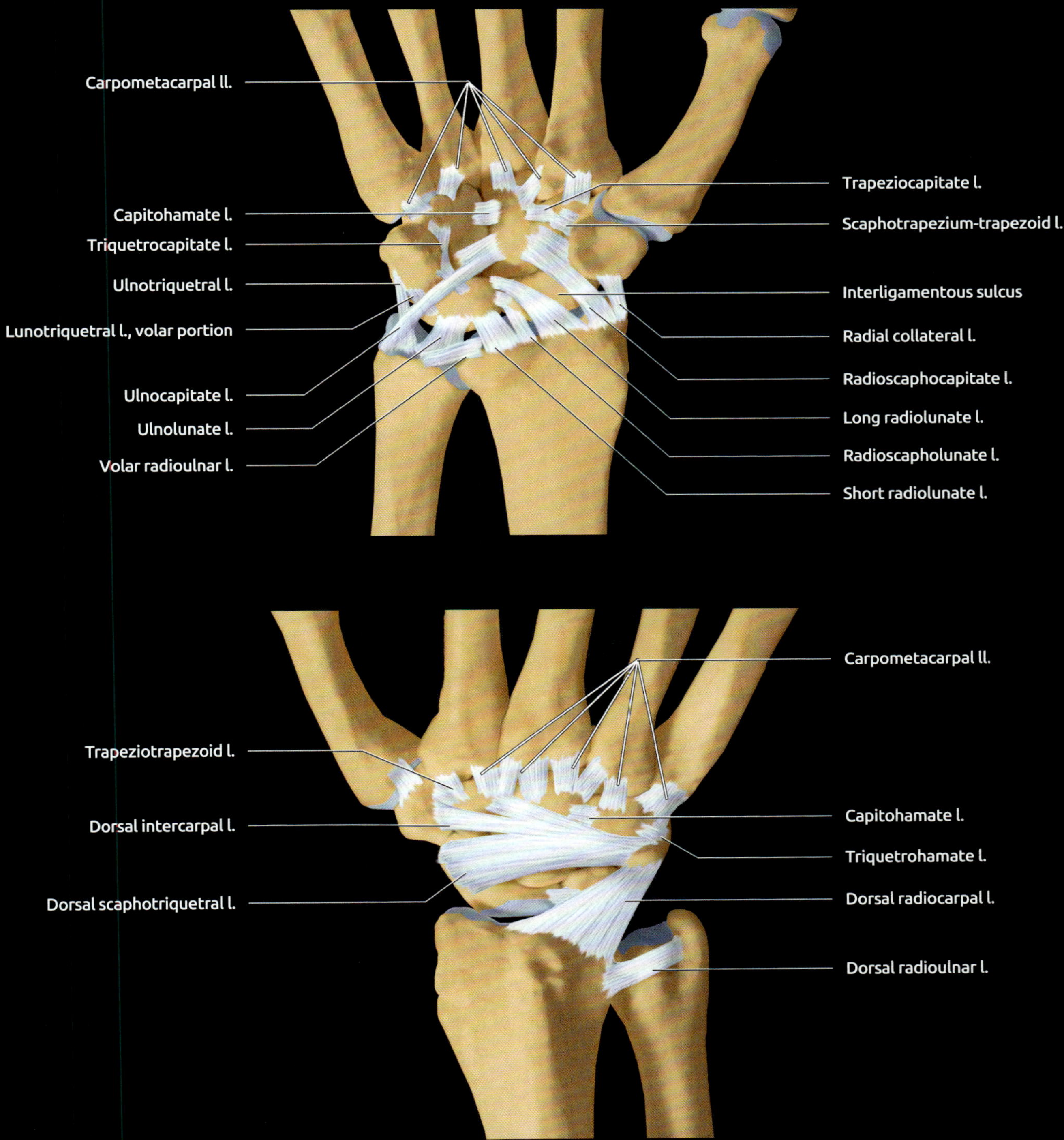

(Top) *Graphic shows volar intrinsic and extrinsic ligaments. The extrinsic ligaments connect the bones of the forearm (radius and ulna) and those of the carpus. The intrinsic ligaments connect carpal bones to carpal bones.* **(Bottom)** *Dorsal ligaments stabilize and restrict motion but are less critical to the stability of the wrist structures than the volar ligaments. On the volar aspect of the wrist, there is a triangular area of weakness (called the space of Poirier) between the lunate and capitate, which is not covered by any ligaments.*

EXTRINSIC FLEXOR MUSCLES: DIGITS 2-5

Flexor Digitorum Superficialis

- Origin: Common flexor tendon origin (medial humeral epicondyle) & midradius
- Insertion: Volar plate proximal interphalangeal (IP) joints & bases of middle phalanges digits 2-5
- Innervation: Median nerve
- Superficialis tendons split at bases of proximal phalanges into 2 tendon slips
  - Slips pass around flexor digitorum profundus (FDP) tendon before inserting into middle phalangeal bases
  - Flexor digitorum superficialis (FDS) forms "tunnel" for FDP
- Flexes metacarpophalangeal joints (aided by lumbricals & interossei) & proximal IP joints

Flexor Digitorum Profundus

- Origin: Proximal 1/2 of radius & interosseus membrane
- Insertion: Volar plates, distal IP joints, & bases of distal phalanges of digits 2-5
- Innervation: Ulnar & median nerves
- Flexes distal IP joints & proximal IP, metacarpophalangeal (aided by lumbricals & interossei) joints of digits 2-5

INTRINSIC MUSCLES: HYPOTHENAR AREA

Hypothenar Eminence

- Mirror image of thenar eminence
- Muscles (superficial to deep): Abductor digiti minimi (ADM), flexor digiti minimi (FDM), opponens digiti minimi (ODM)
  - Origin: ADM from pisiform; FDM & ODM from flexor retinaculum & hook of hamate
  - Insertion: ADM & FDM → conjoint insertion ulnar aspect base of proximal phalanx little finger
    - ODM to proximal 2/3 metacarpal shaft little finger
  - Innervation: All by ulnar nerve

INTRINSIC MUSCLES: THENAR AREA

Thenar Eminence

- Muscles: Abductor pollicis brevis (APB), flexor pollicis brevis (FPB), opponens pollicis (OP)
  - Origin: Flexor retinaculum & tubercle of trapezium
  - Insertion: APB, FPB conjoint insertion at base of proximal phalanx radial aspect
    - OP to volar aspect 1st metacarpal shaft
- Innervation: All by median nerve except FPB, which has also partially supplied by ulnar nerve

Adductor Pollicis Muscle

- Oblique (proximal) & transverse (distal) heads
- Origin: Capitate, trapezoid, 2nd & 3rd metacarpals
- Insertion: Ulnar aspect base of proximal phalanx of thumb, IP joint, volar plate of thumb
  - Also contributes fibers to extensor hood of thumb, which forms adductor aponeurosis
- Innervation: Ulnar nerve
- Actions: Adducts thumb toward 3rd digit & aids extension of IP joint
- Separated from thenar muscles by fascial plane & innervated by ulnar rather than median nerve

INTRINSIC MUSCLES: PALMAR AREA

Lumbricals

- Numbered 1-4 from radial to ulnar
- Origin: Radial aspect FDP tendon index to little fingers
- Insertion: Radial aspect extensor tendon expansion index to little fingers
- Innervation: Most radial 2 lumbricals by median nerve, most ulnar 2 lumbricals by ulnar nerve
- Actions: Flex metacarpophalangeal joints & extend IP joints

Palmar Interossei

- Numbered 1-3 from radial to ulnar
- Origin: Mesial palmar aspect diaphyses of 2nd, 4th, & 5th metacarpals
- Insertion: Mesial lateral bands & mesial aspect bases of proximal phalanges of same digit as origin
- Innervation: Ulnar nerve
- Actions: Adduct digits & assist lumbricals with flexion of metacarpophalangeal joints & extension of IP joints of digits 2, 4, & 5

Dorsal Interossei

- Numbered 1-4 from radial to ulnar
- Origin: FDP tendons just distal to carpal tunnel
- Insertion: Radial lateral bands of digits 2-5
- Innervation: Ulnar nerve
- Actions: Extends IP joints 2-5 & abducts digits 2-5

TENDONS

Tendon Sheaths

- Common flexor sheath (a.k.a. ulnar bursa)
  - Contains FDS & FDP tendons
  - Begins just proximal to carpal tunnel
  - Ends just beyond carpal tunnel for digits 2-4
  - Encloses 5th digit flexor tendons over entire course to distal IP joint
- Flexor pollicis longus tendon sheath (a.k.a. radial bursa)
  - Encloses flexor pollicis longus tendon from just proximal to carpal tunnel to tendon insertion at base of distal phalanx thumb
  - ± communication with ulnar bursa at level of carpal tunnel in ~ 50% of normal population
    - Potential route for spread of infection
- Individual flexor digital sheaths (digits 2-4)
  - Encloses flexor tendons from level of metacarpal necks to bases of distal phalanges

Extensor Expansion (or Extensor Hood)

- Expansions of extensor digitorum, extensor indicis, & extensor digiti minimi tendons on dorsum of fingers
  - Help prevent lateral translation of tendon
- 3 bands of extensor expansion
  - Single, wide middle band passes over dorsum of proximal phalanx & proximal 1/2 of middle phalanx
  - 2 cord-like lateral bands extend from proximal phalanx to distal phalanx
    - Lumbricals, palmar & dorsal interossei insert into these lateral bands

PALMAR ASPECT

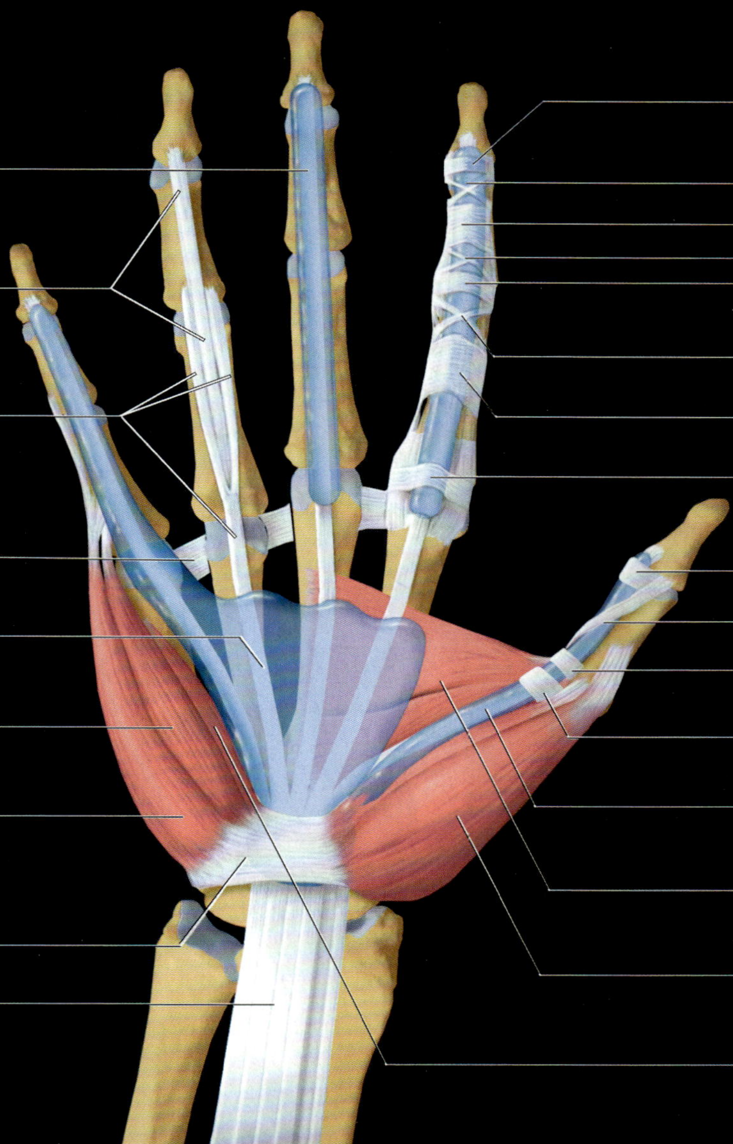

Pulley system for the 3rd-5th digits is identical to that demonstrated for the 2nd digit. Common digital sheath of the 4th digit is removed to show the relationship of flexor digitorum superficialis (FDS) and flexor digitorum profundus (FDP) tendons. Deep transverse metacarpal ligaments connect the volar plates (not shown) of digits 2-5. Although there is overlap of the radial and ulnar bursae in this image, these structures normally do not invariably communicate. The radial and ulnar bursae may communicate as a normal variation in ~ 50% of the population. Similarly, any 1 or more of the common digital sheaths may communicate with the ulnar bursa in up to 10% of the normal population. These different bursal communications are important as they provide routes for more extensive spread of infection.

LUMBRICAL AND INTEROSSEI MUSCLES

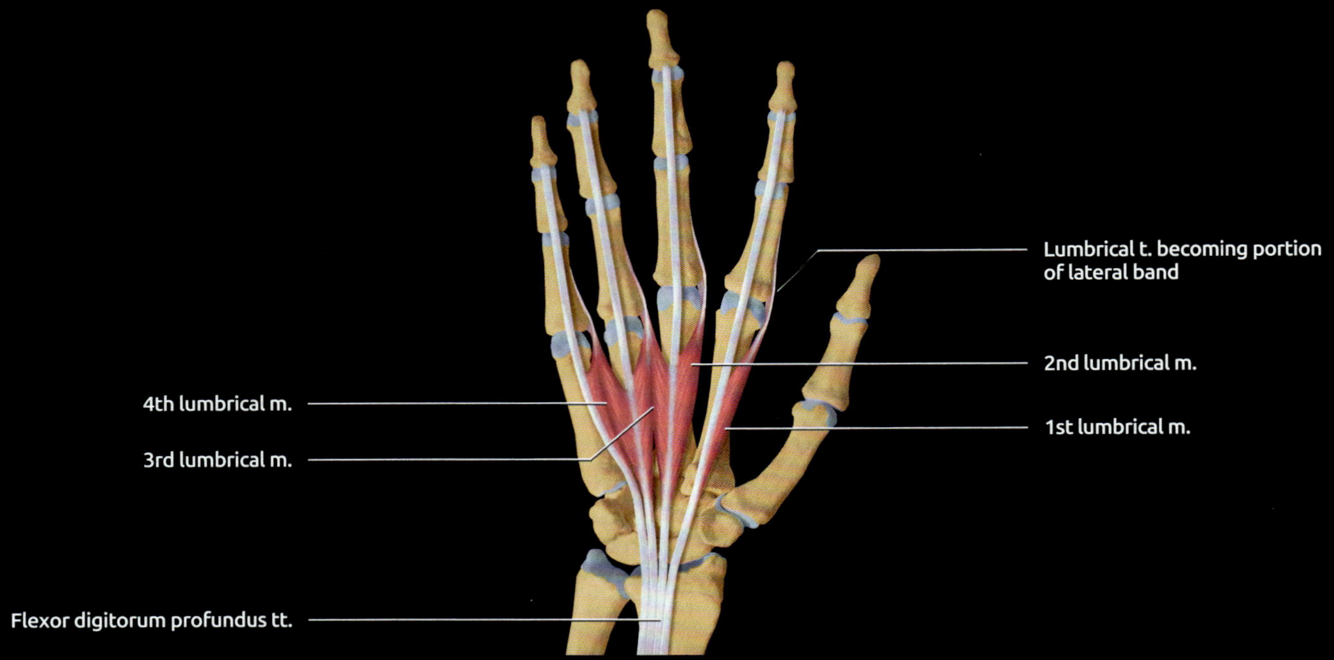

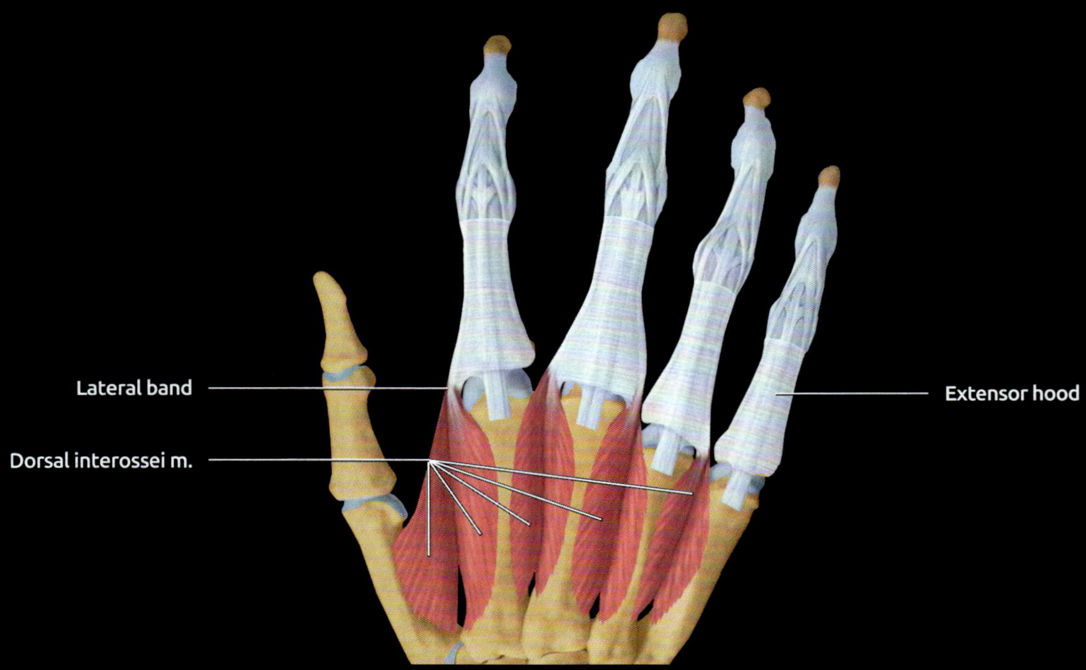

(Top) *Lumbricals originate from the FDP tendons as shown. Note how the 1st and 2nd lumbricals arise only from tendons to the 2nd and 3rd digits, respectively (unipennate), whereas the 3rd and 4th lumbrical arise from the tendons to both the 3rd and 4th and the 4th and 5th digits, respectively (bipennate).* **(Bottom)** *Graphic of the dorsal surface of the extensor mechanism of the hand and digits 2-5 demonstrates the complex relationship of the various fiber bands. As in imaging, distinction of individual fiber bands is often difficult and must be inferred by knowledge of where the structure "should" be with respect to more easily identifiable structures (such as bones and joints).*

PALMAR INTEROSSEI

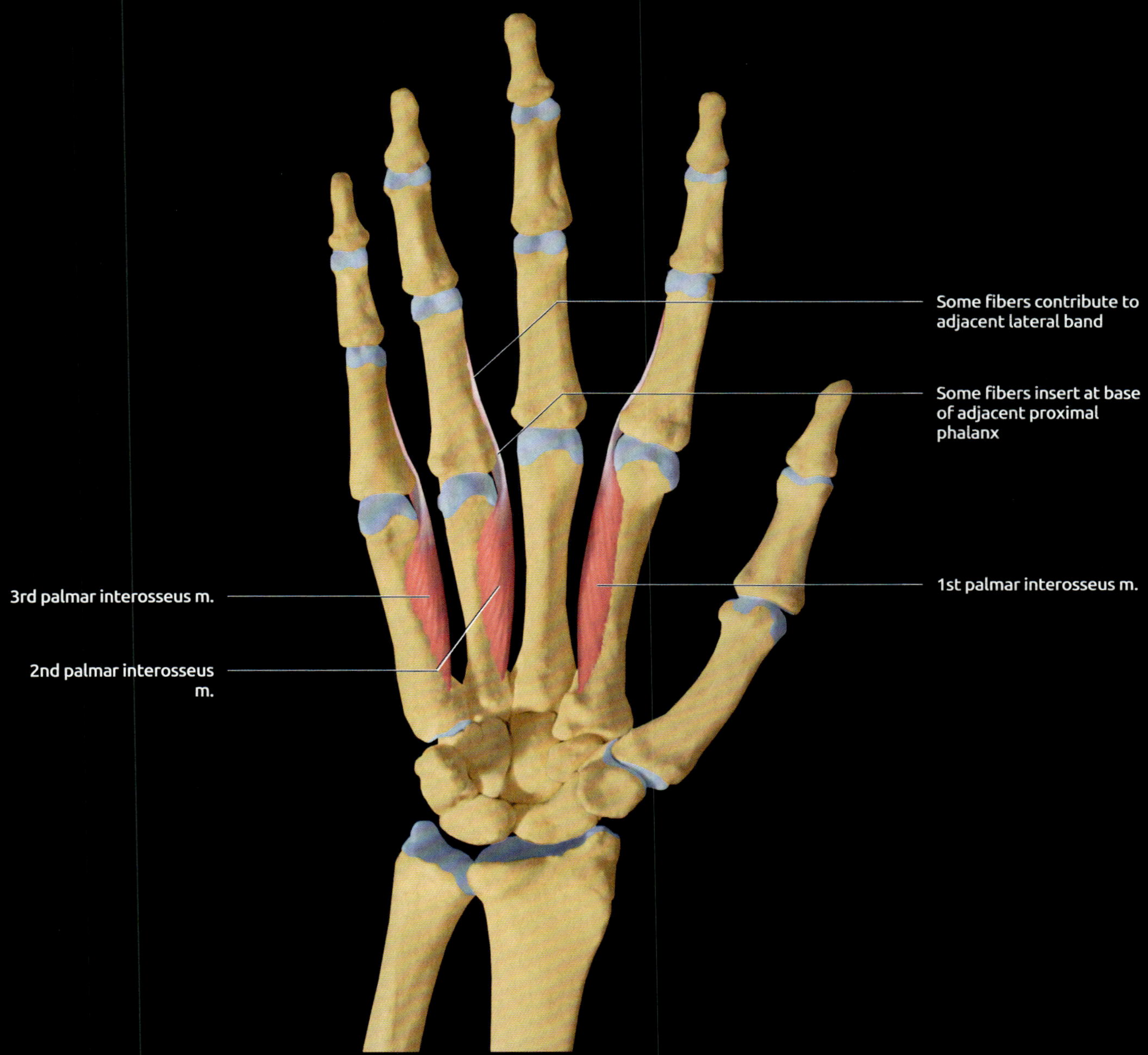

Palmar interossei insert at both bases of adjacent proximal phalanges as well as adjacent lateral bands. The 2nd and 3rd palmar interossei contribute to the radial lateral bands of the 4th and 5th digits, respectively, whereas the 1st palmar interosseus contributes to the ulnar lateral band of the 2nd digit.

TRANSVERSE US, PALM

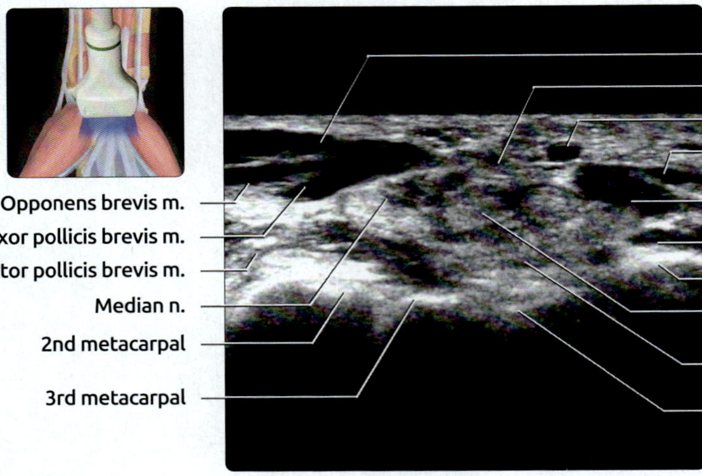

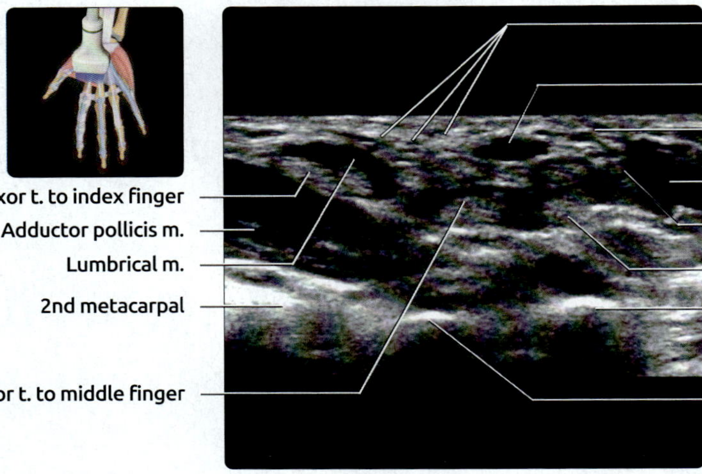

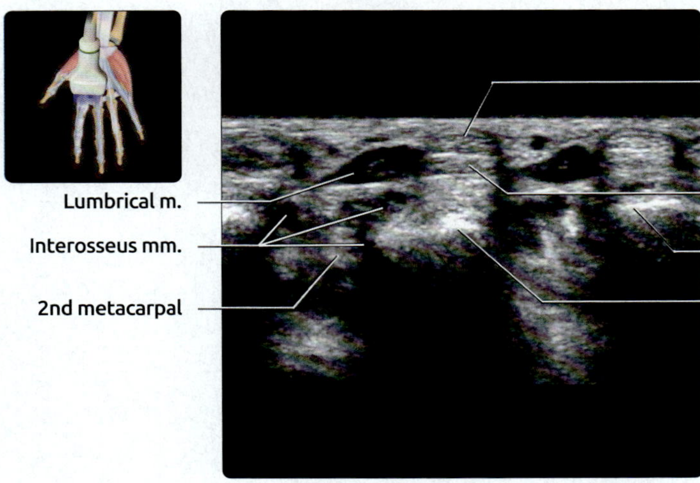

(Top) *Transverse grayscale US shows the proximal aspect of the palm. The palmar aponeurosis is thin and difficult to demonstrate in the mid- to distal palm especially.* **(Middle)** *Transverse grayscale US shows the midportion of the palm. Beneath the aponeurosis lies the flexor tendons, the lumbrical muscles, the muscles of the thenar and hypothenar eminences, and, most deeply, the interossei. There are 3 palmar interossei and 4 dorsal interossei muscles.* **(Bottom)** *Transverse grayscale US shows the distal portion of the palm.*

THENAR-HYPOTHENAR EMINENCE

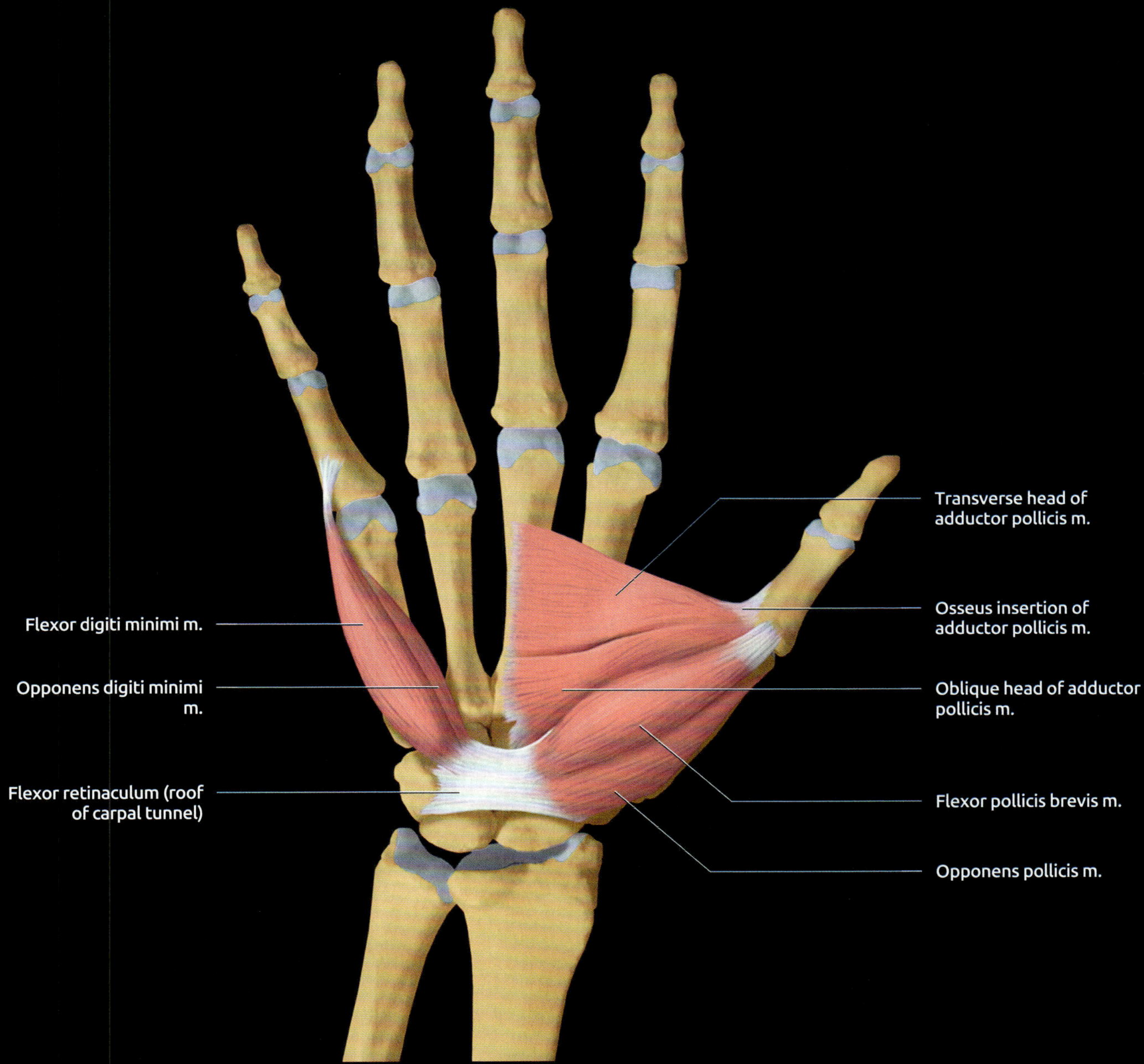

Graphic shows the thenar and hypothenar muscles of the hand. Strictly speaking, the adductor pollicis is not one of the thenar group of muscles since it has a separate fascia and separate nerve supply.

TRANSVERSE US, THENAR EMINENCE

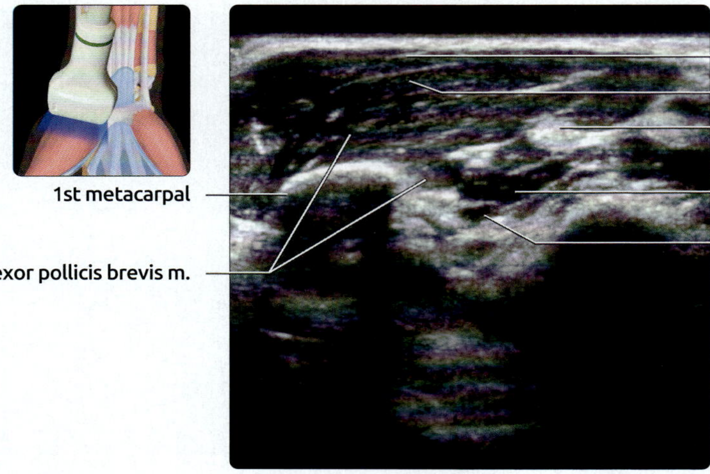

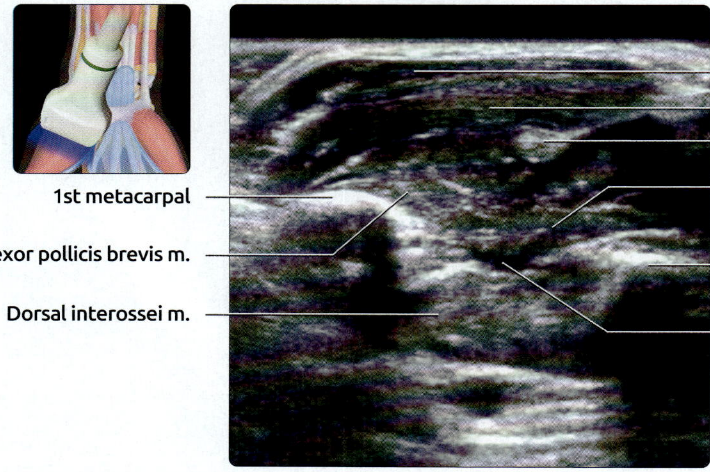

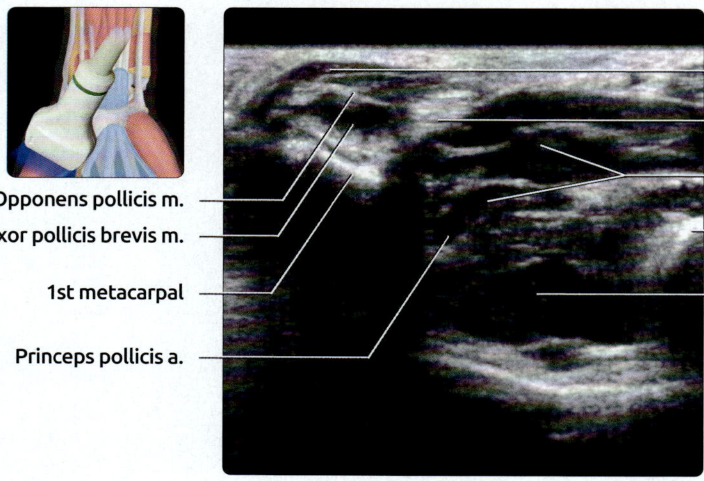

(Top) *Transverse grayscale US shows the proximal aspect of the thenar eminence. The thenar eminence comprises 4 muscles: Abductor, opponens, flexor, and adductor pollicis from above downward. The adductor muscle is, strictly speaking, not part of the thenar eminence since it is separated by a fascial layer and has a separate innervation.* **(Middle)** *Transverse grayscale US shows the midportion of the thenar eminence. The flexor pollicis longus tendon is easily recognized as an echogenic tendon alongside the flexor pollicis brevis, deep to the opponens pollicis and superficial to the adductor pollicis muscle.* **(Bottom)** *Transverse grayscale US shows the distal portion of the thenar eminence.*

LONGITUDINAL US, THENAR EMINENCE

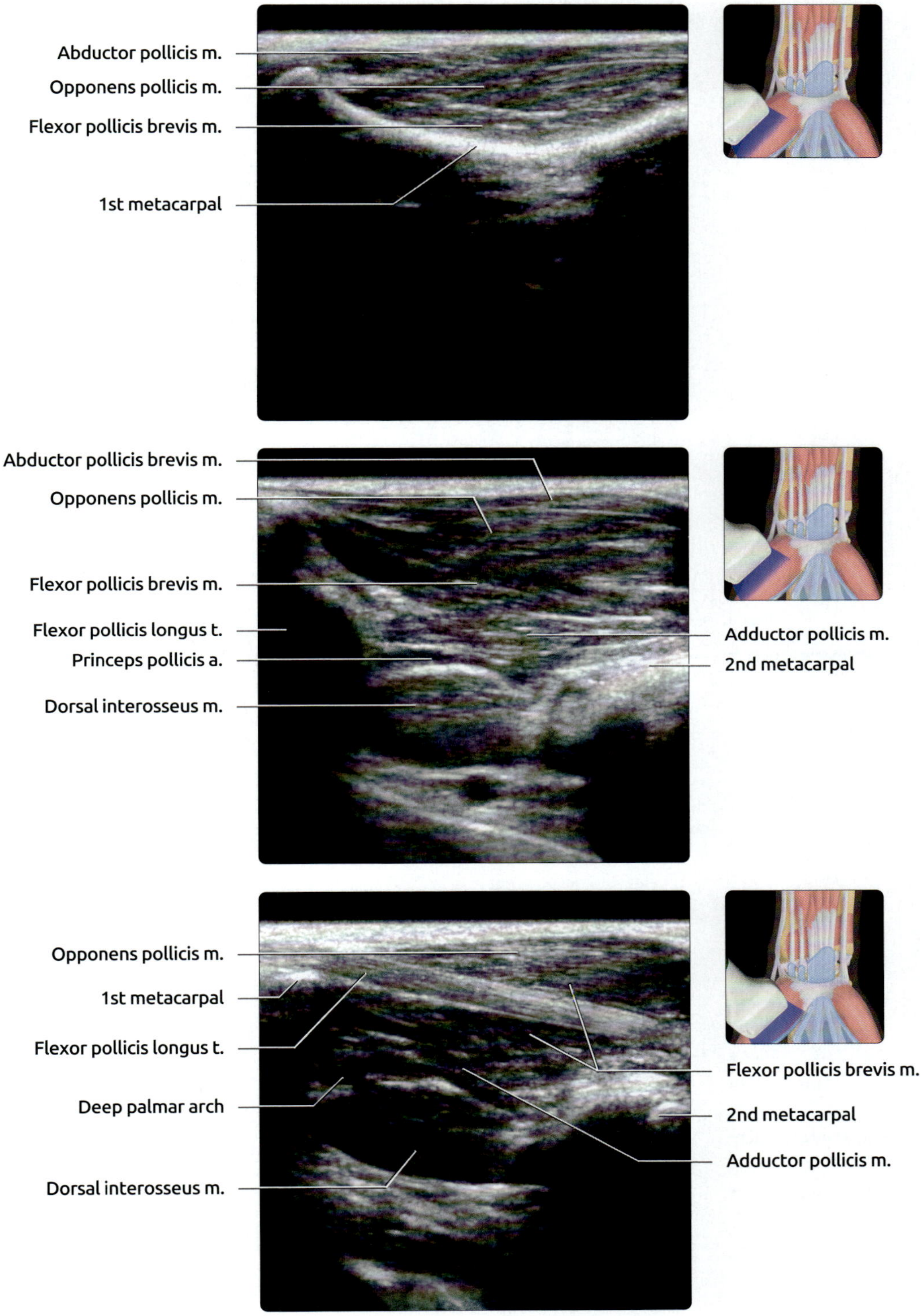

(Top) *Longitudinal grayscale US shows the proximal aspect of the thenar eminence. Use the mnemonic abductor pollicis brevis, opponens pollicis, and flexor pollicis brevis (ABOF) to remember the muscles of the thenar eminence.* **(Middle)** *Longitudinal grayscale US shows midthenar eminence. The ABOF lie superficial to deep in that order.* **(Bottom)** *Longitudinal grayscale US shows the distal aspect of the thenar eminence. The flexor pollicis longus tendon passes through the more distal aspect of the thenar eminence.*

TRANSVERSE US, HYPOTHENAR EMINENCE

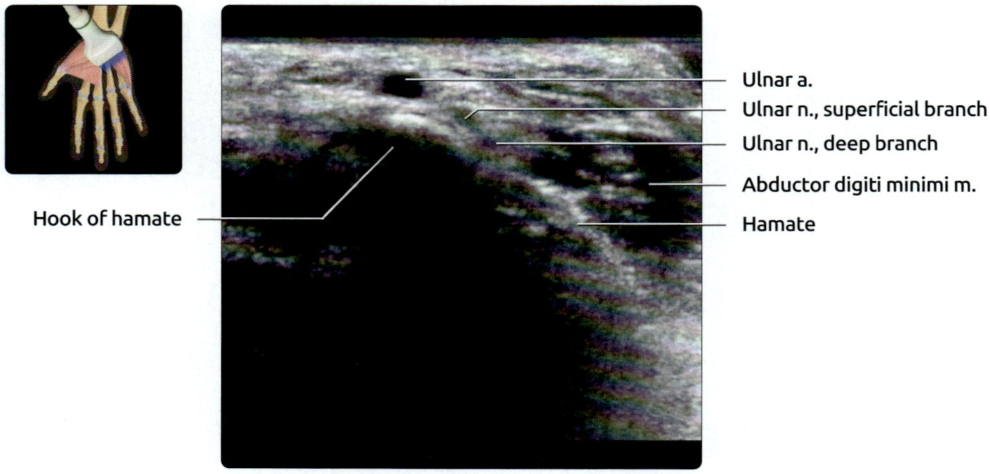

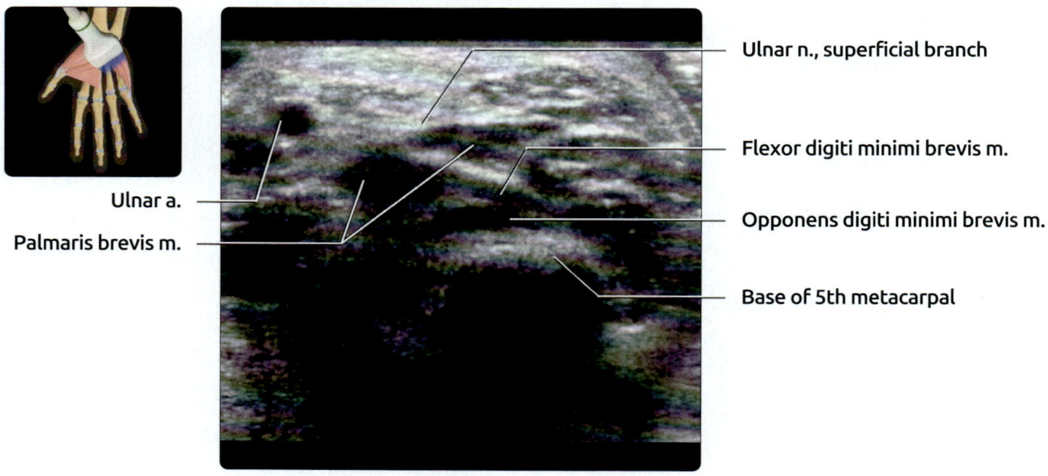

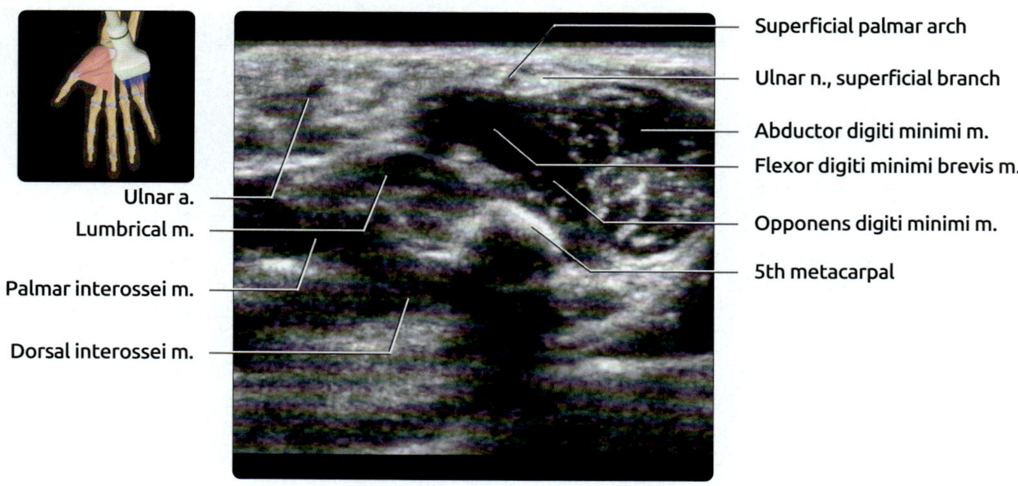

(Top) *Transverse grayscale US shows the proximal 1/3 of the hypothenar eminence. The hypothenar eminence along with the other muscles of the hand, except for the thenar, are supplied by the ulnar nerve.* **(Middle)** *Transverse grayscale US shows the middle 1/3 of the hypothenar eminence. Hypothenar muscle atrophy can occur with injury to the deep branch of the ulnar nerve just distal to Guyon canal.* **(Bottom)** *Transverse grayscale US shows the distal 1/3 of the hypothenar eminence. The order of muscles in the hypothenar eminence is different from the thenar eminence with abductor digiti minimi, flexor digiti minimi, and opponens digiti minimi from superficial to deep.*

DORSAL ASPECT OF HAND

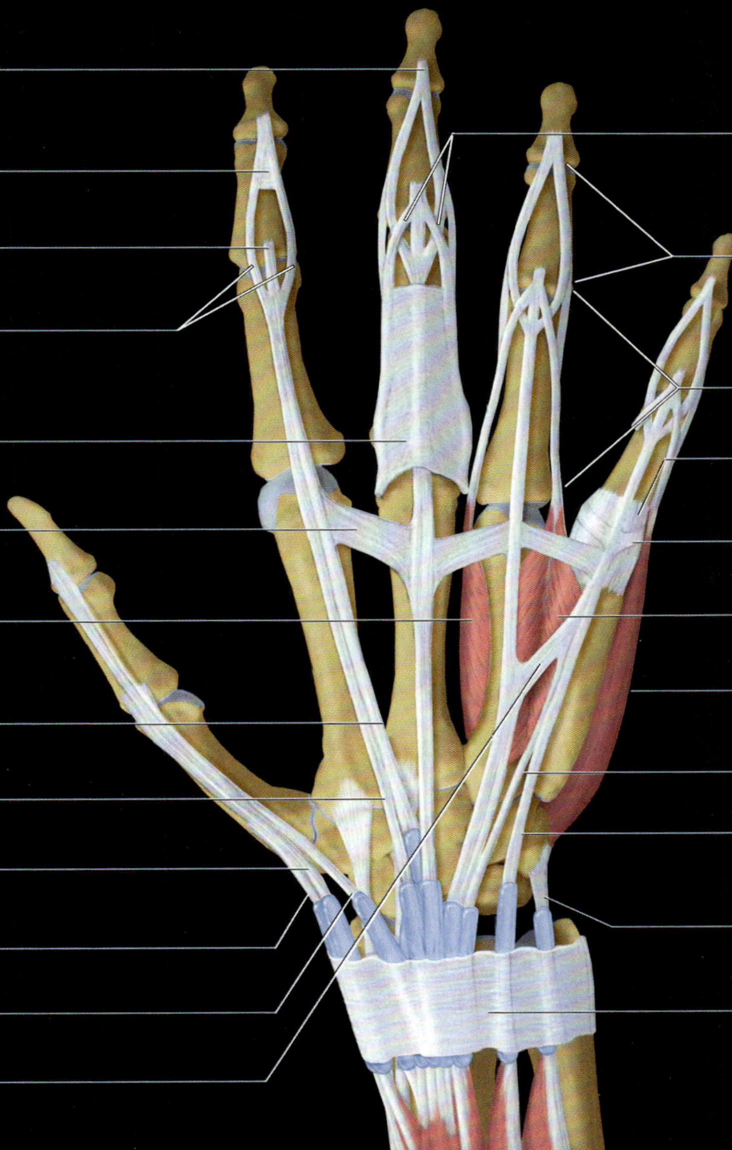

Extensor mechanism composite shows different components of the extensor mechanism on different digits. With the exception of the extensor indicis, extensor digiti minimi, abductor digiti minimi, and 4th extensor digitorum communis (EDC) tendon contribution to the 5th EDC tendon, any structure on any of the 2nd-5th digits can be extrapolated to any and all of the other 2nd-5th digits. Note that the 2nd extensor retinaculum compartment and its contents (extensor carpi radialis longus and brevis tendons) are not included in this graphic. Although the abductor pollicis longus tendon travels in the 1st compartment of the extensor retinaculum, its insertion (not shown) is actually on the radial aspect of the volar surface of the base of the 1st metacarpal.

DORSAL INTEROSSEI

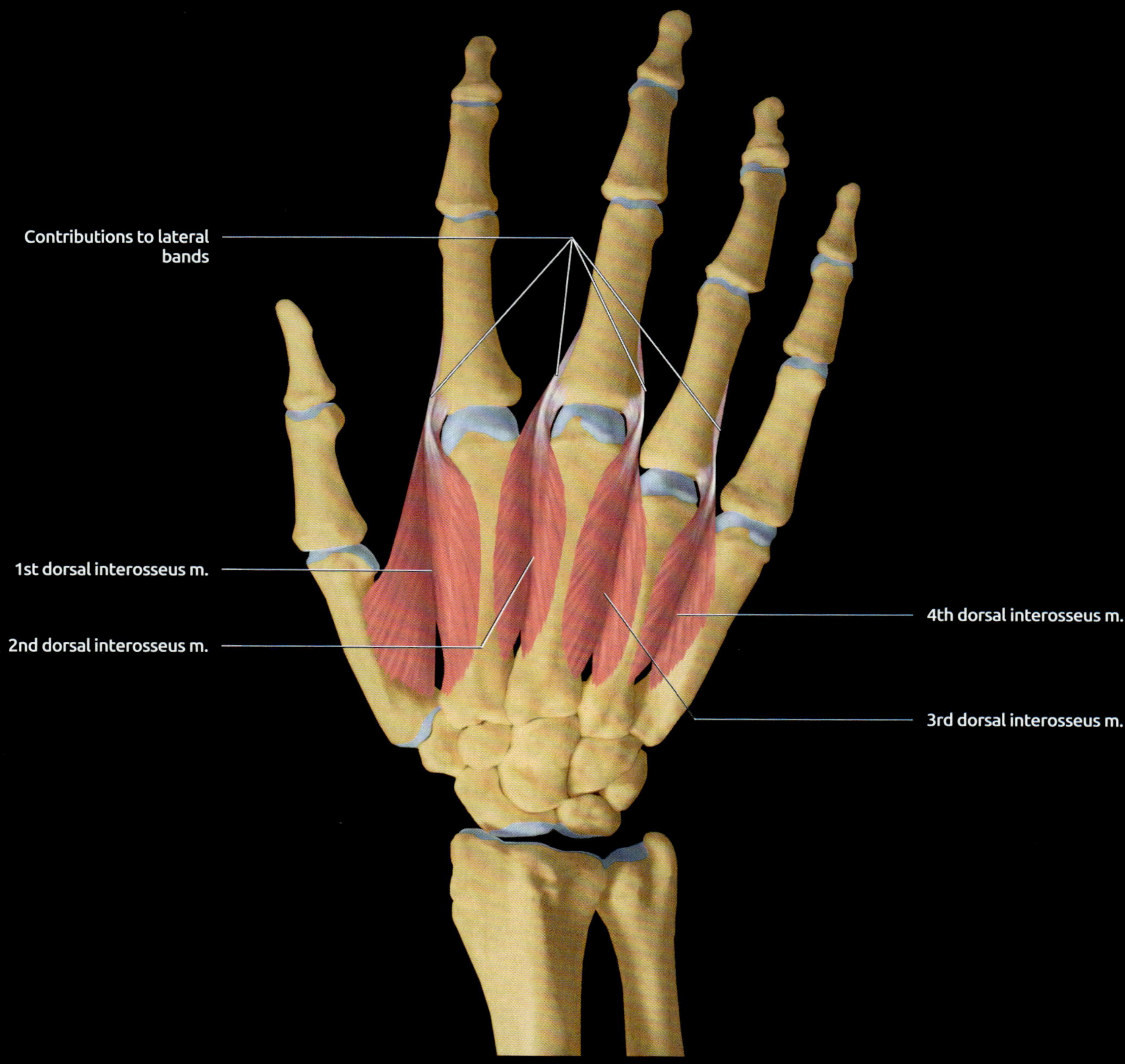

Tendons of the dorsal interossei muscle (along with tendons of the palmar interossei and tendons of the lumbricals) help form lateral bands. Note how each interosseus muscle only contributes fibers to the mesial-most adjacent lateral band. The dorsal interossei do not contribute any lateral band fibers to the 1st or 5th digits.

TRANSVERSE US, DORSUM HAND

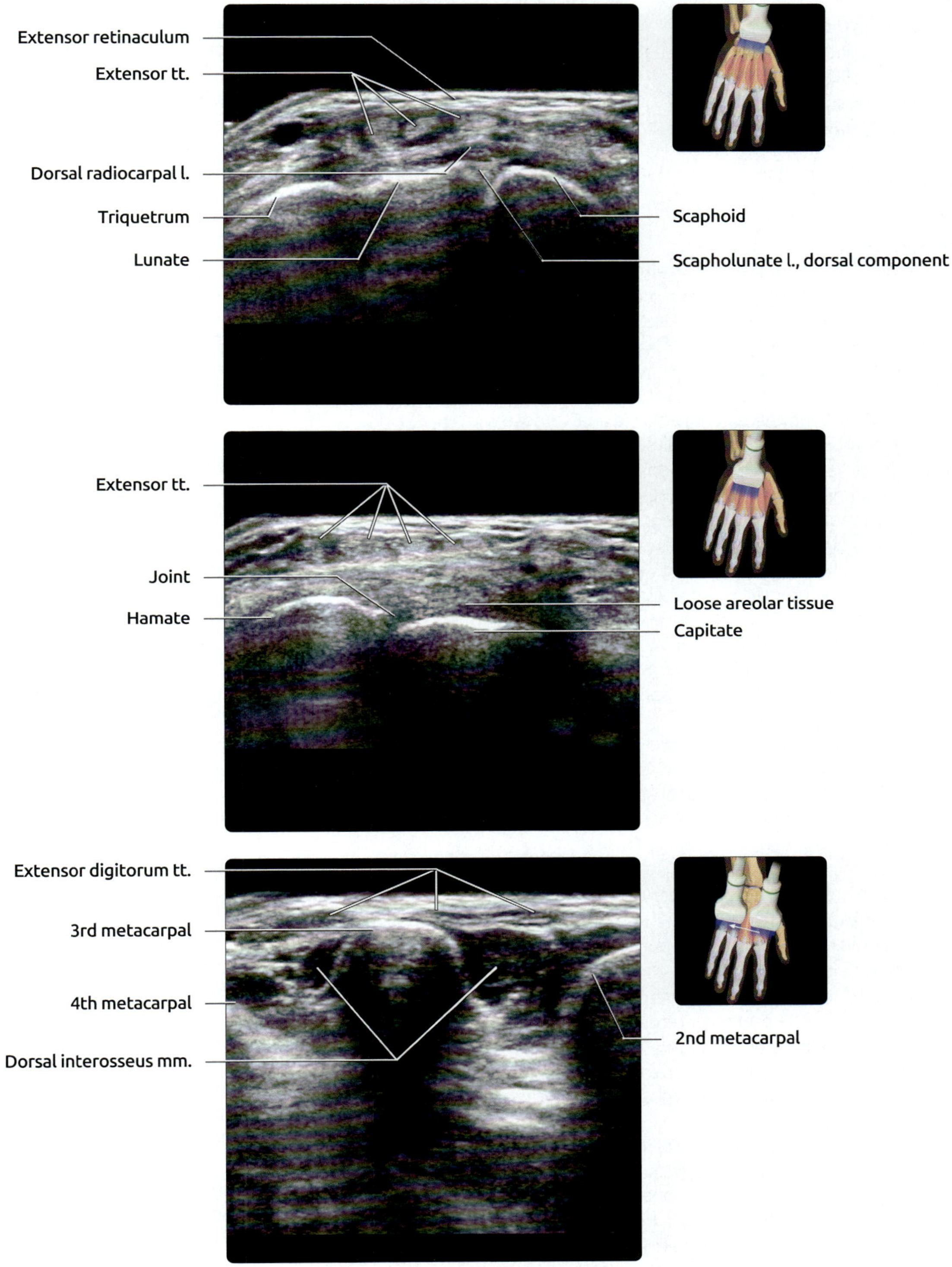

(Top) *Transverse grayscale US shows the proximal aspect of the dorsum hand. The tissues on the dorsum hand are less tightly bound than the tissues on the palmar aspect of the hand, and this part of the hand tends to swell much more with inflammatory or edematous processes.* **(Middle)** *Transverse grayscale US shows the mid aspect of the dorsum of hand. Loose areolar tissue separates the extensor tendons from the metacarpal bones. The extensor tendons do not have a complete tendon sheath from the level of the extensor retinaculum to the insertion.* **(Bottom)** *Transverse grayscale US shows the distal aspect of the dorsal hand. There are 4 dorsal interossei, 1 for each metacarpal interspace.*

TERMINOLOGY

Abbreviations

- Abductor pollicis longus (APL)
- Extensor pollicis longus (EPL)
- Extensor pollicis brevis (EPB)
- Flexor retinaculum (FR)
- Pronator quadratus (PQ)

GROSS ANATOMY

Arteries

- Radial artery
  - Terminal branch of brachial artery
  - Located superficial to PQ, then continues dorsally around radial styloid process, then passes deep to APL and EPB, across anatomic snuffbox and deep to EPL
  - Branches of radial artery
    - Superficial palmar branch continues distally superficial to FR through thenar eminence to anastomose with ulnar artery as superficial palmar arch
  - Deep palmar arch is continuation of radial artery in hand
    - Main radial artery continues dorsally at radial styloid across anatomic snuffbox deep to EPB and APL
    - Passes over dorsal aspect of scaphoid bone
    - Passes between heads of 1st dorsal interosseus muscle to enter palm; joins ulnar artery to form deep palmar arch
    - Deep palmar arch is located at level of metacarpal bases 1 cm proximal to superficial palmar arch
    - Lies deep to long flexor tendons and superficial to interossei
    - Deep branch of ulnar nerve lies on convexity of deep palmar arch
    - Princeps pollicis artery, radial artery of index finger, and palmar metacarpal arteries arise from deep palmar arch
- Ulnar artery
  - Terminal branch of brachial artery
  - Located superficial to PQ, then runs between flexor carpi ulnaris and flexor digitorum superficialis tendons
  - Branches of ulnar artery
  - Common interosseus artery branches in forearm divide into anterior and posterior interosseus arteries
    - Anterior interosseus artery travels distally; at proximal PQ, gives small branch that pierces membrane, enters dorsal forearm, and anastomoses with posterior interosseus artery
    - Median artery (branch of anterior interosseus artery) arises in forearm and may persist accompanying median nerve into carpal tunnel, contributing to superficial palmar arch
    - Posterior interosseus artery travels distally between deep and superficial extensors to anastomose with anterior interosseus artery
  - Superficial palmar arch is continuation of ulnar artery in hand
    - Located along line extended in arch across palm from thumb web space
    - Similar to metacarpal shaft level ~ 1 cm distal to deep palmar arch
    - Located deep to palmar aponeurosis but superficial to long flexor tendons and digital nerves
    - Common palmar digital arteries arise from superficial palmar arch
    - Run distally on surface of 2nd, 3rd, and 4th lumbrical muscles
    - Near metacarpophalangeal joints, each common palmar digital artery divides into 2 proper digital arteries
    - Run in subcutaneous fat of digits (lateral aspects)

Veins

- Dorsal venous network of hand and network formed by dorsal metacarpal veins
  - Found on dorsum of hand
  - Unite to form cephalic vein (radial side of wrist) and basilic vein (ulnar side of wrist)
    - Basilic and cephalic veins travel proximally in superficial tissues
    - Both veins unite at antecubital fossa via median cubital vein
    - Continues proximally as cephalic vein until it joins brachial vein in axilla to form axillary vein
  - Layout of superficial veins in hand and forearm is highly variable
- Causes of deep venous thrombosis in upper limb extremity are different from that of lower limb
  - Related to intravenous cannulations, infusions, radiotherapy, effort-induced thrombosis, and malignant obstruction

IMAGING ANATOMY

Overview

- Extensive volar and dorsal anastomoses typically exist

ANATOMY IMAGING ISSUES

Imaging Recommendations

- Assessment of caliber of small arteries of hand best performed by digital subtraction angiography as they are close to limit of resolution by US
  - For example, when investigating suspected Raynaud disease or Buerger disease

Imaging Pitfalls

- Vessel visualization on standard imaging is limited
  - Radial and ulnar artery: 2-3 mm, typically visualized
  - Vascular arches: 1 mm, inconsistently visualized
  - Multiple branches and anastomoses: < 1 mm, rarely visualized
- Entry points of nutrient vessels into bone should not be mistaken for erosions
- If angle of insonation of US beam is at or near 90° to vessel, Doppler and color sensitivity will ↓
- Optimize gate size for small-caliber arteries and position at center of arterial lumen to detect peak flow velocity

HAND VASCULATURE

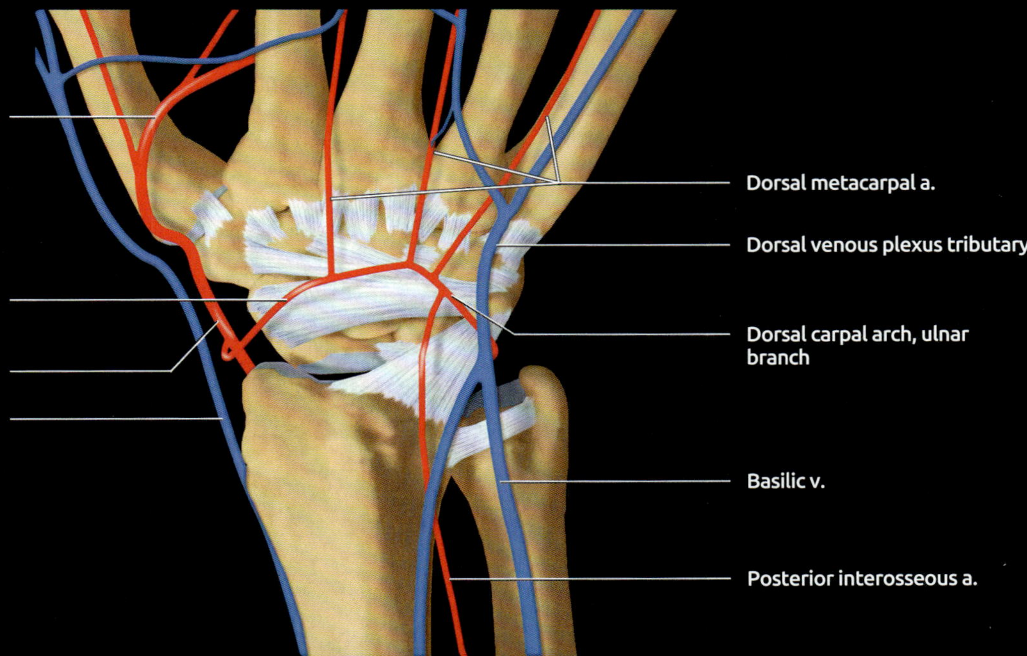

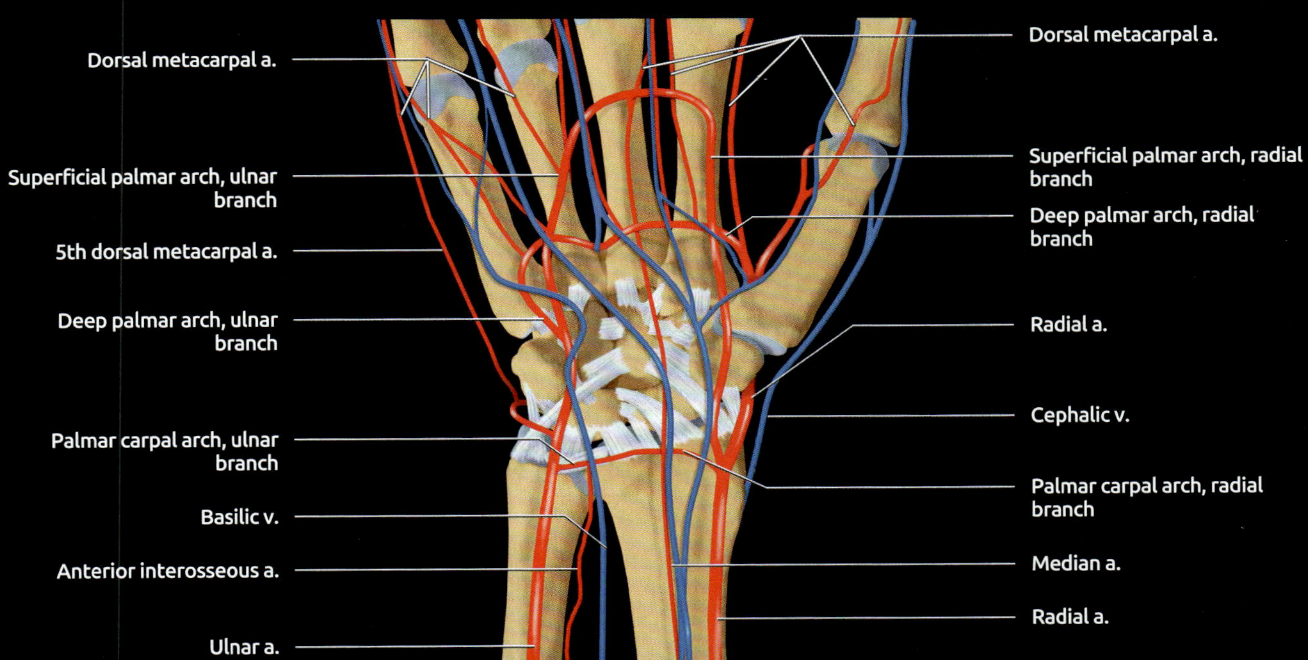

(Top) *Vasculature of the dorsal wrist is shown. The dorsal carpal arch supplies the distal radius, distal carpal row, and lateral proximal carpal row. The venous plexus drains into 2 main venous systems, the cephalic and basilic, with multiple anastomotic communications.* **(Bottom)** *Vasculature of the volar wrist is shown. Three major arterial arches are contributed to by the radial, ulnar, and interosseous arteries: Palmar (volar) carpal, deep, and superficial palmar arches. A dorsal venous plexus drains into the cephalic and basilic veins.*

TRANSVERSE US, WRIST AND PALM

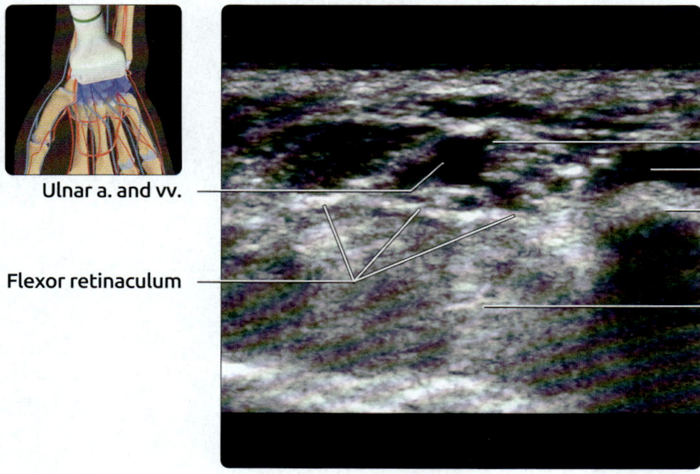

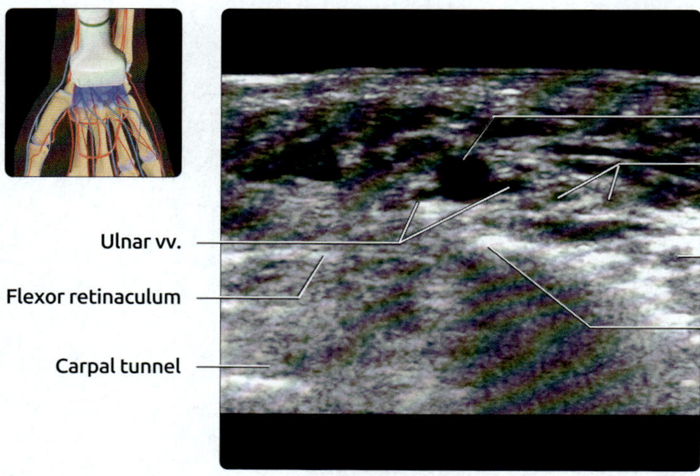

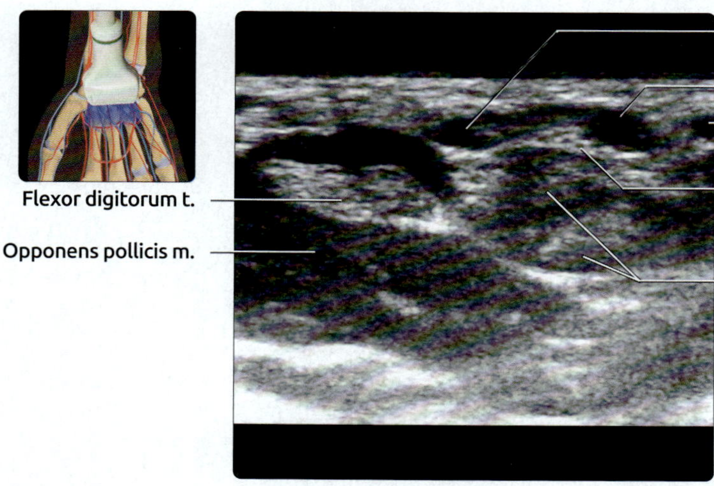

(Top) *Transverse grayscale US shows the ulnar artery at the level of the pisiform bone within Guyon canal. The ulnar artery at the wrist lies on the radial side of the pisiform. Between the ulnar artery and the pisiform lies the ulnar nerve. The ulnar artery divides either within or just beyond the Guyon canal.* **(Middle)** *Transverse grayscale US shows the ulnar artery at the level of the hook of hamate. Here, it is prone to compression injury (hypothenar hammer syndrome), which may damage its walls leading to either thrombosis, occlusion, or aneurysms. It is usually the superficial branch of the ulnar artery that is most prone to hypothenar hammer syndrome.* **(Bottom)** *Transverse grayscale US shows the ulnar artery at the level of the midpalm. The superficial branch of the ulnar artery contributes to the superficial palmar arch.*

TRANSVERSE US, DORSAL ASPECT

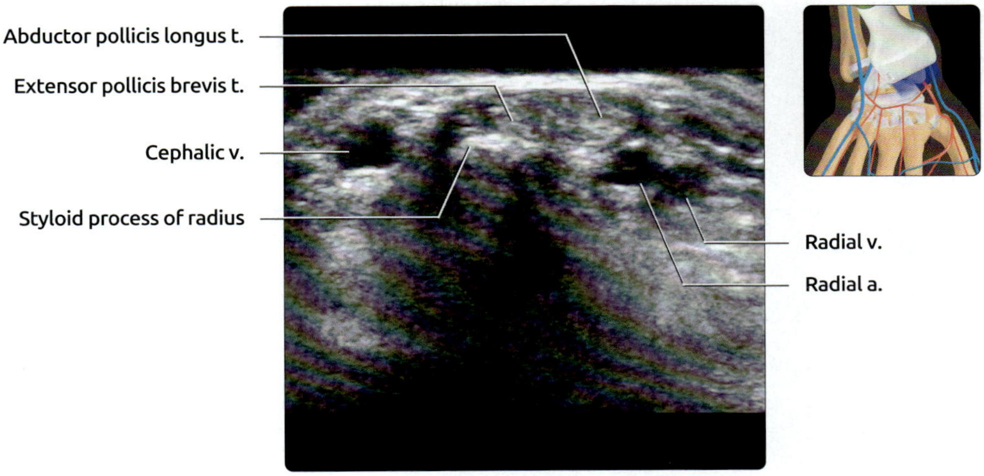

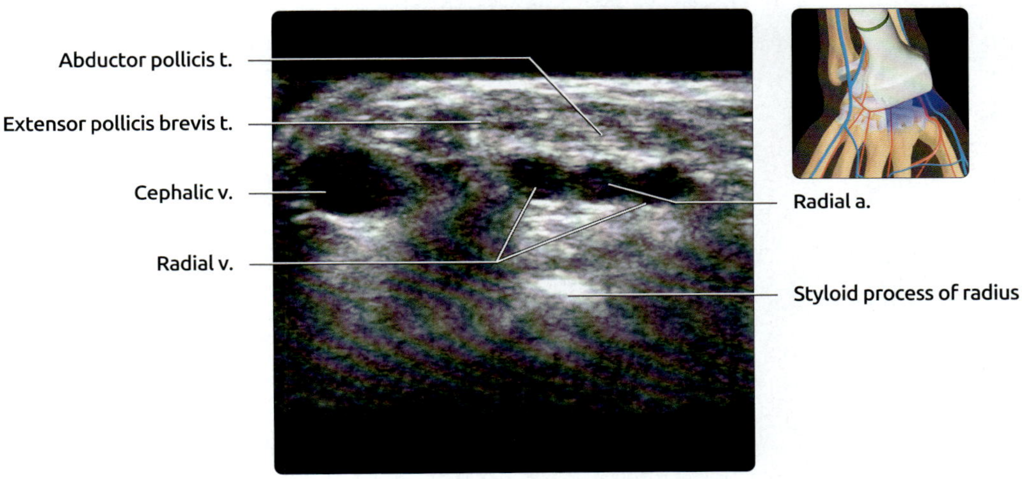

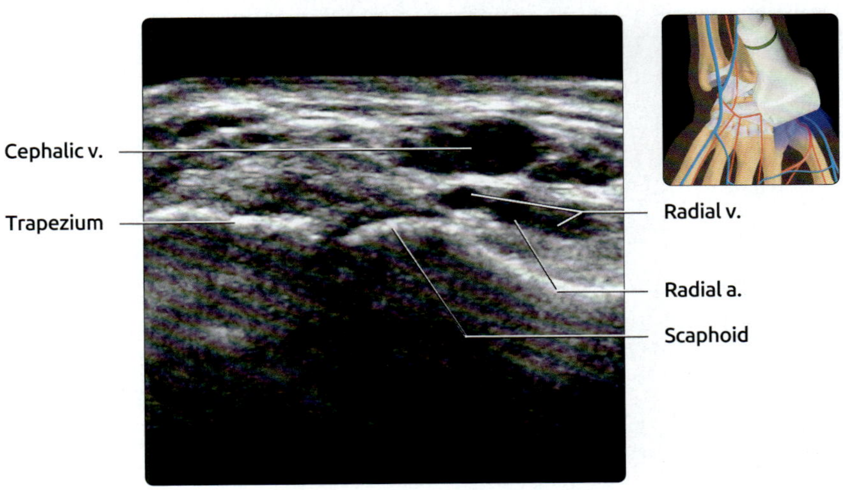

(Top) *Transverse grayscale US shows the radial artery just proximal to the wrist. The radial artery runs a superficial course parallel with the shaft of the radius in the distal forearm. At the wrist, it lies alongside the 1st extensor compartment, comprising abductor pollicis and extensor pollicis brevis tendons.* **(Middle)** *Transverse grayscale US shows the radial artery at the wrist joint level. The radial artery runs deep to the 1st extensor compartment tendons. Just before it winds around the wrist, it gives off the superficial palmar branch, which helps form the superficial palmar arch.* **(Bottom)** *Transverse grayscale US shows the radial artery at the dorsum of the hand. The deep branch of the radial artery runs on the dorsal surface of the scaphoid and trapezium bones and then passes deep between the oblique and transverse heads of the adductor pollicis muscle to enter the palm to form the deep palmar arch along with the deep branch of the ulnar artery.*

TRANSVERSE US, PALMAR ASPECT

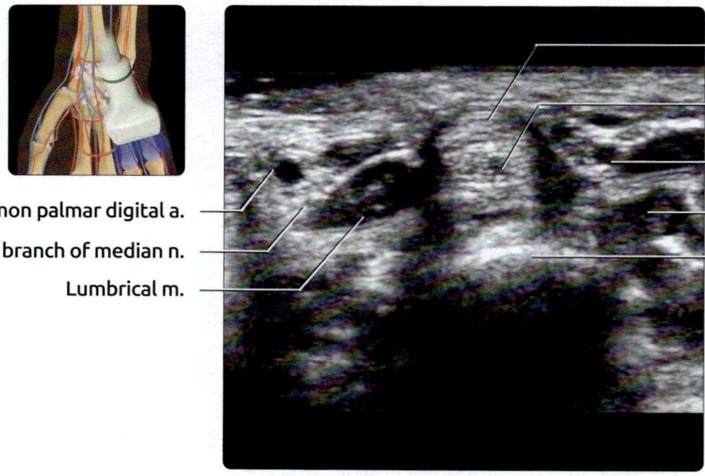

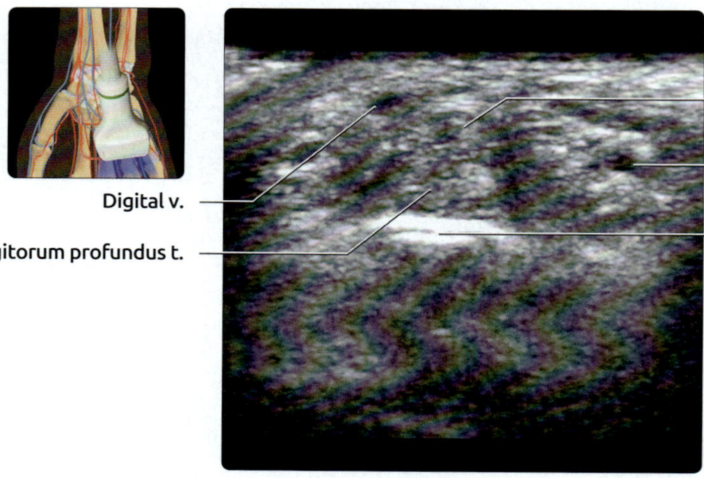

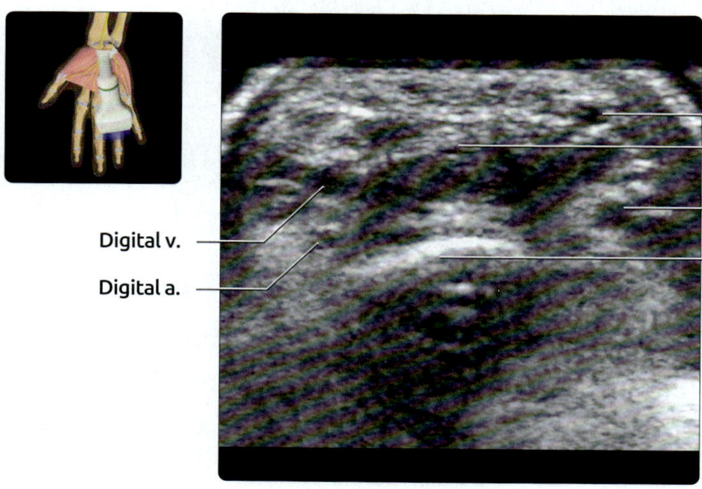

(Top) *Transverse grayscale US shows the vessels of the midpalm at the midmetacarpal shaft level. The common palmar digital arteries arise from the superficial palmar arch and divide at the finger bases into proper digital arteries. Each finger has 4 (2 volar, 2 dorsal) proper digital arteries.* **(Middle)** *Transverse grayscale US shows the digital vessels at the proximal phalanx level. The palmar digital arteries lie dorsal to the digital nerves. The palmar digital arteries are usually bigger than the dorsal digital arteries. Ulnopalmar arteries are 2x as big as the radiopalmar arteries (1.8 mm vs. 1.1 mm). Peak systolic velocity in hand arteries is variable, ranging from 30-70 cm/sec.* **(Bottom)** *Transverse grayscale US shows vessels at the middle phalanx level. The veins of the fingers and the hand are variable in number and location. Dorsal veins are of variable size, ranging from 1.0-1.5 mm.*

HAND ARTERIES

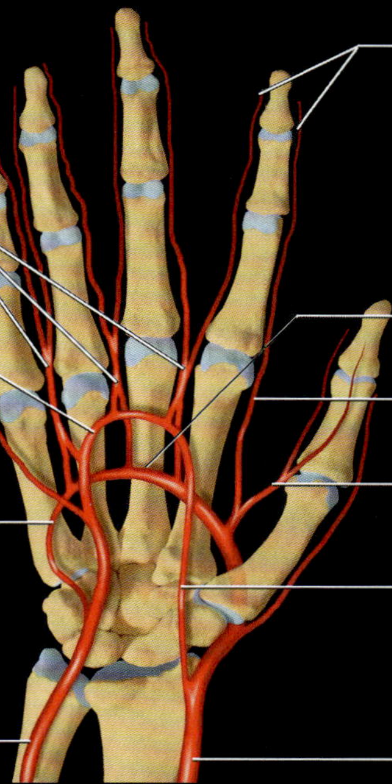

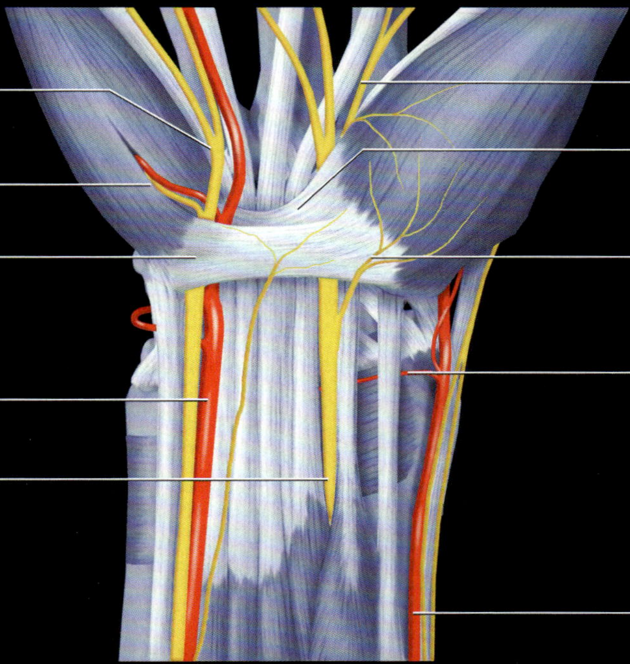

(Top) *Graphic shows the hand arteries. After giving off its contribution to the superficial palmar arch, the radial artery travels around the radial aspect of the wrist to the dorsum of the hand where it travels deep in the anatomic snuffbox. It then dives into the 1st interspace between the heads of the 1st dorsal interosseus muscle as well as between the transverse and oblique heads of the adductor pollicis muscle before forming the deep palmar arch.* **(Bottom)** *Graphic shows the relationship of the nerves and arteries to the volar carpal ligament (superficial fibers of flexor retinaculum that form the roof of Guyon canal) and flexor retinaculum (which forms the roof of the carpal tunnel). The ulnar nerve and accompanying artery pass deep to the volar ligament before branching into the deep and superficial branches.*

GROSS ANATOMY

Muscles

- Thenar muscle
  - Composed of 4 muscles, mnemonic: "One for all and" (ABOF)
    - **Ab**ductor pollicis brevis (APB)
    - **O**pponens pollicis (OP)
    - **F**lexor pollicis brevis (FPB)
    - Adductor pollicis (AP)
  - Origin: All originate from flexor retinaculum and tubercle of trapezium
    - APB also originates from tubercle of scaphoid
  - Insertion: APB and FPB share combined insertion at base of proximal phalanx radial aspect
    - OP inserts onto volar aspect of 1st metacarpal
  - AP is often grouped with thenar muscles but is anatomically distinct from them
    - AP is separated from thenar muscles by fascial plane and innervated by ulnar nerve rather than median nerve
- Extensor muscles
  - Extensor pollicis longus
    - Midulna and interosseus membrane → dorsal aspect base of distal phalanx thumb
    - Extends both metacarpophalangeal (MCP) and interphalangeal joints of thumb
  - Extensor pollicis brevis
    - Distal radius and interosseus membrane → dorsal aspect base of proximal phalanx thumb
    - Extends MCP joint of thumb
- Long flexor muscle
  - Flexor pollicis longus (FPL)
    - Midradius and interosseus membrane → base distal phalanx of thumb
    - Flexes MCP and interphalangeal joints of thumb
  - FPL tendon sheath (i.e., radial bursa)
    - Encompasses FPL tendon from just proximal to carpal tunnel to its insertion (distal phalanx of thumb)
    - Radial bursa occasionally communicates with ulnar bursa (common flexor sheath) in 50% of subjects
- Extensor expansion
  - Formed by fibers from AP on ulnar side and APB on radial side
  - Aids extension of interphalangeal joint
  - AP contribution to extensor expansion is known as adductor aponeurosis

Tendon Pulley System

- More variable than finger tendon pulley system
- In addition to standard A1 and A2 pulleys, have oblique and variable annular pulleys (Av)
- A1 pulley: At MCP joint level
- Av pulley: Along proximal 1/2 of proximal phalanx, variable in position, runs either obliquely or transversely across proximal phalanx
- Oblique pulley: Runs obliquely just proximal to A2 pulley
- A2 pulley: At interphalangeal joint level

IMAGING ANATOMY

Overview

- On dorsal transverse imaging, extensor expansions are seen as thin, regular, hypoechoic bands arising from edges of extensor tendon
- On palmar transverse imaging, annular pulleys are seen as thin, ring-shaped structures covering flexor tendons and inserting in palmar plate

ANATOMY IMAGING ISSUES

Imaging Recommendations

- Thenar muscles can be identified reliably on US
- For suspected ulnar collateral ligament (UCL) injury of thumb, first assess contralateral unaffected side to appreciate normal thickness of UCL
  - Check for avulsion fracture from base of proximal phalanx (uncommon)
  - Hyperabduction of thumb during US examination is usually neither necessary nor feasible (due to patient pain)
  - Best to perform US soon after injury as scar tissue may blur anatomical delineation in later stages

CLINICAL IMPLICATIONS

Clinical Importance

- Radial bursa often communicates with ulnar bursa at level of carpal tunnel allowing spread of infection
- UCL of MCP joint of thumb is especially prone to tearing in forced abduction and extension
  - Referred to as gamekeeper thumb or skier thumb
  - Nearly all tears occur at base of proximal phalanx
  - If UCL is torn, position of proximal retracted end with respect to adductor aponeurosis is of importance
    - If proximal portion is displaced superficial to adductor aponeurosis, it is referred to as Stener lesion
      - 75% of complete UCL tears are Stener lesions
    - This interposition of adductor aponeurosis prevents UCL from reuniting with proximal phalanx
      - Surgical repair is usually required
    - If proximal portion remains deep to adductor aponeurosis, conservative management may suffice
- A1 pulley implicated in pathogenesis of trigger thumb
  - Normal thickness of A1 pulley < 0.5 mm
  - In trigger thumb, A1 pulley is thickened to ~ 1 mm or greater
  - ± restricted movement of FPL tendon under A1 pulley during passive flexion on real-time US imaging

DEEP DISSECTION OF HAND

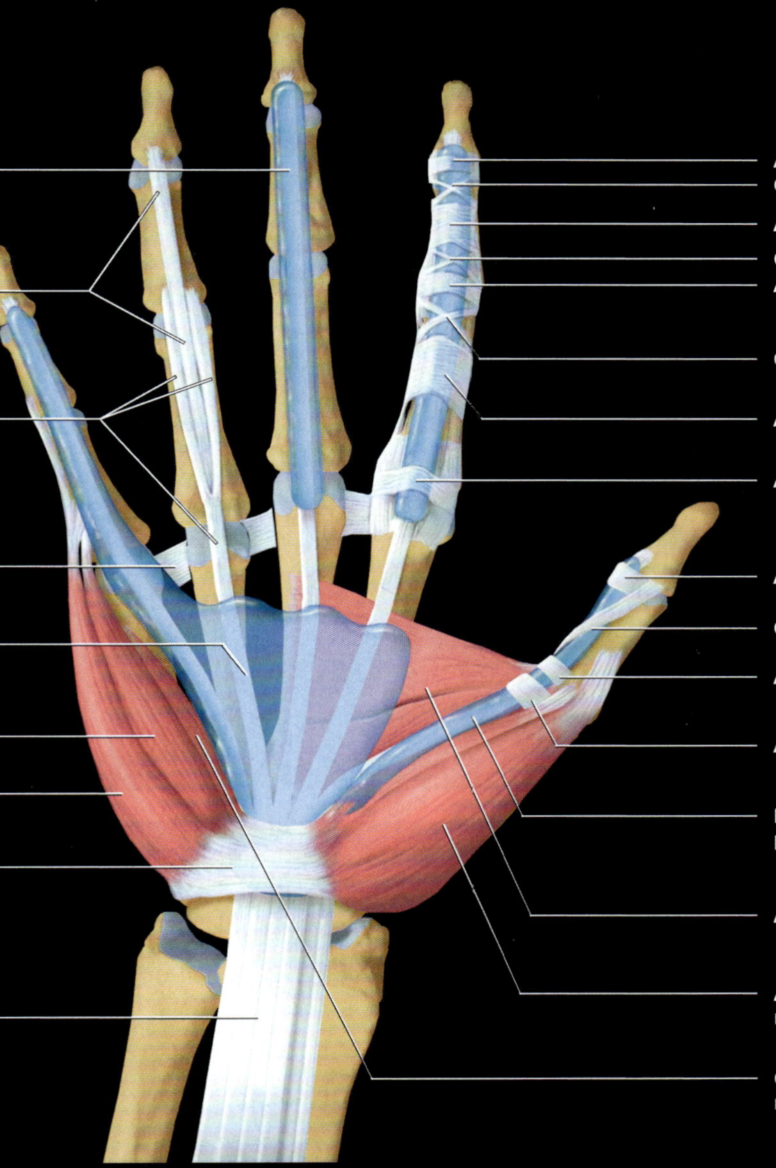

Graphic shows the pulley system of the 2nd digit. The pulley system for the 3rd-5th digits, which are identical to that demonstrated for the 2nd digit, are removed in this graphic. The common digital sheath of the 4th digit is removed to show the relationship of the flexor digitorum superficialis and flexor digitorum profundus tendons. Deep transverse metacarpal ligaments connect volar plates (not shown) of digits 2-5. Although there is overlap of radial and ulnar bursae in this image, these structures normally do not communicate. It is, however, important to know that radial and ulnar bursae may communicate as a normal variation in a small percentage of the population. Similarly, 1 or more of the common digital sheaths may communicate with radial bursa in up to 10% of the normal population. These normal variant bursal communications are important, as they can provide routes for more extensive spread of infection.

US, CARPOMETACARPAL JOINT

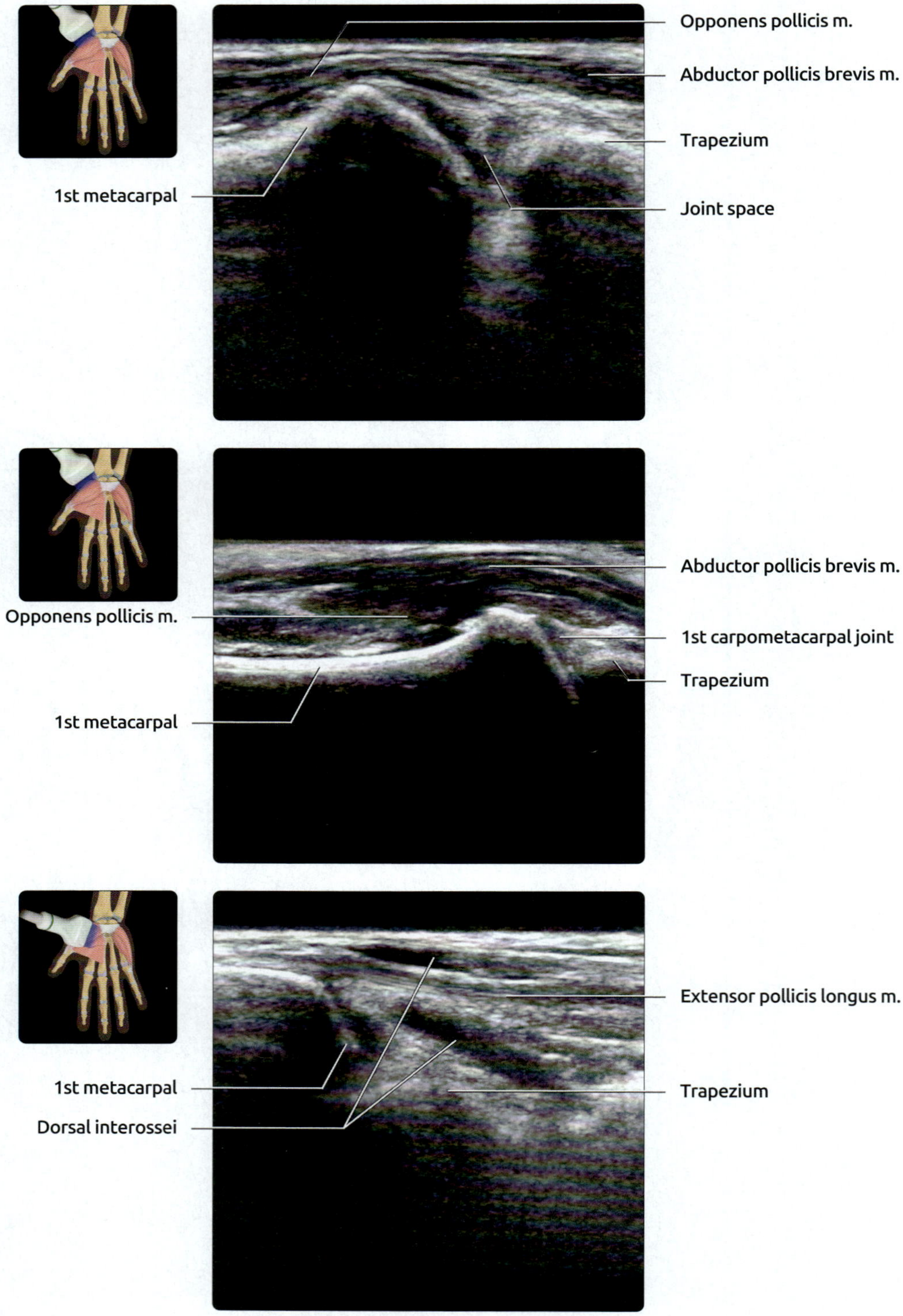

(Top) *Longitudinal grayscale US shows the palmar aspect of the carpometacarpal joint with radial deviation of the thumb. The carpometacarpal joint of the thumb is by far the most mobile of the carpometacarpal joints. It is best viewed by radially deviating the thumb.* **(Middle)** *Longitudinal grayscale US shows the palmar aspect of the carpometacarpal joint of the thumb. Because of its mobility, the carpometacarpal joint of the thumb is the only carpometacarpal joint prone to osteoarthritis. This gives rise to deep pain in the thenar eminence.* **(Bottom)** *Longitudinal grayscale US shows the dorsal aspect of the carpometacarpal joint of the thumb. The large dorsal component of the 1st metacarpal base relative to the articulating surface on the trapezium leads to a normal step-off on the dorsal aspect of this joint.*

US, METACARPOPHALANGEAL JOINT

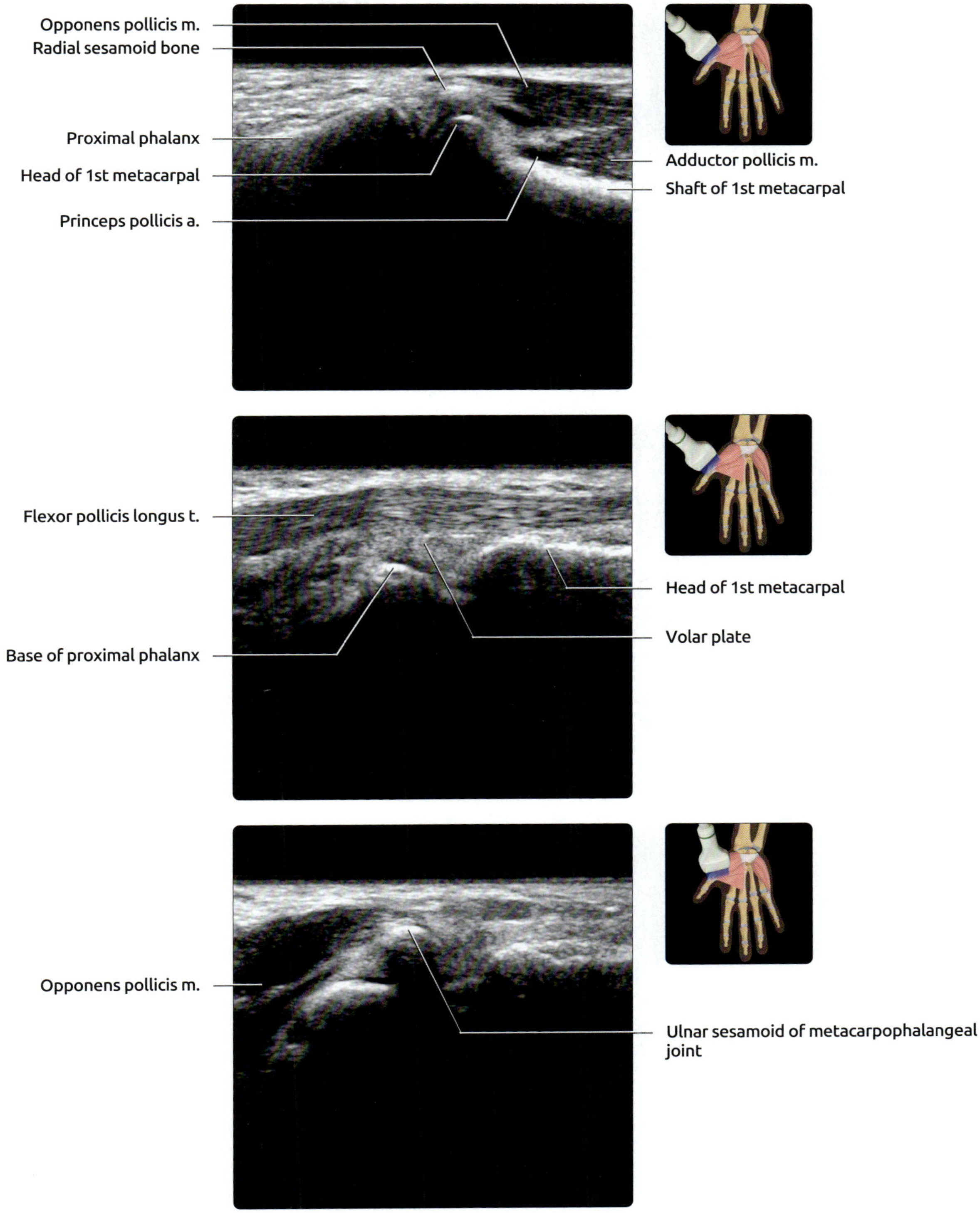

(Top) *Longitudinal grayscale US shows the metacarpophalangeal joint of the thumb. The collateral ligament complex makes up most of the lateral aspects of the metacarpophalangeal and interphalangeal joint capsules. The ulnar collateral ligament of the 1st metacarpophalangeal joint is especially prone to tearing during forced abduction and extension. Such an injury is commonly referred to as gamekeeper thumb or skier thumb.* **(Middle)** *Longitudinal grayscale US shows the metacarpophalangeal joint of the thumb. The volar plate constitutes the majority of the volar aspect of the joint capsule of the metacarpophalangeal joint. The volar plate may be avulsed with forced extension.* **(Bottom)** *Transverse grayscale US shows the 1st metacarpophalangeal joint. The metacarpal head is usually involved in erosions before the proximal phalanx. The metacarpophalangeal joint is an area prone to erosions in inflammatory arthritis.*

ULNAR COLLATERAL LIGAMENT, THUMB

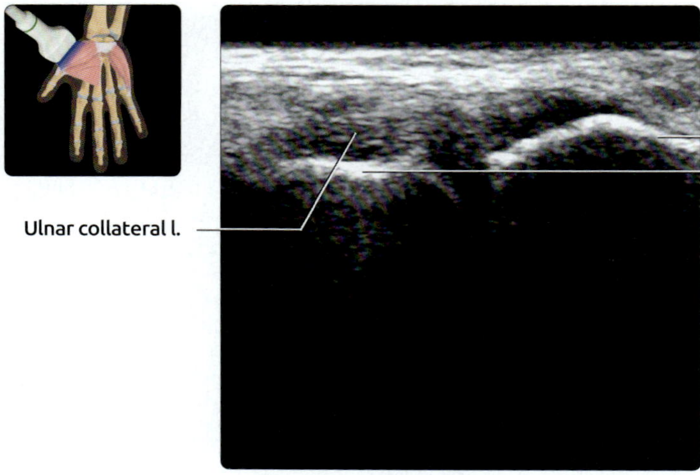

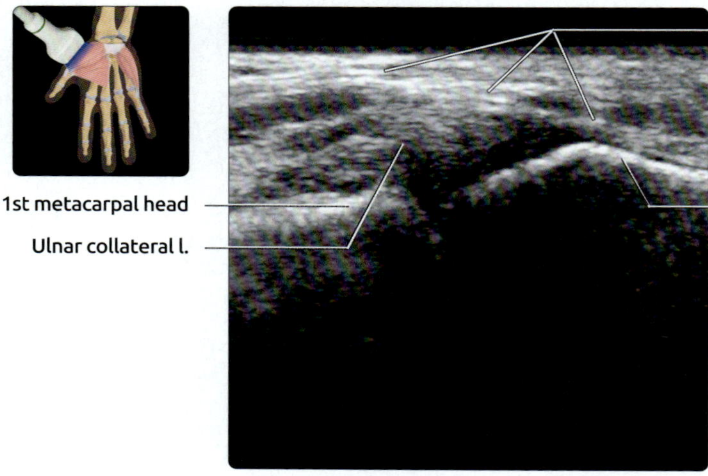

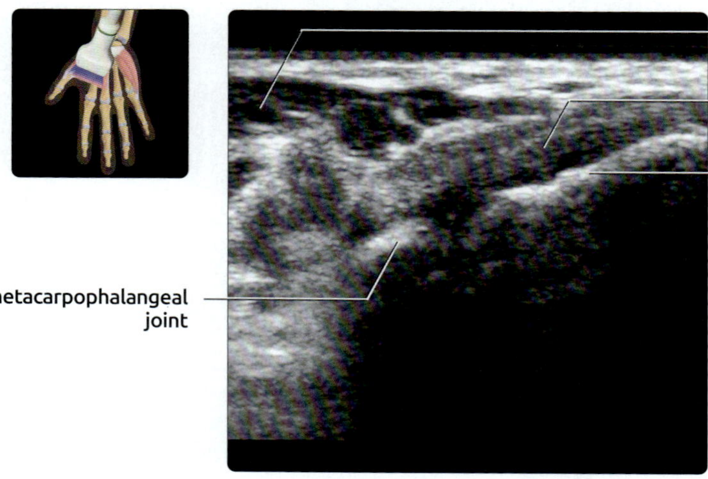

(Top) *Longitudinal grayscale US along the palmar aspect of the ulnar collateral ligament of the thumb is shown. The ulnar collateral ligament of the metacarpophalangeal joint is commonly injured by forced abduction of the thumb. Avulsion normally occurs at the distal (i.e., proximal phalangeal) attachment.* **(Middle)** *Longitudinal grayscale US shows the ulnar collateral ligament at the midportion. The adductor pollicis aponeurosis overlies the ulnar collateral ligament. The thin adductor aponeurosis is attached to the medial side of the proximal phalanx base and the ulnar sesamoid bone. If the torn ulnar collateral ligament displaces above the adductor aponeurosis (Stener lesion), this is an indication for operative repair.* **(Bottom)** *Transverse grayscale US shows the metacarpophalangeal joint of the thumb. This shows the normal thick ulnar collateral ligament in transverse section.*

TRANSVERSE US, PALMAR ASPECT, THUMB

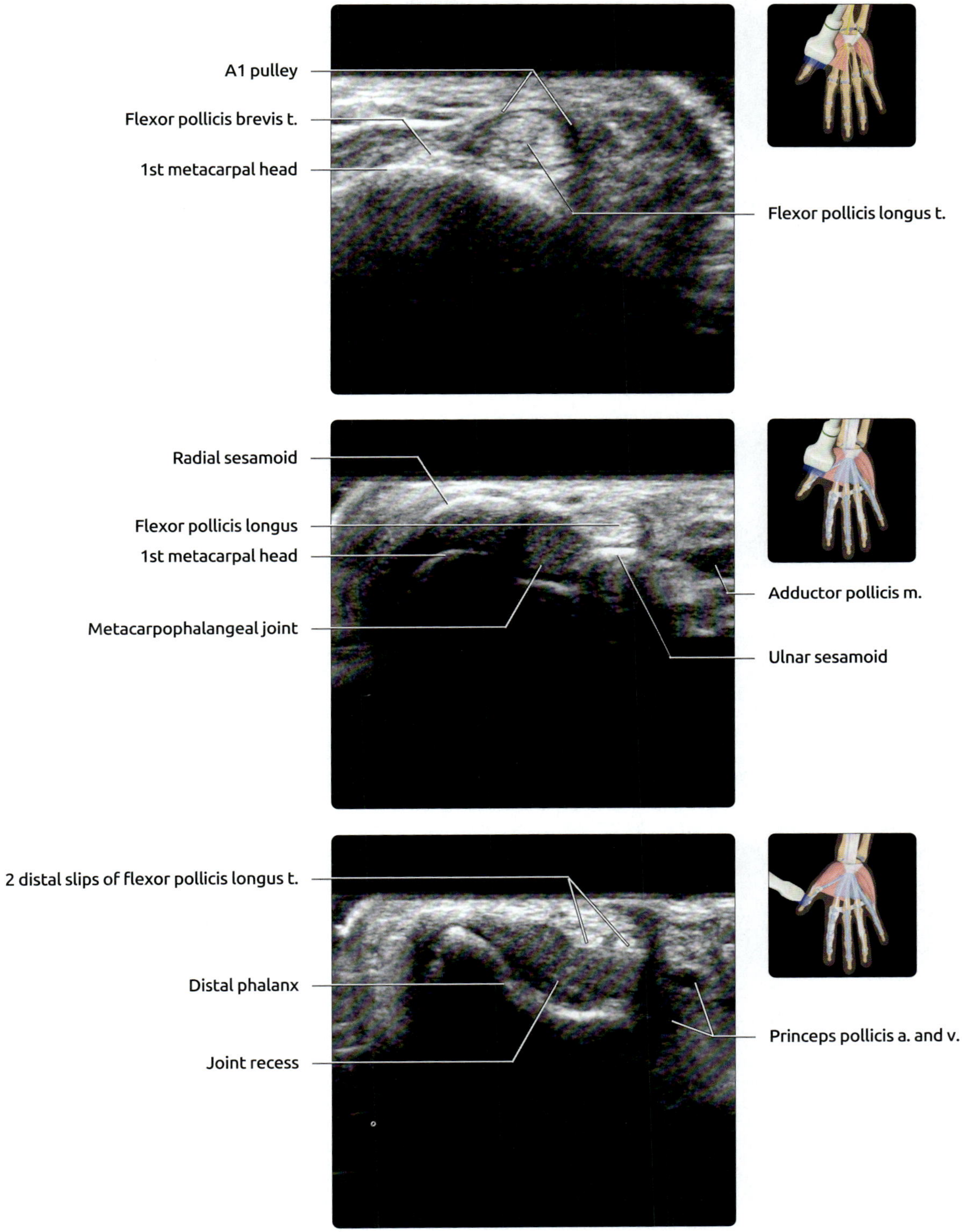

(Top) *Transverse grayscale US along the flexor aspect of the thumb at the metacarpal head level is shown. The flexor pollicis brevis inserts into the radial sesamoid and the base of the proximal phalanx. The flexor pollicis longus inserts into the base of the distal phalanx.* **(Middle)** *Transverse grayscale US along the flexor aspect of the thumb at the metacarpophalangeal joint level is shown. The flexor pollicis longus tendon passes over the ulnar sesamoid bone of the metacarpophalangeal joint.* **(Bottom)** *Transverse grayscale US along the flexor aspect of the thumb at the interphalangeal joint level is shown. The flexor pollicis longus tendon is held in position primarily by the A1 pulley at the metacarpophalangeal joint and the A2 pulley at the interphalangeal joint. The flexor pollicis longus tendon may divide into 2 slips at its insertion.*

EXTENSOR ASPECT, THUMB

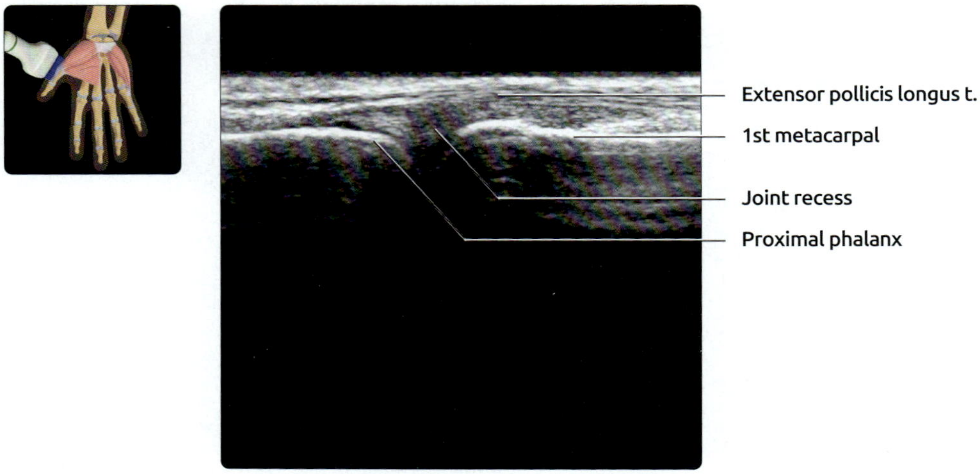

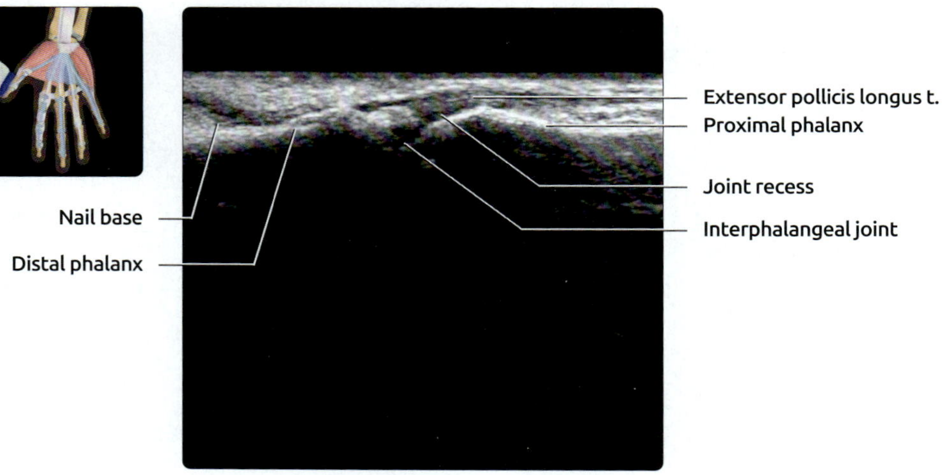

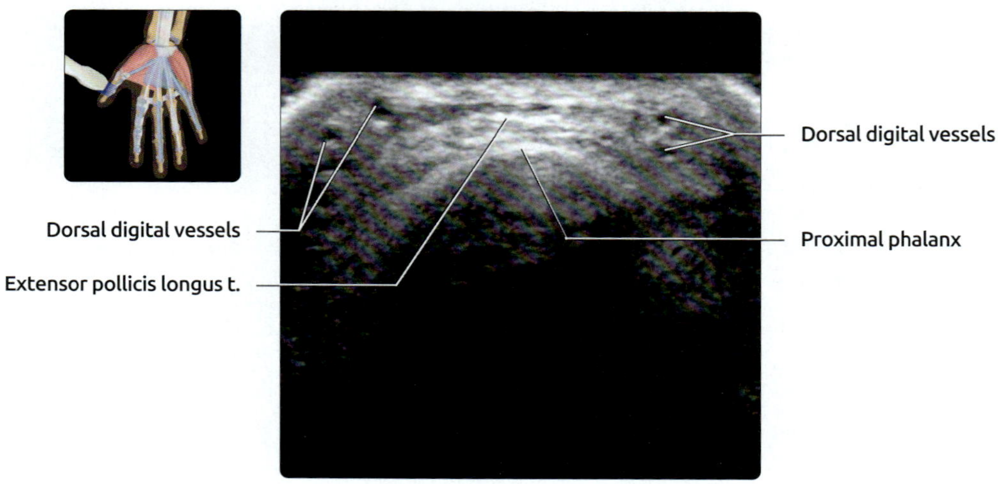

(Top) *Longitudinal grayscale US along the extensor aspect of the thumb at the metacarpophalangeal joint level is shown. The extensor pollicis longus tendon thins as it merges with the extensor expansion on the dorsum of the thumb.* **(Middle)** *Longitudinal grayscale US along the extensor aspect of the thumb at the interphalangeal joint level is shown. The extensor pollicis longus tendon inserts into the distal phalanx.* **(Bottom)** *Transverse grayscale US along the extensor aspect of the thumb at the proximal phalanx level is shown. The thin extensor expansion of the extensor pollicis longus is visible.*

TRANSVERSE US, THENAR EMINENCE

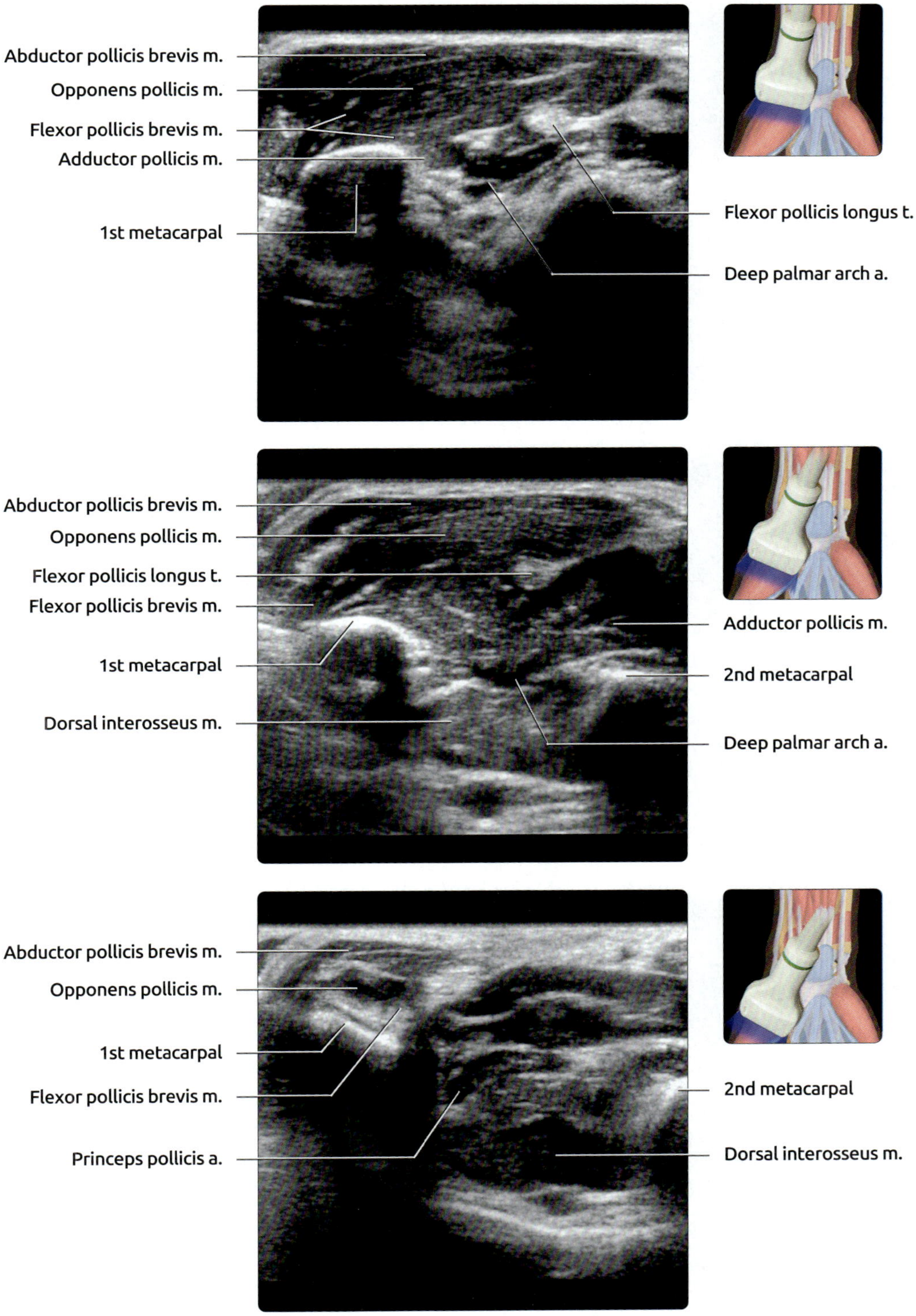

(Top) *Transverse grayscale US along the proximal aspect of the thenar eminence is shown. The thenar eminence is composed of 4 muscles: The abductor, opponens, flexor, and adductor pollicis from superficial to deep. The adductor muscle is, strictly speaking, not part of the thenar eminence since it is separated by a fascial layer and has a separate innervation.* **(Middle)** *Transverse grayscale US shows the midportion of the thenar eminence. The flexor pollicis longus tendon is easily recognized as an echogenic tendon alongside the flexor pollicis brevis, deep to the opponens pollicis and superficial to the adductor pollicis muscle.* **(Bottom)** *Transverse grayscale US shows the distal portion of the thenar eminence.*

LONGITUDINAL US, THENAR EMINENCE

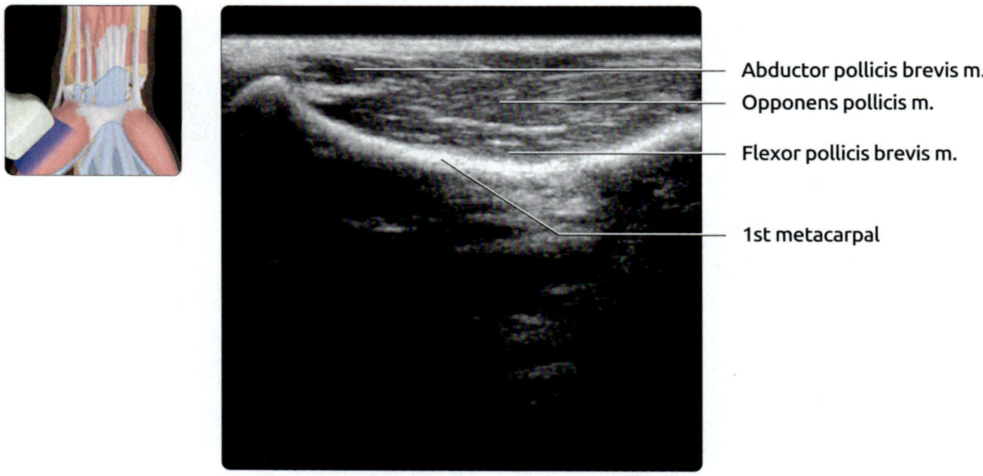

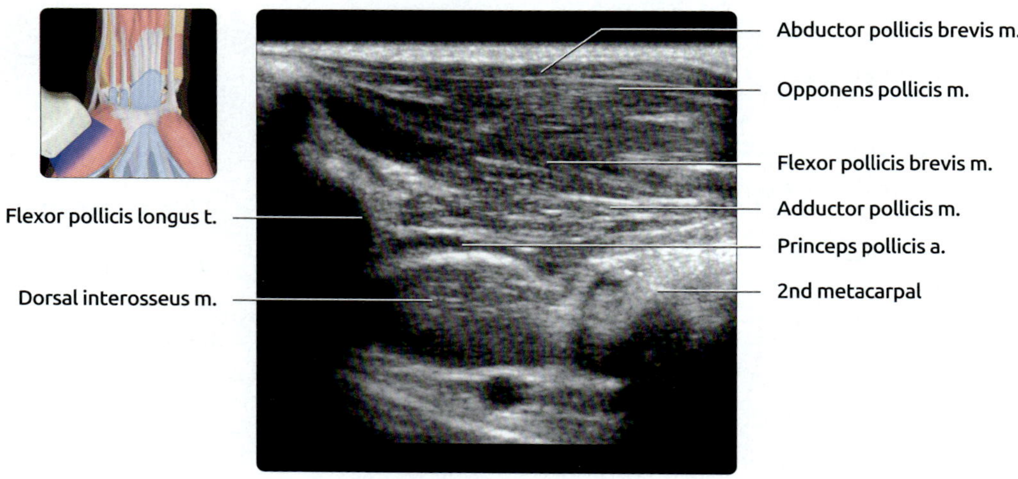

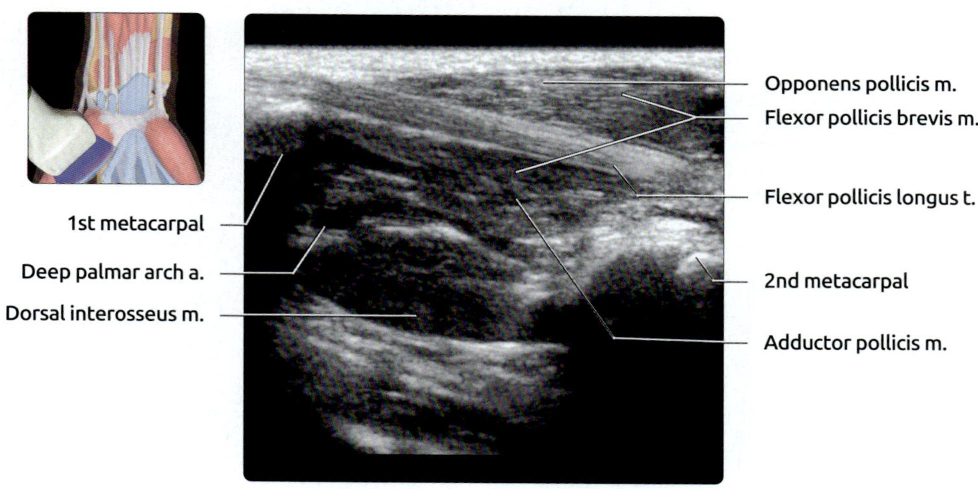

(Top) *Longitudinal grayscale US along the proximal aspect of thenar eminence is shown. ABOF, a mnemonic to remember the muscles of the thenar eminence, stands for abductor pollicis brevis, opponens pollicis, and flexor pollicis brevis, which lie superficial to deep in that order.* **(Middle)** *Longitudinal grayscale US shows the midthenar eminence. The abductor pollicis brevis and flexor pollicis brevis share a combined insertion at the lateral base of the proximal phalanx of the 1st digit.* **(Bottom)** *Longitudinal grayscale US along the ulnar aspect of the thenar eminence is shown. Although often grouped with the thenar muscles due to proximity, the adductor pollicis is distinct from them, as it is separated from the thenar muscles by a fascial plane and innervated by the ulnar nerve as opposed to the median nerve.*

GRAPHICS, STENER LESION

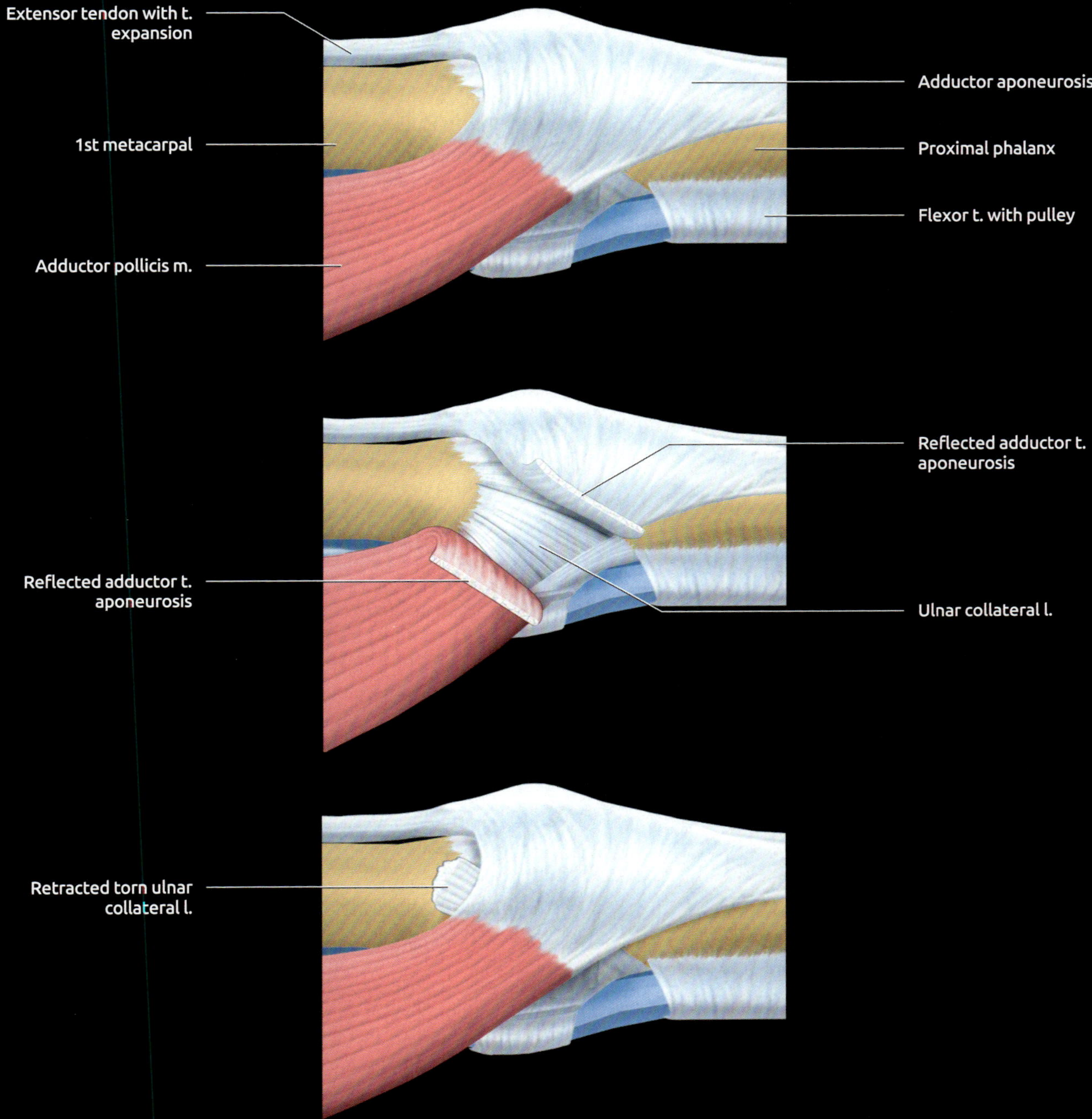

Graphics show the ulnar aspect of the metacarpophalangeal joint of the thumb. On the top image, the adductor tendon aponeurosis attaches to the extensor expansion on the dorsum of the thumb. The middle image shows a reflected adductor aponeurosis with the ulnar collateral ligament beneath. When a complete tear of the ulnar collateral ligament occurs, shown in the bottom image, the proximal end of the ligament may be retracted back over the adductor aponeurosis. This is referred to as a Stener lesion and does require surgical intervention.

TERMINOLOGY

Abbreviations

- Extensor digitorum communis (EDC)
- Extensor digiti minimi (EDM)
- Interphalangeal (IP) joint
- Interphalangeal (MCP) joint

GROSS ANATOMY

Muscles

- Flexor aspect
  - Flexor digitorum superficialis
    - Origin: Common flexor tendon origin and midradius
    - Insertion: Volar plates of proximal IP joints and bases of middle phalanges digits 2-5
    - Superficialis tendon is superficial to profundus tendon in palm until it divides at level of proximal 1/3 of proximal phalanx
    - 2 slips of superficialis tendon pass around profundus tendon and reunite deep to profundus tendon prior to insertion
    - Flexes both MCP joints, aided by lumbricals and interossei and proximal IP joints of digits 2-5
  - Flexor digitorum profundus (FDP)
    - Origin: Proximal radius, midradius, and interosseus membrane
    - Profundus tendon passes through divided superficialis tendon to insert at base of distal phalanx
    - Insertion: Volar plates of distal IP joints and bases of distal phalanges digits 2-5
    - Flexes distal and proximal IP joints as well as MCP joints aided by lumbricals and interossei
- Palmar interossei
  - Denoted 1-3 from radial to ulnar
  - Origin: Mesial palmar diaphyses of 2nd, 4th, and 5th metacarpals
  - Insertion: Mesial lateral bands and proximal phalanges have same base as digit of origin
  - Adducts digits and assists lumbricals with flexion MCP and extension IP joints in digits 2, 4, and 5
- Extensor aspect
  - EDC
    - Origin: Common extensor tendon origin (lateral humeral epicondyle)
    - Insertion: As central slip at bases of middle phalanges and proximal IP joint capsule
    - Separate EDC tendon to 5th digit tendon is present in 50% of general population
  - Extensor indicis
    - Origin: Posterior ulna, distal ulna, and interosseus membrane
    - Insertion: Blends with 2nd digit EDC tendon and extensor hood
  - EDM
    - Origin: Common extensor tendon origin (lateral humeral epicondyle)
    - Insertion: 2 tendons of EDM fuse with one another and with 5th digit EDC tendon prior to insertion at base of proximal phalanx 5th digit
- Lumbricals
  - Numbered 1-4 from radial to ulnar
  - Origin: FDP tendons, just distal to carpal tunnel
  - Insertion: Radial lateral bands of digits 2-5
    - Extend IP and flex MCP joints of digits 2-5
- Dorsal interossei
  - Numbered 1-4 from radial to ulnar
  - Origin: Dorsolateral metacarpal diaphyses
  - Insertion: Adjacent lateral bands of digits 2-4
    - Extend 2nd-5th IP joints and abduct digits 2-5

Tendon Pulley System

- Annular pulley
  - Retinaculum that retain flexor tendons at 5 points = annular pulleys
  - Prevent tendon bowstringing during finger flexion
  - **Odd**-numbered pulleys at **oval** parts (i.e., joints)
  - **Even**-numbered pulleys at **elegant** waists (i.e., shafts)
  - A2 and A4 are functionally most important pulleys to prevent bowstringing
    - A1: MCP joint and base of proximal phalanx
    - A2: Middle and distal 3rds of proximal phalanx
    - A3: Proximal IP joint
    - A4: Midshaft of middle phalanx
    - A5: Distal IP joint
  - Also cruciform pulleys, C1-C3
- Extensor expansion
  - Begins just proximal to MCP joints and terminates just proximal to proximal IP joint
  - Dorsal expansion of fibers oriented perpendicular to long axis of extensor tendons
  - Fibers of extensor expansion interdigitate with EDC tendons to prevent lateral translation

Tendon Sheaths

- 2nd-4th flexor synovial sheaths extend from metacarpal neck to bases of distal phalanges
- Different for 5th digit flexor tendons, which are covered by synovial sheath over their entire course (to level of distal IP joint)

ANATOMY IMAGING ISSUES

Imaging Approaches

- Extensor tendons in digits often seen best on longitudinal imaging
- Pulleys are seen as focal hypoechoic thickening overlying tendon sheath
  - As pulleys are widened at base, thickness of pulleys should be measured at 10 or 2 o'clock positions
  - Do not misdiagnose as tendon sheath thickening
  - A1 and A2 pulley can be depicted in most cases
  - Injury inferred by bowstringing of flexor tendons
  - Compare with normal fingers

CLINICAL IMPLICATIONS

Clinical Importance

- Digital sheaths may connect to ulnar bursa in up to 10% of population, providing route for spread of infection from digits 2-4 to common flexor sheath and vice versa
- Pulley injury inferred by bowstringing of flexor tendons

GRAPHIC, TENDON SHEATHS

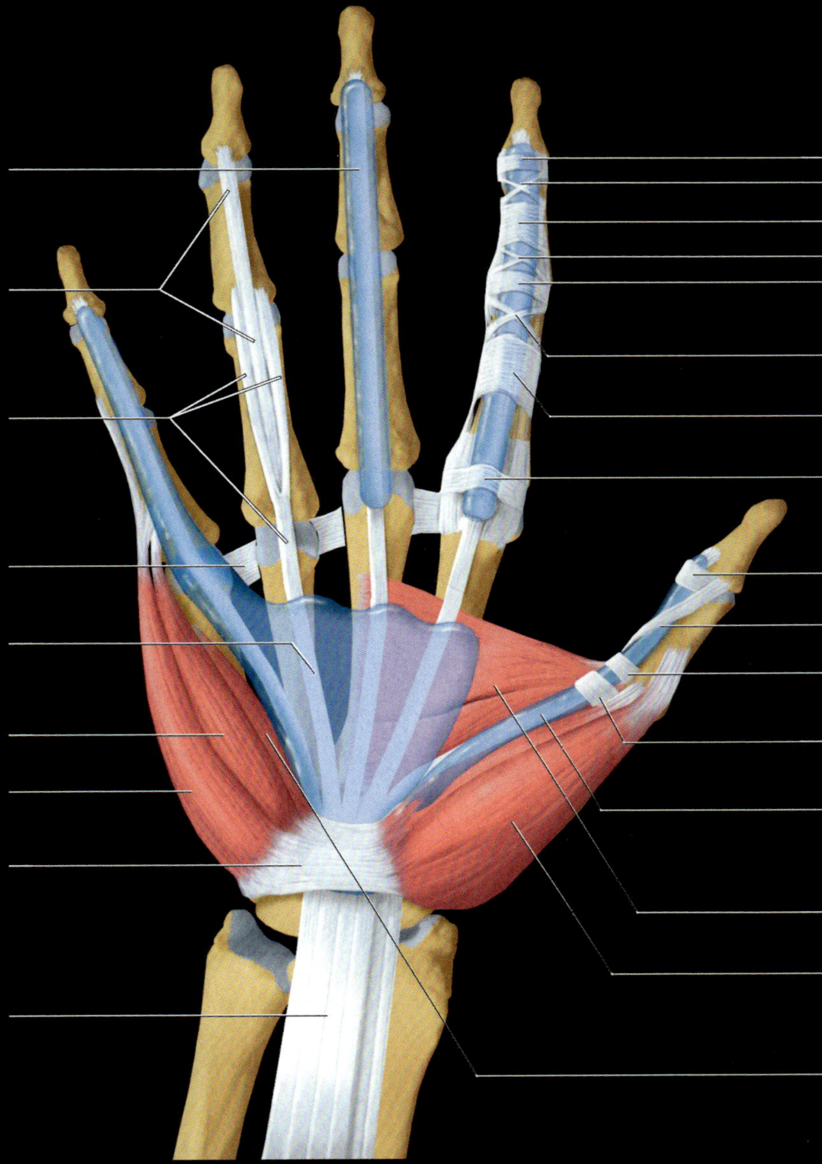

Graphic of pulley system on the palmar side of the 3rd-5th digits is identical to the pulley system demonstrated for the 2nd digit. The common digital sheath of the 4th digit is removed to show the relationship of the flexor digitorum superficialis and flexor digitorum profundus tendons. Deep transverse metacarpal ligaments connect the volar plates (not shown) of digits 2-5. Although there is overlap of the radial and ulnar bursae in this image, these structures normally do not communicate. It is, however, important to know that the radial and ulnar bursae may communicate as a normal variation in a small percentage of the population. Similarly, any one or more of the digital sheaths may communicate with the radial bursa in up to 10% of the normal population. These normal variant bursal communications are important as they can provide routes for more extensive spread of infection.

GRAPHIC, LIGAMENTS AND PULLEYS

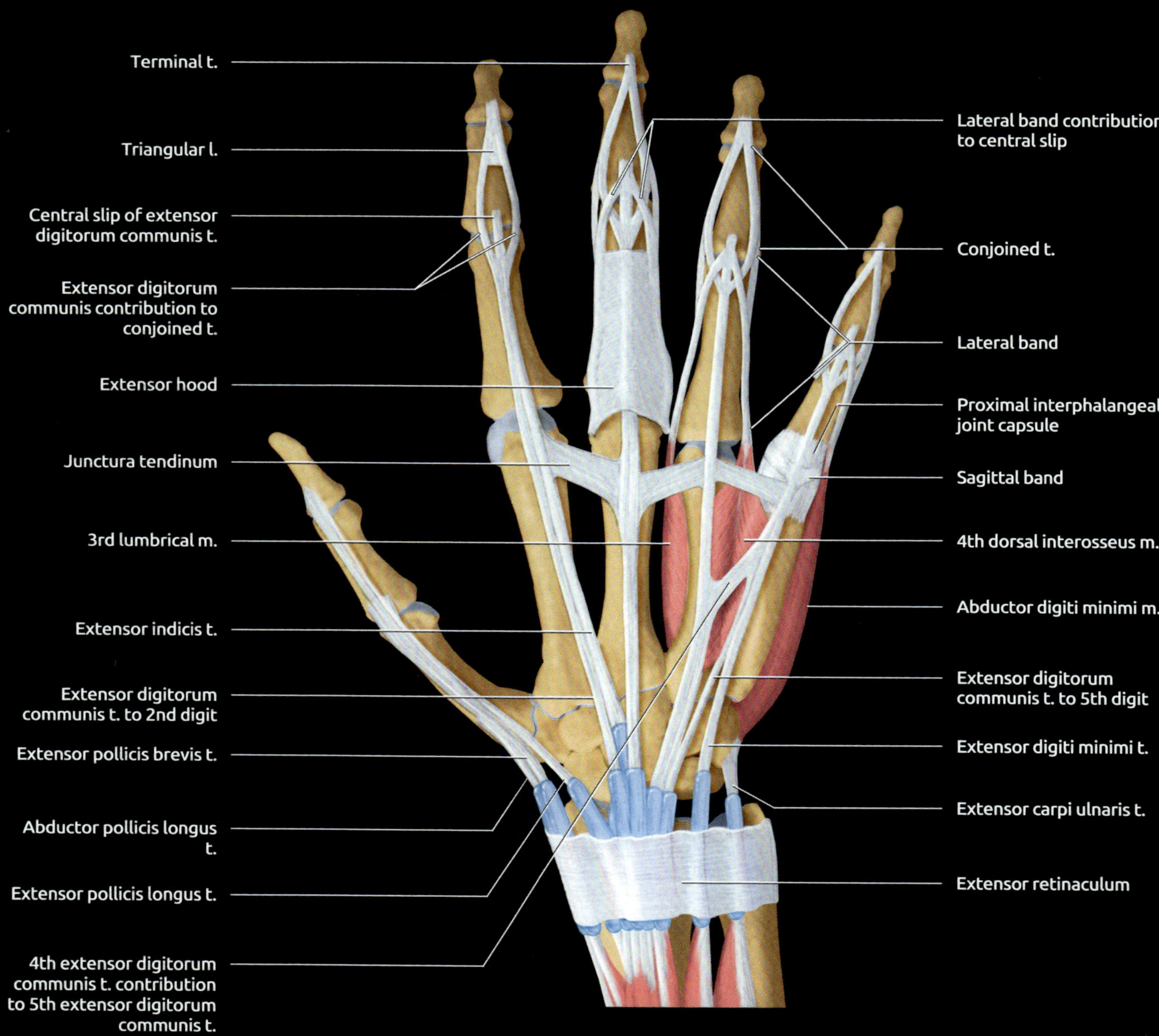

Graphic of the dorsum hand and finger extensor mechanism shows different components of the extensor mechanism on different digits. With the exception of the extensor indicis, extensor digiti minimi, abductor digiti minimi, and 4th extensor digitorum communis tendon contribution to the 5th extensor digitorum communis tendon, any structure on any of the 2nd-5th digits can be extrapolated to any and all of the other 2nd-5th digits. Note that the 2nd extensor retinaculum compartment and its contents (extensor carpi radialis longus and brevis tendons) are not included in this graphic. Although the abductor pollicis longus tendon travels in the 1st compartment of the extensor retinaculum, its insertion (not shown) is actually on the radial aspect of the volar surface of the base of the 1st metacarpal.

GRAPHIC AND LONGITUDINAL US, METACARPOPHALANGEAL JOINT

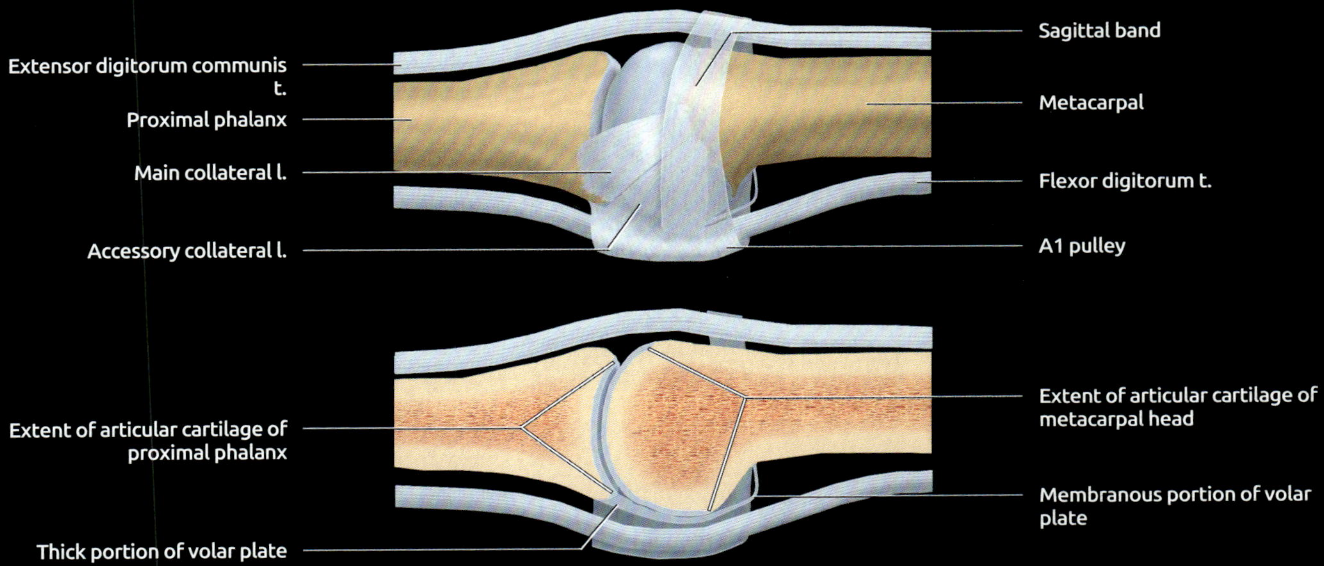

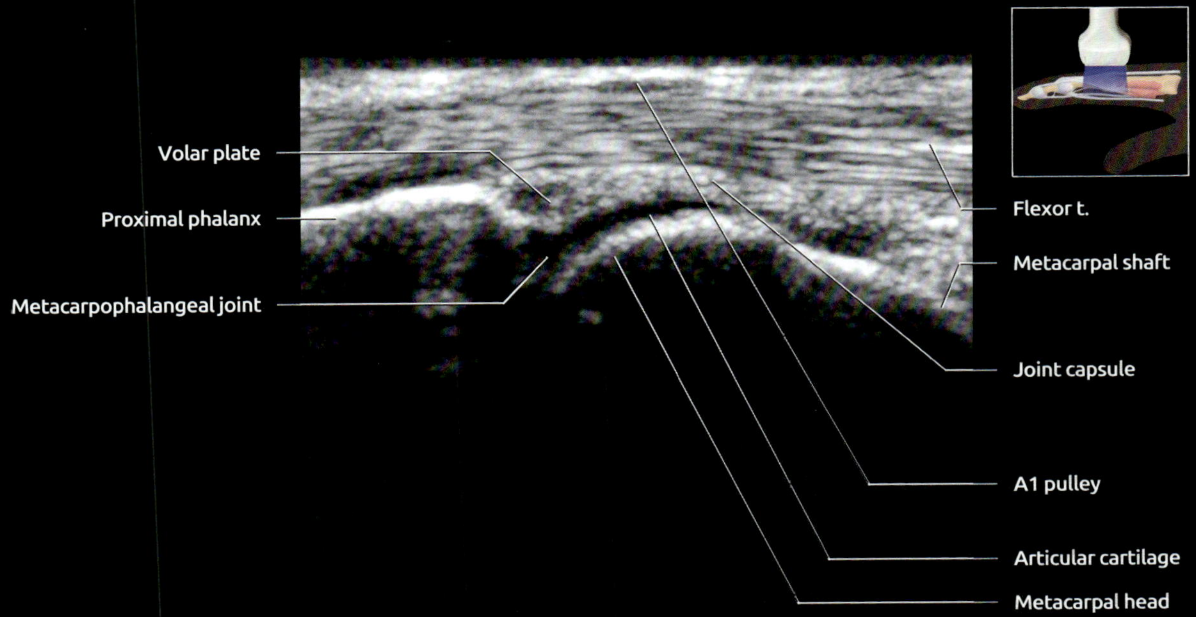

(Top) *Surface and cut-section graphics show the metacarpophalangeal (MCP) joint. The collateral ligament complex is composed of 2 separate fiber bands. The main collateral ligament inserts on the base of the adjacent phalanx. The accessory collateral ligament inserts on the volar plate. The volar plate is thick distally and thin and redundant proximally. With the exception of the sagittal band (present only at MCP joints), anatomy in this graphic is duplicated in all MCP and interphalangeal (IP) joints in the hand.* **(Bottom)** *Longitudinal grayscale US shows the middle finger at the MCP joint.*

LONGITUDINAL US

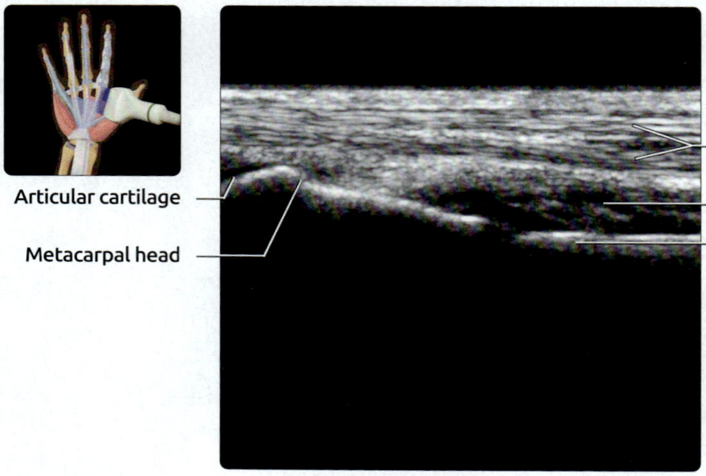

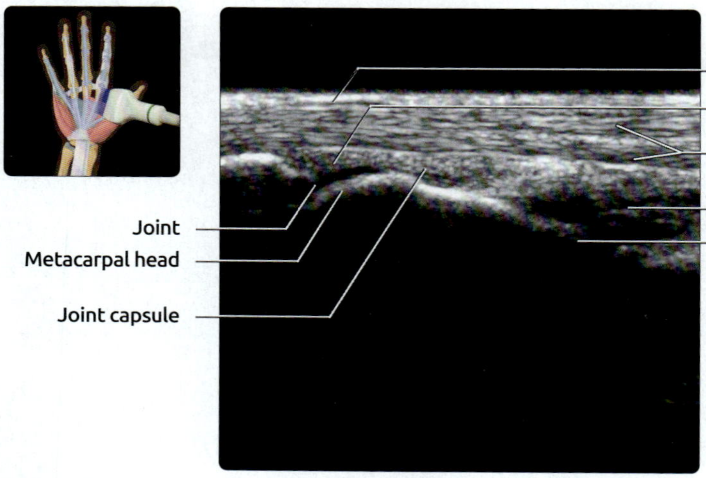

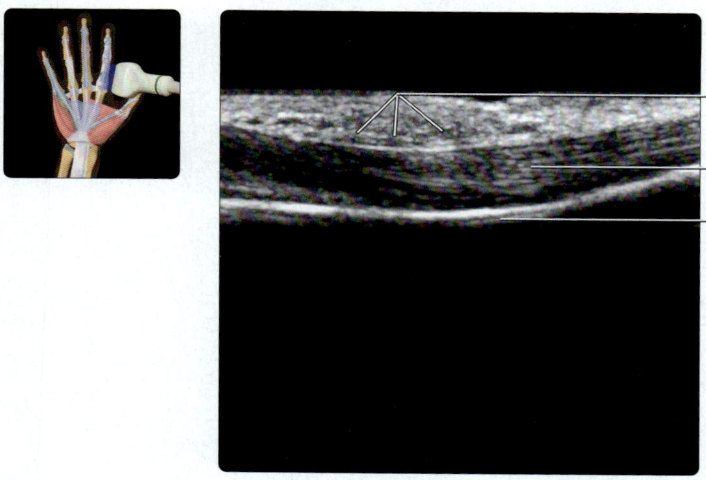

(Top) *Longitudinal grayscale US shows the finger at the metacarpal shaft level. From the base of the metacarpals (carpal tunnel outlet) to the neck of the metacarpals, the flexor tendons of the 2nd-4th fingers (i.e., index, middle, and ring fingers) are devoid of flexor tendon sheaths. A normal flexor tendon sheath cannot be seen on US and can only be seen when it is distended. Do not confuse the flexor tendon pulleys with tendon sheath thickening.* **(Middle)** *Longitudinal grayscale US shows fingers at the MCP joint level. MCP joints are narrowed dorsally and flare in the volar direction, increasing contact with the proximal phalanx as the joint is flexed. The capsule of the MCP joint extends from the metacarpal neck level to the base of proximal phalangeal joint.* **(Bottom)** *Longitudinal grayscale US shows the A2 pulley of the middle finger at the proximal phalanx.*

LONGITUDINAL US

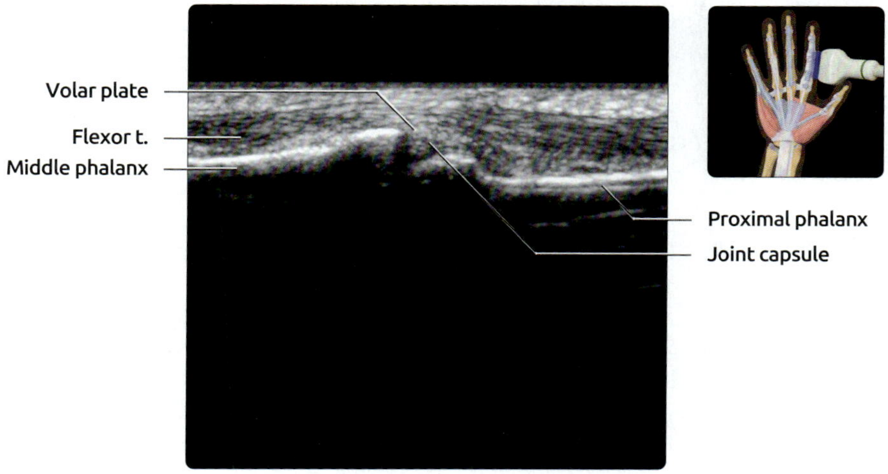

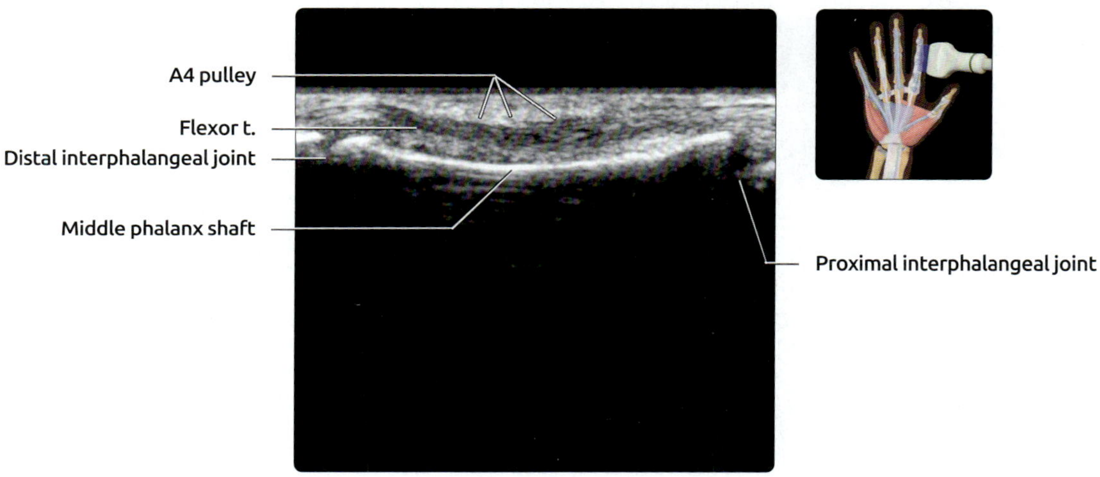

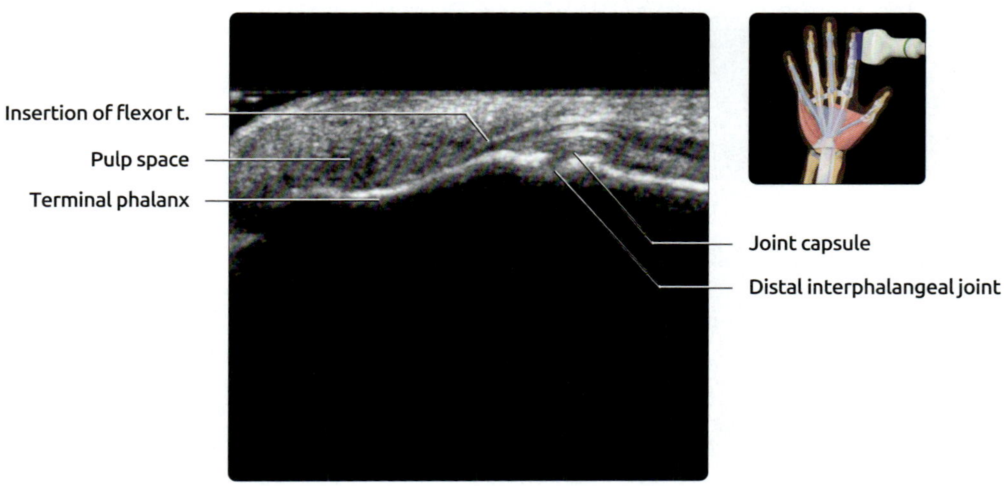

(Top) *Longitudinal grayscale US shows the proximal IP joint of the middle finger. The volar plate is a fibrocartilaginous thickening on the volar aspect of the MCP joint and proximal IP joints. At the proximal IP joint, it is firmly attached to the base of the middle phalanx and has a thinner insertion onto the distal end of the proximal phalanx. It may become avulsed during hyperextension injuries often associated with a small avulsion fracture. There is an inability to fully extend and flex the joint, a condition that usually improves with time unless the joint is locked.* **(Middle)** *Longitudinal grayscale US shows the middle finger at the middle phalanx. The A4 pulley prevents bowstringing of flexor tendons at the middle phalanx.* **(Bottom)** *Longitudinal grayscale US shows the distal IP joint of the middle finger. An avulsion of the flexor tendon from the distal phalangeal base is usually accompanied by a small avulsion fracture.*

TRANSVERSE US

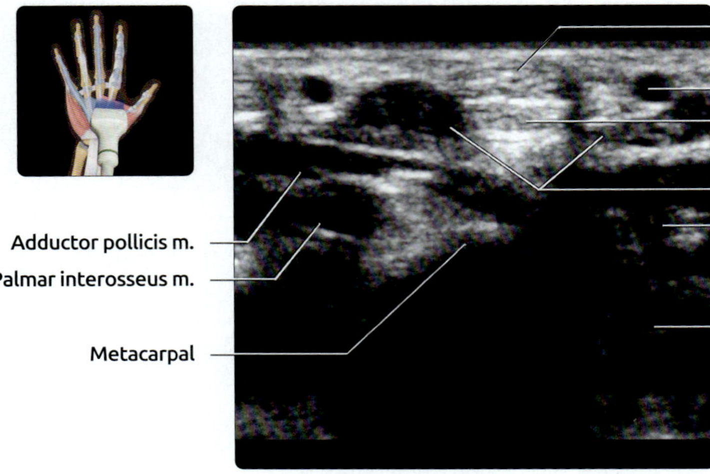

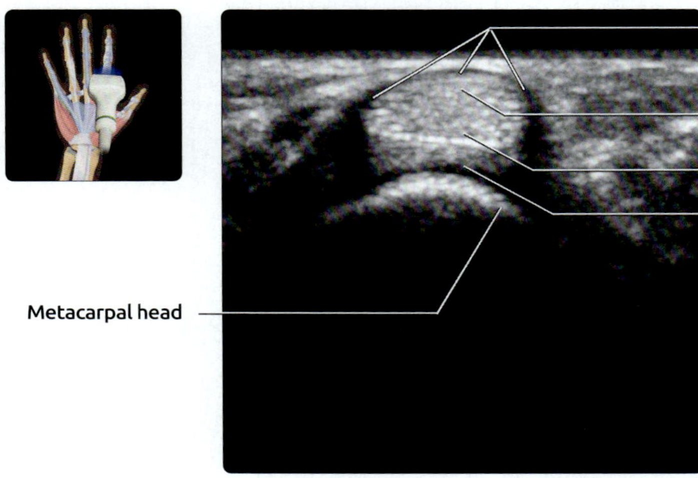

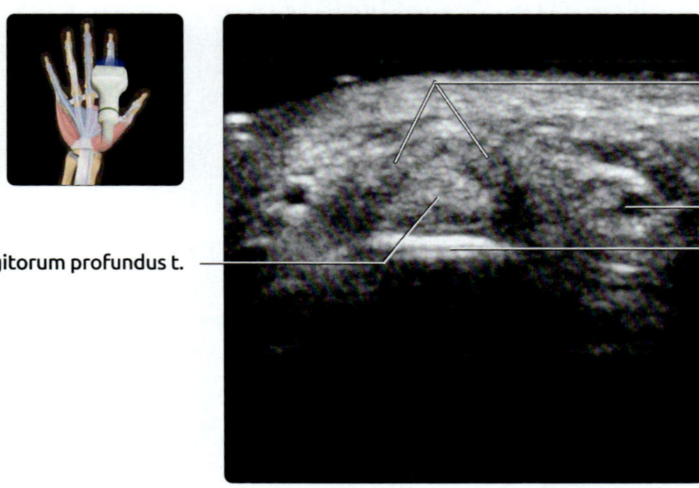

(Top) *Transverse grayscale US shows the middle finger at the metacarpal shaft. There are 4 lumbrical muscles in the hand. These are unusual muscles in that they do not originate from the skeleton but instead originate from the flexor digitorum tendons and insert into the extensor expansions.* **(Middle)** *Transverse grayscale US shows the middle finger at the metacarpal head. This image is of the A1 pulley level. The A1 pulley extends from the metacarpal head to the proximal phalanx base. It is seen as a thin, hypoechoic to hyperechoic rim (depending on the beam angulation) around the flexor tendons. It broadens considerably at the base, which cannot be clearly depicted.* **(Bottom)** *Transverse grayscale US shows the middle finger at the proximal phalanx. The flexor digitorum superficialis tendon splits just proximal to the insertion, enabling the flexor digitorum profundus tendon to push through the gap.*

TRANSVERSE US

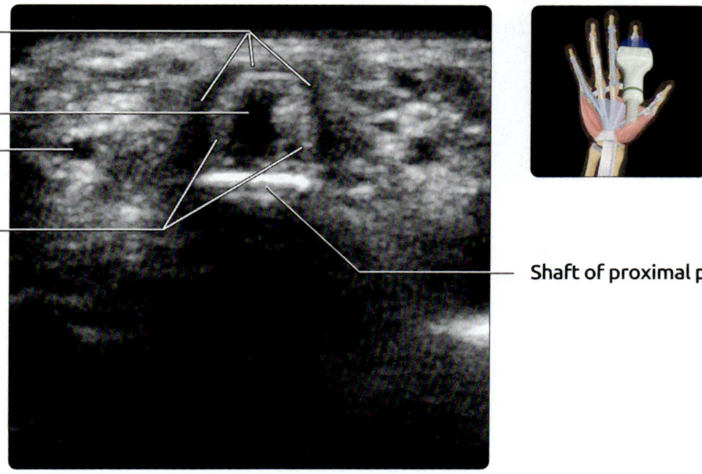

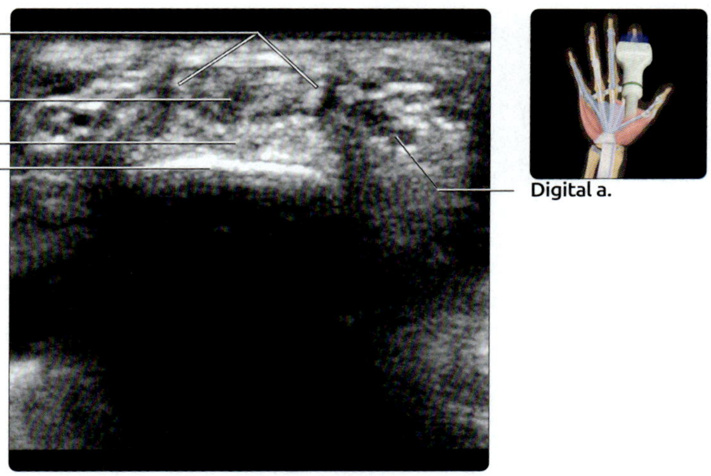

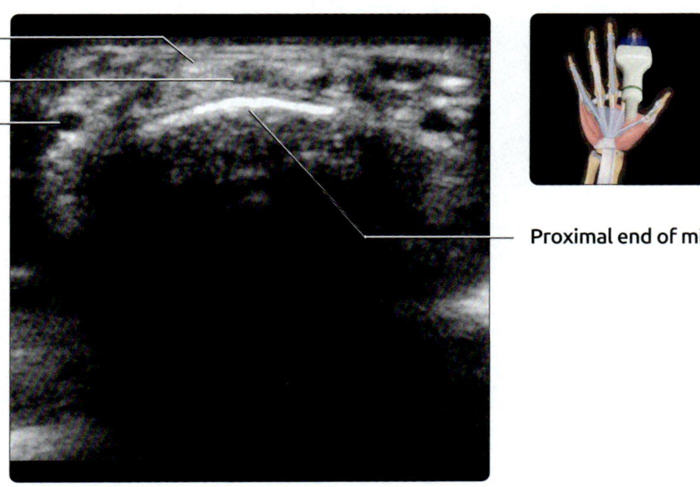

(Top) *Transverse grayscale US shows the middle finger shaft of the proximal phalanx. Do not confuse thickening of the A2 pulley with tendon sheath thickening.* **(Middle)** *Transverse grayscale US shows the middle finger proximal phalangeal head region just distal to the A2 pulley. The small slips of the flexor digitorum superficialis tendon descend on either side of the flexor digitorum profundus tendon to insert onto the volar aspect of the base of the middle phalanx.* **(Bottom)** *Transverse grayscale US shows the middle finger at the proximal end of the middle phalanx. The insertions of the slips of the flexor digitorum superficialis are small. As such, particularly in the presence of edema, it may be difficult to confirm isolated avulsion of 1 slip on US. In this situation, the flexor digitorum superficialis tendon will not be retracted.*

TRANSVERSE US AND GRAPHIC

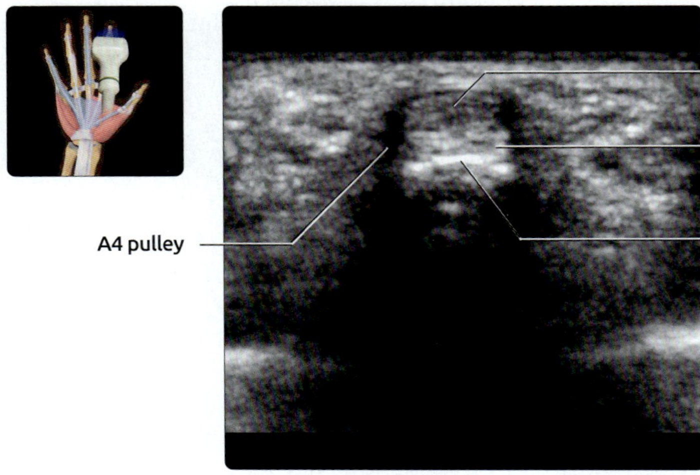

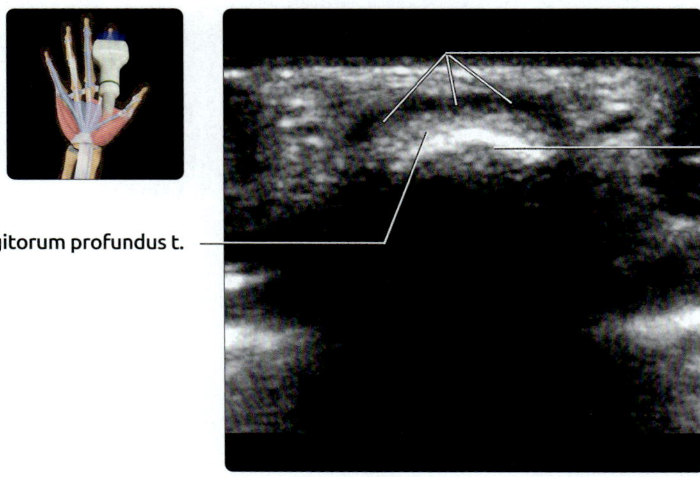

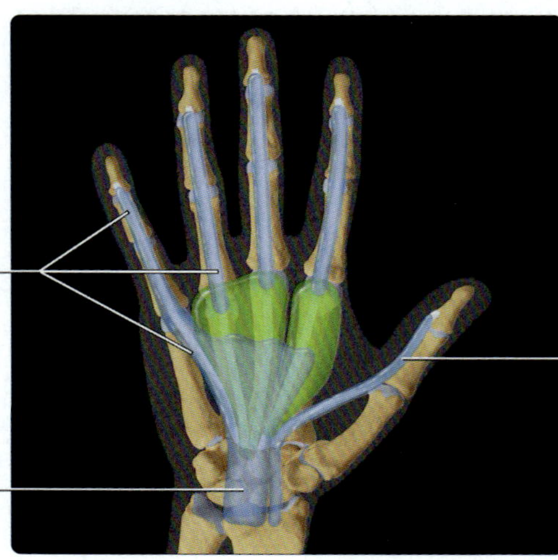

(Top) *Transverse grayscale US shows the middle finger at the shaft of the middle phalanx. Do not confuse the subtendinous connective tissue as part of the tendon.* **(Middle)** *Transverse grayscale US shows the middle finger at the base of the distal phalanx. The flexor digitorum profundus tendon gains broad attachment to the base of the distal phalanx and is supported by the A5 pulley just proximal to the insertion.* **(Bottom)** *Graphic shows the tendon sheaths of the fingers (blue). Digital sheaths may connect to the ulnar bursa (green) in up to 10% of the population, providing a route for the spread of infection from digits 2-4 to the common flexor sheath and vice versa.*

LONGITUDINAL US, EXTENSOR ASPECT

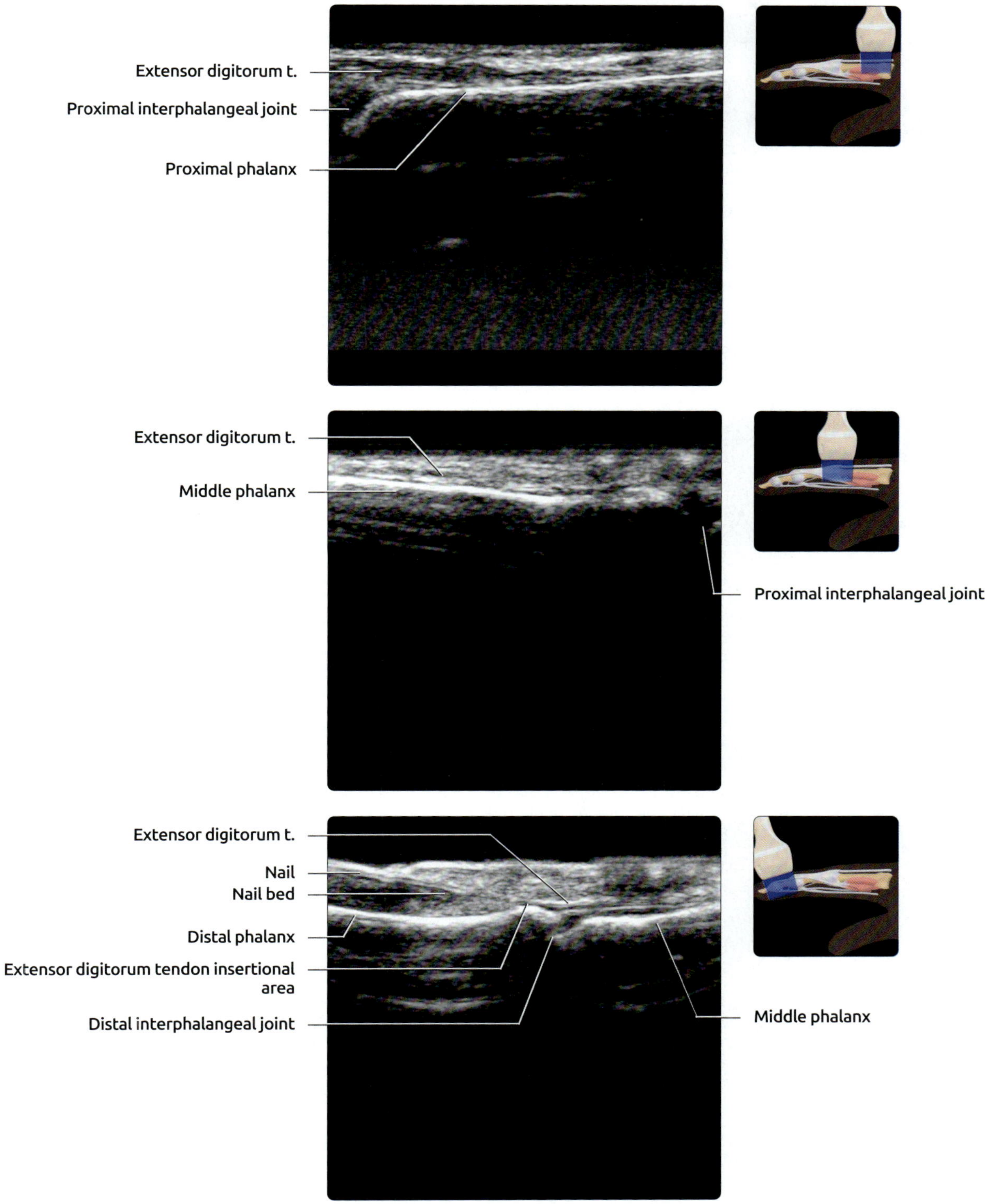

(Top) *Longitudinal grayscale US shows the dorsal aspect of the proximal phalanx. The extensor expansion on the dorsum of the fingers is thin and flat and does not have a synovial sheath.* **(Middle)** *Longitudinal grayscale US shows the dorsal aspect of the middle phalanx. No pulleys are present on the extensor aspect of the fingers. The extensor expansion (or hood) are attached to the phalanges and the lateral bands.* **(Bottom)** *Longitudinal grayscale US shows the dorsal aspect of the distal phalanx. The extensor digitorum is attached to the base of the distal phalanx. This is a common site of avulsion, often with a small avulsion fracture.*

TRANSVERSE US, EXTENSOR ASPECT

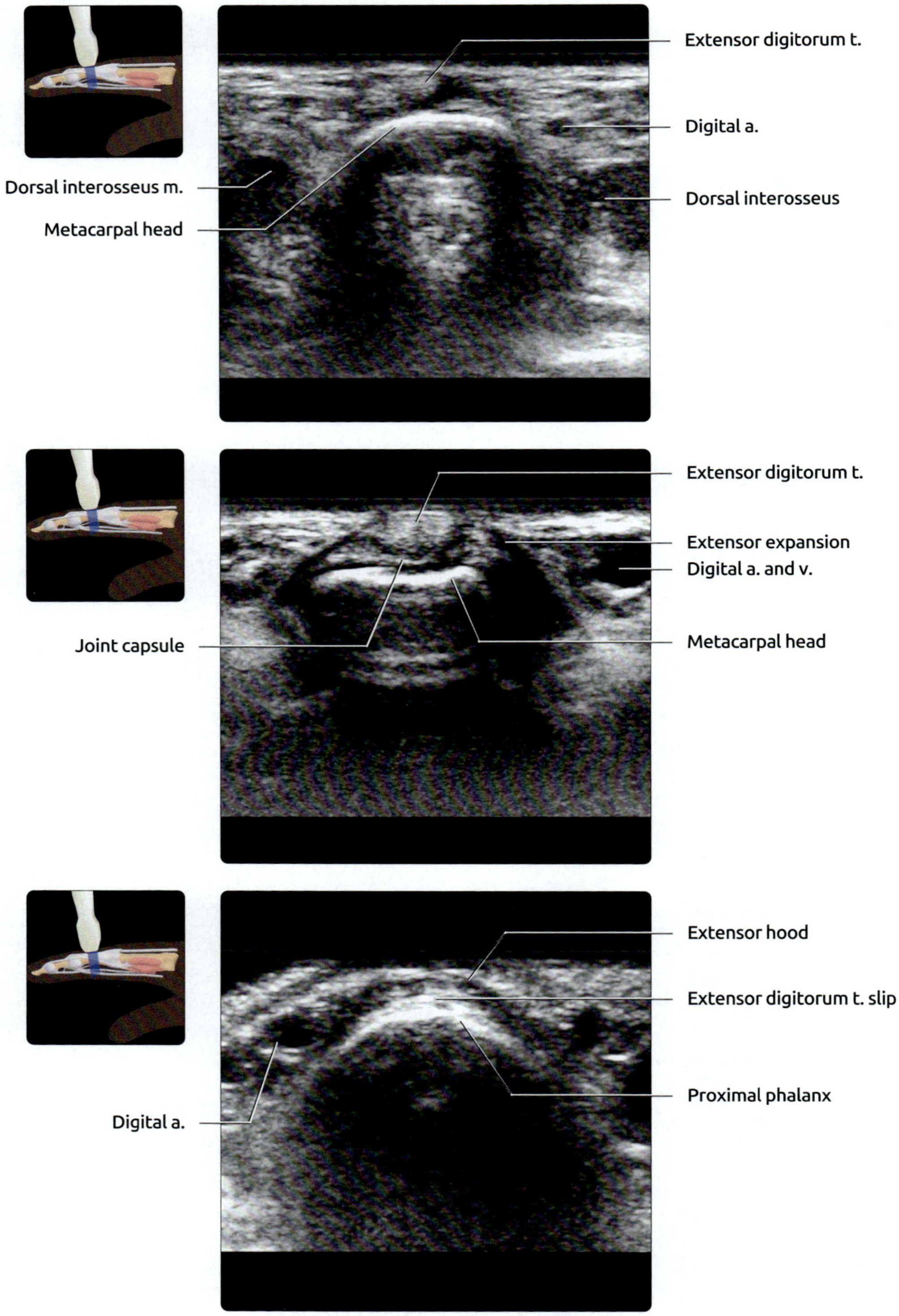

(Top) *Transverse grayscale US shows the metacarpal head/neck region of the extensor aspect of the fingers. The extensor tendons of the hand do not have a complete synovial sheath. They are covered by thin paratenon. The tendons have an ovoid configuration.* **(Middle)** *Transverse grayscale US shows the extensor aspect of the fingers at the metacarpal head region. At the proximal phalanx extending over the middle phalanx, the extensor tendons spread out and connect to the extensor expansion or hood.* **(Bottom)** *Transverse grayscale US shows the extensor aspect of the fingers at the proximal phalanx level. With expansion of the extensor tendon, it becomes very thin such that the normal tendon is difficult to illustrate on the dorsum of the fingers. If injured, the tendon swells and is usually more readily seen.*

GRAPHICS, AXIAL FINGER AND HAND DERMATOMES

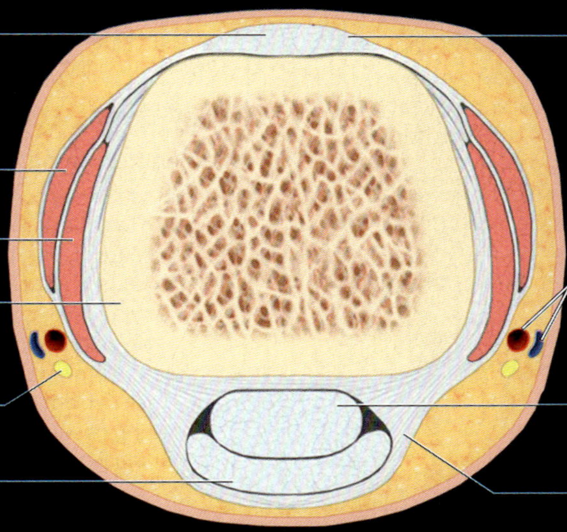

(Top) *Graphic shows the transverse section of a finger depicting the relationships of the digital nerve, vessels, and tendons to the phalanx. On the dorsal aspect, the extensor expansion interdigitates with the extensor digitorum communis tendons to prevent lateral translation. On the palmar side, the collateral ligaments extend from the palmar aponeurosis in the palm to fuse with the fibrous digital sheaths. Other thickenings of the fibrous digital sheaths form annular pulleys and cruciform bands. The common digital artery and the accompanying nerve intervene between the lumbrical muscle and the tendons.* **(Bottom)** *Graphic shows dermatomes of the hand and wrist correlated to corresponding cervical or thoracic nerves. It is necessary for a physician to test the integrity of spinal cord segment C5 to T1 in the upper limb. Usually, physicians will test the C6 to C8 dermatomes in the hand, i.e., the thumb represents C6 dermatome, the middle finger tip represents C7 dermatome, and the little finger represents C8 dermatome.*

GROSS ANATOMY

Arm

- Radial nerve is largest branch of posterior cord of brachial plexus
  - Receives contributions from cervical roots C5-C8
  - Contains motor and sensory components that supply extensor muscles of arm and forearm
  - Lies posterior and deep to axillary artery in axilla
- Lies between coracobrachialis and teres major muscles and then between bellies of medial and lateral head of triceps
- Runs deep to triceps in spiral groove and then deep to brachioradialis
- 10 cm proximal to lateral epicondyle of humerus, radial nerve penetrates lateral intermuscular septum to enter anterior compartment of arm
- Area between lateral intermuscular septum and supinator inlet known as radial tunnel
- Just anterior to lateral epicondyle, nerve bifurcates into sensory (superficial) and motor (deep) branches

Forearm

- Branches of radial nerve
  - Superficial branch of radial nerve (purely sensory)
    - Courses distally, deep to brachioradialis, to extensor compartment of distal forearm and then to dorsal wrist
    - Cephalic vein crosses over superficial branch in distal forearm
      - Nerve may be injured during venous cannulation
    - Divides into lateral and medial branches 5 cm proximal to radial styloid
    - Lateral branch supplies radial wrist and thumb skin; medial branch (supplies mid and ulnar wrist skin)
      - Dorsal digital nerves supply ulnar thumb, index, middle, and radial ring fingers
  - Deep branch of radial nerve (purely motor)
    - Deep branch of radial nerve becomes PIN at entry into supinator muscle (supinator arch or arcade of Fröhse)
    - PIN passes between superficial and deep heads of supinator muscle (supinator canal)
      - Then exits distally and runs posteriorly as PIN
    - Supplies extensor compartment muscles of forearm
    - 5 potential anatomical causes of PIN compression near elbow
      - Fibrous band of radial head, which is continuous with extensor carpi radialis brevis (ECRBr) and superficial head of supinator muscle
      - Radial recurrent vessels (leash of Henry) at level of radial head
      - Tendinous margin of ECRBr
      - Aponeurotic proximal margin of supinator (arcade of Fröhse) = most common site of compression
      - Distal margin of supinator muscle
    - After exiting from supinator canal, PIN travels between superficial and deep extensor muscles until it reaches interosseous membrane distally
      - Travels 1st with posterior interosseous artery and then anterior interosseous artery
    - At wrist, PIN lies deep to radial side of 4th extensor compartments where it can be focally enlarged (pseudoganglion)
      - Thereafter, PIN divides into 2-4 terminal branches

Hand: Branches of Radial Nerve

- Not usually seen on routine imaging
- No motor innervation in hand
- Sensory to back of hand and dorsal aspect of radial 3.5 digits

ANATOMY IMAGING ISSUES

Imaging Recommendations

- Radial nerve is best identified at lateral aspect of distal arm deep to brachioradialis muscle
  - Alternatively, identify nerve deep to triceps in spiral groove midhumeral level
  - Or in proximal aspect of arm medially
    - Lying just deep to axillary and brachial arteries
- Trace medium-sized branches by following their course as they branch from parent nerve
- Examining nerves in transverse plane is preferable to longitudinal plane is it allows easier, continuous tracing of nerves
  - If one loses sight of small nerve, continue moving transducer slowly along expected path of nerve and it will often be identified again after a few cm

Imaging Approaches

- Most common site of entrapment of PIN is at supinator arch (arcade of Fröhse)
  - Nerve will be focally thickened to > 2 mm thick (normal = 1 mm thick)

Imaging Pitfalls

- Smaller nerves and their branches (measuring ~ 1 mm wide) can be difficult to identify and separate from vessels
  - Color Doppler US helps
  - Identify nerves at target areas and follow distally or proximally
- Functional compression is not as readily appreciated as structural compression

CLINICAL IMPLICATIONS

Clinical Importance

- More common sites of radial nerve entrapment or injury
  - Lateral to long head of triceps due to fibrous arch
  - Spiral groove between medial and lateral heads of triceps muscle due to fibrous arch, fractures, tourniquet
  - Between brachialis and brachioradialis muscles at antecubital fossa due to muscle (body builders)
  - Recurrent radial artery (leash of Henry) proximal to supinator muscle
  - Arcade of Fröhse (supinator inlet) compressing PIN
  - Between deep and superficial heads of supinator muscle compressing PIN
  - Superficial branch of radial nerve in mid to distal forearm (Wartenberg syndrome)
  - PIN deep to 4th extensor compartment at wrist by ganglion, during wrist injection, arthroscopy, or radial bone fixation

RADIAL NERVE

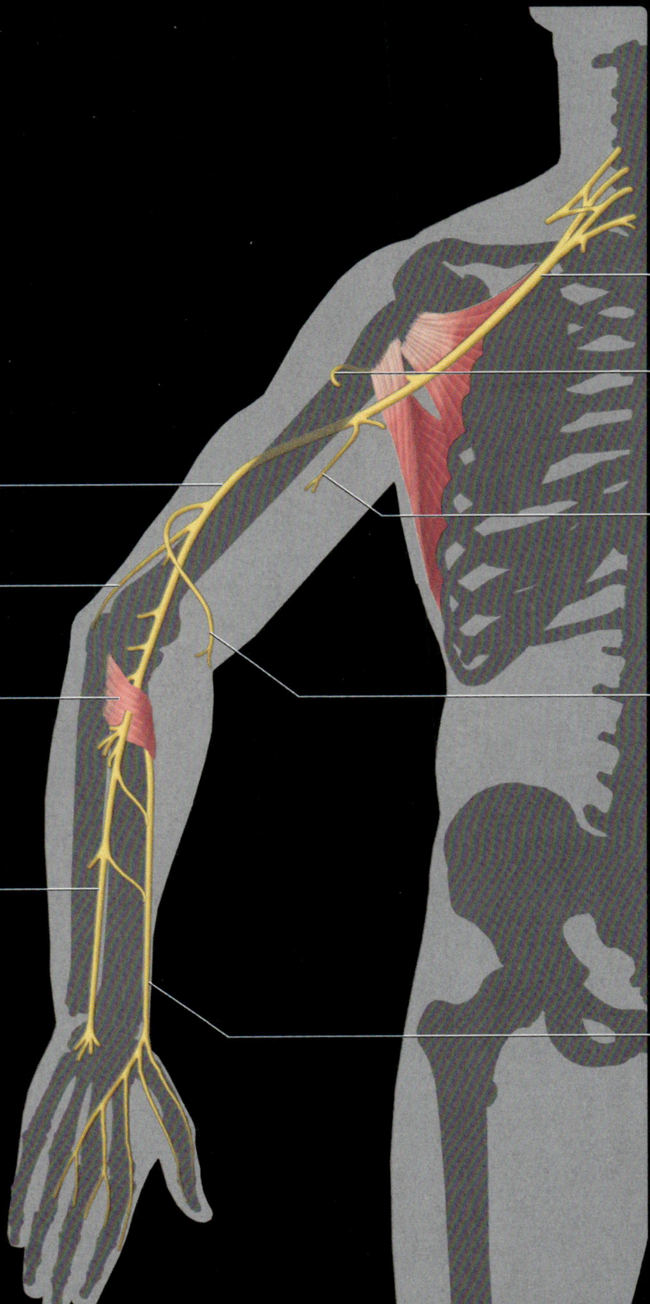

Graphic shows the course and distribution of the radial nerve. The radial nerve arises from the posterior cord of the brachial plexus (C5-C8, T1). It spirals posterolaterally around the humerus with the deep brachial artery located anterolateral between the brachialis and the brachioradialis. It gives off the posterior cutaneous nerve of the forearm, which passes posterior to the lateral condyle and supplies the posterior forearm. It divides into deep and superficial branches at the lateral epicondyle. The deep branch is purely motor and pierces the supinator muscle and winds around the lateral aspect of the radial neck. The superficial branch is purely sensory and is located in the anterolateral aspect of the forearm, superficial to the supinator and pronator teres muscles.

POSTERIOR ARM

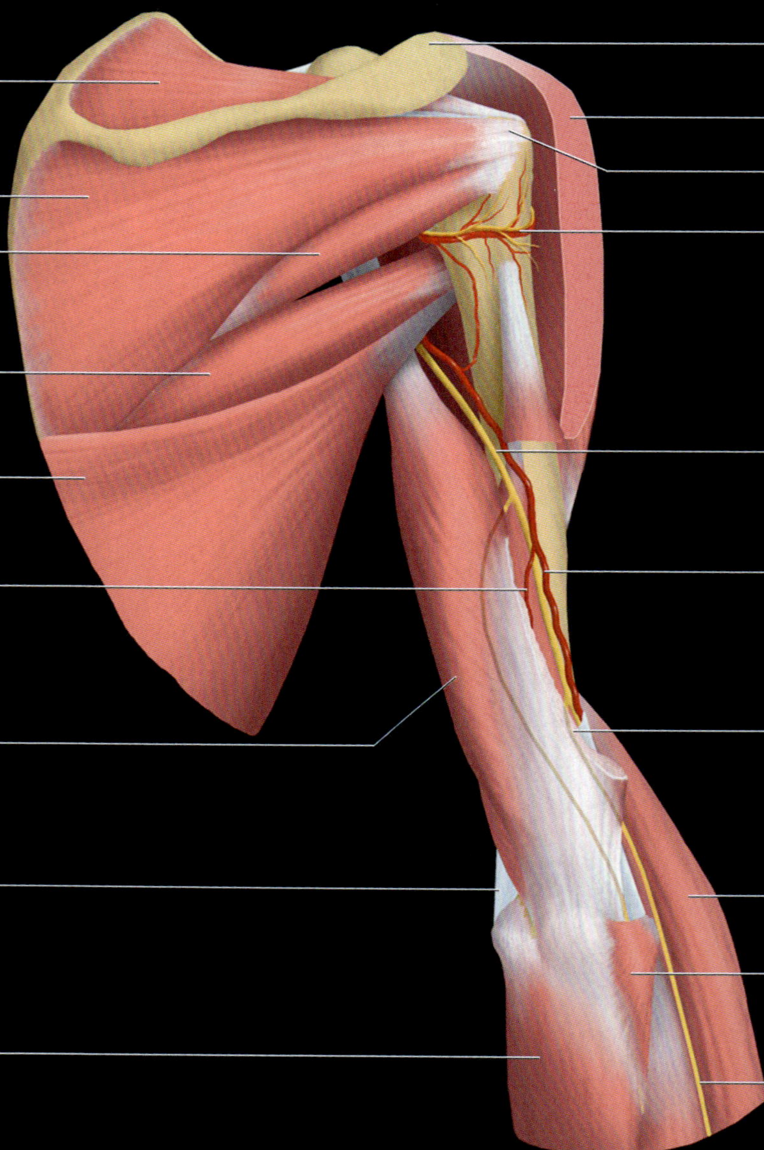

Graphic shows the course of the radial nerve in the posterior aspect of the arm. The radial nerve spirals posterolaterally around the humerus with the deep brachial artery. It gives off the posterior cutaneous nerve of the forearm, which passes posterior to the lateral condyle and supplies the posterior forearm. It is located anterolaterally, between the brachialis and brachioradialis, where it supplies the triceps, anconeus, brachioradialis, and the lateral portion of the brachialis.

TRANSVERSE US: ARM

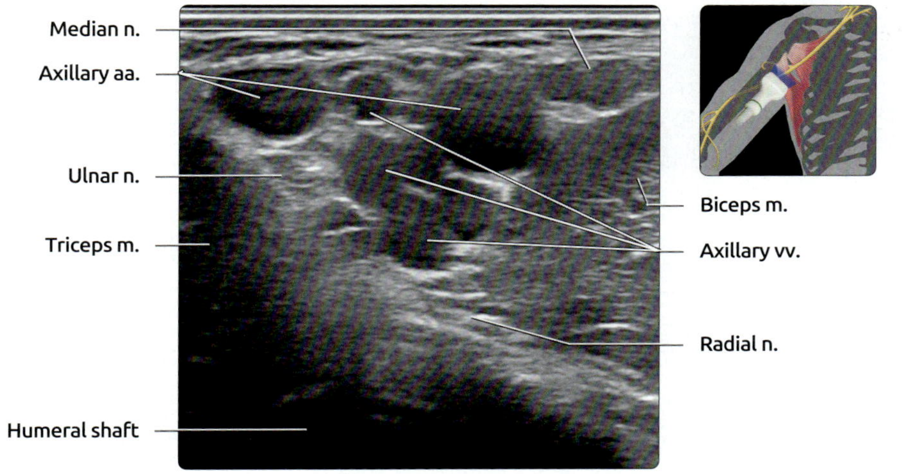

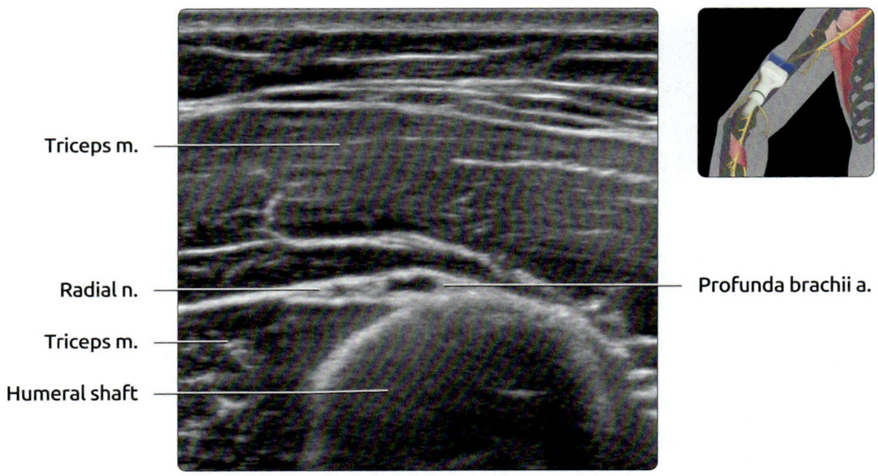

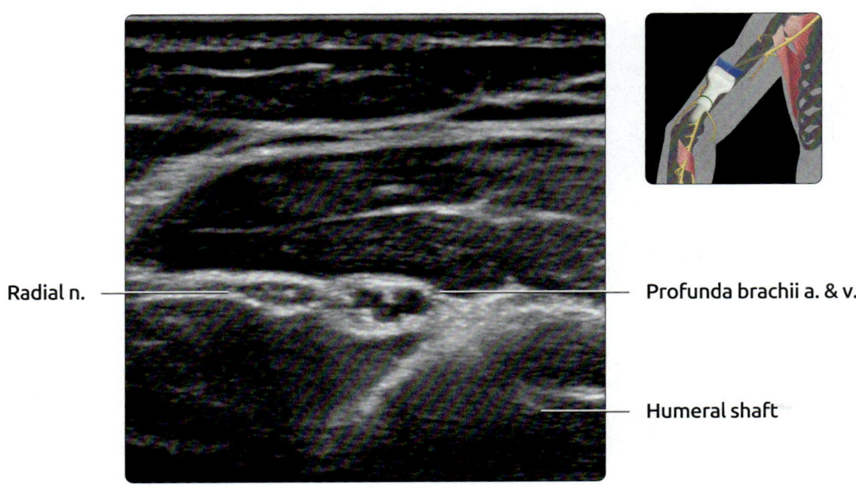

(Top) *Transverse US shows the radial nerve in the axilla. The radial nerve is a continuation of the posterior cord of the brachial plexus. It supplies the triceps and the supinator. In the proximal arm, it lies deep to the axillary artery and vein. There is commonly at least 2 axillary arteries and 3 axillary veins.* **(Middle)** *Transverse US shows the radial nerve at the mid 1/3 of the arm. The radial nerve may be injured as it passes around the posterior aspect of the mid to distal 1/3 of the humeral shaft. This injury may occur at the time of fracture, during fracture manipulation or fixation, or due to postoperative fibrosis.* **(Bottom)** *More localized view of the radial nerve and profunda brachii artery and veins are seen close to the posterior aspect of the humeral shaft in the midarm region.*

TRANSVERSE US: SPIRAL GROOVE

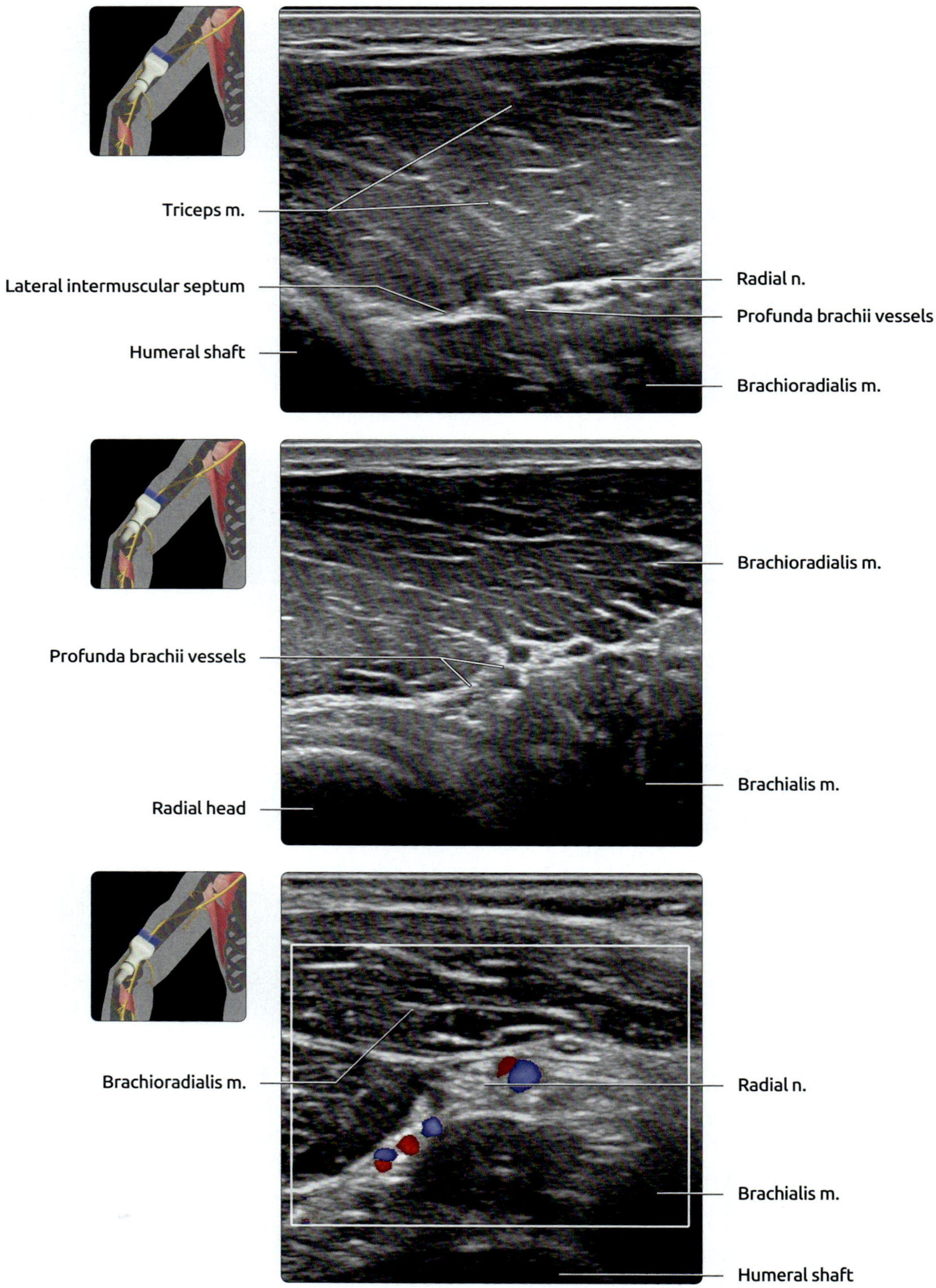

(Top) *Transverse US shows the radial nerve on the volar aspect of the distal 1/3 of the arm. The radial nerve passes from the posterior to the anterior compartment of the arm by piercing the lateral intermuscular septum at the lateral end of the distal humerus. Compression of the radial nerve at the lateral intermuscular septum may occur.* **(Middle)** *Transverse US shows the radial nerve in the radial tunnel. The radial tunnel is the area between the lateral intermuscular septum and the supinator muscle inlet, deep to the brachioradialis muscle. The radial nerve can be readily identified in this area. The radial nerve divides into superficial and deep branches in the radial tunnel area.* **(Bottom)** *Transverse color Doppler US can help identify the radial nerve from adjacent vessels in the radial canal. Rarely, vascular compression of the radial nerve can occur in the radial tunnel area from a mesh of profunda brachii vessels (leash of Henry), leading to radial tunnel syndrome.*

TRANSVERSE US: POSTERIOR FOREARM

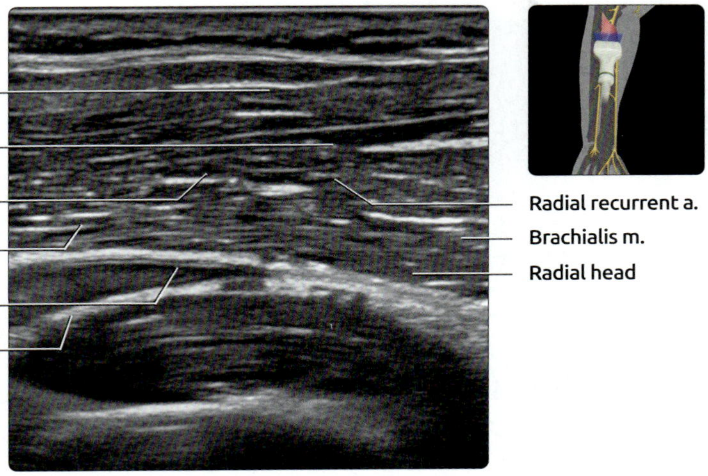

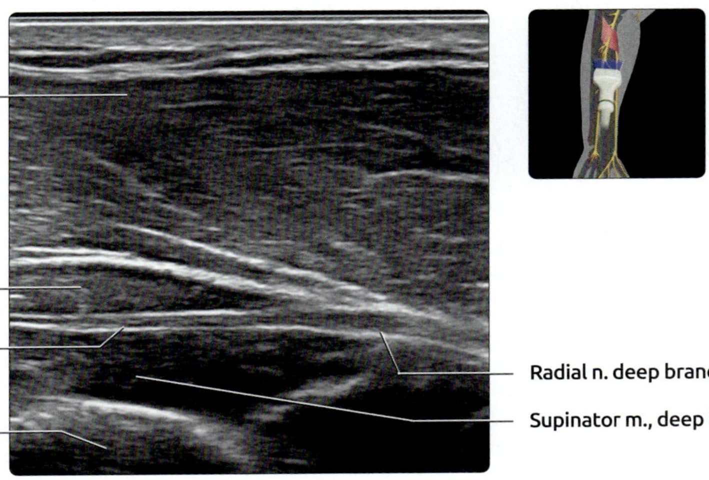

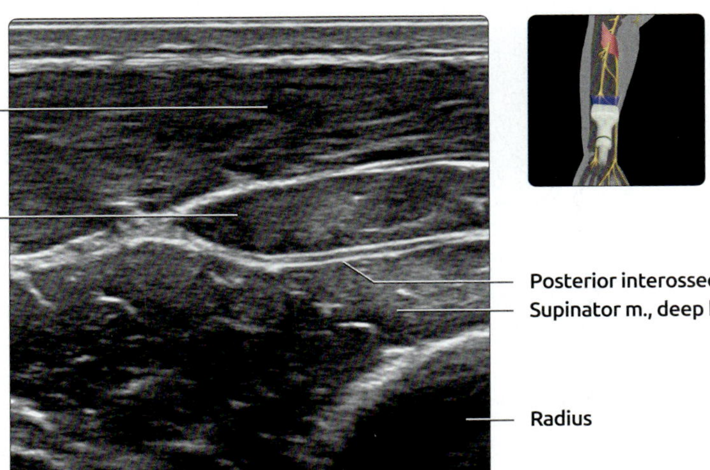

(Top) *Transverse US of the proximal forearm shows the deep branch of the radial nerve entering the supinator canal course in the proximal forearm. The inlet of the supinator tunnel, known as the arcade of Fröhse, is the most common site of radial nerve compression. Compression of the posterior interosseous nerve at this location leads to supinator syndrome.* **(Middle)** *Longitudinal US of the proximal forearm shows the deep branch of the radial nerve in the radial tunnel. Once the deep branch of the radial nerve passes into the supinator inlet (arcade of Fröhse), it is known as the posterior interosseous nerve. The supinator canal lies between the superficial and deep heads of the supinator muscle. The deep branch of the radial nerve normally slightly thickens just before entering the supinator canal. More pronounced thickening (> 2 mm width) just proximal to the supinator inlet would suggest compression. This compression can lead to supinator syndrome.* **(Bottom)** *Longitudinal US of the proximal forearm shows the posterior interosseus nerve within and exiting the supinator tunnel. This tunnel lies between the superficial and deep heads of the supinator muscle.*

TRANSVERSE US: FOREARM

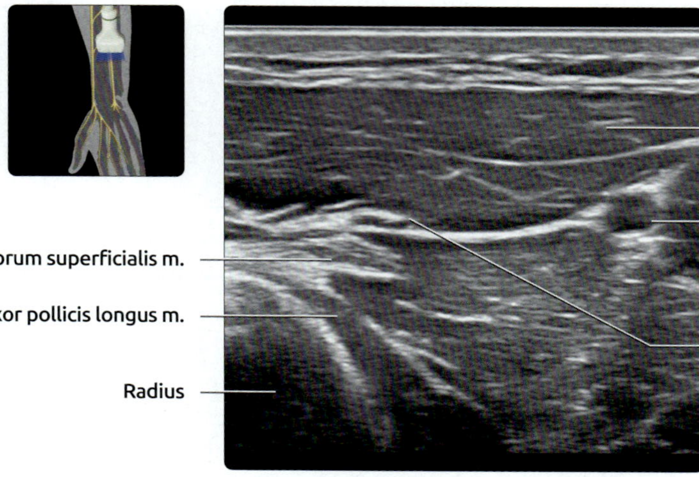

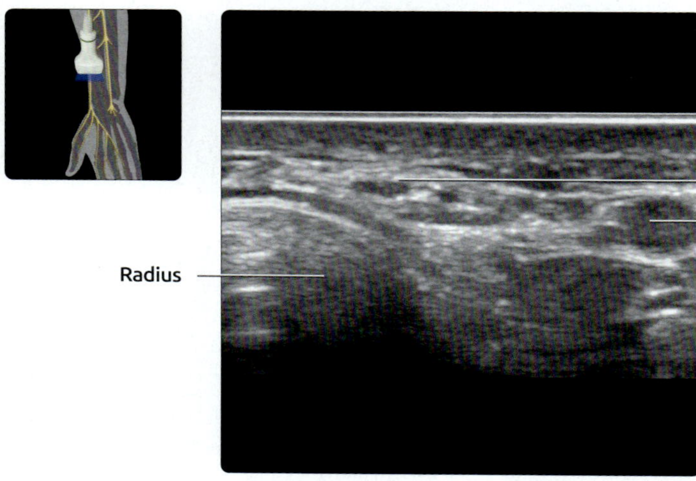

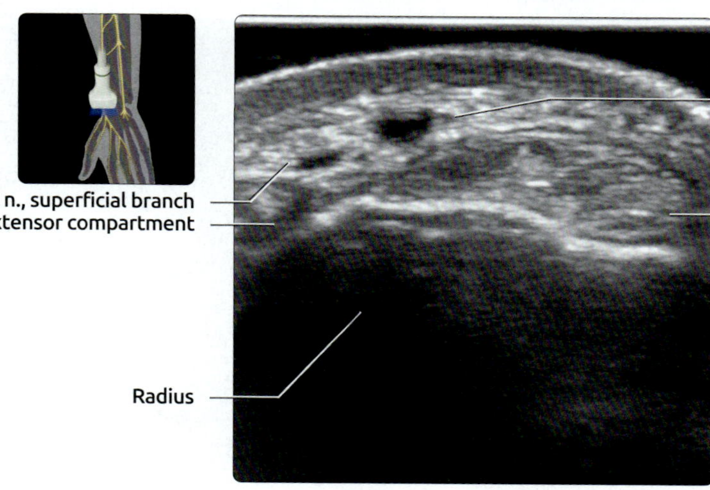

(Top) *Transverse US shows the superficial branch of the radial nerve on the volar aspect of the midforearm. The superficial branch of the radial nerve moves from the deep to the superficial tissues.* **(Middle)** *Transverse US shows the superficial branch of the radial nerve on the volar aspect of the distal forearm. About 8 cm proximal to the radial styloid, the superficial branch of the radial nerve emerges from underneath the brachioradialis muscle to lie superficially. Entrapment or injury to the superficial branch of the radial nerve is called Wartenberg syndrome and can be due to a tight wrist watch, plaster of Paris cast, internal fixation of the radius, or cephalic vein cannulation.* **(Bottom)** *Transverse US shows the superficial branch of the radial nerve on the radial side of the distal forearm traversing the 2nd and 1st extensor tendon compartments. The superficial branch of the radial nerve can be injured during cephalic vein cannulation or percutaneous injection of these extensor compartments.*

ELBOW AND WRIST

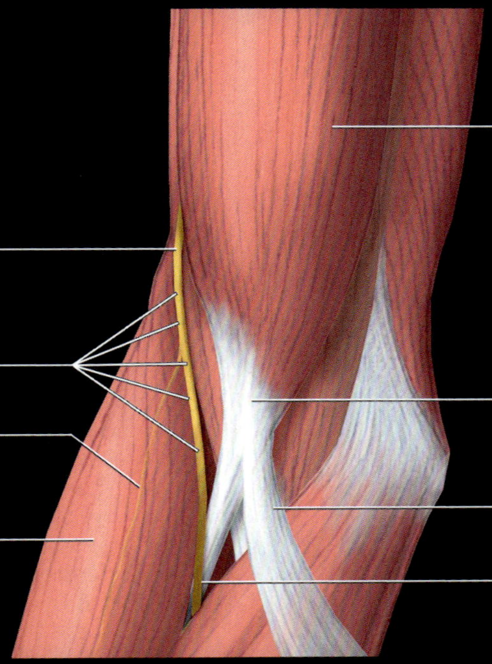

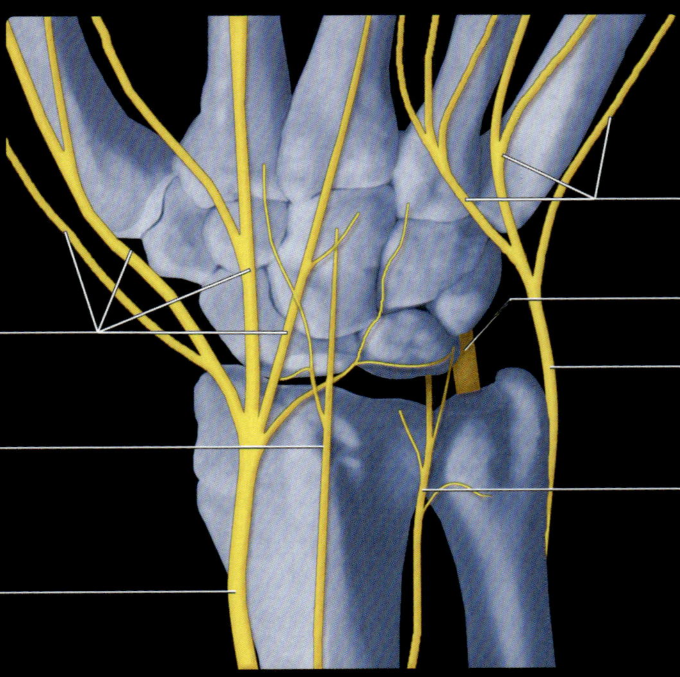

(Top) *Graphic shows the radial nerve course at the anterolateral aspect of the elbow. The deep branch of the radial nerve is potentially prone to impingement at the superficial proximal margin of the supinator muscle (arcade of Fröhse), resulting in posterior interosseus nerve syndrome.* **(Bottom)** *Graphic shows the nerves of the dorsal wrist. The radial nerve branches in the forearm with the superficial, posterior cutaneous, and posterior interosseus branches supplying the wrist and hand. The ulnar nerve provides branches to both the dorsal and volar wrist.*

TERMINOLOGY

Abbreviations

- Flexor digitorum profundus (FDP)
- Flexor digitorum superficialis (FDS)
- Flexor carpi ulnaris (FCU)
- Flexor pollicis longus (FPL)
- Pronator quadratus (PQ)
- Palmaris longus (PL)

GROSS ANATOMY

Arm: Median Nerve

- Arises from medial and lateral cords of brachial plexus (C6-C8, T1)
- In axilla, lies superolateral to artery close to short head of biceps muscle
- Lies medial to brachial artery and vein
- Located deep to bicipital aponeurosis in cubital fossa
- Supplies pronator teres, PQ, and flexors of anterior compartment forearm (except FCU and medial 1/2 of FDP, which are supplied by ulnar nerve)

Forearm: Median Nerve and Branches

- Enters forearm by passing between humeral and ulnar heads of pronator teres; located in forearm between FDP and FDS
  - Courses distally and attached to deep surface of FDS by fascial sheath
  - Anterior interosseous nerve
    - Purely motor nerve
    - Arises from median nerve at level of pronator teres
    - Located in forearm anterior to interosseous membrane, between FPL and FDP, alongside anterior interosseous artery
- At wrist, median nerve emerges from lateral side of FDS, becomes more superficial, and runs deep to PL tendon toward carpal tunnel
  - Supplies FPL, PQ, and lateral 1/2 of FDP

Wrist and Hand: Median Nerve and Branches

- Median nerve lies along axis of ring finger, i.e., to ulnar side of midline
- Lies deep to antebrachial fascia and PL before dipping deeply to enter carpal tunnel
- At distal radioulnar joint: Nerve is rounded, deep to PL, medial and superficial to FCR and FPL, and lateral and superficial to FDS
- Carpal tunnel is fibroosseous tunnel located immediately distal to volar wrist crease
  - Formed by carpal bones and flexor retinaculum (transverse carpal ligament)
  - Flexor retinaculum attaches to pisiform, hook of hamate, scaphoid, and trapezium
  - 9 structures pass through carpal tunnel
    - 4 FDS tendons, 4 FDP tendons, FPL tendon, and median nerve
    - Median nerve is located superficially in carpal tunnel just deep to retinaculum
  - Tunnel outlet is narrowest part of carpal tunnel
- Branches
  - Motor to thenar and 1st and 2nd lumbrical muscles
  - Sensory to radial 1/2 of palm, radial 3.5 digits
    - Palmar cutaneous branch typically arises proximal to carpal tunnel
  - Median nerve divides just distal to carpal tunnel

ANATOMY IMAGING ISSUES

Imaging Recommendations

- Examination of nerve in transverse plane is preferable to longitudinal scanning since it allows nerve to be followed continuously throughout limb
- Use proximal and distal borders of retinaculum as markers of carpal tunnel

Imaging Approaches

- Use high-frequency 12- to 17-MHz transducer

Imaging Pitfalls

- Median nerve may be bifid
  - If quantifying size of nerve, both components should be measured and summated
- Keep transducer at right angles to nerve as angulation will ↑ nerve caliber
  - May need to make use of anisotropy to best differentiate nerve from tendons in carpal tunnel
- Avoid too much transducer pressure as this will ↓ nerve caliber and ↑ flattening

CLINICAL IMPLICATIONS

Clinical Importance

- Normal cross-sectional area (CSA) of median nerve ranges from $< 10\ mm^2$
  - Nerve swelling is considered if CSA $> 10\ mm^2$
    - Between 10-13 mm^2 questionable swelling
    - $> 13\ mm^2$ nearly always abnormal
  - Larger median nerve, greater likelihood of carpal tunnel syndrome
  - US nerve CSA measurements 20% smaller than on MR as US measurements do not usually include echogenic epineurium
- Potential sites of median nerve compression
  - Deep to ligament of Struthers
    - Ligament between medial epicondyle and anomalous bony spur arising from distal humeral diaphysis (present in 1% of population)
    - May compress median nerve, brachial artery, or both
  - Between brachialis and bicipital aponeurosis
    - Bicipital aponeurosis = medial fascial expansion of biceps tendon that extends superficial to brachial artery and median nerve and merges with investing fascia of forearm
  - Between heads of pronator teres
    - Either median nerve or anterior interosseus nerve may be compressed
  - Carpal tunnel syndrome
    - Most common compressive neuropathy
    - Nerve swollen proximal to inlet ± at inlet ± at outlet ± distal to outlet
    - Do not use opposite side as comparative standard as subclinical compression common

MEDIAN NERVE

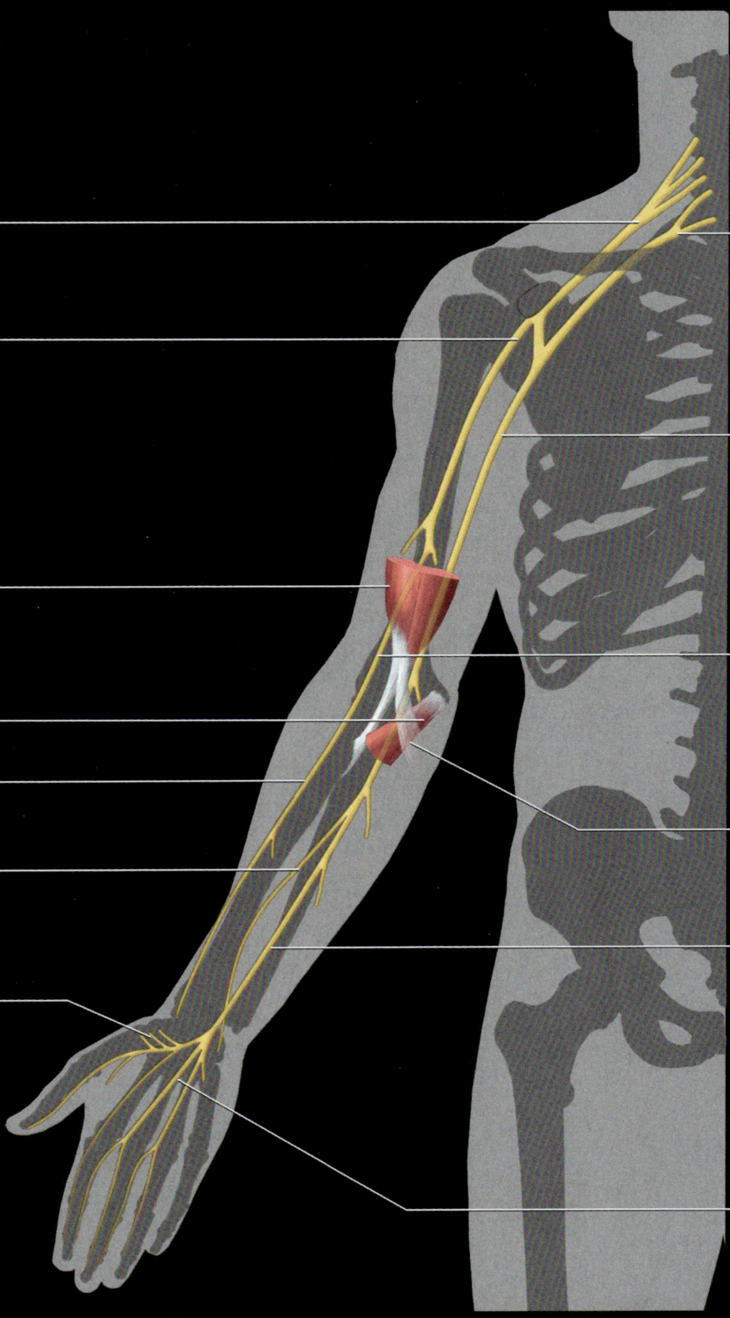

Graphic shows the course and distribution of the median and musculocutaneous nerves. The median nerve arises from the medial and lateral cords of the brachial plexus (C6-C8, T1). It travels together with other nerves of the arm. In the elbow, it is located deep to the biceps aponeurosis, enters the forearm by passing between the heads of the pronator teres, and is located in the forearm between the flexor digitorum superficialis and profundus muscles. The musculocutaneous nerve arises from the lateral cord of the brachial plexus (C5, C6, C7). It lies between the brachialis and biceps brachii muscles and supplies both of them. It then becomes superficial at the elbow and continues laterally as the lateral cutaneous nerve of the forearm, which innervates the skin of the lateral side of the forearm.

CUBITAL FOSSA

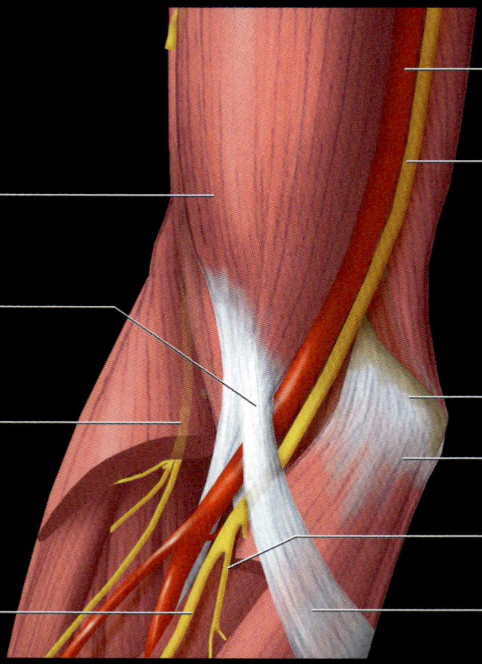

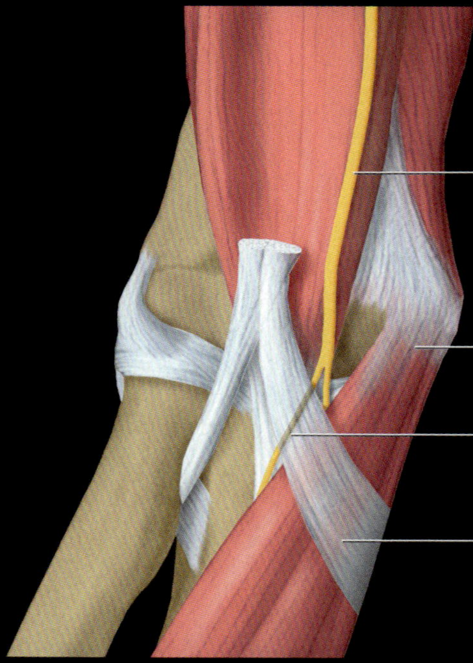

(Top) *Graphic of the anterior view of the cubital fossa shows the median nerve and brachial artery passing underneath the biceps aponeurosis. The anterior interosseus nerve arises from the median nerve as the median nerve passes between 2 heads of the pronator teres muscle.* **(Bottom)** *The median nerve may be entrapped between 2 heads of the pronator teres muscle or by the overlying biceps aponeurosis. The median nerve gives rise to the articular branches to the elbow joint. It supplies the pronator teres, pronator quadratus, and flexors of the anterior compartment of the forearm (except the flexor carpi ulnaris and medial 1/2 of the flexor digitorum profundus, which are supplied by the ulnar nerve).*

TRANSVERSE US

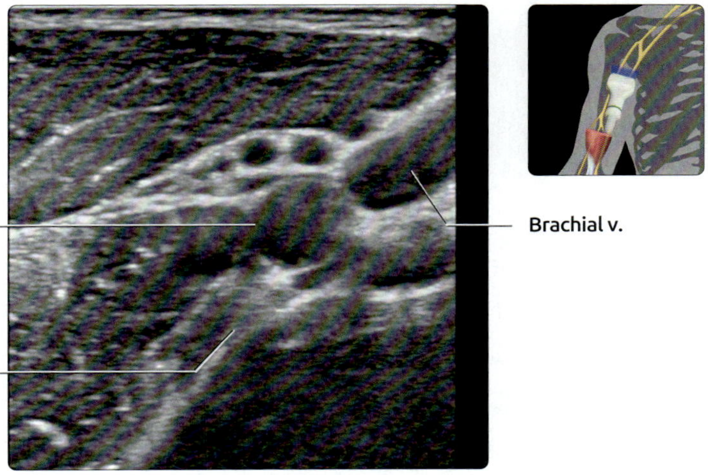

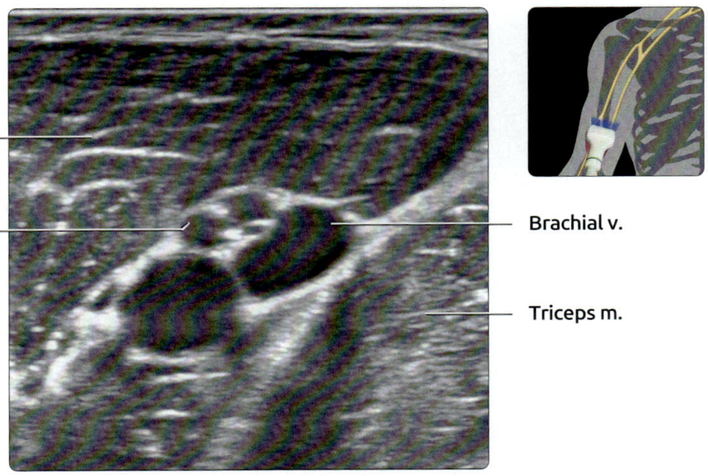

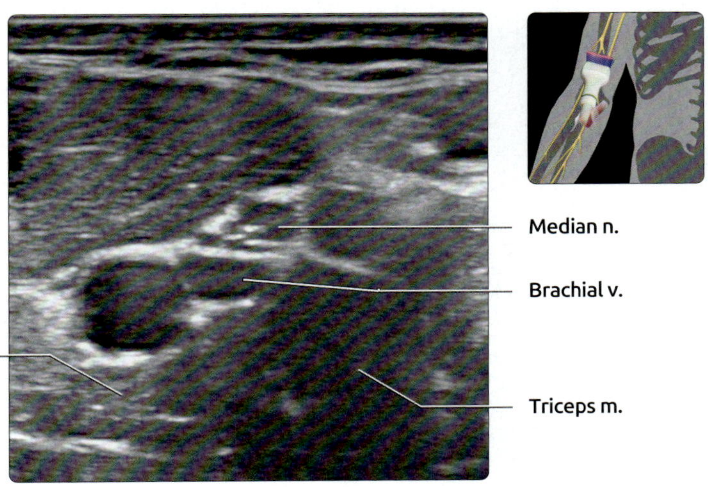

(Top) *Transverse US shows the median nerve at the proximal arm. The median nerve arises from the lateral and medial cords of the brachial plexus. In the proximal arm, it lies between the biceps muscle and the brachial artery/vein.* **(Middle)** *Transverse grayscale US shows the median nerve at the midarm. Little happens to the median nerve in the upper to midarm. It gives off no branches and is not prone to compression by normal structures.* **(Bottom)** *Transverse US shows the median nerve of the distal arm. The median nerve passes over the brachial artery and vein from lateral to medial. Occasionally, there is a ligament (of Struthers) in the distal arm with a bony spur attached to the humerus ~ 5 cm proximal to the medial epicondyle. The pronator teres gains additional attachment from this ligament. The median nerve may be compressed deep to this ligament.*

TRANSVERSE US

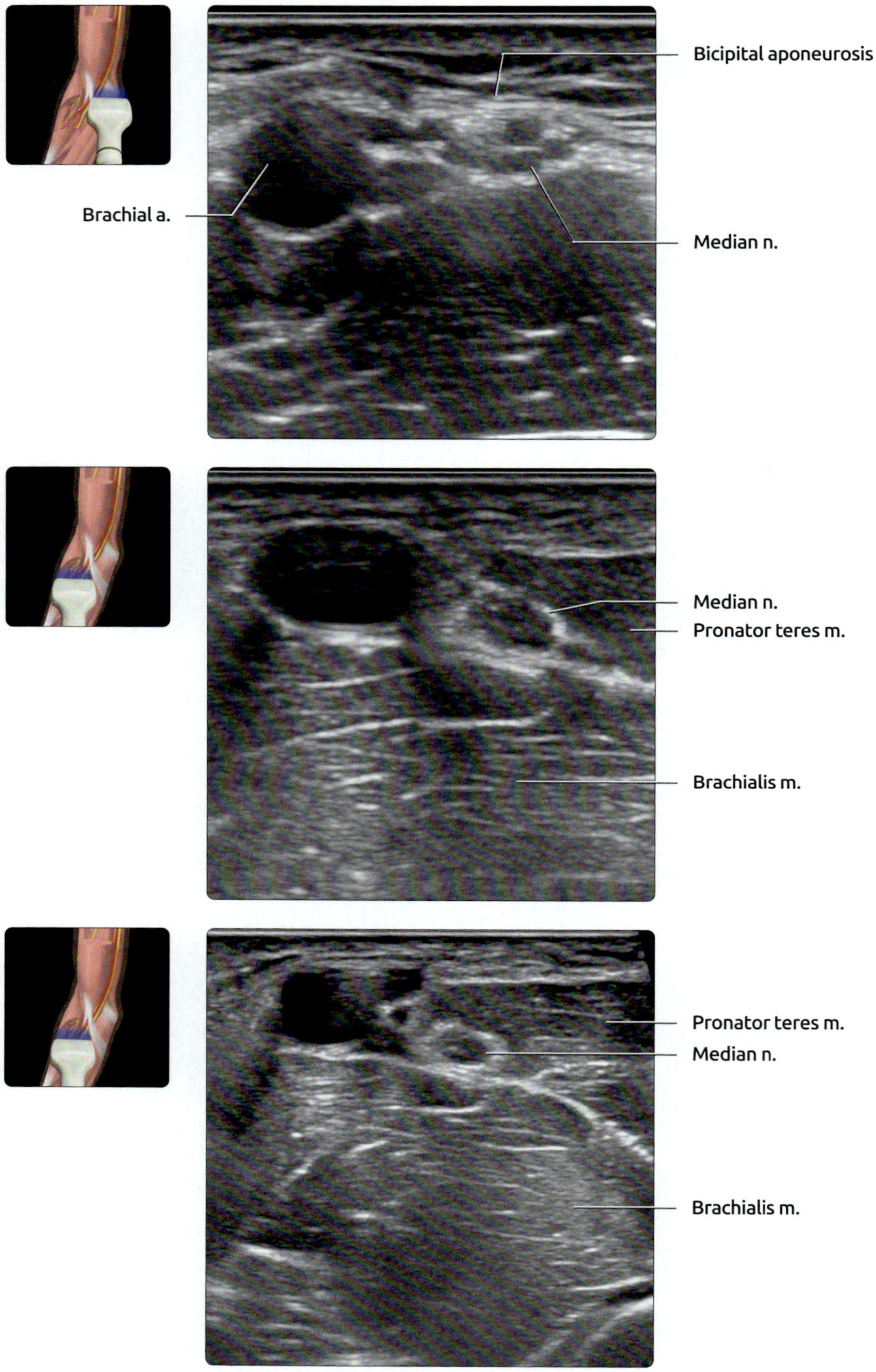

(Top) *Transverse US shows the median nerve at the distal arm. The median nerve travels deep to the bicipital aponeurosis. A fascial aponeurosis (a.k.a. lacertus fibrosis) arises from the biceps musculotendinous junction and inserts into the antebrachial fascia. The median nerve may be compressed deep to this fascia.* **(Middle)** *Transverse US shows the median nerve at the elbow region.* **(Bottom)** *Transverse US shows the median nerve at the proximal forearm. The median nerve is also prone to compression between the 2 heads of the pronator teres muscle (i.e., common flexor origin head and coronoid process head). The median nerve branches as it passes through these heads, giving rise to the anterior interosseus nerve, and may be compressed at this location. Pronator teres syndrome is characterized by pain in the forearm or hand with either prolonged forearm pronation or recurrent pronation/supination of the forearm.*

TRANSVERSE US

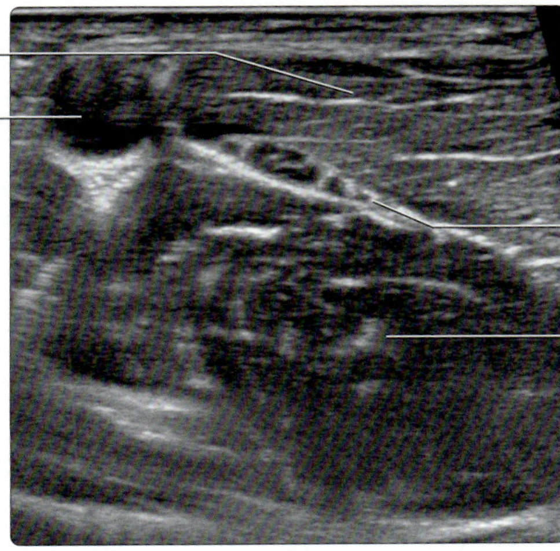

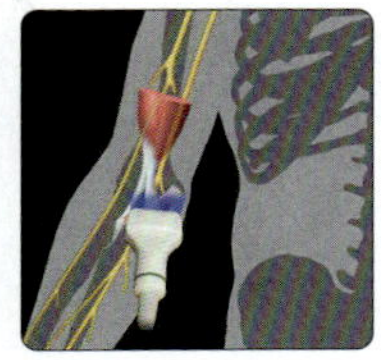

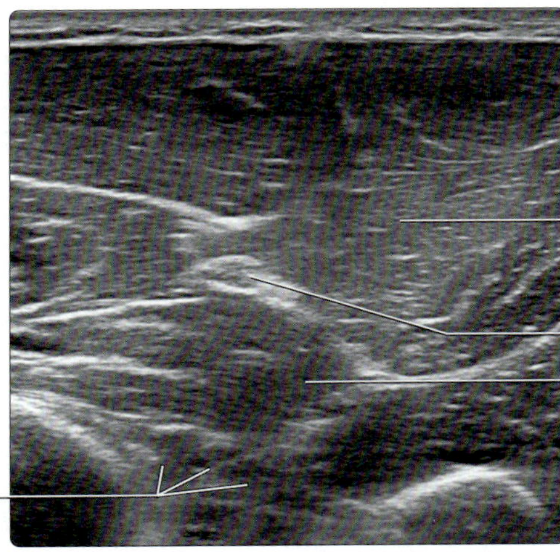

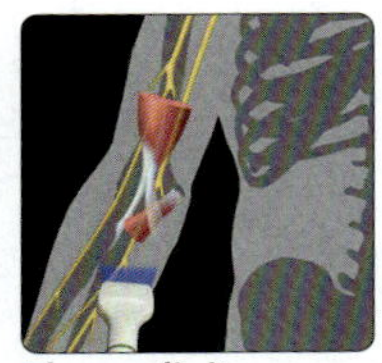

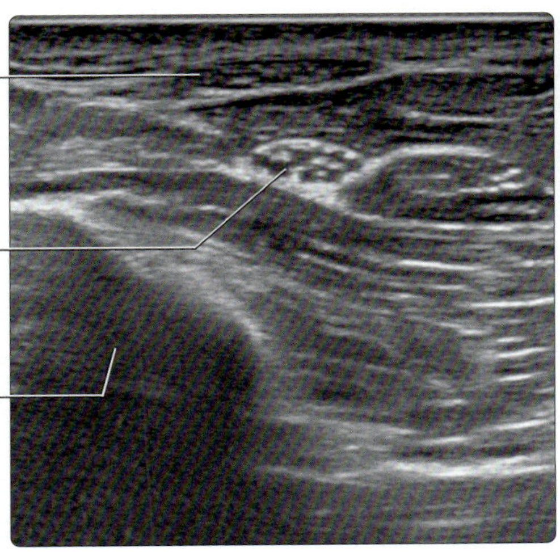

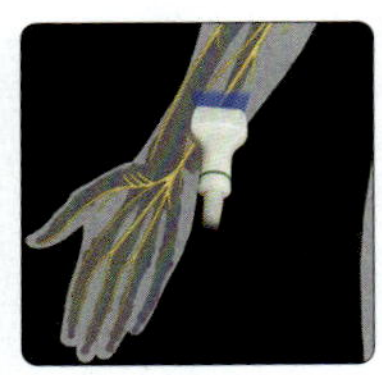

(Top) *Transverse US shows the median nerve at the proximal forearm. The anterior interosseus nerve supplies all the muscles of the anterior compartment of the forearm except the ulnar 1/2 of the flexor digitorum profundus muscle. It may be injured with fracture of the radial shaft.* **(Middle)** *Transverse US shows the median nerve at the midforearm. The median nerve is clearly identified in the mid-distal forearm as it travels between the flexor digitorum superficialis and profundus muscles. The anterior interosseous nerve is more difficult to see clearly, though its location can be inferred through identification of the adjacent anterior interosseous artery and veins.* **(Bottom)** *Transverse US shows the median nerve at the distal forearm.*

TRANSVERSE US: WRIST

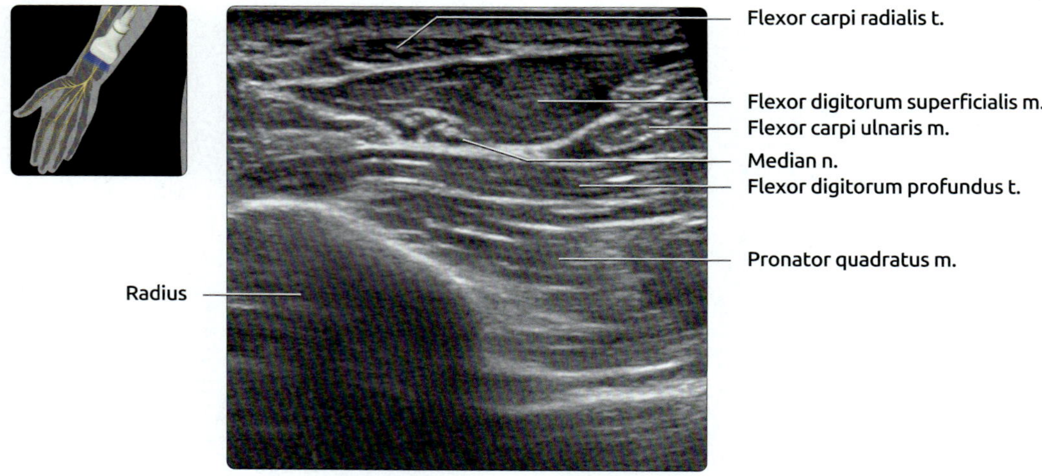

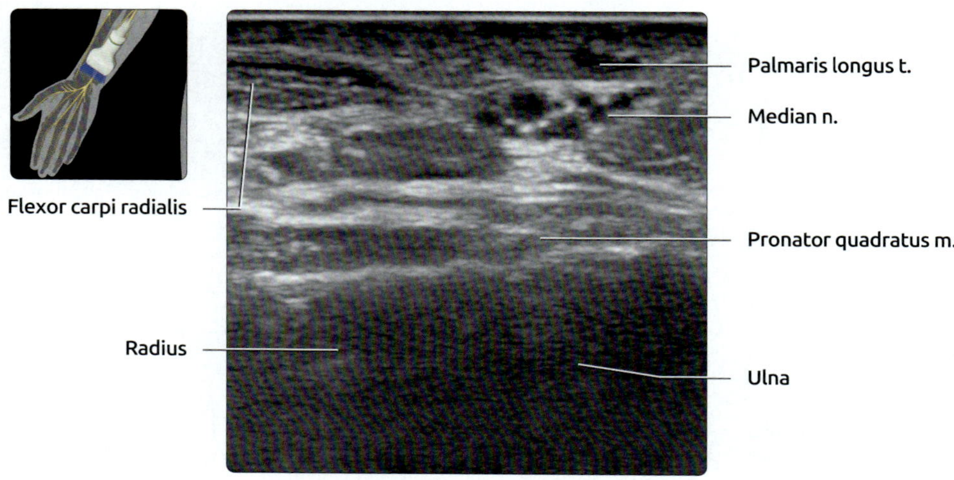

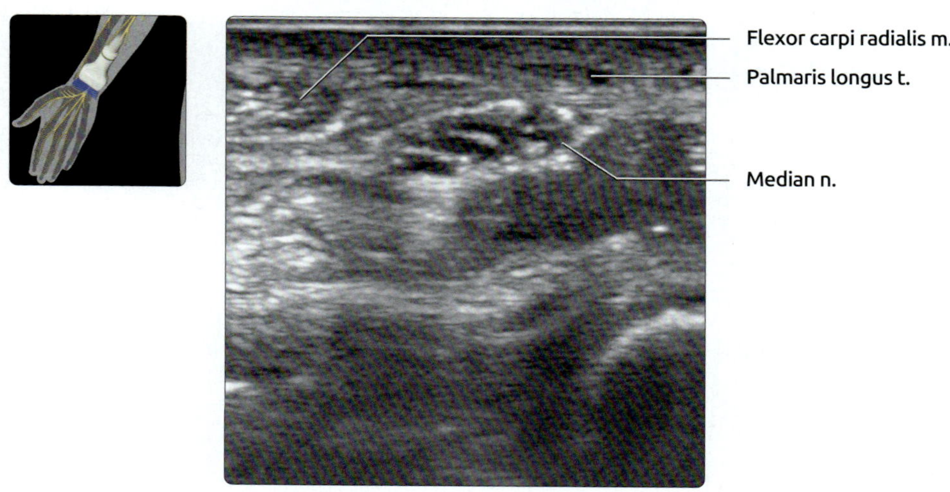

(Top) *Transverse US shows the median nerve in the distal forearm.* **(Middle)** *Transverse US shows the median nerve just proximal to the wrist. The median nerve gives off the palmar cutaneous branch in the distal forearm. This branch innervates the skin on the palm of the hand.* **(Bottom)** *Transverse US shows the median nerve just proximal to the carpal tunnel; the median nerve ascends from a deeper position (between the flexor digitorum profundus and superficialis) to lie just deep to palmaris longus tendon. Enlargement of the median nerve in patients with carpal tunnel syndrome may present clinically as a mass-like lesion in this area. In suspected carpal tunnel syndrome, the caliber of the nerve should be measured at this location.*

TRANSVERSE US: CARPAL TUNNEL

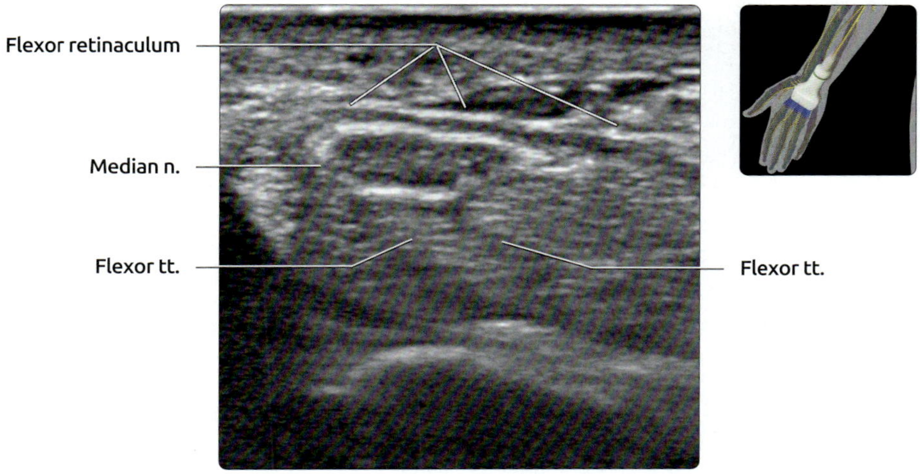

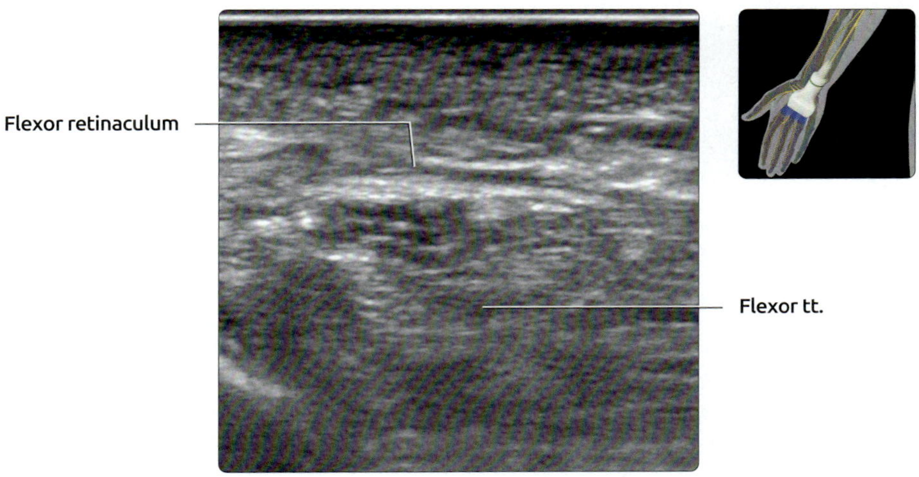

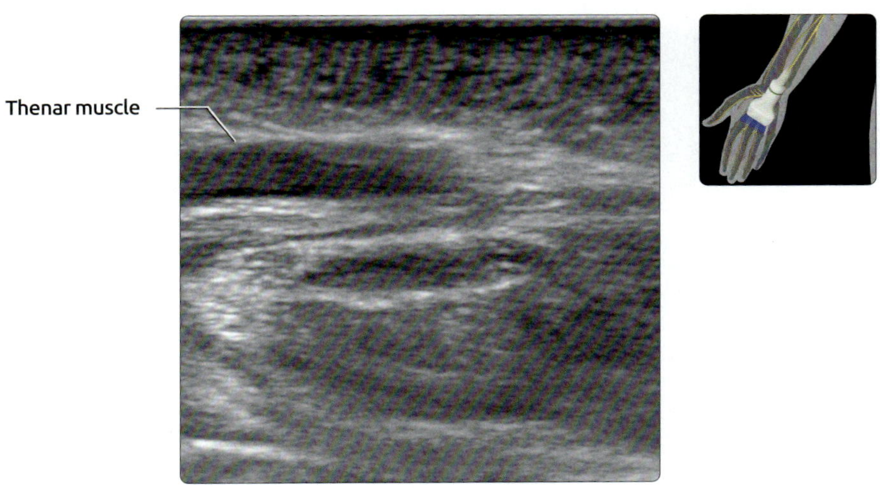

(Top) *Transverse US shows the median nerve at the carpal tunnel inlet. The best way to identify the carpal tunnel is to identify the leading and trailing edges of the flexor retinaculum. The median nerve dips deep to the retinaculum. The cross-sectional area should be measured at this location in suspected carpal tunnel syndrome. The echogenic epineurium around the nerve is not included in US cross-sectional area measurements.* **(Middle)** *Transverse US shows the median nerve at the midcarpal tunnel. The median nerve lies just deep to the retinaculum. It is variable in shape and changes with flexor tendon movement.* **(Bottom)** *Transverse US shows the median nerve at the carpal tunnel outlet. The cross-sectional area of the median nerve should also be measured at this location in patients with suspected carpal tunnel syndrome. This is the narrowest part of the carpal tunnel.*

TRANSVERSE AND LONGITUDINAL US

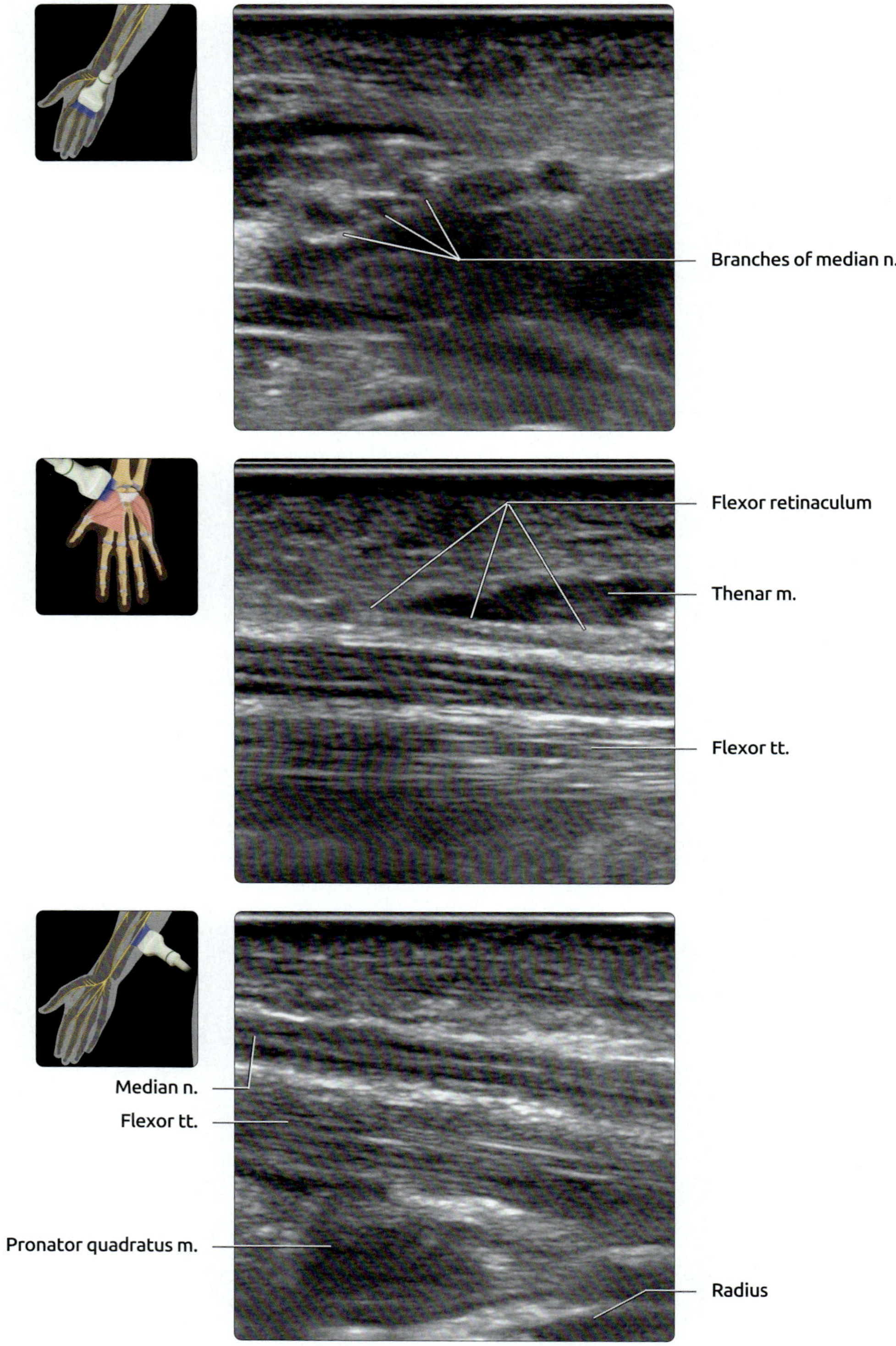

(Top) *Transverse US shows the median nerve just beyond the carpal tunnel outlet. The median nerve fairly consistently divides into its digital branches just beyond the tunnel outlet. If measured beyond the outlet, the median nerve cross-sectional area should be measured before the nerve divides into its digital branches.* **(Middle)** *Longitudinal US shows the median nerve in the carpal tunnel.* **(Bottom)** *Longitudinal US shows the median nerve in the distal forearm.*

PALMAR NERVES

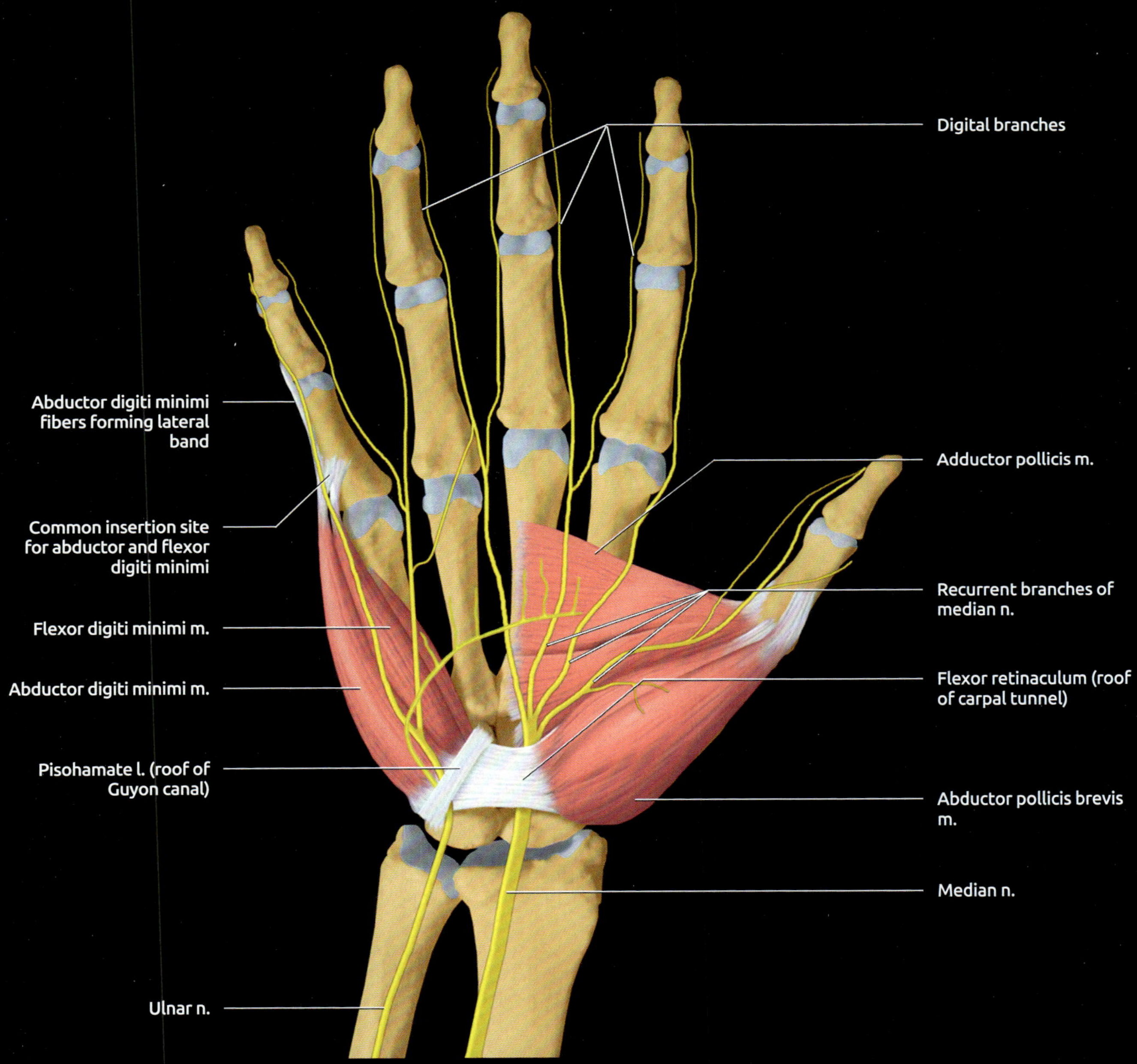

The median nerve travels deep to the flexor retinaculum (within the carpal tunnel). The ulnar nerve travels superficial to the flexor retinaculum and deep to the pisohamate ligament (within the Guyon canal). Both are prone to compression syndromes within these fibroosseous tunnels. Motor supply of the median nerve is to the thenar muscles (via the recurrent branch) and the 1st and 2nd lumbricals. The ulnar nerve motor supply is to all intrinsic muscles of the hand, which are not supplied by the median nerve. Muscle atrophy should prompt an evaluation of the innervating nerve.

GROSS ANATOMY

Arm: Ulnar Nerve

- Arises from medial cord of brachial plexus (C8, T1)
- In axilla, located superficial and medial to axillary artery
- In arm, located medial to brachial artery on coracobrachialis muscle in anterior compartment
  - Enters posterior compartment by piercing medial intermuscular septum before running distally with superior ulnar collateral artery
  - Courses posterior to medial humeral epicondyle in condylar groove (cubital tunnel)
- Cubital tunnel: Fibroosseous tunnel formed by medial epicondyle, olecranon, and Osborne fascia
  - Cubital tunnel syndrome (CUTS)
- Ulnar nerve may sublux over medial epicondyle in 20% of normal subjects during flexion
  - Recurrent subluxation can lead to friction neuritis
- Anconeus epitrochlearis is accessory muscle extending from medial part of olecranon to medial epicondyle deep to Osborne ligament
  - Present in 10% normal subjects, bilateral in 25%
  - Osborne ligament may be vestigial remnant of anconeus epitrochlearis muscle

Forearm: Ulnar Nerve and Branches

- On exiting cubital tunnel, enters forearm by passing deep to arcuate ligament between humeral and ulnar heads of flexor carpi ulnaris (FCU)
- Courses distally between FCU and flexor digitorum profundus (FDP)
  - Supplies FCU, medial 1/2 of FDP
- Palmar cutaneous branch
  - Arises in midforearm and supplies skin on ulnar 1/2 of palm
- Dorsal cutaneous branch
  - Arises in distal forearm and supplies skin on ulnar side of dorsum hand
- Distally, becomes superficial, comes closer to ulnar artery and passes into wrist superficial to flexor retinaculum

Hand: Ulnar Nerve and Branches

- Enters Guyon canal; small, triangular-shaped fibroosseous tunnel
  - Floor = flexor retinaculum; ulnar wall = pisiform and hook of hamate; roof = fascia continuous with flexor retinaculum (superficial palmar ligament)
  - Contains ulnar artery and ulnar nerve
  - Ulnar nerve lies between ulnar artery and pisiform or hook of hamate
  - Within tunnel, proximal to hook of hamate, it divides into superficial sensory and deep motor branches
    - Superficial branch passes over hook of hamate before bifurcating into digital nerves
      - Provides sensory innervation to skin of hypothenar eminence, little finger, and ulnar 1/2 of ring finger
    - Deep branch passes ulnar to hook of hamate
      - Provides motor innervation to hypothenar muscles, all interossei, 3rd and 4th lumbricals, and adductor pollicis

ANATOMY IMAGING ISSUES

Imaging Recommendations

- Knowledge of nerve location and anatomic relationships is essential for US identification
- Examination of nerves in transverse plane is preferable to longitudinal plane as it allows nerves to be followed continuously throughout limb
- Identify nerves at target area and follow proximally and distally
- Target areas for ulnar nerve = cubital tunnel, distal forearm alongside ulnar artery, and Guyon canal

Imaging Approaches

- Ulnar nerve prone to compression in cubital tunnel, between humeral and ulnar origins of FCU, and in Guyon canal
- CUTS = pain, weakness, and numbness in ulnar nerve distribution due to compression of ulnar nerve in cubital tunnel
  - Due to primary compression, secondary compression, or combination
  - Common causes of secondary cubital tunnel compression
    - Bone (osteophytosis)
    - Capsule (synovitis, capsular thickening)
    - Muscle (anconeus epitrochlearis, medial belly triceps)
    - Tumor (ganglion, nerve sheath tumor)
  - If cross-sectional area (CSA) ulnar nerve at cubital tunnel > 10 mm^2, sensitivity 76%, specificity 61% for CUTS
  - Alternatively, nerve CSA within tunnel should not be > 1.5x CSA of nerve in distal arm
  - The more swollen the ulnar nerve in cubital tunnel, the greater the likelihood of CUTS
- Recurrent subluxation of ulnar nerve can lead to friction neuritis and symptoms of CUTS
- Guyon canal compression is more often structural
  - Ganglion cyst, fracture hook of hamate, nerve sheath tumor, or other tumor
  - Repetitive microtrauma ("handlebar" palsy)

Imaging Pitfalls

- Smaller nerves (< 1 mm) are difficult to identify, though their location can be inferred by adjacent anatomy
  - Color Doppler helpful to differentiate between vessels and nerve
- Fixed structural causes of nerve compression are more easy to identify than functional compression
- Ulnar nerve normally enlarges slightly in cubital tunnel, though not to same degree as seen in CUTS

CLINICAL IMPLICATIONS

Clinical Importance

- Identification of cause of ulnar neuropathy is key to successful treatment
- Cubital tunnel compression and cubital tunnel neuritis can give similar symptoms
- Compression at Guyon canal spares palmar and dorsal cutaneous branches, which arise in distal forearm
  - Should have less pronounced sensory denervation than CUTS

ULNAR NERVE

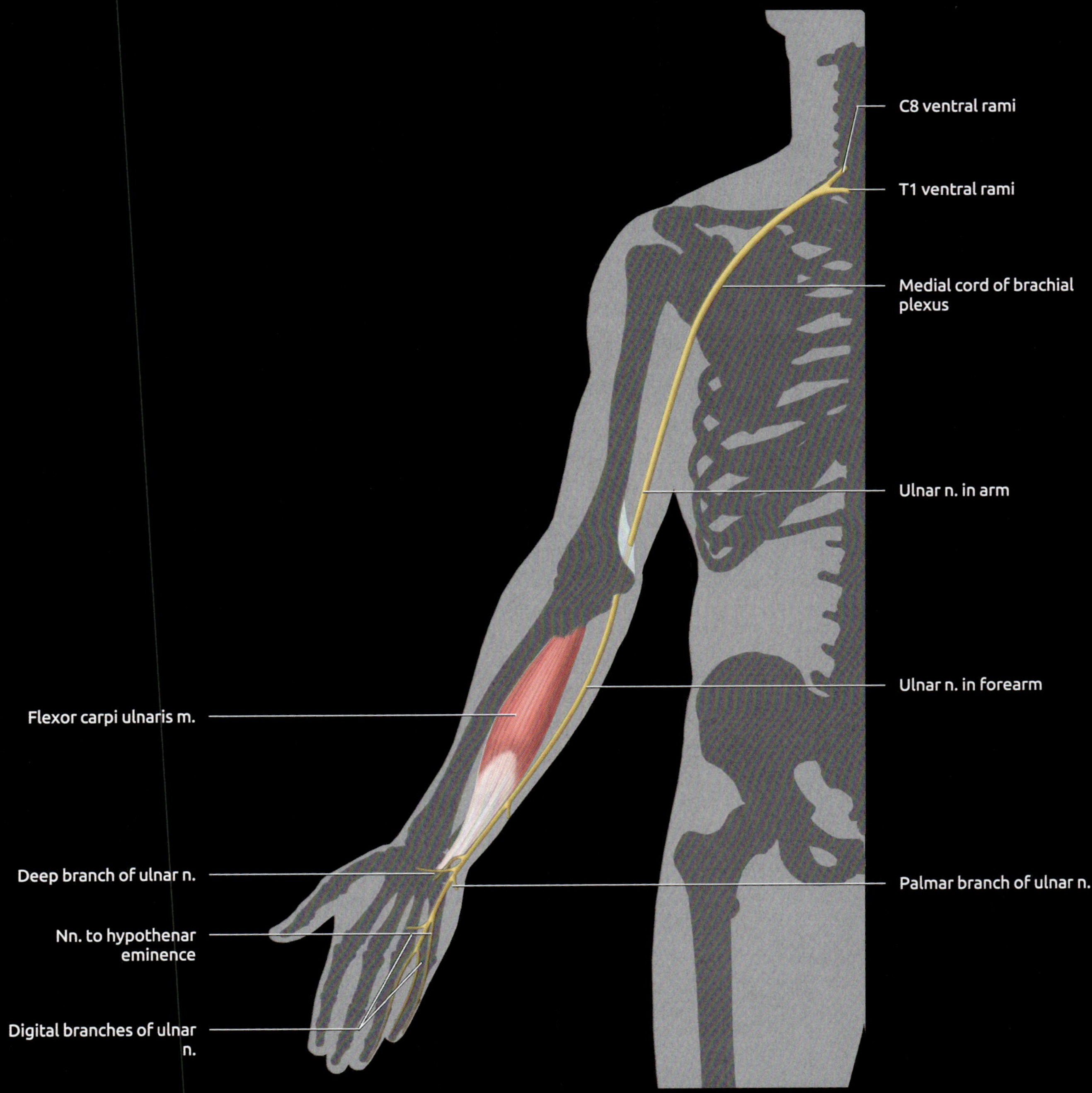

Graphic shows the course of the ulnar nerve. The ulnar nerve arises from the medial cord of the brachial plexus (C8, T1). It is located posteromedially, deep to the triceps muscles in the distal arm, where it passes posterior to medial epicondyle in the cubital tunnel (cubital syndrome due to compression of ulnar nerve in cubital tunnel). The ulnar nerve may sublux anterior to the medial epicondyle in ~ 15% of normal people, usually during flexion. The ulnar nerve passes behind the medial epicondyle, entering the forearm between the humeral and ulnar heads of flexor carpi ulnaris. The ulnar nerve then courses distally between flexor carpi ulnaris and the flexor digitorum profundus muscles, supplying the flexor carpi ulnaris muscle and medial 1/2 of the flexor digitorum profundus muscle.

CUBITAL TUNNEL

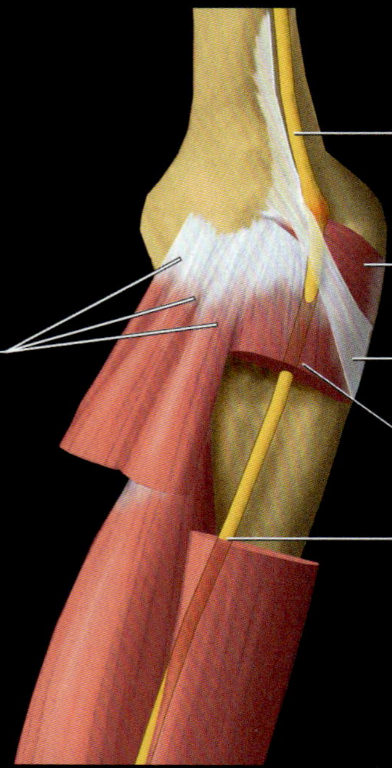

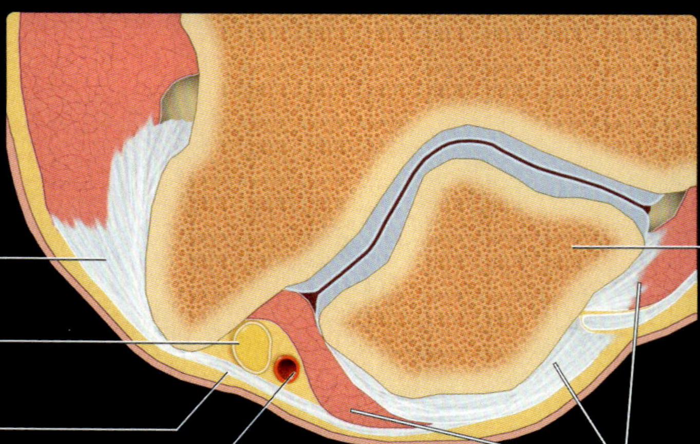

(Top) *Medial graphic shows the cubital tunnel. The anconeus epitrochlearis is an inconstant accessory muscle that lies superficial to the ulnar nerve and may compress the nerve against the medial epicondyle.* **(Bottom)** *Axial graphic shows the cubital tunnel. The ulnar nerve may be compressed within the cubital tunnel (cubital tunnel syndrome) by a mass, synovitis, posttraumatic osseous deformity, or aneurysm of the recurrent ulnar artery. The ulnar nerve may sublux out of the cubital tunnel aided by the adjacent medial head of the triceps. Recurrent subluxation, which can be found in ~ 15% of asymptomatic subjects, can lead to ulnar neuritis, which can produce similar symptoms to ulnar nerve compression.*

TRANSVERSE US: ARM

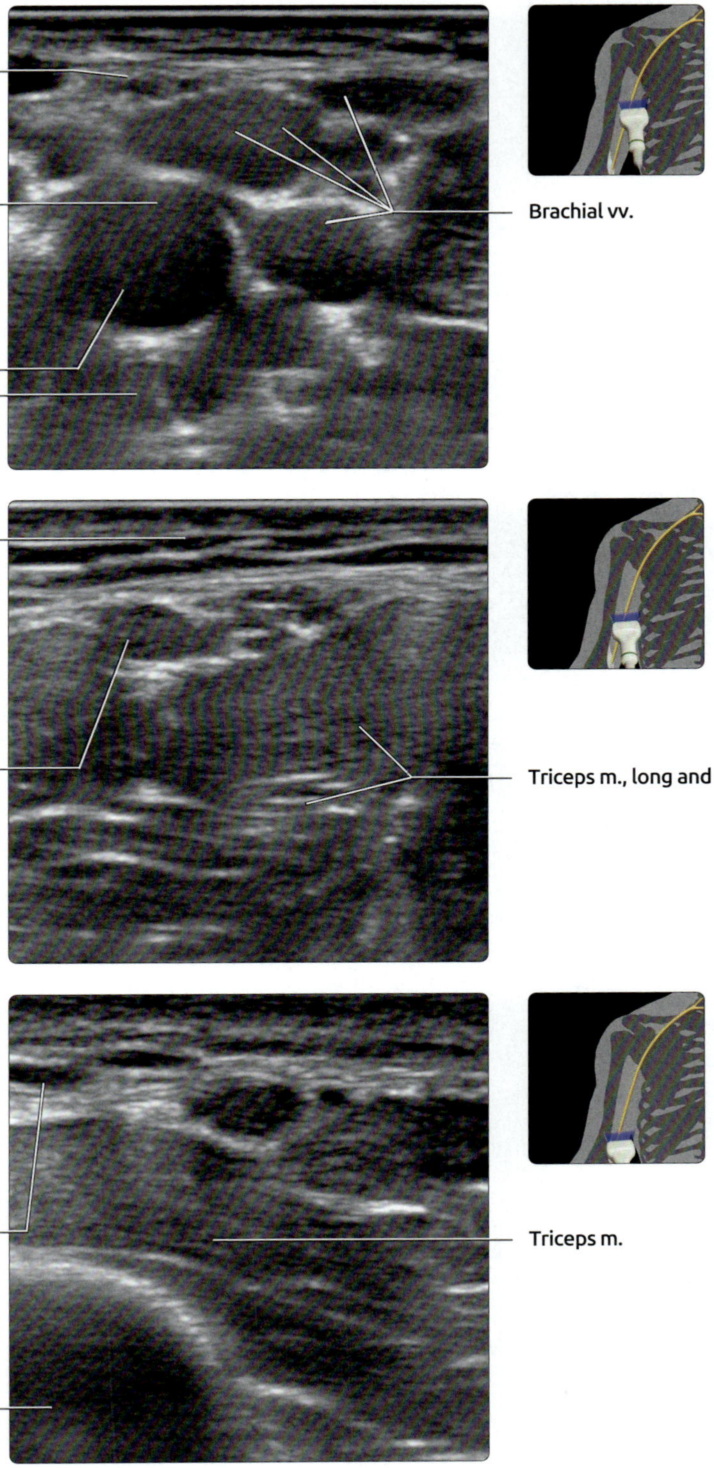

(Top) *Transverse US shows the ulnar nerve at the proximal arm. The ulnar nerve arises from the medial cord of the brachial plexus and lies just medial to the brachial vein.* **(Middle)** *Transverse US shows the ulnar nerve at the midarm. The ulnar nerve pierces the medial intermuscular septum in the midarm where it may be potentially compressed. The ulnar nerve gives off no branch in the arm.* **(Bottom)** *Transverse US shows the ulnar nerve in the distal arm. The ulnar nerve runs distally on the surface of the triceps muscle to reach the area between the olecranon and the medial epicondyle, i.e., the cubital tunnel.*

TRANSVERSE US: CUBITAL TUNNEL

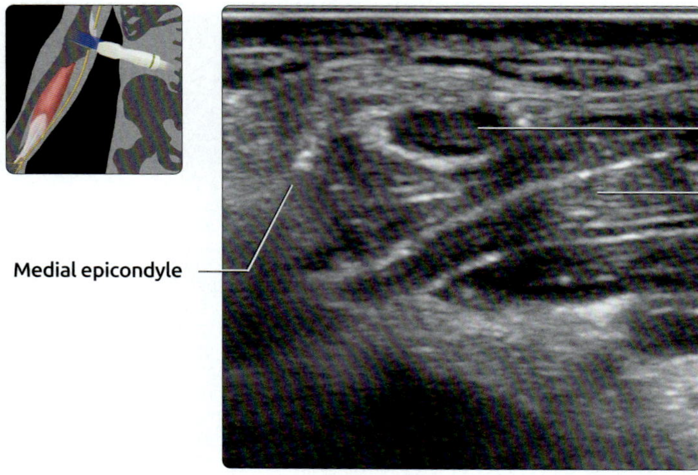

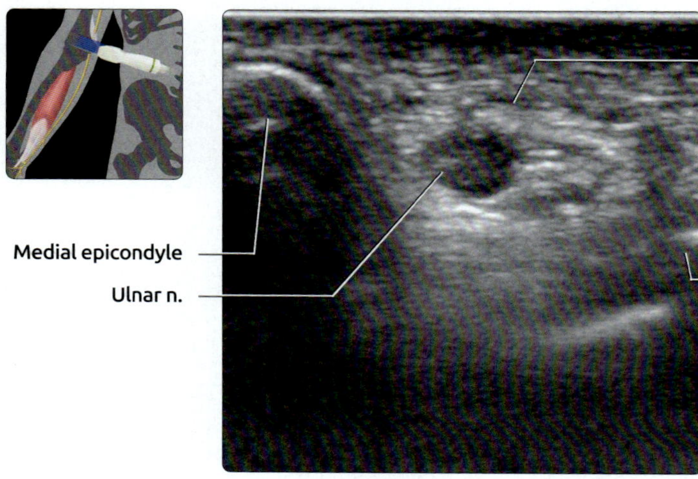

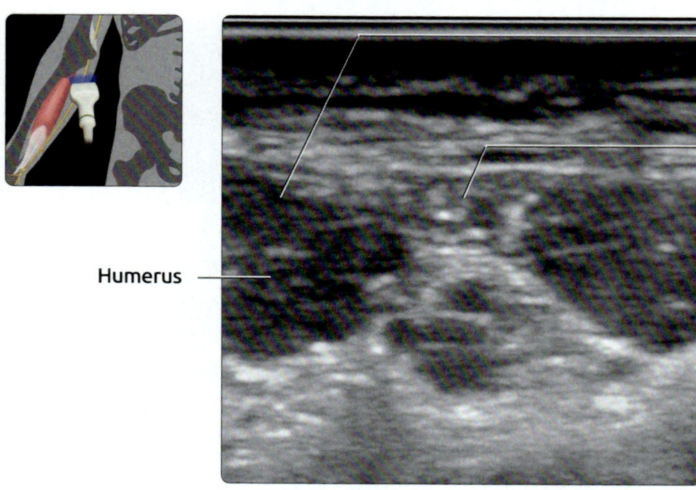

(Top) *Transverse US shows the ulnar nerve just proximal to the cubital tunnel. The cubital tunnel is the most common site of ulnar neuropathy. Neuropathy may take the form of idiopathic compression, bony impingement, synovial or muscular encroachment, recurrent subluxation of the ulnar nerve over the medial epicondyle, or direct injury.* **(Middle)** *Transverse US shows the ulnar nerve at the cubital tunnel. The cubital tunnel retinaculum (formed from an extension of investing fascia of the flexor carpi ulnaris and ligament of Osborne) forms the roof of the cubital tunnel. The retinaculum is 4 mm wide and difficult to depict on US. The floor and wall of cubital tunnel are formed by containing bones, elbow capsule, and the medial collateral ligament. The ulnar nerve normally enlarges slightly within the cubital tunnel.* **(Bottom)** *Transverse US shows the ulnar nerve at the distal end of the cubital tunnel where the ulnar nerve runs deep to the arcuate ligament traversing between the humeral and ulnar heads of the flexor carpi ulnaris muscle. Less frequently, the ulnar nerve may be compressed at this location. This area is sometimes referred to as the carpal tunnel proper.*

TRANSVERSE US: FOREARM

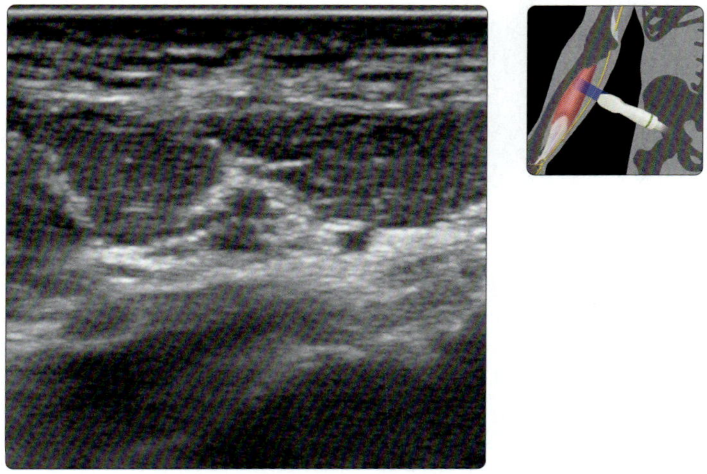

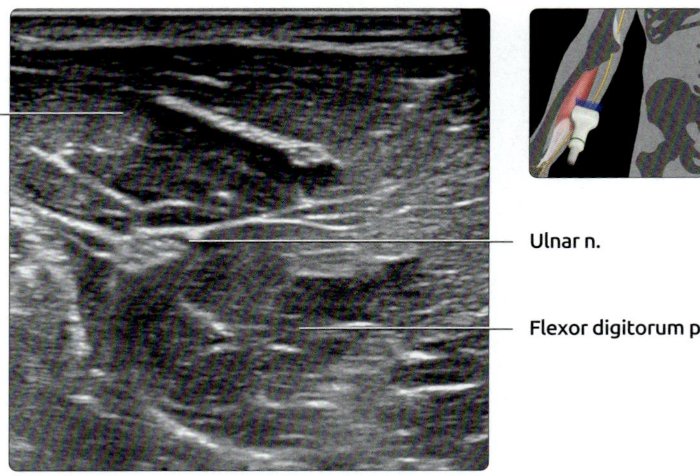

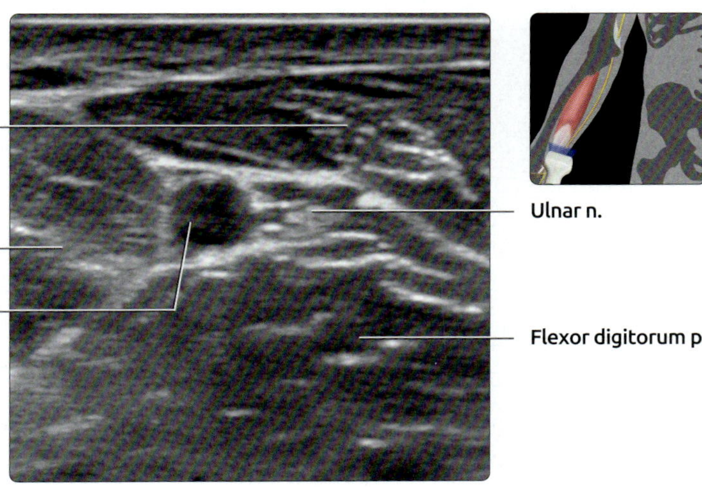

(Top) *Transverse US shows the ulnar nerve just beyond the cubital tunnel, where the ulnar nerve passes between the 2 heads (humeral and ulnar) of the flexor carpi ulnaris tendon. This is another potential site of compression.* **(Middle)** *Transverse US shows the ulnar nerve in the midforearm. The ulnar nerve pierces the intermuscular septum, dividing the flexor and extensor compartments of the forearm to descend in the flexor compartment between flexor carpi ulnaris and flexor digitorum profundus muscles.* **(Bottom)** *Transverse US shows the ulnar nerve in the distal forearm. The ulnar nerve is joined on its radial side by the ulnar artery in the distal forearm, and the 2 structures travel to the wrist, deep to the flexor carpi ulnaris muscle.*

TRANSVERSE AND LONGITUDINAL US: WRIST AND FOREARM

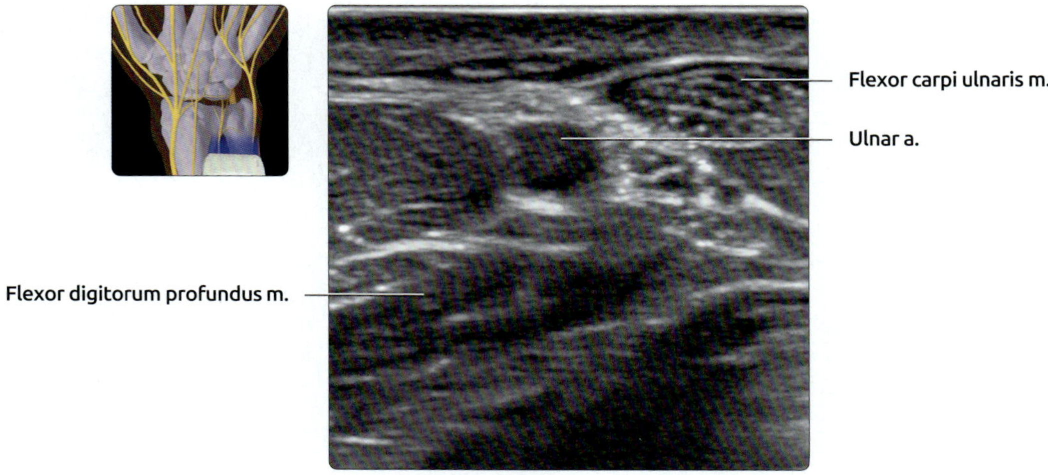

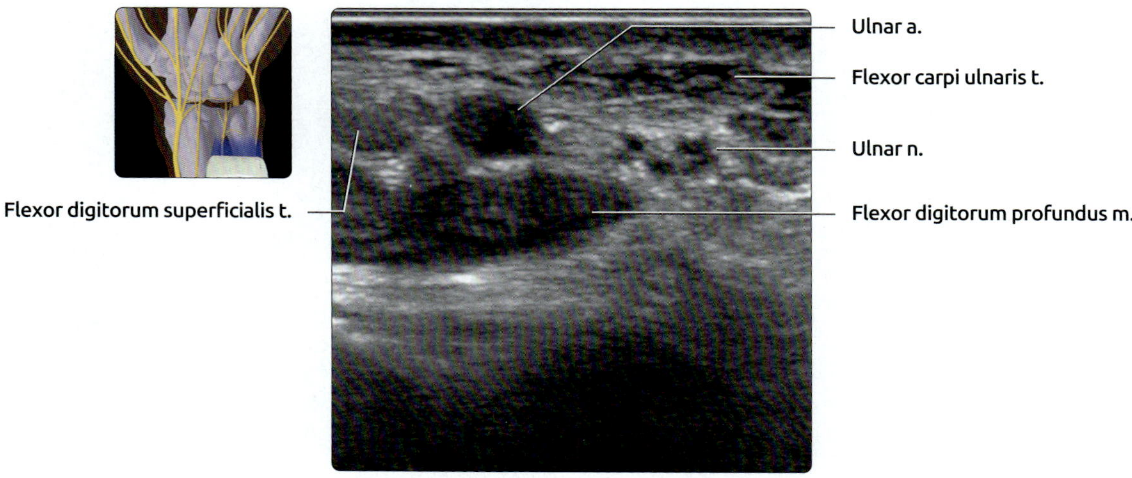

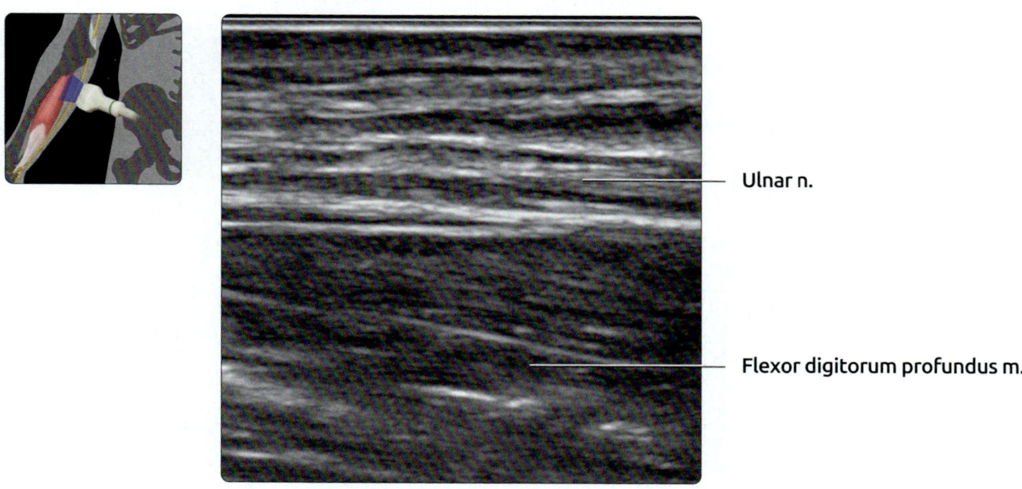

(Top) *Transverse US shows the ulnar nerve at the distal forearm. The ulnar nerve can be identified by first visualizing the ulnar artery and then identifying the nerve just ulnar to the artery.* **(Middle)** *Transverse US shows the ulnar nerve just proximal to the wrist. At this location, the ulnar nerve is prone to injury from trauma, such as dog bite, other penetrating injury, and nonaccidental self-injury.* **(Bottom)** *Longitudinal US shows the ulnar nerve at the distal 1/3 of the forearm. Note the fibrillar pattern of the nerve with a mixed hypoechoic and hyperechoic pattern due to hypoechoic nerve fascicles separated by hyperechoic perineurium.*

TRANSVERSE US: GUYON CANAL

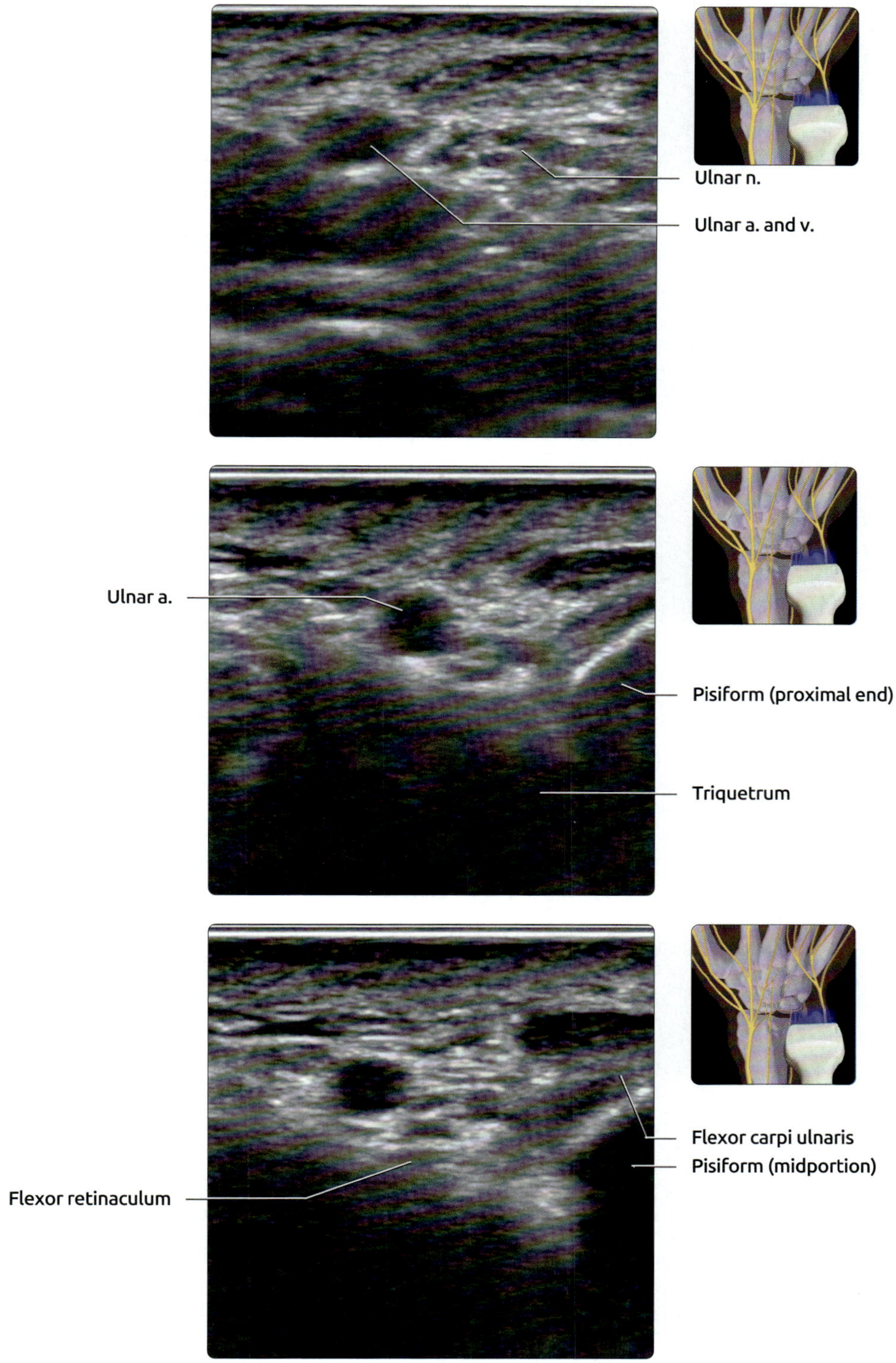

(Top) *Transverse US shows the ulnar nerve just proximal to the Guyon canal.* **(Middle)** *Transverse US shows the ulnar nerve at the proximal end of the Guyon canal. The Guyon canal is the 2nd most common site of ulnar neuropathy.* **(Bottom)** *Transverse US shows the ulnar nerve at the proximal end of the Guyon canal. Ganglia arising from the pisiform-triquetral articulation are the most common cause of ulnar nerve compression in Guyon canal.*

TRANSVERSE US: DISTAL TO GUYON CANAL

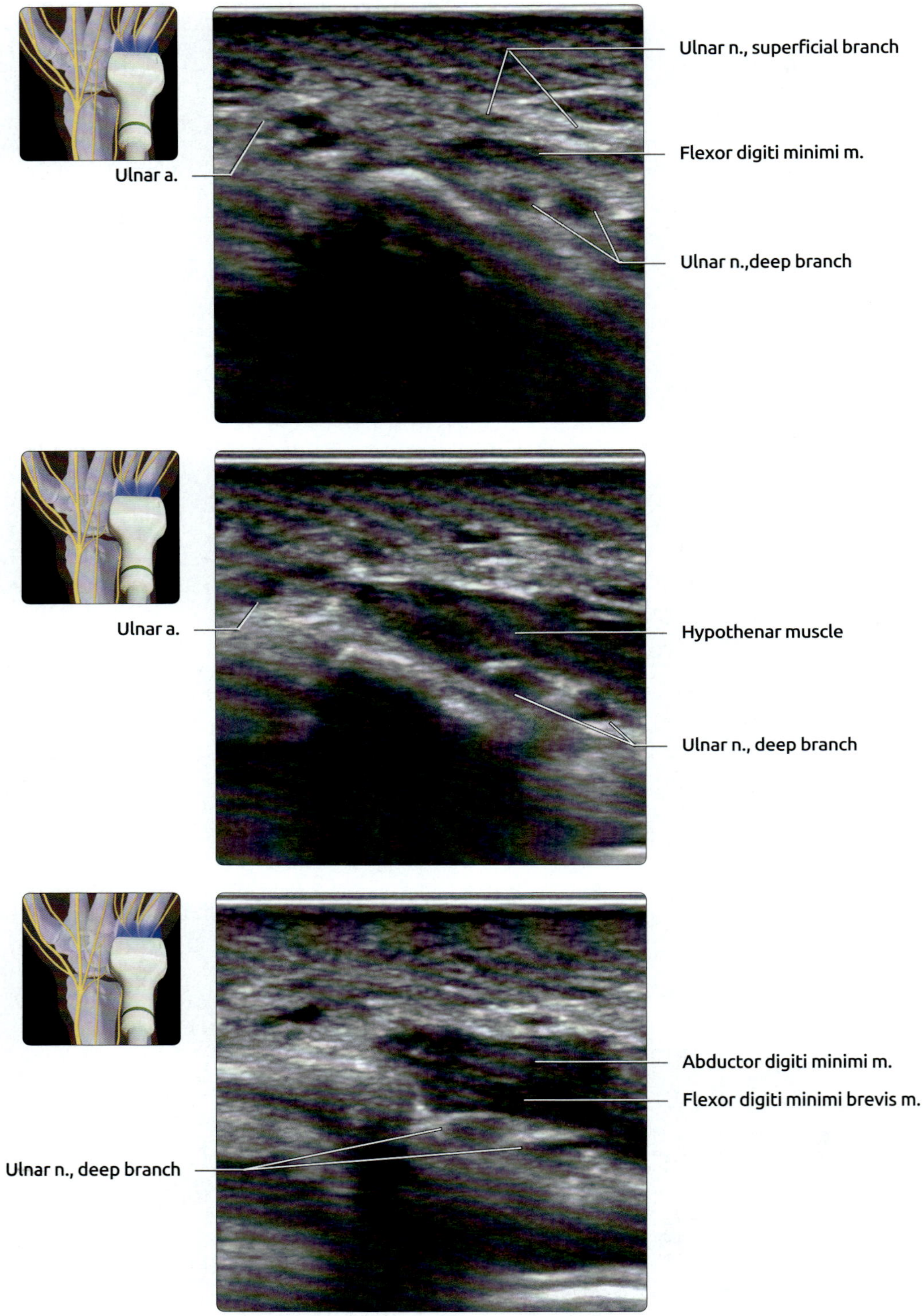

(Top) *Transverse US shows the ulnar nerve at the distal end of the Guyon canal where the ulnar nerve divides into superficial and deep branches. These pass close to the top of (superficial branch) and outer side of (deep branch) the hook of hamate.* **(Middle)** *Transverse US shows the ulnar nerve at the distal end of the hook of hamate level. Hook of hamate fractures (common in golfers and baseball players) and repeated blunt trauma ("handlebar" palsy) may injure the ulnar nerve as it passes alongside the hook of hamate. The ulnar artery can be injured at this location as well (hypothenar hammer syndrome).* **(Bottom)** *Transverse US shows the ulnar nerve at the base of the hypothenar muscles. The superficial branch of the ulnar nerve stays superficial to supply the skin, whereas the deep branch stays deep to supply the hypothenar muscles and the 3rd and 4th lumbrical muscles.*

PROXIMAL AND DISTAL ULNAR NERVE

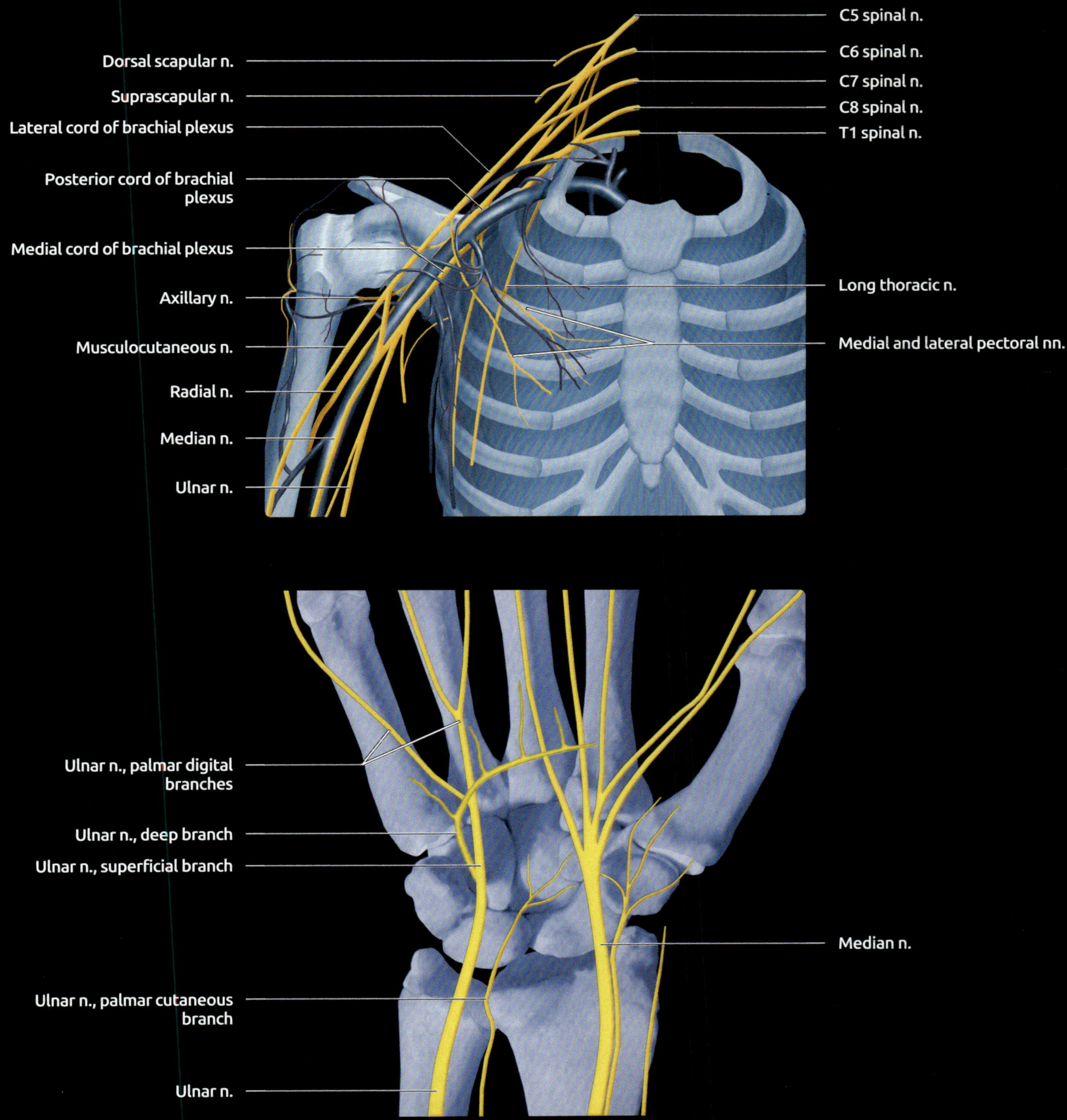

(Top) *Anterior graphic shows the ulnar nerve arising from the medial cord of the brachial plexus.* **(Bottom)** *Ulnar nerve supplies the ulnar wrist and hand. In addition to the branches shown, a dorsal branch of the ulnar nerve passes between the flexor carpi ulnaris and ulna ~ 5 cm proximal to the wrist to supply the dorsal ulnar wrist and the little finger. Deep and superficial branches supply motor and sensory branches to the hypothenar eminence.*

IMAGING ANATOMY

Overview

- Ball and socket joint with wide range of motion 2nd only to glenohumeral joint

Acetabulum

- Formed by pubis, ilium, ischium
- Oriented anterior, inferior, lateral
- Covers > 50% of femoral head
- **Anterior and posterior rims**: Osseous margins of acetabulum
- **Medial wall**: Quadrilateral plate ilium

Femoral Head

- 2/3 of sphere
- Covered by articular cartilage
  - Cartilage thickest superiorly
  - Cartilage thins at head/neck junction

Labrum

- Fibrocartilage lip deepens acetabular rim
- Joins transverse ligament at margins of acetabular notch
- Thickest posteriorly and superiorly
- Widest anteriorly and superiorly
- Vascular supply: Branches of obturator, superior, and inferior gluteal arteries
  - Mainly capsular surface
- Shape
  - Triangular in 66-94%
  - Variants: Rounded/blunt or absent
  - Absent labra: Constellation absent anterior labrum and small remnant superiorly
    - In 10-14% of asymptomatic individuals
- Function
  - Protect cartilage: Distributes forces by maintaining synovial fluid layer between articular surfaces
  - Prevents lateral translation femoral head
- Labral tear
  - Frequently has associated paralabral ganglion/cyst
  - Easier to detect in presence of joint effusion/arthrographic contrast
  - More common anterosuperiorly and anteriorly

Joint Capsule

- 2 layers: Internal synovial and external fibrous (not usually separable on imaging)
  - External layer forms capsular ligaments
- Attachments
  - Acetabulum
    - Base of labrum anteriorly and posteriorly
    - Several millimeters above labrum superiorly
  - Femur
    - Anterior: Intertrochanteric line
    - Posterior: Proximal to intertrochanteric crest
    - Anterior attachment more lateral than posterior attachment
  - **Perilabral recess**: Between labrum and capsule
    - Smaller anteriorly and posteriorly
    - Larger superiorly

Vascular Supply

- Branches of medial and lateral circumflex femoral, deep division of superior gluteal, inferior gluteal arteries, artery of ligamentum teres (branch of obturator artery)

Innervation

- Branches from nerve to rectus femoris, nerve to quadratus femoris, anterior division of obturator nerve, accessory obturator nerve, superior gluteal nerve

Major Adjacent Structures

- **Greater trochanter**
  - Insertion
    - Gluteus minimus and medius muscles
    - Gluteus minimus to anterior facet
    - Gluteus medius to lateral and posterosuperior facets
  - Separated from overlying iliotibial tract by thin trochanteric bursa (normally barely perceptible on imaging)
- **Lesser trochanter**
  - Insertion
    - Iliopsoas muscle
- **Anterior inferior iliac spine**
  - Origin
    - Rectus femoris muscle
- **Iliopsoas muscle**
  - Run inferomedially across anterior surface of hip joint
- **Femoral vessels**
  - Separated from hip joint by iliopsoas muscle

Neonatal Hip

- High success rate in treatment of developmental dysplasia demands early imaging diagnosis (with US)
- Established measurements of bony and cartilaginous components for comparison with normal reference
- Main measurements reflect depth of acetabulum and acetabular coverage of femoral head

ANATOMY IMAGING ISSUES

Imaging Recommendations

- Radiographs for assessing osseous alignment
- US for joint effusion and surrounding soft tissue
  - Also used for real-time guided procedures: Hip joint or abscess aspiration/biopsy; intraarticular contrast injection
  - Ideal in neonates for suspected developmental hip dysplasia
- MR arthrography preferred to assess intraarticular structures

Imaging Sweet Spots

- Small joint effusions are best detected on US by hypoechoic thickening/layer over anterior surface of femoral neck
  - Best site for aspiration/injection
- Paralabral ganglion cysts should alert to presence of labral tear

Imaging Pitfalls

- Synovial proliferation usually shows no evidence of vascularity on color Doppler scanning and is therefore difficult to distinguish from effusion

AXIAL AND SAGITTAL HIP SECTIONS

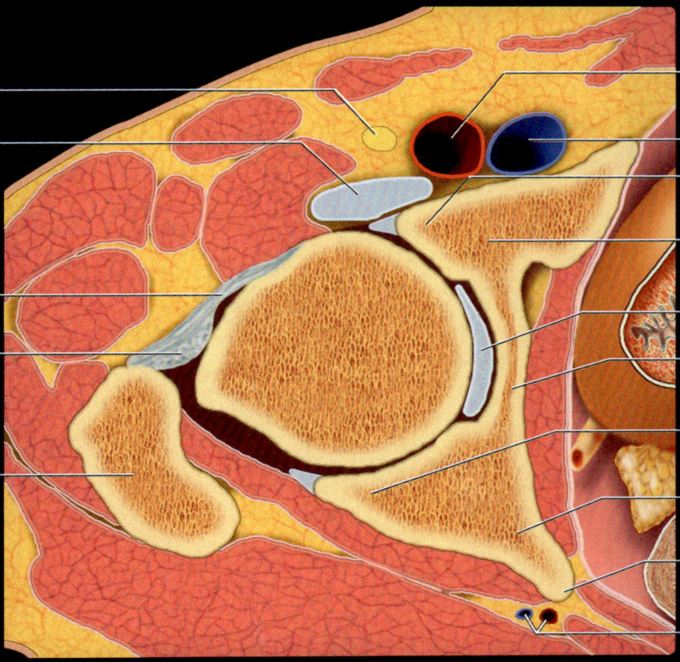

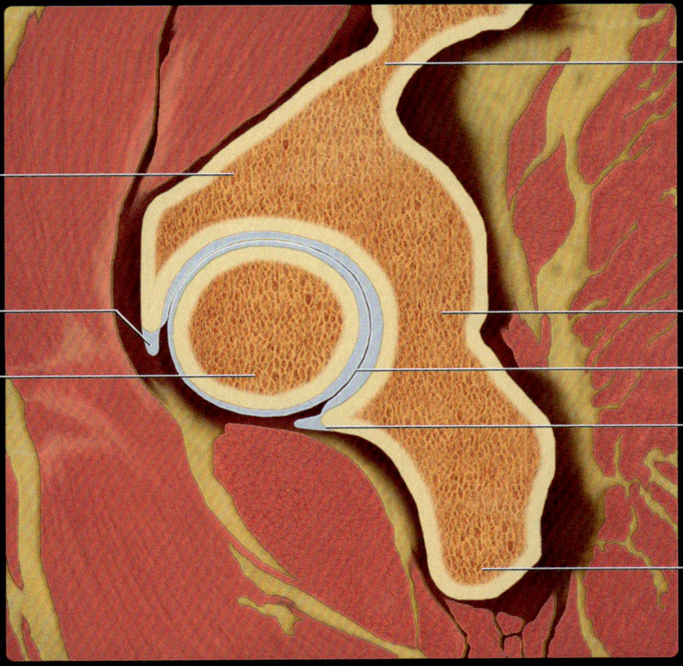

(Top) *Axial graphic shows the hip joint. The iliopsoas bursa is closely approximated to the joint anteriorly. The ligamentum teres is seen in profile as a flattened structure. The 2 layers of the external capsule, the more superficial longitudinal fibers and deep circular fibers, are visible.* **(Bottom)** *Sagittal representation of the hip joint nicely demonstrates the inverted Y of the acetabulum. The ilium is the stem, and the limbs are the anterior and posterior acetabular columns.*

CORONAL HIP SECTION AND LIGAMENTS

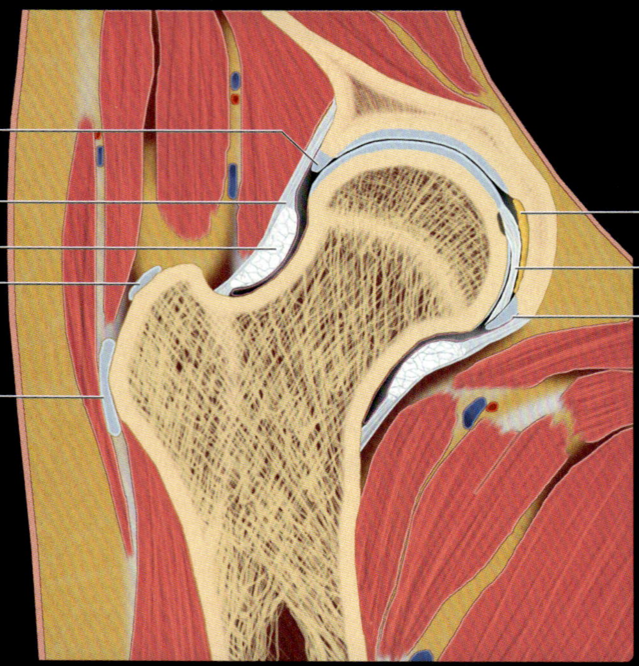

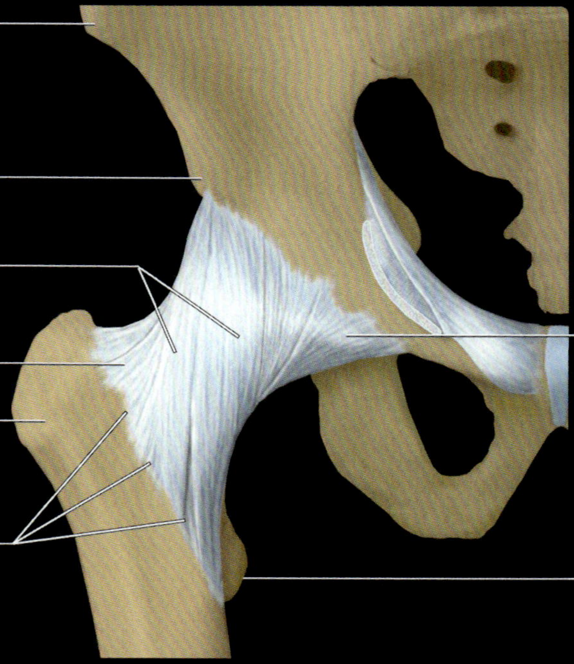

(Top) *Coronal graphic demonstrates many of the important structures of the hip joint. The longitudinally oriented fibers of the ligaments of the external joint capsule and the deeper, circularly oriented fibers of the zona orbicularis are visible. Note the long axis of the ligamentum teres and its insertion onto the transverse ligament. The pulvinar fills the acetabular fossa.* **(Bottom)** *Anterior ligaments are shown. The longitudinal spiral of the iliofemoral and pubofemoral ligaments is well seen with attachments to the anterior inferior iliac spine and pubic aspect of the obturator foramen, respectively. The ischiofemoral ligament wraps over the superior aspect of the femoral neck to attach to the anterior femoral neck.*

HIP OVERVIEW

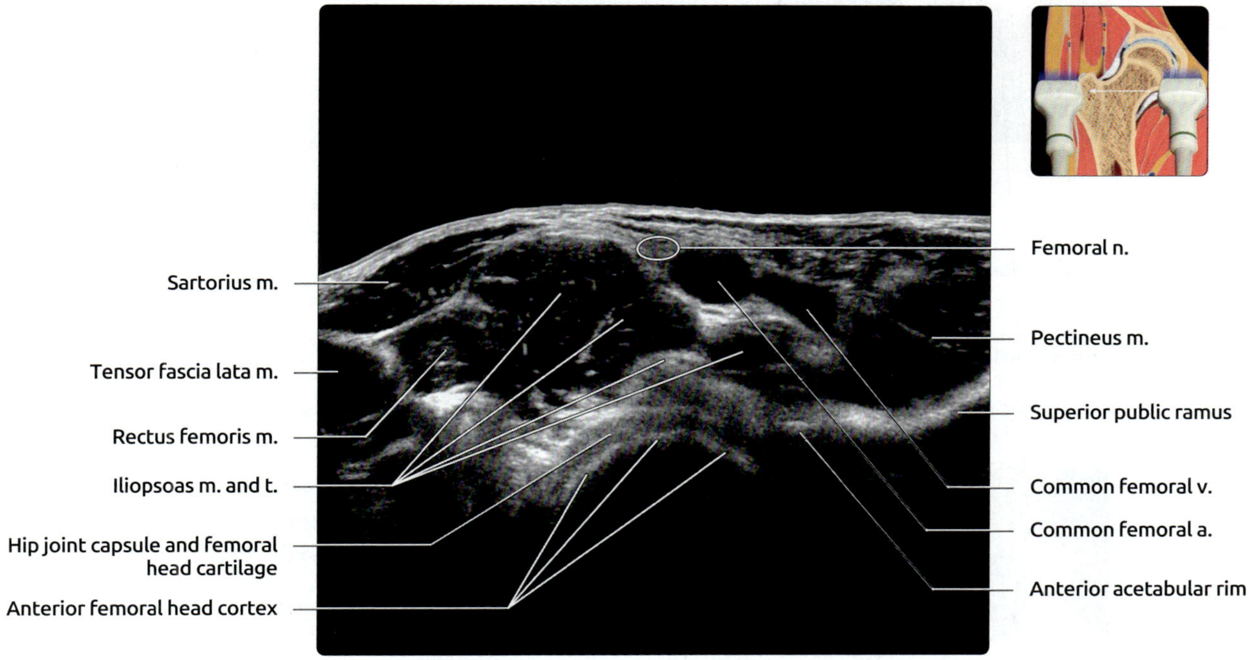

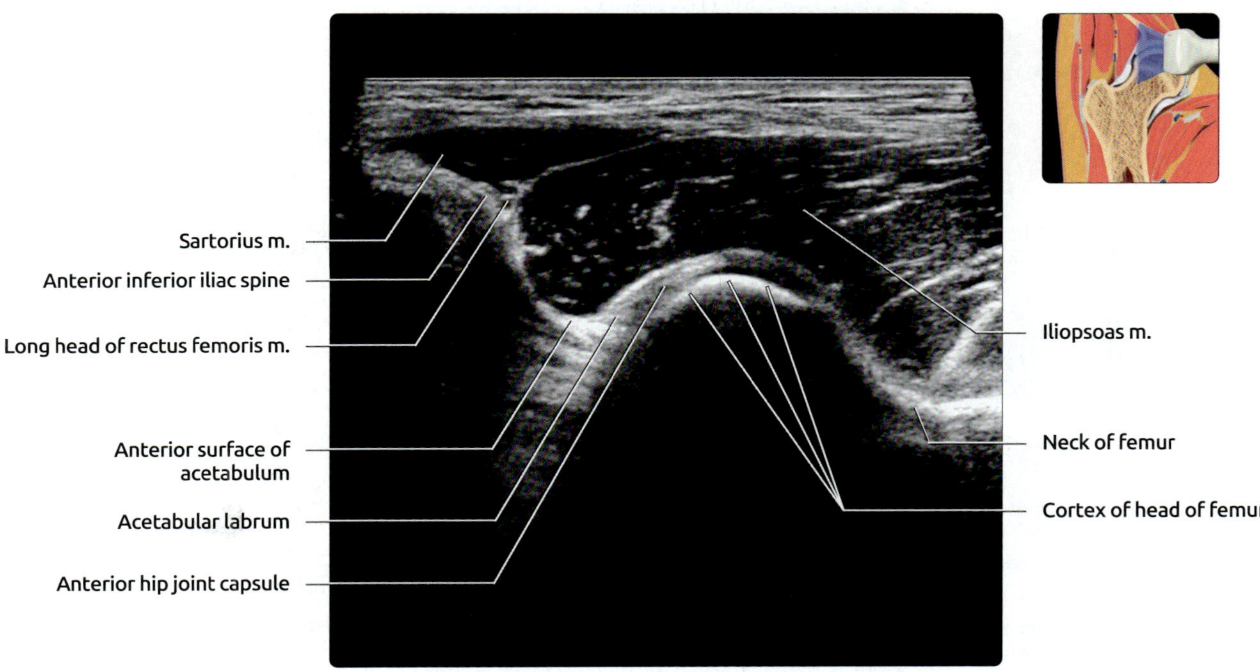

(Top) *Transverse panoramic US shows the femoral head. Medially and superiorly, the femoral head is covered by bony acetabulum and not visible with US. Laterally, the hip joint lies deep to the iliopsoas muscle and femoral triangle. Further laterally, the hip joint is associated with the gluteal muscles.* **(Bottom)** *Sagittal oblique US shows hip joint and anterior inferior iliac spine. The acetabular labrum, which is thickest and widest anteriorly, can be seen between the acetabular rim and joint capsule. Small joint effusions are 1st detected, not over the femoral head, but over the femoral neck as a hypoechoic layer. The femoral neck's relatively safe access makes it a suitable site for guided aspiration/injection.*

ACETABULUM

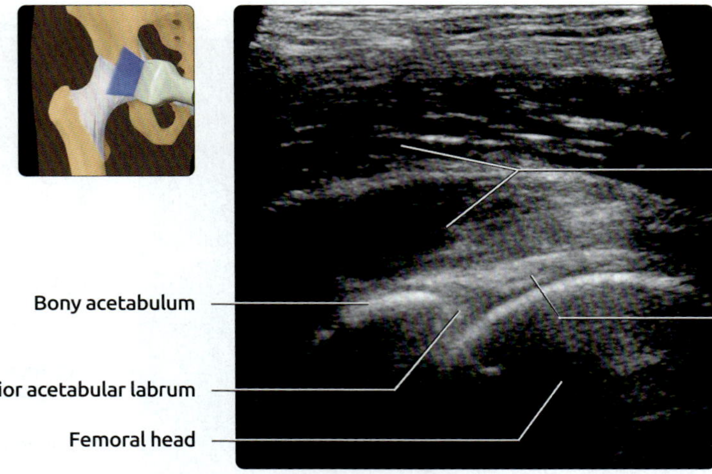

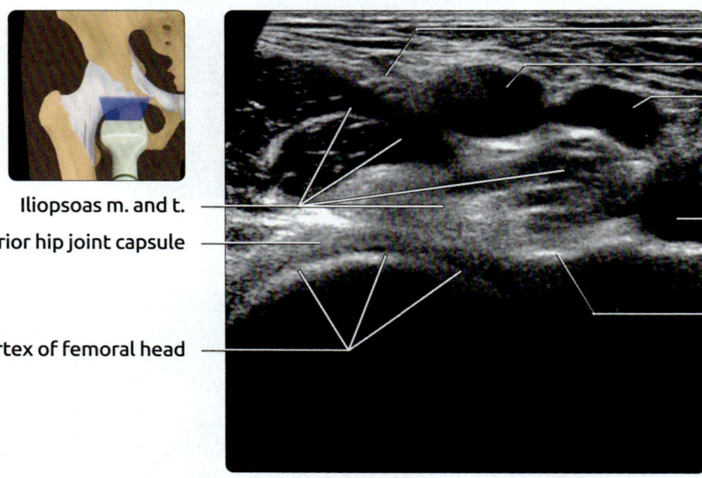

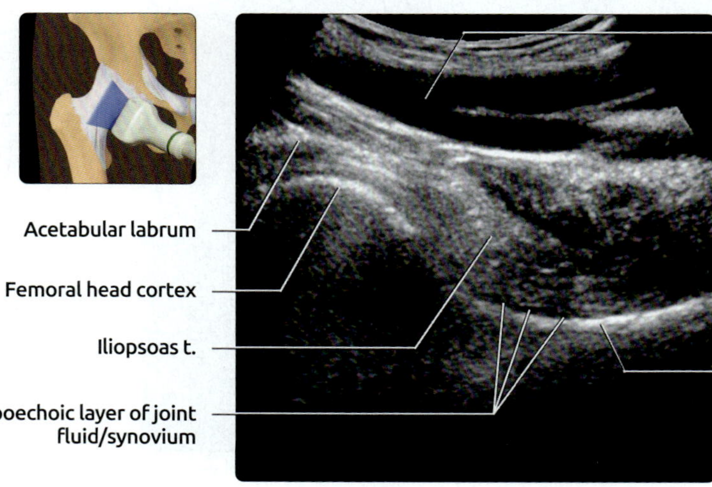

(Top) *Sagittal oblique US along the axis of the femoral neck shows the triangular cross section of anterior acetabular labrum. Joint effusion/contrast tends to allow better delineation of its contour and lesions.* **(Middle)** *Transverse US is shown at the level of the hip joint. Anterior femoral cortex/cartilage, hip joint capsule, and acetabular rim are well visualized. The contents of the femoral triangle are separated from the hip joint by the iliopsoas muscle and tendon, which explains why paralabral cysts may tract along the iliopsoas muscle.* **(Bottom)** *Effusions are 1st detected, not over the femoral head, but over the femoral neck as an increase in the normally minimal hypoechoic layer of joint fluid and synovium. This is also the safest site for joint injections (intraarticular contrast or medication).*

ILIOPSOAS INSERTION

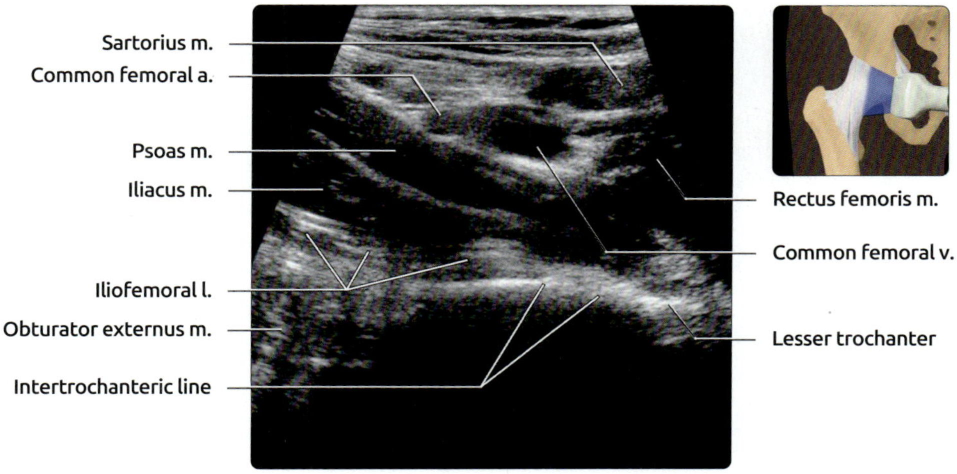

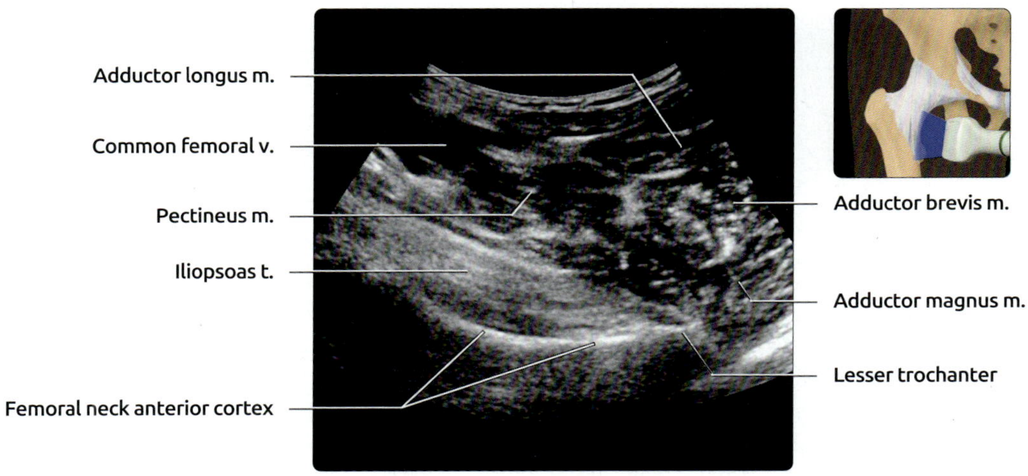

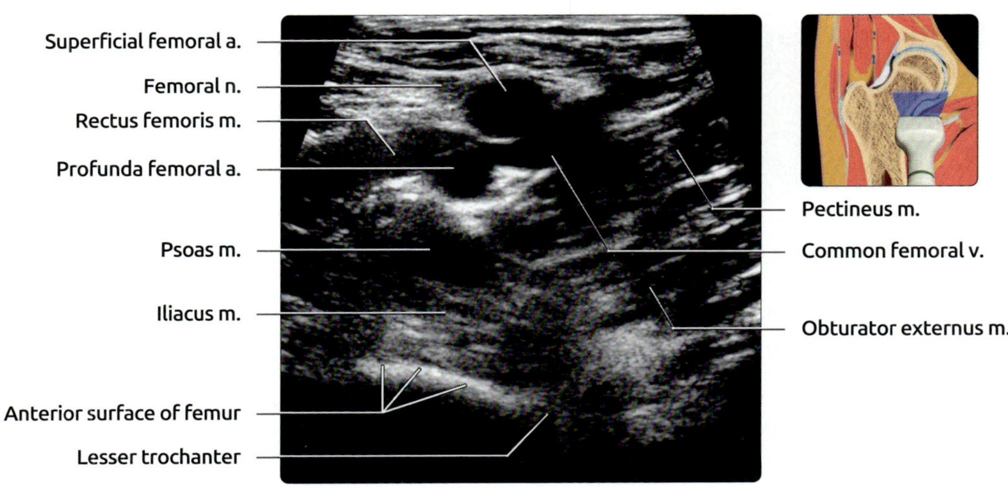

(Top) *Sagittal oblique US along the course of the iliopsoas muscle is shown. Resolution is reduced due to depth of structures, which may be partially improved by externally rotating the hip to bring deep structures closer to the surface.* **(Middle)** *Sagittal oblique US is shown slightly more distal. Insertion of iliopsoas tendon is at the lesser trochanter, which may be avulsed in injury. Notice the close relationship between these structures and the adductor compartment.* **(Bottom)** *Transverse US shows distal iliopsoas tendon just before insertion. A functional division is present between the hip flexor (iliopsoas) and hip adductors. This is also the apex level of the femoral triangle where superficial and profunda vessels bifurcate.*

ANTERIOR INFERIOR ILIAC SPINE

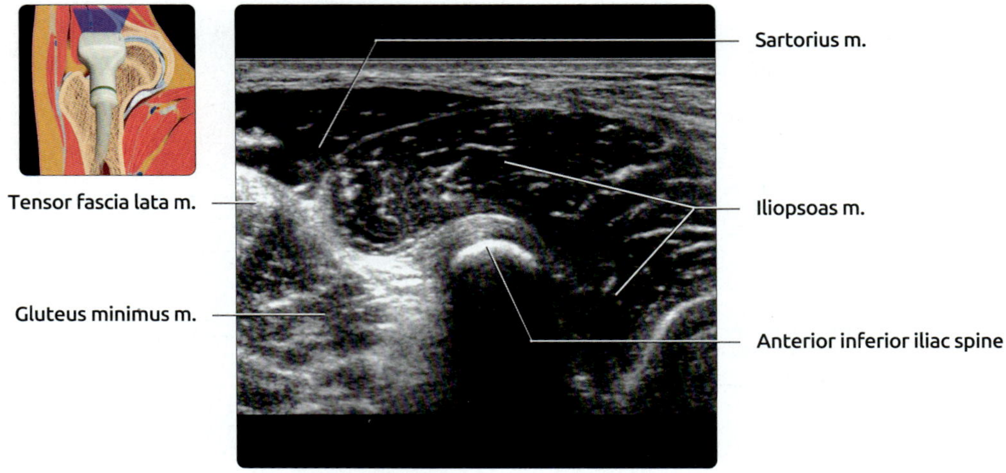

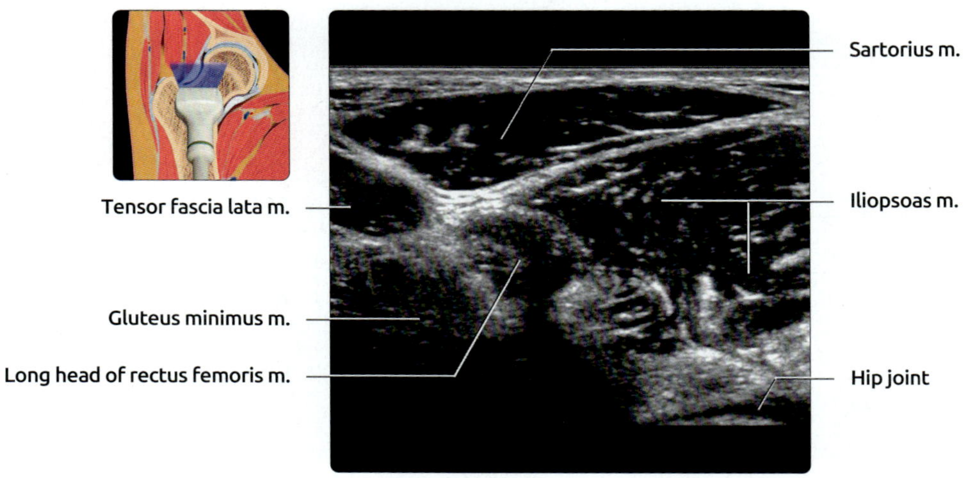

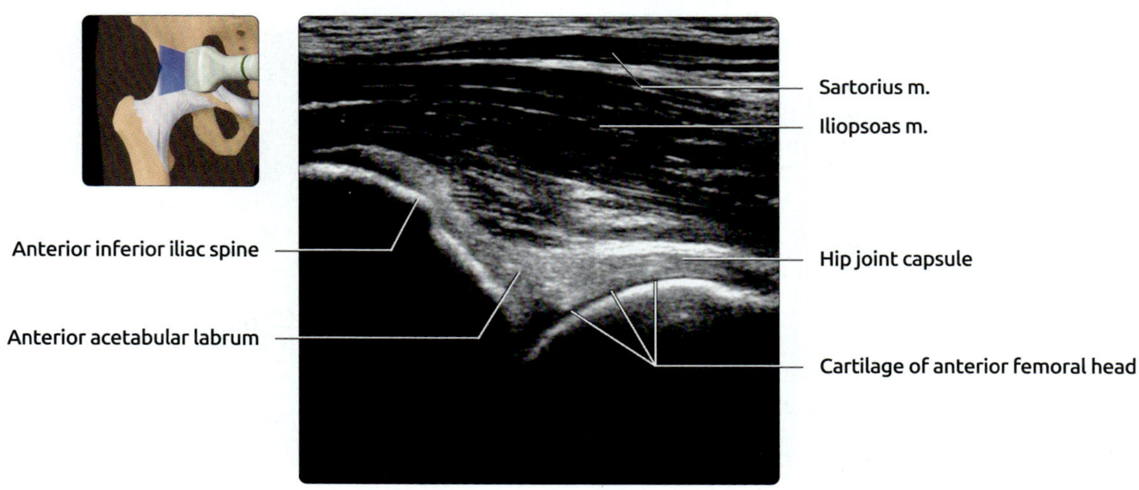

(Top) *Transverse US through the anterior inferior iliac spine is shown. The rectus femoris muscle, together with the sartorius muscle (which originates from anterior superior iliac spine) and the iliopsoas muscle (which originates from iliac fossa and lumbar transverse processes), cross the hip joint and form the hip flexors.* **(Middle)** *Transverse US more inferior shows the long head of the rectus femoris muscle running deep to the sartorius muscle and lateral to the iliopsoas muscle. Note that the rectus muscle is the only quadriceps muscle that crosses the hip joint and thus acts as both hip flexor and knee extensor.* **(Bottom)** *Oblique US along the sartorius muscle at the hip joint level shows it to be roughly parallel with the iliopsoas muscle. The rectus femoris lies in a more lateral plane.*

GREATER TROCHANTER

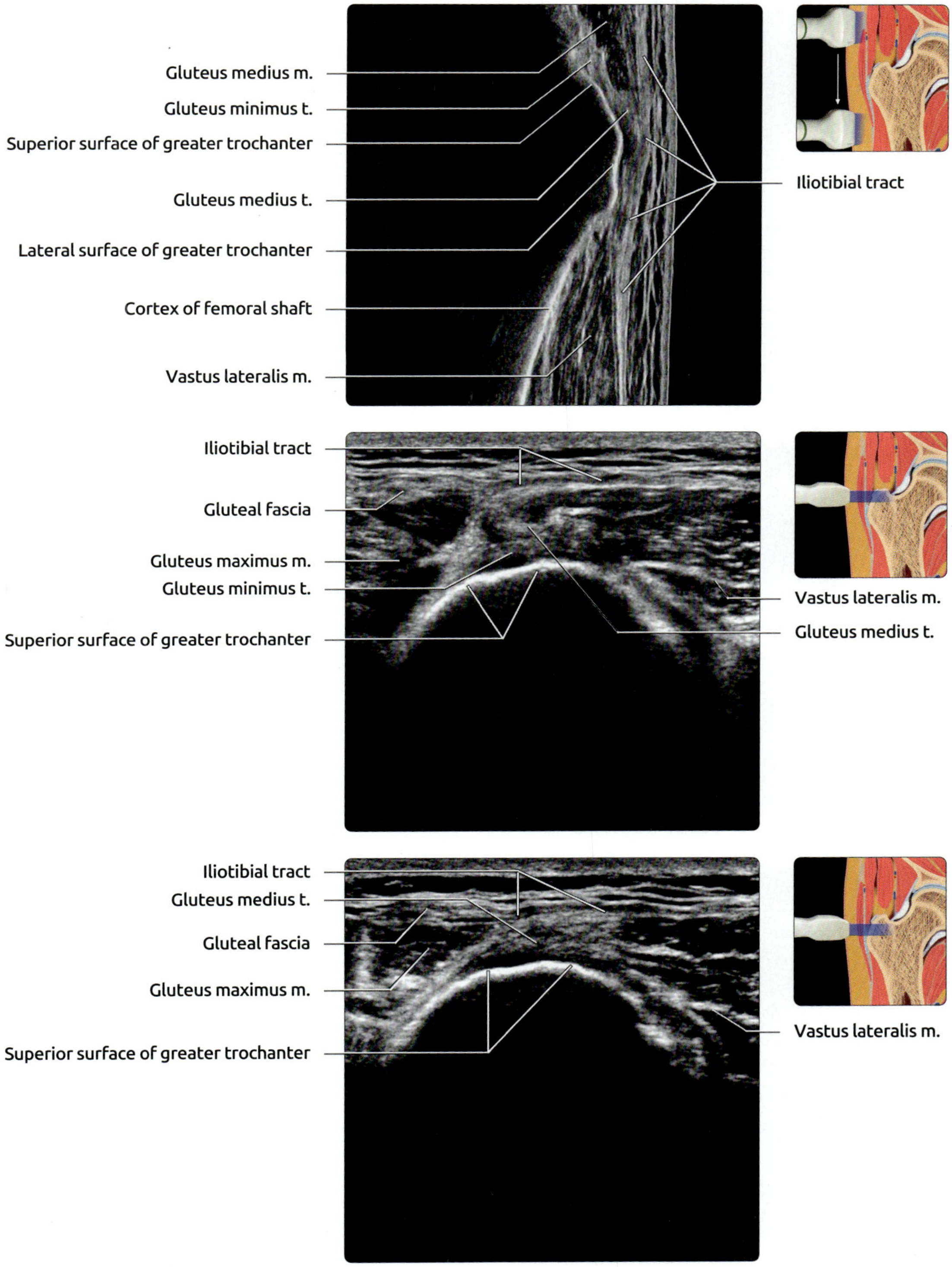

(Top) *Coronal panoramic US shows the greater trochanter. The gluteus minimus and medius tendons insert on the anterior, lateral, and superoposterior facets of the greater tuberosity. The iliotibial tract originates from the anterior aspect of the iliac crest (as the tensor fascia lata muscle) and runs inferiorly, passing superficial to the lateral surface of the greater trochanter.* **(Middle)** *Transverse US through the anterior facet of the greater trochanter shows insertion of the gluteus minimus tendon. The gluteus medius tendon and iliotibial tract pass superficial to this.* **(Bottom)** *Transverse US of the lateral facet of the greater trochanter shows the insertion of the gluteus medius tendon. The iliotibial tract is closest to the trochanter at this point and may be irritated, resulting in the formation of bursitis (trochanteric bursitis).*

PEDIATRIC HIP

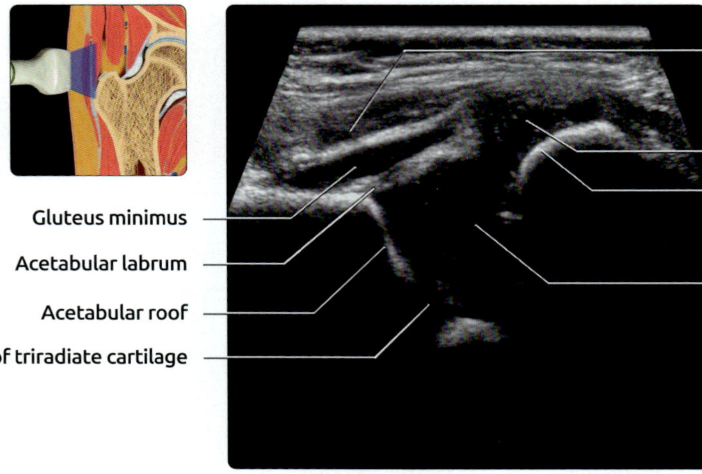

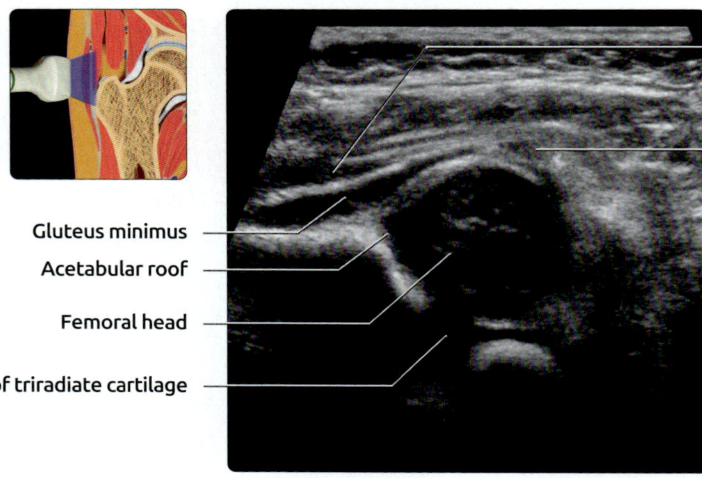

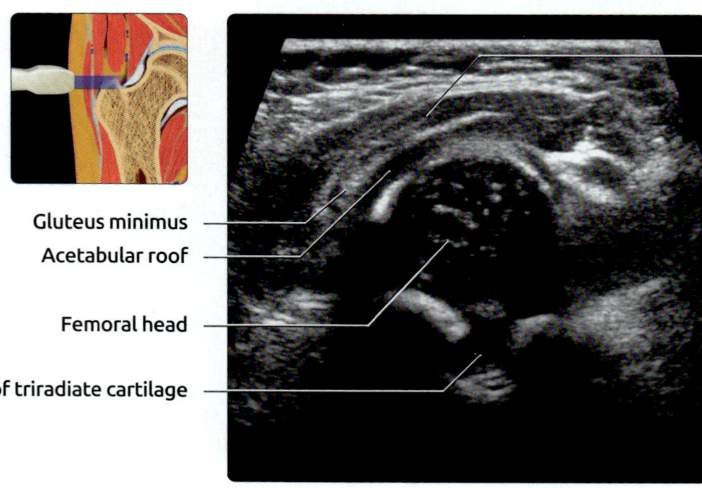

(Top) *Coronal US shows the hip of a 1-month-old neonate with the hip in extension. The femoral head and the greater trochanter are purely cartilaginous. This allows US transmission and thus visualization of the acetabular fossa.* **(Middle)** *Coronal US shows the hip of a 1-month-old neonate with the hip in flexion, which produces an axial section of the femoral head. The femoral head should appear round in this section. Both the femoral head and the greater trochanter are purely cartilaginous, allowing US transmission and thus better visualization of the depth of the acetabular fossa.* **(Bottom)** *Transverse US of the femoral head with hip in extension also shows the central position of a normal femoral head on the triradiate cartilage (golf-ball-on-tee appearance).*

PEDIATRIC HIP

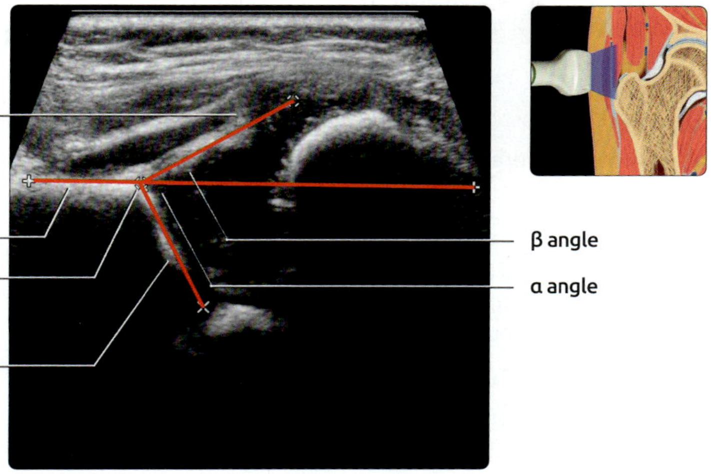

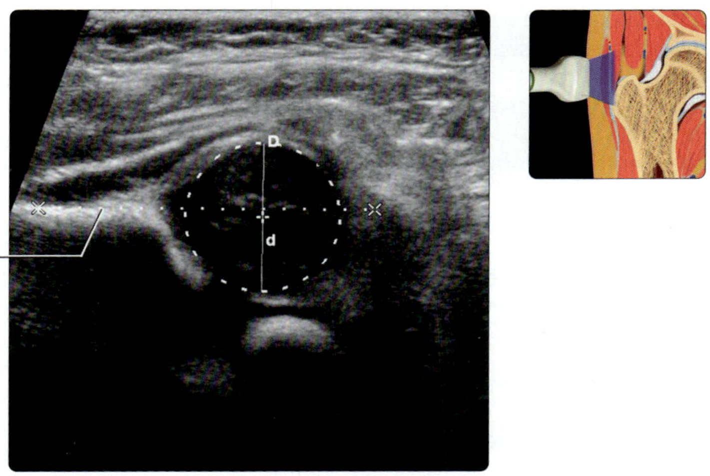

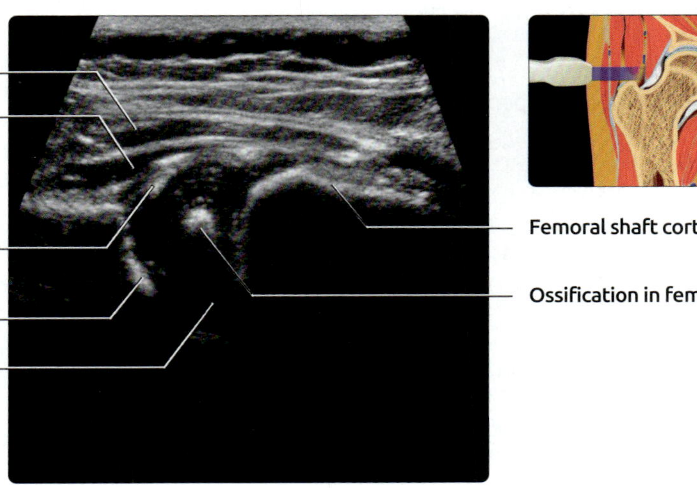

(Top) *Three lines are drawn on the standard coronal image of the pediatric hip: (1) Along the straight contour of the iliac cortex, (2) from promontory along the acetabular roof to describe the α angle, and (3) from promontory to the tip of the labrum for the β angle. Normal ranges are α angle > 60° and β angle < 55°. Measurements outside of these ranges are suggestive of developmental hip dysplasia.* **(Middle)** *The proportion of the femoral head contained within the acetabulum is calculated by dividing the depth (d) of femoral head below line a by the diameter of the femoral head (D). In normal subjects, > 50% of the femoral head should be below line a.* **(Bottom)** *Coronal US shows the hip of a 4-month-old child with the hip in extension. The center of the femoral head has begun to ossify, showing up as an echogenic focus with posterior acoustic shadowing.*

TERMINOLOGY

Abbreviations

- Function (F), insertion (I), origin (O)

IMAGING ANATOMY

Compartment Anatomy

- **Compartmental anatomy is different from functional groupings**
- Thigh divided into anterior, medial, and posterior compartments
  - **Anterior compartment**: Iliotibial tract, tensor fascia lata muscle, quadriceps muscles, sartorius muscle
  - **Medial compartment**: Gracilis muscle, adductor muscles
  - **Posterior compartment**: Hamstring muscles, short head of biceps femoris muscle
- Additional muscles at junction pelvis/thigh
  - Pectineus, iliopsoas, obturator externus, lateral femoral muscles
- Extensions from fascia lata divide compartments
  - Medial intermuscular septum separates anterior from medial compartment
  - Lateral intermuscular septum separates anterior from lateral compartment
  - Thin fascia separates medial from posterior compartment
- Compartment anatomy critical to tumor staging and biopsy planning
  - When performing percutaneous biopsy, try not to cross between affected and nonaffected compartment

Medial Compartment Muscles

- Anterior division: Adductor brevis and longus, gracilis muscles
- Posterior division: Adductor magnus muscle
- Common function: Hip adduction; assist hip flexion, internal rotation (except obturator externus muscle)
- **Adductor brevis**
  - O: Inferior pubic ramus
  - I: Inferior 2/3 pectineal line, superior 1/2 medial lip of linea aspera
- **Adductor longus**
  - O: Pubic body inferior to crest
  - I: Medial lip of linea aspera
- **Adductor magnus**: 2 separate muscles with different innervations and functions
  - Adductor portion
    - O: Ischiopubic ramus
    - I: Gluteal tuberosity, medial lip linea aspera, medial supracondylar line
  - Adductor hiatus between 2 divisions of muscle in distal thigh
- **Gracilis**
  - O: Inferior pubic ramus, symphysis pubis
  - I: Medial proximal tibia (pes anserine)
  - F: Also assists knee flexion
- **Obturator externus**
  - O: External margins of obturator foramen membrane
  - I: Piriformis fossa
  - F: Hip external rotation only
- **Pectineus**
  - O: Superior pubic ramus, pecten
  - I: Pectineal line femur

Anterior Compartment Muscles

- Common function: Knee extension (except sartorius)
- **Quadriceps femoris**: Rectus femoris, vastus lateralis, vastus medialis, vastus intermedius muscles
  - Common tendon of insertion onto superior, lateral, medial patella
- **Rectus femoris**
  - O: Straight head = anterior inferior iliac spine; reflected head = groove above acetabulum
  - I: Superior patella, tibial tuberosity
  - F: Also hip flexion
  - Crosses 2 joints
- **Vastus lateralis**
  - O: Superior intertrochanteric line femur, anterior and inferior greater trochanter, gluteal tuberosity, lateral lip of linea aspera, lateral intermuscular septum
  - I: Lateral tibial condyle (lateral patellar retinaculum), superolateral patella (quadriceps tendon)
  - Largest quadriceps muscle
- **Vastus medialis**
  - O: Entire medial lip of linea aspera, inferior intertrochanteric line, medial intermuscular septum
  - I: Tendon of rectus femoris muscle, superomedial patella (quadriceps tendon), medial condyle tibia (medial patellar retinaculum)
- **Vastus intermedius**
  - O: Anterior and lateral femoral shaft, inferior lateral lip of linea aspera, lateral intermuscular septum
  - I: Blends along deep aspect of rectus femoris, vastus medialis, vastus lateralis muscles
- **Sartorius**: Tailor's muscle
  - O: Anterior superior iliac spine, notch below
  - I: Proximal medial tibia (pes anserine)
  - F: Hip flexion, abduction, external rotation; knee flexion
  - Crosses 2 joints
  - Longest muscle in body
  - Separate fascial covering
- **Tensor Fascia Lata Muscle**
  - **Iliotibial tract/band**: Lateral thickening fascia lata
    - O: Tubercle iliac crest
    - I: Anterolateral proximal tibia (Gerdy tubercle)

Posterior Compartment Muscles

- Common functions: Hip extension, knee flexion
- **Hamstrings**: Long head biceps femoris, semimembranous, semitendinosus, ischiocondylar portion of adductor magnus muscles
  - Does not include short head biceps femoris muscle
- **Pes anserine**
  - Common aponeurosis for insertion gracilis, semitendinosus, sartorius tendons
  - Pes anserine bursa between tendons and tibia
- **Biceps femoris**
  - Long head O: Ischial tuberosity (inferior, medial)
    - Common tendon with semitendinosus muscle
  - Short head O: Lateral lip of linea aspera femur, lateral supracondylar line, lateral intermuscular septum

    - Not part of hamstring muscles
  - I: Fibular head, lateral condyle tibia
  - F: Also external rotation flexed knee
- **Semimembranosus**
  - O: Ischial tuberosity (superior, lateral)
  - I: Posterior medial condyle tibia, popliteal fascia
    - Some fibers extend to form oblique popliteal ligament
  - F: Also internal rotation flexed knee
  - Membranous in upper thigh
- **Semitendinosus**
  - O: Ischial tuberosity (inferior, medial)
    - Common tendon of long head biceps femoris muscle
  - I: Medial proximal tibia (pes anserine)
  - F: Internal rotation flexed knee
  - Entirely tendinous in distal thigh
- **Ischiocondylar portion of adductor magnus**
  - O: Ischial tuberosity
  - I: Adductor tubercle
  - Medial-most aspect of adductor magnus muscle

Hip Flexors

- **Iliopsoas** I: Lesser trochanter
  - Iliacus
    - O: Iliac crest, iliac fossa, sacral ala, sacroiliac joint capsule
  - Psoas major
    - O: Lateral vertebral body and intervertebral discs T12-L5, all lumbar transverse processes
  - Psoas minor
    - O: Lateral vertebral body T12, L1, and T12-L1 intervertebral disc
- Sartorius and pectineus

Femoral Triangle

- Superior border: Inguinal ligament
- Lateral border: Sartorius muscle
- Medial border: Adductor longus muscle
- Floor: Adductor longus and pectineus muscles (medial), iliopsoas muscle (lateral)
- Apex: Crossing of adductor longus and sartorius muscles
- Roof: Fascia lata
- Contents: Femoral nerve and branches, femoral vessels, lymph node (node of Cloquet), femoral sheath
  - Structures medial to lateral at entrance
    - Mnemonic is VAN: Vein, artery, nerve
- Femoral artery/vein relationships
  - Entrance: Artery lateral
  - Apex: Artery anterior
- Femoral nerve branches within triangle
  - Saphenous nerve and nerve to vastus medialis are only branches to exit triangle
- **Femoral sheath**: Transversalis fascia covers vessels proximally
  - 3 compartments
    - Medial: Lymph node (femoral canal)
    - Middle: Vein
    - Lateral: Artery
- **Femoral canal**: Medial compartment of femoral sheath
  - Anterior border: Inguinal ligament
  - Posterior border: Pubic bone
  - Medial border: Lacunar ligament
  - Lateral border: Femoral vein
  - Entrance: **Femoral ring**
    - Anterior border: Medial inguinal ligament
    - Posterior border: Superior pubic ramus
    - Medial border: Lacunar ligament
    - Lateral border: Femoral vein
    - Open to peritoneal cavity
  - Contents: Lymphatic vessels and nodes (**node of Cloquet**), fat, connective tissue
    - Groin nodes divided into superficial and deep groups
    - Superficial nodes are situated around superficial veins
    - Deep nodes lie medial to femoral vein
    - Node of Cloquet in most medial of deep group; located medial to femoral vein at femoral triangle
      - Large oblong node with prominent fatty hilum
  - **Femoral hernia**
    - Lie lateral to and distal to pubic tubercle
    - Deep and distal to inguinal ligament
    - Travels in femoral ring to femoral canal to saphenous opening to subcutaneous tissue

Adductor (Subsartorial or Hunter) Canal

- Fascial passageway for vessels in midthigh
  - Boundaries are fascial surfaces of adjacent muscles
- Anteromedial border: Sartorius muscle
- Anterolateral border: Vastus medialis muscle
- Posterior border: Adductor longus and magnus muscles
- Entrance: Apex femoral triangle
- Exit: **Adductor hiatus**
  - Gap adductor magnus muscle between adductor portion and ischiocondylar portion of distal thigh
- Vessel passageway from thigh to popliteal fossa
- Contents: Femoral artery and vein, saphenous nerve
  - Nerve initially anterior to artery then medial
  - Artery anterior to vein
  - Descending geniculate artery arises in canal

ANATOMY IMAGING ISSUES

Imaging Recommendations

- US
  - Excellent for detecting and characterizing smaller lesions or fine anatomy, e.g., small soft tissue tumors, especially if superficial and internal vascularity of lesions
  - Real-time interrogation allows evaluation of muscle and tendon movement during dynamic maneuvers, e.g., for snapping hips, nerve impingement, etc.
  - Guides percutaneous biopsy or intervention
  - Limited by depth of penetration, in assessment of muscle edema, and assessment of large or diffuse lesions
- CT: Best for evaluating osseous integrity
  - Soft tissue assessment inferior to US and MR
- MR: Highest contrast resolution especially when using fluid-sensitive or contrast-enhanced sequences
  - Most sensitive modality to detect tissue edema
  - Best method to assess extent of diffuse lesions, such as large avascular malformation

ANTERIOR THIGH

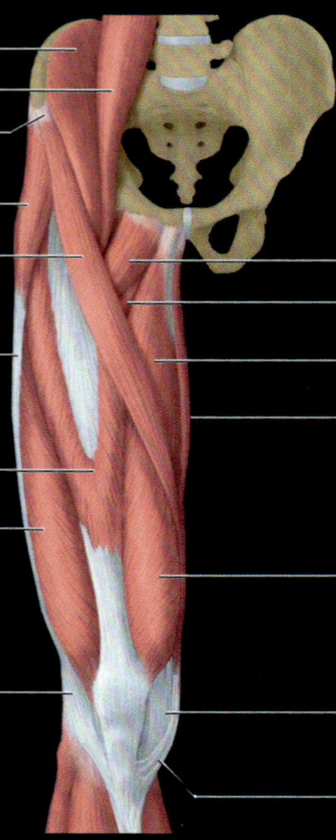

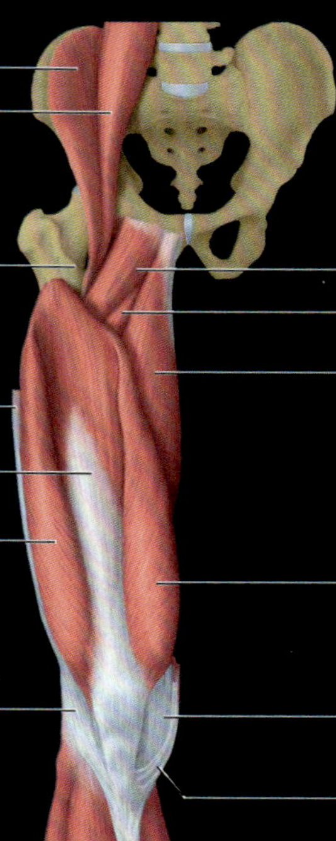

(Top) *Superficial muscles of the anterior thigh are shown. The oblique course of the sartorius muscle is easily appreciated. The adductor brevis muscle is deep to the adductor longus and pectineus muscles. The most medial muscle is the gracilis. The vastus lateralis and medialis muscles continue to their insertions as the lateral and medial patellar retinacula, respectively.* **(Bottom)** *Deep muscles of the anterior thigh are shown. The vastus intermedius muscle is deep to the rectus femoris muscle. The vastus intermedius muscle tendon blends with the rectus femoris tendon along its deep surface. The iliopsoas muscle dives deep after crossing over the pelvic brim. With removal of the sartorius muscle, a little more of the adductor brevis muscle is visible. The pes anserine tendons, the gracilis, sartorius, and semitendinosus have a conjoined insertion into the proximal medial tibia.*

OVERVIEW, ANTERIOR THIGH

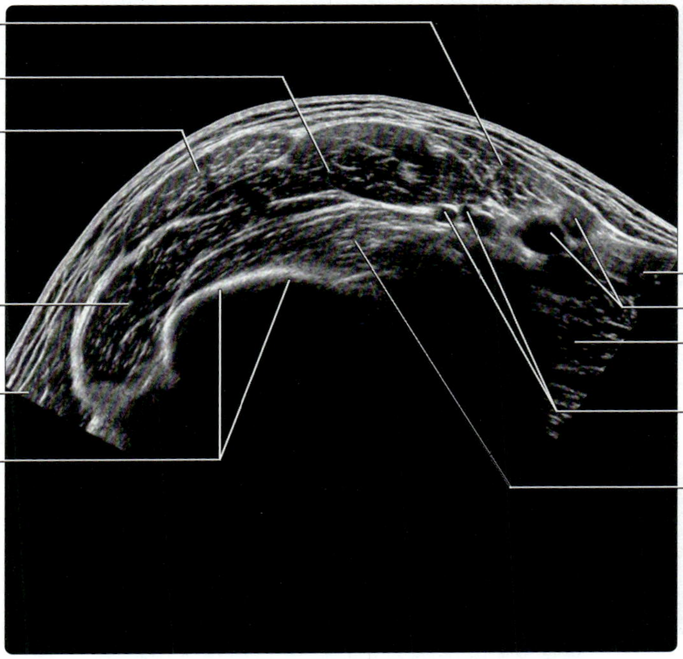

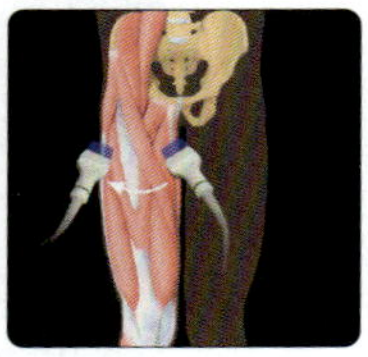

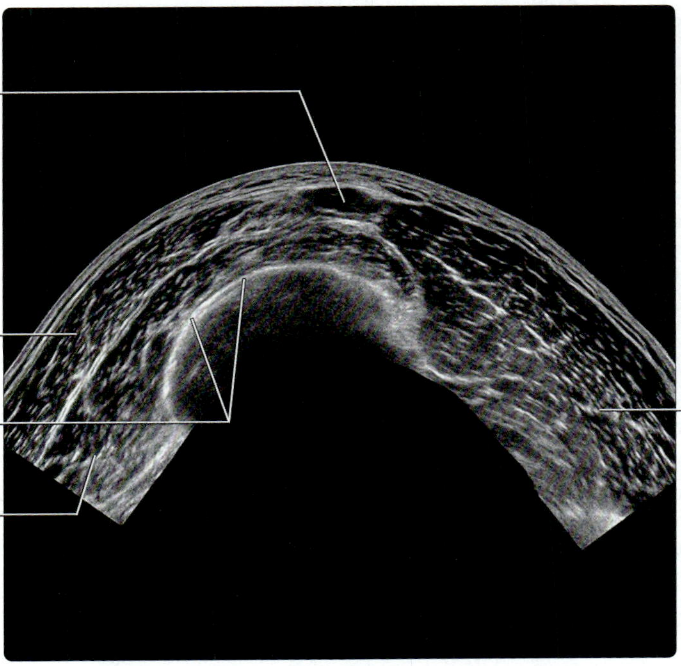

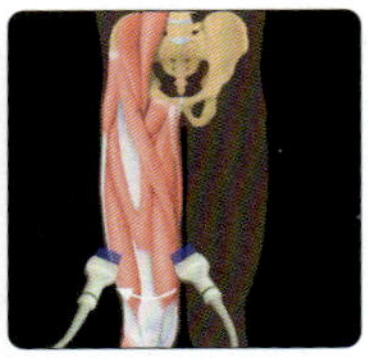

(Top) *Transverse panoramic view shows the anterior compartment of the upper thigh. The anterior compartment is composed of the quadriceps muscle, tensor fascia lata/iliotibial tract, and sartorius muscle. Medially, it is bordered by the femoral neurovascular bundle and adductor muscles. Laterally, it is bordered by the gluteal and hamstring muscles.* **(Bottom)** *In the lower thigh, the rectus femoris is much reduced in size and will continue inferiorly as the quadriceps tendon. The rest of the quadriceps continue as bulky muscles down to the suprapatellar level. They contribute directly to the quadriceps femoris tendon. The major neurovascular bundle has moved to the posterior compartment, and the only anterior compartment vessels are branches of the profunda femoris artery and vein.*

ANTERIOR COMPARTMENT, CENTRAL

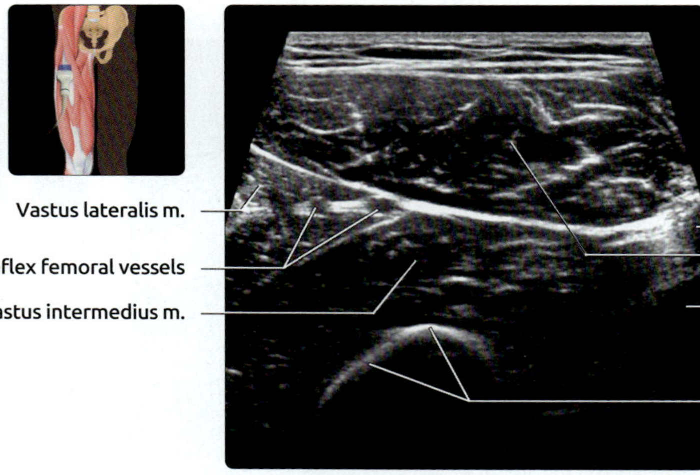

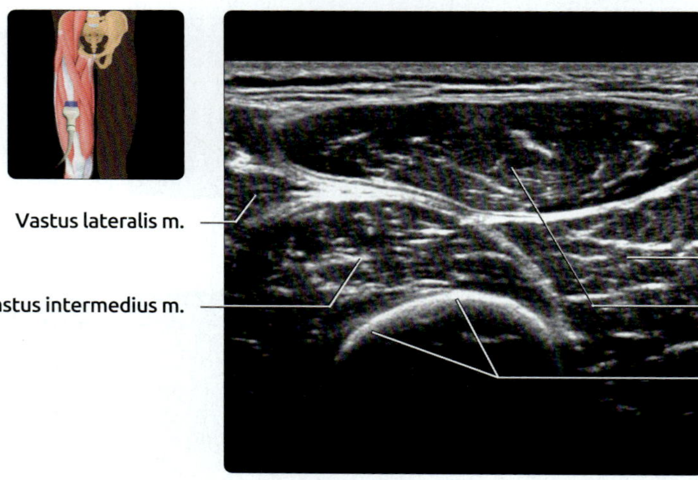

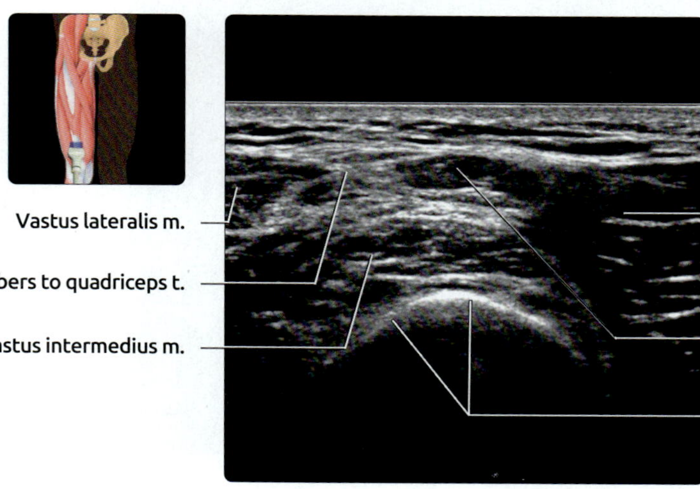

(Top) *Series of 3 transverse scans shows the central part of the anterior compartment. At the proximal level, the dominant (in terms of bulk) quadriceps muscle is the rectus femoris muscle. It is easily identified by its central position; it is lateral to the femoral neurovascular bundle (the femoral triangle). The lateral circumflex artery is important because together with the medial circumflex, its posterior counterpart, they supply the head of the femur.* **(Middle)** *Transverse scan is shown through the midthigh. The vastus medialis and intermedius have originated, and at this level, all 4 quadriceps muscles can be seen.* **(Bottom)** *Transverse scan shows the distal thigh above the level of the suprapatellar pouch. The rectus femoris has become a central tendon, which receives contributions from the vasti muscles.*

ANTERIOR COMPARTMENT, LATERAL

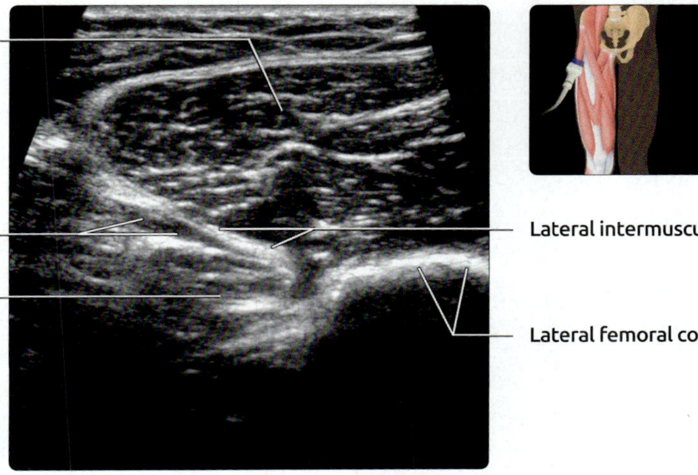

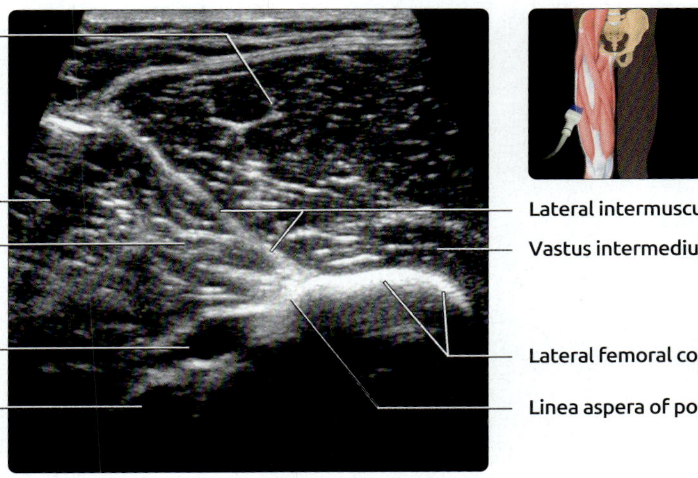

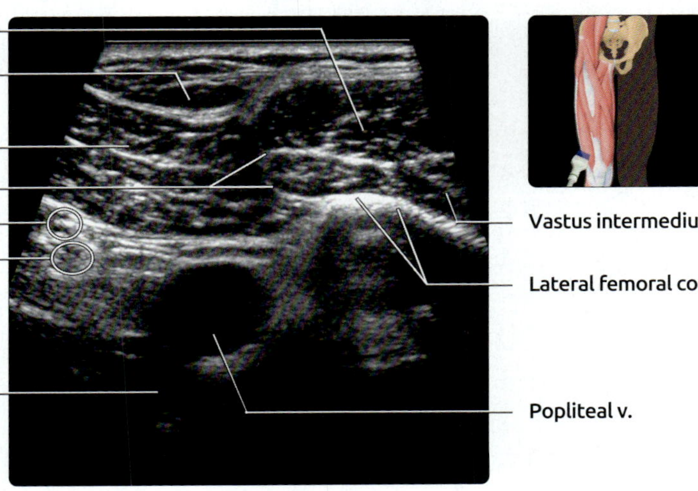

(Top) *A series of transverse scans through the lateral aspect of the thigh is shown. Proximally, the vastus lateralis muscle is dominant and borders with the gluteus maximus muscle (which inserts onto the posterior surface of the femur). The lateral intermuscular septum separates anterior from posterior compartments laterally and also acts as a conduit for vessels.* **(Middle)** *Further inferiorly, the vastus lateralis borders the biceps femoris. The linea aspera is a distinct protrusion from the posterior femoral shaft where many muscles originate and insert, including the lateral intermuscular septum. On US, it is consistently seen and therefore serves as a useful bony landmark.* **(Bottom)** *At the superior popliteal level, the vastus lateralis muscle becomes less bulky (the reverse occurs with the vastus intermedius muscle), as does the bordering long head of the biceps muscle. The sciatic nerve divides into tibial and common peroneal branches.*

MEDIAL AND POSTERIOR THIGH

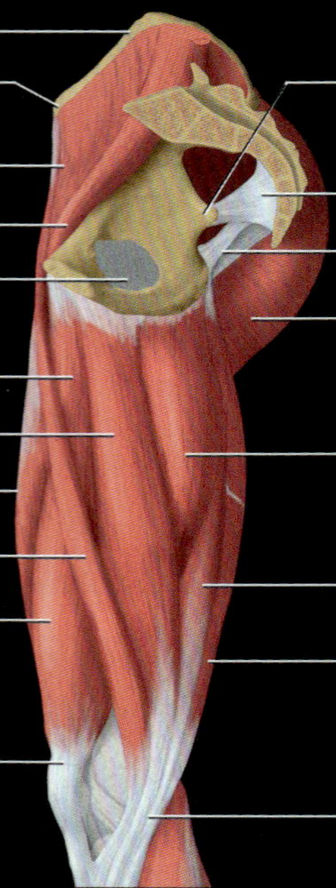

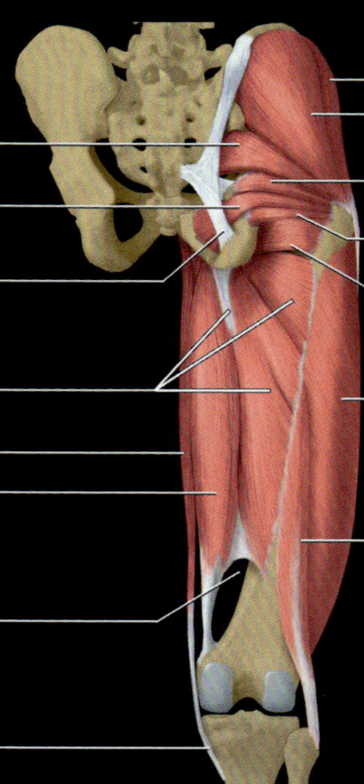

(Top) *Medial muscles of the thigh are shown. The gracilis muscle has a thin profile when viewed from the front; however, when viewed from the side, it is quite broad. The semimembranosus muscle runs along the deep surface of the semitendinous muscle and inserts onto the tibia posterior to the pes anserine tendons. Its insertion is hidden on this image. The iliopsoas muscle courses over the pelvic brim on its course to the lesser trochanter.* **(Bottom)** *Deep muscles of the posterior thigh are shown. With removal of the hamstring muscles, the expansive adductor magnus muscle is visible. The separation of its 2 heads in the distal thigh forms the adductor hiatus.*

OVERVIEW, MEDIAL THIGH

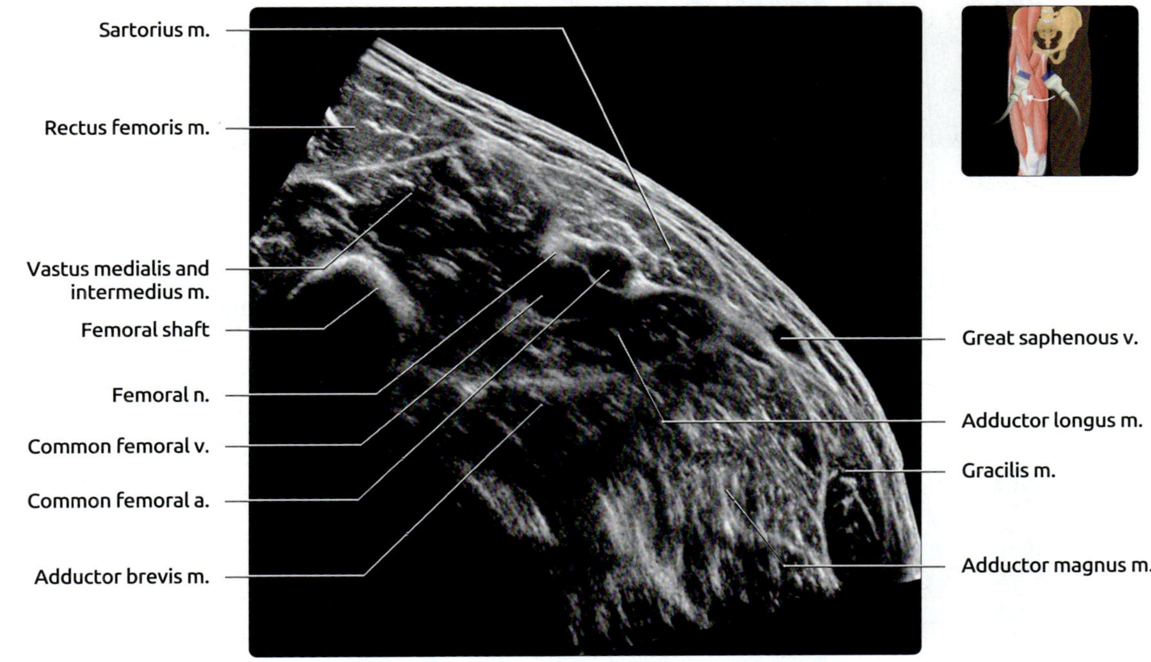

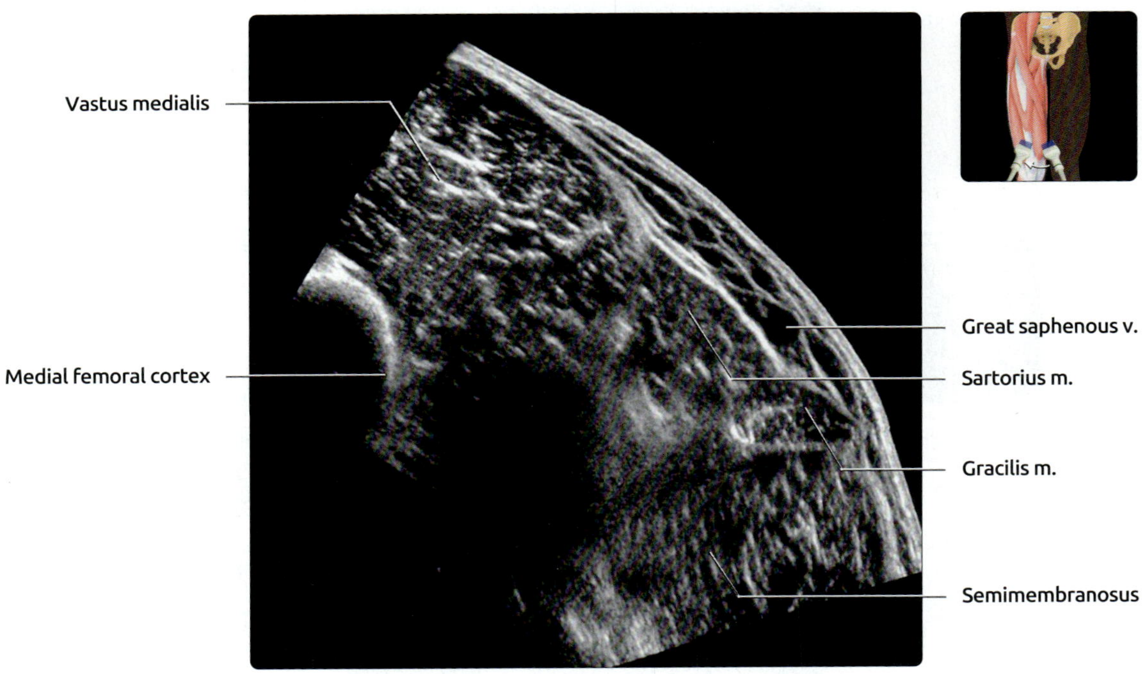

(Top) *Transverse panoramic view shows the medial compartment at the upper thigh level. At this level, all 3 adductor muscles can be seen, together with the gracilis muscle. This compartment is bordered anteriorly by the femoral triangle, which continues inferiorly as the adductor canal. Posteriorly, the medial compartment is bordered by the semimembranosus muscle.* **(Bottom)** *Transverse panoramic scan shows the medial aspect of the lower thigh. At this level, the adductor muscles have inserted superiorly; therefore, the medial compartment is much smaller and composed mainly of the gracilis muscle. The adductor hiatus, which is above this level, provides an entrance for the femoral vessels into the popliteal fossa.*

MEDIAL THIGH, ADDUCTOR CANAL

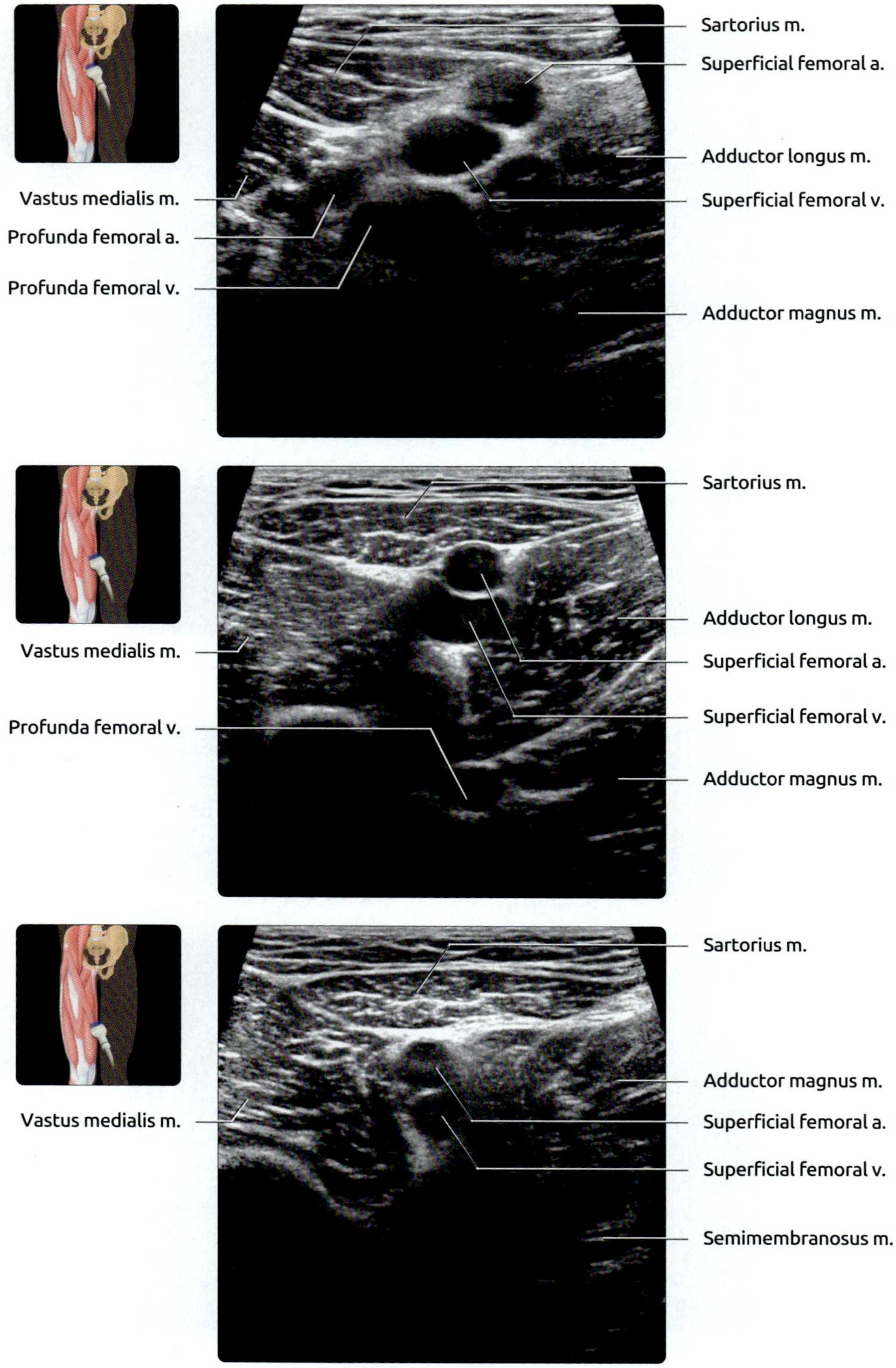

(Top) *Transverse scan shows the anterior part of the medial thigh. Above the adductor canal, the apex of the femoral triangle contains bifurcation of the common femoral vessels. The profunda will run deep to supply the thigh muscles and femur. The superficial femoral vessels continue into the adductor canal.* **(Middle)** *Transverse scan more inferiorly shows that the superficial femoral vessels are now under the cover of the sartorius muscle, indicating that they are in the adductor canal.* **(Bottom)** *Transverse scan shows the medial aspect of the midthigh. The common adductor canal continues medially and more deeply. At this level, the medial wall of the adductor canal changes from the adductor longus muscle to the adductor part of the adductor magnus muscle.*

MEDIAL THIGH

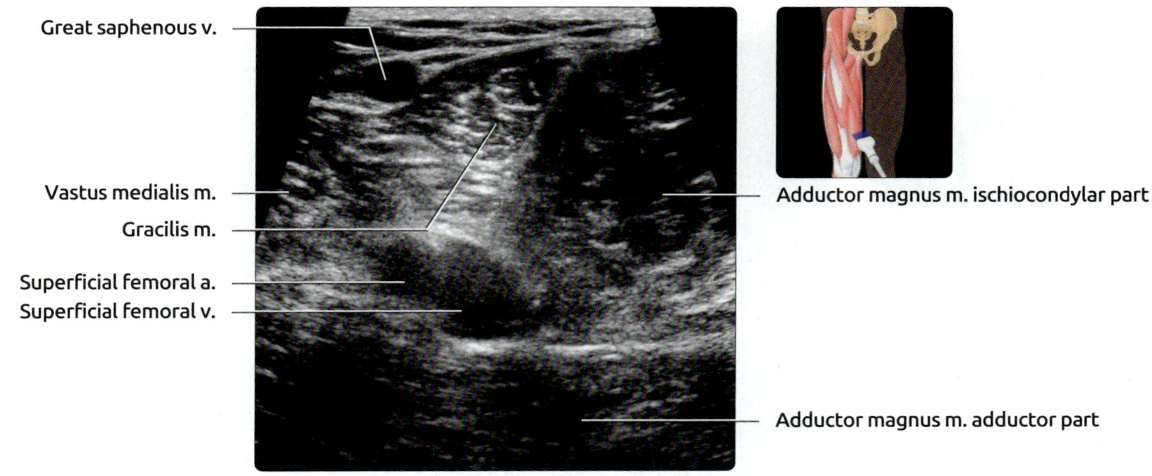

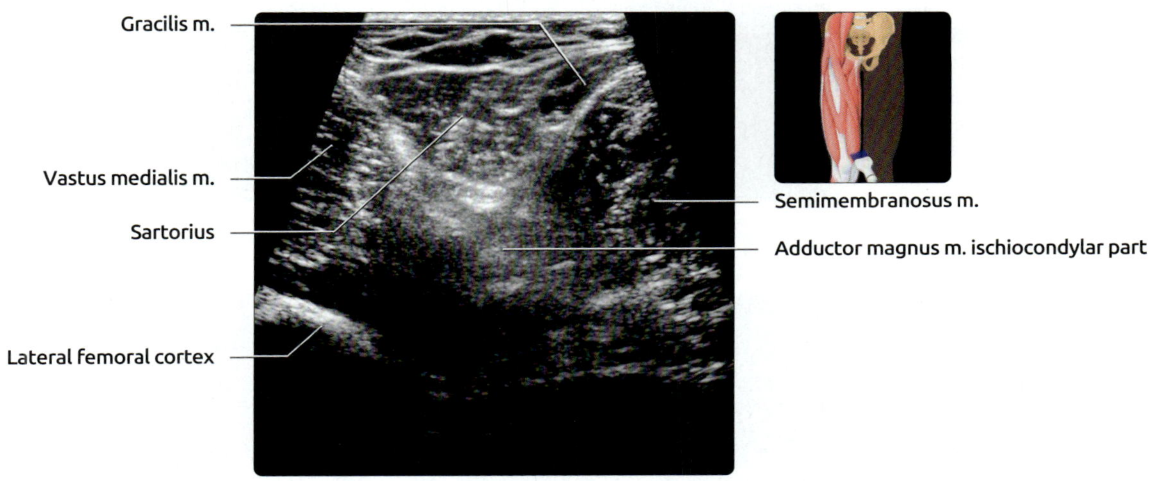

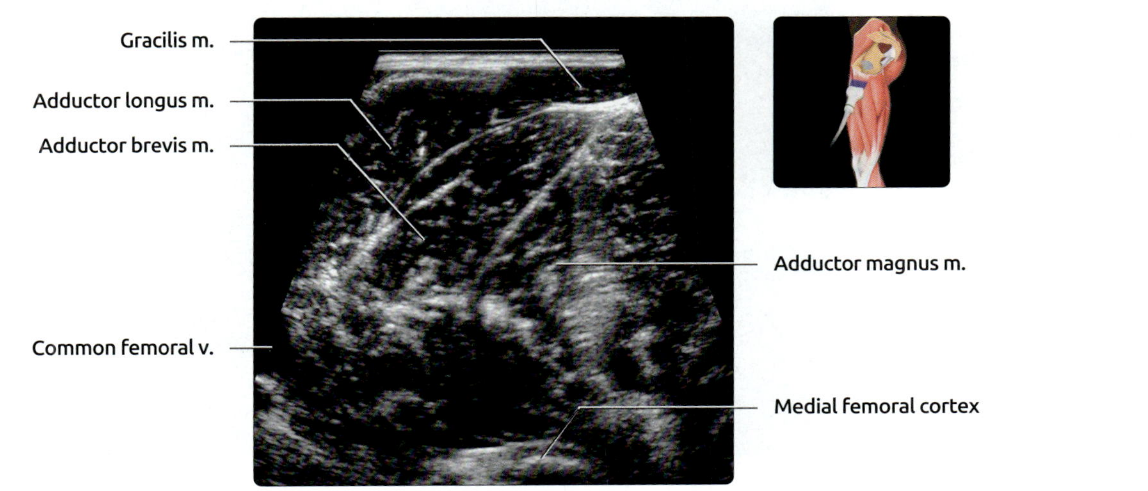

(Top) *Transverse scan using an anteromedial approach is shown. At this level, the superficial femoral vessels enter the popliteal fossa via the adductor hiatus, which is a gap in the adductor magnus muscle, between the adductor and ischiocondylar parts before their insertions (on the medial supracondylar line and adductor tubercle, respectively).* **(Middle)** *More inferiorly, only the gracilis muscle and the ischiocondylar part of the adductor magnus (acting like a hamstring) are left in the medial compartment. This space is encroached on by the vastus medialis muscle anteriorly and semimembranosus muscle posteriorly.* **(Bottom)** *Using a medial approach (from the inner thigh), the deeper muscles can be demonstrated. The adductor longus and brevis muscle plus the gracilis muscle form an anterior division, while the bulky adductor magnus muscle forms the posterior division.*

OVERVIEW, POSTERIOR

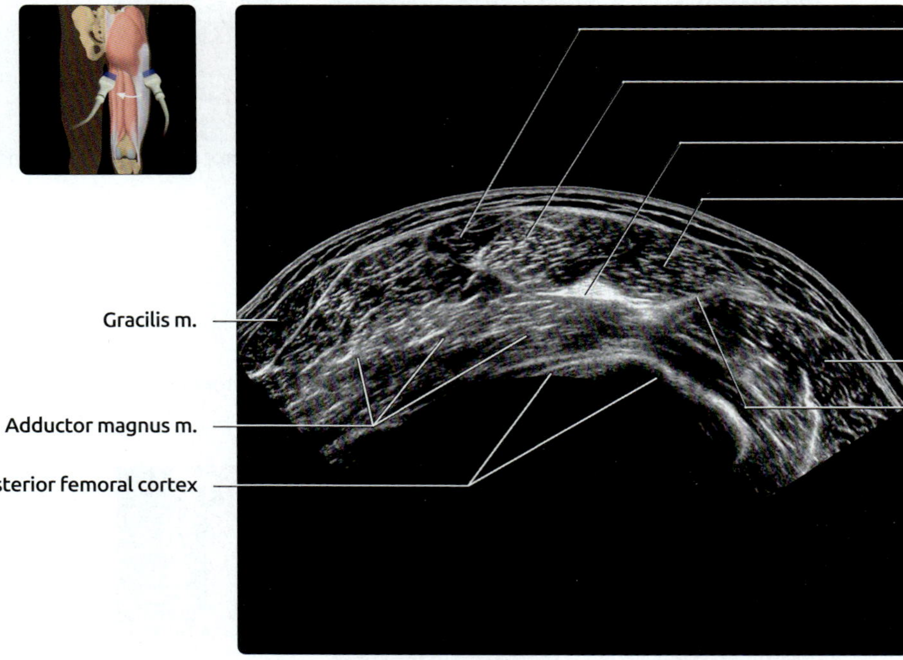

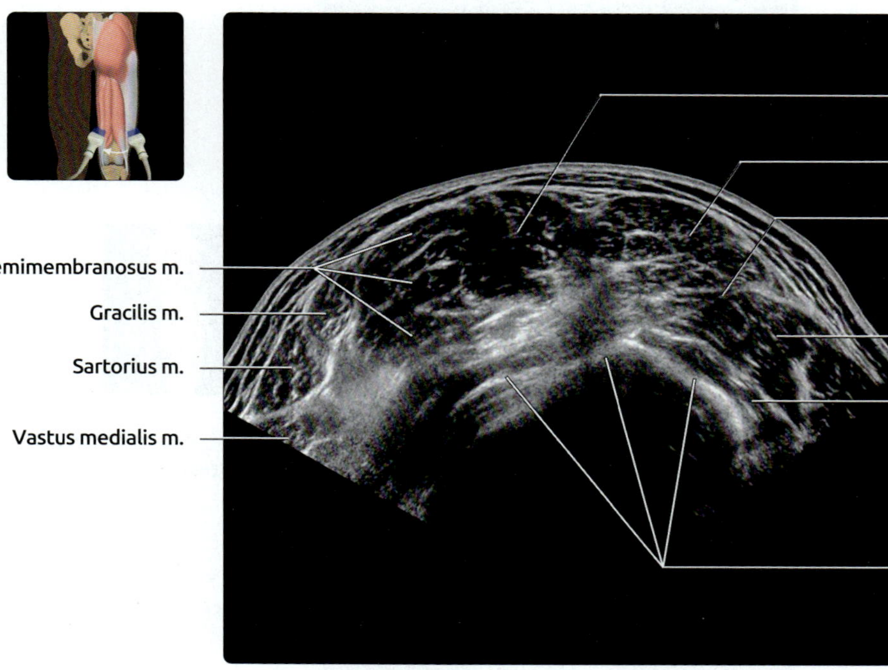

(Top) *Transverse panoramic scan shows the posterior compartment at the upper thigh level. At this level, the hamstring muscles form a small proportion of the posterior muscle bulk. The main contributions come from adductors (the medial compartment) and vastus lateralis (the anterior compartment). The sciatic nerve runs under the cover of the biceps femoris muscle.* **(Bottom)** *Transverse panoramic scan shows the posterior aspect of the distal thigh. At this more inferior level, the hamstrings form a much larger proportion of the posterior muscle bulk. In particular, the semimembranosus becomes the dominant muscle on the posteromedial side, as most of the adductors have inserted onto the femur above this level.*

POSTERIOR THIGH

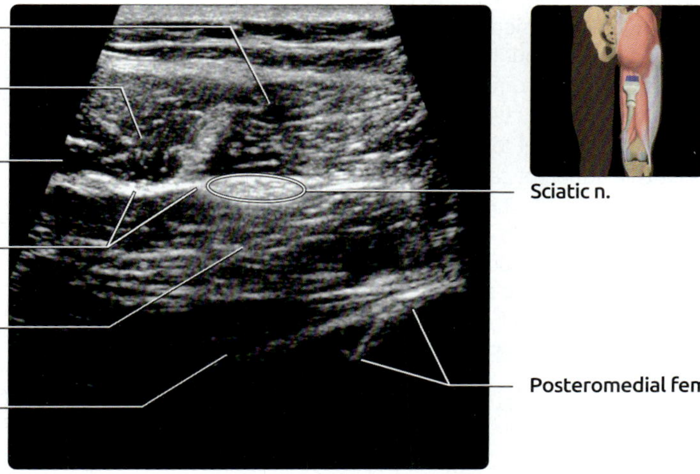

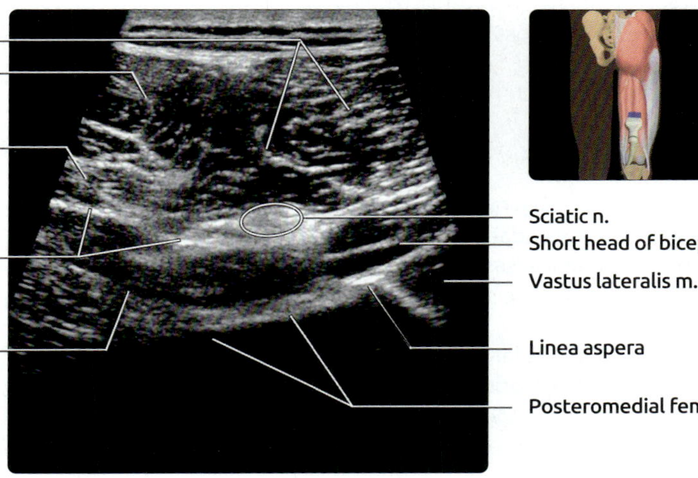

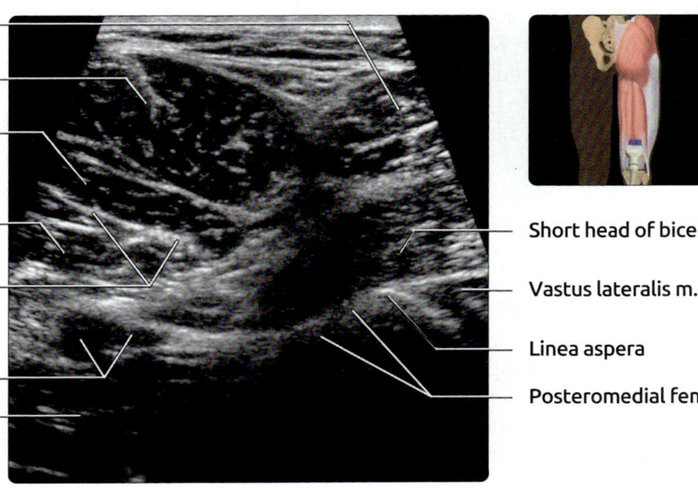

(Top) *A series of 3 transverse scans shows the central portion of the posterior thigh. After their origin from the ischial tuberosity, the semitendinosus and semimembranosus muscles run on the medial side, and the biceps femoris muscle runs on the lateral side. The sciatic nerve is under the cover of the biceps femoris for most of its way to the popliteal fossa.* **(Middle)** *At the midthigh level, the hamstrings are bulkier and the adductors are smaller. The 2 groups are divided by the posteromedial intermuscular septum.* **(Bottom)** *Further inferiorly, the popliteal vessels enter via the adductor hiatus into the popliteal fossa. The superior aspect of the popliteal fossa is lined by the semimembranosus and semitendinosus muscle medially and the biceps femoris laterally.*

IMAGING ANATOMY

Femoral Vessels

- Enter thigh deep to inguinal ligament midway between anterior superior iliac spine and pubic symphysis
  - External iliac vessels become common femoral vessels
- Upper thigh: Vessels within femoral triangle
  - Vein, artery, nerve from medial to lateral
- Midthigh: Vessels within adductor canal
  - Adductor canal: 15-cm long, triangular-shaped canal running from femoral triangle to adductor hiatus
    - Bordered by vastus medialis, adductor longus/adductor magnus, and sartorius
  - Artery anterior to vein
    - Saphenous nerve accompanies femoral vessels in adductor canal
- Distal thigh: Exits adductor canal via adductor hiatus and enters popliteal fossa
  - Adductor hiatus is gap between adductor magnus muscle and femoral shaft 10 cm proximal to knee
    - Saphenous nerve does not pass through adductor hiatus
  - Superficial femoral vessels become popliteal artery and vein
- **Common femoral artery**
  - Branches
    - **Superficial epigastric, superficial circumflex iliac, superficial external pudendal** arise anteriorly
    - **Deep external pudendal** arises medially
  - Divides into superficial and deep branches
  - **Superficial femoral artery**
    - Branch: **Descending genicular**
  - **Deep femoral (profunda femoris)**
    - Arises laterally in femoral triangle
    - Dives between pectineus and adductor longus muscles
    - Medial to femur, deep to adductor longus muscle
    - Branches in femoral triangle: **Medial circumflex femoral** (supplies femoral head), **lateral circumflex femoral**, muscular branches
    - Branches in adductor canal: 3 perforating branches, descending geniculate
- **Femoral vein**: Travels with artery
  - Tributaries: Similar to arteries and long saphenous vein
  - **Long saphenous vein**
    - Longest vein in body
    - Toes → anterior to medial malleolus → medial leg → posteromedial knee → proximal thigh dives deeply through cribriform fascia → saphenofemoral junction → common femoral vein
      - Saphenous nerve accompanies long saphenous vein in leg and superficial femoral artery in thigh

Femoral Nerve

- Nerve contributions from L2 to L5
- Travels in groove between psoas and iliacus muscles
- Exits pelvis beneath inguinal ligament, lateral to femoral vessels where it enters femoral triangle
- Multiple muscular and sensory branches in femoral triangle
  - Saphenous nerve is largest cutaneous branch of femoral nerve
    - Enters adductor canal until 10 cm proximal to knee where it pierces fascia between gracilis and sartorius muscles to become subcutaneous and join long saphenous vein

Sciatic Nerve

- Nerve contributions from L4 to S3
  - 2 nerves in 1 sheath
    - Tibial nerve (medial) and common peroneal nerve (lateral)
    - Separate in popliteal fossa
- Course
  - Exits pelvis inferior to piriformis muscle
    - Anomalous relationship in 10%
  - Deep to gluteus maximus muscle, more distally deep to biceps femoris muscle
    - Crosses over superior gemellus, obturator internus, inferior gemellus, quadratus femoris muscles
    - Midway between ischial tuberosity and greater trochanter in transverse plane
  - Runs deep to long head of biceps femoris muscle after exiting gluteal region
- **Tibial nerve**: Larger division of sciatic nerve
  - Descends through popliteal fossa
- **Common peroneal nerve**
  - Runs obliquely to fibular head

Lateral Cutaneous Nerve of Thigh

- Nerve contributions from L2 and L3
  - Sensory innervation to anterior/lateral aspects of thigh
- Runs from medial to lateral superficial to proximal end of sartorius muscle
- Passes either under or through split in inguinal ligament close to anterior superior iliac spine to enter thigh
  - Runs distally lateral to sartorius before dividing into many small branches

ANATOMY IMAGING ISSUES

Imaging Recommendations

- **Sciatic nerve**
  - Best located lateral to ischial tuberosity or else at gluteal fold deep to long head of biceps femoris and lateral to semimembranosus muscle
    - Alternatively, tibial nerve can be identified at proximal end of popliteal fossa and followed proximally
- **Femoral nerve**
  - Best located in femoral triangle lateral to superficial femoral artery and medial to sartorius muscle
    - Initially large (cross-sectional area 23 mm^2) but then divides into many small (< 1-mm) branches
- **Saphenous nerve**
  - Best located lateral to artery in femoral triangle, between gracilis and sartorius muscles just proximal to knee and alongside long saphenous vein just distal to knee
- **Lateral cutaneous nerve of thigh**
  - Best located in transverse plane as it runs medial to lateral superficial to sartorius muscle just distal to inguinal ligament

THIGH ARTERIES

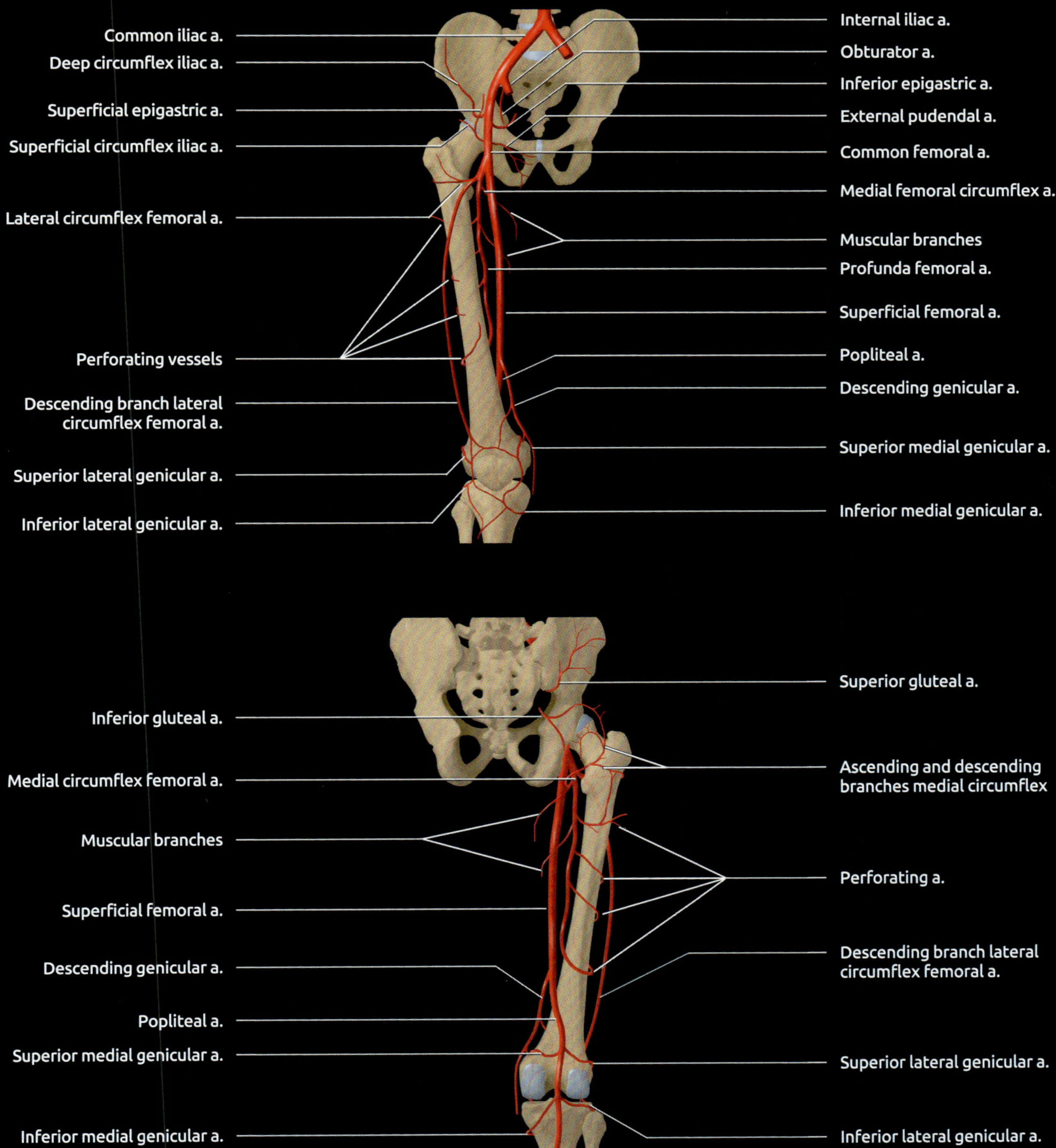

(Top) *Anterior view shows the external iliac, femoral, and popliteal arteries and their branches.* **(Bottom)** *Posterior view shows the femoral and popliteal arteries and the superior and inferior gluteal arteries of the greater sciatic notch. All of the main arteries and veins can be readily evaluated with US. CT angiography or MR angiography are the best investigations to provide an imaging roadmap of the main arteries in the lower limb, and these investigations have replaced standard angiography in this respect. Standard DSA is still the best investigation to demonstrate the smaller arteries of the lower limb (such as those around the feet or around the hip).*

THIGH VEINS

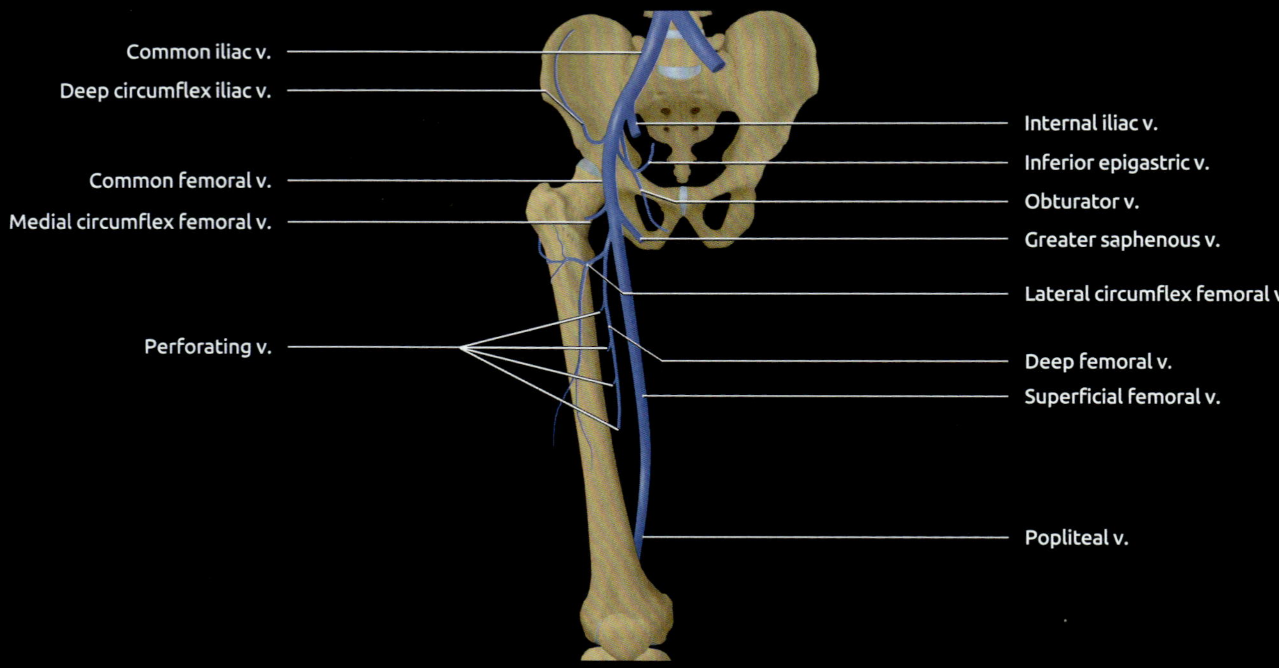

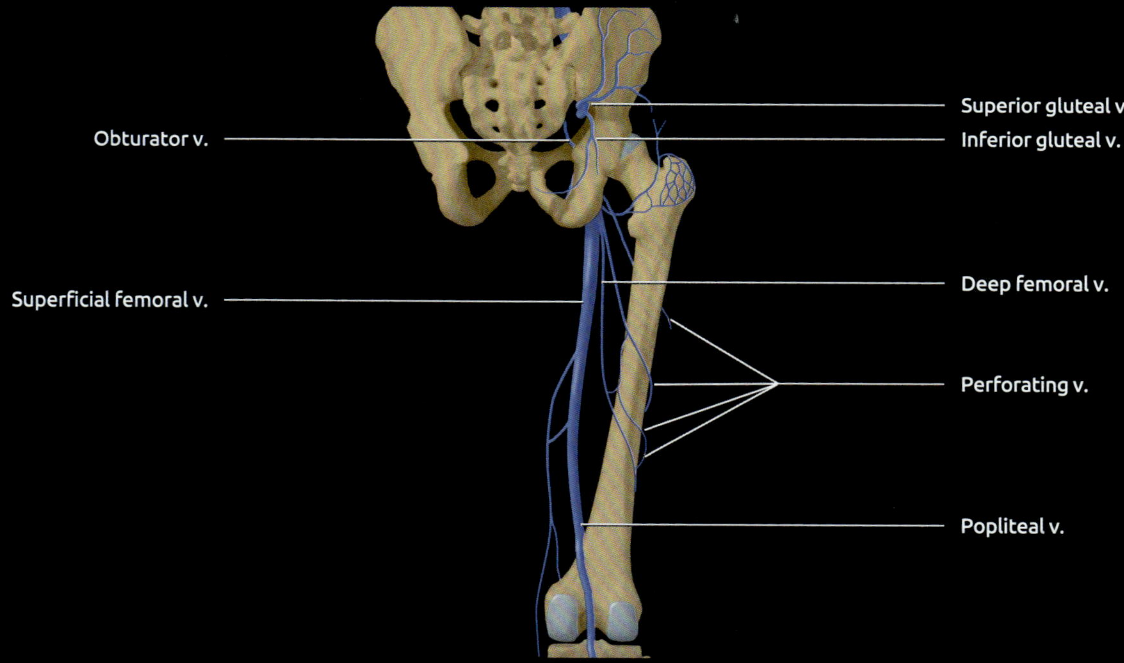

(Top) *Anterior view shows the veins of the thigh. The veins typically follow the arterial tree. The main venous drainage is the superficial femoral vein, which is actually a deep venous structure. Note the entrance of the greater saphenous vein into the femoral vein; the saphenous venous system has no accompanying artery.* **(Bottom)** *Posterior view shows the veins of the thigh and buttocks. The veins generally follow the arterial tree, although more variability is seen within the venous system.*

THIGH NERVES

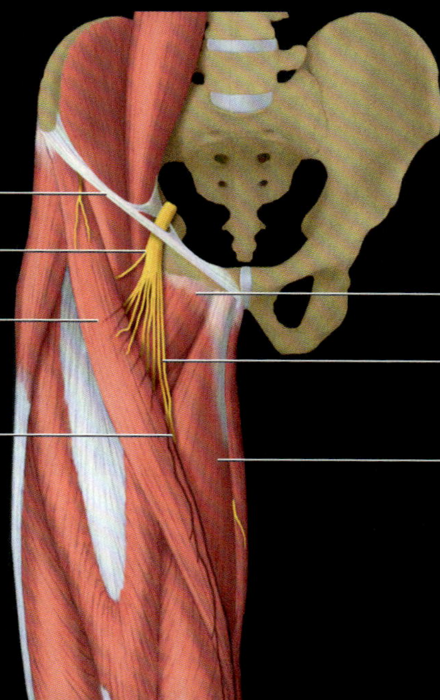

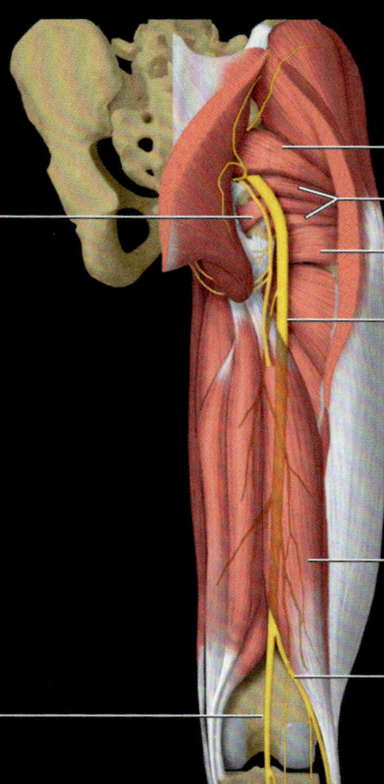

(Top) *Femoral nerve is shown. After emerging from the groove between the psoas and iliacus muscles, the nerve passes under the inguinal ligament to enter the femoral triangle. It immediately branches into muscular, cutaneous, and articular nerves. The only nerve to pass through the femoral triangle into the adductor canal is the saphenous nerve.* **(Bottom)** *The sciatic nerve enters the lower extremity by passing under the inferior border of the piriformis muscle. The sciatic nerve passes posterior to the external rotator muscles and then courses deep to the biceps femoris muscle. It separates into the tibial and common peroneal nerves in the distal thigh. The tibial nerve bisects the popliteal fossa. The common peroneal nerve follows the biceps femoris muscle around the fibular head.*

LONGITUDINAL THIGH VESSELS

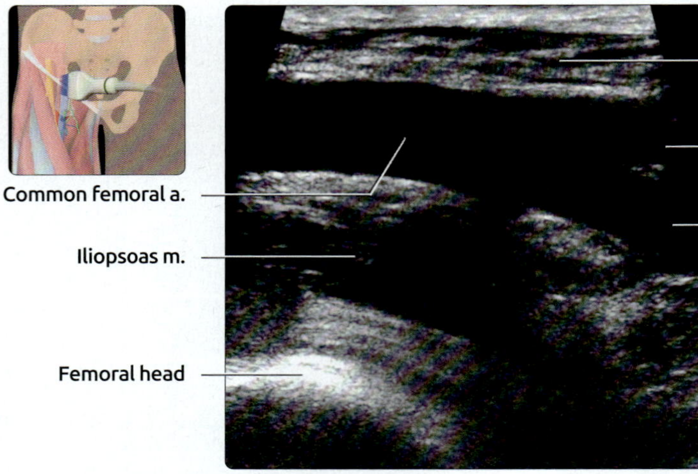

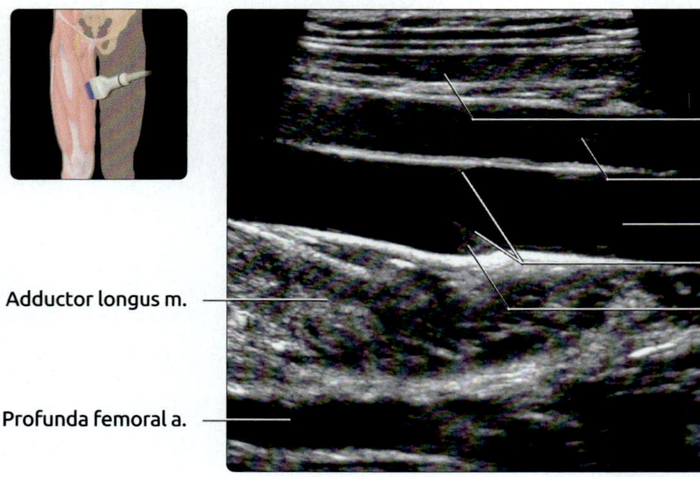

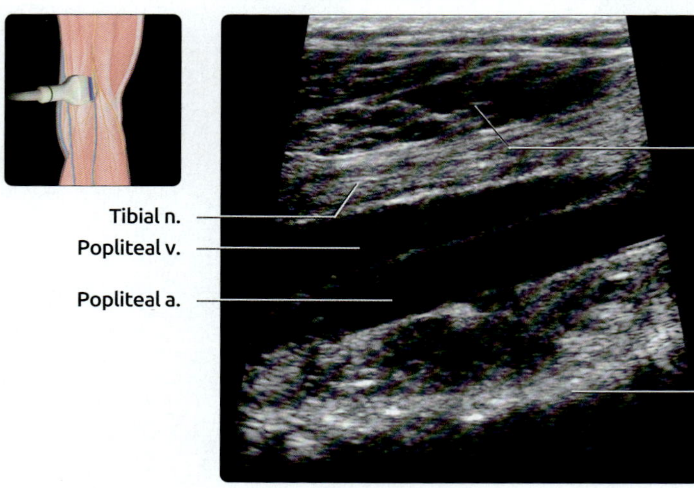

(Top) *Sagittal scan shows the common femoral artery and its bifurcation into the superficial femoral artery and the profunda femoral artery. The superficial femoral artery continues into the femoral triangle and adductor canal before ending at the adductor hiatus to become the popliteal artery. The profunda femoris artery supplies the muscles of the thigh.* **(Middle)** *Sagittal scan shows the upper adductor canal. The superficial femoral vessels are covered by the sartorius muscle. The high spatial resolution of US allows venous valves and sluggish flow to be visualized (the latter as increased echogenicity).* **(Bottom)** *Sagittal scan shows the popliteal vessels using a posterior approach. Note the reversed relationship between the vessels: The vein now lies more superficially than the artery.*

TRANSVERSE THIGH VESSELS

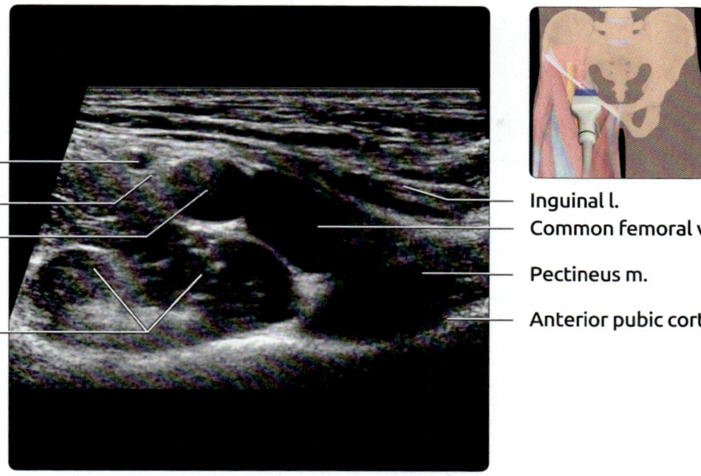

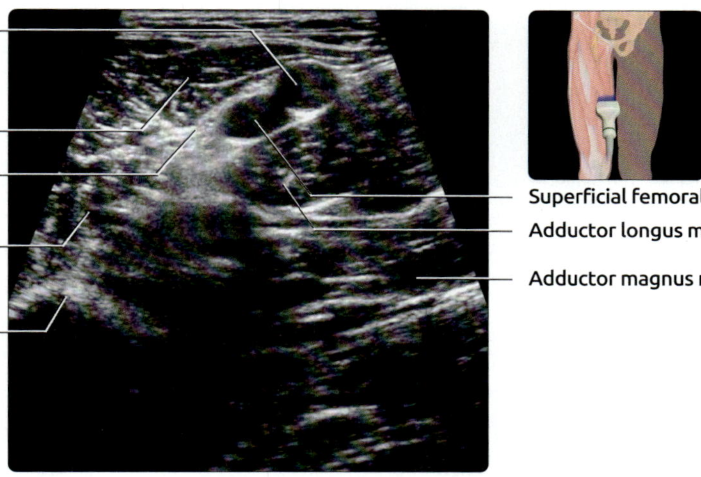

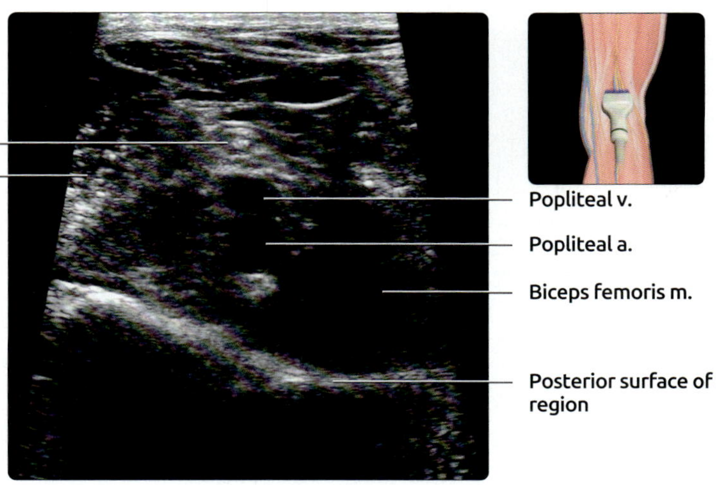

(Top) *Transverse scan shows the common femoral vessels emerging from under the inguinal ligament. The femoral nerve lies lateral to the common femoral artery. The iliopsoas muscle and pectineus muscle form the floor.* **(Middle)** *Transverse scan shows the lower aspect of the femoral triangle. The sartorius begins to cover the femoral vessels. The vastus medialis and the adductor longus muscles form the side walls of the femoral triangle. More inferiorly, the common femoral vessels bifurcate into the superficial and profunda branches.* **(Bottom)** *Transverse scan shows the posterior thigh at the upper level of the popliteal fossa. After passing through the adductor hiatus, the superficial femoral vessels enter the popliteal fossa. The hamstring muscles form the side walls of this fossa.*

COLOR DOPPLER THIGH ARTERY

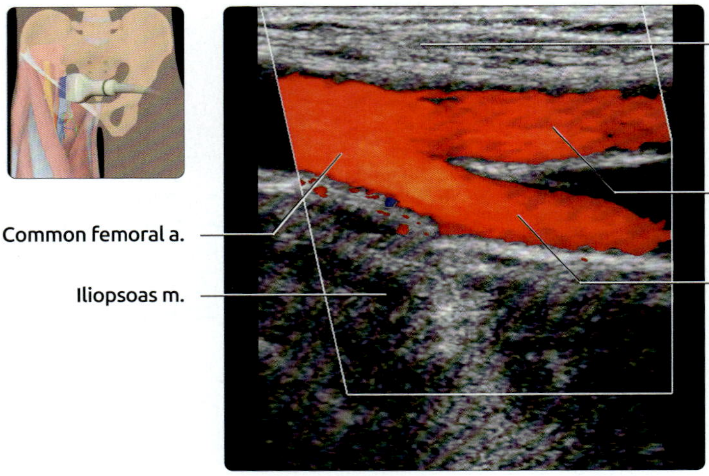

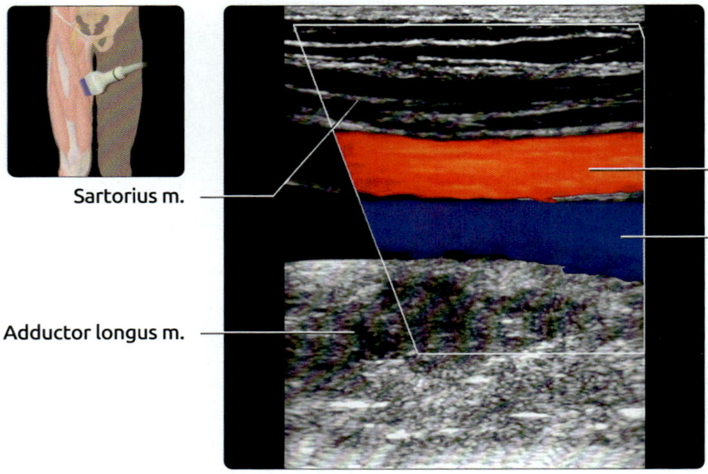

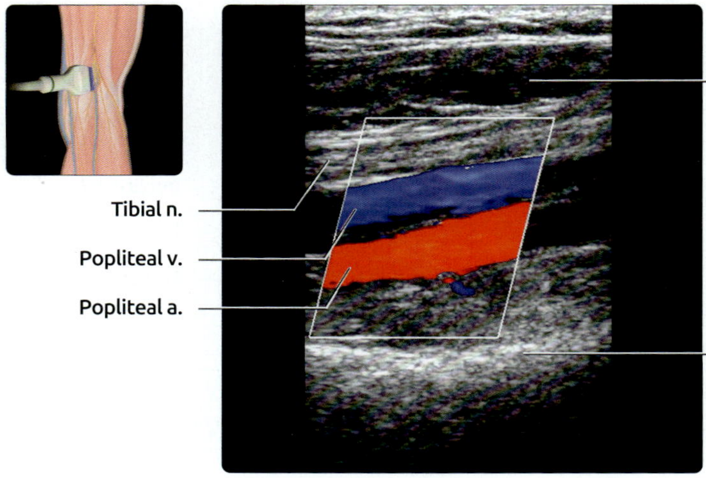

(Top) *Sagittal color Doppler scan shows the common femoral bifurcation. Color, representing flow, should fill the entire lumen with appropriate scan settings (as seen in this image).* **(Middle)** *Sagittal color Doppler scan shows the superficial femoral vessels in the upper adductor canal. The color of flow is direction dependent (the Doppler effect) and here demonstrates the opposite direction of flow between the artery (red, away from the trunk) and vein (blue, toward the trunk). Again, the color should fill the entire lumen (red superficial to blue).* **(Bottom)** *Sagittal color Doppler scan shows the popliteal vessels using a posterior approach. The vein now lies superficial to the artery (blue superficial to red).*

SPECTRAL DOPPLER THIGH ARTERIES

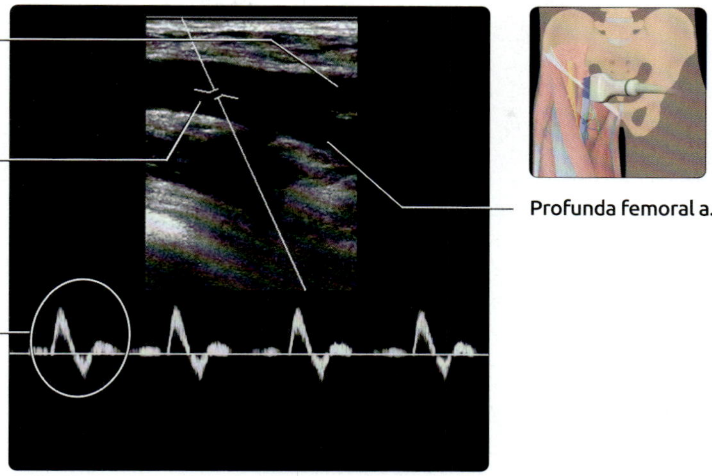

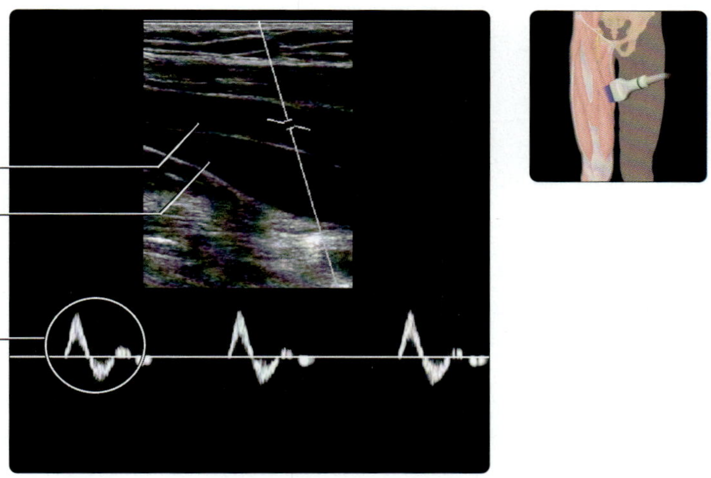

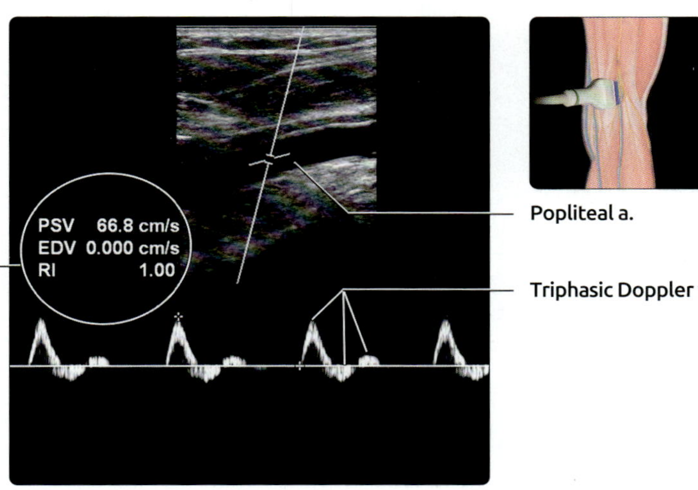

(Top) *Spectral Doppler trace at the common femoral artery is shown. The typical waveform for an artery should be triphasic: Forward sharp upstroke of forward flow (systolic), then reverse flow (early diastolic), and finally slow forward flow (late diastolic), as seen in this trace.* **(Middle)** *Spectral Doppler trace performed at the superficial femoral artery shows a similar waveform and slight decrease in the peak systolic flow. In general, as one moves further away from the heart, the peak velocity decreases.* **(Bottom)** *The typical arterial waveform is again seen in a spectral Doppler trace of the popliteal artery.*

SAPHENOFEMORAL JUNCTION

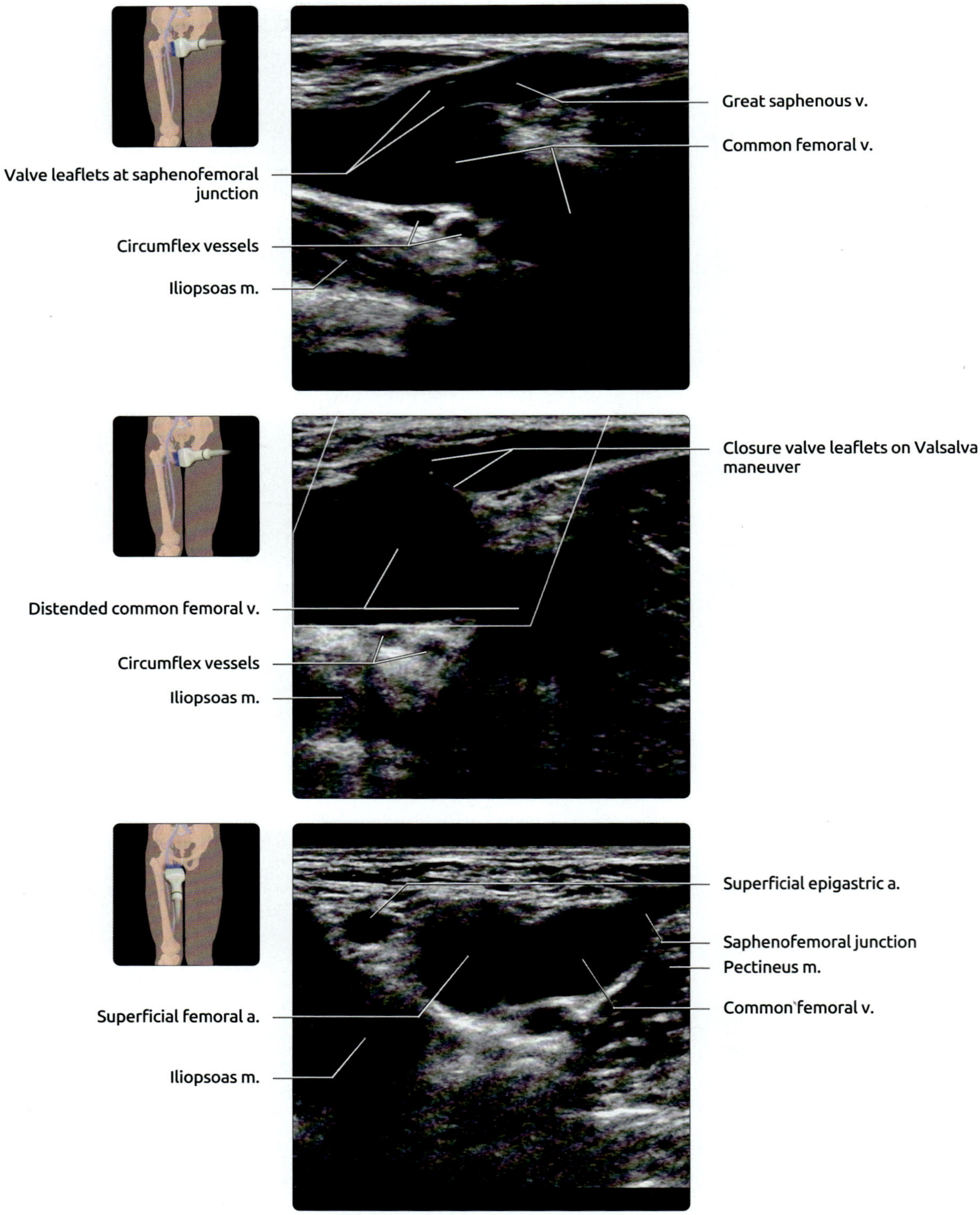

(Top) *Sagittal scan at the confluence of the great saphenous vein with the common femoral vein is shown. On this image, the valve at the saphenofemoral junction is open to allow blood from the great saphenous vein to drain into the common femoral vein.* **(Middle)** *Sagittal color Doppler scan shows the saphenofemoral junction during a Valsalva maneuver. The increased abdominal pressure is transmitted in the common femoral vein. This closes the valve at the saphenofemoral junction to prevent reflux into the great saphenous vein and also distends the common femoral vein (seen here as an increase in caliber). The lack of color flow across the valve indicates that the valve is competent.* **(Bottom)** *Transverse scan shows the saphenofemoral junction. The great saphenous vein penetrates the femoral sheath to enter the common femoral vein.*

DOPPLER SAPHENOUS VEIN

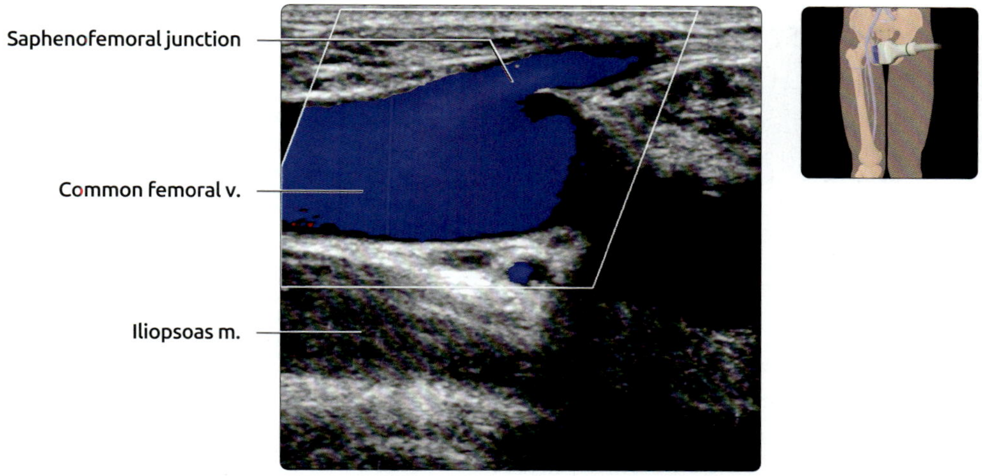

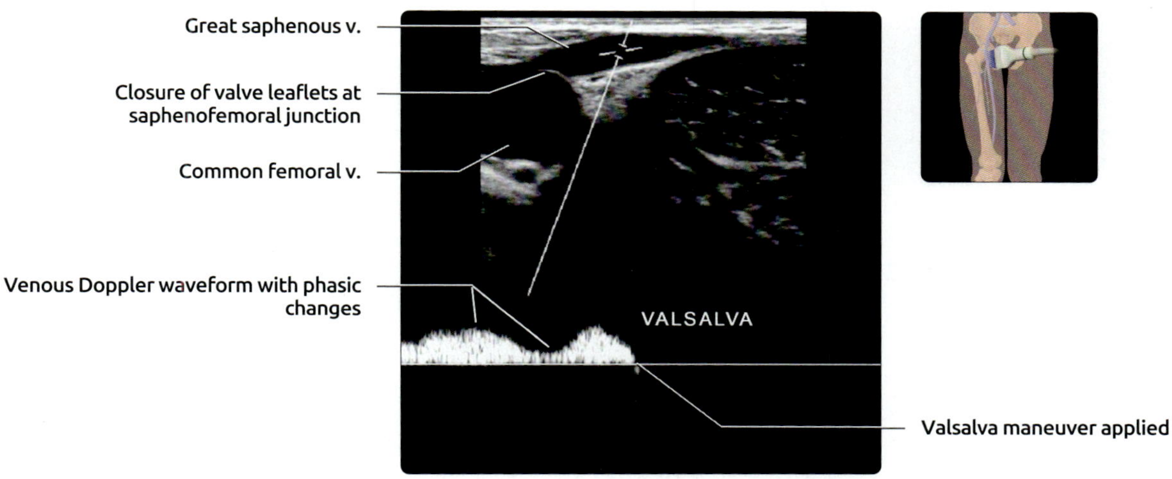

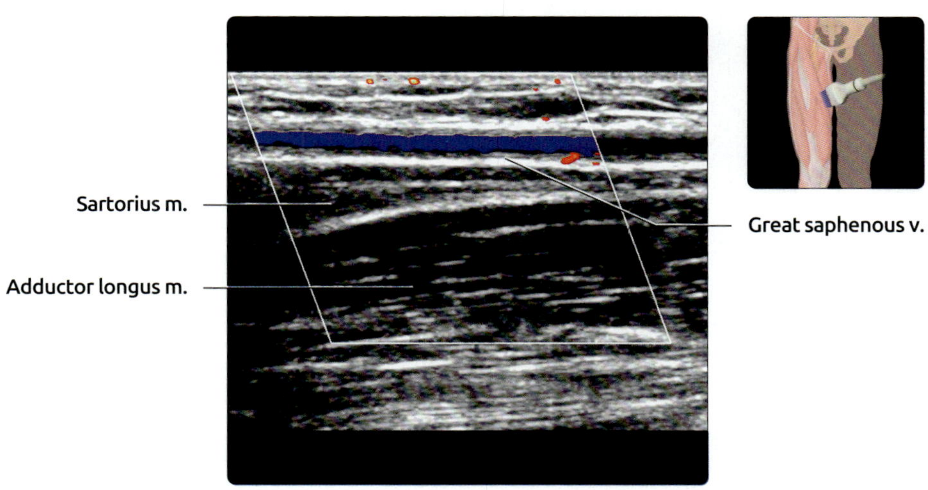

(Top) *Sagittal color Doppler scan shows the saphenofemoral junction. The color flow in both the great saphenous vein and the common femoral vein is toward the trunk.* **(Middle)** *Spectral Doppler trace of the great saphenous vein before and during a Valsalva maneuver is shown. The normal venous Doppler spectrum is biphasic due to respiratory variations. Applying a Valsalva maneuver, the valve closes abruptly at the saphenofemoral junction and stops flow in the great saphenous vein, which is seen as a zero-flow velocity on the trace.* **(Bottom)** *Coronal oblique color Doppler scan of the great saphenous vein in the medial thigh shows blood flow toward the trunk (correct direction).*

THIGH VEIN

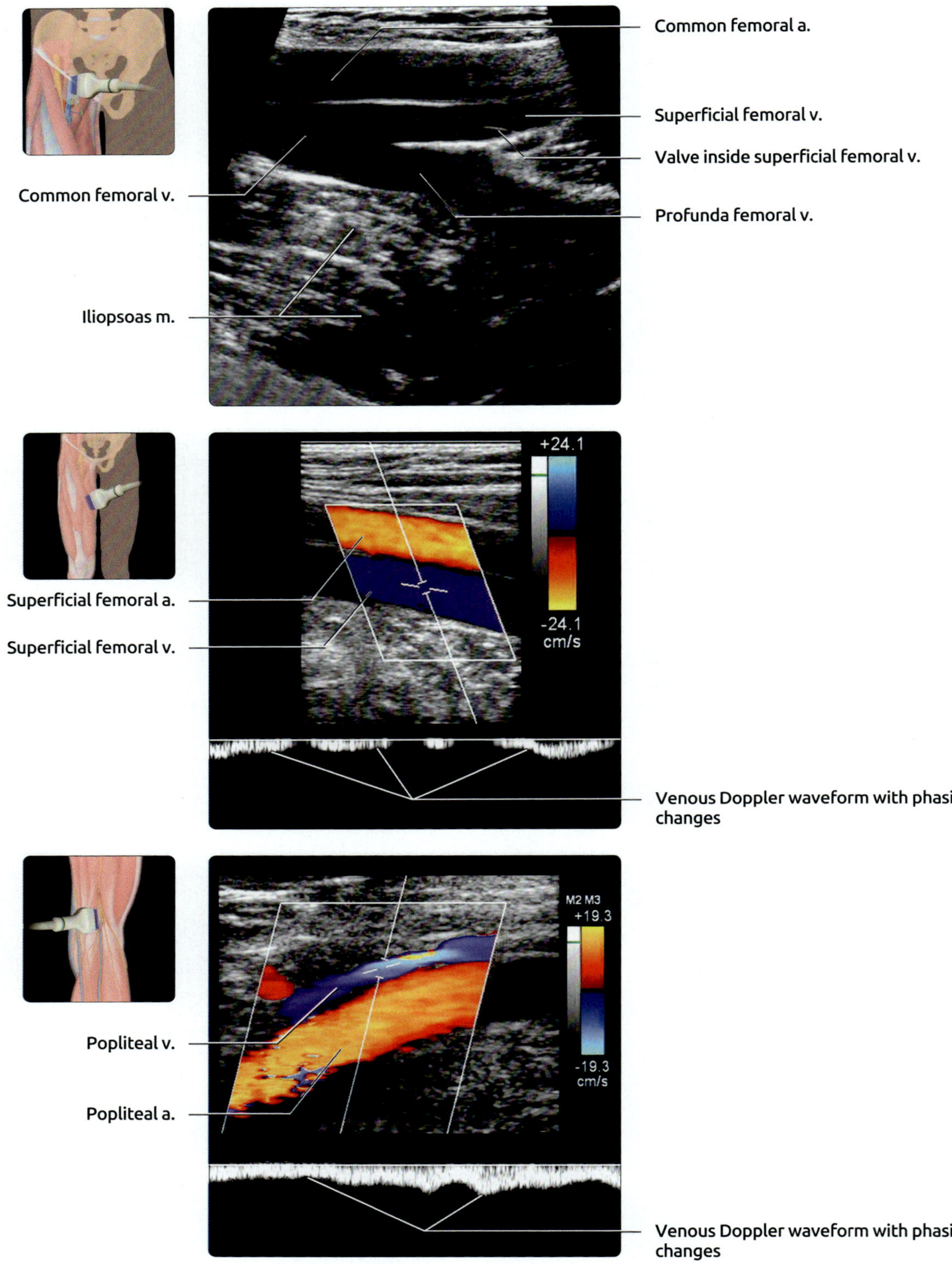

(Top) *Sagittal scan shows the confluence of the common femoral vein. The common femoral vein is formed by its main tributaries: The superficial femoral vein and the profunda femoral vein (both accompanying the course of the arteries of the same name).* **(Middle)** *Spectral Doppler trace shows the superficial femoral vein. The normal venous Doppler waveform is biphasic and varies with the phase of respiration. The flow velocity is also much slower than that in the artery.* **(Bottom)** *Spectral Doppler trace of the popliteal vein shows a similar venous waveform with respiratory phasic changes.*

FEMORAL VEIN: COMPRESSION

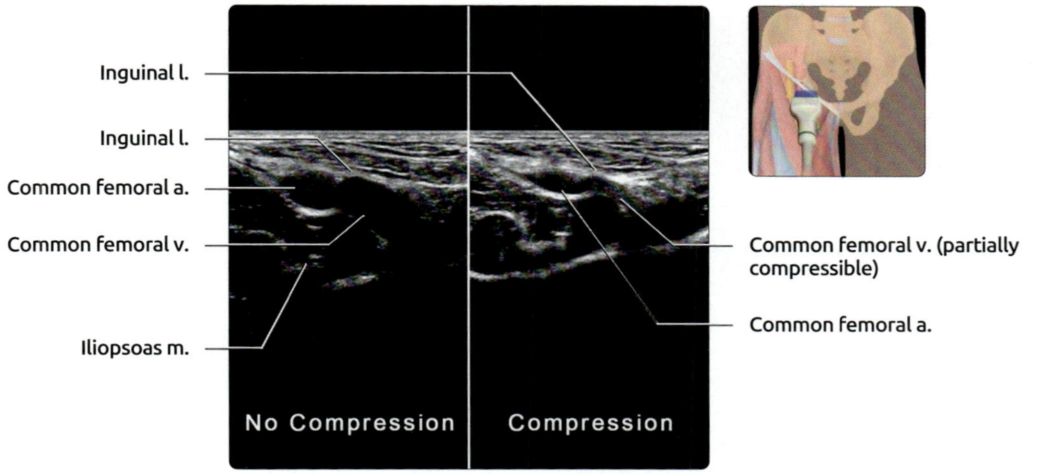

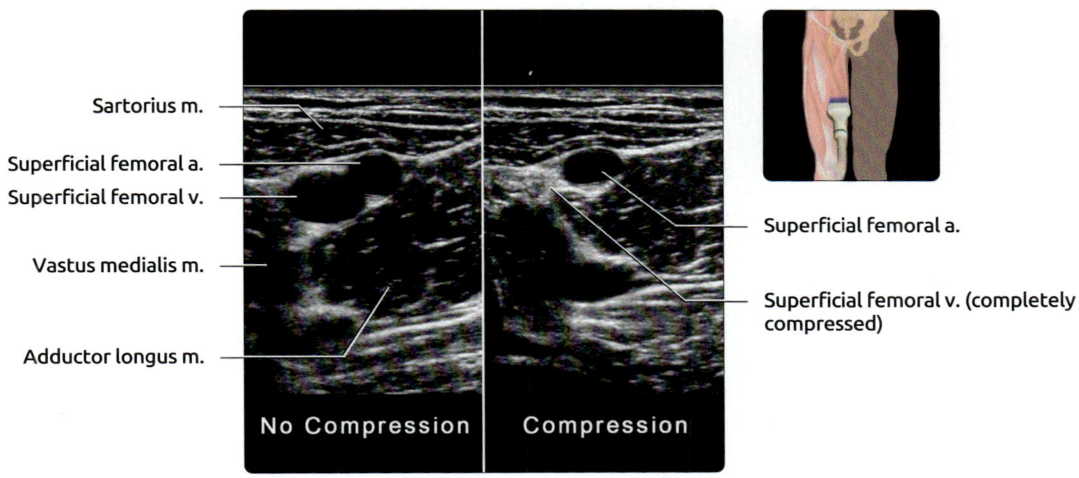

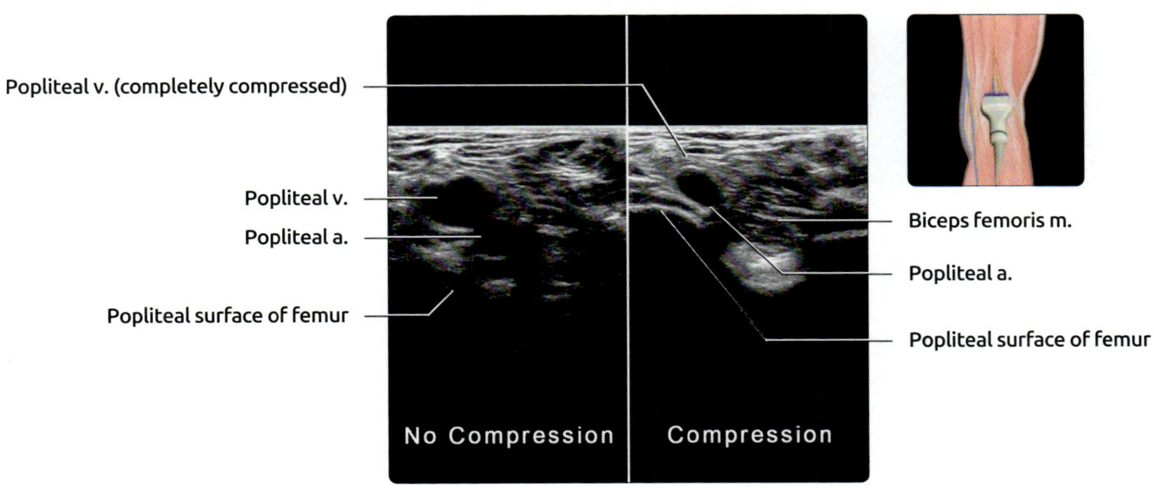

(Top) *Without and with compression split screen transverse scan shows the common femoral vessels. The left 1/2 of the image is obtained by applying normal gentle transducer pressure; the right 1/2 is scanned with firm compression. Patent veins are normally compressible with firm transducer pressure, with the exception demonstrated here due to the inguinal ligament. This is not a sign of venous thrombosis.* **(Middle)** *Without and with compression split screen transverse scan in the adductor canal shows the normal compressible superficial femoral vein. With firm pressure, the superficial artery is distorted, but the lumen is not obliterated.* **(Bottom)** *Without and with compression split screen transverse scan of the popliteal vessels shows the same response: Obliterated venous lumen and distorted arterial lumen.*

SCIATIC NERVE

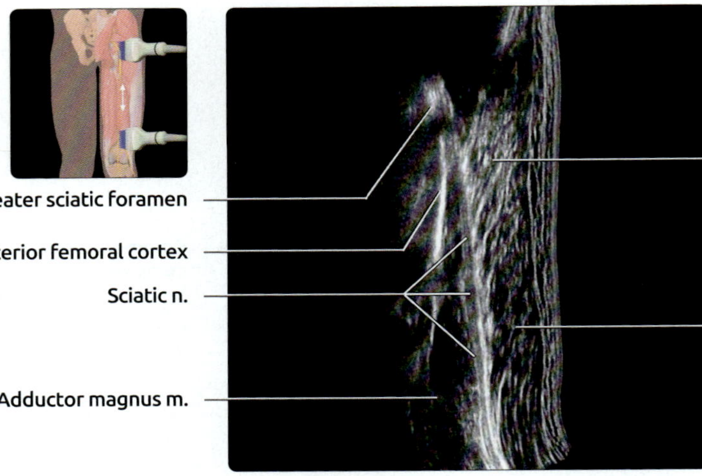

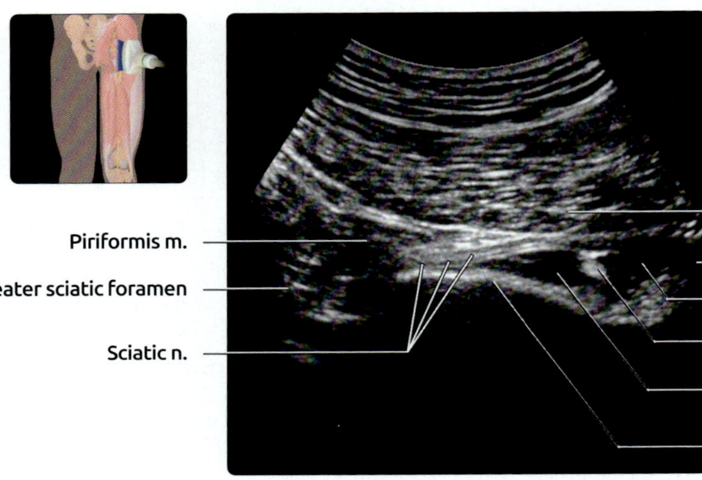

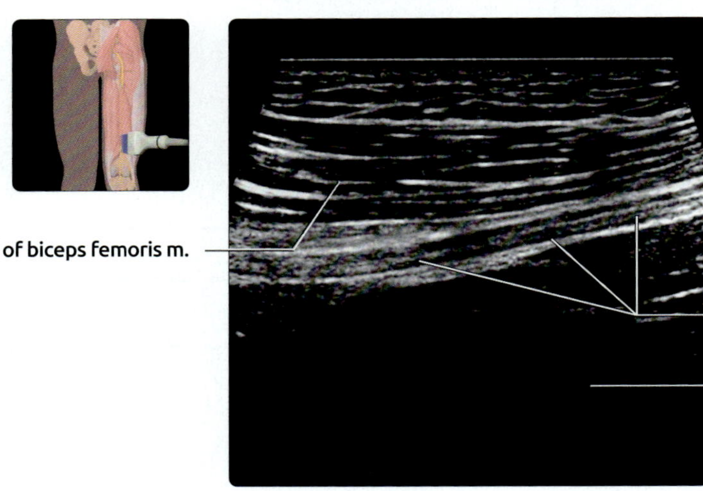

(Top) *Longitudinal panoramic view of the sciatic nerve shows its course after it has exited the greater sciatic foramen. It runs distally deep to the gluteus maximus and then deep to the biceps femoris to the popliteal fossa.* **(Middle)** *Sagittal scan shows the greater sciatic foramen using the ischial tuberosity as a landmark. The sciatic nerve and its relationship with the small external rotators of the hip is shown here. The sciatic nerve emerges between the piriformis and superior gemellus muscles after passing through the greater sciatic foramen.* **(Bottom)** *This longitudinal view of the sciatic nerve demonstrates the parallel and uniform hypoechoic tracts (representing nerve bundles). Any disruption of the uniformity or parallel course should be considered abnormal.*

TRANSVERSE SCIATIC NERVE

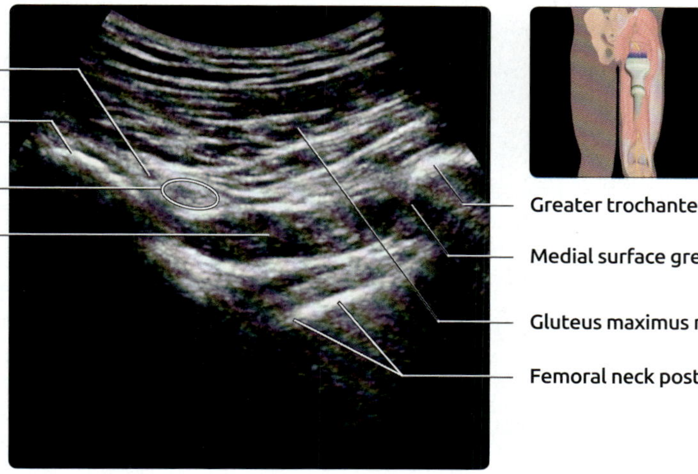

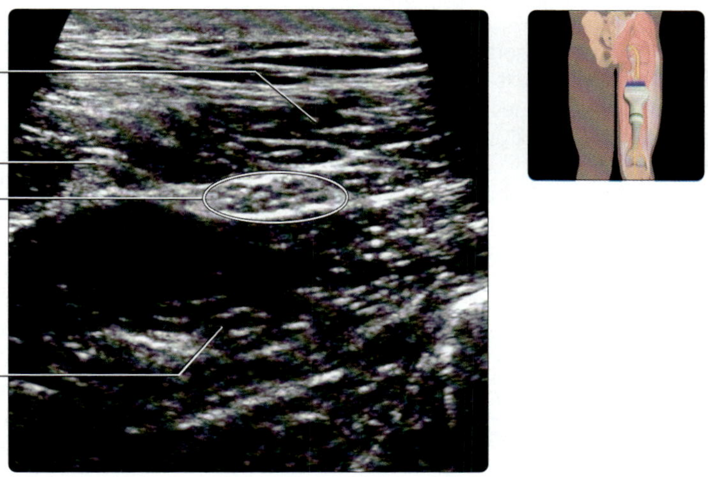

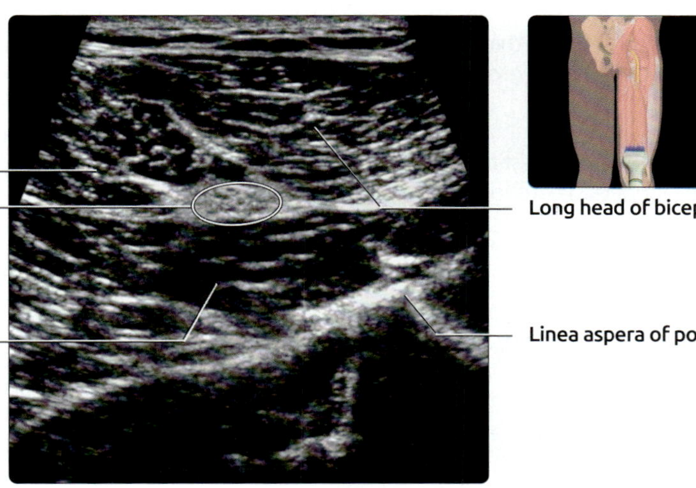

(Top) *Transverse scans at 3 different levels show the sciatic nerve. The sciatic nerve, after it has exited the greater sciatic foramen, passes superficial to the small external rotators of the hip joint (obturator internus, superior and inferior gemelli, and quadratus femoris). The inferior gluteal artery lies medial to the nerve.* **(Middle)** *At the upper thigh level, the sciatic nerve is covered by the long head of the biceps femoris muscle and is relatively superficial. The high spatial resolution of US allows individual nerve bundles to be identified as black dots in a bright matrix of epineurium.* **(Bottom)** *The same relationship with the long head of biceps femoris muscle continues inferiorly down the thigh until the sciatic nerve divides into the peroneal nerves.*

TERMINOLOGY

Abbreviations

- Origin (O), insertion (I), function (F)

GROSS ANATOMY

Overview

- Largest and most complex joint
    - Hinge joint throughout its main range of motion
    - Stability is maintained by ligaments, tendons, capsule, and menisci
- **Motion** of knee and relationship of osseous structures
    - In full flexion
        - Supporting ligaments are not taut and rotation of leg is allowed
    - During knee extension
        - Patella slides upward on femur, engaging with femoral trochlea until near full extension
        - Femoral condyles roll forward on tibial condyles and menisci
        - Lateral femoral condyle is shorter anteroposteriorly than medial so reaches full extension earlier
        - Medial tibial condyle continues to slide after lateral stops rotating slightly laterally on femur ("screw home" mechanism)
        - Tightens cruciate ligaments and stabilizes knee in extension
    - Initiating flexion from fully extended knee
        - Requires slight medial rotation of tibia, produced by popliteus muscle
        - Unlocks joint, allowing remainder of motion to take place

IMAGING ANATOMY

Osseous Anatomy

- **Distal femur**
    - Medial femoral condyle larger than lateral
    - Anteriorly, V-shaped trochlear groove accommodates patella
- **Proximal tibia**
    - Gerdy tubercle anterolateral just distal to joint
    - Tibial tubercle (apophysis) anterior and slightly lateral, several cm distal to joint
- **Proximal fibula**
    - Posterolateral relative to tibia
- **Tibiofibular joint**
    - True synovial joint; subject to any arthritic process
    - Connects to knee joint in 10%
- **Patella**
    - Triangular sesamoid bone wider at base superiorly than at apex inferiorly
    - Nonarticular outer surface may develop prominent enthesopathy where quadriceps tendon I blends into O of inferior patellar tendon
    - Bipartite (multipartite) patella: Always upper outer quadrant; osseous fragments may not appear to "match," but cartilage is continuous over defect

Articular Capsule

- Complex, noncontiguous structure

Extensor Mechanism

- **Quadriceps**
    - Quadriceps tendon inserts onto anterosuperior surface of patella
    - Tendon contributions from rectus femoris, vastus lateralis, medialis, and intermedius muscles
    - 3-layered structure but may have 1-5 laminae separated by fibrofatty connective tissue
        - Rectus femoris → anterior lamina, which continues over patella as prepatellar quadriceps continuation
        - Vastus lateralis and medialis → middle lamina
        - Vastus intermedius → posterior lamina
    - Tendons contribute not only to their own lamina but also to other lamina
- **Patellar tendon**
    - Runs from inferior pole of patellar to tibial tuberosity
        - Rectus femoris fibers that course over patella strengthen continuity with quadriceps tendon
    - Fibers of vastus lateralis and medialis contribute to lateral and medial patellar retinacula, respectively
- **Medial retinaculum**
    - 3-layered fascial sheet between patella and medial femoral condyle
        - Superficial layer: Investing fascia of thigh (crural fascia)
        - Middle layer: Medial collateral ligament (MCL) and medial patellofemoral ligament (MPFL)
        - Deep layer: Knee joint capsule to which synovium is attached
    - MPFL is strongest component of medial retinaculum
        - Contributes 60% of medial patellar stability
        - Extends from superomedial aspect patella to adductor tubercle at attachment of MCL
        - Femoral attachment of MPFL has transverse band (outer fibers) and oblique band (inner fibers)
            - Transverse fibers cross superficial to MCL
            - Oblique fibers attach superficially onto anterior fibers of MCL
- **Lateral retinaculum** (rarely injured)
    - 3-layered fascial sheet between patella and lateral femoral condyle
        - Superficial layer: Investing fascia of thigh (crural fascia)
        - Intermediate layer: Longitudinal fibers of iliotibial band and more transverse fibers from iliotibial band (iliotibial band-patella fibers)
        - Deep layer: Lateral patellofemoral ligament (indistinct thickening of joint capsule)

Fat Pads

- Suprapatellar: Anterior and posterior (prefemoral) to suprapatellar bursa
- Infrapatellar (Hoffa fat pad): Distal to patella, deep to capsule, and superficial to synovium
- Posterior: Posterior joint margin extending into intercondylar region

Medial Supporting Structures

- Superficial (layer 1)
    - Pes anserinus: Anteromedial tibial I
        - Sartorius embedded in crural fascia
        - Gracilis immediately deep to sartorius
        - Semitendinosus immediately deep to gracilis

- All insert onto anteromedial aspect tibia
- Semimembranosus: Posteromedial I
  - Oblique popliteal arm → oblique popliteal ligament (OPL) along posterior capsule knee
  - Direct arm → groove at posteromedial aspect tibia
  - Anterior arm → medial aspect of tibia deep to superficial component MCL
  - Capsular arm → posterior oblique ligament (POL), which is capsular thickening of posteromedial corner of knee
- Middle (layer 2)
  - Superficial MCL (longitudinal and oblique components)
    - Runs slightly anteromedially from medial epicondyle to insert onto tibia 5 cm distal to joint line
    - Anteriorly, longitudinal component fascia blends with superficial layer
    - Posteriorly, oblique component blends with deep layer as POL
- Deep (layer 3)
  - Capsular layers (deep fibers of MCL) at midportion of knee
    - Meniscofemoral ligament: From outer superior aspect of body of meniscus to either superficial MCL or femur
  - More posteriorly, superficial MCL blends with capsular layers of MCL
  - POL arises from oblique fibers of superficial MCL extending posteriorly
    - Blends with posteromedial meniscus
    - Receives fibers from semimembranosus tendon
    - Envelops posteromedial aspect of femoral condyle
    - Blends with OPL along posterior capsule of knee

Lateral Supporting Structures

- Superficial (layer 1)
  - Iliotibial band anteriorly, inserts onto Gerdy tubercle
    - Fibers of iliotibial band and anterior oblique band of lateral collateral ligament (termed anterolateral ligament) → tibia between Gerdy tubercle and fibular head
      - Segond fractures occur at this more posterior I
      - Anterior arm of short head of biceps femoris tendon inserts just distal to this point
  - Biceps femoris short and long heads
    - Long head: Ischial tuberosity
    - Short head: Posterior aspect of distal femoral shaft
      - Both heads insert onto fibular head (direct head) and lateral tibia just distal to anterolateral ligament I (anterior arm)
- Middle (layer 2)
  - Quadriceps retinaculum anteriorly
  - Posteriorly, 2 ligamentous thickenings, which originate from lateral patella
    - Proximal: Terminates at lateral intermuscular septum on femur
    - Distal: Terminates at femoral I of posterolateral capsule and lateral head of gastrocnemius
- Deep (layer 3)
  - Several thickenings in lateral part of capsule F as discrete structures
    - Lateral (fibular) collateral ligament originates from lateral femoral epicondyle, extends posterolaterally to insert on lateral fibular head
    - Arcuate ligament originates from styloid process of fibular head, interdigitates with popliteus, and inserts into posterior capsule near OPL

Internal Structures

- **Menisci**
  - Fibrocartilage cushion to stabilize knee
  - Only peripheral portion visualized by US
  - Medial attached to capsule throughout extent of knee
  - Lateral attached to capsule anteriorly and far posteriorly, but by fascicles to popliteus at body and posterior horn
- **Cruciate ligaments**
  - Major stabilizing structures to anteroposterior motion
  - Not well visualized by US

Synovial Recesses of Knee Joint

- Suprapatellar, medial patellar, lateral patellar
  - Increase in synovial fluid 1st accumulates in either medial or lateral recess and then suprapatellar recess
  - Small amount of fluid in knee joint is normal but should not get more than small amount
  - Synovial proliferation can also be best assessed at medial and lateral recesses
- Baker cyst is distension of gastrocnemius-semimembranous bursa
  - Slit-like opening occasionally connects knee joint to Baker cyst
  - Synovial fluid can be expressed from knee joint into Baker cyst if this interconnection is present

ANATOMY IMAGING ISSUES

Imaging Approaches

- US
  - Very good for assessing extraarticular abnormalities, e.g., tendinosis, ligament disruption, bursitis, masses, etc.
  - Given size and complexity, targeted approach to knee US examination is best
    - e.g., posterior examination for popliteal mass, lateral examination for iliotibial band syndrome, etc.
  - In acutely inflamed joint, US is most effective as screening tool for effusion, facilitating image-guided aspiration or synovial biopsy
  - For assessment of ligaments and tendons, longitudinal examination is more helpful in confirming integrity
    - Slightly flex knee for examination of anterior structures ± as adjunct to examination of other areas
    - Flexion tenses many tendons and ligaments and accentuates straight course of fibers

SELECTED REFERENCES

1. De Maeseneer M et al: Ultrasound of the knee with emphasis on the detailed anatomy of anterior, medial, and lateral structures. Skeletal Radiol. 43(8):1025-39, 2014

EXTENSOR MECHANISM

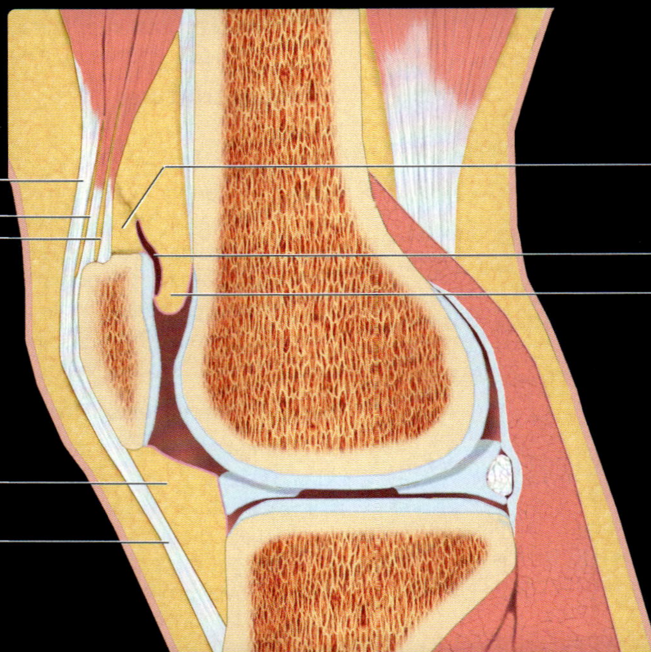

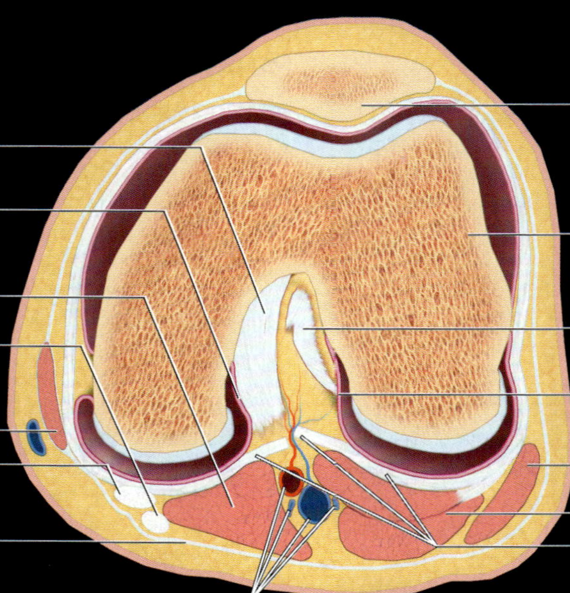

(Top) *Graphic shows trilaminar configuration of the quadriceps tendon. The superficial portion is the rectus femoris, the middle portion is the aponeurosis of the vastus medialis and lateralis, and the deep portion is the vastus intermedius tendon.* **(Bottom)** *Graphic shows the relationships of the posterior capsule and overlying structures. The most superficial layer is the crural fascia, which envelops the sartorius and otherwise confines all of the structures. The posterior capsule (white) is continuous with the synovium (pink thin line) both posteromedially and posterolaterally. Superficial to the capsule are the plantaris and gastrocnemius muscles, hamstring muscle and tendons, and the popliteal vessels.*

EXTENSOR MECHANISM

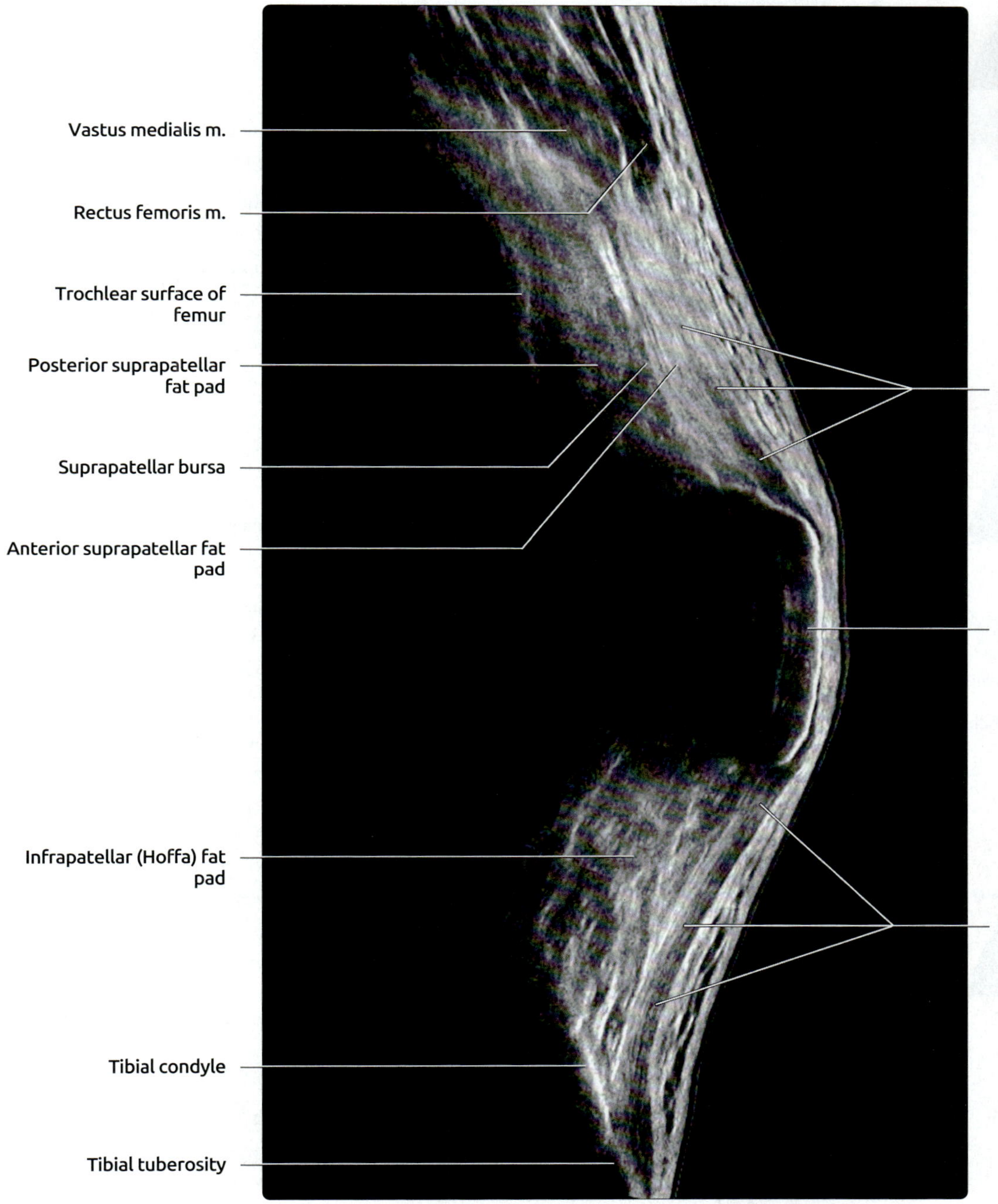

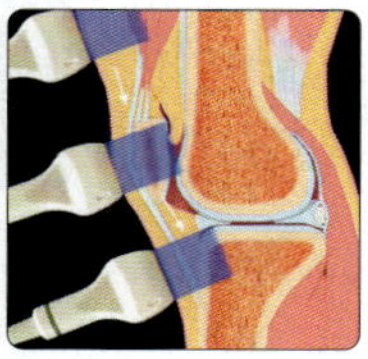

Sagittal panoramic scan of the anterior knee shows the main components of the extensor mechanism: Quadriceps muscle and tendon, patella, and patellar tendon. The scan performed here is with the knee in extension. More focused views of individual components of the anterior knee can be performed with the knee in flexion (with a pillow to support the popliteal region). Flexion increases the tension of the extensor mechanism, thus straightening the tendons. This will help highlight abnormalities in terms of tendon alignment and continuity. However, the amount of suprapatellar fluid may be decreased (due to displacement) in flexion, and assessment of the suprapatellar region in extension may be more sensitive in this respect.

EXTENSOR MECHANISM

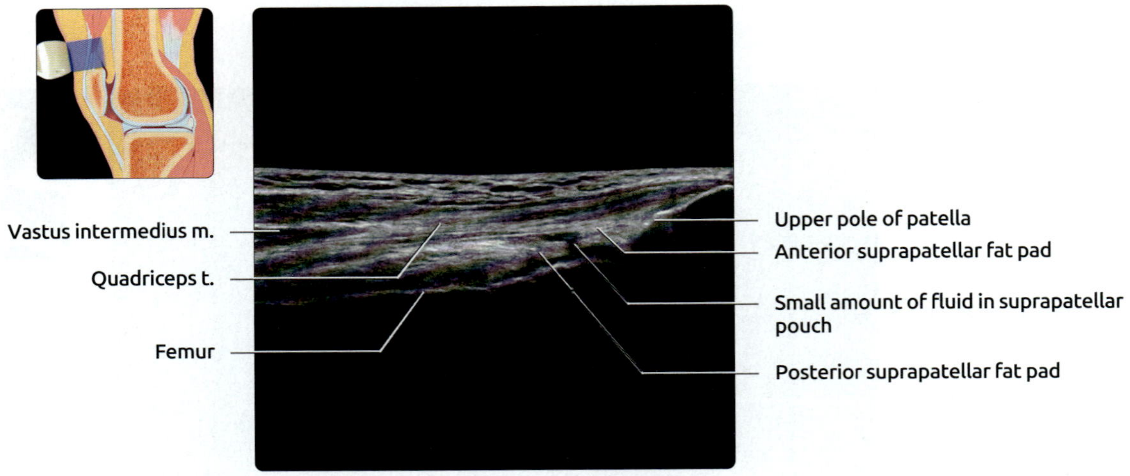

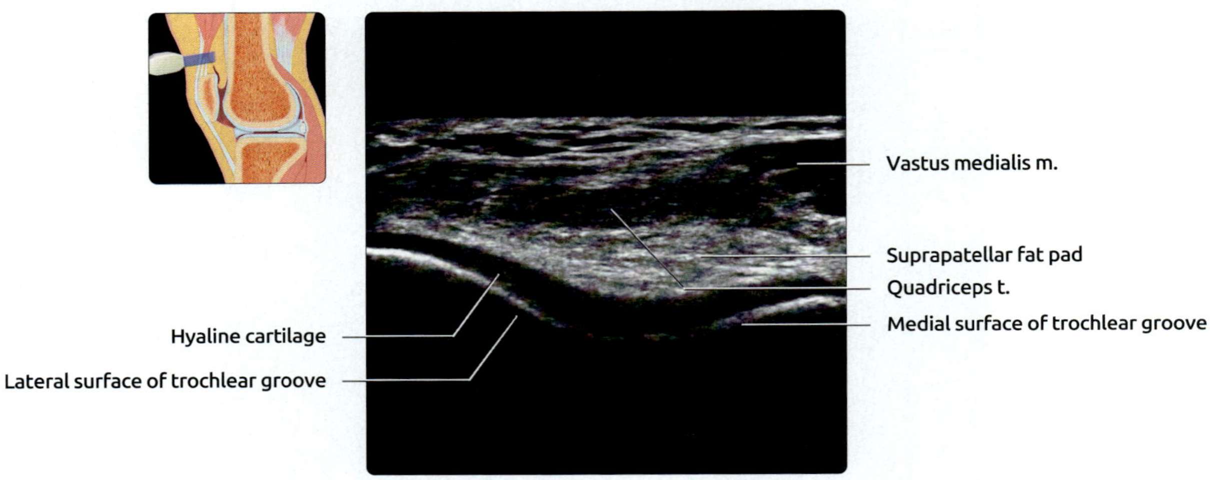

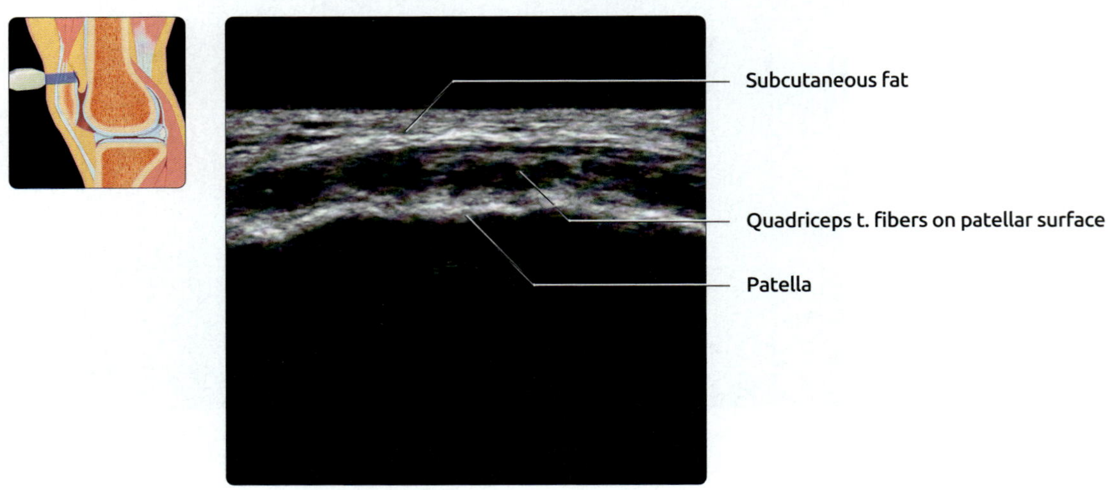

(Top) *Sagittal US shows the quadriceps tendon. The quadriceps tendon is the direct continuation of the rectus femoris with contributions medially and laterally from the vastus medialis and the vastus lateralis, respectively, and posteriorly from vastus intermedius. The tendon inserts on the superoanterior surface of the patella. In this scan, the suprapatellar pouch can be seen. It may normally contain a small amount of fluid (several millimeters thick).* **(Middle)** *Transverse US shows quadriceps tendon with the knee flexed. The central position of the quadriceps tendon can be appreciated here, receiving contributions from the vastus muscles. The articular cartilage of the trochlear groove can also be seen with the patella out of the way during flexion.* **(Bottom)** *Transverse US of the anterior surface of the patella shows tendon fibers inserting or running on the patellar surface.*

EXTENSOR MECHANISM

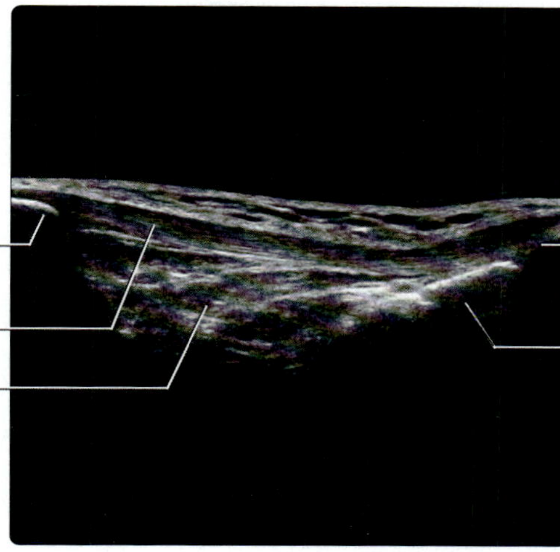

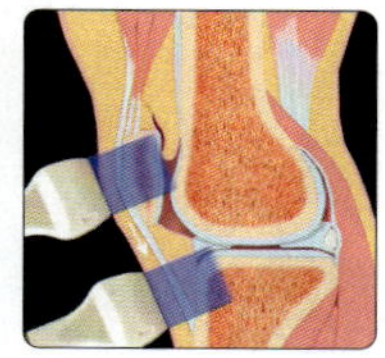

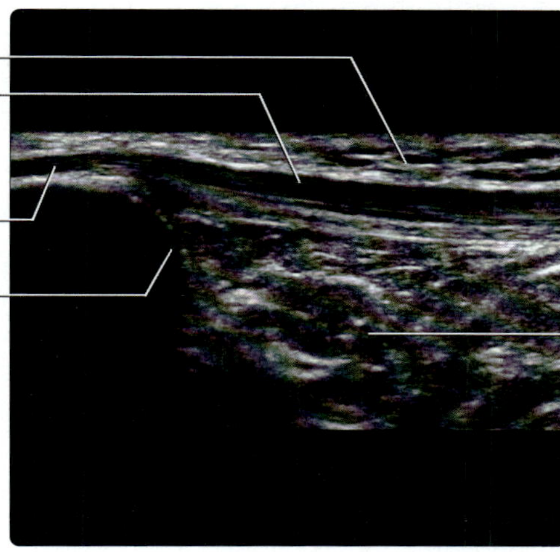

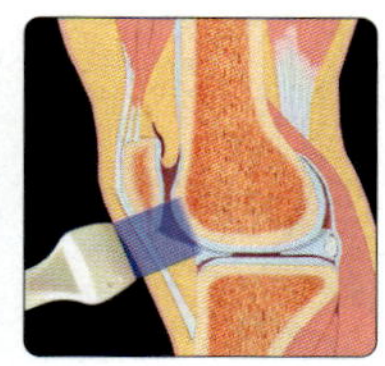

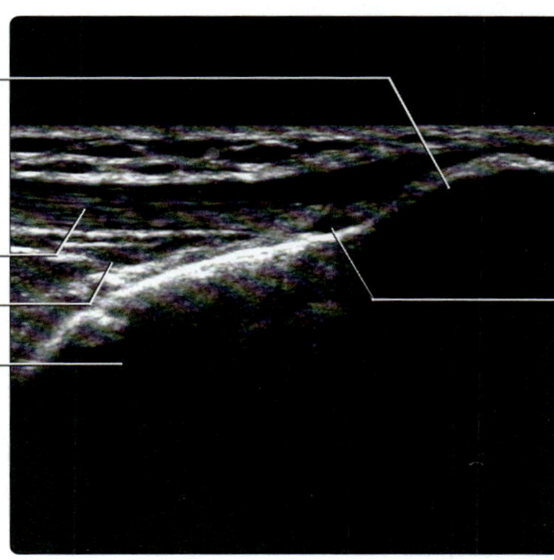

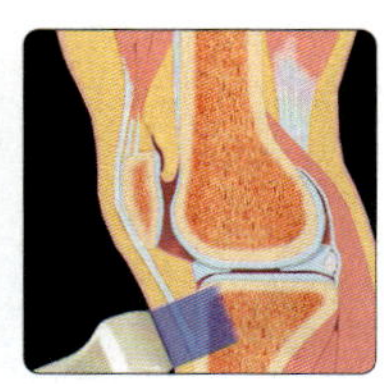

(Top) *Sagittal panoramic view shows the patellar tendon. Most of the fibers of the tendon are continuations of quadriceps tendon fibers.* **(Middle)** *Focused sagittal view of the apex of the patella shows quadriceps fibers running on the anterior surface of the patella, continuing distally as the patellar tendon. For this reason, many consider the patella a sesamoid bone. Also seen here are fibers originating from the apex of the patella itself.* **(Bottom)** *Focused sagittal scan shows the tibial tuberosity. The distal fibers of the patellar tendon are separated from the anterior surface of the tibial condyle by the infrapatellar bursa. A small amount of fluid (2-3 mm thick) may be normally present in this bursa.*

EXTENSOR MECHANISM

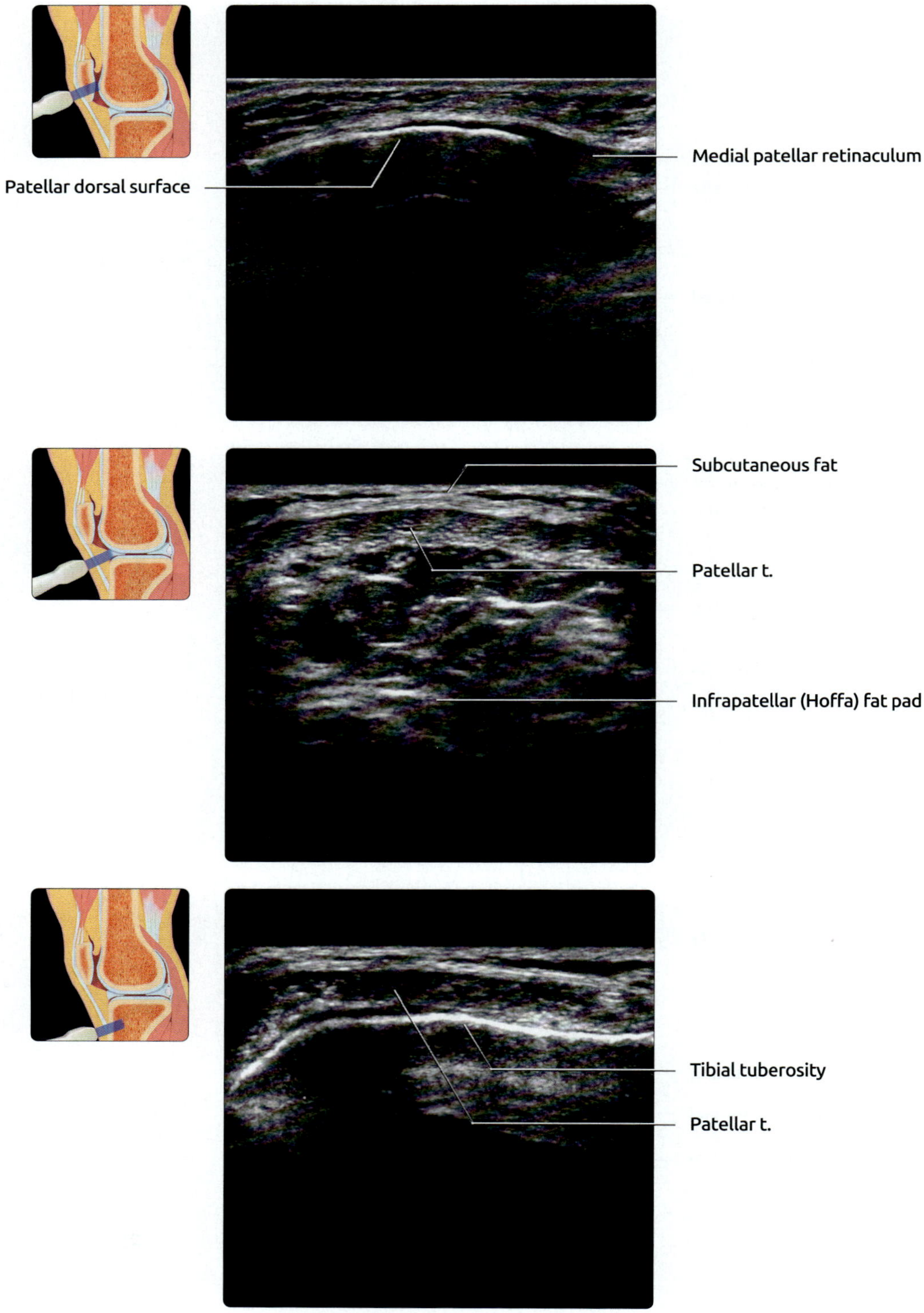

(Top) *Series of 3 transverse scans through the patellar tendon is shown. Proximally, fibers from the quadriceps tendon continue over the anterior patellar surface to contribute to the patellar tendon. On both sides, fibers from the medial and lateral patellar retinaculum also insert onto the patella and blend with the patellar tendon fibers.* **(Middle)** *At the midsubstance of the patellar tendon, the tendon assumes a flattened appearance. The fibers within the tendon can be seen as tiny hyperechoic dots. A disruption to the uniformity of these dots or a change in thickness (focal or general) of the patellar tendon suggests tendinosis.* **(Bottom)** *More inferiorly, a transverse scan shows the patellar tendon inserting onto the tibial tuberosity. The patellar tendon paratenon is seen as thin echogenic line on the anterior surface and side walls of the tendon. It becomes thickened in patellar paratenonitis. This is an infrequently encountered condition.*

EXTENSOR MECHANISM

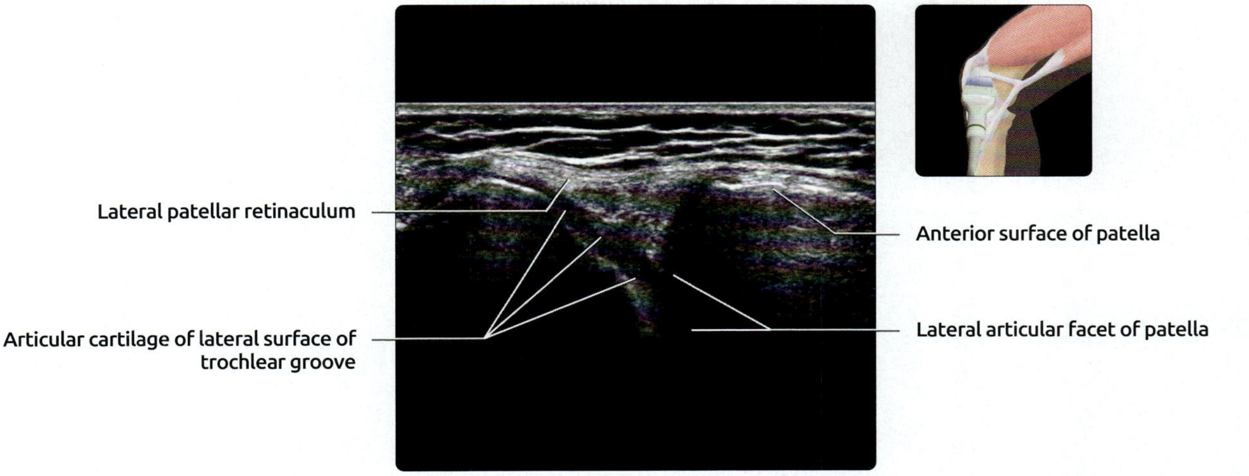

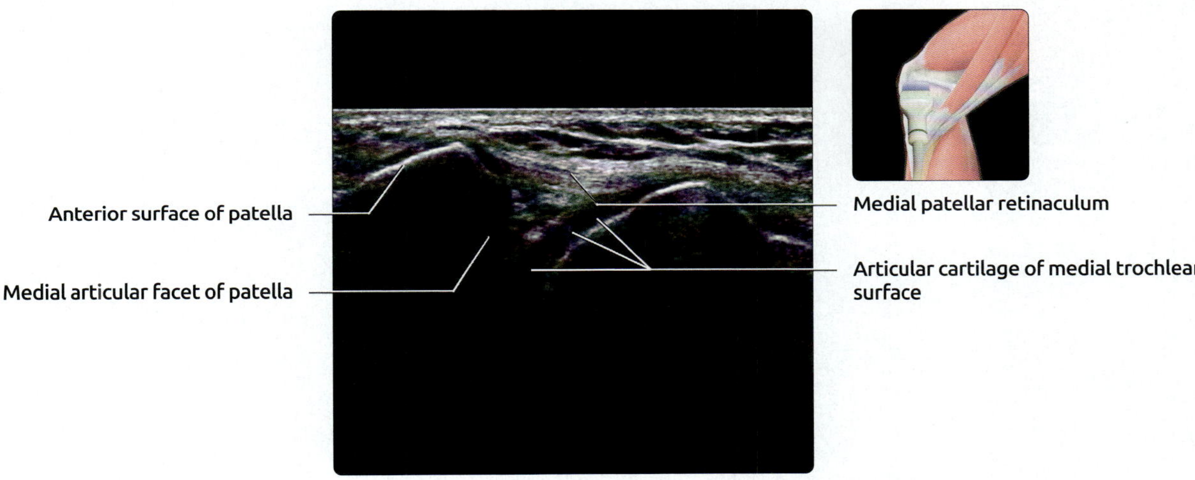

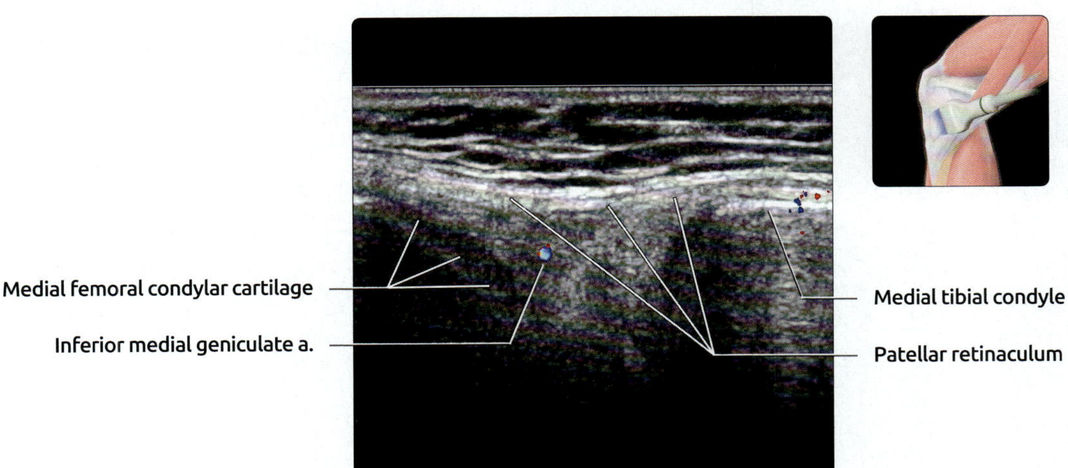

(Top) *Transverse US at the lateral margin of the patella shows the articulation between the lateral patella and the lateral surface of the trochlear groove. This articular plane is steeper (closer to sagittal) than the medial side for mechanical advantage. As a result of the obliquity of the course of the patellar tendon with respect to the tibial axis, the patella is prone to lateral dislocation or subluxation. The steeper lateral articular plane helps prevents this.* **(Middle)** *Transverse US at the medial margin of the patella shows the articulation between the patella and the medial surface of the trochlear groove, showing the patellar retinaculum as a thick structure.* **(Bottom)** *Sagittal color Doppler US of the medial infrapatellar region shows the patellar retinaculum inserting onto the tibia. At this level, the retinaculum has fanned out, becoming much thinner.*

MEDIAL KNEE

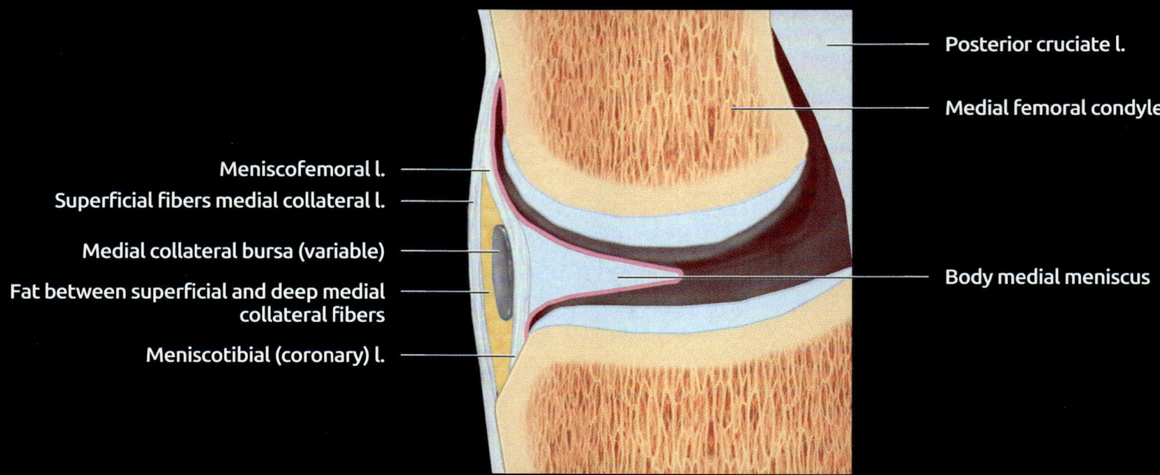

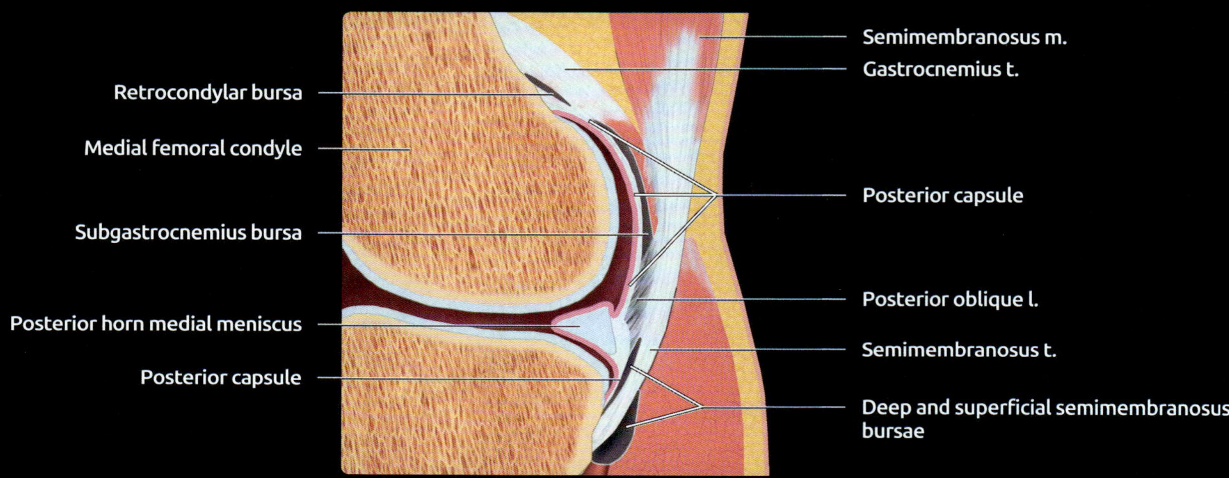

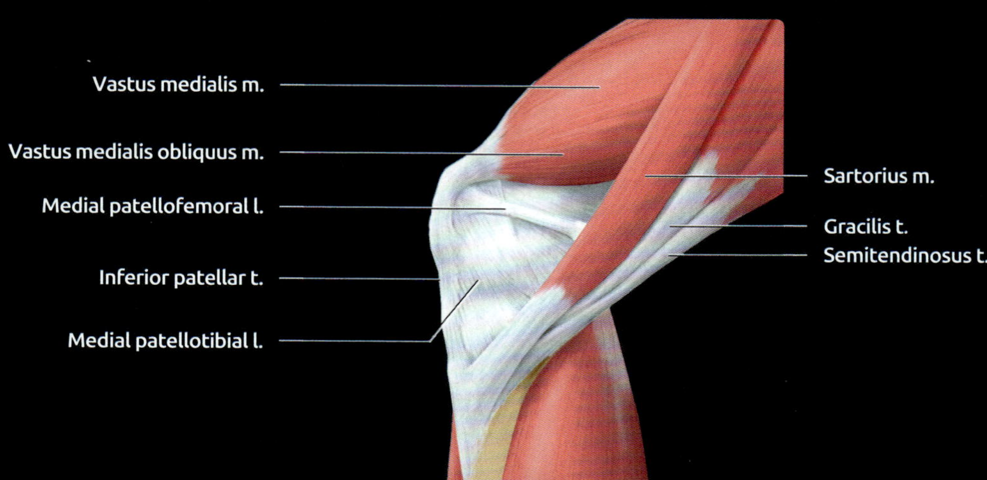

(Top) *Coronal graphic at the midjoint shows superficial fibers extending from the adductor tubercle to ~ 5 cm below the joint line. Deep fibers are shorter. The meniscotibial (coronary) ligament extends from the meniscus to the tibia adjacent to the joint line and the meniscofemoral ligament extends from the meniscus to the femur or superficial medial collateral ligament.* **(Middle)** *Sagittal graphic shows the far medial portion of the posterior capsule. Superiorly, the capsule attaches to the cortex of the posterior femoral condyle and fuses with fibers of the medial gastrocnemius.* **(Bottom)** *Graphic shows the anteromedial attachment of pes anserinus tendons. The sartorius is the most anterior and superficial. The gracilis inserts directly adjacent and deep to the sartorius, and the semitendinosus inserts directly posterior and inferior to the gracilis. These tendons and their crural fascia comprise the superficial layer of the medial capsuloligamentous complex.*

MEDIAL KNEE

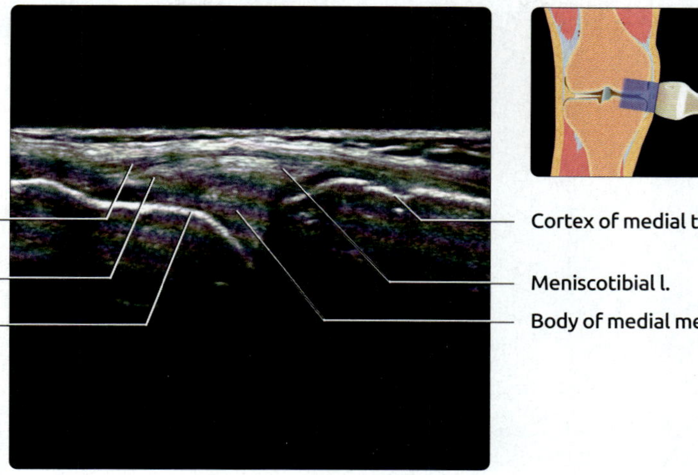

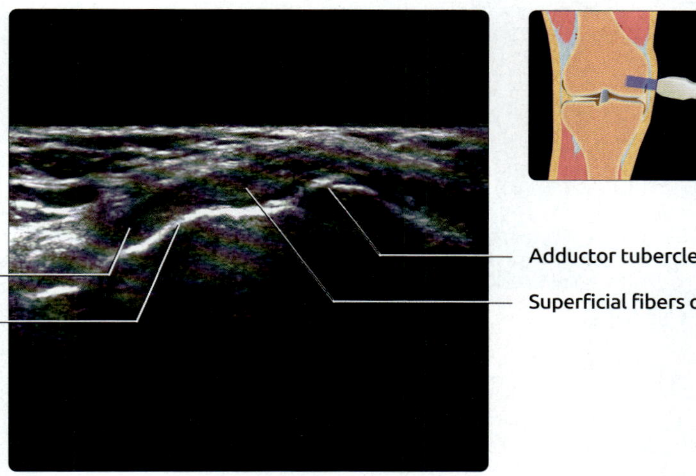

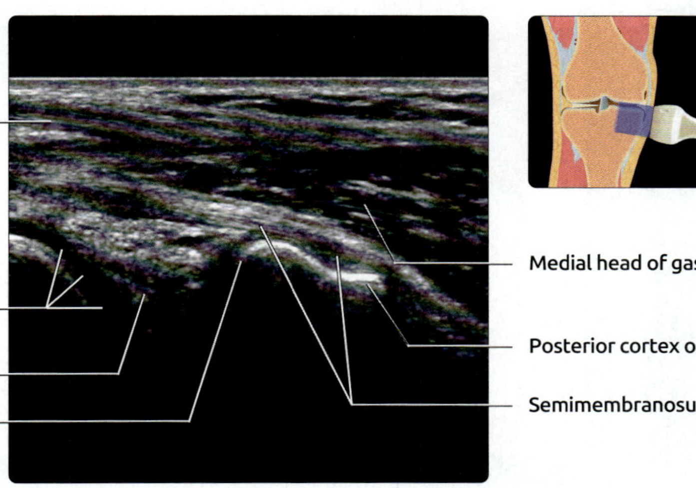

(Top) *Coronal US shows the medial joint line. The medial collateral ligament is composed of 2 layers: Superficial and deep. The superficial fibers run from the medial femoral condyle to the tibial condyle 5 cm below the joint line. The deep layer consists of the meniscofemoral (femoral condyle to body of meniscus) and meniscotibial (body of meniscus to tibia just distal to joint line).* **(Middle)** *Transverse US shows the medial condylar region. The superficial fibers of the medial collateral ligament arise just distal to the adductor tubercle. The popliteal oblique ligament arises adjacent to this and runs posterior to the posterior joint capsule.* **(Bottom)** *Sagittal US shows the posteromedial joint line. The semimembranosus tendon inserts onto the posterior surface of the medial tibial condyle, just below the joint margin. Only the outer part of the meniscus is visible.*

LATERAL KNEE

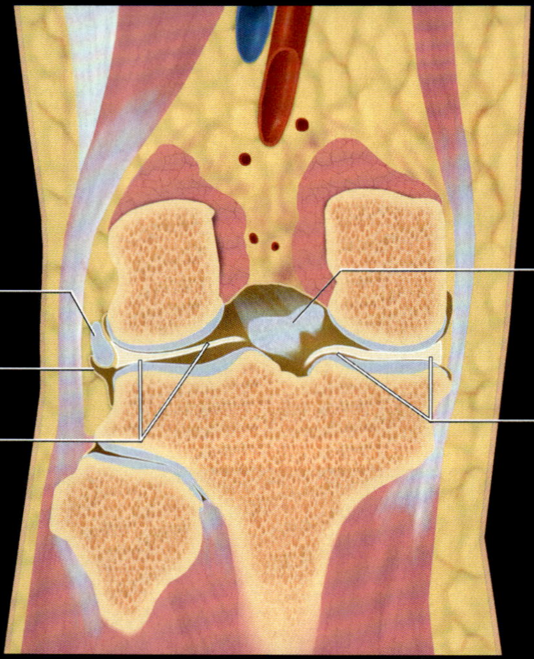

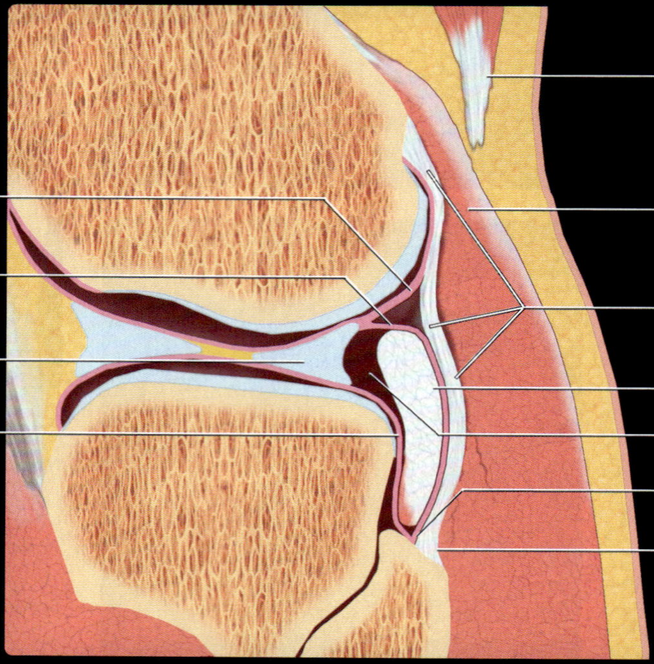

(Top) *Coronal graphic shows the roots of the posterior horns of both the medial and lateral menisci. The popliteus tendon courses through the hiatus adjacent to the body/posterior horn of the lateral meniscus.* **(Bottom)** *Sagittal graphic of the lateral part of the posterior capsule shows that it is intimately attached to the lateral gastrocnemius muscle. The popliteus tendon is intraarticular but extrasynovial and is attached to the capsule. Fibers from the posterolateral corner and arcuate ligament originating from the fibular head contribute to the lateral portion of the posterior capsule.*

LATERAL KNEE

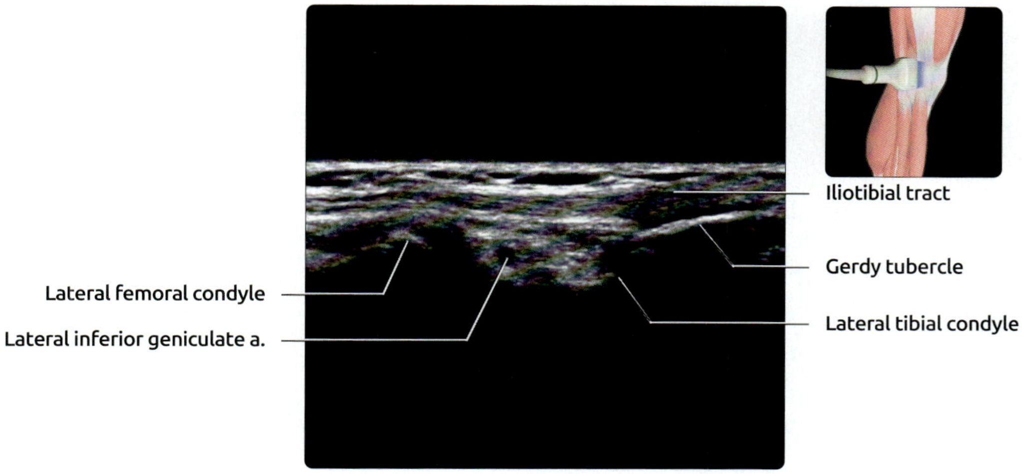

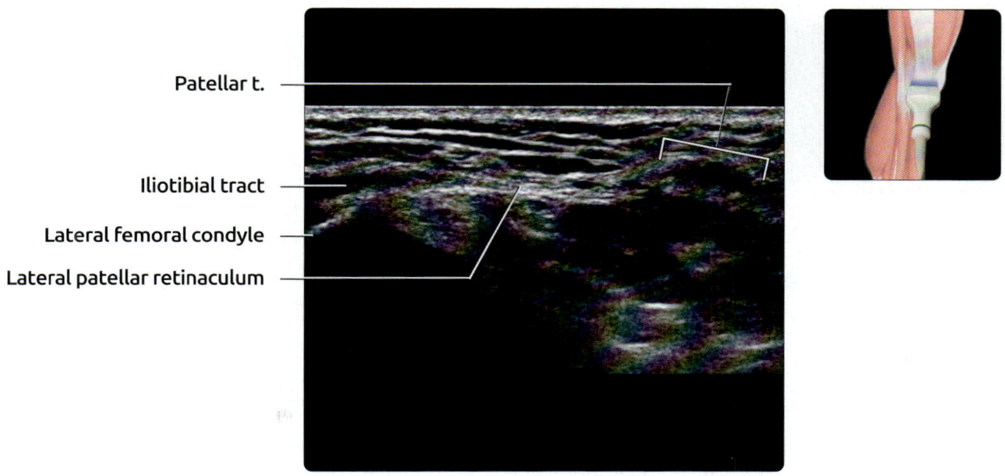

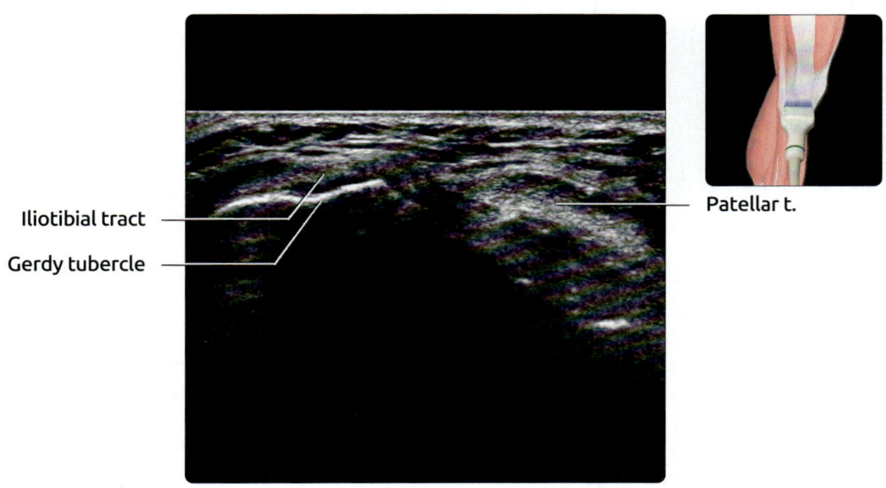

(Top) *Oblique sagittal US shows the lateral aspect of the joint line using the Gerdy tubercle as a landmark. The iliotibial tract is shown here, passing over the edge of the lateral femoral condyle [where it is subject to impingement (iliotibial tract syndrome)] before inserting on the Gerdy tubercle on the anterior surface of the lateral tibial condyle.* **(Middle)** *Transverse US of the anterolateral knee at the level of the femoral condyle shows that the iliotibial tract and the patellar tendon are connected by fibers of the lateral patellar retinaculum.* **(Bottom)** *More inferiorly, the 2 tendons have moved closer to each other. This transverse US shows insertion of the iliotibial tract onto the Gerdy tubercle. The patellar tendon is also approaching its tibial insertion.*

LATERAL KNEE

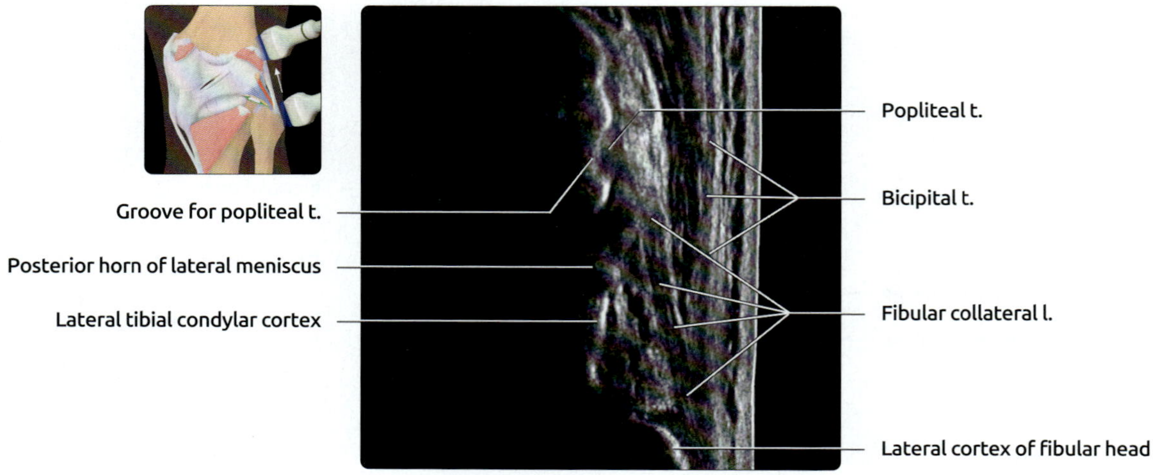

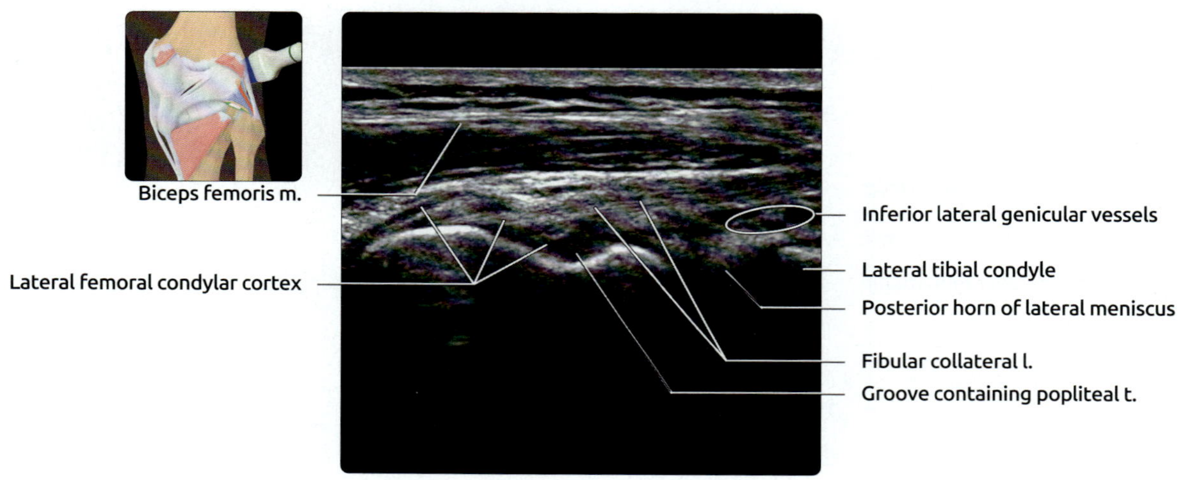

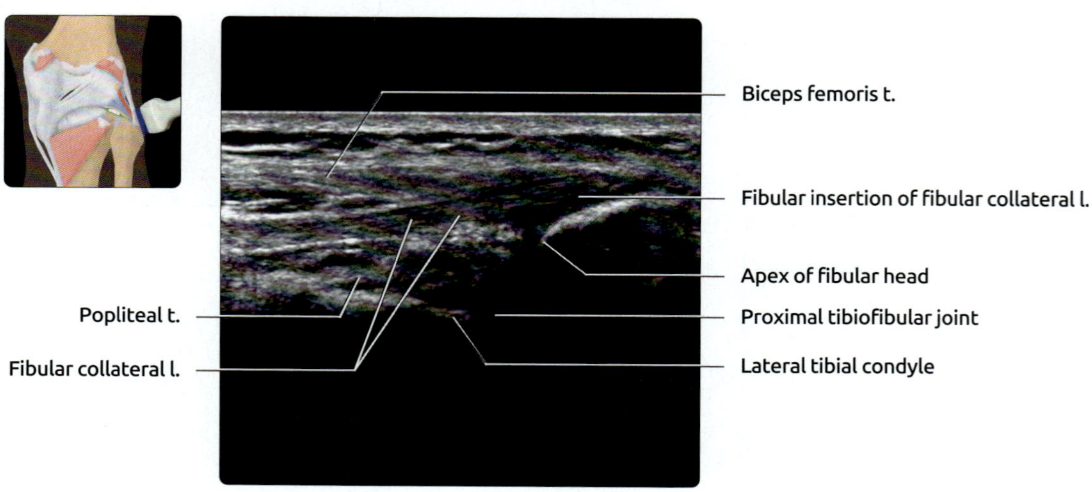

(Top) *Coronal panoramic US shows the lateral collateral complex. The biceps femoris muscle is the dominant muscle on this side of the knee. It runs from the ischial tuberosity laterally and inferiorly to insert onto the lateral aspect of the head of the fibula.* **(Middle)** *Coronal US of the lateral femoral condyle shows the insertion of the fibular collateral ligament. Just under this insertion is the groove for the popliteal tendon. The fibular collateral ligament is cord-like compared to the superficial component of the lateral collateral ligament, which is a flatter strap.* **(Bottom)** *The fibular collateral ligament inserts onto the lateral wall of the head of the fibula. The tendon of the biceps femoris inserts at and around the apex of the fibular head.*

POSTERIOR KNEE

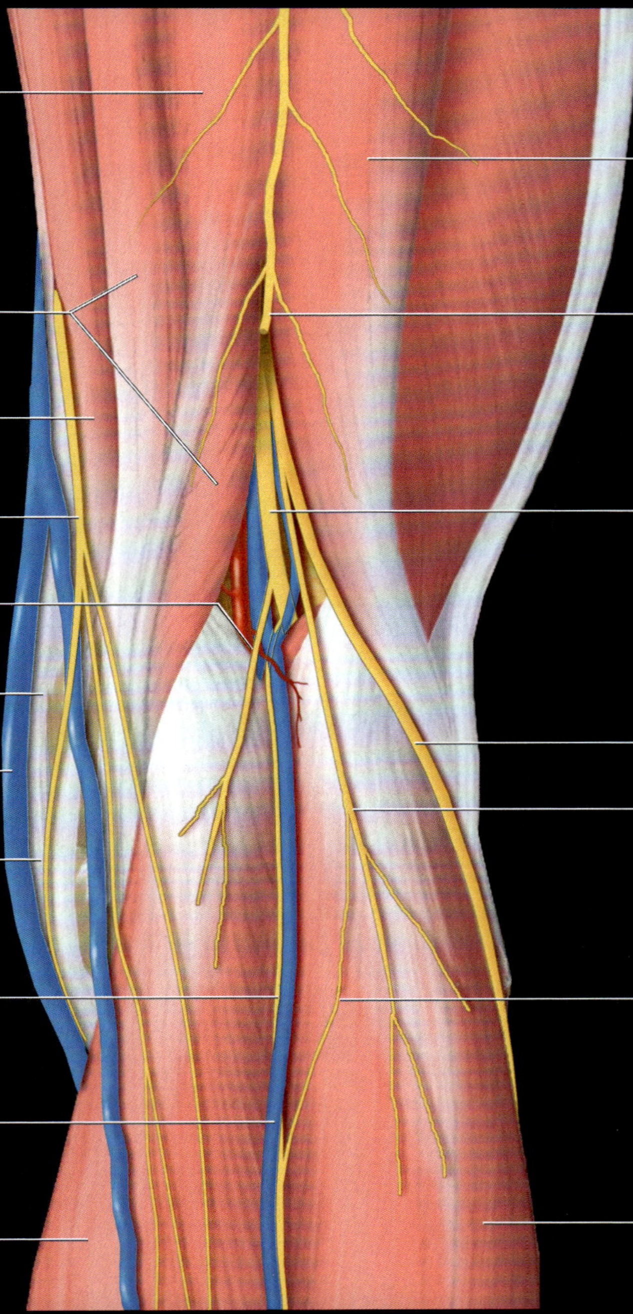

Graphic shows the superficial posterior muscles and nerves. The popliteal fossa is a diamond-shaped fossa bordered superiorly by the hamstring muscles and inferiorly by the head of the gastrocnemius muscle. Superiorly, the medial margin is represented by the semimembranosus and semitendinosus muscles and laterally by the biceps femoris muscle. Inferiorly, the medial and lateral borders are the medial and lateral heads of the gastrocnemius muscle. This fossa contains the popliteal vessels, the sciatic nerve, and its branches. Note that the common peroneal nerve is part of the deeper nerve system, but it travels more superficially, following the posterior biceps femoris muscle and tendon until it wraps around the fibular neck.

POSTERIOR KNEE

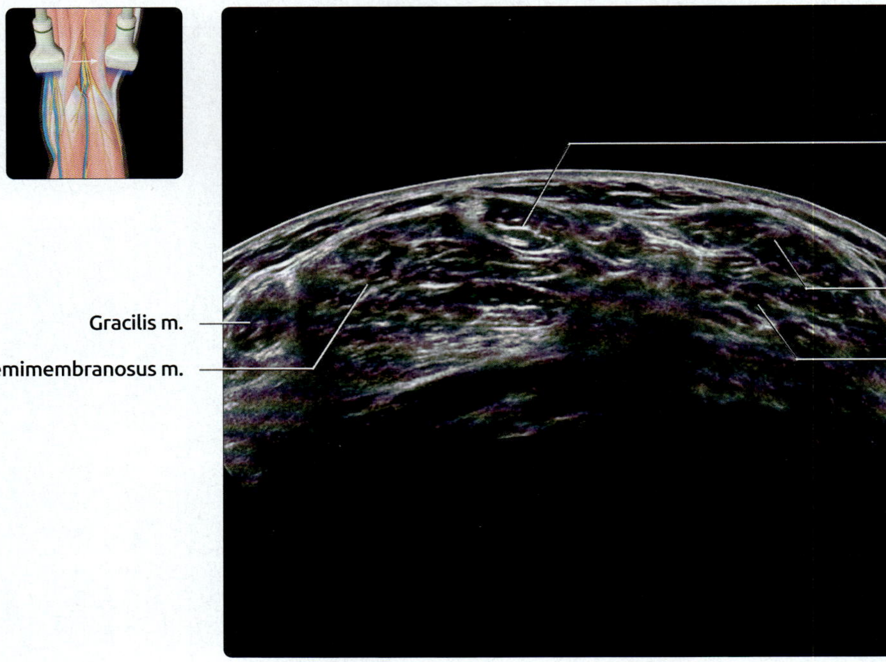

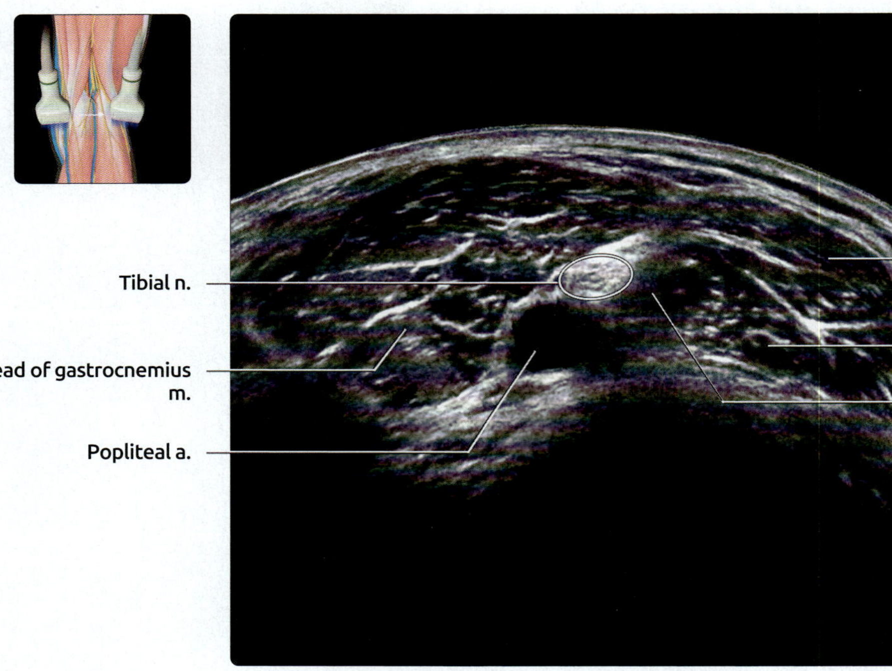

(Top) *Panoramic transverse US of the posterior knee shows the upper popliteal fossa. At this level, the hamstring muscles form the superior boundaries (medial and lateral) of the popliteal fossa. The semimembranosus and semitendinosus muscles form the medial margin, while the biceps femoris muscle forms the lateral margin.* **(Bottom)** *Panoramic view of the posterior knee shows the lower aspect of the popliteal fossa. At this level, the boundaries of the popliteal fossa are the heads of the gastrocnemius muscle, the lateral head of the gastrocnemius being the lateral margin and the medial head being the medial margin. The popliteal vessels are a continuation of the superficial femoral vessels after they have emerged through the adductor hiatus.*

POSTERIOR KNEE

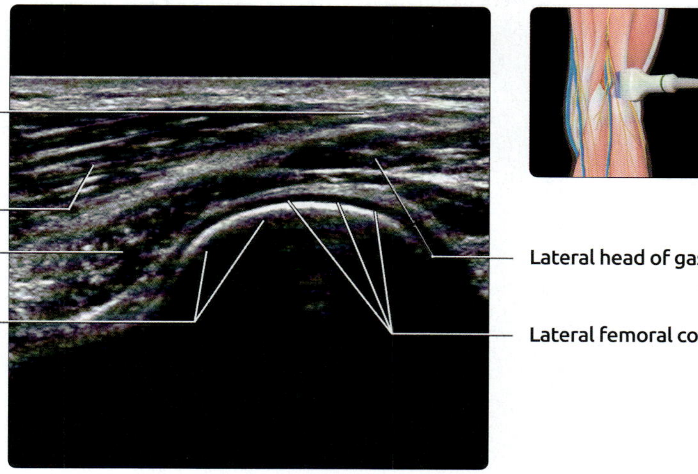

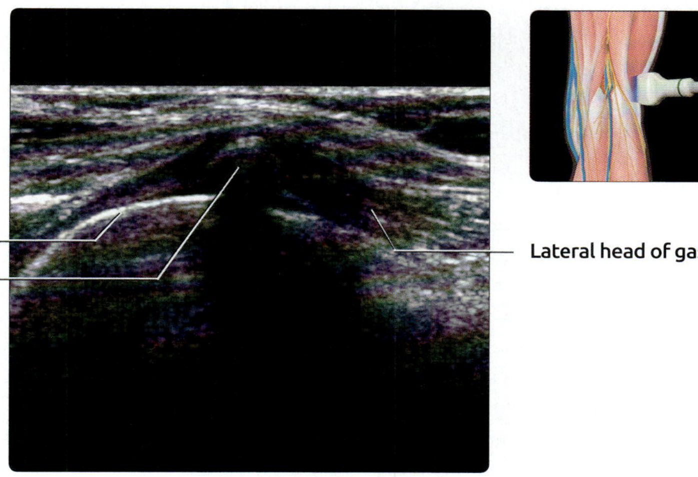

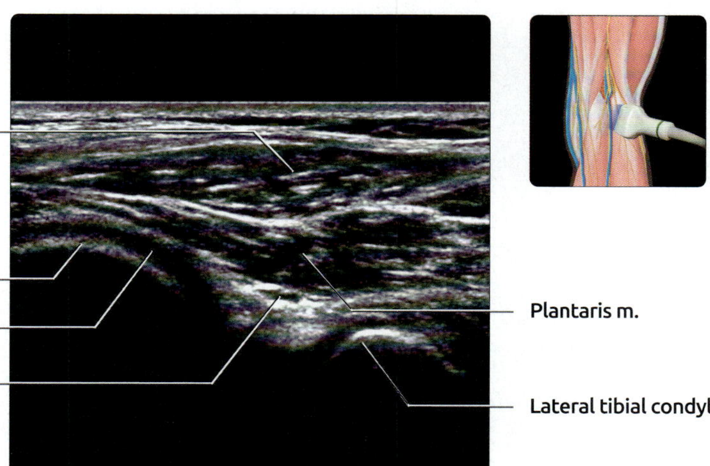

(Top) *Sagittal US shows the lateral femoral condyle. The plantaris muscle originates slightly more superior and medial to the lateral head of the gastrocnemius. The plantaris runs inferomedially across the thigh to lie on the medial side of the Achilles tendon. More superficially, the bicipital tendon runs laterally toward its fibular head insertion.* **(Middle)** *A more lateral sagittal view shows the sesamoid bone within the lateral head of the gastrocnemius muscle, the fabella.* **(Bottom)** *A more inferior oblique US shows both the plantaris and lateral heads of the gastrocnemius muscle running over the posterior surface of the knee joint. Further distally, the plantaris muscle becomes a thin tendon and is difficult to discern until it reaches the medial side of the Achilles tendon in the distal calf.*

POSTERIOR KNEE

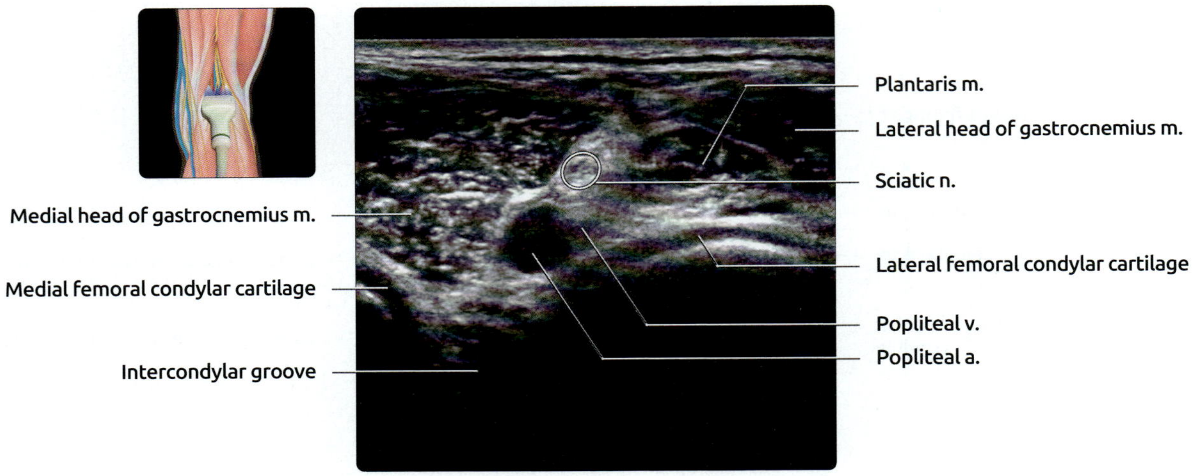

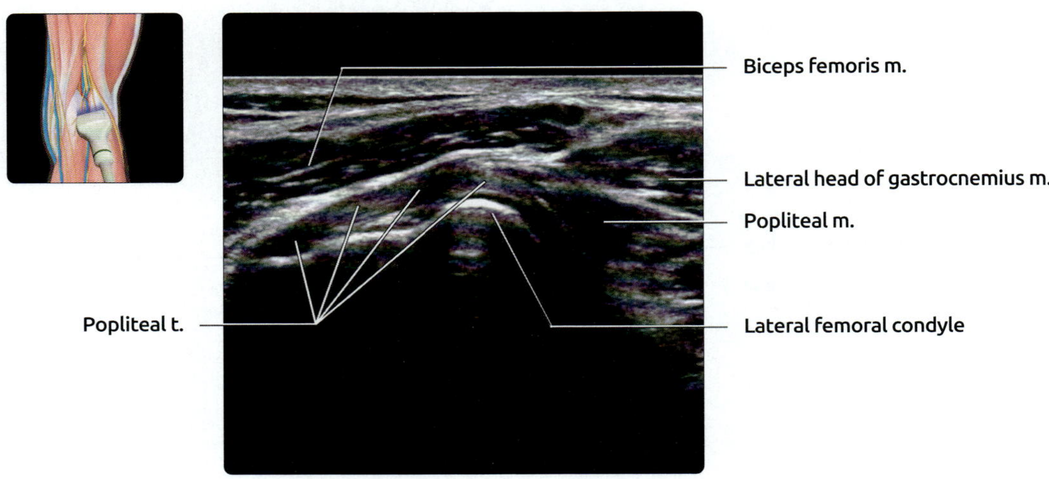

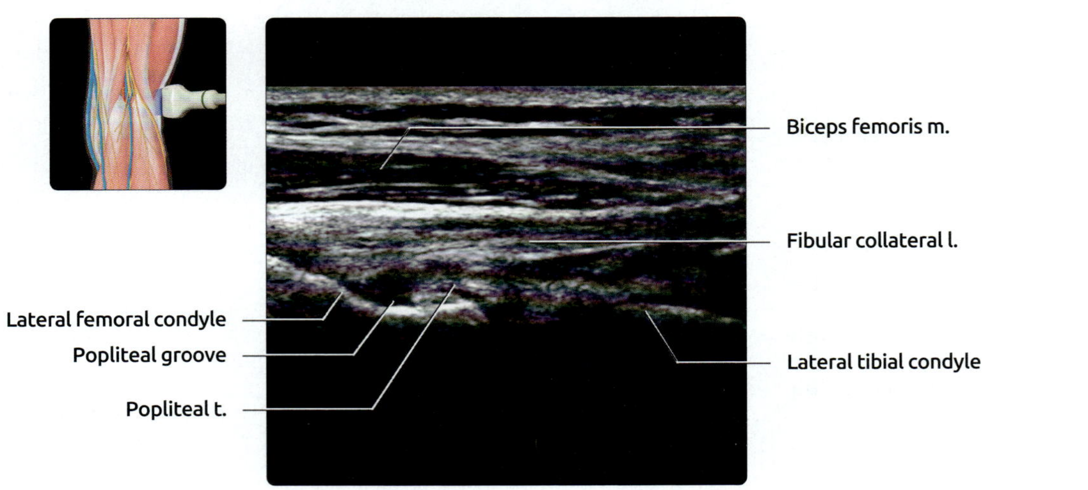

(Top) *Transverse US shows the posterolateral aspect of the knee joint at the level of the inferior popliteal fossa. The sciatic nerve is the largest nerve in the body. In cross section, nerve fibrils are seen as hypoechoic dots separated by hyperechoic perineurium.* **(Middle)** *Scanning more laterally and inferiorly, the popliteal muscle and its tendon can be demonstrated. The popliteal tendon originates from the proximal posterior tibial surface and runs superolaterally under the lateral head of the gastrocnemius muscle to insert onto the lateral cortex of the lateral femoral condyle.* **(Bottom)** *Coronal US shows the popliteal groove on the lateral femoral condyle and the popliteal tendon entering this groove as it heads toward its insertion.*

POSTERIOR KNEE

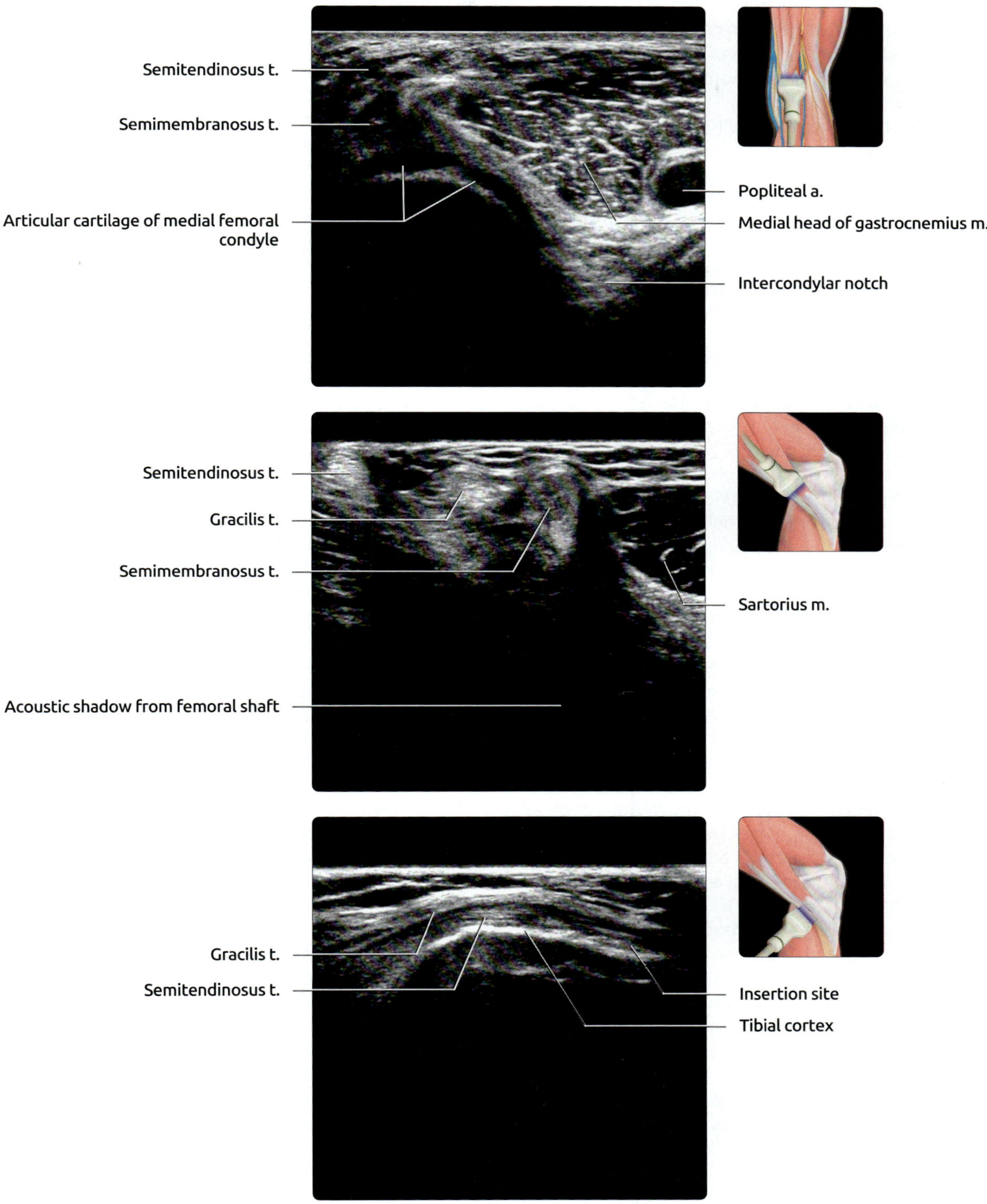

(Top) *Transverse US shows the posteromedial knee at the level of the femoral condyles. The semimembranosus tendon can be seen running medial to the medial head of the gastrocnemius muscle and deep to the semitendinosus tendon. The semimembranosus tendon will continue inferiorly to its posterior tibial insertion, whereas the semitendinosus tendon will be joined by others to insert on the anteromedial surface of the tibia.* **(Middle)** *More superiorly, transverse US of the posteromedial aspect of the supracondylar region shows the medial tendons and muscles. Three of these, the sartorius, gracilis, and semitendinosus, will continue together inferiorly and insert adjacent to each other as the pes anserine on the anteromedial aspect of the proximal tibial shaft.* **(Bottom)** *Oblique transverse US along the course of the pes anserine tendons shows their insertion on the anteromedial tibial cortex.*

GROSS ANATOMY

Interosseous Membrane

- Stretches across interval between tibia and fibula
- Greatly extends surface for origin of muscles
- Strong, oblique fibers run downward and laterally from tibia to fibula
- In upper part, below lateral condyle of tibia, there is opening for passage of anterior tibial vessels
- Distally, opening allows passage of perforating branch of peroneal artery
- Tibialis posterior and flexor hallucis longus take partial origin from back of membrane
- Tibialis anterior, long extensors of toes, and peroneus tertius take partial origin from front of membrane

Muscles of Leg

- Compartments separated by deep fascia, which give partial origin to several muscles
- **Posterior compartment: Superficial muscles**
  - **Gastrocnemius muscle**
    - Origin: Medial from posterior femoral metaphysis; lateral from posterior edge of lateral epicondyle
    - Heads separated from posterior capsule by bursa
    - 2 heads unite to form main bulk of muscle
    - Join in thin aponeurotic tendon near midleg
    - Joins soleus aponeurosis to form Achilles tendon; concave in cross section; musculotendinous junction is 5 cm above calcaneal insertion
    - Action: Plantar flexor of ankle and flexor of knee
  - **Accessory medial and lateral gastrocnemius slips**
    - Accessory slips of medial or lateral bellies can result in popliteal artery entrapment syndrome (PAES)
    - 5 different types of PAES
  - **Plantaris muscle**
    - Origin: Superior and medial to lateral head of gastrocnemius origin as well as from oblique popliteal ligament
    - Continues deep to lateral head of gastrocnemius
    - Myotendinous junction at level of origin of soleus (muscle is 5-10 cm long)
    - Tendon then lies between medial head of gastrocnemius and soleus
    - Follows medial side of Achilles to insert either anteromedially on Achilles or on calcaneus
    - Plantaris absent in 7-10% of general population
    - Action: Acts with gastrocnemius
  - **Soleus muscle**
    - Origin: Extensive, from back of fibular head and upper 1/3 of posterior surface of shaft of fibula, from soleal line and middle 1/3 of medial border of tibia, and from tendinous arch joining these across popliteal vessels
    - Flat, thick, powerful muscle that ends in strong tendon
    - Joins with tendon of gastrocnemius to form Achilles tendon
    - Action: Stabilizes ankle in standing, plantarflexes ankle
  - **Achilles tendon**
    - Formed from conjoined tendon of soleus and gastrocnemius muscles
    - Soleus → deep and medial fibers; gastrocnemius → superficial fibers
      - These fibers rotate 90° distally such that at calcaneum, soleus attaches superficially & medially while gastrocnemius attaches deep and laterally
    - Paratenon covers sides and superficial margin of Achilles tendon
  - **Accessory soleus muscle**
    - Accessory muscle with prevalence of < 5%
    - Originates from posterior aspects of tibia ± fibula and inserts into Achilles or calcaneum anteromedially to Achilles tendon
    - Inserts into Achilles or onto calcaneum anteromedially to Achilles where it can present as soft tissue mass
      - Hypertrophied accessory soleus muscle may compress tibial nerve before nerve enters tarsal tunnel
  - **Tibiocalcaneus internus muscle**
    - Accessory muscle with prevalence of < 1%
    - Originates from distal tibia and passes down deep to flexor retinaculum
      - Implicated in tarsal tunnel syndrome
    - Inserts onto medial border of calcaneum 1-2 cm anterior to Achilles tendon insertion
      - Similar to accessory soleus in being located anteromedial to Achilles tendon
      - Different to accessory soleus in extending deep to flexor retinaculum
- **Posterior compartment: Deep muscles**
  - **Popliteus muscle**
    - Origin: Tendon from popliteal groove of lateral femoral condyle
    - Passes through popliteal hiatus posteriorly and medially, pierces posterior capsule of knee
    - Muscle fibers directed medially and downward to insert on posterior surface of tibia above soleal line
    - Action: Flexes knee and medially rotates tibia with respect to femur at onset of flexion (unlocking extension "screw home" mechanism)
  - **Tibialis posterior muscle**
    - Origin: Interosseous membrane and adjoining parts of posterior surfaces of tibia and fibula
    - Superior end bifid; anterior tibial vessels pass forward between 2 attachments
    - Distally, it inclines medially, under flexor digitorum longus
    - Grooves and curves around medial malleolus
    - Action: Plantarflexes and inverts foot
  - **Flexor digitorum longus muscle**
    - Origin: Posterior surface of tibia, below popliteus, and medial to vertical ridge
    - Crosses superficial to distal part of tibialis posterior
    - Tendon grooves lower end of tibia lateral to that of tibialis posterior, passes around medial malleolus to foot
    - Action: Flexes interphalangeal and metatarsophalangeal joints of lateral 4 toes; plantarflexes and inverts foot
  - **Flexor hallucis longus muscle**
    - Origin: Posterior surface of fibula, below origin of soleus

- Passes medially, descends down posterior to midtibia
- Associated with os trigonum posterior to talus
- Tendon occupies deep groove on posterior surface of talus, passes around medial malleolus, under sustentaculum tali, to great toe
- Action: Flexes interphalangeal and metatarsophalangeal joints of great toe; plantarflexes foot

- **Flexor digitorum accessorius longus muscle**
  - Accessory muscle with prevalence of < 10%
  - Originates from distal tibia or fibula posteriorly and passes down into tarsal tunnel
    - May contain fleshy fibers in tarsal tunnel, which is distinguishing feature
    - ± implicated in tarsal tunnel syndrome
  - Inserts into quadratus plantae or flexor digitorum longus tendon
- **Peroneocalcaneus internus muscle**
  - Accessory muscle with prevalence of 1%
  - Originates from distal fibula beyond origin of flexor hallucis longus, passes through tarsal tunnel and deep to sustentaculum tali
    - Fully tendinous before it enters tarsal tunnel
    - Not implicated in tarsal tunnel syndrome
  - Inserts onto anterior aspect of calcaneum

- **Lateral compartment**
  - Peroneals separated from extensors by anterior intermuscular septum and from posterior muscles by posterior septum
  - **Peroneus longus muscle**
    - Origin: Upper 2/3 of lateral surface of fibula and intermuscular septa and adjacent muscular fascia
    - Becomes tendinous few cm above lateral malleolus
    - Curves forward behind lateral malleolus, posterior to peroneus brevis
    - Action: Everts foot and secondarily plantarflexes foot
  - **Peroneus brevis muscle**
    - Origin: Lower 2/3 of lateral surface of fibula and intermuscular septa and adjacent muscular fascia
    - Muscle is medial to peroneus longus at origin but overlaps peroneus longus in middle 1/3
    - Tendon curves forward behind lateral malleolus, in front of peroneus longus tendon
    - Action: Everts foot and secondarily plantarflexes foot
  - Synovial sheath for peroneals begins 5 cm above tip of lateral malleolus and envelops both tendons; divides into 2 sheaths at level of calcaneus
  - **Peroneus quartus muscle**
    - Accessory muscle with prevalence of 10%
    - Originates from distal leg, frequently from peroneal muscles, with variable insertion sites at foot, usually to calcaneum (either peroneal tubercle or retrotrochlear eminence)
    - At level of malleolus, located medial or posterior to both peroneal tendons
  - **Peroneus digiti minimi muscle**
    - Accessory muscle with prevalence of 25%
    - Originates from peroneus brevis muscle and extends around lateral malleolus to foot
    - Tiny tendinous slips insert into metatarsal and proximal phalanx of little toe
- **Anterior compartment**
  - **Tibialis anterior muscle**
    - Origin: Upper 1/2 of lateral surface of tibia and interosseous membrane
    - Action: Dorsiflexor and invertor of foot
  - **Extensor digitorum longus muscle**
    - Origin: From upper 3/4 of anterior surface fibula
    - Descends behind extensor retinacula to ankle
    - Action: Extends interphalangeal and metatarsophalangeal joints of lateral 4 toes, dorsiflexes foot
  - **Peroneus tertius muscle**
    - Small muscle present in 90% of subjects
    - Origin: Continuous with extensor digitorum longus, arising from distal 1/4 anterior surface of fibula and interosseous membrane
      - Extends distally, usually with extensor digitorum longus
    - Inserts into dorsal surface base of 5th metatarsal
    - Action: Dorsiflexes ankle and everts foot
  - **Extensor hallucis muscle**
    - Thin muscle hidden between tibialis anterior and extensor digitorum longus
    - Origin: Middle 1/2 of anterior surface of fibula and interosseous membrane
    - Tendon passes deep to retinacula to great toe
    - Action: Extends phalanges of great toe and dorsiflexes foot

ANATOMY IMAGING ISSUES

Imaging Recommendations

- Owing to its smaller size, compared to thigh, leg muscles and vessels are well demonstrated by US throughout their course
  - Relationship of different structures in posterior compartment tends to change along their courses
    - If in doubt, identify tendons at ankle level and trace structures proximally

Imaging Sweet Spots

- Common indication of leg US is suspected muscle tear
  - Most common site of muscle tear is at distal myofascial junction medial belly of gastrocnemius
  - Look carefully for preservation of arrowhead appearance of myofascial junction at this location
  - With plantaris tendon tear, this junction is intact, and there is more extensive hemorrhage deep to medial belly of gastrocnemius
- There are more accessory muscles in leg than any other body region
  - These accessory muscles are usually innocuous but can be cause of
    - Neural compression (flexor digitorum accessorius longus, accessory soleus, tibiocalcaneus internus muscles)
    - Vascular compression, PAES (accessory medial and lateral gastrocnemius slips)
    - Clinical soft tissue mass (accessory soleus muscle)
    - Peroneal tendon compression (accessory peroneal muscles)

ANTEROLATERAL LEG

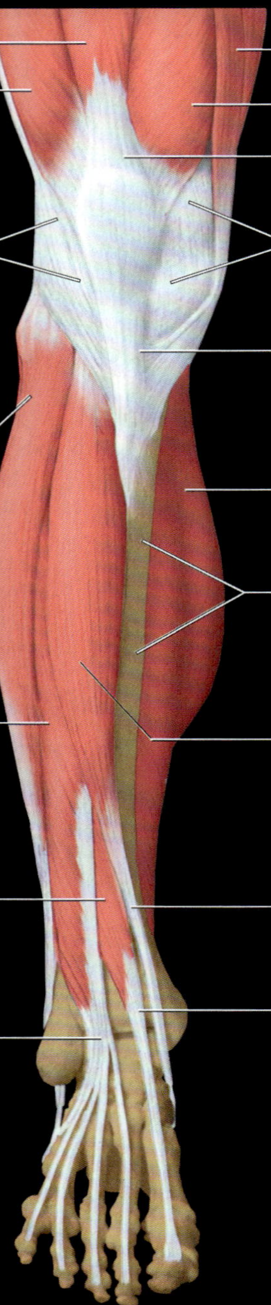

Graphic shows the anterior leg muscles. Note that nearly the entire anteromedial tibia is bare of muscle. Thus, it is poorly vascularized, resulting in slow fracture healing. The tibialis anterior has an extensive origin from both the tibia and interosseous membrane and is the most substantial muscle of the anterior compartment. The extensor digitorum longus originates from the fibula. The extensor hallucis longus originates between the 2, from the fibula and the interosseous membrane. These 3 muscles retain the same orientation as they become tendinous anterior to the ankle. The mnemonic "Tom, Harry, and Dick" applies to the tendon order, from medial to lateral (tibialis anterior, extensor hallucis longus, extensor digitorum).

PANORAMIC ANTEROLATERAL LEG

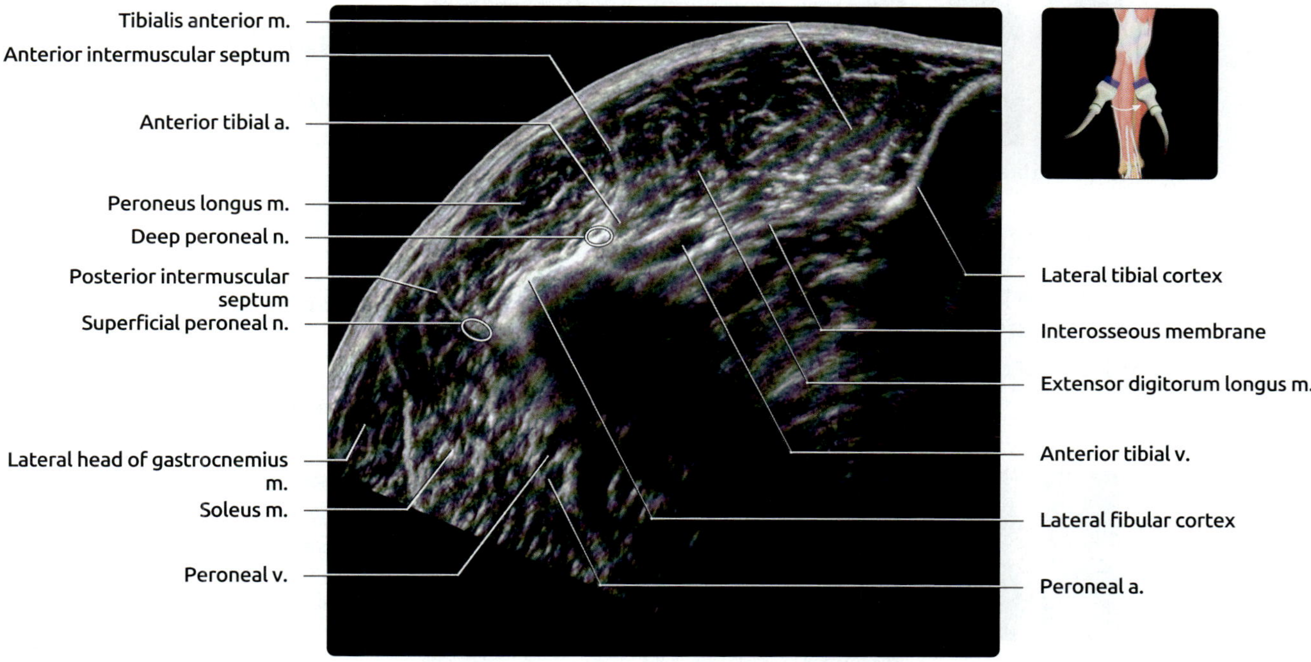

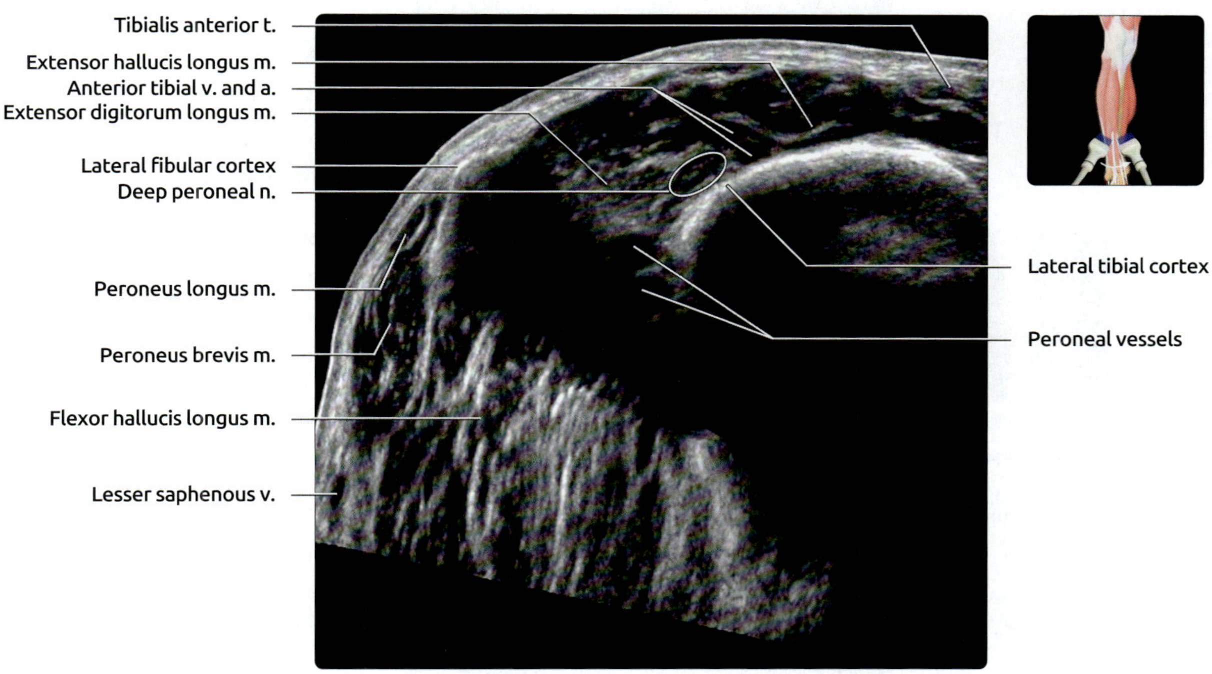

(Top) *Transverse panoramic US shows the anterior and lateral compartments of the upper leg. The crural compartments are divided by the tibia and the 2 intermuscular septa (anterior and posterior) into the anterior (extensor), lateral (peroneal), and posterior (flexor) compartments. The intermuscular septum separates the anterior from the posterior compartments.* **(Bottom)** *Transverse panoramic US shows the anterior and lateral compartments of the inferior leg. All muscles are smaller in caliber compared to those seen superiorly. An interosseous gap is still present between the tibia and fibula, but US interrogation through this gap is no longer possible and the interosseous membrane cannot be demonstrated.*

ANTERIOR COMPARTMENT

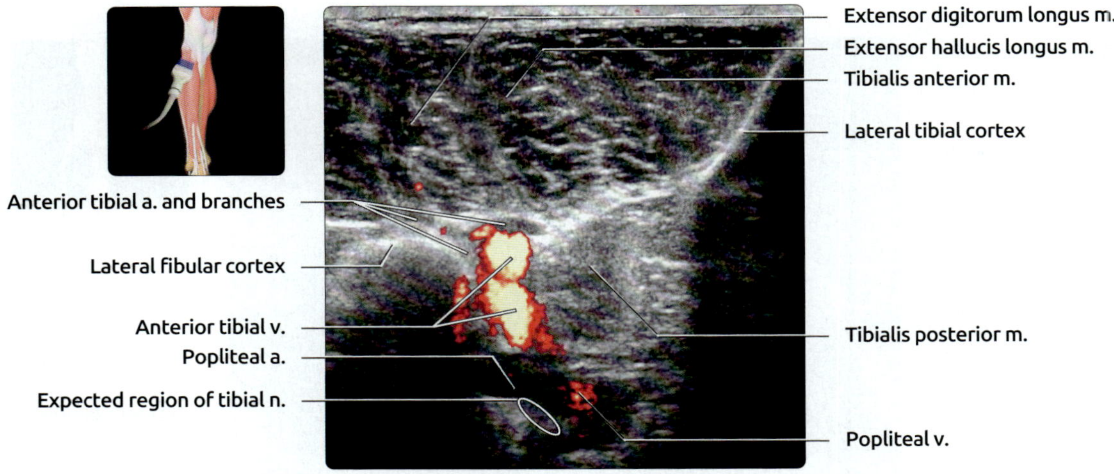

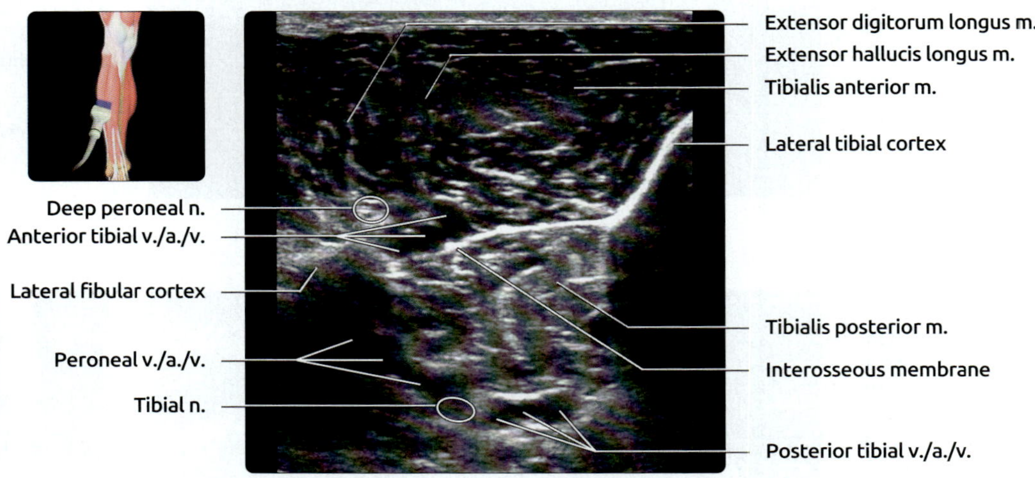

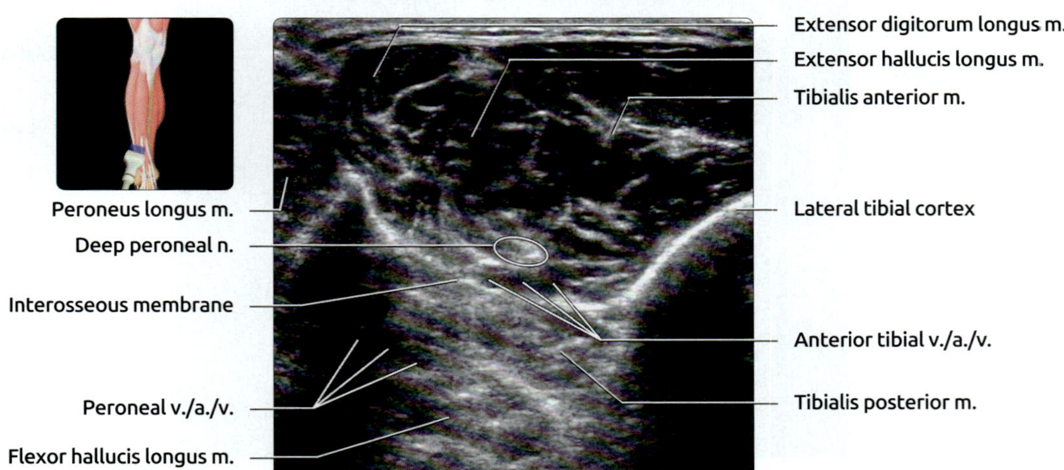

(Top) *Transverse color Doppler US shows the anterior compartment at the upper leg level slightly below the proximal tibiofibular joint. At this level (above the superior margin of the interosseous membrane), the anterior tibial artery and vein (branches of the popliteal artery and vein) run anteriorly to supply the anterior compartment.* **(Middle)** *Transverse US shows the anterior compartment at a slightly lower level. The tibialis anterior, extensor hallucis longus, and extensor digitorum longus muscles form the muscular component of this compartment. The deep peroneal nerve has penetrated laterally to supply this compartment. The anterior tibial vessels run on the surface of the interosseous membrane.* **(Bottom)** *Transverse US more inferiorly shows a decrease in size of the extensor muscles and gradual migration of nerve and vessels more medially and superficially.*

ANTERIOR COMPARTMENT

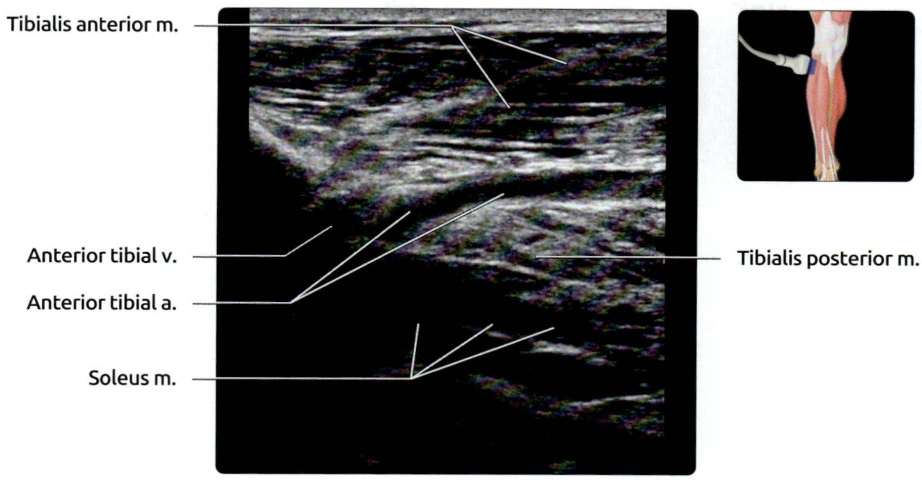

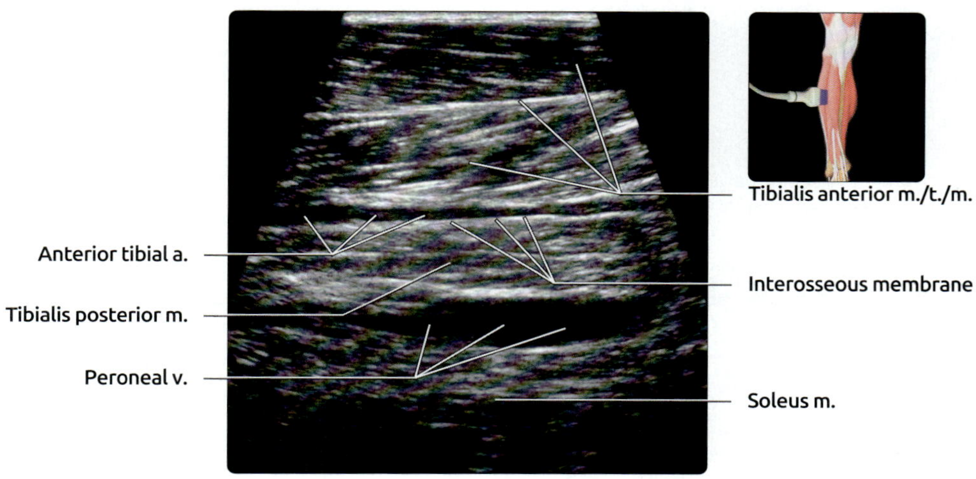

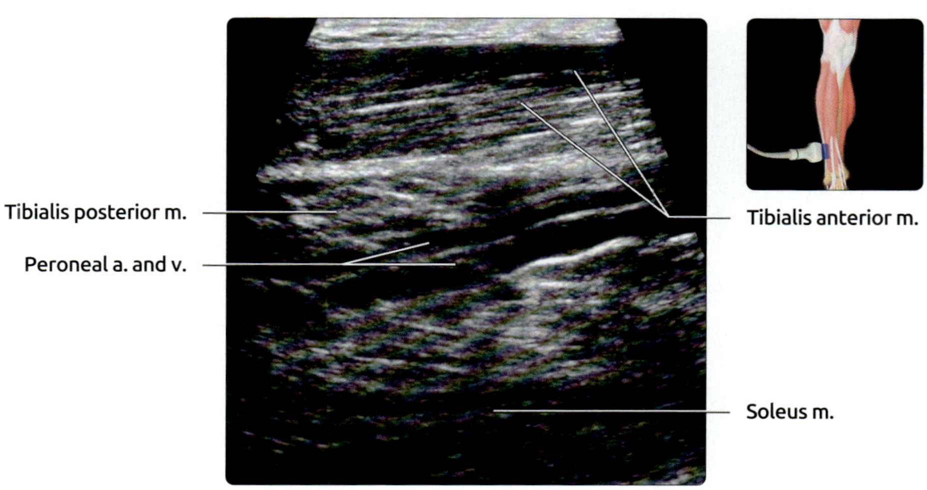

(Top) *Sagittal US shows the anterior compartment at the upper leg level. The anterior tibial vessels pass above the superior margin of the interosseous membrane to enter the anterior compartment and supply the extensor muscles.* **(Middle)** *Sagittal US of the anterior and posterior compartments at the midleg level shows that the interosseous gap between the tibia and fibula can be used as a window to interrogate the deep posterior structures.* **(Bottom)** *Sagittal US shows the anterior compartment at a lower level. Within the posterior compartment, the peroneal vessels run anteriorly and just below the inferior margin of the interosseous membrane.*

LATERAL COMPARTMENT

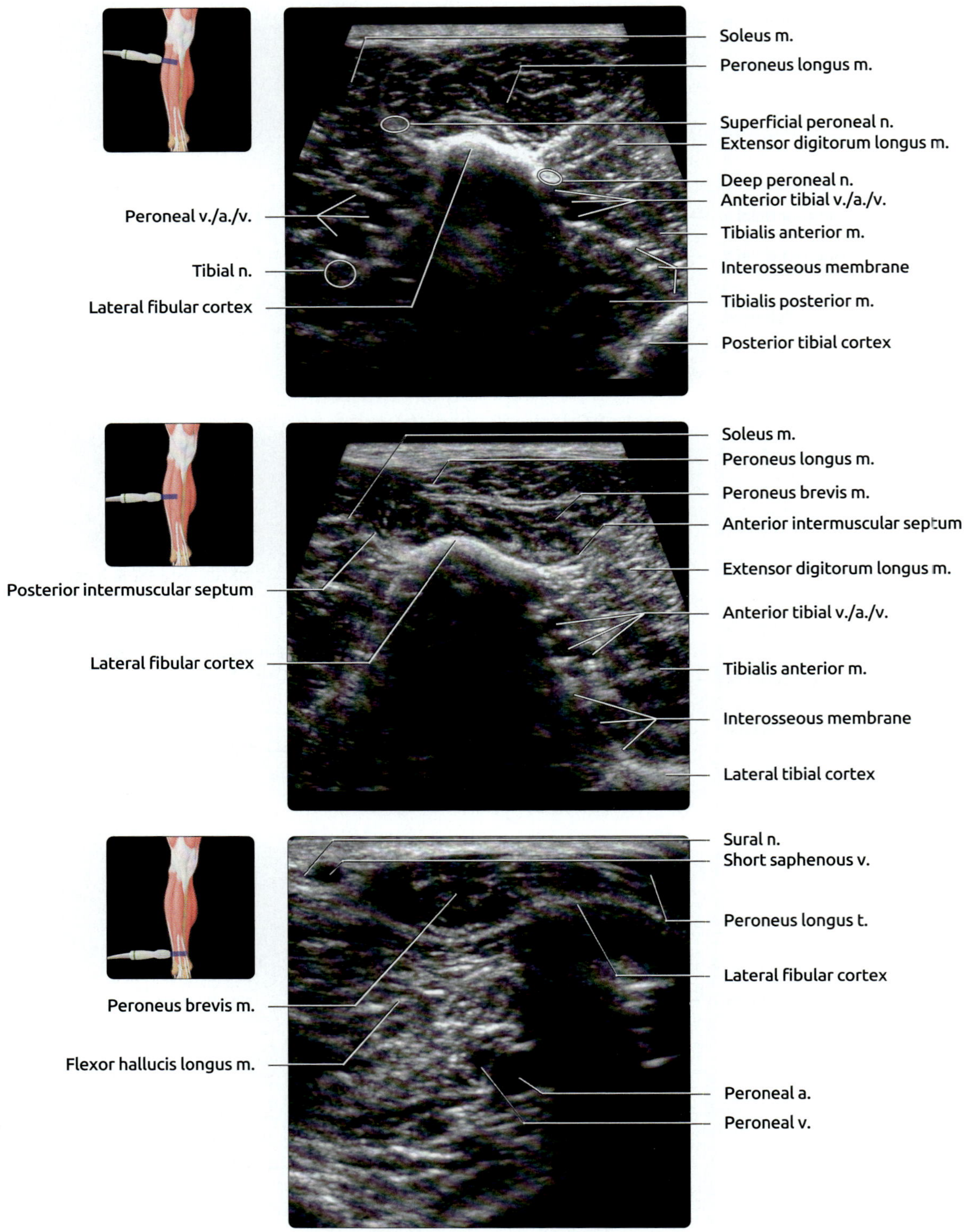

(Top) *Transverse US shows the lateral compartment at a high level. At this level, only the peroneus longus muscle has originated and is the only muscle occupying this compartment. The superficial peroneal nerve lies on the posterior surface of the peroneus longus muscle. The deep peroneal nerve has already entered the anterior compartment after curving around the fibular neck (above this scan level).* **(Middle)** *Transverse US at the midleg level shows that the peroneus brevis muscle has now originated and resides anterior and deep to the peroneus longus muscle. More inferiorly, the peroneus longus muscle will become smaller and turn into a tendon while the peroneus brevis will still have muscle fibers down to the level of the ankle joint.* **(Bottom)** *Transverse US shows the level of the lateral malleolus. The peroneal tendons will run posterior to the lateral malleolus and curve around its tip to run toward the lateral foot.*

LATERAL COMPARTMENT

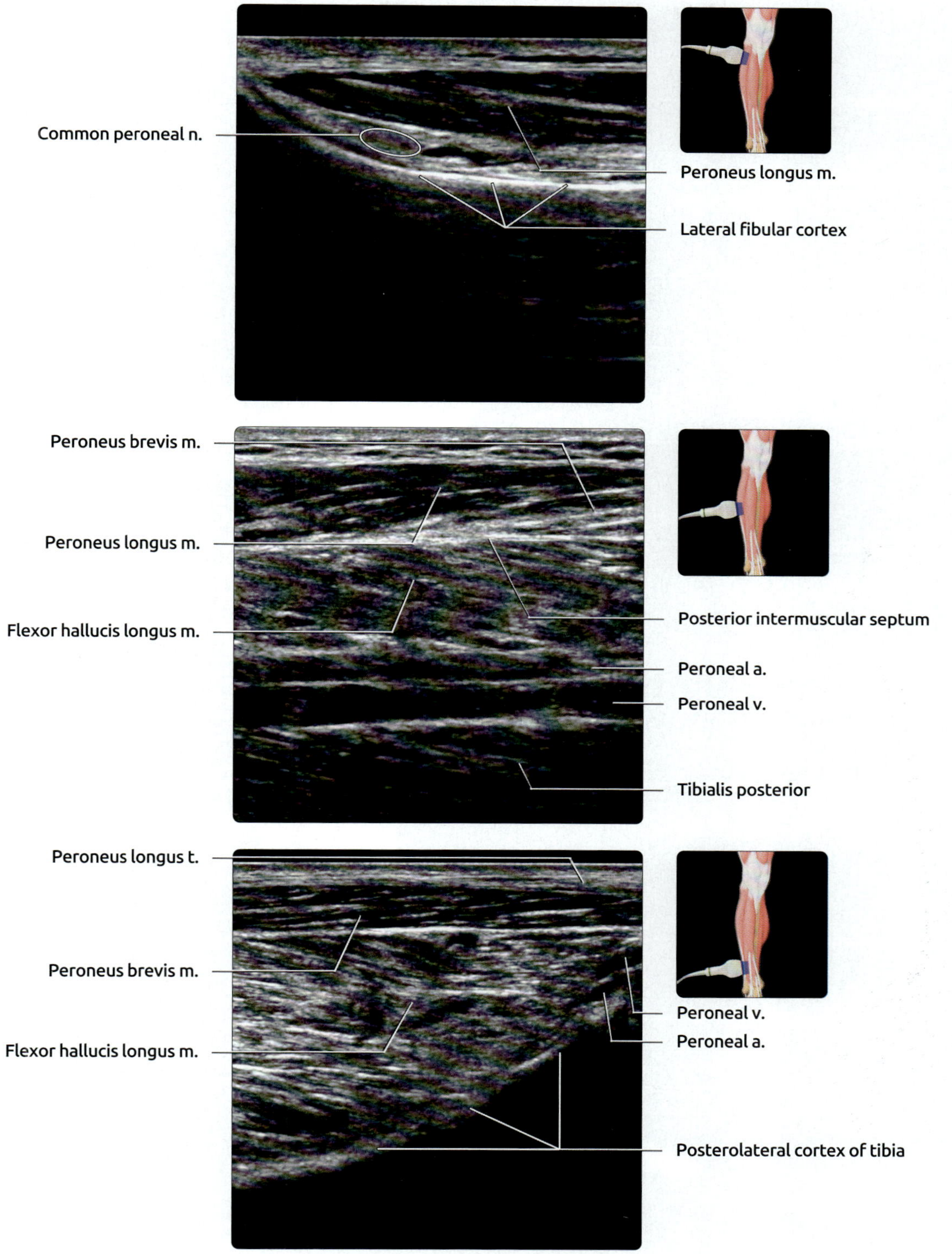

(Top) *Coronal US shows the lateral compartment at the level of the fibular neck. The common peroneal nerve runs together with the biceps femoris tendon, then curves anteriorly around the neck of the fibula to enter the lateral and then the anterior compartments.* **(Middle)** *Coronal US of the lateral compartment at the midleg level shows the transition between the peroneus longus and peroneus brevis muscles. These are separated from the posterior compartment by the posterior intermuscular septum.* **(Bottom)** *Coronal US further inferiorly shows that the peroneus longus muscle is now only a tendon, compared to the muscular peroneus brevis. Similarly, in the posterior compartment, the flexor hallucis longus muscle is the only muscle at this level, as the others have also become tendons.*

LATERAL LEG

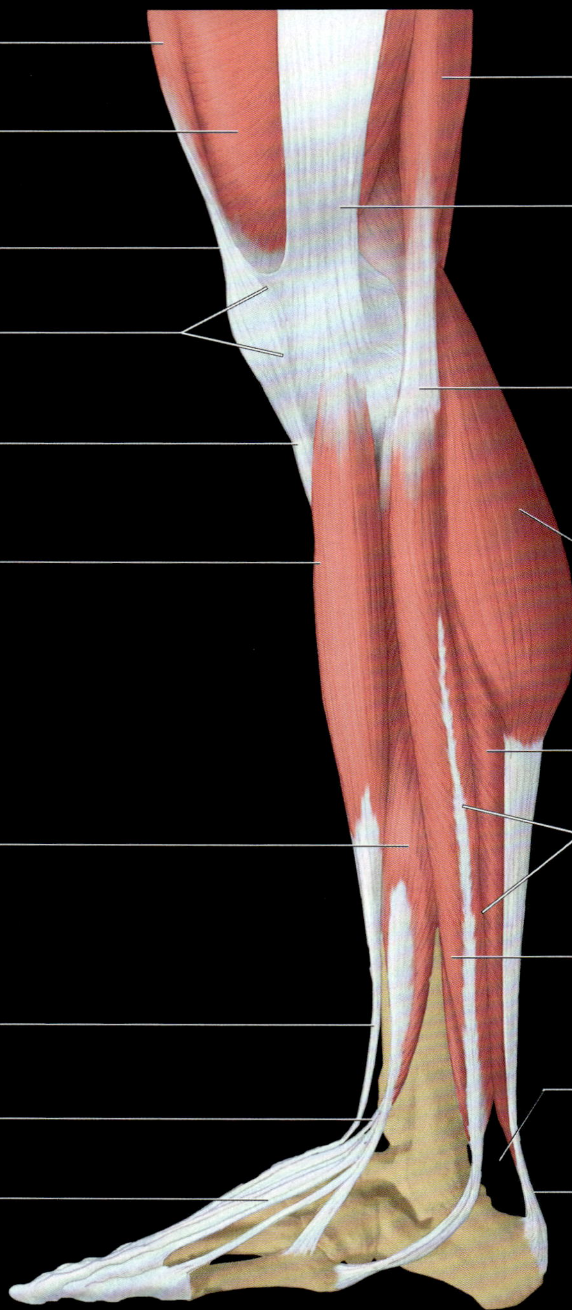

Graphic of the lateral leg shows the anterior compartment (extensors), lateral compartment (peroneals), and superficial muscles of the posterior compartment. Of the ankle tendons, tendinosis most commonly affects the Achilles tendon, the posterior tibialis tendon, and the peroneal tendons. The tibialis anterior tendon is also affected, though less commonly. As the peroneus brevis lies closest to the bone, it is more prone to trauma, particularly in the retromalleolar region. Typically, this tear consists of a longitudinal split in which the overlying peroneus longus tendon pushes into the peroneus brevis tendon, splitting it into 2 sections. One may then see 3 rather than 2 tendons behind the lateral malleolus. Another reason to see 3 tendons behind the lateral malleolus is when there is a quartus tendon present.

POSTERIOR LEG

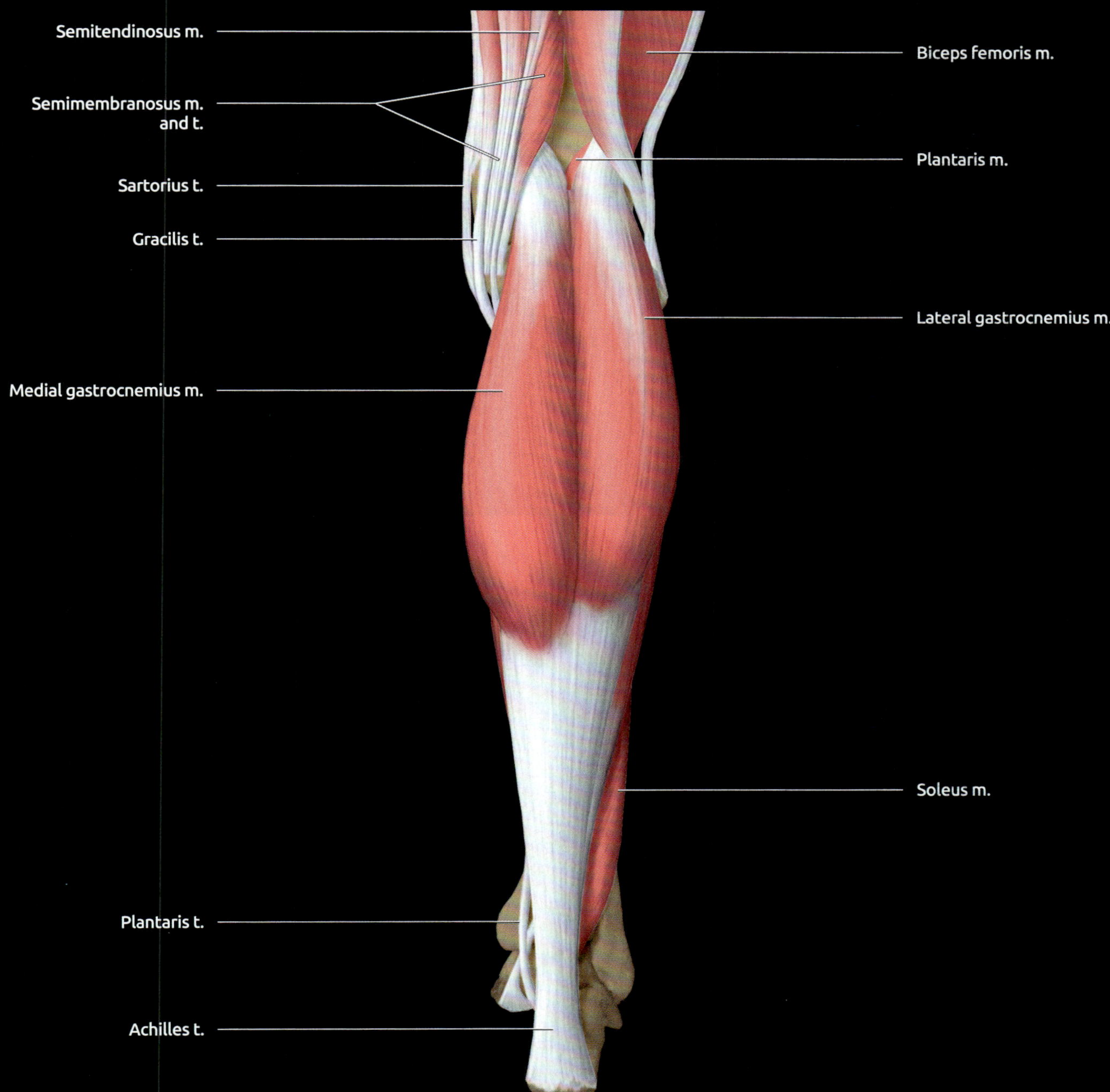

Posterior view shows superficial muscles and tendons of the leg. The gastrocnemius muscles are bulky in the proximal 1/2 of the leg and taper to an aponeurosis, which blends with the soleus more distally to become the Achilles tendon. The plantaris muscle is superficial only at its origin from the lateral femoral metaphysis, just medial to the origin of the lateral head of gastrocnemius. The muscle extends only a few centimeters before becoming tendinous, extending distally between the soleus and medial gastrocnemius. Eventually, the plantaris tendon merges with the medial side of the Achilles or inserts on the calcaneus, medial to the Achilles insertion.

PANORAMIC POSTERIOR LEG

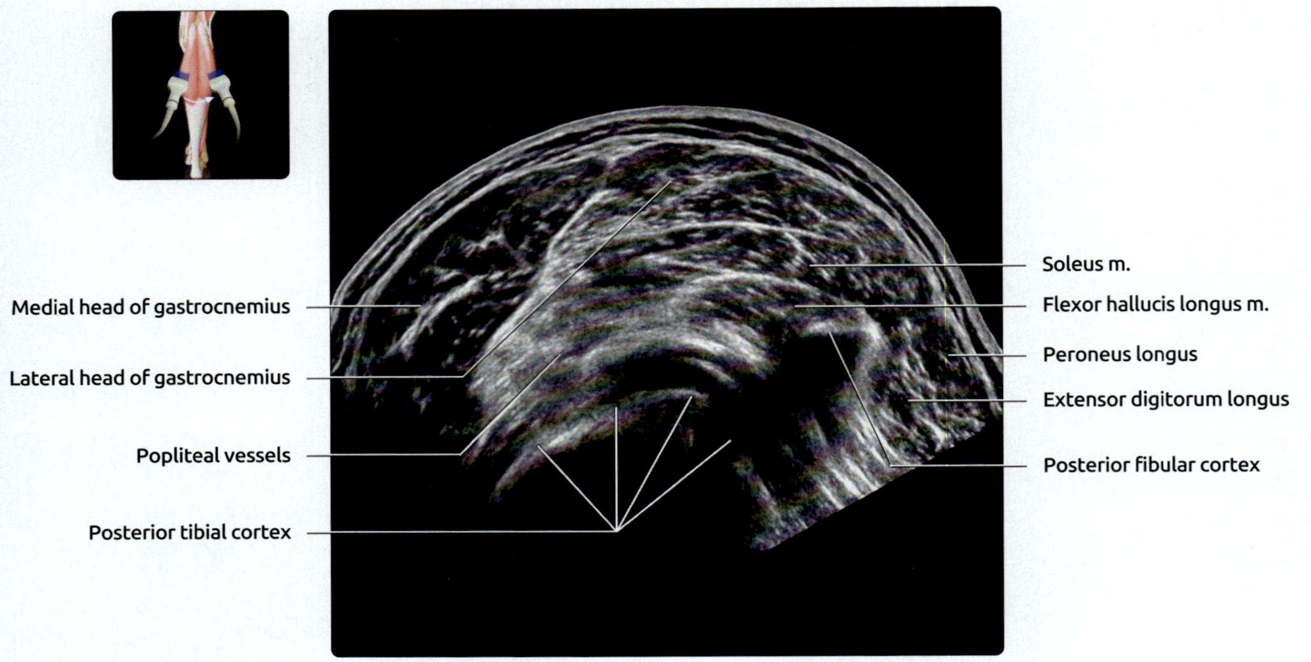

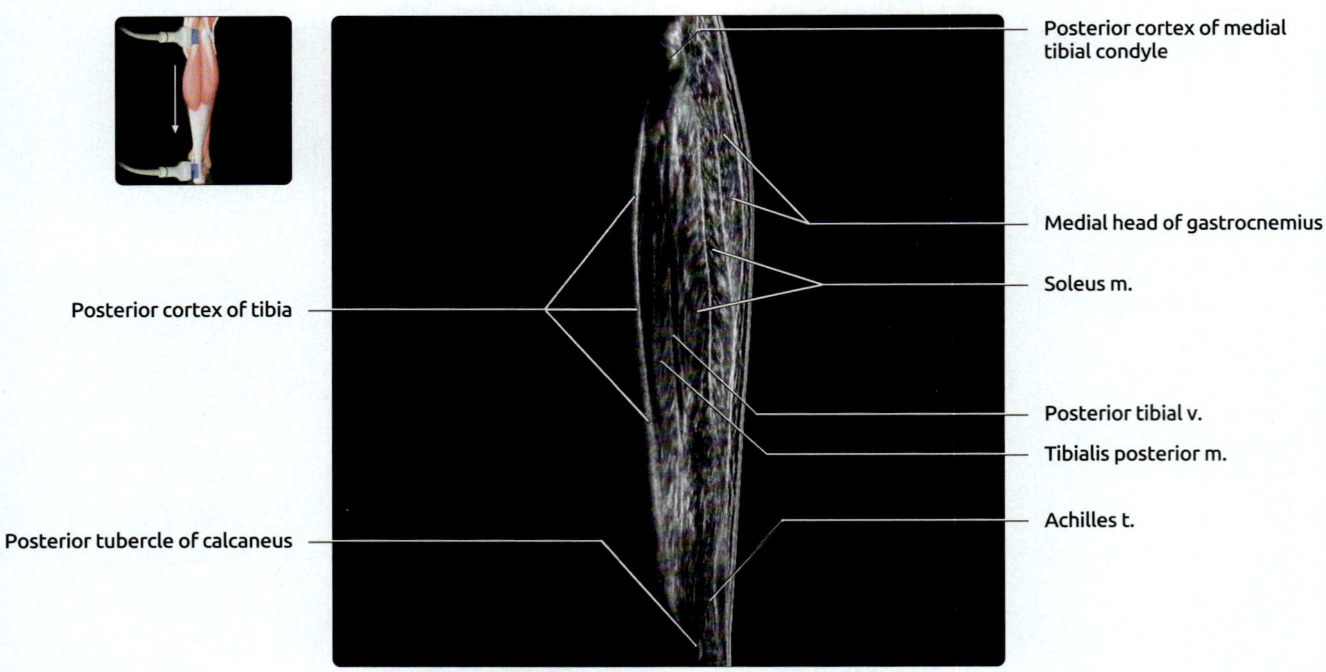

(Top) *Transverse panoramic US shows the posterior compartment at the level of the midleg. The medial head of the gastrocnemius muscle is larger than the lateral head throughout its course. From this midcalf level downward, the soleus muscle will assume relatively more bulk as the gastrocnemius turns tendinous.* **(Bottom)** *Sagittal panoramic scan shows the posterior compartment. The gastrocnemius is muscular down to approximately the level of the midcalf. Below this level, it forms the Achilles tendon, receiving contributions from the soleus muscle. For the deep layer, the tibialis posterior muscle lies on the surface of the tibial shaft and interosseous membrane.*

TRANSVERSE POSTERIOR LEG

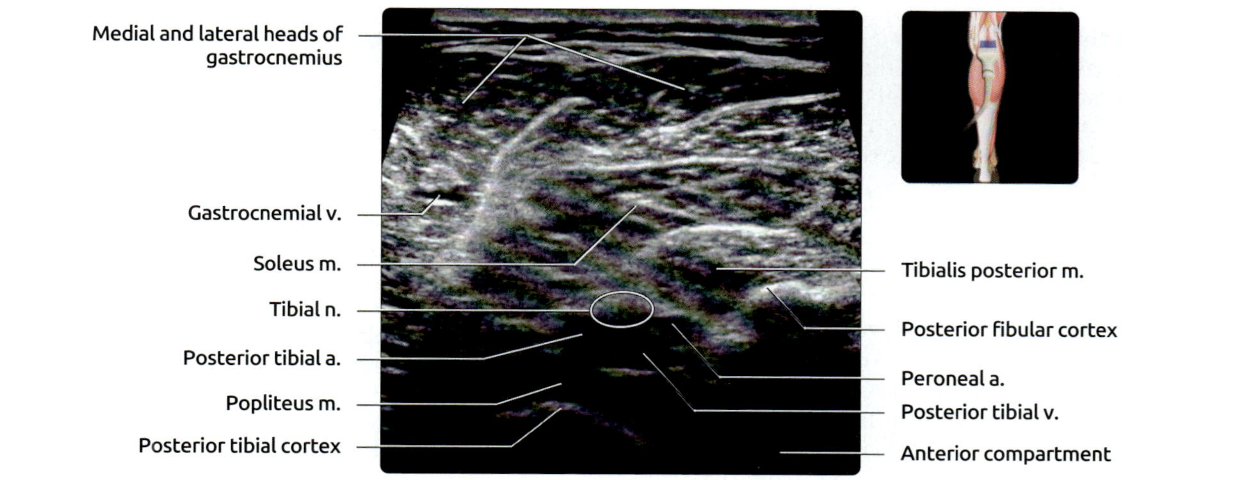

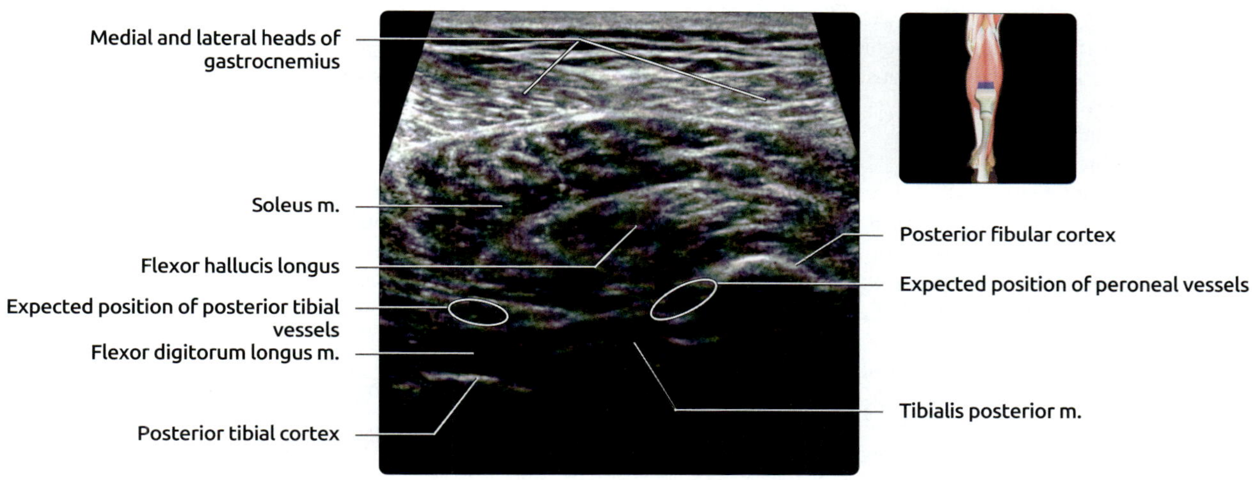

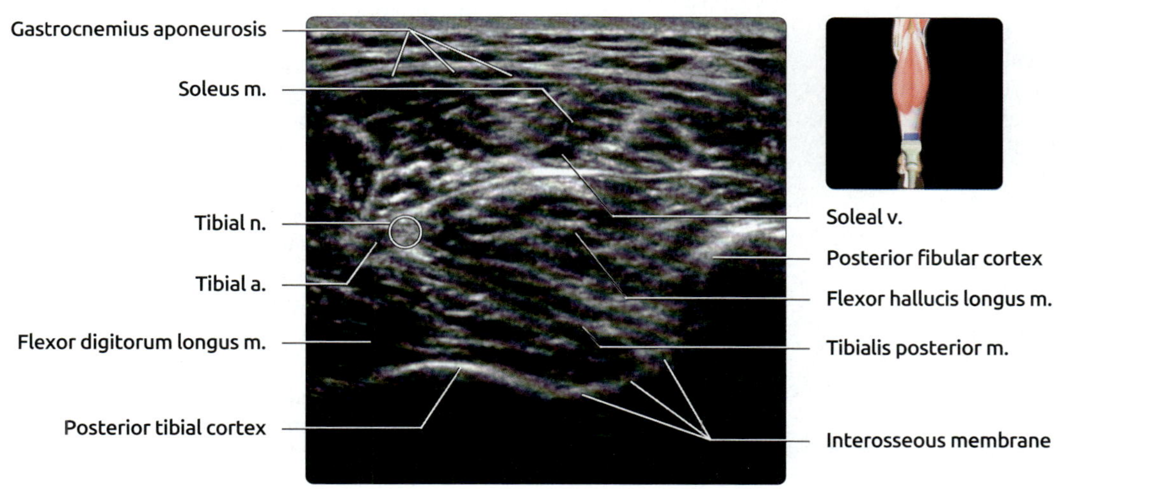

(Top) *Transverse US is seen through the posterior compartment at the upper-calf level. For this compartment, the superficial layer consists of the gastrocnemius, plantaris, and soleus muscles. The deep layer consists of the popliteus, tibialis posterior, flexor digitorum longus, and flexor hallucis longus muscles.* **(Middle)** *Transverse US shows the midcalf level. The tibial vessels have bifurcated into their peroneal and posterior tibial branches, which delineates a separation between the tibialis posterior and flexor digitorum longus from the rest of the calf muscles. Given the thickness of the gastrocnemius and soleus muscle, these deep muscles may be better seen using the anterior compartment as a window.* **(Bottom)** *Further inferiorly, the gastrocnemius has become a thin aponeurosis, and together with the tendon from the soleus muscle, will form the Achilles tendon.*

GROSS ANATOMY

Arteries

- **Popliteal artery**
  - Begins as continuation of superficial femoral artery after it has passed through adductor hiatus (inferior end of adductor canal)
  - Runs through fat of popliteal fossa
  - Relationships
    - Deep to artery: Femoral shaft, knee joint capsule, popliteus fascia
    - Superficial to artery: Semimembranosus and gastrocnemius muscles
    - **Relation to popliteal vein is variable**
      - Artery lies just deep to vein in 40%
      - Deep and medial to vein in 40%
      - Deep and lateral to vein in 10%
      - Just medial or lateral to vein in 10%
  - Gives off 4 named geniculate branches around knee
    - Superior medial and lateral geniculate arteries
    - Inferior medial and lateral geniculate arteries
  - Ends at distal border of popliteus muscle and divides into 2 branches
    - Anterior tibial artery
    - Tibioperoneal trunk
- **Anterior tibial artery**
  - Smaller of 2 terminal branches of popliteal artery
  - Origin in back of leg, at distal border of popliteus muscle
  - Passes through upper part of interosseous membrane
  - Straight course down front of leg to become **dorsalis pedis artery**
  - Runs on anterior surface of interosseous membrane, deep to extensor muscles
  - Muscular branches along length
  - Malleolar branches ramify over malleoli
    - Lateral branch anastomoses with perforating branch of peroneal artery
- **Tibioperoneal trunk**
  - Larger of 2 terminal branches of popliteal artery
    - Divides after short course into posterior tibial and peroneal arteries
  - Branches
    - Circumflex fibular (may arise from anterior tibial)
      - Runs laterally around neck of fibula
    - Nutrient artery to tibia
    - Muscular branches
- **Posterior tibial artery**
  - Main blood supply to foot
  - Passes downward and slightly medially along with tibial nerve
  - Divides into lateral and medial plantar arteries in tarsal tunnel behind medial malleolus
  - Divides behind medial malleolus into medial and lateral plantar arteries
- **Peroneal artery**
  - Largest branch of posterior tibial artery
  - Runs obliquely downward and laterally beneath soleus muscle to fibula
  - Descends deep to flexor hallucis longus

Veins

- **Deep leg veins**
  - Paired venae comitantes follow arteries of same name
    - Anterior, posterior, and peroneal veins
  - **Popliteal vein**
    - Runs superficial to artery
    - Usually begins at inferior border of popliteus muscle from anterior and posterior tibial veins
    - Crosses from medial to lateral side of popliteal artery as it runs in popliteal fossa
    - Ends at adductor hiatus by becoming superficial femoral vein
    - Also receives small saphenous vein in popliteal fossa
- **Superficial leg veins**
  - **Great saphenous vein**
    - Longest vein in human body
    - Begins at medial border of foot and ascends anterior to medial malleolus
    - Passes obliquely upward and backward across medial surface of distal 1/3 of tibia
    - Passes vertically upward along medial border of tibia and over posterior border of medial femoral epicondyle
    - Courses anteriorly along thigh
    - Passes through saphenous opening in fascia lata to join common femoral vein
  - **Small saphenous vein**
    - Extends behind lateral malleolus, ascends lateral to Achilles tendon
    - At midline of calf in lower popliteal region, pierces popliteal fascia and terminates in popliteal vein

ANATOMY IMAGING ISSUES

Imaging Recommendations

- Examination of lower limb vessels requires use of **morphologic and functional techniques**
  - Morphology: Documentation of areas of stenosis or occlusion, sites of venous valves, aberrant branches, aneurysm formation, etc.
  - Functional: Combination of color and spectral Doppler examination
- **Arteries**
  - Require use of spectral Doppler scanning to show their phasicity; normal waveform is triphasic
    - Sharp forward flow upstroke (systolic phase)
    - Small reverse flow (early diastolic phase)
    - Final smaller forward flow (late diastolic phase)
- **Veins**
  - Require use of dynamic maneuvers
    - **Compression**: Normal vein should compress with transducer pressure (thrombus will not compress)
    - **Augmentation**: To increase venous return (to demonstrate venous flow and patency) by manually squeezing calf or by gently moving toes
      - Frequently required to demonstrate flow in deep calf veins (which normally show no color Doppler flow)
    - **Valsalva**: To increase abdominal pressure and accentuate reverse flow and incompetent venous valves

LEG ARTERIES, POSTERIOR

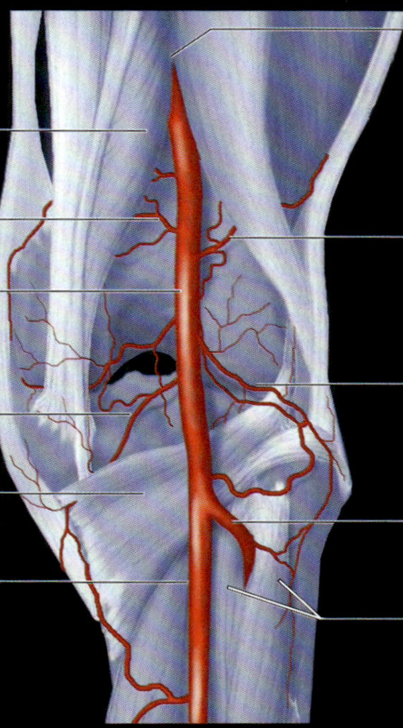

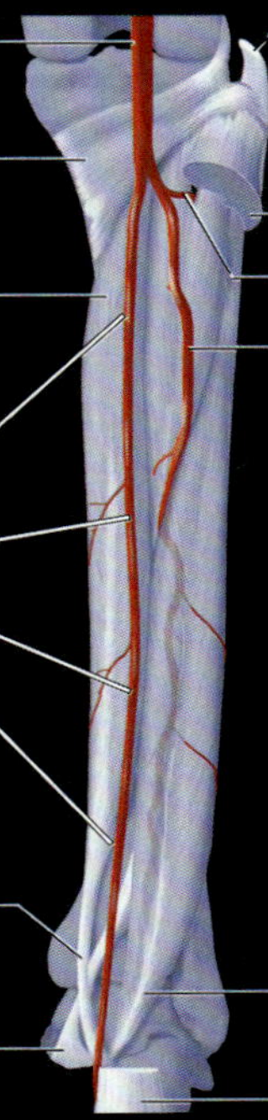

(Top) *Graphic shows the popliteal artery from the hiatus in adductor magnus proximally to the lower border of popliteus distally. At this point, the popliteal artery bifurcates into anterior and posterior tibial arteries. The anterior tibial is the smaller branch that extends through a slit in the tibialis posterior muscle and on through the interosseous membrane to descend along the interosseous membrane down the anterior compartment. There are 4 named geniculate branches, a superior and inferior both medially and laterally.* **(Bottom)** *The posterior tibial artery passes downward and slightly medially adjacent to the tibial nerve to end in the space between the medial malleolus and calcaneus. The largest branch of the posterior tibial artery is the peroneal artery, which runs obliquely downward and laterally beneath the soleus muscle to the fibula.*

LEG ARTERIES, ANTERIOR AND ANKLE

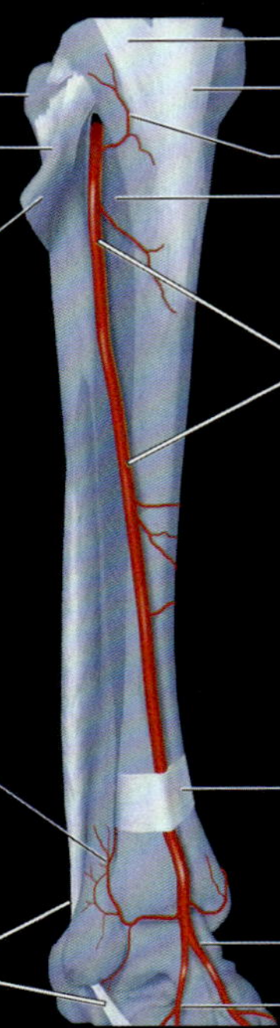

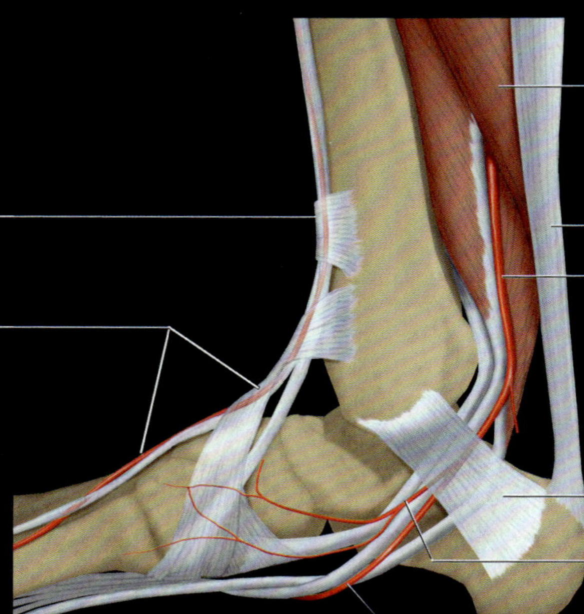

(Top) *Graphic of the anterior leg shows the anterior tibial artery perforating the interosseous septum proximally and descending along this membrane down the front of the leg to terminate as the dorsalis pedis. Distally, a perforating branch of the peroneal artery is seen, which, in a variant situation, may provide the major blood supply to the dorsum of the foot.* **(Bottom)** *Graphic of the medial aspect of the ankle shows the distal portion of both the posterior and anterior tibial arteries. The posterior tibial artery passes between the medial malleolus and calcaneus and divides into the lateral and medial plantar arteries. The dorsal pedis artery is a direct continuation of the anterior tibial artery and supplies the dorsum of the foot.*

LEG VEINS

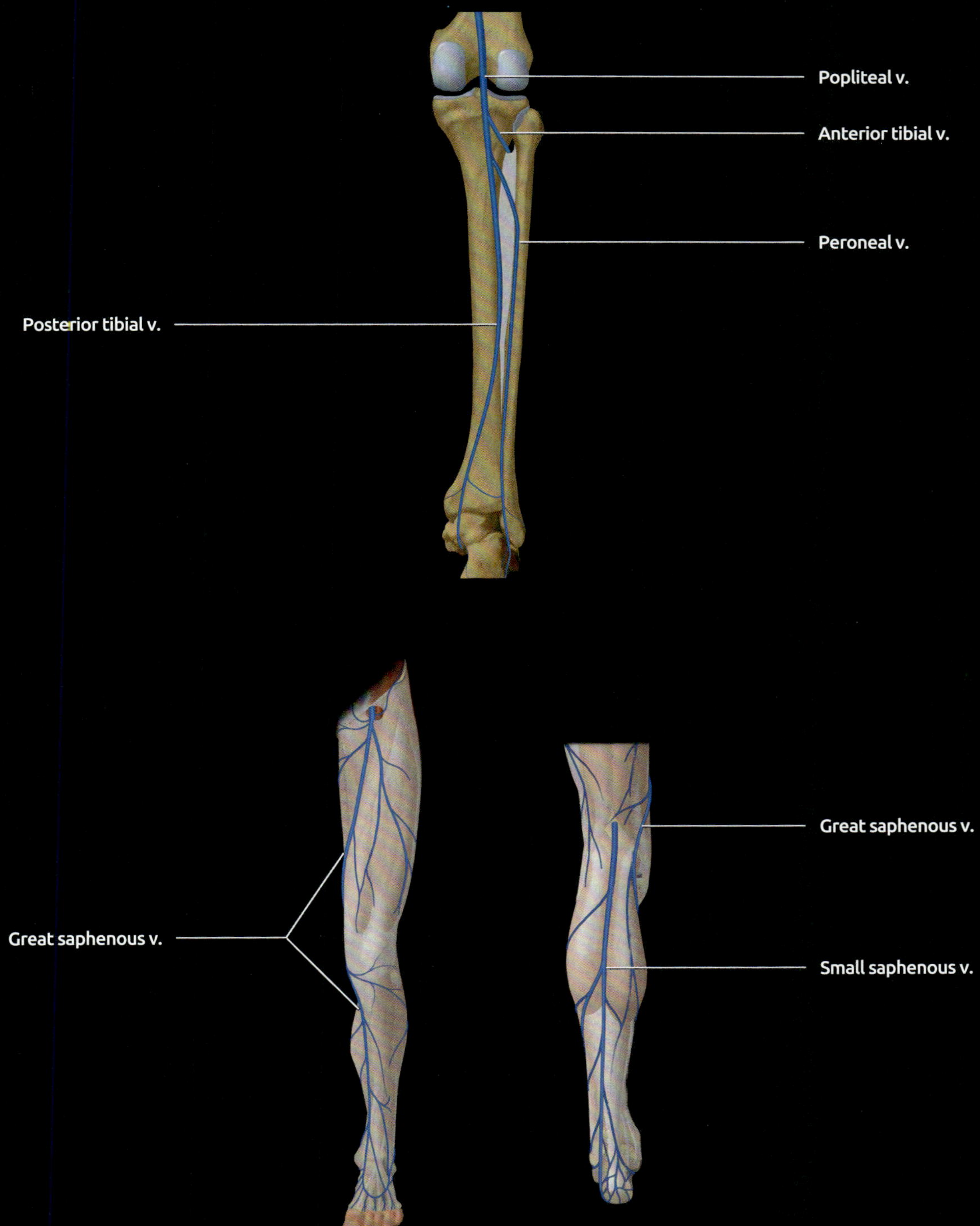

(Top) *This graphic shows the deep veins of the leg, which follow their accompanying artery of the same name and are often paired.* **(Bottom)** *This graphic shows the superficial leg veins. The great saphenous vein is the longest vein in the human body. It begins in the dorsomedial area of the foot. It passes obliquely upward over the posterior medial femoral epicondyle, then anteriorly across the thigh to finally join with the common femoral vein. The small saphenous vein drains from the lateral malleolus and up the posterior calf to join with the popliteal vein in the popliteal fossa.*

POPLITEAL VESSELS

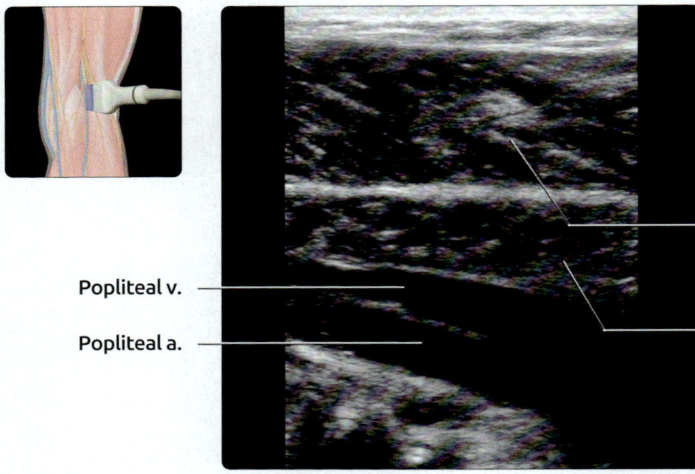

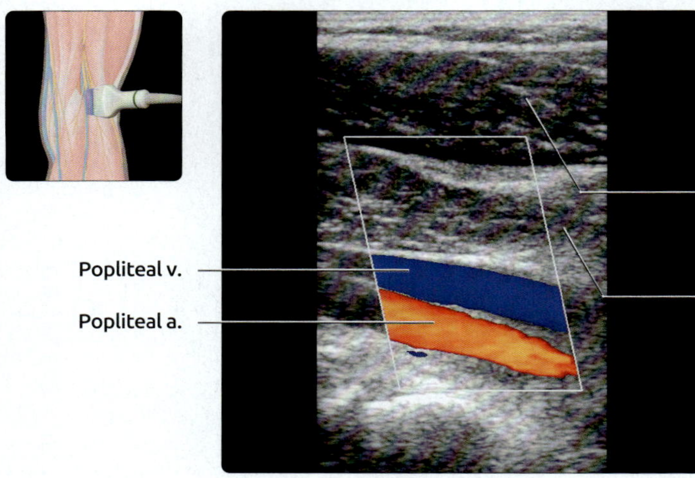

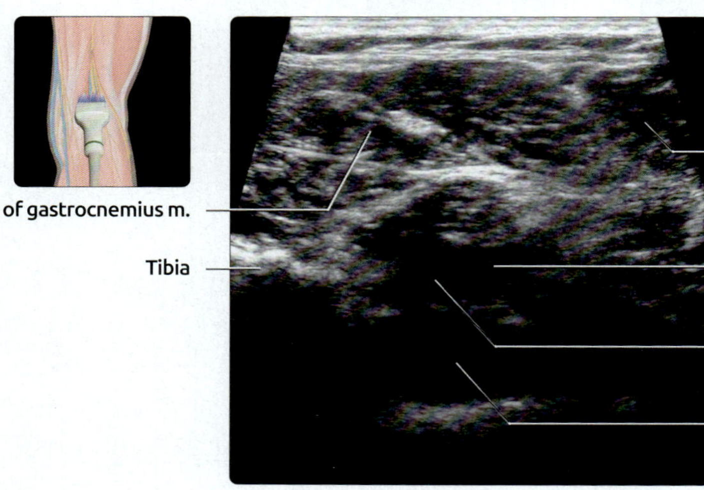

(Top) *Sagittal US shows the lower part of the popliteal vessels. The popliteal vessels are continuations of the superficial femoral vessels after they have exited the adductor canal through the adductor hiatus. The popliteal vein lies superficial to the popliteal artery, a reverse of the relationship between the superficial femoral vein and artery. The popliteal vessels pass inferolaterally through the fat of the popliteal fossa.* **(Middle)** *Sagittal color Doppler US of the popliteal vessels shows complete color filling of the lumina (excluding luminal thrombus) and normal flow direction: Toward the trunk (blue) for the vein and away from the trunk (red) for the artery.* **(Bottom)** *Transverse US shows the lower part of the popliteal vessels. The popliteal vein runs superficial to the popliteal artery and usually bifurcates at the inferior margin of the popliteus muscle.*

POPLITEAL VESSEL BIFURCATION

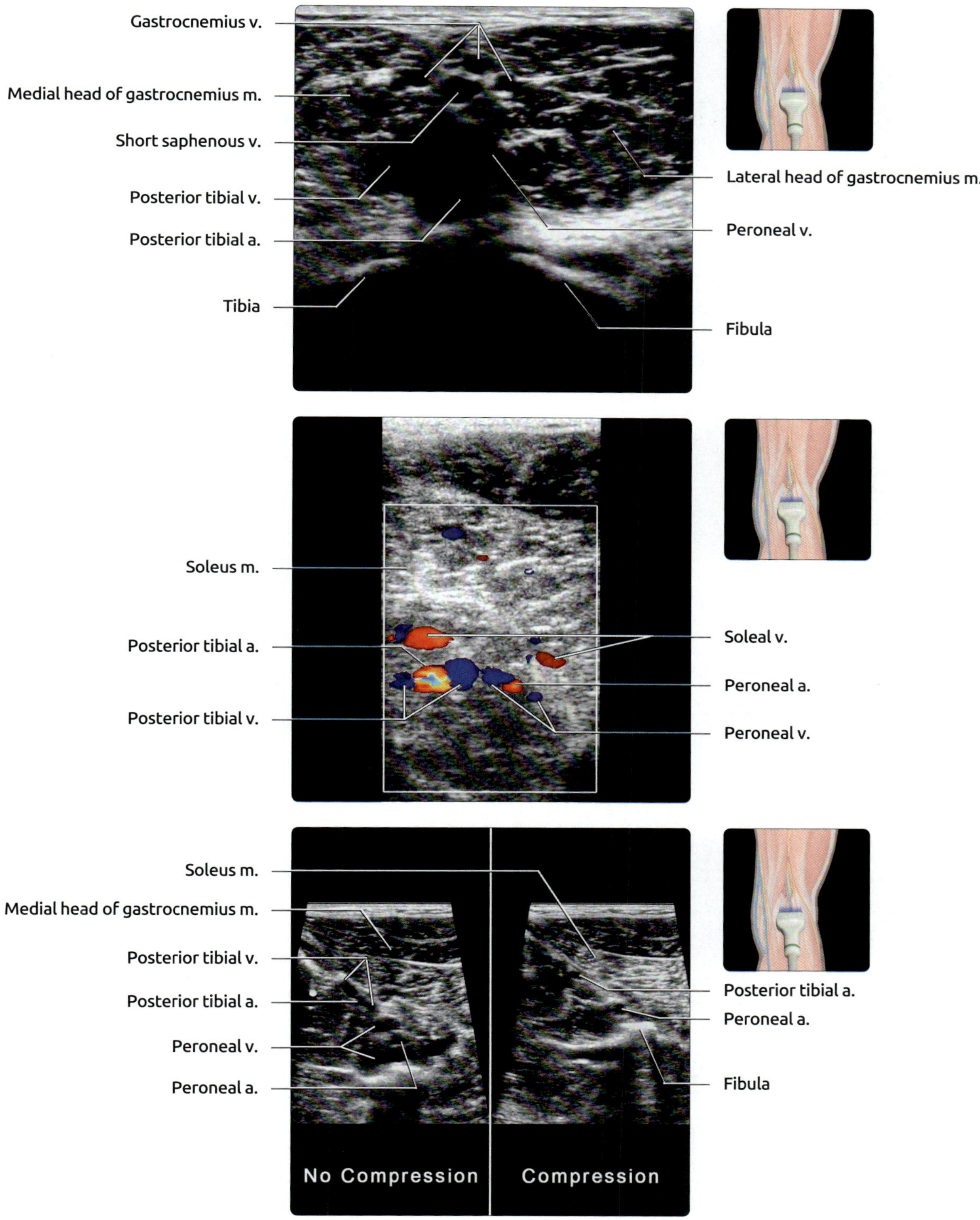

(Top) *Transverse US below the bifurcation of the popliteal artery is shown. The popliteal artery bifurcates into the anterior and posterior tibial arteries. The posterior tibial artery is the larger of the 2 terminal branches of the popliteal artery and passes inferiorly for a short distance before giving out its largest branch, the peroneal artery.* **(Middle)** *Transverse color Doppler US shows the posterior compartment vessels of the proximal calf. Color flow highlights the paired venae comitantes nature of the posterior tibial and peroneal veins, lying on both sides of the corresponding artery.* **(Bottom)** *Transverse US at the midproximal calf scanning from the medial side is shown. The lumina of the peroneal and tibial veins can be compressed by firm transducer pressure (right 1/2 of image), indicating the lack of thrombus.*

TRANSVERSE POSTERIOR TIBIAL VESSELS

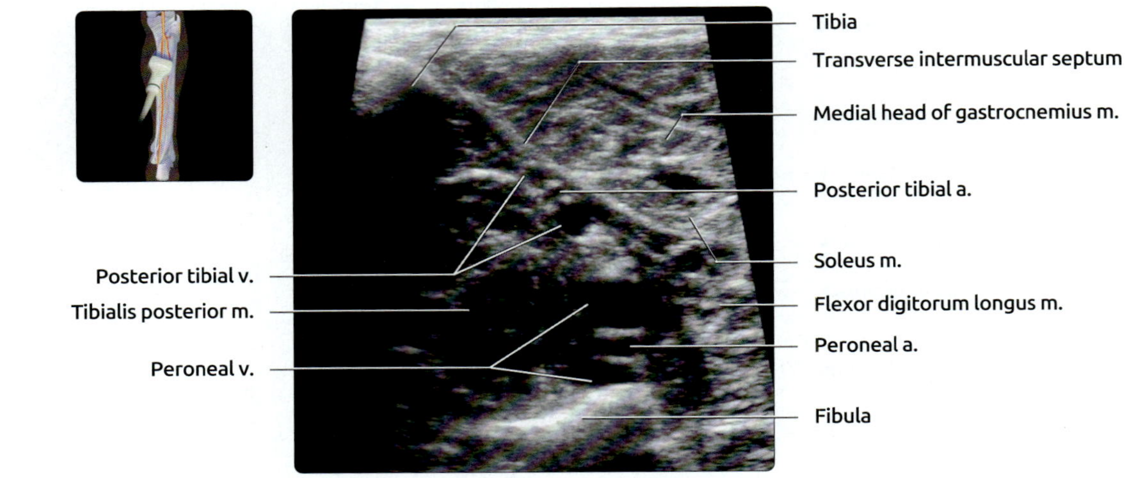

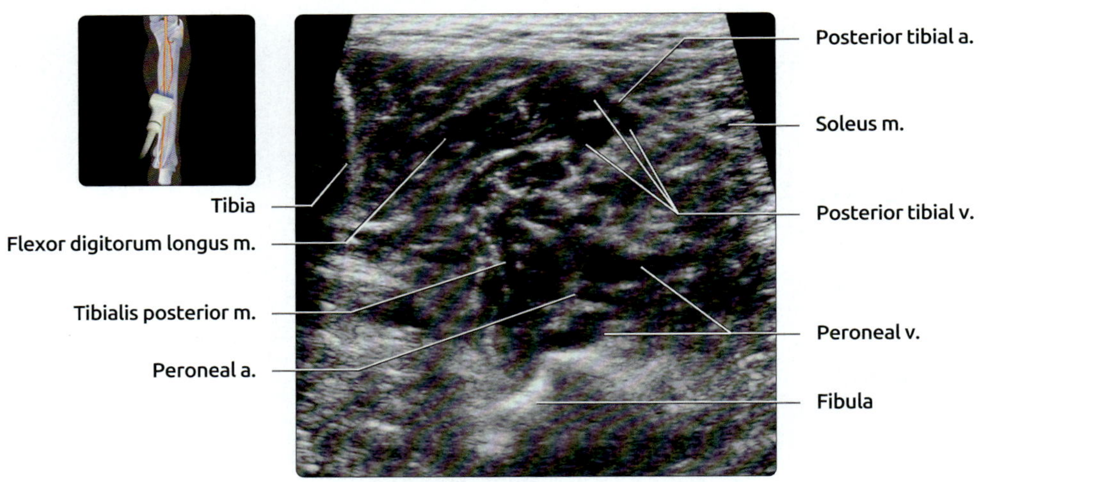

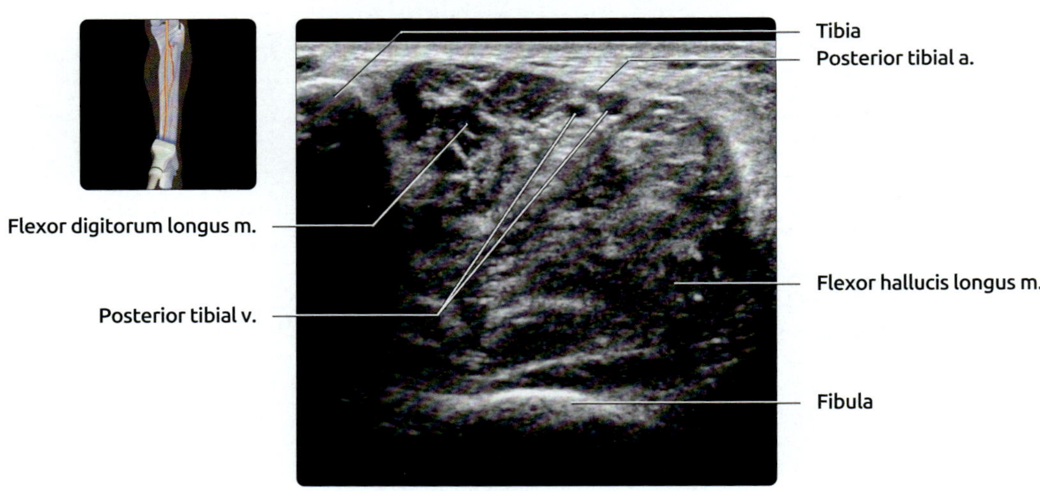

(Top) *This is the 1st of 3 transverse scans that show the posterior tibial vessels using a medial scanning approach. Superiorly, the posterior tibial vessels give off their main branches, the common peroneal vessels. The posterior tibial artery lies just deep to the transverse intermuscular septum while the peroneal arteries are deep to the flexor digitorum muscle.* **(Middle)** *At the midcalf level, the posterior tibial vessels no longer line up side by side (3 in a line). Anatomical variations occur as seen here with 3 posterior tibial veins accompanying 1 artery. The peroneal vessels continue their deeper course, running on the surface of the tibialis posterior, which separates them from the anterior leg compartment.* **(Bottom)** *Further inferiorly, the posterior tibial vessels emerge superficially as they prepare to enter the tarsal tunnel behind the medial malleolus.*

LONGITUDINAL POSTERIOR TIBIAL VESSELS

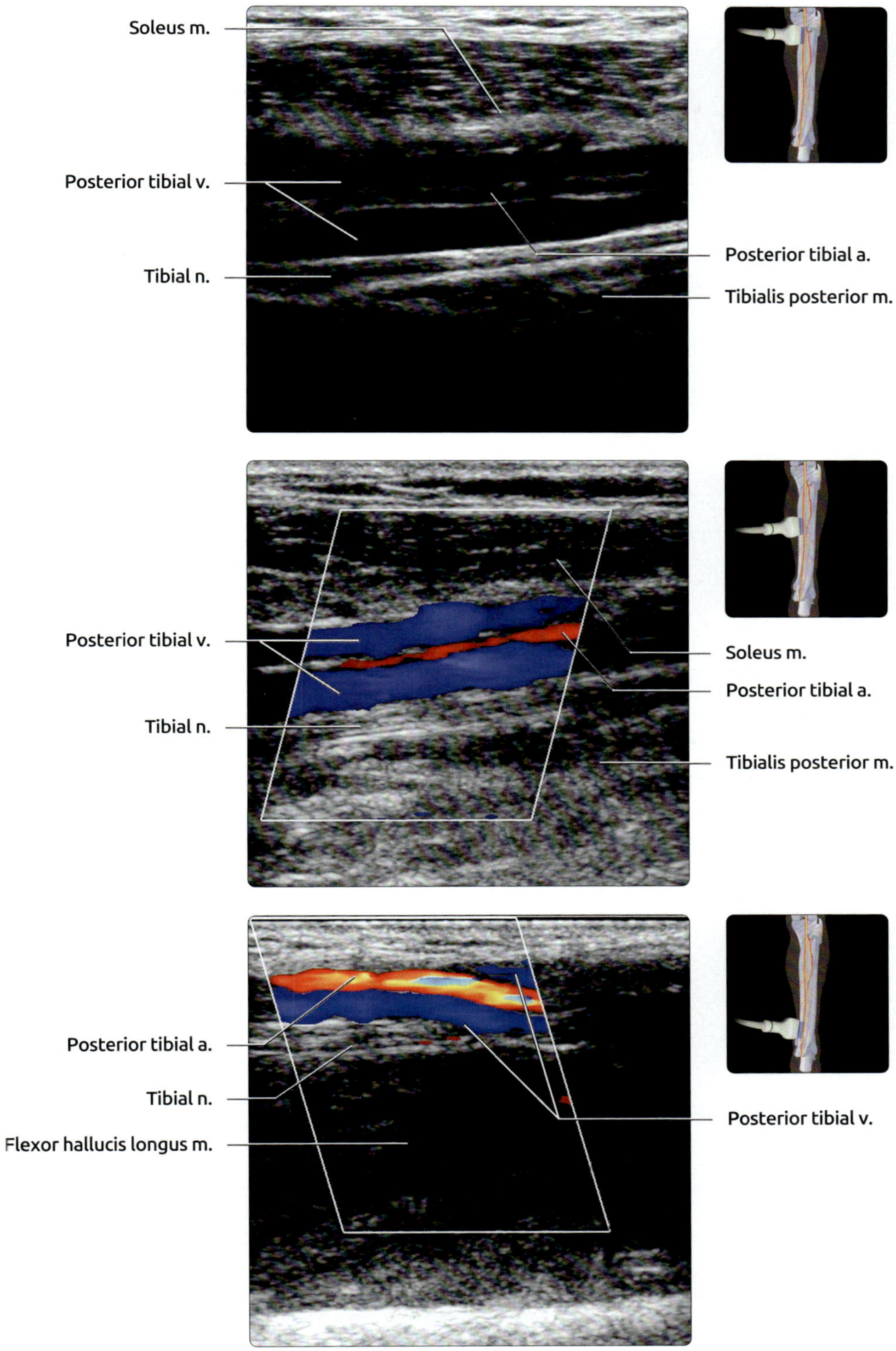

(Top) *Oblique coronal US shows the posterior tibial vessels using a medial approach. The posterior tibial artery is accompanied by paired venae comitantes (posterior tibial veins) and the tibial nerve. They run deep to the transverse intermuscular septum of the leg.* **(Middle)** *Color Doppler US shows the posterior tibial vessels. Notice the marked difference in caliber between the artery and the veins, all of which show normal, complete color filling and normal direction of flow.* **(Bottom)** *Further inferiorly, oblique coronal US of the lower calf shows the distal part of the posterior tibial vessels running superficial to the flexor hallucis muscle and getting ready to curve under the medial malleolus. The posterior tibial vessels and tibial nerve run in the tarsal tunnel together with the tendon flexors of the foot.*

TRANSVERSE PERONEAL VESSELS

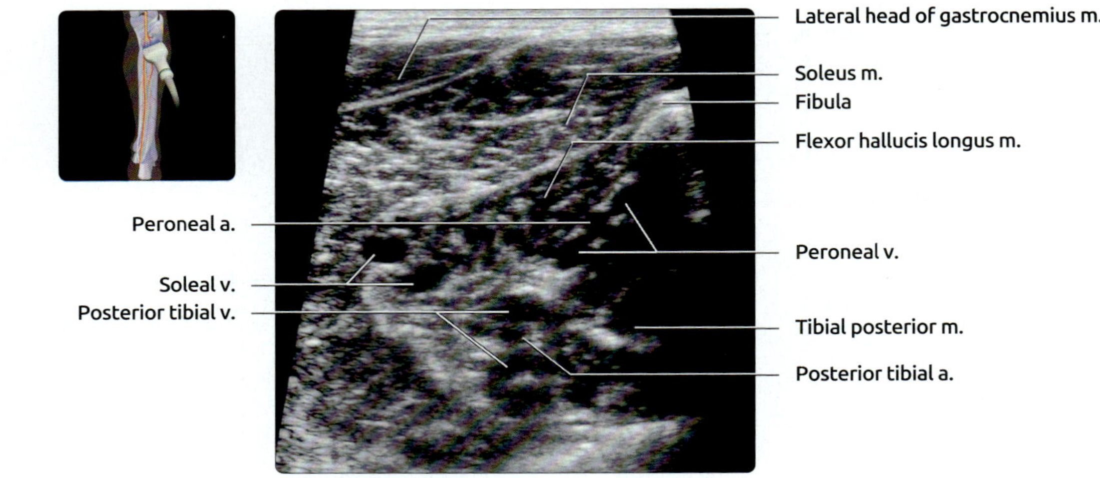

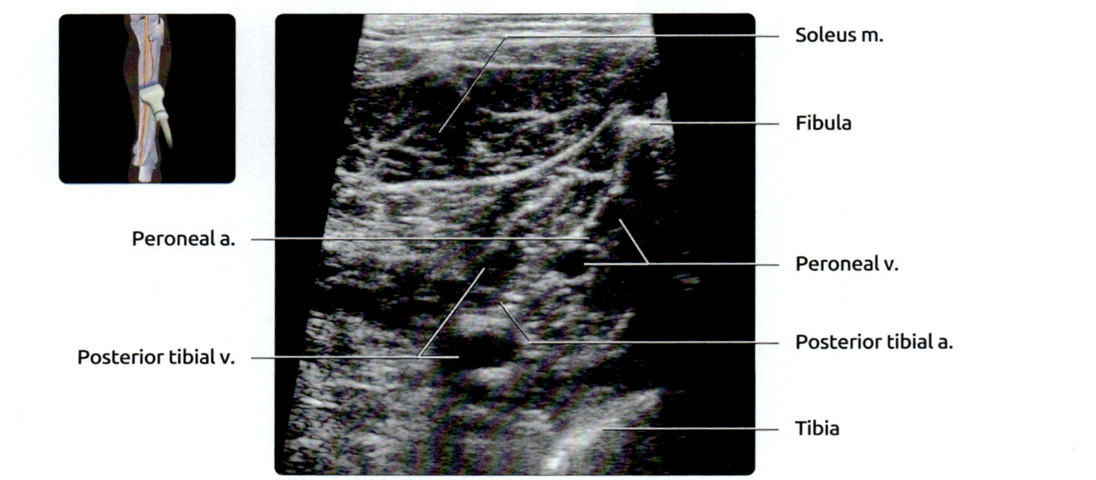

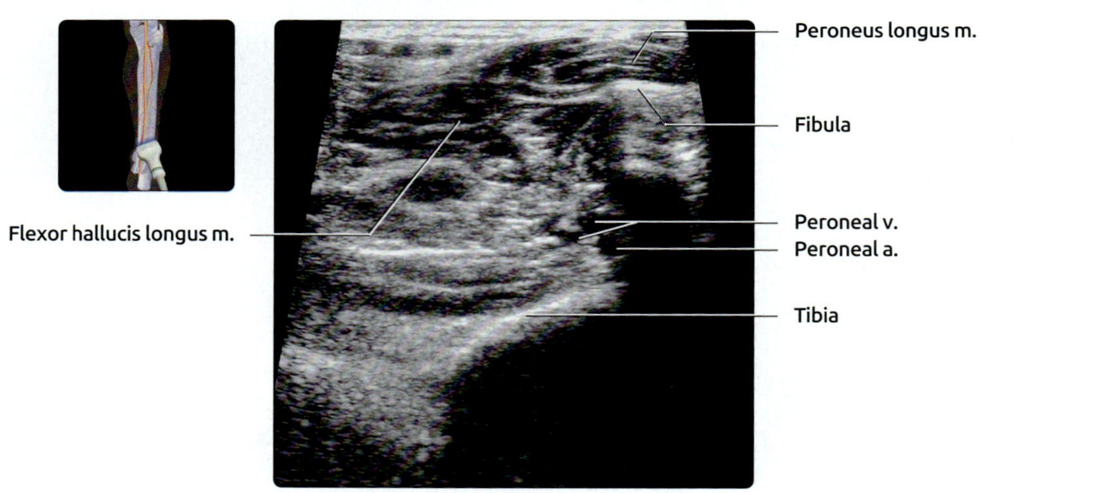

(Top) *First of 3 transverse scans of the calf using a lateral scanning approach is shown. The peroneal artery is a branch of the posterior tibial artery. It runs obliquely toward the fibula and then runs along the medial surface of the fibula distally, sandwiched between the flexor hallucis longus muscle and tibialis posterior muscle.* **(Middle)** *More distally, the relationship between the peroneal vessels and the fibula and flexor muscles remains the same. Nutrient vessels branch out to supply the fibula and the adjacent muscles.* **(Bottom)** *Most distally, the peroneal artery runs behind the tibiofibular syndesmosis before dividing into the lateral calcaneal branches, which run over the lateral and posterior surfaces of the calcaneum.*

LONGITUDINAL PERONEAL VESSELS

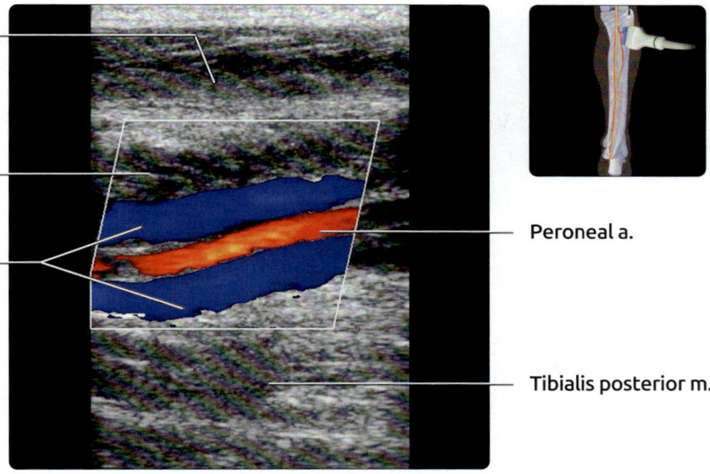

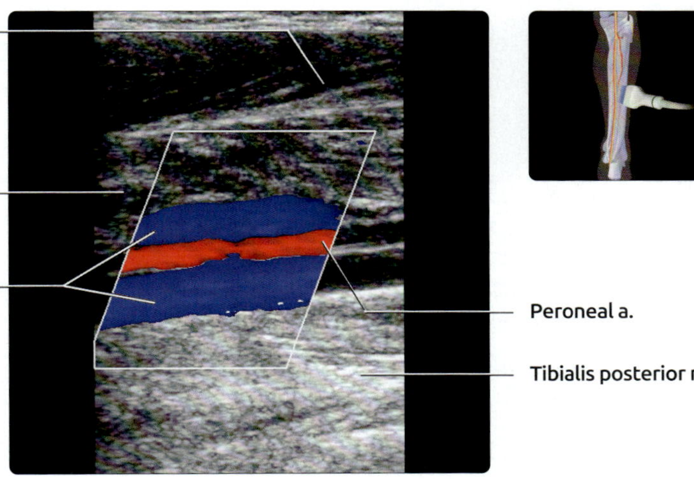

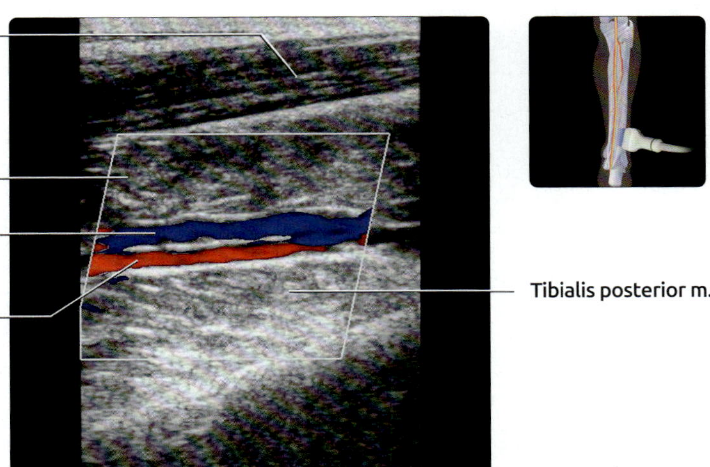

(Top) *Oblique coronal US shows the peroneal vessels using a lateral approach. The peroneal artery, like the posterior tibial, is also accompanied by venae comitantes (peroneal veins). The difference is that there is no peroneal nerve to accompany the vessels. The peroneal nerve runs in the lateral leg compartment after curving around the neck of the fibula.* **(Middle)** *More inferiorly, oblique coronal US shows the peroneal vessels lie deep to the flexor hallucis longus and soleus muscles while running superficial to the tibialis posterior muscle.* **(Bottom)** *Further inferiorly, the peroneal vessels are much smaller in caliber after giving out many branches. They pass posterior to the tibiofibular syndesmosis.*

TRANSVERSE ANTERIOR TIBIAL VESSELS

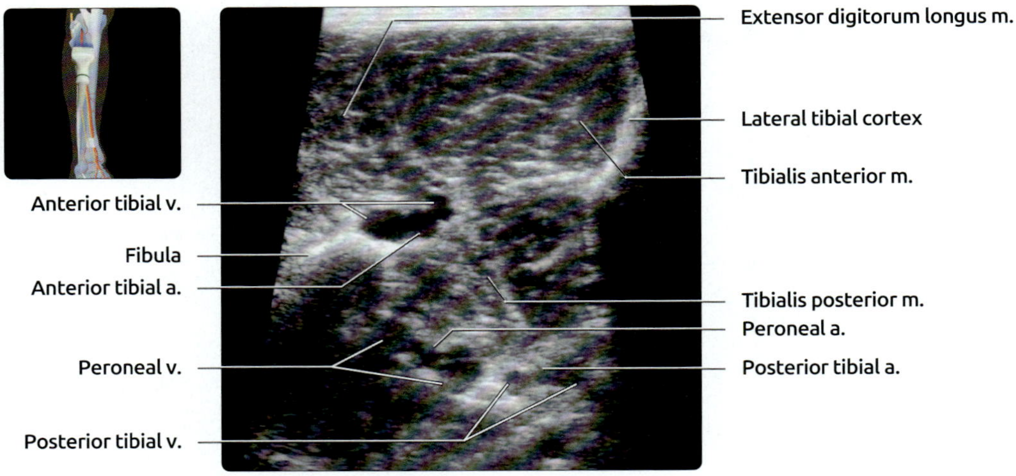

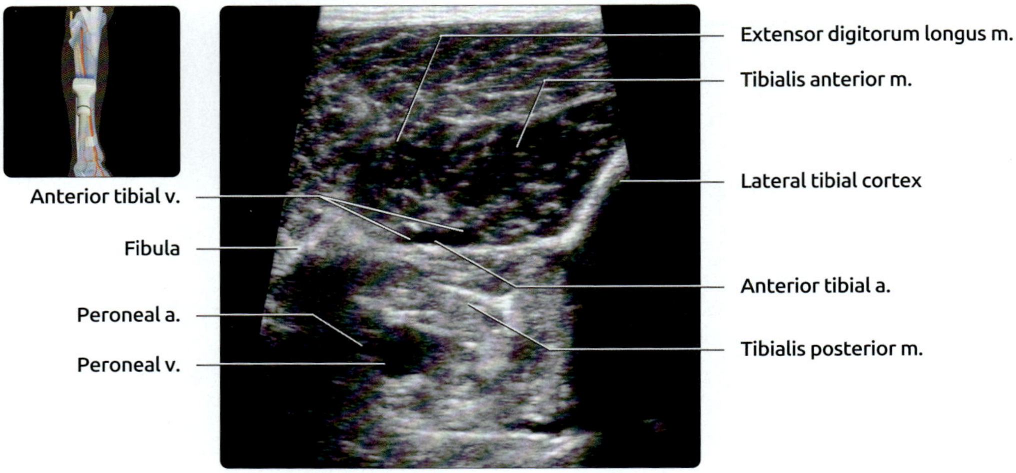

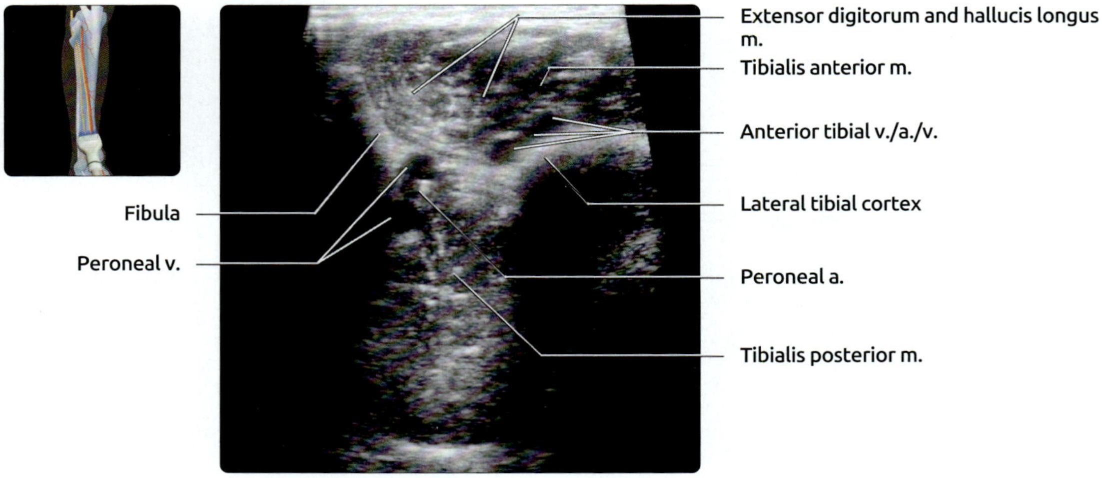

(Top) *This is the 1st of 3 transverse scans that show the anterior tibial vessels using an anterior scanning approach. The anterior tibial vessels are the smaller terminal branches of the popliteal vessels (smaller compared to the posterior tibial vessels at the bifurcation). After its origin, the anterior tibial vessels run anterior through the interosseous membrane and then pass inferiorly on the anterior surface of this membrane.* **(Middle)** *At the midleg level, the anterior tibial vessels continue to run in a deep position on the surface of the interosseous membrane. Through the window between the tibia and fibula, the peroneal vessels and tibialis posterior muscle are seen.* **(Bottom)** *In the most distal part of the leg, the anterior tibial vessels move medially to lie on the surface of the tibia. At this level, the peroneal artery branches run anteriorly to penetrate the interosseous membrane.*

LONGITUDINAL ANTERIOR TIBIAL VESSELS

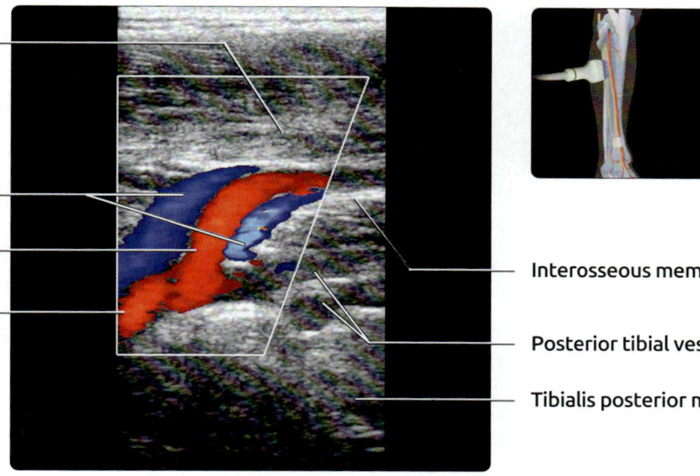

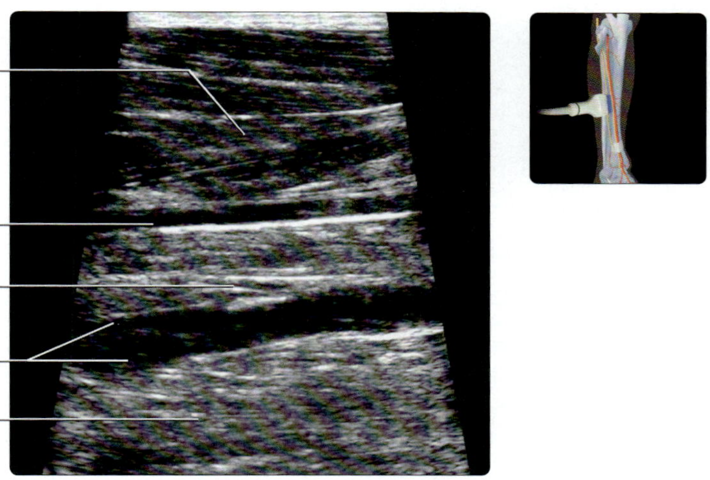

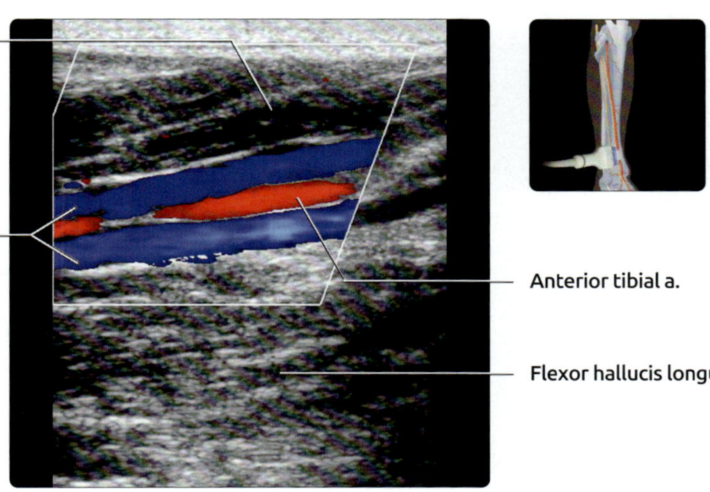

(Top) *Oblique sagittal color Doppler US shows the anterior tibial vessels as they bifurcate from the popliteal vessels. The anterior tibial vessels run anteriorly to penetrate the interosseous membrane into the anterior compartment.* **(Middle)** *Sagittal US shows the anterior tibial artery in the midleg level. The anterior tibial artery runs deep within the anterior compartment and on the surface of the interosseous membrane, being covered by the tibialis anterior muscle.* **(Bottom)** *Sagittal color Doppler US shows the lower aspect of the anterior tibial vessels. The vessels begin to run more superficially, and in the foot, the anterior tibial artery continues as the dorsalis pedis artery.*

SPECTRAL DOPPLER OF LEG ARTERIES

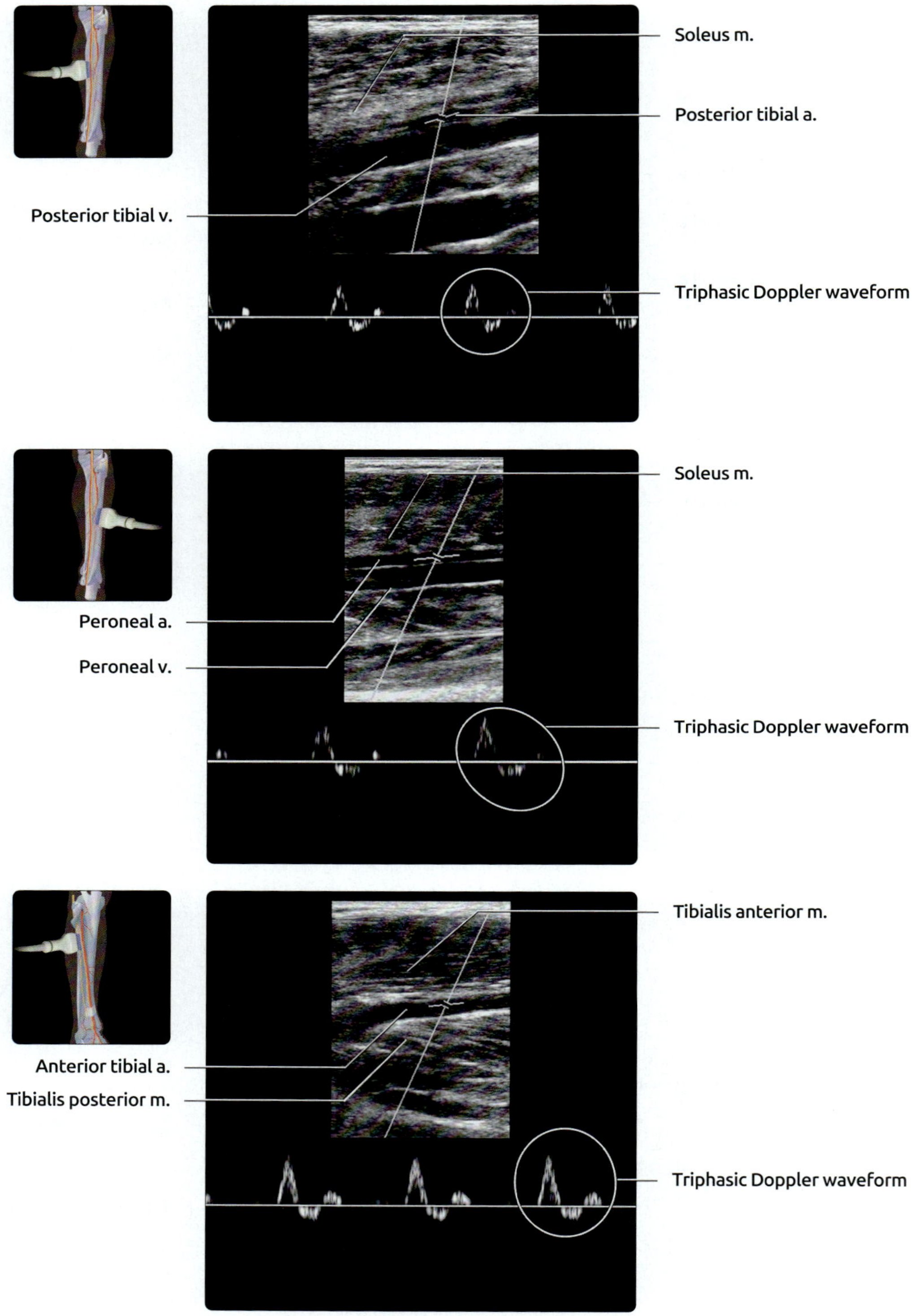

(Top) *Spectral Doppler US of the posterior tibial artery demonstrates the typical triphasic pattern seen in normal arteries. Loss of the reverse flow in the diastolic phase is termed a biphasic pattern and is suggestive of stenosis proximal to the point of interrogation. With proximal stenosis, the peak flow velocity is also decreased.* **(Middle)** *Similarly, the peroneal artery shows a triphasic waveform on spectral Doppler US.* **(Bottom)** *Oblique coronal spectral Doppler US of the anterior tibial artery in the upper calf level shows a normal triphasic pattern. Abnormal increase in reverse flow (negative flow in early diastolic phase) suggests increased distal resistance from stenosis.*

SPECTRAL DOPPLER OF LEG VEINS

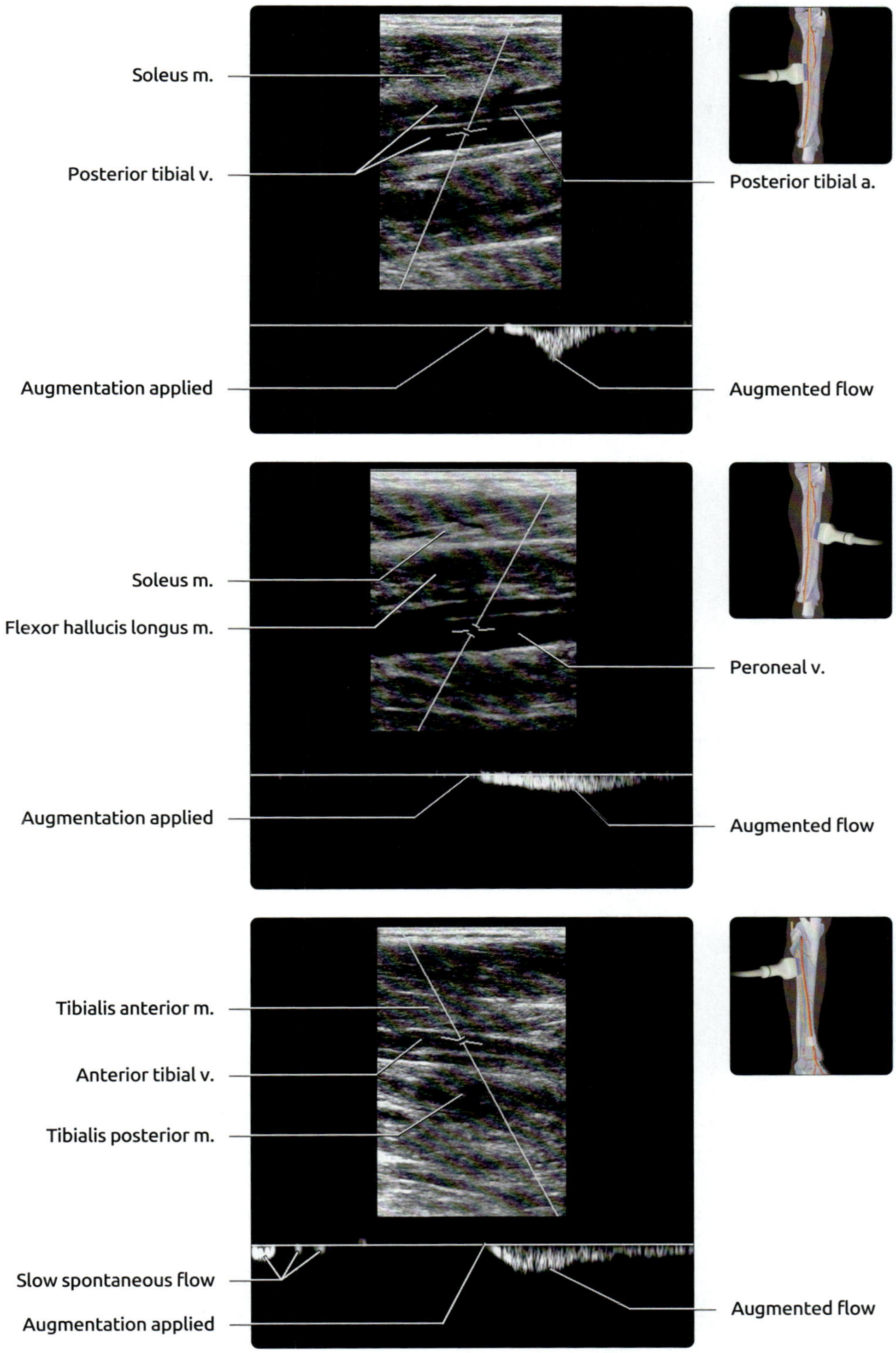

(Top) *Oblique coronal spectral Doppler US of the posterior tibial vein shows absence of spontaneous flow at rest (left 1/2 of the spectral trace). After augmentation (squeezing calf muscles distal to transducer), a surge of blood flow toward the trunk (negative values) can be demonstrated and indicates luminal patency.* **(Middle)** *Oblique coronal spectral Doppler US of the peroneal vein before and after augmentation shows a similar sudden surge of blood flow.* **(Bottom)** *Oblique coronal spectral Doppler US of the anterior tibial vein demonstrates slow spontaneous flow at rest. After augmentation, a sustained surge of flow is demonstrated.*

SOLEAL VEINS

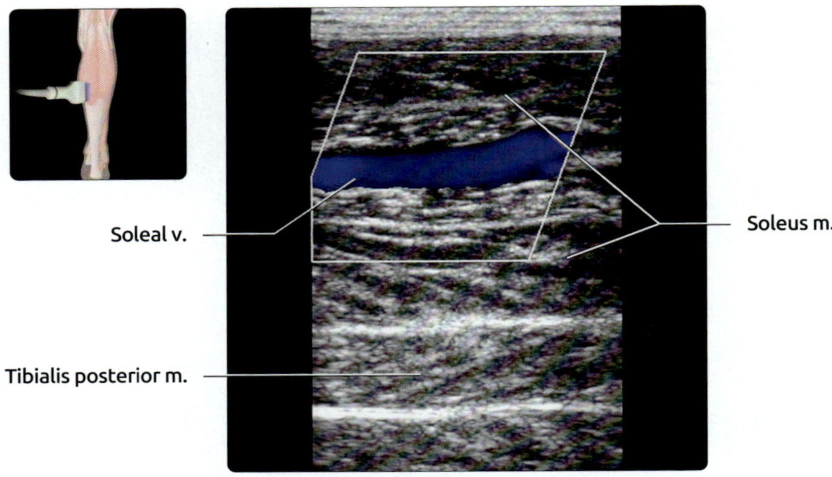

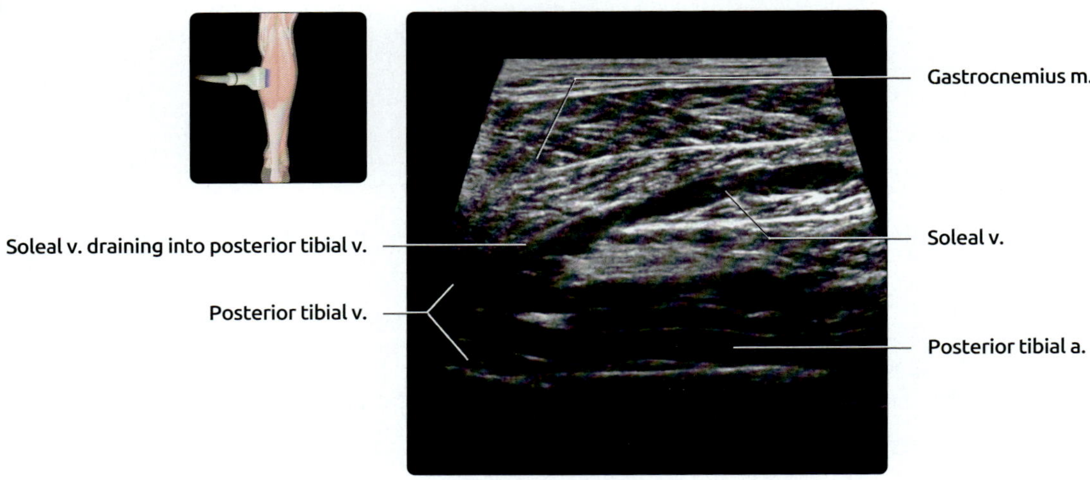

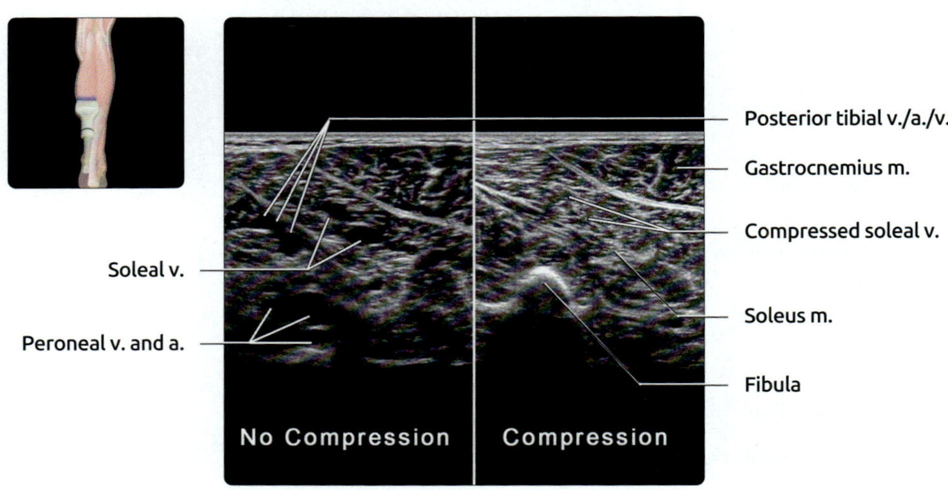

(Top) *Oblique sagittal color Doppler US shows the posterior compartment in the midcalf level. The soleal vein shows complete color filling of its lumen, indicating the absence of thrombus. Soleal veins are the most common veins to develop thrombosis and should be assessed when scanning for deep venous thrombosis.* **(Middle)** *Oblique sagittal US of the posterior compartment shows the soleal vein running proximally and joining the posterior tibial vein. This pathway allows a soleal vein thrombus to extend into the posterior tibial vein and propagate proximally.* **(Bottom)** *Dual transverse images show the posterior compartment using a medial approach, before (left image) and during (right image) firm transducer compression. The soleal veins are completely compressed by transducer pressure, indicating an absence of thrombus.*

GASTROCNEMIUS VEINS

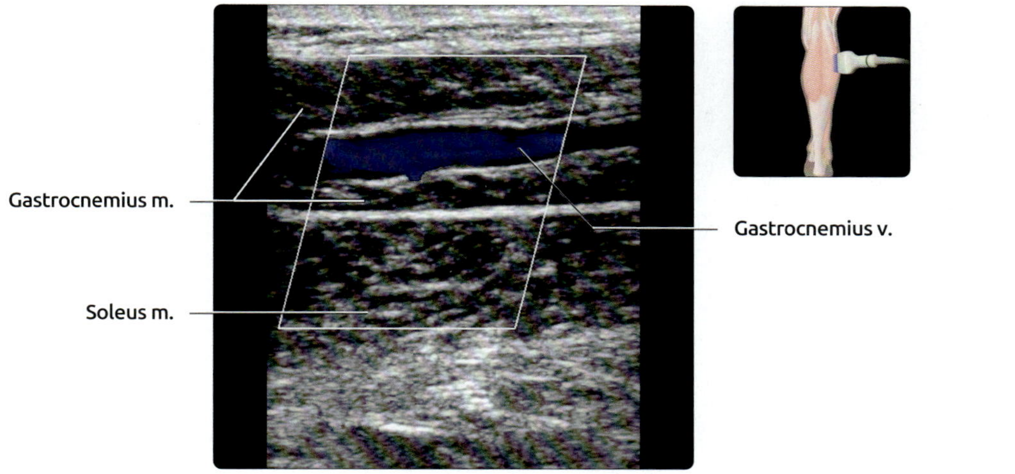

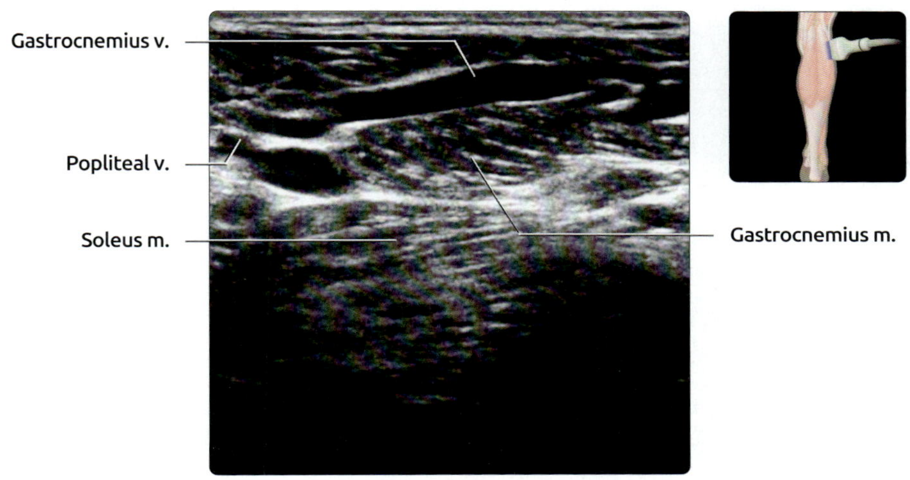

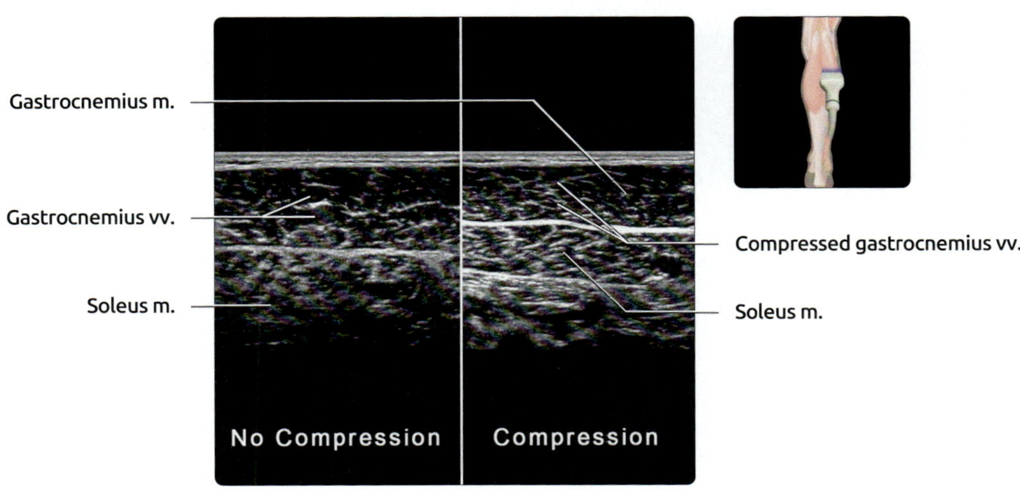

(Top) *Oblique sagittal color Doppler US shows the posterior leg compartment. The gastrocnemius vein shows complete color filling, indicating absence of thrombus. The gastrocnemius veins are the 2nd most common site for venous thrombosis.* **(Middle)** *Oblique sagittal US of the posterior compartment of the upper calf shows the gastrocnemius vein joining the popliteal vein, allowing a gastrocnemius vein thrombus to extend proximally to the popliteal vein.* **(Bottom)** *Dual transverse images show gastrocnemius veins before (left image) and during (right image) compression. The gastrocnemius veins are completely compressed during compression, indicating an absence of thrombus. Compression is regarded as the most reliable technique to rule out venous thrombosis.*

SMALL SAPHENOUS VEIN

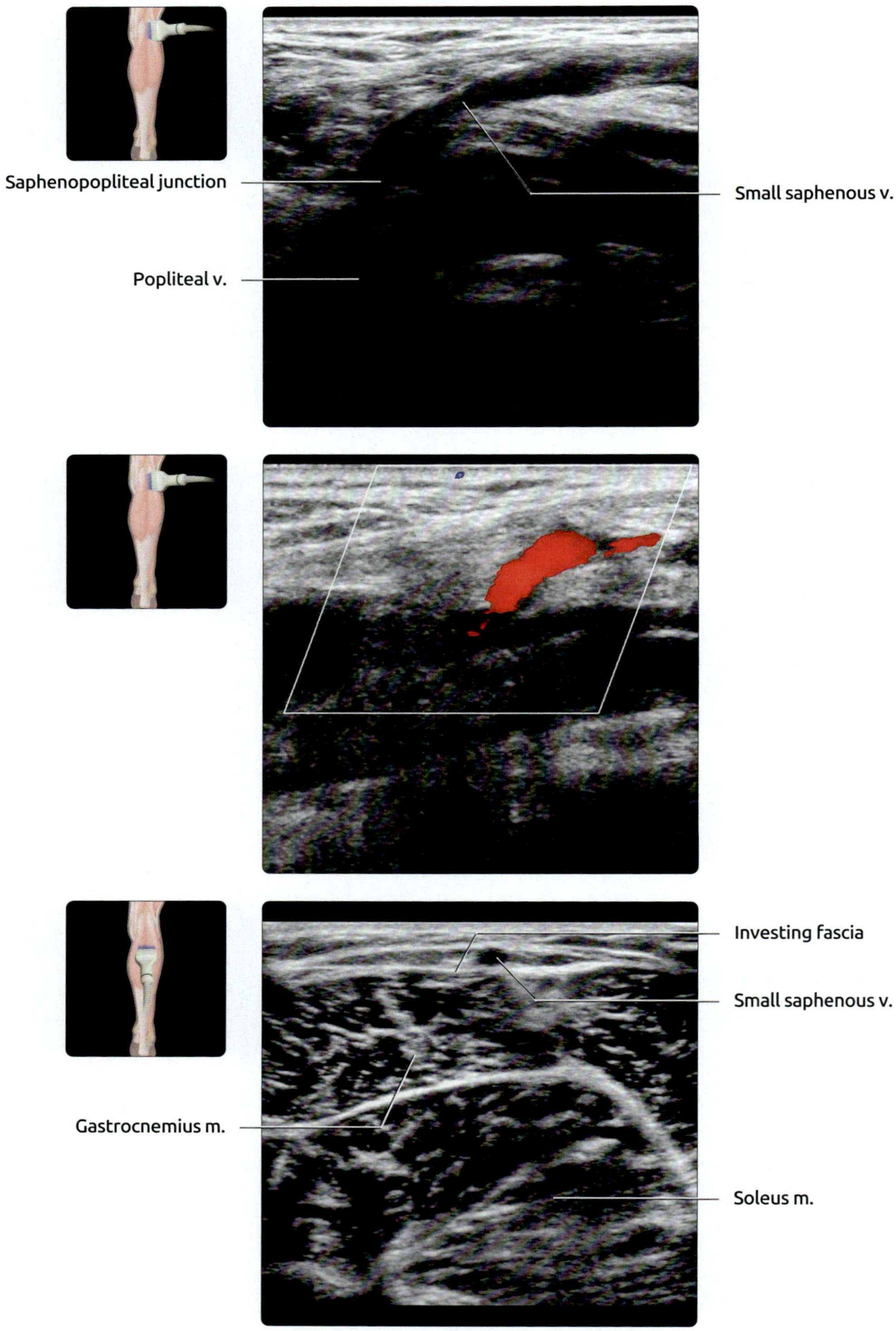

(Top) *Sagittal US shows central popliteal fossa. The small saphenous vein running from the lateral surface of the calf turns medially and penetrates the popliteal fascia to drain into the popliteal vein. Varicose veins arise as a result of reflux and stasis due to incompetent valves. Varicose veins on the lateral side of the calf suggest an incompetent small saphenous system.* **(Middle)** *Sagittal color Doppler US shows the central popliteal fossa with augmentation. Flow toward the popliteal vein (red) is demonstrated at the saphenopopliteal junction. Reverse flow into the small saphenous vein would suggest valvular incompetence.* **(Bottom)** *Transverse US shows the posterior surface of the upper calf. The small saphenous vein is sandwiched between 2 layers of investing fascia. Minimal probe pressure is required to avoid inadvertent compression of the small saphenous vein.*

GREAT SAPHENOUS VEINS

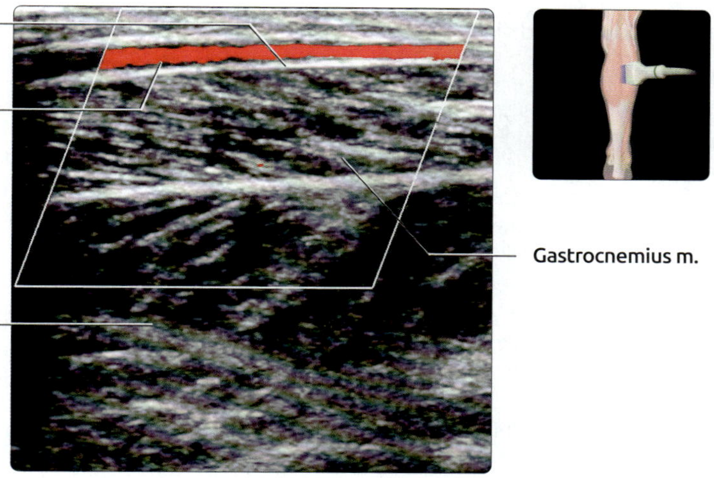

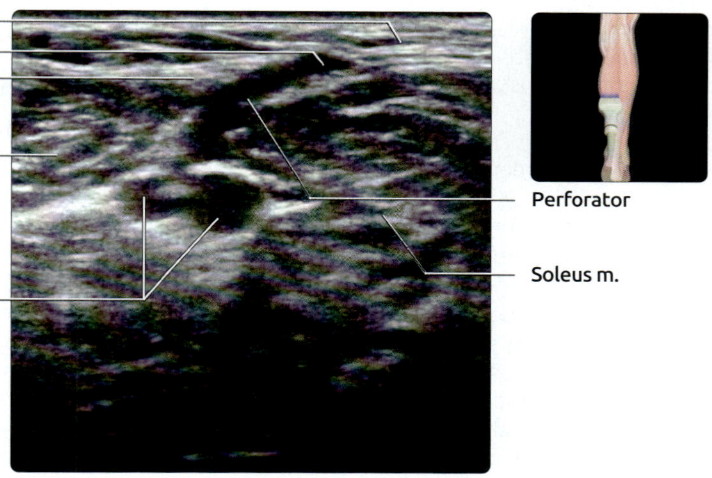

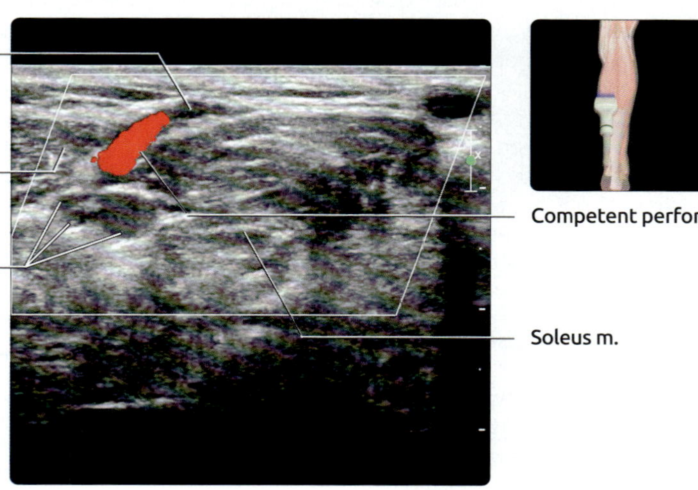

(Top) *Sagittal color Doppler US shows upper midcalf with augmentation. The complete filling of the venous lumen of the small saphenous vein is demonstrated, indicating the absence of a superficial vein thrombus. If flow toward the trunk is demonstrated, valvular incompetence can be excluded.* **(Middle)** *Transverse US shows the medial aspect of the lower 1/3 of the calf. A perforator arising from the great saphenous vein is seen penetrating through the investing fascia to join the posterior tibial vein.* **(Bottom)** *Transverse color Doppler US shows the medial aspect of the lower 1/3 of the calf. With augmentation, color filling of the perforator lumen is demonstrated in the perforator between the great saphenous vein and the posterior tibial vein, indicating patency of vessel. If flow toward the posterior tibial vein can be documented on Doppler study, incompetence can be excluded.*

GROSS ANATOMY

Sciatic Nerve

- Comprises common peroneal and tibial nerves

Common Peroneal Nerve

- Smaller of 2 terminal divisions of sciatic nerve
- Separates from tibial nerve midthigh, at superior angle of popliteal fossa
- Leaves popliteal fossa by passing posterior to head of fibula
- Winds in subcutaneous tissues around lateral aspect of fibular neck
  - Passes into fibular tunnel between fibular neck and peroneus longus aponeurosis
  - After exiting fibular tunnel, nerve trifurcates into 3 terminal branches (articular, deep, and superficial)
    - **Recurrent articular branch**
      - Extends to proximal tibiofibular joint
      - Acts as intraneural conduit for transmission of ganglion arising from joint
    - **Deep peroneal nerve**
      - Pierces anterior intermuscular septum and extensor digitorum longus to enter anterior compartment
      - Descends between tibialis anterior and long extensors supplying anterior compartment muscles
      - 1.5 cm proximal to ankle, divides into medial sensory and lateral motor branches
      - Passes deep to extensor retinaculum and is crossed by extensor hallucis longus tendon
      - Traverses ankle alongside dorsalis pedis artery
      - Medial sensory branch runs alongside dorsalis pedis artery to 1st web space; crossed by extensor hallucis brevis tendon in midfoot
      - Lateral motor branch runs alongside lateral tarsal artery to supply extensor digitorum brevis, tarsal joints, and 2nd-4th metatarsophalangeal joints
    - **Superficial peroneal nerve**
      - Descends in substance of peroneus longus until it reaches peroneus brevis
      - Passes obliquely over anterior border of brevis to descend in groove between peroneus brevis and extensor digitorum longus
      - 10 cm proximal to fibular tip, it pierces deep fascia to enter subcutaneous tissues and divide into intermediate and medial dorsal cutaneous nerves

Tibial Nerve

- Runs along midaxis of popliteal fossa between 2 heads of gastrocnemius
- Descends under fascial septum between deep and superficial posterior muscle compartments of leg
  - In upper 2/3 of leg, lies on fascia of tibialis posterior and flexor digitorum longus
  - In lower 1/3 of leg, located midway between Achilles tendon and medial border of tibia
- Crosses posterior surface of ankle joint alongside posterior tibial vessels
- At medial ankle, under flexor retinaculum within tarsal tunnel, divides into lateral and medial plantar nerves
  - 1st branch of lateral plantar nerve is inferior calcaneal nerve (Baxter nerve)
    - Neuritis of this nerve, which supplies calcaneal periosteum, can simulate plantar fasciitis
  - Medial plantar nerve may compress between navicular bone and abductor hallucis muscle (jogger's foot)

Saphenous Nerve

- Longest branch of femoral nerve, arising 2 cm below inguinal ligament and descends with superficial femoral artery via adductor canal
- Passes posterior to sartorius posteromedial to knee before piercing deep fascia to subcutaneous tissues
- In leg, accompanies long saphenous vein

Sural Nerve

- Purely sensory nerve arises in midcalf with contributions from tibial and common peroneal nerves
- Runs in posterior midline subcutaneous fat with short saphenous vein
- 12 cm proximal to heel, it passes to lateral side of Achilles tendon and descends behind lateral malleolus
- Provides sensory innervation to lateral aspect of ankle and lateral aspect of foot

ANATOMY IMAGING ISSUES

Imaging Approaches

- **Common peroneal nerve**
  - Best located posterior and lateral to fibular head and neck
  - Most frequently entrapped around fibular neck
    - Thick arch of peroneus longus aponeurosis, fracture, trauma, osteophytosis, and ganglia
      - Compressed nerve may appear enlarged proximally
      - Compression may be dynamic
- **Tibial nerve**
  - Best located between 2 heads of gastrocnemius in proximal calf, alongside posterior tibial artery in mid to distal calf, and beneath medial malleolus at ankle
  - Entrapped in tarsal tunnel
    - Talocalcaneal hyperostosis, synovitis, ganglia, or accessory muscles
      - Encroachment onto tibial nerve seen but tibial nerve caliber usually normal
- **Deep peroneal nerve**
  - Best located in lower leg between anterior tibial artery and tibialis anterior muscle
  - Entrapped deep to extensor retinaculum, by extensor hallucis longus tendon, by midfoot osteophytosis, or by extensor hallucis brevis tendon
- **Superficial peroneal nerve**
  - Best located in fat plane between anterior and lateral compartments of distal leg
  - Entrapped passing through deep fascia of distal leg
    - Stretching injury to fascia during ankle trauma
      - Fascial thickening ± nerve thickening
- **Saphenous nerve**
  - Best located near knee alongside long saphenous vein 10 cm proximal to posteromedial knee or near ankle between medial malleolus and anterior tibialis tendon
- **Sural nerve**
  - Best located subcutaneously alongside short saphenous vein between lateral malleolus and Achilles tendon

GRAPHICS: ANTERIOR AND POSTERIOR LEG NERVES

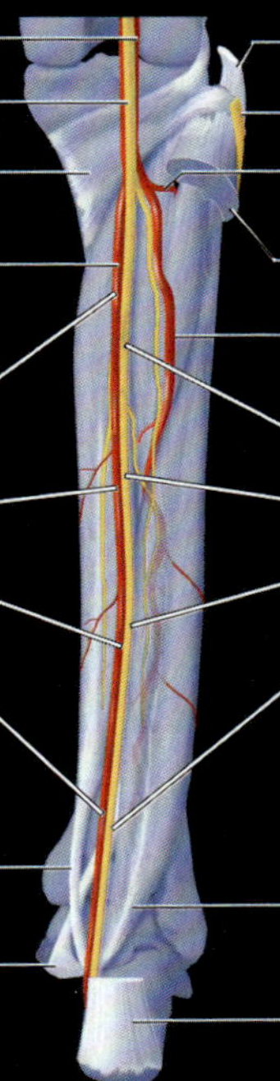

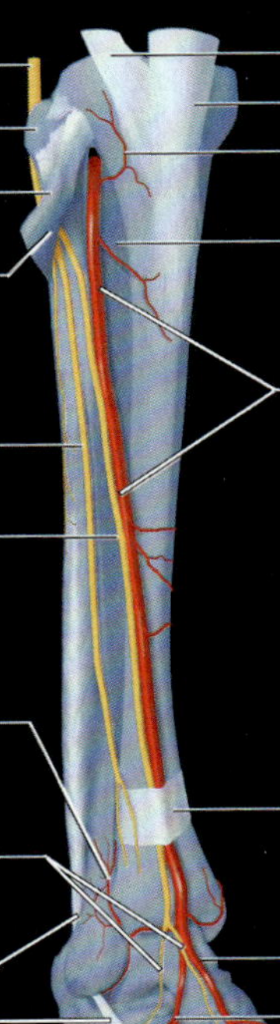

(Top) *Graphic shows posterior leg nerves and arteries. The sciatic nerve divides into common peroneal and tibial branches. The common peroneal nerve then follows the biceps femoris around the lateral fibular neck. The tibial nerve descends between the deep and superficial posterior leg muscles, paralleling the posterior tibial artery.* **(Bottom)** *Graphic of the anterior leg shows the common peroneal nerve extending through the peroneal tunnel (between the peroneus longus tendon and the fibular neck), branching into deep and superficial components. Both have muscular branches along their length; the deep peroneal nerve parallels the anterior tibial artery and terminates in the medial and lateral branches to the dorsum of the foot and ankle, respectively.*

TIBIAL NERVE

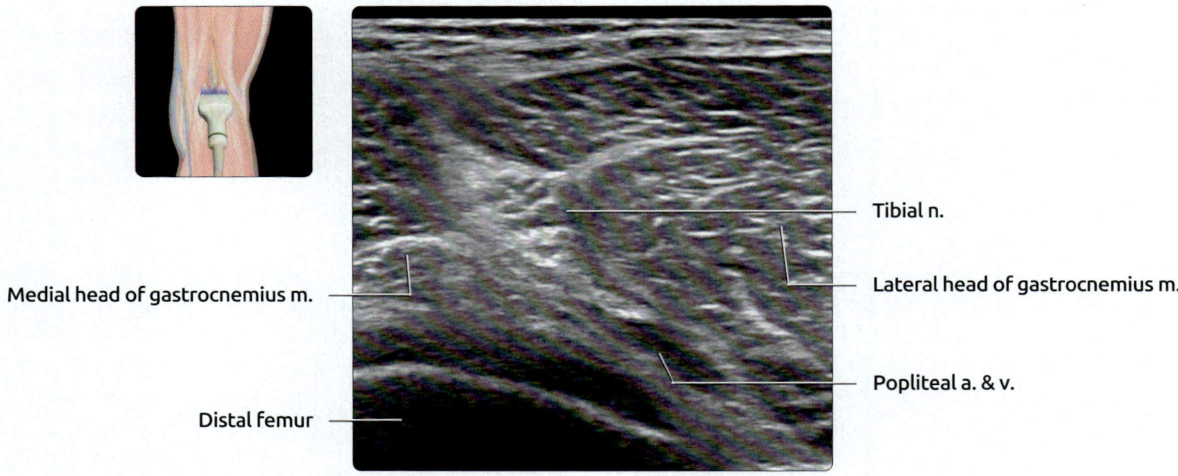

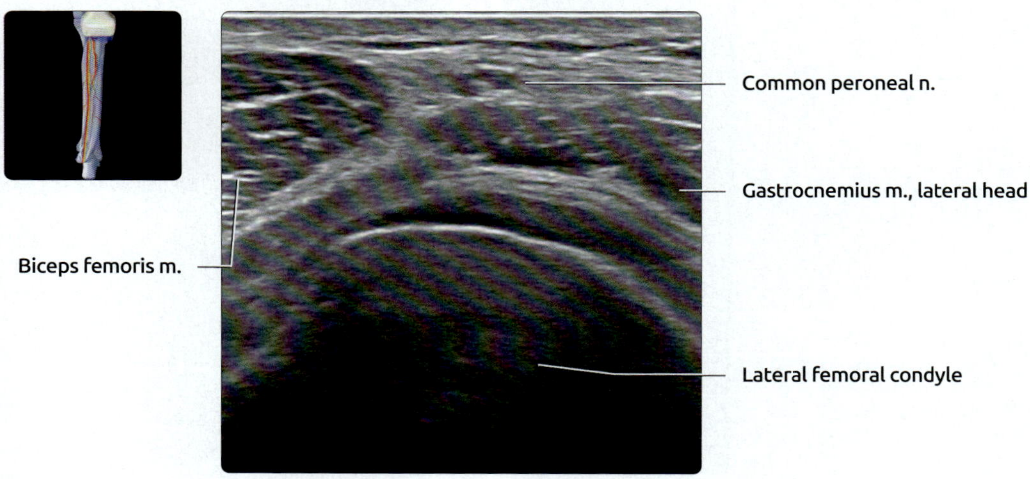

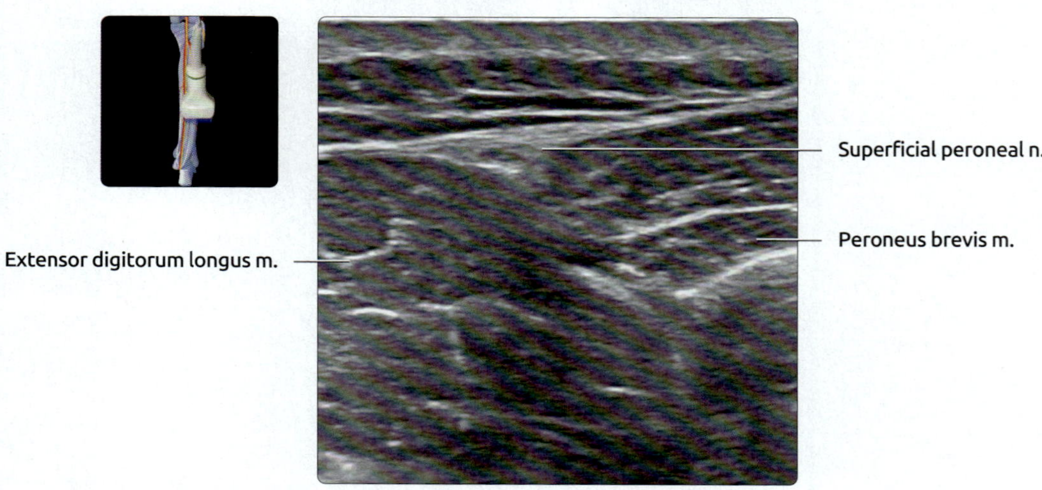

(Top) *Transverse US shows the proximal aspect of the popliteal fossa. The tibial nerve demonstrates the typical neural signature of small black dots within a bright matrix. It lies superficial to the popliteal artery and vein, which are partially effaced.* **(Middle)** *Transverse US at the proximal end of the popliteal fossa shows the common peroneal nerve, having branched from the tibial nerve, alongside the biceps femoris muscle as it moves laterally toward the fibular head. At the fibular head area, it will divide into superficial and deep branches. It also gives rise to an articular branch to the proximal tibiofibular joint.* **(Bottom)** *Transverse US of the lateral aspect of the leg ~ 10 cm proximal to the lateral malleolus shows the superficial peroneal nerve just deep to the crural fascia at the interspace between the peroneus brevis and the extensor digitorum longus muscles.*

PERONEAL NERVE

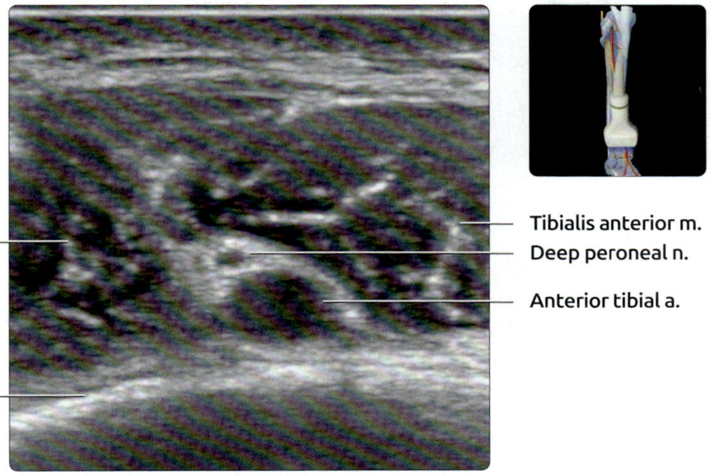

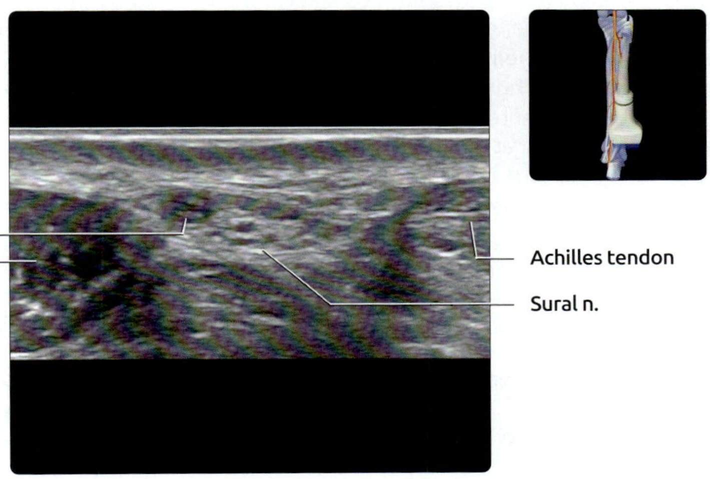

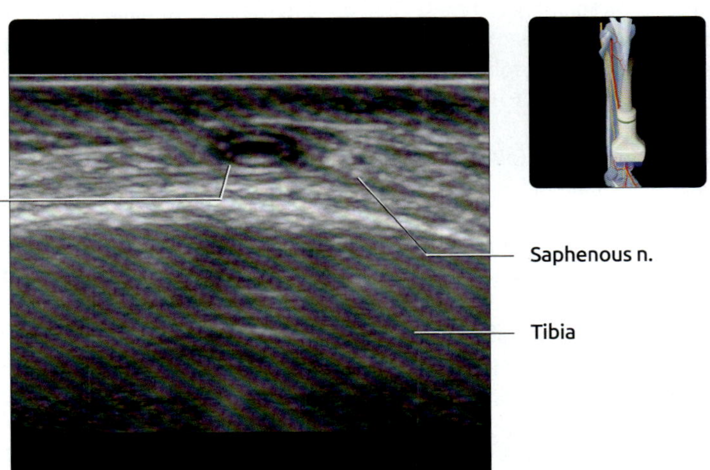

(Top) *Transverse US of the leg just proximal to the ankle joint shows the deep peroneal nerve passing superficial to the anterior tibial artery. The deep peroneal nerve passes from the medial to the lateral side of the anterior tibial artery as it approaches the ankle joint.* **(Middle)** *Transverse US of the distal leg posterolaterally shows the sural nerve lateral to the small saphenous vein. This vein, which lies between the peroneus brevis muscle and the Achilles tendon, acts as a good marker for the sural nerve. The sural nerve provides sensory sensation to the lateral ankle region and lateral aspect of the foot.* **(Bottom)** *Transverse US at the anteromedial aspect of the ankle shows the saphenous nerve medial to the great saphenous vein. This nerve, which is difficult to see clearly with US, lies between the medial malleolus and the tibialis anterior muscle.*

GROSS ANATOMY

Osseous Anatomy

- **Ankle (talocrural) joint**
  - Tibia, fibula, and talus form synovial joint
  - Supported by lateral and medial collateral ligaments
  - Mainly uniaxial hinge joint, dorsiflexion, plantar flexion, also dynamic shift of axis of rotation during dorsiflexion and plantar flexion
- **Distal tibiofibular joint**
  - Fibrous joint
  - Supported by syndesmotic ligaments
  - Synovial recess from ankle joint extends into joint
  - May have articular cartilage far distally
  - Minimal stretch ("give") during dorsiflexion: Allows increase in malleolar gap and slight fibular lateral rotation

Ligaments

- 3 sets bind ankle: Distal tibiofibular syndesmotic complex, lateral collateral ligaments, and deltoid ligaments
- **4 tibiofibular syndesmotic ligaments**
  - **Anterior and posterior tibiofibular ligaments**
    - Extend obliquely between anterior and posterior tibial and fibular tubercles, widest at tibial attachment
    - Distal anterior tibiofibular ligament: Bassett ligament
  - **Inferior transverse ligament**: Distal part of posterior tibiofibular ligament
  - **Interosseous ligament**: Distal thickening of syndesmotic membrane
- **3 lateral collateral ligaments**
  - **Anterior talofibular ligament**
    - Originates anterior aspect lateral malleolar tip; inserts on talar neck, 2 mm thick
    - Stabilizes talus against anterior displacement, internal rotation, and inversion
    - Weakest, 1st lateral ligament to tear
  - **Calcaneofibular ligament**
    - Originates from lateral malleolar tip; inserts on calcaneal trochlear eminence
    - Deep to peroneal tendons
    - Lateral restraint of subtalar joint, often tears with anterior talofibular ligament
  - **Posterior talofibular ligament**
    - Extends from lateral malleolar fossa to lateral talar tubercle
    - Strongest, rarely tears
- **Deltoid ligament (medial collateral ligament)**
  - Fan-shaped, originates from anterior, apex, and posterior medial malleolus; inserts on talus, sustentaculum tali, spring ligament, and navicular
  - Deep: Posterior and anterior tibiotalar bands
  - Superficial: Tibiocalcaneal, tibiospring, tibionavicular, and posterior tibiotalar (variable) bands
- **Spring ligament (plantar calcaneonavicular ligament)**
  - Binds calcaneus to navicular, 3 components
    - Superomedial = **origin**: Sustentaculum tali; **insertion**: Superomedial navicular, tibiospring band of deltoid
    - Medioplantar oblique = **origin**: Calcaneal coronoid fossa; **insertion**: Plantar navicular
    - Inferoplantar longitudinal = **origin**: Coronoid fossa; **insertion**: Navicular beak

Retinacula

- Focal thickening of deep fascia
- Prevents bowstringing, binds tendons down
- **Superior extensor retinaculum**
  - Few cm above ankle joint
  - Attaches to anterior fibula laterally, tibia medially
  - Proximally continues with fascia cruris
  - Distally attaches to inferior extensor retinaculum
  - Binds down anterior compartment muscles
- **Inferior extensor retinaculum**
  - At ankle joint, Y-shaped, stem laterally, proximal and distal bands medially
  - Stem attaches laterally to upper calcaneus
    - Loops around extensor tendons
    - Roots extend into sinus tarsi
  - Proximal medial band has deep and superficial layers, loops around extensor hallucis longus tendon and occasionally tibialis anterior tendon
  - Distal medial band superficial to extensor hallucis longus and tibialis anterior tendons
    - Attaches to plantar aponeurosis
  - Dorsalis pedis vessels, deep peroneal nerve: Deep to all layers of inferior extensor retinaculum
- **Flexor retinaculum**
  - Attaches to medial malleolus
  - Proximally continuous with deep fascia of leg
  - Distally continuous with plantar aponeurosis
  - Abductor hallucis partly attached to it
  - Binds deep flexor tendons to tibial and calcaneal grooves
  - Lateral border of tarsal tunnel
- **Superior peroneal retinaculum**
  - Origin: Lateral malleolus, insertions vary, most commonly to deep fascia of leg and calcaneus
  - Binds peroneal tendons into retrofibular groove
- **Inferior peroneal retinaculum**
  - Continuous with inferior extensor retinaculum
  - Inserts on lateral calcaneus, peroneal tubercle (trochlea)
  - Binds peroneus brevis, peroneus longus tendons to calcaneus

Tendons

- **Anterior (extensor) compartment**
  - **Tibialis anterior tendon**
    - Most medial and largest tendon in anterior compartment
    - Inserts on medial cuneiform, base of 1st metatarsal
    - Dorsiflexes ankle, inverts foot, tightens plantar aponeurosis
    - Supports medial longitudinal arch during walking
  - **Extensor hallucis longus tendon**
    - Inserts on dorsal base of 1st distal phalanx
    - Extends 1st phalanges, dorsiflexes foot
  - **Extensor digitorum longus tendon**
    - Divides into 4 slips on dorsum of foot
    - Slips receive tendinous contributions from extensor digitorum brevis, lumbricals, and interosseous muscles

- Each slip divides into 3: Central slip inserts on dorsal base of middle phalanx and 2 collateral slips, which reunite and insert on bases of 2nd-5th distal phalanges
- Dorsiflexes ankle, extends toes, tightens plantar aponeurosis

- **Peroneus tertius tendon**
  - Typically part of extensor digitorum longus tendon
  - Inserts on dorsal base of 5th metatarsal
- **Lateral compartment**
  - **Peroneus longus tendon**
    - Posterolateral to peroneus brevis tendon in retrofibular groove, deep to superior peroneal retinaculum
    - Proximally has common tendon sheath with peroneus brevis
    - Separates tendon sheath at sole of foot
    - Descends inferior to peroneal tubercle, deep to inferior peroneal retinaculum
    - Curves in cuboid groove deep to long plantar ligament
    - Inserts on plantar base of 1st metatarsal, medial cuneiform
    - Plantarflexes ankle, everts foot, supports longitudinal and transverse arches during walking
    - Os peroneum is in ~ 20% of individuals
  - **Peroneus brevis tendon**
    - Anteromedial to peroneus longus tendon in retrofibular groove, deep to superior peroneal retinaculum
    - Descends superior to peroneal tubercle of calcaneus, deep to inferior peroneal retinaculum
    - Inserts into base of 5th metatarsal
    - Everts foot, limits foot inversion
- **Superficial posterior compartment**
  - **Achilles tendon**
    - Largest and strongest tendon in body
    - Most common tendon to rupture
    - Conjoint tendon of medial and lateral gastrocnemius and soleus muscles
    - ~ 15 cm long
    - Lacks tendon sheath partially enclosed by paratenon
    - Inserts into posterior calcaneal tuberosity
    - Crescent-shaped insertion with medial and lateral projections
    - Retrocalcaneal bursa between distal tendon and calcaneal tuberosity
    - Main plantar flexor of ankle and foot
    - Midsection of Achilles tendon is supplied by peroneal artery
    - Proximal and distal portions supplied by posterior tibial artery
- **Plantaris tendon**
  - Vestigial, slender tendon, medial to Achilles tendon
  - Inserts on or medial to Achilles tendon
- **Deep posterior (flexor) compartment**
  - **Tibialis posterior tendon**
    - Crosses flexor digitorum longus tendon above ankle joint to become most posteromedial tendon
    - Shares tibial groove with flexor digitorum longus tendon
    - Inserts on navicular tuberosity, cuneiforms, bases of 2nd-4th metatarsals
    - Main invertor of foot, aids in plantar flexion
    - Supports medial longitudinal arch
  - **Flexor digitorum longus tendon**
    - Lateral to tibialis posterior tendon in tibial groove
    - Crosses flexor hallucis longus tendon at knot of Henry (where there is also exchange of tendon fibers at this crossing of 2 tendons)
    - Divides into 4 slips, which give origin to lumbricals
    - Slips pass through openings in corresponding tendons of flexor digitorum brevis
    - Slips insert on bases of 2nd-5th distal phalanges
    - Flexes distal phalanges, assists in plantar flexion of ankle
    - When foot on ground: Maintains pads of toes on ground
    - When foot off ground: Plantar flexes 2nd-5th phalanges, aids in maintaining longitudinal arches
  - **Flexor hallucis longus tendon**
    - Passes 3 fibroosseous tunnels: (1) Between medial and lateral talar tubercles, (2) under sustentaculum tali, and (3) between 1st medial and lateral sesamoids
    - Crosses and sends slip to flexor digitorum longus at knot of Henry
    - Inserts on base of 1st distal phalanx
    - When foot on ground: Maintains pad of 1st toe on ground
    - When foot off ground: Plantar flexes 1st phalanges, aids in maintaining medial longitudinal arch
    - Weak plantar flexor of ankle
    - Innervated by tibial nerve

ANATOMY IMAGING ISSUES

Imaging Recommendations

- **US**
  - US best investigation to assess ankle ligaments and tendons
    - As accurate as MR in revealing pathology
  - Assess in both long and short axes
  - Most tendons best assessed in short axis in supra-, retro- and inframalleolar regions
    - Preinsertional portion of posterior tibialis tendon best assessed in long axis
      - This part of tendon is prone to anisotropy
  - Follow tendons one by one around malleoli
    - Do not follow more than 1 tendon at a time
    - Look for tendinosis, tears, tenosynovitis, and peritendinitis
  - Achilles tendon best assessed in long axis
    - If tear found, mark position of tear on skin as this is helpful for surgical planning
  - Ligaments best assessed in long axis
    - Look for continuity, swelling, avulsion
    - Previously torn ligaments can remain swollen for years
    - Reparative periligament fibrosis → ± impingement
  - For joint pathology and large soft tissue masses, MR can provide additional information
  - Radiography and CT best to assess complex fractures and fracture healing

ANTERIOR ANKLE

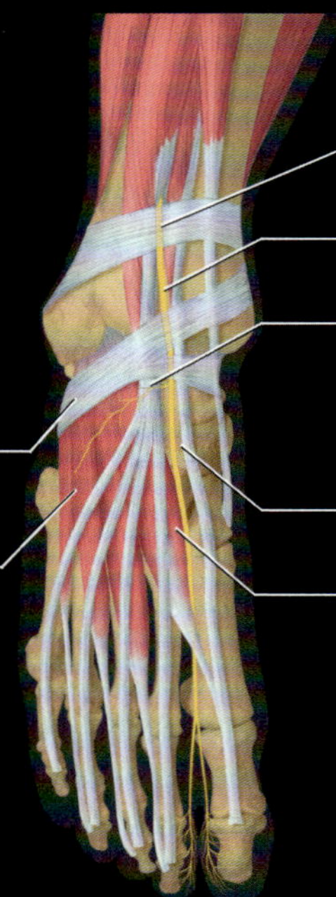

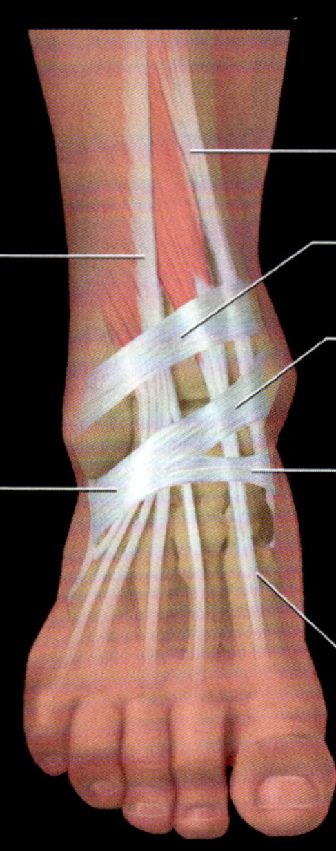

(Top) *The deep peroneal nerve travels deep to the extensor retinacula in the anterior tarsal tunnel and gives a lateral motor branch to the extensor digitorum brevis muscle. The medial branch continues dorsal to the talonavicular joint, middle cuneiform, and in between 1st and 2nd metatarsal to provide mostly sensory, but some motor, supply to the 1st web space.* **(Bottom)** *The peroneus tertius and tibialis anterior tendon insert proximally on the base of the 5th metatarsal and on the medial cuneiform and 1st metatarsal, respectively. The rest of the extensor tendons continue distally toward their insertion sites on the digits. The extensor tendons are held in place by the superior and inferior extensor retinacula.*

ANTERIOR ANKLE

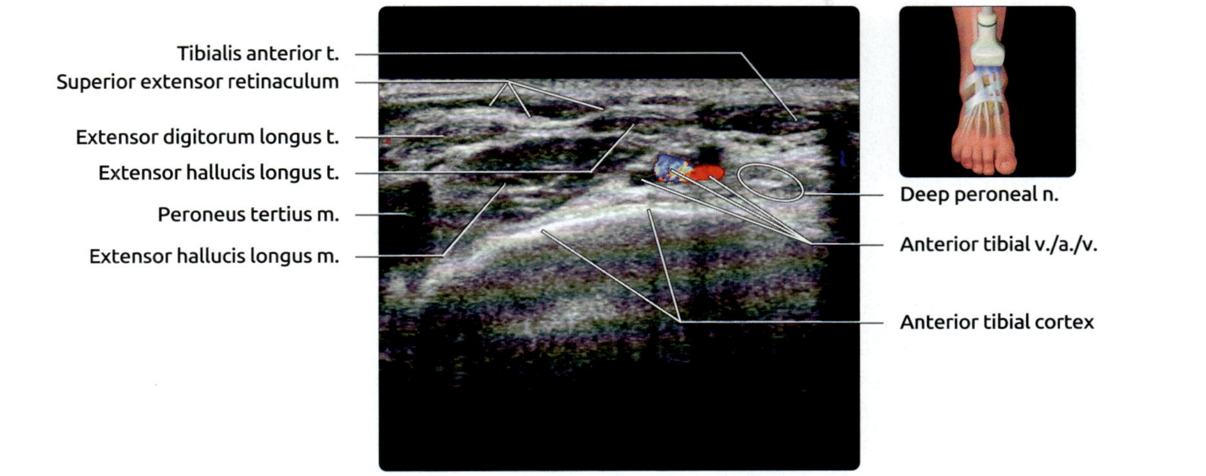

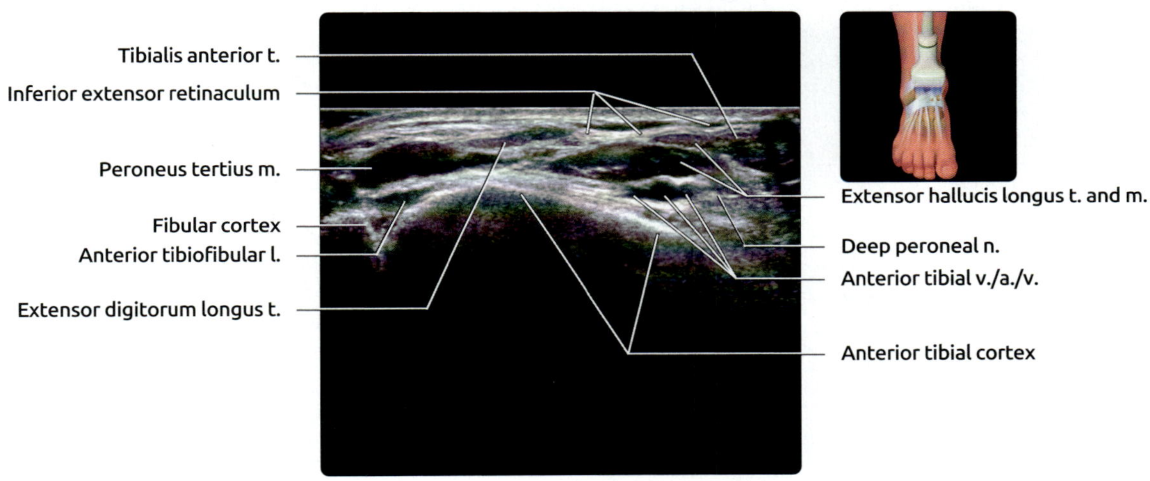

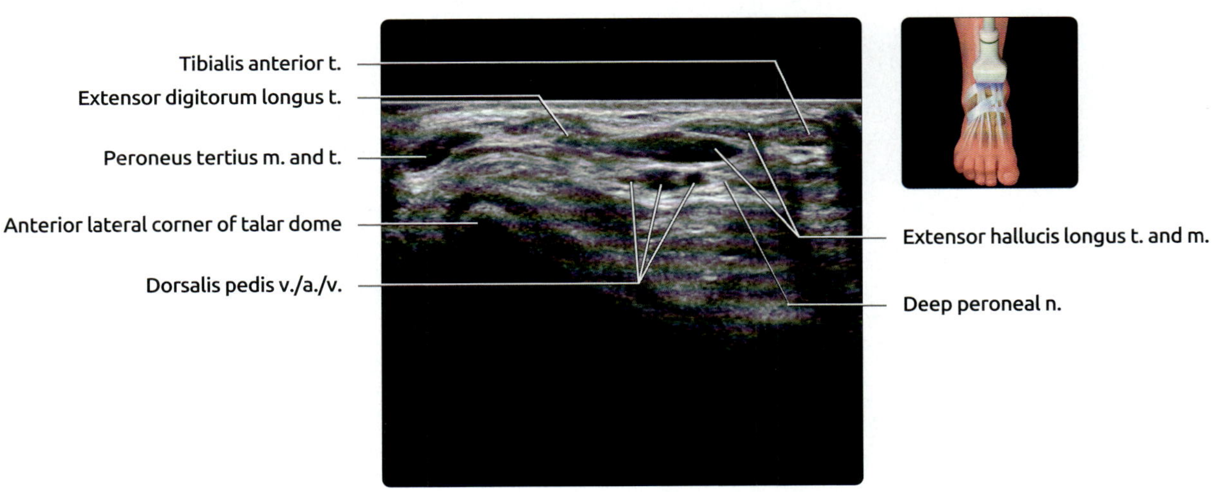

(Top) *Anterior transverse color Doppler US above the level of the ankle joint is shown. The basic relationship of the main structures of the anterior ankle hold true here and down through the ankle joint. From medial to lateral, the structures are: Tibialis anterior, extensor hallucis longus, nerve and vessels, and extensor digitorum longus (mnemonic: Tom, Harry, and Dick).* **(Middle)** *Transverse US of the ankle joint at the level of the anterior talofibular ligament is shown. The anterior talofibular ligament, like most ankle ligaments, runs an oblique course and therefore may appear irregular or interrupted on a straight axial scan. This can be corrected by scanning obliquely along the long axis of the ligament.* **(Bottom)** *Transverse US more inferiorly shows the asymmetrical soft tissue thickness anterior to the talus (thicker medially) due to different orientation axes of talus and foot.*

ANTERIOR ANKLE

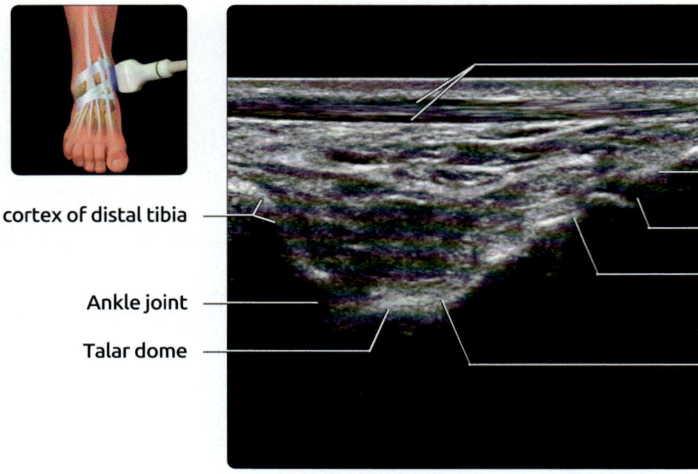

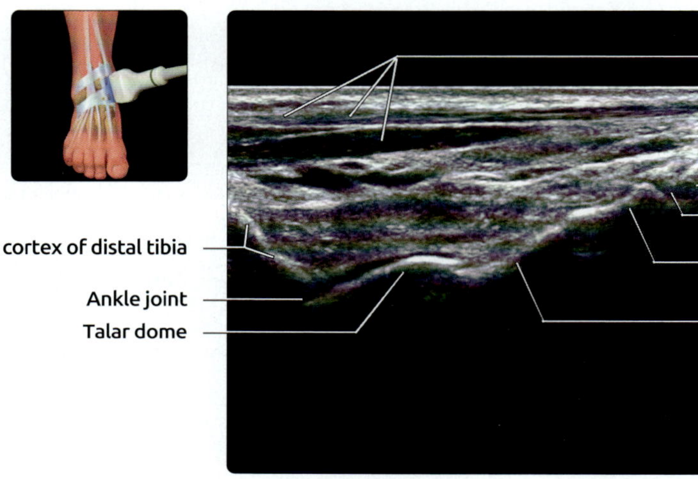

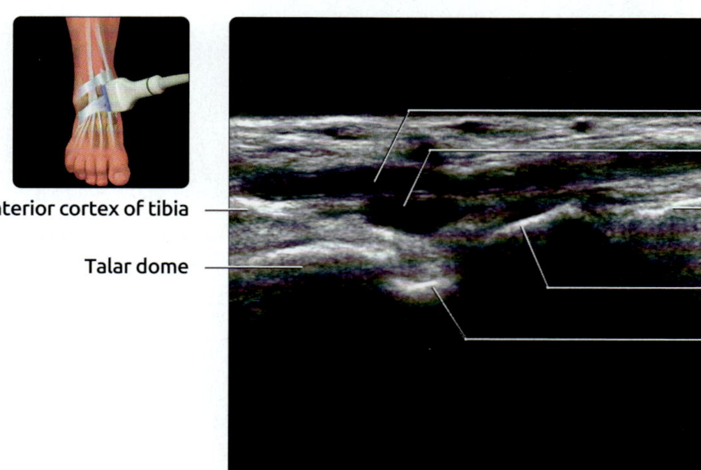

(Top) *Longitudinal US of the lateral aspect of the ankle shows the course of the tibialis anterior tendon. A thin hypoechoic synovial layer can be seen enveloping the tendon.* **(Middle)** *Slightly more laterally, sagittal US shows the extensor hallucis longus tendon and muscle, the latter distinguishing itself from the tibialis anterior tendon.* **(Bottom)** *Further laterally, the dorsalis pedis artery and the accompanying vein(s) come into view. At this point, the subcutaneous tissue is thinnest, a result of the transverse arch of the midfoot with the apex roughly coinciding with the site of this artery. This creates a shallow bony backing for palpating the artery.*

MEDIAL ANKLE

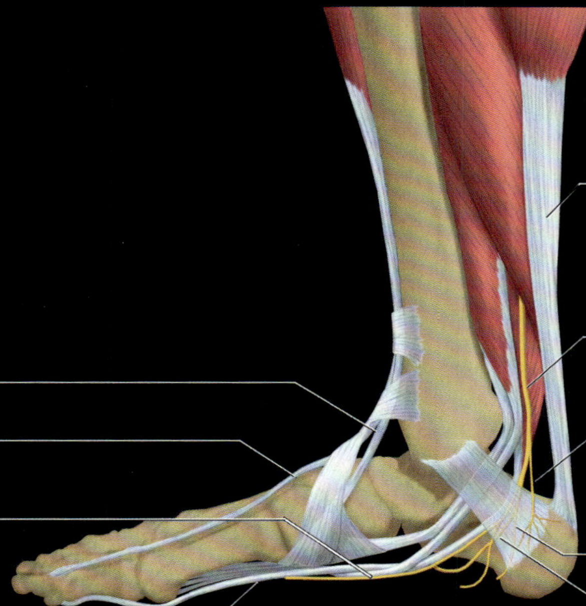

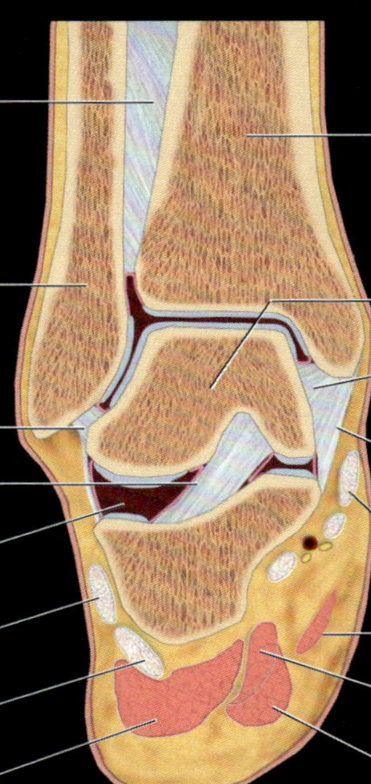

(Top) *Tarsal tunnel is seen accommodating the posterior tibial neurovascular bundle. The medial calcaneal nerve supplies the skin of the medial heel. The medial plantar nerve supplies the flexor hallucis brevis, abductor hallucis, flexor digitorum brevis, and 1st lumbrical. It carries sensation from the medial 2/3 of the plantar foot. The lateral plantar nerve supplies all other plantar muscles, including 2nd-4th lumbricals and all interossei, plus sensation from the lateral 1/3 of the mid- and forefoot.* **(Bottom)** *Illustration of coronal section through the tibia, fibula, talus, and calcaneum shows the ankle ligaments. On the medial side, the medial malleolus is connected via the deep and superficial deltoid ligaments to the medial tarsal bones. The tibialis posterior tendon passes over the deltoid ligament as it curves below the medial malleolus. The talocalcaneal interosseous ligament, which has at least 4 discrete components, maintains apposition of the talus and calcaneus during hindfoot eversion and inversion.*

MEDIAL ANKLE

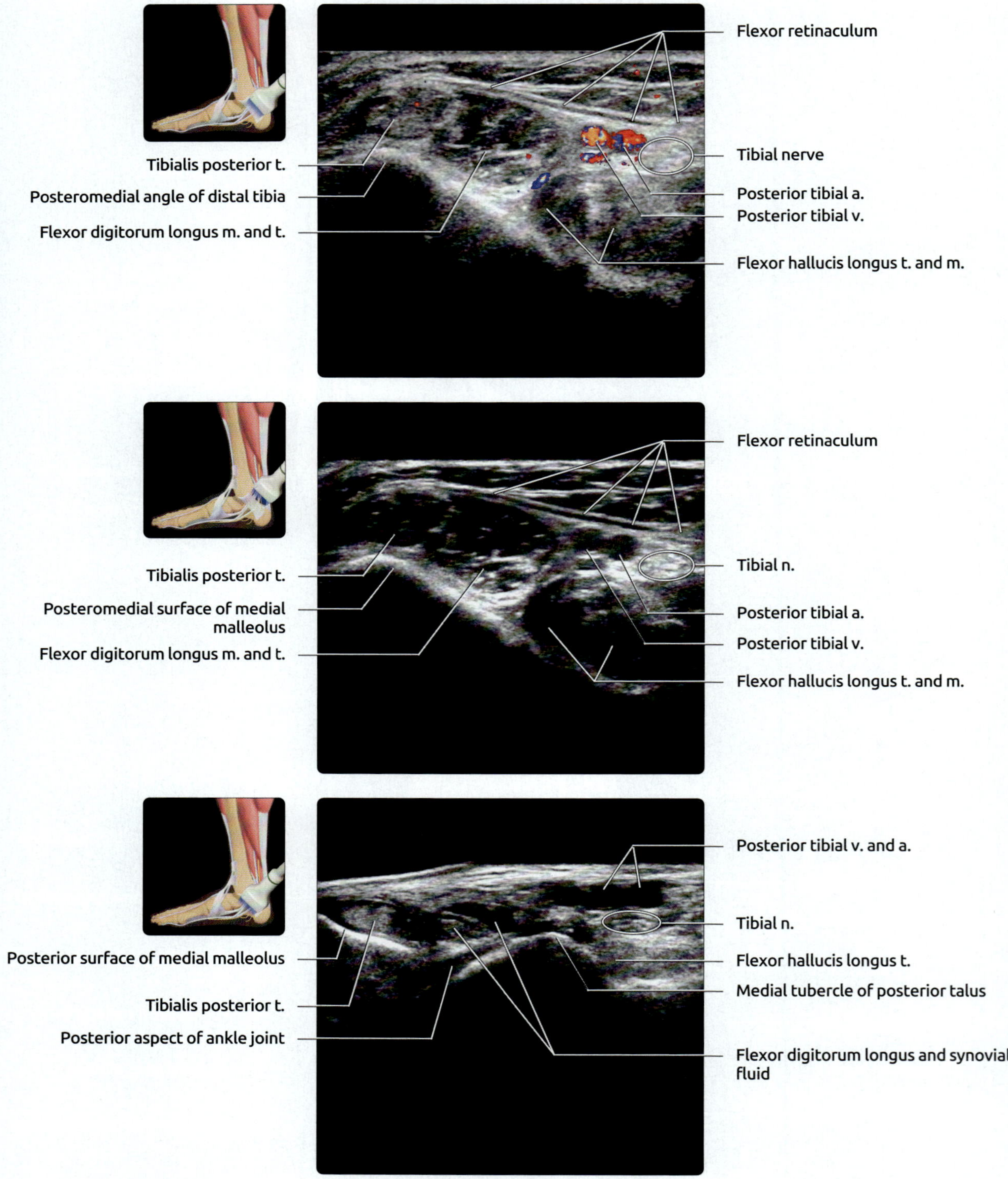

(Top) *Transverse color Doppler US shows the posterior aspect of the distal tibia just above the medial malleolus. This area is held down by the flexor retinaculum and termed the tarsal tunnel. Within this tunnel are tibialis posterior, flexor digitorum longus, vessels and nerve, and flexor hallucis longus (mnemonic: Tom, Dick, and Harry).* **(Middle)** *Corresponding grayscale transverse US just above the medial malleolus is shown. Image is less grainy without simultaneous color Doppler interrogation, and the nerve fibers (hypoechoic dots) give the typical neural signature.* **(Bottom)** *At the level of the talus, the tibial nerve moves anteriorly to run deep to the posterior tibial vessels and superficial to the flexor hallucis longus tendon. The flexor hallucis longus tendon runs in a groove on the posterior surface of the talus.*

MEDIAL ANKLE

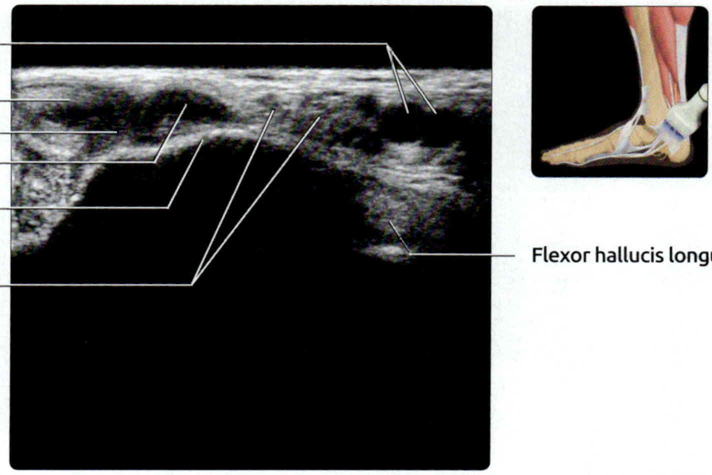

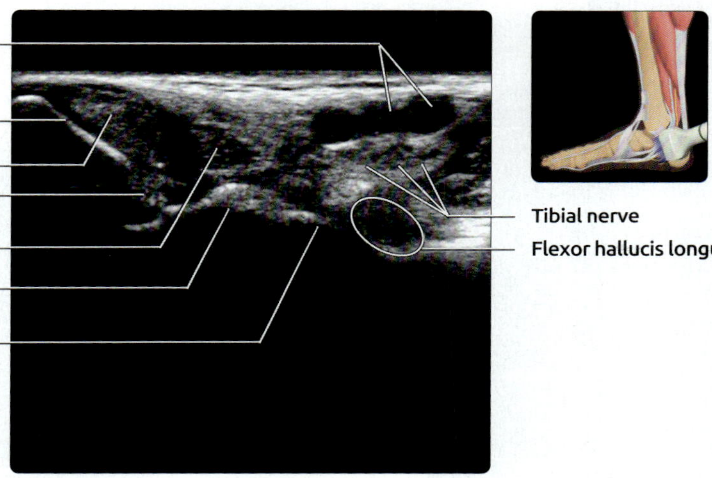

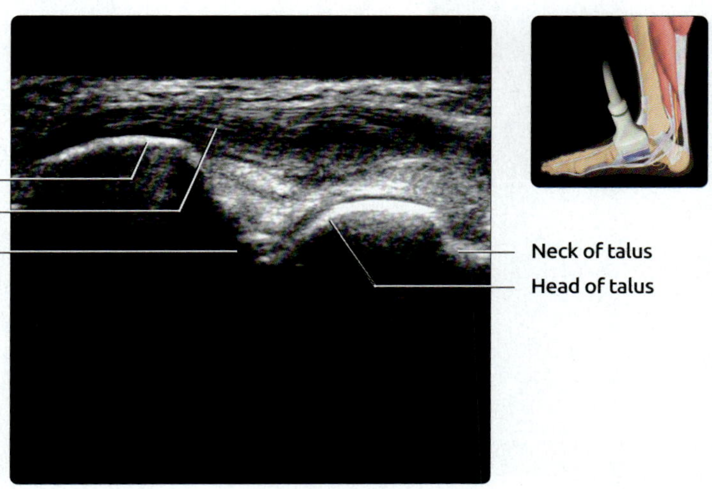

(Top) *More inferiorly, oblique transverse US through the sustentaculum tali of the calcaneus shows that the mnemonic of Tom, Dick, and Harry still holds (tibialis posterior anteriorly and flexor hallucis longus posteriorly). The flexor hallucis longus tendon now runs in a groove behind the sustentaculum tali.* **(Middle)** *Coronal US shows the tarsal tunnel. All structures within the tunnel curve around the medial malleolus to run forward to the midfoot. The tibialis posterior tendon runs on the surface of the deltoid ligament, flexor digitorum longus on the talus, and flexor hallucis longus on the calcaneus under the sustentaculum tali.* **(Bottom)** *Transverse US shows the long axis of the tibialis posterior tendon inserting onto the medial pole of the navicular. This insertion is sometimes accompanied by an accessory navicular bone, which may take on different shapes.*

MEDIAL AND POSTERIOR LIGAMENTS

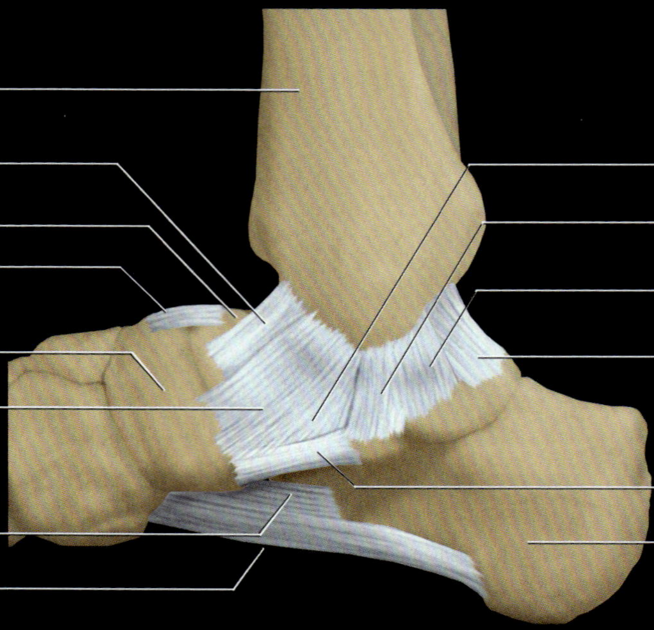

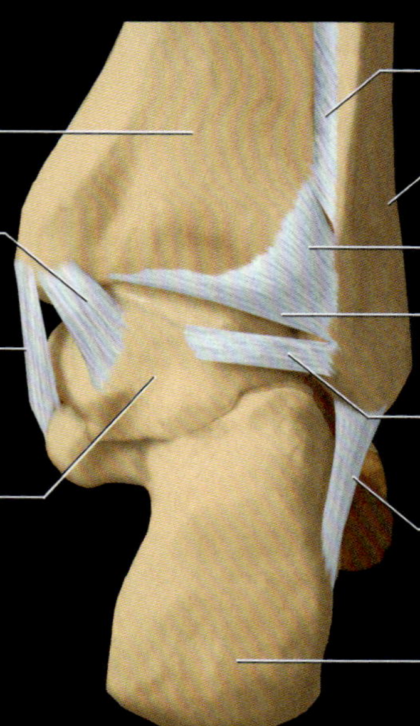

(Top) *Medial view shows the ankle. The deltoid ligament is the major supporter of the ankle. It has many variable components, but a commonly accepted division includes the deep anterior and posterior tibiotalar and superficial anterior and posterior tibiotalar, tibiocalcaneal, and tibiospring and tibionavicular bands. Note that the superficial deltoid is band-like, and distinction between its various components relies on the sites of insertion.* **(Bottom)** *Posterior view shows the ankle. The tibiofibular ligaments are obliquely oriented and their fibular origin is above the fibular fossa. The inferior transverse ligament, which is the inferior aspect of the posterior tibiofibular ligament, extends distal to the tibial posterior surface.*

DELTOID LIGAMENT

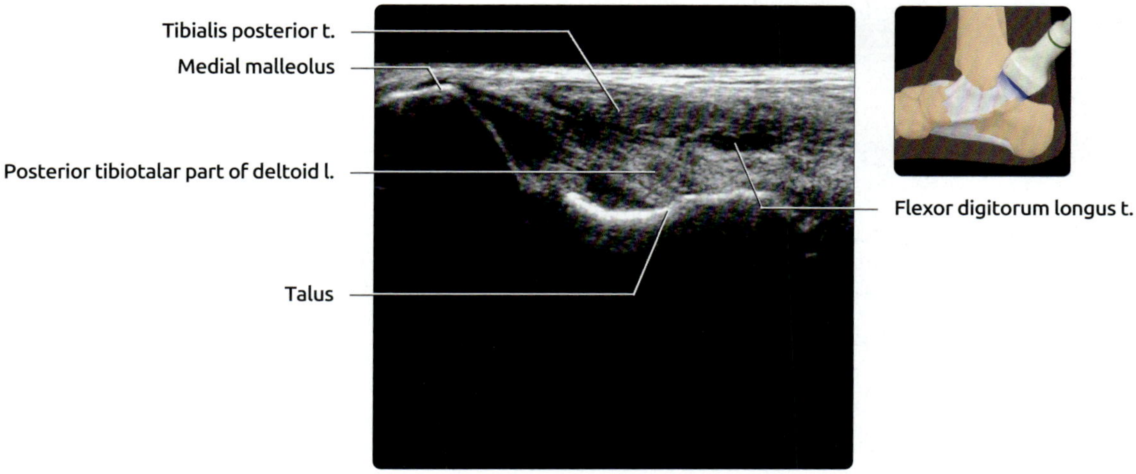

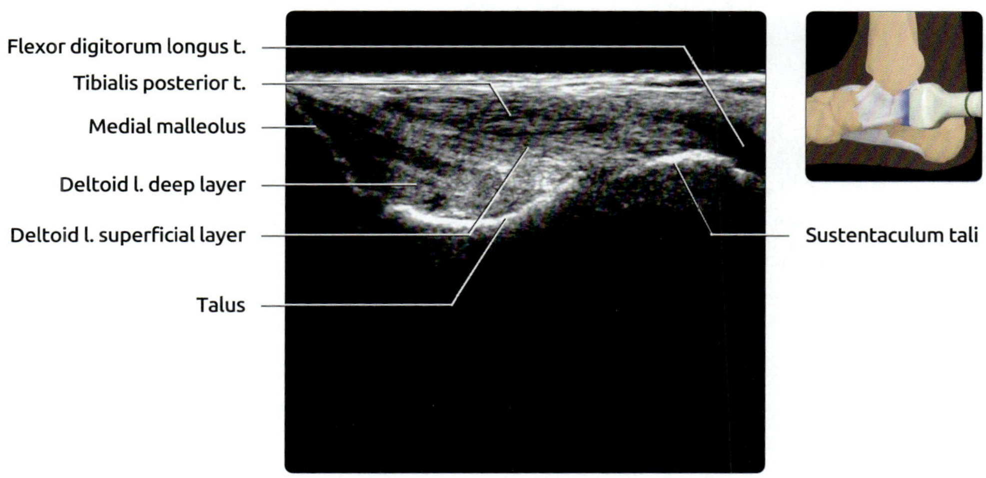

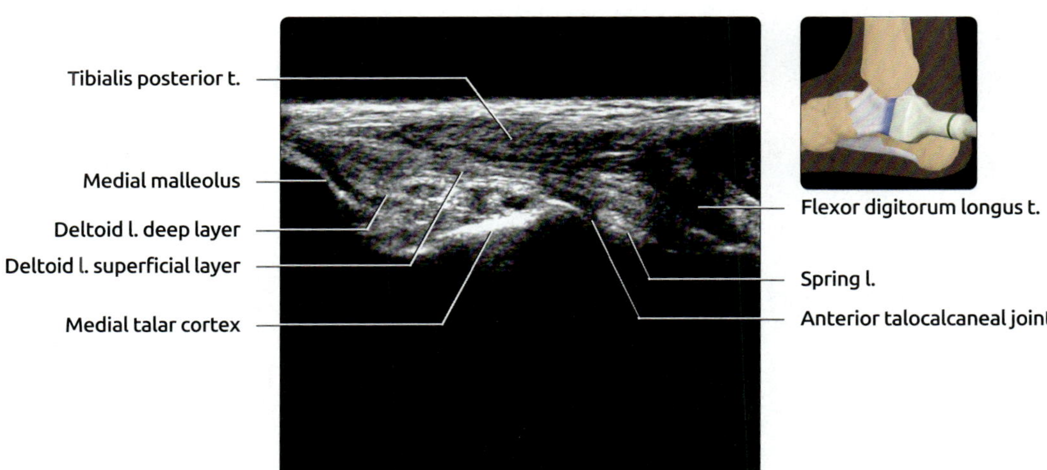

(Top) *Coronal US is seen through the posterior (posterior tibiotalar) portion of the deltoid ligament, which runs from the medial malleolus to the talus. The tibialis posterior and flexor digitorum longus tendons run superficial to this ligament.* **(Middle)** *Coronal US through a more anterior portion of the deltoid ligament is shown. The superficial layer (tibiocalcaneal) runs from the malleolus to the sustentaculum tali, the deep component (posterior tibiotalar) to the talus.* **(Bottom)** *Coronal US is seen through the midportion of the deltoid ligament. The superficial layer (tibiospring) runs from the malleolus to the spring ligament, the deep component to the talus.*

DELTOID LIGAMENT

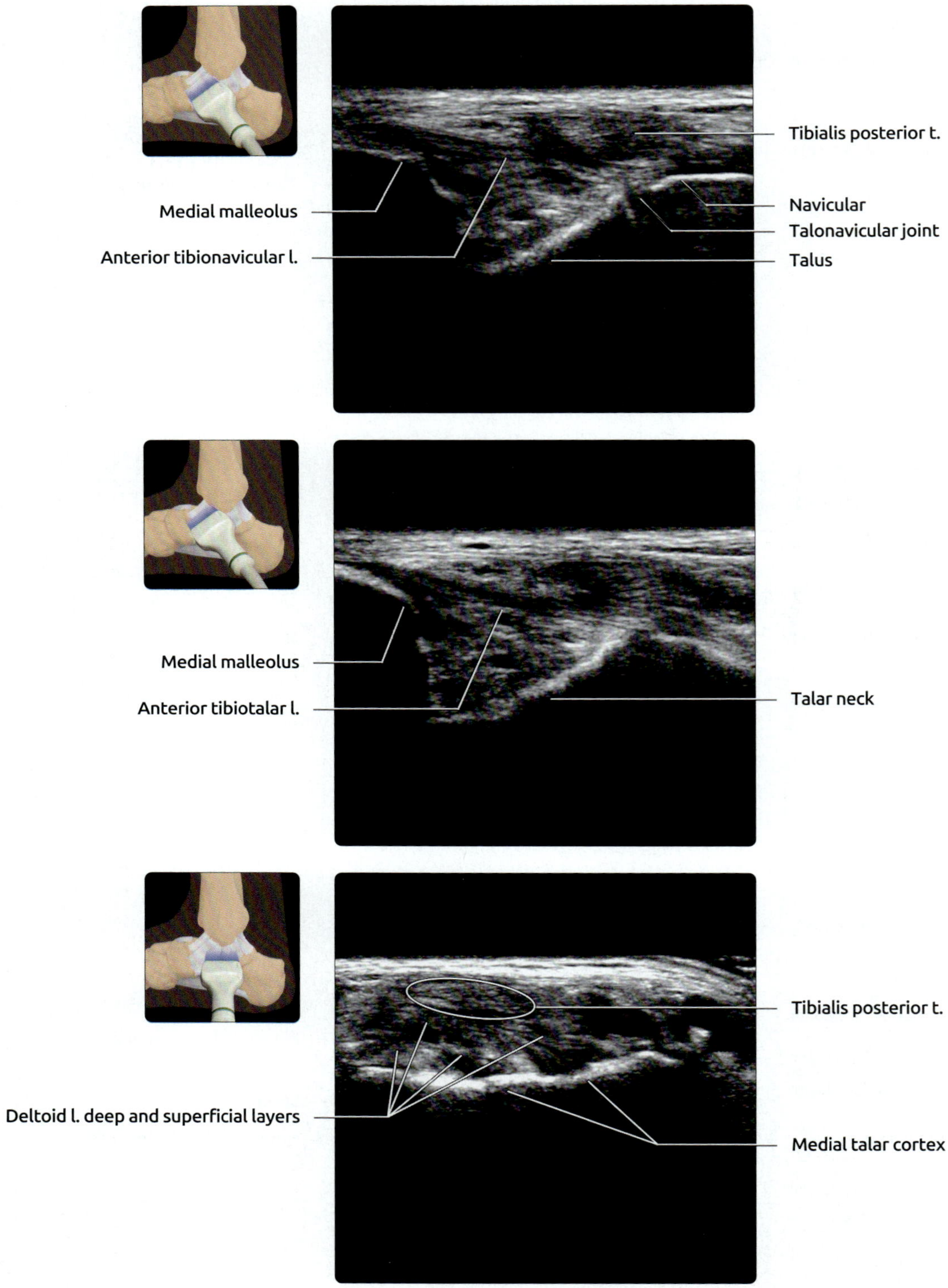

(Top) *Coronal US is seen through the anterior portion of deltoid ligament; a superficial layer (talonavicular) runs from the malleolus to the navicular.* **(Middle)** *Oblique sagittal US is seen through the most anterior portion of the deltoid ligament; fibers (anterior tibiotalar) run from the malleolus to the talar neck. From front to back, these superficial and deep components of the deltoid create a strong restraint against ankle eversion.* **(Bottom)** *Transverse US through the medial wall of the talus shows deep and superficial components of the deltoid ligament. The tibialis posterior tendon runs superficial to this ligament.*

LATERAL ANKLE

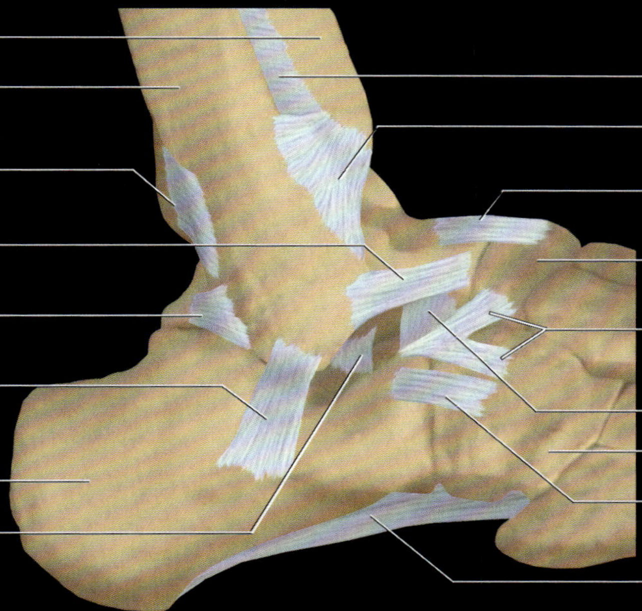

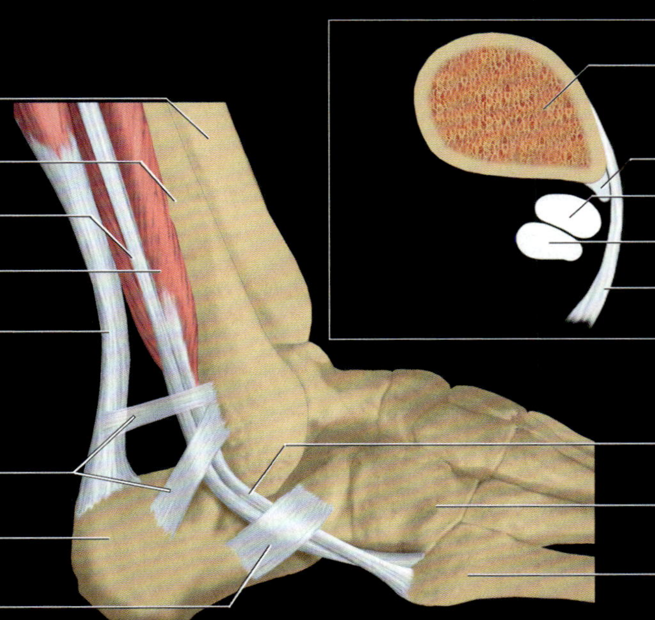

(Top) *Lateral view shows the ankle. Two groups of lateral ligaments support the ankle: (1) The syndesmotic ligaments, including the anterior tibiofibular, posterior tibiofibular, and syndesmotic ligaments, and (2) the lateral collateral ligaments, including the anterior talofibular, posterior talofibular, and calcaneofibular ligaments. Other ligaments that bind the lateral hindfoot include the talocalcaneal interosseous ligament and cervical ligament in the sinus tarsi. Note the bifurcate ligament, which extends from the calcaneus to the navicular and cuboid.* **(Bottom)** *Peroneal retinacula is shown. The superior peroneal retinaculum holds the peroneal tendons in the retrofibular groove. The retinaculum has variable insertions but typically inserts on deep fascia of the leg and on the calcaneus. The inferior peroneal retinaculum holds the peroneal tendons against the calcaneus.*

LATERAL ANKLE

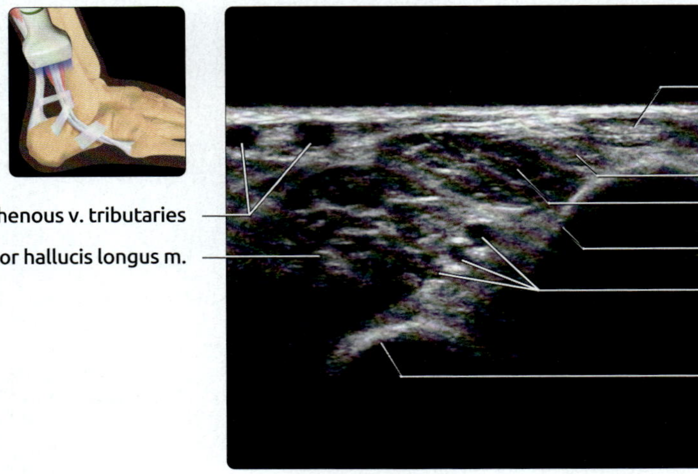

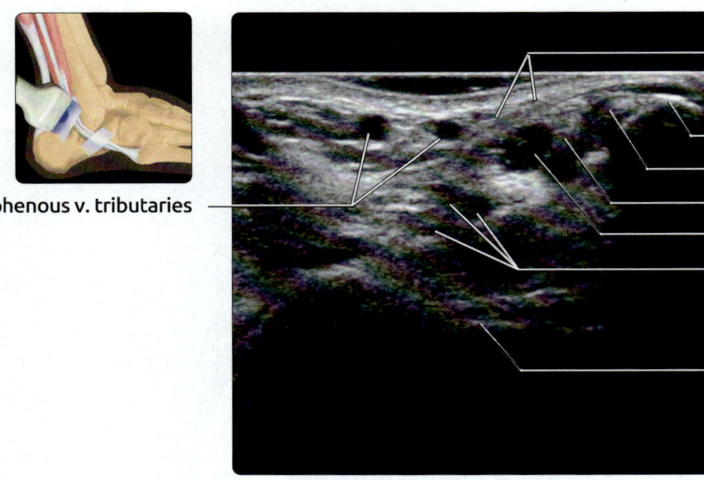

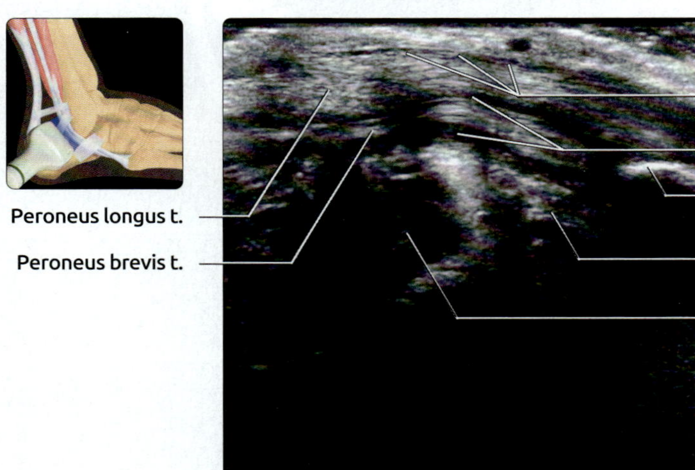

(Top) *Transverse US of the posterior aspect of the distal fibula shows the distal peroneal muscles. The peroneus longus tendon runs anterior to the peroneus brevis. The latter is easily identified by its muscle component, which continues to just above the tip of the lateral malleolus.* **(Middle)** *Transverse US is shown at the level just above the tip of the lateral malleolus. The peroneal tendons are now held down by the superior peroneal retinaculum, which prevents tendon subluxation (bowstringing). There is still a residual component of the peroneus brevis muscle.* **(Bottom)** *Oblique coronal US of the posterior aspect of the lateral malleolus shows the longitudinal course of the peroneal tendons. Both tendons run inferior onto the lateral surface of the calcaneus, passing over the calcaneal fibular ligament on the way.*

LATERAL ANKLE

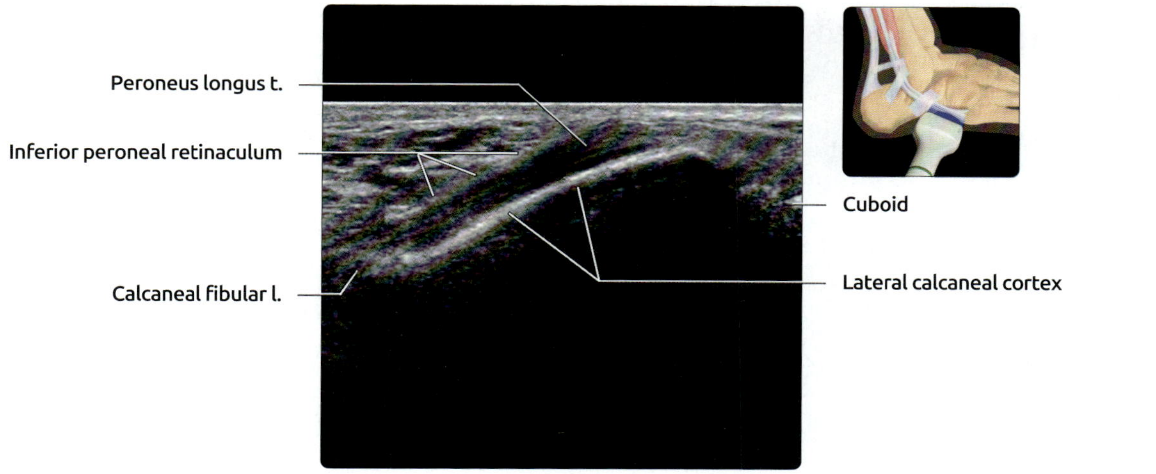

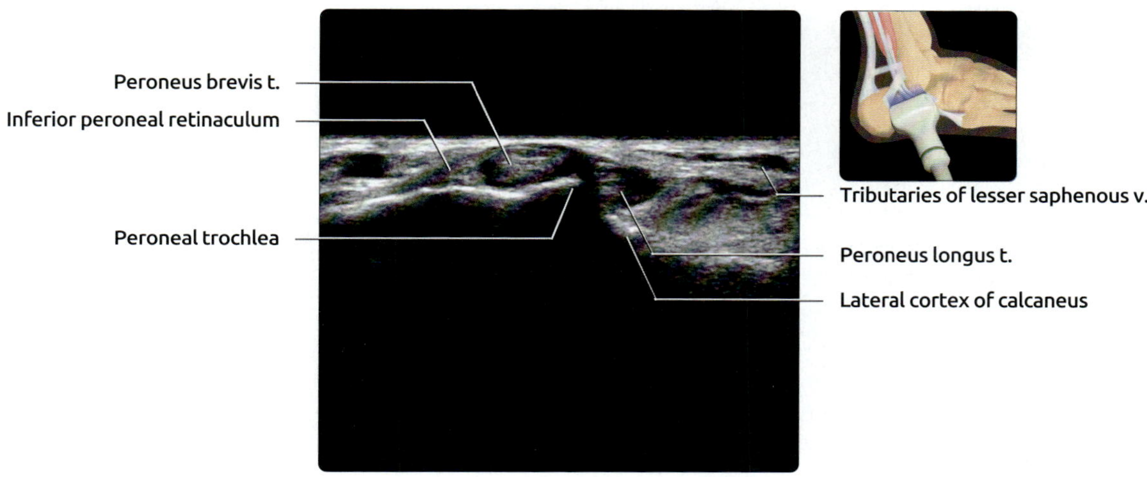

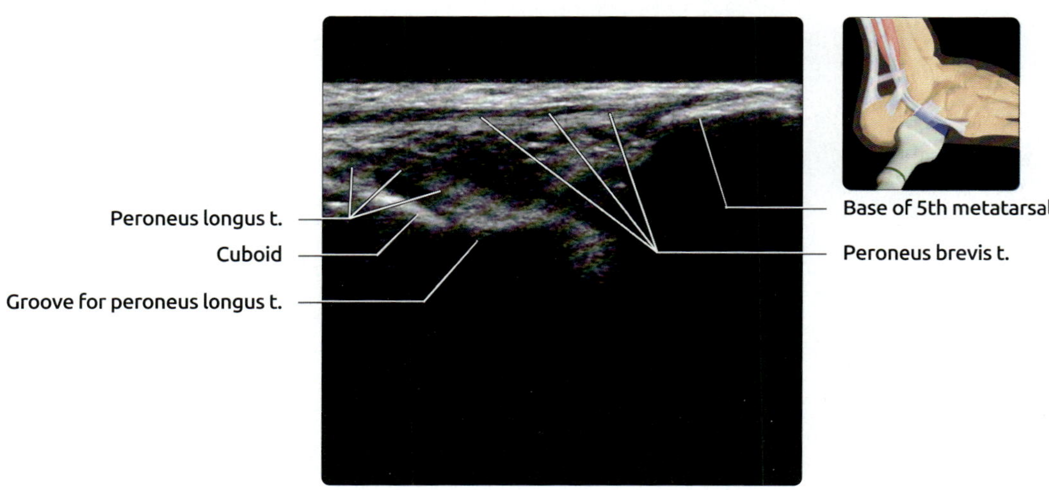

(Top) *Oblique transverse US shows the lateral aspect of the calcaneus along the course of the peroneus longus tendon. Close to the anterior end of the calcaneus, the peroneus longus turns medially and inferiorly to enter the plantar region.* **(Middle)** *Coronal US of the lateral surface of the calcaneus shows the peroneal tendons held down by the inferior peroneal retinaculum but partially separated by the peroneal tubercle, a protuberance on the lateral calcaneal cortex. The peroneus brevis tendon lies more superior than the peroneus longus tendon at this level.* **(Bottom)** *Transverse US of the lateral aspect of the foot shows insertion of the peroneus brevis tendon onto the base of the 5th metatarsal bone. Also shown here is the peroneus longus tendon turning inferiorly on the surface of the cuboid into a groove solely for this tendon to enter the plantar aspect of the foot.*

FIBULAR LIGAMENTS

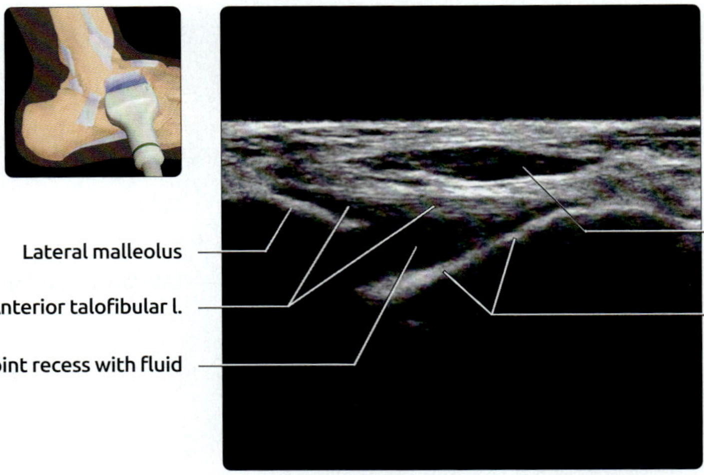

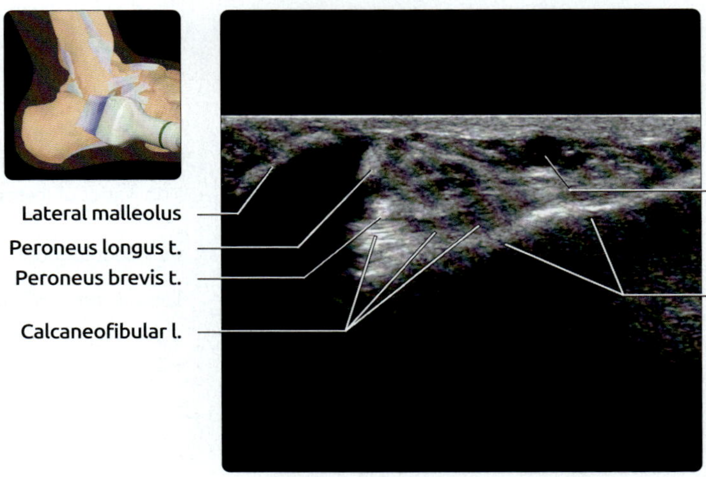

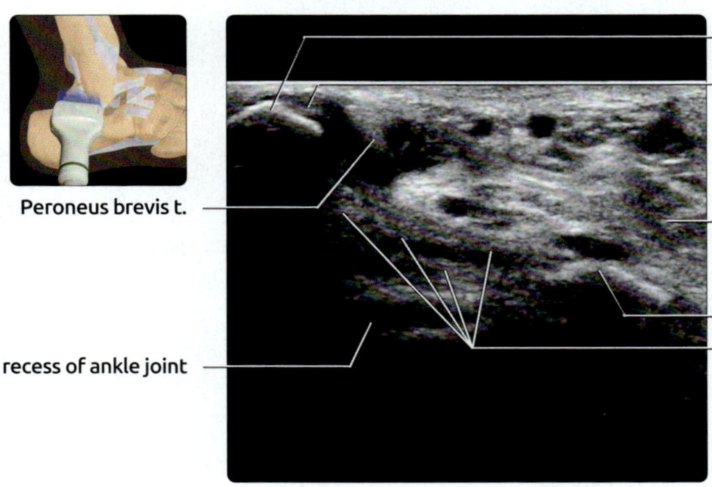

(Top) *Oblique transverse US shows the lateral ankle along the course of the anterior talofibular ligament. An anterolateral ankle joint recess exists deep to this ligament, making it a good area to look for small effusions. The lateral fibers of the extensor digitorum brevis muscle run on the surface of this ligament.* **(Middle)** *Oblique coronal US along the long axis of the calcaneofibular ligament is shown. The peroneal tendons can be seen curving under the lateral malleolus and running partly on the surface of this ligament.* **(Bottom)** *Transverse US of the posterior aspect of the lateral malleolus shows the posterior talofibular ligament. Deep to this ligament is the posterior recess of the ankle joint.*

POSTERIOR ANKLE PANORAMA

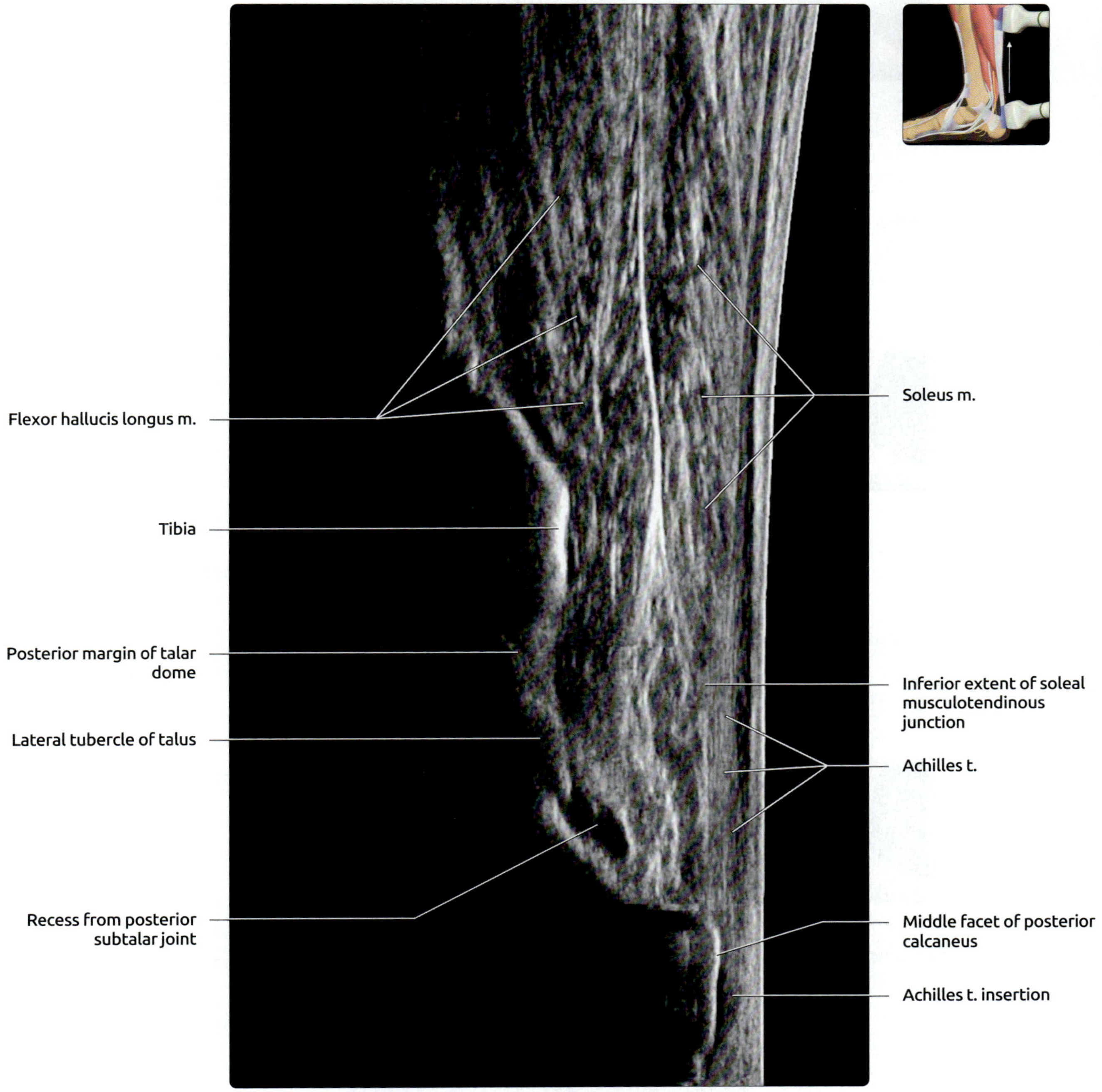

Sagittal panoramic view shows the posterior ankle. The dominant structure in the posterior ankle is the Achilles tendon, the largest tendon in the body. This runs craniocaudally just under the skin and is well demonstrated by US. The ankle should be in dorsiflexion to tighten this tendon during examination so that the organization of its fibers and continuity can be assessed. More superiorly, its contributing muscles, the soleus and gastrocnemius, can be traced. Deep to the Achilles tendon is the flexor hallucis longus muscle. An accessory soleus muscle may be seen inferomedial to the Achilles tendon. It occurs in 30% of the normal population and may be symptomatic in athletes with low-grade pain and stiffness after exercise.

ACHILLES TENDON, LONGITUDINAL

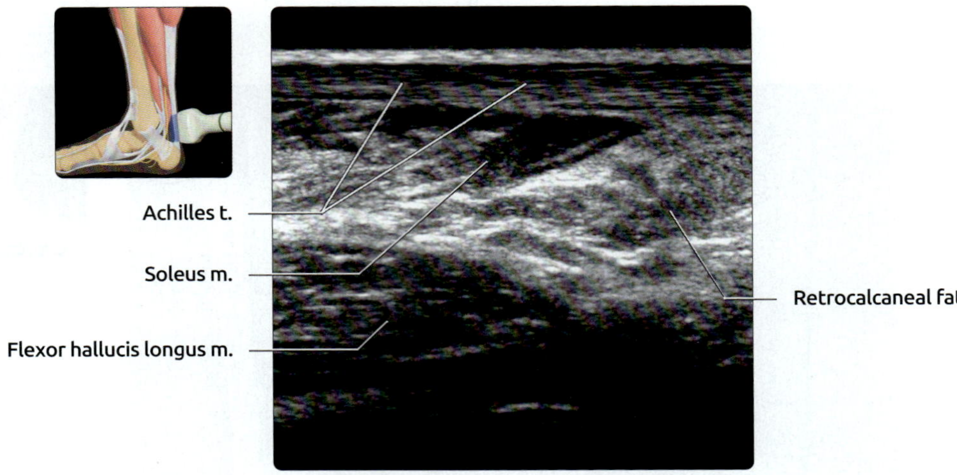

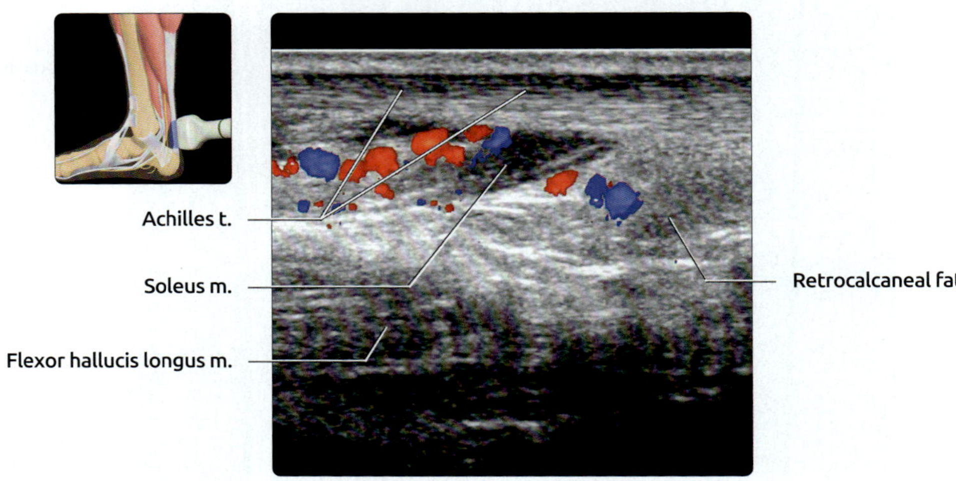

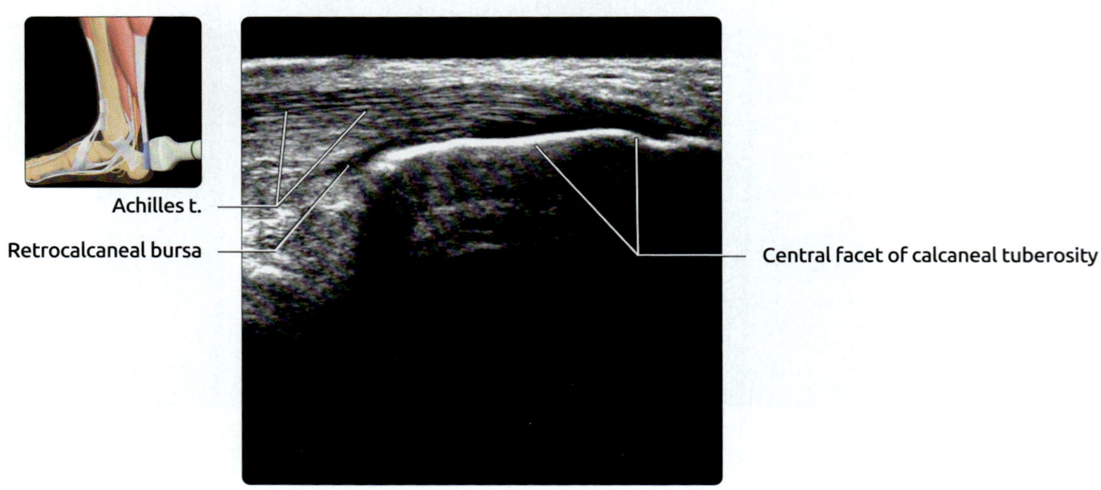

(Top) *Sagittal US of the lower posterior leg shows the musculotendinous junction between the soleus muscle and the Achilles tendon. These structures join at an angle with a crisp border. A disruption of this sharp transition (usually due to hemorrhage and rupture of inserting muscle fibers) is suggestive of injury to the musculotendinous junction.* **(Middle)** *Color Doppler US of the same region shows the rich blood supply to this region and the cause for extensive bruising in tendon damage at this site.* **(Bottom)** *Sagittal US more distally shows the insertion of the Achilles tendon onto the middle facet of the posterior calcaneal surface. The retrocalcaneal bursa separates the Achilles tendon from the superior facet of the posterior calcaneal surface. This bursa usually contains a tiny amount of fluid detectable on US.*

ACHILLES TENDON, AXIAL

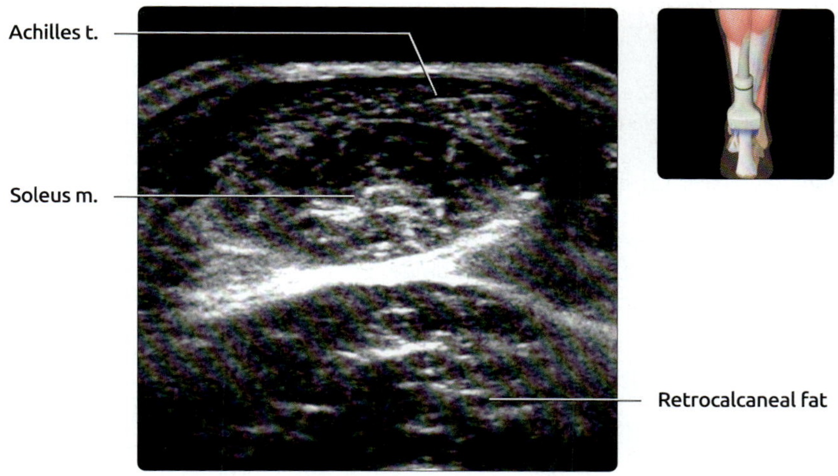

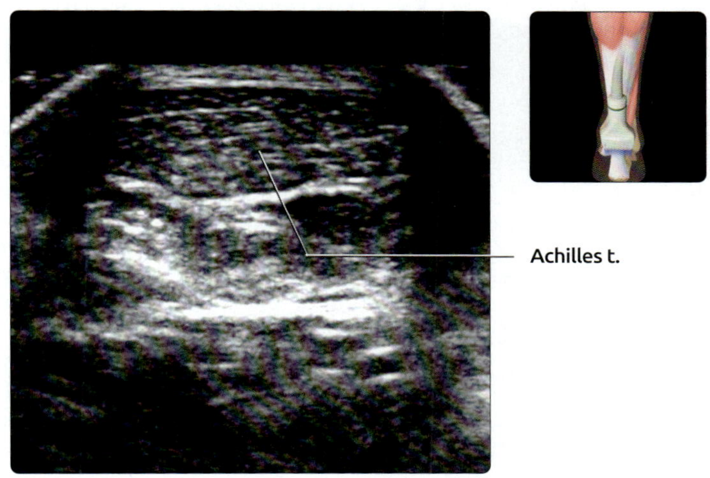

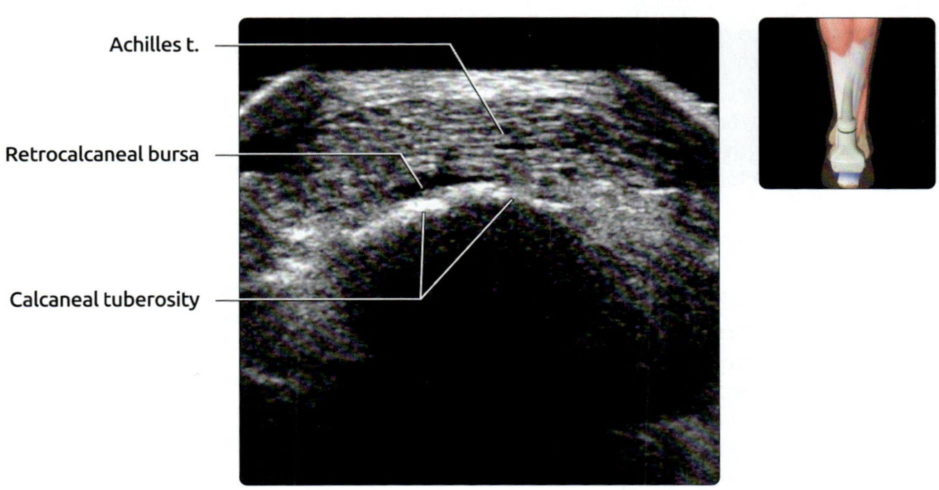

(Top) *A series of transverse scans through the Achilles tendon is shown. At a superior level, the soleus muscle can be seen deep to the Achilles tendon. The Achilles tendon begins superiorly as a continuation of the gastrocnemius aponeurosis, which then joins with the soleus tendon to form the Achilles tendon. At the level of this scan, the tendon is crescent-shaped to accommodate the soleus muscle.* **(Middle)** *Transverse US shows the Achilles tendon at its midportion. The tendon now assumes an oval shape. The fibrils within the tendon produce a coarsely dotted appearance. Focal hypoechogenicity or change in shape of this tendon should suggest tear/rupture or tendinosis.* **(Bottom)** *At a lower level before its insertion onto the calcaneal tuberosity, the Achilles tendon becomes flattened. A small amount of fluid is frequently seen in the retrocalcaneal bursa deep to the tendon.*

TERMINOLOGY

Definitions

- **3 major divisions**
  - **Hindfoot**: Calcaneus and talus
  - **Midfoot**: Navicular, cuneiforms, and cuboid
  - **Forefoot**: Metatarsals and phalanges
- **2 columns**
  - **Medial column**: Talus, navicular, cuneiforms 1-3, digits 1-3
  - **Lateral column**: Calcaneus, cuboid, digits 4 and 5
- Some authors use 2 columns in hind and midfoot but divide forefoot into 3 columns
  - Medial column: 1st toe
  - Middle column: 2nd-4th toes
  - Lateral column: 5th toe

IMAGING ANATOMY

Overview

- Alignment of foot can only be assessed on weight-bearing radiographs

Arches of Foot

- Foot is arched from posterior to anterior and from medial to lateral
- **Transverse arch of foot**
  - Cuneiform bones form keystone of arch due to triangular shape
  - Major supporting structures of transverse arch
    - Spring ligament
    - Lisfranc ligament and intermetatarsal ligaments
    - Intertarsal ligaments
- **Longitudinal arch of foot**
  - From posterior process calcaneus to metatarsal heads
  - Medial side is higher than lateral
  - Apex of arch is at navicular and cuneiforms
  - Metatarsals slant downward from apex of arch to metatarsophalangeal (MTP) joint
    - This is called inclination angle
    - Inclination angle decreases from 20° at 1st to 5° at 5th metatarsal
  - Major supporting structures of longitudinal arch
    - Plantar fascia
    - Long and short plantar ligaments
    - Spring ligament
    - Posterior tibial tendon, peroneus longus tendon
- **Radiographic assessment of normal longitudinal arch**
  - Evaluate talometatarsal alignment
    - Normal: Axis of talus continues along axis of 1st metatarsal
    - Pes planus: Axis of talus points below axis of 1st metacarpal due to flattened arch
    - Pes cavus: Axis of talus points above axis of 1st metacarpal due to exaggerated arch

Distribution of Weight Bearing

- 50% of weight is borne on subtalar joint and calcaneus
- Remainder is transmitted via arch anteriorly to MTP joints, greatest weight on 1st toe

Bony Anatomy

- **Cuboid bone**
  - Roughly cuboidal shape
  - 1 ossification center: Ossifies between 9th fetal month and 6 months of age
  - Articulates with calcaneus, navicular, 3rd cuneiform, and 4th and 5th metatarsals, rarely head of talus
    - Dorsal ligaments (calcaneocuboid, cubonavicular, cuneocuboid, cubometatarsal) strengthen each of these articulations
    - Short and long plantar ligaments attach to plantar surface
  - Sulcus at lateral margin, under which passes peroneus longus tendon
  - 5th metatarsal base extends beyond lateral margin
- **Navicular bone**
  - Curved shape, concave proximally and convex distally
  - 1 ossification center: Ossifies in 3rd year of life
  - Articulates with talus, cuboid, cuneiforms
    - Dorsal ligaments strengthen each of these articulations
    - Single facet proximally for articulation with head of talus
    - 3 facets distally for cuneiform articulations
    - 1 facet laterally for articulation with cuboid
    - Connected to anterior process of calcaneus by bifurcate ligament
    - Connected to sustentaculum tali by spring ligament
  - Large median eminence for attachment of posterior tibial tendon is located more plantar than main body of navicular
- **Cuneiform bones**
  - Wedge-shaped with base of wedge at dorsal surface of 2nd and 3rd cuneiforms, dorsomedial surface of 1st cuneiform
  - In combination, form arch
  - 1st cuneiform (medial cuneiform)
    - Articulates with navicular, 2nd cuneiform, 1st metatarsal
    - 1 or 2 ossifications centers: Ossify in 2nd year of life
  - 2nd cuneiform (middle or intermediate cuneiform)
    - Articulates with navicular, 1st and 3rd cuneiforms, 2nd metatarsal
    - Smallest of cuneiforms
    - 1 ossification center: Ossifies in 3rd year of life
  - 3rd cuneiform (lateral cuneiform)
    - Articulates with navicular, 2nd cuneiform, cuboid, 3rd metatarsal
- **Metatarsal bones**
  - 2 ossifications centers: Shaft ossifies in 9th prenatal week, epiphysis in 3rd-4th years of life
    - 1st metatarsal has epiphysis at proximal end, others at distal end
  - 2nd-5th metatarsals have articulations at bases with adjacent metatarsals
  - 1st metatarsal
    - Largest of metatarsals
    - Articulates with 1st cuneiform, 1st proximal phalanx, sesamoids of metatarsal head
    - Variable articulation with 2nd metatarsal base

- 2nd-3rd metatarsals
  - Articulate with respective cuneiforms and proximal phalanges
  - 2nd metatarsal base recessed relative to 1st and usually 3rd
- 4th-5th metatarsals
  - Articulate with cuboid and respective proximal phalanges
  - Styloid process of 5th metatarsal extends lateral to cuboid
- **Phalanges**
  - 1st toe is biphalangeal, other toes are triphalangeal
  - 5th toe sometimes has failure of segmentation of middle and distal phalanges
  - 2 ossification centers: Shaft ossifies 9th-15th prenatal weeks, epiphysis 2nd-8th years

Musculature

- **Plantar muscles**: 4 muscle layers, numbered from superficial to deep
  - 1st layer: Abductor hallucis, flexor digitorum brevis, abductor digiti minimi
  - 2nd layer: Quadratus plantae (flexor accessorius), flexor digitorum and hallucis longus tendons, lumbricals
  - 3rd layer: Flexor hallucis brevis, adductor hallucis, flexor digiti minimi brevis, tibialis posterior tendon
  - 4th layer: Plantar interossei (3), dorsal interossei (4)
  - Peroneus longus tendon courses across all layers, from superficial plantar laterally to deep plantar medially
- **Dorsal muscles**: 2 muscle layers
  - Superficial layer: Tibialis anterior tendon, extensor hallucis longus tendon, extensor digitorum longus tendon, peroneus tertius tendon
  - Deep layer: Extensor hallucis brevis, extensor digitorum brevis
  - In forefoot, long and short extensors run side by side in single layer

Compartments

- **Plantar compartments: 4 plantar compartments are located** deep to plantar fascia
  - Divided by fascial layers
  - Lateral and medial intermuscular septa determine major compartment divisions
  - **Medial plantar compartment**
    - Contains abductor hallucis, flexor hallucis longus, and flexor hallucis brevis tendons
  - **Central plantar compartment**
    - Superficial subcompartment: Contains flexor digitorum brevis, distal portion of flexor digitorum longus tendons
    - Intermediary subcompartment: Contains proximal plantar portion of flexor digitorum longus tendon, quadratus plantae, lumbricals
    - Deep subcompartment: Limited to forefoot, contains adductor hallucis
  - **Lateral plantar compartment**
    - Contains abductor and flexor digiti minimi
  - **Interosseous compartment**
    - Contains plantar and dorsal interosseous muscles
- **Dorsal compartment**
  - Superficial layer: Extrinsic extensor tendons
  - Deep layer: Intrinsic extensor muscles

Major Ligaments

- **Plantar fascia (aponeurosis): 3 portions** extend from tuberosity of calcaneus to transverse metatarsal ligaments of toes
  - **Medial band**: Thin structure superficial to abductor hallucis muscle
  - **Central band**: Thick, strong structure superficial to flexor digitorum brevis
    - Divides into separate bands to each toe; these are linked by transverse bands
    - Distally sends septa superficially into subcutaneous fat and deep to MTP joints
  - **Lateral band**: Thin structure superficial to abductor digiti minimi
  - Medial and lateral bands sometimes terminate at level of midmetatarsals
- Long plantar ligament
  - Originates at calcaneal tuberosity, inserts at cuboid and bases of 2nd-4th metatarsals
  - Forms retinaculum for peroneus longus tendon as it courses medially on plantar aspect of foot
- Short plantar (plantar calcaneocuboid) ligament
  - Deep to long ligament, inserts more proximally on cuboid
- Plantar calcaneocuboid (spring) ligament
  - Originates on sustentaculum tali, inserts on plantar aspect navicular
- Bifurcate ligament: Originates on anterior process of calcaneus dorsally, inserts on navicular and cuboid
- Lisfranc ligament: Originates at 1st cuneiform, inserts at base of 2nd metatarsal
- Intermetatarsal ligaments: Dorsal and plantar ligaments between 2nd-5th metatarsal bases
- Transverse metatarsal ligaments: Superficial and deep ligaments between metatarsal heads

Nerves

- **Tibial nerve** divides into medial and lateral plantar branches at level of tarsal tunnel
  - **Medial plantar nerve**
    - Between 1st and 2nd muscle layers, accompanies medial plantar artery
    - Motor branches: Abductor hallucis, flexor digitorum and hallucis brevis, 1st lumbrical
    - Plantar digital nerves to 1st-3rd toes, medial aspect of 4th toe
  - **Lateral plantar nerve**: Has deep and superficial divisions
    - Motor branches: Flexor digiti minimi brevis, lumbricals, interossei, adductor hallucis
    - Superficial lateral plantar nerve: Between 1st and 2nd muscle layers
    - Plantar digital nerves to 5th toe, lateral aspect of 4th toe
    - Deep lateral plantar nerve: Between 3rd and 4th muscle layers; accompanies lateral plantar artery
- **Deep peroneal nerve**: Extends along dorsum of foot, between tibialis anterior and extensor hallucis longus
  - Motor branch: Extensor digitorum brevis

- **Superficial peroneal nerve**: Divides into medial and lateral branches at dorsum of foot
  - Sensory branches to dorsal foot
- **Sural nerve**: Lateral, superficial branch of tibial nerve
  - Extends along lateral margin of foot
  - Sensory branches to lateral foot

Arteries

- Posterior tibial artery divides into medial and lateral plantar arteries at level of tarsal tunnel
  - Plantar arteries accompany medial and deep lateral plantar nerves
- Peroneal artery accompanies superficial peroneal nerve down anterolateral aspect of ankle
  - May join or replace posterior tibial artery
- Anterior tibial artery continues into foot as dorsalis pedis artery, deep to extensor retinaculum
  - Divides into multiple branches in midfoot, forming arcade

Bursae

- Extensor digitorum brevis: Between muscle and 2nd cuneiform and metatarsal bases
- Extensor hallucis longus: Between tendon and 1st cuneiform and metatarsal bases
- Abductor digiti minimi: Between muscle and tuberosity of 5th metatarsal
- MTP joints: Dorsally between metatarsal heads and medial to 1st metatarsal head

ANATOMY IMAGING ISSUES

Imaging Recommendations

- Radiographs
  - Weight bearing, when possible
  - Standard views: Anteroposterior (dorsoplantar), lateral, oblique
- US
  - Offers highest spatial resolution for muscle and tendons
  - Particularly useful given limited depth of most soft tissue structures in foot
  - As result of transverse arch of foot, interrogation should be performed separately for medial and lateral sides for best visualization
  - Due to complex soft tissue anatomy, panoramic (extended field-of-view) images provide better depiction of location and extent of lesion and course of different structures
  - Dorsal foot structures tend to be very superficial and thin (tendons) or compressible (vessels)
    - To better visualize these structures, minimal compression and thick layer of coupling gel may help
- CT
  - Multidetector 1-mm images with sagittal and coronal reformations
  - 3D reconstructions often useful for analyzing relationship between bones and bone fragments
- MR
  - Better images obtained when field of view limited to area of concern, not entire foot
  - Demonstration of bone marrow edema aside from soft tissue lesions

Imaging Sweet Spots

- Tendon movement can be easily demonstrated with US by moving bone onto which it inserts
  - Helps with confirming tendon location and also with establishing tendon integrity
- Tarsal ligaments usually seen as hypoechoic thickenings on cortices and between bones
  - Orientation of transducer along long axis makes ligament stand out and easier to identify

Imaging Pitfalls

- Alignment can only be reliably assessed on weight-bearing radiographs

CLINICAL IMPLICATIONS

Foot Motion

- **Supination**: Elevation of medial arch of foot
  - Combination of inversion and adduction
- **Pronation**: Depression of medial arch of foot
  - Combination of eversion and abduction
- Complex motions at multiple joints
  - Chopart (calcaneocuboid and talonavicular) joint
    - These 2 joints move together on oblique axis to produce compound motions
    - Pronation-abduction-extension to supination-adduction-flexion
  - Tarsometatarsal joints
    - Dorsiflexion and plantar flexion
    - 2nd and 3rd tarsometatarsal joints relatively immobile
    - Slight abduction of 1st tarsometatarsal joint
  - MTP joints
    - Dorsiflexion and plantar flexion
    - Abduction and adduction at 1st MTP joint

Alignment

- Normal weight bearing and gait depend on normal foot alignment
- Evaluated initially with anteroposterior and lateral weight-bearing radiographs

Malalignment

- Forefoot adductus: Medial angulation of metatarsals from axis of hindfoot
- Forefoot varus: Inversion of metatarsals resulting in shift of weight bearing to 5th metatarsal from 1st metatarsal
- Metatarsus primus varus: Medial deviation of 1st metatarsal axis relative to 2nd
- Hallux valgus (hallux abductus): Lateral deviation of 1st proximal phalanx relative to axis of 1st metatarsal
  - Valgus refers to angular deformity in vertical plane, where apex points medially
  - Abductus refers to angular deformity in horizontal plane, where apex points medially
  - Hallux valgus is misnomer, but it remains term commonly used for hallux abductus

DORSAL FOOT

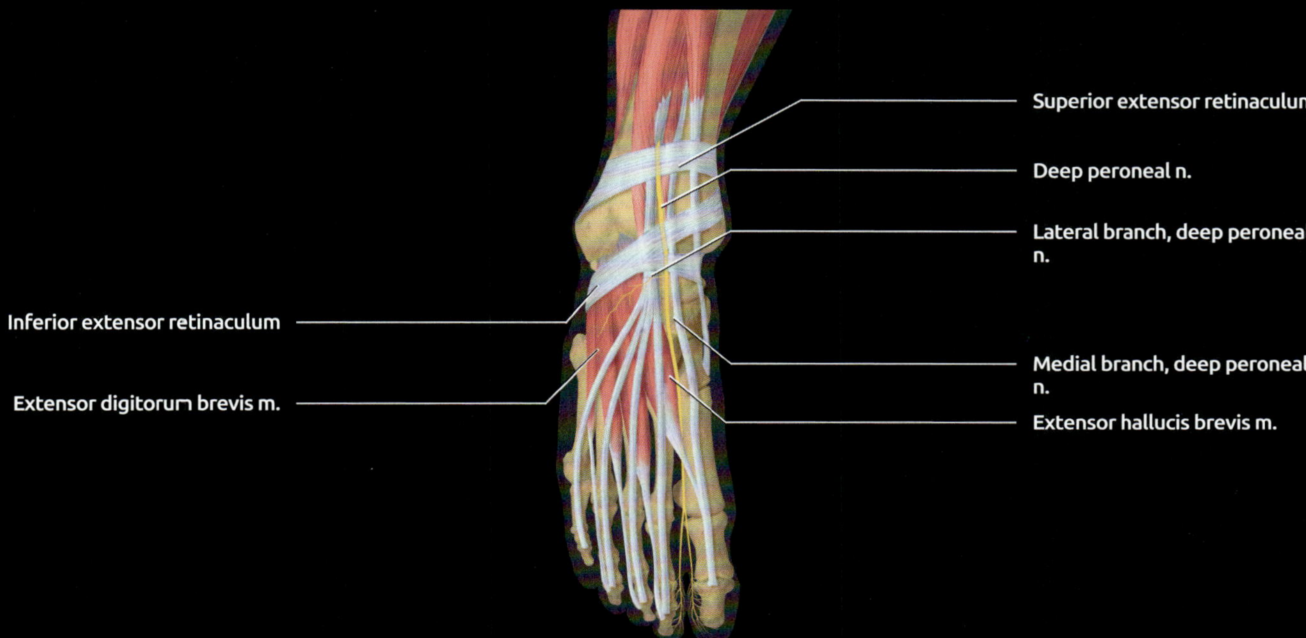

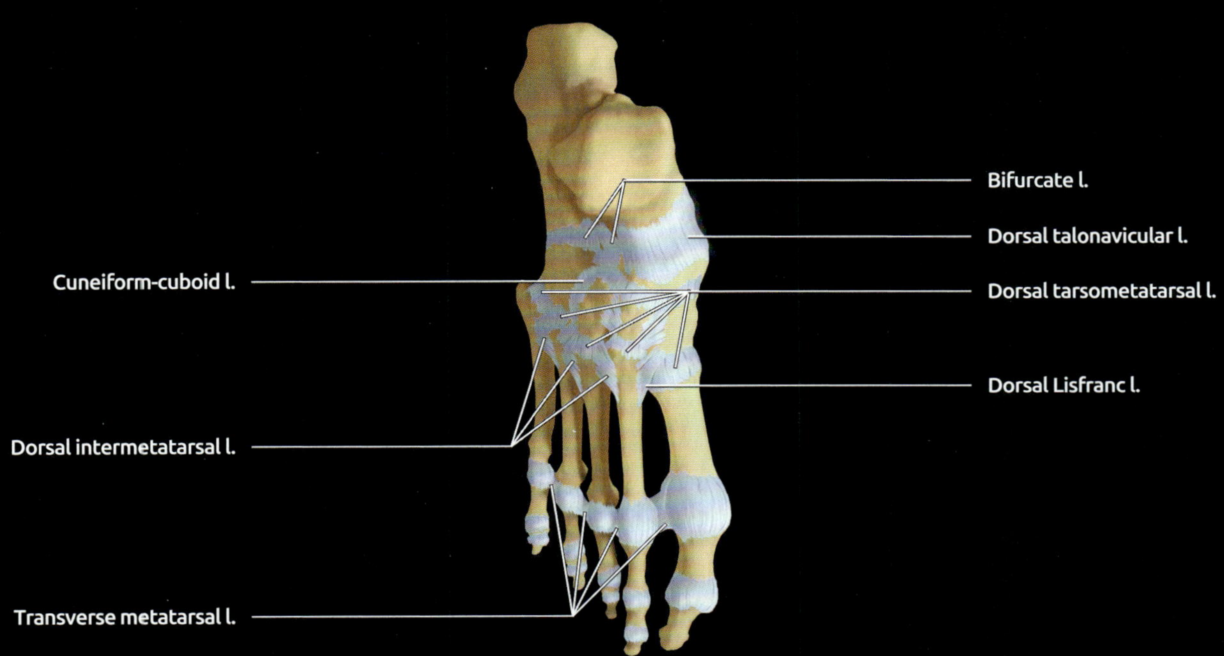

(Top) *The deep peroneal nerve travels deep to the extensor retinacula, in the anterior tarsal tunnel, and gives a lateral motor branch to the extensor digitorum brevis muscle. The medial branch continues dorsal to the talonavicular joint, middle cuneiform, and in between the 1st and 2nd metatarsals to provide mostly sensory, but some motor, supply to the 1st web space.* **(Bottom)** *Graphic of the dorsal ligaments shows dense ligaments between the tarsal bones and between tarsals and metatarsals. Ligaments are generally named for the bones they bridge. Exceptions are the bifurcate ligament, which extends from the anterior process of the calcaneus to the cuboid and the navicular, and the dorsal Lisfranc ligament, from the 1st cuneiform to the 2nd metatarsal.*

DORSAL FOOT

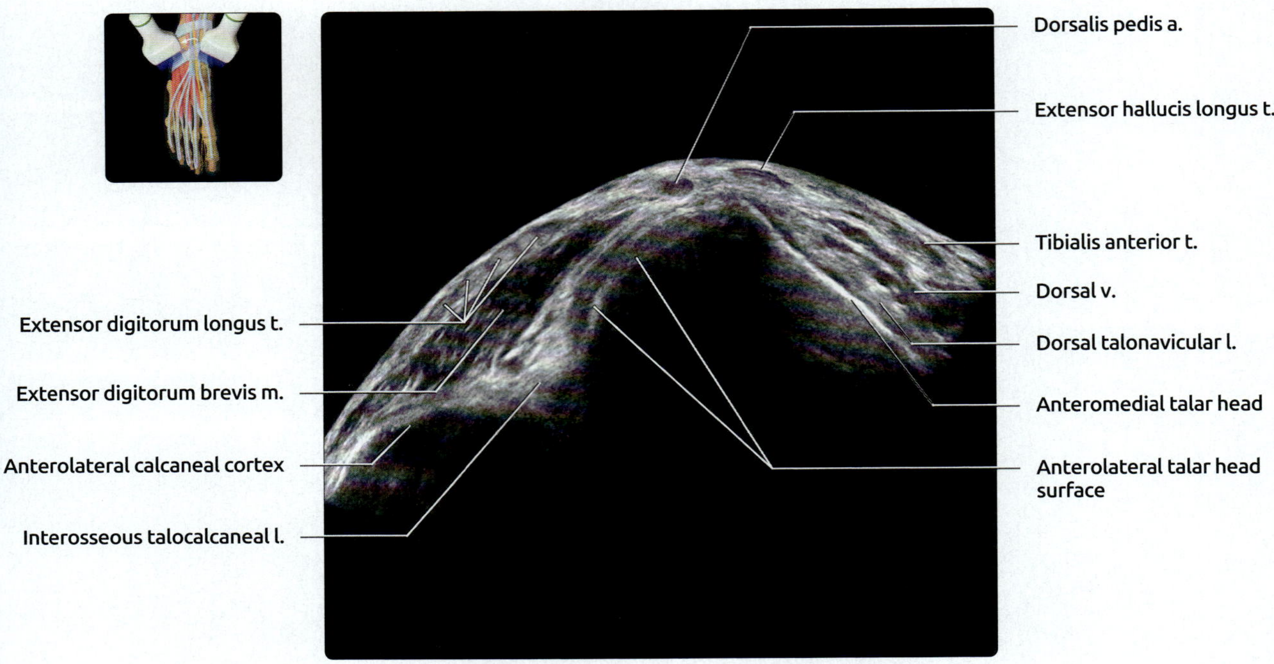

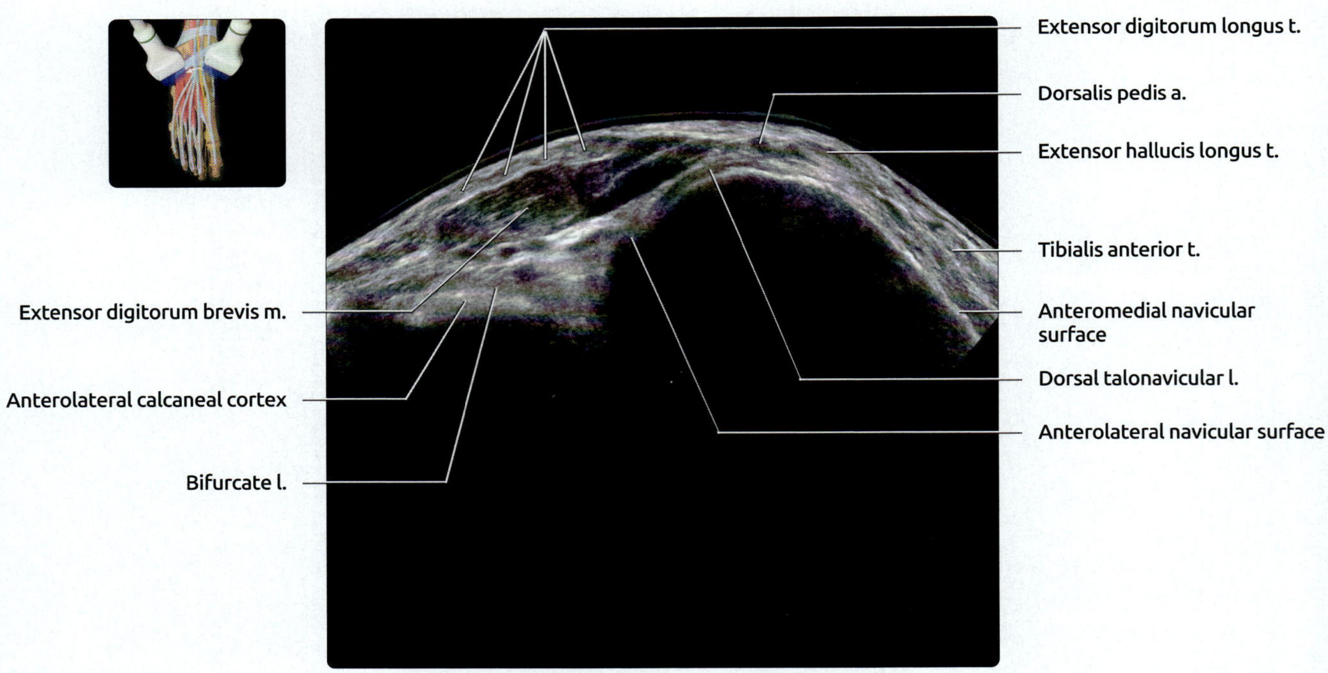

(Top) *Coronal panoramic US shows the dorsal foot at the talocalcaneus level. The foot can be divided into 2 columns: The medial column (talus, navicular, and cuneiforms) and lateral column (calcaneus and cuboid) form the anteromedial and anterolateral surfaces of the transverse midfoot arch. The dorsalis pedis artery runs at the apex of this arch.* **(Bottom)** *Coronal panoramic US shows the dorsal foot through the navicular-calcaneal level. The transverse arch of the dorsal foot creates 2 surfaces: Anteromedial and anterolateral, over which extensor tendons and small muscles lie. More focal examination of these structures requires scanning along each surface separately. Panoramic scans give an overview of anatomy. However, high-resolution imaging of the site of interest provides more details, and both should be performed as complementary techniques.*

DORSAL FOOT

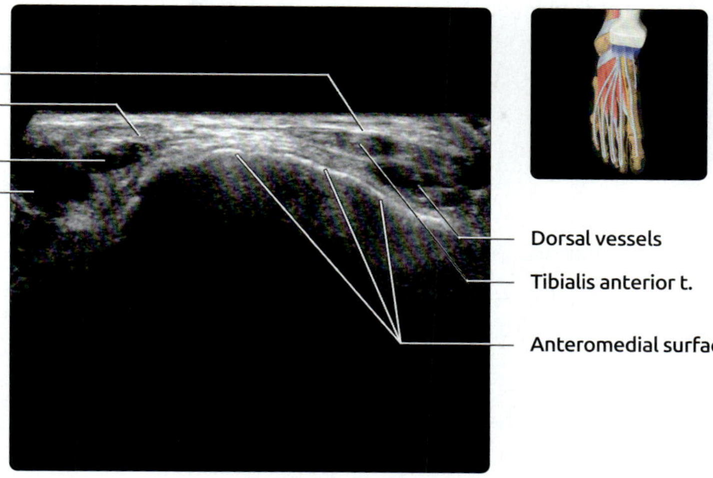

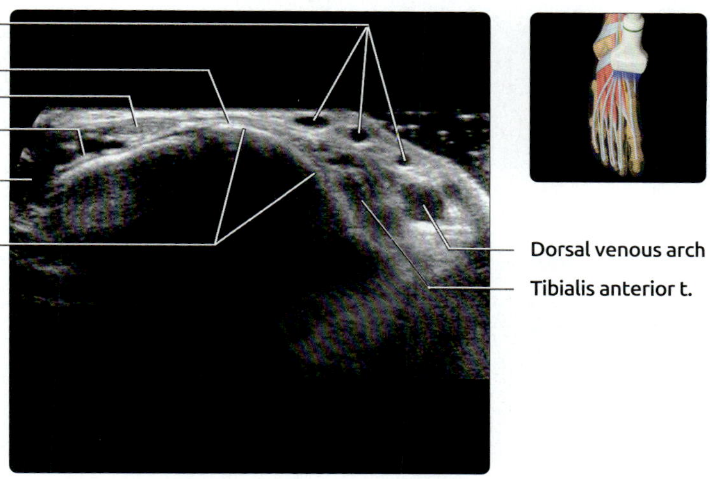

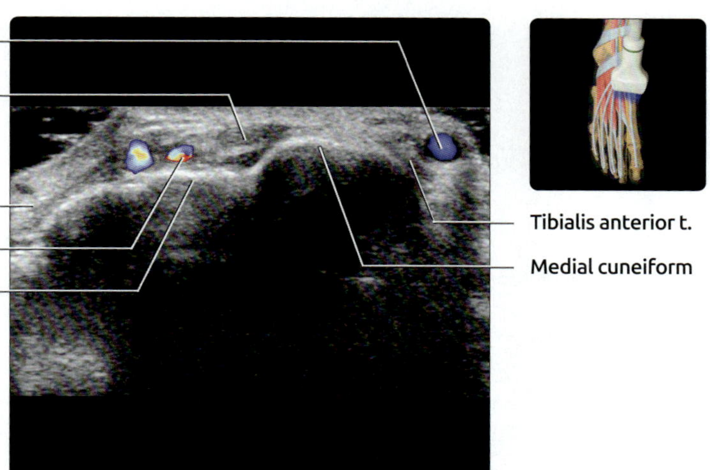

(Top) *Series of 3 coronal scans shows the dorsal medial surface of the foot. At this most proximal level, the talus is the underlying bone. The extensor hallucis can be seen partially covering the dorsalis pedis artery. More medially, the tibialis anterior tendon can be found.* **(Middle)** *At the midtarsal level, coronal US shows the anteromedial surface of the navicular. The tibialis anterior tendon begins to turn acutely medially to curve around to the medial edge of the foot, eventually inserting on the medial cuneiform. As seen here, the use of light pressure and copious amounts of gel will allow visualization of small superficial veins, which are numerous on the dorsum of the foot.* **(Bottom)** *Coronal color Doppler US shows the level of the cuneiforms. The intimate relationship between the dorsalis pedis artery and the extensor hallucis longus tendon continues.*

DORSAL FOOT

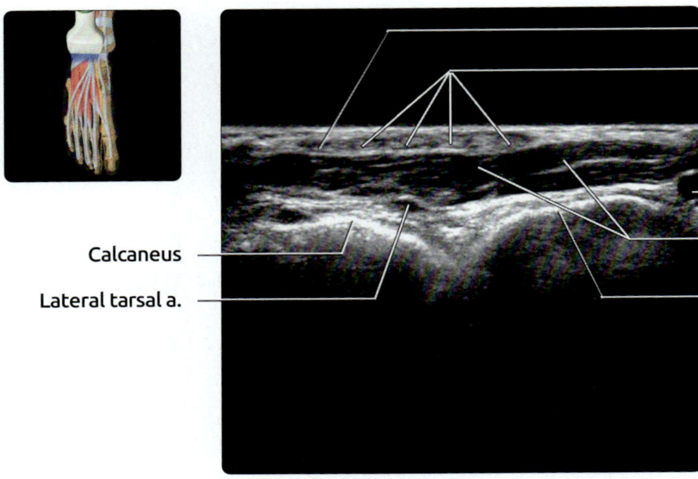

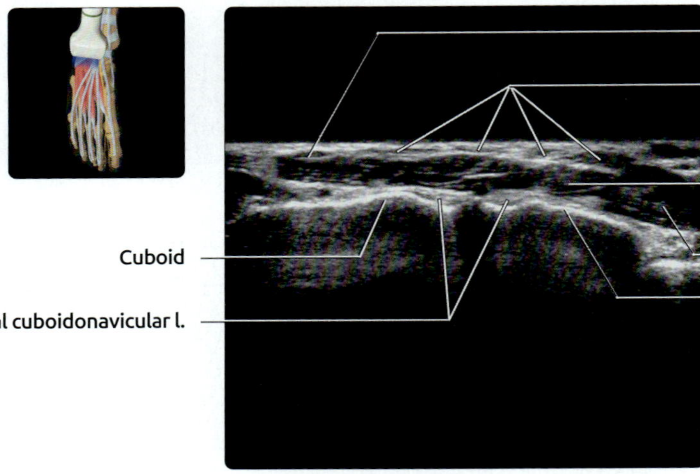

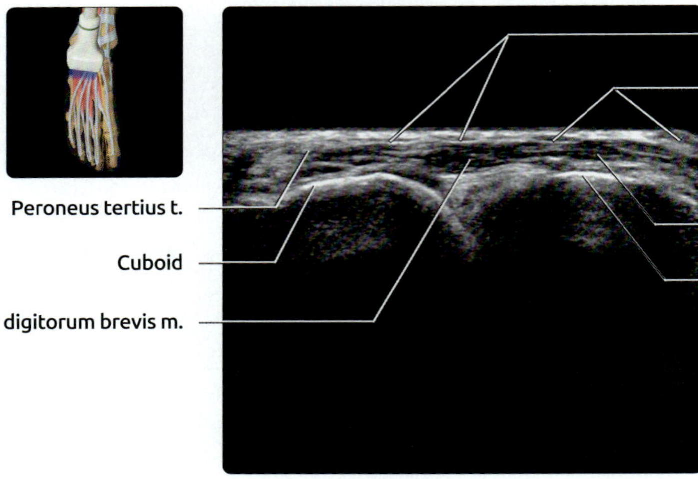

(Top) *Series of 3 coronal scans shows the dorsal lateral surface of the tarsus. At this proximal level, the lateral surface of the calcaneus and the talus are the underlying bones. The extensor digitorum longus tendons are close to each other and share a common synovial sheath. The extensor digitorum brevis muscle lies deep to the longus tendons. The lateral tarsal artery runs deeper and is a branch of the dorsalis pedis artery.* **(Middle)** *At the level of the midtarsus, the extensor digitorum longus tendons for individual toes begin to spread out and also flatten (the latter making them just discernible on US). At this level, the extensor hallucis brevis muscle can also be seen.* **(Bottom)** *At a more distal level, the extensor digitorum brevis muscle and the extensor hallucis brevis muscle are much thinner and will turn into tendons distally.*

DORSAL FOOT

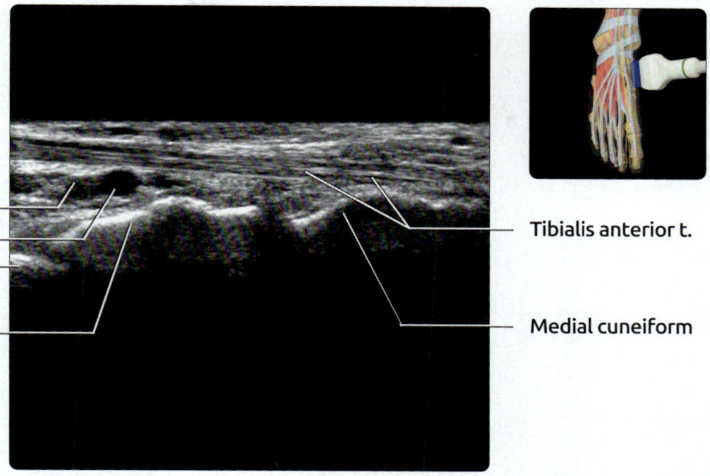

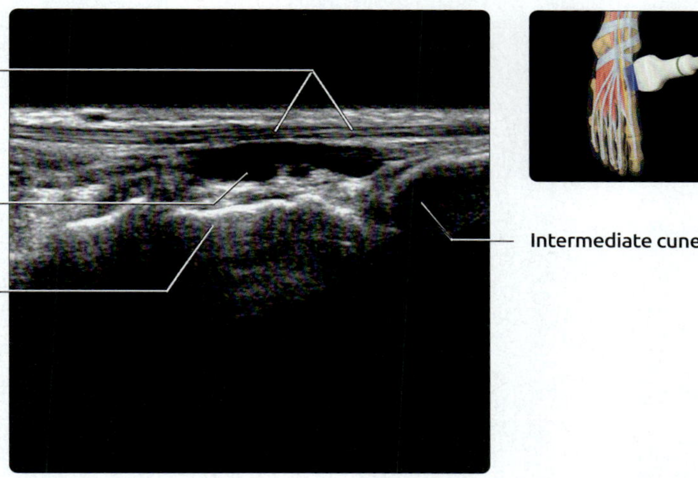

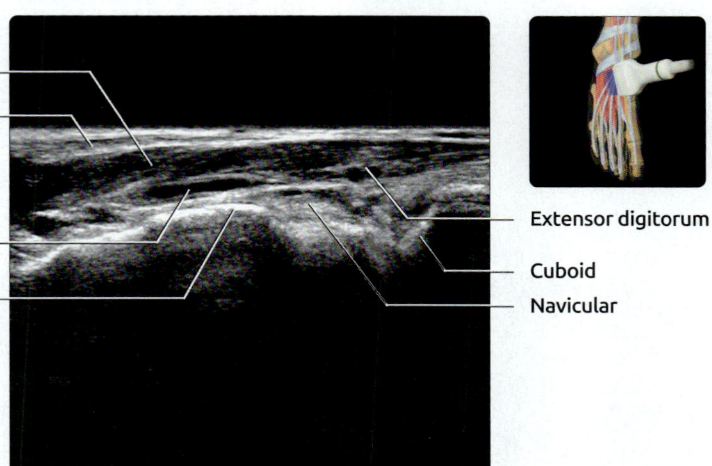

(Top) *Series of sagittal scans is shown through the tarsus. Medially, the tibialis anterior is seen running over the medial tarsal vessels (branch of dorsalis pedis) to insert on the medial and inferior surfaces of the base of the medial cuneiform.* **(Middle)** *Slightly laterally, the extensor hallucis longus tendon runs toward the big toe, passing over the intermediate cuneiform bone.* **(Bottom)** *More laterally, the extensor digitorum longus tendon and the extensor digitorum brevis muscle come into view. They lie on the surface of the calcaneus with the lateral tarsal artery coursing between the muscle and bone. The gradual tapering of the extensor digitorum brevis muscle distally can be appreciated here.*

PLANTAR MUSCLES, SUPERFICIAL LAYERS

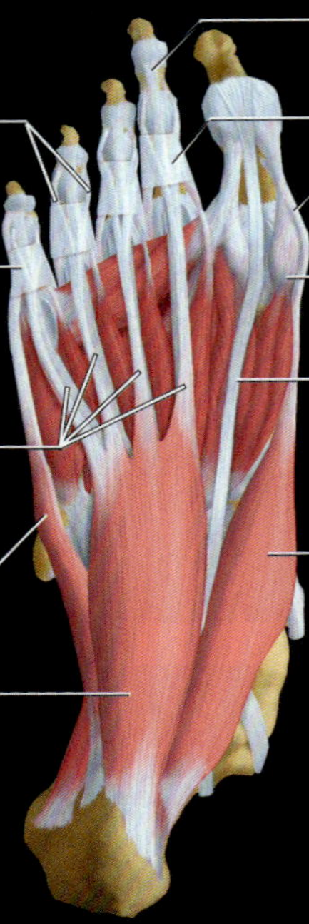

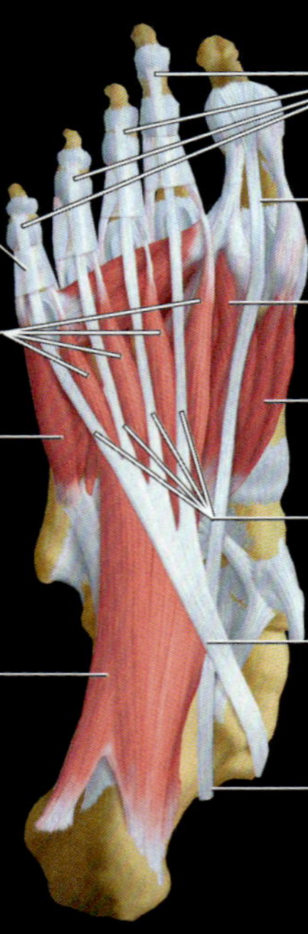

(Top) *Graphic shows the superficial layer of the plantar foot muscles: Abductor digiti minimi, flexor digitorum brevis, and abductor hallucis. The deep muscles are partially visible deep to the superficial layer. The flexor digitorum brevis tendons split in each digit (4th digit labeled), attaching at lateral aspects of the middle phalangeal bases. The flexor tendon sheaths hold the flexor mechanism in close proximity to the phalanges.* **(Bottom)** *Graphic shows the 2nd layer of plantar muscles. The interesting arrangement of interactions between flexors is well seen: Quadratus plantae inserting on flexor digitorum longus tendon and lumbricals arising from individual flexor digitorum longus tendon slips. In the toes, the flexor tendons are contained within fibrous tendon sheaths.*

PLANTAR MUSCLES, DEEP LAYERS

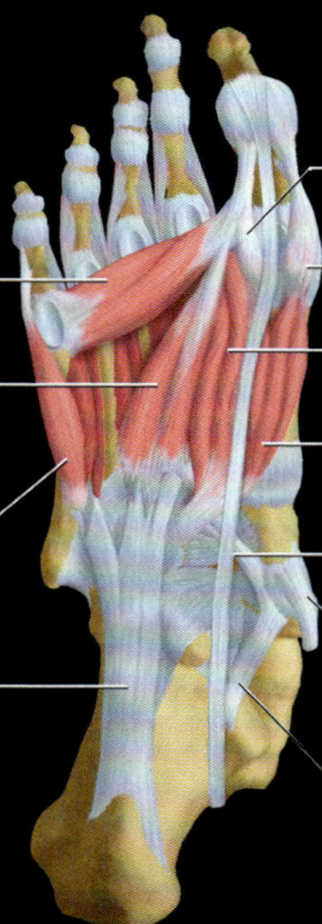

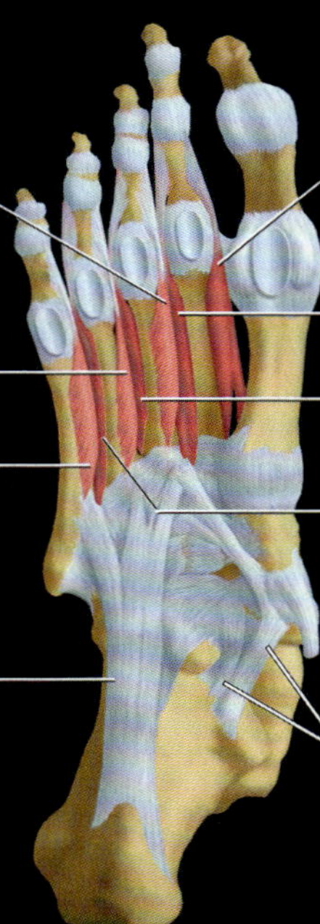

(Top) *Graphic shows the 3rd layer of plantar muscles. This layer contains the flexor hallucis brevis and flexor digiti minimi as well as the adductor hallucis muscle. The oblique head of the adductor hallucis is thick and broad, whereas the transverse head is thin.* **(Bottom)** *Graphic shows the 4th layer of plantar muscles, the interosseous muscles. There are 4 dorsal interossei. They have bipennate origins from 2 adjacent metatarsals. There are 3 plantar interossei. The 1st plantar interosseous muscle originates from the medial side of the 3rd metatarsal and inserts on the medial side of the 3rd proximal phalanx. The 2nd and 3rd plantar interossei originate on the medial sides of the 4th and 5th metatarsals, respectively, and insert on the medial side of the corresponding proximal phalanx.*

PLANTAR APONEUROSIS

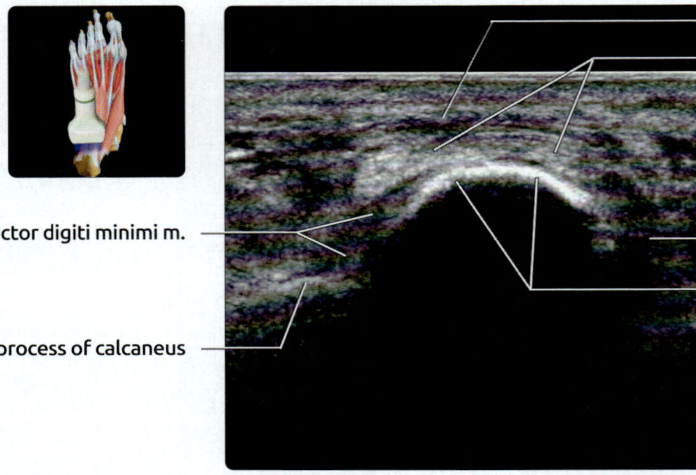

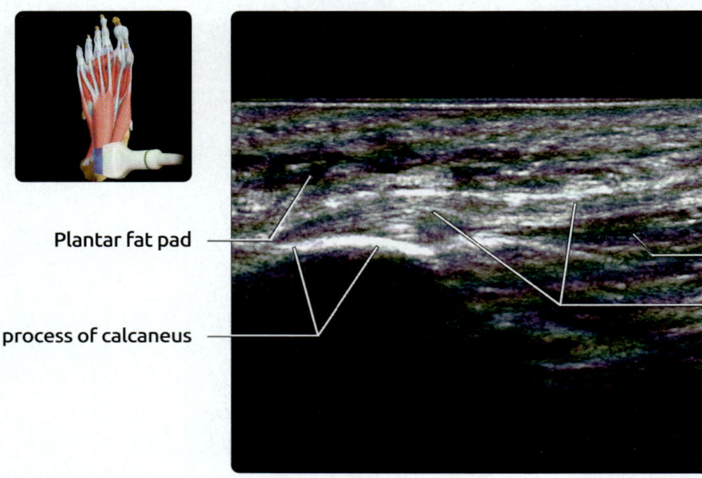

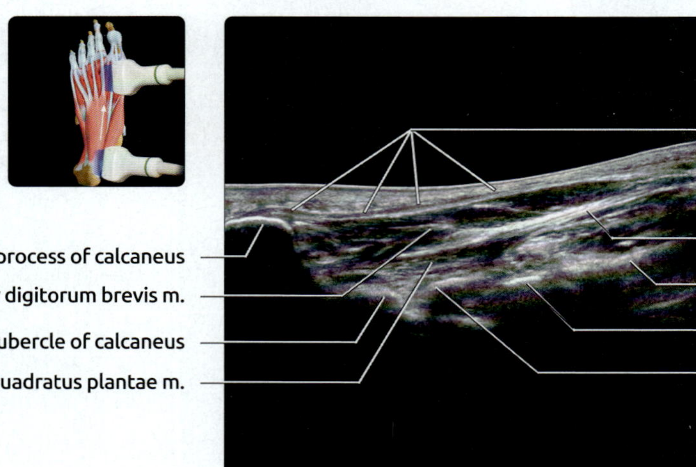

(Top) *Transverse US of the calcaneus shows the origin of the plantar aponeurosis. The plantar aponeurosis fascia consists of 3 parts: The strong, thick central band over the flexor digitorum brevis muscle, the lateral band over the abductor digiti muscle, and the thin medial band over the abductor hallucis muscle.* **(Middle)** *Sagittal US shows the origin of the central band of the plantar aponeurosis. This arises from the medial process of the calcaneus and runs anteriorly deep to the plantar subcutaneous fat.* **(Bottom)** *Sagittal panoramic US shows the central band of plantar aponeurosis. The central band is the thickest and most easily identified component of the plantar aponeurosis. Deep to this are the flexor digitorum brevis and quadratus plantae muscles, which form the bulk of the central tarsal plantar compartment.*

SAGITTAL FOOT OVERVIEW

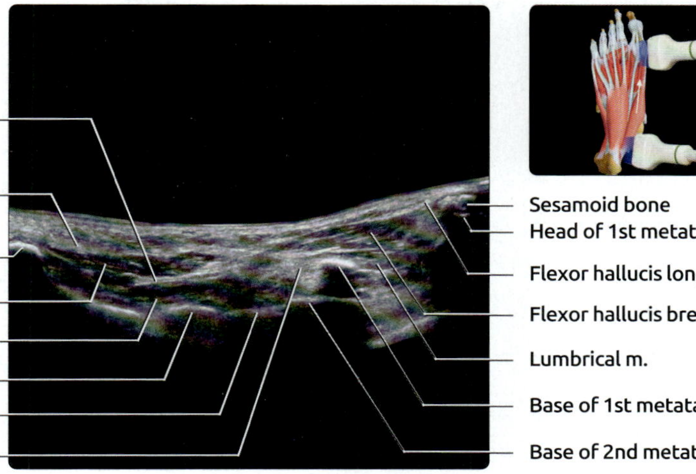

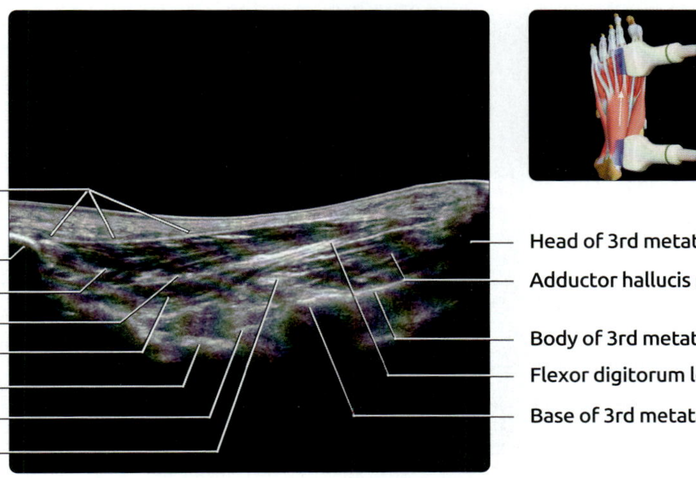

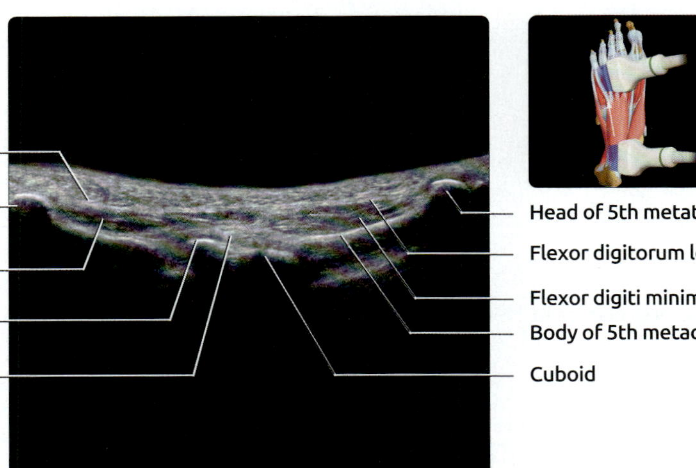

(Top) *Series of 3 panoramic sagittal scans is shown through the plantar aspect of the foot. This scan shows the content of the medial compartment: The abductor hallucis muscle, flexor hallucis longus tendon, and flexor hallucis brevis muscle, together with smaller structures.* **(Middle)** *In the central plantar compartment, the contents can be divided into layers. Superficially, there is the flexor digitorum brevis muscle and the flexor digitorum longus tendon. Deep to this layer are the quadratus plantae muscle, the oblique head of the adductor hallucis muscle, and the lumbricals.* **(Bottom)** *Scan through the lateral compartment shows the abductor digiti minimi muscle and the flexor digiti minimi muscle.*

MEDIAL TARSUS

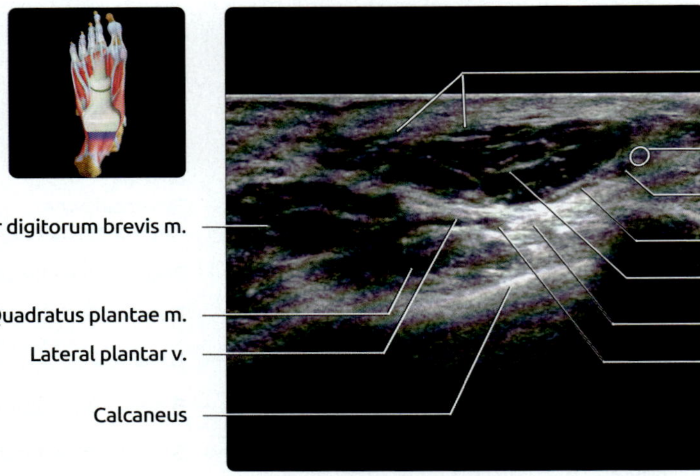

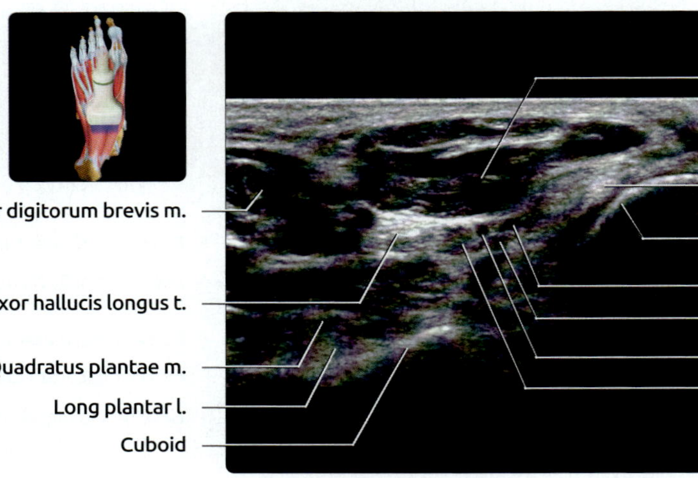

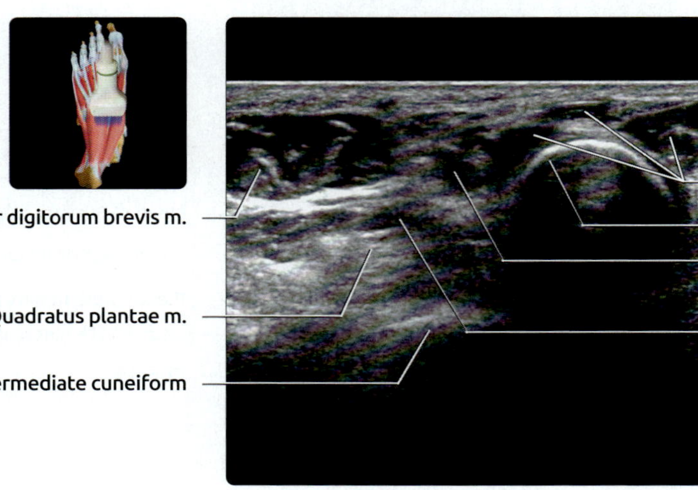

(Top) *Series of 3 coronal scans is shown through the medial 1/2 of the plantar aspect of the foot. At this proximal level, the 2 main muscles (flexor digitorum brevis and quadratus plantae) of the central compartment can be seen lying superficial to the calcaneus. The medial compartment muscle (abductor hallucis) can also be seen medially.* **(Middle)** *At the level of the midtarsus, the knot of Henry can be seen. This is where tendons of the flexor digitorum longus and flexor hallucis longus tendons cross and exchange fibers. The medial plantar neurovascular bundle lies in close proximity to this.* **(Bottom)** *More distally, the flexor hallucis longus and digitorum longus tendons separate and head toward their respective insertions.*

LATERAL TARSUS

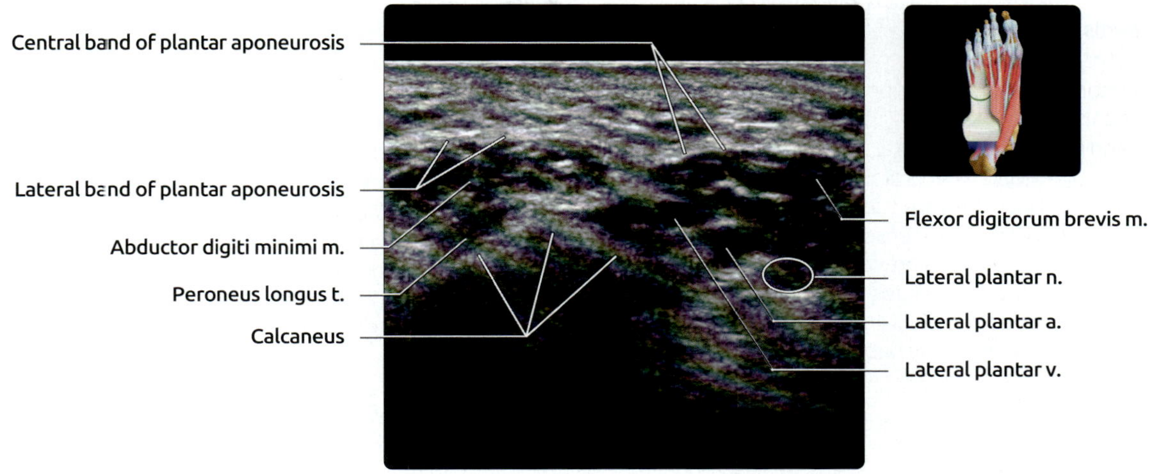

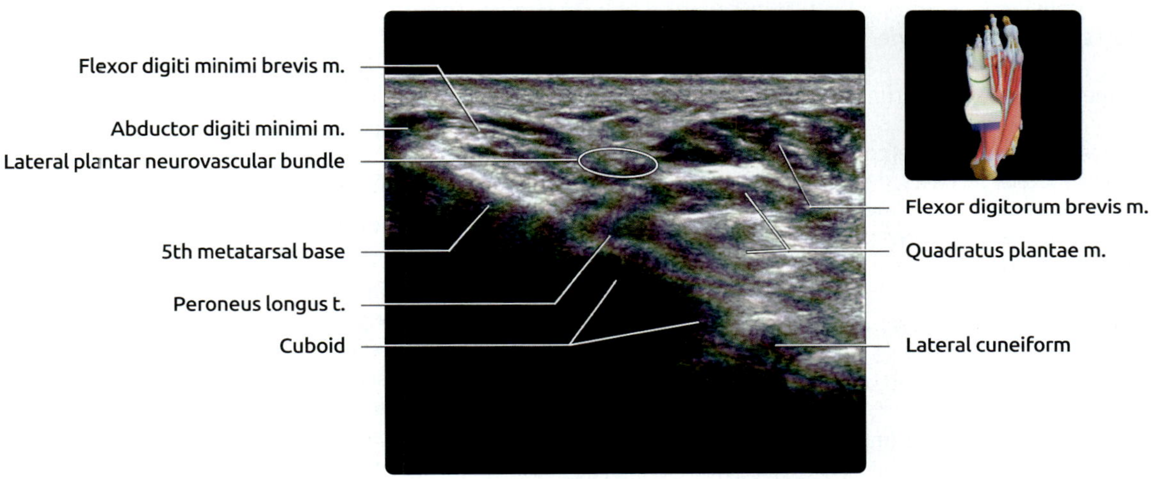

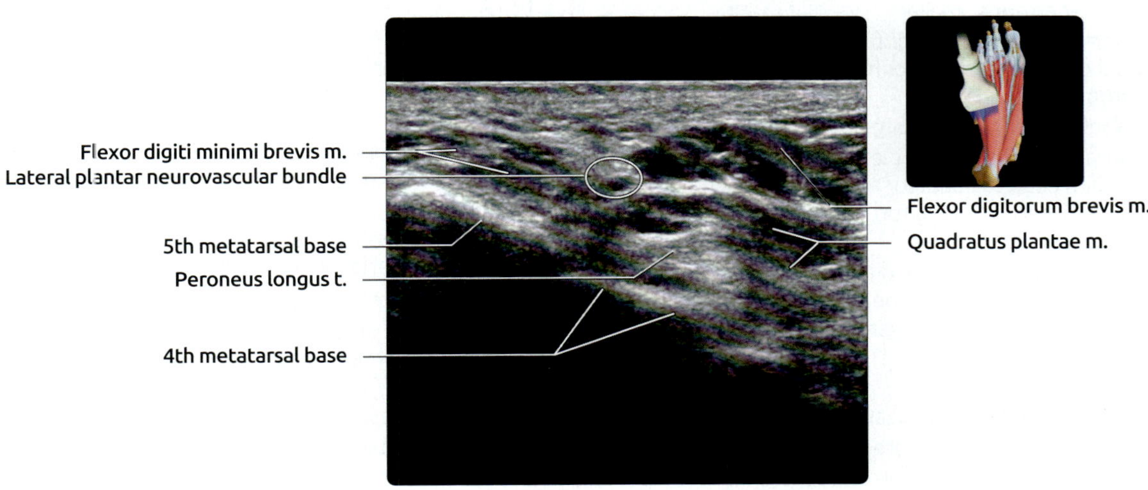

(Top) *Series of 3 coronal scans is shown through the lateral 1/2 of the plantar aspect of the foot. The lateral compartment (abductor digiti minimi) and 1/2 of the central compartment can be seen here, separated by the lateral plantar neurovascular bundle. The calcaneus forms the bony support of all structures at this level.* **(Middle)** *At the level of the midtarsus, the separation between the lateral and the central compartment by the neurovascular bundle continues. Within the lateral compartment, the abductor digiti minimi muscle moves laterally to give way to the other component of the lateral compartment, the flexor digiti minimi brevis muscle.* **(Bottom)** *Most distally, the abductor digiti minimi muscle runs on the lateral border, and only the flexor digiti minimi brevis muscle is seen on the plantar aspect of the foot.*

IMAGING ANATOMY

Arteries

- **Dorsalis pedis artery**
  - Dorsal artery of foot
  - Direct continuation of anterior tibial artery, changing name at ankle joint
  - Course and relations
    - Crosses superficial to talocrural articular capsule, talus, navicular, and intermediate cuneiform bones
    - Passes deep to inferior extensor retinaculum and distal to extensor hallucis brevis
    - Lies lateral to extensor hallucis longus tendon
    - Passes to proximal end of intermetatarsal space and curves deeply to enter sole between heads of 1st dorsal interosseous muscle
    - Forms part of plantar arch
  - Main branches
    - Medial and lateral tarsal arteries: Arise at level of navicular bone
    - Lateral tarsal artery runs under extensor digitorum brevis
    - 2 or 3 medial tarsal arteries run on medial surface of tarsus
  - Arcuate artery: Arises at medial cuneiform level and runs laterally over metatarsal bases, deep to extensor tendons; gives off branches to metatarsals and toes
  - 1st dorsal metatarsal artery: Arises just before dorsalis pedis dives deep into sole
- **Posterior tibial artery**
  - Continuation of calf vessel by same name, branch of popliteal artery
  - Runs in tarsal tunnel behind and below medial malleolus accompanied by tibial nerve
  - Divides into terminal branches (medial and plantar arteries) in tarsal tunnel
  - Medial plantar artery is smaller
    - Runs in medial aspect of foot medial to medial plantar nerve, then deep to abductor hallucis and flexor digitorum brevis muscles
  - Lateral plantar artery is much larger than medial
    - Runs in lateral aspect of foot lateral to lateral plantar nerve, then deep to abductor hallucis muscle and distally between flexor digitorum brevis and quadratus plantae muscles
    - Distally runs between flexor digitorum brevis and abductor digiti minimi, then between base of 1st and 2nd metatarsals, and finally joins dorsalis pedis to form plantar arch
- **Plantar arch**
  - Arterial arch at level of metatarsals
  - Formed by dorsalis pedis and lateral plantar arteries
  - Gives off branches to metatarsals and toes
- **Peroneal artery**
  - Accompanies superficial peroneal nerve down anterolateral aspect of ankle and sends off perforating branch that joins communicating branch from anterior tibial artery
  - May join or replace posterior tibial artery

Veins

- Superficial veins
  - Great saphenous vein
    - Arises from medial part of dorsal venous arch of foot
    - Runs subcutaneously over muscles and tendons to pass anterior to medial malleolus
    - Passes superiorly on medial aspect of calf and thigh to join common femoral vein at femoral triangle
  - Small saphenous vein
    - Arises from lateral part of dorsal venous arch
    - Runs subcutaneously over muscles and tendons to pass posterior to lateral malleolus
    - Passes superiorly on lateral and posterior side of calf to join popliteal vein in popliteal fossa
- Deep veins
  - Plantar venous arch: Receives venous drainage from dorsal and plantar aspects of toes and metatarsals
  - Medial and lateral plantar veins: Arise from plantar venous arch to accompany same named arteries
  - Communication exists with great and small saphenous veins
  - Medial and lateral plantar veins form posterior tibial veins in tarsal tunnel and run proximally to calf, accompanying posterior tibial artery

ANATOMY IMAGING ISSUES

Imaging Recommendations

- US
  - Provides most information on foot vasculature of different imaging modalities available
  - Morphological and functional data provided
  - Color Doppler scan allows assessment of vessel occlusion and thrombosis
  - Spectral Doppler scans allow quantitative and semiquantitative documentation of flow velocity, resistance, waveform, etc.
  - Augmentation of foot veins usually only requires small degrees of flexion/extension of toes to produce enough venous flow for detection (unlike need to squeeze muscles in calf)
- CT or MR angiography
  - Provides very high-quality images of large arteries of abdomen, pelvis, and lower limbs to midfoot region
  - Limited in assessment of smaller arteries of foot or even arteries of distal leg and ankle if arterial flow poor
  - Morphological > functional data provided
- Digital subtraction angiography
  - Usefulness in peripheral arterial disease largely surpassed by CT or MR angiography
  - Helpful in assessing vascular malformation
  - Recently available, 2D color-coded DSA allows quantitative measurement of blood flow
    - Arrival time of contrast material, wash-in rate, time to peak contrast intensity, and area under curve

Imaging Pitfalls

- Superficial veins are easily compressed even with lightest transducer pressure
  - Use thick layer of gel to minimize pressure

DORSAL AND PLANTAR ARTERIES AND NERVES

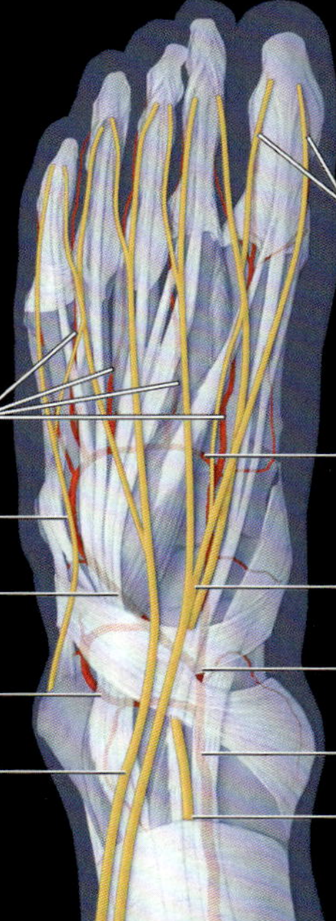

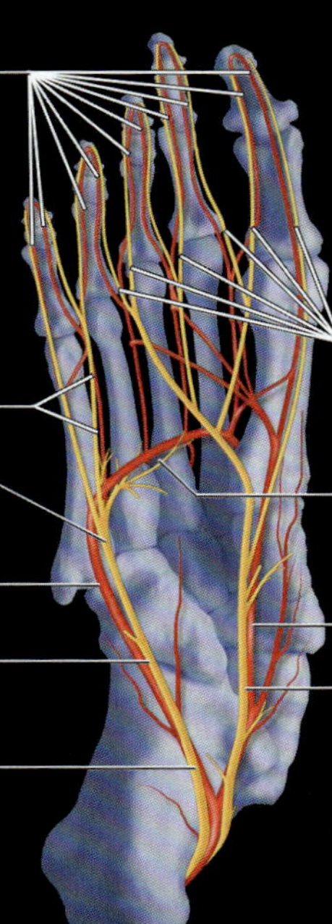

(Top) *Graphic shows arteries and nerves at the dorsal aspect of the foot. The deep peroneal nerve accompanies the anterior tibial artery. The superficial peroneal nerve divides into medial and lateral branches in distal 1/3 of leg. Anterior tibial artery is called the dorsalis pedis in the foot. It terminates in arcuate artery, which communicates with plantar arterial arch, and in digital vessels.* **(Bottom)** *Graphic shows arteries and nerve at the plantar aspect of the foot. The posterior tibial nerve and artery divide into medial and lateral plantar branches. These divide further into digital branches to the medial and lateral aspects of each toe. The plantar arch sends vessels dorsally as well as to the plantar aspect of the toes.*

DORSALIS PEDIS ARTERY

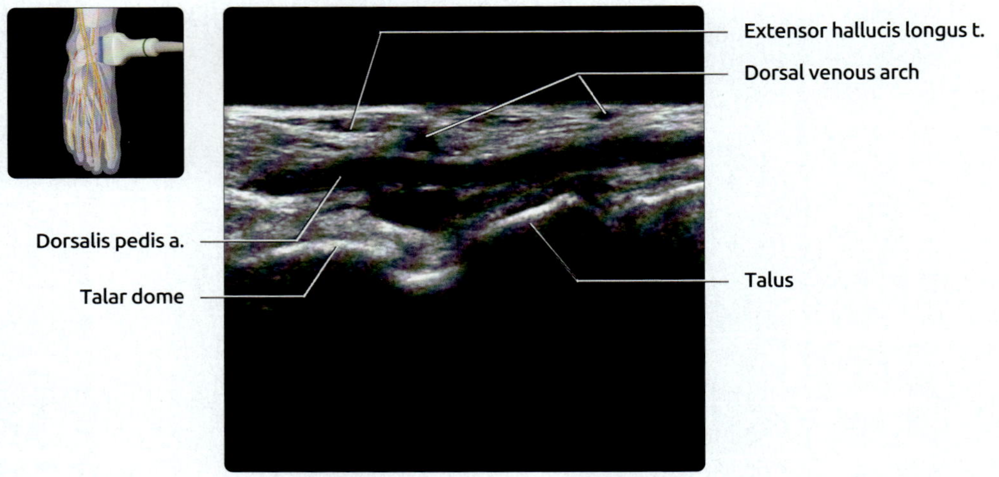

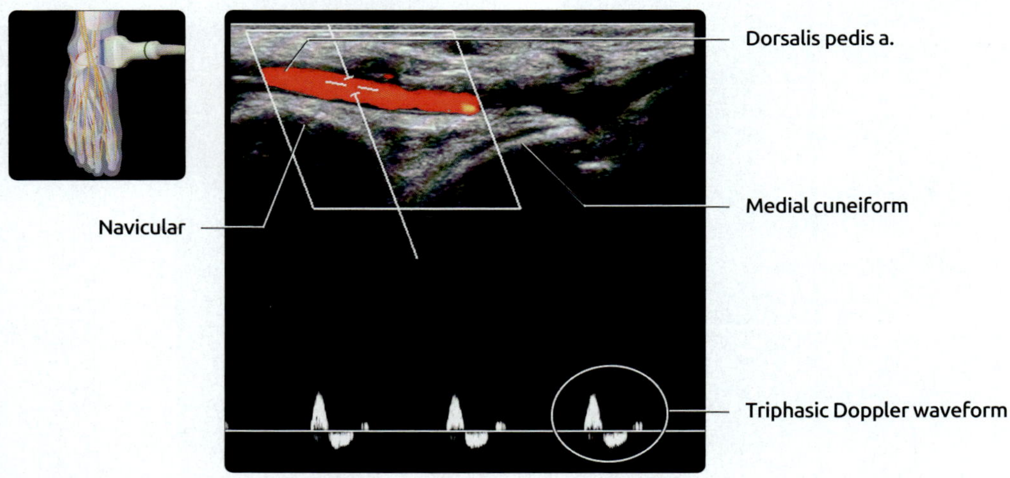

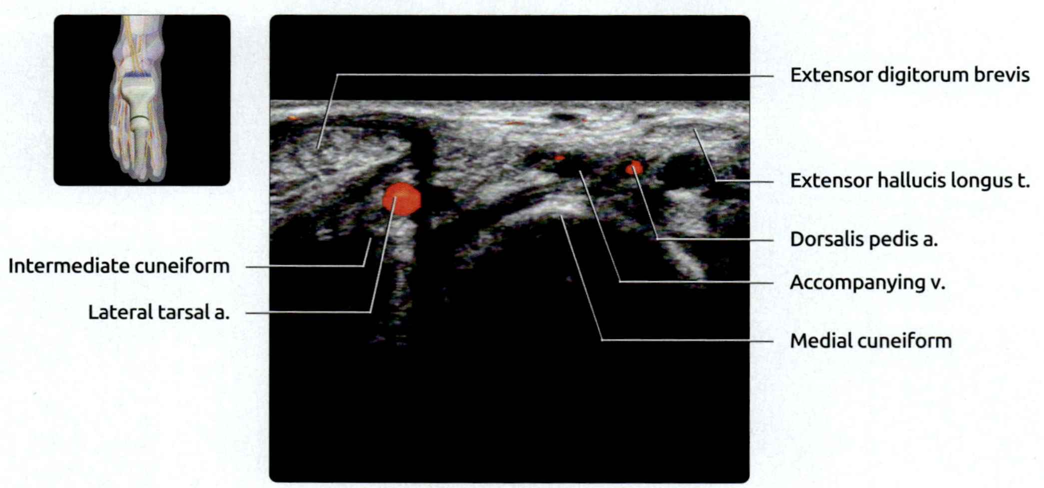

(Top) *Sagittal scan shows the dorsum of the foot. The dorsalis pedis artery runs at the anterior aspect of the transverse arch formed by the tarsal bones. It is a continuation of the anterior tibial artery and runs lateral to the extensor hallucis longus tendon.* **(Middle)** *Sagittal spectral Doppler scan of the dorsalis pedis artery shows the typical triphasic artery flow.* **(Bottom)** *Coronal color Doppler scan shows the dorsum of the foot at the level of the cuneiform. The dorsalis pedis artery gives rise to medial and lateral tarsal arteries at the level of the navicular. The dorsalis pedis artery remains lateral to the extensor hallucis longus tendon, even at this distal level.*

METATARSAL AND DIGITAL ARTERIES

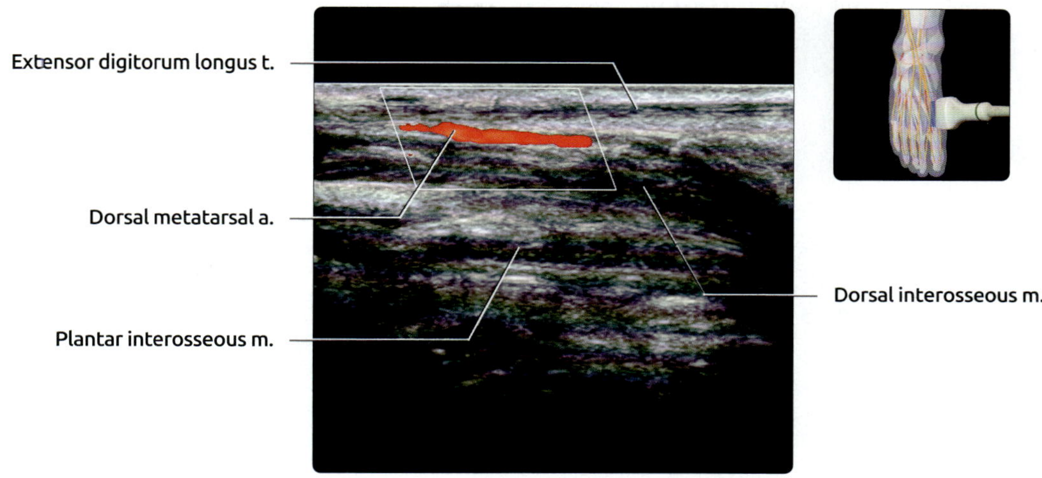

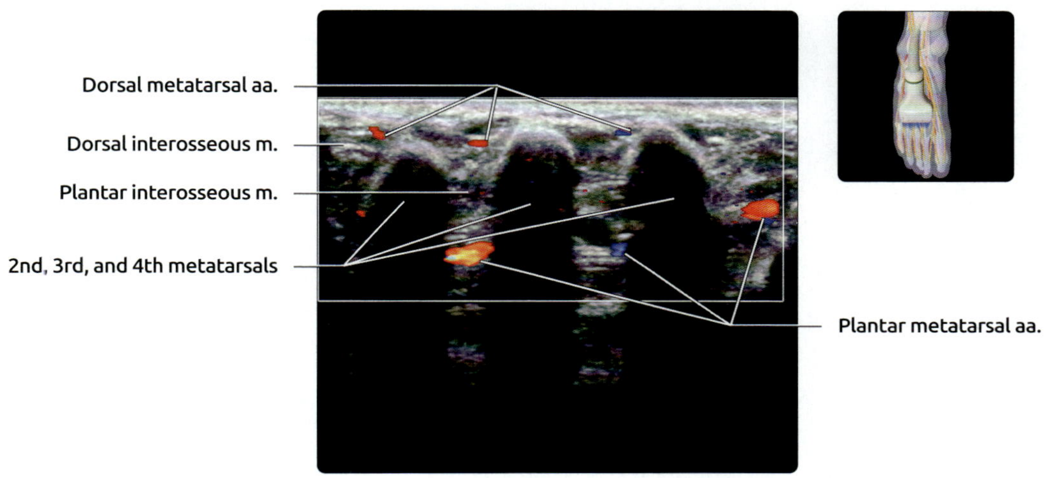

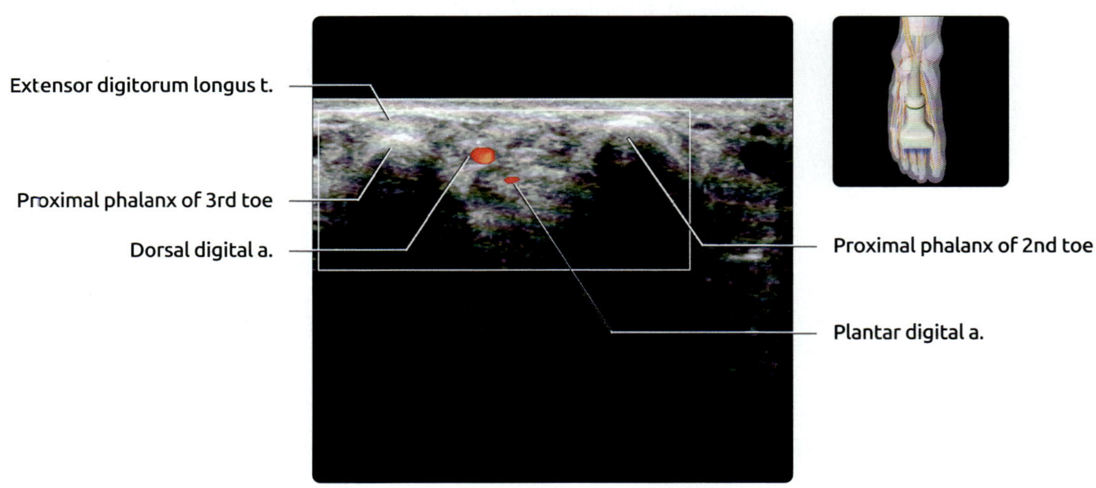

(Top) *Sagittal color Doppler shows the 2nd interosseous space. The arcuate artery gives rise to dorsal metatarsal arteries that run in the interosseous spaces to supply the metatarsals and toes. The 2 layers of interosseous muscles can be appreciated here.* **(Middle)** *Transverse color Doppler scan shows the metatarsal bodies. Both the dorsal and plantar interosseous muscles can be seen between the bones. The dorsal and plantar metatarsal arteries are also demonstrable by the use of color Doppler. The plantar metatarsal arteries arise from the plantar arch.* **(Bottom)** *Coronal color Doppler scan shows the 2nd web space. The dorsal and plantar metatarsal arteries continue distally as the digital arteries, which will divide further to run on the sides of the phalanges.*

PLANTAR ARTERIES

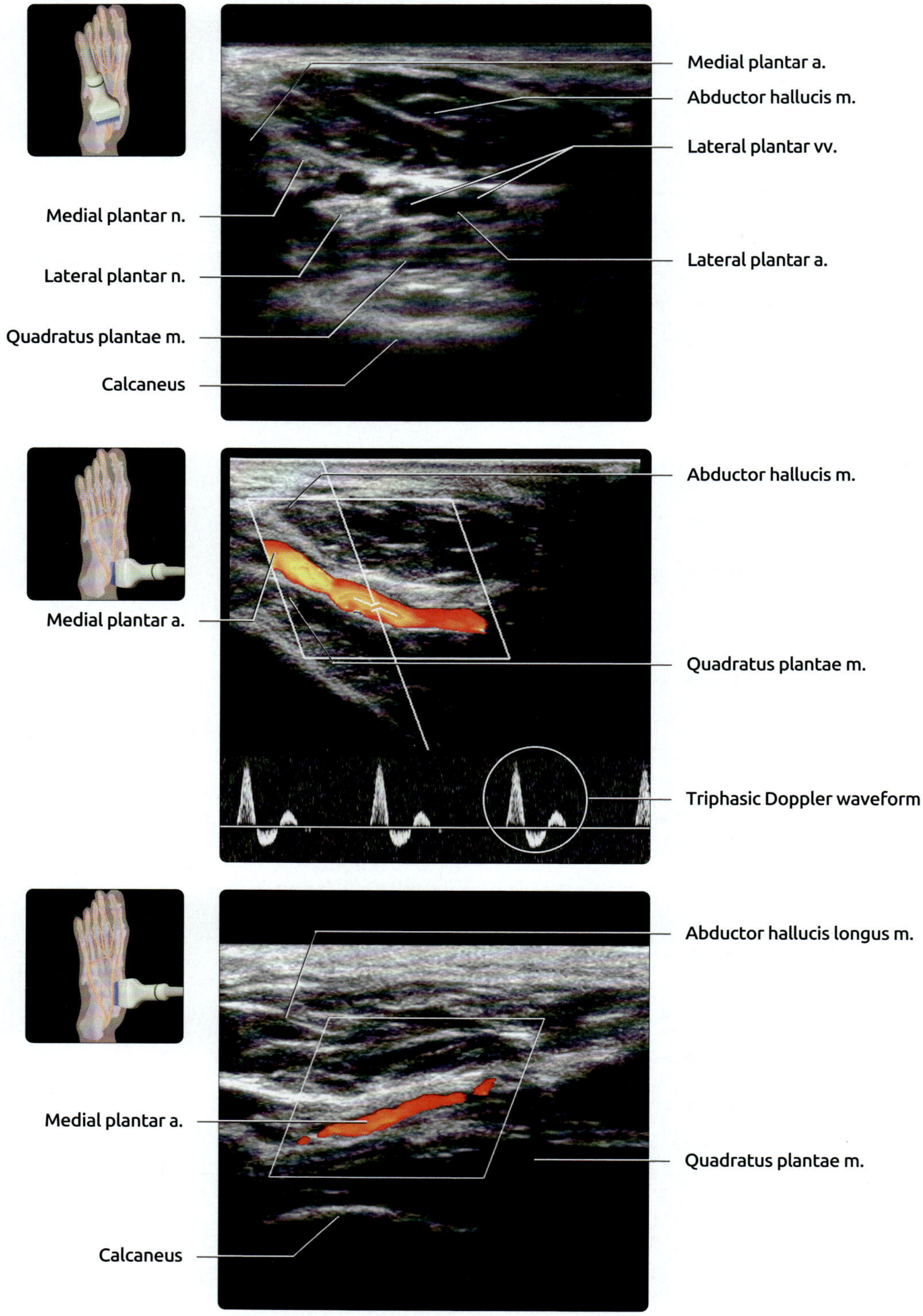

(Top) *Coronal scan shows the medial plantar compartment. The tibial nerve has divided into the medial and lateral plantar nerves, and, similarly, the posterior tibial vessels have divided into medial and lateral plantar vessels. Note that this is a scan of the left foot, and the lateral side is on the right of the image.* **(Middle)** *Spectral Doppler scan of the medial plantar artery shows a typical triphasic arterial waveform.* **(Bottom)** *Oblique sagittal color Doppler scan of the medial compartment shows the medial plantar artery running between the abductor hallucis muscle and the quadratus plantae muscle.*

PLANTAR ARTERIES

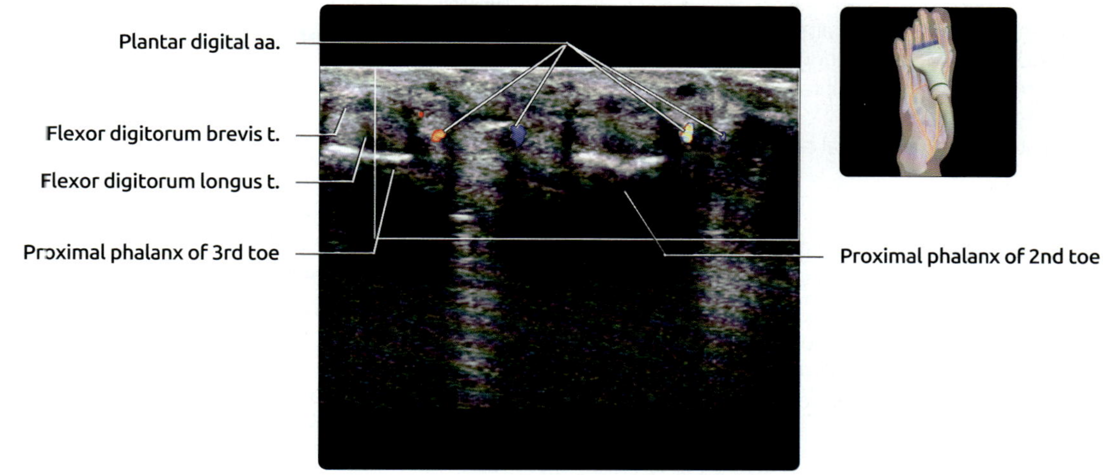

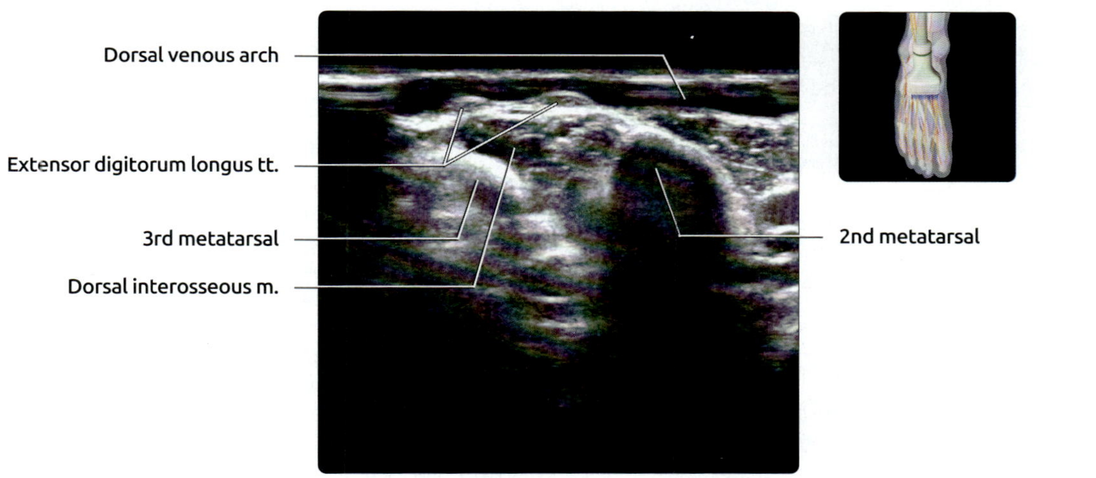

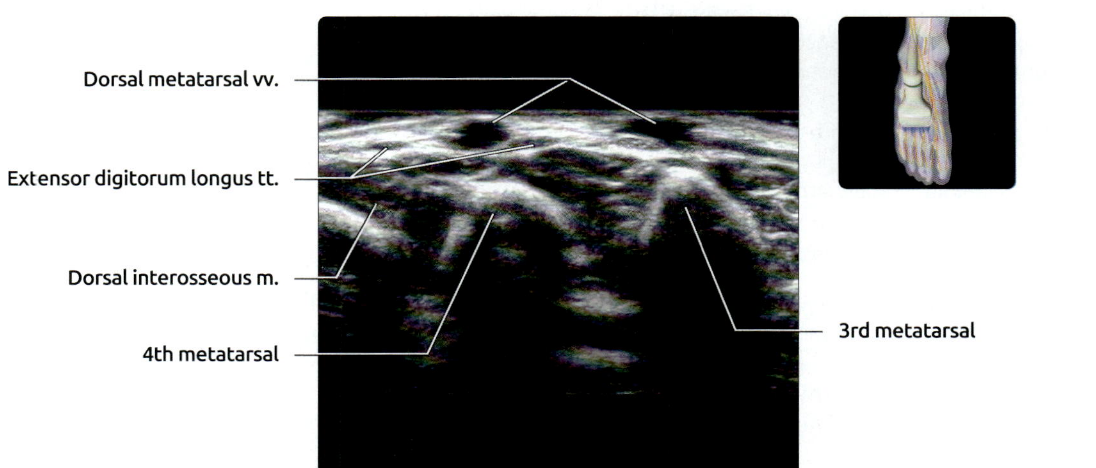

(Top) *Coronal color Doppler scan shows the plantar aspect of the toes. The digital arteries can be seen on both sides of the phalanges as they travel more distally.* **(Middle)** *Coronal scan shows the dorsum of the foot at the metatarsal level. Minimal pressure was used during this scan to prevent compression of the superficial (subcutaneous) veins. Here, the transversely running dorsal venous arch is demonstrated. The extensor tendons and the interossei run deep to this arch.* **(Bottom)** *More distally, a coronal scan of the dorsum of the foot shows the tributaries of the dorsal venous arch, the dorsal metatarsal veins.*

IMAGING ANATOMY

Overview

- In normal weight-bearing stance, all metatarsal heads are at same level and all are weight bearing
  - Metatarsophalangeal (MTP) joints are slightly extended in standing position
- Each MTP joint is separate synovial cavity

Metatarsal Bones

- 1st metatarsal
  - Largest metatarsal
  - Articulates with 1st cuneiform, 1st proximal phalanx, sesamoids
  - Medially receives attachments from tibialis anterior tendon; plantar side receives peroneus longus tendon
  - Origin of medial head of 1st dorsal interosseous muscle on its lateral surface
- 2nd-4th metatarsals
  - Gradually smaller and shorter moving laterally
  - Receive interosseous muscle on medial and lateral sides
- 5th metatarsal
  - Insertion of peroneus tertius and brevis

Phalanges

- Distal phalanx is insertion site for tendons from flexor and extensor digitorum longus
- Middle phalanx is insertion site for tendons from flexor and extensor digitorum brevis
- 2nd-4th proximal phalanges are attachments of interosseous muscles on both sides and lumbricals on medial side

1st Metatarsophalangeal Joint

- Dorsiflexion of toe important in push-off phase of gait
- Metatarsal head has 2 concave facets at plantar surface, 1 for each sesamoid, separated by ridge (crista)
- Distal articular surface of metatarsal head may be flat, rounded, or have central prominence
- Base of proximal phalanx has concave contour
- **Sesamoids**
  - Either sesamoid may be bipartite or tripartite, medial > lateral
    - Bipartite sesamoid typically larger than single sesamoid with irregular opposing contour
  - Alternatively, either sesamoid may be hypoplastic or aplastic
  - Medial sesamoid in medial head of flexor hallucis brevis and abductor hallucis
  - Lateral sesamoid in lateral head of flexor hallucis brevis and adductor hallucis and deep metatarsal ligament
  - Medial and lateral sesamoids joined by intersesamoid ligament
    - Intersesamoid ligament is floor of canal in which flexor hallucis longus tendon runs
  - Both are partially embedded in plantar plate of joint
  - **Sesamophalangeal apparatus**
    - Sesamoids fixed in position relative to 1st proximal phalanx, move relative to 1st metatarsal
    - Therefore, displaced laterally in hallux valgus
- **Plantar plate**
  - Fibrocartilaginous plantar plate extending from metatarsal neck to base of proximal phalanx
  - Prevents hyperextension

Other Metatarsophalangeal Joints

- Convex articular surface of metatarsal head articulates with concave articular surface of proximal phalangeal base
- Plantar aspect of metatarsal head has rounded contour
- Dorsal aspect of metatarsal head is smaller than plantar aspect
  - Has concave or notched contour along medial and lateral margins
- Sesamoids variably present, most commonly at 5th toe
- **Phalangeal apparatus** is combination of plantar plate and proximal phalanx
- **Plantar plate**
  - Fibrocartilaginous plate thickening extending from metatarsal neck to base of proximal phalanx
  - Attached to capsule, deep transverse metatarsal ligament, plantar fascia and flexor tendon sheath, medial and lateral collateral ligaments
  - Instability may mimic Morton neuroma

ANATOMY IMAGING ISSUES

Imaging Recommendations

- US
  - Good for assessing tendon integrity and joint alignment
    - Tendon integrity recognized by morphology (continuity of tendon and densely packed appearance of tendon fibers) and by behavior during passive movement
    - Joints can also be assessed dynamically during flexion and extension
  - Limited use in assessing for global bony alignment

CLINICAL IMPLICATIONS

Clinical Importance

- Instability of MTP joints results in pain and deformity of forefoot

Stability of 1st Metatarsophalangeal Joint

- Collateral ligaments
- Flexor and extensor hallucis brevis muscles
- Flexor and extensor hallucis longus have smaller contribution to stability

Stability of Lateral Metatarsophalangeal Joints

- Collateral ligaments
- Plantar plate
  - Rupture of plantar plate results in dorsal subluxation of MTP joint and hammer toe deformity

Short 1st Metatarsal (Morton Foot)

- Normal variant but increases stress on 2nd metatarsal
- Predisposes to osteonecrosis of 2nd metatarsal head (Freiberg infraction)

1ST METATARSAL HEAD SECTION AND TOES

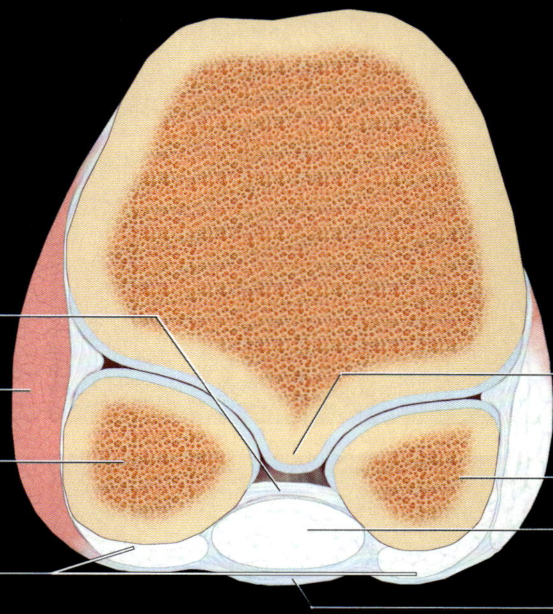

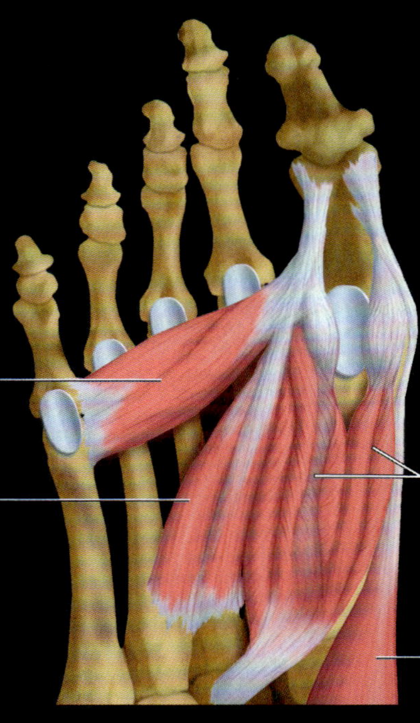

(Top) *Coronal (perpendicular to long axis of metatarsal bone) graphic shows the sesamoid bone located on the plantar surface of the 1st metatarsal head, separated by the median crest of the metatarsal bone. The sesamoids are united by the intersesamoid ligament. The intersesamoid ligament forms the floor of the groove between the sesamoids; the flexor hallucis longus tendon runs along this groove.* **(Bottom)** *Axial graphic shows muscle attachments to sesamoids of the 1st metatarsophalangeal joint. The abductor hallucis and the medial head of the flexor hallucis brevis attach to the medial sesamoid. The lateral head of the flexor hallucis brevis and adductor hallucis attach to the lateral sesamoid.*

DORSUM METATARSAL

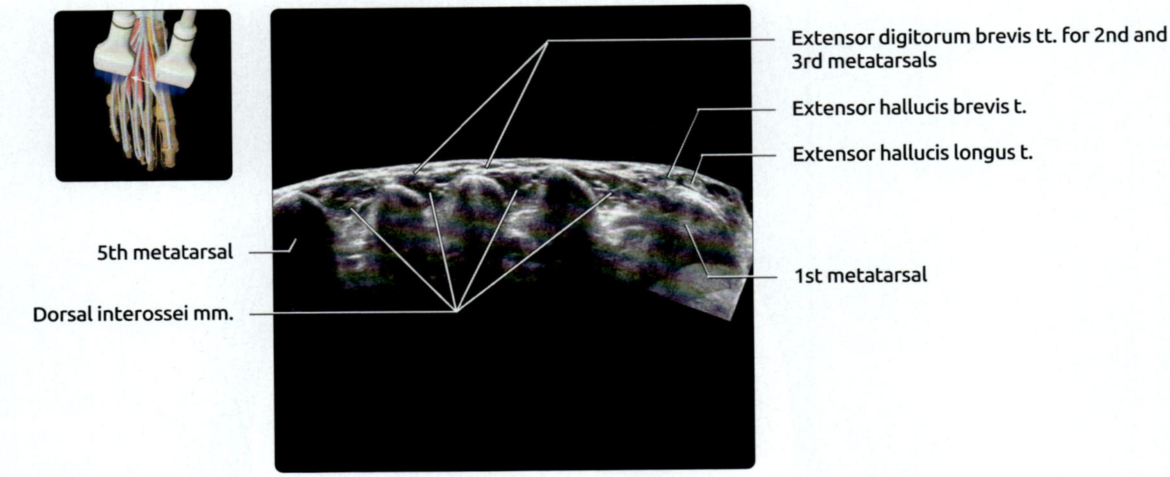

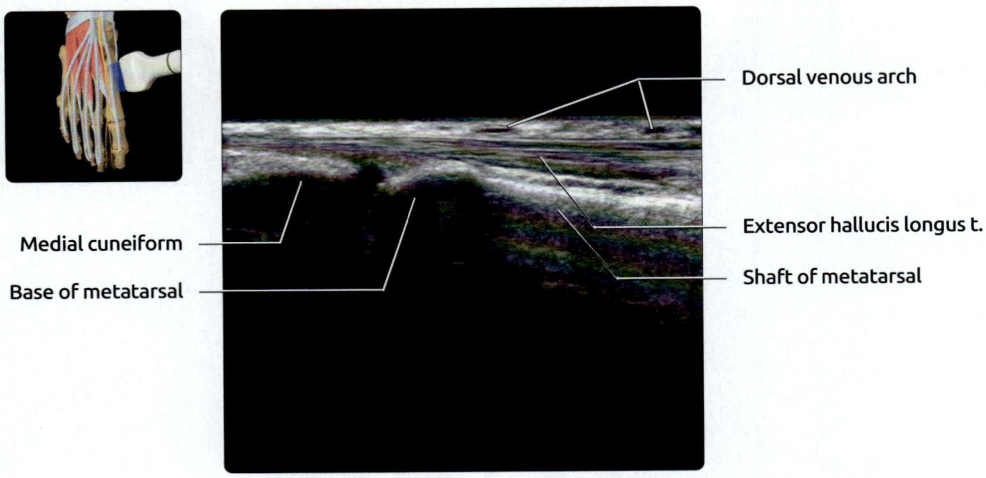

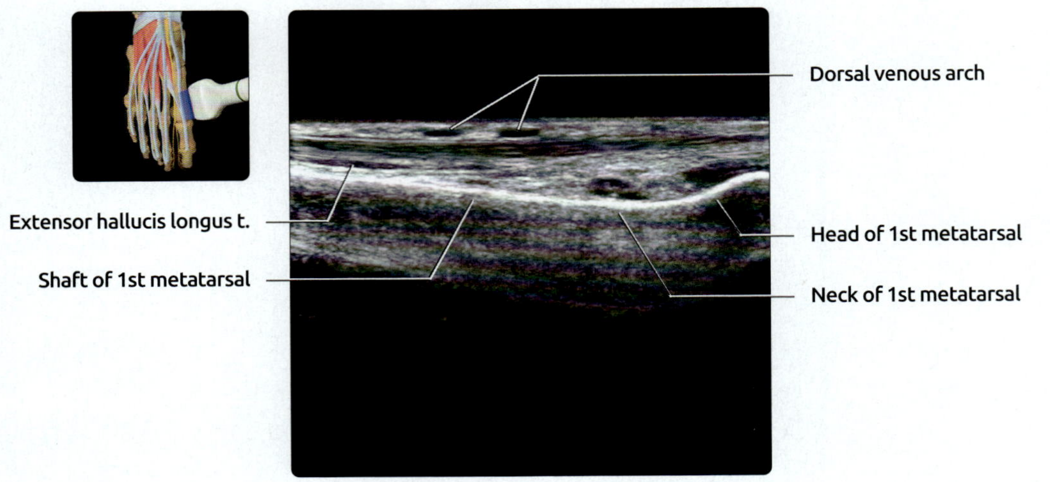

(Top) *Coronal panoramic US shows the dorsal surface of the bodies of the metatarsals. The dorsal interosseous muscles lie between the bodies of the metatarsals. Their action is to abduct the toes. The plantar interosseous muscles act to adduct the toes. The extensor digitorum longus and extensor hallucis longus tendons receive assistance from their brevis tendons, which are deeper and approach from a lateral aspect.* **(Middle)** *Sagittal US shows the dorsum of the big toe at the level of the tarsometatarsal joint. The extensor hallucis longus tendon is in close proximity to the bones around the joint. Note the densely packed and parallel echogenic tendon fibers.* **(Bottom)** *Sagittal US shows the head of the 1st metatarsal. The extensor hallucis longus tendon is separated from the neck and most of the body of the metatarsal by fat.*

PLANTAR METATARSAL

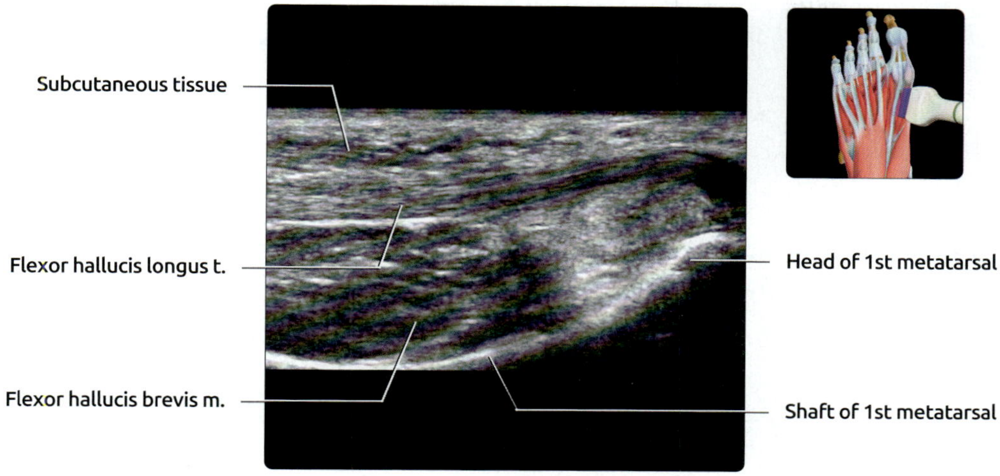

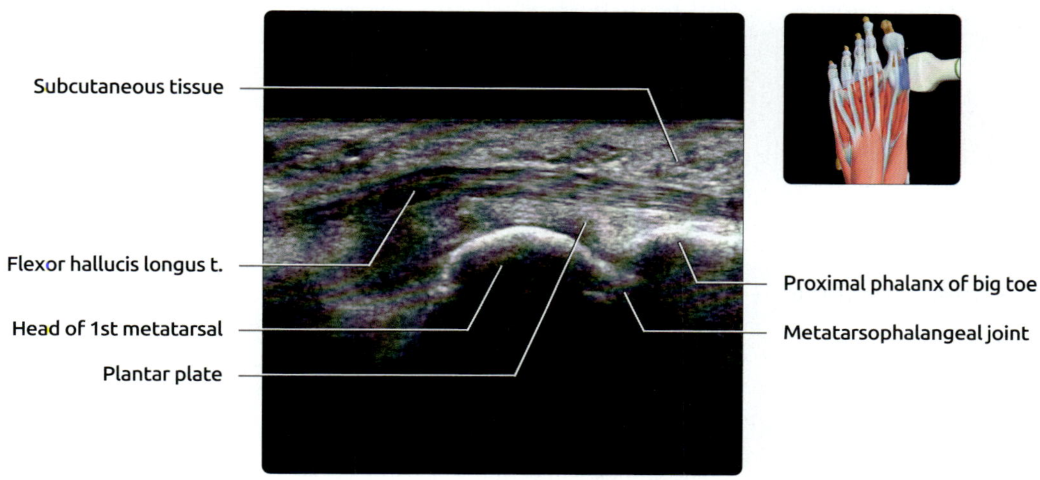

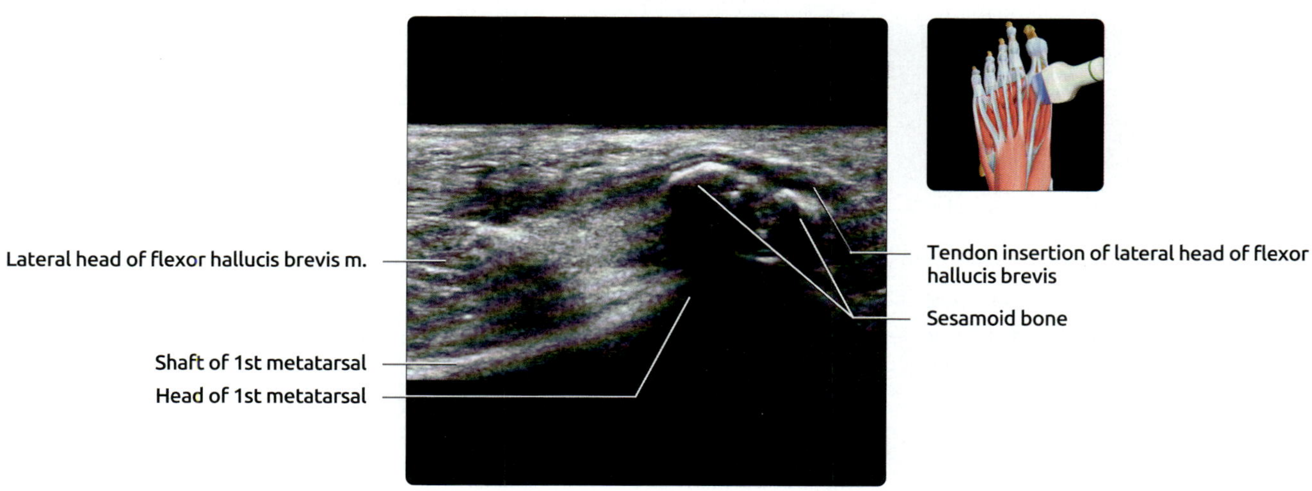

(Top) *Sagittal US shows the plantar aspect of the head of the 1st metatarsal. The flexor hallucis longus muscle originates in the calf, while the brevis muscle arises from the cuboid and lateral cuneiform bones and tendon of the tibialis posterior.* **(Middle)** *Sagittal US shows the center of the 1st metatarsophalangeal joint. The flexor hallucis longus tendon passes over the plantar plate of this joint to reach its insertion at the distal phalanx.* **(Bottom)** *Oblique sagittal US shows the 1st metatarsophalangeal joint just lateral to the center. The flexor hallucis brevis tendons insert onto both sides of the base of the proximal phalanx via 2 tendons. Each tendon contains a sesamoid bone. Sesamoid bones may be bipartite as seen here, especially for the medial sesamoid.*

TRANSVERSE METATARSALS

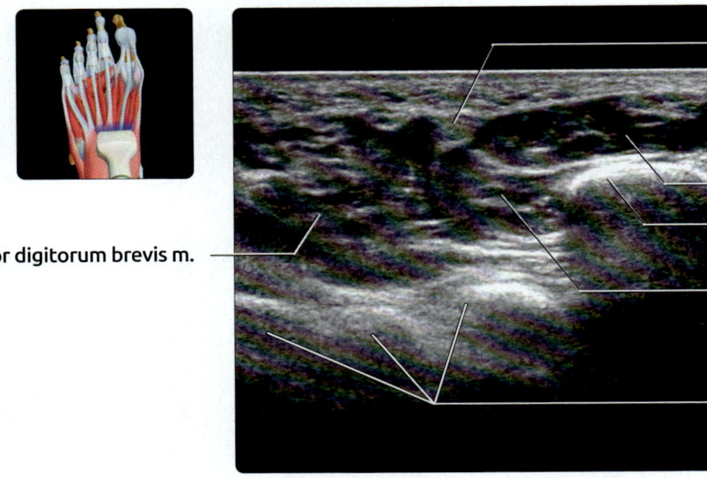

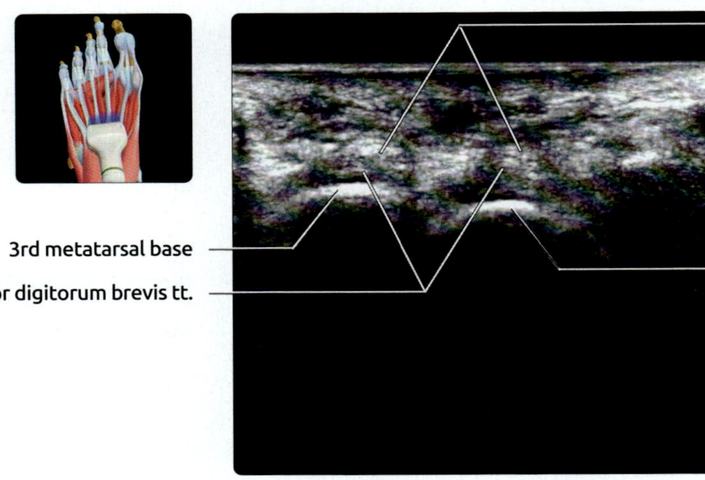

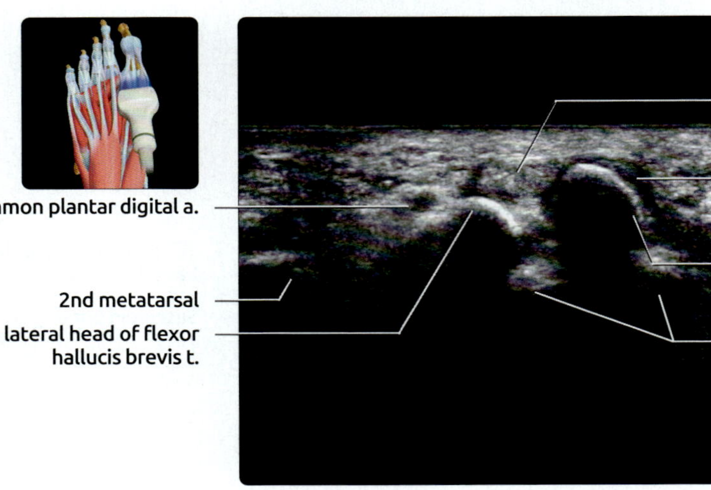

(Top) *Transverse US shows the plantar aspect of the bases of the medial metatarsals. The abductor hallucis muscle is superficial to the 1st and 2nd metatarsals. The base of the 1st metatarsal is uniquely large compared to the rest of the metatarsals and is easily identifiable. The flexor hallucis longus tendon runs superficial to the brevis muscle.* **(Middle)** *Transverse US shows the plantar surface of the base of the 2nd and 3rd metatarsals. The base of the metatarsals form an arch with plantar concavity. The flexor digitorum longus tendons run centrally on the plantar surfaces of the metatarsals.* **(Bottom)** *Transverse US shows the plantar surface of the 1st metatarsal. The flexor hallucis longus runs between the 2 sesamoid bones on the plantar aspect of the 1st metatarsal. The sesamoids adhere to the medial and lateral heads of the flexor hallucis brevis tendons.*

TOE, PLANTAR AND DORSUM

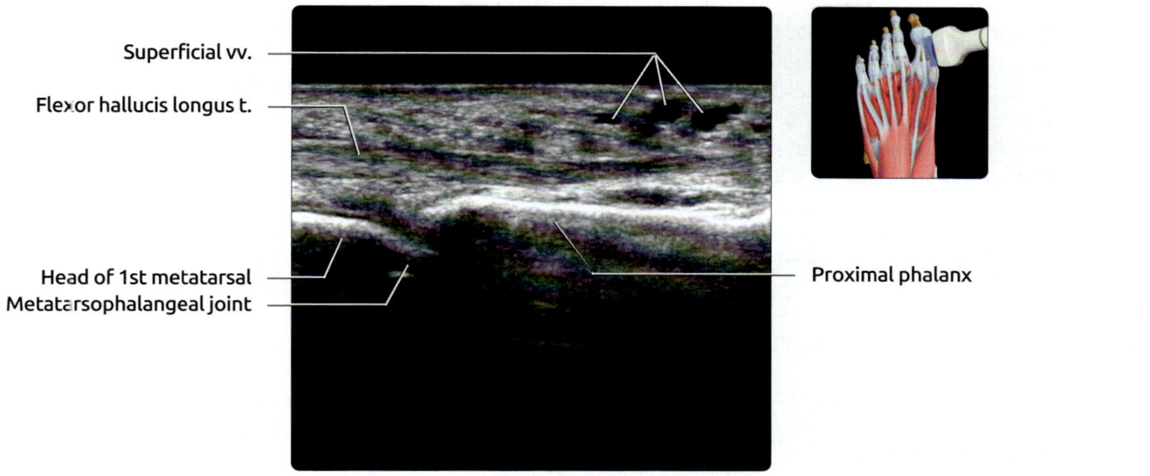

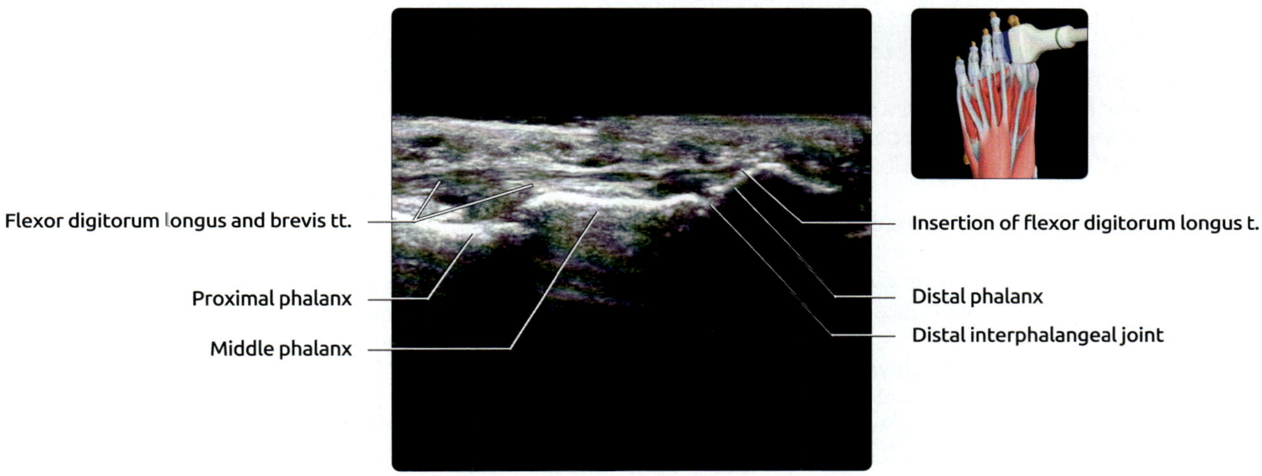

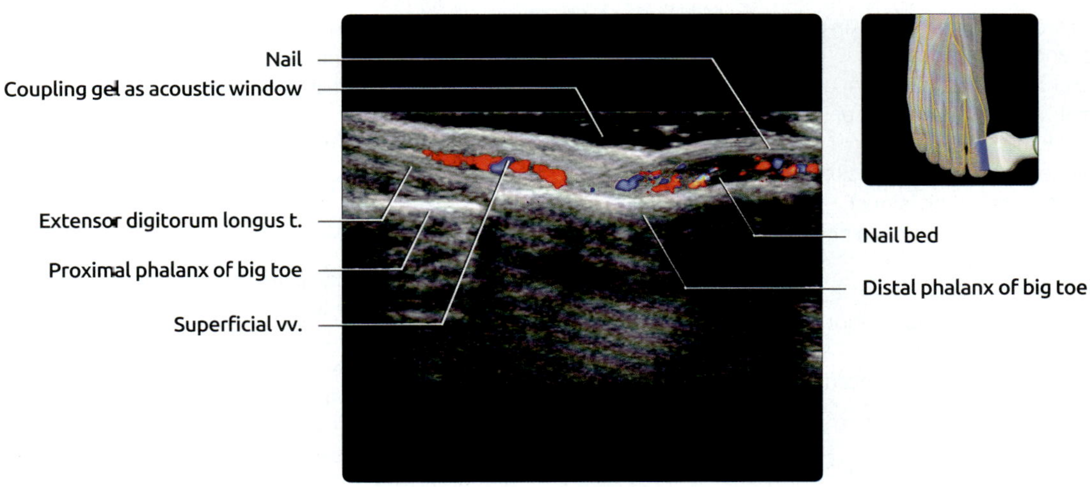

(Top) *Sagittal US shows the plantar surface of the 1st tarsometatarsal joint. The flexor hallucis tendon runs close to the plantar surface of the phalanges and inserts at the base of the distal phalanx.* **(Middle)** *Longitudinal US shows the plantar surface of the 2nd toe. The flexor digitorum longus tendon inserts centrally at the base of the distal phalanx. The flexor digitorum brevis inserts on the sides of the base of the middle phalanx. Passive flexion of only the distal interphalangeal joint allows isolated evaluation of the flexor digitorum longus tendon movement (and integrity).* **(Bottom)** *Sagittal color Doppler US shows the dorsal surface of the distal phalanx of the big toe. The extensor digitorum longus tendon inserts on the dorsal surface of the base of the distal phalanx. The nail bed has a rich vascular supply as demonstrated by color flow.*

GROSS ANATOMY

Overview

- Brachial plexus is neural highway to upper limbs
  - Formed from C5-T1 roots combining to form 3 trunks, 6 divisions, 3 cords, and 5 terminal branches (classic description)
    - 5 terminal branches are ulnar, median, musculocutaneous, axillary, and radial nerves
  - Many variations in this classic anatomy exist
    - Cannot rely on having definite number of trunks, divisions, or cords at any particular location

US Anatomy

- 5 specific areas should be assessed to ensure complete evaluation
  - Extraforaminal, interscalene, supraclavicular, infraclavicular, and axillary areas
- **Extraforaminal area (5 roots)**
  - Single large anterior ramus emerging from spinal cord levels C5-T1
    - Seen between anterior and posterior tubercles of C5 and C6 cervical vertebrae
    - Anterior tubercle of C7 transverse process is small, so root is seen lying anterior to posterior tubercle
    - C8 and T1 roots more difficult to see due to their deep location
- **Interscalene area (5 roots → 3 trunks)**
  - Between scalenus anterior muscle anteromedially and scalenus medius muscle posterolaterally
  - Roots unite to form trunks in most distal interscalene area
    - **Upper (C5-6), middle (C7), lower (C8, T1) trunks**
    - Average cross-sectional area of these trunks is 7 mm^2 (upper), 9 mm^2 (middle), and 9 mm^2 (lower)
    - Upper (C5-6) trunk or part of it may run through or anterior to anterior scalene muscle
    - Dorsal scapular artery may run posteriorly between upper and middle trunks
  - Suprascapular nerve arises from upper trunk
    - Normal cross-sectional area of scapular nerve is 2.8 mm ± 0.8 mm
  - Interscalene triangle common site for nerve block
- **Supraclavicular area (6 divisions)**
  - Trunks divide into anterior and posterior divisions in supraclavicular fossa
    - 3 trunks → 6 divisions
  - Lie posterior and superoposterior to subclavian artery ("bunch of grapes")
  - Anterior divisions innervate anterior (flexor) muscles
  - Posterior divisions innervate posterior (extensor) muscles
  - Transverse cervical artery often loops over divisions of brachial plexus
  - Costoclavicular space is between clavicle and 1st rib
    - Narrowest passageway for brachial plexus
    - Can be compressed with fracture of clavicle or 1st rib
    - Not possible to see costoclavicular space with US
- **Infraclavicular area (3 cords)**
  - Passageway between supraclavicular fossa and apex of axilla
  - Also called retropectoralis minor space or subcoracoid space
  - 6 divisions → 3 cords, which position themselves lateral, medial, and posterior to axillary artery
    - Lateral and medial cords innervate flexor muscles
    - Posterior cord innervates extensor muscles
  - Cords tend to be more hyperechoic than hypoechoic trunks or divisions
    - Brachial plexus becomes more fascicular and contains more connective tissue as it goes distally
      - Roots are monofascicular and hypoechoic, while terminal branches are polyfascicular and hyperechoic
  - Pectoralis major overlies cords medially, pectoralis minor laterally, while serratus anterior covers chest wall medially
- **Axillary area (5 terminal branches)**
  - Pyramidal space with apex superomedially
  - **Lateral cord** → median nerve and musculocutaneous nerve (C5-6)
  - **Medial cord** → median nerve and ulnar nerve (C8-T1)
  - **Posterior cord** → axillary nerve (C5-6), radial nerve (C5-T1), thoracodorsal nerve (C6-8), as well as upper (C6-7) and lower (C5-6) subscapular nerves
  - Both lateral and medial cords contribute to median nerve
  - **Nerves are arranged around axillary artery** with ulnar nerve at 12-3 o'clock, radial nerve at 4-6 o'clock, and median nerve at 9-12 o'clock
    - Medial nerve lies closest to muscles of proximal arm (short head of biceps and coracobrachialis)
    - This neural arrangement is quite constant, seen in about 80% of patients
  - In most cases, musculocutaneous nerve branches early from lateral cord
    - Continues distally with median nerve in 30% of patients
    - Musculocutaneous nerve runs medially between coracobrachialis and biceps muscles before piercing coracobrachialis muscle

ANATOMY IMAGING ISSUES

Imaging Approaches

- US is straightforward and efficient way of assessing brachial plexus
  - Excellent first-line investigation for many pathologies
- Becoming familiar with anatomy will allow you to focus on pathology
- Routinely evaluate extraforaminal (roots), interscalene (trunks), supraclavicular (divisions), infraclavicular (cords), and axillary (terminal branches) areas
- Realize that considerable anatomical variation exists in neural and vascular anatomy of brachial plexus
- Remember that brachial plexus is not isolated group of nerves
  - Many inflammatory neural pathologies (inflammatory polyneuropathy, neuralgic amyotrophy) also affect nerves well beyond brachial plexus

BRACHIAL PLEXUS

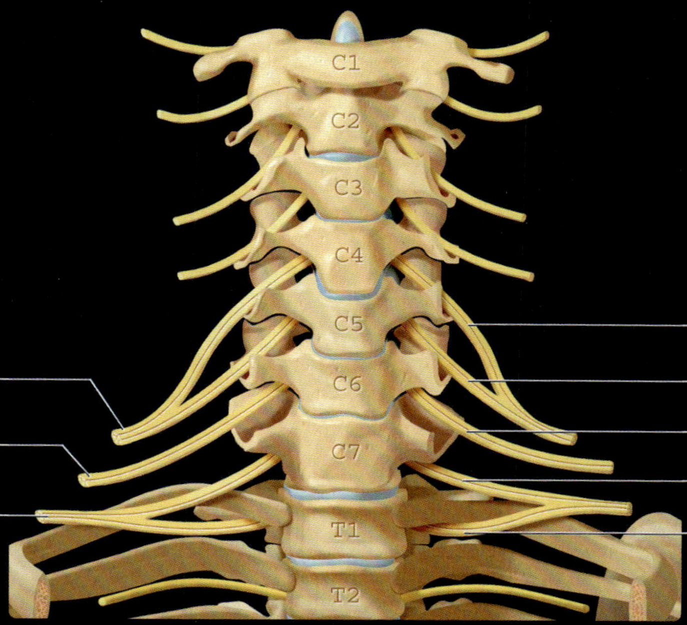

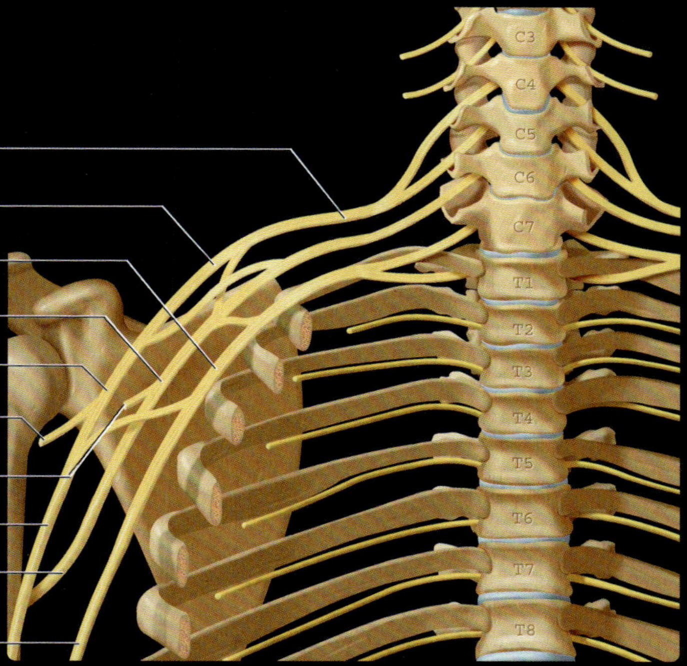

(Top) *Coronal graphic of the cervical spine and supraclavicular brachial plexus demonstrates the cervical ventral primary rami combining to form the brachial plexus. The C1-C7 roots exit above the same numbered pedicle, C8 exits above the T1 pedicle, and more caudal roots exit below their numbered pedicle. The trunks are visualized on US mainly in the interscalene area.* **(Bottom)** *Coronal graphic of the brachial plexus demonstrates the more distal plexus elements extending into the axilla. The trunks recombine into posterior and anterior divisions that form the cords. The posterior cord forms the radial and axillary nerves. The medial cord forms the ulnar nerve, while the lateral cord forms the musculocutaneous nerve. The median nerve is formed from branches of both the lateral and medial cords.*

BRACHIAL PLEXUS AND PERIPHERAL NERVES

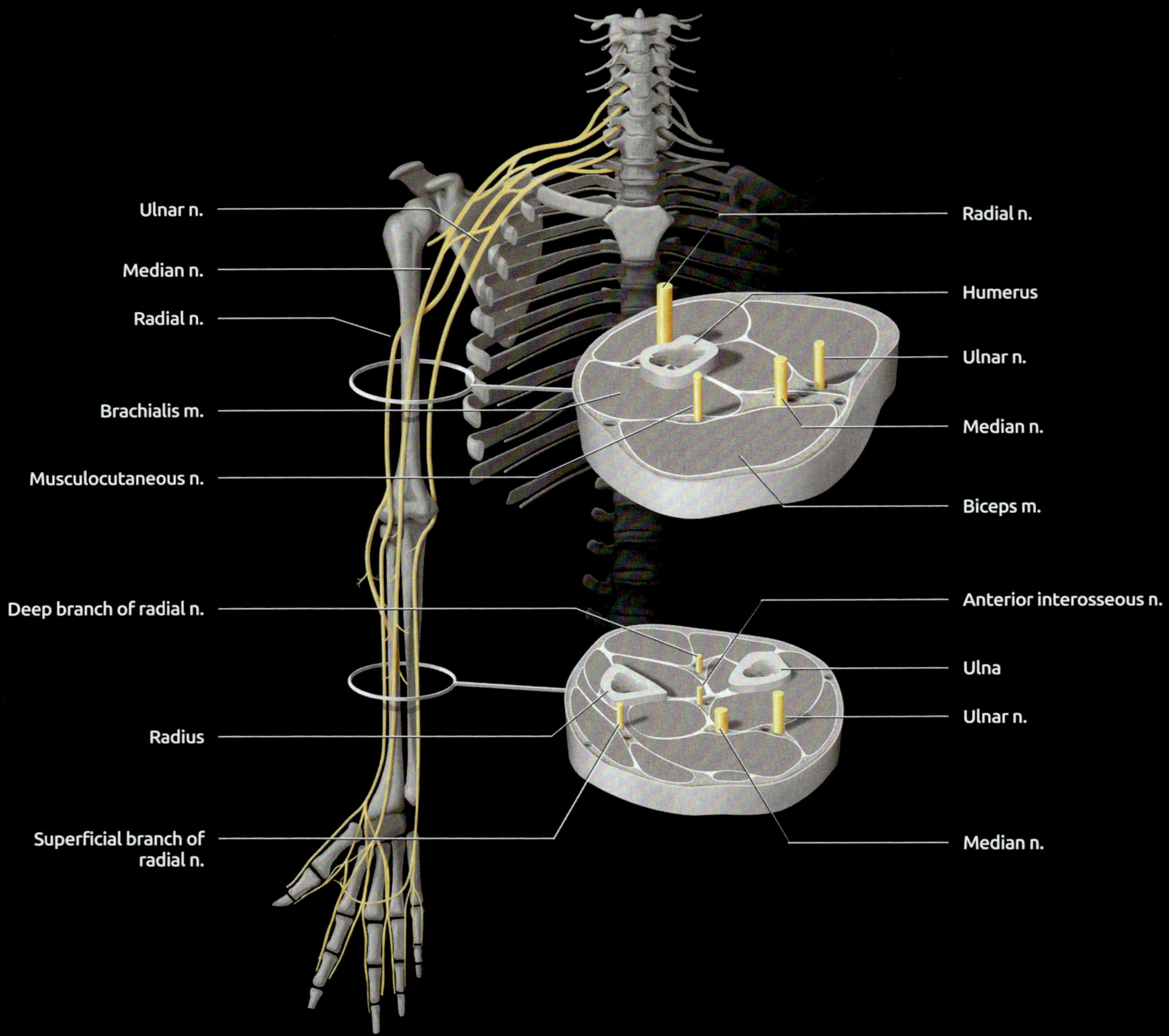

This graphic shows 3 of the 5 terminal branches of the brachial plexus (median, radial, and ulnar nerves) as they course down the arm. The other 2 terminal branches are the musculocutaneous and axillary nerves. The axillary nerve innervates the teres minor and deltoid muscles, while the musculocutaneous nerve passes through the coracobrachialis muscle and descends between the biceps and brachialis muscles.

TRANSVERSE US

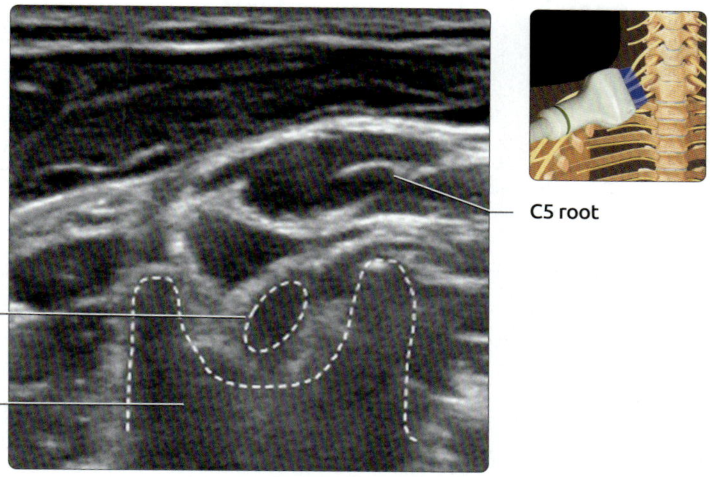

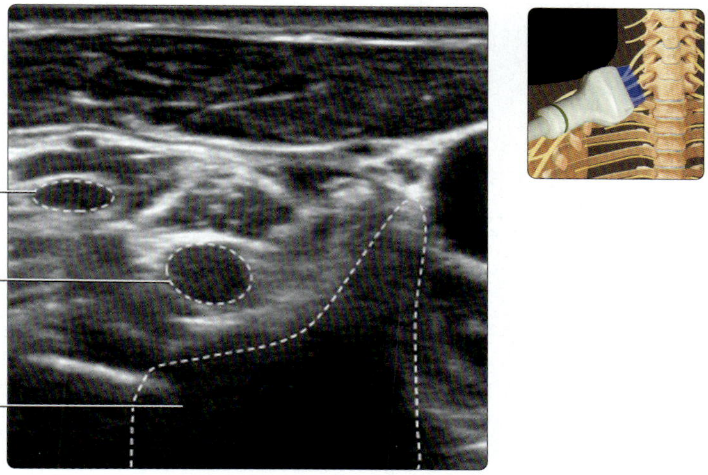

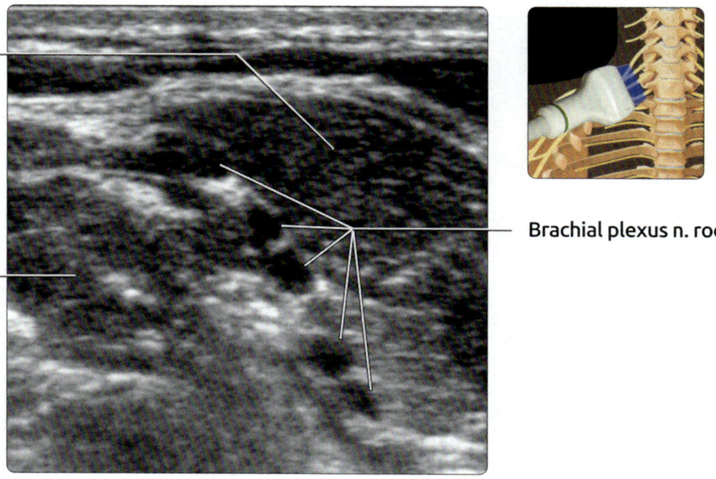

(Top) *Transverse US at the level of C6 shows the C6 nerve root (outlined by dotted lines) between the anterior and posterior tubercles of the C6 transverse process. The C6 nerve root exits from the spinal canal at the C5/6 exit foramen cephalad to this level. The C5 nerve root is located more laterally.* **(Middle)** *Transverse US at the level of C7 shows the C7 nerve root (outlined by dotted line) in front of the anterior tubercles of the C7 transverse process (outlined by dotted line). The C6 nerve root (outlined by dotted line) is located more laterally.* **(Bottom)** *Transverse US at the entry into the interscalene area shows 5 hypoechoic roots of the brachial plexus between the scalenus anterior and medius muscles. The hypoechoic roots are clearly seen against the adjacent hyperechoic intermuscular fat.*

TRANSVERSE AND LONGITUDINAL US

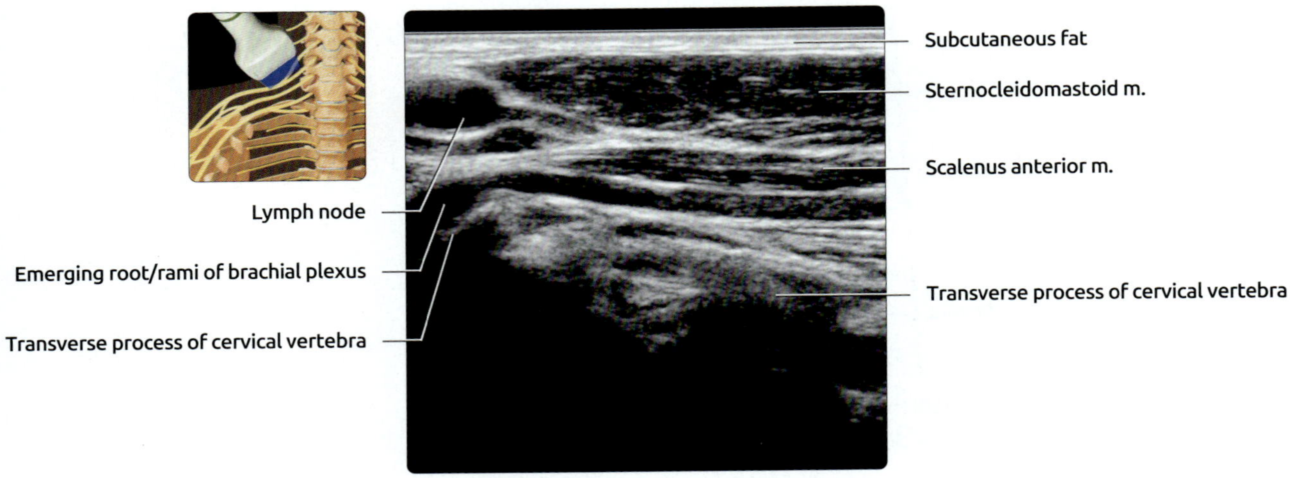

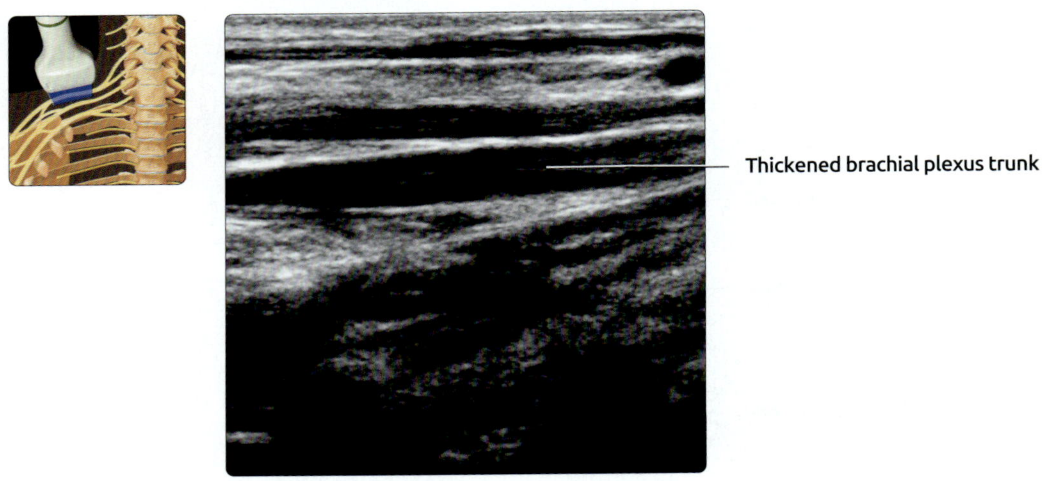

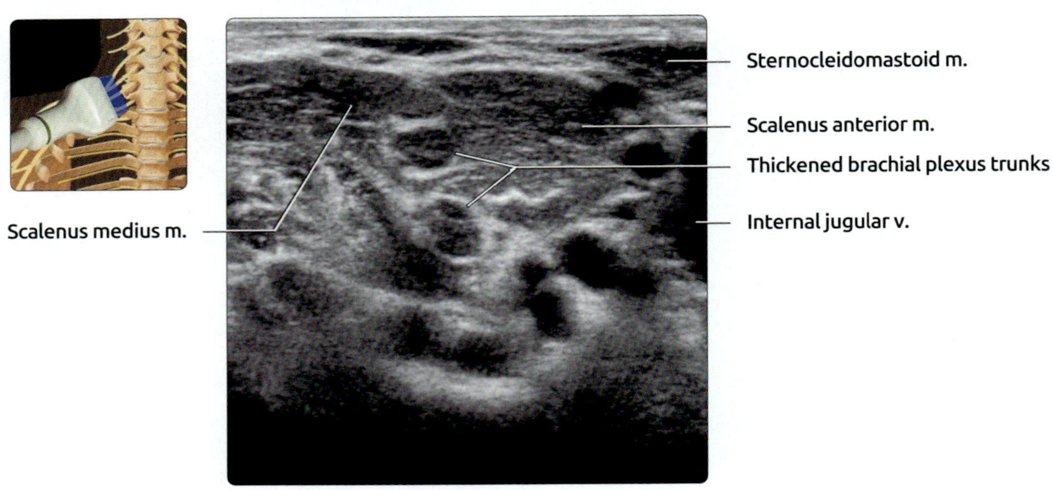

(Top) *Longitudinal US of the posterior triangle of the neck shows the root and trunk of the brachial plexus, which appears as a thin, tubular, hypoechoic structure related superficially to the scalenus anterior muscle and deeply to the cervical vertebrae.* **(Middle)** *Longitudinal US of the right posterior triangle/supraclavicular fossa shows the elongated linear, hypoechoic trunks of the brachial plexus. The patient had a past history of neck irradiation for metastatic neck nodes. The thickening middle trunk may be due to previous radiotherapy.* **(Bottom)** *Transverse US of the right lower posterior triangle/supraclavicular fossa shows round smooth hypoechoic trunks between the scalenus anterior and medius muscles. Mild, relative thickening of the middle and lower trunks may be due to previous radiotherapy.*

TRANSVERSE AND LONGITUDINAL US

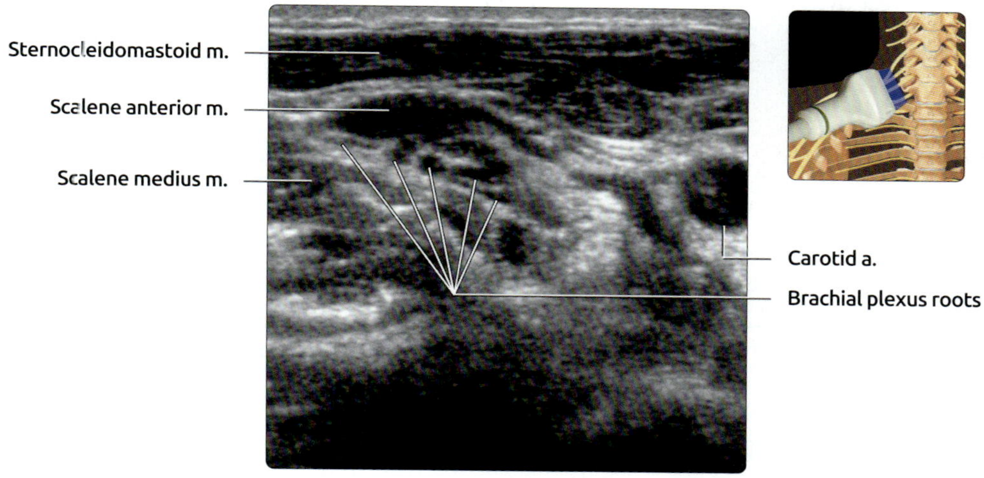

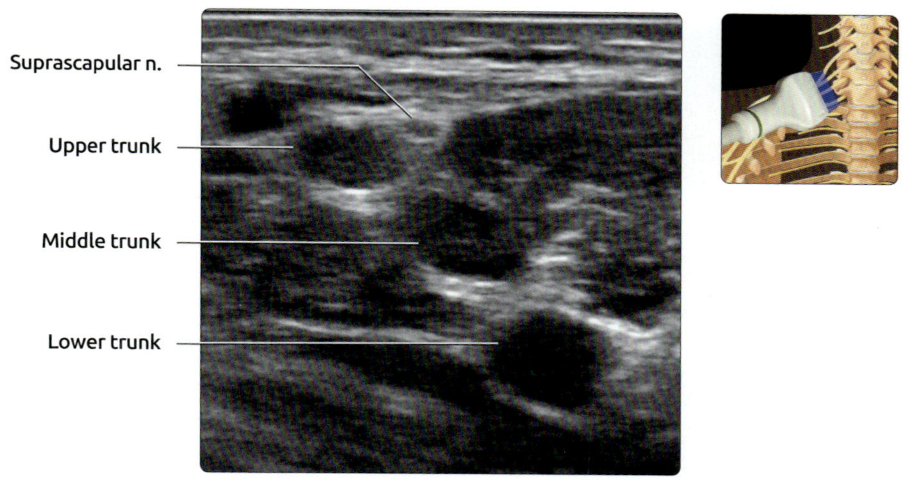

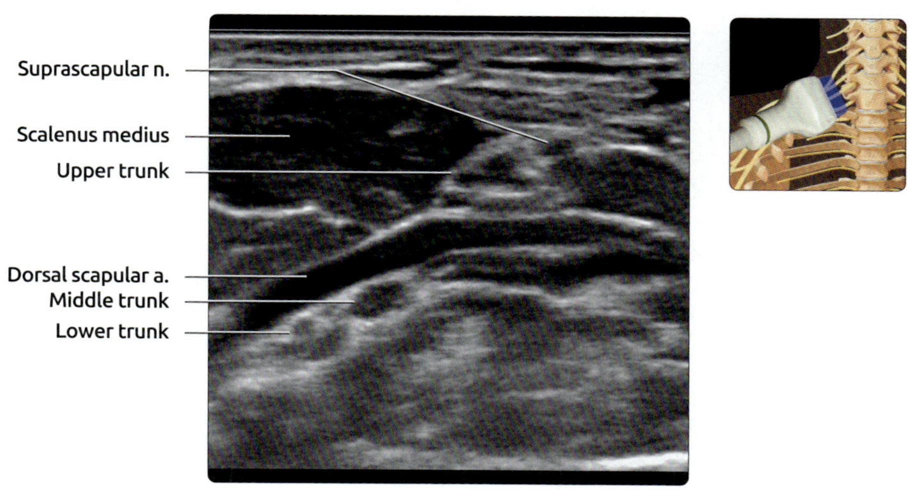

(Top) *Transverse US of the anterolateral lower neck shows the hypoechoic roots of the brachial plexus as they exit the neural foramina and move to the interscalene area.* **(Middle)** *Transverse US at the interscalene level shows the upper, middle, and lower trunks of the brachial plexus between the anterior and medius scalene muscles. Note the diffusely hypoechoic appearance of the monofascicular trunks. The suprascapular nerve, which arises from the upper trunk, is also shown. There is considerable variability in the size of the trunks in different patients.* **(Bottom)** *Transverse US at the outlet of the interscalene triangle is shown. The dorsal scapular artery is running between the upper and middle trunks. This is a common finding. The small suprascapular nerve is also seen.*

MR AND US

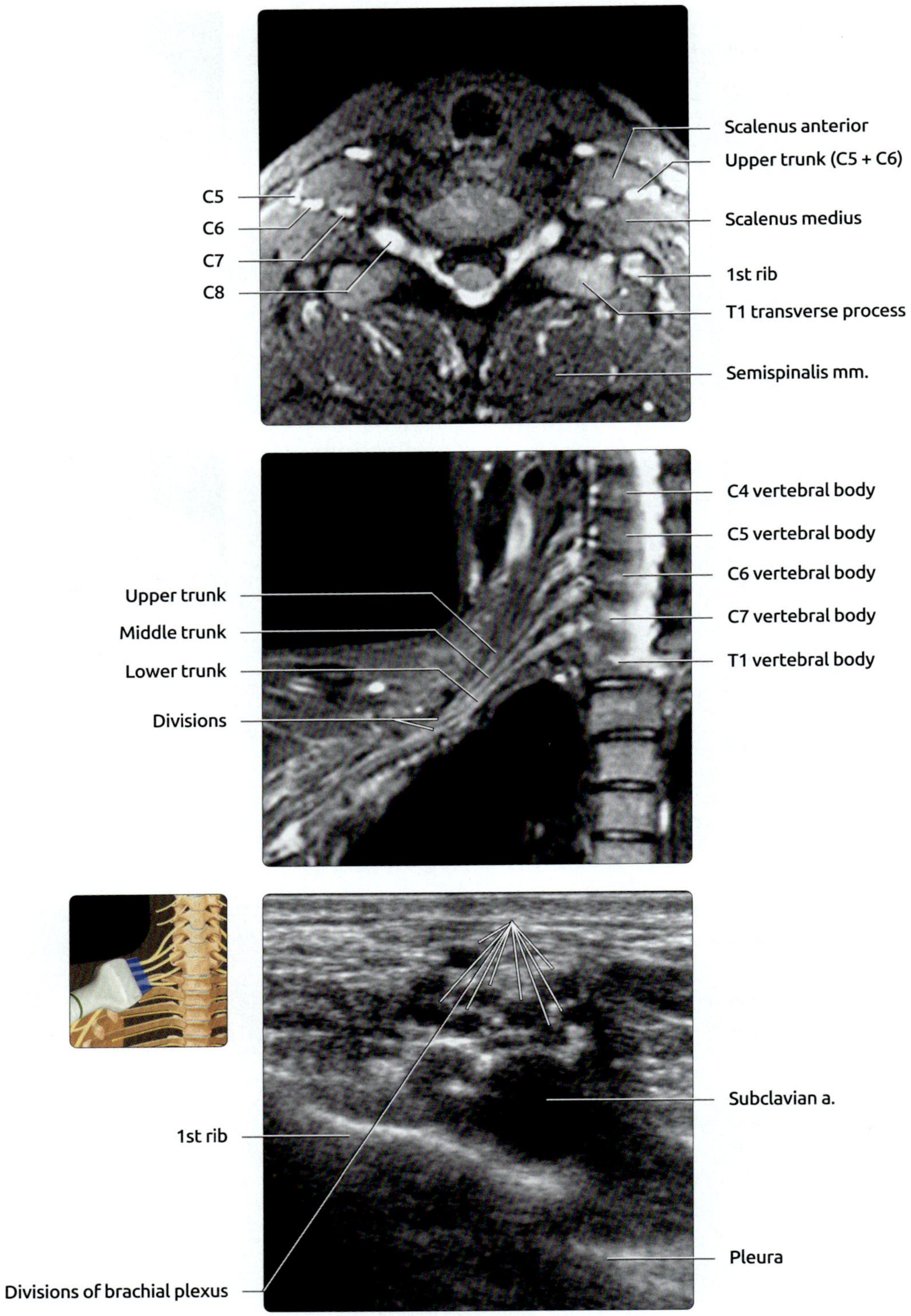

(Top) *Axial STIR MR at the C7/T1 level depicts the linear alignment of the C5 through C8 ventral primary rami. C5 and C6 are closely approximated and form the left upper trunk.* **(Middle)** *Reformatted oblique coronal T2WI MR of the brachial plexus shows that the trunks and divisions of the brachial plexus continue from the cervical nerve roots to the axilla.* **(Bottom)** *Transverse US at the supraclavicular area shows hypoechoic divisions of the brachial plexus located dorsocranial to the subclavian artery, akin to a "cluster of grapes." In a small percentage of patients, some of the divisions may lie caudal to the subclavian artery.*

TRANSVERSE US

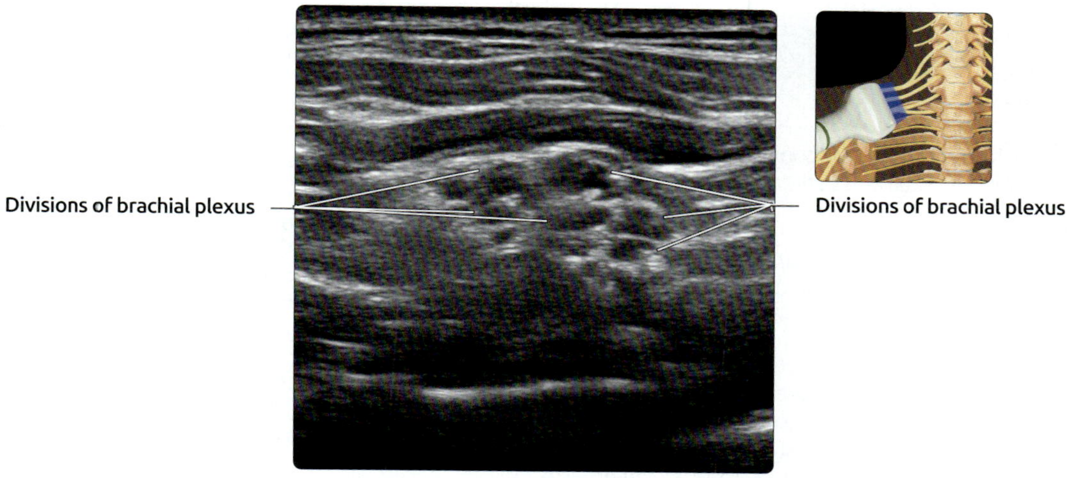

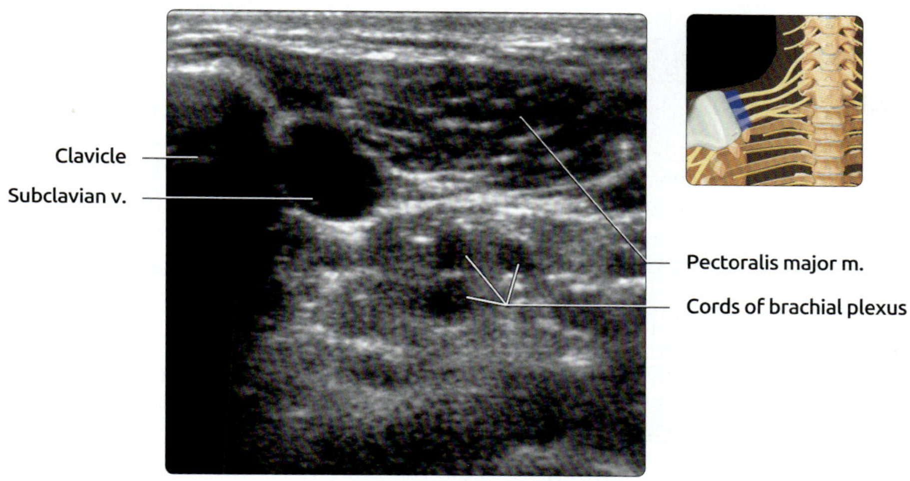

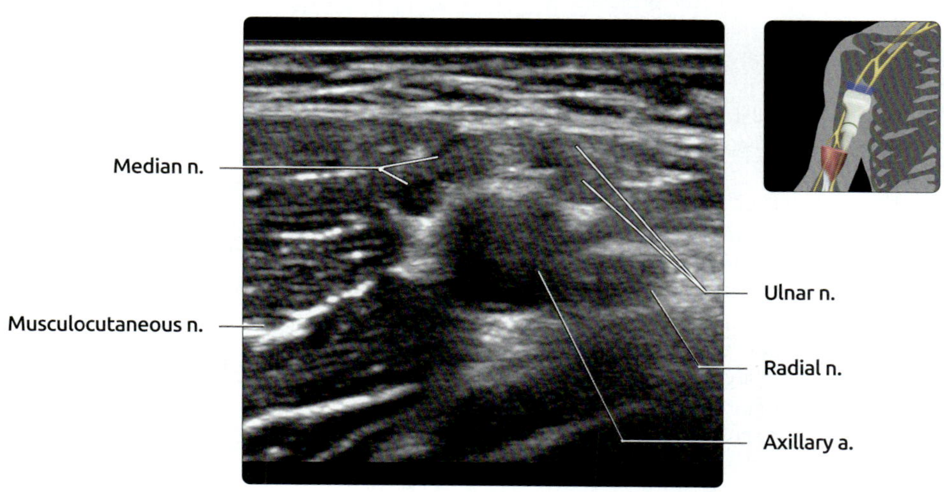

(Top) *Transverse US of the supraclavicular portion of the brachial plexus beyond the scalene muscles shows the divisions of the brachial plexus. The upper, middle, and lower trunks divide into 6 divisions. Note the normal hyperechoic fat between the divisions. Following radiotherapy, this perineural fat can become more hypoechoic.* **(Middle)** *Transverse US at the infraclavicular area shows hypoechoic cords of the brachial plexus located dorsocaudal to the subclavian artery. More medially, the cords are covered by the pectoralis major muscle, while more laterally, they are covered also by the pectoralis minor muscle.* **(Bottom)** *Transverse US of the axilla with the arm abducted shows the median, ulnar, and radial nerves located around the axillary artery. The axillary veins are compressed. The musculocutaneous nerve lies between the biceps muscle anteriorly and the coracobrachialis muscle posteriorly.*

GENERAL ANATOMY AND FUNCTION

Chest Wall Anatomy

- Skin, subcutaneous fat
- Blood vessels, lymphatics, nerves
- Bone, cartilage
- Muscles
- Endothoracic fascia: Fibroelastic connective tissue between inner aspect of chest wall and parietal pleura

Function

- Musculoskeletal cage: Surrounds cardiorespiratory system; affects respiration by expanding and contracting during ventilation

Surface Landmarks

- Suprasternal (jugular) notch: At superior manubrium of sternum; between sternal ends of clavicles
- Sternal angle: Landmark for internal thoracic anatomy; anterior projection at level of costal cartilage of 2nd rib
- Costal margin: Inferior margins of lowest ribs and costal cartilages

SKELETAL STRUCTURES

Sternum

- Flat, broad bone that forms anterior thoracic wall; 3 parts
  - Manubrium forms superior part of sternum
  - Body articulates with manubrium superiorly, xiphoid process inferiorly, bilateral costal cartilages of 2nd-7th ribs
  - Xiphoid process variable size, shape, ossification; articulates with body of sternum superiorly

Ribs

- 12 pairs, symmetrically arrayed; numbered in accordance with attached vertebral body
- True ribs (1-7) attach to sternum by costal cartilages (synovial joints)
- False ribs (8-10) articulate by costal cartilages with costal cartilage of 7th rib
- Floating ribs (11-12) do not articulate with sternum or rib costal cartilages; short costal cartilages terminate in abdominal wall muscle
- Head articulates with hemifacets of 2 adjacent vertebral bodies; neck is located between head and tubercle of each rib; tubercle articulates with vertebral transverse process
- Body: Longest part of each rib
- Angle: Most posterior part
- Ribs can be broadly categorized as midclavicular, anterior axillary, midaxillary, posterior axillary, and dorsal sections
- Costal groove on inner surface of inferior border, which accommodates intercostal neurovascular bundle

MUSCLES

Pectoral

- Pectoralis major: Largest muscle in pectoral region; extends from anterior chest wall, sternum, and clavicle to lateral lip of bicipital groove
  - Adducts, flexes, and medially rotates arm
  - Prone to hypoplasia or aplasia: Sternocostal > clavicle portion > whole muscle
- Pectoralis minor: Deep to pectoralis major; extends from chest wall to coracoid process of scapula
  - Stabilizes scapula
- Sternalis: Anatomical variant (in 5% of cadavers), runs parallel to sternum and connects sternocleidomastoid to rectus abdominis

Intercostal

- External: Contained within 11 intercostal spaces; extends from tubercle of ribs to costochondral junction
- Internal: Middle layer; occupies 11 intercostal spaces; extends from border of sternum to angle of ribs
- Innermost: Forms inner layer of chest wall muscles with subcostales and transversus thoracis muscles

Serratus Anterior

- Thin muscular sheet; overlies lateral thoracic cage and intercostal muscles; arises from upper 8 ribs; wraps around rib cage; inserts along medial anterior surface of scapula

VESSELS AND NERVES

Arteries

- Internal thoracic (internal mammary): Branch of subclavian artery; descends posterior to first 6 costal cartilages; supplies upper anterior chest wall
  - Supplies anterior intercostal arteries to first 6 intercostal spaces

Veins

- Azygos vein receives drainage from posterior intercostal veins, hemiazygos, and accessory hemiazygos veins

Nerves

- Anterior rami of thoracic spinal nerves (T1-T11) form intercostal nerves
- Intercostal nerves run in costal groove between internal and innermost intercostal muscles
- T1-T2 supply upper limb, T3-T6 supply chest wall, T7-T11 supply chest and abdominal wall, T12 supplies abdomen and groin

IMAGING

Radiography

- Limited capability for chest wall abnormalities; may detect congenital deformities and other bony abnormality
  - Poor sensitivity to rib fracture detection
  - Even localized rib views will reveal < 20% of rib fractures

US

- Allows detailed examination of intercostal spaces, ribs, and costal cartilages
- Intercostal muscles are seen as hypoechoic structures
- Intercostal membrane is seen as hyperechoic layer
  - Intercostal structures are often thin, making it difficult to separate them into distinct layers
- Only anterior cortex of ribs evaluated
- Costal cartilage and costochondral junction clearly seen

Computed Tomography

- Multidetector CT with multiplanar reformations provides excellent visualization of chest wall lesions

GRAPHICS, RIBS AND INTERCOSTAL SPACES

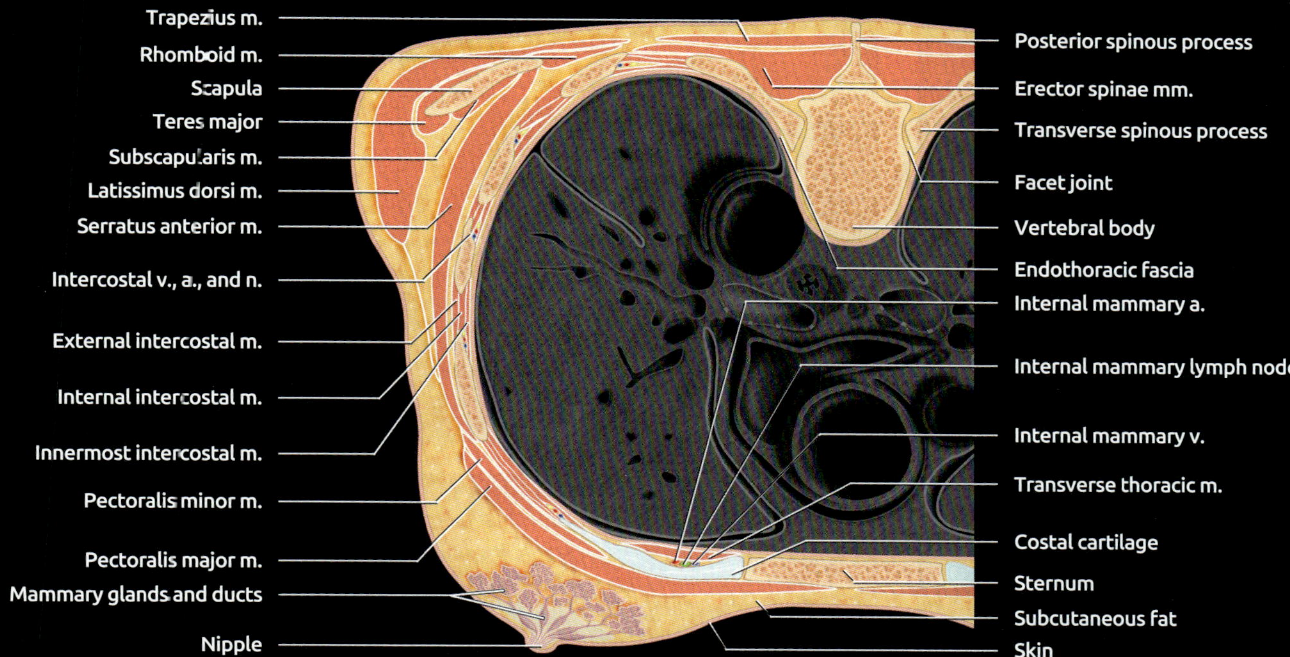

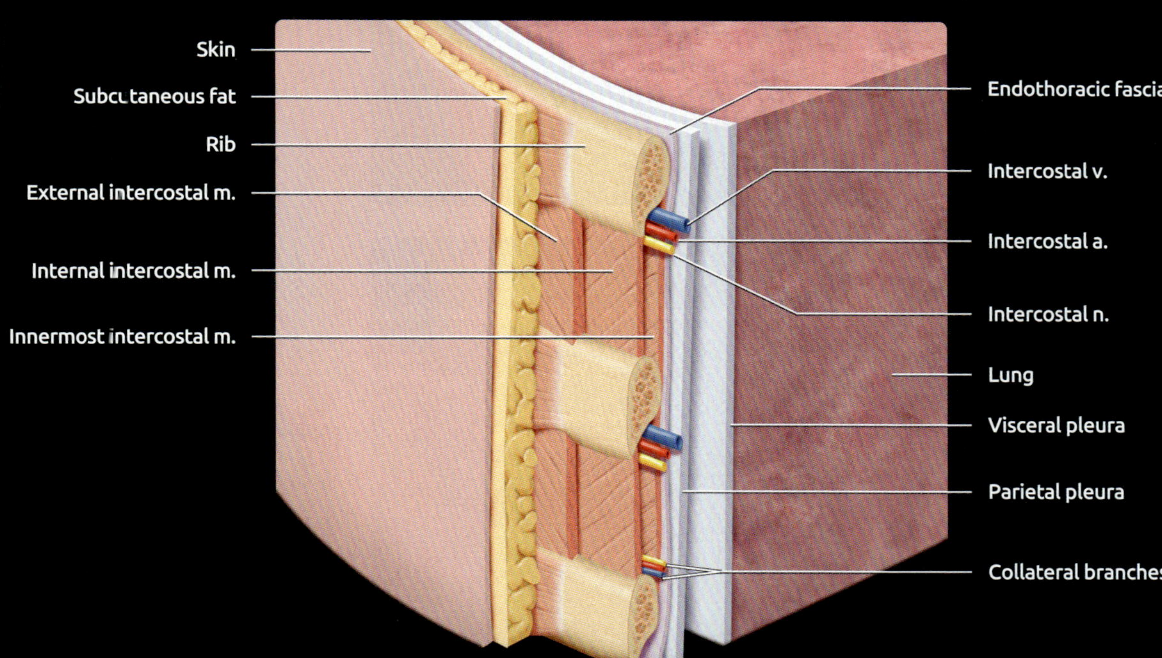

(Top) *Graphic depicts the chest wall layers as visualized in the axial plane, including skin, subcutaneous fat, blood vessels, lymphatics, and musculoskeletal structures. The innermost layer, the endothoracic fascia, is a fibroelastic connective tissue layer between the inner aspect of the chest wall and the pleura.* **(Bottom)** *Graphic demonstrates details of the intercostal region, showing 3 layers of intercostal muscles (external, internal, and innermost) between the ribs. The costal groove along the inferomedial aspect of each rib accommodates the intercostal neurovascular bundle (vein, artery, and nerve). Small collateral branches of the major intercostal vessels and nerves may be present above the body of the subjacent rib. The endothoracic fascia forms a connective tissue layer between the inner aspect of the chest wall and the costal parietal pleura.*

RIBS AND INTERCOSTAL SPACES

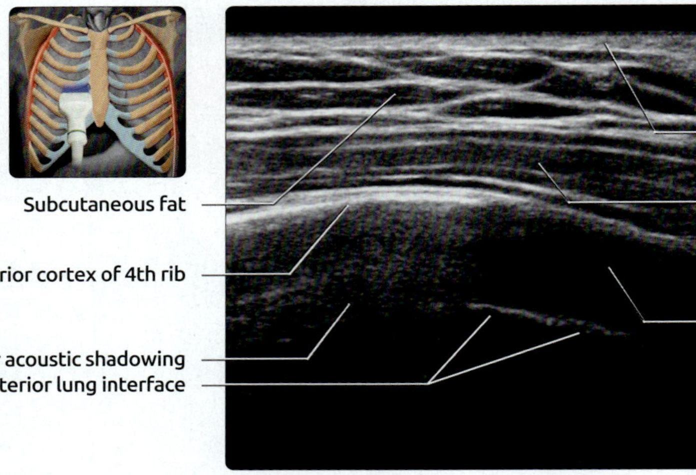

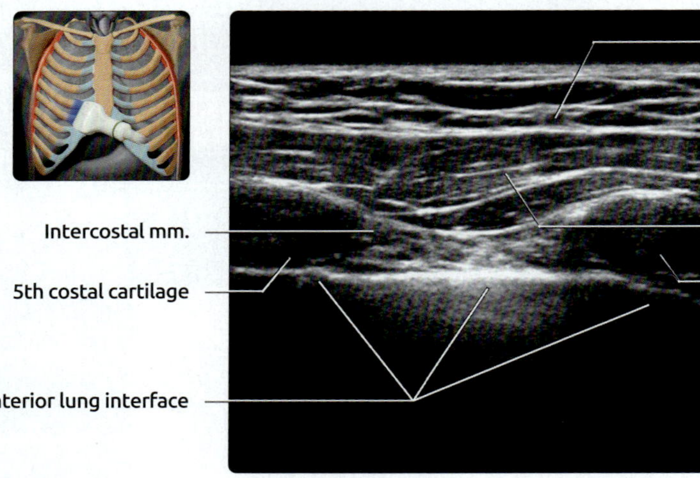

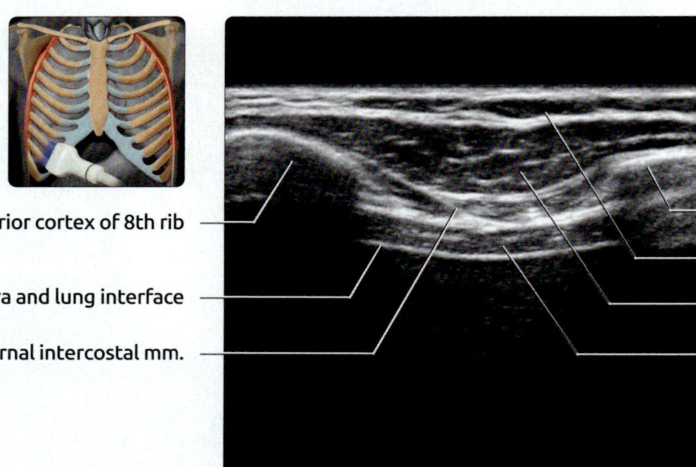

(Top) *Transverse view shows the costochondral junction of the 4th rib. The costal cartilage appears homogeneously hypoechoic and does not produce significant posterior acoustic shadowing, thus allowing the underlying pleural surface to be seen. The cortex of the 4th rib casts a strong posterior shadow, obscuring the underlying pleura.* **(Middle)** *Oblique sagittal view shows intercostal space between the costal cartilages in the anterior chest wall. The costal cartilages are well defined and oval-shaped on cross section. Intercostal muscles between the costal cartilages and overlying pectoralis muscle can be clearly visualized.* **(Bottom)** *More laterally, oblique sagittal view shows intercostal spaces between the ribs. Areas behind the ribs are obscured by acoustic shadowing. The intercostal muscles laterally are better differentiated compared to those located more anteriorly.*

RIBS AND INTERCOSTAL SPACES

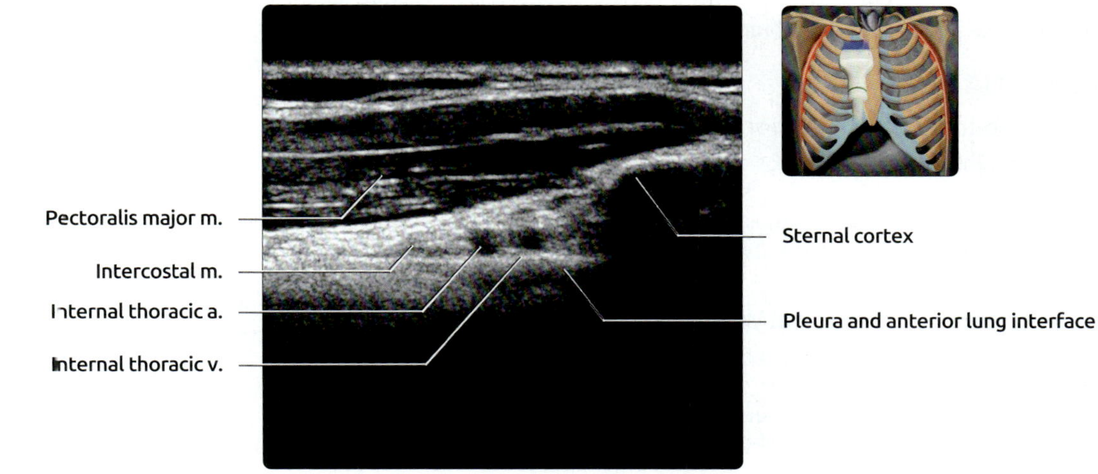

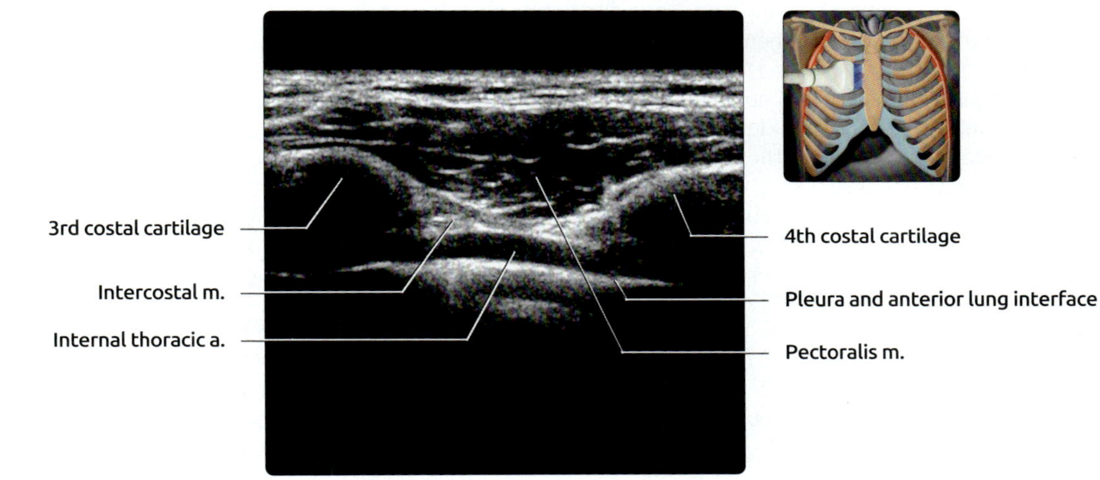

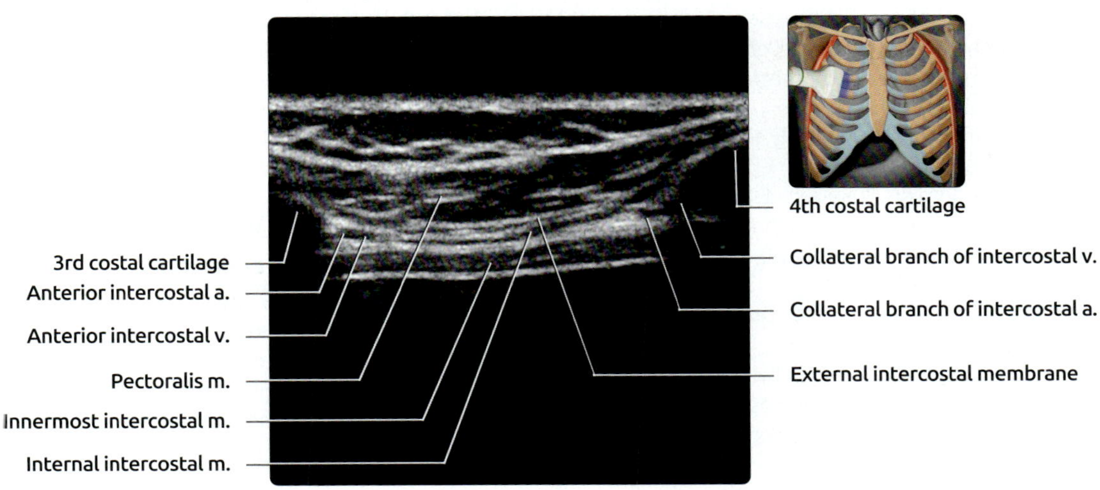

(Top) *Transverse view shows the anterior intercostal space just lateral to the sternum. The internal thoracic artery and vein next to the sternum are well visualized with this approach, as are the overlying pectoralis muscle and underlying pleura.* **(Middle)** *Sagittal view of the anterior intercostal space shows the internal thoracic artery. The long axis of the internal thoracic artery can be well delineated by scanning along the lateral margin of the sternum.* **(Bottom)** *Sagittal view shows anterior intercostal space and the anterior intercostal vessels. The anterior intercostal artery and anterior intercostal vein are identified just inferior to the rib above. Smaller collateral branches of intercostal artery and vein can be identified just superior to the rib above. At the anterior intercostal space, the external intercostal muscle is seen as a thin membrane.*

TERMINOLOGY

Definitions

- Abdomen: Region between diaphragm and pelvis

GROSS ANATOMY

Anatomic Boundaries of Anterior Abdominal Wall

- Superiorly: Xiphoid process and costal cartilages of 7th-10th ribs
- Inferiorly: Iliac crest, iliac spine, inguinal ligament, and pubis
- Inguinal ligament is inferior edge of aponeurosis of external oblique muscle

Muscles of Anterior Abdominal Wall

- Consist of 3 flat muscles (external oblique, internal oblique, and transverse abdominal) and 1 strap-like muscle (rectus)
- Combination of muscles and aponeuroses (sheet-like tendons) act as corset to confine and protect abdominal viscera
- Linea alba is fibrous raphe stretching from xiphoid to pubis
  - Forms central anterior attachment for abdominal wall muscles
  - Formed by interlacing fibers of aponeuroses of oblique and transverse abdominal muscles
  - Rectus sheath also formed by these aponeuroses as they surround rectus muscle
- Linea semilunaris is vertical fibrous band at lateral edge of rectus sheath bilaterally
  - Aponeuroses of internal and transversus abdominis join in linea semilunaris before forming rectus sheath
- External oblique muscle
  - Largest and most superficial of 3 flat abdominal muscles
  - Origin: External surfaces of ribs 5-12
  - Insertion: Linea alba, iliac crest, pubis via broad aponeurosis
- Internal oblique muscle
  - Middle of 3 flat abdominal muscles
  - Runs at right angles to external oblique
  - Origin: Posterior layer of thoracolumbar fascia, iliac crest, and inguinal ligament
  - Insertion: Ribs 10-12 posteriorly, linea alba via broad aponeurosis, pubis
- Transversus abdominis (transversalis) muscle
  - Innermost of 3 flat abdominal muscles
  - Origin: Lowest 6 costal cartilages, thoracolumbar fascia, iliac crest, inguinal ligament
  - Insertion: Linea alba via broad aponeurosis, pubis
- Rectus abdominis muscle
  - Origin: Pubic symphysis and pubic crest
  - Insertion: Xiphoid process and costal cartilages 5-7
  - Rectus sheath: Strong, fibrous compartment that envelops each rectus muscle
    - Contains superior and inferior epigastric vessels
- Actions of anterior abdominal wall muscles
  - Support and protect abdominal viscera
  - Help flex and twist trunk, maintain posture
  - Increase intraabdominal pressure for defecation, micturition, and childbirth
  - Stabilize pelvis during walking, sitting up
- Transversalis fascia
  - Lies deep to abdominal wall muscles and lines entire abdominal wall
  - Separated from parietal peritoneum by layer of extraperitoneal fat

Muscles of Posterior Abdominal Wall

- Consist of psoas (major and minor), iliacus, and quadratus lumborum
- Psoas: Long, thick, fusiform muscle lying lateral to vertebral column
  - Origin: Transverse processes and bodies of vertebrae T12-L5
  - Insertion: Lesser trochanter of femur (passing behind inguinal ligament)
  - Action: Flexes thigh at hip joint; bends vertebral column laterally
- Iliacus: Large, triangular sheet of muscle lying along lateral side of psoas
  - Origin: Superior part of iliac fossa
  - Insertion: Lesser trochanter of femur (after joining with psoas tendon)
  - Action: Iliopsoas muscle flexes thigh
- Quadratus lumborum: Thick sheet of muscle lying adjacent to transverse processes of lumbar vertebrae
  - Invested by lumbodorsal fascia
  - Origin: Iliac crest and transverse processes of lumbar vertebrae
  - Insertion: 12th rib
  - Actions: Stabilizes position of thorax and pelvis during respiration, walking; bends trunk to side

Paraspinal Muscles

- Also called erector spinae muscles
  - Invested by lumbodorsal fascia
- Composed of 3 columns
  - Iliocostalis: Lateral
  - Longissimus: Intermediate
  - Spinalis: Medial
- Origins: Sacrum, ilium, and spines of lumbar and 11th-12th thoracic vertebrae
- Insertions: Ribs and vertebrae with additional muscle slips joining columns at successively higher levels
- Action: Extends vertebral column

ANATOMY IMAGING ISSUES

Imaging Recommendations

- High-frequency (5-12 MHz) linear transducer for anterior abdominal wall and paraspinal muscles
- 3-5 MHz for posterior abdominal wall muscles
- Compare with contralateral side to check for symmetry
- Routine image during Valsalva maneuver if suspected hernia
  - Particularly if symptoms are aggravated by weight bearing, also examine in standing position if supine examination is negative for hernia
- Pay particular attention to deep margin of any mass
  - Critical to ensure that any mass is not emerging from abdominal cavity
- Perform color Doppler imaging during breathhold
  - Even movement from quiet respiration can artifactually simulate color flow

ANTERIOR ABDOMINAL WALL

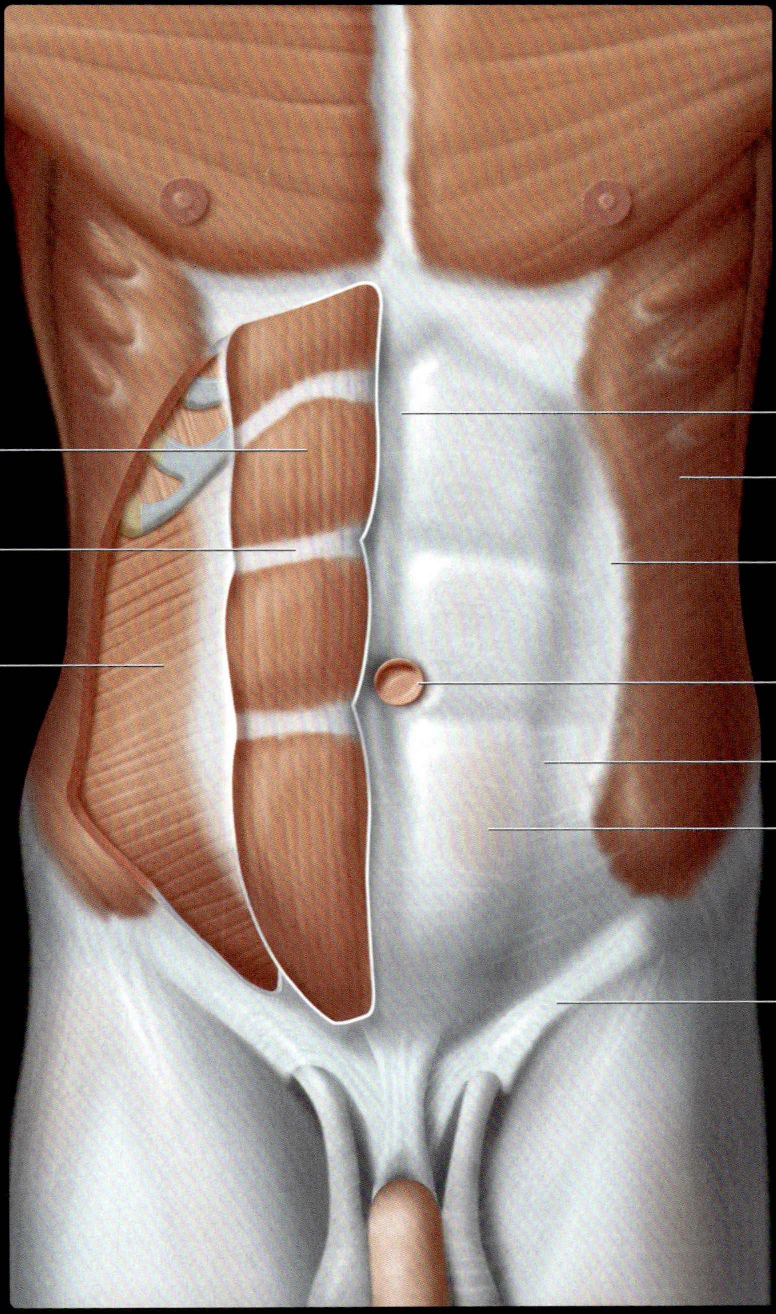

Graphic shows the aponeuroses of the internal and external oblique and transverse abdominal muscles are 2 layered and interweave with each other, covering the rectus muscle, constituting the rectus sheath and linea alba. About midway between the umbilicus and symphysis, at the arcuate line, the posterior rectus sheath ends (arcuate line), and the transversalis fascia is the only structure between the rectus muscle and parietal peritoneum.

POSTERIOR ABDOMINAL WALL

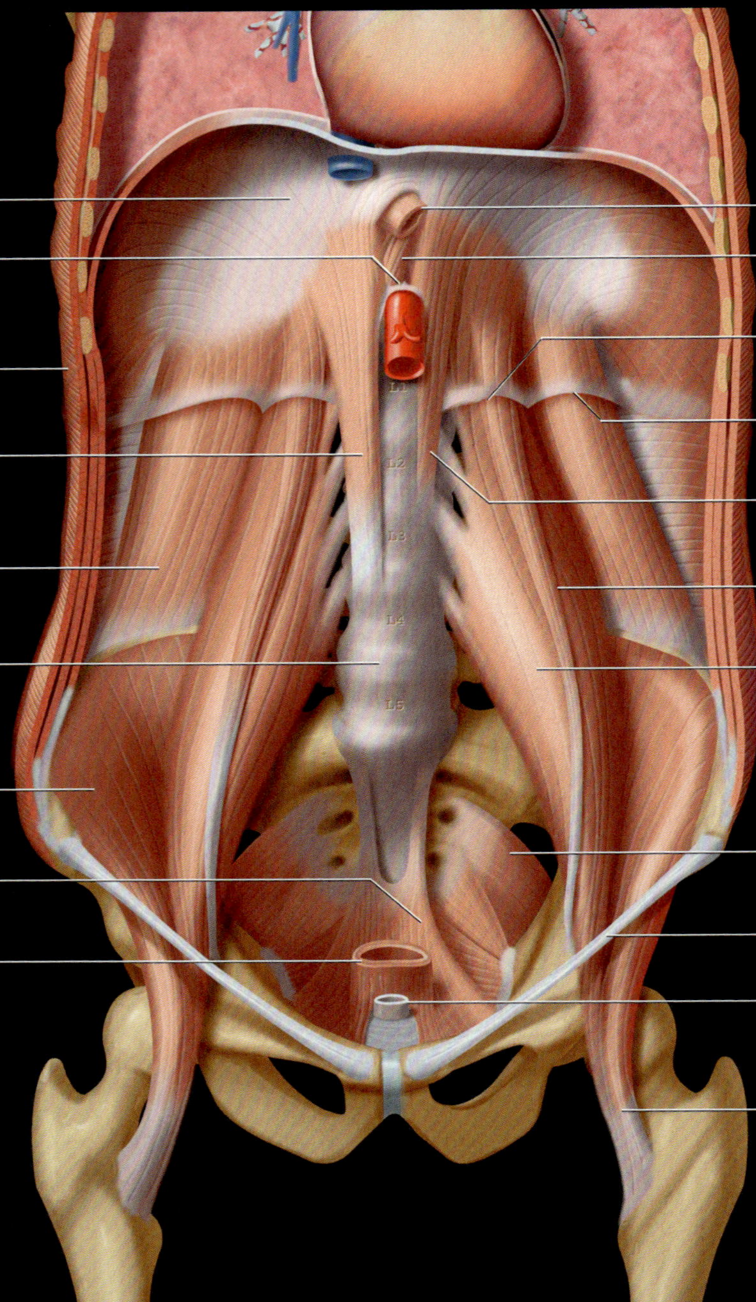

Graphic shows the lumbar vertebrae are covered and attached by the anterior longitudinal ligament, and the diaphragmatic crura are closely attached to it, as are the origins of the psoas muscles, which also arise from the transverse processes. The iliacus muscle arises from the iliac fossa of the pelvis and inserts into the tendon of the psoas major, constituting the iliopsoas muscle, which inserts onto the lesser trochanter. The quadratus lumborum arises from the iliac crest and inserts onto the 12th rib and transverse processes of the lumbar vertebrae. The diaphragmatic and transverse abdominal fibers interlace. The psoas and quadratus lumborum pass behind the diaphragm under medial and lateral arcuate ligaments.

MUSCLES OF BACK IN SITU

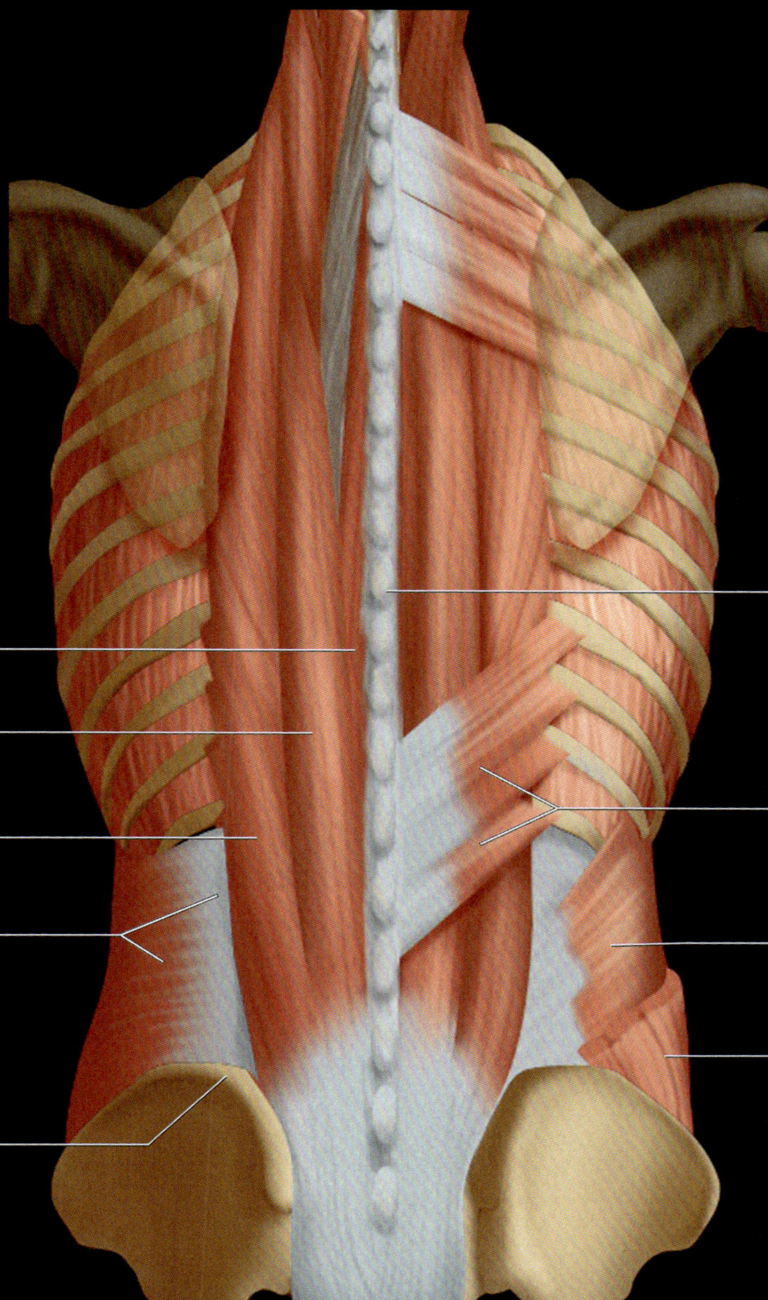

Graphic shows the paraspinal muscles and muscles of the back. The latissimus dorsi muscles are not included. The erector spinae have thick, tendinous origins from the sacral, iliac crests, the lumbar, and 11th-12th thoracic spinous processes. Superiorly, the muscle becomes fleshy and in the upper lumbar region subdivides to become the iliocostalis, longissimus, and spinalis muscles (from lateral to medial), tapering as they insert into the vertebrae and ribs. The erector muscles flank the spinous processes and span the length of the posterior thorax and abdomen. They are responsible for extension of the vertebral column.

ANTERIOR ABDOMINAL WALL

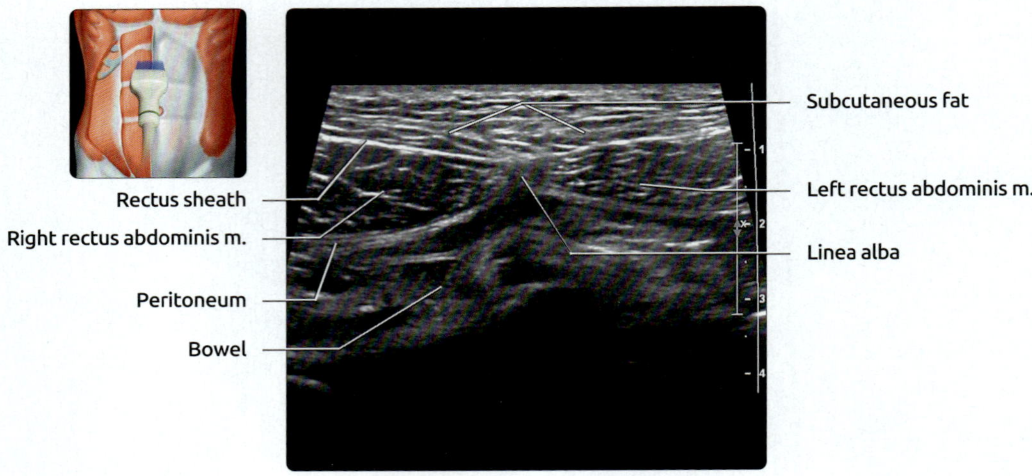

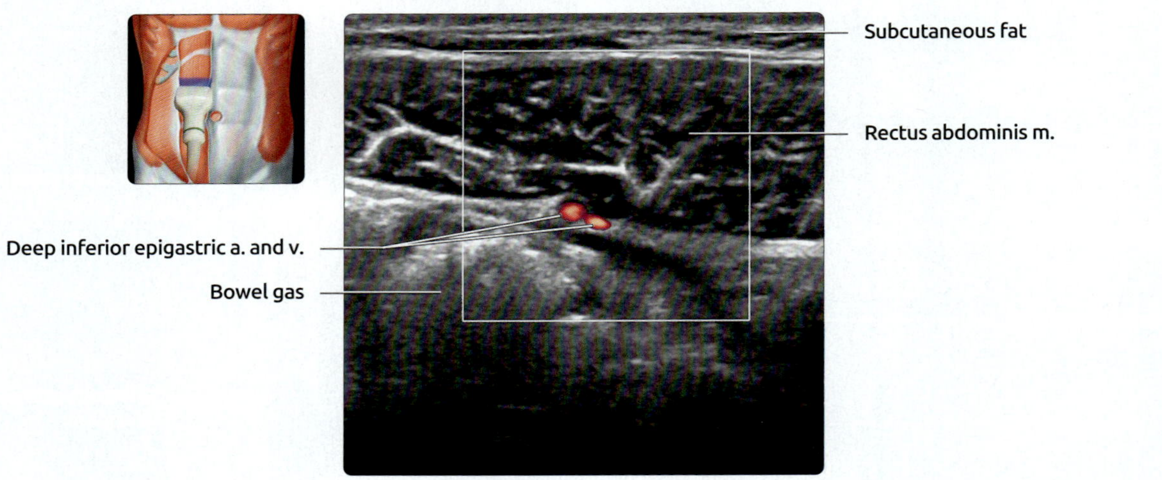

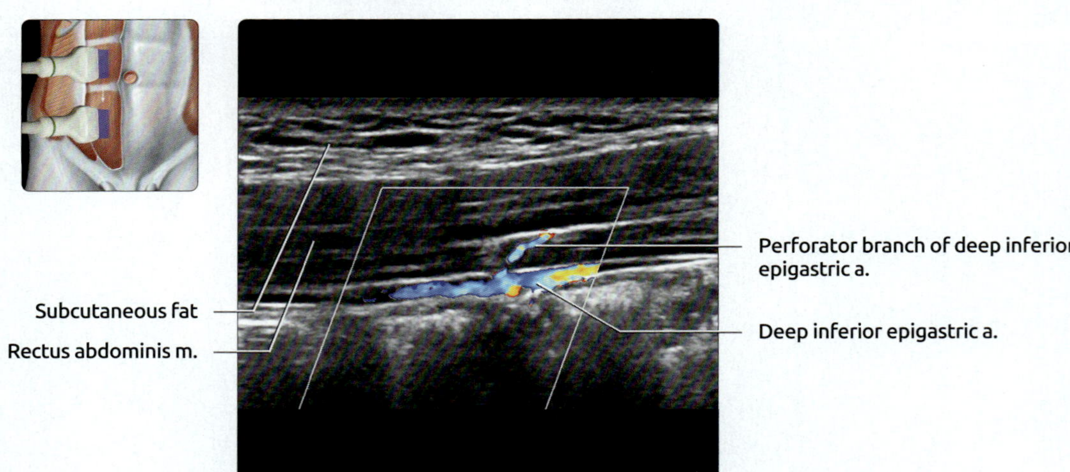

(Top) *Transverse grayscale US of the midline anterior abdominal wall shows the paired rectus abdominis muscles separated by the linea alba. The rectus abdominis muscles are comparable in echogenicity and thickness. The surrounding rectus sheath is seen as a fine, thin, echogenic structure around the muscles.* **(Middle)** *Transverse power Doppler US of a rectus abdominis muscle in the lower abdomen shows the deep inferior epigastric artery and vein. Branches of the superior epigastric artery that arise in the upper abdomen anastomose with branches of the inferior epigastric artery at the umbilicus.* **(Bottom)** *Longitudinal color Doppler US shows a perforating branch of the deep inferior epigastric artery extending into the rectus muscle. These perforators are important for breast reconstruction with abdominal wall flaps.*

ANTEROLATERAL ABDOMINAL WALL

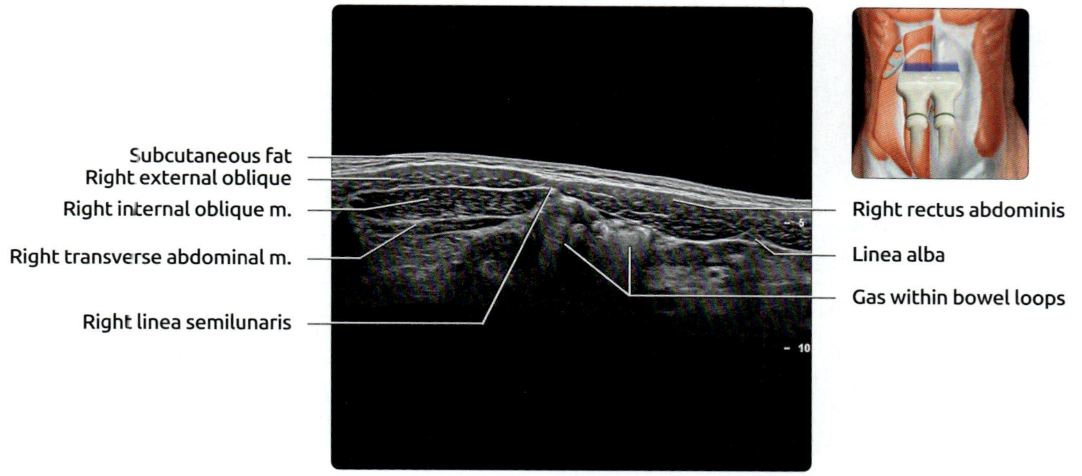

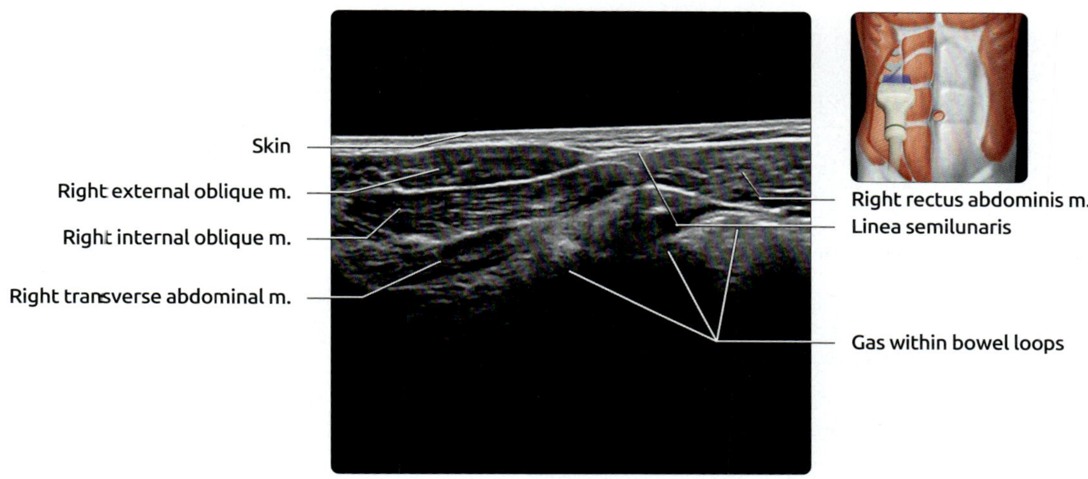

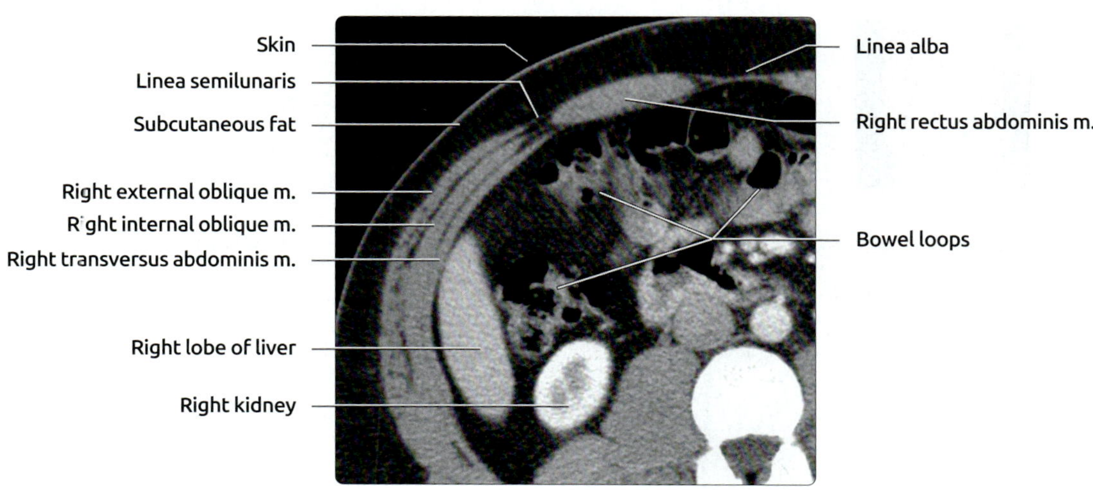

(Top) *Transverse extended FOV grayscale US shows the relationship of the medially located rectus abdominis and the laterally located oblique and transverse abdominal muscles. Medially, the external and internal oblique and the transversus abdominal muscles form aponeuroses that comprise the rectus sheath with the muscles thinning at the linea semilunaris. The linea alba is thin in the lower abdomen.* **(Middle)** *Transverse grayscale US at the right anterolateral abdominal wall shows the relationship of the lateral abdominal wall muscles in better detail. Note the oblique and transverse abdominal muscles taper medially as they become aponeuroses.* **(Bottom)** *Correlative axial CECT illustrates the muscles of the abdominal wall. The rectus abdominis muscle in the anterior abdominal wall is shown, along with the oblique and transverse abdominal muscles in the anterolateral abdominal wall and their aponeuroses.*

POSTERIOR ABDOMINAL WALL

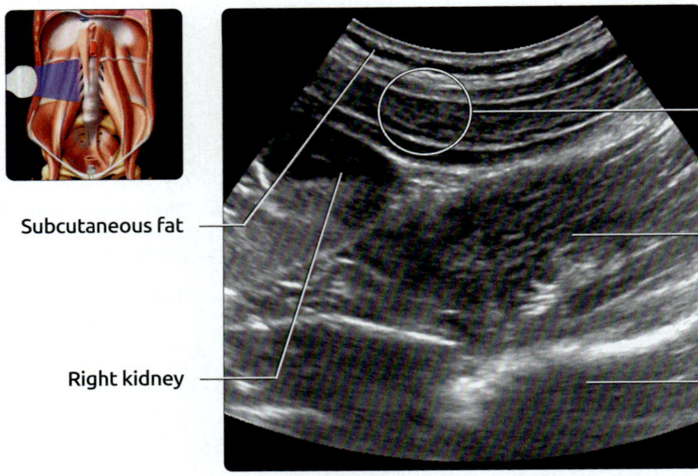

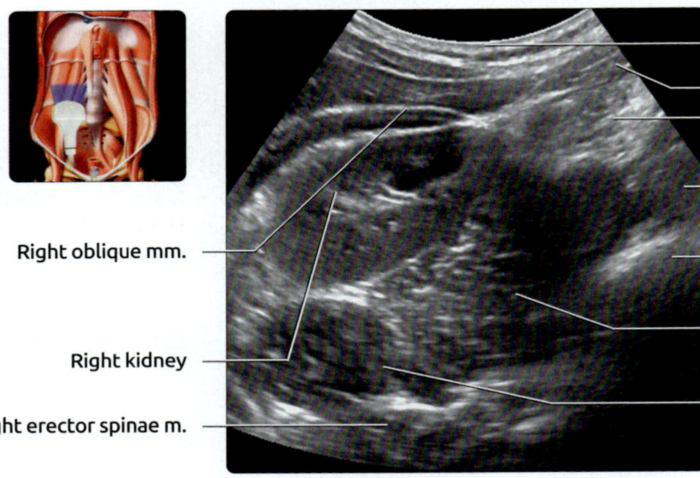

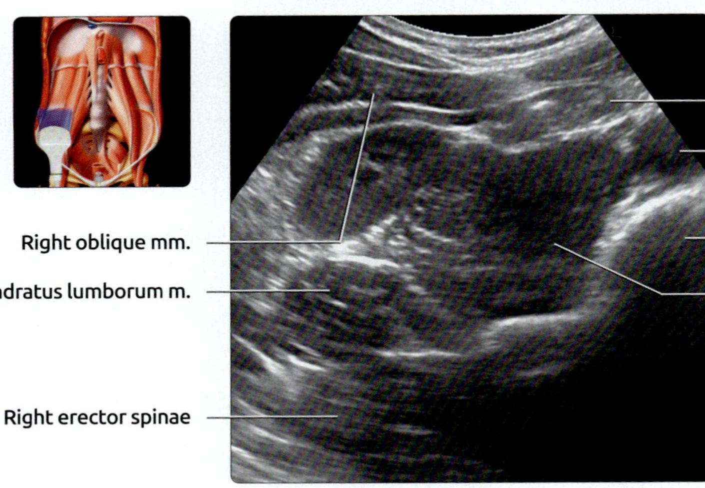

(Top) *Longitudinal oblique grayscale US through the lower right abdomen shows the right psoas muscle, which originates from the lumbar spine and inserts into the proximal femur.* **(Middle)** *Transverse grayscale US of right mid abdomen using the kidney as an acoustic window is shown. The kidney is anterior and lateral to the psoas and anterior to the quadratus lumborum. The psoas runs along the paravertebral region in its entire abdominal course. The quadratus lumborum originates from the iliolumbar ligament and iliac crest to insert into the last rib and lumbar transverse processes. It is easily identified as the muscle on which the kidney rests.* **(Bottom)** *Transverse grayscale US of the right upper abdomen, continuing the scan inferiorly, shows the relationship of the posterior abdominal wall muscles is maintained.*

POSTERIOR ABDOMINAL WALL, CT CORRELATION

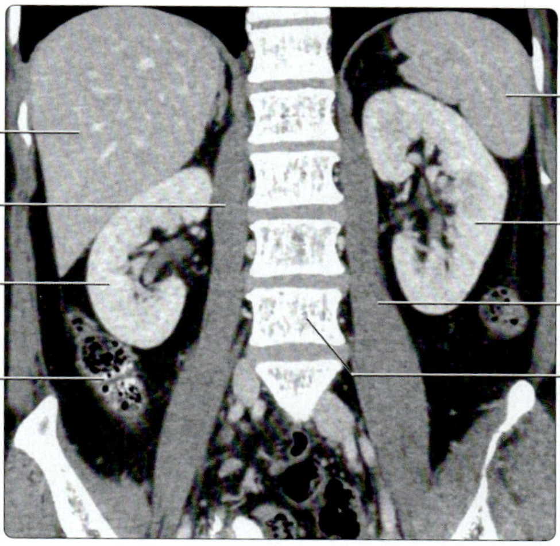

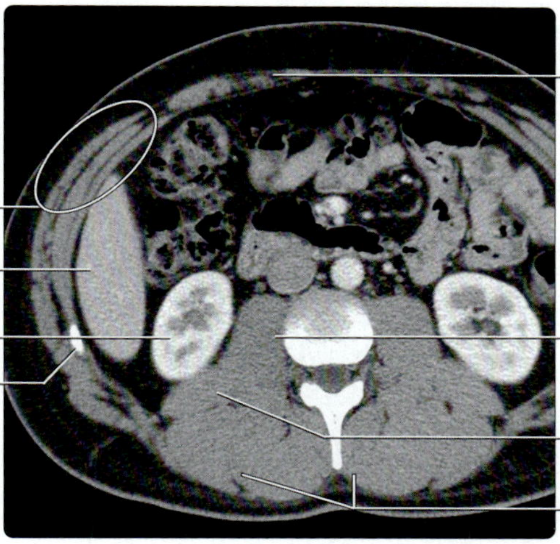

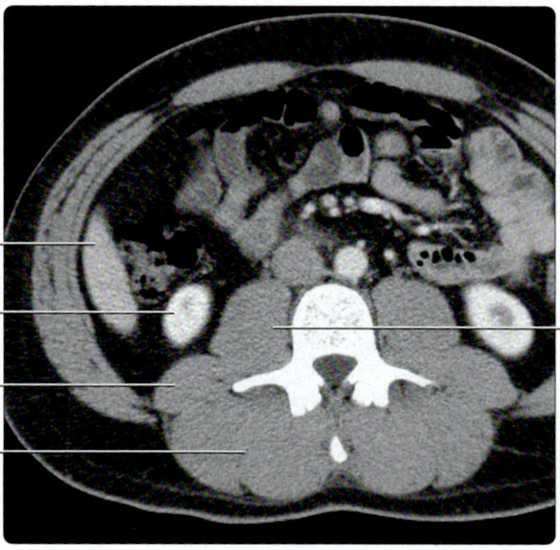

(Top) *Correlative coronal CECT shows the paralumbar location of the psoas muscles and their medial location relative to the kidneys. The psoas muscles originate from the lumbar and 12th thoracic vertebral bodies and their transverse processes and run past the pelvic brim, where they course inferolaterally to be joined by the iliacus muscle.* **(Middle)** *Correlative axial CECT better illustrates the anatomic relationships of the kidney with the posterior abdominal wall muscles. The kidney is lateral to the psoas muscle and rests upon the quadratus lumborum muscle. The erector spinae muscles are immediately posterior to the quadratus lumborum, and the 2 muscles are invested by the lumbodorsal fascia.* **(Bottom)** *Correlative axial CECT at the level of the inferior pole of the right kidney is shown. The psoas muscle and quadratus lumborum muscles, seen in their midsections, are now thicker.*

POSTERIOR ABDOMINAL WALL

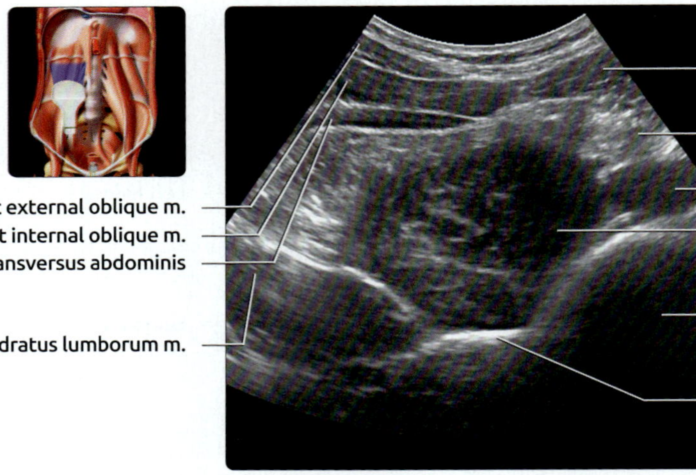

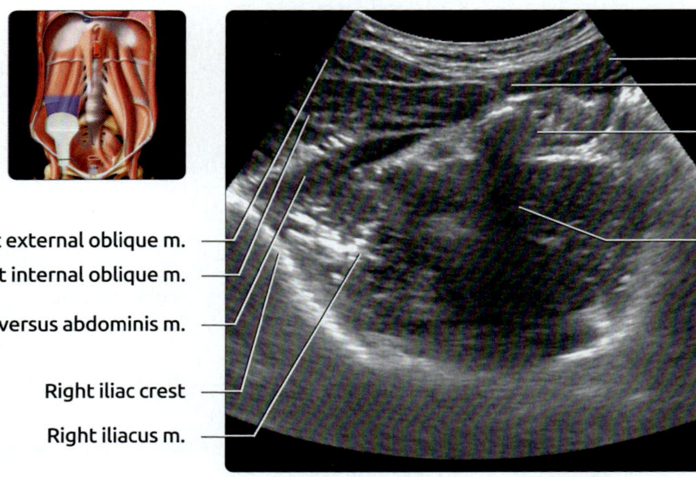

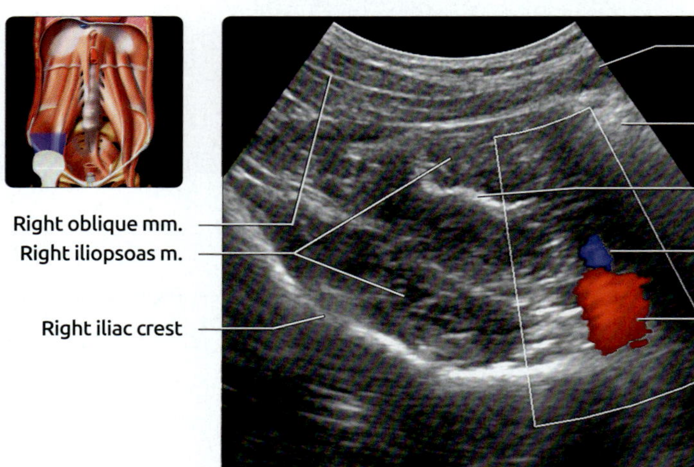

(Top) *Transverse grayscale US in the lower abdominal region shows the right psoas muscle, composed of the psoas minor, which rests upon the psoas major. The 2 muscles cannot be separated clearly on US. Because of their depth, the paraspinal muscles cannot be demonstrated in detail.* **(Middle)** *Transverse grayscale US of the right lower abdomen in the same patient shows that the distal psoas muscle has diminished in size. It rests on the medial portion of the iliacus muscle; the latter is a flat muscle that fills the iliac fossa. Both continue inferiorly together.* **(Bottom)** *Distally, the fibers from the iliacus muscle converge and insert into the lateral side of the psoas muscle to form the iliopsoas muscle. Common iliac vessels can be seen medially.*

POSTERIOR ABDOMINAL WALL, CT CORRELATION

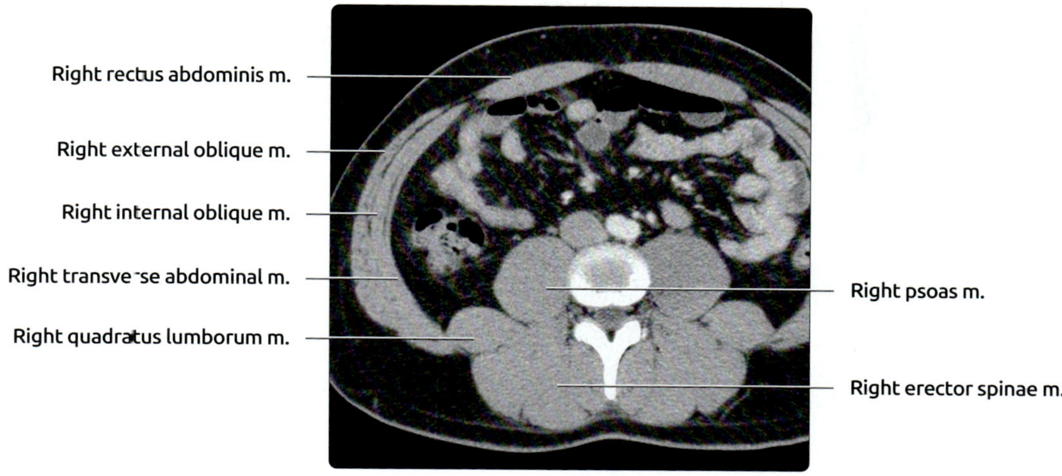

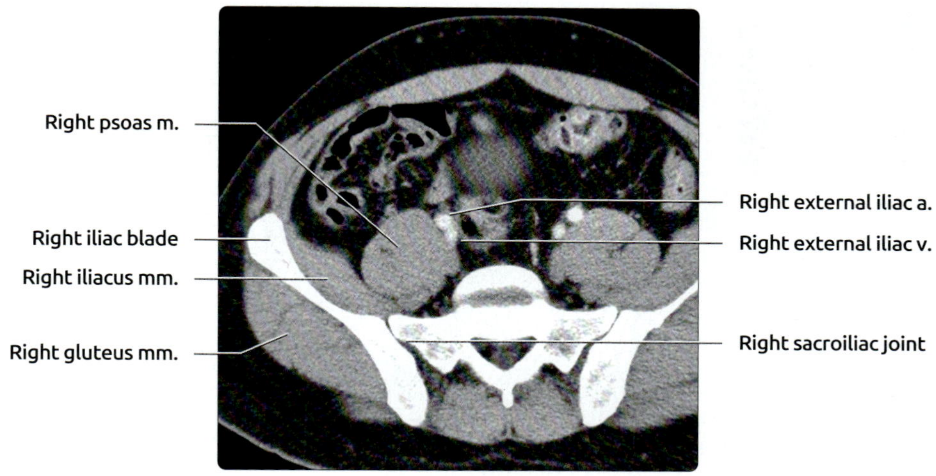

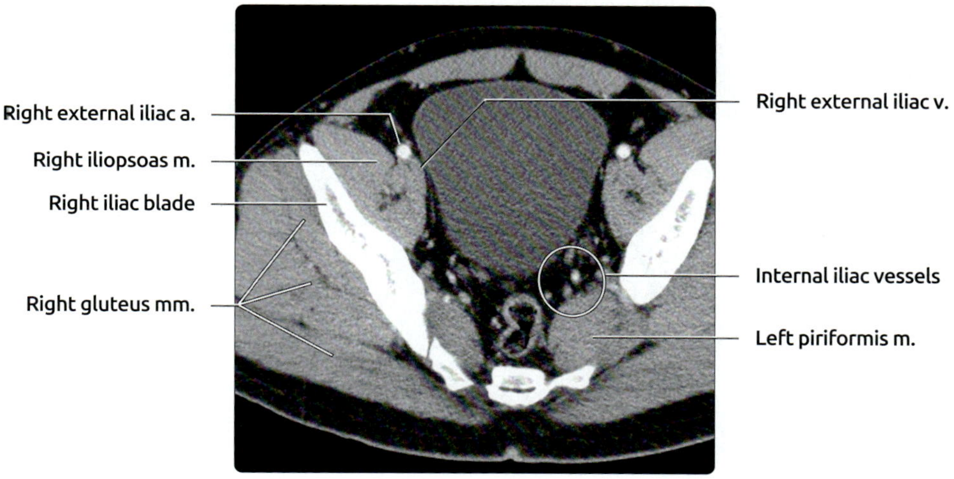

(Top) *Correlative axial CECT below the kidneys shows the quadratus lumborum muscle is more laterally located, and the psoas muscle is directly anterior to the erector spinae muscle.* **(Middle)** *Axial correlative CECT shows the psoas muscle has begun its dorsolateral course and is now anterior to the iliacus muscle. The iliacus muscle is easily identified as a flat muscle filling the iliac fossa, arising from the upper 2/3 of the iliac fossa, inner lip of the iliac crest, anterior sacroiliac and the iliolumbar ligaments, and base of the sacrum.* **(Bottom)** *Correlative axial CECT shows the psoas and iliacus muscles have converged and are now indistinguishable from one another. The resultant iliopsoas muscle passes beneath the inguinal ligament and becomes tendinous as it inserts into the lesser trochanter of the femur.*

PARASPINAL MUSCLES

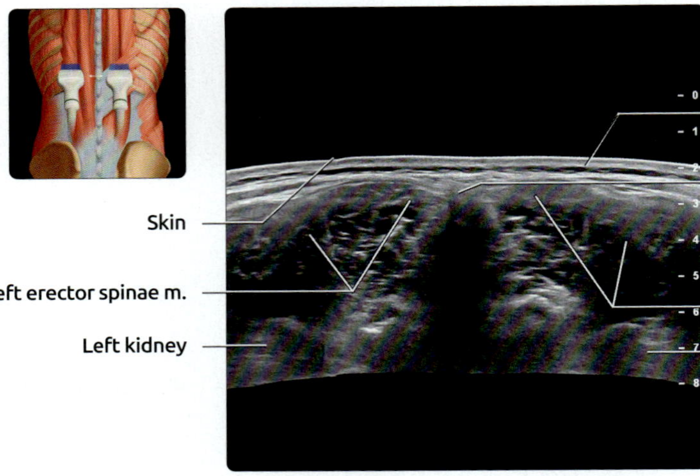

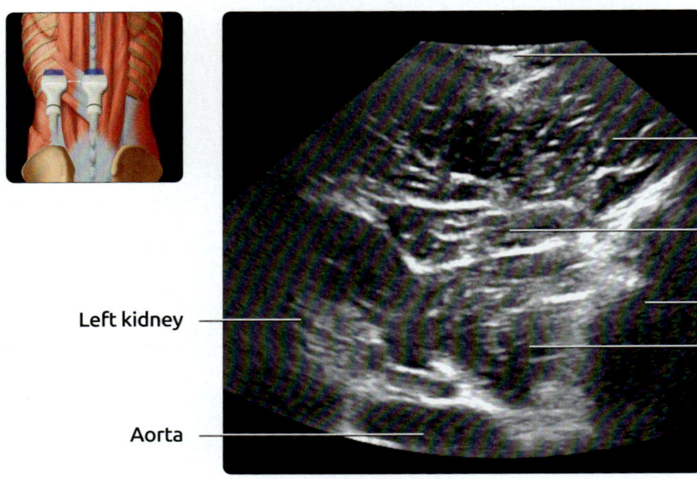

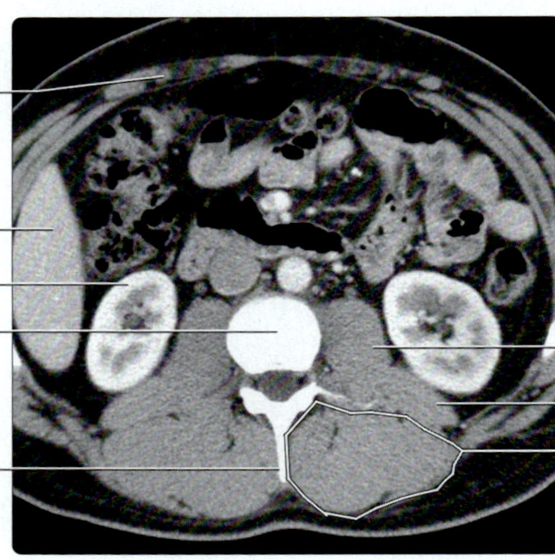

(Top) *Transverse extended FOV grayscale US of the back (with the patient prone) shows the erector spinae muscles flanking the spinous process. They are invested by the lumbodorsal fascia, which also invests the anteriorly located quadratus lumborum muscle. The kidneys are partially demonstrated.* **(Middle)** *Transverse oblique grayscale US shows the left erector spinae muscle (with the patient prone). The 3 columns (iliocostalis, longissimus, and spinalis muscles, from lateral to medial) comprising the erector spinae are not clearly separated from one another on US. They are identified collectively as a thick, fleshy muscle lateral to the spinous process.* **(Bottom)** *Correlative axial CECT of the paraspinal muscles at the level of the kidneys shows the erector spinae muscles originate from a broad and thick tendon, which originates from the sacrum and iliac crest, lumbar, and 11th and 12th thoracic spinous processes.*

MR

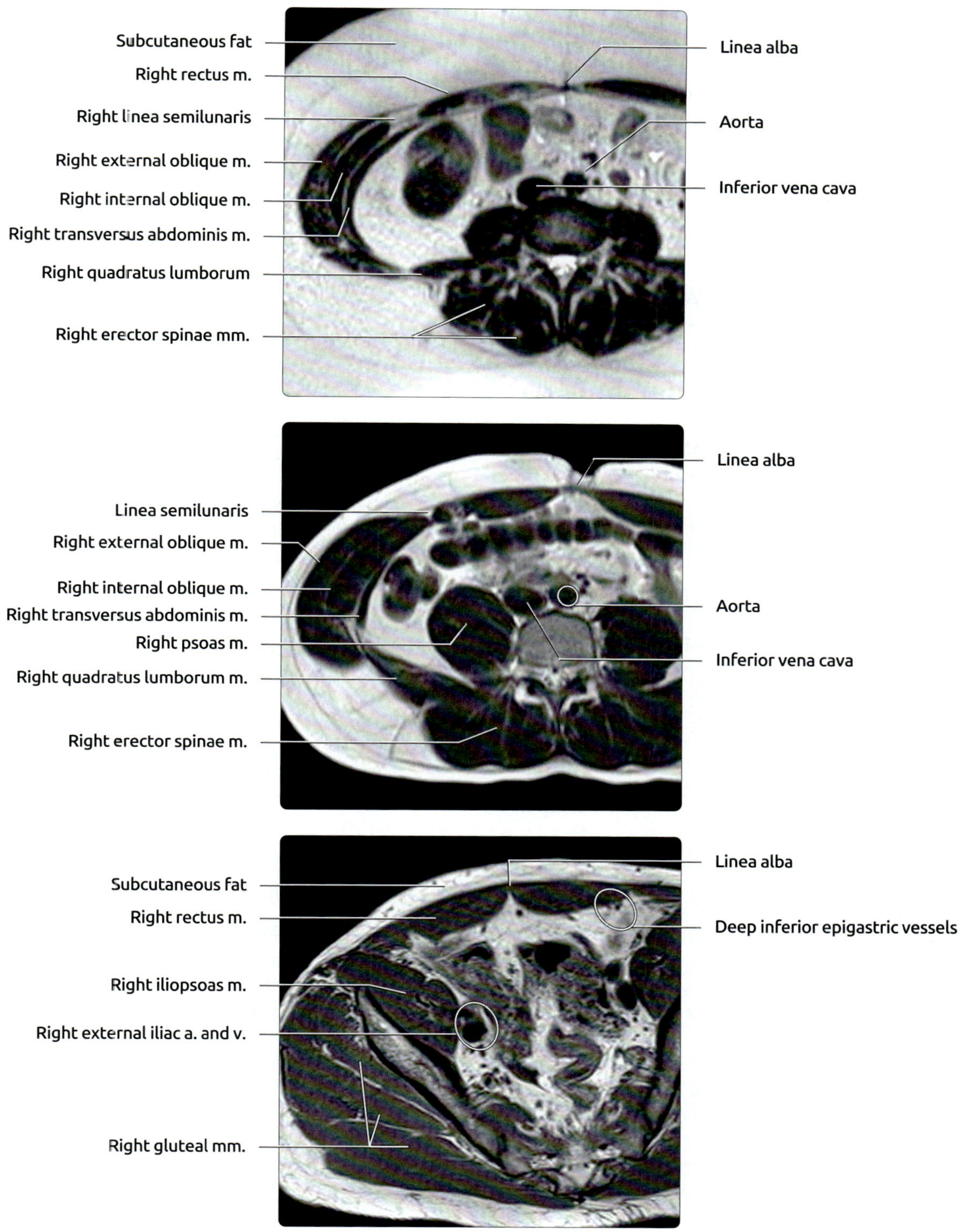

(Top) *Axial T2 HASTE MR in an older patient with muscle atrophy shows fat in between the individual muscles of the anterior and posterior abdominal wall.* **(Middle)** *Axial T2 HASTE MR in a younger male patient shows more bulky abdominal wall musculature with little intermuscular fat.* **(Bottom)** *Axial T1 MR at a lower level shows the iliopsoas as 1 muscle bundle.*

TERMINOLOGY

Definitions

- Abdomen: Region between diaphragm and pelvis

GROSS ANATOMY

Anatomic Boundaries

- Anterior abdominal wall is bounded superiorly by xiphoid process and costal cartilages of 7th-10th ribs
- Anterior wall is bounded inferiorly by iliac crest, iliac spine, inguinal ligament, and pubis
- Inguinal ligament is inferior edge of aponeurosis of external oblique muscle

Muscles of Anterior Abdominal Wall

- Consist of 3 flat muscles (external oblique, internal oblique, and transverse abdominal) and 1 strap-like muscle (rectus)
- Combination of muscles and aponeuroses (sheet-like tendons) act as corset to confine and protect abdominal viscera
- Linea alba is fibrous raphe stretching from xiphoid to pubis
  - Forms central anterior attachment for abdominal wall muscles
  - Formed by interlacing fibers of aponeuroses of oblique and transverse abdominal muscles
  - Rectus sheath is also formed by these aponeuroses as they surround rectus muscle
- External oblique muscle
  - Largest and most superficial of 3 flat abdominal muscles
  - Origin: External surfaces of ribs 5-12
  - Insertion: Linea alba, iliac crest, pubis via broad aponeurosis
- Internal oblique muscle
  - Middle of 3 flat abdominal muscles
  - Runs at right angles to external oblique
  - Origin: Posterior layer of thoracolumbar fascia, iliac crest, and inguinal ligament
  - Insertion: Ribs 10-12 posteriorly, linea alba via broad aponeurosis, pubis
- Transverse abdominal muscle
  - Innermost of 3 flat abdominal muscles
  - Origin: Lowest 6 costal cartilages, thoracolumbar fascia, iliac crest, inguinal ligament
  - Insertion: Linea alba via broad aponeurosis, pubis
- Rectus abdominis muscle
  - Origin: Pubic symphysis and pubic crest
  - Insertion: Xiphoid process and costal cartilages 5-7
  - Rectus sheath: Strong fibrous compartment that envelops each rectus muscle
    - Contains superior and inferior epigastric vessels
  - 3 tendinous intersections divide rectus sheath into distinct muscle bellies
    - This gives rise to 6-pack appearance of well-developed rectus abdominis musculature
  - Arcuate line lies about midway between umbilicus and pubic crest
    - Above arcuate line, internal oblique aponeurosis splits to enclose rectus muscle
    - Below arcuate line, internal oblique passes wholly in front of rectus muscle so that rectus muscle lies directly on transversalis fascia
- Actions of anterior abdominal wall muscles
  - Support and protect abdominal viscera
  - Help flex and twist trunk, maintain posture
  - Increase intraabdominal pressure for defecation, micturition, and childbirth
  - Stabilize pelvis during walking, sitting up
- Transversalis fascia
  - Lies deep to abdominal wall muscles and lines entire abdominal wall
  - Separated from parietal peritoneum by layer of extraperitoneal fat

Muscles of Posterior Abdominal Wall

- Consist of psoas (major and minor), iliacus, and quadratus lumborum
- Psoas: Long, thick, fusiform muscle lying lateral to vertebral column
  - Origin: Transverse processes and bodies of vertebrae T12-L5
  - Insertion: Lesser trochanter of femur (passing behind inguinal ligament)
  - Action: Flexes thigh at hip joint; bends vertebral column laterally
- Iliacus: Large, triangular sheet of muscle lying along lateral side of psoas
  - Origin: Superior part of iliac fossa
  - Insertion: Lesser trochanter of femur (after joining with psoas tendon)
  - Action: Iliopsoas muscle flexes thigh
- Quadratus lumborum: Thick sheet of muscle lying adjacent to transverse processes of lumbar vertebrae
  - Invested by lumbodorsal fascia
  - Origin: Iliac crest and transverse processes of lumbar vertebrae
  - Insertion: 12th rib
  - Actions: Stabilizes position of thorax and pelvis during respiration, walking, and bending trunk to side

Paraspinal Muscles

- Also called erector spinae muscles
  - Invested by lumbodorsal fascia
- Composed of 3 columns from lateral to medial: Iliocostalis, longissimus, spinalis
- Origins: Sacrum, ilium, and spines of lumbar and 11th-12th thoracic vertebrae
- Insertions: Ribs and vertebrae with additional muscle slips joining columns at successively higher levels
- Action: Extends vertebral column

ANATOMY IMAGING ISSUES

Imaging Recommendations

- High-frequency (7-10 MHz) linear transducer for anterior abdominal wall and paraspinal muscles
  - 3-5 MHz for posterior abdominal wall muscles
- Patient may be asked to perform Valsalva maneuver to increase abdominal pressure and exaggerate abdominal wall hernia for better visualization
- Comparison with contralateral side to check for symmetry

ANTERIOR ABDOMINAL WALL

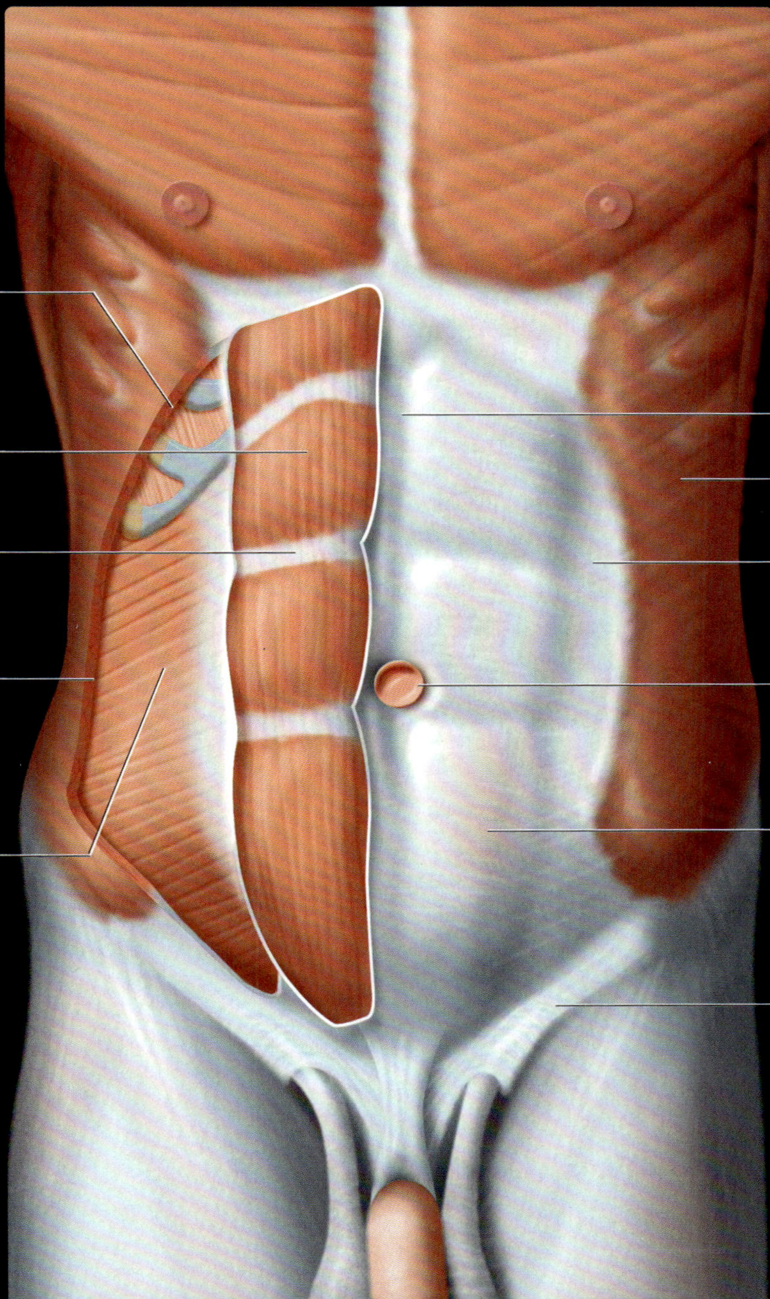

Graphic shows the aponeuroses of the internal and external oblique and transverse abdominal muscles are 2 layered and interweave with each other, covering the rectus muscle and constituting the rectus sheath and linea alba. About midway between the umbilicus and symphysis, at the arcuate line, the posterior rectus sheath ends. The transversalis fascia is the only structure between the rectus muscle and parietal peritoneum below the arcuate line. The arcuate line is also the location where the inferior epigastric artery enters the rectus sheath.

POSTERIOR ABDOMINAL WALL

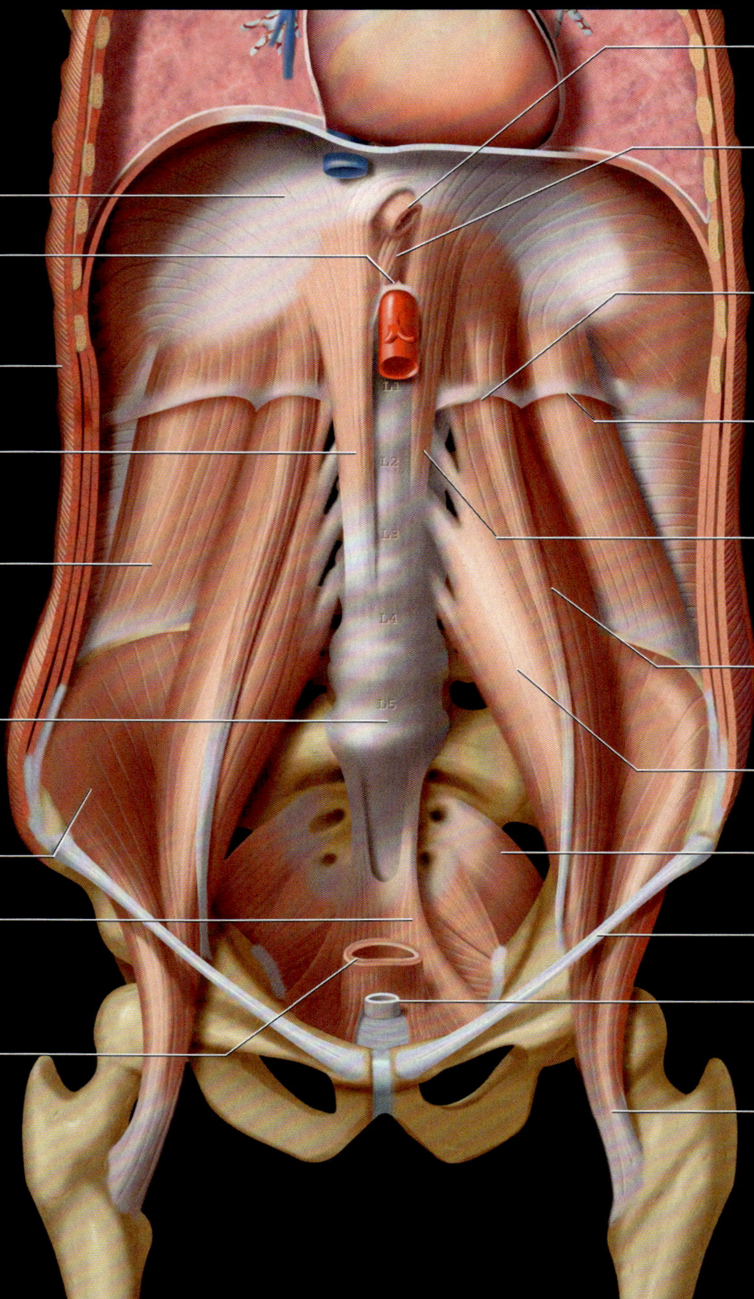

Graphic shows the lumbar vertebrae are covered and attached by the anterior longitudinal ligament, and the diaphragmatic crura are closely attached to it, as are the origins of the psoas muscles, which also arise from the transverse processes. The iliacus muscle arises from the iliac fossa of the pelvis and inserts into the tendon of the psoas major, constituting the iliopsoas tendon, which inserts into the lesser trochanter. The quadratus lumborum arises from the iliac crest and inserts into the 12th rib and transverse processes of the lumbar vertebrae. Diaphragmatic and transverse abdominal fibers interlace. The psoas and quadratus lumborum pass behind the diaphragm under the medial and lateral arcuate ligaments.

MUSCLES OF BACK IN SITU

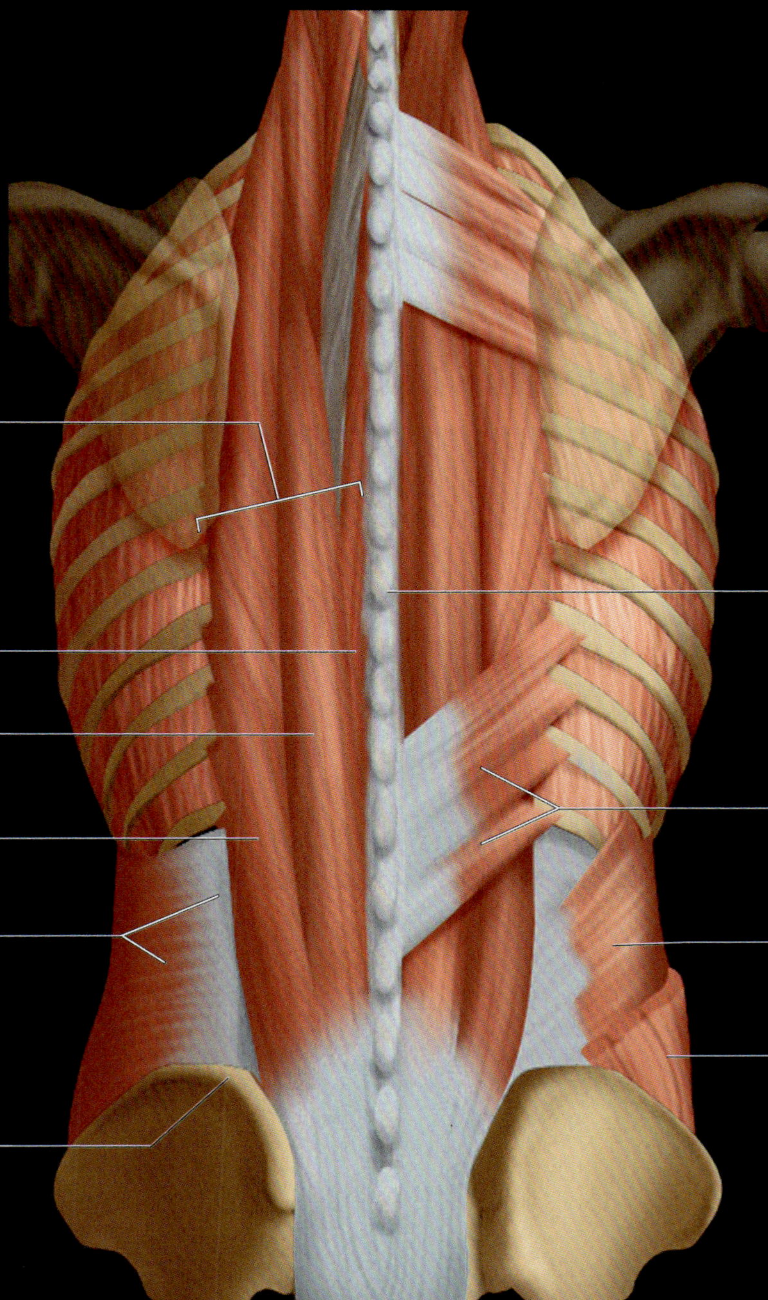

Graphic shows the paraspinal muscles and muscles of the back. The latissimus dorsi muscles are not included. The erector spinae have thick, tendinous origins from the sacral and iliac crests and the lumbar and 11th-12th thoracic spinous processes. Superiorly, the muscle becomes fleshy and subdivides in the upper lumbar region to become the iliocostalis, longissimus, and spinalis muscles (from lateral to medial), tapering as they insert into the vertebrae and ribs. The erector muscles flank the spinous processes and span the length of the posterior thorax and abdomen. They are responsible for extension of the vertebral column.

ANTERIOR ABDOMINAL WALL

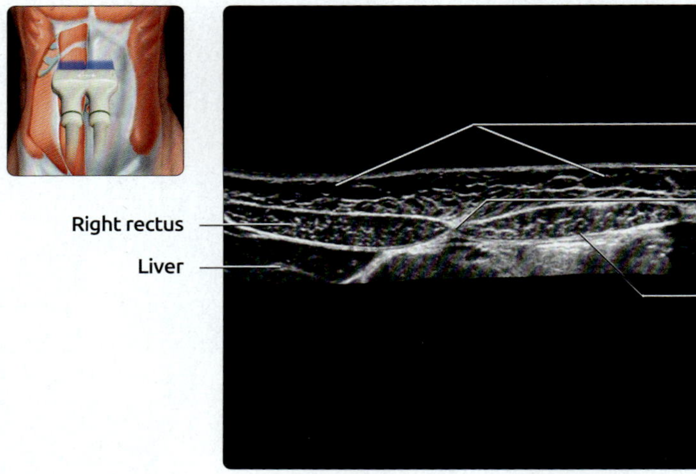

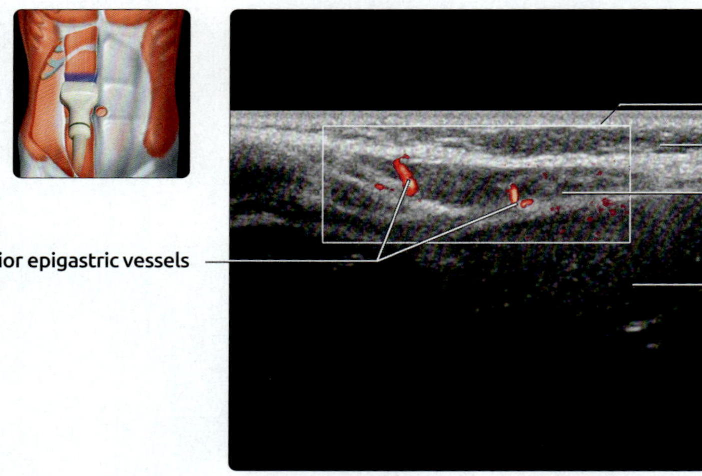

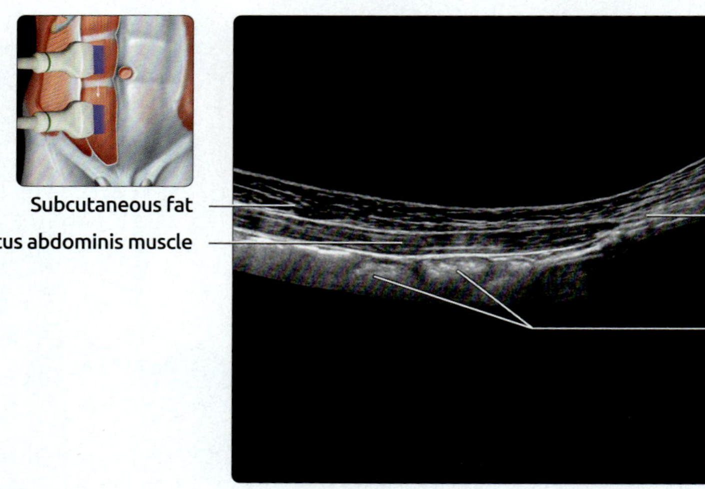

(Top) *Transverse extended FOV grayscale US of the midline anterior abdominal wall shows the paired rectus abdominis muscles separated by the linea alba. The rectus abdominis muscles are comparable in echogenicity and thickness. The surrounding rectus sheath is seen as a fine, thin, echogenic structure around the muscles.* **(Middle)** *Transverse color Doppler US of a rectus abdominis muscle shows arteries within the muscle. At the upper abdomen, these are branches of the superior epigastric artery and, at the lower abdomen, comprise branches of the inferior epigastric arteries. These anastomose at the umbilicus.* **(Bottom)** *Longitudinal extended FOV grayscale US shows the distal rectus abdominis muscle and its tendinous insertion into the symphysis pubis. Note how the rectus abdominis muscle tapers distally.*

ANTEROLATERAL ABDOMINAL WALL

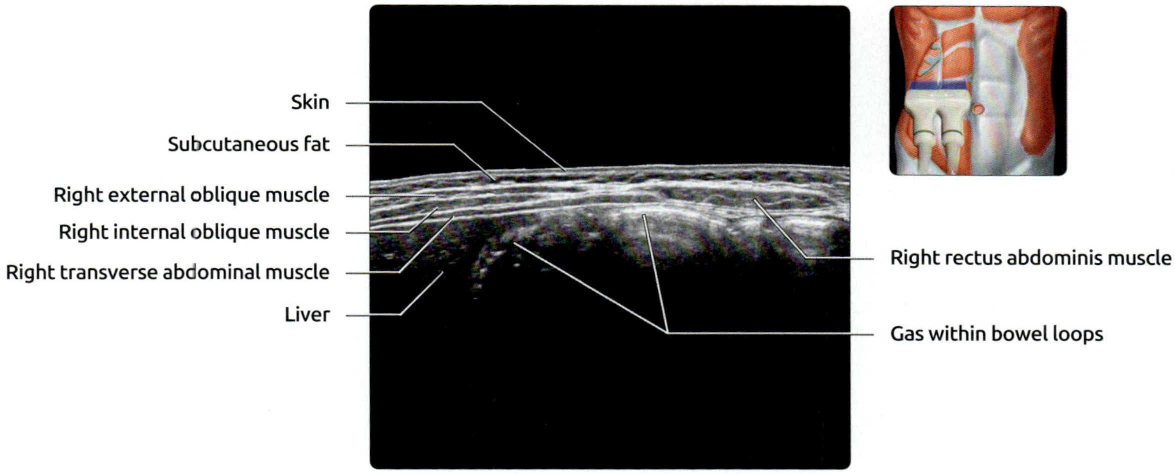

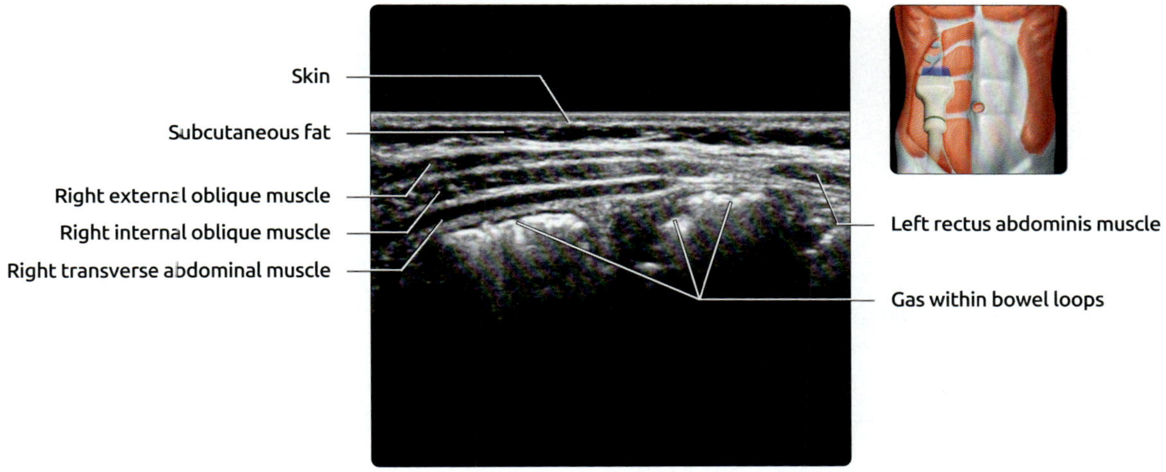

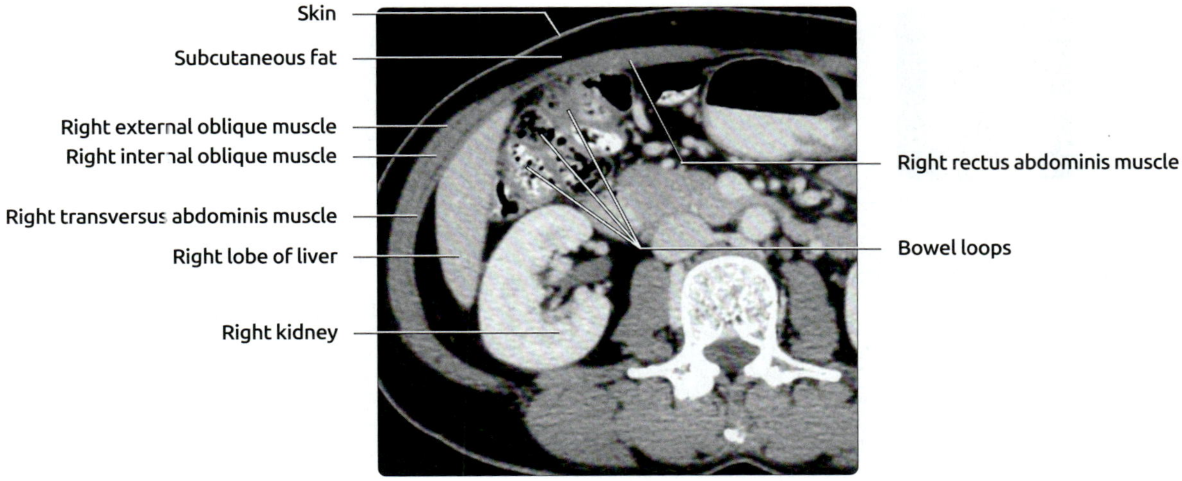

(Top) *Transverse extended FOV grayscale US shows the relationship of the medially located rectus abdominis and the laterally located oblique and transverse abdominal muscles. Medially, the external and internal oblique and transversus abdominal muscles form aponeuroses that compose the rectus sheath.* **(Middle)** *Transverse grayscale US at the right anterolateral abdominal wall shows the relationship of the lateral abdominal wall muscles in better detail. Note the oblique and transverse abdominal muscles taper medially as they become aponeuroses.* **(Bottom)** *Correlative axial CECT illustrates the muscles of the abdominal wall. The rectus abdominis muscle in the anterior abdominal wall and the oblique and transverse abdominal muscles in the anterolateral abdominal wall and their aponeuroses are shown.*

POSTERIOR ABDOMINAL WALL

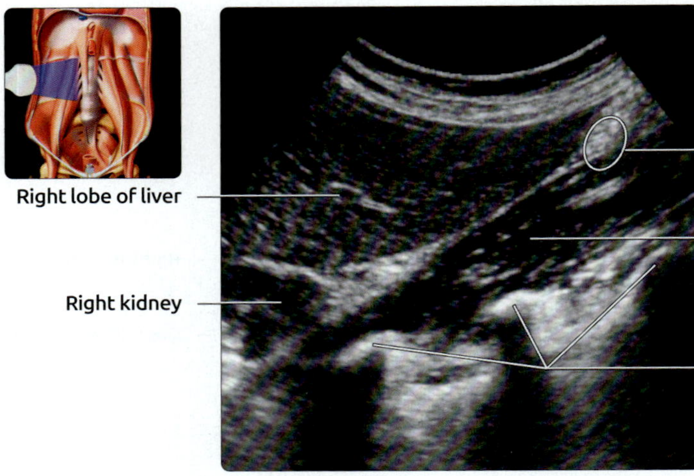

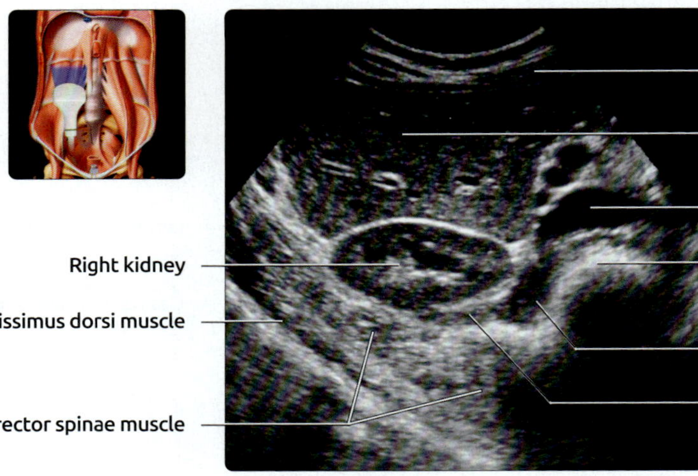

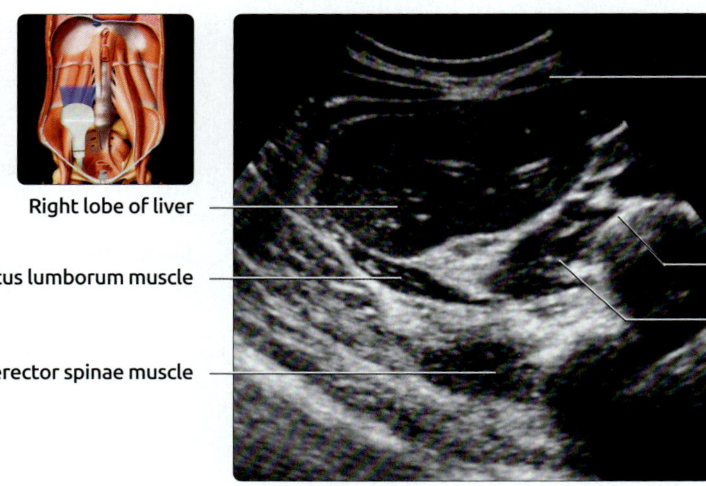

(Top) *Coronal oblique grayscale US using the liver and right kidney as acoustic windows shows the right psoas muscle that originates from the lumbar spine and inserts into the proximal femur.* **(Middle)** *Transverse grayscale US shows the right upper abdomen using the liver and kidney as acoustic windows. The kidney is identified lateral to the psoas and anterior to the quadratus lumborum. The psoas is identified along the paravertebral region in its entire abdominal course. The quadratus lumborum originates from the iliolumbar ligament and iliac crest to insert into the last rib and lumbar transverse processes. It is easily identified as the muscle on which the kidney rests.* **(Bottom)** *Transverse grayscale US shows the right upper abdomen. Continuing the scan inferiorly, the relationship of the posterior abdominal wall musculature would be maintained.*

POSTERIOR ABDOMINAL WALL, CT CORRELATION

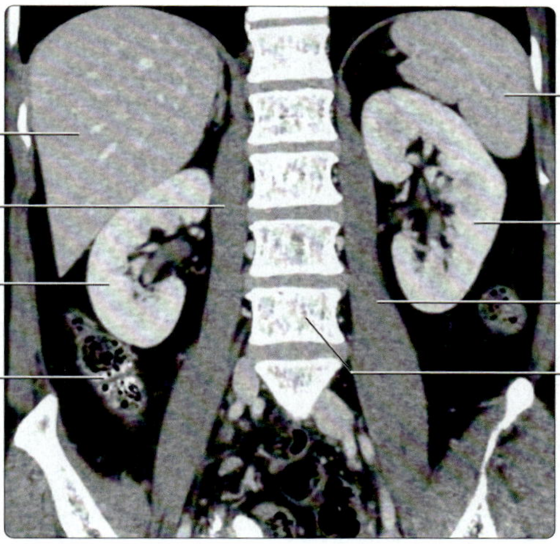

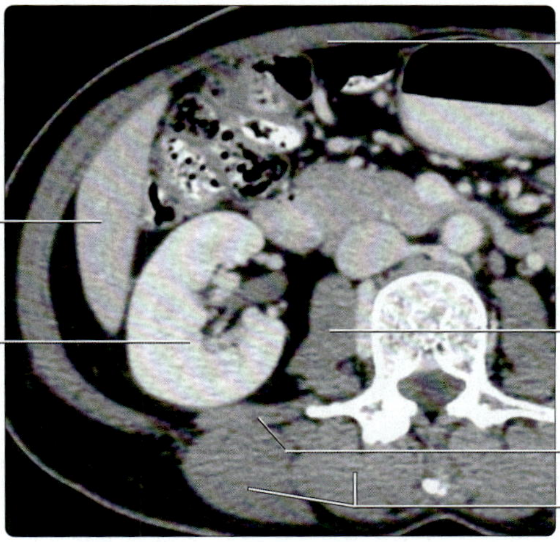

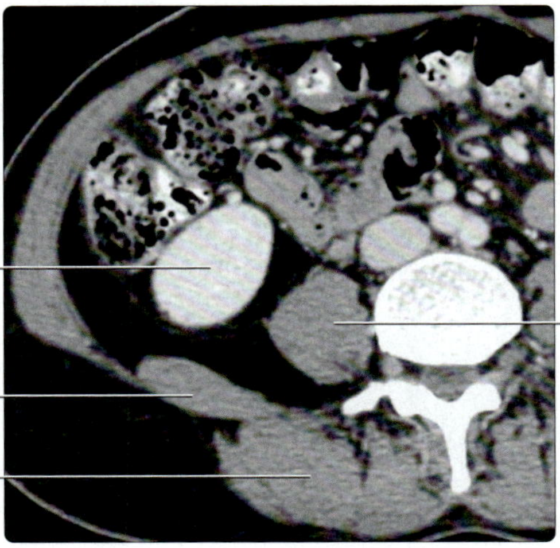

(Top) *Correlative coronal CECT shows the paralumbar location of the psoas muscles and their medial location relative to the kidneys. The psoas muscles originate from the lumbar and 12th thoracic vertebral bodies and their transverse processes, running past the pelvic brim, where they course inferolaterally to be joined by the iliacus muscle.* **(Middle)** *Correlative axial CECT better illustrates the anatomic relationships of the kidney with the posterior abdominal wall muscles. The kidney is lateral to the psoas muscle and rests upon the quadratus lumborum muscle. The erector spinae muscles are immediately posterior to the quadratus lumborum, and the 2 muscles are invested by lumbodorsal fascia.* **(Bottom)** *Correlative axial CECT shows the level of the inferior pole of the right kidney. The psoas muscle and quadratus lumborum muscles, seen in their midsections, are now thicker.*

POSTERIOR ABDOMINAL WALL

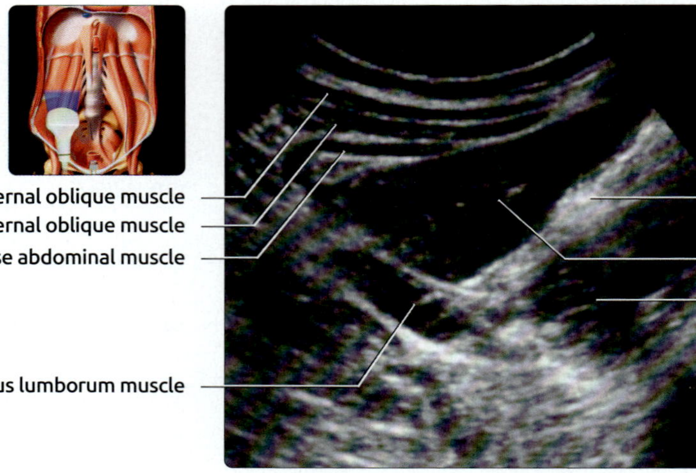

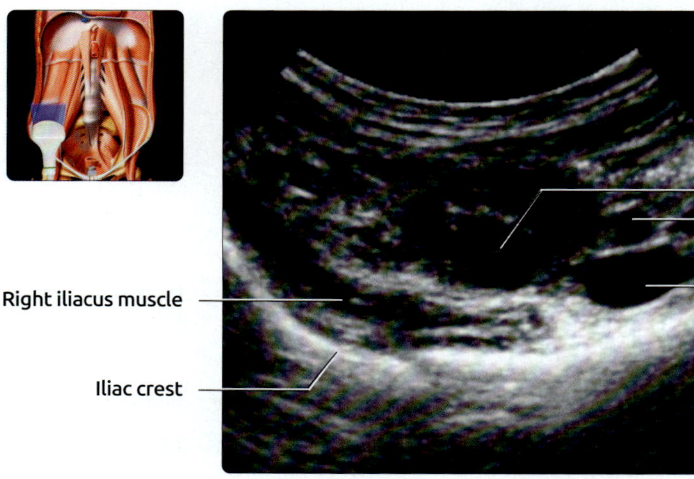

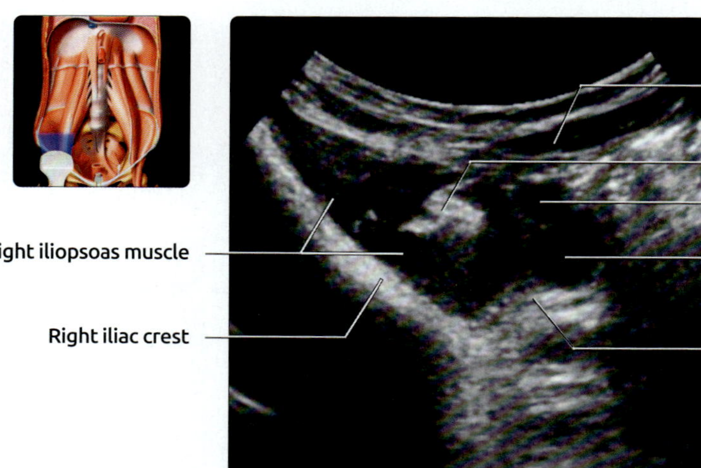

(Top) *Transverse grayscale US in the lower abdominal region shows the right psoas muscle, composed of the psoas minor resting upon the psoas major muscle. The 2 muscles cannot be separated clearly on US. Owing to their depth, the paraspinal muscles cannot be demonstrated in detail.* **(Middle)** *Transverse grayscale US of the same patient shows the right lower abdomen. The distal psoas muscle has diminished in size. It rests on the medial portion of the iliacus muscle. The iliacus muscle is a flat muscle that fills the iliac fossa. Both the psoas and iliacus muscles continue inferiorly together.* **(Bottom)** *Distally, the fibers from the iliacus muscle converge and insert into the lateral side of the psoas muscle to form the iliopsoas muscle.*

POSTERIOR ABDOMINAL WALL, CT CORRELATION

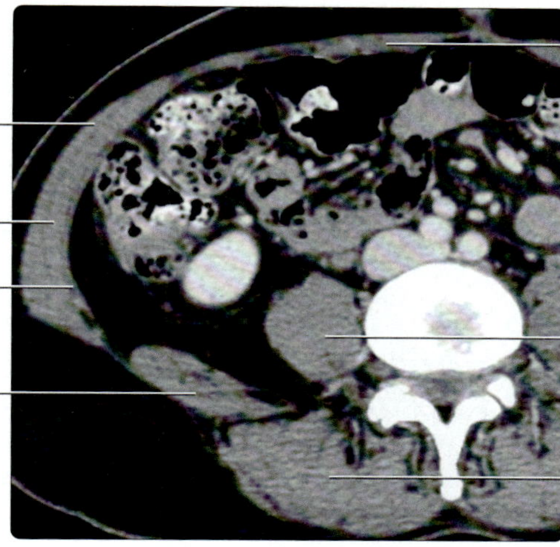

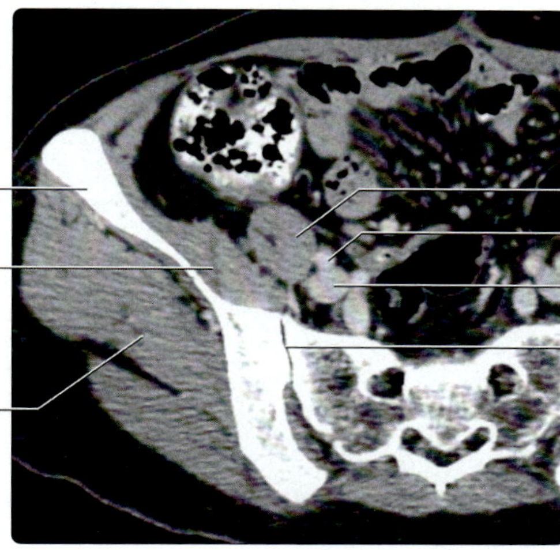

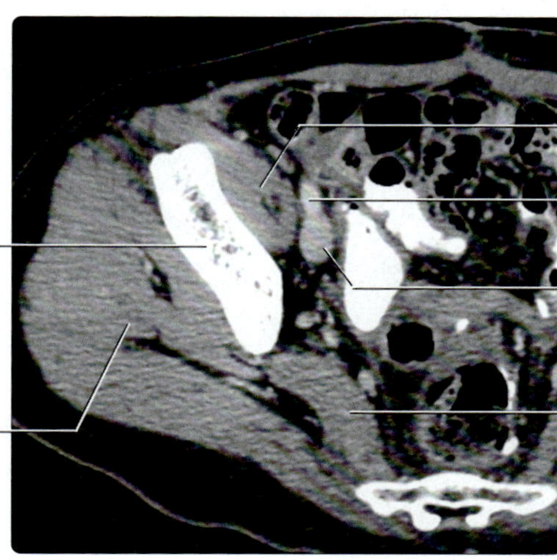

(Top) *Correlative axial CECT shows the level of the inferior pole of the kidney. The quadratus lumborum muscle is more laterally located, and the psoas muscle is directly anterior to the erector spinae muscle.* **(Middle)** *Correlative axial CECT shows that the psoas muscle has begun its dorsolateral course and is now anterior to the iliacus muscle. The iliacus muscle is easily identified as a flat muscle filling the iliac fossa, arising from the upper 2/3 of the iliac fossa, inner lip of the iliac crest, anterior sacroiliac and the iliolumbar ligaments, and base of the sacrum.* **(Bottom)** *Axial correlative CECT shows the psoas and iliacus muscles have converged and are now indistinguishable. The resultant iliopsoas muscle passes beneath the inguinal ligament and becomes tendinous as it inserts into the lesser trochanter of the femur.*

PARASPINAL MUSCLES

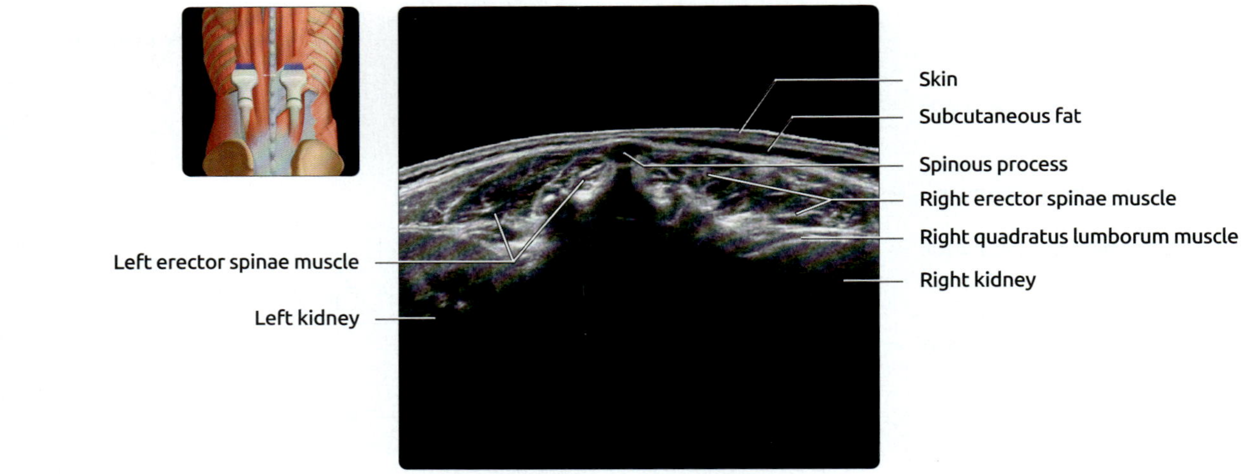

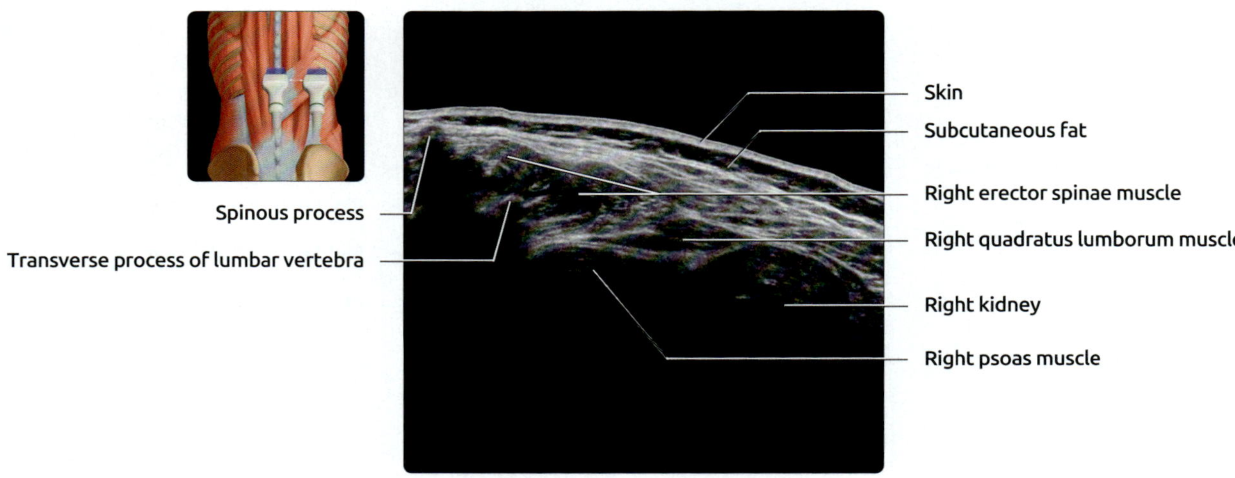

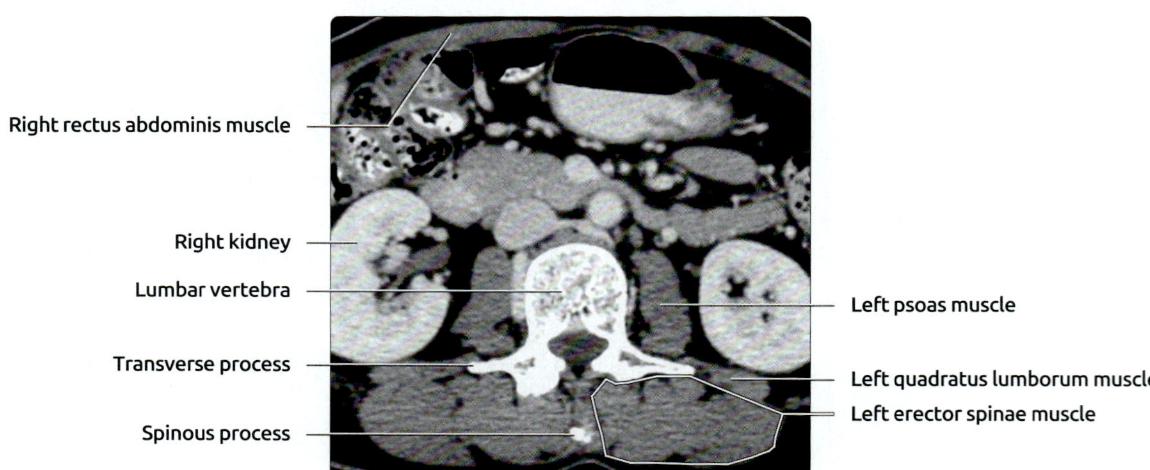

(Top) *Transverse extended FOV grayscale US of the back (with patient prone) shows the erector spinae muscles flanking the spinous process. They are invested by lumbodorsal fascia, which also invests the anteriorly located quadratus lumborum muscle. The kidneys are partially demonstrated.* **(Middle)** *Transverse extended FOV grayscale US shows the right erector spinae muscle (with patient prone). The 3 columns (iliocostalis, longissimus, and spinalis muscles, from lateral to medial) composing the erector spinae are not clearly separated on US. They are identified collectively as a thick, fleshy muscle lateral to the spinous process.* **(Bottom)** *Correlative axial CECT shows the paraspinal muscles at the level of the kidneys. The erector spinae muscles originate from a broad and thick tendon that is attached to the sacrum and iliac crest distally as well as the lumbar and lower 2 thoracic spinous processes more proximally.*

PATHOLOGY

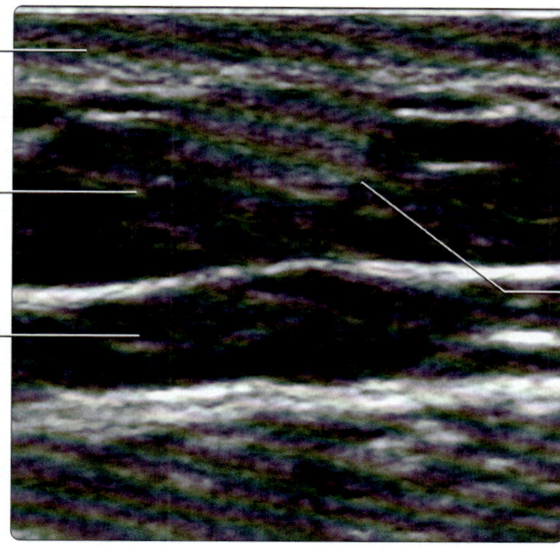

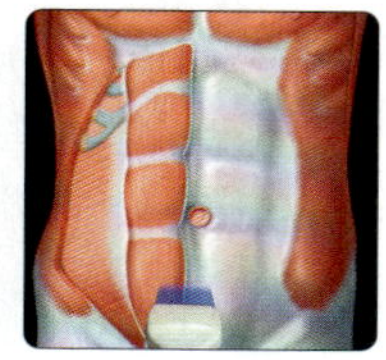

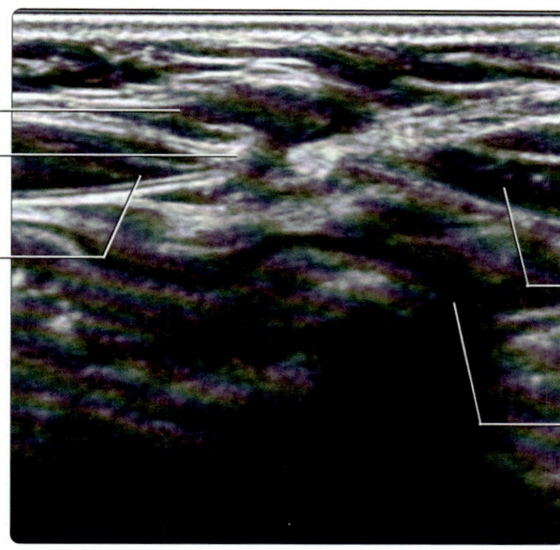

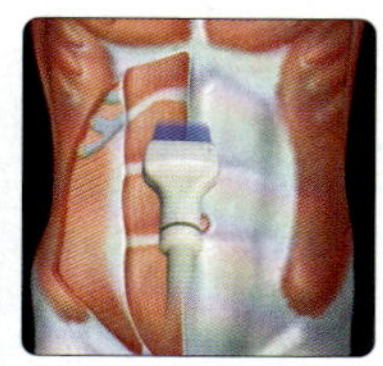

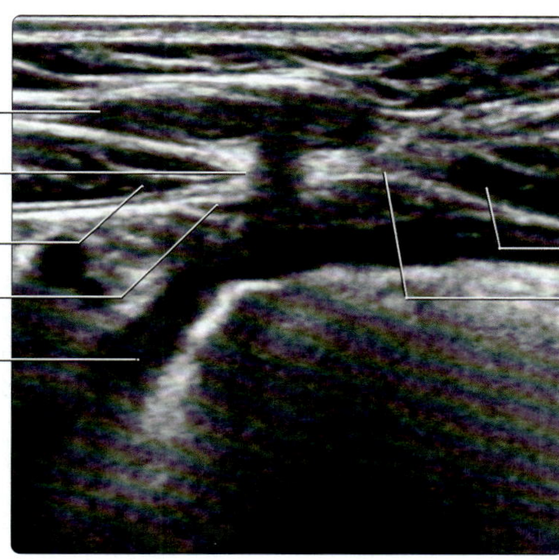

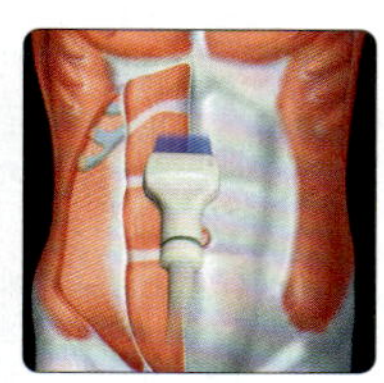

(Top) *Transverse grayscale US shows a small hyperechoic nodule in the subcutaneous fat of the anterior abdominal wall. Mild intrinsic vascularity on color Doppler imaging was present (not shown). Histology following surgical excision revealed an angiolipoma. No deep extension is present.* **(Middle)** *Transverse grayscale US shows a well-defined hypoechoic mass in the anterior abdominal wall. There is a small fascial defect of the linea alba.* **(Bottom)** *Transverse US of the same patient shows widening of the fascial defect and increased herniation of omental fat with Valsalva maneuver. The intraperitoneal communication is much more conspicuous with Valsalva maneuver. Fatty hernias of the linea alba usually occur in the epigastrium and are typically small and painful. The predilection for the epigastrium may be related to extra tension on the linea alba at this location due to pull from the diaphragm.*

TERMINOLOGY

Abbreviations

- Anterior superior iliac spine (ASIS)

Definitions

- Junctional region between thigh and trunk

IMAGING ANATOMY

Overview

- Incorporates lower abdominal wall, inguinal canal, femoral triangle, and femoral adductor muscles

Osseous Anatomy

- Pubic bone
  - **Pubic tubercle**: Small protuberance, lateral border pubic crest
    - Attachment: Inguinal ligament
  - **Superior and inferior rami**: Extend from pubis
  - **Pecten**: Ridge, posterior aspect of superior pubis
    - Origin (O) of pectineus muscle
    - Insertion (I) of conjoint tendon (internal oblique and transverse abdominis)
  - **Pubic crest**: Superior surface, anterior aspect of pubic body
    - O of rectus abdominis muscle
    - I of transversus abdominis and external oblique muscles
  - **Pubic symphysis**: Cartilaginous joint between pubic bodies
  - **Superior pubic ligament**: Laterally extends to pubic tubercles

Anterior Abdominal Wall

- Abdominal wall muscles
  - **Rectus abdominis**: Paired midline muscles
    - O: Superior pubic ramus, pubic crest
    - I: Xiphoid process, costal cartilages 5-7
  - **Linea alba**: Aponeurotic junction of rectus abdominis, transverse abdominis, and internal and external oblique muscles
  - **External oblique**: Most superficial
    - O: Ribs 5-12
    - I: Pubic crest, anterior iliac crest, linea alba
    - Lower border of aponeurosis contributes to inguinal ligament
  - **Internal oblique**: Between external oblique, transversus abdominis
    - O: Lateral inguinal ligament, iliac crest, thoracolumbar fascia
    - I: Pecten (conjoined tendon), pubic crest, inferior aspect of ribs 10-12, linea alba
    - O: Anterior to deep inguinal ring
    - I: Lateral to rectus abdominis muscle, posterior and medial to superficial inguinal ring
    - Arches over inguinal canal, forming roof
  - **Transversus abdominis**: Deepest
    - O: Iliac crest, posterior aspect of lateral inguinal ligament, thoracolumbar fascia
    - I: Pubic crest, pecten (conjoined tendon), linea alba
    - Remains posterior to inguinal canal

Inguinal Ligament

- Thickening inferior border of external oblique aponeurosis
- Attachments: ASIS and pubic tubercle
- Separates lower extremity from pelvis
- Fascia lata attaches to inferior border
- **Subinguinal space**: Deep to inguinal ligament
  - Passageway for femoral vessels and nerve, iliopsoas muscle into femoral triangle
  - External iliac vessels become femoral vessels upon entering this space

Inguinal Canal

- Entrance: **Deep inguinal ring**
  - Located midpoint of inguinal ligament
  - Opening of evaginated transversalis fascia through which spermatic cord/round ligament pass
- Exit: **Superficial inguinal ring**
  - Division of external oblique aponeurosis lateral to pubic tubercle
- Contents: Ilioinguinal nerve; male: Spermatic cord; female: Round ligament; associated vessels
  - Covered by evaginated transversalis fascia

Adductor Musculature

- **Adductor longus muscle**: Thin tendon arises from medial superior pubic ramus
  - Overlies os of gracilis, adductor brevis, and magnus muscles
- **Gracilis muscle**: O from anterior aspect of symphysis pubis and medial aspect of inferior pubic ramus
  - O medial to adductor brevis muscle, deep to adductor longus muscle

Groin Nodes

- 2 groups of groin lymph nodes
- Superficial inguinal nodes
  - Lie in subcutaneous tissues below inguinal ligament and alongside great saphenous vein
  - Subdivided into 3 groups of lymph nodes
    - Inferior group is located distal to saphenous opening and receives drainage from lower limb
    - Superolateral group is located lateral to saphenous opening and receives drainage from lateral buttock and lower anterior abdominal wall regions
    - Superomedial group is located medial to saphenous opening and receives drainage from perineum and external genitalia
- Deep inguinal nodes
  - 3-5 nodes located in femoral triangle just medial to femoral vein
  - Cloquet node is most proximal of deep inguinal lymph nodes located just below inguinal ligament

ANATOMY IMAGING ISSUES

Imaging Recommendations

- Radiography and US for assessing groin pathology are complementary to each other
- If this does not provide necessary information, MR is next best investigation for most groin pathology

GRAPHICS, GROIN

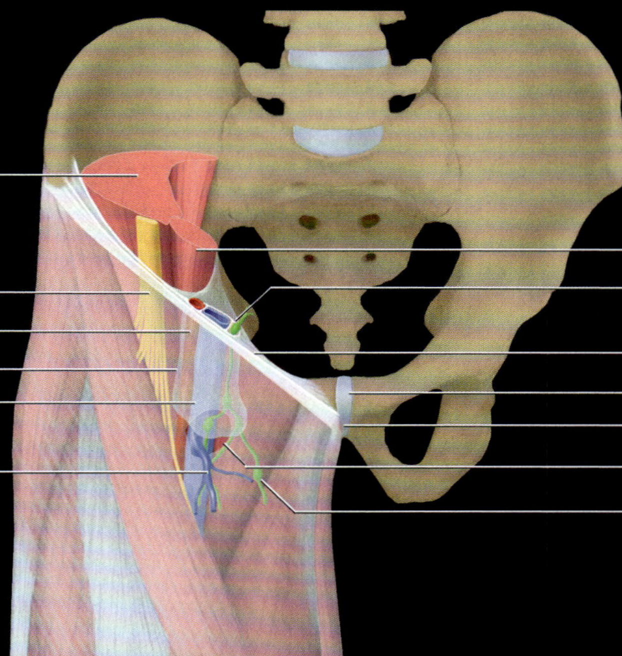

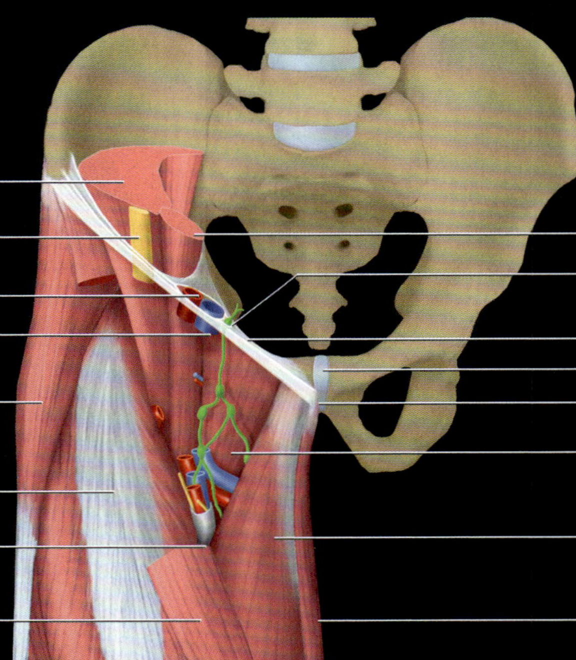

(Top) *The boundaries of the femoral triangle can be remembered by the mnemonic SAIL for sartorius, adductor longus, and inguinal ligament. The contents of the femoral triangle from lateral to medial (remembered by the mnemonic NAVeL) are the femoral nerve, femoral artery and vein, and lymphatics. The nerve lies superficial to the iliopsoas muscle. The fascia lata encases the structures of the thigh. The femoral sheath is the fascial covering over the proximal vessels. Note the septa dividing the sheath into compartments at the cut-away proximal boundary. The femoral canal is the medial compartment.* **(Bottom)** *Femoral triangle is shown after removal of the fascia lata, sartorius muscle, and the vessels. The apex of the triangle is at the crossing of the sartorius and adductor longus muscles. The pectineus, adductor longus, and iliopsoas muscles form the floor of the triangle.*

GRAPHICS, GROIN MUSCLES

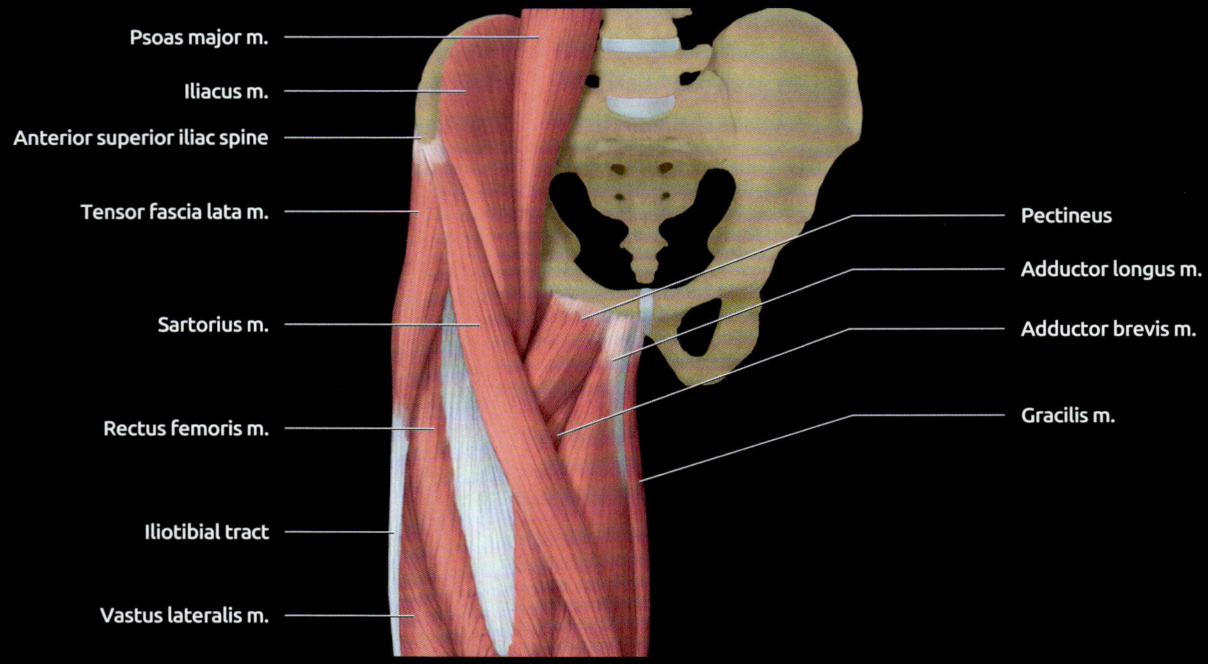

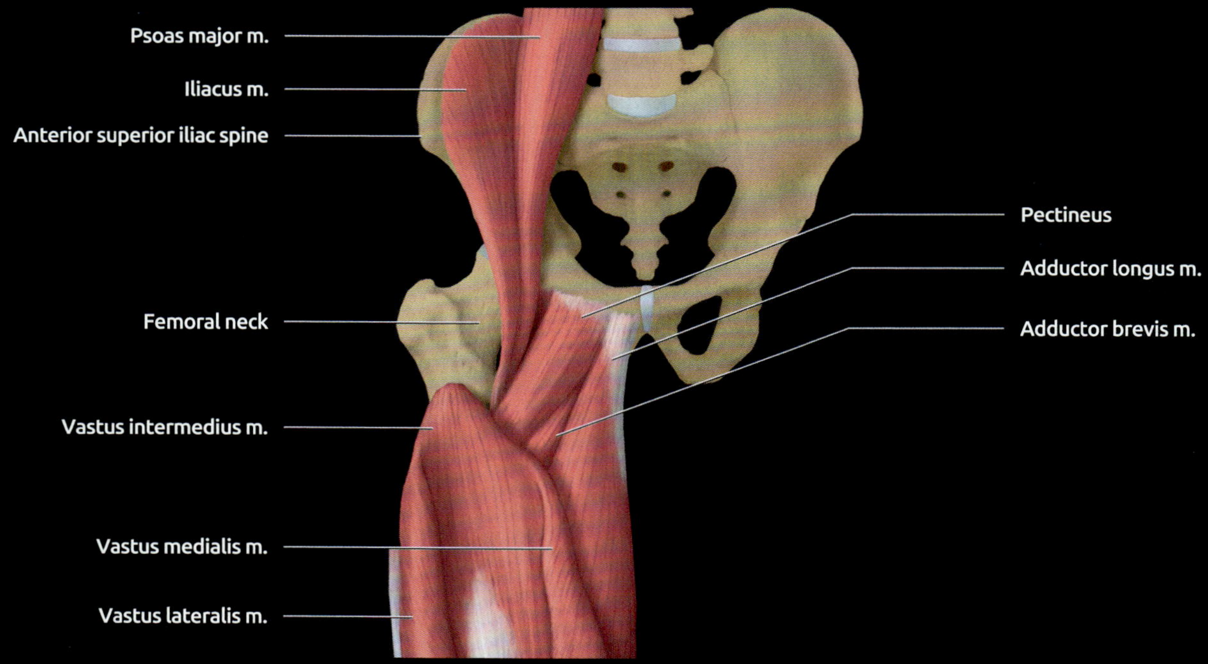

(Top) *Superficial muscles of the groin are shown. The most lateral muscle is the sartorius; its oblique course is easily appreciated. The psoas major and iliacus muscle exit from the pelvis toward the lesser trochanter. On the medial side, the most medial muscle is the gracilis. The adductor brevis muscle is deep to the adductor longus and pectineus muscles.* **(Bottom)** *After removing the superficial layer of muscles, the floor of the femoral triangle is better appreciated. The iliopsoas muscle dives deeply to its trochanteric insertion. The adductors fan out laterally from their pubic origin and constitute the major medial component at a midthigh level. The adductor magnus is in a deeper and more posterior position.*

NERVES IN GROIN

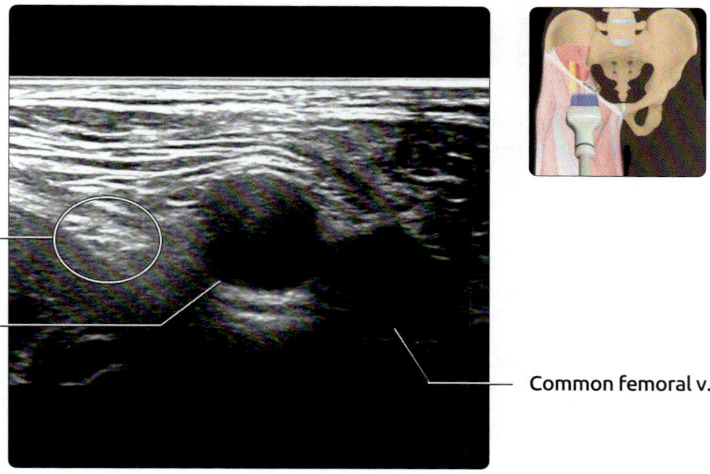

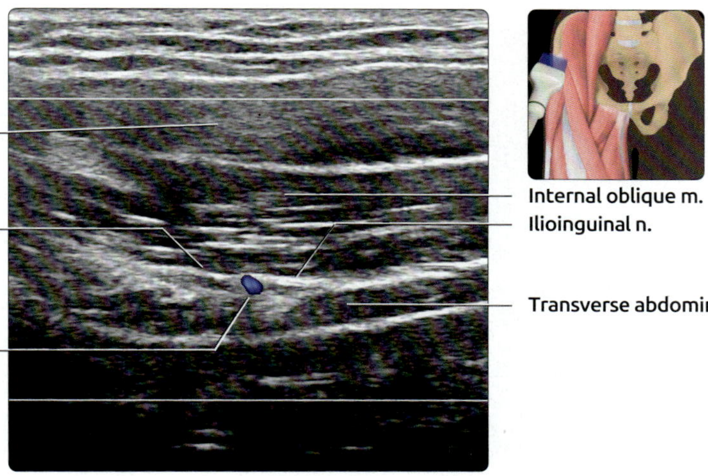

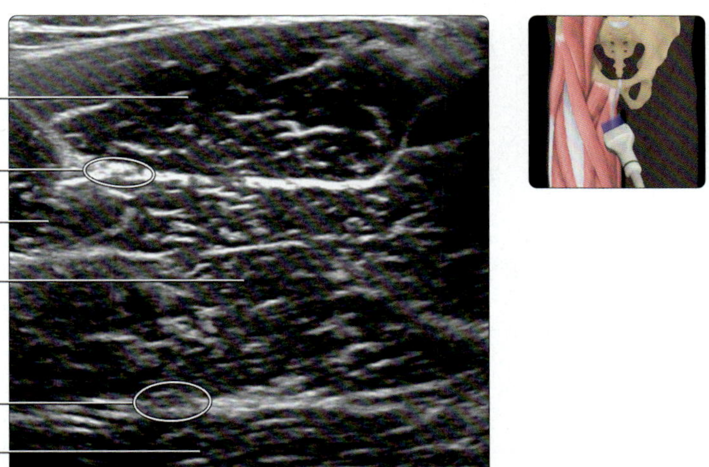

(Top) *The nerve enters the femoral triangle by passing beneath the inguinal ligament, just lateral to the femoral artery. In the groin, the femoral nerve is situated on the lateral side of the common femoral artery. From medial to lateral is the common femoral vein, common femoral artery, and femoral nerve.* **(Middle)** *Transducer is placed transversely along the abdominal wall at the level of the anterior superior iliac spine. The ilioinguinal and iliohypogastric nerves are located between the internal oblique and transverse abdominis muscle. The deep circumflex iliac artery is located close to these nerves.* **(Bottom)** *The obturator nerve is divided into an anterior and posterior branch in the medial groin region. The anterior branch courses in the fascial plane between the adductor longus, adductor brevis, and pectineus muscles, while the posterior branch courses in the fascial plane between the adductor brevis and magnus muscles. The anterior branch provides motor innervation to the adductor muscles and sensory cutaneous branches to the medial aspect of the thigh. The posterior branch provides motor innervation nerve for the adductor muscles.*

MIDLINE GROIN

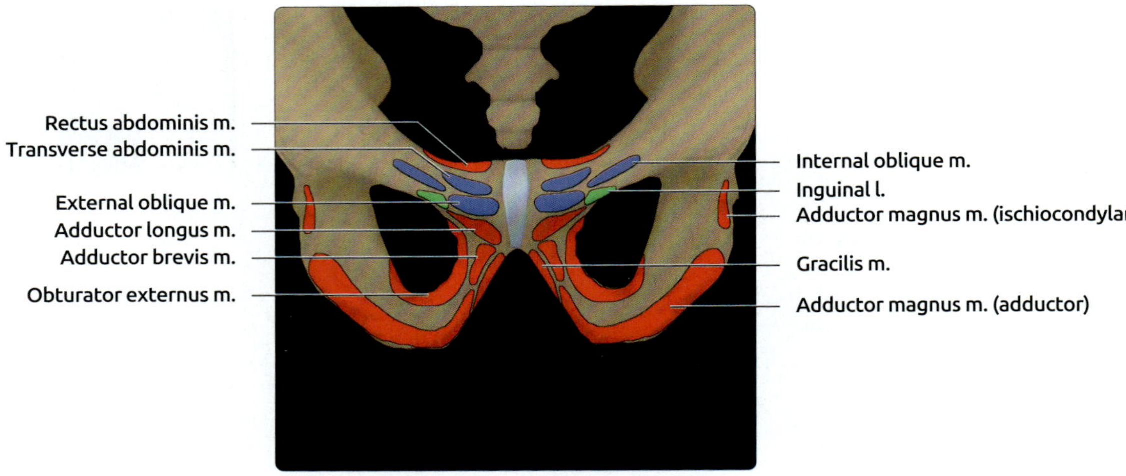

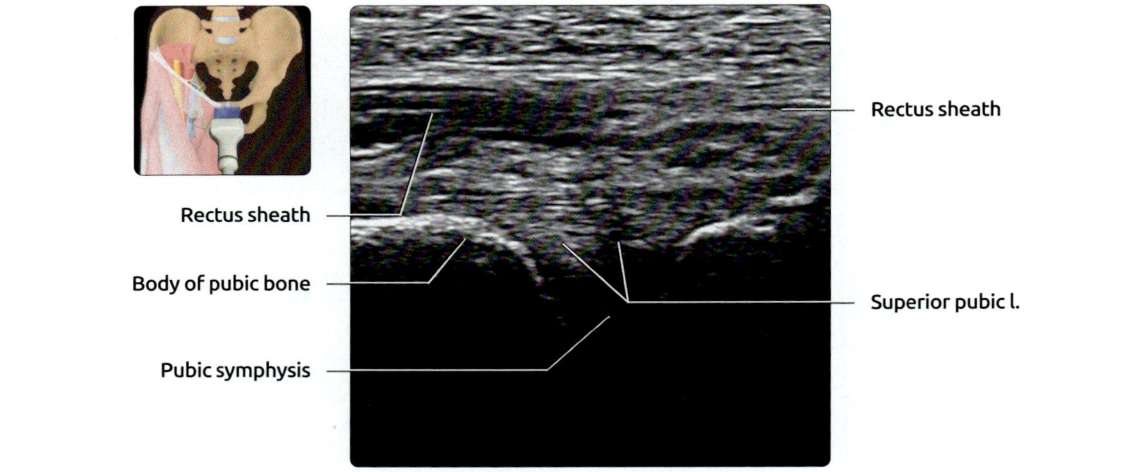

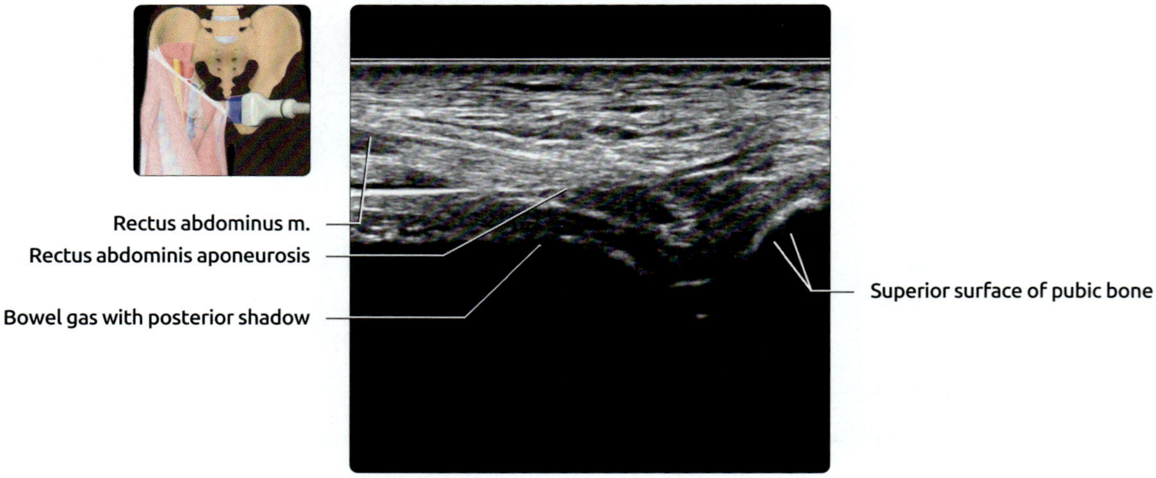

(Top) *Anterior view shows the anterior pelvis and the associated muscle attachments. A number of muscles take origin or insert onto the anterior aspect of the pelvis. These muscles aid in movement and stabilization of the trunk as well as movement and stabilization of the leg and form the medial components of the groin.* **(Middle)** *Transverse scan at the superior aspect shows pubic symphysis. The superior pubic ligament runs horizontally across the symphysis, deep to the rectus abdominis muscle origin.* **(Bottom)** *Sagittal scan shows the lower rectus abdominis originating from the superior surface of pubic bone (pubic crest). Deep to the rectus abdominis muscle is the peritoneal cavity and bowel. Bowel gas produces posterior shadowing artifact, which obscures deeper structures.*

MEDIAL GROIN

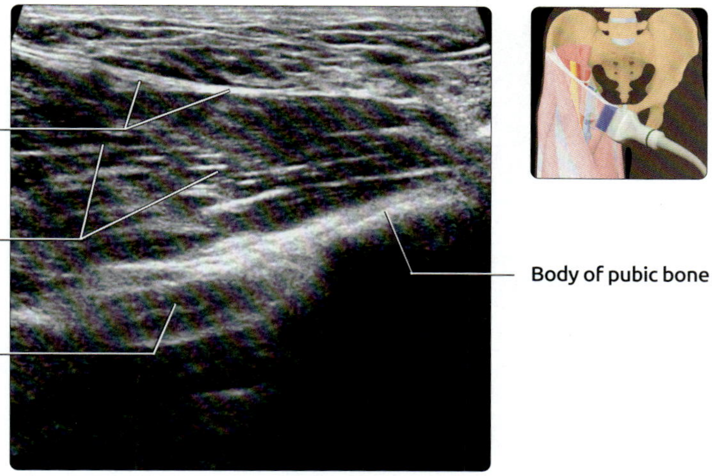

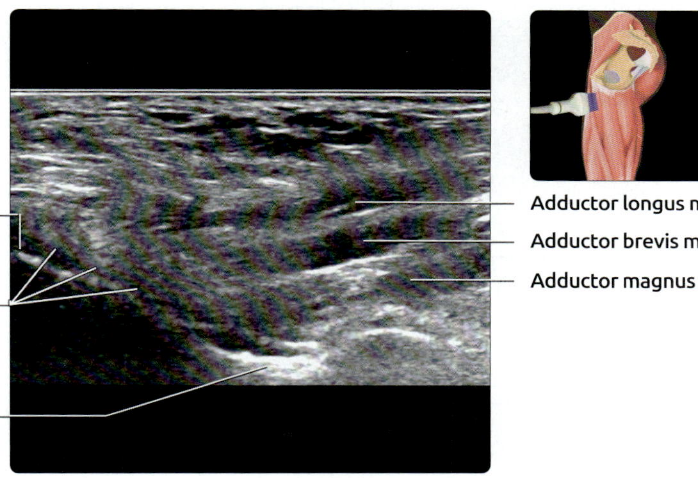

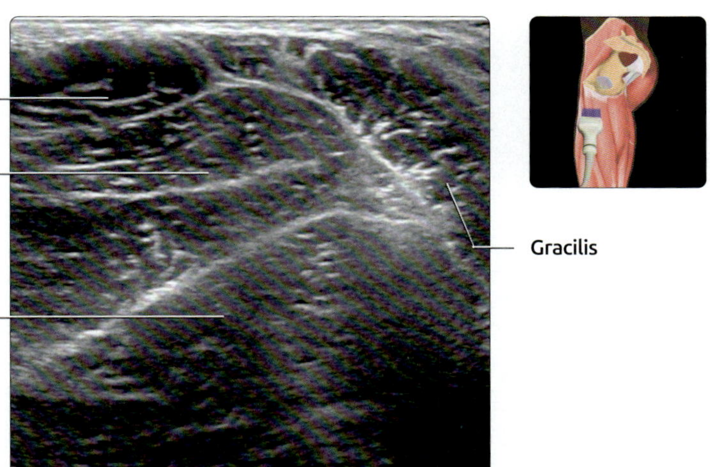

(Top) *Series of (oblique) transverse scans through the adductor muscles are shown. Superiorly, the pectineus muscle originates from the pectineal line (which is a ridge on the superior ramus of the pubic bone) and runs laterally to insert on the pectineal line of the proximal femur. The pectineal line of the proximal femur lies between the lesser trochanter and the linea aspera. The pectineus muscle runs lateral to the adductor longus muscle and superficial to the adductor brevis muscle.* **(Middle)** *Slightly inferiorly and scanning more obliquely, the adductor longus, brevis, and magnus can be seen originating from the pubic body and inferior ramus. Scanning this region requires the thigh abductor and balancing the transducer on the body landmark. From here, individual muscles can be traced inferolaterally.* **(Bottom)** *Oblique transverse scan of adductors using a medial approach is shown. Adductor longus and gracilis muscles form a superficial layer; adductor brevis and magnus muscles run deeper. All muscles run inferolaterally to their femoral insertions.*

CENTRAL GROIN, FEMORAL TRIANGLE

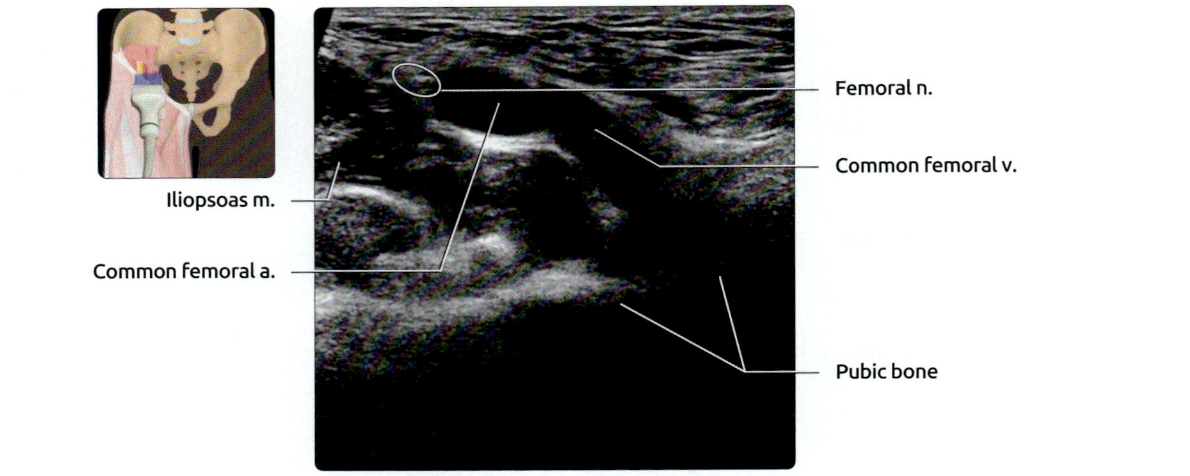

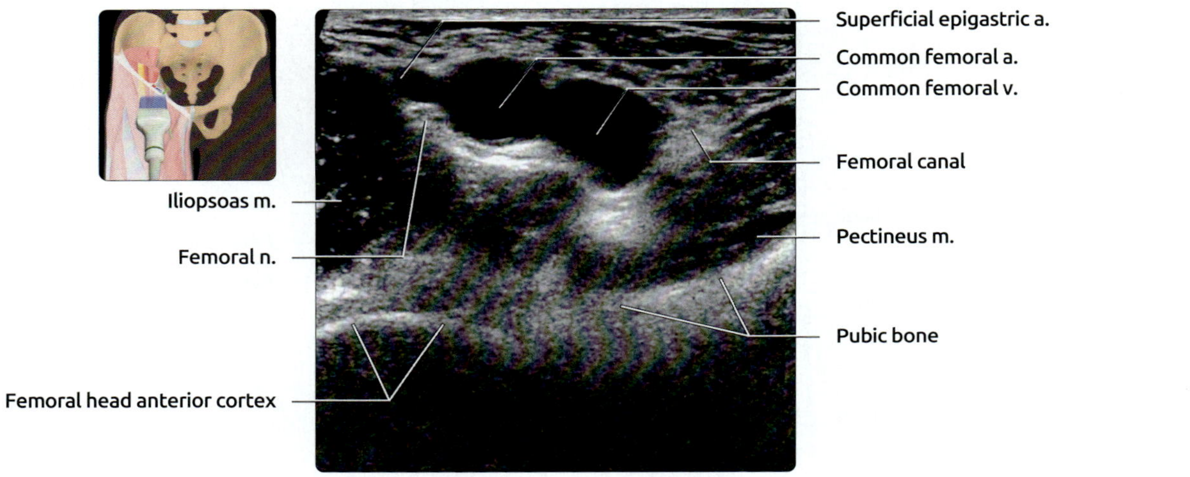

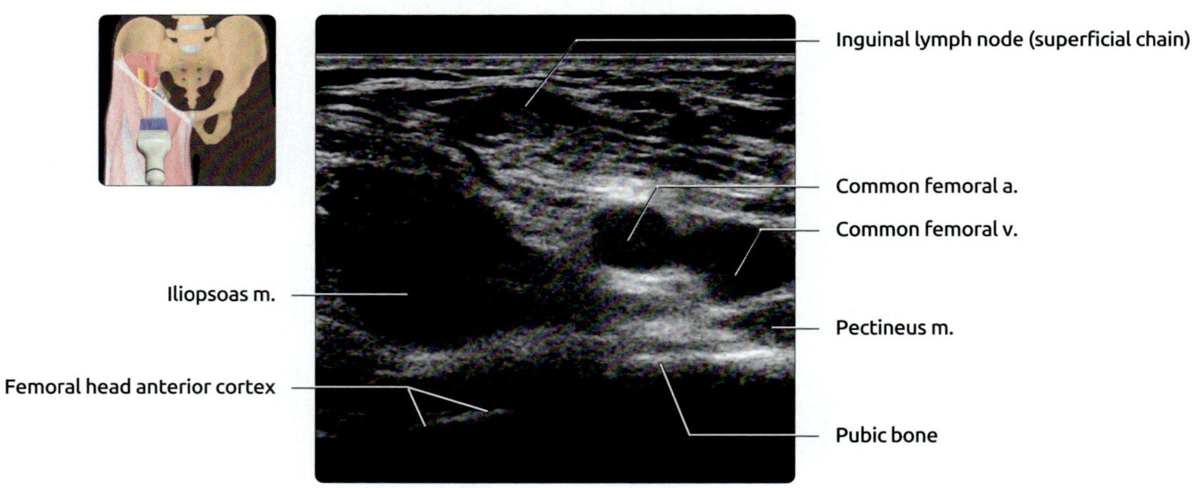

(Top) *Series of transverse scans through the femoral triangle are shown. Above the inguinal ligament, the femoral nerve, artery, and vein run medial to the iliopsoas tendon. The pubic bone acts as a hard support for the iliopsoas and pectineus muscular floor.* **(Middle)** *Slightly inferiorly, below the inguinal ligament, the femoral triangle commences. The mnemonic for the contents is NAVeL, which stands for femoral nerve, common femoral artery, common femoral vein, and lymph nodes (within femoral canal). The pectineus muscle forms the floor of femoral triangle. This floor is supported by the pubic bone and hip joint.* **(Bottom)** *In addition to lymph nodes in the femoral canal, lymph nodes are frequently seen superficially (the superficial inguinal chain) in the subcutaneous tissue.*

CENTRAL GROIN, FEMORAL TRIANGLE

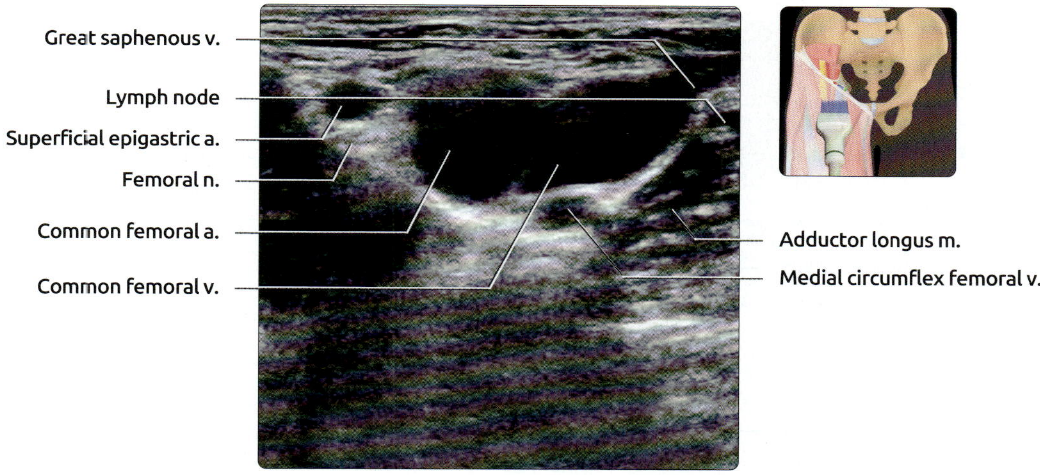

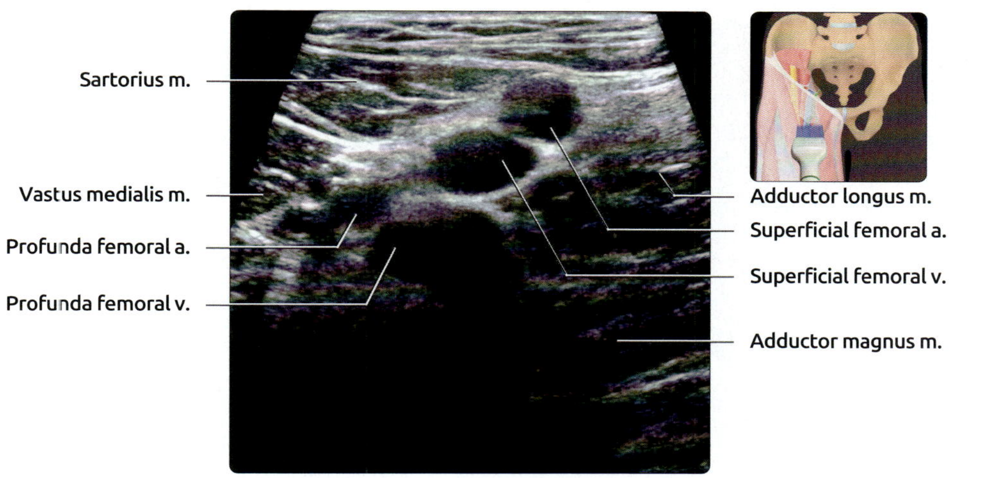

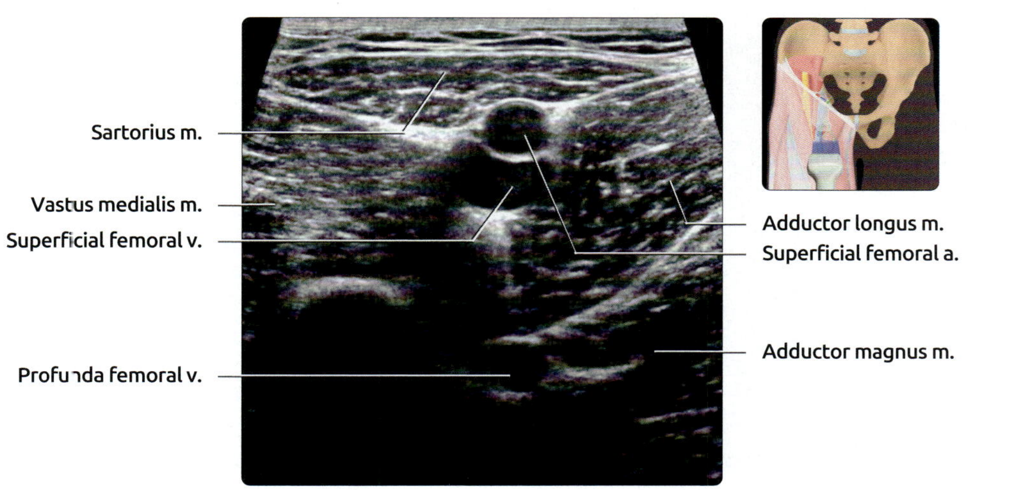

(Top) *Further inferiorly and still within the femoral triangle, the great saphenous vein perforates the anterior wall of the femoral sheath to enter the femoral vein. Also, the floor changes from the pectineus to adductor longus muscle with no bony support.* **(Middle)** *Transverse scan is shown at the apex of the femoral triangle, defined as where the sartorius muscle becomes the roof of the femoral triangle. Just before exiting the femoral triangle, the femoral vessels divide into superficial and profunda branches.* **(Bottom)** *Transverse scan shows the adductor canal. Distal to the femoral triangle, the superficial femoral vessels continue within a muscular conduit, the adductor canal. The groin muscles (the adductor longus and magnus muscles) form the medial wall and floor of this canal.*

LATERAL GROIN, ANTERIOR SUPERIOR ILIAC SPINE

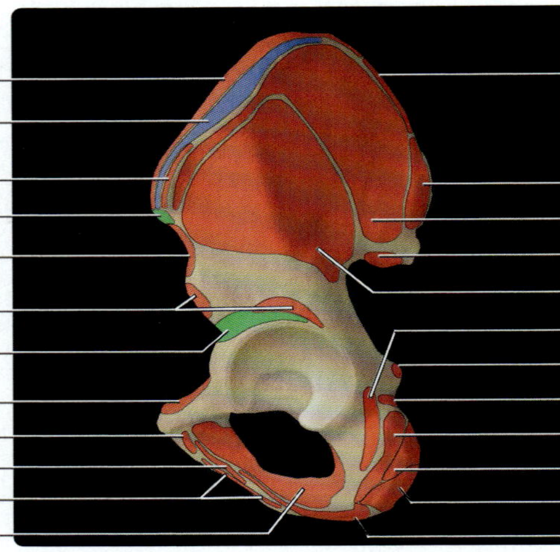

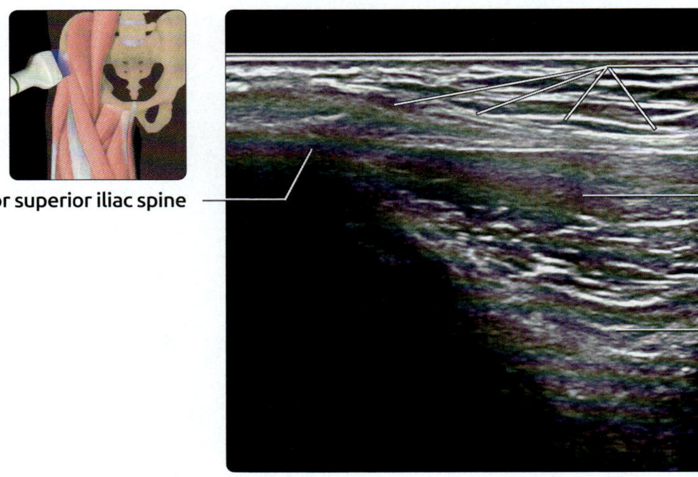

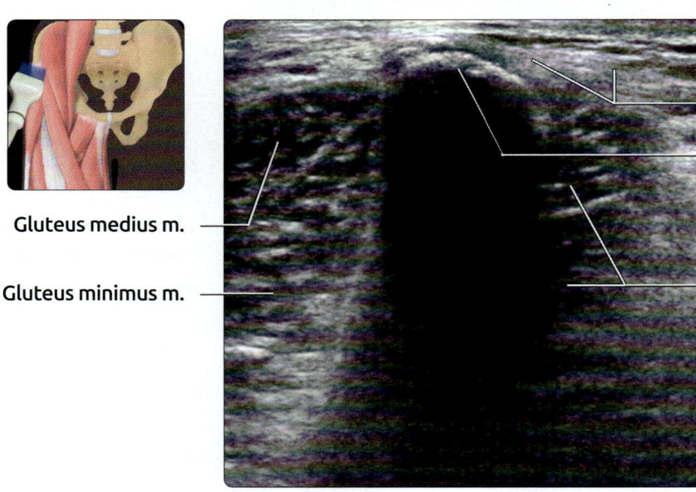

(Top) *Muscle and ligament attachments to the external surface of the pelvis are shown. Note the relatively small origin of the tendon of the adductor longus muscle. The adductor brevis muscle is just deep to the longus muscle. The origin of the gracilis muscle is lateral to the adductor brevis muscle. The adductor magnus muscle has a broad origin with its posterior fibers in close proximity to the hamstring tendon origins.* **(Middle)** *Oblique sagittal scan of the anterior superior iliac spine shows the origin of the inguinal ligament and sartorius muscle (the longest muscle in body). This slender muscle runs inferomedially over the iliopsoas muscle, covers the adductor canal, and inserts at the proximal tibia as part of the pes anserinus.* **(Bottom)** *Coronal scan of the anterior superior iliac spine shows division of muscle groups by this bone. Gluteal muscles lie lateral and hip flexors medial. The inguinal ligament originates here and runs inferomedially to the pubic tubercle.*

ACROSS GROIN, INGUINAL LIGAMENT

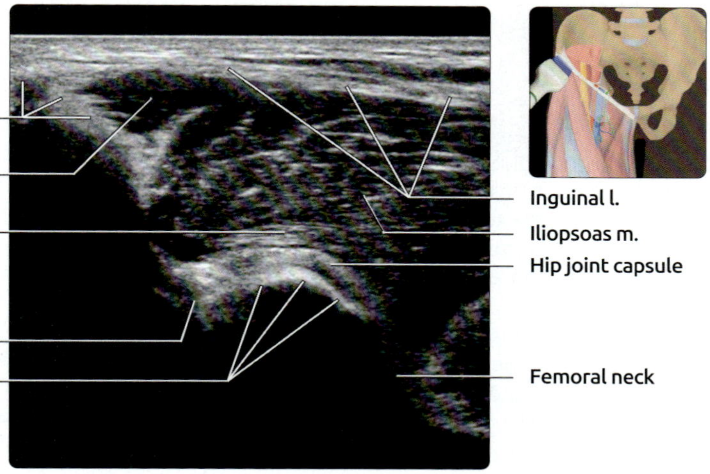

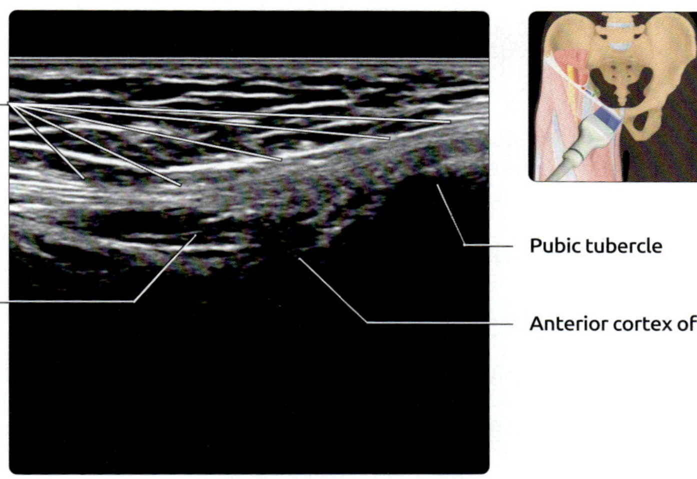

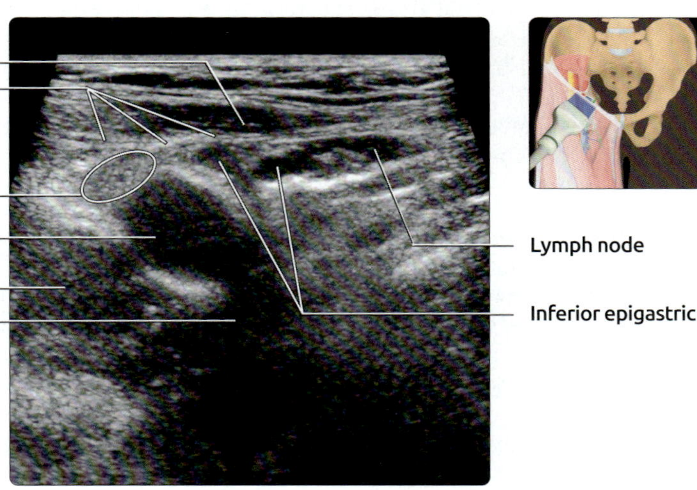

(Top) *Oblique transverse scan of the upper groin shows the inguinal ligament attachment on the anterior superior iliac spine with the sartorius muscle originating immediately inferior. The inguinal ligament runs inferomedially and covers the iliopsoas muscle.* **(Middle)** *Oblique transverse scan along the medial aspect of the inguinal ligament is shown. It runs over the pectineus muscle to insert on the pubic tubercle.* **(Bottom)** *Superficially focused oblique transverse scan over the upper femoral triangle shows the inguinal ligament forming its roof. Inferior epigastric vessels branch out from the femoral vessels and run superomedially. Clinically, the inferior epigastric vessels define the lateral boundary of direct inguinal hernias. Inguinal hernias that emerge lateral to the inferior epigastric vessels are considered indirect.*

TERMINOLOGY

Abbreviations

- Anterior superior iliac spine (ASIS)

IMAGING ANATOMY

Gluteal Muscles

- Components: Gluteus maximus, gluteus medius, gluteus minimus, tensor fascia lata, piriformis, superior and inferior gemellus, obturator internus and externus, quadratus femoris
- Common functions: Hip abduction, external rotation
- **Gluteus maximus**
  - Origin (O): Posterior gluteal line (ilium), posterior sacrum and coccyx, sacrotuberous ligament
    - Passes superficial to bare posterior facet of greater trochanter with interposed trochanteric bursa
  - Insertion (I): Iliotibial tract and proximal aspect of linea aspera on posterior aspect proximal femur
    - Inserts via short tendon
  - Function (F): Hip extension, abduction, and external rotation
- **Gluteus medius**
  - O: Between anterior and posterior gluteal lines (ilium)
  - I: Lateral and posterosuperior facets of greater trochanter with interposed subgluteus medius bursa
    - Inserts via short tendon
  - F: Hip abduction and internal rotation
- **Gluteus minimus**
  - O: External ilium between anterior and inferior gluteal lines
  - I: Anterior facet of greater trochanter with interposed subgluteus minimus bursa
  - F: Hip abduction and internal rotation
- **Tensor fascia lata**
  - O: External lip anterior iliac crest, external ASIS
  - I: Iliotibial tract
    - Iliotibial tract extends to Gerdy tubercle on anterolateral aspect tibia
  - F: Hip flexion, abduction, and weak internal rotation
  - Tensor fascia lata muscle is prone to hypertrophy or pseudohypertrophy
    - Hypertrophy is when there is true increase in muscle size > 50% more than contralateral side
    - Pseudohypertrophy is when fatty infiltration and paradoxical enlargement of muscle occurs due to lumbar radiculopathy or other denervation syndromes
- **Piriformis**
  - O: Anterior sacrum → passes through greater sciatic foramen
  - I: Piriformis fossa at greater trochanter proximal femur (may fuse with obturator internus and gemellus muscles)
    - Piriformis fossa between posterior femoral neck and posterior medial surface greater trochanter
      - Insertion site for piriformis, superior and inferior gemelli, obturator internus
  - F: Hip external rotation, assists abduction
    - Variation in sciatic nerve position through piriformis muscle seen in ~ 20% of patients undergoing hip MR
      - This variation is of questionable significance regarding piriformis syndrome
- **Superior gemellus**
  - O: Ischial spine
  - I: Piriformis fossa
  - F: Hip external rotation, weak abduction
- **Inferior gemellus**
  - O: Ischial tuberosity
  - I: Piriformis fossa
  - F: Hip external rotation, weak abduction
- **Obturator internus**
  - O: Internal surface of obturator foramen and membrane
  - I: Piriformis fossa (joins with gemelli tendons)
  - F: Hip external rotation, weak abduction
- **Obturator externus**
  - O: Edge of obturator foramen and obturator membrane
  - I: Posterior aspect of femur in trochanteric fossa
  - F: Lateral rotation of hip
- **Quadratus femoris**
  - O: Lateral ischial tuberosity
  - I: Quadrate line, intertrochanteric crest of femur
  - F: Strong hip external rotation

Rotator Cuff of Hip

- Gluteus medius and minimus tendons, trochanteric bursa, subgluteus medius and minimus bursae
  - Iliotibial tract, which is conglomeration of facial coverings of gluteus maximus and tensor fascia lata, overlies gluteus medius tendon insertion into lateral facet

Bursa

- Greater trochanter
  - Trochanteric (deep to gluteus maximus), subgluteus medius, and subgluteus minimus
- Ischiogluteal
  - Between ischial tuberosity and gluteus maximus muscle
- Gluteofemoral
  - Between iliotibial tract and vastus lateralis muscle
- Obturator internus
  - Between muscle and ischium
- Obturator externus
  - Beneath inferior border of muscle

Ischial Tuberosity

- Covered by gluteus maximus muscle when hip extended, uncovered in hip flexion
- Origin of hamstring tendons on ischial tuberosity
  - Semitendinosus and biceps femoris posteromedially
  - Semimembranosus laterally
- Sciatic nerve passes just posterolateral to ischial tuberosity
- Ischiofemoral impingement: Compression of quadratus femoris muscle between ischial tuberosity and lesser trochanter
  - Narrowing of ischiofemoral space and quadratus femoris space

Sacroiliac Joint

- Anteroinferiorly synovial, posterosuperiorly syndesmotic
- Weight transfer axial to appendicular skeleton
  - Limited gliding and rotation
- **Sacroiliac ligaments**

- **Interosseous sacroiliac**: Extremely strong within syndesmotic joint
- **Anterior sacroiliac** Weak ligaments
- **Posterior sacroiliac**: Posterior fibers of interosseous ligament
  - **Short dorsal sacroiliac**: Horizontal; lateral sacral crest S1 and S2 to posterior iliac crest
  - **Long dorsal sacroiliac**: Vertical oblique; posterior superior iliac spine to lateral sacral crest S3, S4

Posterior Sacroiliac Ligaments of Pelvis

- Primary stabilizers of pelvis
  - **Sacrotuberous**: Superior and inferior posterior iliac spines, sacrum, and coccyx to ischial tuberosity
  - **Sacrospinous**: Lateral sacrum and coccyx to ischial spine
  - **Iliolumbar**: Tip of L5 transverse process to iliac crest

Greater Sciatic Foramen

- Greater sciatic notch converted to foramen by sacrospinous ligament
  - Contents: Nerves, vessels, piriformis muscle
  - Passageway from pelvis to thigh
    - Structures identified as exiting pelvis above or below piriformis muscle
      - Above: Superior gluteal artery, vein, and nerve
      - Below: Pudendal nerve, internal pudendal artery and vein, nerve to obturator internus, sciatic nerve, inferior gluteal artery, vein, and nerve, posterior cutaneous nerve of thigh, nerve to quadratus femoris muscle
    - Relationship of structures
      - Most medial: Pudendal nerve
      - Most lateral: Sciatic nerve
      - Nerve to quadratus femoris muscle deep to sciatic nerve
- **Inferior gluteal artery**
  - Larger terminal branch of anterior division of internal iliac artery
  - Exits pelvis inferior to piriformis muscle
  - Posteromedial to sciatic nerve
  - **Inferior gluteal vein** travels with artery, drains into internal iliac vein
- **Inferior gluteal nerve**
  - Exits pelvis posterior to sciatic nerve
- **Superior gluteal artery**
  - Continuation of posterior division of internal iliac artery
  - Exits pelvis superior to piriformis muscle
  - **Superior gluteal vein** travels with artery, drains into internal iliac vein
- **Superior gluteal nerve**
  - Exits pelvis superior to piriformis muscle
- **Sciatic nerve**
  - Exits pelvis inferior to piriformis muscle (typically)
    - Most lateral structure
  - Beaton and Anson classification
    - Type A: Sciatic nerve below piriformis (normal)
    - Type B: Sciatic nerve divisions through and below piriformis
    - Type C: Sciatic nerve divisions above and below piriformis
    - Type D: Sciatic nerve through piriformis
    - Type E: Sciatic nerve divisions through and above piriformis
    - Type F: Sciatic nerve above piriformis
- **Pudendal nerve and internal pudendal vessels**
  - Exit inferior to piriformis muscle
  - Nerve most medial structure
- **Nerve to obturator internus muscle**
  - Exits pelvis inferior to piriformis muscle

Lesser Sciatic Foramen

- Lesser sciatic notch converted to foramen by intersection (crossing) with sacrotuberous and sacrospinous ligaments
- Contents
  - **Pudendal nerve and internal pudendal vessels**
    - Pass over sacrospinous ligament
    - Exit buttock, enter perineum
  - **Obturator internus muscle**

ANATOMY IMAGING ISSUES

Imaging Recommendations

- US
  - Subcutaneous fat in gluteal region can be thick and mistaken for muscle
  - Look for investing fascia and fiber orientation to identify muscle layer
  - Anisotropy gives gluteal muscles varying echogenicity due to different orientation of fibers
    - Allows clear distinction between muscles
    - Lesions may be better demonstrated by examining along muscle fibers to make lesion more conspicuous
  - Iliac crest, greater trochanter, and ischial tuberosity are important bony landmarks for orientation
  - Use of panoramic (extended field-of-view) images will help referring clinicians to understand location of lesion
    - Panoramic images should be performed in standard anatomic planes to help interpretation
    - Landmarks, especially osseous (e.g., iliac crest, sacrum, ischial tuberosity), also help interpretation of panoramic image
    - Panoramic images do not provide accurate measurements
- MR
  - Diagnose masses, anatomic anomalies, other causes of nerve compression
  - Preferred modality for sacroiliitis
- CT
  - Mainly used for assessing bone and joint (sacroiliac) abnormalities

Imaging Pitfalls

- Lower frequency transducers (5 MHz or below) should be used initially to exclude deep-seated lesions
- Mistaking thick subcutaneous fat of gluteal region as muscle layer
- Deep tendons around hip region more hypoechoic than most superficial tendons

GRAPHICS, GLUTEAL MUSCLES

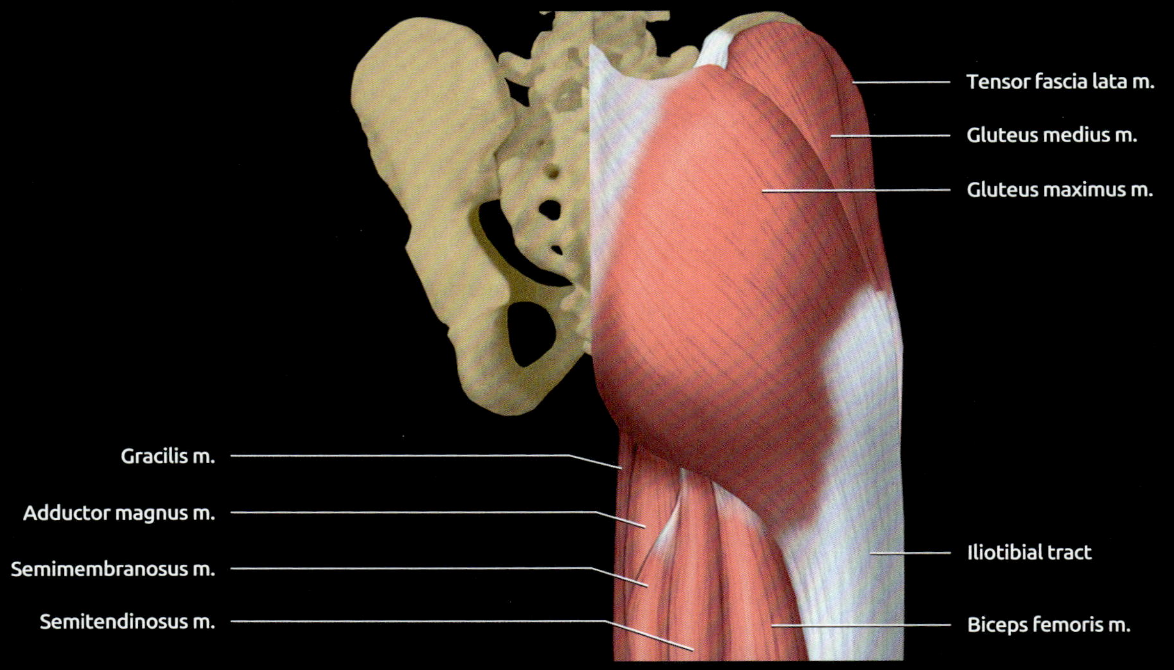

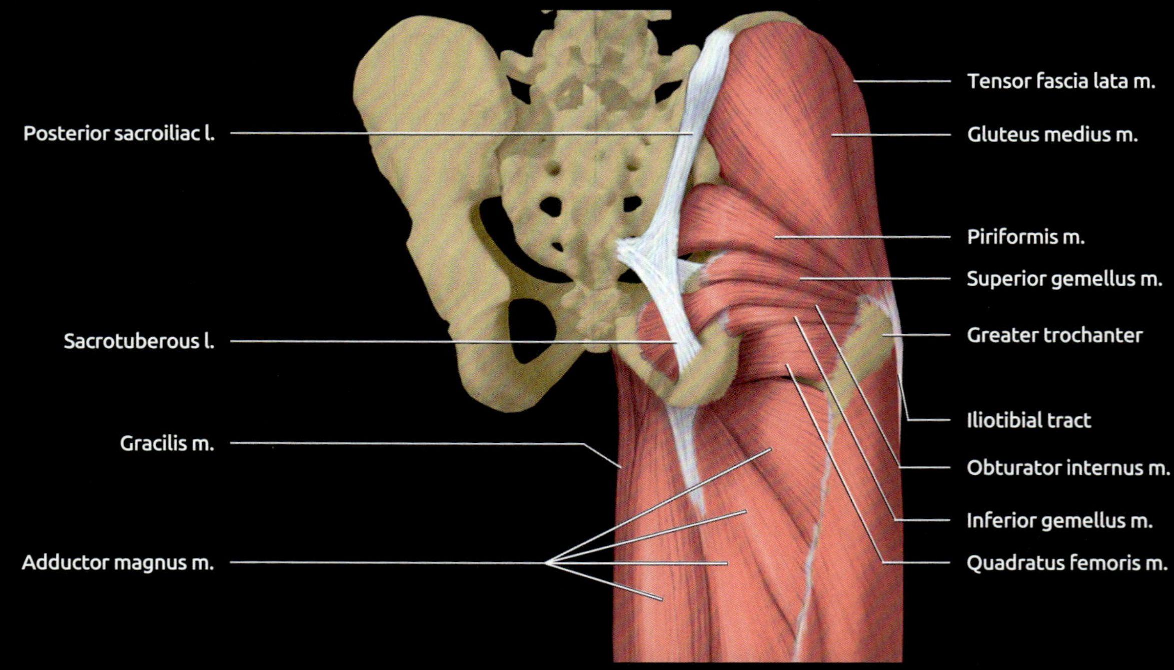

(Top) *The gluteus muscles from anterior to posterior and deep to superficial are the gluteus minimus, medius, and maximus. These muscles form the bulk of the gluteal region. The gluteus maximus, in particular, covers most of the deep gluteal structures.* **(Bottom)** *After removing the gluteus maximus, the deep muscles are visible: Piriformis, superior gemellus, obturator internus, inferior gemellus, and quadratus femoris. These all function as external rotators of the thigh. Note the piriformis muscle, which exits via the greater sciatic foramen, and the obturator internus muscle via the lesser sciatic foramen.*

GRAPHIC, SCIATIC NERVE AND OVERVIEW, GLUTEAL

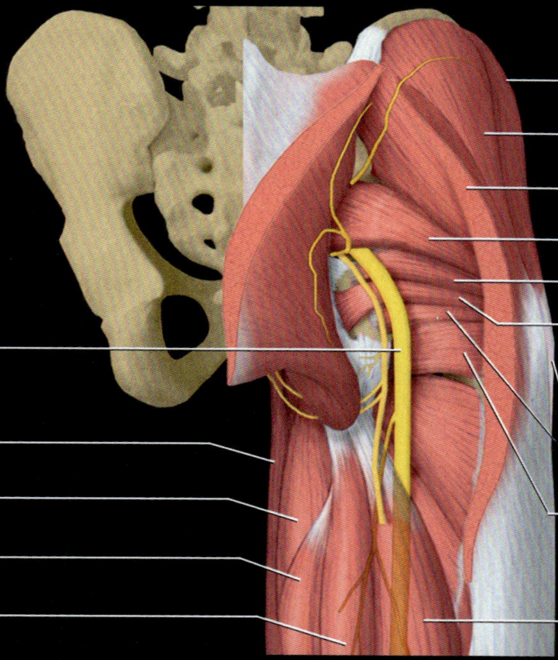

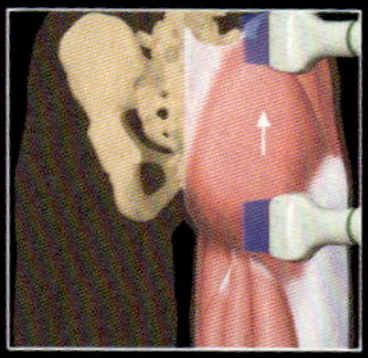

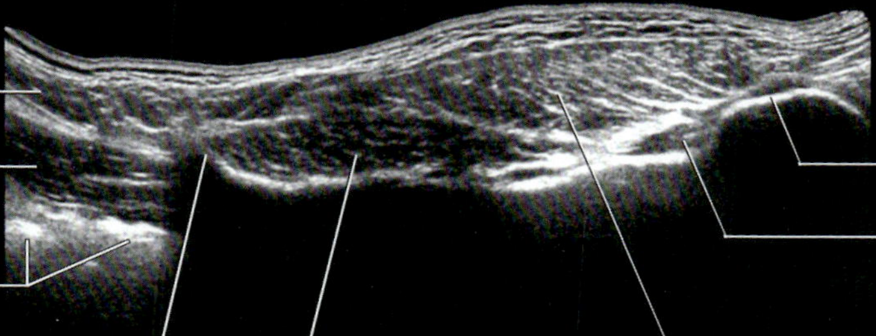

(Top) *The deep external rotators of the hip share an intimate relationship with the sciatic nerve, which emerges between the piriformis and superior gemellus muscles and then between the gluteus maximus and the rest of the deep external rotators. More distally, the sciatic nerve is covered by the biceps femoris muscle.* **(Bottom)** *Sagittal panoramic view shows the medial aspect of the gluteal region. The iliac crest separates the back muscles from the glutei. At this level, the gluteus maximus muscle forms the main muscle bulk. Inferiorly, the gluteus maximus runs across the surface of the ischial tuberosity in this extended (hip joint) position. Coverage is lost (gluteus maximus moves superiorly) when the hip joint is flexed and the ischial tuberosity is exposed.*

OVERVIEW, GLUTEAL REGION

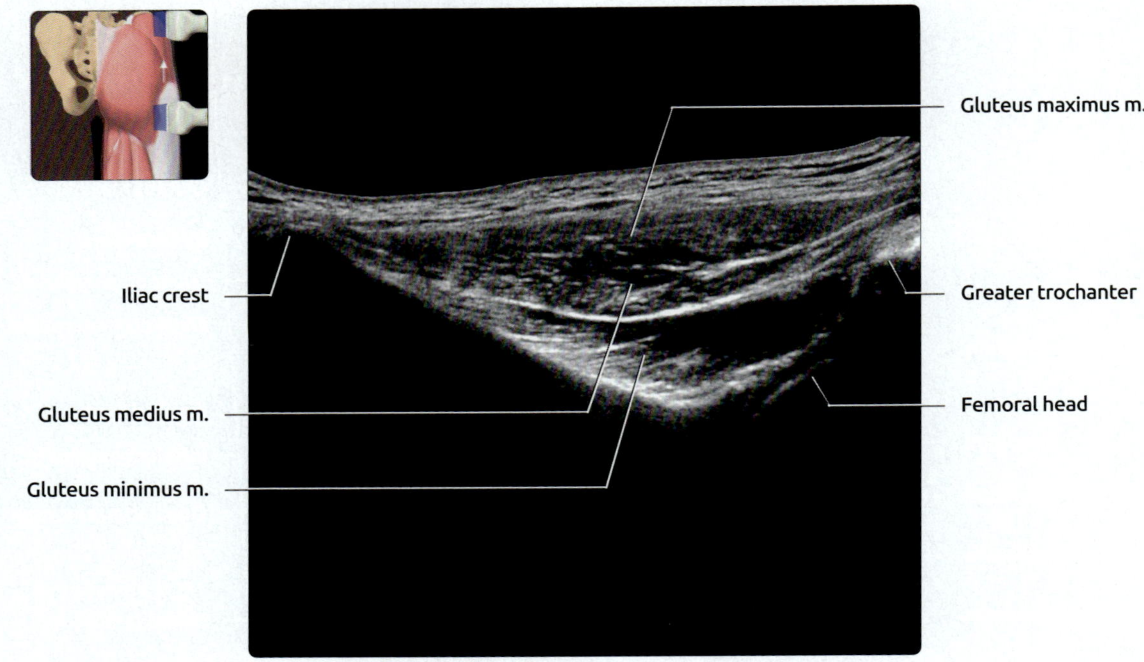

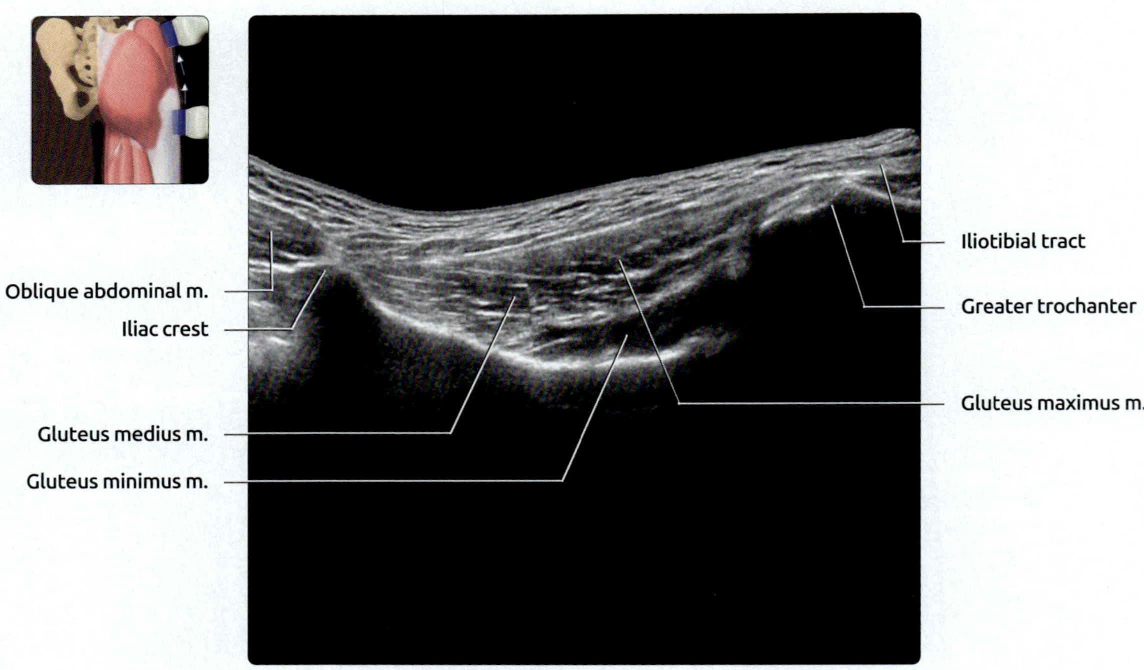

(Top) *Oblique sagittal panoramic view shows the lateral 1/2 of the gluteal region. The close relationship between the gluteus minimus muscle and the posterior aspect of the hip joint can be appreciated here. In this region, all 3 gluteal muscles are present and can be distinguished by their different echogenicity. The gluteus maximus muscle fibers are running approximately horizontally along the scan plane and, being most superficial, produce the highest echogenicity. The gluteus medius is running obliquely (lateral inferiorly) and thus shows short echogenic fibers. The gluteus minimus runs obliquely and almost horizontally toward the anterior facet of the greater trochanter. It is the deepest gluteal muscle and usually appears hypoechoic.* **(Bottom)** *Coronal panoramic view shows the gluteus medius and minimus muscles and their insertion on the greater trochanter. The iliotibial tract runs on the surface of the greater trochanter.*

SAGITTAL GLUTEAL

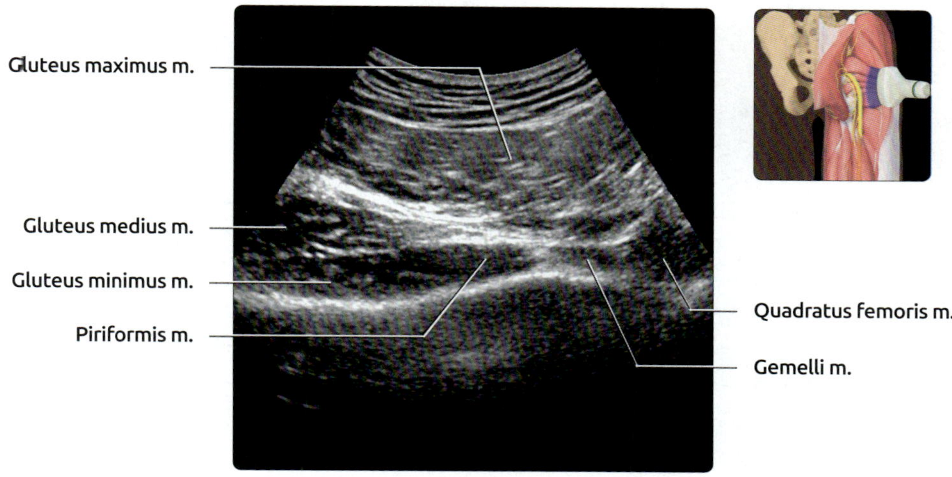

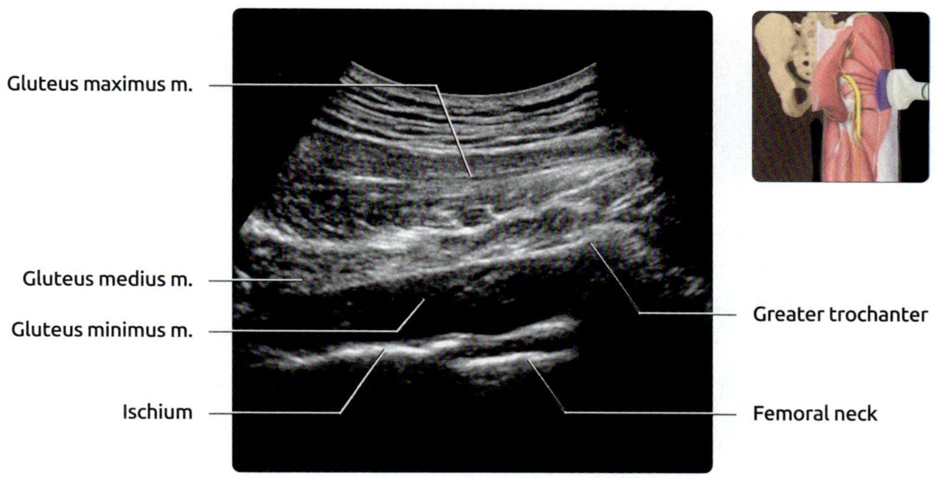

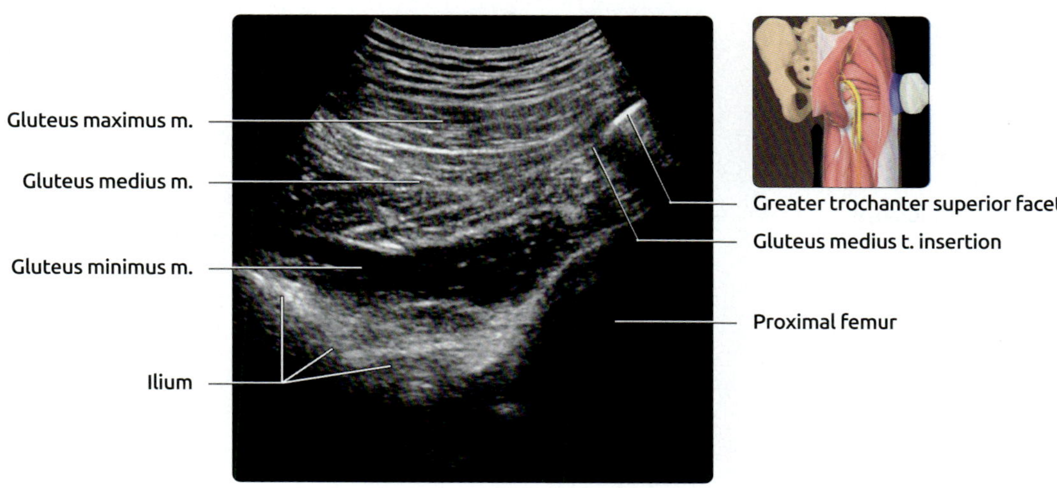

(Top) *Oblique sagittal US in the lateral 1/2 of the gluteal region shows the 3 gluteal muscles and their relationship with the deeper, small external rotators: Piriformis, superior and inferior gemellus, and quadratus femoris muscles.* **(Middle)** *Further laterally, the femoral neck can be seen under the gluteus minimus. Again, the different echogenicity and echotexture of the 3 gluteal muscles can be appreciated.* **(Bottom)** *Further laterally, the gluteus medius tendon fibers insert onto the lateral and posterosuperior facets of the greater trochanter. The gluteus minimus tendon runs anteriorly under the gluteus medius to insert on the anterior facet of greater trochanter.*

TRANSVERSE GLUTEAL

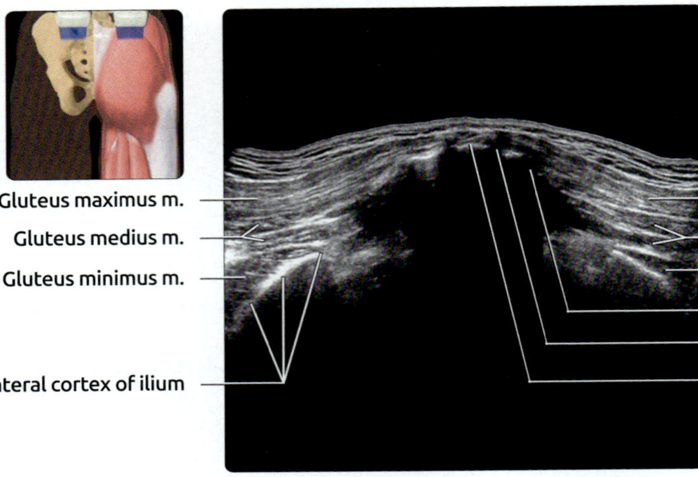

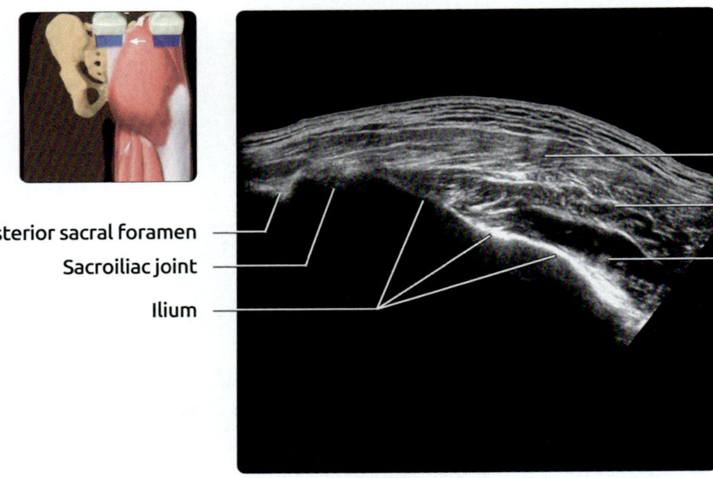

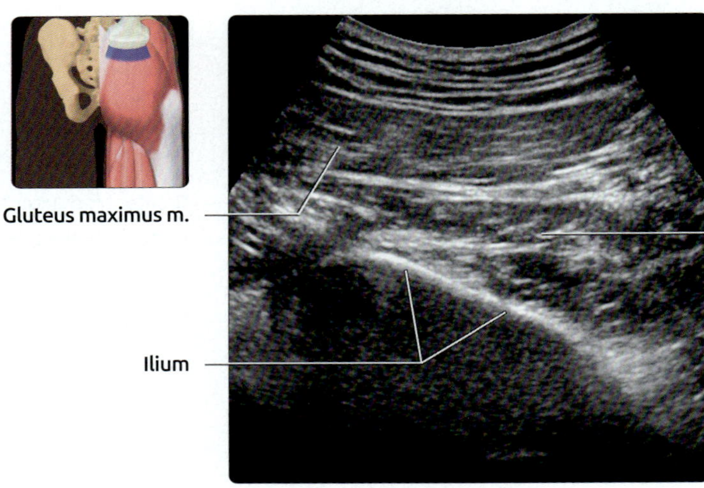

(Top) *Transverse panoramic US is seen across the dorsal surface of the sacrum. This is the medial origin of the gluteus maximus muscle. Deep to this muscle, the gluteus medius and gluteus minimus muscles can be seen originating from the iliac surface.* **(Middle)** *Transverse panoramic view shows the right gluteal region. The 3 gluteal muscles show their signature difference in echogenicity and echotexture even in the horizontal plane.* **(Bottom)** *Transverse focused view of the medial gluteal region shows the fiber orientation of the different muscles. Those of the gluteus maximus are longer, as they run roughly horizontally compared to those of the gluteus medius. This orderly fiber pattern is helpful in detecting intra- and intermuscular lesions in the gluteal region (orderly fiber alignment tends to be disrupted in disease).*

ISCHIAL REGION

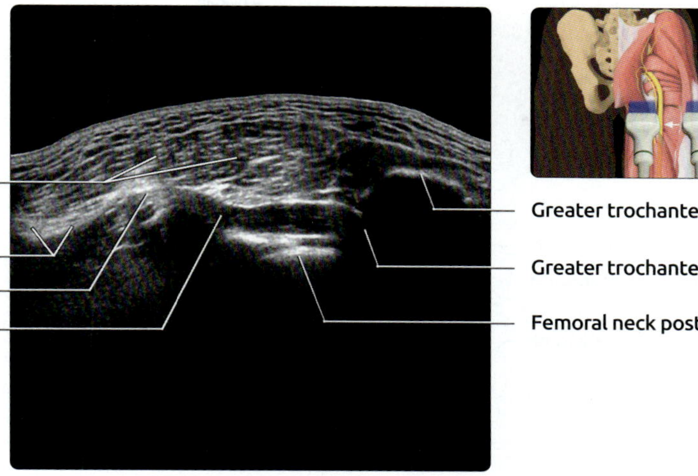

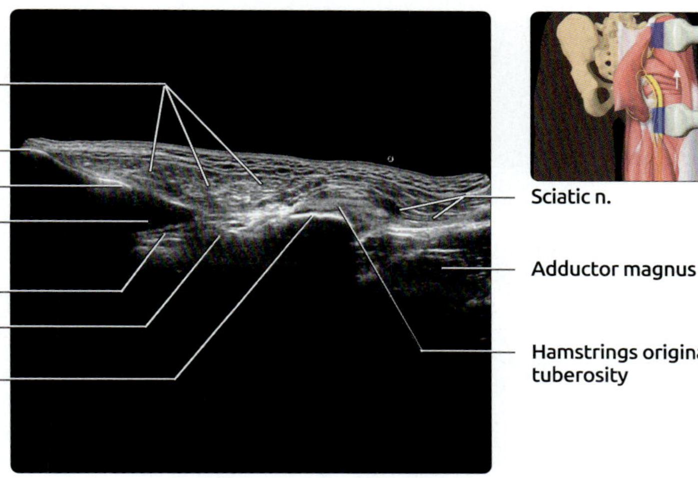

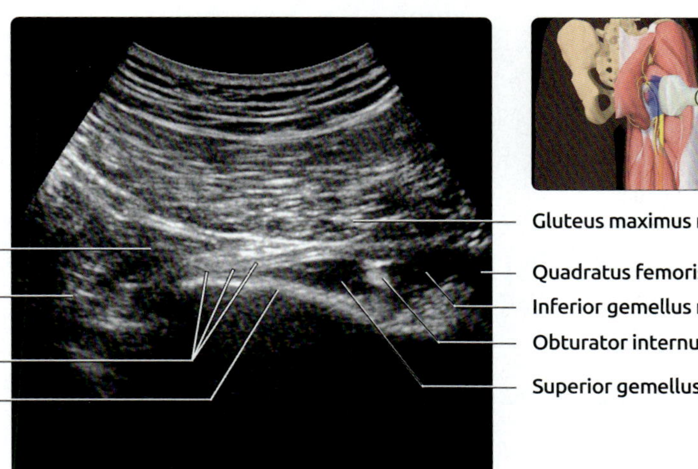

(Top) *Transverse panoramic view shows the ischial region. The gluteus maximus forms most of the bulk at this lower gluteal level. Deep to this, small external rotators of the hip can be seen (in this scan, the quadratus femoris muscle).* **(Middle)** *Sagittal panoramic US is seen from the lower iliac bone to below the ischial tuberosity. The greater sciatic foramen is demonstrated in this scan. The piriformis and sciatic nerve exit the pelvis via this foramen and run inferolaterally, the former to insert on the greater tuberosity and the latter inferiorly to supply the leg.* **(Bottom)** *Sagittal view shows the greater sciatic foramen using the ischial tuberosity as a landmark. The sciatic nerve and its relationship with the small external rotators of the hip are shown. The sciatic nerve emerges between the piriformis and superior gemellus muscles after exiting the pelvis through the greater sciatic foramen.*

TRANSVERSE ISCHIAL

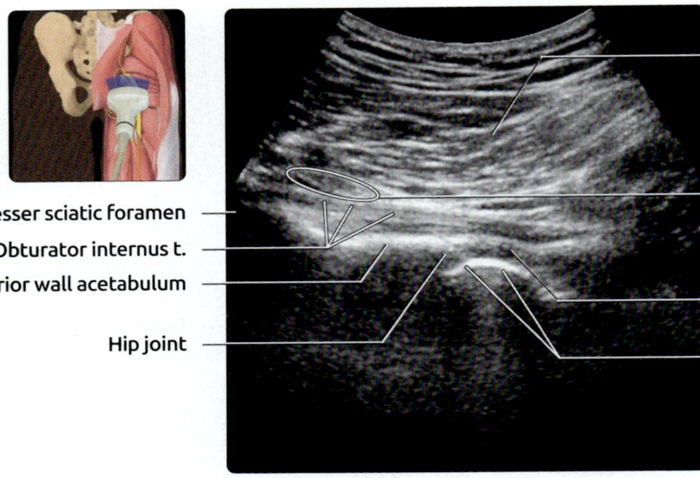

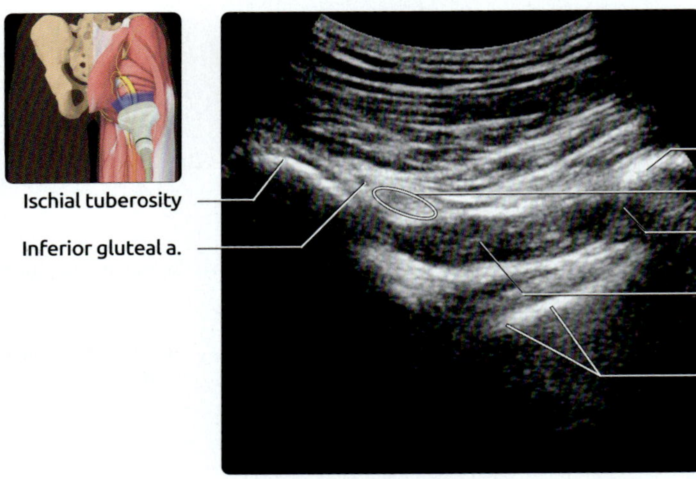

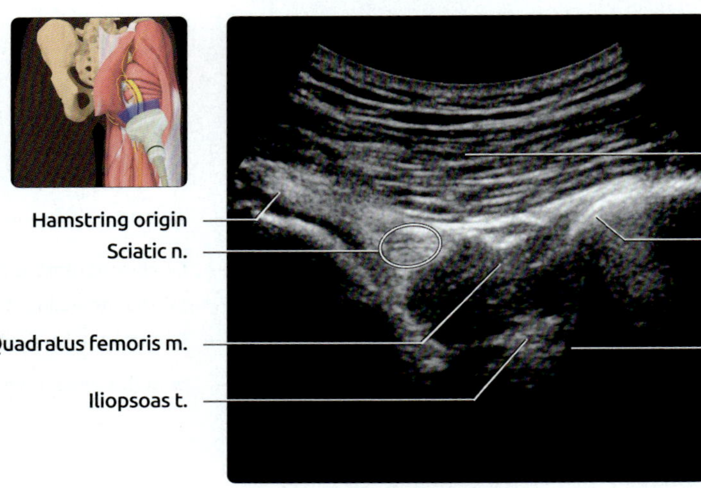

(Top) *Series of 3 transverse scans show the inferomedial aspect of the gluteal region. This image shows the obturator internus tendon exiting the lesser sciatic foramen, running laterally toward the lower aspect of the hip joint. The sciatic nerve runs superficial to this tendon.* **(Middle)** *Slightly inferiorly, this image shows the inferior gemellus muscle arising from the ischial tuberosity, running laterally over the neck of the femur and inserting at the piriformis fossa on the medial surface of the greater trochanter. The sciatic nerve runs superficial to this muscle.* **(Bottom)** *Further inferiorly, this image shows the quadratus femoris muscle arising from the lateral border of the ischial tuberosity and running laterally to insert on the quadrate tubercle of the femur. The sciatic nerve also runs superficial to this tendon.*

ISCHIAL TUBEROSITY

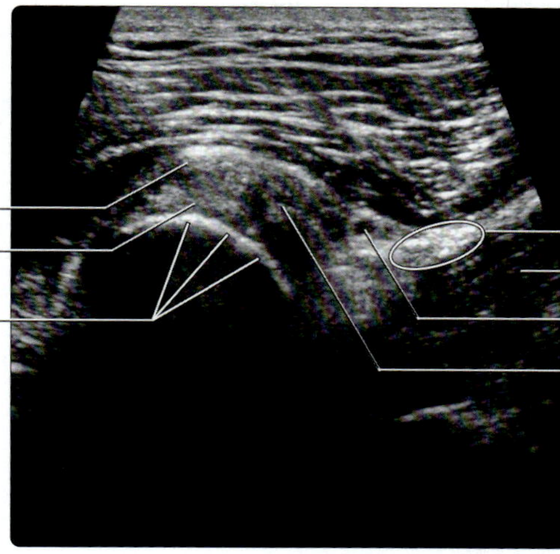

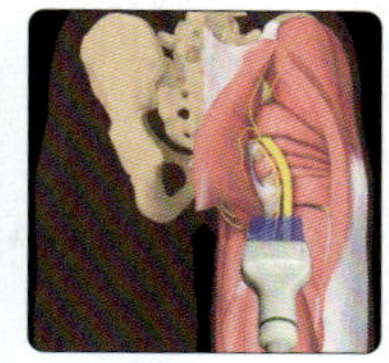

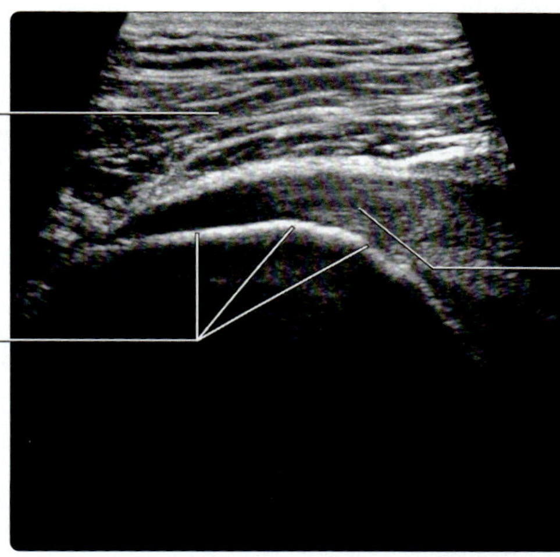

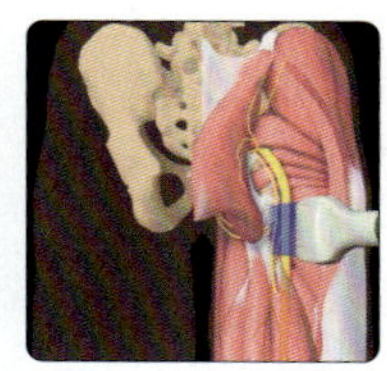

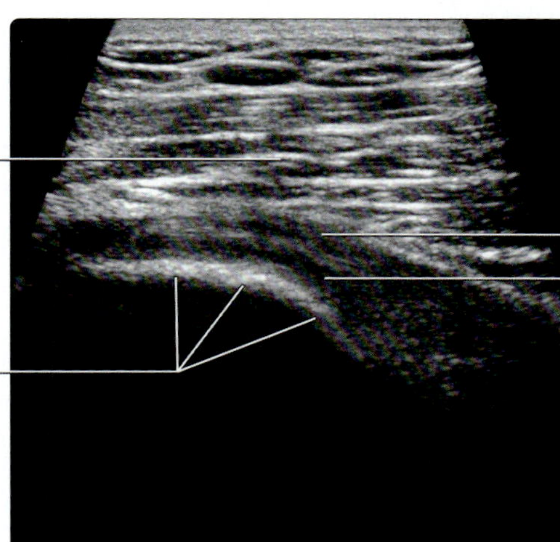

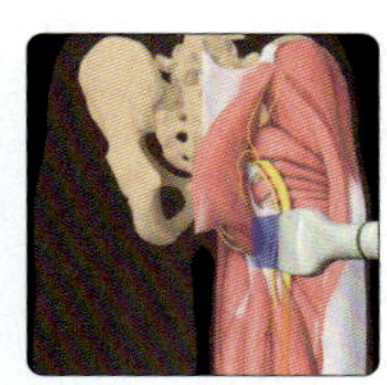

(Top) *Transverse scan shows the ischial tuberosity. The biceps femoris tendon originates lateral to the semimembranosus and semitendinosus tendons. The sciatic nerve and inferior gluteal vessels run just lateral to the ischial tuberosity. Inflammation in this region (e.g., ischiogluteal bursitis) may cause irritation of the sciatic nerve. More importantly, aspiration or injection to treat ischiogluteal bursitis should be performed with image guidance to avoid damaging the sciatic nerve.* **(Middle)** *Sagittal US of the lateral aspect of the ischial tuberosity shows the origin of the biceps femoris tendon.* **(Bottom)** *Sagittal US more medially on the ischial tuberosity shows the insertion of the semitendinosus and semimembranosus tendons. The semitendinosus muscle originates and runs for most of its course on dorsal surface of the semimembranosus tendon.*

SECTION 2

Technique

Approach to Musculoskeletal Ultrasound	360
Musculoskeletal Ultrasound Artifacts	364
Shoulder Ultrasound	372
Elbow Ultrasound	380
Wrist Ultrasound	388
Hand Ultrasound	396
Groin Hernia Ultrasound	400
Hip Ultrasound	408
Knee Ultrasound	416
Ankle Ultrasound	424
Foot Ultrasound	432
Writing an Ultrasound Report	438

Ultrasound Machine

To perform musculoskeletal US, you first need access to a US machine equipped with a high-resolution transducer. Second, you need to understand, at least, the very basics of US technology. High-resolution transducers range from 10-22 MHz and are ideal for examining the superficial soft tissues. Similar to MR and increasing field strength, the relationship between increasing US transducer frequency and resolution is not linear. Most musculoskeletal US can be performed with a linear transducer in the 10-12 MHz range. Moving from 12-17 MHz will improve resolution to a mild degree but usually does not significantly alter diagnostic capability. In other words, access to one high-frequency transducer is usually sufficient to begin musculoskeletal US. A disadvantage of high-resolution transducers is limited depth penetration; for deeper areas (gluteal region, proximal thigh) and large masses, a lower frequency transducer, such as a 5 MHz or 7.5 MHz transducer, is needed. The transducer is a delicate (and expensive) piece of equipment and must be handled with care. The transducer contains an interlinked array of piezoelectric crystals. These crystals convert an electrical signal into a US wave and vice versa. The same array of piezoelectric crystals emits and receives the US signal. This is like the transducer shouting out into tissues and then listening for a returning sound or echo. This "shouting and listening" happens many thousands of times a second (pulse repetition frequency). To improve transmission of sound waves, US gel is applied between the skin and the transducer. Sound waves are transmitted from the transducer through the US gel to the skin and deeper tissues. The ability of tissues to transmit sound waves is known as acoustic impedance. At tissue interfaces where there is a large difference in acoustic impedance, most sound waves are reflected back to the transducer, producing a bright echo. At interfaces where there is little or no difference in acoustic impedance, most sound waves are transmitted and not reflected back to the transducer, producing a dark echo. As expected, most sound waves that contact an interface parallel to the transducer face will be reflected back to the transducer. Sound waves impacting an irregular surface will be retracted away from the transducer (refraction). Sound wave attenuation also results from acoustic absorption, which is greater in tissues of higher viscosity.

Getting Started

The best and only way to learn musculoskeletal US is to get started. In general, the more US examinations you perform, the better you should become. Be conscious of working posture, etc., at the outset. Ensure that both you and the patient are in a comfortable and the most effective position to examine the area of interest. Take a brief clinical history from the patient and palpate any lumps or swelling, if present. This is tremendously helpful when interpreting the subsequent US findings. Warm the US gel using a dedicated US gel warmer, especially if examining the chest or abdomen. Hold the transducer between the thumb and finger. You should have either your little finger or the hypothenar eminence in contact with the patient's skin. This helps to anchor the transducer and ensure sufficient but not excessive transducer pressure on the skin. When assessing superficial structures, and especially when assessing color flow in superficial tissues, one should actively almost lift the transducer from the skin using plenty of intervening US gel to avoid distorting the tissues or effacing the vascularity. Develop a systematic examination procedure for each joint or region that includes all critical anatomical structures. Start your examination at the most symptomatic area and at least include the immediate surrounding area in your examination. For time efficiency, it is not necessary to complete the whole joint examination in every patient. For example, if the patient has medial ankle pain, concentrate on examining the medial and ankle joint proper; there is little to be gained from routinely examining the lateral or posterior ankle structures if there are no symptoms related to these areas. In children, a full examination may not be possible before the child becomes restless, so again concentrate on the most critical area. Sedation is not usually necessary or beneficial in children.

Machine Setup and Scanning

Understanding proper US machine setup is crucial to conducting high-quality US. US is a more operator-dependent examination than CT or MR. One should be pedantic about trying to obtain the best possible quality US examination in every patient. There should be no such thing as a "quick US examination"; this is a recipe for missing significant pathology. Optimize image quality before and during scanning. It is helpful to begin by examining the contralateral normal side, as this allows you to both familiarize yourself with normal US anatomy and also set up the machine properly for that body part. The initial setup will often need frequent readjustment throughout the US scan. It is better to take a few high-quality images than many suboptimal images. Orient the image so that the proximal end is on the left side of the image and the distal end is on the right. Make sure there is no shadowing artifact from too little gel at the sides of the image, adjust the depth to the minimum needed to see all structures of interest, adjust the focal position and zone, gain (overall image gain) and time gain compensation to optimize image quality. During scanning, tissues can be compressed by the transducer to broadly gauge their stiffness. This dynamic capability of US is also helpful when differentiating between fluid and soft tissue in suspected abscess or hematoma, or between synovial proliferation and effusion. The dynamic capability of US is also helpful when assessing tendon tears, tendon subluxation, or shoulder impingement. Color or power Doppler is used routinely in the assessment of soft tissue inflammation, synovitis, and soft tissue masses. The relative sensitivity of color or power Doppler does vary from machine to machine. Color Doppler is flow-direction sensitive, whereas power Doppler is not flow-direction sensitive, though it is considered more sensitive overall to detecting slow flow and flow in small vessels. Low-frequency transducers reveal slow velocity flow better. Reducing the pulse repetition frequency and lowering the wall filter will also improve detection of slow vascular flow.

Supplementary Imaging Techniques

With conventional US, the transducer emits and receives a sound pulse of a specific frequency at a single angle of insonation. The returning signal is of the same frequency though weaker with the transducer "listening" specifically for that frequency. Several techniques over and above standard grayscale imaging may enhance image resolution, depiction, and interrogation. **Spatial compound US** amalgamates sonographic data from several different angles of insonation into a single image. By averaging image data from multiple angles of insonation, spatial compound imaging improves definition of soft tissue planes, reduces speckle and other noise, and improves image detail. With **tissue harmonic**

imaging, the transducer listens for not only returning signals of the transmitted frequency, but also returning signals of harmonic frequencies (generated by passage of the sound wave through tissue) that are twice the transmitted frequency. Benefits of tissue harmonic imaging are most apparent in the midfield region, helping to improve tissue contrast and edge demarcation of soft tissue masses or tendon tears, though with loss of some spatial resolution. **Extended field-of-view imaging** allows panoramic view, which is helpful for measuring large masses, comparing tissue echogenicity in confluent areas, and perceiving anatomical relationships.

Recognizing Pathology

Good-quality musculoskeletal US is dependent on good understanding of musculoskeletal anatomy. You do not want your US technique to be hampered by trying to recall relevant anatomy. It is helpful to familiarize yourself with normal anatomy on the asymptomatic side before examining the symptomatic side. Be aware, however, of using the normal side as an internal reference of normality. Tendinosis and nerve entrapment, in particular, are often asymptomatic, and US often shows abnormal findings (to a comparable or lesser degree) on the asymptomatic normal side. Therefore, you need a good idea as to what the normal appearances are before proceeding to diagnose tendon or nerve pathology. Another prerequisite is knowledge of the likely pathologies in specific areas. Most musculoskeletal pathology is site specific. "You only find what you look for; you only look for what you know." Most tissues (tendons, nerves, ligaments) have the same appearance, irrespective of location. Also, tissue-specific pathology has a similar appearance, irrespective of location. For example, tendinosis, tendon tears, or tenosynovitis appear similar irrespective of location. Tendons, nerves, and ligaments are prone to anisotropy. This is the most common artifact you will encounter, and it can often be put to good use. To overcome this artifact, frequent transducer realignment to examine different areas of curved structures is necessary. For example, anterior and posterior aspects of the supraspinatus tendon need to be assessed separately on transverse imaging. For soft tissue masses, evaluate location, extent, internal structure, consistency, vascularity, relationship to adjacent critical structures, surrounding tissues, and regional nodes.

Clinical Implications

US is the best 1st-line investigation for most soft tissue pathology. An explanation for the patient symptoms can usually be found. Musculoskeletal US is most useful when symptoms are localized to a specific area, e.g., if the patient has point tenderness. US is much less useful when symptoms are not well localized, e.g., pain related to the whole knee or ankle joint. Try to grade every abnormality encountered; there is little benefit in saying that there is supraspinatus tendinosis present. You need to know if this tendinosis is minimal, mild, moderate, or severe. Most superficial soft tissue masses can be accurately diagnosed with US. US diagnosis is formulated not on a basis of 1 or 2 US signs, but on the relative strength of multiple recognized US signs interpreted in conjunction with the clinical presentation. When following tendon or nerves along the limb or around a joint, follow each nerve or tendon in turn rather than together. In other words, do not try to follow 2 tendons, such as the peroneal brevis and longus tendons, at the same time. If the US findings are indeterminate, you can either follow-up with repeat US, proceed to an alternative investigation such as MR, or proceed to US-guided biopsy (or fine-needle aspiration for cytology). Similarly, if a satisfactory explanation for nontumoral symptoms is not found on US, it is usually best to proceed to further investigation, usually MR. There is usually no benefit to be gained from repeating the US examination. MR has particular advantages over US, and in these situations it is usually best to proceed directly to MR. MR is useful for examining areas not accessible to US, e.g., intraarticular or intraosseous abnormalities. MR allows improved perception of large objects (particularly deep, ill-defined lesions and large tumors) and provides a clear anatomical surgical roadmap. MR allows a more reliable quantification of many structures (e.g., synovial volume) and is more sensitive in depicting soft tissue edema, as in myositis. US should ideally precede MR for most soft tissue problems. MR examination, if necessary, can then be tailored to answer only those specific questions not answered by US. For example, there is no need to examine regional lymph nodes if US assessment has already been performed, nor is there a need to give intravenous contrast if tumor vascularity has always been determined by US. If deemed necessary, review additional imaging such as radiography, CT, or MR either before, during, or at the end of the US examination. This too, in some instances, can greatly help with US interpretation. At the end of the examination, explain the findings to the patient.

Newer Ultrasound Techniques

3D US acquires volumetric US data, which can be reconstructed along any plane. This has been used to evaluate rotator cuff tears, soft tissues masses, and synovial proliferation. **Fusion imaging** allows real-time US imaging to be superimposed onto preloaded CT or MR imaging and viewed simultaneously on the US screen. Fusion imaging has been used in rheumatoid arthritis, bone biopsy, and sacroiliac joint injection. **US elastography** measures and displays tissue elasticity and is divided into 2 types. **Strain elastography** involves gentle manual transducer compression of the tissues to measure relative tissue displacement (strain). This strain is displayed as a unitless color-coded map (elastogram) as strain is dimensionless. Stiff areas appear as low-strain areas on elastograms. Operators may compare 2 regions of interest on elastograms to compute strain ratios. **Shear wave elastography** employs a specially modified transducer that produces high-intensity acoustic pulses (push pulses). These induce weak tissue displacements termed shear waves, whose velocities are then tracked by diagnostic impulses. Shear waves travel faster in stiff tissues. Shear wave elastograms have a quantitative output (m/s or kilopascal). Elastography has been studied in tendons, muscles, and soft tissue masses and may be able to detect tendinosis before grayscale imaging. Although elastography seems promising, it requires standardization and precise identification of those areas where it can provide added value over standard grayscale imaging. **Microbubble-enhanced US** improves the detection and quantification of vascularity in, for example, synovitis.

(Left) *Schematic shows a US transducer with and without external casing. An electrical impulse stimulates the piezoelectric crystal, which emits a sound wave . The returning, weaker sound wave of similar frequency is detected by the same crystal stimulating an electrical pulse.* **(Right)** *Extended field-of-view US shows an Achilles tendon following repair . Thickening of the whole tendon can be readily perceived. Thickening is a typical postoperative finding and can persist for years.*

Ultrasound Transducer

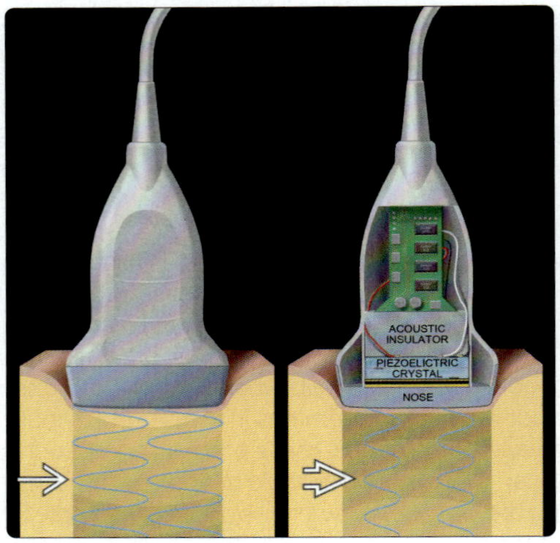

Extended Field-of-View Imaging

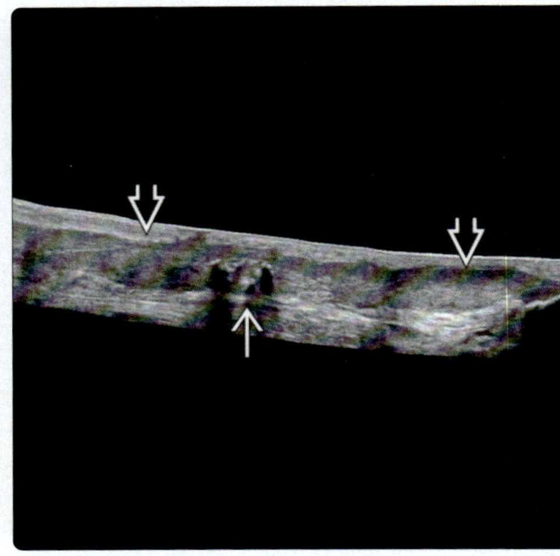

(Left) *Transverse standard grayscale US shows a healing rib fracture with callus .* **(Right)** *Transverse US with compound imaging shows early ossification of the callus more clearly than with standard US imaging. The demarcation between the rib and costal cartilage is also better appreciated on compound imaging, as are the soft tissue planes .*

Standard Ultrasound Imaging

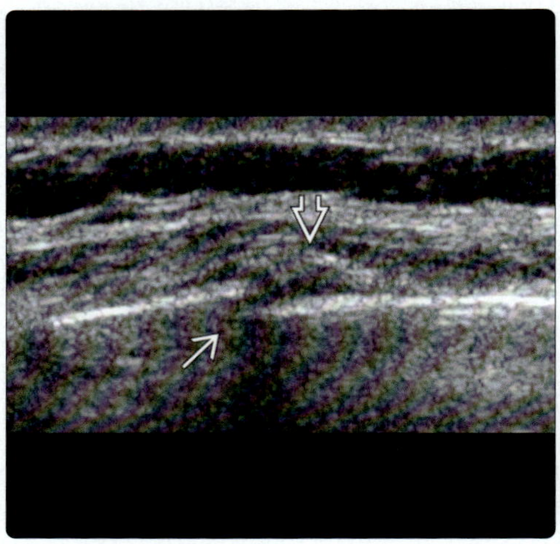

Compound Imaging

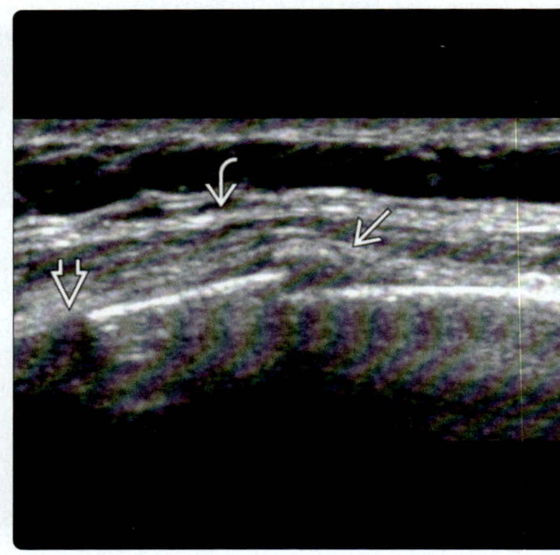

(Left) *Longitudinal color Doppler US shows the radial artery passing deep to a wrist ganglion . Arterial blood flowing towards the midpoint of the transducer head is colored red , while flow away from the midpoint of the transducer head is colored blue .* **(Right)** *Longitudinal power Doppler US shows characteristic irregular chaotic flow within a nerve sheath tumor of the medial nerve. No color direction is present on power Doppler imaging.*

Color Doppler Imaging

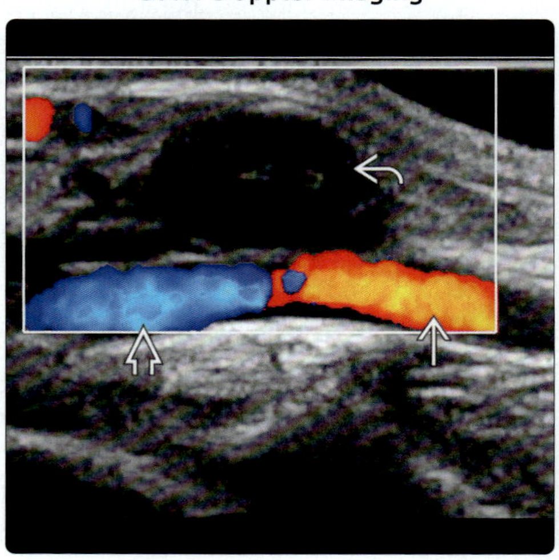

Power Doppler Imaging

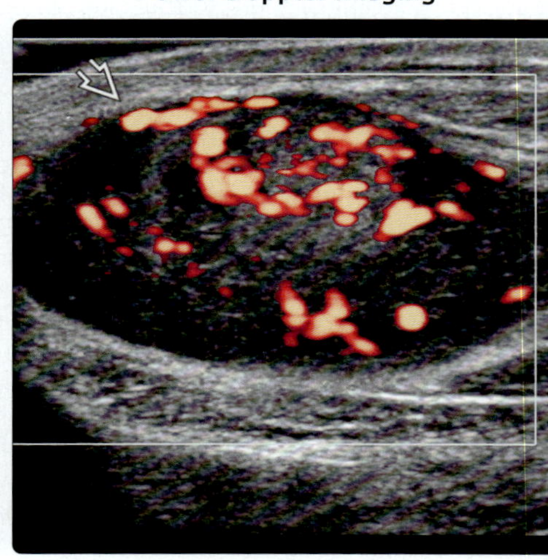

Spectral Analysis

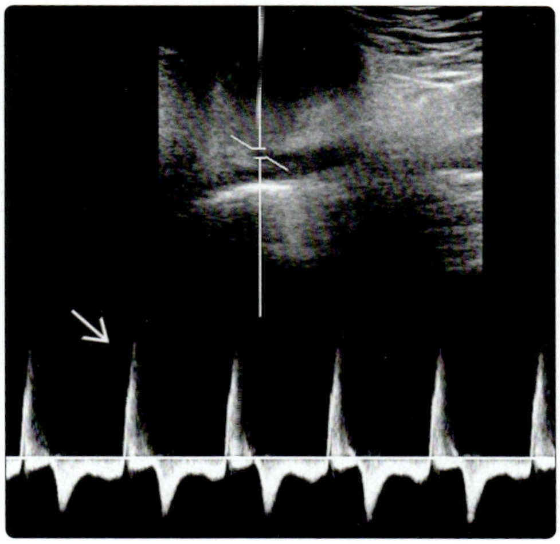

3D Ultrasound

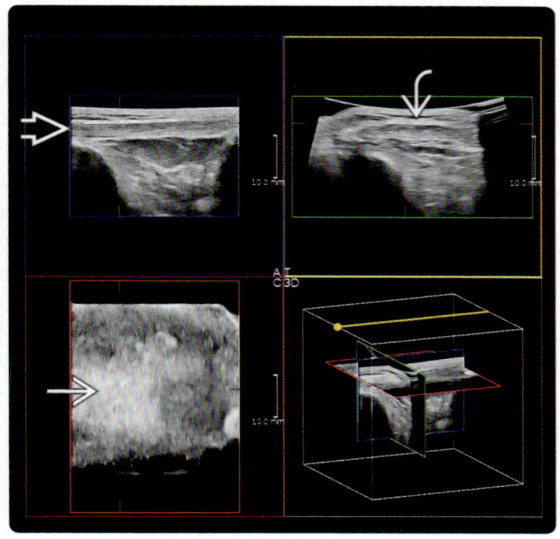

(Left) *Longitudinal pulsed Doppler US shows characteristic to-and-fro* → *spectral waveform pattern at the neck of a femoral artery pseudoaneurysm.* **(Right)** *Longitudinal 3D US of the patellar tendon shows how longitudinal* → *and transverse* → *images can be reconstructed from the volumetric dataset acquired with a large, wide transducer placed on the patellar tendon. The coronal image* → *is not useful in this case.*

Strain Elastography

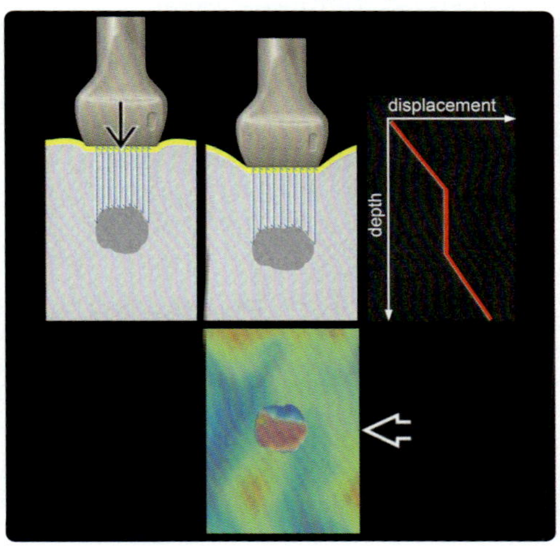

Shear Wave Elastography

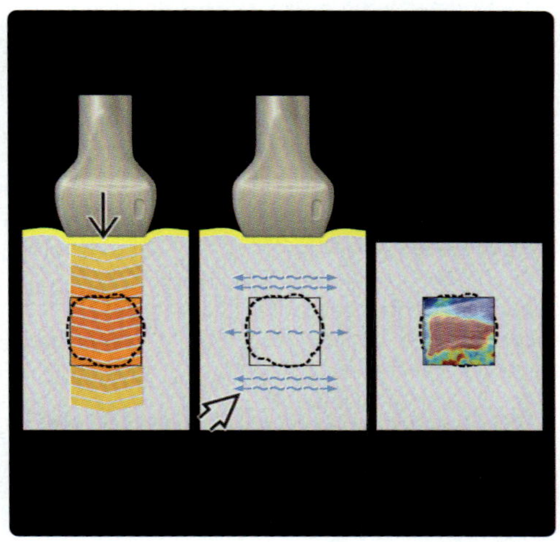

(Left) *Strain elastography schematic shows probe pressure* → *causing axial tissue displacement. Echoes during compression are used to measure relative displacement (strain)* → *and are displayed as color maps without units. Stiff tissue has relatively lower strain.* **(Right)** *Shear wave elastography schematic shows a modified probe sending push pulses* → *into tissues, inducing shear waves* → *that are tracked by diagnostic images. Color maps have quantitative units (m/s or kilopascal). (Courtesy K. Bhatia, MD.)*

Shear Wave Elastography

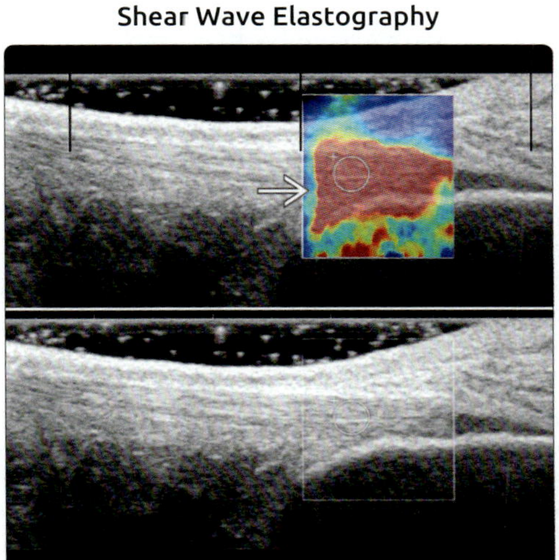

Shear Wave Elastography

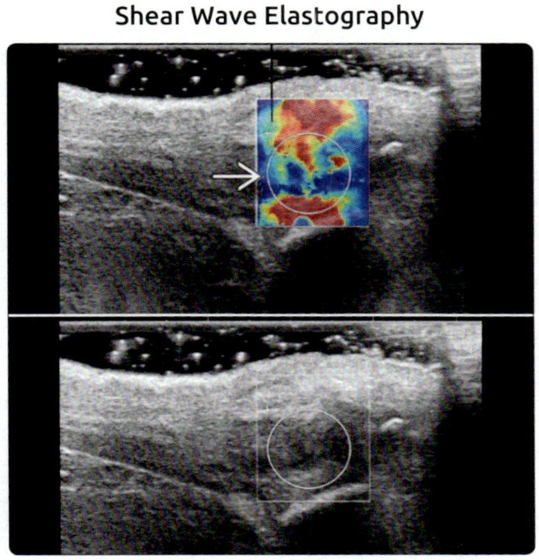

(Left) *Longitudinal shear elastography US shows a normal Achilles tendon. The tissue is mainly hard (expressed as red)* → *with elasticity measured at 497 kilopascal.* **(Right)** *Longitudinal shear elastography US shows severe Achilles tendinosis. The tissue is more heterogeneous and soft (expressed as blue color)* → *with elasticity measured at 133 kilopascal.*

Definition

A US artifact is a feature seen on an image that is not representative of the tissue being scanned. These may be either technology or technique related and can take 1 of 4 forms.

- Feature seen on image does not exist
- Feature that should be seen on image is missing
- Feature is misplaced (misregistered) on image
- Feature seen is of incorrect brightness, shape, or size

Technological Aspects

US technology works on some basic assumptions. These assumptions are not always correct and can lead to artifact formation. Not all US artifacts will be encountered in musculoskeletal US.

- Speed of sound propagation is same for all tissues (1,540 m/sec)
- US beam travels in straight line
- Attenuation of sound is uniform
- All echoes detected by transducer have primarily arisen from transducer
- Time taken for echo to return to transducer is directly related to distance of reflecting interface from transducer

Artifacts Due to Beam Characteristics

Anisotropy

Anisotropy is one of the most common artifacts encountered in musculoskeletal US. The term anisotropy refers to a property being directionally dependent, as opposed to isotropy, which implies identical properties in all directions. Anisotropy can result in a structure appearing hypoechoic if the US beam is not directly perpendicular to the structure. It is seen particularly in tendons, ligaments, nerves, and, to a lesser degree, muscle due to their oblique course and parallel reflective surfaces. Anisotropy can occur when the US beam is angled as little as 5° off perpendicular. It can mimic pathology, such as tendinosis or tendon tear, or may make a biopsy needle less conspicuous. It is particularly noticeable when the structure being imaged is curving, e.g., the ankle tendons curving around the malleoli or the supraspinatus tendon fibers curving to their insertional area. Anisotropy is eliminated by angulating (toggling) the transducer head so that the US beam is aligned perpendicular to the area of interest. It is important to evaluate only the part of the structure that is perpendicular to the US beam and not be tempted to diagnose pathology in other adjacent parts. The anisotropic effect can frequently be used to more clearly demarcate a structure of interest, such as the median nerve in the carpal tunnel or the ankle ligaments.

Off-Axis Lobe Artifact

Off-axis lobe artifact includes side lobe and grating lobe artifacts. Side lobes are multiple beams of low amplitude that project radially from the primary beam axis and are mainly seen with linear transducers. Grating lobes are additional off-axis beams that are stronger than side lobes. Both side and grating lobes are weaker than the primary lobe and, as such, do not normally give echoes of sufficient strength to be displayed. However, if a side lobe or a grating lobe encounters a more highly reflective structure outside the primary beam, their echoes may be incorrectly displayed on the image along the path of the main beam. Off-axis lobe artifacts in musculoskeletal imaging are usually seen as spurious echoes within an expected anechoic structure, such as a cyst or an effusion. Side lobes artifacts will be seen at the edge of Baker cysts. Much less commonly, off-lobe artifacts may give rise to multiple needle paths during US-guided intervention.

Beam-Width Artifact

This artifact is due to incorrect focal zone positioning too superficial to the area of interest. A strong reflector located within the widened beam beyond the focal zone will generate detectable echoes. These echoes are assumed to have originated from the narrow primary beam and are displayed on the primary beam image even though they have not arisen with the primary beam area. Beam-width artifact is typically seen either as spurious echoes displayed at the edges of an anechoic structure or as reduced contrast at the borders of a lesion. Although the distinction can be very difficult, Beam-width artifact produces a more confined artifact than off-axis lobe and, of course, can be improved by optimizing the focal zone. Off-lobe axis artifact will not improve with optimization of the focal zone.

Slice-Thickness Artifact

Slice-thickness artifact is similar to beam-width artifact but instead of being predominantly in a medial-lateral direction, it occurs due to an aberration in the anteroposterior direction. Slice-thickness artifact occurs due to the thickness of the beam, which is 90° to scan plane and equates to the thickness of the transducer head. The transducer will receive echoes from in front of and behind the assumed plane of origin, which will be included in the displayed image. Increasing the thickness of the transducer head will increase the propensity to develop slice-thickness artifact.

Artifacts Due to Multiple Echoes

Reverberation Artifact

Multiple reflections (reverberations) can occur between any 2 strong reflectors parallel to each other or between the transducer and a strong parallel reflector. As a result of these strong reflectors, echoes are reflected back and forth repeatedly. This leads to spuriously reflected echoes being displayed beneath the real reflector at intervals equal to either the distance between the transducer and real reflector or the distance between the 2 strong reflectors. Each subsequent reflection is weaker than the previous one due to sound beam attenuation. Reverberation artifact in musculoskeletal imaging is most commonly seen when the US beam encounters a strongly reflecting surface, e.g., a biopsy needle or a metal prosthesis. An advantage of seeing this artifact is that it ensures your transducer is at a right angle to the structure being imaged. Comet-tail artifacts are a particular type of reverberation artifact usually seen in echo-free regions and are composed of a dense tapering trail of echoes behind a highly reflective small structure, such as calcification, crystal aggregate in gout, or colloid aggregate in a ganglion cyst.

Ring-Down (Resonance) Artifact

Ring-down (resonance) artifact is similar in its appearance to comet-tail artifact but is generated by a different mechanism. When a transmitted US beam passes through small gas locules or fluid bubbles, the US energy leads to vibration of these bubbles, which emit and return sound waves back to the transducer. The effect is a larger artifact than expected.

Mirror-Image Artifact

Mirror-image artifact occurs when an object is located directly in front of a highly reflective smooth surface. The US beam is totally reflected by the strong reflector back toward the object. When the beam hits the object, part of the energy is reflected back to the strong reflector, which then redirects this echo back toward the transducer. True and false images are created equidistant from but on opposite sides of the strong reflector. In musculoskeletal imaging, a mirror-image artifact most commonly manifests as an extraosseous structure being mirrored within bone. The mirror image is inverted and, rather than being a exact copy, is often a distorted view of the real image.

Artifact Due to Velocity Errors

Velocity-Error Artifact

The speed of sound varies within different tissues, though US technology assumes that all tissues have a constant speed of 1,540 m/sec. If the speed of sound through tissues is > 1,540 m/sec, the returning echo will return faster to the transducer leading to the structure being displayed more superficially on the image than expected. Conversely, if the average speed through the tissues is < 1,540 m/sec, the structure will appear deeper than it is in reality. Sound travels ~ 10% slower through fat, which has a sound transmission speed of 1,450 m/sec as opposed to lean tissue (1,540 m/sec).

Refraction Artifact

Refraction of the US beam at a boundary between 2 media of different acoustic impedance will result in image misregistration. In refraction artifact, a nonperpendicular incident US beam will refract and change direction at a boundary between 2 materials of different acoustic impedance. Since US display technology assumes that the sound beam travels in a straight line, it will misplace any returning refracted echoes to the side of the true object location. Refraction artifact may result in a structure appearing longer than expected or a misplaced duplication of a structure (ghost image).

Defocusing Artifact (Edge-Shadowing Artifact)

In defocusing artifact (edge-shadowing artifact), refraction produces shadowing at the edges of a relatively large (compared to the US beam width) curved structure. Refraction occurs at the edge of the structure due to the curved edge and also differences in acoustic impedance between the 2 tissue types involved. The US beam at the edge of a structure, such as a cyst or soft tissue mass, is deviated from its original path. This is commonly seen in musculoskeletal imaging. Edge-shadowing artifact is also the reason for shadowing seen commonly in full-thickness Achilles tendon tear.

Artifact Due to Attenuation Error

Acoustic-Shadowing Artifact (Posterior Acoustic Shadowing)

This is the most commonly encountered artifact in musculoskeletal US and is the reason why US, for example, cannot look deep into joints. Acoustic shadowing artifact is caused by severe attenuation of the US beam at an interface, either by reflection, absorption, or, less commonly, refraction, resulting in reduced sound being transmitted beyond that interface. This leads to an area of hypoechogenicity behind a strongly reflecting interface or strongly attenuating tissue. In musculoskeletal imaging, this is seen as shadowing posterior to calcification, bone, foreign bodies, or dense fibrous tissue. The main implication is that the tissues behind that structure are not visible on US. This can cause problems. For example, in examination of the knee, it may be difficult to tell whether or not a large osseous structure is attached to the bone. It is also an advantageous sign. For example, acoustic shadowing behind noncalcified tissue would make it very likely that this tissue is composed of dense fibrous tissue. Acoustic shadowing, in the absence of calcification, is a good clue to the presence of fibrosis. Objects with a small radius or a rough surface are more inclined to produce a clean posterior shadow, whereas objects with a large radius or smooth surface are more inclined to produce a dirty shadow as a result of reverberation echoes superimposed on the acoustic shadow.

Acoustic-Enhancement Artifact (Increased Through Transmission)

This is caused by low-attenuating structures, such as fluid, resulting in no weakening of the US beam as it passes through these tissues relative to adjacent tissues. This leads to an area of high-amplitude echoes behind the area of low-attenuating tissue. This artifact is seen when examining fluid, myxoid tissue, or associated low-attenuating tissue structures, such as ganglion cysts, nerve sheath tumors, or giant cell tumor of tendon sheath. Tissues deep to the low-attenuation structure appear hyperechoic. Similar to acoustic shadowing, acoustic enhancement is a useful diagnostic sign and can be used, for example, to increase diagnostic accuracy of nerve sheath tumors that typically show mild to moderate acoustic enhancement.

Artifact Due to Improper Machine Setting

Range Ambiguity

US technology assumes that each pulse of echoes is received by the transmitter before the next pulse is emitted. If returning 1st pulse echoes arrive after the 2nd pulse has been transmitted, returning 1st pulse echoes are interpreted incorrectly as having originated from the 2nd pulse and will be incorrectly placed nearer to the transducer. Range ambiguity artifact is not commonly seen now, as the US machine automatically reduces pulse repetition frequency during imaging of deeper structures.

Equipment-Generated Artifact

Incorrect use of equipment controls, such as gain or time-gain compensation, can result in the visualized echoes being either too bright or too dark. To avoid these artifacts, the machine should be properly setup at the start of each examination and repeatedly readjusted during the course of the examination.

Artifact Seen During Doppler Imaging

Transducer Pressure

Too much transducer pressure may obliterate vascular flow, particularly in soft superficial vascular structures. To avoid this happening, you should use a large amount of US gel and actively lift the transducer off the skin when assessing tissue vascularity. Noticing the effect of transducer pressure can be helpful for tissue characterization. For example, the vascularity of nerve sheath tumors typically blanches with only slight transducer pressure.

Noise Artifact

Noise is caused by increased color or power Doppler settings. Noise can be recognized as small, spurious, inconsistent color dots. Color gain settings should be adjusted to minimize noise. To achieve maximum color sensitivity, increase color gain until noise pixels are clearly seen in the Doppler box, then slowly decrease gain until nearly all of the noise has disappeared.

Aliasing Artifact

Aliasing artifact occurs when the velocity range exceeds the scale available to display and is only seen with color and spectral Doppler imaging. Aliasing is not seen with power Doppler, as it does not provide flow-directional information. The sample rate on color or pulsed Doppler is equal to the pulse repetition frequency, which limits maximum velocity scale. The Nyquist limit is the upper limit of Doppler shift that can be detected accurately and is equal to 1/2 the pulse repetition frequency. If Doppler shift signal is higher than the Nyquist limit, aliasing artifact occurs with inaccurate display of color or spectral Doppler velocity. In color Doppler, the display wraps around the scale and overwrites exiting data while in spectral Doppler, velocity peak is cut off at the top end of the scale and is displayed at the lower end of the scale, i.e., below baseline. Aliasing can be overcome by increasing the velocity scale (pulse repetition frequency). Increasing the pulse repetition frequency is disadvantageous in that low velocities cannot be accurately measured.

Range-Ambiguity Artifact

As expected, range-ambiguity artifact can occur in Doppler imaging with increasing pulse repetition frequency used to overcome aliasing artifact. This should be identified in order to avoid detection of frequency shifts deeper than the target object.

Directional Ambiguity/Indeterminate-Flow-Direction Artifact

This artifact leads to the spectral waveform being displayed with nearly equal amplitude above and below spectral analysis baseline in a mirror-image pattern. It is caused by the US beam intercepting the vessel at a 90° angle, particularly for small vessels moving in and out of the imaging plane. True bidirectional flow is a rare occurrence and is only seen in the neck of a pseudoaneurysm or when there is reversed diastolic flow near a high-resistance vascular stenosis. True bidirectional flow is never simultaneously symmetrical above and below baseline and may be seen to vary across the cardiac cycle.

Improper Doppler Angle With No-Flow Artifact

Since Doppler imaging is angle dependent, the angle formed between the interrogating sound beam and vascular flow should be optimized. The angle can be changed by electronic steering. When the angle between the US beam and the vessel being studied approaches 90°, no Doppler shift is detected, and color may not present even though flow is present. The color box should be steered to make the insonation angle to flow closer to 0° or 180° and not close to 90°. This Doppler angle effect artifact is seen in color, power, and spectral Doppler imaging. The ideal Doppler imaging angle is either 0° or 180° but is generally not anatomically achievable when imaging vasculature. A Doppler angle of 60° has traditionally been specified as a standard, though with improvements in US technology, a better criterion would be an angle of 45-55°, striving to measure velocities at 50° whenever possible.

Blooming Artifact ("Color Bleed")

This artifact is usually caused by abnormally high Doppler gain settings, leading to color signal extending beyond the true grayscale vessel margin. It leads to color display outside a vessel wall, making vessels appear larger, and also potentially masking luminal detail, such as the presence of a thrombus. Decreasing color Doppler gain will minimize blooming artifact, though it will also reduce sensitivity to flow detection in small vessels.

Flash Artifact

Movement of either the patient or the transducer during Doppler imaging gives rise to a random color mosaic that obscures the grayscale image. Power Doppler is most susceptible to this artifact.

Mirror (or Ghost)-Image Artifact

This is caused by the same mechanism as that seen with grayscale imaging and is a feature of color, spectral, or power Doppler imaging. It most commonly occurs adjacent to highly reflective surfaces, such as the lung when, for example, imaging the supraclavicular region where one may see duplication of the subclavian artery or vein with the false images being displayed deeper than the real image.

Pseudoflow Artifact

The presence of fluid flow in nonvascular cystic collections can mimic real blood flow in color or power Doppler imaging. However, no arterial or venous spectral waveform will be seen if this apparent fluid flow is subjected to spectral Doppler analysis.

Twinkling Artifact

Twinkling artifact can be seen behind any strongly reflecting surface, such as calcification, bone, or foreign body. It is seen as a mosaic of rapidly changing colors, mimicking turbulent flow, and presents deep to the strongly reflecting surface, such as a calcified nodule, on both color and power Doppler. It is most likely caused by narrow bandwidth noise induced by "phase jitter" in the Doppler circuitry of the US system.

Edge Artifact

Occurs at the margin of a strong, smooth reflector (such as a calcified mass or cortical bone), resulting in persistent color along the rim of a strong reflector. Spectral analysis in this case will show no vascular flow pattern. Edge artifact is seen more commonly with power Doppler than color Doppler imaging.

Motion Artifact

Even a small amount of motion will give rise to noise on color or power Doppler imaging. Noise appears as random dot-like flashes on color or power Doppler imaging. It is caused by low-frequency Doppler shifts and can be minimized by high-pass wall filters, though these filters will reduce sensitivity to detect slow-flowing blood. Very slow-flowing blood is best detected on grayscale imaging alone when moving red cell aggregates will be seen as slow-moving intravascular echoes.

Selected References

1. Serafin-Król M et al: Diagnostic errors in musculoskeletal ultrasound imaging and how to avoid them. J Ultrason. 17(70):188-196, 2017
2. Taljanovic MS et al: Artifacts in musculoskeletal ultrasonography. Semin Musculoskelet Radiol. 18(1):3-11, 2014

Ultrasound Beam

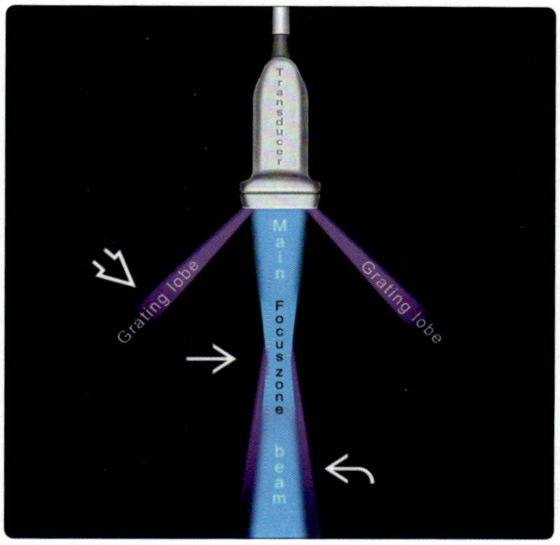

Anisotropy

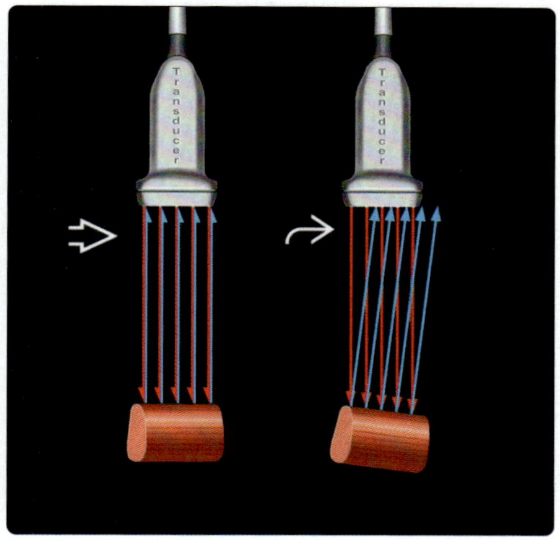

(Left) *Schematic of a US beam shows the primary (main) beam narrowing at the focal zone and then diverging. Grating lobes and slide lobes are both off-axis lobes.* **(Right)** *Schematic shows anisotropy effect. If the structure scanned is parallel to the transducer, all of the reflected echoes can be detected. When the structure is not parallel to the transducer, only some or none of the reflected echoes will be detected, leading to the structure becoming hypoechoic.*

Anisotropy

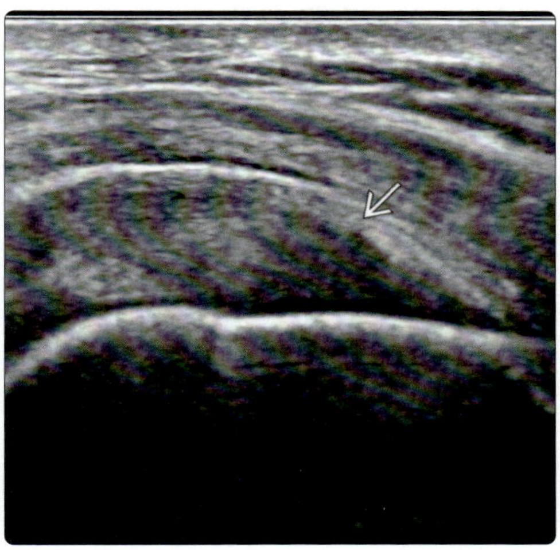

Anisotropy

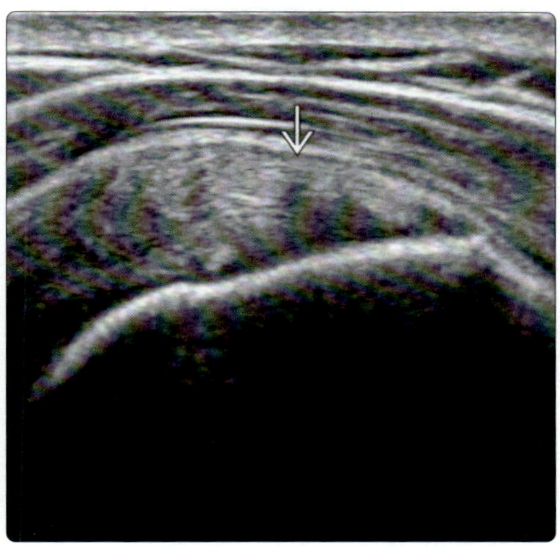

(Left) *Longitudinal US shows moderate hypoechogenicity of the distal fibers of the supraspinatus tendon, which could simulate a tendon tear.* **(Right)** *Longitudinal US of the same tendon following angulation of the transducer so that the fibers lie more parallel to the transducer head shows reduction in hypoechoic anisotropic effect. No tear is present.*

Beam-Width Artifact

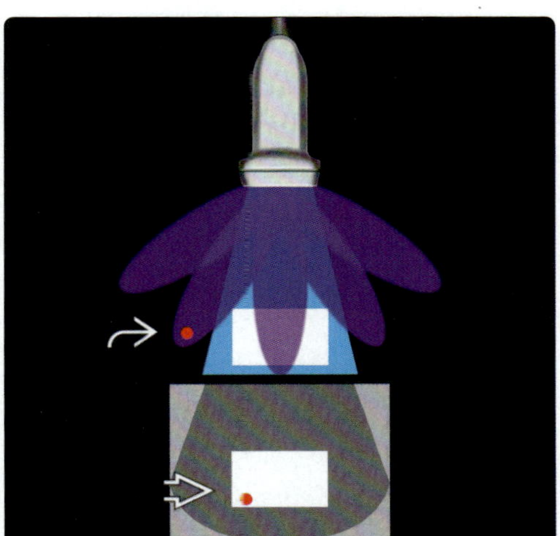

Off-Lobe Axis Artifact

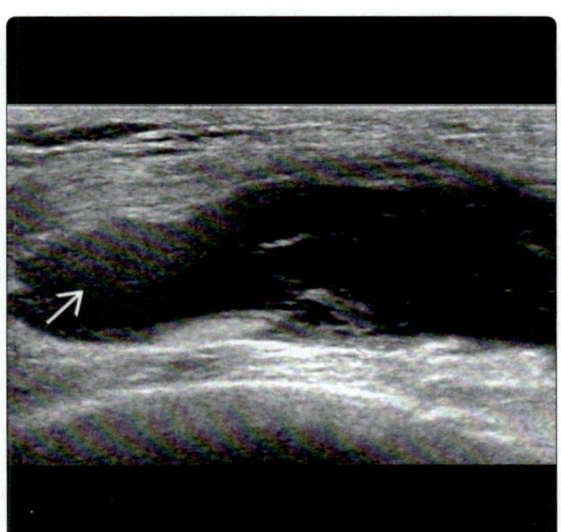

(Left) *Schematic shows beam-width artifact. The purple beams represent multiple side lobes, which encounter an object (red dot). The display, assuming that all returning echoes came from the main beam, will misplace and duplicate the structure as shown.* **(Right)** *Transverse US shows off-axis lobe artifact resulting in spurious echoes from subcutaneous fat appearing in knee joint effusion. This is commonly seen at the edges of cystic structures and can simulate fluid debris.*

(Left) *Schematic shows beam-width artifact. The blue beams represent multiple side lobes, which encounter an object (red dot) ➔. The display, assuming all returning echoes came from the main beam (dotted lines), will misplace the structure ➔. This artifact is corrected by correct positioning of focal zone ➔.* **(Right)** *Transverse US shows blurring of the lateral margin of a lipoma and adjacent tissue ➔ due to either beam-width or off-axis lobe artifact. Optimization of the focal zone will help reduce the former, but not the latter, artifact.*

Beam-Width Artifact

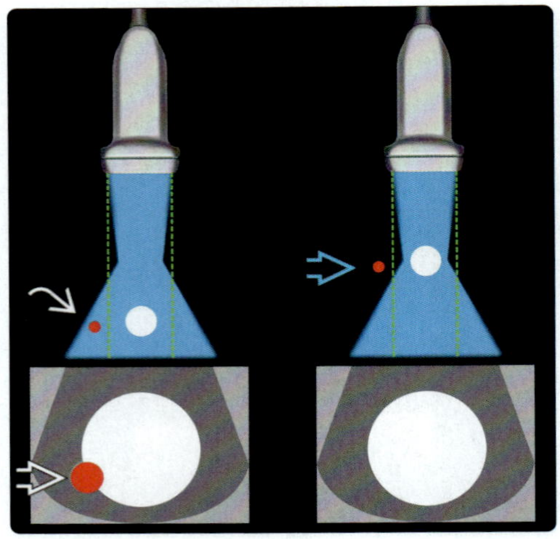

Beam-Width Artifact

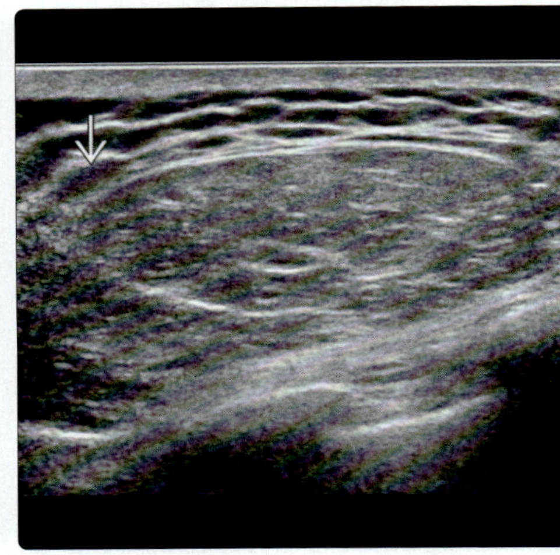

(Left) *Schematic shows ring-down artifact. If the main beam encounters air bubbles, it can induce vibrations in these bubbles ➔, causing a continuous source of sound energy transmitted back to the transducer ➔. Therefore, the artifact appears larger than expected.* **(Right)** *Longitudinal US shows increased echogenicity of gas bubbles ➔ due to ring-down artifact within hypoechoic granulation tissues as a result of low-grade soft tissue infection.*

Ring-Down Artifact

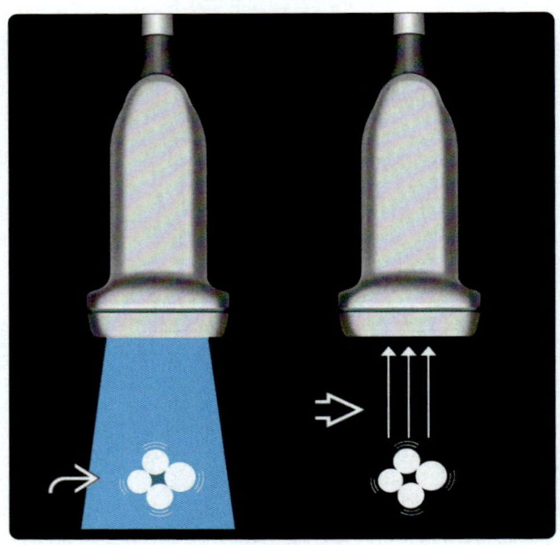

Ring-Down Artifact

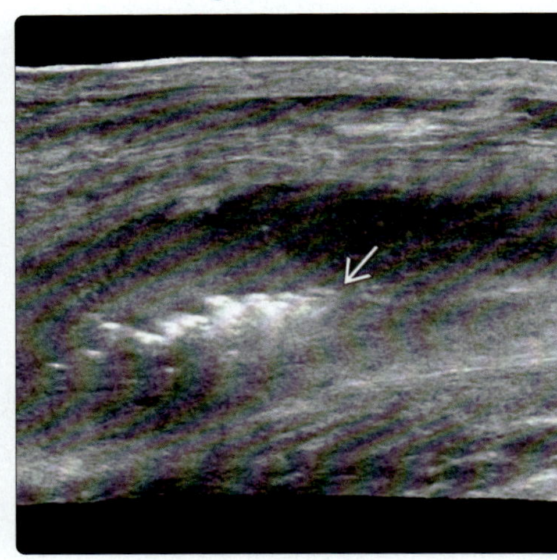

(Left) *Schematic shows reverberation artifact. Repeated reflections ➔ between the transducer face and a very strong reflector causes progressively diminishing reverberations ➔ to occur deep to the real echoes. This artifact can also occur between 2 strongly reflecting surfaces.* **(Right)** *Longitudinal US shows reverberations ➔ deep to the strongly reflecting biopsy needle ➔. This artifact occurs from reverberation between the outer coaxial needle and the inner biopsy needle.*

Reverberation Artifact

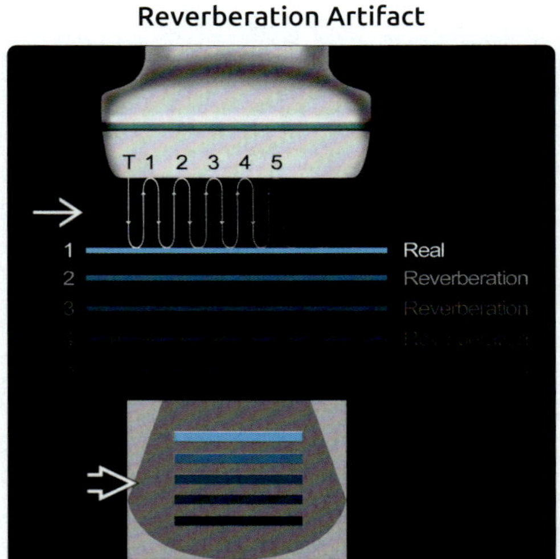

Reverberation Artifact

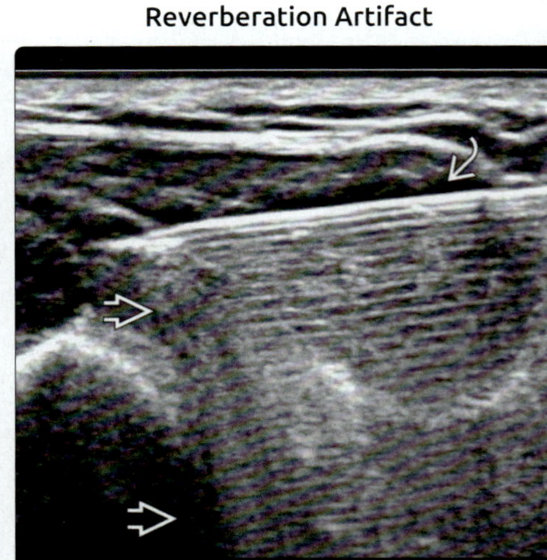

Reverberation Artifact

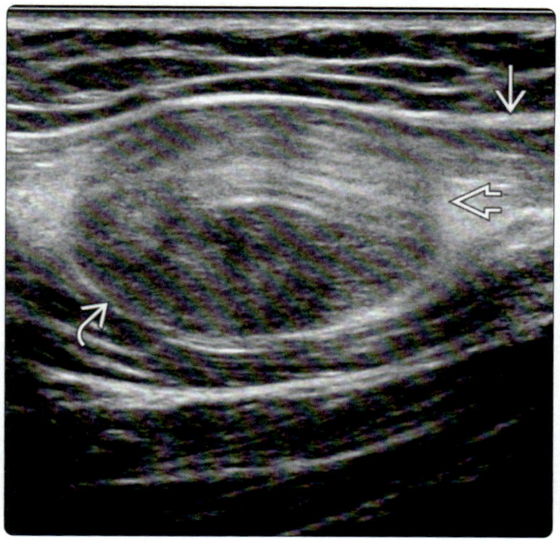

Comet-Tail Artifact

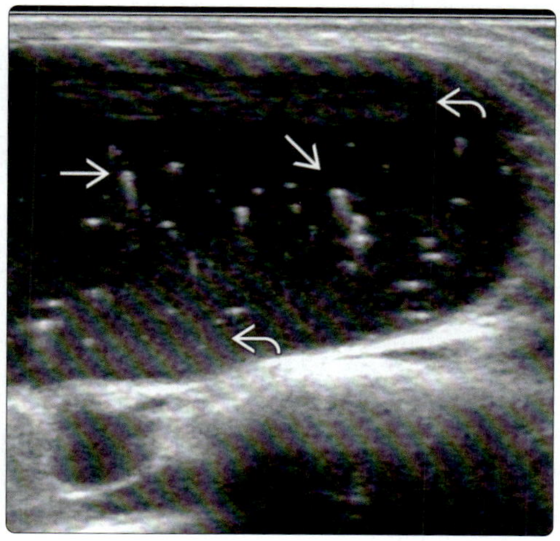

(Left) *Longitudinal US shows linear echoes within a nerve sheath tumor due to reverberations from an overlying strongly reflective investing fascia.* **(Right)** *Transverse US of a ganglion cyst shows comet-tail artifacts due to colloid aggregates within the mucopolysaccharide fluid content. There is also some slice-thickness artifact simulating echogenic fluid content on this image.*

Propagation Speed Artifact

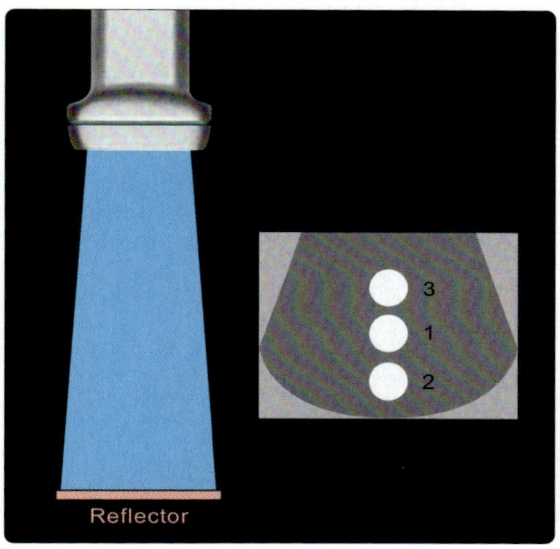

Propagation Speed Artifact

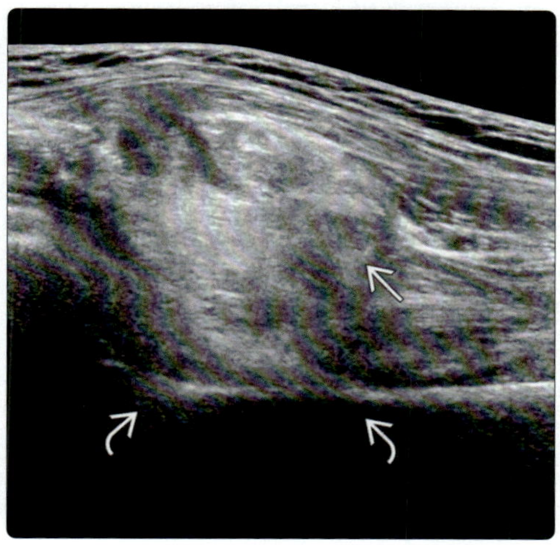

(Left) *Schematic shows propagation speed artifact. The reflector is in position 1. If the actual propagation speed is more than expected, the reflector will appear in position 3. If the actual propagation speed is less than assumed, the reflector will appear in position 2.* **(Right)** *Longitudinal US of the thigh shows ill definition of the proximal femur cortex due to propagation speed artifact from intramuscular soft tissue sarcoma on the anterolateral aspect of the thigh.*

Mirror-Image Artifact

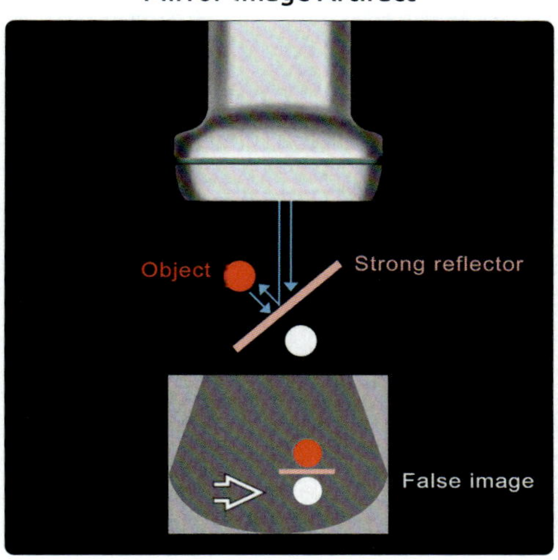

Mirror-Image Artifact

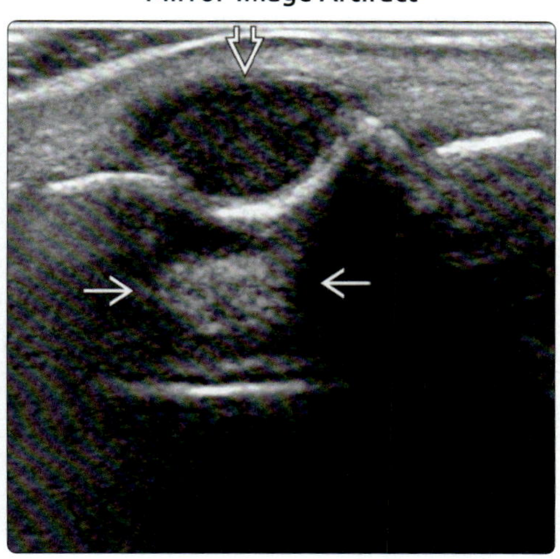

(Left) *Schematic shows mirror-image artifact. Reflections from a strong reflector encounter another object and are reflected first back to the highly reflective surface before returning to the transducer. This produces an inverted mirror image of the object.* **(Right)** *Longitudinal US shows inverted, distorted, and incompletely portrayed mirror-image artifact of a juxtacortical giant cell tumor due to strong cortical bone reflection.*

(Left) *Schematic shows acoustic enhancement artifact posterior to a low-attenuating tumor (e.g., cyst, nerve sheath tumor, giant cell tumor, lipoma).* **(Right)** *Longitudinal US shows characteristic acoustic enhancement ➡ posterior to a nerve sheath tumor ➡ in the thigh.*

Acoustic Enhancement Artifact

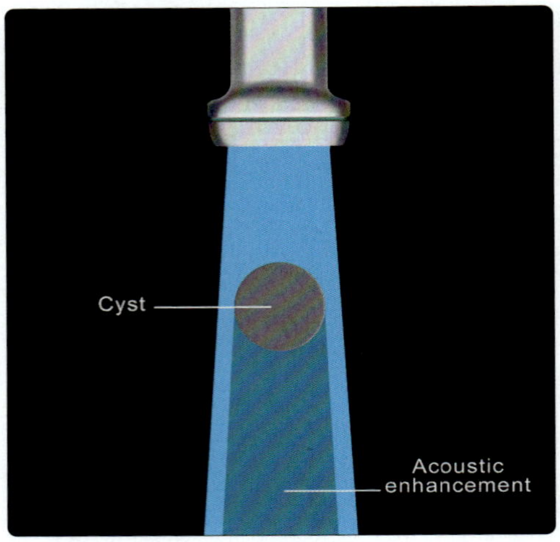

Acoustic Enhancement Artifact

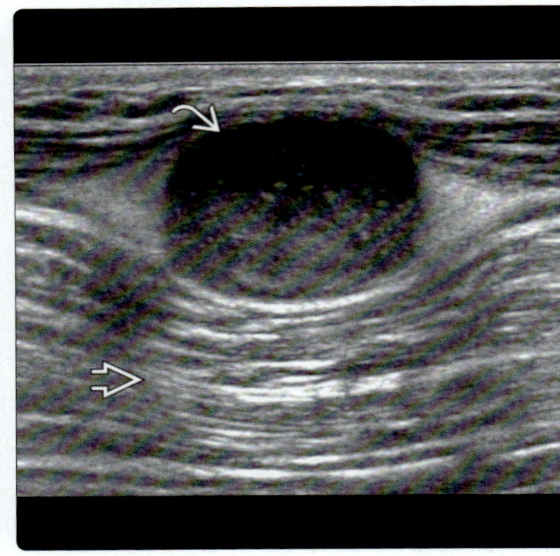

(Left) *Schematic shows acoustic shadowing artifact. A strong reflector, such as a calcified nodule, stops all through transmission, leading to posterior acoustic shadowing.* **(Right)** *Longitudinal US of the thigh shows acoustic shadowing ➡ posterior to a small area of subcutaneous calcification ➡.*

Acoustic Shadowing Artifact

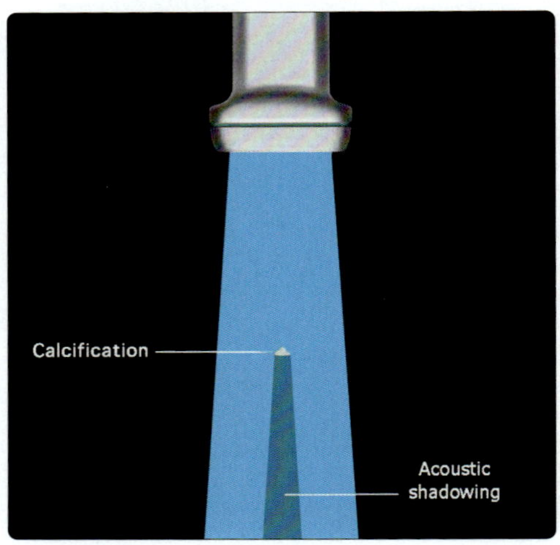

Acoustic Shadowing Artifact

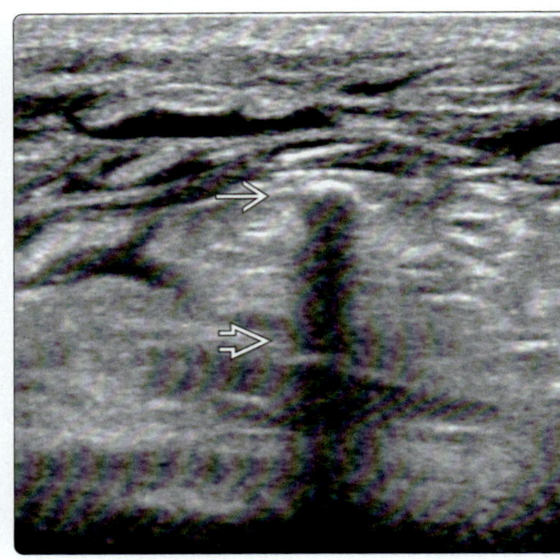

(Left) *Longitudinal US of the gluteal region shows dense acoustic shadowing ➡ deep to the surface of a heavily calcified granuloma ➡.* **(Right)** *Transverse US shows range-ambiguity artifact ➡ with alternating echogenic bands at the sides of a cystic structure due to soft tissue echoes returning from deeper tissues. Increased through transmission ➡ mainly in the central part of the cyst is seen.*

Acoustic Shadowing Artifact

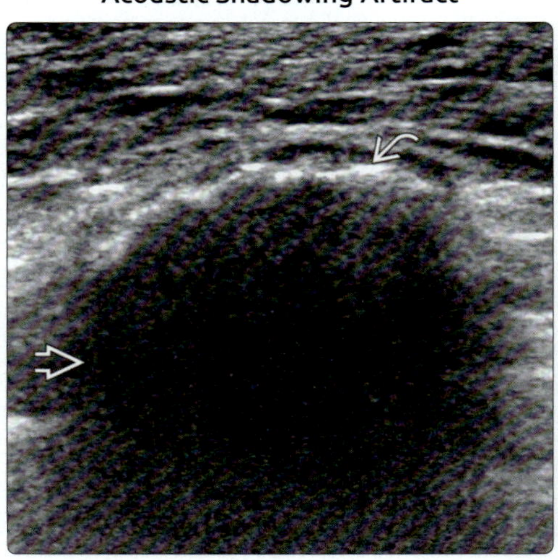

Range-Ambiguity Artifact

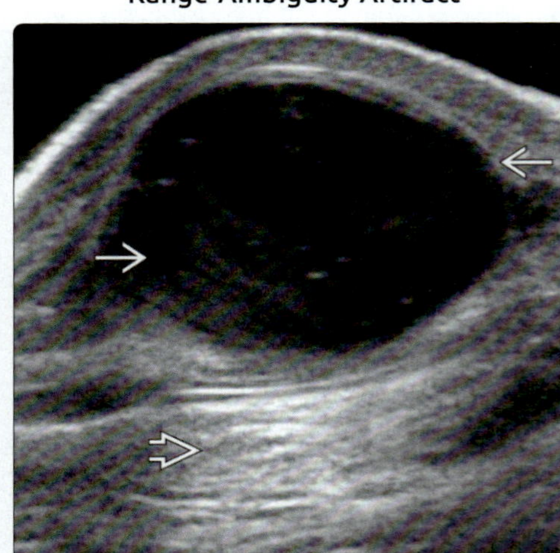

Transducer Pressure Artifact

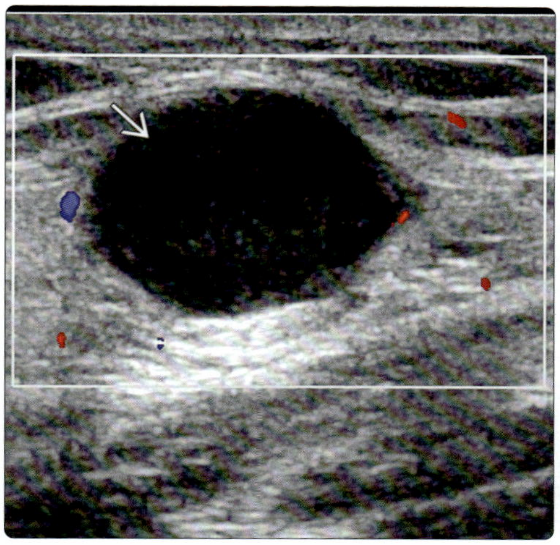

Transducer Pressure Artifact

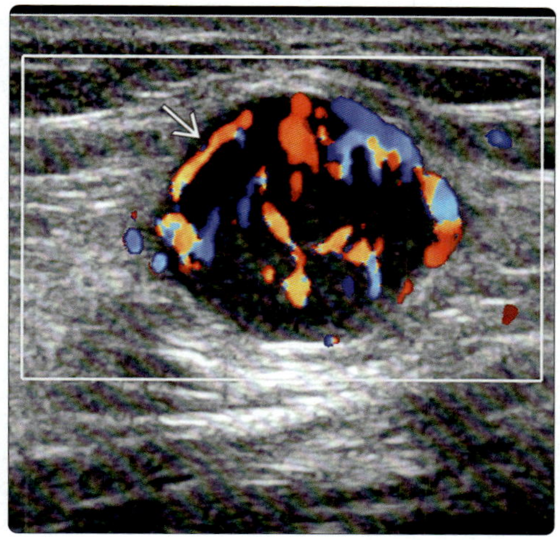

(Left) *Longitudinal color Doppler US shows a subcutaneous nerve sheath tumor with moderate transducer pressure. No intratumoral hyperemia is evident* ➡. **(Right)** *Longitudinal US of same tumor with minimal transducer pressure now shows moderate intratumoral hyperemia* ➡. *Transducer pressure should be kept to a minimum, especially when assessing vascularity of superficial structures.*

Blooming Artifact

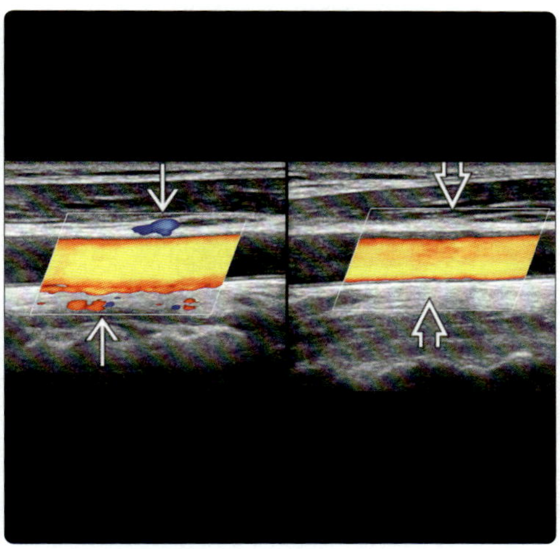

Directional Ambiguity Artifact

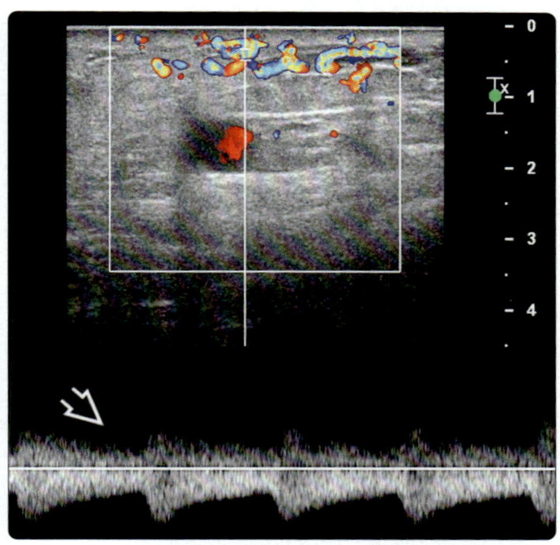

(Left) *Longitudinal color Doppler US shows blooming artifact on the left side* ➡ *("color bleed"). After lowering the color Doppler gain settings, the blooming artifact disappears as seen in the right-sided image* ➡. **(Right)** *Longitudinal color Doppler spectral US of a venous vascular malformation shows directional ambiguity artifact* ➡ *in which the waveform is displayed with nearly equal amplitude above and below the baseline in a mirror-image pattern.*

Noise Artifact

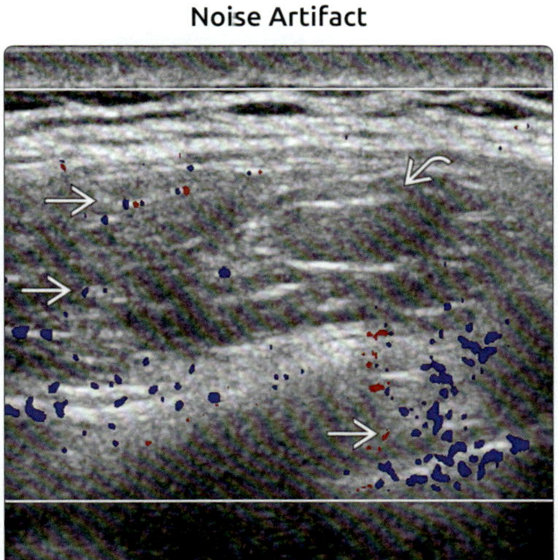

Doppler Edge Artifact

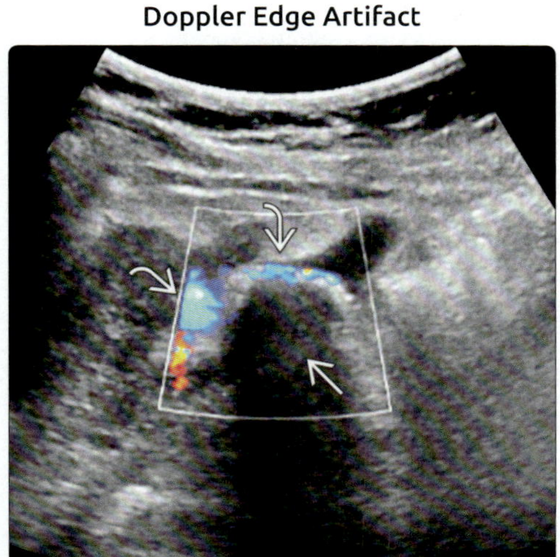

(Left) *Transverse color Doppler US of a subcutaneous lipoma* ➡ *shows noise artifact* ➡ *as a result of too high of a color Doppler gain setting. When the gain was turned down, no increased "vascularity" was present.* **(Right)** *Transverse color Doppler US shows Doppler edge artifact with persistent color* ➡ *signal around the rim of the strong reflector* ➡ *(calcified structure).*

KEY FACTS

GENERAL CONSIDERATIONS

- Common specific indications are rotator cuff and long head of biceps as well as bursal, glenohumeral pathology and diagnostic/therapeutic US-guided intervention
- Less common specific indications are acromioclavicular joint, rotator cuff muscle pathology, suspected paralabral cysts, and labral pathology

TECHNIQUE

- Perform full standardized examination in all patients
- **Supraspinatus**
  - Crass position
  - Modified Crass position
- **Infraspinatus and teres minor**
  - Hand placed on contralateral shoulder or supinated resting on contralateral thigh
- **Biceps tendon**
  - Arm by side, elbow flexed, and examiner holding wrist and finely rotating arm to bring bicipital groove into view
- **Rotator interval**
  - Arm extended and externally rotated (i.e., elbow tucked in) (jogger position)
- **Subscapularis**
  - Arm externally rotated with flexed elbow placed against ipsilateral iliac crest
  - As subscapularis tendon is orientated from inferior to superior, scan in long and short axis along this plane
- **Glenohumeral joint recesses**
  - Lateral decubitus for posterior joint recess
  - Supine with arm externally rotated for anterior joint recess
- **Dynamic scanning for subacromial impingement**
  - Aim to recreate patient's pain
  - Place transducer in long axis either at lateral edge of acromion or over coracoacromial ligament
  - Raise pronated arm at angle ~ 45° forward from shoulder and assess symptoms/degree of impingement

Crass Position

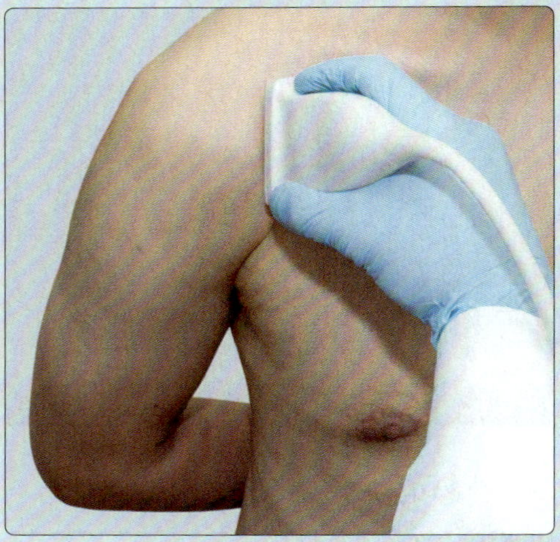

Modified Crass Position

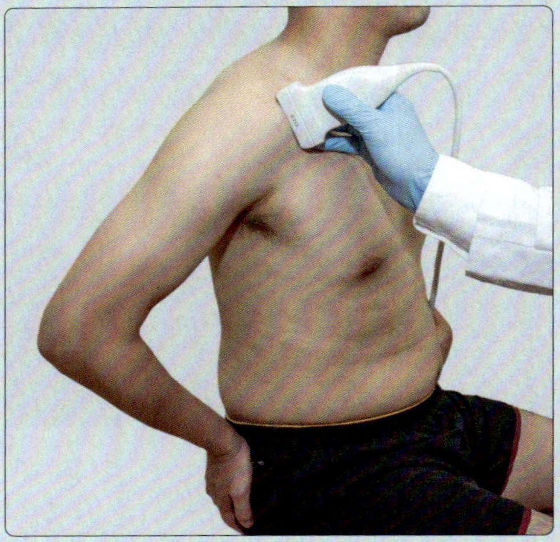

(Left) *Clinical photograph shows Crass position during examination of supraspinatus tendon on long-axis view. Back of the hand is placed across the back, which internally rotates the shoulder, bringing the tendon out from underneath the acromion.* **(Right)** *Clinical photograph shows the modified Crass position. The hand is placed on the ipsilateral back pocket, producing less internal rotation than the Crass position. The anterior edge of supraspinatus can be more clearly seen, especially on short-axis views.*

Supraspinatus Tendon: Crass

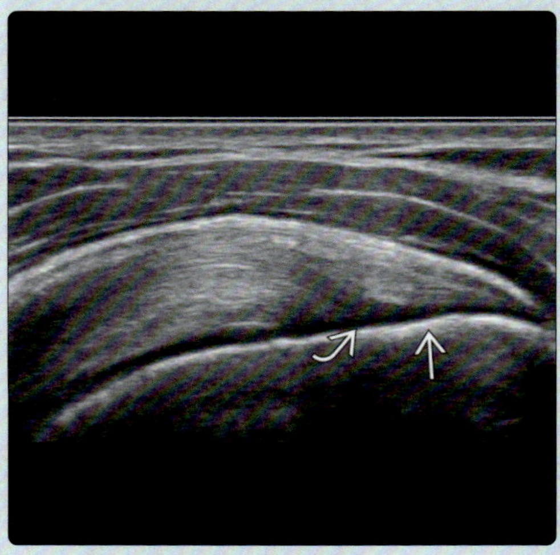

Supraspinatus Tendon: Modified Crass

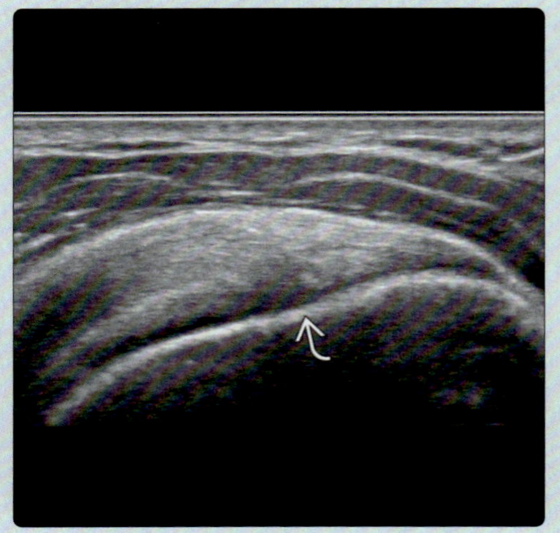

(Left) *Longitudinal US along the supraspinatus tendon shows anterior fibers inserting into the anterior facet ➡ of the greater tuberosity. Focal anisotropy ➡ due to fiber orientation is present. This is a common site for anisotropy.* **(Right)** *Longitudinal US of the supraspinatus tendon in the same location shows alteration in the tendon fiber orientation with lessening of anisotropic effect ➡. Insonation angle can be changed by angling the transducer or altering the patient's position.*

GENERAL CONSIDERATIONS

Clinical Indications for Shoulder US

- Most commonly examined joint with US
- Common specific indications are rotator cuff and long head of biceps as well as bursal, glenohumeral pathology and US-guided intervention
- Less common specific indications are acromioclavicular (AC) joint, rotator cuff muscle pathology, suspected paralabral cysts, and labral pathology
- Patients prefer US to MR

TECHNIQUE

Full Standardized Examination

- Necessary in all patients
- Shoulder pain is poorly localized as large subacromial-subdeltoid bursa is main pain initiator
- Subclinical rotator cuff tendinosis is common; therefore, do not use contralateral asymptomatic shoulder as internal reference of normality

Patient Position and Approach

- Anterior approach with examiner sitting on high seat facing patient who is sitting on low stool
- Posterior approach with patient sitting on couch or revolving stool with back to examiner
- Each tendon is examined from its musculotendinous junction to its insertion with tendon on stretch and moved away from overlying bone that may limit transducer access
- For all tendons, need to continually toggle transducer or move arm position to minimize anisotropy

Supraspinatus Tendon

- **Best position**
  - At least 4 different positions can be used to examine supraspinatus tendon
    - Crass position
    - Modified Crass position
    - Partial internal rotation ("shrug" position)
    - Neutral position
  - In practice, consider scanning supraspinatus long axis in Crass position and short axis in modified Crass position
- **Scan technique**
  - Align transducer longitudinally over supraspinatus tendon
    - Scan from anterior-most portion (i.e., biceps tendon margin) to posterior where supraspinatus tendon imperceptibly merges with infraspinatus
    - Focus primarily on those fibers deep to transducer
  - Turn transducer 90° so that it is aligned transverse to tendon and scan from medial to lateral using biceps tendon to mark anterior edge of supraspinatus tendon
    - Scan anterior, middle, and posterior fibers in turn angling transducer over insertional area to main correct insonation angle

Infraspinatus Tendon

- **Best position**
  - Hand placed on contralateral shoulder or supinated resting on contralateral thigh
- **Scan technique**
  - Align transducer in longitudinal plane along infraspinatus tendon using musculotendinous junction as guide
    - Scan infraspinatus to its most lateral attachment into middle facet
    - Cortical defects of humeral head deep to infraspinatus tendon are common and usually of no clinical significance
  - Turn transducer 90° to align transverse to infraspinatus tendon and scan from myotendinous junction to insertion

Teres Minor Tendon

- **Best position**: Similar to infraspinatus
- **Scan technique**
  - Rotate transducer more vertically than infraspinatus to align along teres minor tendon
    - Teres minor has more vertical course than infraspinatus as it arises on inferolateral side of infraspinatus
    - ± turn transducer 90° to align transverse to teres minor tendon and scan from myotendinous junction to insertion

Long Head of Biceps Tendon

- **Best position**: Arm by side, elbow flexed, and finely rotate arm to bring bicipital groove into view
- **Scan technique**
  - Align transducer transversely across bicipital groove
  - Scan proximally to uppermost aspect of bicipital groove and distally to where pectoralis major tendon passes over biceps tendon
  - Most biceps pathology is localized to proximal-most segment known as "genu" of long head of biceps tendon
  - Slightly extending arm will allow one to see more of proximal biceps tendon
  - ± internally/externally rotate arm to test for biceps subluxation
  - Rotate transducer 90° to align longitudinal to biceps tendon, focusing particularly on proximal end of tendon

Rotator Interval

- **Best position**
  - Arm extended and externally rotated (i.e., elbow tucked in) (jogger position)
  - Mild changes in extension and external rotation can improve visibility
- **Scan technique**
  - Scan transversely across biceps tendon at and proximal to uppermost portion of bicipital groove
  - Proximal, mid-, and distal portions of rotator interval should be examined
  - Uppermost part of bicipital groove equates to distal end of rotator interval
  - Coracohumeral ligament (CHL) seen as echogenic band, ~ 1.5 mm thick, superficial to biceps tendon
  - CHL joins with superior glenohumeral ligament medially forming bicipital sling
    - This sling, located between biceps and subscapularis, helps prevent medial dislocation of biceps tendon
  - Distally, superficial subscapularis fibers passing over biceps tendon to lateral lip of bicipital groove support lateral border of biceps tendon

Subscapularis Tendon

- **Best position**
  - Arm externally rotated with flexed elbow placed against ipsilateral iliac crest
    - Limited external rotation is sign of adhesive capsulitis
  - As subscapularis tendon is orientated from inferior to superior, scan in long and short axis along this plane
  - ± helpful to scan upper fibers with arm more extended
- **Scan technique**
  - On longitudinal scanning, obverse narrow tendon insertion into lesser tuberosity and medial part of surgical neck
    - Toggle transducer or rotate arm slightly to reduce anisotropy
  - Sweep transducer proximally and distally to see full width of tendon
    - Superior 2/3 of tendon is tendinous → lesser tuberosity
    - Inferior 1/3 of tendon is muscular → lesser tuberosity and adjacent part of humeral surgical neck
  - Cortical defects deep to tendon are common
  - Also scan medially deep to coracoid ± check for subcoracoid impingement by rotating arm
  - On transverse scanning, 5-6 tendon slips from multipennate muscle converge onto tendon insertion
    - Do not confuse alternating hypoechoic/hyperechoic pattern with tears or tendinosis
  - Tendon tears of upper fibers more common
  - Uppermost fibers merge with supraspinatus/CHL forming roof and floor of rotator interval over and under biceps tendon

Subacromial-Subdeltoid and Subcoracoid Bursae

- **Best position**
  - Same as for supraspinatus tendon bursa and subscapularis bursa
- **Scan technique**
  - Scan in longitudinal or transverse planes with tissues in relatively relaxed position
  - Supraspinatus tendon bursa is seen between deltoid muscle and supraspinatus tendon
    - Seen as thin (< 0.8 mm) hypoechoic line sandwiched between layers of peribursal fat (< 2 mm thick normally)
  - Subscapularis bursa located between coracoid process and short head of biceps tendon superficial to subscapularis tendon and muscle
    - Should not be confused with subcoracoid recess, which protrudes over upper end of subscapularis tendon

Dynamic Assessment of Subacromial Impingement

- **Best position**
  - Aim to recreate patient's pain
- **Scan technique**
  - Place transducer in long axis either at lateral edge of acromion or over coracoacromial ligament
  - Raise pronated arm at angle ~ 45° forward from shoulder and assess symptoms/degree of impingement
    - Grade 0: No painful arc; no demonstrable impingement
    - Grade 1: Painful arc; no demonstrable impingement
    - Grade 2: Painful arc; demonstrable soft tissue impingement
    - Grade 3: Painful arc; demonstrable humeral impingement

Acromioclavicular Joint

- **Best position**
  - Neutral
- **Scan technique**
  - Scan transversely over superior aspect of AC joint
  - Capsule normally bulges slightly superficially, which is increased in AC joint osteoarthritis
    - Downward transducer pressure inducing pain may help confirm AC joint as source of pain
  - Normal AC joint width is 4 mm
  - AC joint widened > 7 mm at midportion ± > 3 mm wider than opposite side is considered abnormal
  - Dynamic scanning can improve detection of AC joint laxity
    - Moving ipsilateral hand from ipsilateral knee to contralateral shoulder
    - Pull down ipsilateral arm by examiner whilst patient resists and compare to contralateral side
    - Stress imaging with patient holding weights akin to radiographic stress examination

Rotator Cuff Musculature and Suprascapular/Spinoglenoid Notches

- **Best position**
  - Neutral
- **Scan technique**
  - Scan in longitudinal and transverse plane along axis of muscle
  - Assess muscle bulk and also compare echogenicity to central tendon as well as trapezius muscle
  - Align transducer along tendon and trace toward tendon insertion
  - Suprascapular notch seen by scanning in transverse plane deep to supraspinatus muscle at upper margin of scapula
  - Spinoglenoid notch seen by scanning in transverse plane located between posterior glenohumeral joint and scapular spine

Glenohumeral Joint and Glenoid Labra

- **Best position**
  - Lateral decubitus for posterior joint recess
  - Supine with arm externally rotated for anterior joint recess
- **Scan technique**
  - Posterior recess: Scan longitudinally over infraspinatus close to musculotendinous junction
  - Anterior recess: Scan transversely deep to subscapularis tendon, though more difficult to assess
  - Visibility of joint recesses improved by slowly, passively, externally, and internally rotating patient's arm

Symptomatic Area

- Examine symptomatic area if removed from shoulder, such as midarm or supraclavicular region, to avoid missing unexpected pathology

Scan Position for Short-Axis View of Supraspinatus Tendon

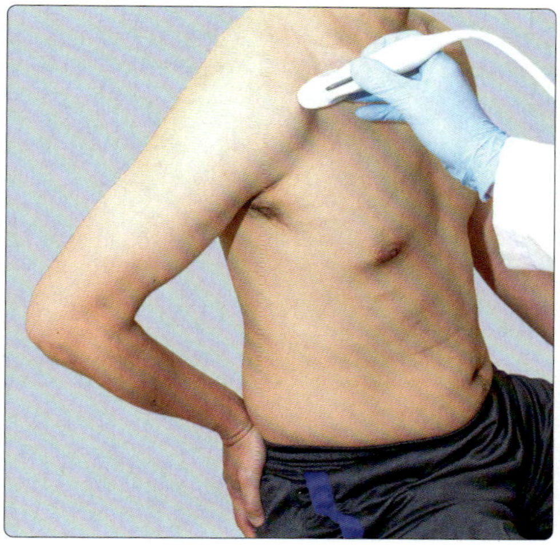

Supraspinatus Tendon: Short Axis

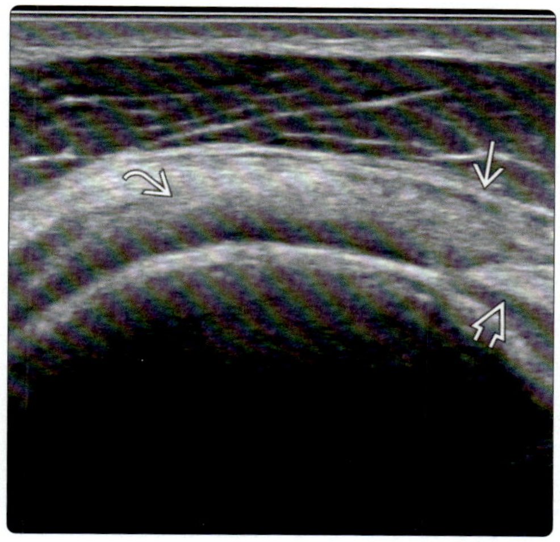

(Left) *Clinical photograph shows modified Crass position with transducer location for short-axis view of the supraspinatus. The modified Crass position produces less internal rotation of the shoulder, improving visibility of the anterior edge of the supraspinatus and the biceps tendon.* **(Right)** *Transverse US of supraspinatus tendon at its insertion is shown. The biceps tendon ➡ marks the anterior edge ➡ of supraspinatus tendon. Mild anisotropy ➡ is due to fiber orientation and intermingling of supraspinatus and infraspinatus tendons.*

Scan Position for Infraspinatus Tendon

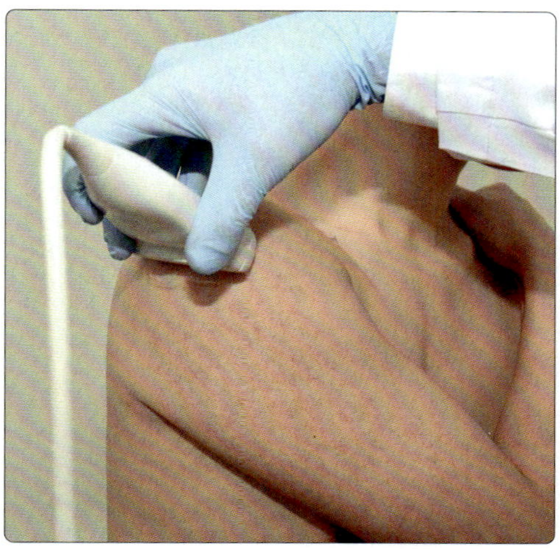

Infraspinatus Tendon: Long Axis

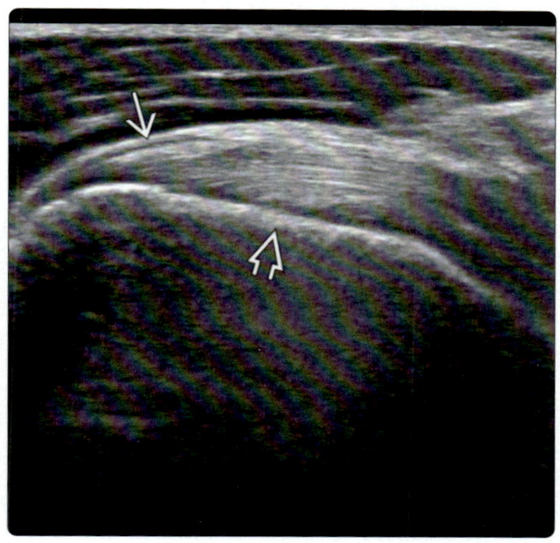

(Left) *Longitudinal US shows the patient and transducer position for examination of the infraspinatus tendon in long axis. The hand is placed on the contralateral shoulder. Alternatively, the hand can be placed on the contralateral thigh with the hand facing upward.* **(Right)** *Longitudinal US shows the long-axis view of the infraspinatus tendon inserted into the middle facet of the greater tuberosity ➡. Peribursal fat overlies the tendon. The thin, hypoechoic bursal line ➡ can be appreciated more laterally.*

Teres Minor: Long Axis

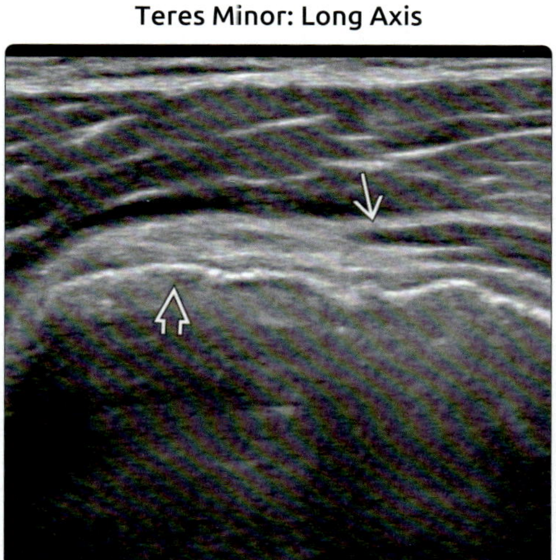

Infraspinatus and Teres Minor Tendons: Short Axis

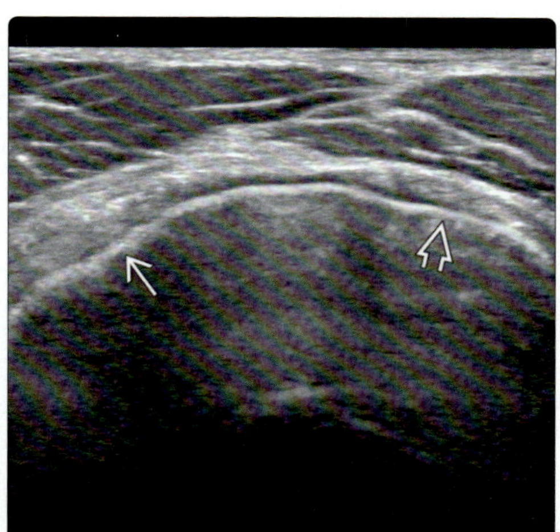

(Left) *Longitudinal US shows the long-axis view of teres minor insertion into the inferior facet of greater tuberosity ➡. Teres minor musculotendinous junction ➡ can also be appreciated.* **(Right)** *Transverse US along the posterosuperior aspect of the humeral head shows the infraspinatus ➡ and teres minor ➡ tendons inserting into the middle and inferior facets of the greater tuberosity, respectively.*

Scan Position for Rotator Interval

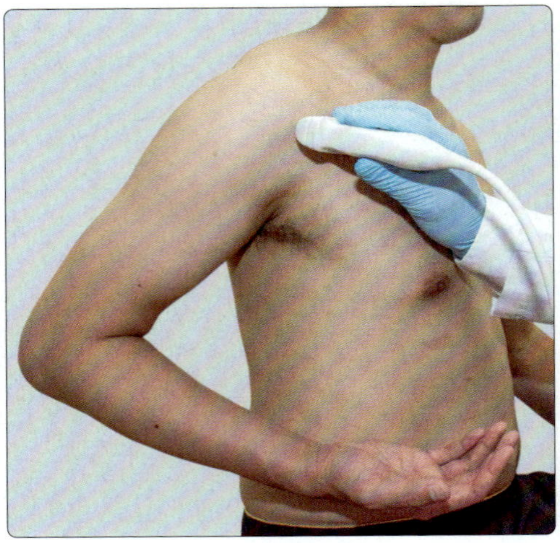

Rotator Interval: Short Axis

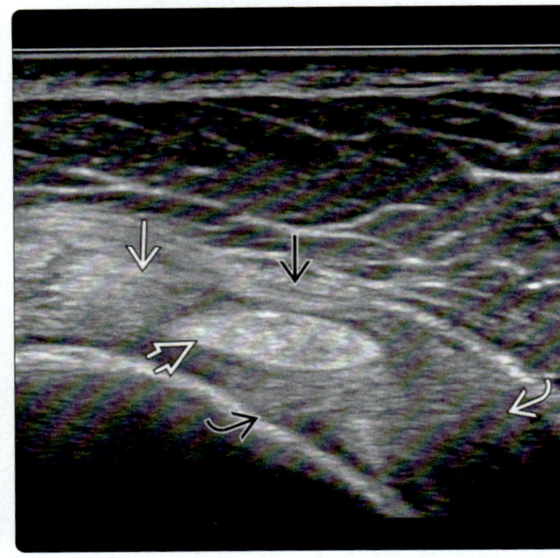

(Left) *Clinical photograph shows the patient and transducer position for rotator interval (jogger position), which is similar to the modified Crass position except that the hand is forward and upward and the elbow is more tucked in.* **(Right)** *Transverse US across the rotator interval shows the anterior edge of supraspinatus ➡, long head of biceps tendon (LHBT) ➡, and superior edge of subscapularis ➡. The coracohumeral ligament ⇨ lies superficial, and superior glenohumeral ligament/bicipital sling ⇨ is medial and inferior to LHBT.*

Scan Position for Biceps Tendon

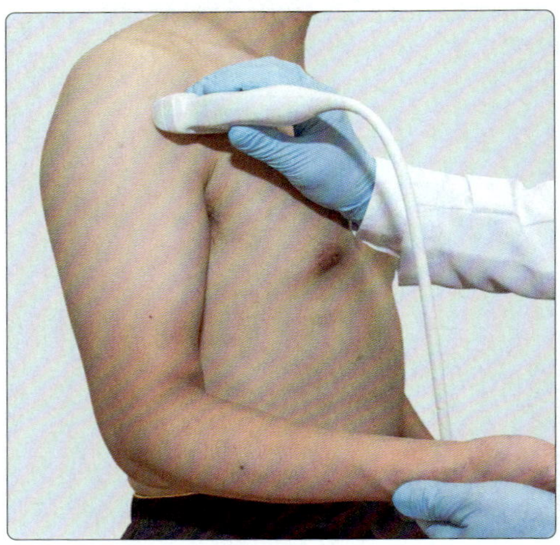

Biceps Tendon: Proximal Portion

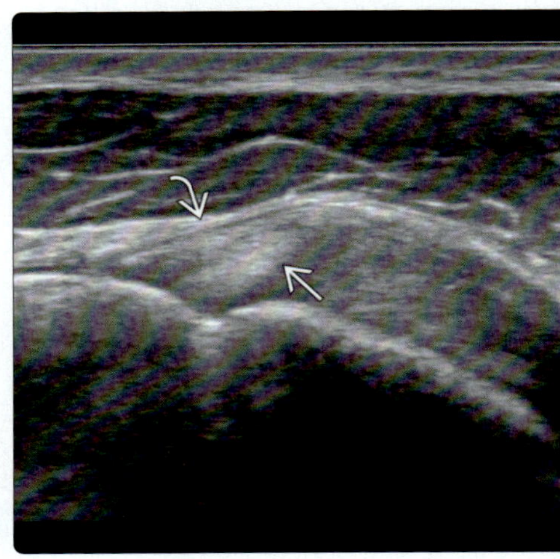

(Left) *Clinical photograph shows the scan position for the biceps tendon. The arm should be rotated slightly to align the bicipital groove and LHBT anteriorly. Greater degree of rotation of the arm allows assessment of LHBT laxity.* **(Right)** *Transverse US shows the normal oblong shape of LHBT ➡ at the upper end of the bicipital groove. This portion of the tendon, known as the genu, is most prone to tendinosis and tear. The superficial fibers ➡ of subscapularis pass over LHBT to insert onto the lateral lip of the bicipital groove.*

Biceps Tendon: Midportion

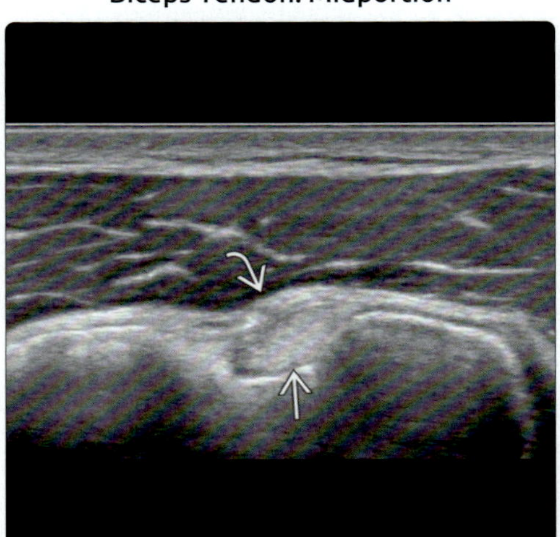

Biceps Tendon: Distal Portion

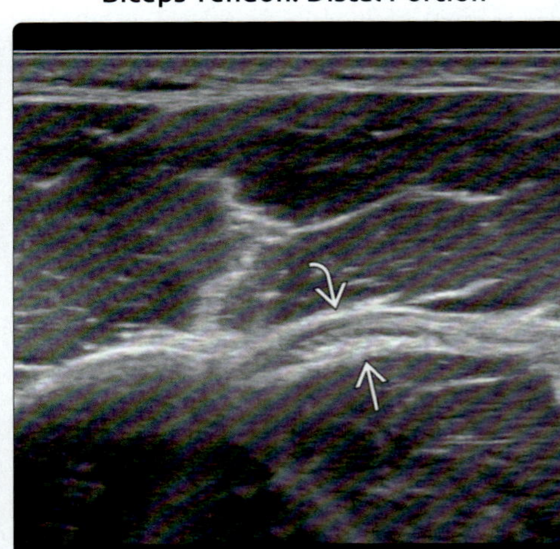

(Left) *Transverse US shows a more rounded appearance of LHBT ➡ within the midportion of the bicipital groove. Traversing superficial fibers ➡ from the subscapularis help retain LHBT in position.* **(Right)** *Transverse US shows a much more flattened, distal LHBT ➡ deep to the pectoralis major insertion ➡ into the lateral lip of the bicipital groove.*

Biceps Tendon: Long Axis

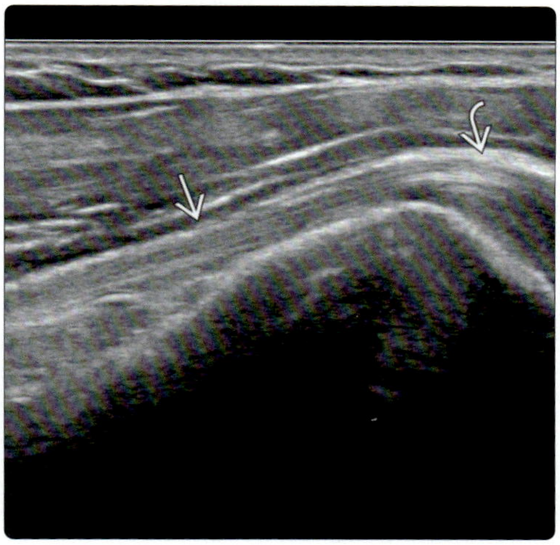

Scan Position for Long-Axis View of Subscapularis Tendon

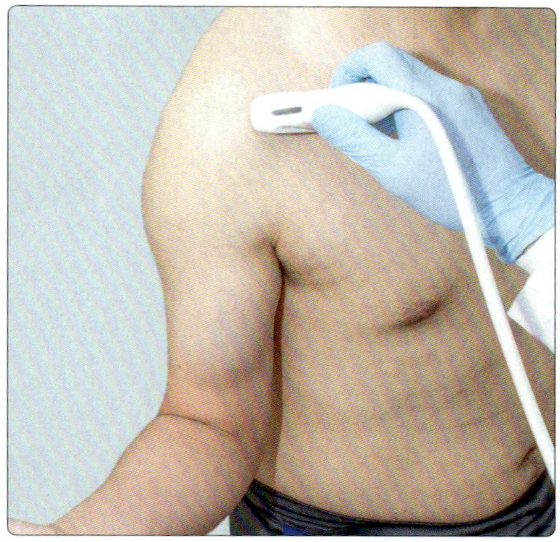

(Left) *Longitudinal US shows the longitudinal view of the mid- to proximal end of LHBT. Slight adduction of the arm may help with visualization. Mild arm extension can help increase visibility of the intraarticular portion of the LHBT.* **(Right)** *Clinical photograph shows the arm and probe position for the long-axis subscapularis tendon. Note how long axis of the transducer should be angled slightly upward as the subscapularis tendon moves from inferior to superior toward its insertion.*

Subscapularis Tendon: Long Axis

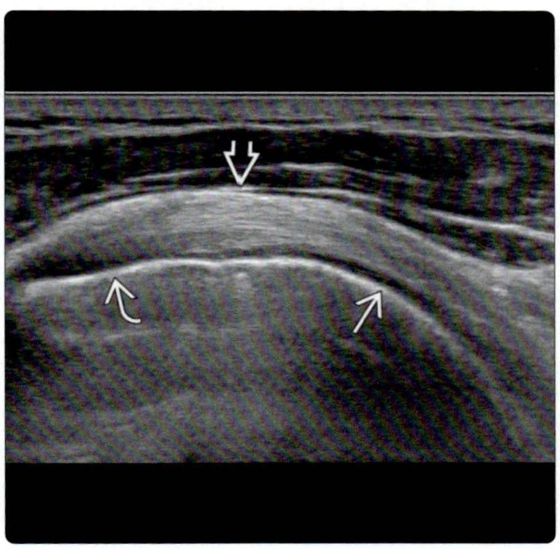

Subscapularis Tendon: Short Axis

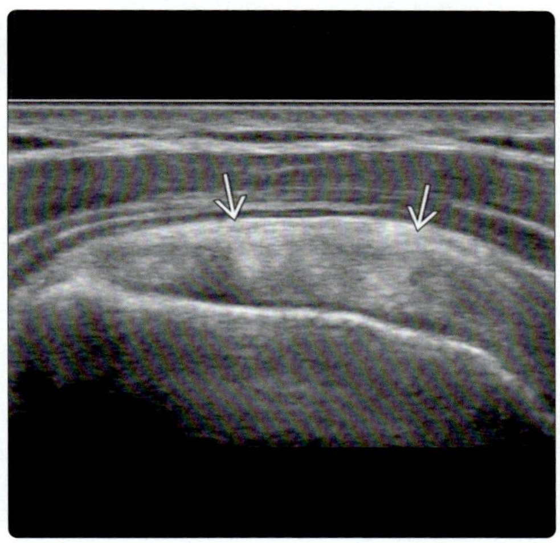

(Left) *Transverse US shows the long-axis view of the subscapularis tendon inserting into the lesser tuberosity. Hypoechoic articular cartilage is visible deep to the tendon medially.* **(Right)** *Longitudinal US shows the transverse section of the subscapularis tendon. The multipennate structure of the subscapularis muscle produces several tendon slips, which lead to the normal alternating hyperechoic/hypoechoic appearance of this tendon.*

Scan Position for Glenohumeral Joint: Posterior

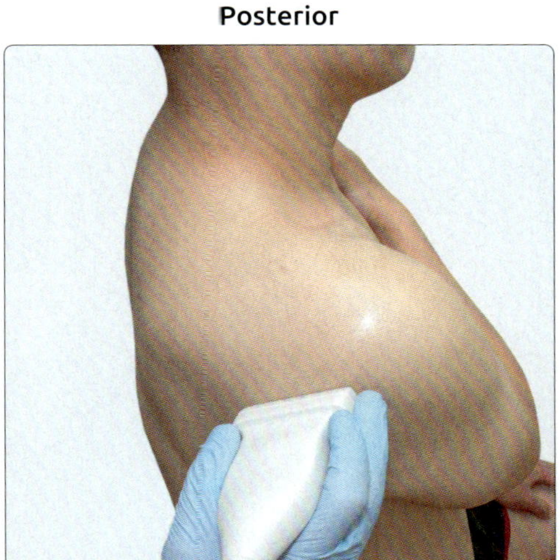

Glenohumeral Joint: Posterior

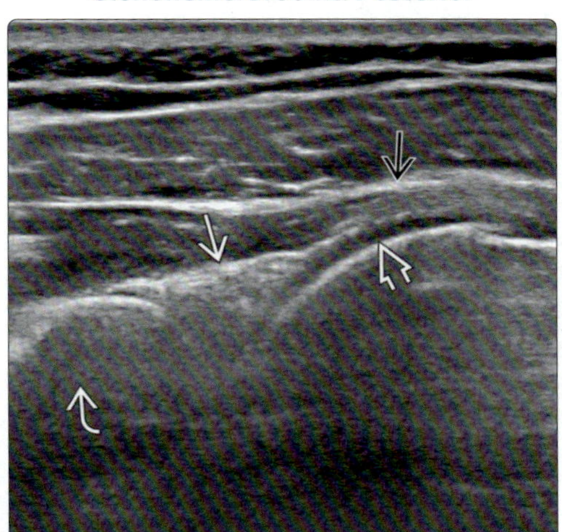

(Left) *Clinical photograph shows the patient and transducer position for examination of the posterior recess of the glenohumeral joint in short axis.* **(Right)** *Transverse US of the posterior aspect of the glenohumeral joint is shown. The humeral head is covered with hypoechoic articular cartilage, posterior glenoid rim, posterior labrum, and overlying infraspinatus tendon. The posterior recess of the glenohumeral joint is effaced and not distended.*

(Left) *Clinical photograph shows the transducer position for assessment of the acromioclavicular joint (AC). Dynamic assessment for AC joint laxity can be performed by either specific arm movement or axial traction on the arm.* (Right) *Transverse US shows a normal AC between the acromion and clavicle. Normal capsuloligamentous thickening is visible superiorly. The intraarticular disc is normally not seen. The clavicle usually lies higher than the acromion. AC joint width should be < 7 mm.*

Scan Position for Acromioclavicular Joint

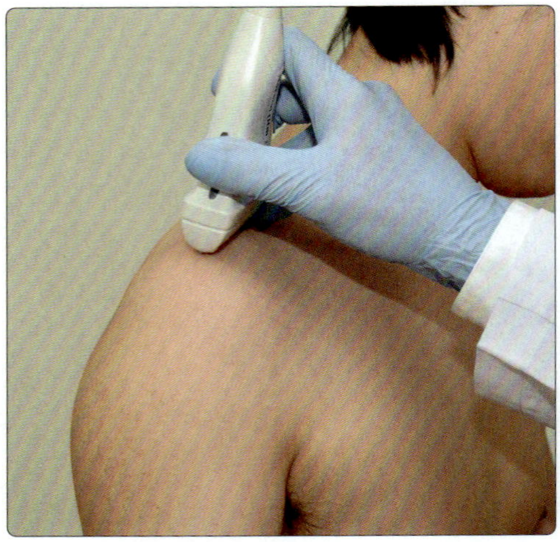

Acromioclavicular Joint

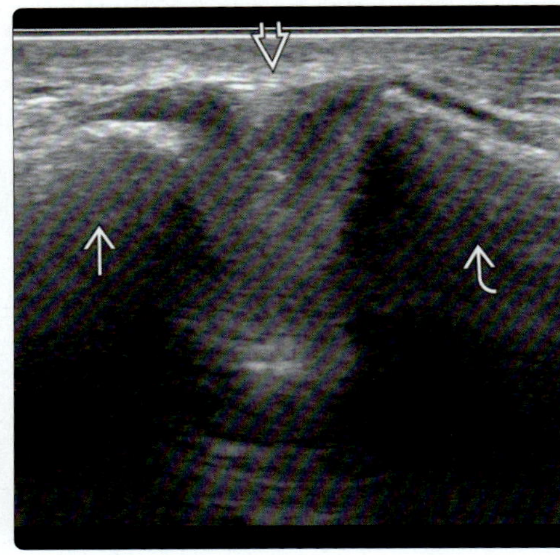

(Left) *Clinical photograph shows the patient and transducer position for dynamic assessment of subacromial or subcoracoacromial ligament impingement. Starting from a neutral position, the arm is abducted to try & recreate the patient's symptoms.* (Right) *Longitudinal US along the length of the coracoacromial ligament is shown stretching between acromion & coracoid process. The supraspinatus muscle lies deep to the coracoacromial ligament & the deltoid muscle is superficial.*

Scan Position for Subacromial Impingement

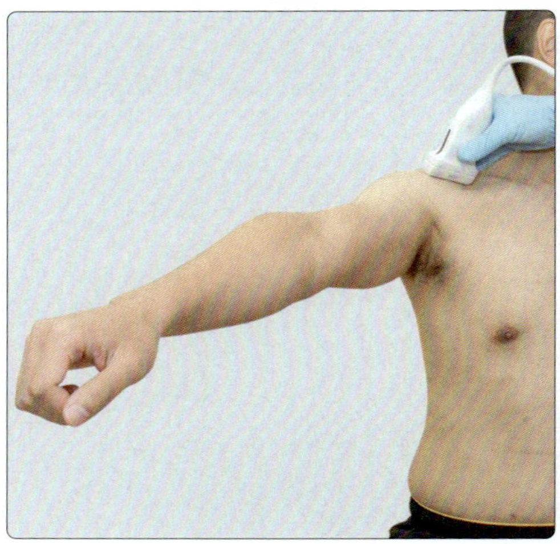

Coracoacromial Ligament: Long Axis

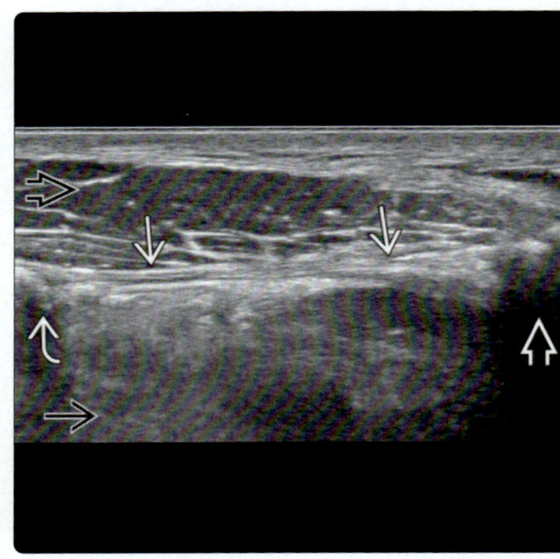

(Left) *Longitudinal US of the normal supraspinatus tendon shows the rotator cable at the mid- to part of the supraspinatus tendon. The alignment is different than that of the remaining supraspinatus fibers. The hypoechoic line superiorly is due to a normal, nondistended subacromial-subdeltoid bursa.* (Right) *Transverse US shows altered supraspinatus tendon echogenicity due to the rotator cable in short axis. This should not be misinterpreted as tendinosis. Biceps tendon lies medially.*

Rotator Cable: Long Axis

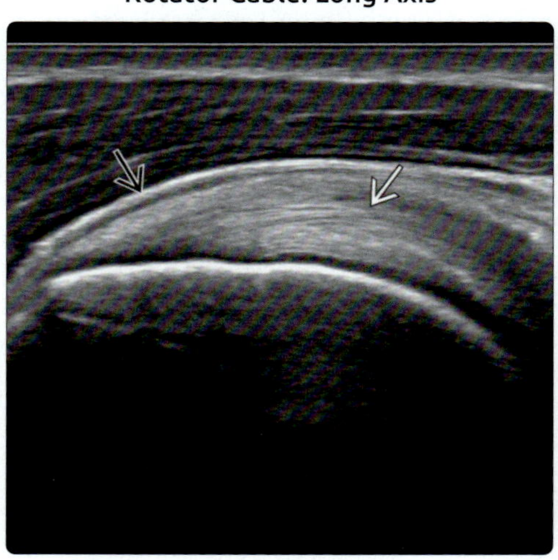

Rotator Cable: Short Axis

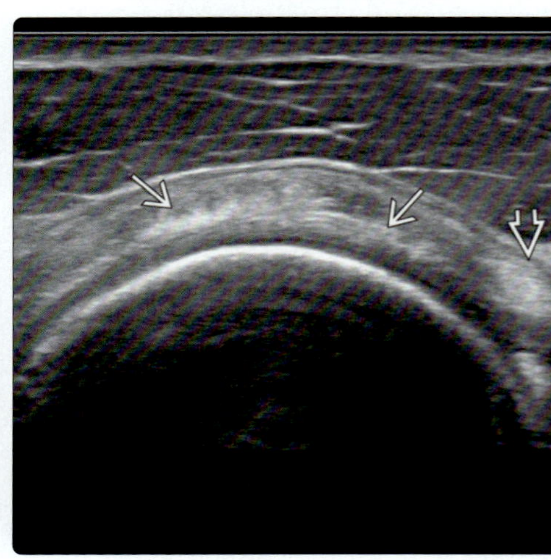

Scan Position for Dorsal Scapular Region

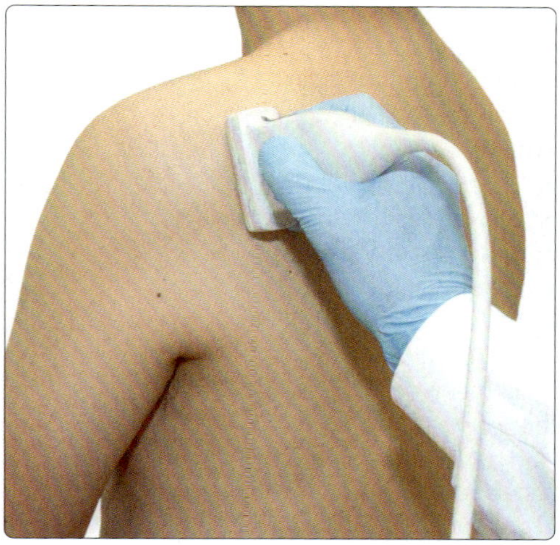

Supraspinatus Muscle: Short Axis

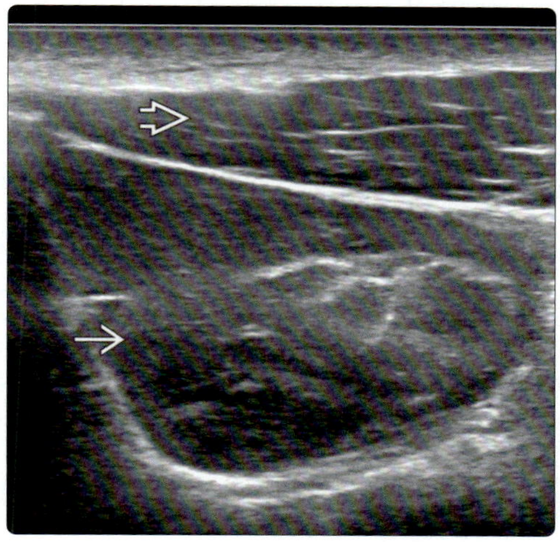

(Left) *Clinical photograph shows the patient position to assess the parascapular muscle. These muscles can be viewed in short axis or long axis along their orientation. The suprascapular notch and spinoglenoid notch are usually scanned in the transverse plane.* **(Right)** *Longitudinal US shows the transverse section of the normal supraspinatus muscle ➡ filling the supraspinatus fossa. Echogenicity can be compared to the overlying trapezius muscle ➡ and to the central tendon of the supraspinatus muscle.*

Infraspinatus Muscle: Short Axis

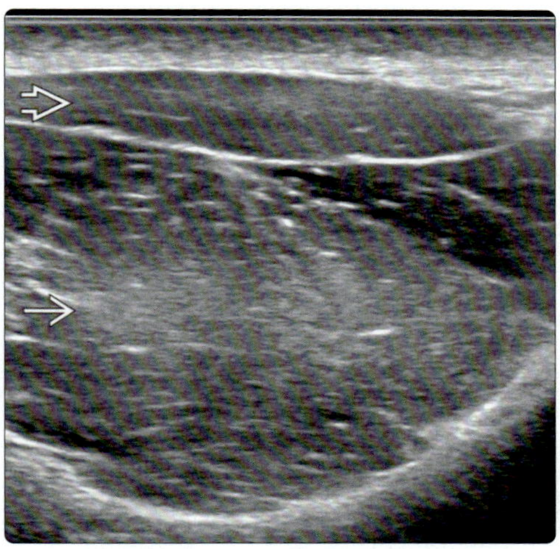

Teres Minor Muscle: Short Axis

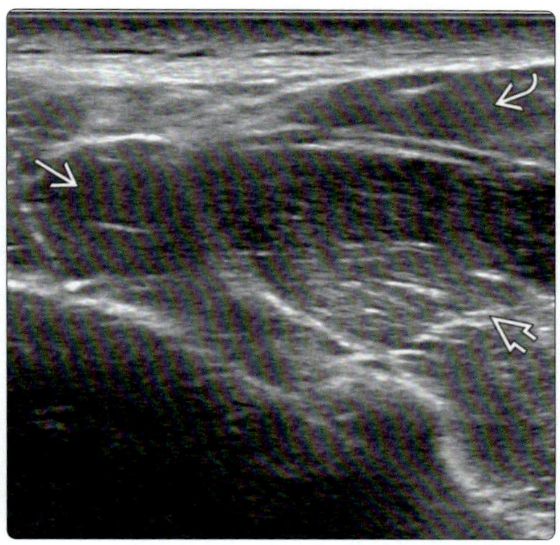

(Left) *Longitudinal US shows the transverse section of the normal infraspinatus muscle ➡ filling the infraspinatus fossa. Muscle echogenicity can be compared to the overlying trapezius muscle ➡ and the central tendon of the infraspinatus muscle.* **(Right)** *Longitudinal US shows the transverse section of the normal teres minor muscle ➡ at the inferomedial tip of the infraspinatus muscle ➡. Teres minor muscle bulk may be spared when the other rotator cuff muscles are atrophied. Trapezius lies superiorly ➡.*

Suprascapular Notch

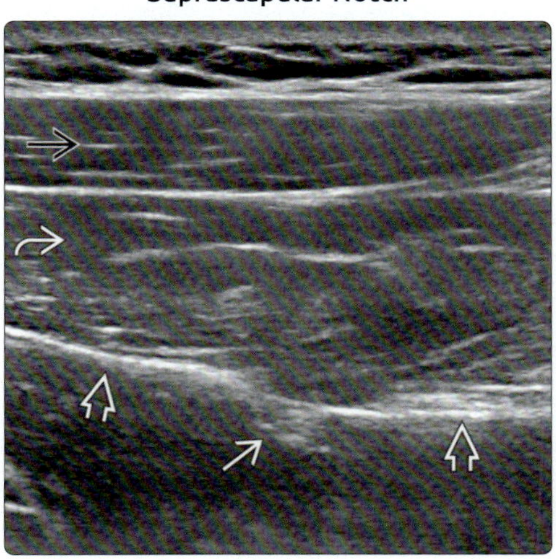

Spinoglenoid Notch

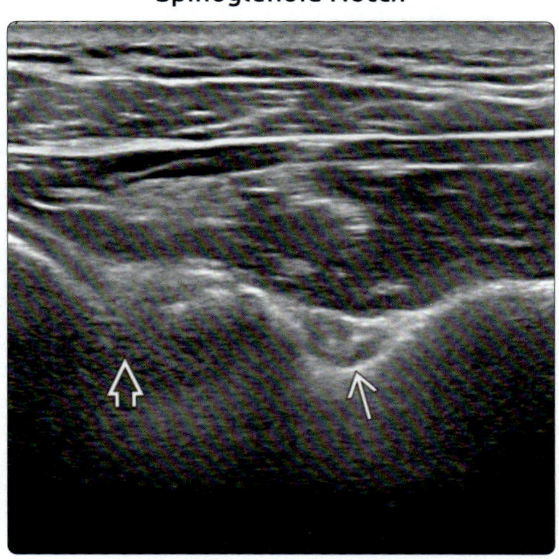

(Left) *Transverse US shows the suprascapular notch ➡ at the superior margin of the scapula bone ➡. The suprascapular notch is very variable in shape and size and often may not be seen as a distinct notch. The supraspinatus ➡ and trapezius ➡ muscles lie superiorly.* **(Right)** *Transverse US shows the spinoglenoid notch ➡ located between the lateral end of the scapular spine and posterior margin of glenohumeral joint ➡. At this location, the suprascapular nerve may be compressed by paralabral cysts originating from the posterior labrum.*

KEY FACTS

GENERAL CONSIDERATIONS

- Common specific indications
  - Suspected joint effusion, synovitis, epicondylitis, and cubital tunnel syndrome
- Less common specific indications
  - Olecranon bursitis, distal biceps and triceps pathology, supinator and radial tunnel syndromes

TECHNIQUE

- **Anterior elbow**
  - Seated opposite examiner with elbow extended on examination couch
  - Increase height of examination table so that elbow is almost horizontal
    - Anterior aspect elbow joint
    - Distal biceps tendon
    - Brachialis
    - Median nerve
    - Radial nerve
- **Medial elbow**
  - Patient facing examiner with elbow extended and forearm externally rotated
    - Common flexor tendon origin
    - Medial collateral ligament complex
    - Cubital tunnel
- **Lateral elbow**
  - Elbow flexed toward examiner and hand pronated
    - Common extensor tendon origin
    - Radial collateral ligament
    - Lateral ulnar collateral ligament
    - Synovial fringe
    - Annular ligament
- **Posterior elbow**
  - Elbow flexed toward examiner with palm resting on examination couch ("crawling crab" position)
    - Triceps tendon
    - Posterior aspect of elbow joint

Scan Position for Anterior Elbow

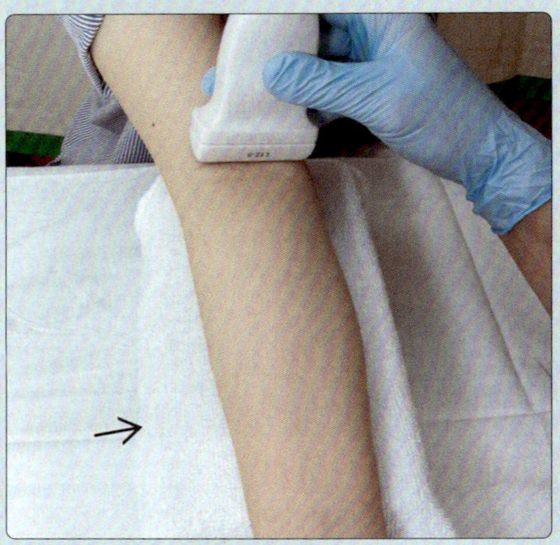

Distal Biceps Tendon

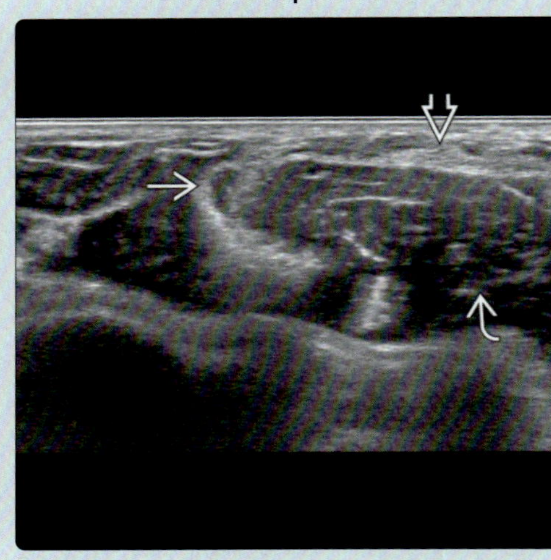

(Left) *Clinical photograph shows the scan position for the anterior elbow. The elbow can be placed on a firm pad → to bring it into a more horizontal orientation. This helps the examination of the anterior elbow region.* **(Right)** *Transverse US shows the biceps tendon → just proximal to the elbow joint located between the subcutaneous tissues and the brachialis muscle →. The bicipital aponeurosis → extending over the brachial artery and median nerve helps maintain the position of the distal biceps tendon.*

Distal Biceps Tendon

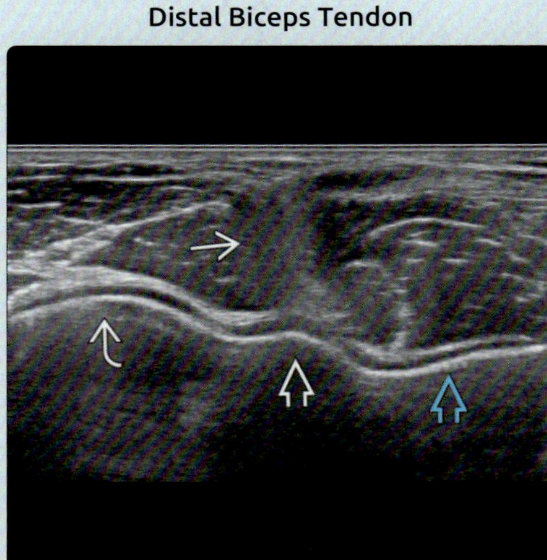

Brachialis Muscle Insertion

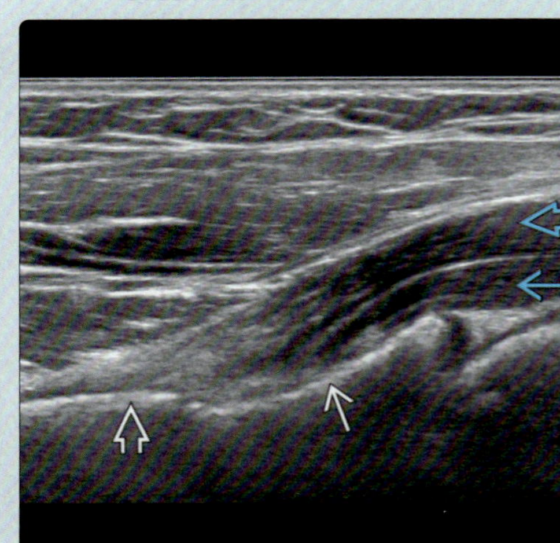

(Left) *Transverse US at the proximal aspect of elbow joint shows the hypoechoic artifact → from the distal biceps tendon as it descends toward the radial tuberosity. The hypoechoic hyaline cartilage overlying the capitellum → and lateral → and medial → pillars of the trochlea is seen.* **(Right)** *Longitudinal US of the anterior elbow shows the broad musculoaponeurotic insertion of the deep head of brachialis muscle → onto the coronoid process → and the smaller tendinous insertion of the superficial head → onto the ulnar tuberosity →.*

GENERAL CONSIDERATIONS

Clinical Indications for Elbow US

- Nearly all elbow conditions are best imaged with US
- Common indications: Suspected joint effusion, synovitis, epicondylitis, and cubital tunnel syndrome
- Less common indications: Olecranon bursitis, distal biceps pathology, radial tunnel syndrome, and triceps pathology

Targeted Approach Most Efficient

- Take brief history regarding site/duration of symptoms, trauma history, etc.
- Check for joint effusion/synovitis and then focus on symptomatic area

TECHNIQUE: ANTERIOR ELBOW

Patient Position

- Seated opposite examiner with elbow extended on examination couch
  - ± place pad under elbow for full extension
- Increase height of examination table so that elbow is almost horizontal

Specifically Examine

- **Anterior aspect of elbow joint**
  - Saddle-shaped trochlea covered with anechoic articular cartilage articulating with pointed triangular coronoid process; medial ridge larger than lateral ridge
  - More rounded capitellum laterally articulates with square-shaped radial head
    - Capitellum faces more anteriorly and is not visible using posterior approach
- **Distal biceps tendon**
  - Seen between subcutaneous tissue and brachialis at distal end of humerus
    - Medial nerve lies medial to brachial artery, while biceps tendon lies lateral
      - Bicipital aponeurosis (lacertus fibrosis) passes from biceps over artery and nerve → pronator teres
  - Tendon is hypoechoic on transverse imaging and best seen on longitudinal imaging
  - Long and short heads are barely separable
    - Long head attached to radial tuberosity and is thus a stronger supinator of elbow
    - Short head attached more medial and distal to radial tuberosity and is mainly elbow flexor
    - Normal bicipitoradial bursa located between biceps and tuberosity not visible
  - To most, readily see distal biceps tendon, hypersupinate forearm, which moves radial tuberosity forward, and angle transducer longitudinally (heel-toe) parallel to tendon
  - Alternatively
    - Medial approach: Elbow flexed to 25° and forearm supinated
      - Scan longitudinally along proximal radius
      - When distal biceps tendon comes into view, minimally adjust forearm and probe position to optimize visibility
    - Lateral approach: Elbow flexed to 90° and forearm pronated
      - Scan longitudinally along distal humerus similar to flexed abducted supinated (FABS) MR view
      - As one moves transducer slowly anteriorly, biceps tendon will come into view parallel to brachial artery
    - Cobra position: Elbow flexed to 90° with forearm pointing upward and pronated
      - Scan in transverse plane over proximal radius at distal end of supinator muscle
      - Distal biceps identified during pronation; hyperflexion of elbow will accentuate biceps tendon visibility
- **Brachialis**
  - Larger superficial head inserts → ulnar tuberosity as short tendon over small area
  - Smaller deep head inserts → ulnar tuberosity and anterior surface coronoid process as musculoaponeurotic attachment over broad area
- **Median nerve**
  - Medial to brachial artery
    - Usually (80%) just alongside artery but occasionally (20%) separate from artery
  - Trace median nerve transversely deep to pronator teres and flexor digitorum superficialis (FDS)
    - Potential sites of compression are between 2 heads of pronator teres and deep to fibrous arcade proximal end of FDS
    - Gives off anterior interosseous branch (purely motor) between pronator teres heads
- **Radial nerve**
  - Best identified deep to brachioradialis
    - Trace along radial tunnel, which is located between radiocapitellar joint and proximal border of supinator
      - Uncommon site of compression by vascular arcade ("leash of Henry")
      - Recurrent radial artery traveling with nerve can be identified by color Doppler
    - Identify radial nerve dividing into superficial and deep branches
  - Deep branch → enters supinator muscle as posterior interosseous nerve (PIN)
    - Gently pronate/supinate forearm to help see nerve passing through and exiting supinator tunnel
    - Proximal end of supinator tunnel (arcade of Frohse) is common site of PIN compression
      - Focal angulation and flattening of PIN at arcade of Frohse is common, normal finding
    - Normal size of deep branch of radial nerve is 2.3 mm^2
  - Superficial branch → ascends to subcutaneous tissues

TECHNIQUE: MEDIAL ELBOW

Patient Position

- Seated opposite examiner with extended elbow and forearm externally rotated

Specifically Examine

- **Common flexor tendon origin (CFTO)**
  - Place finger on medial epicondyle and then place transducer along imaginary line of flexor tendons converging on epicondyle
  - Individual components of CFTO cannot be discriminated

- Pronator teres, flexor carpi radialis longus and brevis, palmaris longus, flexor carpi ulnaris, flexor digitorum superfiicalis
- CFTO is short and forward facing
- **Medial collateral ligament complex**
  - Anterior band is most important and easiest to see
  - Anterior band best seen with elbow in long axis with elbow in flexion and slight valgus stretch
    - Degree of flexion varies but may be up to full flexion
  - Extends from medial epicondyle → sublime tubercle of ulna slightly anterior to CFTO
- **Cubital tunnel**
  - 2 components to cubital tunnel
    - Proximal portion in retrocondylar groove is most common site of compression
    - Distal portion between ulnar and humeral heads of flexor carpi ulnaris ("cubital tunnel proper")
  - Elbow placed on pad and externally rotated with patient leaning away, enabling transducer access to posteromedial elbow
  - Alternatively, examine cubital tunnel with arm extended over edge of examination table and elbow facing upward
    - Dynamic scanning of tunnel more difficult for this position
  - Transverse imaging of cubital tunnel is preferable
    - Measure cross-sectional area of ulnar nerve
    - Ulnar nerve is usually more hypoechoic within cubital tunnel with ↑ size by ~ 15% compared to distal arm
    - Normal size of ulnar nerve in cubital tunnel is 8 mm^2
    - Accompanied by recurrent ulnar artery and veins
  - Continue scanning nerve in transverse plane distally through and beyond distal cubital tunnel
  - Dynamic assessment of ulnar nerve movement
    - Lightly place transducer across cubital tunnel while slowly, passively flexing elbow
    - Continual adjustment of transducer position is required to ensure it remains positioned over medial epicondyle
    - See if ulnar nerve subluxes or dislocates ± subluxation of medial head of triceps ("double snap")

TECHNIQUE: LATERAL ELBOW

Patient Position

- Elbow flexed toward examiner with hand pronated
  - Alternatively, both elbows extended with thumbs upward and hands together ("prayer" position)

Specifically Examine

- **Common extensor tendon origin (CETO)**
  - Place finger on lateral epicondyle and then place transducer along imaginary line of extensor tendons converging on epicondyle
  - Individual components of CETO cannot be discriminated
  - Extensor carpi radialis brevis (ECRB) lies most deeply and inseparable from radial collateral ligament (RCL)
    - ECRB = most commonly torn CETO tendon
  - ECRB, extensor carpi ulnaris, extensor digiti minimi, extensor digitorum (ED); ED is most superficial
  - Examine in longitudinal and transverse planes
    - On longitudinal imaging, normal CETO has straight or slightly concave superficial contour
    - On transverse imaging, CETO has oblong shape over flat area of lateral epicondyle
- **Radial collateral ligament**
  - Best examined in longitudinal plane with transducer angled slightly anterior at anterior edge of CETO tendons
  - Extends from lateral epicondyle → annular ligament overlying radial head
- **Lateral ulnar collateral ligament (LUCL)**
  - For proximal aspect, place transducer on lateral epicondyle of flexed elbow and identify LUCL lateral to radial head
    - Then move probe distally so that proximal end on radial head and rotate distal end of probes 80-90° to scan lateral aspect of proximal ulna in region of supinator crest
    - Supinator crest = bone ridge along radial aspect of proximal ulna to which supinator muscle is attached
  - For distal portion, scan in transverse plane posterolaterally close to anterolateral corner of anconeus muscle
    - LUCL seen deep to anconeus muscle
- **Synovial Fold**
  - Seen on long axis as triangular hyperechoic tissue at posterolateral periphery of radiocapitellar joint
    - Occur posteriorly (95%), anteriorly (70%), and laterally (15%) at radiocapitellar joint
- **Annular ligament**
  - Seen as belt-like hyperechoic band overlying radial head
  - Passive rotation of forearm during scanning allows differentiation of annular ligament from radial head

TECHNIQUE: POSTERIOR ELBOW

Patient Position

- Elbow flexed toward examiner with palm resting on examination couch ("crawling crab" position)
- Or patient supine with elbow flexed across abdomen
  - This position is easier for right elbow
- Or patient prone with posterior aspect of elbow facing examiner

Specifically Examine

- **Posterior aspect of elbow joint**
  - Posterior fat pad deep to distal part of triceps filling olecranon fossa of distal humerus
  - Mild flexion of elbow joint will encourage fluid into posterior aspect of joint
- **Triceps tendon**
  - Examine in both transverse and longitudinal planes
    - Triceps tendon tends to be located slightly medial to midline
  - On longitudinal imaging, triceps tendons show striations as they widen to their olecranon insertion
    - Lateral and long heads contribute mainly to posterior insertion
    - Medial head is inserted mainly onto ulnar side of olecranon process
  - Normal olecranon bursa cannot be appreciated

Anteromedial Aspect of Elbow

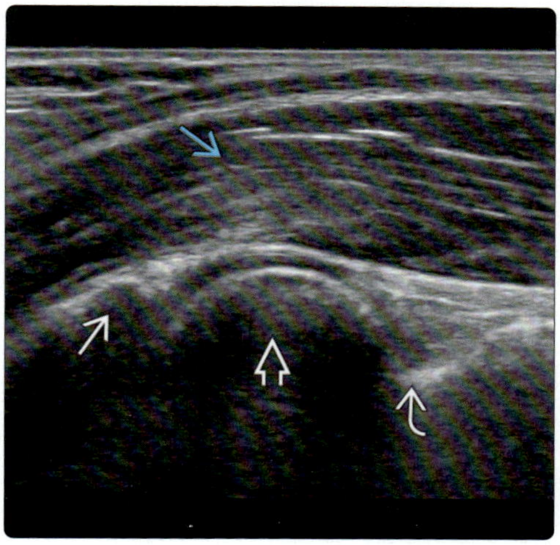

Anterolateral Aspect of Elbow

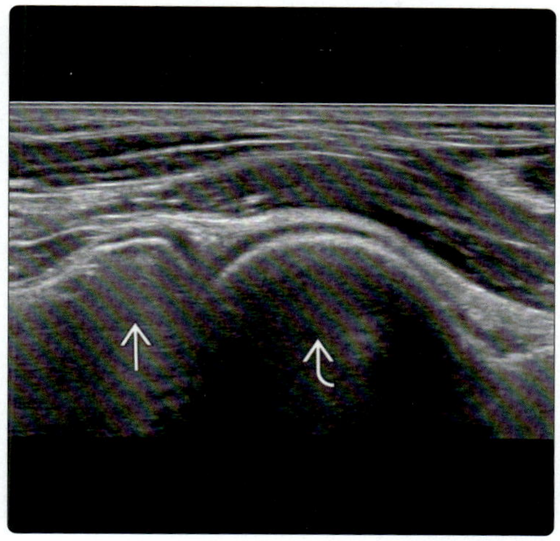

(Left) *Longitudinal US of the anteromedial aspect of the elbow shows the triangular anterior tip of the coronoid process and rounded trochlea process. The anterior fat pad filling the coronoid fossa of the humerus is seen with the brachialis muscle located anteriorly.* **(Right)** *Longitudinal US in the anterolateral aspect of the elbow shows the square-shaped radial head articulating with the rounded capitellum. Hypoechoic cartilage covers the forward-facing capitellum and the sides of the radial head.*

Annular Ligament

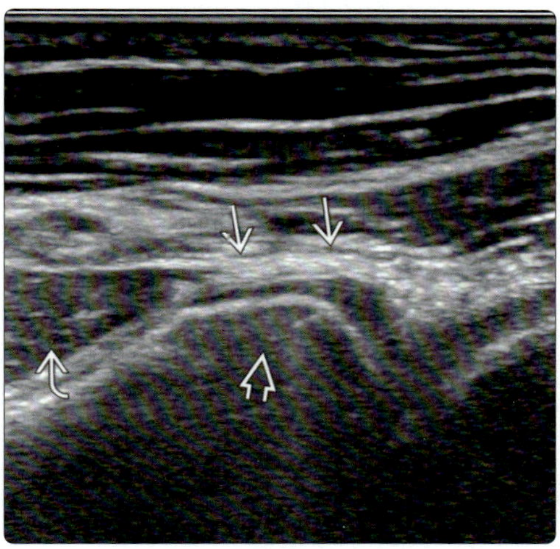

Distal Biceps Tendon: Anterior Approach

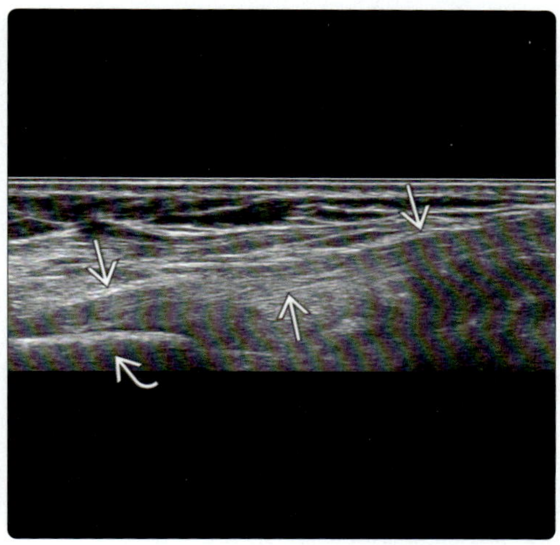

(Left) *Longitudinal US in the anterolateral aspect of the elbow shows an echogenic annular ligament surrounding the radial head. There is hypoechoic hyaline cartilage deep to the ligament for articulation at the proximal radioulnar joint. More distally is the supinator muscle.* **(Right)** *Longitudinal US shows the distal end of the biceps tendon. The transducer is angled heel to toe to bring it more parallel to the biceps tendon, and the forearm is hypersupinated to bring the radial tuberosity forward.*

Scan Position for Biceps Medial Approach

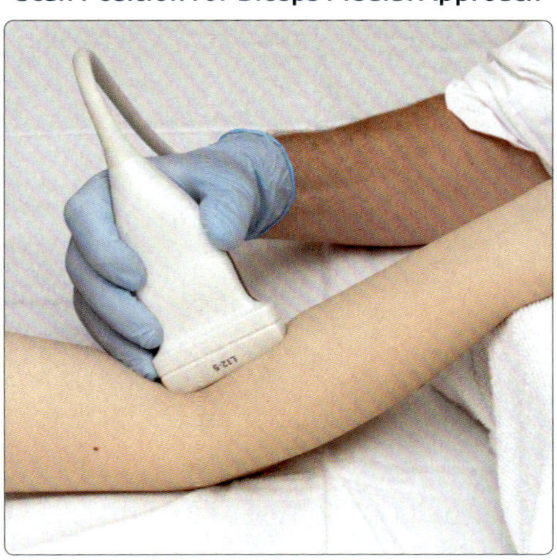

Distal Biceps Tendon

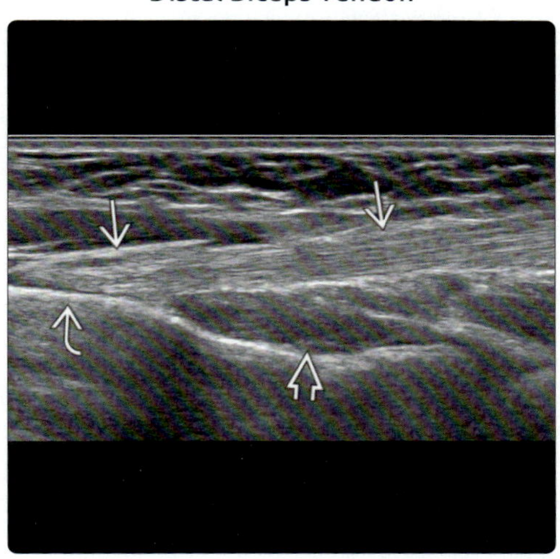

(Left) *Clinical photograph shows the scan position for the distal end of the biceps tendon using a medial approach. The elbow is flexed to about 25°. From a medial approach, identify the radial head and neck. Then angle the transducer very slightly to bring the biceps tendon into view.* **(Right)** *The medial approach provides the best overall view of the distal biceps tendon inserting into the radial tuberosity. The supinator muscle lies just proximal to this insertion.*

Scan Position for Biceps Lateral Approach

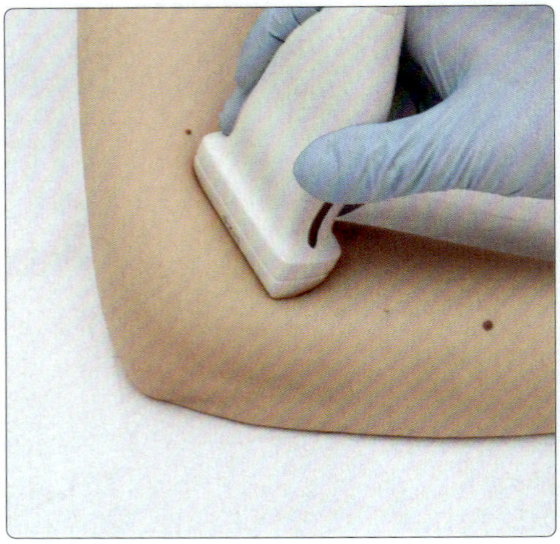

Distal Biceps Tendon: Lateral Approach

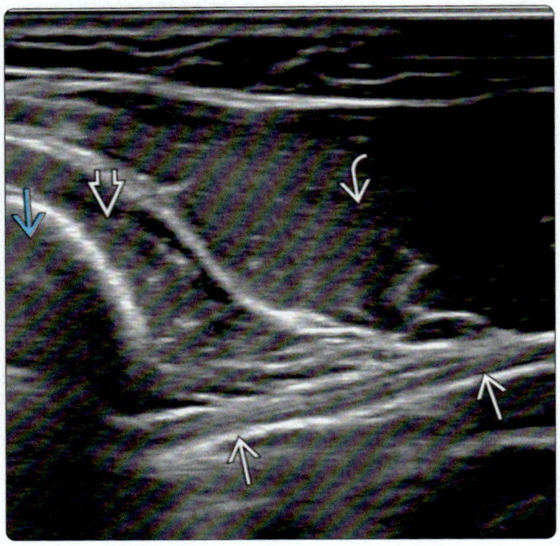

(Left) *Clinical photograph shows the scan position for the lateral approach to the distal biceps. This is similar to the flexed abducted supinated (FABS) view described for MR examination with the elbow flexed, the shoulder abducted, and the wrist supinated. The transducer is aligned almost parallel to the humeral shaft and then angulated more laterally.* **(Right)** *Longitudinal US from the lateral approach shows the distal part of the biceps tendon lying deep to the proximal radius, supinator, and brachialis muscles.*

Scan Position for Biceps Tendon Insertion

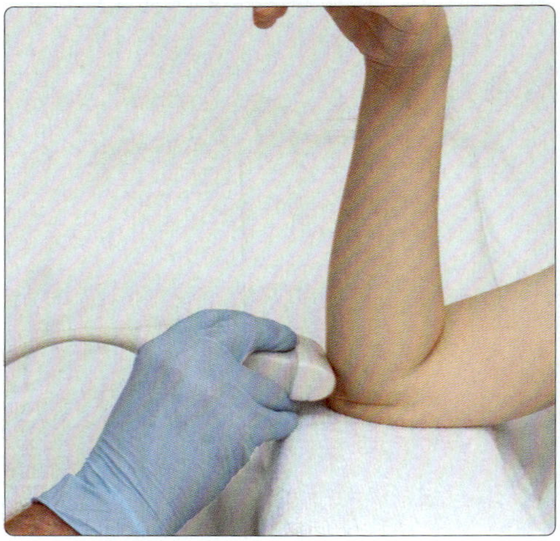

Biceps Tendon Insertion: Cobra Position

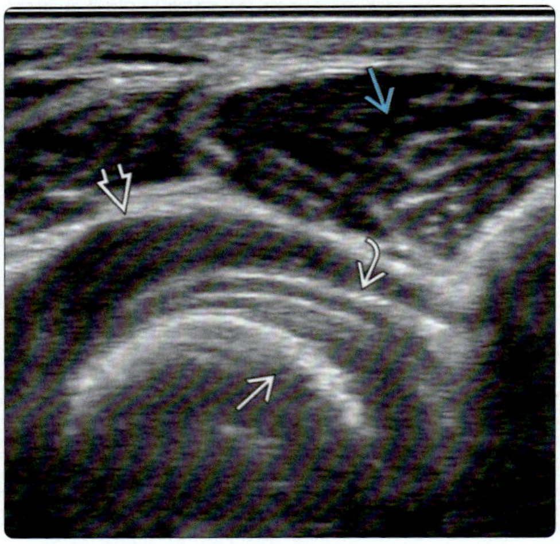

(Left) *Clinical photograph shows the cobra position for examining the radial insertion of the biceps tendon. The transducer is placed over the proximal radius, and the forearm is pronated to bring the distal biceps insertion into view.* **(Right)** *Transverse US shows the biceps tendon insertion into the radial tuberosity. Distal fibers of the supinator muscle overlie the insertion. Visibility of the biceps insertion can be further accentuated by mild hyperflexion of the elbow. Proximal ulna and anconeus muscle are also seen.*

Scan Position for Common Flexor Tendon Origin

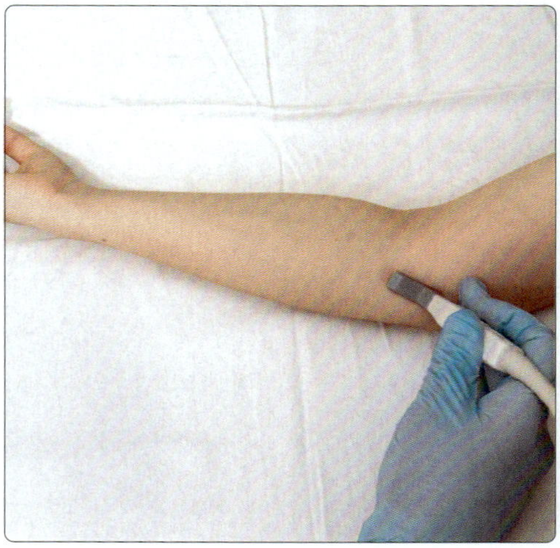

Common Flexor Tendon Origin

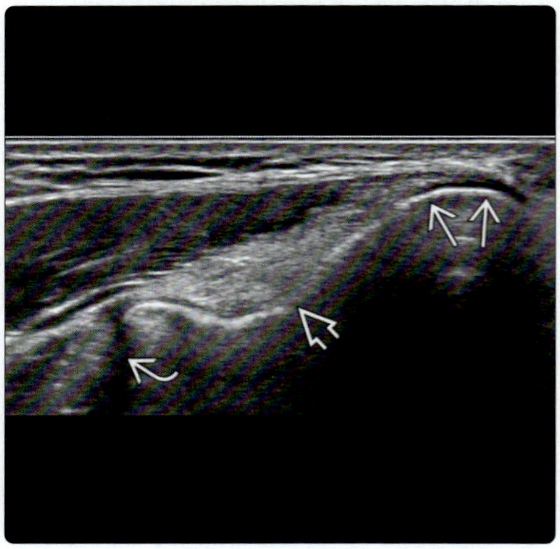

(Left) *Clinical photograph shows the scan position for the common flexor tendon origin (CFTO). The elbow is slightly flexed and rotated externally. The probe is first placed on the medial epicondyle along the assumed path of the flexor tendons and then, using US, is aligned properly along this path.* **(Right)** *Longitudinal US of the medial epicondyle shows the CFTO. The anterior band of the ulnar collateral ligament (UCL) is present deep to the CFTO. The ulnar aspect of the ulnohumeral joint is seen.*

Ulnar Collateral Ligament Complex

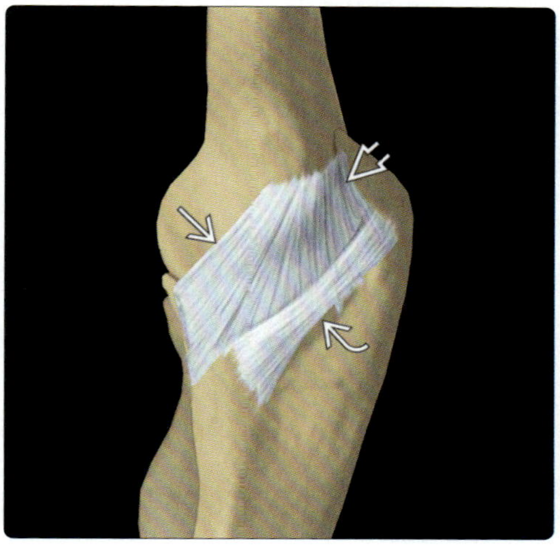

Anterior Band of Ulnar Collateral Ligament

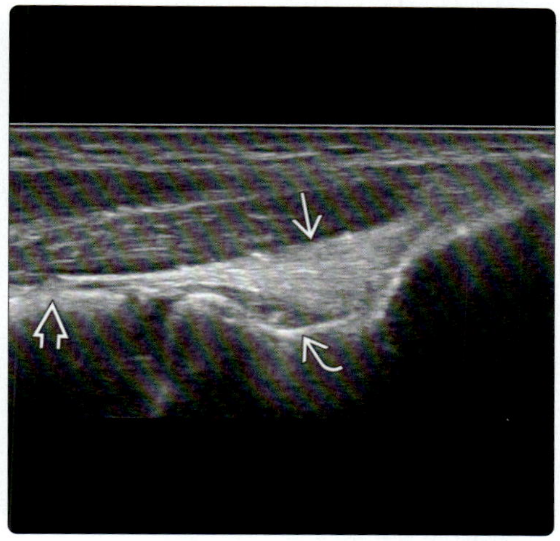

(Left) *Graphic shows the anterior (strongest), posterior, and transverse (weakest) bands of the UCL complex.* **(Right)** *Longitudinal US obtained at the anterior edge of the common extensor tendon origin (CETO) in flexion and mild valgus stretch shows the anterior band of the UCL, which passes between the sublime tubercle of the ulna and the medial epicondyle of the humerus. A small joint recess is present deep to the ligament.*

Scan Position for Cubital Tunnel

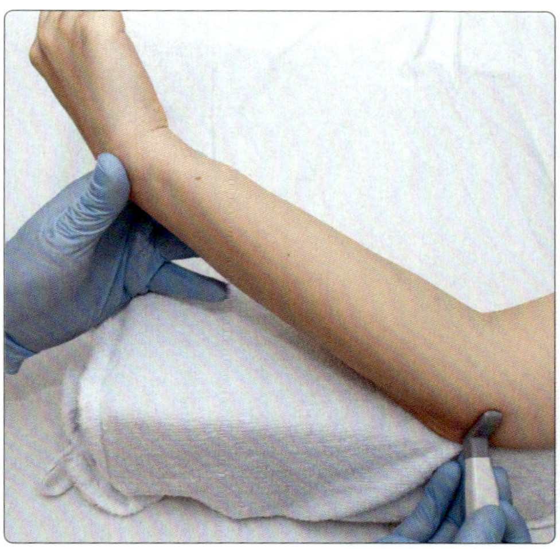

Cubital Tunnel: Proximal Aspect

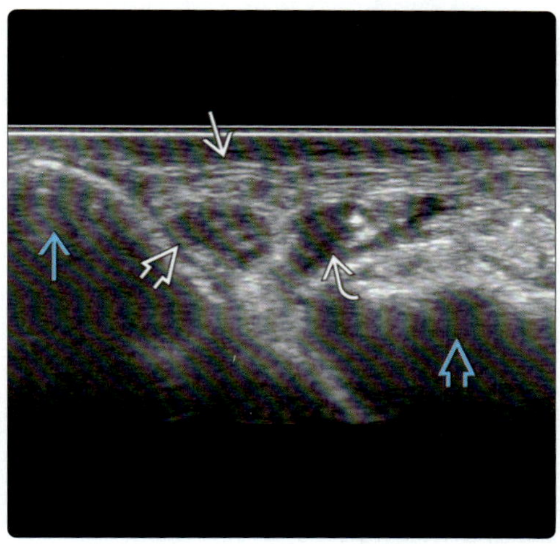

(Left) *Clinical photo shows scan position for cubital tunnel. The elbow is externally rotated and placed at edge of firm pad. This facilitates transducer access to the cubital fossa as well as dynamic scanning of the ulnar nerve during elbow flexion.* **(Right)** *Transverse US in the posteromedial aspect of the elbow shows the cubital tunnel located between the medial epicondyle and the olecranon process. The tunnel, which is covered by the Osborne ligament, contains ulnar nerve & medial belly of triceps muscle slips.*

Cubital Tunnel: Distal Aspect

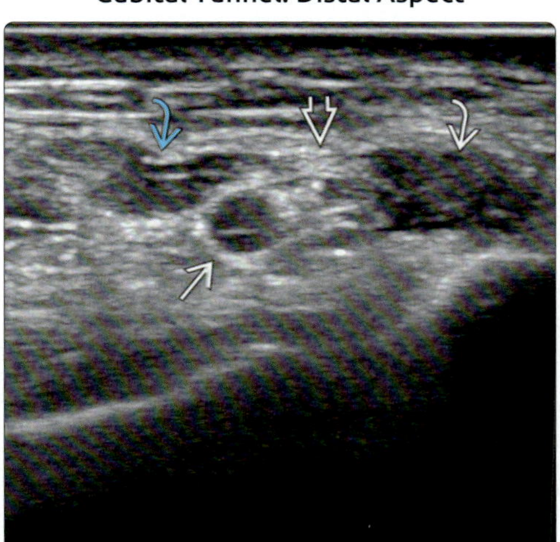

Cubital Tunnel During Elbow Flexion

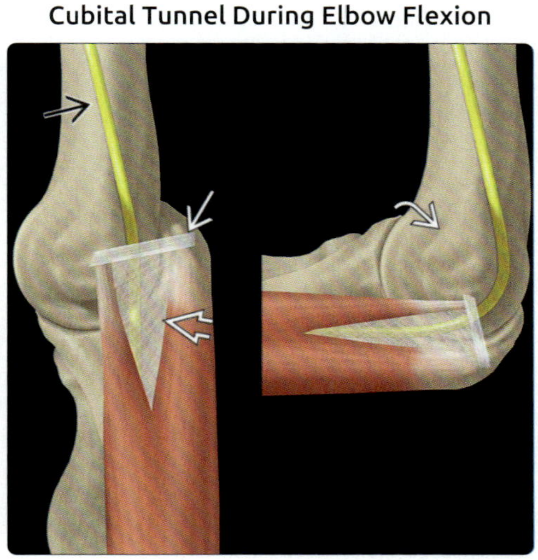

(Left) *Transverse US shows the ulnar nerve deep to the arcuate ligament between the humeral and ulnar heads of the flexor carpi ulnaris. The ulnar nerve may also be compressed at this location.* **(Right)** *Schematic diagram shows how the Osborne and the arcuate ligaments move distally relative to the medial epicondyle during elbow flexion. The resultant undercoverage of the ulnar nerve facilitates subluxation/dislocation over the medial epicondyle during elbow flexion.*

(Left) *Clinical photograph shows the scan position for the CETO. The elbow is flexed. The probe is first placed on the lateral epicondyle along the assumed path of the extensor tendons and then, using US, is aligned properly along this path.* **(Right)** *Longitudinal US shows the CETO attached to the lateral epicondyle. The lateral epicondyle is flatter and more lateral facing than the medial epicondyle. Notice how the superficial contour of the normal CETO is flat or concave.*

Scan Position for Common Extensor Tendon Origin

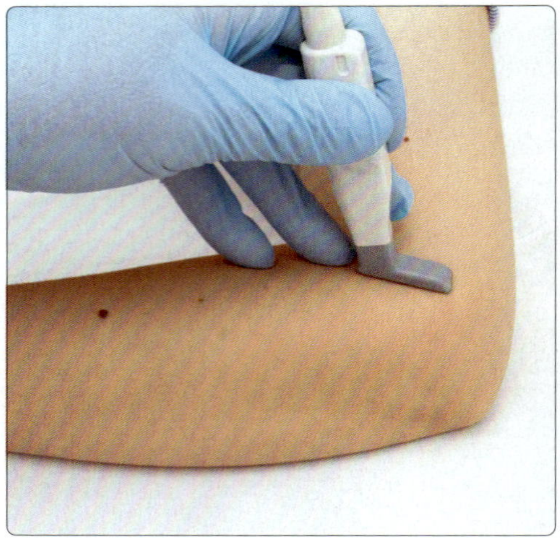

Common Extensor Tendon Origin: Long Axis

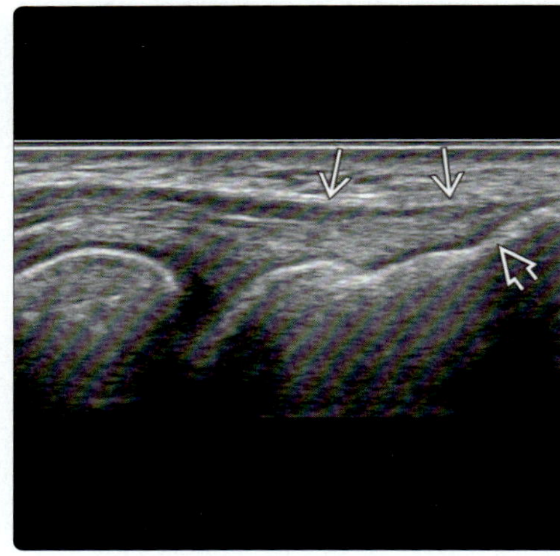

(Left) *Transverse US shows the oblong-type configuration of the CETO at the lateral epicondyle immediately proximal to the insertion. Extensor digitorum and flexor carpi ulnaris muscles are seen.* **(Right)** *Graphic shows the radial collateral ligament (RCL) complex. This comprises the RCL, which inserts into the annular ligament, and the lateral ulnar collateral ligament (LUCL), which inserts into the supinator crest of the proximal ulna.*

Common Extensor Tendon Origin: Short Axis

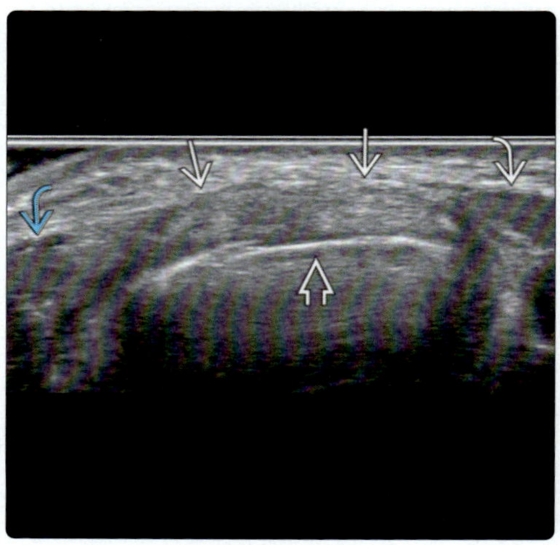

Radial Collateral Ligament Complex

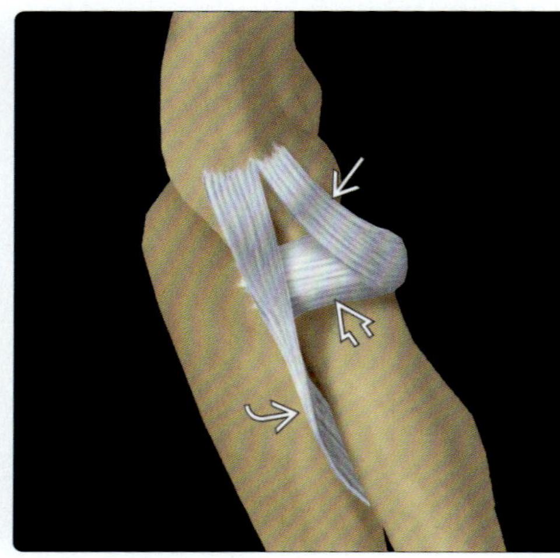

(Left) *Longitudinal US shows the RCL passing to the annular ligament. The RCL fibers are deep to and barely separable from the CETO fibers. The deeper 1/3 of the CETO/RCL complex fibers comprise the RCL. A radiocapitellar synovial flange is present.* **(Right)** *Transverse US in the posterolateral aspect of the elbow shows the mid- to distal portion of the LUCL inserting into supinator crest of proximal ulna. Anconeus muscle is superficial to the ligament, while the radial head is deep.*

Radial Collateral Ligament

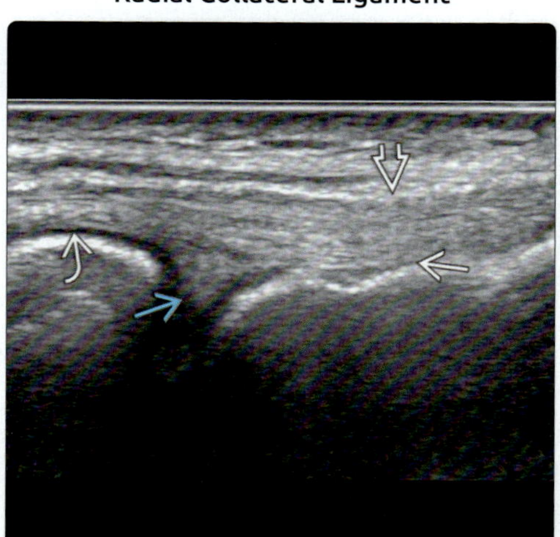

Lateral Ulnar Collateral Ligament

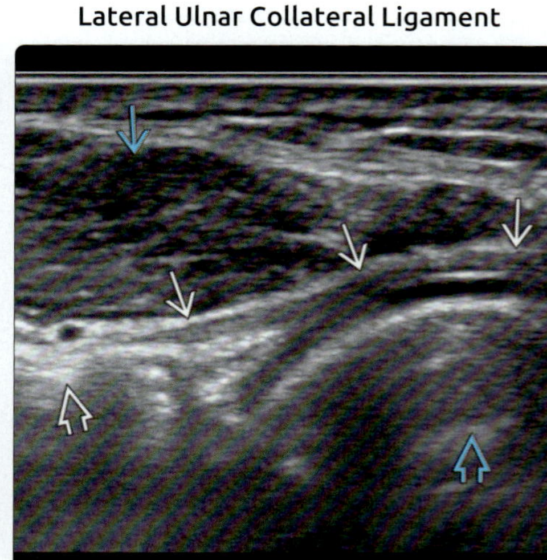

Radial Tunnel: Short Axis

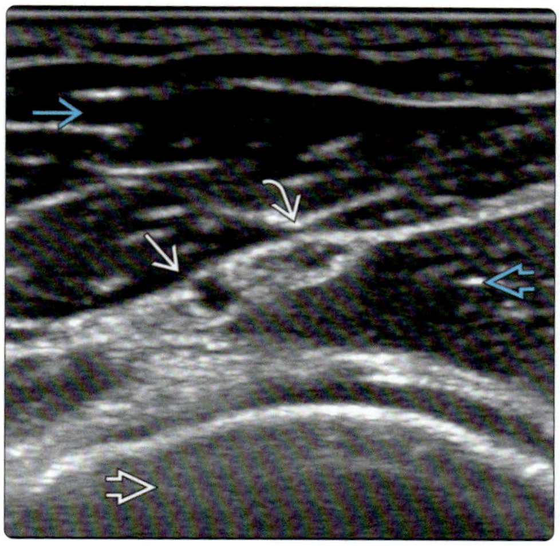

Supinator Tunnel: Short Axis

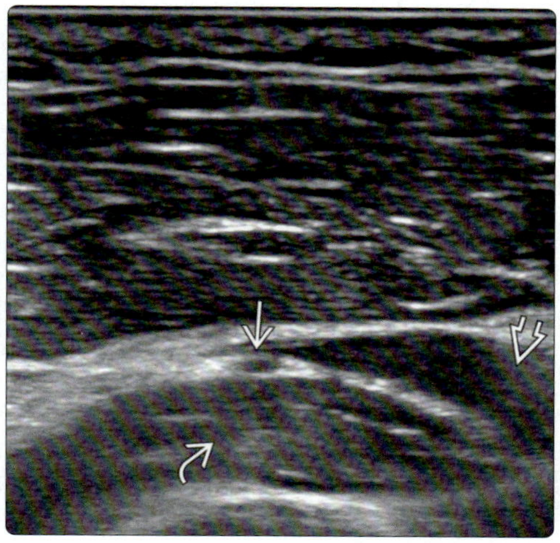

(Left) *Transverse US of the anterolateral elbow shows the radial tunnel with the radial nerve having divided into its deep and superficial branches just superficial to the radial head. The nerves lie between the brachioradialis and the brachialis muscles.* **(Right)** *Transverse US in the anterolateral aspect of the elbow shows the inlet to the supinator tunnel with the deep branch of the radial nerve, now known as the posterior interosseous nerve, between the superficial and deep fibers of the supinator muscle.*

Scan Position for Posterior Elbow

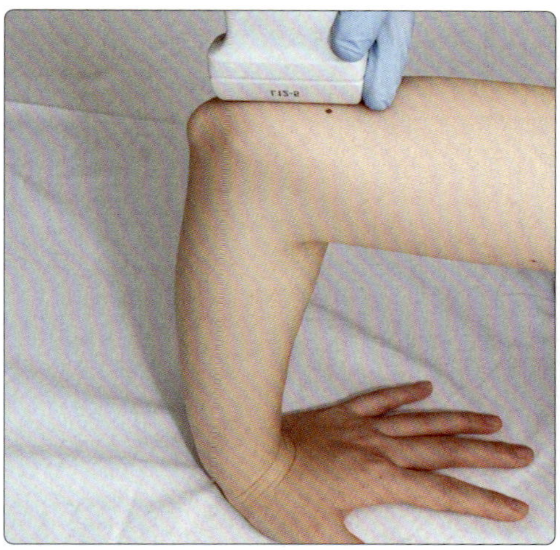

Olecranon Recess

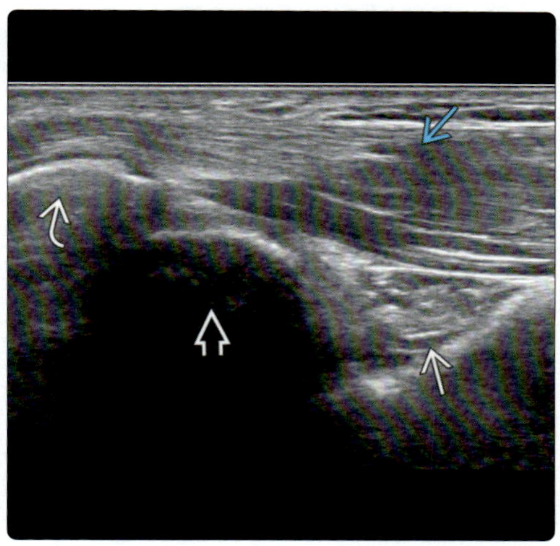

(Left) *Clinical photograph shows the crawling crab position used to examine the posterior aspect of the elbow.* **(Right)** *Longitudinal US along the posterior aspect of the elbow joint shows the posterior fat pad filling the olecranon recess of the distal humerus. The trochlea, olecranon process, and triceps are also visible. The fat pad is intraarticular but extrasynovial. An effusion will displace the fat pad posteriorly.*

Longitudinal Triceps Tendon Insertion

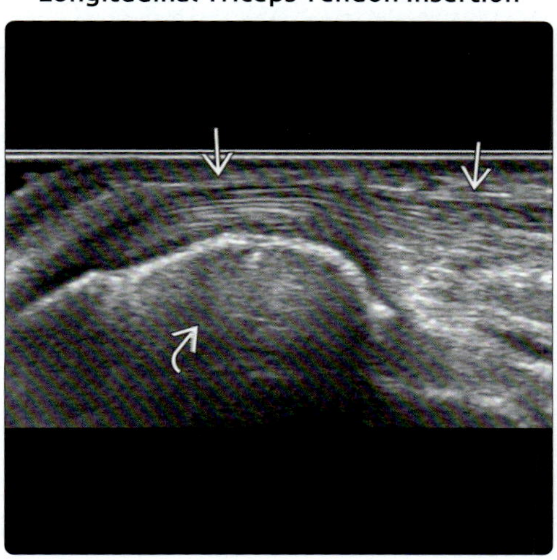

Transverse Triceps Tendon

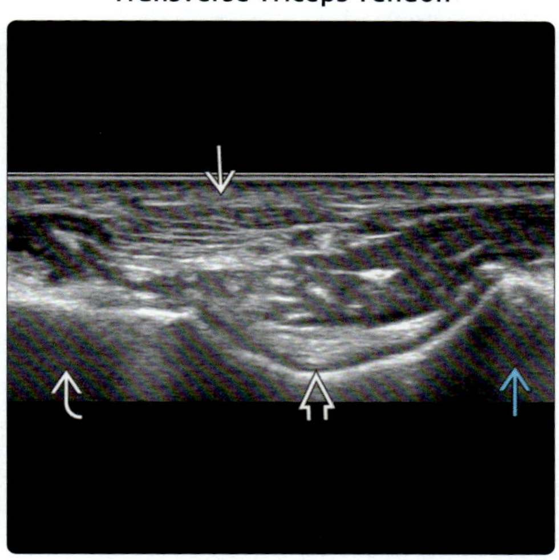

(Left) *Longitudinal US in the posterior aspect of the elbow shows the triceps tendon inserting onto the olecranon process. The long and lateral heads of the triceps muscles insert into the posterior aspect of the olecranon process 1.1 cm from the tip. The medial belly insertion is more muscular and into the medial aspect of the olecranon process.* **(Right)** *Transverse US in the posterior aspect of the elbow shows the triceps tendon. This is overlying the capitellum, posterior fat pad, and trochlea.*

KEY FACTS

GENERAL CONSIDERATIONS

- Common specific indications are suspected ganglia, joint effusion, synovitis, septic arthritis, tendon disorders, carpal tunnel syndrome
- Less common specific indications are Guyon canal syndrome, scaphoid and other fractures, other neural pathology, and wrist ligaments

TECHNIQUE

- **Dorsal wrist**
  - Patient seated across examiner with wrist resting prone on adjustable table
  - ± small pad under hypothenar eminence to align wrist and carpus in more horizontal position
  - Wrist in lateral position ± ulnar deviation to scan scaphoid/1st extensor compartment
  - Wrist in lateral position ± radial deviation to scan ulnocarpal articulation/6th extensor compartment
    - Dorsal wrist joint
    - 1st extensor compartment (abductor pollicis longus and extensor pollicis brevis)
    - 2nd extensor compartment (extensor carpi radialis longus and extensor carpi radialis brevis)
    - 3rd extensor compartment (extensor pollicis longus)
    - 4th (extensor digitorum, extensor indicis proprius) and 5th (extensor digiti minimi) extensor compartments
    - 6th extensor compartment (extensor carpi ulnaris)
    - Posterior interosseous nerve
    - Dorsal ulnar cutaneous nerve
- **Volar wrist**
  - Patient seated across examiner with wrist resting supine on adjustable table
  - ± scan in wrist flexion to see insertion of flexor carpi radialis tendon
    - Volar wrist joint, tendons outside carpal tunnel, median nerve and carpal tunnel, ulnar nerve and Guyon tunnel, superficial radial nerve, palmar ulnar cutaneous nerve

Scan Position for Dorsal Wrist

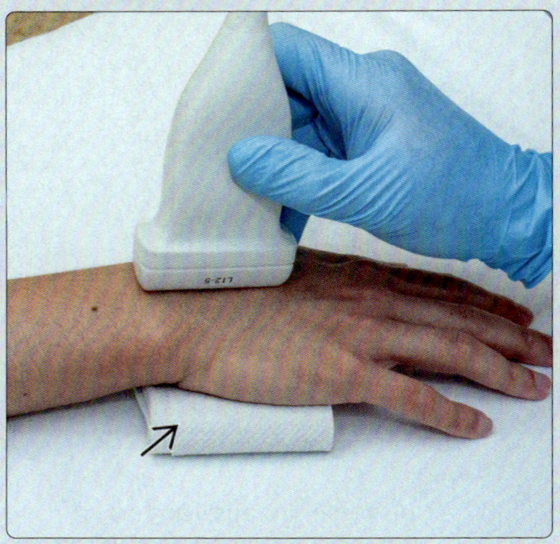

Dorsal Wrist Midline

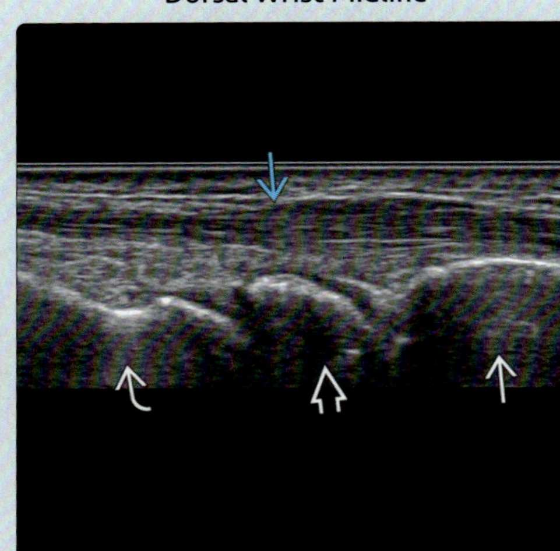

(Left) *Clinical photograph shows the scan position for the dorsal wrist. A small pad ➔ can be placed under the ulnar aspect of the wrist to align the dorsal wrist more horizontally.* **(Right)** *Longitudinal US shows the distal radius ➔, lunate ➔, and capitate ➔ all aligned. The soft tissue interposed between the carpus and extensor digitorum tendon ➔ is composed of joint synovium, extrinsic ligaments, and loose areolar tissue. These individual elements cannot be discerned.*

Dorsal Wrist Scaphoid

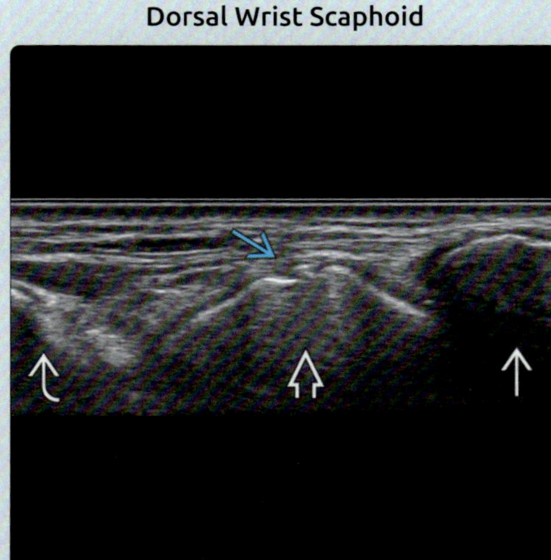

MR of Proximal Wrist

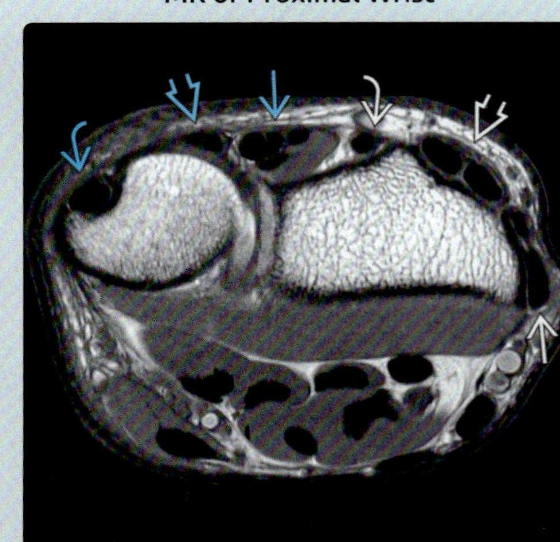

(Left) *Longitudinal US shows the distal radius ➔, scaphoid ➔, and trapezoid ➔ all aligned. The bony irregularity ➔ on the dorsum of the scaphoid is normal. Such irregularity of normal bone contour does somewhat limit US assessment of a carpal fracture.* **(Right)** *Axial intermediate-weighted MR of the wrist at the level of the distal radioulnar joint shows the 1st ➔, 2nd ➔, 3rd ➔, 4th ➔, 5th ➔, and 6th ➔ extensor compartments.*

GENERAL CONSIDERATIONS

Clinical Indications for Wrist US

- US is best investigation for suspected joint effusion/synovitis and all nonarticular pathology
- Common specific indications are suspected ganglia, joint effusion, synovitis, septic arthritis, tendon disorders, and carpal tunnel syndrome (CTS)
- Less common specific indications are Guyon canal syndrome, scaphoid and other fractures, other neural pathology, and wrist ligaments
- As wrist symptoms are usually well localized, exam can usually be targeted to symptomatic area
- Very different causes of radial-sided, central, and ulnar-sided wrist pain exist

TECHNIQUE: DORSAL WRIST

Patient Position

- Patient seated across examiner with pronated wrist resting on adjustable table
  - ± small pad under hypothenar eminence to align wrist and carpus in more horizontal position
- Place wrist in lateral position ± ulnar deviation to examine scaphoid/1st extensor compartment
- Place wrist in lateral position ± radial deviation to examine ulnocarpal articulation/6th extensor compartment

Specifically Examine

- **Dorsal wrist joint**
  - Composed of 3 distinct compartments
    - Distal radioulnar joint (DRUJ), radiocarpal and ulnocarpal joints, midcarpal joint
  - All 3 joints are best viewed using dorsal approach as joints are more superficial
  - Scan transversely over distal radius/ulna to view DRUJ
    - Located deep to 5th extensor compartment tendon
    - Mild passive pronation and supination helps assessment of DRUJ content
  - Rotate transducer into longitudinal plane to view radiocarpal/ulnocarpal and midcarpal joints
    - Echogenic lip of distal radius and rounded proximal contours of lunate and capitate longitudinally aligned provide good starting landmark
    - Then, identify other wrist bones by scanning from landmark in both longitudinal or transverse planes
  - Note joint distension, hypoechoic/hyperechoic synovial proliferation and joint fluid deep to echogenic capsule
  - Color Doppler imaging and especially, to lesser degree, compression can help differentiate fluid from synovitis
    - For hyperemia, optimize settings, use minimal transducer pressure, and record as mild, moderate, severe
- **Dorsal wrist tendons**
  - Best scanned primarily in transverse plane moving along length of tendon
  - Lister tubercle on dorsal aspect of distal radius is useful landmark
- **1st extensor compartment [abductor pollicis longus (APL) and extensor pollicis brevis (EPB)]**
  - Place ulnar border of wrist on table ± resting on pad used to help ulnar deviate wrist
  - Note larger APL and smaller EPB
  - Both tendons rest on shallow groove in radial styloid covered by echogenic extensor retinaculum (< 0.5 mm thick)
  - 1/3 of wrists have septum (90% complete, 10% partial) between APL and EPB
    - This septum is always associated with intervening osseous ridge on underlying radial styloid process
    - Therefore, osseous ridge provides good clue as to likelihood of 1st extensor compartment septum
  - 95% of APL tendons have multiple slips while < 5% of EPB tendons have > 1 slip
    - EPB may occasionally (5%) be absent and replaced with ligamentous structure
  - Superficial radial nerve branches pass over ± under 1st extensor compartment from volar to dorsal wrist
- **2nd extensor compartment (extensor carpi radialis longus and extensor carpi radialis brevis)**
  - Isolated pathology of this compartment uncommon
  - Just proximal to wrist, 1st compartment tendons pass superficial to 2nd compartment tendons (intersection syndrome)
  - Dorsal scapholunate ligament seen deep to 2nd compartment tendons
    - Side-to-side discrepancy in scapholunate joint space should not be > 2 mm
- **3rd extensor compartment [extensor pollicis longus (EPL)]**
  - EPL uses Lister tubercle as pulley
  - Passes over 2nd extensor compartment tendons (distal intersection) to reach base of distal phalanx of thumb
  - Follow in transverse plane, rotating transducer to ensure it is maintained at right angles to tendon
- **4th [extensor digitorum (ED), extensor indicis proprius (EIP)] and 5th [extensor digiti minimi (EDM)] extensor compartments**
  - 4th compartment contains 4 ED tendons (1 for each finger) and EIP tendon
  - 4th compartment of extensor retinaculum is thickest
  - Passive finger movement can help identify tendons
  - 5th extensor compartment is fibrous rather than fibroosseous and overlies DRUJ
  - Visibility of small EDM can be improved by moving little finger; EDM typically joins ED tendon just proximal to metacarpophalangeal joint
- **6th extensor compartment [extensor carpi ulnaris (ECU)]**
  - Best examined with wrist fully pronated & ulna facing up
  - Observe ECU within groove on ulnar head covered by fibrous retinaculum (ECU subsheath)
    - ECU is only extensor tendon with its own retinaculum, serving to keep ECU in its groove
  - For dynamic assessment of ECU, flex elbow with ulnar head facing examiner
    - Using light transducer pressure over ulnar groove
    - First supinate forearm and then allow wrist to drop passively into flexion and ulnar deviation
    - % of ECU displacement from ulnar groove: < 50% = normal; > 50% = subluxation; 100% = dislocation
    - If abnormal, compare with opposite side to exclude physiological laxity of subsheath
- **Posterior interosseous nerve**

- Sensorimotor branch of radial nerve; average size on US is 1.5 mm^2
  - Located deep to radial aspect of 4th extensor compartment at distal radius
    - Enlarges to form small "pseudoganglion" before dividing into its terminal branches
- **Dorsal ulnar cutaneous nerve**
  - Superficial sensory nerve; average size on US is 0.3 mm^2
  - Arises from ulnar side of ulnar nerve (UN) 5-6 cm proximal to ulnar styloid process
  - Passes close to ulnar styloid when extending from radial → dorsal aspect of wrist
    - Innervates mediodorsal hand and ulnar 1 & 1/2 digits

TECHNIQUE: VOLAR WRIST

Patient Position

- Patient seated across examiner with supinated wrist resting on adjustable table
- ± scan in wrist flexion to see insertion of flexor carpi radialis (FCR) tendon

Specifically Examine

- **Volar wrist joint**: Best scanned in longitudinal plane, noting distal radius and ulna and bones of proximal carpal row
- **Tendons outside carpal tunnel**
  - Scan in transverse plane over distal radius and ulna, noting pronator quadratus, flexor tendons, and radial/ulnar arteries
  - FCR → 2nd metacarpal base
  - FCR has its own synovial sheath and runs in separate fibroosseous tunnel covered by extension of transverse carpal ligament
    - Floor of FCR tunnel is distal scaphoid, trapezium, and trapezoid
    - Angle probe & flex wrist to best see distal part of FCR
  - Flexor carpi ulnaris (FCU) → pisiform
    - Some fibers continue to hook of hamate (as pisohamate ligament) and 4th metacarpal base (as pisometacarpal ligament)
    - FCU has no tendon sheath
- **Median nerve (MN) and carpal tunnel**
  - MN best scanned in transverse section from midforearm through carpal tunnel
  - In distal forearm, MN becomes more superficial and is located in midline
    - Palmar cutaneous nerve is small (0.5-0.7 mm^2) branch of MN seen in 80% of wrists with high-resolution US
    - Arises from MN at point where it crosses between FCR and radial border of flexor digitorum superficialis muscle, 5 cm proximal to wrist
      - Remains closely bound to MN for first 2 cm before traveling between palmaris longus and FCR
  - Just proximal to wrist crease, MN lies in more superficial position over deep flexors and deep to palmaris longus
  - MN then dips down to enter carpal tunnel by passing below transverse carpal ligament
    - Recognition of leading edge of transverse carpal ligament is best way to recognize tunnel inlet
  - Carpal tunnel contains 4 pairs of tendons (1 for each finger), 1 unpaired tendon for thumb [flexor pollicis longus (FPL)], and MN
    - FPL has its own tendon sheath (radial bursa)
    - Finger flexor tendons encased by 3 layers of synovium in "E" shape (ulnar bursa)
  - MN normally flatter, more hypoechoic, and more difficult to see at narrower tunnel outlet
  - MN then rises on exiting carpal tunnel and quickly divides into its digital branches
    - Recognition of trailing edge of transverse carpal ligament is best way to recognize tunnel inlet
  - Measure MN cross-sectional area (CSA) of distal forearm, proximal to tunnel, at tunnel inlet, at tunnel outlet, and just distal to tunnel
  - Check hyperemia at most swollen level
  - Bifid MN seen in 20% of population and usually, but not always, accompanied by persistent median artery
    - In this case, measure CSA of both components and summate applying usual diagnostic criteria for CTS
    - Report presence of persistent median artery as may be injured during arthroscopic release of transverse carpal ligament
- **UN and Guyon tunnel**
  - UN best scanned in transverse section from midforearm to Guyon tunnel
    - UN located on medial aspect of distal forearm between FCR and ulnar artery
    - Guyon canal contains UN, ulnar artery, and accompanying veins ± abductor digiti minimi muscle slips
  - At distal end of Guyon canal beyond pisiform, UN divides into superficial sensory branch and deep motor branch
    - Superficial sensory branch runs with ulnar artery over hook of hamate
    - Deep motor branch courses alongside outer margin of hook of hamate
- **Superficial radial nerve**
  - Superficial sensory nerve; average size on US is 3.9 mm^2
  - Prone to iatrogenic injury at wrist in vascular cannulation and fracture fixation
  - Superficial radial nerve divides into 2-3 branches 4.5-6.0 cm proximal to radial styloid tip
    - In 90% of subjects, superficial radial branch passes dorsal to 1st extensor compartment when traveling from volar to dorsal aspect of waist and hand
      - Provides sensory innervation to skin on dorsal aspect of thumb and dorsoradial sides of hand and fingers
    - In smaller percentage, superficial radial branch passes dorsal to 1st extensor compartment when traveling from volar to dorsal aspect of waist and hand
- **Palmar ulnar cutaneous nerve**
  - Superficial sensory nerve; average size on US is 0.3 mm^2
  - Arises from UN radially in midforearm and travels distally deep to FCU
    - Innervates skin of hypothenar eminence
- **Lateral antebrachial cutaneous nerve**
  - Terminal branch of musculocutaneous nerve; another nerve that can be potentially injured at wrist
  - Difficult to see on US of wrist as at this level, nerve is very small, almost intradermal, and quite variable in position

1st Extensor Compartment

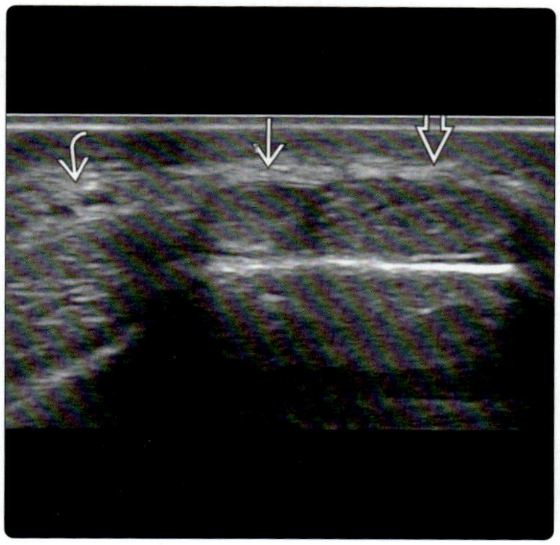

2nd Extensor Compartment

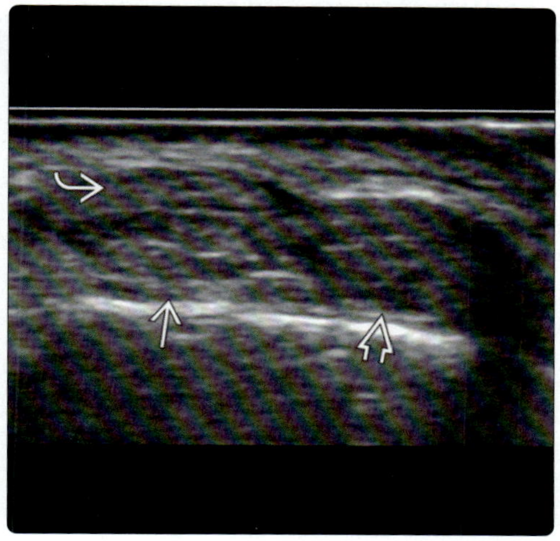

(Left) *Transverse US of the 1st extensor compartment shows smaller extensor pollicis brevis alongside larger abductor pollicis longus. There is no intervening septum. A thin extensor retinaculum overlies the tendons. The lateral branch of the superficial radial nerve passes dorsal to the 1st extensor compartment.* **(Right)** *Transverse US shows 2nd extensor compartment. Extensor pollicis longus (EPL) is passing over flexor carpi radialis brevis tendon, which is closely apposed to the extensor carpi radialis longus tendon.*

3rd Extensor Compartment

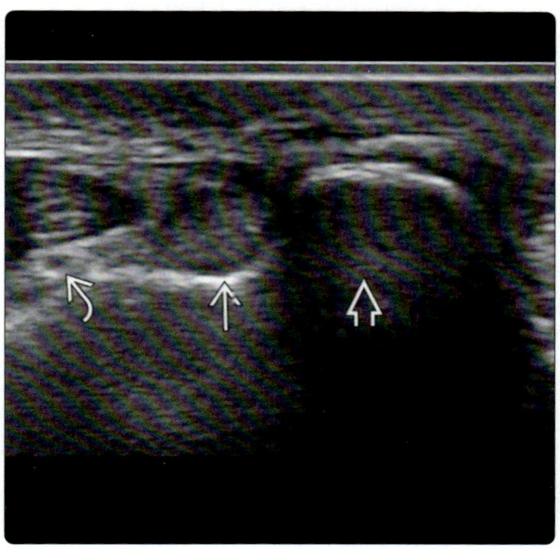

4th Extensor Compartment

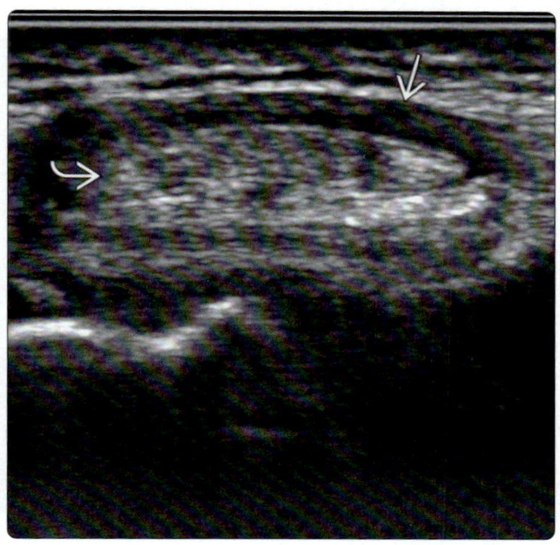

(Left) *Transverse US of the 3rd extensor compartment shows the EPL tendon lying alongside the Lister tubercle. This tubercle acts as a pulley for the EPL. Tears of the EPL at this location are quite common. The posterior interosseous nerve lies nearby.* **(Right)** *Transverse US of the 4th extensor compartment shows the closely packed extensor digitorum tendons lying deep to the extensor retinaculum. This extensor retinaculum is thickest over the extensor digitorum tendons.*

5th Extensor Compartment

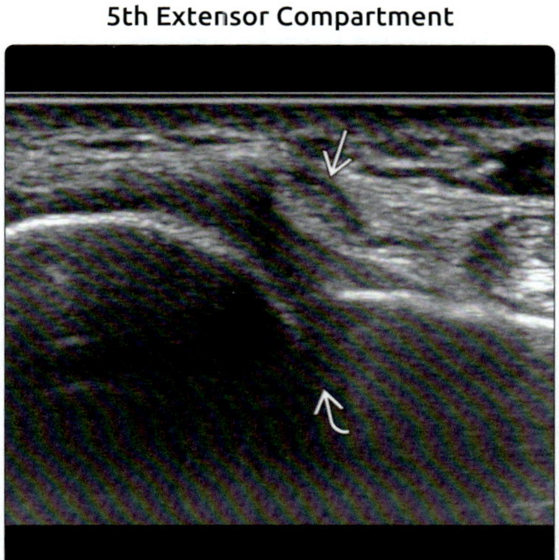

6th Extensor Compartment

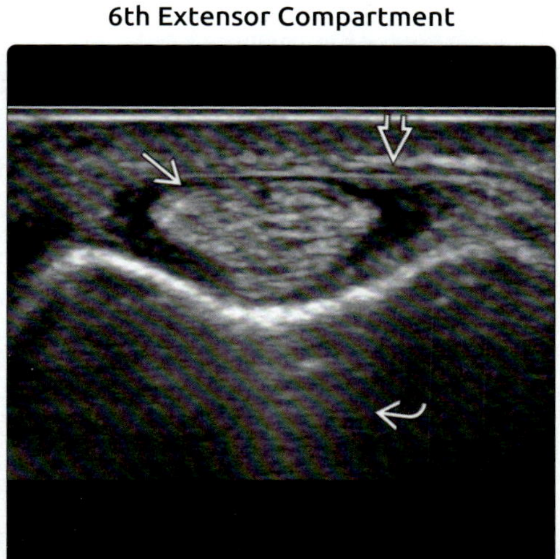

(Left) *Transverse US shows the 5th extensor compartment tendon, extensor digiti minimi, lying on the dorsal aspect of the distal radioulnar joint. The 5th compartment is a purely fibrous compartment that is not in direct contact with the underlying bone.* **(Right)** *Transverse US of the 6th extensor compartment shows the extensor carpi ulnaris (ECU) tendon lying within a groove on the dorsoulnar aspect of the ulnar head. The ECU has its own restraining subsheath in addition to the extensor retinaculum.*

(**Left**) *Clinical photograph during US assessment for ECU subluxation from the ulnar groove of the ulnar head is shown. The ECU will tend to sublux most when the wrist is supinated, flexed, and ulnar deviated.* (**Right**) *Transverse US during dynamic testing of ECU stability shows slight movement of the ECU toward the side of the ulnar groove . This degree of laxity is normal. The hypoechoic line within the ECU is also normal, representing physiological splitting of the more distal ECU fibers.*

Extensor Carpi Ulnaris Subluxation Test

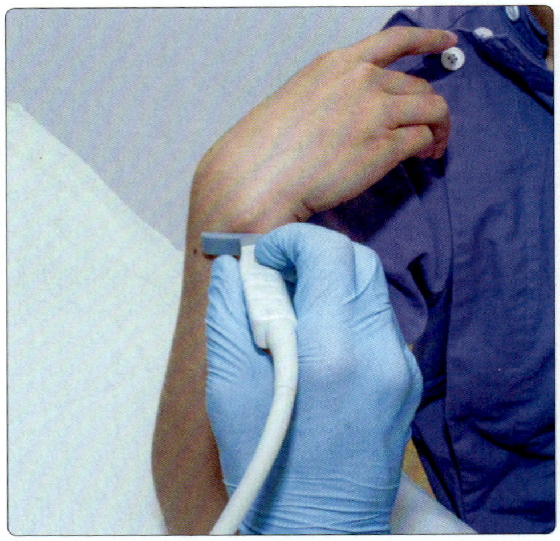

Dynamic Testing of Extensor Carpi Ulnaris

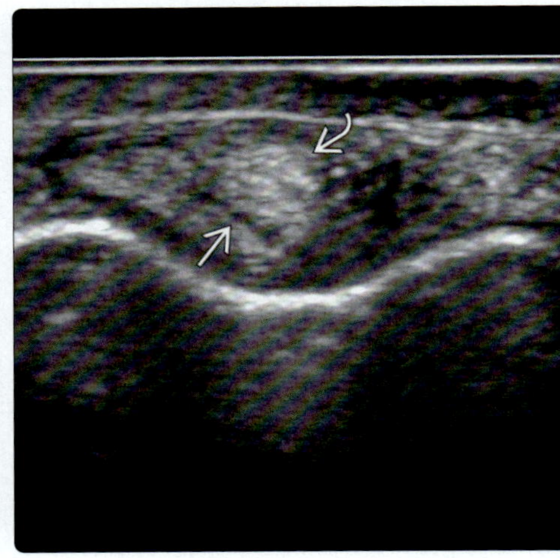

(**Left**) *Transverse US of the wrist just beyond the radiocarpal joint shows the diverging extensor digitorum tendons on the dorsum of the carpus. Deep to the extensor tendons is a dorsal extrinsic ligament .* (**Right**) *Transverse US shows the dorsal component of the strong intrinsic scapholunate ligament traversing between the dorsal aspects of the scaphoid and lunate . The volar component of the scapholunate ligament is less readily appreciated on the volar aspect of the wrist.*

Extensor Digitorum Tendons

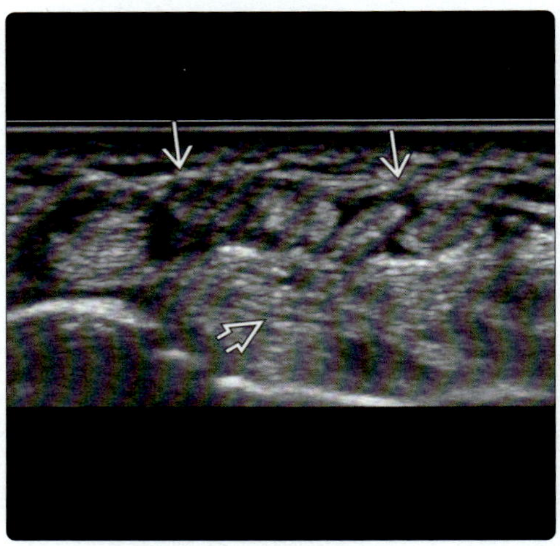

Dorsal Scapholunate Ligament

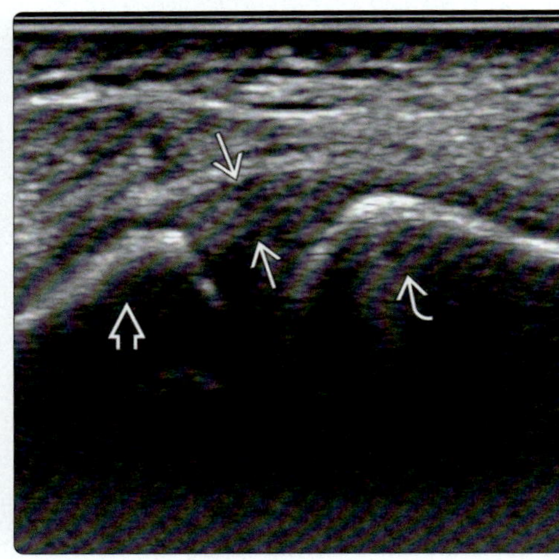

(**Left**) *Axial intermediate-weighted MR of the proximal carpal tunnel shows the transverse ligament covering the carpal tunnel. The carpal tunnel contains 9 tendons (4 paired tendons for each finger and 1 tendon to the thumb) as well as the median nerve .* (**Right**) *Clinical photograph shows the scan position for the volar wrist. The wrist should be horizontal and neither flexed nor extended.*

MR of Carpal Tunnel

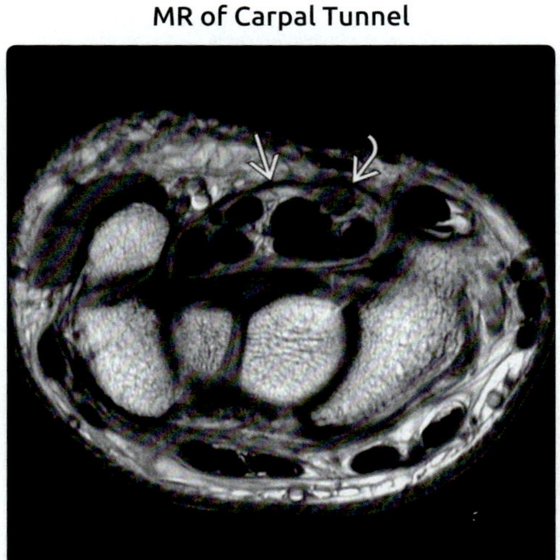

Scan Position for Volar Wrist

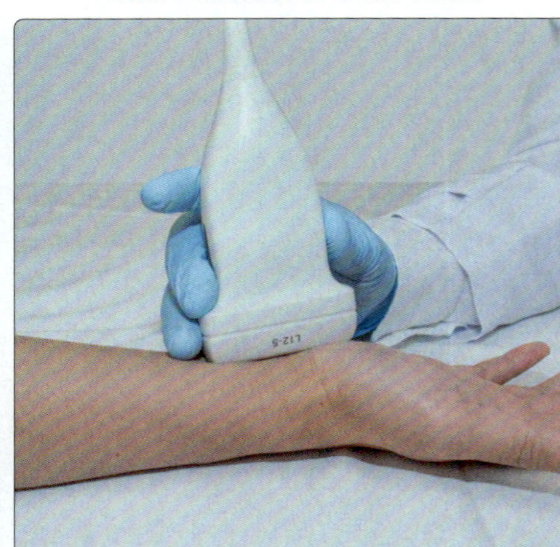

Volar Aspect of Carpus

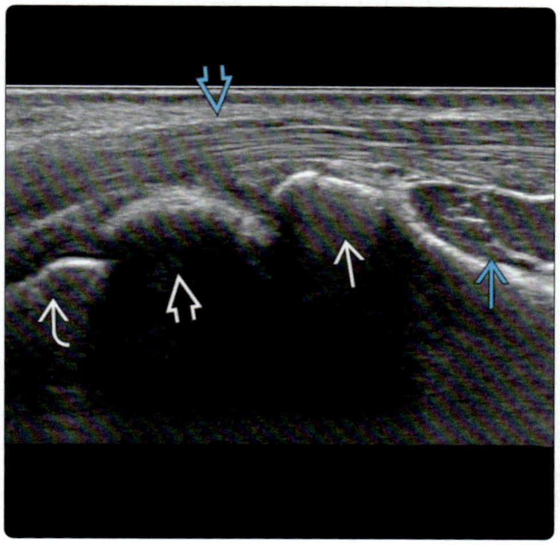

Volar Aspect of Radial Side of Wrist

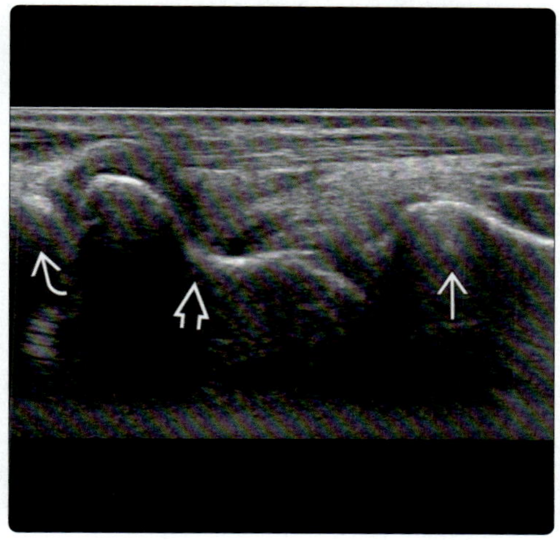

(Left) *Longitudinal US shows the volar aspect of the carpus in the midline with the distal radius, lunate, and capitate all aligned. The pronator quadratus muscle is visible proximally. The flexor tendons lie superficially.* **(Right)** *Longitudinal US shows the volar aspect of the radial side of the wrist with distal radius, scaphoid, and trapezoid aligned. The wrist has been ulnar deviated to show more of the volar surface of the scaphoid. US is commonly used to diagnose scaphoid fractures.*

Flexor Carpi Radialis of Neutral Wrist

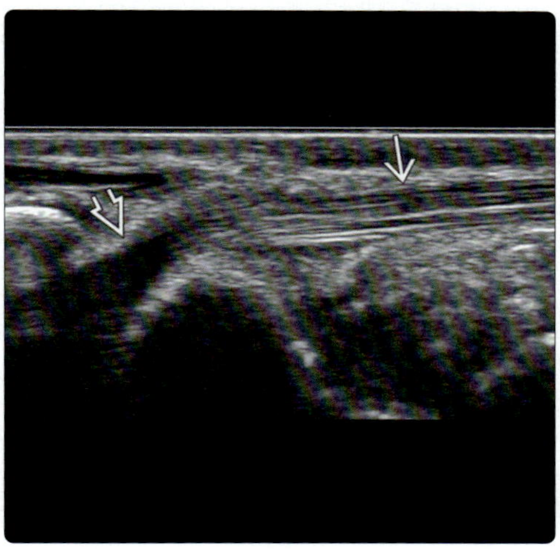

Flexor Carpi Radialis of Flexed Wrist

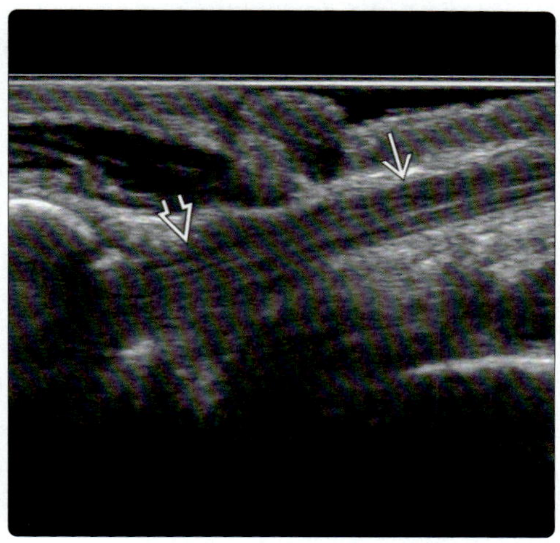

(Left) *Longitudinal US with the wrist in a neutral position shows the hypoechoic appearance of the normal flexor carpi radialis tendon due to anisotropy as it descends to the base of the 2nd metatarsal.* **(Right)** *Longitudinal US with the wrist in a flexed position shows how the distal end of the normal flexor carpi radialis tendon can be more readily appreciated than when the wrist is in a neutral position.*

Flexor Carpi Ulnaris Tendon

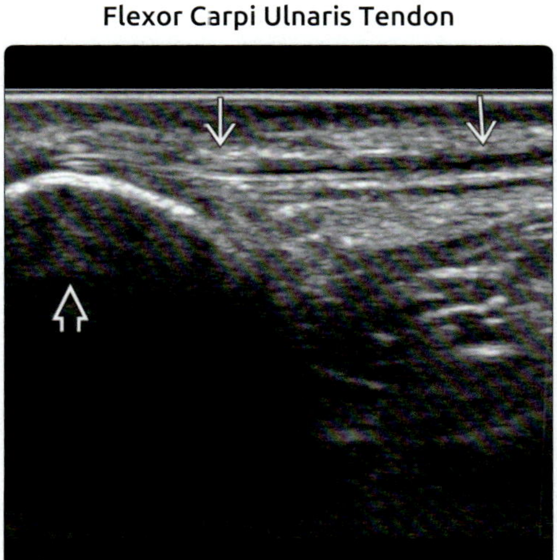

Ulnocarpal Articulation

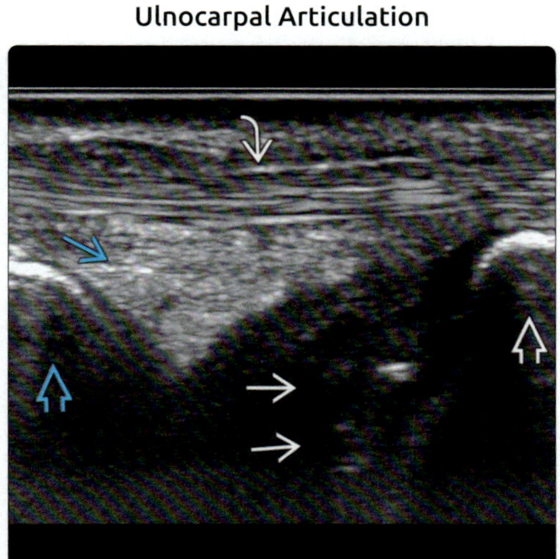

(Left) *Longitudinal US shows flexor carpi ulnaris (FCU) tendon inserted into volar aspect of pisiform bone. Some FCU fibers continue as the restraining ligaments of the pisiform bone. FCU tendon does not have a tendon sheath.* **(Right)** *Longitudinal US of the ulnar aspect of the wrist joint shows the ulnar head, ECU tendon, meniscal homologue, and triquetrum. Articular disc of triangular fibrocartilaginous complex (TFCC) is barely discernible. TFCC injury is best imaged by MR.*

(Left) *Transverse US shows the median nerve* ➡ *alongside the flexor digitorum superficialis muscles and tendons* ➡ *on the volar aspect of the distal forearm. The flexor digitorum* ➡ *and pronator quadratus muscle* ➡ *lies deep to the nerve.* **(Right)** *Transverse US shows the median nerve* ➡ *just proximal to the tunnel inlet. The median nerve at this level is covered by the thin antebrachial fascia* ➡, *which is normally slightly thickened just proximal to the tunnel inlet. This should not be mistaken for the transverse carpal ligament.*

Median Nerve of Distal Forearm

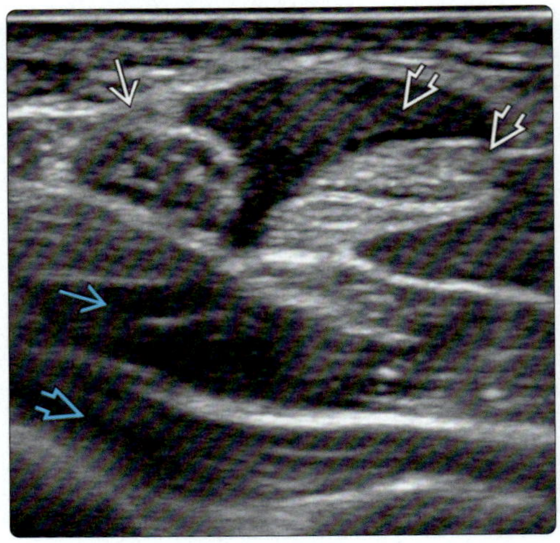

Median Nerve Proximal to Tunnel Inlet

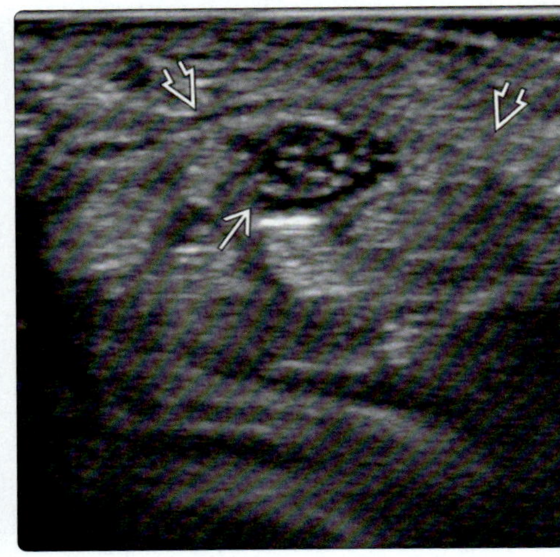

(Left) *Transverse US shows the median nerve* ➡ *at the tunnel inlet just deep to the thick transverse carpal ligament* ➡. *The median nerve dives deep to enter the carpal tunnel inlet. The flexor tendons* ➡ *are crowded together within the carpal tunnel.* **(Right)** *Transverse US at the tunnel outlet shows the normally more flattened median nerve* ➡ *just deep to the thick distal edge of the transverse carpal ligament* ➡.

Median Nerve at Tunnel Inlet

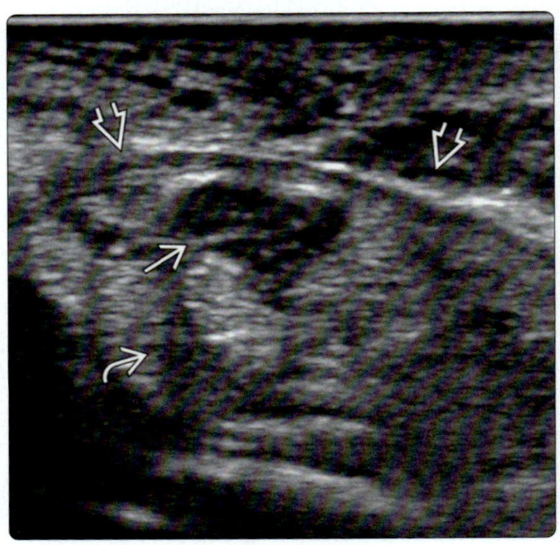

Median Nerve at Tunnel Outlet

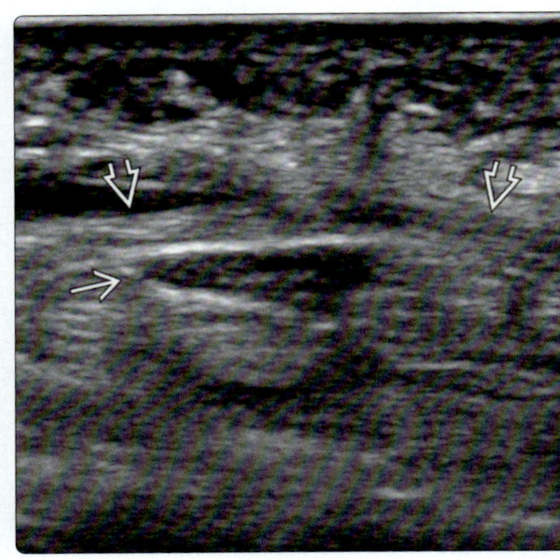

(Left) *Transverse US just distal to the tunnel outlet shows the median nerve dividing into its digital branches* ➡. *This branching usually occurs immediately distal to the tunnel outlet. Part of the thenar muscle is seen* ➡. *The diverging flexor tendons* ➡ *lie deep to the digital nerves.* **(Right)** *Longitudinal US shows median nerve* ➡ *passing deep to transverse carpal ligament* ➡. *Flexor tendons* ➡ *lie mainly deep to the nerve. The coarse fibrillar pattern of the nerve compared to the fine fibrillar pattern of the tendons can be appreciated.*

Distal to Carpal Tunnel

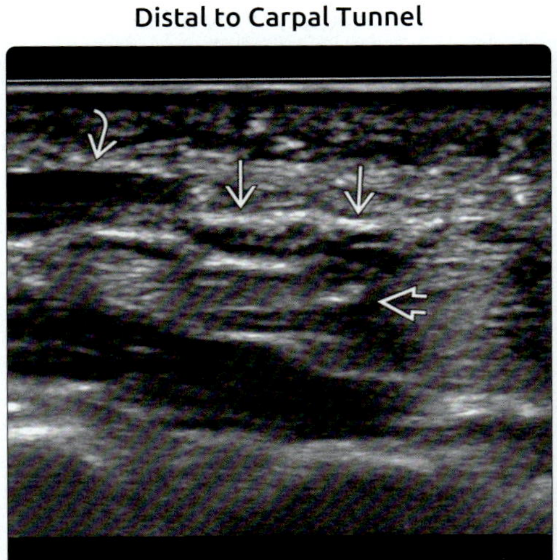

Median Nerve to Carpal Tunnel: Long Axis

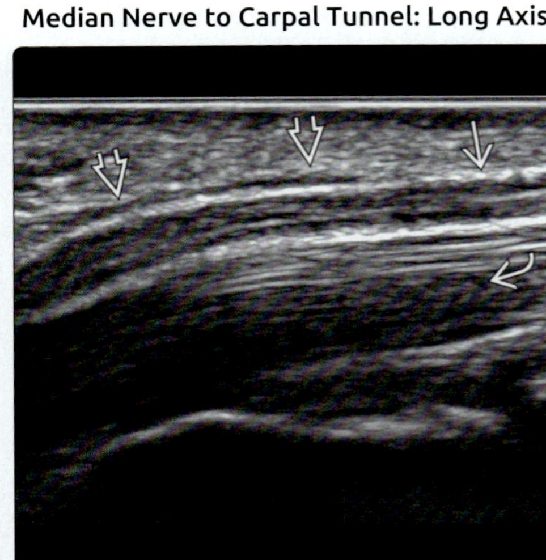

Bifid Median Nerve

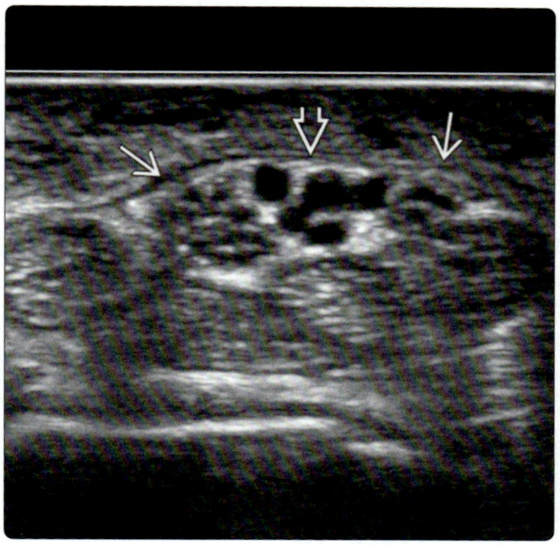

Ulnar Nerve and Palmar Ulnar Cutaneous Nerve

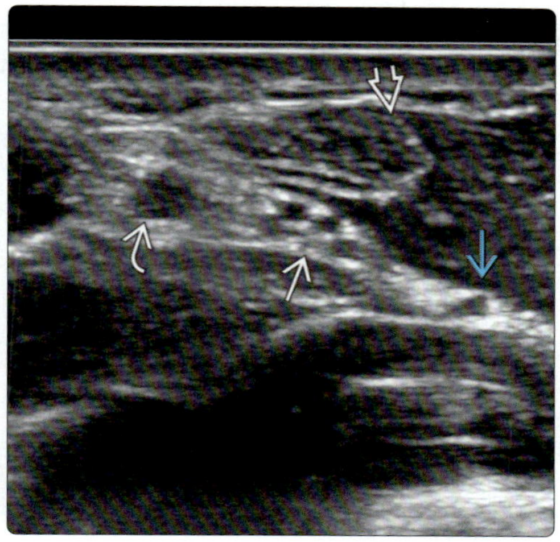

(Left) *Transverse US just proximal to the carpal tunnel shows the bifid median nerve with several vessels comprising arteries and veins separating both median nerve bundles.* **(Right)** *Transverse US at the distal forearm levels shows the ulnar nerve alongside the ulnar artery deep to the FCU tendon and muscle. The palmar ulnar cutaneous nerve has just branched off the parent ulnar nerve and will travel to the hand deep to the FCU.*

Proximal Aspect of Guyon Canal

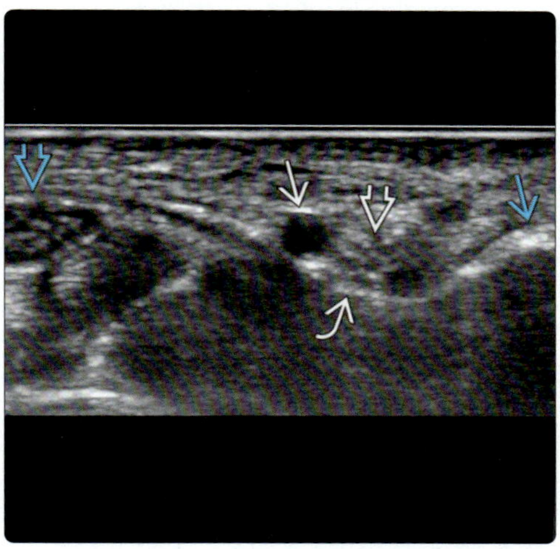

Proximal Aspect of Guyon Canal MR

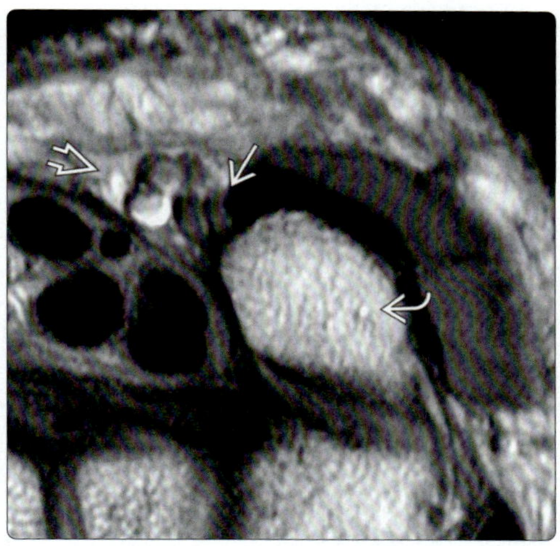

(Left) *Transverse US shows the ulnar artery and the ulnar nerve at the proximal end of the Guyon canal alongside the pisiform bone. The transverse carpal ligament lies deep to the canal. The median nerve within the carpal tunnel is seen.* **(Right)** *Axial PD MR shows proximal end of the Guyon canal alongside pisiform bone. The ulnar nerve is on the ulnar side of the ulnar artery. Note how thin the palmar carpal ligament is for the roof of Guyon canal compared to the thicker transverse carpal ligament deeply.*

Distal Aspect of Guyon Canal

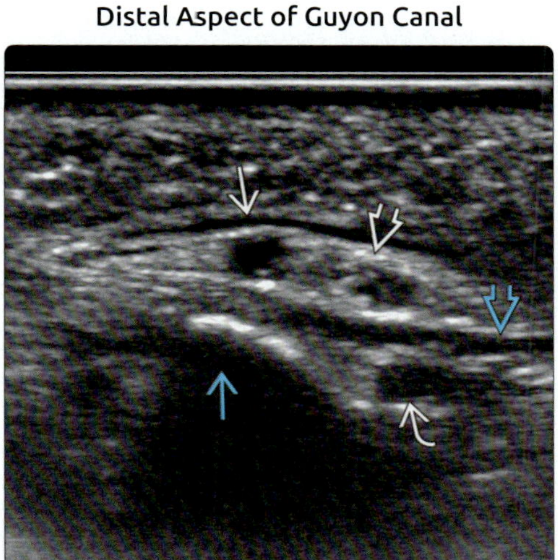

Distal Aspect of Guyon Canal MR

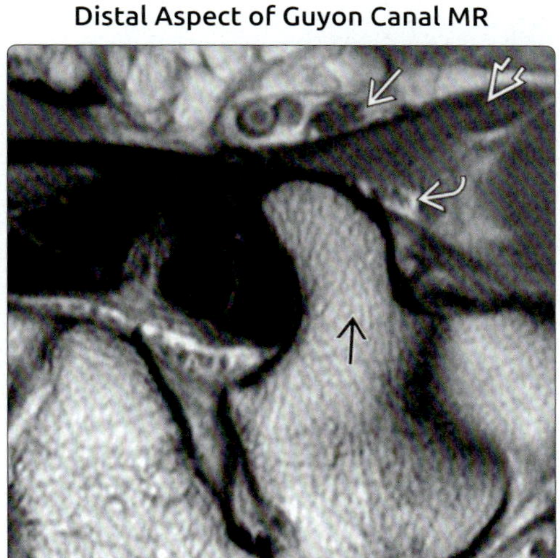

(Left) *Transverse US shows the ulnar artery and superficial and deep branches of the ulnar nerve at the distal end of the Guyon canal alongside the hook of hamate. These nerves are separated by a flexor digiti minimi muscle slip.* **(Right)** *Axial PD MR shows the distal end of the Guyon canal alongside the hook of hamate. The superficial and deep branches of the ulnar nerve are separated by the flexor digiti minimi muscle slip.*

Hand Ultrasound

KEY FACTS

GENERAL CONSIDERATIONS

- Common specific indications: Tenosynovitis, tendinosis, trigger finger, tendon rupture and other tendon disorders, joint inflammation, & ulnar collateral ligament (UCL) injury
- Left common specific indications: 1st carpometacarpal joint osteoarthritis, sagittal band injury, pulley rupture, specific small tumors of hand, tendon xanthoma

TECHNIQUE: DORSAL HAND

- Palm flat on examination table or couch with wrist in neutral position, fingers relaxed
  - Extensor tendons of fingers
    - To assess sagittal band integrity, scan dorsal hood at MCP joint during flexion and extension
  - Extensor tendon of thumb
  - UCL of metacarpophalangeal (MCP) joint of thumb
    - Scan in longitudinal plane
    - Flexion at interphalangeal (IP) joint will help identification of adductor aponeurosis
    - Careful thumb abduction during scanning will stress UCL and widen ulnar aspect of MCP joint

TECHNIQUE: VOLAR HAND

- Palm back of hand on examination table or couch with wrist in neutral position, fingers extended
  - Flexor tendon pulleys of fingers
    - Passively flex and extend of finger while observing movement of flexor tendons deep to pulley
    - For bowstringing, examine finger in resting extended position and during forced flexion
  - Flexor tendon pulleys of thumb
  - Flexor tendons of fingers
  - Flexor tendon of thumb
  - MCP and IP joints
    - Collateral ligaments on radial and ulnar sides, examine in flexion and extension
  - Carpometacarpal joint of thumb

Extensor Digitorum Tendons

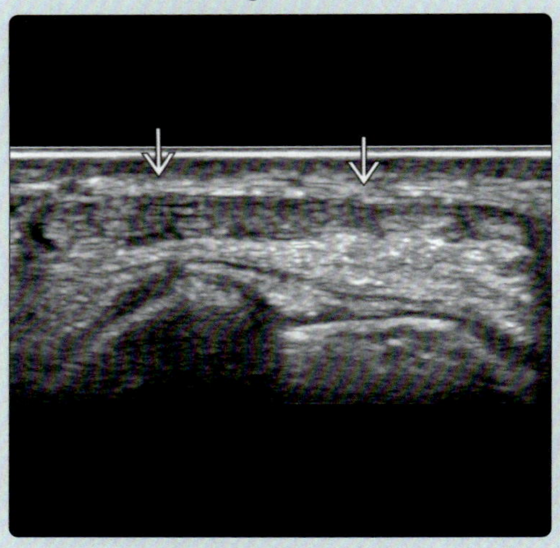

Extensor Digitorum Tendons

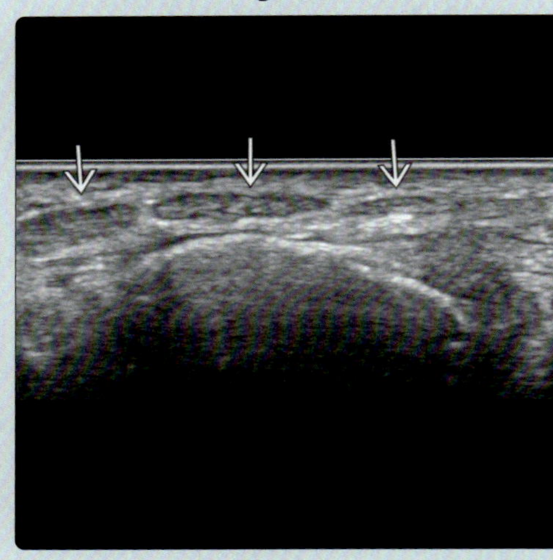

(Left) *Transverse US shows the tightly packed extensor digitorum tendons on the dorsum of the carpus.* **(Right)** *Transverse US of the dorsal aspect of the metacarpal bones shows the extensor digitorum tendons diverging to the individual fingers. Each tendon should be followed individually in the transverse plane initially.*

Extensor Tendons of Fingers

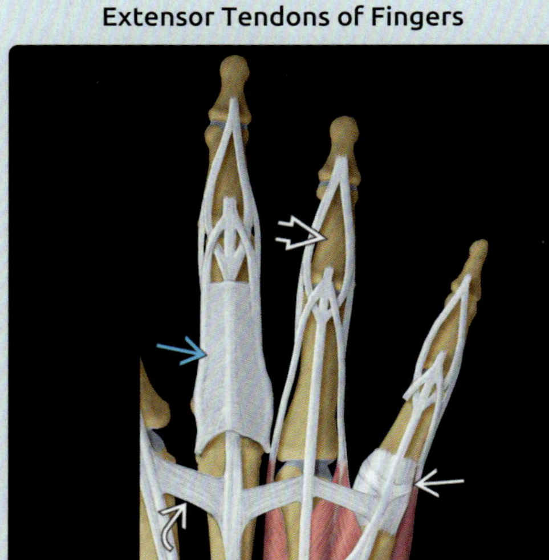

Sagittal Bands: Short Axis

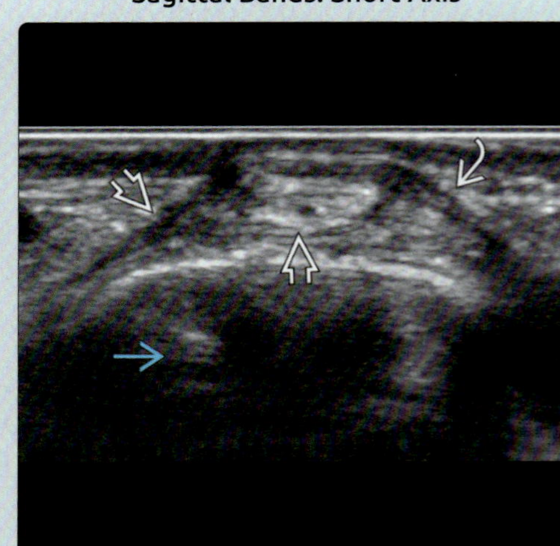

(Left) *Graphic shows the arrangement of the extensor tendons on dorsum of fingers. Note radial and ulnar sagittal bands, deep transverse metacarpal ligaments, extensor hood, and absence of central extensor tendon on proximal aspect of the middle phalanx.* **(Right)** *Transverse US of the dorsum of the metacarpal head of middle finger shows radial and ulnar sagittal bands overlying extensor tendon. The sagittal bands attach to the volar plates on palmar aspects of the metacarpophalangeal (MCP) joints.*

GENERAL CONSIDERATIONS

Clinical Indications for Hand US

- US is excellent modality at assessing all suspected soft tissue and joint abnormalities of hand
- Common specific indications: Tenosynovitis, tendinosis, trigger finger, tendon rupture and other tendon disorders, joint inflammation, and ulnar collateral ligament (UCL) injury
- Left common specific indications: 1st carpometacarpal joint osteoarthritis, sagittal band injury, pulley rupture, small tumors of hand, tendon xanthoma

TECHNIQUE: DORSAL HAND

Patient Position

- Palm flat on examination table or couch with wrist in neutral position, fingers extended

Specifically Examine

- **Extensor tendons of fingers**
  - Ring and middle fingers have 1 extensor tendon, remaining fingers have 2 extensor tendons
  - Radial and ulnar sagittal bands (dorsal hood) help maintain position of extensor tendons on dorsum of metacarpophalangeal (MCP) joint
    - More distally, extensor expansion provides stability with contributions from lumbricals and interossei
  - Extensor tendons on dorsum of fingers do not have tendon sheath
    - Cannot get tenosynovitis or 1° tenosynovial tumors on dorsum of fingers
- Transverse scanning best to identify individual tendons on dorsum of hand and follow them proximally/distally
  - Follow each tendon individually rather than following 2 tendons simultaneously
  - When abnormality found, tendon can be examined in longitudinal plane
    - Long axis good for assessing extensor tendons on dorsum of fingers, including tendon insertion, dynamic movement, and tendon morphology
  - To assess sagittal band integrity, scan dorsal hood during flexion and extension
- **Extensor tendon of thumb**
  - Extensor pollicis longus inserts in base of distal phalanx
    - Has extensor hood similar to fingers
- **UCL of MCP joint of thumb**
  - Scan in longitudinal plane
  - Extends from innominate tubercle and groove at distal end of 1st metacarpal bone → groove base of proximal phalanx
  - Covered by adductor pollicis aponeurosis, which is seen as thin, echogenic band overlying UCL
    - Interphalangeal (IP) joint flexion helps identify adductor aponeurosis
    - On dorsal aspect of 1st web space, adductor pollicis muscle seen as 1st muscle deep to thick dorsal interosseous muscle
  - Careful thumb abduction during scanning will stress UCL and widen ulnar aspect of MCP joint
    - Do not overdo as can potentially induce Stener lesion

TECHNIQUE: VOLAR HAND

Patient Position

- Dorsum of hand flat on examination table or couch with wrist in neutral position, fingers extended

Flexor Tendon Pulleys

- A1-A5 annular pulleys with A1 being most clinically important (trigger finger)
- A1, A3, and A5 seen as hyperechoic or hypoechoic bands looping over flexor tendons at MCP joint, proximal and distal IP joint level
  - Fuse with volar plates & bases of more distal phalanges
  - A1 pulley is thickest and should be < 0.5 mm thick, measured at 11 o'clock or 1 o'clock position
    - A2 is longest and strongest pulley
  - Passively flex and extend finger while observing movement of flexor tendons deep to pulley
  - Normally, tendon movement should be unrestricted and pulley should not have to lift up to allow flexor tendons to glide underneath
- A2 and A4 pulleys occur at proximal and middle phalangeal shaft level; A2 and A4 are particularly relevant to climbing injuries as they prevent bowstringing
  - For bowstringing, examine finger in resting extended position and during forced flexion
    - Forced flexion of IP joints against pressure to fingertips with MCP joint extended
    - Distance between flexor tendons and phalangeal shafts should not increase with forced flexion

Flexor Tendons of Thumb

- Only 1 long flexor tendon (flexor pollicis longus)
  - Has continuous tendon sheath (radial bursa of wrist), which starts 3 cm proximal to radial styloid
  - Best identified passing through thenar muscle
  - Has A1 and A2 pulleys and also Y-shaped pulley across proximal phalanx

Metacarpophalangeal and Interphalangeal Joints

- Examined from volar and dorsal aspects and, if symptomatic, radial and ulnar aspects
- On dorsal side, observe proximal capsular recess, hypoechoic articular cartilage, joint space, bony margins, extensor tendon ± sagittal bands
- On volar side, observe proximal capsular recess, fibrocartilaginous volar plate, hypoechoic articular cartilage, joint space, bony margins, and overlying flexor tendons and A pulley
- Collateral ligaments on radial and ulnar sides
  - Collateral ligament proper runs roughly in line with joint and is taut in flexion
  - Accessory collateral ligament runs obliquely across joint from dorsal to volar, fusing with volar plate distally, and is taut in extension
  - MCP joint collateral ligament assessment is limited except for thumb, radial aspect of 2nd MCP, and ulnar aspect of 5th MCP joint

Carpometacarpal Joint of Thumb

- Examined from volar and dorsal aspects
- Assess triscaphe (scaphotrapeziotrapezoid) joint at same time as it is closely related clinically

(Left) *Near the MCP joint, the extensor tendon divides into 2 lateral slips and a central portion. Central portion inserts into base of the middle phalanx* ➡. *Lateral slips converge at the distal end of middle phalanx and insert into base of distal phalanx* ➡. *No central tendon is present on the proximal part of the shaft of the middle phalanx.* **(Right)** *Transverse US shows flattened central tendon* ➡ *on the dorsal aspect of the proximal phalanx* ➡ *of the middle finger. The extensor hood is too thin to clearly depict with US.*

Extensor Mechanism of Fingers

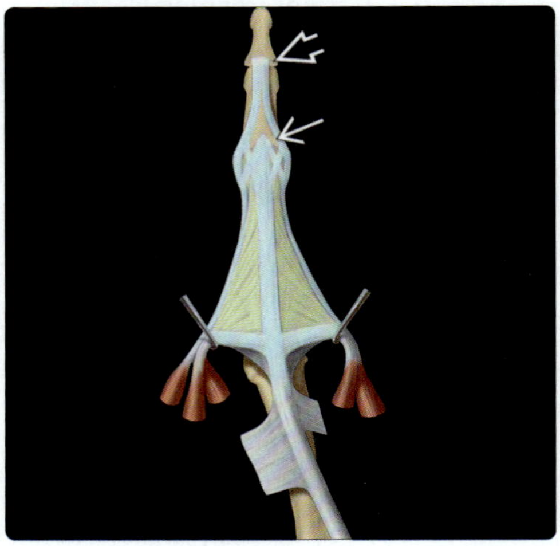

Extensor Digitorum of Middle Finger

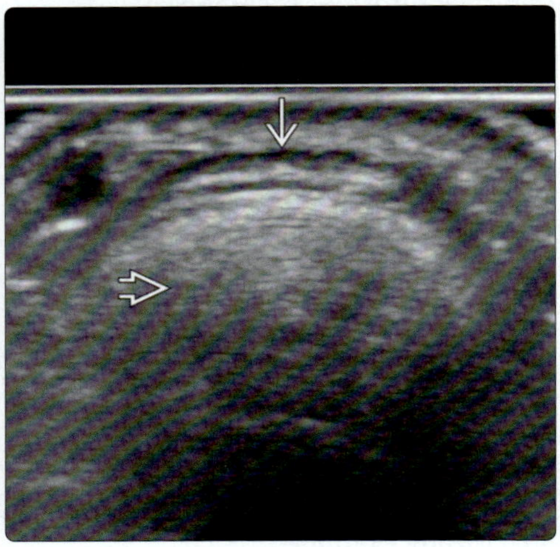

(Left) *Longitudinal US shows a thin extensor digitorum tendon* ➡ *inserting into the base* ➡ *of the distal phalanx. The extensor digitorum tendon in people with small fingers can be very thin. The base of the nail* ➡ *is shown.* **(Right)** *Graphic shows the ulnar aspect of the MCP joint of the thumb. The adductor pollicis aponeurosis* ➡ *attaches to the extensor expansion on the dorsum of the MCP joint. The lower image shows a reflected adductor aponeurosis with the ulnar collateral ligament (UCL)* ➡ *beneath.*

Extensor Digitorum Insertion

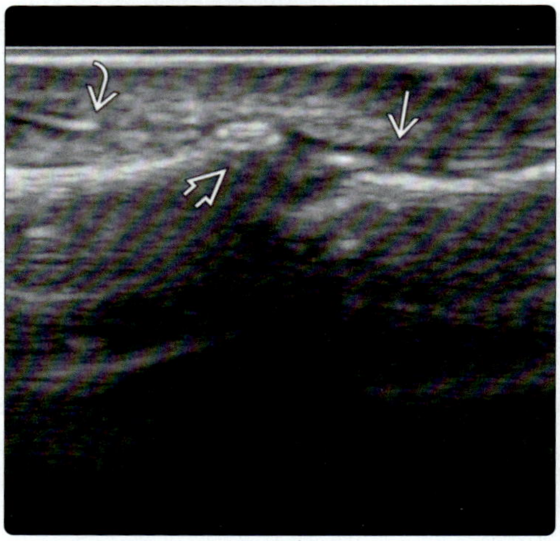

Ulnar Collateral Ligament of Metacarpophalangeal Joint of Thumb

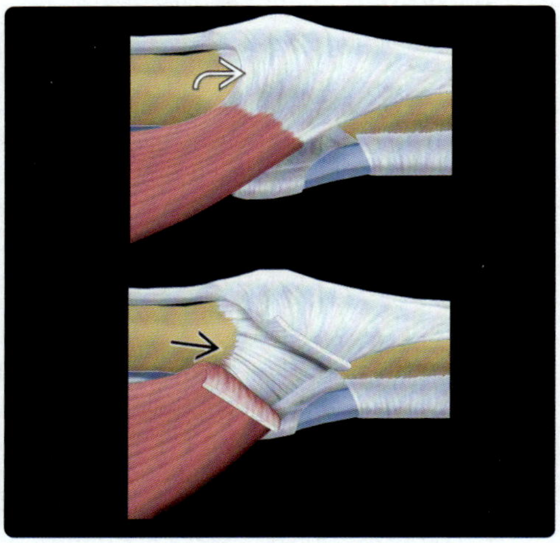

(Left) *Clinical photograph shows the scan position for the UCL of the thumb. The ligament can either be examined with the thumb fully abducted with the hand flat or placed on a pad as shown.* **(Right)** *Longitudinal US shows the hypoechoic UCL* ➡ *at the MCP joint of the thumb. The thin, echogenic line* ➡ *superficial to the ligament represents the adductor aponeurosis. The UCL runs from the metacarpal bone* ➡ *to the proximal phalanx* ➡.

Scan Position for Ulnar Collateral Ligament of Thumb

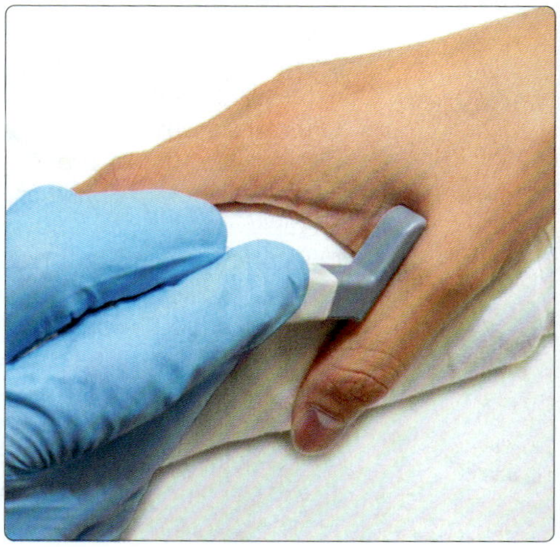

Ulnar Collateral Ligament of Metacarpophalangeal Joint of Thumb

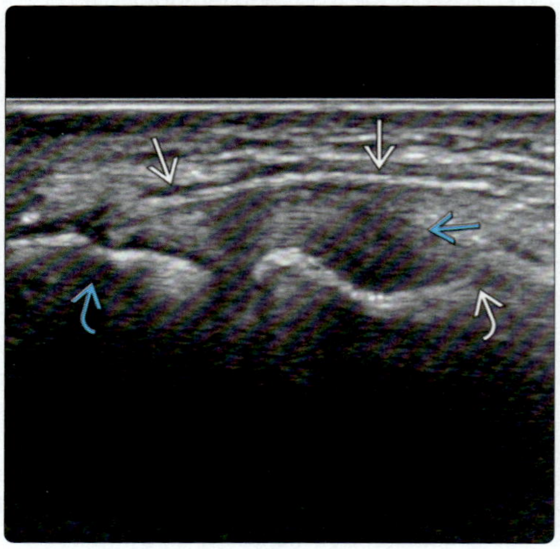

Flexor Pulley System

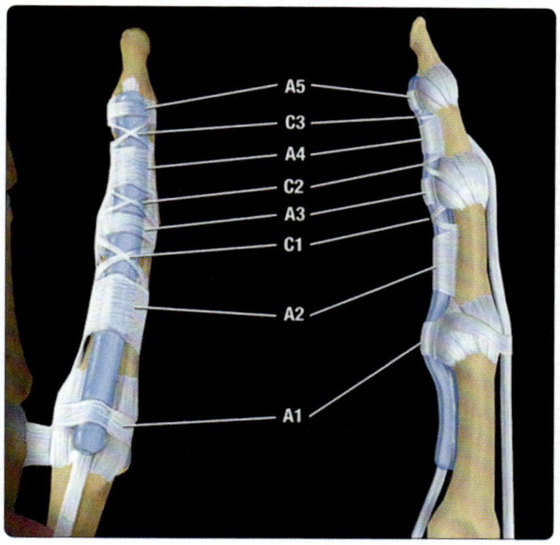

A1 Pulley: Short Axis

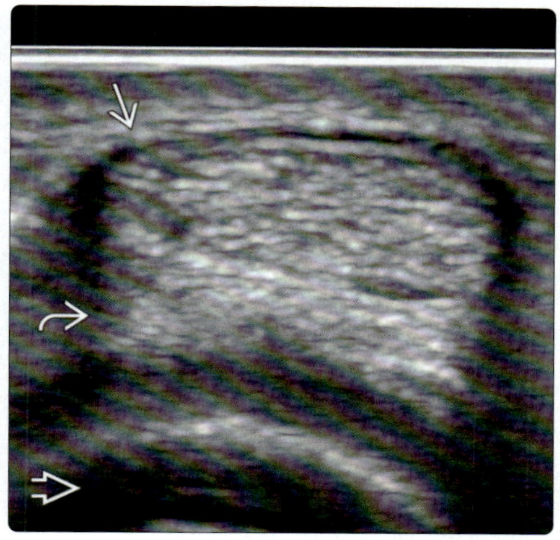

(Left) *Graphic shows the flexor pulley system of the fingers comprising annular (A1-5) and cruciate (C1-C3) pulleys. The most important pulley is the A1 pulley (trigger finger) followed by the A2 and A4 pulleys (bowstringing).* **(Right)** *Transverse US of the volar aspect of the metacarpal head ➡ shows the flexor digitorum superficialis (FDS) and flexor digitorum profundus (FDP) tendons deep to A1 pulley ➡. Deep to the tendons is the MCP joint volar plate ➡. Normal A1 pulley thickness, measured at either the 11 or 1 o'clock position, is < 0.5 mm.*

A1 Pulley: Long Axis

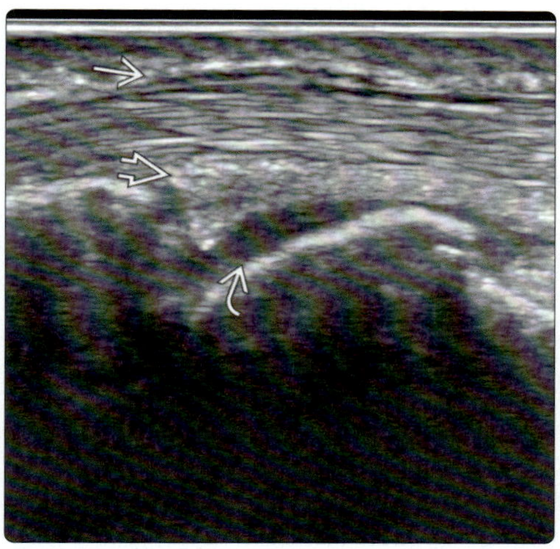

A2 Pulley: Short Axis

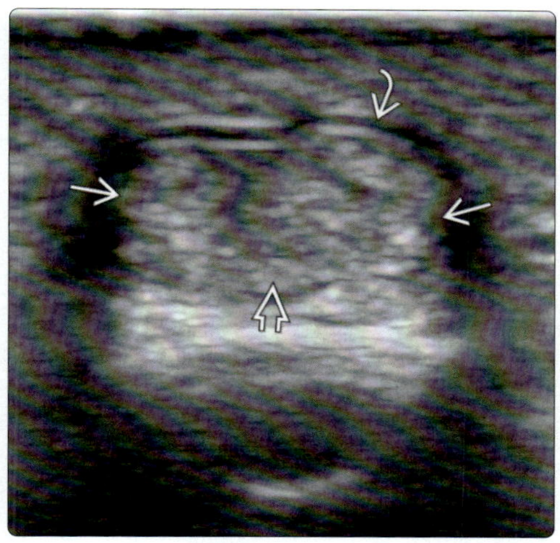

(Left) *Longitudinal US at the MCP joint level shows the A1 pulley ➡ overlying the flexor tendons. The fibrocartilaginous volar plate ➡ is deep to the tendons. The hyaline cartilage ➡ of the metacarpal head is deep to the volar plate.* **(Right)** *Transverse US at the level of the proximal phalangeal head shows the A2 pulley ➡ covering the 2 FDS tendons ➡ and the central FDP tendon ➡.*

Flexor Digitorum Profundus Tendon Insertion

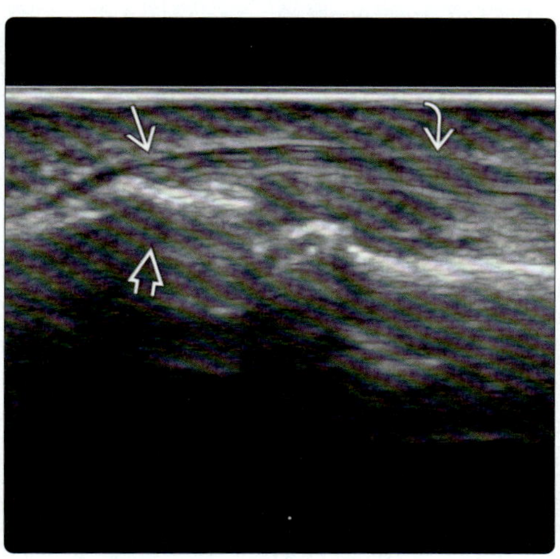

1st Carpometacarpal Joint

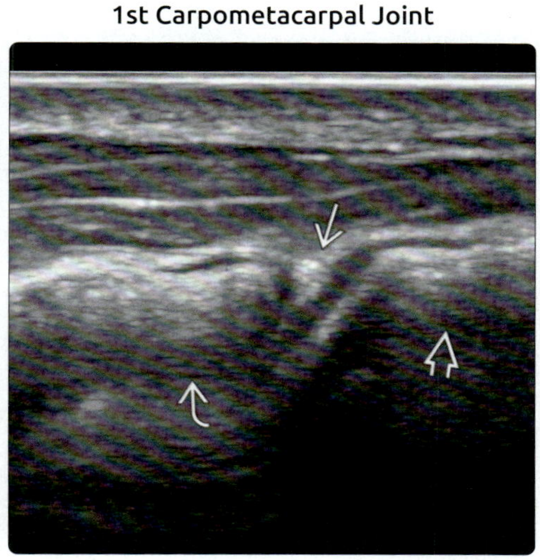

(Left) *Longitudinal US of the volar aspect of the finger shows the insertion ➡ of the flexor digitorum profundus tendon ➡ into the base of the distal phalanx ➡.* **(Right)** *Longitudinal US shows the 1st carpometacarpal joint ➡ between the trapezium ➡ and the base of the 1st metacarpal bone ➡. The hypoechoic articular cartilage can be appreciated. The echogenic line within the joint is due to either gas or a synovial fold. This joint does not have an intraarticular disc.*

Groin Hernia Ultrasound

KEY FACTS

TECHNIQUE

- Adequate exposure of lower abdominal wall from umbilicus to groin
- Specifically examine deep inguinal ring, Hesselbach triangle, and femoral canal

IDENTIFICATION OF IMPORTANT LANDMARKS

- Deep inguinal ring: Situated just proximal and lateral to confluence of inferior epigastric vessels with external iliac vessels; spermatic cord passes through deep ring into peritoneal cavity in males
- Inferior epigastric vessels: Trace transversely from lateral aspect of rectus abdominis muscle distally to external iliac vessels
- Spermatic cord: Trace transversely from scrotum, along inguinal canal before it dips into peritoneal cavity through deep ring
- Inguinal ligament: Trace longitudinally from pubic tubercle to anterior superior iliac spine

TECHNIQUE IN CLASSIFICATION OF GROIN HERNIA

- Inguinal hernia: Neck of sac above inguinal ligament, extends along inguinal canal
- Indirect: Neck of sac above inguinal ligament, extends along inguinal canal
- Direct: Neck of sac is wide and medial to inferior epigastric vessels in Hesselbach triangle
- Femoral hernia: Neck of sac below inguinal ligament

PROTOCOL

- Identify inferior epigastric vessels, inguinal ligament, spermatic cord (in males)
- Examine deep ring, Hesselbach triangle, femoral canal
- In each area, ask patient to perform Valsalva maneuver
- If present, pay particular attention to edges of surgical mesh
- Reexamine patient in standing position, particularly if this clinically accentuates symptoms

Landmarks of Inguinal Canal

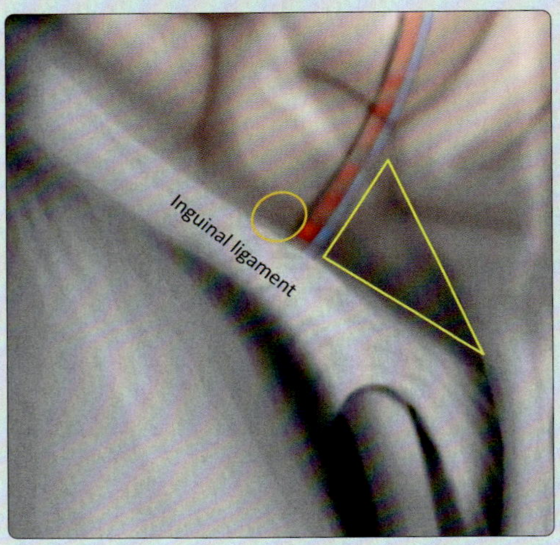

Landmarks of Inguinal Hernia

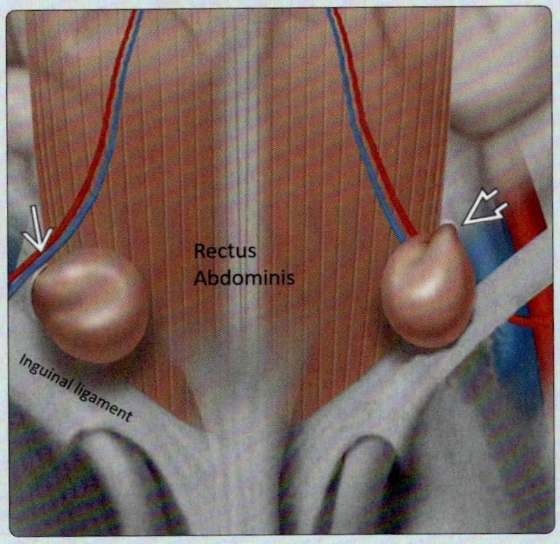

(Left) *Schematic drawing shows relevant groin anatomy for depicting groin hernia. The inguinal ligament extends from pubic tubercle (PT) to the anterior superior iliac spine (ASIS). Hesselbach triangle (yellow triangle) lies medial to inferior epigastric vessels and lateral to rectus abdominis above inguinal ligament. Deep ring (DR) (orange circle) lies lateral to inferior epigastric vessels.* **(Right)** *Direct inguinal hernia* ➡ *tends to arise with a wide neck through Hesselbach triangle. Indirect inguinal hernia* ➡ *passes through the deep inguinal ring.*

Landmarks of Femoral Canal

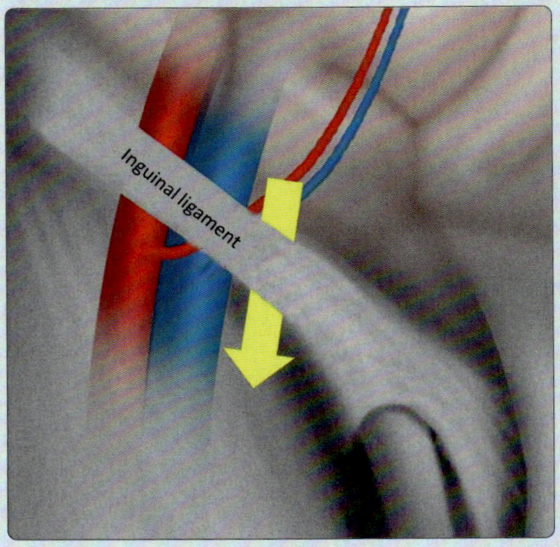

Landmarks of Femoral Hernia

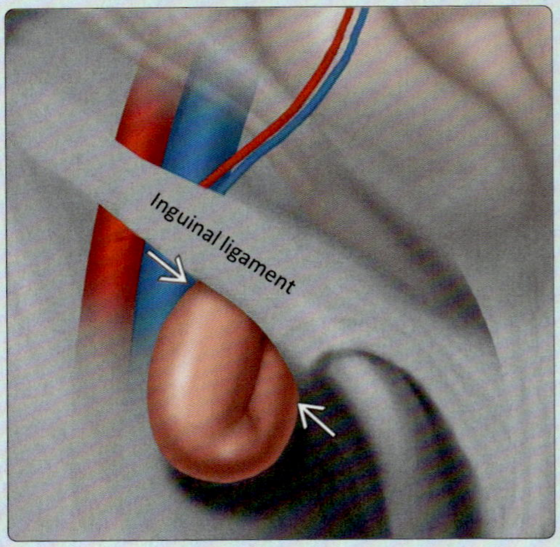

(Left) *Schematic drawing shows the femoral canal (yellow arrow) lying under the inguinal ligament medial to the femoral vein.* **(Right)** *The neck of femoral hernia is along the femoral canal* ➡ *under the inguinal ligament. The common femoral vessels are usually slightly displaced laterally.*

IMAGING ANATOMY

Critical Anatomic Structures

- Inguinal ligament
  - Thickened inferior border of external oblique aponeurosis
  - Runs from pubic tubercle → anterior superior iliac spine (ASIS)
- Inferior epigastric vessels
  - From external iliac artery, just above inguinal ligament
  - Ascends to lateral border of rectus abdominis muscle
- Inguinal canal
  - 4 cm long from deep → superficial inguinal ring
    - Deep inguinal ring
      - Opening in transversalis fascia through which spermatic cord/round ligament passes
      - 1 cm above midpoint inguinal ligament and 1 cm lateral to inferior epigastric vessels
    - Superficial inguinal ring
      - Opening in external oblique aponeurosis just lateral to pubic tubercle
      - Contains ilioinguinal nerve, spermatic cord (male), round ligament (female), and associated vessels
- Spermatic cord
  - Originates from vas deferens in scrotum
  - Enters superficial ring and passes along inguinal canal
  - Passes into peritoneal cavity through deep ring
  - Contains pampiniform plexus, testicular artery, cremasteric artery, artery of vas, lymphatics, sympathetic nerves, and vas deferens

TECHNIQUE

Standard Examination

- Patient supine with legs extended
- Expose lower abdomen from umbilicus to groin
- First identify (1) inguinal ligament, (2) inferior epigastric vessels, and (3) spermatic cord (in males)
  - Normal round ligament in females not seen on US
- Inguinal ligament
  - Trace in longitudinal plane from pubic tubercle → ASIS
  - Hyperechoic structure with linear striated appearance
    - Most inferior edge of external aponeurosis
    - Deep ring at midpoint of inguinal ligament (i.e., halfway between pubic tubercle and ASIS)
      - Deep ring lies between inferior epigastric and external iliac vessels
    - Femoral vessels deep to midinguinal point (i.e., halfway between pubic symphysis and ASIS)
- Inferior epigastric vessels
  - Trace in transverse plane from rectus abdominis muscle → external iliac vessels
    - Artery, 2 veins seen as 3 small, hypoechoic dots
    - Color Doppler aids identification in obese patients
- Spermatic cord (in males)
  - Trace in either longitudinal or transverse plane from upper scrotum → inguinal canal → deep ring
  - Reasonably distinct echogenic, rounded, tubular structure containing arteries and vas deferens
- Then, specifically perform Valsalva maneuver at deep inguinal ring, Hesselbach triangle, and femoral canal
  - Deep inguinal ring
    - 1 cm lateral to inferior epigastric vessels
    - Site of origin of indirect inguinal hernia
  - Hesselbach triangle
    - Medial to inferior epigastric vessels above inguinal ligament
    - Site of origin of direct inguinal hernia
  - Femoral canal
    - Medial to femoral vein below inguinal ligament
    - Site of origin of femoral hernia
- Use only mild transducer pressure to avoid impeding emergence of hernia
- Examine in erect position if no hernia visible in supine position and symptoms worsen when standing upright

Points to Note

- **If hernia visible**
  - Confirm hernia emerges through abdominal wall
    - Determine whether hernia is indirect inguinal, direct inguinal, or femoral in type
      - By assessing relationship of neck to inferior epigastric vessels and inguinal ligament
    - Examine hernia at rest and during Valsalva
      - Best to perform Valsalva by asking patient to blow onto back of hand
    - Assess contents of hernia (mesenteric fat ± bowel) and maximum size, extent
    - Measure width of fascial defect
    - Check whether hernia reduces fully during relaxation
    - If hernia is found, still examine other locations as double hernia may occur
- **If recurrent hernia present following surgical mesh repair**
  - Identify surgical mesh, which is usually seen as thin echogenic interface with posterior acoustic shadowing
    - Mesh may be wavy in appearance and can be differentiated from bowel by lack of peristalsis
  - Inspect and perform Valsalva at margins of mesh
- **If no hernia present**
  - Assess whether painful or symptomatic area is located at deep ring, Hesselbach triangle, or femoral canal
  - See if any abnormality, such as scarring at these locations
  - See if any undue bulging of fascia at these locations during Valsalva, indicating laxity
  - Reexamine with patient standing if uncertain about diagnosis and symptoms worsen when standing
- **If mass other than hernia present**
  - Discriminate other masses, such as hematoma, lipoma, seroma, cysts, canal of Nuck hydrocele
    - Canal of Nuck hydrocele or cyst due to incomplete obliteration of canal (patent processus vaginalis)
- **If unsure regarding US findings**
  - Arrange either CT or MR of groin region

Protocol

- Identify inferior epigastric vessels, spermatic cord, inguinal ligament
- Examine deep ring, Hesselbach triangle, femoral canal
- In each area, ask patient to perform Valsalva maneuver
- Reexamine patient in standing position, particularly if this clinically accentuates symptoms

(Left) *Schematic drawing shows the probe position at the umbilical level to identify inferior epigastric vessels (red and blue lines). The inferior epigastric vessels are located under the lateral border of the rectus abdominis muscle. Inguinal ligament (black line) extends from the ASIS to the PT.* **(Right)** *Corresponding transverse US shows the inferior epigastric artery accompanied by the inferior epigastric veins under the lateral border of the rectus abdominis muscle.*

Identify Inferior Epigastric Vessels

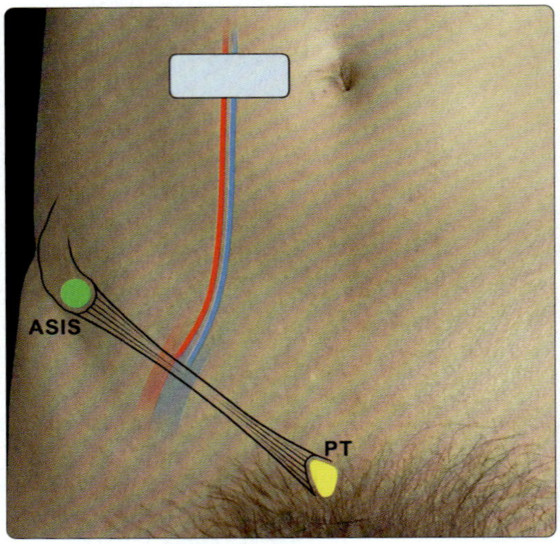

Identify Inferior Epigastric Vessels

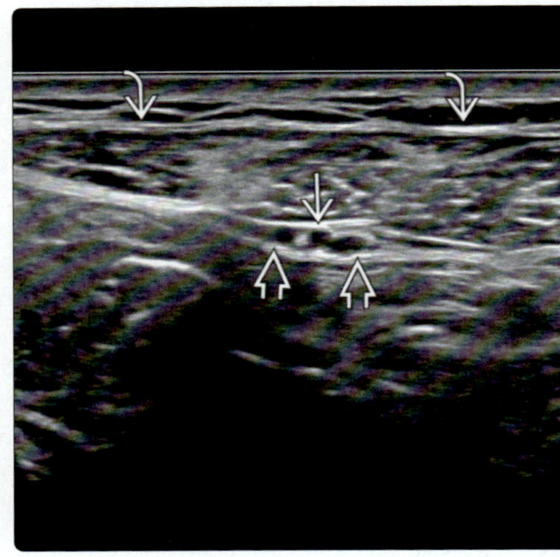

(Left) *Schematic drawing shows the probe position moving down to the right lower quadrant. The inferior epigastric vessels (red and blue lines) are located beneath the lateral border of the rectus abdominis muscle.* **(Right)** *Corresponding color Doppler US shows the inferior epigastric artery accompanied by inferior epigastric veins under the more lateral border of the rectus abdominis muscle. Color Doppler US is useful to identify the inferior epigastric vessels.*

Identify Inferior Epigastric Vessels

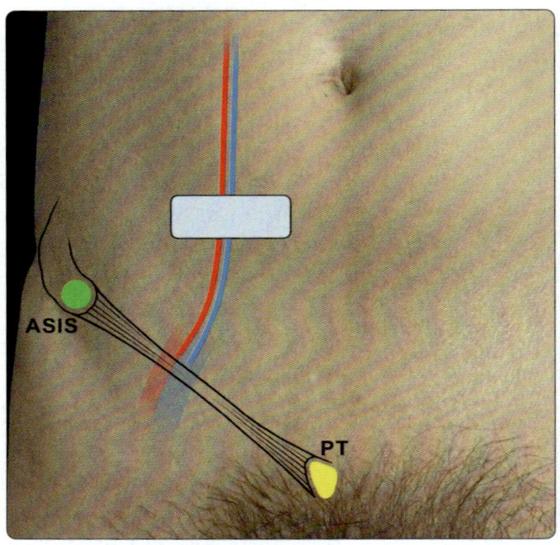

Identify Inferior Epigastric Vessels

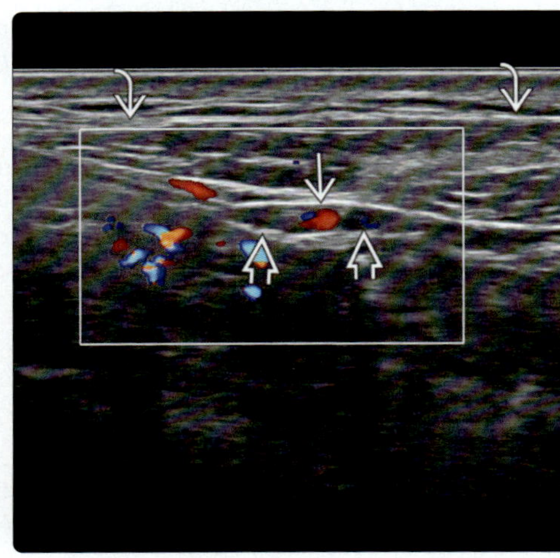

(Left) *Schematic drawing shows the probe position (blue box) at the medial groin region. The inferior epigastric vessels (red and blue lines) lie above the inguinal ligament (black line) just before they join the external iliac vessels. The DR lies lateral to the inferior epigastric vessels at this point.* **(Right)** *Corresponding transverse US shows the inferior epigastric vessels immediately before joining the external iliac artery and vein. The DR is a potential space between the external iliac and inferior epigastric vessels.*

Identify Deep Ring of Inguinal Canal

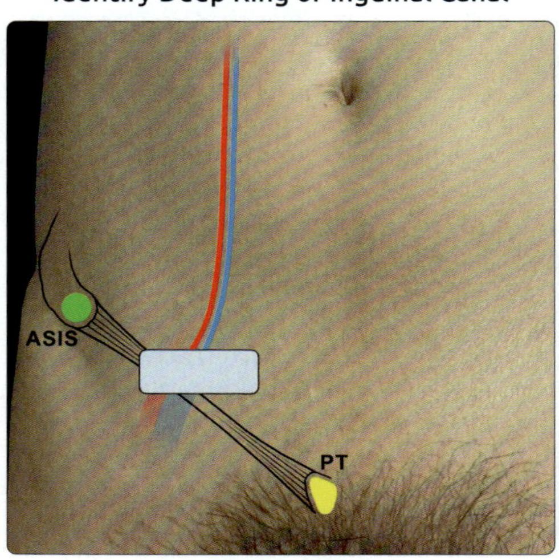

Identify Deep Ring of Inguinal Canal

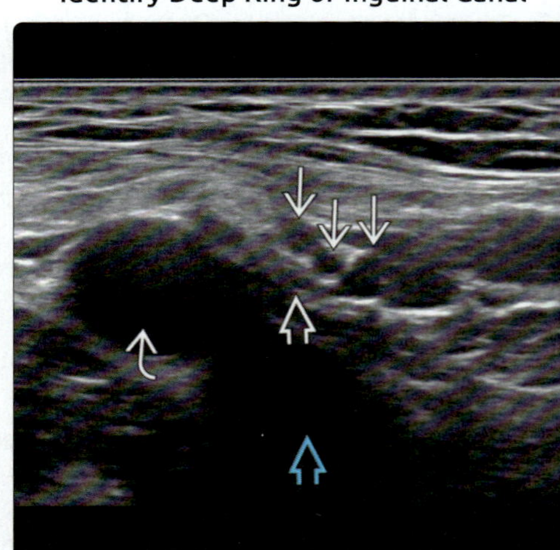

Identify Spermatic Cord

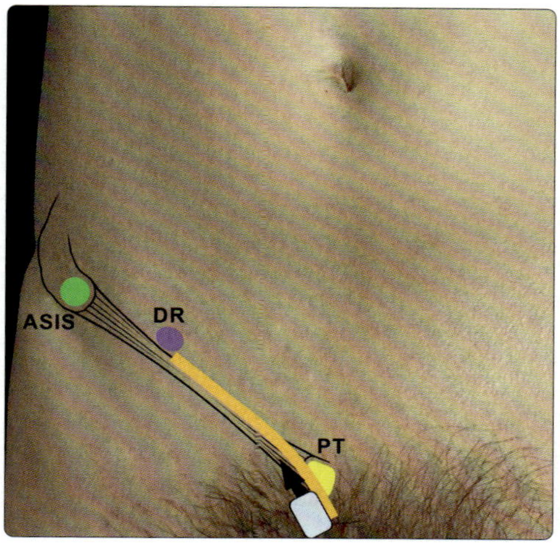

Identify Spermatic Cord

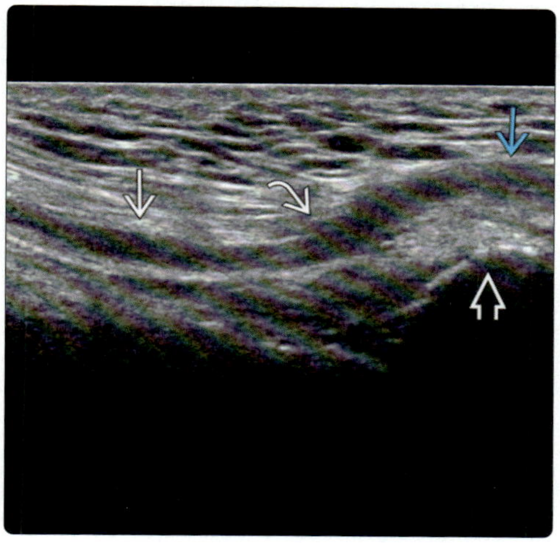

(Left) *Schematic drawing shows the probe position (blue box) at the pubic region. The spermatic cord (orange line) is traced from the high scrotal region below the PT along the inguinal canal before it dips through the DR into the peritoneal cavity.* **(Right)** *Oblique longitudinal US shows the spermatic cord coursing from the high scrotal region over the PT before extending through the superficial inguinal ring into the inguinal canal.*

Identify Spermatic Cord

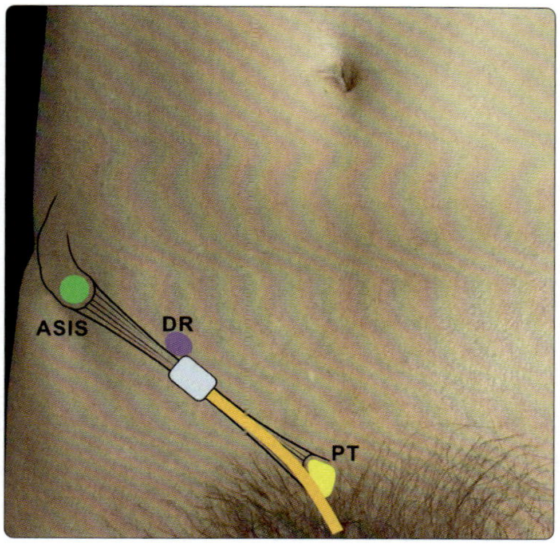

Identify Spermatic Cord

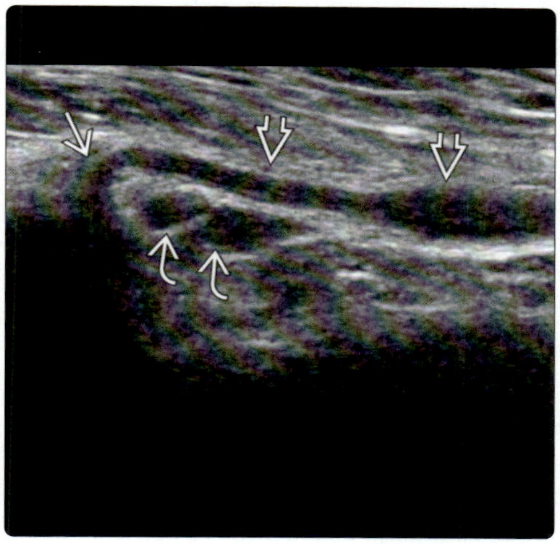

(Left) *Schematic drawing shows the probe position (blue box) at the midinguinal region. At this region, the spermatic cord (orange line) descends through the DR into the peritoneal cavity.* **(Right)** *Oblique longitudinal US shows the spermatic cord coursing along the inguinal canal before dipping through the DR into the peritoneal cavity. The DR lies immediately lateral to the inferior epigastric vessels.*

Identify Inguinal Ligament

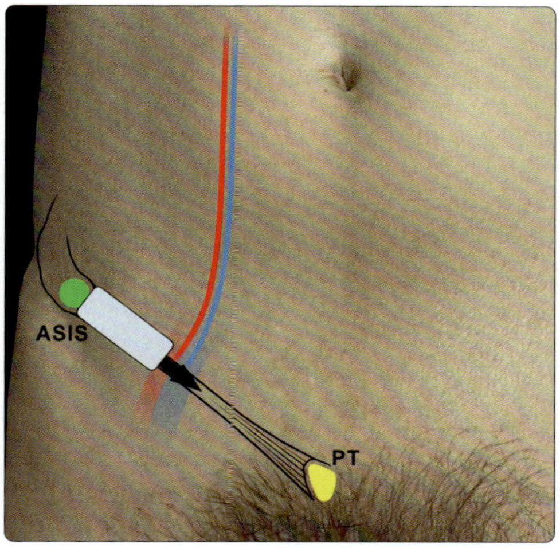

Identify Inguinal Ligament

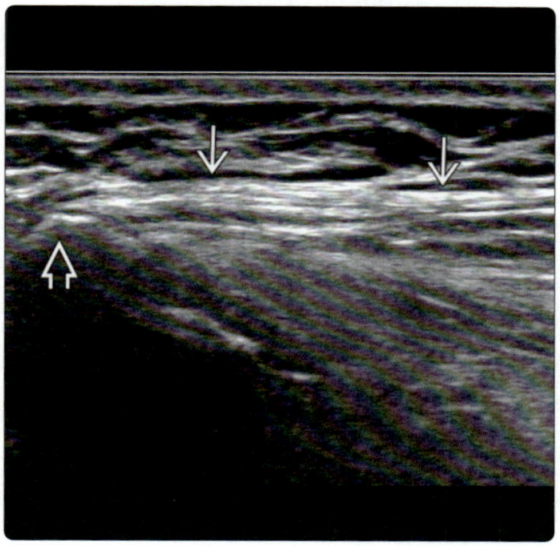

(Left) *Schematic drawing shows the probe position (blue box) at the ASIS orientated inferomedially along the inguinal ligament toward the PT.* **(Right)** *Oblique longitudinal US shows the lateral 1/3 of the inguinal ligament, which is an echogenic striated structure extending from the ASIS to the PT. The inguinal ligament is the free-rolled inferior border of the external oblique aponeurosis.*

(Left) *Schematic drawing shows the probe position (blue box) on the inguinal ligament (black line) at the midinguinal point, which is midway between the pubic symphysis ➡ and the ASIS. The femoral vessels lie deep to the midinguinal point.* (Right) *Oblique longitudinal US at the midinguinal point shows the echogenic striated inguinal ligament ➡ lying immediately superficial to the common femoral artery ➡ and vein ➡.*

Identify Inguinal Ligament

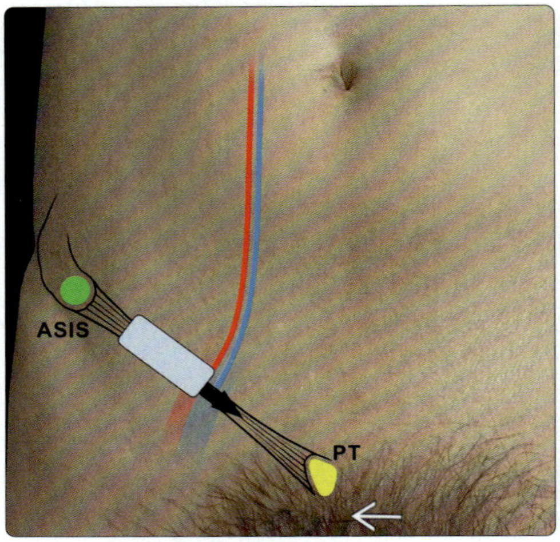

Identify Inguinal Ligament

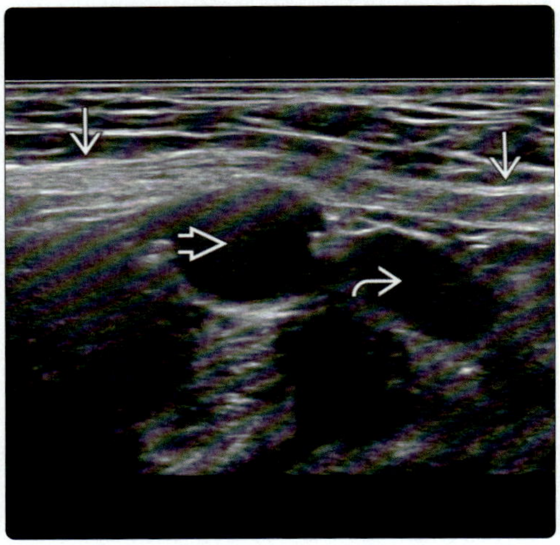

(Left) *Schematic drawing shows the probe position (blue box) moved to the lower groin region overlying the medial 1/3 of the inguinal ligament (black line).* (Right) *Oblique longitudinal US shows the medial 1/3 of the inguinal ligament ➡ seen as a thick, echogenic striated structure attached to the PT ➡.*

Identify Inguinal Ligament

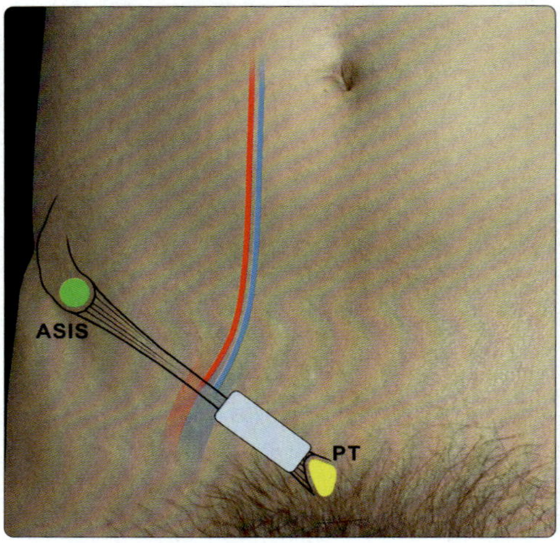

Identify Inguinal Ligament

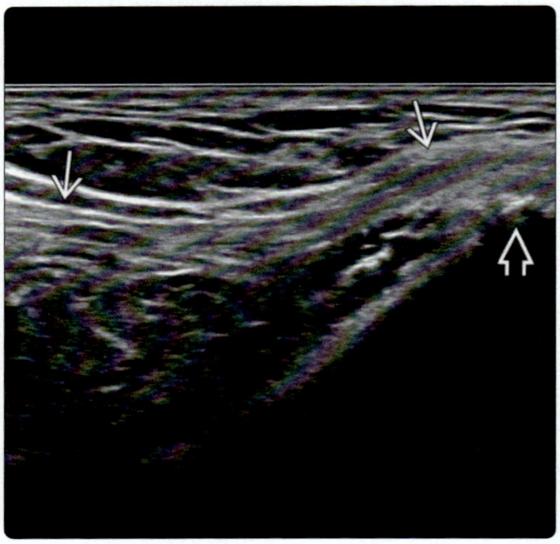

(Left) *Transverse US shows an echogenic mass ➡ present in the right groin region lateral to the inferior epigastric vessel ➡ and medial to the external iliac artery ➡ and vein ➡. This is an indirect inguinal hernia.* (Right) *Corresponding US during Valsalva maneuver shows how the mass is now more bulging ➡, helping to confirm an indirect inguinal hernia. The hernia is between the femoral artery ➡, femoral vein ➡, and the partially effaced inferior epigastric vessels ➡.*

Indirect Inguinal Hernia At Rest

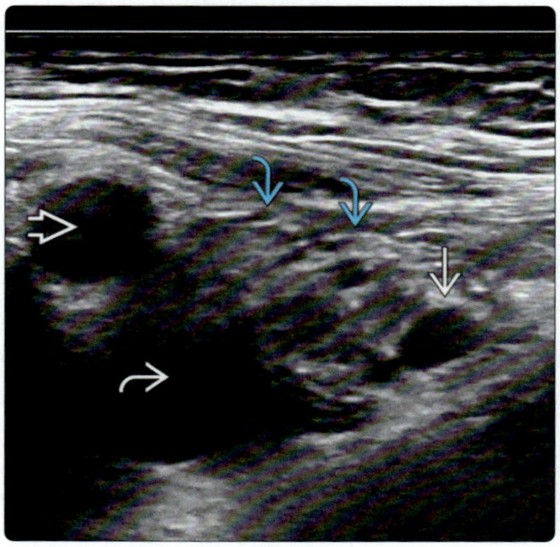

Indirect Inguinal Hernia During Valsalva

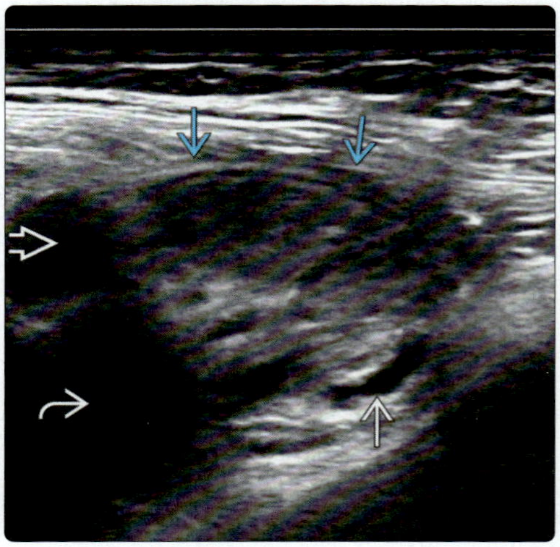

Identify Indirect Inguinal Hernia

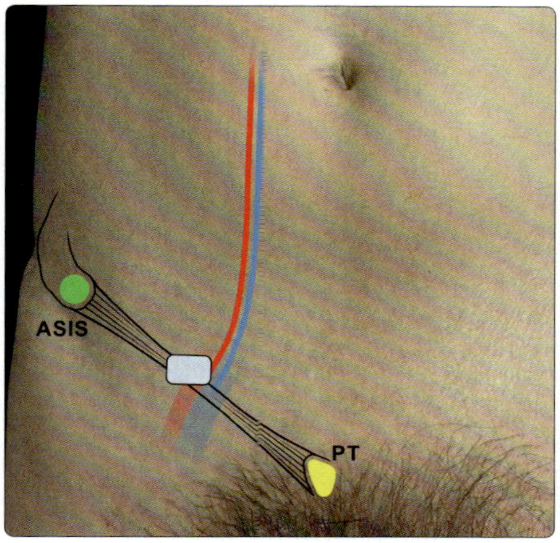

Indirect Inguinal Hernia

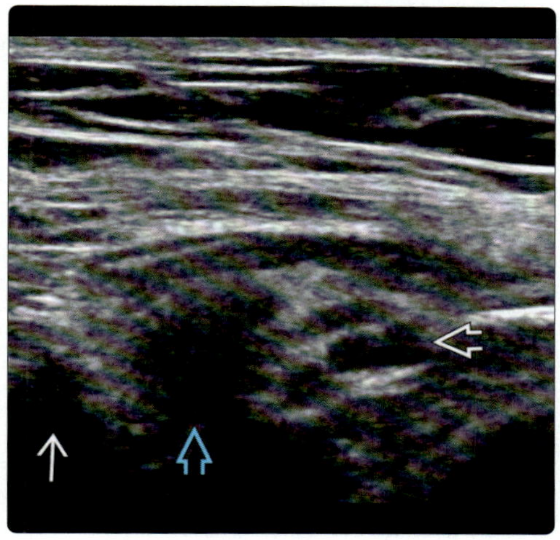

(Left) *Schematic drawing shows the probe position (blue box) at the DR, the point of origin of indirect inguinal hernia. The DR is located at the midpoint of the inguinal ligament, i.e., midway between the PT and ASIS. The neck of an indirect hernia is lateral to the inferior epigastric vessels and above the inguinal ligament (black line).* **(Right)** *Transverse US shows the neck of an indirect inguinal hernia → present at the DR, i.e., lateral to inferior epigastric vessels → and medial to external iliac vessels →.*

Identify Direct Inguinal Hernia

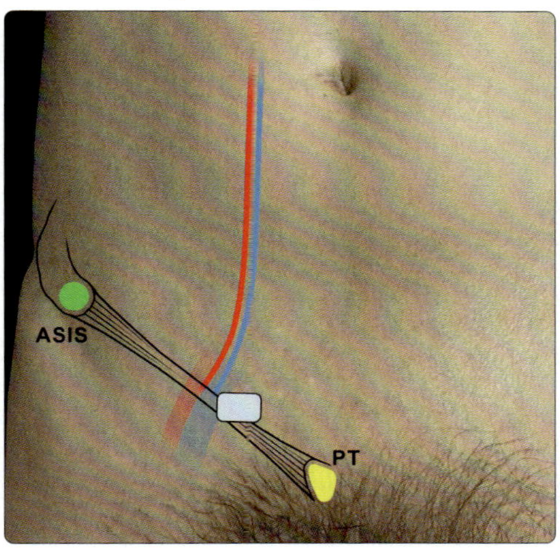

Direct Inguinal Hernia

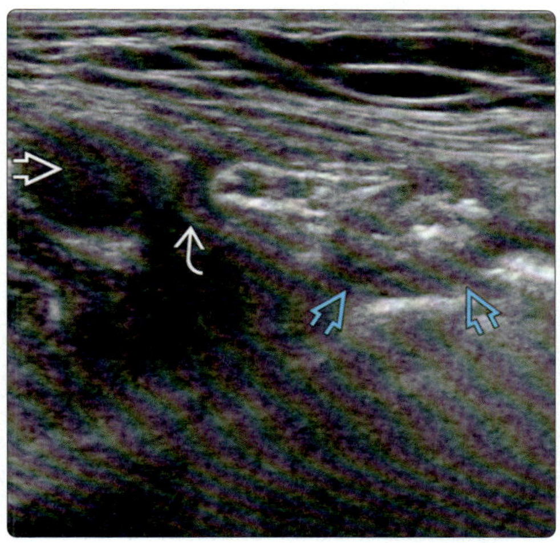

(Left) *Schematic drawing shows the probe (blue box) position at the Hesselbach triangle, i.e., at the site of the origin of a direct inguinal hernia. The neck of a direct inguinal hernia is medial to the inferior epigastric vessels and proximal to the inguinal ligament (black line).* **(Right)** *Oblique transverse US shows the wide direct hernial neck → located medial to the partially effaced inferior epigastric vessels → just before these vessels join the external iliac vessels →.*

Identify Femoral Hernia

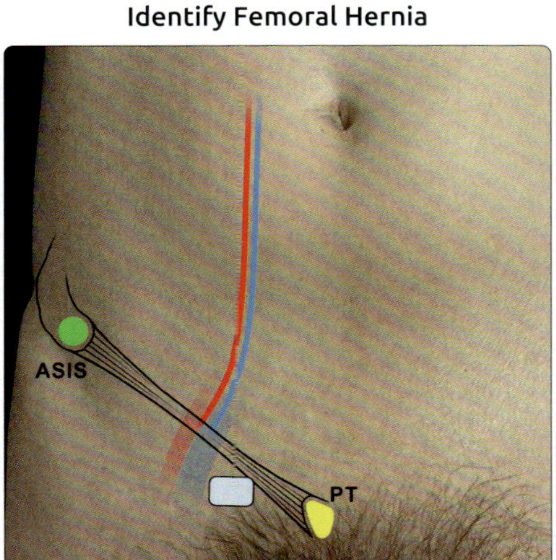

Femoral Hernia

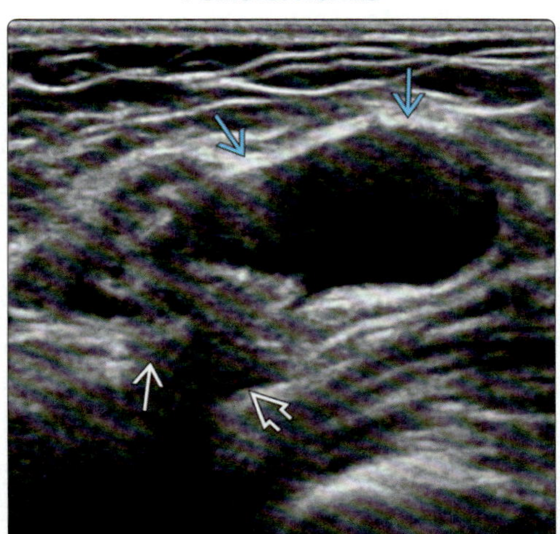

(Left) *Schematic drawing shows the probe (blue box) position at the femoral canal, i.e., the point of origin of a femoral hernia. The neck of the femoral hernia is located medially to the common femoral vein and distally to the inguinal ligament (black line).* **(Right)** *Oblique longitudinal US shows the neck of a femoral hernia → in the femoral canal, i.e., inferior to the inguinal ligament → and medial to the common femoral vein (not shown). The sac of this femoral hernia contains fluid → though no bowel or mesentery.*

(Left) *Oblique transverse US shows an indirect hernial sac containing bowel. The echostructure of bowel is visible with 2 hypoechoic lines: The superficial one representing muscularis propria and the deeper one representing submucosa. The neck of the hernia lies lateral to the inferior epigastric vessels.* **(Right)** *Oblique longitudinal US shows the hernial sac containing gas-filled bowel. This is a very useful sign to indicate a hernia.*

Bowel Wall in Hernial Sac

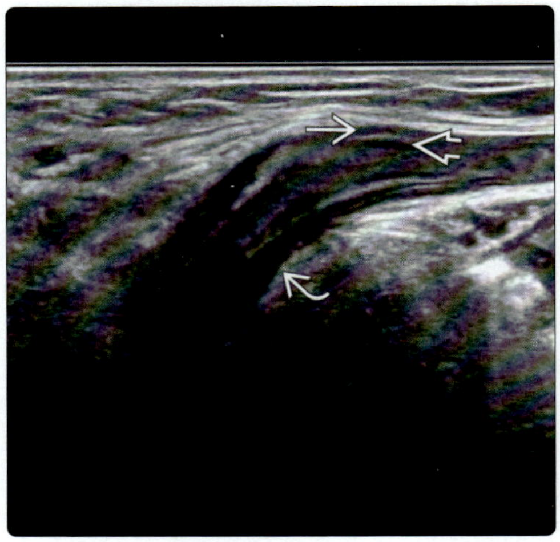

Bowel Gas in Hernial Sac

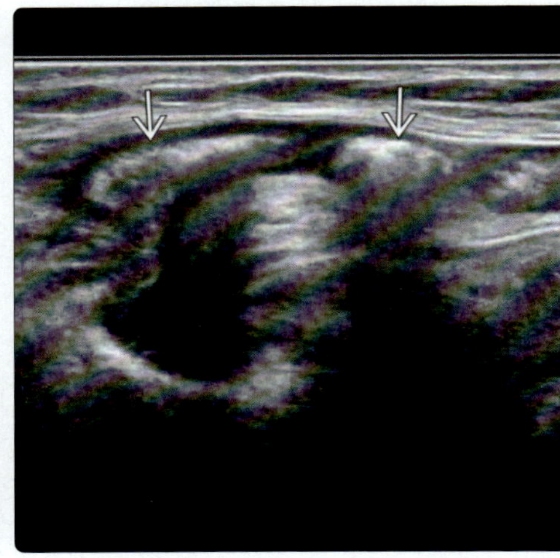

(Left) *Oblique longitudinal US shows the neck of a femoral hernia in the femoral canal lying inferior to the inguinal ligament. The thickened hernial sac contains fluid.* **(Right)** *During Valsalva, bowel passes through the femoral canal into the hernial sac, which helps confirms the presence of a femoral hernia. The femoral canal is narrow and bowel may not reduce back into the peritoneal cavity (i.e., an incarcerated hernia). As expected, incarcerated hernias are more prone to strangulation.*

Femoral Hernia Before Valsalva

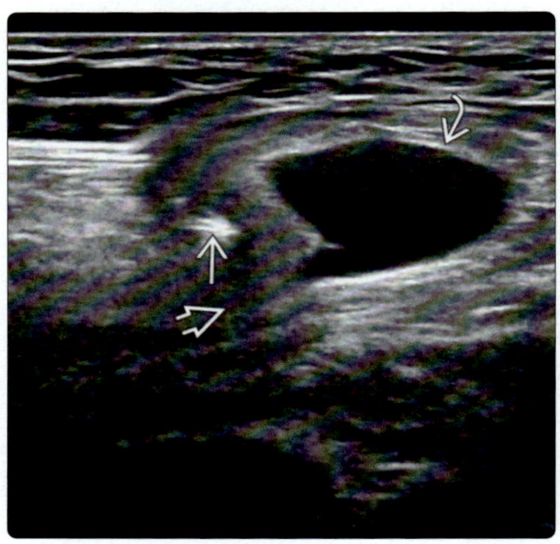

Femoral Hernia With Bowel During Valsalva

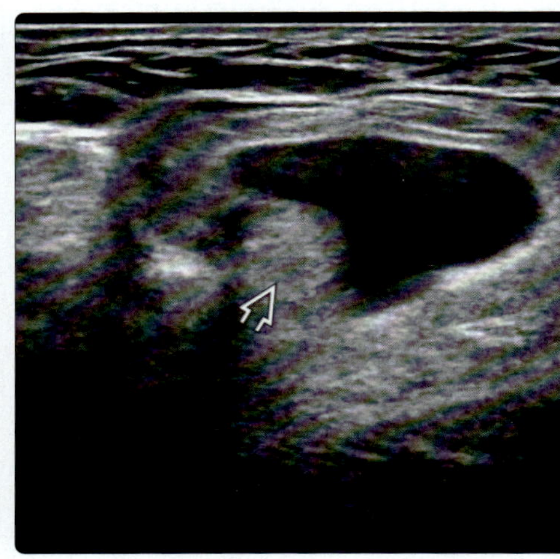

(Left) *Transverse US shows an echogenic mass due to mesentery in the scrotum from an indirect inguinal hernia. A normal testis is present alongside the hernial sac.* **(Right)** *Transverse color Doppler US shows a mixed hyperechoic and hypoechoic soft tissue mass containing scattered spots of arterial and venous-type vascularity reflecting the normal vascular pattern of the mesentery. This helps distinguish herniated mesenteric tissue from a lipoma.*

Mesentery in Scrotum

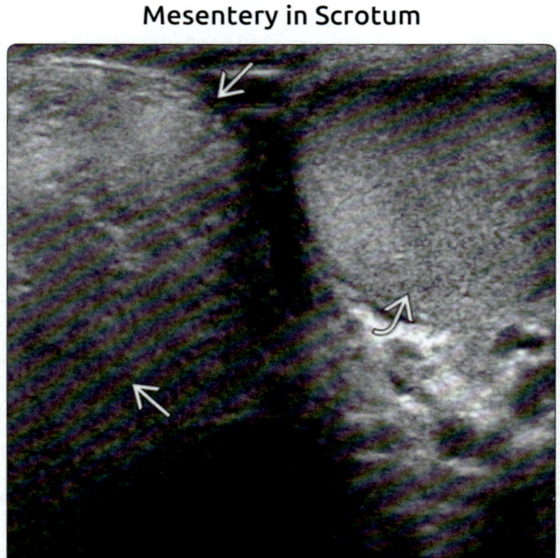

Vascularity Within Mesentery

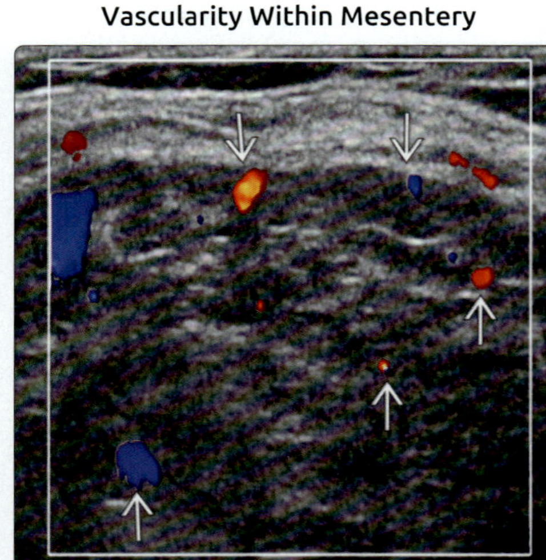

Indirect Inguinal Hernia: Step 1

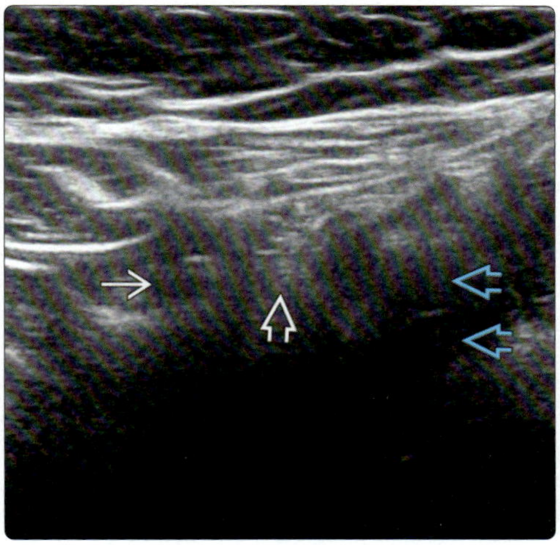

Indirect Inguinal Hernia: Step 2

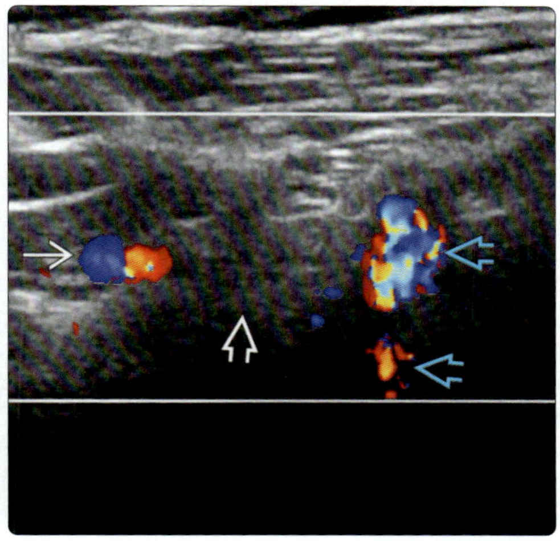

(Left) *Transverse US of a patient with left groin pain is shown. At rest, there is no hernia seen in the region of the deep inguinal ring . The DR is located lateral to the inferior epigastric vessels and medial to the external iliac vessels .* **(Right)** *Color Doppler US is useful to demonstrate the position of the DR of the inguinal canal , which lies between the inferior epigastric vessels and the external iliac vessels .*

Indirect Inguinal Hernia: Step 3

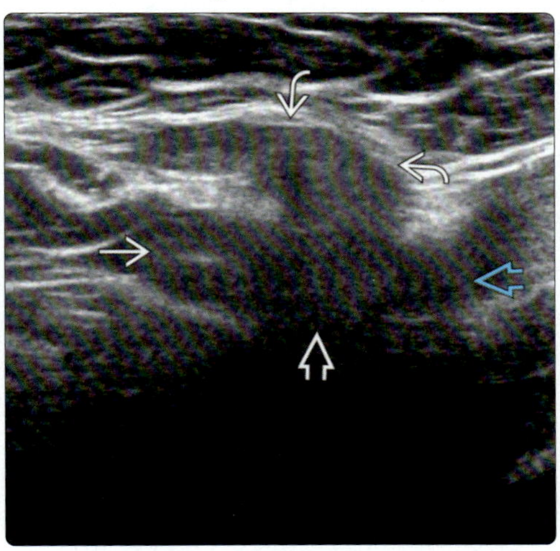

Indirect Inguinal Hernia: Step 4

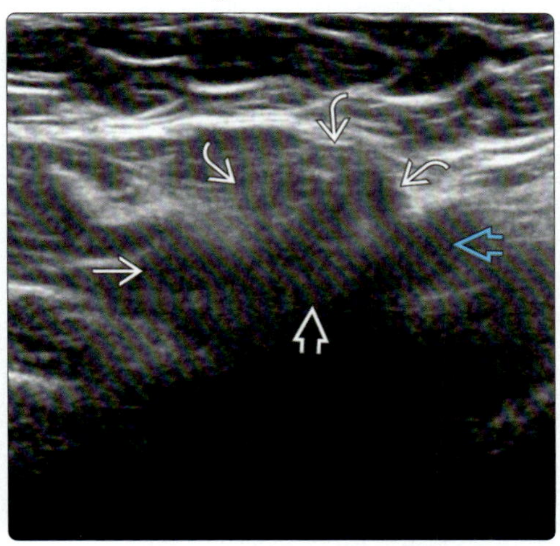

(Left) *Transverse US during Valsalva shows herniation of hypoechoic mesentery from the peritoneal cavity into the inguinal canal through the DR . The neck is located between the inferior epigastric vessels and the external iliac vessels .* **(Right)** *Corresponding US during Valsalva shows the change in configuration and size of the hernia sac . The DR between the inferior epigastric vessels and the external iliac vessels is shown.*

Indirect Inguinal Hernia: Step 5

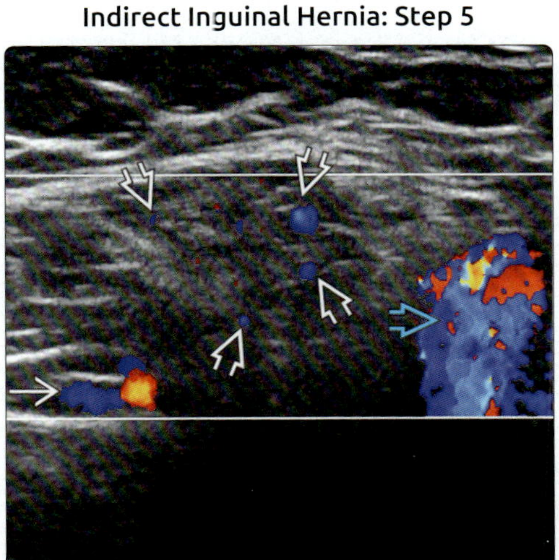

Indirect Inguinal Hernia: Step 6

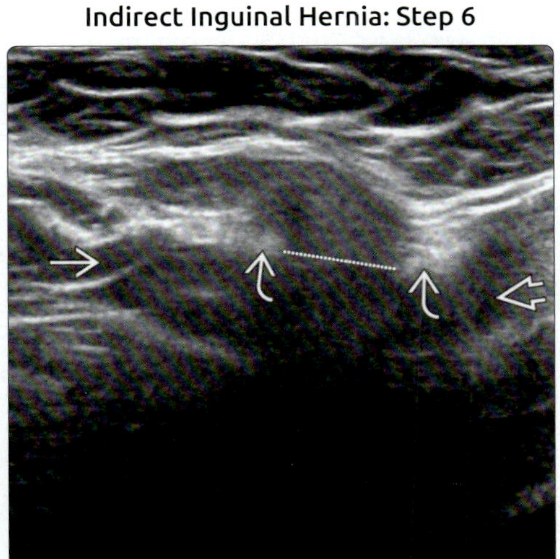

(Left) *Corresponding color Doppler US shows vascularity that is present within the hernial sac, which helps confirm mesentery within the sac. No gas or bowel wall is present. The inferior epigastric vessels and the external iliac vessels are demonstrated.* **(Right)** *The size of the fascial defect (broken line) between the medial and lateral borders of the fascia is measured to guide treatment. The inferior epigastric vessels and external iliac vessels are demonstrated.*

Hip Ultrasound

KEY FACTS

GENERAL CONSIDERATIONS

- US very good for suspected hip effusion, synovitis, some adductor tendon pathology, and greater trochanteric pain
  - Can be supplemented by MR, particularly for deep gluteal pain and small tendon tears around hip
  - Also helpful to guide diagnostic and interventional procedures
- Full examination is preferable in most patients with poorly localized hip pain

ANTERIOR HIP

- Supine on examination couch with leg in neutral position
- Helpful to passively, internally or externally, rotate hip when assessing different structures
  - Anterior aspect hip joint
  - Anterior superior iliac spine region
  - Rectus femoris tendon and insertion
  - Iliopsoas tendon
  - Dynamic examination for snapping iliopsoas tendon

MEDIAL HIP

- Hip abducted to variable degree and externally rotated with knee flexed (frog-leg position)
  - Adductor tendons and muscles

LATERAL HIP

- Lateral decubitus or oblique lateral decubitus
- Flexing and extending hip for suspected iliotibial band impingement
  - Gluteus minimus and medius tendons
  - Abductor tendons and muscles
  - Tensor fascia lata, iliotibial tract, and gluteal muscles

POSTERIOR HIP

- Patient prone with lower limb neutral position
- Low-frequency transducer may be required
  - Posterior thigh musculature just distal to ischial tuberosity; deep gluteal space proximal to ischial tuberosity; ischial tuberosity & hamstring tendon origins

Scan Position for Anterior Hip

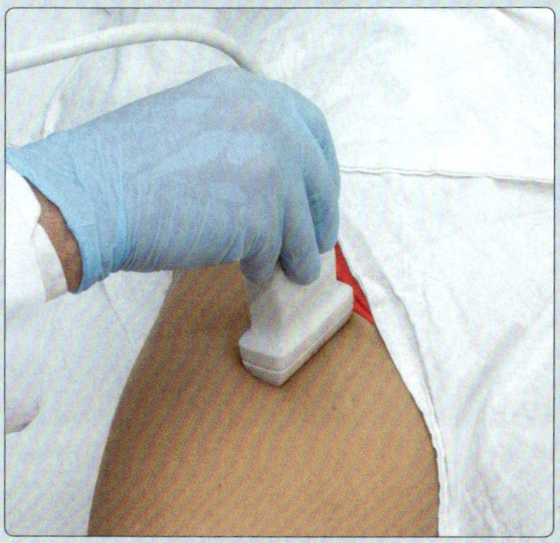

Anterior Recess of Hip Joint

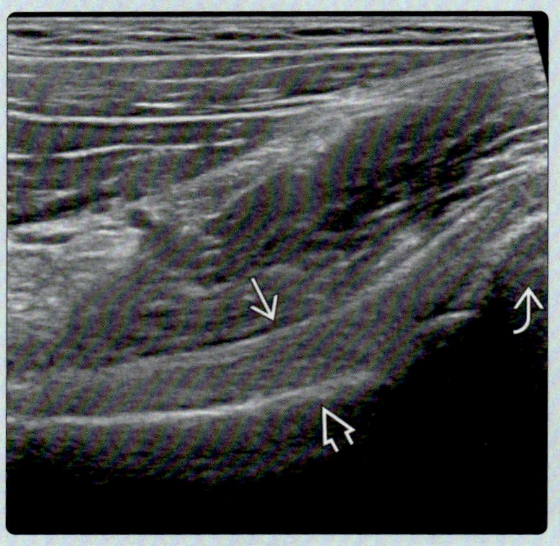

(Left) *Clinical photograph shows the patient and probe position for examination of the anterior hip recess. The patient lies supine with the hip extended. Internal or external rotation of the hip during scanning of the anterior hip structures can be helpful.* **(Right)** *Oblique longitudinal US shows the anterior recess ➡ of the hip joint anterior to the neck ➡ and head ➡ of the proximal femur.*

Acetabular Attachment of Hip Capsule

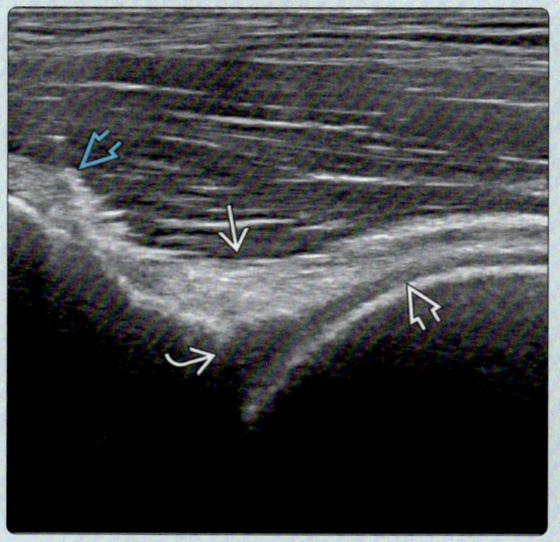

Femoral Head and Iliopsoas Tendon

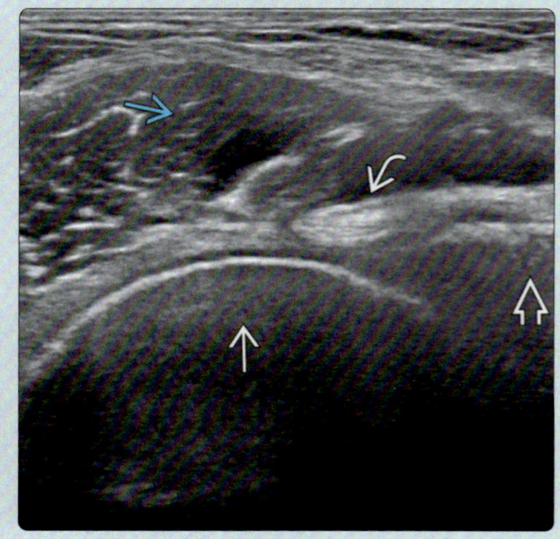

(Left) *Longitudinal US shows the capsular attachment ➡ to the rim of the acetabulum. The labrum ➡ is hypoechoic due to anisotropy. The femoral head of articular cartilage ➡ is clearly seen. Part of the rectus femoris tendon attachment ➡ is visible.* **(Right)** *Transverse US at the level of the femoral head shows the close relationship between the femoral head ➡, the acetabulum ➡, and the iliopsoas tendon ➡ and muscle ➡.*

GENERAL CONSIDERATIONS

Clinical Indications for Hip US

- US is very useful investigation for most hip pathology; especially good for suspected hip effusion, synovitis, some adductor tendon pathology, and greater trochanteric pain
- US can be supplemented by MR, particularly for deep gluteal pain and small tendon tears around hip
- US also helpful to guide interventional procedures
- Full exam is preferable in most patients with poorly localized hip pain

TECHNIQUE: ANTERIOR HIP

Patient Position

- Supine on examination couch with leg extended; helpful to passively, internally or externally, rotate hip when assessing different structures

Specifically Examine

- **Anterior aspect of hip joint**
  - First identify rounded cortex of femoral shaft in midthigh and trace in transverse plane proximally
  - At intertrochanteric region, where bone contour is seen to widen, rotate medial aspect of transducer upward to align longitudinally along femoral neck
  - Anterior joint capsule is seen extending from acetabulum to intertrochanteric line; superficial capsular fibers extend over proximal femoral shaft, while deep fibers reflect backward to head/neck junction
  - Each collagenous layer is seen as 2-4 mm thick echogenic band, while inner synovium is too thin to be seen
  - Normal thickness of anterior recess from outer margin capsule to bone is 6-8 mm
  - If no effusion, both capsular layers are opposed ± thin intervening hypoechoic line (stripe sign) representing anterior hip recess
  - Normal contour of anterior capsule is concave or flat rather than convex in outline
- **Acetabular labral junction**
  - Move transducer proximally along femoral neck to head/acetabular margin region
  - Anterior and superior labrum seen as echogenic triangular structure below echogenic acetabular rim
  - Passively rotating hip can help with delineation
- **Iliopsoas tendon**
  - Seen on transverse imaging over femoral head, located just anterior to hip capsule and deep to femoral neurovascular bundle
    - Tendon located on deep margin of iliopsoas muscle
    - Frog-leg lateral position best for viewing length of iliopsoas tendon and insertion → lesser trochanter
  - Normal iliopsoas bursa between tendon and anterior capsule cannot be seen
- **Dynamic examination for snapping iliopsoas tendon**
  - Transducer in oblique transverse plane just proximal to hip joint
  - Patient is asked to reproduce symptoms or else move hip from flexion-abduction to external rotation to neutral position during real-time scanning in transverse plane
  - Normally, iliopsoas tendon and muscle will move laterally in smooth clockwise fashion
  - Snap usually produced and felt as sudden flip of iliopsoas tendon around iliacus muscle
    - Iliacus muscle fibers become temporarily trapped between tendon and superior pubic ramus
  - Asymptomatic snapping is quite common finding on testing normal subjects
- **Rectus femoris tendon**
  - On transverse and longitudinal scanning proximal to hip joint, characteristic rectus femoris tendon with direct and indirect heads
    - Indirect head arises from supraacetabular ridge, while direct head arises from anterior inferior iliac spine
    - Indirect tendon joins lateral part of direct tendon, causing posterior acoustic shadowing deep to lateral aspect of rectus femoris tendon
  - Evaluate rectus femoris tendon from both medial and lateral sides to see both direct and indirect tendons
  - More distally, within proximal rectus femoris muscle, indirect tendon → central aponeurosis, while direct tendon → superficial aponeurosis
  - Rectus femoris tendon is C-shaped
- **Anterior superior iliac spine region**
  - Observe inguinal ligament, sartorius medially, and tensor fascia lata muscle laterally
  - On transverse scanning, proximal ends of these muscles can have pseudothyroid appearance
  - On longitudinal scanning, short tendons and attachments of these muscles are best seen
  - Observe lateral femoral cutaneous nerve deep to lateral aspect of inguinal ligament
  - Nerve moves from medial to lateral over iliacus, then over sartorius to run between sartorius and tensor fascia lata
  - Sartorius muscle passes distally over femoral neurovascular bundle

TECHNIQUE: MEDIAL HIP

Patient Position

- Hip abducted to variable degree and externally rotated with knee flexed (frog-leg position)

Specifically Examine

- **Adductor tendons and muscles**
  - Scan transversely from pubic symphysis over pectineus muscle
  - Move transducer inferomedially on pectineus to see adductor muscles in short axis
    - Adductor longus (more superficial), adductor brevis, and adductor magnus (more deep) seen
  - Rotate transducer into long axis, maintaining it as parallel to adductor tendons as possible
  - Move transducer obliquely along inferior pubic ramus toward pubic bone and pubic symphysis
    - Observe common adductor tendon originating from pubic body and inferior pubic ramus
      - Adductor longus originates from pubic body just distal to pubic tubercle
      - Adductor brevis originates from medial aspect inferior pubic ramus
      - Adductor magnus originates from lateral aspect of inferior pubic ramus extending to ischial tuberosity

- Moving distally from origin, note 3 adductor muscles in long axis
  - Anterior and posterior obturator nerve branches lie between longus/brevis and brevis/magnus muscles, respectively
- Gracilis muscle arises just medial to adductor longus and descends more vertically in thigh

TECHNIQUE: LATERAL HIP

Patient Position

- Lateral decubitus or oblique lateral decubitus; flexing and extending hip for suspected iliotibial band impingement

Specifically Examine

- **Gluteus minimus and medius tendons**
  - On transverse scanning, move transducer proximally along femoral cortex to firefighter helmet-shaped greater trochanter
  - Gluteus minimus tendon → anterior facet of greater trochanter
    - For longitudinal scanning, align transducer toward middle 1/3 of iliac crest along gluteus minimus tendon
      - Normal subgluteus minimus bursa not seen
  - Gluteus medius tendon → larger, triangular, lateral facet and smaller, flatter, posterosuperior facet of greater trochanter
    - For longitudinal scanning, align transducer toward posterior 1/3 of iliac crest along gluteus medius tendon; normal subgluteus medius bursa not seen
    - Move transducer proximally and posteriorly to view posterosuperior attachment
  - Rounded posterior facet of greater trochanter is bare
- **Tensor fascia lata, iliotibial tract, and gluteal muscles**
  - On transverse plane, identify tensor fascia lata and iliotibial band at anterior superior iliac spine region
    - Iliotibial band = focal thickening of fascia lata from posterior part of tensor fascia lata muscle overlying anterior part of gluteus maximus muscle
  - Fascia lata descends over anterior aspect of gluteus maximus, lateral aspect of gluteus medius muscle, and greater trochanter
    - Scanning transversely while slightly extending hip, observe normal smooth movement of iliotibial tract over trochanter
    - Normal trochanteric bursa at posterolateral aspect of greater trochanter is not seen
  - Also, on transverse scanning, identify gluteus minimus, medius, and maximus muscles
    - Anterior margins of gluteus minimus and medius merge so that these muscles are more clearly defined posteriorly
    - Anterior fibers of gluteus maximus muscle cover posterior aspect of gluteus medius tendon
    - Gluteus maximus passes over greater trochanter and inserts → posterior aspect of proximal femoral shaft

TECHNIQUE: POSTERIOR HIP AND DEEP GLUTEAL SPACE

Patient Position

- Prone: Low-frequency transducer may be required
  - ± internally rotate hip to put short hip rotators on stretch
  - Firm downward transducer pressure usually required to examine tissues proximal to ischial tuberosity

Specifically Examine

- **Ischial tuberosity and hamstring tendon origins**
  - On transverse scanning, hamstring tendon origin area can be seen posterolaterally
  - Semimembranosus tendon lies deep and lateral
  - Semitendinosus/biceps femoris tendon lies superficially
  - Normal ischial tuberosity bursa deep to gluteus maximus is not seen
  - Sciatic nerve lies lateral to ischial tuberosity with posterior cutaneous nerve of thigh and inferior gluteal artery/nerve
- **Posterior thigh musculature just distal to ischial tuberosity**
  - Slowly scan in transverse plane down from ischial tuberosity
  - Identify semimembranosus tendon descending medially as large, thin, sheet-like, membranous aponeurosis
    - Tadpole-shaped membranous aponeurosis gives semimembranosus its name, lying deep to muscular portion of semitendinosus
    - No semimembranosus muscle fibers in proximal thigh
      - Muscle fibers arise from anterior aspect of aponeurosis more distally
  - Identify comma-shaped conjoint semitendinosus/biceps tendon lying more superficially
    - Semitendinosus muscle arises medially and slightly more proximally to that of biceps femoris, which arises laterally from conjoint tendon
    - Oblique central semitendinosus tendon has veil-like appearance
  - Identify fatty triangle formed by conjoint tendon, semimembranosus aponeurosis, sciatic nerve with adductor magnus at base
  - Identify sciatic nerve lying deep to biceps femoris muscle in midline posterior thigh
  - 2 heads to biceps femoris muscle; 2nd belly seen in distal thigh arising from posterior aspect of femoral cortex and joining biceps laterally
- **Deep gluteal space proximal to ischial tuberosity**
  - Scan transversely between more distal end of ischial tuberosity and posterior facet of greater trochanter (ischiofemoral space)
    - Observe quadratus femoris muscle with sciatic nerve superficially
  - Scanning slowly in transverse plane more proximally, one can observe
    - Sciatic nerve deep to gluteus maximus
    - Posterior aspect of hip joint
    - Twinned gemelli muscles with prominent tendon and short muscle belly and intervening obturator internus tendon
    - Sciatic nerve emerging from pelvis through greater sciatic notch just distal to piriformis muscle
  - On longitudinal scanning, long axis of sciatic nerve can be appreciated; moving probe proximally, sciatic nerve can be traced to greater sciatic foramen, where it emerges from pelvis; passes deep to gluteus maximus and superficial to short hip rotators

Anterior Aspect of Hip Joint

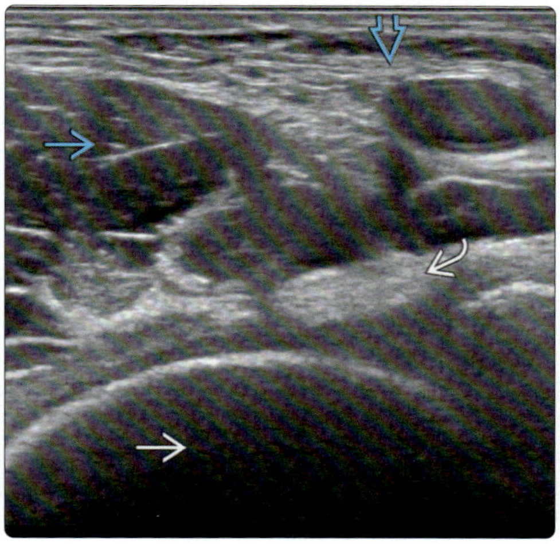

Femoral Nerve and Artery

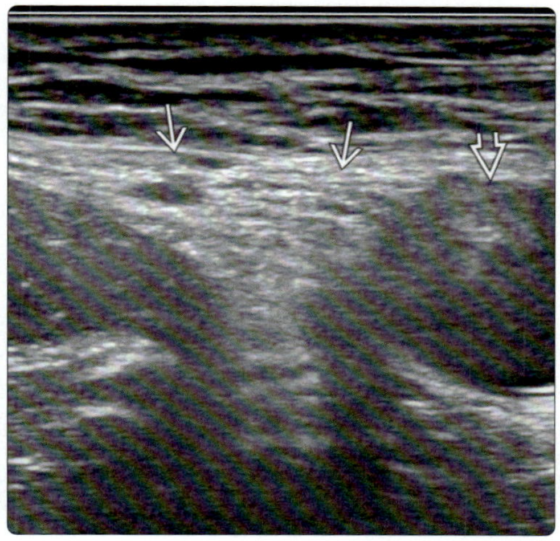

(Left) *Transverse US at the level of the femoral head shows the close relationship between the femoral head, the iliopsoas tendon and muscle, and the femoral neurovascular bundle.* **(Right)** *Transverse US shows how the femoral nerve, which lies lateral to the femoral artery, is indistinct and not as clearly outlined as most nerves. The femoral nerve divides into its many branches (pectineus, sartorius, quadriceps muscles as well as medial cutaneous and saphenous) 2-4 cm distal to the inguinal ligament.*

Rectus Femoris Tendon: Short Axis

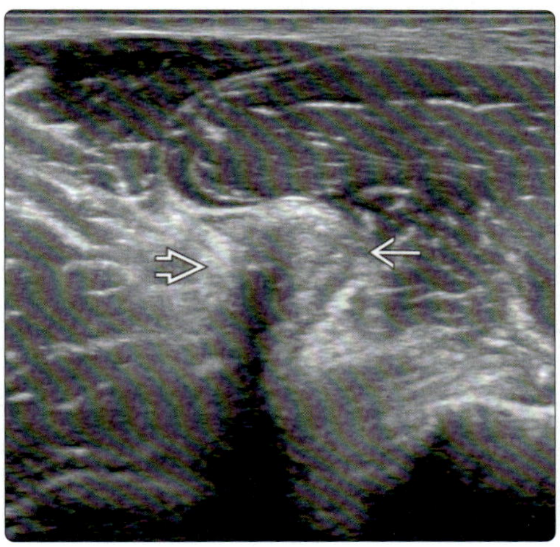

Direct Head of Rectus Femoris Tendon

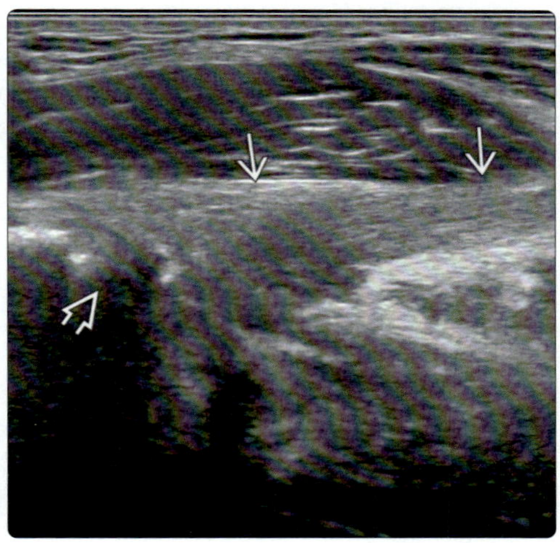

(Left) *Transverse US shows the direct and reflected heads of the rectus femoris tendon. The reflected head, which lies laterally, is hypoechoic due to anisotropy.* **(Right)** *Longitudinal US shows the direct head of the rectus femoris tendon inserted into the anterior inferior iliac spine. Scanning in a more oblique coronal plane allows one to reduce the effect of anisotropy from the reflected head.*

Indirect Head of Rectus Femoris Tendon

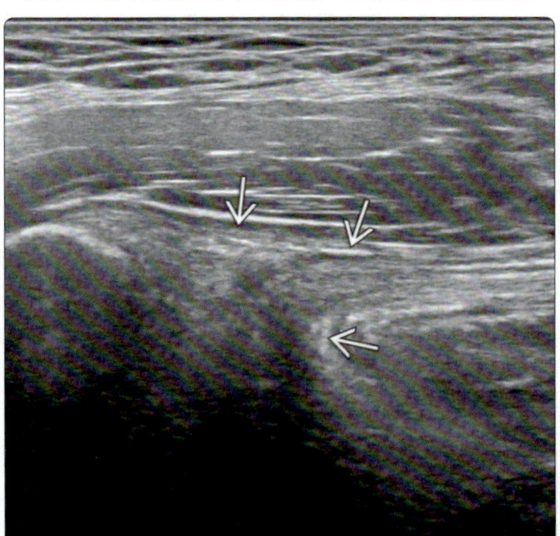

Superior Aspect of Hip Joint

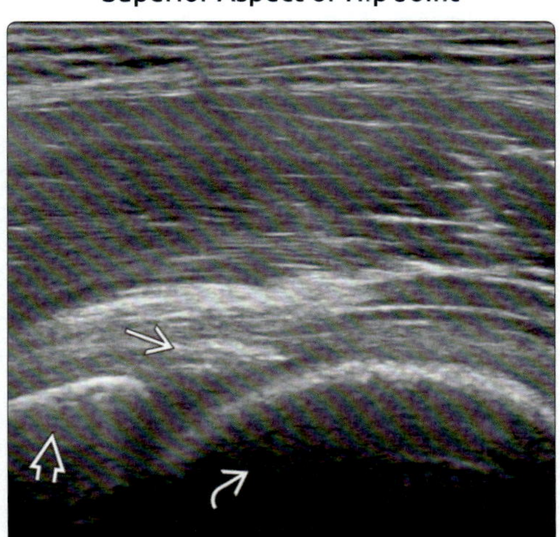

(Left) *Longitudinal US shows the indirect head of the rectus femoris tendon, which is hypoechoic due to fiber obliquity. These fibers are oblique to the scan plane, as they descend deeply to insert into the supraacetabular region.* **(Right)** *Longitudinal US just distal to the prior image shows the superior aspect of the hip joint with the acetabulum and acetabular labrum alongside the femoral head.*

(Left) *Transverse US close to the anterior superior iliac spine shows the lateral cutaneous nerve of the thigh between the inguinal ligament and the iliacus muscle.* **(Right)** *Transverse US just distal to the prior image shows the lateral cutaneous nerve of the thigh between the fascia lata and the sartorius muscle.*

Lateral Cutaneous Nerve of Thigh

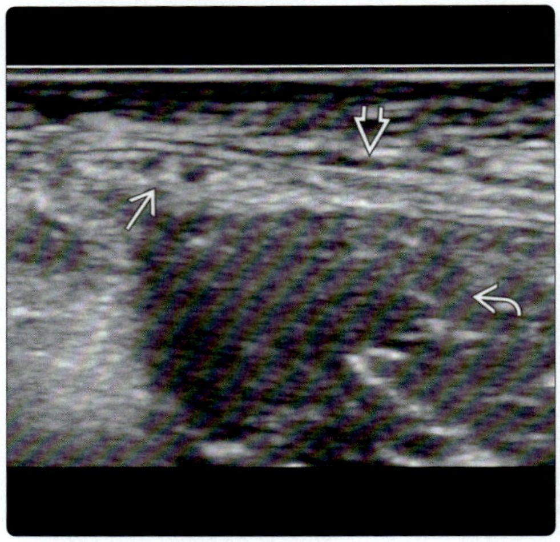

Lateral Cutaneous Nerve of Thigh

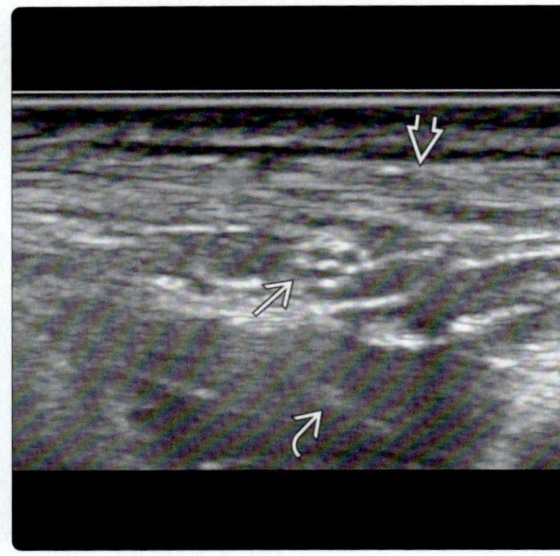

(Left) *Oblique longitudinal US shows a close relationship between the capsule overlying the femoral head and the iliopsoas tendon.* **(Right)** *Oblique longitudinal US shows the iliopsoas tendon insertion into the lesser tuberosity of the proximal femur. The normal iliopsoas bursa between the tendon and the hip joint is not visible.*

Iliopsoas Tendon and Hip Joint

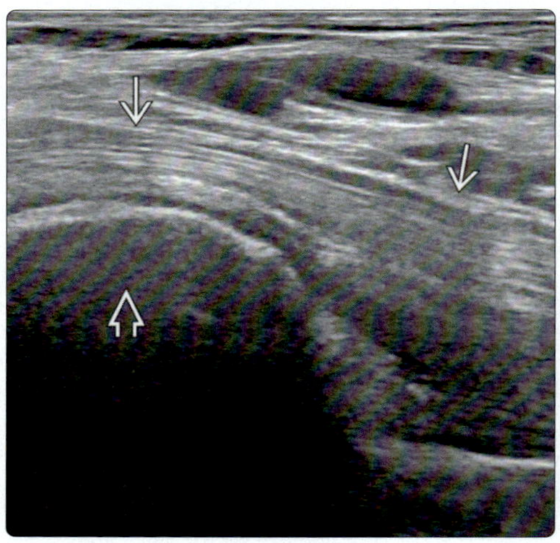

Iliopsoas Tendon Insertion

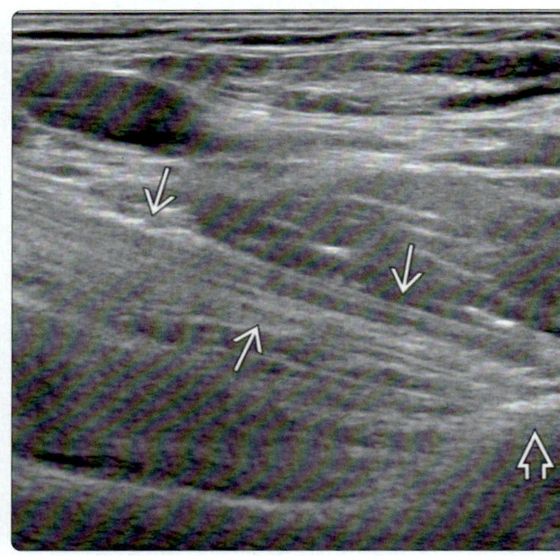

(Left) *Clinical photograph shows the patient and probe position for examination of the lateral hip. The patient lies in a decubitus position. Flexing and extending the limb under examination can be helpful to demonstrate iliotibial band impingement against the greater trochanter.* **(Right)** *Graphic shows facets of the greater trochanter of the proximal femur. The gluteus minimus is inserted into the anterior facet, while the gluteus medius is inserted into the lateral and posterosuperior facets. The posterior facet is bare.*

Scan Position for Lateral Hip

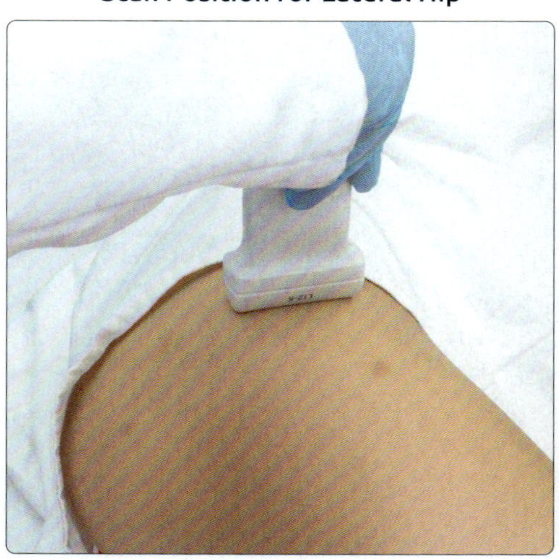

Facets of Greater Trochanter

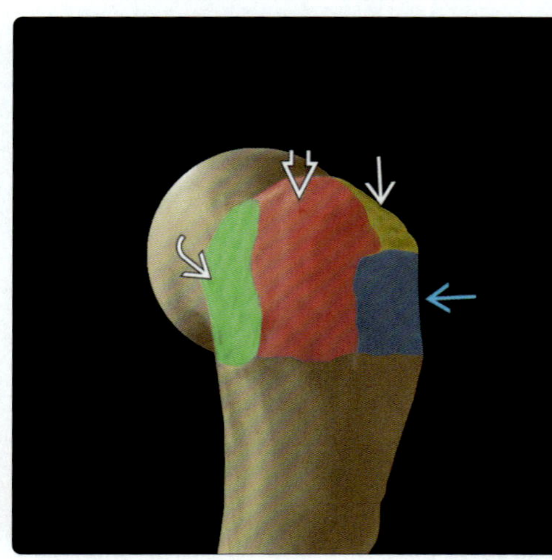

Facets of Greater Trochanter

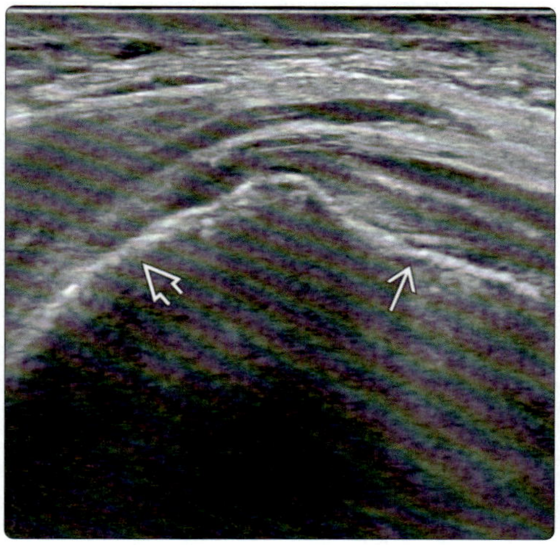

Gluteus Minimus Insertion

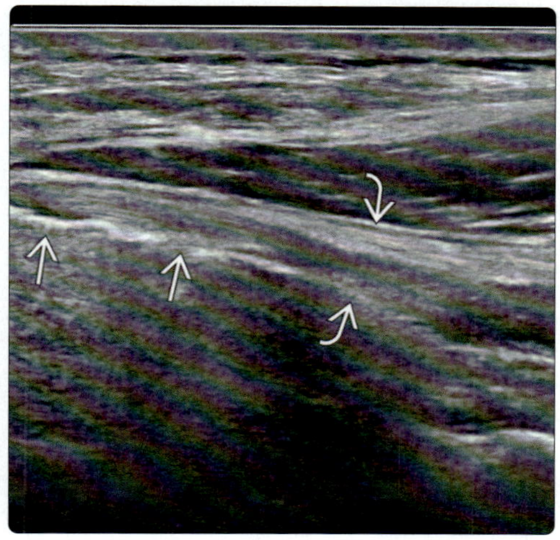

(Left) *Transverse US shows the firefighter helmet configuration of the anterior and lateral facets of the greater trochanter.* **(Right)** *Oblique longitudinal US shows the gluteus minimus musculotendinous junction and tendon insertion into the anterior facet of the greater trochanter.*

Gluteus Medius Insertion

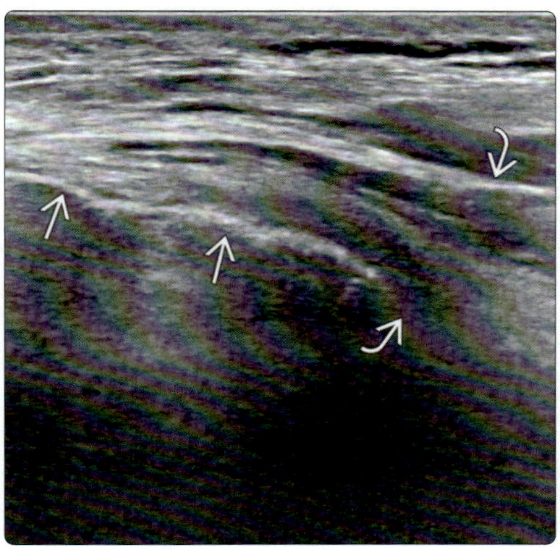

Bare Posterior Facet

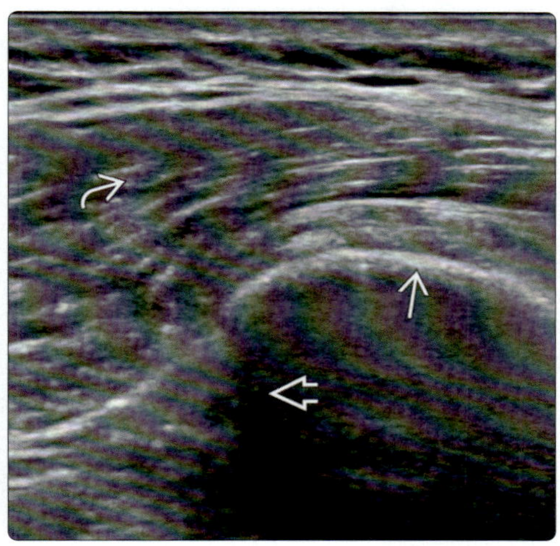

(Left) *Oblique longitudinal US shows gluteus medius musculotendinous junction and tendon insertion into the lateral facet of the greater trochanter.* **(Right)** *Transverse US shows the bare posterior facet of the greater trochanter. The gluteus medius tendon inserted into the lateral facet is shown. This insertion is covered by the gluteus maximus muscle.*

Gluteus Medius Insertion

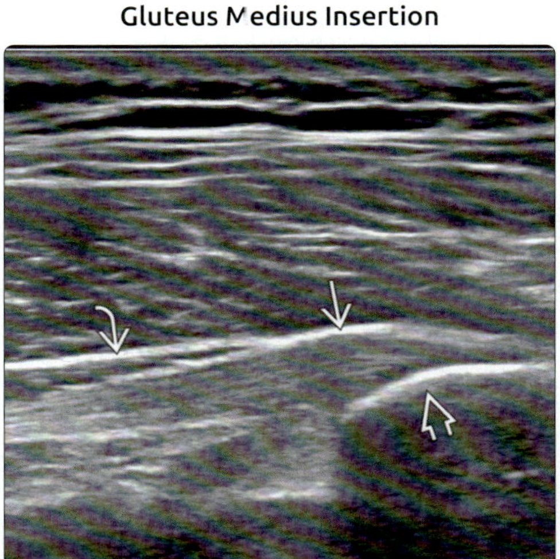

Iliotibial Tract

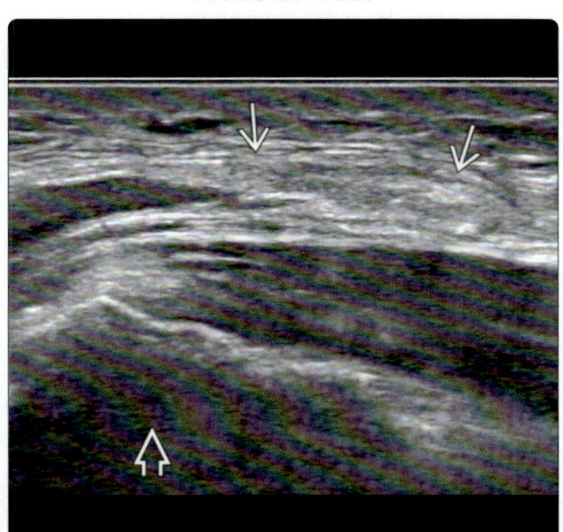

(Left) *Longitudinal US shows the gluteus medius musculotendinous junction and tendon insertion into the posterosuperior facet of the greater trochanter. The gluteus medius has 2 tendinous insertions, a large one into the lateral facet and a smaller one into the posterosuperior facet.* **(Right)** *Transverse US at the lateral facet of the greater trochanter shows the thick iliotibial band overlying the greater trochanter. The terms iliotibial tract and iliotibial band can be used synonymously.*

(Left) *Clinical photograph shows the scan position for the medial hip. The limb is placed in an abducted, externally rotated position. The adductor muscle is scanned in both transverse and longitudinal planes. This is also a good position for scanning the iliopsoas tendon.* **(Right)** *Transverse US shows the 3 adductor muscles lateral to the pectineus muscle. These are the adductor longus, adductor brevis, and adductor magnus from superficial to deep.*

Scan Position for Medial Hip

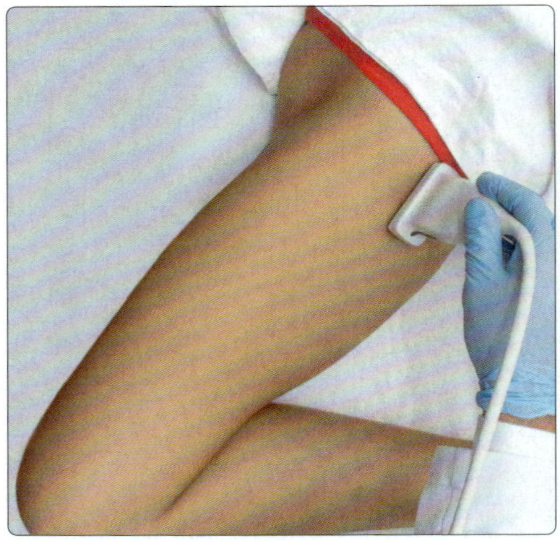

Pectineus and Adductor Muscles: Short Axis

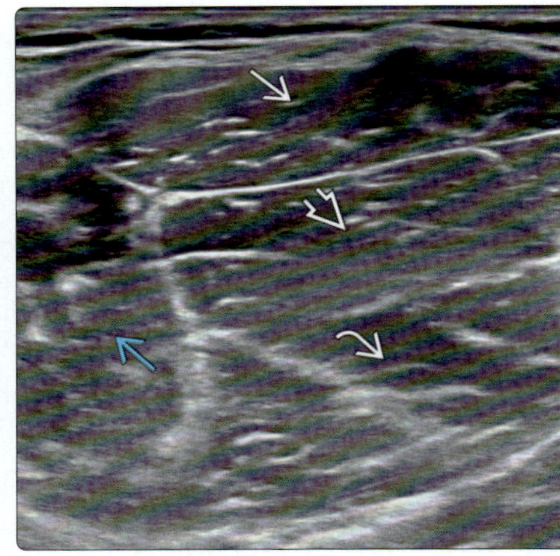

(Left) *Longitudinal US shows adductor muscles diverging from their origin on the pubic body and inferior pubic ramus.* **(Right)** *Longitudinal US shows the adductor muscles diverging from their origin. The anterior branch of the obturator nerve lies between the adductor longus and the adductor brevis, while the posterior branch lies between the adductor brevis and the adductor magnus.*

Adductor Muscle Origin: Long Axis

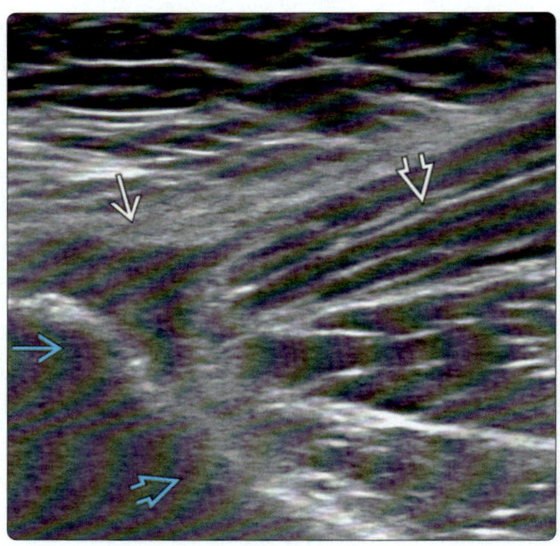

Adductor Muscles: Long Axis

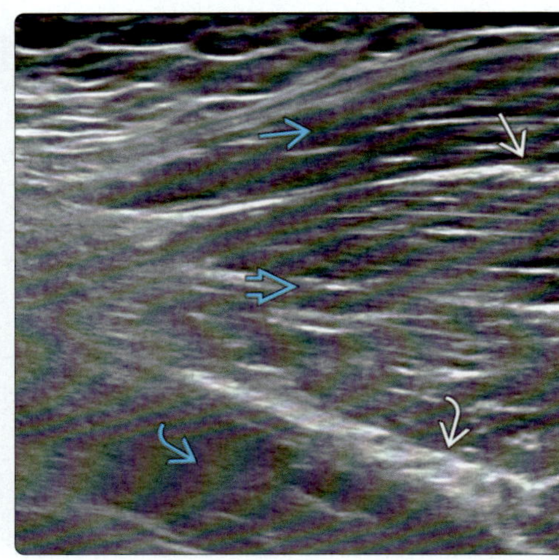

(Left) *Transverse US of the medial thigh shows the adductor longus, adductor brevis, and adductor magnus muscles. The gracilis muscle lies more superficially.* **(Right)** *Clinical photograph shows the patient and probe position for the posterior hip region. The patient lies prone. Especially in larger patients, deep pressure on the transducer is necessary to see the posterior hip structures more proximal to the ischial tuberosity.*

Adductor Muscles

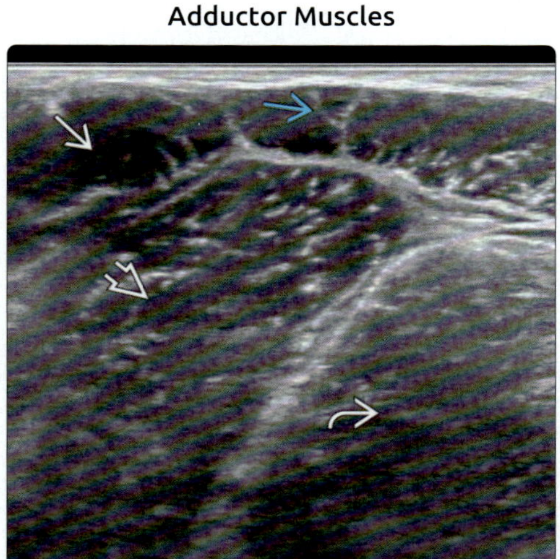

Scan Position for Posterior Hip

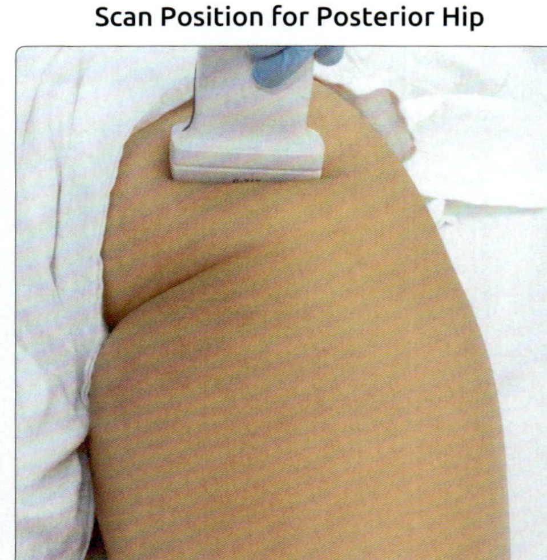

Sciatic Nerve

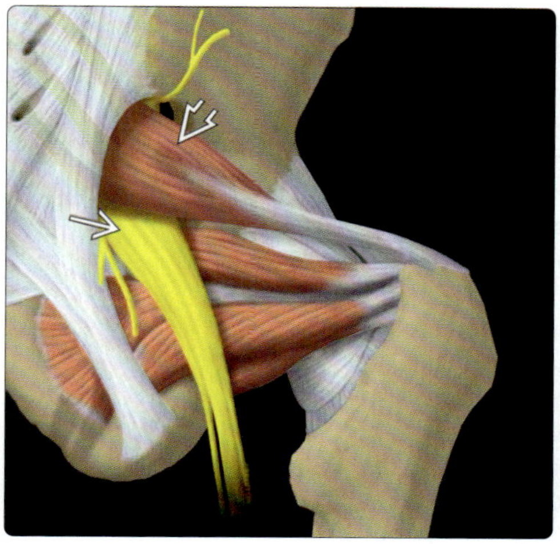

Sciatic Nerve

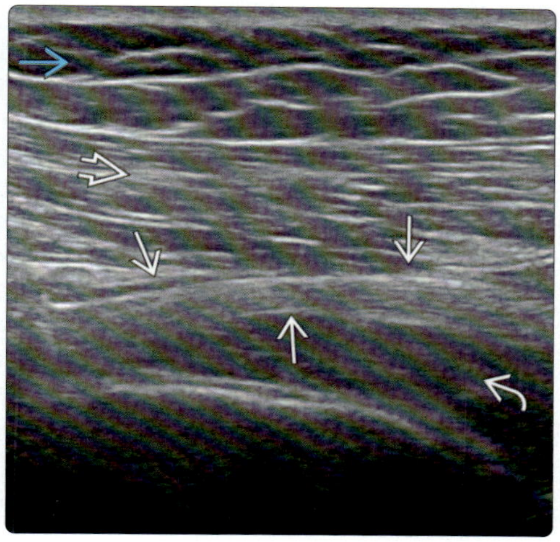

(Left) *Schematic shows deep gluteal region anatomy with the sciatic nerve ➡ exiting the pelvis below the piriformis muscle ➡. The sciatic nerve passes superficial to the short hip rotators (gemelli, obturator internus, and, more distally, quadratus femoris).* **(Right)** *Longitudinal US of the buttock region shows sciatic nerve ➡ emerging from sciatic foramen to descend the thigh. Gluteus maximus ➡ and the thick subcutaneous fat of the buttock ➡ lie superficially. The short rotators ➡ of the hip lie deep to the sciatic nerve.*

Deep Gluteal Region

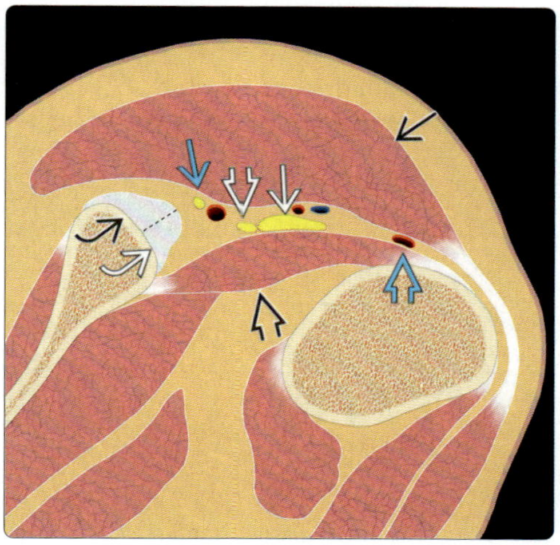

Ischiofemoral Space

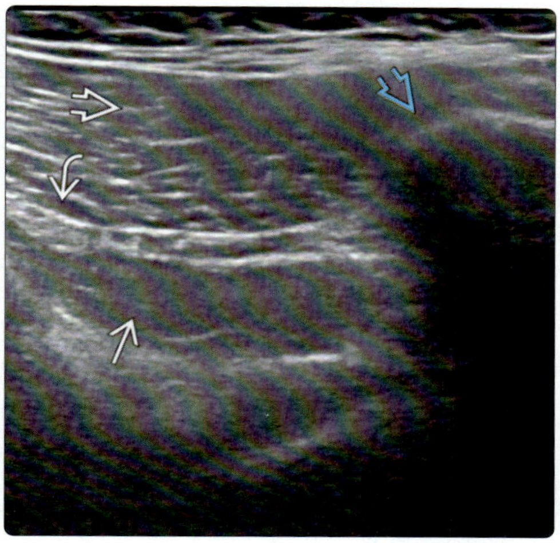

(Left) *Graphic shows deep gluteal anatomy with gluteus maximus ➡ and quadratus femoris muscles ➡, sciatic nerve ➡, posterior cutaneous nerve of thigh ➡, inferior gluteal nerve and artery ➡, ascending circumflex femoral artery ➡, biceps-semitendinosus ➡, and semimembranosus ➡ ischial tuberosity attachment.* **(Right)** *Transverse US of posterior thigh proximal to ischial tuberosity shows quadratus femoris ➡ & gluteus maximus muscles ➡ with intervening sciatic nerve ➡. Greater trochanter ➡ lies laterally.*

Ischial Tuberosity

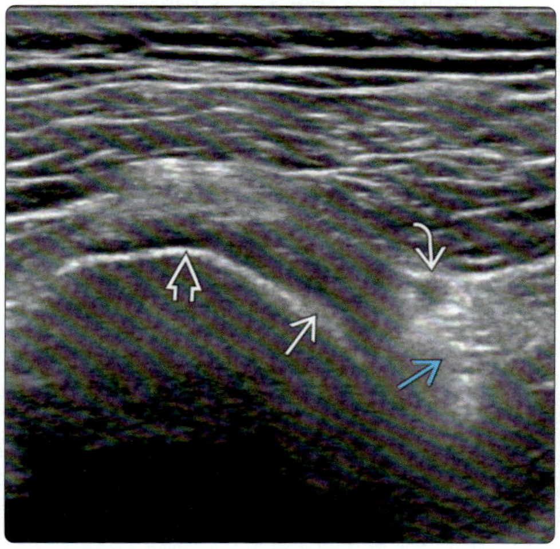

Proximal Thigh Musculature

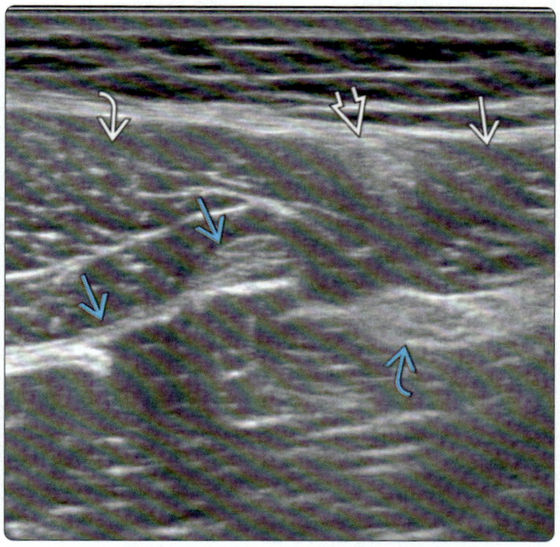

(Left) *Transverse US shows tendon attachment to ischial tuberosity. Conjoint semitendinous/biceps tendon ➡ attaches posteriorly, while semimembranosus tendon ➡ attaches laterally. Sciatic nerve ➡ and inferior gluteal artery ➡ are located lateral to tuberosity.* **(Right)** *Posterior aspect of proximal thigh shows conjoint biceps/semitendinous tendon ➡ with biceps muscle ➡ laterally and semitendinous muscle ➡ medially. The sciatic nerve ➡ and semimembranosus tendon ➡ lie deep to these muscles.*

KEY FACTS

GENERAL CONSIDERATIONS

- Clinical indications for knee US
  - Knee effusion, synovitis, Baker cyst, medial collateral ligament (MCL) injury, parameniscal cyst, paraarticular ganglion, quadriceps/patellar tendinosis/tear
- Less common specific indications
  - Pes anserinus and other bursitis, tendinosis/tear of other tendons besides patella and quadriceps tendons, gout, lateral collateral ligament injury, and peroneal nerve pathology

TECHNIQUE

- **Anterior knee**
  - Patient lying supine on examination table ± slight flexion
    - Anterior knee joint
    - Quadriceps tendon
    - Patellar bone and patellar tendon
    - Patellar retinacula
    - Anterior cruciate ligament
- **Medial knee**
  - Supine with knee externally rotated
    - MCL
    - Body of medial meniscus
    - Semimembranosus tendon
    - Pes anserinus tendons (semitendinosus, gracilis, and sartorius)
- **Lateral knee**
  - Lateral decubitus with knee extended
    - Distal iliotibial band
    - Lateral collateral ligament
    - Popliteus tendon
    - Biceps femoris tendon
- **Posterior knee**
  - Prone with knee extended
    - Gastrocnemius-semimembranosus bursa (Baker cyst)
    - Tibial and peroneal nerves
    - Posterior cruciate ligament

Scan Position for Anterior Knee

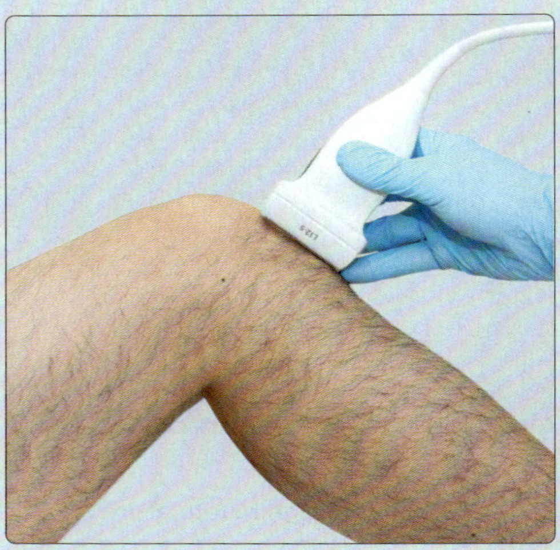

Quadriceps Tendon

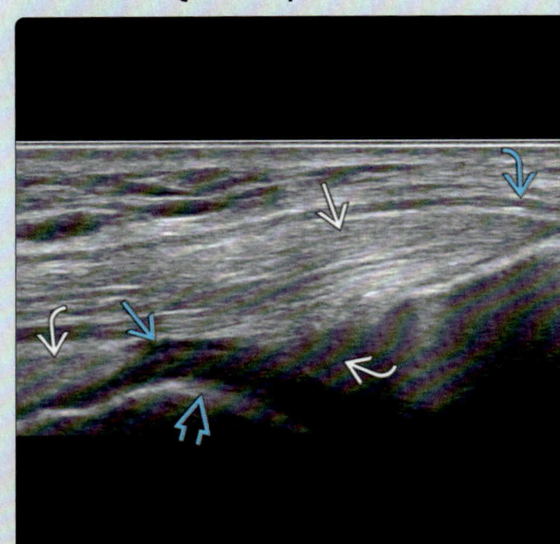

(Left) *Clinical photograph shows scan position for anterior knee, especially for patellar and quadriceps tendons. To assess fluid in patellar recesses, the knee is placed flat.* **(Right)** *Longitudinal US shows multilaminated quadriceps tendon inserting into superior pole of patella. Suprapatellar and prefemoral fat pads are visible as is trochlear cartilage. Suprapatellar recess is not distended. Most superficial lamina passes superficial to patella as the prepatellar quadriceps continuation.*

Quadriceps Tendon: Short Axis

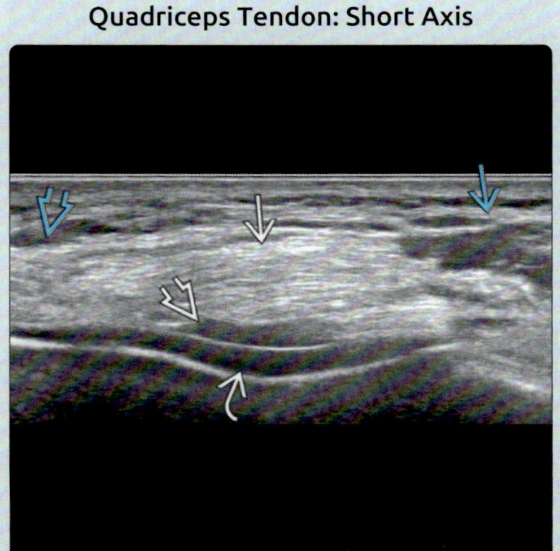

Prepatellar Quadriceps Continuation

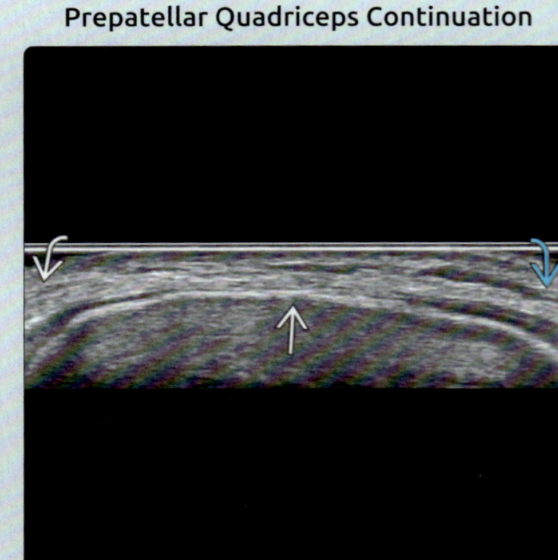

(Left) *Transverse US shows the quadriceps tendon, suprapatellar recess containing a small amount of fluid, and trochlear hyaline cartilage. The vastus medialis extends more distally than the vastus lateralis, which is tendinous at this level.* **(Right)** *Longitudinal US anterior aspect of patella shows how superficial fibers of the quadriceps tendon proximally pass superficial to the patella as the "prepatellar quadriceps continuation" to connect with the superficial fibers of the patellar tendon distally.*

GENERAL CONSIDERATIONS

Clinical Indications for Knee US

- US is preferred investigation for all suspected extraarticular knee pathology as well as joint effusion and synovial proliferation
- Common specific indications are knee effusion, synovitis, Baker cyst, medial collateral ligament (MCL) injury, parameniscal cyst, paraarticular ganglion, quadriceps/patellar tendinosis/tear
- Less common specific indications are pes anserinus and other bursitis, tendinosis/tear of other tendons besides patella and quadriceps tendons, lateral collateral ligament (LCL) injury, and peroneal nerve pathology
- Guides diagnostic and therapeutic intervention

TECHNIQUE: ANTERIOR KNEE

Patient Position

- Patient lying supine on examination table
  - Slight flexion (30°) helps minimize anisotropy of quadriceps and patellar tendons
    - More flexion (45-60°) is required to see anterior cruciate ligament (ACL)
    - Full flexion best for seeing femoral trochlear cartilage
  - Best to assess color flow in relaxed, extended position
  - Standing can accentuate meniscal extrusion and joint space narrowing
  - Standing or decubitus positions can help appreciate mobility of intraarticular bodies
  - Squatting may accentuate small parameniscal cysts of anterior horn lateral meniscus
  - Full extension of relaxed knee allows medial subluxation of patella to view medial retropatellar articular cartilage

Specifically Examine

- **Anterior knee joint**
  - Transverse scanning at medial and lateral aspects of patella to view medial and lateral patellar recess
    - Assess joint fluid and synovial thickening
      - Normal synovium is too thin to see clearly
  - Longitudinal scanning of suprapatellar region to view suprapatellar recess
    - Notice suprapatellar and prefemoral fat pads
      - Joint recess lies between these 2 pads
    - Nondistended suprapatellar recess is seen as thin, hypoechoic line
    - If distended, suprapatellar plica may be seen
  - Fluid tends to collect more in parapatellar recesses rather than suprapatellar recess
    - Mild fluid distension of 1 recess is normal while fluid seen in ≥ 2 recesses indicates effusion
- **Quadriceps tendon**
  - Scan longitudinally and transversely
  - Quadriceps tendon is taut with knee slightly flexed and lax when knee is extended
    - Tendon has multilayered (3-4 layers) appearance though often only superficial layer is distinct
    - Rectus femoris → superficial layer
      - Extends over surface of patella as "prepatellar quadriceps continuation"
    - Vastus medialis and lateralis → intermediate layer
      - These 2 muscles also insert into superomedial and superolateral aspects of patella
      - Musculotendinous junction of vastus medialis lower than that of vastus lateralis
    - Vastus intermedius → deep layer
- **Patellar bone and patellar tendon**
  - Examine in longitudinal and transverse planes
  - Proximal and distal ends of patellar tendon expand slightly
  - More superficial fibers of patellar tendon are continuous "prepatellar quadriceps continuation"
  - Hoffa fat pad seen deep to patellar tendon
  - Below distal end of patellar tendon is deep infrapatellar bursa
    - Small amount of fluid in this bursa is normal
  - Normal prepatellar and superficial infrapatellar bursa cannot be seen
- **Patellar retinacula**
  - Scan in transverse plane over medial and lateral patellar recesses to see bilaminar patellar retinacula
  - Follow posteriorly to insertion into MCL medially and iliotibial band laterally
  - Medial patellofemoral ligament (MPFL) is most important medial stabilizer of patellofemoral joint
    - Contributes > 50% of medial patellar stability
    - Extends from superomedial aspect patella → Nomura point between medial epicondyle and adductor tubercle
    - Wraps over femoral attachment of MCL on way to Nomura point
    - Visible on US as 1- to 3-layered structure that fuse posteriorly
    - Patellar portion is thicker than tapering femoral portion
- **Femoral trochlear cartilage**
  - Knee fully flexed to see femoral trochlear cartilage in transverse and longitudinal planes
  - Knee extended, relaxed, and patella moved medially to show periphery of medial retropatellar cartilage
- **Anterior cruciate ligament**
  - Scanning longitudinally with knee flexed to 45-60°
  - Place transducer longitudinally below patella and rotate toward medial aspect lateral femoral condyle
  - Distal portion of ACL can be seen but is not routinely assessed with US

TECHNIQUE: MEDIAL KNEE

Patient Position

- Supine with knee externally rotated
  - Placing knee on firm pad can help valgus stretch for medial knee ligaments and joint space

Specifically Examine

- **Medial collateral ligament**
  - Examined in both longitudinal and transverse planes
  - 3-layered structure, but uppermost layer (deep fascia) is not visible on US
    - Superficial layer extends from medial femoral epicondyle → tibia 5 cm below medial joint line
      - This is 2-layered hyperechoic band that is most characteristically recognized as MCL on US

- Posteriorly, this layer merges with posterior oblique ligament
- Proximally, MPFL wraps over femoral attachment of MCL
  - Deep layer comprises meniscofemoral and meniscotibial bands
    - Both bands are tightly attached to medial meniscal body but are loosely attached to superficial layer of MCL
- **Body of medial meniscus**
  - Normal medial meniscus is separated from MCL by loose hypoechoic connective tissue, which is usually echo poor
    - Meniscal degeneration, deformation, and extrusion can be appreciated and accentuated on weight bearing
- **Semimembranosus tendon**
  - Scan in transverse plane over posteromedial knee to identify large semimembranosus tendon
  - Follow tendon distally to see direct arm inserting into shallow groove at posteromedial corner tibia
    - 3 other insertions of semimembranosus are anterior arm, capsular arm, and oblique arm
      - Anterior arm → medial tibia deep to MCL
      - Capsular arm → posterior oblique ligament
      - Oblique arm → oblique popliteal ligament
    - All of these arms can be seen on US but not routinely assessed with US
- **Pes anserinus tendons (semitendinosus, gracilis, and sartorius)**
  - Best scanned in transverse plane
  - Identify semitendinosus at posteromedial aspect knee
    - Seen as "cherry on pie" with semitendinosus tendon sitting atop semimembranosus muscle
  - Follow transversely to insertion at anteromedial aspect proximal tibia
  - Identify other 2 pes anserinus tendons in transverse plane and follow distally to insertion
  - Alternatively, follow MCL in longitudinal plane distally
    - At distal end of tibial MCL insertion, move transducer anteriorly to see pes anserinus tendons
  - Pes anserinus tendons are arranged linearly with fleshy sartorius anteriorly and semitendinosus posteriorly
    - Slender gracilis tendon lies between sartorius and semitendinosus tendons

TECHNIQUE: LATERAL KNEE

Patient Position

- Lateral decubitus with knee extended ± firm pad under knee to help varus stretch for lateral knee ligament and joint space
- Supine and internally rotated for anterolateral structures
- Prone and externally rotated for posterolateral structures

Specifically Examine

- **Distal iliotibial band**
  - Best examined mainly in longitudinal plane
  - Thin, tubular structure inserts into Gerdy tubercle anterolateral aspect proximal tibia
    - Iliotibial band fans out before insertion
  - Proximally, passes close to lateral femoral condyle
- **Lateral collateral ligament**
  - Scan longitudinally with knee slightly flexed (30°), as this stretches ligament
  - Place transducer over fibular head and then rotate proximal end of transducer ~ 20° to align along LCL
    - LCL is thicker, more rounded than MCL and fans out at both femoral and fibular attachments
- **Popliteus tendon**
  - Located using longitudinal plane within popliteal sulcus deep and just anterior to femoral attachment LCL
  - Mild knee flexion can improve visibility of proximal end of popliteus tendon
- **Biceps femoris tendon**
  - Scan longitudinally posterolateral aspect of knee
  - Biceps femoris shares common insertion with LCL to posterolateral aspect of proximal fibula
    - Short and long heads of biceps femoris each have 2 tendon insertions (direct and anterior heads)
    - Direct heads of both tendons → fibular head

TECHNIQUE: POSTERIOR KNEE

Patient Position

- Prone with knee extended
- Flexion of knee may improve visibility of Baker cyst pedicle

Specifically Examine

- **Gastrocnemius-semimembranosus (Baker cyst)**
  - Identify medial margin of medial belly gastrocnemius in calf
    - Scan proximally along this medial margin in transverse plane to junction with semimembranosus tendon medially
    - Pedicle of Baker cyst, if present, will be located between these 2 structures
    - Anisotropy is common at gastrocnemius muscle-semimembranosus tendon interspace
      - Do not confuse this with small Baker cyst
- **Tibial and peroneal nerves**
  - Tibial nerve located superficial to popliteal artery and vein
    - Peroneal nerve branches off lateral side of tibial nerve, usually in distal thigh
  - Peroneal nerve best identified on transverse scanning, either arising from parent nerve or deep to distal part of biceps femoris muscle
    - Follow peroneal nerve transversely around fibular head where it divides into deep and superficial branches
- **Posterior cruciate ligament (PCL)**
  - Scan in oblique longitudinal plane with proximal end of transducer aligned 30° toward medial side
  - Distal 2/3 of PCL can be seen with US
    - Seen as thick, hypoechoic, cord-like structure inserted into posterior part of proximal tibia
- **Posterolateral corner structures**
  - Fabellofibular ligament
  - Popliteofibular ligament
  - Arcuate ligament
  - All of these small structures can be seen but are not routinely assessed with US

Scan Position for Medial Patellar Recess

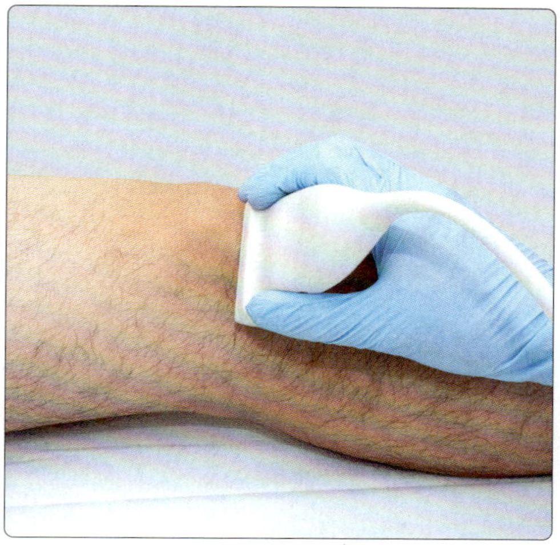

Lateral Patellar Recess

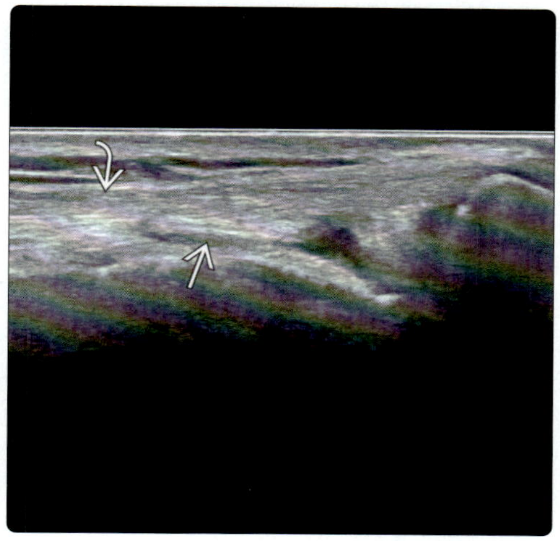

(Left) *Clinical photograph shows scan position for the medial patellar recess. The medial and lateral patellar recesses are best examined when the knee is extended or only slightly (10°) flexed.* **(Right)** *Transverse US shows the lateral patellar recess deep to the lateral retinaculum. The slightly hypoechoic normal synovium is seen along with a very small amount of joint fluid. This is normal. The medial patellar recess will have a similar appearance.*

Medial Patellofemoral Ligament

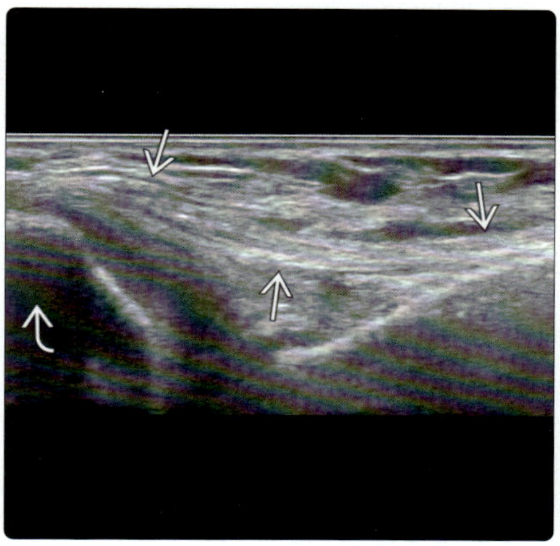

Patellar Tendon: Proximal Portion

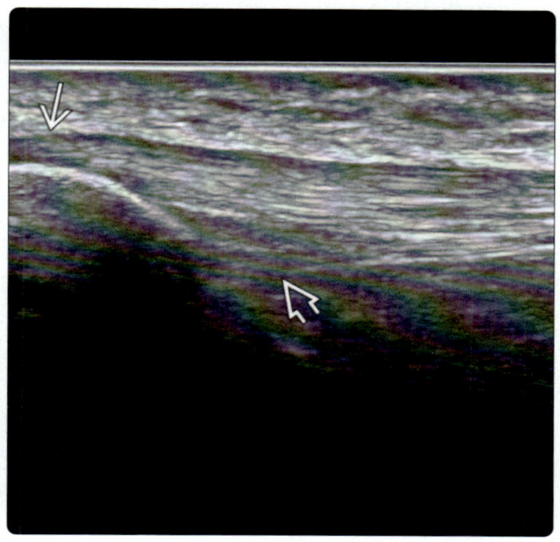

(Left) *Transverse US shows medial patellofemoral ligament (MPFL) extending from proximal medial aspect of patellar bone toward femoral attachment of medial collateral ligament.* **(Right)** *Longitudinal US shows normal proximal patellar tendon. Note how the superficial fibers extend over anterior surface of patella, while deep fibers are hypoechoic due to anisotropy. This is a normal appearance. Typical appearance and absence of hyperemia helps distinguish this hypoechogenic area from tendinosis.*

Patellar Tendon: Distal Portion

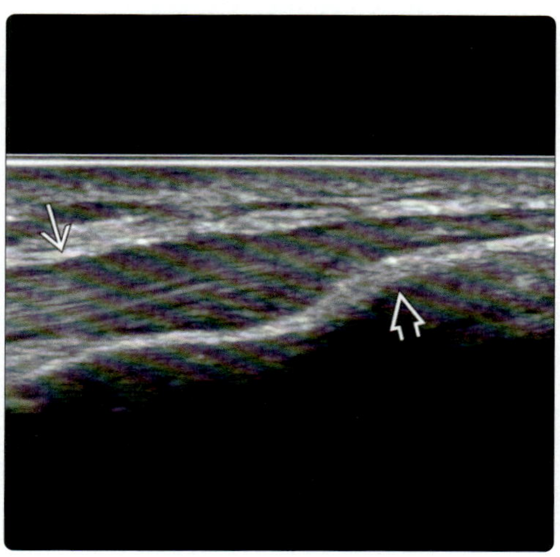

Patellar Tendon: Short Axis

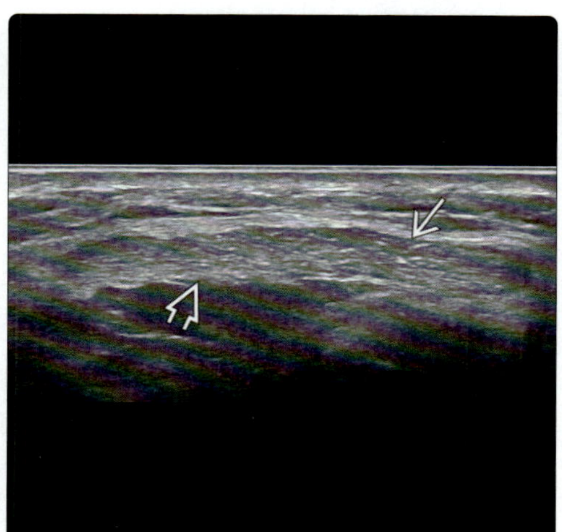

(Left) *Longitudinal US shows the distal part of the patellar tendon inserting into the anterior tibial tuberosity. The patellar tendon does normally widen at this level.* **(Right)** *Transverse US shows the transverse section of the midportion of the patellar tendon. The normal paratenon is seen as an echogenic layer anteriorly.*

(Left) *Transverse US shows normal femoral trochlea hyaline cartilage ➡. The quadriceps tendon ➡ lies close to the femoral trochlea.* **(Right)** *Longitudinal US obtained with the knee in full flexion shows the tibial portion of the anterior cruciate ligament (ACL) ➡. The probe is angled toward the lateral femoral condyle. US only provides a limited view of the ACL. This is not assessed in routine clinical practice.*

Femoral Trochlear Articular Cartilage

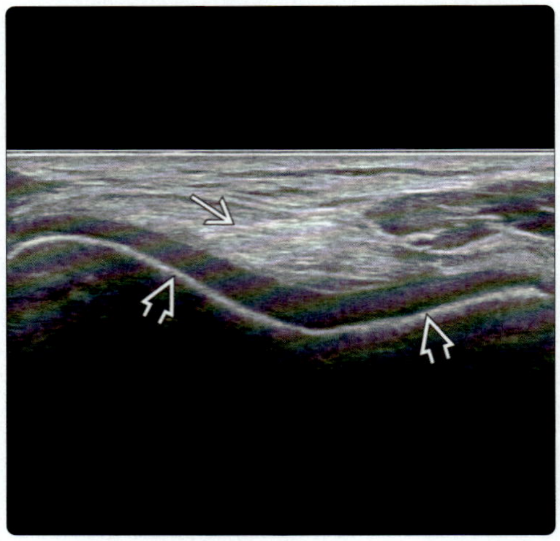

Anterior Cruciate Ligament

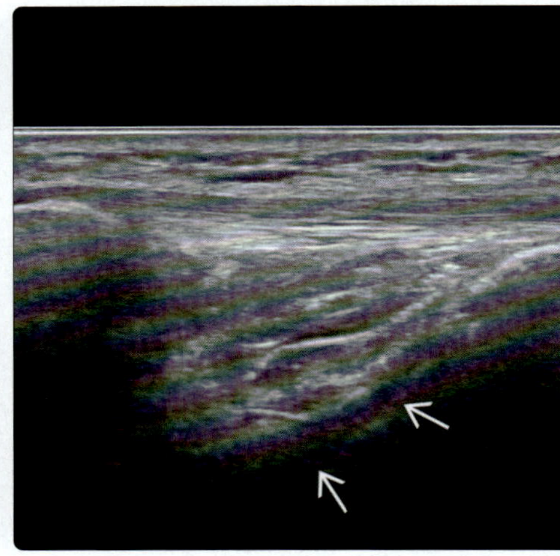

(Left) *Clinical photograph shows scan position for the medial knee, which is best examined in a decubitus position with the knee extended. From this position, rotate the knee posteriorly for anteromedial aspect and anteriorly for posteromedial corner.* **(Right)** *Longitudinal US shows the periphery of the body of the medial meniscus ➡ deep to the medial collateral ligament (MCL) ➡. Extrusion and deformation of the meniscus and present can be appreciated and accentuated by weight bearing.*

Scan Position for Medial Knee

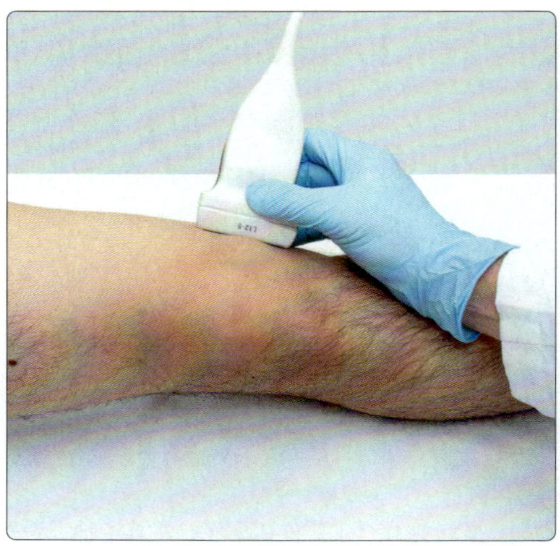

Medial Collateral Ligament

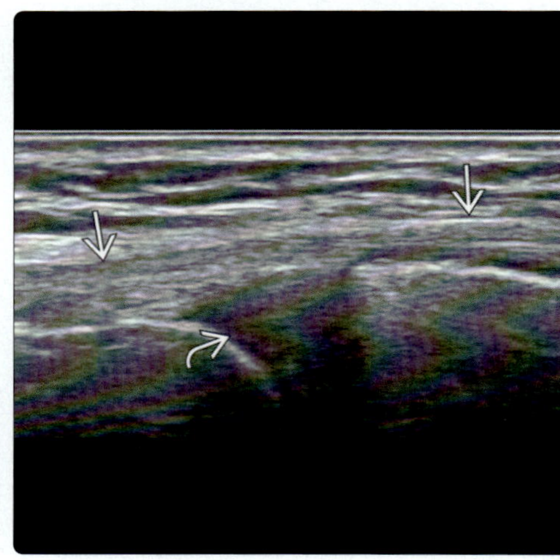

(Left) *Longitudinal US shows the femoral attachment ➡ of the MCL. This is a main side of MCL tear, especially the anterior part of the insertion. The meniscofemoral and meniscotibial bands are seen ➡ deep to the superficial component of the MCL ➡.* **(Right)** *Transverse US shows the MPFL ➡ passing superficial to the femoral attachment of the MCL ➡. The posterior oblique ligament ➡ is continuous with the posterior margin of the MCL attachment.*

Femoral Attachment of Medial Collateral Ligament

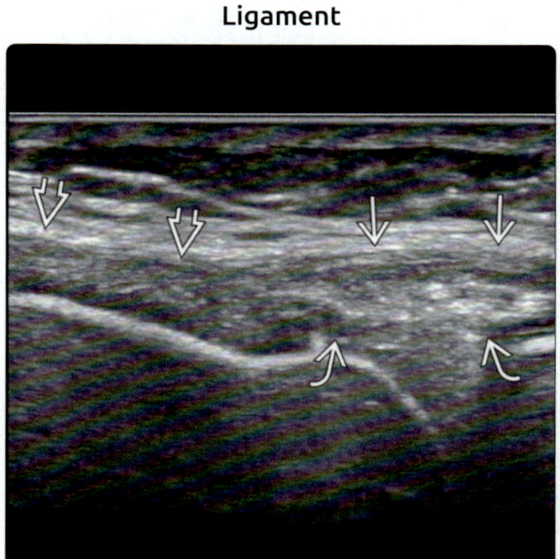

Femoral Attachment of Medial Collateral Ligament

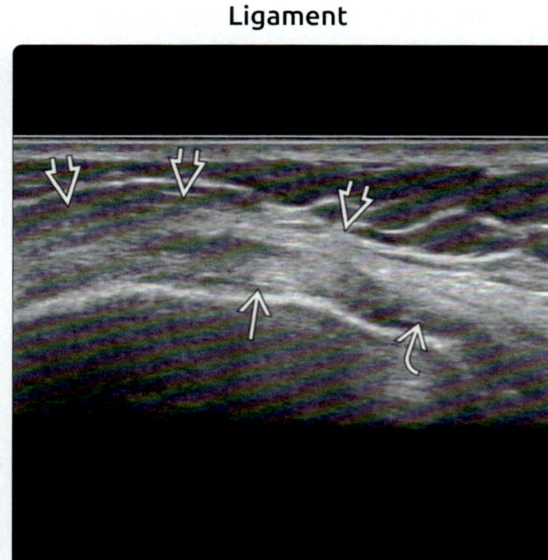

Distal Semimembranous Tendon

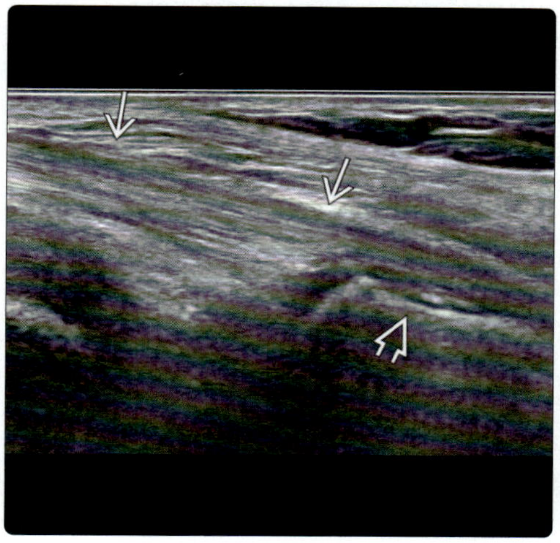

Pes Anserinus Tendons

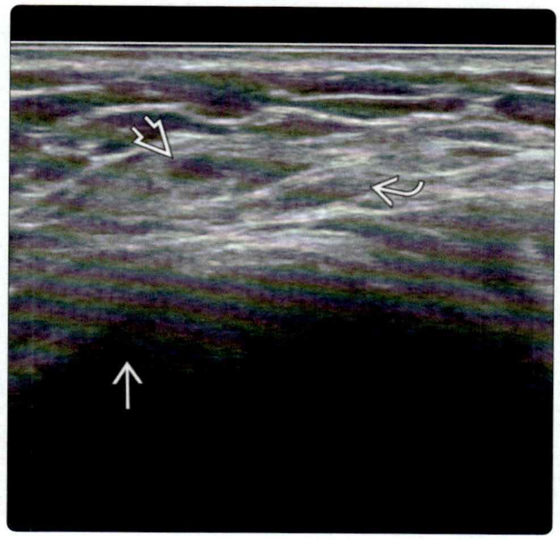

(Left) *Longitudinal US at posteromedial aspect of knee shows the direct arm of the semimembranosus tendon inserted into a shallow groove at the posteromedial aspect of the proximal tibia. In addition to the direct arm, the semimembranosus insertion has anterior, capsular, and oblique arms.* **(Right)** *Transverse US posteromedial corner of the knee shows the gracilis and sartorius tendons. These pes anserinus tendons, together with the semitendinosus, insert into the anteromedial aspect of the proximal tibia.*

Scan Position for Lateral Knee

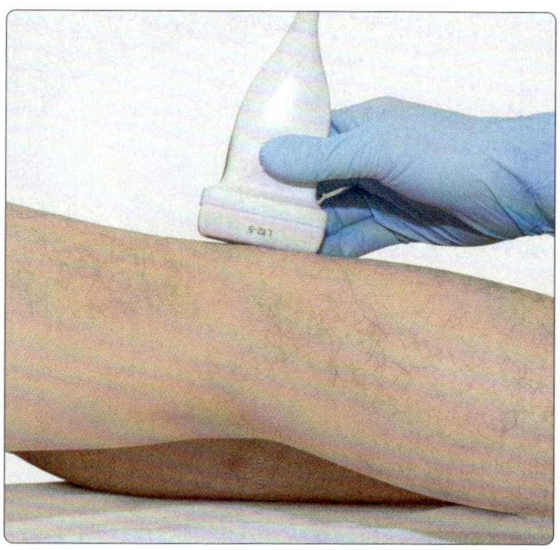

Iliotibial Band

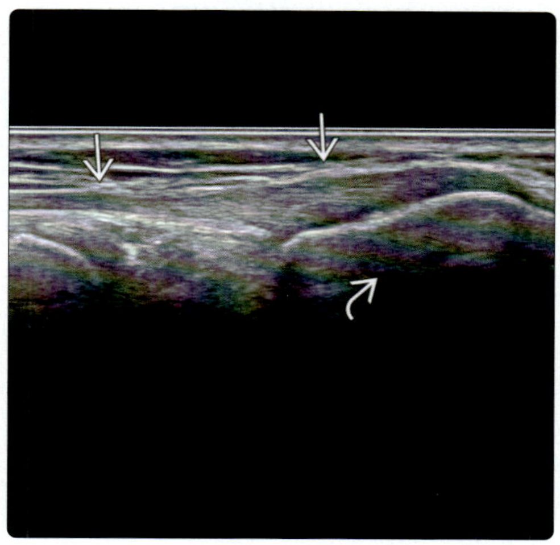

(Left) *Clinical photograph shows scan position for the lateral knee which is best examined in a decubitus position with the knee extended. From this position, rotate the knee posteriorly for anterolateral aspect and anteriorly for posterolateral corner.* **(Right)** *Longitudinal US shows a normal, thickened distal portion of the iliotibial band inserting into the anterolateral aspect of the proximal tibia (Gerdy tubercle).*

Iliotibial Band

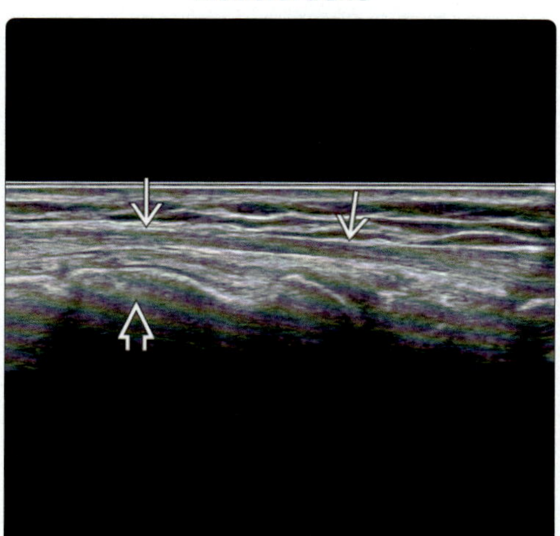

Lateral Collateral Ligament: Fibular Attachment

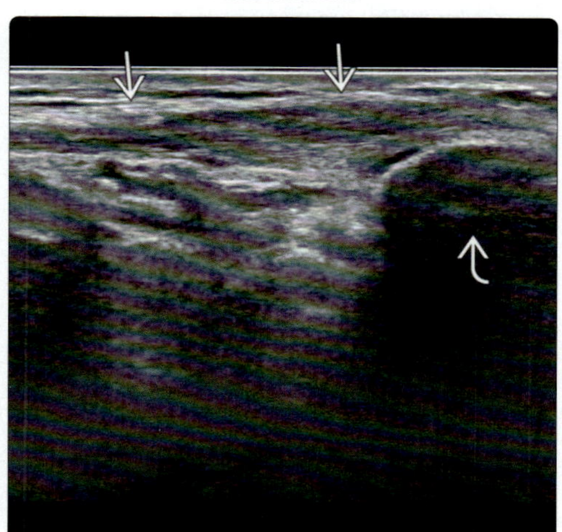

(Left) *Longitudinal US more proximal shows the iliotibial band passing alongside the lateral aspect of the lateral femoral condyle. Friction between the iliotibial band the femoral condyle can lead to iliotibial band friction syndrome.* **(Right)** *Longitudinal US shows a normal, thickened distal portion of lateral collateral ligament (LCL) attached to the proximal fibula. The proximal end of the transducer is angled ~ 20° anteriorly to align with the LCL.*

(Left) *Longitudinal US shows the proximal attachment of the LCL ➡ to the lateral aspect of the femoral condyle ➡ just proximal to the sulcus ➡ for the popliteus tendon attachment. The normal LCL fans out slightly at proximal and distal ends.* **(Right)** *Longitudinal US shows the proximal portion of the popliteus tendon ➡ attached to a groove ➡ in the lateral aspect of the femur just distal to the attachment of the LCL.*

Lateral Collateral Ligament: Femoral Attachment

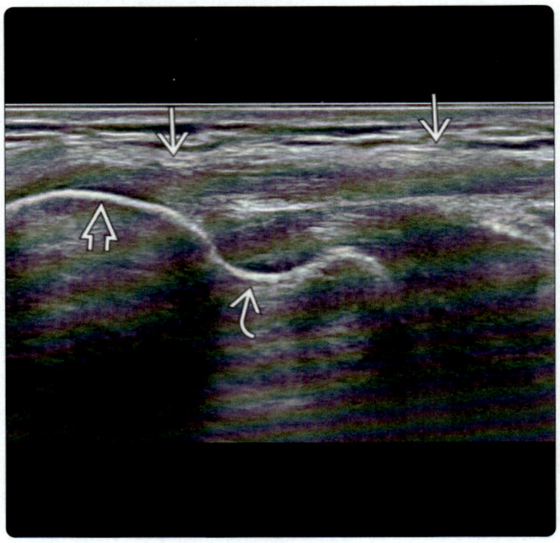

Popliteus Tendon

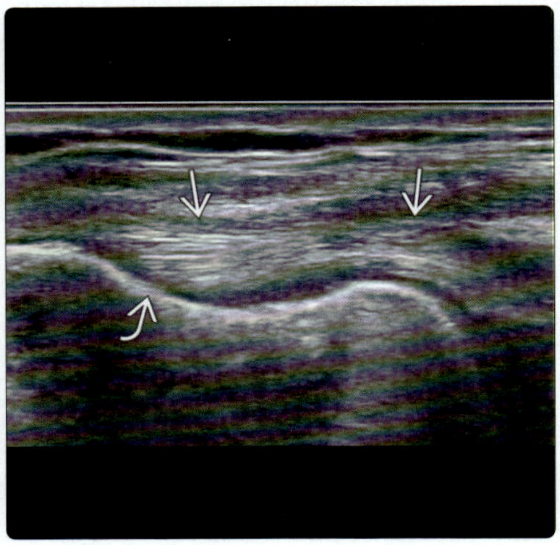

(Left) *Longitudinal US shows the broad attachment of the direct arm of the biceps femoris tendon ➡ to the posterolateral aspect of the proximal fibula ➡. This tendon is aligned vertically as opposed to the LCL, which aligns toward the lateral femoral condyle.* **(Right)** *Clinical photograph shows scan position for the posterior knee. This is best examined with the patient lying prone and the knee extended.*

Biceps Femoris Insertion

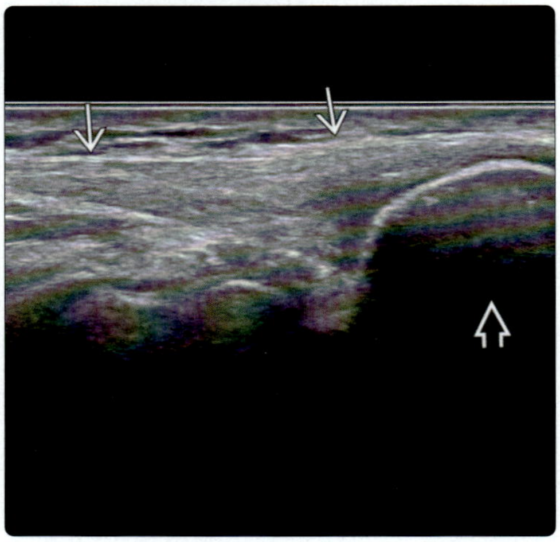

Scan Position for Posterior Knee

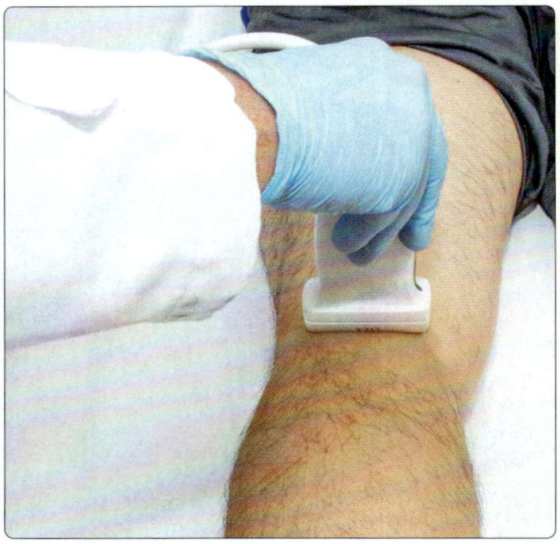

(Left) *Transverse US posteromedial aspect of the knee shows the potential space ➡ between the medial belly of the gastrocnemius ➡ and the semimembranosus ➡ and semitendinosus ➡ tendons through which Baker cysts extend.* **(Right)** *Transverse US shows the tibial nerve ➡ superficial to the popliteal artery and vein ➡ at the proximal end of the popliteal fossa. The peroneal nerve ➡ has already arisen from the lateral side of the tibial nerve.*

Gastrocnemius-Semimembranosus Recess

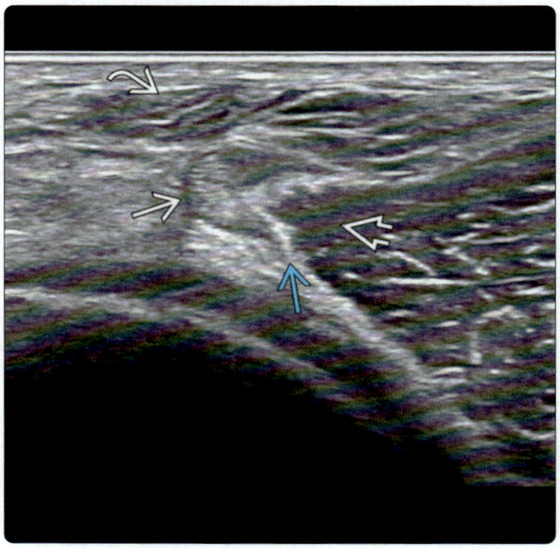

Tibial Nerve

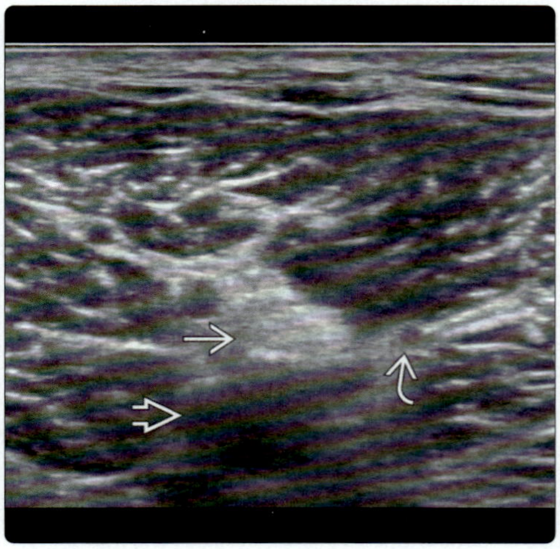

Tibial Nerve

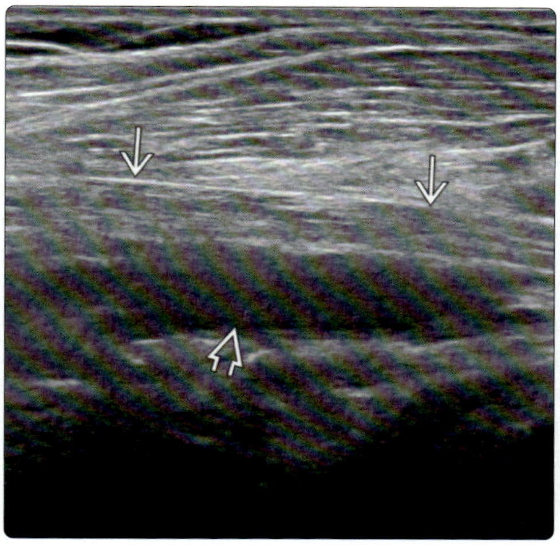

Peroneal Nerve

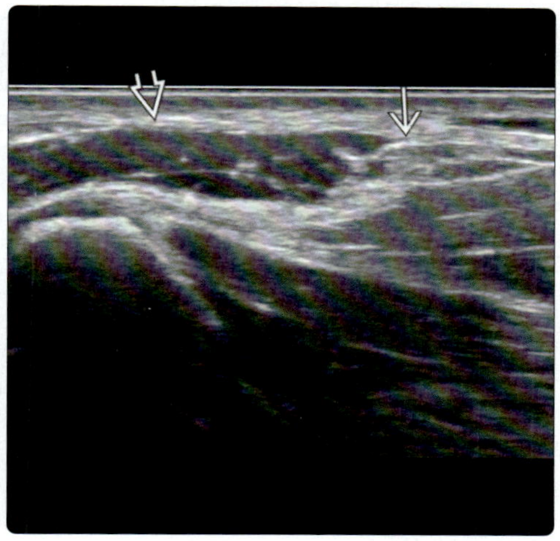

(Left) *Longitudinal US shows the tibial nerve ➡ located posterior to the popliteal artery and vein ➡. The vein is effaced from transducer pressure and is usually located either alongside the artery or between the artery and nerve.* **(Right)** *Transverse US shows the peroneal nerve ➡ seen passing toward the fibular head alongside the biceps femoris muscle ➡.*

Peroneal Nerve

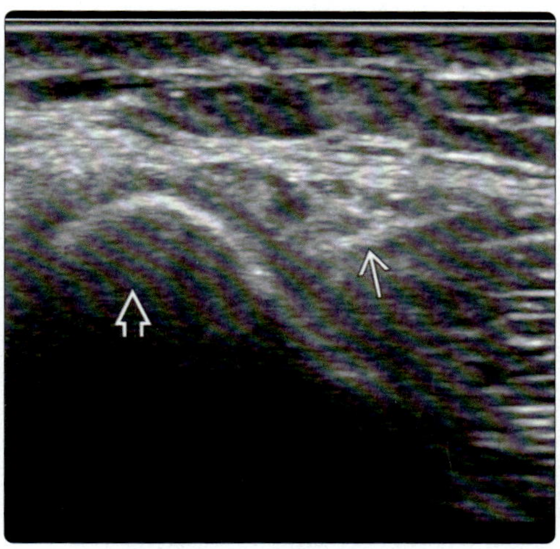

Peroneal Nerve

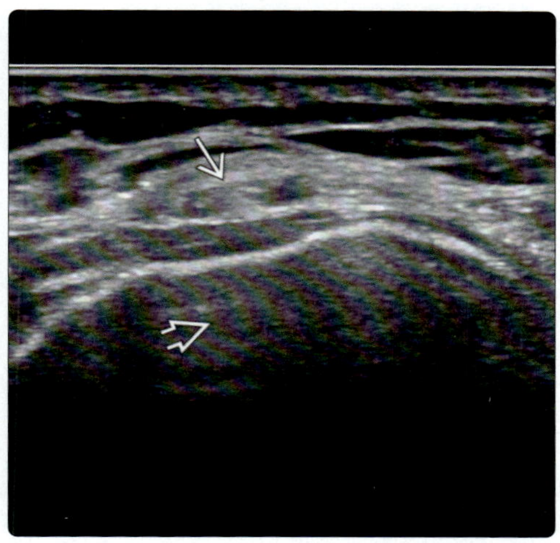

(Left) *US shows the peroneal nerve ➡ just alongside the fibular head ➡.* **(Right)** *Transverse US shows the peroneal nerve ➡ passing posterior to the fibular head ➡.*

Peroneal Nerve

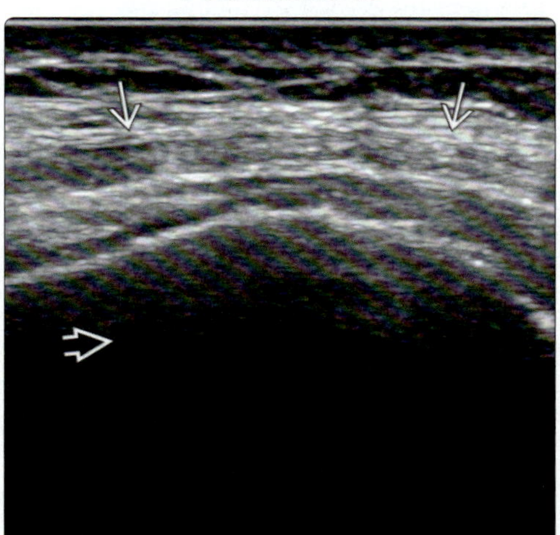

Posterior Cruciate Ligament

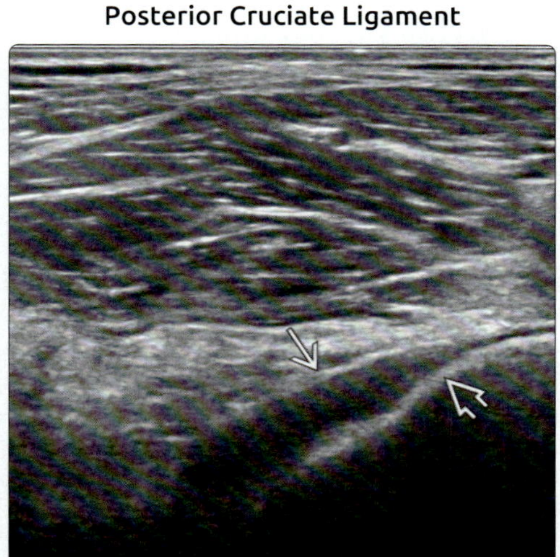

(Left) *Oblique longitudinal US shows the long axis of the peroneal nerve ➡ passing posterior to the fibular head ➡. The nerve will soon divide into its superficial and deep peroneal branches.* **(Right)** *Longitudinal US of the posterior knee shows the tibial portion of the posterior cruciate ligament ➡, which is attached to the posterior aspect of the tibia ➡ in the midline. To obtain this view, the proximal end of the probe is angled toward the inner aspect of the medial femoral condyle.*

KEY FACTS

GENERAL CONSIDERATIONS

- Common specific indications: Ligament tear, tendon disorders, and suspected joint infection and inflammation
- Less common specific indications: Suspected tarsal tunnel syndrome, impingement, and ganglia

TECHNIQUE

- **Anterior ankle**
  - Patient supine with knee flexed so that sole of foot is flat on examination couch
    - Anterior recess ankle joint
    - Anterior ankle tendons
    - Tibialis anterior tendon
    - Extensor digitorum longus tendon
    - Superior and inferior extensor retinacula
    - Deep peroneal nerve
- **Medial ankle**
  - Patient supine with hip abducted and knee flexed
    - Tibialis posterior tendon
    - Flexor digitorum tendon
    - Flexor hallucis longus tendon
    - Deltoid ligament
    - Tibial nerve and tarsal tunnel
- **Lateral ankle**
  - Rotate foot internally ± place firm pad under heel for examination of retromalleolar region
    - Anterior tibiofibular ligament
    - Anterior talofibular ligament
    - Calcaneofibular ligament
    - Peroneal tendons
    - Superficial peroneal nerve
- **Posterior ankle**
  - Patient prone with foot hanging over end of examination couch
    - Achilles tendon
    - Plantaris tendon
    - Sural nerve

Scan Position for Anterior Ankle

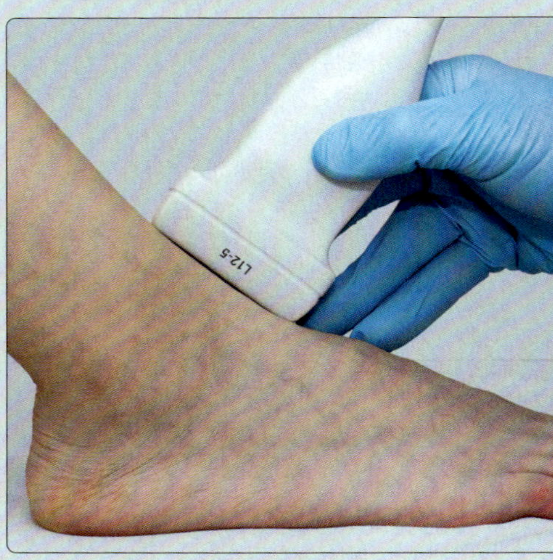

Anterior Ankle Recess

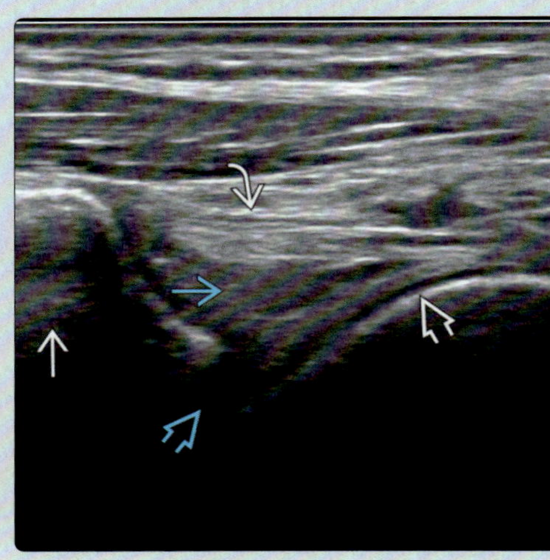

(Left) *Clinical photograph shows the patient position for scanning the anterior ankle. The patient lies on the examination couch with the knee flexed and the sole of the foot resting on the couch.* **(Right)** *Longitudinal US of the midline anterior aspect of the ankle joint shows the distal tibia, anterior capsule, intraarticular extra synovial fat pad, and articular cartilage covering the talar dome. The anterior ankle joint should be scanned from medial to lateral.*

Tibialis Anterior and Extensor Hallucis Longus

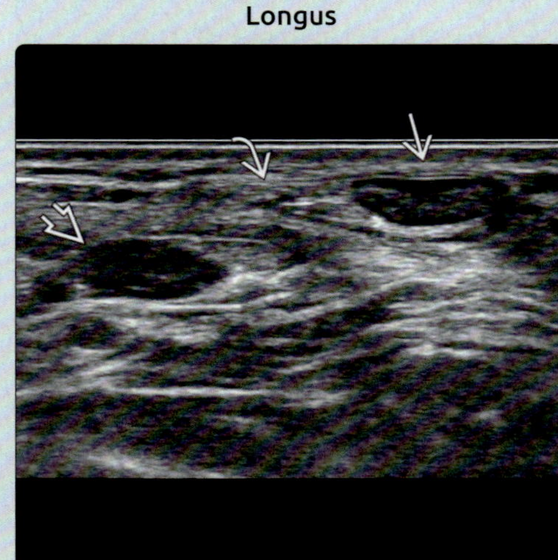

Extensor Digitorum Longus

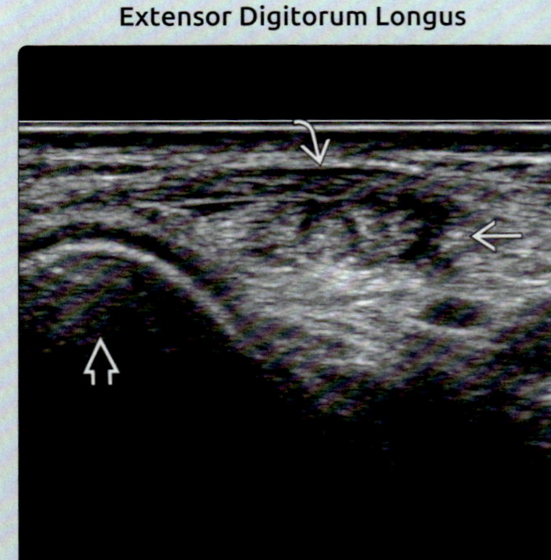

(Left) *Transverse US just proximal to ankle joint shows the tibialis anterior and extensor hallucis longus tendons lying deep to the superior band of the extensor retinaculum.* **(Right)** *Transverse US shows the extensor digitorum longus tendon at the level of the ankle joint just deep to the extensor retinaculum. The tendon has just divided into its digital slips. Note the proximity of the lateral aspect of the ankle to the skin. This makes the lateral ankle a preferred site for intraarticular injection.*

GENERAL CONSIDERATIONS

Clinical Indications for Ankle US

- Nearly all common conditions of ankle region are amiable to full assessment by US
- Common specific indications are ligament tear, tendon disorders, including Achilles tendon (AT), suspected joint infection or inflammation
- Less common specific indications are suspected tarsal tunnel syndrome, impingement, and ganglia

TECHNIQUE: ANTERIOR ANKLE

Patient Position

- Supine with knee flexed so that sole of foot is flat on examination couch

Specifically Examine

- **Anterior recess ankle joint**
  - Much more accessible than posterior recess and routinely examined
  - Absence of joint effusion makes significant intraarticular pathology very unlikely
  - Scan in longitudinal plane from most medial side to most lateral side
    - Capsule attached to tibia 10 mm proximal to articular margin and to talar body/neck junction distally
    - Extrasynovial anterior fat pad lies deep to capsule
    - Joint to skin distance is less lateral than medial; hence, lateral approach for injection is preferable
- **Anterior ankle tendons**
  - Mnemonic is "Tom [tibialis anterior (TA)], Harry (extensor hallucis), and Dick (extensor digitorum)" from medial → lateral
  - Anterior ankle tendons less frequently affected by disease or injury than other ankle tendons
  - Scan primarily in transverse section from distal leg to insertion
- **Tibialis anterior tendon**
  - Seen on anteromedial aspect of ankle to be ~ 2x as large as other extensor tendons; ovoid-shaped as opposed to more flattened shape of other extensor tendons
  - Proximally, TA tendon has tenosynovium whereas distally it has paratenon
  - In 25% of patients, superior extensor retinaculum splits into 2 to form separate TA tunnel
    - TA passes deep to or within superior retinaculum but superficial to inferior retinaculum
  - Scan distal insertion of TA tendon in longitudinal plane with foot slightly everted
    - Inserts into dorsomedial aspect of medial cuneiform
- **Extensor hallucis longus tendon**
  - Seen as thin tendon that moves on passive flexion/extension of big toe
    - Inserts as aponeurotic sheath onto dorsal aspect of base of distal phalanx big toe
- **Extensor digitorum longus (EDL) tendon**
  - Seen on anterolateral aspect of ankle
    - Splints into 4 tendon slips after passing deep to extensor retinacula → middle and distal phalanges of toes
- **Superior and inferior extensor retinacula**
  - Seen as thin (< 1.5 mm), echogenic bands covering extensor tendons
  - Superior retinaculum is rectangular-shaped and located (6-9 mm) proximal to ankle joint
    - From distal fibula laterally to distal tibia medially
  - Inferior retinaculum seen just distal to ankle joint and is Y-shaped
    - Stem attached to lateral hindfoot → 2 limbs
      - Oblique superomedial limb → medial malleolus
      - Oblique inferomedial limb → cuneonavicular joint
- **Deep peroneal nerve (DPN)**
  - Seen on transverse scanning of tibial artery at ankle where nerve crosses from medial → lateral side of artery
    - Usually best seen on medial side of artery just proximal to ankle joint
    - DPN compression deep to inferior retinaculum termed "anterior tarsal tunnel syndrome"

TECHNIQUE: MEDIAL ANKLE

Patient Position

- Patient supine with hip abducted and knee flexed (frog-leg position)

Specifically Examine

- **Tibial nerve and tarsal tunnel**
  - Lies posterior and inferoposterior to medial malleolus bounded superficially by flexor retinaculum
  - Contains "Tom (tibialis posterior), Dick [flexor digitorum (FD)], and a very nervous (artery, vein, tibial nerve) Harry [flexor hallucis longus (FHL)]"
  - Tibial nerve lies alongside posterior tibial artery and veins between FD and FHL
  - Compression of tibial nerve within confines of tarsal tunnel leads to tarsal tunnel syndrome
  - Small medial calcaneal nerve arises from tibial nerve or lateral plantar nerve proximal to, within, or distal to tarsal tunnel
    - Highly variable in position & number & difficult to see
  - In 80% of patients, tibial nerve divides in tarsal tunnel into medial and lateral plantar nerves
    - Inferior calcaneal nerve is 1st branch of lateral plantar nerve (supplies abductor digiti minimi muscle; compression mimics plantar fasciitis, Baxter neuropathy)
- **Tibialis posterior (TP) tendon**
  - Scan primarily in transverse plane from supramalleolar region to insertion
  - Inspect supra-, retro-, infra-, and premalleolar segments keeping transducer at right angles to tendon
  - TP tendon lies in shallow groove behind medial malleolus, which acts as pulley
  - Covered by thick flexor retinaculum with hyaline cartilage more deeply (pulley enthesis)
  - Normally, TP is ~ 2x size of FD tendon
  - Small TP tendon sheath effusion is normal
  - Insertional area best seen on longitudinal scanning
    - Just before insertion, TP widens and angles deeply, making it prone to anisotropy
    - Plantar flexion of foot can increase visibility of insertional fibers

- Superomedial calcaneonavicular ligament (CNL) component of spring ligament between talus and distal TP tendon
  - Extends from sustentaculum tali → navicular bone
  - Normally < 4 mm thick
  - "Gliding layer" of loose areolar tissue superficial to and articular fibrocartilage deep to CNL
- Accessory navicular bone is common finding and strongly associated with pes planus
- 3 types of accessory navicular bone
  - Small accessory bone not directly attached to navicular bone
  - Large accessory bone with synchondrotic attachment to navicular bone
  - Large accessory bone with osseous fusion to navicular bone
- TP tendon partially attaches to all 3 types
- Cornuate configuration with elongated medial pole of navicular bone is also common finding

- **Flexor digitorum tendon**
  - Located posterolateral to TP and usually not affected by tendinosis; hence, FD used as reference for TP tendon size; passes superficial to sustentaculum tali
- **Flexor hallucis longus tendon**
  - Scan longitudinally at posterior ankle deep and medial to AT
  - Scan transversely at posterior ankle to see FHL passing between medial and lateral posterior processes of talus
  - Scan in coronal plane inframalleolar region to see FHL deep to sustentaculum tali
    - Passive flexion and extension of big toe can increase FHL visibility
  - As FHL tendon sheath commonly communicates with ankle, FHL tendon sheath effusions are common
- **Deltoid ligament**
  - Deep (anterior and posterior tibiotalar) and superficial (tibionavicular, tibiospring, tibiocalcaneal) components
  - Scan in longitudinal plane with proximal end of transducer on medial malleolus ± foot dorsiflexion to tense ligament
  - Superficial component seen as thinner, vertical band from medial malleolus → sustentaculum tali, spring ligament, and navicular bone
  - Deep component seen as triangular, striated, echogenic tissue with wide portion attached to talus
    - Anisotropy at talar aspect due to fiber obliquely is normal finding

TECHNIQUE: LATERAL ANKLE

Patient Position

- Rotate foot internally ± place firm pad under heel for examination of retromalleolar region

Specifically Examine

- **Anterior tibiofibular ligament**
  - Scan at distal end of tibia in oblique transverse plane
  - Align transducer along line extending from distal end of lateral malleolus to supramalleolar area of medial malleolus
    - Seen as thin, striated band connecting tibia and fibula; wider at tibial attachment
- **Anterior talofibular ligament (ATFL)**: Place transducer on tip on lateral malleolus and scan transversely or slightly downward along length of ATFL
  - ATFL seen as linear, striated band, 2 mm thick, between fibular and talus with anterolateral gutter deep to ligament; small artery and vein superficial to ATFL
- **Calcaneofibular ligament (CFL)**
  - Seen deep on transverse scanning of peroneal tendon
  - Place transducer along line from anterior tip of lateral malleolus to posteroinferior corner of heel
  - Dorsiflexion tenses CFL and elevates peroneal tendons
  - Angle transducer to reduce anisotropy at fibular insertion
- **Peroneal tendons**
  - Scan in transverse plane from supramalleolar region → midfoot
  - In retromalleolar region, peroneal brevis tendon seen deep to peroneus longus tendon
    - Peroneus brevis tendon is < 1/2 size of peroneal longus and is more oblong in shape
  - Note retinaculum and fibrocartilaginous anchor attached to posterolateral corner fibula; ± dorsiflex & invert ankle to see if peroneal tendons sublux (external or internal)
  - When following tendons around lateral malleolus in transverse plane, follow one tendon at a time
  - Peroneal tubercle (should be < 5 mm) may separate tendons at lateral part of calcaneum
  - Peroneus brevis → 5th metatarsal base
  - Peroneal longus → cuboid groove → medial cuneiform/1st metatarsal base; ± os peroneum (20% of subjects) within tendon just proximal to groove
- **Superficial peroneal nerve**: Located alongside lateral intermuscular septum
  - Emerges at interspace between peroneus muscles and EDL muscle
  - Pierces crural fascia on anterolateral aspect of leg 9 cm (6-12 cm) proximal to fibular tip; can be stretched by fascia giving rise to lateral ankle pain

TECHNIQUE: POSTERIOR ANKLE

Patient Position

- Prone with foot hanging over end of examination couch

Specifically Examine

- **Achilles tendon**: Scanned in longitudinal and transverse planes from musculotendinous junction to calcaneal insertion; should be < 7 mm thick
  - Surrounded by paratenon rather than complete synovial sheath; Kager fat pad and retrocalcaneal bursa lie deeply
  - Inserts into posterior calcaneal tuberosity over 1 cm length with thin entheseal fibrocartilage between tendon and bone
- **Plantaris tendon**: Scan transversely on medial aspect of AT to see small, hypoechoic, oval-shaped plantaris tendon
  - Proximally, seen deep to medial belly of gastrocnemius
  - Distally, usually blends with AT
- **Sural nerve**: Scan transversely between AT and peroneal tendons; sural nerve seen just medial or lateral to short saphenous vein, which acts as marker for this nerve
  - Passes behind lateral malleolus deep to peroneal tendons

Tibialis Anterior Tendon Insertion

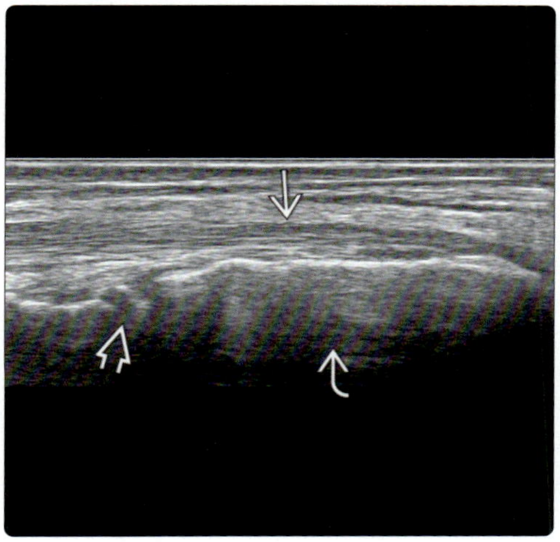

Deep Peroneal Nerve

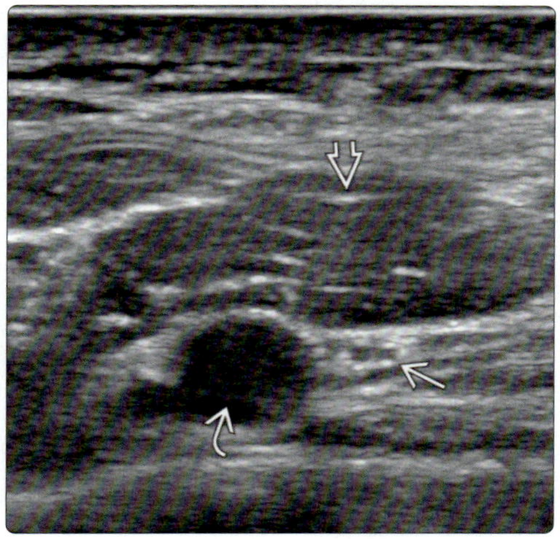

(Left) *Longitudinal US at the dorsomedial aspect of the midfoot shows the tibialis anterior tendon inserted into the medial cuneiform. The navicular-cuneiform articulation is also shown.* **(Right)** *Transverse US just proximal to the ankle joint shows the deep peroneal nerve just medial to the anterior tibial artery and deep to the extensor hallucis longus muscle. Slightly more distal to this location, the nerve passes to the lateral aspect of the anterior tibial artery.*

Anterior Tibiofibular Ligament

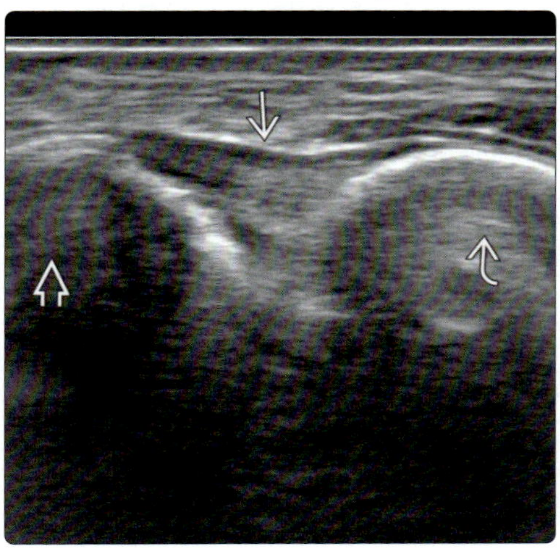

Anterior Talofibular Ligament

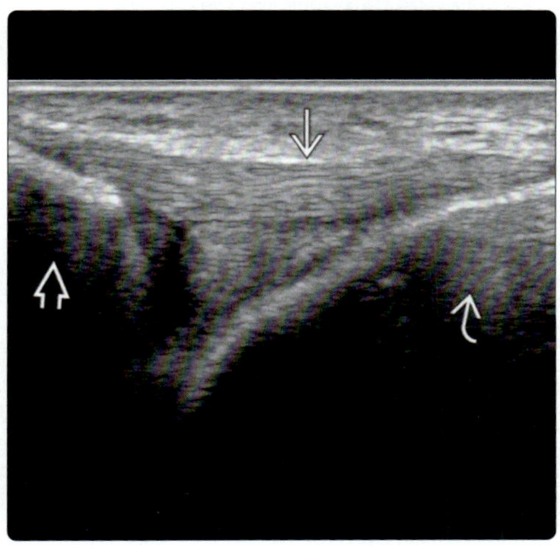

(Left) *Transverse oblique US immediately proximal to the ankle joint, with the medial side of the transducer angled ~ 45° proximally, shows the triangular-shaped anterior tibiofibular ligament between the tibia and fibula. The anterior tibiofibular ligament is wider at its tibial attachment.* **(Right)** *Transverse US at the level of the fibular tip shows the anterior talofibular ligament extending between the fibular tip and the talus.*

Scan Position for Lateral Ankle

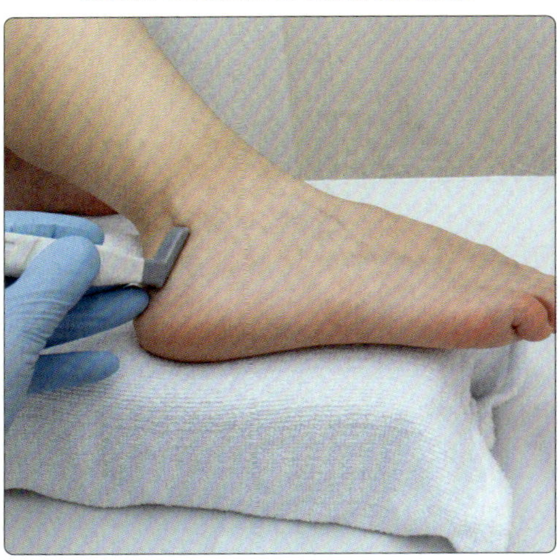

Peroneal Tendons in Retromalleolar Area

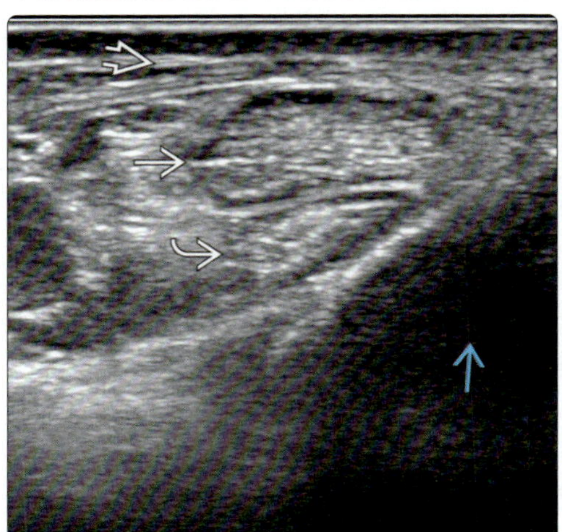

(Left) *Clinical photograph shows the position used for examination of the lateral ankle region. The heel can be placed in a firm pad. This allows easier transducer access to the retromalleolar region. Alternatively, the heel can be placed at the edge of the examination couch.* **(Right)** *Transverse US shows the peroneus longus and brevis tendons covered by the superior peroneal retinaculum posterior to the distal fibula.*

(Left) *Transverse US at the more distal retromalleolar region shows the peroneus longus and peroneus brevis tendons posterior to the fibular tip and overlying the calcaneofibular ligament close to its fibular attachment.* **(Right)** *Transverse US shows the peroneus longus and brevis tendons overlying the lateral aspect of the calcaneum at the inframalleolar region.*

Peroneal Tendons in Retromalleolar Region

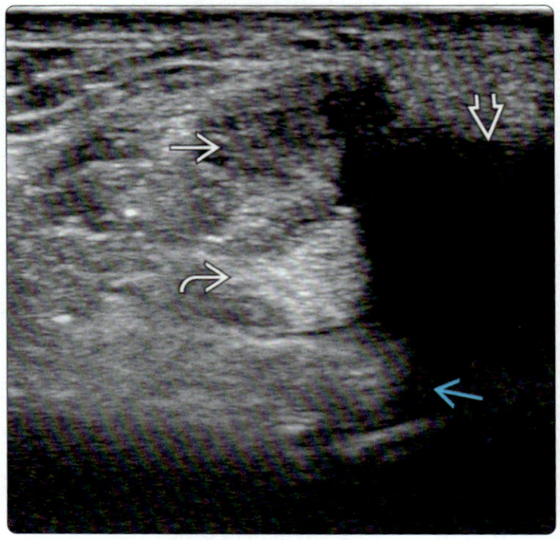

Peroneal Tendons in Inframalleolar Region

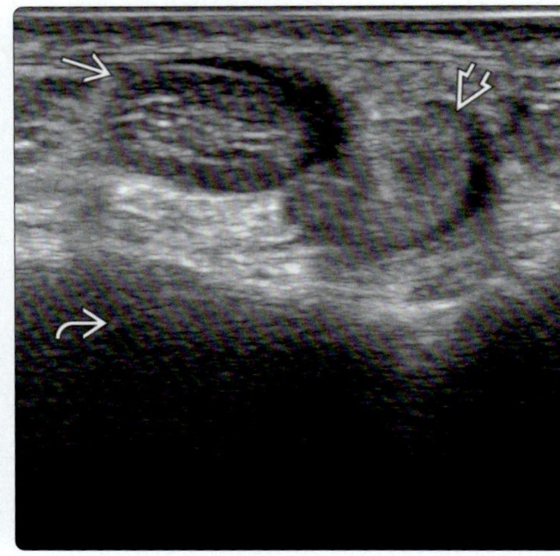

(Left) *Longitudinal US at the posteroinferior aspect of the fibular tip shows the distal part of the calcaneofibular ligament lying deep to the peroneal tendons and inserting into the lateral aspect of the calcaneum.* **(Right)** *Longitudinal US along the proximal part of the calcaneofibular ligament shows the ligament below the peroneal tendons. The fibular attachment is difficult to see due to acoustic shadowing. Integrity can be tested by dorsiflexing the ankle. Posterior talocalcaneal articulation is shown.*

Calcaneofibular Ligament

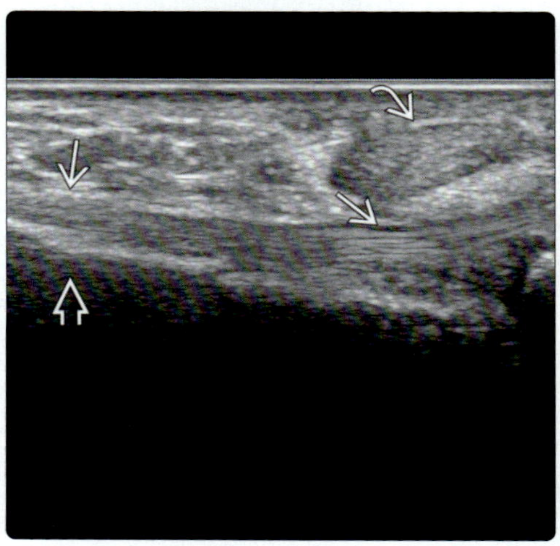

Calcaneofibular Ligament

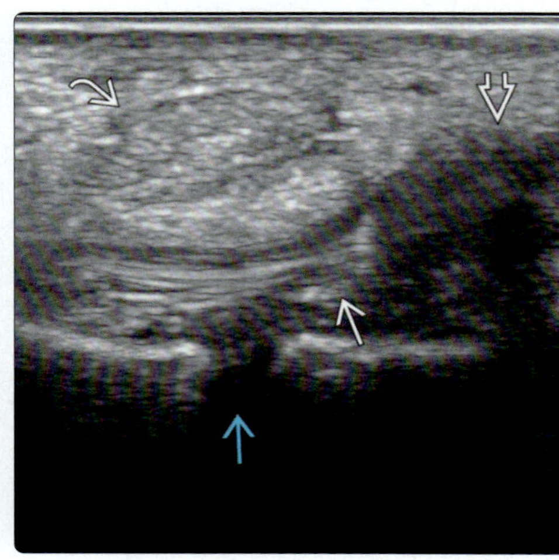

(Left) *Transverse US at mid- to lower leg laterally shows the superficial peroneal nerve emerging through the crural fascia at the interspace between the peroneal brevis and extensor digitorum longus muscles. The nerve can be injured indirectly at this location giving rise to lateral ankle pain.* **(Right)** *Clinical photograph shows the patient position for scanning the medial ankle. The patient lies on the examination couch with the hip adducted, knee flexed, and the side of the ankle and foot resting on the couch (frog-leg position).*

Superficial Peroneal Nerve

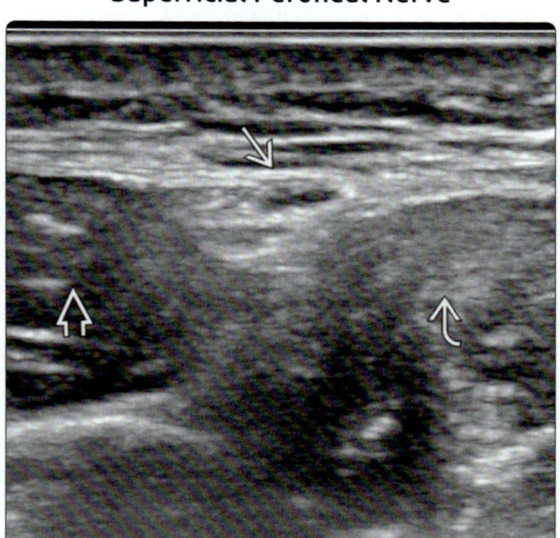

Scan Position for Medial Ankle

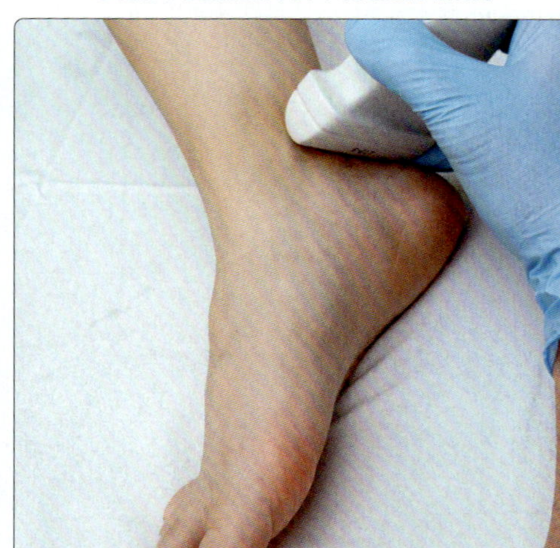

Tarsal Tunnel

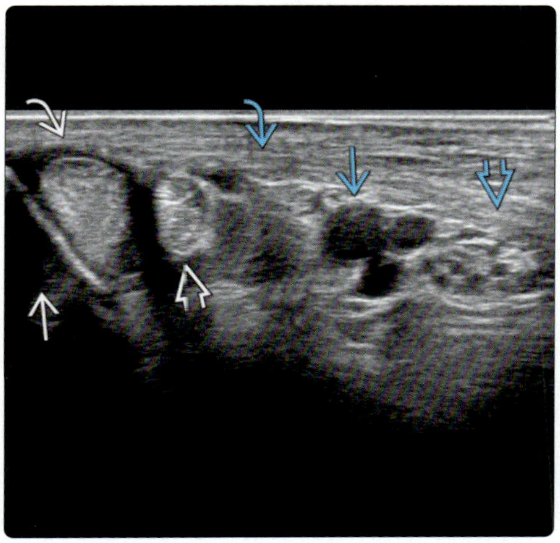

Retromalleolar Tibialis Posterior

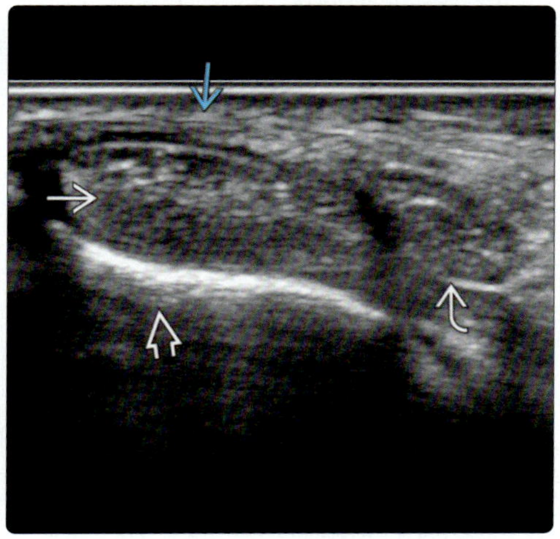

(Left) *. Transverse US posterior to medial malleolus shows the posterior tibialis tendon, flexor digitorum longus, posterior tibial artery and veins, and tibial nerve deep to the flexor retinaculum.* **(Right)** *Transverse US at the medial retromalleolar region shows the posterior tibialis and flexor digitorum longus tendons covered by the flexor retinaculum.*

Superomedial Fibers of Calcaneofibular Ligament

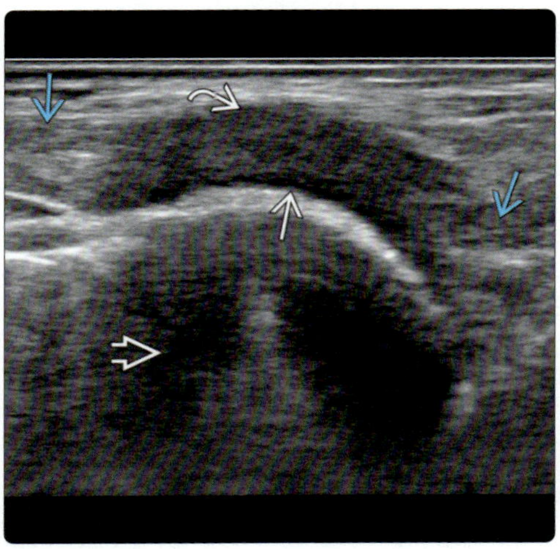

Tibialis Posterior Insertion

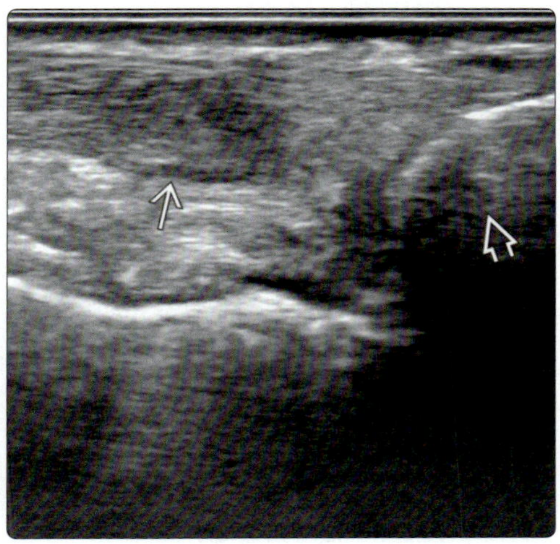

(Left) *Longitudinal US along superomedial fibers of the calcaneonavicular ligament, lying medial to talar head and just dorsal and deep to the tibialis posterior tendon is shown. There is a thin layer of fibrocartilage between the fibers and bone. Fibers running to the tibia proximally and the navicular bone distally can be appreciated.* **(Right)** *Longitudinal US along the posterior tibialis tendon shows fibers inserting into medial pole of navicular bone. The tendon also inserts deeply into the cuneiform bones.*

Tibiospring Ligament

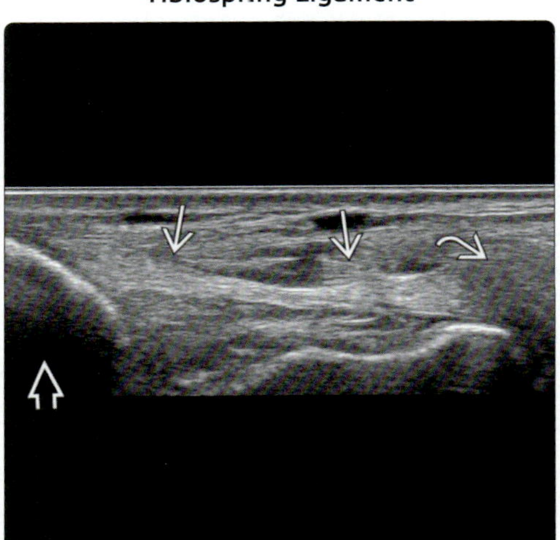

Tibiotalar Ligament

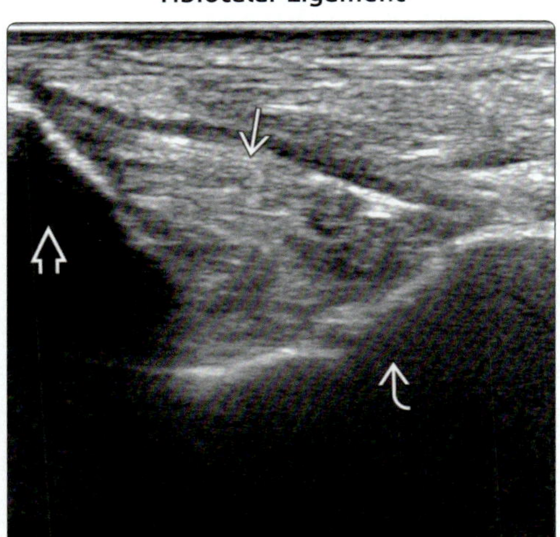

(Left) *Longitudinal US shows the tibiospring ligament extending from the medial malleolus to the superomedial fibers of the spring ligament.* **(Right)** *Longitudinal US shows the triangular-shaped posterior tibiotalar fibers extending between the medial malleolus and the talus.*

(Left) *Clinical photograph shows the scan position for the posterior ankle region. The patient lies prone with the feet hanging freely over the end of the examination couch.* **(Right)** *Longitudinal US shows the musculotendinous junction of the Achilles tendon. The soleus muscle ➔ is attached to the undersurface of the Achilles tendon ➔. The gastrocnemius muscle is attached much more proximally on the dorsal surface. Kager fat pad ➔ and flexor hallucis longus muscle ➔ lie deep to the Achilles tendon.*

Scan Position for Posterior Ankle

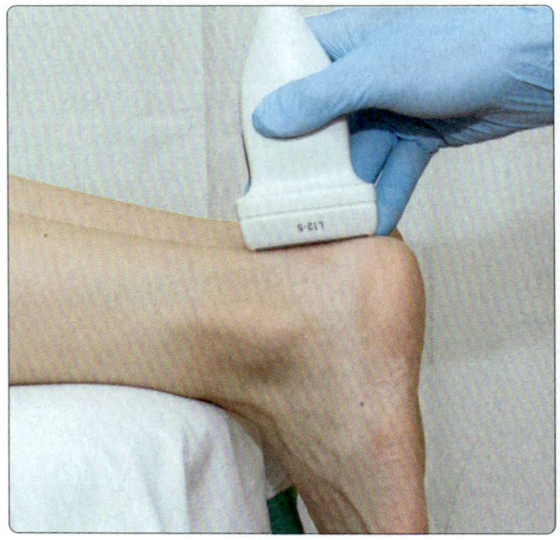

Longitudinal Musculotendinous Junction of Achilles Tendon

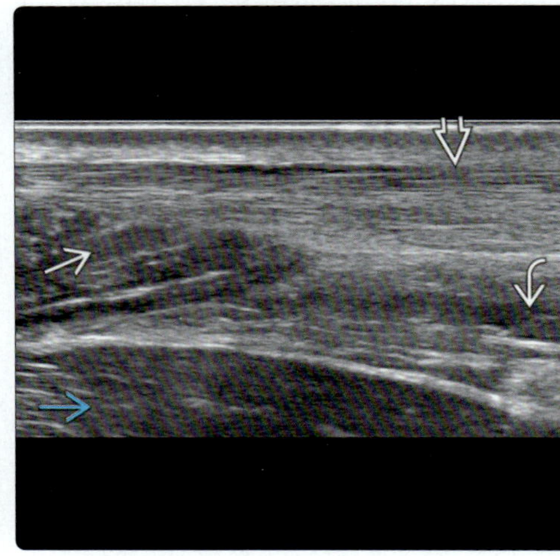

(Left) *Longitudinal US shows the Achilles tendon ➔ insertion to the posterior aspect of the calcaneum. There is hypoechoic entheseal fibrocartilage ➔ at the interface between tendon and bone. Mild retrocalcaneal bursa distension ➔ is normal.* **(Right)** *Transverse US shows the Achilles tendon ➔ at the musculotendinous junction. The soleus muscle ➔ lies deep to the Achilles tendon.*

Longitudinal Achilles Tendon Insertion

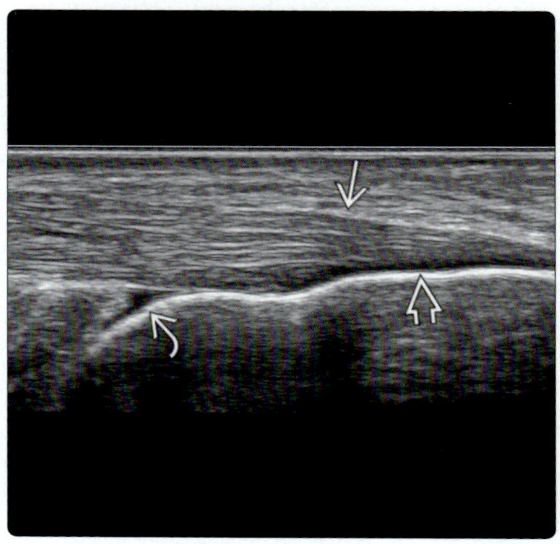

Transverse Achilles Tendon: Proximal Aspect

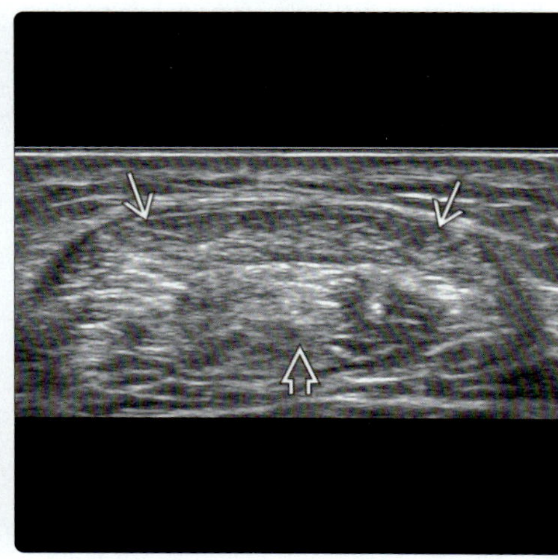

(Left) *Transverse US shows the midportion of the Achilles tendon ➔. The normal echogenic, nondistended paratenon ➔ is barely visible. Kager fat pad ➔ lies deep to the Achilles tendon.* **(Right)** *Transverse US shows the Achilles tendon at the calcaneal insertion. The Achilles tendon ➔ broadens out toward its insertion on the posterior aspect of the calcaneum. Copious gel should be applied to the sides of the transducer to reduce air artifact.*

Transverse Midportion of Achilles Tendon

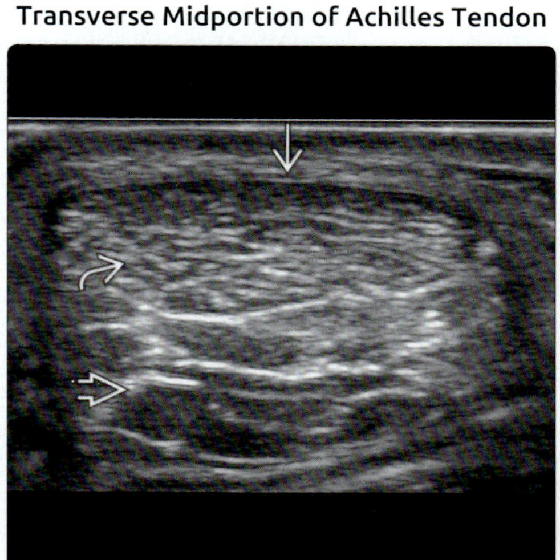

Transverse Achilles Tendon Insertion

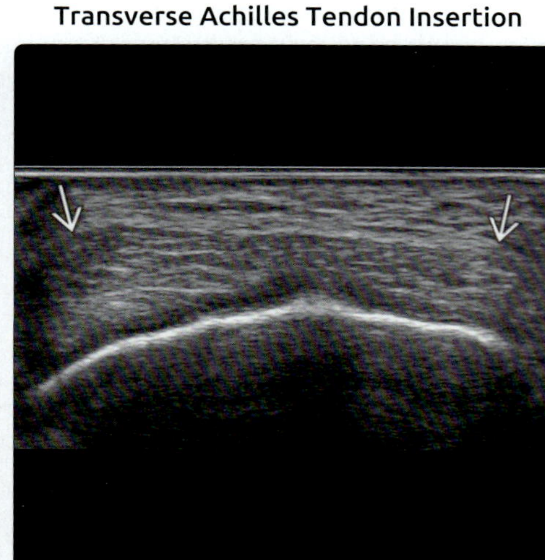

Plantaris Tendon: Midleg

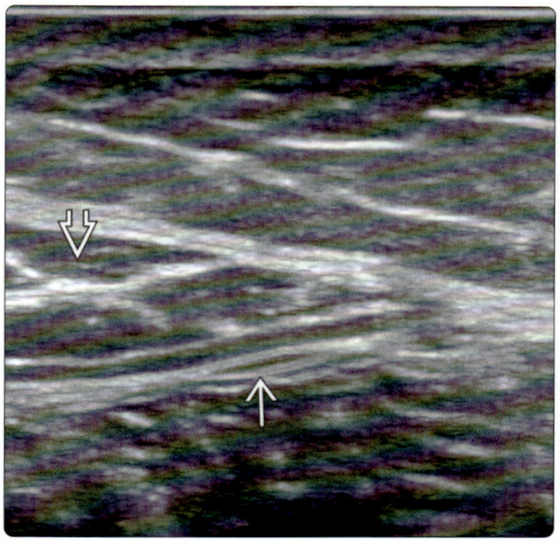

Plantaris Tendon: Distal Leg

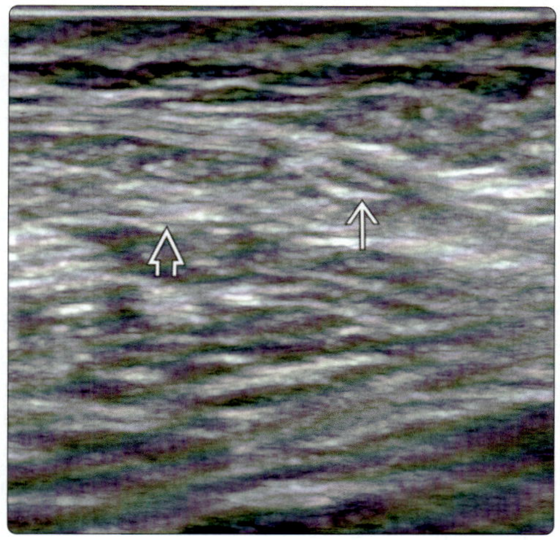

(Left) *Transverse US shows the plantaris tendon deep to the medial belly of the gastrocnemius muscle. The plantaris tendon is best scanned proximally and distally in a transverse plane.* **(Right)** *Transverse US shows the plantaris tendon on the medial side of Achilles tendon. Distally, the plantaris tendon is inserted into either the Achilles tendon or the medial aspect of the calcaneum.*

Posterior Ankle Recess

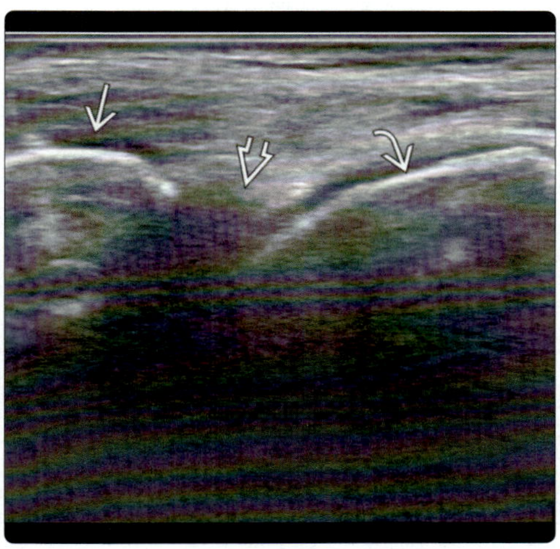

Flexor Hallucis Longus

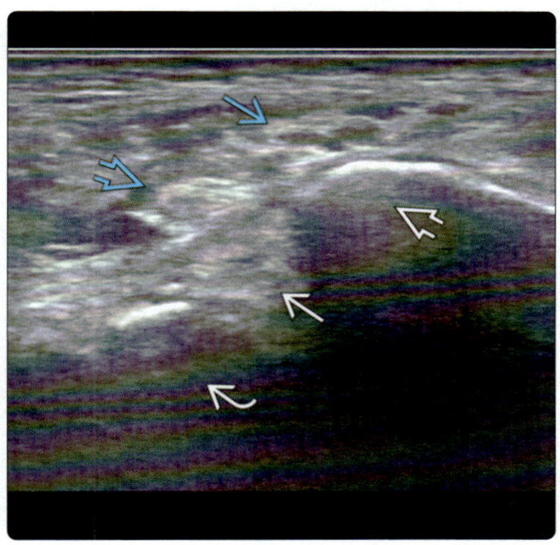

(Left) *Longitudinal US shows posterior recess of the ankle joint located between the posterior aspect of the distal tibia and talus. Visibility of this recess can be improved by ankle dorsiflexion.* **(Right)** *Transverse US of the posteromedial aspect of ankle shows flexor hallucis longus (FHL) between lateral and medial tubercles of the posterior talus. FHL is closely applied to the ankle joint with its tendon sometimes in direct communication with the posterior recesses. The tibial nerve and posterior tibial vessels lie posteriorly.*

Neural Anatomy of Ankle Level

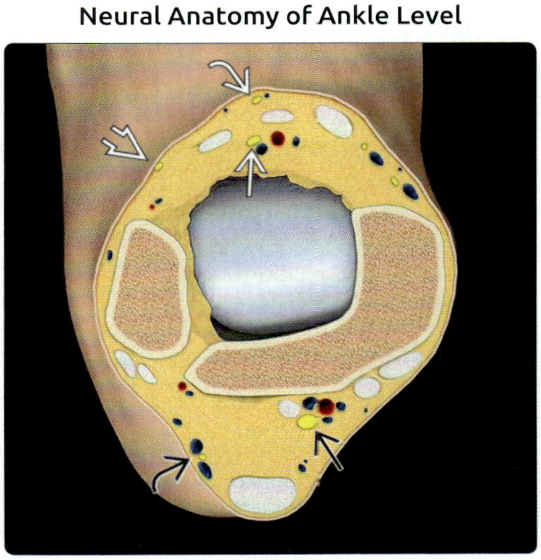

Sural Nerve

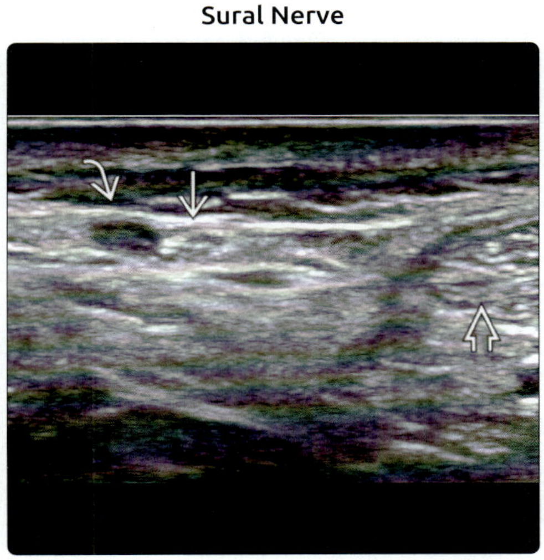

(Left) *Transverse graphic at the level of the ankle joint shows the position of the deep peroneal nerve, medial dorsal cutaneous nerve, intermediate dorsal cutaneous nerve, tibial nerve, and sural nerve.* **(Right)** *Transverse US at the midportion of the Achilles tendon shows the sural nerve lying alongside the short saphenous vein. The sural nerve may lie medial or lateral to the vein. The vein is located between the Achilles tendon and the peroneal muscles/tendons.*

Foot Ultrasound

KEY FACTS

GENERAL CONSIDERATIONS

- Common specific indications: Suspected plantar fasciitis and fibroma, Morton neuroma, ganglia, osteoarthritis, gout, and other joint disorders
- Less common specific indications: Suspected ligament injury, muscle swelling, plantar fascial tear, plantar plate injury, stress fracture

TECHNIQUE

- **Dorsal foot**
  - Patient position: Supine with knee flexed and ankle plantar flexed so that sole of foot flat on examination couch
  - Bones and joints of foot
  - Subtalar joints
    - Anterior and posterior subtalar joints, which do not communicate
    - 4 different US approaches to subtalar joints
  - 1st metatarsophalangeal (MTP) joint
    - Most frequently symptomatic foot joint
  - Calcaneocuboid ligament
  - Dorsal talonavicular ligament
    - Along with calcaneocuboid ligament, helps support midtarsal joint
  - Lisfranc ligament
    - Dorsal component readily evaluated with US
  - Extensor tendons
  - Muscles
- **Plantar foot**
  - Patient position: Prone with foot dorsiflexed and toes resting on examination couch
  - Plantar fascia: Medial, central, and lateral bands
  - Plantar muscles
  - Flexor tendons
  - Toe web spaces
  - Hallucal-sesamoid complex 1st MTP joint
    - Readily evaluated with US

Scan Position for Dorsum of Foot

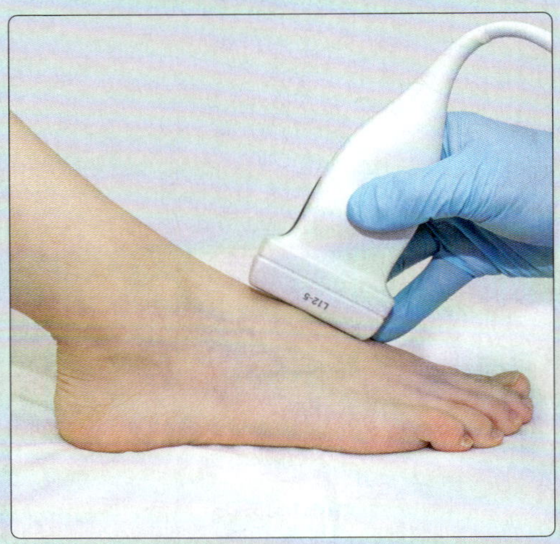

Dorsal Foot Bones

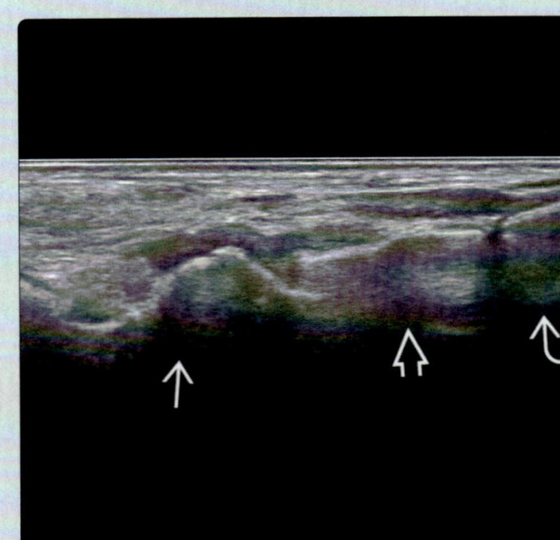

(Left) *Clinical photograph shows the probe and patient position for the dorsum of the foot.* **(Right)** *Longitudinal US shows the bones of the forefoot and midfoot with the talus ➔, navicular ➔, and medial cuneiform ➔. The location and shape of the foot bones makes them quite readily recognizable.*

Dorsal Foot Bones

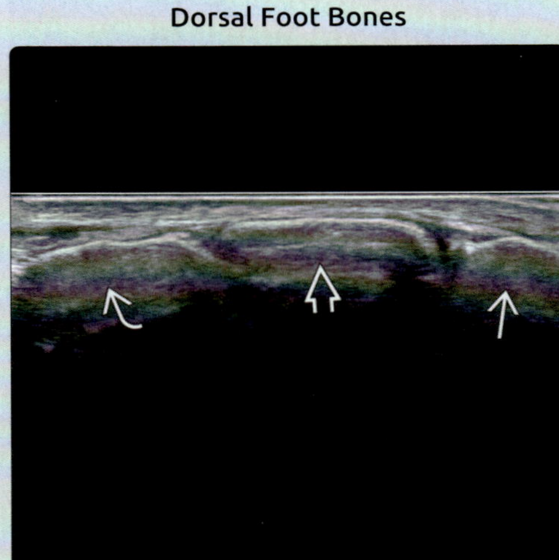

Dorsal Foot Soft Tissues

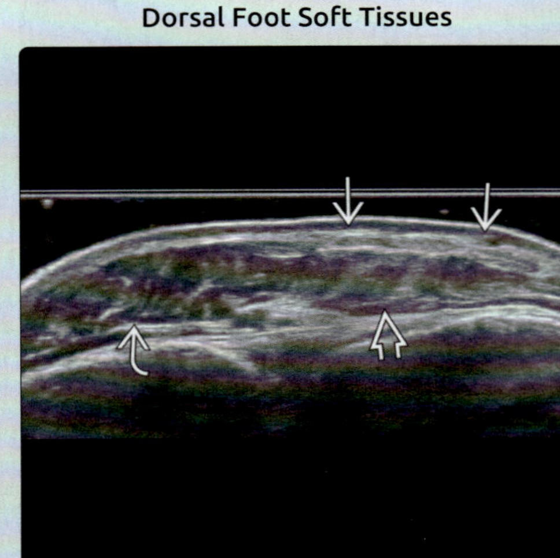

(Left) *Longitudinal US in medial aspect of tarsus shows the base of the 1st metatarsal ➔, medial cuneiform ➔, and navicular ➔ bones.* **(Right)** *Transverse US of the diverging extensor digitorum tendon slips ➔ overlying the extensor hallucis brevis ➔ and extensor digitorum brevis ➔ muscles is shown.*

GENERAL CONSIDERATIONS

Clinical Indications for Foot US

- While closely integrated with ankle pathology, many foot conditions can be fully evaluated with US
- Common specific indications: Suspected plantar fasciitis & fibroma, Morton neuroma, ganglia, osteoarthritis, gout, and other joint disorders
- Less common specific indications: Suspected ligament injury, muscle swelling, plantar fascial tear, plantar plate injury, stress fracture

TECHNIQUE: DORSAL FOOT

Patient Position

- Supine with knee flexed and sole resting on examination couch

Specifically Examine Bones and Joints

- Easily recognizable bones or bone areas are talar head and neck, navicular bone, and metatarsal bases
- **Subtalar joints**
  - Comprises anterior (ASTJ) and posterior subtalar joint (PSTJ), which do no communicate with each other
    - ASTJ does communicate with talonavicular joint ("talocalcaneonavicular joint")
      - ASTJ seen on US from medial approach
    - PSTJ does communicate with posterior recess of ankle joint
      - PSTJ seen on US from posteromedial, posterolateral, and lateral approaches
- **1st metatarsophalangeal joint**
  - Evaluated from both dorsal and plantar aspects, primarily in longitudinal plane
  - Passive flexion and extension can help in assessment
  - On plantar aspect, assess plantar plate, paired sesamoid bones, and intervening flexor hallucis longus (FHL)
    - Sesamoid bones with intervening ligament and plantar plate may be bipartite, hypoplastic, or aplastic
    - Bipartite sesamoid is overall larger than expected sesamoid size with smooth bony margins
    - Both sesamoids are located within tendons of medial and lateral heads of flexor hallucis brevis muscle

Specifically Examine Ligaments

- **Calcaneocuboid ligament**
  - Evident on lateral side of calcaneocuboid articulation
    - Smaller calcaneonavicular ligament also arises at this calcaneal location
- **Dorsal talonavicular ligament**
  - Along with calcaneocuboid ligament, helps support Chopart joint (midtarsal joint)
    - Injured in up to 20% of ankle sprains
- **Lisfranc ligament**
  - 3 components: Dorsal, interosseous (strongest), and plantar
  - Interosseous and plantar components difficult to see
    - Plantar component is Y-shaped with medial cuneiform stem → 2nd and 3rd metatarsal bases
- **Extensor tendons**
  - Tibialis anterior → dorsomedial medial cuneiform
  - Extensor hallucis longus → base of distal phalanx hallux
  - Extensor digitorum longus → base of middle and distal phalanges of lesser digits
- **Dorsal muscles**
  - Extensor hallucis brevis and extensor digitorum brevis muscles inseparable on transverse scanning of dorsocentral midfoot
    - Lie deep to extensor digitorum tendons

TECHNIQUE: PLANTAR FOOT

Patient Position

- Patient prone with knee slightly flexed and ankle dorsiflexed with toes on examination couch

Specifically Examine

- **Plantar fascia**
  - Assessed in both longitudinal (primarily) and transverse planes
  - Composed of medial, central, and lateral bands
    - Medial band not present in hindfoot
    - Central band is largest and extends from medial calcaneal tuberosity to metatarsal heads
      - < 4.3 mm thick at leading edge of calcaneum
    - Lateral band extends from calcaneum to 5th metatarsal base
- **Plantar muscles**
  - 4 muscle layers are identified on both transverse and longitudinal scanning
  - Infrequently diseased and usually assessed with MR
  - 1st layer: Flexor digitorum brevis located deep to plantar fascia with abductor hallucis and abductor digiti minimi on either side
  - 2nd layer: Quadratus plantae and lumbrical muscles
    - Quadratus plantae extends from calcaneum to flexor digitorum longus (FDL) tendon
    - Lumbricals arise from flexor digitorum tendon slips
  - 3rd layer: Peroneus longus tendon traverses plantar aspect of foot from cuboid groove to medial cuneiform and 1st metatarsal base
    - More distally, flexor hallucis brevis and adductor hallucis medially and flexor digiti minima laterally
  - 4th layer: Interossei and muscles on dorsum of foot
- **Flexor tendons**
  - On longitudinal scanning on medial aspect of hindfoot, identify sustentaculum tali
  - FDL passes superficial to this bony landmark while FHL passes deep to it
  - FDL and FHL can be traced on transverse scanning to where they cross over in midfoot (knot of Henry)
    - FDL crosses more superficially and receives insertion of quadratus plantae just beyond this point
    - FHL passes between sesamoid bones of 1st metatarsophalangeal joint
- **Toe web spaces**
  - Best if scanned in neutral position with heel on examination couch
  - Can be assessed from either plantar or dorsal approach
    - Apply finger pressure from either dorsal or plantar side to bring tissues of interest closer to transducer
  - Transverse metatarsal ligament difficult to see

(Left) *Longitudinal US shows the dorsal talonavicular ligament* → *between the distal part of the talus* → *and the proximal part of the navicular bone* →. **(Right)** *Longitudinal US shows the calcaneocuboid ligament* → *between the distal part of the calcaneum* → *and the proximal part of the cuboid bone* →. *The calcaneonavicular ligament (not shown) arises from the same part of the calcaneum, extending medially toward the navicular bone.*

Talonavicular Ligament

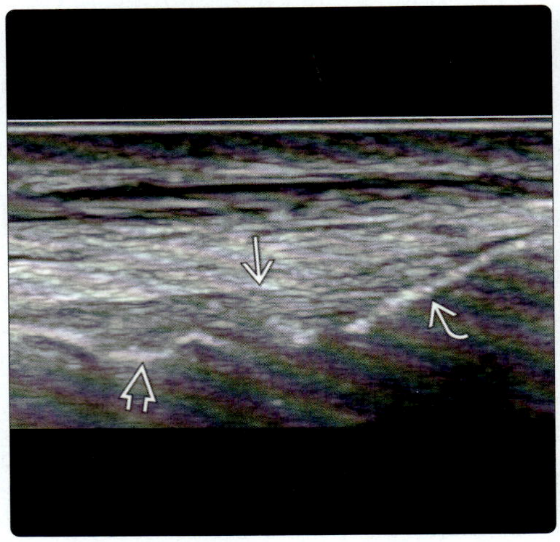

Calcaneocuboid Ligament

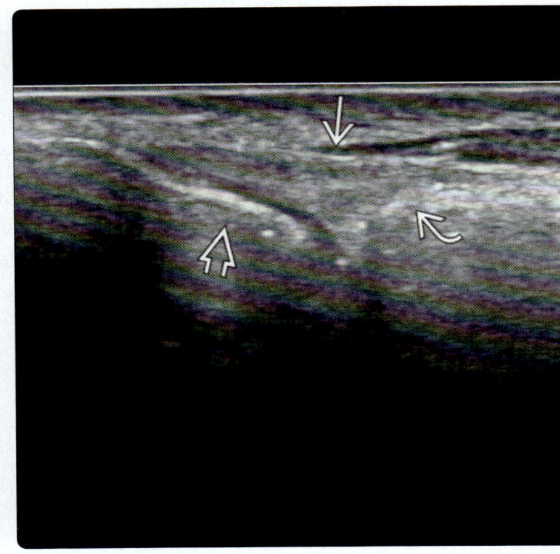

(Left) *Transverse US of the midfoot shows the dorsal component of the Lisfranc ligament* →, *extending between the medial cuneiform* → *and the base of the 2nd metatarsal bone* →. *This strong ligament is aligned almost transversely.* **(Right)** *Transverse US of the midfoot shows the flexor digitorum tendon* → *crossing the flexor hallucis longus tendon* →. *Medial plantar nerve* → *is also visible deep to the abductor hallucis longus muscle.*

Lisfranc Ligament

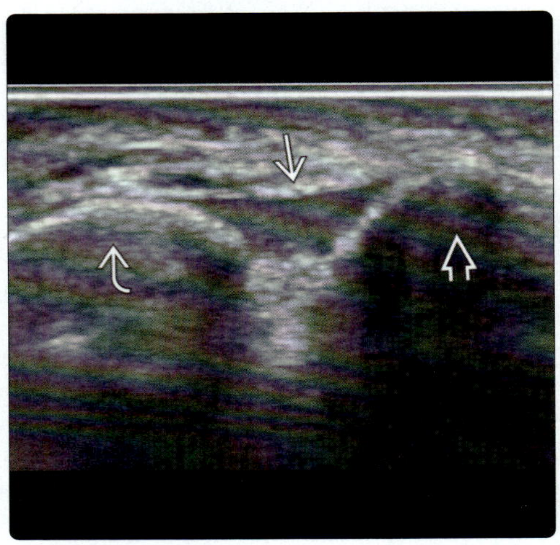

Knot of Henry

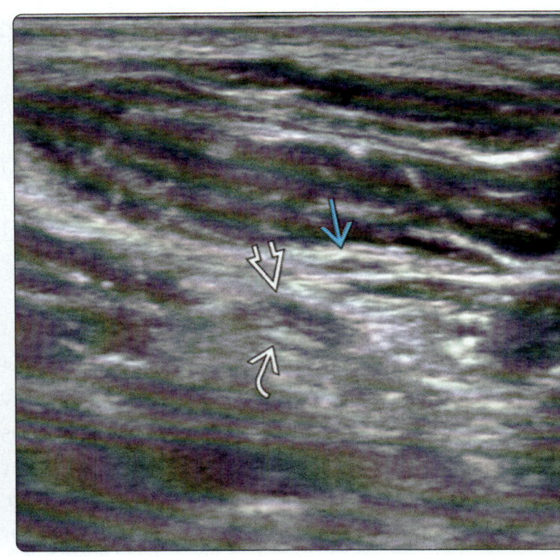

(Left) *Oblique transverse US shows the anterior subtalar joint location. The orientation is from the medial malleolus to the superomedial fibers of the spring ligament. Varus strain on the midfoot can help accentuate the anterior subtalar joint.* **(Right)** *Oblique transverse US shows the talus* → *and the sustentaculum tali* → *with the intervening anterior subtalar joint* →. *Overlying this area are the superomedial fibers of the spring ligament* → *and the tibialis posterior tendon* →.

Scan Position for Anterior Subtalar Joint

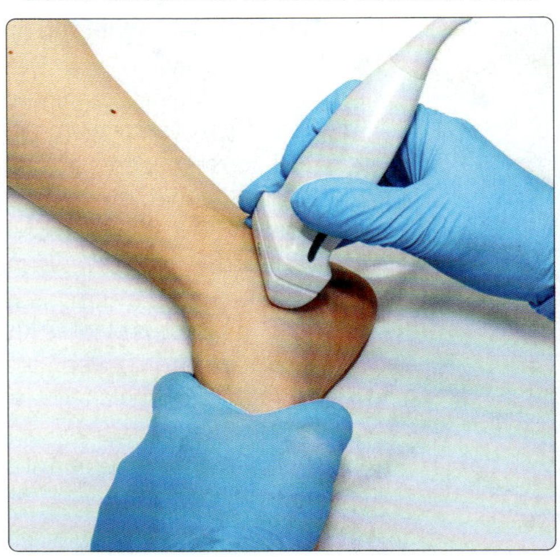

Anterior Subtalar Joint

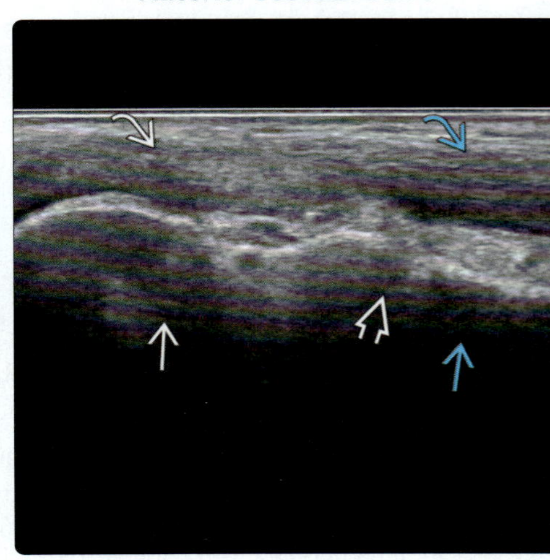

Scan Position (1) for Posterior Subtalar Joint

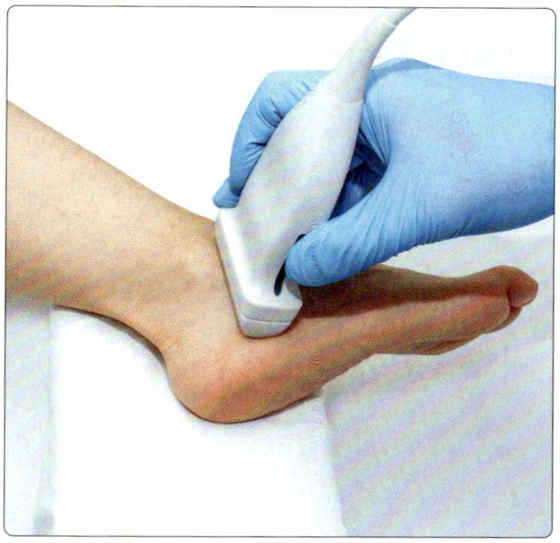

Lateral View of Posterior Subtalar Joint

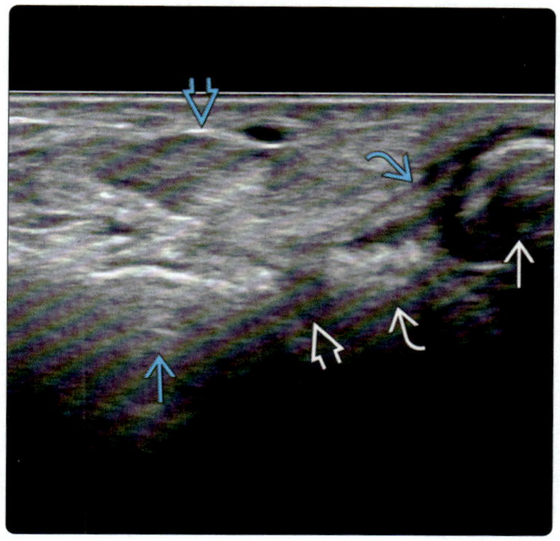

(Left) *Clinical photograph shows the foot and probe position for the posterior subtalar joint (PSTJ), lateral approach. The foot is placed on a pad to aid valgus strain. The orientation of the probe is from the anterior aspect of the lateral malleolus to the midcalcaneal region.* **(Right)** *Longitudinal US shows the lateral view of the PSTJ . The lateral malleolus , talus , and calcaneum are seen. Overlying this joint are the peroneal tendons and part of the calcaneofibular ligament .*

Scan Position (2) for Posterior Subtalar Joint

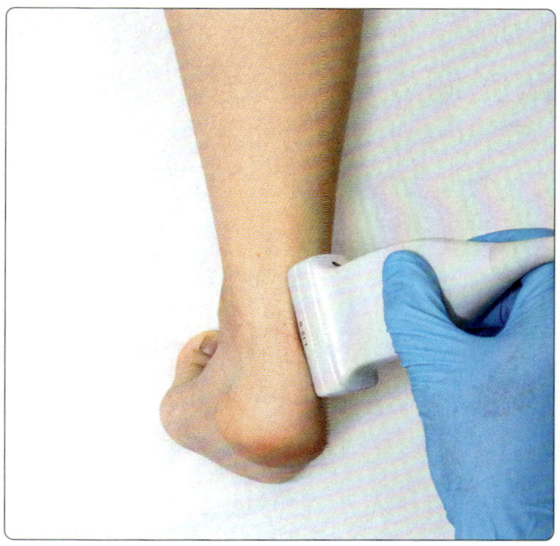

Posteromedial View of Posterior Subtalar Joint

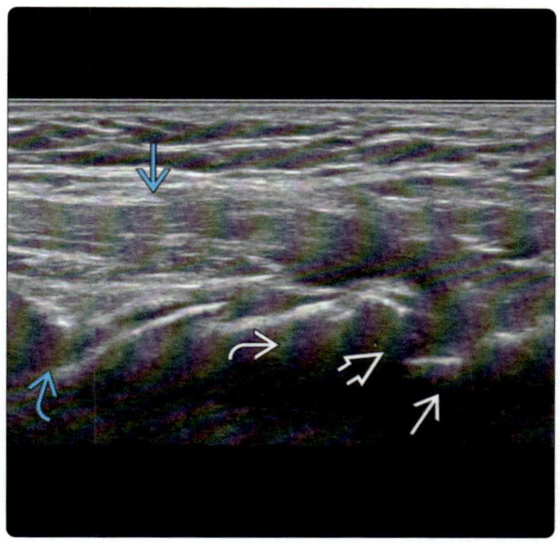

(Left) *Clinical photograph shows foot and probe position for the PSTJ, posteromedial view. The probe is oriented along the flexor hallucis longus tendon at the medial border of the Achilles tendon. The lower part of the probe is moved medially to view the PSTJ.* **(Right)** *Longitudinal US shows posteromedial aspect of the PSTJ . PSTJ lies between the posterior process of talus and calcaneum . Overlying this joint is the flexor hallucis longus tendon . PSTJ space is widened by dorsiflexion and foot eversion. Ankle joint is shown.*

Scan Position (3) for Posterior Subtalar Joint

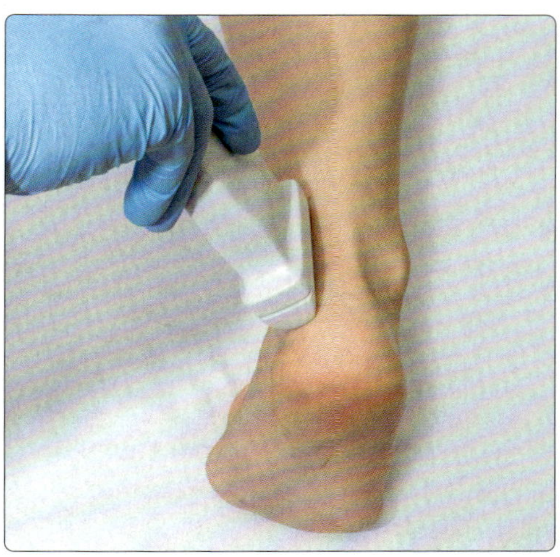

Posterolateral View of Posterior Subtalar Joint

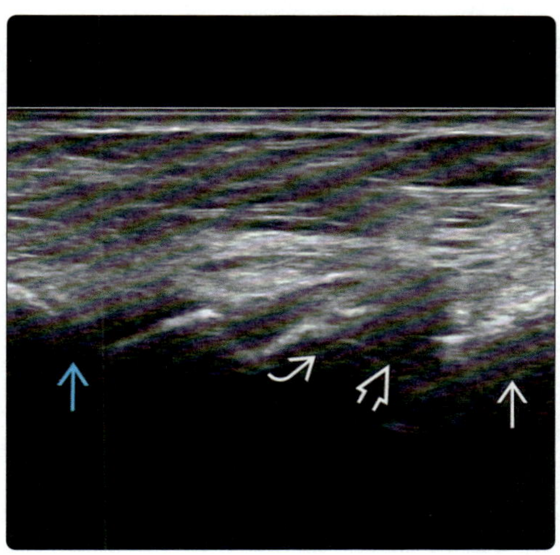

(Left) *Clinical photograph shows the foot and probe position for the PSTJ, posterolateral view. The probe is oriented along the lateral border of the Achilles tendon, and the lower part of the probe is angled medially over the PSTJ.* **(Right)** *Longitudinal US shows the posterolateral view of the PSTJ . The joint lies between the posterior process of talus and the calcaneum . PSTJ space can be widened by dorsiflexion and foot inversion. Posterior recess of the ankle is also shown.*

(Left) *Longitudinal US shows the scan position for the examination of the plantar foot. An advantage of this position is that toe flexion causes stretching and possibly aids visibility of structures, such as the plantar fascia.* **(Right)** *Longitudinal US shows the plantar fascia insertion into the anteromedial aspect of the medial calcaneal tuberosity. Plantar fascia thickness should be measured at the leading edge of the calcaneum. Superficially, the heel fat pad is seen.*

Scan Position for Plantar Foot

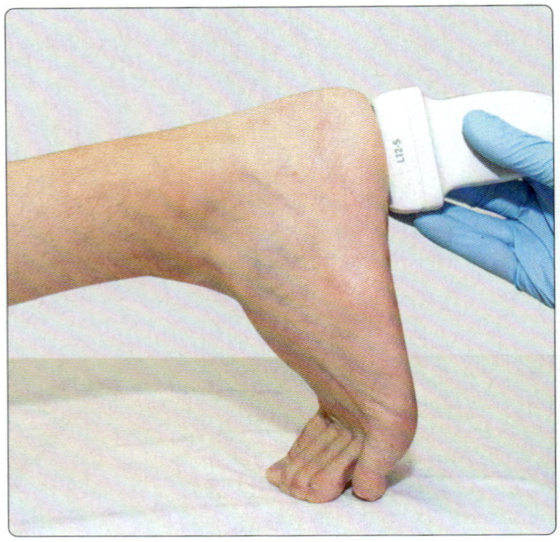

Plantar Fascia Insertion

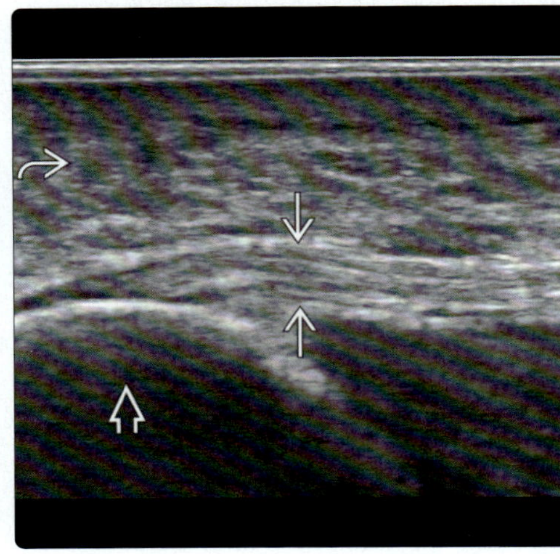

(Left) *Longitudinal US shows the central band of the plantar fascia at the midfoot. The flexor digitorum brevis muscle lies deep to the proximal end of the central band of the plantar fascia. Abductor digiti minimi lies deep to the proximal end of the lateral band of the plantar fascia (not shown).* **(Right)** *Longitudinal US shows the lateral band of the plantar fascia inserting into the inferior aspect of the base of the 5th metatarsal bone.*

Plantar Fascia Midfoot

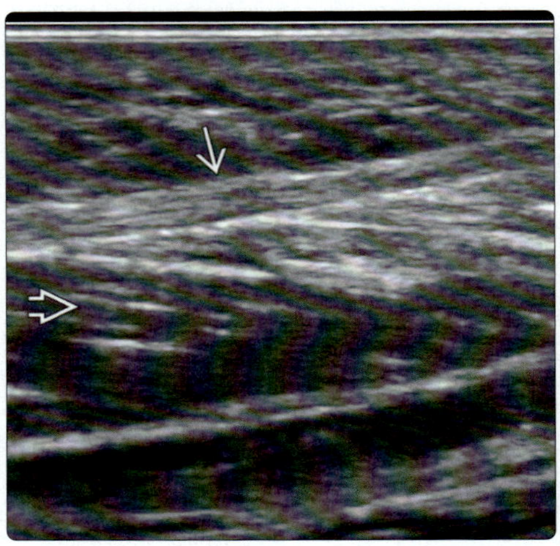

Lateral Band of Plantar Fascia

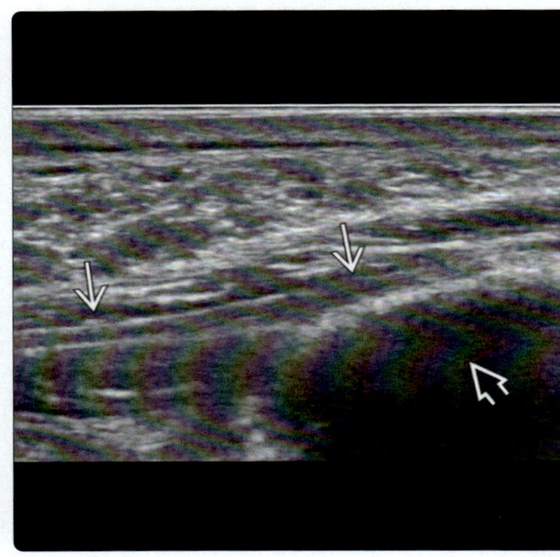

(Left) *Longitudinal US shows the peroneus brevis tendon insertion into the lateral aspect of the base of the 5th metatarsal bone.* **(Right)** *Longitudinal US shows the peroneus longus tendon on the lateral aspect of the midfoot. The peroneus longus passes along the cuboid to the cuboid groove and then along the 3rd layer of plantar foot muscles to insert into the medial cuneiform and 1st metatarsal base. In 20% of subjects, there is an os peroneum located within the peroneus longus tendon just proximal to the groove.*

Peroneus Brevis Tendon Insertion

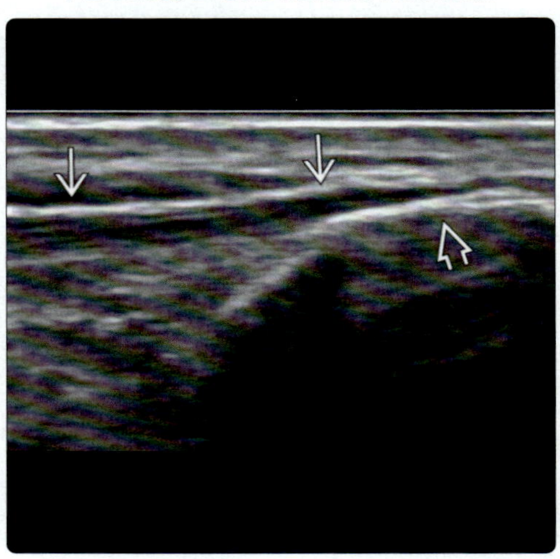

Peroneus Longus Tendon

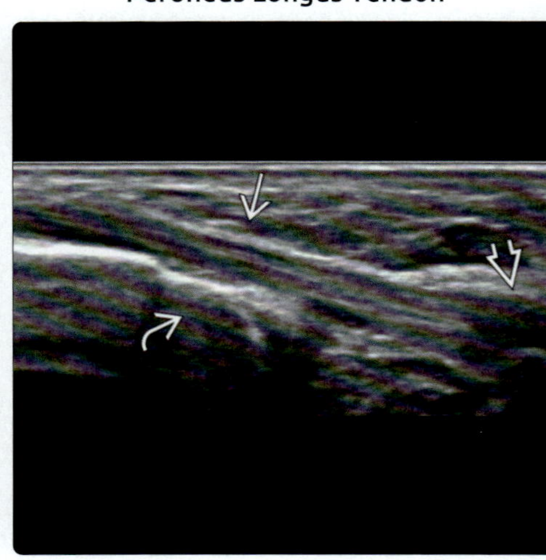

1st Metatarsophalangeal Joint

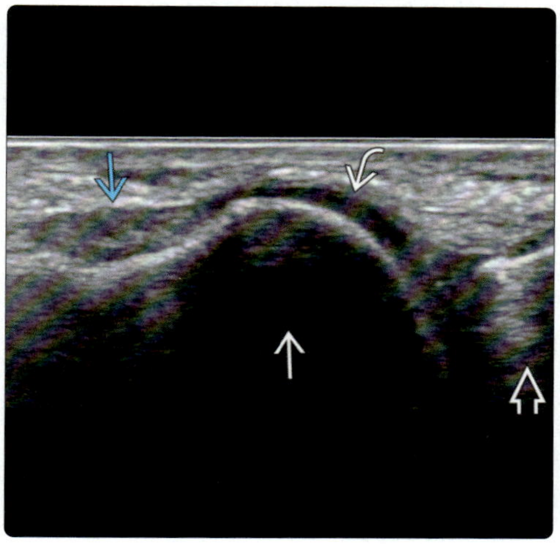

Extensor Hallucis Longus Insertion

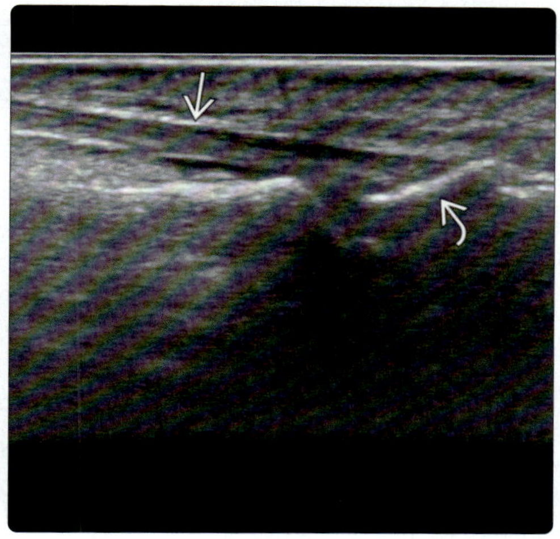

(Left) *Longitudinal US shows the 1st metatarsophalangeal (MTP) joint with a rounded metatarsal head and rather flat base of proximal phalanx . The hyaline cartilage of the metatarsal head is visible along with the dorsal joint recess . This degree of joint distension is normal.* **(Right)** *Longitudinal US shows the normal extensor hallucis longus (EHL) tendon insertion to base of distal phalanx of toe. Variability in position and number of the EHL insertion exists, which may have etiological bearing on hallux valgus.*

Flexor Hallucis Longus

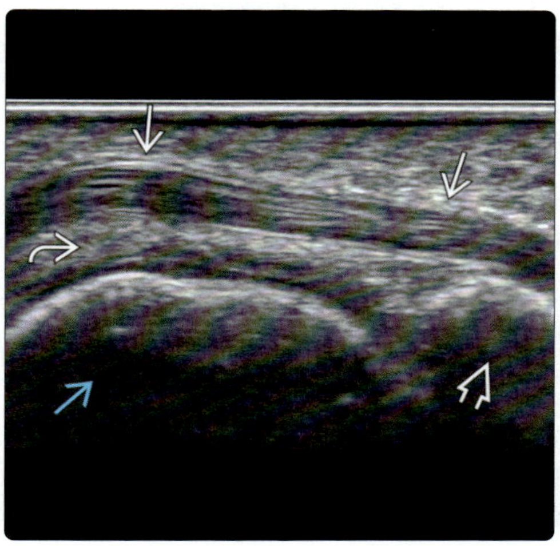

Hallucal-Sesamoid Complex: Short Axis

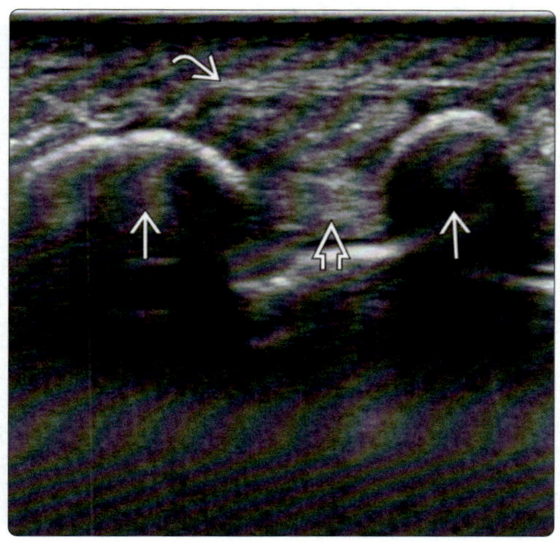

(Left) *Longitudinal US shows flexor hallucis longus tendon overlying the plantar plate of 1st MTP joint. The plantar plate, overlying the metatarsal head , is firmly adherent to the base of the proximal phalanx .* **(Right)** *Transverse US in the plantar aspect of the 1st MTP joint shows the paired sesamoid bones , larger on the medial side. The plantar plate lies deeply between these bones with the flexor hallucis longus tendon more superficially. The intersesamoid ligament is indistinguishable from the plantar plate.*

Hallucal-Sesamoid Complex: Long Axis

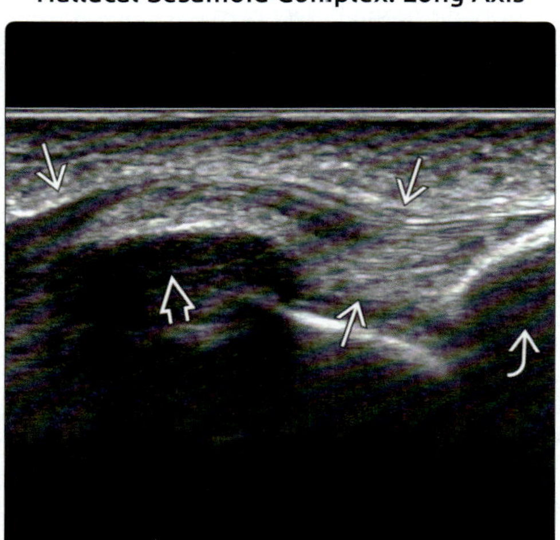

Scan Position for Interdigital Web Space

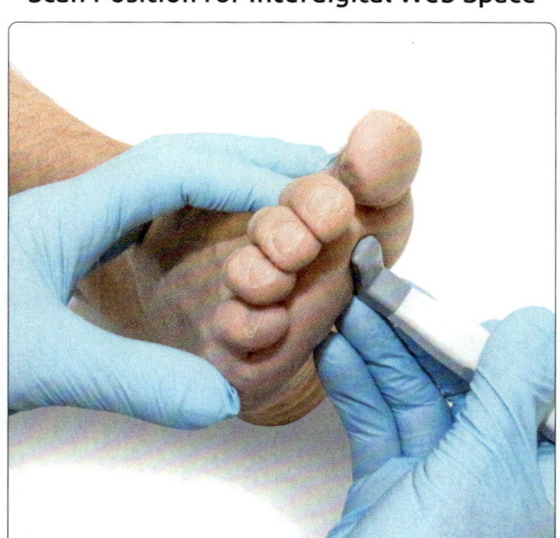

(Left) *Longitudinal US of the base of the 1st MTP joint shows the medial sesamoid embedded in the medial head of the flexor hallucis brevis tendon , which is inserted to the base of the proximal phalanx .* **(Right)** *Clinical photograph shows scan position for examining the interdigital web space. The transducer is placed on the plantar surface with finger pressure applied to the dorsal surface. Conversely, the transducer can be placed dorsally with plantar finger pressure.*

Key General Principles

The operator dependency on US means that the US report is even more relevant for conveying information than for other imaging studies.

Write the report as soon as possible after finishing the US examination.

Address and answer the clinical question.

Make the report brief, clear, precise and pertinent.
- The longer the report, the more likely it will not be read in full
- Report length has been considered inversely proportional to confidence of report writer

Minimize reporting of largely irrelevant pathology, such as small humeral head cortical defects.

Report clinically relevant, but not clinically irrelevant, negative findings.
- Rather than report exhausted list of negative findings, best to say something like: "Otherwise unremarkable appearances"

Avoid ambiguous terminology such as prominent, pronounced, or significant.
- Best to use more precise qualitative descriptors such as mild, moderate, severe or small, medium, large

Avoid use of US descriptors, such as describing a "hypoechoic mass with posterior enhancement though no flow on color Doppler imaging."
- Decide on likely pathology, and then describe depending on how certain you are with respect to diagnosis

If you are certain about the US findings, then be definite in your report.
- "There is medium-sized subcutaneous lipoma present in..."
- "There is small nerve sheath tumor located eccentric to radial nerve in..."
- "There is medium-sized (mm mediolateral x mm anteroposterior) avulsive tear of anterior fiber supraspinatus tendon"

If you are not sure, then you can use terminology such as
- "There is small lipomatous-type mass present in..."
- "There is large irregular area of fibrotic-type tissue present in..."
- "There is probable nerve sheath tumor..." or "This is most likely..."

If you have no idea as to the nature of the lesion, then you can use terminology such as
- "There is abnormal soft tissue mass in...; appearances are nonspecific, although most likely represent..."

If you can completely confirm or exclude a pathology based on your US examination, then be definite in your report and use the term present.
- e.g., "there is no mass present"

If you cannot completely exclude a pathology based on your US findings, then use the term evident rather than present.
- "There is no myositis evident"

Avoid more ambiguous terminology, such as "there is no mass noted" or "there is no mass seen," as this suggests that there may be a mass present but you have not seen it.

Try to minimize use of abbreviations, and only use after abbreviation has first been spelled out in the report.
- "There is severe partial tear of anterior talofibular ligament (ATFL); torn fibers of ATFL have retracted..."

Grading Severity of Disease

Every abnormality mentioned in your report should be graded.
- If qualitative, use minimal, mild, moderate, severe; or use minimal to mild, moderate to severe
- If quantitative, use small, medium sized, large; or use small to medium sized, medium sized to large

There is little benefit in stating, for example, that "there is supraspinatus tendinosis present," as that can mean anything from a minuscule to a severe degree of tendinosis.

Measurements

Do not use mm for one measurement and cm for another measurement in the same report.

Insert a table, whenever possible, if multiple measurements are being provided as, for example, in carpal tunnel syndrome assessment.

If providing measurements as a measure of pathology, always provide the normal range.
- "Plantar fascia thickness, near medial calcaneal tuberosity, was 6.4 mm (normal < 4.3 mm)"

Terminology Best Avoided

"Malignancy cannot be excluded."
- This term should preferably not be used, as, of course, malignancy cannot be fully excluded in any situation

"Clinical correlation is required."
- This states obvious because clinical correlation should always be done, and person doing US examination should be sufficiently well versed to make a clinical judgment

Differential Diagnosis

If appropriate, provide a short differential diagnosis, usually up to 3 possibilities.

Providing a long list of differential diagnoses is not helpful, as it adds little to clinical assessment.

State the relative likelihood of each differential diagnosis, and suggest alternative tests.
- "This most likely represents nerve sheath tumor; an organizing hematoma is another less likely possibility; sarcoma is considered unlikely; US-guided biopsy is recommended"

Conclusion

It should be precise and address the clinical question.

List specific abnormalities found in order of clinical relevance.

Selected References

1. Edwards H et al: What makes a good ultrasound report? Ultrasound. 22(1):57-60, 2014

Pretibial Myxoedema

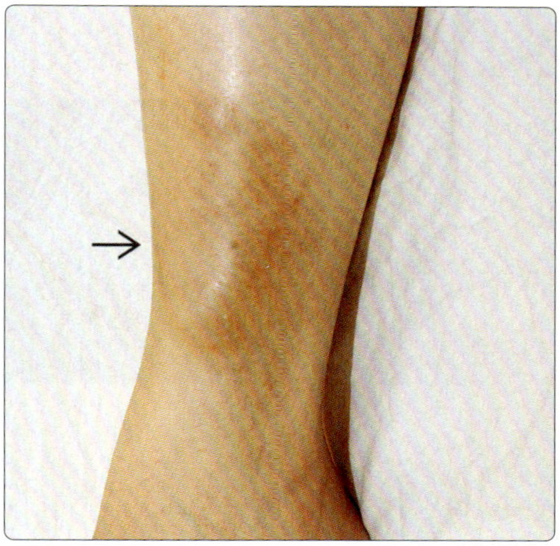

Pretibial Myxoedema

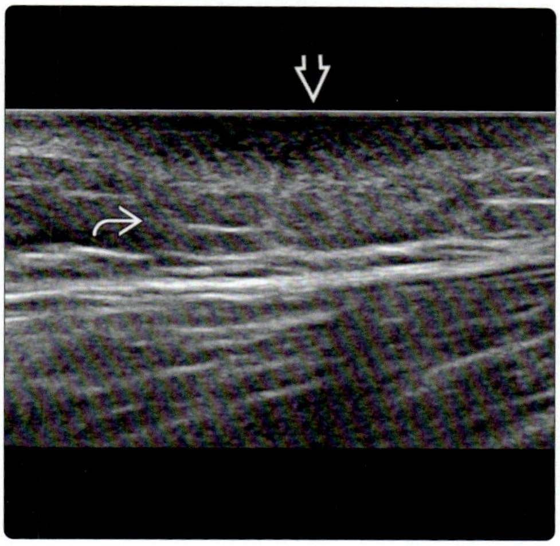

(Left) *Clinical photograph shows a 61-year-old woman with pretibial skin rash* → *and swelling for 2 years.* **(Right)** *Longitudinal US shows localized hypoechoic dermal* ⇨ *and subdermal thickening displacing the mildly edematous subcutaneous fat* ↷ *deeply. On direct questioning, the patient was being treated for hyperthyroidism. In this clinical context, the appearances are consistent with pretibial myxoedema. No further investigation was undertaken.*

Venous Vascular Malformation

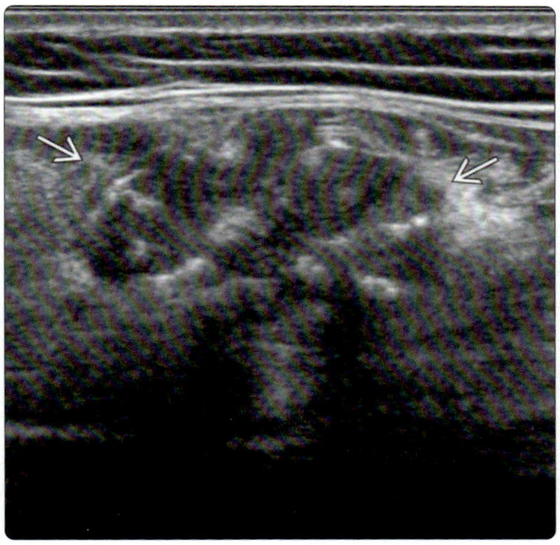

Venous Vascular Malformation

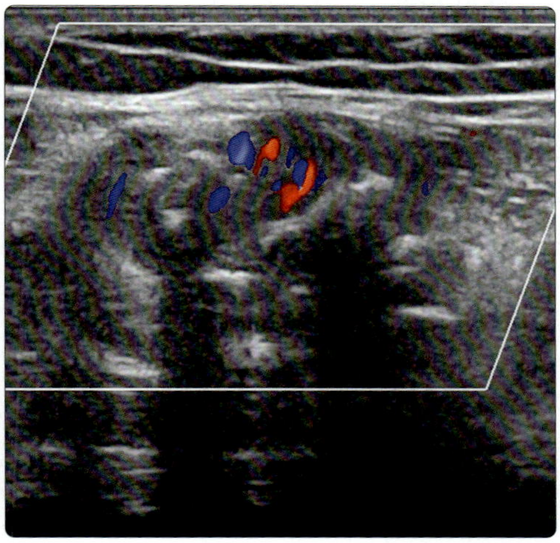

(Left) *Longitudinal US shows the calf in a 28-year-old woman with calf muscle swelling for several years slowly increasing in size. There is a quite well-defined, partially calcified mass* → *within the soleus muscle.* **(Right)** *Color Doppler imaging shows mild hyperemia. Overall a venous vascular malformation was considered the most likely diagnosis. Other possibilities included an organizing chronic hematoma, previous myositis, or synovial sarcoma. Biopsy confirmed a vascular malformation.*

Myositis Ossificans

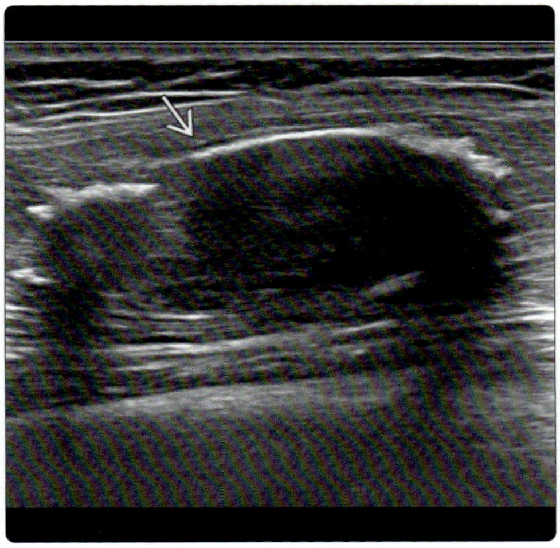

Seroma

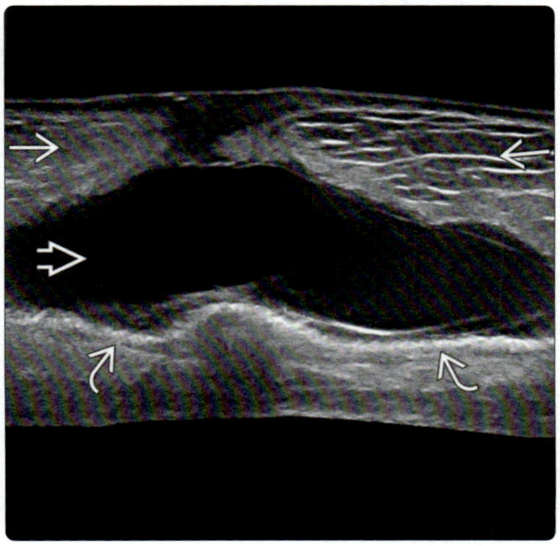

(Left) *Longitudinal US in a 51-year-old woman with thigh swelling shows a well-demarcated mass in the vastus medialis muscle with solid rim-like calcification* →*. Myositis ossificans was the most likely diagnosis. Radiographs confirmed progressive ossification of the mass. No further investigation was undertaken.* **(Right)** *Transverse panoramic US in an 88-year-old woman 1 month after paraumbilical incisional hernia repair shows a large seroma* ⇨ *located between the surgical mesh* ↷ *and the subcutaneous tissue* →*.*

SECTION 3

Diagnoses

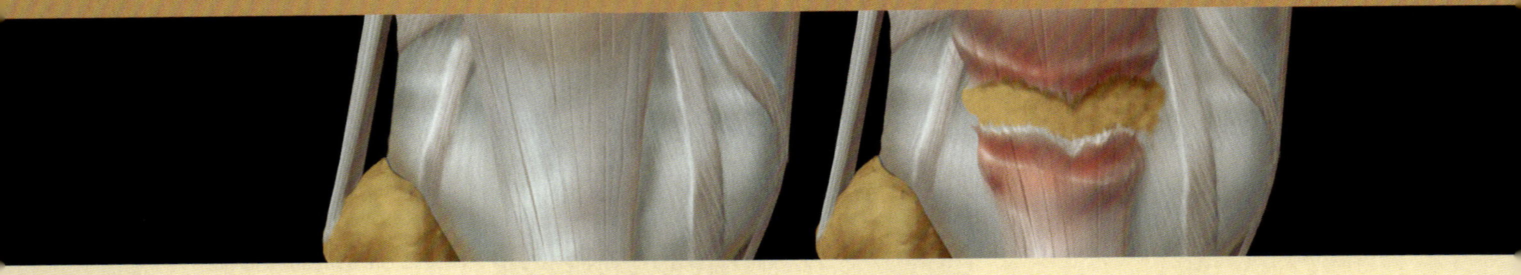

Tendon Disorders

Rotator Cuff/Biceps Tendinosis 442
Rotator Cuff/Biceps Tendon Tear 454
Nonrotator Cuff Tendinosis 462
Nonrotator Cuff Tendon Tears 472
Tenosynovitis 484
Elbow Epicondylitis 494

Soft Tissue, Bone, and Joint Injury

Fat Injury 502
Muscle Infarction 510
Muscle Injury 514
Hematoma/Seroma 524
Ligament Injury 532
Bone Fracture 540

Arthropathies

Osteoarthritis 550
Inflammatory Arthritis 558
Gout and Pseudogout 568
Developmental Hip Dysplasia 576

Neurovascular Abnormalities

Nerve Injury 582
Nerve Sheath Tumors 590
Carpal Tunnel Syndrome 602
Cubital Tunnel Syndrome 610
Tarsal Tunnel Syndrome 614
Vascular Dilatation or Inflammation 620

Infection

Soft Tissue Infection 628
Bone Infection 636
Joint Infection 646
Postoperative Infection 652

Articular and Paraarticular Masses

Hemarthrosis and Lipohemarthrosis 658
Baker Cyst 662
Bursitis 668
Ganglion Cyst 680
Parameniscal Cyst 690
Synovial Tumor 694

Soft Tissue and Bone Tumors

Plantar Fasciitis and Fibromatosis 706
Lipoma 712
Epidermoid Cyst 720
Pilomatricoma 726
Dermatofibrosarcoma Protuberans 730
Vascular Leiomyoma 734
Superficial Metastases, Lymphoma, and Melanoma 738
Vascular Anomaly 744
Foreign Body and Injection Granuloma 752
Lymph Node Abnormality 760
Soft Tissue Sarcoma 766
Bone Tumor 774
Local Tumor Recurrence 786

Hernia

Abdominal Wall Hernia 796
Groin Hernia 804

Rotator Cuff/Biceps Tendinosis

KEY FACTS

TERMINOLOGY

- Collagenous degeneration of rotator cuff ± biceps tendons with proteoglycan deposition

IMAGING

- Supraspinatus > subscapularis, biceps and infraspinatus
- Subclinical tendinosis very common and usually bilateral
  - Do not rely on asymptomatic contralateral side as normal internal reference
- US better than MR for assessing tendinosis as tendon detail, such as fibrillar pattern, better appreciated
- Thickened tendon with diffuse hypoechogenicity and indistinct fibrillar pattern
  - Grade as mild, moderate, or severe
- ± cortical irregularity of tendon insertional area
- Tendon hyperemia unusual in rotator cuff tendinosis
- Strong association between degree of tendinosis and tendon tears
  - The more severe the tendinosis, the greater the likelihood of tendon tear
- Focal hypoechoic area of proteoglycan accumulation within tendinosis can mimic intrasubstance tear
  - Tears are more linear, more hypoechoic, sharper in outline, ± associated with tendon volume loss
- Calcific tendinosis: Formative, latent (resting), and resorptive phases
  - Focal echogenic area within rotator cuff tendon with acoustic shadowing
  - Surrounding tendon hyperemia in resorptive phase
- Biceps tendinosis usually accompanies rotator cuff tendinosis
  - Proximal end ("genu") of bicipital groove most common site of biceps tendinosis

TOP DIFFERENTIAL DIAGNOSES

- Artifact from anisotropy or rotator cable
- Intrasubstance tendon tear

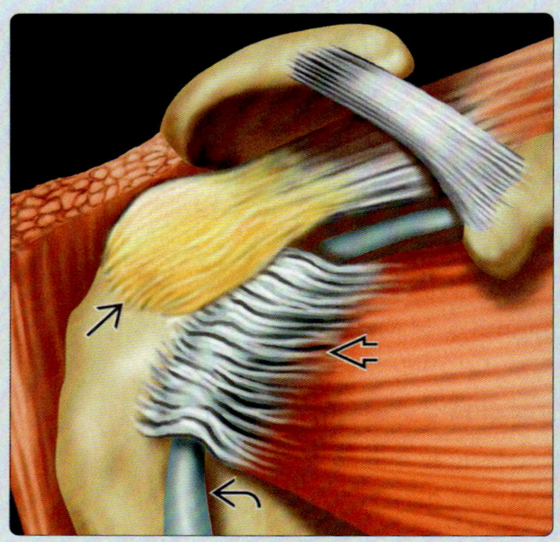

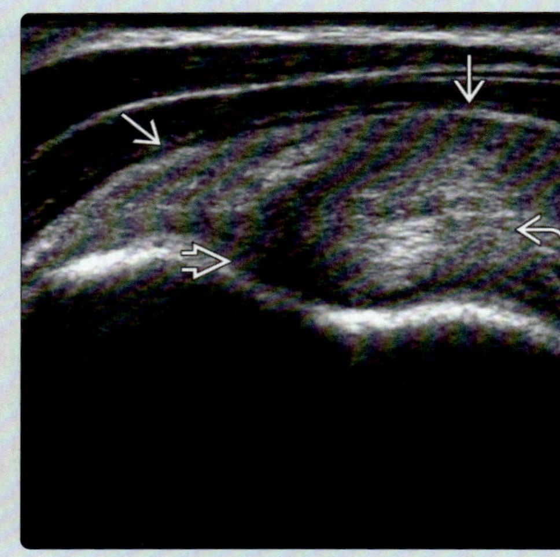

(Left) *Oblique graphic shows a diffusely thickened supraspinatus tendon → indicative of tendinosis. A normal subscapularis tendon → and biceps tendon → are also shown.* **(Right)** *Longitudinal US shows a diffusely thickened and hypoechoic supraspinatus tendon → indicative of moderate to severe tendinosis. A barely discernible fibrillary pattern → is present. Note the hypoechoic area → close to the insertion due to a combination of anisotropy and rotator cable effects.*

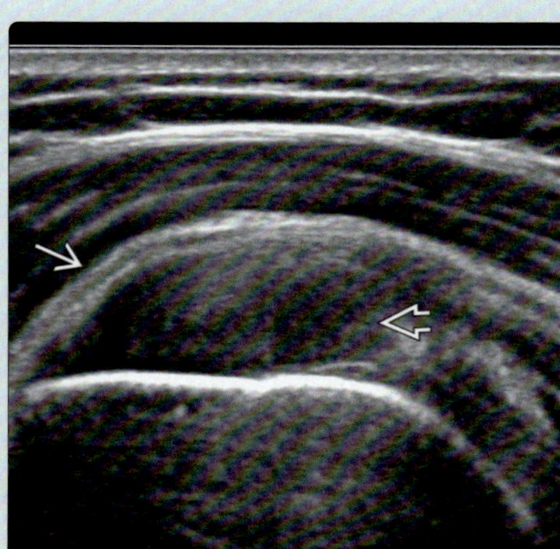

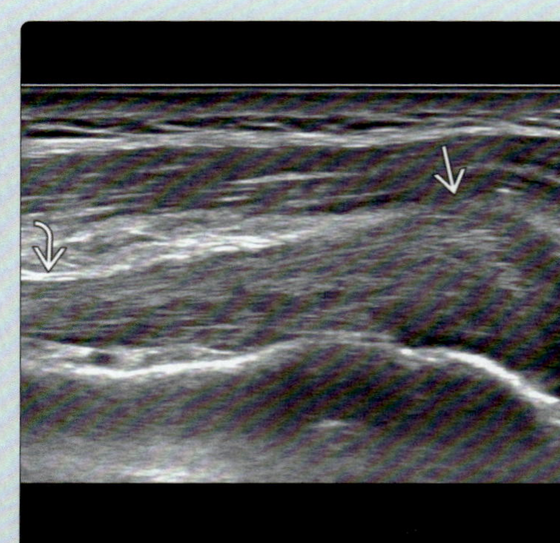

(Left) *Longitudinal US shows a diffusely thickened and hypoechoic supraspinatus tendon →. Multiple, poorly demarcated, hypoechoic areas → with loss of distinct fibrillary structure are present. Features are indicative of severe supraspinatus tendinosis.* **(Right)** *Longitudinal US shows severe tendinosis of long head of biceps tendon → at proximal end of bicipital groove at "genu" between horizontal and vertical components. The more distal tendon is relatively normal →. Tendinosis usually occurs at this location.*

TERMINOLOGY

Synonyms

- Rotator cuff tendinopathy

Definitions

- Collagenous degeneration of rotator cuff ± biceps tendons with proteoglycan deposition

IMAGING

General Features

- Best diagnostic clue
  - Thickened tendon with diffuse hypoechogenicity and indistinct fibrillar pattern
- Location
  - Supraspinatus > subscapularis, biceps and infraspinatus
  - Slight preponderance in dominant shoulder
- Morphology
  - Thickened, inhomogeneous tendon ± surface fraying
    - Tendon torn or partially torn in advanced cases with fluid entering defect
    - Strong association between tendinosis severity and likelihood of tendon tear

Radiographic Findings

- Radiography
  - Flat-undersurface acromion, anteriorly hooked acromion, anterior or lateral downsloping acromion, os acromiale
  - Acromioclavicular joint osteophytes, acromial spurs
  - Surface irregularity, sclerosis, or subcortical cysts of greater tuberosity
  - Tendon calcification in calcific tendinosis

MR Findings

- T1WI
  - Thickened tendon with intermediate signal intensity
- T2WI
  - Increased signal intensity of tendon on PD FSE, FS PD FSE, STIR & T2*, and T2 GRE

Ultrasonographic Findings

- Thickened, hypoechoic tendon with patchy and linear areas of more pronounced hypoechogenicity
  - Graded as mild, moderate, or severe based on degree of thickening, hypoechogenicity, and loss of normal fibrillar pattern
    - Measurement of tendon thickness is generally not helpful as there are no distinct landmarks and thickness is dependent on patient size
  - Focal hypoechoic areas of proteoglycan deposition can mimic intrasubstance tears
    - Tears are more linear, more hypoechoic, and sharper in outline, ± associated with tendon volume loss
  - Nonintrasubstance tears are discernible through changes in tendon contour (contour retraction, loss of volume, fluid gap)
- Cortical irregularity of tendon insertional area occurs with more severe degrees of tendinosis
- Tendinosis generally affects all rotator cuff tendons to some degree with supraspinatus most affected
  - Occasionally, only a single tendon affected
  - Biceps tendinosis usually accompanies rotator cuff tendinosis
    - Normal biceps tendon has oblong cross-sectional contour at proximal end of bicipital groove ("genu"), which is usual site of biceps tendinosis
    - Biceps tendon becomes thickened, more rounded, and more indistinct with increasing severity of tendinosis
- Likelihood of tendon tear increases as degree of tendinosis increases
  - Tendinosis and tendon tears usually occur in same areas
  - Tears uncommonly occur in normal tendons
  - Likelihood of successful repair ↓ as severity of tendinosis ↑
- Tendon hyperemia very common in rotator cuff tendons
  - Tendon hyperemia is much more common in tendinosis at other sites
- US allows ready comparison with opposite side
- Subclinical tendinosis of similar severity is often present on opposite asymptomatic side
  - Be careful about using opposite asymptomatic side as internal reference of normality
- **Calcific tendinosis**
  - Formative, latent (resting), and resorptive phases
    - Focal echogenic mass within rotator cuff tendon with acoustic shadowing
      - Variable size and multiplicity
      - During formative and resorptive stages, calcification can be less echogenic with less acoustic shadowing
      - Tendon hyperemia may be seen in resorptive phase due to intense inflammation and vascularization of tissue around calcium deposit

Imaging Recommendations

- Best imaging tool
  - US is better than MR for assessing tendinosis as tendon detail, such as fibrillar pattern, better appreciated
    - Patients tend to prefer US
- Protocol advice
  - Systematically examine tendons, taut and nontaut, in longitudinal and transverse planes with patient in sitting position
  - Arm rotated into different positions to optimize visibility of each tendon in turn
  - Important to always grade severity of tendinosis as mild, moderate, or severe for each tendon individually
    - Can also grade as "mild to moderate" or "moderate to severe" in intermediate in severity

DIFFERENTIAL DIAGNOSIS

Intrasubstance Tendon Tears

- Focal areas of proteoglycan deposition may mimic intrasubstance tears and vice versa
  - Tears are more hypoechoic, more sharply defined, and more linear, ± contour flattening or retraction of tendon

Anisotropic Effect

- Tendon appears hypoechoic when US beam is not perpendicular to tendon fibers
- Realign transducer to eliminate anisotropic effect

Rotator Cable

- Band-like tendinous thickening at medial aspect of rotator crescent
- Running perpendicular to beneath supraspinatus and infraspinatus tendon
- Strengthens rotator cuff tendons
- More prevalent on US with background tendinosis
- Cable can be preferentially spared when remaining tendon shows tendinosis
  - Uncommonly, reverse may happen

PATHOLOGY

General Features

- Etiology
  - Combination of intrinsic, extrinsic, biomechanical, and genetic factors
    - Intrinsic theory: Overuse, degeneration, and microtears
    - Extrinsic theory: Secondary to repeated impingement, steroid injections
    - Biomechanical: Eccentric tensile overload of rotator cuff tendons
- Genetics
  - Familial predisposition

Staging, Grading, & Classification

- Grade arbitrarily into mild, moderate, or severe, depending on degree of thickening, hypoechogenicity, and loss of normal fibrillar pattern
  - Severity of tendinosis has implications regarding likelihood for tear and surgical outcome

Gross Pathologic & Surgical Features

- Thickened grayish or yellow-brown tendon rather than normal white color

Microscopic Features

- Collagen degeneration and disruption without inflammatory cell infiltrate
  - Normal architectural arrangement of collagen is impaired
  - ↓ in type I collagen content with ↑ in weaker type III collagen content
  - Tenocyte apoptosis (cell death) in some areas
  - Increase in tenocyte number in other areas with more rounded nuclei
  - Increase in extracellular proteoglycan ground substance with high concentration of glycosaminoglycan chains
- **Calcific tendinosis**
  - Deposition of hydroxyapatite crystals into matrix crystals vesicles
    - These vesicles coalesce to form foci of calcification
      - Chalk-like during formative and latent phases
      - Paste-like during resorptive phase

CLINICAL ISSUES

Presentation

- Most common signs/symptoms
  - Often subclinical; usually bilateral in older patients
  - Slowly progressive, poorly localized shoulder pain with decreased range of motion
  - Relationship between pain and severity of tendinosis is unclear
    - Tendinosis per se probably does not induce pain
      - Normal tendon does not contain neurons
      - High prevalence of asymptomatic tendinosis
    - Pain most likely related to associated bursitis ± bone change or impingement
  - **Calcific tendinosis**
    - Resorptive phase is characterized by severe pain and limitation of movement
    - Shedding of crystals into bursa, into joint or into bone causes pain
- Clinical profile
  - In young athletes, precipitating factor, such as os acromiale, overhead sporting activity is usually present
  - In older athletes or older nonathletic population, onset is more insidious ± structural abnormality

Demographics

- Age
  - Uncommon < 30 years, gradually increasing prevalence and severity with increasing age
- Gender
  - Slight female predominance

Treatment

- Physiotherapy
- Corticosteroids and long-acting local anesthetic injection to subacromial-subdeltoid bursa
  - Multiple steroid injections best avoided as may accelerate tendon degeneration
- Subacromial decompression
- **Calcific tendinosis**
  - Usually self limiting
  - US-guided dry needling and aspiration
  - US-guided lavage and aspiration
    - 1- or 2-needle technique
    - Warm saline (42°C) lavage seems better at calcium dissolution than room temperature saline
      - Decreases procedure time and likelihood of postprocedural bursitis
  - Extracorporeal shock wave therapy

DIAGNOSTIC CHECKLIST

Consider

- Small tear if no abnormality on 1st inspection in patient with tendinosis and sudden deterioration in symptoms; check again
- Look especially at leading edge of tendon for small avulsive-type tear
- Tears are very common in more severe degrees of tendinosis, particularly intrasubstance tears
  - Intrasubstance tears are almost integral part of severe tendinosis; expect to see them

SELECTED REFERENCES

1. Lee MH et al: Comprehensive shoulder US examination: a standardized approach with multimodality correlation for common shoulder disease. Radiographics. 36(6):1606-1627, 2016

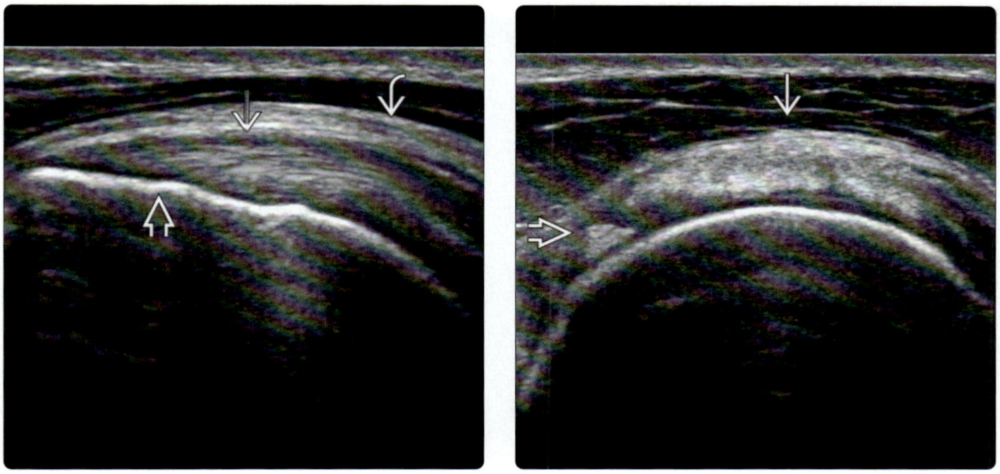

(Left) *Longitudinal US shows a normal, hyperechoic supraspinatus tendon ➔ of normal size and with a distinct fibrillary pattern. Note the smooth insertional area on the greater tuberosity of the humerus ➔. The peribursal fat ➔ deep to the deltoid muscle is also shown.* **(Right)** *Transverse US of the same patient shows the normal, hyperechoic supraspinatus tendon ➔ in transverse section. Note the tendon of the long head ➔ of biceps within the rotator interval.*

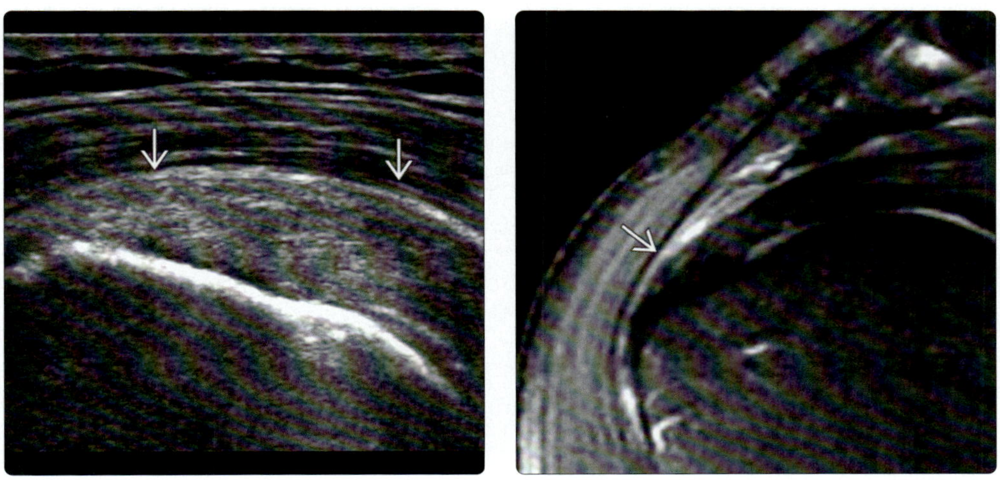

(Left) *Longitudinal US shows mild supraspinatus tendinosis. The tendon ➔ is mildly thickened and hypoechoic. Most of the hyperechoic fibers can still be identified.* **(Right)** *Oblique coronal T2 SPIR MR shows mild supraspinatus tendinosis with mild tendon hyperintensity ➔ over and above that due to magic angle effect and mild tendon thickening. The fibrillar pattern of the tendon is better visualized by US than MR.*

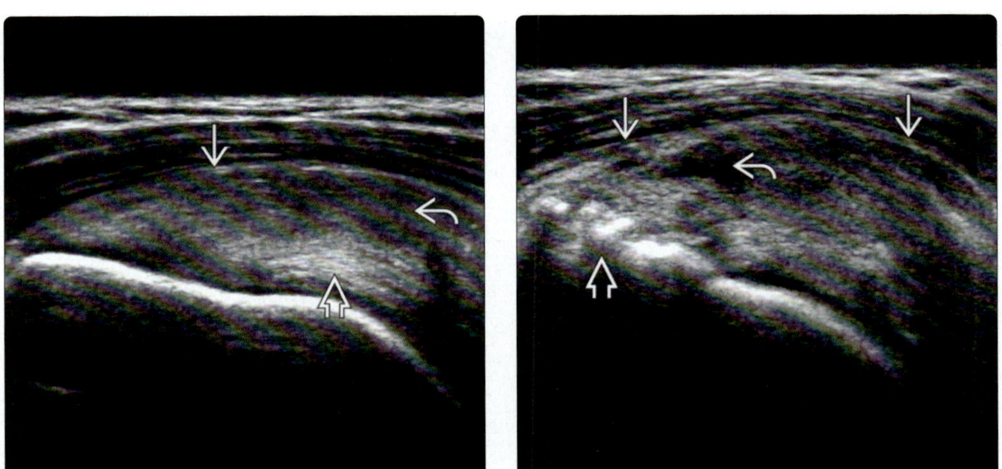

(Left) *Transverse US shows moderate supraspinatus tendinosis. The tendon ➔ is more thickened and heterogeneous. There are ill-defined areas of hypoechogenicity ➔ interspersed with areas of preserved fibrillar pattern ➔.* **(Right)** *Longitudinal US shows a markedly thickened and hypoechoic supraspinatus tendon ➔ due to severe tendinosis. Normal fibrillar pattern is largely lost. There is also severe cortical irregularity at the insertional area ➔, and there are small intrasubstance tears ➔.*

(Left) *Frontal radiograph of the shoulder shows a subacromial hyperostotic spur, which narrows the subacromial space, predisposing to tendon impingement and tendinosis. Note mild cortical irregularity and hyperostosis of the greater tuberosity as a feature of tendinosis.* **(Right)** *Longitudinal US in the same patient shows mild supraspinatus tendinosis. The hypoechoic area is due to a different orientation between rotator cable fibers and remaining supraspinatus fibers.*

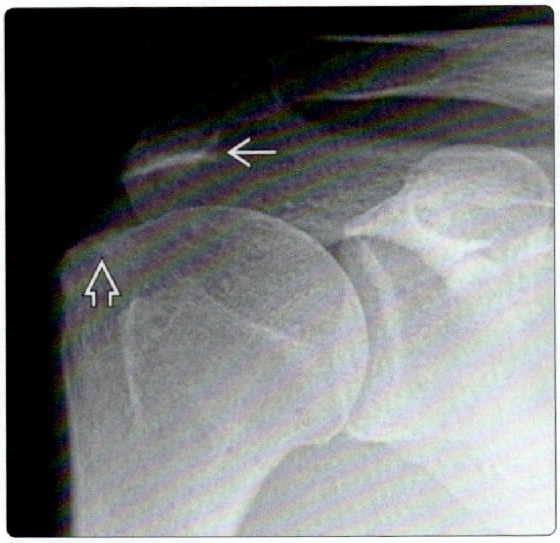

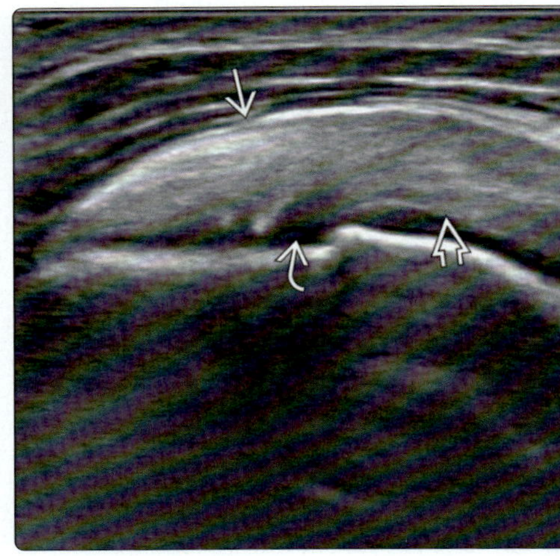

(Left) *Longitudinal US shows tendinosis in the main supraspinatus fibers and no tendinosis in the rotator cable fibers. This accentuates the distinction between both fiber groups more than normal. The supraspinatus distal to the cable is the crescent.* **(Right)** *Oblique graphic shows fiber condensation of the rotator cable, which extends perpendicular and deep to the supraspinatus and infraspinatus tendons. The semicircular area between the rotator cable and the tendon insertion is the rotator crescent.*

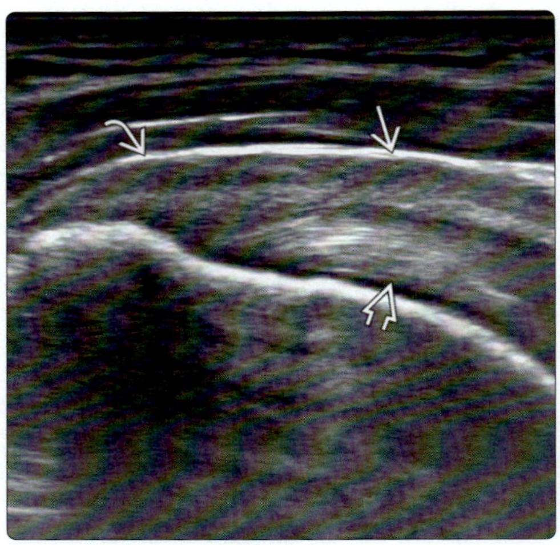

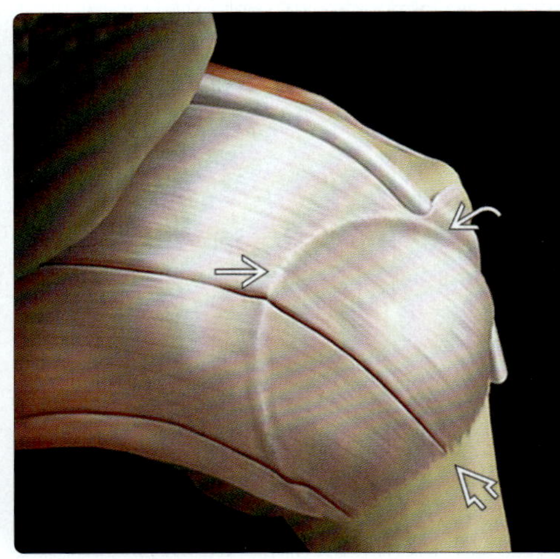

(Left) *Longitudinal US shows mild supraspinatus tendinosis. The tendon is mildly and diffusely thickened. Its normal hyperechogenicity and fibrillar pattern are largely preserved. It is important to grade tendinosis as mild, moderate, or severe.* **(Right)** *Transverse US of the same patient shows minimal distension of the subacromial subdeltoid bursa associated with supraspinatus tendinosis.*

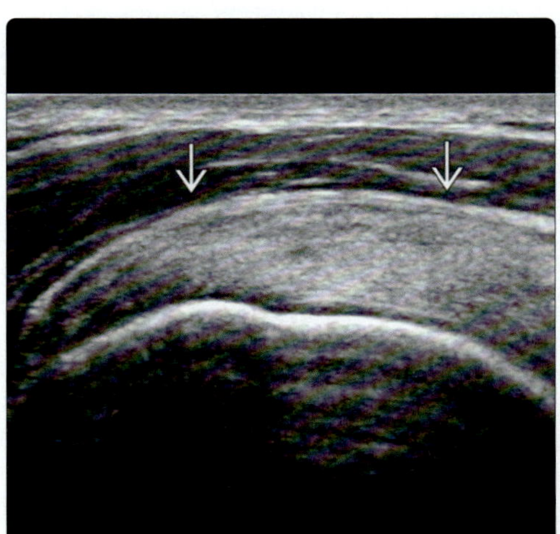

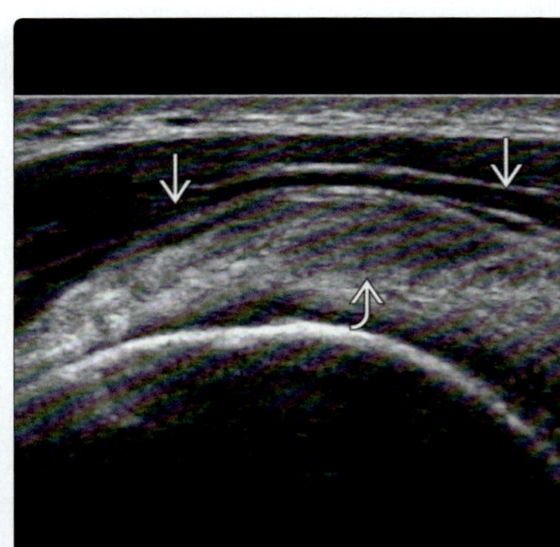

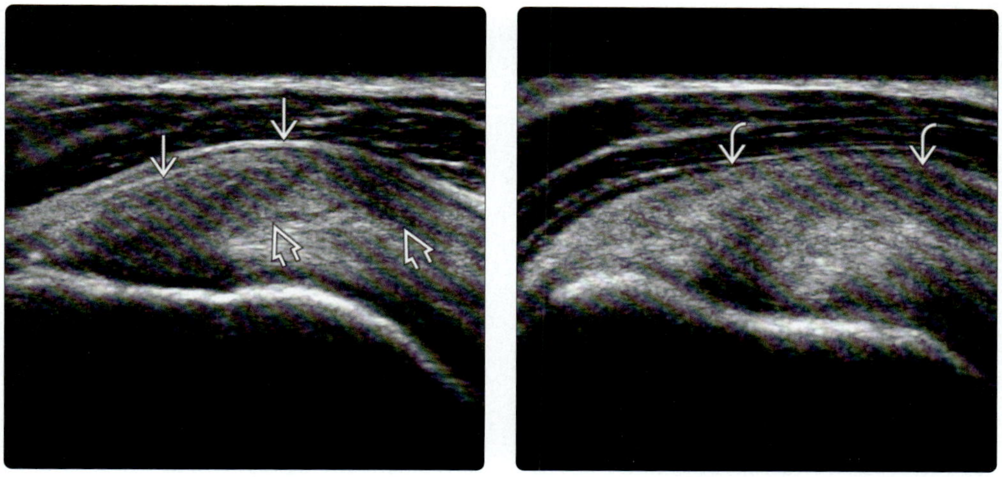

(Left) *Longitudinal US shows moderately severe supraspinatus tendinosis →. The tendon is thickened, and the fibrillar structure is distinct with ill-defined hypoechoic areas →. One should use all 3 criteria to grade the severity of tendinosis.* **(Right)** *Transverse US of the same patient with moderate supraspinatus tendinosis shows mild bursal surface fibers fraying →. This is a common feature in more severe degrees of tendinosis.*

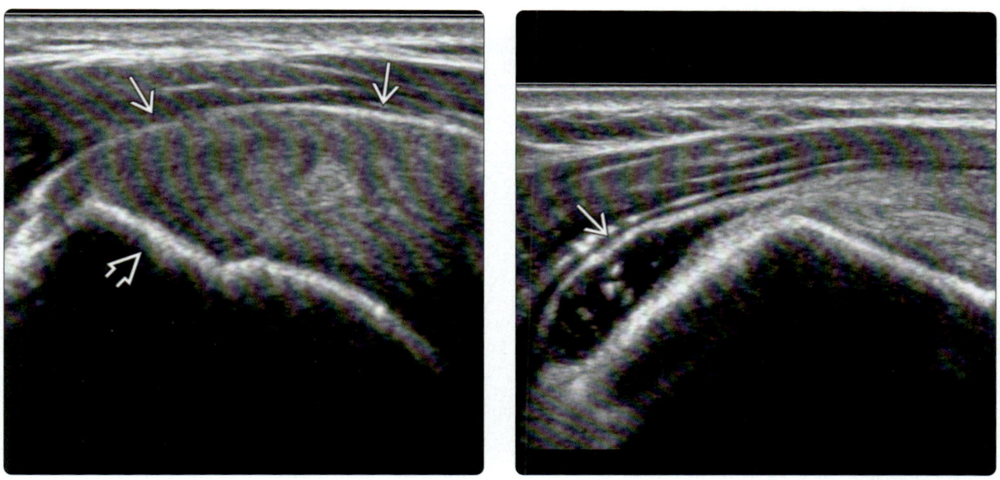

(Left) *Longitudinal US shows a diffusely thickened and hypoechoic supraspinatus tendon →. The loss of distinct fibrillary structure is present. Features are indicative of moderate to severe supraspinatus tendinosis. Note the supraspinatus insertional footprint →.* **(Right)** *Longitudinal US of the same patient at a slightly more lateral portion shows mild distension of the subacromial-subdeltoid bursa →, indicative of mild bursitis. In this situation, one should search carefully for a tendon tear.*

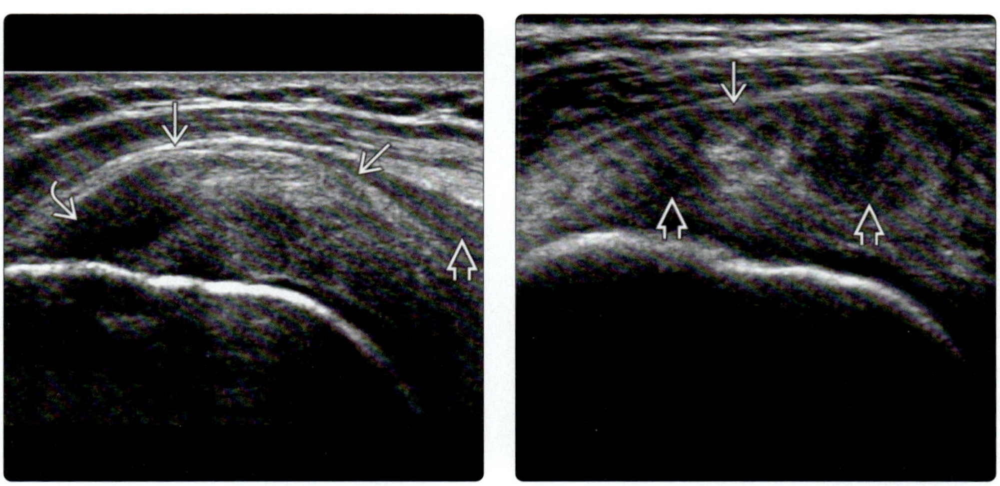

(Left) *Longitudinal US shows moderate-severity supraspinatus tendinosis →. Note the hypoechoic anisotropic effect →. There is distension of the subacromial-subdeltoid bursa → with echogenic fluid probably due to proteinaceous content.* **(Right)** *Longitudinal US shows severe supraspinatus tendinosis →. The tendon is more thickened and heterogenous. There are indistinct areas of hypoechogenicity → due to proteoglycan accumulation.*

(Left) *Transverse US of the same patient shows diffuse thickening and heterogeneous echogenicity of the supraspinatus tendon, associating with mild bursal surface fraying and mild subacromial-subdeltoid bursitis.* **(Right)** *Sagittal T2 FS MR shows moderate supraspinatus tendinosis within moderate tendon thickening and hyperintensity. Moderate fluid distension in the subacromial-subdeltoid bursa is present. Note the normal subscapularis and infraspinatus tendons.*

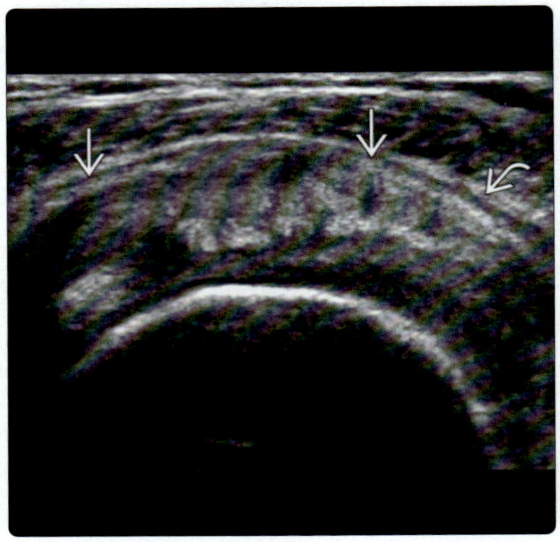

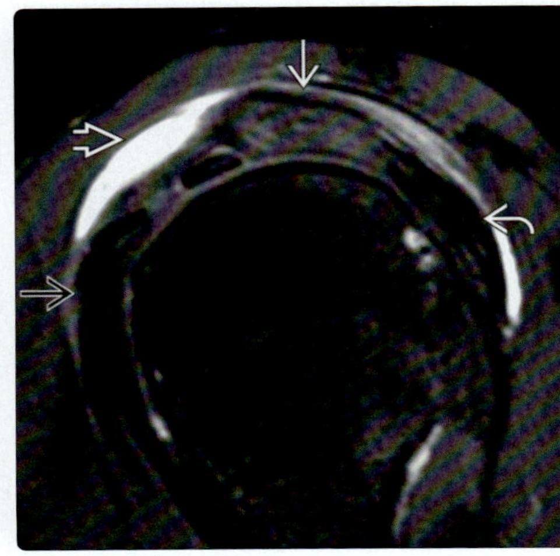

(Left) *Longitudinal US shows severe supraspinatus tendinosis with a well-demarcated, linear, hypoechoic intrasubstance tear. The other more globular hypoechoic area within the tendon is a feature of tendinosis.* **(Right)** *Longitudinal US shows a partial-thickness bursal-surface tear of the supraspinatus tendon at the insertional area. The tear is sharply demarcated against the background of moderate tendinosis. Note the cortical irregularity at the greater tuberosity.*

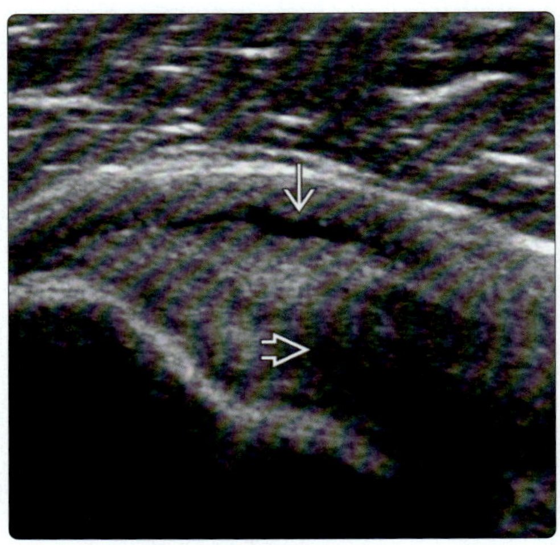

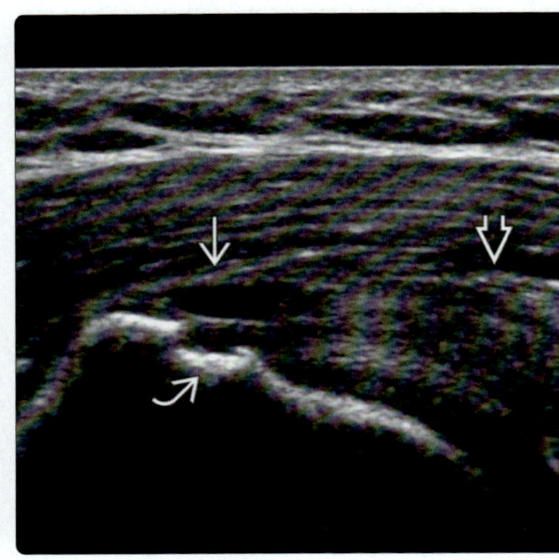

(Left) *Longitudinal US shows severe supraspinatus tendinosis. There is subtle bursal surface contour undulation present on this scan plane. Any contour deformity in tendinosis raises the suspicion of an underlying tendon tear.* **(Right)** *Transverse US of the same patient shows a full-thickness partial tear of the midfibers of the supraspinatus. Note the severe thickening of the anterior and posterior fibers. There is a strong relationship between severity of tendinosis and tendon tear.*

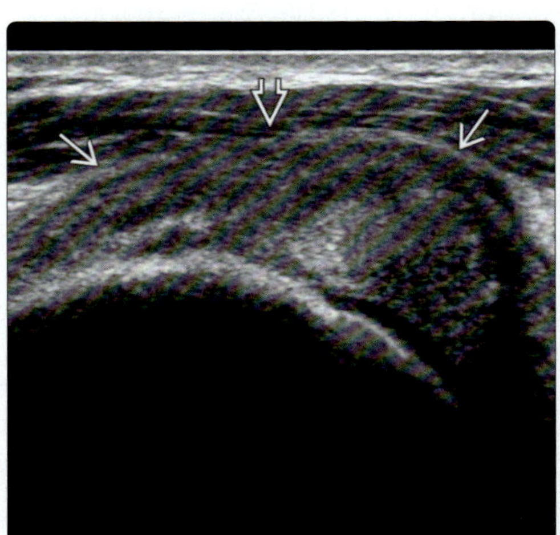

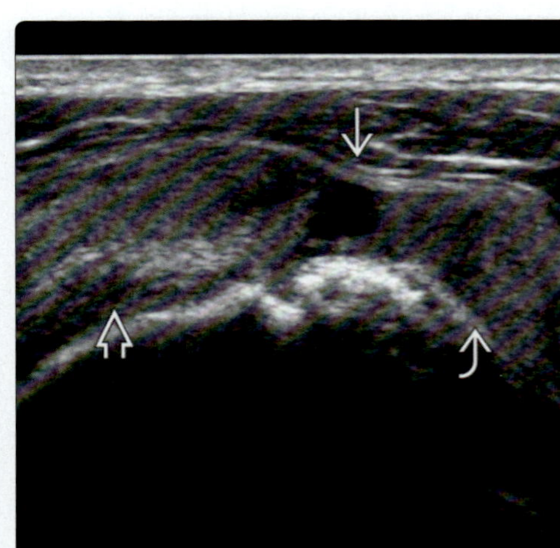

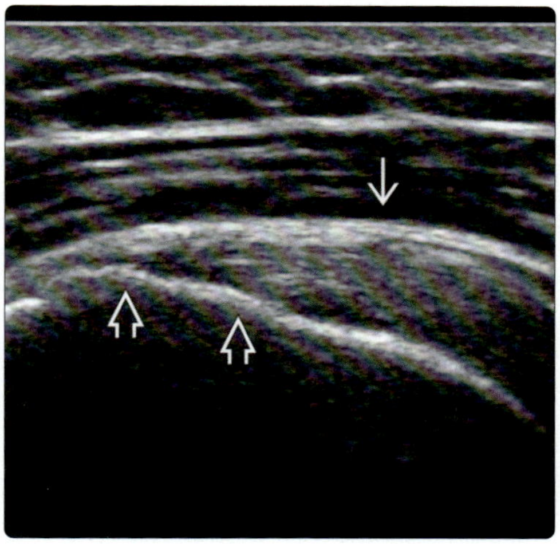

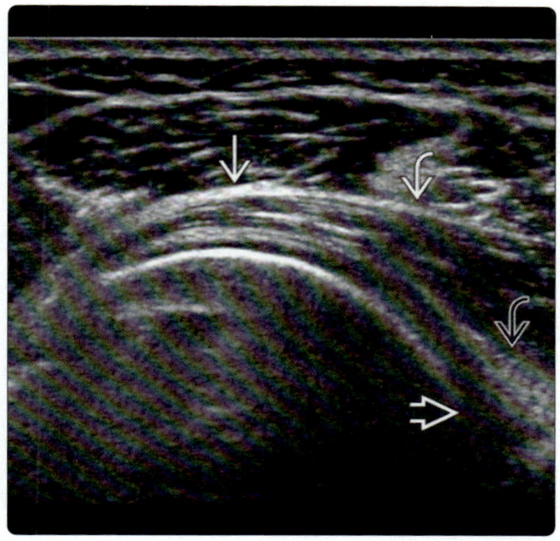

(Left) *Longitudinal US shows a normal infraspinatus tendon inserting into the middle facet of the greater tuberosity. The infraspinatus, supraspinatus, and teres minor are inseparable at insertion.* **(Right)** *Longitudinal US shows the infraspinatus tendon at a slightly more proximal portion. In this position, the myotendinous junction as well as the posterior glenohumeral joint can be visualized. Note the posterior glenoid labrum.*

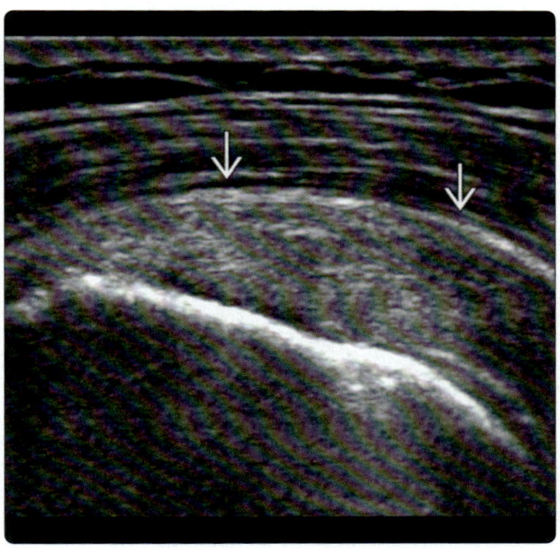

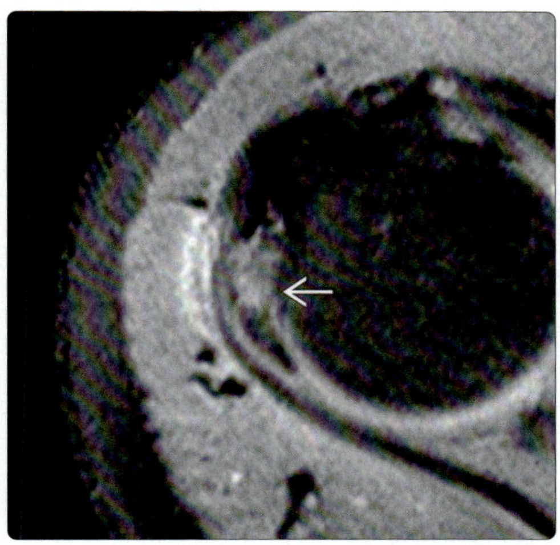

(Left) *Longitudinal US shows mild thickening and hypoechogenicity of the infraspinatus tendon indicative of mild tendinosis.* **(Right)** *Axial PD FSE FS MR shows thickening and hyperintense signal of the insertional fibers of the infraspinatus tendon consistent with moderate tendinosis. You can appreciate how MR gives a global picture, though it does not provide the same fine tendon detail as an US.*

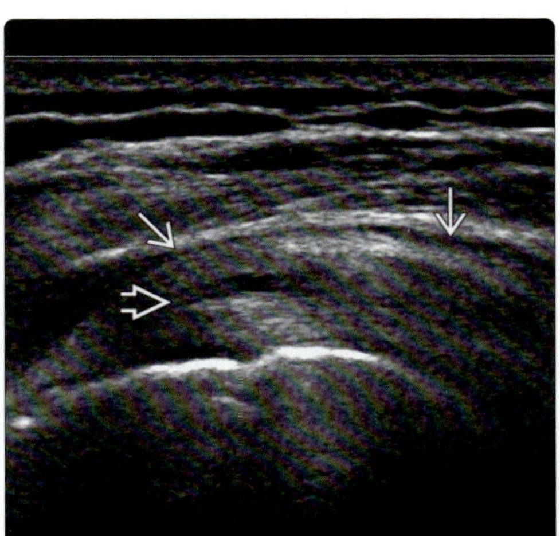

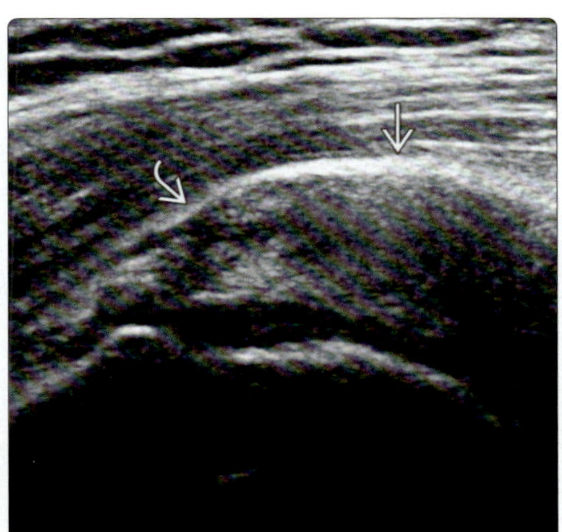

(Left) *Longitudinal US shows moderate-severity infraspinatus tendinosis. The tendon is moderately thickened and hypoechoic. There is an intrasubstance tear without extension to either the bursal or articular surface present.* **(Right)** *Longitudinal US shows a diffusely thickened and hypoechoic infraspinatus tendon indicative of severe infraspinatus tendinosis. There is a small, irregular bursal surface tear near the insertion.*

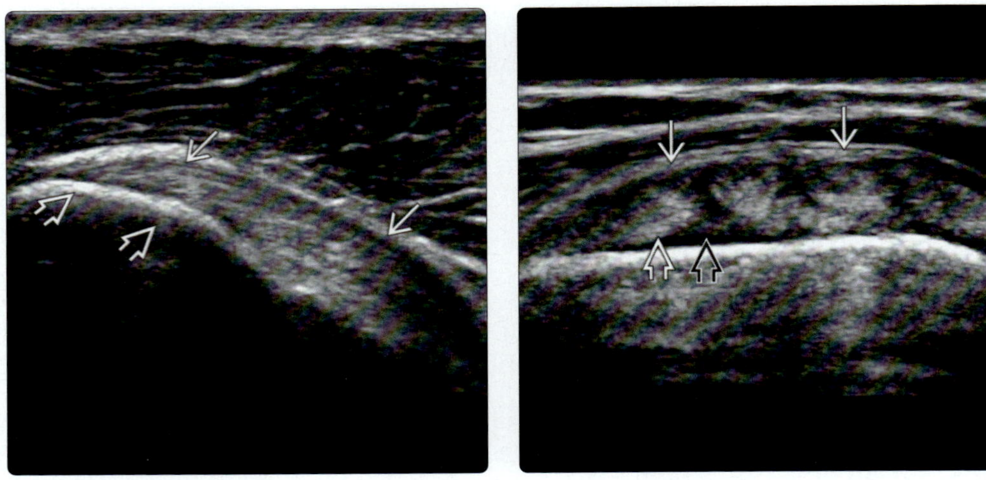

(Left) *Longitudinal US shows a normal subscapularis tendon ➡ inserting onto the lesser tuberosity ➡ of the humerus. The tendon is normal in size with distinct hyperechoic fibers.* **(Right)** *Transverse US shows normal appearances of the subscapularis tendon ➡ in the transverse section. The distinctive bundle heterogeneity is due to the juxtaposition of hyperechoic tendon bundles ➡ with hypoechoic muscle ⇨.*

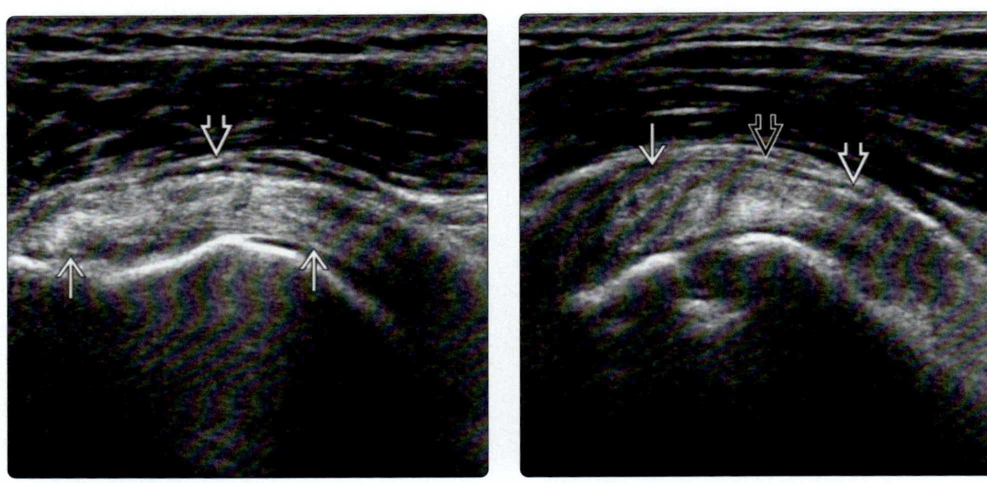

(Left) *Longitudinal US shows mild to moderate subscapularis tendinosis ➡ associated with mild fluid distension within the subcoracoid recess ➡.* **(Right)** *Transverse US of the same patient shows the superior subscapularis fibers ➡ are thickened to a greater degree than the mid ⇨ and inferior ➡ fibers. Tendinosis preferentially affects the superior subscapularis fibers.*

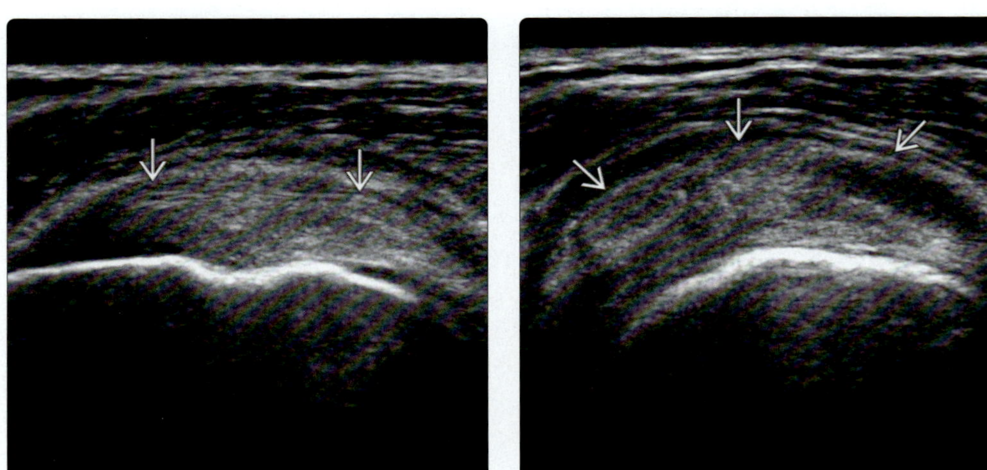

(Left) *Longitudinal US shows moderate to severe subscapularis tendinosis ➡.* **(Right)** *Transverse US of the same patient shows diffuse thickening of superior, mid, and inferior fibers of the subscapularis tendon ➡. Distinction between the tendon fibers and muscle fibers is diminished as the severity of tendinosis increases. The appearances are those of moderate to severe tendinosis.*

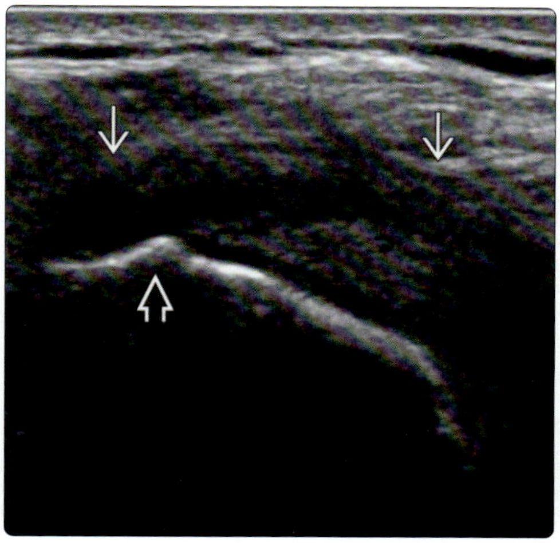

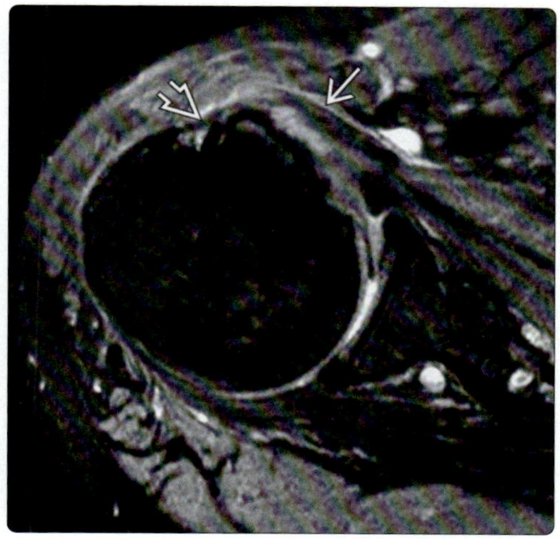

(Left) *Longitudinal US shows severe subscapularis tendinosis with a severely thickened and diffusely hypoechoic subscapularis tendon ➡ associated with mild cortical irregularity of the lesser tuberosity ⮕.* **(Right)** *Axial PD FSE FS MR of the same patient shows severe subscapularis tendinosis ➡ with hyperintense signal within the thickened tendon. Note the normal long head of biceps tendon within the bicipital groove ⮕.*

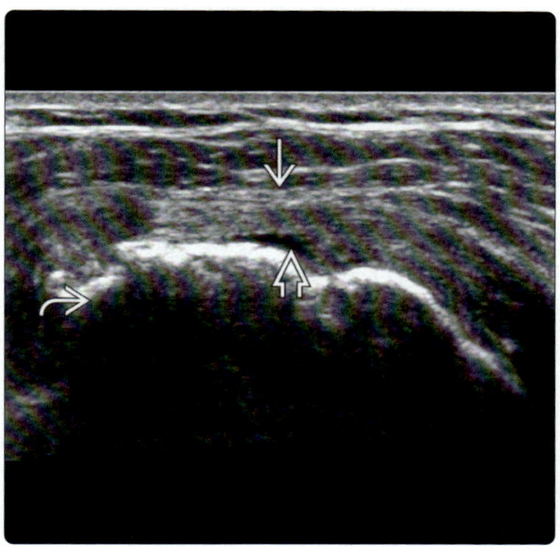

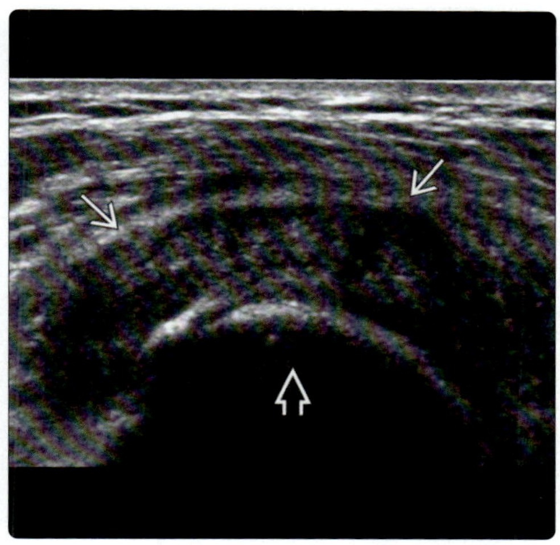

(Left) *Longitudinal US shows severe cortical irregularity of the lesser tuberosity ⮕. There is subscapularis tendinosis associated with articular surface avulsive-type tear ➡ near the insertion. Note flattening of the tendon surface contour ➡ due to the tendon tear.* **(Right)** *Transverse US of the same subscapularis tendon ➡ proximal to the tear shows severe background tendinosis with diffuse thickening and hypoechogenicity. Note the lesser tuberosity ⮕.*

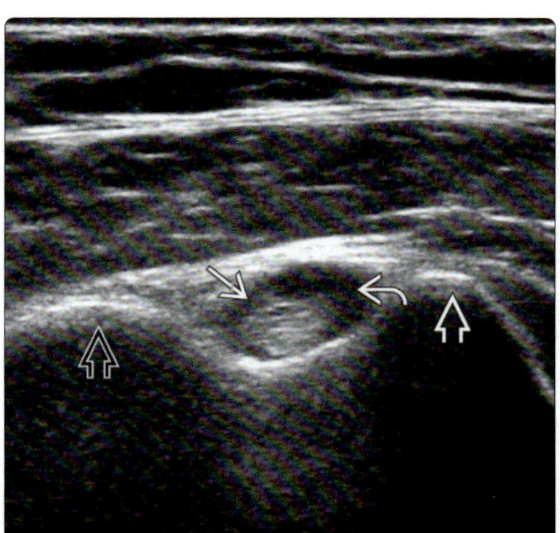

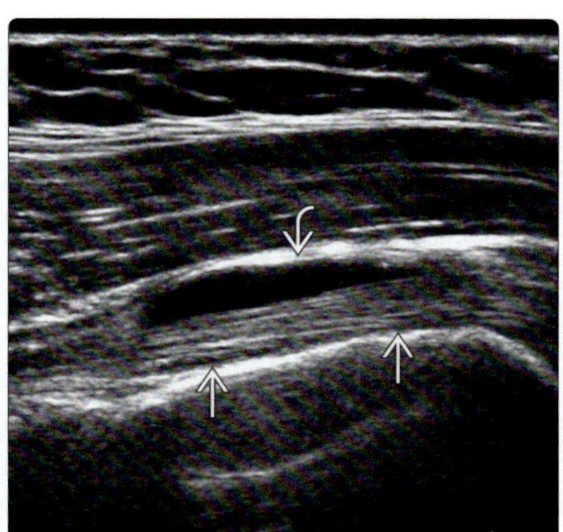

(Left) *Transverse US shows a normal biceps tendon ➡ within the bicipital groove. There is a small amount of fluid ➡ within the biceps tendon sheath. Note the medial ⮕ and lateral ⮕ lips of the bicipital groove.* **(Right)** *Longitudinal US of the same patient shows the normal long head of biceps tendon ➡. As the biceps tendon sheath ➡ communicates with the glenohumeral joint, fluid in one compartment mirrors that in the other compartment.*

(Left) *Transverse US shows the normal oblong configuration of the long head of the biceps tendon at the uppermost portion of the bicipital groove.* **(Right)** *Transverse US shows mild tendinosis of the long head of the biceps tendon with mild tendon swelling leading to a slightly ovoid, cross-sectional contour. Tendinosis of the biceps long head most commonly affects the tendon at the uppermost portion of the bicipital groove.*

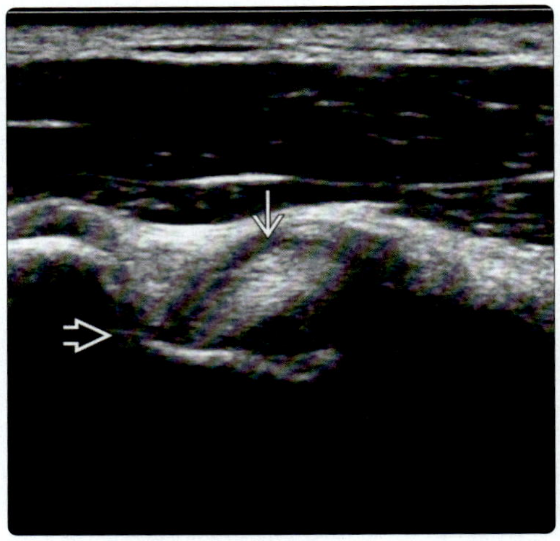

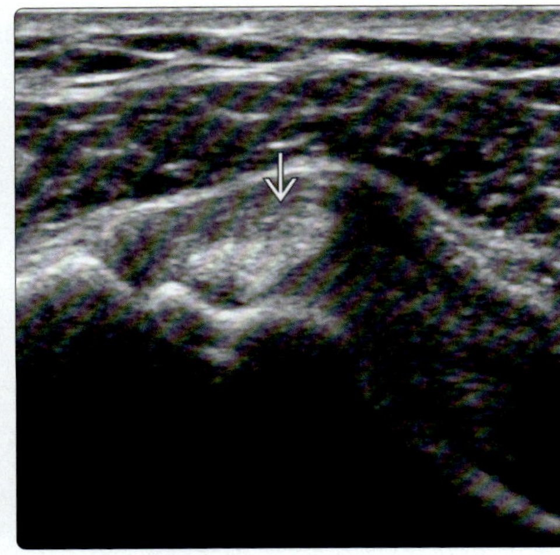

(Left) *Transverse US scan shows moderate biceps tendinosis. The tendon is more hypoechoic and more thickened than normal. The appearances are those of moderate biceps tendinosis.* **(Right)** *Transverse US shows severe biceps tendinosis at the upper end of the bicipital groove. The tendon is moderately swollen and heterogeneous in echotexture with several longitudinal splits. Longitudinal split tears extending for ~ 1 cm are common in severe biceps tendinosis.*

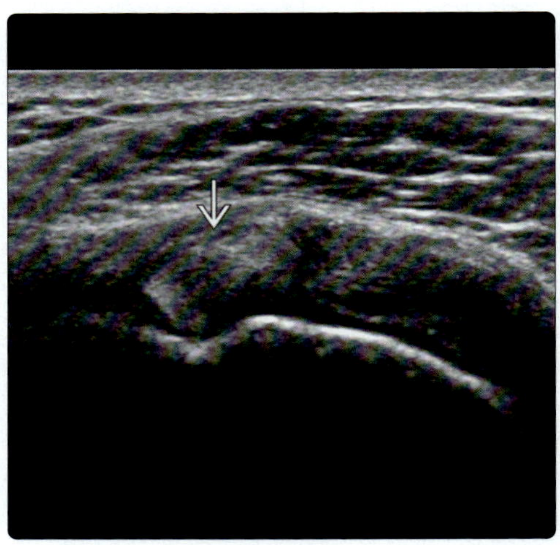

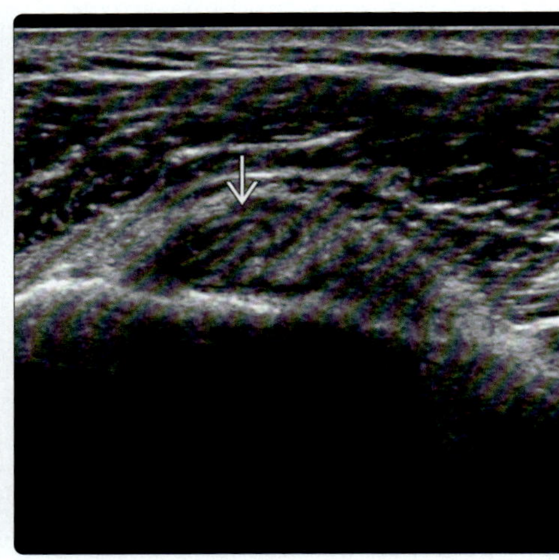

(Left) *Transverse US shows severe biceps tendinosis. The tendon is rounded and heterogeneous with loss of fibrillar pattern. Small longitudinal split tears are almost an inevitable feature of severe biceps tendinosis. Note mild synovial sheath thickening and tendon sheath fluid.* **(Right)** *Transverse US shows very severe biceps tendinosis at the proximal end of the bicipital groove. The tendon is severely thickened with several small longitudinal split tears.*

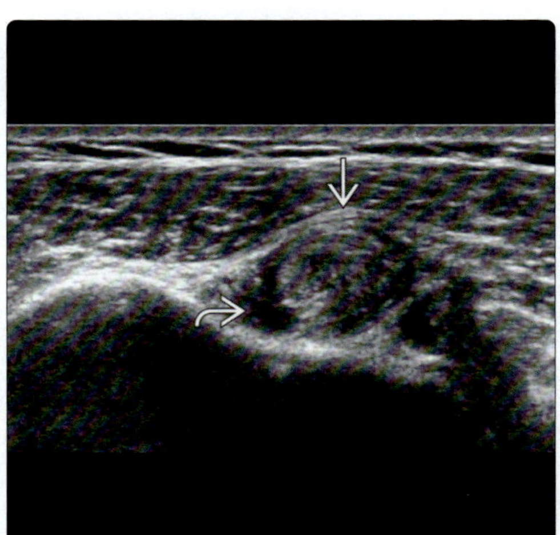

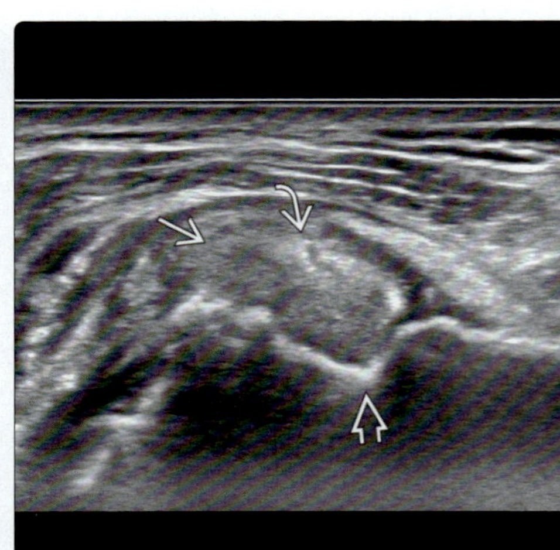

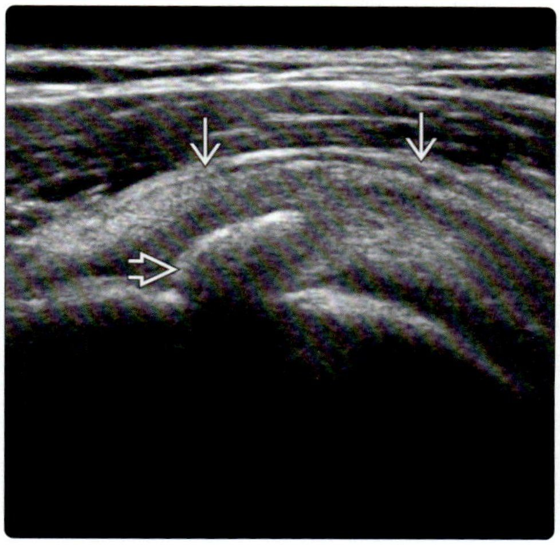

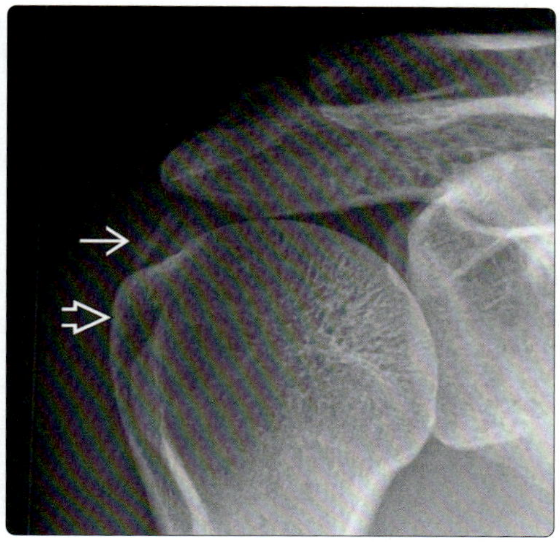

(Left) *Longitudinal US shows a medium-sized area of calcific deposition ➔ with posterior acoustic shadowing at the supraspinatus insertional area. There is mild background tendinosis ➔. This can be summarized as mild supraspinatus tendinosis with a medium-sized area of calcific deposition.* **(Right)** *Frontal radiograph of the same patient shows the calcific focus ➔ close to the greater tuberosity ➔. US is more sensitive than radiography at depicting calcific deposition within tendons.*

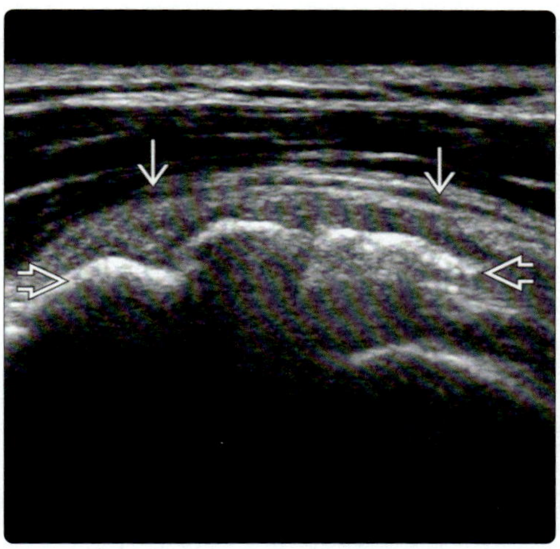

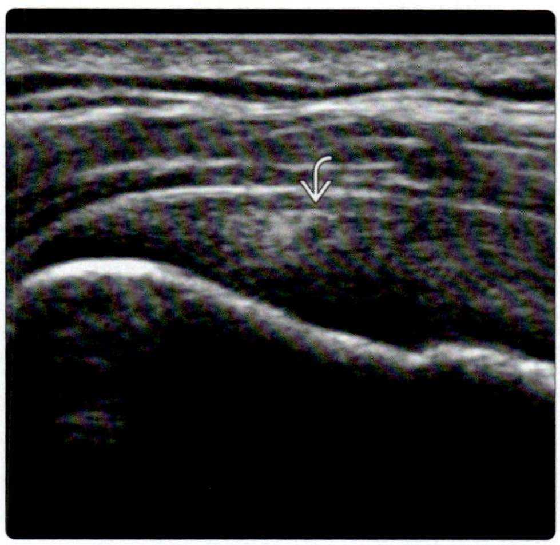

(Left) *Longitudinal US shows mild supraspinatus tendinosis ➔ with large clump-like areas of calcification deposition ➔ within the tendon substance. This can be summarized as mild supraspinatus tendinosis with severe calcific deposition.* **(Right)** *Longitudinal US shows a small, poorly defined area of faint calcification ➔ within the supraspinatus tendon. Only mild background tendinosis is present.*

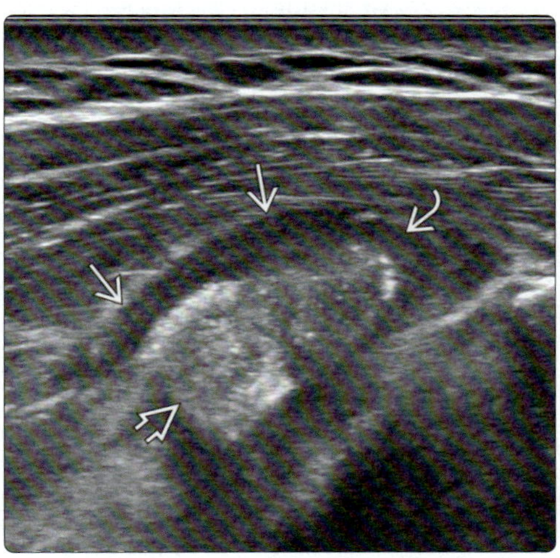

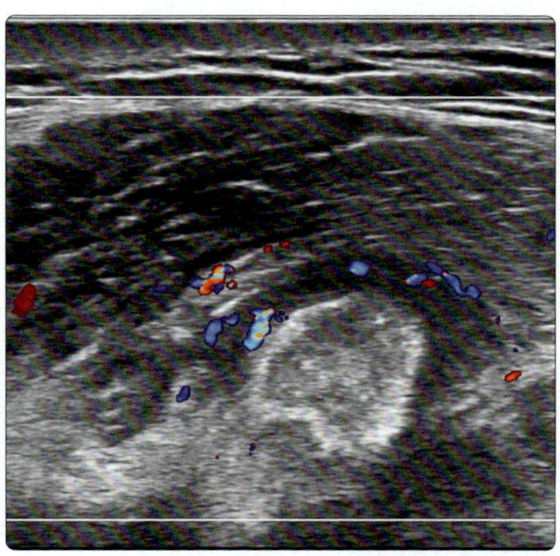

(Left) *Longitudinal US shows severe thickening of the subscapularis tendon ➔ with a large area of paste-like calcium deposition ➔ within the tendon substance. There is also an edematous rim surrounding this area of calcium deposition ➔.* **(Right)** *Transverse color Doppler US of the same patient shows moderate hyperemia in the tendon substance and adjacent soft tissue indicative of an inflammatory response associated with the resorptive phase of calcific tendinosis.*

Rotator Cuff/Biceps Tendon Tear

KEY FACTS

TERMINOLOGY

- Partial-thickness, full-thickness, or complete rotator cuff ± biceps tendon tear

IMAGING

- US and MR have comparable accuracy for rotator cuff and biceps tendon tear detection
- Supraspinatus tendon tears most common close to or at insertional area, especially anterior fibers
  - Most tears are avulsive in type with retraction from insertional area
  - Pay particular attention to leading edge of supraspinatus tendon
- Isolated infraspinatus tendon tear is less common
- Tears in normal tendons are uncommon
- Biceps tendon tears: Either split partial tears or complete transverse tears
  - Most biceps tears occur at upper part of bicipital groove (genu of tendon)

DIAGNOSTIC CHECKLIST

- Discontinuity of tendon and gap filled with anechoic fluid
  - Focal flattening or concavity of tendon surface
- Complete tear with retraction → ± nonvisualization of tendon
  - ± cartilage interface sign, ± superior migration humeral head
- For partial tears, measure anteroposterior and mediolateral dimensions of tear
  - Note tear location in tendon (anterior, middle, or posterior fibers supraspinatus tendon)
- For partial-thickness tears, note % of depth of tear relative to tendon thickness
  - Note bursal sided, articular sided, or intrasubstance
- Grade severity of tendinosis (as mild, moderate, or severe)
- Recognition of intrasubstance tear can be more challenging in moderate to severe tendinosis
  - Intrasubstance tear: Linear, irregular, sharp, hypoechoic area ± volume loss

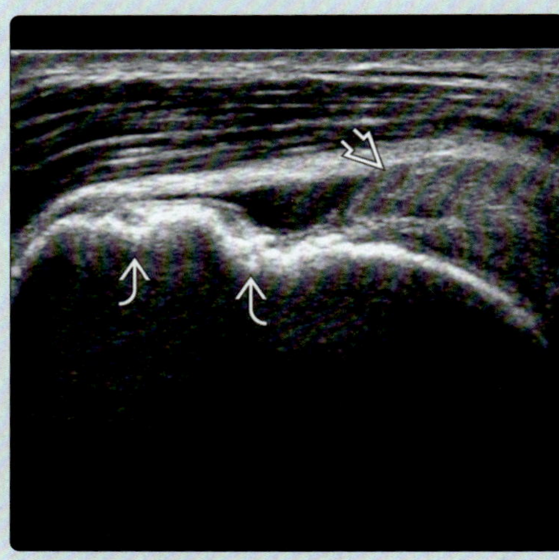

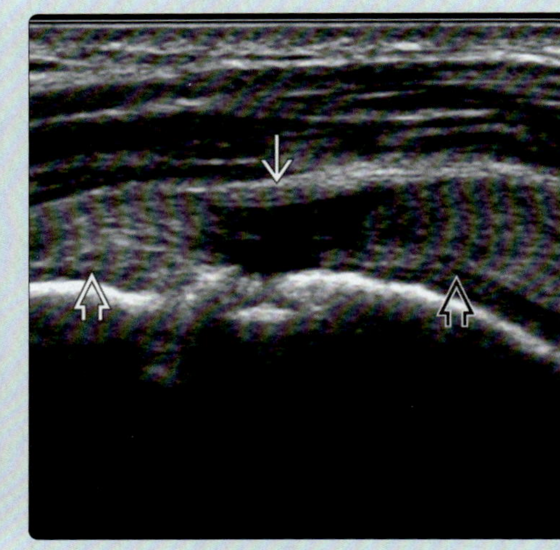

(Left) *Longitudinal US shows full-thickness tear of supraspinatus tendon midfibers. Gap is filled with anechoic fluid with fibers retracted medially. Note cortical irregularity of the insertional area.* **(Right)** *Transverse US of the same patient shows full-thickness tear of the supraspinatus tendon midfibers. Anterior fibers and posterior fibers remain intact. Note the loss of volume with bursal surface contour depression.*

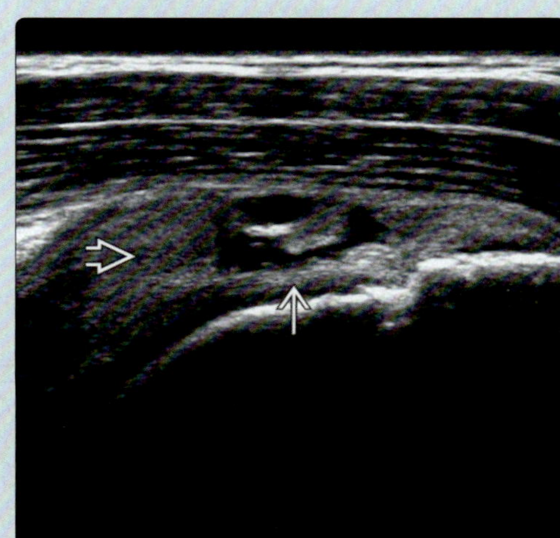

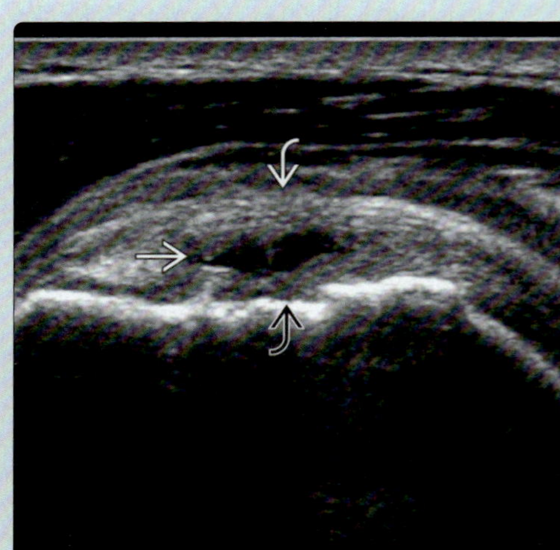

(Left) *Longitudinal US shows a near full-thickness partial tear of the supraspinatus tendon. The gap is filled with anechoic fluid. There are a few intact fibers close to the articular surface with moderate severity supraspinatus tendinosis.* **(Right)** *Longitudinal US shows an irregular intrasubstance tear within the supraspinatus tendon midfibers. The tear does not extend to the bursal or articular surface.*

TERMINOLOGY

Definitions

- Complete tear: All fibers of tendon are torn
- Partial tear: Some fibers of tendon are torn
  - Full-thickness tear: Tear involves only part of tendon (e.g., anterior, mid-, or posterior supraspinatus fibers)
  - Partial-thickness tear: Tear involves only part of tendon depth (e.g., bursal surface, intrasubstance, articular surface tear)
    - Intrasubstance tear does not extend to bursal or articular surface (not seen at arthroscopy)
- Massive tear: Complete tear of supraspinatus and infraspinatus tendons

IMAGING

General Features

- Best diagnostic clue
  - Discontinuity of tendon or tendon torn from its insertional area
- Location
  - Most rotator cuff tendon tears involve supraspinatus
    - Anterior fibers most commonly torn, particularly at leading edge
    - Tear may extend posteriorly to infraspinatus or anteriorly to biceps pulley and subscapularis
      - Biceps pulley and transverse ligament tear → biceps tendon subluxation
    - Most supraspinatus and infraspinatus tears are avulsive in type (i.e., tendon stripped from Sharpey fibers at insertional area)
  - Isolated tear of infraspinatus tendon uncommon
    - Supra- and infraspinatus closely applied at insertion; not easy to distinguish posterior supraspinatus fibers from anterior infraspinatus fibers
    - Maximum supraspinatus anteroposterior length onto superior and middle facets greater tuberosity is 25 mm
      - Tear extending posteriorly > 25 mm from biceps likely to involve infraspinatus tendon
    - Maximum infraspinatus anteroposterior length onto middle facet greater tuberosity facets is 28 mm

Radiographic Findings

- Radiography
  - Features of chronic tendinosis: Cortical irregularity, subcortical sclerosis, and cysts of greater tuberosity
  - Superior migration humeral head with ↓ subacromial space (< 7 mm) in massive rotator cuff tears
  - Acromial osteophytic spur, hooked (type III) acromion, acromioclavicular osteoarthrosis (on supraspinatus outlet view, modified Y view)
  - Rotator cuff arthropathy
    - Accelerated glenohumeral osteoarthropathy from glenohumeral decentering secondary to loss of dynamic stabilizers in large rotator cuff tears

MR Findings

- T1WI
  - Structural abnormalities predisposing to tear (better seen on MR than US)
    - Acromial osteophytic spurs or hook, thickened coracoacromial ligament, acromioclavicular osteoarthrosis
  - Rotator cuff muscle atrophy
- T2WI
  - Fluid-filling complete or incomplete gap in tendon
    - ± fluid within subacromial bursa (more consistently seen on MR than US)
  - MR arthrography is possibly more sensitive for articular surface tear

Ultrasonographic Findings

- Discontinuity or gap within tendon filled with clear hypoechoic fluid (primary sign)
  - Gap filled with blood may appear as isoechoic or hypoechoic
  - Most tears are avulsive type with retraction from insertional area
  - ± focal flattening or concavity of convex bursal tendon surface, undulation of smooth tendon contour
  - ± gas locules with comet-tail artifacts in acute tear
- Intrasubstance tear: Linear, irregular, sharp hypoechoic area within tendon substance
- Complete tear with retraction: Nonvisualization of tendon
  - Cartilage interface sign with clear visualization of humeral head cartilage
  - Superior migration of humeral head
- Features of tendinosis; tears in normal tendons are uncommon
- Fluid within subacromial-subdeltoid bursa
- Cortical irregularity of tendon insertional area
- Muscle atrophy
- Biceps tendon tears: Either split tears or complete transverse tears
  - Most tears preferentially affect tendon at uppermost part of bicipital groove (genu)
  - Typically occur on background of preexisting biceps tendinosis
    - Tendinosis particularly affects uppermost part of biceps tendon in bicipital groove at junction between horizontal and vertical components
      - ± hyperostosis of bicipital groove margins
  - Complete tears seen as empty bicipital groove
    - Collapsed synovial sheath ± synovial hypertrophy may mimic attenuated biceps tendon
    - Retracted long head biceps muscle belly in arm may mimic soft tissue mass

Imaging Recommendations

- Best imaging tool
  - US
    - Comparable accuracy to MR for detecting rotator cuff or biceps tendon tears
      - Patients prefer US to MR
- Protocol advice
  - High-frequency transducer; systematic examination
    - Pay particular attention to anterior leading edge of supraspinatus tendon, particularly if unexplained bursal fluid
    - Make sure all insertional area is covered with tendon

- For partial tears, measure anteroposterior and mediolateral dimension of tear
  - Determine whether tear is located in anterior, middle, or posterior fibers
- For partial-thickness tears, note % of depth of tear relative to tendon thickness
  - Assess whether bursal sided, articular sided, or intrasubstance
- Be wary of diagnosing supraspinatus or infraspinatus musculotendinous junction tears with US
  - Tear uncommon in these areas
- Grade tendinosis severity (mild, moderate, or severe)
  - Difficult in acute tears as tendon is often swollen
- Recognition of intrasubstance tear is more difficult in moderate to severe tendinosis
  - Intrasubstance tear: Linear, irregular, sharp, hypoechoic area ± volume loss
  - Small intrasubstance tears are almost expected feature of severe tendinosis
- Muscle atrophy is assessed on longitudinal scanning over posterior aspect of scapula
  - ↓ volume, ↑ echogenicity (should be = trapezius or teres minor)
  - Also compare with central tendon echogenicity
- Always examine uppermost portion of biceps tendon in bicipital groove
  - Grade biceps tendinosis as mild, moderate, or severe
  - Assess for synovial proliferation around uppermost part of biceps tendon
  - If empty bicipital groove, check for medial subluxation of biceps tendon
  - Examine mid to upper arm to confirm retracted long head of biceps muscle belly

DIFFERENTIAL DIAGNOSIS

Bursitis

- Fluid distension of entire subacromial-subdeltoid bursa
- Fluid first accumulates under and adjacent to acromion, lateral aspect of upper arm, and medial to coracoid process

Rotator Cuff and Biceps Tendinosis

- Thickened tendon with loss of fibrillar pattern and increasing hypoechogenicity
- Frequent coexistence of biceps and rotator cuff tendinosis

Calcific Tendinitis

- Calcium hydroxyapatite deposition
  - Echogenic foci within tendon ± acoustic shadowing

Rotator Cable

- Band-like fibrous thickening located at medial aspect of rotator crescent
- Running perpendicular and beneath supraspinatus and infraspinatus tendons

PATHOLOGY

General Features

- Etiology
  - Rare in patients < 40 years old
  - Partial- or full-thickness rotator cuff tear in 60% of symptomatic patients > 60 years old
    - Often spontaneous on background of tendinosis
  - Tears in normal tendons are usually due to macrotrauma
- Associated abnormalities
  - Superior labrum anterior to posterior tears
  - Posterosuperior internal impingement
  - Rotator cuff arthropathy

Gross Pathologic & Surgical Features

- Thickened, indurated tendon edges
- Hemorrhage in interstitial tears
- Associated subacromial-subdeltoid bursitis

Microscopic Features

- Collagen degeneration without inflammatory cells

CLINICAL ISSUES

Presentation

- Most common signs/symptoms
  - Pain and impingement
    - Insidious onset of pain in adult patient with impingement syndrome
    - Persistence of pain following acute trauma
    - Pain during descent of abducted arm (painful arc)
    - Partial tears are more painful than full-thickness tears
  - Weakness of abduction
    - Weakness of external rotation: Infraspinatus tear
- Clinical profile
  - Athlete; patient > 40 years old with impingement

Natural History & Prognosis

- Many partial-thickness tears progress to full-thickness tears within 2 years
- Many symptomatic tears can become asymptomatic
- Smaller tears may heal

Treatment

- Autologous blood injection for partial tears
  - Physiotherapy for minor partial tears
- Arthroscopic debridement, suture anchoring, and subacromial decompression for larger tears
  - Muscle atrophy can improve with successful repair

DIAGNOSTIC CHECKLIST

Consider

- Small supraspinatus surface tear or leading edge tear if unexplained bursal fluid is seen
  - Pay particular attention to tendon contour and anterior edge of supraspinatus at insertional area

Image Interpretation Pearls

- Systematic examination for secondary signs as well as primary sign of tear
  - Cortical irregularity insertional area
  - Bursal fluid
  - Flattened tendon contour (flagging peribursal fat sign)

SELECTED REFERENCES

1. Lee MH et al: Comprehensive shoulder US examination: a standardized approach with multimodality correlation for common shoulder disease. Radiographics. 36(6):1606-1627, 2016

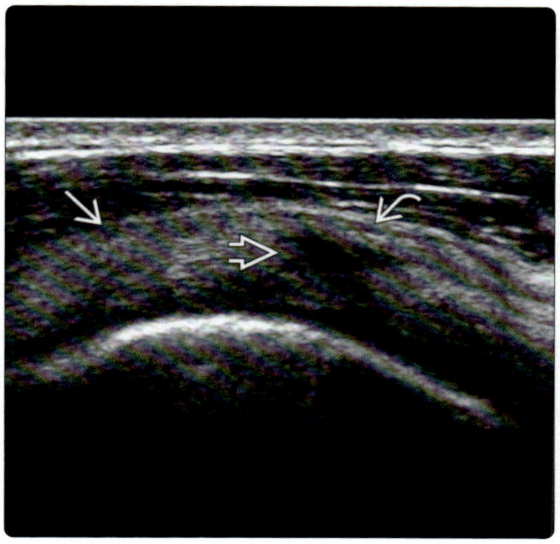

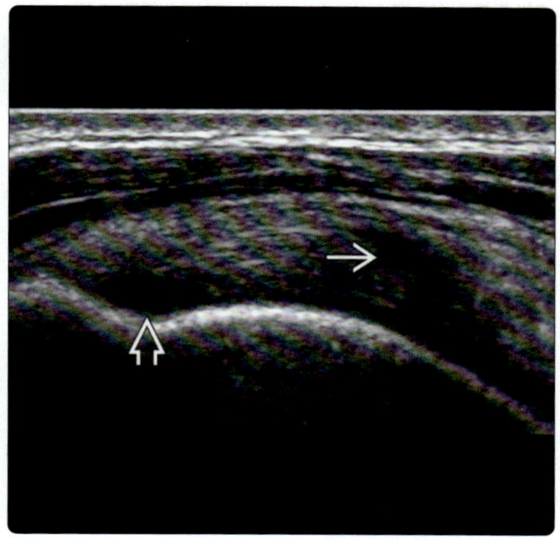

(Left) *Transverse US shows mild supraspinatus tendinosis with a well-defined irregular hypoechoic area in the anterior fibers due to an intrasubstance tear. Note the mild associated bursal surface contour flattening.* **(Right)** *Longitudinal US of the same patient shows an intrasubstance tear just proximal to the insertional area. All rotator cuff tendon tears should be confirmed in 2 planes. The hypoechoic area near the greater tuberosity is due to anisotropy.*

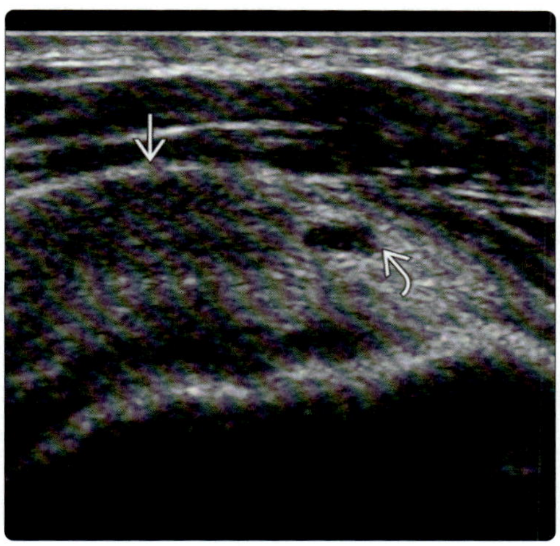

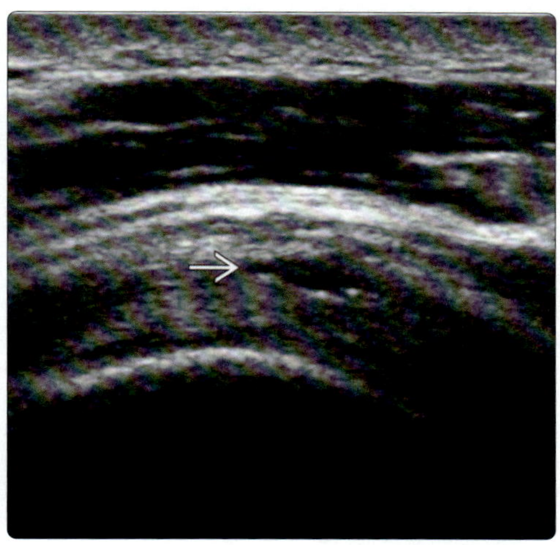

(Left) *Longitudinal US shows moderate to severe supraspinatus tendinosis. There is a more discrete small, partial-thickness, intrasubstance tear present. The fibrillar echotexture of the supraspinatus tendon is indistinct.* **(Right)** *Longitudinal US shows a partial intrasubstance tear involving 25% of the infraspinatus tendon thickness close to the myotendinous junction. This tear does not extend to either the bursal or articular surfaces and would not be seen on arthroscopy.*

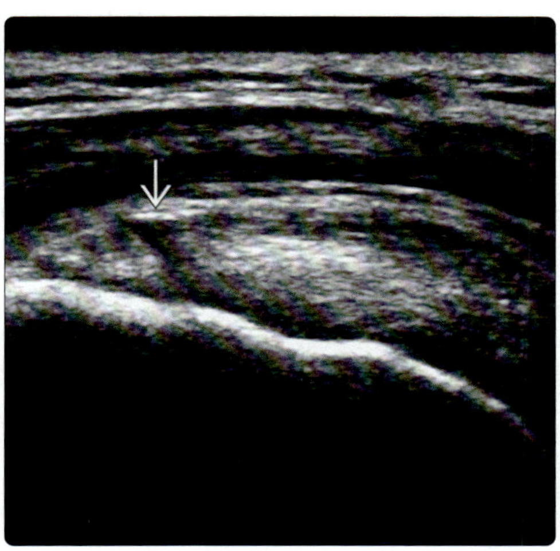

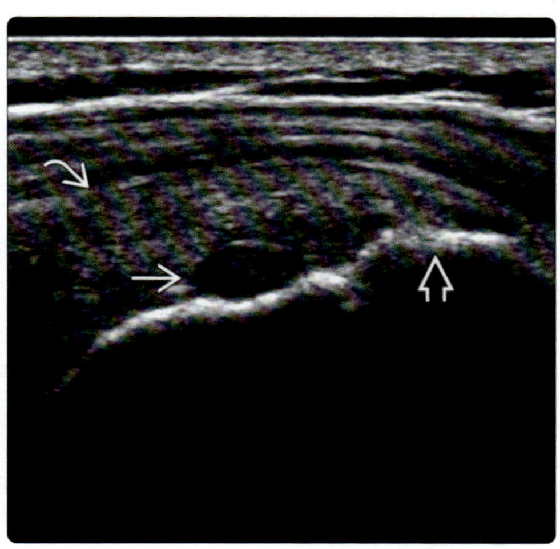

(Left) *Longitudinal US shows a well-defined, hypoechoic, bursal surface tear overlying the insertional area of the supraspinatus tendon. No bursal fluid is present.* **(Right)** *Longitudinal US shows a small, well-defined, hypoechoic partial-thickness tear at the articular surface of the supraspinatus tendon. Note the mild supraspinatus tendinosis and cortical irregularity at the greater tuberosity.*

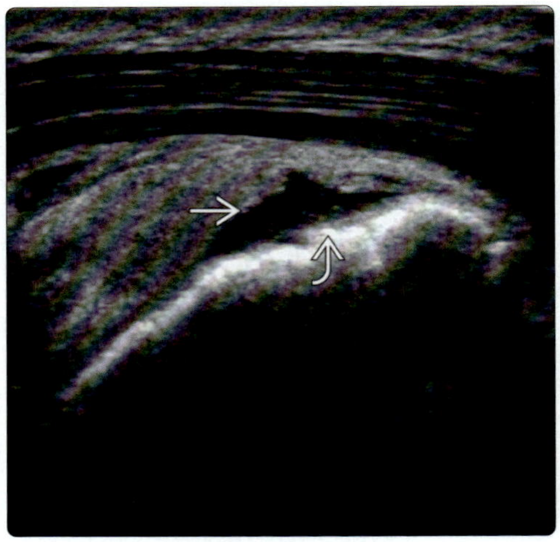

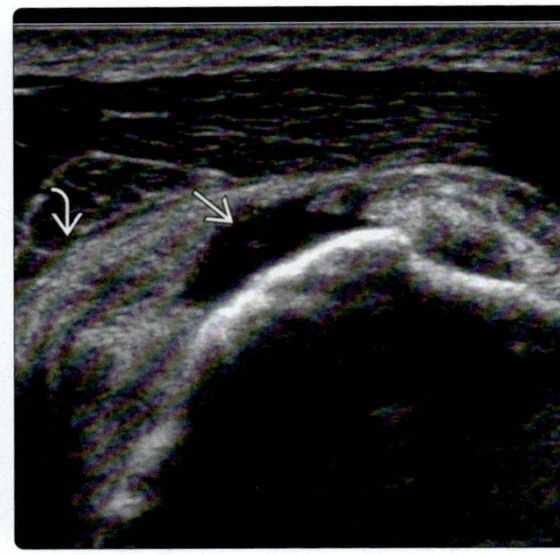

(Left) *Longitudinal US shows a full-thickness avulsive-type tear* → *of the supraspinatus tendon. The fibers have stripped or retracted from the insertional area* ↪. **(Right)** *Transverse US shows the same tear* → *at the anterior fibers of the supraspinatus tendon. The gap is filled with hypoechoic fluid. Moderate tendinosis of the remaining supraspinatus tendon is present* ↪. *The anteroposterior and mediolateral dimensions of the tear should be measured.*

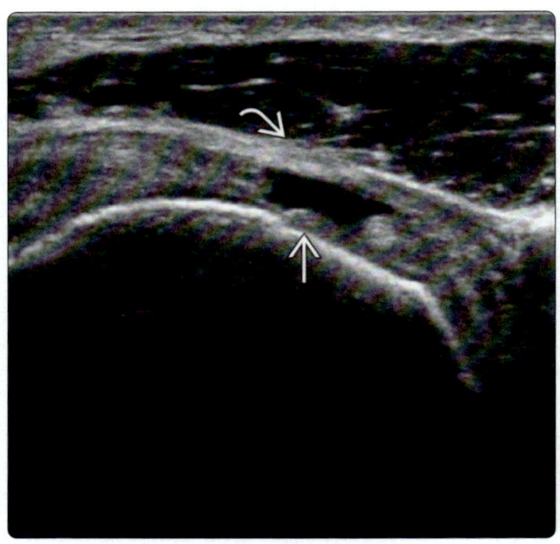

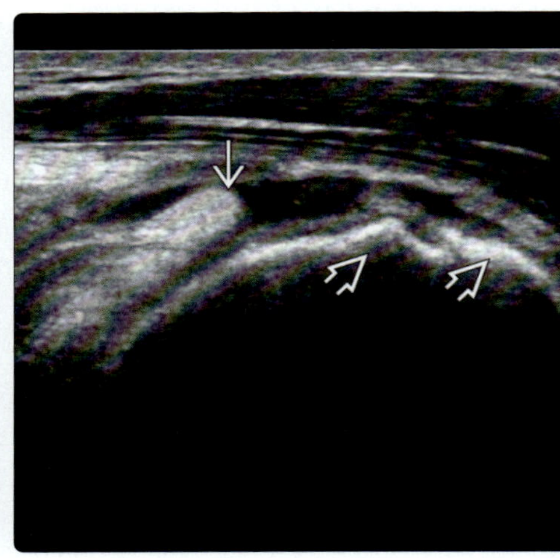

(Left) *Transverse US shows full-thickness tear* → *of the supraspinatus tendon midfibers. The gap is filled with anechoic fluid. The overlying peribursal fat* ↪ *is made more prominent.* **(Right)** *Longitudinal US of the same patient shows how the torn fibers* → *are retracted medially from the insertional area in the greater tuberosity. The insertional area at the greater tuberosity is bare and irregular* ⇨. *The retraction gap is filled with fluid.*

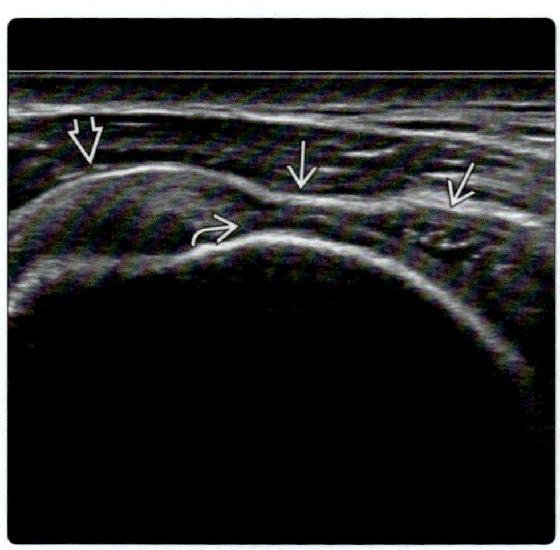

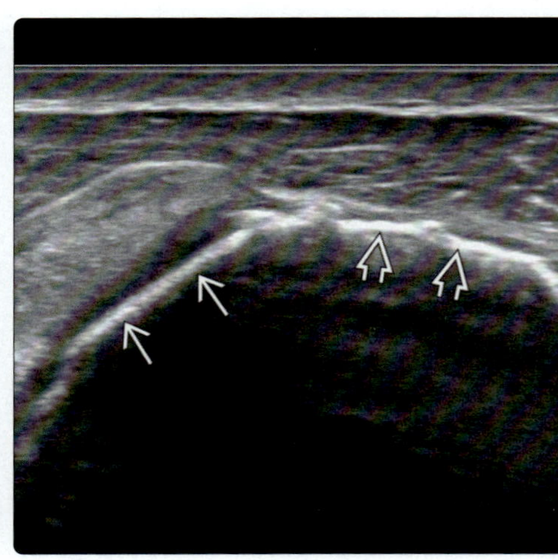

(Left) *Longitudinal US shows a full-thickness tear* → *of the mid and posterior fibers of supraspinatus. The retraction gap is filled with fluid and echogenic foci. The humeral head cartilage* ↪ *underneath the tendon gap is clearly visualized. Note the intact anterior fibers* ⇨. **(Right)** *Transverse US of the same patient shows a bare area in the middle facet* → *of the greater tuberosity due to full-thickness tear of the mid and posterior fibers. Note the intact insertional fibers at the superior facet* ⇨.

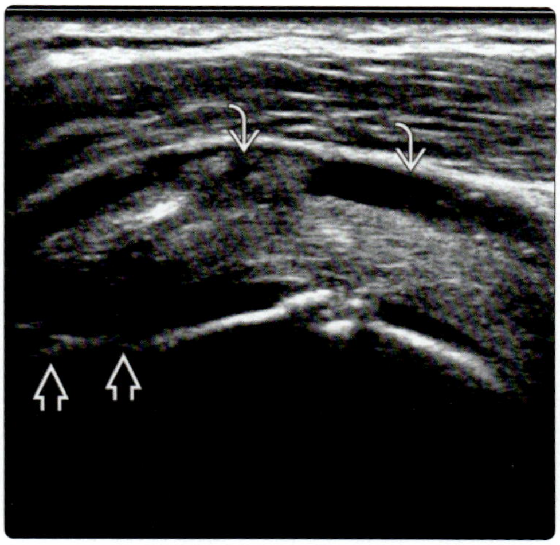

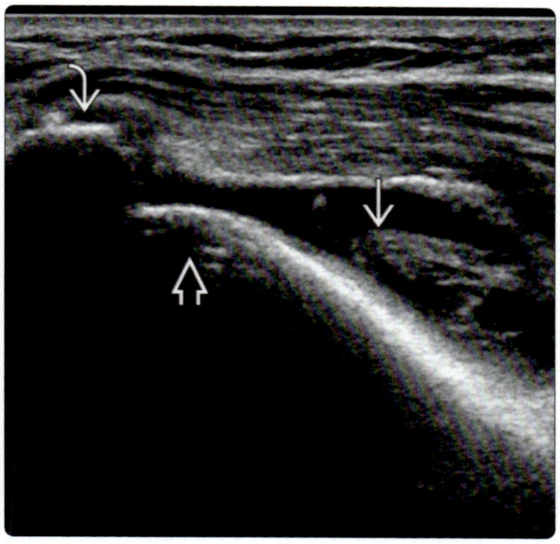

(Left) *Longitudinal US shows moderate supraspinatus tendinosis with a complete avulsive-type tear from the insertional area. There is moderate subacromial-subdeltoid bursal distension containing some echogenic material consistent with hemorrhage.* **(Right)** *Longitudinal US shows a complete tear of the supraspinatus tendon. The torn tendon end is retracted proximally. The humeral head has superiorly migrated and now lies close to the acromion.*

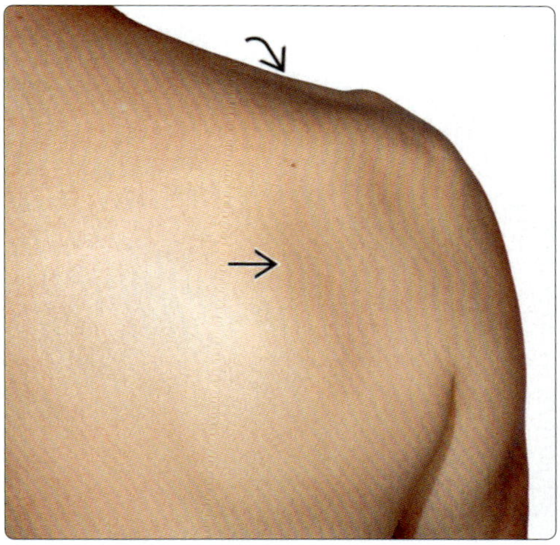

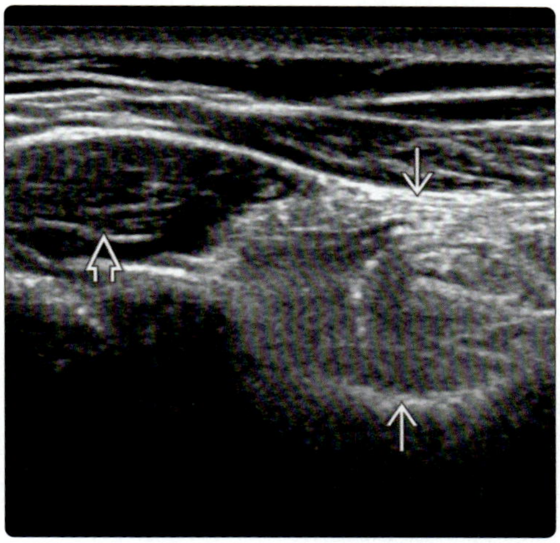

(Left) *Clinical photograph of a patient with rotator cuff tear shows muscle wasting at supraspinatus and infraspinatus muscle areas. Muscle atrophy is more readily appreciated in thin patients.* **(Right)** *Longitudinal US shows fatty atrophy of the infraspinatus muscle in a patient with massive rotator cuff tear. The hyperechoic infraspinatus muscle shows a depressed contour due to volume loss compared to the normal appearance of the teres minor muscle.*

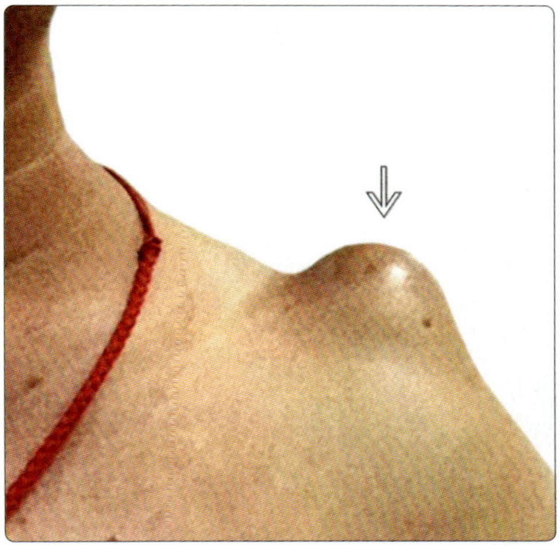

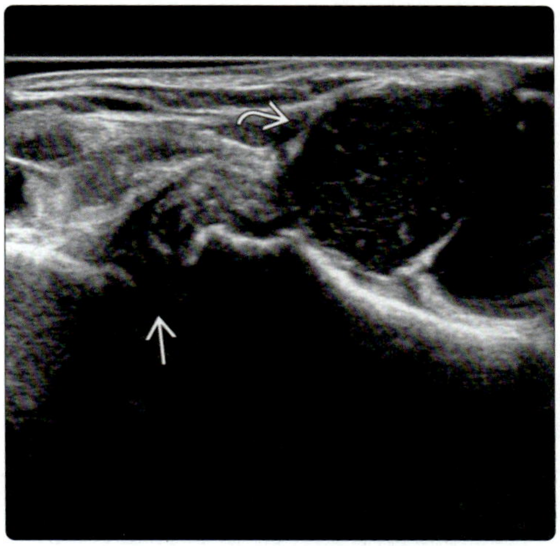

(Left) *Clinical photograph shows a middle-aged man with insidious onset of cystic shoulder swelling.* **(Right)** *Transverse US of the same patient shows a large ganglion overlying the lateral end of the clavicle. This ganglion has arisen as a result of a large supraspinatus tear, followed by humeral head superior migration, acromioclavicular capsule attrition, and leakage of glenohumeral fluid through the joint to the ganglion. This is characteristic of rotator cuff arthropathy and geyser phenomenon.*

(Left) *Transverse US shows a large partial-thickness tear ➔ of the midfibers of the subscapularis tendon just extending to the superficial margin of the tendon ➔.* **(Right)** *Longitudinal US shows an avulsive-type tear of the subscapularis tendon from the insertional area ➔ on the medial lip of the bicipital groove. The tendon fibers are retracted medially ➔ with fluid filling the retraction gap.*

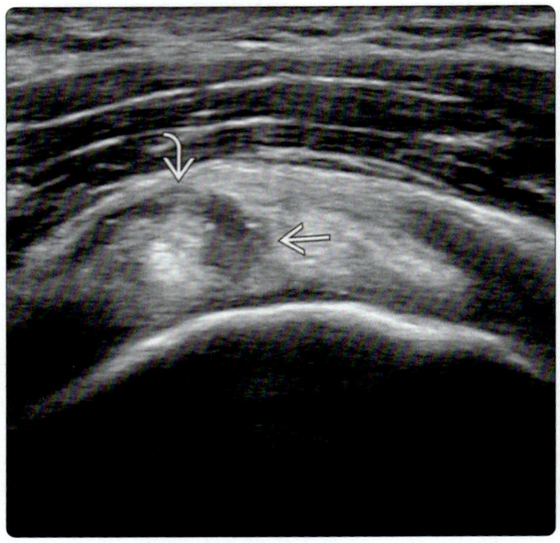

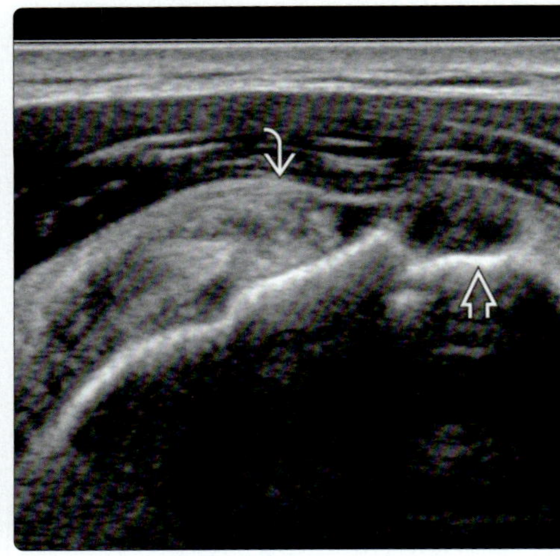

(Left) *Transverse US shows a longitudinal split tear ➔ in the subscapularis tendon ➔, involving the biceps pulley. The biceps long head tendon ➔ is medially subluxed and located at the superior aspect of the lesser tuberosity ➔ rather than within the bicipital groove ➔.* **(Right)** *Transverse US of the same patient at a slightly more cephalad location shows the medial subluxation of the thickened biceps long head tendon ➔ into the subscapularis tendon substance ➔.*

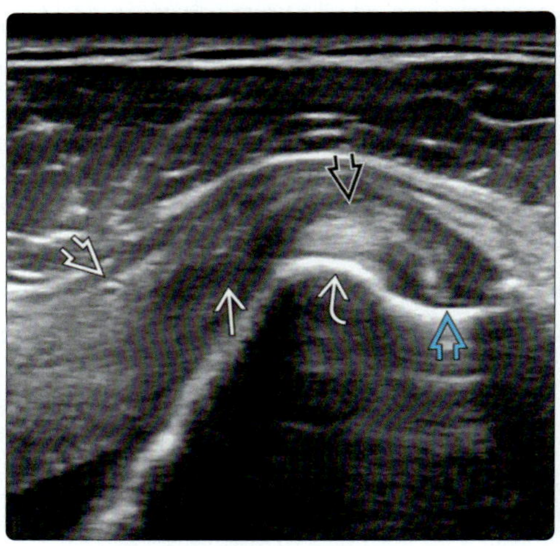

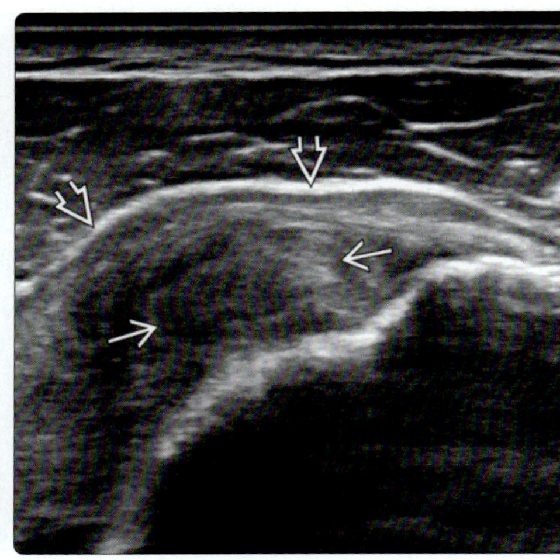

(Left) *Transverse US shows moderate biceps tendinosis with mild synovial proliferation within the surrounding tendon sheath ➔. There is a wide intrasubstance tear ➔ within the deeper aspect of the biceps tendon.* **(Right)** *Longitudinal US of the same patient shows moderate biceps tendinosis ➔ near the proximal end of the bicipital groove with an irregular, longitudinal, split-like tear ➔ of the deeper fibers. Nearly all partial-thickness tears of the biceps tendon are longitudinal split tears.*

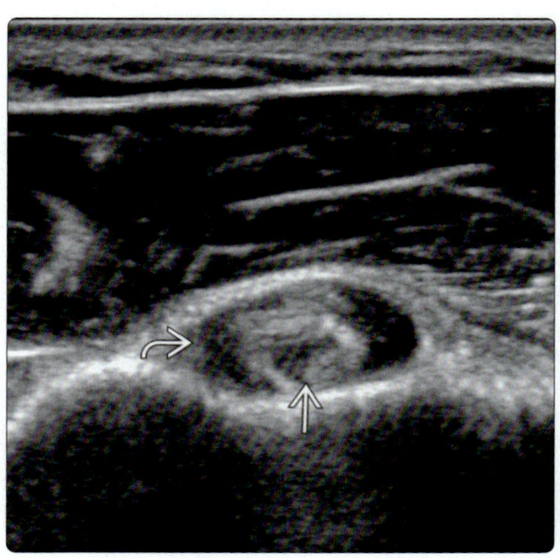

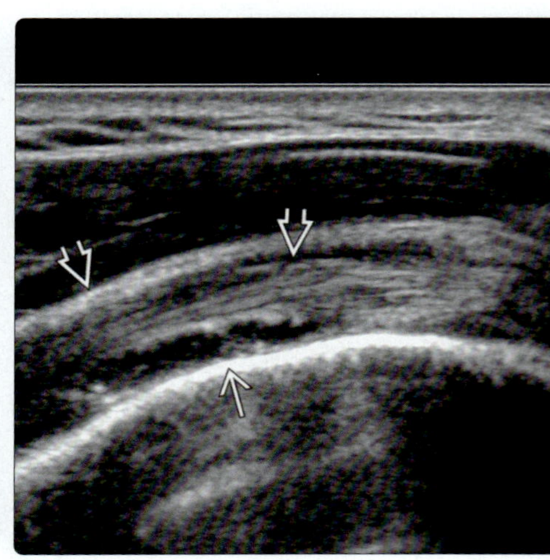

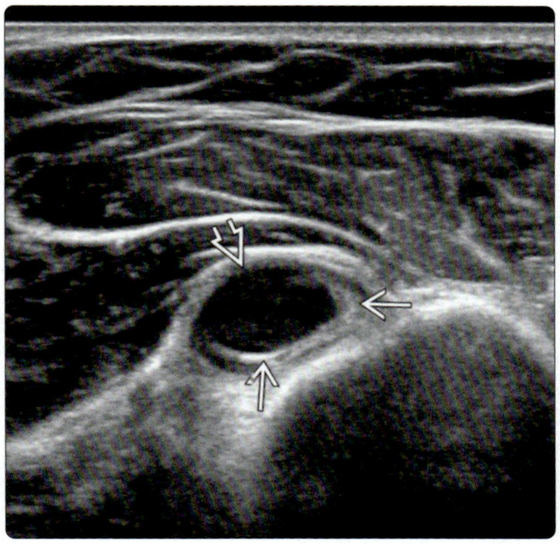
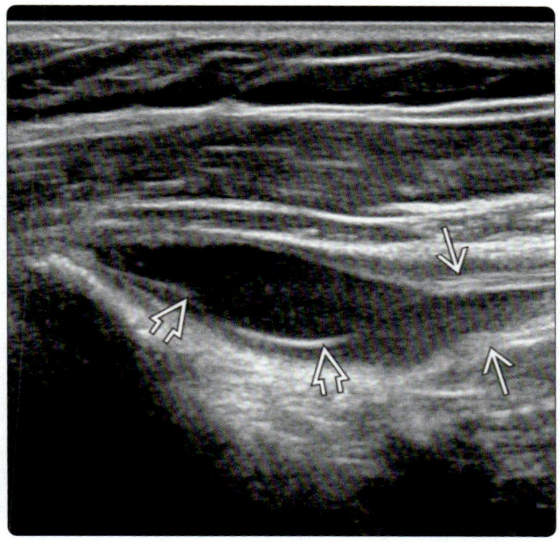

(Left) *Transverse US shows an intrasubstance ganglion cyst within the biceps tendon. Note the eccentric displacement of tendon fibers by the ganglion cyst.* **(Right)** *Longitudinal US of the same patients shows a large intrasubstance ganglion cyst within the biceps tendon. The formation of such a ganglion cyst is due to encysted fluid accumulation with a longitudinal split tear. Note the intact fibers of the biceps tendon.*

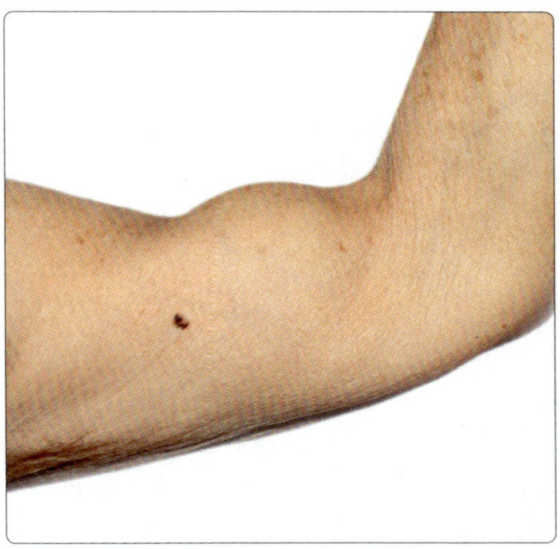
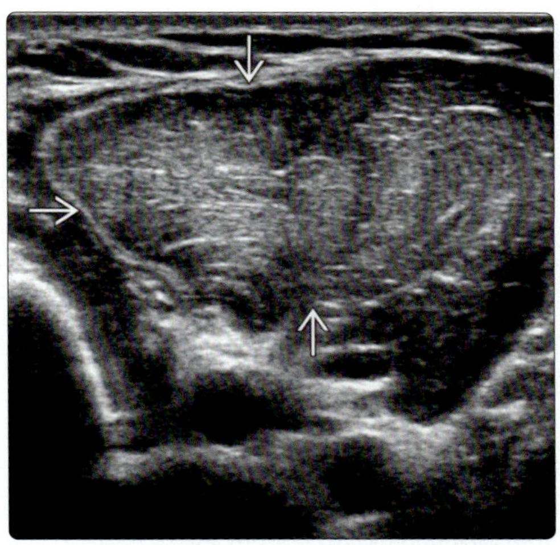

(Left) *Clinical photograph of the same patient shows a roundish swelling on the flexor aspect of arm, which became more noticeable with attempted biceps contraction. This swelling is due to the retracted biceps muscle belly (Popeye sign).* **(Right)** *Transverse US of the arm of the same patient shows a retracted bulbous long head of biceps muscle belly. This can be mistaken for a soft tissue mass if one is not aware that the long head of the biceps has torn.*

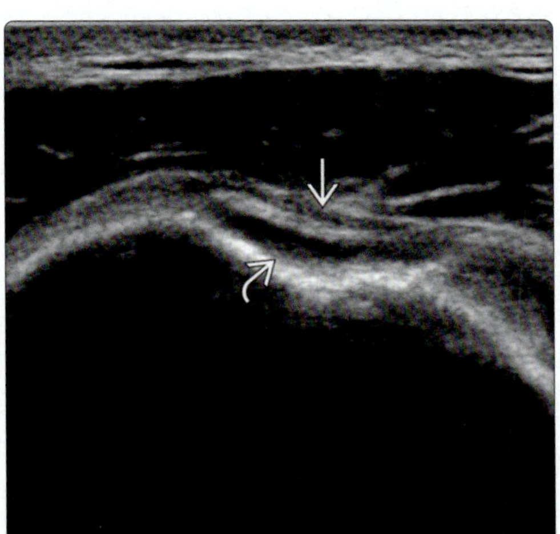
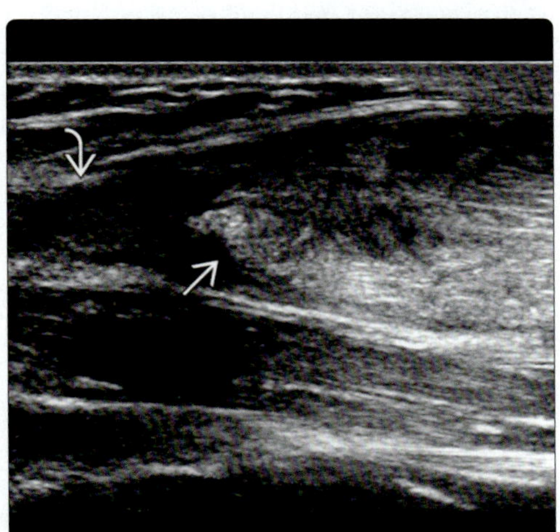

(Left) *Transverse US shows a complete rupture of the long head of the biceps tendon. There are no tendon fibers within the bicipital groove. Some fibers of the subscapularis tendon transverse the groove as the transverse ligament to insert onto the lateral lip of the bicipital groove.* **(Right)** *Longitudinal US shows a retracted myotendinous junction of the long head of the biceps due to a complete tendon tear. The tendon sheath located proximally seems to contain blood.*

Nonrotator Cuff Tendinosis

KEY FACTS

TERMINOLOGY

- Tendinosis: Degenerative disease of tendon with variable degree of peritendinitis

IMAGING

- Some tendons (rotator cuff, biceps, common extensor tendon origin, 1st extensor compartment wrist, Achilles, patellar, posterior tibialis) are particularly susceptible to tendinosis
- Focal, diffuse, or nodular tendon enlargement with hypoechogenicity & loss of normal fibrillar pattern
- Hyperemia correlates with activity of tendinopathic process
- Tendinosis predisposes to tendon tears (tendon tears uncommon in normal tendons)
- Partial tears are seen as sharp, linear, hypoechoic defects within area of tendinosis
  - Usually split tears
  - Should not be confused with dilated vascular channels within hyperemic tendon
- Variable peritendinitis: Swelling, edema, & hyperemia of tissues around tendon
  - Peritendinitis is important feature to assess as it will respond to antiinflammatory medication
- Additional secondary signs
  - Tendon calcification
  - Tendon sheath effusion
  - Bursitis
  - Cortical erosion or hyperostosis at insertional area

DIAGNOSTIC CHECKLIST

- Grade level of tendinosis and, where applicable, peritendinitis, as mild, moderate, or severe
- Routinely examine with color Doppler imaging
- Be careful about using contralateral asymptomatic side as normal reference as subclinical tendinosis is common
- Scrutinize tendon for small tendon tears, especially tendon edges for small avulsive-type tears

(Left) *Longitudinal US of distal Achilles tendon shows moderate diffuse thickening ➔ close to calcaneal insertion ➔ consistent with moderate insertional tendinosis. There is mild distension of the retrocalcaneal bursa ➔ with a localized avulsive-type tear of the calcaneal insertion ➔. There is no calcaneal surface irregularity.* **(Right)** *Longitudinal color Doppler US of the same area shows mild hyperemia of the tendon ➔. Only mild activity is present.*

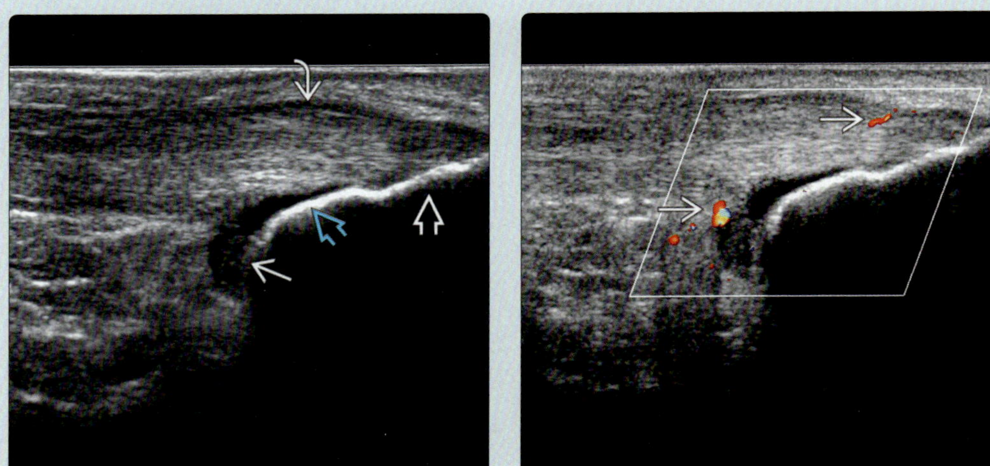

(Left) *Longitudinal US of the distal portion of the Achilles tendon shows diffuse severe thickening with tendon hypoechogenicity ➔. Small foci of dystrophic calcification ➔ are present. Prominent calcaneal surface irregularity and bone spur ➔ are also present.* **(Right)** *Longitudinal color Doppler US of the same area shows moderate hyperemia within the tendon ➔. Mild peritendinous hyperemia is also present close to the insertion ➔.*

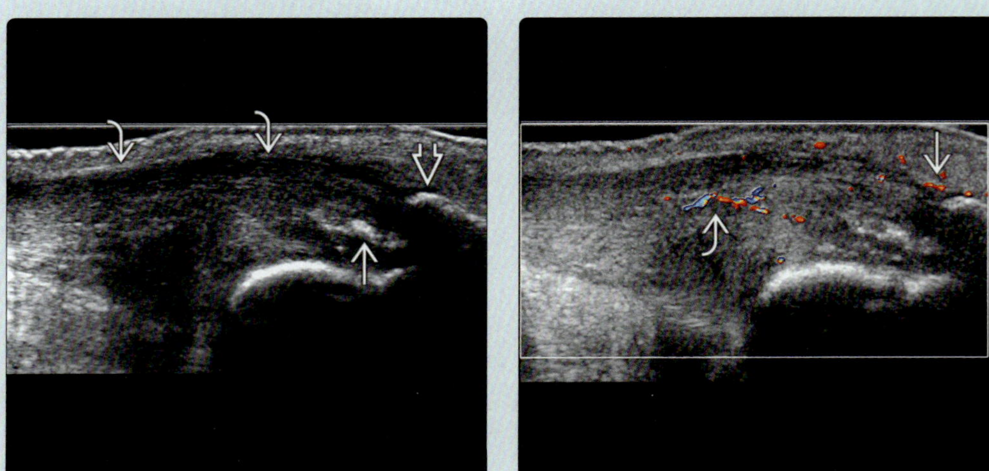

TERMINOLOGY

Synonyms

- Tendinosis or tendinopathy

Definitions

- Primarily disease of tendon with variable degree of peritendinitis
  - Peritendinitis is inflammation of soft tissues surrounding tendon
- Collagenous tendon degeneration with proteoglycan deposition

IMAGING

General Features

- Best diagnostic clue
  - Focal, diffuse, or nodular tendon enlargement with loss of normal fibrillar echotexture
- Location
  - Some tendons are more susceptible to tendinosis
    - Rotator cuff & biceps tendons
    - Achilles: Calcaneal insertion > midportion of tendon
    - Patellar: Proximal > distal; deep > superficial
    - Posterior tibialis: Insertion > retro- or inframalleolar areas
    - Common extensor > common flexor tendon origin of elbow
    - Gluteus minimus & medius insertion at greater trochanter
    - Ischial attachment of hamstring tendons
  - Susceptible areas
    - Tendon insertion or areas where tendons curve (e.g., genu of biceps tendon at cephalad end of bicipital groove)
- Morphology
  - Anterior convexity of middle 1/3 of Achilles tendon
  - Normal cross-sectional area of Achilles tendon
    - 0.54 ± 0.11 cm^2 at musculotendinous junction; 0.54 ± 0.12 cm^2 at mid-level; 0.74 ± 0.15 cm^2 at calcaneal insertion
  - Normal size of posterior tibialis tendon
    - ~ 2x size of flexor digitorum longus tendon
    - < 0.16 cm^2 at retromalleolar; < 0.18 cm^2 at inframalleolar; < 0.27 cm^2 at insertion
  - Peroneus longus tendon normally slightly larger than peroneus brevis tendon

Ultrasonographic Findings

- Increased tendon thickening with hypoechogenicity & variable loss of normal fibrillar pattern
  - Cross-sectional area of tendon measured at specific locations
    - Useful for follow-up rather than diagnosis; abnormal tendon morphology slow to change
- Proteoglycan accumulation: Focal hypoechoic area within tendon that may occasionally bulge beyond tendon margins
- Hyperemia within or around tendon is common feature
  - Level of hyperemia correlates with disease activity
    - Hyperemia, though not seen in rotator cuff tendinosis
- Partial tear seen as sharp, linear hypoechoic defect within area of tendinosis
  - Occurs with moderate to severe degrees of tendinosis; usually runs longitudinally in tendon (split tears)
  - May be difficult to distinguish from areas of focal proteoglycan deposition
    - Tears tend to be more linear, sharper, & hypoechoic
  - Should not be confused with dilated vascular channels within hyperemic tendon (use color Doppler US)
- Secondary signs of tendinosis are more common in moderate-to-severe tendinosis
  - Tendon sheath effusion (if tendon sheath present)
  - Dystrophic calcification within area of tendinosis (particularly for Achilles, quadriceps, & patellar tendons)
  - Reactive cortical erosion or hyperostosis at insertional area
  - Adjacent bursitis (Achilles)
  - Paratenonitis (Achilles, patella)
- Secondary signs that are commonly seen in normal asymptomatic subjects
  - Focal calcification of Achilles tendon (21% normal subjects)
  - Fluid in retrocalcaneal bursa (35% normal subjects)
  - Calcaneal bony contour abnormality (63% normal subjects)
- Variable peritendinitis of surrounding soft tissues
  - Seen as swelling, edema, & hyperemia on color Doppler imaging
  - Grade as mild, moderate, or severe
  - Peritendinitis important feature to note as it will respond to antiinflammatory medication
  - Tendinosis does not have inflammatory component per se
- Elastography to diagnose & grade tendinosis
  - Shows promise in early stages of tendinosis before grayscale features become apparent

Radiographic Findings

- Radiography
  - Sclerosis, cortical irregularity, hyperostosis of insertional tendinosis
  - Intratendinous dystrophic calcification
  - Accessory navicular bone associated with posterior tibialis tendinosis

MR Findings

- Increased T1W, T2W signal within enlarged tendon
- Hyperintense inflammatory fluid around tendon
- Partial tears hyperintense on FS PD FSE or STIR images
- Reactive marrow change

Imaging Recommendations

- Best imaging tool
  - US, as nearly all tendons are readily accessible
- Protocol advice
  - Examine both symptomatic & contralateral side
    - Be careful about using asymptomatic contralateral side as normal standard as subclinical tendinosis is common
    - Always evaluate level of peritendinitis
  - Use minimal transducer pressure to avoid blanching color Doppler signal

- Anisotropy can be used to differentiate tendon from surrounding fat
- Anisotropy can mimic tendinosis; minor angulation of probe can make anisotropy disappear
- Regularly toggle transducer when examining tendons

DIFFERENTIAL DIAGNOSIS

Xanthoma

- Affects Achilles > patellar tendon
- Affected areas tend to be more nodular & less hyperemic than tendinosis
- Features of severe hypercholesterolemia
  - ± thickening of extensor tendon hoods over metacarpophalangeal joints

Tenosynovitis

- Primarily disease of tendon sheath
- Affects tendon with complete synovial sheath
- Tenosynovial inflammation rather than tendon thickening predominates
- Often may get elements of tenosynovitis & tendinosis coexisting

PATHOLOGY

General Features

- Etiology
  - Repetitive overuse → microtrauma with collagen bundle microfracture & incomplete healing
  - Achilles: Hypovascular watershed zone 2-6 cm proximal to calcaneal insertion
    - Eccentric loading of fatigued muscle-tendon unit or overtraining
    - Runners susceptible to acceleration (sprinting) & deceleration (eccentric contraction) injuries
  - Posterior tibialis: Flatfoot & accessory navicular bone
    - Reactive thickening superior calcaneonavicular fibers spring ligament portion on US
  - Patellar tendon: Running & jumping
    - Microtrauma between patellar bone tip & central deep fibers of adjacent patellar tendon
- Relevant anatomy
  - Achilles & patellar tendons have incomplete synovial sheath
- Posterior tibialis inserts into accessory navicular bone (when present), medial pole of navicular bone, & cuneiform bones
  - Cornuate navicular is elongated medial pole of navicular bone

Gross Pathologic & Surgical Features

- Tendon is gray instead of glistening white
- Nodular thickening, calcification, & partial tendon tears
- Inflamed tissues around tendon

Microscopic Features

- Features of failed healing response
  - Noninflammatory intratendinous collagen degeneration
  - Fiber degeneration (fibrinoid, mucoid) & thinning with partial tears
  - Hypercellularity with scattered vascular ingrowth
  - Increased interfibrillar glycosaminoglycans (chondroitin sulfate)
- Inflammation of peritendinous tissues

CLINICAL ISSUES

Presentation

- Most common signs/symptoms
  - Pain with stretching of tendon
- Other signs/symptoms
  - Tendinosis is often visible on US with minimal or no symptoms (subclinical)
    - Subclinical disease is common on contralateral side in patients with unilateral symptoms
- Clinical profile
  - Chronic dull to sharp pain at onset of exercise
    - Gradually subsides with more exercise & returns following cessation of exercise

Demographics

- Age
  - Young middle-aged adults (most common), young athletes, & older sedentary populations
- Epidemiology
  - Usually associated with either physical or physiological overuse

Natural History & Prognosis

- Severe US features bode for chronic relapsing & remitting symptomatology
- Mild US features bode for favorable outcome
- Tendinosis predisposes to tendon tears
  - Tears uncommon in normal tendons
  - Likelihood of tendon tear increases with increasing severity of tendinosis

Treatment

- Conservative
  - Rest & immobilization
  - Controlled, slow stretching exercise
  - Antiinflammatory medication if peritendinitis
- Nonconservative
  - US-guided dry needling, autologous blood injection (ABI), platelet-rich plasma (PRP) injection, barbotage, sclerotherapy
    - No evidence that PRP is better than dry needling or ABI for treating tendinosis
  - US-guided steroid injection for peritendinitis
  - Focal proteoglycan accumulation treated with debridement
  - Suturing of large tears

DIAGNOSTIC CHECKLIST

Consider

- Grade tendinosis as mild, moderate, or severe
- Grade peritendinitis as mild, moderate, or severe
- Routinely examine with color Doppler imaging
- Look for secondary features & tendon tears
- Measurement of tendon caliber is useful for follow-up

SELECTED REFERENCES

1. Robinson P: Sonography of common tendon injuries. AJR Am J Roentgenol. 193(3):607-18, 2009

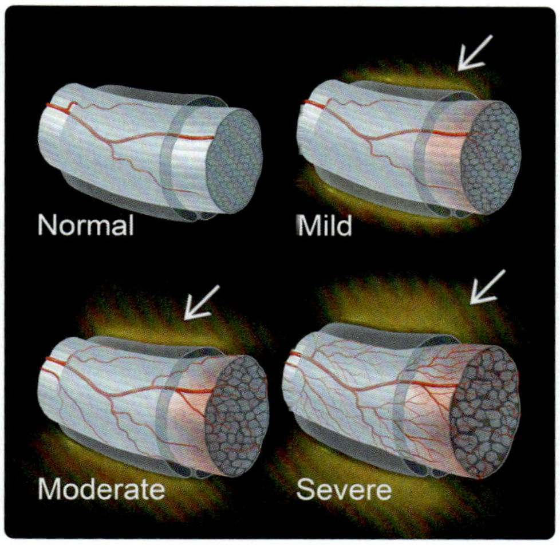

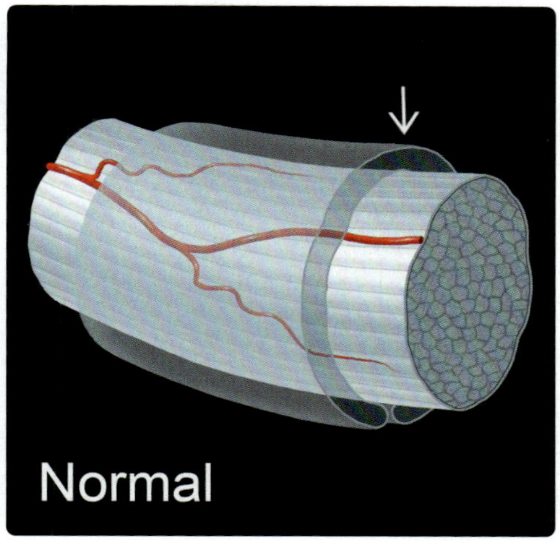

(Left) *Graphic shows grades of tendinosis. The severity of tendinosis and degree of surrounding peritendinitis ➔ should be independently graded. The more severe the tendinosis, the greater the likelihood of tendon tear. The more severe the peritendinitis, the greater the likelihood of clinical response to antiinflammatory medication.* **(Right)** *Graphic shows a normal tendon structure with a tendon sheath ➔ shown for illustrative purposes. Most tendons do not have a tendon sheath.*

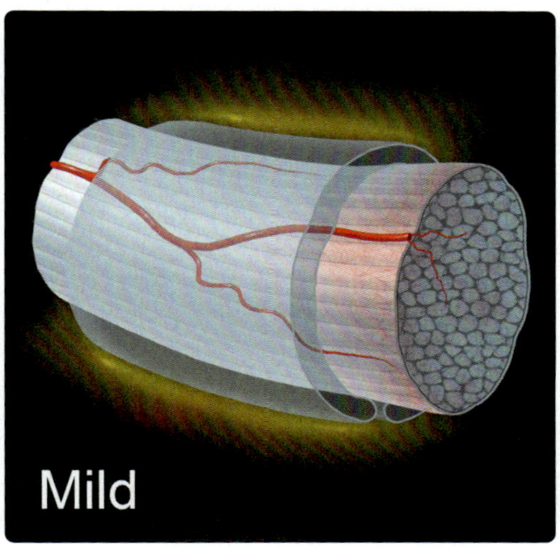

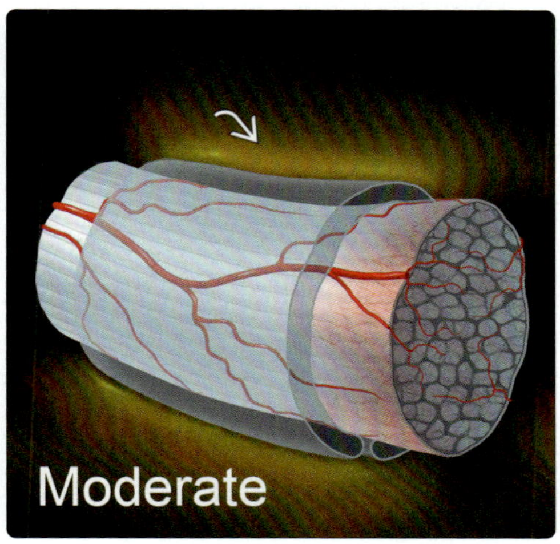

(Left) *Graphic shows mild tendinosis with mild tendon swelling (due to proteoglycan accumulation) and mild tendon hyperemia. Intrinsic vascularity is not a feature of normal tendons.* **(Right)** *Moderate tendinosis with more severe tendon swelling, proteoglycan deposition, and tendon hyperemia is shown. Although tendinosis and peritendinitis (seen as peritendinous edema and hyperemia) ➔ severity are broadly interrelated, both features should be independently graded as mild, moderate, or severe.*

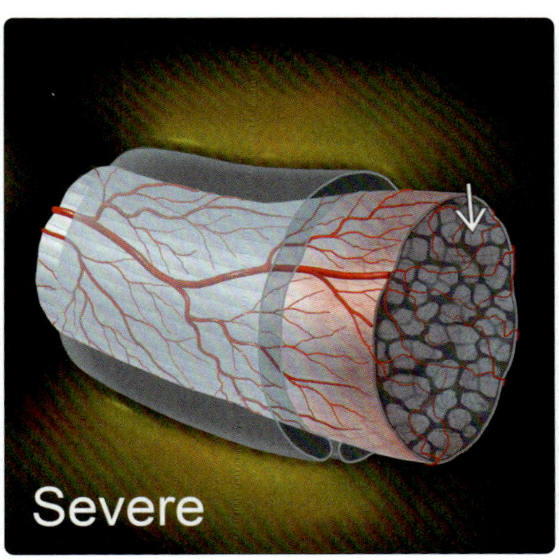

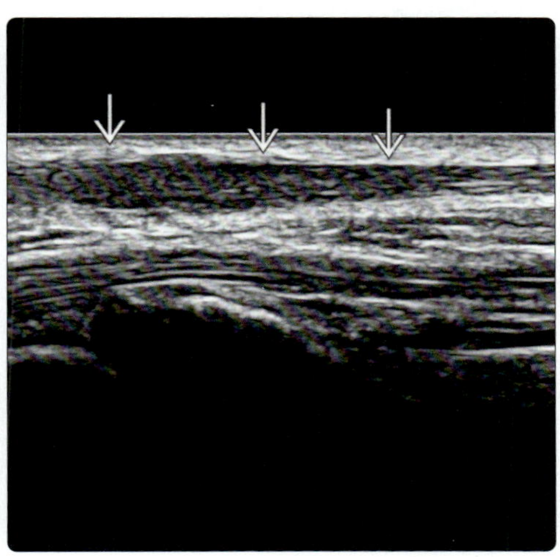

(Left) *Graphic shows severe tendinosis with more severe tendon swelling and tendon hyperemia ➔ (manifest on US as tendon swelling, tendon hypoechogenicity, loss of fibrillar echo pattern, and increasing intratendinous hyperemia). Mild tendon sheath swelling is often seen and is a feature of peritendinitis rather than a primary tendon sheath inflammation.* **(Right)** *Longitudinal US of palmaris longus tendon shows how tendinosis is often localized to that part of a tendon ➔ under most physical stress.*

(Left) *Transverse US in the same patient shows moderate severity palmaris longus tendinosis* ➡ *with tendon thickening, hypoechogenicity, and loss of the normal fibrillar pattern. The adjacent median nerve* ➡ *is mildly indented.* **(Right)** *Transverse color Doppler US shows moderate peritendinous* ➡ *and mild intrinsic hyperemia* ➡ *(reflective of disease activity). The overall appearances are those of moderate-severity palmaris longus tendinosis with moderate peritendinitis and mild disease activity.*

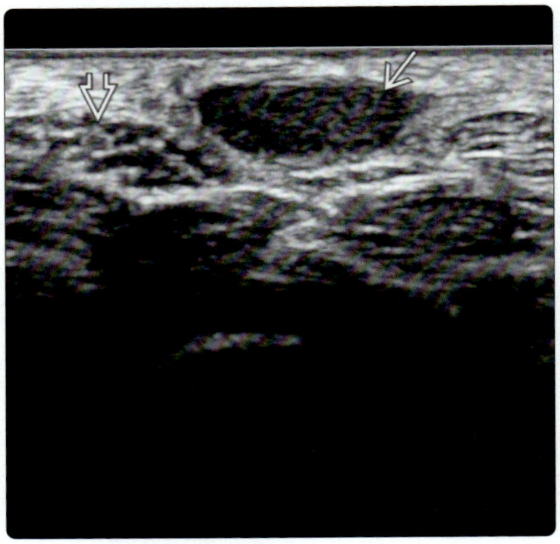

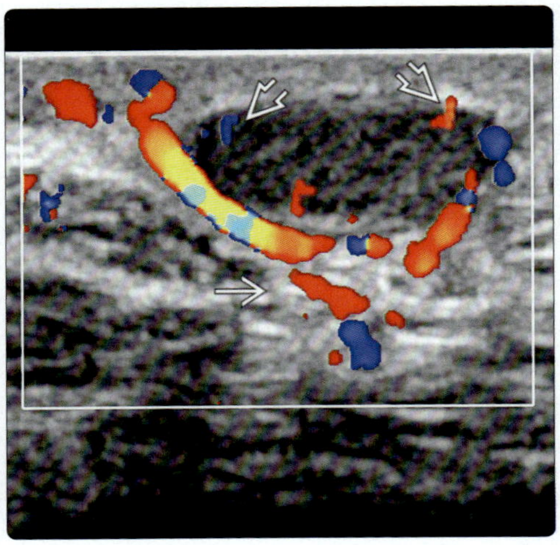

(Left) *Clinical photograph of a patient with severe swelling of the Achilles tendon region* ➡ *is shown. There was moderate localized tenderness.* **(Right)** *Transverse US shows a severely thickened Achilles tendon with mildly thickened paratenon* ➡*. Small curvilinear hypoechoic areas* ➡ *may represent partial tears or prominent intratendinous vessels. These can be differentiated with color Doppler imaging. Small intrasubstance tears are almost an expected feature of severe tendinosis.*

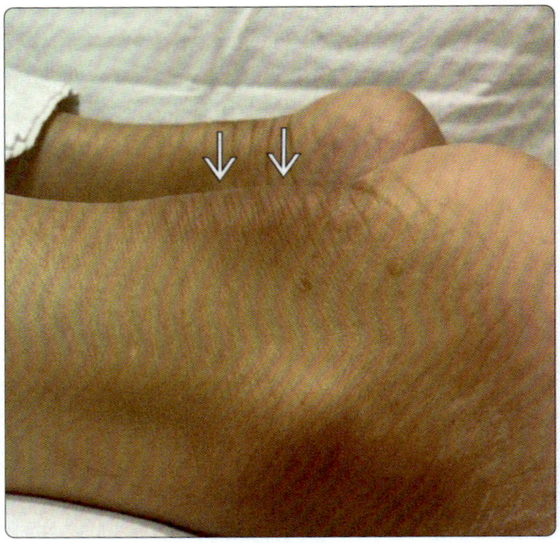

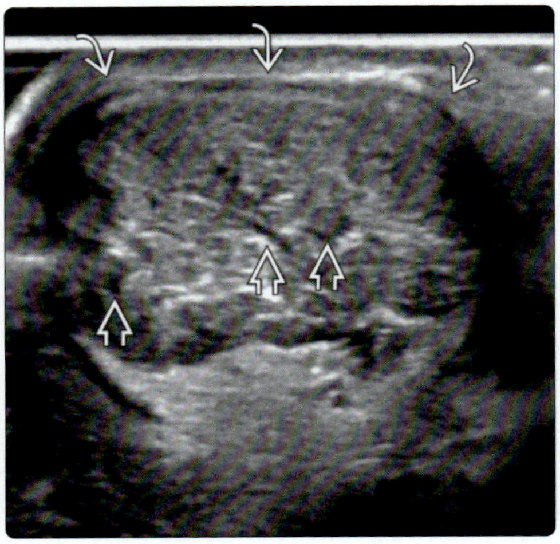

(Left) *Longitudinal US of the midportion of Achilles tendon of the same patient shows severe diffuse thickening* ➡ *due to severe Achilles tendinosis. Small linear hypoechoic areas* ➡ *are present, most likely due to small tears. These small tears are a common feature in severe tendinosis.* **(Right)** *Longitudinal color Doppler US of the same area shows severe intratendinous and peritendinous hyperemia indicating an active tendinopathic process. Color flow is best assessed with the tendon relaxed.*

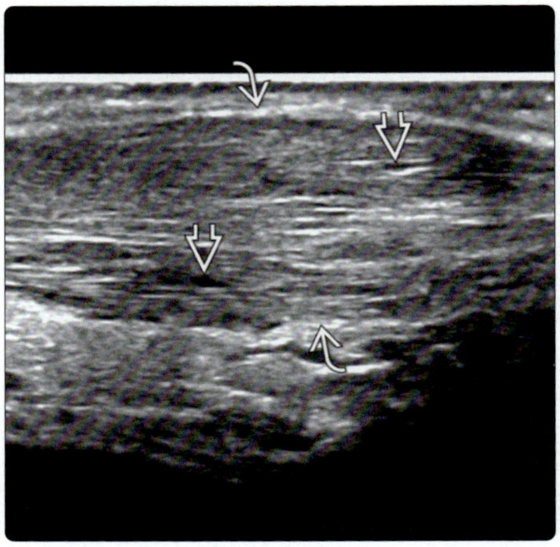

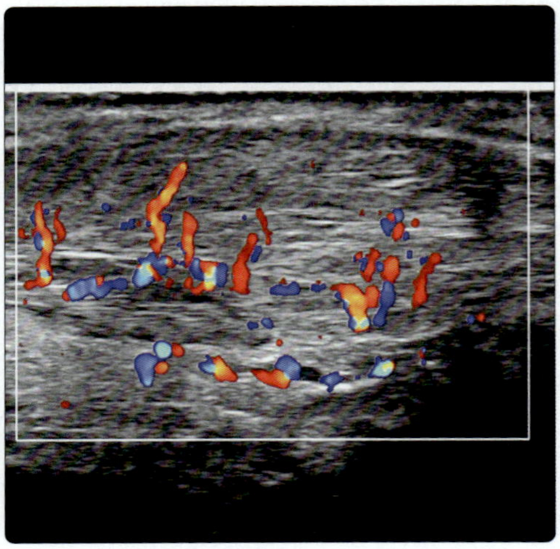

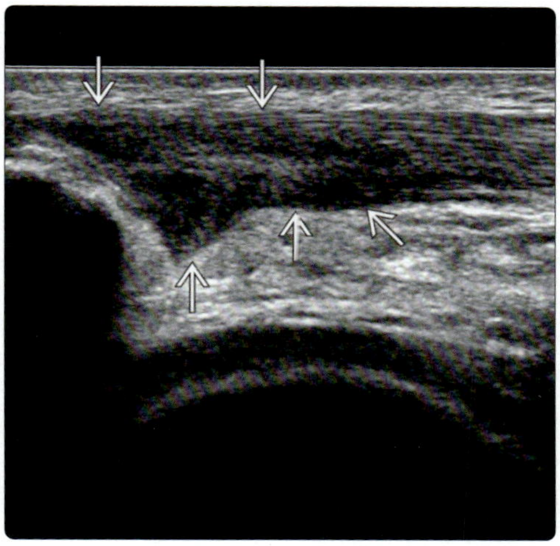

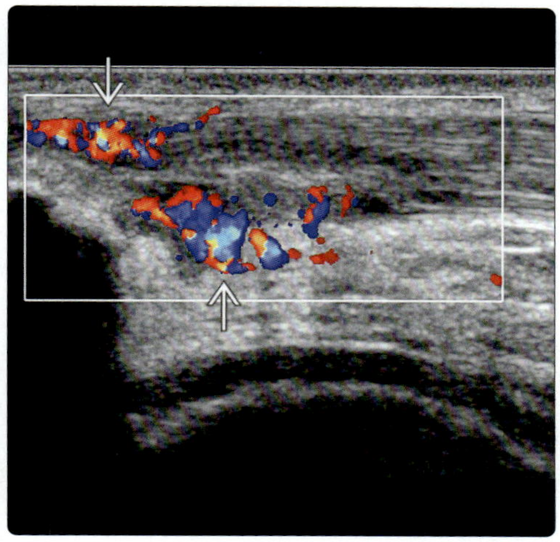

(Left) *Longitudinal US of a patellar tendon shows a severely thickened and heterogeneous hypoechoic tendon with loss of normal fibrillar pattern, particularly of the deep fibers* ➔. *There is no associated tendon calcification. Features are consistent with severe tendinosis.* **(Right)** *Longitudinal US of the same area shows severe intratendinous hyperemia, particularly of the deep fibers* ➔. *This is consistent with severe disease activity.*

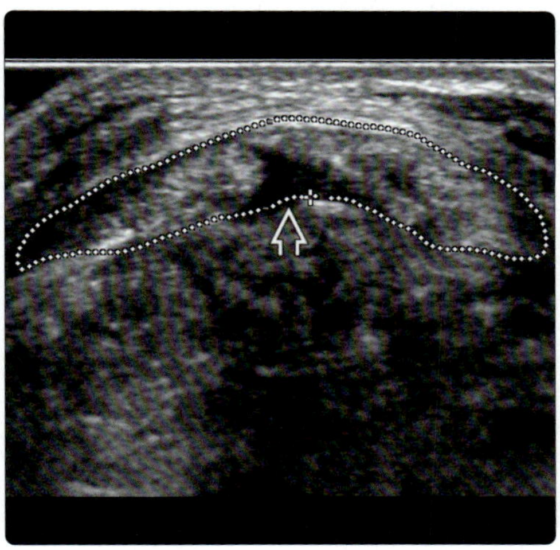

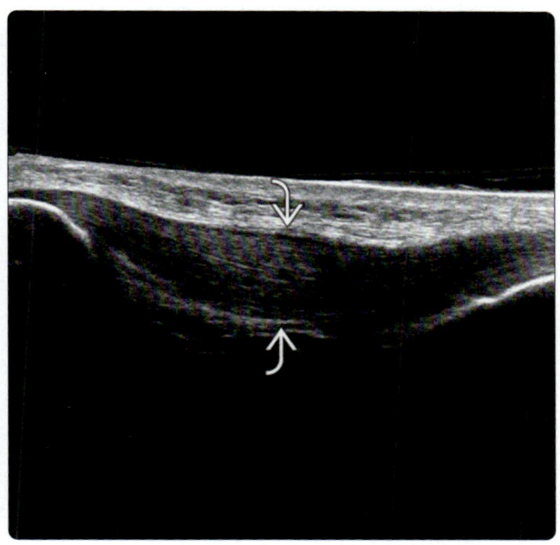

(Left) *Transverse US shows a cross-sectional area measurement of a slightly thickened proximal patellar tendon (dotted lines). There is a discrete area of proteoglycan accumulation centrally* ➔. **(Right)** *Longitudinal panoramic US view of a patellar tendon shows diffuse thickening, particularly of the midportion* ➔. *The whole tendon is more hypoechoic than usual.*

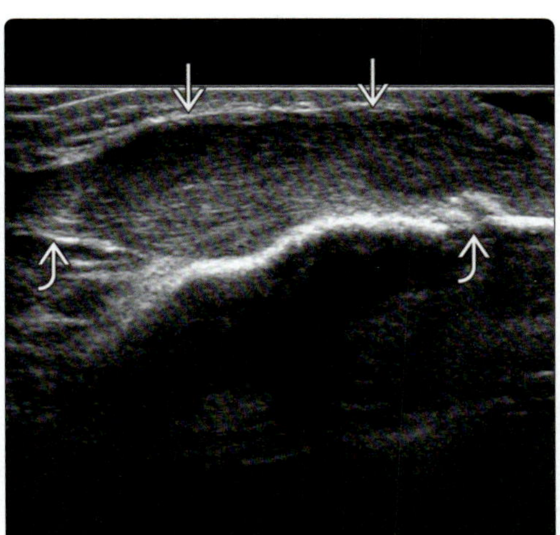

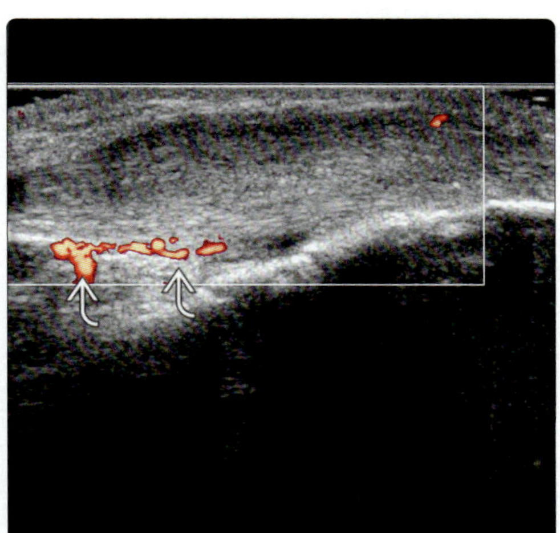

(Left) *Longitudinal US of a patellar tendon close to the tibial insertion shows severe focal thickening with hypoechogenicity and loss of normal fibrillar pattern* ➔ *consistent with severe tendinosis. Mild cortical irregularity is present* ➔. **(Right)** *Longitudinal power Doppler US at the same location shows peritendinous hyperemia* ➔ *consistent with inflammation.*

(Left) *Longitudinal color Doppler US of a quadriceps tendon shows mild thickening at the patellar insertion. Mild intrasubstance dystrophic calcification is present. There is no significant hyperemia present, which indicates quiescent disease.* **(Right)** *Sagittal proton-density MR of the knee in the same patient shows a moderately thickened quadriceps tendon with minimal increased tendon signal intensity corresponding to the US findings of tendinosis.*

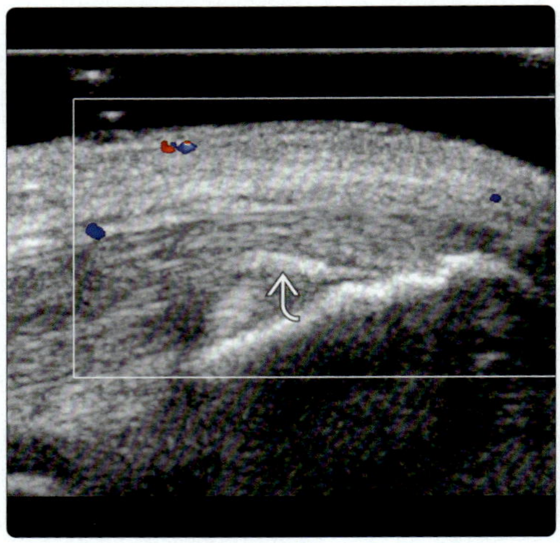

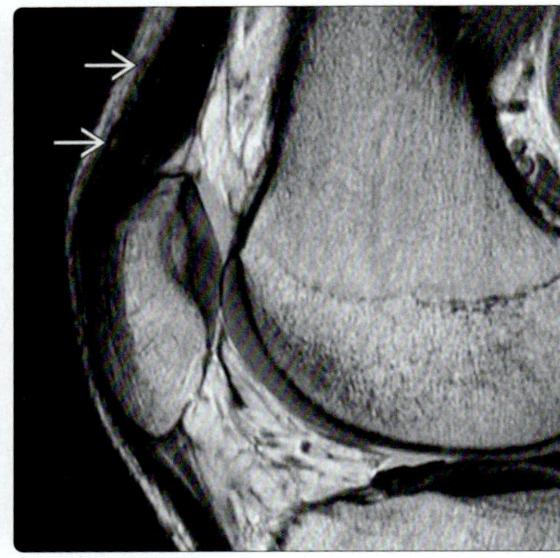

(Left) *Longitudinal US of the quadriceps tendon shows moderate tendon thickening with severe intratendinous dystrophic calcification. Clinical symptoms are helpful to distinguish this from calcific tendinitis.* **(Right)** *Transverse US of the same patient shows several calcified plaques within the thickened quadriceps tendon. This is severe dystrophic calcification within an area of tendinosis rather than calcific tendinitis. The latter is much more painful, particularly in the resorptive phase.*

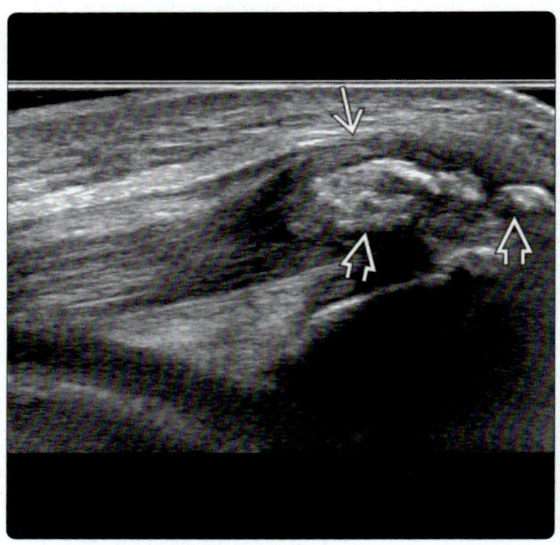

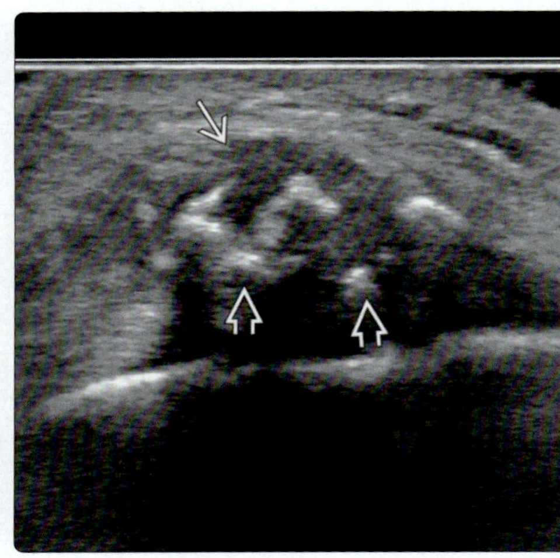

(Left) *Longitudinal color Doppler US of the same patient shows no hyperemia within or around the quadriceps tendon. This favors quiescent disease.* **(Right)** *Longitudinal US of common flexor tendon origin (CFTO) tendinosis shows moderate tendon thickening and hypoechogenicity with a plaque of dystrophic calcification. Dystrophic calcification is common in Achilles and quadriceps tendinosis and, to a lesser degree, tendinosis of the CFTO and common extensor tendon origin (CETO) at the elbow.*

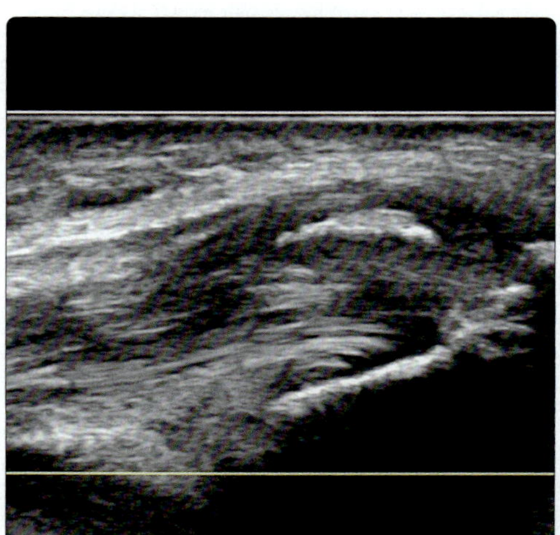

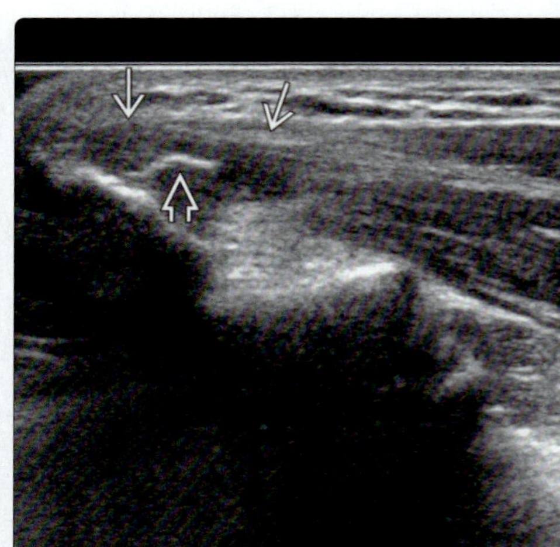

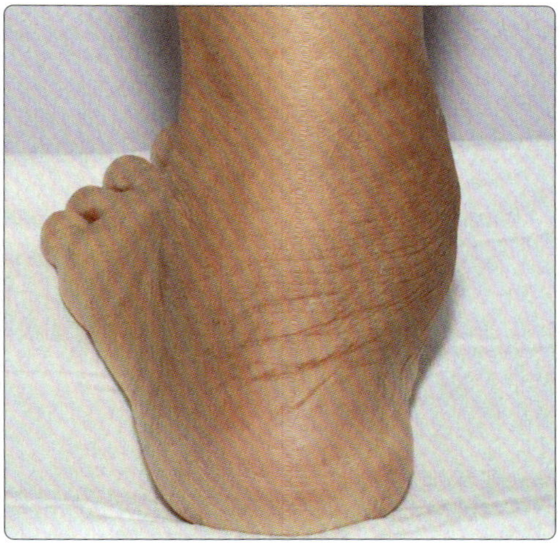

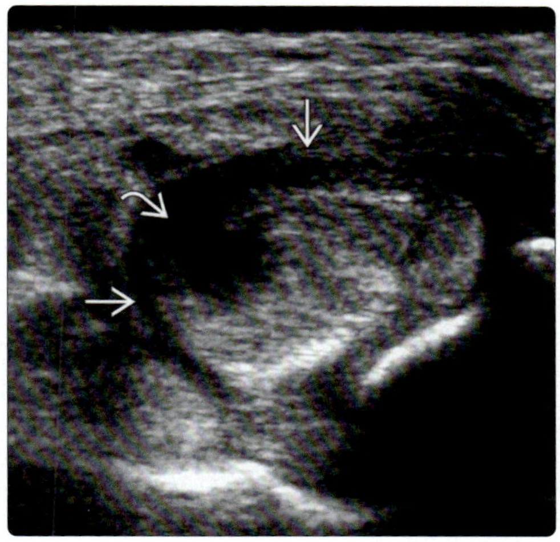

(Left) *Clinical photograph shows a patient who presented with flat foot for 1 year. There is collapse of the longitudinal arches with moderate heel valgus deformity. Abnormal stresses are placed on the medial side of the ankle and foot.* **(Right)** *Transverse US of the retromalleolar tibialis posterior tendon in the same patient shows severe tendinosis with a thickened heterogeneous tendon →. A small hypoechoic area ↪ with ill-defined margin is consistent with an area of proteoglycan accumulation.*

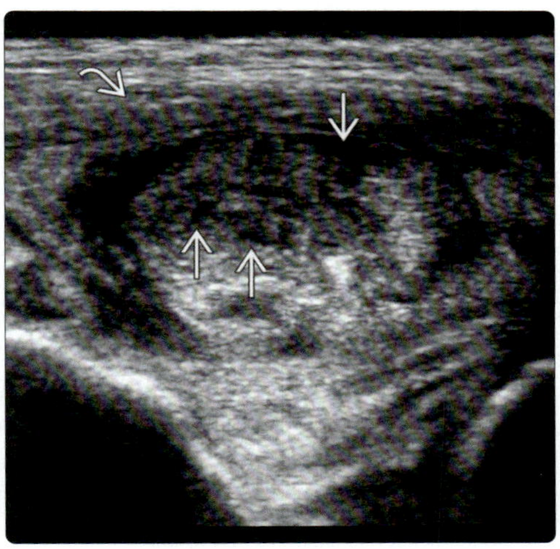

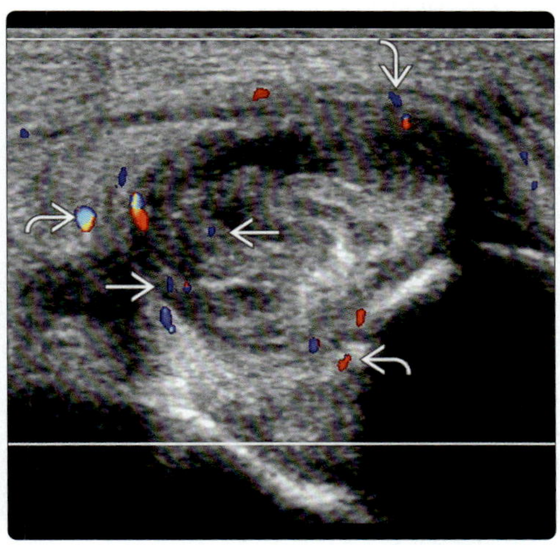

(Left) *Transverse US of the same tibialis posterior tendon at the inframalleolar level shows severe tendinosis with severe small hypoechoic areas → consistent with small intrasubstance tears. There is moderate thickening of the overlying flexor retinaculum ↪.* **(Right)** *Transverse color Doppler US at the same location shows moderate intratendinous → and peritendinous hyperemia ↪ consistent with moderate disease activity and moderate peritendinitis.*

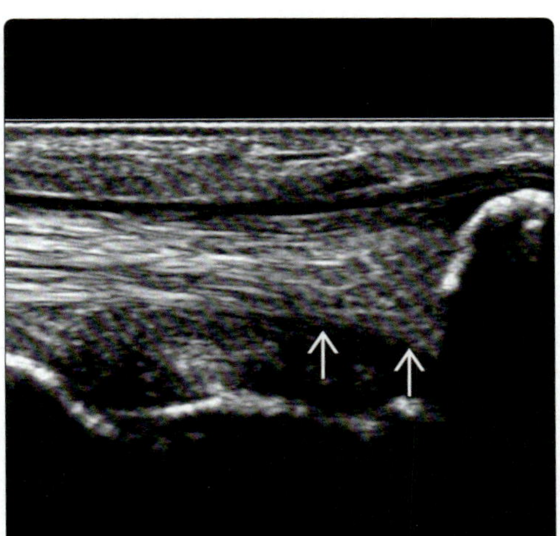

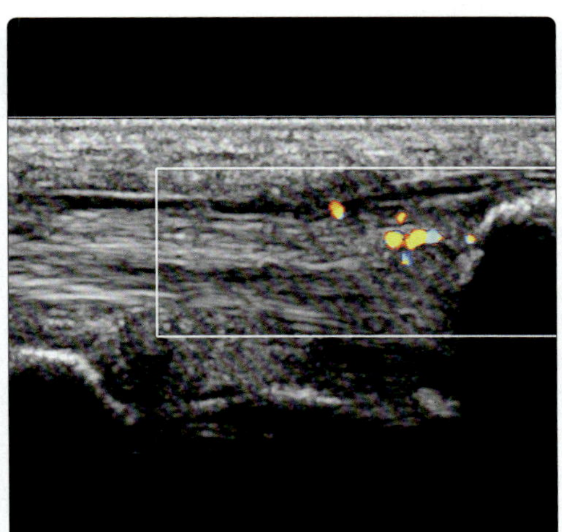

(Left) *Longitudinal US of the tibialis tendon at the navicular bone insertion shows mild focal tendon thickening →. The normal tendon shows this degree of expansion at its insertion.* **(Right)** *Longitudinal color Doppler US at same location shows mild hyperemia within the tendon substance indicative of mild tendinosis. Normal tendons do not display flow on color Doppler imaging. Hyperemia helps confirm the presence of tendinosis.*

(Left) *Longitudinal US of an elbow shows normal common extensor tendon origin at the lateral epicondyle. It is not thickened and has a normal fibrillar pattern & no calcification or tear. Note the tendons coursing over the radiohumeral articulation.* **(Right)** *Longitudinal US of the contralateral abnormal CETO shows severe tendon thickening and an intrasubstance partial tear of the midfibers.*

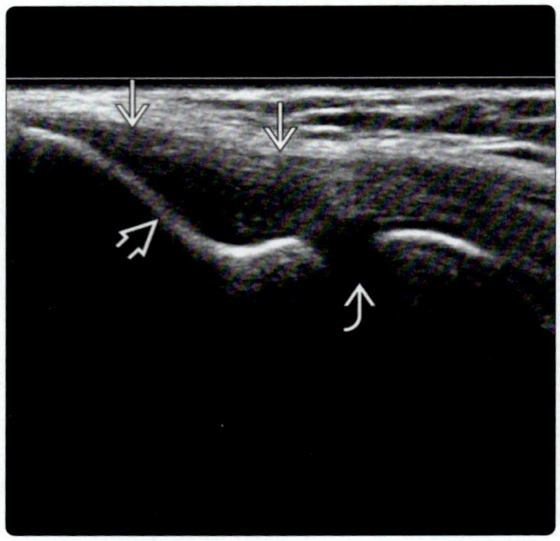

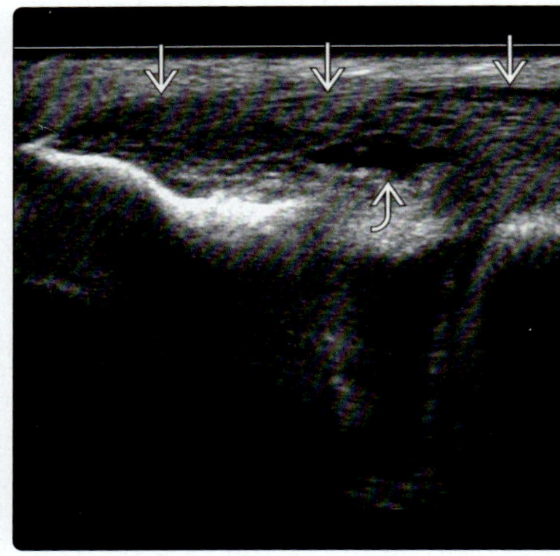

(Left) *Longitudinal color Doppler US of the same location shows severe hyperemia within the thickened tendon, consistent with severe disease activity.* **(Right)** *Transverse color Doppler US of another patient shows a severely thickened CETO with several small echogenic foci due to dystrophic calcification. Small discrete tears are present at the insertion. Minimal hyperemia is present.*

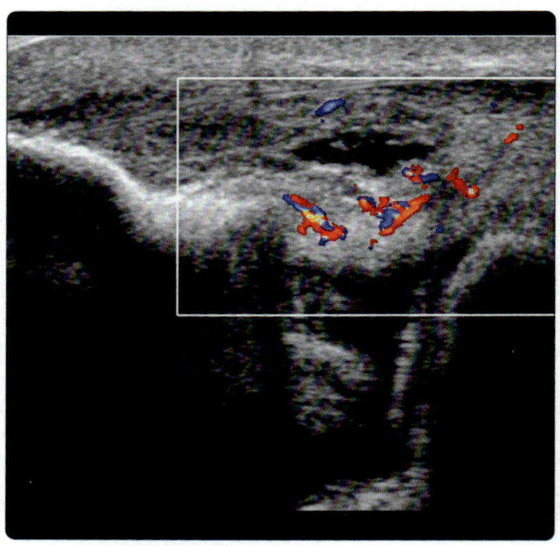

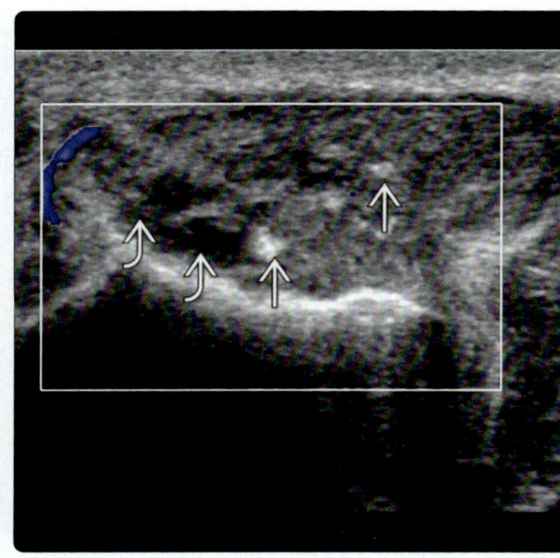

(Left) *Longitudinal US of the adductor longus tendon shows multiple foci of tendon calcification close to the adductor longus origin at the pubic ramus. The tendon is moderately swollen and hypoechoic, consistent with tendinosis.* **(Right)** *Transverse US at the same location shows a moderately thickened adductor longus tendon with foci of dystrophic calcification close to the pubic ramus.*

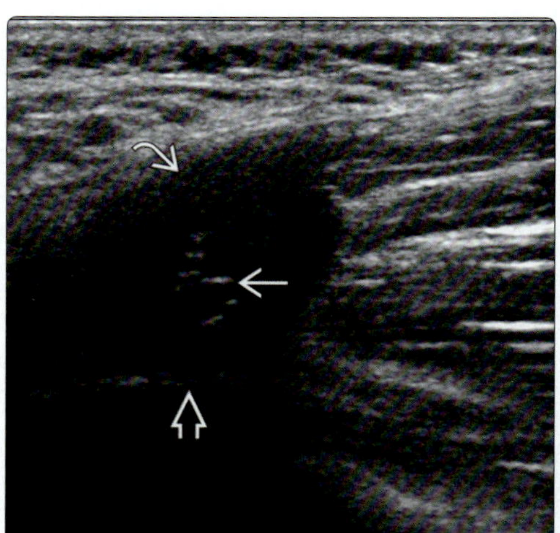

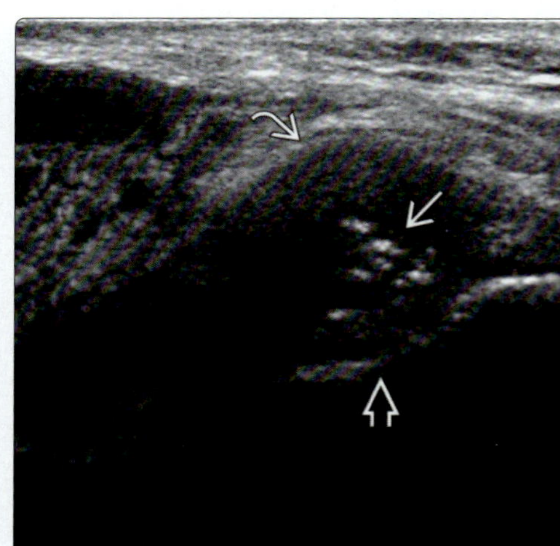

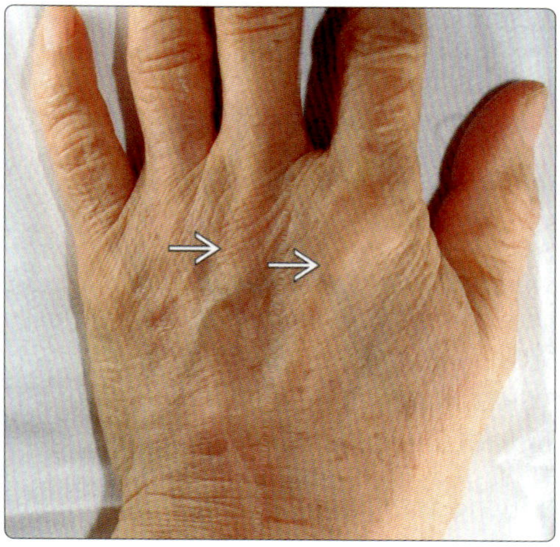

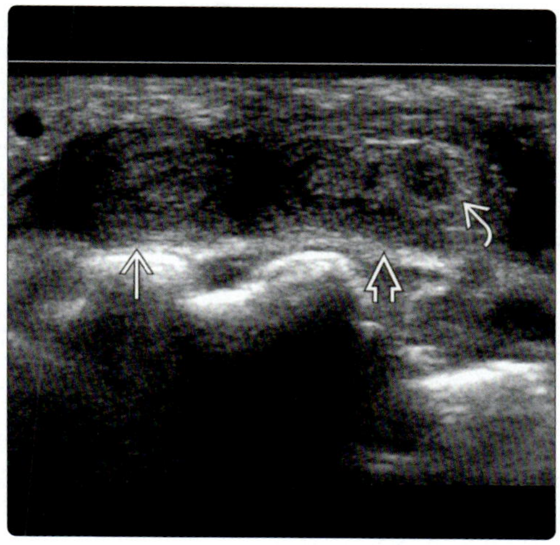

(Left) *Clinical photograph shows a patient with dorsal hand pain and swelling. Note swelling of the extensor digitorum (ED) tendons to the middle and index fingers* ➔. **(Right)** *Transverse US of ED of the index* ➔ *and middle* ➔ *fingers (same patient) shows marked thickening of the tendons and loss of the normal fibrillar pattern. Note the focal hypoechoic area* ➔ *within the tendon, consistent with myxoid accumulation.*

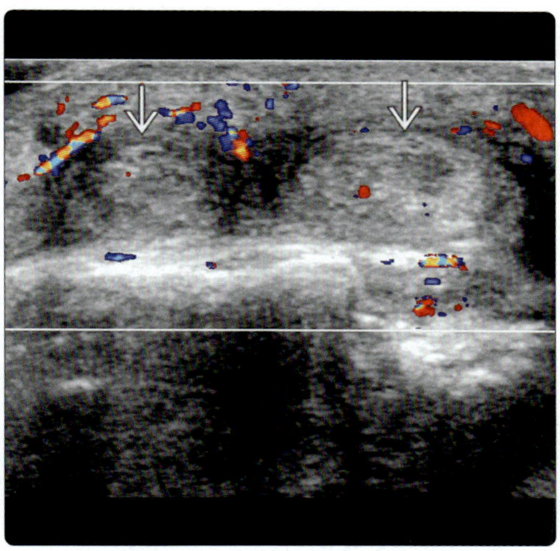

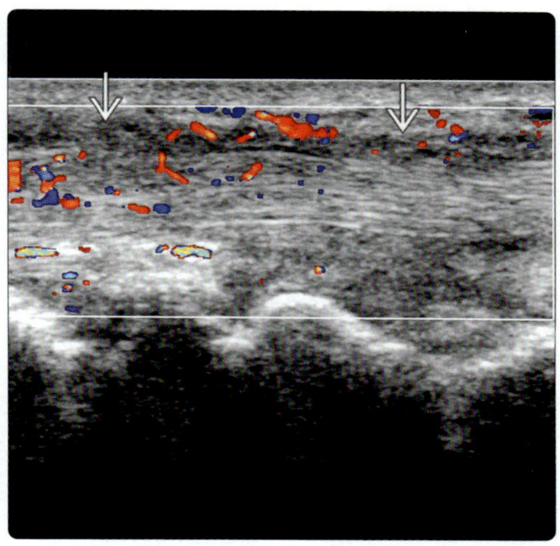

(Left) *Transverse color Doppler US of the same patient shows moderate hyperemia around the thickened ED tendons* ➔, *consistent with moderate peritendinitis. The tendon sheath is not present at this level.* **(Right)** *Longitudinal color Doppler US of the same location shows ED tendon* ➔ *thickening with intratendinous hyperemia, consistent with moderate disease activity.*

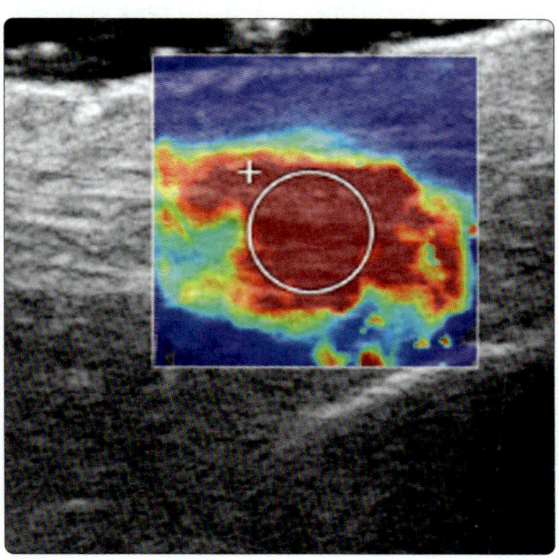

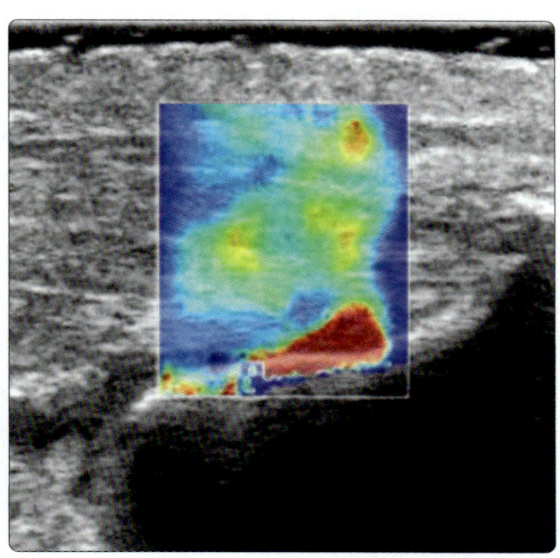

(Left) *Longitudinal elastography shows the normal stiff appearance of the Achilles tendon at its insertion as indicated by the red color. The elasticity spectrum varies between red (hard tissue) and blue (soft tissue). The normal tendon is red in color.* **(Right)** *Longitudinal US elastography shows softening of the Achilles tendon at its insertion, compatible with severe tendinosis, as indicated by the blue coloring.*

Nonrotator Cuff Tendon Tears

KEY FACTS

TERMINOLOGY

- Tendon tear = discontinuity either within tendon substance or at attachment
  - Partial thickness, full thickness, or complete

IMAGING

- Location and prevalence of tendon tears mirrors that of tendinosis
  - Tendinosis is strong prerequisite to tendon tear
  - Tendon tears occur in those tendon areas prone to tendinosis
  - The more severe the tendinosis, the greater the likelihood of tendon tear
  - Tendon tears relatively uncommon in normal tendons, except in severe trauma
- US allows recognition of tendon tears with high accuracy
  - For small tendons, examine 1st in transverse plane
    - Swollen tendon ends make tear more conspicuous on US
  - For larger tendons, examine 1st in longitudinal plane
  - Passive tendon movement can help in identifying tear
- Gap within tendon is either filled with blood, serous fluid, or edematous fat ± gas
- Tendon swelling following acute tear impedes assessment of preexisting tendinosis
- Collapsed tenosynovial sheath can mimic attenuated tendon
- US is accurate at detecting retear and defining cause of limited movement following repair
  - Limited movement may be due to adhesions
  - Tendon swelling can remain for years following repair

DIAGNOSTIC CHECKLIST

- Follow tendon from musculotendinous junction to insertion
- Minimize anisotropy, which can mimic tendon tears

(Left) *Clinical photograph shows both heels hanging freely over the edge of a bed with the patient lying prone. There is loss of plantar flexion on the affected left side with swelling of the Achilles region.* **(Right)** *Longitudinal US of the same patient shows a complete tear of the proximal 1/3 of the Achilles tendon. The proximal and distal stumps are retracted. The soleus muscle deep to the tear is normal.*

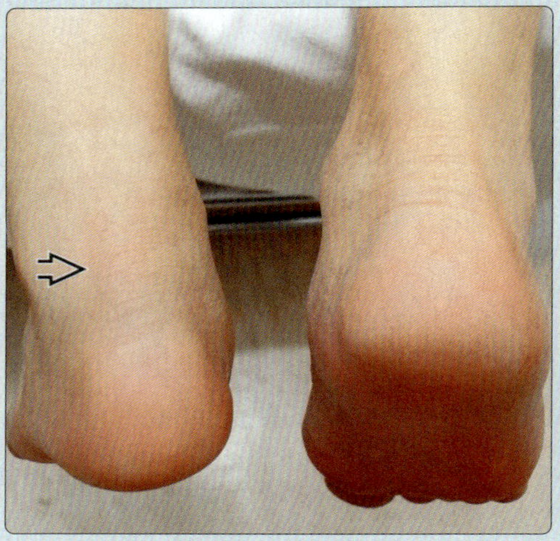

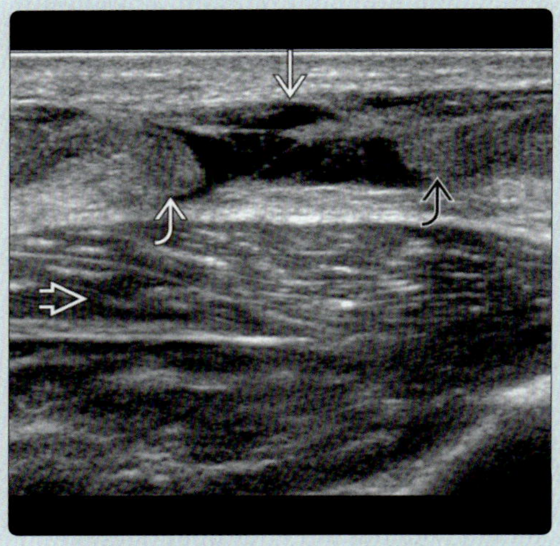

(Left) *Longitudinal US of the distal leg shows a complete tear in the proximal 1/3 of the Achilles tendon with hematoma partially filling the retraction gap. Note the swollen proximal and distal tendon ends.* **(Right)** *Longitudinal US shows a complete Achilles tendon tear just distal to the musculotendinous junction. This is a common site of an Achilles tendon tear. The proximal and distal tendon ends are retracted. The plantaris tendon is intact.*

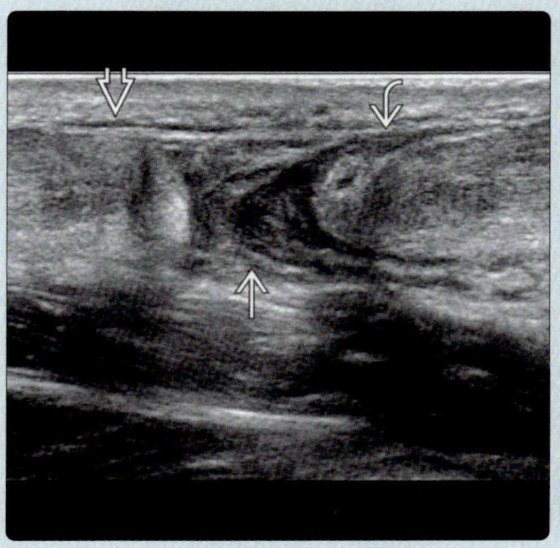

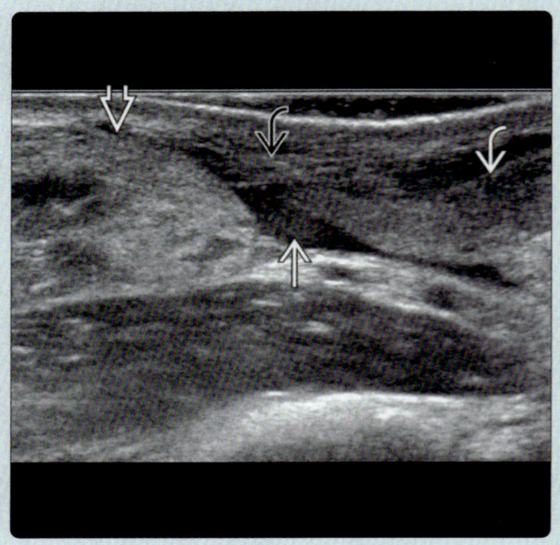

TERMINOLOGY

Synonyms

- Tendon rupture, tendon tear, tendon sprain, tendon strain
  - Tendon sprain and strain are clinical rather than imaging terms used to describe partial-thickness tendon tears

Definitions

- Tendon tear = discontinuity either within tendon substance or at attachment
  - Partial-thickness tear: Some of tendon fibers are torn, but this does not involve full thickness of tendon
  - Full-thickness tear: Some of tendon fibers are torn, and this does involve full thickness of tendon
  - Complete tear: All of tendon fibers are torn
  - High-grade partial tear: Partial tendon tear involving > 50% of tendon fibers

IMAGING

General Features

- Best diagnostic clue
  - Fluid-filled gap in tendon
  - Discontinuity of tendon fibers
- Location
  - In most patients, occurrence and location of tendon tears closely mirror that of tendinosis
    - Tendinosis is strong prerequisite of tendon tear
    - Tendon areas prone to tendinosis are also those prone to tendon tears
      - Biceps tendon at proximal end bicipital groove ("genu")
      - Common extensor tendon origin at lateral humeral epicondylar area
      - Extensor pollicis longus tendon at level of Lister tubercle
      - Extensor carpi ulnaris just distal to ulnar head
      - Hamstring tendons at ischial tuberosity
      - Adductor tendons at pelvic attachment
      - Gluteus medius/minimus tendons at greater trochanter
      - Quadriceps tendon at patellar insertion
      - Patellar tendon at patellar origin
      - Achilles tendon close to musculotendinous junction and at calcaneal insertion
      - Posterior tibialis tendon around medial malleolus or near insertion
      - Peroneal tendons around lateral malleolus, peroneal tubercle, or os peroneum (peroneus longus)
      - Tibialis anterior tendon close to insertion
- Morphology
  - Some tendons preferentially tear transversely
    - Adductor, quadriceps, patellar, Achilles, tibialis anterior, and wrist extensors and flexors
  - Some tendons preferentially tear longitudinally
    - Hamstrings, posterior tibialis, peroneal tendons, and biceps
  - Larger tendons tend to tear transversely
    - Smaller tendons tend to tear either longitudinally or transversely

Ultrasonographic Findings

- Fluid-filled gap within tendon
  - Acutely, tendon retraction gap may be filled with blood, serous fluid, fat ± gas locules, making tendon ends difficult to see
  - Irregular tendon ends
- Often swollen tendon ends
  - Swollen tendon ends make small tendon tears more conspicuous on US
- Collapsed tendon sheath may be mistaken for attenuated tendon
  - Applies to smaller tendons (i.e., finger flexors, extensor pollicis longus tendon, and biceps tendon)
- Partial tear
  - Focal discontinuity of tendon not involving all tendon fibers
  - Lack of complete tendon retraction
- Partial intrasubstance tear
  - Tear is linear, sharp, and hypoechoic within substance of tendon
  - Can be difficult to distinguish from focal areas of proteoglycan accumulation in severe tendinosis
    - Focal proteoglycan accumulation is rounded, more irregular, and less hypoechoic
- Postrupture tendon swelling
  - Can hinder evaluation of presence and severity of preexisting tendinosis
    - Intratendinous calcification and hyperemia are good clues to preexisting tendinosis
    - Examining contralateral tendon may help as comparable degree of tendinosis often exists on contralateral side
- Some tendon tears, especially those occurring spontaneously due to tendinosis, may present late
  - Partial gap in reparative hypoechoic granulation tissue may be seen in chronic tears
  - Mild hyperemia of granulation tissue appreciated on color Doppler imaging
- Tendon subluxation may lead to restricted range of movement without tendon tear
- US is accurate at detecting retear or defining cause of limitation of movement following repair
  - Suture material is often seen traversing gap
  - Limitation in movement of tendons may be due to tendon adhesion and not retear
- Tendon remains swollen and hyperemic long after tendon integrity and strength have returned
  - Rehabilitation schedule best guided by clinical assessment rather than serial US examination

MR Findings

- T1WI
  - Assess tendon enlargement + retraction
- T2WI
  - Disruption with discontinuity ± wavy retracted tendon
  - Enlarged tendon with intratendinous fluid or scar may be seen for many years post surgery

Imaging Recommendations

- Best imaging tool

- US allows recognition of tendon tears with high accuracy and precision
- Protocol advice
  - Start by examining unaffected side to become familiar with normal anatomical appearances
  - For small tendons, examine 1st in transverse plane along length of tendon
    - Examine tendon slowly from distal to proximal, looking for any gap in tendon
  - Continually realign transducer to avoid anisotropy, which may mimic tears
  - Passive movement of affected tendon aids in identification of tendon ends
  - If bicipital groove or extensor carpi ulnaris groove is empty, check that tendon has not subluxed
  - For Achilles tendon examination, patient should be prone with foot dorsiflexed over end of bed
    - Measure distance of tear to musculotendinous junction and to calcaneal insertion
    - ± examine in full plantar flexion to determine tendon end apposition with plantar flexion, if conservative treatment being contemplated
  - Mark retracted tendon ends on skin to aid surgical localization

DIFFERENTIAL DIAGNOSIS

Scarred Tissue

- Post surgery
- May limit normal tendon excursion during passive movement

Tendinosis With Peritendinitis

- Usually associated with small tendon tears
  - Microtears be too small to resolve

PATHOLOGY

General Features

- Etiology
  - Young to middle-aged patients tear large or medium-sized tendons during modest to high exertion
    - Typically, torn tendon has minimal or mild tendinosis
  - Elderly patients tear medium-sized or small tendons spontaneously or with minimal exertion
    - Typically, torn tendon has moderate to severe tendinosis
  - Other common etiological factors
    - Inflammatory arthropathy, particularly rheumatoid arthritis
      - Commonly affects extensor carpi ulnaris and extensor tendons of 4th and 5th fingers
      - Due to chronic friction from subluxed ulnar head ± chronic tenosynovitis
    - Malaligned fracture, particularly distal radial fracture ± fixation plate
      - Usually affects flexor carpi radialis, flexor pollicis longus, and long flexor tendons of index finger
    - Chronic joint deformity, penetrating injury, and possibly repeated steroid injection

Gross Pathologic & Surgical Features

- Achilles tendon
  - Plantaris tendon usually remains intact
- Quadriceps tendon
  - Composed of 3 lamina: Superficial (from rectus femoris), intermediate (vastus lateralis and medialis), and deep (vastus intermedius)
  - Partial tears usually involve superficial ± intermediate lamina

Microscopic Features

- Degeneration with decreased collagen cross-linking
  - Loss of viscoelasticity and increased stiffness

CLINICAL ISSUES

Presentation

- Most common signs/symptoms
  - Acute onset of pain ± swelling during exertion
  - Loss of tendon function
  - Clinical assessment incorrect in up to 25%
- Clinical profile
  - Exertional activity in young/middle-aged subjects
  - Nonexertional activity in elderly subjects
  - In Achilles tendon exam, squeezing calf muscles normally causes plantar flexion (Thompson test)

Demographics

- Age
  - Most common after 30 years
- Gender
  - M:F = 5:1

Natural History & Prognosis

- Achilles tendon
  - Nonsurgical rate of rerupture: 21%
  - Surgical rate of rerupture: 2%

Treatment

- Complete Achilles tendon tear
  - Cast immobilization
    - Above-knee cast with plantar flexion for 4 weeks followed by below-knee cast with decreased plantar flexion
    - Indicated for acute tears and when US shows good approximation of tendon ends with plantar flexion
  - Surgical repair is usually required
    - Either end-to-end anastomosis or tendon graft flap depending on length of tear
- Other tendons: Surgical repair

DIAGNOSTIC CHECKLIST

Image Interpretation Pearls

- Utilize passive or active tendon movement during US examination
- Minimize anisotropic effect

SELECTED REFERENCES

1. Dams OC et al: Imaging modalities in the diagnosis and monitoring of Achilles tendon ruptures: a systematic review. Injury. 48(11):2383-2399, 2017

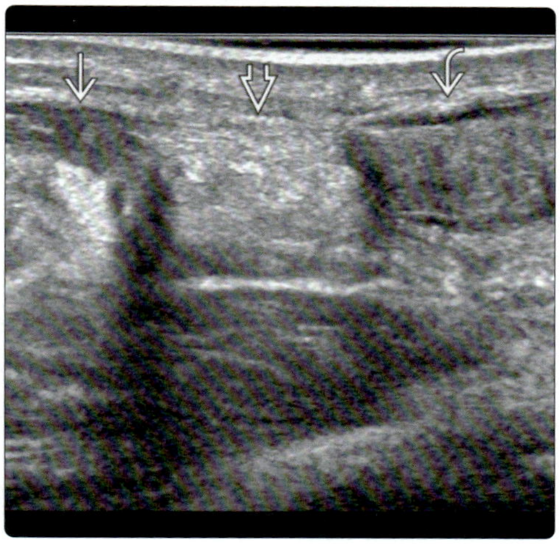

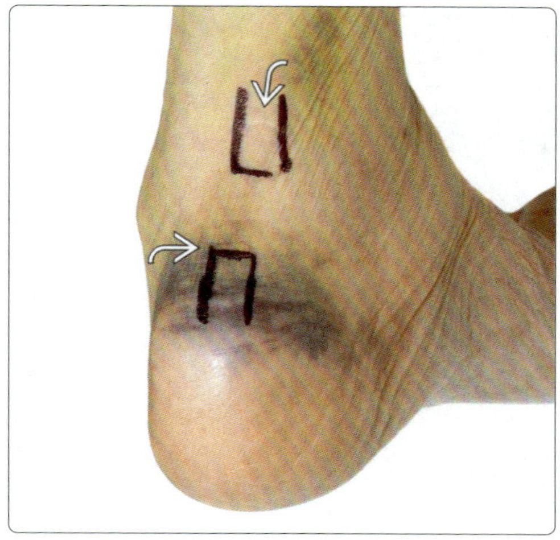

(Left) *Longitudinal US of an acute full-thickness Achilles tendon tear shows a gap filled with echogenic edematous fat . The swollen proximal and distal ends of the tendon are shown. The retraction gap can be filled with hematoma, gas, fat, or reparative tissue.* **(Right)** *Clinical photograph shows proximal and distal ends of the tendon tear marked on the skin. The shallow, rounded skin depressions are made with an upturned marker cap before the gel is removed, and the tendon ends are marked with a felt pen.*

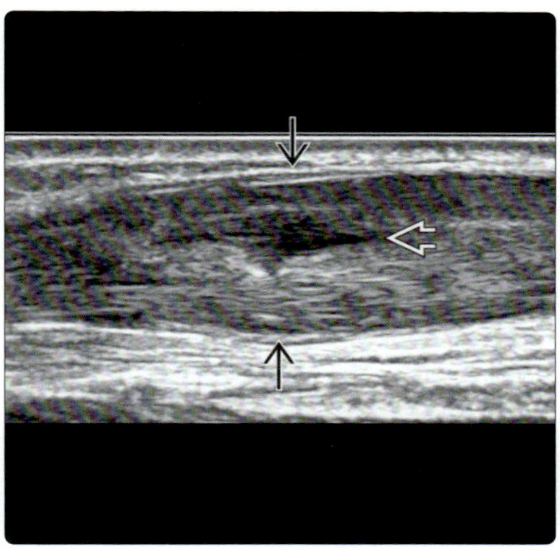

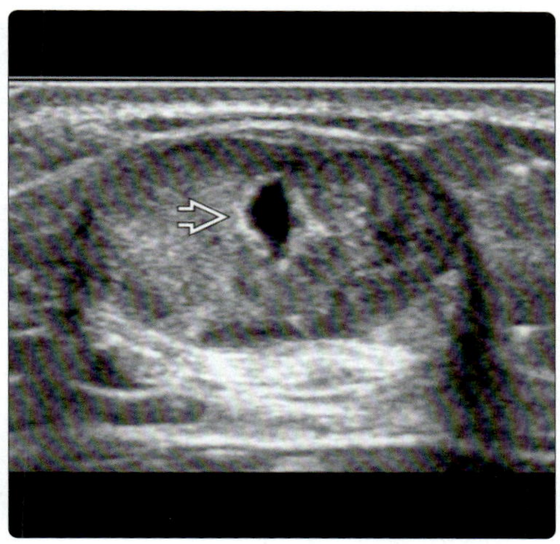

(Left) *Longitudinal US shows tendinosis and fusiform thickening of the midsegment of the Achilles tendon . A small longitudinal intrasubstance partial tear is located within the central portion of the tendon .* **(Right)** *Transverse US at the same location shows the intrasubstance tear of the Achilles tendon. Complete tears of the Achilles tendon are much more common than partial tears.*

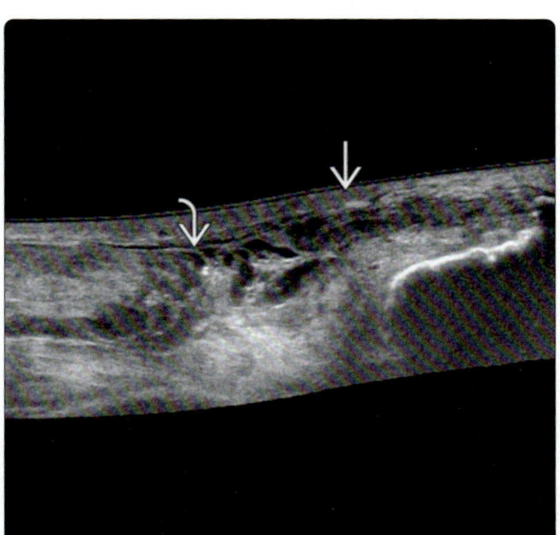

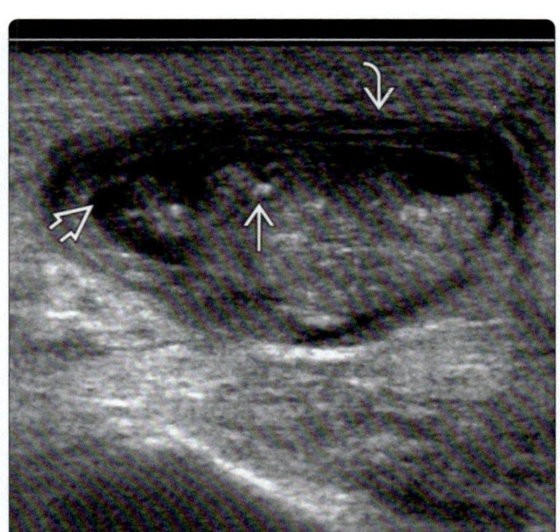

(Left) *Longitudinal US shows an acute complete Achilles tendon tear. The gap is filled with echogenic hematoma . The proximal end is swollen and irregular. Accurately determining the extent of preexisting tendinosis in the presence of acute tears is difficult as the tendon ends will often be swollen.* **(Right)** *Transverse US of the same tear shows small calcifications in the hematoma partially filling the tendon gap. Fluid is also present. The Achilles paratenon is moderately swollen.*

(Left) *Longitudinal US of an Achilles tendon shows a complete tear (avulsion) from the calcaneal insertion. There is a small sliver of echogenic bone at the distal end of the retracted tendon. Most avulsive-type Achilles tendon tears from the calcaneum will have a thin rim of bone.* **(Right)** *Longitudinal US shows a complete chronic rupture of the Achilles tendon. There is a large retraction gap. The chronic tear is filled with echogenic fibrous tissue. The underlying soleus muscle is mildly atrophic.*

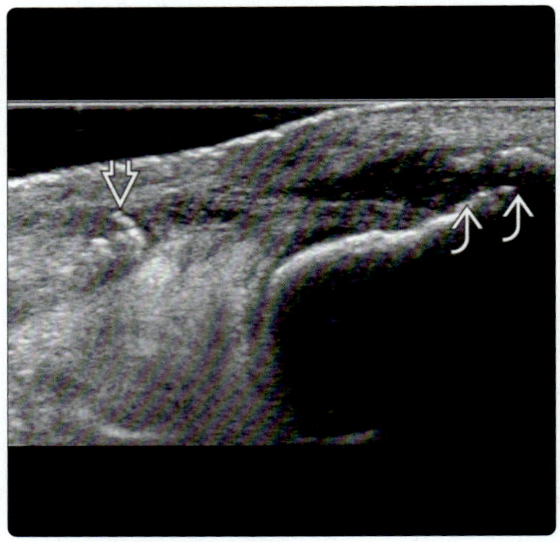

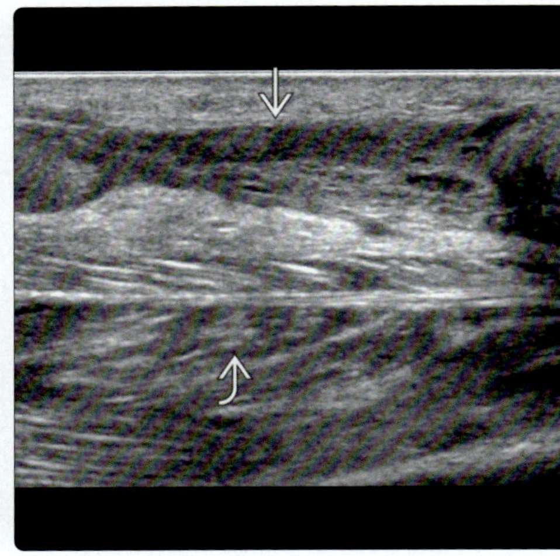

(Left) *Longitudinal US of an Achilles tendon in a patient with heel pain for 9 months and sudden complete tendon tear 2 months ago shows that the retracted proximal stump is severely swollen and heterogeneously hypoechoic with small flecks of calcification, suggestive of severe tendinosis.* **(Right)** *Transverse power Doppler US of the proximal stump of the Achilles tendon in the same patient shows moderate hyperemia and small flecks of calcification within the tendon.*

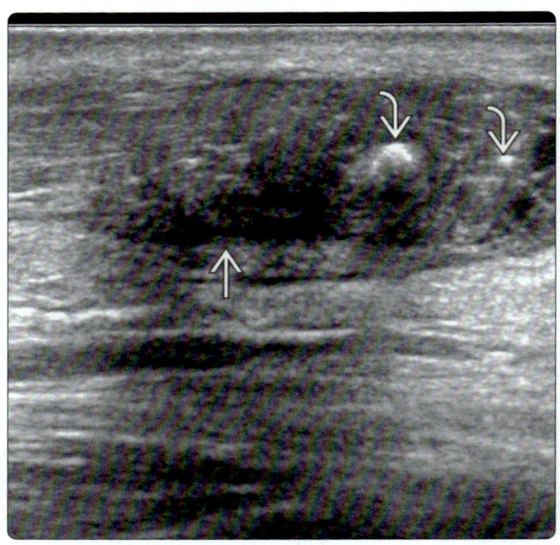

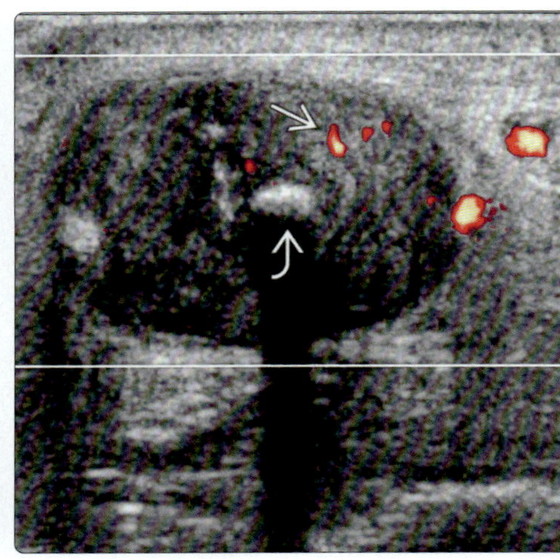

(Left) *Longitudinal US shows an acute tear of the adductor longus tendon close to the inferior pubic ramus attachment. The tendon fibers are retracted distally. The gap is partially filled with hematoma.* **(Right)** *Longitudinal US of the same patient close to the pubic tubercle shows a torn tendon with cortical irregularity of the insertional area. Tendon fibers are displaced distally.*

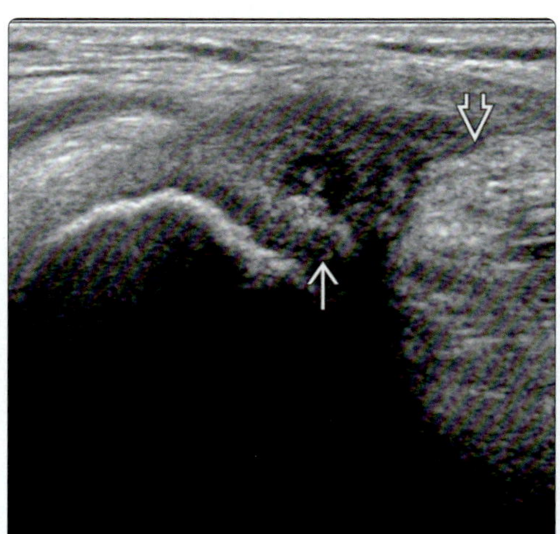

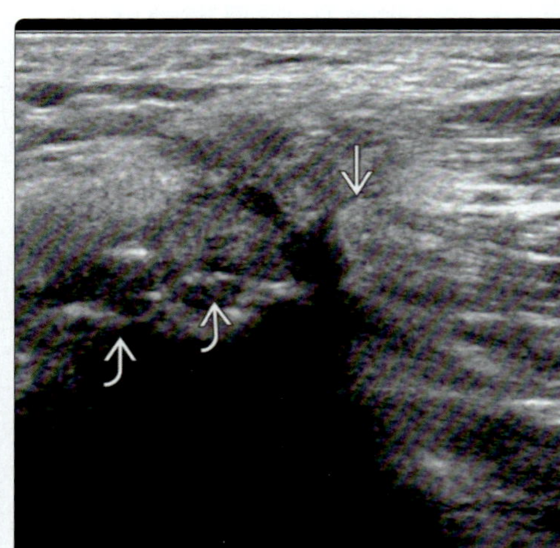

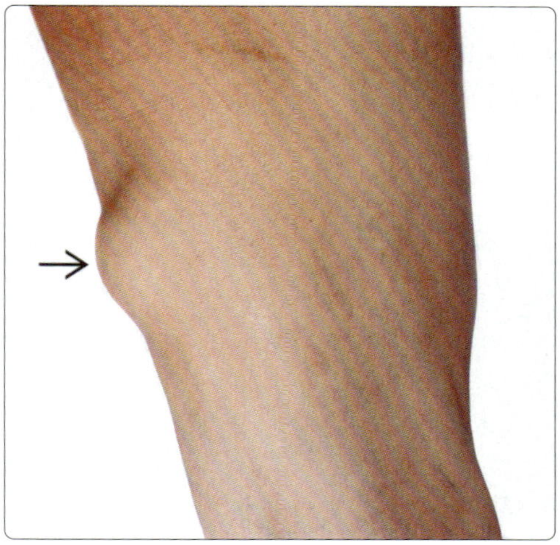

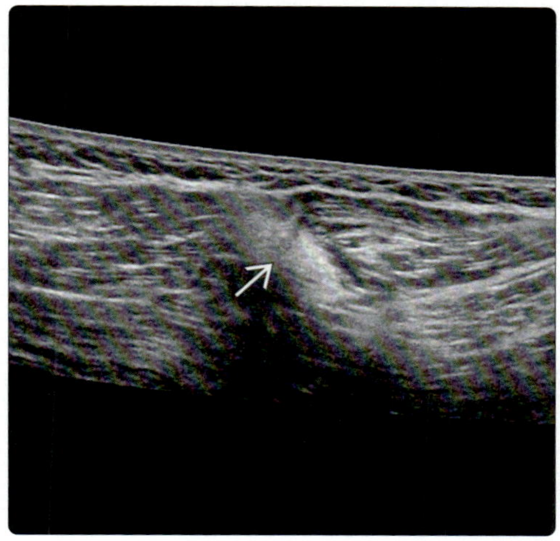

(Left) *Clinical photograph shows a mass on the medial aspect of the distal thigh. The mass had been present for several years. It moves more distally and becomes more apparent with muscle contraction, and it almost disappears with muscle relaxation.* **(Right)** *Longitudinal US of the thigh of the same patient shows an incomplete rupture of the semimembranosus tendon from its proximal insertion and retraction to the midthigh region. There is a fibrotic mass at the site of the retracted torn tendon.*

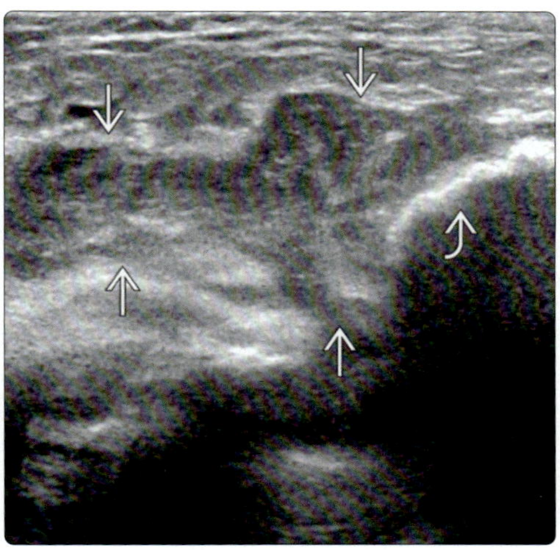

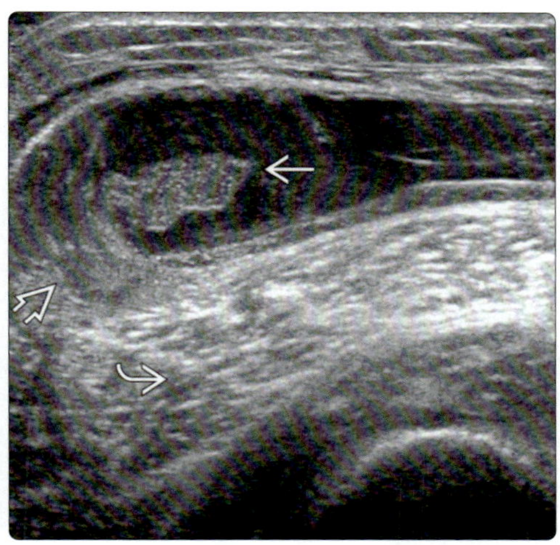

(Left) *Longitudinal US shows a complete quadriceps tendon tear at the superior pole of the patella. The proximal end of the tendon has retracted and is not visible on this image. The tendon gap is filled up with hemorrhagic fluid.* **(Right)** *Transverse US shows hemorrhage with blood clot within the tendon gap. Intact but edematous vastus medialis fibers are present peripheral to the hematoma. The vastus intermedius muscle deep to the hemorrhagic area is normal.*

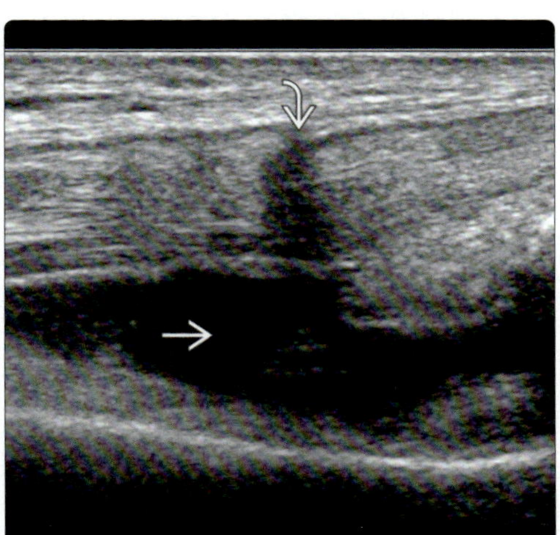

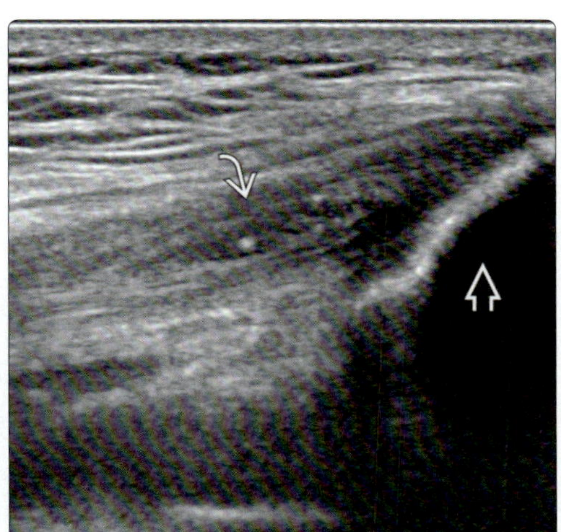

(Left) *Longitudinal US shows a localized quadriceps tendon tear of the lateral fibers with moderate suprapatellar pouch distension. The medial quadriceps fibers were intact.* **(Right)** *Longitudinal US shows a moderate-severity intrasubstance quadriceps tendon tear. This tear involves the central, i.e., intermedius/rectus femoris, fibers just proximal to the patellar insertion. About 20-30% of the tendon fibers are torn. The vastus medialis and lateralis components of the quadriceps tendon were intact.*

(Left) *Graphic shows a patellar tendon tear ➡ just distal to the patella, which is also stabilized by the medial ➡ and lateral ➡ patellofemoral ligaments. The less functionally relevant lateral ➡ and medial ➡ patellotibial ligaments, as well as the patellomeniscal ➡ ligament, are also shown.* **(Right)** *Longitudinal extended-field-of-view US of the knee shows a complete tear (avulsion) of the patellar tendon ➡ from the inferior pole of the patella ➡. The tendon gap ➡ is filled with blood, serous fluid, and fat.*

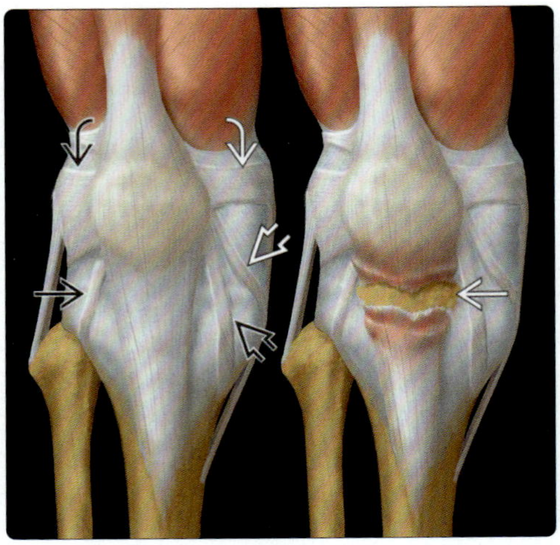

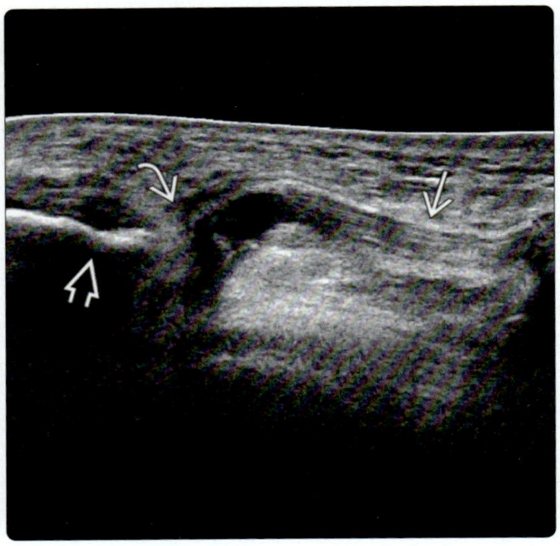

(Left) *Lateral radiograph in a patient with acute knee trauma shows a high-riding patella ("patella alta") with a small osseous fragment ➡, which seems to be attached to the patellar tendon ➡, suggestive of a complete patellar tendon tear.* **(Right)** *Longitudinal US in the same patient shows a complete avulsion of the patellar tendon ➡ from the patellar bone ➡. The small osseous fragment ➡ attached to the avulsed patellar tendon end is seen. The tendon retraction gap ➡ is filled with blood, serous fluid, and fat.*

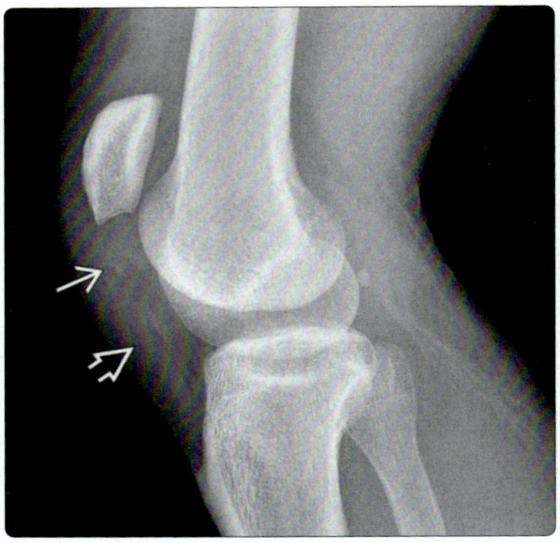

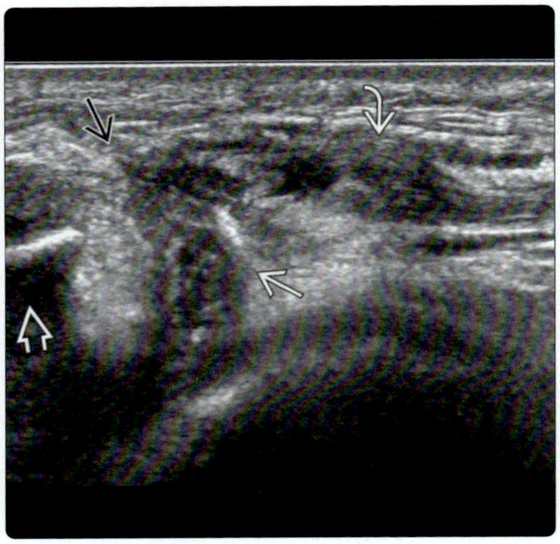

(Left) *Longitudinal US shows a chronic, deep-seated, intrasubstance semimembranous tendon tear ➡. The superficial fibers ➡ are intact. The tear is sharp, irregular, and hypoechoic and likely to be filled with organizing hematoma and fibrotic tissue. Note the medial femoral condyle ➡.* **(Right)** *Longitudinal US of the contralateral tendon in the same patient shows a normal fibrillar pattern ➡ of the semimembranous tendon coursing over the medial femoral condyle ➡.*

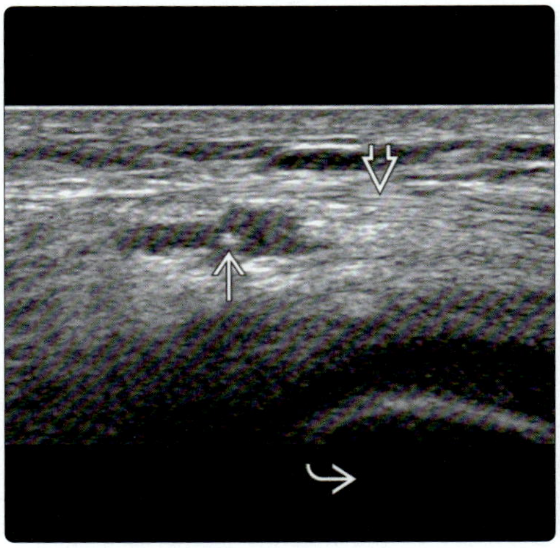

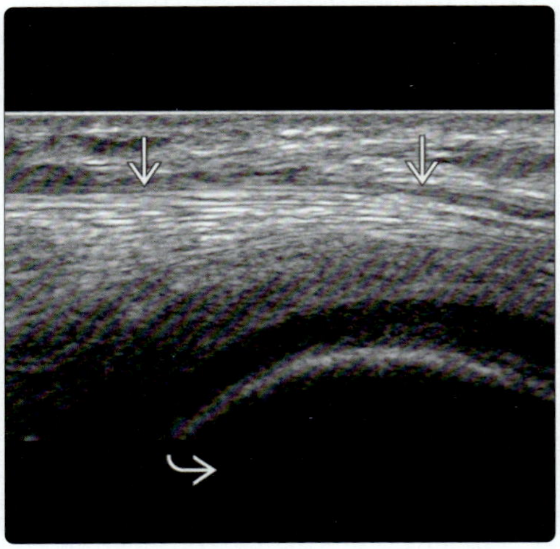

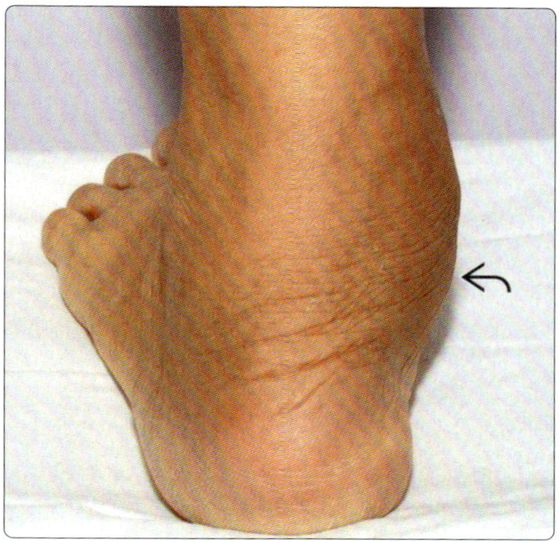

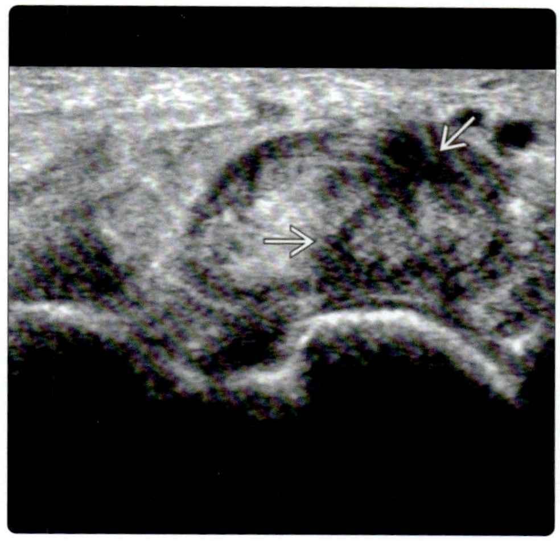

(Left) *Clinical photograph shows quite severe pes planus with loss of the medial longitudinal arch leading to heel valgus deformity ➔.* **(Right)** *Transverse US of the medial retromalleolar region shows severe posterior tibialis tendinosis with several irregular, linear, hypoechoic areas ➔ within the tendon substance, consistent with longitudinal split tears. Longitudinal spit tears, which typically extend over 2-3 cm, are quite common in moderate to severe tendinosis of the tendons around the ankle.*

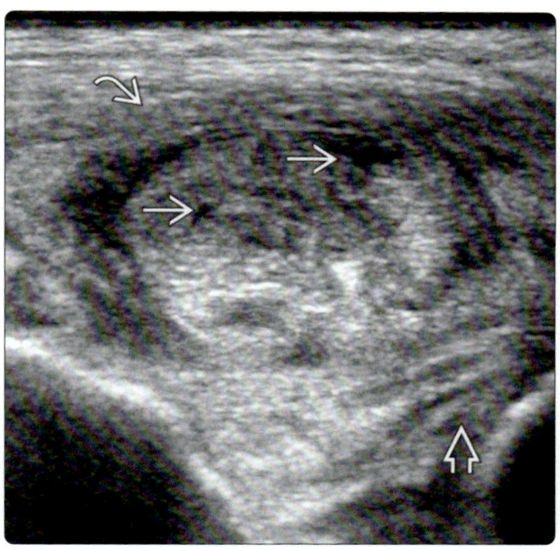

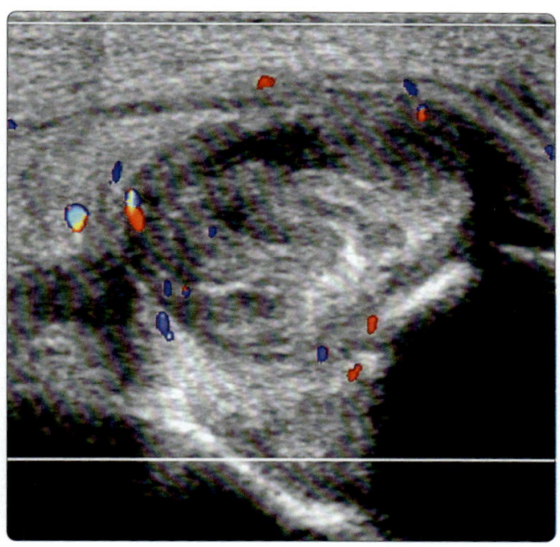

(Left) *Transverse US of the inframalleolar region shows several small, linear, hypoechoic split tears ➔ in the posterior tibialis tendon. Moderate background tendinosis is present. Linearity is the main feature that helps differentiate tendon tears from focal proteoglycan accumulation. Note the thickened flexor retinaculum ➔. The deltoid ligament ➔ is present deep to the tendon.* **(Right)** *Transverse color Doppler US of the same tendon shows mild peritendinous and intrasubstance hyperemia.*

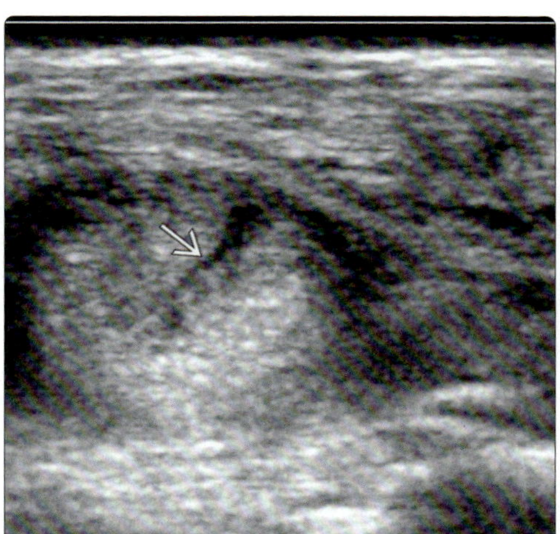

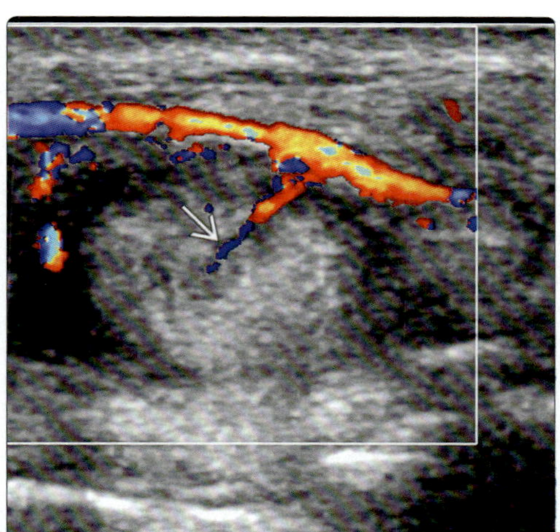

(Left) *Transverse US of the posterior tibialis tendon shows a linear, discrete, hypoechoic line extending from the surface to the center of the tendon ➔, suspicious for a partial-thickness tear. Always check any suspected partial tears with color Doppler US.* **(Right)** *Transverse color Doppler US of the corresponding region shows the linear, discrete, hypoechoic line represents a vascular channel ➔. This may be due to vascularized granulation tissue with a healing tear.*

(Left) *Transverse US of the inframalleolar region shows a normal peroneus longus and a moderately attenuated peroneus brevis tendon.* **(Right)** *Transverse US at a site just proximal shows absence of the peroneus brevis tendon due to a complete tear. The peroneus longus tendon is unchanged. Tears in the small tendons around the ankle and wrist are best appreciated 1st on transverse scanning along the tendon length.*

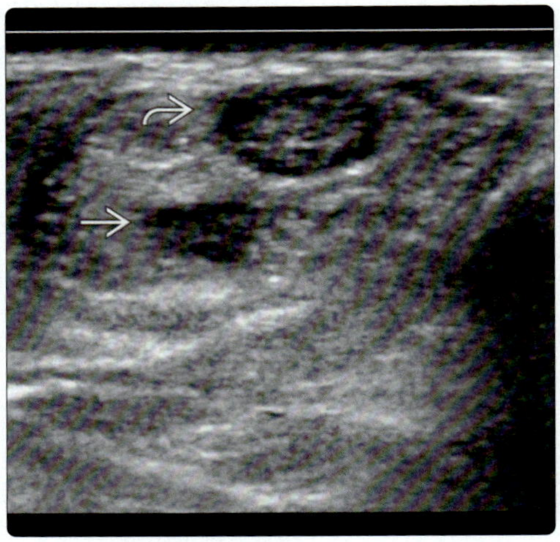

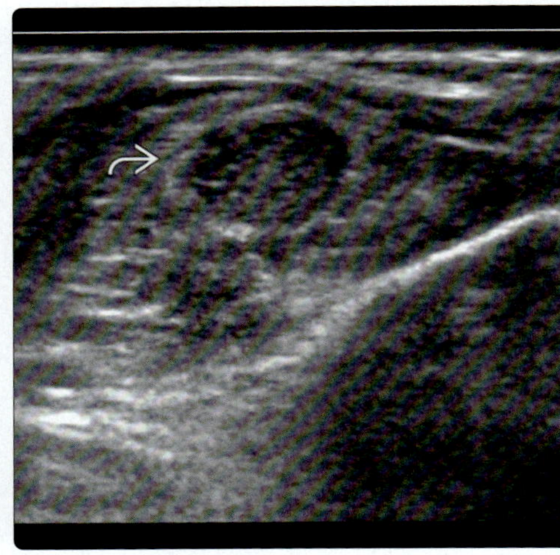

(Left) *Longitudinal US in the lateral retromalleolar area of the same complete tear of the peroneus brevis tendon shows proximal retraction of the swollen tendon end to the supramalleolar region. The peroneus longus tendon is intact.* **(Right)** *Transverse US of the lateral retromalleolar region shows several full-thickness longitudinal (split) tears of the peroneus longus tendon. There is mild peroneus brevis tendinosis with a small tendon sheath effusion, though no tendon tear.*

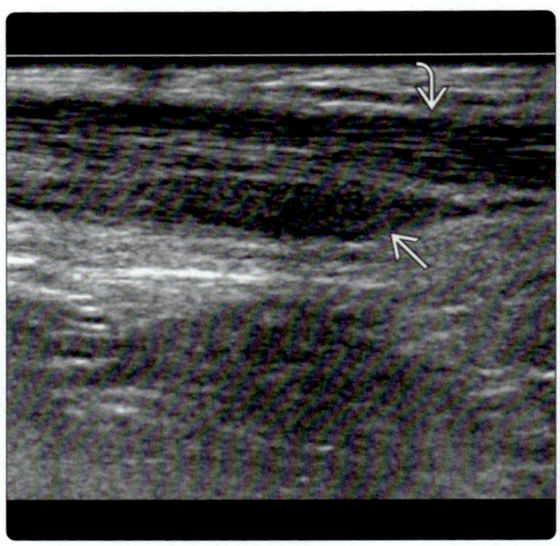

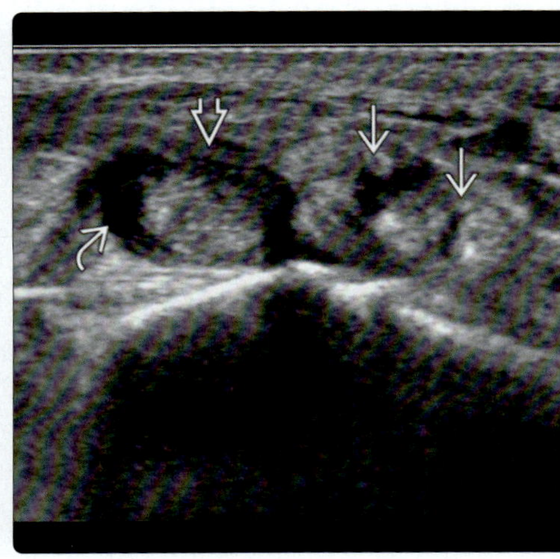

(Left) *Clinical photograph of a patient with foot trauma and impaired big toe dorsiflexion shows the location of the distal and proximal stumps of the torn extensor hallucis longus drawn on the skin. Skin marking is helpful for subsequent surgical repair.* **(Right)** *Longitudinal US in the same patient shows the complete tear of the extensor hallucis longus with proximal retraction to the level of the navicular bone. There is hematoma filling the tendon gap overlying the 1st metatarsal shaft.*

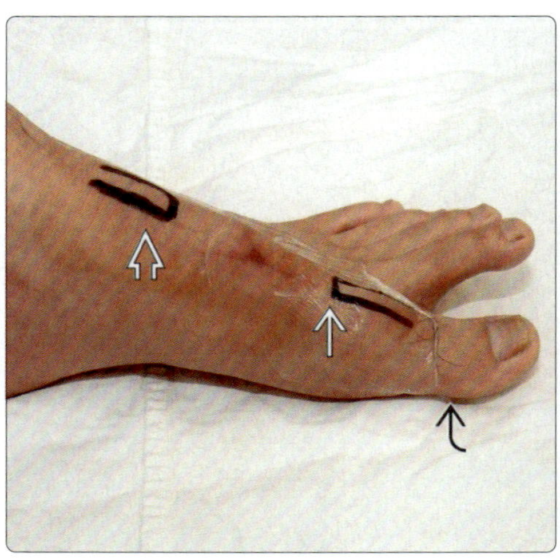

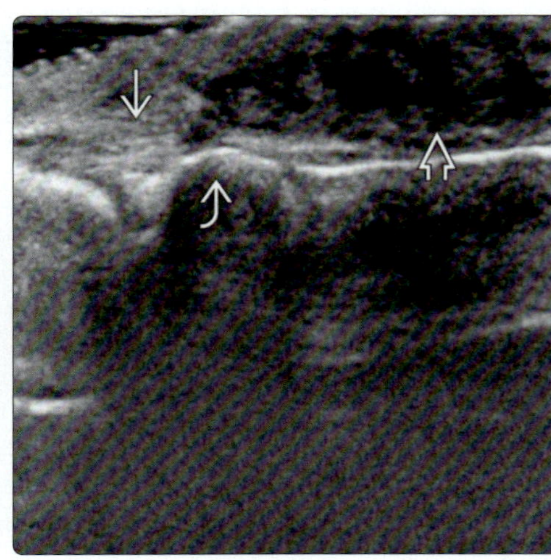

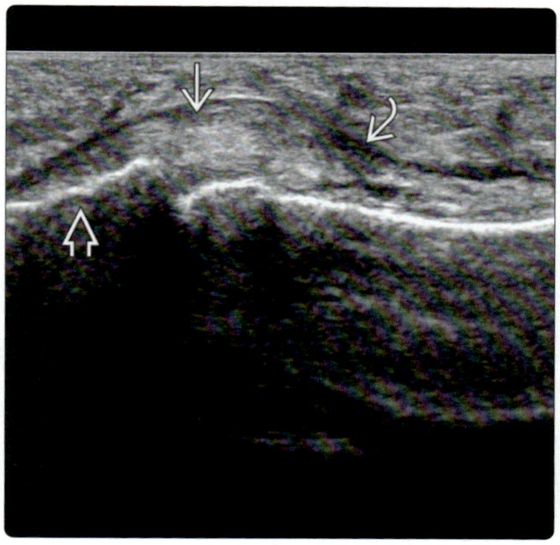

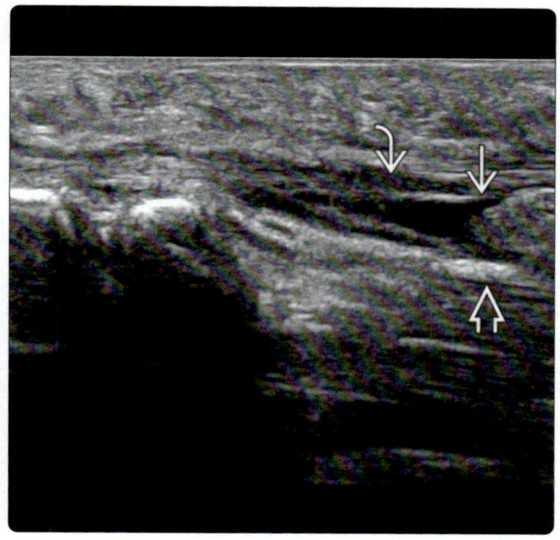

(Left) *Longitudinal US shows a complete tear of the flexor pollicis longus tendon following direct trauma to the thumb. A swollen distal tendon stump is inserted into the base of the distal phalanx.* **(Right)** *The end of the proximal stump is retracted to the proximal 1/3 of the middle phalanx. The partially collapsed tendon sheath should not be misdiagnosed as intact tendon fibers.*

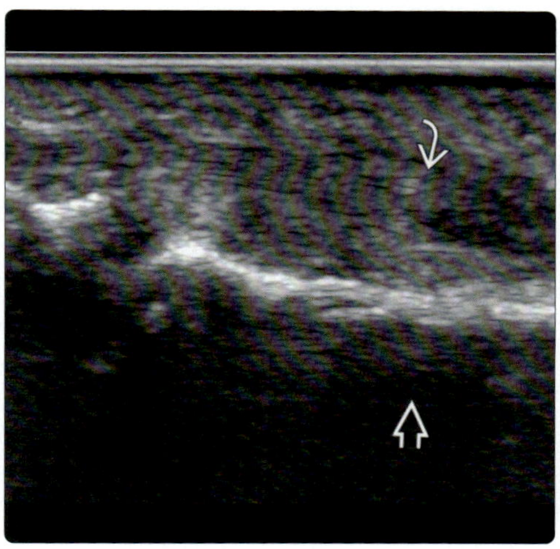

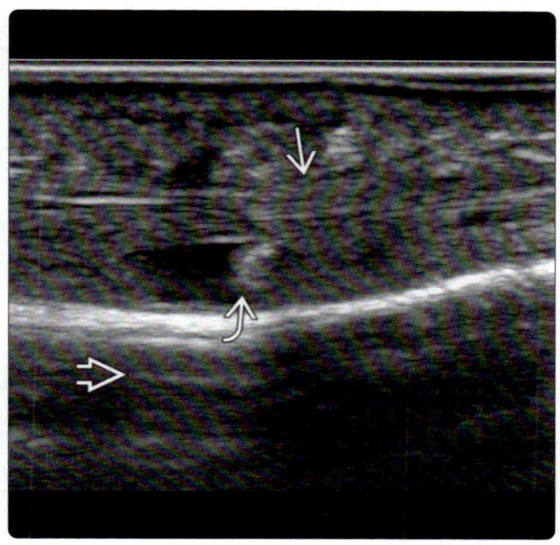

(Left) *Longitudinal US shows a complete transection of the flexor digitorum profundus (FDP) tendon to the little finger. The distal insertion of the FDP tendon (not shown) was intact. The distal end of the transected FDP tendon is located at the level of the middle phalanx.* **(Right)** *Longitudinal US more proximally in the hand shows the proximal end of the transected FDP tendon located at the metacarpal shaft level. The overlying flexor digitorum superficialis tendon is intact.*

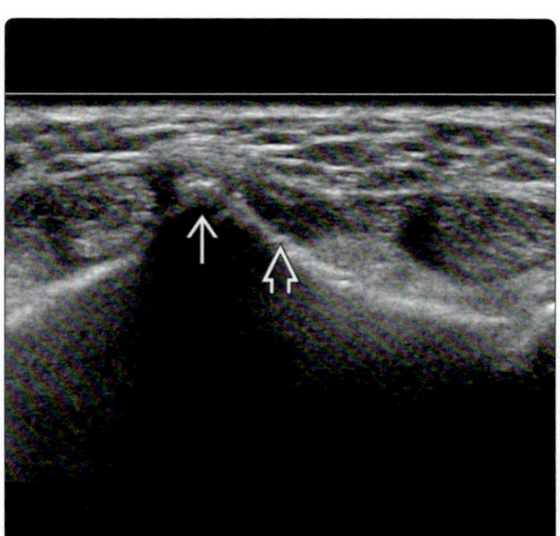

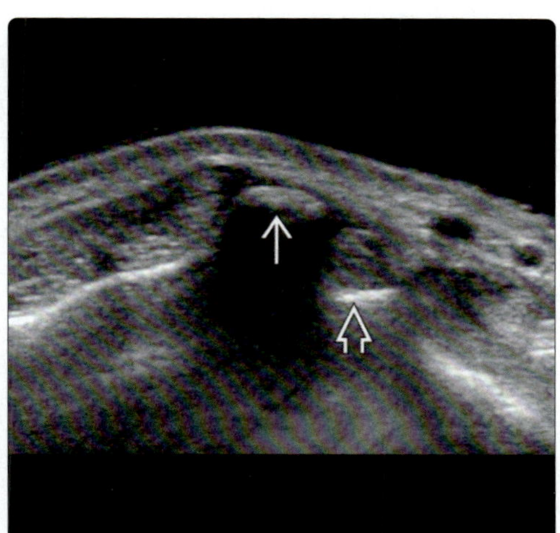

(Left) *Transverse US in a patient with limited thumb movement shows absence of the extensor pollicis longus (EPL) tendon at the distal radius level. The tendon is normally located just ulnar to the Lister tubercle.* **(Right)** *Transverse US of the contralateral wrist shows a normal EPL tendon readily identified adjacent to the Lister tubercle. This is the most common site of EPL tendon rupture with Lister tubercle acting as a fulcrum for the tendons action.*

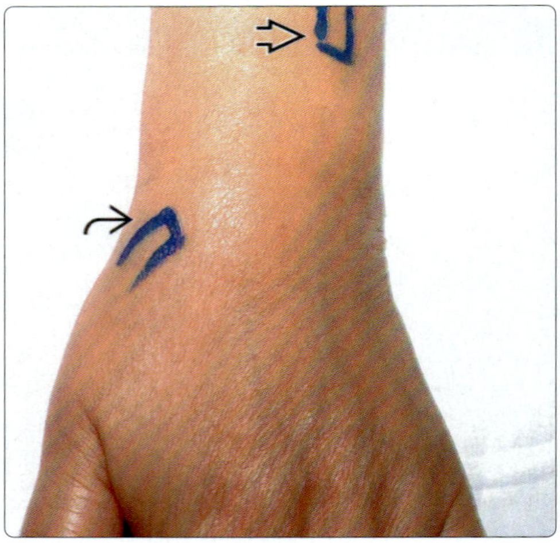

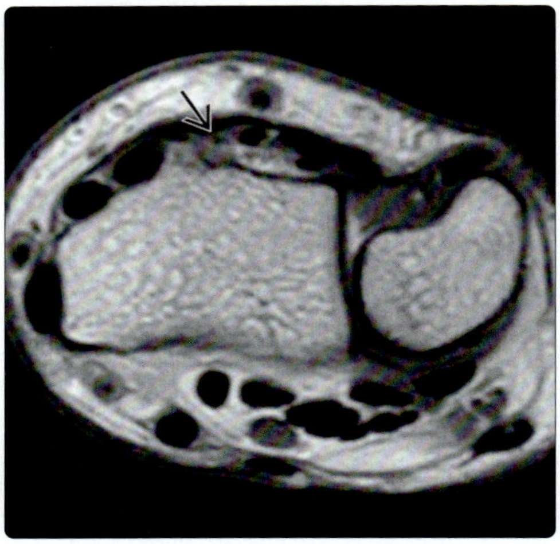

(Left) *Clinical photograph shows the sites of the retracted proximal* ➯ *and distal* ➯ *ends of the torn EPL tendon marked in the patient's skin.* **(Right)** *Transverse PDWI MR shows the wrist with a complete tear of the EPL tendon. EPL tendon tears typically occur at alongside Lister tubercle, deep to the extensor retinaculum. The small, hypointense dot* ➯ *represents the collapsed EPL tendon sheath. The torn, swollen EPL tendon ends (not shown) were more visible proximally and distally.*

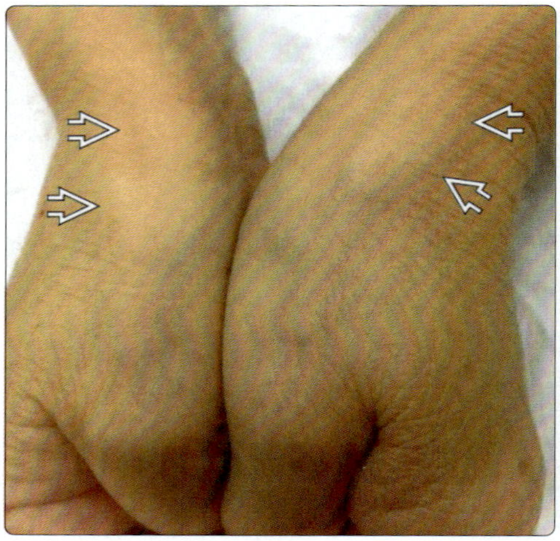

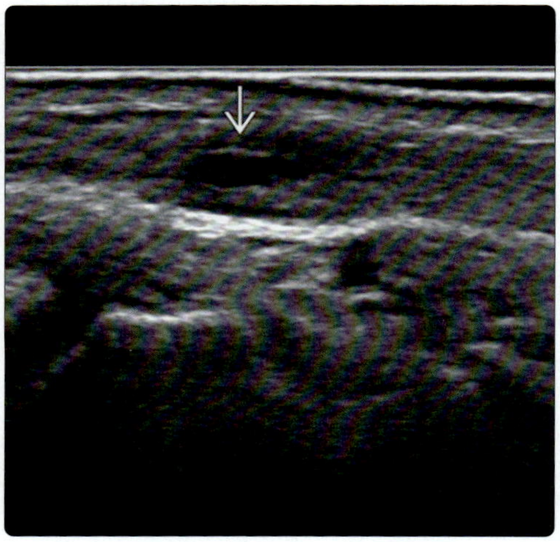

(Left) *Clinical photograph of a patient presenting with pain and swelling over the radial side of both hands* ➯ *is shown.* **(Right)** *Longitudinal US of the left extensor pollicis brevis shows intrasubstance focal split tear with intratendinous cyst formation* ➯.

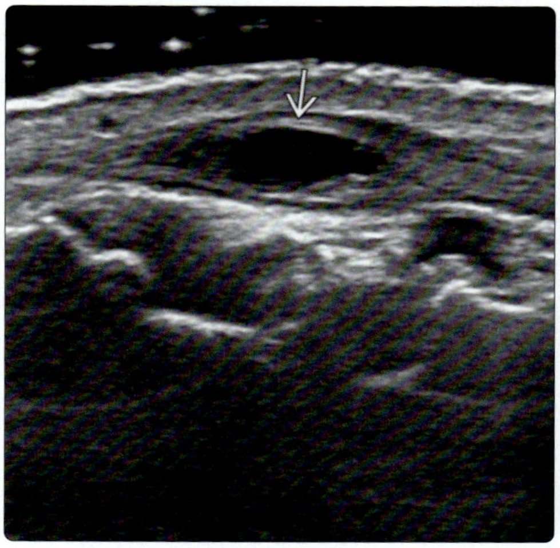

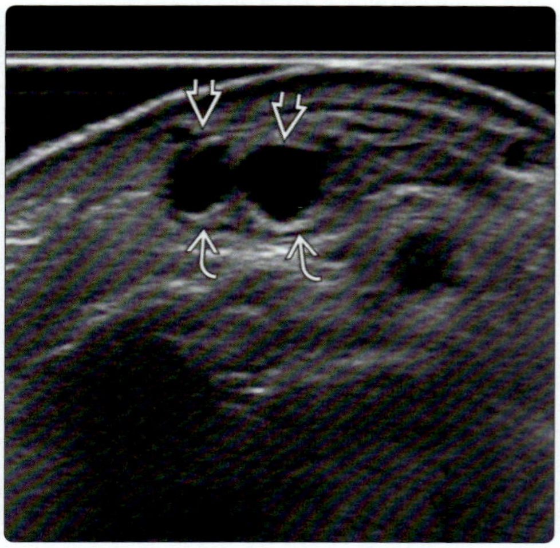

(Left) *Longitudinal US of the contralateral right extensor pollicis brevis shows a larger intrasubstance focal split tear with intratendinous cyst formation* ➯. **(Right)** *Transverse US of the same right extensor pollicis brevis tendon shows a bilobed intratendinous cyst consequent on an focal intrasubstance split tear* ➯. *The normal intact tendon fibers are displaced peripherally* ➯.

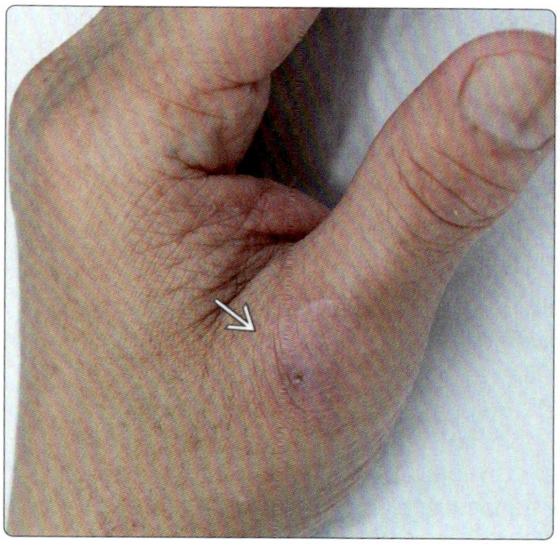

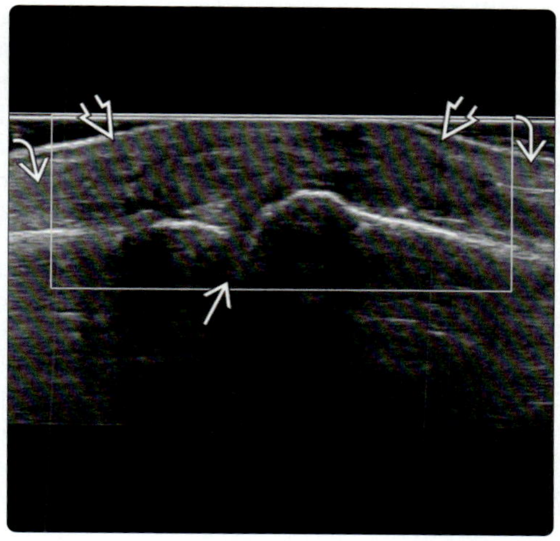

(Left) *A 64-year-old man 4 months following repair* ➡ *of completely torn EPL and brevis tendons to the thumb is shown. He was complaining of restricted thumb movement and pain on the dorsoradial aspect of the wrist.* **(Right)** *Longitudinal US at the metacarpophalangeal joint* ➡ *level shows a fibrotic-type mass* ➡ *surrounding the indistinct extensor tendons* ➡ *of the thumb. No hyperaemia is present. Limited movement of the extensor tendons is present on passive flexion/extension.*

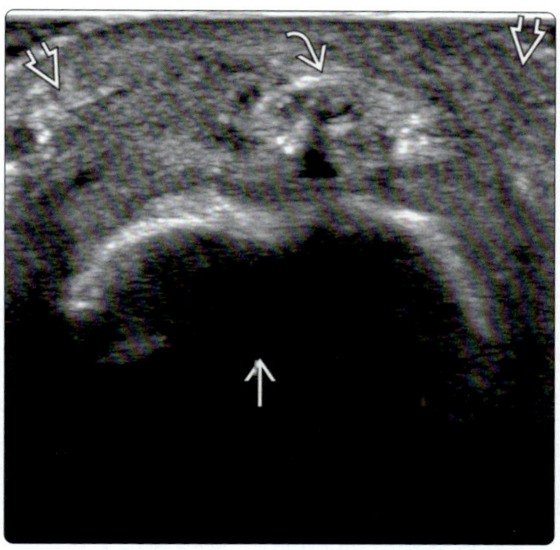

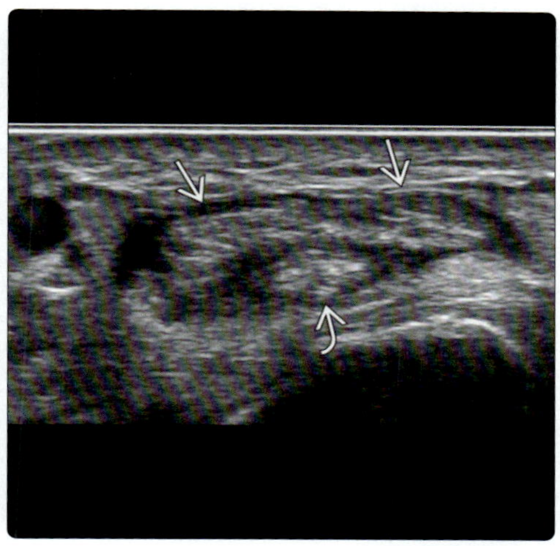

(Left) *Transverse US at the same level shows the disrupted, irregular extensor tendons* ➡ *on the dorsum of the metacarpal shaft* ➡*. There is moderate surrounding fibrotic-type tissue* ➡ *in the area of the sagittal bands. These are fairly typical appearances following extensor tendon repair.* **(Right)** *Transverse US more proximally on the dorsum of wrist shows that the EPL tendon* ➡ *is located deep to, rather than superficial to, the 2nd extensor compartment tendons* ➡*, consequent on previous tear/repair.*

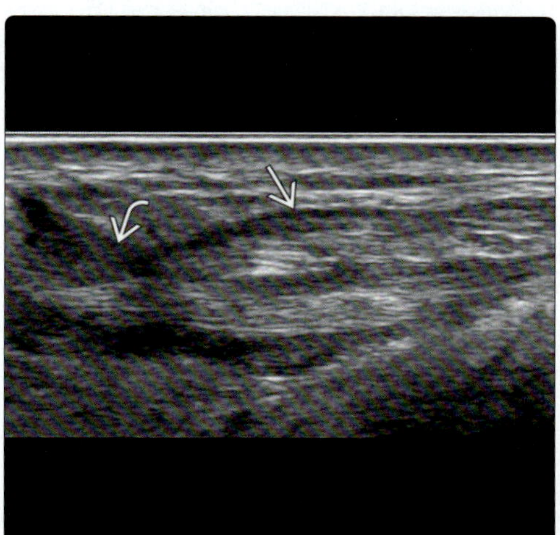

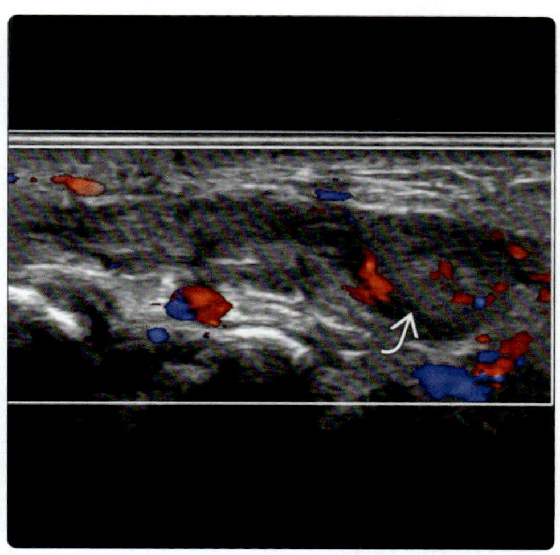

(Left) *Transverse US of the same patient shows mild tendinosis of the malpositioned EPL tendon* ➡ *on the radial side of the 2nd extensor compartment tendons* ➡*.* **(Right)** *Transverse US at this location shows moderate hyperemia around the malpositioned EPL* ➡ *just radial to the 2nd extensor compartment tendons. This localized tendinosis of the EPL was most likely due to malpositioning and was also the most likely the cause of postoperative dorsoradial wrist pain.*

Tenosynovitis

KEY FACTS

IMAGING

- Occurs only in tendons encased by synovium
- Distension of tenosynovium (tendon sheath) due to fluid or synovitis with variable tendon swelling
- **Acute exudative tenosynovitis**
  - Usually due to acute infection or gout
  - Fluid accumulation within sheath with little or no synovial proliferation
  - Fluid may have speckled echogenic appearance due to aggregates of debris or pus
- **Acute nonexudative tenosynovitis**
  - Usually due to infection, inflammatory disease, or gout
  - Synovial proliferation within sheath predominates over fluid accumulation
- **Chronic tenosynovitis (active)**
  - Synovial proliferation within hyperemia with tendon sheath
- **Chronic tenosynovitis (inactive)**
  - Clear fluid accumulation within mildly thickened sheath
  - Hyperemia mostly around, rather than within, tendon sheath
- **Stenosing tenosynovitis**
  - Hypertrophy of retinaculum or pulley component of tendon osseofibrous tunnel
  - Restricts tendon movement (trigger finger and de Quervain disease)

DIAGNOSTIC CHECKLIST

- Tenosynovitis: Disease of tendon sheath with minimal tendon involvement
- Tendinosis: Disease of tendon with variable degree of peritendinitis
- Use color Doppler to distinguish synovial inflammation from synovial effusion
- Use minimal transducer pressure to avoid effacing tendon sheath effusion or vascularity

(Left) *Clinical photograph shows severe swelling dorsum of the wrist in a patient with a 2-week history of increasing wrist pain.* **(Right)** *Transverse US at the metacarpal level shows a severely thickened tendon sheath ➡ around the extensor digitorum communis tendon. Synovial proliferation is present. The extensor digitorum tendons are only mildly swollen.*

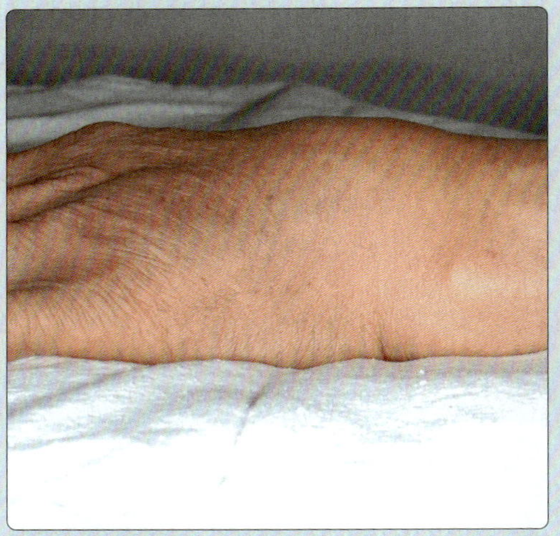

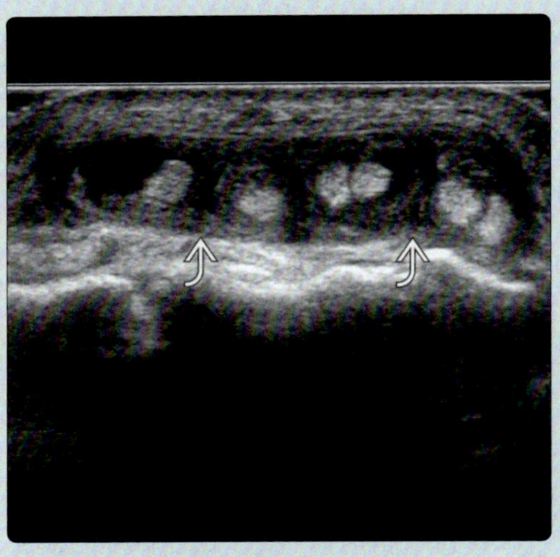

(Left) *Transverse US of the same patient's wrist at the level of the proximal carpal row shows even more severe distension of the extensor digitorum communis tendon sheath with thickened proliferative synovium ➡.* **(Right)** *Transverse color Doppler US of the extensor digitorum communis tendon sheath (same patient) shows severe hyperemia of the thickened synovium, consistent with active chronic tenosynovitis.*

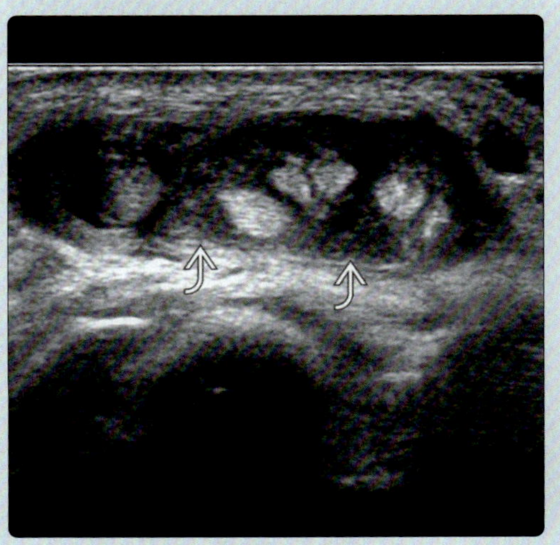

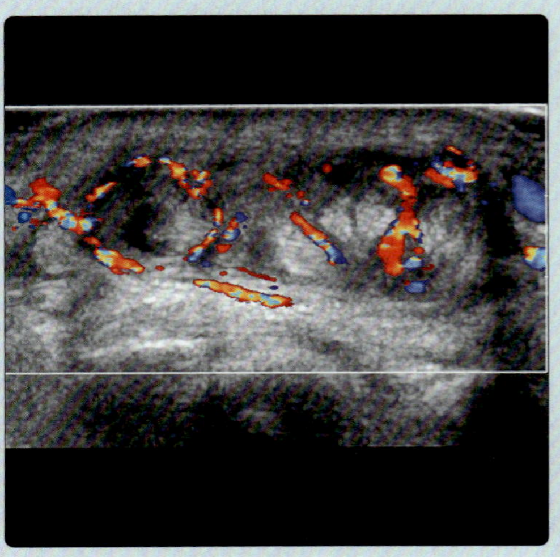

TERMINOLOGY

Synonyms

- Tendinitis or tendonitis are arbitrary terms best avoided

IMAGING

General Features

- Best diagnostic clue
  - Inflammation of tendon sheath (tenosynovium)
  - Thickening of synovial tendon sheath (synovial proliferation)
  - Exudate or fluid within tendon sheath form "halo" around tendon
- Location
  - Can occur only in tendons encased by synovial sheath
  - Wrist extensors > wrist flexors > finger flexors > ankle tendons
    - Tenosynovitis cannot occur if no tendon sheath present
      - For example, flexor carpi ulnaris tendon has no tendon sheath
      - Extensor digitorum tendons of fingers have no tendon sheath
- Morphology
  - Distension of tendon sheath with mild tendon swelling
  - Variable tendon sheath distension and content, depending on cause, severity, duration

Radiographic Findings

- Radiography
  - Signs of underlying cause
    - Calcifications in gouty tophi or calcium pyrophosphate dihydrate deposition
    - Erosive or septic arthropathy
    - Radiopaque foreign body

MR Findings

- T2WI
  - Fluid signal within distended synovial sheath
- T1WI C+ FS
  - Contrast enhancement helps differentiate synovial proliferation from tendon sheath fluid

Ultrasonographic Findings

- Depends on capacity of tendon sheath to distend
- **Acute exudative tenosynovitis**
  - Usually due to infection and gout
    - Fluid accumulation within tendon sheath with little or no synovial proliferation
      - Forms hypoechoic halo around tendon
    - Fluid may have speckled echogenic appearance due to aggregates of debris or pus
      - Echogenic comet-tail artifacts due to gouty crystal aggregates tend to be larger than those due to inflammatory aggregates in infected tenosynovial fluid
    - Hyperemia around but not within tendon sheath
    - Infection may spread from tendon sheath into peritendinous tissues
    - ± foreign body within tendon sheath
- **Acute proliferative tenosynovitis**
  - Due to rheumatological disease, gout, or infection
    - Distension of tendon sheath with minimal or no tendon swelling
    - Synovial proliferation within sheath predominates over fluid accumulation
    - Hyperemia predominantly around rather than within tendon sheath
  - Mild secondary tendon sheath swelling can occur with severe cellulitis or edema in absence of actual tenosynovitis
- **Chronic active tenosynovitis**
  - Due to rheumatological disease, gout, or infection
  - Markedly thickened tendon sheath due to synovial proliferation rather than fluid
    - Synovial proliferation particularly prevalent in tuberculous tenosynovitis
    - Probe pressure can differentiate fluid from tenosynovial thickening
      - Fluid can be squeezed away but tenosynovial thickening is not fully compressible
      - Intrasheath hyperemia confirms tenosynovial proliferation rather than fluid
  - Established hyperemia both within and around tendon sheath
  - ± tendon rupture
    - Chronic inflammation weakens tendon
- **Chronic inactive tenosynovitis**
  - Either synovial proliferation or clear fluid accumulation within mildly thickened sheath
    - Multiple "rice bodies" (small fibrinous nodules) may develop in chronic inflammation (e.g., tuberculosis or rheumatoid arthritis)
    - Adhesions can lead to entrapped tenosynovial fluid with marked distension
  - Little or no hyperemia either around or within tendon sheath
- **Stenosing tenosynovitis**
  - Describes tendon disorders associated with hypertrophy of fibrous component of tendinous osseofibrous tunnel
  - "Tenosynovitis" terminology is contentious as this entity is more related to tendinosis than tenosynovitis
  - Hypertrophy of flexor pulley (trigger finger) or 1st extensor retinaculum (de Quervain disease)
    - For trigger finger, A1 pulley is thickened more frequently than A2 pulley, and any finger may be affected
  - Thickening (1-1.5 mm) of pulley or retinaculum
    - Normal thickness of pulley or retinaculum ≤ 0.5 mm
  - ± hyperemia of retinaculum or tendon thickening
  - Occasional tendon sheath effusion
  - Restriction of tendon movement on dynamic imaging

Imaging Recommendations

- Best imaging tool
  - US: Accurate at revealing presence and extent of tenosynovitis
    - Facilitates fine-needle aspiration (for culture, crystals) or biopsy to determine cause
    - US-guided steroid injection treatment
- Protocol advice

- Examine full length of tendon or group of tendons in transverse plane
- Use minimal transducer pressure to avoid effacing tendon sheath effusion or vascularity
- Always include evaluation with color Doppler
  - To determine activity, distinguish synovial proliferation from fluid and confirm disease chronicity
- Examine nearby joints for synovitis or effusion
- Look for fluid leakage outside tendon sheath
- In stenosing tenosynovitis, measure thickness of retinaculum or pulley
  - Measure retinacular or pulley thickness at 10 o'clock or 2 o'clock positions
  - Perform dynamic assessment to see if tendon movement restricted beneath pulley or retinaculum

DIFFERENTIAL DIAGNOSIS

Tendinosis

- Disease of tendon with variable degree of peritendinitis
- Tendon swelling, disruption of fibrillar echotexture, and intratendinous hyperemia predominate
  - Tendon sheath swelling is less than in tenosynovitis
- Intratendinous neovascularization is not usual feature of tenosynovitis
- Distinction between tenosynovitis and tendinosis is not always clear cut; mixed patterns may exist

Paratendinous Cystic Lesion

- Fluid is located eccentrically to tendon

PATHOLOGY

General Features

- Etiology
  - Overuse, infection (acute or chronic), systemic inflammatory arthropathy, crystal deposition
  - Often no specific cause is found
- Relevant anatomy: Wrist flexor tendons
  - Flexor tendon sheaths of thumb and little finger communicate via radial and ulnar bursae on volar aspect of wrist
  - In 50% of cases, radial and ulnar bursae communicate via intermediate bursa
- Relevant anatomy: Wrist extensor tendons
  - 6 discrete extensor tendon sheaths that do not communicate
  - Extensor tendon expansions on dorsum of fingers do not have tendon sheaths
  - Septum separates 1st extensor compartment tendons in 25% of cases

Gross Pathologic & Surgical Features

- Inflammation of tendon sheath
- Thickening of tendon retinaculum or pulley

Microscopic Features

- Inflammatory cells infiltrate tenosynovium

CLINICAL ISSUES

Presentation

- Most common signs/symptoms
  - Pain, swelling, and tenderness of affected tendons
  - Limitation of movement
- Other signs/symptoms
  - Crepitus, tendon triggering
  - Surrounding cellulitis

Demographics

- Age
  - Most common in middle-aged to elderly people
- Gender
  - Inflammatory tenosynovitis, de Quervain tenosynovitis, and repetitive strain injury have female predominance
- Epidemiology
  - Rheumatological disease, e.g., rheumatoid arthritis, systemic lupus erythematosus
  - Metabolic disease, e.g., gout, pseudogout
  - Repetitive strain injury
  - Infective tenosynovitis more common in immunosuppressed patients and patients prone to penetrating injury
  - Stenosing tenosynovitis of 1st extensor compartment is common post partum (baby wrist), and trigger finger is more common in diabetes patients

Natural History & Prognosis

- Majority (80%) improve with conservative treatment
  - If infective, may need surgical drainage

Treatment

- Noninfective tenosynovitis and chronic infective tenosynovitis
  - Rest, splinting, physiotherapy
    - Antiinflammatory medication ± US-guided steroid injection
    - ± appropriate antibiotics, if infective
    - Synovectomy for failed conservative treatment
- Acute exudative tenosynovitis
  - Surgical drainage, elevation, antibiotics
- Stenosing tenosynovitis
  - US-guided steroid injection between tendon sheath and pulley or retinaculum

DIAGNOSTIC CHECKLIST

Consider

- Tendinosis rather than tenosynovitis if tendon is predominantly swollen with intratendinous hyperemia
- Secondary tenosynovial swelling if there is severe surrounding cellulitis or edema

Image Interpretation Pearls

- Use color Doppler to distinguish synovial inflammation from synovial effusion
- Excessive transducer pressure may displace small amount of tendon sheath fluid

SELECTED REFERENCES

1. Gitto S et al: Sonography of non-neoplastic disorders of the hand and wrist tendons. J Ultrasound Med. 37(1):51-68, 2018

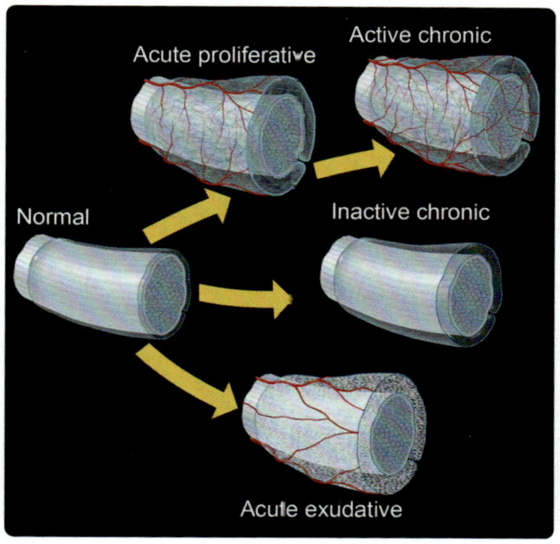

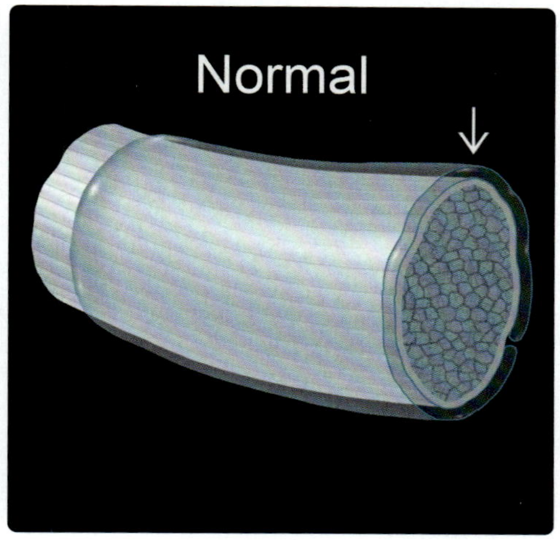

(Left) *Graphic shows different types of tenosynovitis. Tenosynovitis is a disease of the tendon sheath. Unlike tendinosis, which can affect any tendon, tenosynovitis can only affect those tendons with a tendon sheath.* **(Right)** *Graphic shows normal tendon sheath ➡ for comparative purposes. Most, but not all, of the tendons of the wrist/hand and ankle/foot have a tendon sheath. The long head of biceps has a tendon sheath. The patellar and Achilles tendons have a paratenon.*

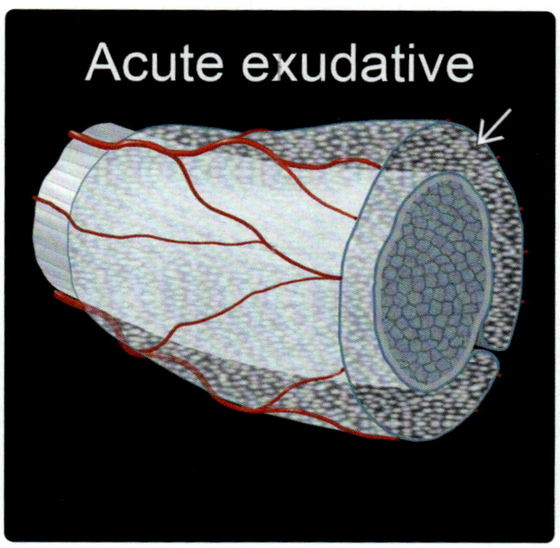

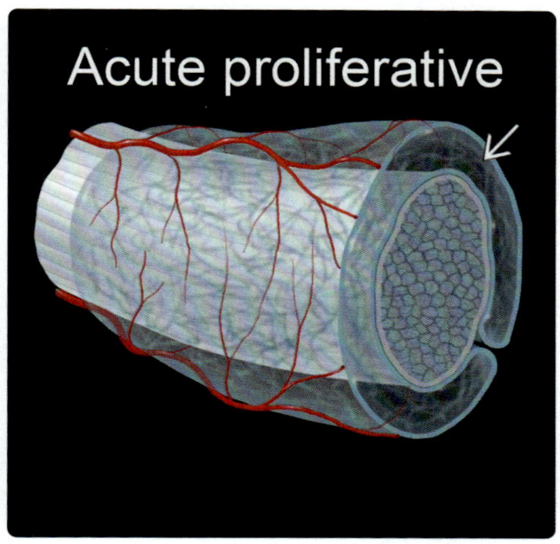

(Left) *Graphic shows acute exudative tenosynovitis ➡, usually secondary to bacterial infection/penetrating trauma and rapidly progressive. Purulent exudate can be aspirated. Hyperemia occurs around, rather than within, the tendon sheath.* **(Right)** *Graphic shows acute proliferative tenosynovitis. This is a proliferative process and usually due to repetitive trauma or a systemic inflammatory disorder. Tendon swelling occurs due to synovial proliferation ➡ with a mild exudate. Hyperemia is limited to the surface.*

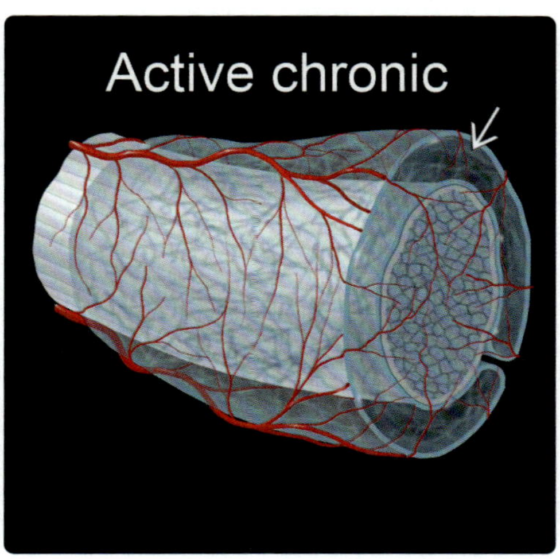

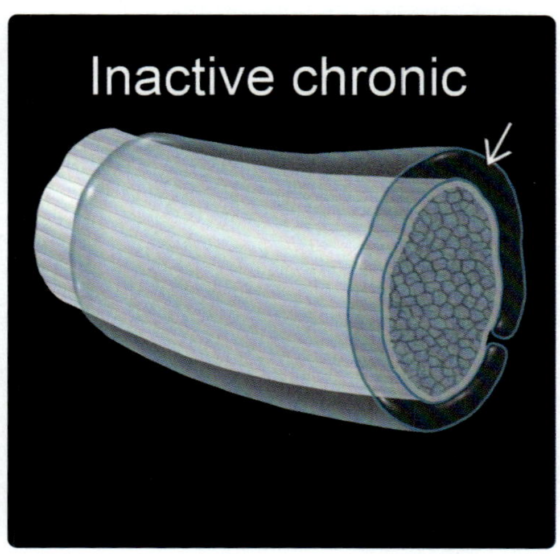

(Left) *Graphic shows active chronic tenosynovitis. Mature hyperemic synovial proliferation ➡ fills the distended tendon sheath. Hyperemia may extend into the tendon, which may be slightly enlarged due to edema. Otherwise, the tendon remains normal in appearance.* **(Right)** *Graphic shows inactive chronic tenosynovitis. The tendon sheath is distended with serous fluid ➡. This distension may be severe and localized due to tendon sheath adhesions. Little or no hyperemia is present.*

(Left) *Transverse US of the dorsal wrist in a patient with a 2-month history of wrist swelling shows moderately severe synovial thickening ➔ with no visible tendon sheath effusion, suggestive of chronic tenosynovitis. Note mild swelling of the extensor digitorum tendons.* **(Right)** *Longitudinal color Doppler US of the ring finger extensor tendon shows synovial thickening extending along the dorsum of the hand with severe hyperemia, consistent with active chronic tenosynovitis.*

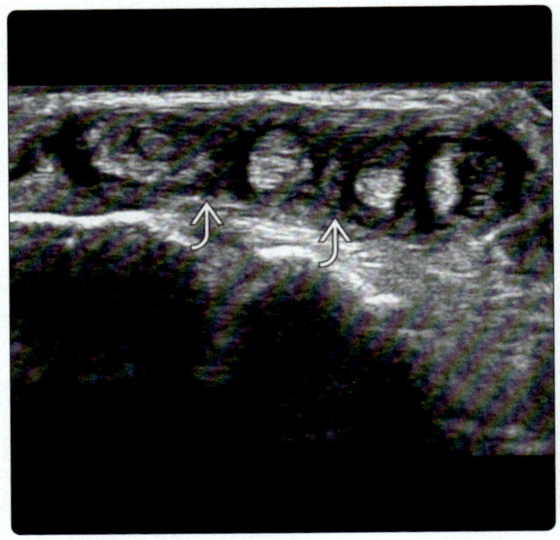

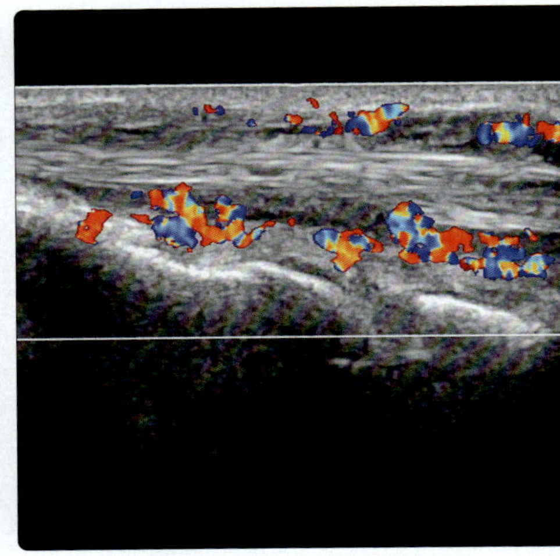

(Left) *Transverse US of the extensor digitorum communis in a patient with a history of gout and progressive wrist swelling for 3 months shows severe synovial thickening ➔ surrounding the mildly swollen tendons, suggestive of chronic tenosynovitis.* **(Right)** *Transverse color Doppler US at the same level shows severe peritendinous hyperemia, suggestive of active chronic tenosynovitis.*

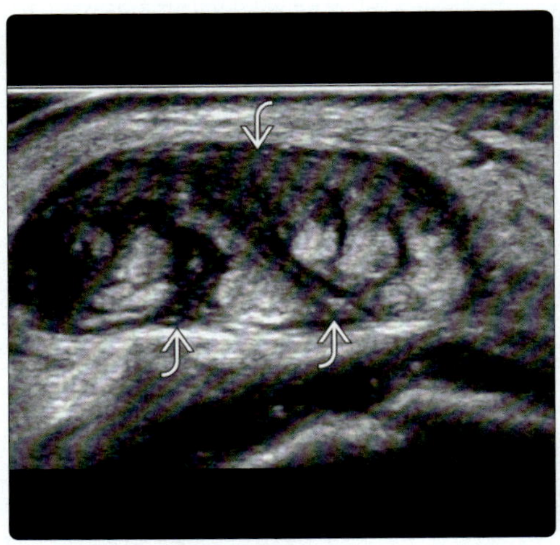

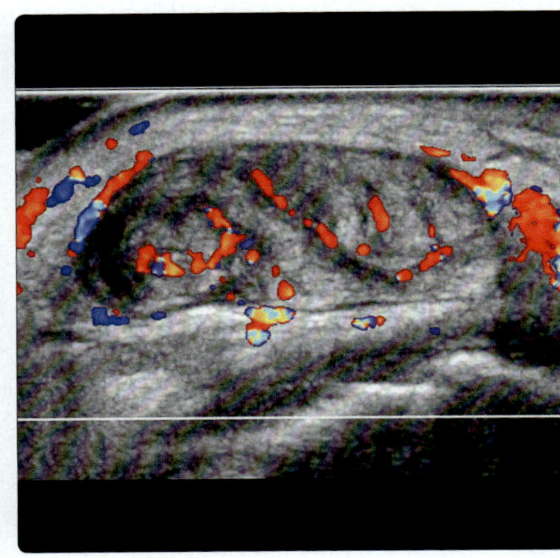

(Left) *Clinical photograph of patient with a history of gout shows marked swelling on the dorsum of the wrist with erythema.* **(Right)** *Longitudinal US of the same wrist shows marked distension of the common extensor digitorum tendon sheath. Small echogenic speckles ➔ within the thickened tenosynovium are consistent with crystal aggregates. These features are compatible with a gouty tenosynovitis.*

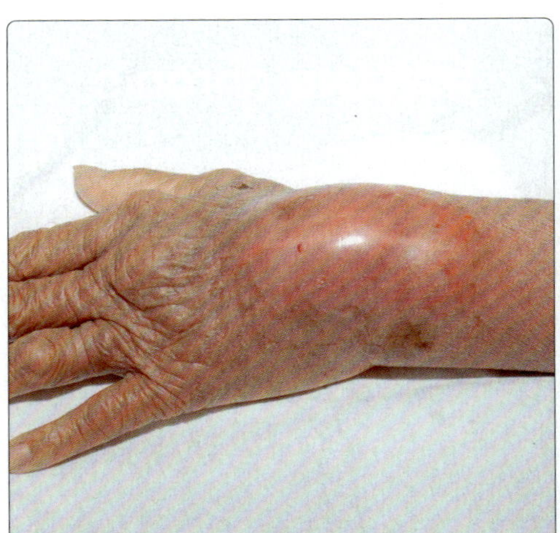

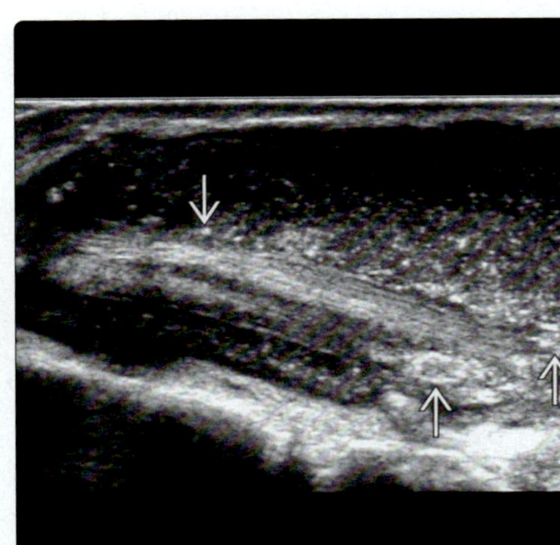

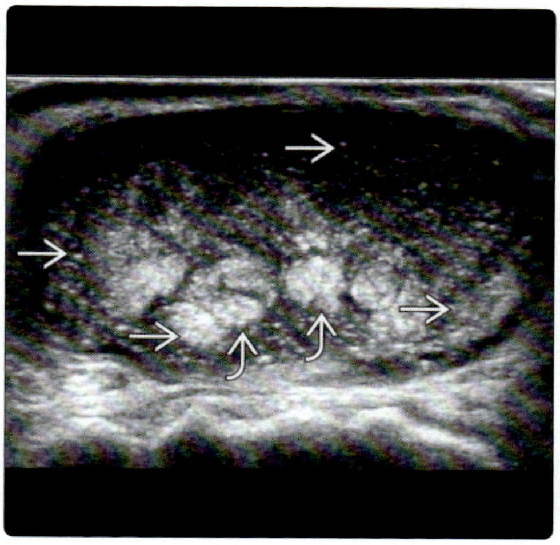

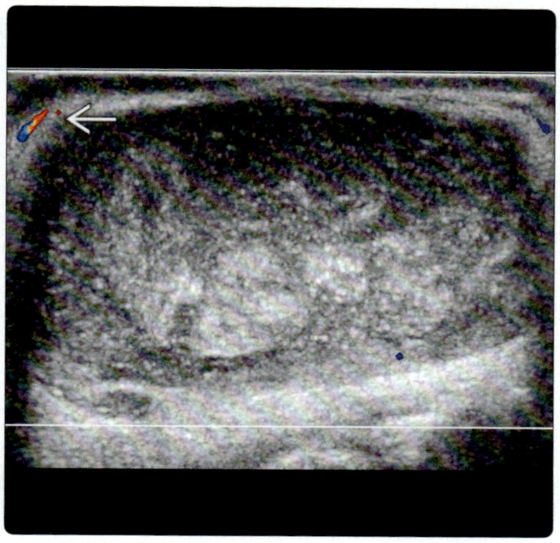

(Left) *Transverse US of the same wrist shows echogenic crystal aggregates ➔, some of which are quite large within the thickened tenosynovium and difficult to separate from the mildly swollen tendons ➔ on this static image.* **(Right)** *Transverse US of the same wrist shows minimal hyperemia ➔ surrounding the extensor tendon sheath, suggestive of inactive chronic tenosynovitis.*

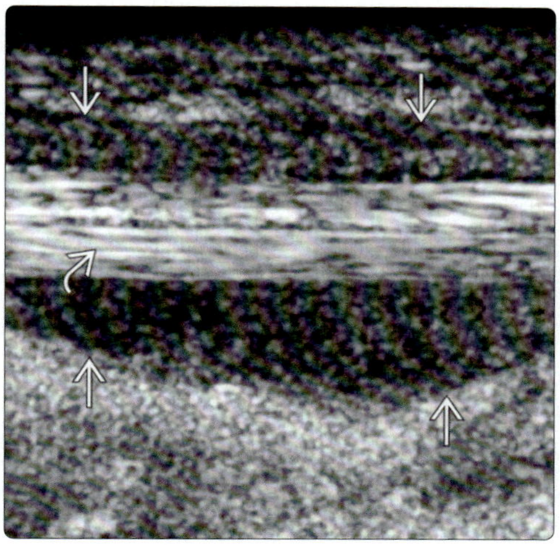

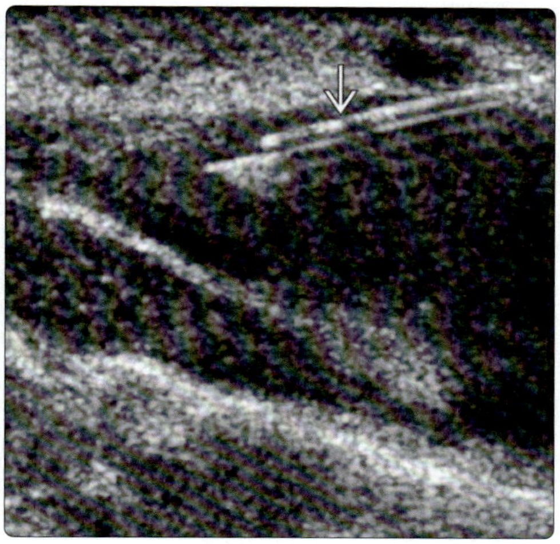

(Left) *Longitudinal US shows acute exudative flexor tenosynovitis of the middle finger with distension of the flexor digitorum tendon sheath ➔ by fluid containing many small echogenic aggregates, which tend to be less echogenic than those seen in gout. The flexor tendons ➔ are slightly swollen.* **(Right)** *Longitudinal US (same patient) of the same distended tendon sheath shows needle during needle aspiration ➔ of purulent tendon sheath fluid. This tendon sheath aspirate grew Staphylococcus aureus.*

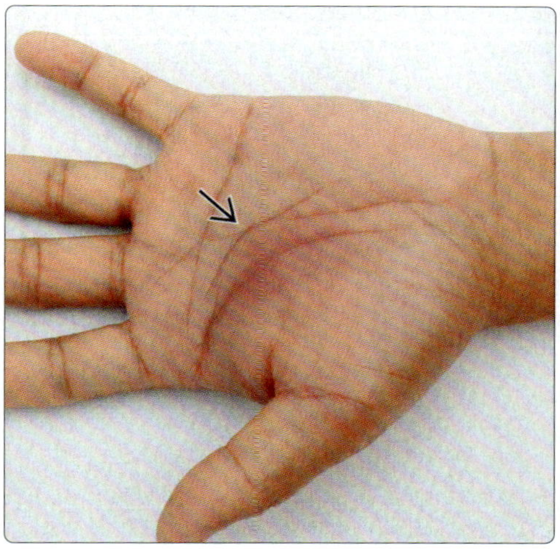

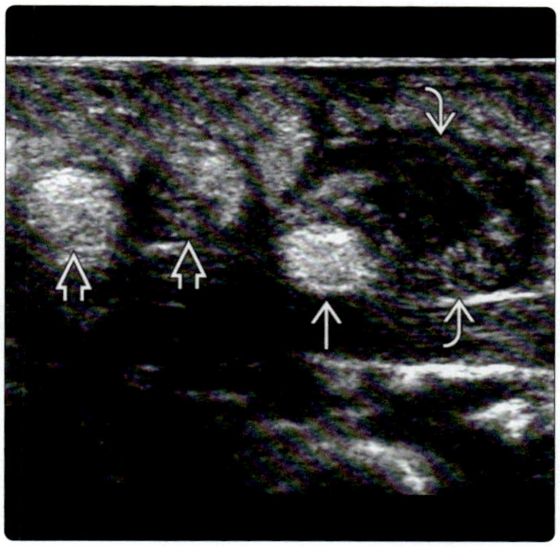

(Left) *Clinical photograph of a palm shows swelling and redness localized to the radial aspect of the palm ➔. US was requested to rule out deep abscess.* **(Right)** *Transverse US of the palm of the same patient shows thickened tenosynovium ➔ with hyperemia (not shown) eccentrically surrounding the index finger flexor tendon ➔ with moderate soft tissue swelling and edema, consistent with active chronic tenosynovitis. The remaining flexor tendons ➔ are not affected.*

(Left) *Transverse US shows moderate hypoechoic tenosynovial proliferation of the extensor digitorum communis tendon sheath on the dorsum of the wrist following a dog bite. The extensor digitorum tendons are only minimally swollen.* **(Right)** *Color Doppler US revealed marked tenosynovial hyperemia. Features are consistent with chronic active proliferative tenosynovitis. Intrasheath hyperemia on color Doppler imaging helps distinguish synovial proliferation from synovial fluid.*

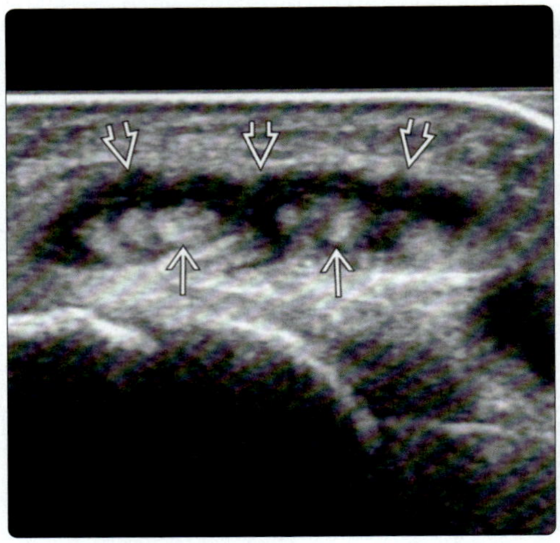

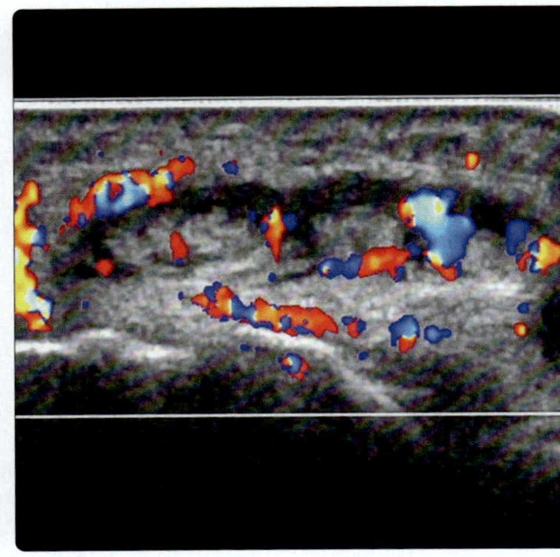

(Left) *Clinical photograph shows moderate swelling of the middle finger. Limited movement was present.* **(Right)** *Longitudinal US of the same finger shows tenosynovitis of the flexor digitorum tendon with a mildly thickened tendon sheath encasing the tendon. The tendon is mildly to moderately thickened.*

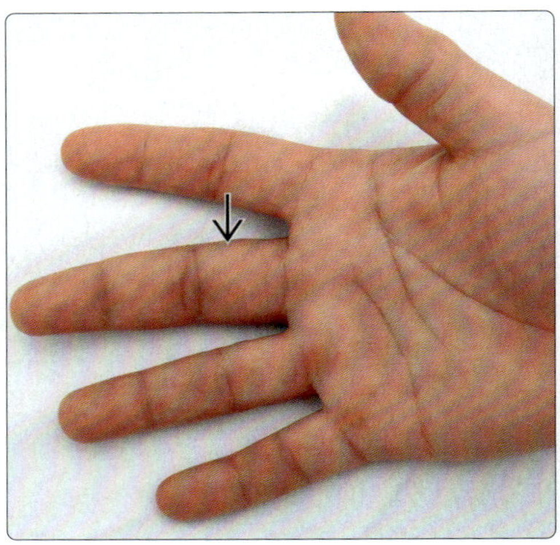

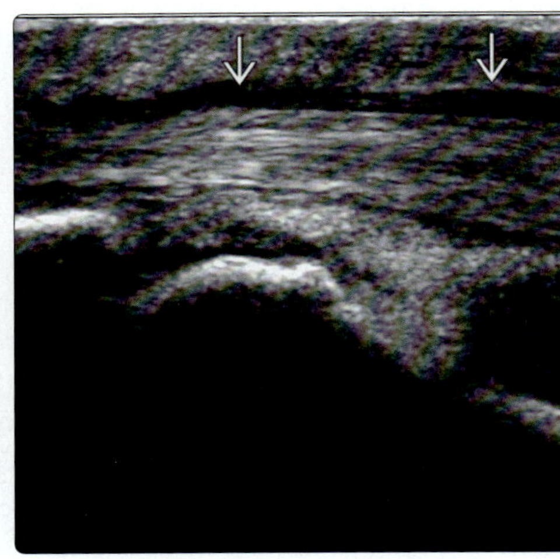

(Left) *Transverse US of the same finger shows moderate tendon sheath thickening encasing the flexor tendons. It is not clear whether this represents fluid or synovial proliferation.* **(Right)** *Transverse color Doppler US at the same location shows moderate tenosynovial hyperemia, indicating that this tendon sheath thickening is due to synovial proliferation and not fluid.*

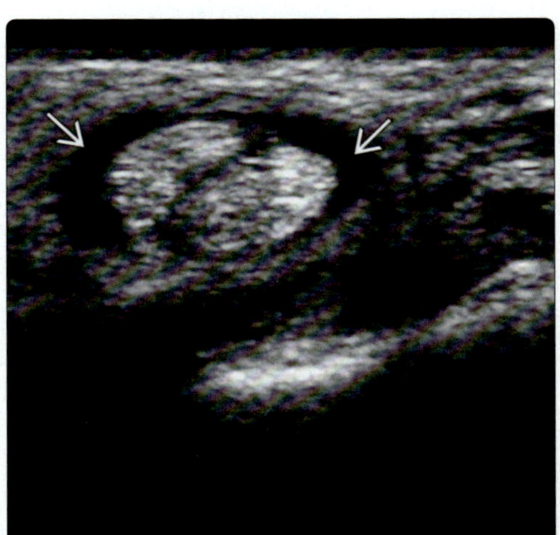

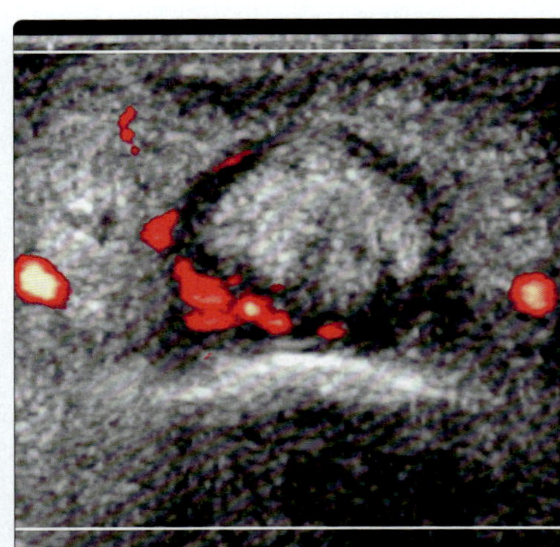

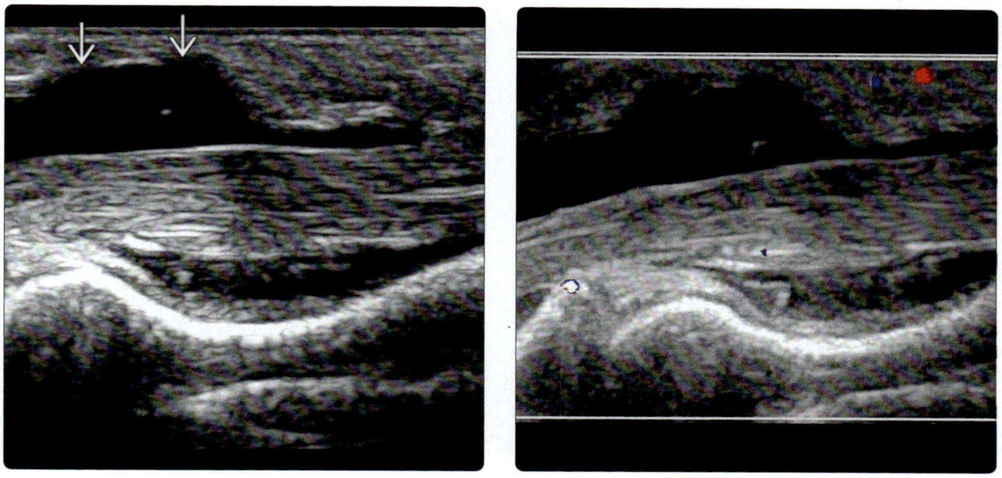

(Left) *Longitudinal US of the flexor digitorum of the ring finger in a patient with lymphoma shows a moderate amount of fluid present within the tendon sheath →. US-guided aspiration yielded gelatinous content, consistent with inactive chronic tenosynovitis.* **(Right)** *Longitudinal color Doppler US at the same location shows no significant hypervascularity, consistent with inactive chronic tenosynovitis.*

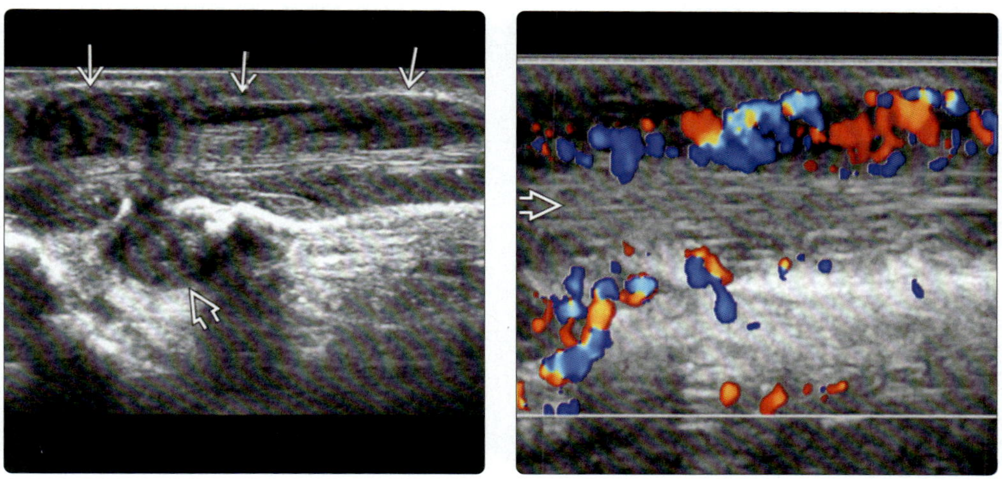

(Left) *Longitudinal US shows severe tenosynovial proliferation → of the extensor carpi ulnaris at the wrist in a patient with systemic lupus erythematosus (SLE). This is associated with moderate-sized erosion of the ulnar head → filled with proliferative granulation tissue (pannus).* **(Right)** *Color Doppler US (same patient) revealed marked tenosynovial hyperemia. The extensor carpi ulnaris tendon → is mildly swollen and of normal echogenicity. There is active chronic extensor carpi ulnaris tenosynovitis due to SLE.*

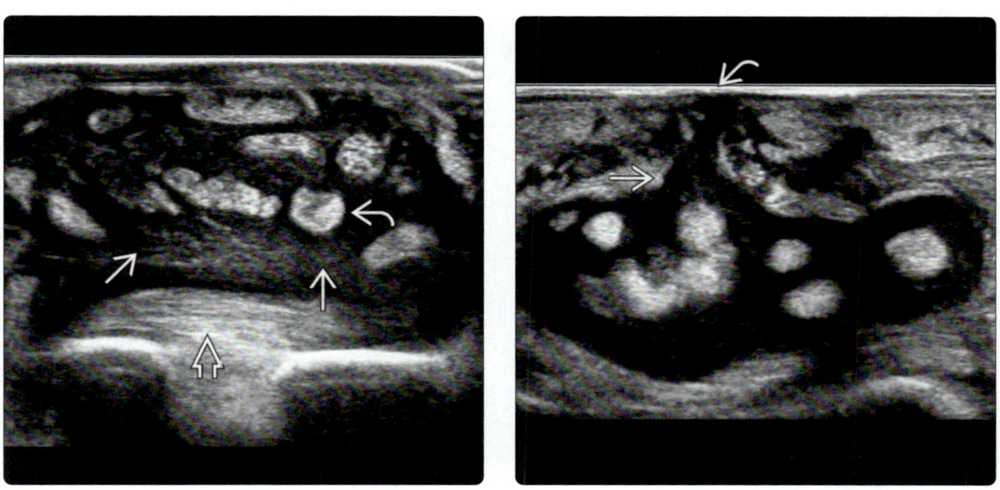

(Left) *Transverse US shows severe tenosynovial proliferation → of the wrist flexor tendons in a patient with known tuberculous (TB) tenosynovitis. Color Doppler imaging (not shown) revealed marked tenosynovial hyperemia. The tendons → are mildly swollen and of normal echogenicity. Pronator quadratus muscle → is normal.* **(Right)** *Transverse US of the same wrist at the more distal level shows inflamed tenosynovium extending along a recent surgical incision tract → toward the skin →.*

(Left) *Transverse US of the same patient shows TB tenosynovitis of the 1st extensor compartment. Both the extensor pollicis brevis and abductor pollicis longus tendons have split longitudinally with interdigitation of inflamed synovium.* **(Right)** *Transverse color Doppler US of the same area shows severe tenosynovial hyperemia. The main radial artery passing deep to the 1st extensor compartment tendons and the superficial palmar branch of the radial artery are shown.*

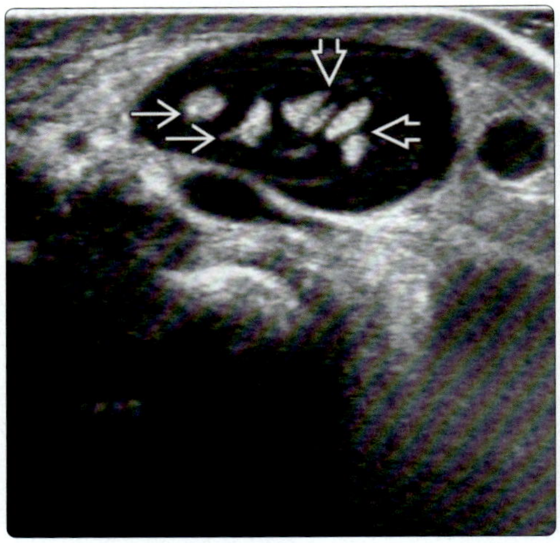

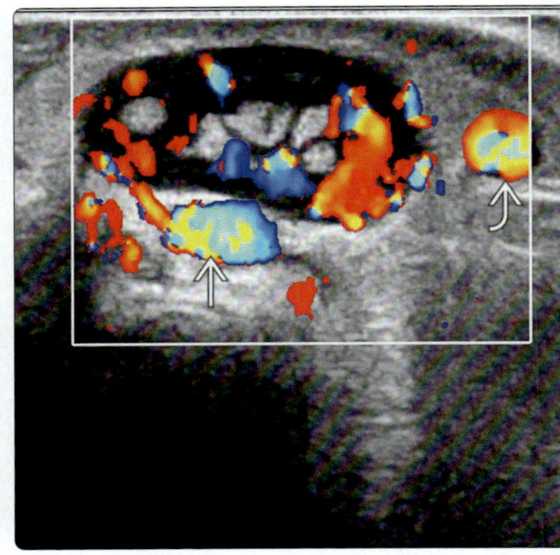

(Left) *Clinical photograph shows moderate swelling of the 1st extensor compartment tendons on the right side.* **(Right)** *Transverse US of the 1st extensor compartment shows a thickened abductor pollicis longus tendon and, to a lesser degree, a thickened extensor pollicis brevis tendon. The overlying extensor retinaculum is also moderately thickened. These features are compatible with de Quervain disease.*

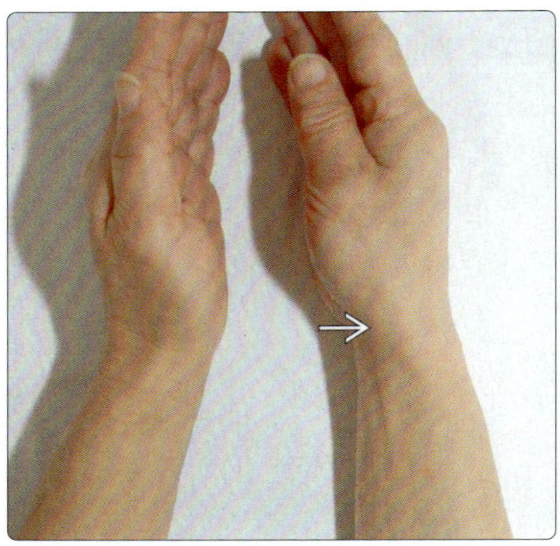

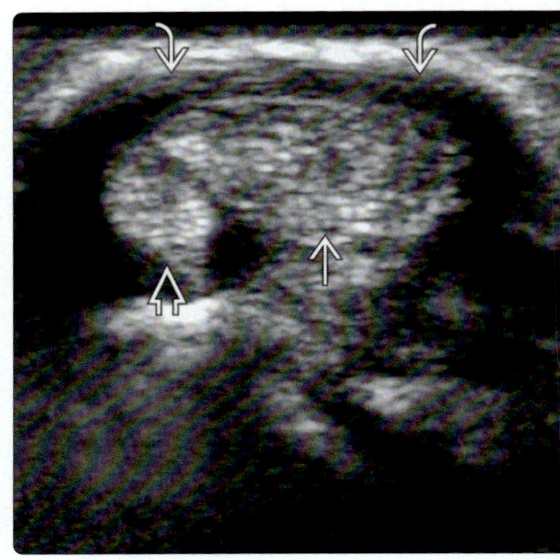

(Left) *Transverse US at the same level shows a cross-sectional area measurement using a continuous trace (dotted lines) of the 1st extensor compartment tendons. These measurements are helpful in monitoring the response to treatment.* **(Right)** *Transverse color Doppler US at the same location shows severe hyperemia of the thickened extensor retinaculum, consistent with active inflammation.*

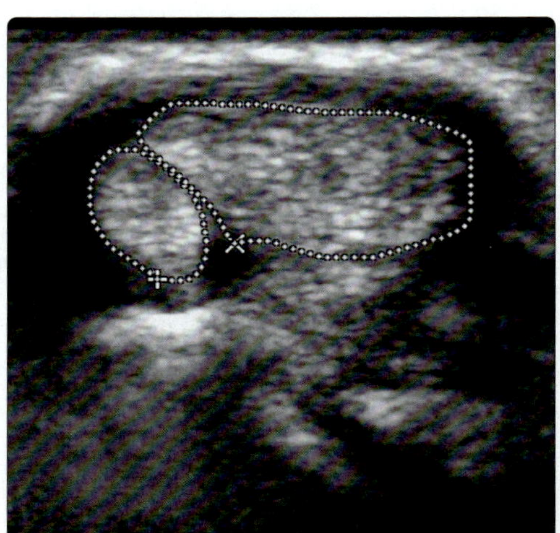

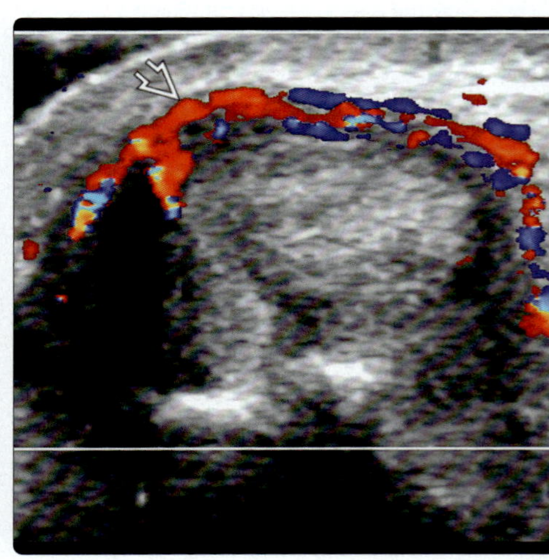

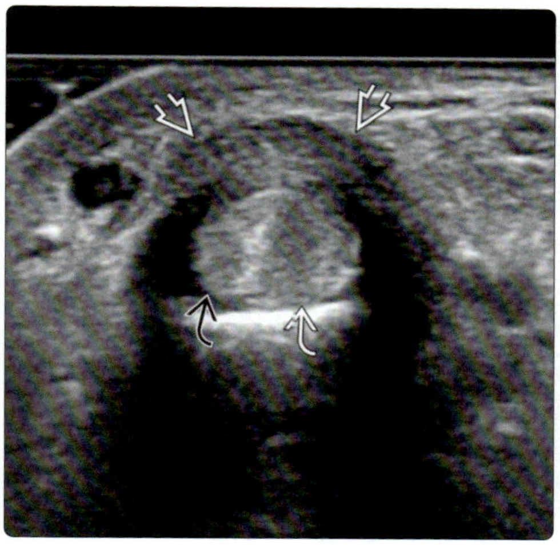

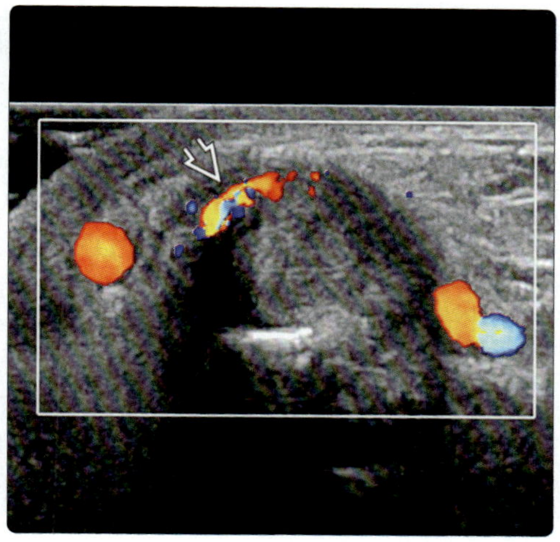

(Left) *Transverse US of the 1st extensor compartment shows a moderately thickened abductor pollicis longus tendon and barely distinguishable mildly thickened extensor pollicis brevis tendon. The overlying extensor retinaculum is severely thickened. Features are compatible with de Quervain disease.* **(Right)** *Transverse color Doppler US at the same location shows moderate hyperemia of the thickened extensor retinaculum due to active inflammation.*

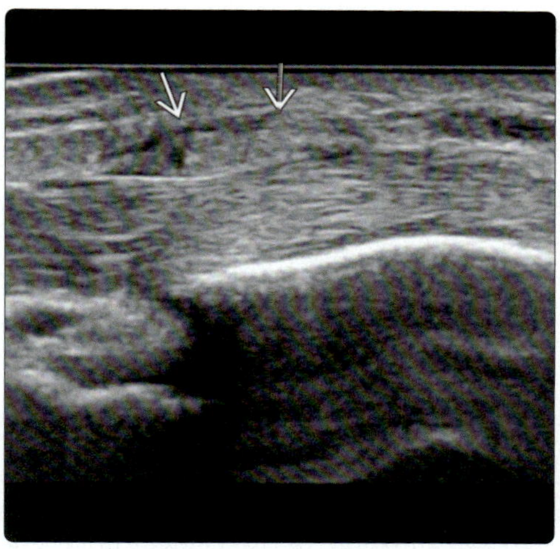

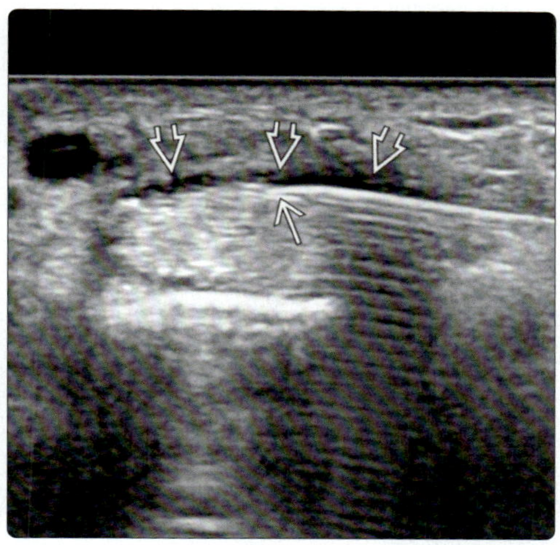

(Left) *Longitudinal US of the same patient at the same location shows severe thickening of the 1st extensor compartment retinaculum. The underlying abductor pollicis longus tendon is moderately indented. There was severe restriction of passive tendon movement on dynamic testing.* **(Right)** *Transverse US during injection of steroid between the tendon sheath and the retinaculum is shown. The needle tip is located at the interface between the tendon and the retinaculum with a thin rim of injected fluid.*

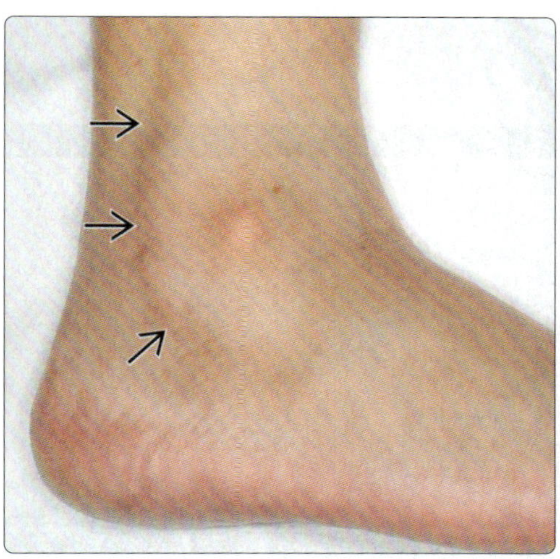

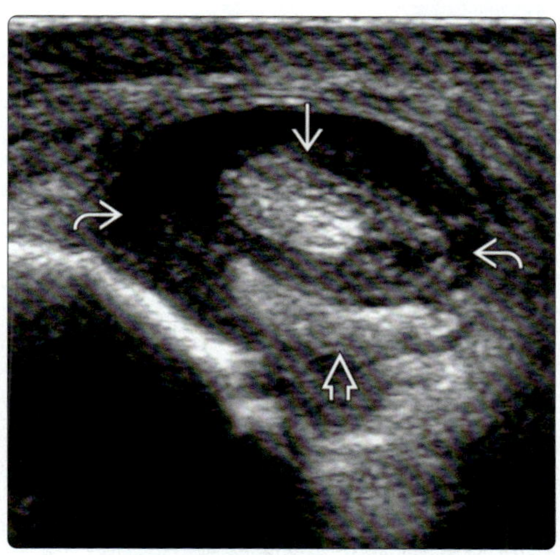

(Left) *Clinical photograph shows lateral retromalleolar ankle swelling with a thickened tubular structure in line with the peroneal tendons.* **(Right)** *Transverse US of the supramalleolar common peroneus tendon sheath shows severe synovial proliferation surrounding the mildly swollen peroneus longus and peroneus brevis tendons. Color Doppler imaging (not shown) revealed diffuse tenosynovial hyperemia indicating chronic active tenosynovitis and no tendon sheath effusion.*

Elbow Epicondylitis

KEY FACTS

TERMINOLOGY

- Tennis elbow (lateral epicondylitis), golfer's elbow (medial epicondylitis)

IMAGING

- Thickening, hypoechogenicity, fibrillar disruption ± hyperemia of common extensor tendon origin (CETO) or common flexor tendon origin (CFTO)
  - Grade tendinosis severity (minimal, mild, moderate, severe); relates to likelihood of tear/long-term prognosis
  - Independently grade level of hyperemia (minimal, mild, moderate, severe); relates to disease activity
- Hypoechoic, roundish areas with internal echoes indicative of focal proteoglycan deposition
- Sharp, linear, hypoechoic areas within tendon indicative of intrasubstance tendon tears
- Deeper, anechoic areas indicative of avulsive-type tear
- CETO tendinosis: Usually first affects extensor carpi radialis brevis followed by anterior fibers of extensor digitorum communis
- CFTO tendinosis: Usually first affects flexor carpi radialis, pronator teres, and palmaris longus tendons

TOP DIFFERENTIAL DIAGNOSES

- Cubital tunnel syndrome

CLINICAL ISSUES

- Lateral epicondylitis most common cause of elbow pain
  - 20x more common than medial epicondylitis

DIAGNOSTIC CHECKLIST

- Ligament injury if posttraumatic onset
- Look to deep margin of tendon for small tendon tears
- Severe pain or acute exacerbation is usually due to tendon tear
- Close association between lateral epicondylitis and ulnar neuritis

(Left) *Longitudinal graphic shows moderately severe common extensor tendon origin (CETO) tendinosis (lateral epicondylitis or tennis elbow) with a small tear (avulsion) at the insertion into the lateral epicondyle. The radial collateral ligament is normal.* **(Right)** *Longitudinal US shows a large tear (avulsion) of the common extensor origin from the lateral epicondyle. Only the superficial fibers remain intact. The radial head is normal.*

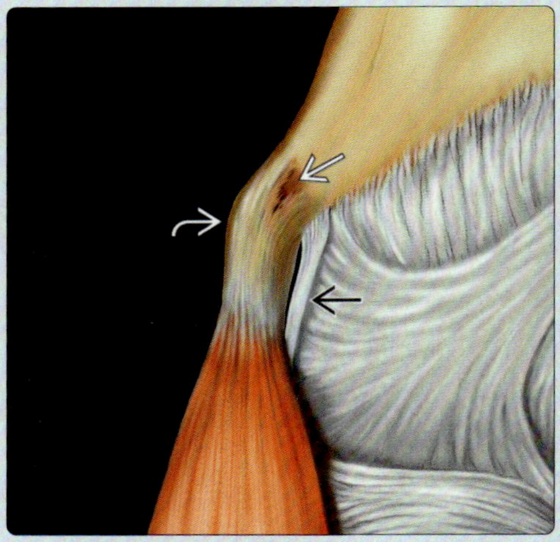

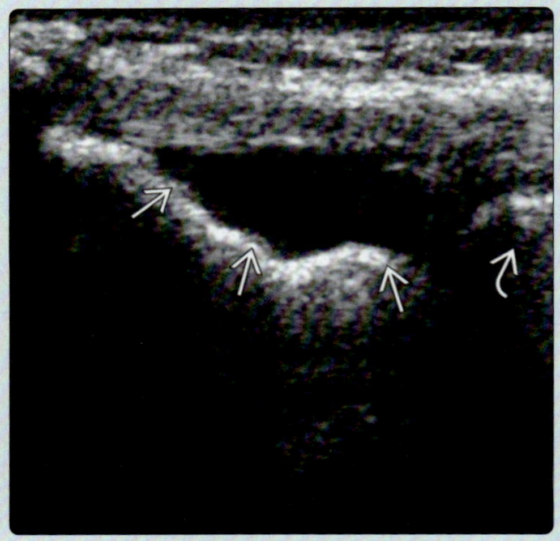

(Left) *Longitudinal US shows moderate tendinosis of the common extensor tendon insertion. There is a small, deep-seated intrasubstance tear. Early dystrophic tendon calcification is present. It is important to grade tendinosis severity as minimal, mild, moderate, or severe as this relates to the likelihood of tear.* **(Right)** *Longitudinal color Doppler US of the same patient shows mild hyperemia indicative of active disease. The small tear is also shown.*

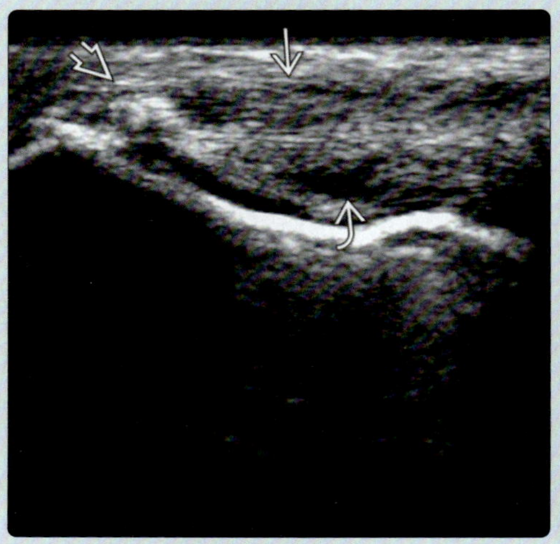

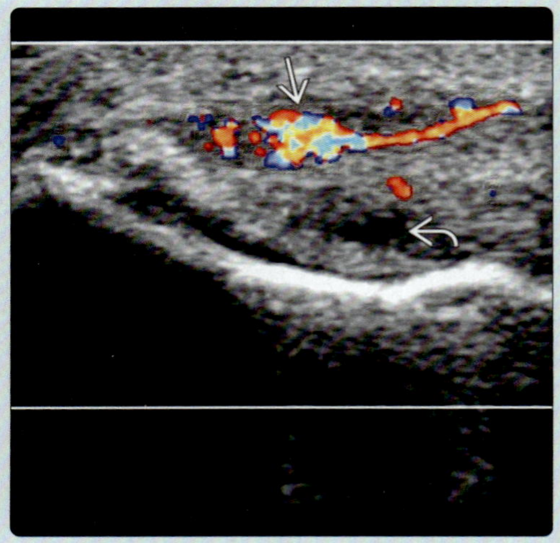

TERMINOLOGY

Synonyms

- Tennis elbow: Lateral epicondylitis, lateral epicondylalgia
- Golfer's elbow: Medial epicondylitis, medial epicondylalgia, medial tennis elbow, flexor-pronator sprain

Definitions

- Tendinosis of common extensor tendon origin (CETO) (lateral epicondylitis)
- Tendinosis of common flexor tendon origin (CFTO) (medial epicondylitis)

IMAGING

General Features

- Best diagnostic clue
    - Rather similar appearances irrespective of whether common extensor or flexor tendon origin is affected
        - Features of tendinosis of tendon origin (hypoechogenicity, thickening, loss of normal fibrillar structure) with hyperemia
            - Severity of tendinosis strongly associated with likelihood of tendon tear
- Location
    - CETO at lateral epicondyle
        - Attached to lateral aspect of lateral epicondyle
        - Tendinosis usually first affects extensor carpi radialis brevis (ECRB) tendon followed by anterior fibers of extensor digitorum communis (EDC) tendon
            - Anterior more than posterior fibers, deeper more than superficial fibers
        - ± tear of underlying radial collateral ligament
    - CFTO at medial epicondyle
        - Attached to undersurface of medial epicondyle
        - Musculotendinous junction is more proximal than lateral side so it has more muscular or fleshy appearance
        - Tendinosis usually first affects flexor carpi radialis (FCR), pronator teres (PT), and palmaris longus (PL) tendons
- Size
    - Varies from fibrillar disruption to tendon thickening, to intrasubstance tear, to complete avulsion
- Morphology
    - Diffuse tendon thickening to tear or avulsion
        - Smaller tears tend to be linear and intrasubstance in type, while larger tears tend to be avulsive in type

Radiographic Findings

- Radiography
    - Cortical irregularity and hyperostosis of common tendon origin
    - Dystrophic calcification within common tendon origin
    - Soft tissue swelling at common tendon origin

MR Findings

- T1WI
    - Thickening and increased signal intensity within common extensor (or flexor) tendon origin at lateral (or medial) epicondyle, respectively
- T2WI FS
    - Thickening and increased signal intensity of common extensor (of flexor) tendon origin
        - Peritendinitis is feature of more severe cases
        - Insertional subcortical bone marrow edema occasionally seen

Ultrasonographic Findings

- Tendinosis (> 90% of symptomatic subjects)
    - Thickened, hypoechogenic tendon with surface bowing and disruption of normal fibrillar pattern
        - Tendon hypoechogenicity usually affects deeper aspects of tendon initially, spreading to more superficial aspects as disease progresses
        - Ill-defined, hypoechoic areas are indicative of focal proteoglycan accumulation
        - Increased cross-sectional area and thickness of tendon
        - For lateral epicondylitis, average (range)
            - Cross-sectional area: 44 mm^2 (35-55 mm^2) for patients vs. 26 mm^2 (17-35 mm^2) for normal subjects; > 32 mm^2 highly predictive of lateral epicondylitis
            - Thickness: 4.9 mm (3.9-6.0 mm) for patients vs. 3.0 mm (2.0-3.7 mm) for normal subjects; > 4.2 mm highly predictive of lateral epicondylitis
- Tendon hyperemia (60% of symptomatic subjects)
    - Normal tendons do not display intrinsic vascularity
- Tendon tears (20% of symptomatic subjects)
    - Sharp, linear, hypoechoic areas within tendon indicative of intrasubstance tendon tears
        - Deeper, anechoic areas indicative of avulsive-type tear
- Secondary signs of tendinosis
    - Dystrophic calcification within tendon substance (30% symptomatic subjects)
    - Cortical irregularity at insertional area (20% symptomatic subjects)
    - Calcific enthesophytes at insertional area (45% symptomatic subjects)
- Radial collateral ligament can be injured with more advanced lateral epicondylitis
- Sonoelastography may be helpful in diagnosis of lateral epicondylitis (100% sensitivity, 89% specificity reported)
- ~ 15% of patients may have characteristic symptoms of epicondylitis but no abnormality on US
    - Abnormality too small to be apparent on US
- Conversely, patients may have no symptoms of epicondylitis and mild to moderate common extensor or flexor tendon origin tendinosis on US

Imaging Recommendations

- Best imaging tool
    - US
- Protocol advice
    - Examination with shoulder elbow flexed 90°; elbow internally rotated for lateral epicondylitis, externally rotated for medial epicondylitis
        - Examine with transducer aligned both parallel to and at right angles to long axis of tendons
        - Measure cross-sectional area and thickness of tendon at flat lateral aspect of lateral epicondyle (for lateral epicondylitis)
    - Always include evaluation with color Doppler and compare with contralateral side

- Grade vascularity subjectively as absent, minimal, mild, moderate, or severe
- Level of hyperemia is good guide to level of disease activity
- Always examine ulnar nerve in medial epicondylitis
  - Check integrity of collateral ligaments
  - Consider MR if US findings do not explain patient symptoms

DIFFERENTIAL DIAGNOSIS

Cubital Tunnel Syndrome

- Compression of ulnar nerve in cubital tunnel
- Inflammation of ulnar nerve may occur with medial epicondylitis

Hypertrophy of Radiohumeral Synovial Flange (Plica)

- Synovial fold projecting into posterolateral aspect of elbow joint
- Fold can hypertrophy with repetitive pronation and supination

Collateral Ligament Injury

- Usually history of moderate to severe trauma

PATHOLOGY

General Features

- Etiology
  - Trauma or overload injury, typically after minor and often unrecognized injury
- **Relevant anatomy: CETO**
  - ECRB, EDC, extensor digiti minimi (EDM), and extensor carpi ulnaris (ECU) (humeral head) all arise from lateral epicondyle
    - ECRB, EDC, and EDM have common origin from superior aspect of lateral epicondyle
    - ECU arises from posteroinferior aspect of lateral epicondyle and lies deep to EDC and EDM tendons
  - ECRL arises from supracondylar ridge separate from CETO
  - Lateral collateral ligament lies deep to extensor tendons and runs in slightly different direction
    - Bursa may exist between extensor tendons and lateral collateral ligament
- **Relevant anatomy: CFTO**
  - PT, FCR, flexor digitorum superficialis (FDS), PL, and flexor carpi ulnaris (FCU) all arise from medial epicondyle
    - FDS lies deep to FCR, PL, and FCU

Staging, Grading, & Classification

- Classify as mild, moderate, or severe, depending on overall severity of tendinosis ± tears

Gross Pathologic & Surgical Features

- Tendinosis with tendon thickening ± macroscopic partial tear or full-thickness tear

Microscopic Features

- Angiofibroblastic hyperplasia with disrupted collagen fibers, mucoid degeneration, neovascularization, and tenocyte proliferation
  - Tendon tears heal with reparative granulation tissue
  - Maturation of fibrous tissue to scar-like tendon tissue
- Epicondylitis is not inflammatory condition, though peritendinitis is feature
- Pathology is similar for both medial and lateral epicondylitis

CLINICAL ISSUES

Presentation

- Most common signs/symptoms
  - Adult patient with lateral or medial elbow pain
    - Pain aggravated by vigorous exercise, gradually lessens with rest over several days
    - Lateral epicondylitis pain aggravated by forced wrist extension; medial epicondylitis pain aggravated by forced wrist flexion
    - Local tenderness at tendon origin

Demographics

- Age
  - Peak incidence: 40-50 years
- Epidemiology
  - Population prevalence ~ 3%
  - Lateral epicondylitis is most common cause of elbow pain
    - 20x more common than medial epicondylitis
      - □ Most patients with tennis elbow do not play tennis
      - □ Tennis elbow is more common in golfers than golfer's elbow

Natural History & Prognosis

- Usually self-limiting; 1/3 will improve by 3 months, 1/2 by 6 months, 3/4 by 1 year
  - Chronic symptoms or relapse more likely if established tendinosis on US

Treatment

- Conservative
  - Most patients respond to rest ± bracing
    - ± physiotherapy, exercise, stretching
    - ± nonsteroidal antiinflammatory medication (oral or topical)
- Many additional treatments for refractory cases
  - Limited evidence currently to recommend any particular method over another
    - Corticosteroid therapy: Short-term benefit but likelihood of long-term harm
      - □ Iontophoresis (drug such as corticosteroid introduced through skin using electric charge)
    - US-guided: Dry needling, autologous blood injection, platelet-rich plasma injection
    - Other treatments: Extracorporeal shock wave therapy, pulsed low-intensity US, pulsed electromagnetic field therapy, low-level laser therapy
- Surgical
  - Anchoring of tendon insertion or tendon release

SELECTED REFERENCES

1. Amin NH et al: Medial epicondylitis: evaluation and management. J Am Acad Orthop Surg. 23(6):348-55, 2015
2. Latham SK et al: The diagnostic test accuracy of ultrasound for the detection of lateral epicondylitis: a systematic review and meta-analysis. Orthop Traumatol Surg Res. 100(3):281-6, 2014

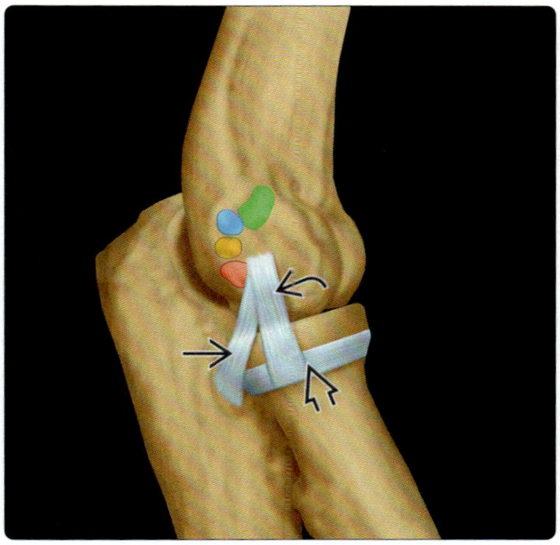

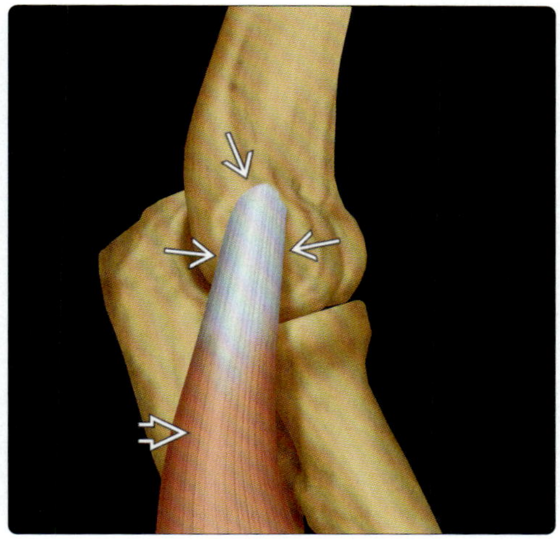

(Left) *Schematic diagram shows the attachment of common extensor tendons at the lateral epicondyle (green = extensor carpi radialis brevis; blue = extensor digitorum; yellow = extensor digiti minimi; red = extensor carpi ulnaris) and lateral collateral ligaments (radial collateral ligament, lateral ulnar collateral ligament, and annular ligament).* **(Right)** *Schematic diagram shows the common extensor attachment at the lateral epicondyle. Note the extensor muscle fibers.*

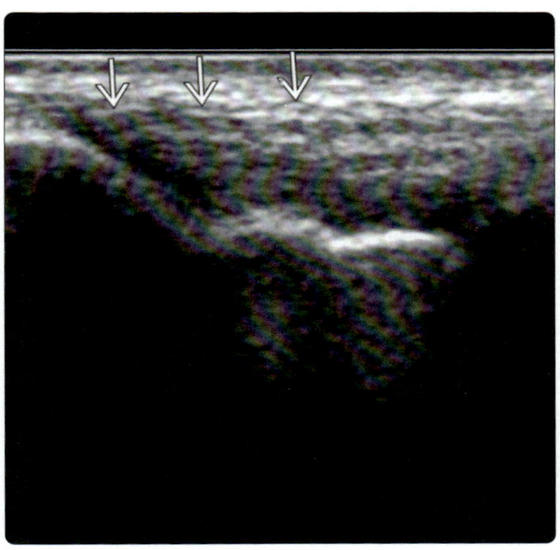

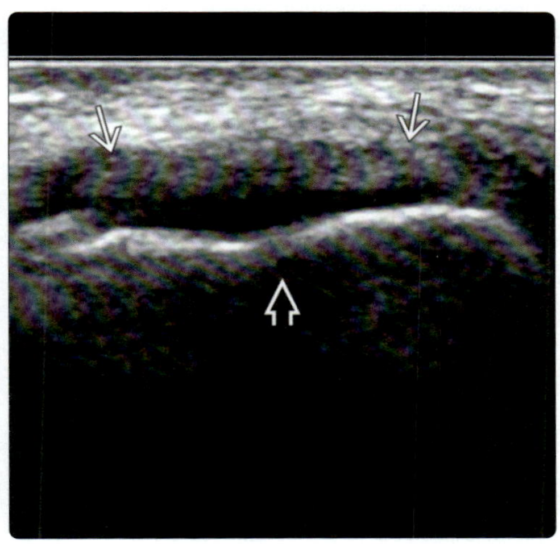

(Left) *Longitudinal US shows the normal CETO at the lateral humeral epicondyle. There is no abnormal thickening or tear. The fibrillar pattern of the tendon is preserved.* **(Right)** *Transverse US of the same patient shows the normal CETO on the flat portion of the lateral humeral condyle. This is the area where the cross section of the common extensor tendons can be measured.*

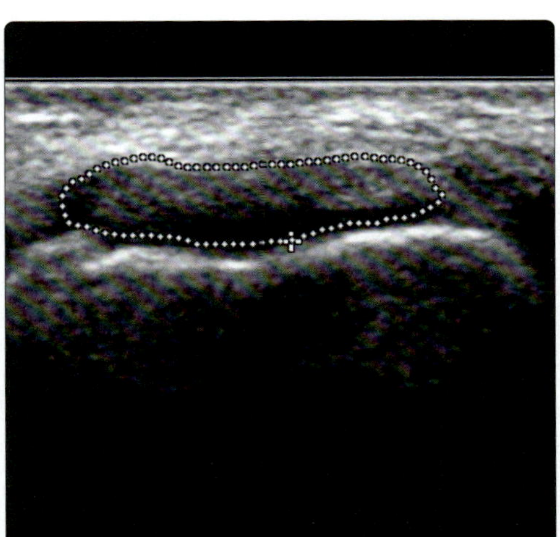

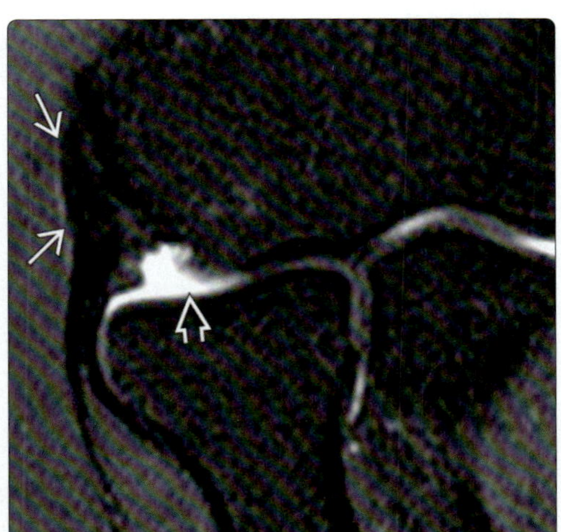

(Left) *Transverse US shows measurement of the cross-sectional area of the common extensor tendon just proximal to the bony attachment. Such measurements are useful for follow-up and for broadly gauging severity of disease.* **(Right)** *Coronal T2WI FS MR shows the normal CETO from the lateral epicondyle. There is no abnormal hyperintense signal or fiber discontinuity (tear). A small amount of normal physiological joint fluid is present at the radiocapitellar articulation.*

(Left) *Longitudinal US shows mild thickening of the CETO* ➡ *at the lateral epicondyle, indicative of mild tendinosis.* **(Right)** *Longitudinal color Doppler US of the same patient shows only minimal hyperemia* ➡*, indicative of minimal disease activity. Tendinosis around the elbow should be routinely assessed with color Doppler imaging as this provides the best indication of disease activity. Normal tendon does not show color flow.*

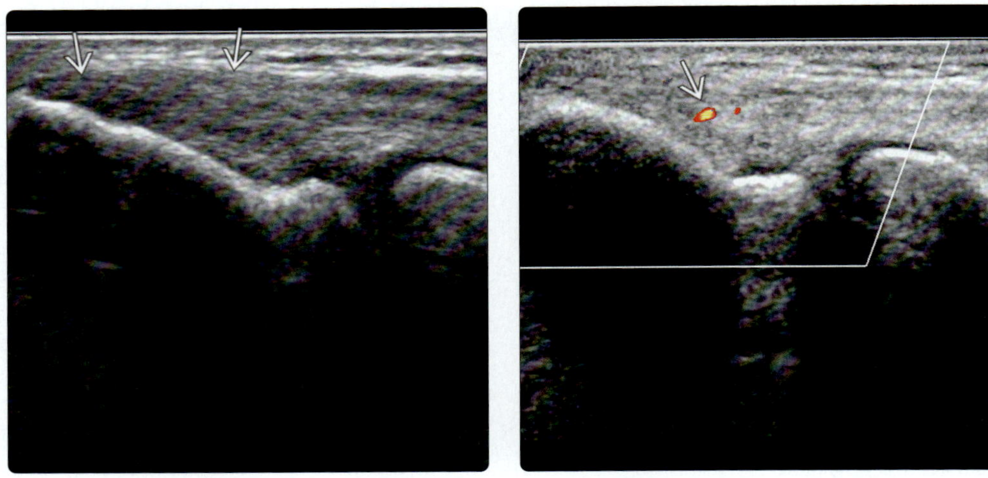

(Left) *Longitudinal US shows mild thickening of the CETO* ➡ *at the lateral epicondyle, indicative of mild tendinosis. The ill-defined area of increased hypoechogenicity* ➡ *is due to proteoglycan accumulation rather than an intrasubstance tear.* **(Right)** *Longitudinal color Doppler US of the same patient shows mild hyperemia* ➡*, indicative of mildly active disease.*

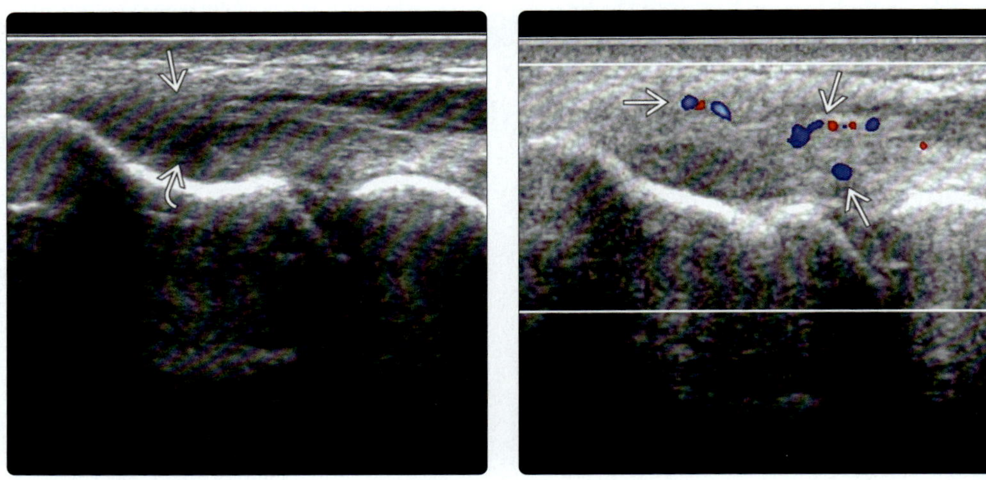

(Left) *Lateral US shows moderate thickening and hypoechogenicity of the CETO* ➡ *at the lateral epicondyle, indicative of moderate tendinosis. There is a medium-sized intrasubstance tendon tear* ➡*. Mild insertional cortical hyperostosis* ➡ *is present.* **(Right)** *Longitudinal color Doppler US of the same patient shows mild to moderate hyperemia* ➡*, indicative of active disease.*

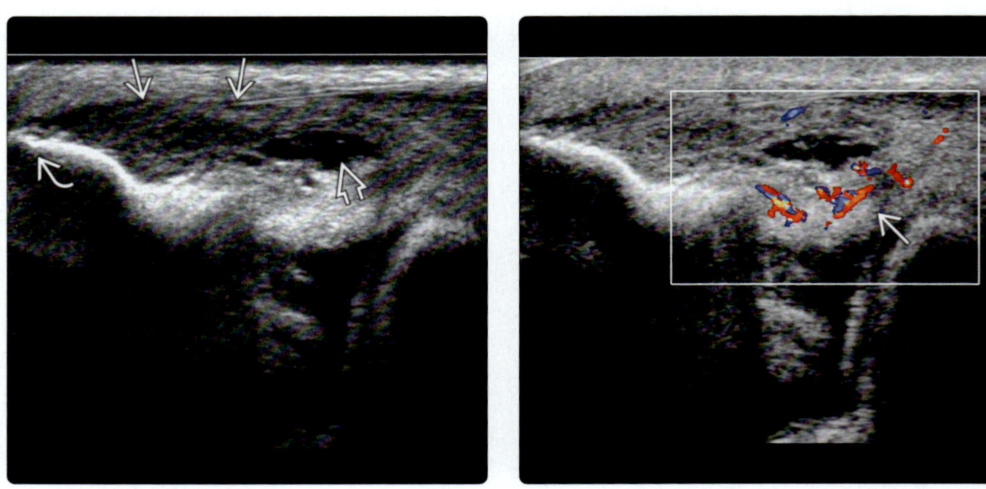

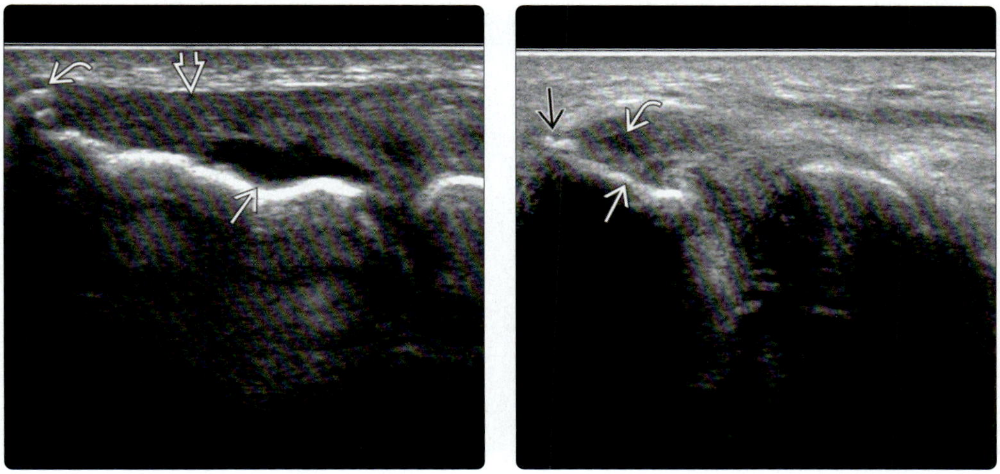

(Left) *Longitudinal US shows mild tendinosis → of the CETO with hypoechogenicity and loss of fibrillar pattern. There is a medium-sized tear (avulsion) of the deeper part of the tendon insertion →. Mild insertional cortical hyperostosis → is present.* **(Right)** *Longitudinal US shows moderate tendinosis of the CETO → associated with a small central tear → as evidenced by a localized area of more pronounced hypoechogenicity. A small area of dystrophic calcification is present →.*

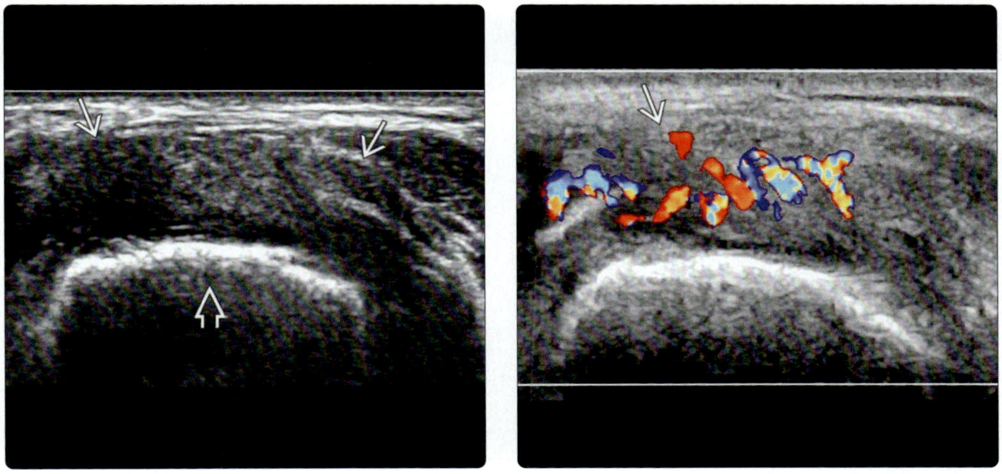

(Left) *Transverse US shows moderate thickening of the CETO → overlying the lateral humeral condyle →. No tear is present.* **(Right)** *Transverse color Doppler US of the same patient shows moderate to severe hyperemia →, indicative of moderate to severe disease activity. Disease activity does not invariably accompany macroscopic tendon tear and vice versa.*

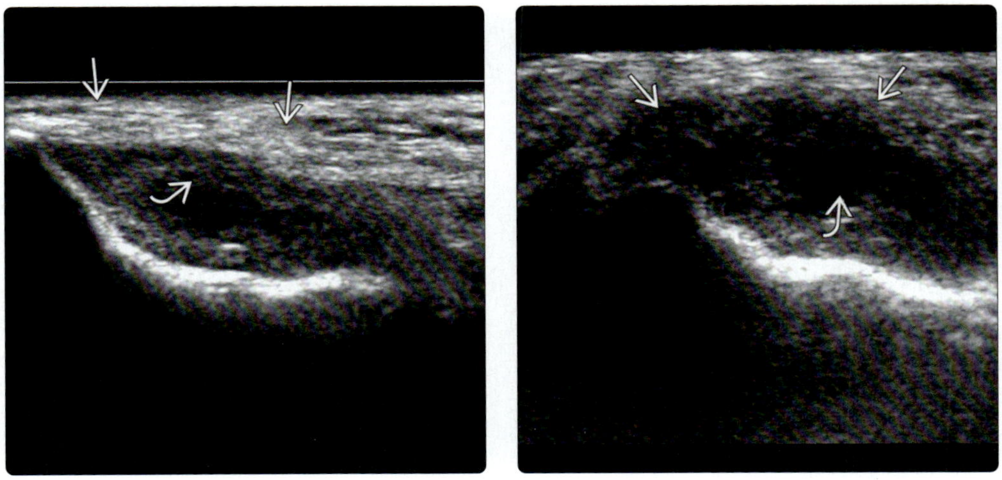

(Left) *Longitudinal US shows moderate tendinosis of the CETO → with tendon thickening and hypoechogenicity associated with a small intrasubstance tear →. Distinguishing between focal severe proteoglycan (mucoid) accumulation and tear is not always easy as both entities may coexist.* **(Right)** *Oblique US of the same patient shows moderate CETO tendinosis → with an intrasubstance tendon tear →.*

(Left) *Lateral US shows moderate tendinosis of the CETO* → *with dystrophic intratendinous calcification* ↷. **(Right)** *Coronal T2WI FS MR of the same patient shows abnormal thickening and hyperintense T2-weighted signal intensity* ↷ *of the CETO, indicative of moderate tendinosis. There is a very small intrasubstance tear present* →. *Unlike insertional tendinosis at other sites, bone marrow edema is not a common feature of CETO tendinosis.*

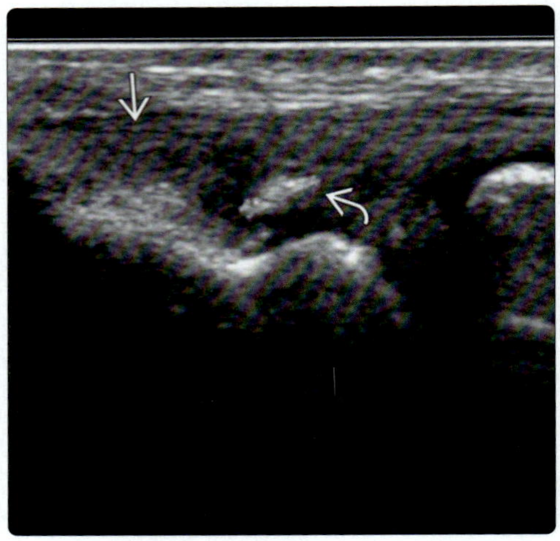

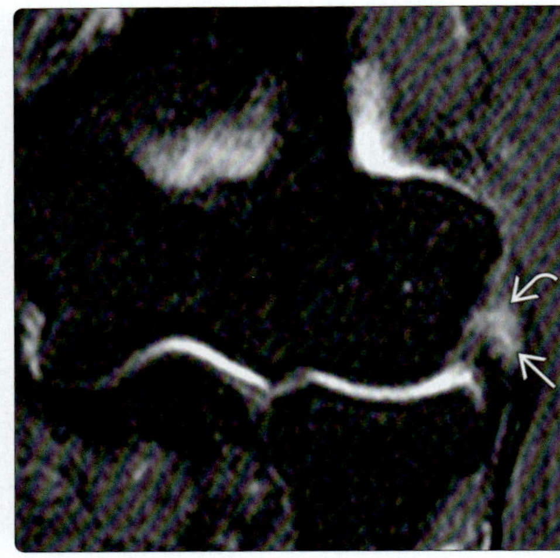

(Left) *Longitudinal US shows severe CETO* → *with tendon thickening, hypoechogenicity, and loss of the normal fibrillar pattern. No tear is present.* **(Right)** *Transverse US of the same patient shows severe CETO tendinosis* → *with dystrophic intratendinous calcification* ⇨. *Mild surface irregularity of the lateral humeral condyle is present* ↷.

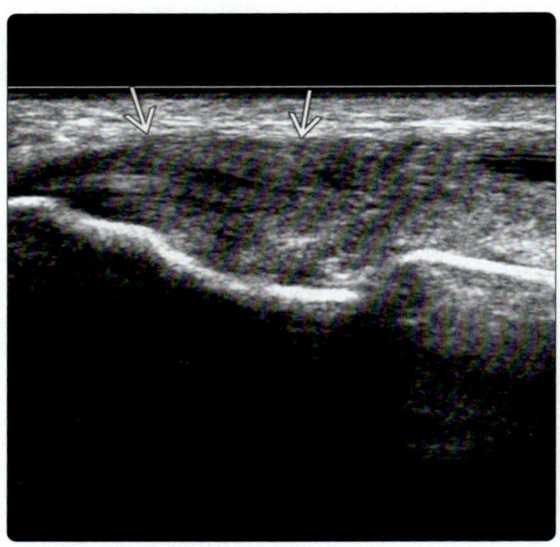

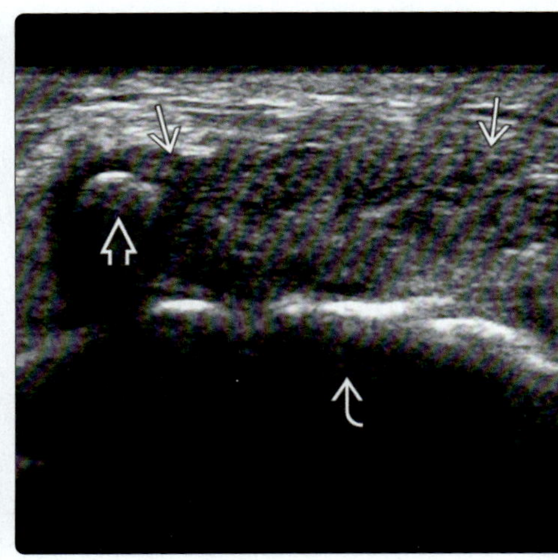

(Left) *Axial T1WI MR of the same patient shows the thickened CETO, which is isointense* → *to adjacent muscle.* **(Right)** *Coronal T2WI FS MR of the same patient shows severe CETO tendinosis* → *with diffuse hyperintensity. No tear is present. Only that portion of the CETO close to the bone is affected. There is very little reactive bone change.*

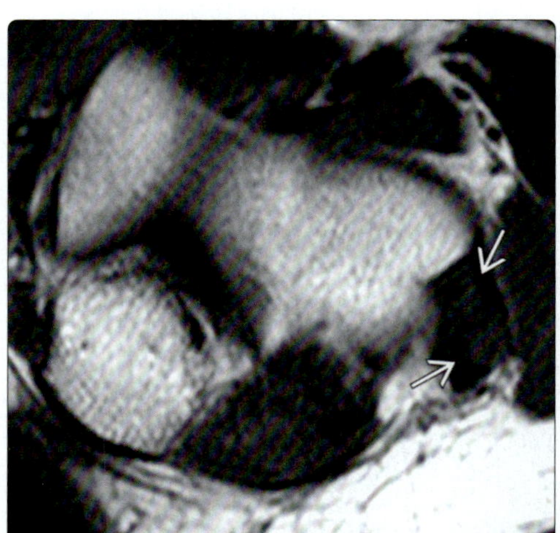

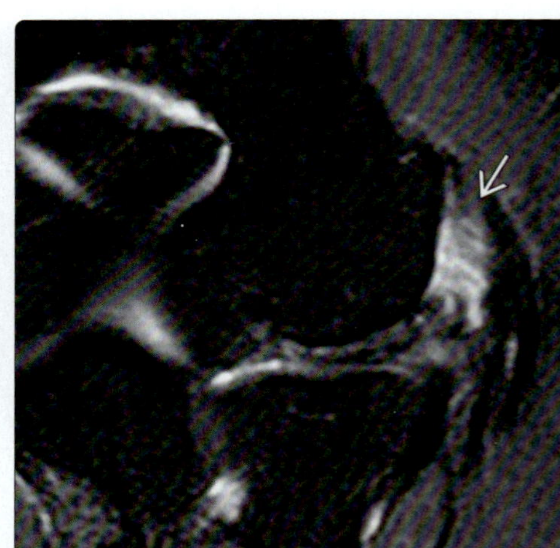

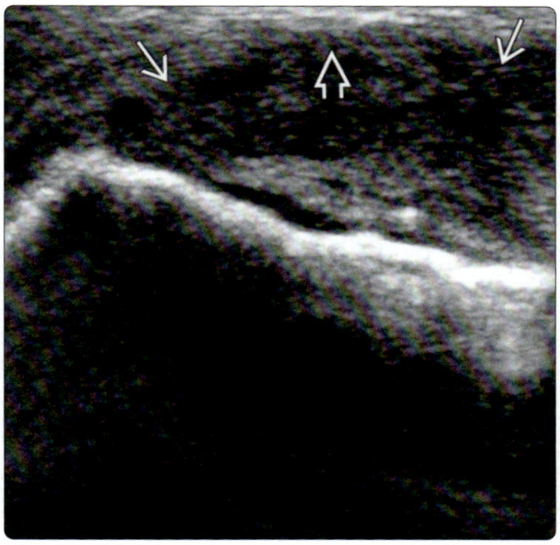
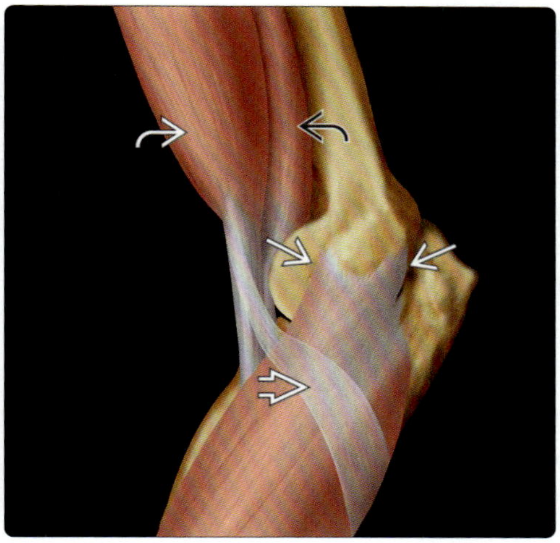

(Left) *Longitudinal US shows severe tendinosis of the CETO with a superficial bulging contour of the tendon surface.* **(Right)** *Schematic diagram shows the attachment of common flexor tendons at the medial epicondyle. Note the bicipital aponeurosis, biceps muscle, and brachialis muscle.*

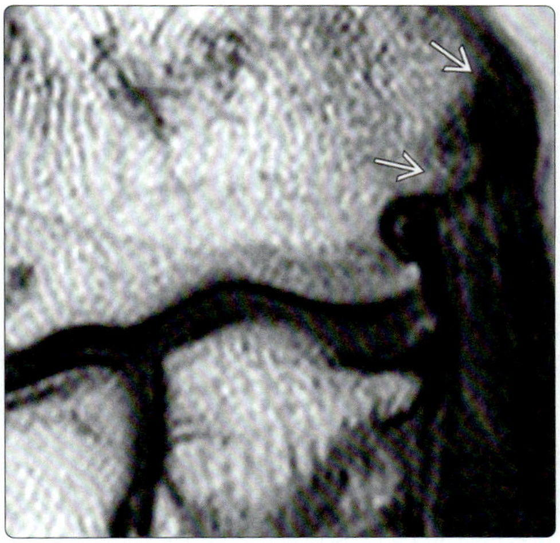
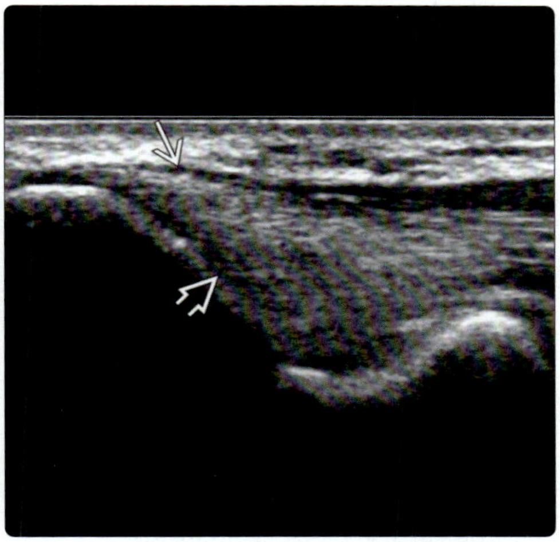

(Left) *Coronal T1WI MR shows the normal common flexor tendon origin (CFTO) at the medial epicondyle. The tendon is hypointense and inserted over a broader area than the CETO on the lateral side.* **(Right)** *Longitudinal US shows the normal CFTO at the medial epicondyle. Note the largely flattened superficial contour.*

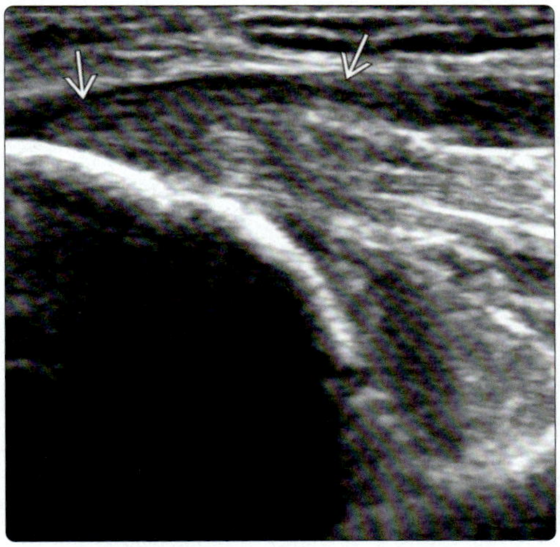
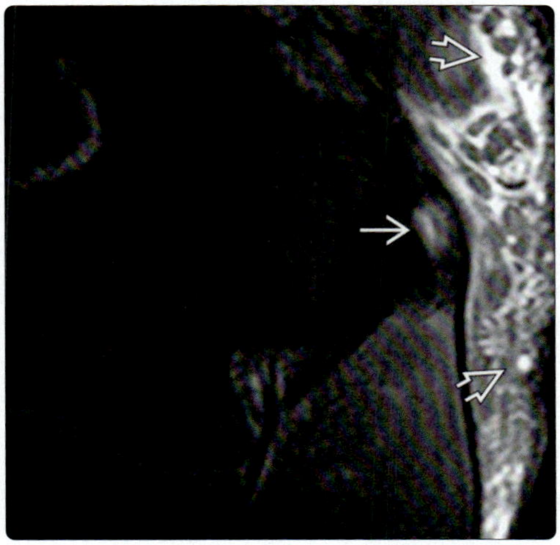

(Left) *Longitudinal US shows mild CFTO tendinosis with hypoechogenicity and a more bulging superficial contour. No tear is apparent.* **(Right)** *Coronal T2WI FS MR of the same patient shows a small intrasubstance tear within the CFTO. There is moderate adjacent subcutaneous soft tissue edema. MR is helpful if US fails to fully explain the patient's symptoms.*

Fat Injury

KEY FACTS

TERMINOLOGY

- Focal inflammation ± necrosis ± atrophy of subcutaneous fat

IMAGING

- Variable appearances depending on chronicity of lesion
- Earlier stages
  - Localized mild swelling and increased echogenicity of subcutaneous fat
  - Loss of normal subcutaneous fat striation
  - Irregular poorly defined margins without discrete mass
  - Hypoechoic stranding extending into surrounding fat
  - ± fluid accumulation within affected fat (liquefaction)
  - ± calcification of affected fat
    - US very sensitive at depicting early calcification
  - ± hypoechoic pseudocapsule
  - Usually mild or no hyperemia
- Usually fascia and muscle deep to affected area are normal
- Later stages
  - Area of fat necrosis becomes better defined and more hypoechoic
  - Loss of volume of affected area
- US allows definitive diagnosis in most cases
  - US is better at depicting associated calcification than MR
- Trauma may be mild and not remembered
- Morel-Lavallée lesion
  - Compressible hypoechoic or anechoic collection between subcutaneous fat and investing fascia
  - Most commonly occurs in anterolateral thigh

TOP DIFFERENTIAL DIAGNOSES

- Subcutaneous lipoma
- Superficial thrombophlebitis
- Subcutaneous panniculitis
- Subcutaneous fat necrosis of newborn
- Subcutaneous fat necrosis associated with pancreatitis

(Left) *Transverse US of the lateral thigh region in a patient with recent trauma shows a quite well-defined area of subcutaneous edema → due to fat injury. Note the mirror image artifact deep to the investing fascia ↪.* **(Right)** *Longitudinal US of the same lesion shows a posttraumatic dermal and subcutaneous injury with edema →. The entire thickness of the subcutaneous fat down to the investing fascia is affected. The signal alteration deep to the investing fascia is artifactual ⇨.*

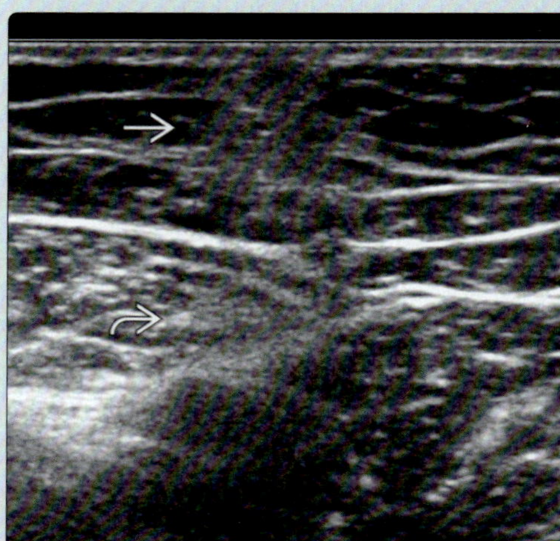

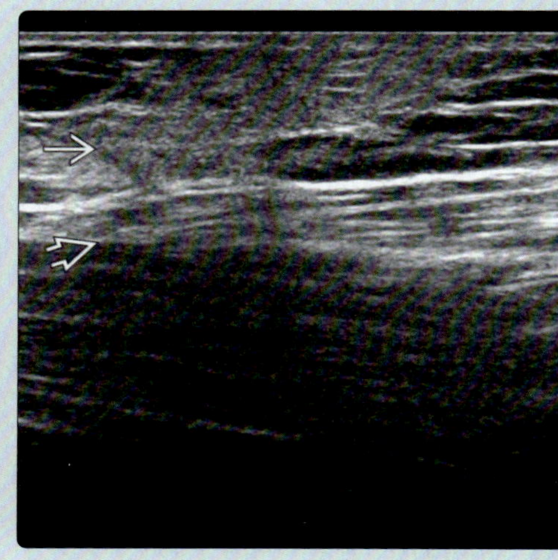

(Left) *Longitudinal US of a patient who developed an anterior abdominal wall mass following trauma 1 year earlier is shown. There is a discrete, irregular, hypoechoic mass → in the subcutaneous fat due to chronic fat injury. Note the mild degree of posterior shadowing ⇨ due to fibrotic tissue.* **(Right)** *Longitudinal color Doppler US of the same lesion shows mild, marginal hyperemia ↪ due to perilesional reparative change. Clinically, such lesions become smaller and firmer over time, possibly with some skin depression.*

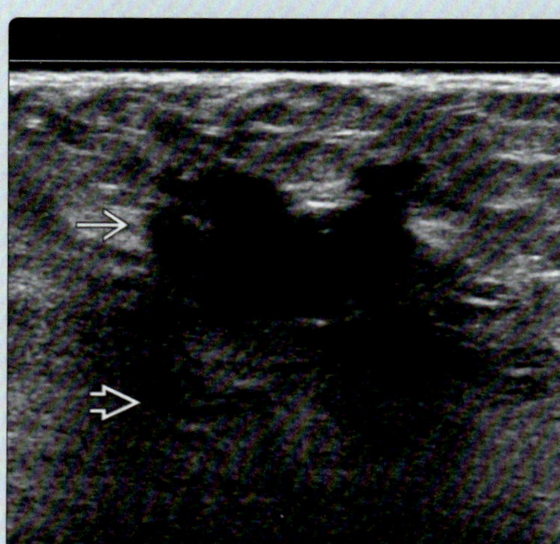

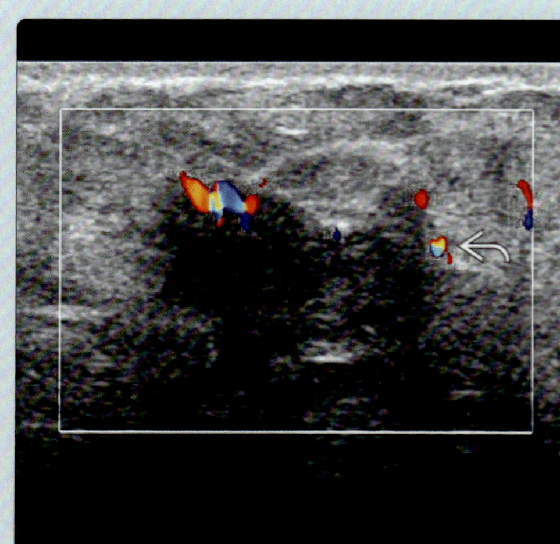

TERMINOLOGY

Definitions

- Focal inflammation ± necrosis ± atrophy of subcutaneous fat
- Morel-Lavallée lesion: Shear injury leading to separation of subcutaneous tissue from underlying investing fascia creating potential space filled with blood, lymph, or other breakdown products

IMAGING

General Features

- Best diagnostic clue
  - Localized altered echogenicity and thickness of subcutaneous fat with loss of normal fat lobulation
- Location
  - Subcutaneous fat usually over bony protuberances
    - Mainly affects gluteal or proximal thighs
- Size
  - From a few mm to 10 cm
  - Commonly full depth of subcutaneous tissue is affected
- Morphology
  - Focal swelling of subcutaneous tissues progressing to induration and atrophy

Radiographic Findings

- Focal swelling or atrophy of subcutaneous soft tissue
- Subcutaneous calcification
- Subcutaneous gas locules after high-pressure trauma

MR Findings

- T1WI
  - Focal hyperintense subcutaneous fat swelling
  - Spiculated areas of linear hypointensity within subcutaneous fat due to reactive fibrosis
    - ± pseudocapsule
  - Focal atrophy of subcutaneous fat
  - Usually investing fascia and muscle are normal
- T2WI FS
  - Focal hyperintense subcutaneous fat swelling with surrounding edema
    - ± nodular areas of subcutaneous hypointensity due to calcification
    - ± larger hyperintense areas within subcutaneous fat due to fluid
- T1WI C+ FS
  - Minimal or no intrinsic contrast enhancement
  - Enhancement of pseudocapsule and surrounding edema

Ultrasonographic Findings

- **Traumatic fat necrosis**
  - Appearances vary depending on chronicity of lesion
  - Earlier stages (within 1 year)
    - Localized swelling and increased echogenicity of edematous subcutaneous fat
      - Poor definition of subcutaneous fat lobulation
      - Irregularly defined margins without discrete mass
      - Lobulation of subcutaneous fat, which becomes better defined and more hypoechoic with increasing chronicity
      - ± pseudocapsule, which is usually hypoechoic though may be hyperechoic
    - ± fluid accumulation within affected fat
      - Linear fluid collections located superficial to investing fascia
      - Larger, more rounded fluid collections in areas with more abundant subcutaneous fat (e.g., buttock)
    - ± calcification of affected subcutaneous fat
      - Seen in subacute or chronic stage of fat necrosis
      - Varies from mild, speckled calcification to dense calcification
      - Increasing calcification with increasing lesion chronicity
      - Shadowing artifact may impede full depiction of necrotic area
      - Aggregations of fibrosis without calcification in chronic lesions can cast less severe shadowing artifact
      - US more sensitive than radiographs at depicting milder degrees of calcification
    - Hyperemia is usually mild or absent
      - Moderate or severe hyperemia suggests secondary infection
  - Later stages
    - Subcutaneous fat necrosis becomes better defined and more hypoechoic with increasing chronicity
      - Subcutaneous fat atrophy may occur with depression of skin surface
  - Usually normal fascia and muscle deep to affected area
    - May rarely show disruption if severe trauma
- **Morel-Lavallée lesion**
  - Uncommon lesion
  - Compressible hypoechoic or anechoic collection between subcutaneous fat and investing fascia
    - Characteristic location at interface between fascia and subcutaneous fat
    - Most commonly occurs in anterolateral thigh
  - Acute/subacute: More heterogeneous and irregular
    - ± echogenic, nondependent fat globules within fluid collection
    - ± signs of secondary infection
  - Chronic: More homogeneous and well defined
    - Partial or complete fibrous capsule

Imaging Recommendations

- Best imaging tool
  - US provides definitive diagnosis in most cases with ready comparison of opposite side
    - Better at depicting associated calcification than MR
    - MR is better at defining extent at markedly edematous subcutaneous tissue
- Protocol advice
  - Minimal transducer pressure to avoid effacement of subcutaneous fat
  - Ample acoustic gel to fill any skin depression
  - Compare with adjacent normal subcutaneous fat on ipsilateral side or normal subcutaneous fat on contralateral side
  - Use color Doppler to assess vascularity

- Extended field-of-view imaging can improve perception of localized subcutaneous fat atrophy and show full length of Morel-Lavallée lesion

DIFFERENTIAL DIAGNOSIS

Superficial Thrombophlebitis

- Localized swelling wall of subcutaneous vein
- ± perivenous edema ± intraluminal thrombosis

Subcutaneous Lipoma

- Discrete, nontender, subcutaneous lipomatous mass
- Fine linear striations parallel to long axis of mass
- Atypical lipoma and angiolipoma mimic fat necrosis

Subcutaneous Panniculitis

- Localized subcutaneous inflammation with edema and hyperemia
    - Erythema nodosum = septal panniculitis
- More common in shin regions and in children
- Many causes of panniculitis
    - Infection
        - *Streptococcus*, tuberculosis, rheumatic fever
    - Systemic inflammatory disorder
        - Polyarteritis nodosa, systemic lupus erythematosus, sarcoidosis, dermatomyositis, morphea, Takayasu arteritis, necrotizing vasculitis
    - Myeloproliferative disorder
        - Lymphoma, leukemia
    - Pancreatic disease
        - Pancreatitis (50%) or pancreatic carcinoma (50%)
        - May precede recognition of pancreatitis in 50%
    - Drug hypersensitivity
    - Weber-Christian disease: Rare inflammatory panniculitis
        - Fever, arthralgia, anemia, leucopenia, deranged liver function, subcutaneous panniculitis

Subcutaneous Fat Necrosis of Newborn

- Seen in otherwise normal newborns
- Single or multiple violaceous plaques or nodules usually on trunk
- Associated with preeclampsia, hypothermia, obstetric trauma, hypoxia, etc.
- Diffuse increased echogenicity of subcutaneous fat in affected areas ± calcification
- Histology shows lobular panniculitis with fat necrosis, granulomas, fibrosis, and calcification

PATHOLOGY

General Features

- Etiology
    - Following trauma
        - ~ 50% of patients report prior trauma
            - Trauma is likely cause even when patients cannot recall antecedent trauma
        - Shearing stress of subcutaneous fat; investing fascia interface with disruption of perforating blood vessels and creation of potential spaces in Morel-Lavallée lesion
    - Following panniculitis

Gross Pathologic & Surgical Features

- Vary due to evolutionary nature of lesion
- Lobular thickening of subcutaneous fat surrounded by fibrous septa
- Later stages may show more dense fibrosis and atrophy of subcutaneous fat

Microscopic Features

- No need to biopsy majority of cases
- Inflammatory infiltrate predominantly involving subcutaneous septa (septal panniculitis) or lobules (lobular panniculitis) ± vasculitis
- Aggregates of foamy phagocytic macrophages
- Reactive fibrous tissue with areas of fat necrosis
- Cytology of cystic aspirate may reveal anuclear fat cells (vacuolated fat cells or "ghost" cells)

CLINICAL ISSUES

Presentation

- Most common signs/symptoms
    - 50% recall trauma (days, weeks, or months)
    - Evolving lesion over weeks or months
    - Tender subcutaneous swelling
        - Progressing to induration and later atrophy
- Other signs/symptoms
    - Pain and tenderness in 50%
    - May be asymptomatic and may present de novo as focal skin depression

Demographics

- Age
    - Young to middle age
- Gender
    - No sex predilection

Natural History & Prognosis

- Resolves over weeks or months with induration ± atrophy
- Indurated subcutaneous tissue either returns to normal, remains slightly indurated, or undergoes progressive fat atrophy leading to focal skin depression

Treatment

- Often no specific treatment
- ± antiinflammatory medication, if lesions are painful
- Percutaneous draining for Morel-Lavallée lesion
    - Usually recurs following percutaneous aspiration
    - Surgical drainage and sclerodesis for chronic lesion

DIAGNOSTIC CHECKLIST

Consider

- Pancreatic disease in patients presenting with unexplained panniculitis or fat necrosis

Image Interpretation Pearls

- Ill-defined, localized echogenicity of subcutaneous fat clinically evolving over weeks or months is usually due to either panniculitis or fat injury

SELECTED REFERENCES

1. Depaoli R et al: Morel-Lavallée lesion of the knee in a soccer player. J Ultrasound. 18(1):87-9, 2015

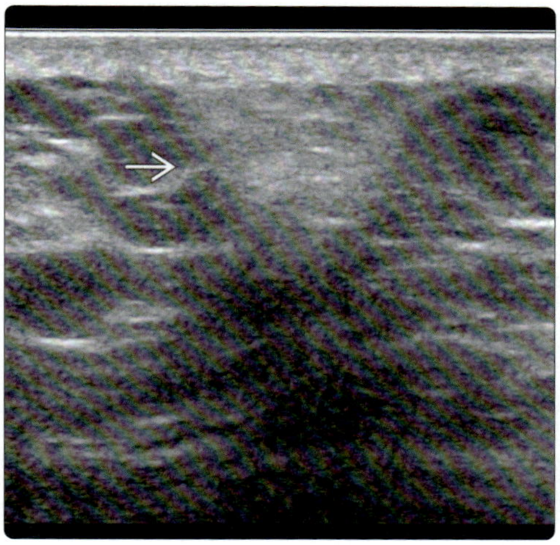
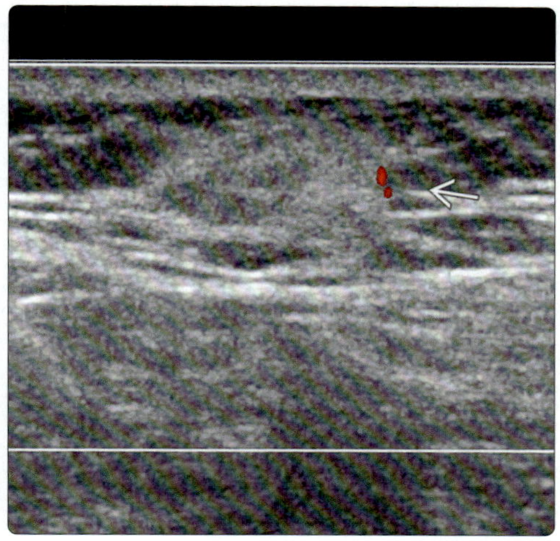

(Left) *Transverse US in a patient with blunt buttock injury 4 months earlier shows a small (2.2 cm long x 1.5 cm wide x 1.4 cm deep) area ➡ of subcutaneous fat necrosis in the upper buttock region. Slight loss of volume is present. There is no liquefaction or calcification.* **(Right)** *Color Doppler US of the same lesion shows minimal peripheral hyperemia ➡. Minimal hyperemia is typical of fat necrosis.*

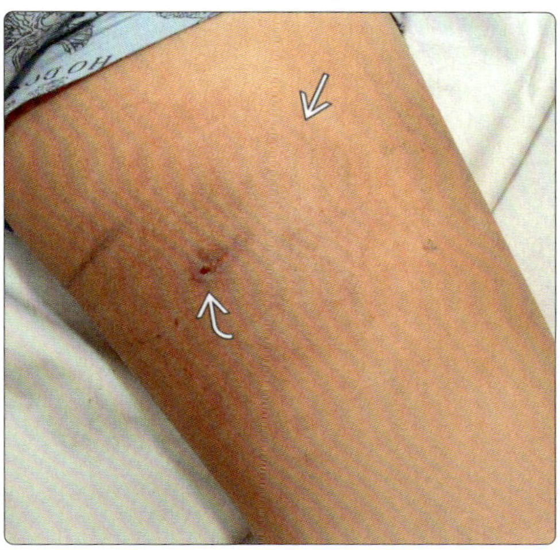
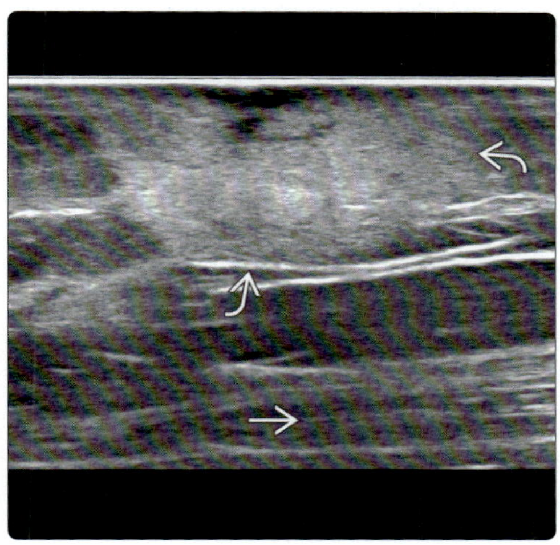

(Left) *Clinical photograph of a 10-year-old boy shows medial thigh swelling ➡ following trauma with a skin penetration ➡ from a bicycle brake handle 4 weeks earlier.* **(Right)** *Longitudinal US of the same region shows a medium-sized (28 mm wide x 37 mm long x 12 mm deep) area of subcutaneous fat necrosis ➡ at the inner thigh. The underlying investing fascia and muscles were normal ➡. The superficial shearing injury is most likely related to the initial penetrating trauma.*

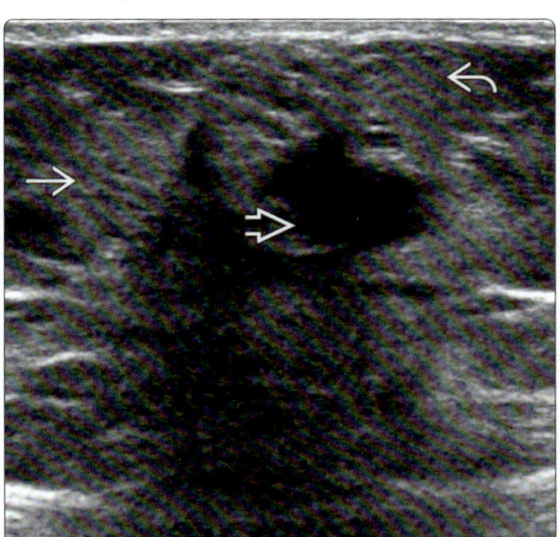
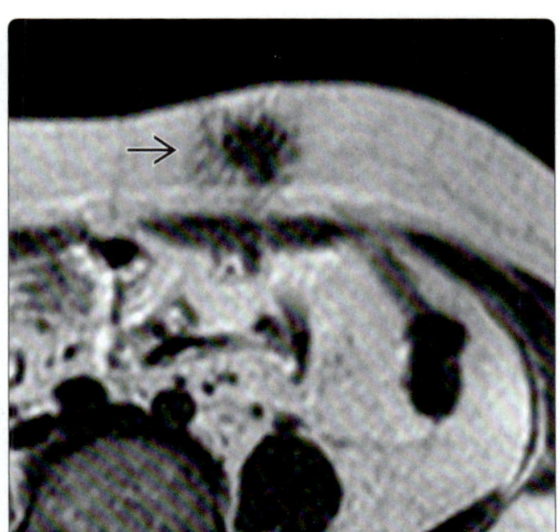

(Left) *Transverse US of a patient with a slowly enlarging posttraumatic mass in the abdominal wall shows a heterogeneous subcutaneous mass ➡. The mass is composed of an ill-defined peripheral hyperechoic component ➡ and a discrete central hypoechoic component ➡. Areas of subcutaneous fat necrosis become better defined and more hypoechoic with increasing chronicity.* **(Right)** *Axial T1 MR of the same lesion shows a spiculated area of subcutaneous fat necrosis ➡.*

(Left) *Transverse US of the calf in a patient with inflammatory arthropathy is shown. There is a 2-cm-wide area of localized dermal ➡ and subcutaneous edema with early fat calcification ➡ in the midcalf. Investing fascia and the muscle are normal. As there was no trauma, panniculitis related to the underlying inflammatory arthropathy was considered more likely than traumatic fat necrosis.* **(Right)** *Color Doppler US shows only minimal hyperemia ➡. Panniculitis can look very similar to fat injury on US.*

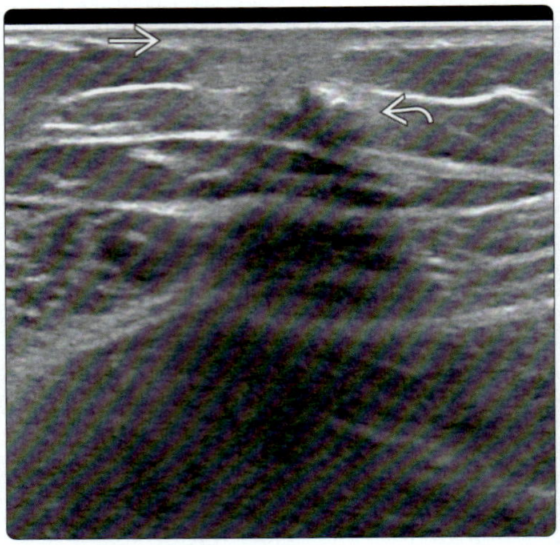

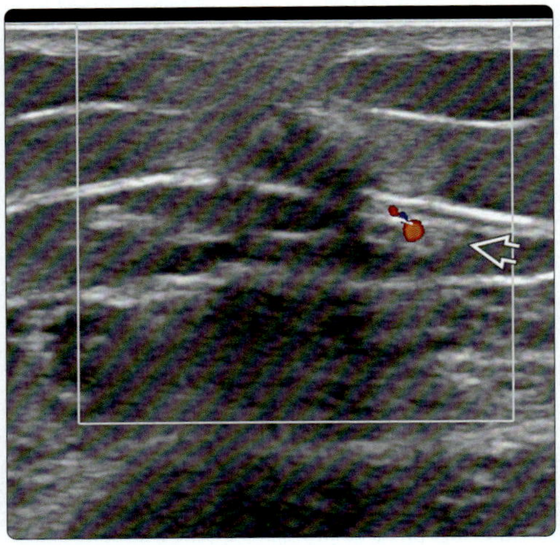

(Left) *Extended field-of-view US of a patient with a painful posttraumatic buttock nodule due to fat necrosis shows an irregular hypoechoic mass ➡ in the subcutaneous fat of the buttock. There is hypoechoic stranding ➡ extending into the surrounding fat with mild posterior acoustic shadowing ➡.* **(Right)** *Axial T2WI FS MR of the same lesion shows a hypointense subcutaneous mass ➡ with a fibrotic hypointense pseudocapsule ➡ and mild perilesional edema ➡.*

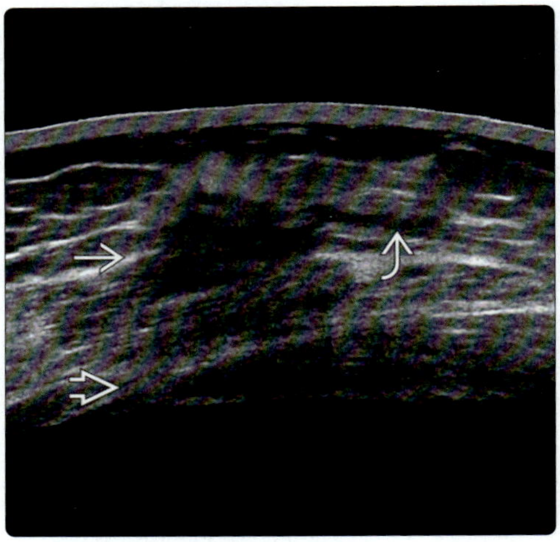

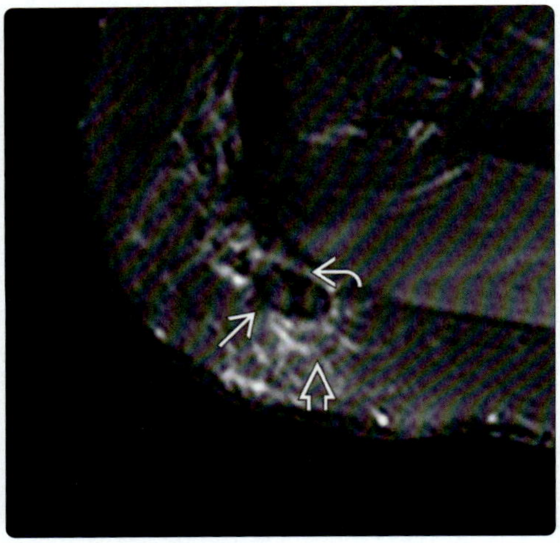

(Left) *Longitudinal US shows a heterogeneous subcutaneous mass in the anterior abdominal wall. The mass is composed of an ill-defined peripheral hyperechoic component ➡, a central hypoechoic fat lobule ➡, and a small area of cystic component ➡.* **(Right)** *Longitudinal color Doppler US of the same lesion shows mild peripheral hyperemia. The degree of hyperemia in noninfected fat necrosis is usually absent or mild.*

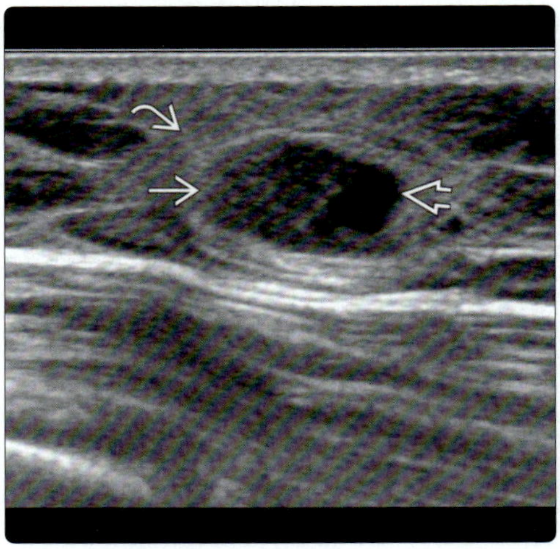

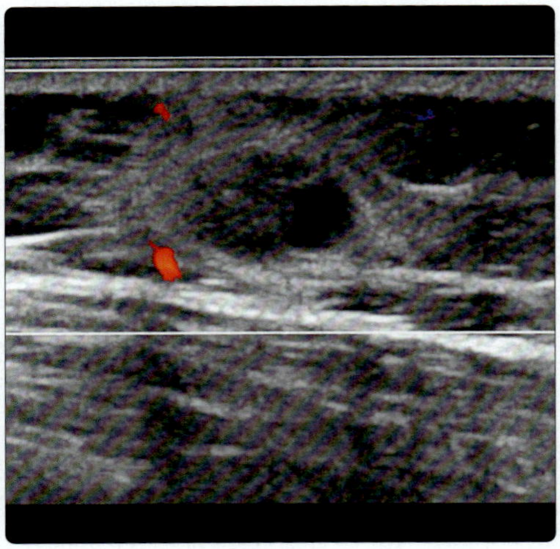

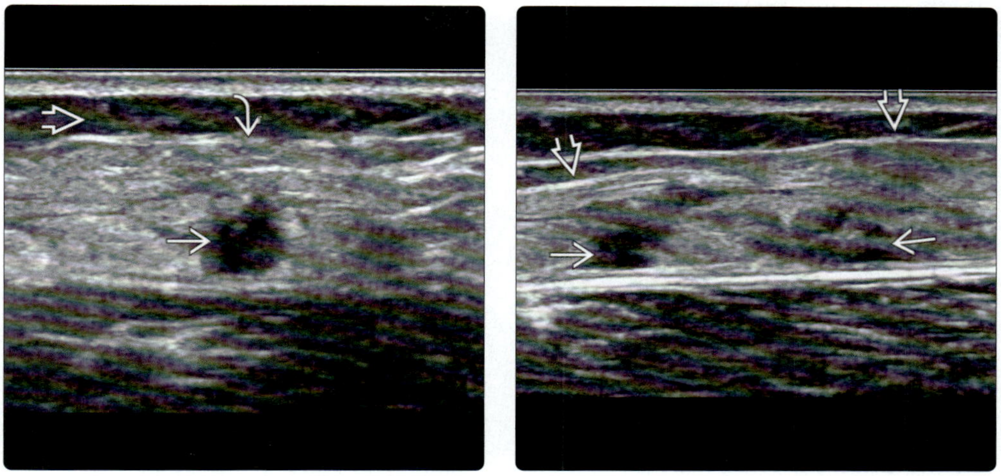

(Left) *Transverse US of the calf in a patient with prior local blunt trauma shows ill-defined hyperechoic subcutaneous fat* ➡ *with a small irregular hypoechoic area* ➡*. Note the relatively normal superficial subcutaneous fat* ➡*.* **(Right)** *Longitudinal US of the same patient shows a large ill-defined area of subcutaneous fat injury* ➡*. The normal lobulation of the subcutaneous fat is indistinct. Multiple small irregular areas of hypoechogenicity* ➡ *representing focal fluid accumulation are present.*

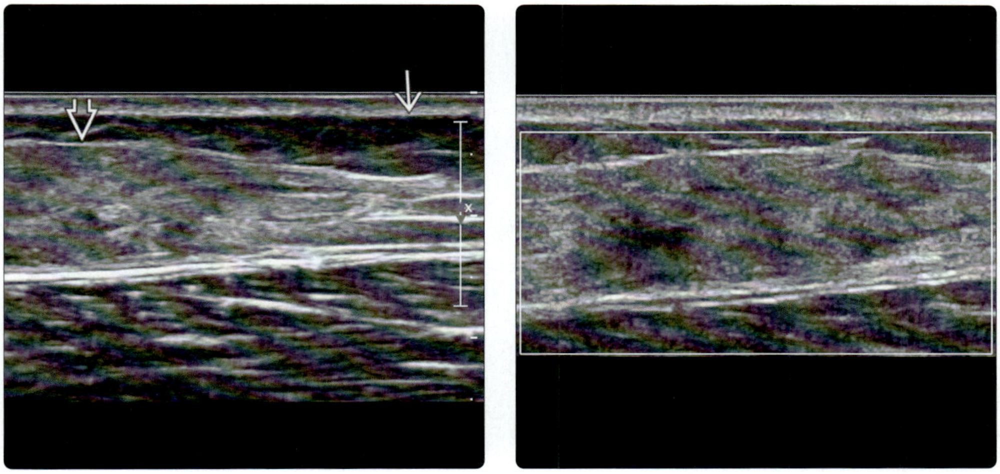

(Left) *Longitudinal US of the same patient taken at a slightly more caudal location shows the distinction between the normal hypoechoic subcutaneous fat* ➡ *and the posttraumatic hyperechoic subcutaneous fat* ➡*.* **(Right)** *Longitudinal US of the same lesion shows no significant hyperemia throughout the whole area of fat injury.*

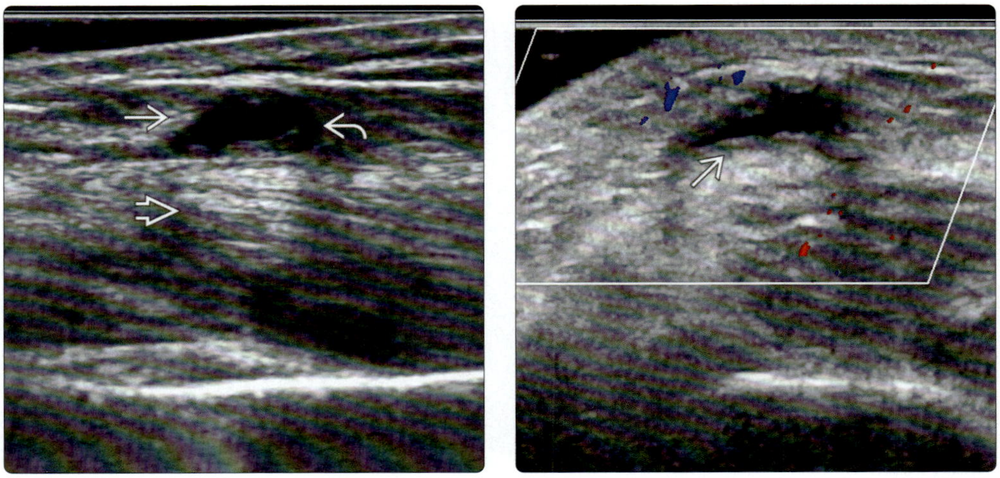

(Left) *Transverse US of the lateral thigh region shows a well-defined, cystic-type, subcutaneous mass* ➡ *with a pseudocapsule* ➡ *and mild posterior acoustic enhancement* ➡*.* **(Right)** *Longitudinal color Doppler US of the same lesion shows no significant hyperemia around this area of cystic, posttraumatic fat necrosis* ➡*. This is a mature lesion with no surrounding inflammation.*

(Left) *Transverse US of a patient with recent history of crush injury to the calf shows an area of subcutaneous fat liquefaction* → *secondary to subcutaneous fat necrosis. The subcutaneous fat is hyperechoic* → *and swollen due to subcutaneous fat edema.* **(Right)** *Longitudinal US of the same lesion shows how the full depth of subcutaneous fat is affected. Mild edema is also present in the medial belly of the gastrocnemius* → *just deep to the investing fascia* →.

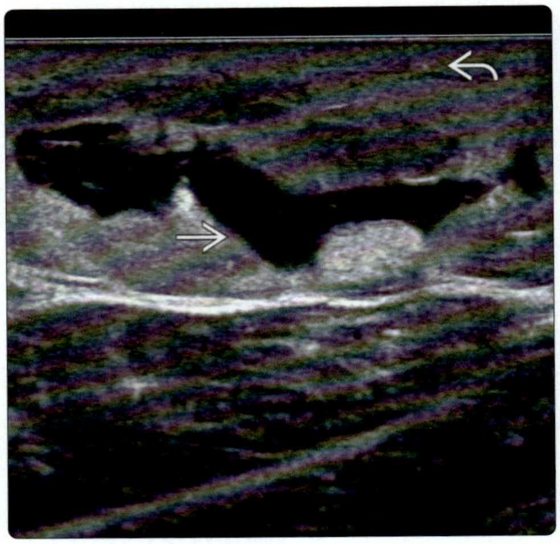

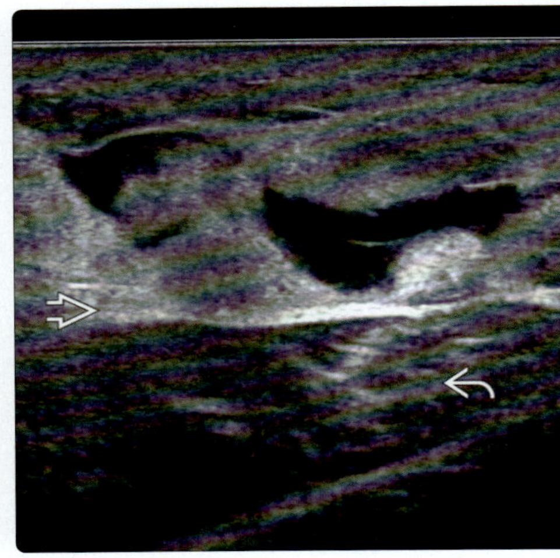

(Left) *Clinical photograph of a patient with previous history of buttock trauma shows severe focal contour depression* → *in the buttock region.* **(Right)** *Extended field-of-view US in the same patient shows a focal area of severe subcutaneous fat atrophy* →. *There is severe thinning of the subcutaneous fat compared to the adjacent normal subcutaneous fat* →. *The underlying ischial tuberosity* → *is normal.*

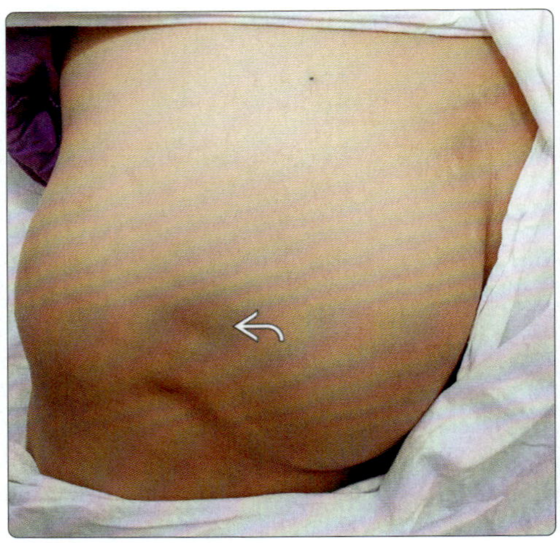

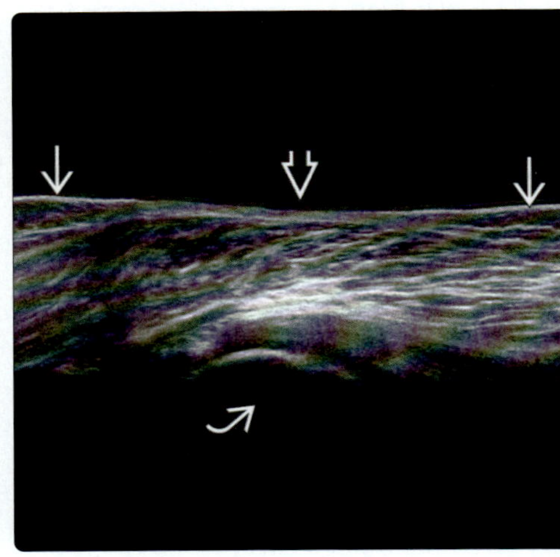

(Left) *Transverse US at the edge of the affected area in the same patient shows moderate thinning of the subcutaneous fat* →. *The muscle layer is normal* →. **(Right)** *Transverse US of the unaffected contralateral buttock in the same patient shows the normal thickness of the subcutaneous fat* → *for comparison. One of the most frequent findings in fat necrosis is localized thinning of otherwise normal subcutaneous fat.*

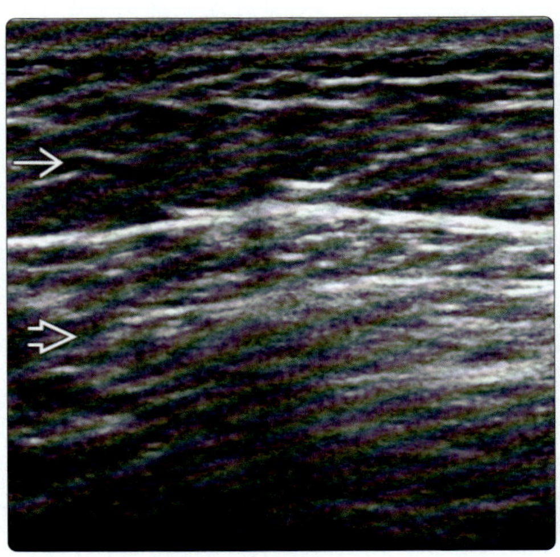

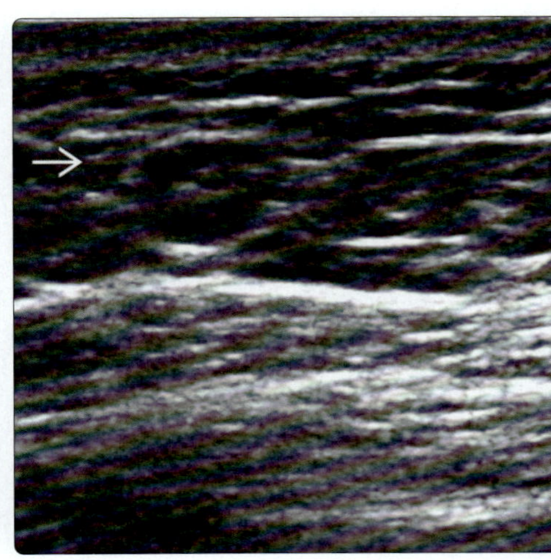

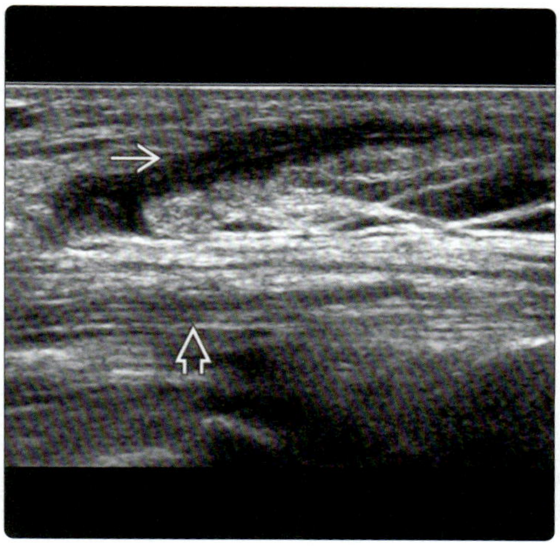

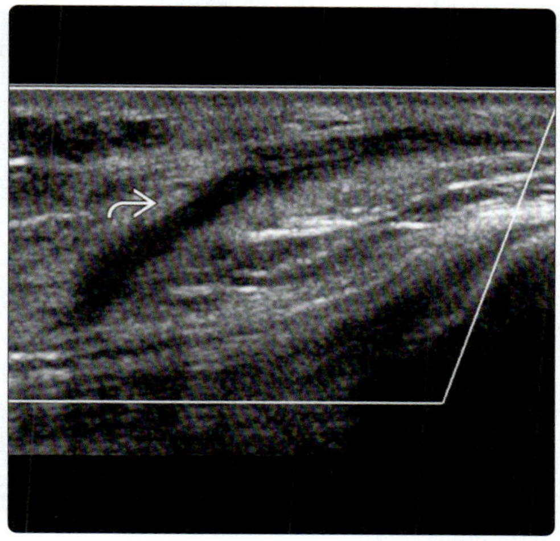

(Left) *Longitudinal US of a patient with subacute trauma to the knee shows a cleavage-type injury to the subcutaneous fat in the suprapatellar region. The quadriceps tendon is normal .* **(Right)** *Longitudinal color Doppler US of the same lesion shows no hyperemia. The linear hypoechoic cleavage space is filled with fluid. This cleavage injury is different to a Morel-Lavallée lesion where subcutaneous fat separates from the investing fascia.*

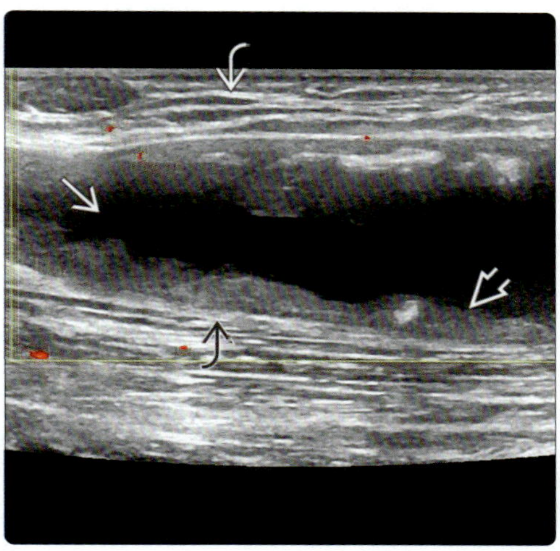

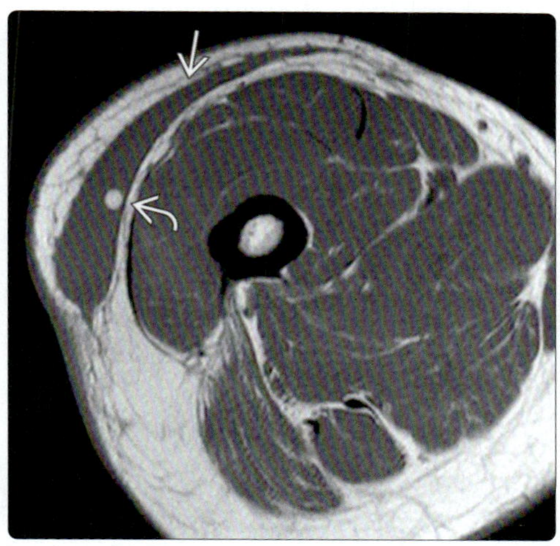

(Left) *Longitudinal color Doppler US of the thigh several weeks after injury shows a Morel-Lavallée lesion with an elongated fluid collection lying between the investing fascia and the subcutaneous fat . The fluid collection has a thick irregular rim of hypoechoic traumatized fat . (Courtesy P. Robinson, MD.)* **(Right)** *Axial T1 MR in a patient with lateral thigh swelling, clinically suspected of sarcoma, shows an elongated subcutaneous fluid collection with a small fat lobule characteristic of a Morel-Lavallée lesion.*

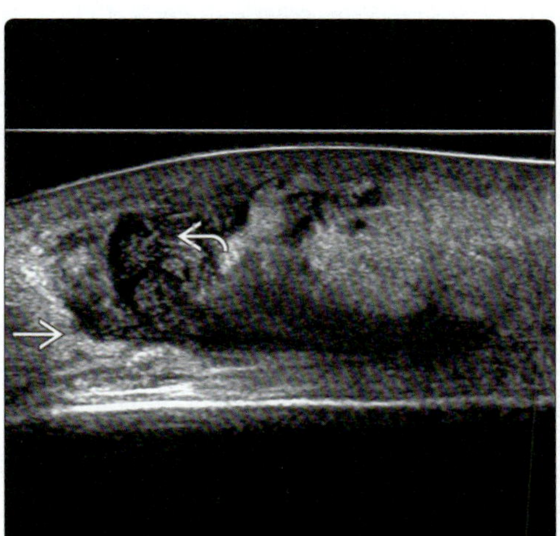

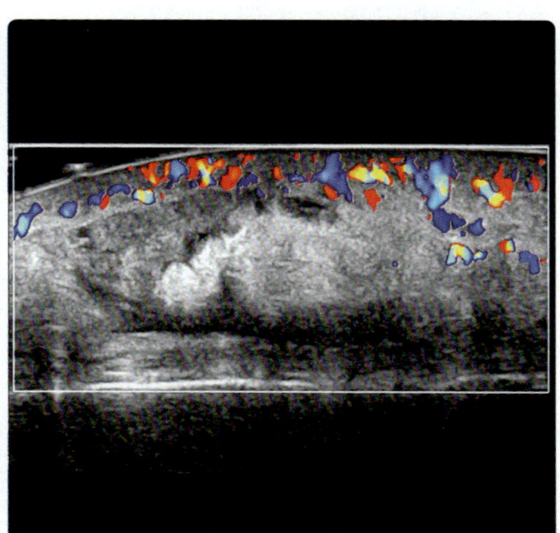

(Left) *Longitudinal US of the shin region in a patient with prior history of trauma to this area, who subsequently developed progressive erythema and swelling, shows a large subcutaneous mass with marked surrounding edema. There are hypoechoic areas within the affected area, but no discrete liquefaction is present.* **(Right)** *Longitudinal color Doppler US of the same lesion shows moderate to severe hyperemia consistent with an infected area of subcutaneous fat necrosis.*

Muscle Infarction

KEY FACTS

IMAGING

- Predilection for involvement of larger muscles in lower limb
  - Thigh (60%) > leg (40%)
  - Thigh: Anterior > posterior > medial compartment
  - Leg: Posterior > anterior = lateral compartment
  - Usually involves several muscles rather than single muscle
  - Bilateral muscle involvement uncommon (10%)
- Focal, ill-defined swelling and hypoechogenicity of affected muscle
  - Effacement and disruption of affected muscular architecture
- ± subfascial fluid
- ± gas within infarcted area
- ± hemorrhage within infarcted area (best seen on MR)
- Subcutaneous tissue edema overlying infarcted area
- Mass-like area appears during reparative phase
  - Muscle echotexture broadly recognizable within mass-like area
- Initially infarcted area iso- or hypovascular relative to surrounding muscle
  - Becomes mildly hypervascular relative to surrounding muscle during reparative phase
- Biopsy for histological confirmation usually not necessary

CLINICAL ISSUES

- Sudden onset of severe localized muscle pain with swelling of affected muscle
- Preexisting advanced diabetes
  - 20% of patients with muscle infarction die within 2 years due to other diabetic complications
- Settles in 5-8 weeks with conservative treatment
- Muscle infarction may also occur in patients with chronic renal failure or severe angiopathy though no diabetes

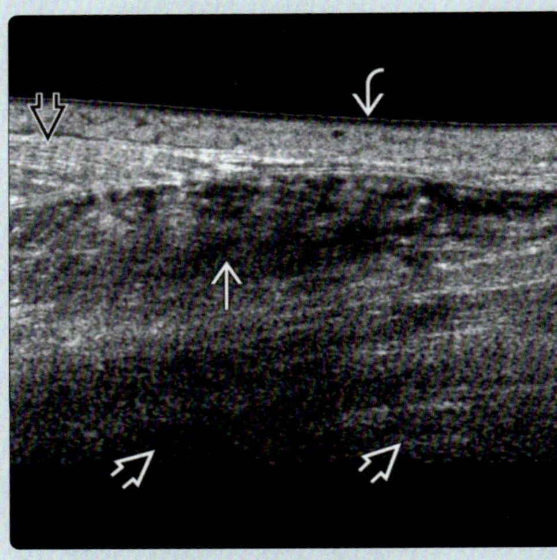

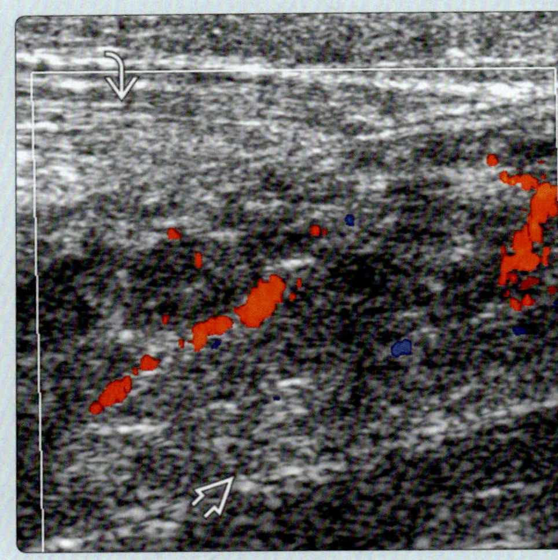

(Left) *Longitudinal US of a diabetic patient shows a swollen soleus muscle ➔ with a more hypoechoic area superficially ➔, indicative of diabetic muscle infarction. There is mild subcutaneous edema ➔. Note the relatively normal gastrocnemius muscle ➔.* **(Right)** *Longitudinal color Doppler US in the same patient shows moderate hyperemia of the affected soleus muscle ➔. The gastrocnemius muscle is not hyperemic ➔.*

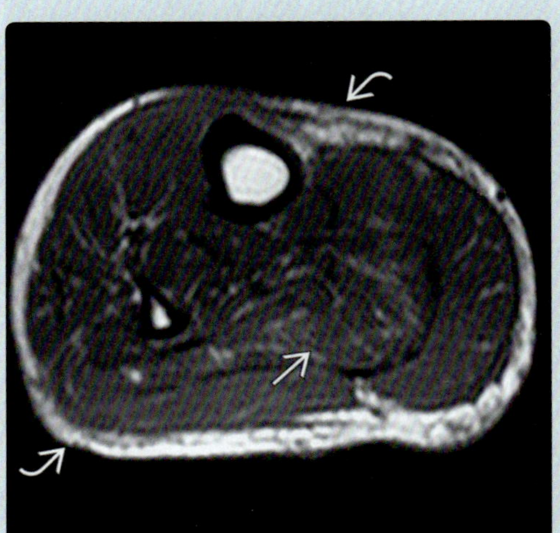

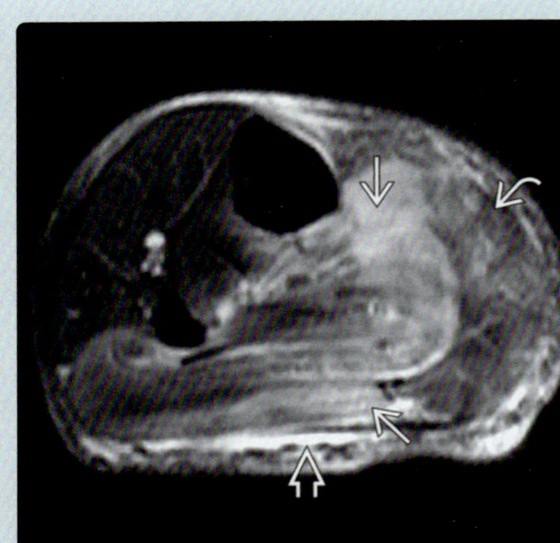

(Left) *Axial T1WI MR in the same patient shows mild, diffuse subcutaneous edema ➔ with mild swelling of the soleus muscle ➔. No muscle hemorrhage is present.* **(Right)** *Axial T2WI FS MR shows moderate to severe edema of the soleus muscle and lateral belly of gastrocnemius ➔. The medial belly of the gastrocnemius muscle is only mildly edematous ➔. Note mild subcutaneous edema with a small subfascial fluid collection ➔. The appearances are compatible with diabetic muscle infarction.*

TERMINOLOGY

Synonyms

- Diabetic muscle infarction, diabetic myonecrosis, aseptic myonecrosis, ischemic myonecrosis

Definitions

- Muscle infarction leading to muscle cell death

IMAGING

General Features

- Best diagnostic clue
  - Focal muscle edema in susceptible diabetic patient
- Location
  - Deep within affected muscle
  - Usually involves several muscles rather than single muscle
  - Particularly larger muscles of lower limb
  - Thigh (60%) > leg (40%)
    - Thigh: Anterior compartment (100%), posterior compartment (65%), medial compartment (55%)
    - Leg: Posterior compartment (90%), anterior compartment (20%), lateral compartment (20%)
  - Gluteal muscle also commonly affected
  - Upper limb & abdominal muscle uncommonly affected
  - Bilateral muscle involvement uncommon (10%)
- Size
  - Variable, though typically 3-5 cm
- Morphology
  - Initially, affected muscle swollen
  - Later, firm muscle mass may develop at affected site

Radiographic Findings

- Radiography
  - Arterial calcification feature of diabetic arteriopathy as well as renal disease and arteriosclerosis

MR Findings

- T1WI
  - Swollen, isointense affected muscle
  - ± focal hyperintensity within affected muscle due to hemorrhage (40%)
    - Hemorrhage within affected area better seen on MR than US
- T2WI FS
  - Focally edematous muscle
    - Degree of muscle edema more pronounced on MR than US
    - ± discrete mass at site of infarction
    - ± subfascial fluid collection (60-100%)
  - Subcutaneous edema (100%)
- T1WI C+
  - Generally not advocated due to risk of nephrogenic systemic fibrosis
  - Diffuse or rim-enhancement pattern
    - Not beneficial in predicting outcome
  - Most diagnostic information available from noncontrast examination

Ultrasonographic Findings

- Grayscale ultrasound
  - Focal, ill-defined swelling and hypoechogenicity of affected muscle
  - Effacement and disruption of affected muscular architecture if severe
    - Intrinsic muscle architecture remains apparent in milder cases
  - ± subfascial fluid superficial to infarcted muscle area
    - Can be gravity dependent so less readily appreciated on US than MR
  - ± gas within infarcted area, seen as echogenic foci with reverberation artifact
  - Subcutaneous tissue edema overlying infarcted area
    - Varies according to extent, severity, and duration of infarction
    - Degree of subcutaneous edema may provide clue to duration of infarction
  - Infarcted area may resolve or evolve into medium-sized, discrete, well-marginated, hypoechoic mass
    - Mass appears during reparative phase; not feature of acute phase
    - Echotexture of muscle broadly preserved within mass-like area
    - Mass slowly resolves, tending to become better defined and firmer with time
  - Calcification not feature
  - US can exclude deep venous thrombosis
- Color Doppler
  - Initially, affected area iso- or hypovascular relative to surrounding muscle
    - Becomes mildly hypervascular relative to surrounding muscle after 2-3 weeks

Imaging Recommendations

- Best imaging tool
  - US to help confirm diagnosis of diabetic muscle infarction and exclude likely differential diagnoses
- Protocol advice
  - Perform early examination followed by serial examinations at intervals of 3-4 weeks for 9-12 weeks to observe changing nature of lesion
  - Routinely compare vascularity of affected muscle with contralateral unaffected muscle
  - Biopsy for histological confirmation usually not necessary
    - Biopsy indicated only for atypical presentation, equivocal findings, or if not settling as expected over time
    - Aspiration of fluid for culture may be helpful if secondary infection suspected

DIFFERENTIAL DIAGNOSIS

Muscle Abscess

- Severe pain and tenderness ± fever
- Predominant fluid component
- Deteriorates and becomes more liquefied with time
- Purulent fluid on aspiration

Infective or Inflammatory Myositis

- Severe pain and tenderness ± fever
- Diffuse muscle edema and hyperemia
- Progresses rather than regresses with time

Deep Venous Thrombosis

- Noncompressibility of affected vein
- Visible thrombus within vein
- Poor phasicity on spectral analysis

PATHOLOGY

General Features

- Etiology
  - Probably caused by microvascular occlusion due to arteriosclerosis obliterans, thrombus, or embolus
  - Hypoxia: Reperfusion injury and hypercoagulable state may also be contributory
- Associated abnormalities
  - Longstanding, insulin-dependent diabetes
  - Poor control of glycemia
  - Presence of microvascular diabetic complications (neuropathy, retinopathy, nephropathy)

Gross Pathologic & Surgical Features

- Swollen, nonhemorrhagic, pale-looking muscle

Microscopic Features

- Muscle necrosis (myonecrosis) and edema
  - Phagocytosis of necrotic muscle fibers
- Acute or chronic inflammatory infiltrate
  - Greater if secondary infection
- Granulation tissue repair with fibrosis
- Small arteries of infarcted area may have thickened, hyalinized walls and be occluded with fibrin clot or calcium fragments

CLINICAL ISSUES

Presentation

- Most common signs/symptoms
  - Sudden severe localized muscle pain and swelling
  - Preexisting advanced diabetes
- Other signs/symptoms
  - Often delay between onset of symptoms and presentation (average 4 weeks)
  - Resolution of swelling with appearance of painful palpable mass (33%) over 3 weeks
  - Fever (10%)
  - Secondary infection of affected muscle
- Clinical profile
  - Mean duration of diabetes prior to diabetic muscle infarction is 15 years
    - Insulin-dependent diabetics (30%), noninsulin-dependent diabetics (30%), noninsulin-dependent diabetics on insulin regime (30%)
  - Accompanying diabetic complications common
    - Nephropathy (70%)
    - Retinopathy (60%)
    - Neuropathy (60%)
    - Peripheral vascular disease (10%)
    - Hypertension (10%)
  - Laboratory tests inconclusive
    - Elevated creatine kinase (50%)
      - Particularly if obtained soon after onset of symptoms
      - Will normalize in ~ 1 week
    - Elevated white cell count (10%)
    - Elevated C-reactive protein (50%)
    - Elevated erythrocyte sedimentation rate (50%)
  - Muscle infarction may also occur in patients with chronic renal failure or severe angiopathy though no diabetes

Demographics

- Age
  - Range: 19-81 years (mean: 42 years)
    - Insulin-dependent diabetics affected at younger age than noninsulin-dependent diabetics
- Gender
  - Slightly more common in diabetic women (60%)
- Epidemiology
  - Diabetes mellitus affects 5% of population
  - Muscle infarction uncommon complication of diabetes

Natural History & Prognosis

- Settles in ~ 5-8 weeks with conservative treatment
- ~ 1/3 of patients have 2nd episode of muscle infarction
  - Most commonly involves different muscle (85%) rather than same muscle (15%)
- Diabetic muscle infarction poor prognostic sign
  - 20% of patients die within 2 years due to other diabetic complications

Treatment

- Conservative
  - Bed rest and non-weight-bearing
    - Leg elevation
  - Analgesia and symptomatic treatment
  - Gradual resumption of activity as pain resolves
  - Careful metabolic control of diabetes
  - Gradually resolves spontaneously
    - Nonsteroidal antiinflammatory agents
    - Antiplatelet therapy if hypercoagulable state
    - Steroids if antiphospholipid antibodies present
- Complications
  - Biopsy may induce bleeding into infarcted area
  - Secondary infection of infarcted muscle (uncommon)

DIAGNOSTIC CHECKLIST

Consider

- Diabetic muscle infarction in any patient with known diabetic complication presenting with muscle pain

Image Interpretation Pearls

- Diagnosis based on recognition of typical US findings in at-risk patient
- Serial examination showing evolving US changes will allow definitive diagnosis without need for percutaneous biopsy

SELECTED REFERENCES

1. Baker JC et al. Diabetic musculoskeletal complications and their imaging mimics. Radiographics. 1959-74; 32(7), 2012

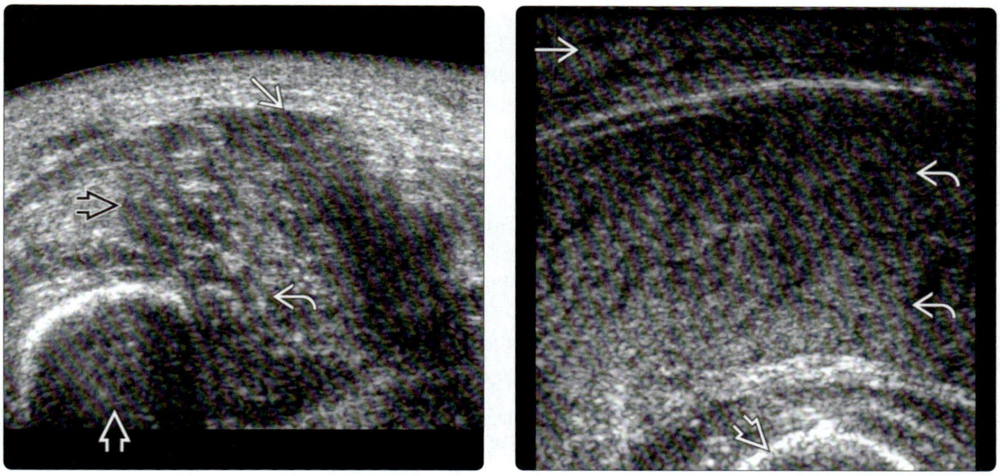

(Left) *Transverse US in a diabetic patient shows a well-defined, swollen, hypoechoic area* → *of vastus lateralis muscle infarction. The small echogenic foci* → *are possibly due to gas. There is a more hypoechoic edematous area laterally* →*. The femoral shaft is normal* →*.* **(Right)** *Transverse US in a diabetic patient shows a diffusely swollen vastus intermedius muscle* → *with subcutaneous edema* →*, consistent with recent muscle infarction. The femoral shaft* → *is shown.*

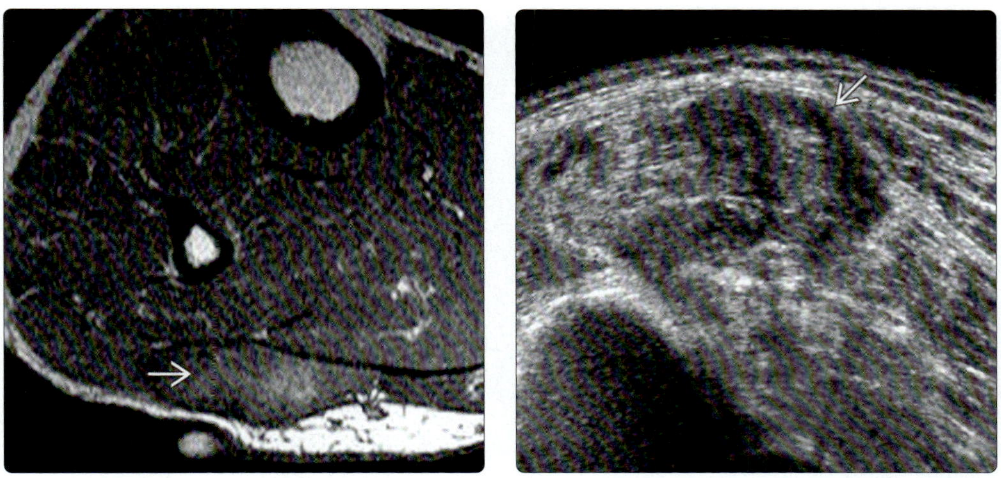

(Left) *Transverse T1WI MR in a diabetic patient shows an ill-defined area of hyperintensity* → *within the mildly swollen lateral belly of the gastrocnemius muscle due to hemorrhage within an area of muscle infarction.* **(Right)** *Transverse US in a diabetic patient shows a well-defined, hypoechoic area* → *due to peroneal muscle infarction. The muscle architecture can still be appreciated within the infarcted area, i.e., no discrete muscle mass is present. The proximal fibula is normal.*

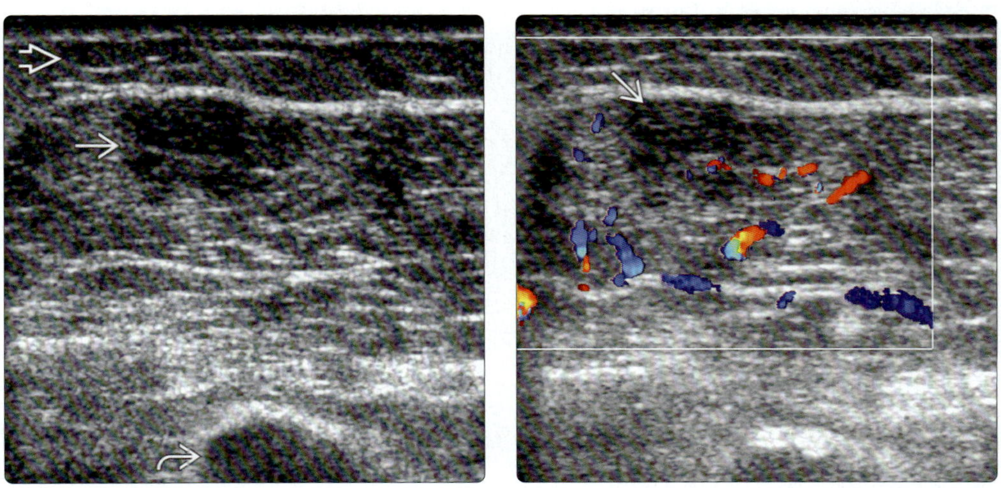

(Left) *Transverse US in a diabetic patient shows a well-defined, hypoechoic area* → *with broadly preserved background muscle architecture due to resolving biceps brachii muscle infarction. There is no subcutaneous edema* →*. The humeral shaft is normal* →*.* **(Right)** *Transverse color Doppler US at the same location (same patient) shows moderate hyperemia around the infarcted hypoechoic area* →*. In a typical patient, these appearances require no additional imaging.*

Muscle Injury

KEY FACTS

TERMINOLOGY

- **Contusion**: Direct blunt compressive trauma to muscle without definable tear
- **Tear**: Discontinuity of muscle fibers typically at myofascial or myotendinous junction
- **Hernia**: Focal protrusion of part of muscle through epimysium
- **DOMS**: Overuse → muscle edema without definable tear
  - Pain peaking 1-3 days after overuse

IMAGING

- **Muscle contusion**
  - Ill-defined hyperechogenicity of muscle or muscles
  - No discrete muscle tear (fiber discontinuity)
- **Muscle tear**
  - Discontinuity ± retraction of muscle fibers
  - Tear filled with hematoma or fluid
  - ± mild muscle edema or hemorrhage
  - Estimate what actual percentage of muscle fibers torn
  - Grade 1: < 5% muscle fibers torn
  - Grade 2: 5-99% muscle fibers torn
  - Grade 3: 100% muscle fibers torn
- **Muscle hematoma**
  - Typically associated with grade 2 or 3 tears
  - Appearance varies with duration of hematoma, severity, and location
- **DOMS**
  - Affected muscle may be normal or diffusely hyperechoic due to diffuse muscle edema
- **Muscle hernia**
  - Focal protrusion of muscle through epimysium
  - Accentuated by muscle contraction or exercise

CLINICAL ISSUES

- Most muscle injuries improve with conservative treatment
  - Serial US examination not shown to be useful for guiding speed of rehabilitation
- Functional recovery precedes structural recovery

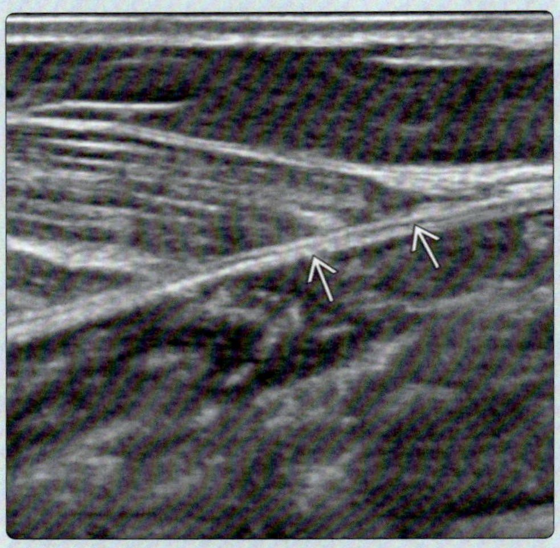

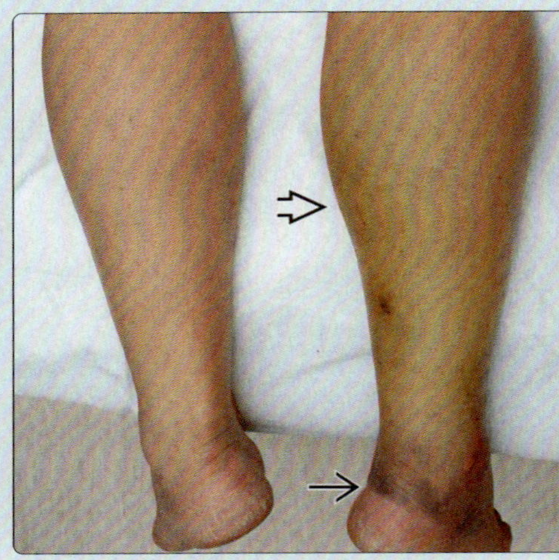

(Left) *Longitudinal US of an unaffected leg shows normal sharp angulation of the myotendinous and myofascial junction ➡ at the distal end of the medial belly of gastrocnemius muscle. This is the most common site of calf muscle tears.* **(Right)** *Clinical photograph shows a patient with a recent calf muscle tear. There is moderate calf swelling and bruising ⇨ with more intense, gravity-related bruising at the ankle region ➡.*

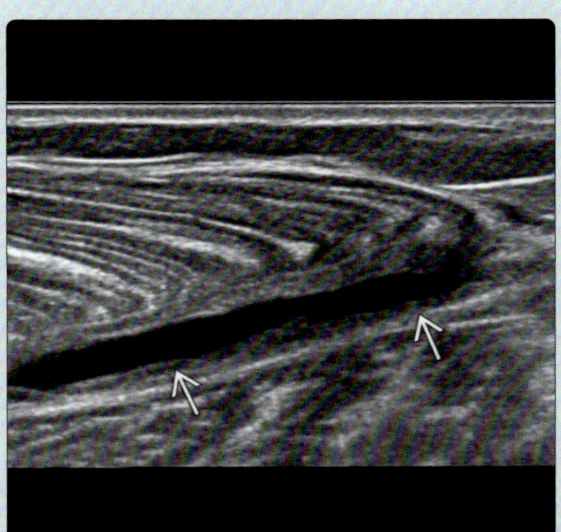

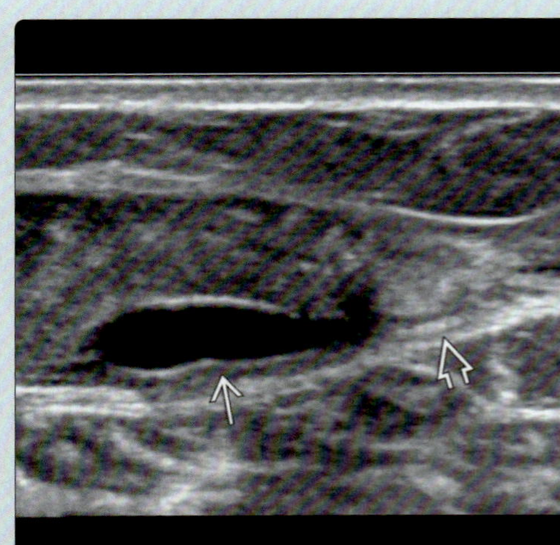

(Left) *Longitudinal US of a symptomatic leg shows a myofascial junction tear ➡ at the distal end of the medial belly of gastrocnemius muscle. The medium-severity tear is filled with fluid. No muscle retraction is present.* **(Right)** *Transverse US of the same area shows the moderate-severity myofascial tear ➡. Overall, ~ 15% of the cross-sectional area (CSA) of the muscle insertion is torn. The plantaris tendon ➡ is normal.*

TERMINOLOGY

Synonyms

- Muscle tear, muscle contusion, muscle hematoma, muscle strain, "pulled muscle," "tennis leg," delayed-onset muscle soreness (DOMS), muscle hernia

Definitions

- Tear: Discontinuity of muscle fibers typically at myofascial or myotendinous junction
- Contusion: Direct blunt compressive trauma to muscle without definable tear
- Hernia: Focal protrusion of part of muscle through epimysium
- DOMS: Overuse causing muscle edema without definable tear
  - Pain peaking 1-3 days after overuse

IMAGING

General Features

- Best diagnostic clue
  - Muscle tear; hypoechoic area within muscle or at myotendinous junction
- Location
  - Usually affects lower limb muscles (gastrocnemius > rectus femoris, adductor longus, biceps femoris, semimembranosus, semitendinosus, and soleus muscles)
  - Typically at myofascial junction within muscle or myotendinous junction
    - These are weak areas at interface between tissues of different elastic property
- Morphology
  - Muscle fiber discontinuity with local hemorrhage and edema

MR Findings

- T1WI
  - Muscle tears
    - Low to intermediate signal intensity hemorrhage
    - ± hyperintense subacute hemorrhage
    - Peripheral hypointense hemosiderin
    - Loss of normal muscle striations ± retracted muscle belly
  - DOMS
    - Muscle swelling of intermediate signal intensity
  - Muscle hernia
    - Focal defect in investing fascia ± muscle protrusion
- T2WI
  - Muscle tear
    - Hyperintense edema in affected muscle
    - Inhomogeneous signal intensity mass in midsubstance tears due to hemorrhage and edema
    - Perimuscular hyperintense fluid
  - DOMS
    - Diffuse homogeneous hyperintensity
- T2* GRE
  - Susceptibility artifact due to hemorrhage

Ultrasonographic Findings

- **Muscle contusion**
  - Direct injury
  - Ill-defined hyperechogenicity of muscle or muscles due to edema ± hemorrhage
  - No discrete muscle tear (fiber discontinuity)
  - Hyperemia during reparative phase
- **Muscle tear**
  - Usually indirect injury
    - Typically at myofascial or myotendinous junction
  - Grade 1 injury (Peetrons 2002)
    - Muscle injury with either no visible tear or small tear involving few muscle fibers (< 5% muscle fibers torn)
    - Muscle either normal, focal, or diffuse hyperechogenicity of muscle
  - Grade 2 injury
    - Partial (5-99%) muscle fibers torn
    - Discontinuity (tear) ± retraction of muscle fibers
      - Tear filled with hematoma or fluid
      - ± mild to moderate muscle edema
  - Grade 3 injury
    - Complete tear of muscle at musculotendinous junction usually with retraction
- **Healing muscle tear**
  - Bridging scar tissue or regenerated muscle at site of tear
  - Clearer definition of injured area
  - Localized hyperemia, which will disappear when healing is complete
- **Muscle hematoma**
  - Typically associated with grade 2 or 3 tears
  - Appearance varies with duration, severity, and location of hematoma
  - During first 48 hours, initially hypoechoic then becomes more hyperechoic as hematoma solidifies
    - ± acoustic enhancement
  - After 72 hours, hematoma can start to liquefy
    - Liquefaction seen as anechoic areas, starting centrally in hematoma
  - Marginal hyperemia during reparative phase (reparative fibrovascular granulation tissue)
    - Peripheral granulation tissue usually hyperechoic and hyperemic
    - Granulation tissue is more hyperechoic than fresh hematoma
- **DOMS**
  - Affected muscle may be normal or, if severe, diffusely hyperechoic due to diffuse muscle edema
  - No discernible muscle tear
- **Muscle hernia**
  - Focal protrusion of muscle fibers through epimysial fascial defect
  - Accentuated by muscle contraction or exercise

Imaging Recommendations

- Best imaging tool
  - US since most muscle tears are accessible
    - MR preferable in pelvitrochanteric muscles, adductor or hamstring insertions as less accessible with US
- Protocol advice
  - Examine with muscle at rest and compare with contralateral side
  - Examine all tears in 2 planes

- Measure dimensions of tear and try to gauge what percentage of muscle or musculotendinous junction is torn
  - US during muscle contraction can help accentuate muscle retraction
- Gastrocnemius muscle tear: Look closely at distal end of myotendinous junction medial belly
  - ± comparison with opposite side
  - Check for integrity of plantaris tendon
    - Seen medial to Achilles tendon and deep to medial belly gastrocnemius muscle
- Muscle hernia: Reexamine after exercise if hernia is not apparent on initial examination

DIFFERENTIAL DIAGNOSIS

Tendon Tear

- Quadriceps, patella, or Achilles tendon

Soft Tissue Tumor

- May mimic muscle tear with palpable mass
- Healing muscle tears have recognizable background muscle architecture and less mass effect
- Peripheral hyperemia of healing tear as opposed to central hyperemia of tumor

PATHOLOGY

General Features

- Muscle tear
  - Extrinsic tear from direct external trauma
    - Commonly affects rectus femoris
  - Intrinsic tear from forced contraction of elongated muscle
    - Usually located at myofascial junction
    - Most common type of muscle injury
- DOMS: Reversible structural damage at cellular level due to unaccustomed severe repetitive exertion

Staging, Grading, & Classification

- Several other grading/classification systems
  - Munich Consensus Statement (2013)
  - British Athletics Muscle Injury Classification (2014)

Microscopic Features

- Destructive phase: 0-3 days
  - Hematoma with area of tear (central zone)
  - Surrounding (2- to 3-mm) myofibril phagocytosis and denervation
- Early remodeling phase: 3-12 days
  - Resorption of hematoma
  - Bridging of tear with fibrous tissue or myofibrils and hyperemia
- Late remodeling phase: 12 days to 9 months
  - Scar formation ± myofibril formation

CLINICAL ISSUES

Presentation

- Most common signs/symptoms
  - Tear and hematoma: Sudden onset of limb pain at time of exertion; "pop" sensation
    - Localized swelling and tenderness ± discrete mass
    - Complete rupture: Palpable defect with retracted mass
  - DOMS: Swelling and pain beginning 1-2 days after exercise, peaking at 3 days, and resolving by day 7
  - Muscle hernia: Chronic pinpoint pain and tenderness
    - Usually has well-developed musculature

Demographics

- Age
  - Young to middle-aged athlete
    - In children, apophyseal fractures predominate
    - In older patients, tendon tears predominate
- Epidemiology
  - Muscle injury: 30% of sports-related injuries
  - Risk factors
    - Eccentric contraction force
    - Muscles that cross more than 1 joint
    - Muscles that comprise mainly type II fibers

Natural History & Prognosis

- Most muscle injuries improve with conservative treatment
- Serial US has not been shown to be useful for guiding speed of rehabilitation
- Grade 1: 1-2 weeks recovery time
- Grade 2: < 50% muscle fibers torn; 3-4 weeks recovery time
- Grade 2: > 50% muscle fibers torn; 6-8 weeks recovery time
- Grade 3: 6-8 weeks or longer recovery time
- Hamstring tears, rectus femoris central tendon tears, and retears at site of previous tear tend to take longer to heal
- Functional recovery precedes structural recovery

Treatment

- Conservative
  - Rest, ice, compression, elevation
  - Physical therapy: Stretching and exercises
  - Nonsteroidal antiinflammatory drugs
  - Gradual increase in activity
- Surgery
  - Decompression for compartment syndrome
  - Hematoma evacuation
  - Suturing of some complete muscle tears
  - Repair of investing fascia defect (if muscle hernia)
  - Resection of fibrotic mass

DIAGNOSTIC CHECKLIST

Consider

- Muscle hernia in any muscular patient with pinpoint muscle pain and tenderness

Image Interpretation Pearls

- Pay particular attention to distal myofascial junction of medial head of gastrocnemius for small tears

SELECTED REFERENCES

1. Guermazi A et al: Imaging of muscle injuries in sports medicine: sports imaging series. Radiology. 282(3):646-663, 2017
2. Patel A et al: British athletics muscle injury classification: a reliability study for a new grading system. Clin Radiol. 70(12):1414-20, 2015
3. Mueller-Wohlfahrt HW et al: Terminology and classification of muscle injuries in sport: the Munich consensus statement. Br J Sports Med. 47(6):342-50, 2013
4. Peetrons P: Ultrasound of muscles. Eur Radiol. 12(1):35-43, 2002

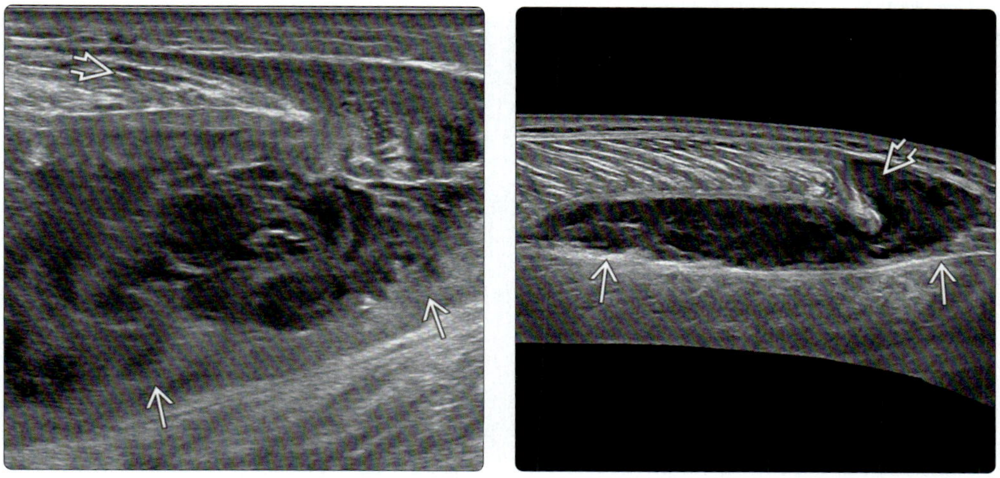

(Left) *Longitudinal US of an adult patient who sustained a calf muscle injury the previous day shows a large tear of the myofascial and myotendinous junction* ➔ *at the distal end of the medial belly of gastrocnemius muscle* ➔. *The muscle is retracted with the retraction gap filled with hematoma.* **(Right)** *Longitudinal extended field-of-view US shows the full extent of the tear* ➔. *Overall, ~ 70% of the CSA of the insertion of the medial belly of gastrocnemius muscle is torn with moderate muscle retraction* ➔.

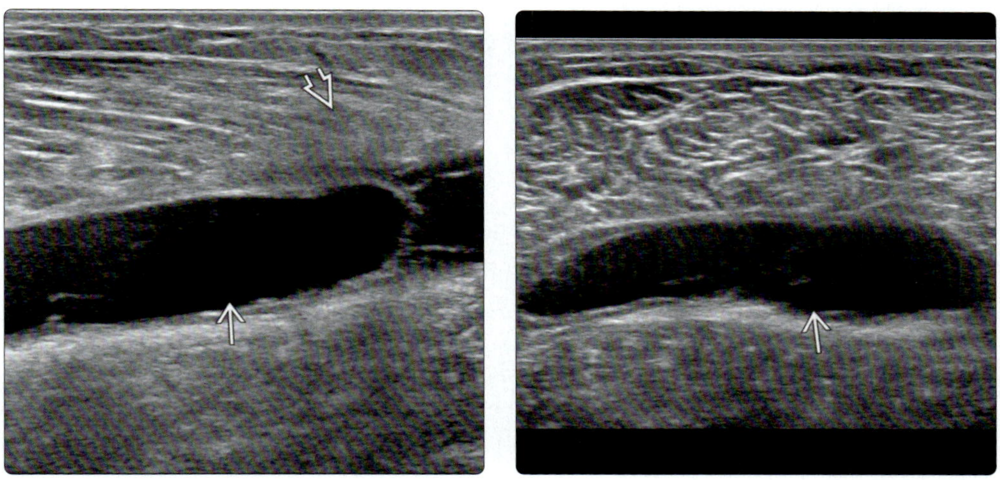

(Left) *Longitudinal US 1 week following calf muscle injury shows a large liquefied tear with anechoic fluid filling the myofascial junction at the distal end of the medial belly of gastrocnemius muscle* ➔. *The muscle is moderately edematous* ➔. **(Right)** *Transverse US of the same area shows the moderate-severity myofascial tear* ➔ *filled with liquefied-type hematoma. Overall, ~ 40% of the CSA of the muscle insertion is torn.*

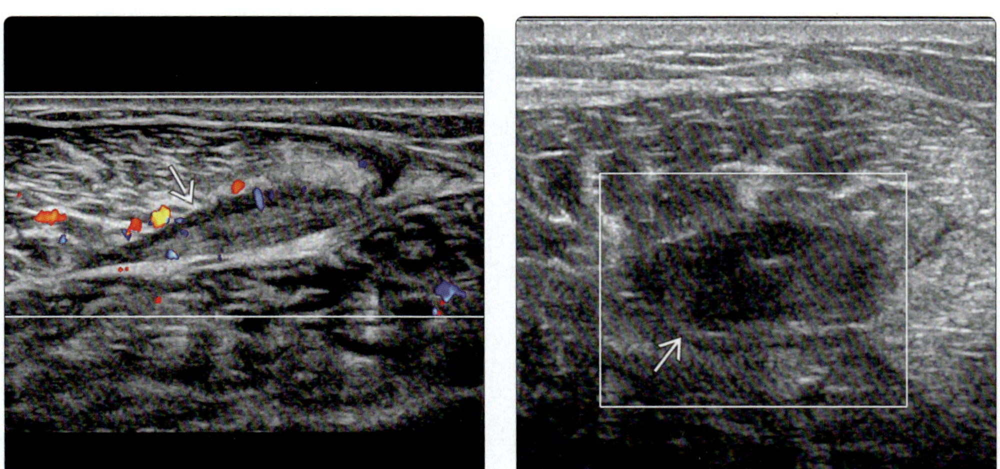

(Left) *Longitudinal color Doppler US shows a healing gastrocnemius myofascial tear with mild to moderate marginal hyperemia* ➔. *Such features of healing can persist for many months following injury and do not necessarily prevent return to play.* **(Right)** *Transverse color Doppler US of a healing gastrocnemius myofascial tear shows how the tear is filled with organizing, slightly echogenic hematoma and is not hyperemic* ➔. *The patient was symptom free. This reparative change can exist for many months following muscle tear.*

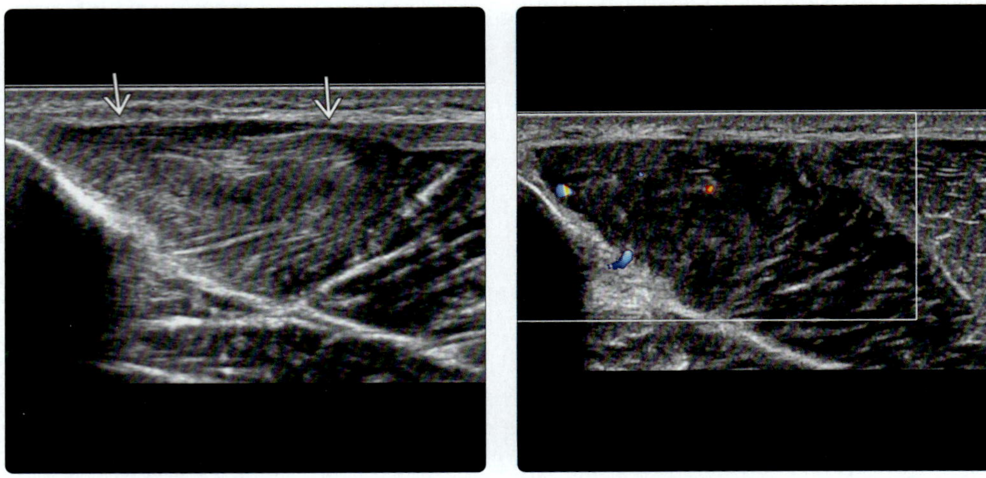

(Left) *Transverse US in a patient with recent blunt trauma to the shin shows a medium-sized area of muscle contusion ➡ with echogenic hemorrhage and edema at the proximal aspect of the tibialis anterior muscle. No discrete muscle tear is evident.* **(Right)** *Transverse color Doppler US of the same area shows mild hyperemia. This is a grade 1 muscle injury.*

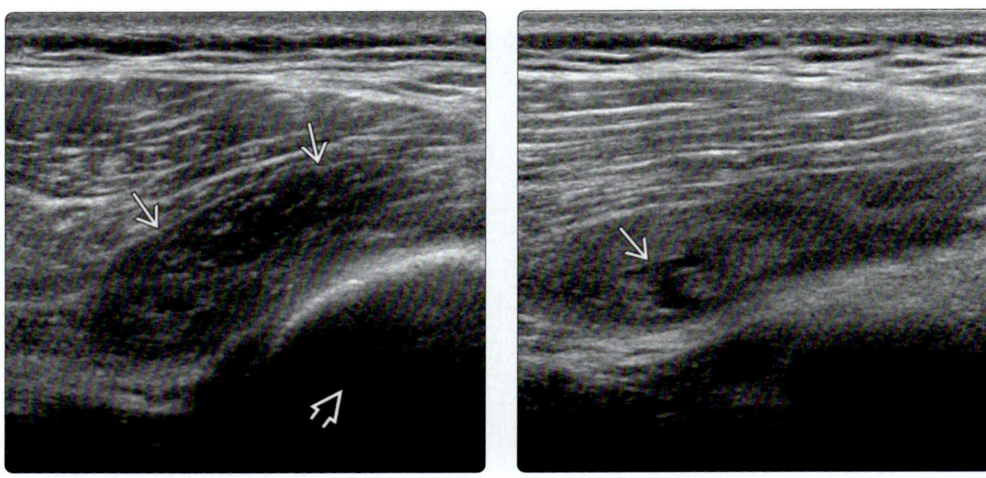

(Left) *Longitudinal US in patient with recent blunt trauma to the lateral hip region shows a medium-sized area of muscle contusion ➡ with edema at the deeper part of the gluteus maximus muscle overlying the posterior edge of the greater trochanter ⇨.* **(Right)** *There is a small tear ➡ within the contused area of the muscle. No hypervascularity is present (not shown). This is consistent with a grade 1 muscle tear.*

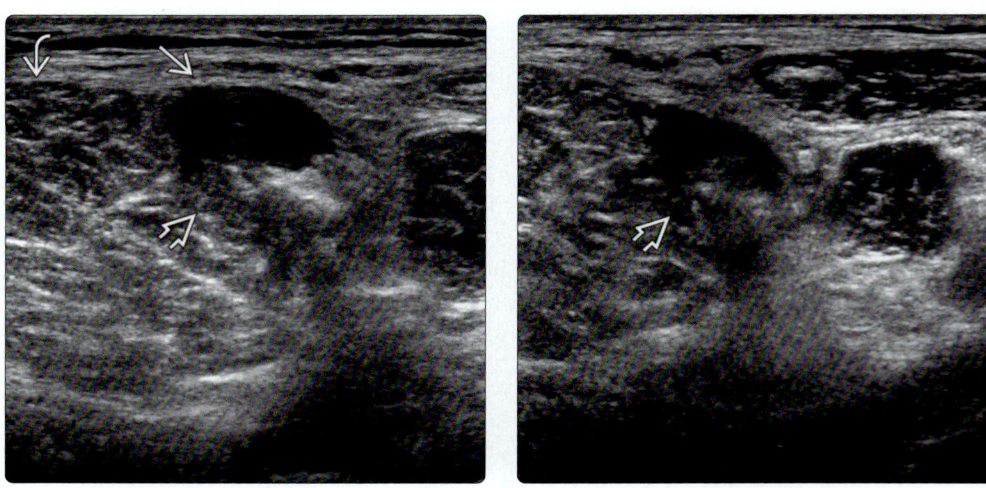

(Left) *Transverse US of the thigh in an elderly patient with blunt thigh trauma and bruising shows a small hematoma ➡ in the proximal medial portion of the tensor fascia lata muscle ⇨ at the anterolateral aspect of the upper thigh. There is adjacent muscle fiber discontinuity due to a small grade 1 muscle tear ⇨.* **(Right)** *Transverse US close to the same location shows a partial tear ⇨ involving only ~ 5% of the tensor fascia lata muscle CSA.*

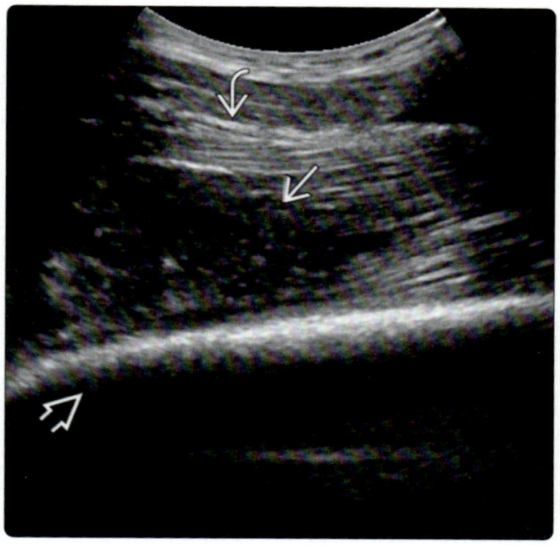

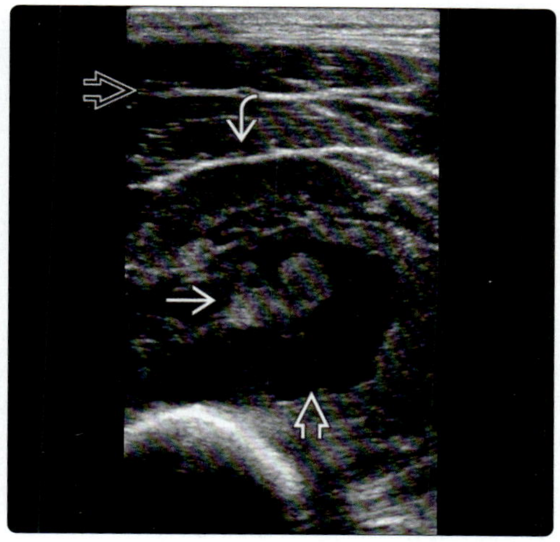

(Left) *Longitudinal US in a patient with recent leg trauma shows a severe tear at the upper to midportion of the vastus intermedius muscle. This grade 2 tear occupied ~ 70% of the muscle cross-sectional area. The femoral shaft is normal.* **(Right)** *Transverse US of the same area shows torn vastus intermedius muscle fibers with a large intramuscular hematoma. The overlying rectus femoris muscle is normal.*

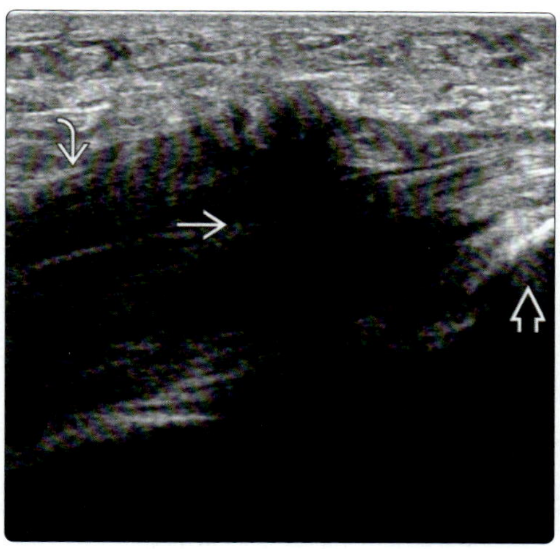

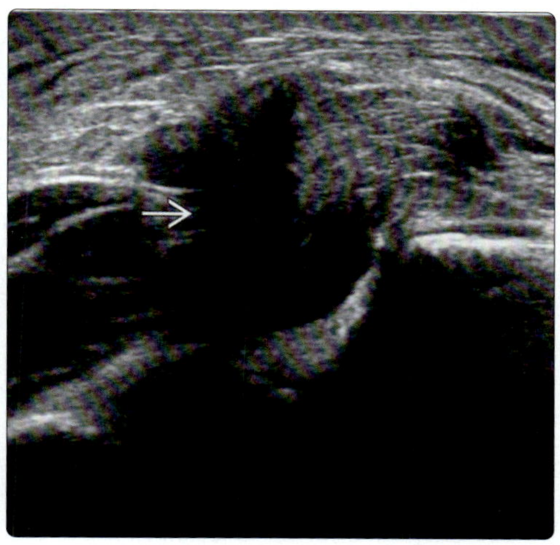

(Left) *Longitudinal US in a patient who developed pain while doing a 1-armed pull-up shows severe swelling of the distal triceps muscle and surrounding soft tissues. There is a full-thickness partial (grade 2) tear of the distal triceps muscle where the lateral fibers insert into the central tendon just proximal to the olecranon attachment.* **(Right)** *Longitudinal US shows the same tear. Overall, ~ 30% of the triceps muscle fibers were torn.*

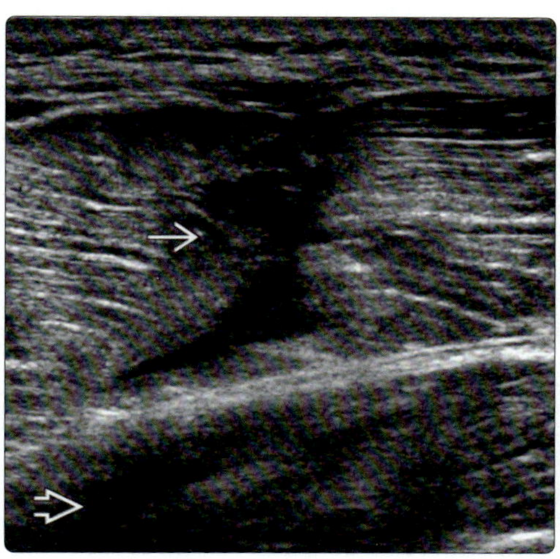

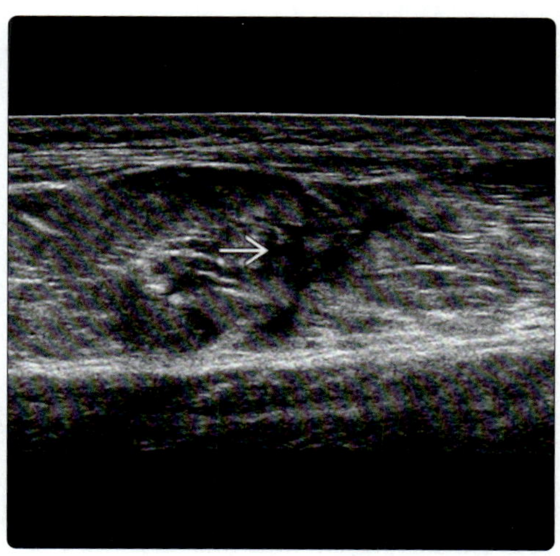

(Left) *Longitudinal US of the thigh in a patient recently kicked by a horse shows a full-thickness tear in the vastus lateralis muscle of the distal thigh. Overall, 90% of the fibers were torn, consistent with a severe grade 2 tear. The tear was associated with a surrounding hematoma and a moderate-sized knee joint effusion (not shown). The adjacent vastus intermedius muscle is normal.* **(Right)** *Extended field-of-view US shows the same full-thickness tear of the vastus lateralis muscle. There is no muscle retraction.*

(Left) *Extended field-of-view US shows severe retraction of the rectus femoris muscle due to a complete (grade 3) tear at the quadriceps tendon junction. The underlying vastus muscles were not torn.* **(Right)** *Longitudinal US of the same patient shows a clapper bell configuration of the completely torn, retracted rectus femoris muscle. The underlying vastus intermedius muscle is not torn.*

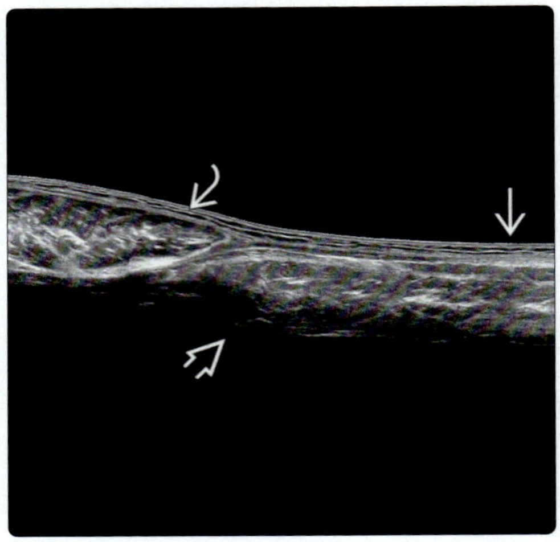

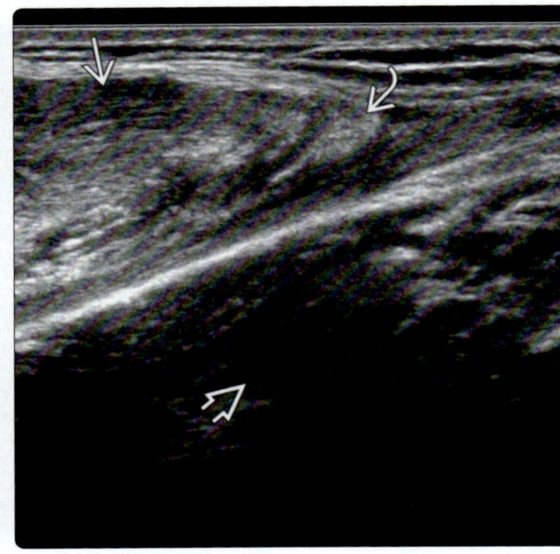

(Left) *Transverse US of the same rectus femoris muscle tear shows an amorphous appearance of a torn rectus femoris muscle around the central muscle tendon compared to the adjacent normal vastus muscle.* **(Right)** *Transverse US of the same patient shows a normal contralateral rectus femoris muscle around the central tendon for comparison.*

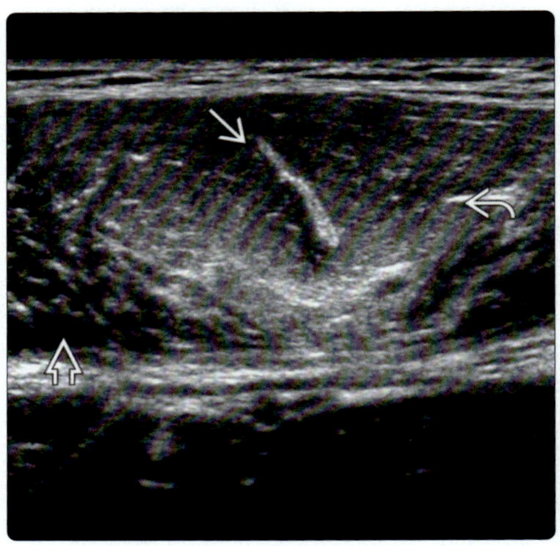

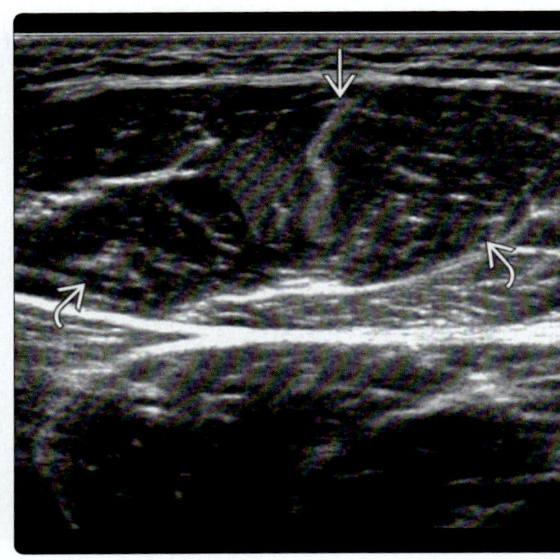

(Left) *Longitudinal US in a patient with blunt calf trauma 3 weeks earlier shows a healing muscle contusion within the peroneus longus muscle. No discrete muscle tear is evident. The fibular shaft is normal.* **(Right)** *Transverse US of the same area shows moderate hyperemia around the contused muscle area. The background muscle architecture is largely preserved. In another clinical context, this US appearance would fit for resolving diabetic muscle infarction.*

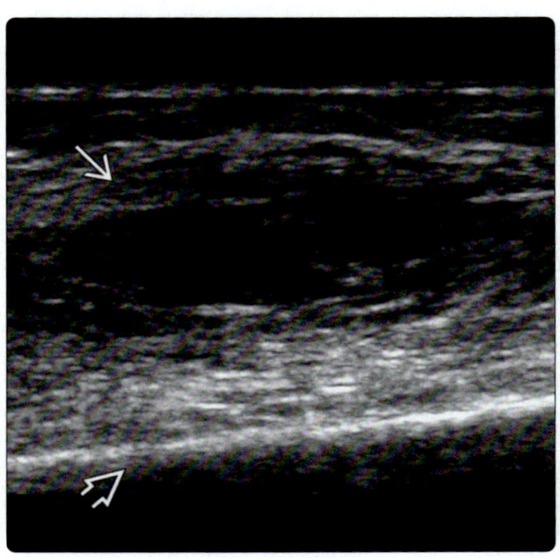

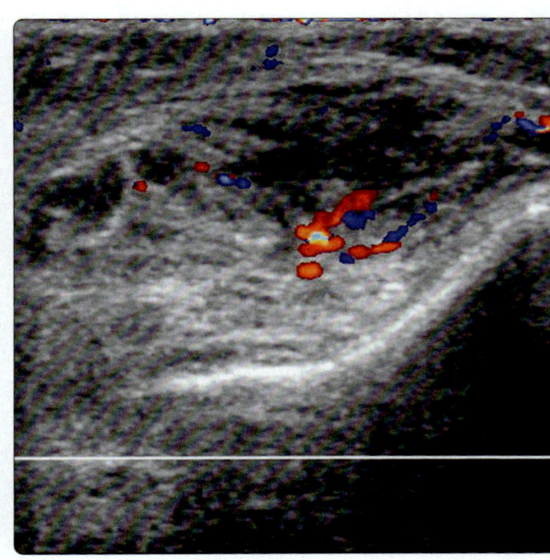

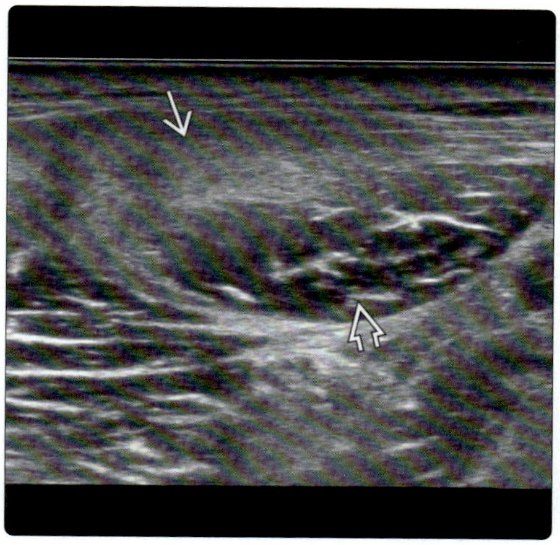

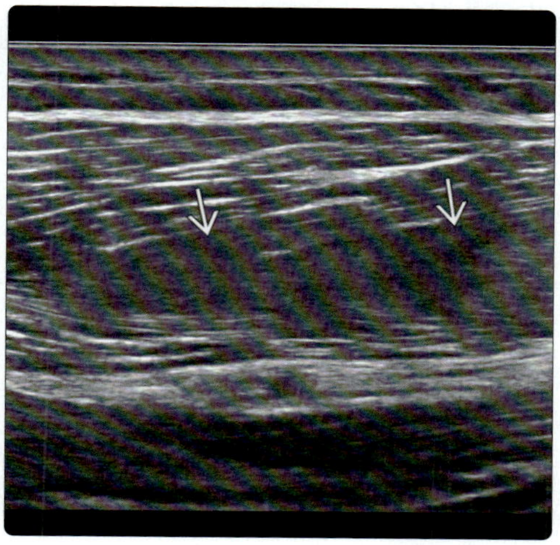

(Left) *Transverse US of the distal thigh shows a healing partial tear ➡ of the long head of biceps femoris muscle ➡ just proximal to the myotendinous junction. This involves 30% of the CSA of the long head. There is fibrosis and granulation tissue filling the tear.* **(Right)** *Longitudinal US of posterior midthigh shows a healing tear ➡ of the semimembranosus muscle. No calcification or muscle retraction is present. There was mild marginal hyperemia. The time since injury also helps in differentiating fibrosis from hematoma.*

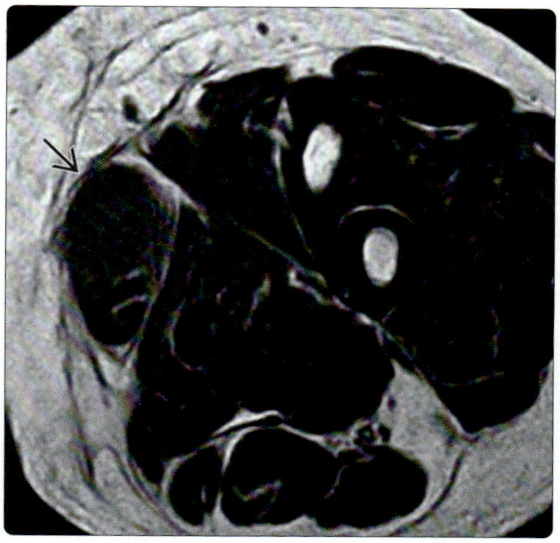

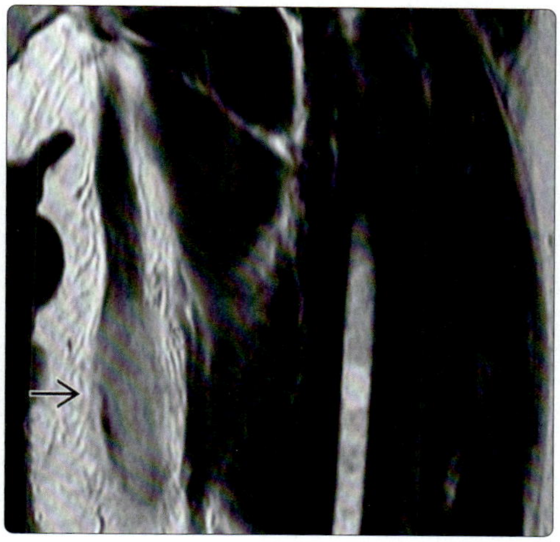

(Left) *Axial T1 MR in a middle-aged patient with vague thigh trauma 2 months earlier shows localized isointense swelling ➡ in the anterior aspect of the gracilis muscle. The remaining thigh musculature is normal.* **(Right)** *Coronal T2 MR of the same lesion shows diffuse muscle edema and swelling within the middle 1/3 of the gracilis muscle ➡.*

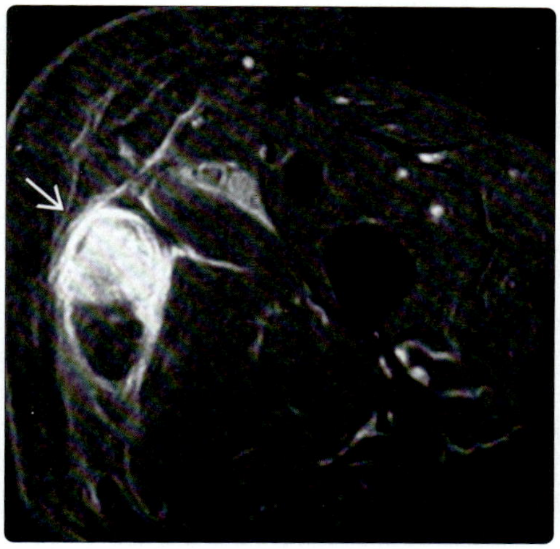

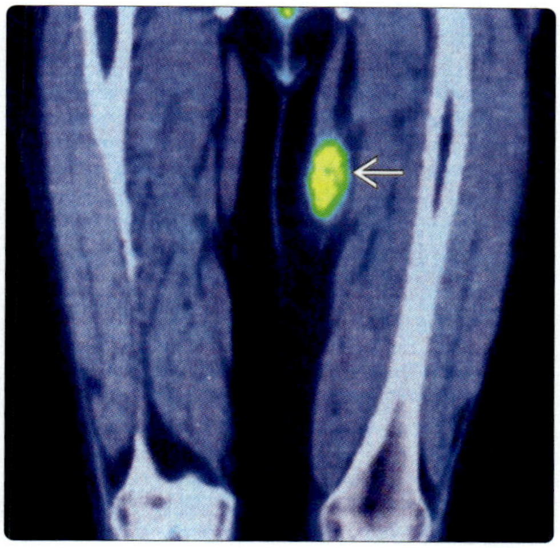

(Left) *Transverse T1 C+ FS MR of the same lesion shows intense contrast enhancement ➡ in and around the affected area of the gracilis muscle.* **(Right)** *Coronal reconstructed FDG PET/CT of the same lesion shows increased metabolic activity (standardized uptake value 4.4) in the swollen gracilis muscle area ➡. Based on the MR and PET/CT appearances, this was reported as a soft tissue sarcoma and a US-guided biopsy was advised.*

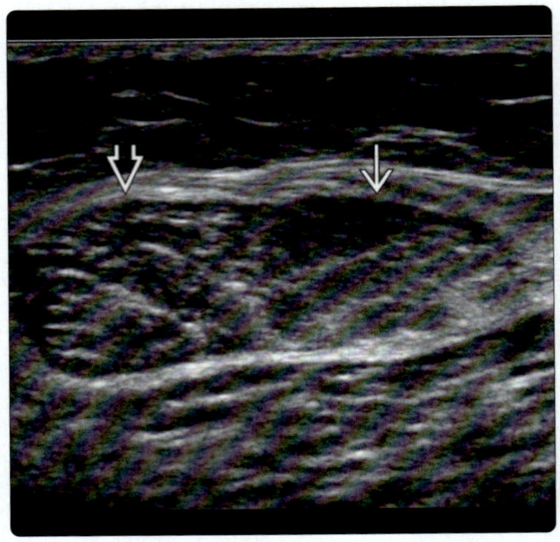

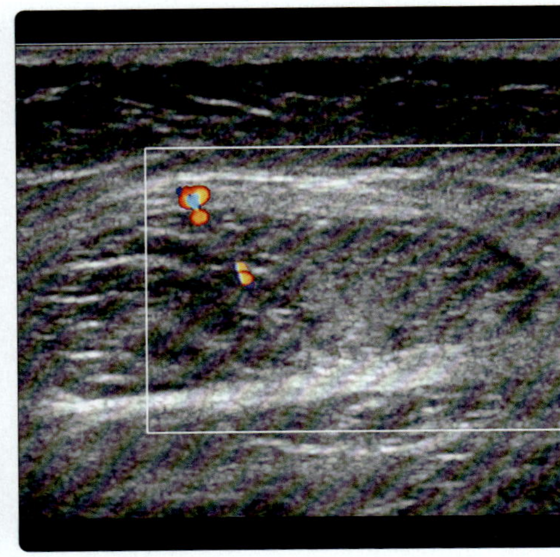

(Left) *Transverse US of the same lesion shows moderate swelling confined to the anterior part* ➔ *of the gracilis muscle* ➔*. The background muscle architecture is preserved. This favored a healing muscle tear with no evidence of tumor.* **(Right)** *Transverse color Doppler US of the same lesion shows mild perilesional hyperemia. In view of the MR and PET findings, a US-guided biopsy for histology was performed, which confirmed a healing muscle tear. US is helpful when other imaging is ambiguous.*

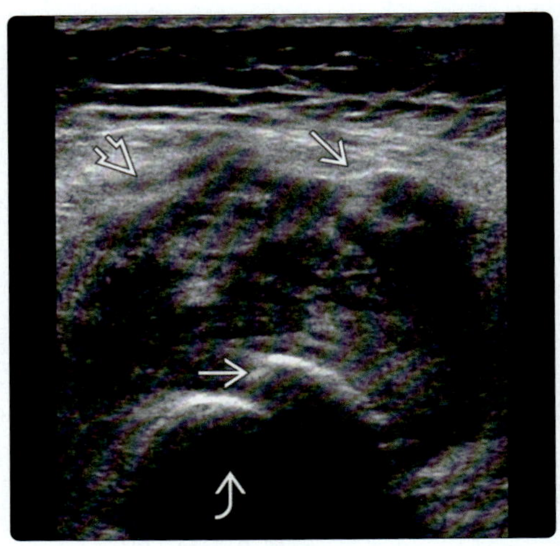

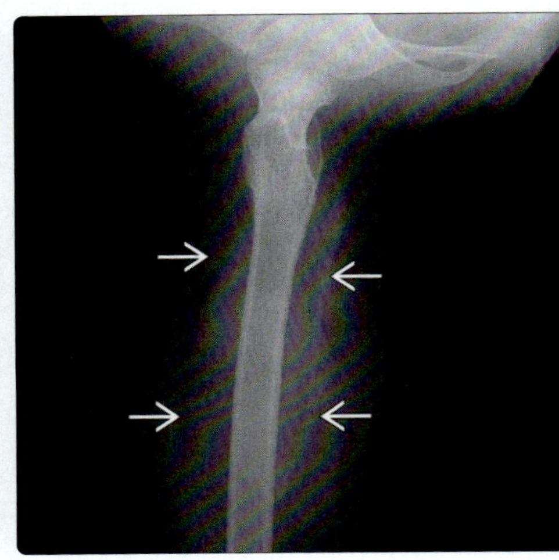

(Left) *Transverse US of the thigh in a patient 4 weeks after severe blunt thigh trauma shows a partially calcified* ➔ *hematoma* ➔ *deep in the quadratus femoris muscle and alongside the femoral shaft* ➔*.* **(Right)** *Lateral radiograph of the same patient shows the partially ossified hematoma* ➔ *surrounding the proximal femoral shaft, consistent with posttraumatic myositis ossificans (MO).*

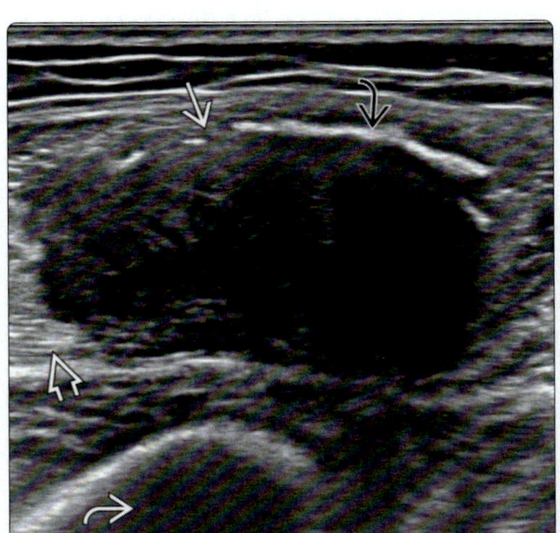

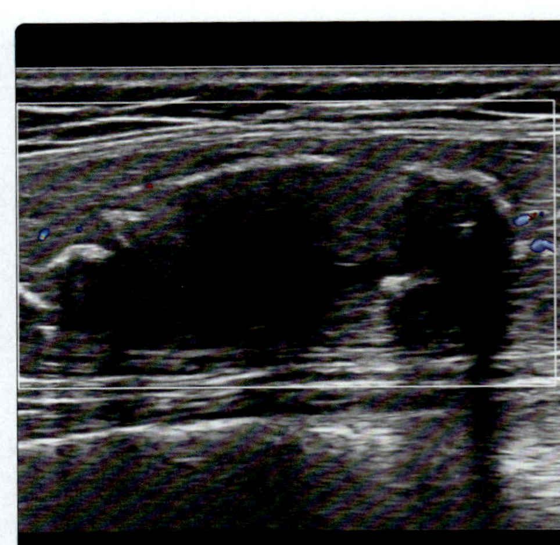

(Left) *Transverse US of the thigh in a patient 10 weeks after blunt thigh trauma shows MO* ➔ *with an oval-shaped, hypoechoic mass within the slightly edematous, hyperechoic vastus lateralis muscle* ➔ *alongside the femoral shaft* ➔*. There is quite mature peripheral ossification* ➔ *present.* **(Right)** *Color Doppler US of the same area shows mild marginal hyperemia. The maturation of MO can be followed with US.*

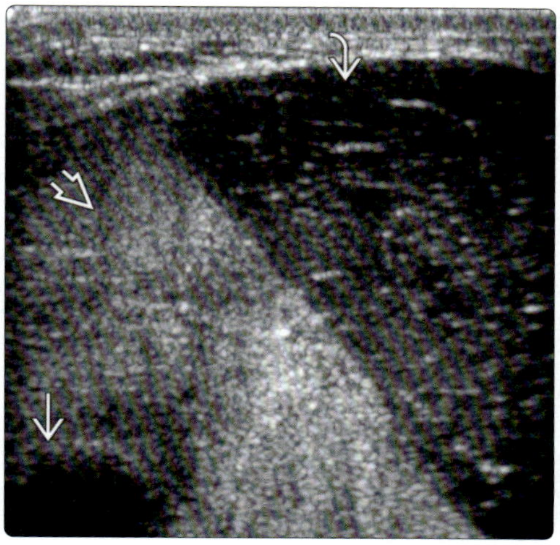

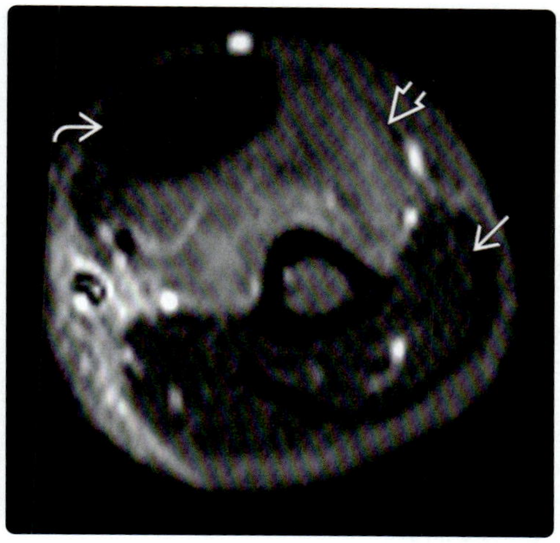

(Left) *Transverse US in a patient with unaccustomed overuse rowing injury and pain onset 1 day later shows severe edema of the brachialis muscle consistent with delayed-onset muscle soreness (DOMS). The overlying biceps muscle and humeral shaft are normal.* **(Right)** *Axial T2 FS MR of the same patient shows severe edema of the brachialis muscle consistent with DOMS, which is a cellular-related muscle injury. The biceps and triceps muscles are normal.*

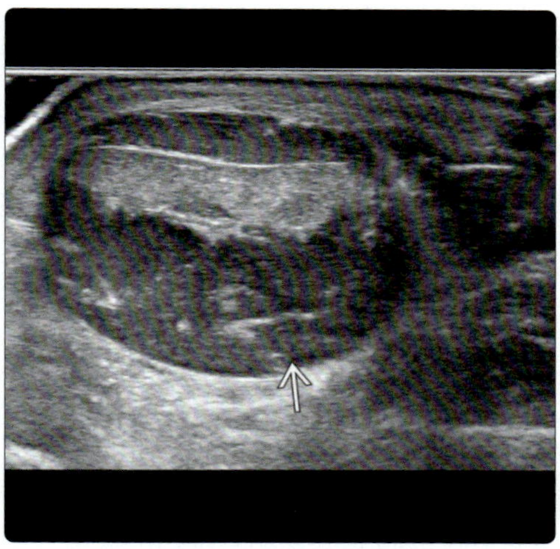

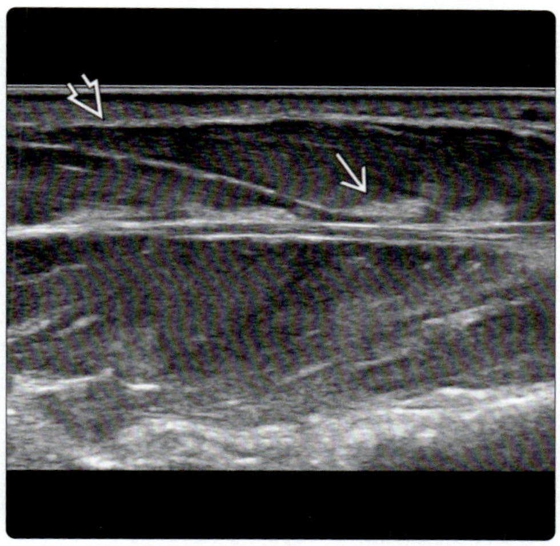

(Left) *Transverse US in a patient with pes planus and severe overuse trekking injury to the foot 1 day prior shows severe diffuse swelling of the abductor hallucis longus muscle, which is swollen to 3x the normal size. Moderate diffuse muscle hyperemia was present. The appearances are due to DOMS.* **(Right)** *Longitudinal US of the same area shows how muscle architecture is maintained. The echogenic area alongside the swollen central tendon is due to intense muscle edema ± hemorrhage.*

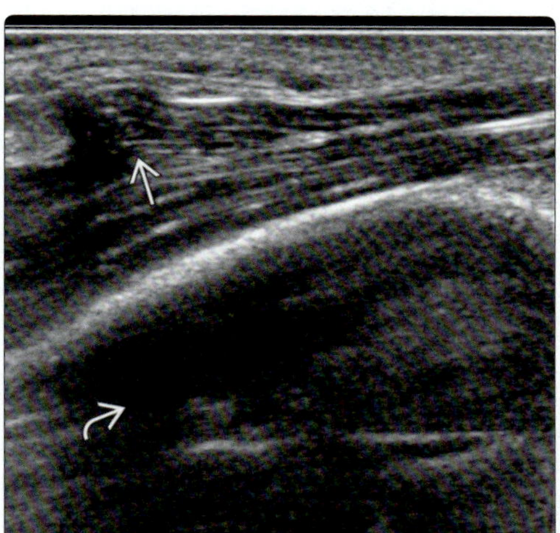

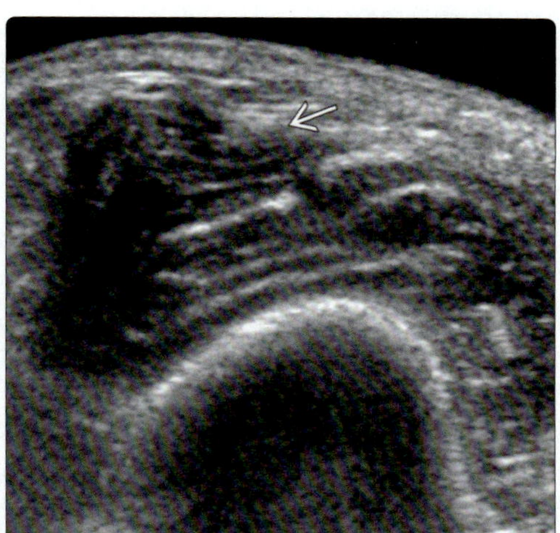

(Left) *Longitudinal oblique US shows a small muscle hernia of the peroneus brevis muscle lateral to fibular shaft. The muscle herniates through a small (7-mm) epimysial fascial defect into the subcutaneous tissue. No muscle entrapment is present.* **(Right)** *Transverse US of the same area shows the muscle herniation. Muscle hernias can be accentuated by muscle contraction or exercise. They are more commonly seen in the lateral and anterior thigh musculature and are treated with surgical repair of the fascial defect.*

Hematoma/Seroma

KEY FACTS

IMAGING

- **Hematoma: Acute**
  - Quite well-defined heterogeneous mass ± echogenic pseudosolid appearance
  - Multilaminated whorled appearance
  - Anechoic cystic areas
  - Echogenic foci due to gas locules
  - Surrounding soft tissue edema
  - Mild surrounding hyperemia
  - No internal hyperemia
  - Underlying cause of hematoma, such as muscle tear or vascular malformation, may be apparent
- **Hematoma: Subacute and chronic**
  - More well-defined margin ± distinct capsule
  - Reduces in size on serial examination
  - Hypoechoic fluid areas due to separation of serum from cellular elements in process of liquefaction
  - Mild marginal hyperemia on color Doppler imaging
  - ± more central hyperemia if of long duration
- **Organizing hematoma**
  - Encapsulated, increasing size due to recurrent bleeding, intrinsic hyperemia
  - May simulate soft tissue tumor
  - May need biopsy to distinguish
- **Seroma**
  - Often close to site of surgery (chest wall, abdomen, groin)
  - Echo-poor or anechoic with strong posterior acoustic enhancement
  - Usually no wall or barely discernible wall
- **Lymphocele**
  - Close to site of vascular surgery
  - Very similar US appearances to seroma
- Serial follow-up helpful to confirm hematoma and to check for underlying cause
- Relate to clinical setting

(Left) *Clinical photograph shows a patient with an acute knee injury associated with prepatellar soft tissue swelling, abrasions, and bruises.* **(Right)** *Transverse US of the same patient shows a medium-sized acute subcutaneous hematoma anterior to the patella. The hematoma shows a heterogeneous, hypoechoic, pseudosolid appearance. A more fluid component is present at the superficial aspect of the hematoma.*

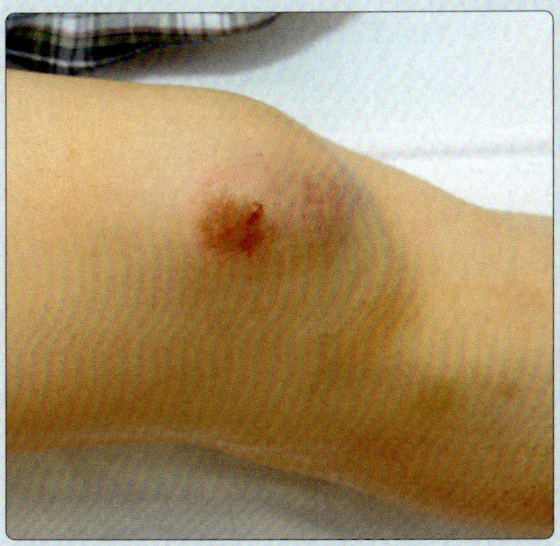

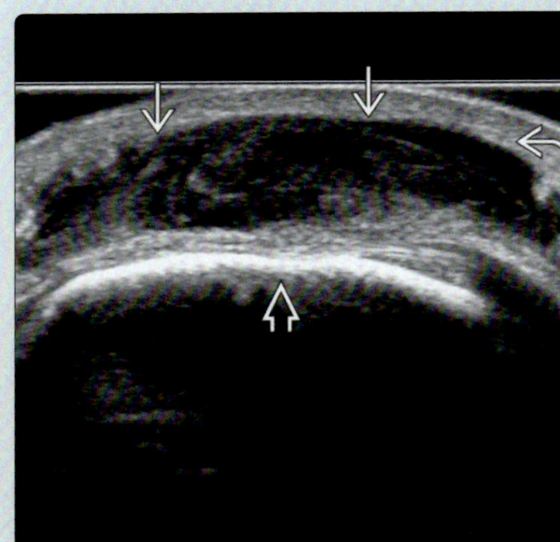

(Left) *Transverse US of the groin 2 days after arterial puncture shows a large, well-defined, oval-shaped hematoma. It shows typical multilaminated whorled appearance with posterior enhancement and a central anechoic fluid component. Another smaller hematoma is present posteriorly. A moderate degree of surrounding edema is present.* **(Right)** *Transverse color Doppler US of the same location shows no internal vascularity. The femoral artery is located posteriorly.*

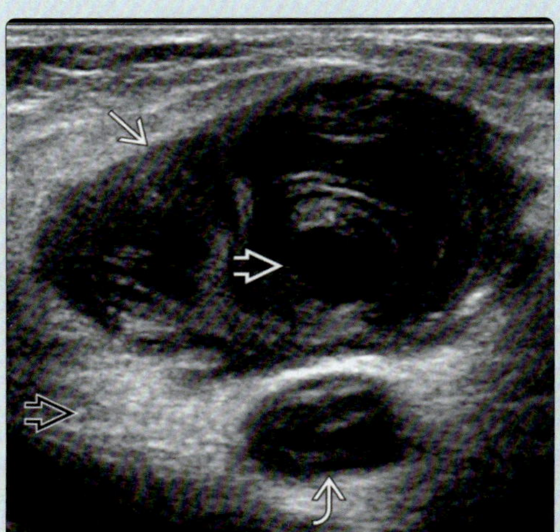

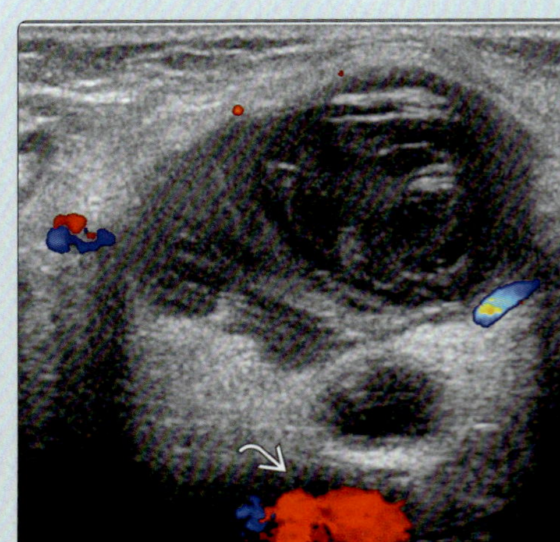

TERMINOLOGY

Definitions

- Hematoma: Localized collection of blood outside of blood vessel
- Seroma: Serous fluid collection within soft tissues
- Lymphocele: Collection of lymphatic fluid due to leakage from lymphatics

IMAGING

General Features

- Best diagnostic clue
  - Hematoma: Largely solid-looking echogenic mass with small cystic areas and no intrinsic hyperemia
    - Often history of trauma
  - Seroma: Anechoic fluid collection ± internal debris; minimal or no wall hyperemia
    - Nearby surgery
  - Lymphocele: Anechoic fluid collection, no internal debris, no wall hyperemia
    - Nearby vascular surgery
    - Difficult to distinguish seroma from lymphocele
- Location
  - Usually at site of trauma or surgery
- Size
  - Variable but usually large in buttock or thigh

Radiographic Findings

- Radiography
  - Mildly hyperdense soft tissue opacity and swelling
    - ± calcification in old hematoma

MR Findings

- T1WI and T2WI
  - Hematoma
    - Acute: Isointense on T1WI; variable hyper- to hypointense on T2WI
    - Subacute/chronic: Hyperintense on T1WI and T2WI
    - T2* GRE: Hemosiderin deposition with susceptibility blooming artifact
  - Seroma/lymphocele
    - Fluid-like collection near site of surgery

Ultrasonographic Findings

- Internal echogenicity depends on degree of blood coagulation, liquefaction, and organization
- Hematoma
  - Appearances depend on chronicity
  - Acute: Fairly well-defined heterogeneous mass in subcutaneous or muscle layer
    - Echogenic pseudosolid appearance
    - Multilaminated whorled appearance
    - Anechoic cystic components
    - Echogenic foci due to gas locules
    - Surrounding soft tissue edema
    - Usually no surrounding hyperemia in hyperacute stage; develops over a few days; no intrinsic hyperemia
    - Mild acoustic enhancement
    - Underlying cause of hematoma, such as muscle tear or vascular malformation, may be apparent
  - Subacute and chronic: Well-defined, variable echogenicity depending on liquefaction and degree of reparative organization
    - More well-defined margin ± distinct capsule
    - Reduces in size on serial examination
    - Hypoechoic fluid areas due to separation of serum from cellular elements during liquefaction process
    - Mild marginal hyperemia; ± more central hyperemia if hematoma is of long duration; may remain avascular
    - ± marginal calcification
    - Acoustic enhancement if liquefaction; acoustic shadowing if calcification or fibrosis
    - Underlying cause may become more apparent
  - Organizing hematoma
    - May simulate soft tissue mass; may require biopsy for confirmation
    - Encapsulated, increasing in size due to recurrent internal bleeding, internal hyperemia
    - Neovascularization, although present, may not show on color Doppler imaging due to small size of neovascular vessels
  - Hemophiliac pseudotumor
    - Organizing hematoma of hemophilia; affecting 1-2% of all hemophiliacs
    - Thick-walled hematoma with internal vascularity
    - Can grow due to repeated bleeding
    - Mass effect with bone scalloping
    - Usually in soft tissues but can be subperiosteal or intraosseous
  - Rectus sheath hematoma
    - Above arcuate line
      - Ovoid in shape with long axis superoinferior
      - Localized to one rectus muscle, does not cross midline
      - Expansile in appearance, mimicking soft tissue tumor
    - Below arcuate line
      - No linea alba, so can cross midline; pad-like with transverse long axis
- Seroma
  - Often close to site of surgery (chest wall, abdomen, groin)
  - US appearances vary with chronicity and content of seroma
  - Echo-poor with strong posterior acoustic enhancement
  - Acute or subacute seroma
    - Usually anechoic because of sterile fluid content
    - Usually no wall or barely discernible wall
    - No adjacent soft tissue edema or hyperemia
    - No internal echogenic component unless complicated by hemorrhage
    - US-guided aspiration yields clear or pale straw-colored fluid
  - Chronic seroma
    - Increasing echogenicity of content ± fibrous septa
    - Increasing wall thickness and mild hyperemia
    - Difficult to distinguish from chronic liquefied hematoma
- Lymphocele
  - Close to site of vascular surgery

- Very similar US appearances to seroma
  - Etiology of seroma and lymphocele is different
- US-guided aspiration yields clear or pale straw-colored fluid
  - Laboratory analysis yields slight increase in lymphocyte count
  - Chylous fluid (↑ triglyceride and protein count) is feature of high retroperitoneal and thoracic lymphoceles only

Imaging Recommendations

- Best imaging tool
  - US can make distinction between hematoma and seroma/lymphocele
  - Serial follow-up helpful to confirm hematoma and to check for underlying cause
    - MR may provide additional help in assessing underlying injury or vascular malformation
  - Percutaneous aspiration can help distinguish seroma/lymphocele and low-grade infection (e.g., fungal infection)
    - Increased lymphocyte count in lymphocele
- Protocol advice
  - Correlate closely with clinical setting
  - Always include evaluation with color Doppler
  - Check for signs of any infection
    - Aspirate if doubt exists as to infection

DIFFERENTIAL DIAGNOSIS

Soft Tissue Mass/Sarcoma

- Predominantly solid in appearance
- Usually more internal vascularity
- Enlarging clinically or on serial examination
- No trauma or surgical history

Low-Grade Abscess

- Thicker walls ± internal septation
- Internal debris with variable amount of internal echoes ± gas locules
- Adjacent soft tissue edema and hyperemia
- ± sinus tract

Aneurysm

- Doppler spectral analysis shows arterial flow

PATHOLOGY

General Features

- Etiology
  - Hematoma: Blood leakage from vasculature
    - Spontaneous hematoma
      - Underlying bleeding tendency (e.g., hemophilia, anticoagulant therapy, aspirin)
      - Unrecognized self trauma, such as severe coughing or sneezing
      - Rectus sheath hematoma due to inferior epigastric vessel constriction during physical exertion
  - Seroma: Serous exudate from serous membrane
    - Generally occurs following surgery to chest, abdomen, or groin

Gross Pathologic & Surgical Features

- Hematoma: Weeks later, more progressive hemolysis and liquefaction
  - Chronic hematoma can have multilaminated concentric appearance due to repeated bleeding
  - Months later, hematoma completely resolves
- Organizing hematoma
  - Central mass of blood, wall of granulation tissue, and dense, fibrous tissue at periphery
  - Expanding nature due to irritant effects of blood products, causing repeated exudation ± bleeding from capillaries

CLINICAL ISSUES

Presentation

- Most common signs/symptoms
  - Localized swelling
  - History of trauma and surgery
  - Hematoma: Bruising or bluish discoloration of skin
- Other signs/symptoms
  - Tenderness, redness, and increased warmth
  - Abnormal clotting profile and thrombocytopenia, if bleeding tendency

Demographics

- Age
  - More common in elderly due to impaired coagulation (hematoma) and increasing likelihood of surgery (seroma/lymphocele)
- Gender
  - Any gender though rectus sheath hematoma is more common in females due to weaker rectus muscles

Natural History & Prognosis

- Hematoma and seroma/lymphocele usually resolve gradually over 2-6 months

Treatment

- Hematoma
  - Rest, ice, compression, and elevation ± analgesia
  - Intravenous fluid resuscitation/blood transfusion
  - Correction of coagulopathy (e.g., stop anticoagulant, infusion of fresh frozen plasma, platelet concentrates)
  - Aspiration if liquefied
- Seroma/lymphocele
  - US-guided aspiration or drainage
  - Pressure bandaging

DIAGNOSTIC CHECKLIST

Consider

- Infection if adjacent soft tissue hyperemia and edema

Image Interpretation Pearls

- Relate to clinical history
- Always evaluate with color Doppler imaging

SELECTED REFERENCES

1. Kummer T et al: Diagnosis of an actively bleeding rectus sheath hematoma by contrast-enhanced ultrasound. Med Ultrason. 19(2):239-240, 2017
2. Lee RK et al: Ultrasound of the abdominal wall and groin. Can Assoc Radiol J. 64(4):295-305, 2013

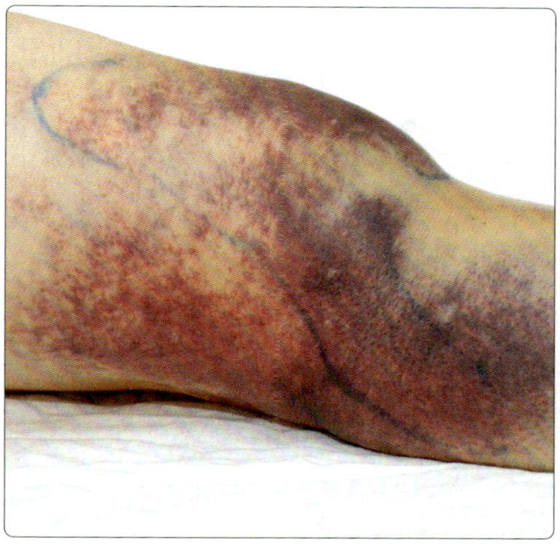

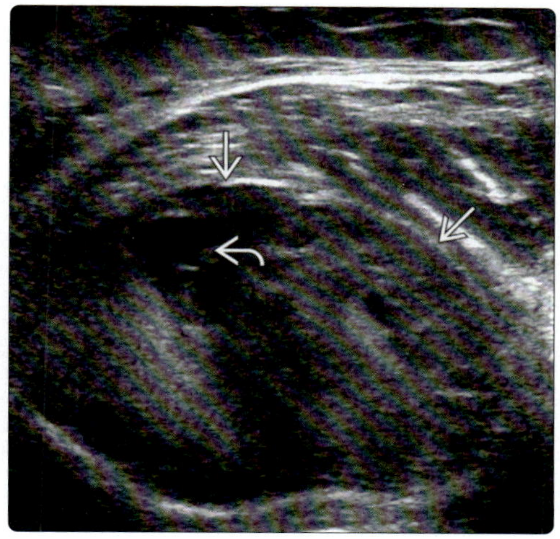

(Left) *Clinical photograph shows a patient on warfarin with spontaneous swelling and bruising on the lateral side of the lower thigh and knee.* **(Right)** *Transverse US of the same patient shows a large, well-defined hematoma within the vastus lateralis muscle. A small, anechoic cystic area due to liquefaction is present more superiorly.*

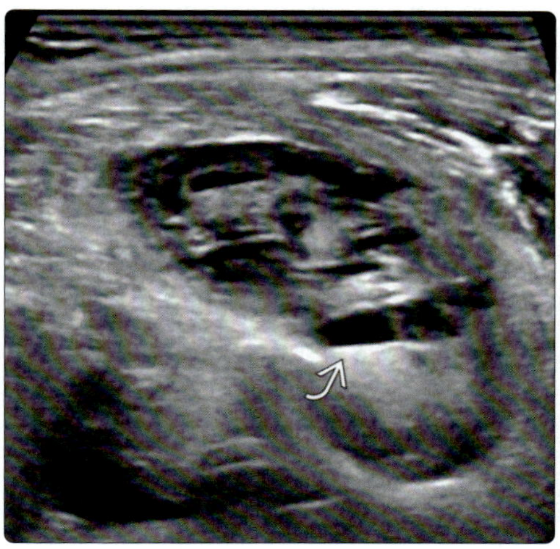

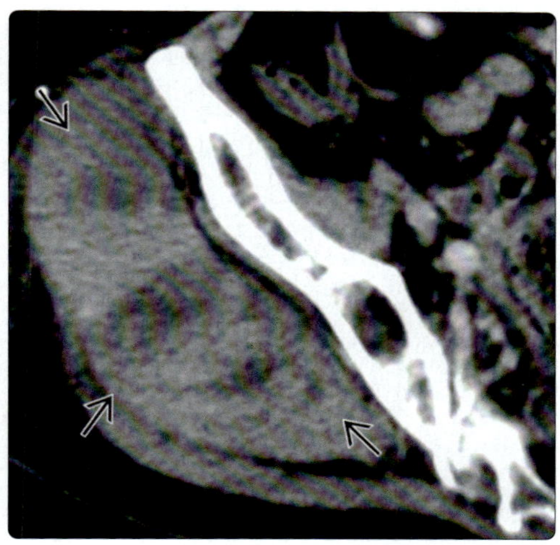

(Left) *Transverse US of the gluteus muscle in a patient who had a fall 3 weeks earlier shows a large, partially solid, partially cystic mass with fluid-fluid levels within the gluteal muscle, consistent with a hematoma.* **(Right)** *CT scan at the corresponding level shows a mixed hypodense and isodense mass with ill-defined margins in the swollen gluteus medius muscle. In this clinical setting, this is entirely consistent with a hematoma. Follow-up US is nevertheless advisable to confirm resolution.*

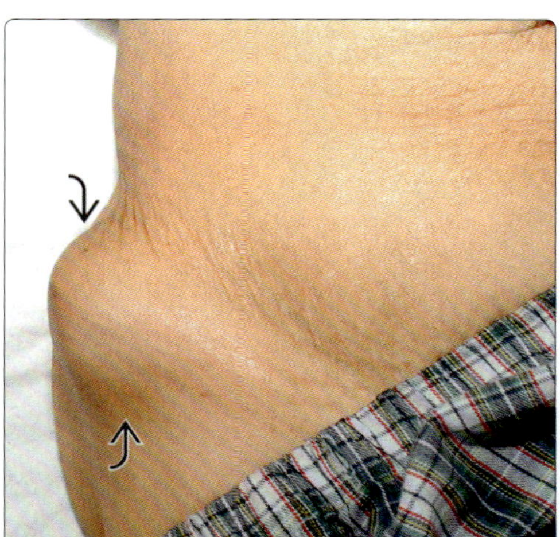

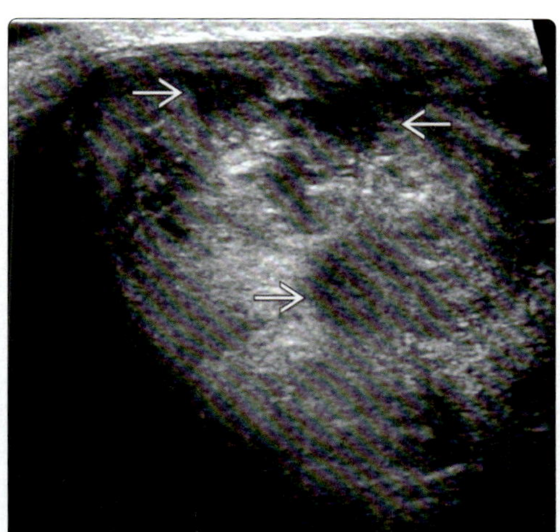

(Left) *Clinical photograph shows a patient with spontaneous painful swelling of the iliac crest region, which had been increasing for 6 months with no history of trauma.* **(Right)** *Transverse US of the same patient shows a largely solid mass in the subcutaneous tissue above the iliac crest. The lesion is well encapsulated with no calcification. Small, hypoechoic areas are present within the mass. This is consistent with an expanding hematoma, which was confirmed by percutaneous biopsy.*

(Left) *Longitudinal US shows the calf in a patient with spontaneous swelling. Note the large, well-defined, oval-shaped mass with multilaminated whorled appearance and mild posterior enhancement, consistent with a hematoma. Small liquefied areas are seen within the hematoma ⇨. No discrete tear or vascular malformation is seen.* **(Right)** *Color Doppler US of the same mass shows increased vascularity → eccentrically within the mass, suggestive of underlying vascular anomaly. Note the surrounding edema ⇨.*

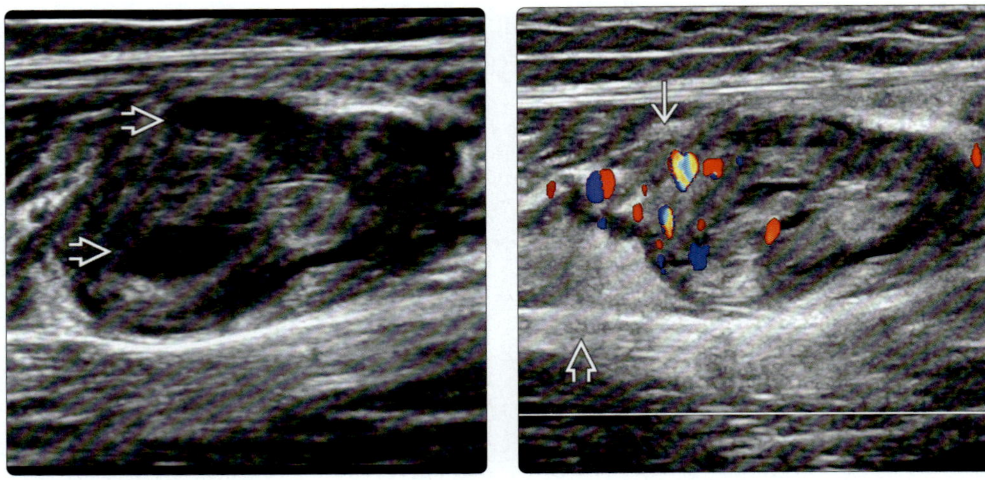

(Left) *Transverse US of the calf 3 days following blunt trauma shows a large hematoma within the soleus muscle belly. Layering → is present within the more liquefied portions.* **(Right)** *Transverse US of the medial aspect of the knee shows a large subcutaneous cystic collection with fluid-fluid levels. This hematoma is composed of an echogenic blood clot ⇨, a clear liquefied serous component →, and a more dependent cellular component ↷.*

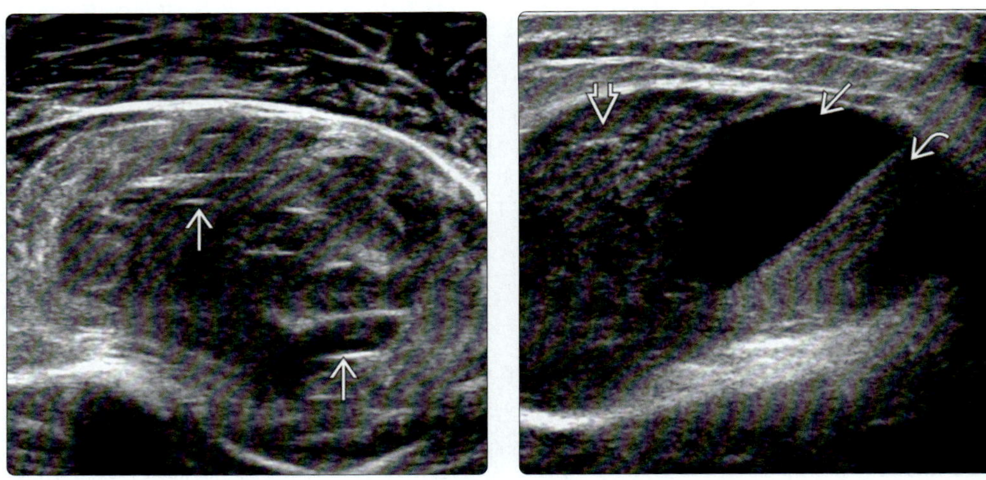

(Left) *Transverse US of a patient with recent-onset thigh pain and swelling when playing football shows a large hematoma → within the rectus femoris muscle ↷. No liquefaction is present.* **(Right)** *Transverse color Doppler US at the same location shows moderate peripheral hyperemia. Although no discrete muscle tear is evident, this is invariably the precursor to traumatic muscle hematoma. No specific treatment is necessary as this hematoma will resolve spontaneously.*

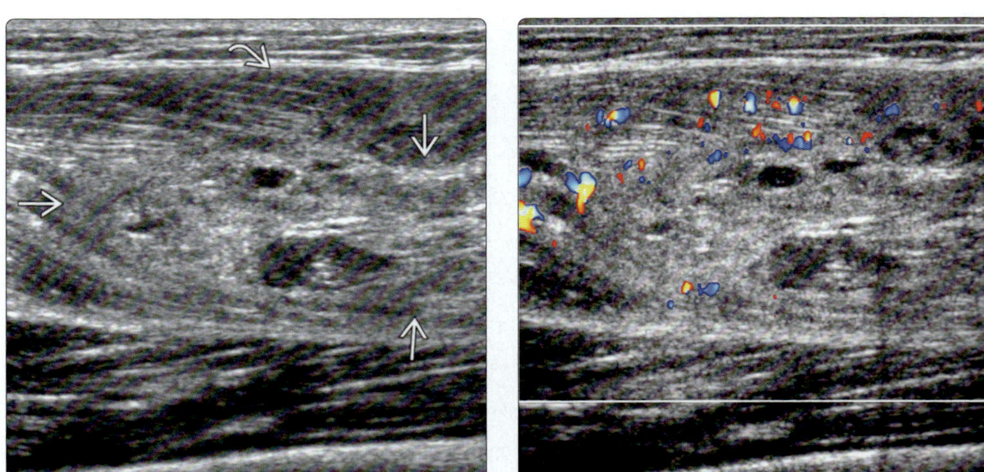

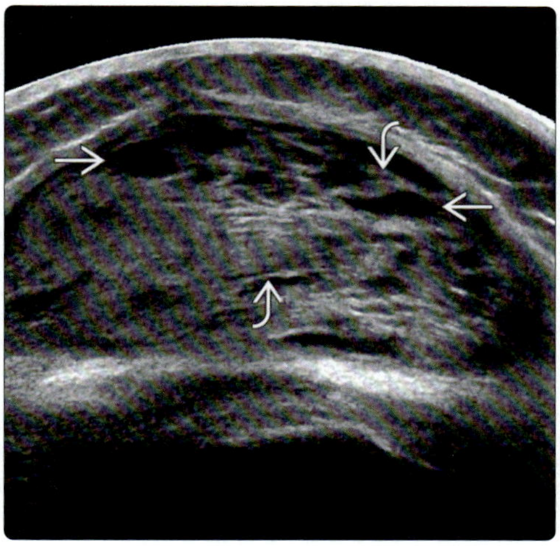

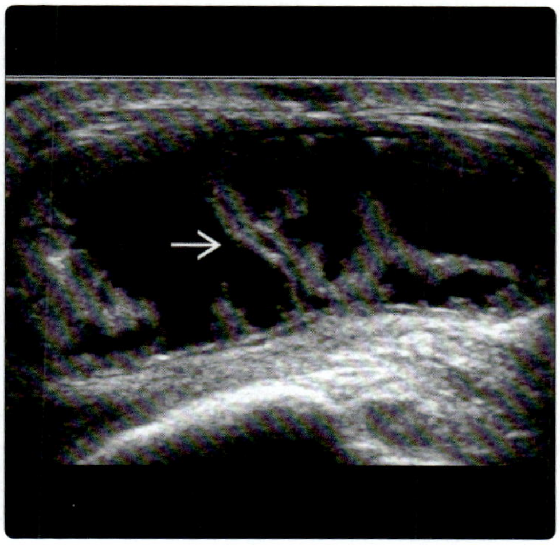

(Left) *Longitudinal US of the proximal thigh after total hip arthroplasty shows a well-defined mass with layering, anechoic liquefied areas ➡ and internal septations ➡. Features are consistent with a resolving large hematoma.* **(Right)** *Transverse US of the same region 1 month later shows increased liquefaction of the hematoma with a greater cystic component, septa ➡, and fibrin clots. Features are consistent with liquefying hematoma.*

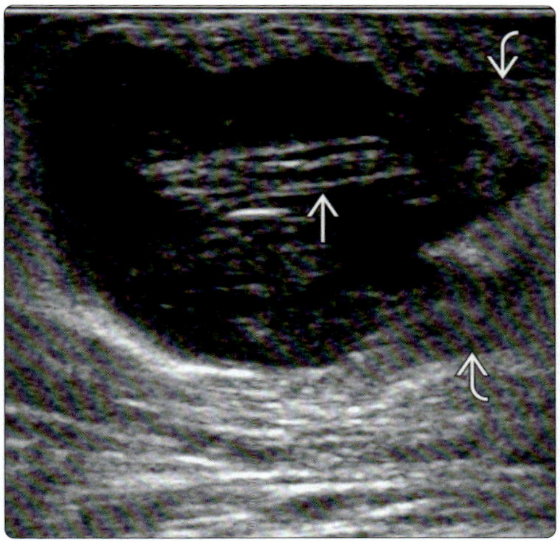

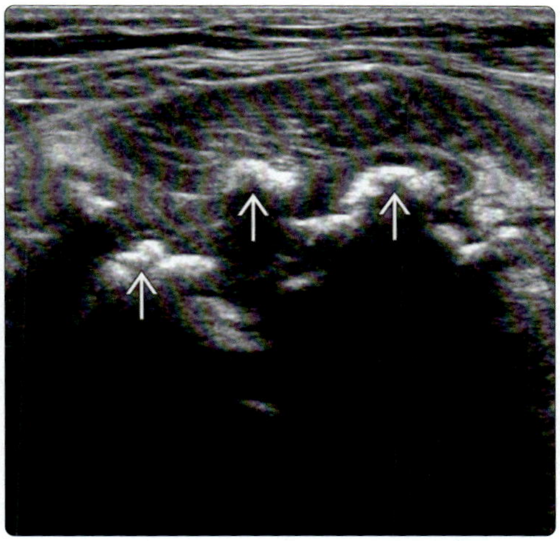

(Left) *Longitudinal US of a calf in an anticoagulated patient with spontaneous swelling shows a medium-sized hematoma with fibrinous-type layering ➡ confined to the subcutaneous tissues. Moderate surrounding soft tissue edema ➡ is present.* **(Right)** *Transverse US of the thigh following a road traffic accident several months previously shows a large, heavily calcified mass with peripheral calcification ➡ obscuring more posterior detail. CT confirmed a large calcified hematoma.*

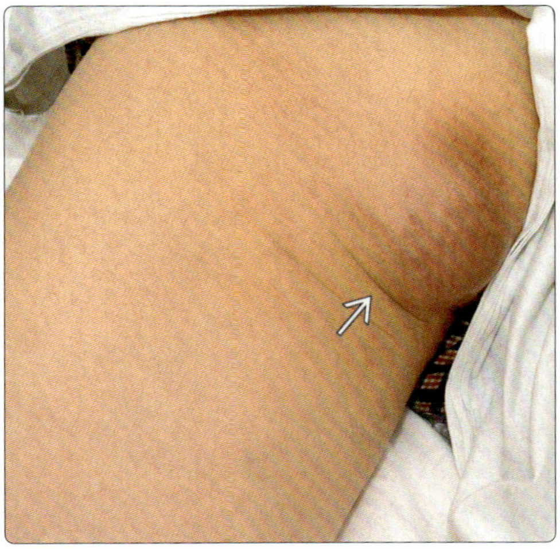

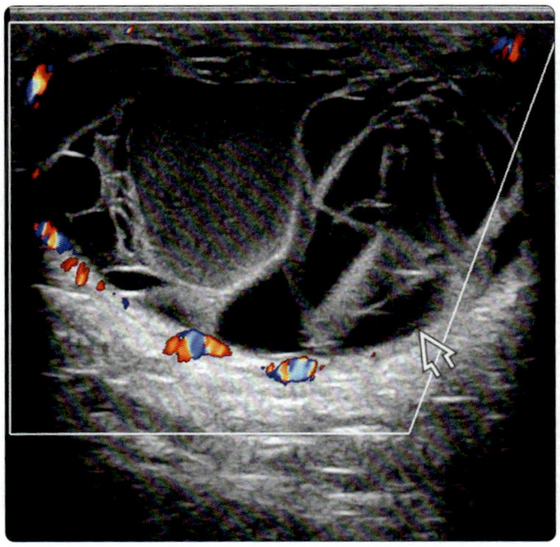

(Left) *Clinical photograph of an elderly patient who presented with painful swelling ➡ of the medial aspect of the proximal thigh is shown. Clinically, an epidermoid cyst was suspected.* **(Right)** *Transverse color Doppler US of the same patient shows a large, well-defined, multiseptated cystic subcutaneous mass ➡ with peripheral hyperemia, consistent with an organizing hematoma. Following aspiration, which yielded old blood, the mass settled and did not recur.*

(Left) *Longitudinal US in a patient with trauma to the parascapular region 10 days earlier, who then developed a painful swelling at this site, shows a hypoechoic, partially liquefied hematoma within the latissimus dorsi muscle superficial to the chest wall. Overall, features were suspicious of hematoma complicated by infection.* **(Right)** *CT scan of the corresponding region reveals a cystic area with a thick rim of contrast enhancement suggestive of an infected hematoma.*

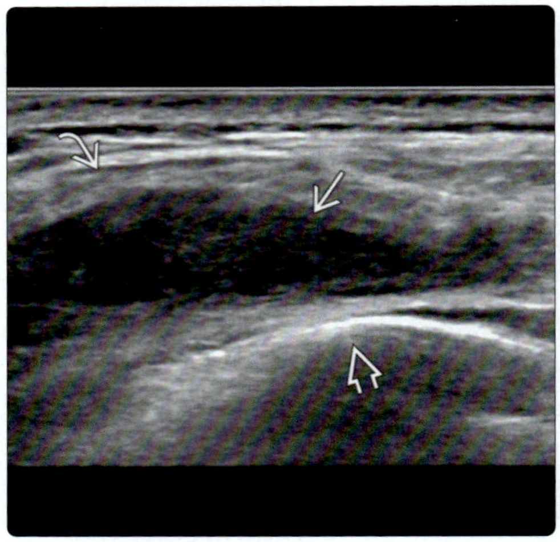

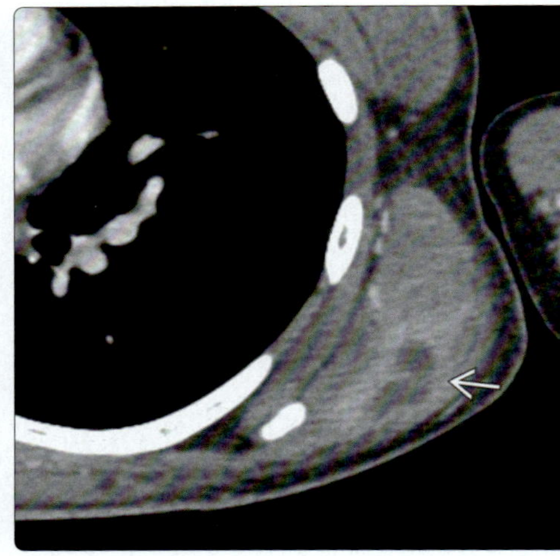

(Left) *Transverse US during needle aspiration of the same patient is shown. A small amount of pus was aspirated, which grew Staphylococcus aureus, confirming the infective nature of this hematoma.* **(Right)** *Longitudinal US shows a large, partially septated liquefied hematoma within the gastrocnemius muscle. The surrounding muscle is moderately edematous. With aspiration, which yielded old blood, the hematoma collapsed and did not recur.*

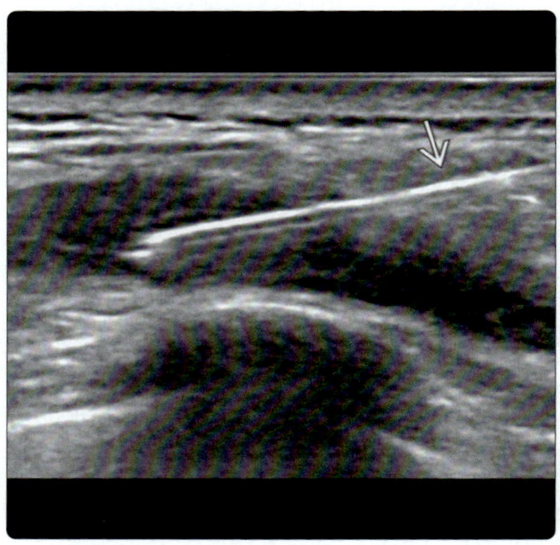

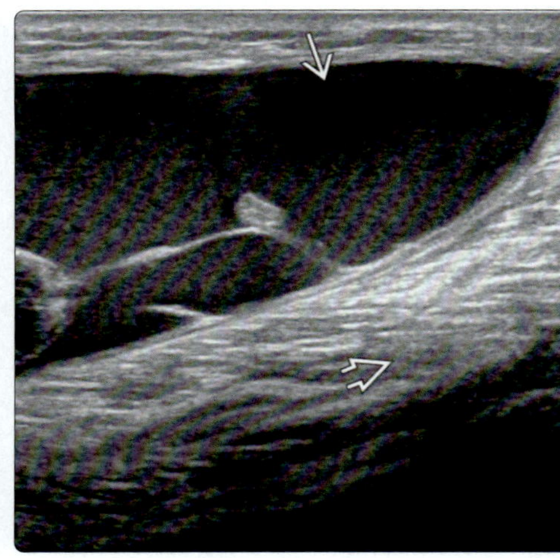

(Left) *Clinical photograph shows a groin swelling at the site of a scar due to inguinal hernia repair 2 months previously.* **(Right)** *Longitudinal oblique US in the same patient shows an oval-shaped, well-defined collection within the inguinal canal. The mass contains predominantly cystic components with multiple thin internal septations and small areas of more solid material. Features are compatible with a chronic seroma.*

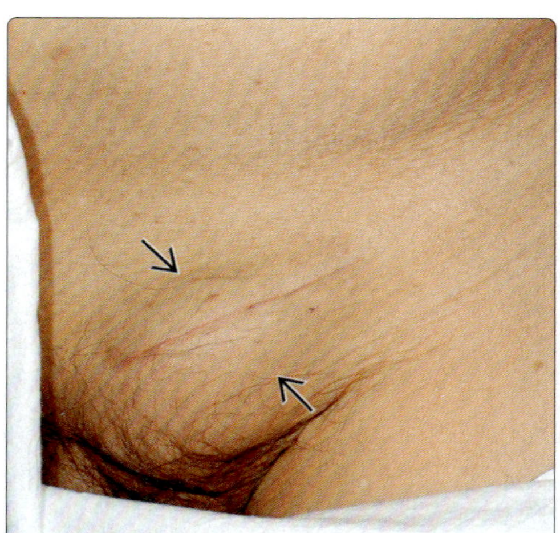

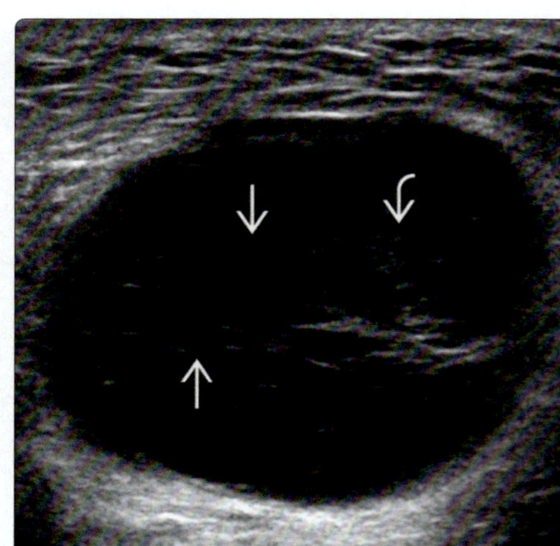

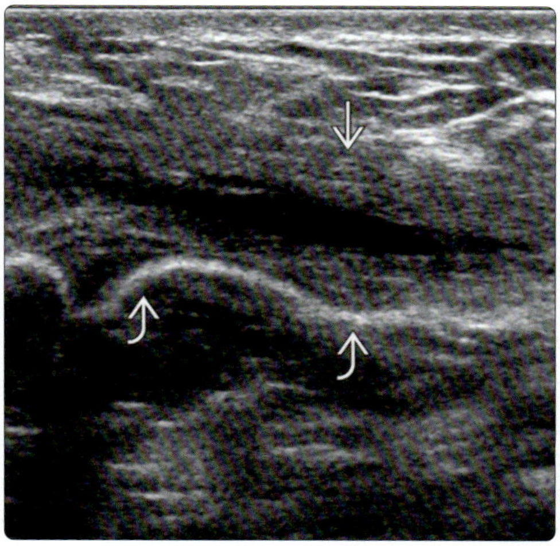

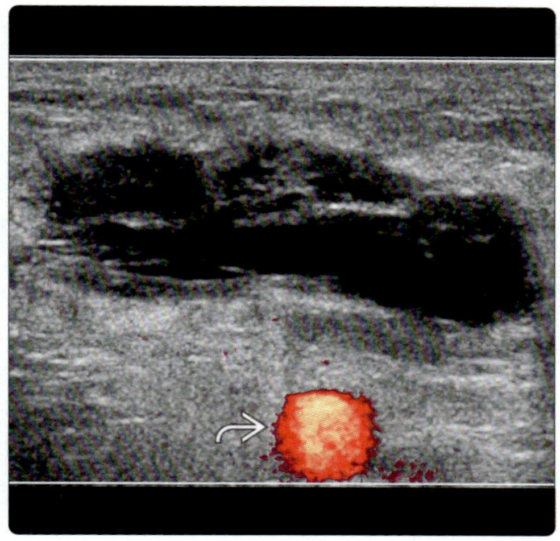

(Left) *Transverse US of the groin in a patient following hernial surgery shows that a thick-walled seroma containing anechoic fluid is located just anterior to the surgical mesh.* **(Right)** *Transverse color Doppler US of the same seroma more proximally shows no vascularity within the lesion or in the adjacent soft tissues. The femoral artery is normal.*

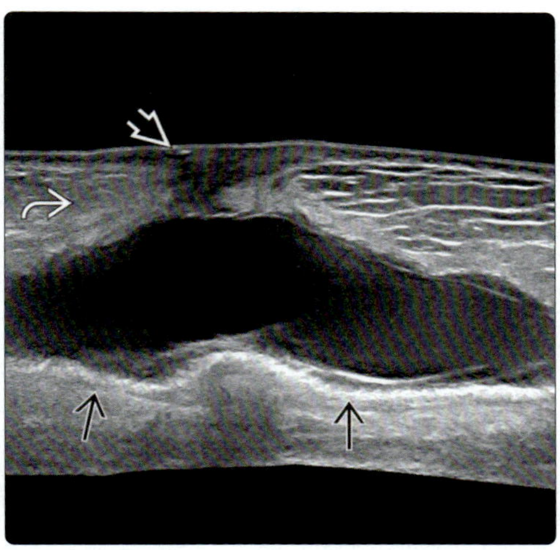

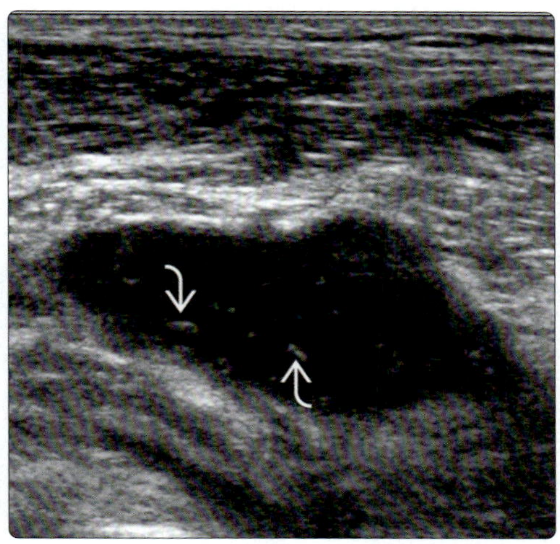

(Left) *Panoramic transverse US shows a large seroma present within the anterior abdominal wall. The seroma lies between the subcutaneous fat with the surgical wound and the underlying surgical mesh.* **(Right)** *Transverse US of the medial aspect of the popliteal fossa 3 months after knee operation shows a well-defined cystic lesion with small, echogenic foci and no discernible wall. Appearances are consistent with a lymphocele or a seroma.*

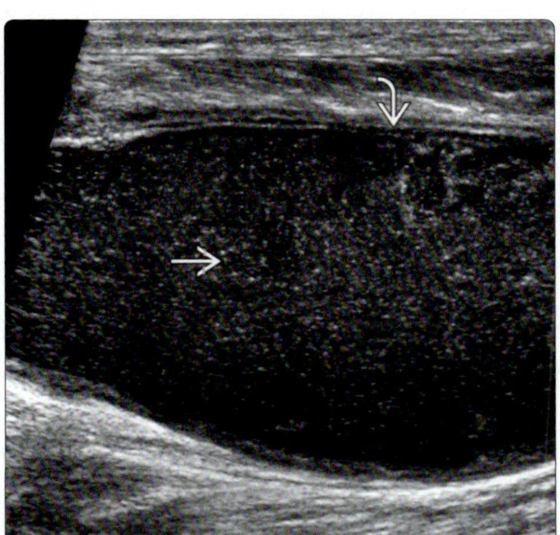

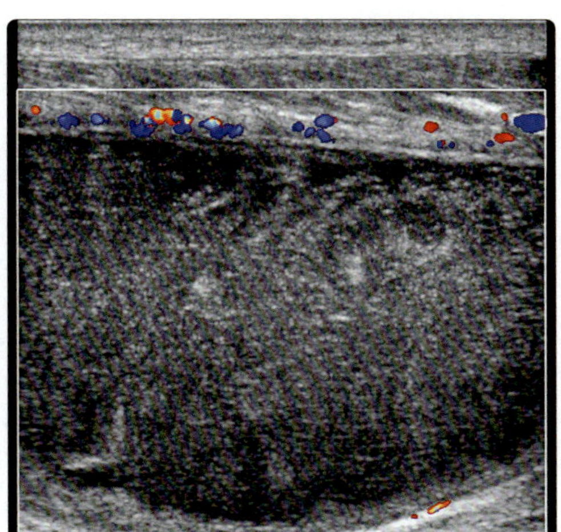

(Left) *Longitudinal US of the calf in a patient with end-stage renal failure shows multiple echogenic speckles and a well-defined wall, consistent with an abscess rather than a seroma, lymphocele, or liquefied hematoma. Aspiration yielded purulent fluid.* **(Right)** *Longitudinal color Doppler US of same lesion shows mild peripheral hyperemia. The size of intrinsic echogenic speckles is indicative of an abscess rather than seroma, lymphocele, or liquefied hematoma.*

Ligament Injury

KEY FACTS

TERMINOLOGY

- Ligament tear: Tear or avulsion of ligament fibers

IMAGING

- Ligament tears have similar appearances irrespective of ligament location
- Acute stage
  - Diffuse or localized ligament swelling
  - Tear can be partial or full thickness
  - Ligament retraction or discontinuity if complete
  - ± wavy, displaced ligament
  - Widening of ligament tear or joint space when ligament is stressed
  - Surrounding soft tissue edema and hemorrhage
- Chronic stage: 4 possible sequelae of ligament tear exist
  - Diffusely thickened continuous ligament
    - Can remain for many years following injury
    - Most common sequela of ligament tear
  - Diffusely thickened discontinuous ligament
  - Ligament is continuous but attenuated compared to normal ligament
    - Ligament laxity with widening of joint space is demonstrated during stress maneuver
  - Absent ligament
    - No ligament visible due to attrition or resorption of ligament fibers

PATHOLOGY

- Grade 1: Mild partial ligament tear
- Grade 2: Moderate-to-severe partial ligament tear
- Grade 3: Complete ligament tear or avulsion

CLINICAL ISSUES

- Most frequent musculoskeletal injury
- Usually does not require surgical intervention
  - Most partial and some complete ligament tears heal spontaneously
- Ligament tears can lead to joint instability or impingement

(Left) *Clinical photograph shows probe orientation for examining the anterior talofibular ligament (ATFL). The ankle is slightly plantar flexed and the probe orientation is transverse.* **(Right)** *Transverse US of the anterolateral ankle shows normal intact ATFL coursing between the fibula and talus. Note how the ligament is echogenic and taut with continuity of fibers. A small amount of physiological fluid fills the anterolateral recess of the ankle deep to the ligament.*

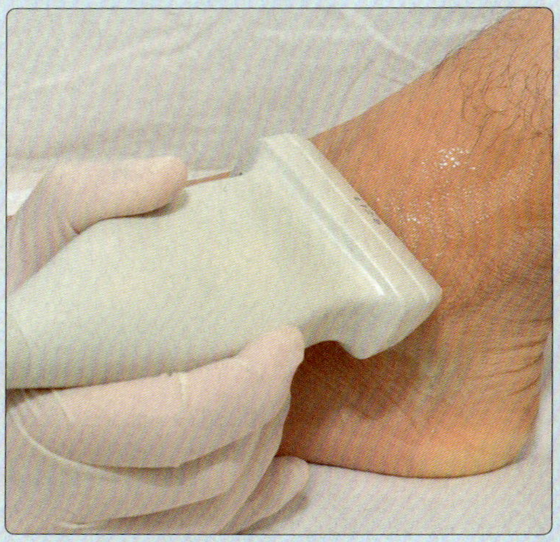

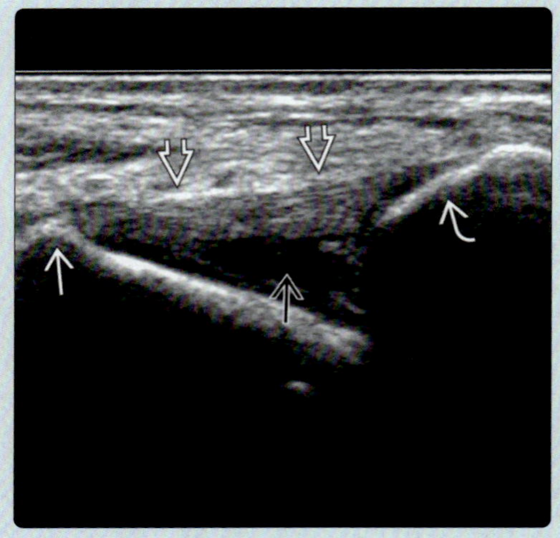

(Left) *Transverse US of the lateral ankle region shows an acute complete (grade 3) tear of the ATFL with loss of fibers in between the talus and the fibula. Hematoma partially fills the gap.* **(Right)** *Transverse US of the lateral ankle region shows an acute complete (grade 3) tear of the ATFL with loss of the normal fibers between the talus and the fibula. The intervening gap is partially filled by hematoma.*

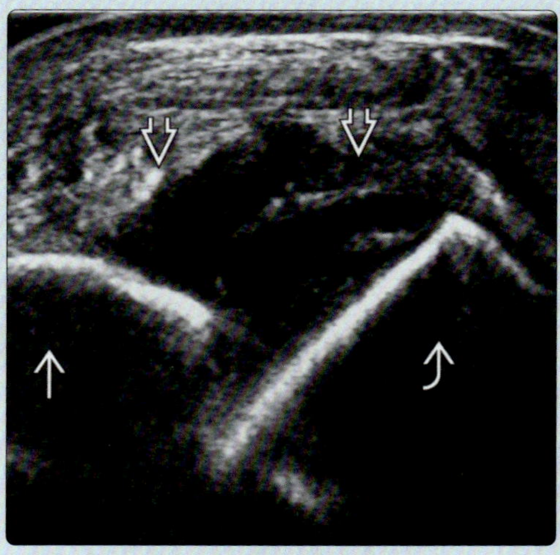

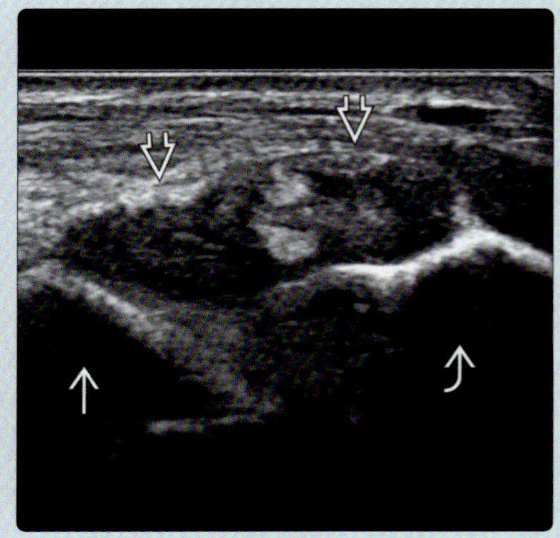

TERMINOLOGY

Synonyms

- Skiers or gamekeepers thumb: Injury of ulnar collateral ligament (UCL) of thumb
- Stener lesion: Proximal retraction of UCL, which then overlies adductor aponeurosis
- Ligament sprain or strain

Definitions

- Ligament: Fibrous band that connects 2 bones and helps stabilize intervening joint
- Ligament tear: Tear or avulsion of ligament fibers
- Avulsion: Some or all ligament fibers are torn away (avulsed) from their bony attachment
  - ± attached avulsed fragment of bone (avulsion fracture)
- Ligament sprain or strain: Partial ligament tear

IMAGING

General Features

- Best diagnostic clue
  - Discontinuity of ligament fibers
  - Swollen ligament with history of trauma
- Location
  - Some ligaments are more susceptible to tear
    - Ankle
      - Anterior talofibular ligament (ATFL)/calcaneofibular ligament (CFL)/anterior tibiofibular ligament (ATibFL): Anywhere along course of ligament
      - ATFL ligament tears before CFL; ATibFL tear is "high ankle sprain" and associated with deltoid ligament tear
      - Deltoid ligament: Particularly anterior deep fibers
    - Knee
      - Medial collateral ligament (MCL): Anterior fibers at femoral attachment
      - Lateral collateral ligament (LCL): Anywhere along course of ligament, usually in association with posterolateral corner injury
    - Thumb
      - UCL of metacarpophalangeal (MCP) joint: Particularly phalangeal attachment
    - Elbow
      - Radial collateral ligament: Particularly at humeral attachment
- Morphology
  - Normal
    - Normal fibrillar pattern and fibers preserved
    - No edema, normal thickness and tautness
  - Partial tear
    - Edema and swelling
    - Loss of fibrillar pattern ± partial discontinuity
    - ± laxity on stress testing
  - Complete tear
    - Discontinuity or complete attenuation of fibers
    - Avulsion of bony attachment

Radiographic Findings

- Soft tissue swelling ± joint effusion
  - ± avulsion fracture at ligament insertional area
  - ± diastasis of opposing bone margins
- Stress radiographs
  - Less frequently used nowadays
  - Mainly indicate complete ligament tear

Ultrasonographic Findings

- All ligament injuries have same appearance irrespective of where ligament is located
- Acute stage
  - Diffuse or localized ligament swelling
  - Ligament fiber discontinuity with hypoechoic gap at site of tear
    - Tear can be partial or full thickness
    - ± avulsion fracture at bony attachment
  - Ligament retraction if complete
    - ± wavy, displaced ligament
    - Widening of ligament tear or joint space when ligament is stressed
  - Surrounding soft tissue edema and hemorrhage
  - Surrounding soft tissue hyperemia on color Doppler
  - Variable size joint effusion
- Chronic stage: 4 possible sequelae of ligament tear exist
  - Diffusely thickened continuous ligament
    - Most common sequela of ligament tear, which can remain unchanged for many years following injury
    - Disorganized reparative tissue undergoing gradual remodeling
    - Even completely torn ligaments can regain full continuity with healing
    - ± ligament calcification
    - Pellegrini Stieda: Calcification at proximal attachment of MCL
      - Due to periosteal elevation, proximal end of medial femoral condyle ± associated with adductor magnus tendon avulsion
  - Diffusely thickened discontinuous ligament
    - Uncommon sequela
    - Stener lesion: Yo-yo sign due to retracted stump of torn ligament lying superficial to adductor aponeurosis
  - Diffusely attenuated ligament
    - Ligament is continuous but reduced in caliber compared to normal ligament
    - Repeated partial tear with fiber resorption
    - Ligament laxity with widening of joint space demonstrable during stress maneuver
  - Absent ligament
    - No ligament visible
    - Torn fibers undergo attrition or resorption during repair process
- Associated findings
  - Ankle ligament injury: Osteochondral injury talar dome
  - MCL tear of knee: Meniscal and cruciate ligament injury
  - LCL tear of knee: Posterolateral corner injury
- Chronic stress
  - Some ligaments react to chronic stress by reactive thickening, particularly superomedial fibers spring ligament

MR Findings

- If acute: Edema and thickening ± some fiber loss

- If chronic: Thickening or attenuation of fibers
- If complete: Fiber discontinuity ± hematoma filling gap

Imaging Recommendations

- Best imaging tool
  - US can assess ligament injuries
    - MR for associated injuries and for visualizing tears in ligaments with limited US access
- Protocol advice
  - Assess noninjured contralateral side
  - Align transducer along long axis of ligament
  - Adjust transducer alignment to minimize anisotropy
  - ± stretch ligament to maximize visibility of tear and determine joint laxity

DIFFERENTIAL DIAGNOSIS

Anisotropic Artifact

- US beam should be oriented perpendicular to ligament to prevent artifact
- Ligaments are best examined under stretch

Traction Apophysitis

- Apophysitis, epicondylitis, or avulsion fracture of medial humeral condyle (little league elbow syndrome)
- More common in adolescents

PATHOLOGY

General Features

- Etiology
  - Trauma with undue stretching of ligament
  - Epidemiology
    - Most frequent musculoskeletal injury and injured tissue around joint
    - Ankle sprain is most common cause of injury in college athletes
      - 85% of ankle ligament tears occur laterally, 15% medially

Staging, Grading, & Classification

- General
  - Grade 1: Mild partial tear/sprain
    - Ligament swollen with no visible ligament discontinuity or avulsion
  - Grade 2: Moderate-to-severe partial tear/sprain
    - Ligament severely swollen with discontinuity of some fibers
  - Grade 3: Complete ligament tear or avulsion

Gross Pathologic & Surgical Features

- Inflammatory phase: Bleeding → platelet-rich fibrin clot → release growth factors → repair stimulation
  - Acute inflammation begins immediately after injury and lasts for ~ 72 hours
- Proliferative phase: Fibroblasts → disorganized scar tissue and proteoglycan filling ligament gap → progressive remodeling
  - Lasts 4-6 weeks
- Reparative phase: Progressive collagen maturation
  - Lasts for months to years

Microscopic Features

- Normal histology
  - Ligament is composed of 2/3 water, 1/3 solid tissue; solid component is mainly (85%) type I collagen
  - Ligaments consist of bundles of collagen fibers aligned parallel along course of ligament
    - Collagen fibers are crosslinked and, if not under tension, undulate along long axis of ligament
    - Fibroblasts and proteoglycans reside in extracellular matrix between collagen fibers
- Remodeled ligament tissue remains morphologically, biochemically, and functionally inferior to normal ligament
  - Normal ligament (type I ligament 85%); healed ligament is mainly composed of type III collagen
  - Remodeled fibers are smaller with less crosslinking
    - Ligaments only regain 40-80% of original stiffness and strength but elasticity recovers to ~ 90%

CLINICAL ISSUES

Presentation

- Most common signs/symptoms
  - Joint pain, especially with apprehension testing
- Other signs/symptoms
  - Pain and swelling after traumatic event and instability

Natural History & Prognosis

- Complications
  - Chronic instability → chondral injury or premature osteoarthritis
  - Impingement due to hyperplastic scar tissue
  - Dystrophic calcification

Treatment

- Conservative
  - Rest, ice, compression, elevation
  - Controlled exercise regimen
    - Remobilization improves ligament remodeling
  - Prolotherapy (i.e., proliferative therapy)
    - US-guided injection of platelet-rich plasma (PRP) to torn ligament
      - Enhances fibroblastic and capillary proliferation
      - No proven clinical benefit of PRP treatment for ligament injury
- Surgical
  - Selected complete tear
    - Many complete tears will heal with conservative treatment
  - Ligament fixation to bone

DIAGNOSTIC CHECKLIST

Image Interpretation Pearls

- Examine contralateral noninjured side for comparison
- Perform dynamic assessment with stress maneuver if pain permits

SELECTED REFERENCES

1. Bica D et al: Reliability and precision of stress sonography of the ulnar collateral ligament. J Ultrasound Med. 34(3):371-6, 2015

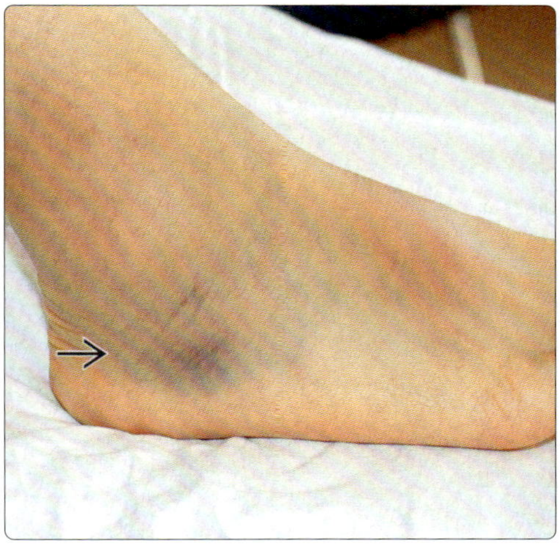

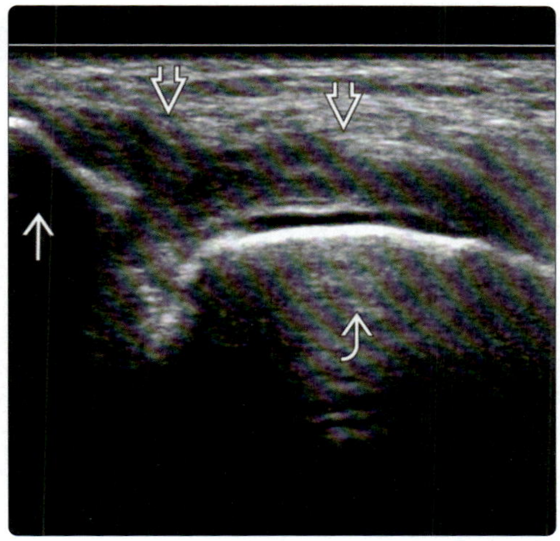

(Left) *Clinical photograph shows lateral ankle swelling with a dependent bruise → at the inframalleolar region 2 weeks following an inversion injury. Clinically, this is consistent with an ATFL tear.* **(Right)** *Longitudinal US of the same patient shows a severely thickened, hypoechoic ATFL → with fiber continuity between the distal fibula → and talus →, compatible with severe (grade 2) partial tear of the ATFL.*

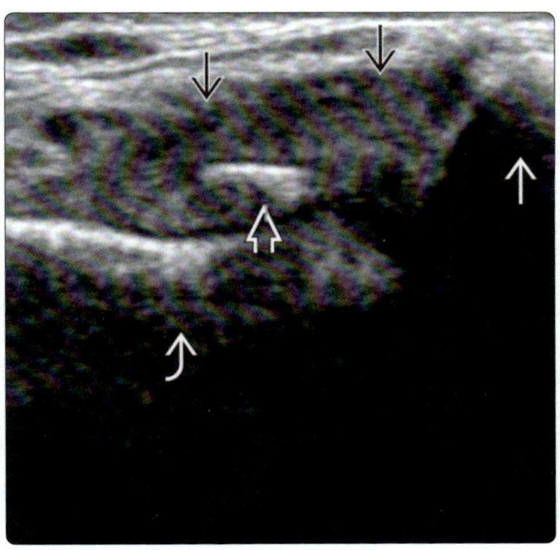

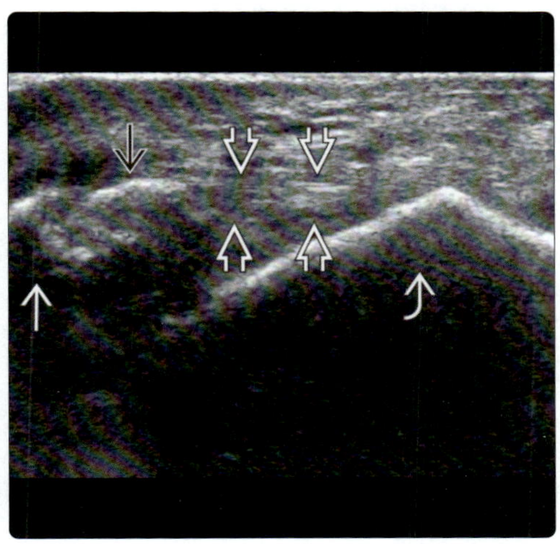

(Left) *Longitudinal US of the lateral ankle region shows an acute, high-grade partial tear of the ATFL → with a thickened, hypoechoic ATFL and loss of the normal fibrillar structure. There is a small, avulsed osseous fragment → from the talus →. No avulsion is present at the fibular attachment →.* **(Right)** *Longitudinal US of the lateral ankle region shows a fracture → of the distal fibula. The intact ATFL →, which attaches the fracture fragment → to the talus →, is barely seen.*

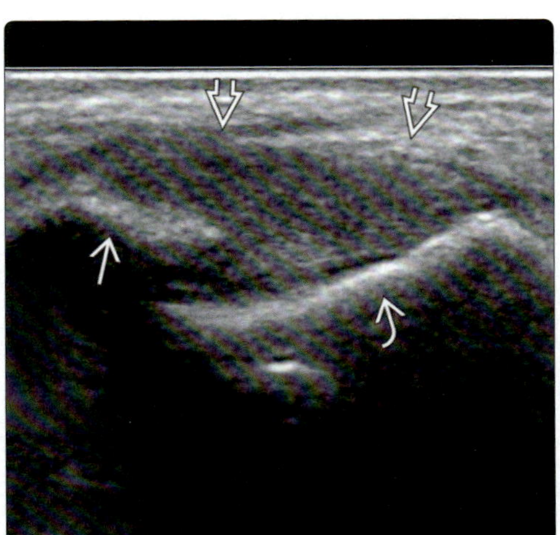

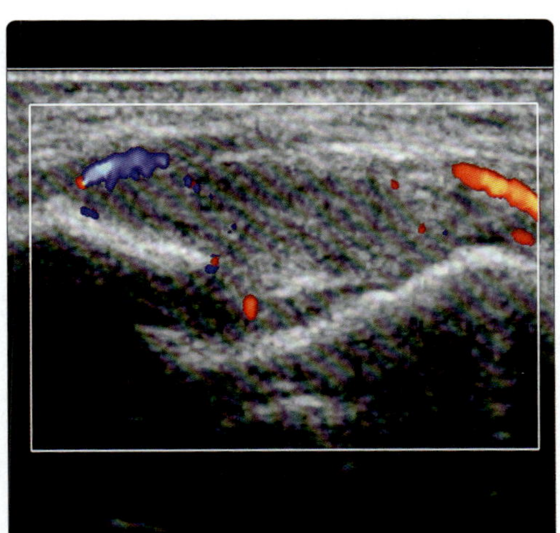

(Left) *Transverse US of the anterolateral ankle region 6 weeks after an inversion injury shows a severely thickened, hypoechoic ATFL → with loss of normal fibrillar pattern between the distal fibula → and the talus →, compatible with a healing partial tear.* **(Right)** *Color Doppler US at the same area shows mild peri- and intraligamentous hyperemia consistent with a healing ligament. The tissues around the ATFL are normally well vascularized.*

(Left) *Clinical photo shows probe orientation for examining the calcaneofibular ligament (CFL). Ligament visibility can sometimes be improved by dorsiflexing the ankle.* **(Right)** *Longitudinal US of the same patient shows the normal CFL coursing between the fibular tip and the calcaneum deep to the peroneal tendons.*

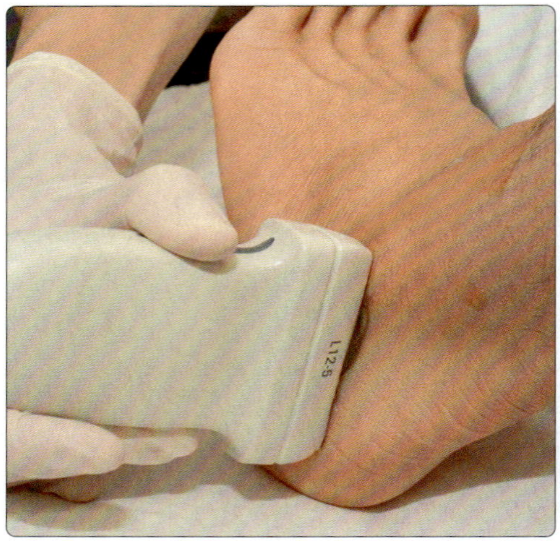

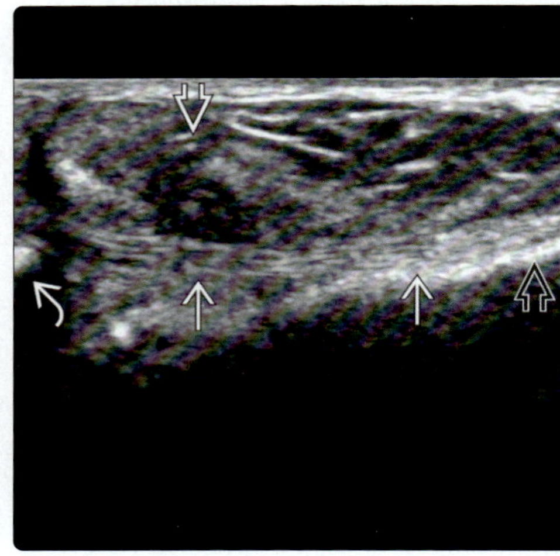

(Left) *Longitudinal US of the ankle shows a complete (grade 3) CFL tear with a severely thickened, hypoechoic ligament overlying the calcaneum and subcutaneous edema. The peroneal tendons are located close to the calcaneum due to loss of the CFL sling effect.* **(Right)** *Axial PD FSE MR of the ankle in a patient with an injury 2 weeks previously shows a severely swollen CFL and the peroneal tendons close to the calcaneum consistent with a complete (grade 3) CFL tear.*

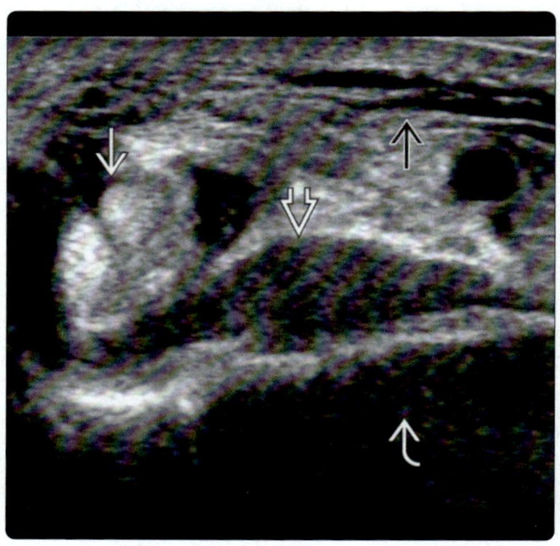

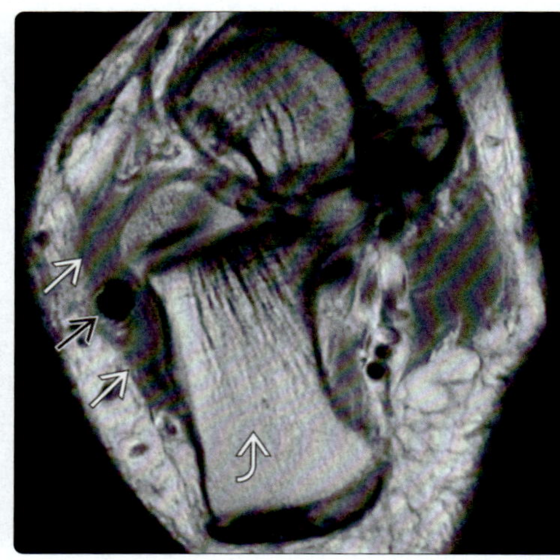

(Left) *Clinical photograph shows the probe orientation for assessing the deltoid ligament. The proximal end of the probe is on the medial malleolus while the distal end sweeps anterior to posterior to view different components of the deltoid ligament.* **(Right)** *Longitudinal US of the medial malleolus shows the normal fan-shaped deltoid ligament deep to the tibialis posterior tendon. The 5 different components of the deltoid ligament are not separable on US.*

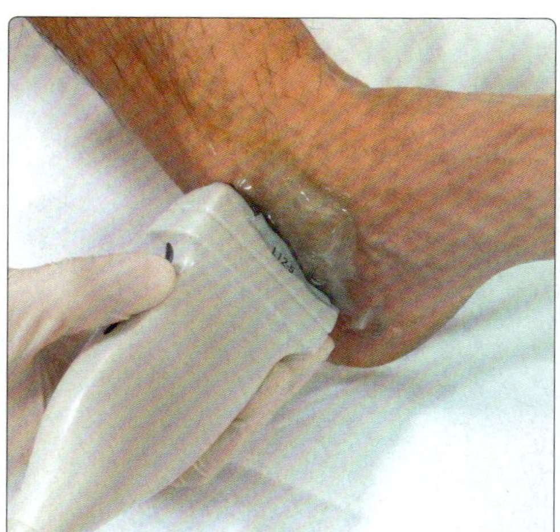

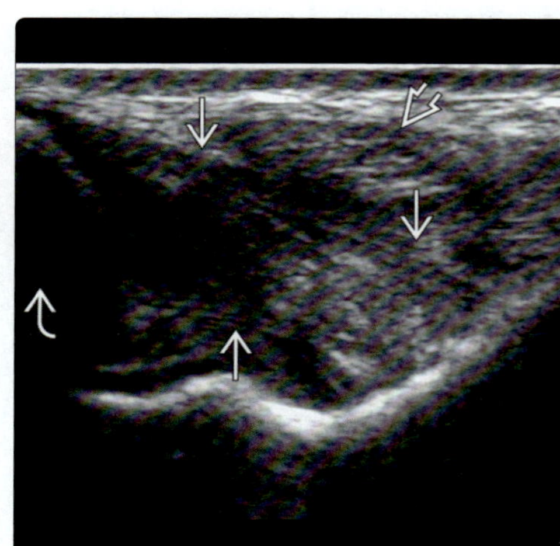

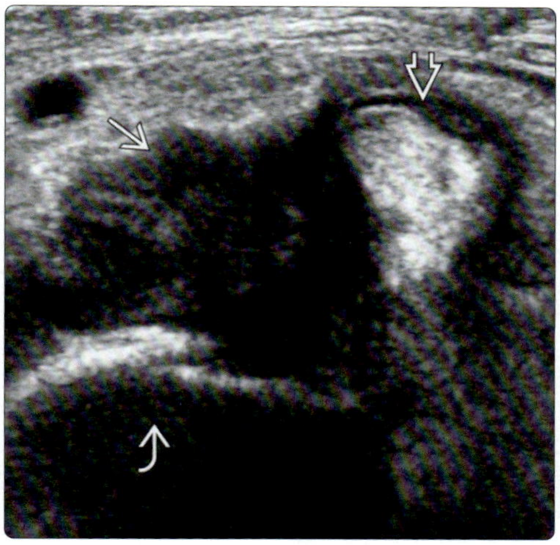

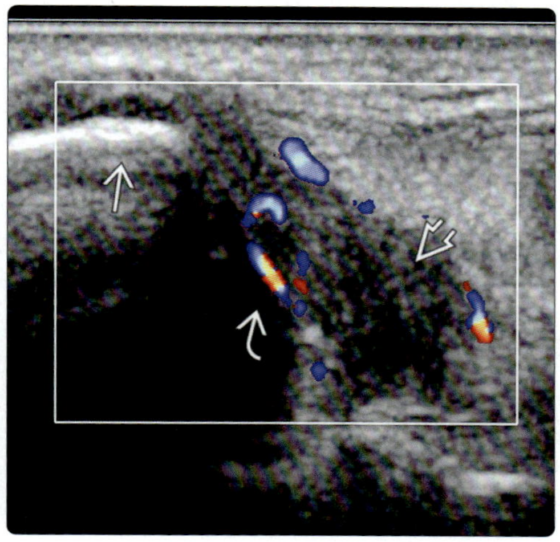

(Left) *Transverse US of the medial inframalleolar region 1 day after injury shows a severe tear of the deltoid ligament with a hypoechoic, severely thickened deltoid ligament located deep to the tibialis posterior tendon.* **(Right)** *Color Doppler US of a patient with ankle trauma 1 month previously shows a healing high-grade partial tear of the deltoid ligament with moderate peripheral vascularity just distal to the tip of the medial malleolus.*

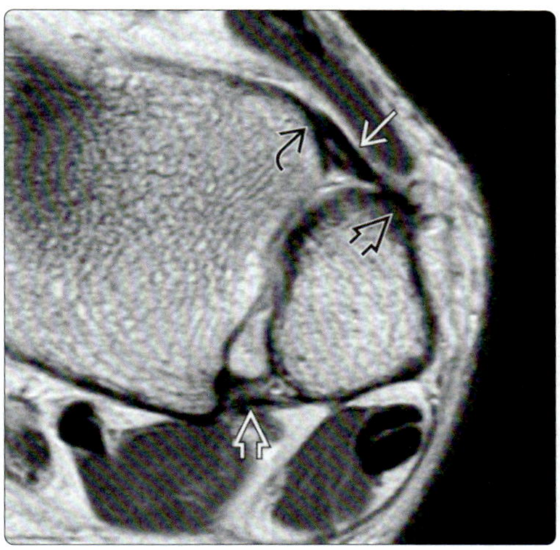

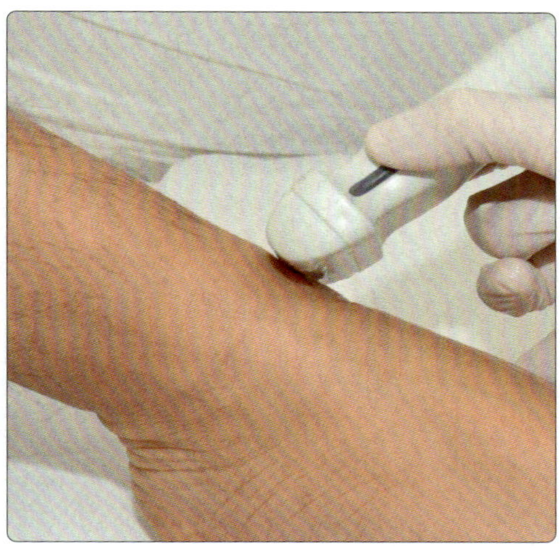

(Left) *Axial PD FSE MR scan of the tibiofibular articulation shows a normal anterior tibiofibular ligament with intact low-signal fibers connecting the anterolateral margin of the tibia to the fibula. The normal posterior tibiofibular ligament with intact low-signal fibers is also seen connecting the posterior tibia and fibula.* **(Right)** *Clinical photograph shows the transducer position for examining the anterior tibiofibular ligament (ATibFL).*

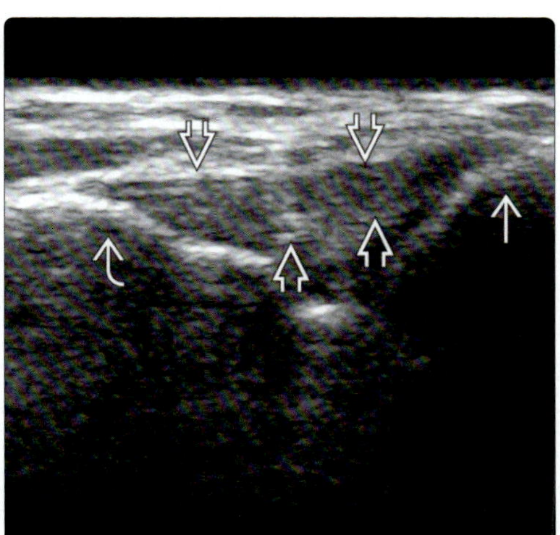

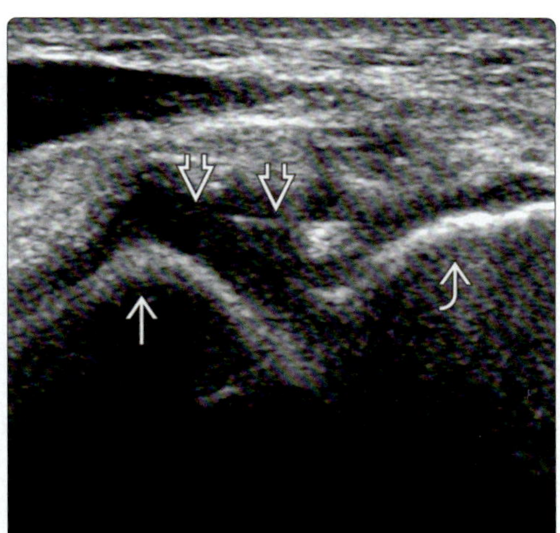

(Left) *Longitudinal US of a normal subject shows intact ATibFL with normal fibrillar structure coursing between the talus and fibula. The ligament is normally wider at the tibial end.* **(Right)** *Transverse US of the anterior leg shows complete tear of the ATibFL coursing between the tibia and fibula with the gap filling up with fluid. There is no widening of the tibiofibular distance.*

(Left) *Clinical photograph of a standing foot shows pes planus with flattening of the medial longitudinal arch .* **(Right)** *Longitudinal oblique US of the medial foot in a patient with pes planus shows severe thickening of the superomedial fibers of the spring ligament. These fibers are located deep to the posterior tibialis tendon and medial to the talar head and navicular bones . The ligament thickens from chronic stress.*

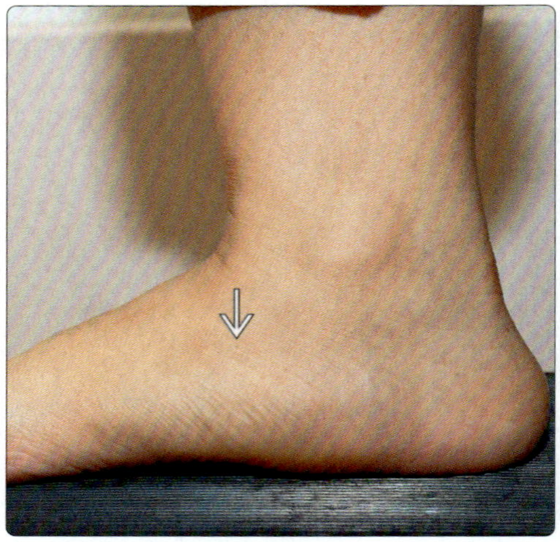

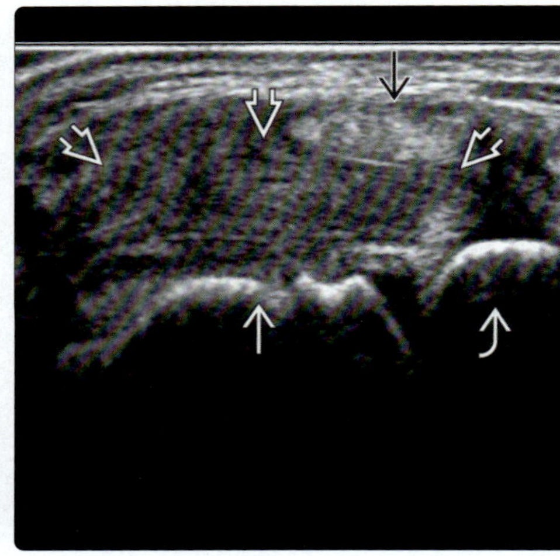

(Left) *Coronal T2WI FS MR shows a severe tear of the anterior fibers medial collateral ligament (MCL) with partial periosteal stripping of adductor magnus insertion and partial tear of the superior coronary ligament of the medial meniscus .* **(Right)** *Longitudinal US shows a complete MCL tear at the femoral attachment with swelling and loss of normal fibrillar echotexture. The distal MCL fibers are normal . MCL tears most commonly affect the anterior fibers at the femoral attachment.*

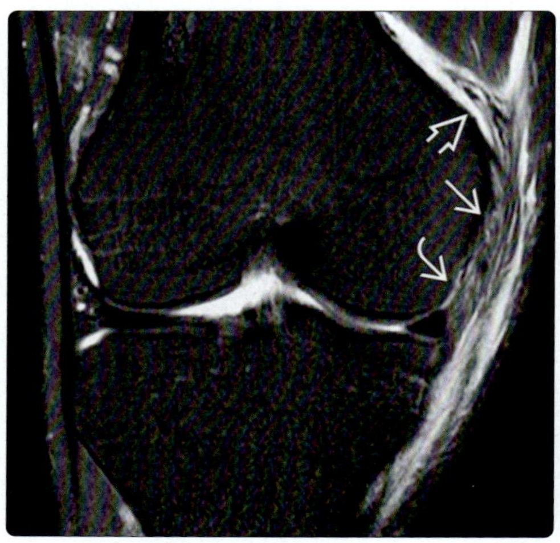

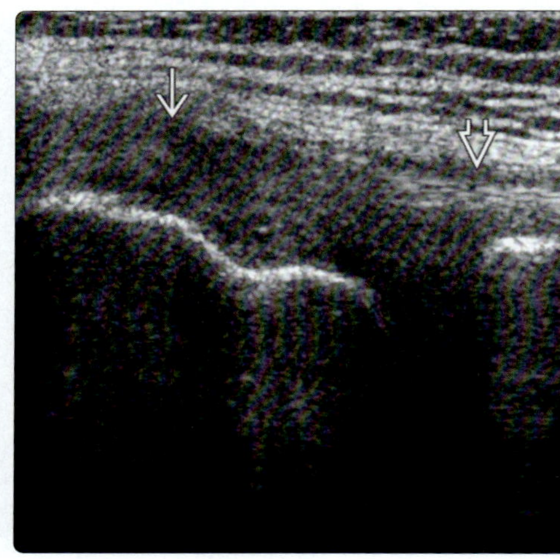

(Left) *Longitudinal US shows moderate thickening of the anterior MCL fibers at the femoral attachment , especially the superficial fibers. Fiber structure and continuity can still be resolved, consistent with a moderate severity partial tear.* **(Right)** *Longitudinal US shows a Pellegrini-Stieda lesion with linear calcification just proximal to a healing MCL tear at the femoral attachment . This dystrophic calcification is due to either adductor magnus tendon insertional injury or femoral periosteal stripping.*

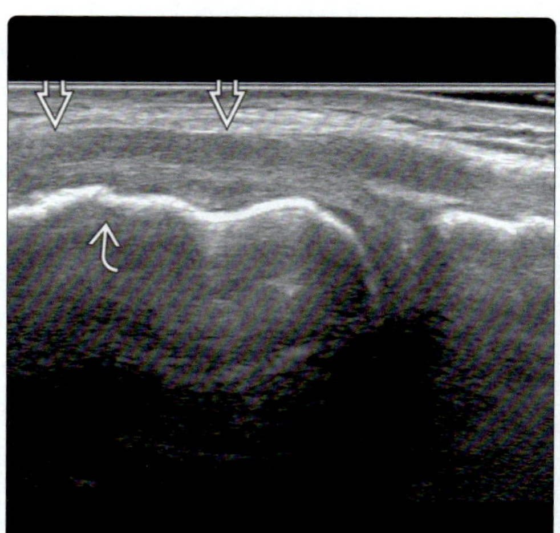

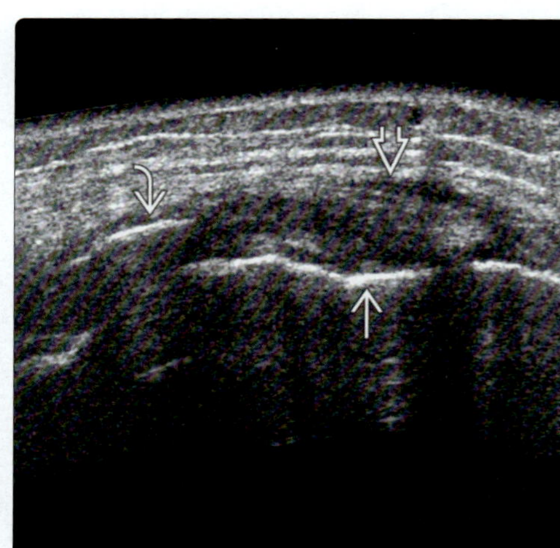

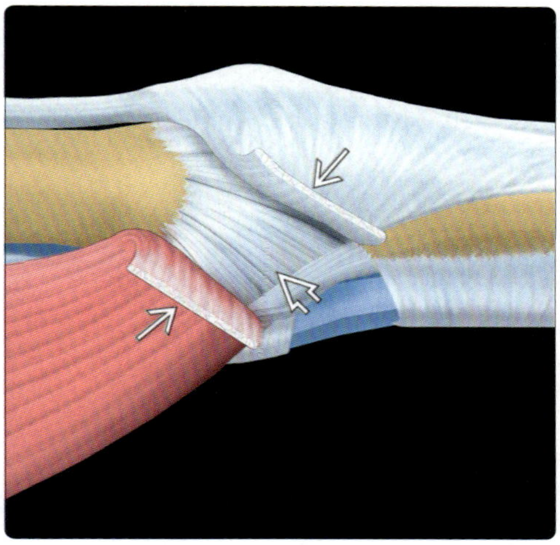

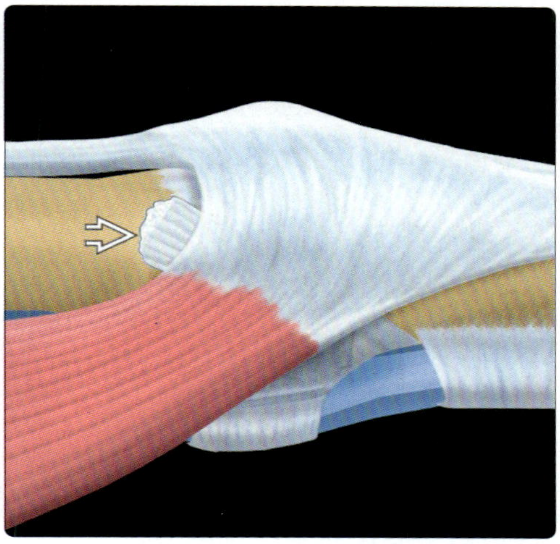

(Left) *Graphic shows the ulnar aspect of the metacarpophalangeal (MCP) joint of the thumb and a reflected adductor aponeurosis ➡ with the ulnar collateral ligament (UCL) ➡ beneath.* **(Right)** *Graphic shows the ulnar aspect of the MCP joint of the thumb. When a complete tear of the UCL occurs, the proximal end of the ligament may retract back ➡ over the adductor aponeurosis. This is referred to as a Stener lesion.*

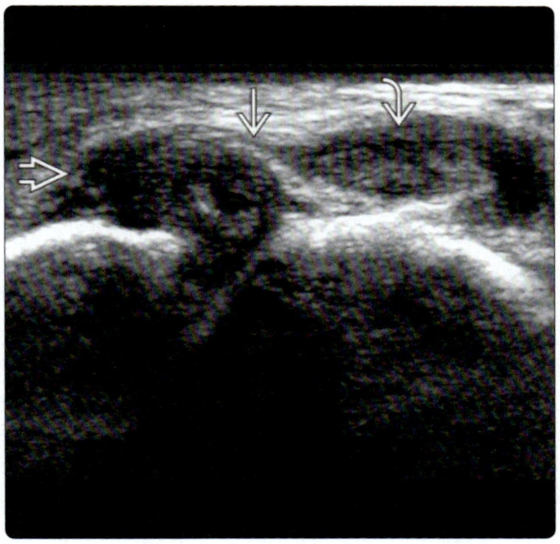

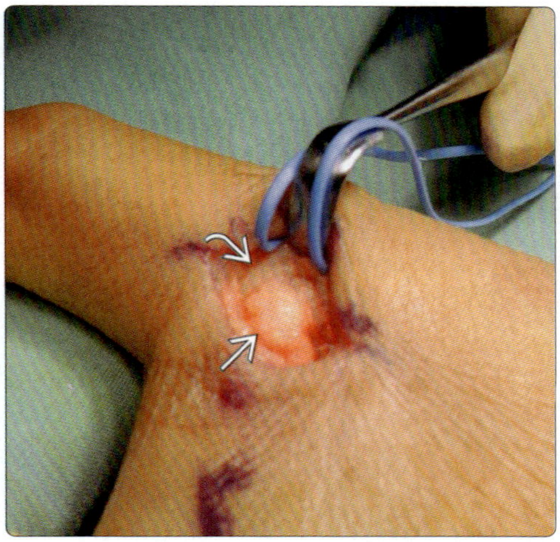

(Left) *Longitudinal US of the ulnar aspect of the MCP joint of the thumb shows a swollen, irregular, and hypoechoic torn distal end of the UCL ➡. The swollen proximal end of the UCL ➡ is displaced proximal to the adductor aponeurosis ➡.* **(Right)** *Clinical intraoperative photograph of the same patient shows a Stener lesion with the torn and retracted UCL ➡ located proximal and slightly superficial to the adductor aponeurosis ➡.*

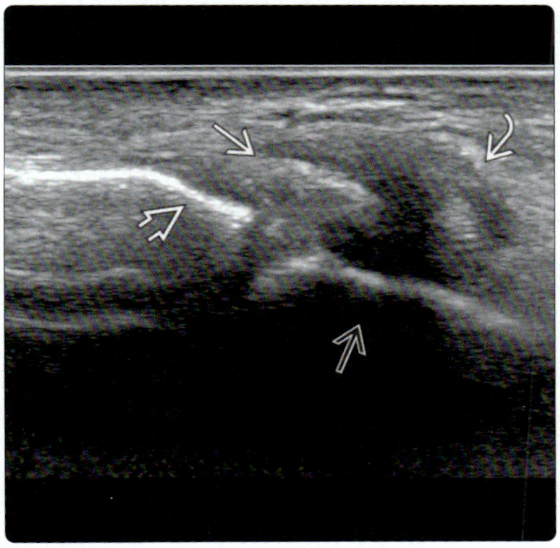

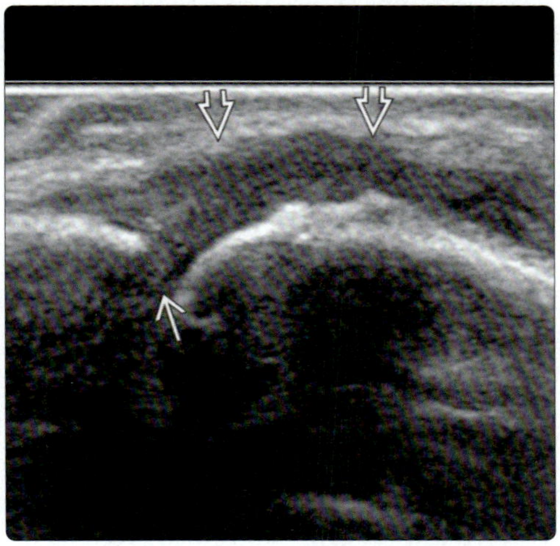

(Left) *Longitudinal US of thumb 1 week following an abduction injury is shown. The swollen ulnar collateral ligament ➡ of the 1st MCP joint is completely torn from the proximal phalanx ➡ and is retracted superficial to the adductor aponeurosis ➡. This is a Stener lesion. This shows the 1st metacarpal bone ➡.* **(Right)** *Longitudinal US of the radial side of the proximal interphalangeal joint of the little finger ➡ shows a severely thickened, hypoechoic radial collateral ligament ➡, compatible with a high-grade partial tear.*

Bone Fracture

KEY FACTS

TERMINOLOGY

- Fracture (fx): Traumatic discontinuity of bone usually involving both cortical and trabecular bone
- Stress fx: Fx occurring in normal bone due to excessive repetitive submaximal loading
- Insufficiency fx: Fx occurring in diffusely weakened bone subjected to normal loading
- Pathologic fx: Fx occurring in focally weakened bone subjected to normal loading

IMAGING

- Standard radiographs will **not** show
  - 85% rib fxs, 50% nasal bone fxs, 15% scaphoid fxs
  - 25% pediatric elbow fxs, 100% chondral fxs
- Fx
  - Focal cortical disruption with soft tissue edema, hematoma, or callus ± reverberation artifact at fx line
  - → fx edge resorption with hypoechoic osteoid deposition across fx line
  - → progressive callus mineralization
- Stress fx
  - Thickened hyperechoic periosteum at symptomatic site
  - Visible on US before radiographs
  - Hyperemic thickened periosteal rim on color Doppler imaging
- Pathologic fx
  - Often wider fx line ± cortical irregularity extending beyond fx margins
  - ± visible tumor protruding from fx gap
  - US-guided biopsy helpful
- Chondral fx
  - Focal disruption of hypoechoic cartilage
  - ± juxtachondral soft tissue edema or hematoma
- Insufficiency fx
  - US insensitive to detection in early stage
  - Focal cortical hyperostosis or exuberant callus in later stages

(Left) *Transverse US shows an acute undisplaced fracture in the anterior aspect of the rib with focal discontinuity of the cortex, minimal soft tissue swelling, and posterior reverberation ("chimney phenomenon"). Note that the costochondral junction and hypoechoic costal cartilage radiograph was normal.* **(Right)** *Transverse US shows a mildly angulated acute fracture in the anterior axillary line with mild overlying soft tissue swelling. Radiograph was normal.*

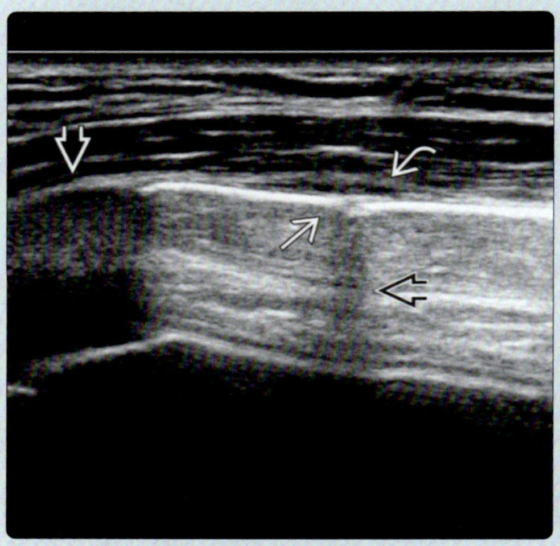

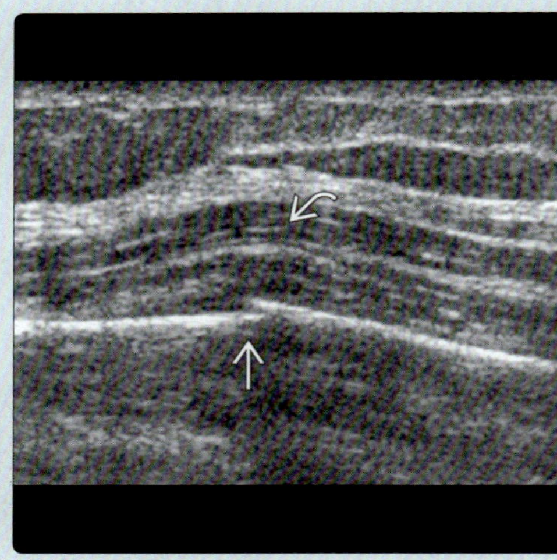

(Left) *Transverse US shows a partially healing fracture in the midaxillary line. There is hypoechoic callus surrounding the fracture with echogenic mineralization more deeply. This stage of fracture healing may no longer be painful.* **(Right)** *Transverse US shows a healing fracture at the anterior aspect of the rib close to the costochondral junction. Most of the callus has mineralized. There is no soft tissue swelling.*

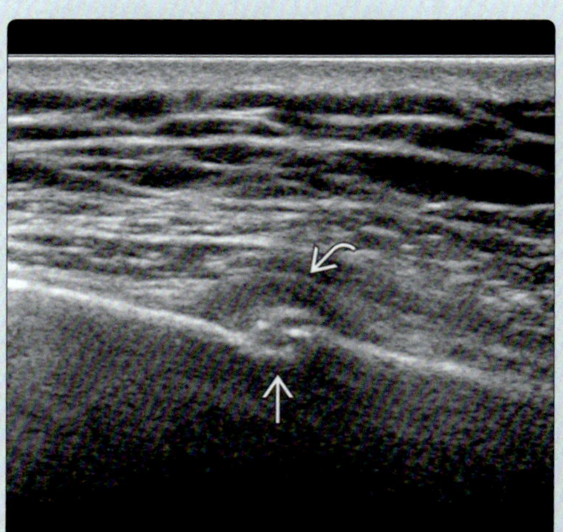

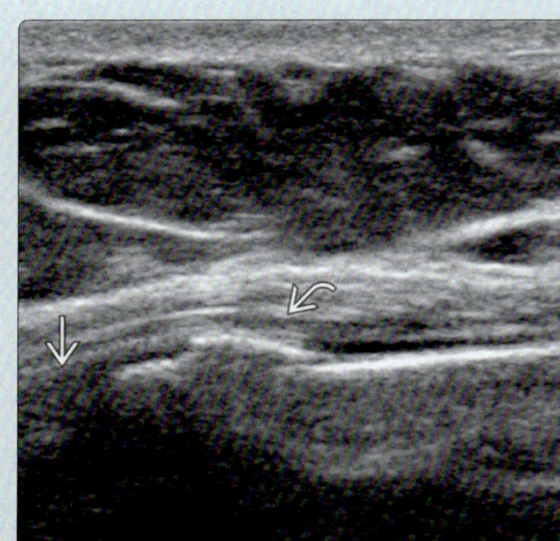

TERMINOLOGY

Definitions

- Fracture (fx): Traumatic discontinuity of bone usually involving both cortical and trabecular bone
- Trabecular microfracture: Fx limited to trabecular component of bone without clear fx line
  - Diffuse bone marrow edema on MR imaging but no discrete fx line
- Stress fx: Fx occurring in normal bone due to excessive repetitive submaximal loading
- Stress reaction of bone: Trabecular microfracture occurring in normal bone due to repetitive excessive submaximal loading
- Insufficiency fx: Fx occurring in diffusely weakened bone subjected to normal loading
  - Usually due to osteoporosis; other causes include renal osteodystrophy and chronic bisphosphonate therapy
- Pathologic fx: Fx occurring in focally weakened bone subjected to normal loading
  - Usually due to metastatic infiltration or primary bone tumor; infection is another cause

IMAGING

General Features

- Best diagnostic clue
  - Focal disruption of bone cortex with juxtacortical soft tissue edema/hemorrhage ± callus formation
- Location
  - Most common in ribs, distal radius, scaphoid, metacarpals, phalanges, metacarpals, nasal bones, skull, and vertebral bodies

Radiographic Findings

- Standard method of fx assessment
  - Shows majority of fx to good effect
  - Limited depiction of some fxs
  - Standard radiographs will not show
    - 85% of rib fxs, 50% of nasal bone fxs, 15% of scaphoid fxs
    - 25% of pediatric elbow fxs, 100% of chondral fxs
    - < 5% of proximal femoral and distal radial fxs in osteopenic patients

MR Findings

- Gold standard for fx detection
  - Detects both cortical and trabecular component of fx as well as stress reaction with high sensitivity, specificity, and negative predictive value

Ultrasonographic Findings

- Fx
  - Focal cortical disruption with adjacent soft tissue edema, hematoma, or callus
    - ± reverberation artifact ("chimney phenomenon") or shadowing artifact at fx site
    - Effusion, hemarthrosis, or lipohemarthrosis if intracapsular surface extension is present
  - Later, fx edge resorption occurs with hypoechoic osteoid deposition across Fx line
  - → progressive callus mineralization
  - → complete callus mineralization → smooth bump on cortical contour
- Stress fx
  - Thickened rim of hyperechoic periosteum at symptomatic site
    - Visible on US before radiographs
  - Hyperemia of thickened periosteum on color Doppler imaging
  - → progressive increasing echogenicity of periosteal thickening due to increasing bone deposition
- Pathologic fx
  - Often wider fx line ± cortical irregularity extending beyond fx margins
  - ± visible tumor protruding from fx gap
- Insufficiency fx
  - Periosteal thickening and hyperemia with adjacent soft tissue edema
  - ± focal cortical irregularity or disruption
  - Overall sensitivity of US for early insufficiency fx is low
    - Propagates from trabecular bone and is therefore not visible on US at early stage
- Chondral fx
  - Focal disruption of hypoechoic cartilage
  - ± juxtachondral soft tissue edema or hematoma
    - Heals by fibrocartilaginous repair
    - No callus formation

Imaging Recommendations

- Best imaging tool
  - Radiography ± US for selected fx detection
  - Increasingly, US is advocated as best initial investigation for suspected rib, sternal, nasal bone, and skull fxs
  - With suitable training, US can detect or exclude fxs in these areas with high (> 90%) sensitivity and specificity
    - US can be used as standalone investigation or as supplement to radiography
  - US assessment of fx is severely limited in
    - Retroscapular area of rib cage and spine
    - Intraarticular areas, carpus or tarsus
    - Reduced transducer access due to open wound, severe abrasion, or external fixation
    - Gas or severe trauma to soft tissues impairing acoustic transmission
    - Assessment of any trabecular fx
    - Unconscious or uncooperative patient
      - Cannot localize painful area
    - Complex fx as overall fx perception is limited
  - In addition to fx detection, US can also
    - Check alignment of fx post reduction
      - More time efficient compared to repeating radiographs
    - Guide hematoma local anesthetic block
      - Significantly ↓ need for systemic analgesics in immediate post-fx period
    - Check for neurovascular bundle injury in at-risk fx
      - For example, displaced pediatric supracondylar fx (Garland type III)
    - ~ 25% have compromised brachial artery flow
    - Due to kinking, spasm, thrombosis ± laceration of brachial artery
- Protocol advice

- Focus examination on area of maximum pain or tenderness
- Align transducer with long axis of bone
- Check for cortical as well as juxtacortical signs of fx
- Determine location, displacement, angulation, comminution, and alignment of fx as well as integrity of adjacent cortex
  - Check juxtacortical soft tissues for edema, hematoma, presence and maturity of callus
- ± finger-palpation test to ensure US abnormality corresponds to point of maximum tenderness
- Check for fx-related complications
- For rib fx
  - Usually examine patient in decubitus position but also in supine or prone positions depending on where painful site is located
  - Ask patient to point to most painful area and start examination at this site
  - Align transducer along long axis of rib
  - Then examine this rib from costosternal junction to costovertebral junction
  - Repeat procedure for all ribs in painful area
  - If rib fx is detected, note site of fx on skin with bleb of US gel
    - Place transducer in oblique longitudinal plane over 12th rib, and count ribs backward toward bleb of gel
    - In addition to rib number, report whether rib fx is located in midscapular line; posterior, mid-, or anterior axillary line; or midclavicular line
  - Only weak correlation exists between degree of rib fx displacement/alignment and time to clinical healing
  - Check for pneumothorax
    - Align transducer within intercostal space parallel to ribs at least dependent portion of affected hemithorax
    - Look for absence of gliding sign (lung gliding at pleural interface on real-time scanning)
    - Look for absence of moving comet-tail artifacts at lung surface (on real-time scanning)
    - US is as sensitive as radiography for pneumothorax detection
  - Check for pleural effusion or hemothorax at most dependent portion of affected hemithorax
  - ± brachial plexus injury in upper rib fx
    - C8 and T1 roots most common injury, though least well seen on US
    - Check for presence of pseudomeningocele
- For nasal bone fx
  - Examine longitudinal and transverse planes midline of nasal bones, right and left lateral nasal bones
    - Evaluate osseous and cartilaginous components
    - Ideally, use "hockey stick" transducer
    - US more sensitive than radiography, though slightly less than CT for nasal bone fx
    - US cannot evaluate nasal septum
  - If nasal bone fx depressed on US, CT can be helpful to further evaluate

DIFFERENTIAL DIAGNOSIS

Physiological Cortical Irregularity or Physis

- No adjacent soft tissue edema, hematoma, or callus

Bone-Allograft or Bone-Prosthetic Junction

- No adjacent soft tissue edema, hematoma, or callus

PATHOLOGY

Staging, Grading, & Classification

- Hematoma at fx site with inflammatory response
  - Production and release of proinflammatory cytokines
  - Peaks at 24 hours; completes after 7 days
  - Local and systemic recruitment of mesenchymal stem cells
  - Fibrin-rich granulation tissue forms and replaces hematoma in and around fx site
- Elevation and stimulation of periosteum → intramembranous ossification at subperiosteal site (hard callus) → bridging woven bone
- Endochondral formation within granulation tissue (soft callus) → progressive endochondral ossification of soft callus
- Bone remodeling → original lamellar structure and external shape

CLINICAL ISSUES

Presentation

- Most common signs/symptoms
  - Acute localized pain and tenderness after trauma
- Other signs/symptoms
  - Swelling, hematoma, ecchymosis, and limitation of movement
  - Stress fx: Pain induced by exercise and relieved by rest
    - Pain increases with increasing intensity of exercise

Natural History & Prognosis

- Severe pain for ~ 1 week; symptom free in 6-8 weeks

Treatment

- Initially, rest, ice, compression, and elevation
- ± fx reduction, ± cast immobilization
- Functional casting or brace
- External or internal metallic fixation

DIAGNOSTIC CHECKLIST

Consider

- Using US to detect fx in those known sites where radiography is limited

Image Interpretation Pearls

- Localized cortical discontinuity at tender spot with adjacent soft tissue injury or reparative change

SELECTED REFERENCES

1. Lee IS et al: Ultrasonography in the diagnosis of nasal bone fractures: a comparison with conventional radiography and computed tomography. Eur Arch Otorhinolaryngol. 273(2):413-8, 2016

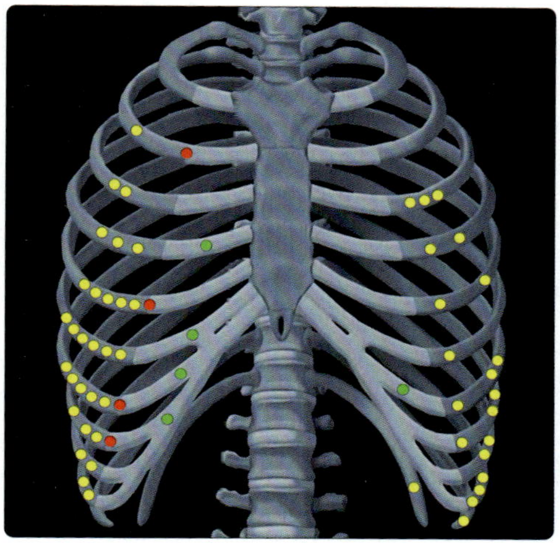

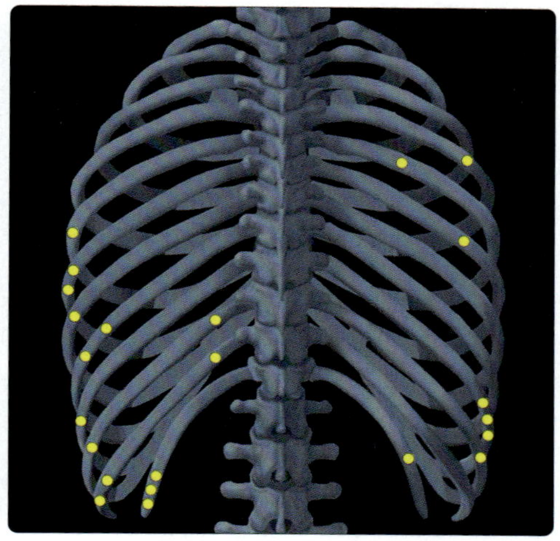

(Left) *Graphic of the anterior chest wall shows the common locations of rib fractures (yellow dots), costochondral junction fractures (red dots), and costal cartilage fractures (green dots).* **(Right)** *Graphic of the posterior chest wall shows the common locations of rib fractures (yellow dots). Note that most chest wall fractures occur anteriorly (preferentially affecting the lower ribs) and are more common on the right side anteriorly and on the left side posteriorly.*

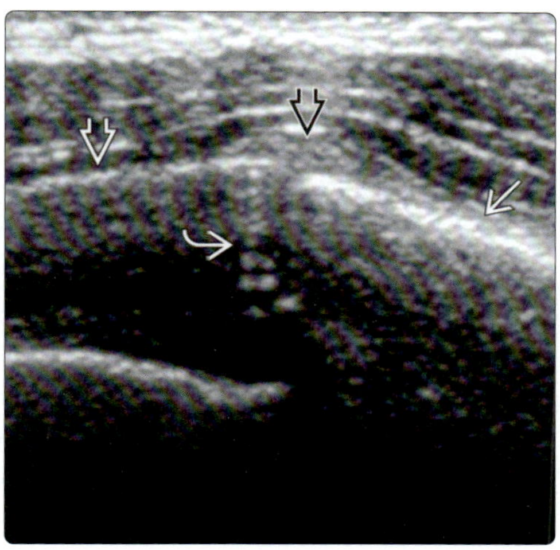

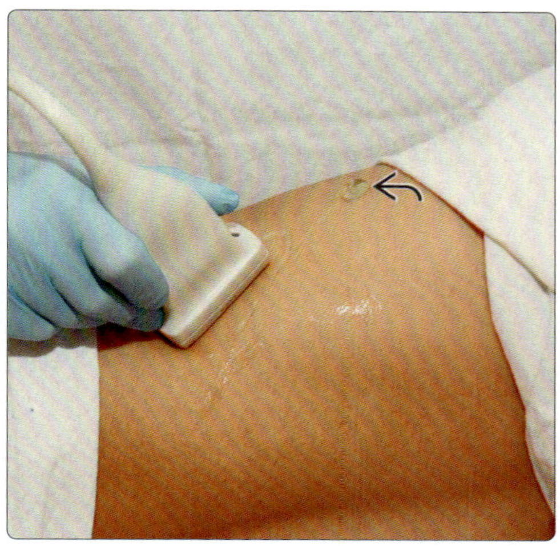

(Left) *Transverse US shows an acute undisplaced costochondral junction fracture with mild soft tissue swelling anterior to the junction between the osseous and chondral portions of the rib. Note reverberation artifact at the fracture site ("chimney phenomenon").* **(Right)** *Photograph shows how to determine which rib is fractured. Mark fracture site with gel. Place transducer transversely over the 12th rib, and move it upward to the fracture site while counting down the ribs as the transducer passes over them.*

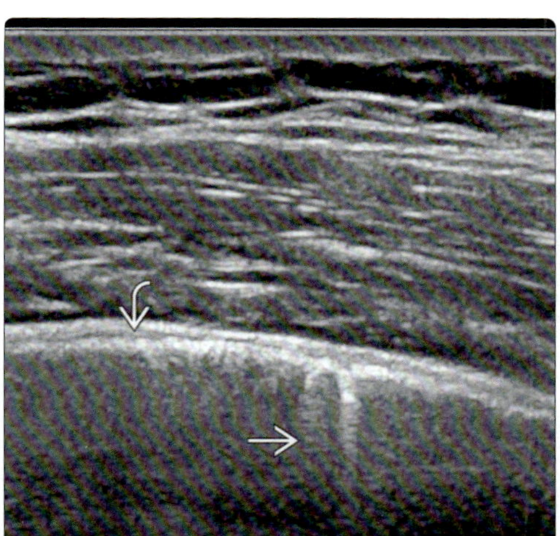

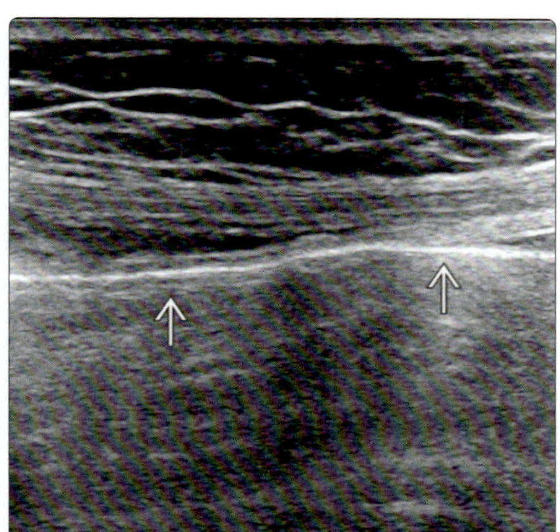

(Left) *Transverse US of a normal lung while scanning in the intercostal space shows normal, thick, irregular echogenic pleural interface with reverberation artifacts, which moves to and fro on real-time grayscale imaging.* **(Right)** *Transverse US shows pneumothorax with a thin, sharp echogenic line on the parietal pleural surface and absence of reverberation artifact. On real-time imaging, no pleural interface to-and-fro movement is present, enabling clear recognition of pneumothorax.*

(Left) *Longitudinal US of an active surf ski paddler shows a stress fracture at the anterior aspect of the right 6th rib. There is hyperechoic juxtacortical hyperostosis and hypoechoic soft tissue swelling.* **(Right)** *Transverse color Doppler US of the right 6th rib at the same location shows a hypoechoic rim of juxtacortical soft tissue swelling with moderate hyperemia. The normal 5th rib is shown.*

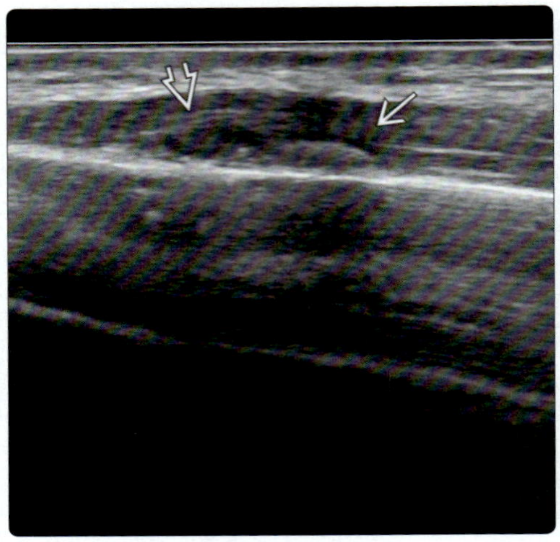

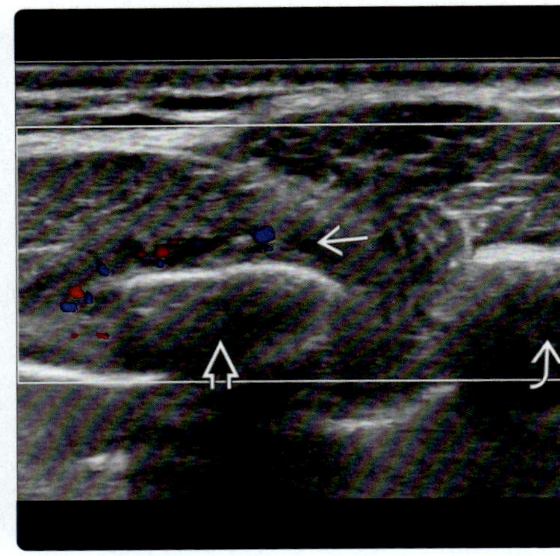

(Left) *Longitudinal US in a patient with anterior chest pain following a fall shows an angulated fracture of the sternal body with mild associated soft tissue swelling. Radiograph was normal.* **(Right)** *Sagittal T1 MR confirms a mildly angulated transverse fracture of the sternal body. A high percentage of sternal fractures are not visible on radiographs. When compared with MR, US will allow correct identification or exclusion of most sternal fractures.*

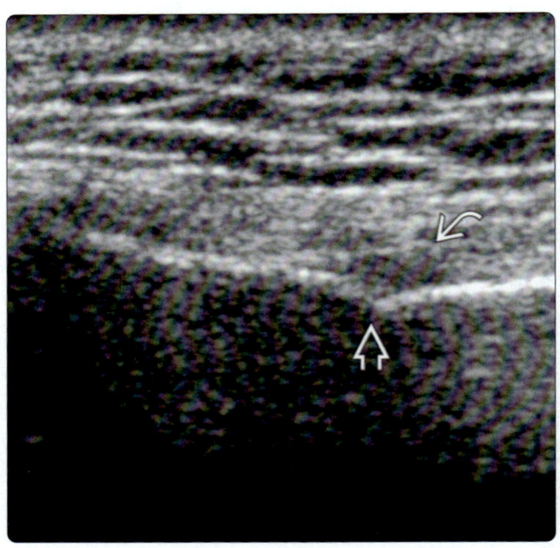

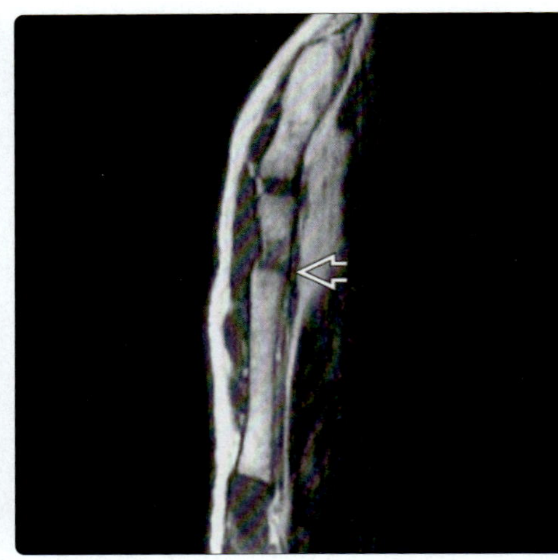

(Left) *Clinical photograph shows depressed fracture of nose. US is almost as sensitive as CT for detecting both lateral nasal wall and midline nasal bone fractures.* **(Right)** *Longitudinal midline US of the same patient shows a very comminuted, midline nasal fracture involving both the osseous and cartilaginous components. US, unlike CT, cannot assess the nasal septum.*

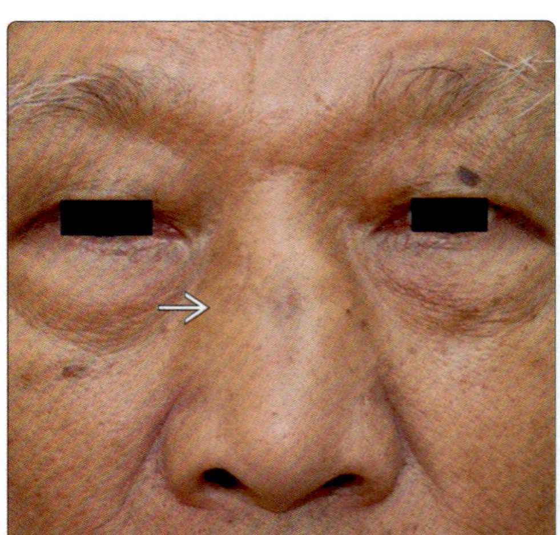

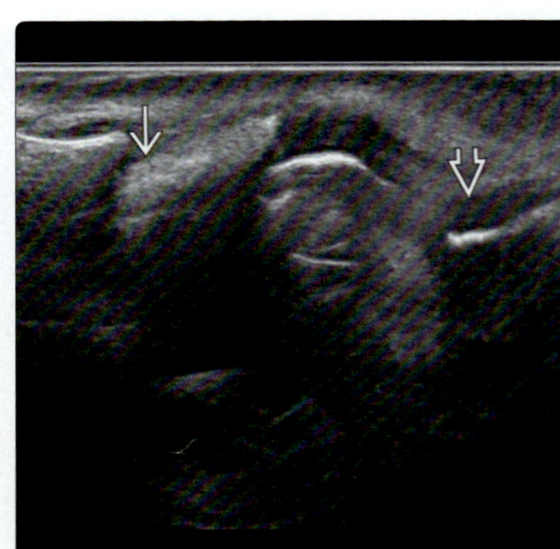

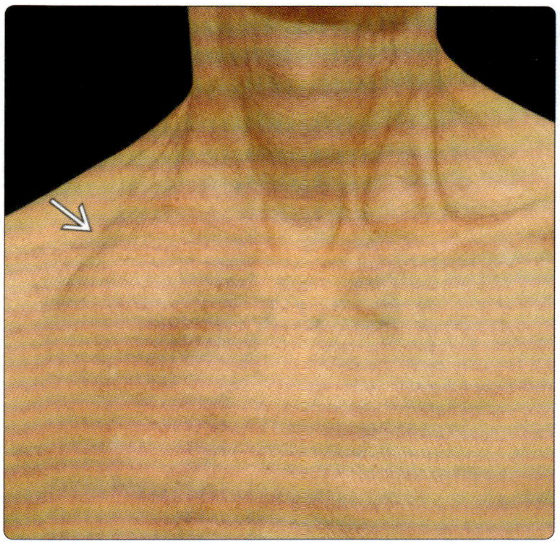

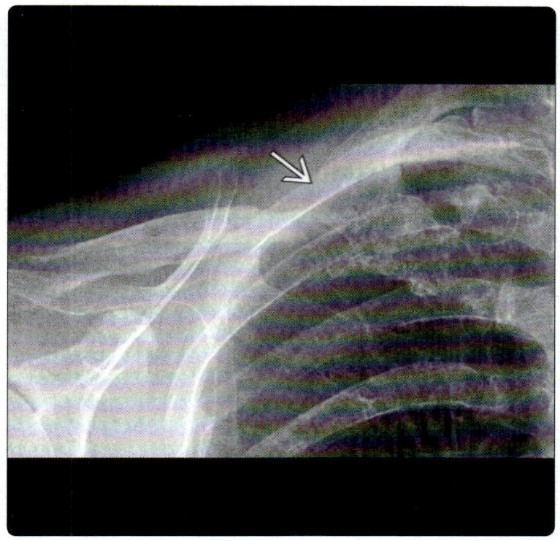

(Left) *Clinical photograph shows a 67-year-old man with right clavicular swelling → 30 years after radiotherapy treatment for nasopharyngeal carcinoma.* **(Right)** *Radiograph shows severe osteolysis → of the medial clavicle. A radiation-induced sarcoma was clinically suspected.*

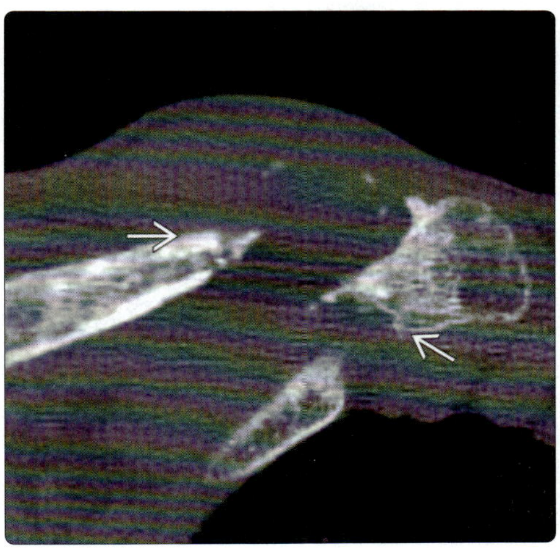

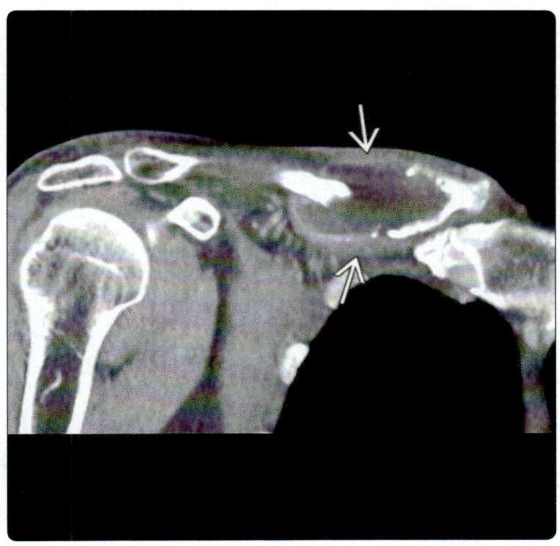

(Left) *CT of the same patient shows severe osteolysis of the medial clavicle with sharply tapered bone ends →, not compatible with malignant osteolysis.* **(Right)** *Soft tissue windowing of the same CT examination shows a large, intervening soft tissue mass with a hyperdense, calcified rim →. The overall appearances were compatible with poor healing of a radiation-induced insufficiency fracture of the medial clavicle.*

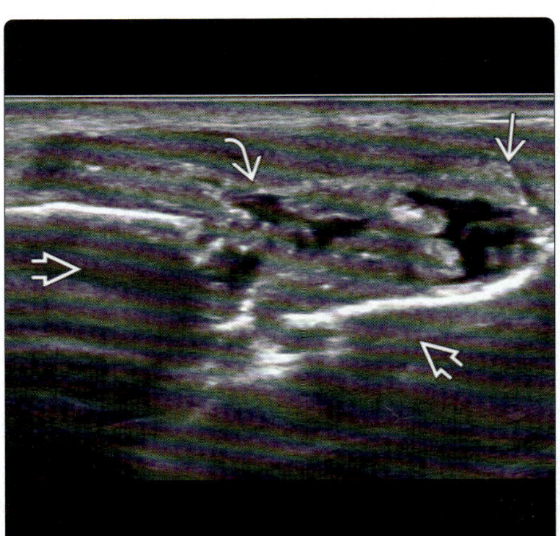

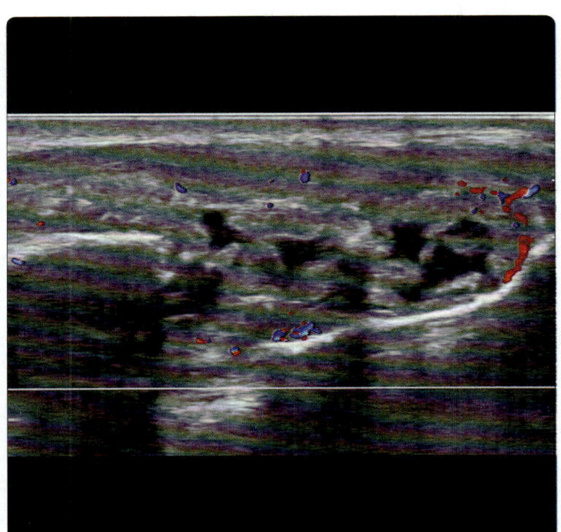

(Left) *Transverse US of the same area shows a large, partially cystic → mass with foci of calcification → between the fractured ends of the clavicle →.* **(Right)** *Color Doppler US at the same location shows mild marginal hyperemia, all compatible with poor healing of an insufficiency fracture, which was confirmed by US-guided biopsy.*

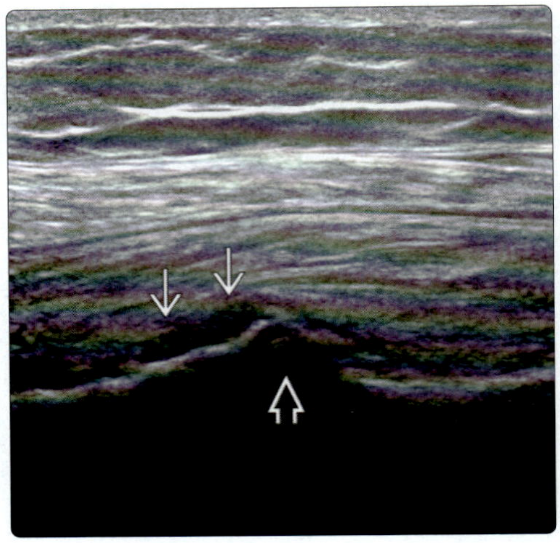

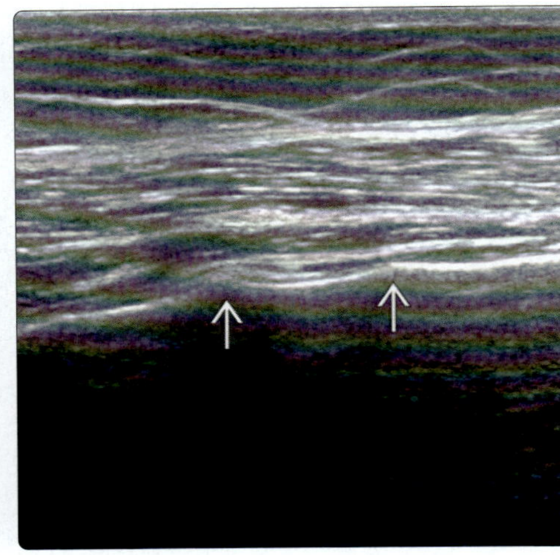

(Left) *Longitudinal US of the lateral aspect of the proximal femoral shaft in an elderly patient with thigh pain shows an osseous spike ➔ with a thin, hypoechoic rim ➔ due to unossified callus. This is consistent with an insufficiency fracture, associated with chronic bisphosphonate therapy and atypical subtrochanteric fracture.* **(Right)** *Longitudinal US of an elderly patient shows 2 similar, though smaller, fractures ➔. These fractures should be further assessed by radiography ± CT.*

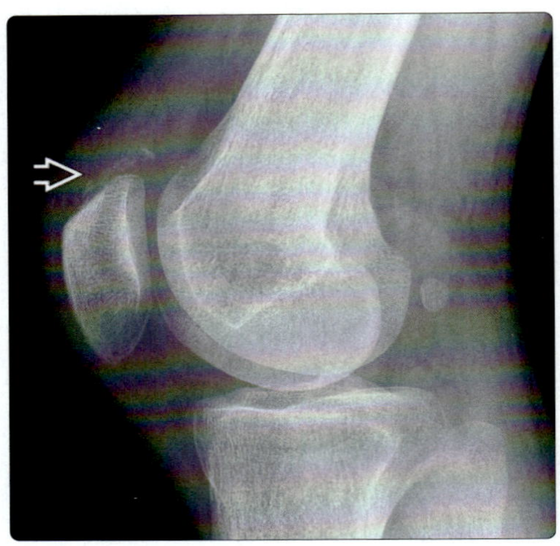

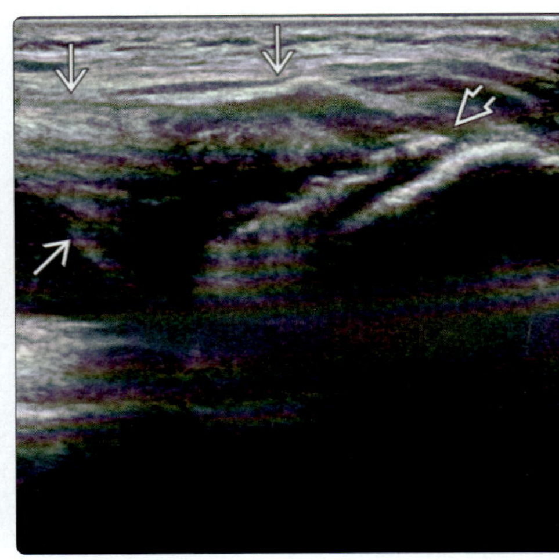

(Left) *Lateral knee radiograph in a young patient with chronic renal failure and renal osteodystrophy shows a faint periosteal avulsion fracture ➔ from the superior pole of the patella. There is radiographic evidence of hyperparathyroidism.* **(Right)** *Longitudinal US of the same patient shows the quadriceps tendon ➔ attached to the thin periosteal sleeve avulsion fracture ➔. This type of fracture is more common in children and in patients with metabolic bone disease.*

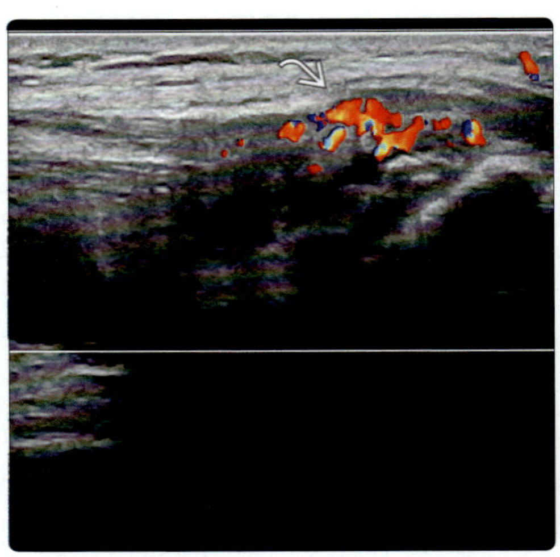

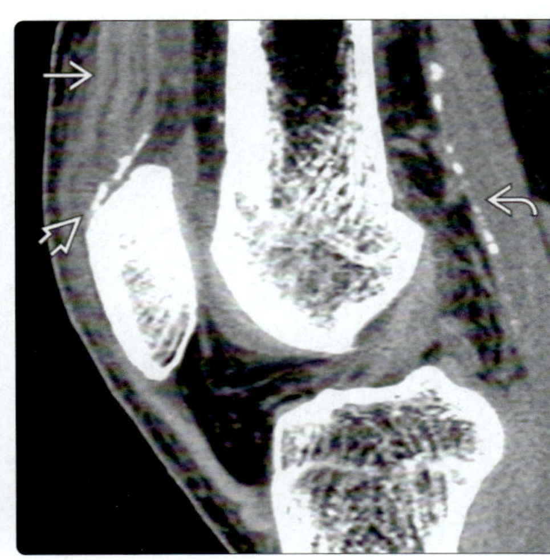

(Left) *Longitudinal color Doppler US of the same patient shows moderate hyperemia ➔ of the fracture site.* **(Right)** *Sagittal reformat CT of the same patient shows a periosteal sleeve fracture ➔, which is attached to the quadriceps tendon ➔ avulsed from the superior pole of the patella. This type of fracture can also occur at the inferior pole of the patella. Note arterial calcification ➔ in this young patient due to chronic renal failure.*

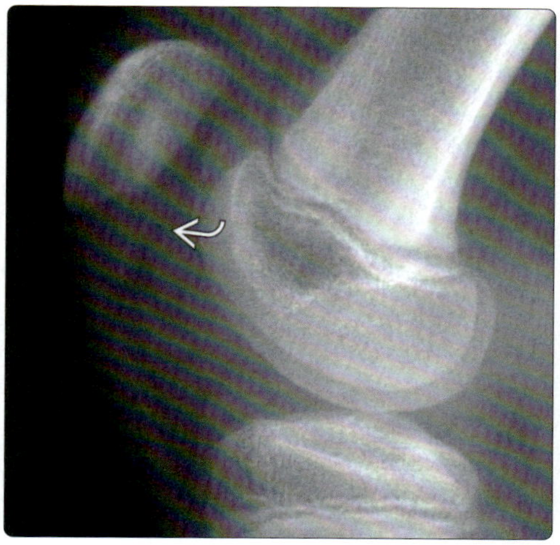

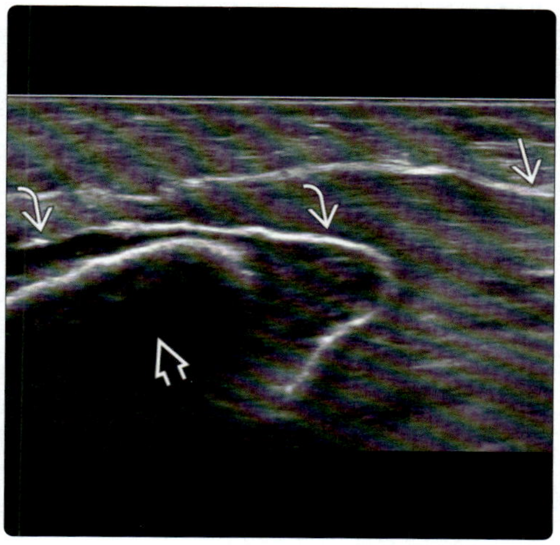

(Left) *Lateral radiograph of the knee in a 12-year-old boy with acute hemarthrosis following a fall is shown. There is a faint opacity at the inferior pole of the patella ➡. The radiograph was interpreted as no fracture.* **(Right)** *Longitudinal US shows a periosteal sleeve avulsion ➡ from the inferior pole of the patella ➡ with an attached patellar tendon ➡.*

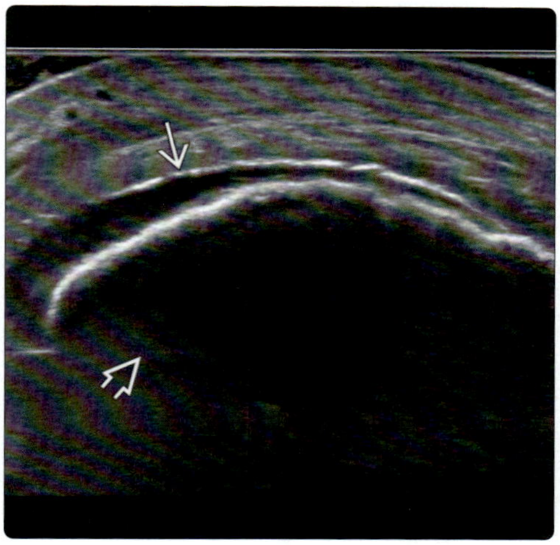

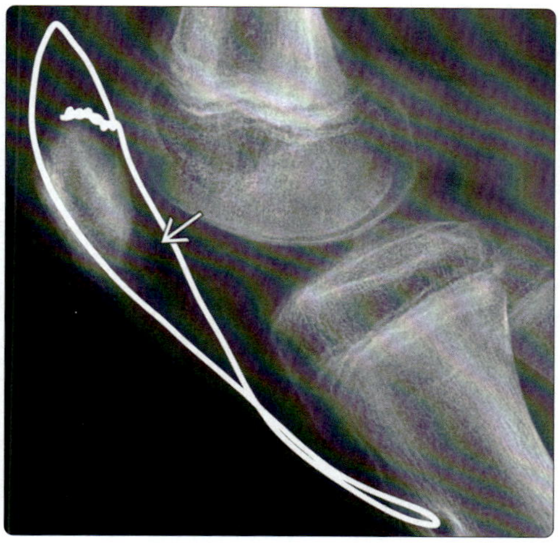

(Left) *Transverse US of the inferior pole of the patella shows the periosteal sleeve ➡ separated from the patella ➡.* **(Right)** *Lateral radiograph 2 months later shows wire fixation of the quadriceps tendon to the tibial tuberosity with reparative new bone formation filling the previously avulsed periosteal sleeve ➡.*

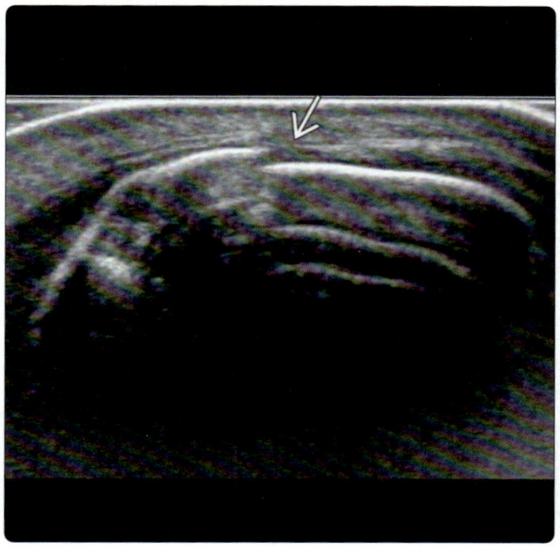

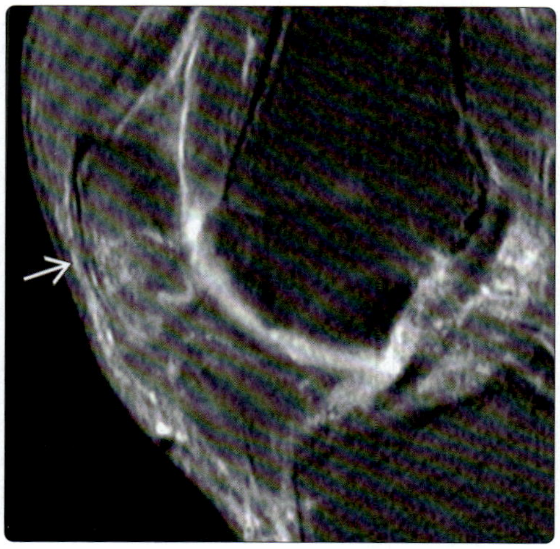

(Left) *Longitudinal US in a 66-year-old woman with knee pain following a fall shows a minimally displaced fracture of the patella ➡, which was not apparent on radiographs (not shown).* **(Right)** *Sagittal T2 FS MR of the knee in the same patient confirms the patellar fracture ➡ with a hyperintense thin fracture line and mild patellar bone marrow edema. There is mild movement artifact present. Radiographically occult patellar fractures are not uncommonly found on US.*

(Left) *Photograph of a young female with bilateral foot pain and swelling for 2 months following commencement of an exercise regimen shows that the left foot is moderately swollen ➡. Pain was initially more severe on the right side though is now more severe on the left side.* **(Right)** *Transverse US shows an echogenic rim of thickened periosteum ➡ around the left 2nd metatarsal shaft. This is consistent with a stress fracture. The normal 3rd metatarsal shaft ➡ is also shown for comparison.*

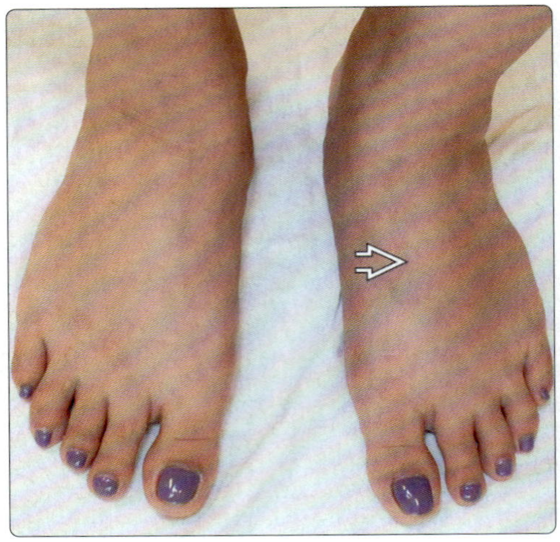

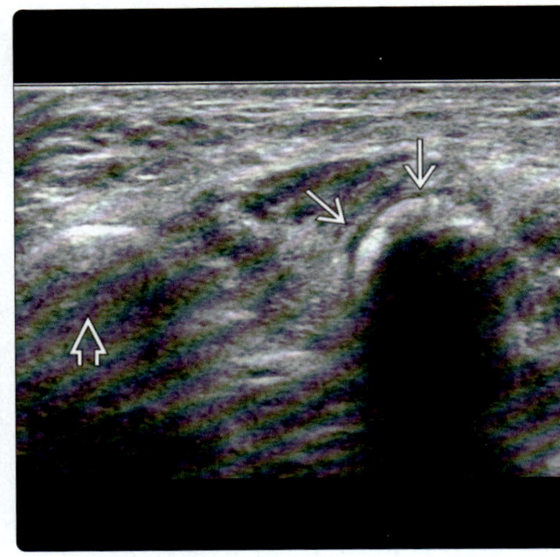

(Left) *Longitudinal US of the left 2nd metatarsal shaft of the same patient shows a focally thickened, echogenic periosteum ➡. Normal nonthickened periosteum is not visible on US.* **(Right)** *Longitudinal color Doppler US of the same patient shows mild hyperemia ➡ around the thickened periosteum. US is more sensitive than radiography at detecting early metatarsal stress fracture.*

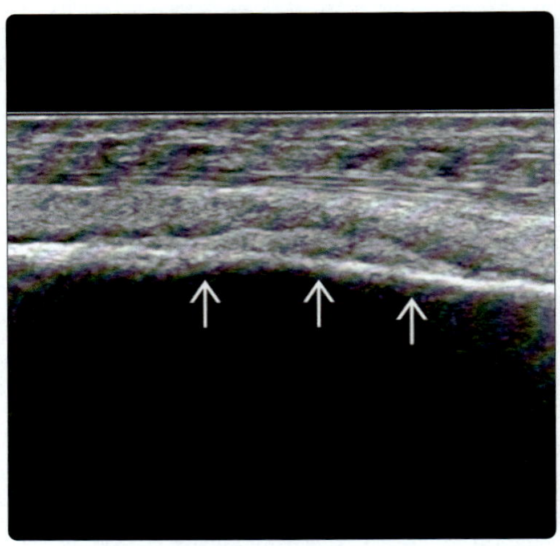

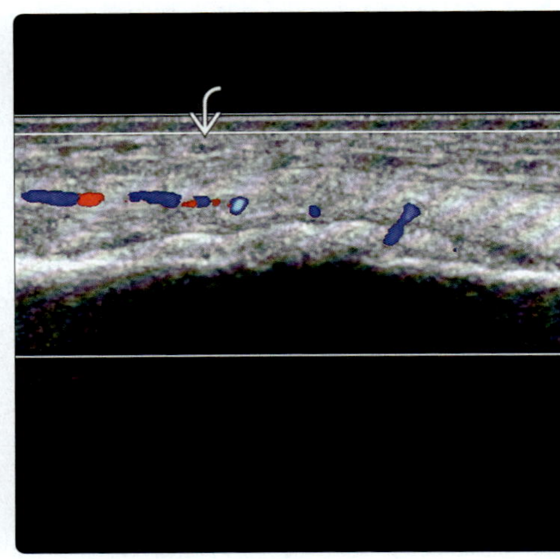

(Left) *Dorsoplantar radiograph of the left foot in the same patient performed at the same time as US examination shows no radiographic evidence of a stress fracture, particularly in the 2nd metatarsal shaft ➡.* **(Right)** *Dorsoplantar radiograph of the same foot 2 months later shows unequivocal focal cortical thickening in the 2nd metatarsal shaft ➡, consistent with a stress fracture. No other imaging was obtained.*

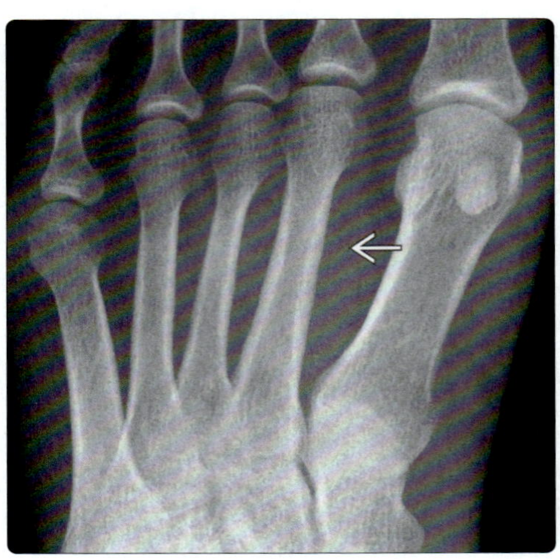

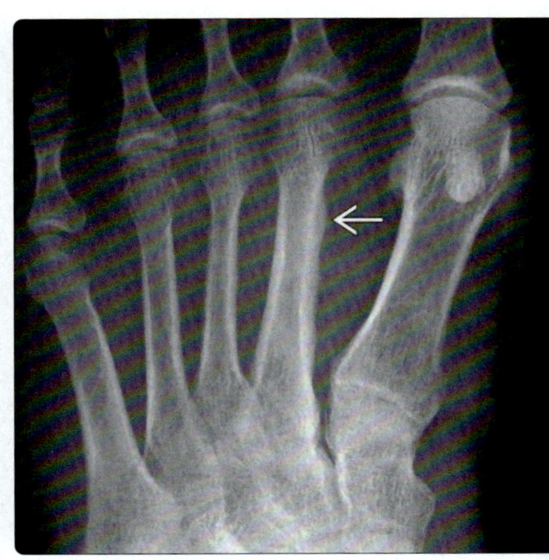

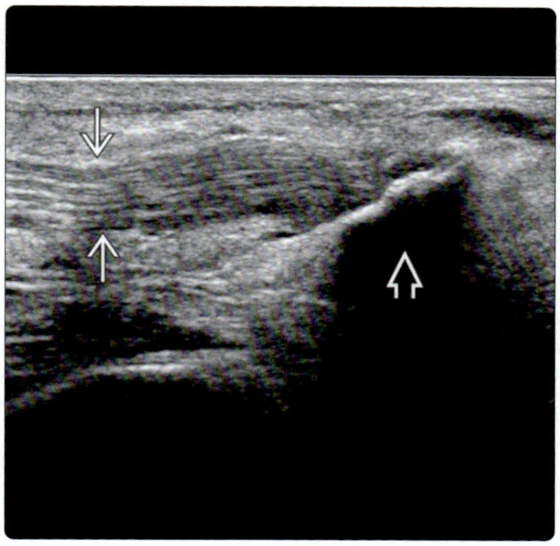

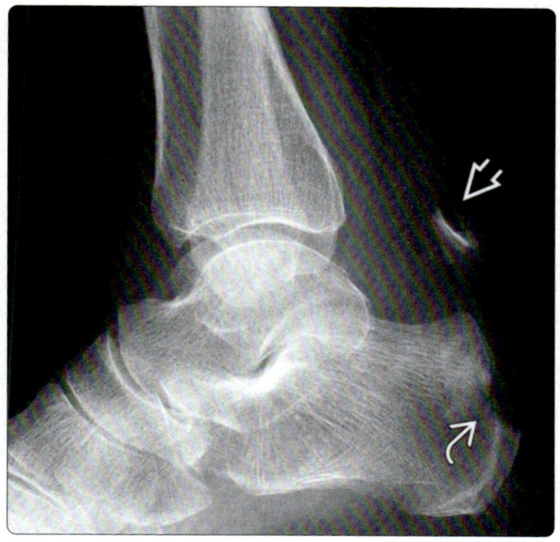

(Left) *Longitudinal US in a middle-aged patient with heel pain shows an avulsion fracture in the posterosuperior aspect of the calcaneum with an attached Achilles tendon.* **(Right)** *Lateral radiograph subsequently obtained in the same patient allows the thickness of the fracture fragment to be better appreciated. The calcaneal bone defect is readily apparent.*

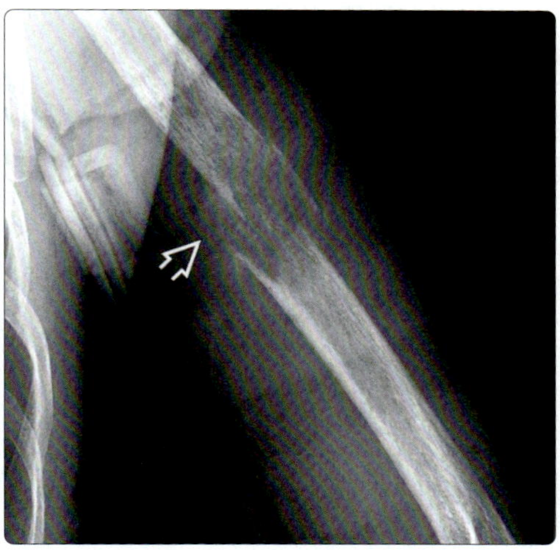

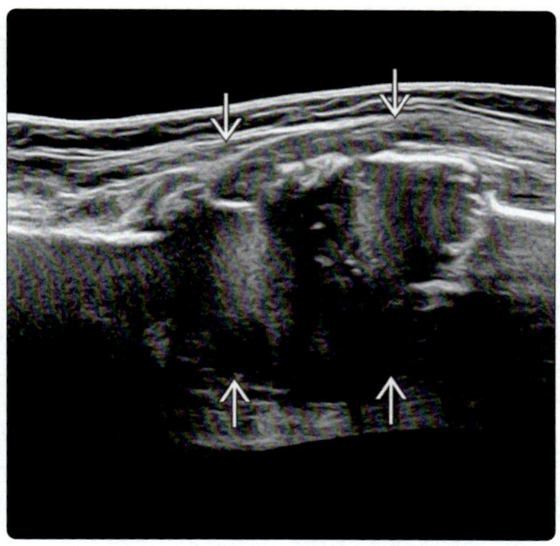

(Left) *Frontal radiograph shows a mildly displaced transverse pathologic fracture of the midhumeral shaft in a patient with no known primary tumor.* **(Right)** *Extended field-of-view US in the same patient shows a large, pathologic-type fracture in the midshaft of the humerus with irregular cortical osteolysis and mild osseous expansion.*

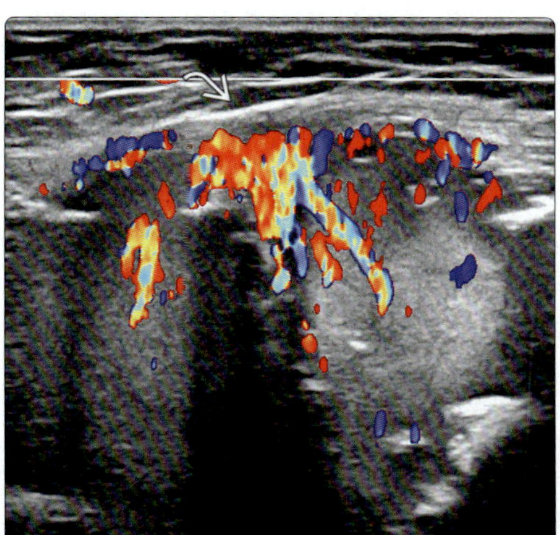

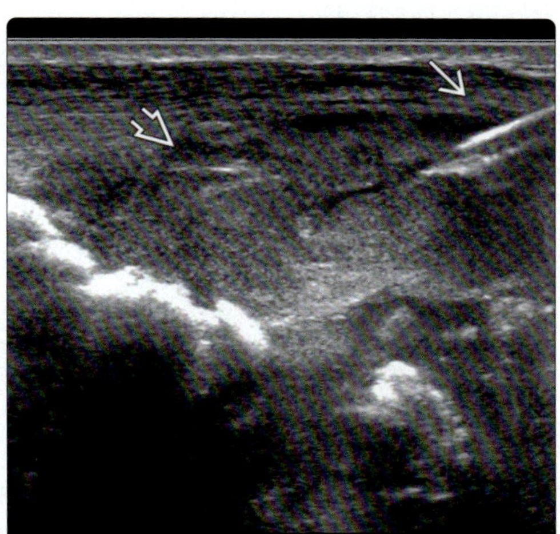

(Left) *Transverse color Doppler US of the same tumor shows marked hyperemia, confirming the presence of a viable tumor rather than a hematoma.* **(Right)** *Longitudinal US shows the same extraosseous tumor component during biopsy. Biopsy revealed metastatic adenocarcinoma, most likely from a lung primary. US-guided biopsy is usually indicated and feasible in most pathologic fractures of the appendicular skeleton if there is no known primary.*

Osteoarthritis

KEY FACTS

TERMINOLOGY

- Osteoarthrosis, osteoarthropathy, degenerative joint disease

IMAGING

- Joint space narrowing, marginal osteophytosis, capsular swelling, synovitis, & effusion are US signs of OA
- Most commonly affects knee, hip, fingers, 1st carpometacarpal, triscaphe, wrist, & acromioclavicular & tarsal joints
- US is increasingly used to determine inflammatory component of OA & differentiate OA from crystal deposition disease or inflammatory arthropathy
- Inflammatory element of OA is characterized by
  - Capsular swelling
  - Joint effusion
  - Synovial proliferation
  - Synovial & pericapsular hyperemia
  - Edema of periarticular soft tissues

CLINICAL ISSUES

- Only modest correlate between US severity of OA & clinical symptoms

DIAGNOSTIC CHECKLIST

- US is best for assessing degree of synovitis
  - Size of effusion, degree of synovial proliferation, & level of hyperemia to gauge inflammatory component of OA
  - Presence of echogenic comet-tail artifacts as indicator of crystal arthropathy
  - Joint aspiration to check for crystals or infection
- In most situations, excellent at guiding needle placement for joint aspiration, therapeutic injection, or synovial biopsy
- Degree of inflammatory component helps differentiate inflammatory-type OA from inflammatory arthropathy
  - Less pronounced in OA & commensurate with severity of underlying OA
  - Inflammatory arthropathy associated with more pronounced synovitis ± synovitis

(Left) *Longitudinal US of medial femorotibial joint line shows mild to moderate osteoarthritis (OA) with mild marginal osteophytosis* ➡ *and mild peripheral extrusion of the medial meniscus* ➡*. The medial collateral ligament* ➡ *is mildly displaced superficially.* **(Right)** *Longitudinal US shows severe peripheral extrusion of the medial meniscal body* ➡ *displacing the medial collateral ligament* ➡*. There is moderate to severe joint space narrowing between the medial femoral* ➡ *and tibial* ➡ *condyles.*

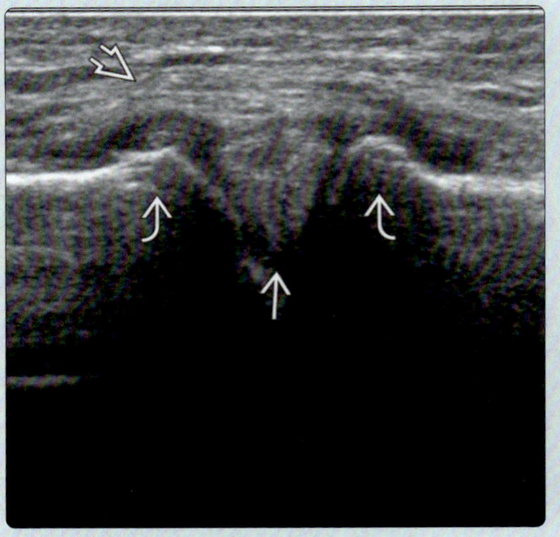

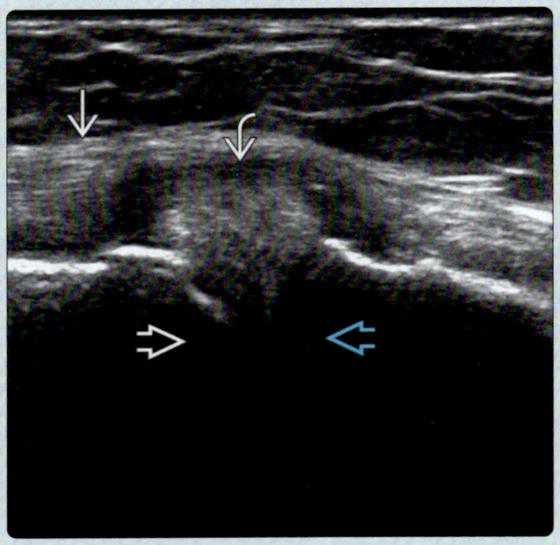

(Left) *Longitudinal US in a patient with knee OA shows moderate marginal osteophytosis* ➡*, severe peripheral meniscal extrusion* ➡*, & moderate to severe medial femorotibial compartment joint space narrowing* ➡*. The body of the medial meniscus is that portion of the meniscus best appreciated on US.* **(Right)** *Transverse US shows moderate distension of medial patellofemoral recess between the patella* ➡ *and femoral condyle* ➡ *with mild synovial proliferation* ➡*. This pattern is common in knee OA.*

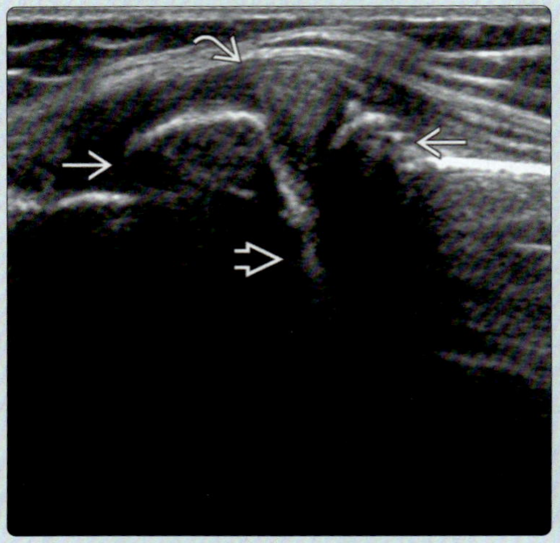

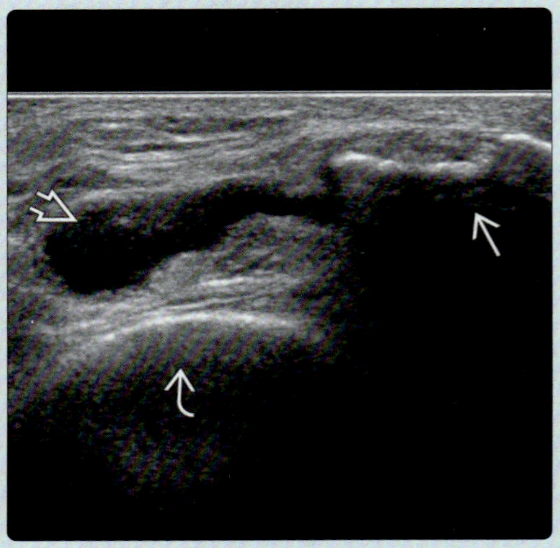

TERMINOLOGY

Abbreviations

- Osteoarthritis (OA)

Definitions

- Degenerative disease with thinning of articular cartilage & associated bone change, soft tissue change, positional change & inflammation
  - Bone change: Marginal osteophytosis, subchondral reactive changes (oedema, cyst formation, sclerosis, attenuation)
  - Soft tissue change: Usually mild synovial proliferation, capsular swelling
  - Positional change: Joint subluxation ± misalignment
  - Inflammation: Variable degree of inflammation present
- Secondary: Accelerated OA due to existing joint disorder or deformity

IMAGING

General Features

- Best diagnostic clue
  - Articular cartilage thinning accompanied by marginal osteophytosis, subchondral sclerosis, cyst formation ± inflammatory component
- Location
  - Most commonly affects knee, hip, fingers, 1st carpometacarpal, scaphotrapezium-trapezoid articulation, wrist, acromioclavicular & tarsal joints
  - May affect only localized area or whole of joint

Radiographic Findings

- Radiography
  - Primary screening modality for OA
  - Joint space narrowing, subchondral sclerosis, subchondral cysts, & osteophyte formation
    - Mild subluxation, especially in small joints
  - Focal cartilage loss may be present with no joint space narrowing radiographically

MR Findings

- T1WI
  - Cartilage thinning, subchondral cysts, marrow fat signal extending into osteophytes
- T2WI
  - Subchondral edema
  - Joint effusion ± intraarticular particulate debris
  - Reactive-type synovitis
  - Periarticular edema

Ultrasonographic Findings

- Increasingly used to determine inflammatory component of OA & differentiate OA from crystal or inflammatory arthropathy
  - Radiography can determine presence of OA but US can reveal inflammatory component
  - One can also detect OA when US performed for other suspected pathology
- Joint space narrowing, marginal osteophytosis, capsular swelling, synovitis, & effusion
  - Focal cartilage thinning or defects cannot be fully evaluated with US
  - Joint space narrowing of some joints (e.g., patellofemoral) may not be readily appreciable
- Joint effusion (may be anechoic or slightly echogenic)
  - Medial & lateral patellar recesses of knee usually distend prior to suprapatellar recess
- Mild synovial proliferation is common finding in OA
- Inflammatory element is characterized by
  - Capsular swelling
  - Joint effusion
  - Synovial proliferation (synovial thickening & fronds)
    - Important sign of inflammatory component present in OA
  - Pericapsular & synovial hyperemia on color Doppler
    - Not very sensitive sign in OA
    - Color Doppler imaging is often negative, even with inflammatory component present in OA
      - Indicative of low-grade inflammation seen usually in OA
    - Capsular-synovial hyperemia is difficult to depict in deeper joints such as hip, shoulder, & even knee
    - Contrast-enhanced US is more sensitive
  - Edema of periarticular tissues (better appreciated with MR than US)
- Clinical picture, radiographic findings, & degree of inflammatory component helps differentiate inflammatory-type OA from inflammatory arthropathy
  - Inflammatory arthropathy has more pronounced synovial proliferation & inflammation
    - ± erosions & other clinical/radiographic features of inflammatory arthropathy
- Particulate debris may be visible within joint fluid
  - US can detect radiographically occult intraarticular debris
    - Though limited in assessing whether this particulate debris is free or attached
  - Central aspects of joint cannot be fully visualized with US (better seen with MR)
- Joint subluxation can be exaggerated on US as dependent on transducer angulation & alignment
- Medial meniscal extrusion is common feature in moderate to severe medial femorotibial OA
- Associated Baker cysts, parameniscal cysts, or ganglia identified by US with high level of accuracy
  - Inflammatory component, crystal aggregates, & synovial debris can extend into Baker cysts & ganglia
- US-guided joint aspiration is helpful to exclude infective, inflammatory arthropathy or crystal deposition disease
  - < 1,000 WBC/μL synovial fluid is indicative of OA (normally ~ 200 WBC/μL)
  - > 5,000 WBC/μL synovial fluid is indicative of acute inflammatory or infective arthropathy
  - Specimen for crystal identification should be sent fresh as alcohol will dissolve crystals

Imaging Recommendations

- Best imaging tool
  - US for assessing degree of synovitis & guiding intraarticular needle placement
  - MR for assessing cartilage thinning & subchondral bone marrow edema
- Protocol advice
  - Examine joints in systematic fashion

- Use varying degree of flexion or internal/external rotation to maximize cartilage visibility
- Use minimal transducer pressure to avoid effacing effusions & blanching color Doppler signal
- Assess joint space narrowing & marginal osteophytosis as well as medial meniscal extrusion (knee)
- Assess degree of synovial effusion & echogenicity, degree of synovial proliferation, & hyperemia
- Assess degree of capsular swelling, periarticular edema, bursa, & cysts
- Can semiquantitatively grade synovial hyperemia on color Doppler imaging
  - Grade 0: No flow
  - Grade 1: Up to 3 color signals
  - Grade 2: > 3 color signals but < 50% of synovial area shows color signal
  - Grade 3: > 50% of synovial area shows color signal
- Can measure degree of medial meniscal extrusion in supine & standing positions
  - Degree of extrusion beyond line joining outer margins of femoral & tibial condyles (exclusive of osteophytes)
    - Normal < 2 mm, increasing to < 3 mm on standing
    - OA often > 4 mm, increasing to > 5 mm on standing
  - Circumferential collagen bundle damage → impaired ability to resist hoop strain (circumferential stress) → meniscal extrusion

DIFFERENTIAL DIAGNOSIS

Inflammatory Arthritis

- Signs of symmetrical disease ± positive serological markers
- Degree of synovitis is more than expected relative of degree of OA ± presence of erosions

Crystal Deposition Disease

- Echogenic foci within synovial fluid, synovium, articular cartilage, or menisci
- Otherwise similar to acute OA exacerbation
  - ± discrete hard or soft tophi
  - US-guided aspiration for crystals

Focal Chondral Injury

- Best assessed with MR

PATHOLOGY

General Features

- Etiology
  - Common risk factors are age, sex, prior joint injury, obesity, genetic predisposition, & mechanical factors such as malalignment
  - Primary & secondary OA may be different diseases with common final pathway
  - Secondary OA often explained by overload paradigm ("wear & tear")
  - Primary OA does not always reconcile with overload paradigm, e.g., common in sedentary individuals
    - Diminished synovial clearance of normal joint enzymes
    - Closed, positive feedback loop modulates joint enzymatic level & activity
  - Meniscal degeneration & extrusion has etiological role in knee OA

Staging, Grading, & Classification

- Articular cartilage damage (stages)
  - I: Edema (chondromalacia)
  - II: Surface fibrillation with superficial defects
  - III: Deep fibrillation with large deep defects
  - IV: Full-thickness defect & subchondral sclerosis, cyst formation ± bony attenuation

CLINICAL ISSUES

Presentation

- Most common signs/symptoms
  - Pain aggravated by movement & joint swelling
  - Only modest correlate between US severity of OA and clinical symptoms
- Clinical profile
  - Catching, locking, or grinding in older patient

Demographics

- Age
  - Typically older patients (> 50 years)
- Gender
  - Female predominance; accelerated joint degeneration occurs after menopause

Natural History & Prognosis

- Progressive, debilitating disease

Treatment

- Conservative
  - Minimize overloading, e.g., weight loss
  - Muscle strengthening
  - Pain management by analgesics or NSAIDs
  - Steroid injection
    - US modest at predicting clinical response
    - In general, more severe arthropathy on US, less sustained clinical response
  - Viscosupplementation via intraarticular injection
  - Glucosamine & chondroitin by oral supplements
- Surgical
  - Arthroscopy (debridement)
  - Resurfacing procedure (hip or knee)
  - Tibial osteotomy to redistribute weight-bearing load
  - Prosthetic joint replacement

DIAGNOSTIC CHECKLIST

Consider

- Size of effusion, degree of synovial proliferation, & level of hyperemia to gauge inflammatory component of OA
- Presence of crystals as indicator of crystal arthropathy
- Joint aspiration or synovial biopsy to check for infection or crystals

SELECTED REFERENCES

1. Beitinger N et al: The value of colour Doppler sonography of the knee joint: a useful tool to discriminate inflammatory from non-inflammatory disease? Rheumatology (Oxford). 52(8):1425-8, 2013

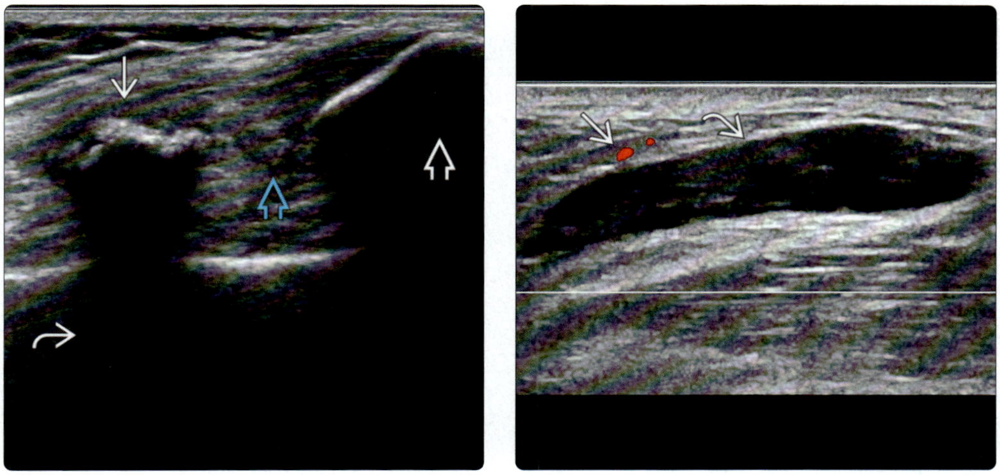

(Left) *Longitudinal US shows a calcified intraarticular body in the suprapatellar recess ➔. There is no suprapatellar pouch effusion; hence, the recess cavity ➔ is effaced. The patellar bone ➔ & femoral shaft ➔ are shown.* **(Right)** *Longitudinal color Doppler US shows synovial proliferation ➔ within a distended Baker cyst in a patient with knee OA. Any knee inflammation will usually manifest in an accompanying Baker cyst. Absence of color flow ➔ within the cyst reflects relatively low-grade inflammation.*

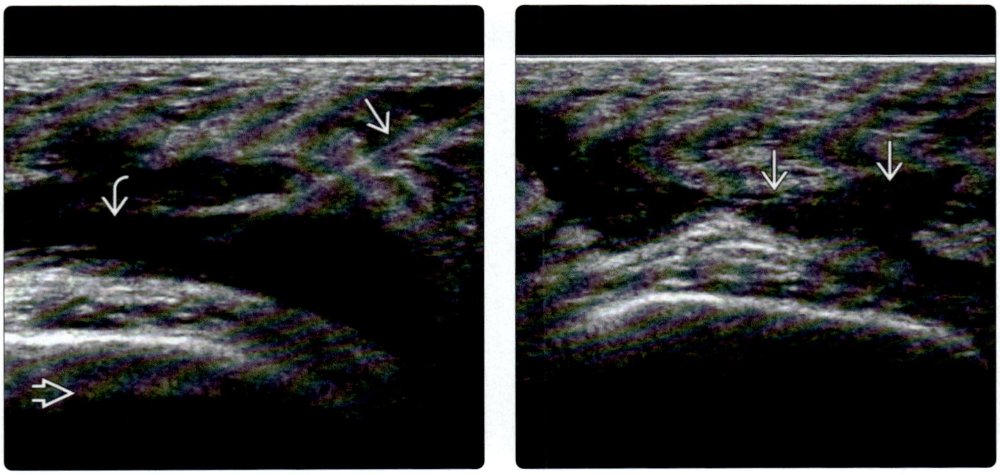

(Left) *Transverse US shows needle ➔ placement in the lateral recess of the knee ➔ for joint aspiration. US ensures proper needle placement for joint aspiration or intraarticular therapeutic injection. Note the lateral femoral condyle ➔.* **(Right)** *Transverse US in the same patient following fluid aspiration shows mild residual synovial proliferation ➔. US is very helpful in determining the relative component of knee swelling due to effusion or synovial proliferation.*

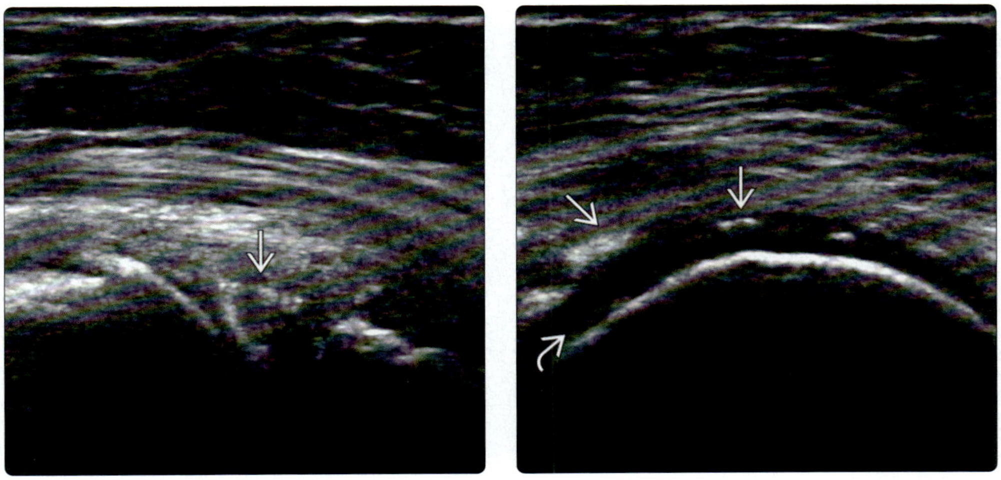

(Left) *Longitudinal US in a patient with knee OA shows rim-like calcification ➔ of the medial meniscus from chondrocalcinosis. Chondrocalcinosis can be seen more easily in the medial meniscus than in the lateral meniscus.* **(Right)** *Longitudinal US shows echogenic calcification both on the surface ➔ and embedded into the articular cartilage ➔ on the medial femoral condyle posteriorly. This distribution favors calcium pyrophosphate deposition over uric acid deposition.*

(Left) *Longitudinal US of the finger shows prominent marginal osteophytosis of the distal interphalangeal joint with mild capsular swelling. Finger flexion may help to better show the distal interphalangeal joint space. The nail overlies the distal phalanx.* **(Right)** *Clinical photograph shows associated swelling (Heberden node) and mild flexion deformity of the distal interphalangeal joint. Bouchard nodes are similar nodes at the proximal interphalangeal joint.*

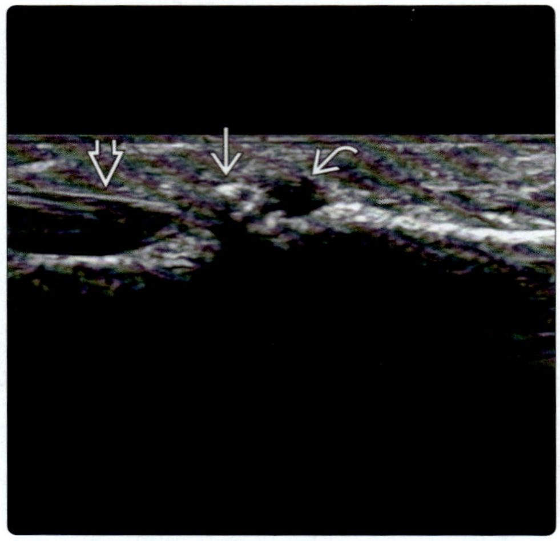

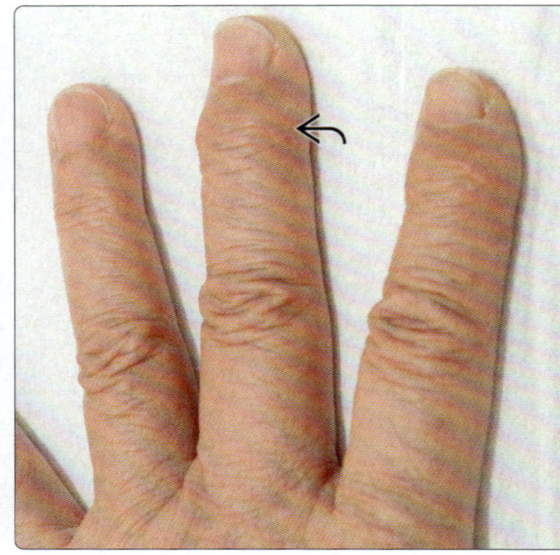

(Left) *Longitudinal US of a patient with recent trauma to the finger shows moderate proximal interphalangeal joint OS with prominent dorsal osteophytosis and a small avulsion fracture, which is attached to the extensor expansion. There is marked dorsal swelling.* **(Right)** *Color Doppler US shows marked periarticular hyperemia dorsally. Osteoarthritic osteophytes can fracture with resultant periarticular inflammation.*

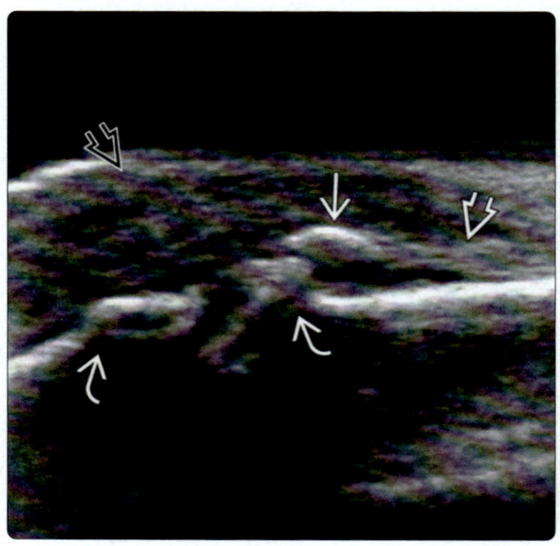

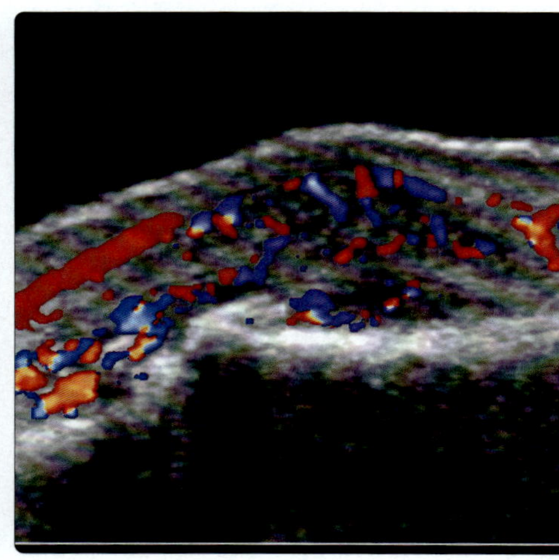

(Left) *Longitudinal US shows only mild joint space narrowing of the 1st carpometacarpal joint between the metacarpal base and the trapezium, consistent with mild osteoarthritis. No periarticular edema is present.* **(Right)** *Longitudinal US shows severe OA of the 1st carpometacarpal with marginal osteophytosis of the trapezium, subluxation of the 1st metacarpal base, and moderate periarticular soft tissue edema.*

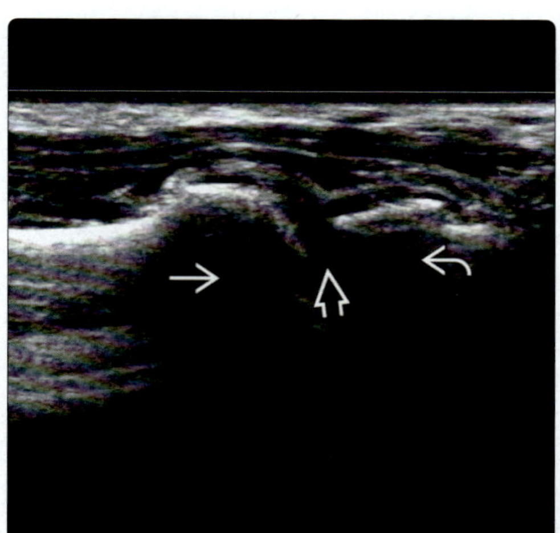

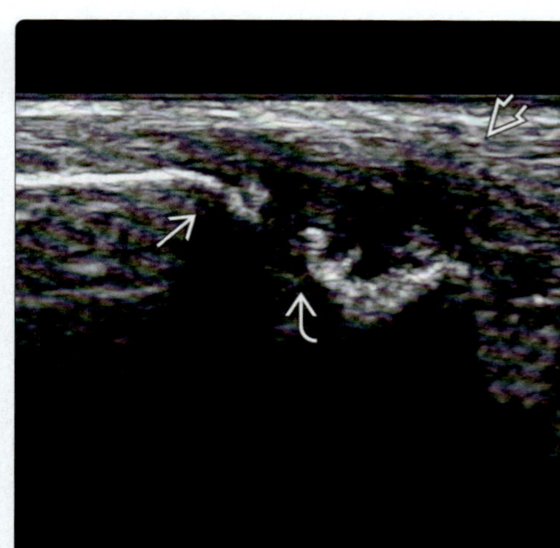

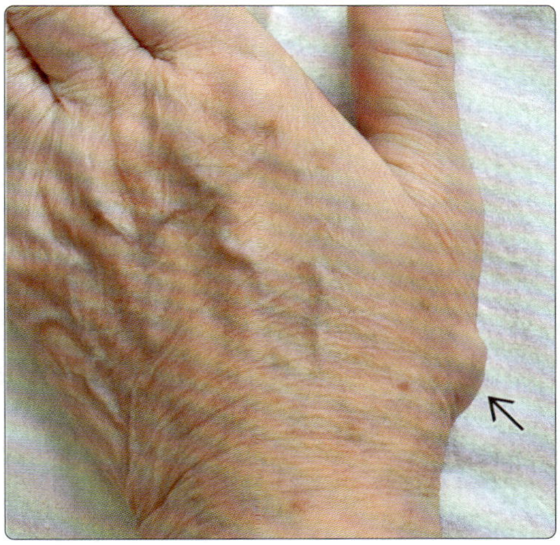

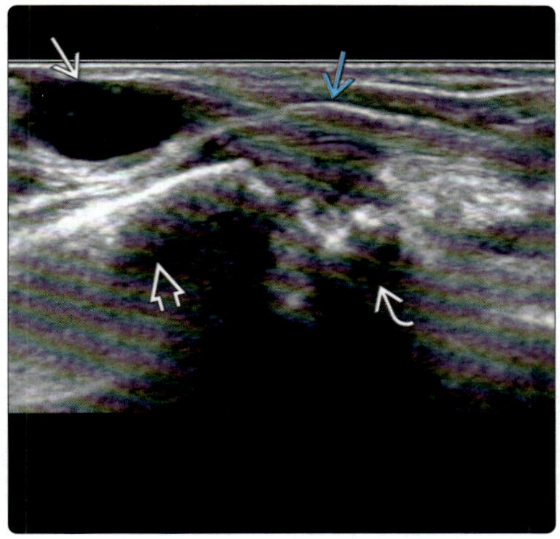

(Left) *This 76-year-old woman has pain and swelling → on the radial border of the hand.* **(Right)** *Longitudinal US of this area shows severe 1st carpometacarpal osteoarthritis between the trapezium → and the 1st metacarpal bone → with severe joint space irregularity. A small ganglion cyst → overlies the base of the 1st metacarpal bone and the thenar muscle →. This ganglion arose from the 1st carpometacarpal joint. Ganglia are more common around osteoarthritic joints.*

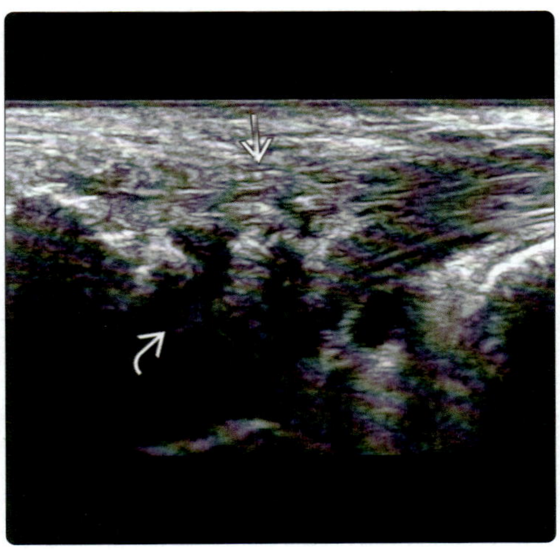

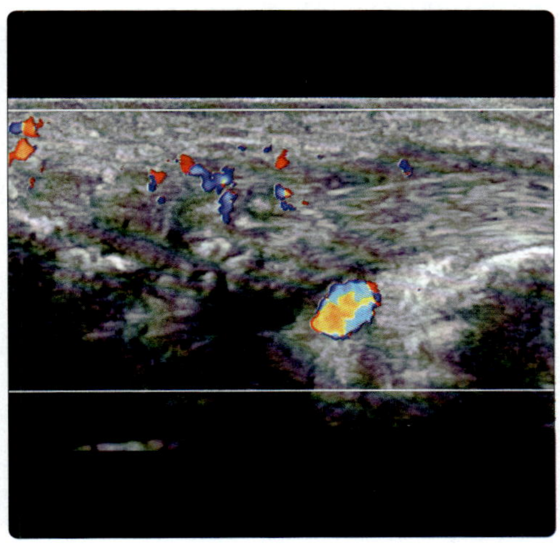

(Left) *Longitudinal US shows severe flexor carpi radialis (FCR) insertional tendinosis → and triscaphe OA →. Individual components of the triscaphe articulation (distal pole scaphoid, trapezium, and trapezoid) cannot be delineated.* **(Right)** *Longitudinal color Doppler US of the same area shows moderate hyperemia of the FCR insertional area. Anatomical proximity is felt to account for the association between triscaphe osteoarthritis, 1st carpometacarpal OA, and FCR tendinosis.*

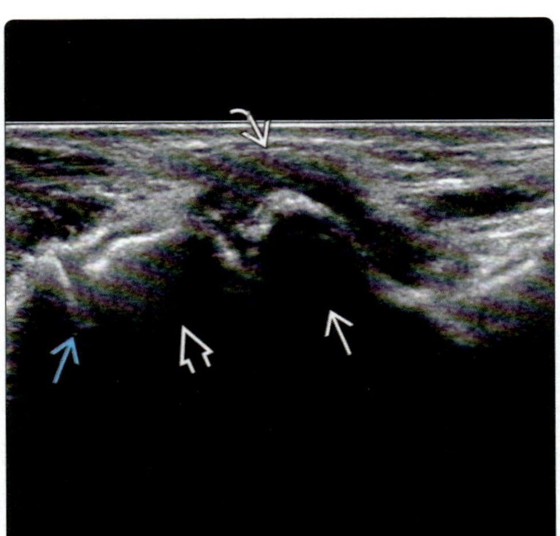

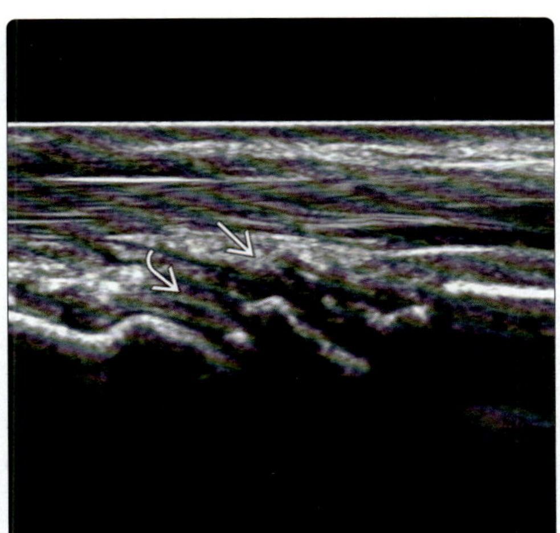

(Left) *Longitudinal US shows moderate to severe OA between the distal pole of the scaphoid → and the trapezium →. There is moderate capsular swelling → as well as periscaphoid inflammation. The 1st carpometacarpal joint → is well preserved.* **(Right)** *Longitudinal US of wrist OA shows severe joint space narrowing at the dorsal aspect of the radiolunate → and lunocapitate → joints with minimal synovial proliferation. These carpal bones are quite severely attenuated.*

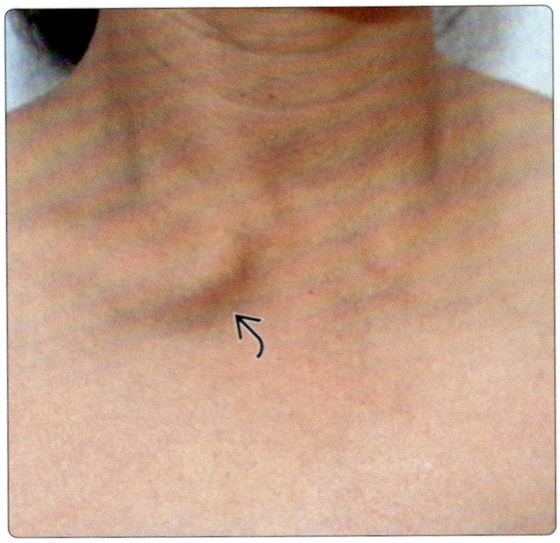

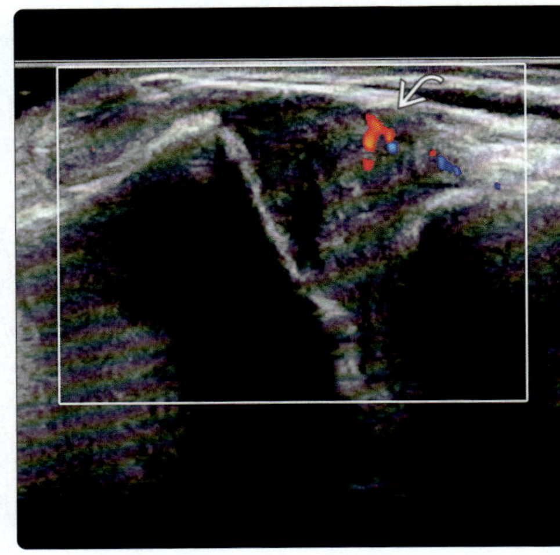

(Left) *Clinical photograph shows moderate swelling of the right sternoclavicular joint →. The superficial location of this joint means that any disfiguration will become readily apparent clinically. This joint is very accessible to US.* **(Right)** *Transverse color Doppler US of the same joint shows mild capsular hyperemia →. There is no erosive component. Distinction from an inflammatory arthropathy is best made on clinical and serological grounds as CT or MR are often not helpful in this context.*

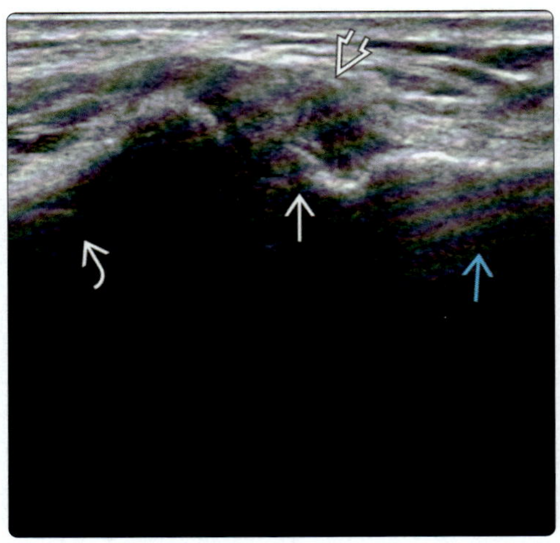

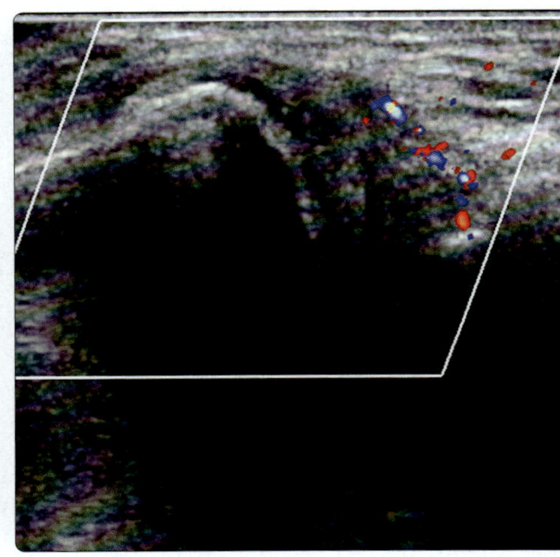

(Left) *Transverse US shows moderate capsular swelling → of the right sternoclavicular joint between the medial end of the clavicle → and the manubrium → due to osteoarthritis. There is mild marginal osteophytosis →.* **(Right)** *Color Doppler imaging of the same joint shows moderate capsular hyperaemia. This is all indicative of an inflammatory-type sternoclavicular joint OA.*

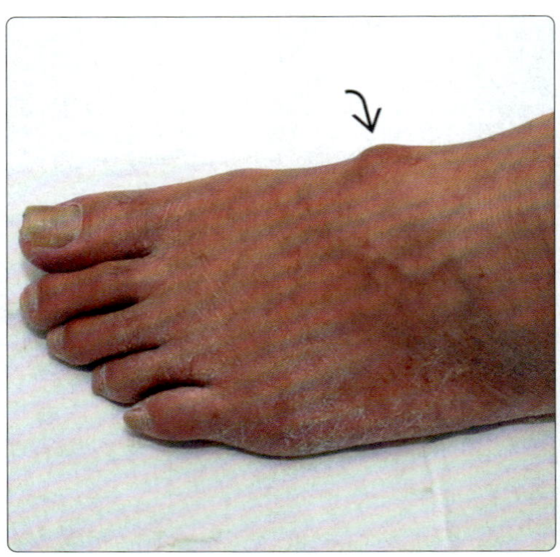

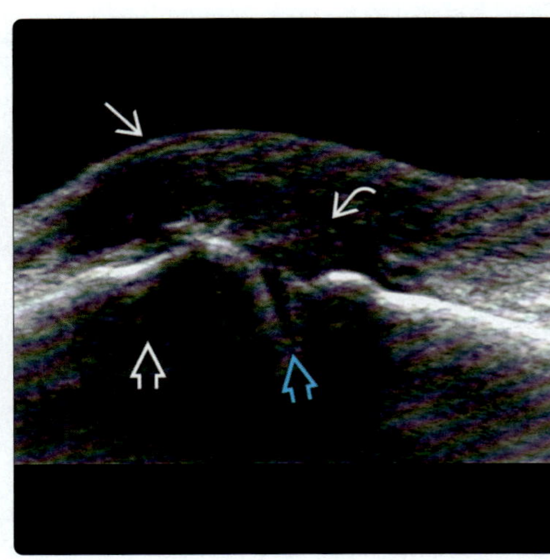

(Left) *Clinical photograph of a patient with foot pain & swelling shows a localized swelling → on mediodorsal aspect of midfoot.* **(Right)** *Longitudinal US of the same area shows moderate 1st tarsometatarsal joint OA between the medial cuneiform → and 1st metatarsal base with joint space narrowing →, capsular thickening →, and periarticular soft tissue edema →. This is consistent with inflammatory-type OA. There is no evidence of gout. Color Doppler US (not shown) showed mild capsular hyperemia.*

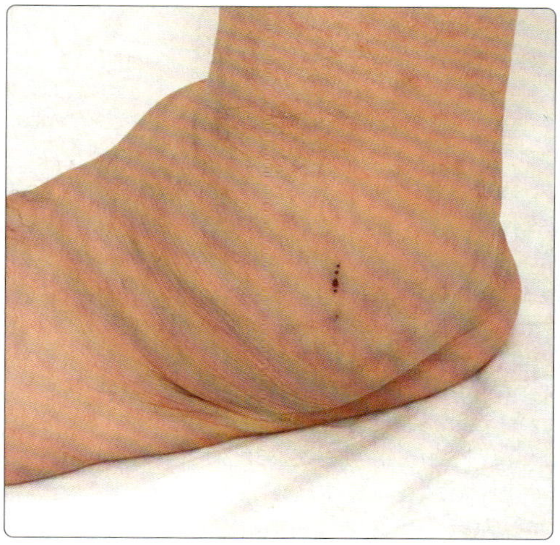

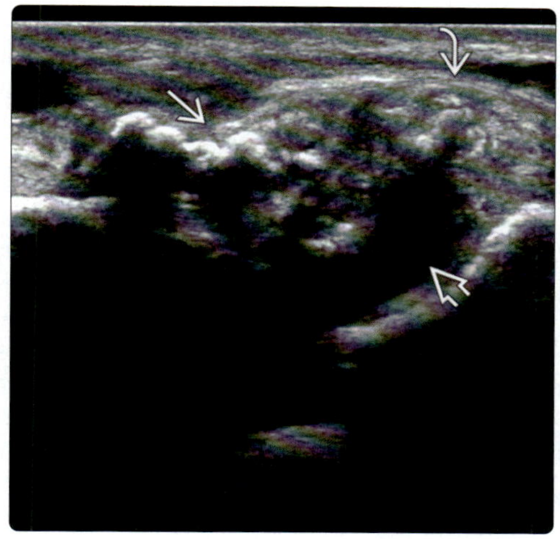

(Left) *Clinical photograph of a diabetic patient with progressive ankle swelling shows severe deformity.* **(Right)** *Longitudinal US in the same patient shows severe ankle arthropathy with chronic synovial thickening, osseous fragmentation, debris, and a small effusion. Possibilities include either a diabetic neuroarthropathy or, less likely, a tuberculous arthropathy. US-guided synovial biopsy revealed acute on chronic inflammation but no evidence of infection.*

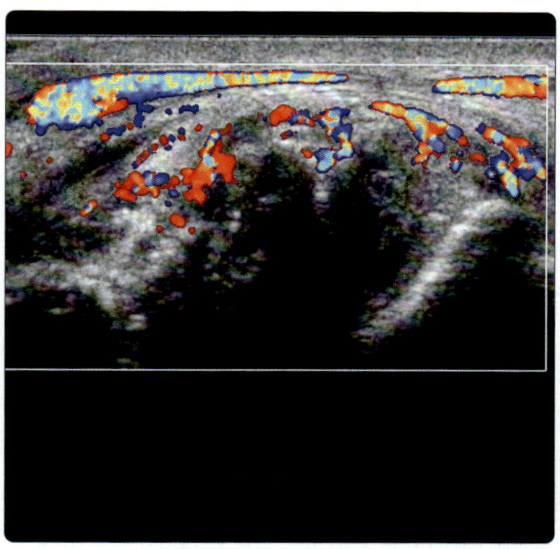

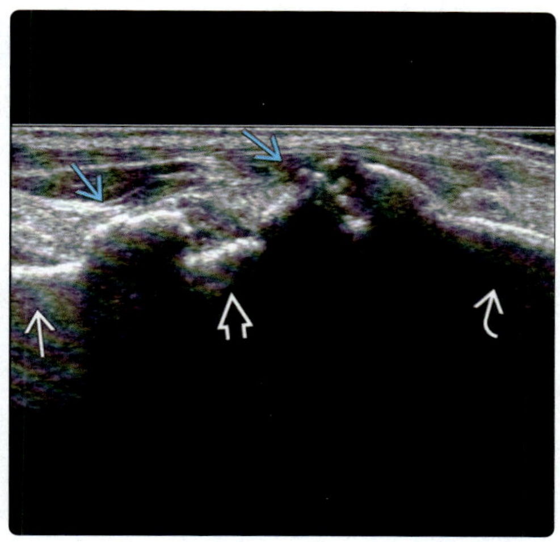

(Left) *Longitudinal color Doppler US in the same patient shows moderate periarticular hyperemia. The appearances are consistent with subacute exacerbation of diabetic neuroarthropathy.* **(Right)** *Longitudinal US in a patient with lateral foot pain shows severe hypertrophic-type OA with large marginal osteophytes at the effaced calcaneum/cuboid joint proximally and the cuboid/5th metatarsal base joint distally. There is little surrounding inflammation.*

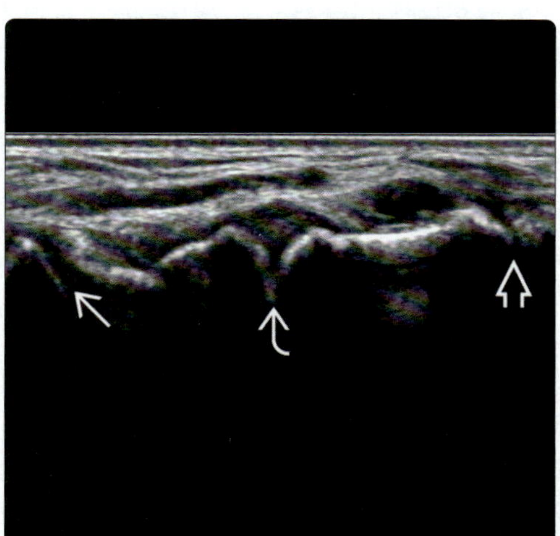

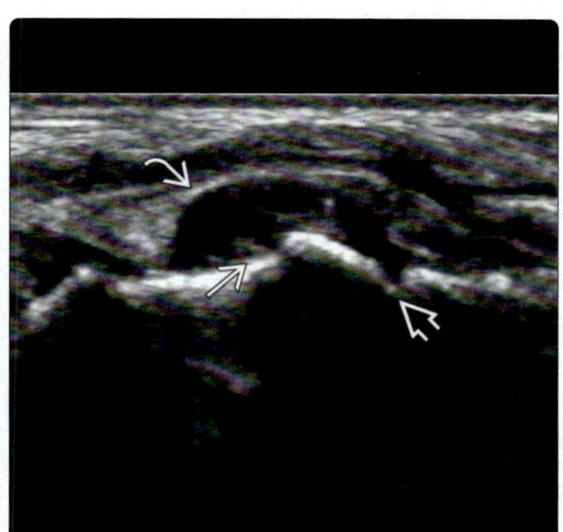

(Left) *Longitudinal US in a patient with medial midfoot pain shows moderate OA with joint space narrowing, capsular swelling, and marginal osteophytosis at the talonavicular, navicular-cuneiform, and 2nd tarsometatarsal articulations.* **(Right)** *Longitudinal US of the same patient shows a medium-sized effusion of the 2nd tarsometatarsal joint with dorsal capsular distension. Synovial proliferation is present, which helps somewhat to distinguish this from a periarticular ganglion.*

Inflammatory Arthritis

KEY FACTS

TERMINOLOGY

- Systemic autoimmune inflammatory disorders of synovium with secondary involvement of articular surface and bone
- Spondyloarthropathy associated with enthesitis

IMAGING

- Thickened hyperemic synovium
- Any synovial-lined joint
  - Wrist joint > small joints of hand and feet
- Mono- to oligo- to polyarthropathy
- **Effusion**
  - Sensitive though nonspecific indicator of synovitis
- **Thickened hypoechoic synovium (pannus) is most common finding**
  - Hypoechoic > hyperechoic
  - Active inflammation → synovial proliferation and hyperemia
  - US is 10x more sensitive than clinical examination at detecting early synovitis
- **Marginal bony erosion**
  - Affects bare areas
  - Focal juxtacortical cortical disruption with adjacent synovitis
  - 6.5x more metacarpophalangeal erosions detected with US than radiography
  - 30% false-positive rate for US detection of erosions
- **Cartilage loss**
  - US good but best assessed with radiography or MR
- **Periarticular soft tissue changes**
  - Tenosynovitis, nerve entrapment, bursitis, rheumatoid nodules, and peripheral enthesopathy
- **Color Doppler helpful in distinguishing synovial proliferation from joint fluid**
  - Allows activity of synovitis to be assessed
  - Active synovitis shows high flow
  - Inactive synovitis → low/absent flow

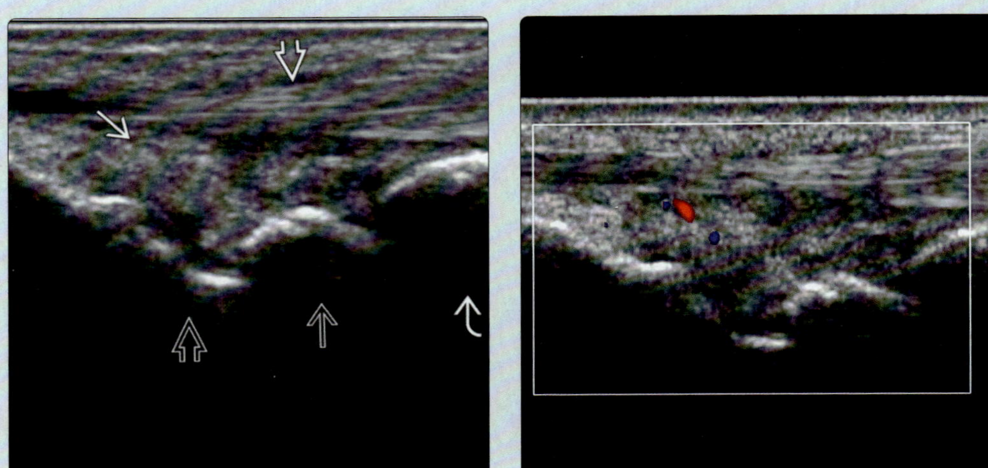

(Left) *Longitudinal US shows moderate-severity echogenic synovial proliferation* ➡ *on the dorsum of the carpus deep to the extensor tendons* ⇨. *The distal end of the radius* ➡, *the lunate* ⇨, *and the capitate* ↪ *are shown.* **(Right)** *Longitudinal color Doppler US of the same area shows mild synovial hyperemia. This is indicative of a low level of synovial inflammatory activity.*

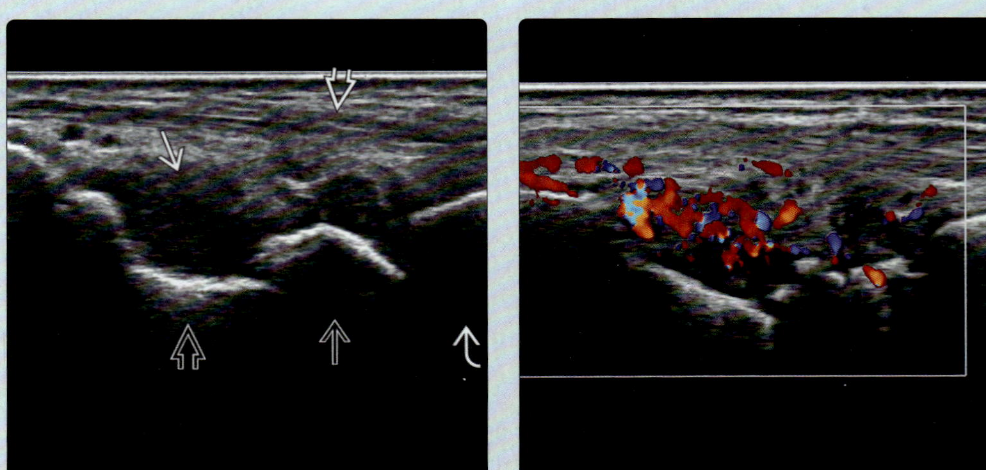

(Left) *Longitudinal US shows quite severe, largely hypoechoic synovial proliferation* ➡ *on the dorsum of the carpus deep to the extensor tendons* ⇨. *The distal end of the radius* ➡, *the lunate* ⇨, *and the capitate* ↪ *are shown. Hypoechoic synovium is more edematous than hyperechoic synovium.* **(Right)** *Longitudinal color Doppler US of the same area shows severe synovial hyperemia. This is indicative of a high level of synovial inflammatory activity.*

TERMINOLOGY

Synonyms

- Rheumatoid arthritis (RA), juvenile inflammatory arthropathy, systemic lupus erythematosus, psoriatic arthropathy, spondyloarthropathy (SpA)

Definitions

- Systemic autoimmune inflammatory disorders primarily affecting either synovium or enthesis with secondary involvement of articular surface and bone

IMAGING

General Features

- Best diagnostic clue
  - Thickened, hyperemic synovium and tenosynovium ± erosions ± joint space narrowing in inflammatory arthropathy
  - Hyperostosis, cortical irregularity, soft tissue thickening, and hyperemia at entheses in peripheral SpA
- Location
  - Any synovial-lined joint or enthesis
  - Wrist joint > carpometacarpal, metacarpophalangeal (MCP), metatarsophalangeal (MTP), and interphalangeal joints
  - Elbow, shoulder, hip, knee, ankle, intertarsal joints
- Size
  - Single (monoarthropathy), few (oligoarthropathy), or many (polyarthropathy) joints involved
- Morphology
  - Synovial thickening, effusion, erosions, joint space narrowing, deformity, subluxation, bony resorption, ankylosis
  - Progressive cortical irregularity and hyperostosis at entheses

Radiographic Findings

- Radiography
  - Often normal in early stages of disease
  - Soft tissue swelling, juxtaarticular osteopenia, joint space narrowing, bone erosion
    - Pattern of disease more important than individual radiographic signs
  - Joint obliteration and ankylosis, particularly of carpus
  - Arthritis mutilans and joint deformities ("swan neck," "boutonniere") less commonly seen with disease-modifying therapy

MR Findings

- T1WI
  - More sensitive at depicting early erosions against background of hyperintense marrow fat
  - Hypertrophied hypointense synovium
- T2WI FS
  - Shows fluid and edematous tissue (synovium, erosions, subchondral marrow, paraarticular soft tissues)
  - Effusions and pannus are both hyperintense and may not be readily distinguishable
  - Joint space narrowing, erosions, bone marrow edema graded semiquantitatively [outcome measures in rheumatology clinical trials (OMERACT) score]
- T1WI C+
  - Effusion fluid will not enhance while pannus and acute or subacute erosions will enhance vigorously
  - Contrast enhancement is necessary to appreciate activity of synovitis
    - Quantification of enhancing synovial volume and degree of synovial perfusion

Ultrasonographic Findings

- **Effusion**
  - Sensitive though nonspecific indicator of synovitis
- **Thickened synovium (pannus) is most common finding**
  - Normal synovium is not visible on US
  - Thickened synovium hypoechoic but may be iso- or hyperechoic
    - The greater the water content, the more hypoechoic the synovium
  - Synovial thickening → diffuse, frond-like, lobulated or nodular proliferation
    - May resemble hypoechoic or echogenic synovial fluid
    - Joint fluid is normally anechoic
  - US is 10x more sensitive than clinical examination in detecting early MTP joint synovitis
- **Marginal bony erosions**
  - Focal, slightly irregular cortical defect ± containing hyperemic pannus
  - 6x more MCP erosions are detected with US than radiography
  - 30% false-positive rate for US detection of erosions
- **Cartilage loss**
  - US good but best assessed with radiography or MR
- **Periarticular soft tissue changes**
  - Tenosynovitis, nerve entrapment, bursitis, rheumatoid nodules, and peripheral enthesopathy all seen well with US
- **Color Doppler helpful in distinguishing synovial proliferation from joint fluid**
  - Allows level of disease activity to be assessed
    - Active synovitis → high flow
    - Inactive synovitis → low/absent flow
    - Grade synovial hyperemia semiquantitatively as absent, mild, moderate, or severe
    - Spectral analysis: Active synovitis associated with ↓ resistive index but not routinely acquired
  - US-guided joint aspiration, biopsy, or injection is useful for both diagnosis and treatment

Imaging Recommendations

- Best imaging tool
  - US as most peripheral joints are accessible, multiple joints can be examined during same examination and diagnostic/therapeutic intervention possible
    - For SpA, US useful for peripheral SpA and, to lesser degree, sacroiliitis
- Protocol advice
  - Joints best examined in long axis, tendons in short axis
  - Schematic examination of each symptomatic joint
    - Dorsal aspects of wrists and small joints of hands and feet are usually most informative
  - If multiple joints symptomatic, examine wrists and 3 most symptomatic other joints

- If no joint symptoms, examination both wrists and 2nd MCP joint of dominant hand for subclinical disease
- Erosions usually associated with surface cortical irregularity and adjacent pannus
  - This helps distinguish normal osseous indentation from erosion ± compare with opposite side
- Use minimal transducer pressure when examining superficial joints, especially for color Doppler imaging

DIFFERENTIAL DIAGNOSIS

Osteoarthritis

- Joint space narrowing with marginal osteophytosis and absence of erosions
  - Synovial proliferation and hyperemia not as pronounced as in inflammatory arthritis
- Inflammatory erosive osteoarthritis is characterized by small joint involvement, erosions, and synovial proliferation similar to other inflammatory arthritides
  - Clinical course, radiography, and laboratory tests are helpful to distinguish

Septic Arthritis

- Often indistinguishable from inflammatory monoarthropathy
- Joint aspiration is required

Crystal Deposition Disease

- Very similar to acute inflammatory mono- or oligoarthropathy
- ± echogenic foci within joint fluid, synovium, or articular cartilage due to crystal aggregates
- Joint aspiration helpful

PATHOLOGY

Gross Pathologic & Surgical Features

- RA is primarily disease of synovial inflammation
- Bilateral, symmetrical joint involvement
- Synovial inflammatory mass (pannus) leads to erosion at bare area between articular cartilage and capsule
- Chronic synovitis/inflammation → capsular, ligamentous laxity, and tendon attrition/rupture
  - Laxity of radioulnar and ulnocarpal ligaments
  - → dorsal subluxation ulna head → extensor carpi ulnaris tendon attrition and tear (caput ulnae syndrome)

Microscopic Features

- Synovial inflammation
  - Hyperplasia of synovial cells with lymphocytic and plasma cell infiltrate
  - Synovial membrane exudes fibrinous exudate
- Rheumatoid factor (RF)
  - Serum IgM antibody present in 70% of RA patients
  - The higher the titer, the greater the likelihood of RA
    - RF titer ≥ 1:40 = 28% likelihood of RA; RF titer ≥ 1:640 = 99% likelihood of RA
  - RF-positive RA patients experience more erosive disease and extraarticular manifestations
  - Anticitrullinated protein antibodies (ACPA)
    - Associated with more severe disease progression

CLINICAL ISSUES

Presentation

- ACR/EULAR Classification Criteria for RA (2010)
  - Accessed via RheumaHelp and used as guide to establish clinical diagnosis
  - Consider the following features of RA
    - Number of large or small joints affected
    - Symptom duration < 6 or ≥ 6 weeks
    - RF and ACPA positivity
    - C-reactive protein and erythrocyte sedimentation rate levels

Demographics

- Age
  - 25-60 years most common
  - Peak incidence: 40-60 years
- Gender
  - M:F = 1:3
- Epidemiology
  - RA incidence = 1% of population

Natural History & Prognosis

- Chronic progressive disease with exacerbation and progressive joint destruction
- Complications
  - Progressive joint disease with secondary osteoarthritis
  - Tendon attrition and rupture
  - Median and ulnar nerve entrapment
  - Rheumatoid nodules at pressure sites
  - Avascular necrosis
  - Pulmonary (granuloma, fibrosis, pleuritis), ocular (episcleritis), stromal (Sjögren syndrome), and renal
  - Lymphadenopathy, splenomegaly (Felty syndrome)

Treatment

- Conservative
  - Nonsteroidal antiinflammatory agents
  - Corticosteroid (systemic or intraarticular)
  - Disease-modifying agents
    - Methotrexate, hydroxychloroquine, sulfasalazine
    - Tumor necrosis factor inhibitors: Etanercept, infliximab
- Surgical
  - Synovectomy and tenosynovectomy
  - Tendon repair
  - Arthroplasty or arthrodesis

DIAGNOSTIC CHECKLIST

Image Interpretation Pearls

- Use color Doppler to determine synovial activity
- Aspirate joint fluid if septic arthritis or crystal deposition disease suspected

SELECTED REFERENCES

1. Plagou A et al: Recommendations of the ESSR arthritis subcommittee on ultrasonography in inflammatory joint disease. Semin Musculoskelet Radiol. 20(5):496-506, 2016

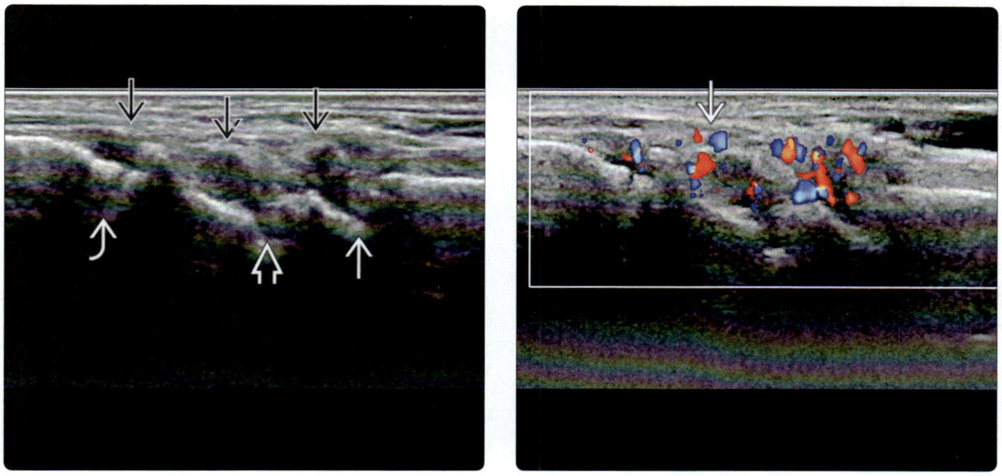

(Left) *Longitudinal US shows moderate carpal synovial proliferation → in a patient with rheumatoid arthritis (RA) with widening secondary to erosion at the radiolunate → and lunocapitate → articulations, but the joint space narrowing → is due to cartilage loss without erosion at the carpometacarpal articulation.* **(Right)** *Longitudinal color Doppler US in the same patient shows moderate-severity synovial hyperemia →, indicative of moderate synovitis.*

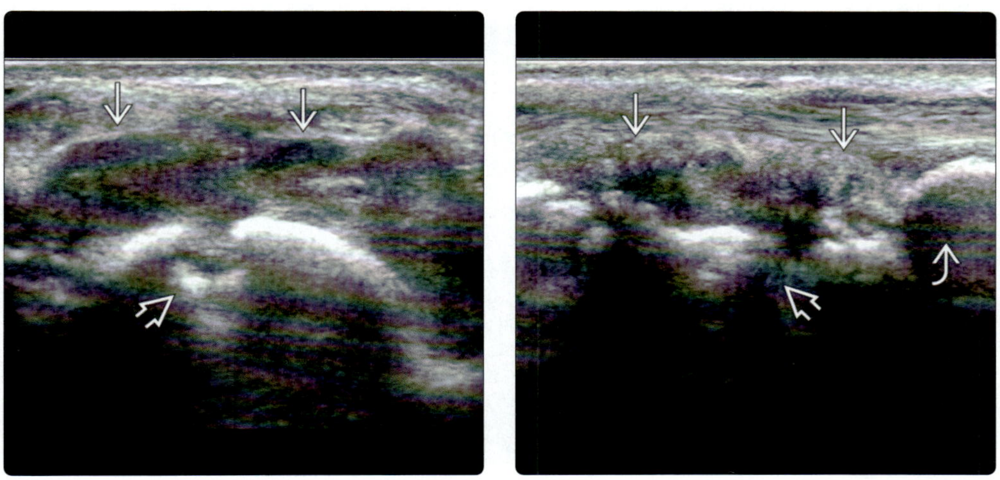

(Left) *Transverse US in a patient with chronic RA and carpal fusion shows severe synovial proliferation → on the dorsum of the wrist joint with a moderate-sized discrete bony erosion → on the dorsal aspect of the fused scapholunate articulation.* **(Right)** *Longitudinal US of same patient shows moderate-severity synovial proliferation → on the dorsal aspect of a largely ankylosed carpus. The midcarpal articulation is still not fused →. The ulnar head is mildly subluxed →.*

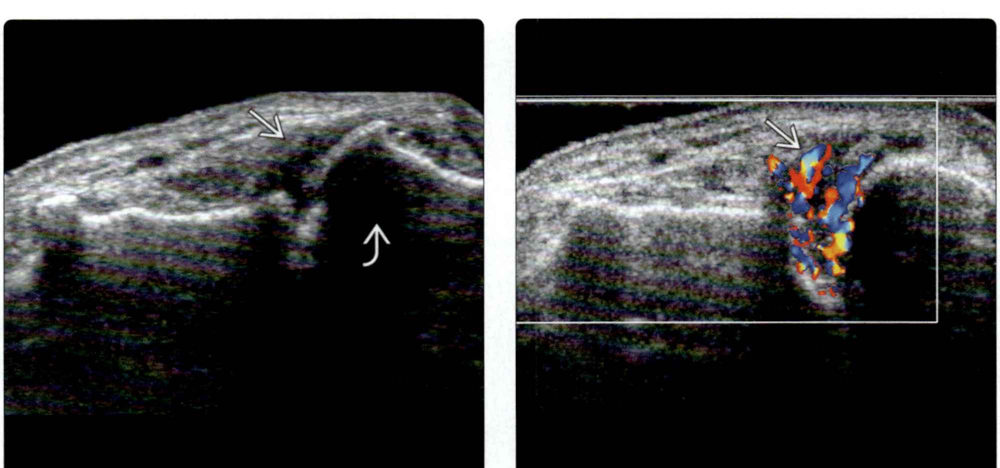

(Left) *Transverse US shows moderate synovial proliferation of the dorsal aspect of the distal radioulnar joint → in a patient with RA. There is moderate dorsal subluxation of the ulnar head →.* **(Right)** *Transverse color Doppler US of the same area shows severe synovial hyperemia →, indicative of active synovitis. Determining the degree of synovial vascularity is the best way to gauge synovial activity.*

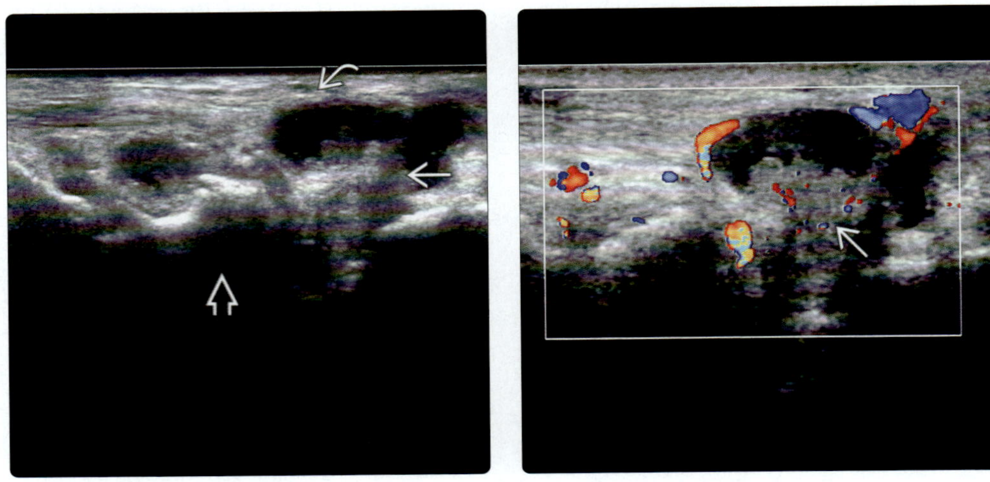

(Left) *Longitudinal US in a patient with end-stage RA shows severe arthropathy of carpus with largely fused carpus ➡. There is outpouching of both thickened synovium ➡ and synovial fluid ➡ dorsally. The thickened synovium shows villous proliferation.* **(Right)** *Longitudinal color Doppler US of the same patient shows moderate periarticular and synovial hyperemia ➡, indicative of moderate-severity disease activity.*

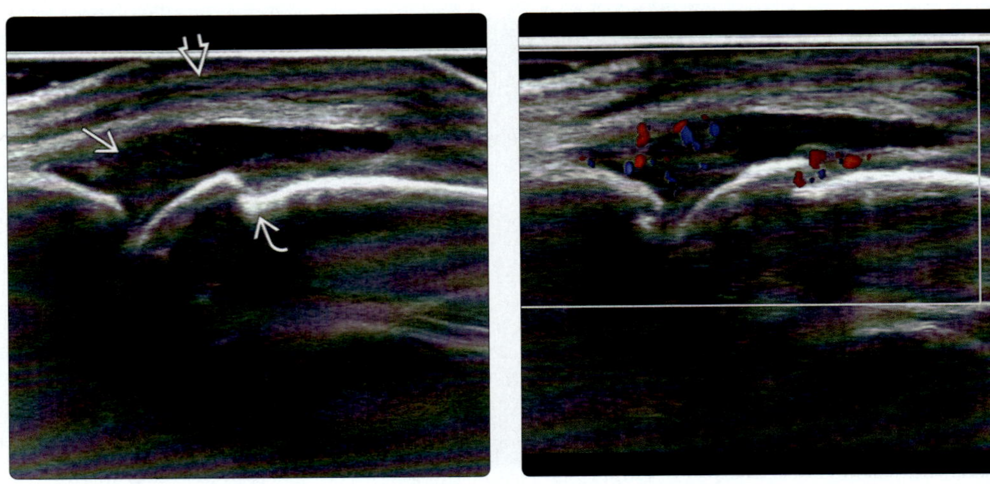

(Left) *Longitudinal US shows moderate synovial proliferation ➡ with an effusion on the dorsum of the 2nd metacarpophalangeal joint deep to the extensor tendon ➡. There is a normal focal cortical depression of the dorsal aspect of the metacarpal neck ➡. This is not an erosion.* **(Right)** *Longitudinal color Doppler US of the same joint shows moderate synovial hyperemia. This is indicative of moderate-severity synovial inflammatory activity.*

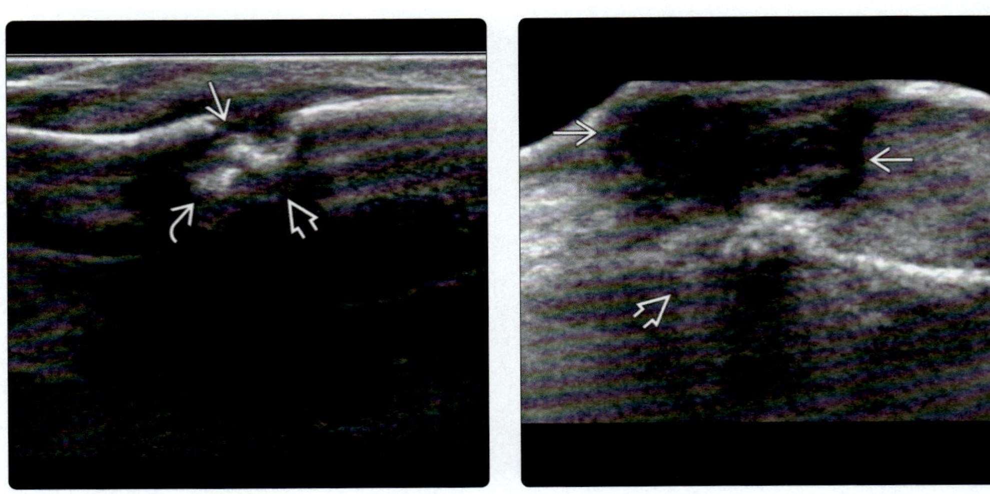

(Left) *Longitudinal US shows mild synovial proliferation ➡ on the dorsum of the middle finger distal interphalangeal joint in a patient with RA. The head of the middle phalanx is eroded ➡. The joint space is irregular and narrowed ➡.* **(Right)** *Longitudinal US of another finger in the same patient shows a well-defined hypoechoic rheumatoid nodule ➡ dorsal to an erosive arthropathy of the proximal interphalangeal joint ➡.*

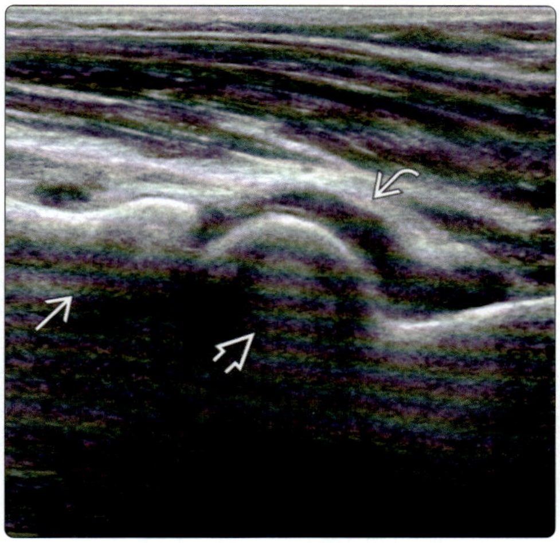

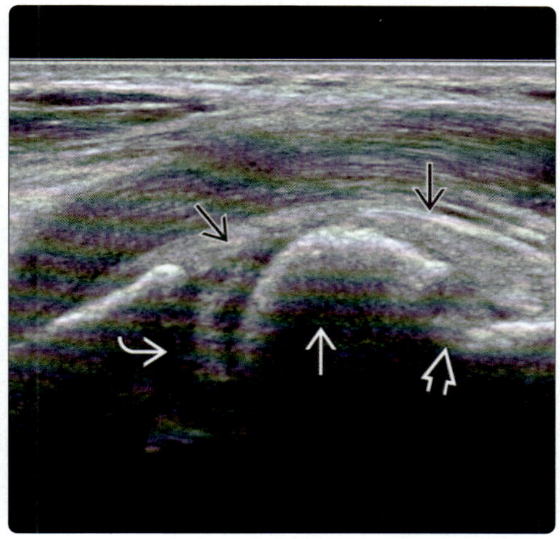

(Left) *Longitudinal US in a patient with early RA shows mild synovial proliferation with moderate effusion ➔ of the elbow joint. There is no erosive change of either the capitellum ➔ or the radial head ➔.* **(Right)** *Longitudinal US in a patient with longstanding RA shows moderate synovial proliferation ➔ on the anterior aspect of the elbow joint with a moderate-sized, discrete erosion ➔ at the proximal end of the capitellum ➔. There is also moderate subchondral resorption of the radial head ➔.*

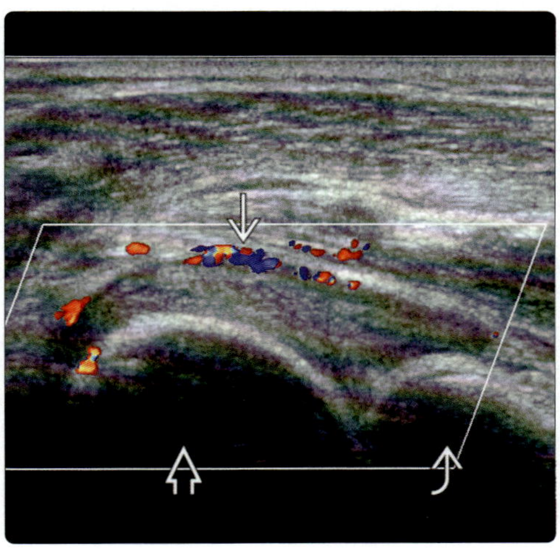

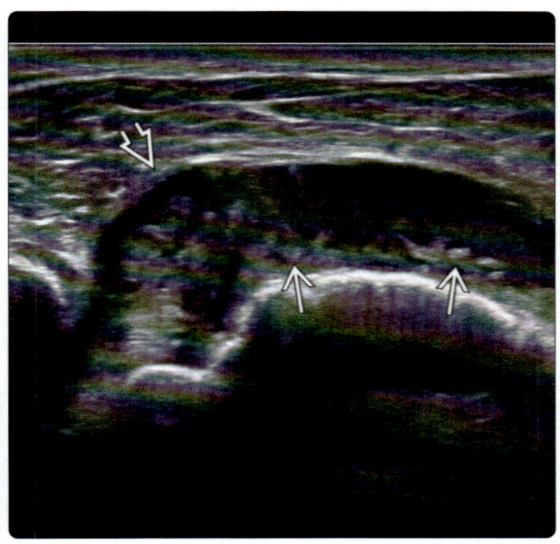

(Left) *Transverse color Doppler US in an RA patient shows moderate synovial proliferation ➔ of the glenohumeral joint posteriorly with moderate-severity synovial hyperemia. Color Doppler signal detection is less sensitive for the deeper joints. No humeral head ➔ or glenoid ➔ erosions are present.* **(Right)** *Transverse US shows severe synovial proliferation with effusion ➔ on the posterior aspect of the glenohumeral joint. The small intraarticular nodules ➔ are due to fibrinous rice bodies.*

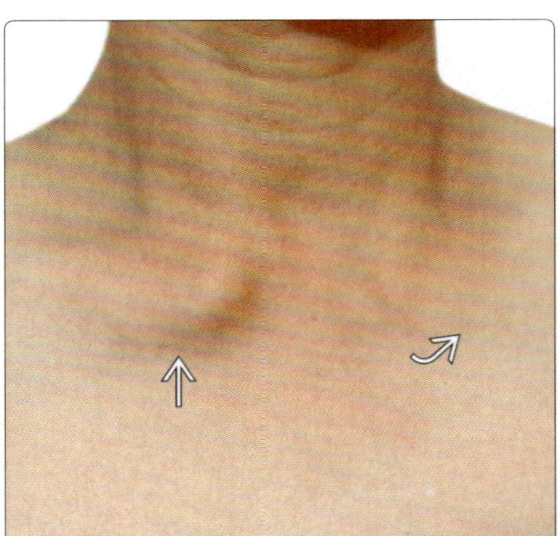

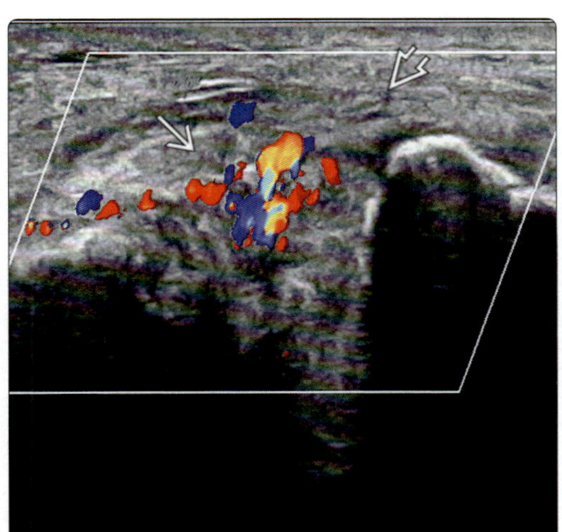

(Left) *Photograph shows an RA patient with a 3-month history of painful, right sternoclavicular joint swelling ➔. There was also pain, though much less swelling, of the left sternoclavicular joint ➔.* **(Right)** *Transverse color Doppler US in the same patient shows moderate synovial hyperemia ➔, indicative of moderately active synovitis. The periarticular soft tissues are moderately edematous ➔.*

(Left) *Longitudinal US in a patient with systemic lupus erythematosus shows a moderate-sized inflammatory joint effusion with mild synovial proliferation. Aspiration yielded straw-colored, nonpurulent, fluid, which had an increased white cell count but was negative on Gram stain and sterile on culture.* **(Right)** *Longitudinal US in a patient with chronic RA and recent hip pain shows moderate-severity synovial proliferation. There is proximal femoral avascular necrosis with femoral head collapse.*

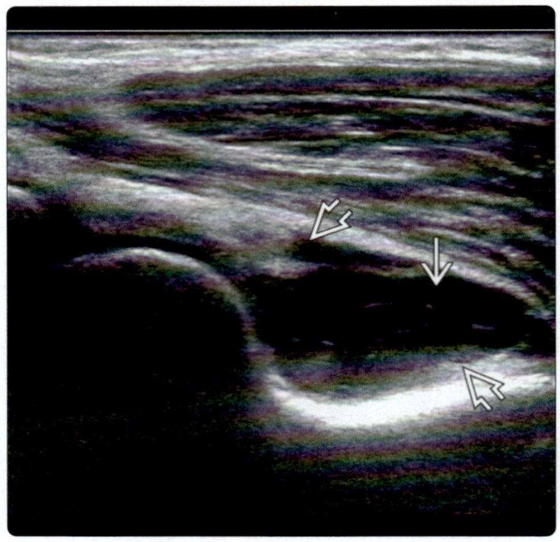

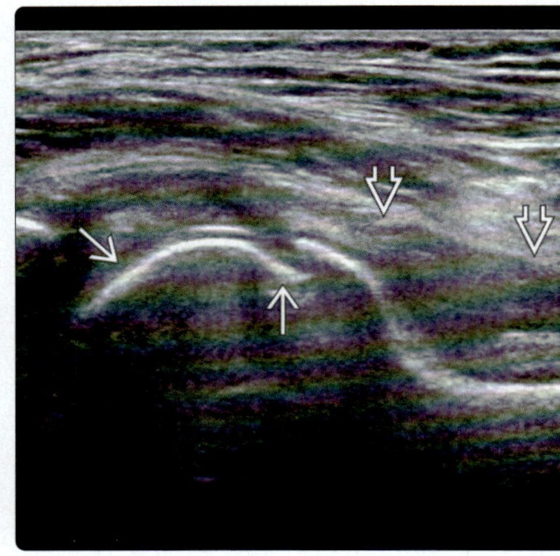

(Left) *Transverse US of the lateral patellar recess in a patient with RA shows mild, frond-like synovial proliferation, which can be readily distinguished from joint fluid.* **(Right)** *Transverse US of the lateral patellar recess in a patient with RA shows more severe, frond-like synovial proliferation.*

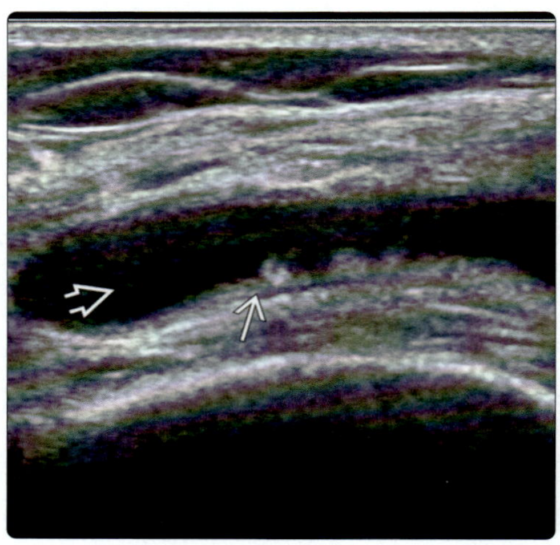

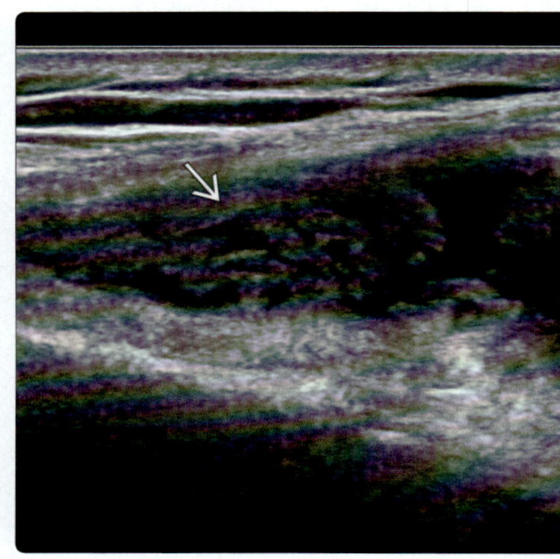

(Left) *Transverse US of the lateral patellar recess in a patient with RA shows more severe, lobulated hyperechoic synovial proliferation, which is readily distinguished from joint fluid.* **(Right)** *Longitudinal US of the suprapatellar pouch shows nodular-like synovial proliferation. There is no known significance to the type of proliferation present (i.e., smooth, frond-like, lobular, or nodular).*

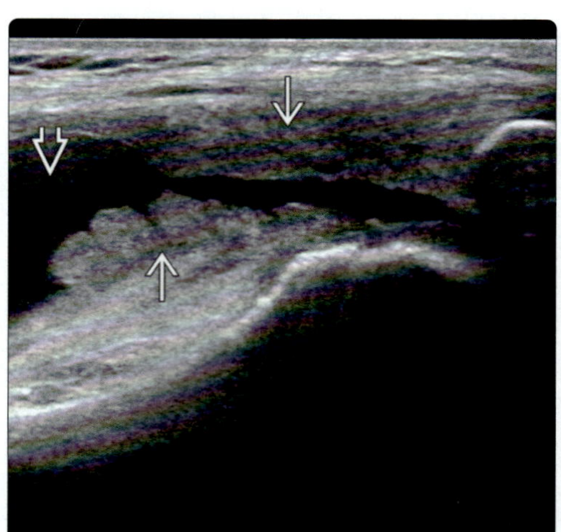

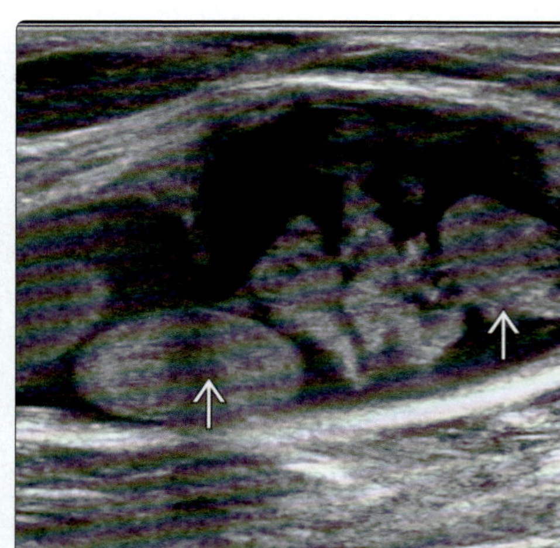

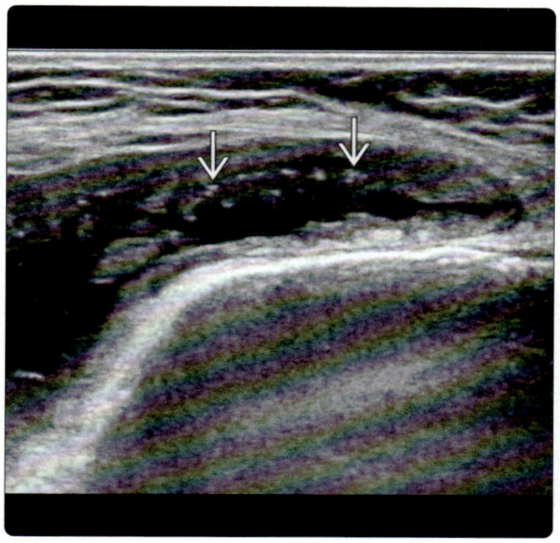

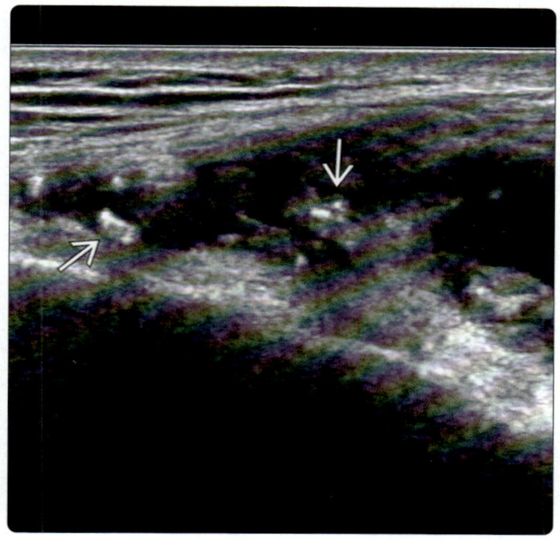

(Left) *Transverse US in a patient with RA shows moderately thickened synovium in the medial patellar recess with several small surface echogenic foci* → *due to ether fibrin aggregates or crystal aggregates. Aspiration would help differentiate.* **(Right)** *Transverse US of the lateral patellar recess in a patient with RA shows more discrete echogenic hyperechoic foci* → *probably due to crystal rather than fibrin aggregates. Aspiration would help to differentiate.*

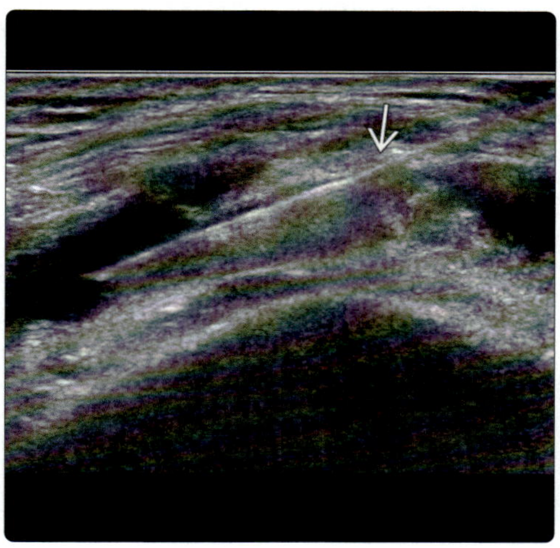

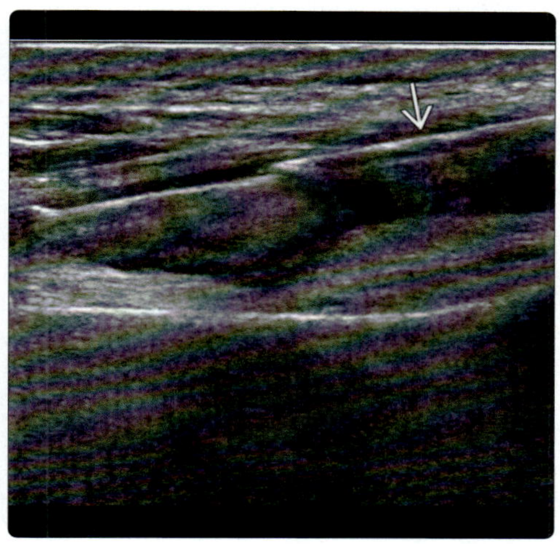

(Left) *Transverse US of the lateral patellar recess is shown in a patient with RA during knee joint needle aspiration* →. **(Right)** *Transverse US of the lateral patellar recess in a patient with RA during synovial biopsy* → *is shown. Both synovial fluid analysis and synovial biopsy are useful at excluding crystal deposition and joint infection in patients either with or suspected of having an inflammatory arthropathy.*

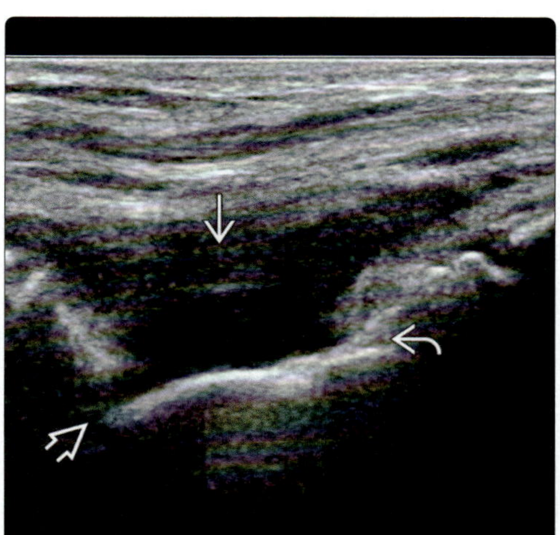

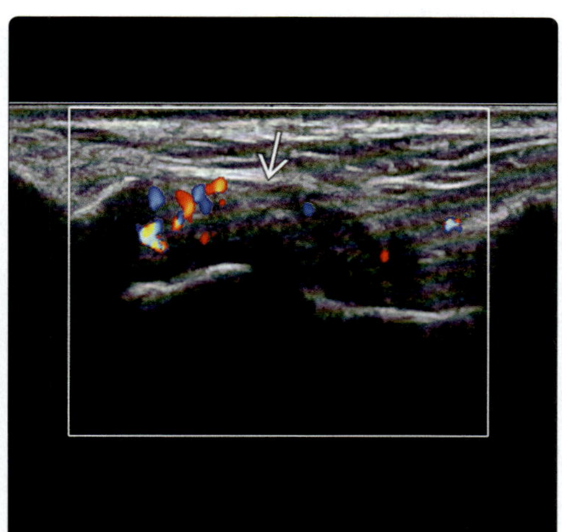

(Left) *Longitudinal US shows moderate ankle joint* → *capsular distension due to a combination of synovial proliferation* → *and joint fluid. The ankle joint capsule is attached distally to the talar neck* →. **(Right)** *Longitudinal color Doppler US of a patient with RA shows moderate ankle joint capsular distension due to synovial proliferation* →. *There is no appreciable joint fluid present. Moderate synovial hyperemia is present.*

(Left) *Longitudinal US in a patient suspected of having RA shows severe synovial proliferation on the dorsum of the cuneiform /2nd metatarsal joint. Small, echogenic foci are embedded in the thickened synovium.* **(Right)** *Longitudinal color Doppler US of the same joint shows severe synovial hyperemia. US-guided synovial biopsy (not shown) helped exclude joint infection and crystal arthropathy.*

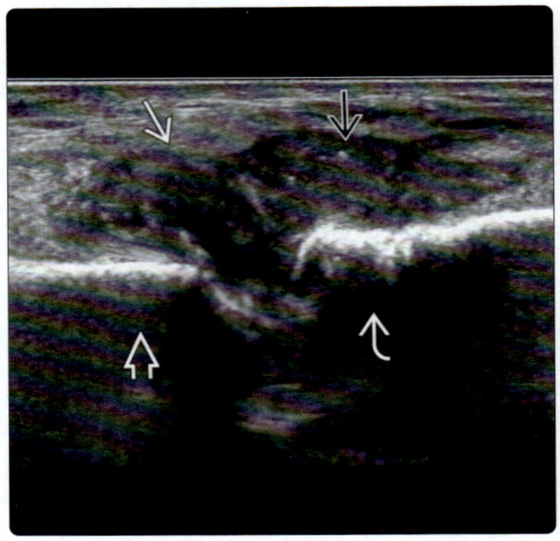

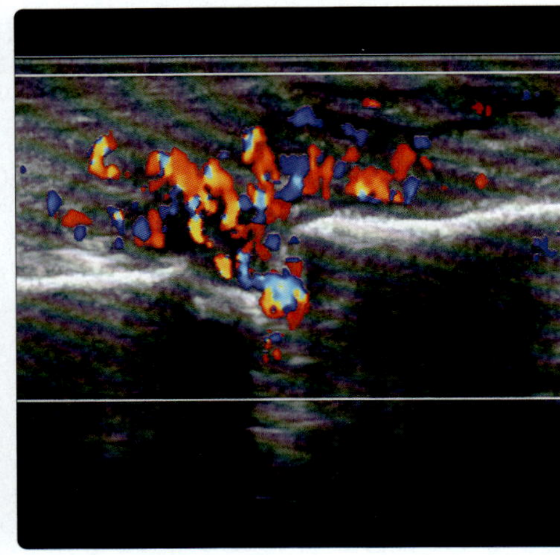

(Left) *Longitudinal color Doppler US in a patient with moderately active RA shows mild synovial hyperemia but moderate synovial proliferation and effusion with joint distension on the dorsum of the hyperextended 2nd metatarsophalangeal joint .* **(Right)** *Transverse US shows a typical hypoechoic rounded appearance and subcutaneous location of a rheumatoid nodule . This is located over the olecranon process . Mild hyperemia was present on color Doppler imaging.*

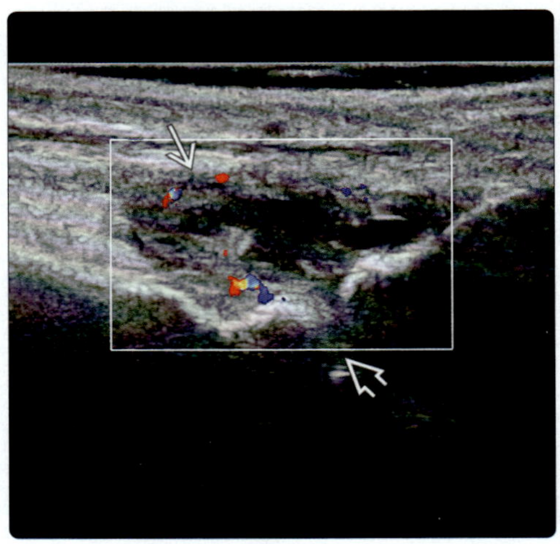

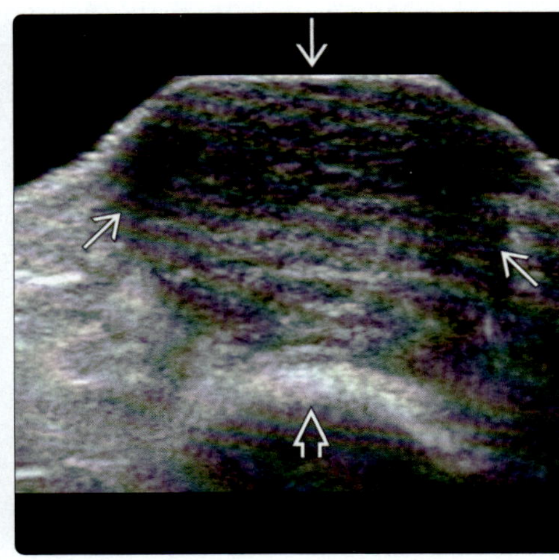

(Left) *Longitudinal US shows a palmar ganglion in an RA patient, which connects via a thin stalk to an inflamed pisiform-triquetral articulation . Inflammatory fibrous-type stranding is present in the ganglion , as inflammatory exudate spreads from the joint to the ganglion.* **(Right)** *Transverse US of a patient with RA-related tenosynovitis shows needle aspiration of the extensor digitorum tendon sheath to help exclude an infective or gouty tenosynovitis. This is helpful only in a minority of patients with tenosynovitis.*

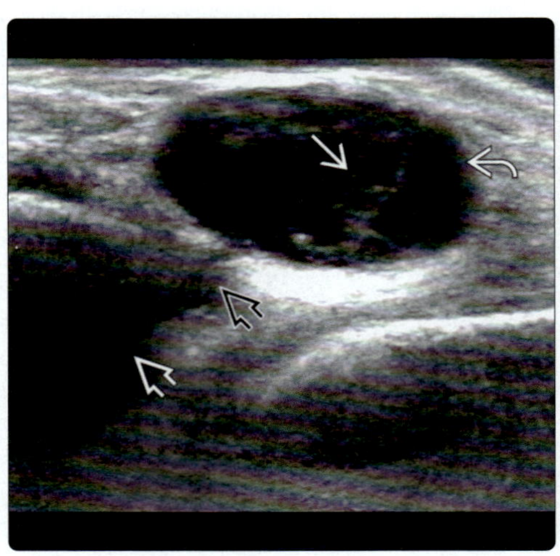

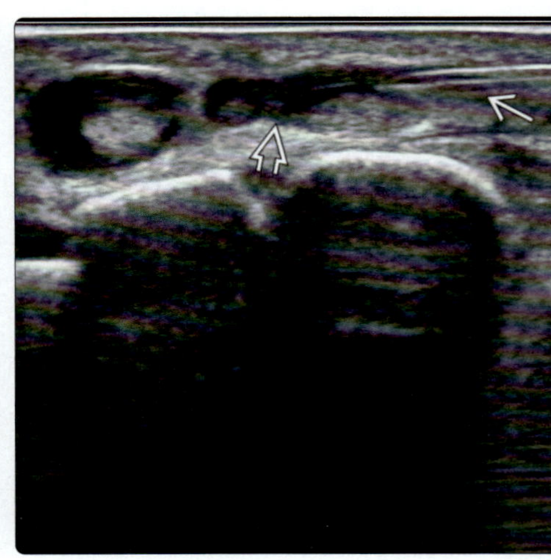

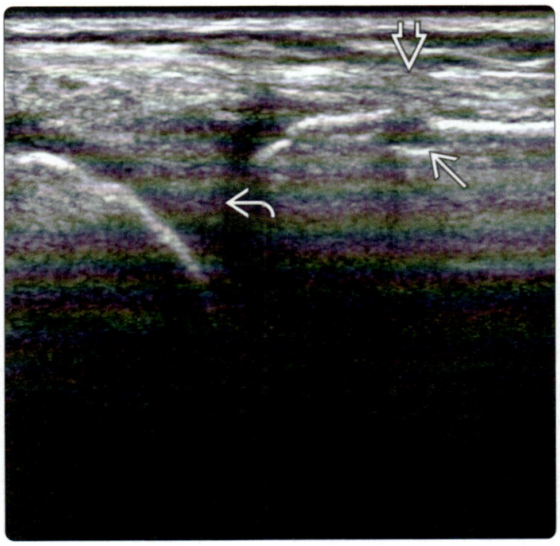

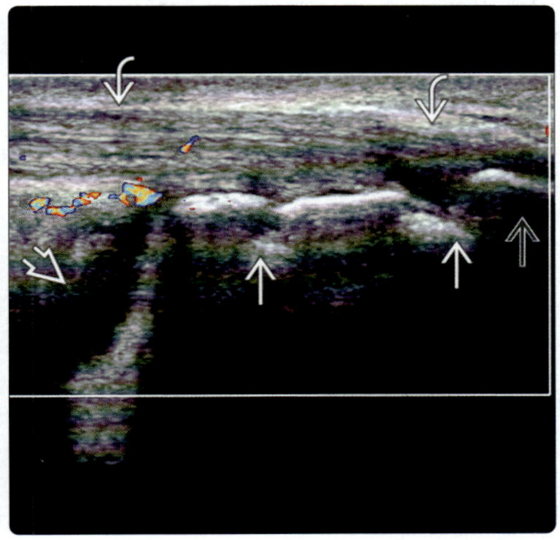

(Left) *Longitudinal US of the medial aspect of the knee joint in a patient with peripheral spondyloarthropathy (SpA) shows a medium-sized periarticular entheseal erosion deep to the normal medial collateral ligament. Medial meniscus is normal.* **(Right)** *Longitudinal color Doppler US (same patient) shows medium-sized entheseal erosions deep to the Achilles tendon. There is mild hyperemia alongside a mildly distended retro-Achilles bursa. Mild enthesophytosis with cortical hyperostosis is present.*

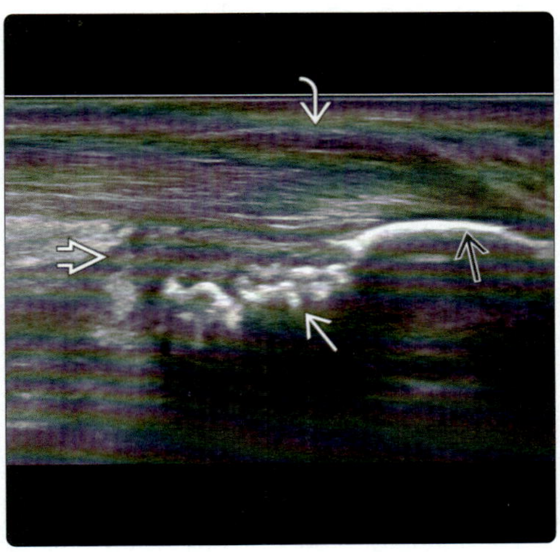

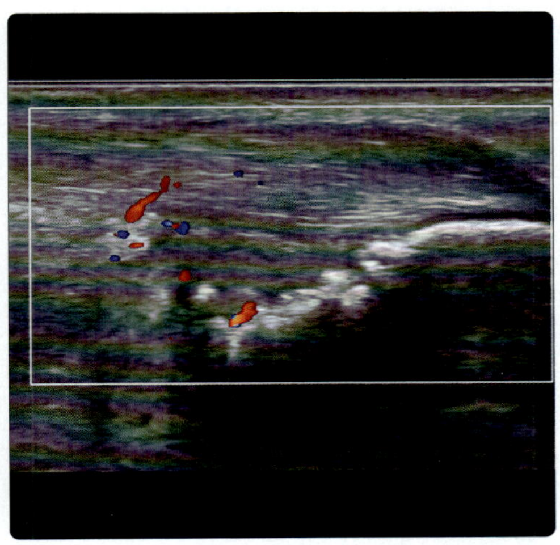

(Left) *Longitudinal US of the posterior heel in a patient with peripheral SpA shows severe irregular cortical erosion deep to the mildly distended retrocalcaneal bursa. The Achilles tendon is slightly thickened while the calcaneal tendon insertion is normal.* **(Right)** *Longitudinal color Doppler US of the same area shows mild marginal hyperemia of the retrocalcaneal bursal area. The retrocalcaneal bursa is an integral part of the Achilles tendon enthesis organ and, in this case, is the main affected area.*

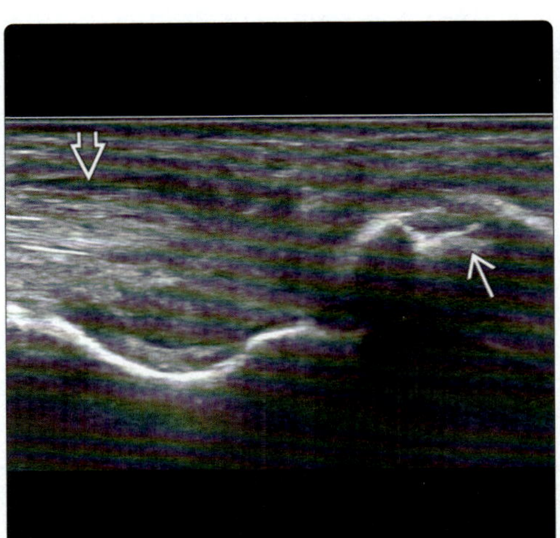

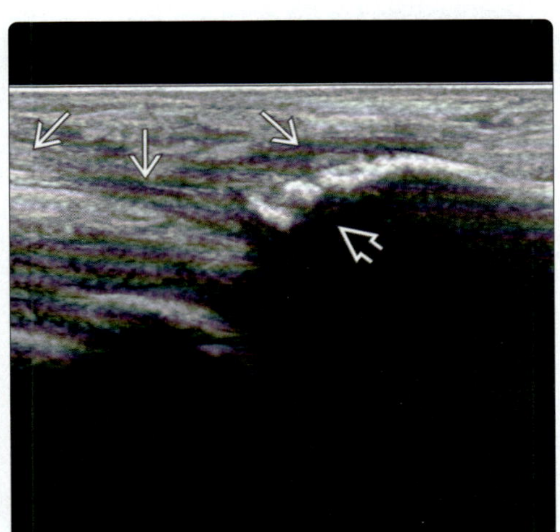

(Left) *Longitudinal US in the same patient shows a quite large entheseal erosion on the medial pole of the navicular bone. This area comprises part of the insertional area of the posterior tibialis tendon.* **(Right)** *Longitudinal US in a patient with peripheral SpA shows moderate thickening of the lateral band plantar fascia at its insertion into the 5th metatarsal base. Moderate cortical irregularity and hyperostosis (enthesophytosis) of the 5th metatarsal base is present.*

Gout and Pseudogout

KEY FACTS

TERMINOLOGY

- Inflammatory arthropathy due to urate crystal deposition (gout) or calcium pyrophosphate dihydrate (CPPD) crystal deposition (pseudogout)

IMAGING

- US signs usually specific enough to make diagnosis without joint aspiration
- **Tophi**
  - Soft tophi (hyperechoic with no acoustic shadowing) → hard tophi (hyperechoic with intense acoustic shadowing)
    - Quite well-defined irregular margins with variable hypoechoic halo ("wet clumps of sugar")
    - Multiple grouped tophi are common
- **Tendons and ligaments**
  - Probably most frequent site of gouty crystal deposition
  - Microdeposits with small hyperechoic foci → large hyperechoic foci with intense acoustic shadowing
- **Synovial fluid**
  - Hyperechoic foci or cloudy areas in synovial fluid
  - Fine punctuate echogenicity ("urate sand")
  - Large echogenic aggregates (snowstorm appearance)
  - Pseudogout: Fewer echogenic foci in synovial fluid
- **Hyaline (articular) cartilage**
  - Gout: Thin echogenic band on surface of cartilage ("urate icing" or double-contour sign)
    - Most sensitive sign for diagnosis of gouty arthritis
  - Pseudogout: Thin echogenic band in mid zone of cartilage (CPPD deposition); hyaline cartilage distribution has high disease specificity
- **Fibrocartilage**
  - Pseudogout: Punctuate echogenic foci within hypoechoic fibrocartilaginous disc (wrist) or menisci (knee) due to CPPD deposition
- **Joint aspiration**
  - Send fresh specimen (since crystals dissolved by water, alcohol, or formalin)

(Left) *Longitudinal US shows typical 1st metatarsophalangeal joint (MTPJ) gouty arthritis. The soft tophus causes marginal erosion of the metatarsal head. A small joint effusion is present in the MTPJ with echogenic gouty crystal aggregates.* **(Right)** *Anteroposterior radiograph of the same patient shows soft tissue swelling and marginal erosions. The tophi are largely noncalcified and not clearly seen.*

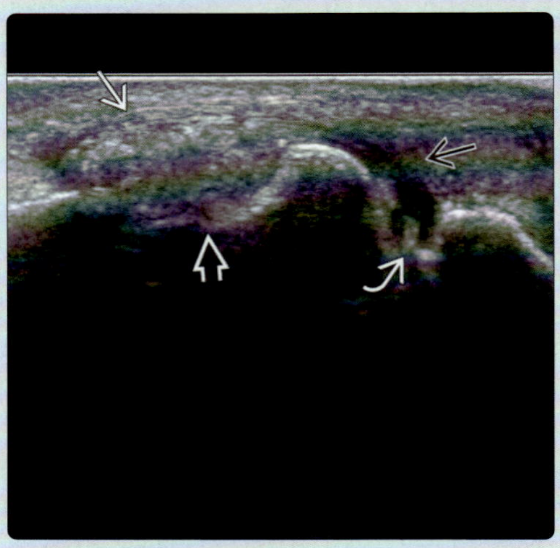

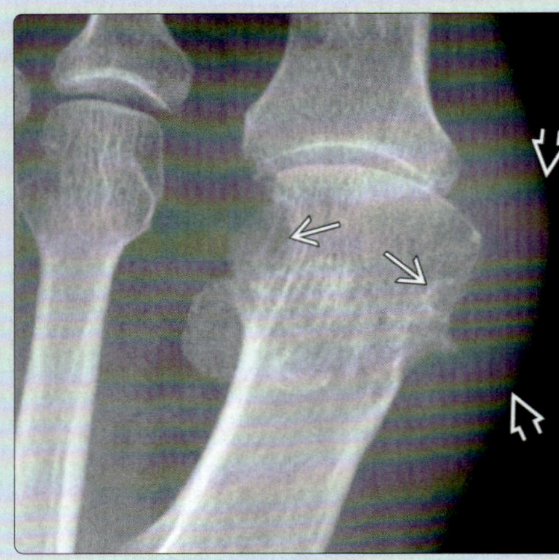

(Left) *Axial NECT of the same patient shows hyperdense calcified tophi. CT is more sensitive than radiography at revealing gouty tophi. There are marginal erosions deep to the tophi.* **(Right)** *Axial T1 MR of the same patient shows the isointense (to muscle) tophi and marginal erosions. Dual-energy CT (if available) is better than MR or US at quantifying gouty tophi volume for monitoring response to treatment.*

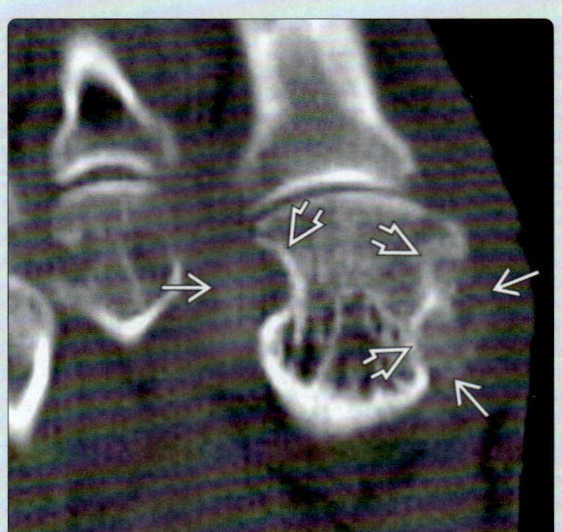

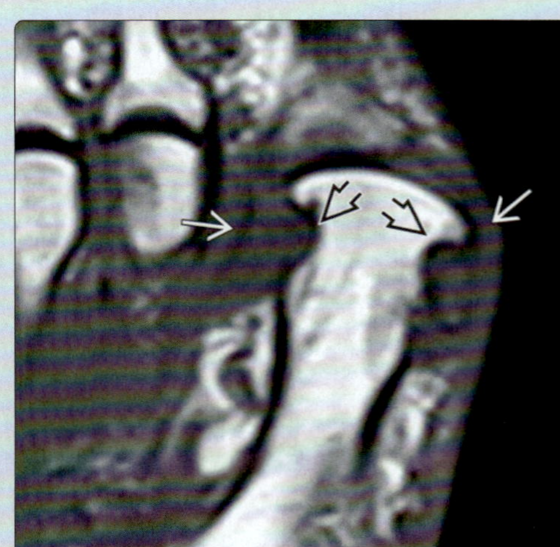

TERMINOLOGY

Synonyms

- Gouty arthropathy, crystalline arthropathy, metabolic arthritis, tophaceous gout, podagra, gouty tophi

Definitions

- Inflammatory arthropathy due to urate crystal deposition (gout) or calcium pyrophosphate dihydrate (CPPD) crystal deposition (pseudogout)

IMAGING

General Features

- Best diagnostic clue
  - Identification of echogenic foci (crystals) in joint or soft tissues
- Location
  - Gout: 1st metatarsophalangeal (MTP) joint (podagra) (50%) > ankle > midfoot > knee > wrist
    - Helix of ear, olecranon bursa, tendons (especially patellar and triceps), knee and ankle ligaments
  - Pseudogout: Any joint, especially knee, wrist, scapho-trapezium-triquetral joints
    - Extrachondral involvement (synovium, tendons, ligaments, soft tissues) is less common in pseudogout than in gout
- Size
  - Gouty tophi range from few mm to 5 cm
- Morphology
  - Monoarticular swelling → deforming polyarthropathy
    - Tophi lead to hard, discrete nodules in subcutaneous, paraarticular, and other tissues

Radiographic Findings

- Eccentric soft tissue hyperdense swelling ± calcific tophi
- Asymmetrical arthropathy
  - Absence of osteopenia
  - Joint space preserved until late
  - Marginal erosions with overhanging edges and thin sclerotic edge
    - Larger than erosions of rheumatoid arthritis

CT Findings

- Dual-energy CT has diagnostic accuracy close to 100% for gout with excellent reader agreement
  - Can automatically quantify urate volume

MR Findings

- T1WI
  - Tophi: Well-defined masses, isointense to muscle
- T2WI
  - Tophi: Low to high signal intensity
    - Variable intensity reflects amount of urate deposition, edema, and maturity of granulation tissue
    - Paucity of bone marrow edema suggests how pathophysiology of gouty erosions differs from other inflammatory arthropathies
- T1WI C+
  - Tophi: Honeycomb-like heterogeneous to near-homogeneous enhancement
    - Honeycomb appearance is due to avascular urate surrounded by vascularized granulation tissue

Ultrasonographic Findings

- **Tophi**
  - Soft tophi (hyperechoic with no acoustic shadowing) → hard tophi (hyperechoic with intense acoustic shadowing)
  - Quite well-defined irregular margins with variable hypoechoic halo ("wet clump of sugar")
  - Multiple grouped tophi are common
- **Tendons (and ligaments)**
  - Probably most frequent site of crystal deposition in gout
  - Ranges from microdeposits with small, hyperechoic foci within tendon to large, hyperechoic foci with intense acoustic shadowing
    - Fibrillar pattern is disrupted, tendon is stretched and displaced
    - ± tendon rupture (uncommon)
    - Acoustic shadowing with larger (> 5-mm) masses
- **Synovial fluid**
  - Gout: Variable appearance of joint fluid
    - Hyperechoic aggregates in synovial fluid
      - ◻ Fine punctuate echogenicity ("urate sand")
      - ◻ Large echogenic aggregates (snowstorm appearance)
  - Pseudogout produces fewer echogenic aggregates in joint fluid
- **Articular cartilage**
  - Gout: Thin echogenic band on surface of cartilage ("urate icing" or double-contour sign)
    - Most sensitive sign for diagnosis of gouty arthritis
  - Pseudogout: Thin echogenic band in mid zone of cartilage (CPPD deposition)
    - No acoustic shadowing
- **Fibrocartilage**
  - Pseudogout: Punctuate echogenic foci within hypoechoic fibrocartilaginous disc (wrist) or menisci (knee) due to CPPD deposition
- **Bone**
  - Marginal erosions
    - Invariably associated with overlying tophus
- **Subcutaneous tissue**
  - Gouty panniculitis, which is lobular panniculitis, particularly of lower extremities

Imaging Recommendations

- Best imaging tool
  - US best 1st-line investigation for gout
  - More accurate than clinical examination or radiography at detecting gout, though less accurate than DECT
  - Conflicting results, though overall sensitivity of US seems to be slightly less than DECT, particularly for extraarticular disease
    - 1st MTP joint, 2nd MTP joint, ankle, knee, wrist, and patellar or triceps tendons all common sites of involvement
      - ◻ Screening 1st MTP joints and either knees or ankles is comparable to screening all areas
  - US signs usually specific enough to make diagnosis without joint aspiration

- Protocol advice
  - Assess all aspects of joint (cartilage, fluid, synovium) and periarticular structures (ligaments, tendons, bone)
  - ± synovial fluid aspiration for analysis by polarizing light microscopy (sensitivity: 84%; specificity: 100%)
    - Only small amount of fluid is required for analysis
    - Send fresh specimen (since crystals are dissolved by water, alcohol, or formalin)
    - Urate crystals: Needle-like, negatively birefringent
    - CPPD crystals: Rod-like, positively birefringent
  - Urate and CPPD crystals may coexist
  - Also send fluid for Gram stain, culture, and cell count

DIFFERENTIAL DIAGNOSIS

Rheumatoid Nodule

- Hypoechoic, well defined ± hyperemia
  - No echogenic foci
  - Typical location (over bony prominences)

Rheumatoid Arthritis

- Symmetrical polyarthropathy
- Proteinaceous aggregates of inflammatory arthropathy less echogenic than crystal aggregates
- Absence of crystals on synovial fluid analysis

Septic Arthritis

- Distinction from acute gout can be difficult
  - Septic arthritis and crystal arthropathy may coexist
  - Synovial fluid analysis helpful

PATHOLOGY

General Features

- Etiology
  - Genetic
  - Diet ("disease of kings")
  - Diabetes, thiazide diuretics, renal impairment
- Associated abnormalities
  - Hypertension, diabetes mellitus, metabolic syndrome, abdominal obesity, hyperlipidemia
  - Renal stones (1,000x increased incidence)
    - Pure uric acid stones (80%) or mixed stones (20%)
- Uric acid = end product of purine catabolism
  - ↓ excretion of uric acid in urine (90% cases)
  - ↑ production of uric acid (10% cases)
- Only fraction of patients with hyperuricemia develop clinical gout
  - US is helpful to diagnose subclinical gouty arthritis in hyperuricemic individuals
- Urate solubility is affected by concentration, cold, pH, tissue composition and turnover
  - Uric acid preferentially deposits in synovium, around joints, tendons, ligaments
- Urate crystals are coated with apolipoprotein E or B
  - Protein coating may stop crystals from triggering inflammation
  - Immunoglobulins bind to bare areas on crystals and initiate phagocytosis by neutrophils
  - Exposure of bare areas on crystals (flux in uric acid level or trauma) may trigger acute inflammation

Gross Pathologic & Surgical Features

- Tophi; multilobulated locules with white, chalky material

Microscopic Features

- Multicentric urate crystals surrounded by vascularized granulation tissue and inflammatory infiltrate

CLINICAL ISSUES

Presentation

- Most common signs/symptoms
  - Acute monarticular arthritis (90%)
    - Severe inflammation reaching peak in 8-12 hours
  - Arthritis usually precedes development of tophi
    - Tophi slowly enlarge over years

Demographics

- Age
  - Peak onset = 30-50 years (males), 50-70 years (females)
  - Earlier with renal insufficiency or genetic predisposition
- Sex
  - M:F = 9:1 in gout
    - Estrogen is mildly uricosuric
    - Gout more common in postmenopausal women
  - M:F = 1.5:1.0 in pseudogout
- Epidemiology
  - 1% of general population

Natural History & Prognosis

- Acute attack resolves spontaneously in < 2 weeks
  - Joint is normal between attacks (early disease)
  - → polyarticular involvement with attacks becoming less intense, more frequent, and lasting longer
  - → chronic polyarthropathy
  - If serum urate levels are kept below 6 mg/dL, ↓ in crystal deposition can be seen on US
- Gouty nephropathy

Treatment

- Limiting acute attack
  - Rest and nonsteroidal antiinflammatory agents
- Prevention or reversal of crystal deposition
  - Diet modification and weight reduction
  - Drug therapy: Allopurinol, probenecid, or colchicine
- Surgery
  - Excision of tophi (cosmetic, chronically discharging)

DIAGNOSTIC CHECKLIST

Consider

- US evaluation should include assessment of synovial fluid, articular cartilage, bone, tendons, ligaments, and soft tissue tophi
  - Distinction between gout and pseudogout rests on site of crystal deposition and synovial fluid analysis

SELECTED REFERENCES

1. Teh J et al: Advanced imaging in the diagnosis of gout and other crystal arthropathies. Semin Musculoskelet Radiol. 22(2):225-236, 2018

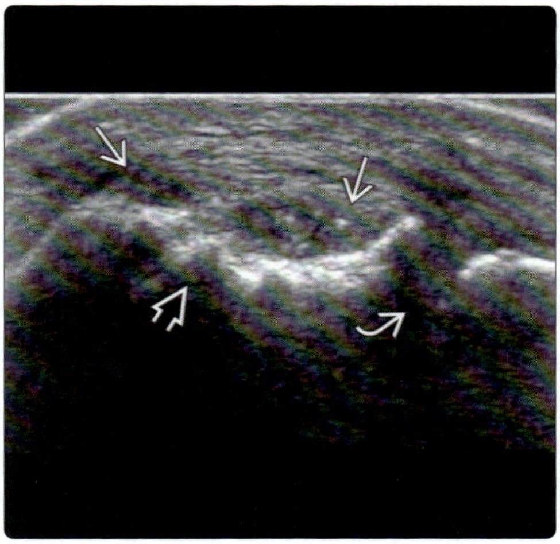

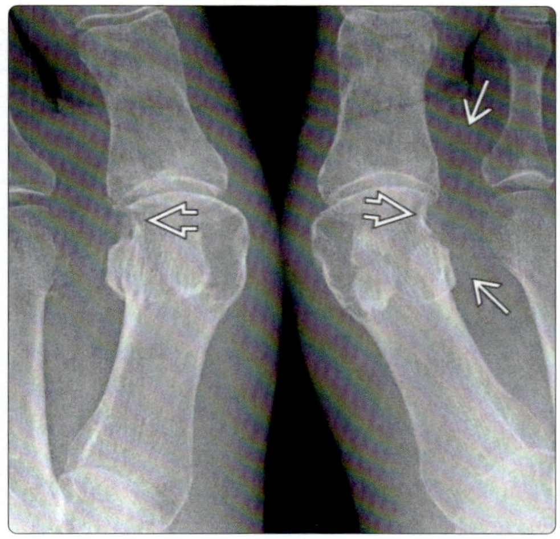

(Left) *Longitudinal US shows a soft tophus near the 1st MTPJ with marginal erosion of the metatarsal head. A small MTPJ effusion is present with tiny echogenic "urate sand" within the synovial fluid.* **(Right)** *Anteroposterior radiograph of the same patient shows soft tissue swelling and marginal erosions of both MTPJs. Symptoms were only present on the right side.*

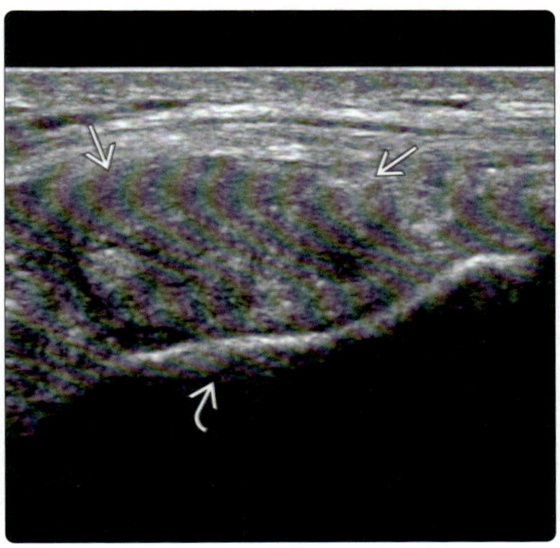

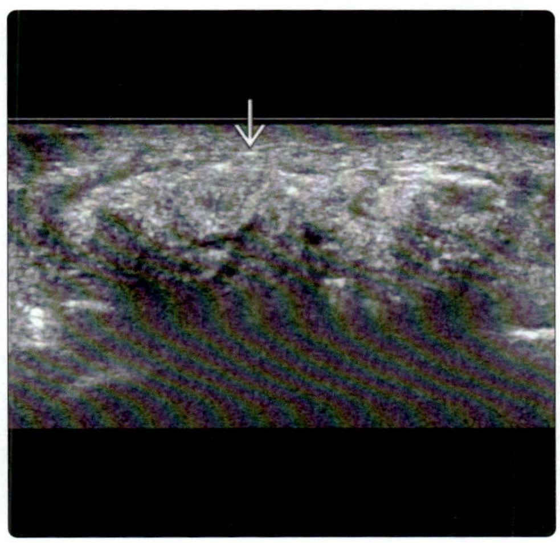

(Left) *Oblique US of the knee shows a large, soft gouty tophus on the lateral aspect of the distal femur. The internal structure of the tophus as well as the bone deep to the tophus can be fully appreciated.* **(Right)** *Transverse US of the dorsum of the foot shows a fairly soft gouty tophus with diffuse increased echogenicity and moderate posterior shadowing. Almost the full depth of the tophus can be seen, but the deeper structures cannot be appreciated.*

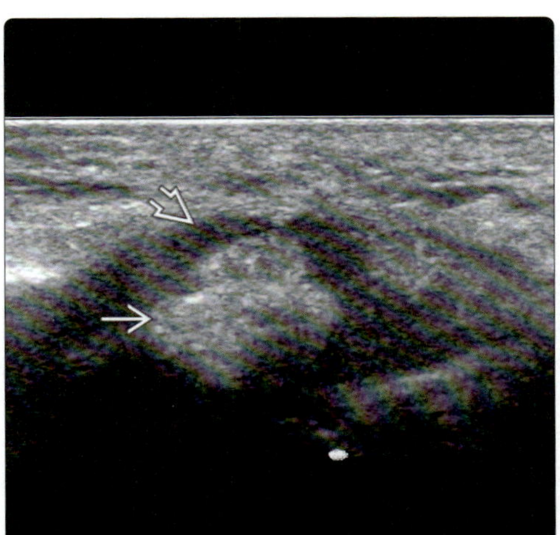

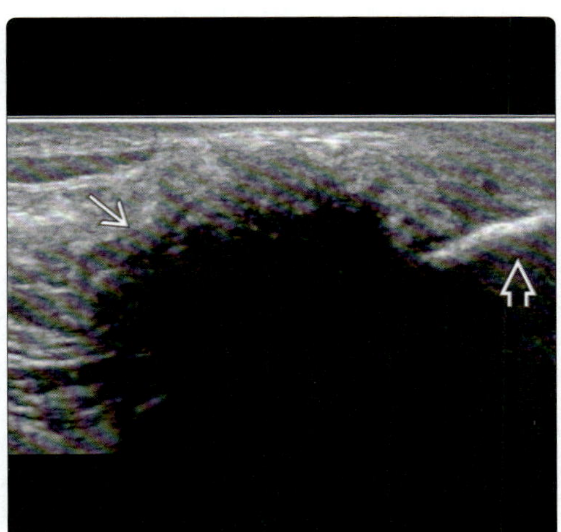

(Left) *Transverse US of the volar aspect of the wrist shows a moderately hard gouty tophus just proximal to the tunnel inlet. There is a hyperechoic central area with a quite well-defined hypoechoic rim of granulation tissue (lump of wet sugar appearance).* **(Right)** *Longitudinal US shows a very hard gouty tophus overlying the anterior tibial tuberosity. Neither the internal composition of the tophus nor the structures deep to the tophus can be appreciated. The tibial shaft is shown.*

(Left) *Three-dimensional reformatted dual-energy CT shows multiple urate deposits (green) in both feet. Total urate volume can be quantified using an automated method. (Courtesy P. Munk, MD.)* **(Right)** *Longitudinal US of the elbow in a 34-year-old man with acute gout shows a small joint effusion with synovial proliferation and microcrystalline deposition . "Urate icing" or double-contour sign is seen on the capitellum. This is a sensitive sign for gouty arthropathy.*

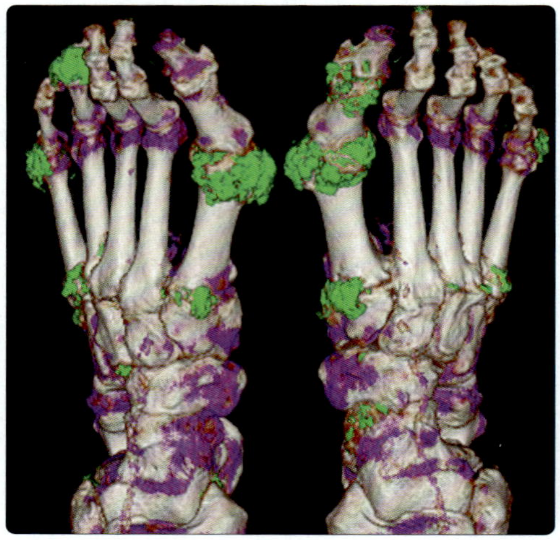

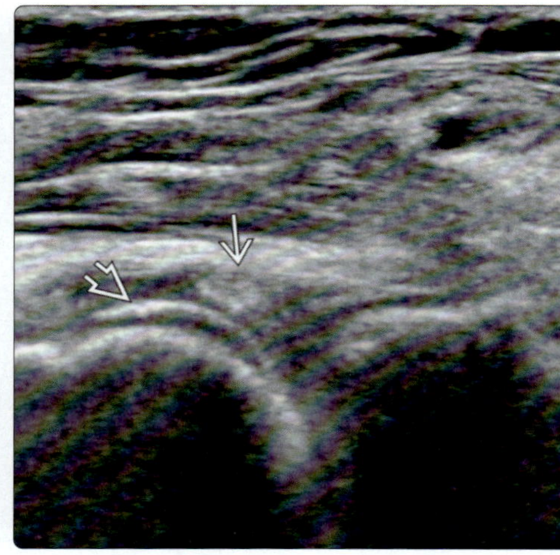

(Left) *Longitudinal US of the knee shows synovial proliferation and fluid on the medial aspect of the knee joint beneath the medial collateral ligament . There are fine punctuate echogenic foci present, compatible with "urate sand."* **(Right)** *Transverse US in the same patient shows joint distension with large gouty crystal aggregates anterior to the articulation between the proximal tibia and the fibula .*

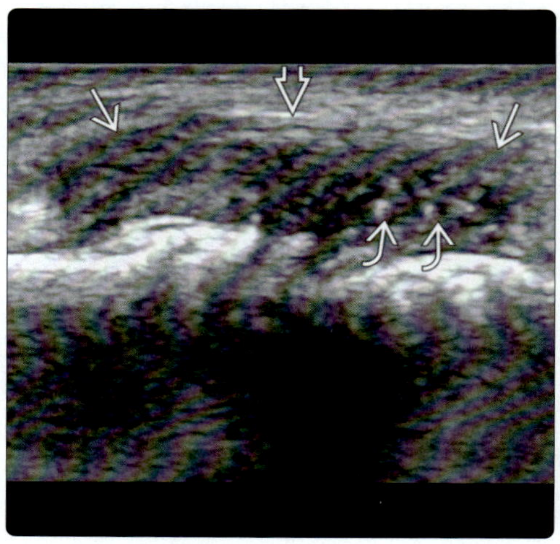

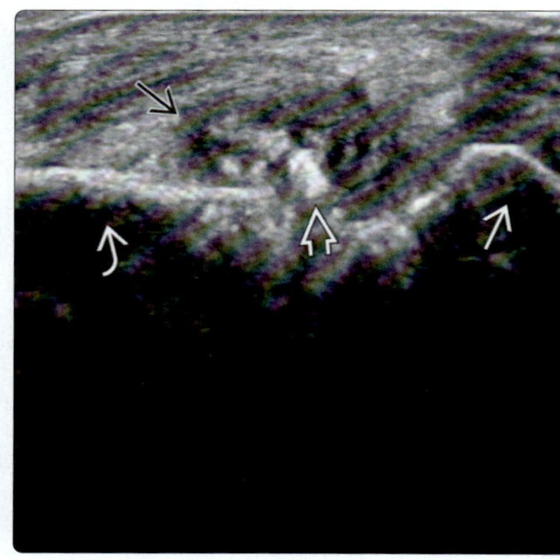

(Left) *Clinical photograph shows a patient with gout with swelling and mild redness on the lateral part of the midfoot/forefoot junction.* **(Right)** *Oblique US of the dorsolateral foot in the same patient shows a soft tophus overlying the 5th metatarsal base at the site of insertion of the peroneus brevis tendon .*

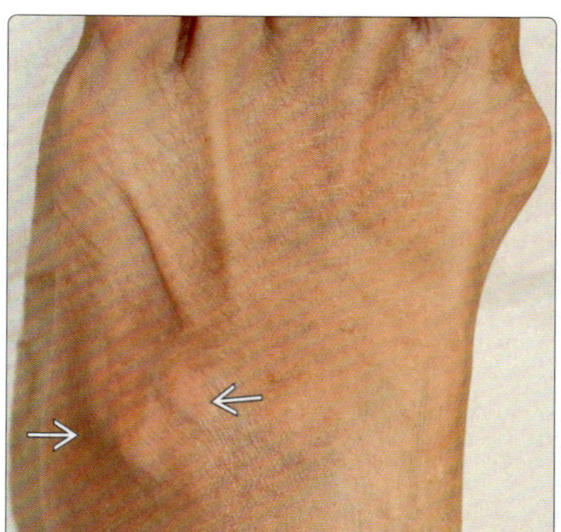

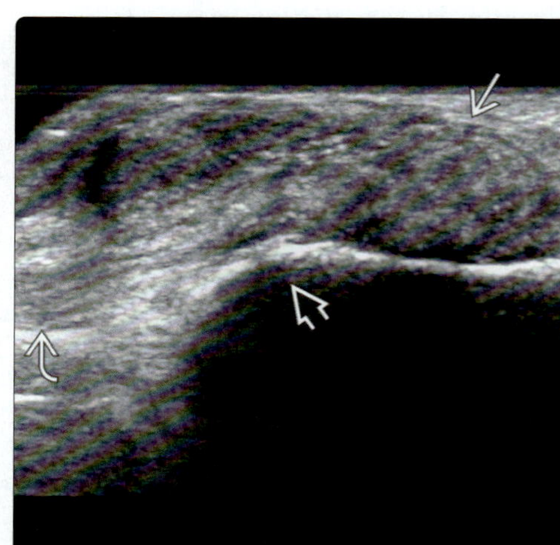

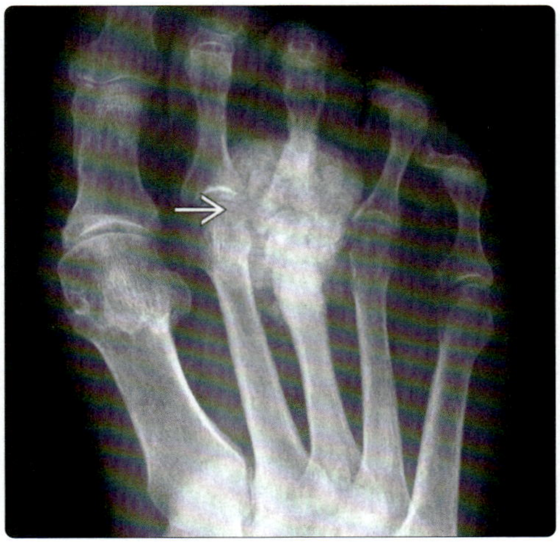

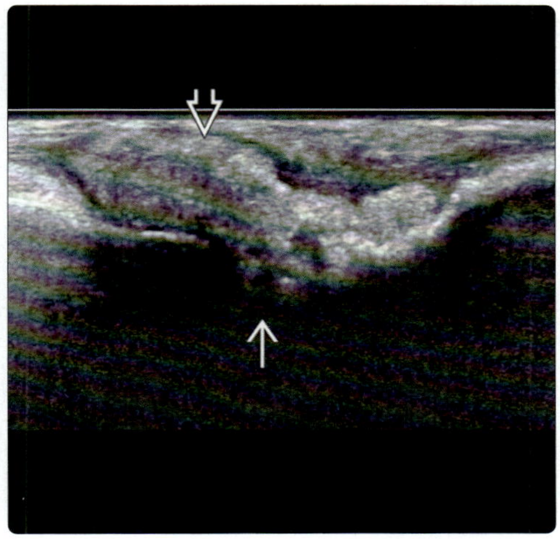

(Left) *Dorsoplantar radiograph shows the foot in an elderly woman with no known gout. Due to recent swelling and radiographic abnormality* ➡, *she was referred with a suspected chondrosarcoma.* **(Right)** *Longitudinal US of the same patient shows severe destruction of the 3rd MTPJ* ➡ *with the joint capsule distended by finely punctuate echogenic fluid ("urate sand")* ➡. *This is entirely consistent with a gouty arthropathy. The appearances are not those of malignancy.*

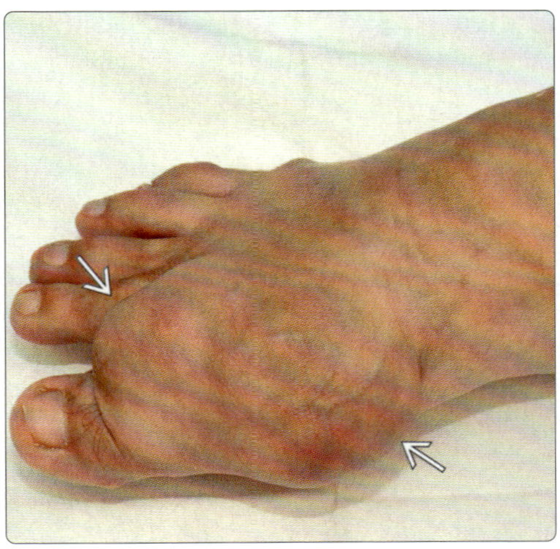

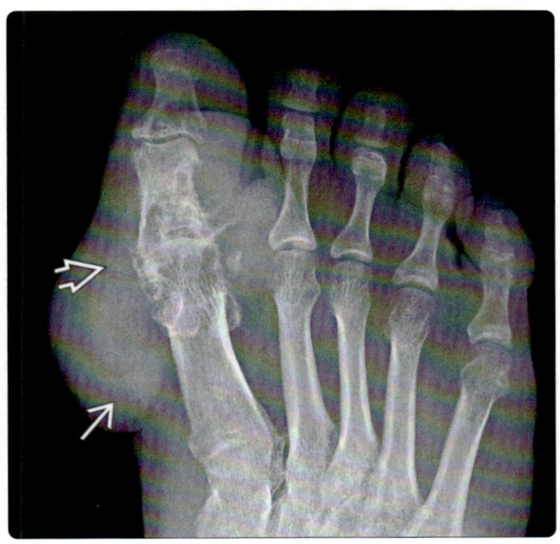

(Left) *Clinical photograph of a 32-year-old man with known gout and progressive forefoot swelling for 2 years shows a large, hard mass on the medial aspect of forefoot* ➡. *A soft tissue sarcoma was suspected clinically.* **(Right)** *Dorsoplantar radiograph of the foot shows a large, partially calcified soft tissue mass* ➡ *with severe bony erosion around the severely narrowed 1st MTPJ* ➡. *The big toe interphalangeal joint is less affected. These appearances are compatible with severe chronic gout.*

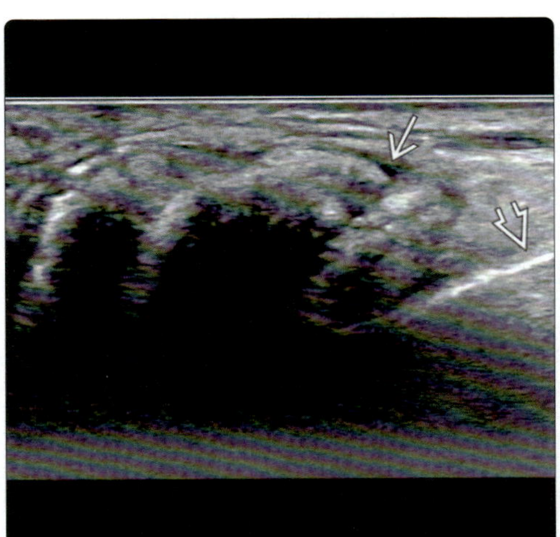

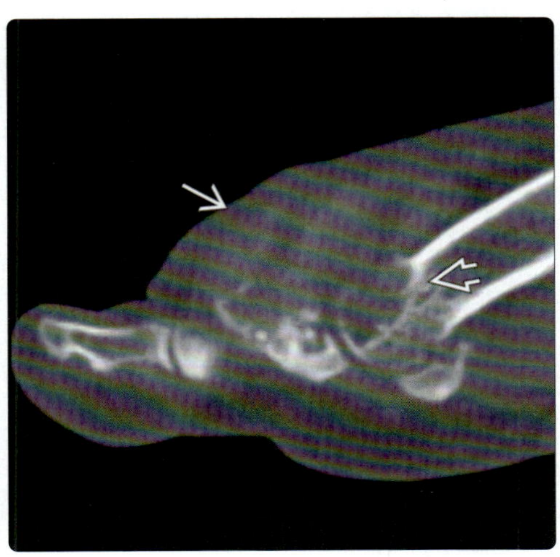

(Left) *Longitudinal US of the same patient shows a moderately hard tophus* ➡ *obscuring the MTPJ distal to the 1st metatarsal shaft* ➡. *No additional soft tissue component was present.* **(Right)** *Sagittal CT reconstruction shows large, well-marginated bony erosions* ➡ *and heavily calcified soft tissue mass* ➡. *Overall appearances are entirely consistent with chronic tophaceous gout. There is no need for percutaneous biopsy in this clinical setting.*

(Left) *Clinical photograph of a patient with known gout shows swelling and redness at the posterior aspect of the elbow, consistent with olecranon bursitis.* **(Right)** *Transverse US of the elbow shows gouty olecranon bursitis superficial to the olecranon. The bursa is distended with finely echogenic fluid ("urate sand").*

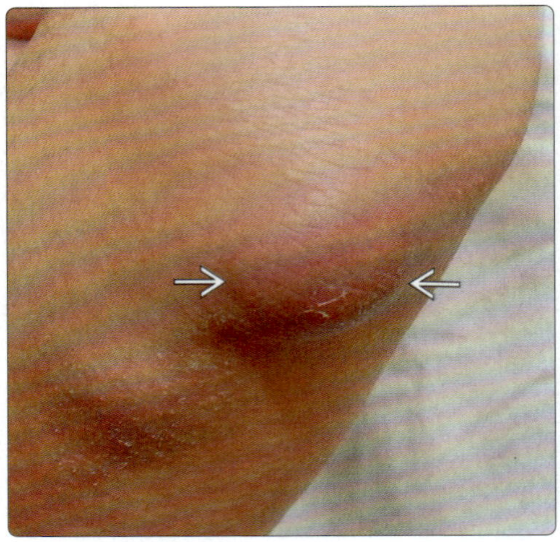

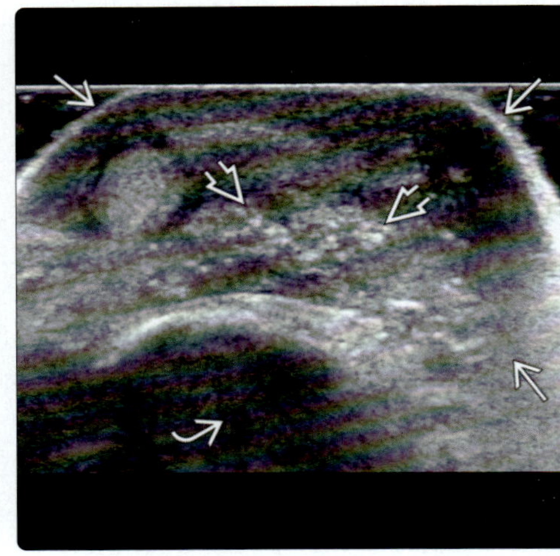

(Left) *Longitudinal US shows gouty arthritis on the dorsoulnar aspect of the wrist joint. There are small echogenic foci in the joint (snowstorm appearance) with urate deposits on the articular disc surface. The ulnar head and extensor carpi ulnaris tendon are normal.* **(Right)** *Longitudinal US of the hand shows severe gouty flexor digiti minimi tenosynovitis. The tendon sheath is distended with fluid and large crystal aggregates. The flexor tendon and metacarpal bone are normal.*

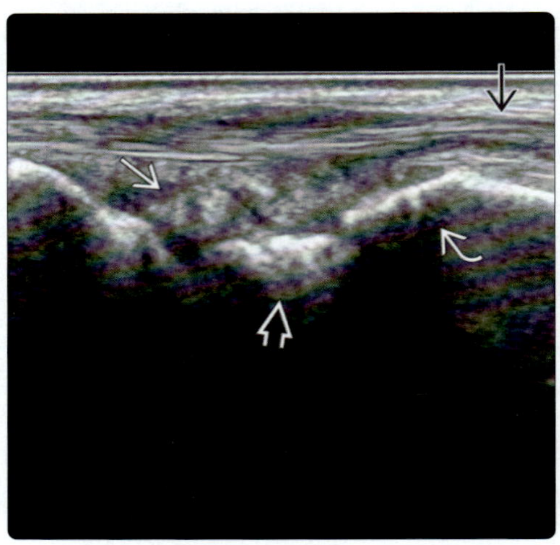

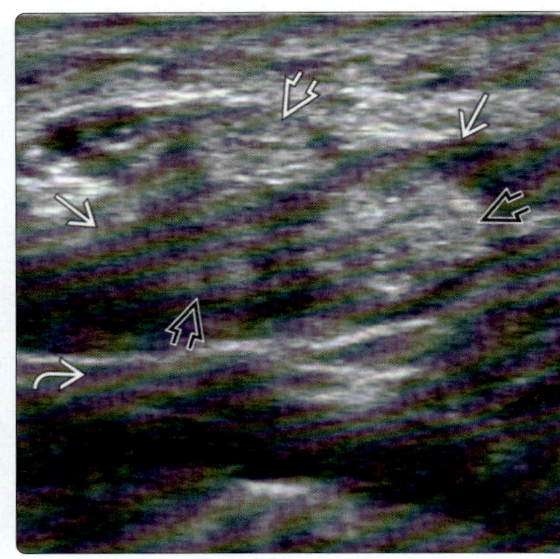

(Left) *Transverse US shows severe gouty flexor tenosynovitis of the wrist. The tendon sheath is severely distended with fine echogenic fluid ("urate sand"). The flexor tendons are poorly defined with features of intratendinous urate deposition.* **(Right)** *Longitudinal US shows anterior tibialis gouty tenosynovitis. Tendon sheath is eccentrically distended with finely echogenic crystals. Tendon is normal. Eccentric deposition favors solid over liquefied forms of urate deposition.*

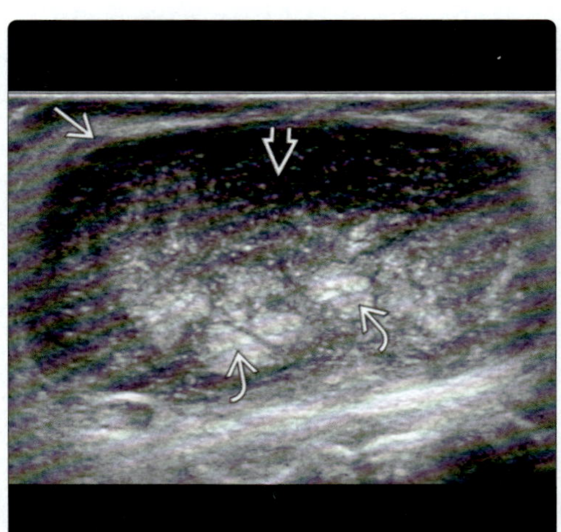

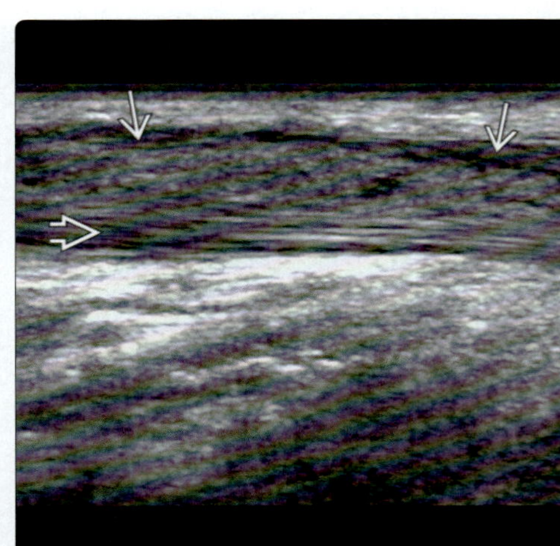

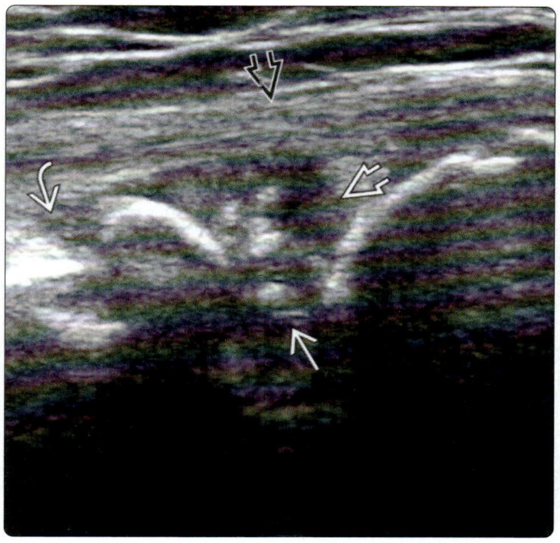

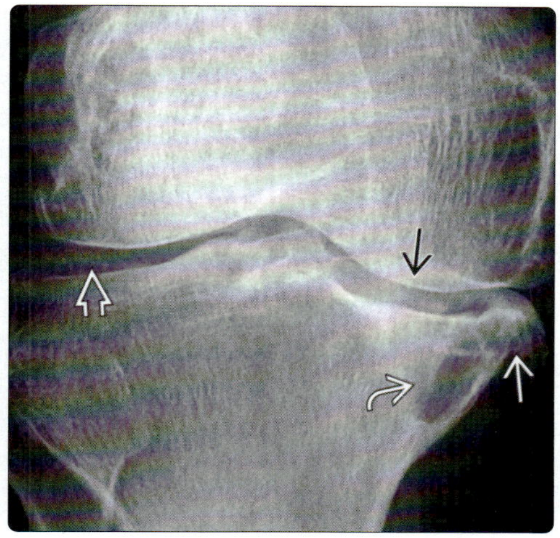

(Left) *Longitudinal US of the knee shows echogenic foci [calcium pyrophosphate dihydrate (CPPD) deposition] within the lateral meniscus. The lateral collateral ligament is normal. There is marginal bony erosion.* **(Right)** *Anteroposterior radiograph of the same patient shows chondrocalcinosis of the lateral meniscus. There are features of quite severe osteoarthritis with marginal osteophytosis, joint space narrowing, and a large subchondral cyst.*

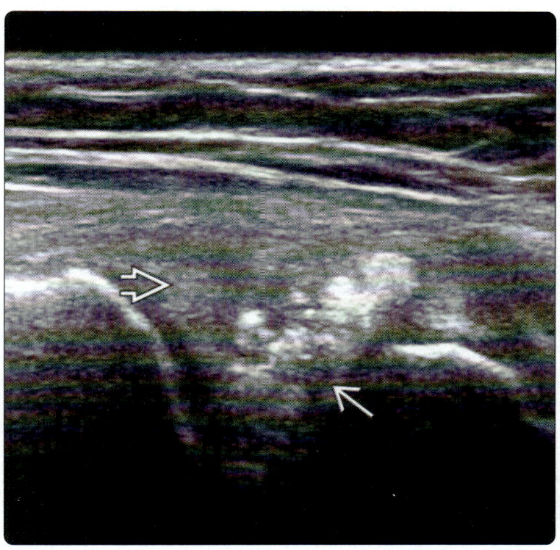

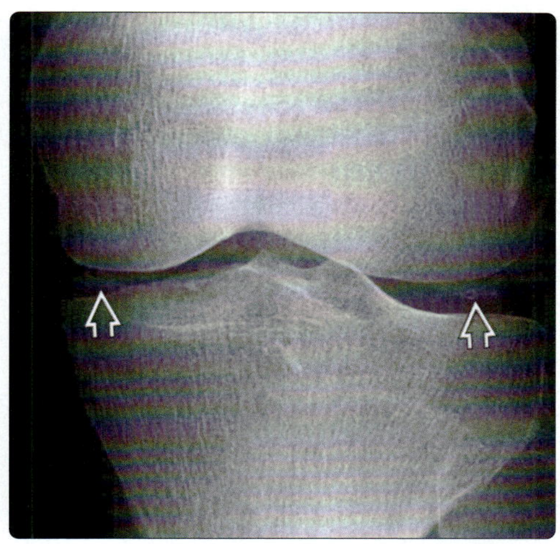

(Left) *Longitudinal US of the knee shows echogenic foci due to CPPD deposition in the medial meniscus.* **(Right)** *Anteroposterior radiograph of the same patient shows moderate chondrocalcinosis of both the medial and lateral menisci. Calcium pyrophosphate deposits centrally in the menisci, while urate deposits initially on the surface of the menisci.*

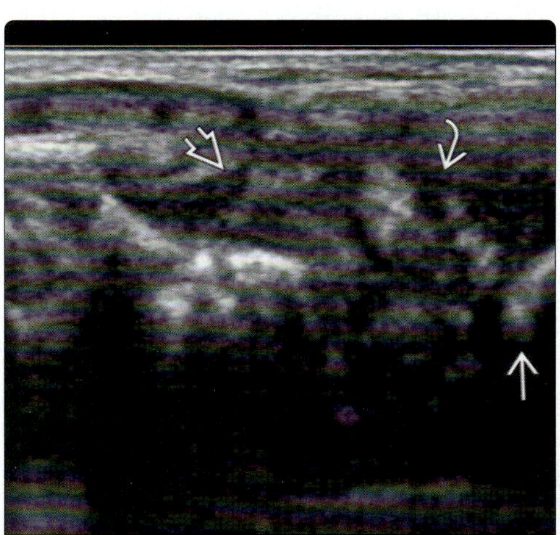

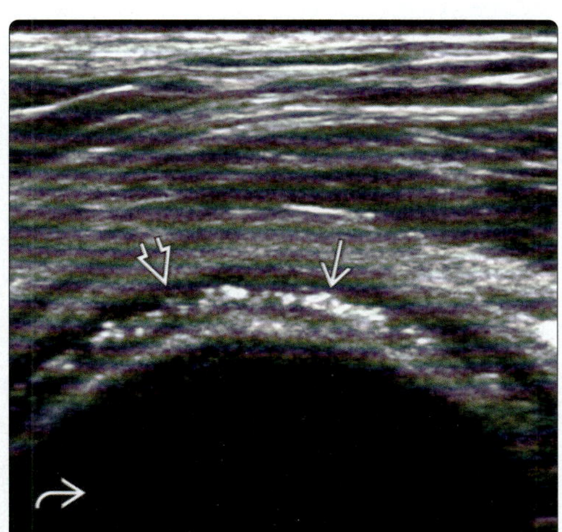

(Left) *Transverse US shows moderate synovitis at the ulnocarpal interval with severe calcification of the articular disc of the triangular fibrocartilaginous complex alongside the ulnar styloid process. This is compatible with CPPD.* **(Right)** *Longitudinal US of the knee shows a quite confluent echogenic band due to CPPD deposition in the mid zone of the articular cartilage of the medial femoral condyle posteriorly. This pattern of crystal deposition is quite specific for CPPD disease.*

Developmental Hip Dysplasia

KEY FACTS

IMAGING

- Shallow acetabulum ± subluxation of femoral head
- US useful before 6 months of age
- Normal hip
  - Capsular laxity: > 50% bony coverage femoral head at rest, with < 50% bone coverage on stress examination or normal movement
- Subluxation: < 50% bony coverage at rest
- Dislocation: Femoral head lies completely outside of bony acetabulum at rest
- Modified Graf staging
  - Type Ia = mature hip, angular bony promontory, $\alpha > 60°$, $\beta < 55°$
  - Type Ib = mature hip, roundish bony promontory, $\alpha > 60°$, $\beta < 55°$
  - Type IIa = physiologic immaturity < 3 months, α = 50-60°, β = 55-77°
  - Type IIb = immaturity > 3 months, α = 50-60°, β = 55-77°
  - Type IIc = critical hip, subluxation, α = 43-49°, $\beta > 77°$
  - Type III = dislocated hip, $\alpha < 43°$
  - Type IV = dislocated hip, inverted labrum, $\alpha < 43°$
- Roundish anterior acetabular rim can be physiological and often seen in immature hips
- Obtain single image showing deepest part of acetabulum, triradiate cartilage, center of femoral head, straight iliac contour

TOP DIFFERENTIAL DIAGNOSES

- Physiologically immature hip in neonate or premature baby

DIAGNOSTIC CHECKLIST

- Key to diagnosis rests on obtaining correct coronal (longitudinal) view
- Be clear with respect to definitions of capsular laxity, subluxation, and dislocation
  - Useful to have these definitions written on US report
- If in doubt, repeat US in 2-4 weeks

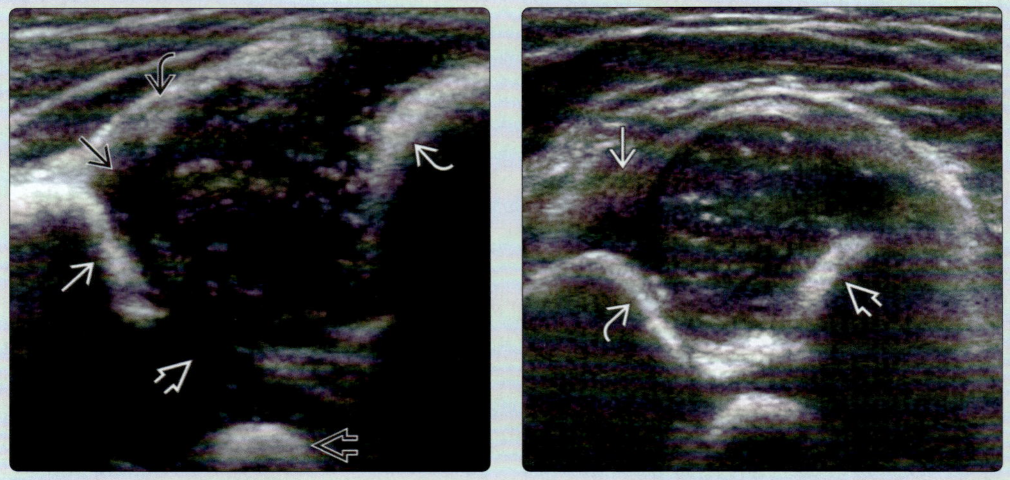

(Left) *Longitudinal coronal US shows a standard view of a normal hip. This view includes the osseous acetabular roof ➔, cartilaginous acetabular roof ➔, acetabular labrum ➔, triradiate cartilage ➔, ischium ➔, and ossified femoral metaphysis ➔.* **(Right)** *Transverse US of a normal hip shows the femoral head well centered within the acetabulum. The anterior pillar of the acetabulum ➔, acetabular labrum ➔, femoral metaphysis ➔, and gluteal muscle are shown.*

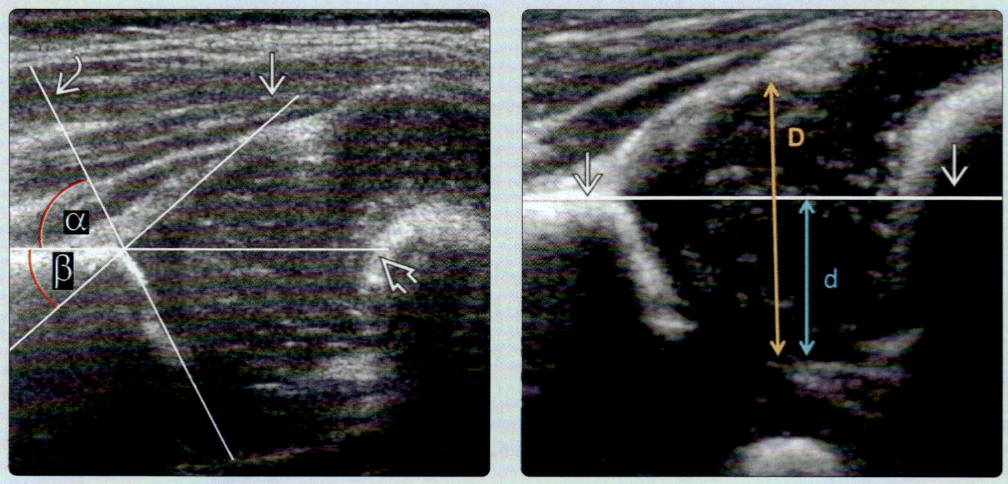

(Left) *Longitudinal coronal US shows a standard view of a normal hip. Note the lines along the ilium ➔, acetabular roof ➔, between the acetabular promontory and the tip of acetabular labrum ➔, and the α and β angles formed by the intersection of these lines.* **(Right)** *Longitudinal US shows a standard coronal view of a normal hip with a well-covered femoral head. The portion of the femoral head diameter (D) covered by bone acetabulum (d) is > 50% (D/d > 50%). Note the iliac line ➔.*

TERMINOLOGY

Abbreviations

- Developmental dysplasia of hip (DDH)

Synonyms

- Congenital dysplasia of hip (CDH) (less appropriate)

Definitions

- Underdevelopment of acetabular component of hip joint

IMAGING

General Features

- Best diagnostic clue
  - Shallow acetabulum ± subluxation of femoral head
- Morphology
  - Ranges from shallow acetabulum to complete dislocation with false acetabulum
    - Secondary hypoplasia of femoral head

Ultrasonographic Findings

- Both morphology (coronal and transverse views) and dynamic evaluation
- Morphology: Graf technique
  - Coronal ("egg in spoon") view
    - Most important image
    - Obtained from lateral approach with hip and knee in slight (~ 20°) flexion; infant in decubitus (or supine) position
    - Align transducer 10-15° obliquely (usually posteriorly) from coronal plane to obtain straight iliac line
    - Obtain single image showing deepest part of acetabulum, triradiate cartilage, center of femoral head, straight iliac contour
      - Position too anterior → false-positive result; too posterior→ false-negative result
    - Normal hip
      - Sharp edge to ossified acetabular roof (promontory)
      - Roundish anterior acetabular rim can be physiological and often seen in immature hips
  - Transverse ("ice cream cone") view
    - Shows femoral head overlying triradiate cartilage
  - 3 lines drawn on coronal view to assess acetabular morphology
    - Line along straight contour of ilium (iliac line)
    - Line along osseous acetabular roof
    - α angle measured at intersection between iliac line and line along osseous acetabular roof
      - α angle is acetabular angle
      - Reflects steepness of acetabular roof
      - Greater the angle, more mature hip; smaller the angle, more dysplasia of hip
    - Line from promontory to tip of labrum
    - β angle measured at intersection between iliac line and line from acetabular promontory to tip of labrum
      - Represents elevation of acetabular labrum
      - Displacement of femoral head; restraint of cartilaginous roof coverage
      - β angle less critical than α angle
  - Modified Graf staging
    - Type Ia = mature hip, angular bony promontory, α > 60°, β < 55°
    - Type Ib = mature hip, roundish bony promontory, α > 60°, β < 55°
    - Type IIa = physiologic immaturity < 3 months, α = 50-60°, β = 55-77°
    - Type IIb = delayed ossification = immaturity > 3 months, α = 50-60°, β = 55-77°
    - Type IIc = very deficient acetabulum = subluxation, α = 43-49°, β > 77°
    - Type III = dislocated hip, α < 43°
    - Type IV = severe dysplasia = dislocated hip, inverted labrum, α < 43°
- Dynamic testing: Apply Barlow maneuver (i.e., posterior force on adducted and flexed hip)
  - Attempt to displace femoral head from acetabulum during real-time imaging
  - Lack of relaxation by infant can limit acquisition of optimal stress view and can produce false-negative examination
  - Acetabular bony coverage
    - Portion of femoral head contained within bony acetabulum
    - Center of femoral head relative to iliac straight line
    - Capsular laxity: > 50% bony coverage femoral head at rest, < 50% bone coverage on stress examination or active movement
    - Subluxation: < 50% bony coverage at rest
      - Some centers define subluxation as < 33% bony coverage and 33-50% as indeterminate
    - Dislocation: Femoral head lies completely outside of bony acetabulum at rest
- Normal hip
  - Should have > 50% of femoral head covered by bony acetabulum ("bony coverage")

Radiographic Findings

- Radiography
  - Hip dysplasia associated with relatively shallow acetabulum with slanting roof and relatively small superolaterally placed femoral head
  - Hilgenreiner line links tops of both triradiate cartilages
    - Femoral metaphysis should lie below Hilgenreiner line
  - Perkin line is perpendicular to Hilgenreiner line through lateral margin of bony acetabulum
    - Femoral capital epiphysis should lie within inner lower quadrant of intersection between Perkin and Hilgenreiner lines
  - Acetabular index is angle between acetabular roof and Hilgenreiner line
    - 30-32° in newborn; ↓ in older children

MR Findings

- T2WI
  - Inverted limbus of acetabular roof best seen on coronal images
    - Hypertrophy of hypointense ligamentum teres and fibrofatty pulvinar
    - Hypointense interposed iliopsoas tendon

Imaging Recommendations

- Best imaging tool

- US useful up to 6 months of age
  - Limited at 6-9 months and not useful thereafter
  - Obscured by femoral capital ossification centers
- Protocol advice
  - 0-3 months: Linear transducer ≥ 10 MHz
    - If fat baby, better to use 7.5 MHz in order to see ischium and posterior femoral head
  - Lateral decubitus (or supine) position with standard orthogonal views
    - Hip and knee in slightly flexed position
  - Ensure that definitions of capsular laxity, subluxation, and dislocation are clear to those interpreting US result
    - Useful to have these definitions written on US report
  - Useful to continue screening hip during active movement if child restless

DIFFERENTIAL DIAGNOSIS

Physiological Immature Hip

- In neonate or premature baby

PATHOLOGY

General Features

- Etiology
  - Laxity of joint capsule
  - Inadequate contact between acetabulum and femoral head
- Relevant anatomy
  - Osseous acetabulum formed by portions of ilium, ischium, and pubis
    - These osseous elements unite at triradiate cartilage (growth cartilage)
  - Echogenic fibrocartilage at edge of cartilaginous roof is true acetabular labrum
  - Ossification centre of femoral head appears on US before it appears radiographically (6th-8th week)

Gross Pathologic & Surgical Features

- Hourglass-shaped joint capsule
- Femoral head flattened medially
- Thickened tight transverse ligament

CLINICAL ISSUES

Presentation

- Most common signs/symptoms
  - Palpable clunk using Ortolani and Barlow maneuvers
- Clinical profile
  - Ortolani maneuver
    - Hip flexed to 90° and gently abducted while thigh lifted anteriorly
    - "Clunk" felt as dislocated femoral head reduces
  - Barlow maneuver
    - Hip flexed to 90° and gently adducted while thigh pushed posteriorly
    - "Clunk" felt as femoral head dislocates

Demographics

- Age
  - Newborns
- Gender
  - M:F = 1:4
  - Left hip 3x more commonly affected than right hip
    - Possibly related to left occiput anterior position of most nonbreech babies in utero
      - In this position, left hip lies against maternal spine, which may limit abduction
- Epidemiology
  - Dysplasia: 0.8% of newborns (0.3% boys, 1.4% girls)
- Risk factors
  - Positive family history, breech presentation, oligohydramnios, torticollis, foot deformity
    - 3% of births breech; 8% of girls with breech birth have developmental hip dysplasia
    - Due of hip flexed, knee extended position of breech babies in utero

Natural History & Prognosis

- Prognosis
  - Physiological hip joint laxity of newborn without acetabular dysplasia resolves spontaneously
  - Early treatment of DDH: Excellent result with harness or splint
    - Acetabulum particularly susceptible to remodeling during first 10 postnatal weeks
    - Contact between femoral head and acetabulum necessary for normal acetabular development
  - Delayed treatment of DDH: Irreversible dysplasia
    - Hip muscle tightness; joint space fills with fibrofatty tissue; ligamentum teres and joint capsule redundant
- Limited range of movement, adductor spasm
- Limb shortening and secondary hip osteoarthritis or pseudoacetabulum

Treatment

- Conservative
  - Pavlik harness ± closed reduction (if dislocated)
- Surgical for dislocation or subluxation failing to respond to conservative treatment
  - Open reduction + spica cast
  - Adductor tenotomy + release of iliopsoas
  - Varus (derotational) vs. reconstructive osteotomy

DIAGNOSTIC CHECKLIST

Consider

- On US examination, incorrect orientation can make deep hip appear shallow
  - Not possible to make shallow hip appear deep

Image Interpretation Pearls

- Key to diagnosis rests on obtaining correct coronal (longitudinal) view
- Be clear with respect to definitions of capsular laxity, subluxation, and dislocation

SELECTED REFERENCES

1. Collins-Sawaragi YC et al: How to use... hip examination and ultrasound in newborns? Arch Dis Child Educ Pract Ed. ePub, 2017

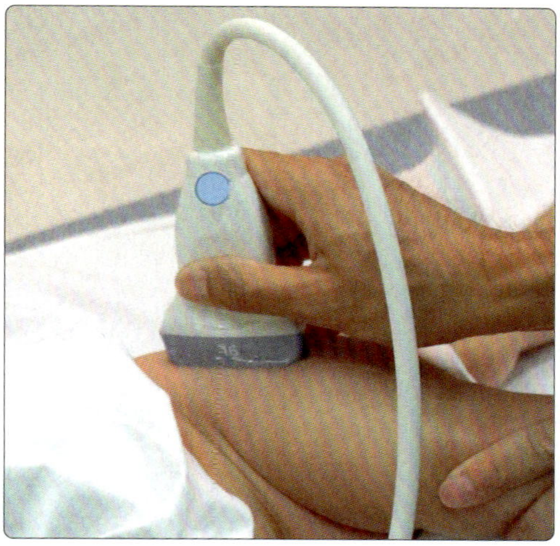

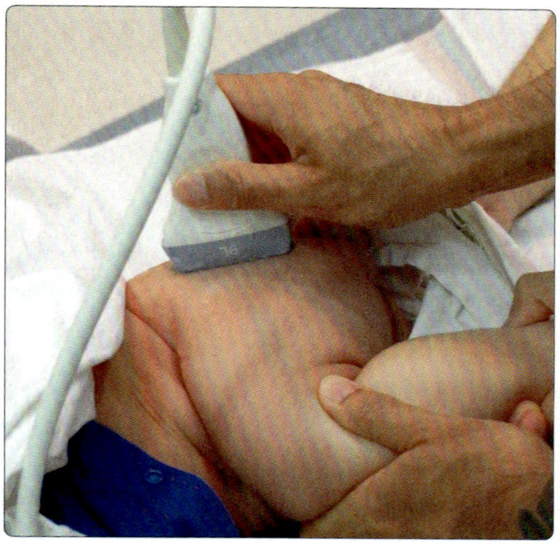

(Left) *Clinical photograph shows the position of a baby during scanning for nonstress views of hip. The baby is in the lateral decubitus position with the hip flexed to only ~ 20°. The transducer is oriented in a nearly coronal position.* **(Right)** *Clinical photograph shows the technique for obtaining stress views of the hip. The baby is in a lateral decubitus position. The hip is adducted and flexed, with a steady applied force being directed posteriorly. The transducer remains orientated in a nearly coronal position.*

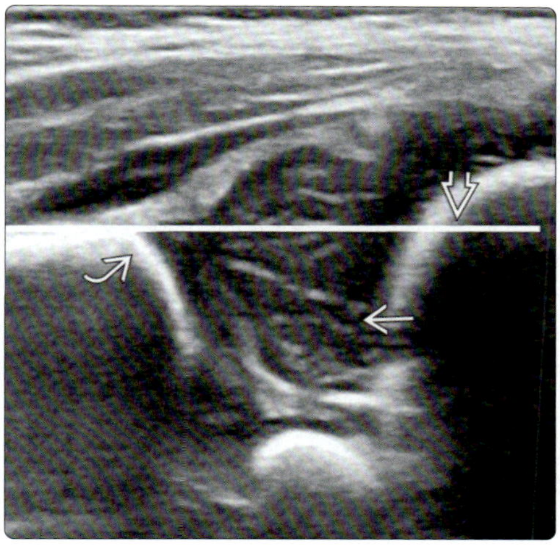

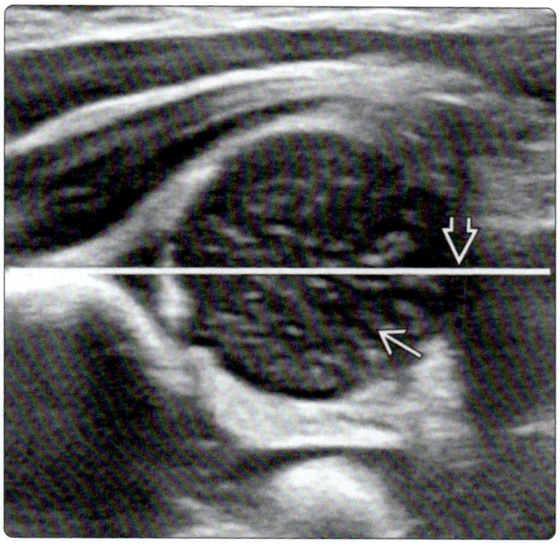

(Left) *Longitudinal US shows an immature acetabulum with a roundish acetabular promontory. The majority of the unossified femoral head lies below the iliac line. There is good bony coverage.* **(Right)** *Longitudinal US of the same patient during stress maneuver shows that most of the femoral head lies slightly above the iliac line. This indicates capsular laxity and is common in newborns and premature infants.*

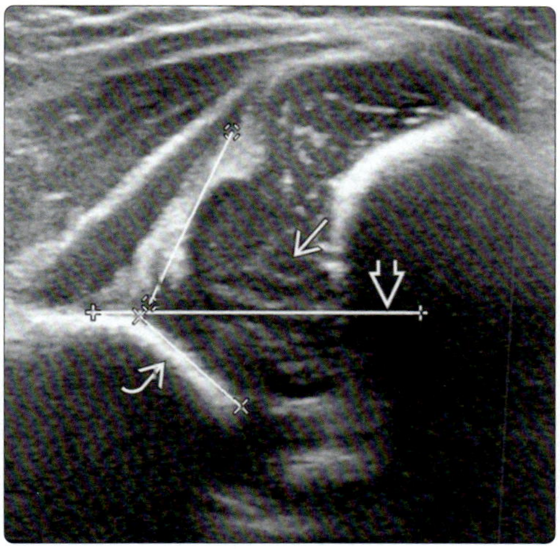

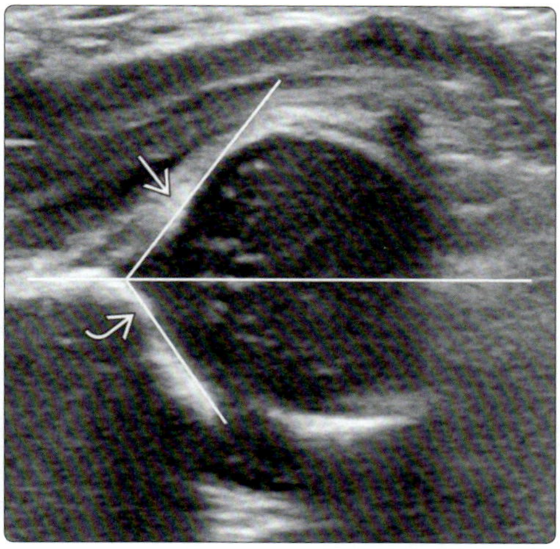

(Left) *Longitudinal coronal US shows an immature shallow acetabulum in a neonate with decreased alpha angle (45°). The femoral head is mildly subluxed and above the iliac line.* **(Right)** *Longitudinal coronal US of the hip obtained on follow-up 3 months later shows that the previously shallow bony acetabulum has now become less shallow (alpha angle = 57°). This is an expected finding with normal maturation of the acetabulum. Note the normal acetabular labrum.*

(Left) *Longitudinal coronal US of the hip shows a roundish acetabular promontory indicative of acetabular immaturity. The acetabular roof is still deep with good bony coverage of the femoral head.* **(Right)** *Longitudinal coronal US of the same hip during stress testing shows moderate capsular laxity with < 50% coverage of the femoral head by the bony acetabulum as indicated by the iliac line. Treatment is indicated with follow-up US in 4 weeks.*

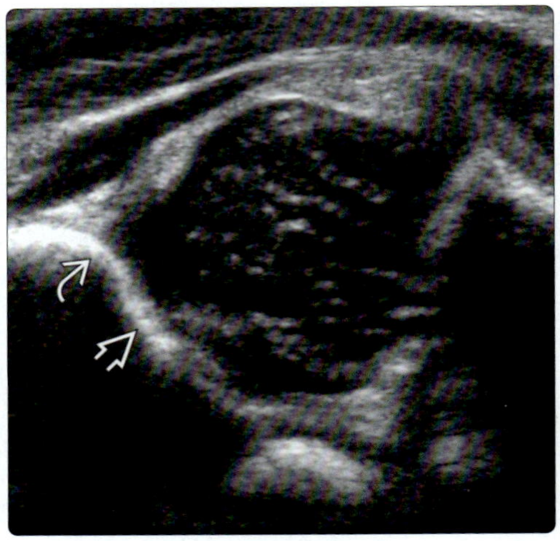

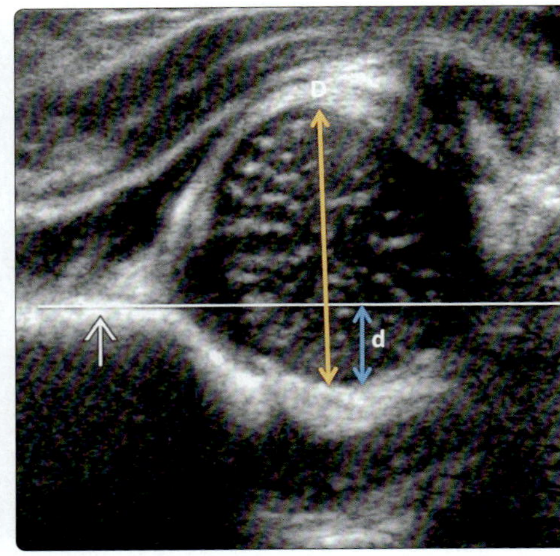

(Left) *Longitudinal coronal US of the same hip 4 weeks later shows that the femoral head is now not subluxed during stress maneuver, i.e., D/d > 50%. The acetabulum promontory is also sharper. Early diagnosis and proper treatment can lead to dramatic improvement.* **(Right)** *Longitudinal US of the contralateral hip in the same patient shows a slightly roundish acetabular promontory. However, the acetabular bony roof is well formed and deep.*

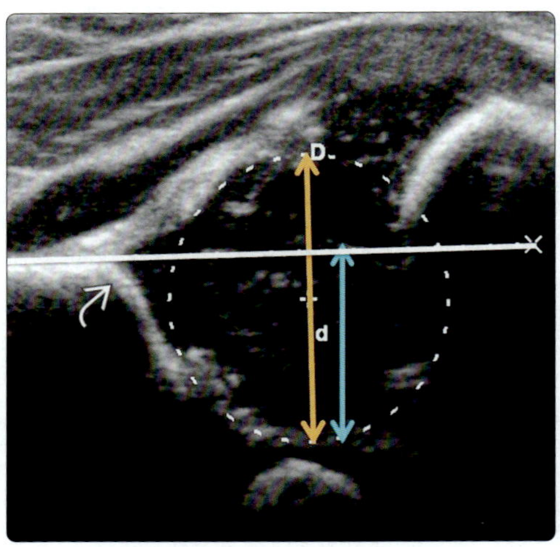

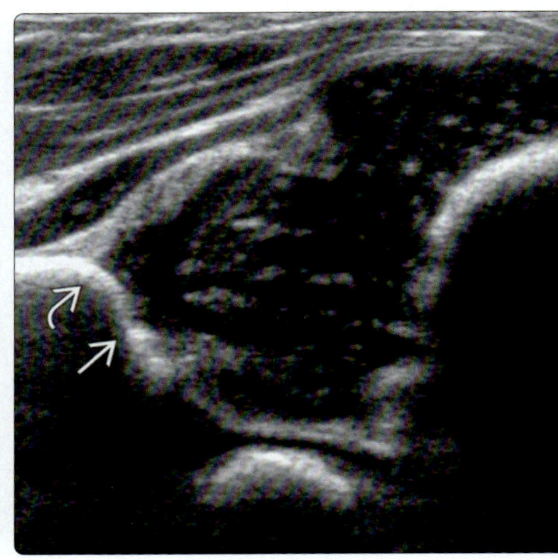

(Left) *Longitudinal coronal US shows a moderately dysplastic hip with a shallow acetabulum. The alpha angle is 44° (type IIc). Moderate femoral head subluxation is present. The acetabular labrum is everted. The femoral head is relatively small.* **(Right)** *Longitudinal US of the hip shows a severely hypoplastic acetabulum with a dislocated femoral head. The labrum is interposed between the bony acetabulum and the femoral head. Features are compatible with type IV DDH.*

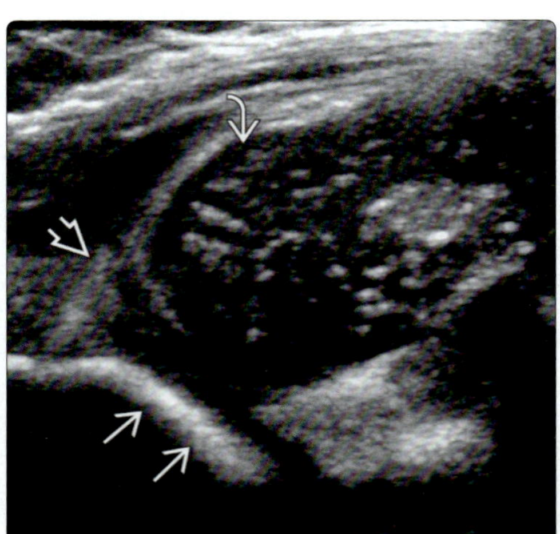

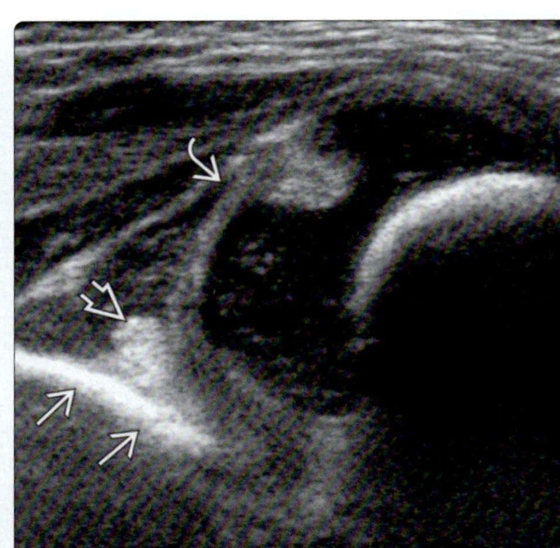

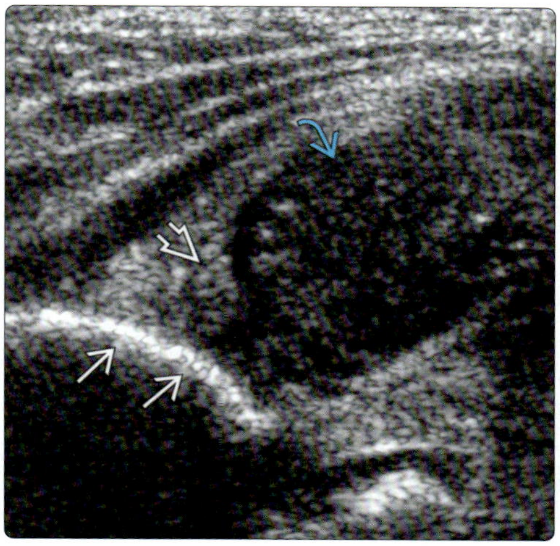

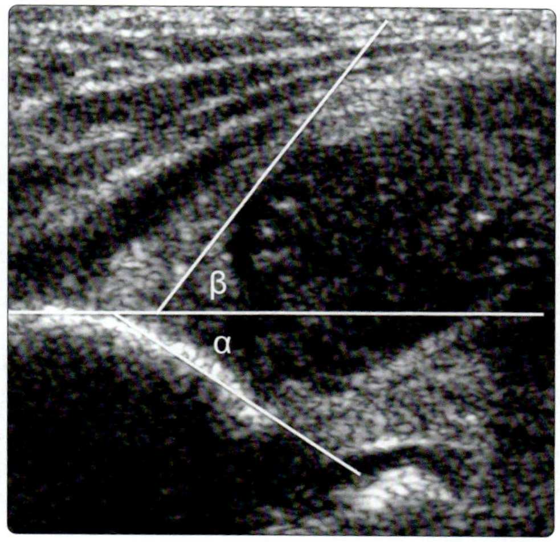

(Left) *Longitudinal US shows a severely dysplastic hip with a rounded shallow acetabulum. The femoral head is severely subluxed, and the acetabular labrum is located between the acetabulum and the femoral head. These features are compatible with type III DDH.* **(Right)** *Longitudinal US of the hip (same patient) shows severe reduction in alpha angle measuring 34°, and increase in beta angle measuring 55°, which is classified as type III DDH.*

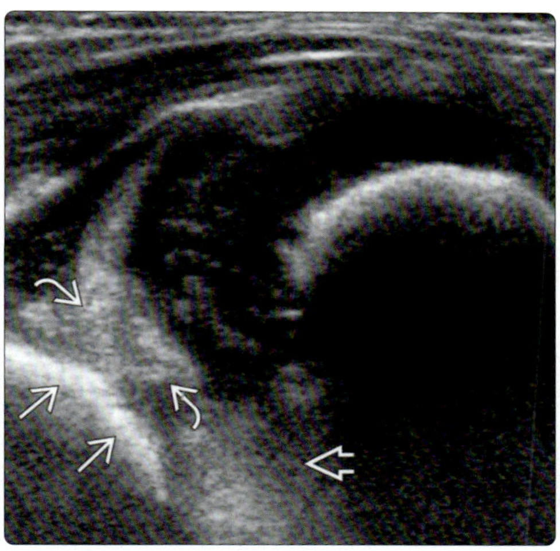

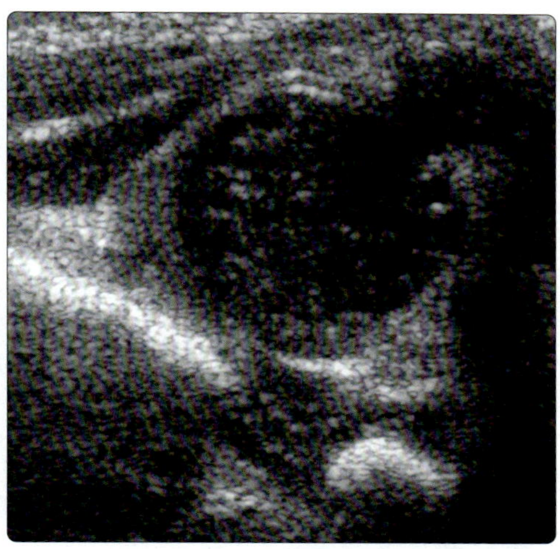

(Left) *Longitudinal US shows a severely dysplastic hip with lateral and cephalad displacement of the femoral head. The labrum is interposed between the femoral head and acetabulum and is thickened. The pulvinar is also thickened. The hip is classified as type IV DDH.* **(Right)** *Longitudinal US shows a severely dysplastic acetabulum with a dislocated hip. The bony acetabulum is not measurable, and the labrum cannot be well delineated. These features are compatible with type IV DDH.*

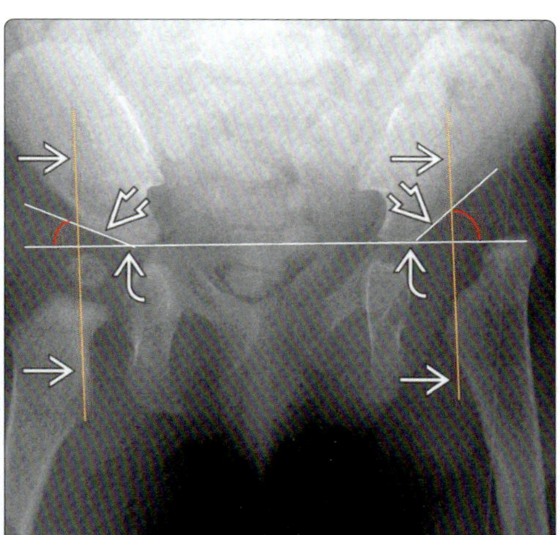

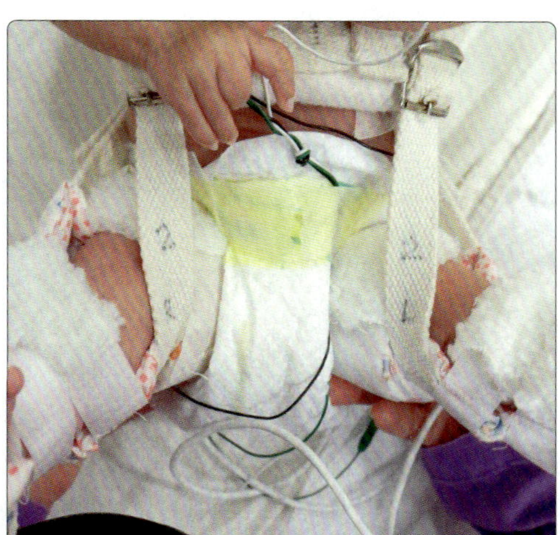

(Left) *Frontal radiograph shows Hilgenreiner, Perkins, and acetabular roof lines as well as acetabular index angles. There is left-sided acetabular dysplasia with dislocation of the femoral head and delayed ossification of the left femoral capital epiphysis.* **(Right)** *Clinical photograph shows a baby wearing a Pavlik harness to treat DDH. The harness helps to keep the hips in an abducted flexed position. This position ensures that the femoral heads remain within the acetabular fossae.*

Nerve Injury

KEY FACTS

TERMINOLOGY

- Neurapraxia → impaired nerve function due to conduction block
- Axonotmesis = "axon cutting" → impaired nerve function due to tear of axon or axons ± sheath
- Neurotmesis = "nerve cutting" → loss of nerve function due to tear of entire nerve

IMAGING

- Disruption of normal fibrillar pattern of nerve + focal nerve swelling
- Digital > median > ulnar > sciatic > radial nerve
- US best shows nerve injury
- MR best shows denervated muscle effect
- Scan transversely along course of nerve
  - Observe caliber, integrity of nerve, echogenicity, and compressibility of perineural fat
- Then align transducer along long axis of nerve
  - Best plane to appreciate continuity and discontinuity of epineurium and fascicles
  - If partial tear, estimate degree of transection (e.g., 10%, 40%, 90%)
- Neurapraxia: Appearance of nerve and surrounding tissues is usually normal
  - ± focal swelling of nerve
- Axonotmesis: Incomplete tear of nerve
- Neurotmesis: Complete tear of nerve ± retraction

DIAGNOSTIC CHECKLIST

- US is very good at localizing site of injury in axonotmesis/neurotmesis, determining severity of injury, and detecting perineural fibrosis/neuroma
- Check for integrity of epineurium followed by integrity of individual fascicles, if visible
- US, in general, probably slightly overestimates rather than underestimates nerve injury

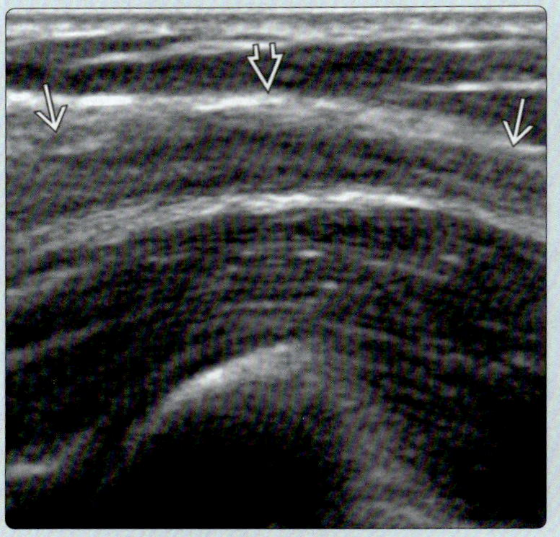

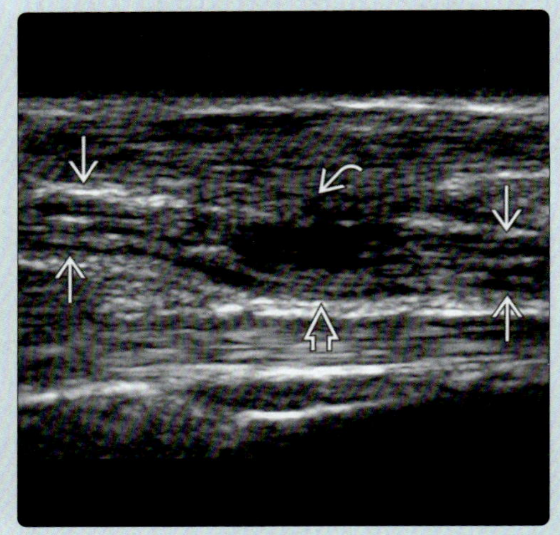

(Left) *Longitudinal US of the ulnar nerve ➡ 1 month following blunt trauma shows mild localized swelling with poor definition of fascicular echo pattern ➡ on the medial aspect of the elbow. No loss of neural continuity is evident. Appearances are consistent with severe neurapraxia.* **(Right)** *Longitudinal US shows axonotmesis ➡ of the median nerve ➡ following a laceration with the tear of the superficial fascicles. 50% of the nerve fascicles are estimated as torn with the deeper fascicles ➡ intact.*

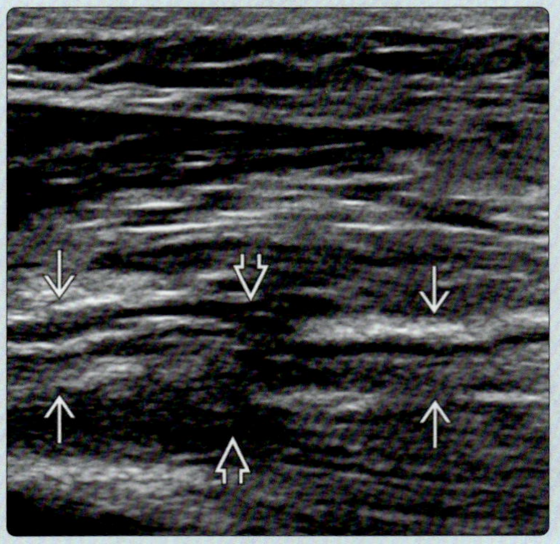

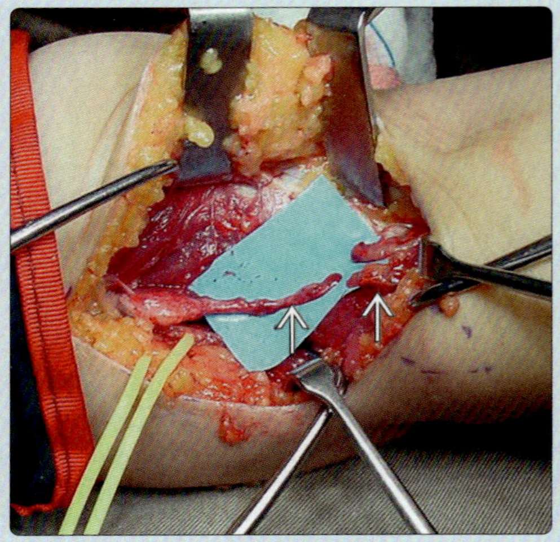

(Left) *Longitudinal US of the posterior arm region following a chop injury shows complete laceration ➡ (neurotmesis) and mild displacement of a swollen radial nerve ➡.* **(Right)** *Clinical photograph shows the complete transection of the median nerve ➡ following a forearm laceration. (Courtesy P. C. Ho, MD.)*

TERMINOLOGY

Synonyms

- Neurapraxia: Saturday-night palsy, honeymoon syndrome, crutch syndrome, wheelchair syndrome

Definitions

- Neurapraxia = impaired nerve function due to conduction block
  - Praxis = function (Greek)
- Axonotmesis = "axon cutting" → impaired nerve function due to tear of axon or axons ± sheath
  - Severity variable depending on number of axons torn
- Neurotmesis = "nerve cutting" → loss of nerve function due to disruption of entire nerve

IMAGING

General Features

- Best diagnostic clue
  - Disruption of normal fibrillar pattern of nerve + focal nerve swelling
- Location
  - Digital > median > ulnar > sciatic > radial nerve
- Size
  - Ranging from marginally swollen nerve up to ~ 4x normal size
    - Chronic denervation → muscle atrophy
- Morphology
  - Ranges from focal nerve swelling to complete discontinuity with retraction ± neuroma
    - Slightly enlarged muscle in acute denervation → severe fatty atrophy with chronic denervation

MR Findings

- T1WI
  - Acute/subacute denervation → ± mild muscle swelling
  - Chronic denervation → muscle atrophy
- T2WI FS
  - Acute/subacute denervation → hyperintense muscle
    - ± mild increase in muscle size
- T1WI C+
  - Acute/subacute denervation → enhancement of denervated muscle
- Muscle edema and enhancement may be seen 24 hours after denervation
  - Increasing muscle edema parallels increase in muscle enhancement
    - Both probably reflect ↑ capillary engorgement and ↑ blood volume within affected muscle
  - Pattern of muscle involvement can predict likely nerve injury
- Muscle edema and enhancement reverse with reinnervation of muscle

Ultrasonographic Findings

- Detection of injury
  - Neurapraxia
    - Appearance of nerve and surrounding tissues is usually normal
    - ± focal swelling of nerve
  - Axonotmesis and neurotmesis
    - Disruption of nerve continuity
    - Disruption of fibrillar pattern
    - ± surrounding hemorrhage or edema
    - Evaluation in acute stage may be limited due to artifact caused by overlying wound
- Localization of injury
  - Nerves divide at inconstant levels
  - US can depict bifurcation level and confirm if main or distal branch is affected
- Type of injury
  - Axonotmesis: Incomplete tear of nerve
  - Neurotmesis: Complete tear of nerve ± retraction
    - Retraction of nerves < tendon retraction
    - Up to few cm
    - Nerve injury overlying mobile joint leads to greater retraction
- Presence of perineural fibrosis
  - Irregular hypoechoic tissue replacing echogenic perineural fat
  - Focal swelling of nerve
  - ± tethering of nerve
  - ± reduced motility of nerve on moving part
- Continuity of nerve postoperatively
  - Continuity of epineurial sheath
    - ± continuity of nerve fascicles
- Presence of neuroma
  - Quite common finding
  - Hypoechoic rounded or ovoid mass at site of disrupted nerve
  - Connected to proximal end of severed nerve
  - Largely avascular
  - May be small and intraneural
    - Small neuromas may be difficult to distinguish from nodular fibrosis
  - Vast majority associated with complete nerve transection though minority can occur with partial nerve transection
- Muscle denervation
  - Mild swelling and increased echogenicity of affected muscle seen after 2 weeks
    - Due to capillary dilatation and muscle edema
  - Later → decreased muscle bulk and increased echogenic fat within and between muscle bundles
    - Due to muscle atrophy and fatty replacement

Imaging Recommendations

- Best imaging tool
  - US best shows nerve injury
    - High resolution
    - Easy orientation along long and short axes of nerve
  - MR best shows denervated muscle effect
- Protocol advice
  - Scan transversely along course of nerve
    - Easiest plane to trace course of nerve
  - Observe caliber, integrity of nerve, echogenicity, and compressibility of perineural fat
  - Align transducer along long axis of nerve
    - Best plane to appreciate continuity and discontinuity of epineurium and fascicles

- If partial tear, estimate degree of transection (e.g., 10%, 40%, 90%)
- Measure retraction gap, if present
- Observe nerve movement when passively moving adjacent joints

DIFFERENTIAL DIAGNOSIS

Tendon Injury

- Sometimes torn in conjunction with nerve
- Discontinuity of tendon fibers

Nerve Sheath Tumor

- May mimic traumatic neuroma
- Entering and exiting nerve
- Traumatic neuromas are less vascular

Nerve Entrapment

- Typical locations
- Nerve swelling proximal to level of entrapment
- → demyelination (neurapraxia) of entrapped nerve with secondary changes in affected muscle

PATHOLOGY

General Features

- Etiology
  - Blunt or penetrating trauma
    - Includes iatrogenic causes: Fracture repair and fixation, arthroscopy, carpal tunnel release, surgical exploration
- Relevant anatomy
  - Axon ± covered by myelin (myelinated or nonmyelinated nerve fiber)
  - Endoneurium = loose connective tissue between nerve fibers
  - Perineurial sheath groups nerve fibers into fascicles
  - Epineurium encircles and runs between fascicles; outer layers of epineurium are condensed into sheath
- Fascicles within large peripheral nerve continually divide and reunite (fascicular plexus)
  - Particularly in lumbar and brachial plexuses

Staging, Grading, & Classification

- Sunderland grading system
  - Grade 1: Neurapraxia
    - Neural tubes intact
      - No axonal regeneration required
  - Grade 2-4: Axonotmesis
    - Grade 2: Axon is disrupted but endo-, peri-, and epineurium intact
      - Regenerating axons always reach original motor and sensory targets
    - Grade 3: Axons and endoneurium are disrupted but peri- and epineurium intact
      - Regenerating axons may not reach original motor and sensory targets
    - Grade 4: Axons, endo- and perineurium are disrupted but epineurium is intact
      - Regenerating axons usually do not reach original motor and sensory targets
  - Grade 5: Neurotmesis
    - Complete transection of all neural tubes
      - Regenerating axons invariably do not reach original motor and sensory targets
    - Classification system is difficult to apply clinically as mixed injury is common

Gross Pathologic & Surgical Features

- Axonal injury → wallerian degeneration → regeneration

Microscopic Features

- Axonal apoptosis followed by phagocytosis
- Schwann cell proliferation
- Sprouting axonal filopodia → growth cone, guided by neurotrophic and neurotropic mediators

CLINICAL ISSUES

Presentation

- Most common signs/symptoms
  - Weakness ± numbness ± pain
- Other signs/symptoms
  - Type of transection, nerve function fails sequentially
  - Motor → proprioception → touch → temperature → pain → autonomic
    - Recovery in reverse order

Demographics

- Age
  - More common in young adults
- Gender
  - M:F = 2.2:1
- Epidemiology
  - 2-3% of patients admitted to level I trauma center
  - More common in dominant limb

Natural History & Prognosis

- Neurapraxia has excellent prognosis
- Axonotmesis and neurotmesis have variable prognosis depending on degree of injury
- Axons regenerate at ~ 1 mm/day (slower in older patients)

Treatment

- Neurapraxia: Supportive treatment
- Axonotmesis or neurotmesis
  - Immediate or delayed repair
    - Immediate repair if clean wound, no crush injury, good vascular supply, and soft tissue coverage
  - End-to-end closure
  - Nerve graft ± vascularization
  - Nerve transfer

DIAGNOSTIC CHECKLIST

Consider

- US is very good at localizing site of injury in axonotmesis/neurotmesis and detecting perineural fibrosis/neuroma
  - In general, US probably slightly over- rather than underestimates nerve injury

SELECTED REFERENCES

1. Hollister AM et al: High frequency ultrasound evaluation of traumatic peripheral nerve injuries. Neurol Res. 34(1):98-103, 2012

Nerve Injury

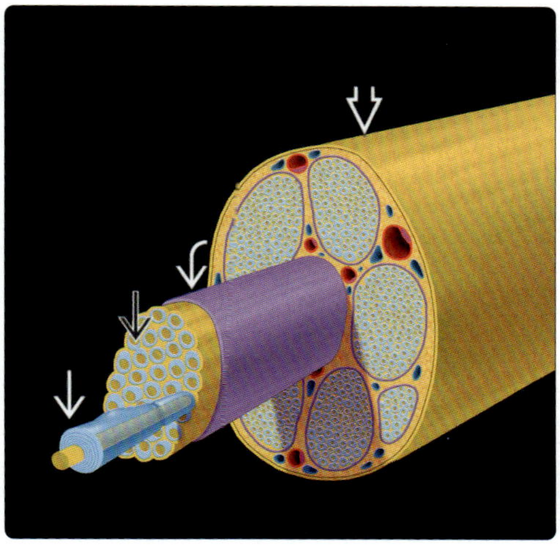

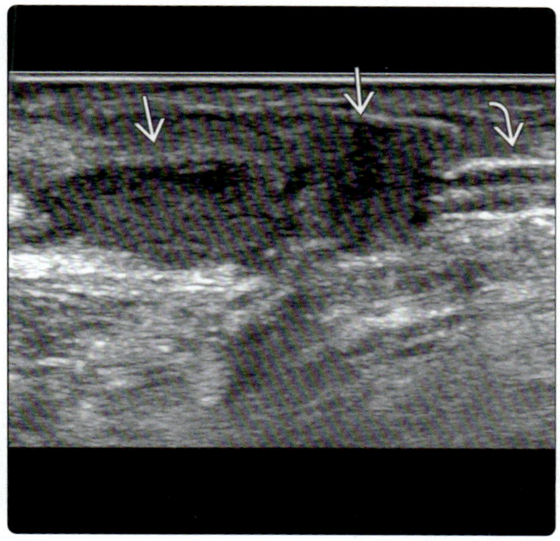

(Left) *Graphic of normal nerve structure shows myelinated axons embedded in endoneurium (seen as yellow). The fascicles are surrounded by perineurium. The nerve is surrounded by epineurium.* **(Right)** *Longitudinal US shows a localized fibrotic-type mass of the deep peroneal nerve at the ankle, 6 months following severe blunt trauma to this area. This most likely represents a completely transected nerve with intervening reparative fibrosis. No surgery was performed in this case.*

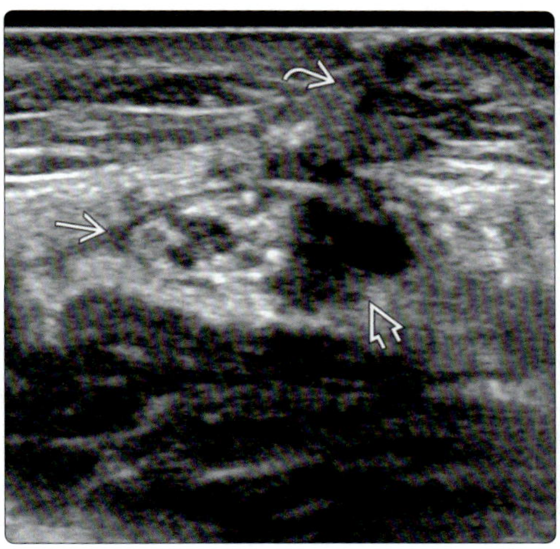

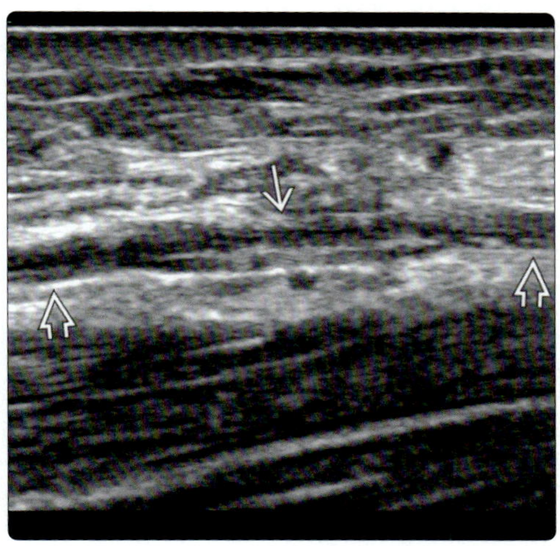

(Left) *Transverse US in patient with suspected median nerve injury following brachial artery cannulation for endovascular procedure is shown. The median nerve is swollen to over 2x its normal size. Surrounding soft tissues are swollen, the brachial artery is irregular, and the endovascular procedure scar is visible.* **(Right)** *Longitudinal US of same area shows moderate swelling and impaired definition of the median nerve but no neural transection. This pattern of nerve injury is consistent with neurapraxia.*

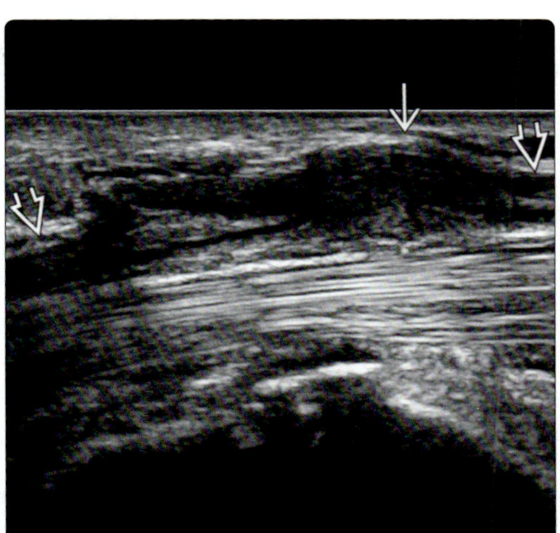

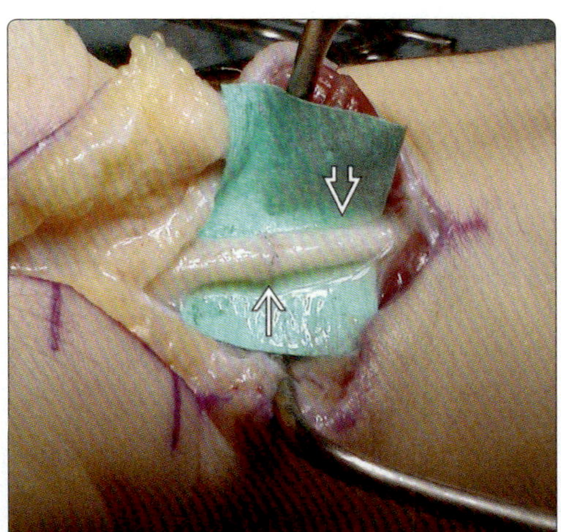

(Left) *Longitudinal US of median nerve following self-induced injury 4 months earlier shows a fibrotic mass replacing median nerve at the wrist. No abnormal neural tissue is evident, but there is no neural discontinuity. At operation, the nerve was completely transected.* **(Right)** *Clinical photograph shows repaired, complete transection (neurotmesis) of the median nerve. Most traumatic nerve injuries are readily apparent clinically and usually do not require preoperative imaging. (Courtesy P. C. Ho, MD.)*

(Left) *Transverse US shows severe diffuse postoperative fibrosis involving nearly all of the infraclavicular brachial plexus following resection of a schwannoma 1 year earlier. Nodular fibrosis particularly involves the medial cord. The subclavian vein and artery are normal.* **(Right)** *Longitudinal US in the same patient shows severe hypoechoic fibrosis involving the infraclavicular brachial plexus. The clavicle is normal.*

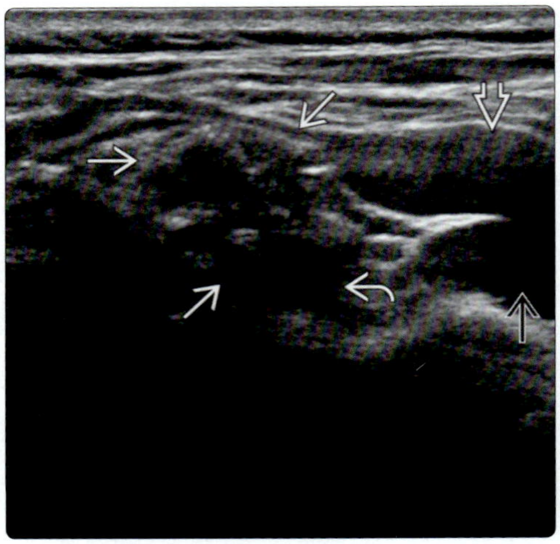

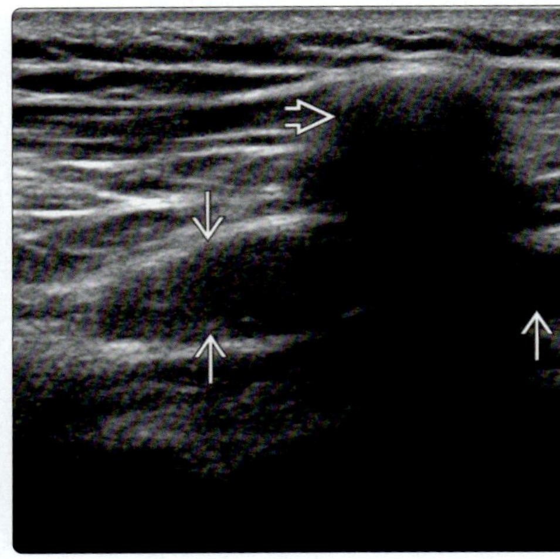

(Left) *Longitudinal US shows neuroma at the distal end of a mildly swollen transected median nerve.* **(Right)** *Longitudinal US immediately distal shows an oval-shaped hypoechoic neuroma of the proximal stump of transected median nerve. There is moderate posterior acoustic enhancement. The underlying flexor tendon is normal.*

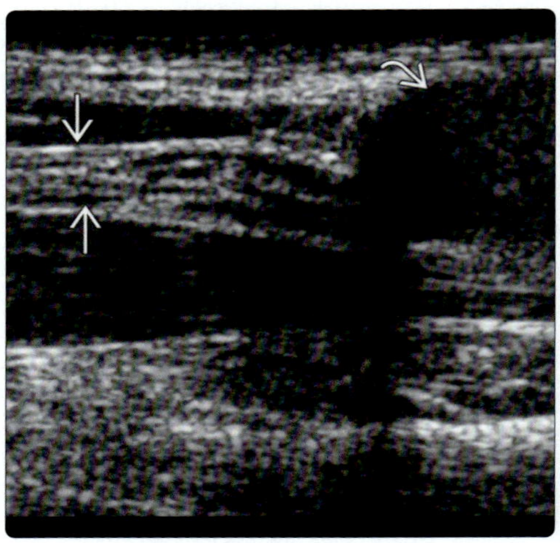

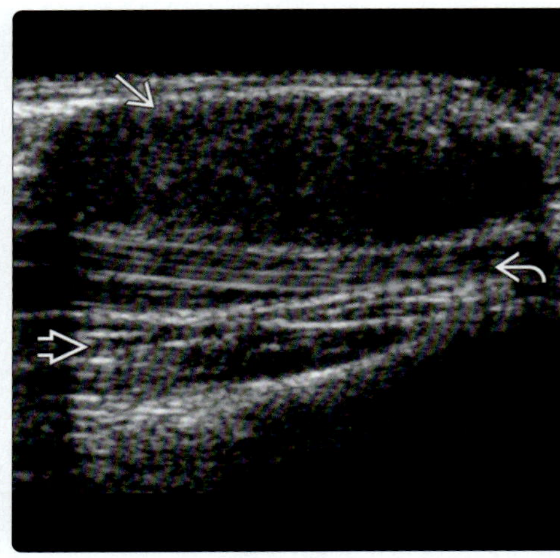

(Left) *Longitudinal T2WI MR of the same patient shows an ovoid posttraumatic neuroma of the medial nerve at the wrist level. The flexor tendon deep to the neuroma is normal.* **(Right)** *Clinical photograph shows a posttraumatic neuroma of the median nerve. The neuroma connects the 2 torn nerve endings. The majority of traumatic neuromas occur following complete nerve transection, though a minority can arise from incomplete nerve transection. (Courtesy P. C. Ho, MD.)*

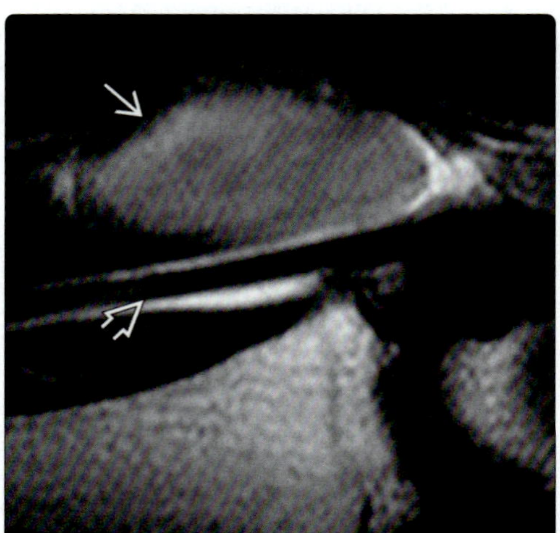

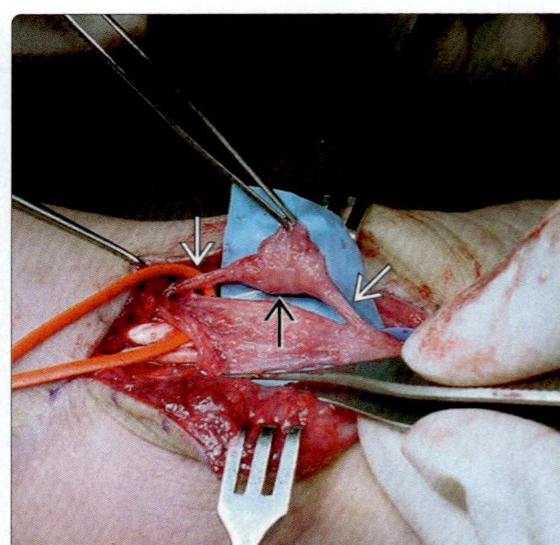

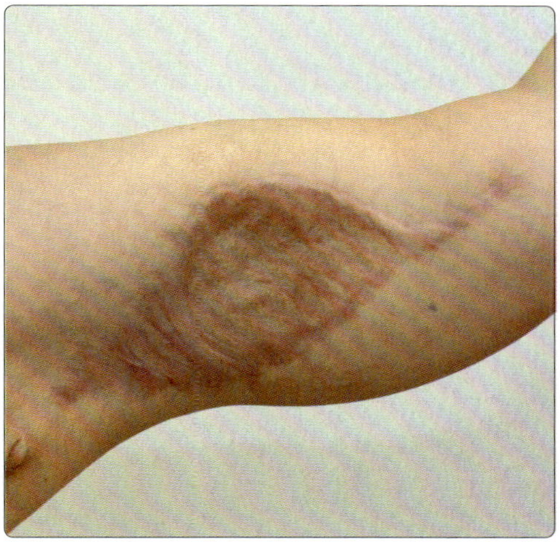

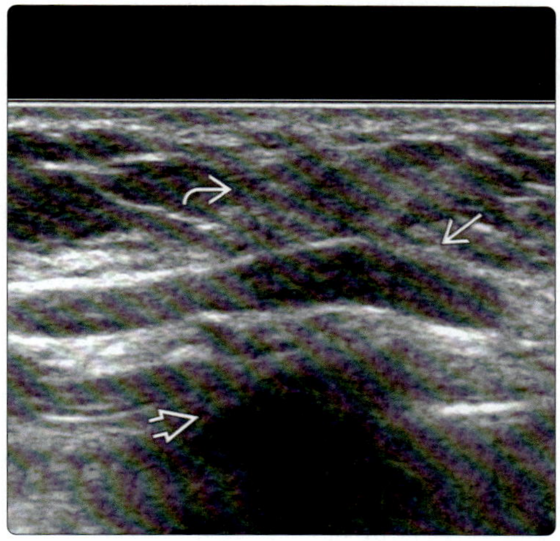

(Left) *Clinical photograph shows large scar following blunt injury to the inner aspect of the arm 8 years previously. The patient was concerned with mild ulnar nerve symptoms.* **(Right)** *Longitudinal US at the same location shows mildly thickened and mildly kinked ulnar nerve ➔ tethered to moderately fibrotic subcutaneous tissues ➔. The underlying humeral shaft ➔ is also slightly irregular.*

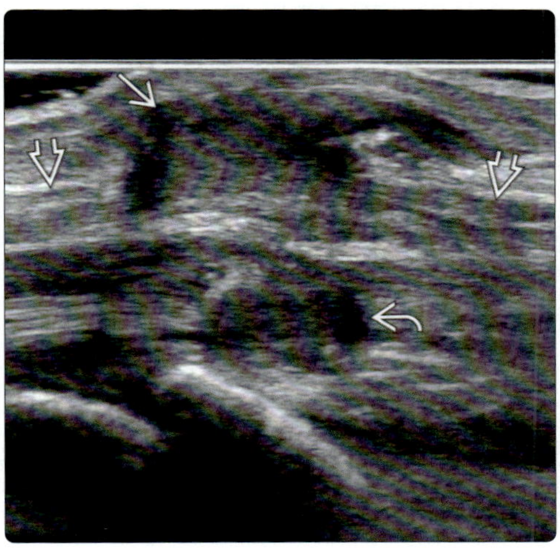

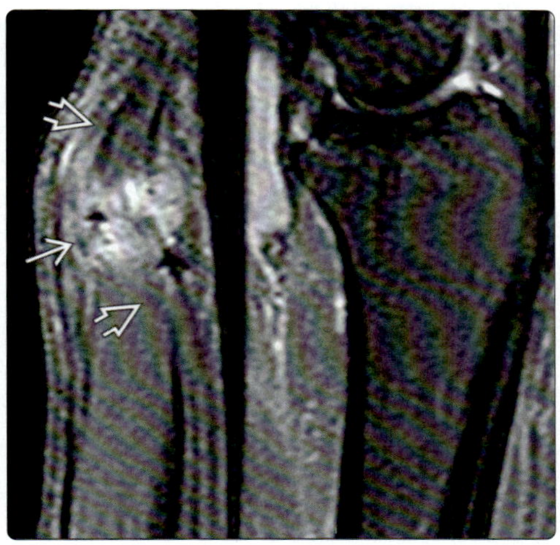

(Left) *Longitudinal US of the median nerve ➔ at the volar wrist in patient with self-inflicted laceration 3 years previously shows exuberant fibrotic tissue ➔ within and superficial to the largely transected median nerve. No median nerve retraction is evident. There is also scarring ➔ of the underlying flexor tendons.* **(Right)** *Longitudinal T2WI FS MR of the same patient shows exuberant hyperintense fibrous tissue ➔ filling the gap of the transected median nerve ➔. The hypointense foci are from previous surgery.*

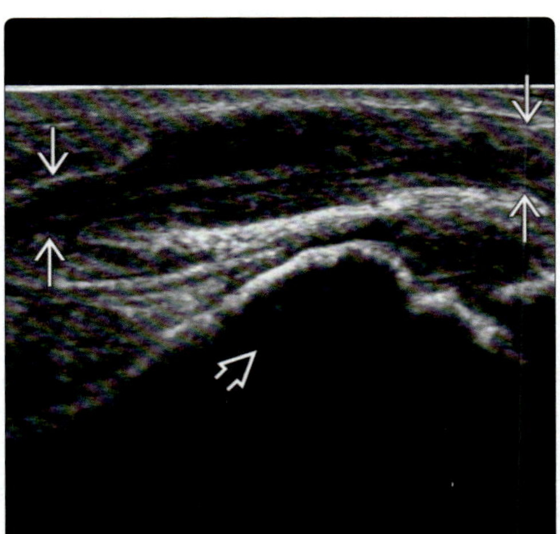

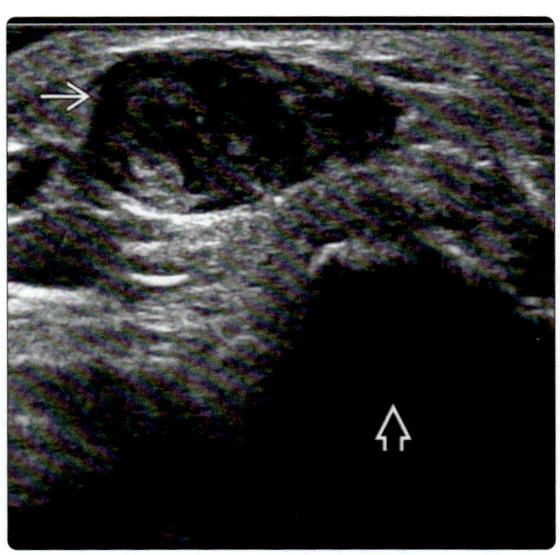

(Left) *Longitudinal US shows a patient with open fracture of the humeral medial epicondyle associated with ulnar nerve contusion and epineural abrasion in whom ulnar nerve transposition was performed. This shows a severely swollen, though continuous, ulnar nerve ➔ transposed anterior to the medial epicondyle ➔.* **(Right)** *Transverse US in the same patient shows the very swollen ulnar nerve ➔, consistent with severe neurapraxia, anterior to the medial epicondyle ➔.*

(Left) *Transverse US of the leg in a young patient with previous deep peroneal nerve injury shows severe fatty atrophy of the anterior compartment muscles (tibialis anterior and extensor digitorum longus). The tibia is normal.* **(Right)** *Transverse US of the contralateral limb in the same patient shows the normal tibialis anterior and extensor digitorum longus muscles for comparison.*

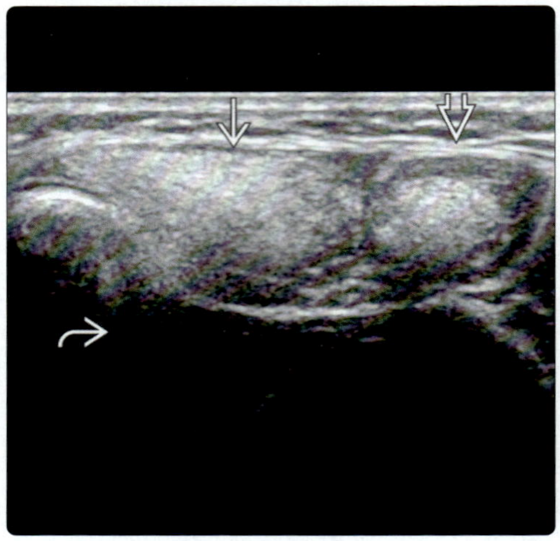

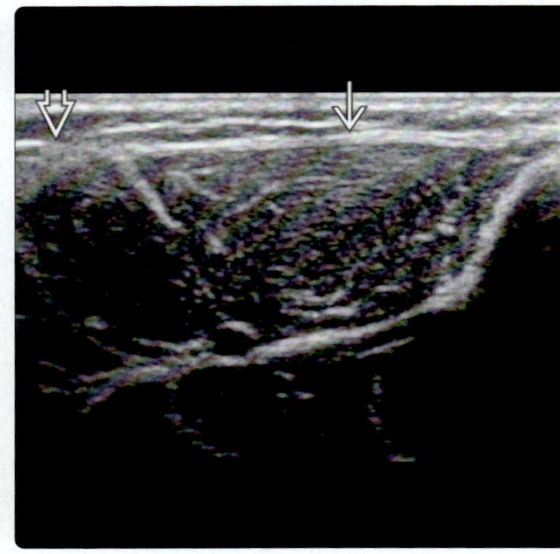

(Left) *Axial T2WI FS MR in a patient with forearm pain and weakness of thumb pincer movement shows moderate edema of the pronator quadratus muscle.* **(Right)** *Axial T2WI FS MR in the same patient shows mild edema of the flexor pollicis longus and flexor digitorum profundus (index and middle) consistent with subacute anterior interosseous syndrome (AIS). Mild muscle edema is more easily appreciated on MR than US.*

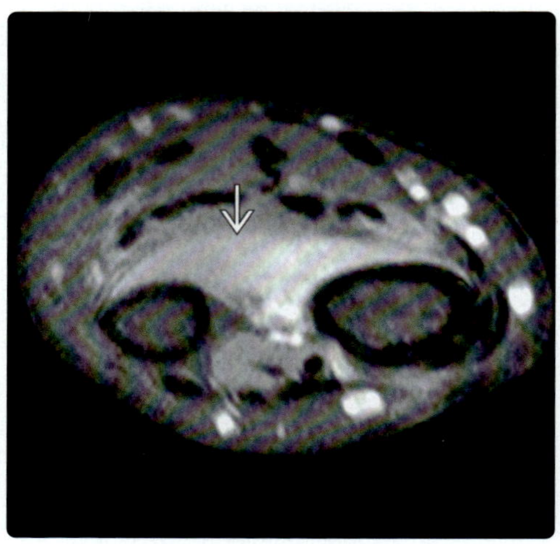

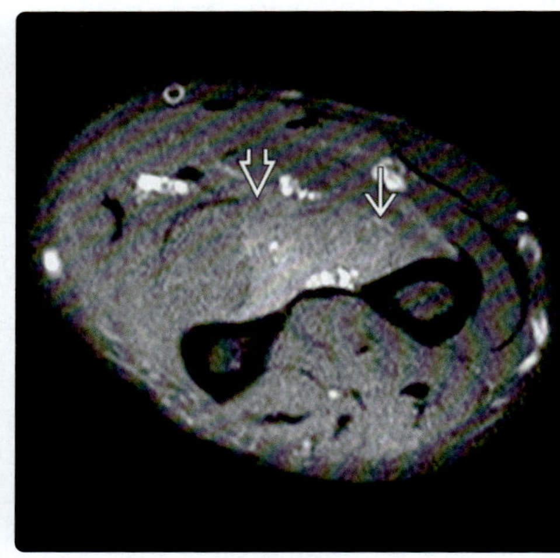

(Left) *Axial T1WI MR shows severe fatty atrophy of the pronatus quadratus muscle due to chronic AIS.* **(Right)** *Axial T1WI MR in the same patient shows severe atrophy of the flexor pollicis longus and flexor digitorum profundus (index and middle) muscles. Most cases of AIS are due to a neuritis (brachial neuritis) with median nerve compression and trauma being less common causes. MR allows better perception of muscle atrophy than US.*

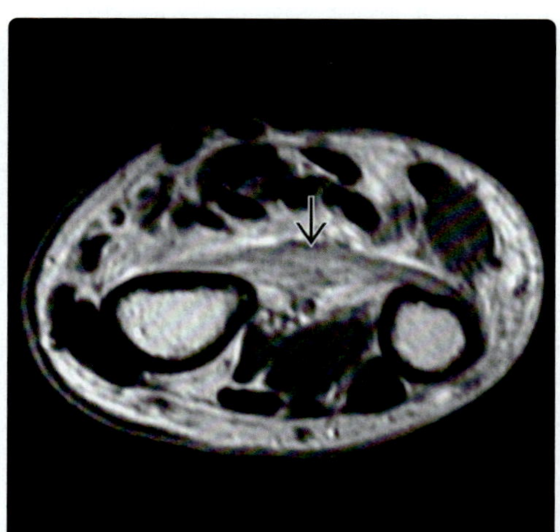

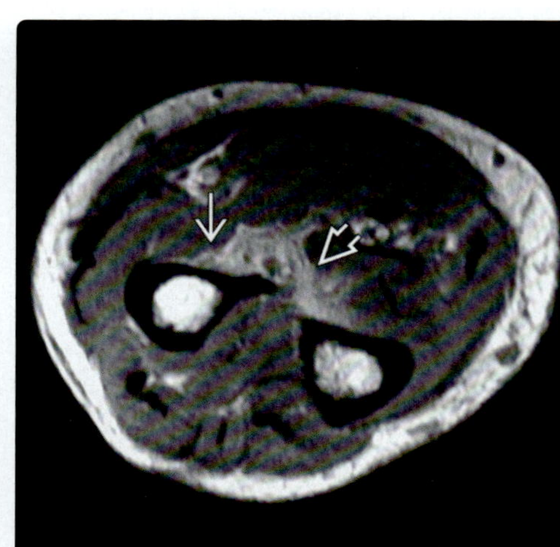

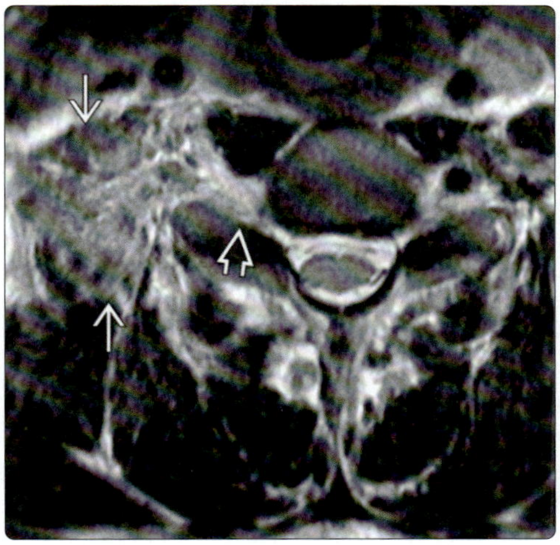

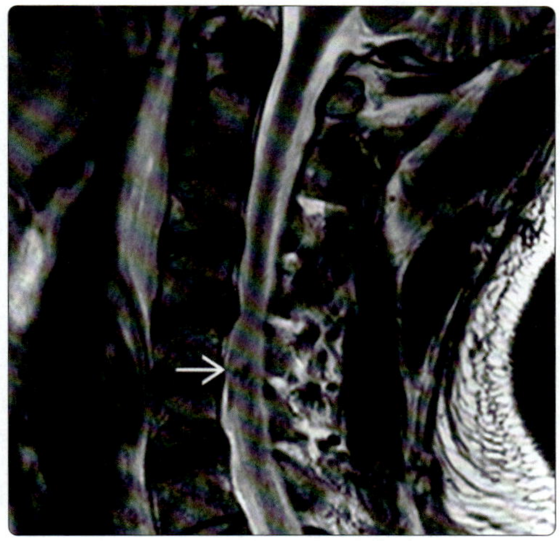

(Left) *Axial T2WI MR shows hematoma within and just outside of the right C6/7 exit foramen in a patient with traction injury to the brachial plexus. This suggests an avulsive injury of the C7 root.* **(Right)** *Sagittal T2WI MR in the same patient shows a small hematoma at the edge of the spinal cord at the site of origin of the right C7, suggestive of an avulsive nerve root injury. MR, rather than US, is recommended to assess preganglionic traction injury of the brachial plexus.*

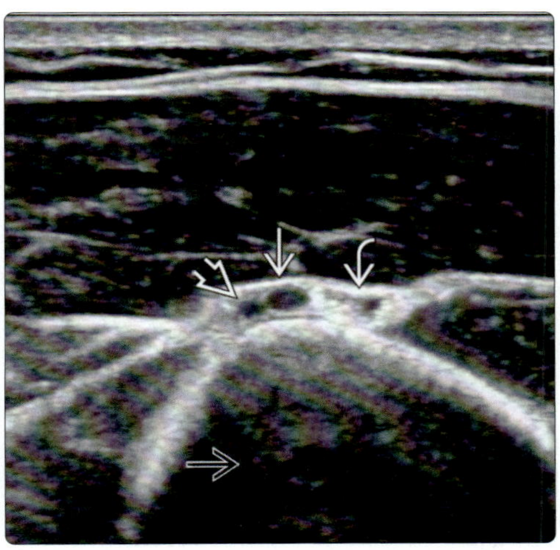

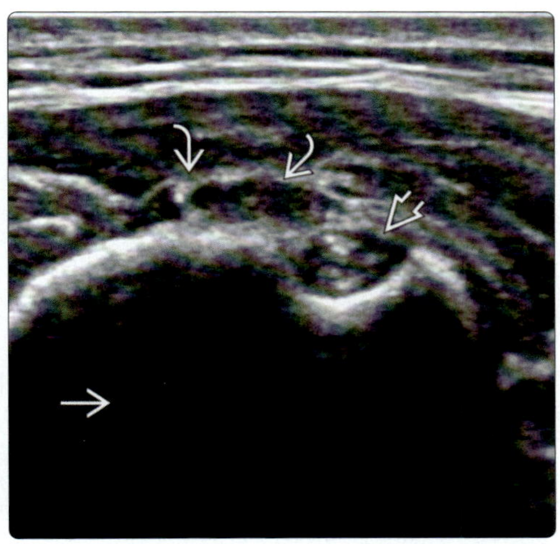

(Left) *Transverse US shows the normal radial nerve at the posterior aspect of humeral shaft. There is an accompanying profunda brachii artery and vein.* **(Right)** *Transverse US in a patient with a healed malaligned humerus shaft fracture and radial nerve palsy shows a thickened radial nerve that has become partially entrapped in a fracture callus at the posterior aspect of the fracture site.*

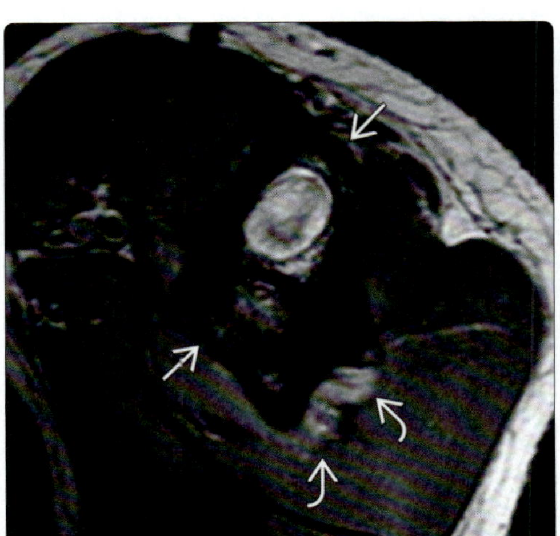

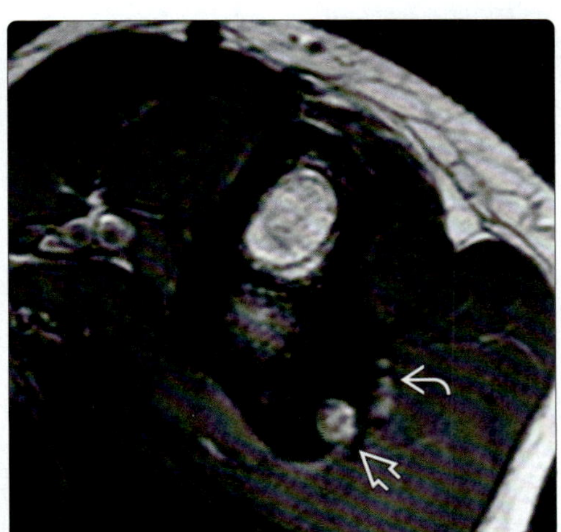

(Left) *Axial PDWI MR in the same patient shows the thickened radial nerve at the posterior aspect of the malaligned humeral shaft.* **(Right)** *Axial PDWI MR in the same patient shows the radial nerve partially entrapped in a healed fracture callus at the posterior aspect of the humeral shaft. Because the radial nerve palsy was gradually improving with conservative treatment, no surgical intervention was undertaken.*

KEY FACTS

TERMINOLOGY

- Tumor arising from nerve; histologically classified as either schwannoma or neurofibroma
- 3 types of neurofibroma: Localized, diffuse, and plexiform

IMAGING

- Any peripheral nerve can be affected
  - Intermuscular > subcutaneous > intramuscular
- Well-defined, largely homogeneous, hypoechoic mass
- Typically fusiform or oblong-shaped and oriented along long axis of nerve
- Nerve sheath tumors align centrally or eccentrically along course of nerve; arise from neurovascular bundle
- Nerve entering or exiting from tumor seen in majority (90%)
- Posterior enhancement common (70%)
- Nerve sheath tumors, especially schwannoma, often contain cystic areas representing myxoid degeneration, hemorrhage, or necrosis
- Vascular pattern varied; usually moderate intrinsic vascularity is present
- US appearances of schwannoma and neurofibroma overlap considerably
- US appearances of benign and malignant nerve sheath tumors overlap considerably
- Accurate diagnosis of nerve sheath tumor is usually feasible based on US findings alone
  - MR better at assessing extent of diffuse neurofibroma or plexiform neurofibroma

DIAGNOSTIC CHECKLIST

- Consider nerve sheath tumor even if parent nerve is not seen, provided other US features are compatible
- Nerve sheath tumors of medium-sized or large nerves can nearly all be diagnosed with confidence based on US appearances alone
  - Specificity for diagnosis of nerve sheath tumors arising from smaller nerves is not as high

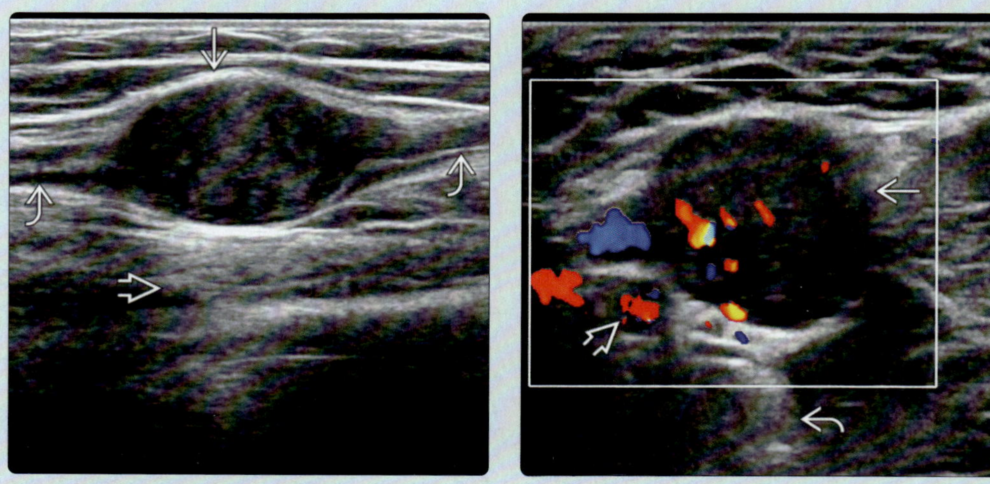

(Left) *Longitudinal US shows a fusiform homogeneous, hypoechoic mass in continuity with the ulnar nerve with posterior acoustic enhancement. Note the entering and exiting components of the ulnar nerve.* **(Right)** *Transverse color Doppler US of the same lesion shows a moderate degree of intralesional hypervascularity. The lesion is arising from the neurovascular bundle and is located alongside the ulnar artery. Note the moderate posterior acoustic enhancement.*

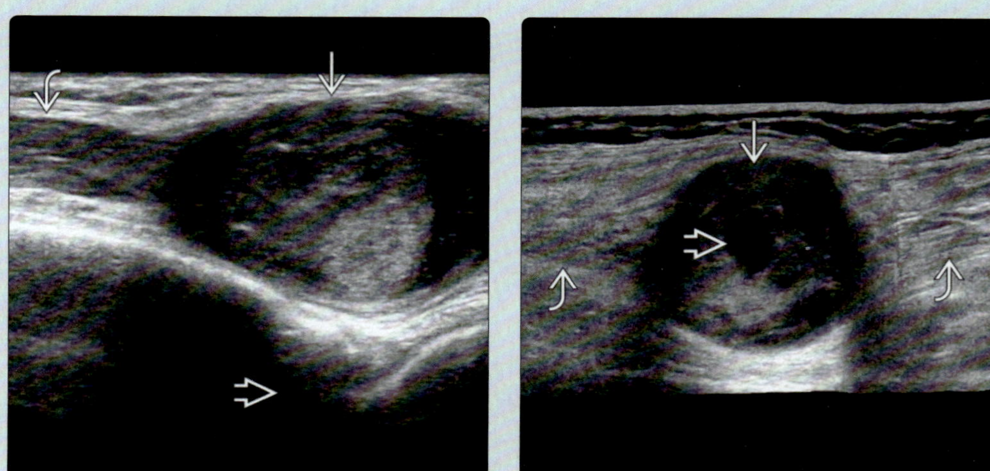

(Left) *Longitudinal US shows a heterogeneous nerve sheath tumor from the deep peroneal nerve. The peroneal nerve is thickened proximally due to tumor extension along the nerve. Note the ankle joint.* **(Right)** *Extended field-of-view US shows a histologically proven schwannoma with a central myxoid component and prominent posterior acoustic enhancement arising from the sciatic nerve. Note the strong posterior acoustic enhancement.*

TERMINOLOGY

Definitions

- Tumor arising from peripheral nerve and histologically classified as schwannoma and neurofibroma

Associated Syndromes

- Neurofibromatosis type 1 (NF1), von Recklinghausen disease

IMAGING

General Features

- Best diagnostic clue
  - Well-defined, largely hypoechoic mass arising from nerve
- Location
  - Any peripheral nerve
    - Intermuscular > subcutaneous > intramuscular
  - Head and neck > upper limbs > lower limbs > > trunk
- Size
  - Usually < 5 cm
    - Larger tumors from large nerves (up to 20 cm)
  - Nerve tumors usually grow slowly until critical mass is reached and then stop growing
- Morphology
  - Soft tissue mass along neurovascular bundle

MR Findings

- T1WI
  - Soft tissue mass isointense to muscle; split-fat sign (intermuscular fat splits to encompass tumor)
- T2WI
  - Well-defined fusiform mass hyperintense to muscle; target sign hypointense center due to collagen and condensed Schwann cells; neural tail sign
- T1WI C+
  - Variable enhancement: May see mainly peripheral enhancement or reverse target sign (enhancement of central fibrous component but no enhancement of peripheral myxoid component)
- MR appearances of schwannoma and neurofibroma overlap considerably

Ultrasonographic Findings

- Well-defined, largely homogeneous, hypoechoic mass; occasionally heterogeneous
  - Heterogeneity due to myxoid component, hemorrhage, fibrosis, or calcification
    - Target sign: Central hyperechogenicity representing fibrocollagenous component surrounded by peripheral hypoechogenicity representing myxoid component; more commonly seen in neurofibroma
  - Intratumoral hemorrhage may lead to sudden ↑ in tumor size
- Arise from nerve
  - Typically fusiform or oblong in shape and oriented along long axis of nerve
    - Nerve sheath tumors align centrally or eccentrically along course of nerve
  - Nerve entering or exiting from tumor is seen in majority (90%)
    - Nerve often shows slight enlargement just proximal and distal to tumor
    - Entering/exiting nerve may not be seen when nerve sheath tumor arises from small peripheral nerves
    - Compressed fascial planes may mimic nerve on longitudinal plane; always confirm on transverse plane
- Nerve sheath tumors often contain cystic areas
  - Degenerative ancient schwannomas are predominantly cystic
  - Cystic areas can be due to myxoid accumulation, previous hemorrhage, or necrosis
  - Cystic areas are more common to schwannoma but also commonly seen in neurofibroma
- Occasionally tumor has echogenic rim due to fibrous capsule, fibrous pseudocapsule, or compression of surrounding fat
- Vascular pattern varied
  - Usually moderately vascular with central irregular vascular pattern
    - Some display predominantly peripheral vascularity
    - Some do not have any demonstrable vascularity on color Doppler imaging
- Diffuse neurofibroma: Thickening of skin and subcutis with retinacular hypoechoic pattern
  - No discrete mass, hyperemic
    - Can mimic soft tissue infection or vascular malformation
- Plexiform neurofibroma: Diffuse nodular enlargement of nerves with "bag of worms"
  - Discrete lobulated mass present with lobulated neural enlargement
- US diagnosis of nerve sheath tumor is based on consortium of features rather than single feature
  - US features usually typical enough to make diagnosis without need for biopsy
    - Biopsy may be very painful
- US appearances of schwannoma and neurofibroma overlap considerably
  - Do not focus on making this histological distinction on imaging
  - Instead report on centric/eccentric location of tumor to parent nerve as this affects surgical outcome
- US appearances of benign and malignant nerve sheath tumors overlap considerably
  - No single feature can reliably differentiate between benign and malignant peripheral nerve sheath tumor
  - Clinical indicators are relatively more reliable: Large size (> 5 cm), rapid growth, and change in neurology
  - Irregular margins, inhomogeneity, internal necrosis, and peritumoral inflammation are also more common in malignant peripheral nerve sheath tumors
  - NF1 is most important risk factor for malignant peripheral nerve sheath tumor
- PET imaging is more accurate in predicting malignancy in plexiform neurofibromas

Imaging Recommendations

- Best imaging tool
  - Diagnosis of nerve sheath tumor is generally feasible based on US findings alone

- MR better at assessing extent of diffuse neurofibroma or plexiform neurofibroma

DIFFERENTIAL DIAGNOSIS

Malignant Lymph Node

- Malignant or lymphomatous node with loss of distinction cortex and hilum may simulate nerve sheath tumor
    - Does not have entering or exiting nerve
    - Occurs at different locations than nerve sheath tumors

Ganglion

- May simulate ancient schwannoma
- Does not have entering or exiting nerve

Vascular Leiomyoma

- Does not have entering or exiting nerve
- May be painful
- Typically has eccentric vascular confluence
- Can mimic small subcutaneous nerve sheath tumors in which entering or exiting nerve not apparent

Glomus Tumor

- Glomus tumor arising in site other than subungual region may simulate nerve sheath tumor
    - Does not have entering or exiting nerve

PATHOLOGY

General Features

- Genetics
    - NF1, von Recklinghausen disease
        - Affects 1 in 3,000 people
        - Autosomal dominant; 50% sporadic
        - Genetic mutation on *NF1* gene chromosome 17

Staging, Grading, & Classification

- Schwannoma
    - Usually solitary; multiplicity associated with NF1
    - Malignant change extremely rare
- Neurofibroma
    - 3 types
        - Localized: Most common type of neurofibroma (90%)
            - Most are solitary and not associated with NF1
        - Diffuse: Most are solitary; 10% associated with NF1
            - Infiltrative retinacular pattern in subcutaneous tissue or plaque-like thickening of skin and subdermal tissues
        - Plexiform: Pathognomonic of NF1
            - Multinodular confluent neural mass with mass effect on bones, lungs, or gastrointestinal tract; "bag of worms"
    - Malignant change occurs in 15% of NF1 patients
        - Common with plexiform neurofibromas (50% risk of malignant transformation), less common with diffuse neurofibroma, and rare in localized neurofibroma

Gross Pathologic & Surgical Features

- Schwannoma
    - Mesenchymal tumors with cellular differentiation to Schwann cells; myelin-producing neuroglial cells covering peripheral axons
    - True capsule composed of epineurium
    - Tumors lie eccentric to nerve with nerve fascicles displaced to periphery of tumor mass
    - Tend to be more prone to myxoid degeneration, hemorrhage, and calcification
- Neurofibroma
    - Mesenchymal tumors with cellular differentiation into various neural and perineural cell elements
    - Neurofibroma may not be covered by epineurium
    - Neurofibroma cannot be separated from parent nerve

Microscopic Features

- Schwannoma
    - Intermixed Antoni A areas: Cellular, arranged in short bundles or interlacing fascicles; Antoni B areas: Less cellular and organized, more myxoid component
    - Immunohistochemistry: S100 protein strongly positive
- Neurofibroma
    - Does not contain Antoni A or B areas
    - Composed of interlacing bundles of multiple cells composed of axonal processes, Schwann cells, fibroblasts, perineural cells, and mast cells; often contains abundant amount of collagen
    - Involvement of nerve fibers
    - Immunohistochemistry: Less extensive expression of S100 protein in localized and diffuse neurofibroma

CLINICAL ISSUES

Presentation

- Most common signs/symptoms
    - Slowly growing painless mass with positive Tinel sign
- Clinical profile
    - NF1: Café au lait spots, axillary or inguinal freckling, optic nerve glioma, Lisch nodules, kyphoscoliosis, tibial pseudoarthrosis, rib deformity, sphenoid wing dysplasia
    - < 10% of patients with neurofibroma have NF1

Demographics

- Age
    - 20-30 years
- Epidemiology
    - Schwannoma and neurofibroma: ~ 15% of all benign soft tissue tumors
        - 6% of soft tissue tumors superficial to investing fascia

Treatment

- Surgical excision of symptomatic lesions
    - Complete resection with sparing of parent nerve
    - Incomplete resection with sparing of parent nerve to minimize neurological impairment
    - Complete resection with excision of parent nerve
        - Autologous nerve graft or biodegradable nerve conduit if defect > 20 mm

DIAGNOSTIC CHECKLIST

Consider

- Nerve sheath tumor, even if parent nerve not seen, provided other US features are compatible

SELECTED REFERENCES

1. James AW et al: Malignant peripheral nerve sheath tumor. Surg Oncol Clin N Am. 25(4):789-802, 2016

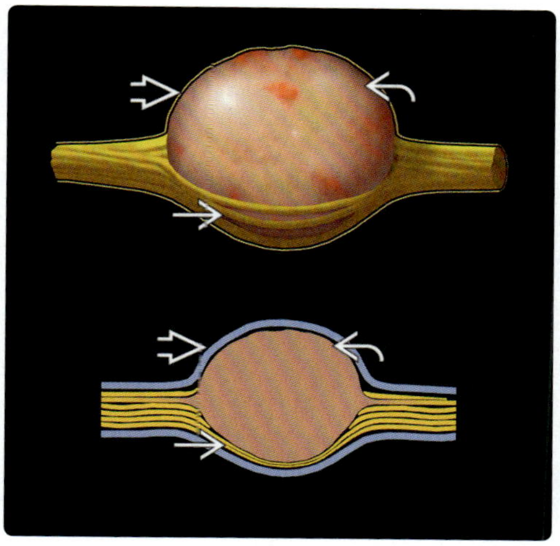

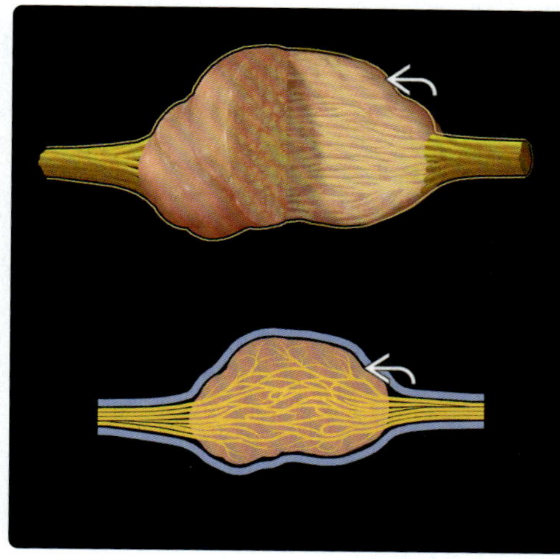

(Left) *Schematic shows a schwannoma. Each nerve is composed of axons arranged into fascicles. Each axon is covered by a myelin sheath composed of Schwann cells from which schwannoma derive. Each fascicle is covered by perineurium. All the fascicles are covered by epineurium.* **(Right)** *Schematic shows a neurofibroma. Neurofibromas arise from cells of origin in the fascicles, which differentiate into different neural and perineural cell elements instead of a Schwann-cell type.*

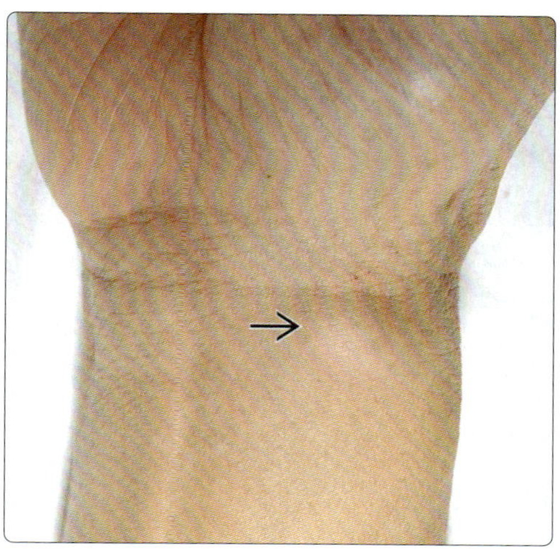

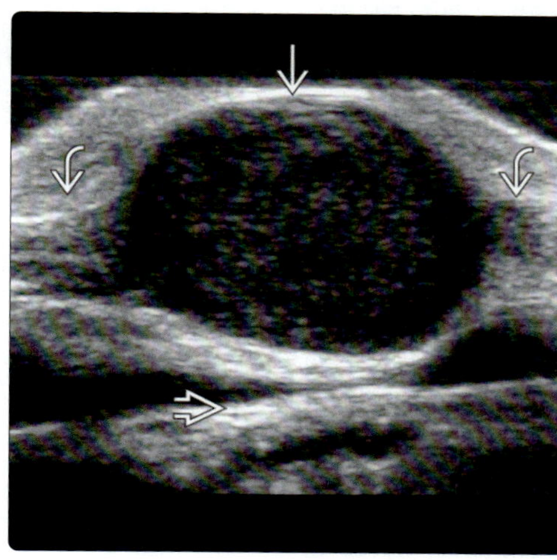

(Left) *Clinical photograph shows a focal swelling at the radial and volar aspect of the wrist region. Tinel sign was positive.* **(Right)** *Longitudinal US of the same patient shows a well-defined, quite homogeneous, ovoid hypoechoic mass continuous with the superficial sensory branch of the radial nerve. There is moderate posterior acoustic enhancement.*

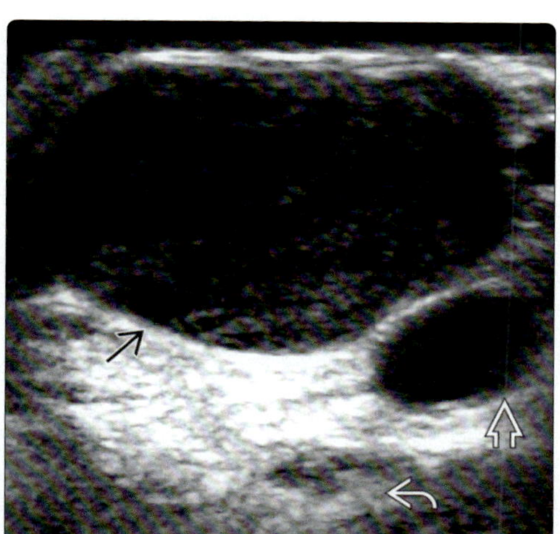

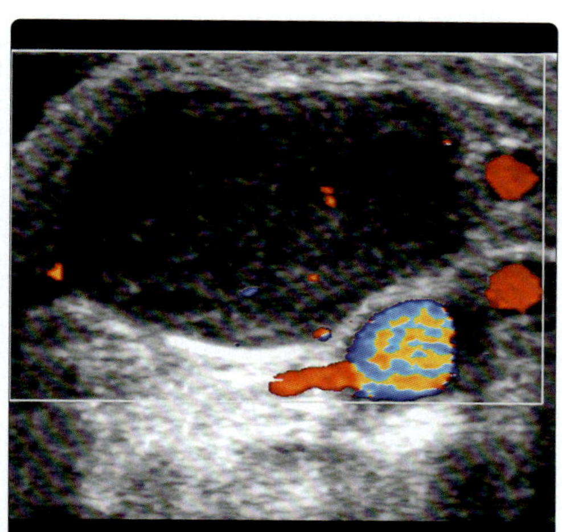

(Left) *Transverse US of the same tumor shows the mass to be in close proximity to the radial artery. Moderate posterior acoustic enhancement is present.* **(Right)** *Transverse color Doppler US of the same tumor shows mild intratumoral vascularity. The mass is arising from the radial neurovascular bundle and abutting the radial artery and veins. These findings are entirely consistent with a nerve sheath tumor of the radial nerve.*

(Left) *Longitudinal US shows an ovoid, homogeneous and hypoechoic mass → in continuity with the median nerve → at the inlet of the carpal tunnel. Note the scaphoid → and distal radius →.* **(Right)** *Transverse color Doppler US of the same patient shows the presence of moderate intrinsic hypervascularity → and moderate posterior acoustic enhancement → deep to the tumor. The tumor is centrally located within the median nerve and displacing the intact fascicles → peripherally.*

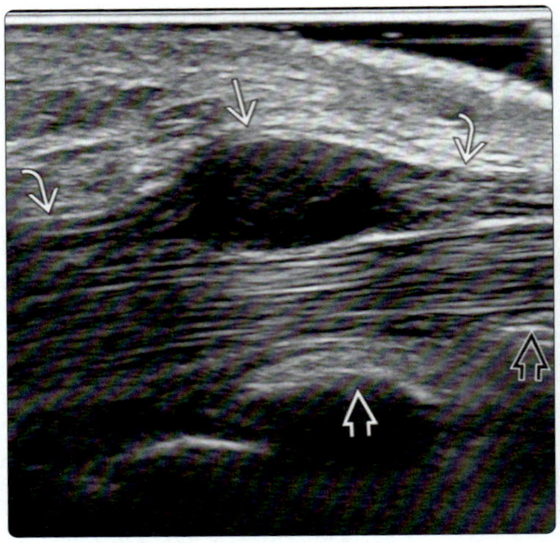

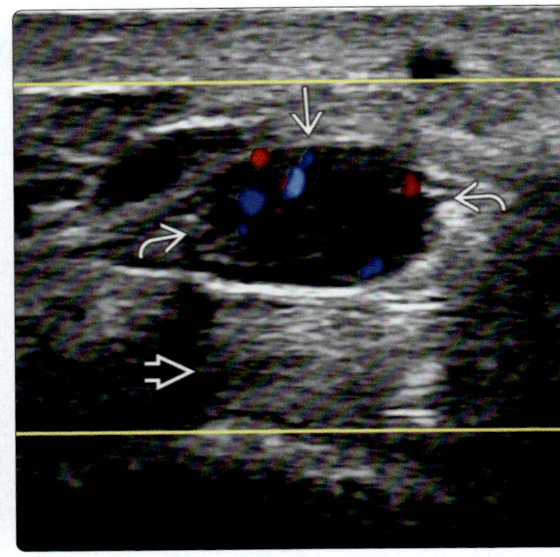

(Left) *Coronal PD MR of the same patient shows that the mildly hyperintense tumor → is centrally located with the median nerve →. Central or eccentric location with respect to the parent nerve should be documented but cannot reliably distinguish between schwannoma or neurofibroma.* **(Right)** *Axial T1 C+ FS MR shows intense tumoral enhancement → in contrast to the mild hyperemia seen on color Doppler imaging. There are mildly enhancing nerve fascicles → surrounding the nerve sheath tumor. Histology revealed a schwannoma.*

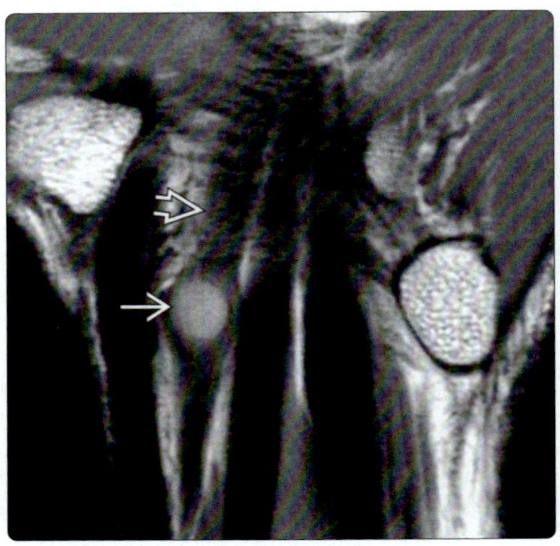

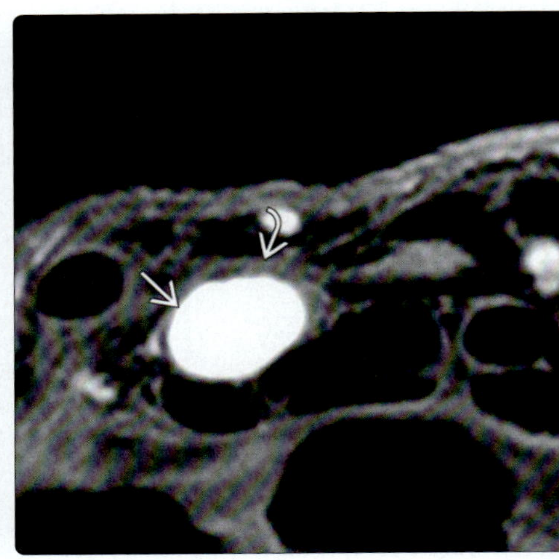

(Left) *Longitudinal US shows a mildly heterogeneous nerve sheath tumor → with moderate posterior acoustic enhancement of the median nerve just proximal to the carpal tunnel. The tumor is arising eccentrically from the radial aspect of the medial nerve → with no extension into the carpal tunnel.* **(Right)** *Longitudinal color Doppler US of the same tumor shows moderate organized hypervascularity →. Organized or chaotic vascular pattern is not helpful in diagnosing nerve sheath tumors.*

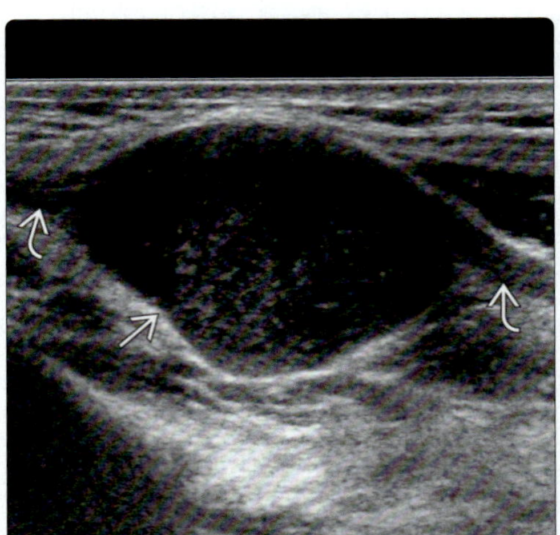

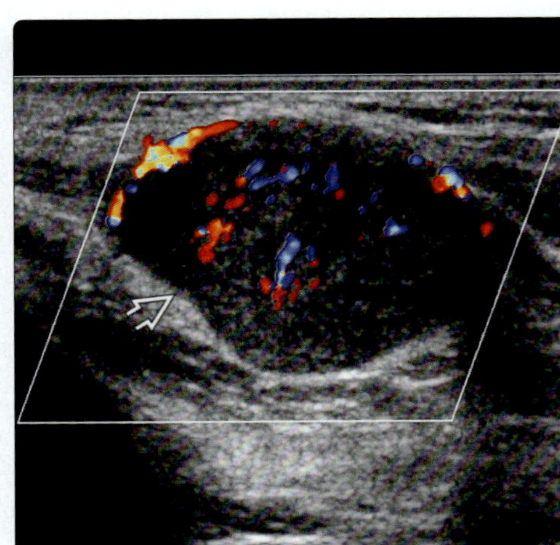

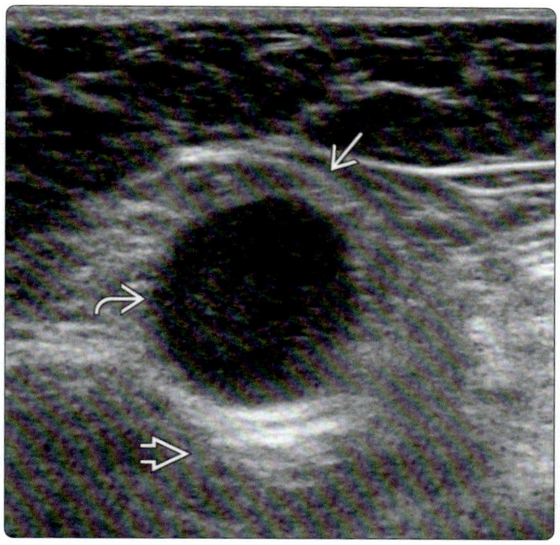

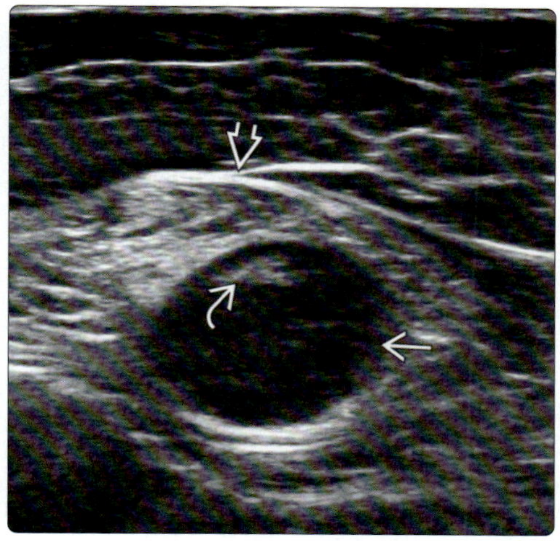

(Left) *Transverse US shows a round, homogenously hypoechoic nerve sheath tumor ➔ within the gracilis muscle ➔. Note the mild posterior acoustic enhancement ➔.* **(Right)** *Follow-up transverse US of the same patient shows mild interval enlargement of the tumor ➔ with the presence of a new hyperechoic area ➔ within the tumor. This new hyperechoic component is most likely due to intralesional hemorrhage. Note mild fatty atrophy of the gracilis muscle ➔.*

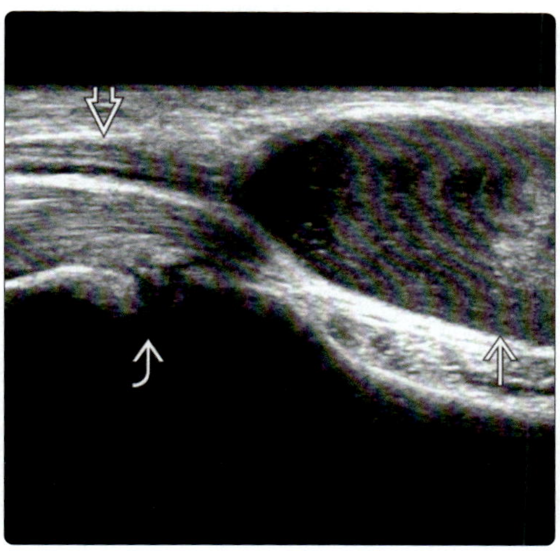

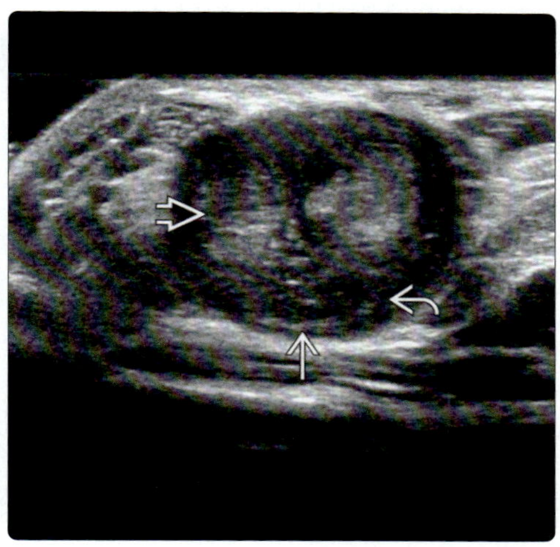

(Left) *Longitudinal US shows a schwannoma ➔ arising from the median nerve proximal to the radiocarpal articulation ➔. The median nerve is slightly thickened ➔.* **(Right)** *Transverse US of the same tumor shows moderate intratumoral heterogeneity ➔. Internal heterogeneity is due to a juxtaposition of hypoechoic fibrocollagenous tissue ➔, hypoechoic myxoid tissue ➔, and possibly previous hemorrhage. This tissue variability can also lead to the variable vascularity seen on color Doppler imaging.*

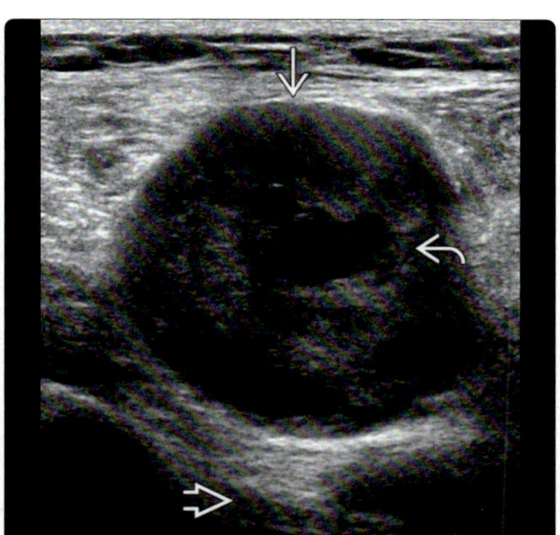

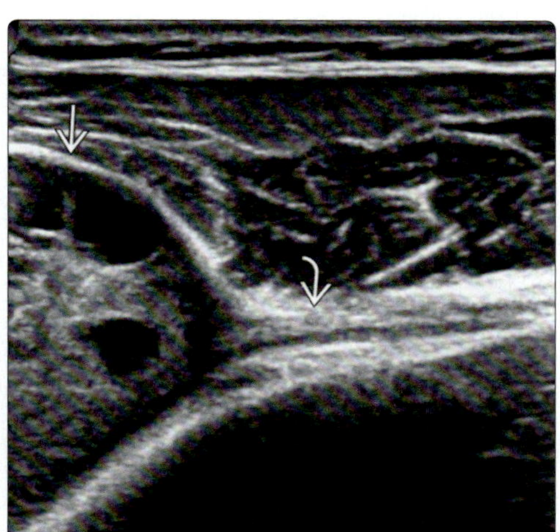

(Left) *Transverse US of the upper thigh region shows a well-defined schwannoma ➔ arising from the sciatic nerve with a small central myxoid-type area ➔. Moderate posterior acoustic enhancement ➔ is present.* **(Right)** *Longitudinal US of another patient shows a mixed solid and cystic tumor ➔ in the posterior aspect of the arm. The tumor is in continuity with the radial nerve ➔ compatible with a nerve sheath tumor of the radial nerve.*

(Left) *Longitudinal US of the same patient at a slightly more cranial location shows that the tumor → is composed of a 30% solid component ↪ and 70% cystic component ⇨. Cystic degeneration is mainly due to myxoid accumulation rather than hemorrhage or necrosis. It is more common in schwannoma.* **(Right)** *Axial T1 C+ FS MR of the same patient shows an enhancing solid component and a nonenhancing cystic component of the radial nerve schwannoma → in the posterior arm.*

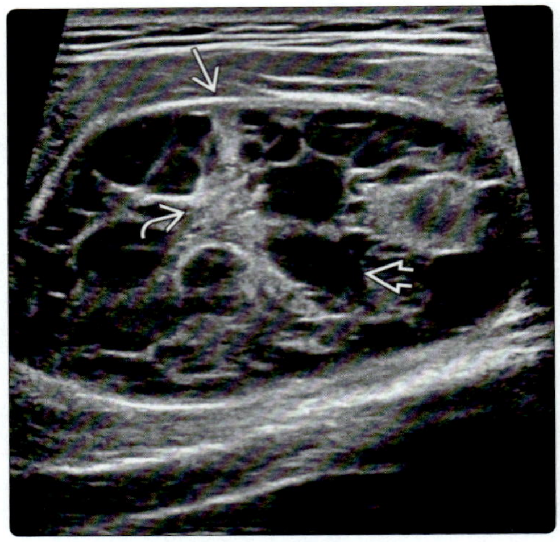

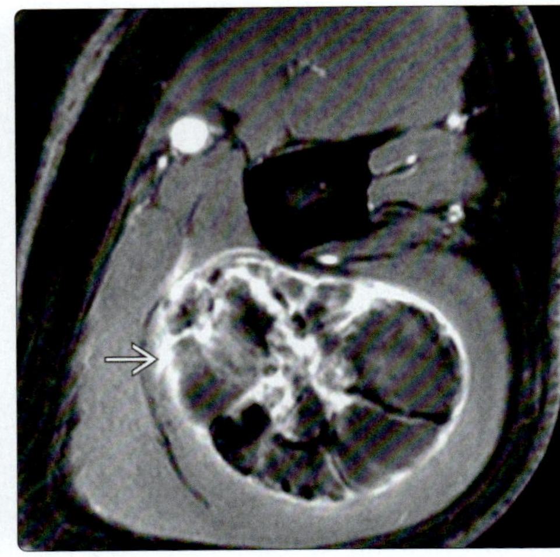

(Left) *Longitudinal US shows a well-defined, multiseptated, and almost entirely cystic tumor → in the axillary region. Note the multiple thin septations ⇨ within the cystic mass. Very little solid tissue is present.* **(Right)** *Coronal oblique T2 FS MR of the same tumor shows the cystic mass → with continuity to the median nerve ⇨. Pathology revealed a cystic schwannoma. This type of cystic schwannoma is sometimes referred to as an ancient schwannoma.*

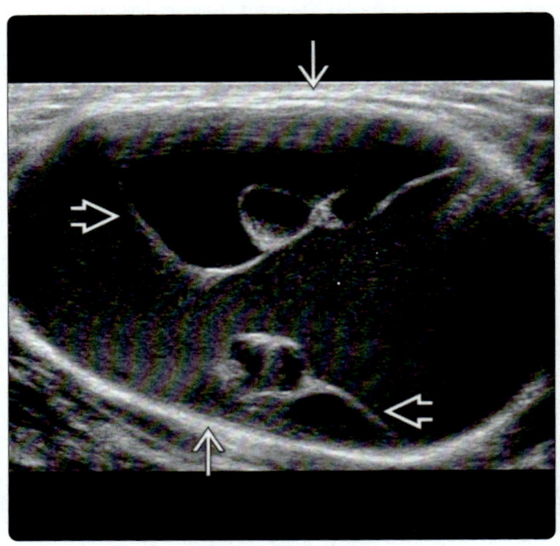

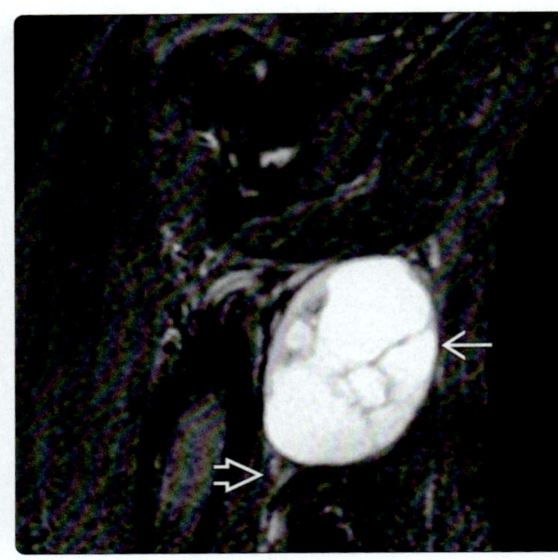

(Left) *Longitudinal US shows a well-defined mixed cystic and solid lesion → in the subcutaneous layer of the midforearm. No neural thickening was present adjacent to this mass.* **(Right)** *Transverse power Doppler US of the same tumor shows minimal tumoral vascularity. Pathology revealed this to be a localized neurofibroma with atypical features. The hypoechoic areas within the tumor are most likely due to dense tumoral fibrosis.*

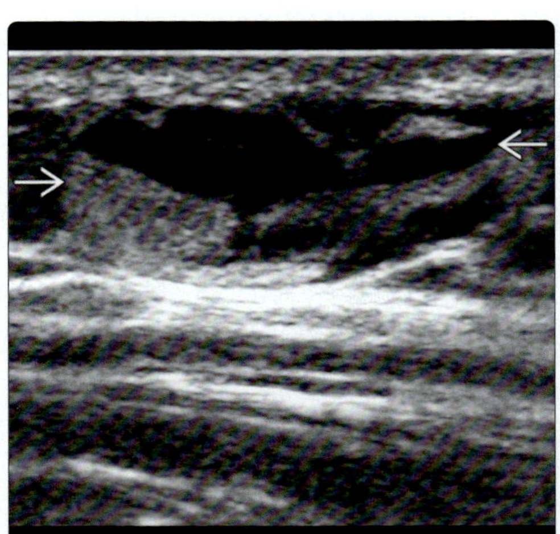

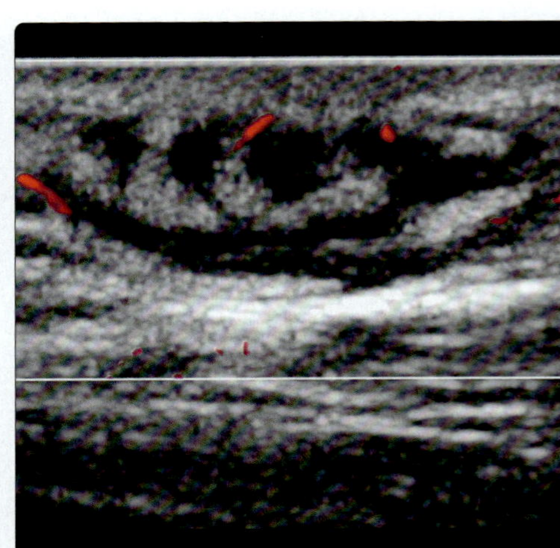

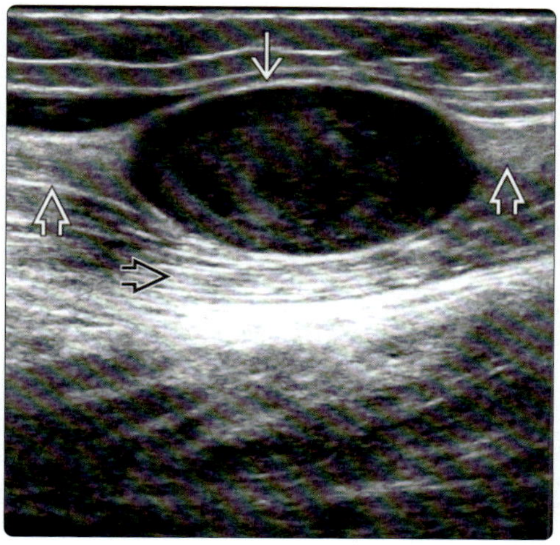

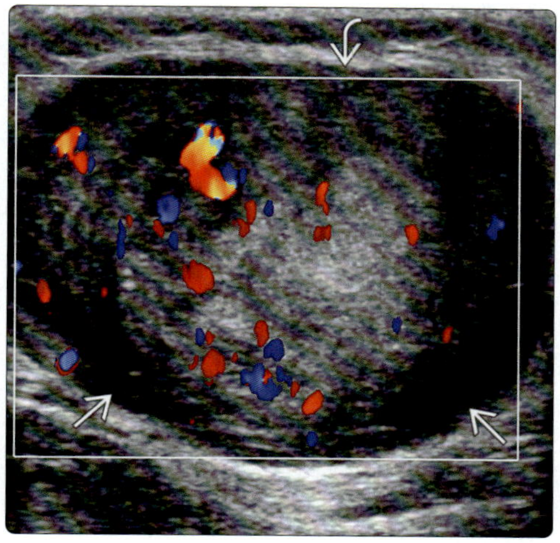

(Left) *Longitudinal US of a neurofibromatosis type 1 (NF1) patient shows nerve sheath tumor with characteristic appearances in the medial thigh region. Continuity with thickened parent nerve & moderate posterior enhancement is present.* **(Right)** *Transverse color Doppler US in the same patient shows nerve sheath tumor located between the vastus medialis and intermedialis muscles. There is moderate intrinsic vascularity and peripheric echogenic rim due to compressed peritumoral fat.*

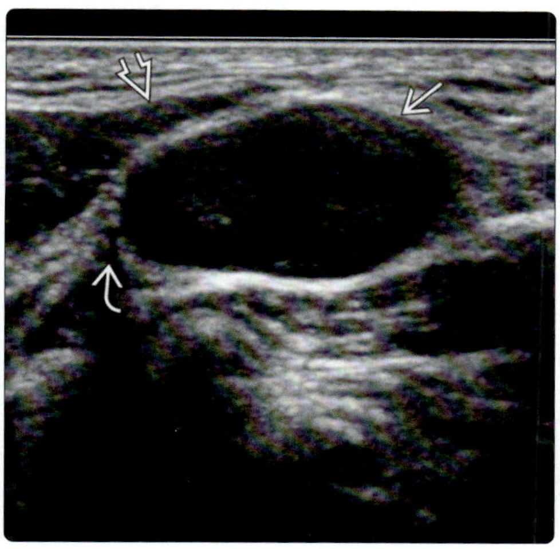

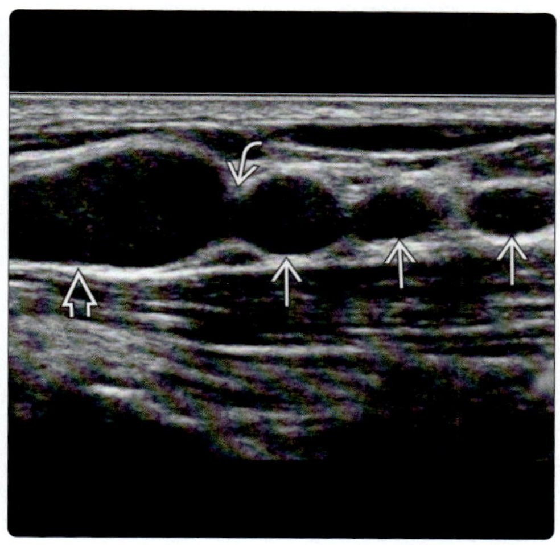

(Left) *Transverse US shows a well-defined nodule in the neck just deep to the posterior edge of sternocleidomastoid muscle. The nodule is arising from the phrenic nerve, which is better shown in the longitudinal plane.* **(Right)** *Longitudinal US of the same patient shows the presence of 3 more smaller nodules caudal to the largest nodule shown in previous image. Overall features are compatible with multiple nerve sheath tumors of the phrenic nerve.*

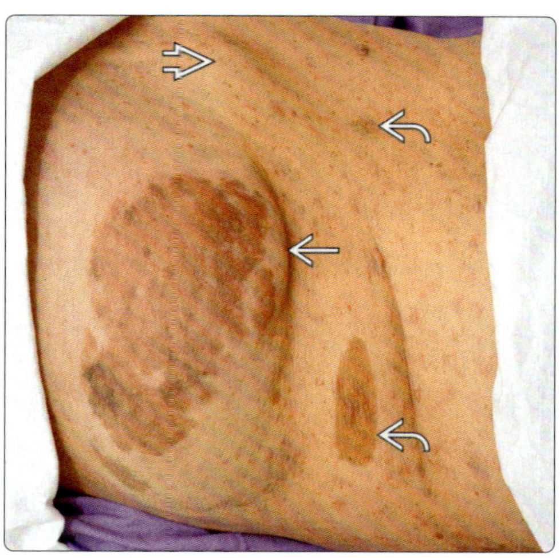

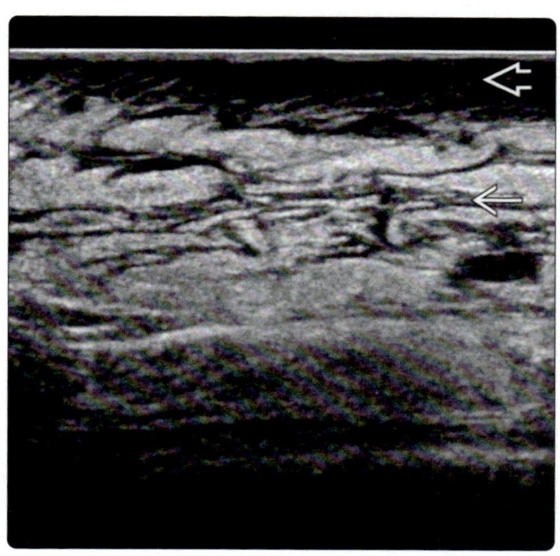

(Left) *Clinical photograph of an NF1 patient lying on his chest shows a large swelling in the posterolateral chest wall and a smaller swelling close to the midline. There are multiple café au lait spots.* **(Right)** *Longitudinal US of the larger swelling shows a large, poorly defined infiltrative area of skin and subcutaneous tissue thickening. No discrete mass was apparent.*

(Left) *Longitudinal color Doppler US of the same lesion shows moderate hyperemia →. The appearances are those of an infiltrative diffuse neurofibroma, confirmed by percutaneous biopsy and histology.* **(Right)** *Longitudinal US of the same area of diffuse neurofibroma shows the quite abrupt zone of transition between normal-appearing hypoechoic subcutaneous fat → and the hyperechoic tissue of diffuse neurofibroma →.*

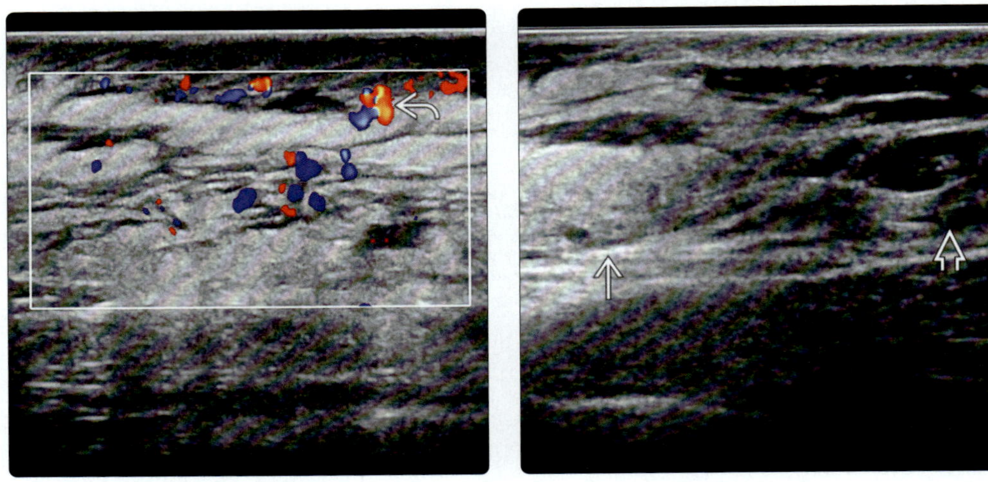

(Left) *Axial T1 MR of the same diffuse neurofibroma shows an area of diffuse cutaneous and subcutaneous tissue thickening → due to diffuse neurofibroma. There is also a discrete T1-hypointense neurofibroma → present in the subcutaneous tissues dorsal to the thoracic spine.* **(Right)** *Axial T1 C+ digital subtraction MR shows avid, reticular-like enhancement in the chest wall lesion → consistent with diffuse neurofibroma. Note the adjacent, moderately enhancing localized neurofibroma →.*

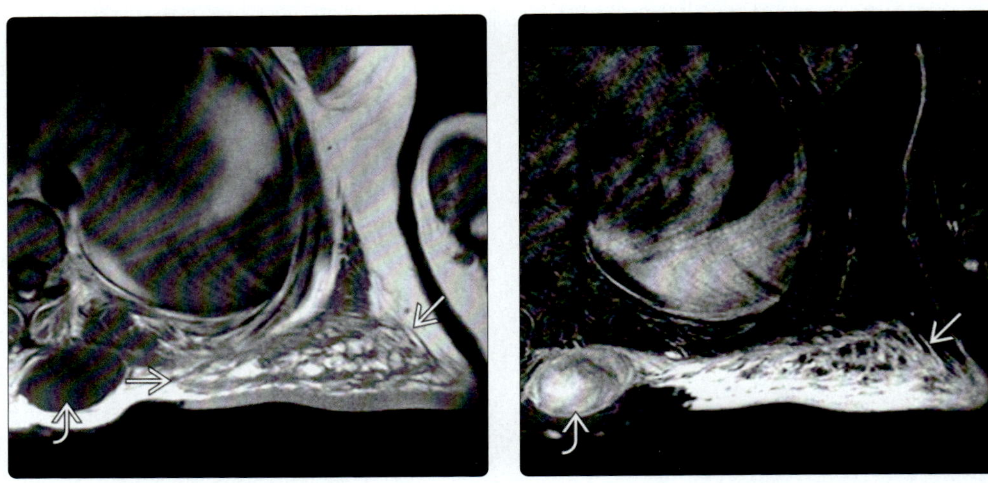

(Left) *Clinical photograph of the foot and ankle of a patient with NF1 shows a localized swelling → on the dorsal mid- to hindfoot. There are are multiple small skin nodules → due to neurofibromatosis.* **(Right)** *Longitudinal US of the same swelling shows a diffuse hypoechoic mass lesion → with bag-of-worms appearances on the dorsal aspect of the midfoot. This is consistent with a plexiform neurofibroma. It is overlying the navicular → and medial cuneiform → bones.*

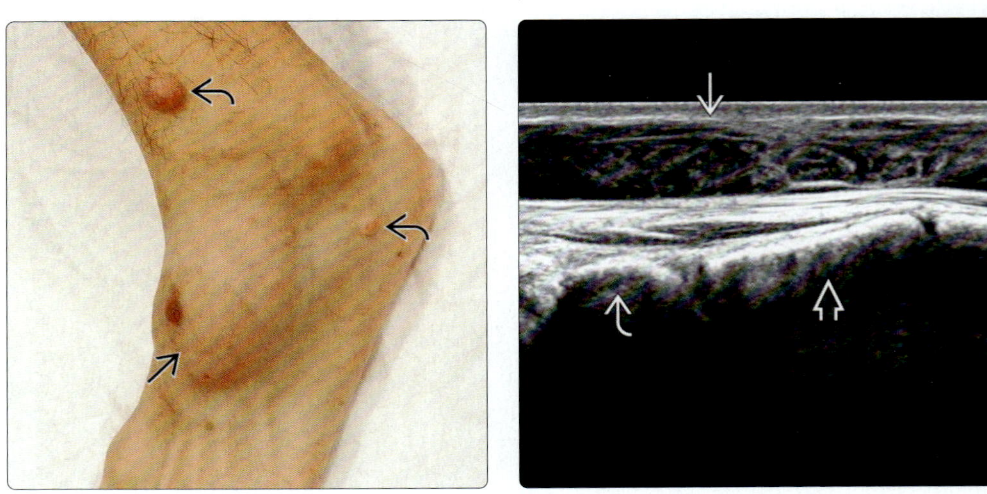

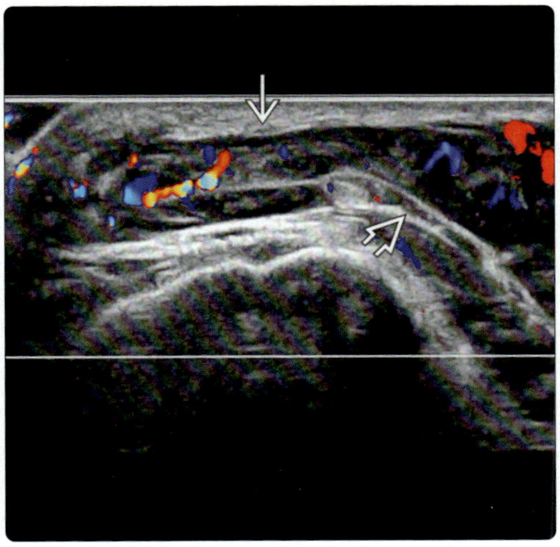

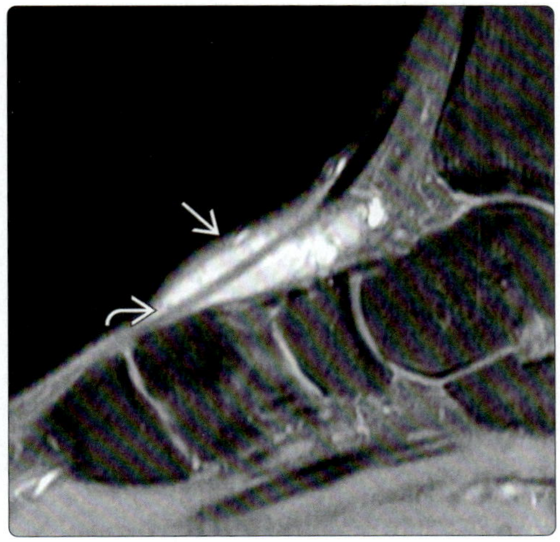

(Left) *Transverse color Doppler US of the same plexiform neurofibroma shows moderate intratumoral hyperemia ➔. The tumor is encasing the extensor hallucis longus tendon ➔.* **(Right)** *Sagittal T1 C+ FS MR of the same plexiform neurofibroma shows avid tumoral enhancement ➔, which encases the extensor tendon ➔. Plexiform neurofibroma is diagnostic of NF1 and associated with a high risk of malignant change.*

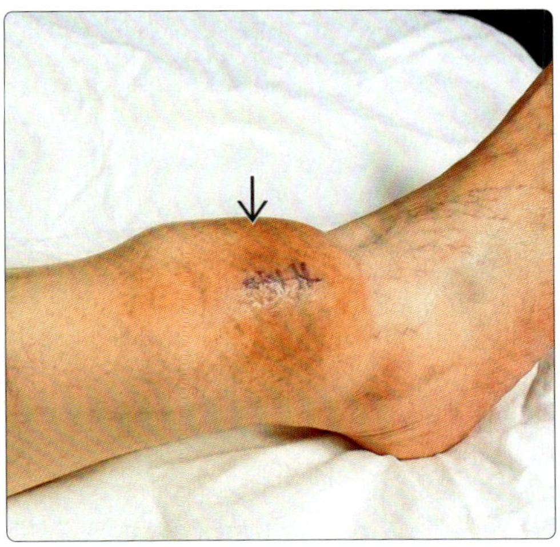

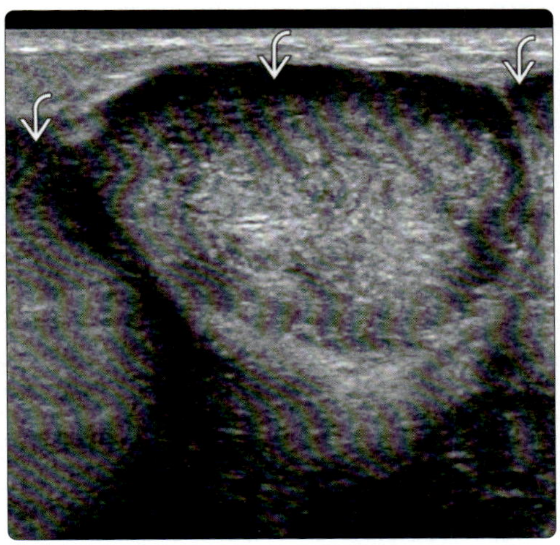

(Left) *Clinical photograph shows a large swelling ➔ on the anterior aspect of the distal leg with associated skin discoloration. This mass had been slowly growing for several years. A suitable site for biopsy has been marked on the skin.* **(Right)** *Transverse US of the same swelling shows part of a large, lobulated, solid hypoechoic mass ➔ in the anterior compartment of the leg. Only a part of the tumor can be depicted at a time by high-resolution US.*

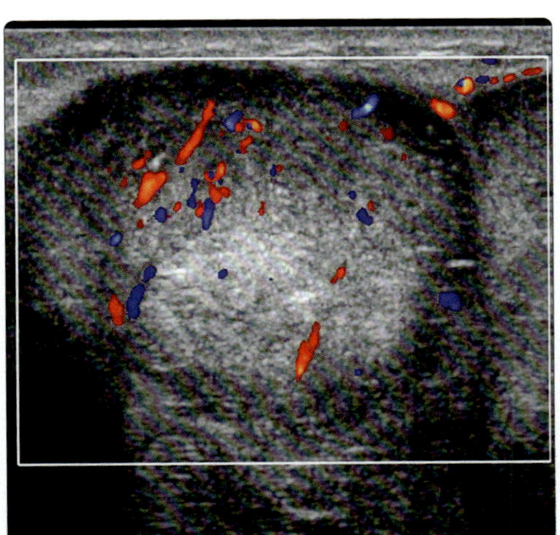

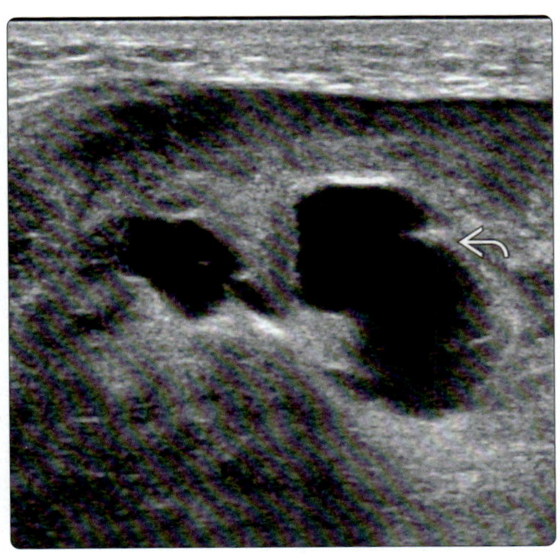

(Left) *Transverse color Doppler US of the same tumor shows moderate intratumoral hyperemia.* **(Right)** *Transverse US of another part of the same tumor shows a large myxoid-like area ➔. Biopsy revealed a cystic schwannoma. As there was concern regarding malignant change, the mass was subsequently excised, and pathology concurred fully with a benign cystic schwannoma.*

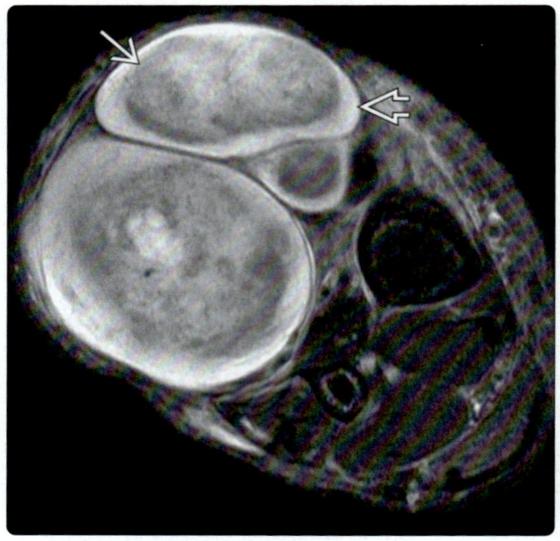

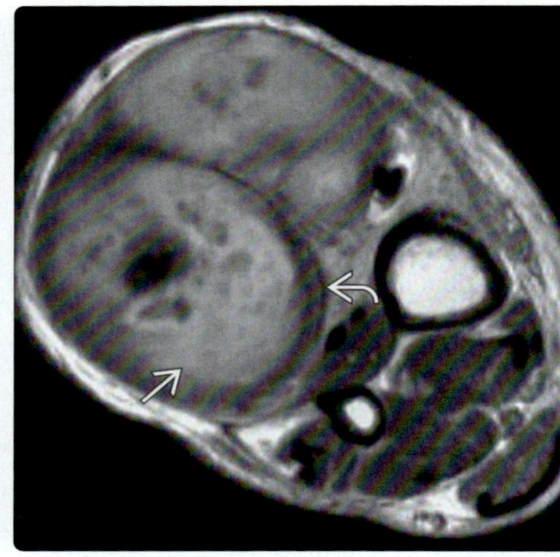

(Left) *Transverse T2 FS MR of the same schwannoma prior to excision shows a large, lobulated mass with each lobule displaying a target sign due to more centralized hypointense fibrous component and a more peripheral hyperintense myxoid component.* **(Right)** *Transverse T1 C+ FS MR of the same tumor shows a reverse-target sign characterized by enhancement of the central fibrous component and lack of enhancement of the surrounding myxoid component.*

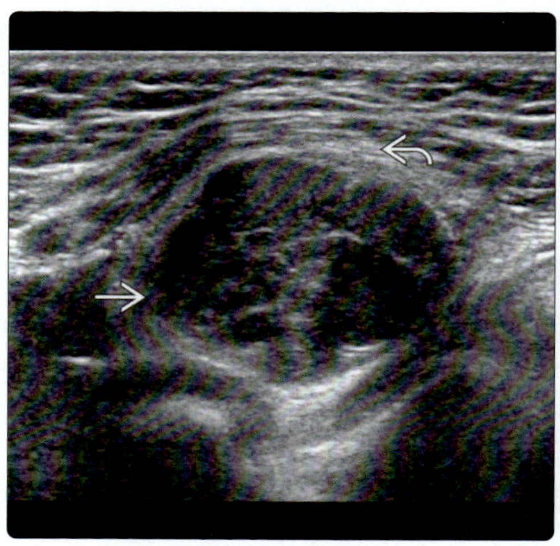

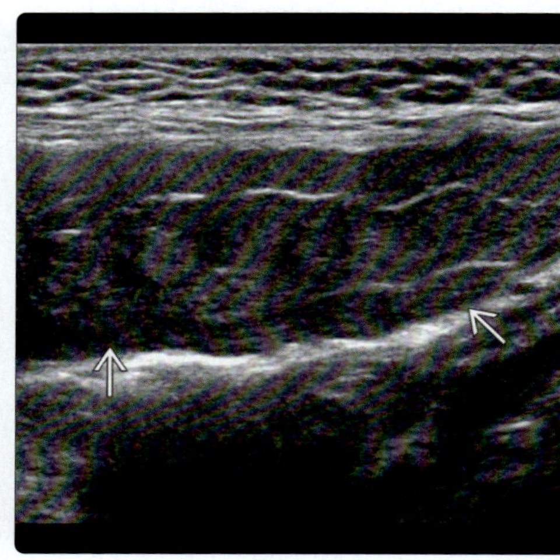

(Left) *Transverse US of the posterior thigh in a 3-year-old girl with a short history of posterior thigh swelling and limping gait shows a multilobulated hypoechoic swelling with posterior enhancement of the sciatic nerve. Note the echogenic rim around the mass due to compressed fat.* **(Right)** *Longitudinal US shows the same sciatic nerve to be diffusely and severely thickened.*

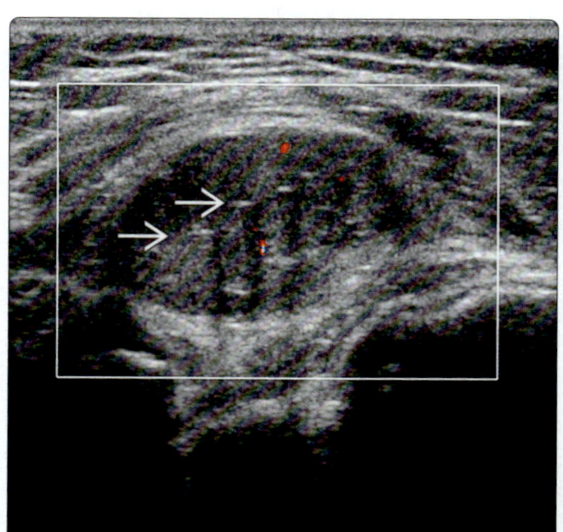

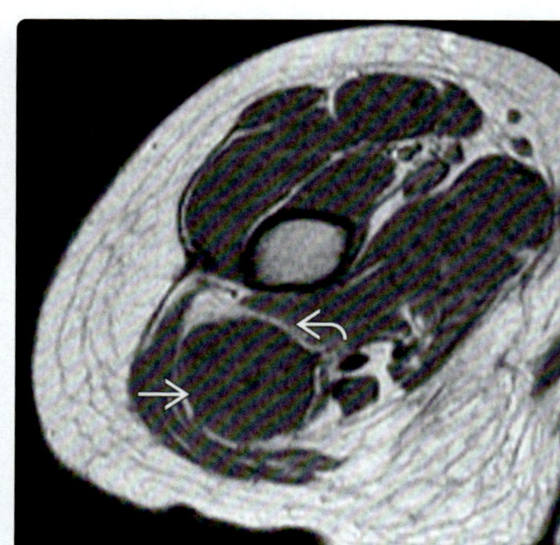

(Left) *Transverse color Doppler US of the same thickened sciatic nerve shows minimal intralesional hyperemia. Several small intratumoral echogenic foci with acoustic shadowing are appreciated on this image, which are most likely due to intratumoral calcification.* **(Right)** *Axial T1 MR of the same lesion shows the thickened isointense sciatic nerve. There is a thin rim of hyperintense fat surrounding the tumor. There is no local invasion. This child was otherwise well with no features of NF1.*

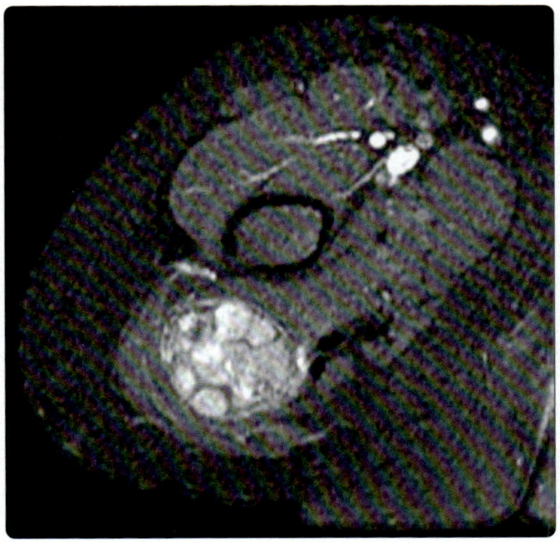

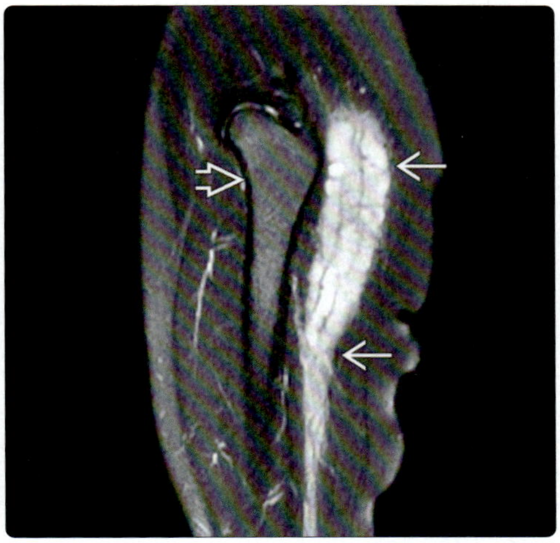

(Left) *Transverse T1 C+ FS MR of the same lesion shows moderate intratumoral contrast enhancement. Note the fascicular appearance of the tumor.* **(Right)** *Sagittal T2 FS MR of the same lesion shows severe, diffuse sciatic nerve enlargement ➡ posterior to the femur ⇨. The appearances are those of a nerve sheath tumor. Pathology revealed a malignant nerve sheath tumor. Benign and malignant nerve sheath tumors share similar imaging features.*

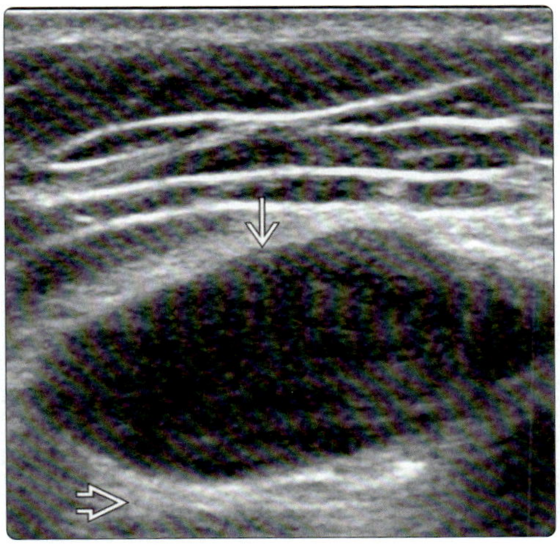

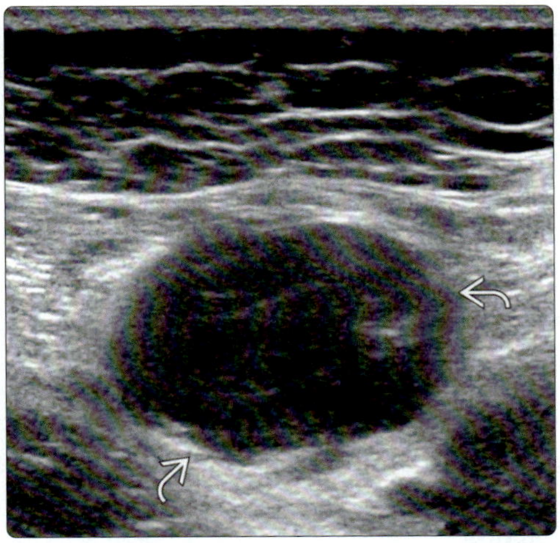

(Left) *Longitudinal US shows a fusiform hypoechoic nerve sheath tumor ➡ along the radial nerve in the posterior arm region. Note the posterior acoustic enhancement ⇨.* **(Right)** *Transverse US of the same tumor shows the slightly irregular tumor border ↪. Clinically, the mass was enlarging, so a follow-up US examination was arranged 6 months later. There were no clinical features of NF1.*

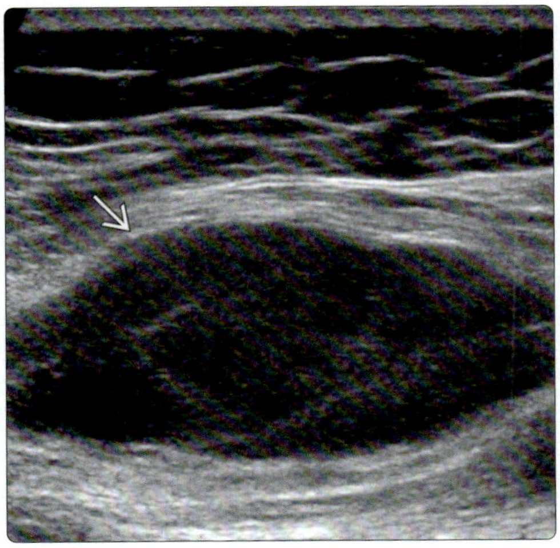

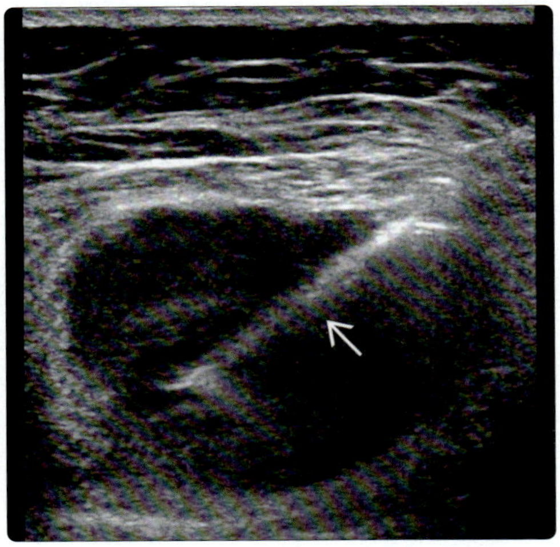

(Left) *Follow-up longitudinal US of the same tumor ➡ 6 months later shows mild to moderate interval increase in tumor size.* **(Right)** *Transverse US of the same tumor shows moderate increase in tumor size, particularly in the transverse plane. Rapid irregular growth is an indicator of malignant transformation of nerve sheath tumor. The tumor was biopsied ➡ and shown to be a malignant nerve sheath tumor.*

Carpal Tunnel Syndrome

KEY FACTS

TERMINOLOGY

- Symptoms secondary to compression of median nerve in carpal tunnel area
- Most common entrapment neuropathy

IMAGING

- More swollen median nerve means more likely possibility of carpal tunnel syndrome (CTS)
- Measure nerve cross-sectional area (CSA) at forearm, immediately proximal to tunnel, tunnel inlet, tunnel outlet, & distal to tunnel outlet (CSAc)
- 2 main measurement methods
- **Largest CSA**
  - Median nerve CSA ≥ 14 mm² at proximal or distal to carpal tunnel: Strong positive predictor of CTS (sensitivity/specificity/accuracy > 88%)
- **Relative increase in nerve CSA**
  - Calculate difference between CSAp & largest carpal tunnel measurement (CSAc)
  - If CSAc & CSAp > 4 mm², positive predictor of CTS
- Additional US signs of CTS have reduced discriminatory power if used in isolation, though may improve discrimination when used with CSA measurements
  - Intraneural hyperemia
  - Retinacular bowing
  - Flattening ratio: Width:depth ratio > 3 of median nerve ↑ with CTS
  - Homogeneously hypoechoic with loss of normal fascicular pattern
  - If either CSA proximal to or distal to carpal tunnel is > 14 mm² or palmar volar bowing of flexor retinaculum at outlet > 1 mm, sensitivity, specificity & accuracy are 100%, 84%, and 93%, respectively

CLINICAL ISSUES

- Night/early morning paresthesia and pain in radial 3.5 digits
- Positive Tinel & Phalen signs
- Thenar muscle wasting

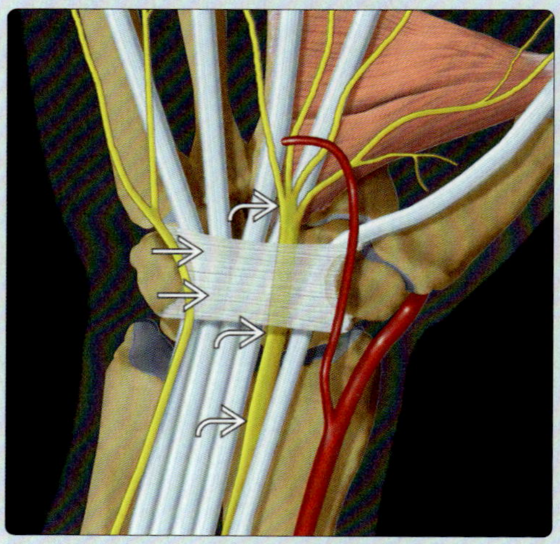

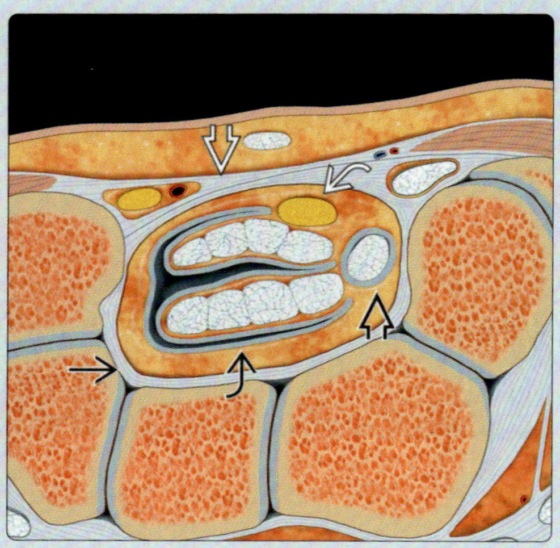

(Left) *Graphic shows the median nerve deep to the flexor retinaculum & superficial to the flexor tendons. The nerve usually divides just beyond the tunnel outlet.* **(Right)** *Schematic axial view of the carpal tunnel shows the median nerve lies immediately below the flexor retinaculum & superficial to the flexor digitorum tendons, which are surrounded by the ulnar bursa & the flexor pollicis longus, which is surrounded by the radial bursa. The extrinsic ligaments lie above the carpal bones.*

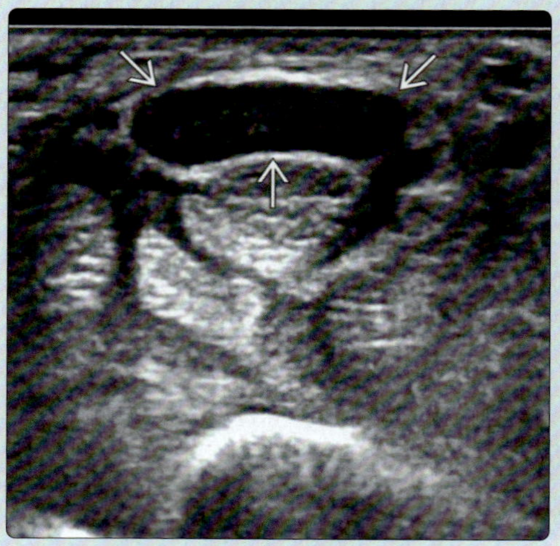

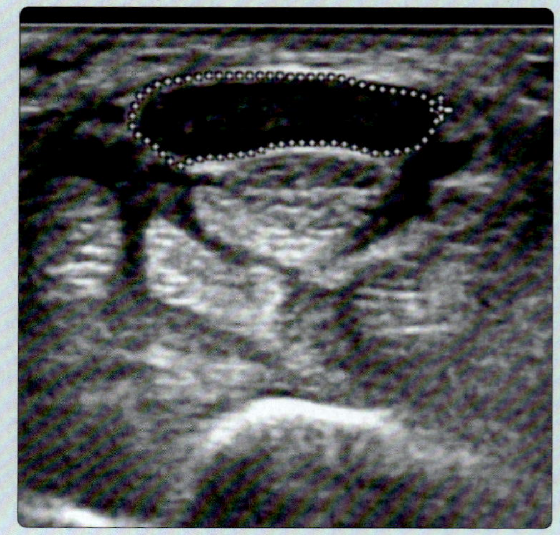

(Left) *Transverse US of the median nerve just proximal to the carpal tunnel inlet shows enlargement (15 mm²). A nerve cross-sectional area (CSA) > 14 mm² is a positive predictor for CTS. Note also the loss of neural fascicular structure, which is an additional sign of CTS.* **(Right)** *Transverse US of the same patient shows CSA measurement by continuous tracing around nerve perimeter. As the nerve is occasionally irregular in outline, electronic tracing can achieve better results than ellipsoid calipers.*

TERMINOLOGY

Abbreviations

- Carpal tunnel syndrome (CTS)
- Transverse carpal ligament (TCL)

Definitions

- Symptoms secondary to compression ± ischemia of median nerve in carpal tunnel area
  - Most common entrapment neuropathy

IMAGING

General Features

- Best diagnostic clue
  - Swollen median nerve immediately proximal or distal to carpal tunnel
- Size
  - Normal nerve CSA varies from 7 mm^2 to ~ 12 mm^2
    - Measured nerve caliber on US is ~ 20% smaller than on MR or anatomy
      - US measurements do not include epineurium
- Morphology
  - Hourglass-type appearance of median nerve due to compression in carpal tunnel with swelling proximally & distally

Ultrasonographic Findings

- Increase in nerve caliber is, by far, most reliable US criterion of CTS
  - Swelling of median nerve either immediately proximal to, within, or immediately distal to carpal tunnel
    - Proximal: Proximal to TCL
    - Inlet: Between scaphoid tubercle and pisiform
    - Outlet: Between trapezium tubercle and hook of hamate
    - Distal to carpal tunnel: Distal to TCL
    - More swollen median nerve means more likely possibility of CTS
    - Rigid criteria not useful in clinical practice as median nerve compression is continuum rather than "all or none" phenomenon
  - Measure CSA by tracing around nerve perimeter
    - Measure around hypoechoic neural fascicles
      - Echogenic epineurium not well delineated from perineural fat on US
  - 2 main measurement methods based on median nerve CSA, either of which can be used
    - **Largest CSA of nerve**
      - Median nerve CSA measured at 4 points: Immediately proximal to tunnel inlet, at tunnel inlet, at tunnel outlet, & distal to carpal tunnel
      - Optimal cutoff point: Median nerve CSA ≥ 14 mm^2
      - Median nerve CSA ≥ 14 mm^2 at proximal or distal to carpal tunnel: Strong positive predictor of CTS (sensitivity/specificity/accuracy > 88%)
    - **Relative increase in nerve CSA**
      - Median nerve CSA measured at proximal 1/3 of pronator quadratus muscle level (CSAf) & immediately proximal to tunnel inlet, at tunnel inlet, & at tunnel outlet (CSAc)
      - Difference between pronator quadratus level measurement (CSAf) & any carpal tunnel measurement (CSAc) should not be > 4 mm^2
      - Difference between CSAc & CSAf (ΔCSA) of > 4 mm^2 is strong positive predictor of CTS
- Additional US signs of CTS
  - Increased intraneural vascularity on color Doppler imaging
    - Hyperemia is marker of neural inflammation & ischemia
    - High specificity but low sensitivity
  - Median nerve edema
    - Nerve becomes homogeneously hypoechoic with loss of normal fascicular pattern
    - Prone to anisotropy & difficult to quantify
  - Altered median nerve movement on dynamic US testing
    - ↓ longitudinal movement or ↑ volar-dorsal movement between flexor tendons during wrist flexion
    - No objective criteria & difficult to judge
  - ↑ retinacular bowing
    - Increased volar convexity of TCL
    - > 1 mm at outlet (specificity > 90%)
  - Flattening ratio
    - Width:depth ratio > 3 of median nerve ↑ with CTS
  - Thenar muscle atrophy
    - Difficult to quantify
- Other US features of CTS may ↑ diagnostic accuracy when combined with CSA measurements, though, in isolation, each has lower discriminatory power
  - If either CSA proximal to or distal to carpal tunnel is > 14 mm^2 or palmar volar bowing of flexor retinaculum at outlet > 1 mm, sensitivity, specificity, & accuracy are 100%, 84%, & 94%, respectively
- Secondary CTS mainly due to mass effect within carpal tunnel
  - Accessory muscle within carpal tunnel
    - Anomalous palmaris longus
      - Duplicated, hypertrophied, or reversed (i.e., fleshy in distal part, tendinous in proximal part)
    - Hypertrophied or bipennate lumbrical muscle
    - Anomalous flexor digitorum superficialis muscle belly
  - Flexor digitorum tenosynovitis
  - Ganglion, lipoma, or gouty tophus
  - Nerve sheath tumor of median nerve
  - Hamartomatous fibrolipoma of median nerve
  - Thrombosed persistent median artery
- Following carpal tunnel decompression
  - Resected bands of retinaculum are wavy, thickened, & discontinuous
  - Flexor retinaculum may reform though in elongated position
  - Abrupt change in nerve caliber if continued compression
  - ± hypoechoic scar tissue encasing median nerve

MR Findings

- Swelling of median nerve
  - Hyperintense T2 signal intensity
  - Swollen fascicles
  - Palmar volar bowing of TCL

Imaging Recommendations

- Best imaging tool
  - US, as it is accurate, quick, comfortable, & relatively inexpensive
- Protocol advice
  - Wrist supported on flat surface with fingers relaxed
    - Hand-wrist exercises should not precede examination as they may increase nerve caliber
  - Measure nerve CSA at distal forearm, proximal to tunnel inlet, at tunnel inlet, at tunnel outlet, & distal to tunnel outlet
  - Try to keep transducer aligned at right angles to nerve rather than oblique
    - However, slight transducer angulation can utilize anisotropy to optimize delineation of nerve boundary
  - Measure median nerve CSA by electronic tracing rather than ellipsoid calipers
  - If median nerve bifid, measure both components separately, excluding any intervening median artery & summation
  - Assess vascularity of median nerve
  - Assess volar bowing of TCL at inlet and outlet levels

DIFFERENTIAL DIAGNOSIS

Tenosynovitis

- Tendon sheath synovial proliferation & hyperemia predominates
  - ± secondary CTS

Inflammatory Synovitis

- Synovial swelling & hyperemia predominates
  - ± secondary CTS

PATHOLOGY

General Features

- Etiology
  - Due to compression ± microvascular ischemia of median nerve in carpal tunnel
  - Majority primary with no specific structural cause found
  - Increased risk in
    - Females, > middle age, ↑ body mass index
    - Repetitive physical work
  - Secondary causes
    - Distal radius or carpal bone injury
    - Inflammatory wrist arthropathy
    - Pregnancy, diabetes, myxoedema
    - Tenosynovitis, anomalous muscles, ganglia, lipoma compressing carpal tunnel

Staging, Grading, & Classification

- No widely used grading scheme

Gross Pathologic & Surgical Features

- Inflammation & edema of median nerve

Microscopic Features

- Venous congestion, intrafascicular edema, endoneural & epineural swelling, Wallerian degeneration, axonal damage

CLINICAL ISSUES

Presentation

- Most common signs/symptoms
  - Night tingling & pain over distribution of median nerve
    - Exacerbated by prolonged strenuous work or exercise
- Other signs/symptoms
  - ± abductor pollicis brevis & opponens pollicis weakness (early finding) followed by atrophy (late finding) = "ape hand"
- Clinical profile
  - Nocturnal exacerbation
    - Wrists tend to flex during sleep
  - Positive Tinel & Phalen signs
  - Electrodiagnostic studies test functional integrity of nerve
    - Median nerve sensory conduction over 8-cm span is compared to ipsilateral ulnar nerve sensory conduction over 8-cm span
    - Main diagnostic parameter is median-ulnar sensory latency delay of > 0.5 ms

Demographics

- Age
  - Peak: 50 years ± 15 years
- Gender
  - M:F = 1:6
- Epidemiology
  - Cumulative lifetime prevalence rate: 10%
    - 0.5 million carpal tunnel decompressions performed yearly in USA

Natural History & Prognosis

- Symptoms usually remain over months or years
  - Tendency to improve clinically & electrophysiologically over time even without specific treatment
    - CTS will continue to progress in minority of patients

Treatment

- Conservative
  - Neutral position wrist splint
  - US-guided steroid injection into carpal tunnel
    - Symptom improvement for up to 12 weeks
- Surgical
  - Flexor retinaculum release; open vs. endoscopic, which may use US guidance

DIAGNOSTIC CHECKLIST

Consider

- CTS in all patients with unexplained hand, wrist, or distal forearm paresthesia ± pain, even though symptoms are atypical for CTS

SELECTED REFERENCES

1. Ng AWH et al: Ultrasound carpal tunnel syndrome: additional criteria for diagnosis. Clin Radiol. 73(2) 214 e11-214 e18, 2018

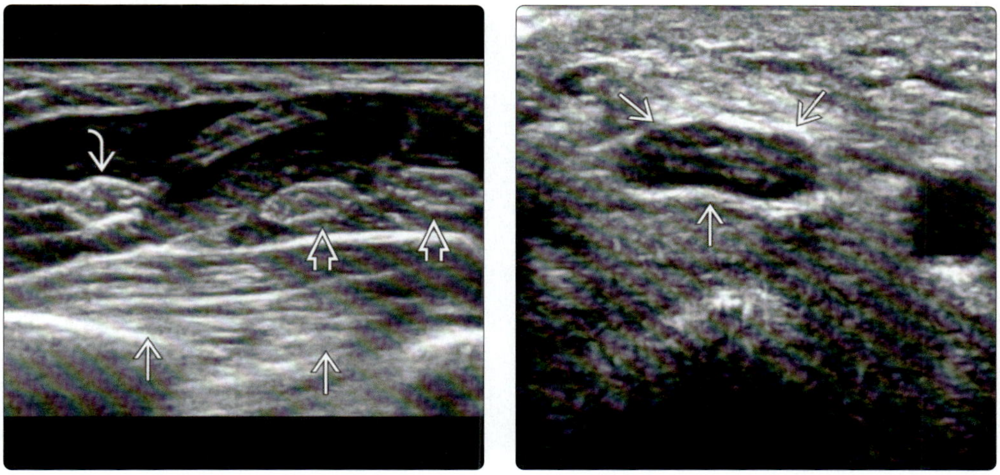

(Left) *Transverse US shows the normal appearance & size of the median nerve at the proximal end of the pronator quadratus muscle. At this location, the median nerve is ovoid in shape & not swollen. Note the adjacent flexor digitorum profundus tendons.* **(Right)** *Transverse US of the median nerve in the same patient just proximal to the carpal tunnel inlet shows the swollen median nerve with loss of normal fascicular nerve echo pattern.*

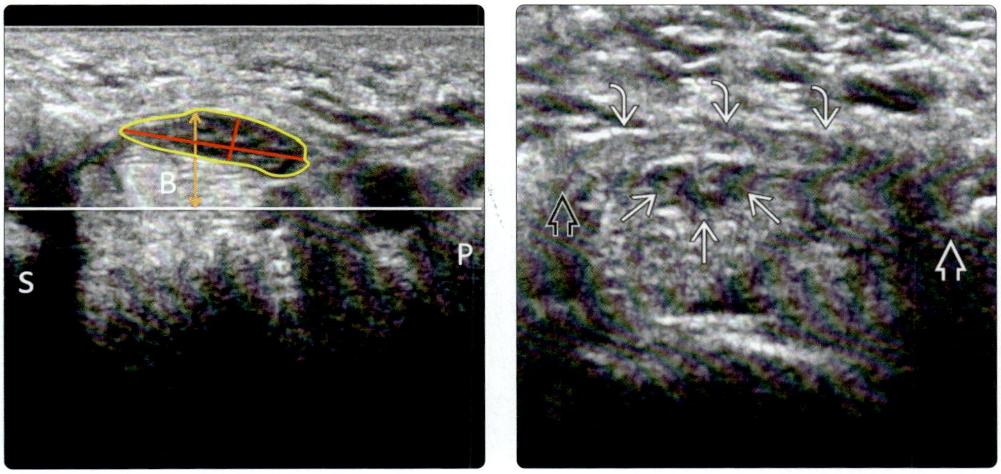

(Left) *Transverse US of the median nerve at the inlet shows features of CTS with severe palmar retinaculum bowing (B) relative to the transverse carpal line drawn between scaphoid tubercle (S) & pisiform (P), high flattening ratio (red lines), & increased CSA of the median nerve (outlined by yellow).* **(Right)** *Transverse US of the carpal tunnel outlet at the level of hamate and trapezium shows flattening of the median nerve with moderate retinaculum thickening.*

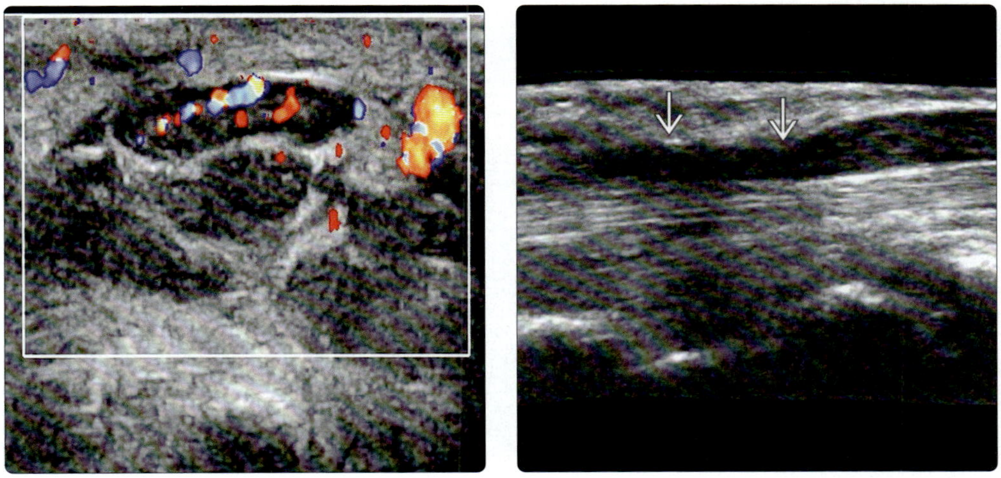

(Left) *Transverse color Doppler US shows moderate intraneural hyperemia proximal to the carpal tunnel in a patient with CTS. Intraneural vascularity should be assessed where the nerve is most swollen.* **(Right)** *Longitudinal US of the median nerve shows severe abrupt change in nerve caliber at the inlet of the carpal tunnel consistent with compression. Note swelling of the medial nerve proximal to the carpal tunnel with loss of the normal fascicular pattern.*

(Left) *Clinical photograph shows electrodiagnostic studies testing median nerve sensory conduction over 8-cm span with stimulating, ground, active recording, and reference electrodes.* **(Right)** *Clinical photograph of a patient with CTS shows more severe, generalized atrophy of the thenar muscles on the right side. The normal contralateral thenar eminence is also shown for comparison. Thenar muscle atrophy is readily assessed clinically with no added benefit from US.*

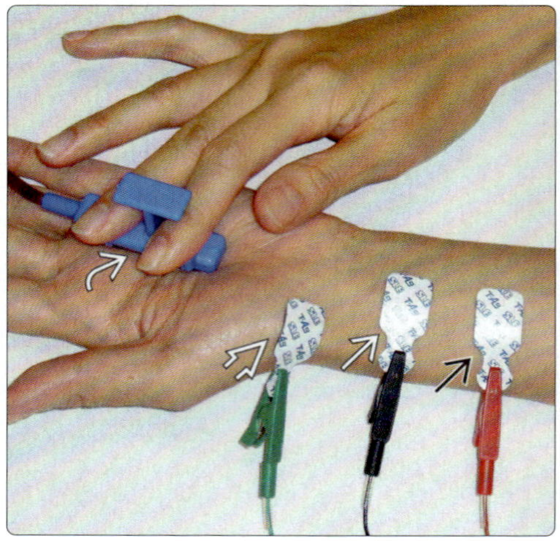

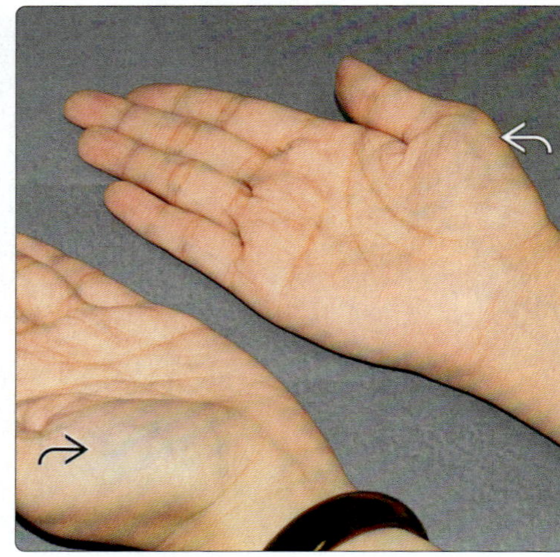

(Left) *Transverse US of a normal subject shows normal bulky thenar muscle, including the abductor pollicis brevis, opponens pollicis, flexor pollicis brevis, & adductor pollicis muscles.* **(Right)** *Transverse US of a patient with CTS shows moderate generalized thenar muscle atrophy, including abductor pollicis brevis opponens pollicis, flexor pollicis brevis, & adductor pollicis muscles. Note flexor pollicis longus tendon. Thenar muscle atrophy grading on US adds little to clinical assessment.*

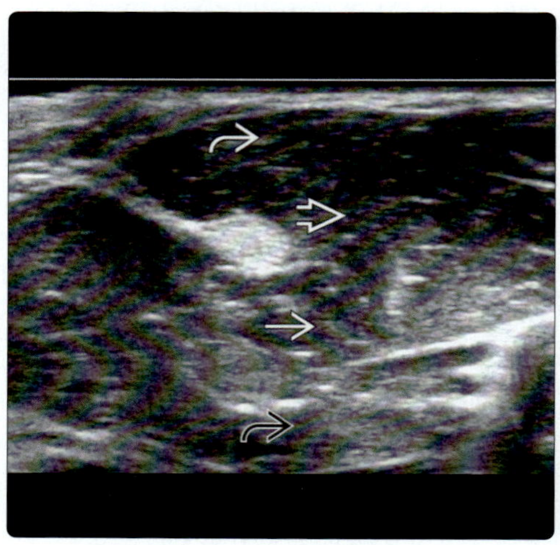

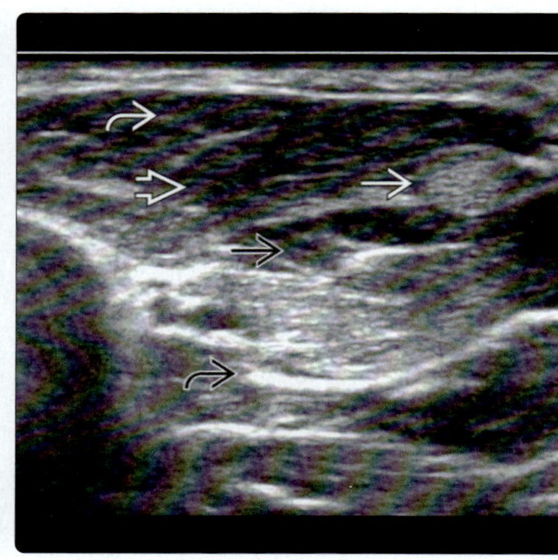

(Left) *Transverse US shows median nerve enlargement (17 mm²) just proximal to the carpal tunnel inlet. Note reduction in neural fascicular structure. The antebrachial fascia overlies the nerve. The palmaris longus tendon is superficial.* **(Right)** *Axial proton density MR shows the same severely swollen median nerve as previous image just proximal to the tunnel inlet. The nerve fascicles are swollen. The antebrachial fascia is shown. The palmaris longus tendon is superficial.*

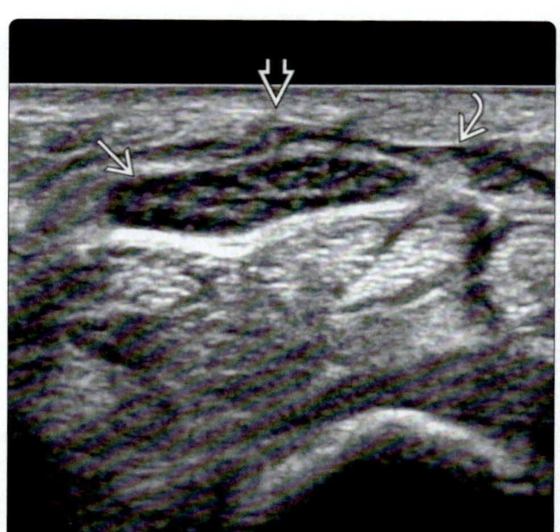

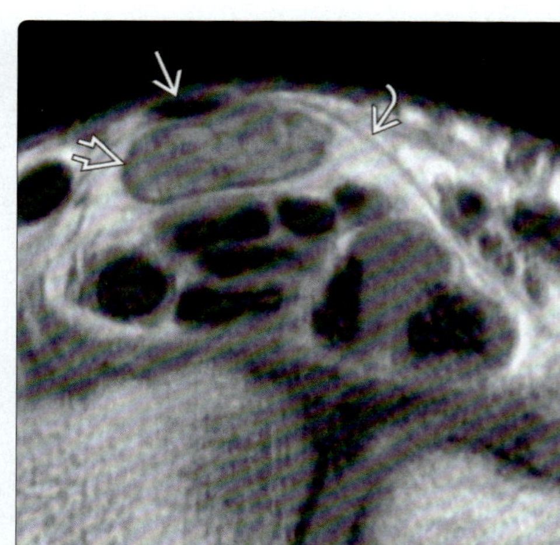

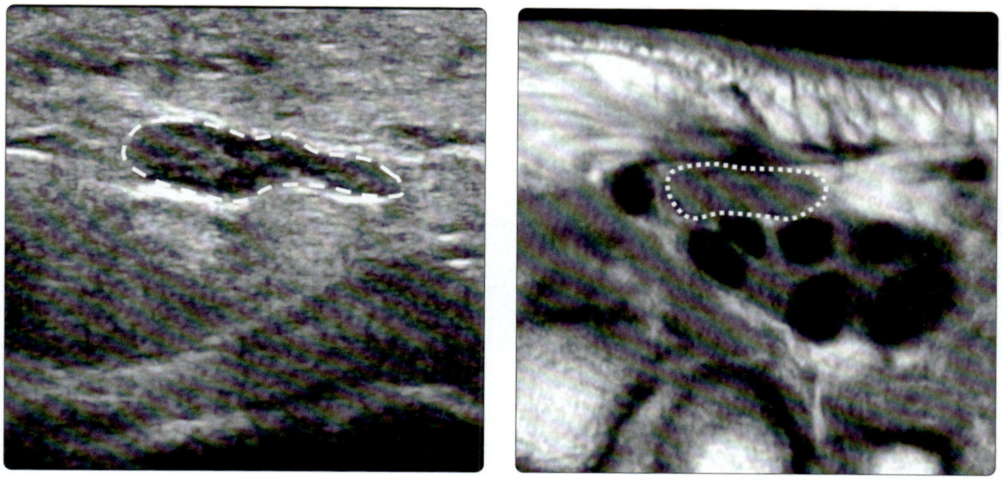

(Left) *Transverse US of the same patient shows severe swelling of the median nerve just distal to the tunnel outlet (broken line). The median nerve has a hypoechoic echo pattern with loss of the normal fascicular pattern.* **(Right)** *Axial proton density MR shows the median nerve (dotted line) of the same patient at a corresponding level just distal to the tunnel outlet. The median nerve is severely swollen.*

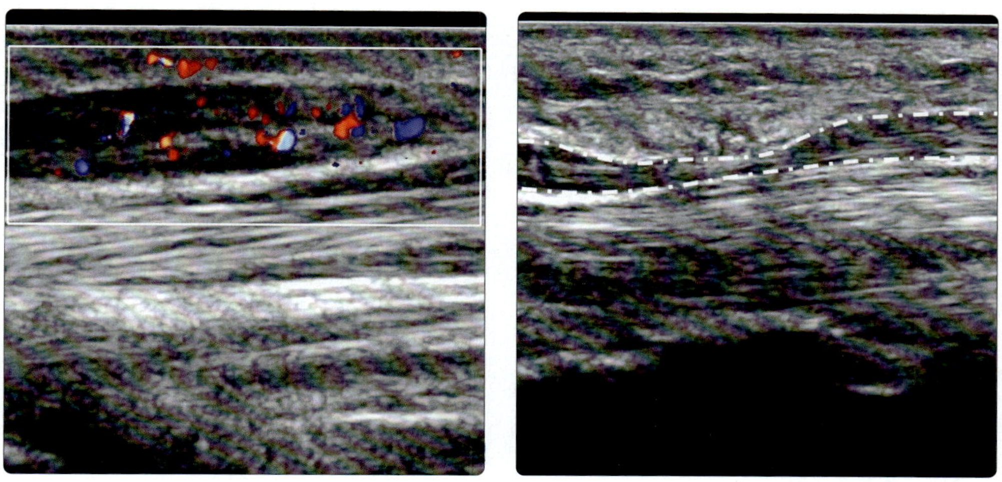

(Left) *Longitudinal color Doppler US of the same patient shows severe intraneural hyperemia proximal to the carpal tunnel. This is likely to be reactive hyperaemia secondary to compression and ischemia more distally.* **(Right)** *Longitudinal US of the same patient shows severe compression of the median nerve (broken lines) within the carpal tunnel. The nerve shows an hourglass configuration to the nerve, swollen both proximal & distal to the compressed site.*

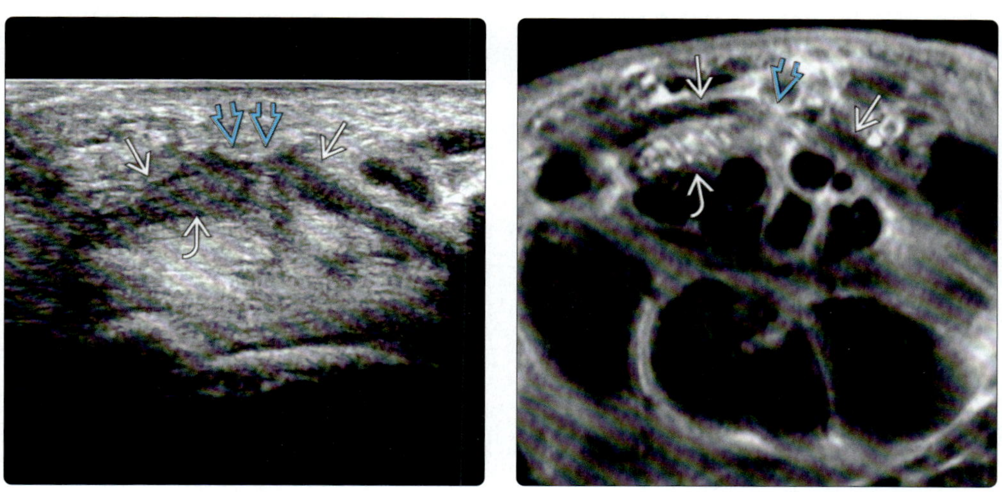

(Left) *Transverse US of the tunnel inlet 3 months postcarpal tunnel release is shown. Retracted ends of the cut transverse carpal ligament ➡ & central defect ⇨ can be appreciated. The median nerve ⮫ lies just under the defect, inseparable from the TCL.* **(Right)** *Axial proton density MR of the tunnel inlet of the same patient shows the retracted ends of the transected TCL ➡ & the resulting central defect ⇨. The median nerve ⮫ lies just under the TCL.*

(Left) *Transverse US in a patient with CTS shows echogenic gouty tophi with posterior acoustic shadow within the carpal tunnel causing compression of the median nerve. The flexor tendons cannot be clearly identified due to gouty crystal deposition.* **(Right)** *Longitudinal US of the same patient shows the median nerve being compressed within the carpal tunnel. The echogenic gouty tophi with posterior acoustic shadow are located deep to the median nerve.*

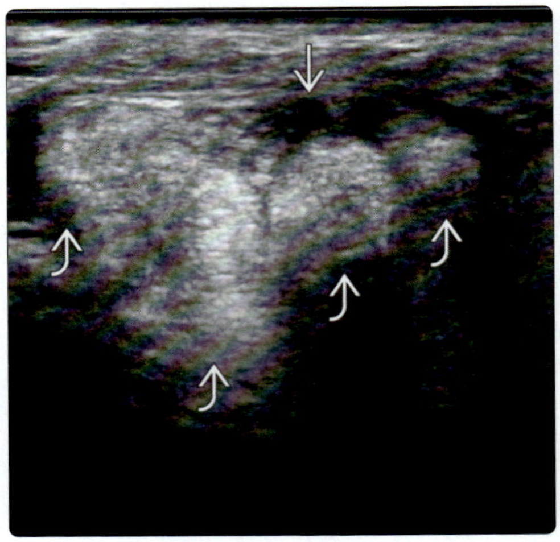

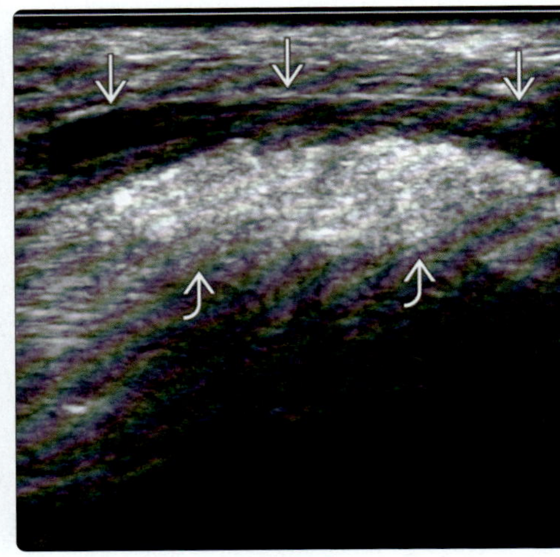

(Left) *Transverse US of the carpal tunnel inlet shows ulnar & radial components of the bifid median nerve associated with persistent median artery coursing in between these components. In this situation, the CSA of both components should be summated.* **(Right)** *Transverse power Doppler US of the carpal tunnel inlet in the same patient shows normal arterial signal of persistent median artery. Thrombosis of this median artery is 1 cause of secondary CTS.*

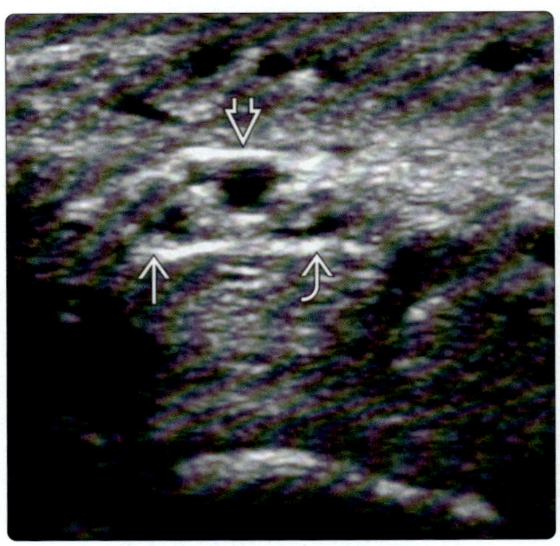

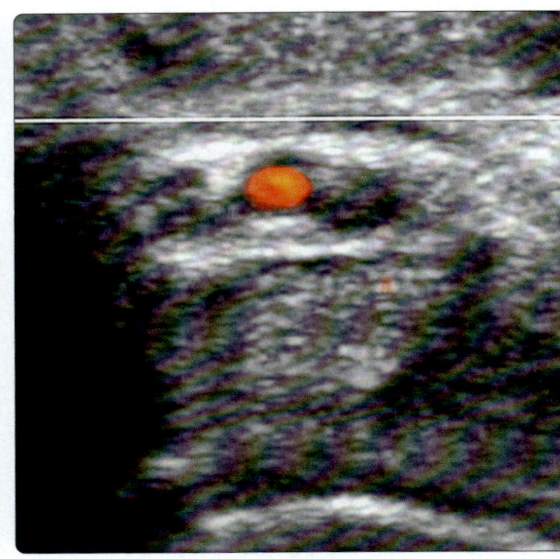

(Left) *Longitudinal US of the median nerve in the same patient shows that the mass is ovoid-shaped with a rather homogeneous echogenicity & is overall consistent with an intraneural lipoma.* **(Right)** *Transverse PD MR at the outlet of the midcarpal tunnel area in the same patient confirms that the mass is of fat signal intensity and arising from the ulnar aspect of the median nerve. The overall appearances are consistent with a lipoma of the median nerve.*

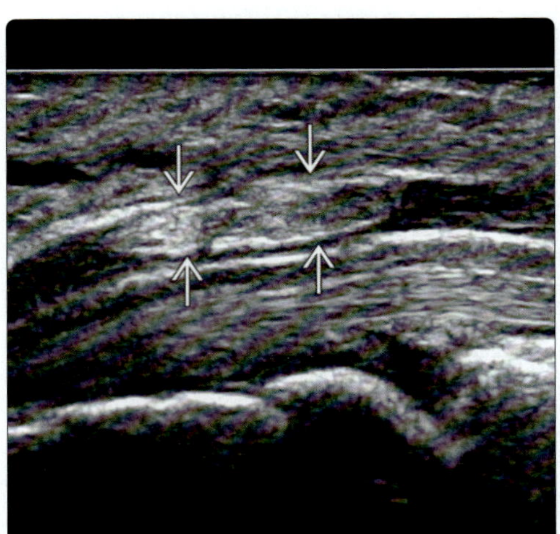

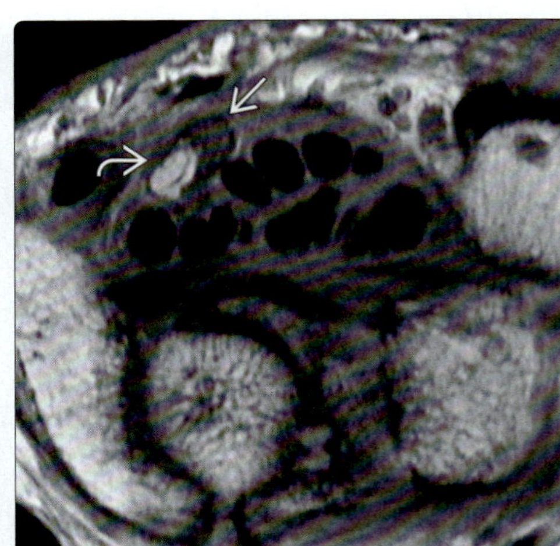

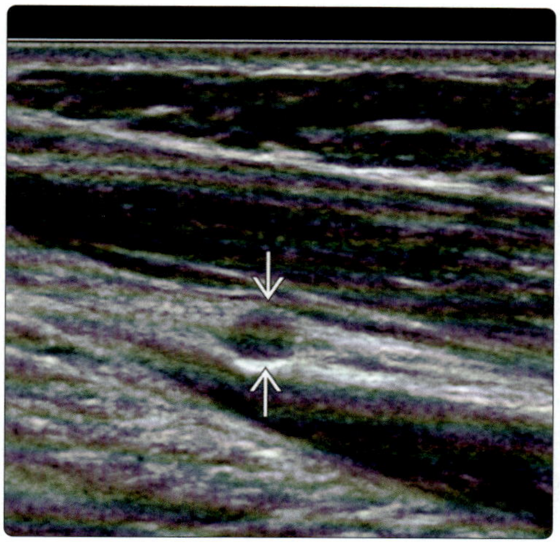

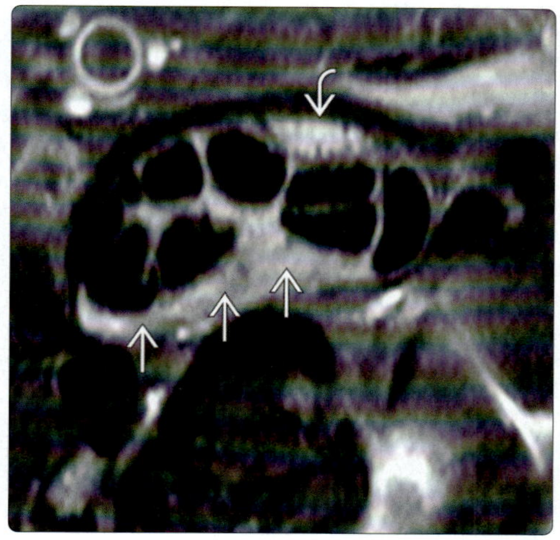

(Left) *Longitudinal US of the median nerve at the distal forearm shows a well-defined hypoechoic, round intraneural lesion, consistent with a small nerve sheath tumor. MR (not shown) revealed an enhancing intraneural lesion compatible with a nerve sheath tumor.* **(Right)** *Transverse PDWI FS MR arthrogram in a patient with CTS symptoms shows median nerve flattening at the tunnel outlet with synovial proliferation of the ulnar flexor synovial sheath, consistent with flexor tenosynovitis.*

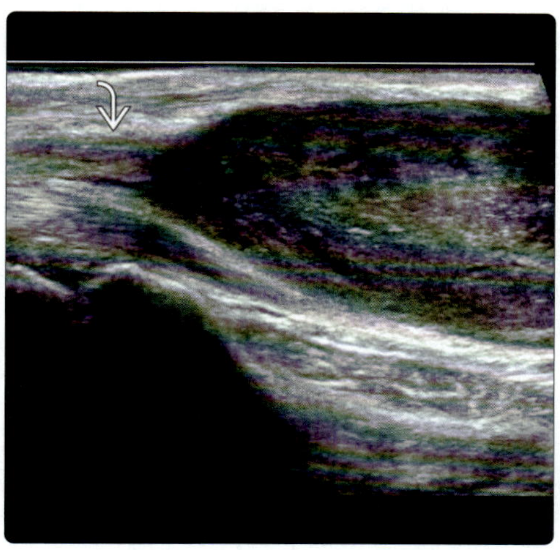

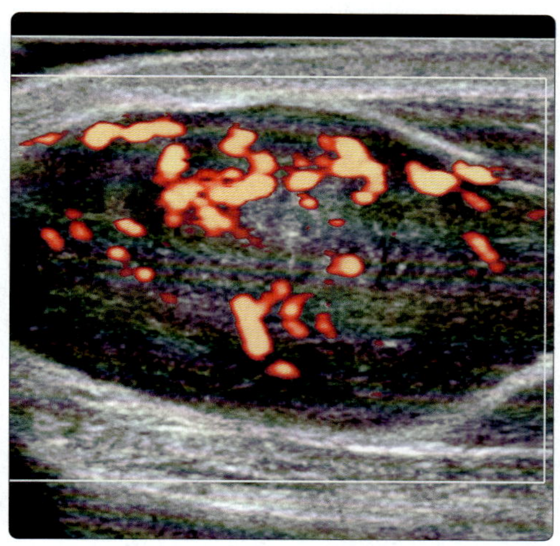

(Left) *Longitudinal US in a patient with CTS symptoms shows a large, heterogeneous, oval-shaped tumor continuous with the median nerve just proximal to the carpal tunnel. The nerve immediately adjacent to the tumor is mildly thickened. Features are compatible with a large median nerve sheath tumor.* **(Right)** *Longitudinal power Doppler US shows moderate tumoral hyperemia with an irregular vascular pattern. This is consistent with a nerve sheath tumor.*

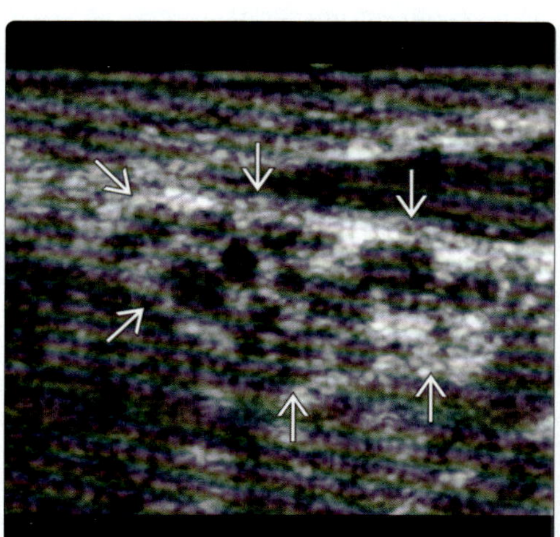

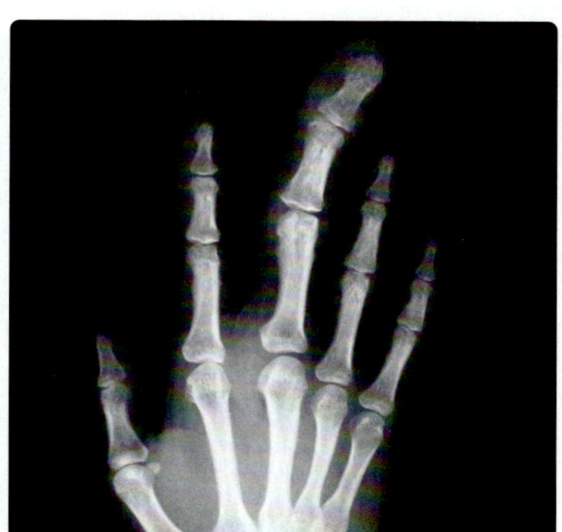

(Left) *Transverse US in a patient with enlarged middle finger & recent onset of CTS symptoms during pregnancy shows a very thickened median nerve at the tunnel inlet with thickened hypoechoic intraneural fascicles. These appearances are consistent with hamartomatous fibrolipoma of the median nerve.* **(Right)** *Frontal radiograph in the same patient shows middle finger macrodactyly, a known association of median nerve fibrolipoma & a more common presenting symptom than CTS.*

Cubital Tunnel Syndrome

KEY FACTS

TERMINOLOGY

- Cubital tunnel syndrome (CUTS)
  - Due to compression of ulnar nerve at cubital tunnel
- Symptoms can also be due to ulnar nerve traction and friction secondary to instability

IMAGING

- Best diagnostic clue: Swollen ulnar nerve within cubital tunnel
- 3 elements of CUTS should be routinely evaluated
  - Space-occupying lesion in cubital tunnel
    - Ganglia, synovitis, osteophytosis, muscle, or tumor
  - Compression of ulnar nerve with swelling
    - Ulnar nerve cross-sectional area measured proximal to tunnel and within tunnel
    - The more swollen the ulnar nerve, the greater the likelihood of CUTS
    - If cross-sectional area of ulnar nerve at cubital tunnel ≥ 10 mm², sensitivity 76%, specificity 61% for CUTS
    - Alternatively, ulnar nerve cross-sectional area in tunnel should not be > 1.5x cross-sectional area in mid- to distal arm region
  - Ulnar nerve instability
    - As elbow flexes, olecranon and roof of cubital tunnel move distally, allowing ulnar nerve to sublux
    - Subluxation or luxation: Present in 25% of CUTS patients; 20% of normal subjects

PATHOLOGY

- 80% of patients with CUTS will have positive findings (either encroachment onto cubital tunnel, nerve swelling, or luxation/subluxation)
- 40% of patients with CUTS will have same positive findings on asymptomatic contralateral side

DIAGNOSTIC CHECKLIST

- More swollen ulnar nerve = more likely possibility of CUTS
- In CUTS due to ulnar nerve instability, cubital tunnel encroachment and neural swelling may be less pronounced

(Left) *Transverse US shows a normal-caliber ulnar nerve (9 mm²) within the cubital tunnel between the medial epicondyle and the olecranon. There is anconeus epitrochlearis accessory muscle within the tunnel.* **(Right)** *Transverse US shows a mildly enlarged (12 mm²) ulnar nerve located within the cubital tunnel just ulnar to the medial epicondyle. An ulnar nerve cross-sectional area of ≥ 10 mm² is an indicator of cubital tunnel syndrome.*

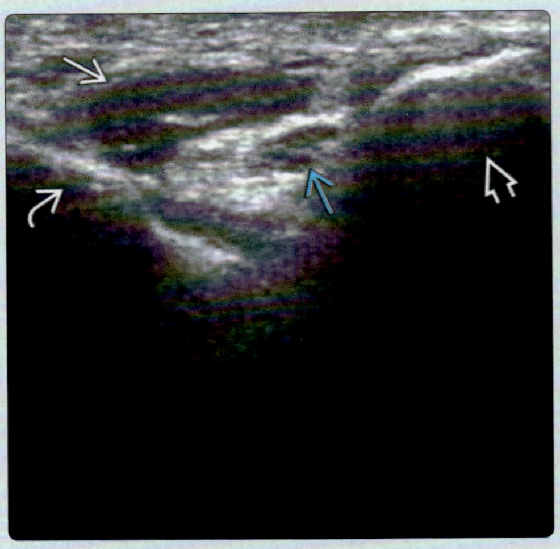

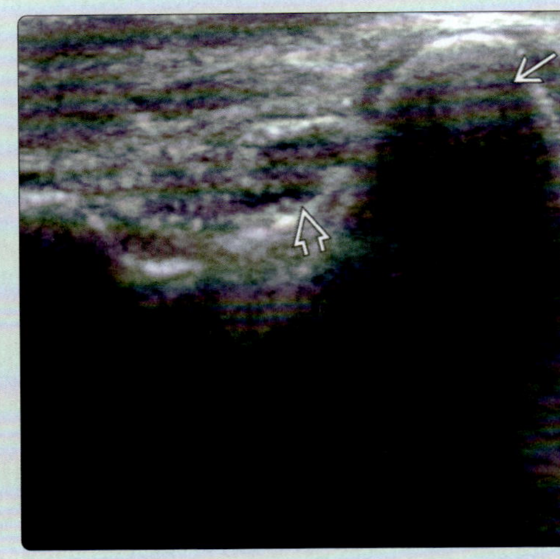

(Left) *Transverse US shows a severely enlarged (22 mm²) ulnar nerve located within the cubital tunnel alongside the medial epicondyle. There is large anconeus epitrochlearis accessory muscle within the tunnel, which may be contributing to ulnar nerve compression.* **(Right)** *Clinical photograph of a patient with chronic cubital tunnel syndrome shows wasting of the small hand muscles and clawing of the ring and little fingers.*

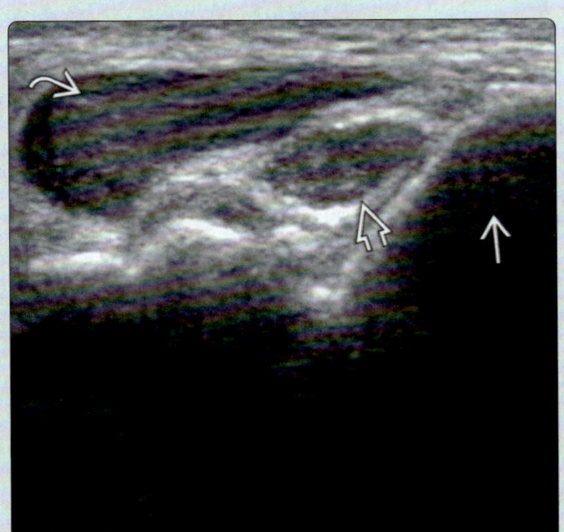

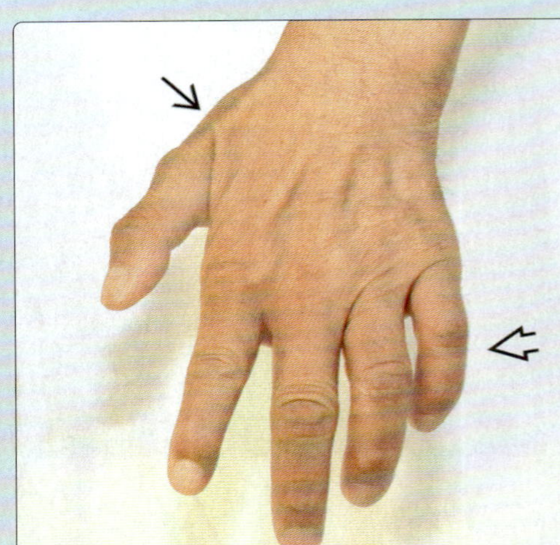

TERMINOLOGY

Abbreviations

- Cubital tunnel syndrome (CUTS)

Synonyms

- Ulnar neuropathy at elbow
- Cubital tunnel
    - "Ulnar neural groove," "retrocondylar groove," or "epicondyle-olecranon groove"

Definitions

- Symptoms secondary to compression or irritation of ulnar nerve at cubital tunnel
    - 2nd most common peripheral entrapment neuropathy after carpal tunnel syndrome

IMAGING

General Features

- Best diagnostic clue
    - Swollen ulnar nerve within cubital tunnel
- Location
    - Between medial epicondyle of humerus and olecranon
- Size
    - Normal nerve cross-sectional area is 6-12 mm²
- Morphology
    - Fusiform appearance on longitudinal scanning due to localized swelling

Radiographic Findings

- Radiography
    - Elbow osteoarthritis with osteophytic impingement into cubital tunnel, especially with hypertrophic osteoarthropathy
    - Elbow deformity from old trauma, chronic inflammatory arthropathy

MR Findings

- T2WI
    - Swelling of ulnar nerve with hyperintensity
- DWI
    - May show nerve compression before electrophysiologic abnormality

Ultrasonographic Findings

- Compression, swelling, and subluxation of ulnar nerve within cubital tunnel
    - Unlike carpal tunnel syndrome, there are 2 elements to CUTS
        - Compressive neuropathy when cubital tunnel encroachment and neural swelling are dominant features
        - Noncompressive neuropathy (i.e., ulnar nerve instability) when recurrent subluxation/luxation results in nerve stretching and friction neuritis
            - In noncompressive neuropathy, cubital tunnel encroachment and neural swelling may be less pronounced
        - Both compressive and noncompressive elements commonly coexist
- Space-occupying masses in cubital tunnel
    - Ganglia, osteophytes, hyperostosis, synovitis, muscle, or tumor
    - Anconeus epitrochlearis muscle present in 25% of CUTS patients; 15% of normal subjects
        - Extends from medial epicondyle → medial border of olecranon superficial to and along path of Osborne ligament
            - When present, contributes to roof of cubital tunnel
        - Osborne ligament considered vestigial remnant of anconeus epitrochlearis
            - Osborne ligament cannot be reliably seen on ultrasound
    - Large medial head triceps (> 50% ulnar groove) present in 15% of CUTS patients, 10% of normal subjects
        - Attached to ulnar side olecranon process
- Enlarged ulnar nerve in cubital tunnel
    - The more swollen the ulnar nerve, the greater the likelihood of CUTS
    - Mean ulnar nerve cross-sectional area in healthy subjects is 9.0 mm² at cubital tunnel
    - For cut-off of 10-mm² cross-sectional area of ulnar nerve at cubital tunnel, sensitivity 76%, specificity 61% for CUTS
    - Swelling ratio criterion
        - Convenient but may not be more accurate than maximum cross-sectional area
            - Compare cross-sectional area of nerve within tunnel with that in middistal arm
            - Cross-sectional area within cubital tunnel should not be > 1.5x than that in mid- to distal arm
- Instability of ulnar nerve during elbow flexion
    - As elbow flexes, olecranon and roof of cubital tunnel move distally, enabling ulnar nerve to sublux
    - Normally with flexion, ulnar nerve shifts superficially in cubital tunnel and flattens slightly
    - Subluxation: Present in 25% of CUTS patients; 10% of normal subjects
    - Luxation: Present in 25% of CUTS patients; 10% of normal subjects

Imaging Recommendations

- Best imaging tool
    - Ultrasound: Quick, comfortable, cost effective, accurate, and dynamic
- Protocol advice
    - Usually evaluate cubital tunnel with elbow extended
    - Trace nerve in transverse plane, from distal arm to proximal forearm
    - Check contents of cubital tunnel
    - Measure nerve caliber just proximal to tunnel and within tunnel
    - Check for ulnar nerve instability
    - Check additional site of possible ulnar nerve compression around elbow
        - Proximally: Arcade of Struthers, medial intermuscular septum
        - Distally: Deep to flexor pronator aponeurosis (humeroulnar arcade) and between ulnar and humeral heads of flexor carpi ulnaris (FCU)
    - General tips

- Try to keep transducer aligned at right angles and not oblique to nerve
- Nevertheless, slight angulation of transducer can utilize anisotropy to improve nerve delineation
- Best to make measurements immediately after freezing image
- Measure cross-sectional area of nerve by trace rather than with ellipsoid calipers
- Use minimal transducer pressure when testing for ulnar nerve instability
- 80% of patients with CUTS will have positive findings (either encroachment onto cubital tunnel, nerve swelling, or luxation/subluxation)
- 40% of patients with CUTS will have same positive findings on asymptomatic contralateral side

DIFFERENTIAL DIAGNOSIS

Common Flexor Tendon Origin Tendinosis With Peritendinitis ± Tear

- May irritate ulnar nerve in cubital tunnel

PATHOLOGY

General Features

- Majority due to compression of ulnar nerve within cubital tunnel
  - Minority due to friction neuritis of ulnar nerve from recurrent subluxation or luxation
- Anatomy of cubital tunnel
  - Roof: Proximally, Osborne ligament, which is thickened fascial band between medial epicondyle and olecranon, is probably vestigial remnant of anconeus epitrochlearis
    - Distally, arcuate ligament, which bridges humeral and ulnar heads of FCU
  - Floor: Elbow joint capsule and olecranon
- Etiology
  - Majority have specific structural cause and are deemed primary CUTS
  - CUTS increased in
    - Cubital tunnel encroachment onto by osteophytes, ganglion, synovial proliferation
    - Recurrent luxation or subluxation of ulnar nerve
    - Large medial triceps muscle (i.e., muscle occupying > 50% ulnar groove)
    - Presence of anconeus epitrochlearis muscle
    - Snapping triceps syndrome
    - Valgus or varus deformity of elbow
- Epidemiology
  - Incidence: ~ 1% in United States

Gross Pathologic & Surgical Features

- Compressed nerve in cubital tunnel

Microscopic Features

- Nerve compression or ischemia leads to fluid leakage/endoneural edema and increase in endoneural pressure
  - Leads to demyelination, inflammation, damming of axoplasmic flow, distal axonal degeneration, fibrosis, growth of new axons, remyelination, and thickening of perineurium and epineurium

CLINICAL ISSUES

Presentation

- Most common signs/symptoms
  - Paraesthesia and numbness of ulnar 1 and 1/2 digits, but frequently difficult to localize
- Other signs/symptoms
  - Pain in medial elbow region
  - Grip problems secondary to intrinsic muscle (lumbrical and interosseous) weakness and atrophy
  - Claw hand: Hyperextension of proximal phalanx with flexion of middle and distal phalanges
- Clinical profile
  - Ultrasound findings supported by combination of history, physical examination, specific provocative tests, and electrophysiologic tests
  - Provocative tests
    - Positive Tinel test, elbow flexion test, pressure provocation test, combined elbow flexion-pressure test, scratch collapse test
  - Electrodiagnostic studies test functional integrity of nerve
    - Ulnar nerve motor conduction velocity across elbow is positive if < 50 m/s
    - Can identify other disease processes, such as upper motor neuron disease or peripheral neuropathy

Demographics

- Age
  - Peak: 45 ± 12 years
- Gender
  - M:F = 1:3

Treatment

- Conservative
  - Antiinflammatory or vitamin B medication, elbow pad activity modification, splints to limit motion, nocturnal splinting, physical therapy
- Surgical
  - In situ decompression, subcutaneous/intramuscular/submuscular transposition, medial epicondylectomy

DIAGNOSTIC CHECKLIST

Consider

- CUTS in all patients with unexplained hand, wrist, or distal forearm paresthesia ± pain on ulnar side

Image Interpretation Pearls

- The more swollen the ulnar nerve, the more likely the possibility of CUTS

SELECTED REFERENCES

1. Chang KV et al: Ulnar nerve cross-sectional area for the diagnosis of cubital tunnel syndrome: a meta-analysis of ultrasonographic measurements. Arch Phys Med Rehabil. 99(4):743-757, 2018
2. Schertz M et al: High-resolution ultrasound in etiological evaluation of ulnar neuropathy at the elbow. Eur J Radiol. 95:111-117, 2017

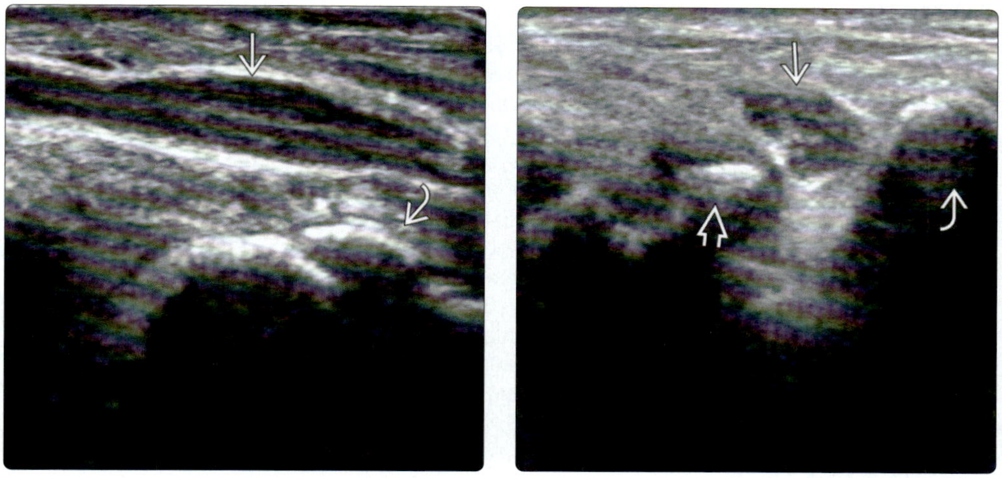

(Left) *Longitudinal US shows a focally enlarged ulnar nerve within and just proximal to the cubital tunnel. Note the medial epicondyle. This exaggerated fusiform swelling of the ulnar nerve is seen in cubital tunnel syndrome.* **(Right)** *Transverse US shows moderate enlargement (18 mm²) of the ulnar nerve alongside the medial epicondyle. There are large osteophytes indenting the base of the cubital tunnel and contributing to the development of cubital tunnel syndrome.*

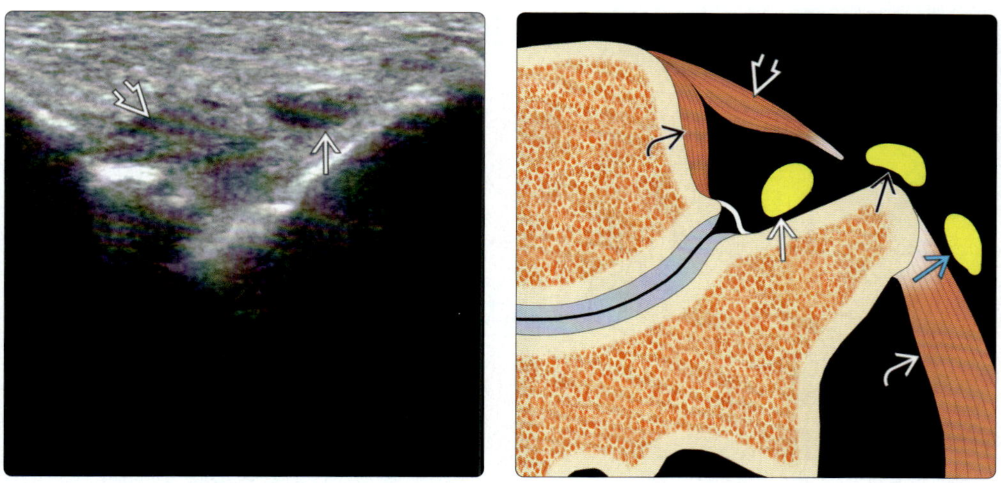

(Left) *Transverse US shows mild enlargement (11 mm²) of the ulnar nerve. There is moderate-severity synovitis indenting the base of the cubital tunnel. This can contribute to the development of cubital tunnel syndrome.* **(Right)** *Schematic diagram shows a normal ulnar nerve position, a subluxed ulnar nerve position, and a dislocated ulnar nerve position. The medial belly of triceps muscle, anconeus epitrochlearis muscle, and common flexor tendon origin are also shown.*

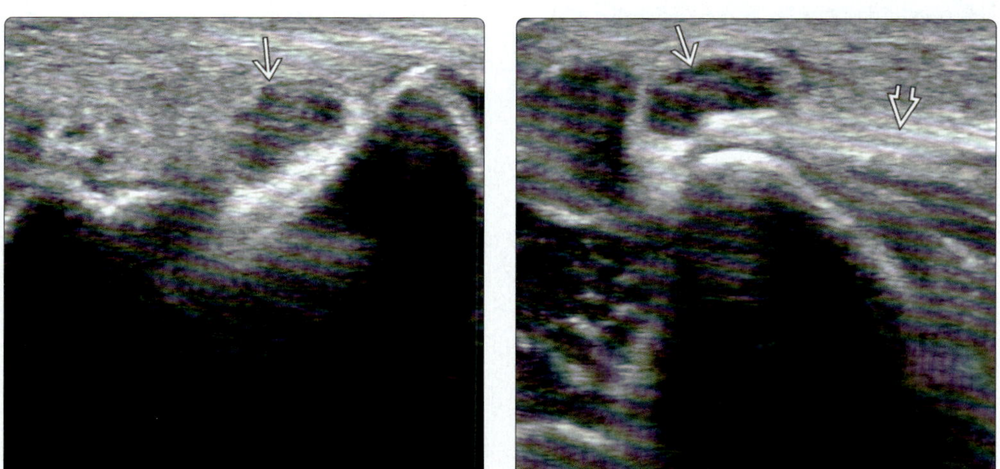

(Left) *Transverse US shows moderate enlargement (16 mm²) of the ulnar nerve within the cubital tunnel in the extended elbow position. These features are compatible with cubital tunnel syndrome.* **(Right)** *Transverse US of the same patient with full elbow flexion shows subluxation of the ulnar nerve, which is now flattened. Note the common flexor tendon origin. Recurrent subluxation or dislocation can lead to friction neuritis of the ulnar nerve and symptoms of cubital tunnel syndrome.*

Tarsal Tunnel Syndrome

KEY FACTS

TERMINOLOGY

- Entrapment neuropathy of tibial nerve and its branches (medial, lateral plantar nerves) on medial aspect of ankle
  - Many causes of tarsal tunnel syndrome (TTS)
- 2 forms of TTS
  - Upper: Compression within tarsal tunnel (classic)
  - Lower: Compression deep and distal to abductor hallucis muscle (variant)
- TTS is much less common than carpal or cubital tunnel syndrome

IMAGING

- Demonstration of structural abnormality within tarsal tunnel is currently of greater diagnostic value than nerve caliber measurements
  - Value and diagnostic capacity of nerve caliber measurements not established
- Mass within tarsal tunnel (ganglion, lipoma, tendon disorder with peritendinitis, varicosities)
- Extrinsic mass encroaching on tarsal tunnel (synovial proliferation, bony overgrowth)
- Intrinsic nerve abnormality (fibrosis, tumor, diffuse enlargement)
- Altered tibial nerve caliber (enlarged proximally, narrowed at site of compression)
- In-tunnel-to-pretunnel cross-sectional area (CSA) ratio is 0.9 ± 0.1 in normal subjects and 1.3 ± 0.1 in TTS patients
  - In-tunnel-to-pretunnel CSA ratio > 1 has sensitivity of 74%, specificity of 100% for TTS although not validated
- Tibial nerve in tarsal tunnel 13.8 ± 4.4 mm^2 in normal subjects, 20.6 ± 8.5 mm^2 in TTS patients
- Side-to-side difference in tibial nerve CSA > 1.8 mm^2 is considered abnormal
- US Tinel sign can be performed by tapping over nerve to see if this elicits symptoms
- Consider examining tarsal tunnel in standing position
  - Weight-bearing will exaggerate heel valgus and distend varicosities

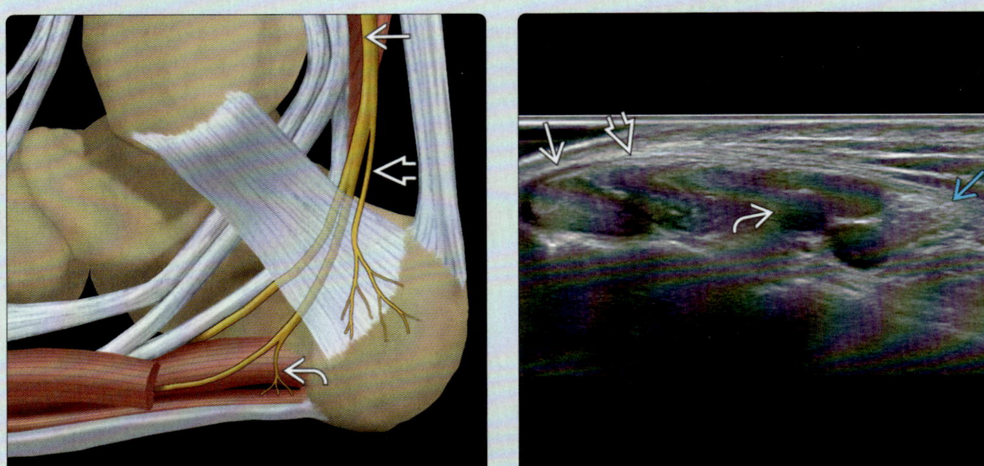

(Left) *Graphic shows the tarsal tunnel with the tibial nerve dividing into the medial and lateral plantar nerves under the flexor retinaculum. The medial calcaneal nerve normally comes off the tibial nerve proximal to the retinaculum. The inferior calcaneal nerve is the 1st branch of the lateral plantar nerve.* **(Right)** *Transverse US shows the normal tarsal tunnel with the tibialis posterior and flexor digitorum longus tendons, the posterior tibial artery and accompanying veins, as well as the tibial nerve.*

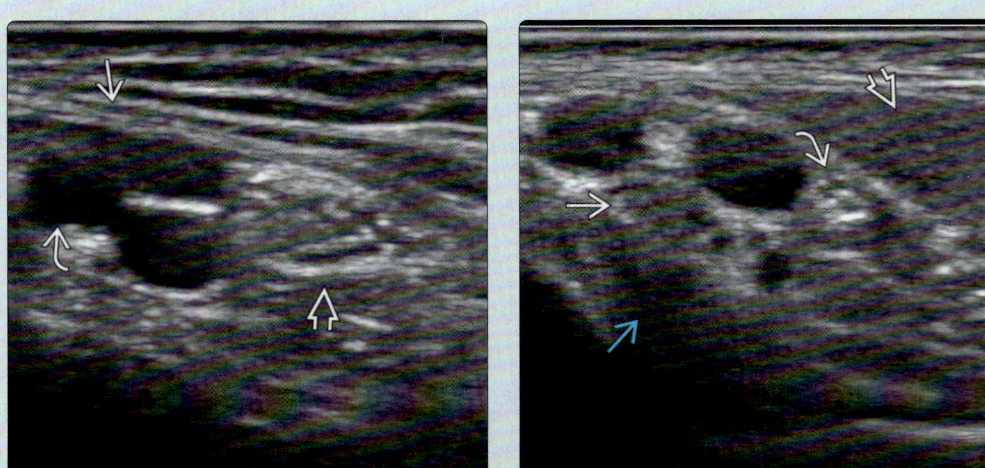

(Left) *Transverse US shows the posterior tibial artery, accompanying veins, and the tibial nerve, which are about to divide into its plantar branches deep to the flexor retinaculum.* **(Right)** *Transverse US shows the medial and lateral plantar nerves in the lower tarsal tunnel with accompanying veins between the abductor hallucis and quadratus plantae muscles.*

TERMINOLOGY

Abbreviations

- Tarsal tunnel syndrome (TTS)

Definitions

- Entrapment neuropathy of tibial nerve and its branches on medial aspect of ankle
- → pain, burning sensation, paraesthesia, and numbness of medial 3 toes, sole of foot, and heel region
- 2 forms of TTS
  - Upper (tibiotalar): Compression of tibial nerve within tarsal tunnel
    - Upper tarsal tunnel extends from retromalleolar region to proximal end of abductor hallucis muscle
  - Lower (tibiocalcaneal): Compression of plantar nerves in their separate tunnels deep and distal to abductor hallucis muscle
    - Lower tarsal tunnel lies deep to and distal to abductor hallucis muscle

IMAGING

General Features

- Location
  - Tarsal tunnel is posteroinferior to medial malleolus
    - Roof: Flexor retinaculum and abductor hallucis muscle
      - Flexor retinaculum at tarsal tunnel is less taut and more distensible than that of carpal tunnel
    - Floor: Calcaneum, talus, posterior talocalcaneal articulation, deltoid ligament, distal tibia
    - Contents "**T**om, **D**ick, **A**nd a **V**ery **N**ervous **H**arry" (**t**ibialis posterior, flexor **d**igitorum longus, posterior tibial **a**rtery, posterior tibial **v**eins, tibial **n**erve, flexor **h**allucis longus)
  - Tibial nerve passes deep to flexor retinaculum and divides into medial and lateral plantar branches
    - In 20% of subjects, tibial nerve divides into its terminal branches just before it enters tarsal tunnel
  - Medial calcaneal nerve is 1st branch of tibial nerve at ankle
    - Usually arises directly from tibial nerve proximal to tunnel or, less frequently, lateral plantar nerve
      - 1/4 subjects have 1, 1/2 have 2, and 1/4 subjects have 3-5 medial calcaneal nerves
    - Supplies skin of heel and medial side sole of foot
  - Medial plantar nerve innervates abductor hallucis, flexor digitorum brevis, flexor hallucis brevis, and 1st lumbrical muscle
    - Larger branch of tibial nerve
    - Runs from anteromedial aspect calcaneum along medial aspect of foot along line of flexor hallucis longus tendon
    - Provides sensation to medial part of sole of foot and medial 3 toes
    - Can be compressed by thickened Knot of Henry, which is flexor hallucis longus/flexor digitorum longus tendon crossover at distal end of sustentaculum tali (jogger's foot)
  - Lateral plantar nerve innervates remaining deep foot muscles, including quadratus plantae and abductor digiti minimi
    - Descends steeply between abductor hallucis and quadratus plantae before running anterolaterally on surface of quadratus plantae, then to base of little toe
    - Provides sensation to lateral part of sole of foot and lateral 2 toes
  - Inferior calcaneal nerve (Baxter nerve) is 1st branch of lateral plantar nerve
    - Arises at lower end of abductor hallucis, then runs between quadratus plantae and flexor digitorum brevis
      - Passes superficial to medial calcaneal tuberosity, then innervates digiti minimi muscle
      - Visible on US deep to flexor digitorum brevis at leading edge of calcaneum
    - Mixed motor and sensory nerve
      - Motor innervation to abductor digiti minimi and part of quadratus plantae
      - Sensory to calcaneal periosteum and long plantar ligament
      - Baxter nerve impingement or neuritis can give symptoms indistinguishable from plantar fasciitis
    - Compression leads to abductor digiti minimi atrophy
      - However, moderate-severity adductor digiti minimi atrophy is seen in 10% of asymptomatic subjects

MR Findings

- T1WI or PD
  - Axial images useful for nerve anatomy, branching variations, assessment of perineural fibrosis, and space-occupying lesions
    - Atrophic fatty atrophy of foot muscles graded as mild (< 25%), moderate (26-50%), severe (51-75%), or very severe (> 76%)
  - Coronal images most useful for assessing osseous or synovial encroachment onto tarsal tunnel
- T2WI FS
  - Best for detection of nerve swelling, hyperintensity, and muscle edema

Ultrasonographic Findings

- Mass lesion within tarsal tunnel
  - Ganglion, lipoma, synovial proliferation, tendon disorder with peritendinitis, varicosities, accessory muscle, or hypertrophied muscle
- Change in tibial nerve caliber (enlarged proximally, narrowed at site of compression)
  - US diagnostic criteria for TTS not firmly established

Imaging Recommendations

- Best imaging tool
  - US because tibial nerve can be consistently seen, as well as compression or abnormality of tibial nerve
    - No established diagnostic US criteria for TTS, unlike carpal tunnel or cubital tunnel syndrome
  - MR is helpful if symptoms not fully explained by US examination
    - Particularly helpful in assessing distal part of lower tarsal tunnel, where thick plantar skin can limit US examination
  - Nerve conduction studies show absent sensory potentials and prolonged motor latencies

- Not specific for TTS as may be seen in polyneuropathy or in elderly patients
- Overall, nerve conduction studies in TTS are of limited value

- Protocol advice
  - Patient lies in lateral decubitus position with medial ankle facing upward
  - Scan transversely over tarsal tunnel (malleolar calcaneal axis) keeping tibial nerve in cross section
  - Diagnosis of TTS depends more on demonstration of structural abnormality than nerve caliber change
  - Check especially for
    - Mass present within tarsal tunnel (lipoma, ganglion, tendon abnormality, varicosity)
    - Mass encroaching onto tarsal tunnel (synovial proliferation, bony overgrowth)
    - Intrinsic abnormality of tibial nerve (fibrosis, tumor, diffuse enlargement)
  - Check for change in tibial nerve caliber between pre- and in-tunnel positions
    - In-tunnel-to-pretunnel cross-sectional area (CSA) ratio is 0.9 ± 0.1 in normal subjects and 1.3 ± 0.1 in TTS patients
      - In-tunnel-to-pretunnel CSA ratio > 1 has sensitivity of 74%, specificity of 100% for TTS, although not validated
  - Tibial nerve in tarsal tunnel 13.8 ± 4.4 mm^2 in normal subjects, 20.6 ± 8.5 mm^2 in TTS patients
  - Side-to-side difference in tibial nerve CSA > 1.8 mm^2 is considered abnormal
    - Side-to-side comparison in patients with unilateral TTS may be more accurate than absolute measurements
  - US Tinel sign can be performed by tapping over nerve to see if this elicits symptoms
  - Consider examining tarsal tunnel in standing position
    - Weight-bearing will exaggerate heel valgus in pes planus and exaggerate venous distension
      - May see sustentaculum tali compressing plantar nerves
      - Dilatation of plantar veins > 5 mm indicates varicosity
      - Clear advantage of US over MR, as MR is usually performed only in non-weight-bearing position
  - Check for atrophy of hindfoot muscles
    - Compare with contralateral side
  - US-guided injection of local anesthetic ± steroid into tarsal tunnel may be of diagnostic and possibly therapeutic benefit

DIFFERENTIAL DIAGNOSIS

Posterior Tibial Disorders

- Thickening, hypoechogenicity, with loss of normal fibrillar pattern of posterior tibialis tendon
  - Many have split tendon tears ± variable degree of peritendinitis
    - May lead to TTS

Tarsal Coalition

- Hypertrophic bone protruding from midtalocalcaneal articulation is strong US sign of subtalar coalition

Plantar Fasciitis

- Thickening (> 4.3 mm) of plantar fascia at leading edge of medial calcaneal tuberosity
  - ± perifascial edema or plantar calcaneal spur
- Identical symptoms to inferior calcaneal neuropathy (Baxter neuritis)

CLINICAL ISSUES

Presentation

- Most common signs/symptoms
  - Medial malleolar pain radiating to medial aspect of foot
    - If symptoms confined to either plantar nerve, distal tarsal tunnel compression may be implicated
- Other signs/symptoms
  - Paraesthesia or burning sensation on plantar aspect of foot ± hyperesthesia on medial aspect of sole
    - Sparing of heel region, which is innervated by medial calcaneal nerve
  - Weakness ± atrophy of foot muscles
    - Weakness is usually asymptomatic, as function is not affected
  - Tenderness behind medial malleolus ± positive Tinel sign
  - Induction of symptoms by passive dorsiflexion and eversion of ankle for 10 seconds

Demographics

- Age
  - Typically young or middle-aged patient
- Sex
  - M:F = 2:1
- Epidemiology
  - Many causes of TTS
  - Mass within tarsal tunnel (ganglion, lipoma, nerve sheath tumor, synovial proliferation, tendon enlargement, venous engorgement)
  - Deformity with heel valgus
  - Osseous impingement from ankle/hindfoot arthropathy, talocalcaneal coalition, calcaneal fracture
  - Accessory muscles
    - Flexor digitorum accessorius muscle (medial aspect of distal tibia to quadratus plantae)
    - Peroneocalcaneus internus muscle (distal fibula to medial aspect of calcaneum)
  - Posttraumatic perineural fibrosis
  - Intrinsic nerve abnormality (chronic inflammatory demyelinating polyneuropathy, nerve sheath tumor)
  - Thickened flexor retinaculum (normal thickness, 0.64 ± 0.03 mm)
  - No cause found in 20% of patients with TTS

Treatment

- Conservative treatment
- Surgical division (open or endoscopic) of flexor retinaculum ± abductor hallucis muscle fasciotomy
  - Leads to symptomatic improvement in 50% of patients

SELECTED REFERENCES

1. De Maeseneer M et al: Normal anatomy and compression areas of nerves of the foot and ankle: US and MR imaging with anatomic correlation. Radiographics. 35(5):1469-82, 2015

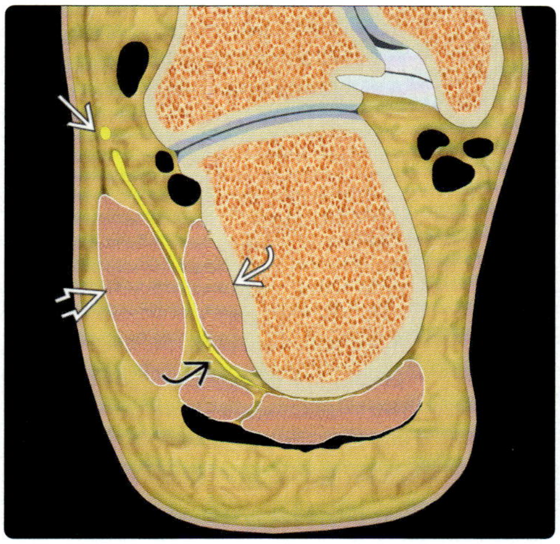

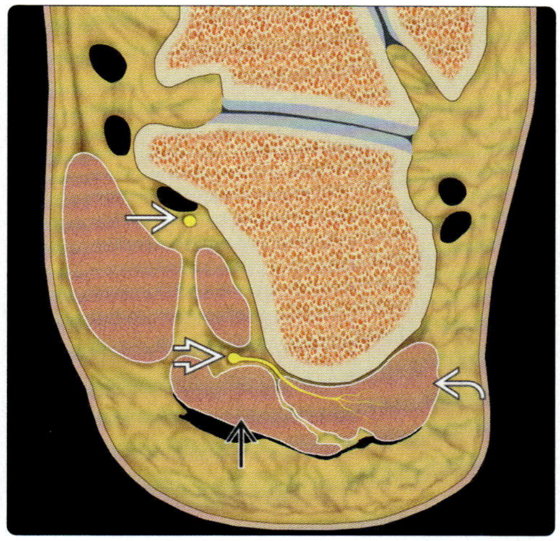

(Left) *Graphic shows the medial and lateral plantar nerves in the proximal portion of the lower tarsal tunnel. Note how the lateral plantar nerve dips deeply between the abductor hallucis and quadratus plantae muscles.* **(Right)** *Graphic of the distal portion of the lower tarsal tunnel shows the medial and lateral plantar nerves. Note how the lateral plantar nerve extends deep to the flexor digitorum brevis muscle before supplying the abductor digiti minimi muscle.*

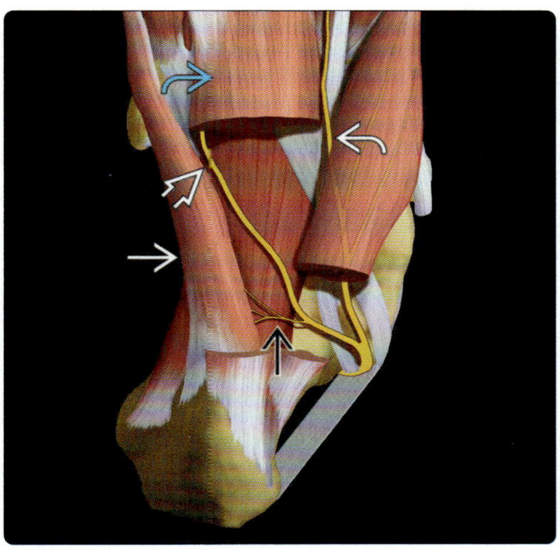

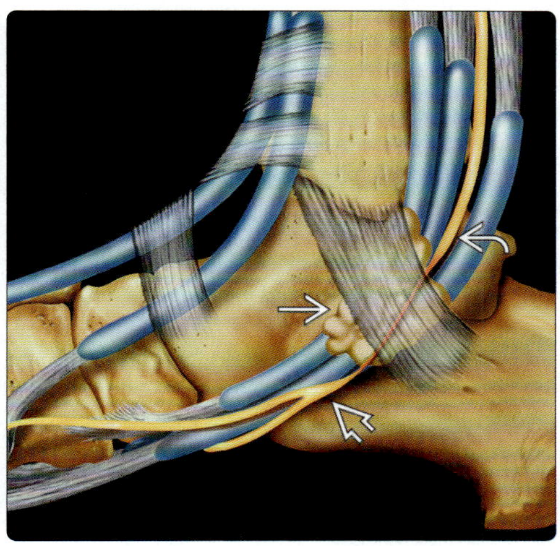

(Left) *Graphic of hindfoot undersurface shows inferior calcaneal nerve passing laterally between the calcaneum and flexor digitorum brevis muscle to supply abductor digiti minimi. The medial and lateral plantar nerves can be seen on US on both sides of the flexor digitorum brevis muscle.* **(Right)** *Graphic shows a ganglion compressing the tibial nerve within the tarsal tunnel. In this schematic, the tibial nerve divides distal to the tarsal tunnel, which is an uncommon anatomic variant.*

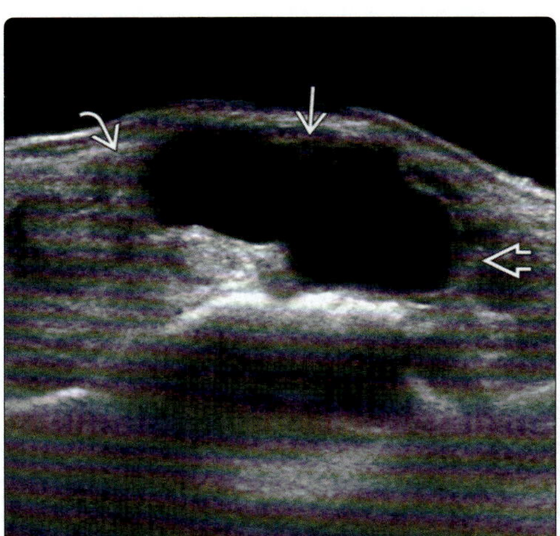

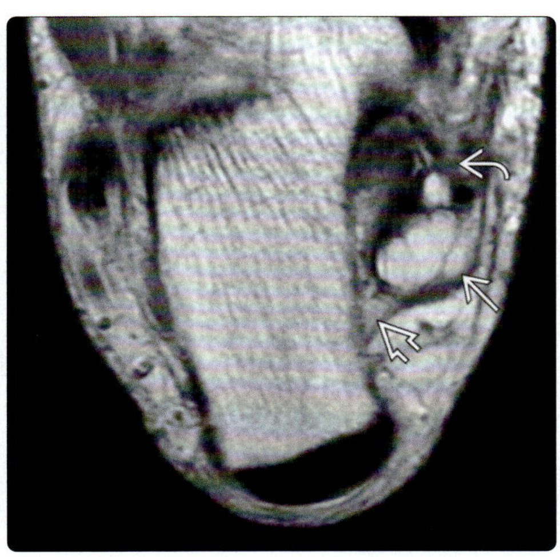

(Left) *Transverse US in a patient with tarsal tunnel syndrome shows a medium-sized ganglion located within the tarsal tunnel between the flexor digitorum longus tendon and posterior tibial neurovascular bundle. The tibial nerve is compressed.* **(Right)** *Axial PDWI MR in a patient with tarsal tunnel syndrome shows a medium-sized ganglion cyst located between and compressing the medial and lateral plantar nerves within the tarsal tunnel.*

(Left) *Transverse US in a 48-year-old man with medial ankle pain shows a medium-sized ganglion ➡ between the flexor digitorum longus ➡ and the tibial neurovascular bundle. The ganglion, which arises from the ankle, mildly compresses the tibial nerve ➡.* **(Right)** *Transverse US in a patient with tarsal tunnel syndrome shows marked synovial proliferation ➡ from the posterior talocalcaneal articulation and ankle joint bulging into the tarsal tunnel with compression of the tibial nerve ➡. The calcaneum ➡ is normal.*

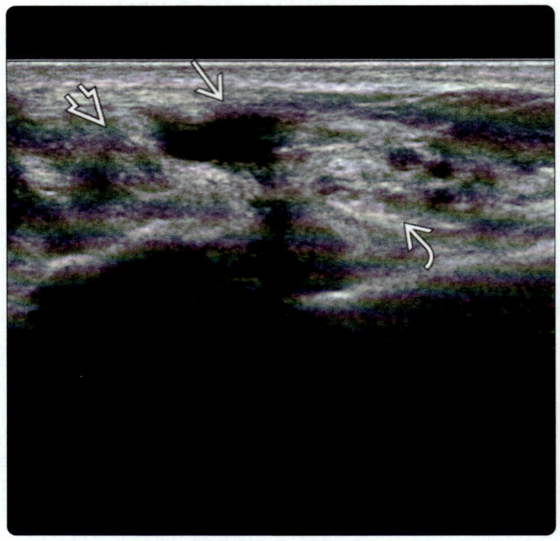

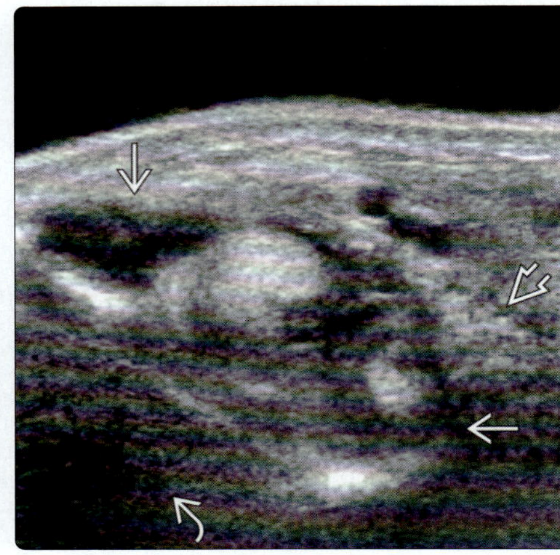

(Left) *Transverse US shows a large nerve sheath tumor ➡ of the tibial nerve within the tarsal tunnel. The tumor is arising eccentric to the tibial nerve ➡. Mild tumoral hyperemia was present on color Doppler imaging.* **(Right)** *Transverse US in a patient with tarsal tunnel syndrome shows severe posterior tibialis tendinosis ➡ with moderate peritendinitis and thickening of the flexor retinaculum ➡. This swelling and edema can lead to tibial nerve compression within the upper tarsal tunnel.*

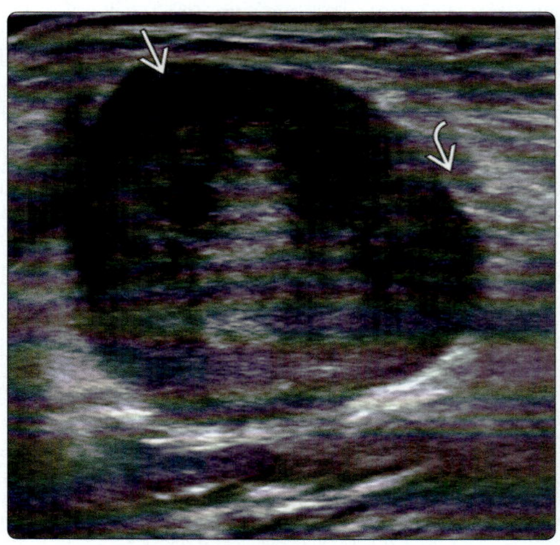

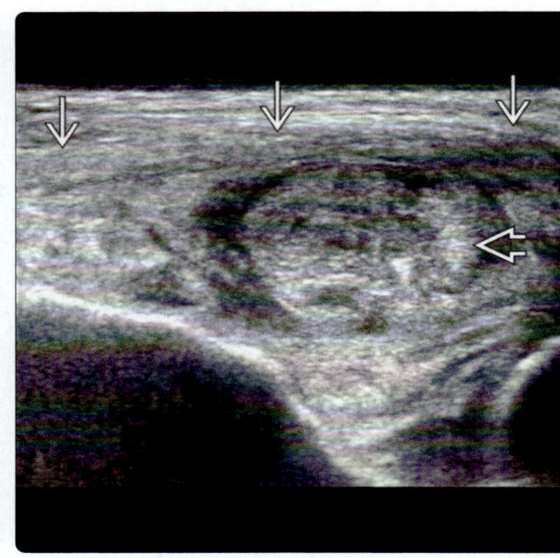

(Left) *Coronal CT reconstruction shows hyperostosis ➡ of the narrowed middle talocalcaneal facet in a patient with fibrous-type talocalcaneal coalition. This hyperostosis can impinge on the tarsal tunnel, giving rise to tarsal tunnel syndrome.* **(Right)** *Axial T1WI MR shows an accessory flexor digitorum muscle ➡, which can extend into the tarsal tunnel and contribute to compression of the tibial nerve. There is a similar, although smaller, accessory muscle ➡ on the contralateral side.*

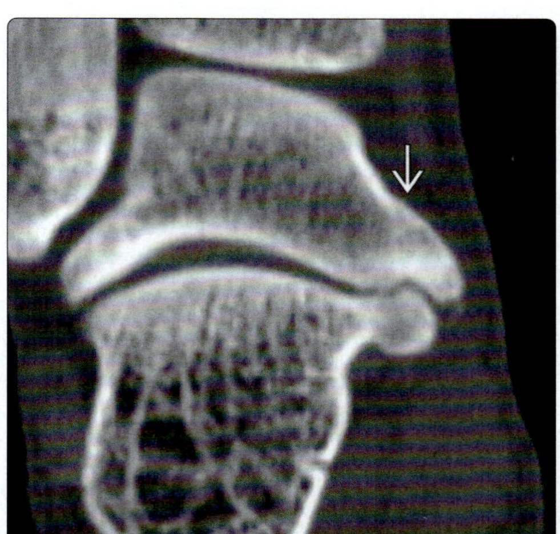

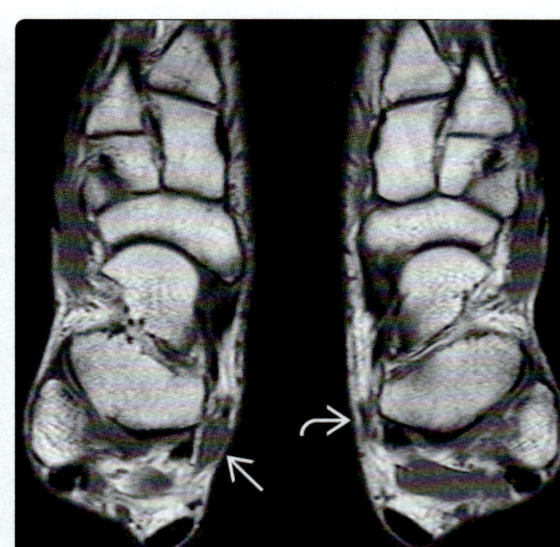

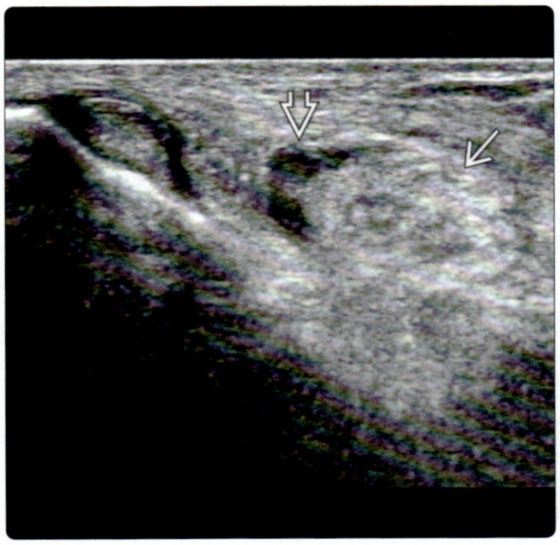

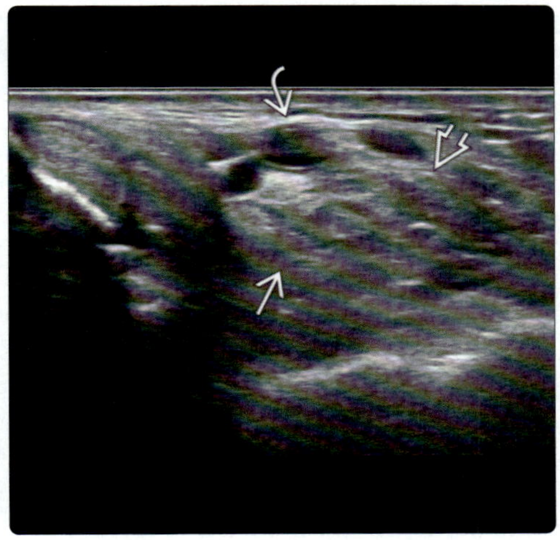

(Left) *Transverse US in a patient with symptoms of tarsal tunnel syndrome and clinically suspected chronic inflammatory demyelinating neuropathy shows an enlarged tibial nerve within the tarsal tunnel. The nerve cross-sectional area was 34 mm².* *The tibial artery and veins are displaced.* **(Right)** *Transverse US in a 65-year-old man with medial ankle pain shows a medium-sized lipoma within the tibial tunnel deep to the posterior tibial artery and veins. The tibial nerve is mildly compressed.*

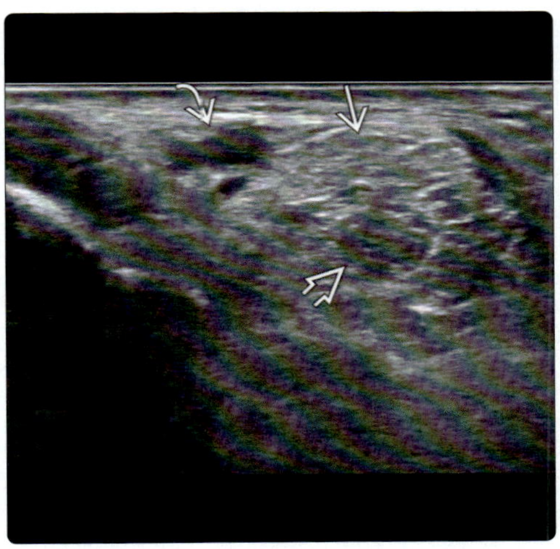

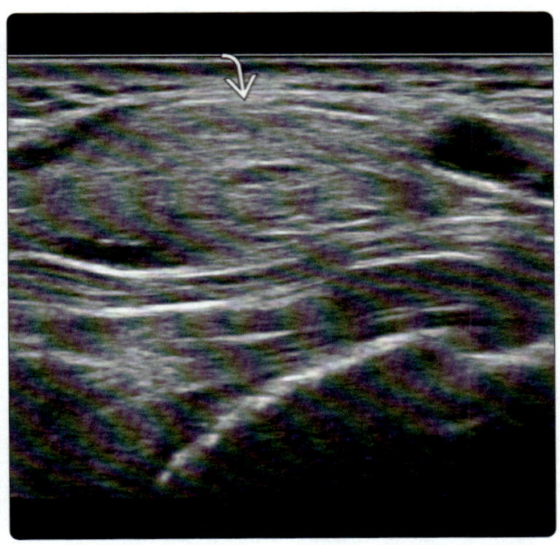

(Left) *Transverse US shows a medium-sized lipoma within the tarsal tunnel alongside the posterior tibial artery and veins. The lipoma is partially encasing the slightly swollen tibial nerve.* **(Right)** *Longitudinal US of the same tarsal tunnel mass shows the typical finely striated appearance of a lipoma. This helps distinguish this tarsal tunnel fatty mass from lipomatosis or lipohypertrophy.*

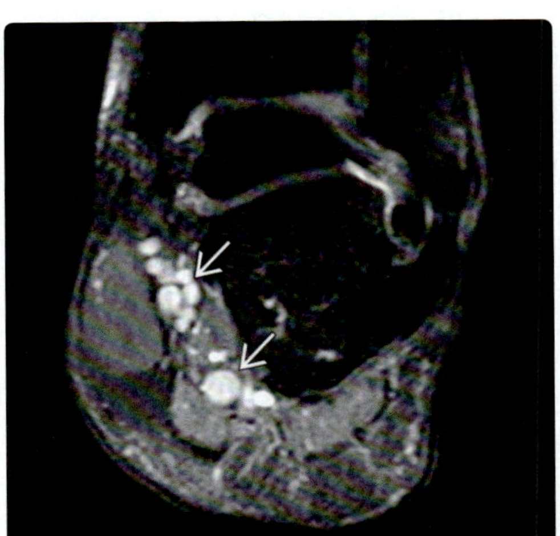

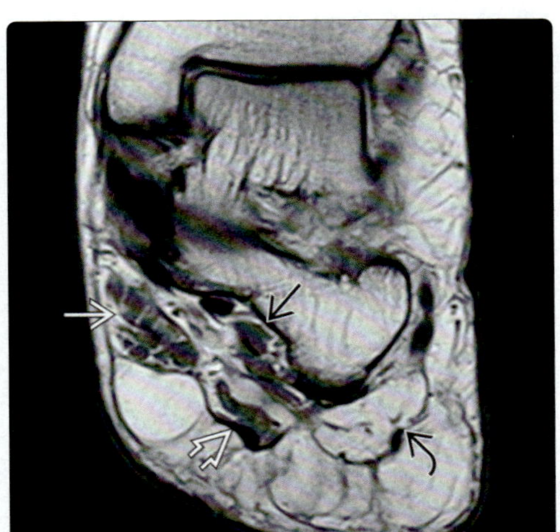

(Left) *Coronal T2WI FS MR in a patient with tarsal tunnel syndrome shows venous dilation of lower tarsal tunnel, which can cause compression of medial and lateral plantar nerves.* **(Right)** *Coronal T1WI MR in another patient with tarsal tunnel syndrome shows moderate abductor hallucis, flexor digitorum brevis, and quadratus plantae atrophy with severe abductor digiti minimi atrophy, due to tarsal tunnel syndrome, affecting inferior calcaneal nerve (Baxter nerve). (Courtesy K. A. Al-Ismail, MD.)*

KEY FACTS

IMAGING

- **Intravascular papillary endothelial hyperplasia**
  - Organizing thrombus with excessive reactive papillary endothelial proliferation
  - Predilection for subcutaneous tissues of extremities, particularly upper limbs
  - Focal vascular dilatation with intraluminal mass
    - Hypoechoic compared with subcutaneous fat
    - Hyperechoic septum (70%)
    - Little or no color flow (30%); moderate (40%); high color flow (30%)
- **Superficial thrombophlebitis**
  - Thrombosed vein surrounded by halo of hyperechoic edematous subcutaneous fat; ± perivascular hyperemia
- **Venous varix**
  - Focal dilatation of vein
- **Intravascular organized hematoma**
  - Focal hypoechoic to hyperechoic lesion within lumen
    - i.e., hematoma-like lesion within vein
    - No flow on color Doppler imaging
- **Pseudoaneurysm**
  - Hypoechoic or anechoic perivascular mass
    - Neck connects pseudoaneurysm lumen to artery
    - Characteristic to-and-fro flow pattern at neck of pseudoaneurysm
    - Characteristic yin-yang sign within pseudoaneurysm sac
    - Thrombus or slow flow within pseudoaneurysm sac also occasionally seen
- **Kimura disease**
  - More common in Asians
  - Fusiform-shaped hypoechoic tumor of arterial wall
    - Variable intratumoral hyperemia

DIAGNOSTIC CHECKLIST

- US is best imaging modality for all focal vascular abnormalities, provided vessel accessible

(Left) *Longitudinal US shows focal dilatation of the short saphenous vein ➡ in the popliteal fossa. There is a mildly hypoechoic mass ➡ within and around the dilated vein.* **(Right)** *Longitudinal color Doppler US of the same patient shows mild internal vascularity ➡ within the mass. These appearances are consistent with intravascular papillary endothelial hyperplasia.*

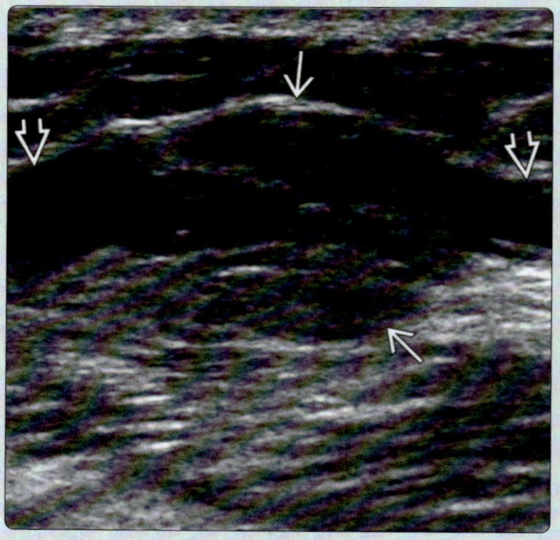

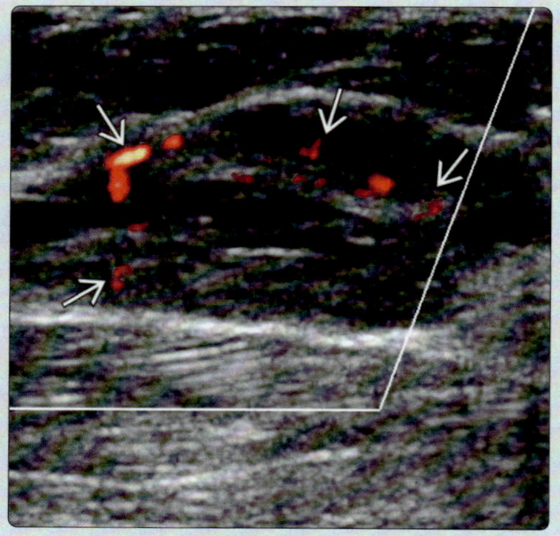

(Left) *Sagittal T1 C+ FS MR of the same patient shows focal dilatation ➡ of the short saphenous vein ➡. The vascular and perivascular tumor component shows heterogeneous contrast enhancement ➡. Excision biopsy revealed Masson tumor (intravascular papillary endothelial hyperplasia).* **(Right)** *Axial T1 C+ FS MR of the same patient shows heterogeneous tumor enhancement ➡. The mass arises eccentrically from the short saphenous vein ➡.*

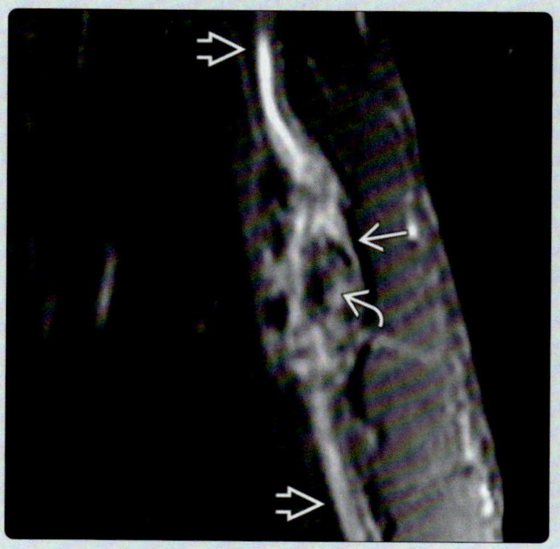

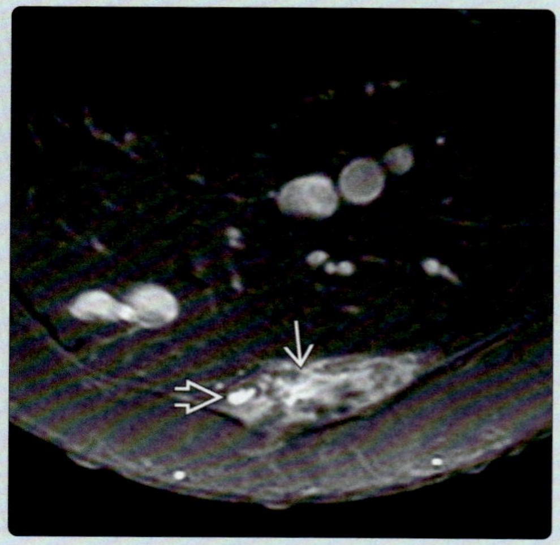

TERMINOLOGY

Synonyms

- **Intravascular papillary endothelial hyperplasia (IPEH)**: Masson tumor, Masson hemangioma, intravascular angiomatosis, vegetant intravascular hemangioendothelioma
- **Pseudoaneurysm**: False aneurysm

Definitions

- **IPEH**
  - Organizing thrombus with excessive reactive papillary endothelial proliferation
- **Superficial thrombophlebitis**
  - Inflammation of superficial vein
- **Venous varix**
  - Focal dilatation of vein
- **Intravascular organized hematoma**
  - Organized hematoma within vessel, usually vein
- **Pseudoaneurysm**
  - Luminal outpouching of artery with
    - Defect in tunica intima and media and preservation of tunica adventitia
    - Or full-thickness wall defect with bleeding contained by hematoma and surrounding soft tissues
- **Kimura disease**
  - Inflammatory response associated with self-limiting allergic or autoimmune reaction to unknown stimulus

IMAGING

General Features

- Location
  - **IPEH**
    - Most commonly within vein; rarely within artery
    - Predilection for subcutaneous tissues of extremities, particularly upper limbs
  - **Superficial thrombophlebitis**
    - Superficial venous system, particularly in lower limbs
  - **Venous varix**
    - Most commonly in superficial vein
  - **Intravascular organized hematoma**
    - Any vessel, most commonly within vein
  - **Pseudoaneurysm**
    - Most commonly affects common femoral artery related to needle puncture for angiography
      - Arises from punctured superficial wall of artery
  - **Kimura disease**
    - Most commonly involves head and neck region
    - Less common in axilla, popliteal region, groin, and forearm

CT Findings

- **Pseudoaneurysm**
  - Arterial enhancement of cavity
  - Contrast extravasation indicates pseudoaneurysm rupture
  - Fresh hematoma around pseudoaneurysm is slightly hyperdense
  - CT reconstruction is useful for delineation of anatomy and assessing suitability for intervention

MR Findings

- **IPEH**
  - T1WI: Isointense mass with node-like hyperintense foci
  - T2WI: Heterogeneous hyperintense mass with node-like hypointense foci
- **Intravascular organized hematoma**
  - Signal intensity depends on age of hematoma
- **Pseudoaneurysm**
  - T1WI
    - Pseudoaneurysm luminal flow void
    - High signal in methemoglobin-containing portions of surrounding hematoma
  - MRA
    - Contrast-enhanced MRA preferred to time-of-flight or phase-contrast technique because of superior signal:noise ratio
    - Double to triple dose gadolinium typically used

Ultrasonographic Findings

- **IPEH**
  - Focal, oblong-shaped vascular dilatation with intraluminal mass
    - Detectable vessel of origin only seen in 40%
  - Mass hypoechoic compared with subcutaneous fat
    - Acoustic enhancement (100%)
    - Hyperechoic septum (70%)
  - Color Doppler imaging
    - Septal and peripheral vascularity
    - Little or no color flow (30%); moderate (40%); high color flow (30%)
- **Superficial thrombophlebitis**
  - Thrombosed vein with thickened vein wall
  - Halo of edematous hyperechoic subcutaneous fat
  - ± perivascular hyperemia
- **Venous varix**
  - Focal, easily compressible, saccular or fusiform dilatation of vein
    - Vein wall of normal caliber with normal perivascular tissues
    - ± low-level internal echoes due to turbulent or slow flow
- **Intravascular organized hematoma**
  - Focal hypoechoic to hyperechoic lesion within vascular lumen (i.e., hematoma-like lesion within vein)
    - Mild to moderate vascular dilatation
    - No color flow on color Doppler imaging
- **Pseudoaneurysm**
  - Hypoechoic or anechoic, perivascular, saccular-shaped mass
  - Neck connects pseudoaneurysm to artery
  - Pulsed Doppler imaging
    - Characteristic to-and-fro flow pattern at neck of pseudoaneurysm
    - ± turbulent high-velocity flow if neck is narrowed
    - Flow can sometimes be detected in pseudoaneurysm neck, though not in lumen
  - Color Doppler imaging
    - Characteristics yin-yang sign within lumen of pseudoaneurysm

- Swirling flow producing different color signals within lesion
- Lumen may be only partially filled with color signal due to partial thrombosis
  - ± slow flow within lumen
- **Kimura disease**
  - Tumor arising from arterial wall
    - Fusiform-shaped, hypoechoic arterial wall thickening
      - Variable flow in thickened arterial wall on color Doppler imaging

Imaging Recommendations

- Best imaging tool
  - US for all focal vascular abnormalities, provided vessel is accessible
- Protocol advice
  - **IPEH**
    - Tourniquet application is useful if arising from vein to maximize venous distension and optimally show intraluminal content
  - **Superficial thrombophlebitis**
    - Valsalva maneuver or compression augmentation can help with flow detection

DIFFERENTIAL DIAGNOSIS

Angiosarcoma or Kaposi Sarcoma

- Rarely confined to vascular wall or lumen

Angiolymphoid Hyperplasia

- Histology similar to Kimura disease
- Eosinophilia less prevalent than in Kimura disease
- IgE level is normal

PATHOLOGY

General Features

- Etiology
  - **IPEH**
    - Macrophage-derived fibroblast growth factor (bFGF) stimulates papillary proliferation of plump endothelial cells that release more bFGF
      - Anomalous thrombus organization rather than primary vascular tumor
      - Less commonly arises in vascular anomaly and rarely in hematoma
      - 0.5-2.0 cm
  - **Superficial thrombophlebitis**
    - Preexisting deep venous thrombosis or cellulitis
    - Recent instrumentation
  - **Venous varix**
    - Previous trauma or surgery, valvular incompetence
  - **Intravascular organized hematoma**
    - Previous trauma
  - **Pseudoaneurysm**
    - Iatrogenic or traumatic needle puncture; infection/inflammation adjacent to artery, septic emboli
  - **Kimura disease**
    - Immunologically mediated disorder

Microscopic Features

- **IPEH**
  - Papillary projections composed of plump endothelial cells within fibrinous matrix and dilated vascular spaces
    - Also microthrombi in various stages of organization
- **Kimura disease**
  - Eosinophilic infiltration with microabscesses
  - Well-developed lymphoid follicles

CLINICAL ISSUES

Presentation

- Most common signs/symptoms
  - **IPEH**
    - Slowly growing palpable lesion, often with bluish or reddish discoloration of overlying skin ± pain
  - **Superficial thrombophlebitis**
    - Swollen, tender area with increased warmth
  - **Venous varix**
    - Palpable, soft, fully compressible mass
  - **Intravascular organized hematoma**
    - Painful palpable lesion along course of vein
  - **Pseudoaneurysm**
    - Increased swelling, which may be pulsatile, at site of prior needle puncture, surgery, or trauma
  - **Kimura disease**
    - Painless subcutaneous mass

Demographics

- Ethnicity
  - **Kimura disease**
    - More common in Asian population

Treatment

- **IPEH**
  - Complete surgical excision
- **Superficial thrombophlebitis**
  - Antiinflammatory agents or heat therapy
- **Venous varix**
  - Pressure garment or surgery
- **Intravascular organized hematoma**
  - Conservative treatment
- **Pseudoaneurysm**
  - US-guided compression
    - Success rate ~ 90%
  - US-guided thrombin (human or bovine) injection
  - Percutaneous transarterial embolization or covered stenting
  - Direct surgical repair or bypass surgery
- **Kimura disease**
  - Surgical excision

DIAGNOSTIC CHECKLIST

Consider

- US best imaging modality for focal vascular abnormalities, provided vessel accessible

SELECTED REFERENCES

1. Kim OH et al: Subcutaneous intravascular papillary endothelial hyperplasia: ultrasound features and pathological correlation. Skeletal Radiol. 45(2):227-33, 2016

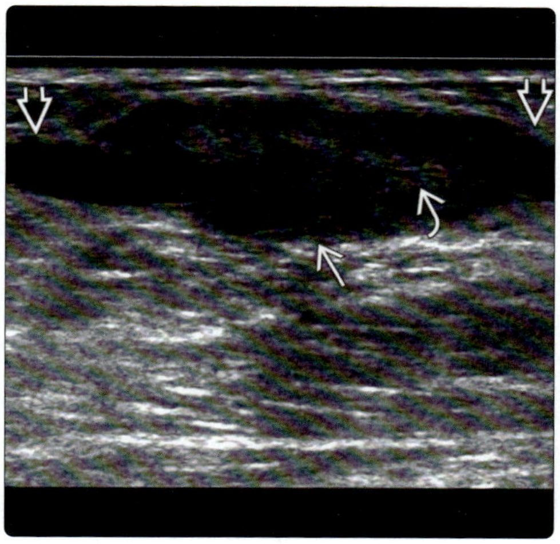

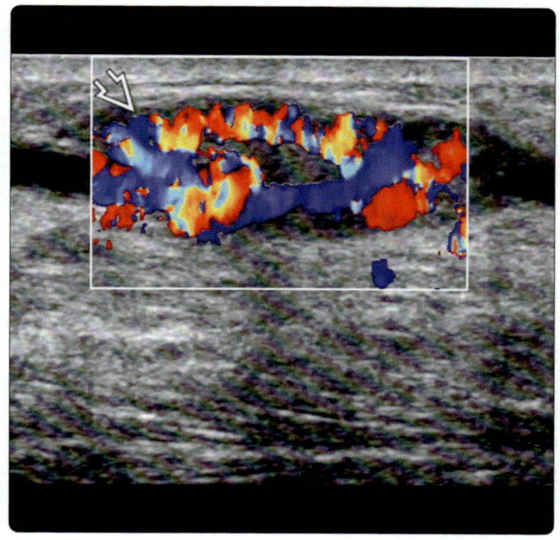

(Left) *Longitudinal US shows a focally dilated, long saphenous vein on the medial aspect of the thigh. There is a mildly hypoechoic tumor within the dilated vein, compatible with histologically proven Masson tumor (intravascular papillary endothelial hyperplasia).* **(Right)** *Longitudinal color Doppler US of the same tumor shows marked tumoral vascularity. This helps to differentiate the intravascular papillary endothelial hyperplasia from intravascular thrombosis.*

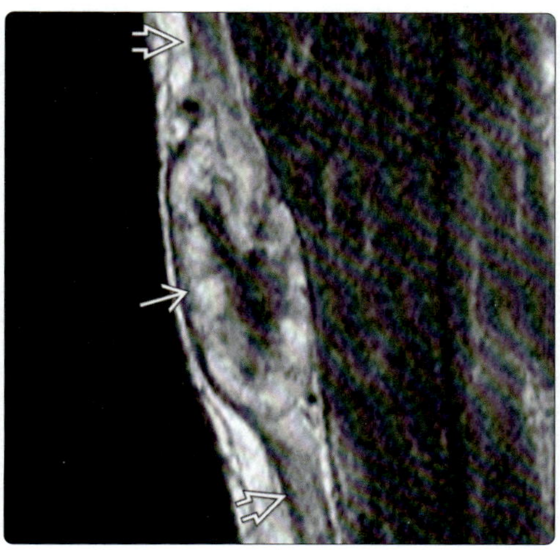

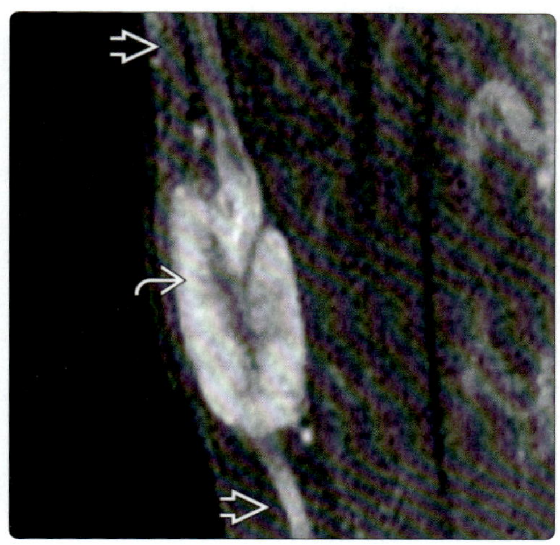

(Left) *Sagittal T2 MR of the same tumor shows the fusiform-shaped, dilated, long saphenous vein filled with heterogeneously hyperintense tumor.* **(Right)** *Sagittal T1 C+ FS MR of the same tumor shows homogeneous enhancement of the Masson tumor within the long saphenous vein. This enhancement is compatible with the hyperemia seen on color Doppler imaging.*

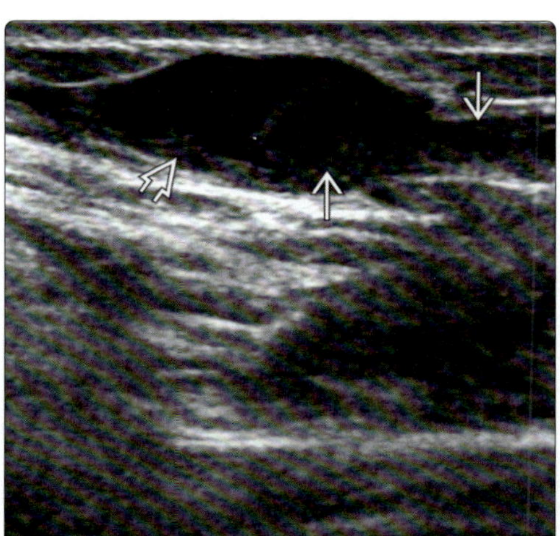

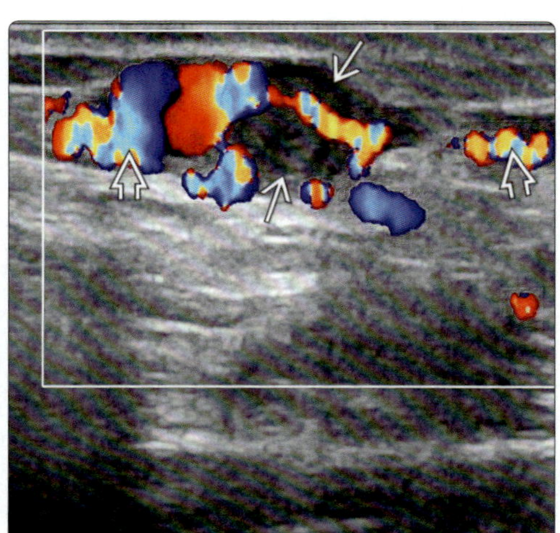

(Left) *Longitudinal US shows focal dilatation of the short saphenous vein on the medial aspect of the knee. The dilated vein and proximal venous segment is partially filled with mildly hypoechoic tissue. Histology confirmed a Masson tumor (intravascular papillary endothelial hyperplasia).* **(Right)** *Longitudinal color Doppler US of the same tumor shows no tumoral vascularity, which is an atypical feature. There is flow within the short saphenous vein.*

(Left) *Transverse US of the forearm shows superficial thrombophlebitis of the basilic vein with venous distension and edema of the perivascular* ➡ *and adjacent soft tissues* ➡. **(Right)** *Transverse color Doppler US at the same location shows marked hyperemia* ➡ *surrounding the distended basilic vein, indicative of perivascular inflammation. There is echogenic thrombus within the distended vein. This basilic vein superficial thrombophlebitis was secondary to prior venous cannulation.*

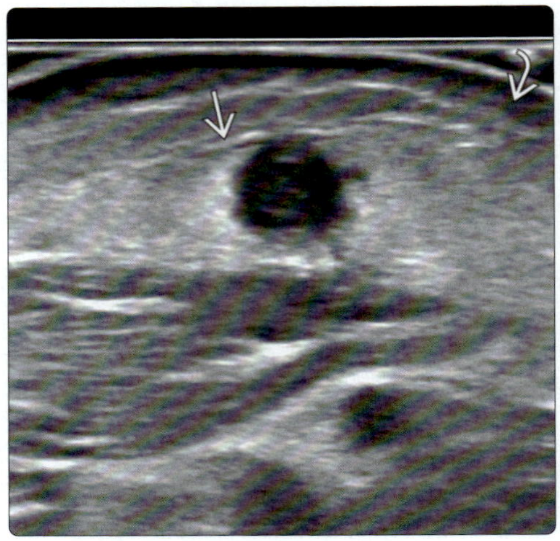

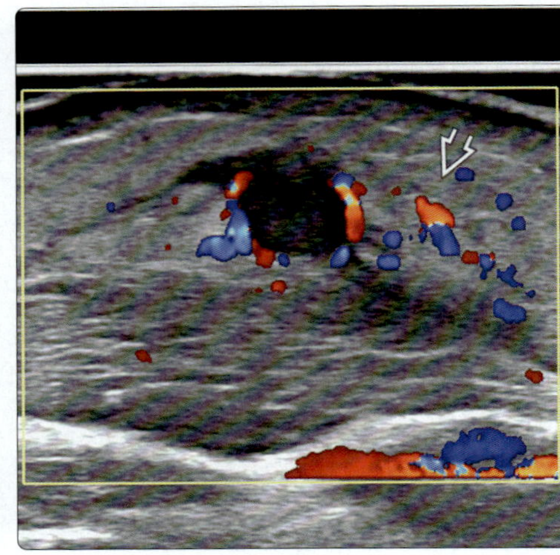

(Left) *Longitudinal US at the same location shows a thickened basilic vein wall* ➡ *with echogenic thrombus largely filling the venous lumen* ➡. *The normal investing fascia* ➡ *and muscle deep to the area of superficial thrombophlebitis is shown.* **(Right)** *Longitudinal US along the edge of the same basilic vein shows marked hyperemia* ➡, *indicative of severe perivascular inflammation. Most superficial thrombophlebitis is sterile.*

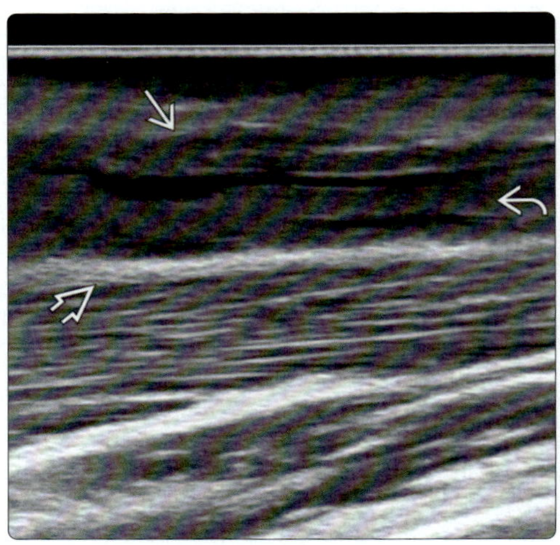

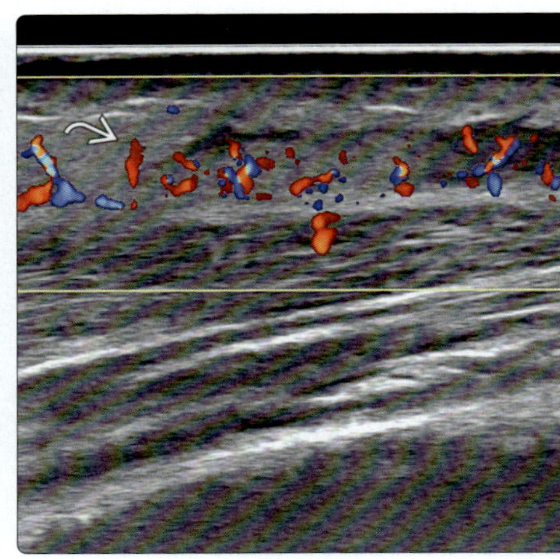

(Left) *Clinical photograph of a leg shows redness and swelling* ➡ *that had developed over the previous week around the site of a subcutaneous lipoma excision* ➡ *performed 1 year earlier.* **(Right)** *Longitudinal US of the same area shows mild subcutaneous vein dilatation* ➡ *with moderate adjacent subcutaneous edema* ➡ *and hyperemia (not shown), indicative of cellulitis. This is primary cellulitis, not centered around the vein, rather than primary superficial thrombophlebitis.*

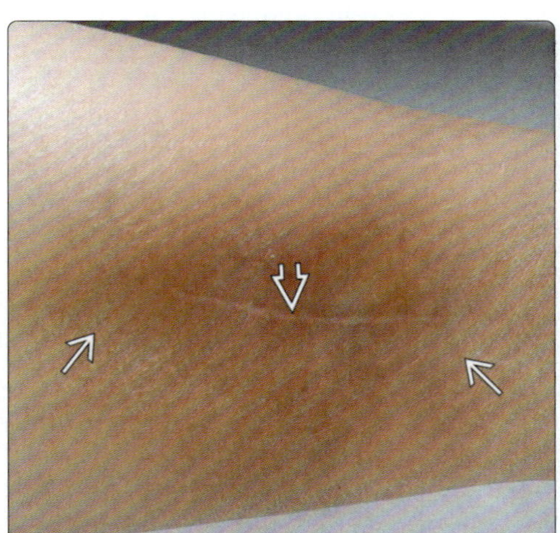

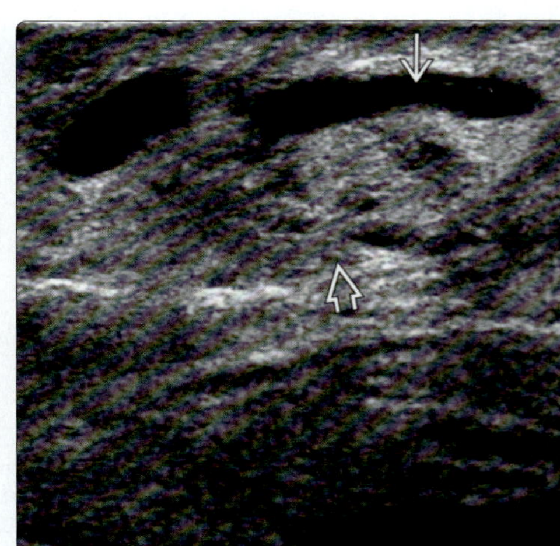

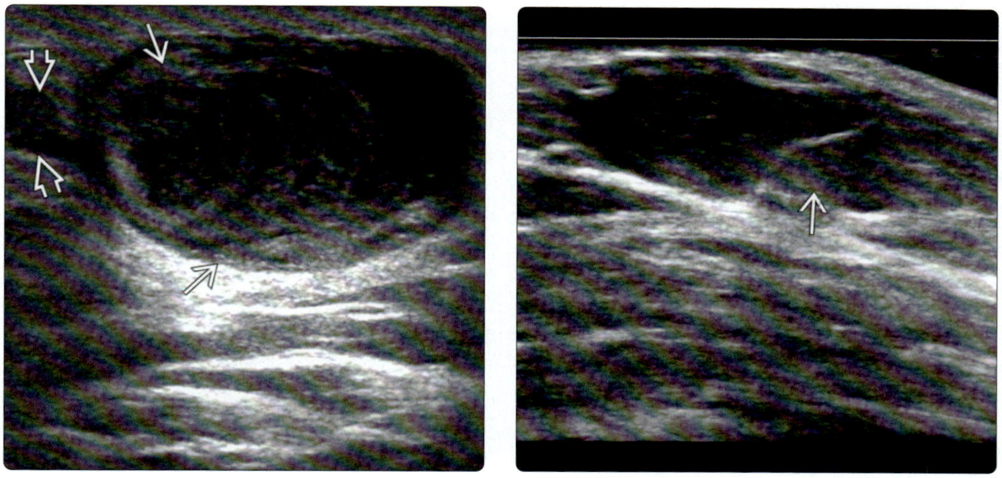

(Left) *Longitudinal US shows an organized intravascular hematoma* → *on the volar aspect of the elbow. The hematoma causes focal dilatation of the superficial vein* →. **(Right)** *Longitudinal color Doppler US shows an intravascular organized hematoma* → *within the long saphenous vein in the thigh. No internal vascularity is present within the hematoma.*

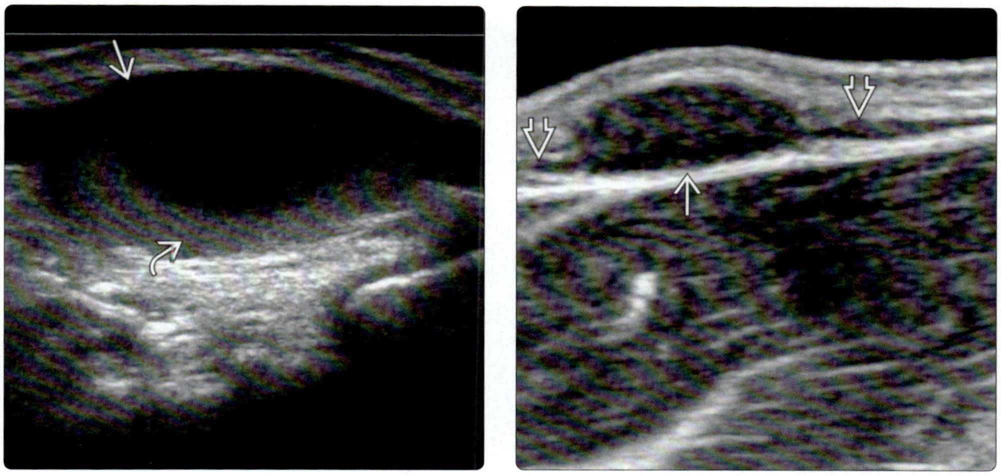

(Left) *Longitudinal US shows a venous varix* → *on the dorsal aspect of the wrist. There is partially stagnant flow* → *within the varix, which mimics a hypoechoic mass. Dynamic US with compression and decompression revealed movement within this stagnant flow area.* **(Right)** *Longitudinal US shows a fusiform-shaped venous varix* → *along a superficial vein* → *on the medial aspect of the elbow. The luminal hyperechogenicity is due to slow-moving, red cell aggregates.*

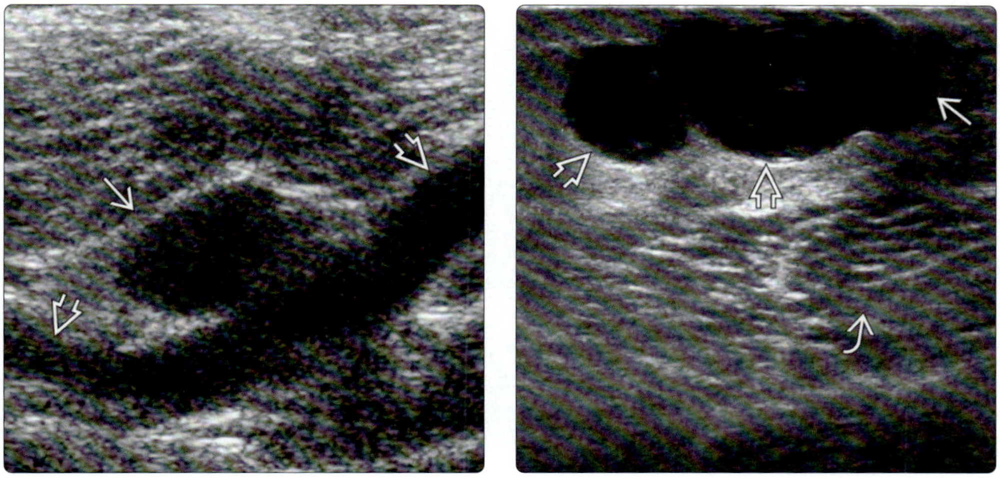

(Left) *Longitudinal US shows a saccular venous outpouching* → *from the popliteal vein* → *in the popliteal fossa, consistent with a venous varix.* **(Right)** *Longitudinal US shows a lobulated venous varix* → *arising from a superficial femoral vein* → *on the medial aspect of the distal thigh. The varix lies superficial to the sartorius muscle* → *and is easily compressible. The echogenic content represents sluggish blood flow with rouleaux formation of red blood cells.*

(Left) *Lateral radiograph shows soft tissue swelling ➡ on the posterior aspect of the distal thigh with an osteochondroma ➡ in the metaphysis of the distal femur. A malignant transformation was suspected clinically.* **(Right)** *Transverse color Doppler US of the same patient shows a large pseudoaneurysm ➡ arising from the superficial femoral artery on the posterolateral aspect of the thigh. There is swirling color flow (yin-yang sign) within the pseudoaneurysm lumen.*

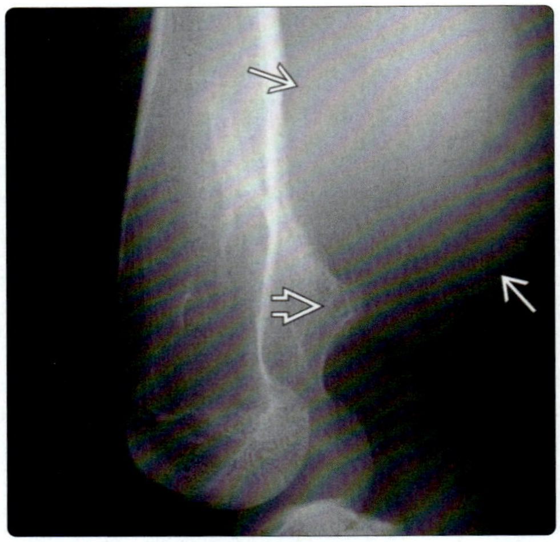

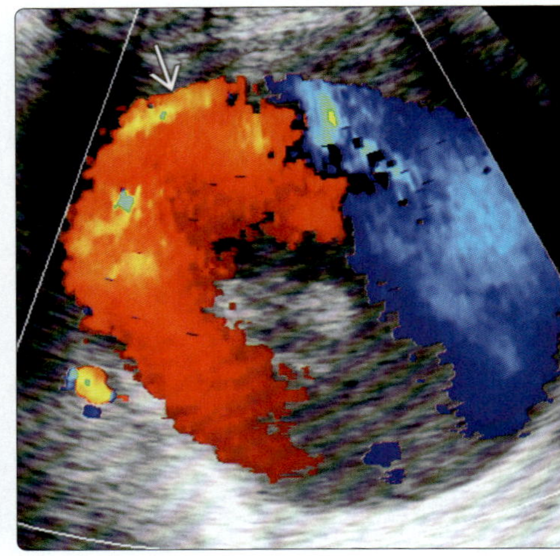

(Left) *Transverse power Doppler US of the same patient shows to-and-fro arterial flow ➡ within the neck of the pseudoaneurysm.* **(Right)** *Sagittal T2 MR of the same lesion shows the large pseudoaneurysm ➡ with flow void similar to the yin-yang sign in the posterior aspect of the thigh. The pseudoaneurysm causes scalloping of the distal femur ➡ as well as truncation of the osteochondroma ➡.*

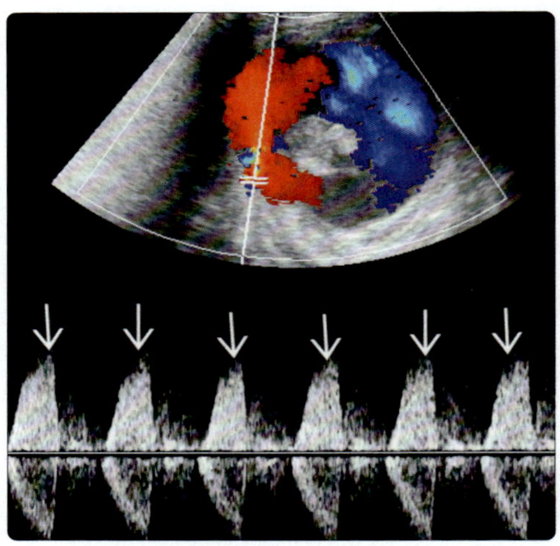

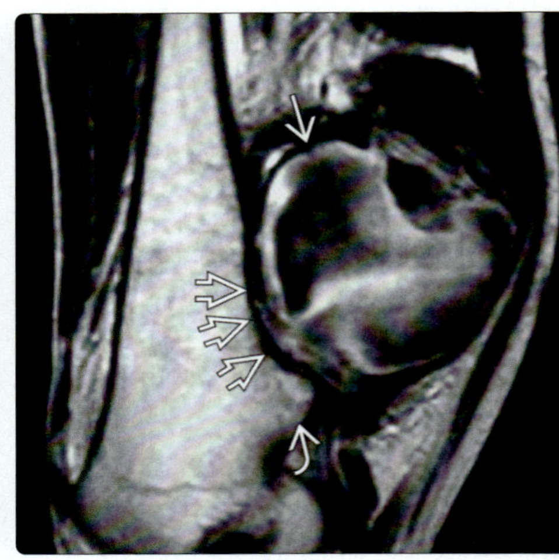

(Left) *Anteroposterior digital subtraction angiography (DSA) in the same patient shows the pseudoaneurysm ➡ arising from a displaced and narrowed superficial femoral artery ➡. Angiographic embolization was subsequently performed (not shown).* **(Right)** *Anteroposterior DSA of the same lesion following embolization shows complete obliteration of the pseudoaneurysm by embolization coils ➡.*

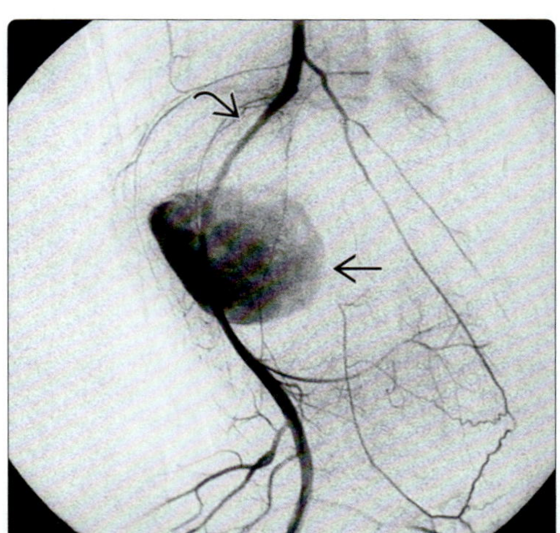

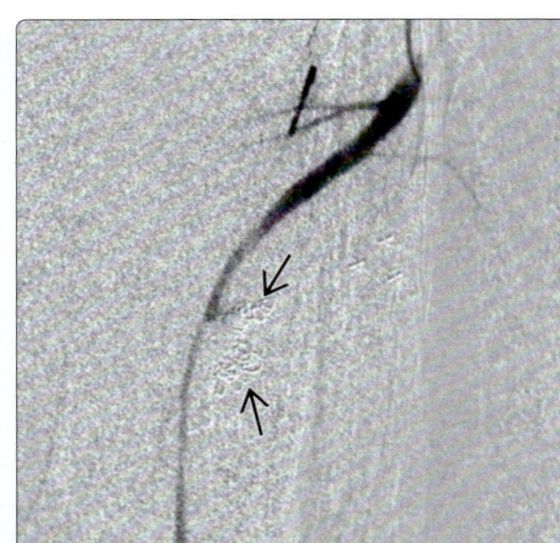

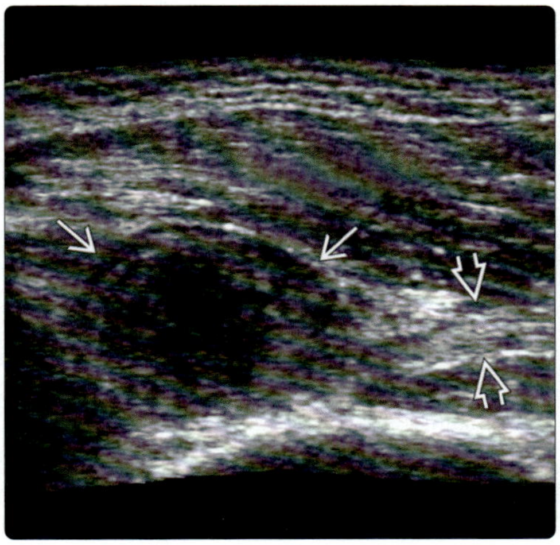

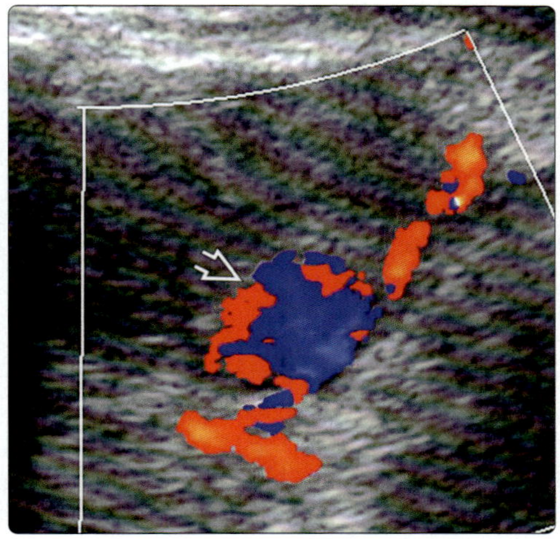

(Left) *Longitudinal US of the buttock in a patient with a recent stab injury to the buttock region shows a pseudoaneurysm of the superior gluteal artery. The adjacent sciatic nerve is moderately compressed by the pseudoaneurysm.* **(Right)** *Color Doppler US of the same patient shows swirling, turbulent flow within the pseudoaneurysm sac. Transarterial embolization was subsequently performed (not shown) with rapid alleviation of sciatica-type symptoms.*

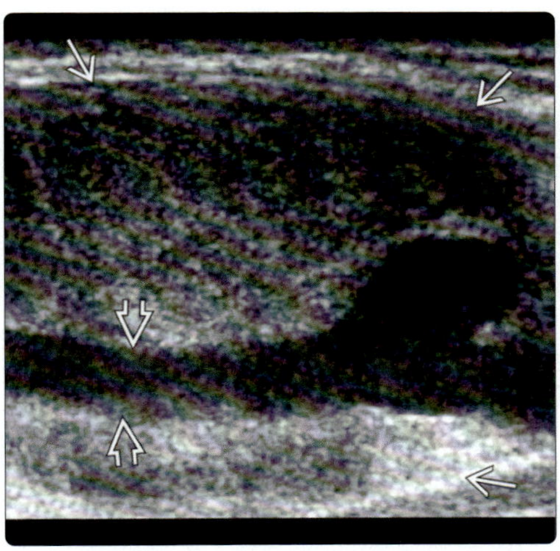

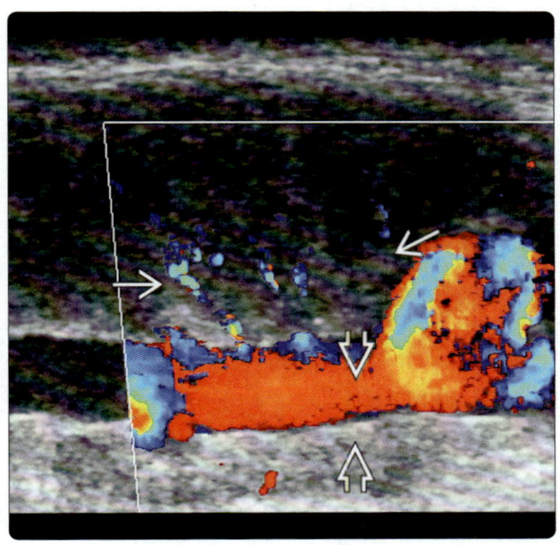

(Left) *Longitudinal oblique US reveals a thick hypoechoic mass arising from the arterial wall of the proximal brachial artery.* **(Right)** *Longitudinal oblique color Doppler US shows mild internal vascularity of the hypoechoic mass supplied by arterioles arising from the brachial artery lumen. US-guided percutaneous biopsy of the arterial wall revealed Kimura disease.*

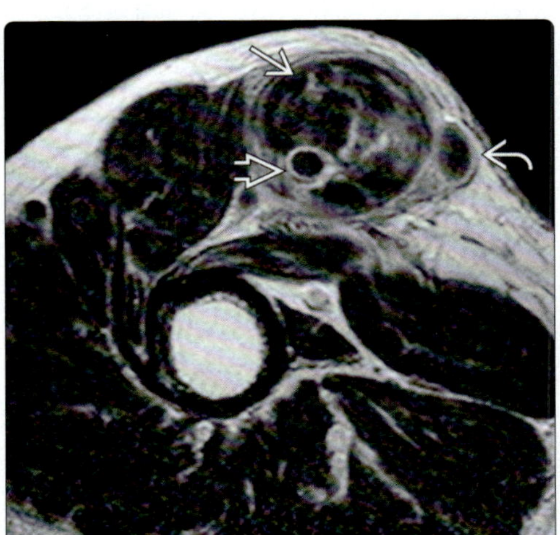

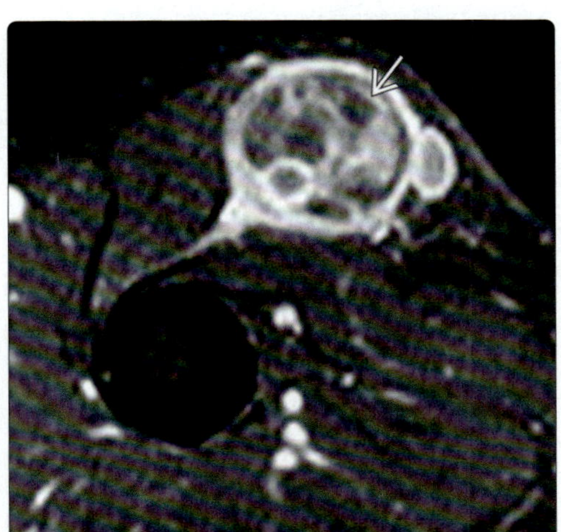

(Left) *Axial T2 MR of the same patient shows a heterogeneous hypointense mass due to Kimura disease arising largely eccentrically from the wall of the brachial artery. The brachial vein is normal.* **(Right)** *Axial T1 C+ FS MR of the same patient shows mild heterogeneous enhancement of the Kimura disease mass. Segmental brachial artery excision was performed followed by arterial repair.*

Soft Tissue Infection

KEY FACTS

TERMINOLOGY

- Infection of soft tissue including cellulitis, infective fasciitis, necrotizing fasciitis, pyomyositis, phlegmon, and abscess

IMAGING

- US excellent for localized soft tissue infection
- Use US ± CT or MR for more generalized soft tissue infection
- US cannot distinguish infective from noninfective fluid
- Always consider secondary soft tissue infection from deeper or other primary source
- **Cellulitis**
    - Edematous hyperechoic subcutaneous fat with hyperemia
        - Hyperemia helps differentiate subcutaneous inflammation from subcutaneous edema
        - Subcutaneous inflammation normally, but not always, equates to infection, i.e., cellulitis
    - Thickened interlobular septa
- **Necrotizing fasciitis**
    - Affects subcutaneous fat, fascia, and muscle
    - Thickened fascia with perifascial fluid (fluid above and below investing fascia)
    - Severe subcutaneous and muscle edema
    - ± fascial focal defect ± muscle necrosis ± gas locules with comet-tail or shadowing artifact
- **Pyomyositis**
    - Diffuse edematous hyperechoic muscle swelling with indistinct fibrillary pattern
    - Hyperemia ± gas locules with comet-tail or shadowing artifact
- **Abscess**
    - Echogenicity of fluid may vary from hypoechoic to isoechoic to hyperechoic
    - Posterior enhancement characteristic
    - Echo movement on real-time imaging
    - Peripheral hyperemia ± septation ± gas locules

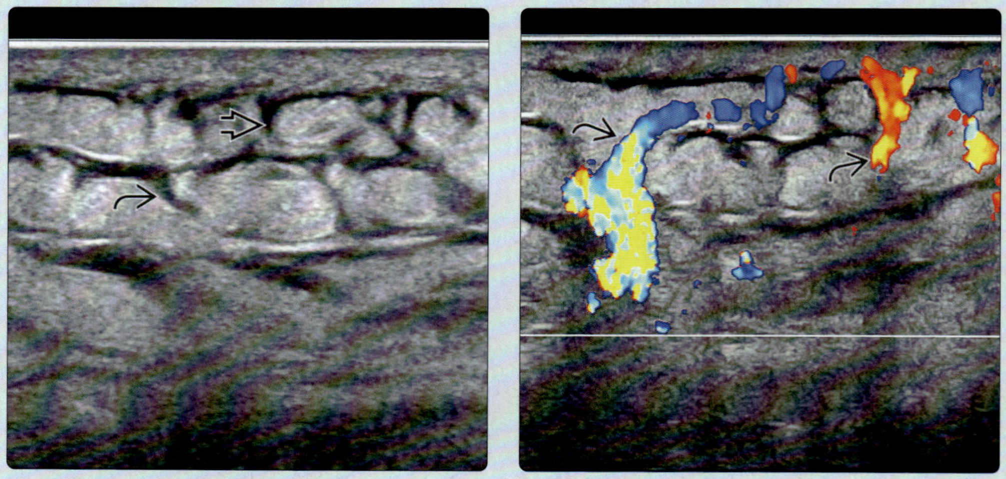

(Left) *Transverse US of calf region shows severe, subcutaneous tissue thickening and edema. Increased subcutaneous fluid leads to thickened interlobular septa ⇨ with intervening, hyperechoic fat lobules ⇨.* **(Right)** *Transverse color Doppler US of the same patient shows moderate-severity hyperemia ⇨. This hyperemia allows distinction between noninflamed and inflamed subcutaneous tissue edema, i.e., cellulitis.*

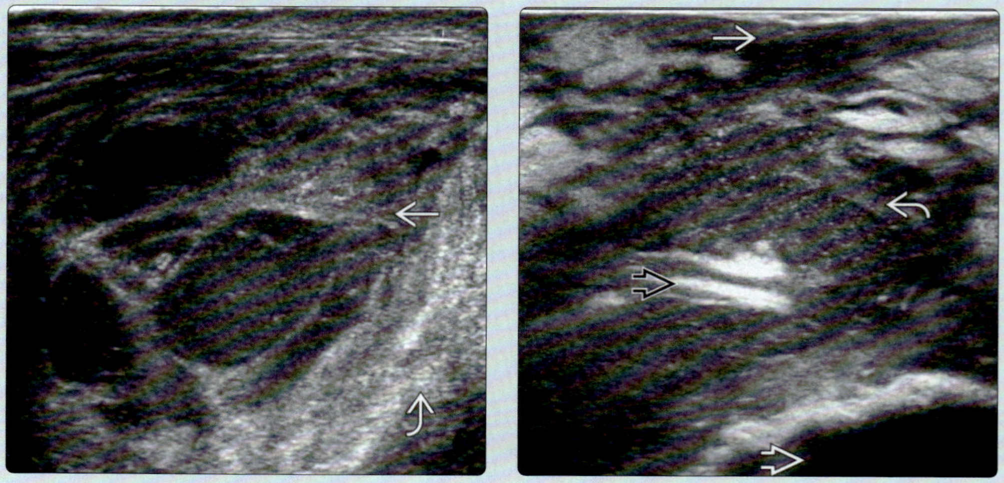

(Left) *Transverse US of the posterior thigh shows a multiloculated intramuscular abscess ➡ within the semimembranosus muscle. This is associated with marked surrounding soft tissue edema ➡.* **(Right)** *Transverse US of the buttock region in a tetraplegic patient shows extensive soft tissue edema with ill-defined abscess formation ➡ overlying the ischial tuberosity ➡ and extending to a cutaneous opening ➡. There is a linear, hyperechoic gas artifact ⇨ present within the abscess.*

TERMINOLOGY

Definitions

- Cellulitis: Localized or diffuse soft tissue inflammation of skin and subcutaneous tissues
- Infective fasciitis: Infection of superficial &/or deep fascia
- Necrotizing fasciitis: Rapidly progressive infection of superficial and deep fascia associating with vascular occlusion and necrosis
- Pyomyositis: Infective myositis = suppurative bacterial infection of muscle
- Phlegmon: Localized, intense inflammation prior to development of abscess or ulceration
- Abscess: Localized collection of pus

IMAGING

General Features

- Best diagnostic clue
  - Edema and hyperemia of affected tissues ± fluid accumulation
- Location
  - Any soft tissue tissue can become infected
  - Most common sites
    - Cellulitis: Dorsum of hands and feet
    - Necrotizing fasciitis: Thigh and leg
    - Pyomyositis: Large muscles of pelvis and lower limb; frequently multifocal
- Size
  - Varies from small localized abscess to multicompartmental disseminated infection
- Morphology
  - Inflammation ± fluid ± abscess

Radiographic Findings

- Often noncontributory; exclude other causes of soft tissue swelling such as fractures
- Soft tissue swelling with obliteration of fascial planes
  - ± gas locules within soft tissues ± ulcers ± radiopaque foreign body ± concomitant osteitis or osteomyelitis

CT Findings

- Increased attenuation of subcutaneous fat
- Swelling of subcutaneous tissue ± muscles
- Fascial thickening or perifascial fluid
- ± gas locules within soft tissues ± foreign body ± concomitant osteitis or osteomyelitis
  - Highest sensitivity for detection of soft tissue gas and foreign body
- Contrast enhancement is useful in detecting fluid collection, abscess, or necrosis

MR Findings

- T2WI FS
  - Cellulitis: Diffuse hyperintensity of subcutaneous tissues
    - ± feathery edema of adjacent tissues
  - Necrotizing fasciitis: Hyperintense thickening of deep fascia with perifascial fluid
    - Edema of adjacent musculature
    - ± fluid areas indicative of necrosis, exudate, or abscess
    - ± hypointense foci due to gas locules
  - Pyomyositis: Diffuse hyperintensity of muscle
    - ± fluid collection
    - ± edematous changes in adjacent subcutaneous tissue > fascial planes > bone and joints
  - Abscess: High signal intensity area with surrounding edema ± hyperemia
    - ± hypointense foci due to gas locules
- T1WI C+
  - Useful in delineating fluid collection, sinus, abscess, or necrosis

Ultrasonographic Findings

- **Cellulitis**
  - Edema and hyperemia of subcutaneous fat
    - Edematous fat is echogenic on US
    - Thickened interlobular septa; characteristic cobblestone appearances
    - ± periseptal fluid and fluid around investing fascia
    - Need combination of edema and hyperemia to diagnose cellulitis
      - Hyperemia is not seen with noninflammatory edema in, for example, venous insufficiency, heart failure, or lymphedema
    - ± secondary tenosynovitis with mild synovial proliferation and tendon thickening
  - Appearance of subcutaneous edema varies according to severity and architecture of subcutaneous fat
    - Varies from amorphous swelling with indistinct lobules to cobblestone appearance with accentuated lobules
  - 1st line to differentiate cellulitis from venous thrombosis, but both conditions may coexist
- **Necrotizing fasciitis**
  - Affects subcutaneous fat, fascia, and muscle
  - Thickened deep fascia (> 3 mm) with perifascial fluid (fluid above and below investing fascia)
    - Extensive involvement of deep intermuscular fascia, not just area contiguous to investing fascia; 3 or more compartments involved
    - ± fascial irregularity or focal defect
  - Severe, associated subcutaneous and muscle edema
  - ± muscle necrosis (difficult to detect)
  - Gas locules with comet-tail artifacts or shadowing artifact are suggestive of necrosis
    - rare finding but high specificity
- **Pyomyositis**
  - Diffuse muscle swelling with edema ± hyperemia
  - Edematous muscle is echogenic on US with indistinct fibrillary pattern of muscle
  - ± focal hypoechoic areas due to serous exudate, abscess, or necrosis
  - Usually involves single muscle; multiple muscles are involved in 40% of cases
- **Abscess**
  - Posterior enhancement characteristic
  - Depending on maturity and content, echogenicity of fluid may vary from hypoechoic → isoechoic → hyperechoic
    - If thick fluid or debris present, may mimic solid mass, so best to have low threshold for aspiration, as fluid is not readily discernible
  - Echo movement on real-time imaging

- ± gas locules with comet-tail artifact or shadowing artifact ± septations ± peripheral hyperemia
- Wall thickness is proportional to chronicity of abscess
- US-guided aspiration of fluid for culture and sensitivity
  - Occasionally, little or no fluid can be aspirated despite visibly moving echoes

Imaging Recommendations

- Best imaging tool
  - US ± US-guided aspiration gives sufficient information to make prompt diagnosis and guide treatment
- Protocol advice
  - Routinely assess with color Doppler imaging
  - Look for focal hypoechoic areas representing fluid collection, abscess, or necrosis
    - If fluid is present, consider aspiration for culture
    - Imaging cannot distinguish between infective and noninfective fluid
  - ± percutaneous biopsy of cellulitis for culture

DIFFERENTIAL DIAGNOSIS

Subcutaneous Edema

- Similar to cellulitis though no hyperemia

Muscle Hematoma

- Similar to muscle abscess but less inflammatory component
- Hypoechoic mass with posterior enhancement, variable liquefaction, and surrounding edema

Muscle Infarction

- Similar to focal pyomyositis but less inflammatory component
- Hypoechoic mass with surrounding edema and hyperemia

Noninfectious Fasciitis

- Eosinophilic fasciitis: Associated with peripheral eosinophilia
  - Usually quite diffuse fascial inflammation and swelling
  - US-guided fascial biopsy shows fasciitis with lymphocytic and eosinophilic infiltrate
- Paraneoplastic fasciitis: Uncommon manifestation of malignancy

PATHOLOGY

General Features

- Etiology
  - Predisposing conditions
    - Old age, immunosuppression, illicit drug abuse, systemic inflammatory disease, peripheral vascular disease, trauma, surgery, obesity, alcoholism, malnutrition
  - Route of infection
    - Loss of skin integrity from trauma, surgery, or diabetic ulceration; hematogenous route is less common
  - Causative organisms
    - Usually polymicrobial with combination of aerobic and anaerobic organisms; fungal and atypical organism are sometimes seen
    - Cellulitis: *Streptococcus pyogenes*, *Staphylococcus aureus*
    - Necrotizing fasciitis: Group A β-hemolytic streptococci
    - Pyomyositis: *S. aureus*, *Mycobacterium tuberculosis*, *Nocardia asteroides*, *S. pyogenes*, Streptococcus viridans, *Cryptococcus neoformans*

Gross Pathologic & Surgical Features

- Cellulitis: Acute inflammatory reaction
- Necrotizing fasciitis: Bacteria spread from subcutaneous tissues to investing and deep fascia and then muscle; induce vascular thrombosis, tissue necrosis, and neural ischemia
- Pyomyositis: Bacterial infection of skeletal muscle ± necrosis or abscess

Microscopic Features

- Cellulitis: Acute inflammatory infiltrate, capillary dilatation ± abscess formation
- Necrotizing fasciitis: Fascial inflammation and necrosis, subcutaneous fat necrosis, arterial thrombi, vasculitis, and myonecrosis
- Pyomyositis: Acute inflammatory infiltrate, capillary dilatation ± abscess formation

CLINICAL ISSUES

Presentation

- Most common signs/symptoms
  - Cellulitis: Redness, heat, swelling, pain, and dysfunction ± lymphangitis, blistering necrosis ± lymphadenopathy
  - Necrotizing fasciitis: Rapidly progressive condition with severe inflammation of tissues ± blistering necrosis, cyanosis, severe tenderness, high fever, tachycardia, hypotension, impaired consciousness
  - Pyomyositis: Deep pain, tenderness ± mass

Demographics

- Age
  - More common in older patients
- Gender
  - M:F = 1:1
- Epidemiology
  - Cellulitis > > pyomyositis > necrotizing fasciitis

Natural History & Prognosis

- Cellulitis: Responds well to treatment
- Necrotizing fasciitis: Mortality up to 29%
- Pyomyositis: Less favorable response in immunocompromised patients

Treatment

- Cellulitis: Elevation, splinting + antibiotic therapy
- Necrotizing fasciitis: Early radical surgical debridement + irrigation + antibiotic therapy
- Pyomyositis: Antibiotic therapy
  - Abscess: Either US-guided percutaneous aspiration catheter drainage ± surgical drainage

SELECTED REFERENCES

1. Hayeri MR et al: Soft-tissue infections and their imaging mimics: from cellulitis to necrotizing fasciitis. Radiographics. 36(6):1888-1910, 2016
2. Park M et al: Sonographic findings of common musculoskeletal diseases in patients with diabetes mellitus. Korean J Radiol. 17(2):245-54, 2016

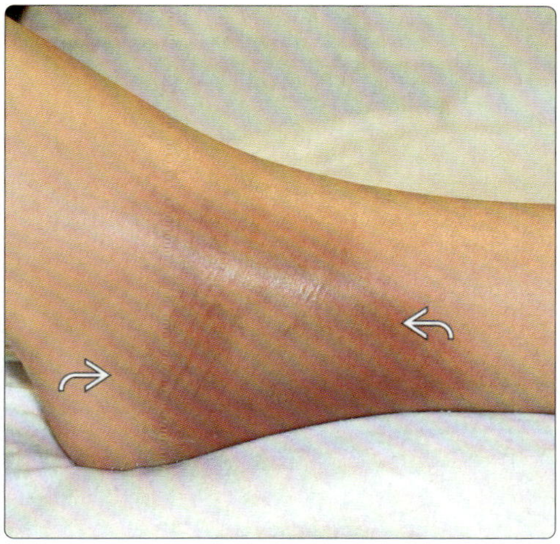

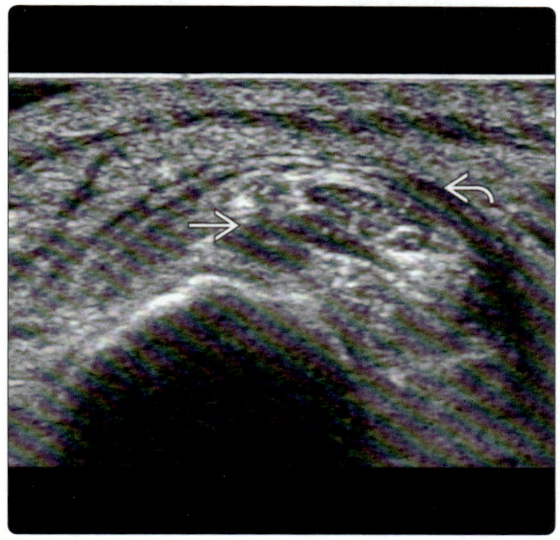

(Left) *Clinical photograph of a patient with ankle swelling and fever shows soft tissue swelling and skin erythema over the lateral aspect of the ankle.* **(Right)** *Transverse US of the same patient shows moderate subcutaneous edema on the lateral aspect of ankle. Mild edematous thickening of the superior peroneal retinaculum, but no subfascial extension, is present. The peroneal tendons are normal. Color Doppler imaging revealed hyperemia, consistent with subcutaneous cellulitis.*

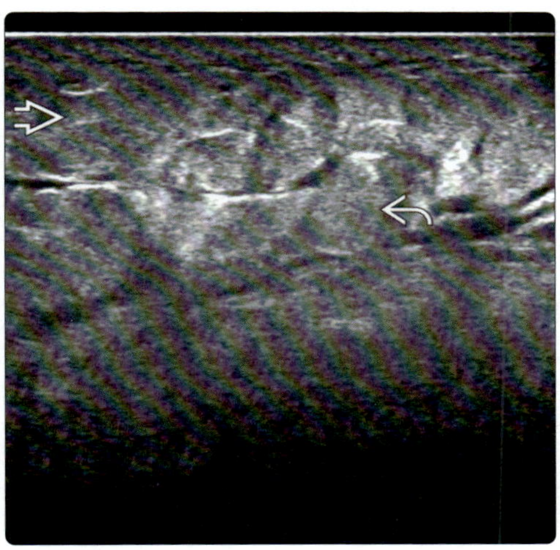

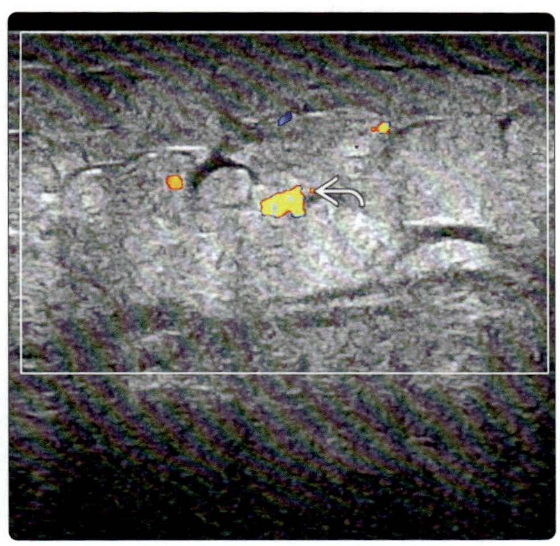

(Left) *Transverse US of the ankle region shows severe, subcutaneous edema with diffuse increase in echogenicity of subcutaneous fat. The fat lobules are indistinct.* **(Right)** *Transverse color Doppler US of the same patient shows mild hyperemia of the subcutaneous tissues consistent with moderate-severity cellulitis. Hyperemia helps differentiate cellulitis from noninfective causes of ankle edema, such as heart failure.*

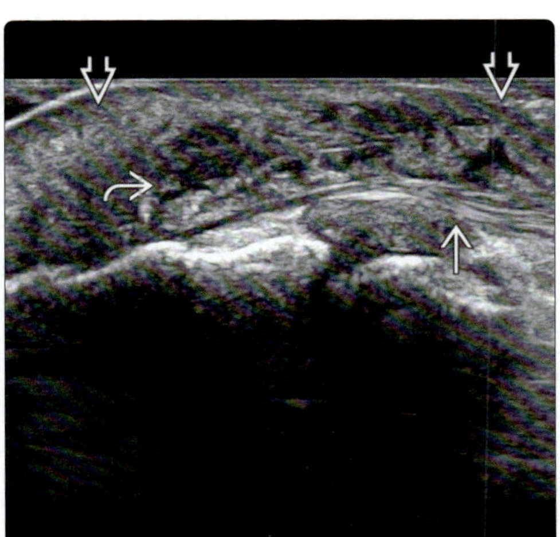

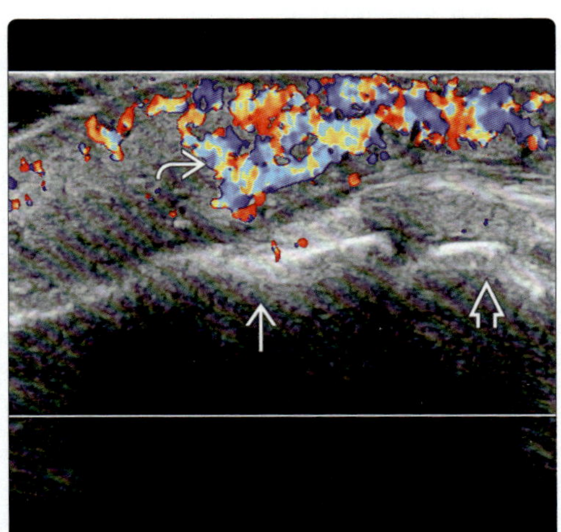

(Left) *Longitudinal US of the palmar aspect of the thumb shows the presence of severe subcutaneous soft tissue thickening and edema. Linear hypoechoic areas representing edematous interlobular septa are present. No focal fluid collection is present. The flexor pollicis longus tendon is normal.* **(Right)** *Longitudinal color Doppler US of the same patient shows severe hyperemia indicative of severe cellulitis. The distal and proximal phalanges are normal.*

(Left) *Transverse US of a patient with extensive hand swelling and erythema shows severe, subcutaneous soft tissue thickening with edematous interlobular septa on the dorsum of the hand. The metacarpal shafts are normal.* **(Right)** *Transverse color Doppler US of the same patient shows severe subcutaneous and peritendinous soft tissue hyperemia indicative of primary severe cellulitis with secondary tenosynovitis.*

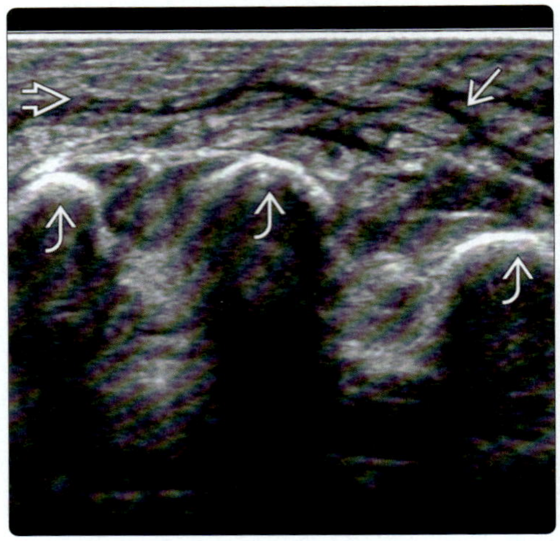

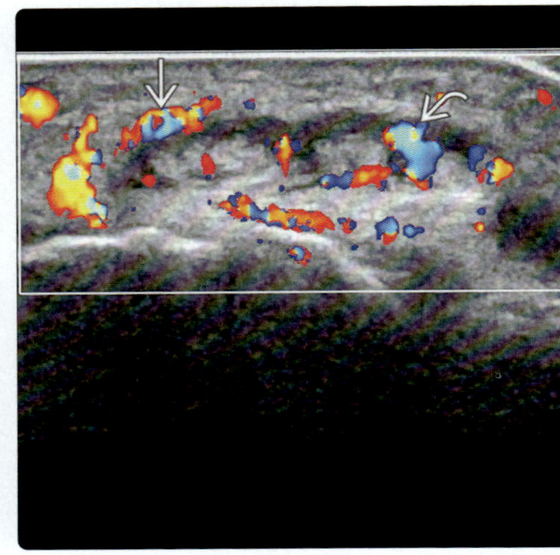

(Left) *Longitudinal US of the index finger in a patient with a recent cat bite to the index finger shows severe cellulitis, mainly superficial to but also deep to the extensor tendon at the metacarpophalangeal joint level.* **(Right)** *Transverse US of the same patient shows the inflammation surrounding the extensor tendon. The extensor tendon is moderately swollen, indicative of cellulitis involving both the subcutaneous tissues and the extensor tendon.*

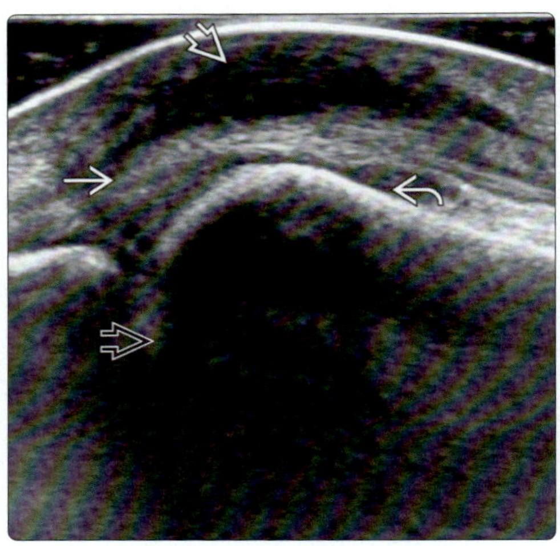

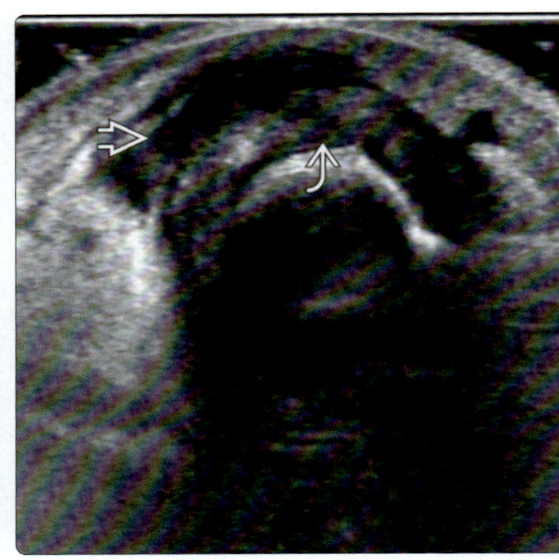

(Left) *Transverse US of a patient with forefoot cellulitis shows diffuse soft tissue edema with a relatively well-defined and localized area of hypoechogenicity present between the metatarsal shafts.* **(Right)** *Transverse color Doppler US in the same patient shows mild hyperemia around the hypoechoic area. The appearances are those of a phlegmonous preabscess inflammatory mass. A follow-up US after a few days would be helpful if symptoms did not settle.*

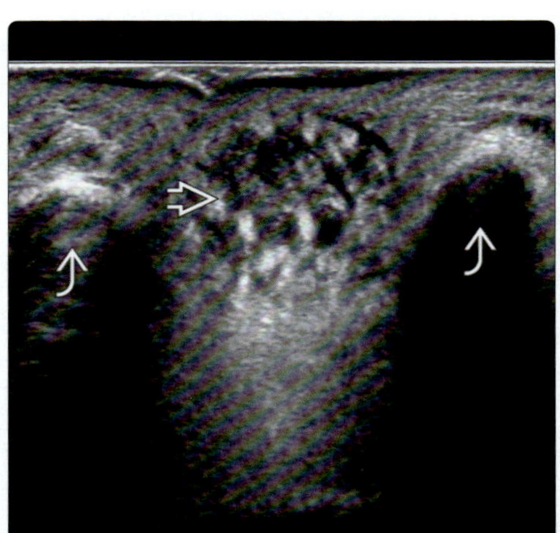

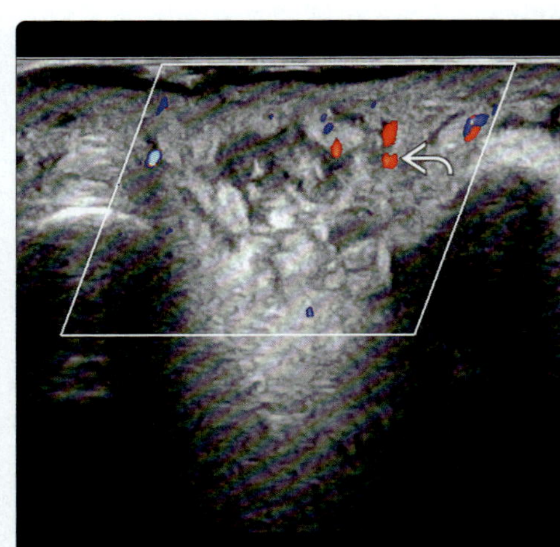

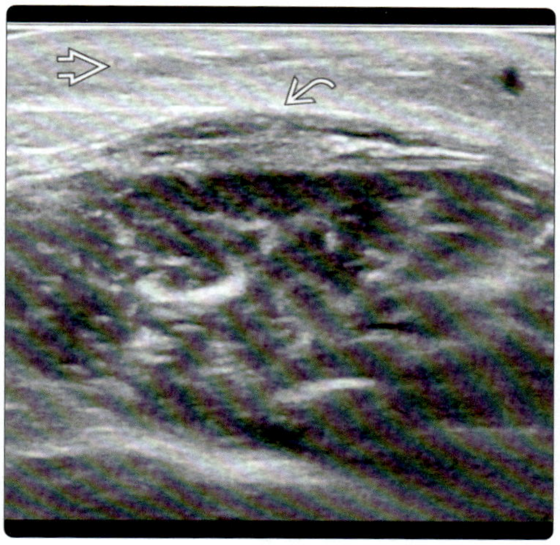

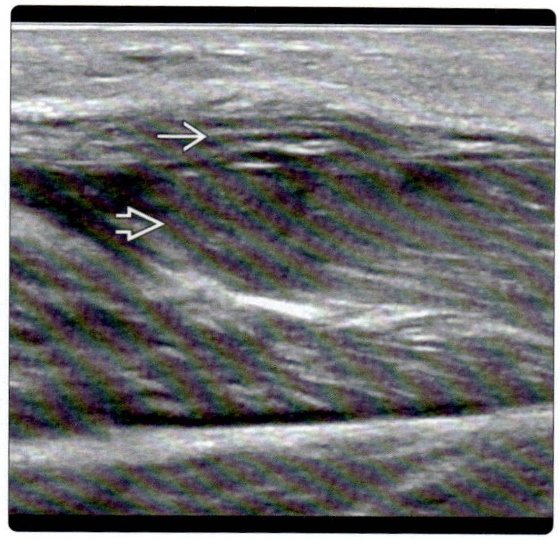

(Left) *Transverse US of the posterior calf shows a localized area of severe thickening of the investing fascia with adjacent hyperechoic edematous subcutaneous tissues.* **(Right)** *Longitudinal US of the same patient shows fascial delamination with intrafascial fluid. The gastrocnemius muscle deep to the investing fascia is moderately edematous with an indistinct fibrillary pattern.*

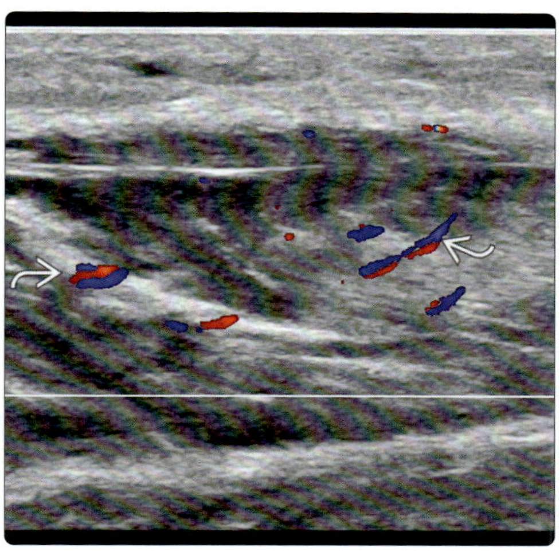

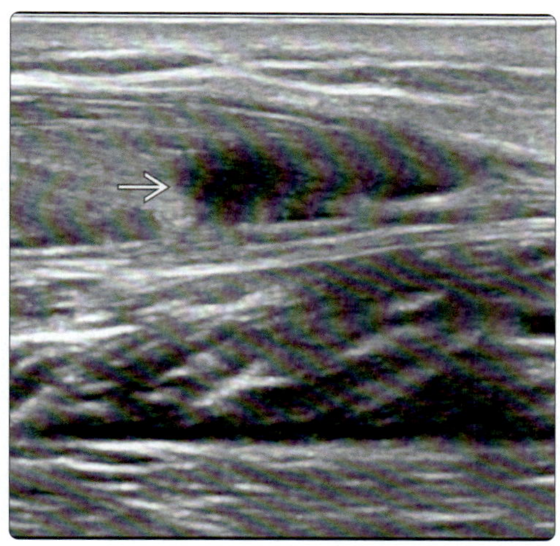

(Left) *Longitudinal color Doppler US of the same area shows fascial and gastrocnemius muscle hyperemia.* **(Right)** *Longitudinal US of the more distal part of the gastrocnemius muscle shows a small intramuscular abscess. The overall appearances are indicative of infective fasciitis with cellulitis and pyomyositis.*

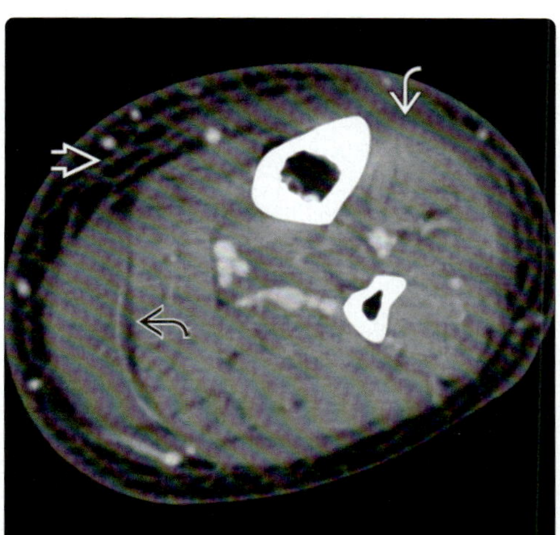

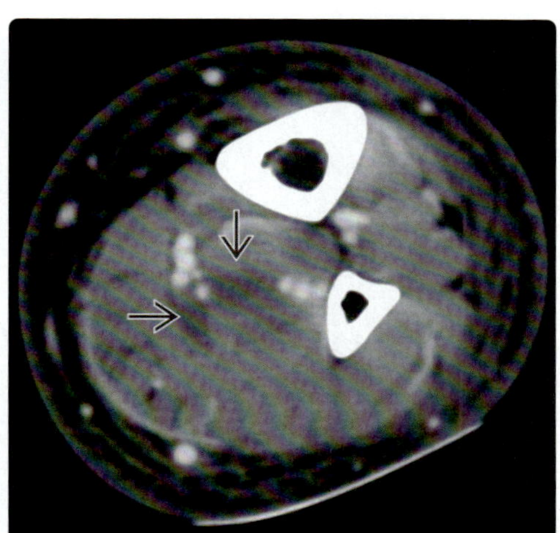

(Left) *Axial CECT of the distal leg in a patient with acute leg swelling and septicemia shows diffuse increased attenuation and stranding of the subcutaneous fat. There is also moderate, generalized edema with fluid tracking along the superficial and deep fascia of the calf.* **(Right)** *Axial CECT of the same patient shows ill-defined, nonenhancing areas of muscle indicative of early muscle necrosis. The overall appearances are consistent with necrotizing fasciitis.*

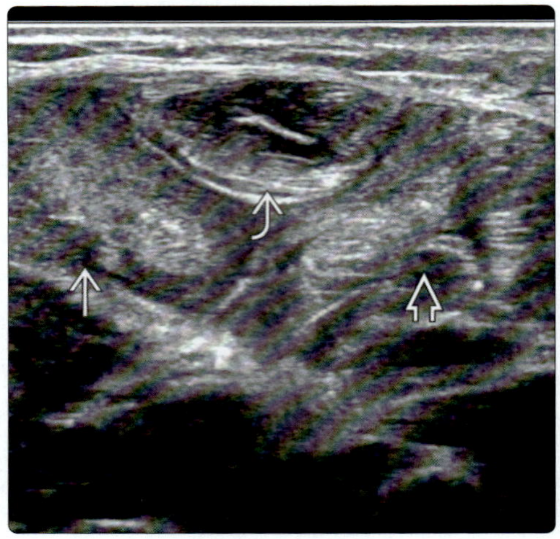

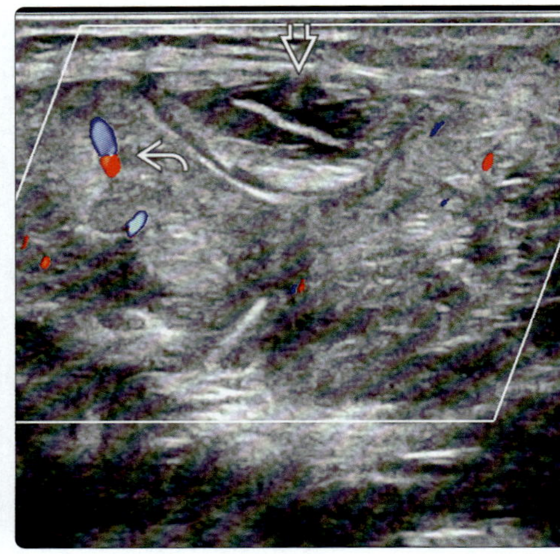

(Left) *Transverse US of the forearm in a patient with clinically progressive cellulitis shows severe edema of the flexor carpi radialis →, palmaris longus →, and flexor digitorum superficialis → muscles. Mild to moderate subcutaneous edema is present. The appearances favor polymyositis.* **(Right)** *Transverse color Doppler US of the same patient shows mild hyperemia → within the flexor muscles. Note the relatively normal appearance of the palmaris longus muscle close to the central tendon →.*

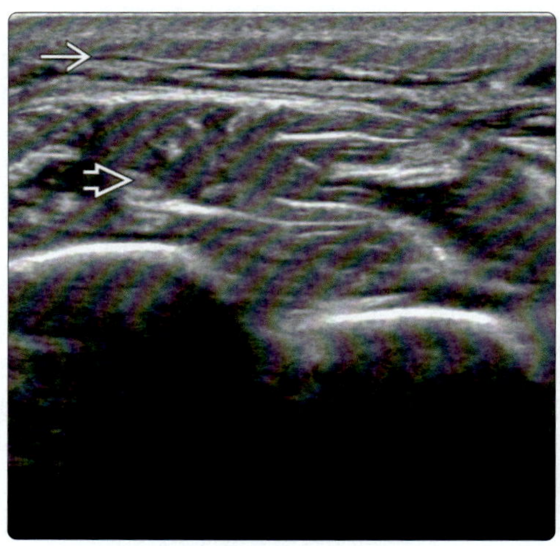

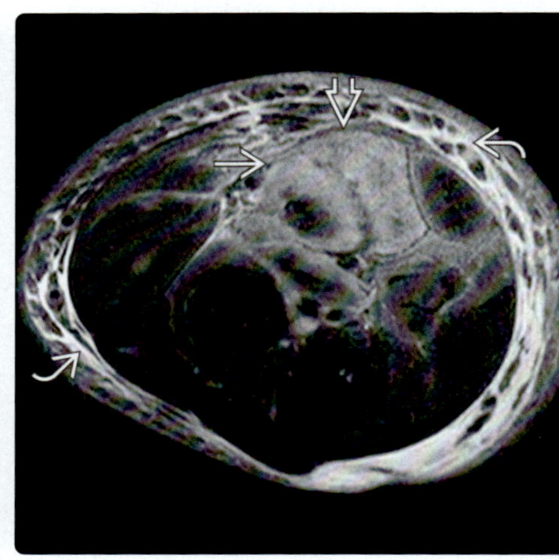

(Left) *Transverse US of the dorsal forearm of the same patient shows normal extensor musculature → with moderate subcutaneous edema →.* **(Right)** *Axial T1 C+ subtraction MR shows increased flexor muscle enhancement →, most severely affecting the flexor carpi radialis muscle →, consistent with pyomyositis. Pyomyositis is typically multifocal. Note the diffuse subcutaneous tissue edema → indicative of associated cellulitis.*

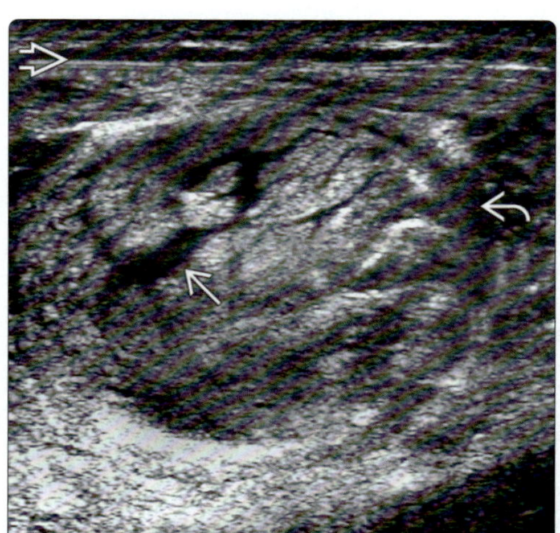

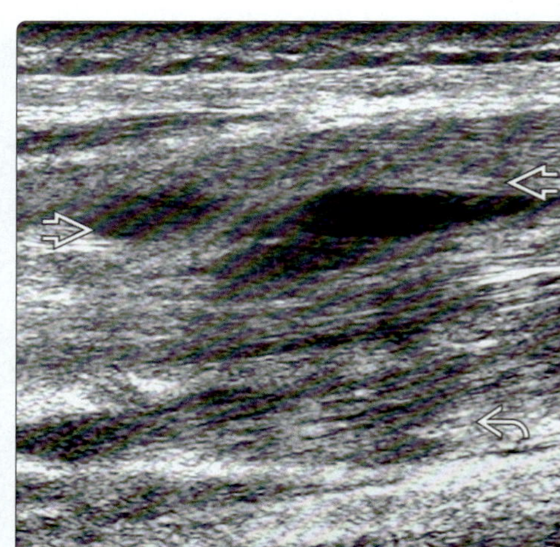

(Left) *Transverse US in an immunocompromised patient shows severe subcutaneous edema → with severe semimembranosus muscle edema. There is fluid tracking between the muscle fibers → and between adjacent muscles →.* **(Right)** *Longitudinal US of the same patient shows the swollen semimembranosus muscle with largely intact muscle fibers →. The appearances represent an infective pyomyositis. A small amount of fluid → is tracking between the muscle fibers.*

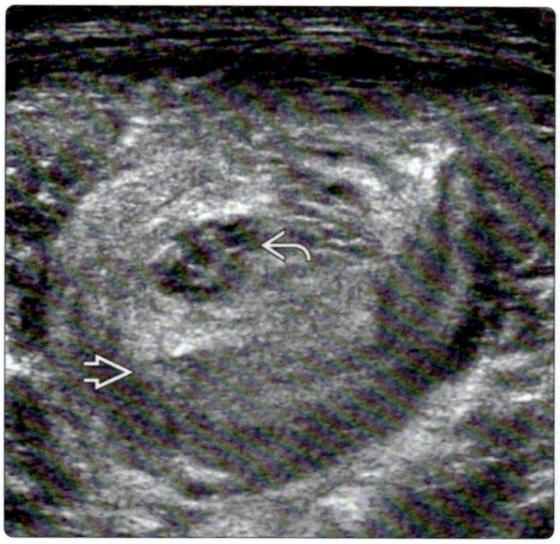

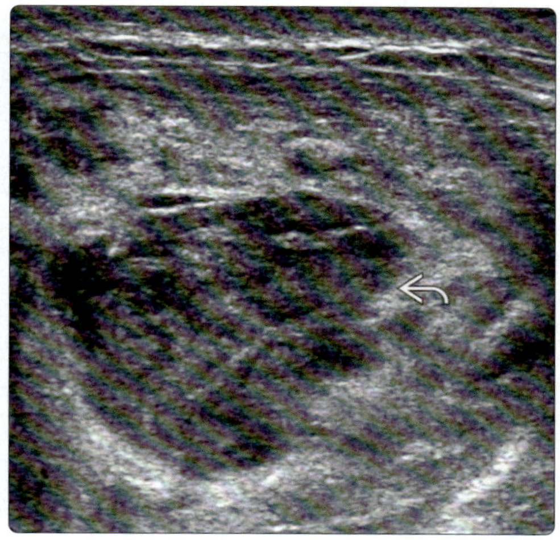

(Left) *Transverse US of the same patient 10 days later shows progressive myositis with loss of the normal fibrillary pattern . A small, hypoechoic intramuscular focus due to early abscess formation is present.* **(Right)** *Transverse US of the same patient 4 days after the prior image shows evolving pyomyositis with an enlarged area of liquefaction due to abscess formation . Aspiration of the abscess yielded Stenotrophomonas maltophilia.*

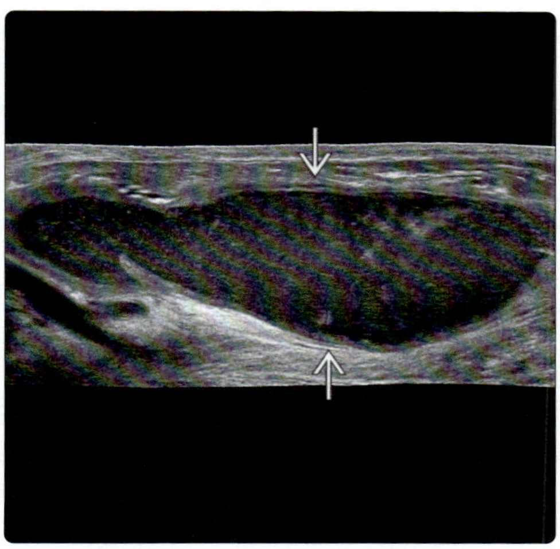

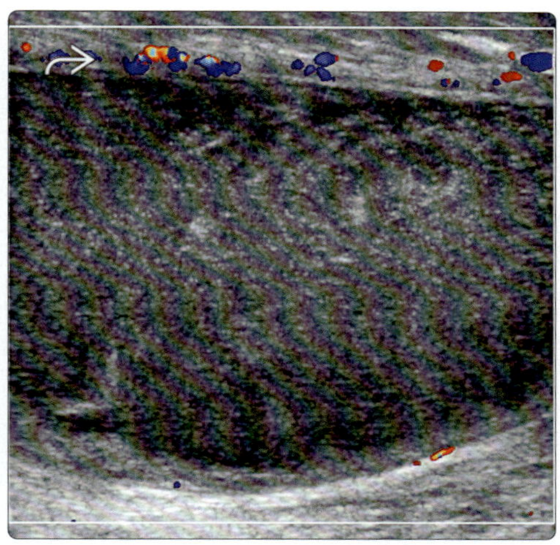

(Left) *Extended field-of-view US in a patient with illicit intravenous drug use shows a large, well-defined, hypoechoic collection within the medial belly of the gastrocnemius. The well-defined, thickened wall of the collection and the lack of surrounding soft tissue edema are indicative of a chronic abscess.* **(Right)** *Transverse color Doppler US of the same lesion shows mild to moderate hyperemia along the wall of the collection. Aspirate grew Staphylococcus aureus.*

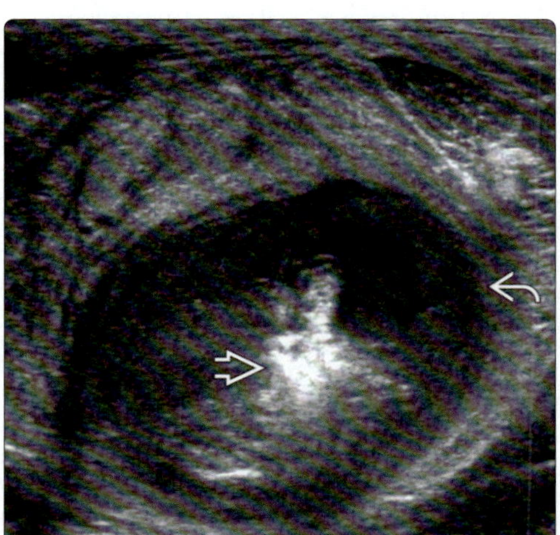

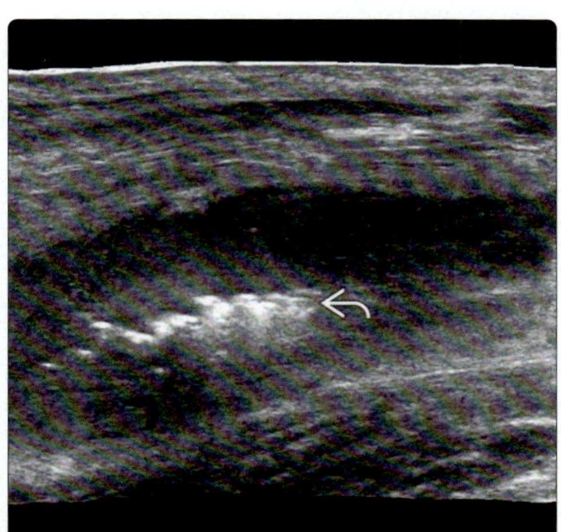

(Left) *Transverse US of the thigh shows severe inflammation of the hamstring muscles with a large, well-defined abscess within the semitendinosus muscle. The hyperechoic foci within the abscess are due to gas.* **(Right)** *Extended field-of-view US of the same patient shows multiple hyperechoic foci with comet-tail artifact within this gas-forming abscess. Aspirate yielded Streptococcus milleri, Staphylococcus aureus, and Klebsiella species.*

Bone Infection

KEY FACTS

IMAGING

- Periosteal thickening and hyperemia with juxtacortical edema
- US can depict outer cortical, periosteal, and extraosseous changes
  - Periosteal thickening
  - ± juxtacortical soft tissue edema and hyperemia
  - ± cortical irregularity
  - ± sympathetic joint effusion
- Severe infection may show
  - Subperiosteal exudate or, less commonly, abscess
  - Soft tissue abscess
  - Cortical disruption ± extraosseous inflammatory mass
- Chronic osteomyelitis may show
  - Sinus tracts
  - Cloaca
  - Sequestra
- US useful in neonates and young infants when clinical signs are often nonspecific
  - High propensity to develop periosteal reaction in children
  - US shows features of acute osteomyelitis several days earlier than radiography
- US useful in postoperative patients when metallic implants limit MR assessment
- US useful in assessment of involvement of adjacent soft tissue and guiding aspiration or biopsy
- MR has highest sensitivity and specificity for both acute and chronic osteomyelitis

TOP DIFFERENTIAL DIAGNOSES

- Soft tissue infection
- Septic arthritis
- Bone sarcoma
- Leukemia or metastatic infiltration

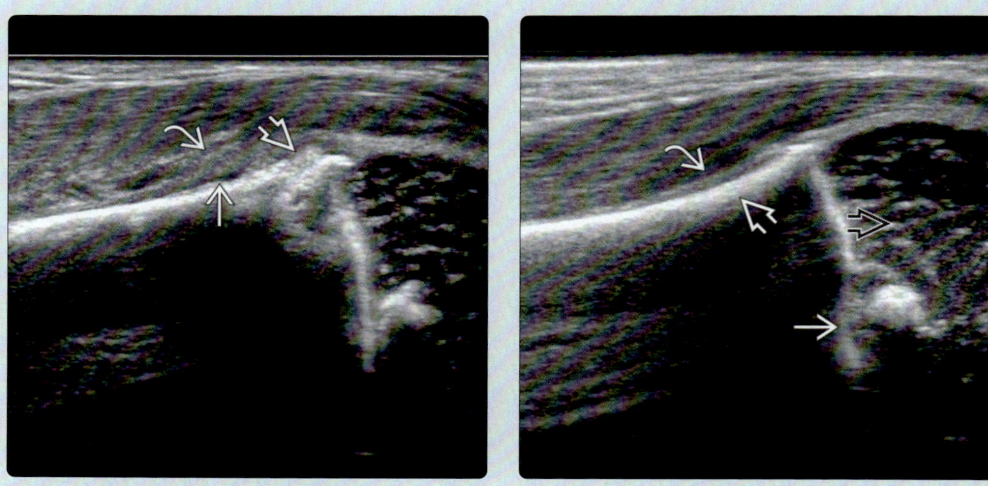

(Left) *Longitudinal US of a neonate with acute osteomyelitis of the femur shows focal cortical irregularity at the distal femoral metaphysis. This is associated with periosteal soft tissue thickening and a small amount of subperiosteal fluid.* **(Right)** *Lateral US of the contralateral femur of the same patient shows the smooth cortex of the femoral metaphysis and normal thin hyperechoic periosteum. The physis and epiphysis are normal.*

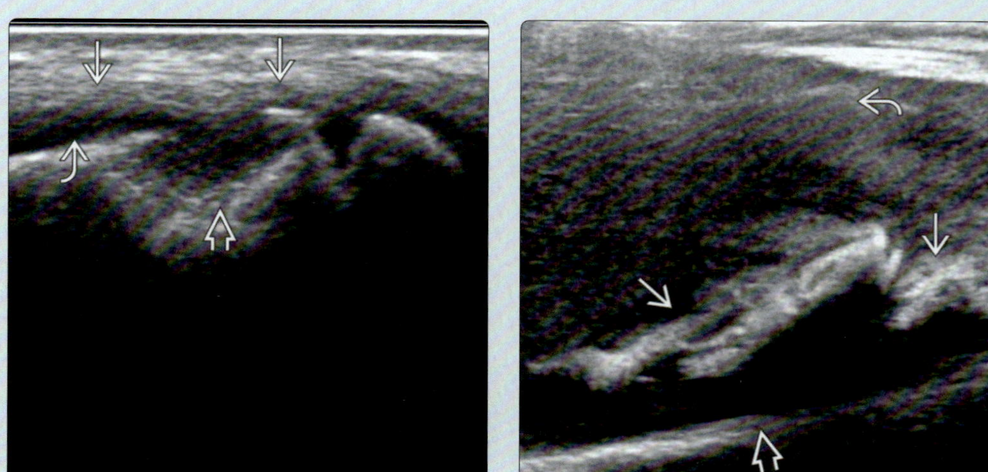

(Left) *Longitudinal US shows acute osteomyelitis with a focal bone erosion filled by hypoechoic inflammatory soft tissue at the distal ulnar metaphysis with associated periosteal thickening and a small amount of subperiosteal exudate.* **(Right)** *Longitudinal US shows partially mineralized, thickened mature periosteal reaction deep to a large collection due to chronic osteomyelitis in the femoral diaphysis. The femoral bony cortex in this section is largely intact.*

TERMINOLOGY

Definitions

- Osteomyelitis: Infection of bone marrow
- Acute osteomyelitis: Symptomatic osteomyelitis < 6 weeks duration
- Chronic osteomyelitis: Symptomatic osteomyelitis > 6 weeks duration
- Infective osteitis: Infection of bone cortex (usually contiguous with juxtacortical infection)
- Sequestrum (meaning "separated"): Fragment of dead bone surrounded by granulation tissue
- Involucrum (meaning "to wrap in"): Layer of living periosteal new bone formed around dead bone
- Cloaca (meaning "drain"): Opening in cortex or involucrum through which sequestra, fluid, or granulation tissue is discharged
- Brodie abscess: Focal area of chronic osteomyelitis with abscess and surrounding sclerosis

IMAGING

General Features

- Best diagnostic clue
  - Periosteal thickening and hyperemia with juxtacortical edema
- Location
  - Acute osteomyelitis: Metaphyses of long tubular bones
    - Metaphyses and epiphyses in infant
  - Chronic osteomyelitis: Diaphyses of long tubular bones
  - Osteitis: Alongside area of chronic soft tissue infection or ulceration
- Size
  - Acute osteomyelitis is usually less extensive than chronic osteomyelitis
- Morphology
  - Acute osteomyelitis: Bony resorption
  - Chronic osteomyelitis: Simultaneous bony resorption, draining infection (sequestra, cloaca, sinus tract), and bony repair (bone deposition, expansion)

Radiographic Findings

- Radiography
  - Insensitive for early acute osteomyelitis
    - Radiographic bony changes occur 10-14 days after symptom onset, depending on severity of infection
  - Sensitive for chronic osteomyelitis
    - Sclerosis, periosteal thickening, sequestra, cloaca, expansion ± bowing
    - Involucrum is rarely seen in modern practice

CT Findings

- NECT
  - Useful when MR is not available
  - Useful in chronic osteomyelitis for depicting cloaca, sequestra, and sinus tracts
    - Sequestra may be difficult to see on MR (hypointense) or radiography (if marked sclerosis and cortical thickening)
- CECT
  - Abscesses and sinus tracts enhance peripherally

MR Findings

- T1WI
  - Hypointense marrow edema ± cortical destruction
- T2WI FS
  - Acute osteomyelitis: Hyperintense marrow edema without discrete mass
    - ± focal cortical destruction, ± periosteal thickening (lamellar, Codman triangle)
    - ± juxtacortical edema, inflammation, or inflammatory mass
    - MR = highly sensitive and specific for acute osteomyelitis
  - Chronic osteomyelitis; additional findings include
    - ± intraosseous, transcortical, or extraosseous sinus tracts
    - ± hypointense sequestra
    - ± periosteal thickening (from lamellar immature to thick mature)
    - Appearances vary with level of inflammatory activity
  - Brodie abscess (intramedullary abscess cavity): Hyperintense + hypointense sclerotic rim
    - Abscess surrounded by reactive sclerotic bone and inflammation
- T1WI C+
  - Differentiates abscess (rim enhancement) from inflammatory tissue (diffuse enhancement)
    - Sequestra, in chronic osteomyelitis, may also enhance peripherally and simulate abscess
    - Enhancing tissue in chronic osteomyelitis may represent reparative or inflammatory tissue

Ultrasonographic Findings

- US useful in neonates and young infants when clinical signs are often nonspecific
  - Shows features of acute osteomyelitis several days earlier than radiography
  - Less sensitive than MR or scintigraphy
- US sensitivity is highest in suspected osteomyelitis of tubular bones in children where propensity to develop periosteal reaction is highest
- US also useful in postoperative patients when metallic implants limit MR assessment
- US usually only depicts outer cortical, periosteal, and extraosseous changes
  - Periosteal thickening is seen as periosteal hyperemic rim, usually with hyperemia
  - ± juxtacortical soft tissue edema and hyperemia
  - ± cortical irregularity
  - ± sympathetic joint effusion
- Severe infection may show
  - Subperiosteal exudate or, less commonly, abscess
  - Soft tissue abscess
  - Cortical disruption ± extraosseous inflammatory mass
    - Visualization of medullary canal if osteolysis of overlying cortex provides acoustic window
  - US-guided aspiration and biopsy for periosteal fluid and inflammatory soft tissue mass, respectively
- Chronic osteomyelitis may show
  - Sinus tracts
  - Cloaca
  - Sequestra

- US appearances of acute bone infection are nonspecific
  - Periosteal thickening ± subperiosteal fluid ± soft tissue edema also seen in
    - Stress reaction of bone
    - Sickle cell disease with medullary infarction
    - Bone sarcoma
- In neonates and young children, pure chondral infection may occur
  - May be difficult to detect on US as hypoechoic fluid is of similar echogenicity to hypoechoic cartilage

Nuclear Medicine Findings

- Increased tracer activity in 3-phase bone scan, gallium scan, white blood cell scan, and PET/CT

Imaging Recommendations

- Best imaging tool
  - MR: High sensitivity and specificity for both acute and chronic osteomyelitis
    - Very high negative predictive value, i.e., normal MR excludes osteomyelitis
  - US: Assess involvement of adjacent soft tissue
    - Guide aspiration or biopsy for histology, microbiology, and treatment
    - Useful in regions with orthopedic instrumentation where infection might not be seen by MR or CT
    - Useful in assessment of soft tissue involvement, especially in chronic osteomyelitis
- Protocol advice
  - In acute osteomyelitis, periosteal reaction is most specific sign
    - Proceed to MR if US is positive, and also if US is negative, and there is strong clinical suspicion of osteomyelitis
  - Check for signs of osteomyelitis if performing US for suspected soft tissue or joint infection

DIFFERENTIAL DIAGNOSIS

Soft Tissue Infection

- Soft tissue edema and inflammation ± inflammatory mass ± abscess
- Check bone surface for signs of osteomyelitis

Septic Arthritis

- Similar clinical picture to acute osteomyelitis
- Medium to large joint effusion ± hyperemia ± positive joint aspiration

Bone Sarcoma

- No clinical signs of infection
- Cortical destruction ± extraosseous soft tissue mass ± spiculated new bone formation

Leukemia or Metastatic Infiltration

- Usually known primary tumor
- Multiple sites of involvement in leukemia

PATHOLOGY

General Features

- Etiology
  - *Staphylococcus aureus*: Most common infecting organism in all age groups
  - Common organisms: *Streptococcus*, *Pseudomonas*, *Haemophilus*, *Enterobacter*, *Mycobacterium*
  - May be due to mixed organisms
    - Especially chronic osteomyelitis

Gross Pathologic & Surgical Features

- Hematogenous seeding
- Direct contamination

Microscopic Features

- Inflammation of marrow
- Vascular compromise
- Necrosis, abscesses, sequestra

CLINICAL ISSUES

Presentation

- Most common signs/symptoms
  - Pain, tenderness, fever with signs of inflammation
- Other signs/symptoms
  - Restricted motion, constitutional symptoms
- Clinical profile
  - Acute osteomyelitis progresses over days to weeks
    - ± positive blood cultures
  - Chronic osteomyelitis usually presents de novo without preexisting acute osteomyelitis
    - Typically established at time of presentation
    - Early disease is subclinical

Demographics

- Age
  - Acute osteomyelitis more common in young children
  - Chronic osteomyelitis more common in older age group
- Epidemiology
  - Both acute and chronic osteomyelitis less common with increasing economic development globally
  - Chronic osteomyelitis in adults usually 2° to trauma

Natural History & Prognosis

- Good prognosis for acute osteomyelitis if treated early
- Chronic osteomyelitis is difficult to fully eradicate; relapsing-remitting clinical course

Treatment

- Conservative
  - Rest, elevation, intravenous antibiotics
- Surgical
  - Debridement ± antibiotic-impregnated beads

DIAGNOSTIC CHECKLIST

Consider

- Bone infection if unexplained periosteal thickening or juxtacortical inflammatory mass is detected even though cortex seems intact

SELECTED REFERENCES

1. DiPietro MA et al: Pediatric musculoskeletal ultrasound. Pediatr Radiol. 47(9):1144-1154, 2017
2. Nguyen JC et al: US evaluation of juvenile idiopathic arthritis and osteoarticular infection. Radiographics. 37(4):1181-1201, 2017

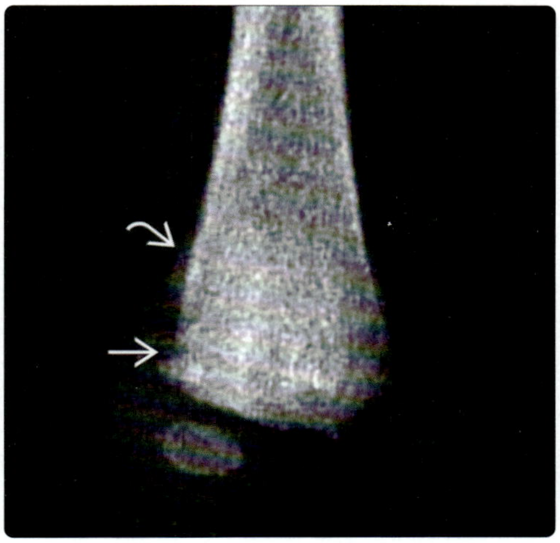
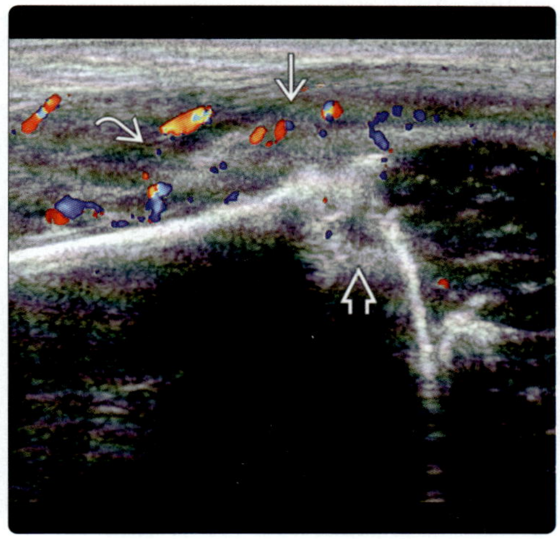

(Left) *Frontal radiograph of a 13-day-old neonate with fever and decreased leg movement shows mild cortical resorption at the distal femoral metaphysis with a fine periosteal reaction.* **(Right)** *Longitudinal color Doppler US of the same patient shows acute osteomyelitis with focal metaphyseal cortical resorption, moderate periosteal soft tissue thickening, and hyperemia. A small amount of fluid is present alongside the periosteal soft tissue.*

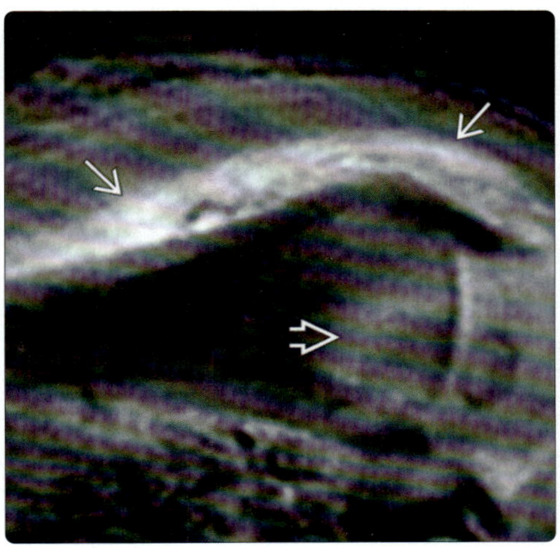
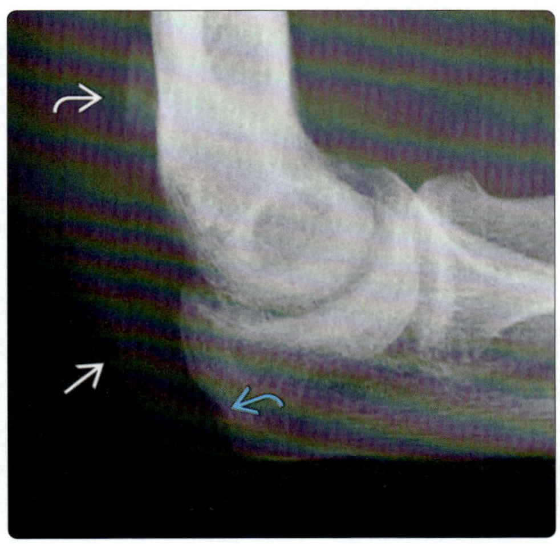

(Left) *Sagittal STIR MR of the same patient shows metaphyseal marrow edema with thickened, hyperintense inflamed periosteal soft tissue consistent with acute osteomyelitis.* **(Right)** *Lateral radiograph of an adult patient with fever and elbow pain shows soft tissue swelling at the posterior aspect of the elbow joint with a focal subtle cortical defect at the tip of the olecranon process. There is also some unrelated focal hyperostosis of the distal humeral shaft.*

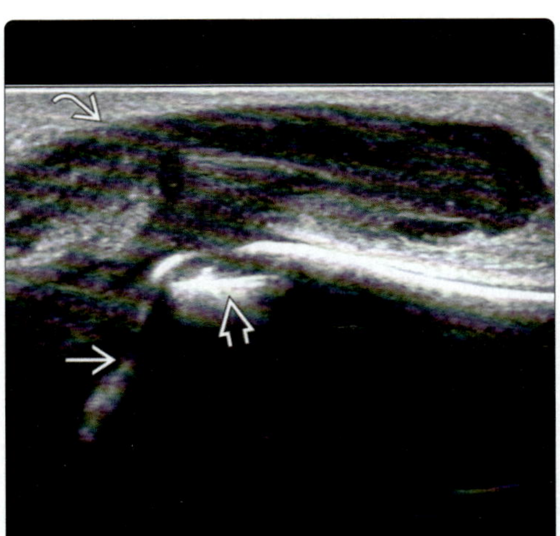
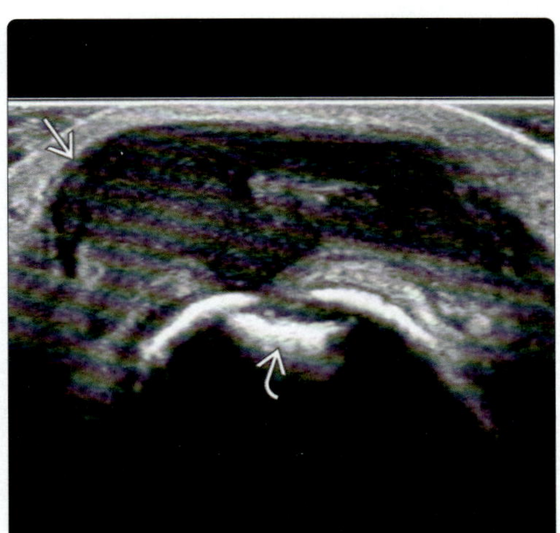

(Left) *Longitudinal US of the same patient shows a localized, partially liquefied abscess at the posterior aspect of the olecranon process and proximal ulna. Note a small cortical defect at the posterior tip of the olecranon process.* **(Right)** *Transverse US of the same region shows the olecranon cortical defect. Osteomyelitis of the olecranon process leads to focal cortical disruption and a secondary juxtacortical abscess. This abscess should not be confused with primary olecranon bursitis.*

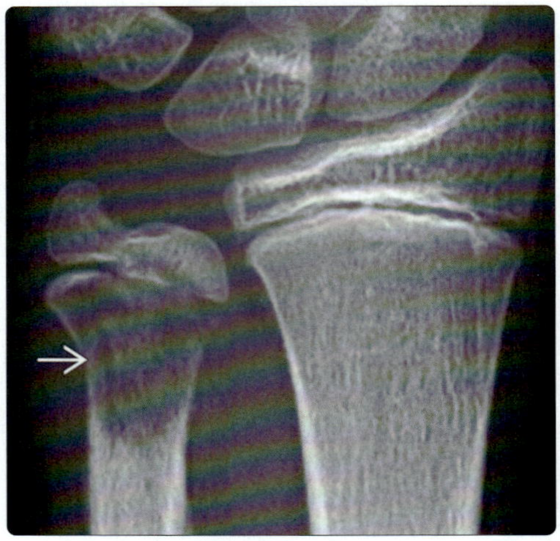

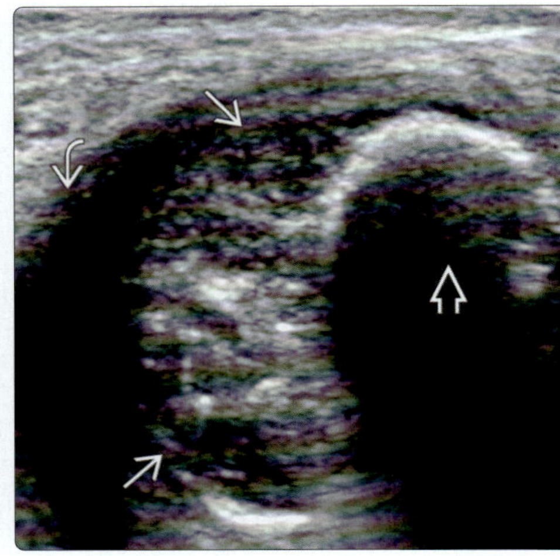

(Left) *Frontal radiograph of a 14 year old who presented with fever and wrist pain shows a lytic lesion → with a wide zone of transition in the distal ulnar metaphysis.* **(Right)** *Transverse US of the same patient shows prominent solid-appearing juxtacortical soft tissue thickening → surrounding the distal ulna →. Echogenic seams represent new bone formation. An outer hypoechoic area → representing fluid or severe edema is present.*

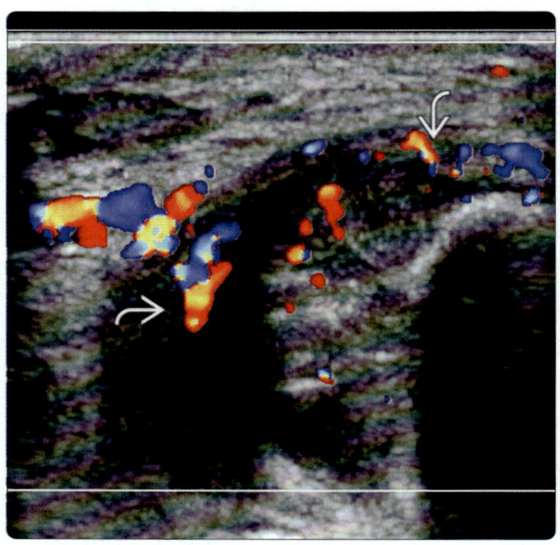

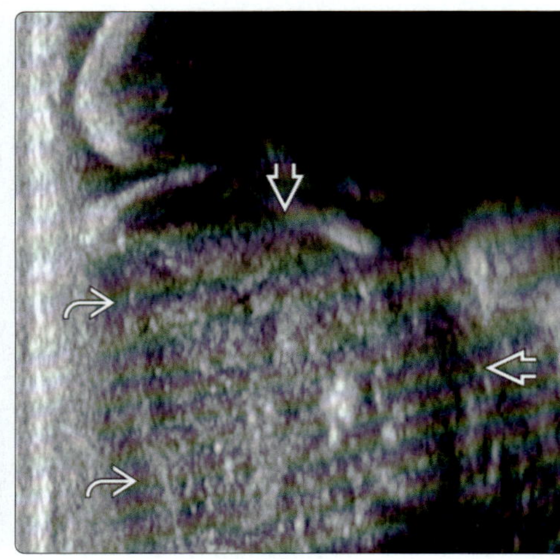

(Left) *Transverse color Doppler US of the same patient shows moderate to severe hyperemia → in the juxtacortical soft tissues. Overall appearances favor either bone sarcoma or infection.* **(Right)** *Longitudinal US of the same patient shows an osteolytic area → filled by heterogeneous inflammatory soft tissue in the distal ulnar metaphysis. The normal hyperechoic outer cortex is lost →.*

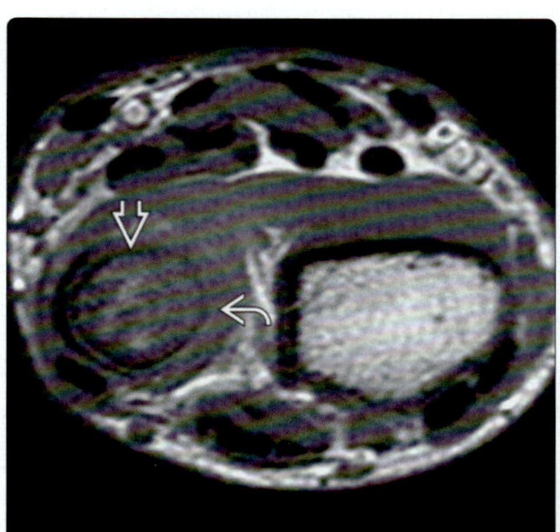

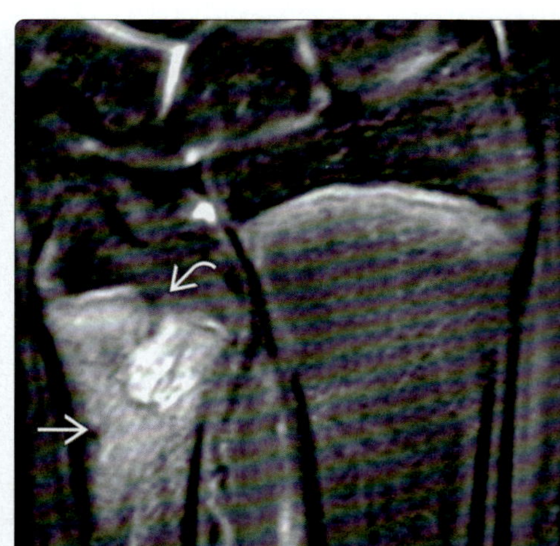

(Left) *Axial T1 MR of the same patient shows a loss of normal fatty marrow signal in the distal ulnar metaphysis → with a surrounding thick rind of hypointense soft tissue →.* **(Right)** *Coronal T2 FS MR shows moderate marrow edema → at the distal ulnar metaphysis with involvement of the physis →. MR is better than US in depicting physeal or transphyseal involvement. Percutaneous biopsy confirmed acute osteomyelitis.*

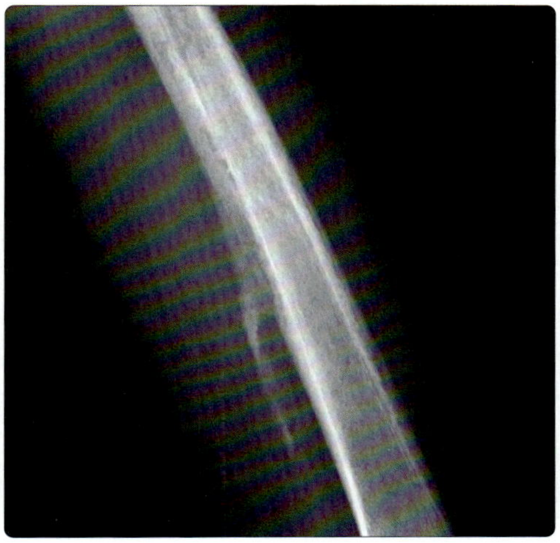

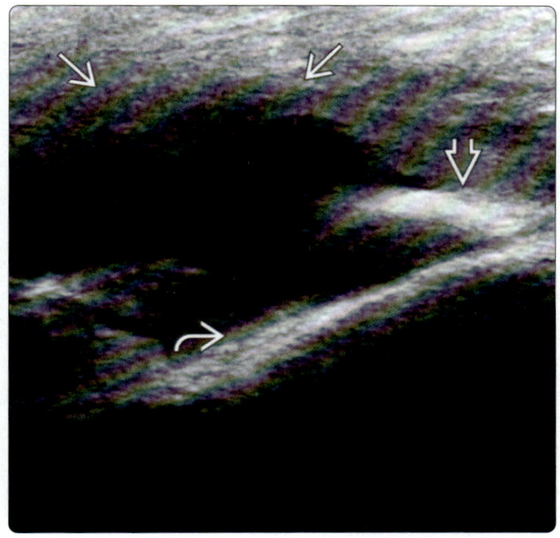

(Left) *Lateral radiograph of an 11-year-old patient with subacute thigh pain shows mild diffuse osteosclerosis with irregular mature-type periosteal thickening along the femoral diaphysis posteriorly.* **(Right)** *Longitudinal US of the same patient shows a hypoechoic extraosseous collection ➔ extending deep to the thickened, ossified periosteal tissue ➔ on the posterior femoral cortex ➔, indicative of a subacute exacerbation of chronic osteomyelitis.*

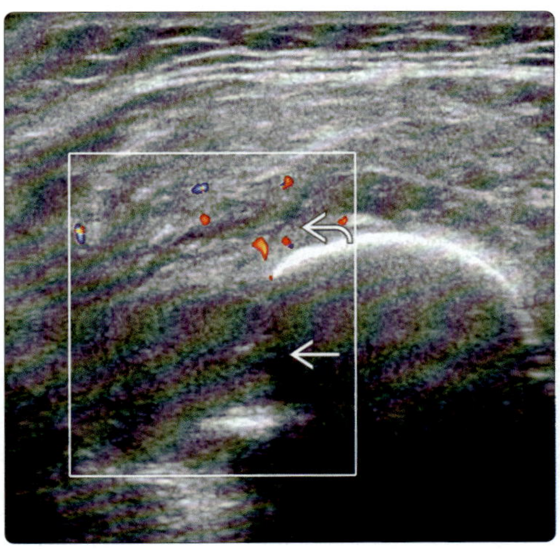

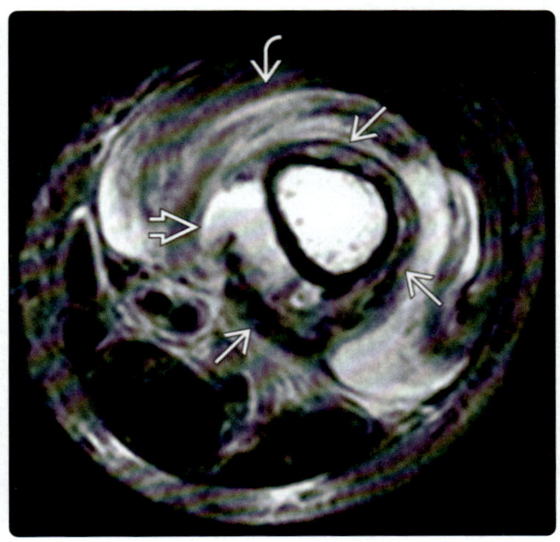

(Left) *Transverse color Doppler US in the same patient shows mild juxtacortical hyperemia ➔ and soft tissue thickening, indicative of moderate active inflammation. The hypoechoic collection ➔ demonstrates no intrinsic vascularity.* **(Right)** *Axial T2 FS MR of the same patient shows bone marrow with a hypointense thickened periosteum ➔ due to ossification (involucrum). A subperiosteal collection ➔ with a fluid level is associated with this subacute infection. There is severe quadriceps muscle edema ➔.*

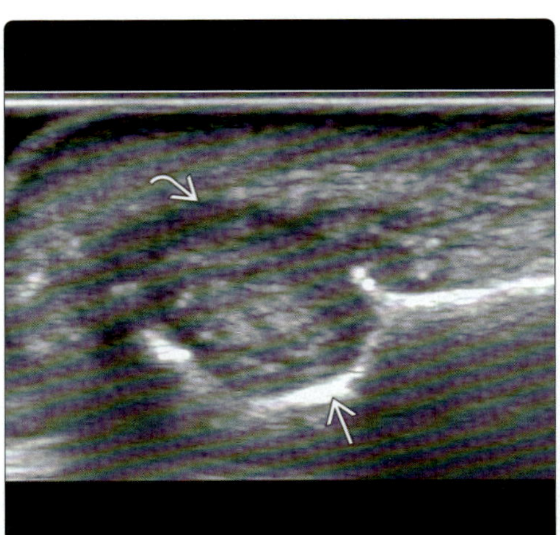

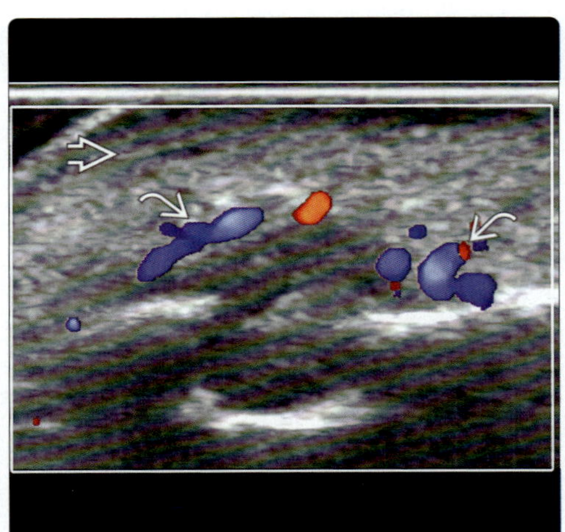

(Left) *Longitudinal US of the index finger in a patient with recent penetrating injury to the nail fold shows a focal cortical erosion ➔ in the distal phalanx. The cortical erosion is filled with hypoechoic inflammatory soft tissue ➔. No foreign body is present.* **(Right)** *Longitudinal color Doppler US of the same region shows moderate hyperemia ➔ around the soft tissue mass, indicating moderate to severe active inflammation. The surrounding subcutaneous soft tissues are moderately edematous ➔.*

(Left) *Frontal radiograph of the same patient shows a well-defined area of osteolysis* ➡ *in the distal phalanx. The soft tissue component is clearly depicted on the previous US, highlighting the mutual benefits of both radiography and US when assessing bone infection.* **(Right)** *Coronal T1 C+ FS MR of the same region shows an enhancing inflammatory mass* ➡ *at the site of osteolysis. There is a small abscess present* ➡*, all consistent with distal phalangeal osteomyelitis.*

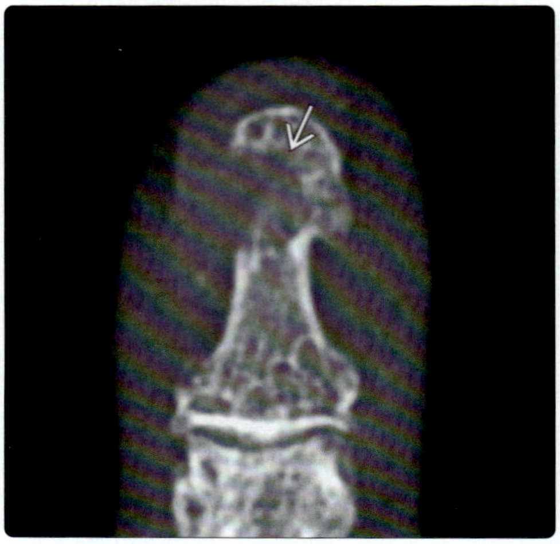

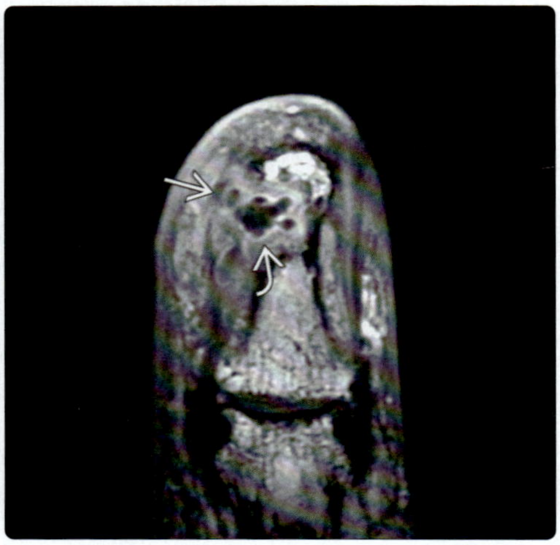

(Left) *Transverse US of an adult with acute fever and subacute thigh pain shows a large hypoechoic collection* ➡ *alongside the femoral diaphyseal cortex* ➡ *associated with a thickened ossified periosteal reaction* ➡*.* **(Right)** *Longitudinal US of the same patient shows minimal hyperemia* ➡ *in the adjacent soft tissues. The ossified periosteal new bone* ➡ *leads to strong acoustic shadowing.*

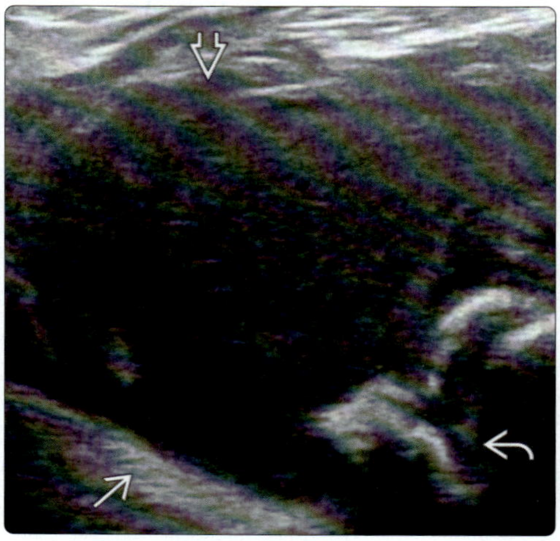

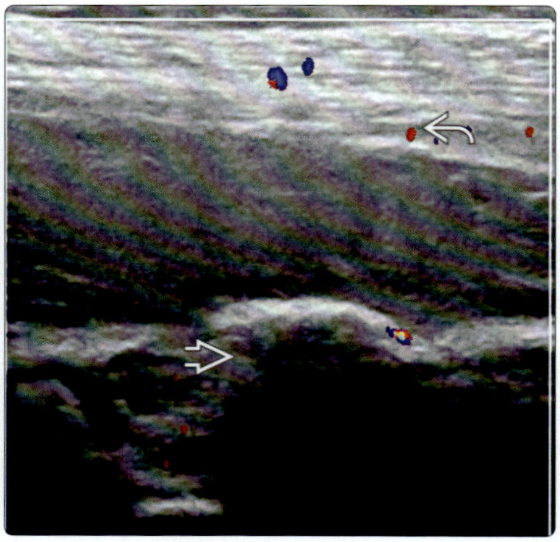

(Left) *Coronal reformatted NECT of the same patient shows focal femoral cortical osteolysis* ➡ *associated with ossified periosteal thickening* ➡*.* **(Right)** *Coronal T1 C+ FS MR of the same patient shows enhancing inflammatory tissue* ➡ *and an internal abscess near the cortical osteolysis* ➡*. This is aspirated under US guidance. Culture grew Streptococcus milleri. The final diagnosis was juxtacortical abscess with established osteitis and early osteomyelitis.*

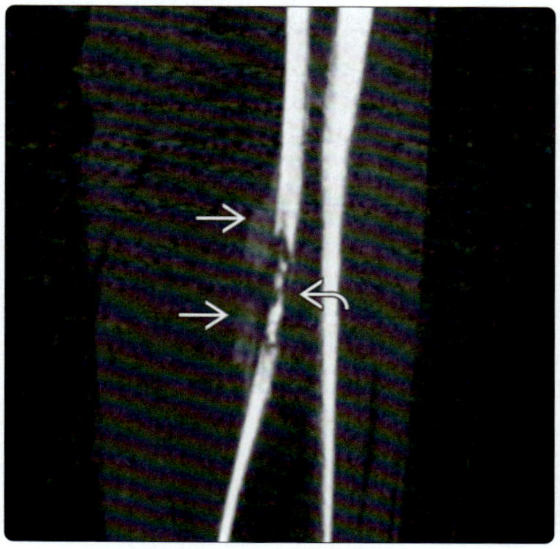

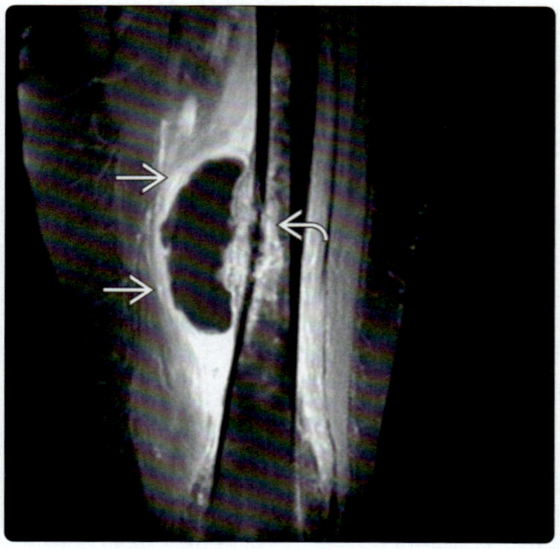

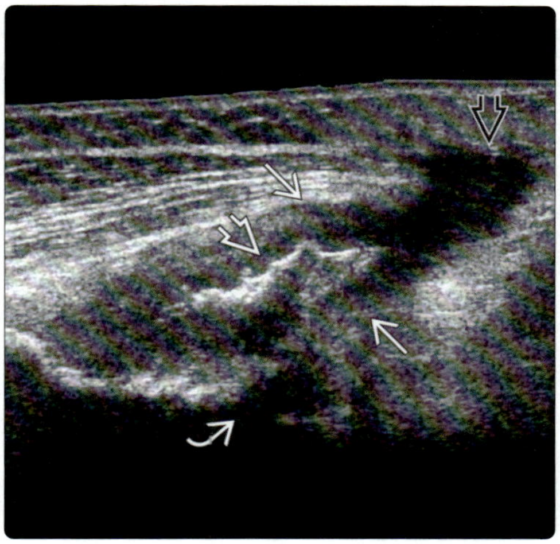

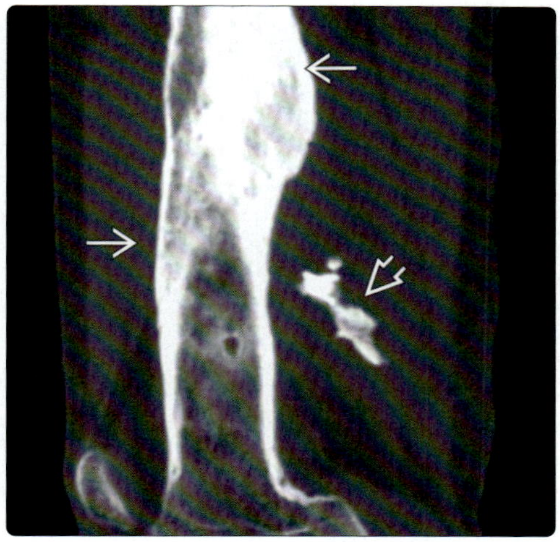

(Left) *Longitudinal US with extended field of view shows a large sequestrum ➡ within a sinus tract ➡ extending from an irregular femoral cortex ➡ to the investing fascia of the thigh ➡.* **(Right)** *Longitudinal NECT with sagittal reconstruction of the same patient shows the sequestrum ➡ and a severely thickened sclerotic femoral diaphysis ➡, all indicative of chronic osteomyelitis.*

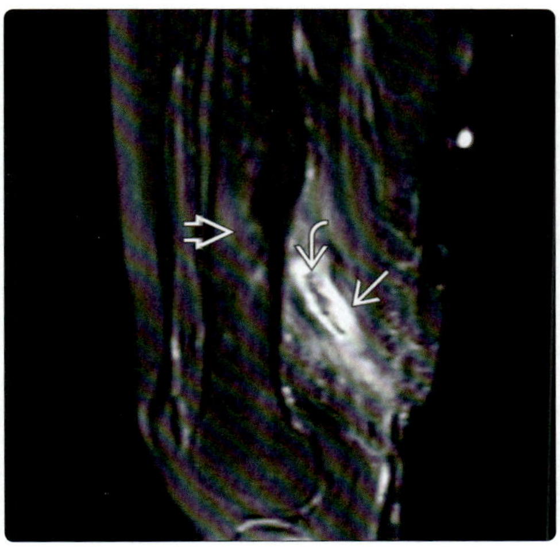

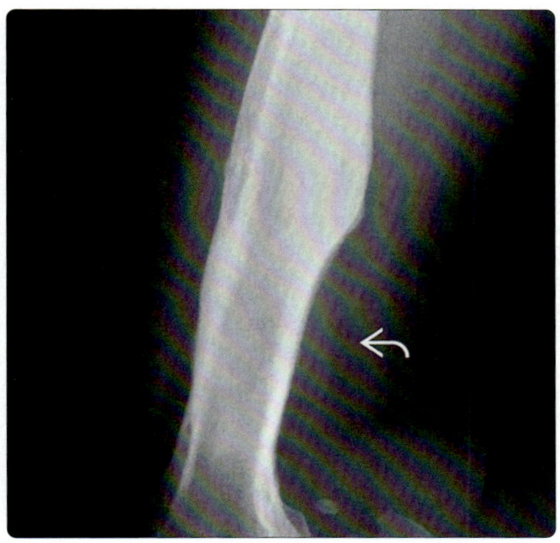

(Left) *Sagittal STIR MR of the same patient shows a hypointense sequestrum ➡ with surrounding hyperintense inflamed granulation tissue ➡ within a thin sinus tract. Mild medullary canal edema ➡ and extraosseous soft tissue indicate low-grade chronic inflammation.* **(Right)** *Lateral radiograph of the femur in the same patient following sequestrectomy and debridement shows clearance ➡ of the sequestrum.*

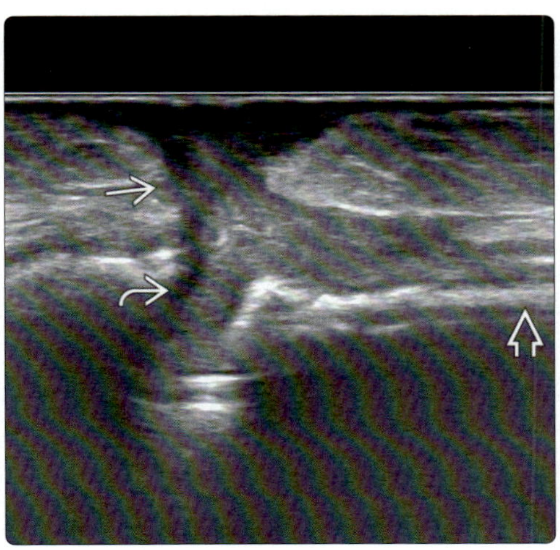

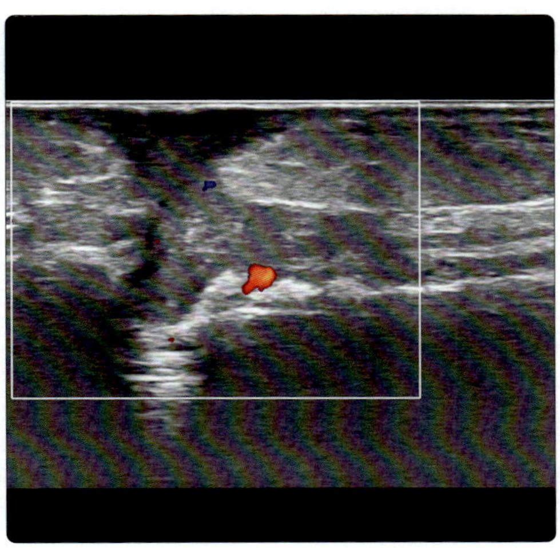

(Left) *Longitudinal US in a patient with previous fracture of the distal tibia is shown. There is a medium-sized sinus tract ➡ filled with inflammatory-type granulation or gelatinous-type tissue extending from the anterior aspect of the distal tibia ➡ to the skin. This tract communicates with a 6-mm-wide bone defect ➡ in the distal tibia at the previous fracture site. Low-grade bone infection is likely.* **(Right)** *Color Doppler US at the same site shows minimal hyperemia, indicative of low-level disease activity.*

(Left) *Frontal radiograph shows mild expansion, mixed sclerosis, and lysis of the midhumeral diaphysis with soft tissue thickening distally. The radiographic appearances are suggestive of sarcoma or chronic osteomyelitis.* **(Right)** *Longitudinal US of the same arm shows focal cortical disruption of the midhumeral diaphysis, a large extramedullary inflammatory-type mass with a small cystic area, indicative of acute exacerbation of chronic osteomyelitis.*

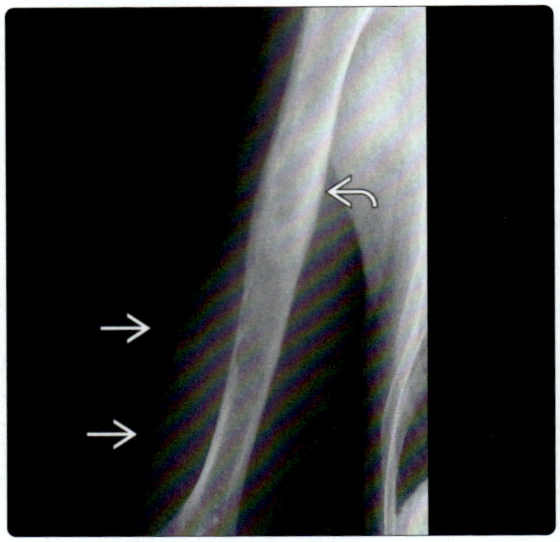

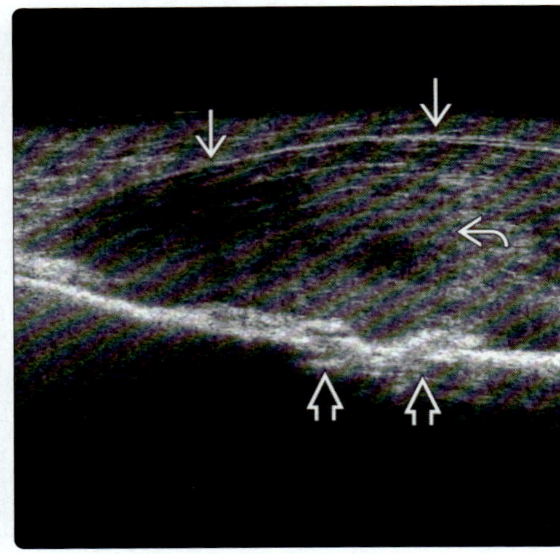

(Left) *Longitudinal T1 C+ MR in the same patient shows several small intramedullary abscesses with a large juxtacortical inflammatory mass distally compatible with acute exacerbation of chronic osteomyelitis.* **(Right)** *CECT in a patient with pain, swelling, and discharge from a recent sternotomy site shows some enhancing soft tissue filling the sternotomy site, which extends to the cutaneous surface.*

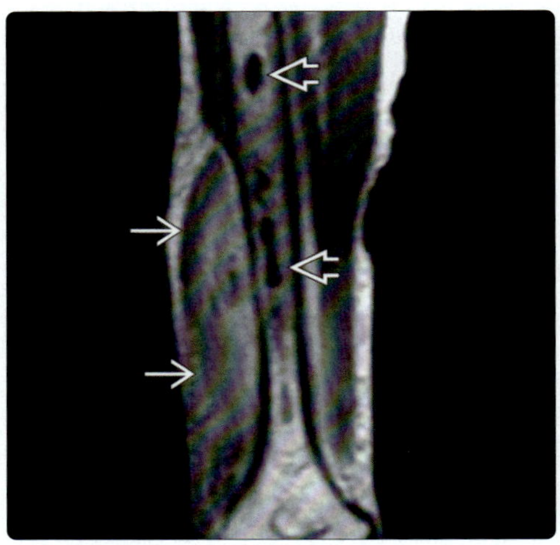

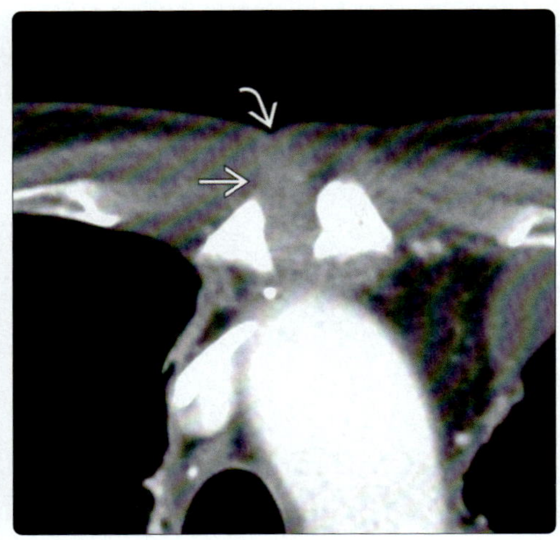

(Left) *Transverse US of the same patient shows the defect in the manubrium. There is a midline sinus tract extending superficially from the manubrium to the skin, associated with diffuse low-grade inflammation of the adjacent soft tissues.* **(Right)** *Longitudinal US of the same region shows the sinus tract filled with hypoechoic granulation tissue. The sinus tract is extending to a cutaneous opening. There is no abscess or collection present.*

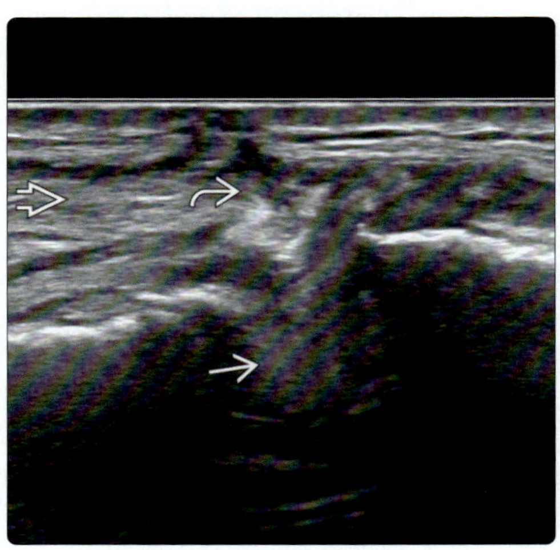

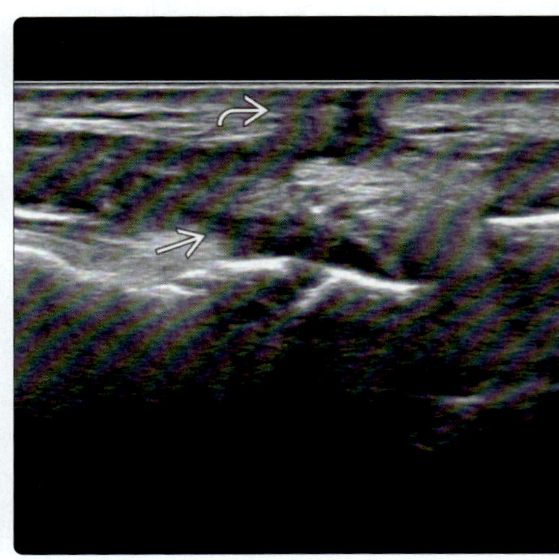

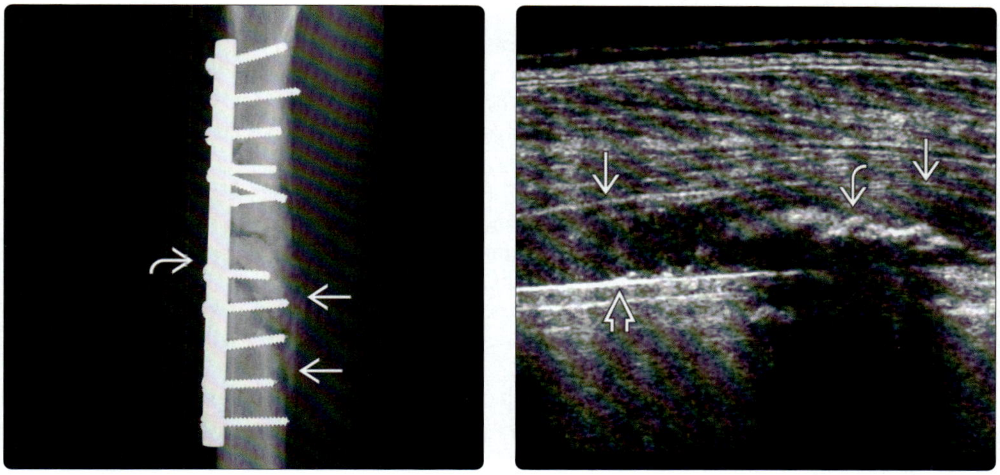

(Left) *Radiograph shows a nonunited fracture of the femoral shaft with plate and screw fixation. There is a mature, interrupted lamellar periosteal new bone suggestive of osteomyelitis.* **(Right)** *Extended field-of-view US of the same patient shows inflammatory-type tissue alongside the fixation plate and periosteal new bone formation, all consistent with postoperative chronic osteomyelitis.*

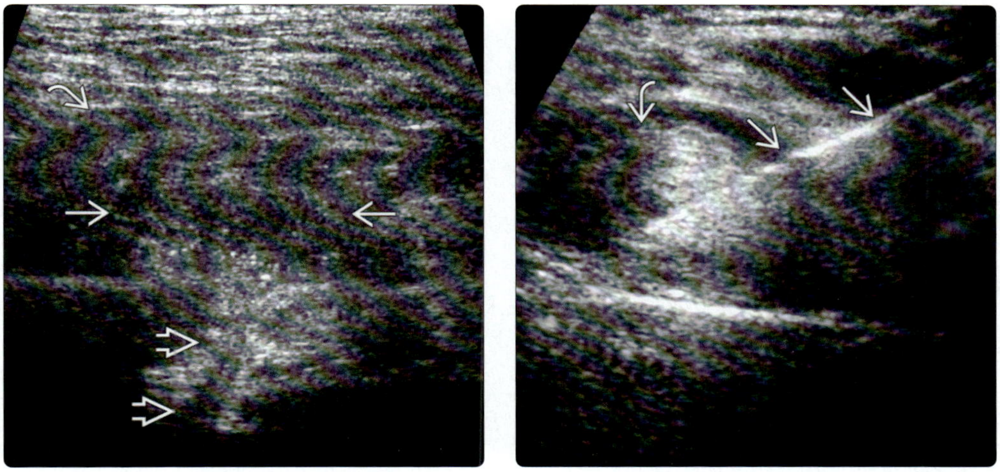

(Left) *Longitudinal US of the same patient shows inflammatory exudate exuding from the fracture site. More generalized edema of the juxtacortical tissue is present.* **(Right)** *Longitudinal US in the same patient shows a percutaneous core biopsy of inflammatory tissue. Culture yielded Staphylococcus aureus. Histology revealed acute on chronic inflammation.*

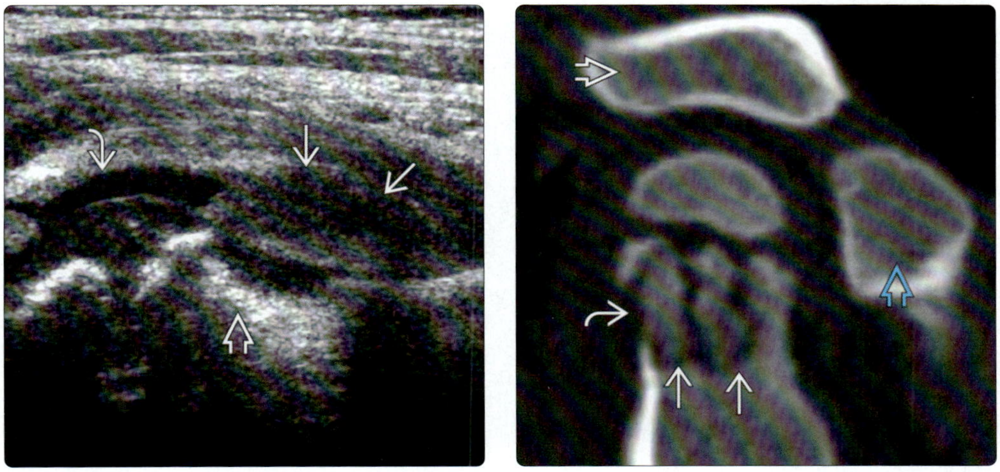

(Left) *Lateral US shows marked synovial thickening of the hip joint, a small effusion, and a large cortical defect on the anterolateral aspect of the proximal femoral metaphysis. Percutaneous biopsy revealed tuberculous infection.* **(Right)** *CECT of the proximal femur in the same patient shows an anterior cortical defect and intramedullary channel-like defects in the femoral metaphysis, indicative of infection (tuberculous osteomyelitis). The iliac and ischial portions of the acetabulum are normal.*

Joint Infection

KEY FACTS

IMAGING

- Echogenic joint effusion with positive aspirate ± synovial/capsular thickening and hyperemia
- Normal US examination excludes septic arthritis
  - Provided joint capsule is intact, then no joint effusion or synovial proliferation = no septic arthritis
- **Joint effusion** is cardinal sign of septic arthritis
  - ± fine echoes within joint fluid ± swirling echoes
  - Effusion is minimal in poorly distensible joints (e.g., acromioclavicular, sternoclavicular, sacroiliac joints)
- Synovial/capsular thickening
  - Chronic infection in prosthetic joints may comprise only synoviocapsular thickening
- Bone and cartilage change
  - Marginal erosion (osteolysis)
  - Progressive joint space narrowing (chondrolysis)
  - Features of osteomyelitis
  - Secondary degenerative change
- Periarticular soft tissue inflammation ± collection
- ± visible capsular rupture
  - ± pericapsular inflammatory mass
  - ± pericapsular abscess
- As chronicity of infection increases
  - Amount of effusion decreases
  - Degree of synovial proliferation increases
  - Degree of bone and cartilage damage increases
- US appearances of infective arthritis and noninfective inflammatory arthritis overlap considerably
  - Patients with inflammatory arthropathy, osteoarthritis, and joint prosthesis more prone to infection
- Joint fluid aspiration and analysis best way to exclude joint infection
- Synovial biopsy is excellent alternative if no synovial fluid present
  - Prosthetic joint infections, in particular, tend to have pronounced synovial proliferation and relatively less fluid

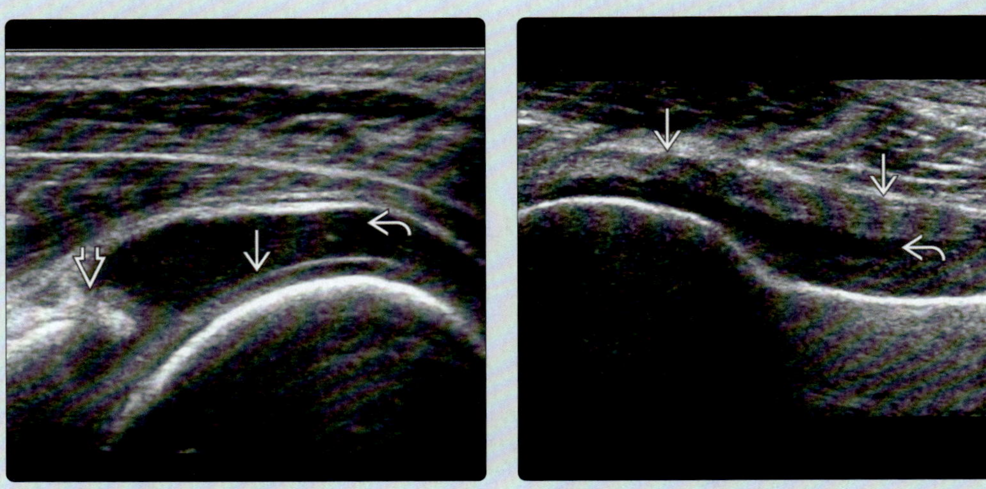

(Left) *Longitudinal US shows a moderate glenohumeral effusion ↪ without synovial proliferation due to acute septic arthritis (Salmonella enteritidis). Note the clear visualization of cartilage → and posterior glenoid labrum ⇨.* **(Right)** *Longitudinal US of the hip shows moderate synovial proliferation → with a small effusion ↪ in a patient with juvenile rheumatoid arthritis. This may be due to either infective or inflammatory arthritis. Joint aspirate yielded Moraxella species.*

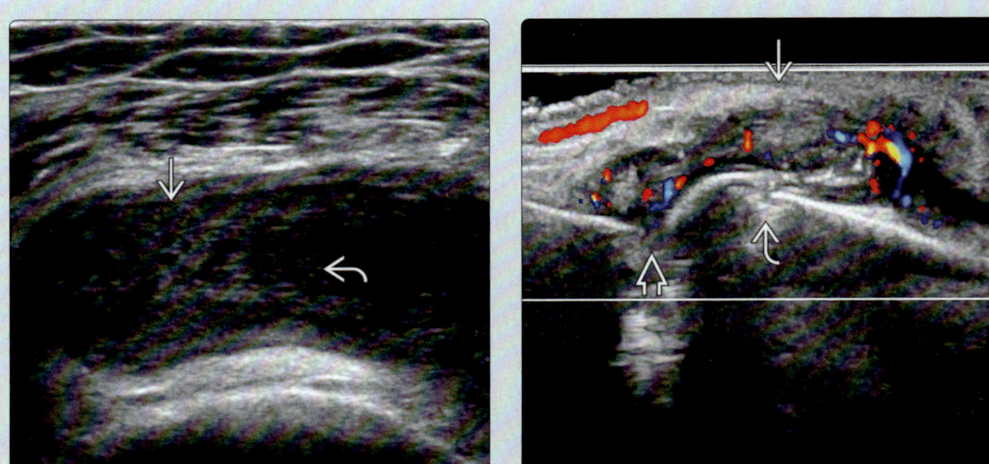

(Left) *Transverse US of the knee shows a small effusion ↪ at the suprapatellar pouch with severe synovial proliferation → due to chronic joint infection with Staphylococcus aureus.* **(Right)** *Longitudinal color Doppler US shows moderately hyperemic synovial proliferation → on the dorsum of the index metacarpophalangeal joint ⇨. There is a small erosion ↪ of the 2nd metacarpal neck. This arthritis was due to tuberculosis confirmed by histology and microbiology following surgical exploration.*

TERMINOLOGY

Synonyms

- Infective arthritis, septic arthritis

Definitions

- Infection of joint

IMAGING

General Features

- Best diagnostic clue
  - Echogenic joint effusion with positive aspirate ± synovial/capsular thickening and hyperemia
- Location
  - Any joint can potentially be infected
    - Knee > hip > shoulder > wrist
      - More common in synovial joints as usually blood-borne infection
    - May be multifocal, especially in at-risk patients
    - Gonococcal arthritis: Predilection for small joints of hands and feet (disseminated gonococcal infection)

Radiographic Findings

- Radiography
  - Early stages
    - Usually normal
    - Later visible joint effusion, joint space widening, periarticular swelling and fat plane obliteration, periarticular osteopenia
  - Late stages
    - ± narrowing of joint space (chondrolysis), marginal erosions (osteolysis), secondary osteoarthritis

MR Findings

- T1WI
  - Hypointense or isointense joint effusion
  - ± hypointense periarticular marrow edema, hypointense periarticular soft tissue extension
- T2WI
  - Hyperintense joint effusion with capsular distension
    - ± synovial proliferation, periarticular edema, inflammation, or abscess
- T1WI C+
  - Enhancement of synovium, capsule, pericapsular tissues

Ultrasonographic Findings

- Normal US examination excludes septic arthritis provided that joint is well seen
- **Joint effusion**
  - Cardinal sign of septic arthritis
    - US can detect 1 ml of joint fluid
  - ± clumps of echoes or multiple fine echoes within joint fluid
    - Swirling or moving echoes on real-time imaging
  - Anechoic joint fluid does not exclude septic arthritis
    - Joint aspiration and analysis of joint fluid is only way to truly exclude joint infection
  - In some nondistensible joints (e.g., acromioclavicular, sternoclavicular, sacroiliac joints), effusion is minimal
  - Amount of joint fluid decreases as chronicity of infection increases
  - Viscosity of joint fluid increases as chronicity of infection increases
- **Synovial/capsular thickening**
  - Synovial proliferation and capsular thickening increase as chronicity of infection increases
    - Chronic infection in prosthetic joints may comprise only synovial/capsular thickening
  - ± hyperemia
    - Hyperemia reflects severity of active inflammation
  - Prominent synovial/capsular thickening and hyperemia in nondistensible joints
- **Bone and cartilage change**
  - Seen in subacute or later phase of infection
  - Progressive joint space narrowing (chondrolysis)
  - Marginal erosions (osteolysis)
  - Secondary degenerative change
  - Features of osteomyelitis
    - Seen as periosteal thickening ± subperiosteal fluid ± cortical resorption
- **Periarticular soft tissue extension**
  - ± visible capsular rupture ± pericapsular inflammatory mass or abscess

Imaging Recommendations

- Best imaging tool
  - US, as nearly all joints are accessible to US examination
    - No metallic artifact from joint prosthesis
- Protocol advice
  - US-guided aspiration for cytology, culture, and crystals (adults)
    - Prior to initiation of antibiotic therapy
  - Cytology: Turbid, low-viscosity fluid with leucocyte count > 50,000 per mm^3
    - ~ 90% polymorphs
  - Crystal: Compensated polarized light microscopy sample should be fresh (i.e., not mixed with water or alcohol, as these will dissolve crystals)
    - Uric acid crystals (negatively birefringent)
    - Calcium pyrophosphate dihydrate crystals (positively birefringent)
    - Simultaneous crystalline and bacterial arthritis may occur
  - Gram stain and culture
    - Send sample for culture immediately
      - Higher yield than if sample is left standing or kept in refrigerator overnight
    - Aerobic, anaerobic, mycobacterial, and fungal cultures
    - Aspirate Gram stain is positive in 50% of infective arthritis
    - Aspirate culture is positive in 90% of infective arthritis
  - US-guided synovial biopsy
    - Indicated if no or only minimal joint fluid is present
    - 100% accuracy for infection of native joints, 78% accuracy for prosthetic joints
      - 3 synovial cores sufficient for native joints
      - More synovial cores (up to 12) required in prosthetic joints
      - ± repeat synovial biopsy if yield negative and clinical suspicion of infection is high

DIFFERENTIAL DIAGNOSIS

Transient Synovitis of Hip

- Self-limiting and predominantly in children
- Anechoic or mildly echogenic fluid ± synovial/capsular thickening
  - Often subclinical synovitis of contralateral hip

Crystal Arthropathy

- Synovitis ± effusion ± crystalline particles (seen as comet-tail artifacts)
  - ± periarticular tophi, ± elevated serum uric acid

Hemarthrosis

- Traumatic or spontaneous onset ± underlying coagulopathy
- Finely echogenic fluid similar to acute septic arthritis
  - Synovial/capsular thickening from repeated hemarthrosis

PATHOLOGY

General Features

- Etiology
  - Most commonly due to bacteremic seeding of synovium
  - Less common causes
    - Arthroplasty, spread from adjacent osteomyelitis, needle inoculation, penetrating trauma
- Organisms
  - *Staphylococcus aureus*
    - Most common organism, especially in diabetic and rheumatoid arthritis patients
    - Late (> 3 months) infection of prosthesis
    - Early infection of prosthesis related to surgical contamination and is more commonly due to *Staphylococcus epidermidis*
  - *Streptococcus pyogenes*
  - Gram-negative organisms (*Pseudomonas aeruginosa, Escherichia coli, Neisseria gonorrhoeae*)
    - Especially intravenous (IV) drug users, extremes of age, and immunosuppressed patient
    - *Neisseria gonorrhoeae* may lead to disseminated gonococcal infection
    - Gonococcal infection also causes reactive arthritis

Gross Pathologic & Surgical Features

- No limiting basement plate under well-vascularized synovium; facilitates bacterial entry to joint
  - → bacterial adherence and colonization
  - → bacterial proliferation → acute inflammatory response
  - → cytokine and acute phase protein release → tissue damage

CLINICAL ISSUES

Presentation

- Most common signs/symptoms
  - Monoarticular pain, swelling, and limitation of movement
- Other signs/symptoms
  - Fever, malaise

Demographics

- Age
  - Extremes of age
- Gender
  - Nongonococcal arthritis: M:F = 1:1
  - Gonococcal arthritis: M:F = 1:10
- Epidemiology
  - Incidence: 2-10 per 100,000 per year
  - More common in groups of patients that follow
    - Patients over 60 years old
      - Due to higher prevalence of degenerative joint disease and reduced immunity
    - Patients with rheumatoid arthritis (10x more common than general population)
    - Patients with diabetes, poor general health, long-term corticosteroid usage
    - Patients with joint prosthesis

Natural History & Prognosis

- Depends on immunocompetency of host, virulence of organism, rapidity of detection, treatment response
  - Nongonococcal septic arthritis
    - Prognosis is good if detected and treated early (≤ 7 days after onset of symptoms)
    - Prognosis is poor if > 6 days to make joint (clinically) sterile after onset of treatment
- Morbidity: 5-20% due to bacteremia
  - Septic oligoarthritis (30% mortality)
  - Septic oligoarthritis + rheumatoid arthritis or *Staphylococcus aureus* infection = 50% mortality
- Gonococcal septic arthritis: Good prognosis
  - Symptoms ↓ rapidly without sequelae

Treatment

- Conservative
  - IV antibiotics
  - Splinting to prevent contractures
  - ± repeated US-guided percutaneous aspiration
  - ± percutaneous US-guided tidal irrigation
- Arthrotomy, arthroscopic debridement, wound irrigation
- Infected prosthesis usually requires replacement
  - Removal and debridement (stage 1)
  - Reimplantation (stage 2)

DIAGNOSTIC CHECKLIST

Consider

- Ultimate diagnosis rests on isolating pathogens from aspirated joint fluid
  - Early diagnosis and treatment is key to success
- Normal US examination excludes septic arthritis
- US and aspiration establishes diagnosis; MR assesses damage

SELECTED REFERENCES

1. DiPietro MA et al: Pediatric musculoskeletal ultrasound. Pediatr Radiol. 47(9):1144-1154, 2017
2. Gaigneux E et al: Ultrasound abnormalities in septic arthritis are associated with functional outcomes. Joint Bone Spine. 84(5):599-604, 2017
3. Nguyen JC et al: US evaluation of juvenile idiopathic arthritis and osteoarticular infection. Radiographics. 37(4):1181-1201, 2017
4. Sitt JC et al: Ultrasound-guided synovial Tru-cut biopsy: indications, technique, and outcome in 111 cases. Eur Radiol. 27(5):2002-2010, 2017

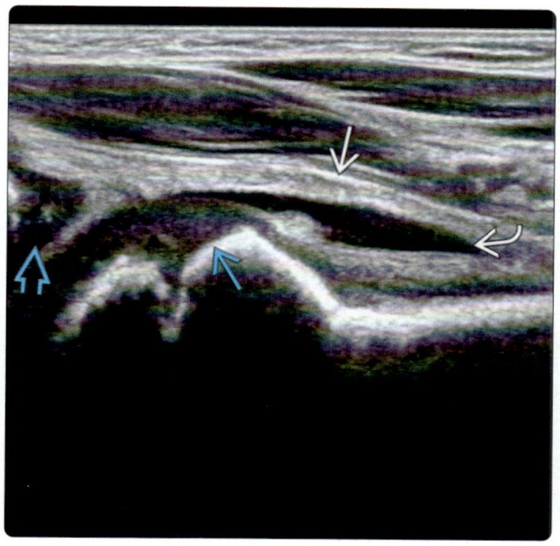

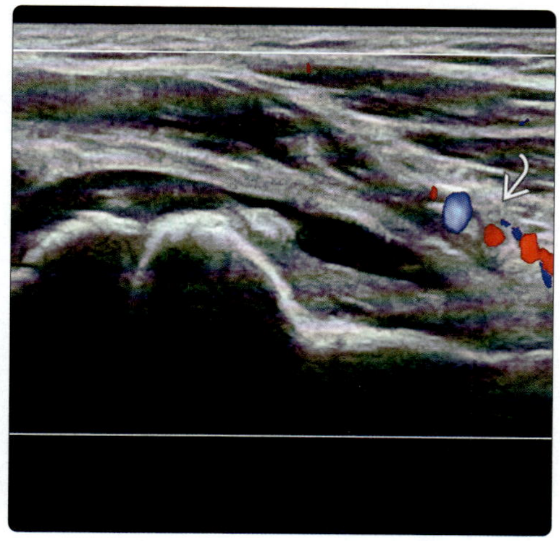

(Left) *Longitudinal US in a child with acute hip pain shows mild synoviocapsular thickening with a small effusion. Note the normal, incompletely ossified cartilage of the femoral head and acetabulum.* **(Right)** *Longitudinal color Doppler US shows mild hyperemia of the periarticular soft tissues. The US appearances of acute septic arthritis are similar to those of transient synovitis of the hip. Blood culture yielded Group D Salmonella indicative of septic arthritis.*

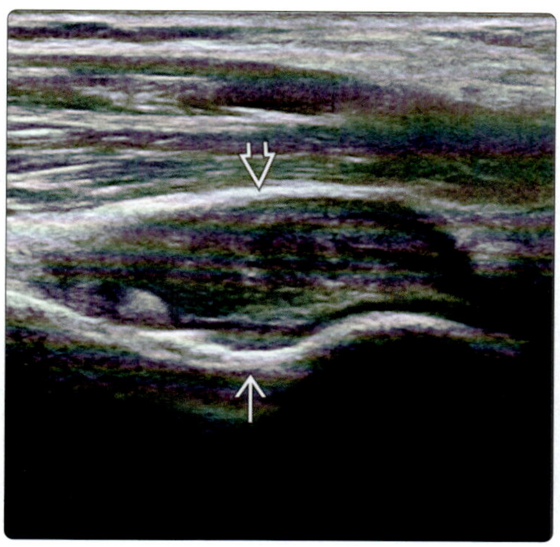

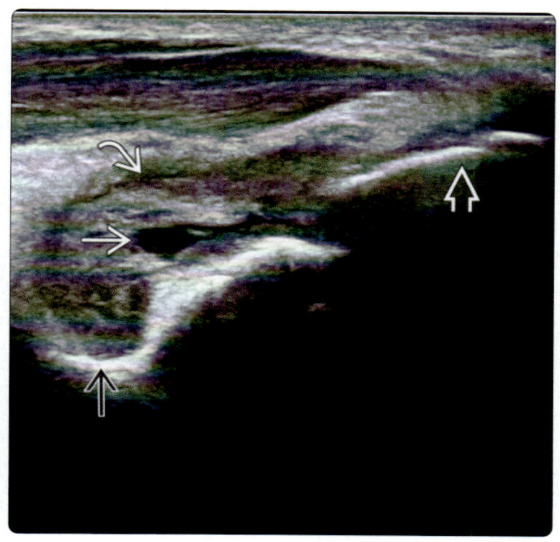

(Left) *Longitudinal US of the anterior elbow recess in a patient with severe elbow pain, fever, and raised inflammatory markers shows severe synovial proliferation. Note the coronoid fossa of the distal humerus.* **(Right)** *Longitudinal US of the posterior elbow recess of the same patient as previous image shows severe synovial proliferation with a small joint effusion. Note the olecranon fossa and the olecranon. Blood culture yielded Streptococcus agalactiae.*

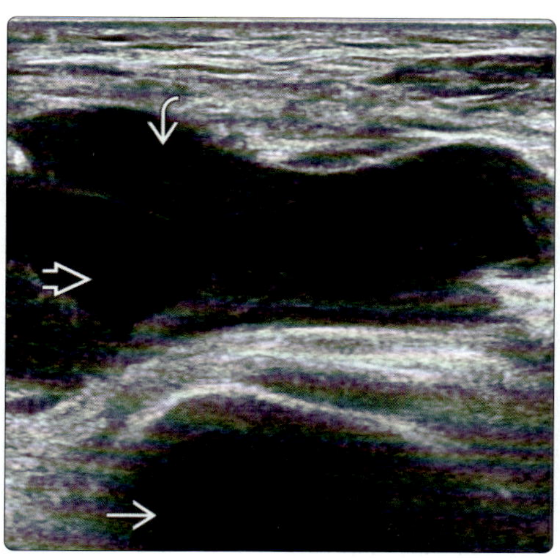

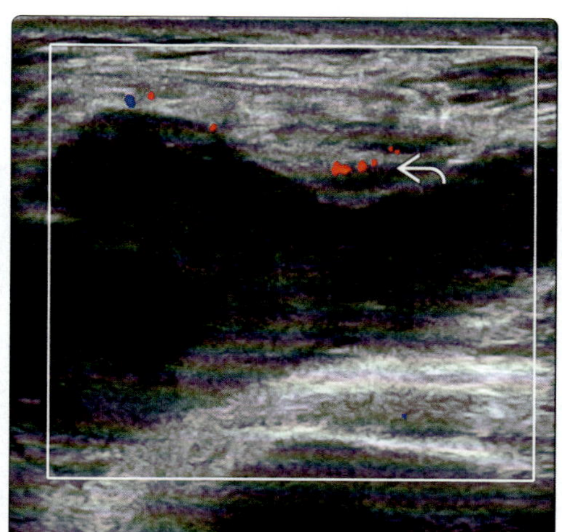

(Left) *Transverse US of a patient with chronic septic arthritis of the knee shows moderate to severe synovial proliferation with only a relatively small effusion. The distal femur is normal.* **(Right)** *Transverse color Doppler US of the same patient shows only mild peripheral hyperemia, indicative of low-grade active inflammation. Aspirated joint fluid grew group G Streptococcus.*

(Left) *Longitudinal US of the shoulder in a patient with subacute shoulder pain and swelling shows marked distension of both the subacromial-subdeltoid bursa ➡ and the shoulder joint associated with asymmetrical synovial thickening ➡.* **(Right)** *Longitudinal US of the same patient shows multiple areas of nodular synovial proliferation ➡ within the distended subacromial-subdeltoid bursa.*

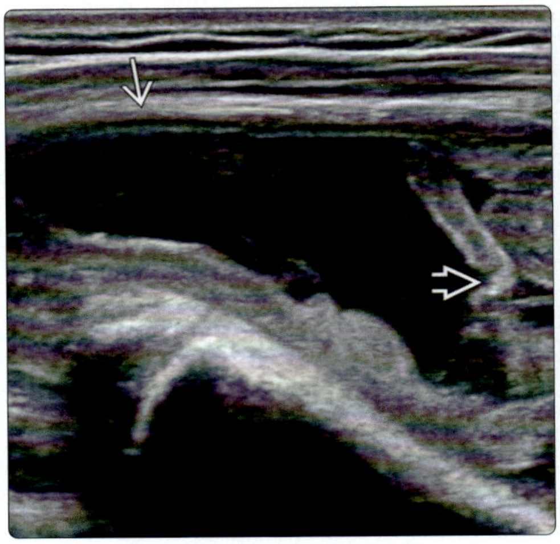

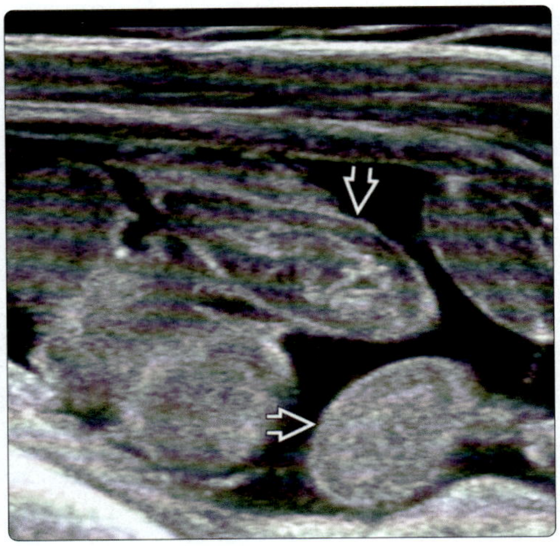

(Left) *Transverse color Doppler US of the same patient shows severe synovial proliferation ➡ at the posterior aspect of the glenohumeral joint. Mild hyperemia ➡ is present indicating active inflammation.* **(Right)** *US-guided biopsy of the same patient as previous image shows satisfactory position of the biopsy needle ➡ within the thickened synovial tissue ➡. The causative organism was shown to be Microbacterium tuberculosis.*

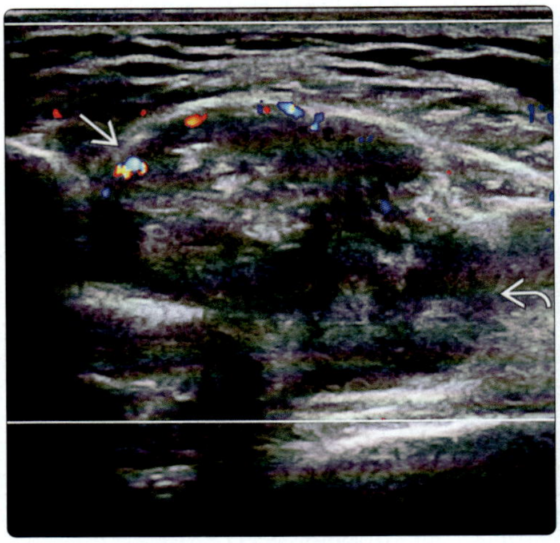

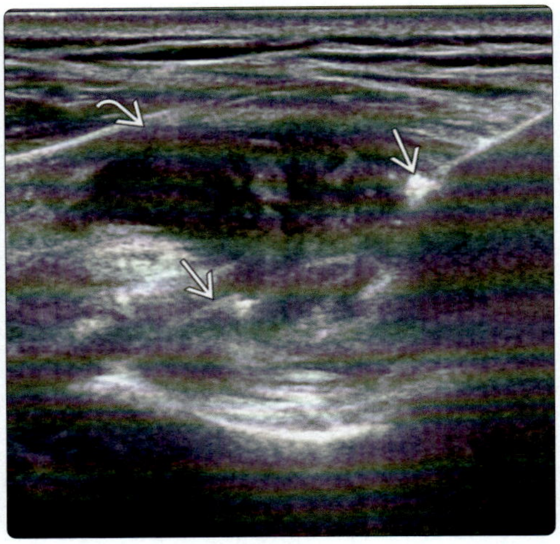

(Left) *Clinical photo of a patient with a slow-growing hand mass shows a focal swelling ➡ with skin discoloration on the dorsoradial aspect of the hand.* **(Right)** *Longitudinal US of the same patient shows a large heterogeneous mass ➡ with internal calcification ➡ and moving echoes on real-time imaging. This was due to a tuberculous cold abscess extending from the metacarpophalangeal joint ➡ into the subcutaneous soft tissues.*

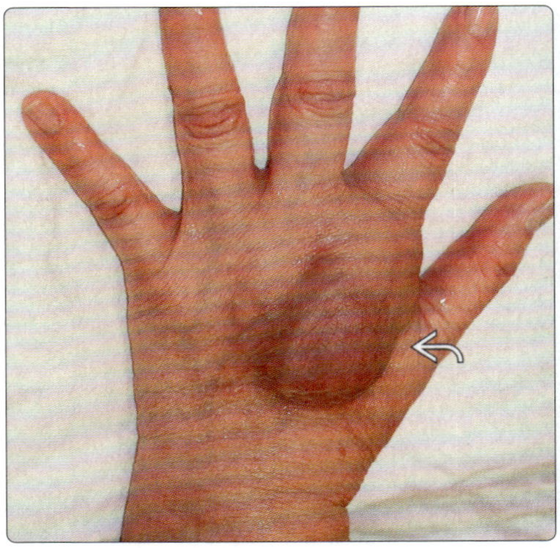

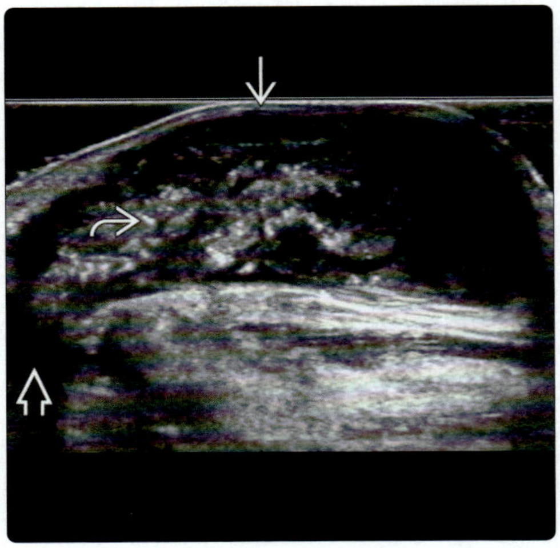

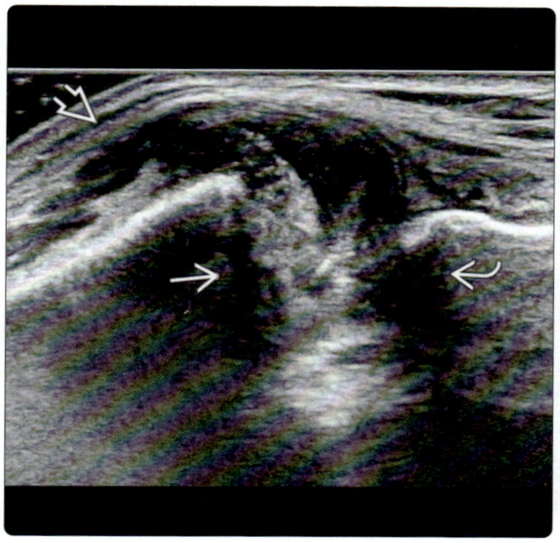

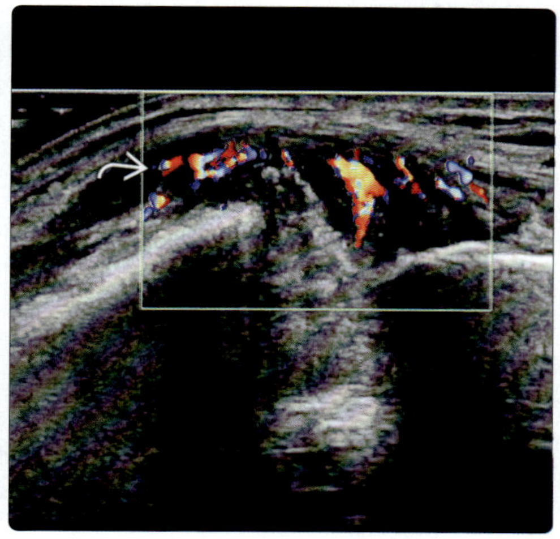

(Left) *Transverse US shows a widened sternoclavicular joint between the medial end of clavicle and the sternum with capsular disruption superficially leading to a medium-sized periarticular inflammatory mass.* **(Right)** *Transverse color Doppler US showing hyperemia indicates joint distension by inflammatory tissue rather than fluid. This patient had concomitant streptococcal septic arthritis of the knee. The sternoclavicular joint septic arthritis settled with antibiotics.*

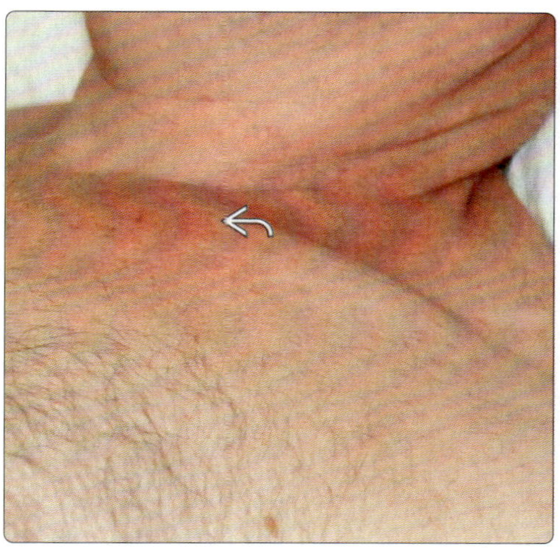

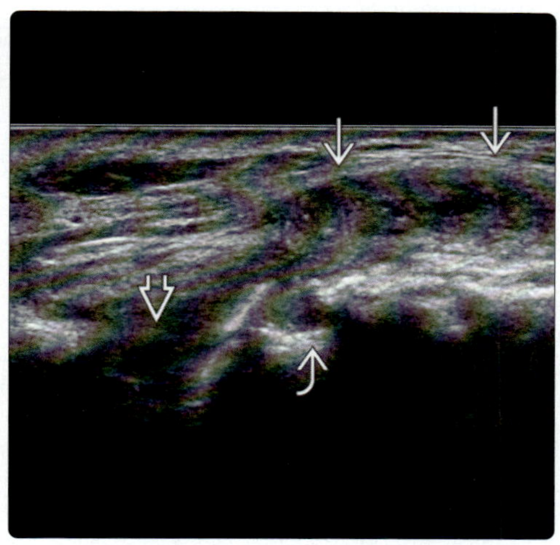

(Left) *Clinical photo of another patient shows focal swelling with overlying skin erythema at the sternoclavicular joint region.* **(Right)** *Longitudinal US of the same patient shows widening of the sternoclavicular joint, bone erosion, and cortical irregularity of the medial clavicle with a moderate-sized periarticular inflammatory mass. No effusion is evident. US-guided biopsy of the inflammatory mass yielded Staphylococcus aureus.*

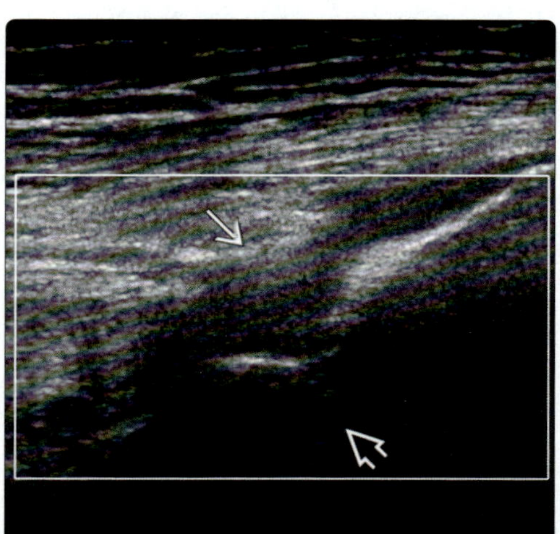

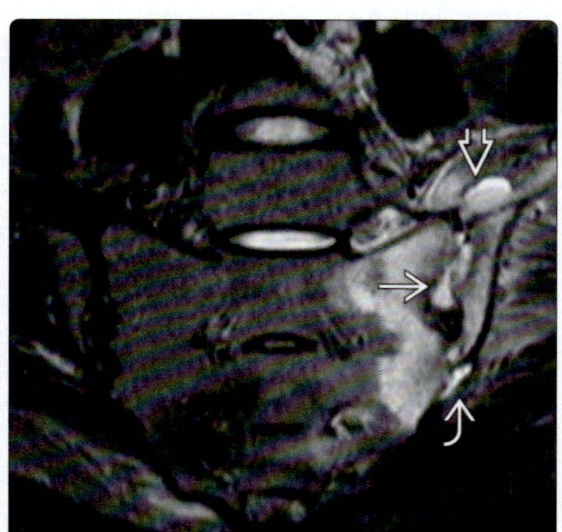

(Left) *Transverse color Doppler US of an intravenous drug user with Staphylococcus septicemia shows a small amount of inflammatory hypoechoic soft tissue posterior to the sacroiliac joint. No hyperemia is evident.* **(Right)** *Oblique coronal T2WI FS MR of the same patient shows sacroiliac joint widening with cortical erosion and periarticular marrow edema. Small fluid collections are present at the superior and inferior aspects of the joint. This was due to an infective sacroiliitis.*

Postoperative Infection

KEY FACTS

TERMINOLOGY

- Infection at surgical site within 1 month of surgery or within 1 year of prosthetic implantation

IMAGING

- Mild soft tissue edema, fluid, & hyperemia are very common signs in early postoperative period
- Intensity & pattern of edema, fluid, & hyperemia helps differentiate infectious from noninfectious inflammation
  - Inflammation from surgical disruption of tissues & subsequent repair follows expected pattern & is usually mild in degree
  - Inflammation from postoperative infection is more severe & does not follow expected pattern
- Edematous tissue = hyperechoic if mildly edematous, hypoechoic if more intensely edematous
- Fluid may collect as discrete collection or as accumulation at fascial or tissue interface
  - Variable echogenicity
  - Cannot reliably differentiate between infected collection & noninfected collection
  - Check gravity-dependent areas around surgical site as fluid will tend to collect here
- US is not prone to metallic artifact & is more suited to assessment of postoperative patient
- US is not sensitive at detecting osteomyelitis

TOP DIFFERENTIAL DIAGNOSES

- Postoperative inflammation
- Hematoma
- Deep venous thrombosis

DIAGNOSTIC CHECKLIST

- US helps to define site of soft tissue infection, isolate fluid, & provide aspirate for culture
- ± US-guided soft tissue biopsy if no fluid collection is present
- US-guided synovial biopsy helpful in confirming prosthetic joint infection if no fluid for aspiration

(Left) *Longitudinal US shows mild distension of the hip joint → with a small amount of internal gas ↩. This is a normal appearance 1 week after hip implant surgery.* **(Right)** *Longitudinal US of a patient with hip surgery 2 months previously shows a moderately thick band of granulation tissue ⇨ extending from the distended hip joint → to the cutaneous wound ↩. Color Doppler US (not shown) revealed no significant hyperemia. This is a normal postoperative appearance.*

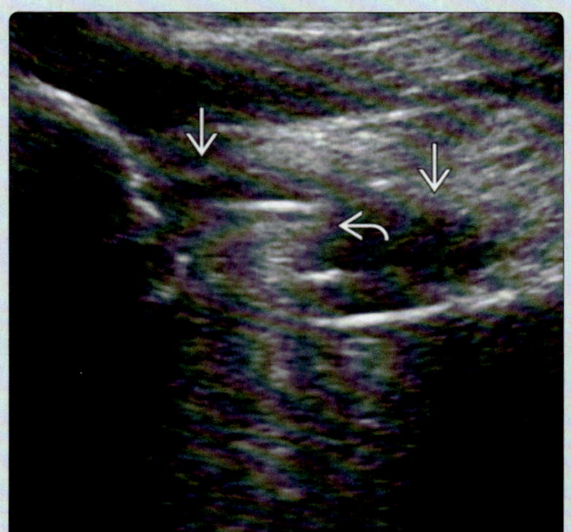

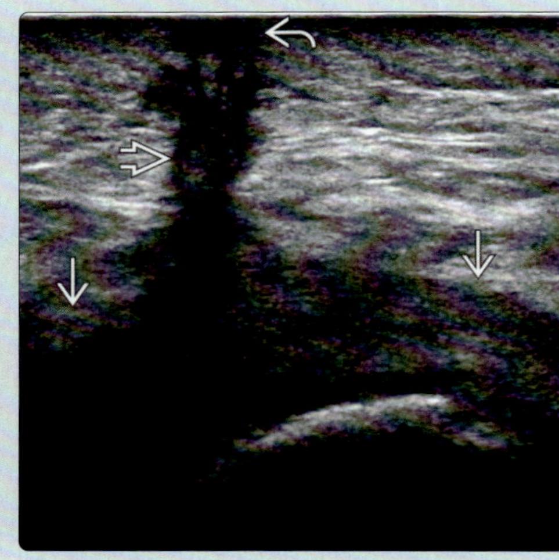

(Left) *Transverse US shows an infected hypoechoic collection → in the lateral aspect of the proximal femur ⇨ in a patient with revision of hip implant 2 weeks previously. There is moderate surrounding deep soft tissue edema ↪ and subcutaneous soft tissue edema ↪.* **(Right)** *Transverse US of a patient with an infected elbow implant shows a moderate degree of inflammatory soft tissue → and edema surrounding the implant ↪.*

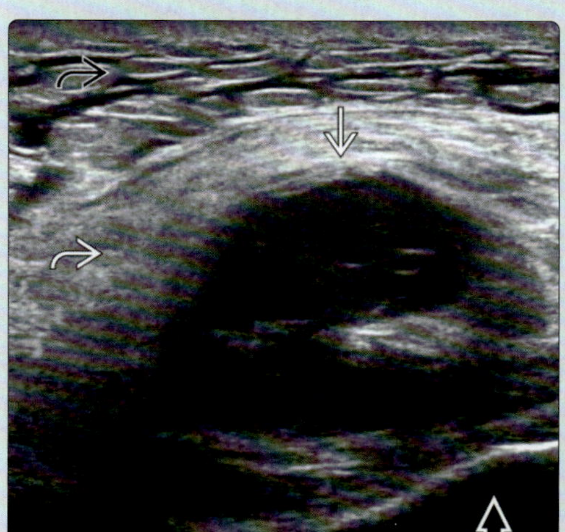

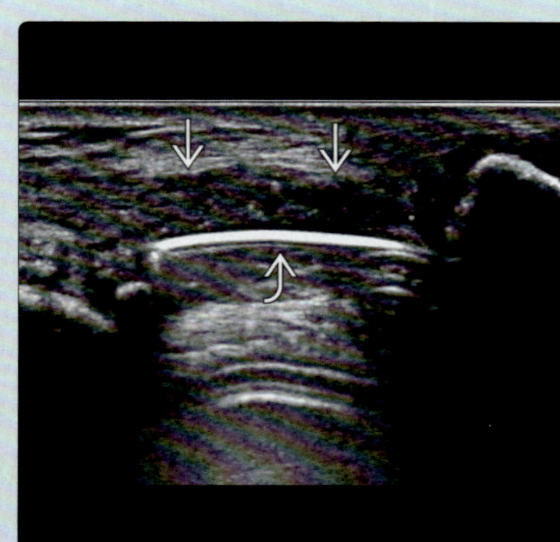

TERMINOLOGY

Definitions

- Infection at surgical site within 1 month of surgery or within 1 year of surgical implant

IMAGING

General Features

- Best diagnostic clue
  - Inflammation ± collection persisting > 1 week after surgery
- Location
  - Soft tissue, bone, or joint
  - Hip, knee, spine, or any operative site
- Morphology
  - From localized wound infection to acute osteomyelitis

Radiographic Findings

- Best initial investigation
  - Aids interpretation of subsequent US and other investigations
  - Progressive osteolysis is key to recognizing infection
  - Radiographs are normal in 50% of periprosthetic implant infections

CT Findings

- CECT
  - Multidetector CT is less prone to metallic artifact than MR
    - CT (or MR) provide most useful assessment of infection after implant has been removed
  - Delineates bony abnormality and soft tissue infection
    - ± CT-guided aspiration or drainage

MR Findings

- T1WI
  - MR imaging is often severely limited by metallic susceptibility artifact
  - Hypointense soft tissue swelling
- T2WI FS
  - Hyperintense soft tissue swelling ± hyperintense fluid or fluid collection
- T1WI C+
  - Infective, noninfective inflammatory, and reparative tissue all enhance
    - Generally, the greater the enhancement, the more likely the infection
  - Useful at delineating fluid collections
    - Degree of wall enhancement is greater with abscesses than with sterile collections

Ultrasonographic Findings

- US is not prone to metallic artifact
  - Good visualization of juxtacortical tissues
- US can access most operative areas
- Helps localize site of infection (superficial or deep), identify drainable collection, and guide fluid aspiration or tissue biopsy for diagnosis
- Mild soft tissue edema, joint fluid and capsular hyperemia are common early postoperative findings
- Intensity and pattern of edema, fluid, and hyperemia helps differentiate infectious from noninfectious inflammation
  - Inflammation from surgical disruption of tissues and subsequent repair follows expected pattern and is usually mild in degree
  - Inflammation from postoperative infection is more severe and does not follow expected pattern
- The greater the degree of soft tissue edema and inflammation, the greater the likelihood of postoperative infection
- Edematous tissue is hyperechoic if mildly edematous, hypoechoic if more intensely edematous
- Fluid may collect as discrete collection or along fascial and tissue interfaces
  - Usually accumulates at most dependent areas
  - Anechoic or hypoechoic or hyperechoic fluid collection
  - ± echogenic debris
  - Usually unable to distinguish between infected collection, noninfected seroma, or liquefied hematoma
    - Need to aspirate fluid for microbiological, cell count, and biochemical analysis
    - Synovial fluid leukocyte count, percentage of polymorphonucleocytes, C-reactive protein, and other synovial fluid biomarkers are predictive of periprosthetic joint infection
- Synovial/capsular thickening is common after surgery
  - Joint capsule disruption during surgery or arthroplasty
    - Limits development of postoperative effusion
      - If capsule has been disrupted, there may be little, if any, joint effusion even with infection as joint fluid drains into periarticular tissues
  - Synovial thickening does not imply infection
    - Synovial biopsy has 78% accuracy for injection of prosthetic joint when no joint fluid present to aspirate
    - Need to take 12 cores of synovial tissue for culture as bacterial load in synovium typically low
- ± gas around operative site
  - May be present normally days after surgery or longer with open wound or drain inserted
- Limitations of US
  - Cannot differentiate mild postoperative infection from noninfective postoperative inflammation
  - Transducer access is occasionally limited by external fixators, bandaging, or cutaneous wound
    - Some areas (such as spine) have very limited transducer access
  - Not sensitive at detecting osteomyelitis

Nuclear Medicine Findings

- Bone scan
  - Used for diagnosis of periprosthetic infection and best if combined with labeled leukocyte scintigraphy or gallium scintigraphy
  - No periprosthetic uptake = no infection
  - Periprosthetic uptake indicates loosening or infection
    - Uptake at same site on labeled leukocyte scintigraphy or gallium scintigraphy makes infection most likely
    - No uptake at same site on labeled leukocyte scintigraphy or gallium scintigraphy makes loosening most unlikely
  - High sensitivity with moderate specificity

- Repair can simulate infection
- PET/CT
  - Combines high sensitivity of PET imaging with high resolution of CT imaging
    - ↓ specificity, particularly in postoperative cases
    - Glucolysis also feature of noninfectious inflammation and repair
  - Activity around shaft and tip of hip prosthesis increases likelihood of infection compared to activity around head and neck of prosthesis

Imaging Recommendations

- Best imaging tool
  - US
- Protocol advice
  - Remove bandaging; cover suture line with thin polyurethane adhesive dressing
  - Copious gel to maximize transducer contact at surgical incision
  - Examine gravity-dependent areas carefully since collections will preferentially accumulate here
    - e.g., following hip arthroplasty, roll patient to lateral decubitus position to examine gluteal and posterior thigh regions
  - Perform US-guided aspiration of joint or fluid collection ± US-guided biopsy of synovium or inflammatory soft tissue
    - If performing aspiration, ensure aseptic technique, as damaged tissues are particularly susceptible to bacterial inoculation

DIFFERENTIAL DIAGNOSIS

Postoperative Inflammation

- Difficult to distinguish from postoperative infection
- Pattern and intensity of edema, fluid accumulation, and degree of hyperemia are relevant to making this distinction

Hematoma

- Hypoechoic avascular mass with fixed fine internal echoes

Deep Venous Thrombosis

- May coexist with postoperative infection
- Noncompressible vein

PATHOLOGY

General Features

- Etiology
  - Incidence of infection following knee or hip arthroplasty = 0.5-2.0%
    - Early infection (< 3 months after surgery) is mostly due to surgical contamination
    - Late infection (within 1 year of surgery) is mostly due to hematogenous seeding of operative site
    - Prostheses: Disrupt integrity of tissues, affect blood supply, and provide surface for bacterial adherence
  - *Staphylococcus aureus*
    - Responsible organism in 1/2 of cases
  - Gram-negative bacilli (*Bacteroides*, *Escherichia coli*, *Pseudomonas*, *Proteus*, *Klebsiella*, *Enterococcus*)

Gross Pathologic & Surgical Features

- Cellulitis, pyomyositis, abscess

Microscopic Features

- Acute or acute on chronic inflammatory infiltrate, capillary dilatation ± abscess formation

CLINICAL ISSUES

Presentation

- Most common signs/symptoms
  - Pain and swelling
- Other signs/symptoms
  - Wound oozing, inflammation, fever
- Clinical profile
  - Leukocytosis, raised inflammatory markers

Demographics

- Age
  - Less common in young patients
- Epidemiology
  - Risks related to surgery
    - Long operation time (> 3 hours), high blood loss (> 1,000 mL), instrumentation, multiple surgical sites (e.g., fixation + bone grafting), staged operation or repeat operation
  - Risks related to patient
    - Frailty, coexistent morbidity, coexistent remote site infection, obesity, smoking

Natural History & Prognosis

- Superficial infection can usually be treated successfully with antibiotic therapy
- Established deep infection is often difficult to eradicate without removing implant

Treatment

- Antibiotic therapy
- Aspiration or drainage (US-guided) of collection(s)
- ± debridement
- ± implant removal and revision

DIAGNOSTIC CHECKLIST

Consider

- Accurate and timely recognition is key to successful eradication of postoperative infection
- Most postoperative infections are clinically evident
  - US helps to define site of soft tissue infection, isolate fluid, and provide aspirate for culture
  - US-guided aspiration and synovial biopsy becoming much more frequently used to diagnose infection in prosthetic joints
- Integrity of recently operated bone is impaired
  - If juxtacortical infection present, high probability that bone is also infected

SELECTED REFERENCES

1. Sitt JC et al: Ultrasound-guided synovial Tru-cut biopsy: indications, technique, and outcome in 111 cases. Eur Radiol. 27(5):2002-2010, 2017
2. Chun KA et al: Postoperative ultrasonography of the musculoskeletal system. Ultrasonography. 34(3):195-205, 2015

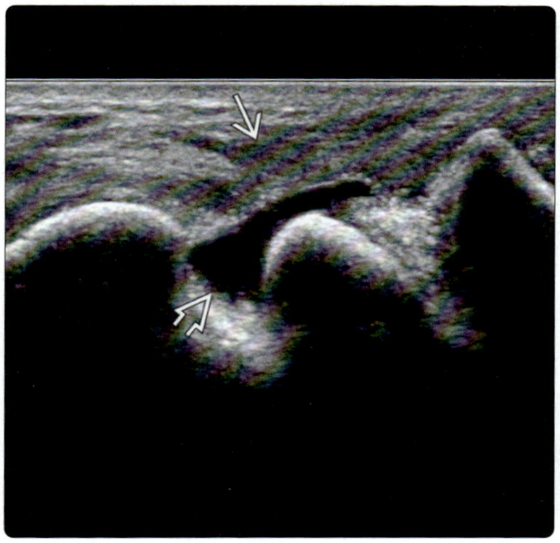

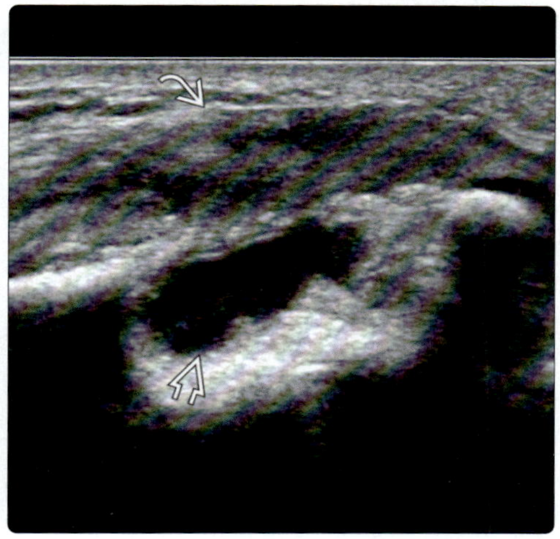

(Left) *Longitudinal US of a patient after elbow replacement shows mild fluid distension of the elbow joint with marked capsular thickening and periarticular soft tissue edema.* **(Right)** *Transverse US of the same patient again shows a small effusion but more severe capsular thickening. Capsular thickening is common in the early post operative status and does not necessarily imply infection.*

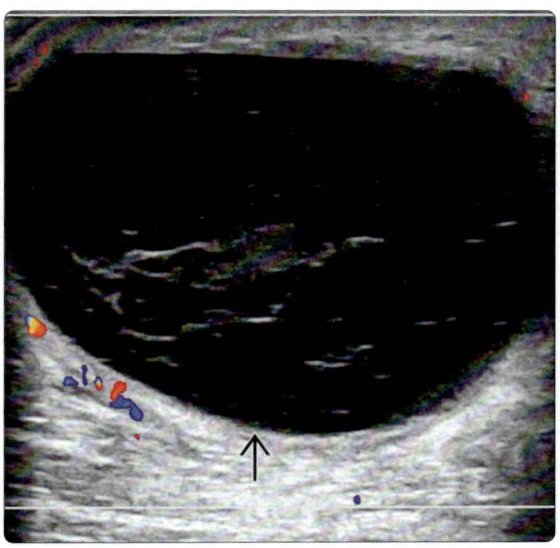

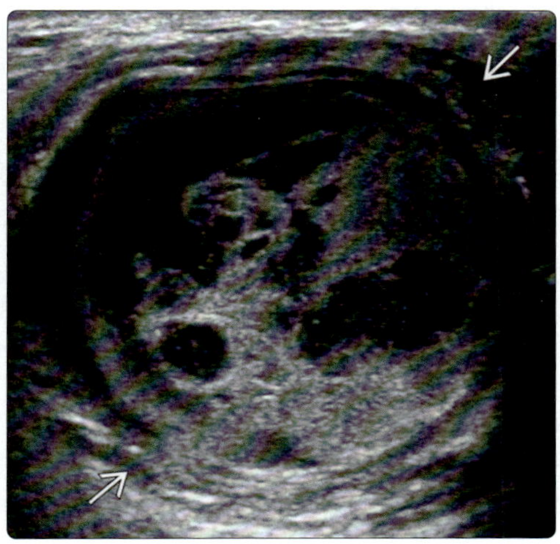

(Left) *Transverse color Doppler US of a patient after groin surgery shows a large subcutaneous collection with multiple thin septations and mild marginal hyperemia. This is a seroma. US cannot reliably differentiate infected from noninfected collection. Aspiration is necessary.* **(Right)** *Transverse US of a patient after groin surgery shows an echogenic collection with a thickened wall and multiple fine internal echoes due to an infected organizing hematoma.*

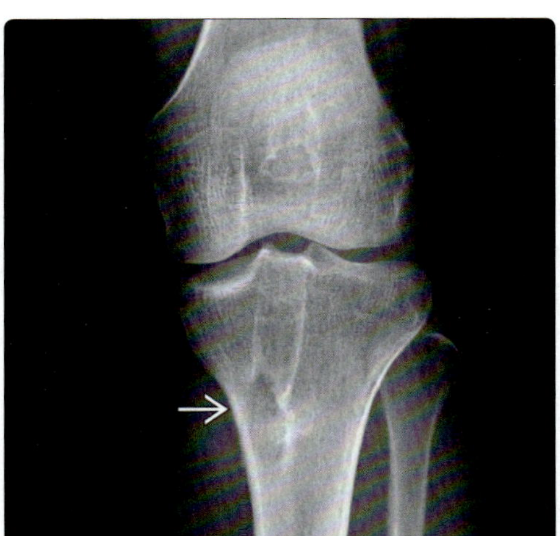

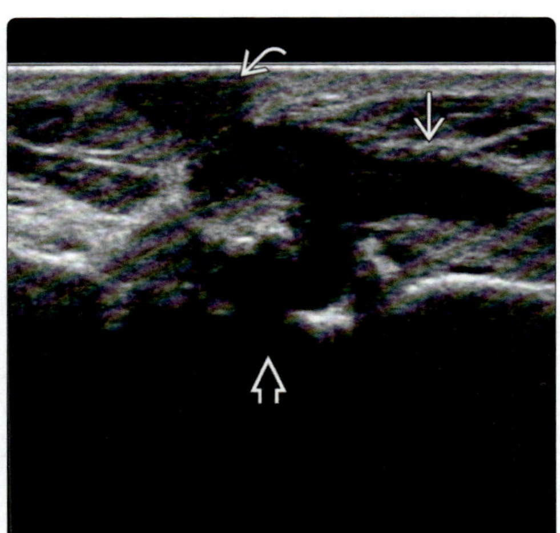

(Left) *Frontal radiograph of a patient with previous anterior cruciate ligament grafting shows an area of osteolysis with a partially sclerotic border at the proximal end of the tibial tunnel.* **(Right)** *Longitudinal US of the same patient shows hypoechoic inflammatory tissue adjacent to the tibial tunnel cortical defect extending to the skin. This radiographic and US picture is consistent with postoperative infection of the tibial tunnel.*

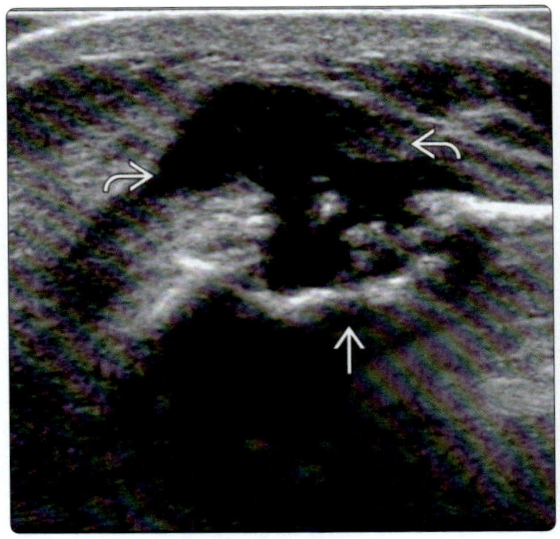

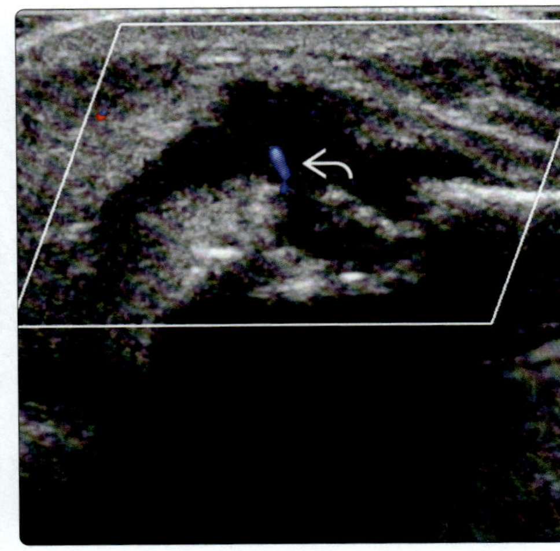

(Left) *Transverse US of the same patient shows moderate cortical irregularity → around the tibia tunnel opening with juxtacortical hypoechoic inflammatory tissue →.* **(Right)** *Transverse color Doppler US of the same patient shows a mild degree of hyperemia → within the hypoechoic inflammatory soft tissue. This hyperemia is helpful at distinguishing hypoechoic inflammatory tissue from fluid.*

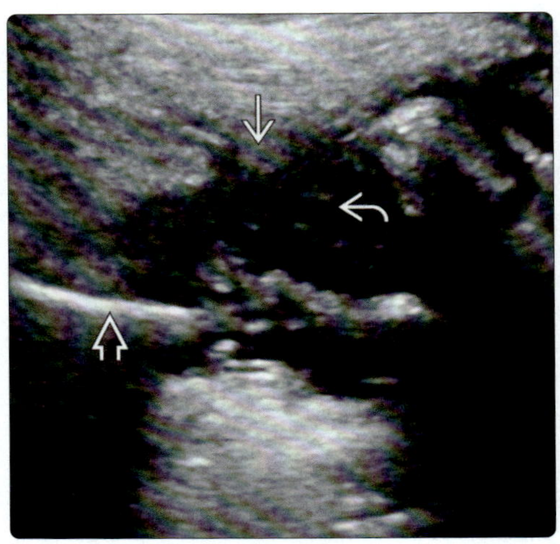

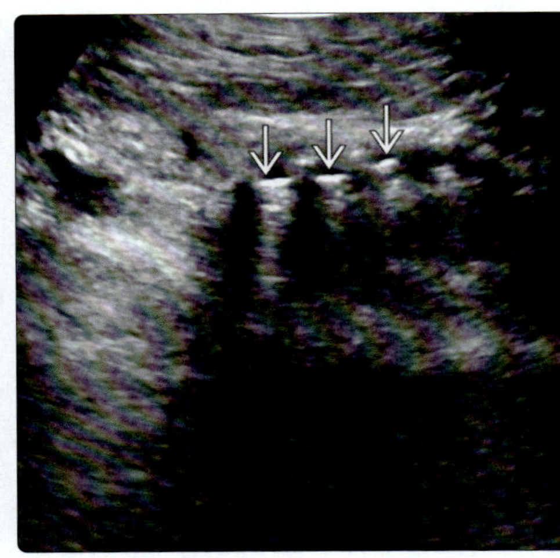

(Left) *Longitudinal US shows a periprosthetic infection of the hip with a moderate degree of hypoechoic inflammatory soft tissue → adjacent to the prosthesis →. There is a small anechoic fluid component → within the inflammatory soft tissue.* **(Right)** *Longitudinal US of the same patient in a slightly more lateral location shows multiple echogenic foci → due to surgically implanted antibiotic beads within the soft tissues.*

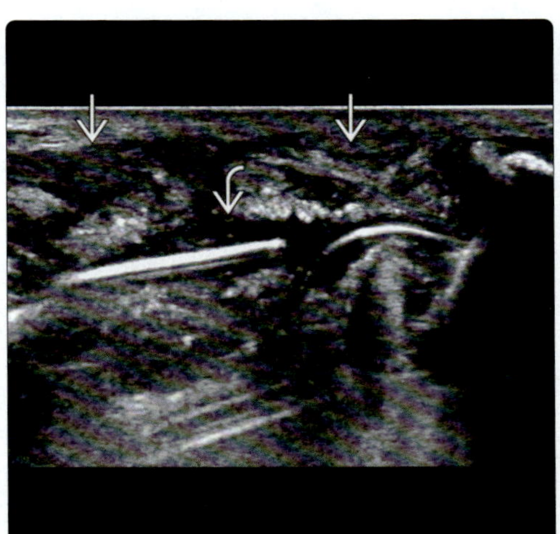

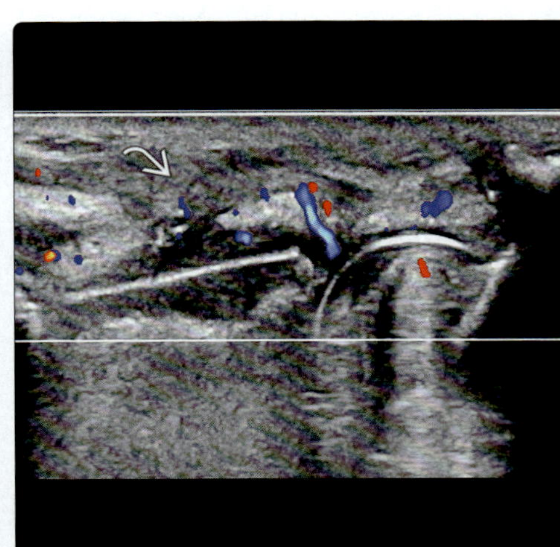

(Left) *Longitudinal US of a patient with an infected elbow replacement shows severe periprosthetic soft tissue edema →. There is a small amount of fluid or very edematous inflammatory tissue → immediately alongside the prosthesis.* **(Right)** *Longitudinal color Doppler US of the same patient shows moderate hyperemia → of the inflamed periprosthetic soft tissues. The periprosthetic hypoechogenicity is due to edematous inflammatory tissue rather than fluid.*

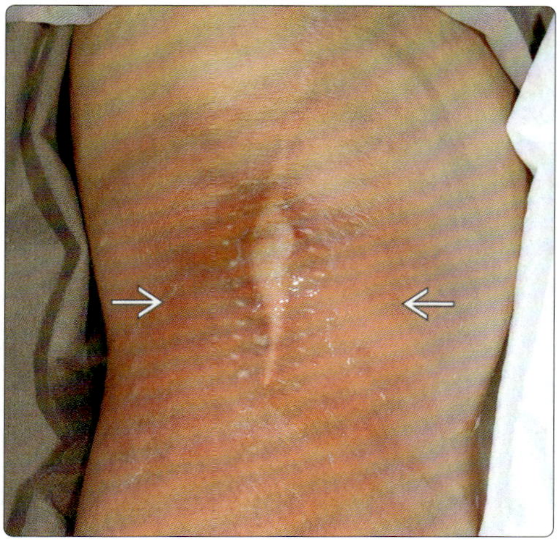

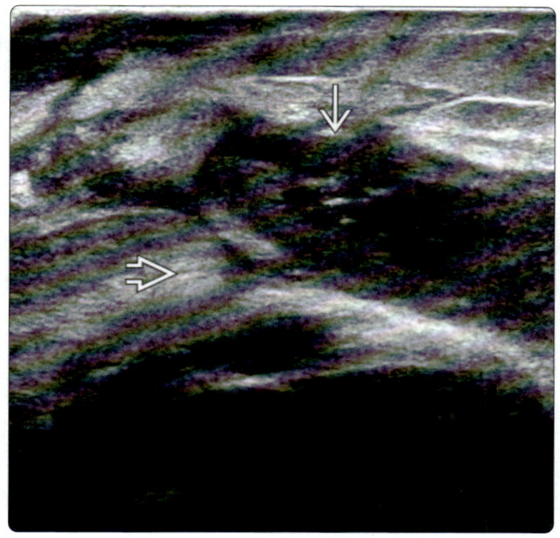

(Left) *Clinical photo of a patient with previous total knee replacement shows swelling and erythema of the skin around the surgical scar.* **(Right)** *Longitudinal US of the same patient shows moderate synovial proliferation of the suprapatellar pouch with moderate surrounding soft tissue edema due to an infected implant.*

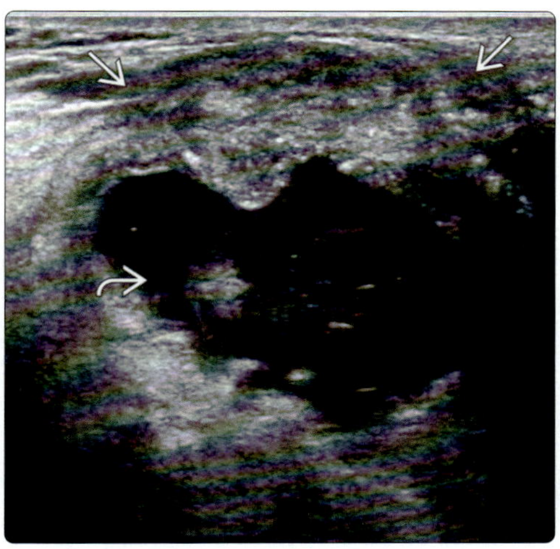

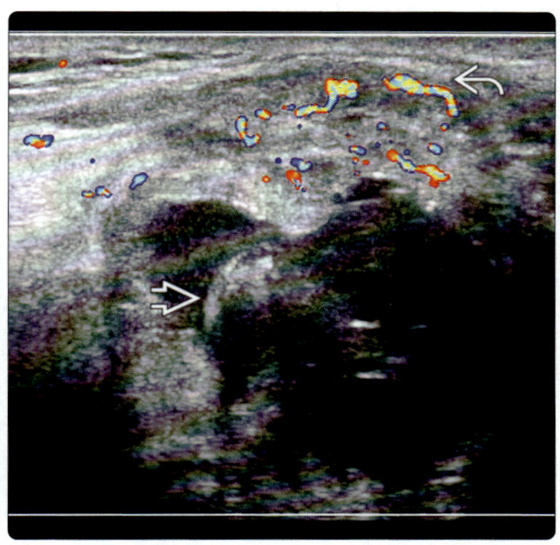

(Left) *Transverse US of a patient with infected total knee replacement for osteoarthritis shows severe synovial proliferation with a moderate amount of mildly echogenic joint fluid.* **(Right)** *Transverse color Doppler US of the same patient shows moderate to severe synovial hyperemia indicative of moderately active infective synovitis. Note the echoes within the joint effusion.*

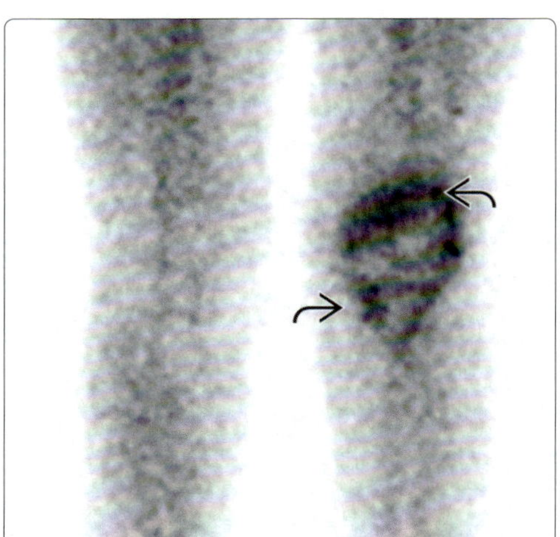

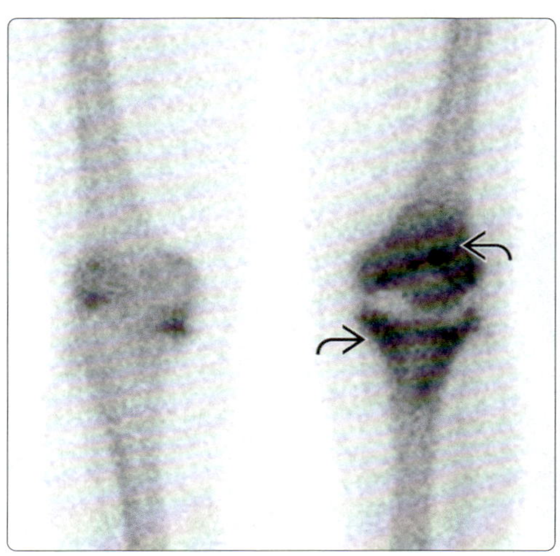

(Left) *Three-phase bone scan of the same patient shows increased periprosthetic activity in the early perfusion phase indicative of hyperemia. There was also increased activity in the blood pool phase (not shown).* **(Right)** *Three-phase bone scan of the same patient shows increased periprosthetic activity in the delayed phase. This can be due to prosthetic loosening or infection. Additional gallium scan revealed increased tracer accumulation at the same sites indicative of infection.*

Hemarthrosis and Lipohemarthrosis

KEY FACTS

TERMINOLOGY

- Hemarthrosis: Blood in joint cavity
- Lipohemarthrosis: Blood and fat in joint cavity

IMAGING

- Layered joint effusion
- **Hemarthrosis initially seen as finely speckled echogenic joint fluid**
  - After ~ 5 minutes, may become 2-layered joint effusion
  - Superficial hypoechoic layer represents supernatant serum
  - Deep hyperechoic layer represents dependent red blood cells (RBCs)
- **Lipohemarthrosis initially seen as finely speckled echogenic joint fluid**
  - After ~ 10 minutes, may become 3-layered joint effusion
  - Superficial hyperechoic layer represents floating fat
  - Central hypoechoic layer represents supernatant serum
  - Deep hyperechoic layer represents dependent RBCs
  - ± hypoechoic areas within floating fat due to trapped blood
- **Hemophilia early arthropathy detection with US score (HEAD-US)**
- Targeted examination of elbow, knee, ankle
- Synovial hypertrophy graded as 0 (absent/minimal), 1 (mild/moderate), or 2 (severe) synovial proliferation
- Subchondral bone graded as 0 (normal), 1 (mild), or 2 (moderate/severe) subchondral bone irregularity ± osteophytes
- Cartilage thinning graded as 0 (normal), 1 (mild), 2 (moderate), 3 (severe), or 4 (complete) thinning

TOP DIFFERENTIAL DIAGNOSES

- Septic arthritis
- Gout
- Inflammatory synovitis

DIAGNOSTIC CHECKLIST

- Lipohemarthrosis is marker of articular surface fracture

(Left) *Longitudinal US of the suprapatellar pouch following knee trauma shows the joint capsule is distended ➔ with echogenic fluid consistent with hemarthrosis. No layering or sedimentation is present.* **(Right)** *Longitudinal US 15 minutes later at the same location in the same patient shows the 3-layered appearance of lipohemarthrosis. The fat layer ➔, serum layer ➔, and cellular [red blood cell (RBC)] layer ➔ are now apparent.*

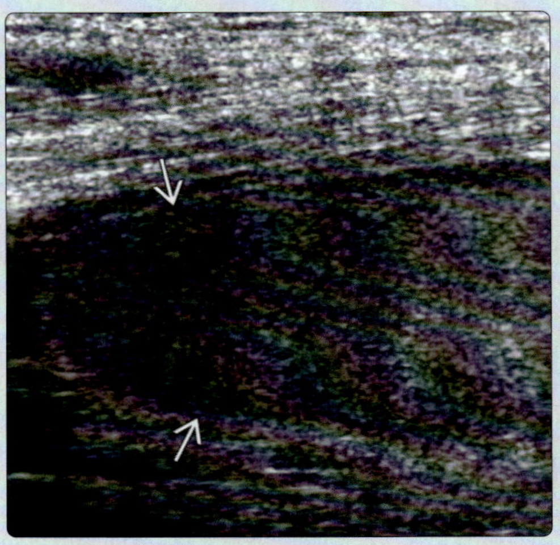

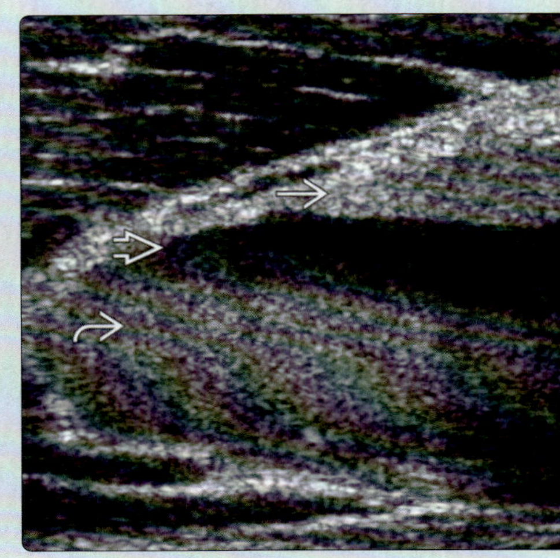

(Left) *Transverse T1 MR in a patient with patellar fracture shows the 3-layered appearance of lipohemarthrosis with fat ➔, serum ➔ and cellular ➔ layers. The fat layer originates from leakage of fatty marrow.* **(Right)** *Transverse US of the posterior aspect glenohumeral joint shows acute hemarthrosis with a distended joint capsule ➔. No layering is present. The glenoid labrum ➔ is shown as is the articular cartilage of the humeral head ➔. Shoulder joint hemarthrosis may accompany massive rotator cuff tear.*

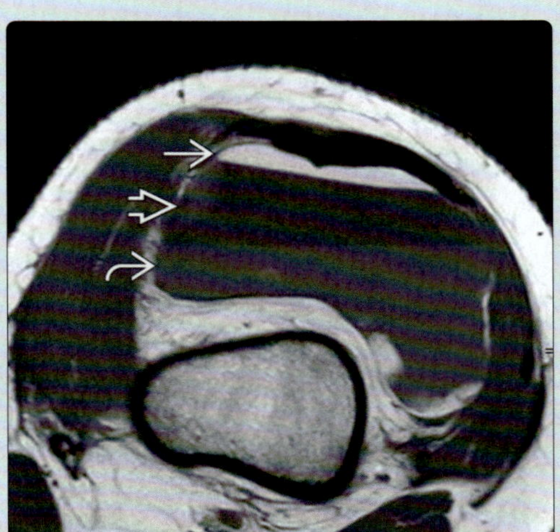

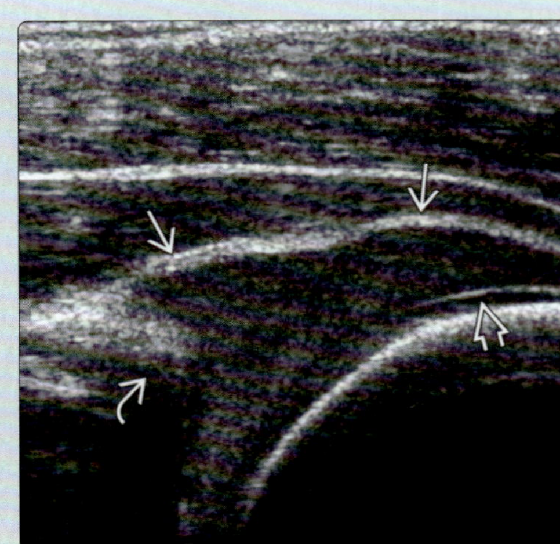

TERMINOLOGY

Synonyms

- Posttraumatic hemarthrosis or lipohemarthrosis
- Spontaneous hemarthrosis
- Hemophiliac arthropathy

Definitions

- Hemarthrosis: Blood in joint cavity
- Lipohemarthrosis: Blood and fat in joint cavity
  - Very specific for intraarticular fracture
- Spontaneous hemarthrosis: Hemarthrosis developing in absence of predisposing trauma, synovial tumor, or hemophilia
- Hemophiliac arthropathy: Arthropathy secondary to repeated hemarthrosis due to inherited clotting disorder

IMAGING

General Features

- Best diagnostic clue
  - Layered joint effusion
- Location
  - Knee > ankle > elbow > shoulder > hip

Radiographic Findings

- Hyperdense joint effusion
  - Density of hemorrhage > simple synovial fluid
- Horizontal beam radiography for lipohemarthrosis: Fat-fluid level due to attenuation differences in fat and blood components
  - Fat is less dense and floats on blood surface
  - Lipohemarthrosis is present in 35% of knee intraarticular fractures
    - High specificity but low sensitivity for intraarticular fractures

CT Findings

- NECT
  - Hemarthrosis: 2 distinct layers
    - Superficial isodense layer [~ 30 Hounsfield units (HU)] = serum
    - Deep hyperdense layer (~ 70 HU) = red blood cells (RBCs)
  - Lipohemarthrosis: 3 distinct layers
    - Superficial hypodense layer (~ -70 HU) = fat
    - Central intermediate layer (~ 20 HU) = serum
    - Deep hyperdense layer (~ 90 HU) = RBCs

MR Findings

- T2WI
  - Hemarthrosis: 2 distinct signal bands
    - Superficial hyperintense band = serum
    - Deep isodense band = dependent RBCs
  - Lipohemarthrosis: 3 distinct signal bands
    - Superficial hyperintense band = floating fat
    - Central hyperintense band = serum
    - Deep isodense band = dependent RBCs
  - Low signal in synovium due to magnetic susceptibility effect of hemosiderin

Ultrasonographic Findings

- Acute stage
  - Large effusion
    - Finely speckled echogenic joint fluid
    - Minimal or no synovial thickening or hyperemia
  - ± ↑ echogenicity from more hypoechoic (superficial) to more hyperechoic (deep)
    - Due to early sedimentation of cellular elements
  - ± layering of different fluid components
    - Specific for hemarthrosis or lipohemarthrosis
    - Seen after lying supine for few minutes to allow time to sediment
  - **Hemarthrosis is initially seen as uniformly finely speckled echogenic joint fluid**
    - After ~ 5 minutes, may become 2-layered joint effusion
      - □ Superficial hypoechoic layer representing supernatant serum
      - □ Deep hyperechoic layer representing dependent RBCs
  - **Lipohemarthrosis initially seen as uniformly, finely speckled echogenic joint fluid**
    - After ~ 10 minutes, may become 3-layered joint effusion
      - □ Superficial hyperechoic layer = floating fat
      - □ Central hypoechoic layer = supernatant serum
      - □ Deep hyperechoic layer = dependent RBCs
    - ± hypoechoic areas within floating fat due to trapped blood
- Subacute stage
  - Blood clots: Discrete echogenic foci within joint
    - Can be difficult to distinguish from synovial proliferation
    - Aspirate if in doubt
  - Joint fluid: More anechoic
  - ± fibrinous adhesions
- Repeated hemarthrosis
  - Early hemophiliac arthropathy: ↑ joint fluid, mild to moderate synovial proliferation
  - Late hemophiliac arthropathy: Less joint fluid, severe synovial thickening
    - Hemosiderin deposition within synovium not apparent on US
    - Synovial hyperemia usually not apparent in hemophiliac arthropathy
      - □ Blood flow in hemophiliac arthropathy lies below sensitivity threshold of Doppler systems
  - Secondary osteoarthritis
    - Diffuse cartilage thinning
    - Irregularity of subchondral bone
    - Marginal osteophytosis
  - ± Baker cyst
  - ± hematoma
  - ± pseudotumor (expanding hematoma)
    - Hypoechoic ± hyperechoic areas (recent bleeding) ± anechoic areas (liquefaction)
- **Hemophilia early arthropathy detection with US score (HEAD-US)**
  - Additive score to define joint status in hemophiliac arthropathy

- Targeted examination of elbow, knee, ankle
  - Synovial hypertrophy graded as 0 (absent/minimal), 1 (mild/moderate), or 2 (severe) synovial proliferation
  - Cartilage thinning graded as 0 (normal), 1 (mild), 2 (moderate), 3 (severe), or 4 (complete) thinning
  - Subchondral bone graded as 0 (normal), 1 (mild), or 2 (moderate/severe) subchondral bone irregularity ± osteophytes

Imaging Recommendations

- Best imaging tool
  - US
- Protocol advice
  - Minimize movement of part under examination
  - ± repeat at 5-20 minutes to observe layering
    - Layering only seen in acute hemarthrosis

DIFFERENTIAL DIAGNOSIS

Septic Arthritis

- Gradual-onset joint swelling over 1-2 days
  - More insidious onset if nonpyogenic infection
- Very tender with marked limitation of movement
  - More severe inflammation than hemarthrosis/lipohemarthrosis
  - Aspirate if there is any doubt

Gout

- Severe inflammation with moderate synovial proliferation
- Echogenic crystal aggregates within joint fluid or synovium
  - ± adjacent tophi or erosions

Inflammatory Synovitis

- Moderate to severe synovial proliferation and hyperemia
  - Hyperemia not prominent feature of synovial proliferation in hemophilia
- ± polyarticular ± marginal erosions
- ± clinical/serological features of inflammatory arthropathy

PATHOLOGY

General Features

- Etiology
  - Posttraumatic hemarthrosis/lipohemarthrosis
    - Bleeding from laceration of synovial vessels or leakage from marrow cavity
    - Lipohemarthrosis always signifies intraarticular fracture extension
  - Spontaneous hemarthrosis
    - Due to spontaneous bleeding from synovium
- Genetics
  - Hemophilia A or B: X-linked recessive

Staging, Grading, & Classification

- Arnold-Hilgartner scale for hemophiliac arthropathy of knee
  - Stage 0: Normal joint
  - Stage I: Soft tissue swelling
  - Stage II: Osteoporosis and overgrowth of epiphysis
  - Stage III: Early subchondral bone cysts, squaring of patella, widening of intercondylar notch
  - Stage IV: As per stage III, but joint space is narrowed
  - Stage V: Fibrous joint contracture, loss of joint space, severe enlargement of epiphysis, and substantial disorganization of joint

Gross Pathologic & Surgical Features

- Synovial proliferation ± hemophiliac arthropathy

Microscopic Features

- Acute on chronic inflammation
- Hemosiderin-laden macrophages; fibrosis

CLINICAL ISSUES

Presentation

- Most common signs/symptoms
  - Rapid-onset joint effusion with tense warm joint

Demographics

- Age
  - Posttraumatic hemarthrosis/lipohemarthrosis: Young adults
  - Spontaneous hemarthrosis: Older patients
  - Hemophiliac arthropathy: Throughout life
- Gender
  - Hemarthrosis and lipohemarthrosis: M > F and related to trauma prevalence
  - Hemophilia: Almost exclusively in males
- Epidemiology
  - Relative frequency: Hemophilia A (85%), hemophilia B (14%), other clotting disorders (1%)
    - 1 in 5,000 males; 60% of cases are severe

Natural History & Prognosis

- Acute hemarthrosis resolves over 1-2 weeks
- Hemophiliac arthropathy
  - Repeated hemarthrosis → progressive arthropathy

Treatment

- Rest, ice, and analgesia
  - ± aspiration if very painful (distension)
- Hemophiliac arthropathy with acute hemorrhage
  - Correct coagulation factor deficiency
    - Infusion of factor VIII (hemophilia A) or factor IX (hemophilia B)
- Hemophiliac arthropathy with repeated hemarthroses
  - Chemical synovectomy with intraarticular administration of rifampicin or osmic acid (± US guidance)
  - Radiosynovectomy with intraarticular administration of yttrium, gold, or phosphorus (± US guidance)
  - Intraarticular administration of hyaluronic acid (± US guidance)
  - Surgery: Joint replacement (if severe arthropathy)

DIAGNOSTIC CHECKLIST

Image Interpretation Pearls

- Lipohemarthrosis has single-fluid level (fat-fluid) on radiography but double-fluid level (fat-serum-RBCs) on US, CT, and MR

SELECTED REFERENCES

1. Di Minno MN et al: Assessment of hemophilic arthropathy by ultrasound: where do we stand? Semin Thromb Hemost. 42(5):541-9, 2016

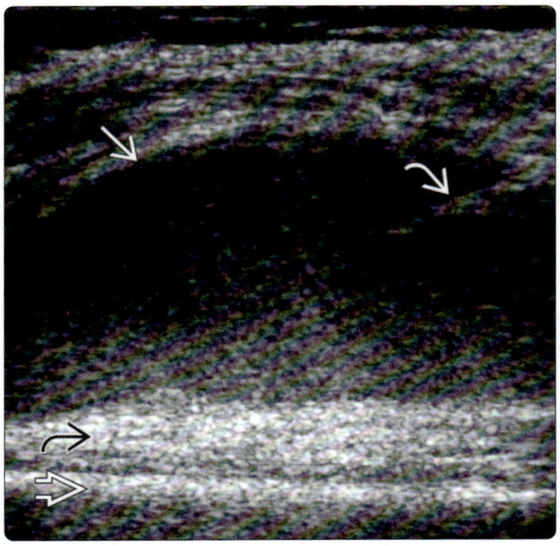

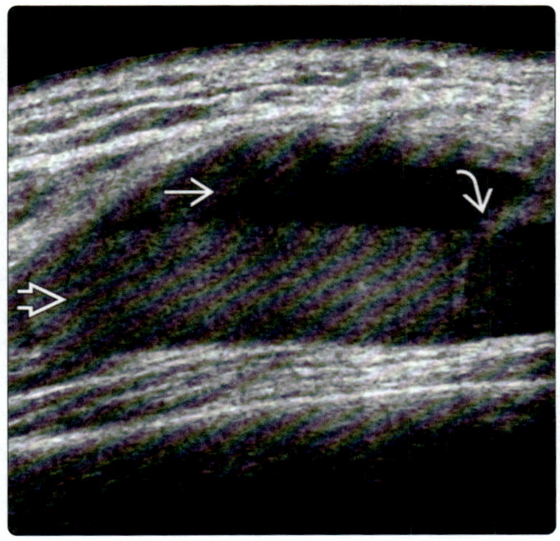

(Left) *Longitudinal US of the suprapatellar pouch in acute hemarthrosis demonstrates sedimentation evidenced by gradually increasing echogenicity from superficial to deep. Note the prefemoral fat , femoral cortex , and synovial fold .* **(Right)** *Longitudinal US 10 minutes later in the same patient shows acute hemarthrosis with a 2-layered appearance: Serum layer and cellular (RBC) layer . Note the synovial plica (fold) traversing the suprapatellar pouch.*

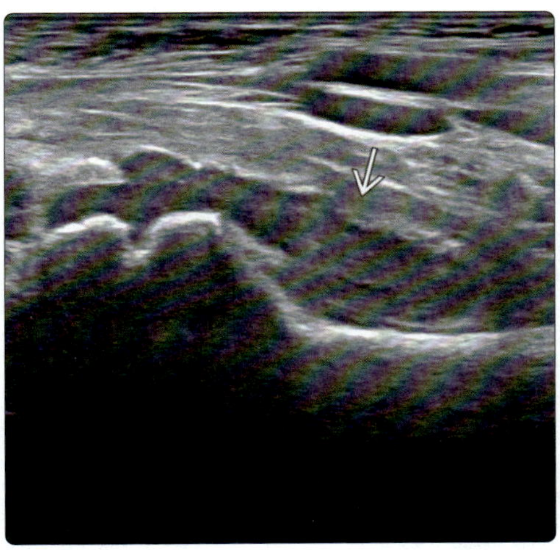

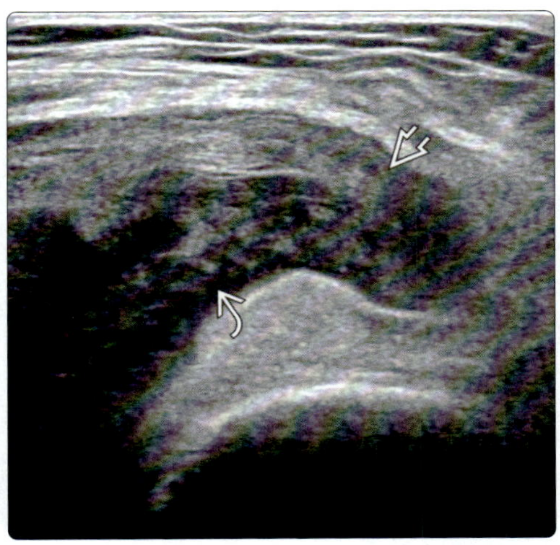

(Left) *Longitudinal hip US in an 8-year-old boy following trauma is shown. No radiographic fracture was present. There is moderate joint distension with echogenic fluid consistent with hemarthrosis. MR, including gradient-echo sequences, showed no occult fracture and confirmed hemarthrosis.* **(Right)** *US lateral patellar recess in a 72-year-old woman on anticoagulants with recent knee swelling shows synovial proliferation and echogenic joint fluid , suggestive of hemarthrosis and confirmed by aspiration.*

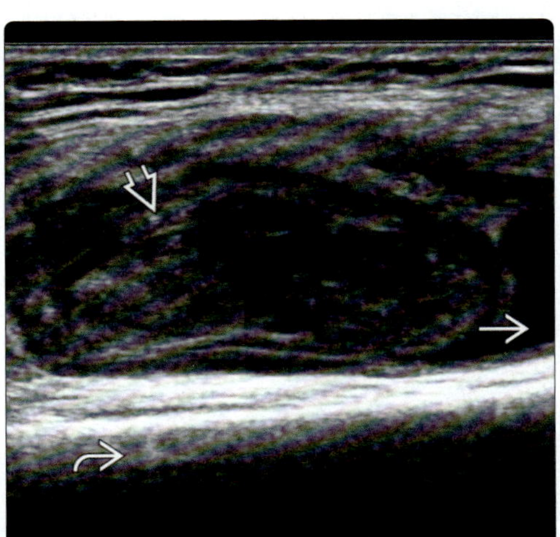

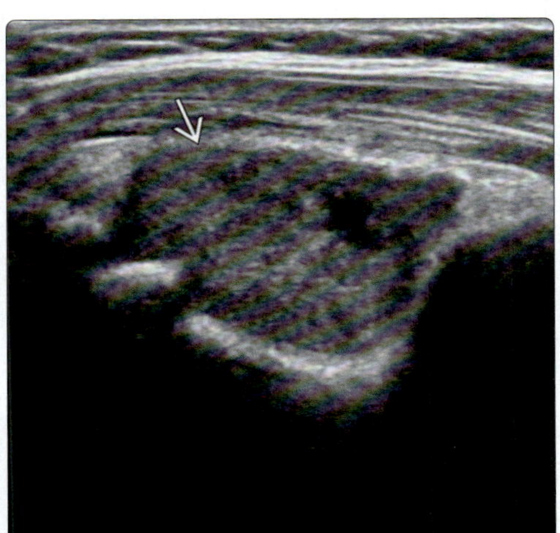

(Left) *Longitudinal US shows an acute hematoma in the synovium of the suprapatellar pouch of a hemophiliac patient. There is an anechoic joint effusion . The distal femur is shown.* **(Right)** *Transverse US shows distension of the ulnar epicondylar-olecranon recess on the posterior aspect of the elbow joint due to acute hemarthrosis in a hemophiliac patient. Of course, the whole joint was distended though distension of only 1 recess.*

Baker Cyst

KEY FACTS

IMAGING

- Fluid-filled sac with neck arising from interspace between medial gastrocnemius muscle & semimembranosus tendon
  - Cyst must have neck emerging from gastrocnemius muscle and semimembranosus tendon interspace
  - Gives "talk-bubble" configuration on transverse scan
- Cysts are typically well defined and thin walled
- As they communicate with knee joint, cyst contents usually mirror those in knee
  - Thick-walled cysts in inflammatory osteoarthritis, inflammatory arthropathy, or with intracystic hemorrhage
  - Thick-walled cysts often have hyperemic walls
- Cyst may be septated
- Contains anechoic synovial fluid; typically gelatinous in consistency
- Free-fluid tracking adjacent to cyst indicates recent leakage
  - Recent leakage is important to note as it can result in localized posteromedial knee pain
- Irregular and multiloculated cysts may result from repeated cyst leakage and collection
- Cysts usually extend distally superficial to medial belly gastrocnemius muscle
- Contralateral subclinical Baker cysts common
- US accurately confirms presence, extent, and content of Baker cyst
  - Examine with patient prone, knees extended
  - Follow medial belly of gastrocnemius proximally and locate interspace between medial gastrocnemius head and semimembranosus tendon
  - Routinely examine contralateral side
  - Can aspirate percutaneously if not too viscous, but high rate of recurrence
    - ± steroid injection, especially if inflammatory component is present (inflammatory-type exudate, thick walled, hyperemia)
    - Not possible to ascertain viscosity of cyst content from US appearance

(Left) *Clinical photograph of the popliteal fossa shows focal swelling → due to Baker cyst. Most patients with Baker cyst just have a palpable swelling rather than a visible focal fullness.* **(Right)** *Longitudinal graphic shows a Baker cyst → located between the semimembranosus tendon → and the medial belly of the gastrocnemius muscle →.*

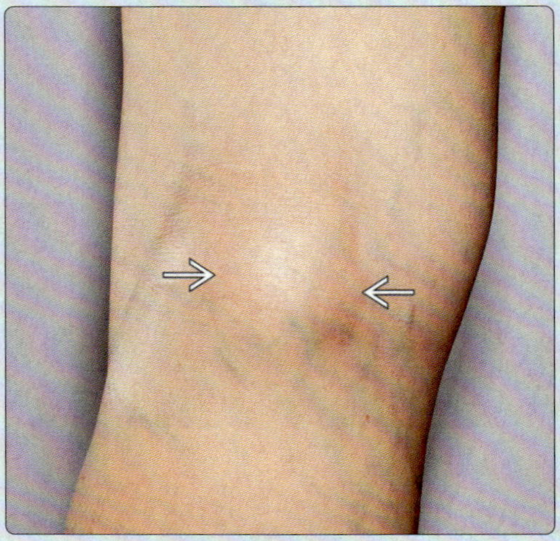

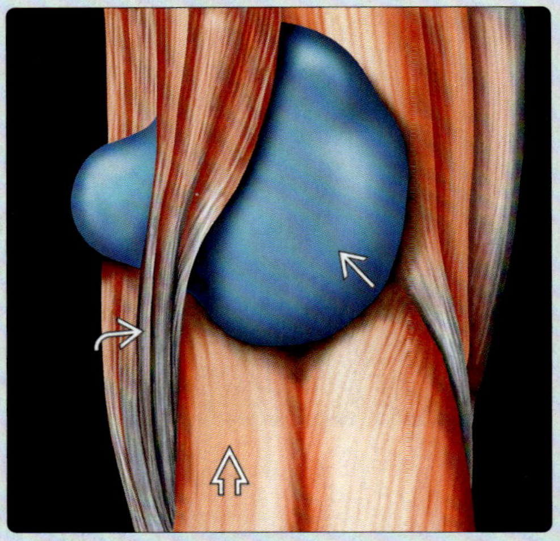

(Left) *Transverse US shows a typical Baker cyst → with a "talk-bubble" configuration. The neck extends between the semimembranosus tendon → and the medial belly of the gastrocnemius muscle →.* **(Right)** *Transverse US shows a typical Baker cyst with the neck arising between the semimembranosus tendon → and the medial belly of the gastrocnemius muscle →. The finely speckled appearance of fluid within the cyst → is due to colloid aggregates suspended in the gelatinous content of the cyst.*

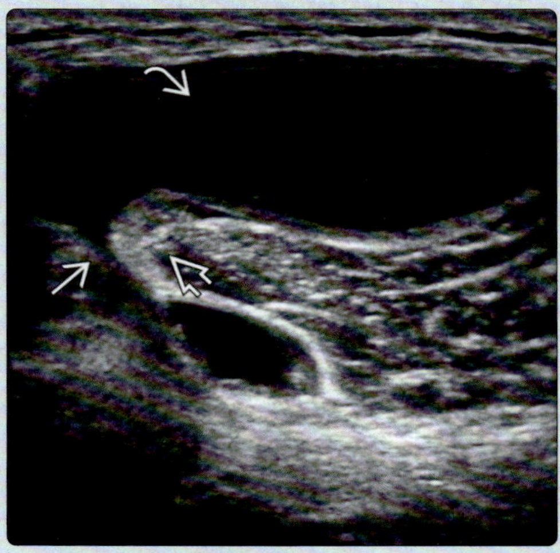

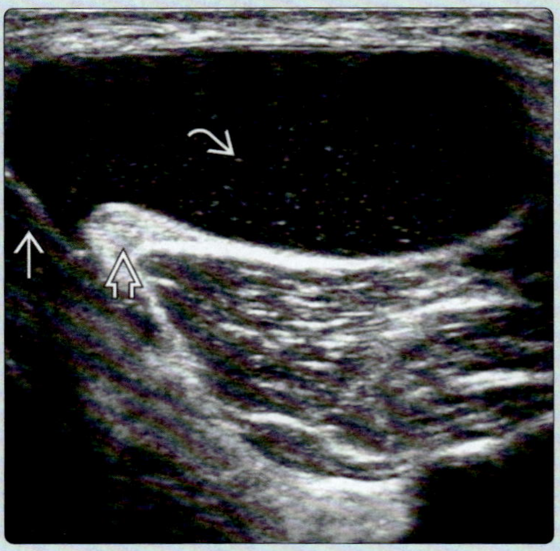

TERMINOLOGY

Synonyms

- Popliteal cyst, popliteal synovial cyst
  - Baker cyst is preferred term since popliteal cyst can also refer to periarticular ganglion cyst or parameniscal cyst

Definitions

- Fluid distension of gastrocnemius-semimembranosus bursa

IMAGING

General Features

- Best diagnostic clue
  - Fluid-filled sac with neck arising from interspace between medial gastrocnemius muscle and semimembranosus tendon
- Location
  - Posterior to medial femoral condyle
- Size
  - Range: 1-30 cm across long axis or greater
    - Larger cysts often seen with inflammatory arthropathy
- Morphology
  - Small unilocular sac to large multiloculated sac extending to mid 1/3 of calf

Radiographic Findings

- Radiography
  - Soft tissue swelling of posteromedial knee
    - ± calcified debris/intraarticular loose bodies
    - ± displacement of calcified popliteal artery

MR Findings

- T1WI
  - Hypointense, well-defined mass in expected location
  - Subacute hemorrhage in cyst seen as areas of hyperintensity
- T2WI
  - Hyperintense mass: Gastrocnemius-semimembranosus bursa
    - ± fluid level, ± hypointense foci (hemosiderin), ± calcified bodies (hypointense), ± thick synovial wall
  - ± surrounding free fluid

Ultrasonographic Findings

- Fluid-filled popliteal fossa mass
  - Anechoic with posterior enhancement
  - Must see neck arising from medial gastrocnemius-semimembranosus tendon interspace to make diagnosis of Baker cyst
    - Characteristic "talk-bubble" configuration on transverse scan
  - Extension through posterior knee capsule not seen
- Cyst is typically well defined and thin walled
- As they communicate with knee joint, cyst contents usually mirror those in knee
- Thick-walled cysts suggest association with inflammatory osteoarthritis, inflammatory knee arthropathy, or intracystic hemorrhage
  - Thick-walled cysts often have hyperemic walls
- Cyst may be septated
  - Usually bilocular or trilocular
- Typically contains anechoic synovial fluid; gelatinous in consistency
  - Swirling fluid artifact is not feature
  - ± fine echogenic speckles due to colloid aggregates
  - ± small echogenic comet-tail artifacts (secondary to colloid aggregates)
  - ± fibrinous-type exudate (particularly with inflammatory arthropathy)
  - ± calcific particulate matter
  - ± blood clots
- Iso- or hyperechoic fluid suggests recent intracystic bleeding or inflammatory exudate
- Free-fluid tracking adjacent to cyst indicates recent leakage
  - Irregular and multiloculated cysts result from repeated cyst leakage and recollection
  - Sharp angular margins (pointed or "stiletto tip") to cyst contour suggest leakage
    - Fluid leaked from Baker cyst will not have discernible wall
    - Free-fluid tracts around posteromedial aspect of knee, into calf, and into distal thigh
      - Recent leakage is important to note as it can result in localized pain
- Cysts usually extend distally superficial to medial belly of gastrocnemius muscle
  - ± extension distally deep to medial belly of gastrocnemius
  - ± extension proximally deep to and around hamstring muscles, especially semimembranous muscle
  - ± intramuscular extension (uncommon)
    - Into medial belly gastrocnemius, vastus medialis
    - Through muscular fascia weakened by trauma or perforating vessels
- Cannot usually ascertain viscosity of cyst content on US
- Contralateral subclinical Baker cysts are common
  - Routinely examine both popliteal fossae

Imaging Recommendations

- Best imaging tool
  - US
    - Confirms presence, extent, content, and recent leakage
- Protocol advice
  - Examine with patient prone, knees extended
  - Follow medial belly of gastrocnemius proximally and locate interspace between medial gastrocnemius head and semimembranosus tendon
    - Baker cyst arises from interspace between these 2 structures
    - Check cyst margins closely looking for fluid leakage, especially most distal margin
  - Routinely examine contralateral side

DIFFERENTIAL DIAGNOSIS

Parameniscal Cyst

- Usually associated with meniscal tear, typically horizontal tear
- Cyst extends to contact or very close to margin of meniscus
- Does not have typical configuration of Baker cyst

- Usually multiloculated

Popliteal Artery Aneurysm

- Saccular or fusiform dilation of popliteal artery
- Color Doppler imaging confirms vascular flow

Cystic Periarticular Tumor

- Cystic synovial sarcoma or myxoma
- Does not have typical configuration of Baker cyst

Pariarticular Ganglion

- Does not arise between medial gastrocnemius tendon and semimembranous

PATHOLOGY

General Features

- Etiology
  - Joint fluid decompressing into cyst
  - Fluid may communicate freely or be restricted with ball valve-like mechanism
- Associated abnormalities
  - Effusion associated with intraarticular pathology in ~ 85-98% of cases in adults
    - Meniscal tears, arthritis (degenerative, inflammatory), chondral defects, cruciate ligament tears
  - In children, usually not associated with intraarticular pathology or joint effusion
    - Possibly related to congenital predisposition
    - Frequently bilateral
- Slit-like capsular opening into gastrocnemius-semimembranosus bursa (Baker cyst) behind posterior horn of medial meniscus
  - Slit = 15-20 mm vertically oriented opening
    - Slit closed in extension, open in flexion
- Capsule unsupported at this location
  - Between expansions of semimembranosus tendon medially and posterior cruciate ligament laterally
- Slit present on cadaveric study in 50% of subjects > 50 years age
  - Prevalence increases with age
  - Probably due to degenerative thinning of joint capsule
- Synovial fluid extruded through slit into cyst
- Wall of cyst composed of fibrous tissue lined by synovium
  - Synovium of cyst in continuity with synovium of knee joint
- Synovial disorders of knee also affect Baker cyst
  - Synovitis, synovial osteochondromatosis, pigmented villonodular synovitis
- Other mechanisms whereby Baker cysts may develop
  - Accumulation of fluid within noncommunicating gastrocnemius-semimembranosus bursa
  - Herniation of posterior joint capsule due to increased intraarticular pressure

Gross Pathologic & Surgical Features

- Fluid-filled cyst in popliteal space

Microscopic Features

- Lined by synovium, similar to knee joint

CLINICAL ISSUES

Presentation

- Most common signs/symptoms
  - Painless mass on medial side of popliteal fossa
  - Symptoms from underlying internal derangement (meniscal tear, osteoarthritis, chondral defect, etc.)
- Other signs/symptoms
  - Sudden onset of calf pain due to
    - Expansion, leakage, intracystic bleeding
  - Anticoagulation for clinically suspected deep venous thrombosis may induce intracystic bleeding
  - Pressure effect on popliteal artery, vein, or tibial nerve is very rare

Demographics

- Age
  - Prevalence increases with age
- Gender
  - More common in males
- Epidemiology
  - Most commonly encountered cyst around knee
    - 15% of general population
    - 5% of pediatric MR knee studies
    - 15% of asymptomatic MR knee studies
    - 50% of elderly MR knee studies

Natural History & Prognosis

- Cyst may regress or enlarge
  - Overall good prognosis, especially in children
- In children, if cysts are treated conservatively, 1/2 will disappear and remainder will regress

Treatment

- Conservative
  - Treatment of associated internal derangement
  - Nonsteroidal antiinflammatory agents, ice
  - Cyst aspiration ± intracystic steroid injection
    - High rate of recurrence
    - Occasionally, contents are too viscous to aspirate
    - Intracystic steroid injection, particularly if inflammatory component to cyst (inflammatory-type exudate, thick walled, hyperemia)
- Surgical
  - Excision ± closure of joint-bursal communication
    - Recurrence is common

DIAGNOSTIC CHECKLIST

Consider

- Intraarticular pathology, as this is very frequently associated with Baker cysts in adults
- Cyst content often mirrors content of knee joint

Image Interpretation Pearls

- Look for characteristic "talk-bubble" configuration of cyst on transverse imaging

SELECTED REFERENCES

1. Shah A et al: A diagnostic approach to popliteal fossa masses. Clin Radiol. 72(4):323-337, 2017

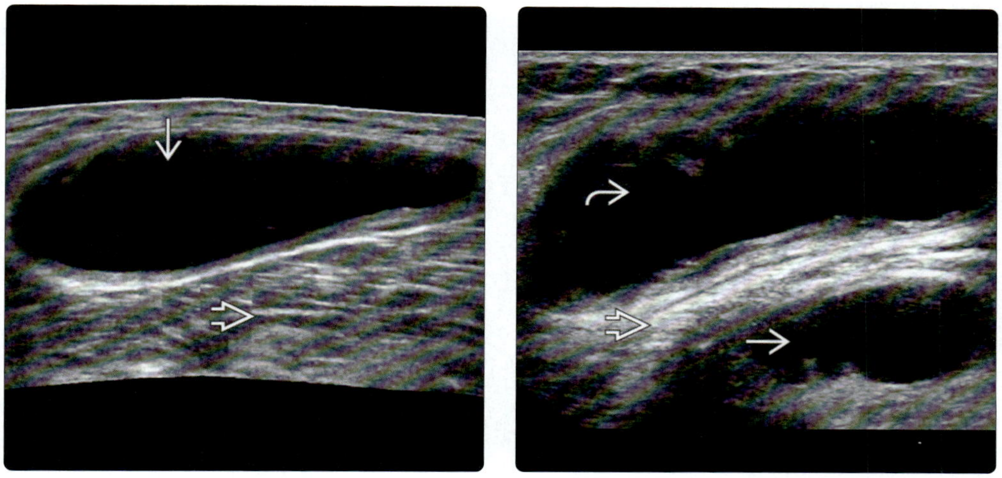

(Left) *Extended field-of-view US shows a thin-walled unilocular Baker cyst ➔ lying superficial to the medial belly of the gastrocnemius muscle ➔.* **(Right)** *Longitudinal US shows a Baker cyst extending distally in the calf both superficial ➔ and deep ➔ to the medial belly of the gastrocnemius muscle ➔. Most cysts just extend superficial to the gastrocnemius muscle.*

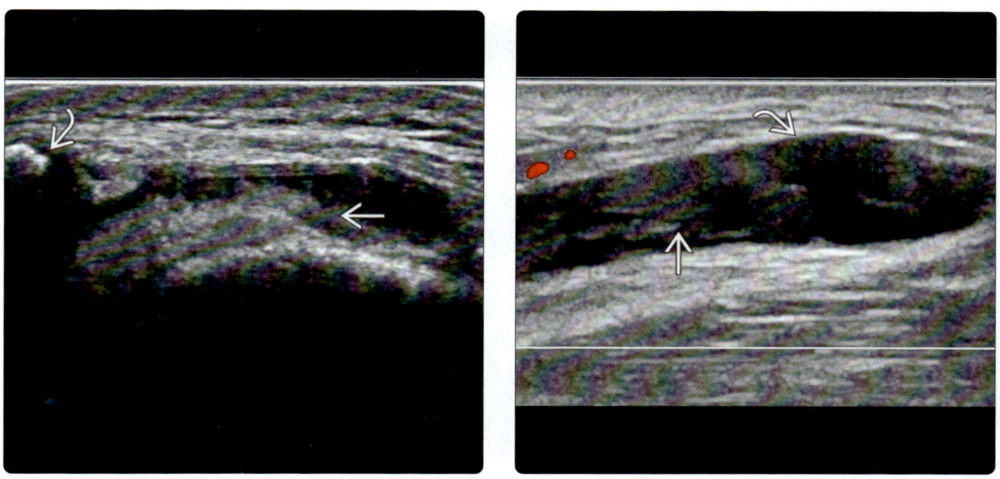

(Left) *Transverse US of the lateral patellar recess ➔ shows synovial proliferation ➔ in this patient with moderate severity, inflammatory-type knee osteoarthritis. The edge of the patella lies medial to the recess.* **(Right)** *Longitudinal color Doppler US of the same patient shows moderate synovial proliferation ➔ within the Baker cyst ➔. However, there is no demonstrable wall hyperemia. The contents of a Baker cyst will often mirror those of the knee joint.*

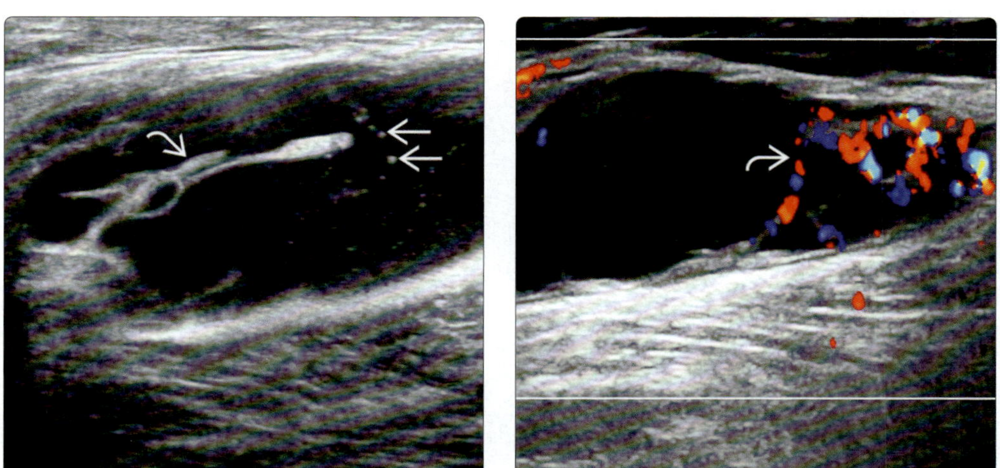

(Left) *Longitudinal US shows a Baker cyst containing thick synovial fronds ➔ and comet-tail artifacts ➔. These comet-tail artifacts are due to colloid aggregates. The fronds are due to fibrous adhesions.* **(Right)** *Longitudinal color Doppler US of the same patient shows marked hyperemia ➔ of the hypertrophied synovial fronds.*

(Left) *Transverse US shows a thick-walled Baker cyst with fibrinous stranding and thick, fibrous-type exudate.* **(Right)** *Longitudinal color Doppler US shows moderate hypervascularity within mass-like synovial proliferation in a patient with an inflammatory arthropathy of the knee joint. One should relate the appearances seen in a Baker cyst to those seen in the knee joint.*

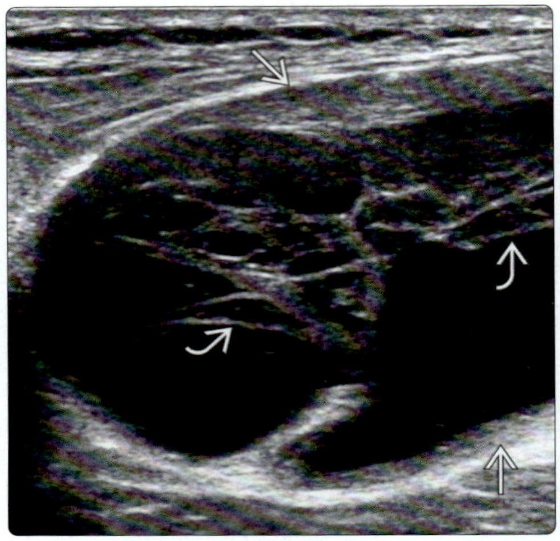

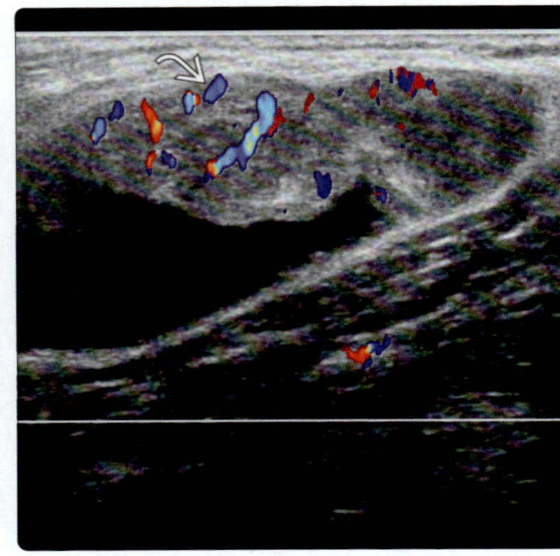

(Left) *Transverse US of a Baker cyst shows small areas of calcified debris within hypertrophied synovium.* **(Right)** *Transverse US of the same patient shows calcified areas within the hypertrophied synovium of a thick-walled Baker cyst.*

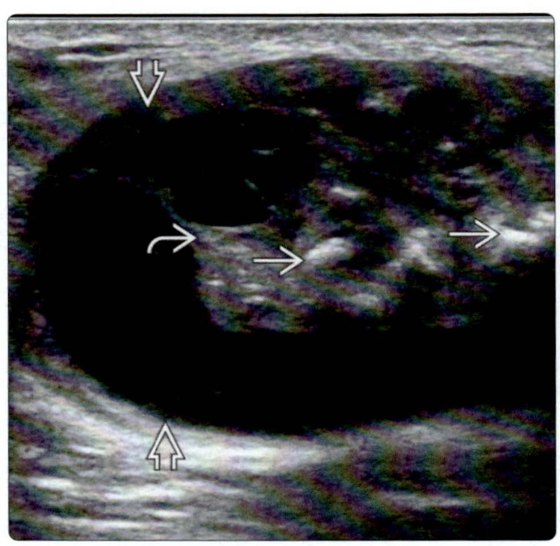

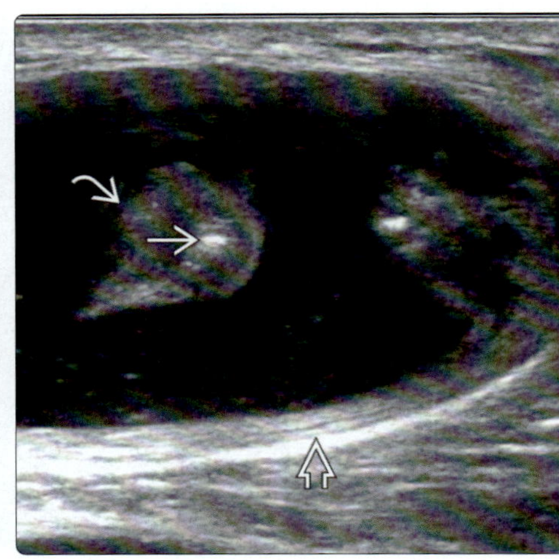

(Left) *Lateral radiograph of a knee shows soft tissue swelling of a Baker cyst containing small amounts of calcified debris.* **(Right)** *Transverse US of the same patient shows a calcified fragment within part of the Baker cyst. Osseous fragments within the knee joint can extend into a Baker cyst. Osseous fragments may also enlarge within Baker cysts. Hence, very large calcified bodies can be seen in Baker cysts.*

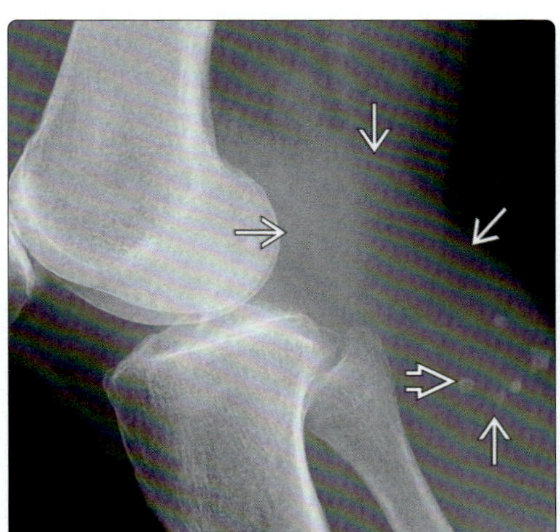

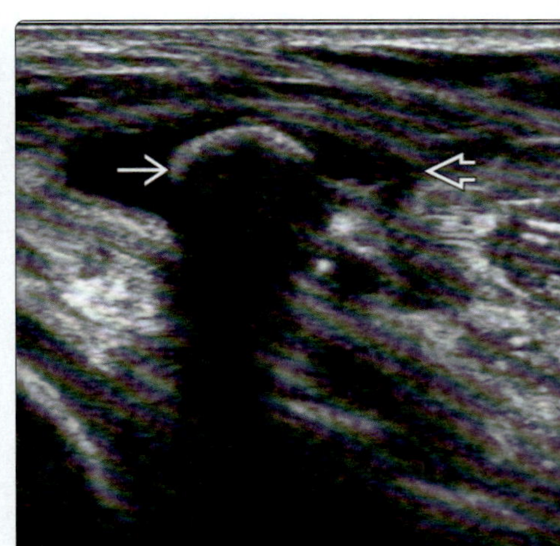

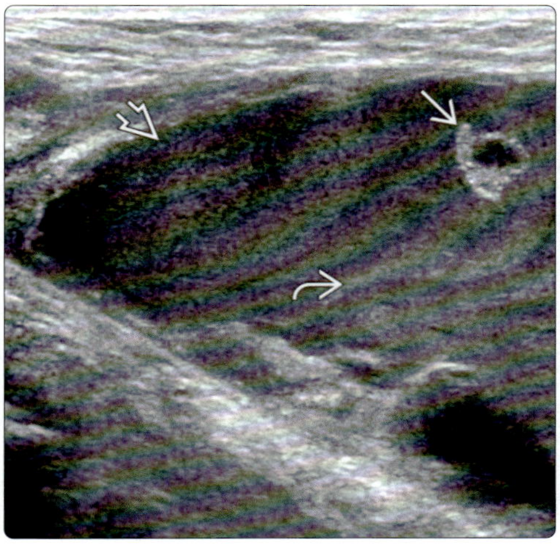

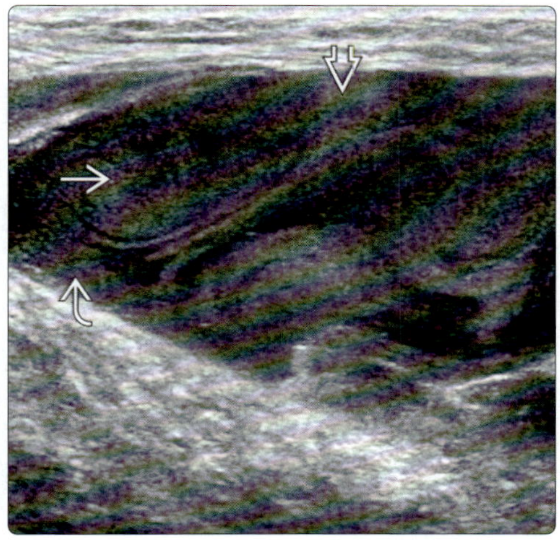

(Left) *Transverse US shows recent hemorrhage within a Baker cyst → with finely speckled echogenic fluid →. The small hyperechoic ring → may be due to a detached synovial fragment.* **(Right)** *Transverse US of the same patient shows recent hemorrhage into the Baker cyst → with echogenic fluid → and early clot → formation within the cyst. The echogenicity and swirling pattern is characteristic of hemorrhage.*

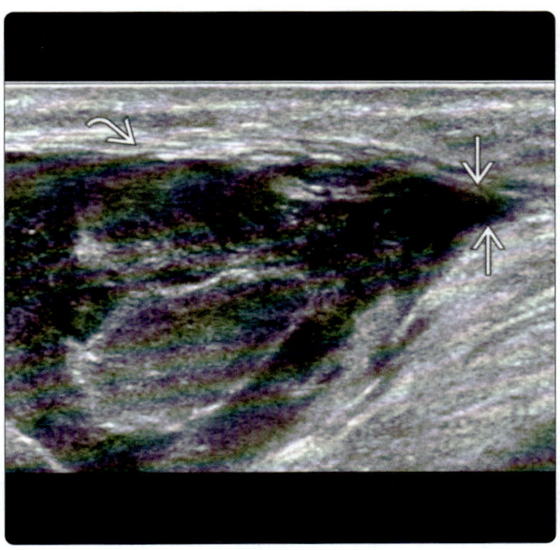

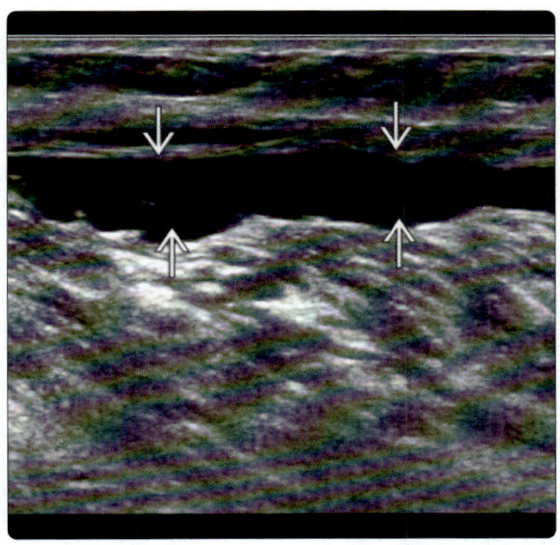

(Left) *Longitudinal US with extended field of view shows localized cyst leakage → as evidenced by a sharp contour to the edge of the Baker cyst → and lack of a discernible wall around the extracystic free fluid. Localized fluid leakage is much more common than frank cyst rupture.* **(Right)** *Longitudinal US of the same patient shows fluid leakage → tracking more distally superficial to the medial belly of the gastrocnemius muscle. Most fluid leakage occurs distally.*

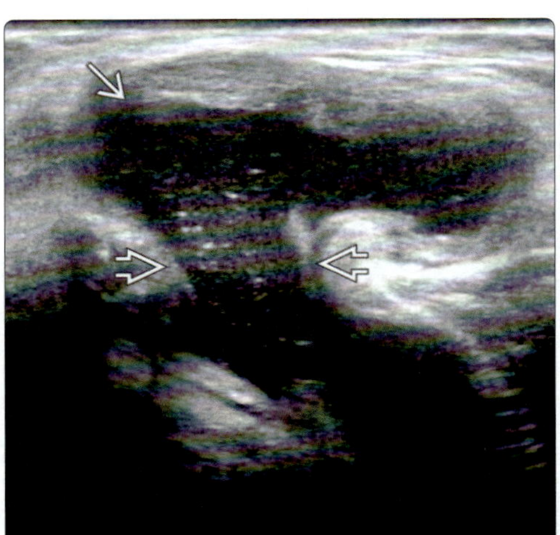

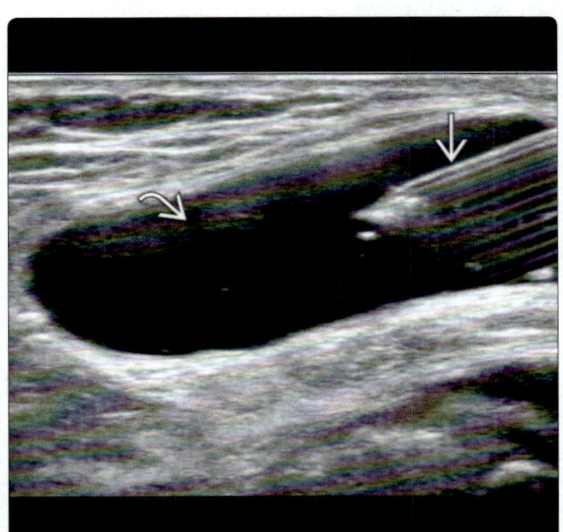

(Left) *Transverse US shows a Baker cyst → with an unusually wide neck →.* **(Right)** *Longitudinal US shows percutaneous aspiration with a 21-g needle → within a Baker cyst →. There is some reverberation artifact deep to the needle. It is difficult to determine the viscosity of Baker cyst contents prior to aspiration. As most Baker cysts tend to reaccumulate following aspiration, only temporary relief of symptoms is obtained. Aspiration is not usually recommended.*

Bursitis

KEY FACTS

TERMINOLOGY

- Bursitis: Inflammation of lining of bursa

IMAGING

- Majority of bursae are not visible on US unless distended with fluid or proliferative synovial tissue
- Acute bursitis: Thin-walled bursa distended with hypoechoic fluid, peribursal hyperemia
- Chronic bursitis: Thickening of bursal wall, more mature synovial proliferation, more echogenic content, & increased likelihood of intrabursal hyperemia
  - Can simulate soft tissue mass or abscess
  - If echogenic foci in distended bursa, consider crystal deposition
- Most bursae are accessible to US examination
  - Use minimal transducer pressure to avoid effacement of superficial bursae
  - Amiable to US-guided aspiration ± injection

TOP DIFFERENTIAL DIAGNOSES

- Cystic collection from underlying structure, e.g., parameniscal cyst
- Abscess or collection
- Cyst adjacent to bursa, e.g., epidermoid or hematoma

CLINICAL ISSUES

- Treatment
  - Rest; antiinflammatory agents: US-guided corticosteroids ± long-acting, local anesthetic injection
  - Excision for resistant cases

DIAGNOSTIC CHECKLIST

- Small amount of fluid is normal in subacromial-subdeltoid bursa, deep infrapatellar bursa, & retrocalcaneal bursa
- Bursitis is common; infective bursitis is uncommon
- Synovial-type bursa are prone to every synovial-based pathology

(Left) *Clinical photo shows a patient with posterior elbow swelling ➡ that has been present for years but recently increased in pain and size.* **(Right)** *Transverse US shows a large olecranon bursitis with a moderately thickened bursal wall ➡. The bursa is distended by fluid and synovial proliferation and overlies the olecranon process ➡. Olecranon bursae frequently have a thickened wall, internal debris, and septa. A noninflamed olecranon bursa is not visible on US.*

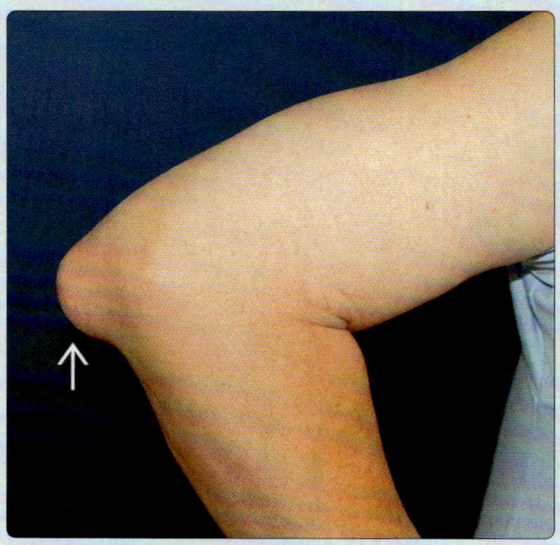

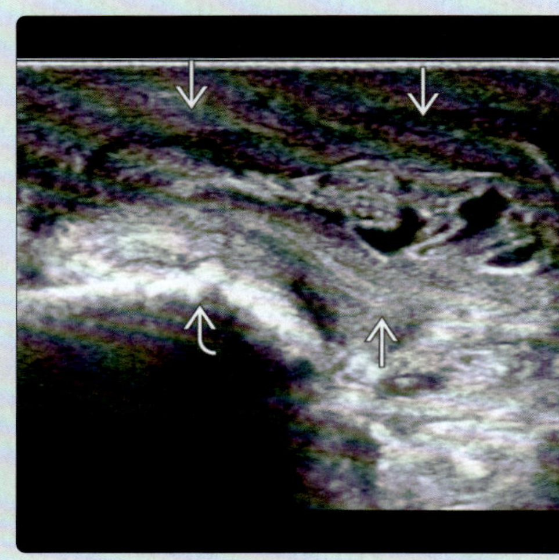

(Left) *Longitudinal US of the same bursa shows the distended, inflamed olecranon bursa with several thick internal septations ➡, giving the bursa a multiloculated appearance. The normal triceps tendon insertion ➡ into the olecranon process is shown.* **(Right)** *Longitudinal color Doppler US shows moderate hyperemia of the thickened bursal wall ➡ and peribursal tissues. The internal content and septations show no hyperemia.*

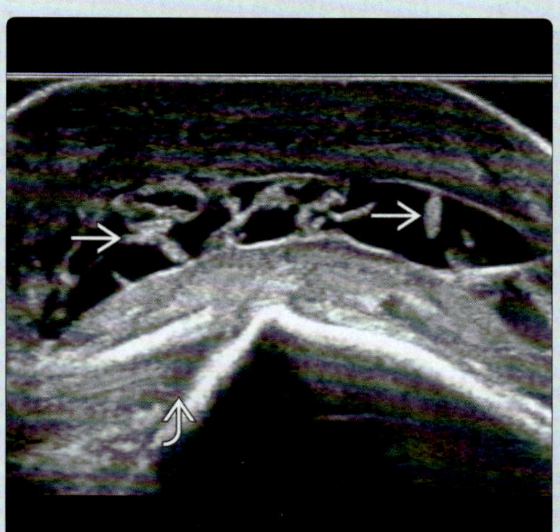

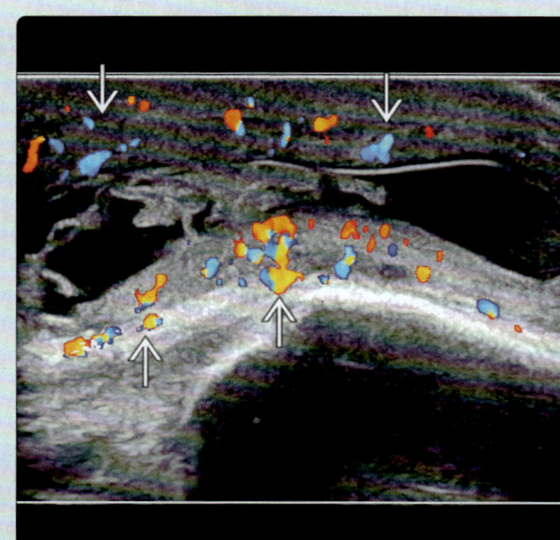

TERMINOLOGY

Synonyms

- Bursopathy, student's elbow (olecranon bursitis), ischial tuberosity (ischiogluteal) bursitis, trochanteric (hip) bursitis, prepatellar bursitis (housemaid's knee), infrapatellar bursitis (preacher's knee)

Definitions

- Bursitis: Inflammation of lining of bursa
  - Bursa: Synovial-lined sac interspersed between tissues at areas of friction
  - 2 types of bursae
    - Synovial bursae: Constant synovial-lined bursae, which develop in utero in defined anatomical locations
    - Adventitious bursae: Nonsynovial-lined bursae, which are acquired post partum due to friction between opposing tissues

IMAGING

General Features

- Best diagnostic clue
  - Fluid-distended sac in typical location
- Location
  - Subacromial-subdeltoid bursa: Large bursa overlying rotator cuff tendons
    - Communicates with glenohumeral joint if full-thickness tear of supraspinatus/infraspinatus tendons
  - Olecranon bursa: Overlies olecranon process & distal end of triceps tendon
  - Ischial tuberosity bursa: Between ischial tuberosity (hamstring tendon attachment) & gluteus maximus muscle
  - Iliopsoas bursa: Large bursa between iliopsoas tendon & anterior part of hip joint
    - Communicates with hip joint in 15% of subjects
  - Trochanteric bursa: Laterally overlies gluteus medius & minimus tendon insertion into greater trochanter femur
  - Subgluteus minimus bursa: Between gluteus minimus tendon & anterior facet greater trochanter
  - Subgluteus medius bursa: Between gluteus medius tendon & posterosuperior facet greater trochanter
  - Semimembranous bursa: Posteromedial to semimembranous musculotendinous junction & tendon posteromedial aspect distal thigh
  - Pes anserine bursa: Between sartorius, gracilis, semitendinosus tendons, & posteromedial tibia
  - Prepatellar bursa: Overlies inferior pole of patella & proximal patellar tendon
  - Infrapatellar bursa: Overlies mid-to-distal aspect of patellar tendon
    - Do not confuse with deep infrapatellar bursa
      - Small amount of fluid in deep infrapatellar bursa is normal
  - Retrocalcaneal bursa: Between calcaneus, Achilles tendon, & Kager fat pad
    - Small amount of fluid in retrocalcaneal bursa is normal
  - Superficial calcaneal bursitis: Overlies distal aspect of Achilles tendon
- Size
  - Few mm to several cm, depending on degree of bursal distension
- Morphology
  - Communicating or noncommunicating bursa
    - Communicating refers to bursa communicating with adjacent joint, e.g., iliopsoas bursa

Radiographic Findings

- Radiography
  - Soft tissue swelling

Ultrasonographic Findings

- General
  - Majority of bursae not visible on US unless distended with fluid or proliferative synovial tissue
  - Distended hypoechoic bursa
    - Thickened wall ± internal septation ± hemorrhage
  - Acute bursitis: Thin-walled bursa distended with hypoechoic fluid, peribursal hyperemia
  - Chronic cases: Greater thickening of bursal wall, more mature synovial proliferation, more echogenic content, & increased likelihood of intrabursal hyperemia
    - May simulate abscess or soft tissue mass
    - Especially for olecranon & ischiorectal bursa
- Subacromial-subdeltoid bursitis
  - Associated with tendinosis, tendon tears, & impingement
  - Fluid collects in dependent positions
    - Lateral aspect of greater tuberosity, over bicipital groove, near coracoid, & deep to acromion
  - Bursal wall thickening & hyperemia generally not featured
    - If present, consider inflammatory arthropathy & check for rice bodies (uncommon)
- Olecranon bursitis
  - Common site for crystal deposition (echogenic foci within distended bursa)
- Ischial tuberosity bursitis: Large, irregular bursa with moderate wall thickening ± internal septa ± debris
- Iliopsoas bursitis: Large bursa beneath musculotendinous portion of iliopsoas muscle, anterior to hip joint capsule
  - Usually associated with hip arthropathy, e.g., inflammatory arthropathy or osteoarthritis
- Trochanteric bursitis: Uncommon
  - Most cases of trochanteric pain syndrome due to gluteus medius, minimus tendinosis, or tears
- Semimembranous bursitis: May become inflamed due to impingement against osteophytes over medial aspect of tibial plateau
- Pes anserinus bursitis: Large, distended bursa indented by 3 tendons (looks like goose foot)
  - Prone to synovial proliferation
- Prepatellar bursitis: Often just subtle, hypoechoic, ill-defined band in prepatellar tissues
  - Usually due to repetitive trauma from kneeling (e.g., housemaid, carpet layer)
  - Occasionally infected (penetrating trauma)
- Infrapatellar bursitis: Similar to prepatellar bursitis but located more distally
- Retrocalcaneal bursitis: Small amount of fluid in retrocalcaneal bursa is seen in 35% of normal subjects

- Larger amount of fluid, bursal wall thickening, synovial proliferation, & peribursal or intrabursal hyperemia are indicative of bursitis
- Nearly always associated with Achilles tendinosis

MR Findings

- T1WI
  - Homogeneous decreased signal intensity mass at typical location
- T2WI
  - Fluid-filled sac
  - ± inflammatory debris (isointense) or, less commonly, calcification (hypointense)
  - ± rice bodies (fibrin nodules, T1 & T2 hypointense)
- T1WI C+
  - ± immediate bursal wall enhancement with delayed enhancement of bursal content

Imaging Recommendations

- Best imaging tool
  - US
  - Most bursae accessible to US
    - Facilitates US-guided aspiration or injection
- Protocol advice
  - High-resolution (> 10 MHz) linear transducer for superficial bursae
    - Lower resolution (~ 7 MHz) for deeper bursae
  - ↓ transducer pressure to avoid bursal effacement of superficial bursa
    - Use copious acoustic gel to lift transducer from skin
  - Transducer palpation or movement of part may allow fluid to be appreciated
  - Angulate transducer to minimize anisotropy & optimize tendon visibility
  - ± compare with opposite side
  - ± aspiration (for cell count, crystals, Gram stain, & culture including fungal, TB culture)
  - Small amount of fluid in some bursae (deep infrapatellar, retrocalcaneal, subacromial-subdeltoid) may be normal
  - If subacromial-subdeltoid bursal fluid & no obvious rotator cuff tear, then pay particular attention to anterior leading edge of supraspinatus tendon for small avulsive-type tear

DIFFERENTIAL DIAGNOSIS

Cystic Collection Adjacent to Bursa

- Ganglion: Loculated cystic collection around tendon
- Sebaceous cyst: Typical appearance with multiple hypo- & hyperechoic foci ± punctum
- Hematoma: History of trauma or bleeding tendency
- Abscess: Signs of infection

Cystic Collection Adjacent to Bursa & Arising Underlying Structure

- Fluid extension to subacromial bursa due to tendon tear at supraspinatus tendon
- Parameniscal cyst due to underlying meniscal tear

Tendinosis

- Insertional tendinosis commonly accompanies bursitis
  - Tendon thickening, hypoechogenicity, & loss of normal fibrillar pattern

PATHOLOGY

General Features

- Etiology
  - Chronic frictional trauma or overuse
  - Acute blunt trauma
  - Crystalline deposition (gout/pseudogout), especially in olecranon bursitis
  - Inflammatory joint disease
  - Infection (± penetrating trauma) is uncommon

Gross Pathologic & Surgical Features

- Bursal sac inflammation with variable synovial-like proliferation
  - Variable amount of fluid

Microscopic Features

- Thickened synovial lining with inflammatory reaction

CLINICAL ISSUES

Presentation

- Most common signs/symptoms
  - Pain ± mass
- Other signs/symptoms
  - Inflammation with warmth, tenderness, & redness
- Clinical profile
  - Overuse phenomenon

Demographics

- Age
  - Typically middle age
- Gender
  - Overall M = F
- Epidemiology
  - Common disorder, especially subacromial-subdeltoid bursitis & olecranon bursitis

Natural History & Prognosis

- Insidious onset
  - Typically resolves with cessation of aggravating activity or removal of aggravating factor

Treatment

- Conservative
  - Rest; antiinflammatory agents
    - US-guided corticosteroid ± long-acting, local anesthetic injection
- Surgical
  - Excision for resistant cases

DIAGNOSTIC CHECKLIST

Consider

- Small amount of fluid is normal in subacromial-subdeltoid bursa, deep infrapatellar bursa, & retrocalcaneal bursa

SELECTED REFERENCES

1. Chang KV et al: An uncommon cause of anterior elbow pain: diagnosis and injection for bicipitoradial bursitis using ultrasonography. Kaohsiung J Med Sci. 33(4):212-213, 2017

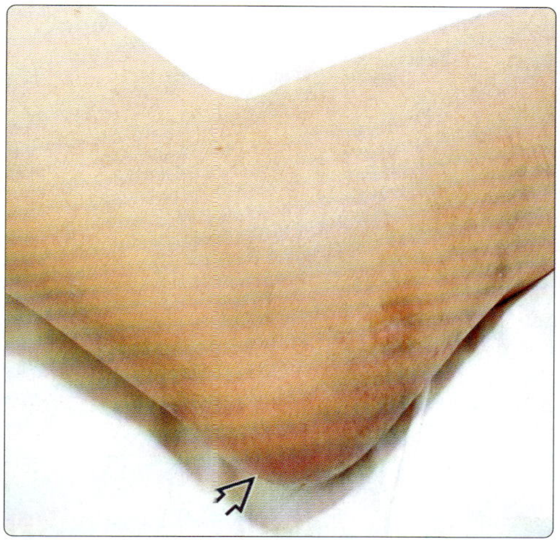

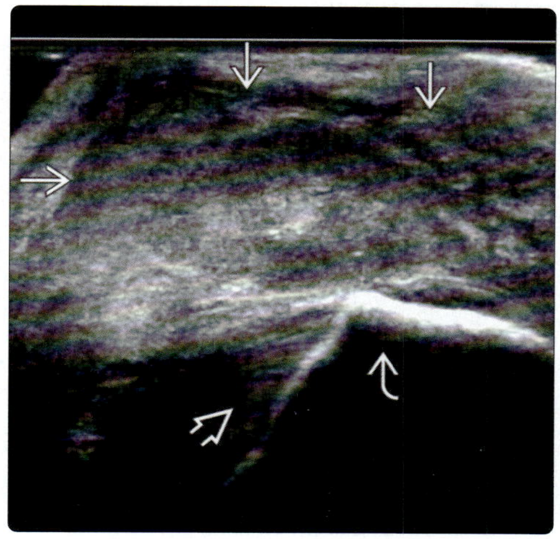

(Left) *Clinical photo shows a patient with posterior elbow swelling →, which had caused pain for several months.* **(Right)** *Longitudinal US shows a large olecranon bursitis → filled with chronic-type synovial proliferative tissue. The olecranon → and triceps tendon → are shown. This solid appearance is quite common for olecranon bursitis, which can still be diagnosed based largely on location and configuration. Biopsy rather than aspiration is usually more helpful in this solid-looking bursitis.*

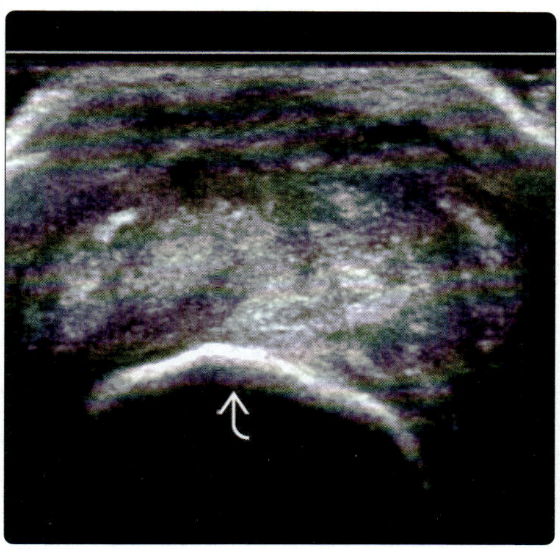

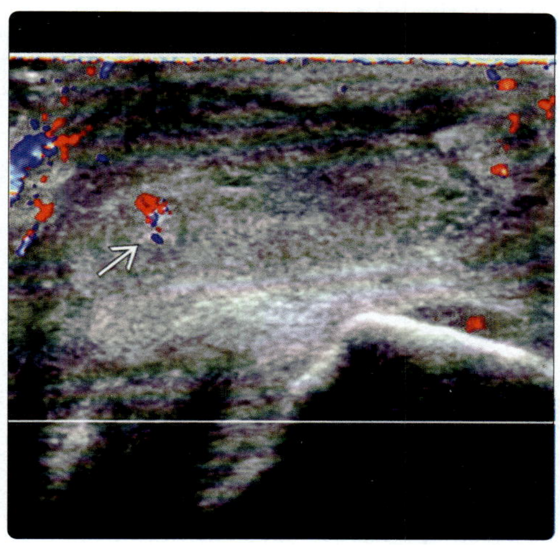

(Left) *Transverse US of the same bursa shows heterogeneous soft tissue within the distended olecranon bursa, located between the olecranon → and the overlying skin.* **(Right)** *Transverse color Doppler US of the same bursa shows mild peribursal hyperemia. There is also mild hyperemia → of the chronic-type synovial proliferation within the olecranon bursa. No percutaneous intervention was undertaken, as it was not likely to be helpful.*

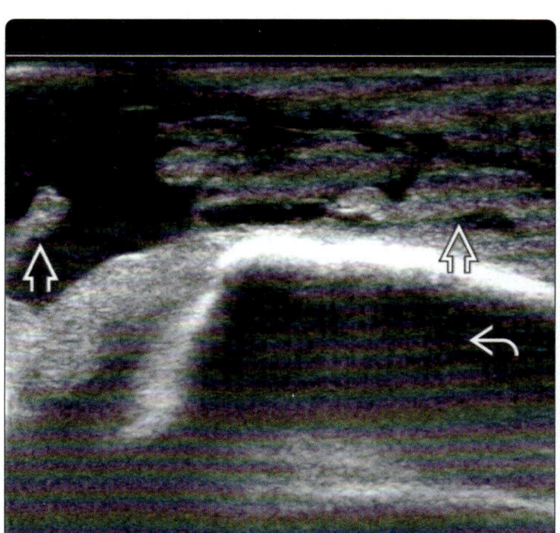

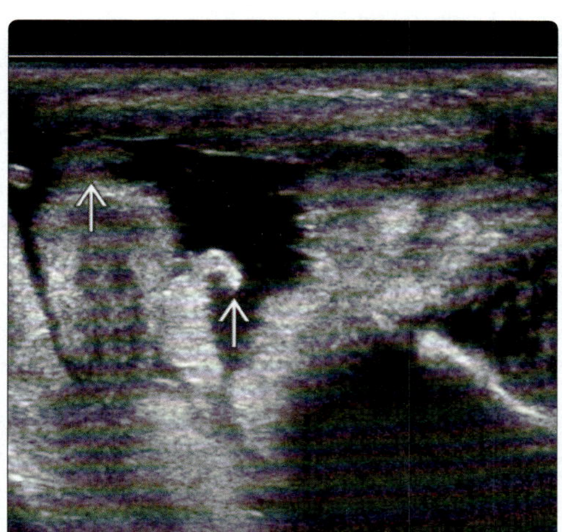

(Left) *Longitudinal US shows severe olecranon bursitis overlying the olecranon process →. The distended bursa shows moderate synovial thickening → with almost anechoic fluid.* **(Right)** *Transverse US of the same olecranon bursitis shows frond-like → synovial proliferation within the distended bursa. Mild vascularity was present in the bursal wall on color Doppler imaging (not shown). The appearances are those of chronic olecranon bursitis.*

(Left) *Clinical photo shows a patient with elbow pain and swelling present for 3-4 months.* **(Right)** *Transverse US of the same swelling shows an olecranon bursa distended with soft tissue containing many echogenic speckles consistent with calcified crystal aggregates. Features are compatible with chronic gouty olecranon bursitis. Chronic trauma, gout, and inflammatory arthropathy are common causes of aseptic bursitis.*

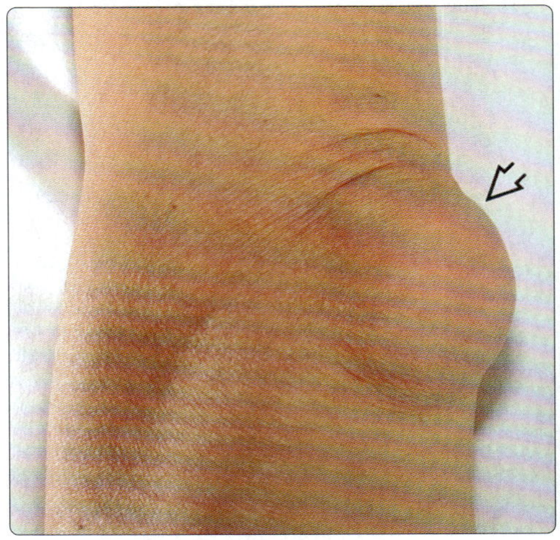

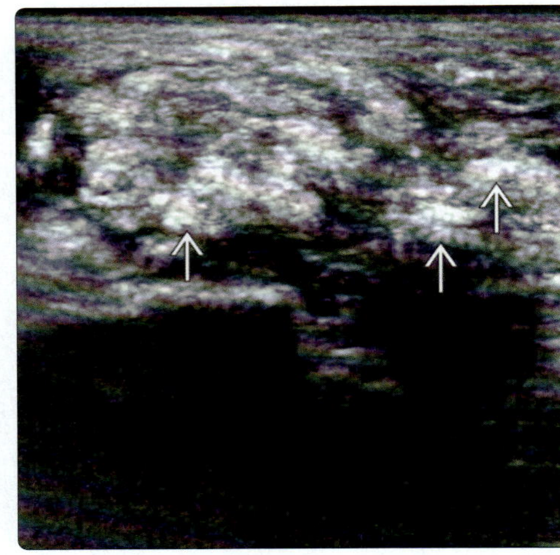

(Left) *Transverse US of the posterior elbow region shows a distended olecranon bursa located between the olecranon and the skin. The bursa is filled with proliferative synovium with prominent posterior acoustic shadowing, a few small echogenic foci, and little demonstrable fluid. Features are compatible with chronic gouty olecranon bursitis.* **(Right)** *Transverse color Doppler US of the same bursa shows only mild marginal hyperemia.*

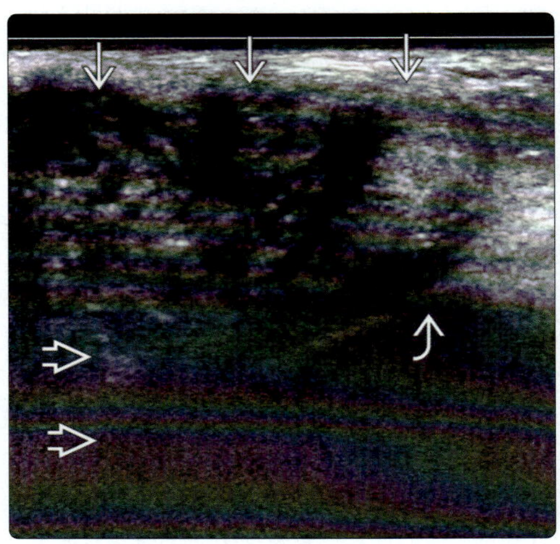

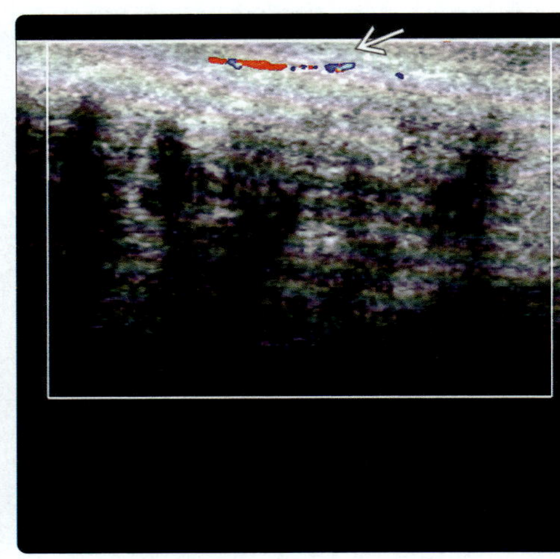

(Left) *Longitudinal oblique US shows a well-defined elbow mass located between the olecranon and the skin, compatible with olecranon bursitis. The bursa is distended by fluid with low-level echoes suggestive of high proteinaceous or hemorrhagic content.* **(Right)** *Transverse US of the same bursa shows how the bursa lies superficial to the triceps tendon and olecranon. One can appreciate the varied appearances of olecranon bursitis.*

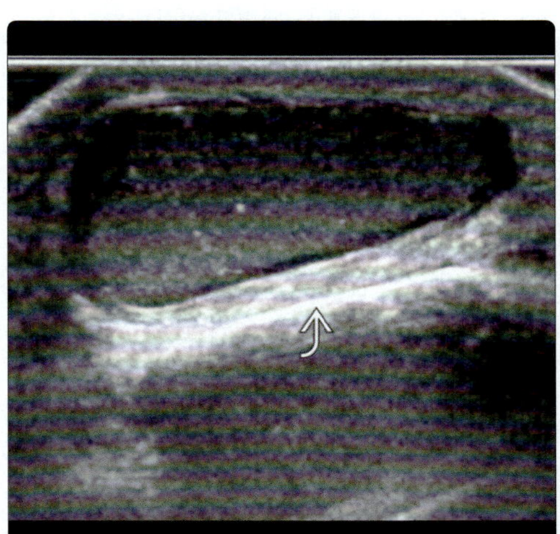

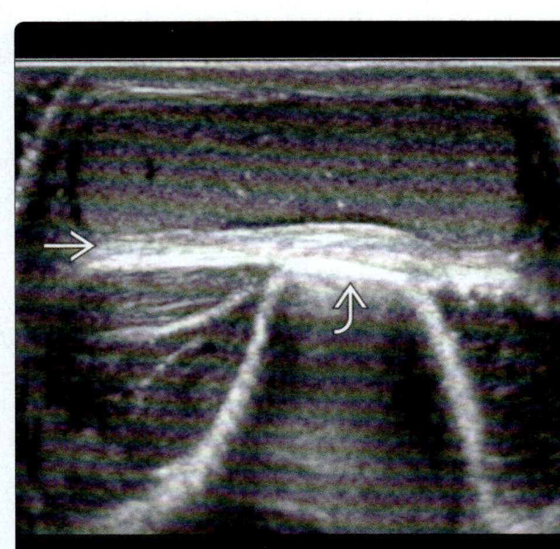

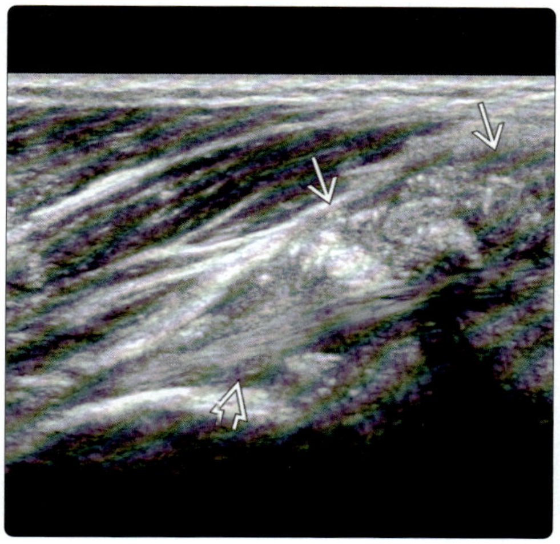

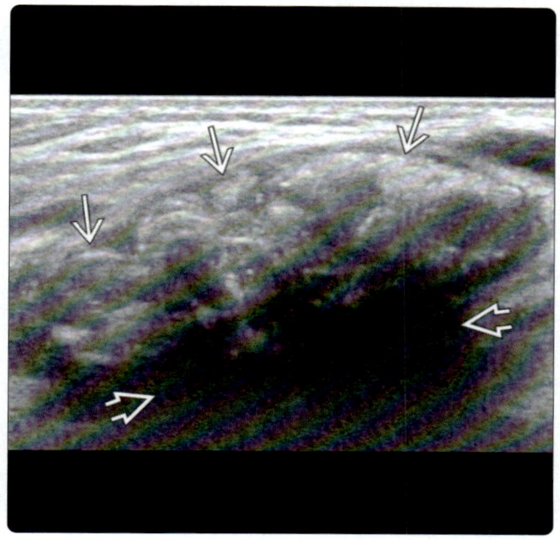

(Left) *Longitudinal US of the anterior elbow in a patient with end-stage renal failure and tertiary hyperparathyroidism is shown. A slightly lobulated mass with multiple fine echogenic foci → is present within the distended bicipitoradial bursa, consistent with crystal aggregates. The biceps tendon → is encased by the lobulated mass.* **(Right)** *Transverse US of the corresponding region shows the bicipitoradial bursa is distended with multiple calcifications → causing significant posterior acoustic shadowing →.*

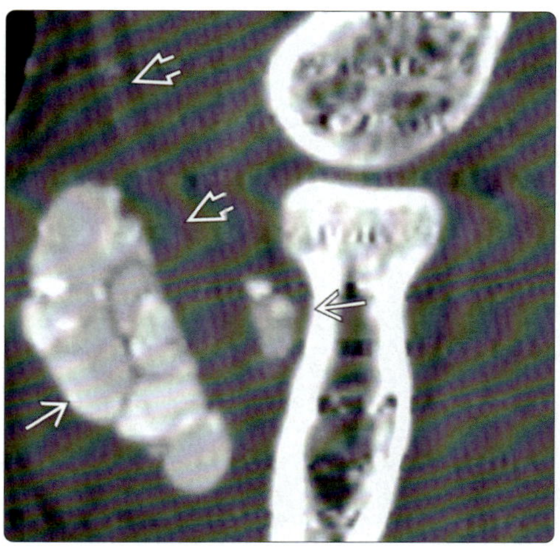

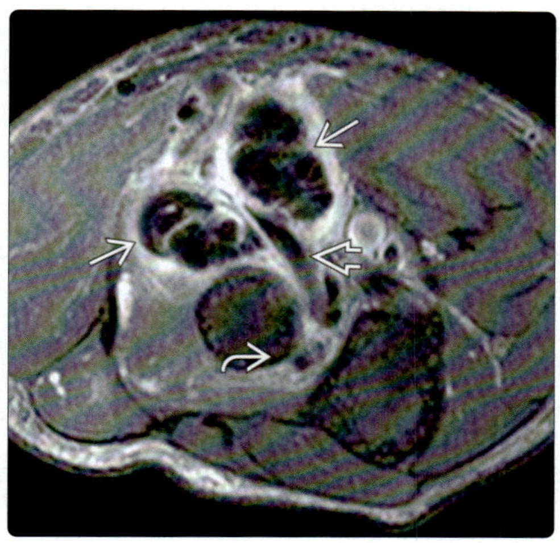

(Left) *Reconstructed sagittal CT of the same patient shows the presence of multiple round calcifications → surrounding the distal biceps tendon →.* **(Right)** *Axial T2 MR of the same patient shows multiple lobulated roundish calcifications → in the bicipitoradial bursa encasing the distal biceps tendon → close to radial insertion →. This is all consistent with bicipitoradial bursal tumoral calcinosis. The bursal distension resolved completely following resection of the parathyroid adenoma.*

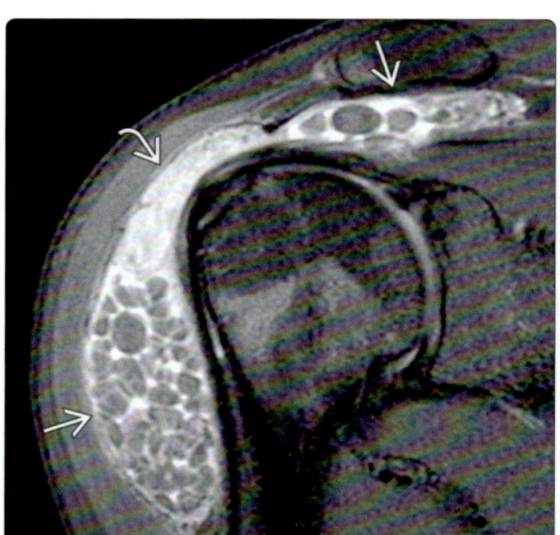

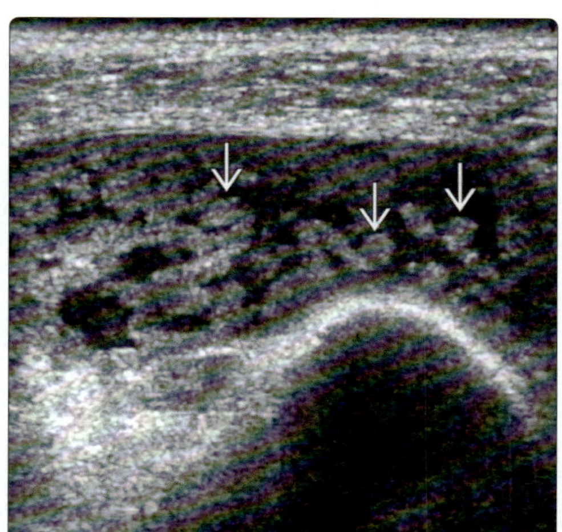

(Left) *Coronal oblique T2 FS MR of the shoulder in a 41-year-old woman with rheumatoid arthritis and shoulder swelling for 6 months shows rice bodies → and synovial proliferation → in distended subacromial-subdeltoid bursa.* **(Right)** *Transverse shoulder US in a rheumatoid arthritis patient shows subacromial-subdeltoid bursa distended with multiple rice bodies →. Rice bodies are more common in inflammatory arthropathy and tuberculosis and are composed of fibrin rather than cartilage.*

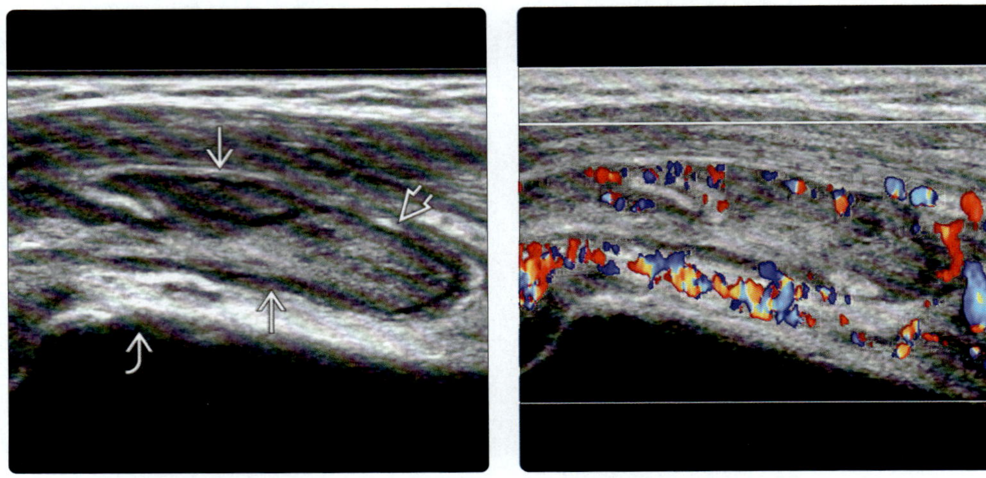

(Left) *Longitudinal US of a shoulder shows subacromial-subdeltoid bursa distended by heterogeneous echogenic tissue* →. *The wall of the bursa is moderately thickened* ⇨. *The humerus is shown* ↪. **(Right)** *Longitudinal color Doppler US of the same area shows severe peripheral hyperemia. Fine-needle aspiration yielded purulent fluid compatible with infective bursitis.*

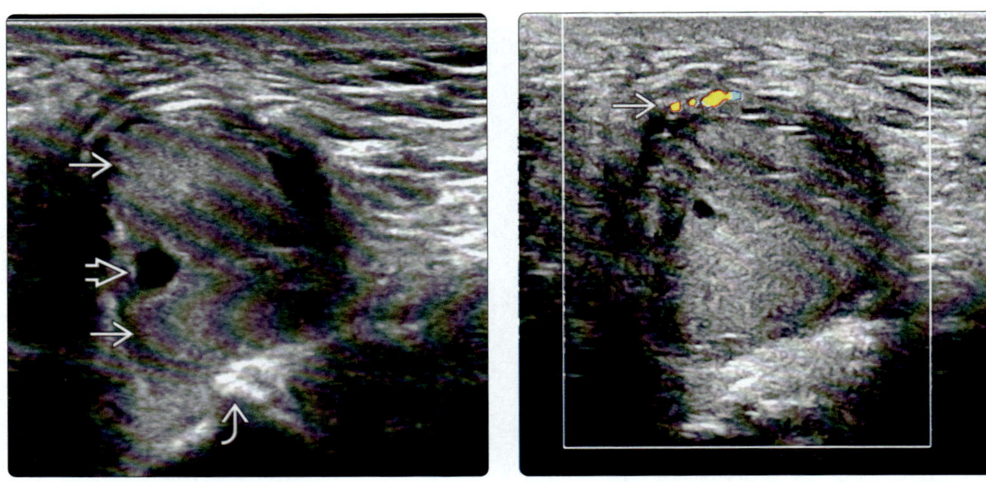

(Left) *Transverse US of the buttock shows ischial bursitis with a rather solid, mass-like, loculated lesion* → *overlying the ischial tuberosity* ↪ *with small, anechoic pockets of fluid* ⇨. **(Right)** *Transverse color Doppler US of the same mass shows only mild peripheral vascularity* → *with no internal hyperemia. This is a fairly typical finding in inflammatory or infective bursitis.*

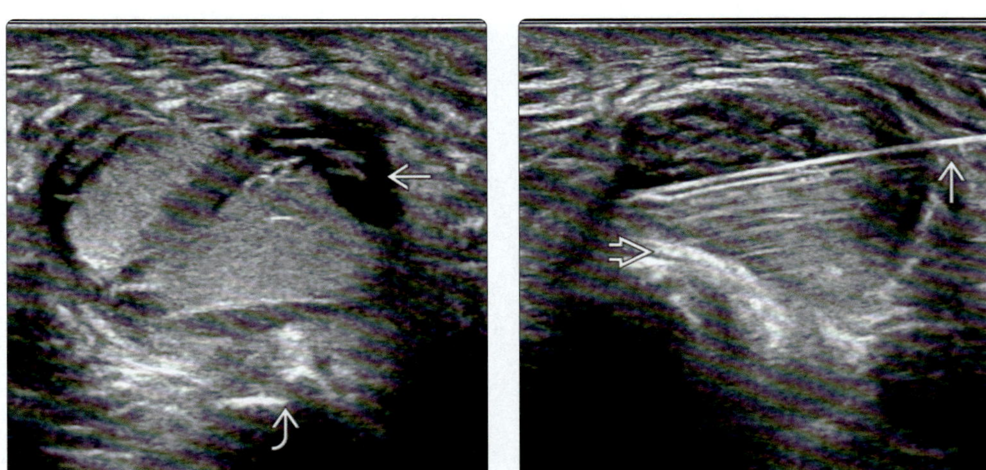

(Left) *Transverse US of the same mass a few minutes later shows more clearly the fluid component* → *to the mass overlying the ischial tuberosity* ↪*, consistent with ischial bursitis. Moderate surrounding soft tissue edema is present.* **(Right)** *Transverse US shows the same mass during fine-needle aspiration* →. *Aspiration is used to relieve bursal pressure and rule out an infective or gouty etiology. Reverberation artifact* ⇨ *is present from the needle.*

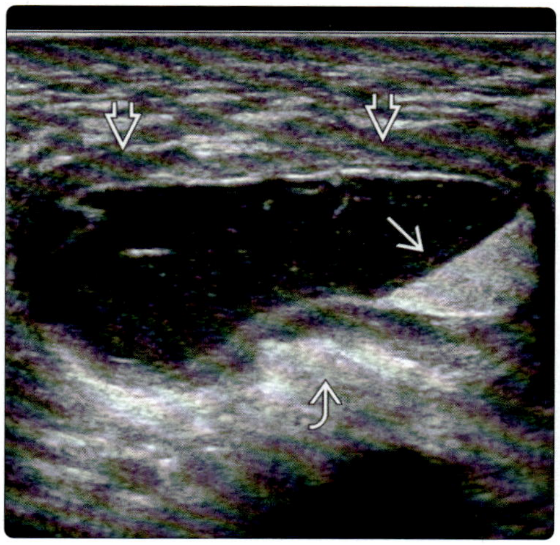

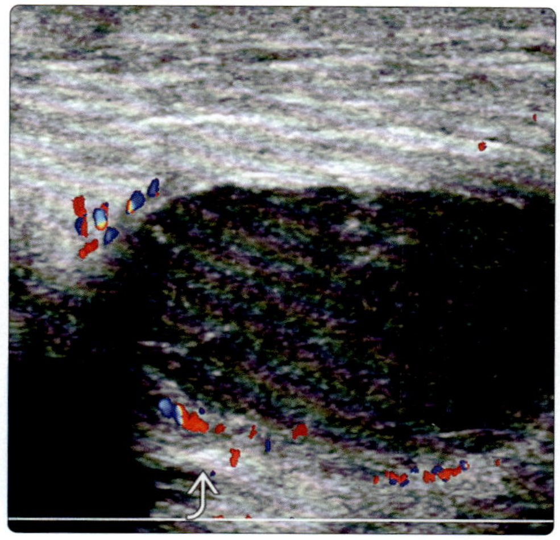

(Left) *Transverse US of the buttock shows a thick-walled, chronic ischial bursitis overlying the ischial tuberosity. There is a fine debris within the bursa, forming a fluid-fluid level.* **(Right)** *Transverse color Doppler US of another patient shows severe ischial bursitis located superficial to the ischial tuberosity with mild marginal hyperemia consistent with chronic bursitis. The bursa is distended with echogenic fluid due to either hemorrhage or chronic inflammation.*

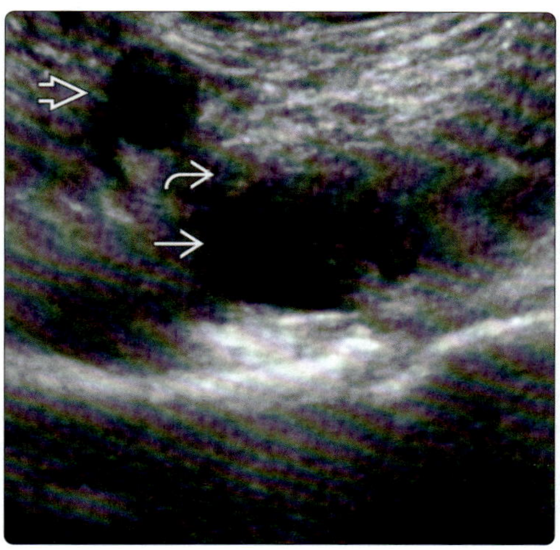

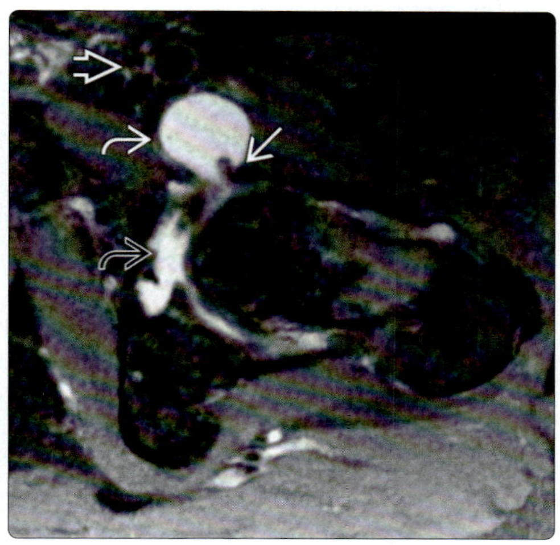

(Left) *Transverse US of the anterior hip region shows a fluid-distended bursa alongside the iliopsoas tendon. Note the anteriorly located common femoral artery.* **(Right)** *Axial T2 FS MR of the same patient shows moderate iliopsoas bursal distension, which communicates via a short tract with the underlying joint. The adjacent iliopsoas tendon and common femoral vessels are shown.*

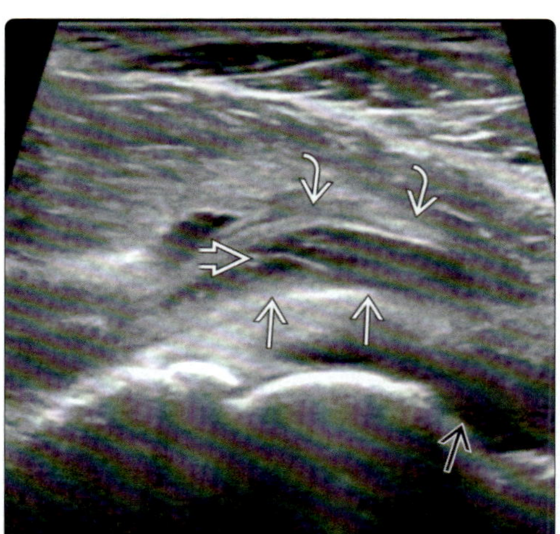

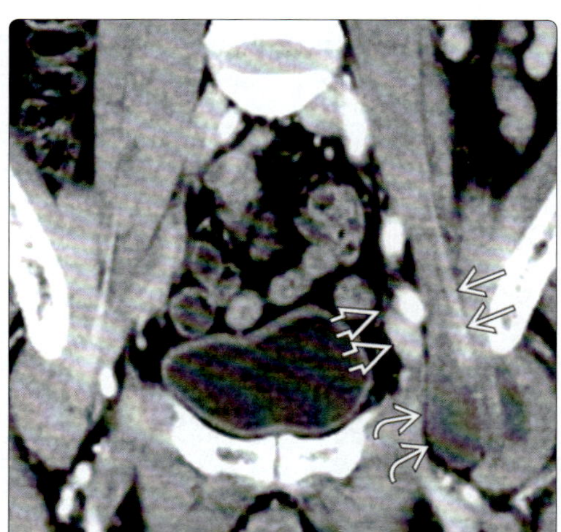

(Left) *A 59-year-old man presented with progressive left hip pain and fever. Longitudinal US of anterior hip region shows a fluid-distended bursa with internal septations under the iliopsoas tendon. Note a small amount of joint effusion in the anterior recess.* **(Right)** *Coronal reformatted CT shows moderate iliopsoas bursal distension. The adjacent iliopsoas tendon and common femoral vessels are shown. Pus was aspirated from the bursa. This pus grew methicillin-resistant Staphylococcus aureus.*

(Left) *Transverse US of the hip region shows moderate iliopsoas bursal distension anterior to the lesser trochanter. Note that the iliopsoas tendon is superficial to the bursal fluid.* **(Right)** *Coronal PD MR of the same patient shows a moderately distended iliopsoas bursa adjacent to the iliopsoas tendon. The bursal wall is better appreciated on MR imaging.*

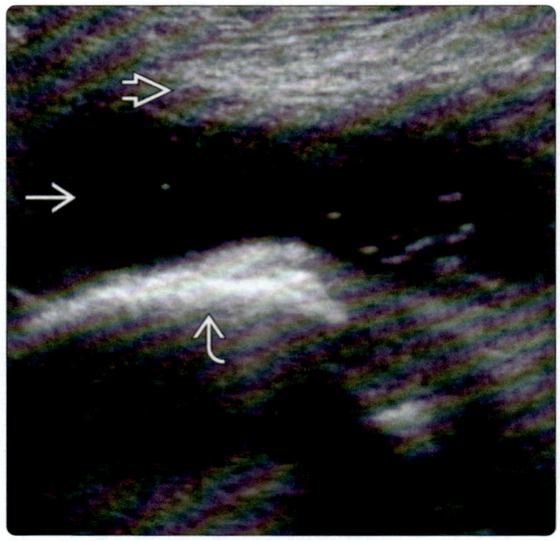

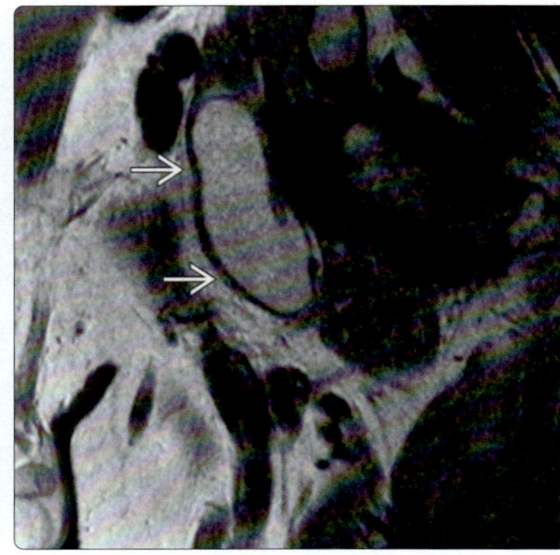

(Left) *Longitudinal US of the hip shows a large, distended iliopsoas bursa superficial to the femoral head. The echogenic strands within the bursa are due to clotted blood. The iliopsoas tendon is not visible on this image.* **(Right)** *Axial NECT of the corresponding level in the same patient shows the distended bursa with clear fluid at the iliopsoas bursa. A small fleck of calcification is present in the bursa wall. The iliopsoas tendon is at the inferomedial aspect of the bursa.*

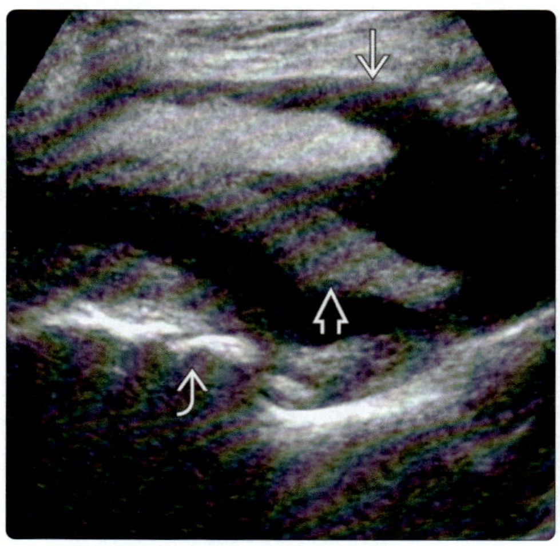

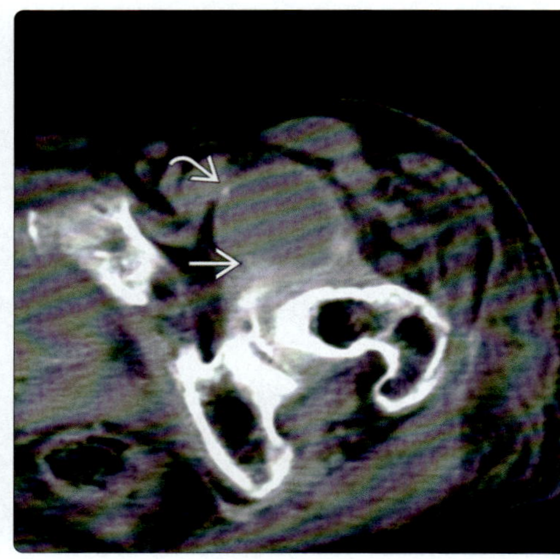

(Left) *Clinical photo shows a patient who had eccentric prepatellar swelling for 2 weeks.* **(Right)** *Longitudinal US of the prepatellar region shows a mild to moderate prepatellar bursitis superficial to the patellar bone. The bursa is distended with fluid, with some thin internal septa.*

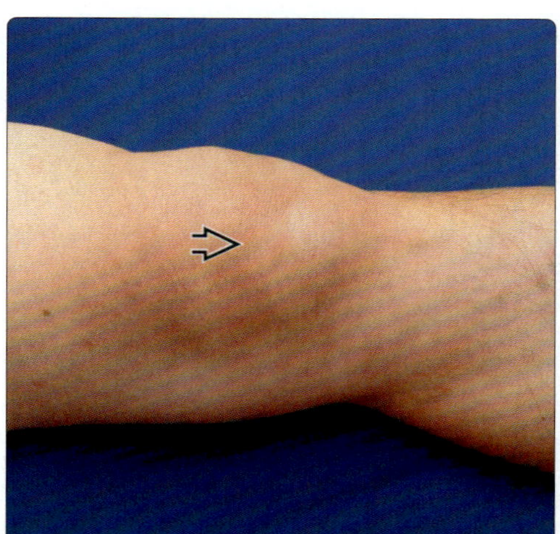

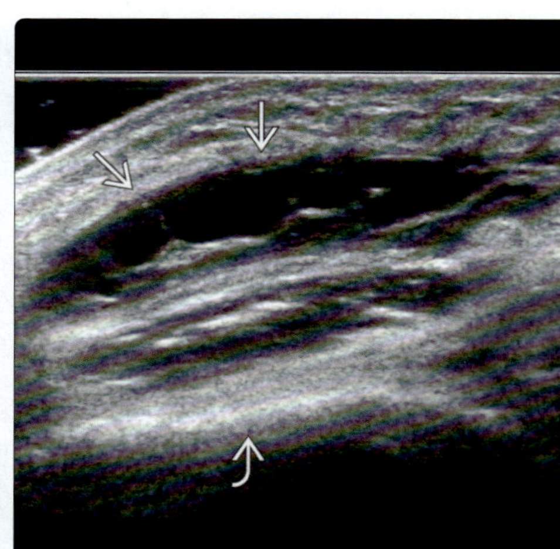

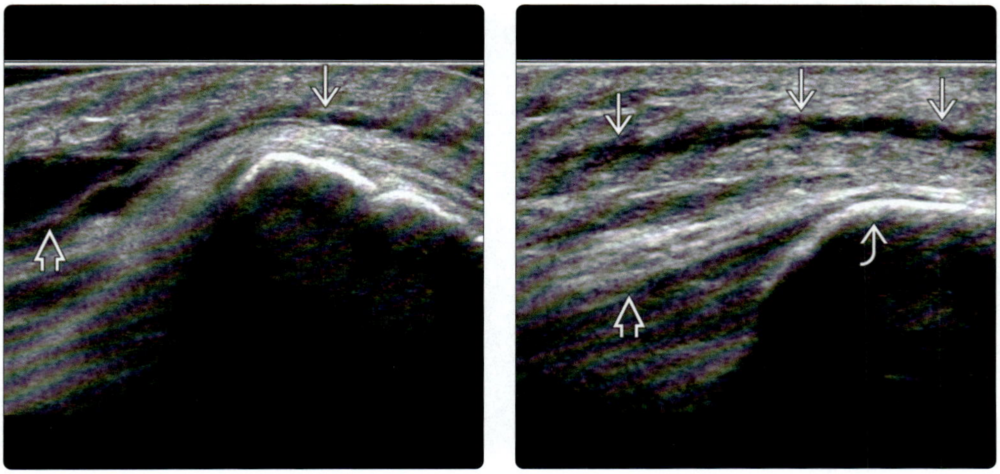

(Left) *Transverse US obtained more medial to the previous image shows the parapatellar collection ➡ communicating with a thin nondistended prepatellar bursitis ➡. Prepatellar bursitis is not just confined to the immediate prepatellar region.* **(Right)** *Longitudinal US of the same bursa shows the thin, hypoechoic prepatellar bursa ➡ located superficial to the patella ➡ and the proximal part of the patellar tendon ➡.*

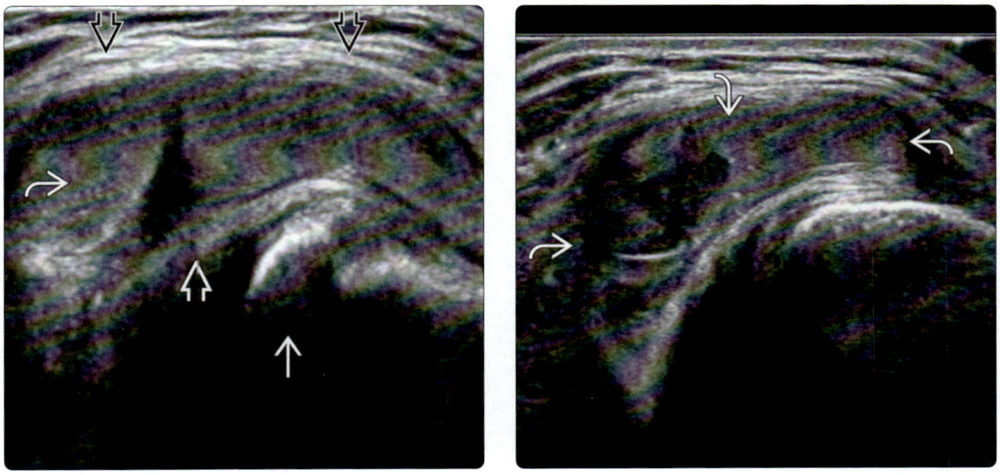

(Left) *Longitudinal oblique US of the knee just medial to the proximal tibia ➡ shows an echogenic fluid collection ➡ immediately superficial to the semitendinosus ➡ and deep to the sartorius tendon ➡, compatible with pes anserinus bursitis.* **(Right)** *Transverse US of the same pes anserinus bursa shows heterogeneous bursal fluid within the distended bursa ➡. The normal pes anserinus bursa is a potential space and cannot be visualized on US.*

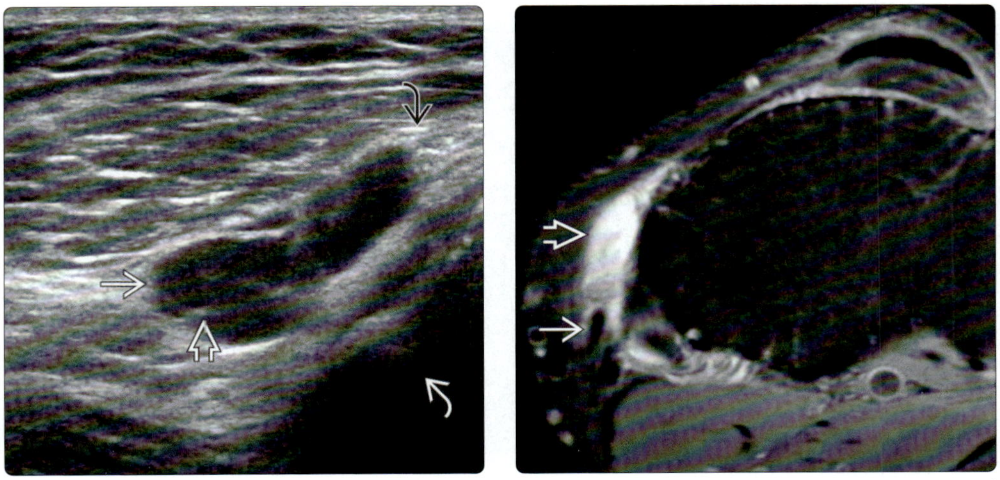

(Left) *Longitudinal oblique US of the knee close to the medial aspect of the proximal tibia ➡ shows a fluid collection ➡ with internal fibrous stranding ➡ immediately adjacent to the pes anserinus tendons ➡, compatible with a pes anserinus bursitis.* **(Right)** *Axial PD FSE FS MR of the same pes anserinus tendon bursitis shows a multiloculated cystic collection ➡ alongside the pes anserinus tendons ➡.*

(Left) *Transverse US at the medial aspect of the popliteal fossa shows an irregular fluid collection → between the medial belly of the gastrocnemius ⇨ and the semitendinosus tendon ↩, compatible with semitendinosus bursitis.* **(Right)** *Transverse color Doppler US of the same patient shows significant hyperemia of the peribursal soft tissues, consistent with acute inflammation.*

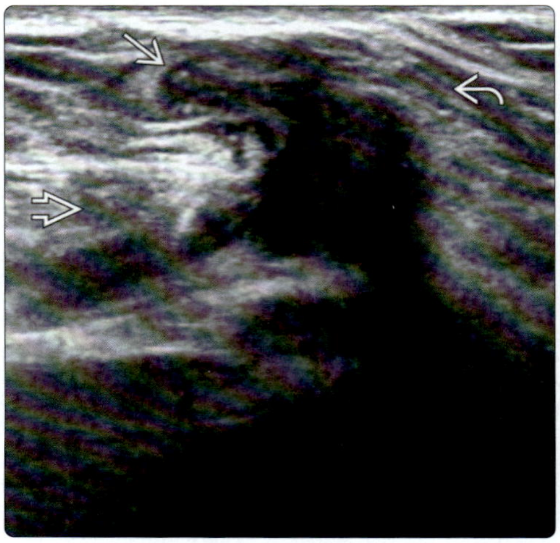

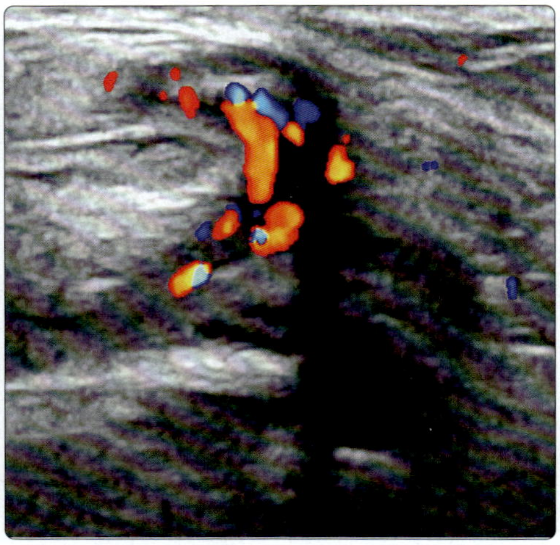

(Left) *Clinical photo shows a patient with a history of posterior ankle swelling → and redness, who had had pain for 3 months.* **(Right)** *Longitudinal color Doppler US of the same area shows a mildly distended retrocalcaneal bursa → with synovial proliferation and mild adjacent hyperemia. Mild distal Achilles tendinosis is present with mild hyperemia ⇨. Mild retrocalcaneal bursal distension is normal, but hyperemia and synovial proliferation are not normal features.*

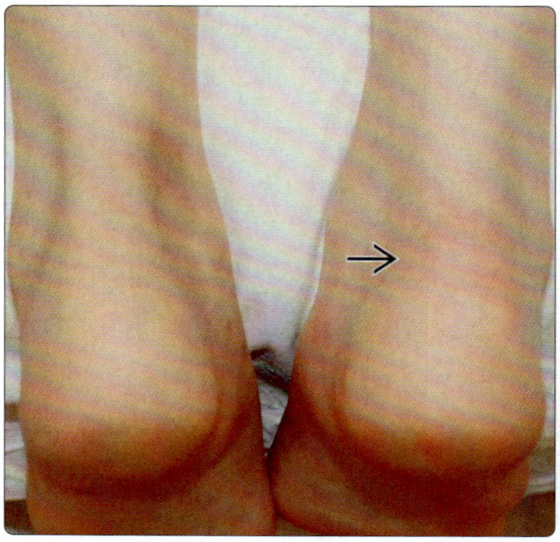

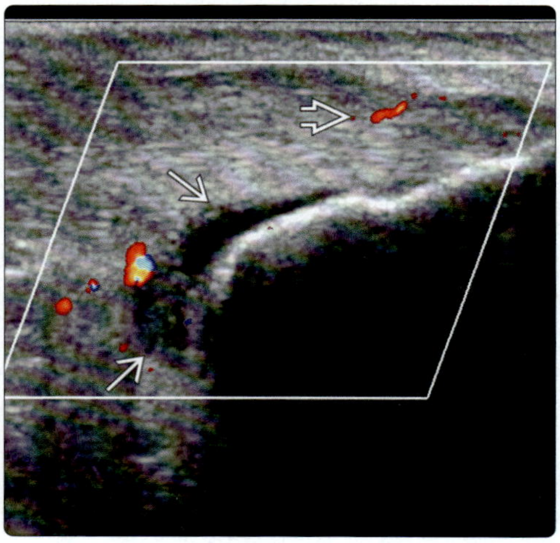

(Left) *Longitudinal US of the plantar aspect of the 1st metatarsophalangeal joint region shows an irregularly distended bursa → with fluid and debris ⇨ overlying the medial sesamoid bone ↩. Features are compatible with chronic bursitis.* **(Right)** *Sagittal PD MR of the same region shows the distended plantar bursa → with a thickened wall ⇨ overlying the medial sesamoid bone ↩. The bursal content can be better appreciated on US than on MR.*

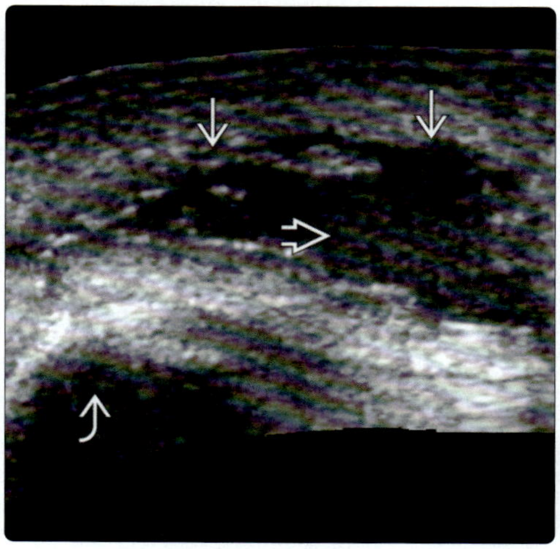

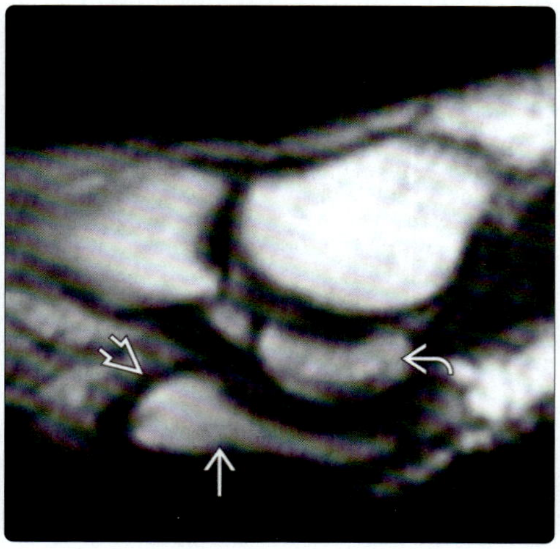

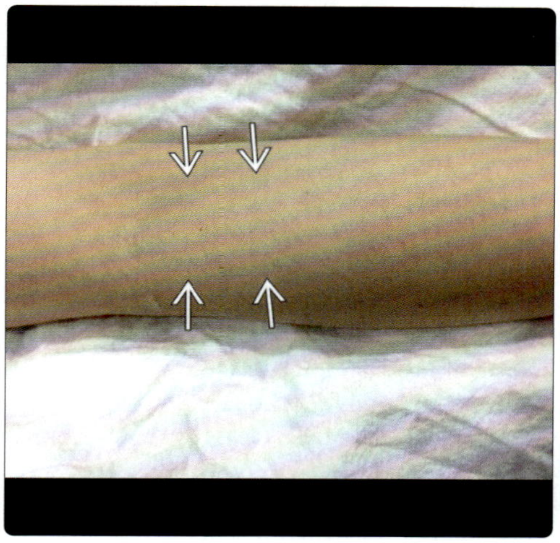

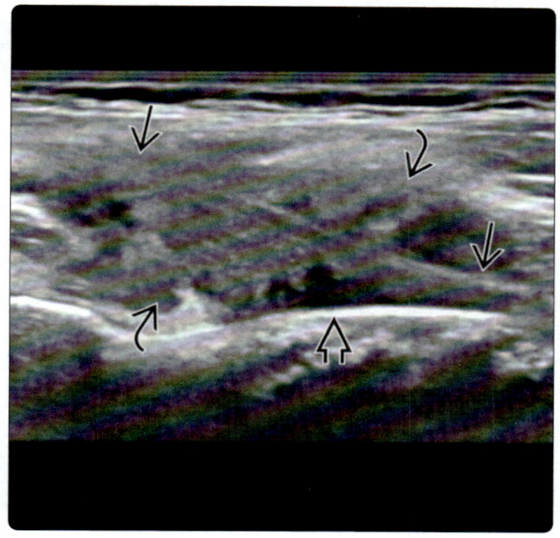

(Left) *A 39-year-old woman presented with ill-defined distal forearm soft tissue swelling clinically suspicious of sarcoma.* **(Right)** *Longitudinal US of the elbow shows the bicipitoradial bursa distended with proliferative synovium and a small amount of fluid. The biceps tendon is encased by this thickened, villonodular-like synovium.*

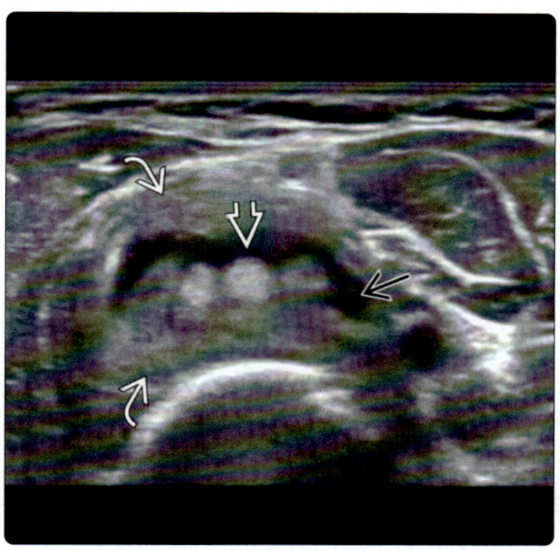

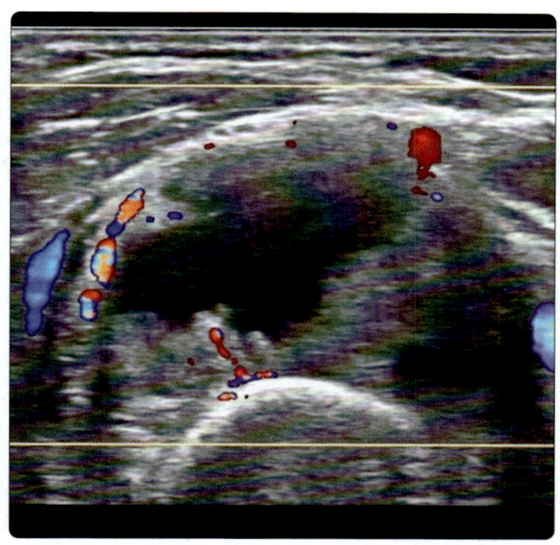

(Left) *Transverse US of the same region as previous imaging shows the biceps tendon surrounded by bicipitoradial bursa distended with synovial proliferation and fluid. Bicipitoradial bursitis often presents clinically as a sarcoma.* **(Right)** *Transverse color Doppler US shows moderate hyperemia of the thickened bursal synovium and peribursal soft tissues, consistent with acute inflammation.*

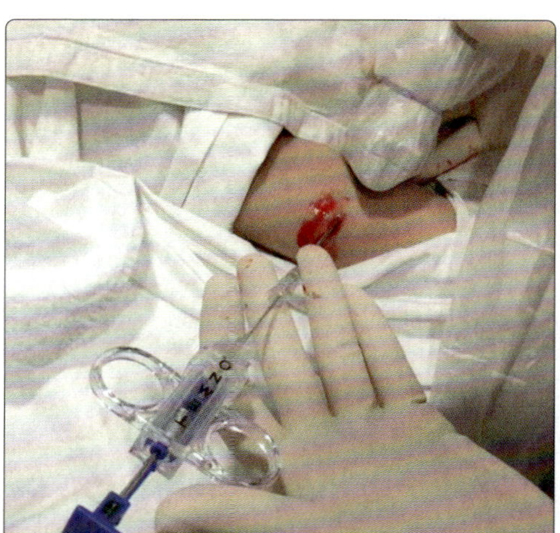

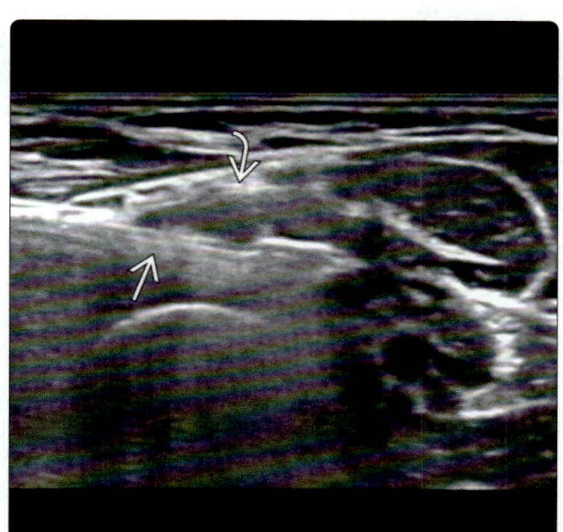

(Left) *US-guided biopsy is performed on the same patient using a coaxial-biopsy needle system. Resultant biopsy specimens were sent for histology and culture to rule out infection.* **(Right)** *Transverse US during biopsy shows synovial tissue filling the slot of the needle prior to firing the biopsy device. Histology and culture revealed inflamed synovial tissue but no evidence of infection, crystal deposition, or malignancy.*

KEY FACTS

IMAGING

- Fluid-filled mass with stalk extending toward joint
- Ovoid or irregular configuration with stalk connecting to joint of origin
  - Elongated neck allows cyst to surface at distance from joint of origin
- Hypoechoic with clear fluid
  - ± septation
  - ± comet-tail artifacts (due to colloid aggregates)
- Not compressible
- No hyperemia except with recent leakage when surrounding tissues may be mildly hyperemic and edematous
  - Leakage of ganglion contents induces soft tissue inflammation
- US will allow diagnosis of most ganglion cysts
  - Multidirectional capability allows stalk to be traced toward joint
- Locate telltale stalk of cyst and trace it back toward joint of origin
- Evaluate with color Doppler to exclude vascular anomaly and nerve sheath tumor

CLINICAL ISSUES

- In adults, up to 50% of wrist ganglia will resolve without treatment within 6 years
- In children, > 90% ganglion cysts will resolve without treatment within 1 year
- Aspiration (US guided)
  - > 70% complete resolution for ganglia alongside flexor tendon sheath (retinacular ganglia)
  - < 50% complete resolution for other ganglia, though they may become smaller and less symptomatic (incomplete resolution)
  - Greater success if ganglion is present for < 1 year
- Surgical excision

(Left) *Graphic shows the typical location of a dorsal wrist ganglion cyst ➡ arising from a defect in the dorsal capsule of the scapholunate ligament ➡. The ligament remains functionally intact.* **(Right)** *Oblique US shows a lobulated ganglion cyst ➡ in the dorsal aspect of the wrist. The stalk of the ganglion ➡ extends toward the articulation between the scaphoid ➡ and the distal radius ➡.*

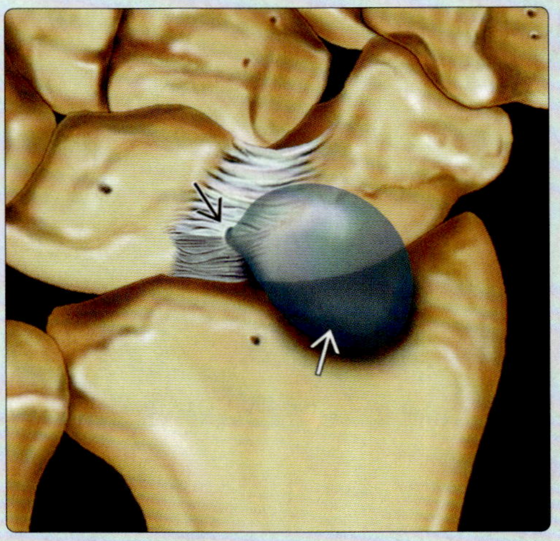

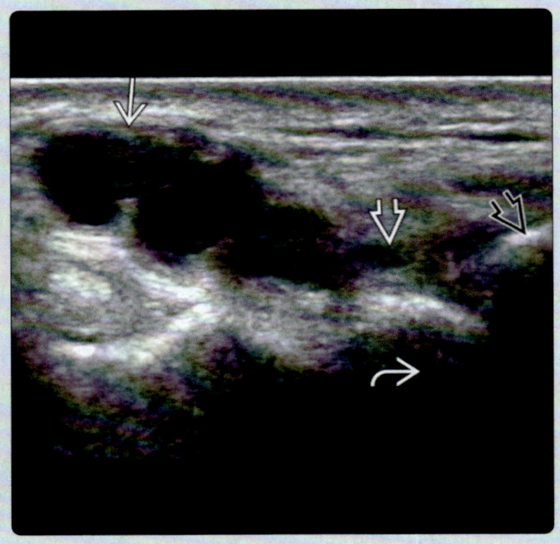

(Left) *Clinical photograph of the wrist of a young patient shows dorsal swelling ➡ suggestive of a ganglion cyst.* **(Right)** *Oblique US of the same patient shows a ganglion ➡ arising from the dorsal aspect of the radioscaphoid articulation. There is mild septation ➡ within the ganglion. The radial epiphysis ➡ and scaphoid ➡ are shown. The small terminal branches of the posterior interosseous nerve, which may be compressed, are not easily visible on US.*

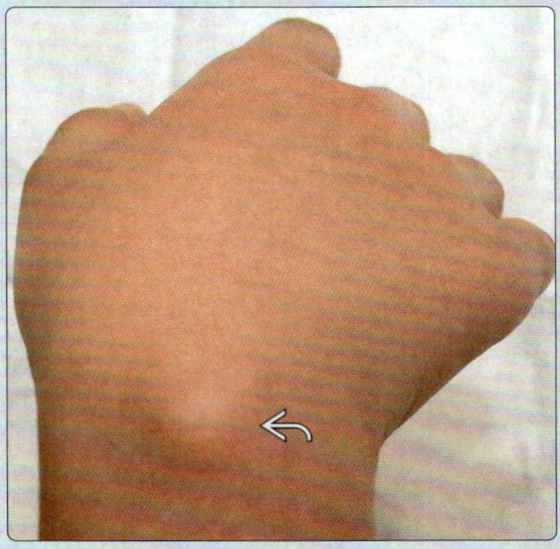

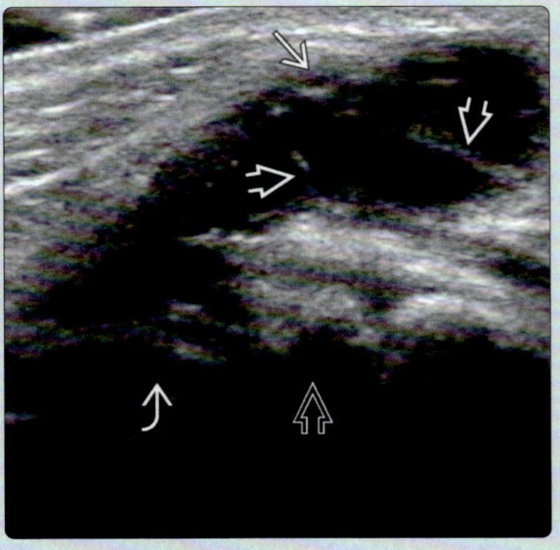

TERMINOLOGY

Synonyms

- Bible bump, Gideon disease, digital mucous cyst

Definitions

- Ganglion: Myxoid tissue encased in fibroblastic collagenous wall with no synovial lining
  - Occur in predictable locations
- Occult ganglion: Ganglion not clinically apparent

IMAGING

General Features

- Best diagnostic clue
  - Fluid-filled mass with stalk extending toward joint
- Location
  - Dorsal wrist ganglia
    - May compress terminal branch of posterior interosseous nerve, 1 branch of which lies near dorsal aspect scapholunate interval
    - Scapholunate joint
      - Typically emerge through or just distal to scapholunate ligament between 3rd and 4th extensor compartments
      - Most common type of dorsal wrist ganglion
      - Functionally intact scapholunate ligament
    - Also commonly arise from radioscaphoid joint, scaphotrapeziotrapezoid joint (triscaphe joint), and ulnocarpal joints
  - Volar wrist ganglia
    - Typically arise in weak spot between radioscaphocapitate & radiolunotriquetral ligaments (space of Poirier) overlying lunocapitate interval
    - Also arise from radioscaphoid joint, scaphotrapeziotrapezoid joint, ulnocarpal joints
    - Pisiform-triquetral joint
      - Can extend into Guyon canal & compress ulnar nerve
      - Do not confuse proximal (larger) or distal (smaller) pisotriquetral recesses with ganglia
  - Finger
    - Flexor tendon sheaths near A1 pulley or A2 pulley
    - Dorsal aspect of distal interphalangeal joint (digital mucous cyst)
  - Acromioclavicular joint
  - Shoulder
    - Posterosuperior, usually associated with labral tear
  - Lumbar facet joints (not usually demonstrable by US)
  - Hip: Anterosuperior, usually associated with labral tear
  - Knee
    - Anterior and posterior cruciate ligaments
    - Paraarticular, deep to supracondylar attachment of medial or lateral belly of gastrocnemius
  - Proximal tibiofibular joint
    - Articular branch of deep peroneal nerve explains propensity of ganglia from this joint to extend intraneurally
    - Ganglia may extend intraneurally into deep peroneal nerve, common peroneal nerve, or even peroneal division of sciatic nerve
  - Distal tibiofibular syndesmosis
  - Foot and ankle: Cuneonavicular joint, intercuneiform joint, subtalar joint, talonavicular joint
- Size
  - Average: 5 mm; ganglia from large joint may reach > 50 mm in maximum dimension
  - May increase with activity
- Morphology
  - Ovoid or irregular configuration with stalk connecting to joint of origin

Radiographic Findings

- Radiography
  - Usually normal ± soft tissue density ± intraosseous carpal ganglion

MR Findings

- T2WI
  - Uniformly hyperintense soft tissue cystic mass
  - Narrow stalk connecting cyst to joint
  - Distension of capsular recess may simulate cyst
    - Capsular recesses occur in expected locations
    - If capsular recess is unusually distended for given amount of joint effusion, consider ganglion
    - Cysts often contain septations
- Very sensitive at detecting occult ganglia

Ultrasonographic Findings

- Well-defined, hypoechoic, fluid-filled mass with stalk extending toward joint
  - Typical location
  - Elongated neck allows cyst to surface at distance from joint of origin
  - ± multiloculated ± thin septations
  - ± small comet-tail artifacts due to colloid aggregates
  - ± inflammatory exudate: Inflammatory content of joint can extend into ganglia
  - Not compressible
  - No hyperemia except with recent leakage when surrounding tissues may be mildly hyperemic and edematous
    - Contents of ganglion cyst can induce short-lived inflammation of adjacent tissues
  - Check for compression of small nerves, such as posterior interosseous nerve or superficial branch of radial nerve
- Stalk may be thin and serpiginous
  - Locating entry site of stalk into joint will help surgical planning
- Ganglia close to A1 pulley have no visible communication with tendon sheath
- Hemorrhagic cyst: Can have pseudosolid appearance
  - Will change appearance on follow-up imaging
- Ruptured cyst or cyst leakage
  - Collapsed cyst with pericystic inflammation

Imaging Recommendations

- Best imaging tool
  - US allows diagnosis of most ganglion cysts
    - Allows stalk to be traced toward joint
- Protocol advice
  - Locate telltale stalk of cyst and trace it back toward joint of origin

- Extension through joint capsule is often not seen
  - Examination of wrist in flexion ± extension may occasionally help
- Note important relationships of cyst (i.e., to tendons and neurovascular bundle)
- Evaluate with color Doppler to exclude vascular anomaly and nerve sheath tumor

DIFFERENTIAL DIAGNOSIS

Giant Cell Tumor of Tendon Sheath

- Soft tissue mass, usually hyperemic
- Eccentrically located to tendon sheath

Vascular Anomaly

- Compressible
- ± phleboliths, ± flow on color Doppler imaging
- Lymphangiomas may appear similar to ganglion cysts
  - Tend to occur in areas different from those of ganglia
  - Do not have stalk communicating with joint

Cystic Schwannoma

- Some nerve sheath tumors may be almost purely cystic
  - Generally occur at locations different from those of ganglia
  - Can see entering/exiting nerve

PATHOLOGY

General Features

- Etiology
  - Typical locations are due to capsular weakness or nerve traversing capsule
    - Degenerative or traumatic tear in capsule
      - Only 15% are related to specific injury
    - Joint fluid pumped 1 way between collagen bundles
    - Posttraumatic ganglia may develop acutely over few hours to days
- Associated abnormalities
  - Intraosseous ganglia are more common in subjects with soft tissue ganglia
  - Single ganglion may simultaneously have both intraosseous and extracapsular components

Gross Pathologic & Surgical Features

- Cystic mucinous fluid is contained in pseudocapsule

Microscopic Features

- Capsular lining = compressed loose areolar tissue = pseudocapsule
- No inner cell lining (not lined by synovial tissue)

CLINICAL ISSUES

Presentation

- Most common signs/symptoms
  - Slowly growing soft tissue mass
- Clinical profile
  - Localized pain ± firm mass ± weakness
    - Typically, pain on attempted full dorsiflexion of wrist
      - e.g., getting out of chair
    - Cyst transilluminates
  - Neurologic findings with extension into carpal tunnel or Guyon canal and nerve compression (uncommon)

Demographics

- Age
  - 20-50 years
- Gender
  - M < F
- Epidemiology
  - 80% of all soft tissue masses of wrist and hand
    - 80% of wrist ganglia presenting clinically are on dorsal aspect
    - 80% of wrist ganglia overall occur on volar aspect (MR evidence)

Natural History & Prognosis

- In adults, up to 50% of wrist ganglia will resolve without treatment within 6 years
  - In children, > 90% of ganglion cysts will resolve without treatment within 1 year

Treatment

- Observe: Quite high rates of spontaneous resolution
- Active treatment
  - Indicated if cyst is painful or fails to resolve after 1 year or if it is cosmetically disfiguring
- Manual rupture by hitting with large book
  - Not recommended (soft tissue injury)
- Aspiration (US guided)
  - > 70% complete resolution for cysts alongside flexor tendon sheath (retinacular ganglia)
  - < 50% complete resolution for other ganglia, though they may become smaller and less symptomatic (incomplete resolution)
  - Greater success if ganglion is present for < 1 year
  - Aspiration ± hyaluronidase injection
    - Hyaluronidase degrades hyaluronic acid, which is macromolecular component of ganglion cyst fluid
- Surgical
  - Open resection
    - Excision of stalk and small portion of joint capsule
      - ≤ 30% may have persistent pain or limitation of movement ± recurrence
      - ~ 5-15% recurrence rate
  - Arthroscopic resection: Less scarring
    - Capsular defect, stalk, and cyst debrided from intraarticular approach
    - Similar recurrence rate to open resection

DIAGNOSTIC CHECKLIST

Image Interpretation Pearls

- Locate stalk of ganglion and trace toward joint of origin
- Check for free fluid or inflammation around cyst
  - These are signs of cyst leakage
- Examine with color Doppler
  - Nerve sheath tumor, giant cell tumor, and vascular anomalies may occasionally simulate cysts

SELECTED REFERENCES

1. Giard MC et al: Ganglion cyst versus synovial cyst? Ultrasound characteristics through a review of the literature. Rheumatol Int. 35(4):597-605, 2015

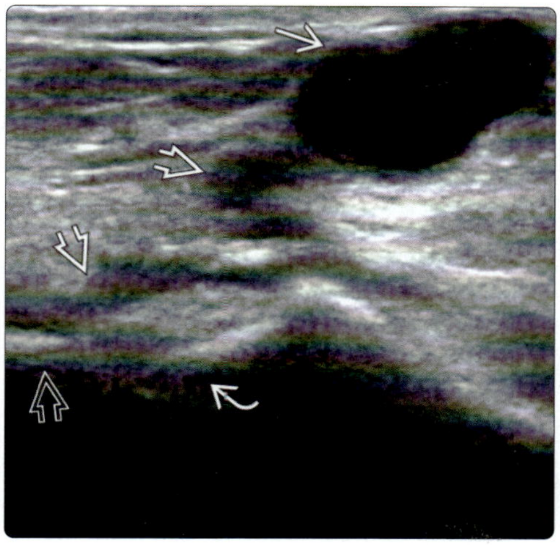
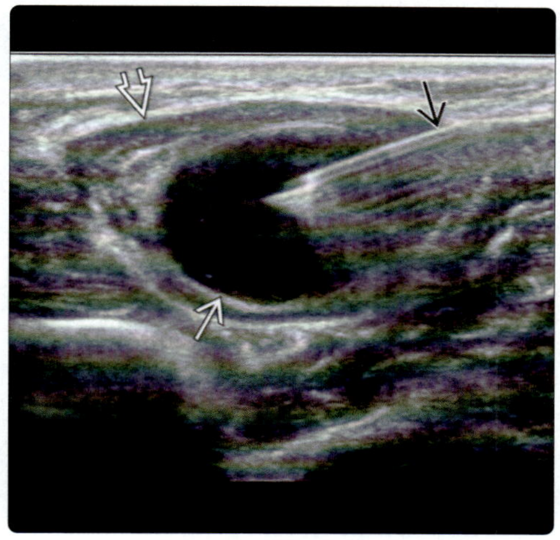

(Left) *Longitudinal US shows a ganglion ➔ arising from the dorsal aspect of the radioscaphoid joint. The serpiginous stalk ➔ of the ganglion extends deeply toward the radiocarpal articulation between the distal radius ➔ and the scaphoid bone ➔.* **(Right)** *Longitudinal US shows a unilocular ganglion ➔ within the flexor carpi ulnaris muscle ➔ at the elbow region. Aspiration was performed using 21-g needle ➔. It is not possible to predict the viscosity of ganglion content on US.*

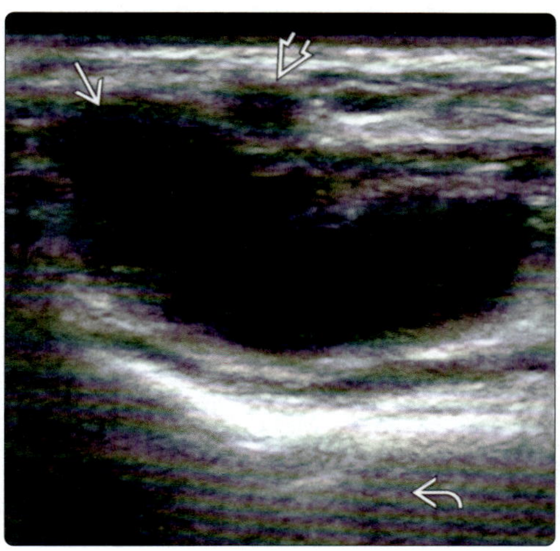
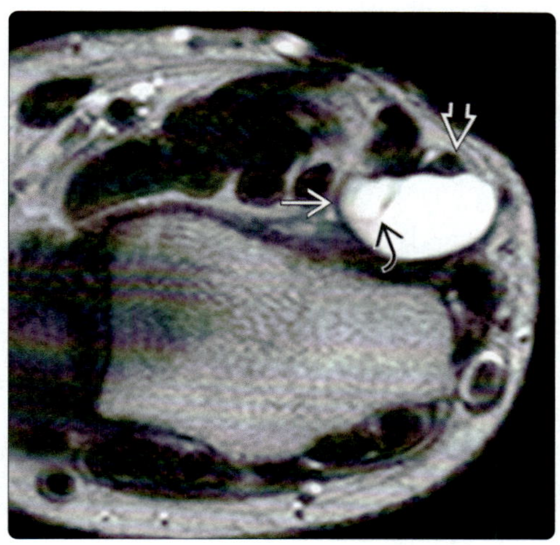

(Left) *Transverse US shows a medium-sized, lobulated ganglion ➔ on the volar aspect of the distal radius ➔. The ganglion lies deep to the radial artery ➔.* **(Right)** *Axial T2 MR of the same patient shows the ganglion ➔ lying deep to the radial artery ➔. There is a thin septum ➔ within the ganglion. It is important to identify any large arteries or nerves lying alongside ganglia.*

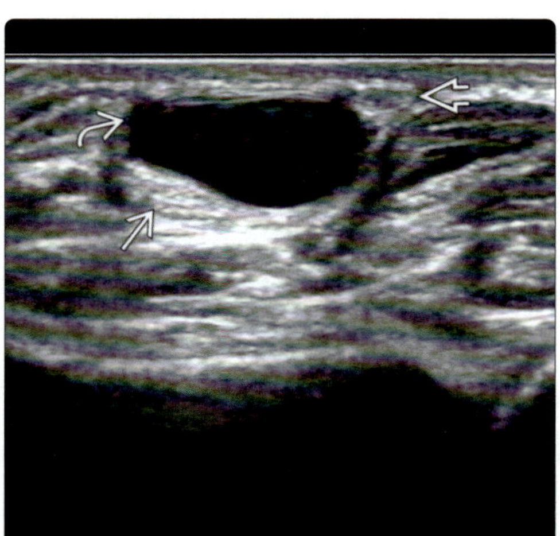
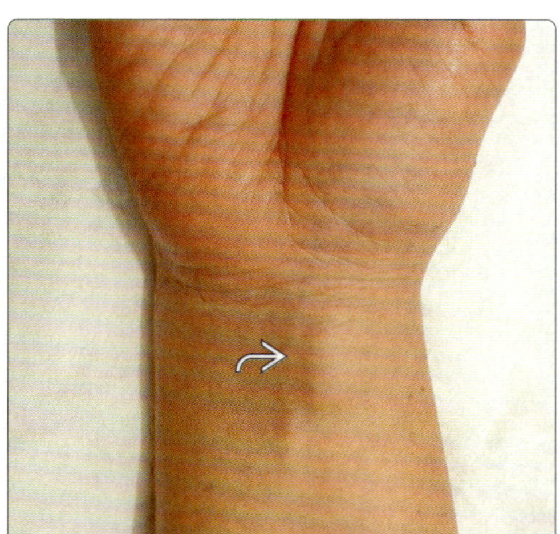

(Left) *Tranverse US shows a small ganglion ➔ on the volar aspect of the wrist. The ganglion lies just radial and deep to flexor carpi radialis tendon ➔. The medial nerve is deep to the ganglion ➔. The ganglion arose from the volar aspect of the radioscaphoid articulation (not shown).* **(Right)** *Clinical photograph of the wrist of the same patient shows a volar swelling ➔ suggestive of a ganglion cyst.*

(Left) *Transverse US shows a small ganglion. The ganglion lies on the radial aspect of the A1 pulley of the middle finger. This type of ganglion is sometimes referred to as a retinacular or pulley ganglion.* **(Right)** *Transverse US shows a small pulley ganglion. The ganglion lies on the ulnar aspect of the A2 pulley of the ring finger. Ganglia are less common around the A2 pulley than around the A1 pulley.*

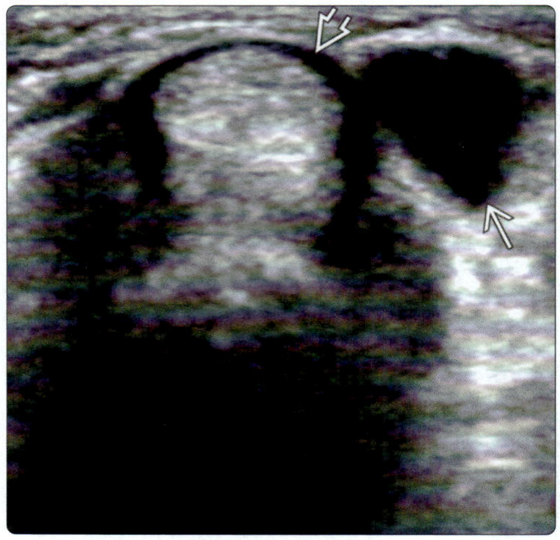

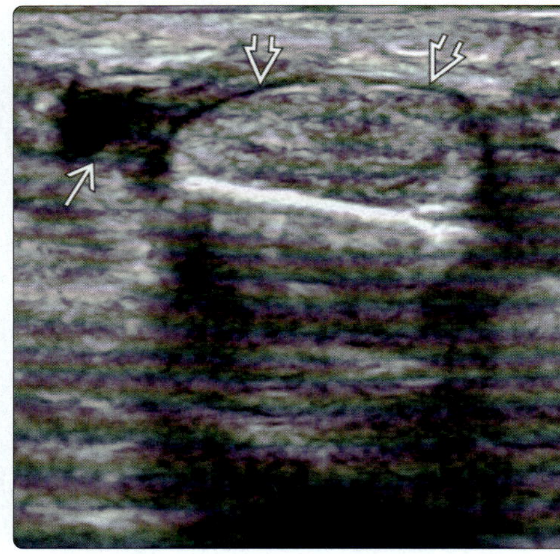

(Left) *Clinical photograph shows swelling consistent with a ganglion on the dorsal aspect of the middle finger. These cysts are covered only by skin and may be translucent.* **(Right)** *Longitudinal US of the same patient shows a subcutaneous unilocular ganglion located dorsal to the extensor tendon of the middle finger. This ganglion most likely arose from the distal interphalangeal joint and is sometimes called a digital mucous cyst.*

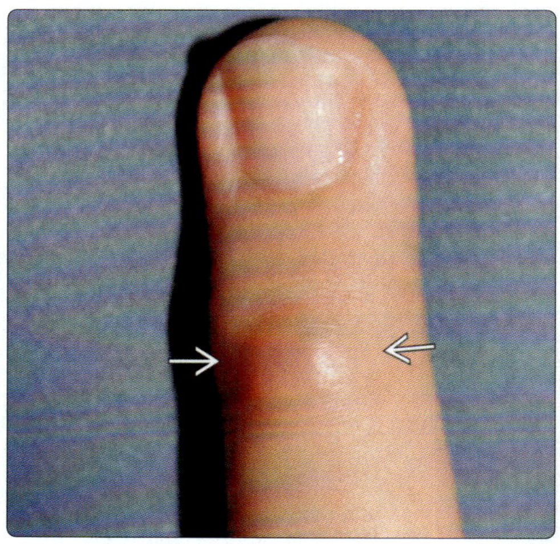

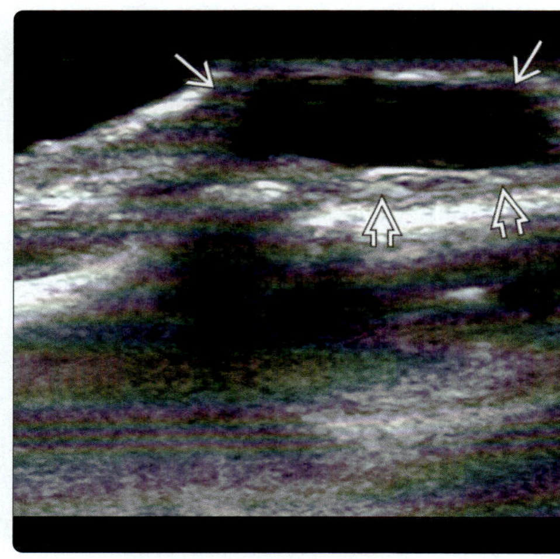

(Left) *Clinical photograph shows a swelling that is suspicious of a ganglion on the ulnovolar aspect of the proximal phalanx close to the proximal interphalangeal joint of the index finger.* **(Right)** *Longitudinal US of the same patient shows an echogenic ganglion just proximal to the proximal interphalangeal joint. No direct communication with the joint is visible, though the thin stalk of the ganglion does point toward the joint.*

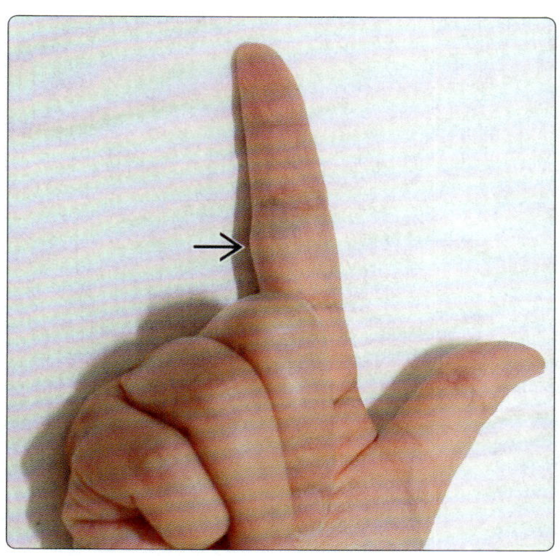

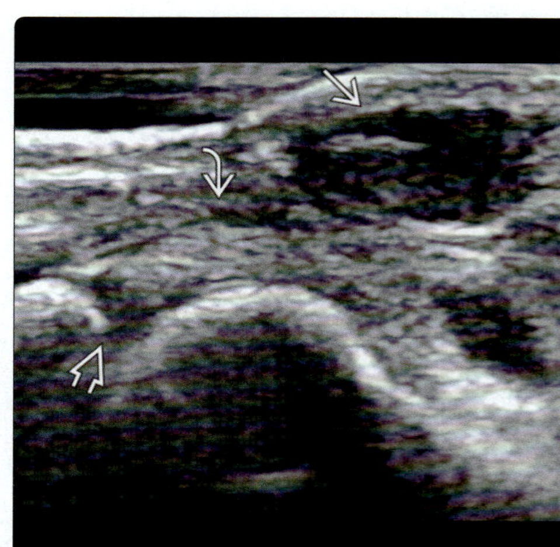

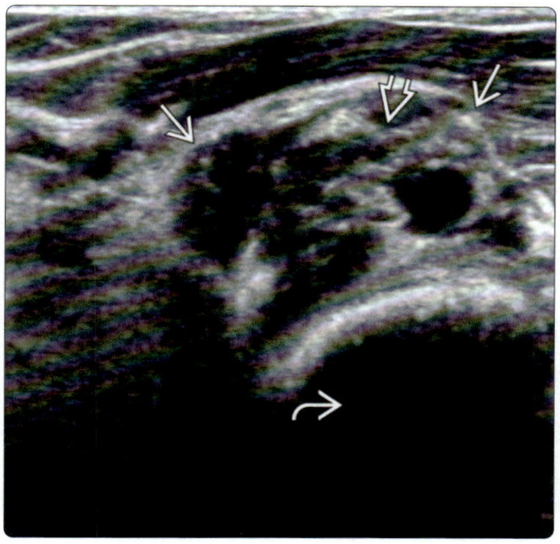

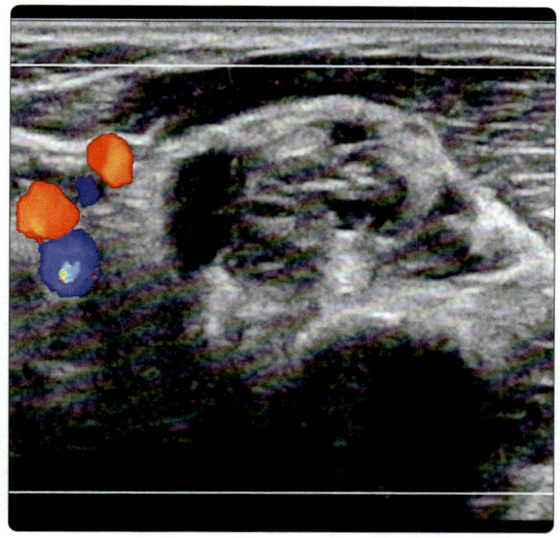

(Left) *Transverse US shows a complex ganglion* ➡ *with thick septations* ➡ *on the anterior aspect of the elbow just superficial to the radial head* ➡*. This ganglion can simulate a solid mass.* **(Right)** *Transverse color Doppler US of the same ganglion shows no vascularity within the septa of the ganglion.*

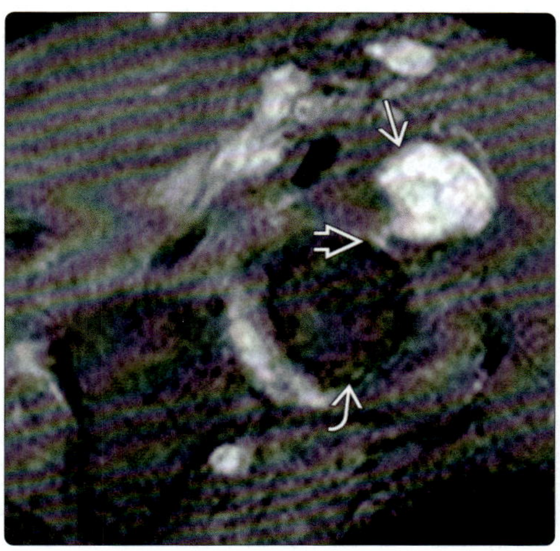

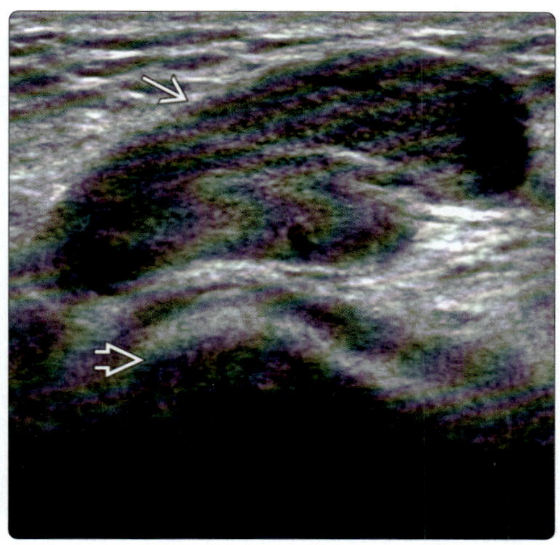

(Left) *Axial T2 FS MR of the same patient shows the ganglion* ➡ *located anterior to the radial head* ➡*. A small tract* ➡ *extending from the ganglion toward the elbow joint is visible.* **(Right)** *Oblique US shows a ganglion* ➡ *located posterior to the olecranon process* ➡*. This should not be confused with an olecranon bursa. A bursa will not point toward the elbow joint.*

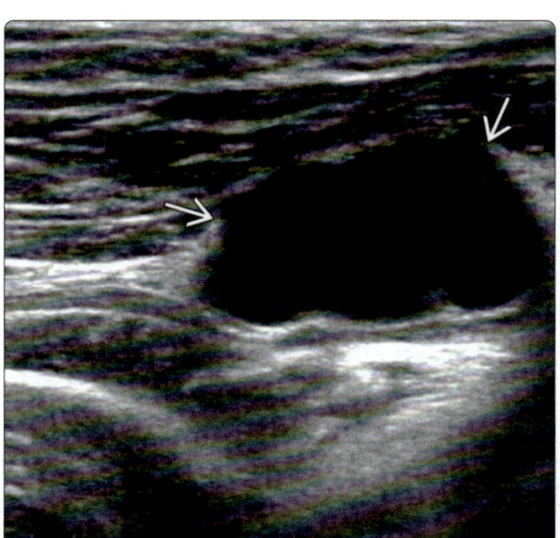

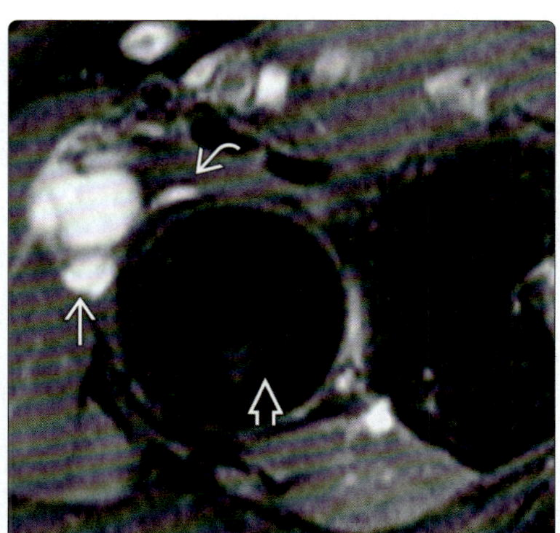

(Left) *Transverse US shows a lobulated ganglion* ➡ *on the lateral aspect of the elbow joint.* **(Right)** *Axial T2 FS MR of the same ganglion shows that the distal aspect of the ganglion* ➡ *is abutting the radial head* ➡ *with a thin tract* ➡ *extending toward the elbow joint.*

(Left) *Transverse US shows a large ganglion with septation and internal echogenic foci due to colloid aggregates superficial to the acromioclavicular joint. The stalk of the ganglion extends toward the acromioclavicular joint.* **(Right)** *Transverse US shows a small ganglion anterior to the acromioclavicular articulation. The thick-walled stalk of the ganglion extends toward the acromioclavicular joint. The clavicle and acromion are shown.*

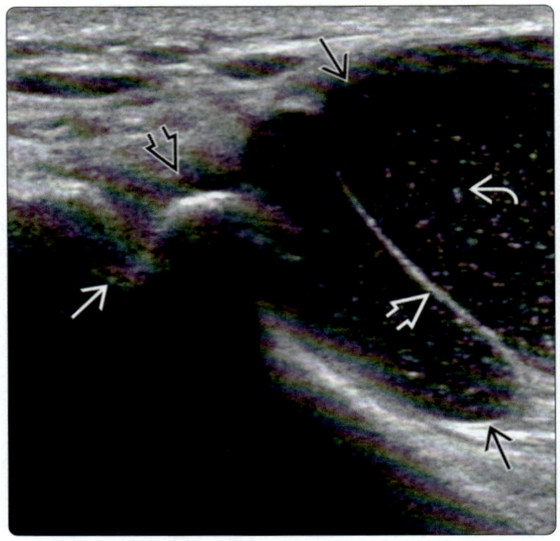

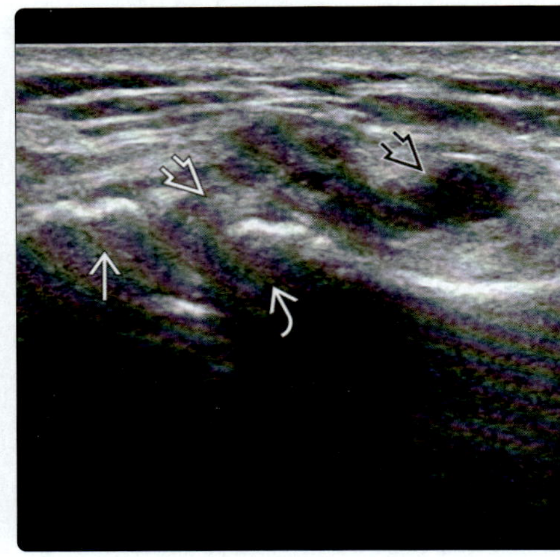

(Left) *Transverse US shows an elongated ganglion anterior to the sternoclavicular joint. No communication with the joint is evident. The clavicular and sternal ends of the joint are shown.* **(Right)** *Axial PD FS MR shows a pseudosolid ganglion in the popliteal fossa alongside the popliteal artery and vein.*

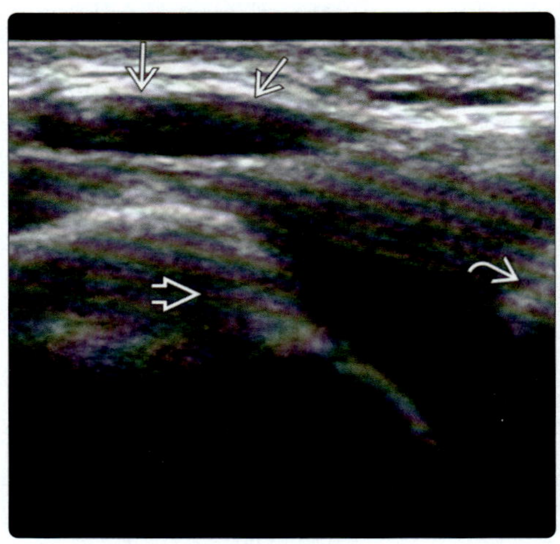

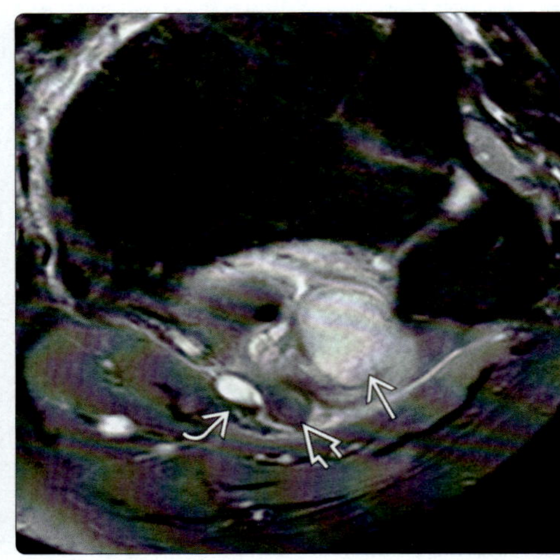

(Left) *Transverse US of the same mass shows that the ganglion lies just posterior to the proximal tibiofibular joint between the tibia and fibula.* **(Right)** *Transverse color Doppler US of the same patient shows that the ganglion lies alongside the popliteal artery and vein. The pseudosolid appearance of this ganglion is due to internal hemorrhage.*

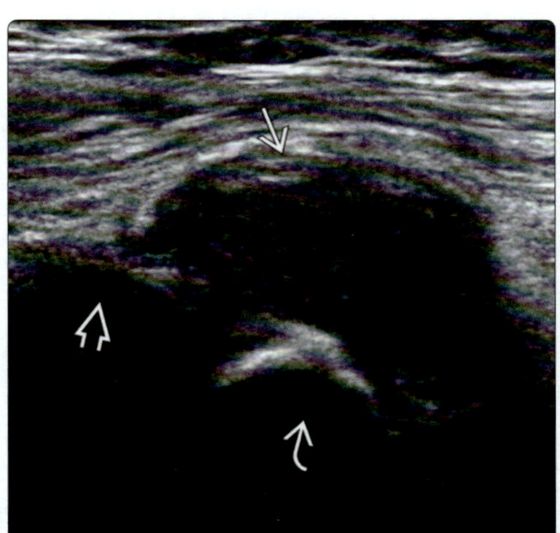

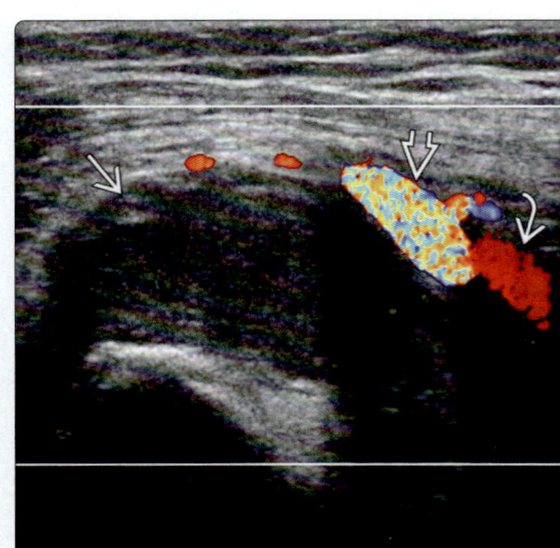

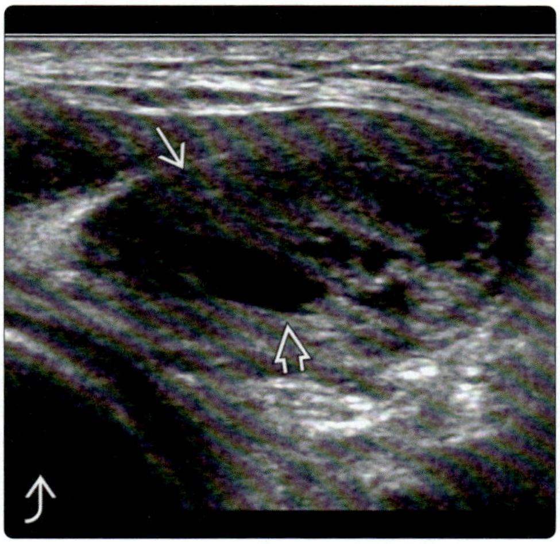

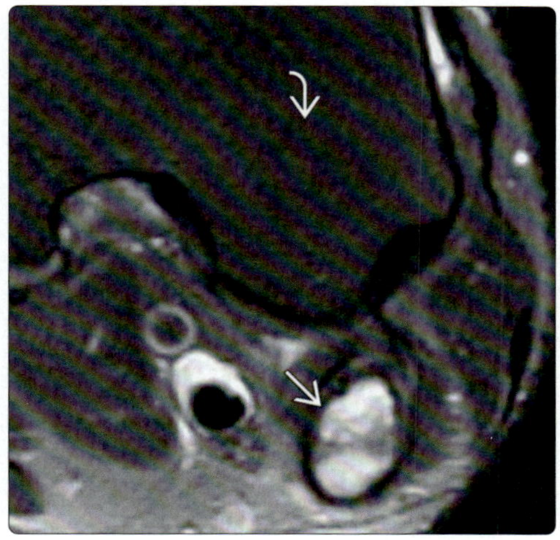

(Left) *Oblique US shows a ganglion on the posteromedial aspect of the knee joint. The ganglion is thick walled with thick septa. This thickening of the cyst wall is likely to be consequent upon repeated intracystic hemorrhage. The medial femoral condyle is shown.* **(Right)** *Axial T2 FS MR of the same patient shows the ganglion lying posterior to the medial femoral condyle. The ganglion has a hypointense wall suggestive of hemosiderin deposition.*

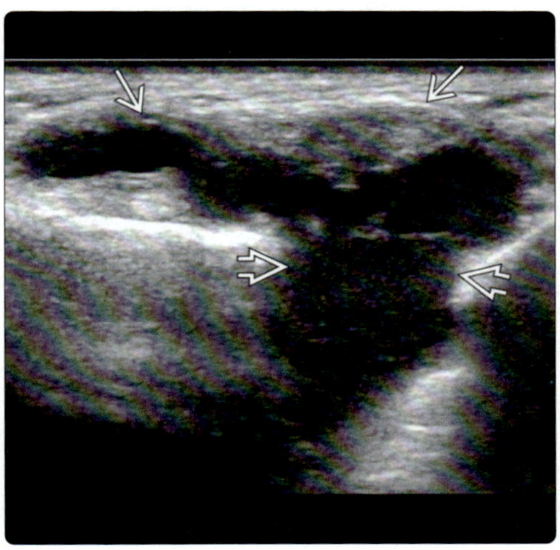

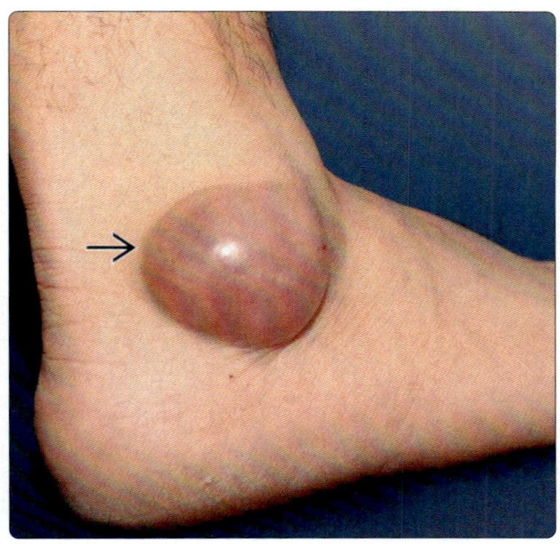

(Left) *Transverse US of the knee region shows a ganglion lying anterior to the external opening of the tibial tunnel for an anterior cruciate ligament graft. This tunnel will communicate with the knee joint, allowing for expression of synovial content.* **(Right)** *Clinical photograph shows a large, cystic swelling on the lateral aspect of the ankle. This swelling occurred 2 days after a relatively minor ankle sprain.*

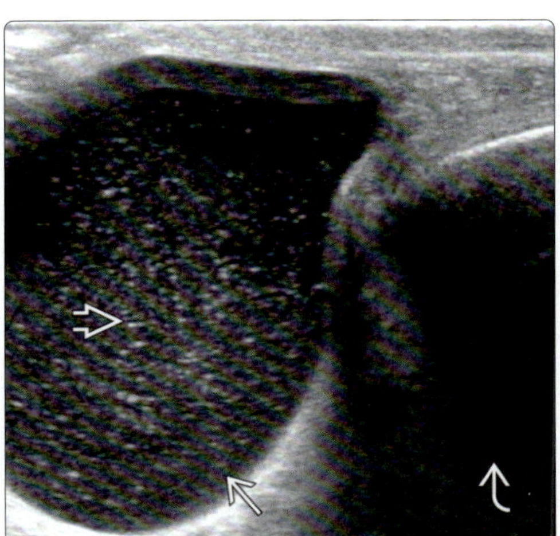

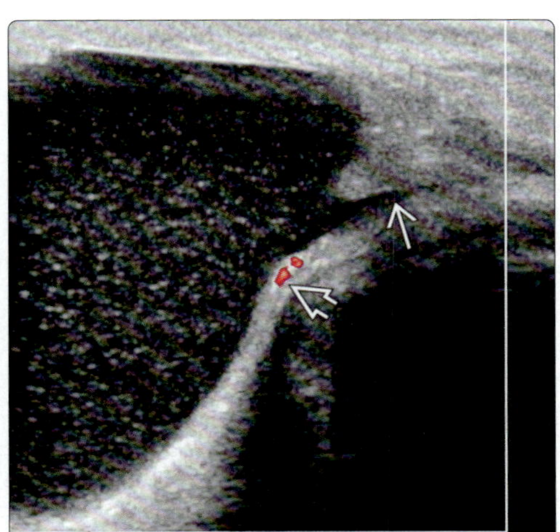

(Left) *Transverse US of the same patient shows a large, acute, traumatic ganglion lateral to the fibula. The tiny echogenic foci within the ganglion cyst are due to colloid aggregates. They are not a sign of chronicity.* **(Right)** *Transverse color Doppler US of the same patient shows a small tract from the anteromedial aspect of the ganglion extending toward the ankle joint with minimal vascularity in the ganglion wall.*

(Left) *Oblique US shows a ganglion* ➡ *on the dorsal aspect of the foot superficial to the articulation between the intermediate cuneiform* ➡ *and the medial cuneiform* ➡*. No artery is visible on this image.* **(Right)** *Oblique color Doppler US of the same area shows that the dorsalis pedis artery* ➡ *lies on the side of the ganglion. This artery could potentially be injured during percutaneous aspiration if no color Doppler imaging was performed beforehand.*

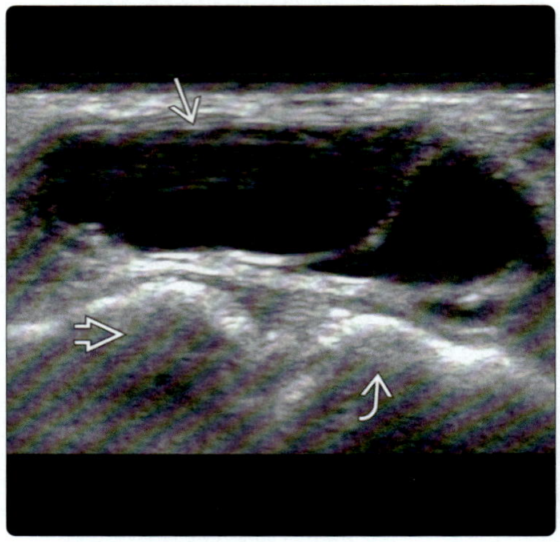

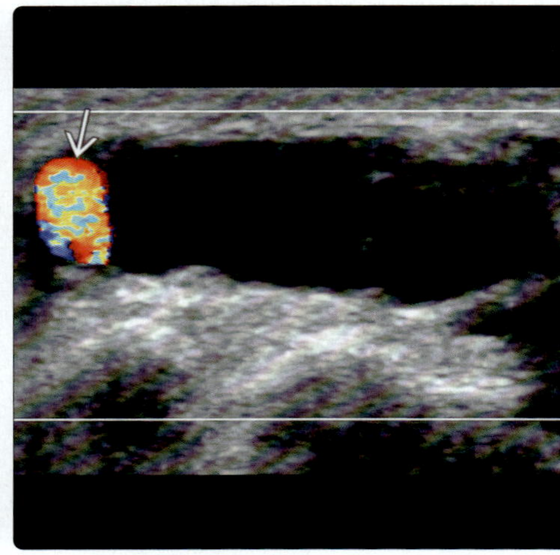

(Left) *Transverse US shows a lobulated ganglion* ➡ *anterior to the distal tibiofibular articulation. The ganglion lies anterior to the anterior tibiofibular ligament* ➡*. No direct communication with the joint is evident.* **(Right)** *Longitudinal US shows a small ganglion* ➡ *on the dorsal aspect of the midfoot. The tortuous stalk* ➡ *of the ganglion is arising from the articulation between the cuboid* ➡ *and the 4th metatarsal base* ➡*.*

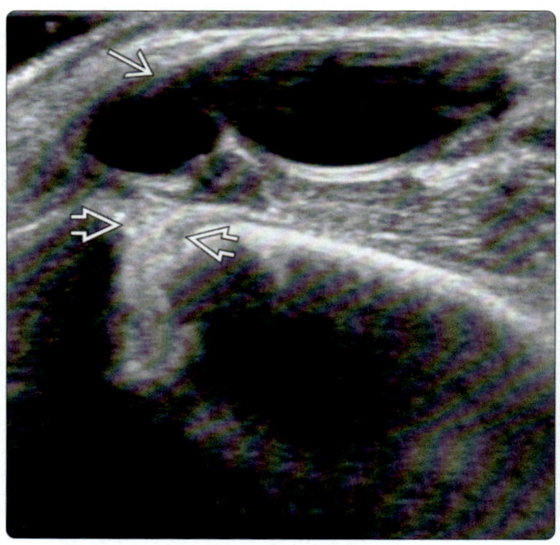

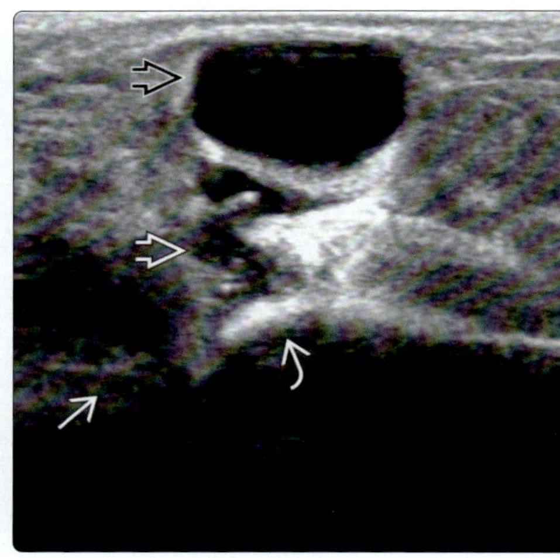

(Left) *Transverse US shows a ganglion* ➡ *on the plantar aspect of the big toe. Ganglion lies medial to the flexor hallucis longus tendon* ➡*, which lies between the medial and lateral sesamoid bones* ➡*.* **(Right)** *Oblique US shows a thick-walled ganglion* ➡ *on the dorsal aspect of the navicular* ➡*/cuneiform* ➡ *joint. This is most likely the result of repeated leakage and recurrent pericyst inflammation. Excessive transducer pressure may efface the fluid component* ➡*.*

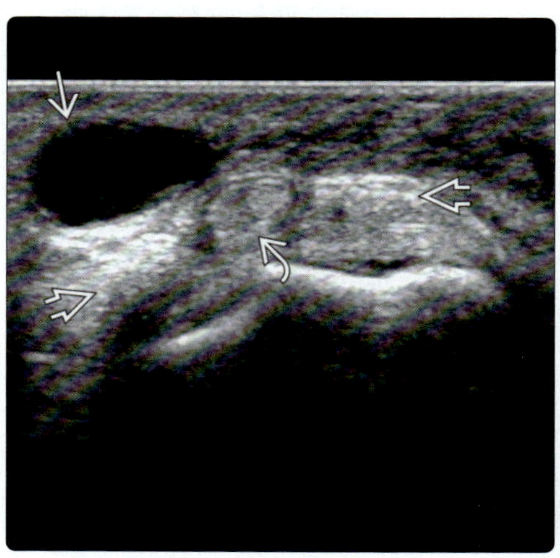

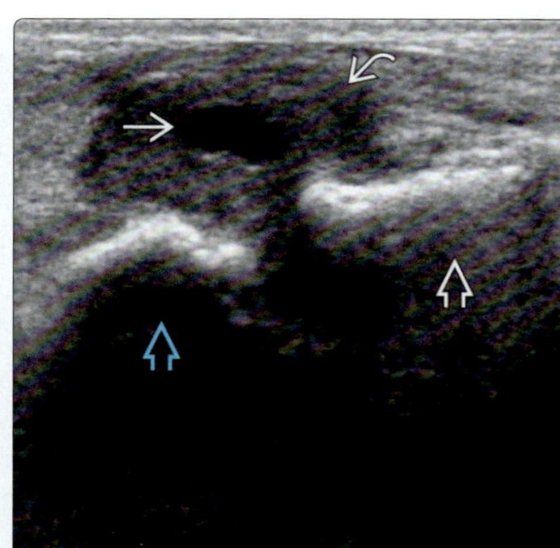

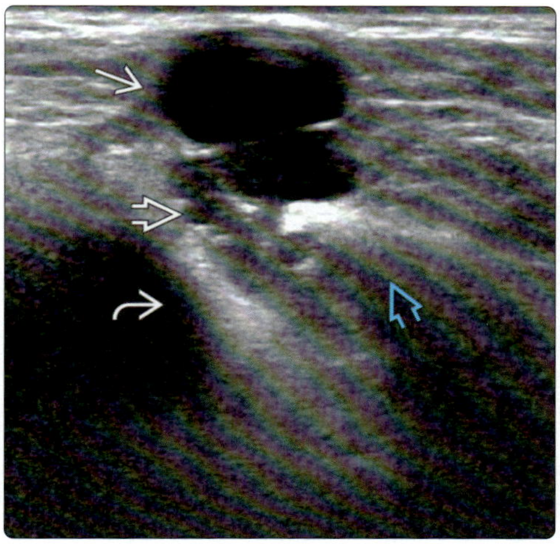
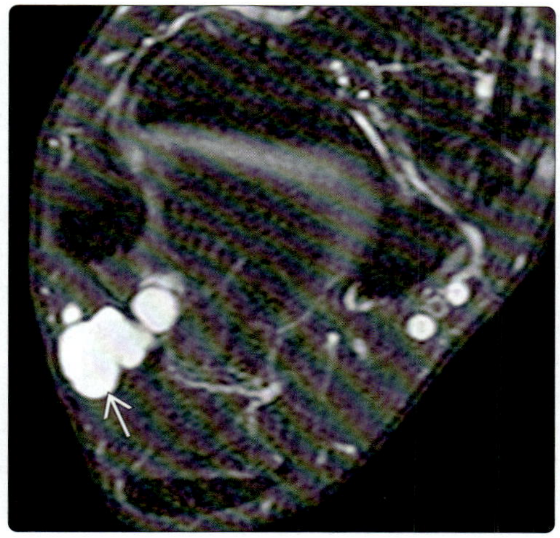

(Left) *Transverse US shows a small, lobulated ganglion arising from the posterolateral aspect of the ankle joint. The stalk of the ganglion extends deep to the articulation between the distal fibula and the talus.* **(Right)** *Axial T2 FS MR of the same patient shows the lobulated ganglion on the posterolateral aspect of the ankle. The ganglion stalk is not clearly seen. Occasionally, an US is superior to MR in this respect.*

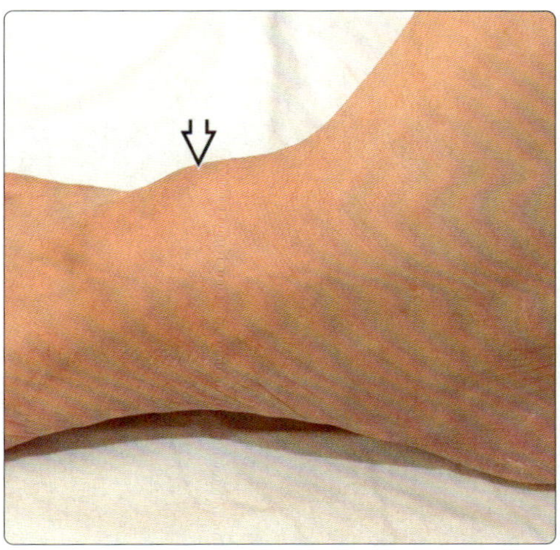
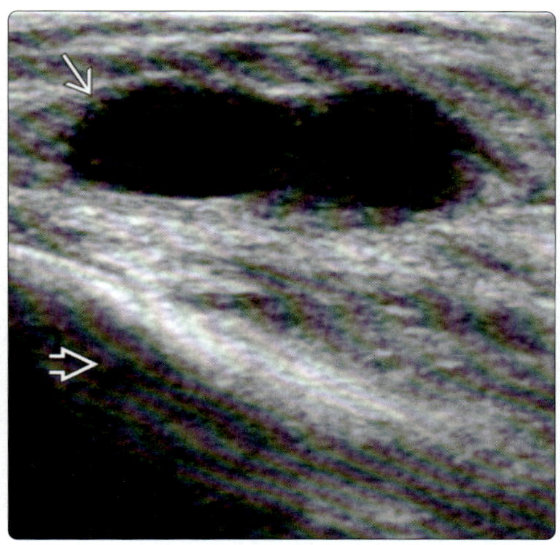

(Left) *Clinical photograph shows a small swelling on the centrodorsal aspect of the midfoot.* **(Right)** *Longitudinal US shows a unilocular uncomplicated ganglion on the dorsum of the midfoot close to the 2nd and 3rd metatarsal bases. Dorsal foot ganglia are much more common than plantar foot ganglia. Additional images (not shown) showed the ganglion neck pointing toward the 2nd tarsometatarsal joint.*

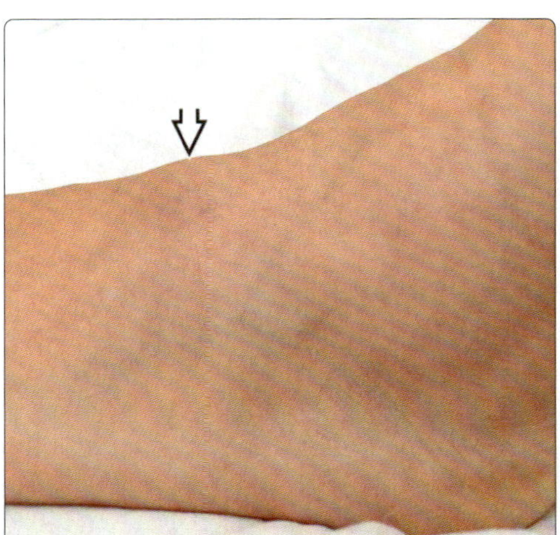
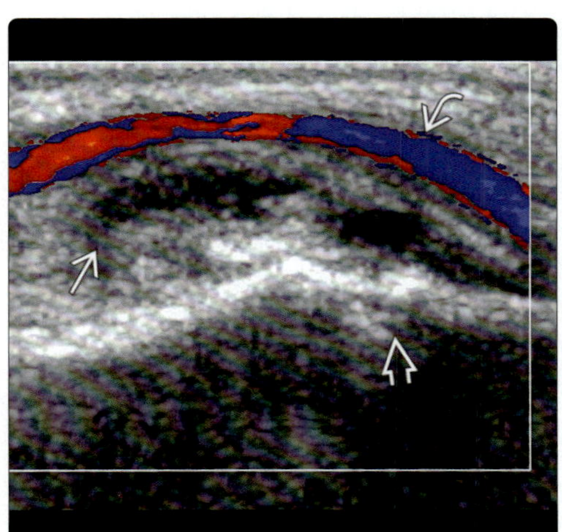

(Left) *Clinical photograph shows a small swelling on the centrodorsal aspect of the midfoot.* **(Right)** *Oblique US shows a thick-walled ganglion partially collapsed on the dorsal aspect of a narrowed talonavicular articulation. The ganglion lies deep to the dorsalis pedis artery. Most dorsal foot ganglia arise from joints of the medial longitudinal arch in the midfoot region.*

Parameniscal Cyst

KEY FACTS

IMAGING

- Cystic mass in continuity with horizontal meniscal tear
  - Medial parameniscal cyst slightly more common than lateral parameniscal cyst
- Meniscal tear forces synovial fluid or mucoid content into parameniscal cyst
- Cyst is usually well defined and filled with anechoic or hypoechoic gelatinous-type fluid
  - Meniscal tear does not always have to be present
    - Tear may have healed
  - Especially with anterior horn lateral meniscus
- Cyst may lie in direct continuity with meniscus or connect via serpiginous connecting stalk
- Colloid content of parameniscal cyst similar to that of ganglia and Baker cyst
  - Diagnosis of parameniscal cyst relies on showing continuity between cyst and meniscus
- Check for pericystic soft tissue inflammation
- Parameniscal cysts, especially medially, can be very large
- Comment on cyst size, loculation, extent, associated meniscal tear, recent cyst leakage ± localized inflammation
  - MR to assess meniscal tear
  - Intracapsular parameniscal cysts may not be visible on US
- US-guided aspiration ± hyaluronidase injection ± steroid injection
  - Provides temporary and often long-term symptom relief
    - Especially for smaller cysts and poor surgical candidates

DIAGNOSTIC CHECKLIST

- Not every paraarticular cyst with pedicle extending to knee joint is parameniscal cyst
  - Need to distinguish from paraarticular ganglion
- Inflammation around parameniscal cyst usually results from recent leakage or extension
  - Leakage common cause of symptom exacerbation

(Left) *Clinical photo shows moderate swelling ➔ on the medial aspect of the knee around the medial joint line due to a parameniscal cyst.* **(Right)** *Longitudinal US of the medial aspect of the knee shows a large parameniscal cyst ➔ located superficial to the medial collateral ligament ➔. The cyst is extending proximally ➔ toward the knee joint.*

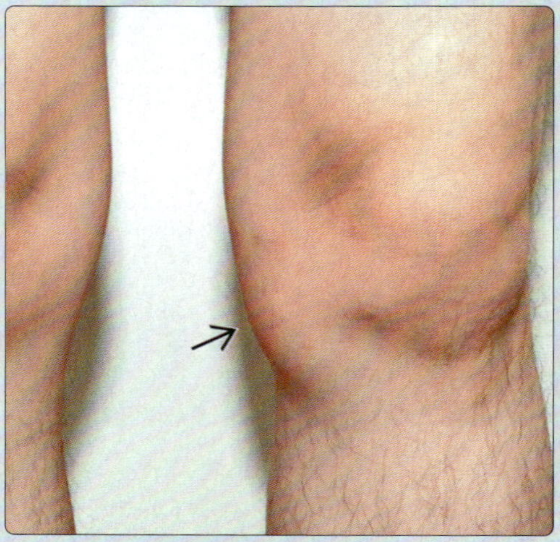

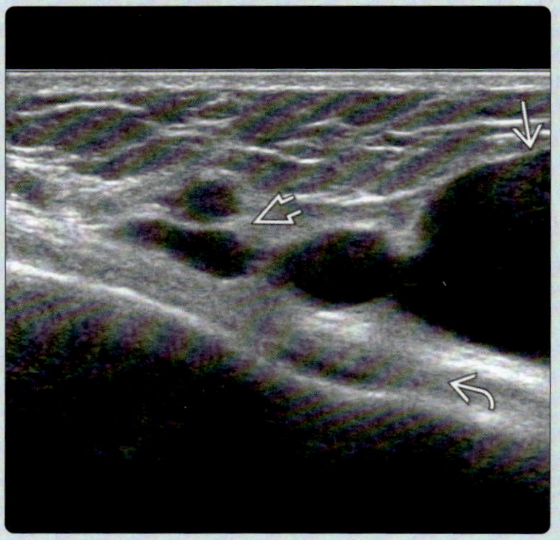

(Left) *Longitudinal US just proximal to the previous image shows a thin tract ➔ extending through the medial collateral ligament ➔ toward the periphery of the thinned degenerate medial meniscus ➔.* **(Right)** *Graphic shows horizontal meniscal tear ("fish mouth" tear) ➔, with meniscal mucoid degeneration ➔ and a small parameniscal cyst ➔ arising from the peripheral aspect of the tear. Meniscal tears are much better depicted on MR than US.*

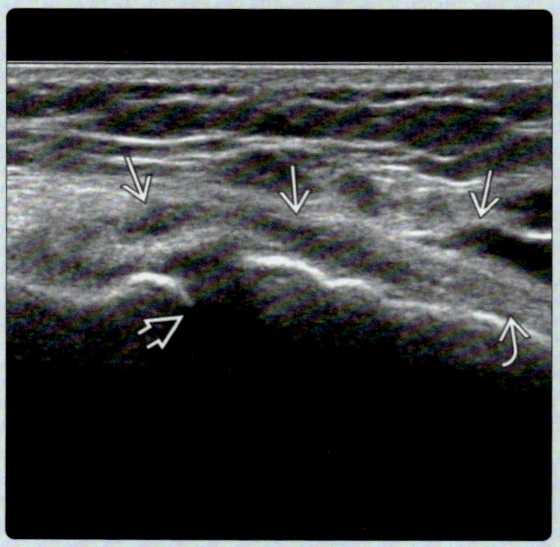

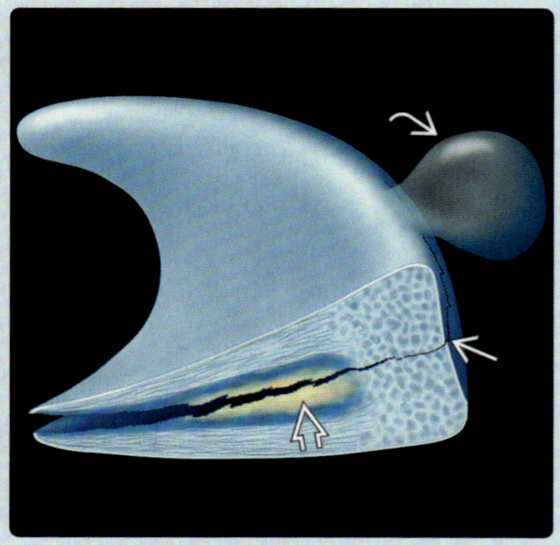

TERMINOLOGY

Synonyms

- Meniscal cyst
  - This term should be reserved for intrameniscal cysts

Definitions

- Cyst extending from meniscal tear
  - Usually horizontal tear (most common) or complex tear (2nd most common)

IMAGING

General Features

- Best diagnostic clue
  - Cystic mass in continuity with horizontal meniscal tear
- Location
  - Can arise from any part of meniscus
    - Medial meniscus (~ 60%), lateral meniscus (~ 40%)
  - Most common locations
    - Junction between body and posterior horn medial meniscus (posterior to medial collateral ligament)
    - Junction between body and anterior horn lateral meniscus (anterior to lateral collateral ligament)
- Size
  - Tiny (few mm) to large extraarticular cyst (> 10 cm)
- Morphology
  - Extension from parameniscal area on medial and lateral aspects of knee
  - Cysts from medial meniscus often extend either
    - Inferiorly deep to semimembranous and pes anserinus tendons
    - Anteriorly superficial to medial collateral ligament or into Hoffa fat pad
    - Posteriorly toward popliteal fossa
  - Cysts from lateral meniscus often extend deep to iliotibial tract or lateral collateral ligament
  - Medial parameniscal cysts in particular can extend quite a distance from meniscal tear
    - Lateral parameniscal cysts tend to remain more localized

MR Findings

- T1WI
  - Hypointense cyst mass
  - If proteinaceous, cyst may be iso- or hyperintense
- T2WI
  - Rounded, hyperintense mass
    - Lobulation and septation
  - Tear = linear hyperintense meniscal signal extending to articular surface
    - Peripheral meniscal tear may not visibly connect with meniscal articular surface (noncommunicating tear)

Ultrasonographic Findings

- Parameniscal cyst ± extraarticular component with communicating stalk
  - Meniscal tear does not always have to be present, especially from anterior horn of lateral meniscus
- Meniscus is seen as diffusely echogenic, sharply defined structure within medial and lateral femorotibial compartments
  - Medial meniscus easier to see on US
    - Only peripheral part of menisci can be seen on US
  - Body of menisci is best seen with limited views of other areas
- Meniscal tear = linear or wedge-shaped hypoechogenicity within meniscus extending to meniscal periphery
  - Be careful because normal meniscus can mimic appearances of horizontal tear
  - Extrusion of medial meniscal body is common accompanying feature
- Parameniscal cyst
  - Usually well defined and filled with anechoic or hypoechoic gelatinous-type fluid, or rarely, heterogeneous hyperechoic appearance
  - ± multiloculation, ± septation, ± comet-tail artifacts due to colloid aggregates
  - ± echogenic mass/debris (due to proteinaceous content or previous hemorrhage)
  - Edema, swelling, and hyperemia around cyst indicates recent leakage, which leads to localized inflammation
    - Check for free fluid around cyst (sign of recent cyst leakage)
- Cyst may lie in direct continuity with meniscus or connect via connecting stalk (pedicle)
  - Small intrameniscal cyst may be visible
- Communicating stalk is usually serpiginous and narrow
  - Cyst may be located at some length from meniscal tear
- Cyst and contents of parameniscal cyst are similar to ganglia and Baker cyst: All 3 contain gelatinous material
  - Main differences
    - Stalk in parameniscal cyst → meniscal tear
    - Stalk in paraarticular ganglion → joint
    - Stalk in Baker cyst → between semimembranous tendon and medial gastrocnemius muscle → joint
- If meniscal communication is not seen, then parameniscal cyst can only be suggested as diagnosis
  - Sometimes difficult to distinguish parameniscal cyst and paraarticular ganglion

Imaging Recommendations

- Best imaging tool
  - US: 1st line of investigation (accuracy: 88%)
    - Very accurate at detecting cystic nature of paraarticular mass
    - Will show meniscal origin in majority of parameniscal cysts
  - MR: Helpful to fully assess meniscal integrity
    - Best to show meniscal tear, articular surface extension of tear, and degree of meniscal mucoid degeneration
- If extraarticular cyst, locate stalk of cyst
  - Trace stalk of cyst to knee joint
  - Determine if stalk originates from meniscal tear
- Note relationship of cyst to adjacent structures
  - Medially: Medial collateral ligament and pes anserinus (sartorius, gracilis, semitendinous) tendons
  - Laterally: Iliotibial band, lateral collateral ligament, and popliteus tendon
- Start examination with knee extended
- Examine with varying degrees of flexion to accentuate visualization of meniscal tear ± stalk (optional)
- Examine anterior horn of menisci with knee flexed

DIFFERENTIAL DIAGNOSIS

Paraarticular Ganglion Cyst

- Can extend from intraarticular to extraarticular compartment
- Does not originate from meniscal tear

Bursitis

- Medial parameniscal cyst mimicked by
  - Semimembranosus bursitis, pes anserinus bursitis, or medial collateral ligament bursitis
- Lateral parameniscal cyst mimicked by
  - Deep infrapatellar bursa (anteriorly located)

Baker Cyst

- Distension of semimembranous-gastrocnemius bursa
  - Popliteal cyst with "talk bubble" configuration
  - Stalk arises through interspace between semimembranous tendon and medial belly gastrocnemius muscle

PATHOLOGY

General Features

- Etiology
  - All have underlying meniscal abnormality
    - > 90% have meniscal tear, invariably with horizontal component ± meniscal degeneration
    - Cysts also arise from complex tears with horizontal component
  - Forcing of synovial joint fluid through horizontal meniscal tear into cyst
    - Compression/decompression of horizontal tear with knee movement forces fluid into cyst
  - 1/3 of tears associated with parameniscal cyst do not extend to meniscal articular surface
  - Possible explanations
    - Articular surface extension is too small to see or may have healed
    - Cyst content is generated solely from meniscal mucoid degeneration and not synovial fluid
  - Anterior horn of lateral meniscus parameniscal cyst typically not associated with meniscal tear
    - Cyst extends to Hoffa fat pad or into anterior meniscal root
  - Horizontal tear and parameniscal cyst associated with discoid meniscus
  - Also associated with meniscocapsular separation

Gross Pathologic & Surgical Features

- Cyst filled with mucinous-type fluid
  - In continuity with horizontal cleavage meniscal tear or flap (oblique) tear with predominately horizontal component

CLINICAL ISSUES

Presentation

- Most common signs/symptoms
  - Palpable mass ± pain/discomfort near joint line ± limitation of movement
  - Size varies with knee position
    - Palpable lateral cyst at 15-30° of flexion, which disappears at full extension and flexion (Pisani sign)
  - Consistency varies from fluctuant to firm or hard
  - Pain tends to settle, though cyst may remain unchanged
    - Symptom exacerbation usually associated with recent cyst leakage or extension

Demographics

- Age
  - Peak: 20-40 years; mean: 41 years
- Sex
  - M > F
- Epidemiology
  - Relatively common: 2-4% of all knee MR exams

Treatment

- US-guided aspiration of cyst
  - ± intracystic percutaneous steroid injection
    - ± prior hyaluronidase injection if cyst content too viscous to aspirate
  - Cyst tends to recur but seems to provide good temporary relief
    - Efficacy of percutaneous treatment not fully evaluated
  - Sometimes percutaneous aspiration and steroid injection provides long-term relief
    - Especially helpful in poor surgical candidates
  - Not possible to predict cyst viscosity of cyst on US
  - Viscosity determined during attempted aspiration
- Arthroscopic or open cystectomy and repair of tear or decompression of cyst
  - Cysts treated with decompression are 6x more likely to recur than those cysts excised

DIAGNOSTIC CHECKLIST

Consider

- Not every paraarticular cyst with stalk extending to knee joint is parameniscal cyst
  - Intraarticular ganglion cyst with paraarticular extension can look very similar
- Demonstration of meniscal tear extending to articular surface is relevant as this will affect operative approach
  - Arthroscopic if communicating meniscal tear present
  - Open if noncommunicating meniscal tear present

Image Interpretation Pearls

- Cyst alongside meniscus in continuity with meniscal tear = parameniscal cyst
  - Meniscal tear usually has horizontal component

Reporting Tips

- Comment on cyst size, loculation, extent, associated meniscal tear, recent cyst leakage, or localized inflammation

SELECTED REFERENCES

1. Botchu R et al: Parameniscal cyst - a rare cause of popliteal artery compression: treatment with ultrasound-guided decompression. West Indian Med J. 64(2), 2014

Parameniscal Cyst

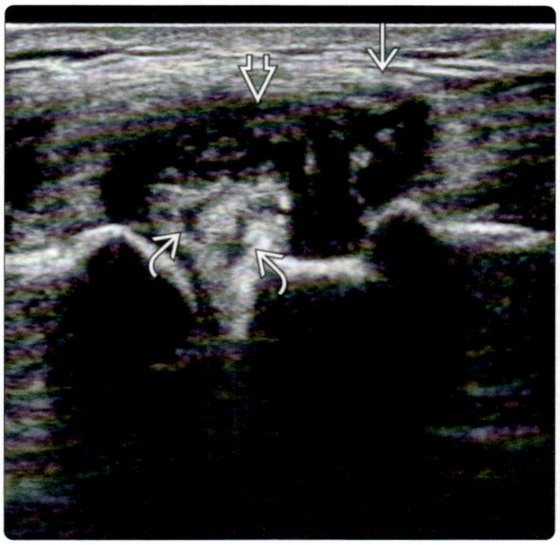

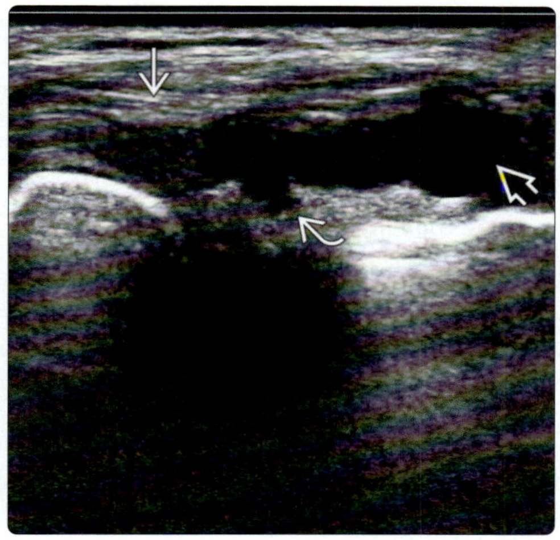

(Left) *Longitudinal US shows a complex tear → at the periphery of the extruded medial meniscus. It is associated with a medium-sized parameniscal cyst → that lies deep to the medial collateral ligament →.* **(Right)** *Longitudinal US shows a parameniscal cyst → deep to the medial collateral ligament →. The cyst connects to a horizontal peripheral tear → of the medial meniscus.*

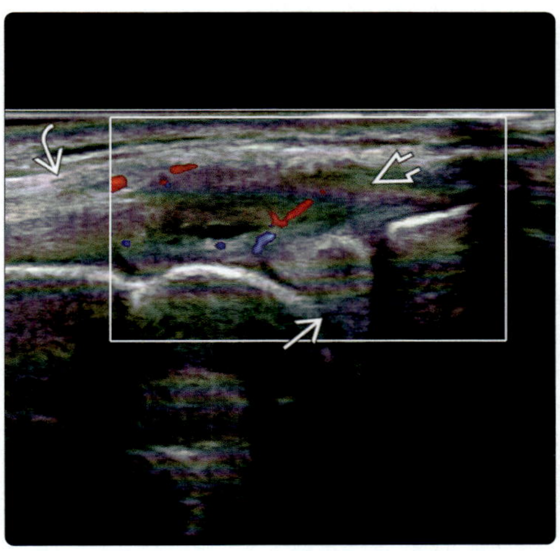

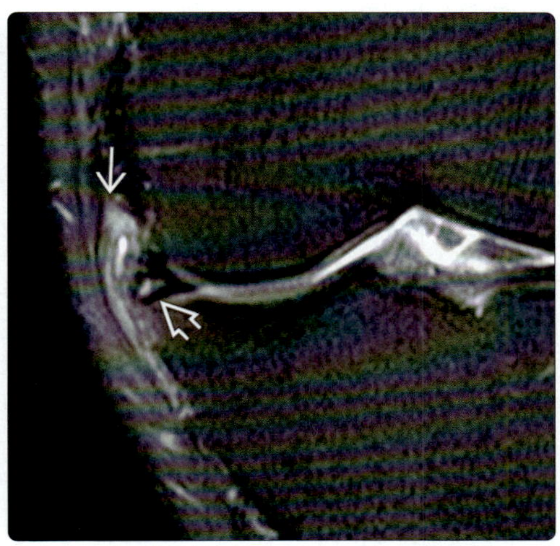

(Left) *Longitudinal US shows a small parameniscal cyst at the periphery of the medial meniscus → and deep to the medial collateral ligament →. There is moderate surrounding soft tissue inflammation → with edema and hyperemia.* **(Right)** *Coronal T2 FS MR at the same location shows a very small parameniscal cyst with surrounding inflammation →. Cyst content leakage is associated with focal inflammation. There is a horizontal meniscal tear → present, which was not visible on US.*

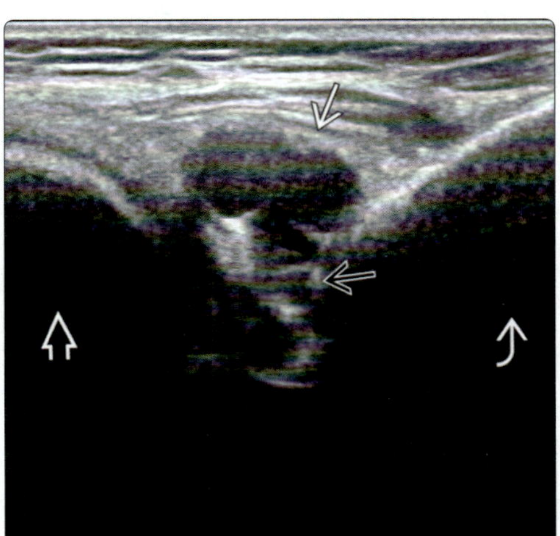

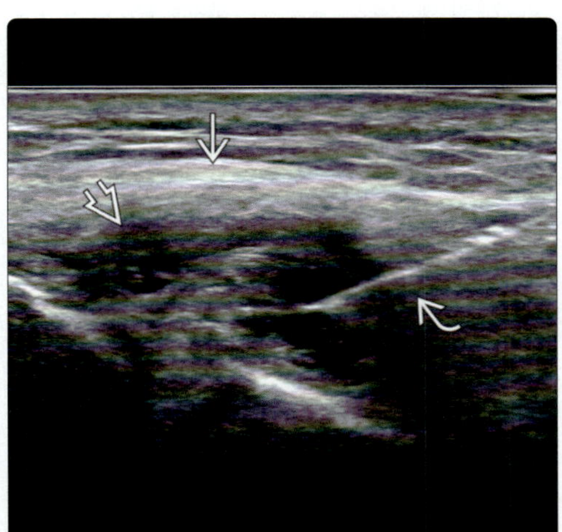

(Left) *Longitudinal US shows a medium-sized parameniscal cyst → on the anterolateral aspect of the knee between the femoral → and tibial → condyles. The tail of the cyst → extends deeply toward the anterior horn of the lateral meniscus.* **(Right)** *Longitudinal US shows a parameniscal cyst → deep to the lateral collateral ligament →. US-guided aspiration and steroid injection with the needle tip → in the cyst is shown. If no fluid can be aspirated, hyaluronidase can be injected and aspiration repeated after 20 minutes.*

Synovial Tumor

KEY FACTS

TERMINOLOGY

- Tenosynovial giant cell tumor (TSGCT)
  - Diffuse intraarticular, localized intraarticular, & tendon sheath forms
- Synovial (osteo)chondromatosis
- Less common synovial tumors
  - Lipoma arborescens, synovial hemangioma, synovial lipoma, synovial fibroma, synovial chondrosarcoma, synovial infiltration (e.g., lymphoma)

IMAGING

- Localized synovial proliferation favors tumor over infection or systemic inflammation
- TSGCT: Tumor-like hypoechoic hypervascular synovium
  - 90% occur in synovial-lined joints, tendon sheaths, & bursae; multicentric occurrence is rare
  - 10% of extraarticular TSGCT may not have demonstrable connection to joint, tendon sheath, or bursa
  - Knee (60%) > hip > ankle > small joints of hand & foot > shoulder & elbow
- TSGCT (diffuse intraarticular)
  - Lobular thickening of hypoechoic synovium; variable synovial hyperemia; hemosiderin not visible on US; ± intraarticular bony scalloping, ± small effusion
- TSGCT (localized intraarticular)
  - Well-defined hypoechoic intraarticular nodule
  - ± mild hyperemia, ± effusion
- TSGCT (tendon sheath): Eccentric nodular mass along tendon sheath ± hyperemia
  - Hand & wrist (80%) > foot & ankle (20%)
- Synovial osteochondromatosis: Thickened synovium with discrete echogenic foci ± hyperemia
  - No acoustic shadowing if pure cartilage nodules
  - Knee > elbow > hip and shoulder; polyarticular (5%)
- Confirm most synovial tumors with US-guided core biopsy as US appearances are usually not specific enough to make definitive diagnosis

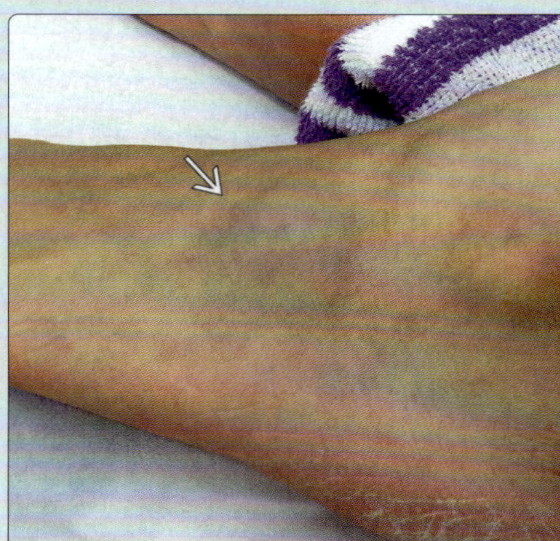

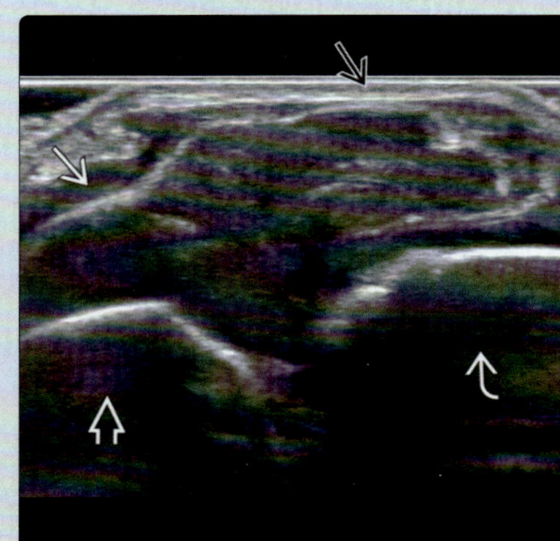

(Left) *Clinical photograph of a 32-year-old woman with a nonpainful dorsal lateral foot mass ➔ for 6 months is shown. One can appreciate the slight bluish discoloration of the overlying skin. Both MR and PET/CT performed prior to referral suggested a sarcoma. The standardized uptake value was 15.3.* **(Right)** *Longitudinal US shows a hypoechoic mass ➔ arising from the junction between the calcaneum, the cuboid ➔, and the lateral cuneiform ➔. The mass lies deep to the extensor digitorum tendons ➔.*

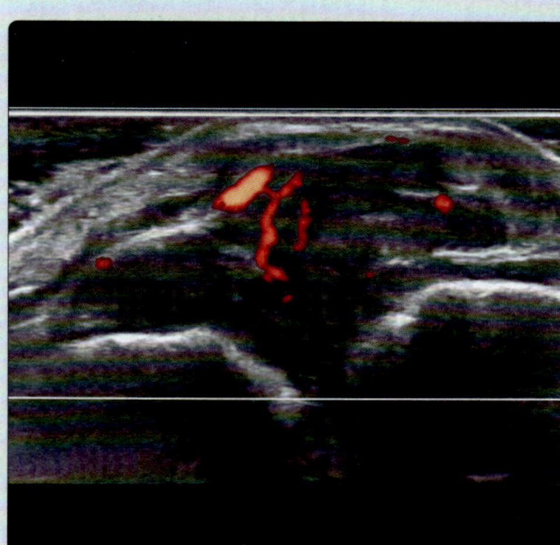

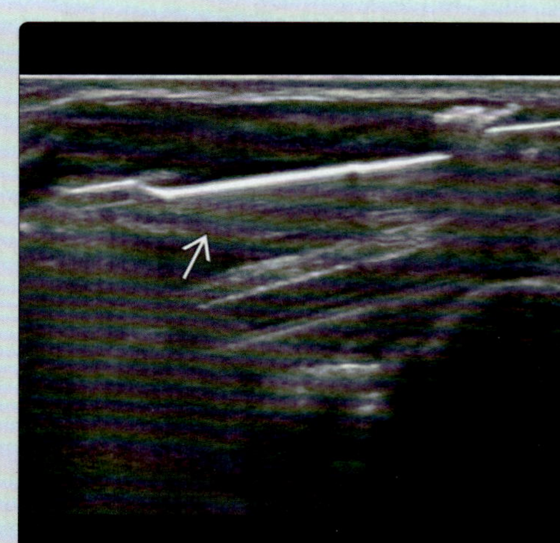

(Left) *Longitudinal power Doppler US shows mild to moderate tumoral hyperemia. The US diagnosis was tenosynovial giant cell tumor (TSGCT) arising from the calcaneo-cuboid-lateral cuneiform articulation.* **(Right)** *Longitudinal US shows percutaneous Tru-Cut needle biopsy ➔. Biopsy revealed cores of fibrous tissue mainly composed of sheets of mononuclear cells with scattered osteoclast-like giant cells and hemosiderin-laden macrophages. This was consistent with TSGCT.*

TERMINOLOGY

Fibrohistiocytic Synovial Tumors

- Tenosynovial giant cell tumor (TSGCT)
  - Diffuse and localized forms of fibrohistiocytic tumors
- Formerly called
  - Pigmented villonodular synovitis if diffuse and intraarticular
  - Focal nodular synovitis if localized and intraarticular
  - Giant cell tumor of tendon sheath if arising from tendon sheath

Other Synovial Tumors

- Synovial osteochondromatosis: Synovial metaplasia characterized by cartilaginous loose bodies that later ossify
- Less common synovial tumors
  - Lipoma arborescens, synovial hemangioma, synovial lipoma, synovial fibroma, synovial chondrosarcoma, synovial infiltration (e.g., lymphoma)
    - Synovial sarcoma is misnomer since not derived from true synovial cells

IMAGING

General Features

- Best diagnostic clue
  - TSGCT: Tumor-like hypoechoic hypervascular synovium or eccentric nodular mass along tendon sheath ± hyperemia
    - 90% occur in synovial-lined joints, tendon sheaths, and bursae; multicentric occurrence is rare
      - □ 10% of extraarticular TSGCT shows no demonstrable connection to joint, tendon sheath, or bursa
      - □ Either original connection may have become disrupted or tumor arose de novo in other tissues
  - Synovial osteochondromatosis: Thickened synovium with small discrete echogenic foci ± hyperemia
- Location
  - TSGCT (diffuse and localized) from joint: Knee (60%) > hip > ankle > small joints of hand and foot > shoulder and elbow
  - TSGCT from tendon sheath: Hand and wrist > foot and ankle
    - TSGCT from tendon sheath should occur only where there is synovial-lined tendon sheath
  - Synovial osteochondromatosis: Knee > elbow > hip and shoulder; polyarticular (5%)
- Size
  - TSGCT: Thickened synovial mass ranging from a few cm to 20 cm
  - Synovial osteochondromatosis: Nodules ranging from a few mm to ~ 20 mm

Radiographic Findings

- Radiography
  - TSGCT
    - Dense soft tissue swelling
      - □ ± intraarticular erosions with sclerotic margins, especially in hip (tight capsule)
    - Preservation of joint space until late stage
    - No juxtaarticular osteopenia
  - Synovial (osteo)chondromatosis
    - Multiple small ossified bodies within joint
    - Cartilage nodules not ossified in 20%
      - □ ± intraarticular erosions, especially in hip
    - Preservation of joint space until late stage
    - No juxtaarticular osteopenia

MR Findings

- T1WI
  - TSGCT
    - Isointense to hypointense mass or masses
- T2WI
  - TSGCT
    - Hypointense to isointense synovial mass secondary to paramagnetic effect of hemosiderin
    - ± intermediate-signal erosions with hypointense margins
- T2* GRE
  - TSGCT
    - Foci of hypointensity due to hemosiderin deposits
    - Hemosiderin is present in 85% of TSGCT cases, i.e., lack of hemosiderin does not exclude diagnosis
    - Localized TSGCT contains less hemosiderin than diffuse TSGCT

Ultrasonographic Findings

- TSGCT (diffuse intraarticular)
  - Lobular thickening of hypoechoic synovium
  - Variable synovial hyperemia
  - Hemosiderin not visible on US
  - ± intracapsular bony erosion ± effusion
- TSGCT (localized intraarticular)
  - Well-defined hypoechoic intraarticular nodule
- TSGCT (tendon sheath)
  - Eccentric, hypoechoic mass along tendon sheath
  - ± mild hyperemia
  - Otherwise normal tendon and tendon sheath
- Synovial osteochondromatosis
  - Mild to moderate synovial thickening with small effusion
  - Multiple small echogenic nodules
  - Nonossified cartilaginous nodules may be difficult to see on US
    - Hypoechoic similar to joint fluid
  - ± mild hyperemia
- Nearly all synovial tumors
  - Use US-guided core biopsy for histologic confirmation
  - US appearances are usually not specific enough to make definitive diagnosis

Imaging Recommendations

- Best imaging tool
  - US for initial assessment
    - Most synovial tumors are readily seen by US
    - If synovium clearly seen in full and is normal, then no synovial tumor
    - Allows US-guided synovial biopsy for definitive histologic confirmation
  - MR for defining extent
    - US can suggest diagnosis of most synovial tumors, though MR is more specific and better defines extent
  - Radiography for synovial osteochondromatosis

- Protocol advice
  - Look carefully for any articular or tendon sheath involvement of extraarticular masses
  - Look for bone erosion or scalloping
  - Always access with color Doppler imaging
    - Use minimal transducer pressure

DIFFERENTIAL DIAGNOSIS

Hemophilic Arthropathy

- Familial (X-linked)
- ± polyarticular, ± hemarthrosis

Synovitis

- Systemic inflammatory disorder
- ± polyarticular + rheumatoid arthritis factor positive

Synovial Hemangioma

- Vascular anomaly involving synovium
- Phleboliths and venous lakes ± hemarthrosis

PATHOLOGY

General Features

- TSGCT
  - Synovial proliferative disorder
    - Localized intraarticular TSGCT seems to be separate entity from diffuse intraarticular TSGCT, although histologically, they are almost identical
- Localized intraarticular TSGCT
  - Localized to 1 area of synovium
  - Does not have frond-like projections
  - Has less hemosiderin deposition
  - Tends to become more pedunculated with growth
- Synovial osteochondromatosis
  - Synovial metaplasia

Gross Pathologic & Surgical Features

- TSGCT (diffuse and localized forms)
  - Yellow-brown cut surface (fat and hemosiderin)
  - ± joint filled with nonclotted blood
- Synovial (osteo)chondromatosis
  - Synovial thickening
  - Multiple small calcified or ossified bodies, both free and attached to synovium

Microscopic Features

- TSGCT
  - Synovial mass with fibrohistiocytic cells, foamy histiocytes, and giant cells with hemosiderin
- Synovial osteochondromatosis
  - Initial active phase of synovial proliferation without loose bodies
    - Progresses to late inactive phase with quiescent synovial disease and loose bodies
  - Multiple nodules of hyaline cartilage formed within subsynovial connective tissue
  - Later ossify (endochondral ossification)
    - Nodules may contain cartilage, cartilage and bone, or mature bone with fatty marrow

CLINICAL ISSUES

Presentation

- Most common signs/symptoms
  - Joint or tendon swelling
- Other signs/symptoms
  - Pain and limitation of movement

Demographics

- Age
  - Typically 30-50 years
- Gender
  - TSGCT: M:F = 2:1
  - Synovial osteochondromatosis: M:F = 2:1

Natural History & Prognosis

- TSGCT (diffuse intraarticular): → progressive arthropathy
- TSGCT (localized intraarticular and of tendon sheath): Good response to excision
- Synovial osteochondromatosis: Active → inactive phase

Treatment

- TSGCT (diffuse intraarticular)
  - Synovectomy
  - Recurrence
    - 20% after open synovectomy
    - 50% after arthroscopic synovectomy
  - Intraarticular yttrium-90 radiation synovectomy
- TSGCT (localized intraarticular and of tendon sheath)
  - Limited synovectomy
- Synovial osteochondromatosis
  - Synovectomy
  - Recurrence in 15%

DIAGNOSTIC CHECKLIST

Consider

- Diffuse and localized intraarticular TSGCT probably represent separate entities

Image Interpretation Pearls

- TSGCT (diffuse intraarticular): Diffuse hypoechoic synovial mass with variable hyperemia
  - T2* GRE identifies hemosiderin (hypointense)
    - Not imperative to see hemosiderin on gradient-echo MR imaging to make diagnosis of diffuse TSGCT
- TSGCT (localized intraarticular): Localized hypoechoic synovial mass with variable hyperemia
- TSGCT (tendon sheath)
  - Eccentric (rather than concentric) tendon sheath mass
- Synovial osteochondromatosis: Calcified bodies not invariably visible in early stages
- US-guided biopsy will confirm diagnosis of most synovial tumors

SELECTED REFERENCES

1. Sitt JC et al: Ultrasound-guided synovial Tru-cut biopsy: indications, technique, and outcome in 111 cases. Eur Radiol. 27(5):2002-2010, 2017
2. Sitt JC et al: Ultrasound-guided synovial biopsy. Br J Radiol. 89(1057):20150363, 2016

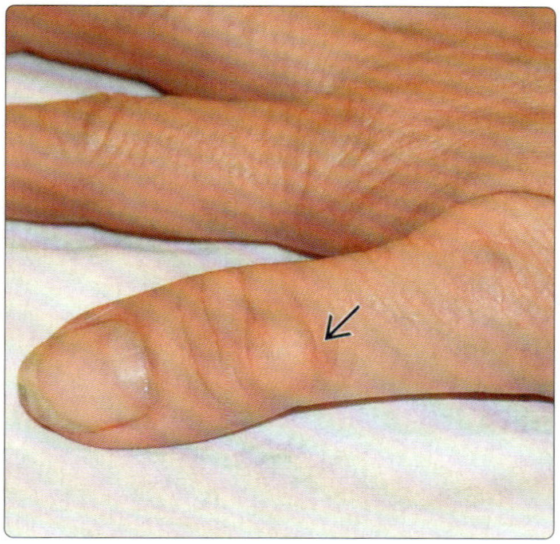

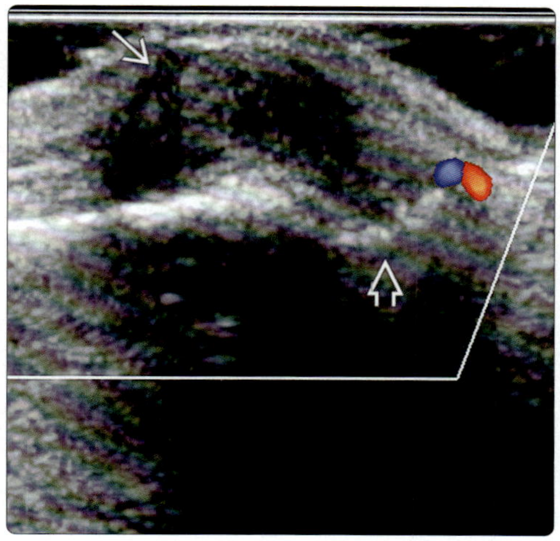

(Left) *Clinical photograph shows a thumb mass → that had been present for 12 months.* **(Right)** *Longitudinal color Doppler US of the same patient shows a medium-sized hypoechoic solid mass → that was in contact with the dorsal capsule of the interphalangeal joint (not shown). No demonstrable tumor hyperemia is present. Overall appearances favor TSGCT, confirmed by histology. The apparent bone scalloping → is spurious due to probe obliquity.*

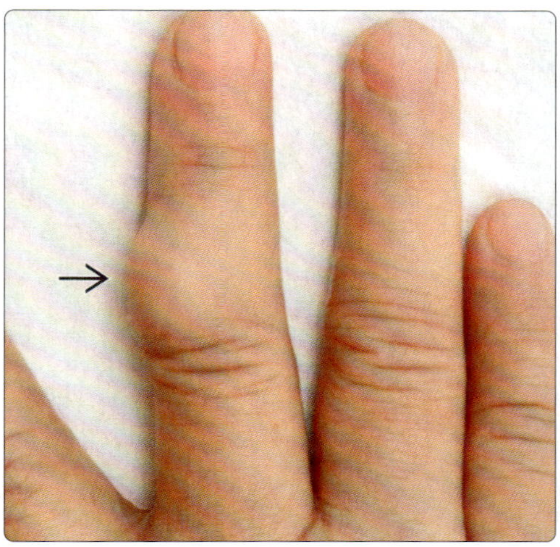

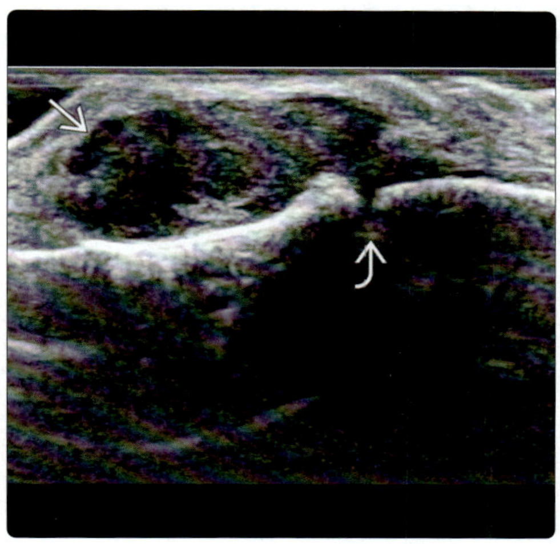

(Left) *Clinical photograph shows a painful middle finger mass → that had been slowly growing for 2 years.* **(Right)** *Longitudinal US of the same patient shows a medium-sized mass → on the dorsoradial aspect of the middle finger at the level of and distal to the proximal interphalangeal joint →. The appearances are consistent with TSGCT arising from the proximal interphalangeal joint.*

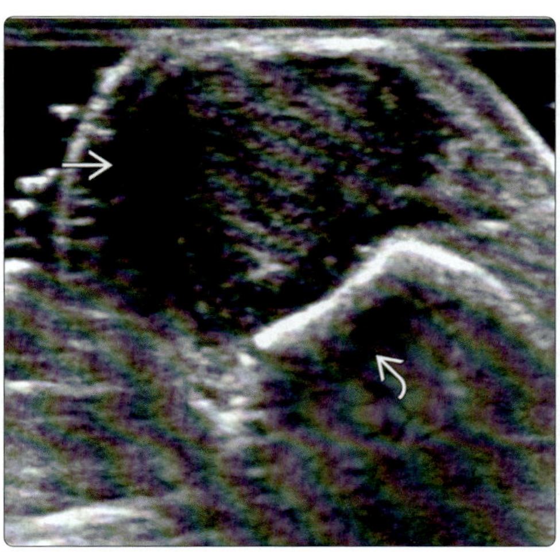

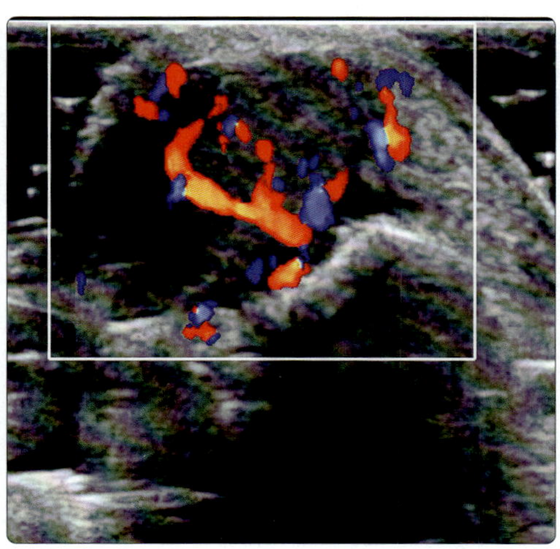

(Left) *Transverse US in the same patient shows the typical hypoechoic appearance of TSGCT → with mild scalloping of the underlying middle phalanx →.* **(Right)** *Transverse color Doppler US of the same patient shows moderate tumoral hyperemia. This level of hyperemia is quite typical for TSGCT, though some TSGCT tumors may be relatively hypovascular.*

(Left) *Longitudinal US shows a large, lobulated mass ➡ present on the dorsum and ulnar aspect of the thumb base. This mass seems to arise from the metacarpophalangeal joint ➡ and encases the extensor pollicis longus tendon ➡.* **(Right)** *Transverse US of the same thumb shows the hypoechoic tumor mass ➡ with moderate scalloping ➡ of the underlying dorsal aspect of the proximal phalanx.*

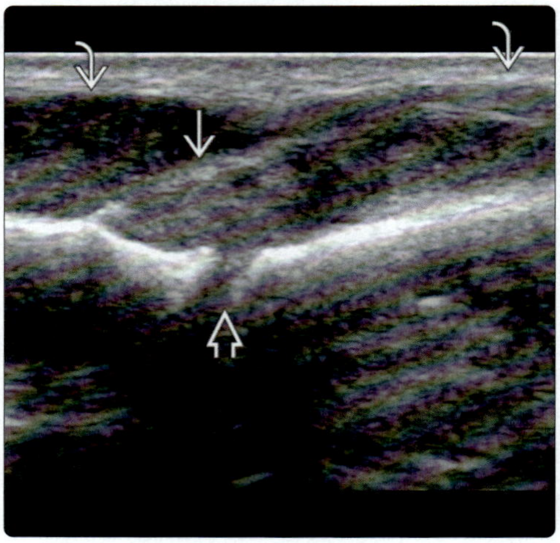

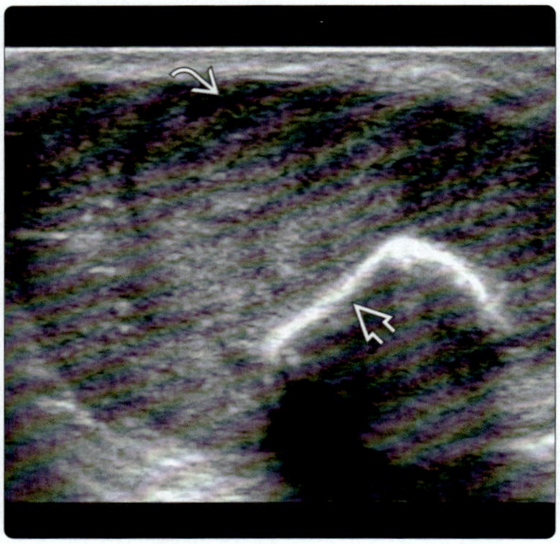

(Left) *Transverse color Doppler US at the same location shows moderate tumoral hyperemia. The appearances are consistent with TSGCT, which was confirmed by US-guided biopsy for histology. Although this tumor probably arose from the metacarpophalangeal joint, the main tumor bulk is extraarticular.* **(Right)** *3D CT angiography of the same tumor shows prominent arterial supply to the tumor mass ➡.*

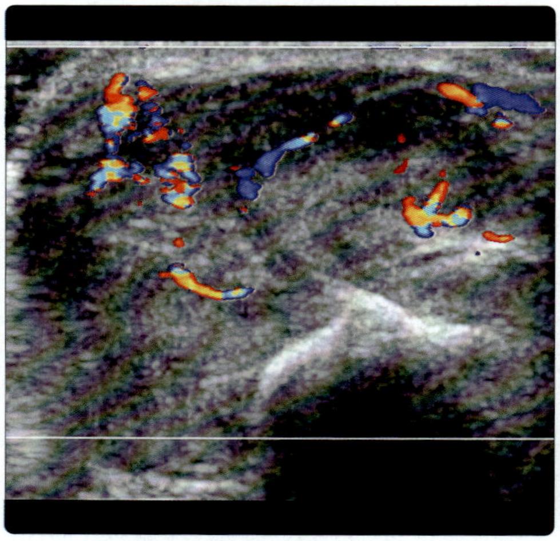

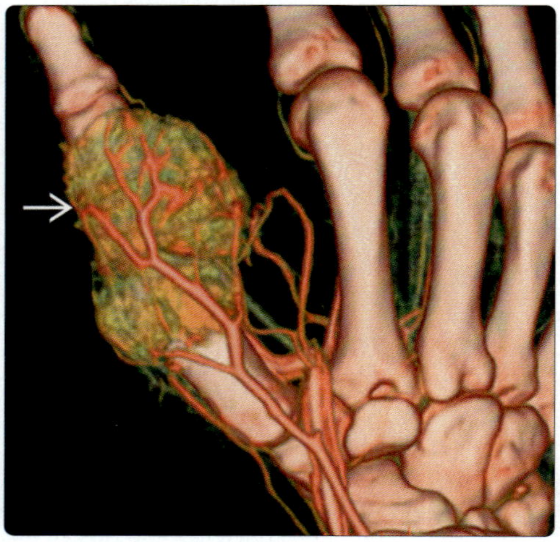

(Left) *Clinical photograph shows a middle finger mass ➡, which had been present for 18 months.* **(Right)** *Transverse US in the same patient shows a medium-sized mass ➡ on the dorsoradial aspect of the proximal phalanx ➡ in contact with the proximal interphalangeal joint ➡. The appearances are consistent with TSGCT, which was confirmed on histology following US-guided biopsy. This is an extraarticular TSGCT, which most likely arose from the proximal interphalangeal joint.*

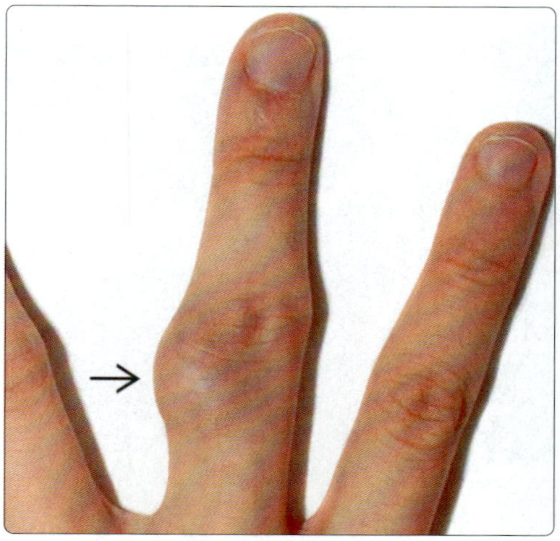

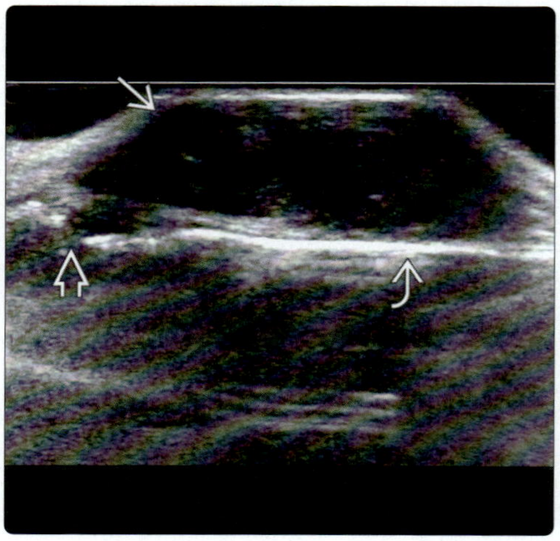

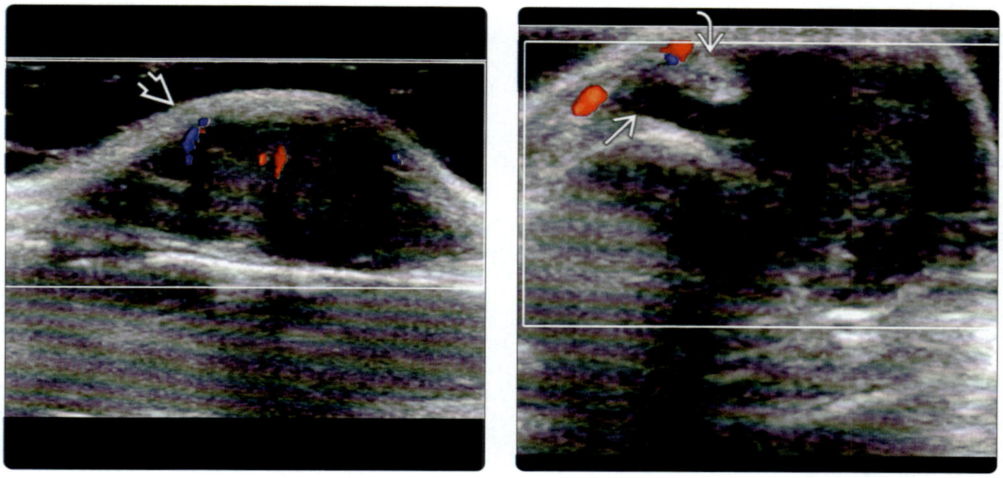

(Left) *Longitudinal color Doppler US of the same tumor shows mild tumoral hyperemia. Note that lifting the transducer from the tumor to assess hyperemia positively alters the tumor shape* ➔. **(Right)** *Transverse color Doppler US of the same tumor shows the tongue of the tumor* ➔ *extending deep to the extensor digitorum tendon* ➔. *This tendon does not have a synovial sheath. Although this tumor is close to the tendon, it most likely arose from the interphalangeal joint.*

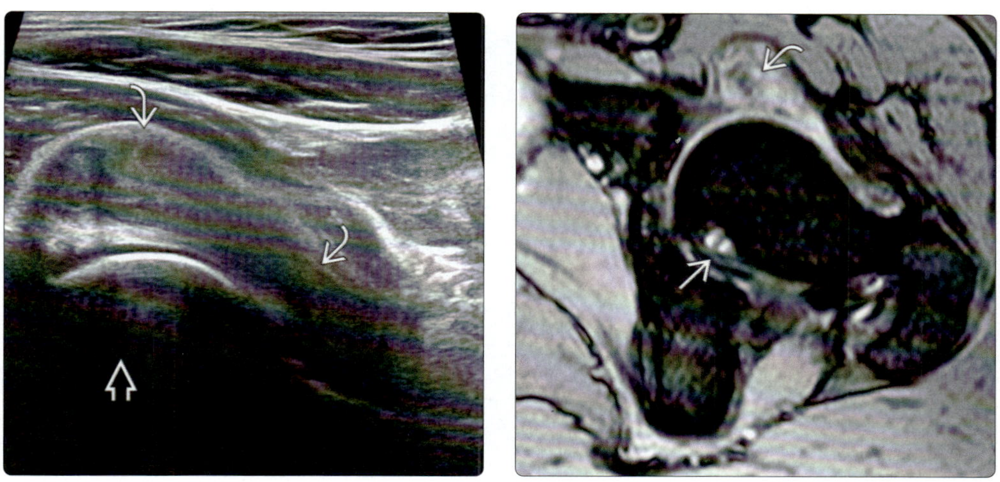

(Left) *Longitudinal US of a patient with hip pain for 15 months shows localized hypoechoic synovial proliferation* ➔ *on the anterior aspect of the hip. The femoral head is normal* ➔. **(Right)** *Axial fat-suppressed T2* GRE MR of the same hip shows minimal hemosiderin deposition within the thickened synovium* ➔. *This is compatible with localized TSGCT, which was confirmed by US-guided biopsy for histology (not shown). There is a small incidental fibrogelatinous cyst posteriorly* ➔.

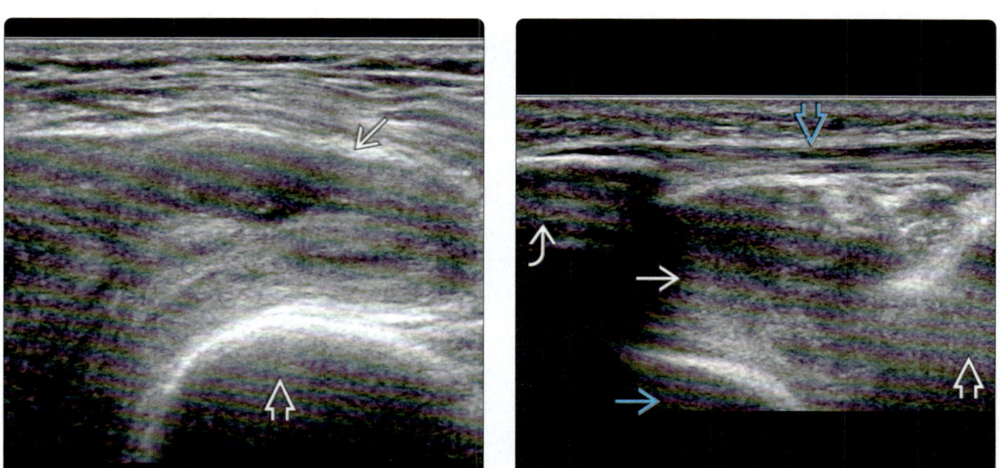

(Left) *Transverse US of the medial patellofemoral recess shows similar severe hypoechoic synovial proliferation* ➔. *The femur is normal* ➔. **(Right)** *Longitudinal US of the same patient shows severe synovial proliferation* ➔ *bulging into the infrapatellar fat pad. The patella* ➔, *patellar tendon* ➔, *distal femur* ➔, *and proximal tibia* ➔ *are normal. The appearances are consistent with diffuse intraarticular TSGCT, which was confirmed on percutaneous biopsy.*

(Left) *Lateral radiograph of the knee in a 69-year-old man with swelling in the suprapatellar region shows moderate distension of the suprapatellar pouch with severe patellofemoral osteoarthritis.* **(Right)** *Longitudinal US of the suprapatellar recess shows severe synovial proliferation with no effusion. No synovial proliferation was present elsewhere in the knee joint. Biopsy of this synovium revealed TSGCT. Localized synovial proliferation favors tumor rather than infection or systemic inflammation.*

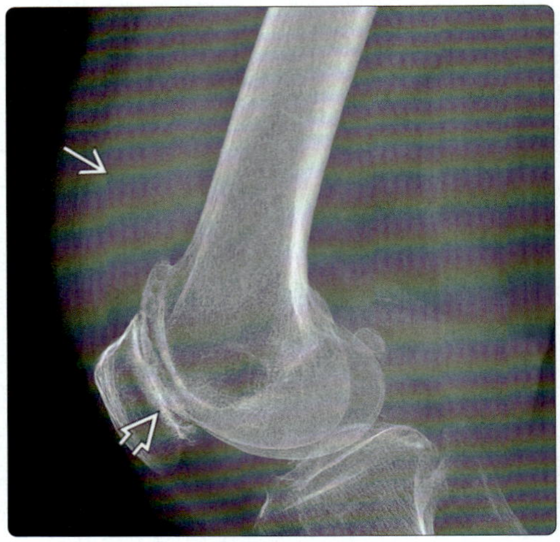

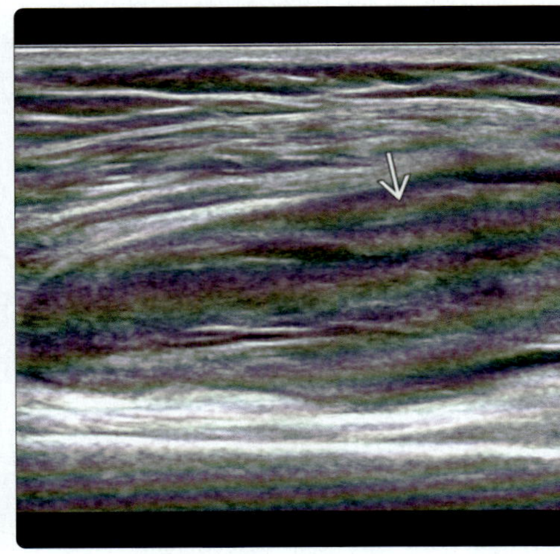

(Left) *Longitudinal US in a 37-year-old man with ankle pain for 4 months shows severe localized synovial thickening at the anterior ankle recess distending the anterior capsule. No bone erosion is present. The talar dome articular cartilage is well preserved.* **(Right)** *Longitudinal color Doppler US shows only minimal tumoral hyperemia. Overall intraarticular TSGCT was considered most likely, but synovial hemangioma and lymphoma were also considered. Biopsy confirmed TSGCT.*

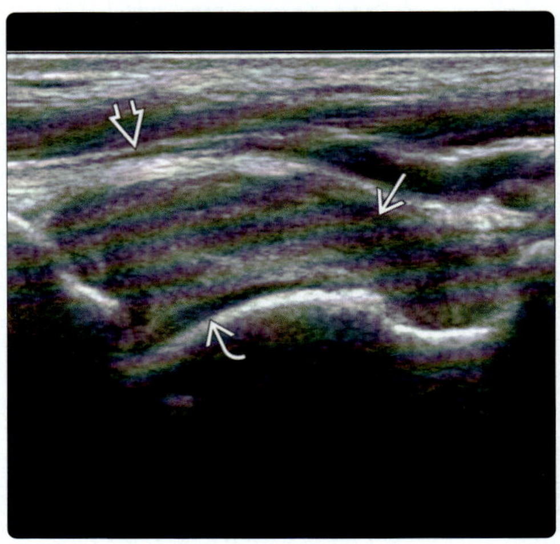

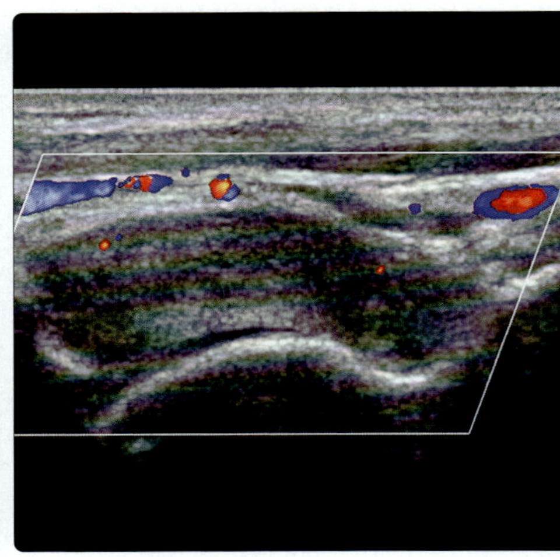

(Left) *Clinical photograph shows a patient with a little toe mass enlarging for 4 years. The skin marker represents the planned line of surgical incision.* **(Right)** *Longitudinal US of the same patient shows a hypoechoic dorsal mass surrounding the extensor digitorum tendon. This tendon does not have a tendon sheath. Histology following percutaneous biopsy revealed TSGCT. This extraarticular TSGCT most likely arose from the metatarsophalangeal joint.*

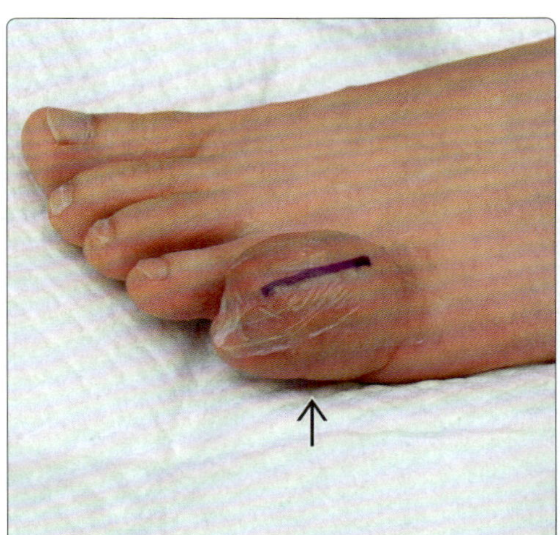

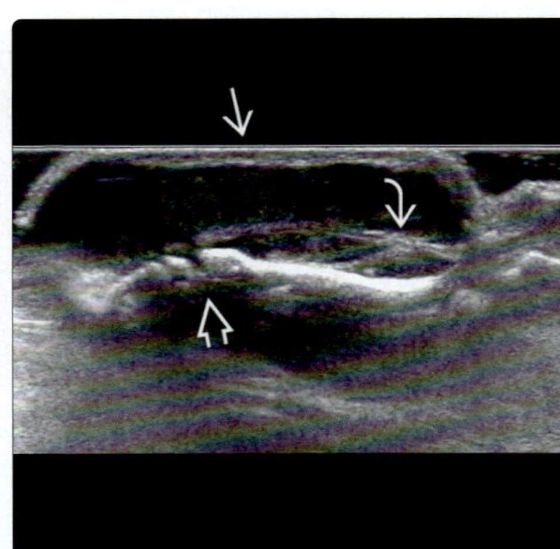

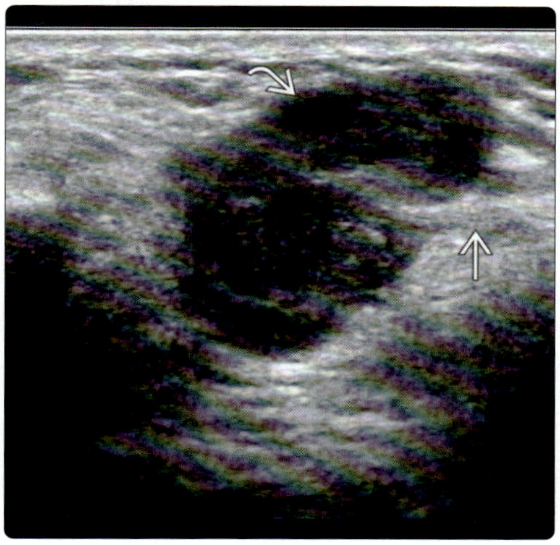

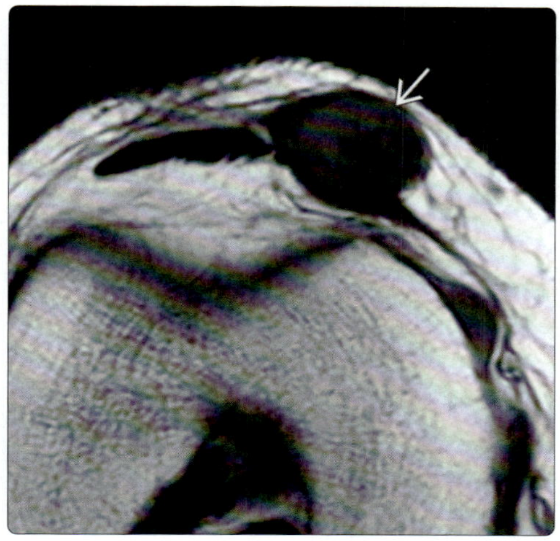

(Left) *Transverse US in a patient with a knee mass present for 6 months shows a medium-sized soft tissue mass ↪ that lies both superficial and deep to the medial patellar retinaculum ➔. Overall appearances favor a focal nodular synovitis.* **(Right)** *Transverse T1 MR of the same patient shows a mass with the characteristic hypointensity of focal nodular synovitis extending through and along the medial retinaculum ➔.*

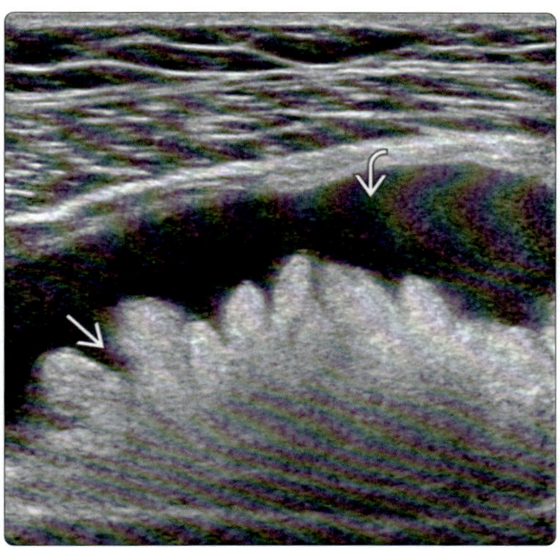

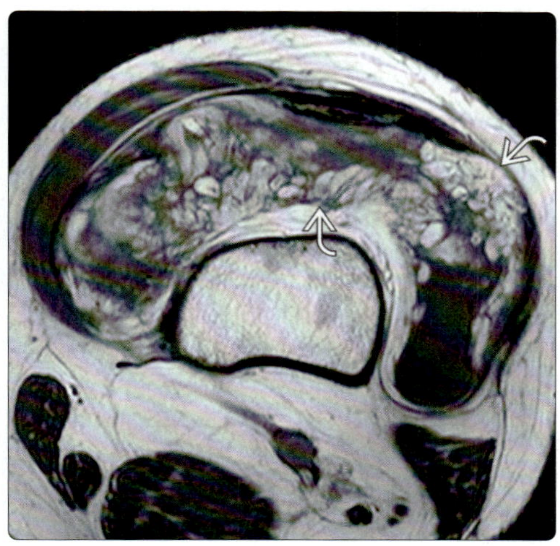

(Left) *Longitudinal US of a patient with chronic knee swelling shows a large effusion ↪ with severe nodular hyperechoic synovial proliferation ➔ on the posterior aspect of the suprapatellar pouch. No free nodules are present.* **(Right)** *Axial T1 MR of the same lesion shows frond-like synovial proliferation ↪ involving much of the suprapatellar pouch. These fronds have a similar intensity to subcutaneous fat.*

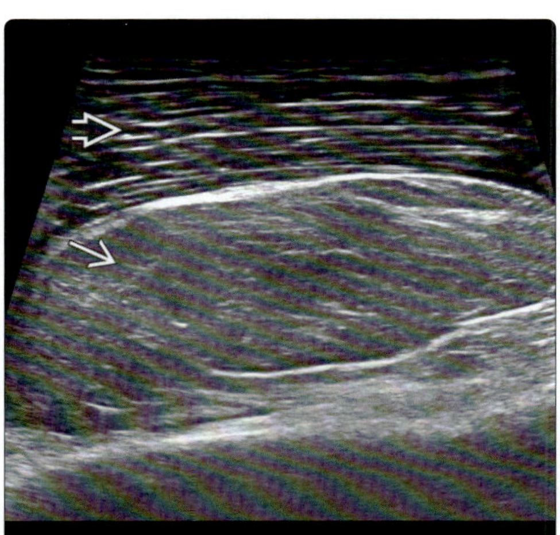

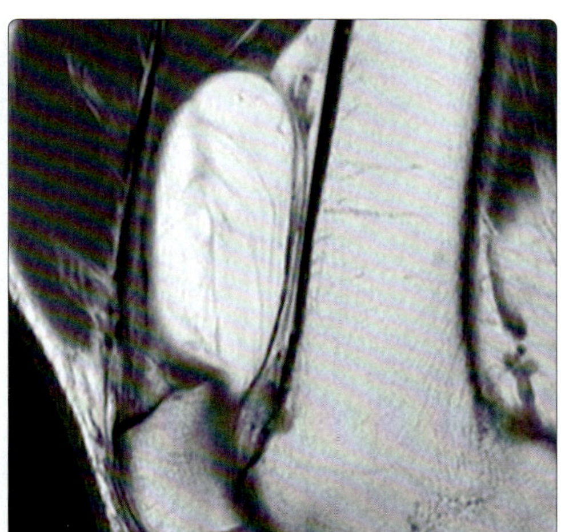

(Left) *Longitudinal US in a 63-year-old man with knee swelling for 4 months is shown. There is a large, intrarticular, synovial-type lipoma ➔ present within the suprapatellar pouch deep to the vastus medialis muscle ➔. The US appearances are similar to those of extraarticular lipomas.* **(Right)** *Sagittal T1 MR shows the synovial lipoma within the suprapatellar recess. Other than the large size and deep location, there are no suspicious features present. In this setting, percutaneous biopsy is not helpful.*

(Left) *Lateral radiograph of the knee in a patient with knee swelling for 20 years shows severe erosion ➡ of the posterior aspect of the lateral femoral condyle and supracondylar area.* **(Right)** *Longitudinal US of the same knee shows a large, lobulated mass ➡ present on the posterior aspect of the knee joint with dorsal bone scalloping ➡. Although some calcific-type foci are present ➡, this was still considered most likely to be pigmented villonodular synovitis (PVNS).*

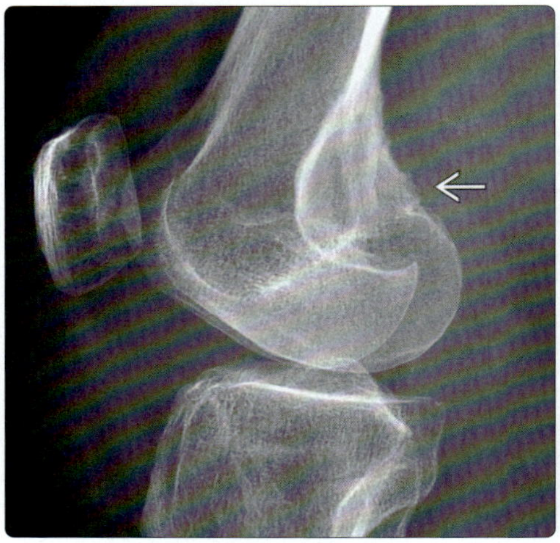

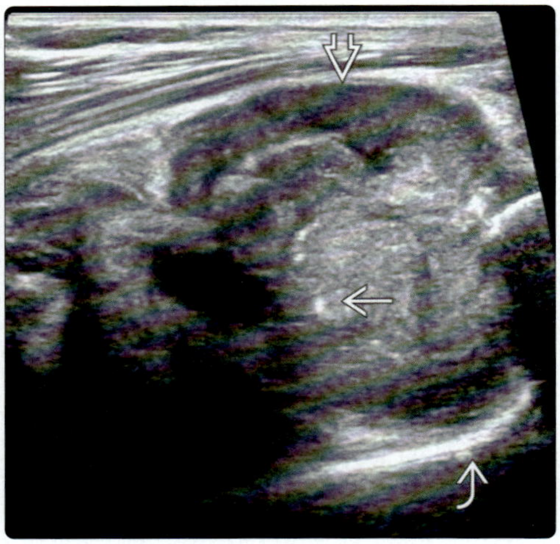

(Left) *Longitudinal US of the same patient shows a separate large, lobulated mass ➡ on the medial aspect of the knee. The medial femoral ⇨ and tibial condyles ➡ are normal. There are possibly some calcific foci ➡ within the mass.* **(Right)** *Transverse color Doppler US of the same patient shows moderate tumoral hyperemia. The US diagnosis was PVNS. However, histology following US-guided biopsy revealed synovial chondrosarcoma.*

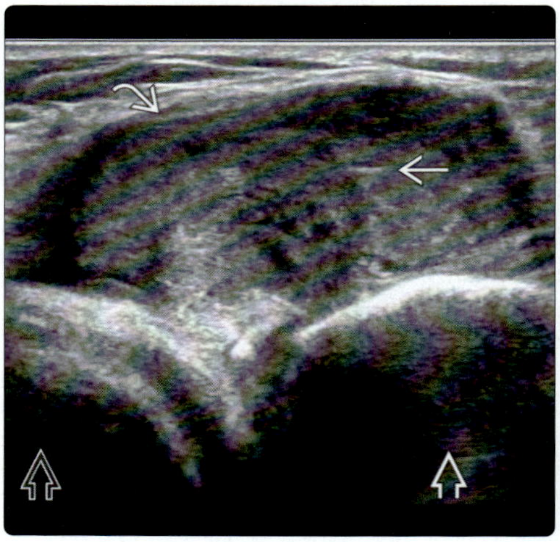

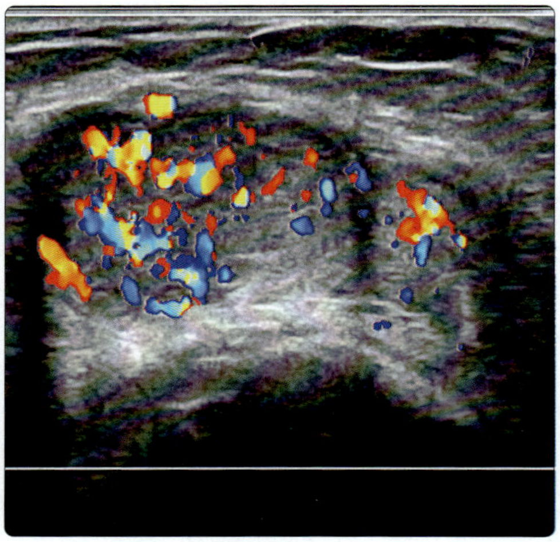

(Left) *Longitudinal US in a patient with knee pain and swelling for 1 year shows thickened, synovial-like tissue ➡ containing small echogenic calcific-like foci ➡ anteromedially, suspicious of synovial osteochondromatosis. US-guided biopsy for histology revealed synovial chondromatosis. The medial femoral ➡ and tibial ➡ condyles are normal.* **(Right)** *Frontal radiograph of the knee of the same patient shows no corresponding soft tissue calcification ➡.*

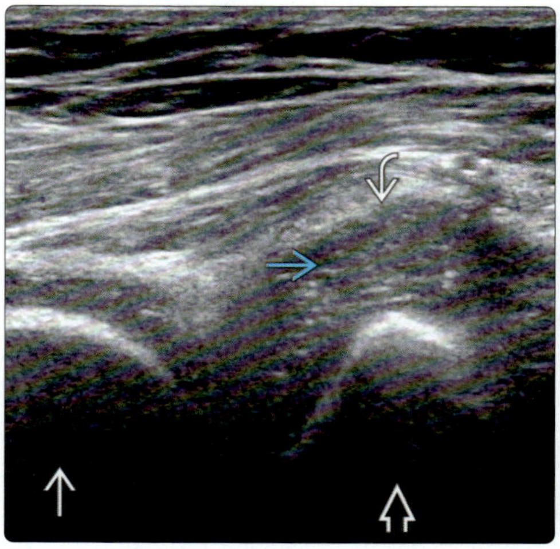

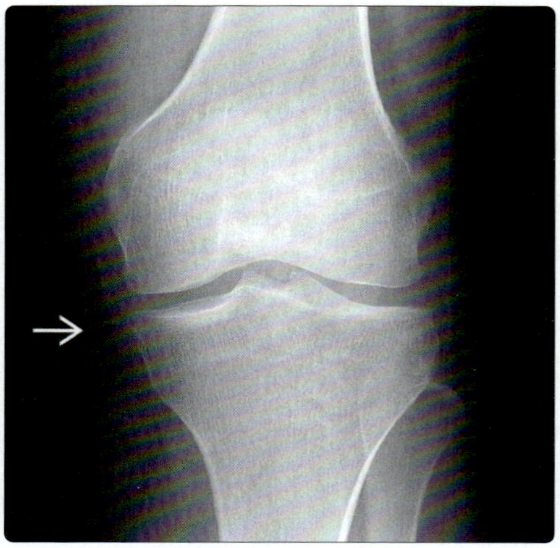

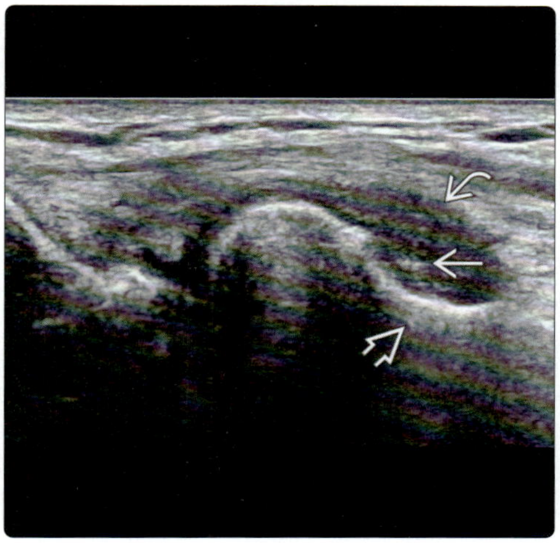

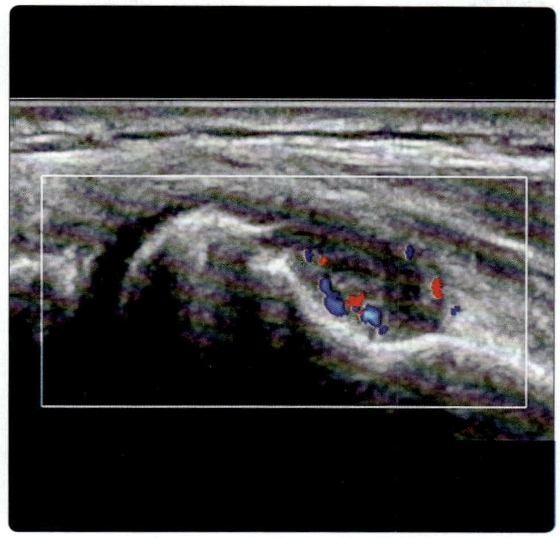

(Left) *Longitudinal US in a patient with elbow pain and limitation of movement for 1 year shows moderate focal synovial proliferation with early scalloping of the radial neck. There is a suspicion of mild synovial calcification.* **(Right)** *Longitudinal color Doppler US of the same area shows mild synovial hyperemia.*

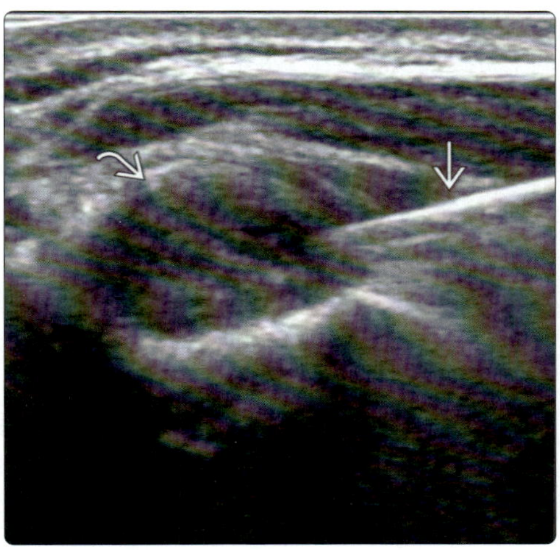

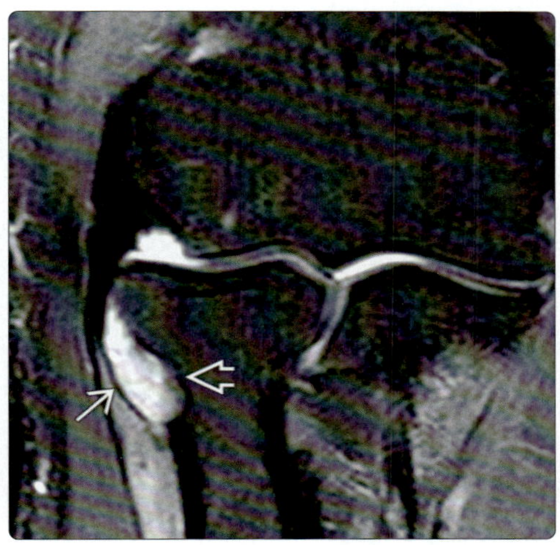

(Left) *Transverse US of the same lesion during US-guided synovial biopsy shows the coaxial needle at the margin of the tumor prior to introduction of the Tru-Cut needle. Histology revealed synovial chondromatosis. As the US appearances are not entirely specific, biopsy is indicated to confirm diagnosis in most synovial tumors.* **(Right)** *Coronal T2 FS MR of the same elbow shows the synovial chondromatosis tumor deep to the radial collateral ligament with moderate scalloping of the radial neck.*

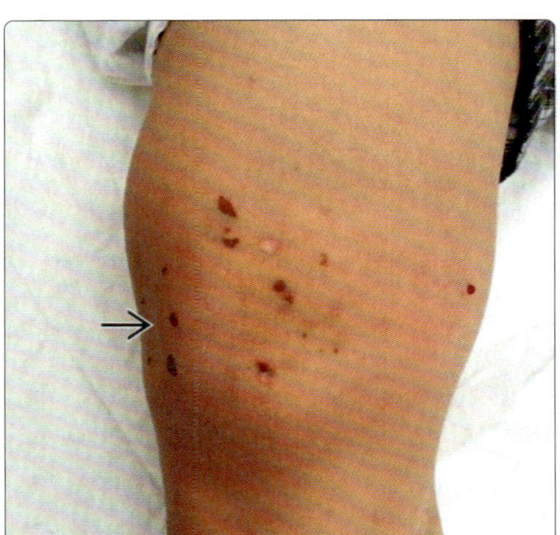

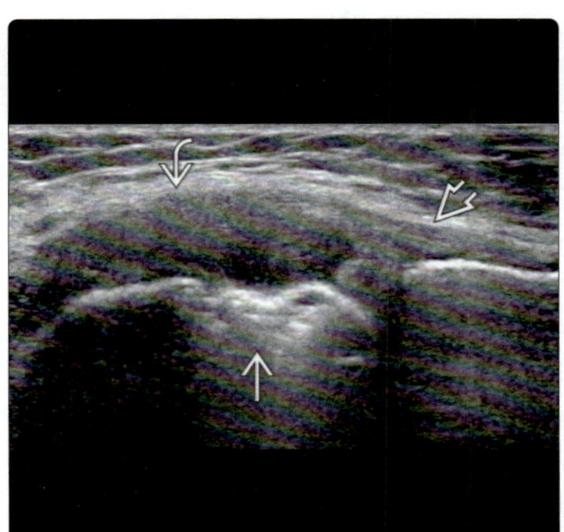

(Left) *Clinical photograph of an otherwise healthy patient with knee pain and swelling after minor sprain 1 month earlier shows some skin petechiae.* **(Right)** *Longitudinal US of the medial aspect of the knee in the same patient shows severe, synovial-like thickening elevating the medial collateral ligament with moderate cortical resorption of the medial femoral condyle.*

(Left) *Transverse US of the same knee shows markedly thickened knee synovium ➡, a small knee joint effusion with additional cortical resorption of the anterior aspect of the medial femoral condyle ➡.* **(Right)** *Axial T2 FS MR of the same knee shows the lymphomatous mass ➡ extending from the medial distal femur with cortical resorption and severe periosteal thickening ➡. There is an enlarged popliteal lymph node ➡.*

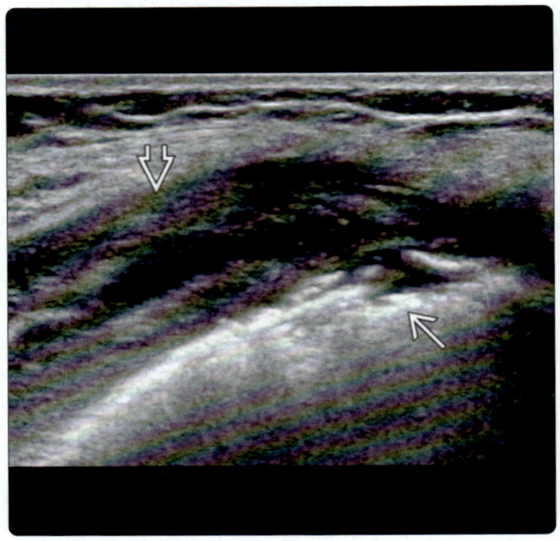

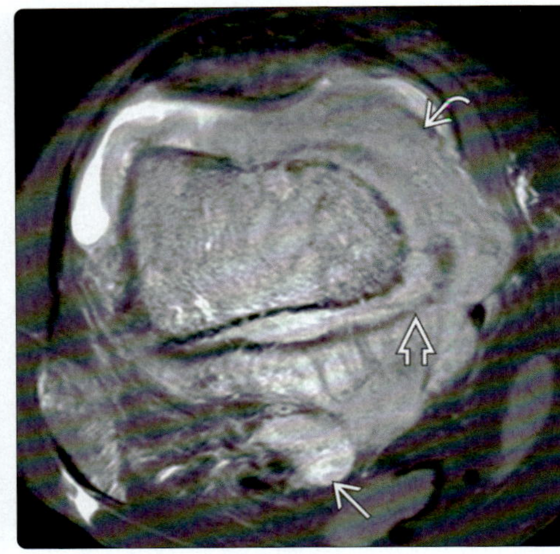

(Left) *Frontal radiograph of the elbow in a patient with elbow pain following mild trauma shows moderate periarticular osteopenia with generalized loss of cortical definition ➡ as well as early marginal erosions ➡.* **(Right)** *Longitudinal US of the anterior elbow of the same patient shows a large, synovial-type mass ➡ on the anteromedial aspect of the elbow. No calcific foci are evident, and there is no joint effusion. The trochlea ➡ and coronoid process ➡ are normal.*

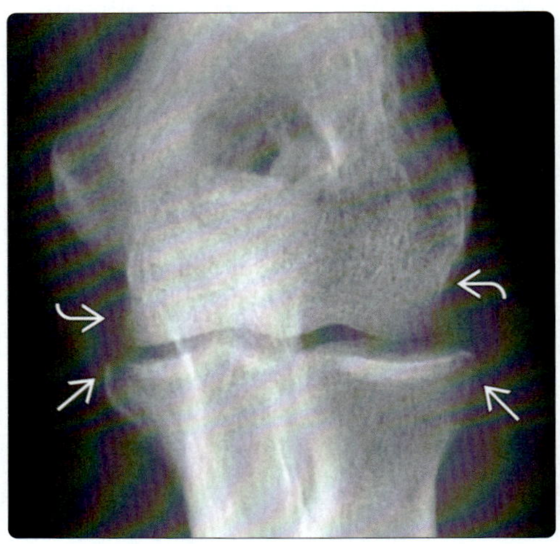

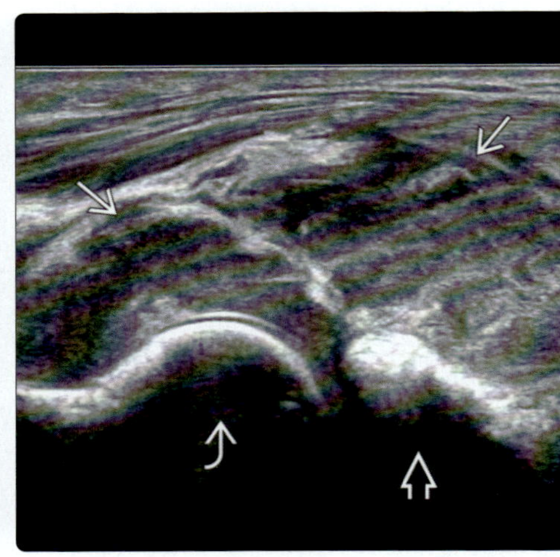

(Left) *Longitudinal US of the posteromedial aspect of the elbow joint of the same patient shows hypoechoic synovial proliferation ➡ deep to the triceps tendon ➡. The distal humerus ➡ is normal.* **(Right)** *Transverse color Doppler US of the posterior aspect of the elbow of the same patient shows severe marginal synovial hyperemia. The appearance would fit a diagnosis of PVNS, but the location of erosions would be atypical. Percutaneous biopsy revealed tuberculosis infection.*

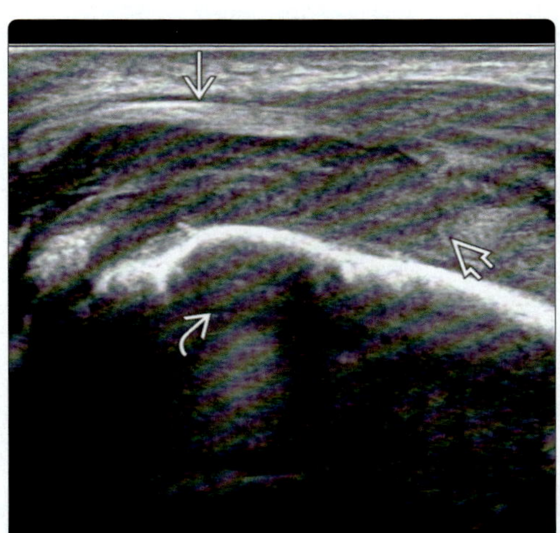

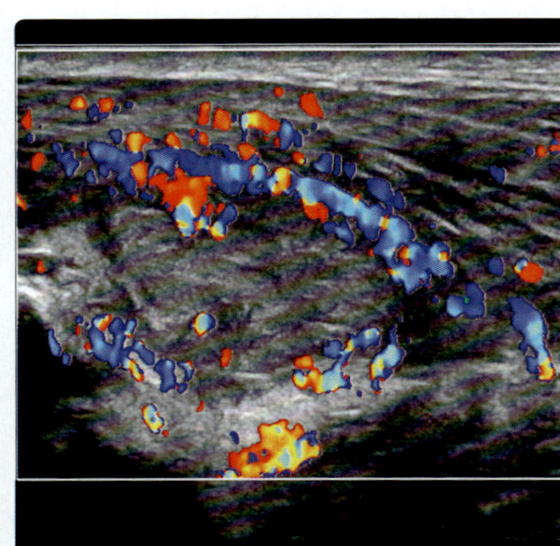

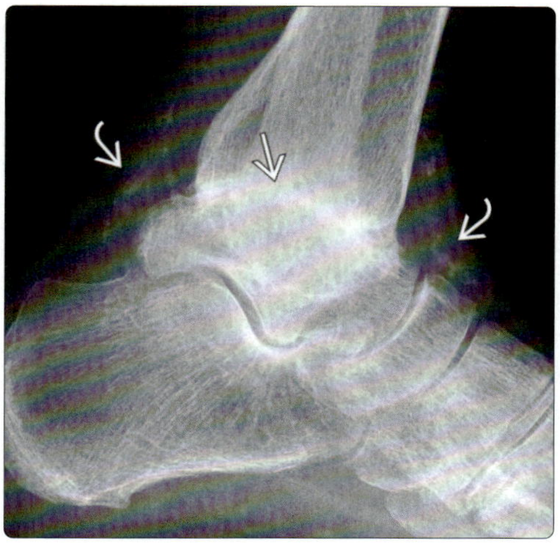

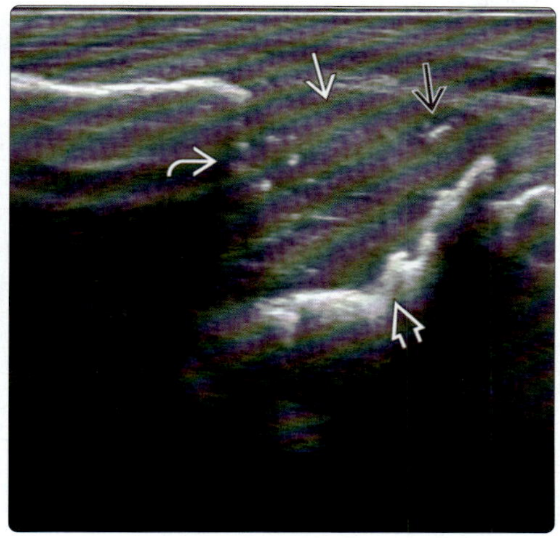

(Left) *Lateral radiograph in a 64-year-old diabetic patient shows severe ankle joint arthropathy ➡ with synovial distension. There is moderate vascular calcification ➡.* **(Right)** *Longitudinal US shows severe synovial proliferation ➡ with severe bony reabsorption of both the distal tibia ➡ and the proximal talus ➡. There is a small amount of osseous debris ➡ embedded in the thickened synovium.*

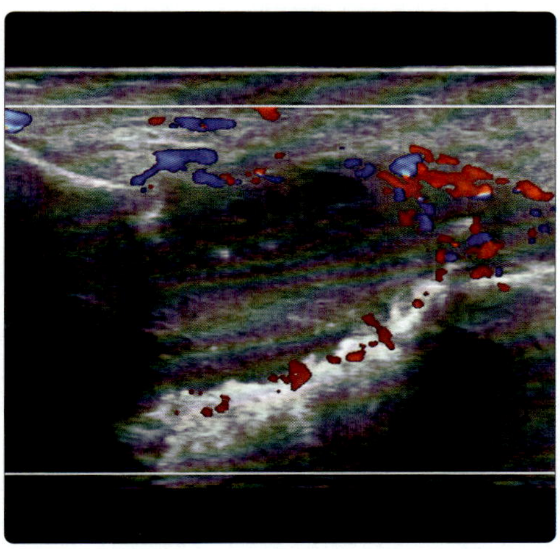

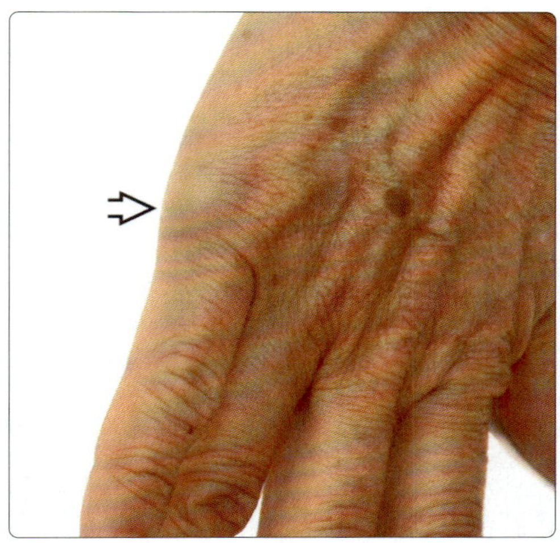

(Left) *Longitudinal color Doppler US shows moderate marginal hyperemia. These appearances with marked arthropathy and bony erosion at initial presentation are more consistent with low-grade septic arthritis than TSGCT. Synovial biopsy showed acute on chronic inflammation with growth of Pseudomonas aeruginosa.* **(Right)** *Clinical photograph of a 62-year-old woman with a 3-month history of nontender swelling ➡ on the lateral aspect of the hand is shown.*

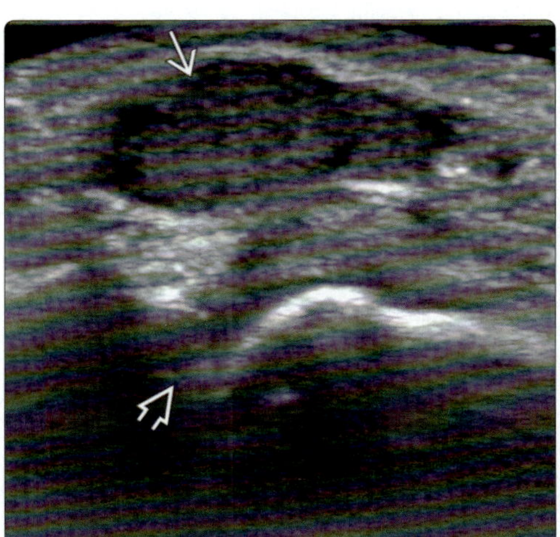

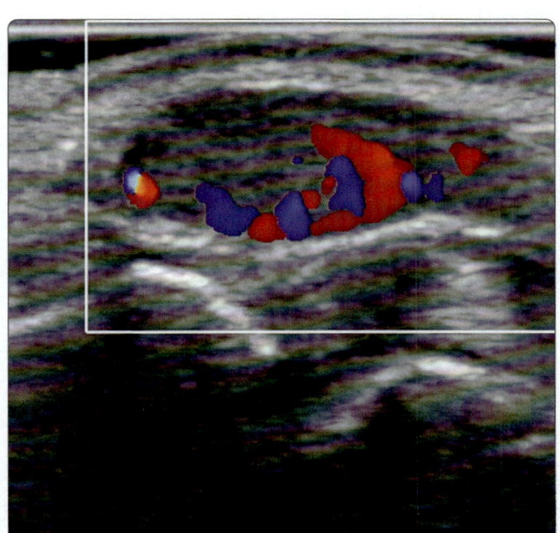

(Left) *Longitudinal US shows a small, well-defined fleshy-type mass ➡ in the subcutaneous tissues on the lateral aspect of the 5th metacarpophalangeal joint ➡. The mass is separate from the joint.* **(Right)** *Color Doppler US shows moderate tumoral hyperemia. Extraarticular TSGCT was considered the most likely possibility, synovial sarcoma less likely, and vascular leiomyoma unlikely. Excisional biopsy, however, revealed lobules of capillary proliferation, congestion, and edema, consistent with a pyogenic granuloma.*

Plantar Fasciitis and Fibromatosis

KEY FACTS

TERMINOLOGY

- Plantar fascia = multilayered fibrous aponeurosis, 1-2 mm thick, with medial, central, and lateral bands

IMAGING

- **Plantar fasciitis**
  - Thickened plantar fascial insertion, especially at medial aspect of calcaneum
  - Plantar fascia thickness > 4.5 mm (most useful sign)
  - ± hypoechogenicity of plantar fascia
  - ± decreased fascial definition
  - ± perifascial edema
  - ± edema, disruption, and atrophy of heel fat pad
  - Spur is located deep to (and not within) fascia
    - Spur may simulate anterior edge of calcaneus
  - Bilateral in 1/3
  - Subclinical disease on opposite side is not uncommon so do not use opposite side as internal normal reference
- **Plantar fibromatosis**
  - Discrete fusiform-shaped nodule of plantar fascia separated from calcaneal insertion
  - Hypoechoic (75%) or isoechoic (25%) to plantar fascia
  - Posterior acoustic enhancement (40%) or shadowing (20%)
  - Intrinsic vascularity (10%)
  - Majority are located within central or medial bands of plantar fascia
  - Bilateral in 1/3 and multiple on symptomatic side in 1/4

DIAGNOSTIC CHECKLIST

- Likelihood of plantar fasciitis increases with increasing thickness of plantar fascia at calcaneal insertion
  - Heel fat pat atrophy is most common differential diagnosis
- Plantar fibromatosis = distinct plantar fascial nodule separate from calcaneal insertion
  - Plantar fascial tear is most similar differential diagnosis

(Left) *Longitudinal US shows a normal central band of the plantar fascia attached to the medial calcaneal tuberosity ➡. The thickness (3.7 mm) of the plantar fascia is measured at the leading edge of the calcaneum ➡. The normal thickness is < 4.3 mm.* **(Right)** *Longitudinal US shows a moderately thickened (5.3 mm) central band of the plantar fascia ➡ at the attachment to the medial calcaneal tuberosity ➡. This is consistent with moderate-severity plantar fasciitis.*

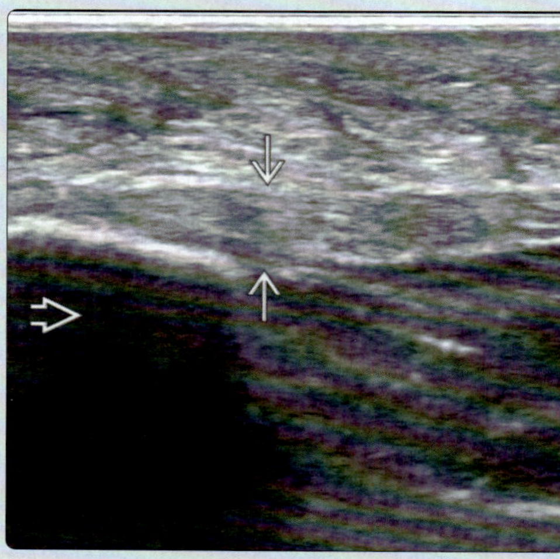

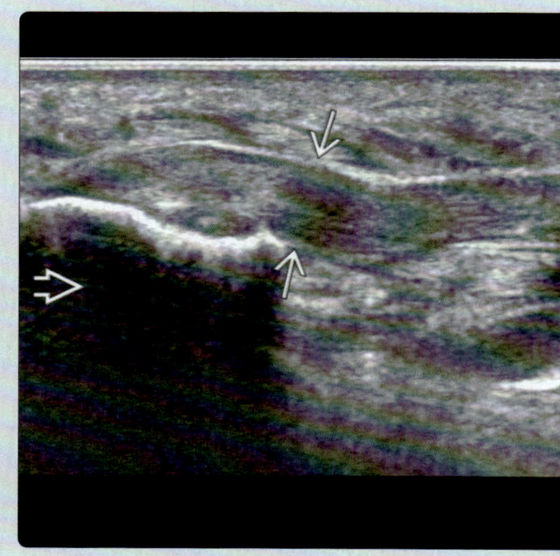

(Left) *Longitudinal US shows a normal plantar fascia ➡ at the midfoot level. The flexor digitorum brevis muscle ➡ is attached to the deep surface of the plantar fascia. The subcutaneous fat of the sole ➡ is superficial.* **(Right)** *Longitudinal US shows a medium-sized (12-mm wide x 4-mm deep x 18-mm long) fibroma ➡ within the cental band of the plantar fascia ➡ at the midfoot region. Moderate posterior enhancement ➡ is present. Additional plantar fibroma was present elsewhere in the feet.*

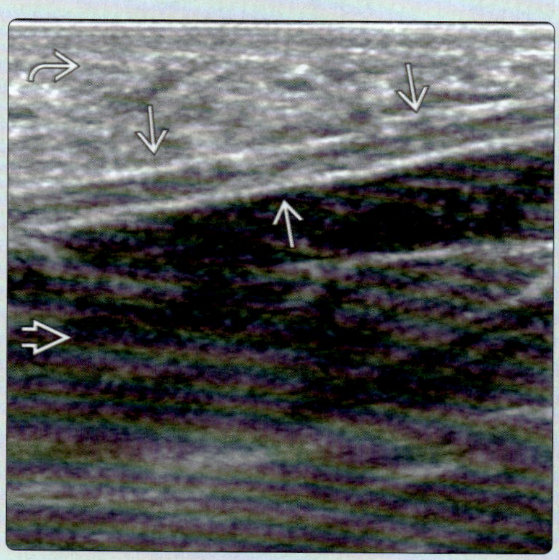

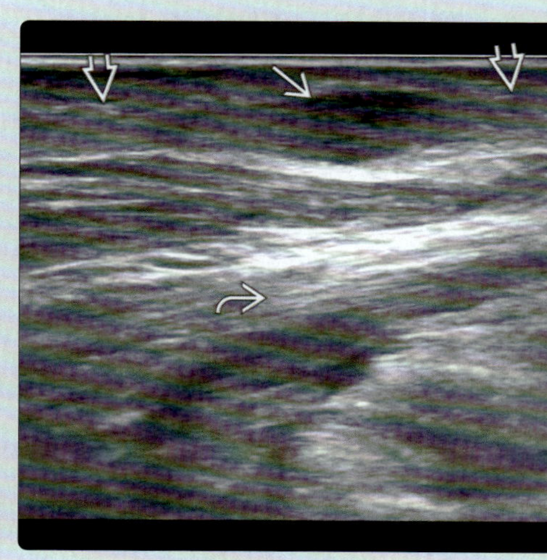

TERMINOLOGY

Synonyms

- Plantar fasciitis: Jogger's heel, tennis heel, policeman's heel, heel spur syndrome, calcaneal spur syndrome, plantar fasciopathy
- Plantar fibromatosis: Ledderhose disease, plantar fibroma

Definitions

- Plantar fascia = multilayered fibrous aponeurosis, 1-2 mm thick, with medial, central, and lateral bands
  - Supports long and short plantar ligaments in maintaining longitudinal arches of foot
    - Innervated by medial calcaneal nerve and 1st branch lateral plantar nerve
      - Entrapment of 1st branch lateral plantar nerve (Baxter neuropathy) can simulate plantar fasciitis
  - Medial band is least significant
    - Inserts into 1st metatarsophalangeal joint
  - Central band is largest; originates from medial tuberosity of calcaneum
    - Divides in forefoot into 5 bands that insert onto plantar plates of metatarsophalangeal joints
  - Lateral band inserts into base of 5th metatarsal bone
- Plantar fasciitis: Degeneration ± fibroblastic proliferation of plantar fascia insertion to medial tuberosity of calcaneus
  - As there is no clear inflammatory component, plantar fasciopathy may be more favorable term
- Plantar fibromatosis: Nodular fibroblastic proliferation of plantar fascia separate from calcaneus

IMAGING

General Features

- Best diagnostic clue
  - Plantar fasciitis: Thickened plantar fascial insertion
  - Plantar fibromatosis: Nodular thickening of plantar fascia
- Location
  - Plantar fasciitis: Medial tuberosity calcaneus
    - Bilateral in 1/3
  - Plantar fibromatosis: Affects medial (60%) or central bands (40%) of plantar fascia
    - Bilateral in 1/3
    - Multiple on symptomatic side in 1/4
- Size
  - Plantar fasciitis: Fascial thickness > 4.5 mm
  - Plantar fibromatosis
    - Average length: 13 mm (range: 4-17 mm)
    - Average width: 10 mm (range: 4-22 mm)
    - Average depth: 4 mm (range: 2-10 mm)
- Morphology
  - Plantar fasciitis: Fascial thickening at calcaneal insertion
  - Plantar fibromatosis: Discrete well-defined plantar fascia nodule

Radiographic Findings

- Radiography
  - Plantar fasciitis: Plantar calcaneal spur present in ~ 50% of patients with plantar fasciitis
    - Also found in asymptomatic subjects
    - Consider enthesopathy when fluffy periostitis, erosion, and sclerosis are present

MR Findings

- T1WI
  - Plantar fasciitis: Hypo - or isointense plantar fascial thickening at calcaneal insertion
  - Plantar fibromatosis: Isointense plantar fascial nodule
- T2WI
  - Plantar fibromatosis: Hypo- to isointense nodule along plantar fascia
- STIR
  - Plantar fasciitis
    - Thickness > 4.5 mm (almost 100%)
    - Subcutaneous or perifascial edema (90%)
    - Insertional bone marrow edema (40%)
    - Markedly insertional marrow edema, erosions, and hyperostosis suggest enthesitis (peripheral spondyloarthropathy)

Ultrasonographic Findings

- **Plantar fasciitis**
  - Plantar fascia thickness > 4.5 mm (most useful sign)
  - ± hypoechogenicity of plantar fascia and loss of normal fibrillar reflective echotexture
  - ± minor dystrophic calcification of fascia
  - ± perifascial edema
  - ± mild fascial hyperemia
  - ± increased stiffness on elastography
  - Calcaneal spur
    - Spur is located deep to (and not within) fascia
    - May simulate anterior edge of calcaneus
- **Plantar fibromatosis**
  - Discrete fusiform-shaped nodule of plantar fascia separated from calcaneal insertion
  - Hypoechoic (75%) or isoechoic (25%) to plantar fascia
  - Posterior acoustic enhancement (40%) or shadowing (20%)
  - Intrinsic vascularity (5%)
  - Majority are located within midsubstance or superficial aspect of plantar fascia
    - If large with infiltrative margins, consider aggressive plantar fibromatosis (relatively rare tumor)
  - Coexistent thickening of ipsilateral plantar fascial calcaneal insertion present in 1/3 of patients without related symptoms

Imaging Recommendations

- Best imaging tool
  - Plantar fasciitis and fibromatosis; US as accurate, quick, and accessible
- Protocol advice
  - Examine fascia in longitudinal and transverse planes
    - Routinely examine both sides as subclinical contralateral disease is common in both diseases
  - Measure thickness of fascia at medial leading edge of calcaneum
  - Routinely access heel fat pad in plantar fasciitis for oedema, internal architecture, and thickness
  - Avoid excessive transducer pressure as this will efface perifascial edema and blanch color Doppler signal

Plantar Fasciitis and Fibromatosis

DIFFERENTIAL DIAGNOSIS

Heel Fat Pad Atrophy

- Focal edema ± atrophy of heel fat pad
  - Often seen in conjunction with plantar fasciitis

Calcaneal Stress or Insufficiency Fracture

- Stress fracture: Abnormal stress on normal bone
- Insufficiency fracture: Normal stress on abnormal bone
- Best assessed by MR

Plantar Bursitis

- Fluid-filled bursa or flat hypoechoic area superficial to plantar fascia
- Typically forefoot region under metatarsal heads

Plantar Fascial Tear

- Trauma history
- Fascial discontinuity ± swelling, edema, or hemorrhage

PATHOLOGY

General Features

- Etiology
  - Plantar fasciitis: Population lifelong risk of 10%
    - Possibly caused by repetitive microtrauma or microvascular injury
  - Plantar fibromatosis: Unknown etiology; not knowingly related to trauma
    - Possibly genetic ± related to collagen profile
- Associated abnormalities
  - May be related to other fibroproliferative diseases (Dupuytren contracture, Peyronie disease, etc.)
    - Onset of other fibromatoses may be delayed for 5-10 years after onset of plantar fibromatosis

Gross Pathologic & Surgical Features

- Plantar fasciitis: Firm thickening of plantar fascia at calcaneal insertion
- Plantar fibromatosis: Firm whitish nodule

Microscopic Features

- Plantar fasciitis: Degeneration of plantar fascia ± fibroblastic proliferation ± mild chronic inflammation
- Plantar fibromatosis: 3 phases of fibroblastic proliferation within plantar fascia
  - Proliferative phase: Nodular fibroblastic proliferation
  - Active phase: Collagen synthesis and deposition
  - Mature phase: Collagen maturation

CLINICAL ISSUES

Presentation

- Most common signs/symptoms
  - Plantar fasciitis: Heel pain ↑ by weightbearing
    - Pain worse on 1st few steps in morning
  - Plantar fibromatosis: Sole pain ± palpable firm lump
    - Pain ↑ by prolonged weightbearing
- Clinical profile
  - Plantar fasciitis: Tenderness on deep palpation just anterior to medial calcaneal tuberosity
  - Plantar fibromatosis: Nontender or mildly tender sole lump attached to plantar fascia

Demographics

- Age
  - Plantar fasciitis: 40-60 years
    - Younger in at risk individuals (runners, military)
  - Plantar fibromatosis: 50 years (range: 30-80 years)
- Gender
  - Plantar fasciitis and fibromatosis: M:F = 1:1
- Epidemiology
  - Plantar fasciitis: Most common cause of heel pain
    - 10% of running injuries
    - Additional risk factors: Mechanical overload, prolonged standing, obesity, pes planus, and tight Achilles tendon
  - Plantar fibromatosis: Less common than plantar fasciitis
    - No specific risk factors identified

Natural History & Prognosis

- Plantar fasciitis
  - Symptoms resolve in 80% within 12 months
- Palmar fibromatosis
  - Similar tendency to improve with time
    - Nodules become smaller and lesser painful

Treatment

- Plantar fasciitis
  - Conservative: Start treatment soon after onset
    - Rest and analgesia
    - ± orthotics splints and shoe inserts
    - ± stretching of plantar fascia
    - ± night splints to keep ankle in neutral position
  - Steroid ± long-acting local anesthetic injection
    - Usually not possible to inject directly into plantar fascia; instead, inject into perifascial tissues
    - Only temporary relief for ~ 4 weeks
  - Extracorporeal shock wave therapy
  - Surgery (for patients with persistent symptoms)
    - Open or endoscopic release of plantar fascia
- Plantar fibromatosis
  - Conservative
    - Shoe inserts
    - ± steroid ± long-acting local anesthetic injection
  - Surgery (for very painful nodules)
    - Recurrence common after surgery

DIAGNOSTIC CHECKLIST

Consider

- Aggressive plantar fibromatosis if plantar fascia nodule has infiltrative superficial or deep margins (rare)

Image Interpretation Pearls

- Likeliness of fasciitis increases with increasing thickness of plantar fascia at calcaneal insertion
- Plantar fibromatosis = distinct plantar fascial nodule separate from calcaneal insertion

SELECTED REFERENCES

1. Draghi F et al: Imaging of plantar fascia disorders: findings on plain radiography, ultrasound and magnetic resonance imaging. Insights Imaging. 8(1):69-78, 2017

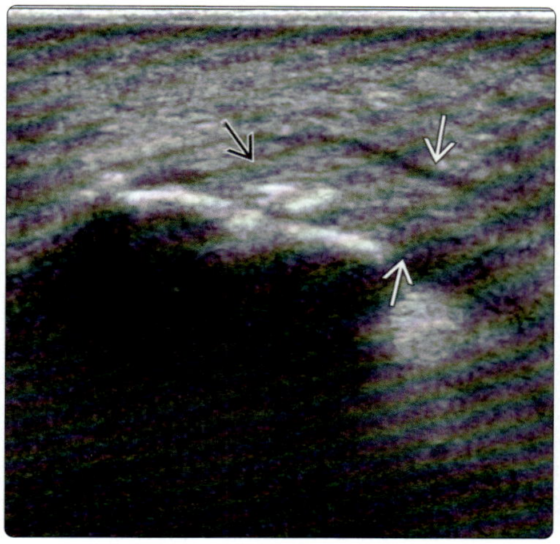

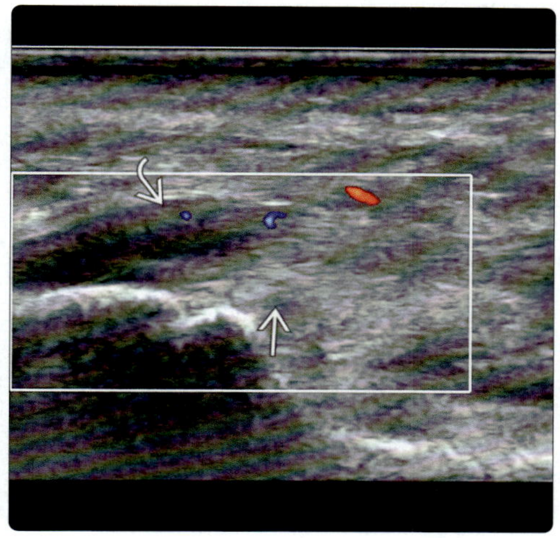

(Left) *Longitudinal US shows a mildly thickened (5.2 mm) plantar fascia* ➡ *at the medial part of the calcaneum. There is moderate localized dystrophic calcification* ➡ *within the plantar fascia at this location.* **(Right)** *Longitudinal color Doppler US shows mild hyperemia* ➡ *present within a moderately thickened (5.7 mm) plantar fascia* ➡ *at the medial calcaneal tuberosity. Plantar fascial hyperemia is uncommon and may be an indication of more active disease.*

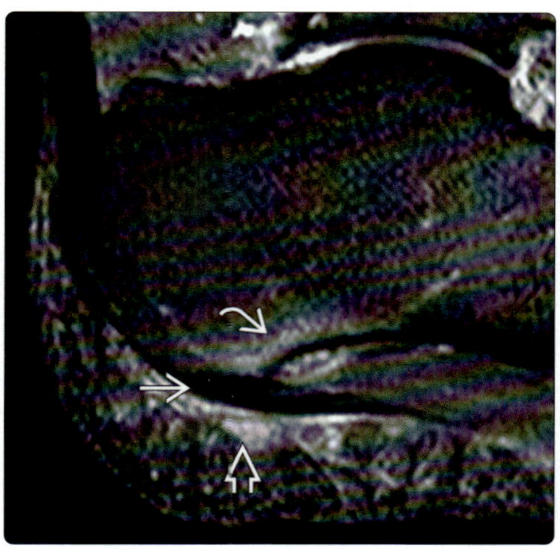

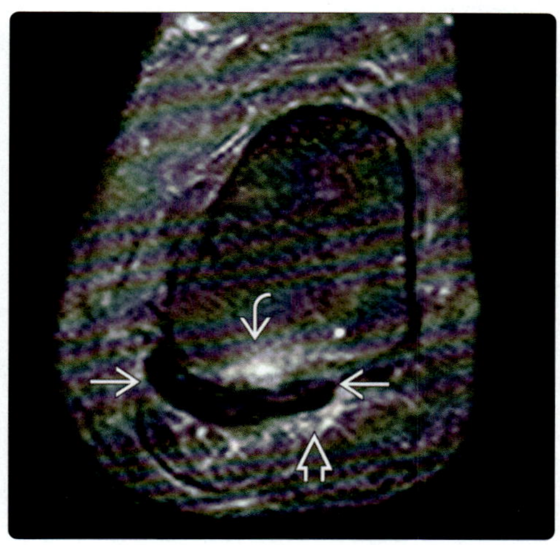

(Left) *Sagittal T2 FS MR shows mild thickening (5 mm) of the plantar fascia* ➡ *at the medial calcaneal insertion. There is mild perifascial edema* ➡ *and mild adjacent subcortical bone marrow edema* ➡*.* **(Right)** *Coronal T2 FS MR of the same patient shows thickening of the plantar fascia* ➡ *at the medial calcaneal insertion. There is mild perifascial edema* ➡ *and mild subfascial bone marrow edema* ➡*.*

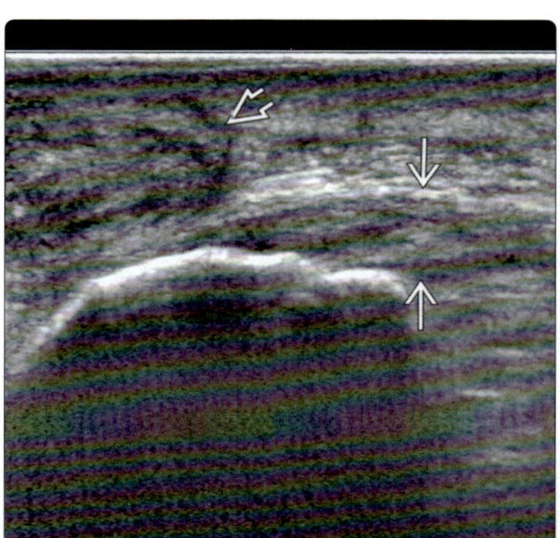

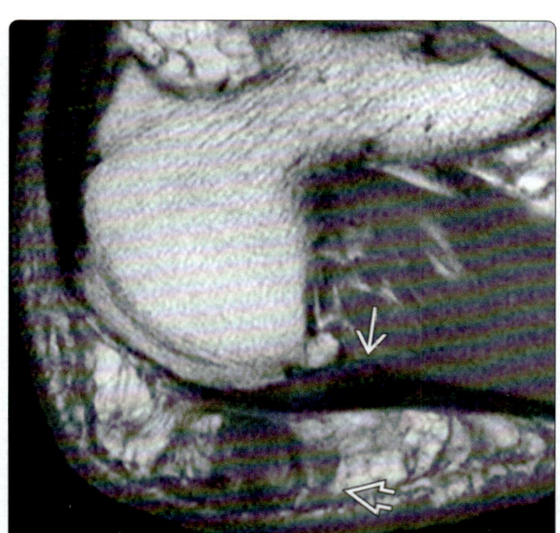

(Left) *Longitudinal US shows a mildly (4.7 mm) thickened plantar fascia* ➡ *at the medial calcaneal tuberosity. There is a nodular mass-like area* ➡ *in the heel fat pad due to fat necrosis and induration. The heel fat pads should also be routinely assessed when examining this area.* **(Right)** *Sagittal T1 MR at almost the same location as previous image shows a moderately thickened plantar fascia* ➡ *with nodular edema and fibrosis of the heel fat pad* ➡*. The normal thick lobular septa of the heel fat pad are disrupted.*

(Left) *Longitudinal US of the base of 5th metatarsal shows bony hyperostosis and thickening at the insertion of the normally thin lateral band of the plantar fascia. This is indicative of an enthesopathy associated with a peripheral spondyloarthropathy.* **(Right)** *Longitudinal color Doppler US at the same location shows moderate hyperaemia indicative of high disease activity. The peroneus brevis tendon is also inserted separately into the base of the 5th metatarsal bone, though was normal in this patient.*

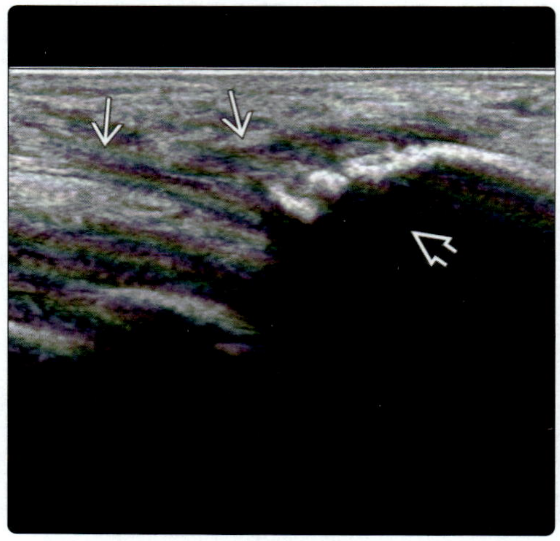

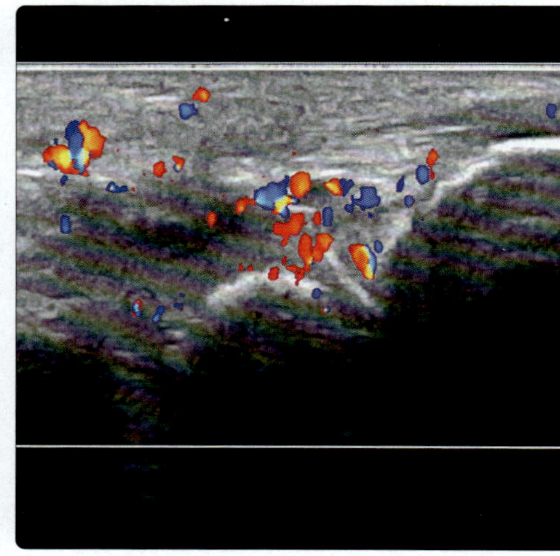

(Left) *Longitudinal graphic shows a large plantar fibroma arising from the plantar fascia.* **(Right)** *Clinical photograph shows a solitary small nodule due to a plantar fibroma on the medial aspect of the midsole. Most plantar fibromas occur in the midfoot region and are often multiple and bilateral.*

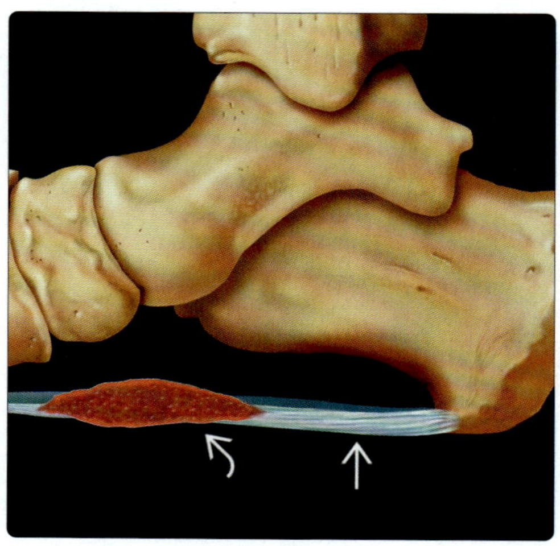

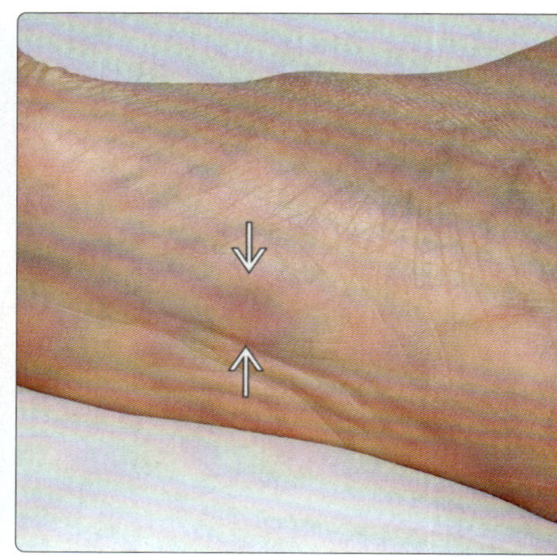

(Left) *Longitudinal US shows a medium-sized (13-mm wide x 3.4-mm deep x 13-mm long) hypoechoic plantar fibroma arising from the plantar fascia on the medial aspect of the midfoot. No hyperaemia was present.* **(Right)** *Longitudinal US shows a small hypoechoic plantar fibroma arising from the plantar fascia. This plantar fibroma primarily arises from the more superficial fibers of the plantar fascia.*

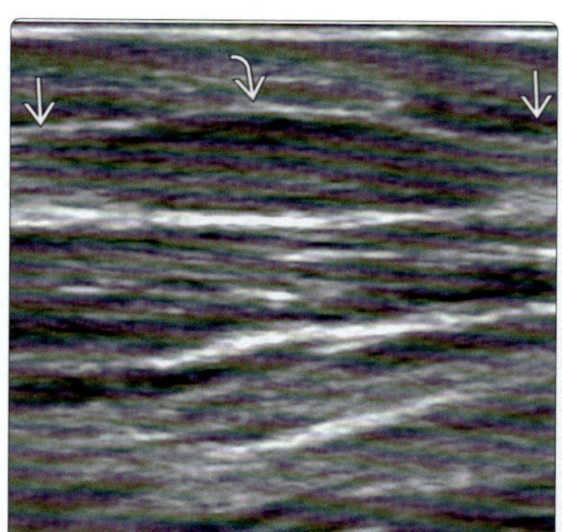

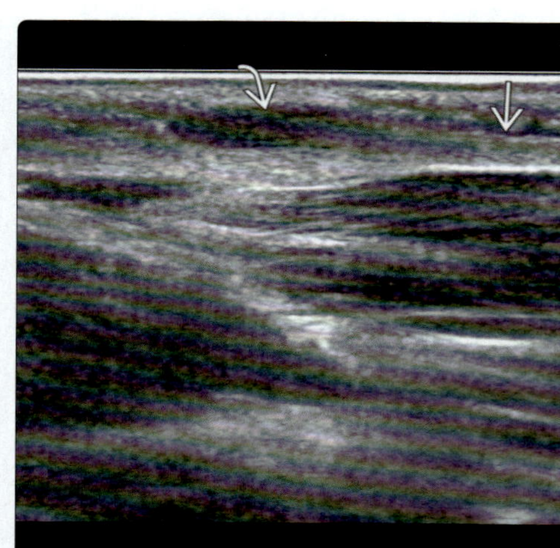

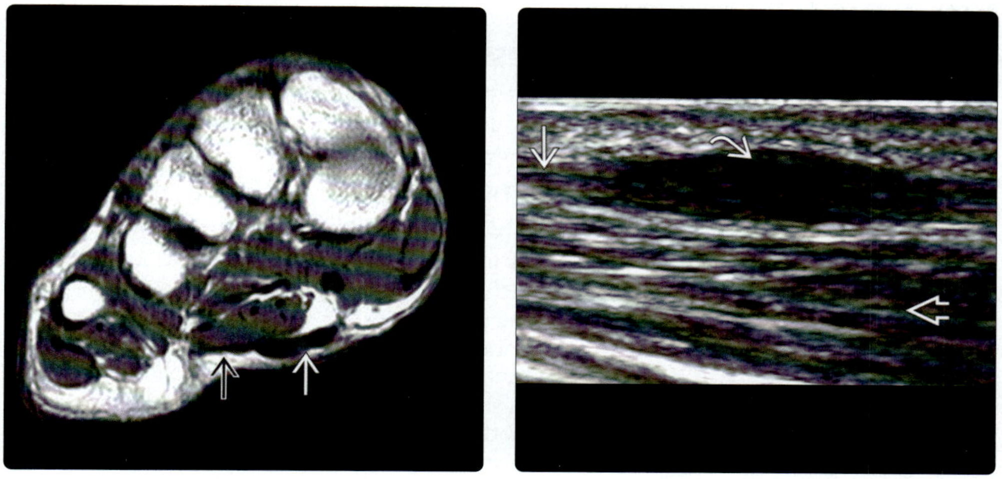

(Left) *Coronal T1 MR shows localized nodular thickening of the medial band of the plantar fascia in the midfoot region consistent with plantar fibromatosis. The lateral band of the plantar fascia is not thickened.* **(Right)** *Longitudinal US of the midfoot region shows a well-defined, fusiform-shaped hypoechoic nodule arising from the plantar fascia consistent with a plantar fibroma. Note the minimal posterior acoustic enhancement deep to the fibroma.*

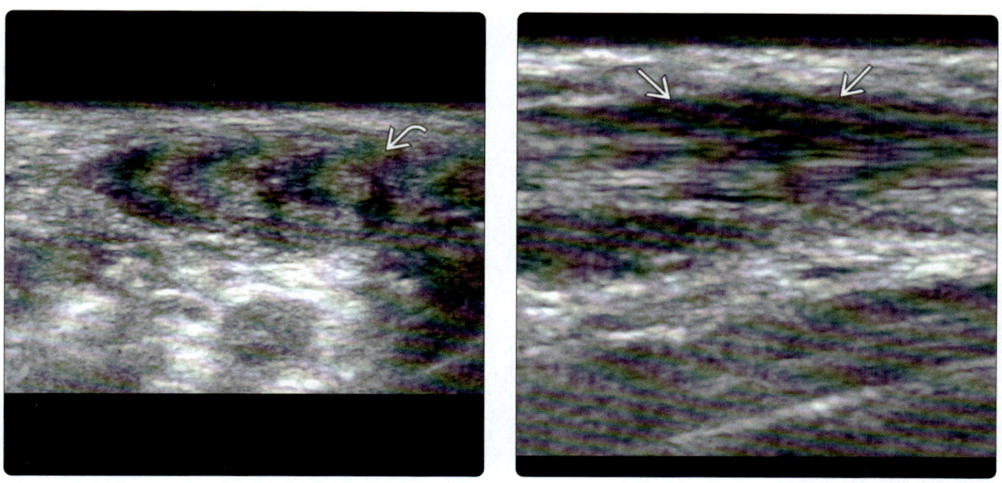

(Left) *Transverse US shows a large oval-shaped, heterogeneous, echogenic plantar fibroma. Although the fibroma is large, it is still well-defined and not infiltrative in nature.* **(Right)** *Longitudinal US shows a small, ill-defined plantar fibroma. The fascial margins of the fibroma are not well-defined. In this case, the more superficial fibers of the plantar fibroma are particularly affected.*

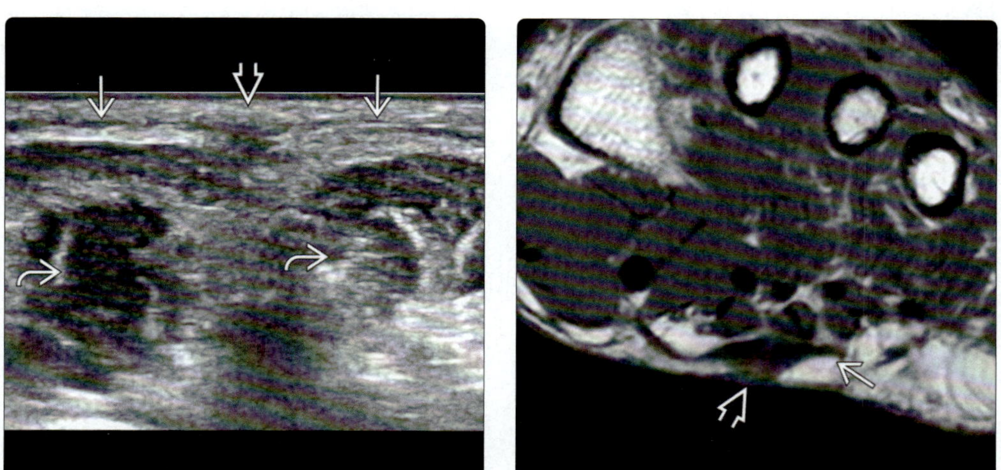

(Left) *Transverse US of the midfoot in a patient with trauma 2 months earlier shows a healing penetrating tear of the midportion of the plantar fascia. This should not be misdiagnosed as a plantar fibroma. The muscle is deep to the plantar fascia.* **(Right)** *Axial T1 MR at the same location shows a healing tear of the midportion of the plantar fascia with reparative fibrous tissue extending from the plantar fascial tear to the skin.*

Lipoma

KEY FACTS

IMAGING

- Subcutaneous, most common
  - Well-defined, oblong-shaped encapsulated mass with fine linear striations parallel to long axis of tumor
    - Of variable echogenicity depending on cellularity and fat-water content
  - Absent or minimal internal vascularity on color Doppler imaging
  - Compressible similar to adjacent fat
- Lipomas in patients with multiple lipomas typically appear different to those in patients with 1 or 2 lipomas
  - Multiple lipomas tend to be more rounded, more echogenic, with less striations, and less well encapsulated
  - Tend to be located in trunk and proximal parts of appendicular skeleton
  - Known as Dercum disease (adiposis dolorosa) if lipomas painful or tender
- Subfascial lipomas are more variable in appearance
  - Nevertheless, appearances usually specific enough to allow confident US diagnosis

CLINICAL ISSUES

- Most common soft tissue tumor
- Soft, well-defined, painless mass enlarging over months, often during period of weight gain

DIAGNOSTIC CHECKLIST

- US findings are usually characteristic
- MR is useful for assessment of large, deep-seated lipomas
- MR is better than US at differentiating between benign or malignant lipomas
  - Large size, deep location, and heterogeneity are best indicators of liposarcoma
  - However, not all heterogeneity within lipoma indicates liposarcoma
  - Calcification, small nonfatty areas, myxoid tissue, and fat necrosis can be features of atypical lipoma and lipoma variant but are not necessarily diagnostic of liposarcoma

(Left) *Clinical photograph shows a large mass on the anterior aspect of the proximal arm.* **(Right)** *Longitudinal US with extended field of view of the same patient shows a large hyperechoic subcutaneous lipoma causing focal contour bulging of the skin. The lipoma contains fine linear striation parallel to the skin. The underlying deltoid muscle is normal.*

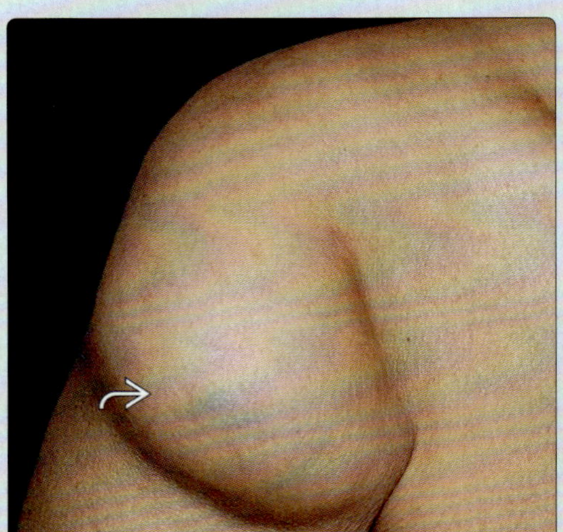

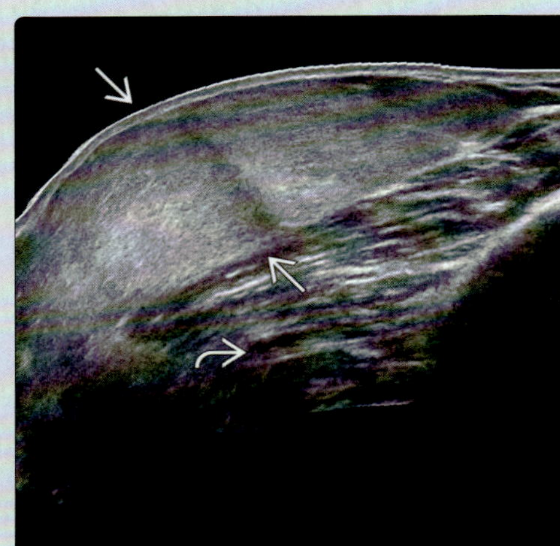

(Left) *Longitudinal US shows a small, well-defined homogeneously hyperechoic subcutaneous lipoma. The fine striations are barely perceivable, though these appearances are still characteristic of a lipoma. Such lipomas are often multiple and tend to be located more centrally than peripherally.* **(Right)** *Longitudinal US shows an intramuscular hyperechoic lipoma within the vastus lateralis muscle. Note the well-defined hyperechoic capsule and the fine linear horizontal striations.*

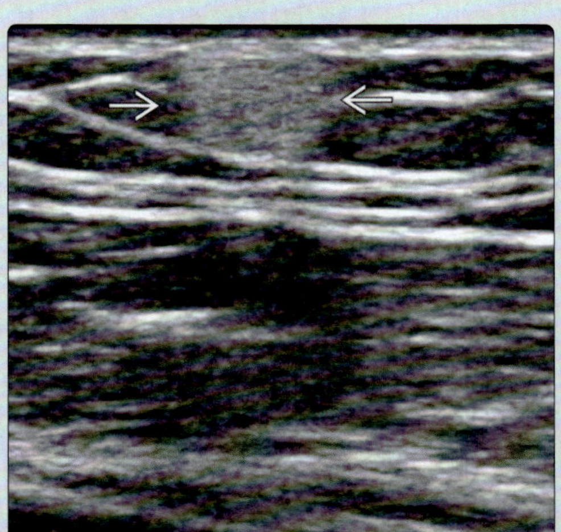

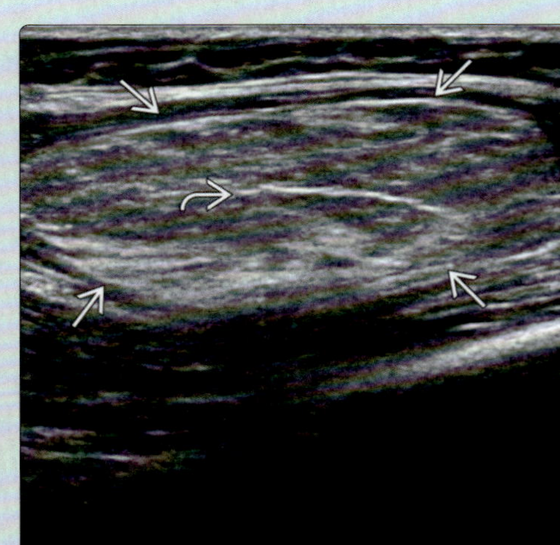

Lipoma

TERMINOLOGY

Synonyms

- Fatty tumor

Definitions

- **Benign**
  - Lipoma: Benign soft tissue tumor composed of mature adipose tissue
  - Angiolipoma: Lipoma in which small capillary-sized vessels containing microthrombi transverse tumor
  - Lipomatosis: Diffuse infiltrative overgrowth of mature adipose tissue, completely or partially encapsulated
  - Lipomatosis of nerve: Median > > > other nerves ± localized mesenchymal overgrowth
  - Spindle cell lipoma: Mixture of fat and spindle cells; more common in head and neck subcutaneous tissues and adult males
  - Lipoblastoma: Admixture of mature and immature adipocytes; more common in extremities and children < 3 years old
  - Hibernoma: Tumor of brown (thermogenic) fat similar to that found in hibernating animals; > subcutaneous tissues of adults, especially lower neck, mediastinum, and upper dorsal region
  - Chondroid lipoma: Adult females, shoulder or pelvic girdle, may calcify
  - Myolipoma: Benign tumor exhibiting features of mature smooth muscles and mature adipose tissue
  - Lipoma arborescens: Subsynovial fatty proliferation, knee > > > ankle or hip; may occur in tendon sheath
- Intermediate
  - Well-differentiated liposarcoma: Slow-growing, usually deep-seated tumor that does not metastasize but is prone to local recurrence; dedifferentiation risk: 6%
- Malignant
  - Dedifferentiated liposarcoma: Deep-seated and large with significant (> 25%) nonfatty component; malignant potential
  - Myxoid liposarcoma: Myxoid tumor either low grade (< 5% round cell component) or high grade (> 5% round cell component)
  - Pleomorphic liposarcoma: Large mass with myxoid, necrotic, fatty, and hemorrhagic components

IMAGING

General Features

- Best diagnostic clue
  - Echogenic mass with fine internal echoes aligned parallel to long axis of tumor
- Location
  - Subcutaneous, most common
    - Posterior trunk, extremities, neck
    - Multifocal: 10%
  - Subfascial (deep to investing fascia)
    - Intramuscular, intermuscular, supramuscular (i.e., subfascial), and submuscular (i.e., juxtacortical) locations
  - Lipomas do not transverse investing fascia
- Size
  - < 5 cm: 80%; > 10 cm: Uncommon
- Morphology
  - Well-defined, homogeneous, fatty soft tissue mass

Radiographic Findings

- Radiography
  - Hypodense soft tissue mass ± calcification or ossification occasionally

CT Findings

- NECT
  - Well-defined, homogeneous soft tissue mass of low attenuation (-65 to -120 HU)

MR Findings

- T1WI
  - Well-defined mass with high signal intensity
  - Fine septa
  - MR is more reliable than US at depicting features suspicious for malignancy
    - Thick (> 2 mm) irregular septa
    - Nodular or globular areas of nonfatty tissue
    - Well-differentiated liposarcoma usually composed of > 75% fatty component
- T2WI
  - Intermediate signal intensity
- STIR
  - Low signal intensity (fat signal suppressed)
- T1WI C+
  - Intravenous contrast is usually not required if no malignant features on noncontrast imaging

Ultrasonographic Findings

- **Solitary lipoma**
  - Variable echogenicity depending on cellularity and fat-water content
  - Well-defined, encapsulated mass, usually iso- to slightly hyperechoic to adjacent fat
    - Variable echogenicity depending on cellularity and fat-water content
  - Distinctive, fine linear striations parallel to skin
  - Typically fusiform or oblong in shape
  - Encapsulated
    - Helps differentiate lipoma from lipohypertrophy
  - Acoustic enhancement is not feature of superficial lipoma
  - Compressible similar to adjacent fat
  - May calcify or ossify infrequently
  - Absent internal vascularity on color Doppler imaging
    - Consider angiolipoma if mild to moderate internal vascularity is present
    - Large lipoma may encase vessels; this feature should not be confused with angiolipoma
- **Multiple lipomas**
  - Lipomas in patients with multiple lipomas typically appear different to those in patients with 1 or 2 lipomas
  - Lipomas tend to be more rounded, more echogenic, with less striations, and less well encapsulated
    - Tend to be located in trunk and proximal parts of appendicular skeleton
    - Known as Dercum disease (adiposis dolorosa) if lipomas painful or tender

- No need for histological confirmation or further imaging for nearly all superficial lipomas
- Subfascial lipomas are more variable in appearance
  - Still have fine striations parallel to long axis of tumor
  - Iso- or hyperechoic to adjacent muscle
  - Usually show acoustic enhancement (acoustic transmission of fat > muscle)
  - ± minimal to mild vascularity, more vascular than subcutaneous lipoma
  - US can usually indicate diagnosis
    - ± MR to exclude features of liposarcoma
      - MR better at helping differentiate lipoma from liposarcoma
  - Large size, increasing size, deep location, and heterogeneity are best indicators of liposarcoma
    - However, not all heterogeneity within tumor necessarily implies liposarcoma
      - Calcification, small nonfatty areas, myxoid tissue, and fat necrosis can be features of atypical lipoma and lipoma variant but are not necessarily diagnostic of liposarcoma

Imaging Recommendations

- Best imaging tool
  - US as diagnostic in most cases
- Protocol advice
  - Avoid transducer pressure but also check compressibility
  - Look for encapsulation and relationship to investing fascia
  - Realize that US appearance of lipomas in patients with multiple lipomas is different
  - Routinely evaluate with color Doppler imaging

DIFFERENTIAL DIAGNOSIS

Lipohypertrophy

- Localized ↑ in subcutaneous fat thickness
  - Nonencapsulated hypertrophied fat gradually merges with normal thickness surrounding fat
  - Similar to lipomatosis except that lipomatosis is more diffuse and ↑ in both superficial and deep fat

Fat Trauma

- Focal subcutaneous fat edema
  - No discrete mass
  - Trauma may be trivial (and forgotten) and precede onset of mass by days to weeks
  - Occurs at sites prone to trauma (buttock, thigh, and shin)

Vascular Malformation

- Often have large fatty stromal component and can be difficult to distinguish from lipoma, especially angiolipoma
  - Not encapsulated and usually more poorly defined and less mass-like than angiolipomas
  - More vascular than angiolipomas, ± phleboliths
  - Lipoma is generally more common in older age group than vascular malformation

PATHOLOGY

General Features

- Etiology
  - Mesenchymal neoplasm
  - More common in overweight subjects and diabetics
    - Clinical emergence of lipoma often coincides with period of weight gain
- Associated abnormalities
  - Familial multiple lipomas
    - Autosomal dominant inheritance
- Lipoma fat is similar to normal body fat
  - Sheltered from systemic metabolism; therefore, lipoma does not change size with weight loss

Gross Pathologic & Surgical Features

- Histologic types
  - Benign: Lipoma, angiolipoma, lipoblastoma, myolipoma, spindle cell lipoma, chondroid lipoma, lipomatosis of nerve, lipoma arborescens, lipomatosis
  - Intermediate: Well-differentiated liposarcoma
  - Malignant: Differentiated liposarcoma, myxoid liposarcoma, pleomorphic liposarcoma

CLINICAL ISSUES

Presentation

- Most common signs/symptoms
  - Compressible painless soft mass slowly growing over weeks to months
- Other signs/symptoms
  - May compress nerves, e.g., carpal tunnel
  - Deep-seated tumors can have firm consistency and simulate sarcoma
    - Deep-seated tumors can become very large before becoming clinically apparent

Demographics

- Age
  - Peak: 40-60 years
- Epidemiology
  - Most common soft tissue tumor
  - 1% of population

Natural History & Prognosis

- Slow-growing mass that becomes latent over time

Treatment

- Excision of large, symptomatic, or suspicious lesions
- Liposuction or lipolysis

DIAGNOSTIC CHECKLIST

Consider

- US usually diagnostic for superficial lipomas
- MR useful for assessment of large, deep-seated lipoma

Image Interpretation Pearls

- Fine internal striations parallel to long axis of tumor = characteristic US finding
- Patients with multiple lipomas may not have this feature

SELECTED REFERENCES

1. Gupta P et al: Spectrum of fat-containing soft-tissue masses at MR imaging: the common, the uncommon, the characteristic, and the sometimes confusing. Radiographics. 36(3):753-66, 2016

Lipoma

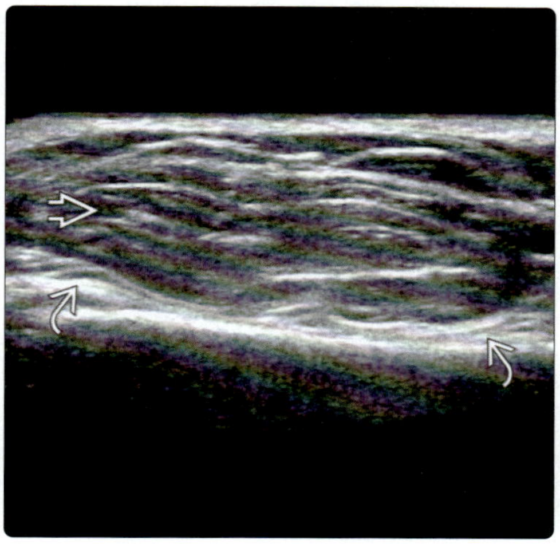

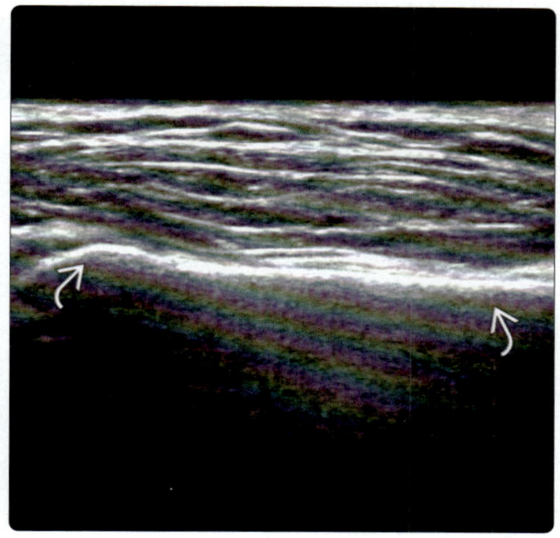

(Left) *Longitudinal US shows a lobulated, oblong-shaped subcutaneous mass isoechoic to adjacent fat with distinct internal horizontal striations aligned parallel to the skin. These appearances are entirely consistent with a subcutaneous lipoma.* **(Right)** *Longitudinal US of the same lesion upon probe compression shows the soft consistency of lipoma, which is easily compressible using mild transducer compression.*

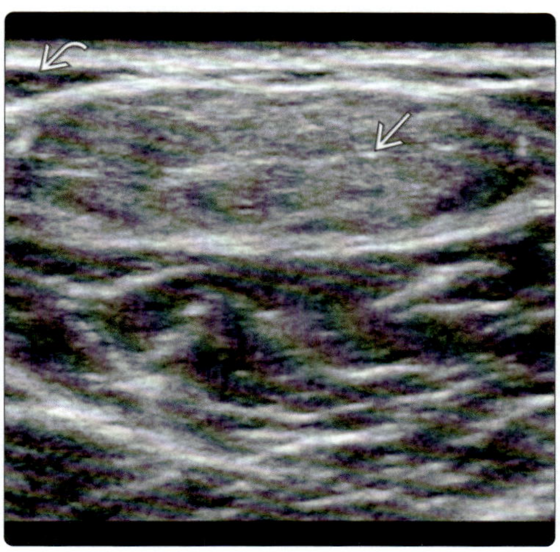

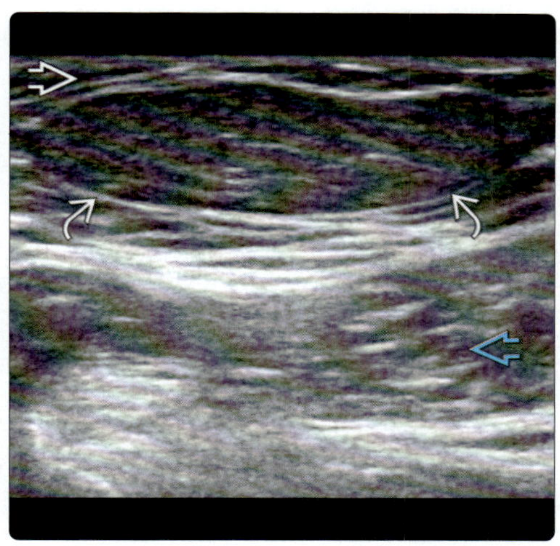

(Left) *Longitudinal US shows another encapsulated lipoma with fine linear internal striations parallel to the long axis of the tumor. This lipoma is hyperechoic to the surrounding subcutaneous fat.* **(Right)** *Longitudinal US shows a subcutaneous lipoma isoechoic to adjacent subcutaneous fat. The echogenicity of lipomas varies and depends on the internal cellularity and fat-water content. Most subcutaneous lipomas tend to be isoechoic to adjacent fat and hypoechoic to muscle.*

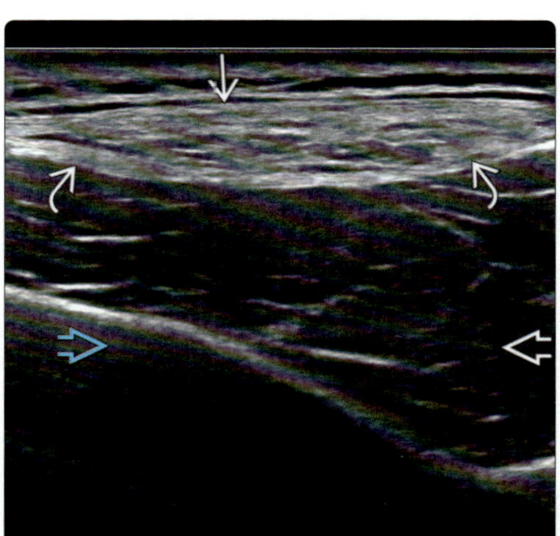

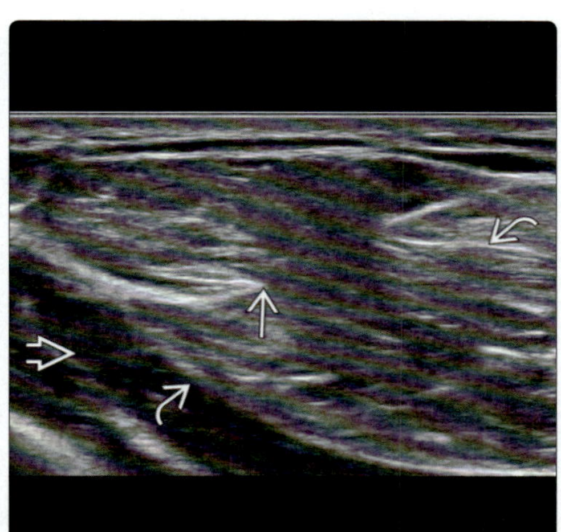

(Left) *Longitudinal US shows a hyperechoic subfascial lipoma on the posterior aspect of the shoulder. The lipoma is located just deep to the investing fascia and indents the underlying infraspinatus muscle. Note the scapula deep to the infraspinatus muscle.* **(Right)** *Longitudinal US shows a subfascial lipoma in the arm herniating into the subcutaneous fat via a fascia defect. Note the deltoid muscle. Such a transfascial herniation is very uncommon.*

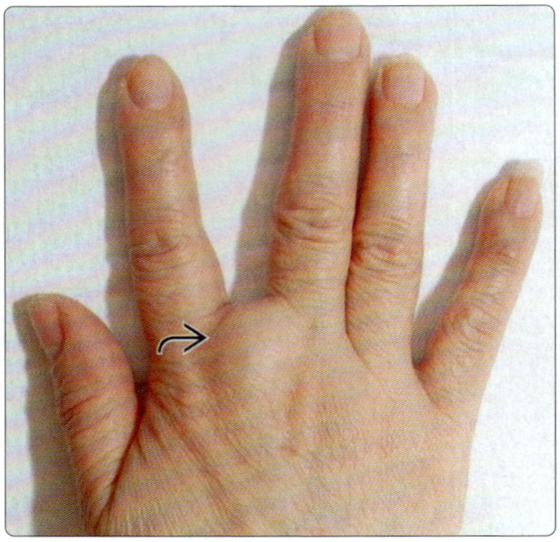

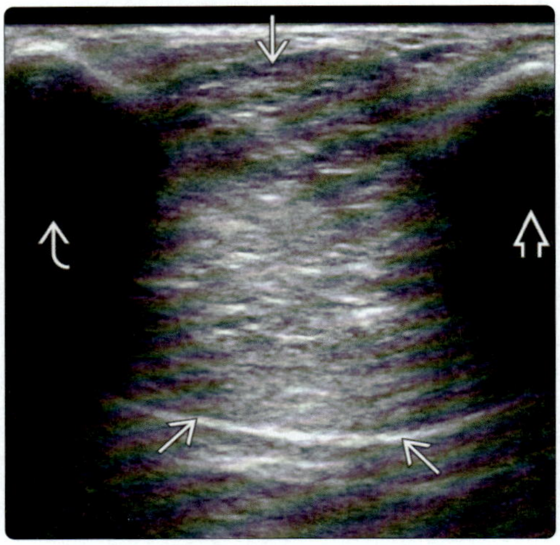

(Left) *Clinical photograph shows a mass ▭ in the 2nd web space between the index and middle fingers.* **(Right)** *Transverse US of the same patient shows a well-defined, dumbbell-shaped lipomatous mass ▭ in the web space between the 2nd ▭ and 3rd ▭ metacarpal bones. Note the fine linear striations parallel to the skin within the tumor, which are indicative of a lipoma.*

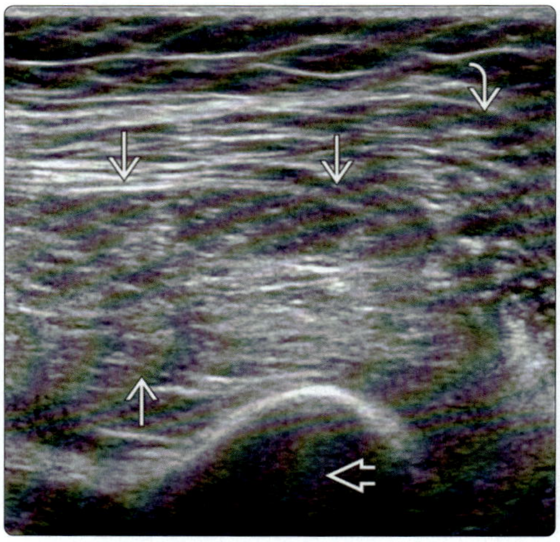

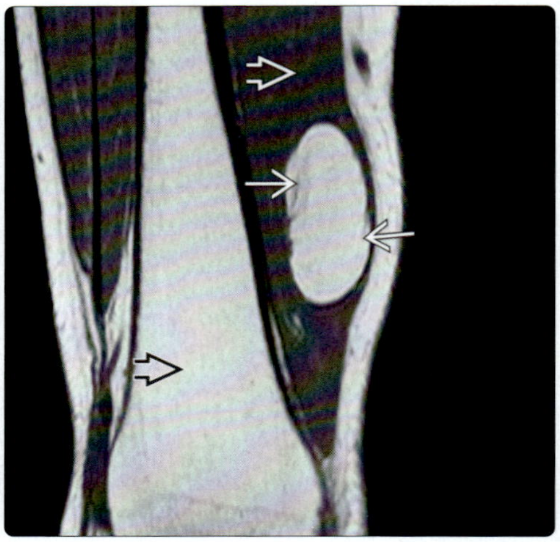

(Left) *Transverse US shows an intramuscular lipoma ▭ within the vastus medialis muscle ▭ close to the distal femur ▭. The lipoma is isoechoic to muscle.* **(Right)** *Coronal T1WI MR of the same patient shows the well-defined lipoma within the vastus medialis muscle ▭. Note the thin internal septations ▭. The appearances are consistent with an intramuscular lipoma. The femur ▭ is normal.*

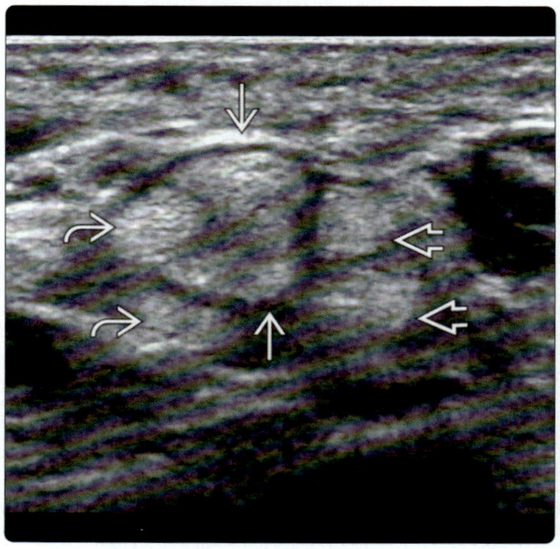

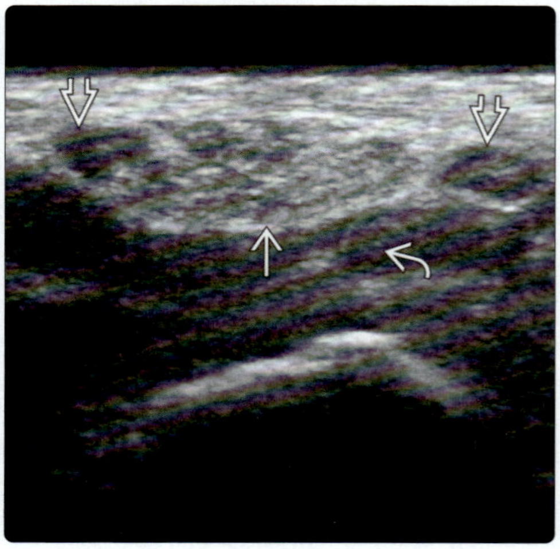

(Left) *Transverse US of the wrist shows a hyperechoic lipoma ▭ located between the flexor tendons of the middle finger ▭ and the ring finger ▭. The superficial flexor tendon of the middle finger is partially encased.* **(Right)** *Transverse US of the wrist shows a lipoma ▭ located between and separating both components ▭ of a bifid median nerve. The lipoma lies superficial to the flexor tendons ▭.*

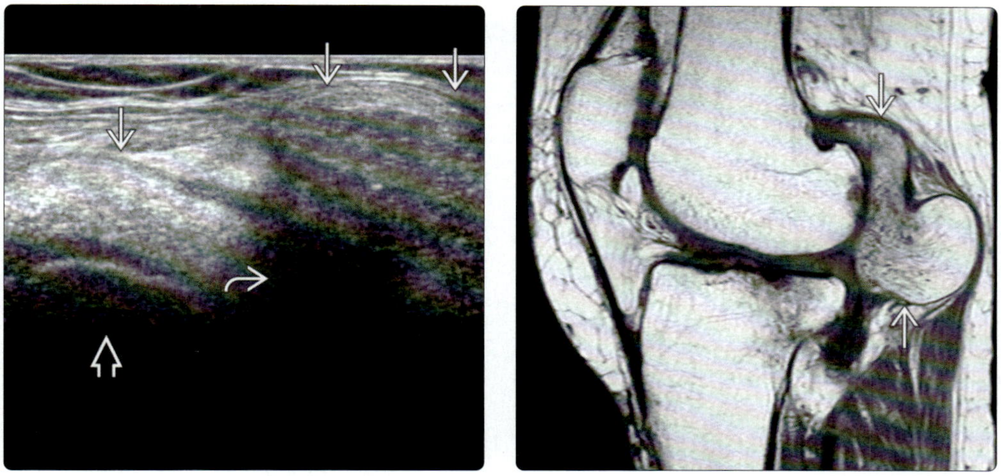

(Left) *Longitudinal US shows a lipomatous-type mass at the posterior aspect of the knee joint. The large size of the mass and the associated acoustic shadowing hinders full assessment of deep or intraarticular involvement. Note the femoral condyle.* **(Right)** *Longitudinal T1 MR of the same lesion shows the mass to be an intraarticular lipoma bulging into the posterior joint capsule. The full extent of this large lipoma is best appreciated on MR imaging.*

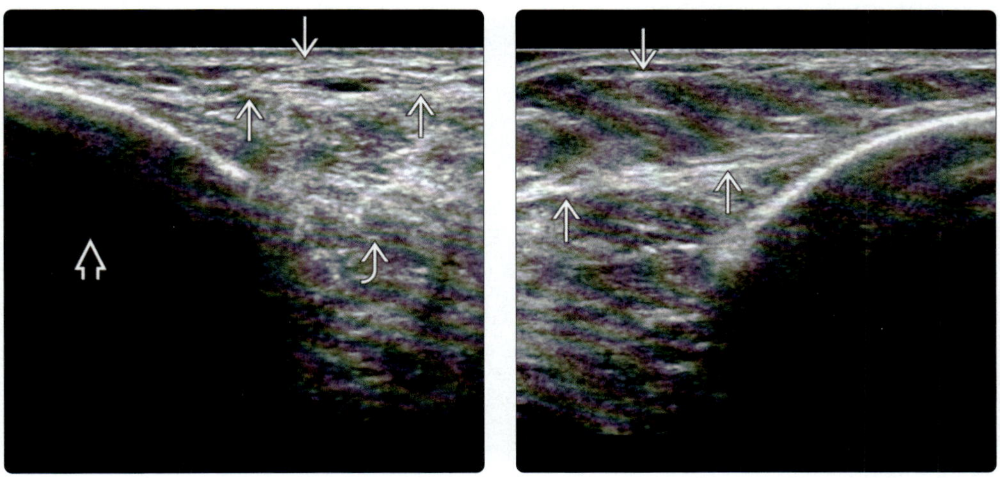

(Left) *Transverse US of the shin shows the normal appearance of the subcutaneous fat layer. Note the tibia and anterior tibialis muscle.* **(Right)** *Transverse US of the contralateral shin of the same patient shows moderate diffuse thickening of the subcutaneous fat without a discrete mass being present. This is consistent with localized subcutaneous lipohypertrophy. Comparison with the same area on the contralateral limb is very helpful in such cases.*

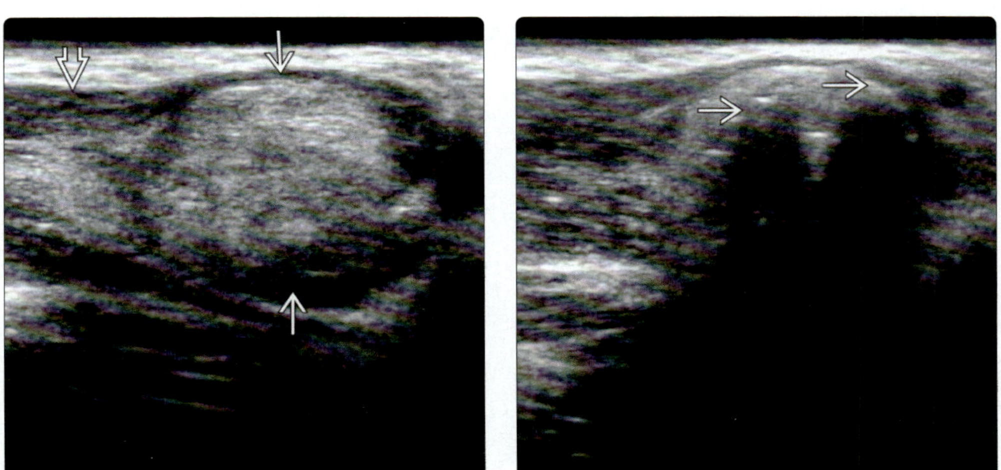

(Left) *Longitudinal US of the hand shows a hyperechoic lipomatous mass within the interosseous muscle between the thumb and index finger metacarpal bones. Note the fine linear striations consistent with an intramuscular lipoma.* **(Right)** *Longitudinal US of the same tumor shows multiple, small, calcific foci within the lipoma with posterior shadowing. The appearances are consistent with a calcified intramuscular lipoma.*

(Left) *Transverse US of the groin region shows a moderately heterogeneous subcutaneous lipomatous mass with linear echogenic foci* →*. The differential diagnosis is atypical lipoma, fibrolipoma, or slow-flow vascular malformation.* **(Right)** *Transverse CT of the same lesion shows a lipomatous mass with a slightly thickened peripheral rim* → *consistent with an atypical lipoma. Histology revealed a lipoma with central fat necrosis, calcification, and fibrinous stranding.*

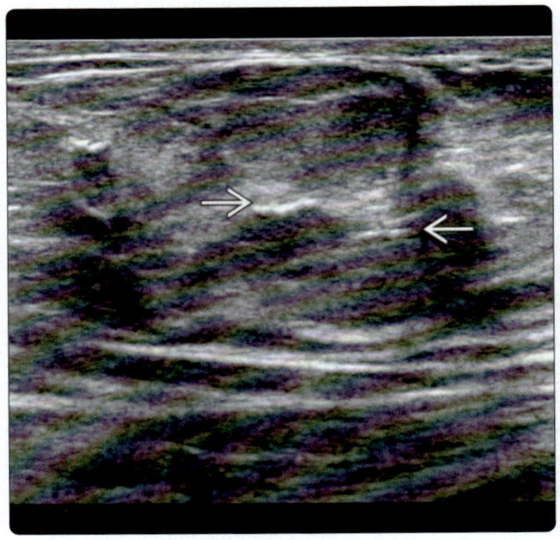

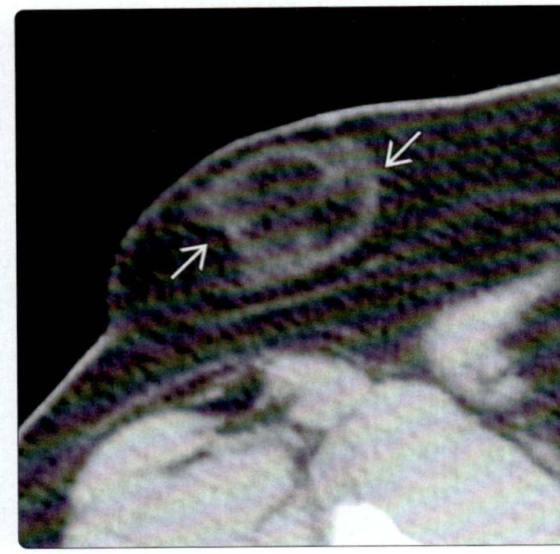

(Left) *Longitudinal color Doppler US shows a subcutaneous hyperechoic mass with moderate intrinsic vascularity, consistent with an angiolipoma. This was confirmed by pathology following excision.* **(Right)** *Longitudinal US of the wrist shows a slightly irregular, heterogeneous, lipomatous subcutaneous mass* → *partially involving the flexor carpi ulnaris muscle* →*. Although lipomatous, the mass is usually heterogeneous with irregular hypoechoic fibrotic-type areas* →*.*

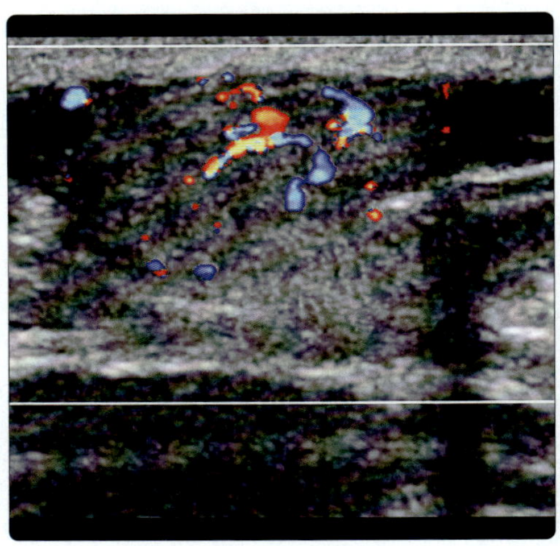

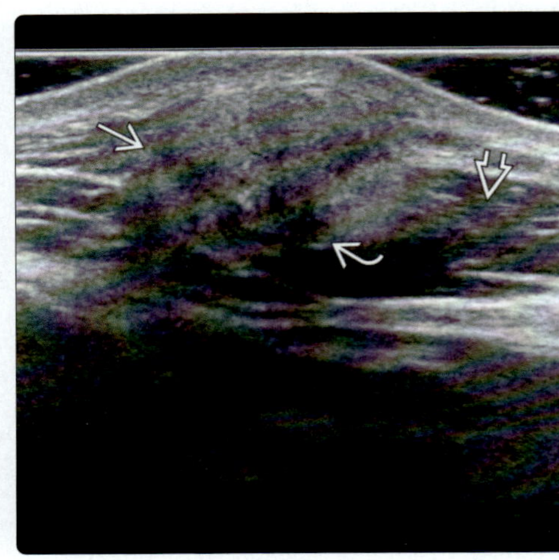

(Left) *Longitudinal color Doppler US of the same lesion reveals no intratumoral hypervascularity.* **(Right)** *Axial T1 MR of the same mass* → *shows it is located in ulnovolar subcutaneous fat of the wrist with intramuscular extension into the flexor carpi ulnaris muscle* →*. There is a small amount of intralesional fat* → *with a considerable nonfatty soft tissue component present. The mass was suspected to be liposarcoma. Final histology revealed spindle cell lipoma.*

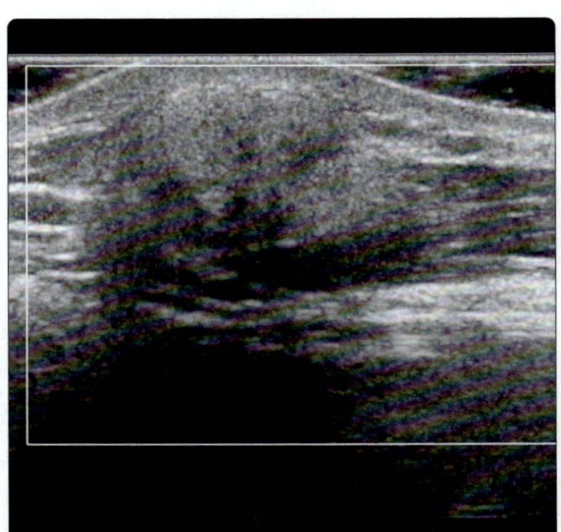

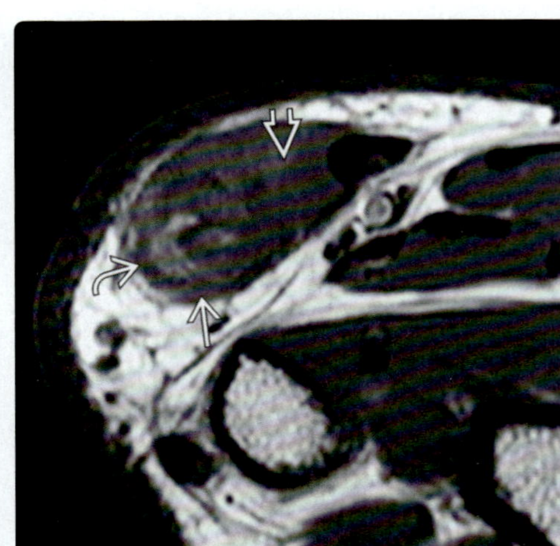

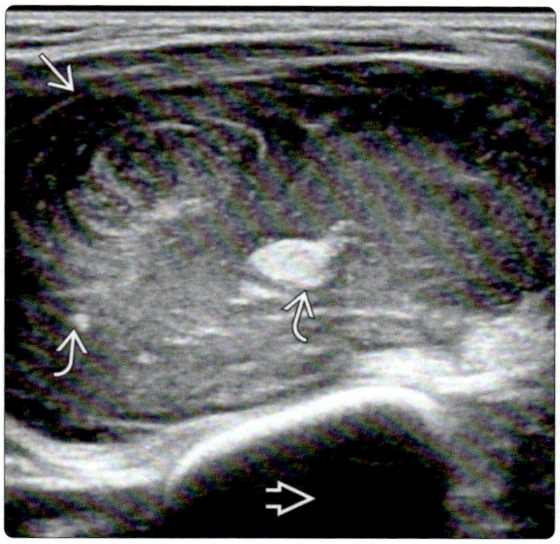

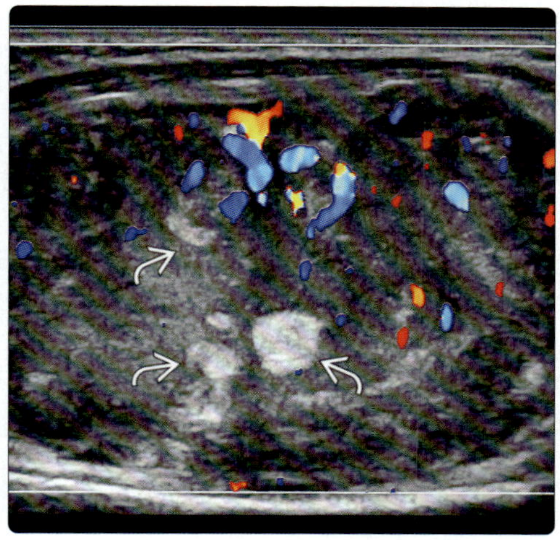

(Left) *Transverse US of a 2 year old with a 1-month history of posterior thigh swelling shows a large, well-circumscribed heterogeneous mass* ➡ *in the posterior thigh superficial to the femur* ➡. *Scattered areas of hyperechoic, lipomatous-type tissue* ➡ *are present.* **(Right)** *Longitudinal color Doppler US of the same lesion shows moderate intratumoral hypervascularity. More foci of echogenic, fatty-type tissue* ➡ *are evident in this scan plane.*

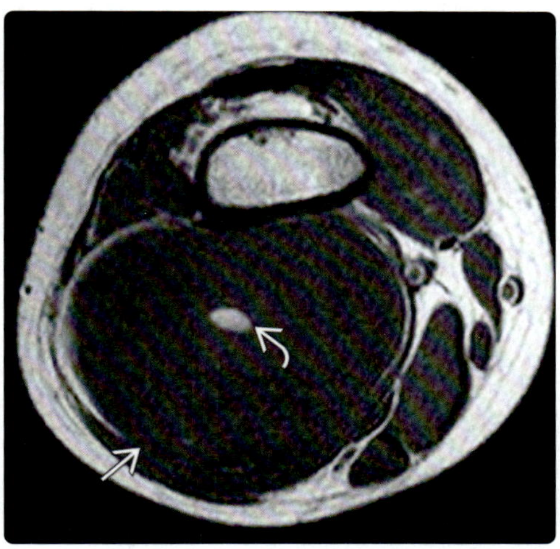

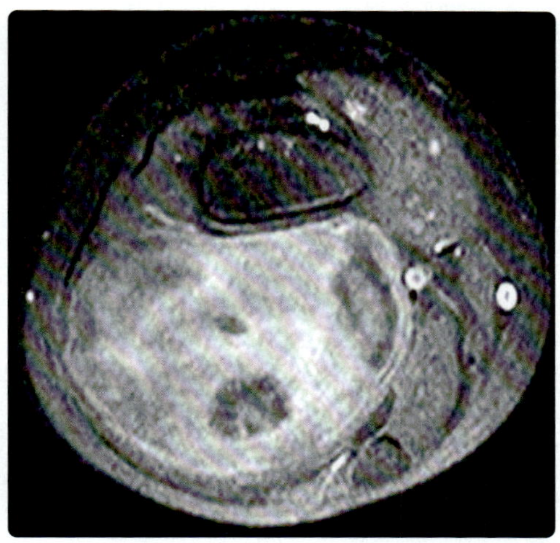

(Left) *Axial T1 MR of the same patient shows a large well-circumscribed mass lesion in the posterior thigh* ➡. *The mass is predominantly nonfatty with a small fat component* ➡ *centrally within the mass, corresponding to the US appearances.* **(Right)** *Postgadolinium axial T1 FS MR of the same patient shows heterogenous enhancement of the mass. Based on the patient's age and the imaging features, the most likely diagnosis was lipoblastoma. This was confirmed on histology.*

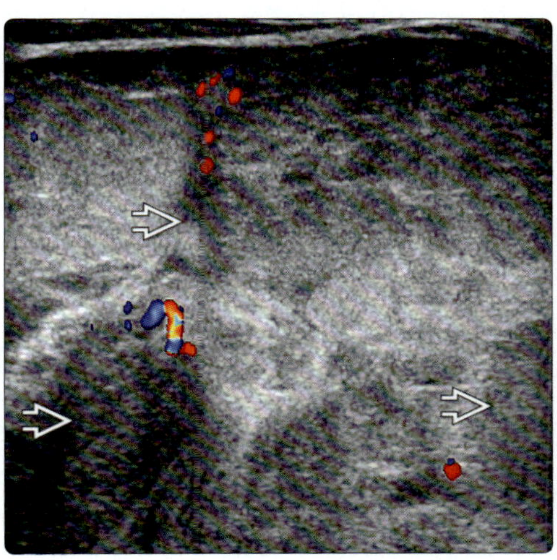

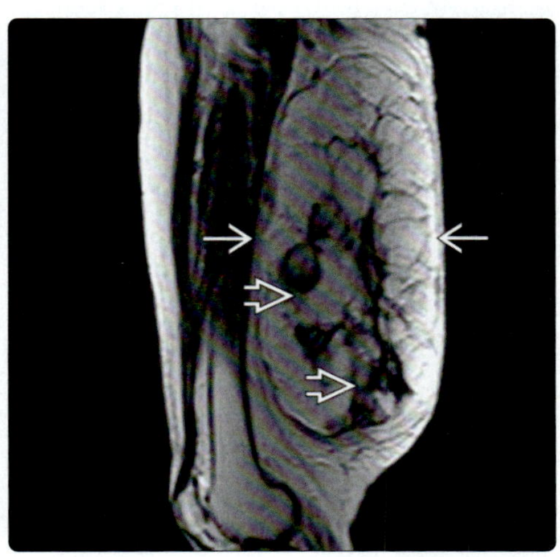

(Left) *Longitudinal US of a pathologically proven myxoid liposarcoma shows a large, intramuscular, heterogeneous lipomatous mass with moderate intrinsic vascularity and areas of nonlipomatous hypoechoic tissue* ➡. **(Right)** *Longitudinal T1 MR of the same lesion shows a large, heterogeneous lipomatous mass* ➡ *in the posterior thigh with medium-sized areas of nonfatty tissue* ➡ *and thickened septa.*

Epidermoid Cyst

KEY FACTS

TERMINOLOGY

- Slow-growing subdermal cyst that contains keratin and is lined by stratified squamous epithelium
  - Cyst wall must not contain dermal elements

IMAGING

- US appearances depend on maturation of cyst and amount and compactness of keratin
  - ± presence of cyst rupture and degree of cyst collapse
- Well-circumscribed ovoid (80%), lobulated (19%), or tubular (1%) configuration
- Increased acoustic transmission (96%)
- Most are heterogeneous and mildly hyperechoic
- Some homogeneous with pseudotestis appearance
- Linear hyperreflective disc-like areas represent layered keratin aggregates ± cholesterol deposition
- Cleft-like or filiform anechoic or hypoechoic areas are due to cyst-like spaces lacking keratin
- Thin, hypoechoic cyst wall
  - Hypoechoic tract extending from superficial wall to dermis (10%) (punctum)
- Contained rupture → lobulated configuration of cyst
- Most epidermoid cysts are avascular
  - Peripheral and pericystic vascularity is feature of cyst rupture and inflammation
- Sensitivity/specificity of US in making diagnosis of epidermoid cyst is 80% and 95%, respectively

TOP DIFFERENTIAL DIAGNOSES

- Lipoma, soft tissue abscess, slow-flow vascular malformation, nerve sheath tumor

DIAGNOSTIC CHECKLIST

- Characteristic US appearances in most cases
- Consider localized leakage of cyst content if inflammation is present around epidermoid cyst

(Left) *Longitudinal US shows a well-circumscribed hypoechoic epidermoid cyst with dermal attachment, multiple internal filiform anechoic areas, linear hyperreflective strands, and posterior acoustic enhancement.* **(Right)** *Longitudinal US shows a slightly hyperechoic epidermoid cyst with dermal protuberance and acoustic enhancement. The epidermoid cyst content is homogeneous with a mildly coarse echotexture giving rise to a pseudotestis appearance.*

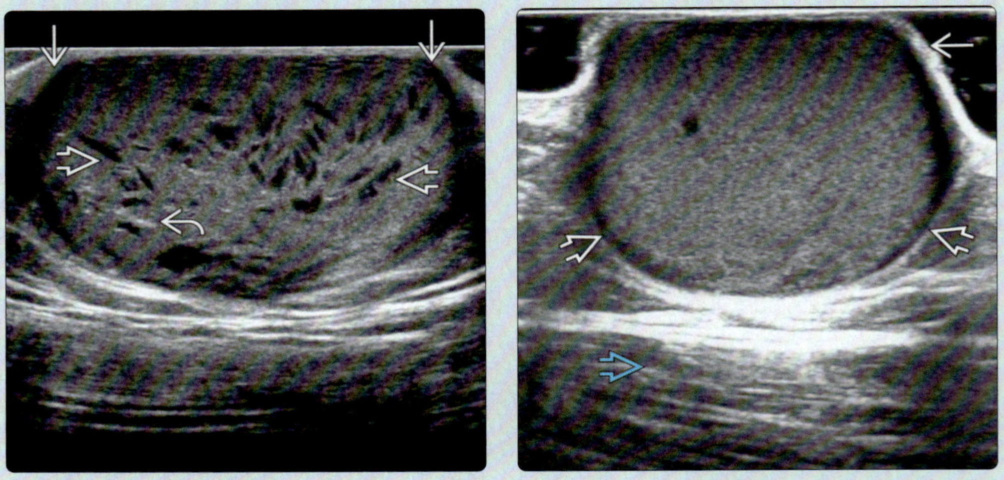

(Left) *Longitudinal US shows a well-circumscribed epidermoid cyst with dermal attachment, heterogeneous echogenic content, alternate layering, multiple floating echogenic foci, and increased through transmission.* **(Right)** *Longitudinal US shows typical appearances with a well-circumscribed ovoid heterogeneous nodule with dermal attachment, internal floating hyperreflective strands, small hypoechoic areas, and a hypoechoic tract extending to a cutaneous punctum.*

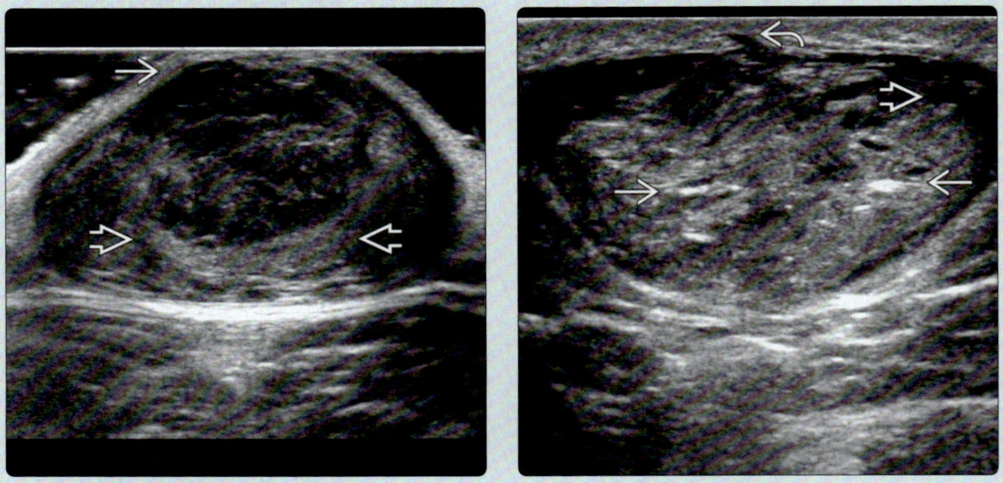

TERMINOLOGY

Synonyms

- Epidermal inclusion cyst, epidermal cyst, infundibular cyst, sebaceous cyst

Definitions

- Slow-growing subdermal cyst that contains keratin and is lined by stratified squamous epithelium
  - Most cysts are due to ductal blockage of hair follicles with downward rupture of obstructed pilosebaceous unit
  - Some cysts are due to traumatic or embryological subdermal sequestration of epithelial tissue
    - Histologically similar and hence grouped under general term "epidermoid cyst"
  - Wall of cyst must not contain dermal elements
    - Presence of dermal elements → designation as dermoid cyst
  - Epidermis = outer layers of skin
    - Contains keratinocytes, corneocytes, and melanocytes
  - Dermis = inner layers of skin that support epidermis
    - Contains hair follicles, nerves, sweat glands, small blood vessels, and sebaceous glands

IMAGING

General Features

- Best diagnostic clue
  - Well-circumscribed, cyst-like mass with thick, linear, seed- or disc-like hyperechoic foci, filiform anechoic areas, and dermal attachment
- Location
  - Predominately in those areas with most body hair
  - Most commonly seen in head and neck
  - Trunk and extremities less commonly found
    - Most cysts occurring in relatively hairless areas, such as palm or sole, are implantation epidermoid cysts
    - Trauma → implantation of epidermal elements to dermis or subcutis, which then survive, grow, and produce keratin
  - Can occur within tendon or bone
- Size
  - < 1 cm to > 10 cm, usually < 5 cm
- Morphology
  - Well-circumscribed subcutaneous or dermal mass

Radiographic Findings

- Soft tissue swelling if large size
- Rarely bony remodeling or lytic bone lesion with sclerotic border

CT Findings

- Noncalcified soft tissue attenuation mass in subdermal layer
  - Low-density, fluid-like attenuation due to high lipid content
    - Rarely hyperdense due to hemorrhage, saponification, or high protein content
- No enhancement or minimal rim enhancement

MR Findings

- T1WI
  - Isointense to slightly hyperintense relative to muscle
- T2WI
  - Hyperintense signal with high or low signal intensity content
    - Low-signal debris represents keratin aggregates
    - High-signal debris represents cholesterol deposition
  - Content may be positionally dependent or layered concentrically
  - T2-hypointense rim due to stratified squamous epithelial wall
- T1WI C+
  - Absent to low level of central enhancement with enhancing capsule

Ultrasonographic Findings

- US appearances depend on maturation of cyst, compactness, and amount of keratin
  - ± presence of cyst rupture and degree of cyst collapse
- Well-circumscribed ovoid (80%), lobulated (19%), or tubular (1%) configuration
- Dermal attachment (i.e., no intervening subcutaneous fat between lesion and dermis)
  - ± dermal protrusion
  - ± cutaneous punctum
    - Feature of blocked duct etiology
    - May not be present if implantation etiology
- Hypoechoic punctum seen between cyst and dermis in 10% of those referred for US examination
  - Many of those with clinically obvious punctum may not be referred for US
- Cyst confined to subcutaneous tissue
  - No subfascial extension
- Most are unilocular
- Most cysts are heterogeneous and mildly hyperechoic
- Some cysts are homogeneous with pseudotestis appearance
  - Linear hyperreflective disc-like areas represent layered keratin aggregates ± cholesterol deposition
  - Cleft-like or filiform anechoic or hypoechoic areas are due to cyst-like spaces lacking keratin
  - ± multilayered appearance with hyperechoic and hypoechoic rings
  - ± target appearance with hyperechoic center surrounded by hypoechoic rim
  - ± "swirling" of mobile cyst contents when compressed
- Increased acoustic transmission (96%)
- Thin, hypoechoic cyst wall
  - Focal discontinuity indicative of cyst rupture
- Irregular cyst contour indicative of cyst rupture
  - Localized contained rupture may give rise to lobulated configuration
- Most epidermoid cysts are avascular
  - Peripheral vascularity, or occasionally central vascularity, is feature of previously ruptured cysts
  - Leakage of cyst contents → pericystic inflammation
  - Twinkle artifact on color Doppler imaging due to strong reflective interfaces arising from keratin layers within cyst

- Soft on strain elastography
- Sensitivity/specificity of US in making diagnosis of epidermoid cyst is 80% and 95%, respectively
  - Most cysts have distinctive US appearance

Imaging Recommendations

- Best imaging tool
  - US
    - Can make correct diagnosis in majority of cases
- Protocol advice
  - Transverse, longitudinal scan, and color Doppler imaging
    - Determine any connection with dermis
    - Ensure continuity of cyst wall
    - Determine any rupture ± surrounding inflammation

DIFFERENTIAL DIAGNOSIS

Lipoma

- Isoechoic to adjacent subcutaneous fat, compressible
- Fixed linear fine horizontal echogenic striations parallel to skin
- More homogeneous than epidermoid cysts

Soft Tissue Abscess

- Can be difficult to differentiate from inflamed, infected, or ruptured epidermoid cyst
- Liquefied component of abscess

Slow Flow Vascular Anomaly

- Vascular anomaly with predominant fatty stroma can mimic epidermoid cyst
- Small vessels may be too small to be depicted by US
- Will have either no or mild acoustic enhancement

Nerve Sheath Tumor

- Occurs along neurovascular bundle
- Usually hypoechoic and hypervascular
- Entering and exiting nerves may be seen

PATHOLOGY

General Features

- Etiology
  - Low-grade inflammation leads to epithelial hyperplasia and obstruction at infundibulum of pilosebaceous unit (composed of hair follicle, sebaceous gland, and arrector pili muscle)
  - Followed by downward rupture of obstructed pilosebaceous unit
  - Alternatively, embryological or traumatic sequestration of epidermal tissue into subdermal tissues

Gross Pathologic & Surgical Features

- Elastic mass with "butter and cheese" content
- Microscopic features
  - Encapsulated cyst with walls of stratified squamous epithelium
    - Walls do not contain dermal elements, such as sebaceous glands or hair follicles
  - Lumen filled with laminated keratinaceous material
- Additional findings associated with cyst rupture
  - Pericystic granulomatous reaction with inflammation and fibrosis
  - Cyst contents excite foreign body reaction

CLINICAL ISSUES

Presentation

- Most common signs/symptoms
  - Most are incidental finding
- Other signs/symptoms
  - May report history of penetrating trauma many months or years earlier
  - Firm, slowly growing, painless mass
  - Mobile or tethered to skin
  - ± punctum
  - ± symptoms and signs of inflammation

Demographics

- Age
  - Any age; more common in 3rd-5th decades
- Gender
  - M > F (slightly)
- Epidemiology
  - 7% of superficial masses referred for US
  - 12% of superficial masses surgically excised

Natural History & Prognosis

- May rupture, become inflamed or infected
  - Infection of epidermoid cyst is probably uncommon
  - No difference in microbiological milieu of infected and noninfected inflamed cysts
  - Most cyst "infections" probably due to inflammation induced by cyst rupture rather than true infection
- No local recurrence if complete surgical excision

Treatment

- Antiinflammatory agents if cyst inflamed
- Antimicrobial agents if cyst infected
- Incision and expression of cyst contents
  - High rate of recurrence
- Incision and expression of cyst contents and removal of cyst wall through same minimal incision
- Total excision of cyst
- Laser therapy for cysts on face and other sensitive areas
- Surgery for previously ruptured epidermoid cysts associated with pericystic scarring is more complicated than surgery for nonruptured cysts

DIAGNOSTIC CHECKLIST

Consider

- Consider localized leakage of cyst content if inflammation present around epidermoid cyst

SELECTED REFERENCES

1. Clarke R et al: Twinkle artefact in the ultrasound diagnosis of superficial epidermoid cysts. Ultrasound. 24(3):147-153, 2016
2. Park HJ et al: Strain elastography features of epidermoid tumours in superficial soft tissue: differences from other benign soft-tissue tumours and malignant tumours. Br J Radiol. 88(1050):20140797, 2015
3. Hung EH et al: Pitfalls in ultrasonography of soft tissue tumors. Semin Musculoskelet Radiol. 18(1):79-85, 2014
4. Hung EH et al: Ultrasound of musculoskeletal soft-tissue tumors superficial to the investing fascia. AJR Am J Roentgenol. 202(6):W532-40, 2014

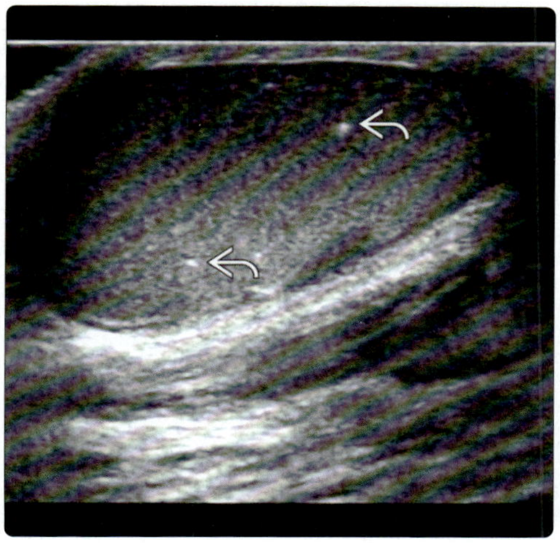

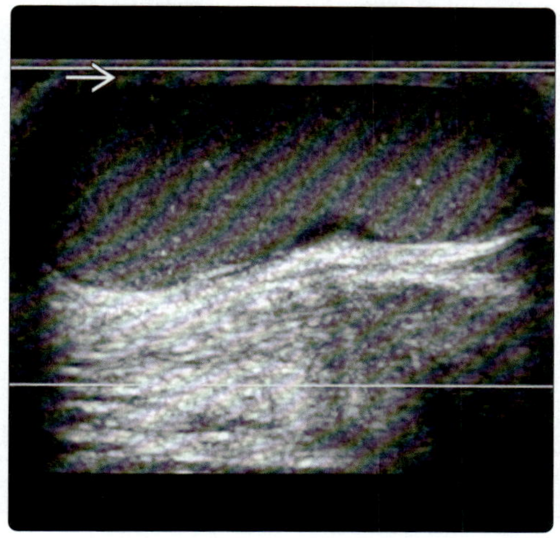

(Left) *Transverse US shows a well-circumscribed, mildly echogenic epidermoid cyst with increased through transmission. The echotexture is homogeneous with a pseudotestis appearance. Note the echogenic foci due to either keratin or cholesterol aggregates.* **(Right)** *Longitudinal color Doppler US of the same lesion reveals no central or peripheral vascularity. Note there is no intervening fat between the epidermoid cyst and the dermis, which indicates dermal attachment.*

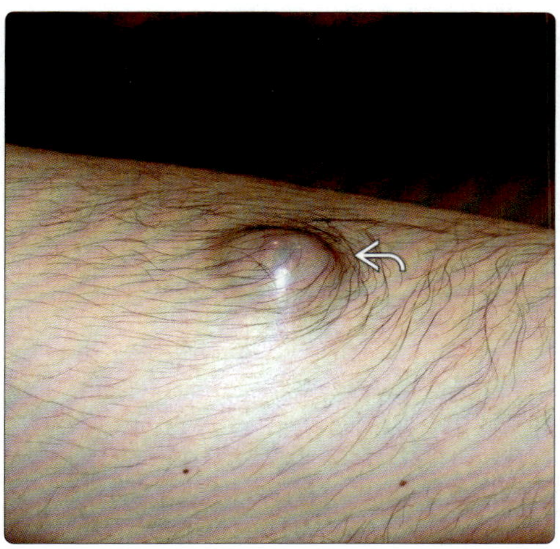

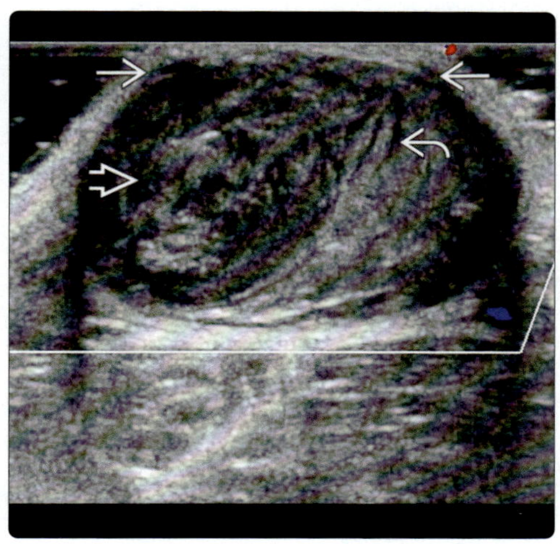

(Left) *Clinical photo shows a protuberant bluish nodule on the medial calf region. No obvious punctum is evident clinically.* **(Right)** *Transverse color Doppler US of the same lesion shows an ovoid heterogeneous epidermoid cyst with dermal attachment and protrusion. Internal filiform hypoechoic areas and a laminated pattern representing compacted keratin layers are shown. There is no intralesional vascularity. Note the mild posterior acoustic enhancement.*

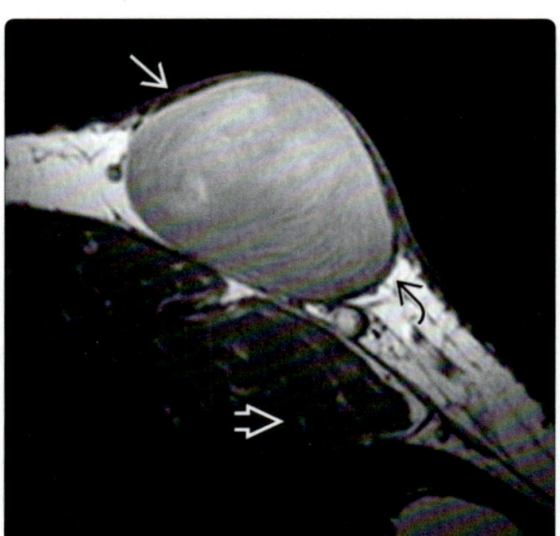

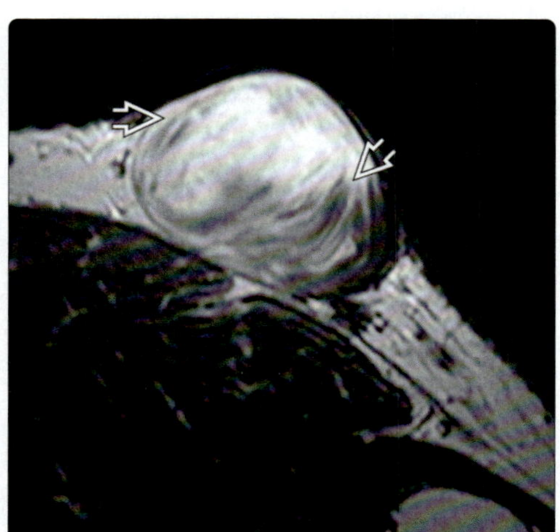

(Left) *Axial T1WI MR of the same well-defined, nonruptured epidermoid cyst shows attachment to the dermis. The content is mildly heterogeneous and moderately hyperintense to muscle. Note the thin yet distinct hypointense wall.* **(Right)** *Axial T2WI MR of the same lesion shows how the cyst content has cyst-like hyperintensity with hypointense lamina due to layers of compacted keratin. These features are very typical of an epidermoid cyst.*

(Left) *Transverse US shows a well-circumscribed epidermoid cyst with increased through transmission* ➡, *internal cleft-like hypoechoic areas* ➡, *echogenic strands, and foci due to keratin and cholesterol aggregates* ➡. **(Right)** *Transverse color Doppler US of the same cyst shows absence of central or marginal vascularity, which is a typical finding in epidermoid cyst without inflammation or rupture.*

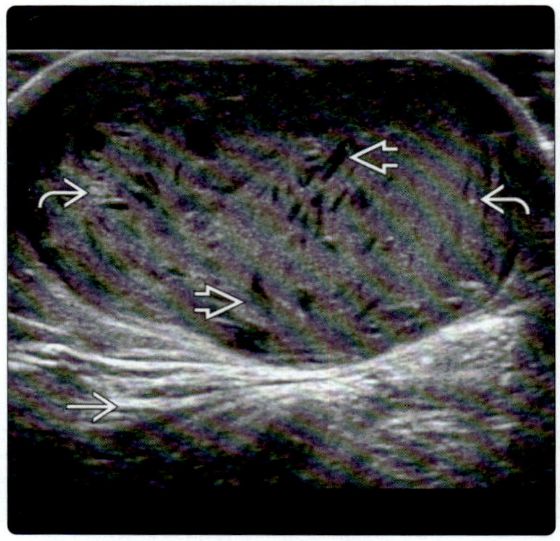

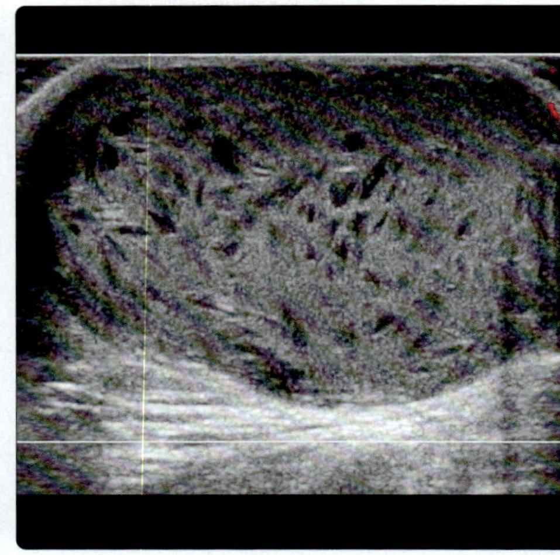

(Left) *Clinical photo shows mild protuberance and mild bluish discoloration* ➡ *due to a small epidermoid cyst on the posterior aspect of the shoulder. Note the central punctum.* **(Right)** *Longitudinal US of the same patient shows an encapsulated epidermoid cyst* ➡ *and the presence of multiple small echogenic foci* ➡ *due to keratin and cholesterol aggregates. There is a superficial punctum* ➡.

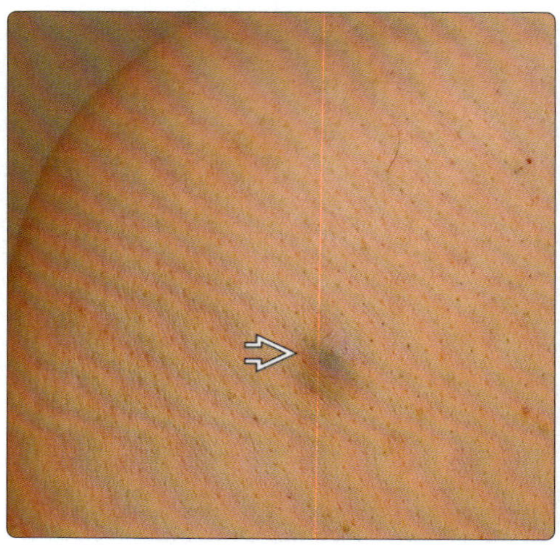

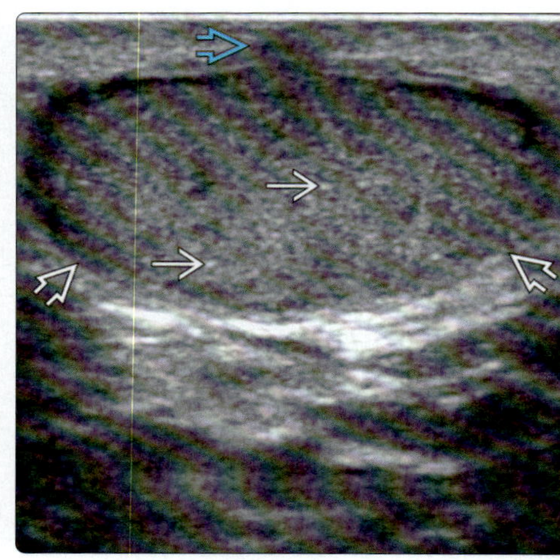

(Left) *Longitudinal US shows a mildly hypoechoic epidermoid cyst* ➡ *with increased acoustic through transmission* ➡. *No hyperemia is present. The epidermoid cyst is small, superficial, and slightly atypical but still sufficient to make a quite definitive diagnosis based on the US appearances alone.* **(Right)** *Transverse color Doppler US of a typical uncomplicated epidermoid cyst* ➡ *shows no intrinsic vascularity.*

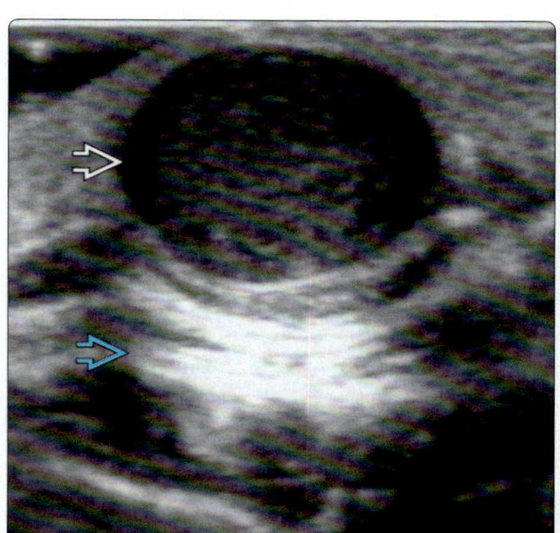

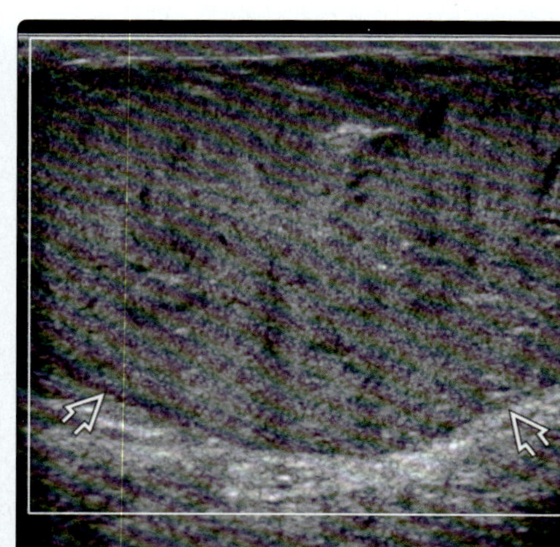

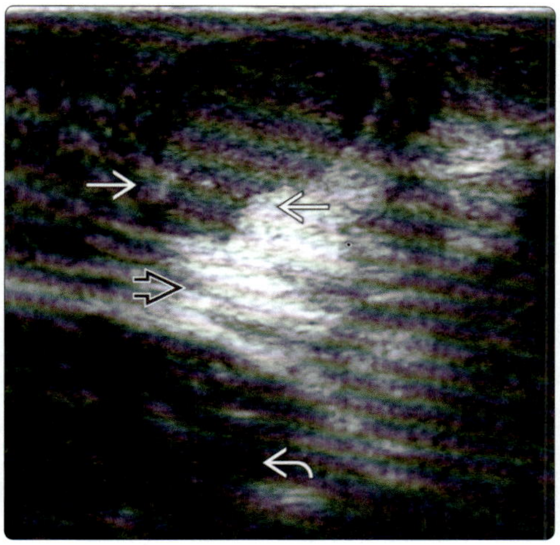

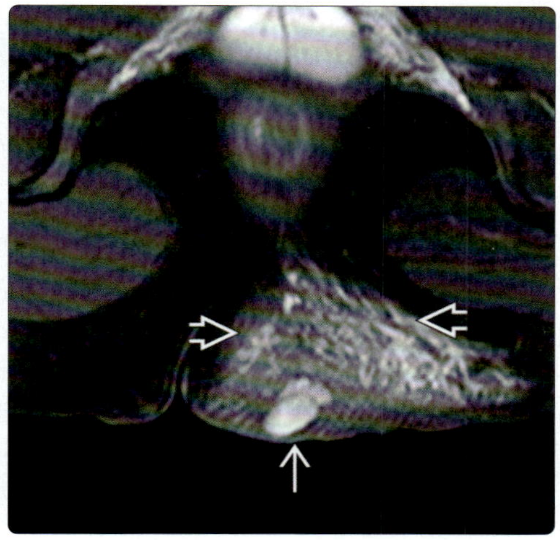

(Left) *Transverse US shows a lobulated cystic subcutaneous nodule with slightly echogenic content and markedly increased through transmission. The irregular border of the epidermoid cyst is a sequelae of recent cyst rupture. Note the gluteal muscle.* **(Right)** *Axial T2WI FS MR of the same patient shows a hyperintense lesion with lobulated contour, hypointense rim, and moderate subcutaneous soft tissue edema, which is consistent with epidermoid cyst rupture.*

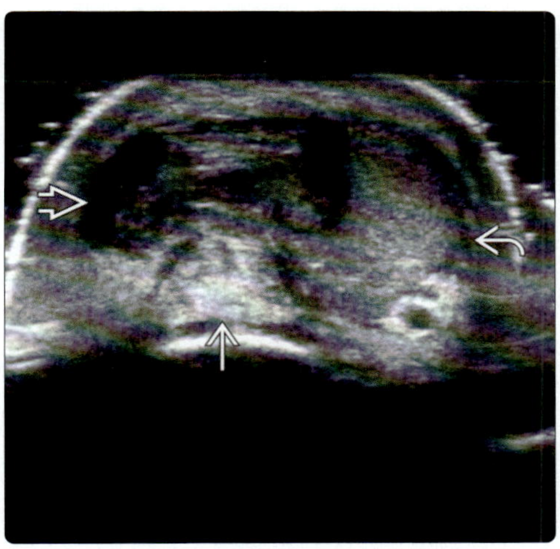

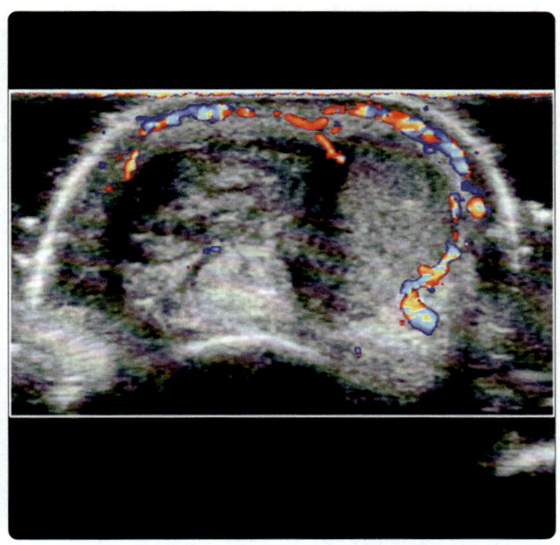

(Left) *Transverse US of a middle finger shows a bilobulated epidermoid cyst superficial to the flexor tendons. The lesion has both a homogeneous echogenic component and a layered heterogeneous component. Moderate pericystic edema is present.* **(Right)** *Transverse color Doppler US of the same lesion shows moderate pericystic soft tissue hyperemia representing inflammation secondary to leakage of cyst contents.*

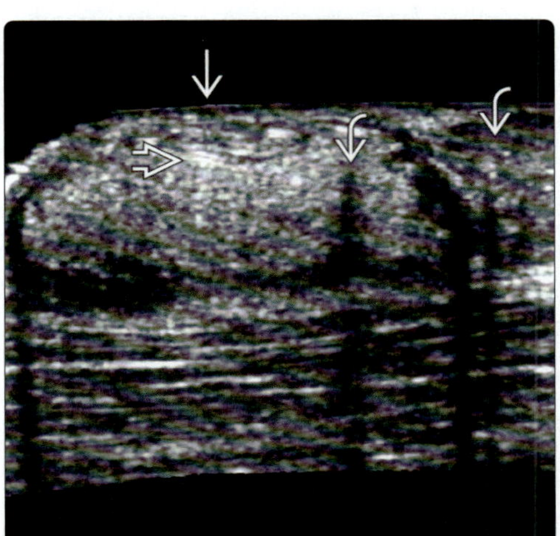

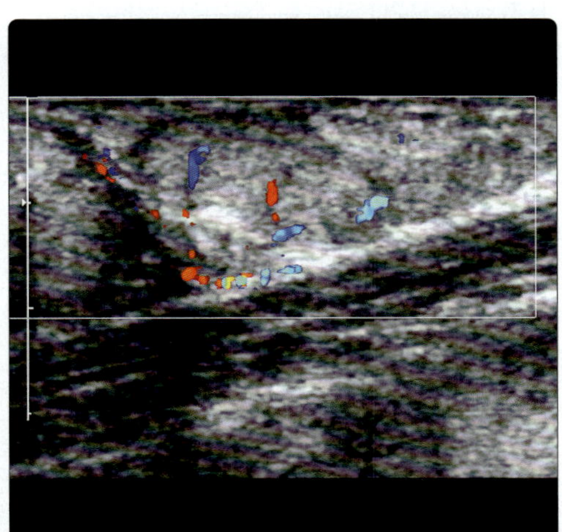

(Left) *Extended field-of-view US shows a lobulated hyperechoic nodule containing hyperreflective foci. Some of these hyperechoic foci are associated with posterior shadowing suggestive of calcific foci.* **(Right)** *Longitudinal color Doppler US of the same lesion shows mild central and peripheral vascularity. The lesion was proven pathologically to be a ruptured epidermoid cyst associated with granulomatous foreign body giant cell reaction.*

Pilomatricoma

KEY FACTS

TERMINOLOGY

- Synonym: Calcifying epithelioma of Malherbe, pilomatrixoma
- Definition: Benign skin tumor arising from cells of hair matrix

IMAGING

- Head and neck most common (60%), followed by upper extremities (25%), trunk (10%), and lower extremities (5%)
- Well-circumscribed, ovoid, heterogeneous, solid subdermal mass containing matrix calcification, intrinsic vascularity, and peripheral hypoechoic rim
- Mass extending from dermal to subcutaneous tissues
- Ovoid (80%), round (10%), or irregular (10%)
- Hypoechoic (70%), isoechoic (15%), or hyperechoic (5%) to subcutaneous fat, heterogeneous (10%)
- Internal hyperechoic foci/reticulation due to matrix calcification (93%)
- Scattered dots (most common), lump- or arc-like calcific deposits (less common), may be completely calcified (uncommon)
- Hypoechoic peripheral rim due to connective tissue capsule (80%)
- Intrinsic vascularity (60%)
- Capsular rupture can lead to peritumoral inflammation
- US appearances are sufficiently specific to make diagnosis in majority of cases

TOP DIFFERENTIAL DIAGNOSES

- Epidermoid cyst
- Hemangioma

CLINICAL ISSUES

- Most present in children but can present at any age
  - 2nd most commonly excised superficial tumor after epidermoid cyst in children
- Treatment is surgical excision

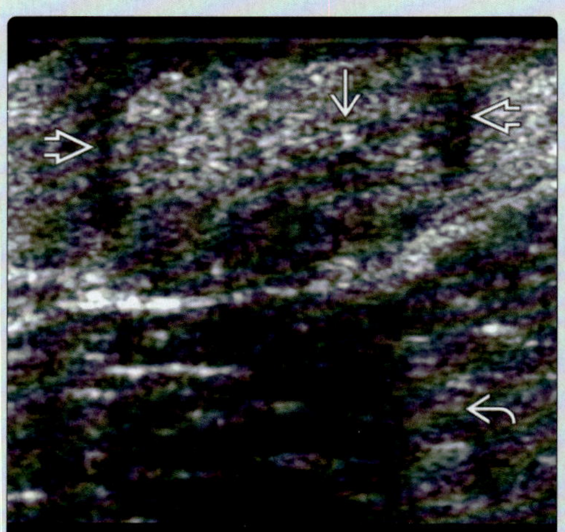

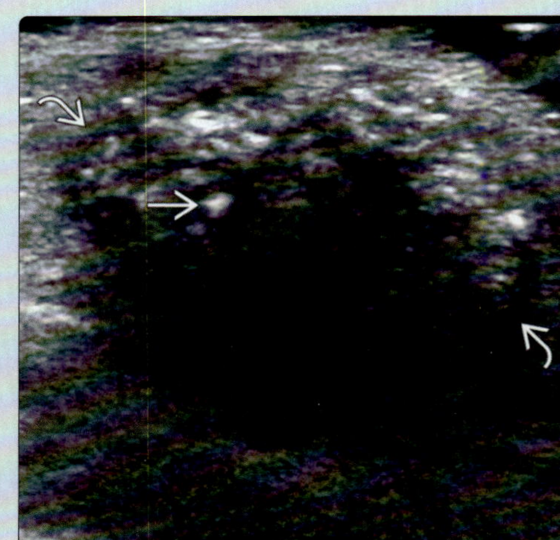

(Left) *Transverse US shows a well-defined, heterogeneous pilomatricoma in the subcutaneous layer. There are multiple scattered fine echogenic foci with mild posterior shadowing due to matrix calcification. Note the peripheral hypoechoic rim. The muscle layer is shown.* **(Right)** *Transverse US shows a pilomatricoma containing multiple coarse, lumpy calcifications associated with moderate posterior shadowing. Note the hypoechoic peripheral rim.*

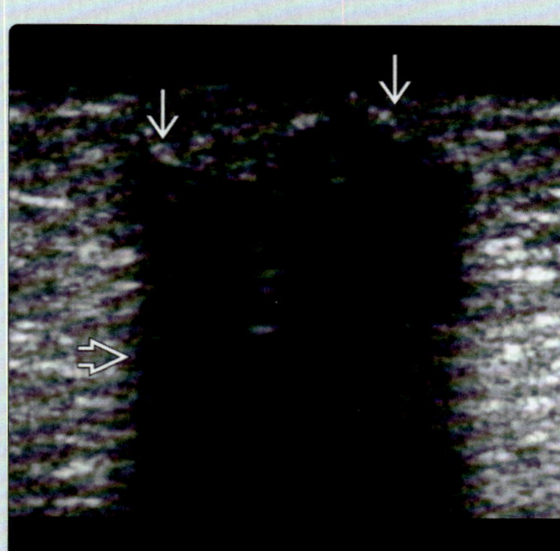

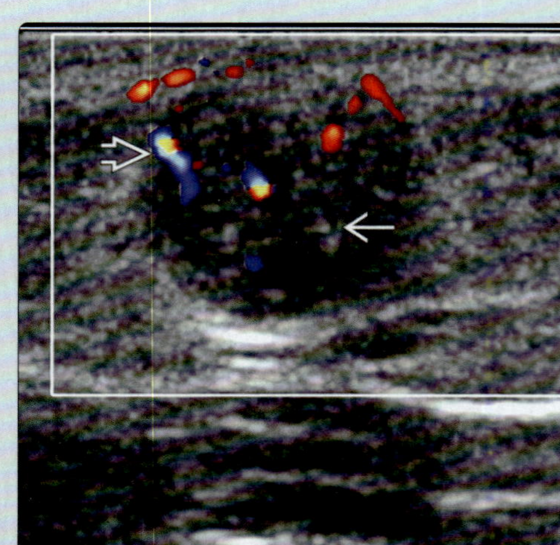

(Left) *Longitudinal US shows a densely calcified pilomatricoma casting strong posterior shadowing and limiting assessment of deeper margin.* **(Right)** *Transverse US shows a well-defined, subcutaneous, hypoechoic pilomatricoma with central and peripheral vascularity containing scattered small calcific foci. Intrinsic vascularity is commonly seen in pilomatricoma.*

Pilomatricoma

TERMINOLOGY

Synonyms

- Calcifying epithelioma of Malherbe
- Pilomatrixoma

Definitions

- Benign skin tumor arising from cells of hair matrix

IMAGING

General Features

- Best diagnostic clue
  - Well-circumscribed subdermal mass lesion with heterogeneous echotexture, internal calcification, vascularity, and peripheral hypoechoic rim
- Location
  - Head and neck (60%), upper extremities (25%), trunk (10%), lower extremities (5%)
- Size
  - < 1 cm to 5 cm mean diameter: 1.4 cm
- Morphology
  - Well-circumscribed, partially calcified mass lying between dermis and subcutaneous tissues

Radiographic Findings

- Visibility dependent on degree of tumoral calcification

CT Findings

- Nonspecific soft tissue attenuation mass in subcutaneous tissue ± calcification ± mild to moderate enhancement

MR Findings

- T1WI intermediate to low, T2WI heterogeneously hyperintense
- Calcification → low-signal foci
- Mild to moderate contrast enhancement
- No restricted diffusion

Ultrasonographic Findings

- Well-circumscribed mass extending from dermis to subcutaneous tissues
- Ovoid (80%), round (10%), or irregular (10%) in shape
- Hypoechoic (70%), isoechoic (15%), hyperechoic (5%) to subcutaneous fat or heterogeneous (10%)
- Internal hyperechoic foci/reticulation due to calcification in stroma or shadow cells (95%)
  - Scattered dots (most common), lump- or arc-like calcific deposits (less common)
- Hypoechoic peripheral rim due to connective tissue capsule (80%)
- Intrinsic vascularity (60%)
  - Peripheral, central, or combined peripheral and central
- ± posterior shadowing
  - Severity depends on degree of matrix calcification
  - Complete calcification may occur uncommonly
- ± surrounding inflammation
  - Capsular rupture can lead to peritumoral inflammation

Imaging Recommendations

- Best imaging tool
  - US
- Protocol advice
  - Assess shape, echotexture, internal calcification, hypoechoic rim, margin, posterior shadowing, and vascularity

DIFFERENTIAL DIAGNOSIS

Epidermoid Cyst

- Linear seed-like internal echogenic strands ± posterior enhancement
- No intrinsic vascularity unless ruptured

Hemangioma

- Fatty stroma with vascular channels ± phleboliths

PATHOLOGY

General Features

- Etiology
  - Usually sporadic, not familial
- Genetics
  - 75% with mutations in β-catenin gene (*CTNNB1*)
  - Associations: Gardner syndrome, familial adenomatous polyposis, myotonic muscular dystrophy, and Turner syndrome

Gross Pathologic & Surgical Features

- Basaloid proliferation of hair matrix cells
- These mature into structureless anucleate eosinophilic cells called shadow cells representing differentiation toward hair cortex
- Calcification occurs in shadow cells or stroma
- ± histiocytic multinucleate infiltrate at site of rupture

CLINICAL ISSUES

Presentation

- Most common signs/symptoms
  - Slow-growing, hard, subcutaneous mass ± localized inflammation ± skin discoloration
- Other signs/symptoms
  - Usually solitary, though multiple in up to 10%

Demographics

- Age
  - Particularly prevalent in children, typically first 2 years of life, though can present at any age
- Gender
  - Slight female predominance
- Epidemiology
  - 2nd most commonly excised superficial lump after epidermoid cyst in children

Natural History & Prognosis

- Spontaneous regression does not occur
- Malignant pilomatricoma (seen in adults) is exceedingly rare

Treatment

- Surgical excision

SELECTED REFERENCES

1. Bulman JC et al: Pilomatricoma of childhood: a common pathologic diagnosis yet a rare radiologic one. AJR Am J Roentgenol. 206(1):182-8, 2016

(Left) *Transverse graphic shows the epidermis, dermis, and subcutaneous fat. The hair follicle, sebaceous gland, and matrix of the hair follicle bulb from which pilomatricoma arise are shown. A sweat gland is also depicted.* **(Right)** *Longitudinal color Doppler US shows a well-defined ovoid pilomatricoma with small echogenic dots, a thin hypoechoic capsule, mild posterior shadowing, and mild, mainly peripheral hyperaemia. Superficially, the lesion contacts the dermis.*

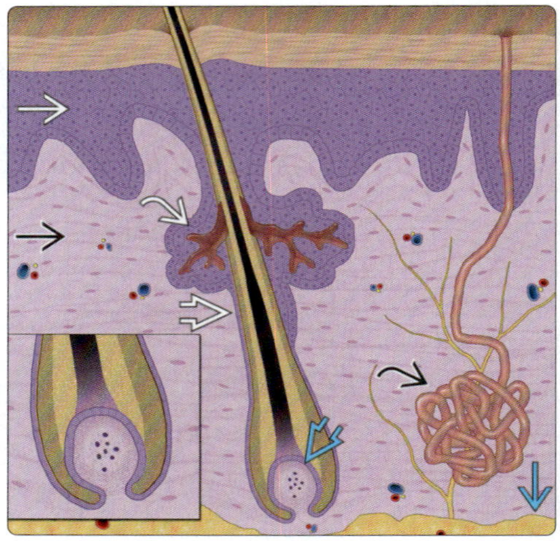

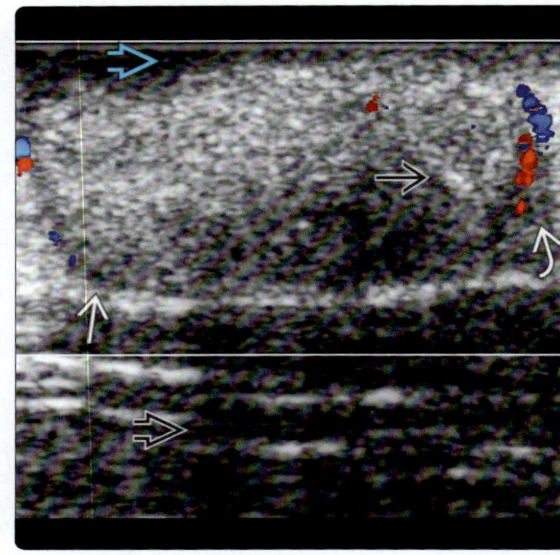

(Left) *Longitudinal US shows a slightly lobulated pilomatricoma with several calcifications, most of which are larger than those usually seen. The upper margin of the pilomatricoma abuts the dermis. Only limited acoustic shadowing is present.* **(Right)** *Transverse US of a 13-year-old boy shows a well-defined, ovoid, subcutaneous pilomatricoma. The mass is hypoechoic to adjacent subcutaneous fat and contains areas of both arc- and lump-like calcifications. Dense posterior acoustic shadowing is present.*

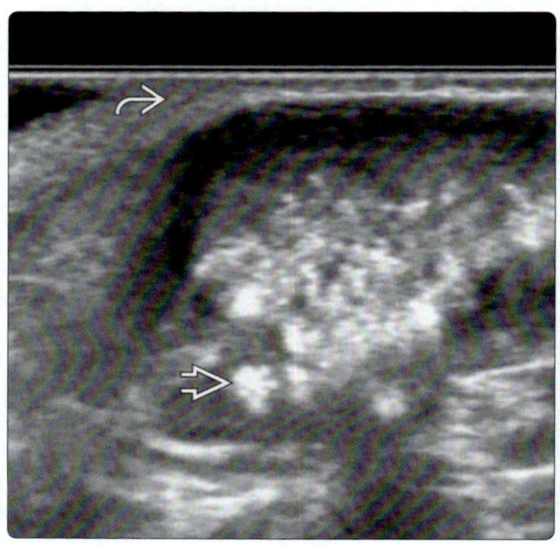

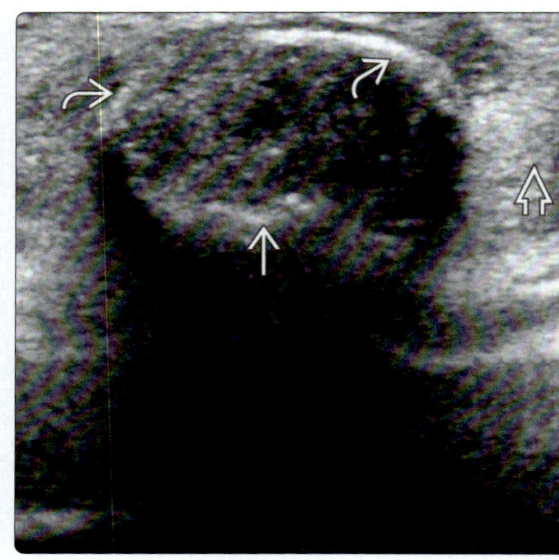

(Left) *Lateral radiograph of the arm of a 10-year-old girl shows a calcified mass due to a pilomatricoma located posterior to the humerus. Radiography is helpful in densely calcified masses with severe acoustic shadowing.* **(Right)** *Axial T2WI MR of a pilomatricoma shows a well-defined subcutaneous mass on the posteromedial aspect of the arm. The lesion is attached to the dermis and contains some low-signal areas due to internal calcification.*

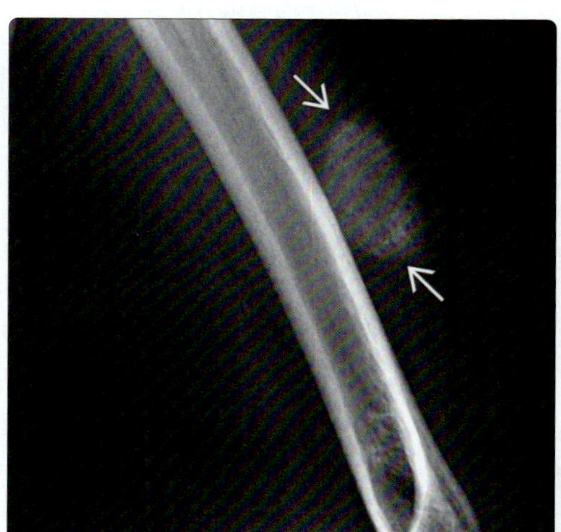

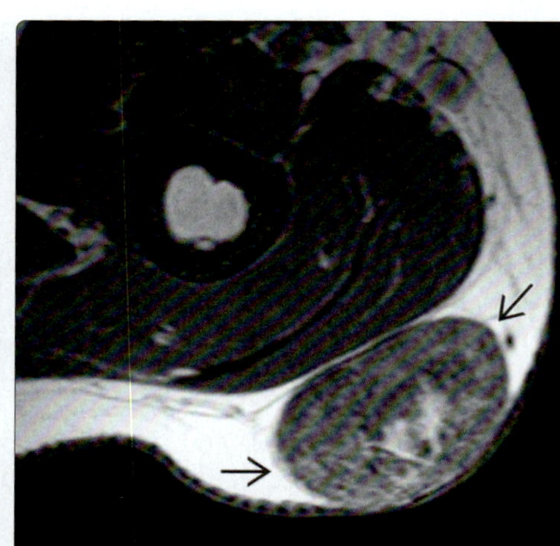

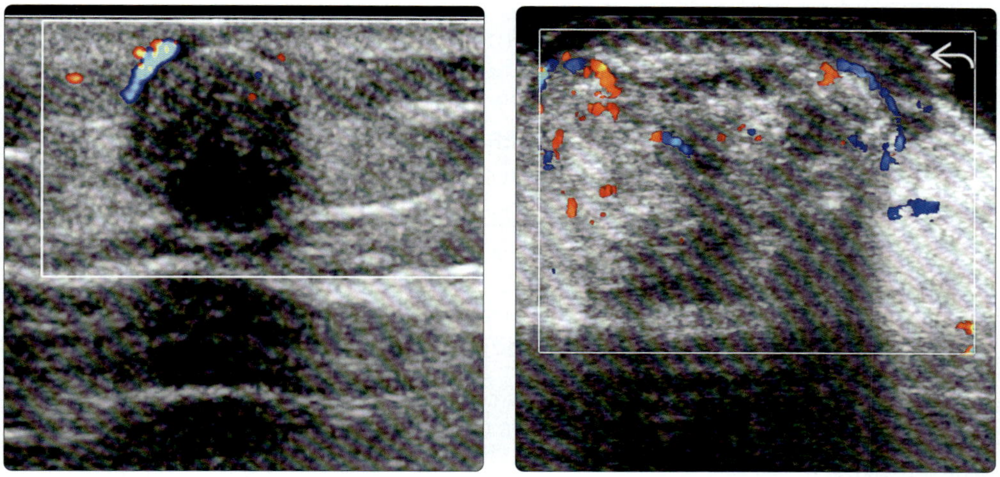

(Left) *Transverse color Doppler US of a young boy shows a round, hypoechoic pilomatricoma with peripheral arc-like calcification and mild peripheral vascularity. Moderate posterior acoustic shadowing is present.* **(Right)** *Longitudinal color Doppler US of an adult man shows a protruding ➔, ovoid, heterogeneous calcified pilomatricoma with dermal attachment in the arm region. Mild central and peripheral vascularity are present. Note the moderate posterior shadowing.*

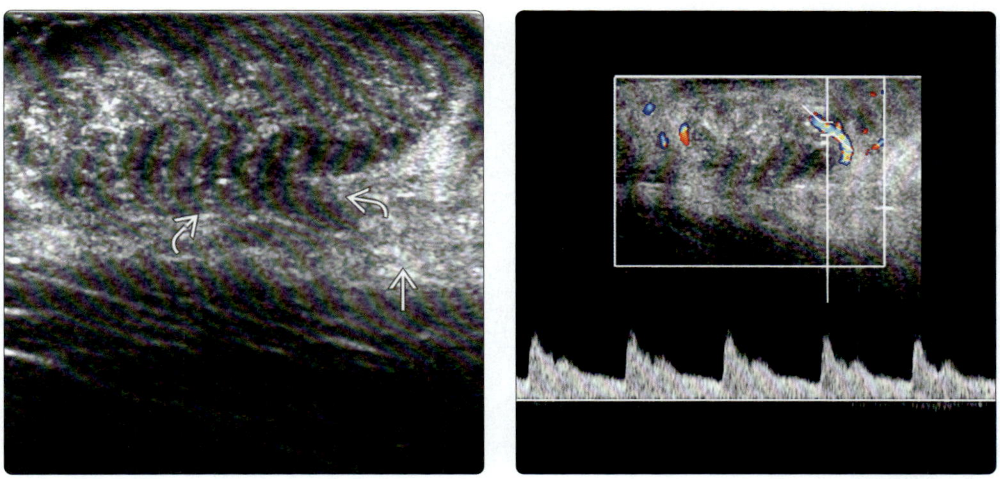

(Left) *Longitudinal US of a 20 year old shows an irregular ovoid pilomatricoma with small internal calcified foci and minimal posterior acoustic shadowing. There seems to be localized rupture at the posterior margin of the tumor ➔. Mild perilesional edema ➔ is present.* **(Right)** *Longitudinal pulsed Doppler US of the same lesion shows intratumoral arterial vascularity. Intrinsic vascularity is present in about 60% of all pilomatricoma tumors.*

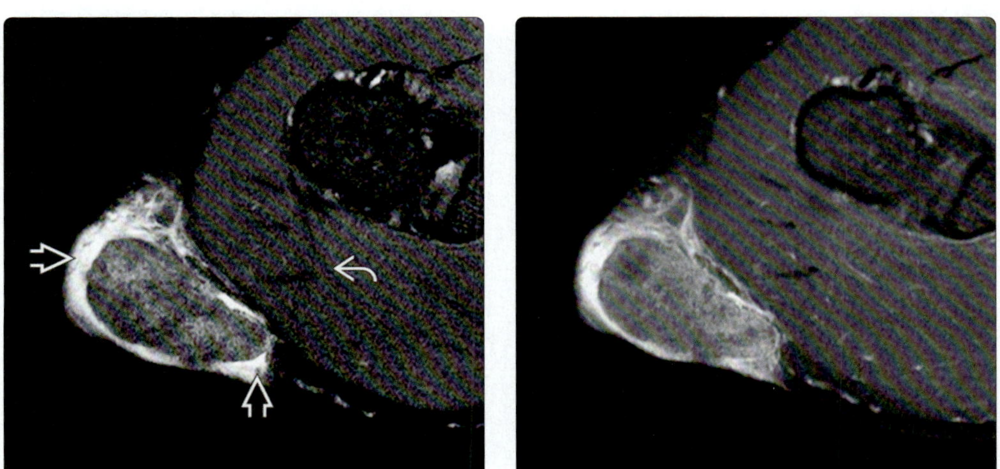

(Left) *Axial T2WI FS MR shows a well-defined, ovoid, predominantly hypointense pilomatricoma in the subcutaneous fat of the posterior deltoid ➔ region. The lesion is hypointense due to matrix calcification. There is a rim of T2 hyperintensity ➔ in the soft tissue, indicative of peritumoral edema.* **(Right)** *Axial T1WI C+ FS MR of the same lesion shows mild tumoral enhancement. Moderate peritumoral enhancement is consistent with peritumoral inflammation, most likely due to capsular leakage.*

Dermatofibrosarcoma Protuberans

KEY FACTS

TERMINOLOGY

- Low-grade spindle cell mesenchymal sarcoma of dermis and subcutis

IMAGING

- Nearly always arising from dermis, then infiltrates subcutaneous tissue
  - Most commonly arises in trunk (~ 50%), particularly chest wall, back, abdominal wall
- Protruding mass arising from skin with bluish or reddish discoloration
- Well-marginated, hypoechoic subcutaneous mass with posterior enhancement
  - Small echogenic speckles and larger discrete hypoechoic cleft-like areas within mass
  - ± discrete cleft-like areas of hypoechogenicity within mass
  - ± claw sign at dermal-subdermal interface
- Color Doppler imaging findings very variable
  - Usually has mild to moderate peripheral > central hyperemia but may have no visible hyperemia or very marked hyperemia
- US helps to characterize tumor type as well as assess local infiltration

TOP DIFFERENTIAL DIAGNOSES

- Malignant skin lesions: Metastasis or cutaneous lymphoma

CLINICAL ISSUES

- Slowly growing, slowly evolving cutaneous mass
- Starts as small nodule, grows into medium erythematous or bluish lesion
- Cutaneous ulceration and satellite nodules in advanced lesions
- Occasionally, plaque-like indurated region progressing to single or multiple protuberant nodules
- Locally infiltrative
- Uncommonly metastasize (3-5%)

(Left) *Clinical photograph of a young man with a protruding mass ➡ with bluish discoloration on the dorsum of the 2nd toe is shown. The mass had been growing slowly for 3 years.* **(Right)** *Transverse US of the same mass shows a well-defined roundish mass ➡ within the subcutaneous tissues in contact with the dermis. The mass contacts the extensor digitorum tendon ➡ over the proximal phalanx ➡. The tendon moved freely on dynamic examination.*

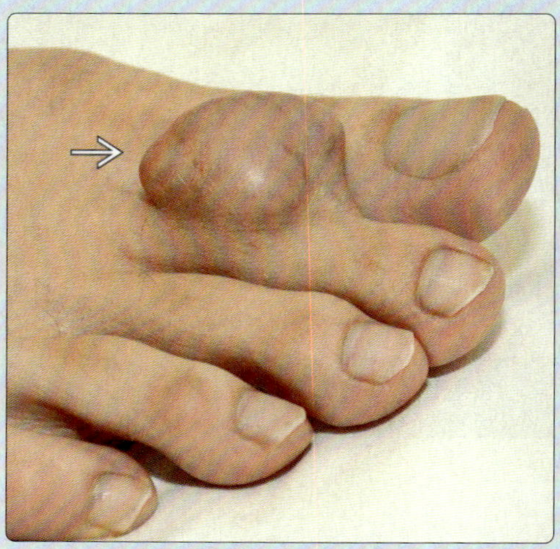

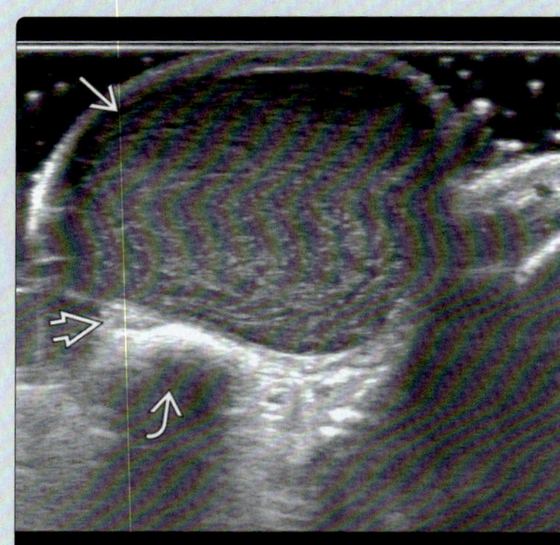

(Left) *Transverse color Doppler US of the same tumor shows marked central and peripheral hyperemia ➡. This level of hyperemia is unusually high for dermatofibrosarcoma protuberans (DFSP).* **(Right)** *Coronal intermediate-weighted MR of the same lesion shows a well-defined dermal/subcutaneous mass ➡ in contact with but separate from the extensor tendon ➡ overlying the proximal phalanx. Histology revealed DFSP.*

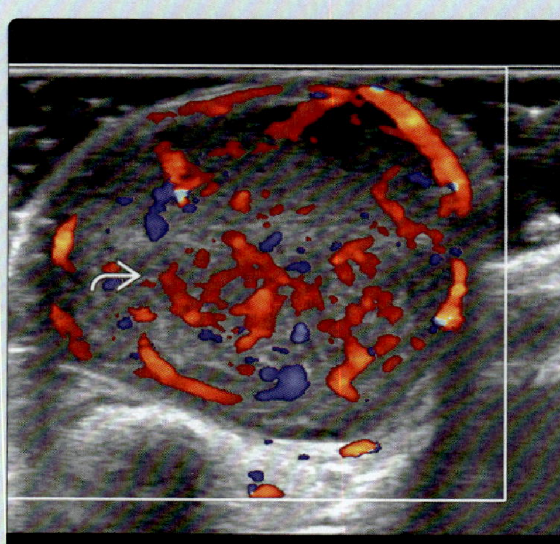

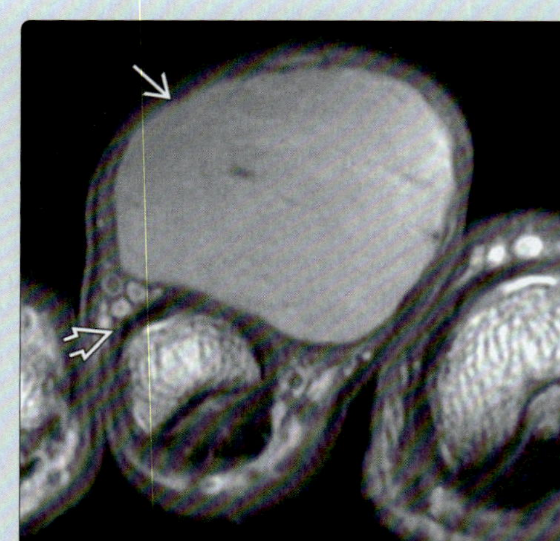

TERMINOLOGY

Abbreviations

- Dermatofibrosarcoma protuberans (DFSP)

Definitions

- Low-grade spindle cell mesenchymal sarcoma of dermis and subdermis

IMAGING

General Features

- Best diagnostic clue
  - Nearly always arises in dermis, then infiltrates subcutaneous tissue
- Location
  - Most commonly arises in trunk (~ 50%), particularly chest wall, back, abdominal wall
  - Next common site is proximal extremities (35-40%)
- Size
  - Ranges from 1 cm to > 25 cm; averages 4 cm at time of excision
- Morphology
  - Protruding mass arising from skin with bluish or reddish discoloration

Ultrasonographic Findings

- Subcutaneous, well marginated, ± lobulated border
  - ± protruding from skin surface
  - Uncommonly arise in deep subfascial tissues (5%)
- Hypoechoic mass with fine echogenic speckles and posterior enhancement (100%)
  - Roundish (70%) > ovoid (30%)
  - Well defined (90%) > ill defined (10%)
  - Calcific foci (100%)
- ± discrete cleft-like areas of hypoechogenicity within mass
- ± claw sign at dermal-subdermal interface
  - Hypoechoic extension along dermal-subdermal interface
- Color Doppler imaging findings very variable
  - Usually has mild to moderate peripheral > central hyperemia but may have no visible hyperemia or very marked hyperemia

CT Findings

- Usually well-defined with distinct lobular or nodular architecture
- Attenuation equal to or slightly higher than that of skeletal muscle
- Central cystic component (myxoid degeneration, necrosis, or hemorrhage) uncommon
- No calcification, moderate contrast enhancement

MR Findings

- Isointense to hyperintense relative to muscle on T1WI
- Hyperintense relative to muscle on fluid-sensitive sequences

Imaging Recommendations

- Best imaging tool
  - US is helpful to characterize tumor type as well as assess degree of local infiltration

DIFFERENTIAL DIAGNOSIS

Malignant Skin Lesions

- Metastasis or cutaneous lymphoma
  - More infiltrative appearance

PATHOLOGY

Gross Pathologic & Surgical Features

- Indurated plaque or nodule involving dermis and subcutaneous layers with firm, gray-white cut surface
  - Tumor infiltration from dermis tends to be asymmetrical with frond-like horizontal or vertical extensions into subcutaneous fat

Microscopic Features

- Histopathology varies within lesion
  - Central regions: Spindle-shaped cells in storiform pattern with myxoid change
  - Peripheral regions: Hypocellular with dermal collagen separating spindle cells
  - Deep regions: Spindle cells expanding fibrous septa and interdigitating with fat
- Morphological variants: Pigmented, myofibroblastic, granular cell, purely or predominantly myxoid tumor
- Focal sarcomatous transformation ~ 10%

CLINICAL ISSUES

Presentation

- Most common signs/symptoms
  - Slowly growing, slowly evolving cutaneous mass
    - Starts as small nodule, grows into medium-sized erythematous or bluish lesion
    - Ulceration and satellite nodules in advanced lesions
    - Occasionally, plaque-like indurated region progressing to single or multiple protuberant nodules

Demographics

- Age
  - Young to middle-aged adults, peak in 3rd-5th decades
- Sex
  - Male predominance
- Epidemiology
  - Most common primary sarcoma in subcutaneous tissues, 6% of all soft tissue sarcomas

Natural History & Prognosis

- Local recurrence in 18-55% depending on completeness of initial excision
- Metastases, usually to lungs, is rare (3-6%) and usually in those cases with focal sarcomatous transformation

Treatment

- En bloc resection with wide surgical excision (≥ 3-cm margin)

SELECTED REFERENCES

1. Sung TH et al: Dermatofibrosarcoma protuberans: a comprehensive review on the spectrum of clinico-radiological presentations. J Med Imaging Radiat Oncol. 61(1):9-17, 2017
2. Lee KL et al: Ultrasound appearances of dermatofibrosarcoma protuberans. J of Med Ultrasound. 21(1):21–8, 2013

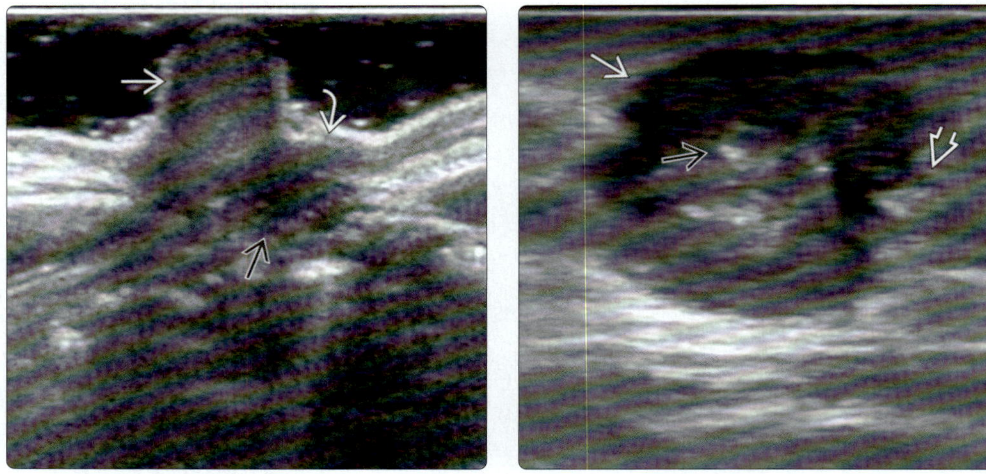

(Left) *Transverse US of a histologically proven DFSP shows a quite well-defined protruding nodule → from the umbilicus with peripheral extension at the dermal-subdermal interface → as well as at the linea alba →.* **(Right)** *Transverse US of the shoulder in a middle-aged woman shows a quite well-defined hypoechoic mass → arising in the subcutaneous layer in contact with the dermis. This histologically proven DFSP is atypical with an irregular margin → and small central punctate calcific foci →.*

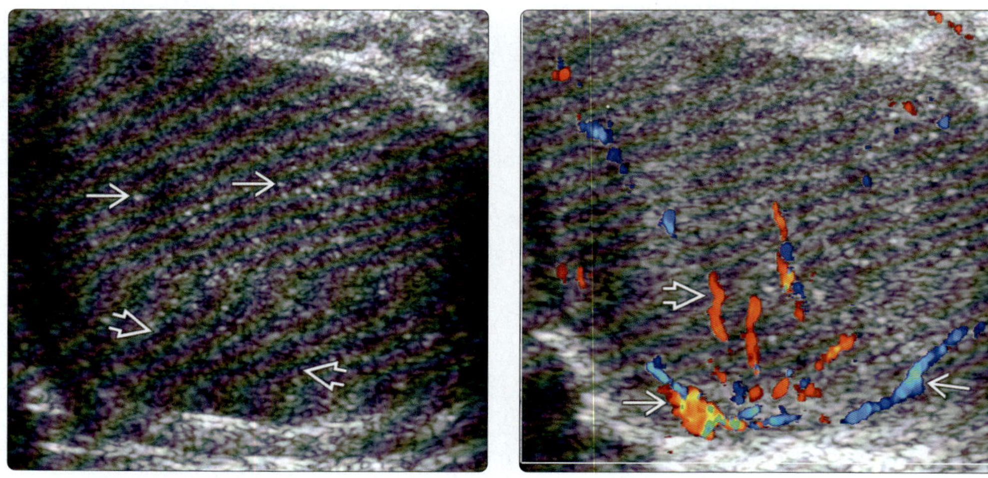

(Left) *Transverse US of DFSP shows typical tiny hyperechoic foci → and central, small irregular cleft-like hypoechoic areas →.* **(Right)** *Transverse color Doppler US of the same tumor shows the typical vascular pattern of DFSP with both peripheral → and inwardly radiating central → vascularity.*

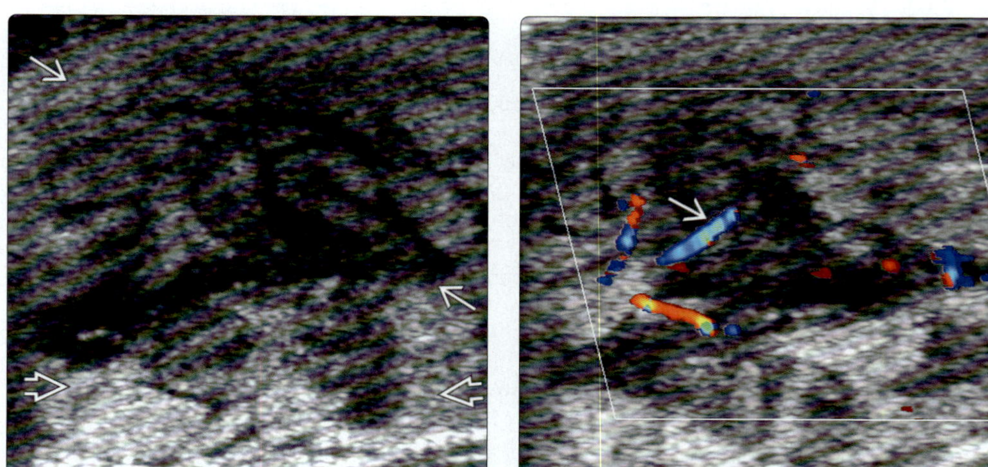

(Left) *Transverse US shows atypical appearances of DFSP → in that it is very ill defined and infiltrating the dermal and subcutaneous regions. The lesion is hypoechoic with posterior acoustic enhancement →.* **(Right)** *Transverse color Doppler US of the same tumor shows predominantly central vascularity →.*

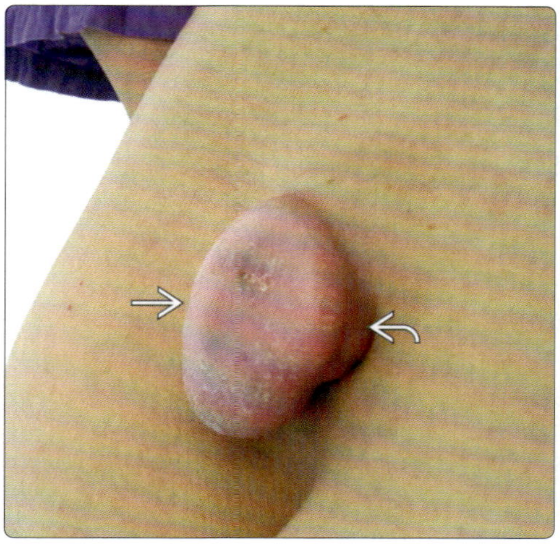

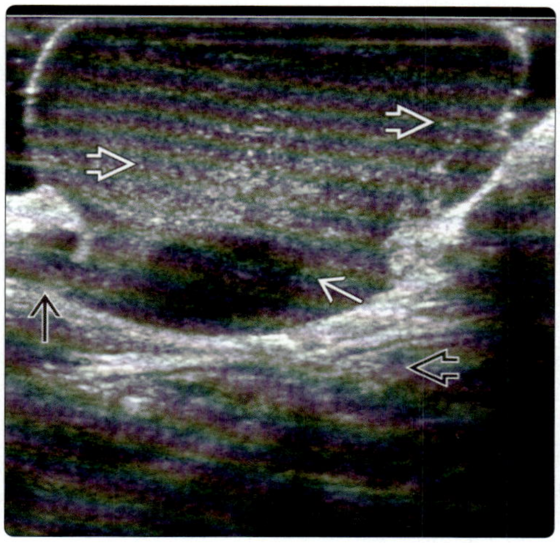

(Left) *Clinical photograph of the same patient shows the typical clinical appearance of DFSP. The reddish mass protrudes from the skin through a narrow stalk. No ulceration is present.* **(Right)** *Transverse US shows typical DFSP as a hypoechoic nodule with small hyperechoic speckles arising from the dermis and superficial subdermis with posterior acoustic enhancement, a lobulated deep margin, an eccentric hypoechoic area, and claw-like extension at the dermal-subdermal interface.*

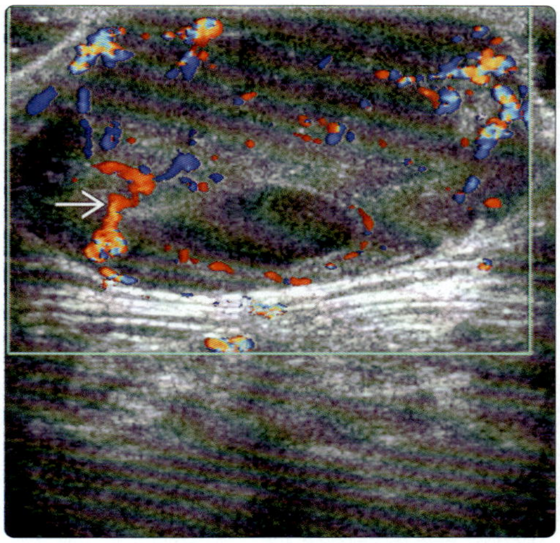

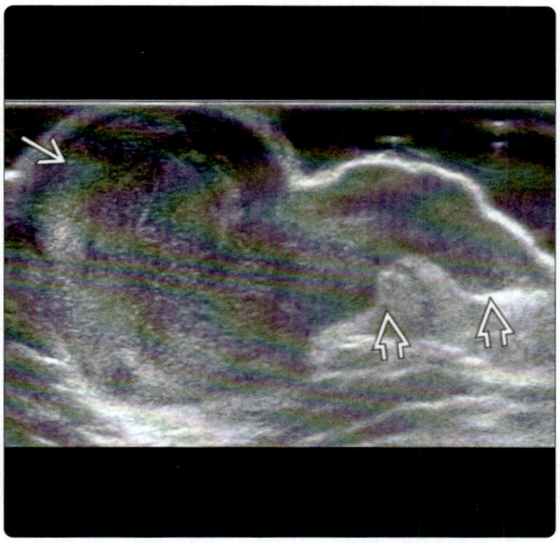

(Left) *Transverse color Doppler US of the same lesion shows typical central and peripheral vascularity.* **(Right)** *Transverse US of a histologically proven DFSP of the anterior abdominal wall in a middle-aged woman shows thick, finger-like tumoral extension at the dermal-subdermal interface.*

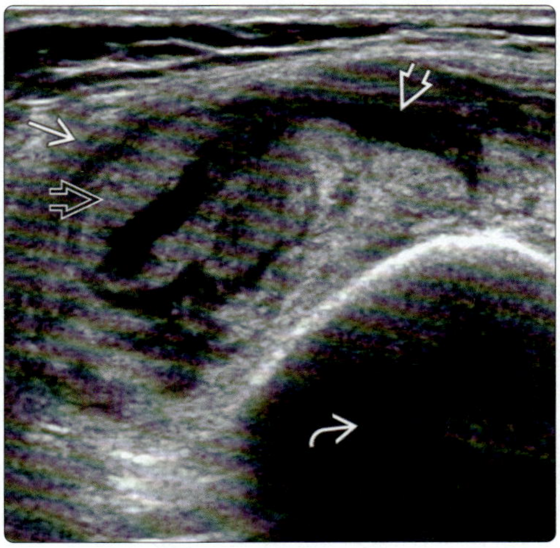

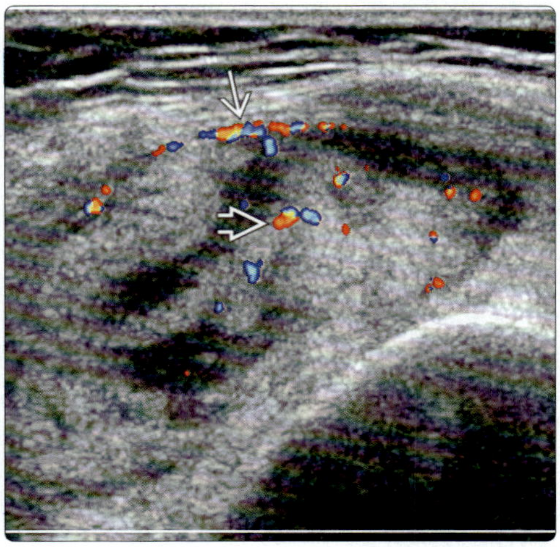

(Left) *Transverse US shows a histologically proven DFSP in the deep subfascial tissues anterior to the tibia. This lesion is atypical in both its deep location and echogenicity containing mixed hypoechoic and hyperechoic areas.* **(Right)** *Transverse color Doppler US of the same tumor shows both central and peripheral vascularity.*

Vascular Leiomyoma

KEY FACTS

TERMINOLOGY

- Benign solitary smooth muscle neoplasm that most commonly occurs in subcutaneous tissues of lower extremities

IMAGING

- 2/3 involve lower leg or ankle
- Majority are < 2 cm in diameter and subcutaneous
- Well-defined smooth margin
- Usually round or ovoid in shape
- Long axis parallel to extremity axis
- Hypoechoic, homogeneous
- No perceivable capsule
- Mild to moderate posterior enhancement
- Occurs in close proximity to artery or vein
- Hypervascular with peripheral vascular convergence
- Atypical appearances are calcification, fat, myxoid area, and hemorrhage

TOP DIFFERENTIAL DIAGNOSES

- Peripheral nerve sheath tumor

PATHOLOGY

- Arises from muscularis media of small arteries or veins

CLINICAL ISSUES

- Often painful or tender
- Most commonly present in middle age
- Slight female predominance

DIAGNOSTIC CHECKLIST

- Very similar to small peripheral nerve sheath tumors, which do not necessarily possess thickened entering or exiting nerves
  - If it looks like nerve sheath tumor, but does not have entering or exiting nerve, consider vascular leiomyoma, particularly if located in lower leg or ankle

(Left) *Transverse US of the thigh shows the typical appearance of a vascular leiomyoma ➡. A well-defined, hypoechoic, oval-shaped mass located in the superficial subcutaneous and deep dermal layers of the lower limb is shown. There is mild posterior acoustic enhancement ➡.* **(Right)** *Transverse color Doppler US of the same patient shows peripheral vascularity ➡ within the lesion with vessels converging to a single point (vessel convergence).*

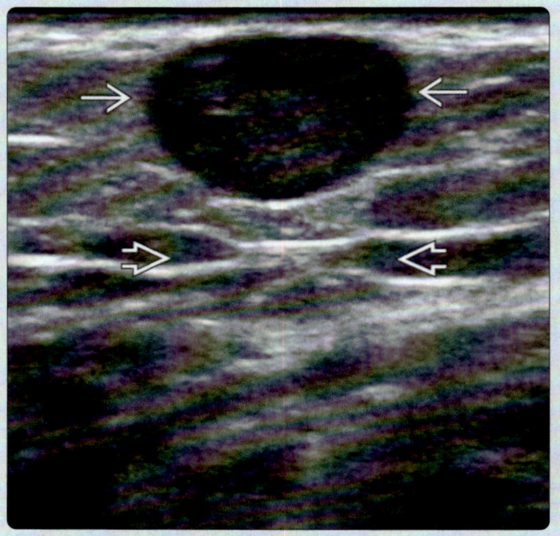

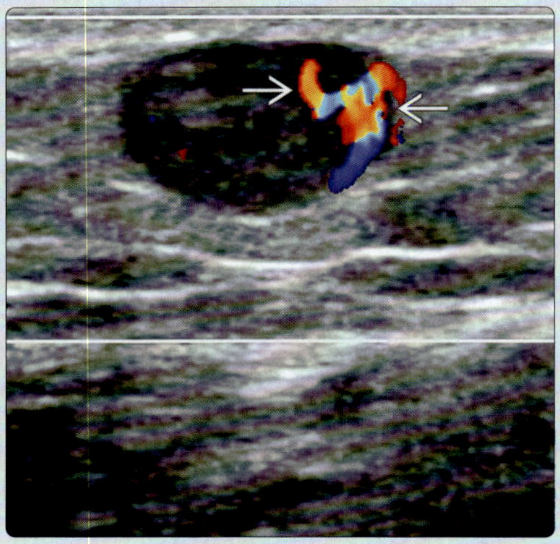

(Left) *Transverse color Doppler US shows a typical vascular leiomyoma ➡. The tumor lies close to a subcutaneous vessel ➡, which is 1 distinguishing feature of vascular leiomyoma. This should not be confused with a thickened entering or exiting nerve.* **(Right)** *Axial T2 FS MR of the heel region shows a well-defined mass ➡ lateral to the calcaneum ➡. Intratumoral hypointense areas ➡ are due to vascular channels. Vascular leiomyomas are typically T1 hypointense, T2 hyperintense, with marked enhancement.*

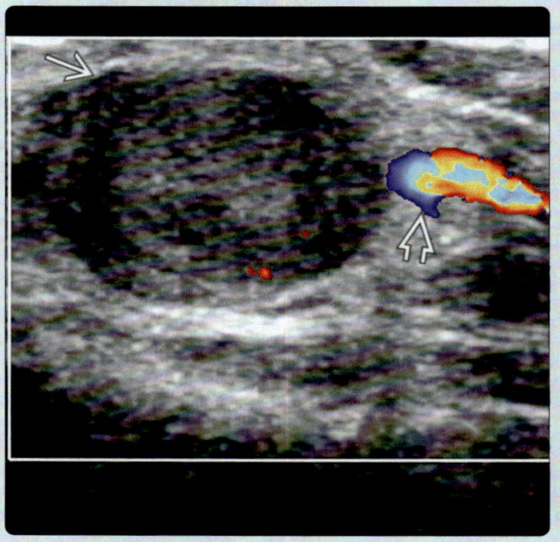

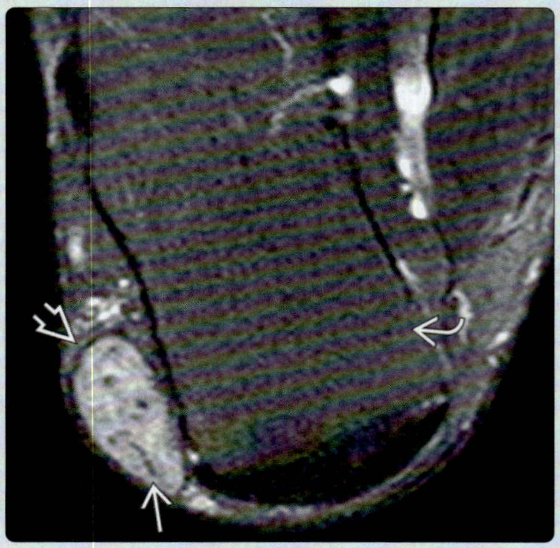

TERMINOLOGY

Synonyms

- Angioleiomyoma, angiomyoma

Definitions

- Benign solitary smooth muscle neoplasm arising from muscular layer of small arteries or veins

IMAGING

General Features

- Location
  - Any location; 2/3 occur around lower leg or ankle
  - Lower leg and ankle > thigh > upper extremity > head and neck > trunk
- Size
  - Majority < 2 cm in diameter
- Morphology
  - Usually roundish or ovoid in shape

Ultrasonographic Findings

- Grayscale ultrasound
  - Small (< 2 cm), roundish or oval-shaped mass
  - Subcutaneous > dermal > subfascial
  - Well-defined smooth margin
  - Long axis parallel to extremity axis
  - Hypoechoic, homogeneous
  - No perceivable capsule
  - Mild to moderate posterior acoustic enhancement
  - Occurs in close proximity to vessels
  - ± calcific foci
  - ± anechoic foci of myxoid accumulation
  - May also have hemorrhagic or lipomatous component
- Color Doppler
  - Hypervascular with high-resistance flow
    - Typically several vessels converge to single point at tumor periphery

CT Findings

- Well-defined mass with similar attenuation to muscle
- ± calcification

MR Findings

- T1WI
  - Isointense to mildly hyperintense to muscle
- T2WI
  - Heterogeneously hyperintense to muscle
- T1WI C+
  - Variable enhancement depending on subtype
  - Located adjacent to vessels
  - Similar appearance to peripheral nerve sheath tumor

Imaging Recommendations

- Best imaging tool
  - US: Can usually either allow specific diagnosis or at least limit list of differential diagnoses

DIFFERENTIAL DIAGNOSIS

Benign Peripheral Nerve Sheath Tumor

- ± thickening of entering or exiting nerve
- More prominent myxoid areas, chaotic vascular pattern

PATHOLOGY

General Features

- Etiology
  - Benign smooth muscle neoplasm that arises from muscular layer of small arteries or veins
- Genetics
  - No consistent chromosomal abnormality

Gross Pathologic & Surgical Features

- Solitary, pseudoencapsulated, gray-white myxoid mass

Microscopic Features

- Mass composed of medium-sized, thin-walled, venous-like vessels and thick-walled, arterial-like vessels that otherwise lack typical features of veins or arteries
  - Walls of these vessels merge imperceptibly with proliferative mass of benign smooth muscle
- Absent or infrequent mitoses
- 3 broad histologic subtypes depending on amount of smooth muscle and type of vessel present (solid, venous, cavernous)
- ± myxoid areas, calcification, hemorrhage, or fat

CLINICAL ISSUES

Presentation

- Most common signs/symptoms
  - Often (> 50%) painful or tender
    - Symptoms are affected by light touch, temperature, or hormonal change
    - Upper extremity lesions are less likely to be painful than lower extremity lesions
  - Grows slowly over several years

Demographics

- Age
  - Any age, but most common in middle-aged adults
- Gender
  - F:M = 1.7:1
- Epidemiology
  - Uncommon, ~ 5% of benign soft tissue tumors

Natural History & Prognosis

- No spontaneous regression

Treatment

- Complete marginal surgical excision

DIAGNOSTIC CHECKLIST

Image Interpretation Pearls

- Well-defined, small, hypoechoic vascular mass within subcutaneous tissue of lower extremity
  - Consider vascular leiomyoma in same differential as small peripheral nerve sheath tumor

SELECTED REFERENCES

1. Park HJ et al: Sonographic appearances of soft tissue angioleiomyomas: differences from other circumscribed soft tissue hypervascular tumors. J Ultrasound Med. 31(10):1589-95, 2012

(Left) *Longitudinal US of a 53-year-old man shows a medium-sized subcutaneous mass → superficial to the Achilles tendon. The mass is mildly hyperechoic with a central hypoechoic area → and prominent acoustic enhancement →. No adjacent neural thickening is present.* **(Right)** *Color Doppler US shows moderate hyperemia. The US diagnosis was either a vascular leiomyoma or nerve sheath tumor. The final histologic diagnosis was a vascular leiomyoma.*

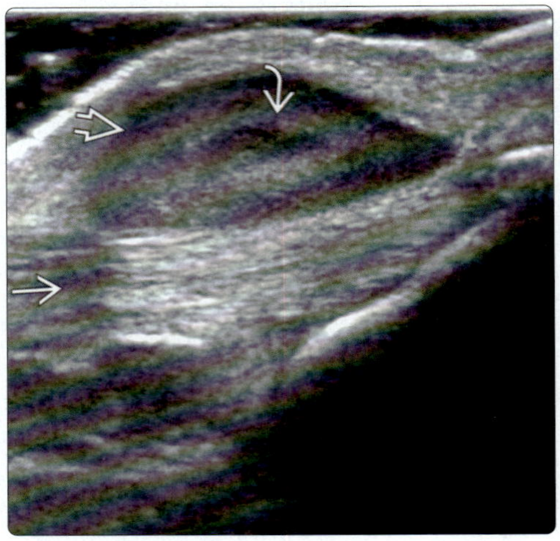

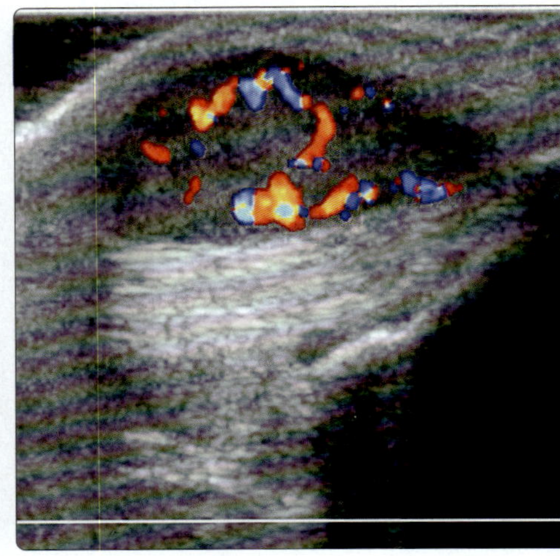

(Left) *Longitudinal US of a 51-year-old man shows a small, oval-shaped, well-defined solid hypoechoic subcutaneous mass → superficial to the Achilles tendon →. Mild acoustic enhancement is present.* **(Right)** *Color Doppler US shows marked color flow. The US diagnosis was either vascular leiomyoma, nerve sheath tumor, or glomus tumor. The final histologic diagnosis was vascular leiomyoma.*

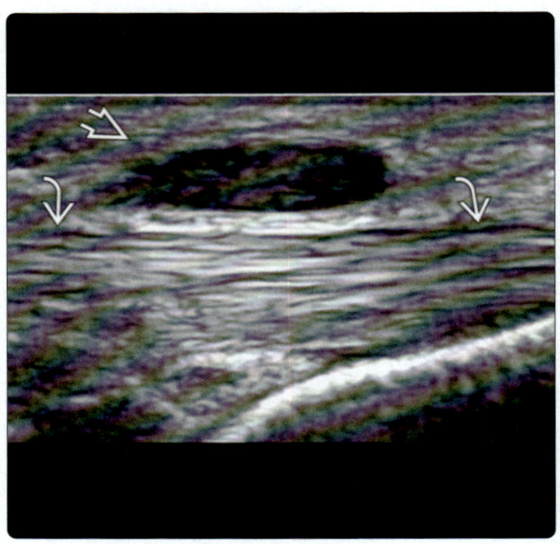

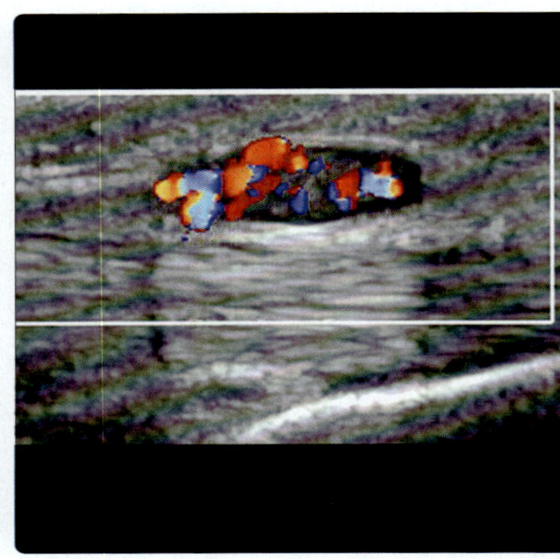

(Left) *Longitudinal US shows a subcutaneous vascular leiomyoma → in the forefoot with a typical hypoechoic echotexture, no visible capsule, and moderate posterior acoustic enhancement →. The head of the 1st metatarsal bone → and flexor hallucis longus tendon → are shown.* **(Right)** *Longitudinal color Doppler US of the same lesion reveals mild vascularity with vessels tending to converge toward the periphery →.*

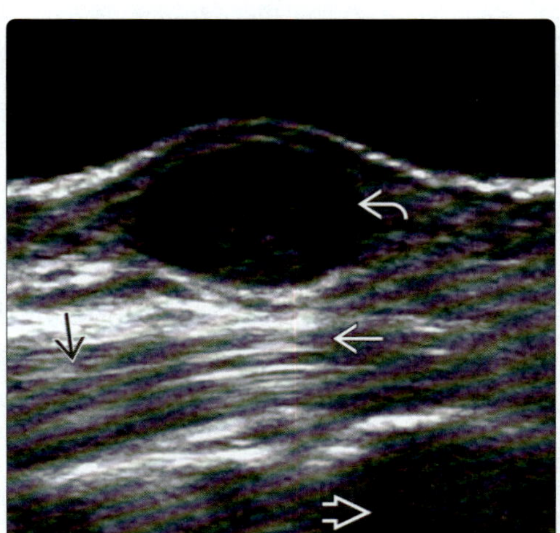

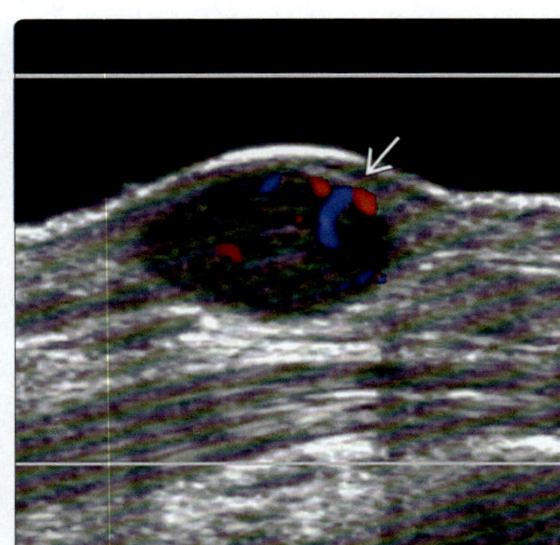

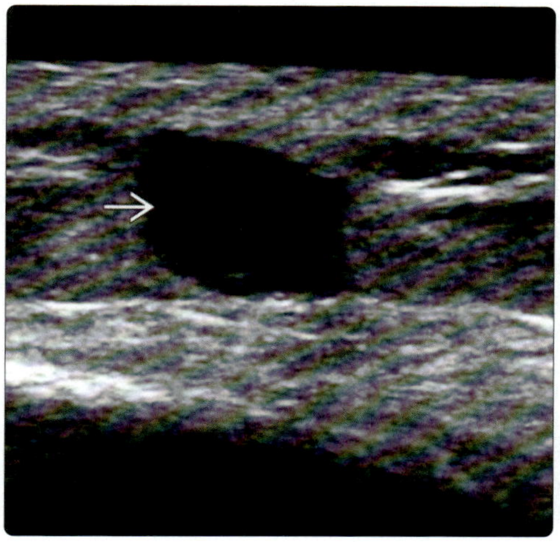

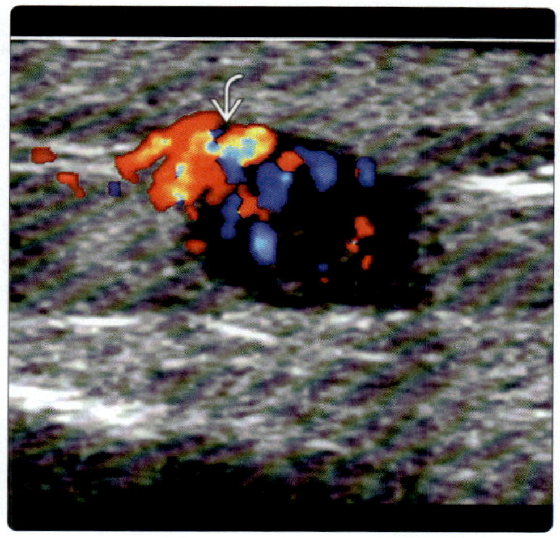

(Left) *Transverse US shows a subcutaneous vascular leiomyoma* ➔ *in the ankle region with a lobulated and mildly angular border. Note the characteristic hypoechoic echotexture of the vascular leiomyoma.* **(Right)** *Transverse color Doppler US of the same tumor reveals moderate peripheral and central vascularity and vascular convergence* ➔ *to a peripheral area.*

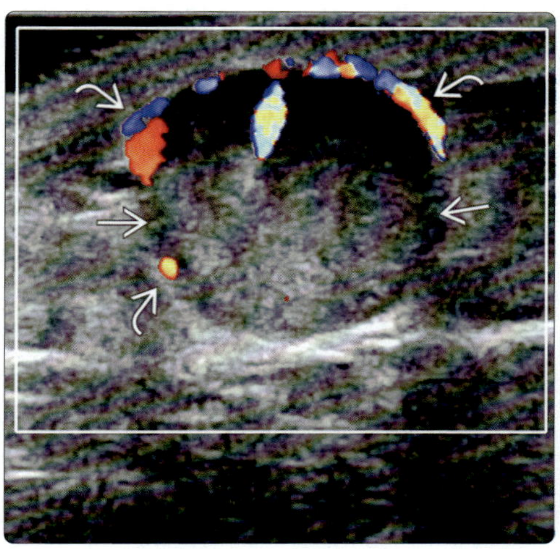

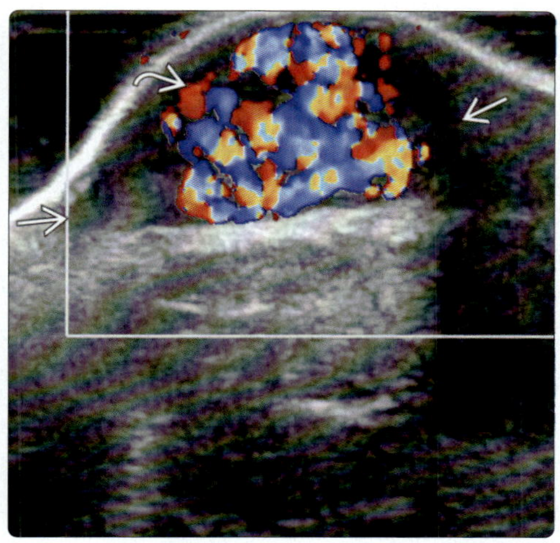

(Left) *Transverse color Doppler US shows a subcutaneous vascular leiomyoma* ➔ *in the ankle region. It shows mild peripheral vascularity* ➔*.* **(Right)** *Transverse color Doppler US shows a subcutaneous vascular leiomyoma* ➔ *in the ankle region. It shows very significant central and peripheral vascularity* ➔*.*

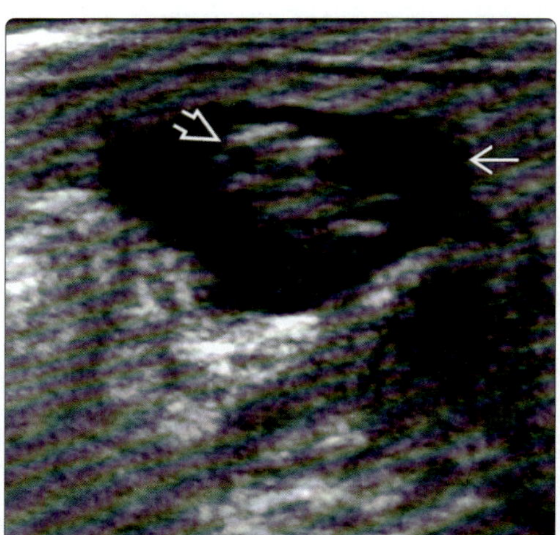

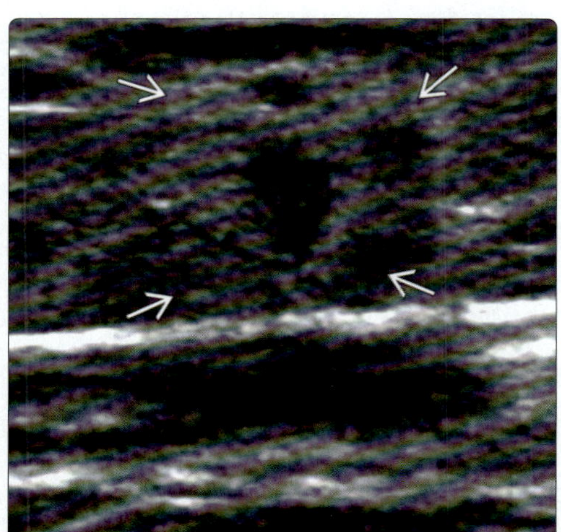

(Left) *Transverse US shows atypical, histologically proven vascular leiomyoma* ➔ *within the subcutaneous tissues of the thigh. The vascular leiomyoma is large (> 2 cm in diameter). It has central calcification* ➔*, which is an uncommon, though recognized, feature of vascular leiomyoma.* **(Right)** *Transverse US shows an atypical subcutaneous vascular leiomyoma* ➔ *in the ankle region. This histologically proven tumor has an ill-defined border with an echogenic lipomatous-like periphery.*

Superficial Metastases, Lymphoma, and Melanoma

KEY FACTS

TERMINOLOGY

- "Superficial" refers to tissues above investing fascia

IMAGING

- Familiarity with imaging characteristics of recognizable superficial soft tissue tumors is key to recognizing malignant tumor
  - 1 of best criterion of malignant superficial tumor (metastases, lymphoma, melanoma, etc.) is that appearance does not conform to recognized appearance of benign superficial tumor (lipoma, vascular anomaly, epidermoid cyst, etc.)
- Well-demarcated, irregular margin
- Melanoma metastases are often markedly hypoechoic ± tail sign ± string sign
  - Melanoma metastases may be satellite, in-transit, interval nodal, regional nodal, or distant

PATHOLOGY

- 5% of superficial tumors are malignant
- 50% of these are metastatic
  - Most common (> 50%) from breast, lung, and melanoma
  - Adenocarcinoma (40%) > squamous cell carcinoma > melanoma
- Remaining 50% are lymphoma, primary sarcoma of superficial tissues
- 30% of lymphomas are extranodal
- Most musculoskeletal involvement is to bone and muscle
- Involvement of superficial tissues is uncommon
- Subcutaneous panniculitis-like lesion is uncommon form of peripheral T-cell lymphoma

CLINICAL ISSUES

- Superficial mass in patient with known malignancy is still more commonly benign than malignant
  - Superficial metastasis is 1st manifestation of malignancy in 0.8%

(Left) *Longitudinal US shows a subcutaneous metastatic nodule → of the anterior chest wall in a patient with disseminated carcinoma of the lungs. The nodule is irregular in outline with a mild degree of perilesional edema →. It abuts the atrophic pectoralis muscles →.* **(Right)** *Transverse power Doppler US of the same tumor shows mild intrinsic vascularity →. Carcinoma of the lungs is 1 of the most common malignancies to metastasize to the superficial tissues.*

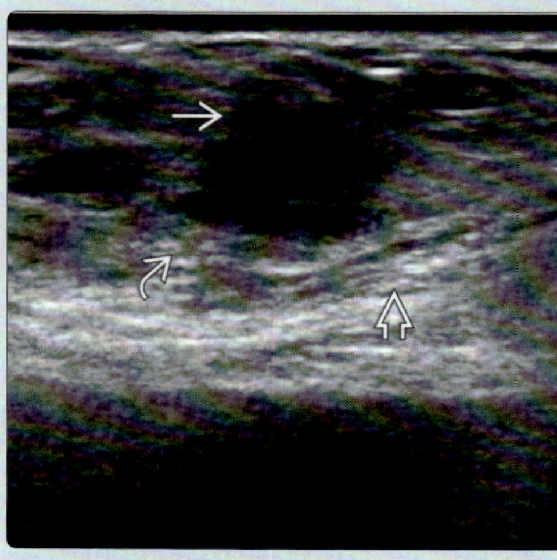

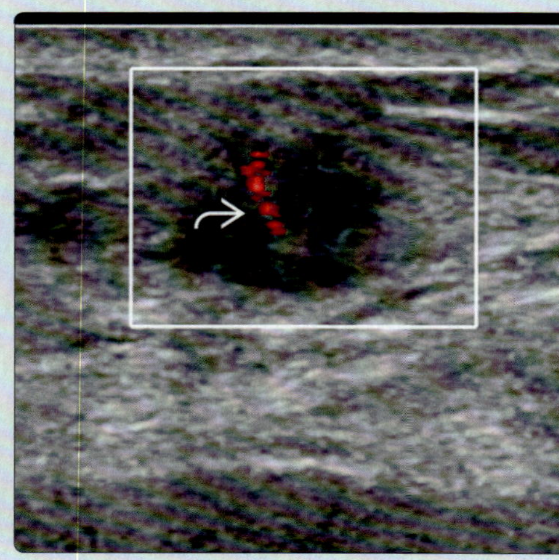

(Left) *Longitudinal US shows a well-defined, oval-shaped, in-transit subcutaneous metastasis → of the midthigh in a patient with foot melanoma. The metastatic lesion is hypoechoic with moderate posterior acoustic enhancement →. Small cystic areas are present →.* **(Right)** *Longitudinal color Doppler US of the same tumor shows moderate intratumoral hypervascularity with a chaotic, mainly peripheral vascular pattern →.*

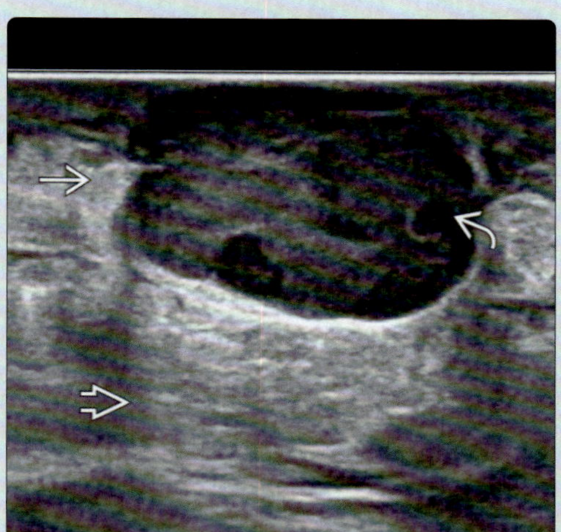

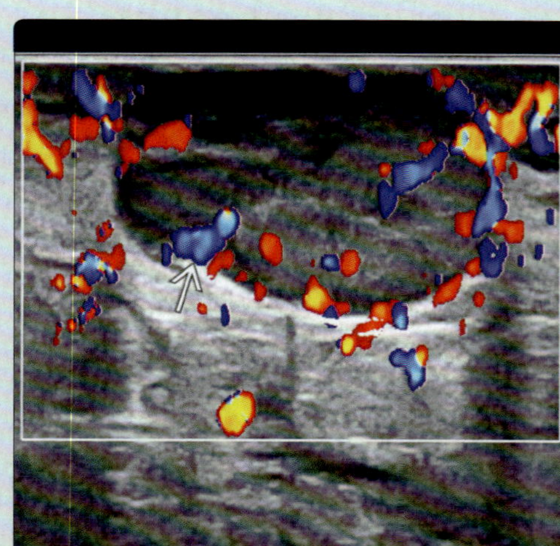

TERMINOLOGY

Synonyms

- Superficial metastasis, skin metastasis, subcutaneous metastasis, secondary deposits, metastatic tumor

Definitions

- Spread of tumor from its primary site to skin or subcutaneous tissues
- "Superficial" refers to tissues above investing fascia (subcutaneous and cutaneous tissues)

IMAGING

General Features

- Best diagnostic clue
  - Hypoechoic, superficial soft tissue mass with intrinsic vascularity
- Location
  - Trunk (> 50%) > > head and neck > extremities
  - Melanoma metastasis within 2 cm of primary tumor is deemed to be satellite metastasis
  - Melanoma metastasis at extranodal site > 2 cm from primary tumor is deemed to be in-transit metastasis
- Size
  - Variable; mean: ~ 2 cm
- Morphology
  - Well-defined to infiltrative mass

Radiographic Findings

- Radiography
  - Soft tissue swelling if large size or mass protuberant

CT Findings

- Isodense to hypodense to muscle
- Variable enhancement

MR Findings

- T1WI
  - Usually isointense to hypointense to muscle
    - Areas of T1 hyperintensity may be seen in melanoma due to paramagnetic effect of melanin and hemorrhage
- T2WI
  - Variable; homogeneous to heterogeneously hyperintense to hypointense to skeletal muscle
- T1WI C+
  - Variable enhancement
    - Homogeneous, heterogeneous, nodular, or ring enhancement

Ultrasonographic Findings

- US appearance does not conform with known appearances of benign superficial tumors (e.g., lipoma, vascular anomaly, nerve sheath tumor, epidermoid cyst, etc.)
- **Shape**
  - Usually round or oval
    - ± elongated shape seen in melanoma due to tumor spreading along lymphatic channels
    - Tail sign and string sign
    - Tail sign = thin, hypoechoic propagation from 1 or both poles of superficial melanoma metastasis
      - Seen in 2.5% melanoma patients
    - String sign = multiple in-transit lesions connected to each other like string of rosary beads
      - Seen in 1% melanoma patients
- **Margin**
  - Usually well-demarcated, irregular margin
- **Echogenicity**
  - Most are hypoechoic
    - Marked hypoechogenicity in melanoma due to poor beam reflectivity of melanotic tissue
- **Internal architecture**
  - Homogeneous or heterogeneous
  - Cystic areas are seen in melanoma, squamous cell metastases, or other cystic/necrotic primary
  - Calcification from calcified or ossified primary tumor, such as osteosarcoma, medullary carcinoma of thyroid, or mucinous adenocarcinoma
- **Vascularity**
  - Mild to moderate intrinsic vascularity is usually present
  - Moderate to high intrinsic vascularity is seen in metastases from hypervascular primary tumors, such as melanoma and renal cell carcinoma
- **Peritumoral change**
  - Mild to moderate peritumoral edema is usually present
  - ± posterior enhancement
  - ± posterior shadowing
- US is useful in assessing locoregional spread of melanoma
  - Primary melanoma tumors are uncommonly imaged
  - Satellite tumor: Tumor arising ≤ 2 cm from primary tumor or its scar
  - In-transit tumor: Tumor arising ≥ 2 cm from primary tumor or scar along lymphatic course toward regional lymphatic nodes
  - Interval lymph node metastases: Metastasis of lymph node along path from primary tumor to regional lymph node station, e.g., epitrochlear node or popliteal node
  - Regional lymph node metastasis: Metastasis of regional nodes, e.g., ipsilateral inguinal group for lower limb melanoma, ipsilateral axillary group for upper limb melanoma
  - Distant cutaneous metastasis: Hematogenous metastasis to cutaneous or subcutaneous tissues

Imaging Recommendations

- Best imaging tool
  - US
    - All superficial tumors are readily accessible to high-resolution US
    - Allows good definition of tumor characteristics
    - Most superficial soft tissue tumors can be diagnosed on imaging characteristics alone without need for histological confirmation
      - However, suspected malignant tumor needs histological confirmation
    - Familiarity with imaging characteristics of recognizable superficial soft tissue tumors is key to recognizing malignant tumor
      - 1 of best criterion of malignant superficial tumor is that appearance does not conform to recognized appearances of any benign tumor

   - US diagnosis of tumor type is based not on 1 or 2 US signs, but on relative strength of multiple recognized US signs interpreted in conjunction with clinical presentation
- Protocol advice
  - US for assessment of superficial masses
    - Apply light transducer pressure to avoid compression or distortion of mass
      - Especially for color Doppler imaging to avoid blanching of tumoral vascularity
    - For melanoma, assess primary tumor site as well as for satellite tumor, in-transit tumor, regional nodes, and (when clinically suspect) distant cutaneous metastases
      - For primary in trunk, examine axillary, supraclavicular, infraclavicular, and deep pectoral nodes bilaterally
      - For primary in upper limb, examine ipsilateral epitrochlear, axillary, supraclavicular, infraclavicular, and deep pectoral nodes
      - For primary in lower limb, examine ipsilateral popliteal and bilateral groin and external iliac nodes
  - US-guided fine-needle aspiration for cytology or biopsy for histology
  - CT, PET/CT, or whole-body MR for systemic staging

DIFFERENTIAL DIAGNOSIS

Soft Tissue Sarcoma

- Very uncommon in superficial tissues
- Usually > 5 cm
- Can look very similar to metastases
  - Epithelioid sarcoma mimics metastatic carcinoma histologically

Nerve Sheath Tumor

- Entering and exiting nerve
- Fusiform or oblong-shaped
- Smooth border

Reactive Lymph Node

- Fatty hilum, hilar vascularity

Hematoma

- Organizing hematoma can have variable hyperemia
- Interval resolution rather than progression

Abscess

- Marked hyperemia and surrounding edema
- Local signs of infection

PATHOLOGY

General Features

- 5% of superficial soft tissue tumors are malignant
  - 50% of these are metastatic
  - Metastatic tumor cell type
    - Most common: Adenocarcinoma (~ 40%)
    - Followed by squamous cell carcinoma and melanoma
    - Less common: Sarcoma or carcinosarcoma
    - Primary tumor
      - Breast (> 50%) > lung > melanoma
  - 50% are nonmetastatic
    - Lymphoma, primary sarcoma of superficial tissues, such as dermatofibrosarcoma protuberans
  - Lymphoid malignancies are classified as T-cell neoplasms, B-cell neoplasms, & Hodgkin disease
    - T-cell neoplasms are divided into precursor T-cell neoplasms and peripheral T-cell neoplasms
    - Subcutaneous panniculitis-like T-cell lymphoma is uncommon type of peripheral T-cell lymphoma
  - Overall, ~ 30% of lymphomas are extranodal (spleen, gastrointestinal, salivary glands, thymus, thyroid, lung, genitourinary, pancreas, liver, adrenal, breast, and musculoskeletal)
    - Most musculoskeletal involvement is to bone and muscle
    - Involvement of superficial tissues is uncommon
    - Most cases of superficial lymphoma have known systemic lymphoma
      - Rarely, lymphoma of superficial tissues may precede diagnosis of systemic lymphoma

CLINICAL ISSUES

Presentation

- Most common signs/symptoms
  - Superficial nodule
  - Often have history of known malignancy
    - Superficial mass in patient with known malignancy is still more commonly benign
    - Superficial metastases is 1st manifestation of occult malignancy in 0.8%
  - Superficial metastases are generally more painful than primary soft tissue sarcoma

Demographics

- Age
  - Any; mean: 53 years

Natural History & Prognosis

- Superficial metastases indicate stage IV disease regardless of primary tumor or location

Treatment

- Highly variable depending on site of primary tumor and degree of metastatic spread
- Primary sarcoma of superficial tissues is treated with wide local excision

DIAGNOSTIC CHECKLIST

Consider

- Superficial malignant tumor if US appearances do not conform to recognized appearances of benign tumor (e.g., lipoma, vascular anomaly, epidermoid cyst)
- Superficial mass in patient with known malignancy is still more commonly benign rather than malignant

SELECTED REFERENCES

1. Corvino A et al: The tail and the string sign: new sonographic features of subcutaneous melanoma metastasis. Ultrasound Med Biol. 43(1):370-374, 2017
2. Hung EH et al: Ultrasound of musculoskeletal soft-tissue tumors superficial to the investing fascia. AJR Am J Roentgenol. 202(6):W532-40, 2014
3. Wong CY et al: Patterns of skin metastases: a review of 25 years' experience at a single cancer center. Int J Dermatol. 53(1):56-60, 2014

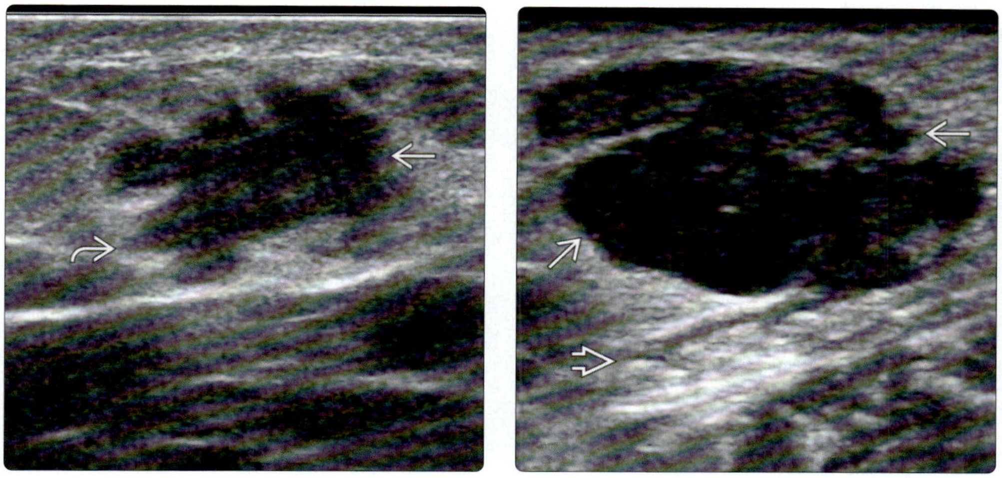

(Left) *Longitudinal US shows an irregular metastatic nodule ➡ in the subcutaneous tissue of the flank from a clear cell sarcoma in the foot. The border is very irregular with moderate peritumoral edema ➡.* **(Right)** *Longitudinal US shows an irregularly lobulated, hypoechoic, subcutaneous metastatic nodule ➡ of the abdominal wall from a malignant nerve sheath tumor in the calf. Mild posterior acoustic enhancement ➡ is present.*

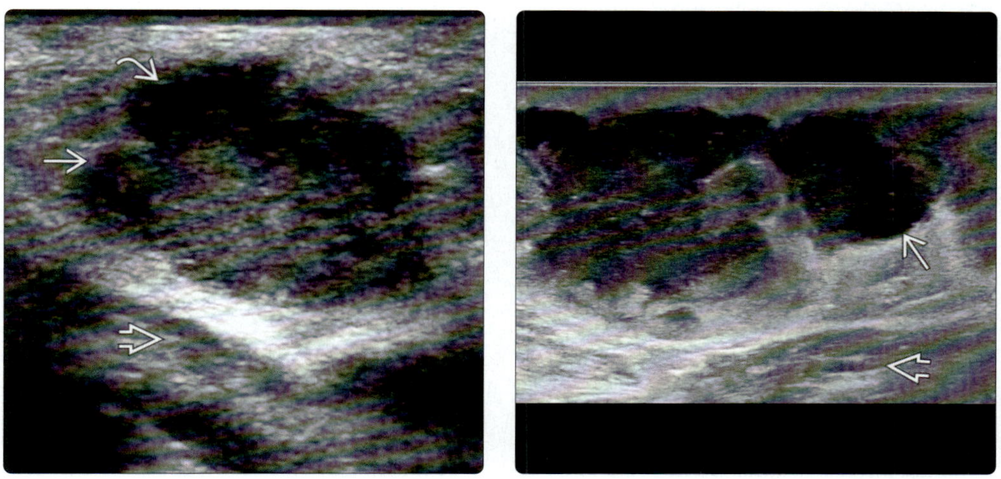

(Left) *Longitudinal US of the axilla in a patient with a high-grade distal femoral osteosarcoma shows an irregular, subcutaneous metastatic nodule ➡. Note anechoic areas ➡ and moderate posterior acoustic enhancement ➡. Unusually, there is no mineralization present.* **(Right)** *Longitudinal US of a patient with a rapidly enlarged superficial soft tissue mass in the lateral arm shows an irregularly lobulated, heterogenous, subcutaneous mass ➡. Mild posterior acoustic enhancement ➡ is present.*

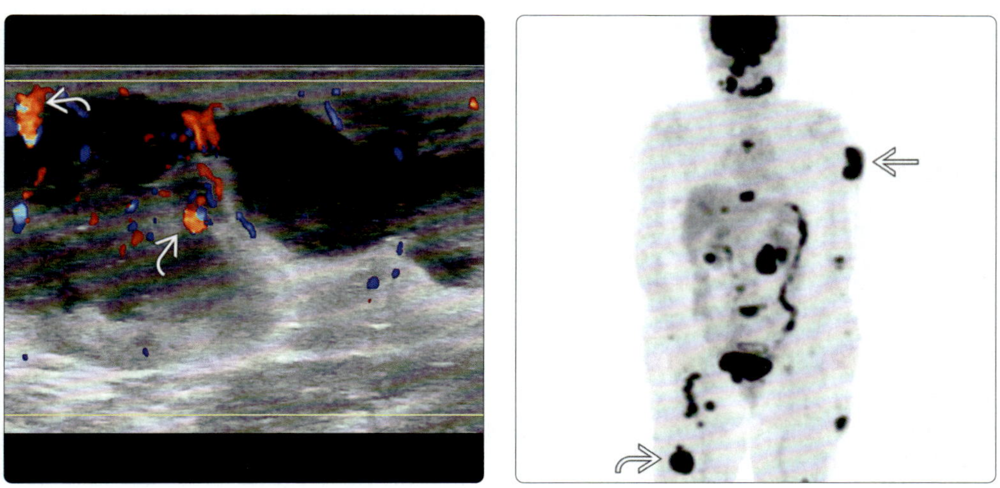

(Left) *Longitudinal color Doppler US of the same tumor shows moderate intratumoral vascularity ➡. Histology following percutaneous biopsy revealed diffuse large B-cell lymphoma.* **(Right)** *Coronal PET of the same patient shows foci of abnormal tracer uptake in neck, thorax abdominal, and pelvis, indicative of stage IV disease. The focus in the left arm ➡ corresponds to the subcutaneous metastasis seen in the previous images. There is another subcutaneous deposit ➡ in the right thigh.*

(Left) *Clinical photograph shows a patient with increasing calf swelling and erythema for 1 month. The patient was otherwise healthy.* **(Right)** *Longitudinal US of the same patient shows a large area of quite well-defined, intense subcutaneous and cutaneous edema and swelling. No discrete tumor mass is present. The edematous swollen subcutaneous tissue is lobulated, while the subdermal tissue is diffusely thickened without lobulation.*

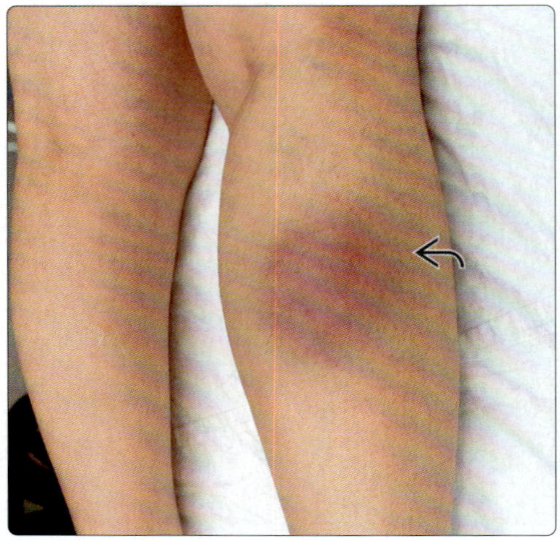

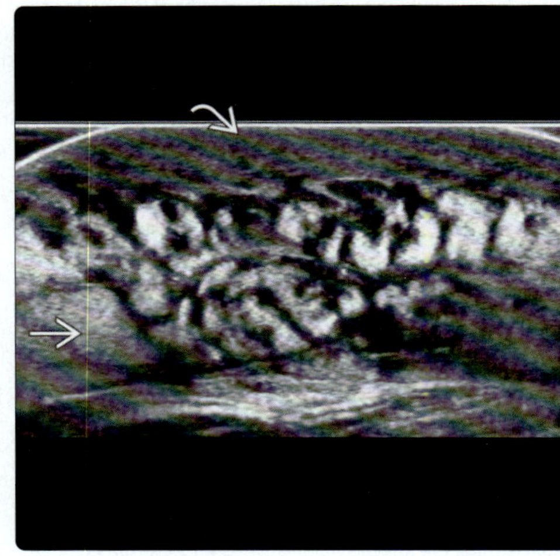

(Left) *Longitudinal color Doppler US of the same patient shows severe lesional hyperemia. Initially, this was thought to represent an unusual infective panniculitis. Biopsy, however, revealed peripheral T-cell malignant lymphoma.* **(Right)** *Longitudinal color Doppler US in the same patient at a site just proximal shows a hyperechoic, mildly vascular subcutaneous and subdermal mass, also due to T-cell lymphoma. This smaller lesion was clinically occult.*

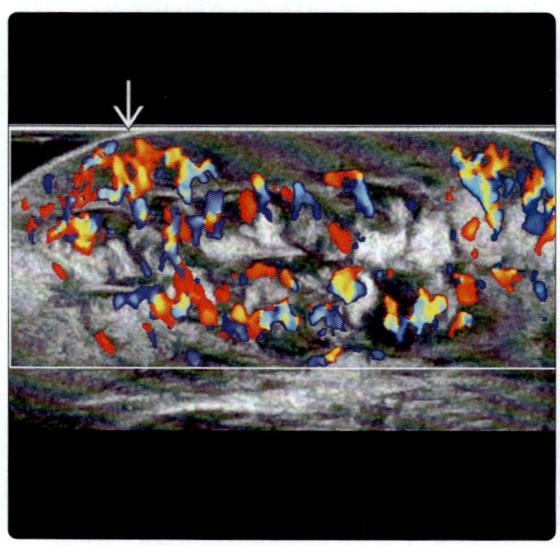

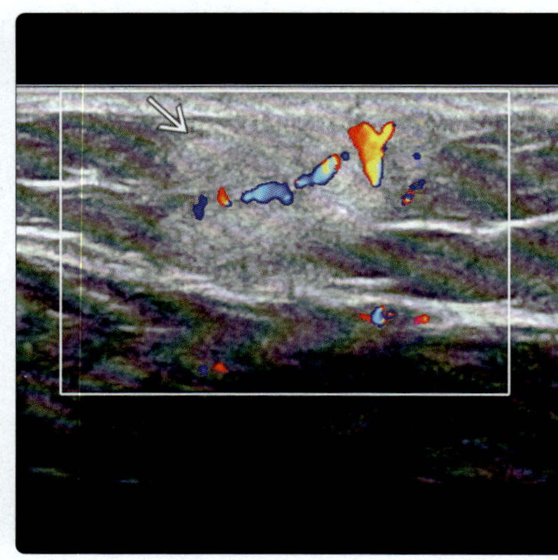

(Left) *Transverse US on the medial aspect of the knee in a patient with malignant melanoma of the ipsilateral heel shows a lobulated heterogeneously hypoechoic subcutaneous mass. Note the severe surrounding soft tissue edema.* **(Right)** *Transverse color Doppler US of the same tumor shows moderate to severe peripheral intratumoral vascularity. There is a small cystic component peripherally due to intralesional haemorrhage.*

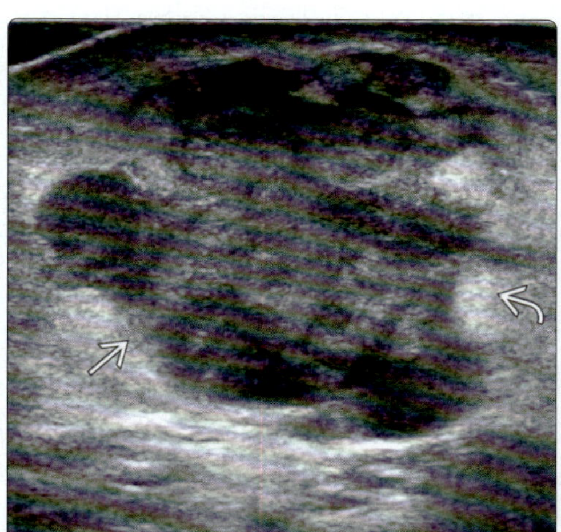

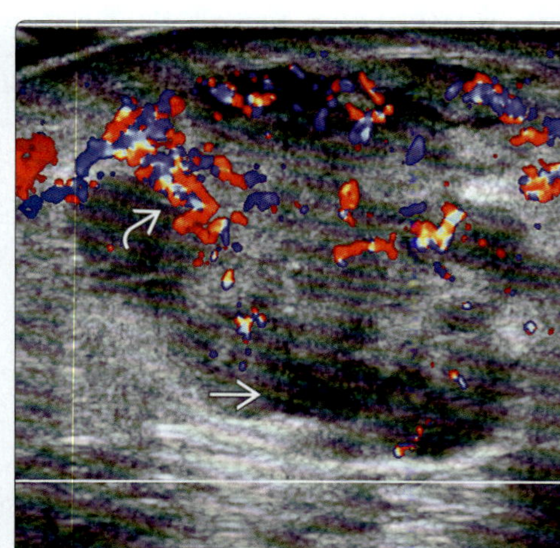

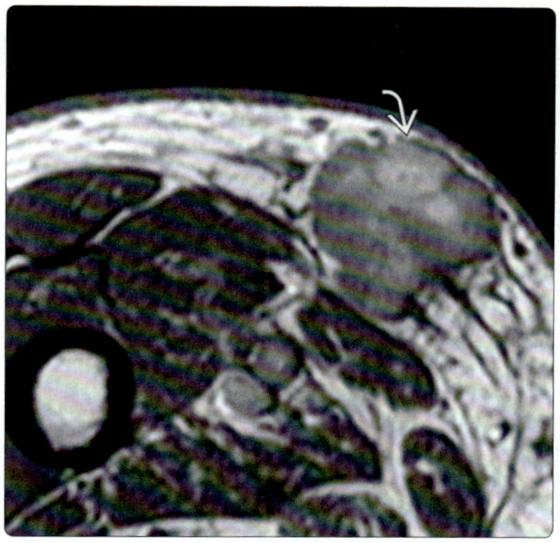

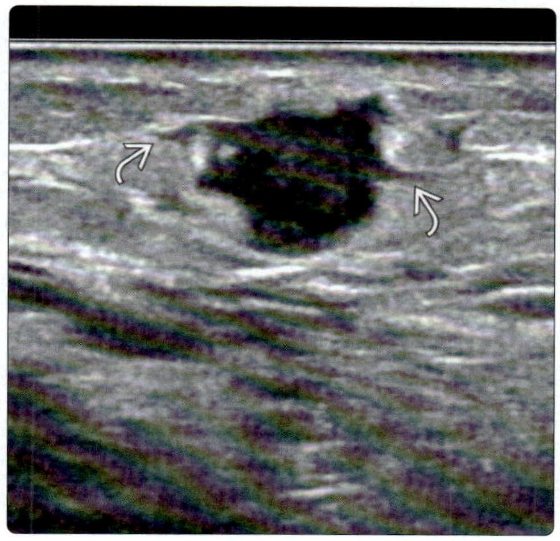

(Left) *Axial T1 MR of the same tumor shows a lobulated subcutaneous mass. The internal T1 hyperintensity can be due to the paramagnetic effect of melanin or subacute hemorrhage.* **(Right)** *Longitudinal US of the same patient shows a smaller in-transit subcutaneous metastasis on the medial thigh region. The tumor is irregular with hypoechoic tails extending from the tumor ends, presenting tumor spread along adjacent lymphatic channels.*

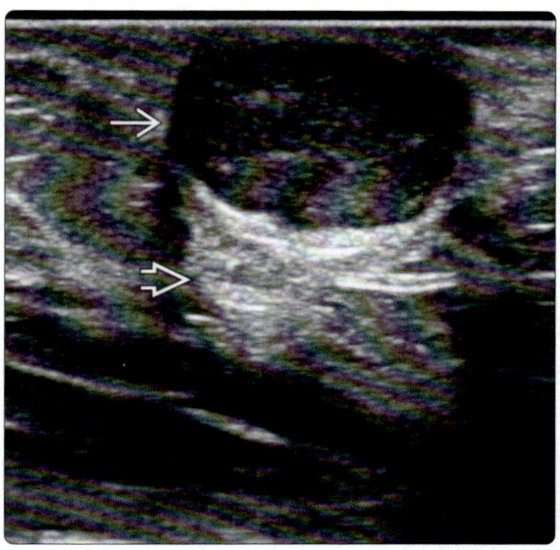

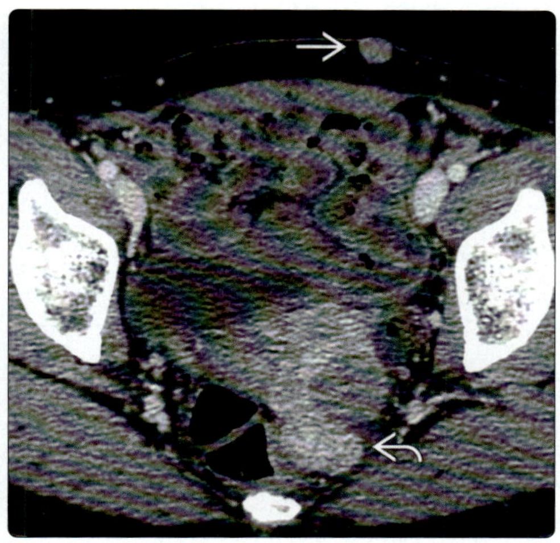

(Left) *Longitudinal US of the abdominal wall in a patient with metastatic melanoma shows a well-defined, hypoechoic subcutaneous nodule with moderate posterior acoustic enhancement. There is no tail sign.* **(Right)** *Axial CECT of the same patient shows the well-defined, mildly enhancing metastatic tumor in the anterior abdominal wall. There is a larger metastatic intraperitoneal deposit in the pouch of Douglas.*

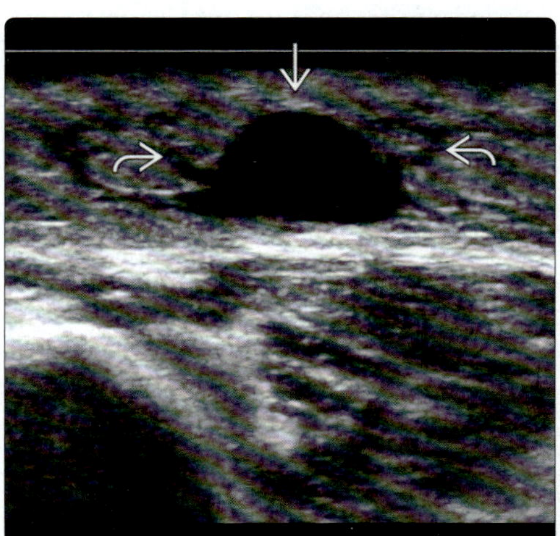

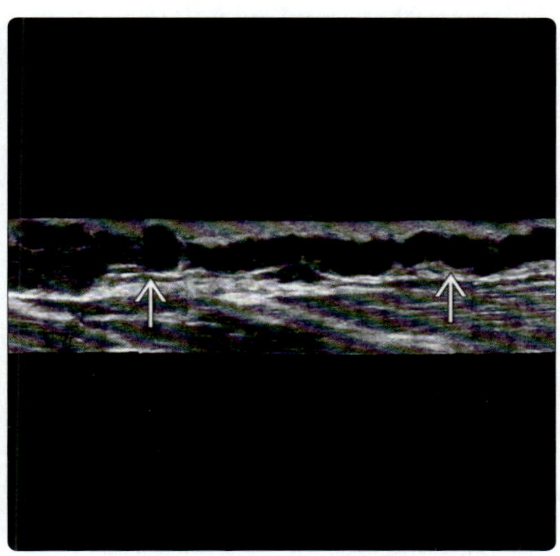

(Left) *Transverse US of a patient with melanoma of the ipsilateral index finger shows an in-transit subcutaneous lesion in the forearm. There is a hypoechoic tail sign at both the proximal and distal ends of the tumor.* **(Right)** *Longitudinal US of the same patient shows another much longer in-transit lesion. The elongated, hypoechoic mass is due to melanotic tumor filling and distending a lymphatic channel between the primary tumor and the regional lymph nodes, giving rise to the string sign.*

Vascular Anomaly

KEY FACTS

TERMINOLOGY

- Term "vascular anomaly" includes both vascular tumors & vascular malformation
- Hemangioma is benign & most common vascular tumor
  - Grows after birth, proliferates, & then involutes
  - Most are low flow
- Vascular malformation is vascular dysplastic lesion
  - Presents at birth, grows commensurately with child & with no proliferation or involution
  - Low flow: Venous malformation, capillary/venous, capillary, lymphatic
  - High flow: Arteriovenous malformation, arteriovenous fistula

IMAGING

- Fairly discrete, usually hypervascular mass ± phleboliths ± medium to large vascular channels
- Variable amount of stromal connective tissue component
  - Stromal component may contain vessels too small to resolve with US
  - Especially in children, assess all lipoma-like masses with color Doppler
- Variable color depending on speed of flow
  - Grayscale is better at detecting very slow flow
- Use US to make diagnosis & assess flow
  - Histological confirmation is not usually necessary
- Use MR to define extent, if lesion is large

DIAGNOSTIC CHECKLIST

- Consider age at onset, behavior, size, extent, vascular vs. connective tissue stromal component, high flow vs. low flow
- Differentiation between hemangioma & vascular malformation is not always clear-cut despite imaging, clinical, & even histological evaluation
- Chronological behavior is best way of differentiating hemangioma from vascular malformation

(Left) *Clinical photo of a 7-month-old baby shows a protuberant mass ➡ on the posterior chest wall with bluish discoloration of the overlying skin.* **(Right)** *Transverse color Doppler US of the same lesion shows a quite discrete mass containing dilated venous vascular spaces ➡. No deep extension is present. The lesion reduced in size on subsequent follow-up. The imaging features and clinical course are consistent with an infantile hemangioma.*

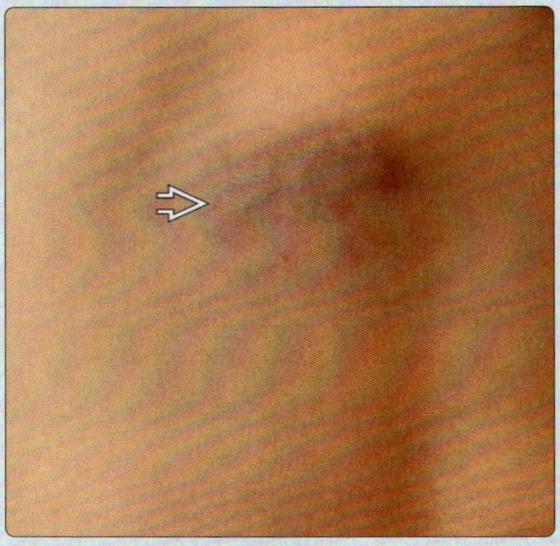

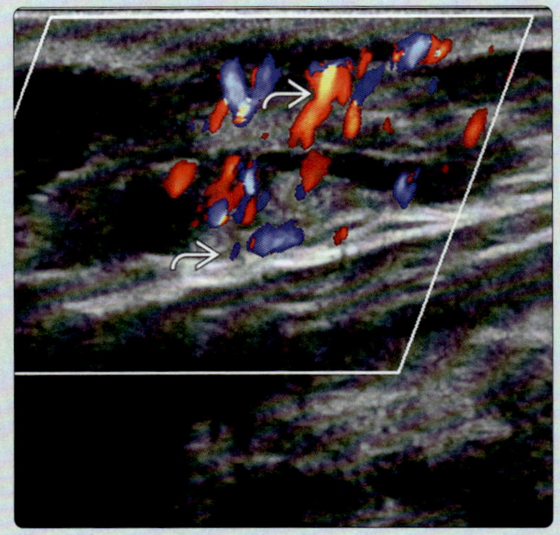

(Left) *Longitudinal US shows an arteriovenous malformation composed mainly of dilated vascular channels ➡ and only a relatively small amount of supporting stroma ➡. Pulsation was visible on grayscale imaging.* **(Right)** *Color Doppler US of the same lesion shows mainly arterial channels indicative of a high-flow arteriovenous malformation rather than a hemangioma. Hemangiomas generally tend not to have such high flow.*

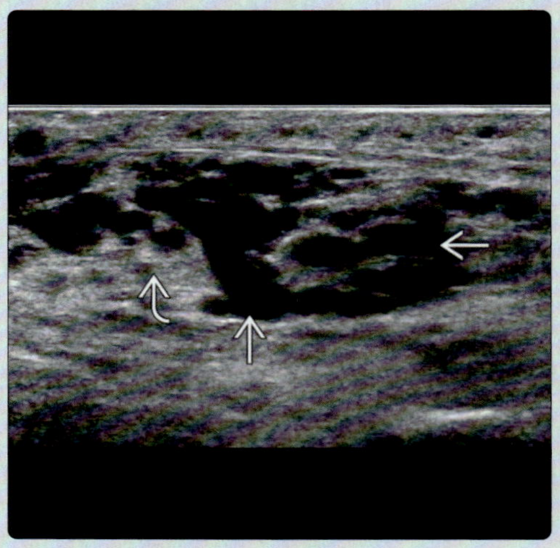

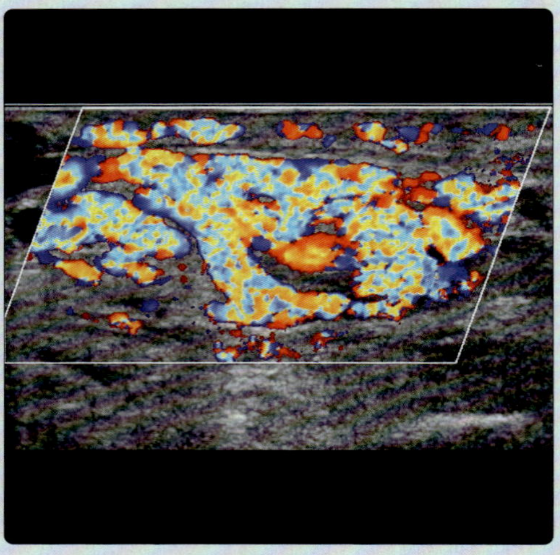

TERMINOLOGY

Definitions

- "Vascular anomaly" comprises both vascular tumors & vascular malformation
  - Differentiation based on cellular turnover, histology, natural history, imaging, & physical findings
- Vascular tumors
  - Features cellular proliferation & hyperplasia
  - Classified as benign (most common), locally aggressive, or borderline & malignant
  - Benign: Infantile hemangioma, congenital hemangioma, pyogenic granuloma (lobular capillary hemangioma)
    - Infantile hemangioma = most common childhood vascular tumor
      - Grows after birth with early proliferative stage & later involuntary stage
  - Local aggressive: Tufted angioma, kaposiform hemangioendothelioma
  - Malignant: Angiosarcoma
- Vascular malformation
  - Vascular dysplasia rather than actual tumor
    - Dysplastic vascular channels with normal endothelial turnover
  - Present at birth & grows commensurately with child without proliferation or involution
  - Subdivided into simple, combined, involving major vessel ± associated with other anomalies
    - Simple type is based on predominant vessel type affected, e.g., capillary, venous, arteriovenous, or lymphatic
      - Venous malformation = most common vascular malformation
    - Combined type: Arteriovenous fistula, capillary/venous, lymphatic/venous, etc.
  - Will be low flow or high flow depending on color Doppler flow dynamics
    - Low-flow vascular malformations: Venous, capillary, lymphatic, or capillary/venous
    - High-flow vascular malformations: Arteriovenous malformation, arteriovenous fistula

IMAGING

General Features

- Best diagnostic clue
  - ↑ vascular channels ± fairly discrete mass ± phleboliths
- Location
  - Head & neck (including intracranial) (40%), trunk (20%), & extremities (40%)
  - Unifocal (97%) or multifocal (3%)
- Morphology
  - Single-vessel ectasia → well-defined mass → large, ill-defined advancing collection of anomalous vessels

Radiographic Findings

- Radiography
  - Soft tissue mass ± phleboliths ± reactive mature periosteal new bone formation

MR Findings

- T2WI FS
  - Extent is well demonstrated on T2 fat-suppressed imaging
  - Low flow: No signal voids or large feeding arteries
  - High flow: Signal voids ± large feeding arteries
- MRA
  - Better at revealing nidus, feeding artery, & draining vein of arteriovenous malformation than US or MR

Ultrasonographic Findings

- Vary greatly, but US diagnosis of vascular anomaly can still be reliably made in most cases
  - Diagnosis is based on compendium of signs rather than only 1 or 2 signs
  - Differentiation between vascular tumor (hemangioma) & vascular malformation can be difficult based on imaging alone
- Abnormal vascular spaces may appear as vascular channels or large cystic spaces
- Variable stromal connective tissue component but usually has fatty-type background
  - Stromal component may contain vessels too small to be resolved with US
- Acoustic shadowing from phleboliths
- Color Doppler imaging varies depending on speed of flow
  - Slow blood flow shows moving echoes on real-time grayscale imaging often with little or no flow on color Doppler imaging
- Infantile hemangioma
  - Solid, usually hyperechoic mass ± lobulation ± intralesional vessels (occasionally)
  - Usually high vascular density, low-flow pattern on color Doppler
- Congenital hemangioma
  - Similar to infantile hemangioma
- Venous malformation
  - Very similar to hemangioma
    - Slow or no flow on color Doppler imaging
    - Can also detect arterial supply in venous malformation
      - 1 or 2 arteries do not imply high-flow lesion
- Capillary malformation
  - Localized skin & subcutaneous echogenic mass
    - Usually minimal or no flow evident
- Lymphatic malformation
  - Largely avascular
  - Macrocystic (i.e., cystic hygroma): Thin-walled, septated, fluid-filled cystic spaces ± internal echoes (hemorrhage or infection)
  - Microcystic: Multiple tiny cysts within echogenic connective tissue matrix
    - Cysts may be too small to be resolved with US
- Arteriovenous malformation
  - Pulsatile mass on grayscale imaging
  - High-velocity flow on spectral & color Doppler imaging
  - Only type of vascular malformation to possess nidus
    - Nidus is clear epicenter of anomaly from which lesion spreads peripherally

Angiographic Findings

- Prelude to sclerotherapy or embolization

Imaging Recommendations

- Best imaging tool
  - Use US to make diagnosis & assess flow
  - Use MR to define extent of large lesions
- Protocol advice
  - Use minimal transducer pressure to avoid compression
  - Determine location, size, extent, multiplicity, type of vessels present, presence/age of clots as well as amount of connective tissue stroma vs. vascular component
  - Assess flow dynamics: High flow or low flow depending on predominant flow pattern
  - Especially in children, assess all lipoma-like masses with color Doppler

DIFFERENTIAL DIAGNOSIS

Alveolar Soft Part Sarcoma

- 2nd or 3rd decade, particularly lower limbs
- Typically, intramuscular hypervascular mass
  - Similar to high-flow vascular malformation
  - Grows rapidly
  - Metastases (nodal, lung) in 20% at presentation

PATHOLOGY

General Features

- Associated abnormalities
  - Maffucci syndrome
    - Enchondromas & low-flow vascular malformation
  - Osler-Weber-Rendu syndrome
    - Arteriovenous malformations in skin, mucosa, lung, & brain
  - Klippel-Trenaunay-Weber syndrome
    - Low-flow vascular malformation ± large aberrant lateral vein ± limb hypertrophy
  - Gorham-Scott syndrome
    - Intraosseous venous & lymphatic malformations

Microscopic Features

- Hemangioma: Plump endothelial cells with hypercellularity, especially during proliferative phase
  - Extracellular matrix contains growth-promoting proteins
- Vascular malformation: Flat endothelial cells with normal cell turnover
  - Patchy deficiency of smooth muscle in wall → dilatation
  - No growth promoting proteins in extracellular matrix
- Histological distinction of vascular anomalies can be difficult

CLINICAL ISSUES

Presentation

- Most common signs/symptoms
  - Pain, swelling
- Other signs/symptoms
  - Bluish discoloration, mass ± engorgement with gravity
  - Discomfort after exercise, edema, hemorrhage, local hypertrophy, ulceration, contractures

Demographics

- Age
  - Congenital hemangioma: Present at birth
  - Infantile hemangioma: Present in early infancy
  - Venous malformation: Birth to early adulthood
  - Capillary or lymphatic malformation: First 2 years of life
  - Arteriovenous malformation: Early teens
- Epidemiology
  - Prevalence: 1.5% of general population
  - Hemangioma: Most common tumor of childhood
  - Vascular malformation: Venous (60%), lymphatic (25%), arteriovenous (10%), mixed (10%), & capillary (5%)

Natural History & Prognosis

- Infantile hemangioma
  - Proliferation (months) → stabilization (months to years) → spontaneous involution
  - Residual cosmetic deformity in ~ 50%
    - Later involution → greater cosmetic deformity
- Congenital hemangioma
  - Rapidly involuting congenital hemangioma (RICH) → involute by 14 months
  - Noninvoluting congenital hemangioma (NICH) → commensurate growth
    - Uncommon, need to see soft tissue mass as well as vessels & GLUT1 antigen positivity
  - Partially involuting congenital hemangioma (PICH) → rapid involution → stabilization → commensurate growth
- Vascular malformation
  - Slow commensurate growth
    - ± more rapid growth precipitated by puberty, pregnancy, infection, trauma, or unknown factor
    - Arteriovenous malformation is unpredictable
      - Threatens limb & life
  - Lymphatic malformation may spontaneously regress

Treatment

- Hemangioma: Wait & see, steroids, propranolol, laser, or surgery
- Vascular malformation
  - Sclerotherapy
    - Angiographic, venographic, or US guidance
      - Ethanol is most commonly used sclerosing agent
    - Contraindicated if high flow (risk of systemic shunting)
  - Coil embolization for high-flow malformation
- Wide surgical excision ± embolization/sclerotherapy

DIAGNOSTIC CHECKLIST

Consider

- Treatment based on clinical & imaging features
  - Classification is not always clear-cut despite imaging, clinical, & even histological evaluation

Image Interpretation Pearls

- Important parameters to assess: Age at onset, behavior, site, size, margination, vascular vs. stromal component, & high flow vs. low flow

SELECTED REFERENCES

1. Johnson CM et al: Clinical and sonographic features of pediatric soft-tissue vascular anomalies part 1: classification, sonographic approach and vascular tumors. Pediatr Radiol. 47(9):1184-1195, 2017
2. Johnson CM et al: Clinical and sonographic features of pediatric soft-tissue vascular anomalies part 2: vascular malformations. Pediatr Radiol. 47(9):1196-1208, 2017

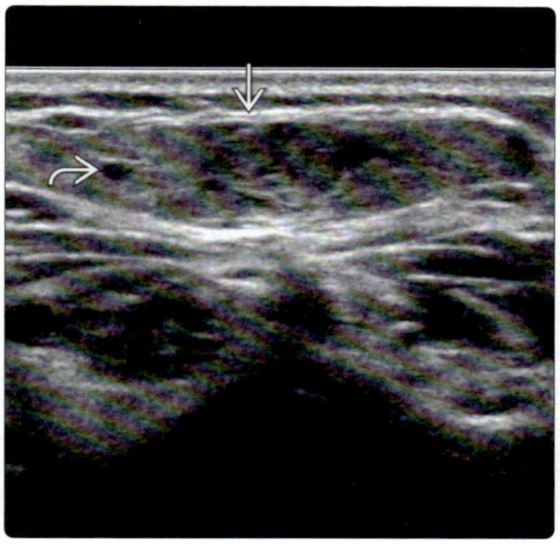

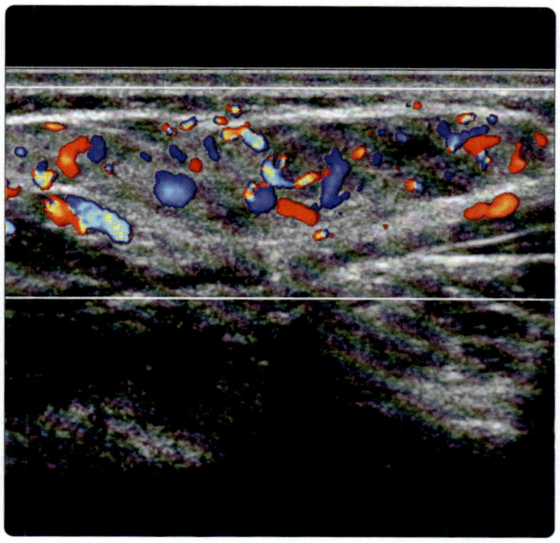

(Left) *Transverse US of the upper back in a 1-year-old child shows a well-defined hyperechoic, fatty-type subcutaneous mass ➡. Multiple, small, hypoechoic vascular channels ➡ are present within the mass. The mass vaguely resembles a lipoma.* **(Right)** *Color Doppler US of the same lesion shows multiple dilated, venous-type vascular channels within the mass, indicative of a slow-flow infantile hemangioma. These lesions are usually followed up by serial US examination.*

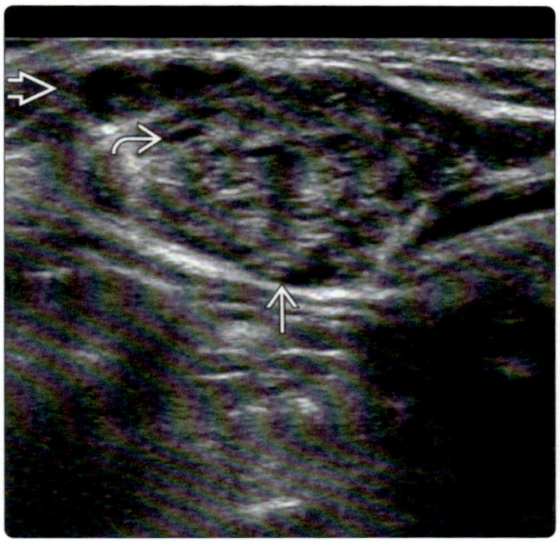

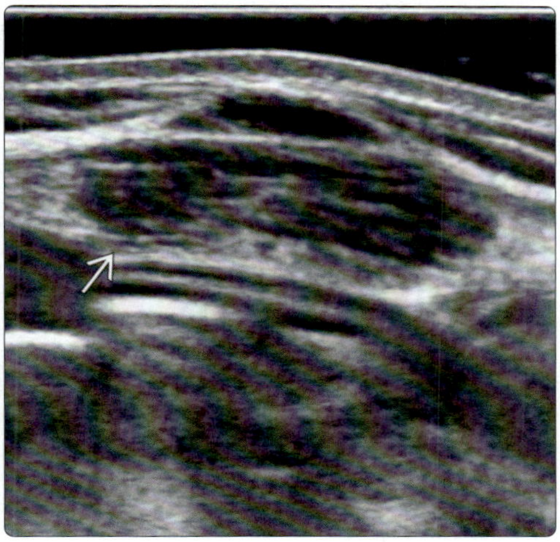

(Left) *Transverse US of a 1-year-old infant with an anterior abdominal mass, which had enlarged over a few months, shows a well-defined, fatty-type mass ➡ within the rectus abdominis muscle, which contains multiple small vascular channels ➡. There is limited subcutaneous extension ➡.* **(Right)** *Follow-up transverse US of the same lesion 2 years later shows a notable reduction in size ➡. Infantile hemangioma typically presents and proliferates in infancy, followed by stabilization and involution.*

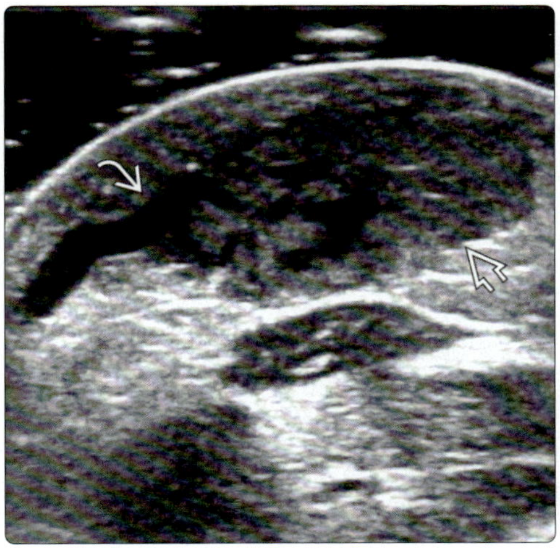

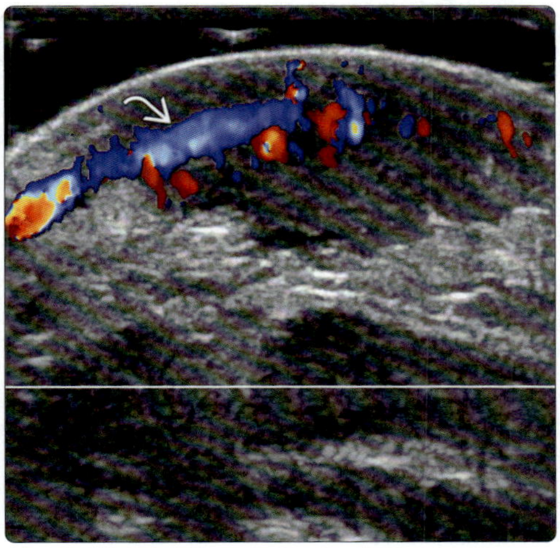

(Left) *Transverse US of a 21-day-old baby with a mass in the anterior thigh since birth shows a congenital hemangioma with a large amount of supporting stroma ➡ and a prominent central vascular channel ➡.* **(Right)** *Longitudinal color Doppler US of the same lesions shows venous signal ➡ within the central vessel. The lesion rapidly decreased in size in the 1st month of life. This clinical course is in keeping with a rapidly involuting congenital hemangioma.*

(Left) *Transverse US of a 17-year-old patient with a pulsatile thigh mass for years shows a discrete mass in the rectus femoris muscle. Several small vascular channels are present in the more solid background stroma of the mass.* **(Right)** *Transverse color Doppler US of the same lesion shows that the mass is highly vascular. This vascular component is mainly composed of arterial flow, though small areas of venous flow are also present.*

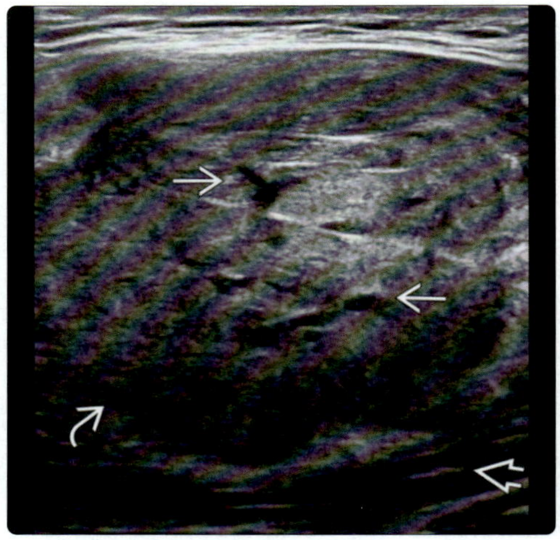

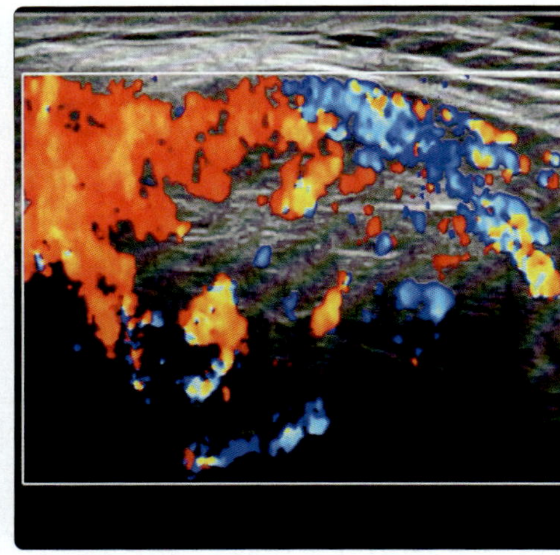

(Left) *Longitudinal spectral Doppler analysis of the same lesion shows a large arterial vascular channel supplying the lesion. Features are indicative of a high-flow noninvoluting hemangioma or high-flow vascular malformation.* **(Right)** *DSA of the common femoral artery shows a hypervascular mass supplied by the profundus femoral artery. The absence of a discrete nidus or early draining veins are more suggestive of a high-flow noninvoluting hemangioma rather than arteriovenous malformation.*

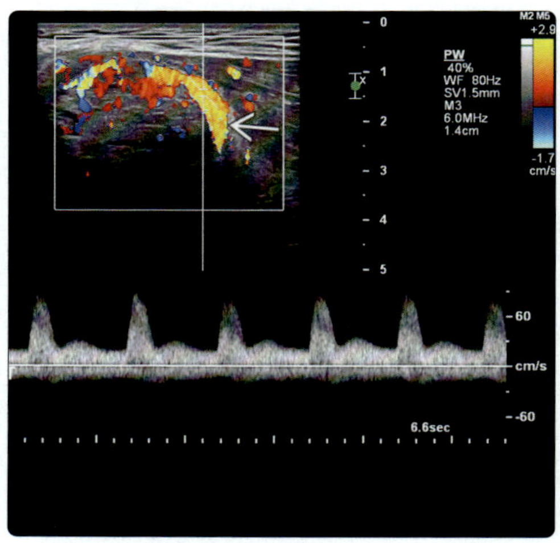

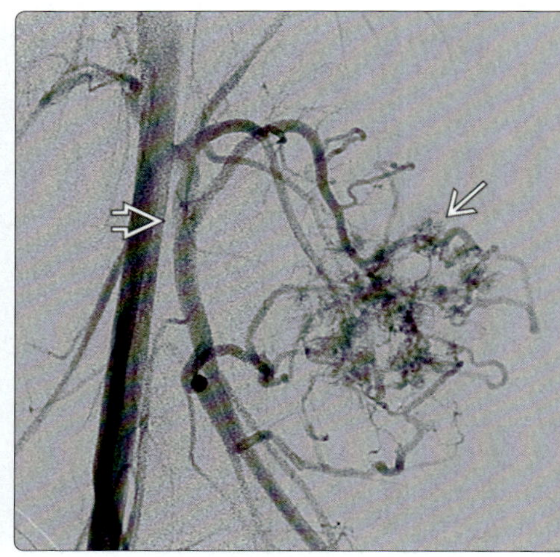

(Left) *Longitudinal US of an adult with an enlarging finger nodule a few weeks after injury shows a well-defined, hypoechoic subcutaneous nodule on the volar aspect of the finger close to the proximal interphalangeal joint.* **(Right)** *Longitudinal color Doppler US of the same lesion shows marked hypervascularity within the nodule and adjacent soft tissues consistent with a pyogenic granuloma. This type of vascular tumor occurs as an overgrowth to irritation, trauma, or hormonal stimulus.*

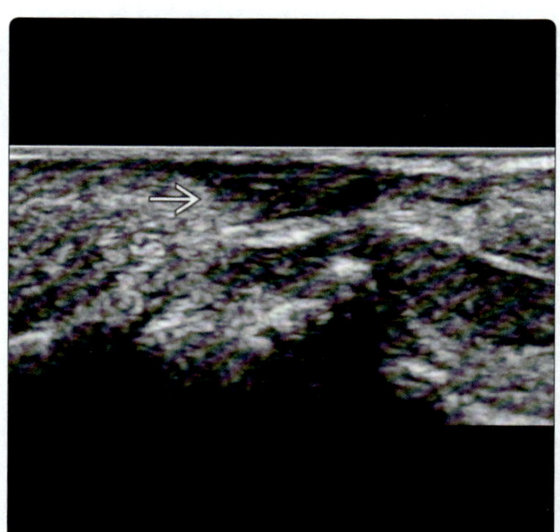

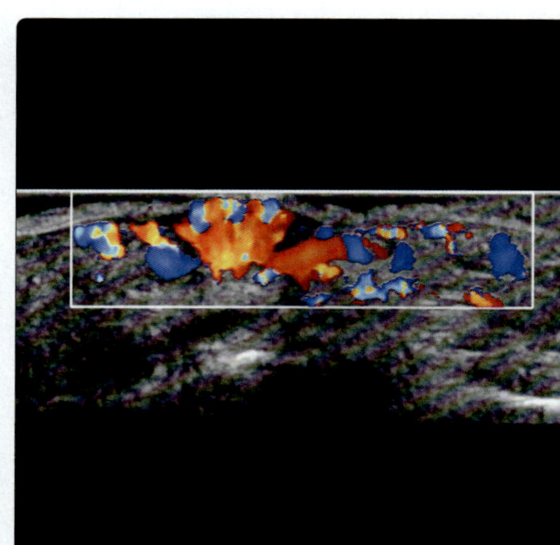

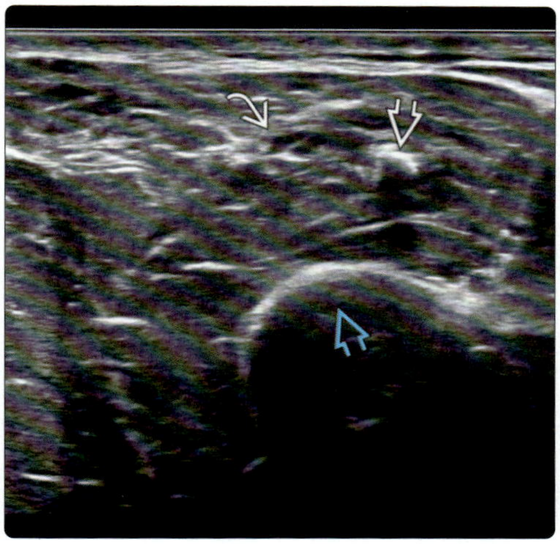

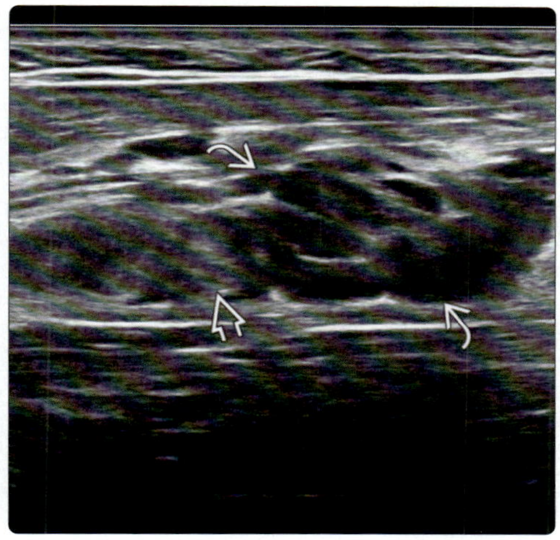

(Left) *Transverse US of the arm in an 18 year old with a 3-year history of arm swelling shows an ill-defined mass in the brachialis muscle alongside the humeral shaft . This venous vascular malformation contains dilated vascular channels and a phlebolith .* **(Right)** *Longitudinal US of the same patient shows that the lesion comprises ~ 70% of dilated vascular channels and ~ 30% of echogenic-supporting stroma . The appearances favor a venous vascular malformation.*

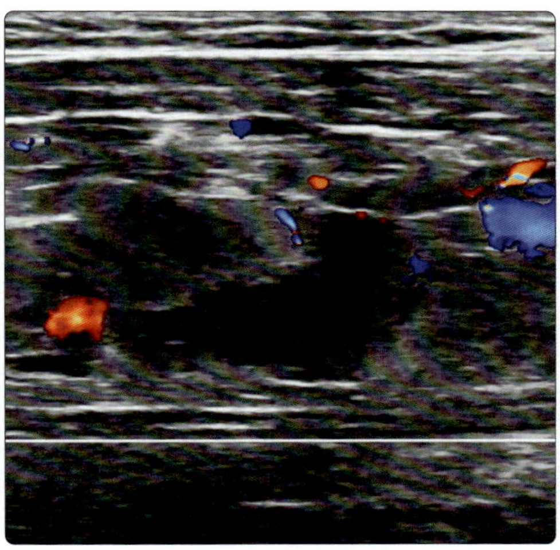

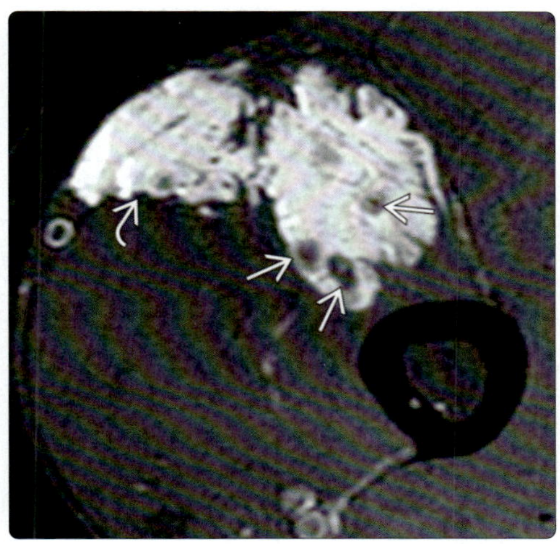

(Left) *Longitudinal color Doppler US of the same lesion shows little vascular flow present in the lesion. Flow within many of the dilated venous channels is too slow to be depicted on color Doppler US.* **(Right)** *Axial T2 FS MR of the same lesion shows a well-defined, lobulated hyperintense mass within the brachialis muscle. The extent of the venous malformation is more readily appreciated on MR imaging. Multiple hypointense foci representing phleboliths are present within the mass.*

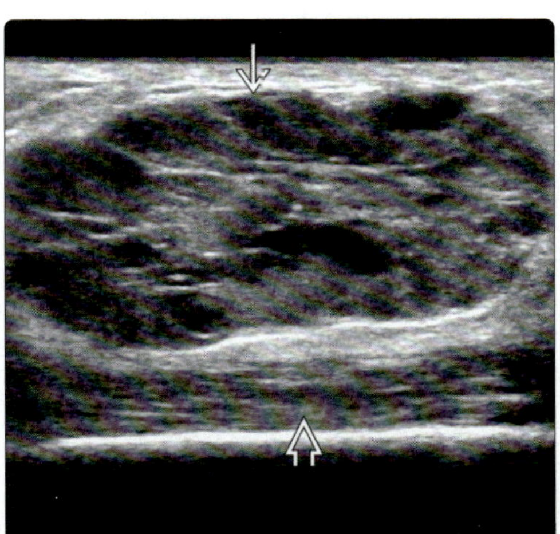

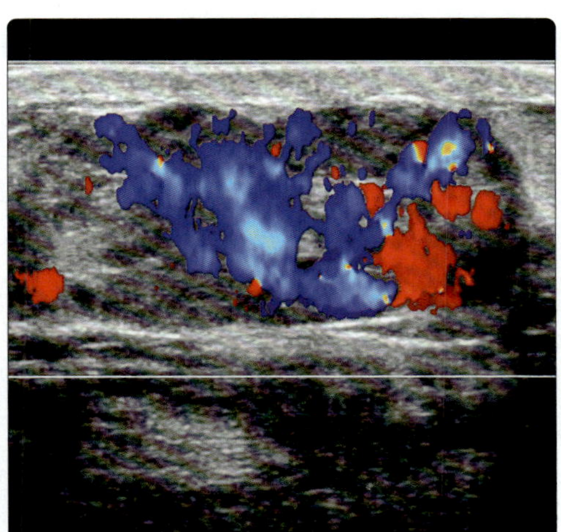

(Left) *Longitudinal US of a 71-year-old patient with a long history of forearm swelling that recently increased in size shows a well-defined subcutaneous lobulated mass without subfascial extension. The muscle layer is intact.* **(Right)** *Longitudinal color Doppler US of the same lesion shows the presence of ~ 60% of mainly venous-type vascular channels and ~ 40% of supporting connective tissue stroma.*

(Left) *Longitudinal spectral Doppler analysis of the same lesion shows a large venous vascular channel supplying the lesion. Features are indicative of a venous vascular malformation. This was confirmed by histology. Change in size of a vascular malformation is usually due to altered consistency and size secondary to thrombosis.* **(Right)** *Clinical photograph of a 41-year-old man shows a cystic-type mass on the dorsum of the hand with bluish discoloration and telangiectasia of the overlying skin.*

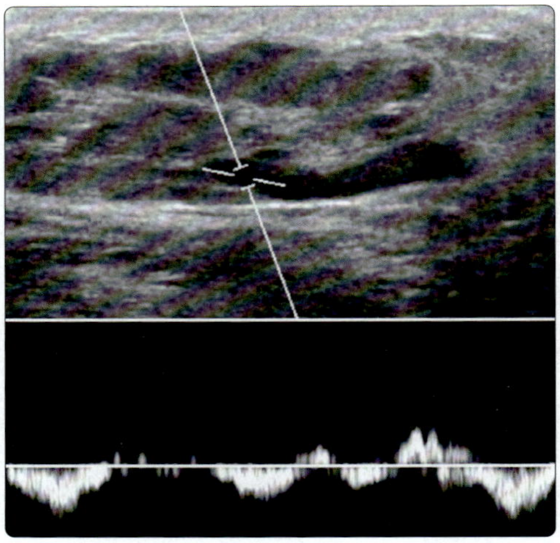

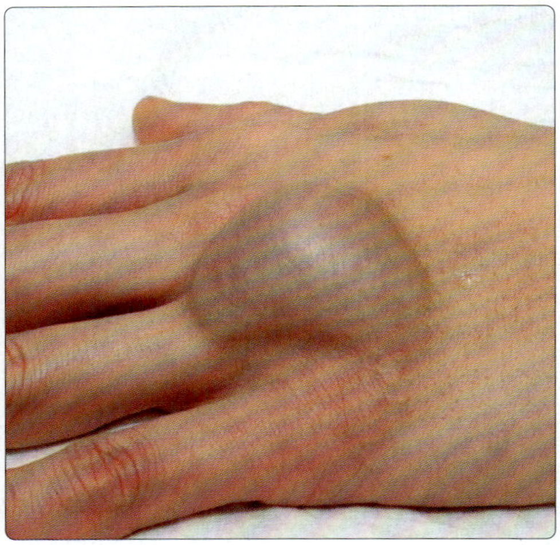

(Left) *Transverse US of the same lesion shows a venous vascular lesion overlying the metacarpal bones ➡. The fluid-fluid interface ⇨ indicates sedimentation of blood product in a slow-flow distended venous channel.* **(Right)** *Longitudinal color Doppler US shows another venous vascular malformation ➡ filled with moving echoes on real-time grayscale imaging. There is proximal and distal vascular supply and drainage evident ⇨. Slow flow is better depicted on grayscale imaging.*

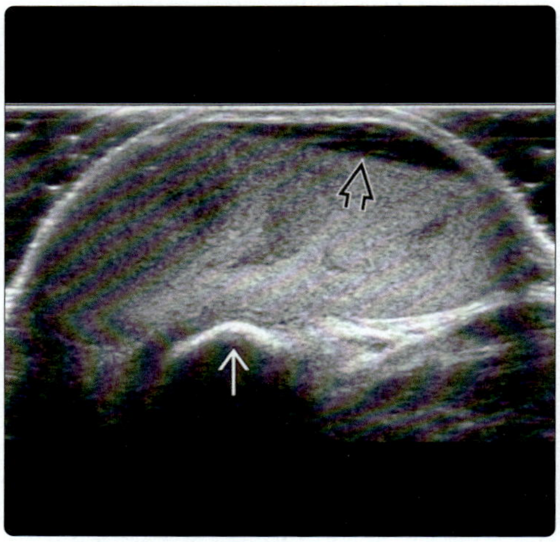

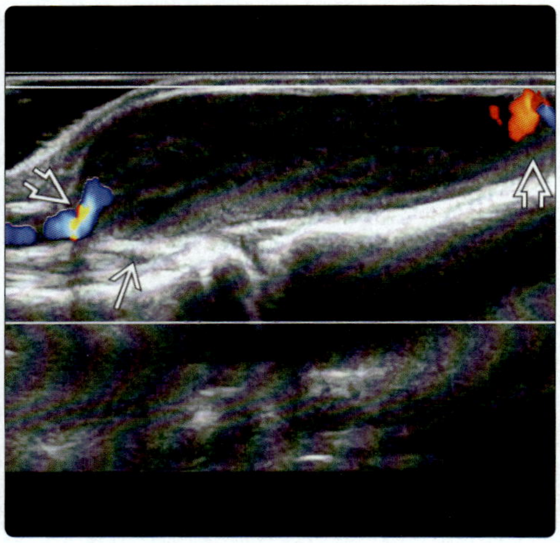

(Left) *Transverse color Doppler US of a 3 year old with a soft tissue mass in the big toe shows an ill-defined hyperechoic mass ➡ in the subcutaneous soft tissue with only a few dilated vascular spaces ➡ overlying the 1st metatarsal bone ➡.* **(Right)** *Follow-up color Doppler US 2 years later shows the static size of the lesion ➡. The apparent reduction in vascularity is due to obliteration of small veins by probe pressure. The appearances are consistent with a capillary venous-type vascular malformation.*

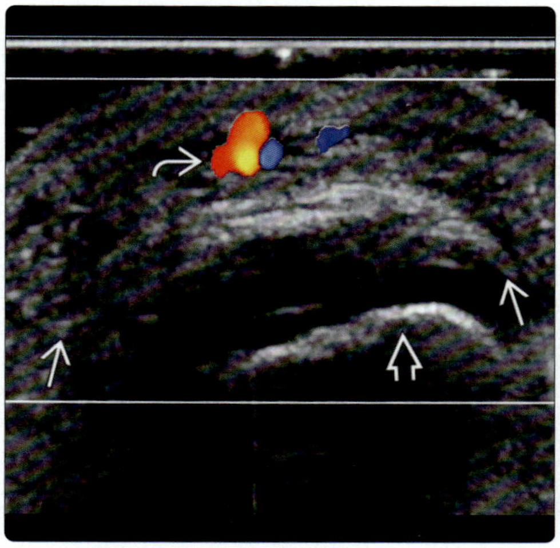

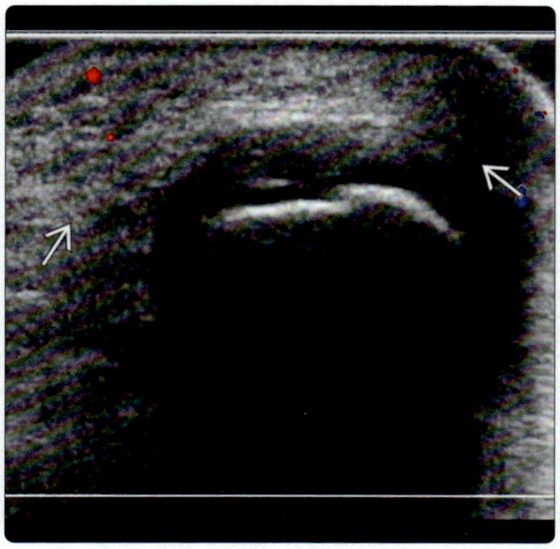

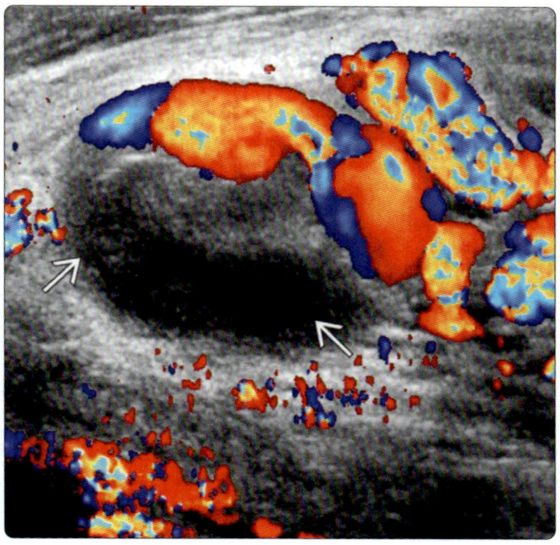
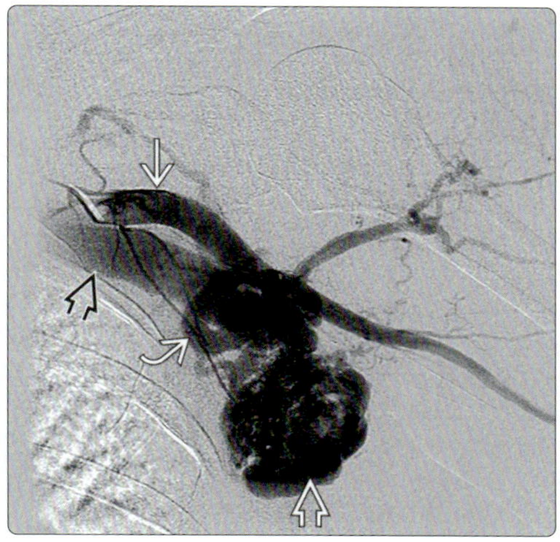

(Left) *Transverse color Doppler US of an arteriovenous malformation shows high flow around the partially thrombosed nidus .* **(Right)** *Selective DSA of the subclavian artery shows an arteriovenous malformation with feeding arteries from the axillary and medial circumflex humeral arteries and with early venous drainage into the subclavian vein . Note the lesion nidus . Angiography is more helpful in high-flow than slow-flow vascular malformations.*

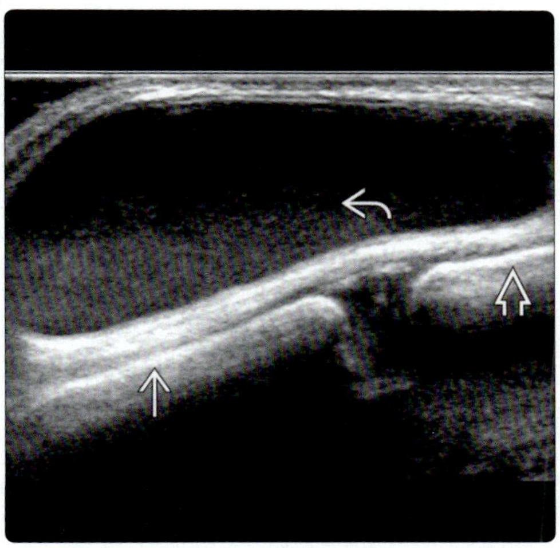
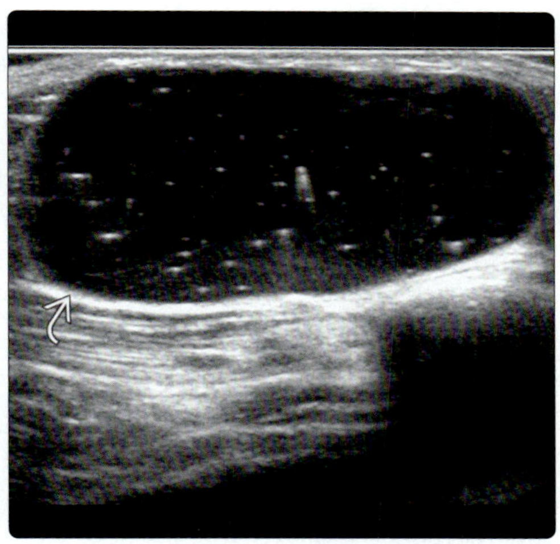

(Left) *Longitudinal US in a young boy shows a well-defined, uniloculated cystic mass overlying the sternum and manubrium . The cystic contents are mildly echogenic . There was no detectable flow on color Doppler imaging. Appearances are consistent with a unilocular lymphatic malformation.* **(Right)** *Longitudinal US in a young child shows a well-defined unilocular lymphatic malformation with multiple internal echogenic foci, including comet-tail artifacts, in the supraclavicular fossa.*

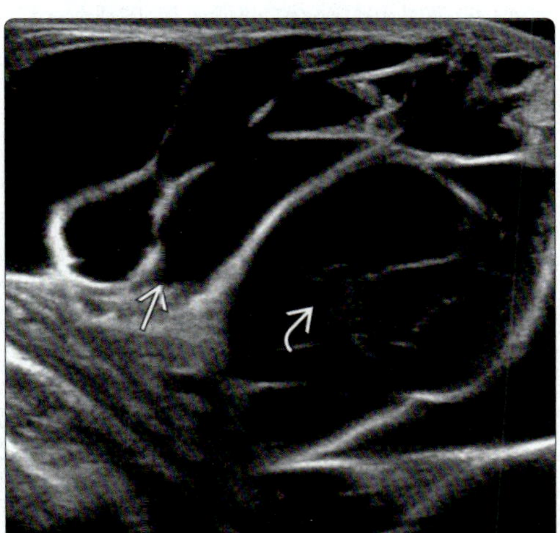
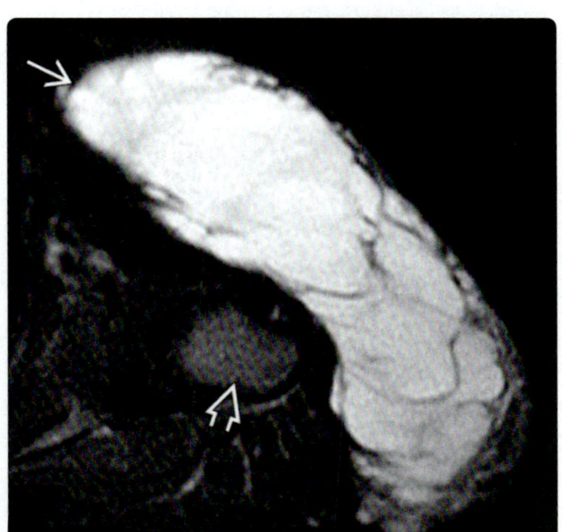

(Left) *Transverse US of a 3 year old shows a multicystic subcutaneous lesion in the proximal arm. Some cysts contain fine echogenic material due to a previous haemorrhage. Color Doppler revealed no vascular flow. The appearances are consistent with a multiloculated lymphatic malformation.* **(Right)** *Axial T2 FS MR of the same lesion shows the subcutaneous lymphatic malformation in the arm laterally in close proximity to the humerus .*

Foreign Body and Injection Granuloma

KEY FACTS

TERMINOLOGY

- Foreign body (FB): Retention of penetrating or iatrogenic object within tissues
- Injection granuloma (IG): Localized tissue reaction to material injected into tissues

IMAGING

- US: Modality of choice to detect radiolucent FBs and diagnose IGs
- **FB**: Nearly 100% are detectable on US
  - Reflectivity of some objects (wood, metal, glass, plastic) exceeds that of others (bamboo)
  - Align probe as parallel to FB as possible; FB may be located some distance from entry wound
  - Echogenic FB ± posterior artifact (reverberation/comet-tail artifact or posterior shadowing)
  - ± hypoechoic rim of edematous inflammatory tissue, granulation tissue, or fibrosis
  - False-positive US with gas or calcification in soft tissues; correlate with radiographs if necessary
  - Repeat US in 2-10 days if clinical history is suspicious and US findings negative or equivocal
    - Bamboo splints may not be visible initially
- **IG**: Focal, hypoechoic subcutaneous mass with variable calcification and posterior acoustic shadowing
  - Especially in buttock and lateral aspect of arm

CLINICAL ISSUES

- Almost 90% of FBs can be successfully removed under US guidance
  - Particularly if injury is recent
  - Chronically retained FB can be firmly adherent
- Hypoechoic rim can be breached by local anesthetic dissection or blunt forceps dissection

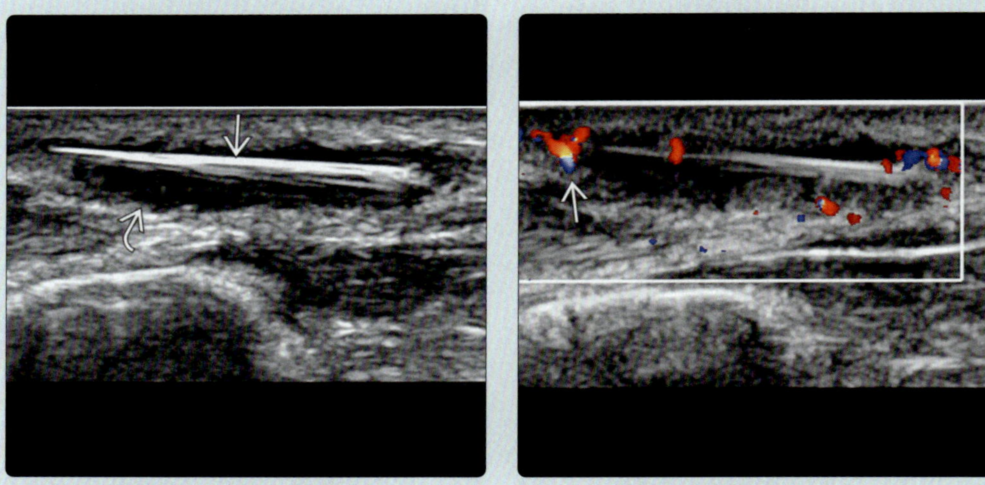

(Left) *Longitudinal oblique US shows an echogenic wooden splinter ➔ in the subcutaneous tissues of the hand 7 days after a penetrating skin injury. A rim of hypoechoic granulation tissue ➔ surrounds the wood splinter.* **(Right)** *Oblique color Doppler US of the same patient shows moderate hypervascularity ➔ in the adjacent granulation tissue and subcutaneous tissues. This foreign body (FB) was successfully removed under US guidance.*

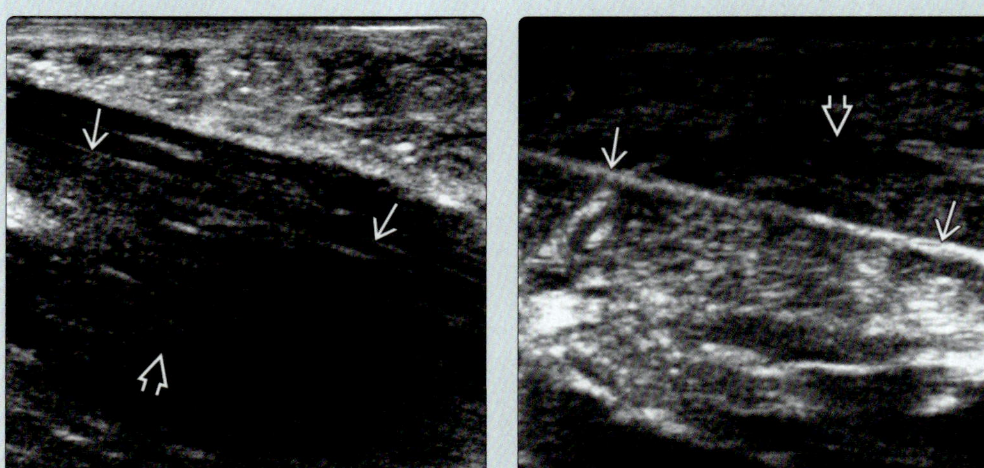

(Left) *Oblique US shows a barely visible bamboo splinter ➔ within the edematous thenar muscle ➔ 1 day after a penetrating injury.* **(Right)** *Oblique US of the same area 4 days later shows that the bamboo splinter is now clearly visible as an echogenic FB ➔. The thenar musculature ➔ is also less hypoechoic than previously. Follow-up US is particularly useful for suspected organic FBs not visible on initial US examination.*

TERMINOLOGY

Abbreviations

- Foreign body (FB)
- Injection granuloma (IG)

Definitions

- FB granuloma: Retention of penetrating or iatrogenic object within tissues with resultant granulomatous reaction
- IG: Localized chronic granulomatous tissue reaction to injected material

IMAGING

General Features

- Best diagnostic clue
  - FB: Echogenic material ± posterior artifact ± hyperemic hypoechoic rim
  - IG: Localized hypoechoic swelling of subcutaneous tissues → ill-defined mass with peripheral hyperemia → well-defined mass ± calcification
- Location
  - FB: Subcutaneous tissues > > deep tissues
    - Hands and wrists > feet
  - IG: Subcutaneous tissue, particularly superolateral aspect of gluteal region and lateral aspect of proximal arm
- Size
  - FB: Usually small (1-3 cm) but can be large (≥ 10 cm)
  - IG: Usually small (2-5 cm)
    - Dependent on amount and dispersion of injected material
- Morphology
  - FB: Depends on size and shape of FB and inflammatory response
    - Wood (45%), glass (25%), needle (10%), metal (10%), and others (e.g., grit, fish fin, pencil lead) (10%)
  - IG: Firm to hard subcutaneous nodule

Radiographic Findings

- FB: Detects radiopaque FBs, such as metal, glass, and some plastics
  - ≤ 15% wood FBs are radiopaque
  - Limited information on exact location of FB and associated soft tissue response
- IG: ± calcified subcutaneous mass

CT Findings

- NECT
  - FB granuloma
    - Typically seen as hyperdense object with variable surrounding inflammation or fibrosis
    - Wood may appear hypodense and simulate air
  - IG: Well-defined subcutaneous mass
    - ± lobulated or rim calcification (often dense)

MR Findings

- Less accurate than US at detecting FBs
  - Good visualization of inflammatory response

Ultrasonographic Findings

- FB
  - Echogenic object ± posterior artifact (reverberation/comet-tail artifact or posterior shadowing)
  - Echogenicity is dependent on acoustic impedance between retained FB and surrounding tissue
    - Reflectivity of some objects (wood, metal, glass, plastic) exceeds that of others (bamboo)
    - Reflectivity also dependent on smoothness of FB and alignment to US beam
  - ± hypoechoic halo of inflammatory edematous tissue, granulation tissue, or fibrosis
    - Presence of halo depends on intensity of inflammatory response and duration of retention of FB
      - May be apparent within 24 hours of retention
    - Improves sensitivity and specificity of US detection
    - Variable hyperemia of hypoechoic rim
- IG
  - Focal swelling and disruption of subcutaneous fat
    - Similar to early stages of fat necrosis
    - ± small areas of liquefaction
    - ± mild peripheral hyperemia
  - Swelling may progress to ill- or well-defined hypoechoic mass
    - ± mild peripheral hyperemia
    - ± dense calcification
      - Limits visibility of deep margin of mass

Nuclear Medicine Findings

- IG may accumulate tracer on FDG PET imaging
  - False-positive for subcutaneous metastatic deposit

Imaging Recommendations

- Best imaging tool
  - US: Modality of choice to detect radiolucent FBs and diagnose IGs
    - Sensitivity and specificity > 90% for radiolucent FB detection
- Protocol advice
  - FB
    - First, check with patient regarding type and size of FB, direction of implantation, and likelihood of FB retention
    - Align probe as parallel to FB as possible
    - Examine beyond entry wound
    - FB may be located some distance from entry wound
      - e.g., hand entry wound may give rise to FB in distal forearm
    - Note size, shape, location, and orientation of FB
      - Relate size and shape to recalled FB
    - Assess presence, thickness, and vascularity of hypoechoic rim
    - Assess adjacent critical structures (neurovascular bundles, tendons, and joints)
    - False-positive US can occur due to gas or calcification in soft tissues
      - Particularly if FB is small, such as glass fragment
      - Correlate with radiographs (calcification is radiopaque)
    - Repeat US in 2-10 days if history suspicious and US findings equivocal

- Bamboo splints particularly may not be clearly visible initially
- IG
  - Assess location, size, and margination of granuloma
  - Assess vascularity and present of calcification

DIFFERENTIAL DIAGNOSIS

Subcutaneous Fat Necrosis

- Localized swelling and ↑ echogenicity and disruption of subcutaneous fat
- ± history of prior blunt trauma

Subcutaneous Fibroma

- Hypoechoic, well-defined subcutaneous nodule
  - No history of injection or injury

Subcutaneous Metastasis

- Hypoechoic, usually noncalcified, vascular nodule
- Usually known disseminated malignancy

PATHOLOGY

General Features

- Etiology
  - FB granuloma
    - Penetrating FB (wood, glass, needle, bamboo, etc.)
    - Retained FB (gauze, cotton swab, etc.)
  - IG
    - Injection of drug into subcutaneous fat
    - Many intended intramuscular injections occur instead in subcutaneous tissue
      - Granulomas are much less frequent with intramuscular injection
    - Tissue reaction is dependent on site, composition, and amount of injected material

Gross Pathologic & Surgical Features

- FB: Variable rim of inflammatory or fibrous tissue
- IG: Tissue necrosis with mass composed of inflammatory or fibrous tissue

Microscopic Features

- IG
  - Necrosis of fat cells
  - Dense fibrous tissue reaction with inflammation
    - Inflammation may persist > 1 year after injection

CLINICAL ISSUES

Presentation

- Most common signs/symptoms
  - FB
    - History of penetrating injury
    - Palpable mass, initially tender but becomes less tender over ensuing weeks
  - IG
    - History of injection
    - Tender area → palpable mass
- Other signs/symptoms
  - Original trauma or injection may be forgotten
    - May present as unexplained mass

Demographics

- Age
  - FB: Young or middle-aged patients
  - IG: Elderly patients

Natural History & Prognosis

- FB
  - Inert material leads to nontender mass
  - Noninert material leads to continuing tissue reaction, which persists as tender FB granuloma
    - Chronic inflammatory reaction ± sinus tract to skin
  - IG
    - Mass becomes less tender, smaller, and harder
    - Usually persists as small, hard nodule

Treatment

- FB
  - Percutaneous US-guided removal
    - ~ 90% of FB can be successfully removed
      - Particularly if injury is recent
      - Chronically retained FBs can be firmly tethered
      - US retrieval may not be feasible if high risk of adjective nerve or vascular injury
    - Superficial FB: Localization → 2- to 3-mm incision usually though can be up to 10 mm
      - Remove under US guidance or mark skin location and remove directly after skin incision
      - Using either small needle as lever under FB or, for larger FB, grab FB with forceps
    - Deep FB: Localization → incision → blunt forceps dissection → removal using US guidance
      - Removal along long axis of FB minimizes likelihood of FB breaking during retrieval
      - Use transducer pressure to fix FB
      - Use crocodile forceps or mosquito forceps
      - Crocodile forceps opened along FB, which is pushed into jaws of forceps with transducer pressure
      - Hard materials are generally easier to remove (positive feedback on contact)
      - Lignocaine injection or needle bevel can be used to free up FB bound by granulation tissue
  - Cutaneous or needle localization aids surgical excision
- IG: Treatment usually not necessary

DIAGNOSTIC CHECKLIST

Consider

- US has dramatically ↑ detection rate and image-guided recovery rate of FBs
- Be careful about diagnosing small, thin (e.g., submillimeter) FBs on US
  - Fine, echogenic lines masquerading as thin FBs are seen in almost any tissue
- US appearances of IGs are specific enough not to warrant further investigation

SELECTED REFERENCES

1. Bradley M: Image-guided soft-tissue foreign body extraction - success and pitfalls. Clin Radiol. 67(6):531-4, 2012

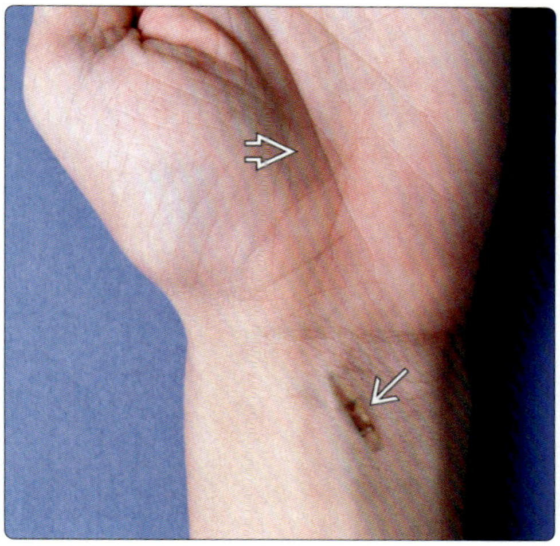

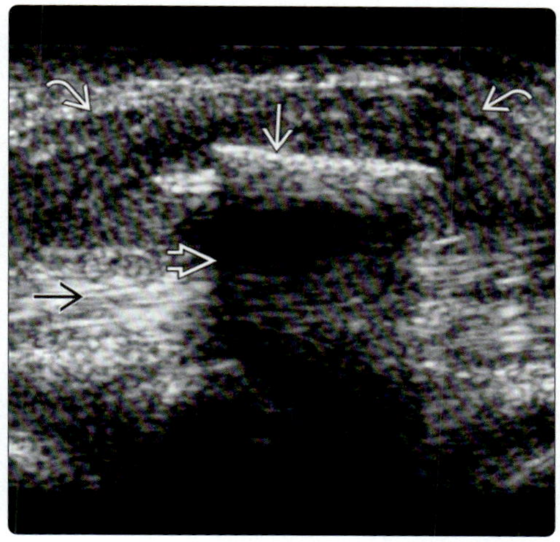

(Left) *Clinical photograph shows an almost healed entry site of a wood splinter in a patient who presented with a palpable mass at the midpalm region 4 weeks following initial injury.* **(Right)** *Oblique US of the palm in the same patient shows the wood splinter with posterior acoustic shadowing on the volar aspect of the carpus, superficial to a flexor tendon . A thick rim of reparative granulation tissue surrounds the FB.*

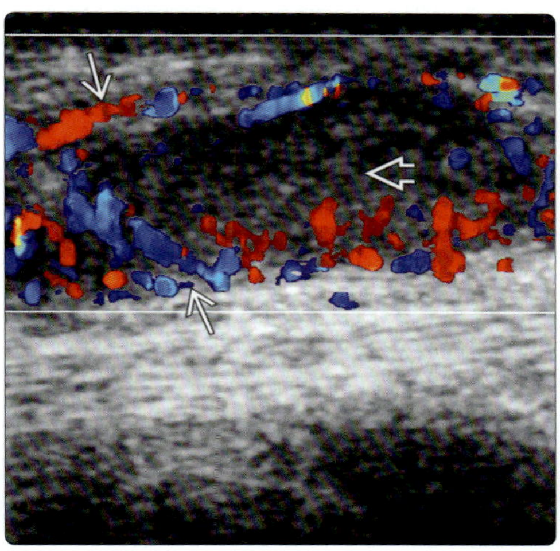

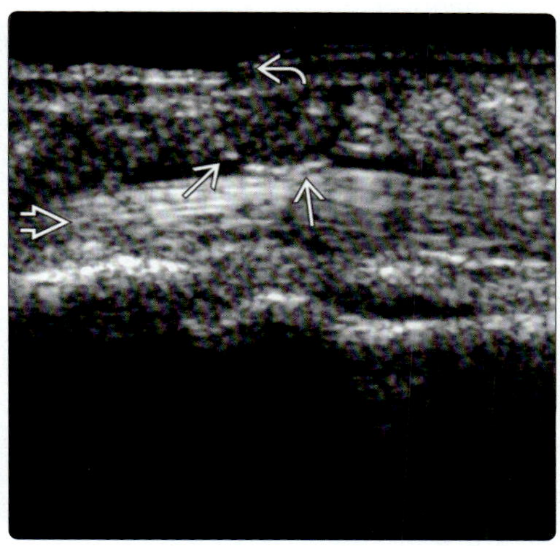

(Left) *Oblique US of the same patient shows marked hypervascularity of the surrounding hypoechoic rim , compatible with immature granulation tissue.* **(Right)** *Longitudinal US of the palm shows a curvilinear, echogenic rim just superficial to a flexor tendon . This echogenic rim moved with transducer pressure and was due to air, not a glass FB. The cutaneous wound is shown. Follow-up US 2 days later showed no FB.*

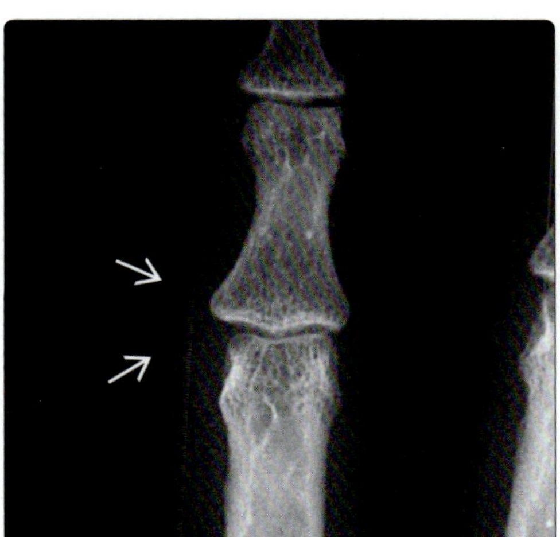

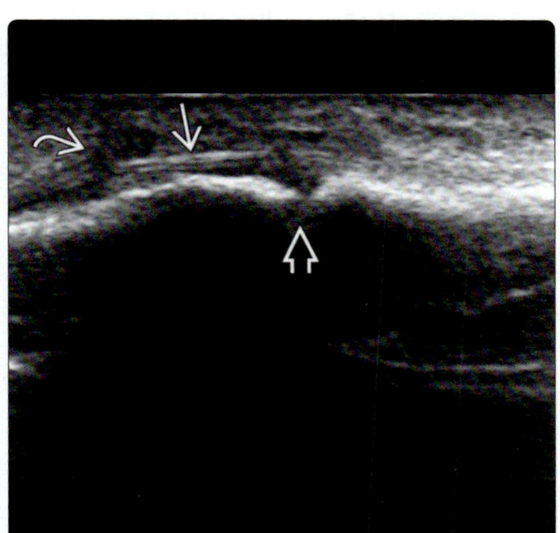

(Left) *Anteroposterior radiograph of a patient with suspected wood splinter in the index finger shows mild focal soft tissue swelling on the radial side of the proximal interphalangeal joint. No visible radiopaque FB is seen.* **(Right)** *Longitudinal US of the same patient shows an echogenic wood splinter on the radial aspect of the proximal interphalangeal joint with a thin, hypoechoic halo of granulation tissue .*

(Left) *Transverse US shows multiple bamboo splinters with surrounding granulation tissue anterior to the tibia. The anterior tibial artery is just deep to this area.* **(Right)** *Clinical photograph of the same patient shows an open entry site wound over the anteromedial aspect of the mid leg. The site of FBs and surrounding granulation tissue is marked for surgical localization. Note how the FBs are located far from the entry site. The path of the anterior tibial artery is also marked to aid surgical planning.*

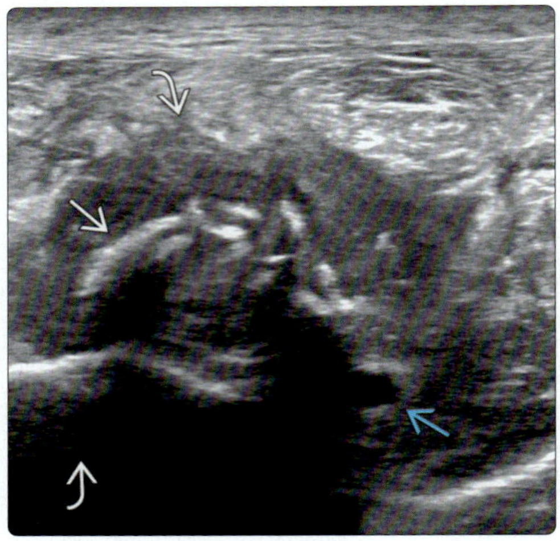

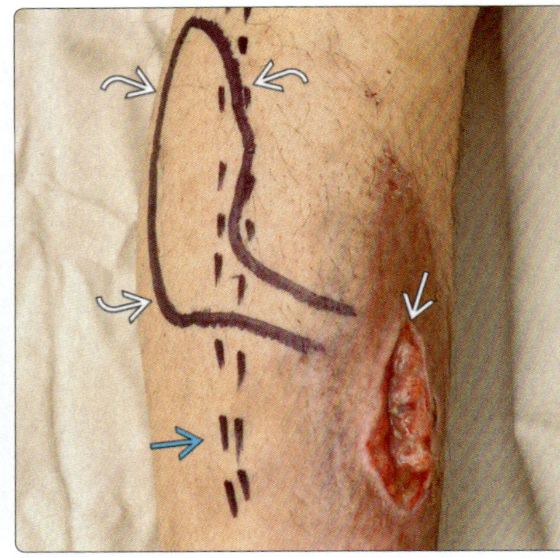

(Left) *Oblique US shows a wood splinter within the thenar muscle. There is a small, hypoechoic area (most likely due to a small fluid collection) at the proximal tip of the splinter.* **(Right)** *Lateral radiograph of the same patient shows only mild soft tissue swelling in the thenar region. No visible radiopaque FB is seen.*

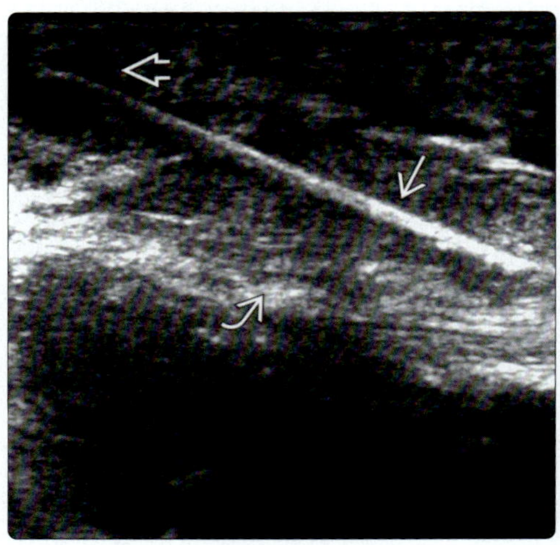

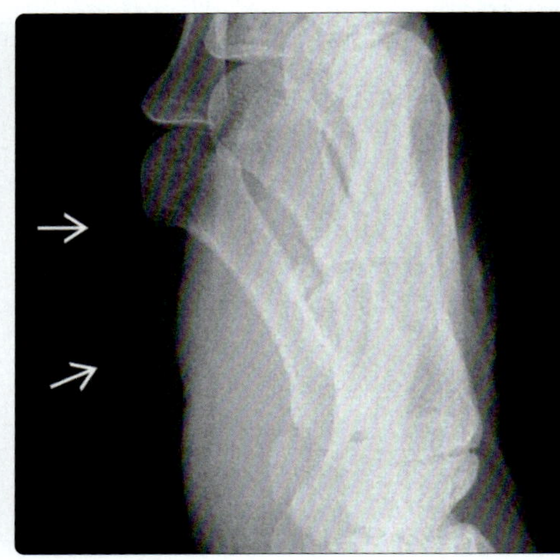

(Left) *Longitudinal US of the arm shows 2 thin, curvilinear subdermal glass fragments with edematous inflammatory tissue. These linear echoes did not move with transducer pressure, helping to distinguish them from air locules.* **(Right)** *Longitudinal US shows wooden splinter on the dorsum of the finger with surrounding fluid and inflammatory tissue. Minimal hyperemia was present. The nail base, distal interphalangeal joint, and extensor tendon are normal.*

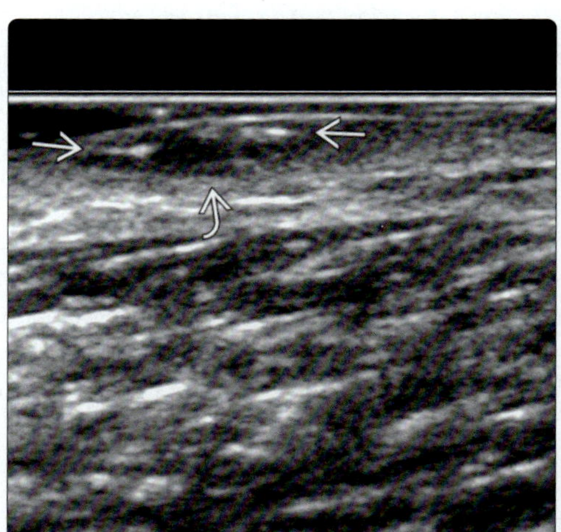

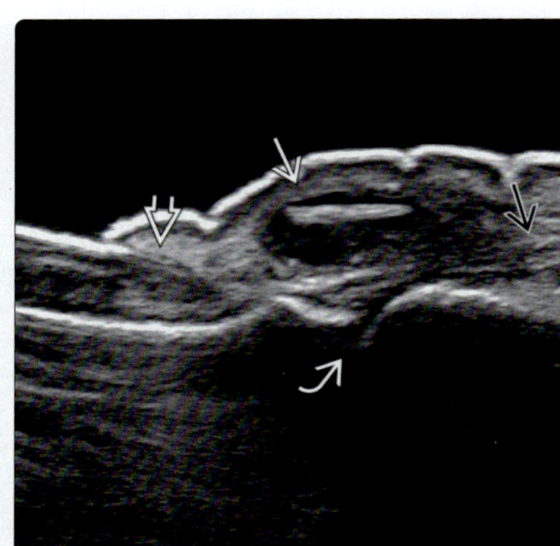

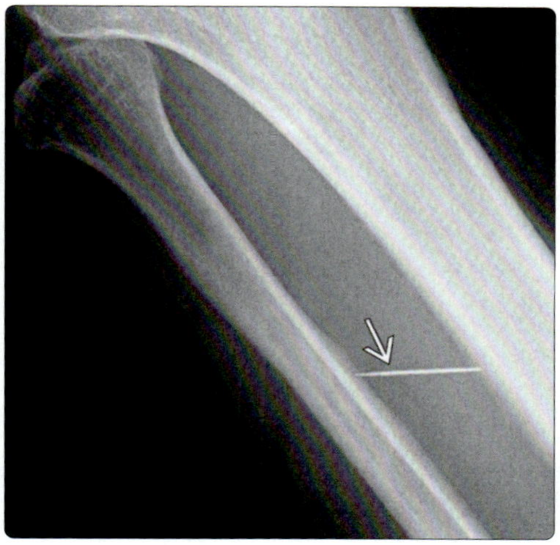

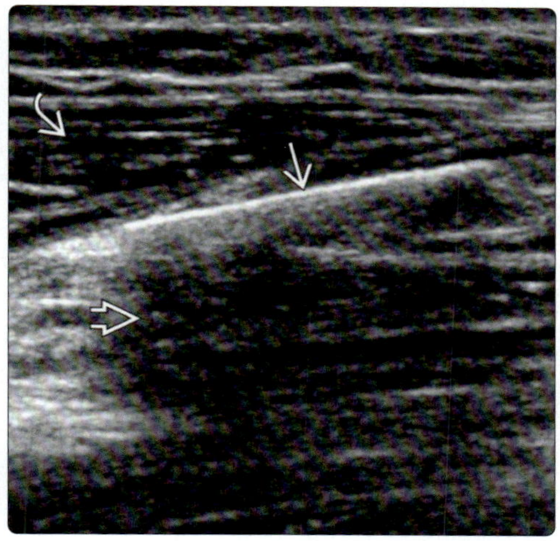

(Left) *Anteroposterior radiograph shows a metallic FB (needle) in the proximal aspect of the leg several hours after a penetrating injury.* **(Right)** *Oblique US of the same patient shows how the echogenic needle is located within the superficial fibers of the soleus muscle. There is mild posterior acoustic shadowing. The overlying gastrocnemius muscle is normal. US allows precise FB localization. This FB was removed under US guidance.*

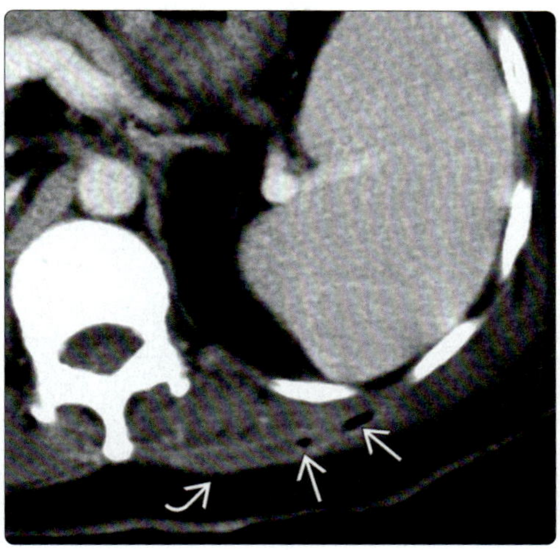

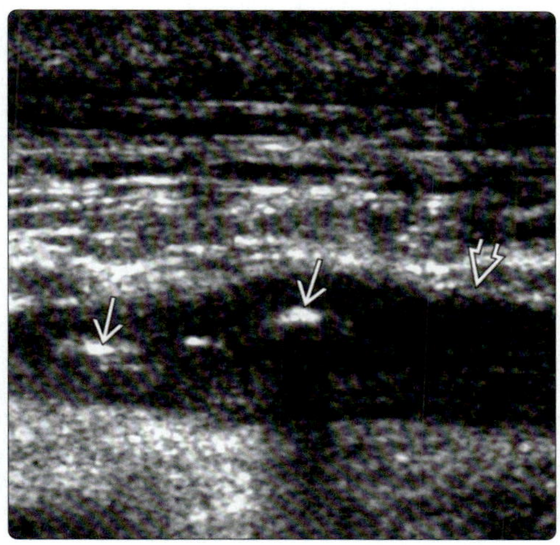

(Left) *Transverse CECT in a patient with penetrating chest trauma shows 2 radiolucent wood splinters in the posterior chest wall with a medium-sized, isodense halo of rim-enhancing granulation tissue.* **(Right)** *Oblique US at the same location (same patient) shows the 2 echogenic wood splinters with a thick, hypoechoic halo of fibrous tissue. Wood FBs can mimic air, being hypodense on CT and echogenic on US.*

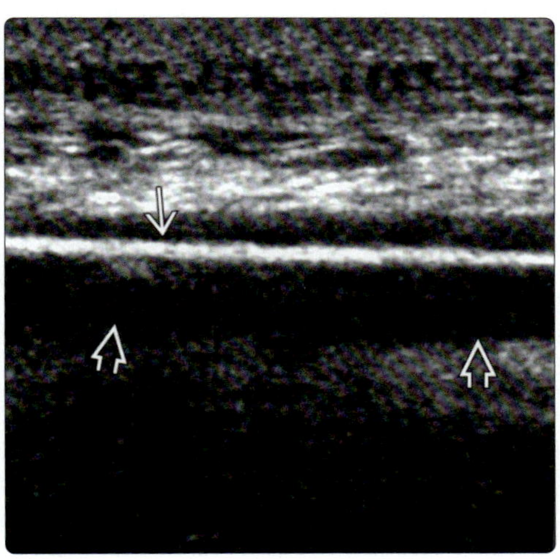

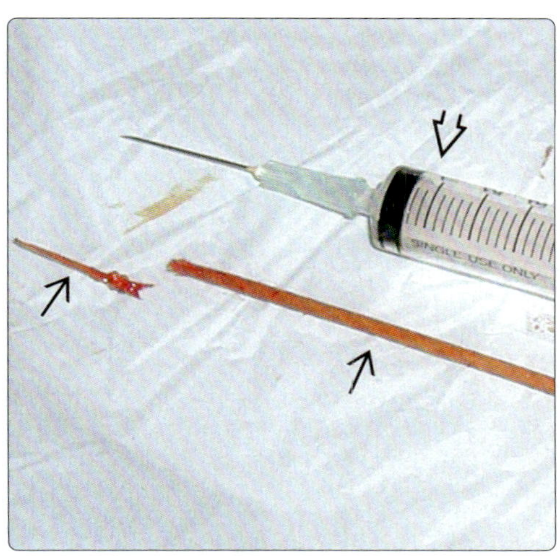

(Left) *Oblique US of the same area (same patient) clearly shows the echogenic wood splinter against a hypoechoic halo of granulomatous-type tissue.* **(Right)** *Clinical photograph of the same patient shows the 2 wood splinters removed using forceps and blunt dissection under US guidance and local anesthesia. A rim of granulation tissue does not preclude image-guided removal as it can be breached by local anesthetic dissection or blunt dissection.*

(Left) *Longitudinal US of the volar aspect of the finger shows a small (3 mm long) subcutaneous metallic FB* ➔ *superficial to the flexor tendons* ➔. *This FB was removed under US guidance.* **(Right)** *Longitudinal US of the thigh shows a small chopstick splinter* ➔ *measuring 10 mm long x 1.5 mm deep x 2 mm wide. There is no surrounding granulation tissue. The splinter lies immediately deep to the skin, deep to and just lateral to the cutaneous wound. This was retrieved with US guidance.*

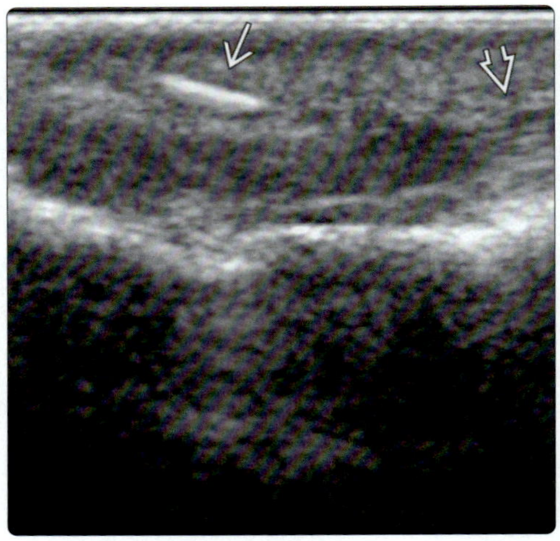

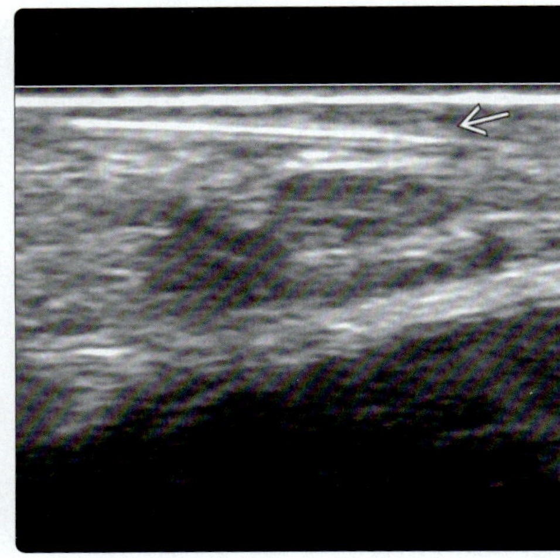

(Left) *Transverse US shows a medium-sized granuloma* ➔ *in the subcutaneous region of the upper outer quadrant of the buttock. Although the granuloma is not calcified, its fibrotic nature still leads to moderate acoustic shadowing* ➔. **(Right)** *Transverse US of the buttock shows a medium-sized subcutaneous granuloma* ➔. *Hypoechoic areas* ➔ *within the granuloma most likely represent areas of more mature fibrous tissue. Mild acoustic shadowing* ➔ *is present.*

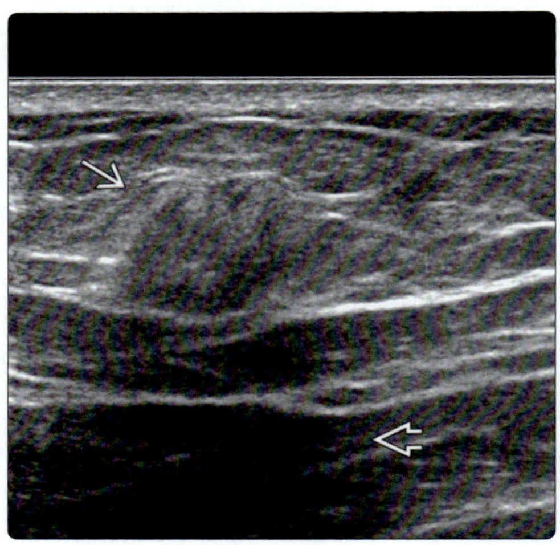

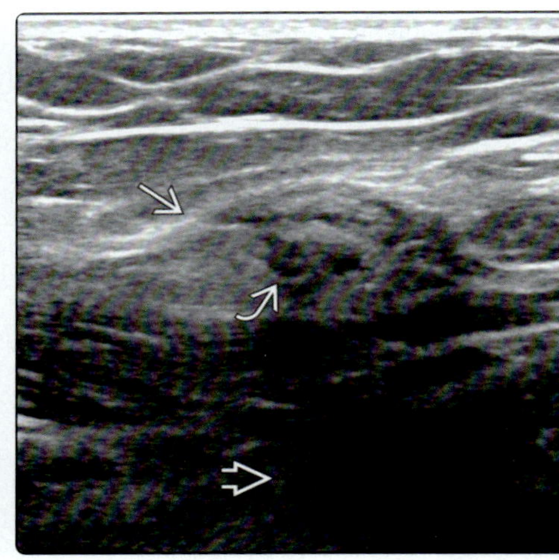

(Left) *Transverse US of the buttock shows a small subcutaneous granuloma* ➔ *in the upper outer quadrant of the buttock with surrounding soft tissue edema* ➔. *This edema may be a feature of immature granulomas, probably related to ongoing tissue reaction to the subcutaneous injectate. Mild acoustic shadowing is present* ➔. **(Right)** *Transverse color Doppler US of the same region shows mild tissue hyperemia, indicative of low-grade inflammation or active tissue repair.*

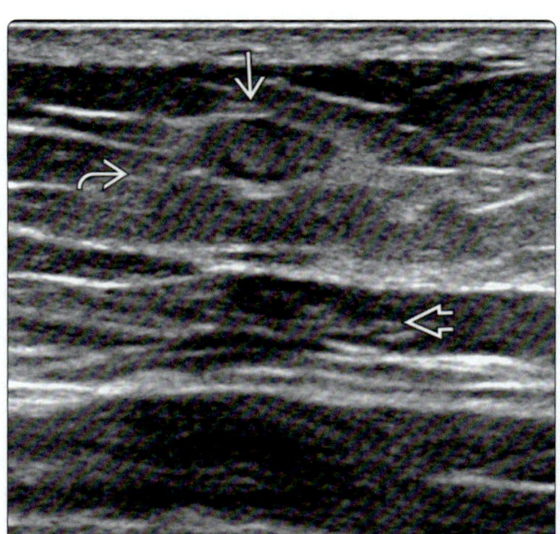

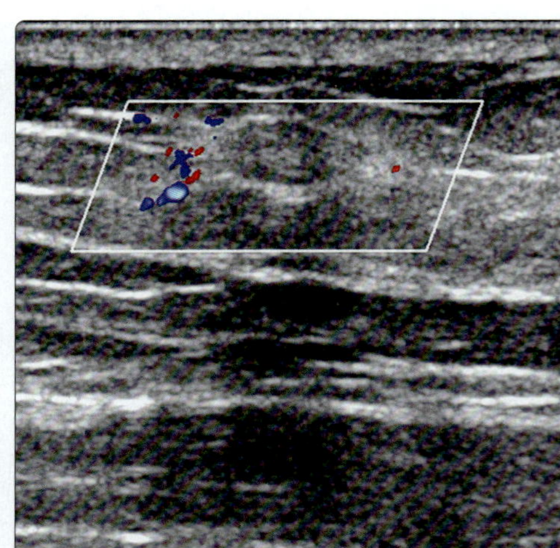

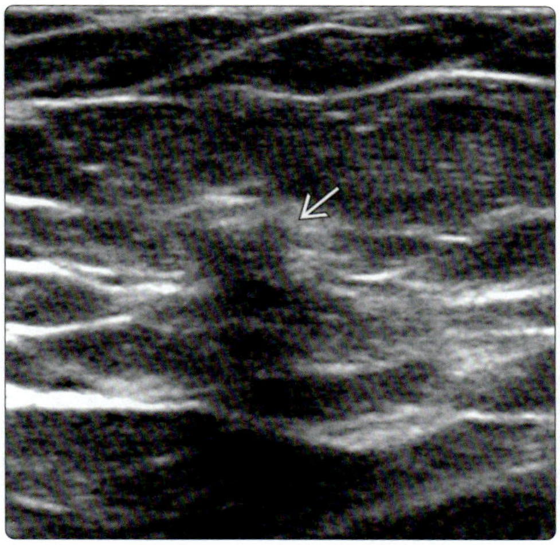

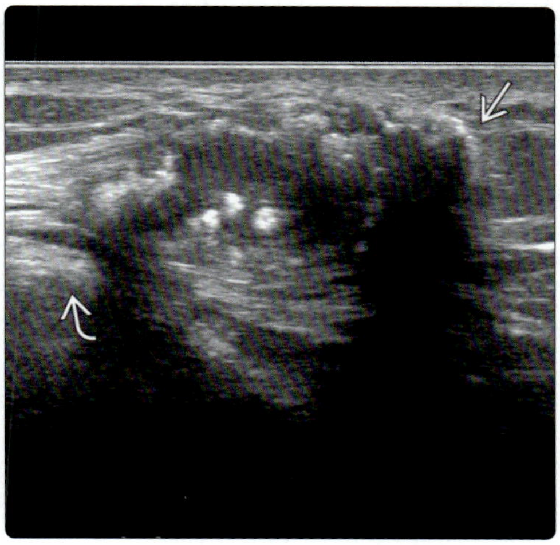

(Left) *Transverse US shows a small, hypoechoic subcutaneous injection granuloma → in the buttock region. Injection granulomas can be small. There is partial calcification of this granuloma.* **(Right)** *Transverse US shows a medium-sized, partially calcified subcutaneous granuloma → overlying the iliac crest ↷. Calcification is both peripheral and central and is punctuate initially but tends to become more solid with increasing chronicity.*

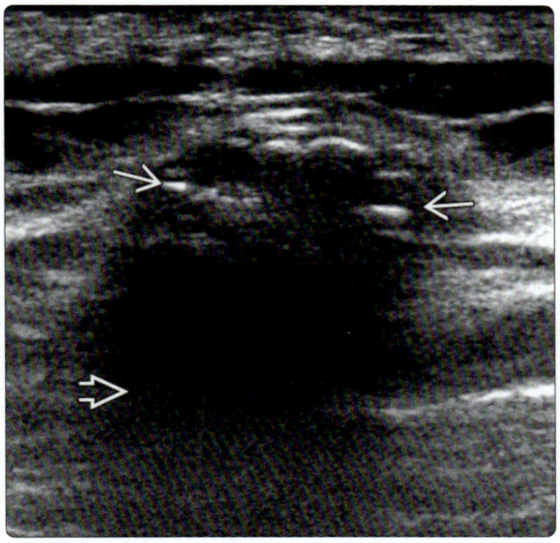

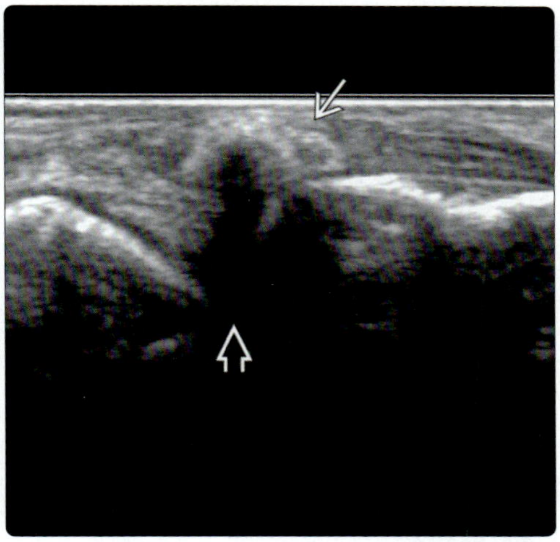

(Left) *Transverse US shows an injection granuloma of the buttock with several medium-sized calcific foci → and strong posterior acoustic shadowing ⇨. Injection granulomas show either no hyperemia or only minimal hyperemia on color Doppler imaging (not shown).* **(Right)** *Transverse US shows a small, heavily calcified granuloma → on the dorsal aspect of the distal radius alongside the distal radioulnar joint ⇨. This granuloma was most likely related to a previous steroid injection at this site.*

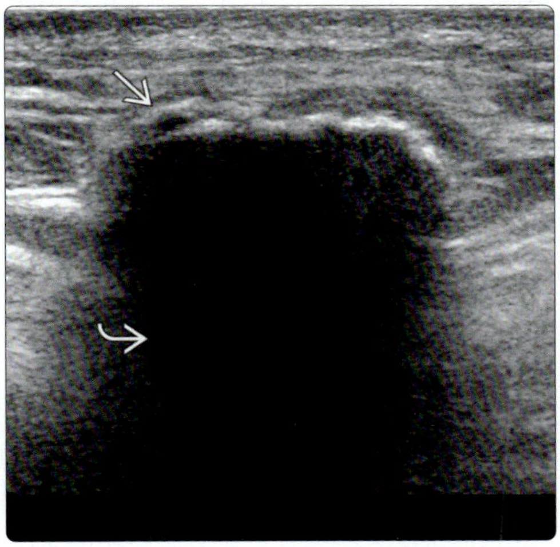

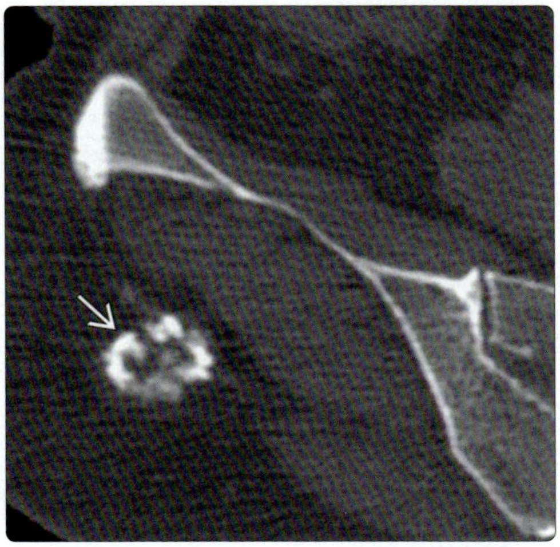

(Left) *Transverse US shows a quite large, heavily calcified granuloma → in the upper outer quadrant of the buttock. There is dense acoustic shadowing ↷.* **(Right)** *Axial NECT shows a calcified granuloma → in the deep subcutaneous region of the buttock. Granulomas are a common incidental finding on radiograph or CT that remain unchanged over years. Clinically, they are either asymptomatic or may present as a mildly painful or painless mass.*

Lymph Node Abnormality

KEY FACTS

IMAGING

- US
  - Accurate at differentiating between benign and malignant nodes
  - More accurate than CT and MR provided that nodes are accessible to US assessment
  - Can guide fine-needle aspiration (FNA) cytology or biopsy
  - Appearances vary depending on whether node is normal, reactive, metastatic, lymphomatous, or infected (tuberculosis or nontuberculosis infection)
  - Important discriminatory criteria: Size, nodal configuration, border, internal architecture, and pattern of vascularity
  - Color Doppler to assess pattern of vascularity
  - Spectral analysis is not routinely helpful
  - US-guided FNA or biopsy for microbiologic, cytologic, or histologic analysis

TOP DIFFERENTIAL DIAGNOSES

- Nerve sheath tumor
- Vascular leiomyoma
- Vascular anomaly

CLINICAL ISSUES

- Lymphadenitis: Very tender
- Reactive adenopathy: Mildly tender
- Metastatic, lymphomatous, and tuberculous nodes: Nontender

DIAGNOSTIC CHECKLIST

- US nodal diagnosis formulated on relative strength of many US signs in conjunction with clinical presentation
  - US is accurate at determining presence and likely etiology of nodal pathology as well as need for FNA cytology or biopsy

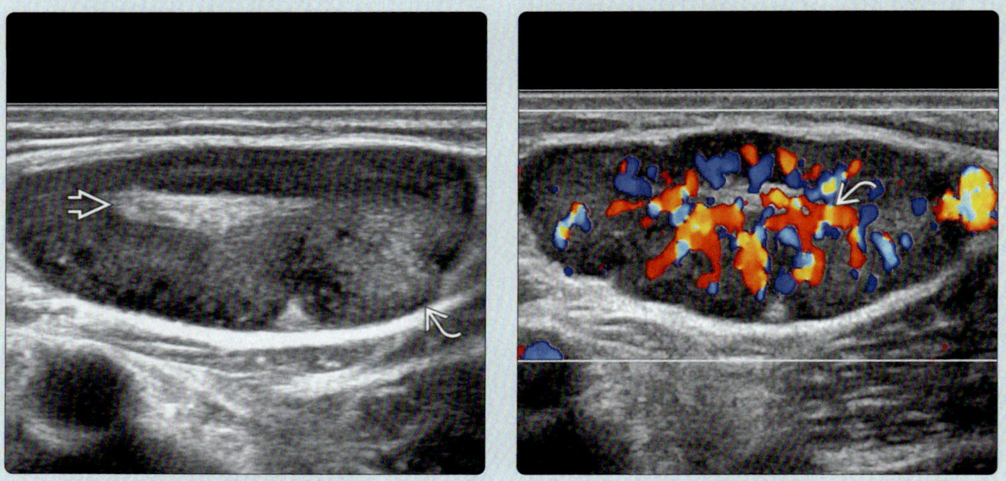

(Left) *Longitudinal US shows a reactive groin node in a patient with ipsilateral calcaneal osteomyelitis. The lymph node is elliptical in shape with moderate cortical thickening and a distinctive central echogenic fatty hilum.* **(Right)** *Longitudinal color Doppler US of the same lymph node shows a normal hilar vascular flow pattern with branching intranodal vessels extending from the hilum into the hypoechoic cortical parenchyma.*

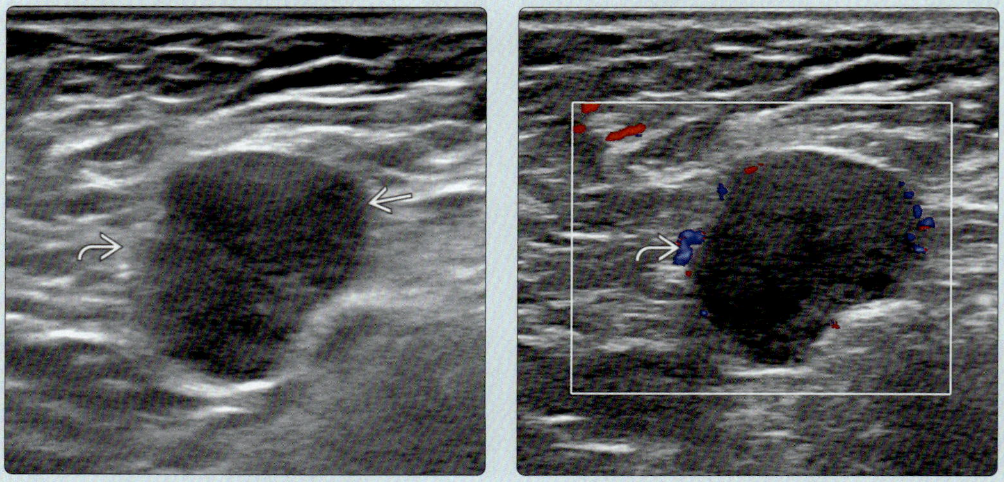

(Left) *Longitudinal US shows a metastatic axillary node in a patient with primary lung cancer. The lymph node is roundish in shape, irregular in contour, and diffusely hypoechoic due to metastatic infiltration with early extracapsular spread.* **(Right)** *Longitudinal color Doppler US of the same node shows loss of normal hilar vascularity. The vascular flow pattern in this metastatic node is peripheral. This peripheral vascularity is highly suggestive of metastasis, which was confirmed on fine-needle aspiration cytology.*

TERMINOLOGY

Synonyms

- Lymphadenopathy, adenopathy, lymph node enlargement, lymphadenitis

Definitions

- Reactive node
  - Benign enlargement of node in response to antigenic stimuli, including infection and inflammation
- Metastatic node
  - Enlargement of node due to metastatic infiltration
- Lymphomatous node
  - Enlargement of node due to lymphoma infiltration
- Tuberculous node
  - Enlargement of node due to tuberculous infection
- Lymphadenitis
  - Enlargement of node due to bacterial or fungal infection

IMAGING

General Features

- Best diagnostic clue
  - Enlargement of lymph node with altered nodal architecture
- Location
  - Any sites along lymphatic drainage
  - Head and neck > groin > axillae
- Size
  - ~ 3 mm to > 10 cm
- Morphology
  - Elliptical, oval, or rounded nodes

CT Findings

- Reactive node: Central fatty hilum with hypodense peripheral cortex of variable thickness
- Malignant node: Solid soft tissue nodule with no identifiable fatty hilum
- ± intranodal cystic change, calcification, matting, or perinodal soft tissue inflammation
- Irregular border indicates extranodal spread

Ultrasonographic Findings

- Accurate at differentiating between benign and malignant nodes
  - Reasonably accurate at identifying other nodal pathologies
  - More accurate than CT and MR if nodes are accessible
  - Guide fine-needle aspiration (FNA) cytology or biopsy
- Size is important, but only one of several discriminatory criteria used
- Nodal configuration, border, internal architecture, and pattern of vascularity are also important
- **Normal and reactive lymph nodes**
  - Normal node: Long axis < 8 mm
  - Reactive node: Long axis usually > 8 mm
  - Elliptical or ovoid in shape
    - Long- to short-axis length ratio > 2
  - Hypoechoic cortex compared with adjacent muscle
    - Variable cortical hyperplasia (thickening)
      - Leading to lobulated cortical hyperplasia
  - Normal echogenic fatty hilum
  - Hilar vascularity on color Doppler imaging
    - Central hilar vessels branch radially into cortical parenchyma of node
- **Metastatic lymph node**
  - Long axis > 8 mm
    - Some malignant nodes have long axes of only 6-8 mm
  - Rounded shape
    - Long- to short-axis length ratio < 2
  - Effacement or replacement of echogenic fatty hilum
  - Thickened, hypoechoic cortex
  - Usually well-demarcated border
    - Irregular, blurred contour suggests extranodal spread
  - ± punctate calcification depending on primary tumor
    - Calcification feature of chondroid, osteosarcoma metastases, papillary, and medullary thyroid carcinoma metastases
  - ± intranodal cystic change
    - Seen with squamous cell carcinoma and melanoma
    - Coagulative necrosis seen as echogenic foci/area
  - ± matting of adjacent nodes
    - Seen in malignancy > tuberculosis > lymphoma
  - Peripheral capsular vascularity or mixed hilar and peripheral vascularity
    - Presence of peripheral vascularity only is highly suspicious of metastases
    - Chaotic rather than regular vascular pattern
    - Impedance values (resistive index, pulsatility index) of limited value due to large overlap with reactive values
  - Shear wave elastography can aid differentiation between malignant and benign lymph nodes
    - Sensitivity, specificity, and accuracy of 81.8%, 80.0%, and 81.0%, respectively, with threshold of 37.9 kPa
- **Lymphomatous node**
  - Usually > 8 cm
    - Can be very large
  - Rounded in shape
    - Focal cortical infiltration → focal cortical thickening
  - Usually sharp border without extracapsular spread
    - Unsharp border suggestive of extracapsular spread and aggressive disease
  - Loss of normal echogenic hilum (75%)
  - Usually very hypoechoic
    - Uniform cellular infiltration within node produces few interfaces → good acoustic transmission → posterior acoustic enhancement
  - Intranodal reticular pattern
    - Thin, echogenic lines traversing hypoechoic node
      - Seen in lymphoma and malignancy
  - Calcification is rare; may be present after treatment
  - Cystic necrosis is uncommon
  - Mixed prominent hilar and peripheral vascularity
    - Peripheral vascularity alone is rare
  - Nodal size ↓ with treatment indicates good response
- **Tuberculous node**
  - Enlarged > 8 cm
  - Oval or rounded
  - Hypoechoic nodes with intranodal necrosis
    - Posterior acoustic enhancement
  - Matting of multiple nodes

- Perinodal soft tissue inflammation (periadenitis)
- Calcification
  - Seen after treatment, usually not seen in acute phase
- Variable vascular pattern
  - Preserved but displaced hilar vascularity (50%)
  - Absent hilar vascularity (19%)
    - Intranodal vessels obliterated by necrotizing granulomatous inflammation
    - Intranodal vessels obliterated by fibrosis and hyalinization in healing phase
  - Perinodal inflammation → capsular vascularity (12%)
- **Lymphadenitis**
  - Suppurative or nonsuppurative nodal infection
  - Enlarged node with preservation of nodal architecture, hyperemia, ± liquefaction and periadenitis

Imaging Recommendations

- Best imaging tool
  - High-resolution US
    - Nodes in neck, supraclavicular fossa, axilla, elbow, groin, and popliteal fossa are all accessible to examination by US
- Protocol advice
  - Grayscale to assess nodal size, shape, border, and internal architecture
    - For axillary nodes, sweep transducer along center and sides of axillary artery
      - Normal axillary nodes have large fatty hilum and can be inconspicuous against axillary fat
    - Routinely examine external iliac nodes when examining inguinal nodes
      - Particularly large subcutaneous node in proximal medial thigh (Cloquet node) is normal finding
    - Normal elbow (epitrochlear) and popliteal fossa nodes are not routinely visible
  - Color Doppler to assess pattern of vascularity
    - Use minimal transducer pressure
    - Spectral analysis not routinely helpful
  - US-guided FNA to obtain microbiologic, cytologic, or histologic diagnosis
    - Apply suction as needle is moved to and fro within node while simultaneously rotating needle

DIFFERENTIAL DIAGNOSIS

Nerve Sheath Tumor

- Thickened entering and exiting nerve
- No echogenic fatty hilum

Vascular Leiomyoma

- Peripheral converging vascularity
- No echogenic fatty hilum

Vascular Anomaly

- Irregular border
- Intrinsic vascular mesh

PATHOLOGY

General Features

- Etiology
  - Reactive node
    - Inflammation, infection, allergy, hemorrhage, inflammatory arthropathy, sarcoidosis
    - Castleman disease: Giant lymph node hyperplasia; lymphoproliferative disorder
      - Localized or multicentric (may be HIV related)
    - Kikuchi disease: Occurs in neck; histiocytic necrotizing lymphadenitis
    - Rosai-Dorfman disease: Typically occurs in neck; sinus histiocytosis
    - Kimura disease: Eosinophilic hyperplastic lymphogranuloma
  - Infection
    - Bacterial > viral > fungal or parasitic
      - *Mycobacterium avium-intracellulare, Mycobacterium scrofulaceum, Mycobacterium bovis*, and *Mycobacterium kansasii*
      - Other causes of granulomatous nodal inflammation are brucellosis and sarcoidosis
    - Cat-scratch disease: Regional adenopathy 7-14 days after cat scratch or bite; infection by *Bartonella henselae*
    - Bacille Calmette-Guérin (BCG) adenitis: Exaggerated expected nodal response → regional lymphadenitis 2 weeks to 2 years after BCG vaccination
      - Most common complication of BCG vaccination
  - Metastases
  - Hematologic
    - Lymphoma (non-Hodgkin > Hodgkin) > leukemia

CLINICAL ISSUES

Presentation

- Most common signs/symptoms
  - Local swelling, tenderness, ± pain
  - Lymphadenitis: Very tender
  - Reactive adenopathy: Mildly tender
  - Metastatic, lymphomatous, and most tuberculous nodes: Nontender

Treatment

- Antiinflammatory agents or antibiotics (including antituberculous therapy) dependent on etiology
- Lymphoma: Systemic chemotherapy
- Metastatic: Surgery ± chemotherapy or irradiation

DIAGNOSTIC CHECKLIST

Consider

- US diagnosis should be formulated on several recognized signs in conjunction with clinical presentation, rather than on just 1 or 2 signs
  - US is overall accurate at determining presence and likely etiology of nodal pathology

SELECTED REFERENCES

1. Tan S et al: Value of shear wave elastography versus contrast-enhanced sonography for differentiating benign and malignant superficial lymphadenopathy unexplained by conventional sonography. J Ultrasound Med. 36(1):189-199, 2017

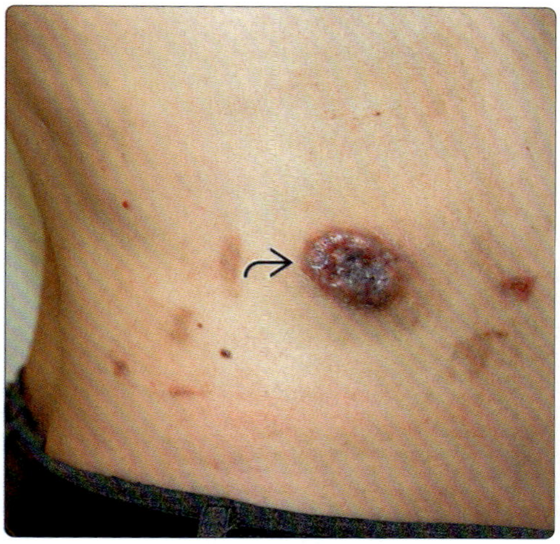

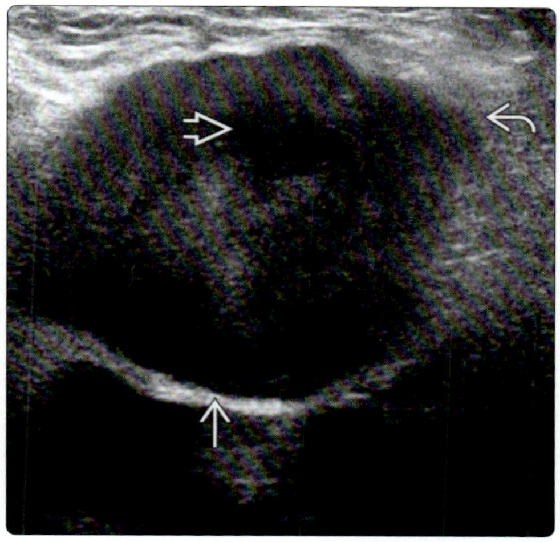

(Left) *Clinical photograph of the flank region shows a malignant-looking skin lesion in a patient with a chronic, nonhealing skin ulcer. Histology revealed a high-grade adenocarcinoma with foci of sweat gland myoepithelial differentiation.* **(Right)** *Longitudinal US of the same patient shows an enlarged metastatic axillary node. The node is round with complete loss of the central fatty hilum. Note the mild intranodal necrosis and localized extracapsular spread.*

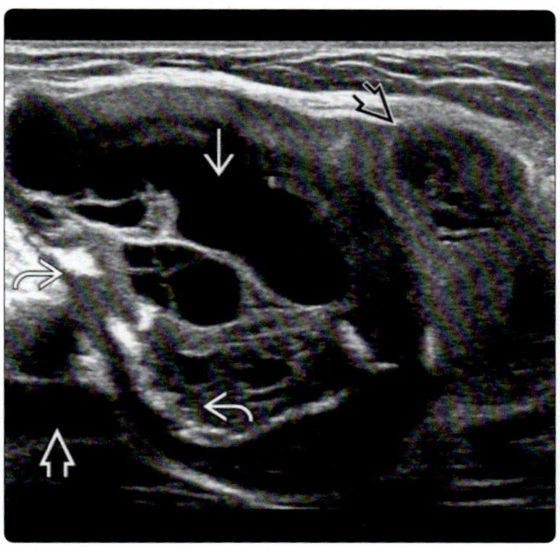

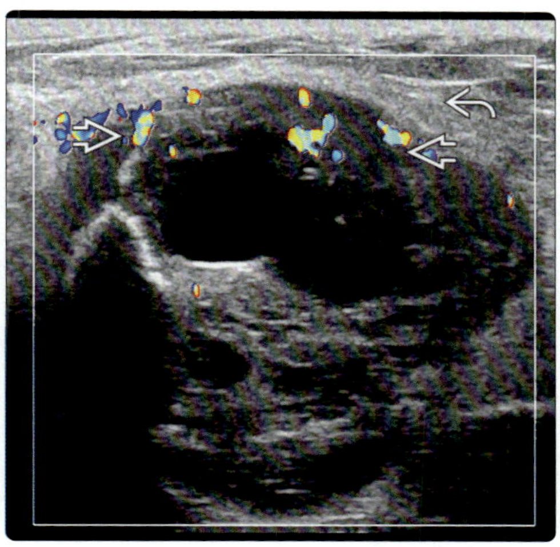

(Left) *Longitudinal US of a patient with telangiectatic-type osteosarcoma shows a metastatic groin node lying lateral to the femoral neurovascular bundle. There is a large cystic element with areas of mineralization and a 2nd discrete metastatic cortical focus.* **(Right)** *Transverse color Doppler US of the same metastatic node shows mild to moderate peripheral nodal vascularity. Mild perinodal soft tissue edema is present.*

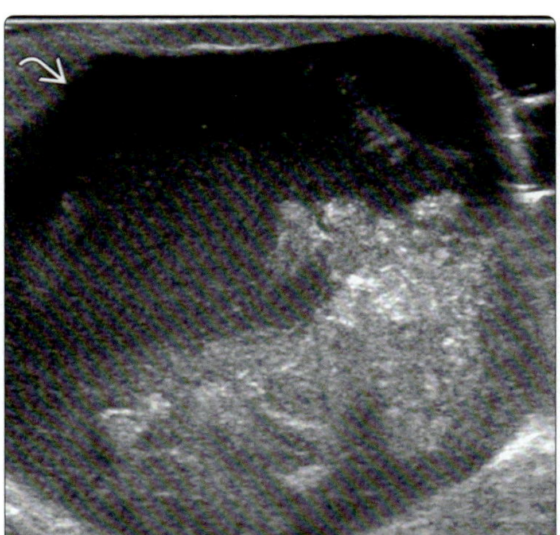

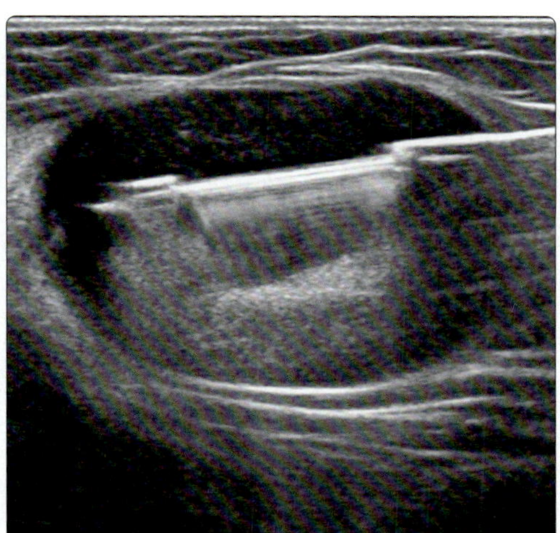

(Left) *Transverse US of the groin in a patient with carcinoma of vulva shows a partially cystic metastatic lymph node. Intranodal cystic change is seen in metastases from squamous cell carcinoma or melanoma.* **(Right)** *Transverse US during US-guided core biopsy of a groin lymph node in a patient with a suspicious ovarian mass is shown. Histology revealed metastatic adenocarcinoma consistent with an ovarian primary. If feasible, biopsy rather than FNAC of suspicious lymph nodes is preferable.*

(Left) *Transverse US shows an enlarged groin lymph node in a patient with non-Hodgkin lymphoma. There is a thickened cortex that has many fine reticulations, moderate posterior acoustic enhancement, and a partially effaced fatty hilum. All of these are features of lymphomatous nodes.* **(Right)** *Transverse color Doppler US of the same lymph node shows the prominent hilar vascular flow pattern with branching vessels extending from the hilum into the thickened cortical parenchyma.*

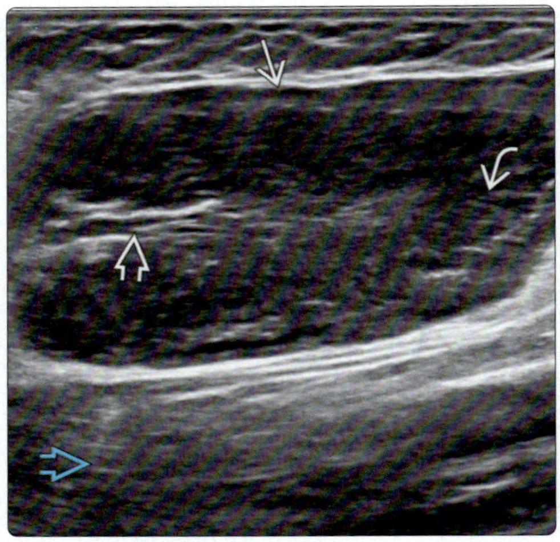

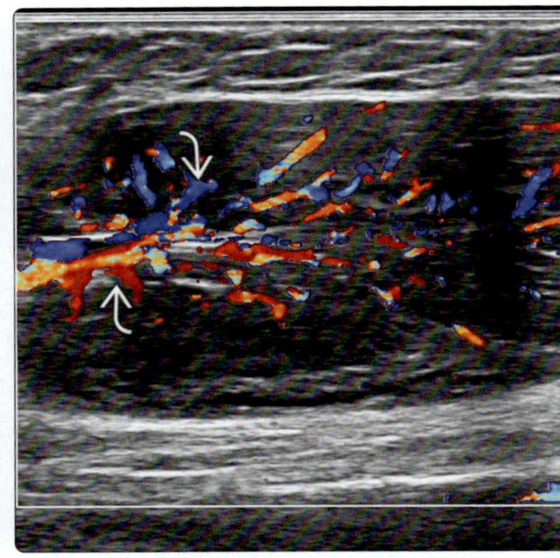

(Left) *Longitudinal color Doppler US of a lymphomatous node shows marked hypoechoic thickening of the cortical parenchyma. Moderate hilar and peripheral vascularity are present. Lymphoma nodes usually have preservation of nodal architecture.* **(Right)** *Longitudinal power Doppler US of a patient with non-Hodgkin lymphoma shows a reticulated thickened cortex with moderate central and peripheral vascularity. Note the extracapsular extension, which indicates aggressive disease.*

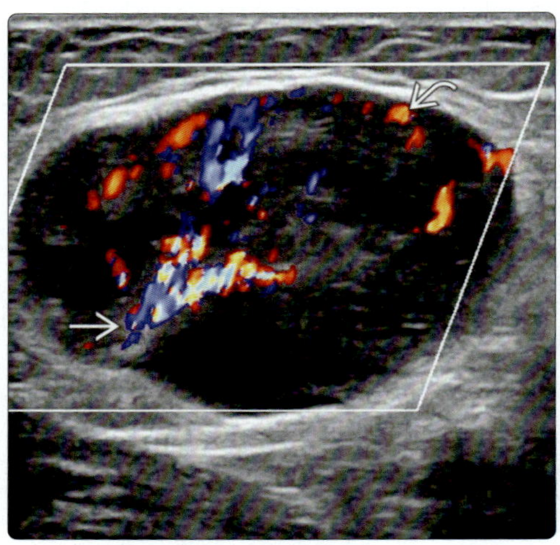

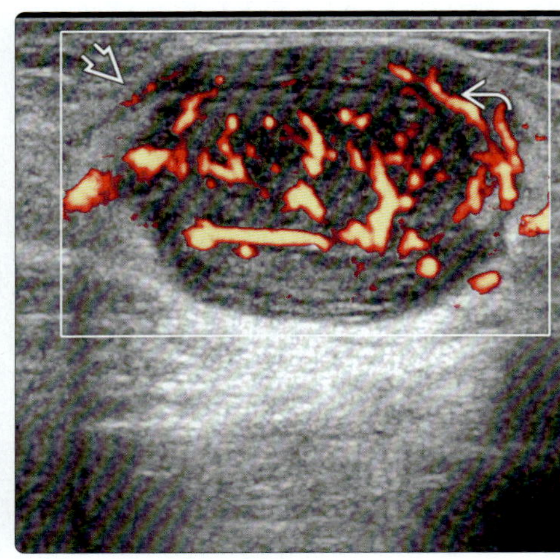

(Left) *Clinical photograph of the ring finger of a patient with a recent cat bite injury shows 2 small wounds (resulting from the cat bite) and associated mild skin erythema.* **(Right)** *Longitudinal US of the elbow region in the same patient shows an enlarged, irregular, oval-shaped epitrochlear lymph node with irregular cortical hyperplasia. The central fatty hilum is still preserved. There is severe perinodal soft tissue edema.*

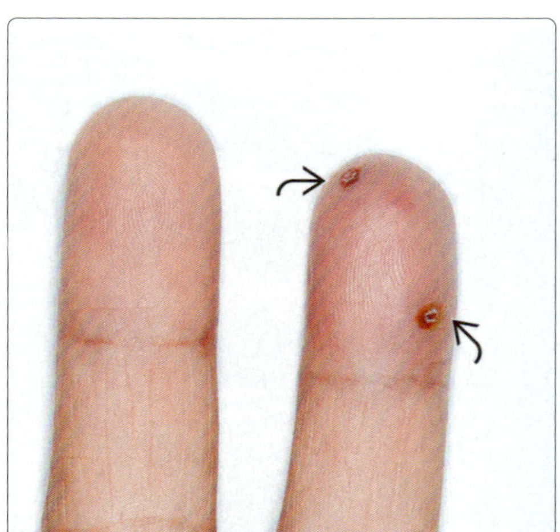

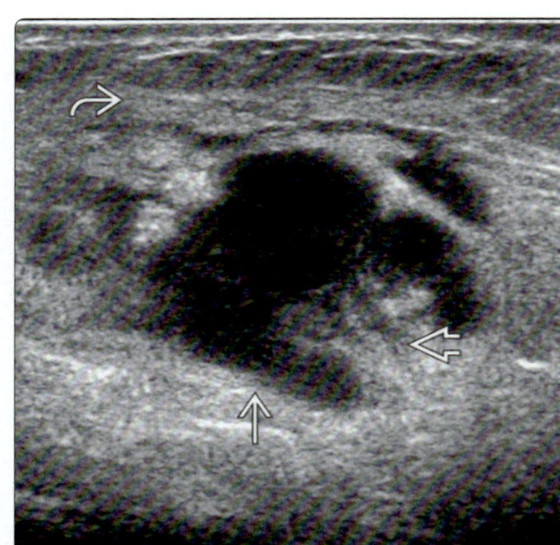

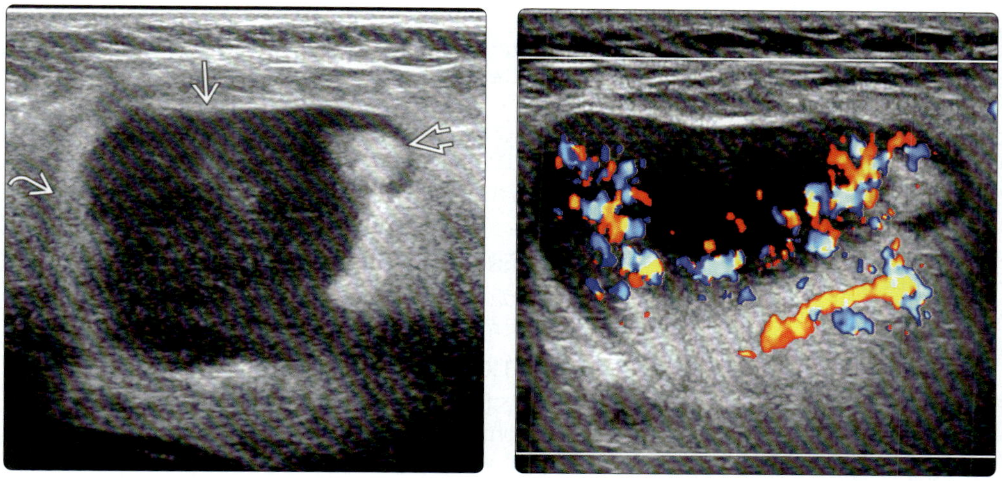

(Left) *Transverse US of the axilla in a patient with ipsilateral soft tissue infection shows a solitary enlarged lymph node in the central part of the axilla. The lymph node is diffusely enlarged with marked cortical hyperplasia with eccentric displacement of fatty hilum. Severe perinodal soft tissue inflammation is present.* **(Right)** *Transverse color Doppler US of the same lymph node shows marked intranodal hyperemia consistent with lymphadenitis. No liquefaction or abscess formation is present.*

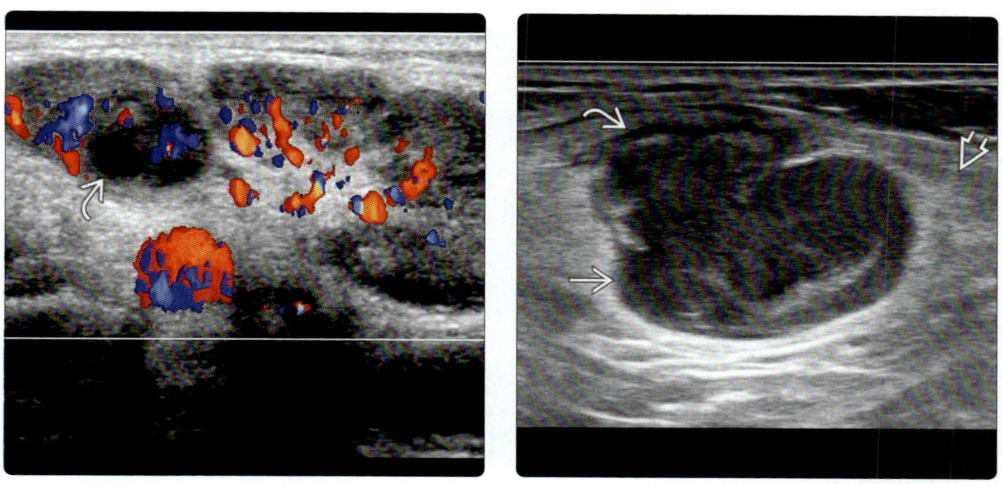

(Left) *Longitudinal color Doppler US shows groin lymphadenitis. A small cortical cystic component consistent with focal suppuration is present. Mild perinodal soft tissue edema is present.* **(Right)** *Longitudinal US shows an active tuberculous groin lymph node with eccentric cortical thickening and irregular extracapsular soft tissue as a result of previous rupture of an intranodal abscess. Moderate perinodal soft tissue edema is present.*

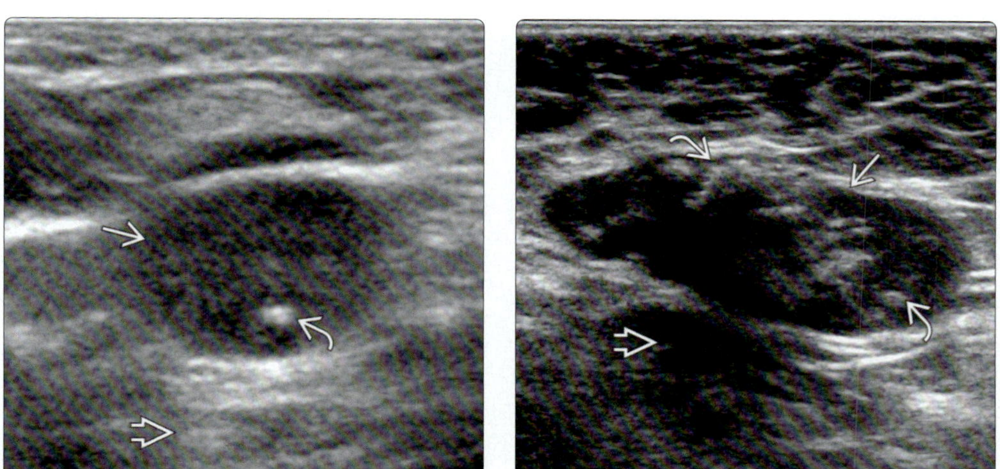

(Left) *Transverse US of the axilla of a 4-year-old boy shows an enlarged roundish node with a small focus of eccentric calcification. Mild posterior acoustic enhancement is present. This is due to healing tuberculous adenitis. Nodal calcification is not seen in acute tuberculous adenitis.* **(Right)** *Longitudinal US shows a healed tuberculous lymph node. The lymph node contains many small foci of intranodal calcification, which led to posterior acoustic shadowing.*

Soft Tissue Sarcoma

KEY FACTS

TERMINOLOGY

- Malignant tumor arising from connective (mesenchymal) tissue other than bone

IMAGING

- Large, heterogeneous, hypoechoic mass
  - Vast majority are located in deep (subfascial) tissues
- Most soft tissue sarcomas (STS) will have well-defined margin
  - As opposed to bone tumor, well-defined margin does not indicate benignity of soft tissue tumor
- ± myxoid tissue, necrosis, hemorrhage, or calcification
- Usually hypervascular with disorganized vascular pattern on color Doppler imaging
  - May be hypovascular or avascular
- Regional nodal metastases (uncommon)
- Neurovascular encasement
- Bone erosion or periosteal reaction (uncommon)
- Determine presence of tumor, US characteristics, and extent
- Identify which compartment is involved
- MR is better than US at defining extent of large, deep tumors
- Perform US-guided biopsy of all suspected sarcoma except for well-differentiated liposarcomas
  - Biopsy margin of tumor deep to capsule and vascularized areas

TOP DIFFERENTIAL DIAGNOSES

- Benign soft tissue tumor
- Muscle hematoma

DIAGNOSTIC CHECKLIST

- Consider STS in all growing, deep soft tissue tumors that do not conform to known benign tumors
- Do not label any tumor as definitively benign unless firm diagnosis can be made based on appearances

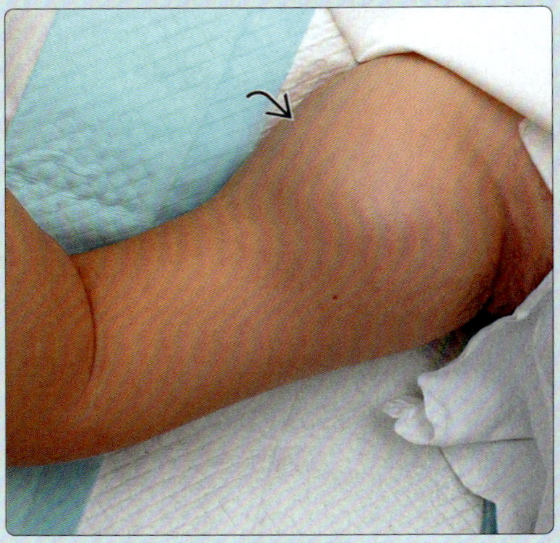
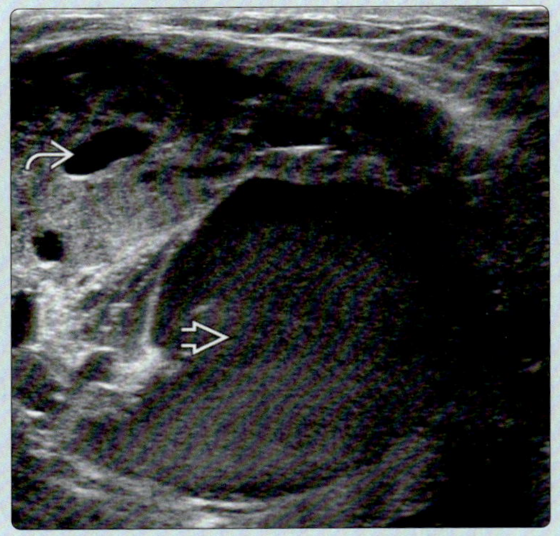

(Left) *Clinical photograph of an adult patient shows an arm mass ➡ that had been slowly growing for 2 years.* **(Right)** *Transverse US of the same patient shows a large, well-defined, mixed cystic/solid tumor within the biceps brachii muscle. Low-level echoes ➡ consistent with hemorrhage are present within some cystic spaces. Some of these cystic spaces are clear ➡. Unlike bone tumor, a well-defined border in a soft tissue tumor does not favor benignity.*

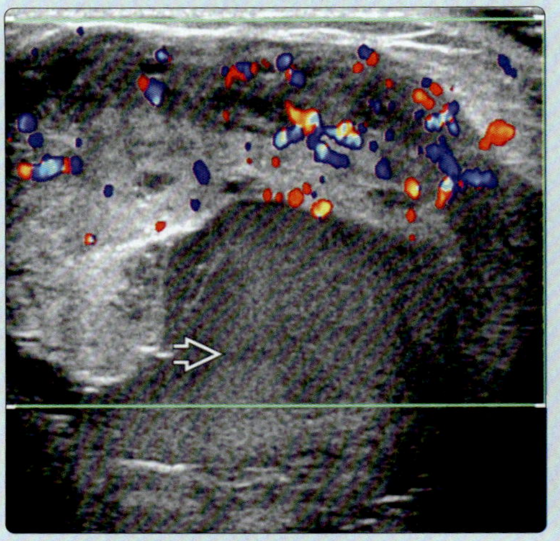
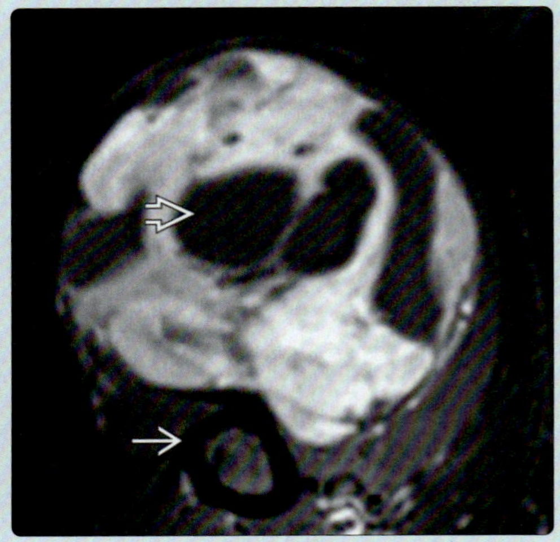

(Left) *Transverse color Doppler US of the same tumor shows moderate hypervascularity in the more solid component of the tumor. The cystic area ➡ is avascular. US-guided biopsy was performed.* **(Right)** *Transverse T1 C+ FS MR of the same tumor shows avid tumoral enhancement with nonenhancing cystic areas ➡. The tumor contacts the humerus ➡. No cortical infiltration is evident. Histology revealed a synovial sarcoma.*

TERMINOLOGY

Abbreviations

- Soft tissue sarcoma (STS)

Synonyms

- Malignant soft tissue tumor

Definitions

- Malignant tumor arising from connective (mesenchymal) tissue other than bone

IMAGING

General Features

- Best diagnostic clue
  - Large, deep soft tissue mass
- Location
  - Lower extremity (45%) > trunk (30%) > upper extremity (20%) > head and neck (5%)
    - Other than dermatofibrosarcoma protuberans, sarcoma arising within subcutaneous tissue is rare
      - Most malignant subcutaneous tumors are metastases or lymphoma
- Size
  - Variable
- Morphology
  - Heterogeneous, usually encapsulated mass

Ultrasonographic Findings

- Large, heterogeneous, hypoechoic mass
  - Usually located in deep (subfascial) tissues
  - Well encapsulated, though pseudocapsule is not usually seen as distinct structure
- Margin: Most STS will have well-defined margin; infiltrative margins suggest aggressive fibromatosis, lymphoma, and inflammatory masses
- Myxoid tissue: Well-defined, intratumoral anechoic or hypoechoic areas
- Necrosis: Poorly defined, more centralized anechoic or hypoechoic areas
- Calcification: Discrete, echogenic foci with posterior acoustic shadowing
  - e.g., liposarcoma (5%), pleomorphic undifferentiated sarcoma (10%), synovial sarcoma (30%)
- Hemorrhage: Ill-defined, intratumoral, echogenic areas without acoustic shadowing; fluid levels
- Usually hypervascular with disorganized vascular pattern on color Doppler imaging
  - May be hypovascular or avascular
    - Large, deep-seated sarcomas often show minimal or no vascularity on color Doppler imaging
  - Lymphoma tends to be hypervascular
- Well-differentiated liposarcoma has US characteristics similar to superficial lipoma though tend to be larger, deep (subfascial), and growing clinically
  - Undertake MR in all deep-seated lipomatous masses to help characterization
- Myxoid liposarcoma can initially resemble cysts and should not be confused with paraarticular ganglia, myxoma, or bursae
  - Internal solid component and vascularity can usually be detected
- Do not conform to recognized US appearances of other tumors, such as lipoma or nerve sheath tumor
  - Particularly true for superficial sarcomas, as most superficial soft tissue tumors can be distinguished on basis of US appearances
- Associated abnormality
  - Adjacent bone erosion or periosteal reaction/scalloping (uncommon)
  - Regional nodal metastasis (not common)
    - Similar US appearances to primary tumor
    - Enlarged, rounded node with loss of hilar echogenicity and peripheral, rather than hilar, vascularity
  - Neurovascular encasement

MR Findings

- Large, well-defined soft tissue mass
- Heterogeneous pattern variably isointense on T1 and hyperintense on T2
- May have high T1 signal due to fat component in liposarcoma or hemorrhage

Imaging Recommendations

- Best imaging tool
  - US followed by MR
- Protocol advice
  - Determine presence of tumor, US characteristics, and extent
    - Identify which compartment is involved
      - If subfascial, identify which muscle compartment is involved
  - Determine relationship to neurovascular bundle and bone
    - Neurovascular bundle may be either separate from, in contact with, or encased (partially or completely) by tumor
  - Evaluate regional nodes
    - Undertake fine-needle aspiration cytology (FNAC) or biopsy, if suspicious
  - MR is better than US at defining extent of large, deep tumors and at characterizing malignant features in large lipomatous masses
  - Final diagnosis depends on histology of biopsy tissue
    - Perform US-guided biopsy of all suspected sarcomas, except for well-differentiated liposarcomas
      - False-negative rate high in well-differentiated liposarcoma
      - If suspected clinically or on imaging, excision is advocated
  - US-guided biopsy for histology
    - Do not contaminate nonaffected compartment
    - Sample solid component
      - Hyperemia on color Doppler imaging helps to identify areas of viable tumor
    - 16-g Tru-Cut core biopsy needle with 15-g coaxial system suitable for tissue typing and histologic staging
      - Obtain at least 3 good cores from noncystic areas of tumor
      - FNAC is not sufficient for initial diagnosis, detecting residual disease, or recurrence

DIFFERENTIAL DIAGNOSIS

Benign Soft Tissue Tumor

- Most benign soft tissue tumors have characteristic US appearances

Muscle Hematoma

- History of trauma is important
  - Some tumors may bleed and present as muscle injury
  - Undertake further investigation (MR or PET imaging), follow-up, or biopsy if suspicious

Fibromatosis

- Soft tissue mass of variable echogenicity, though usually hypoechoic with absent or mild vascularity on color Doppler imaging
- Irregular margins

Muscle Metastasis

- Uncommon
- Smaller and less well-defined than STS ± more peritumoral edema
- Majority will have known primary tumor

PATHOLOGY

General Features

- Etiology
  - Idiopathic
  - Radiation induced
- Classified according to tissue type rather than anatomical origin
  - e.g., fibrosarcoma may not arise from fibrous tissue, but histologically, most resemble fibrous tissue
- Most STS are histologically high grade
- Grow centrifugally from single focus
  - Pseudocapsule formed by compression of adjacent connective tissues
  - Reactive zone peripheral to pseudocapsule may contain tiny extensions (pseudopods) of tumor
    - May detach to give satellite nodules
- STS respects fascial and anatomical boundaries
  - Usually grow within compartment of origin
  - Involvement of > 1 compartment and bone/neurovascular bundle invasion are signs of aggressive tumor

Staging, Grading, & Classification

- Histological grading (as high, intermediate, or low grade) depends on
  - Degree of cellularity, cellular pleomorphism and level of dedifferentiation, mitotic activity, degree of necrosis, and infiltrative growth

Gross Pathologic & Surgical Features

- Myxoid accumulation, hemorrhage, necrosis, pseudocapsule

Microscopic Features

- Liposarcoma
  - Well-differentiated variety: 40%
    - Mainly fat with pathognomic lipoblasts that confer malignancy
    - Thick (> 2-mm) septations, nodular areas of nonlipomatous tissue
  - Myxoid variety (40%), round cell variety (10%), pleomorphic variety (5%), dedifferentiated variety (5%)
- Synovial sarcoma
  - Presence of keratin (epithelial marker) is diagnostic
    - Monophasic: 50-60%; mesenchymal spindle cell predominates
    - Biphasic: 20-30%; mesenchymal spindle cell and epithelial components
    - Poorly differentiated: 15-25%; high mitotic activity with necrosis

CLINICAL ISSUES

Presentation

- Most common signs/symptoms
  - Painless soft tissue mass enlarging over several months
- Clinical profile
  - Often indistinguishable from benign tumor

Demographics

- Age
  - 10-90 years; peak: 50 years
- Epidemiology
  - < 1% of adult cancer; 6% of childhood cancer
  - Pleomorphic undifferentiated sarcoma (25%); liposarcoma (15%); leiomyosarcoma (8%); malignant nerve sheath tumor (5%); dermatofibrosarcoma protuberans (5%); synovial sarcoma (5%); fibrosarcoma (5%); rhabdomyosarcoma (2%); epithelioid sarcoma, alveolar soft part sarcoma, extraskeletal Ewing tumor, lymphoma, and others (15%); undifferentiated (15%)

Natural History & Prognosis

- Depends of histological grade and surgical resectability
  - 50% of high-grade tumors develop metastases despite adequate treatment
    - < 10% of those with metastases survive > 2 years
- Local recurrence rate: ~ 10% for limb or limb-girdle STS
  - 80% of recurrences occur within 2 years of surgery

Treatment

- ~ 80% of adult STS are chemoresistant
- Surgical resection with wide margin
- ± adjuvant radiotherapy to tumor bed

DIAGNOSTIC CHECKLIST

Consider

- STS in any rapidly growing soft tissue mass that does not conform to known benign characteristics

Image Interpretation Pearls

- Nearly all STS are well defined (i.e., sharp border does not equate to benign tumor)
- Do not label any tumor as benign based on US appearances unless sure about diagnosis

SELECTED REFERENCES

1. Morii T et al: Differential diagnosis between benign and malignant soft tissue tumors utilizing ultrasound parameters. J Med Ultrason (2001). 45(1):113-119, 2018

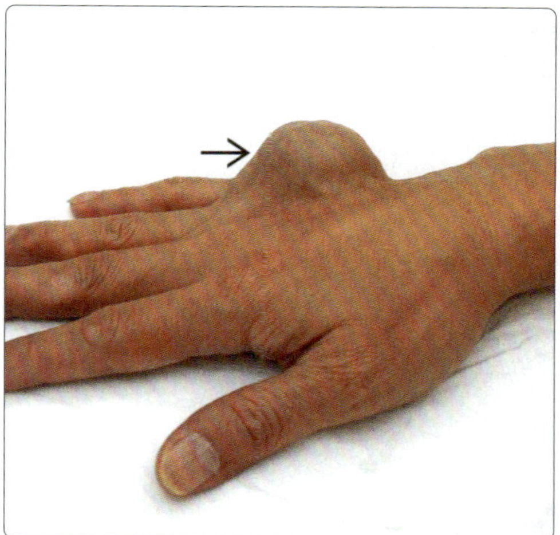

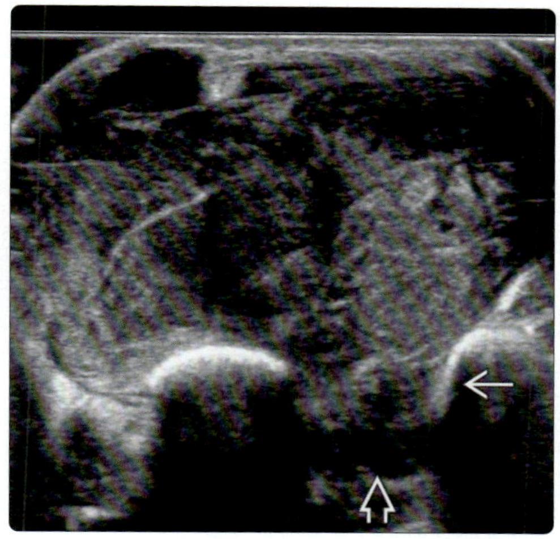

(Left) *Clinical photograph shows a large mass → on the dorsum of the hand. The mass had grown rapidly over the past year.* **(Right)** *Transverse US of the same swelling shows a large soft tissue mass on the dorsum of the hand extending deeply between the 4th and 5th metacarpal bones ⇨. Smooth bony scalloping → is present on the ulnar side of the 4th metacarpal shaft. This scalloping may be due to pressure or bony resorption. It does not necessarily imply malignancy.*

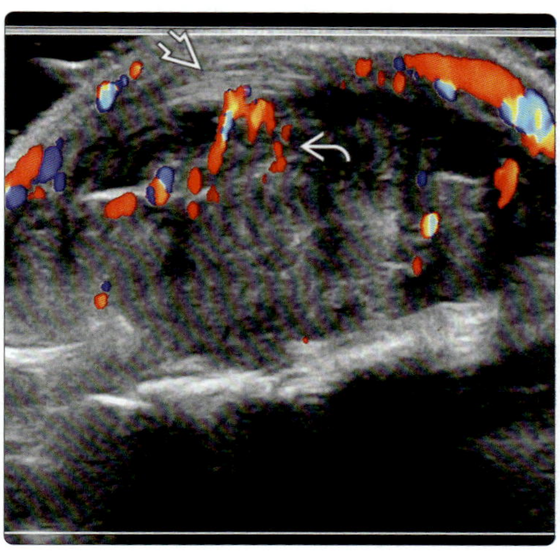

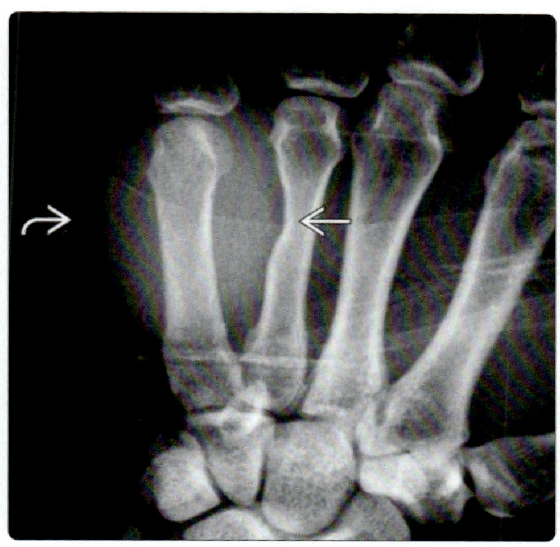

(Left) *Longitudinal color Doppler US of the same tumor shows moderate tumor hyperemia ↪. The mass has arisen deep to the extensor tendons ⇨, which are displaced superficially.* **(Right)** *Frontal radiograph of the hand in the same patient shows a large soft tissue mass ↪ with a long area of smooth, bony scalloping with cortical thinning → of the 4th metacarpal bone deep to the large soft tissue mass. Radiographs show the degree and type of bony erosion better than US.*

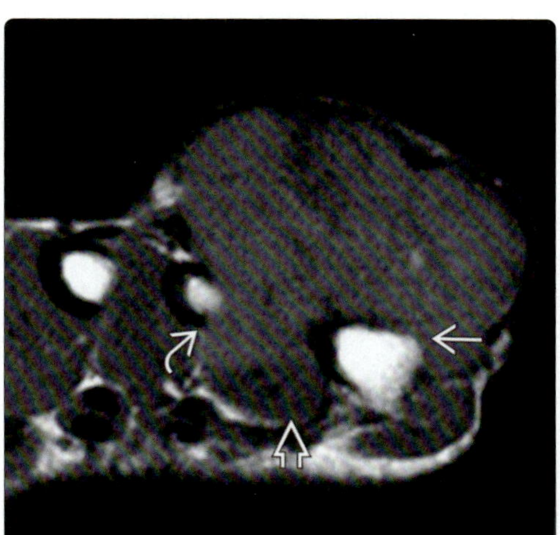

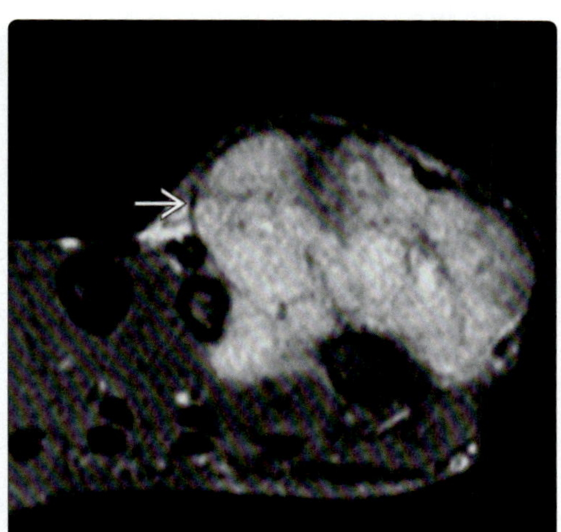

(Left) *Transverse T1 FSE MR of the same tumor shows the soft tissue mass extending deeply ⇨ between the 4th and 5th metacarpal bones. Note how the bony scalloping or erosion of the 4th ↪ and 5th → metacarpal bones is more apparent than on US.* **(Right)** *Transverse T1 C+ FS MR of the same tumor shows significant enhancement of the tumor mass →. Percutaneous biopsy for histology revealed a clear cell sarcoma.*

(Left) *Longitudinal US of a young adult patient with a painless thigh swelling for 1 month shows a well-defined, oval-shaped, homogeneous, hypoechoic soft tissue mass ➔ within the vastus intermedius muscle. No calcification or cystic change is present. The underlying femoral shaft ➔ is normal.* **(Right)** *Longitudinal color Doppler US of the same tumor shows mild, intratumoral hyperemia ➔.*

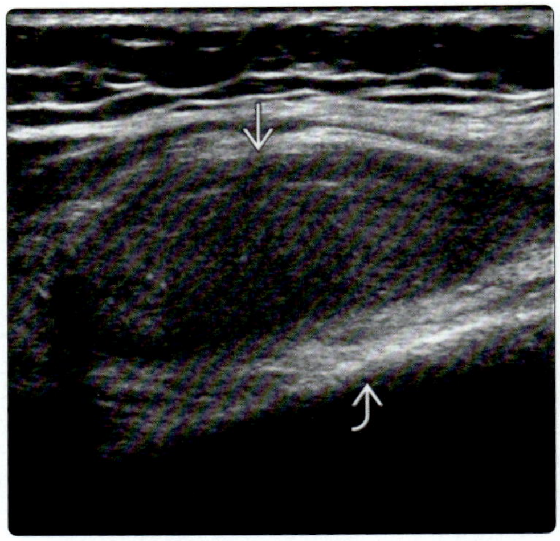

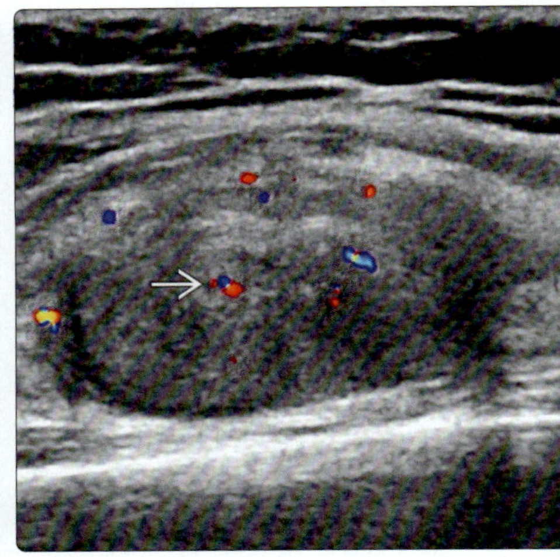

(Left) *Longitudinal US elastography of the same tumor shows variable tumoral elasticity. Red color represents stiff tissue, while blue color represents normal soft tissue. Elastography in assessing soft tissue tumors is currently under investigation. Histology revealed a fibromyxoid tumor of intermediate malignancy.* **(Right)** *Transverse US of a 60-year-old man with a popliteal mass shows a homogeneous echogenic mass ➔ deep to the biceps femoris muscle ➔ and superficial to the popliteal vessels ➔.*

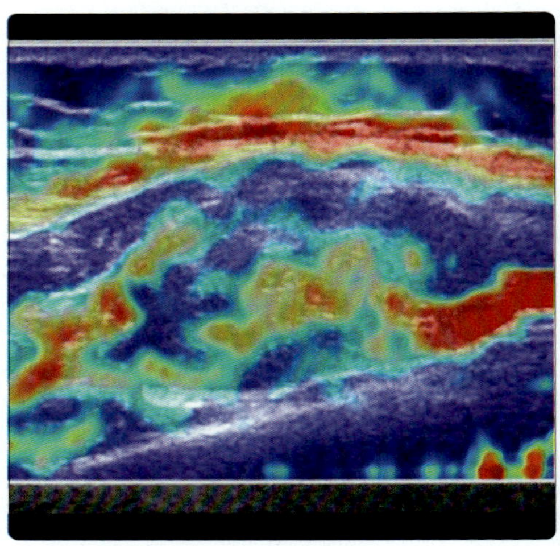

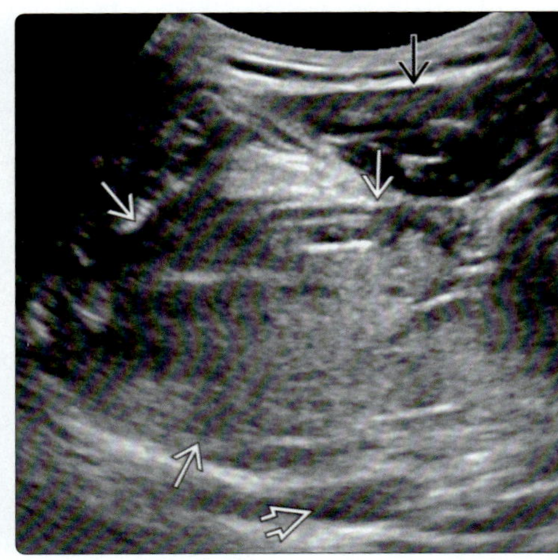

(Left) *Longitudinal US of the same patient shows a largely echogenic lipomatous-type mass ➔, though not with the typical US appearances of a lipoma. The mass is large and deep with infiltrative margins ➔ and suspicious of liposarcoma.* **(Right)** *MR shows the lipomatous tumor ➔ is situated alongside the popliteal vessels ➔ and deep to the biceps femoris muscle ➔. The mass has a large nonlipomatous component ➔. Pathology revealed a liposarcoma (F = femur).*

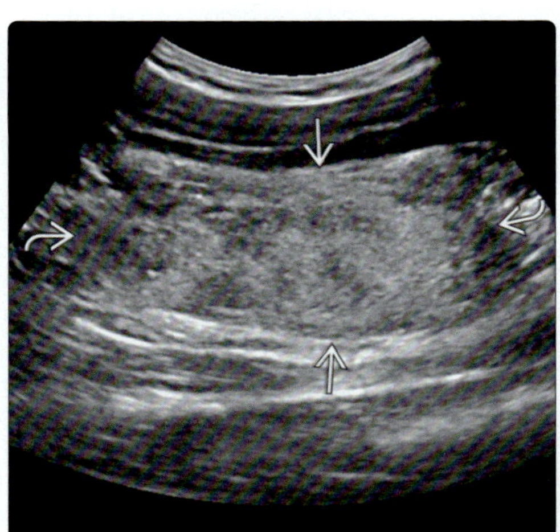

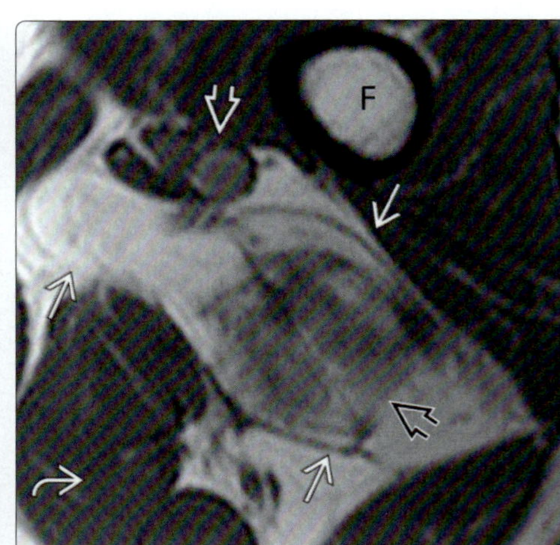

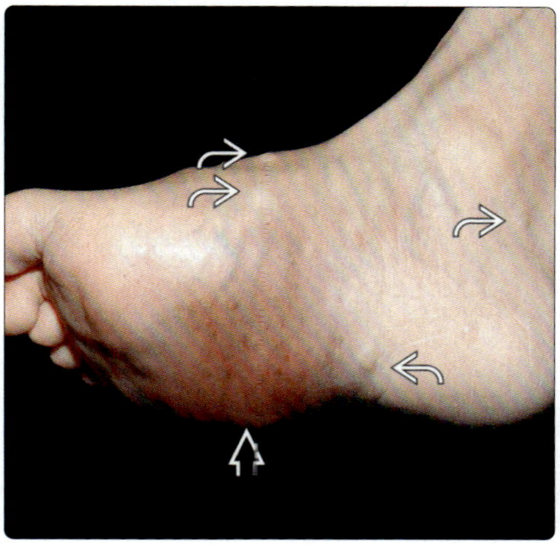

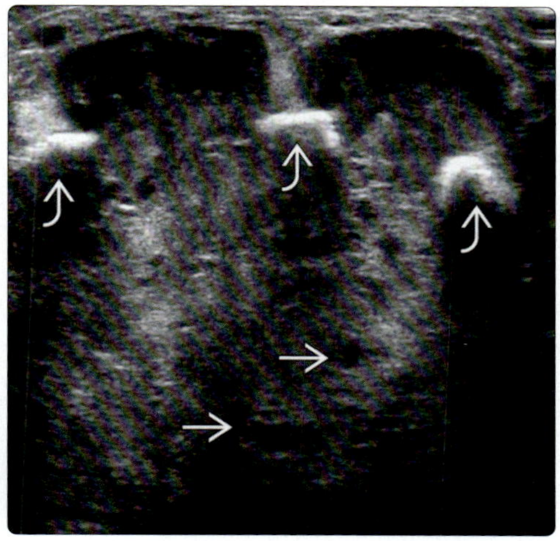

(Left) *Clinical photograph of a patient with known neurofibromatosis shows several small cutaneous neurofibromas . There is a large plantar swelling that had been increasing in size for 6 months.* **(Right)** *Transverse US of the same swelling shows a large, heterogeneous soft tissue tumor mass on the plantar aspect of the foot extending dorsally between the metatarsal bones . Small, intratumoral, cystic spaces are present.*

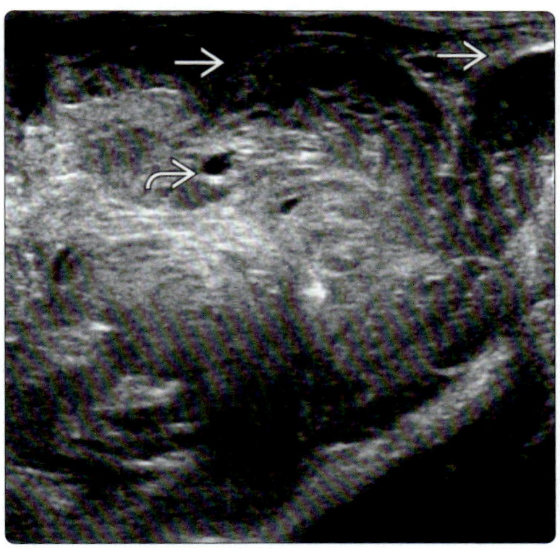

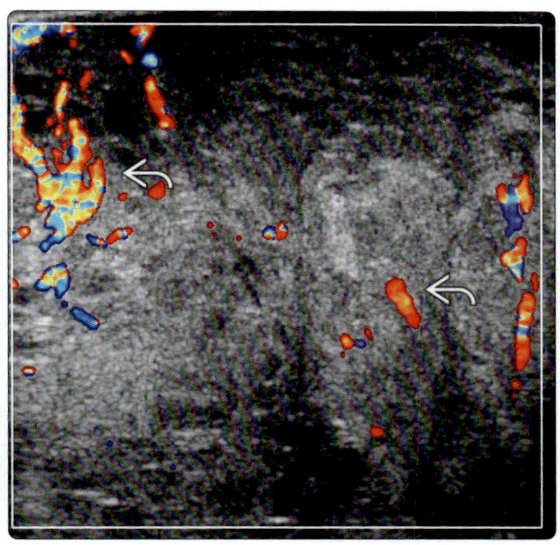

(Left) *Longitudinal US of the same tumor shows the lobulated tumor margin with small, intratumoral, cystic areas .* **(Right)** *Transverse color Doppler US of the same tumor shows moderate hypervascularity throughout the mass. Percutaneous biopsy for histology revealed a malignant peripheral nerve sheath tumor.*

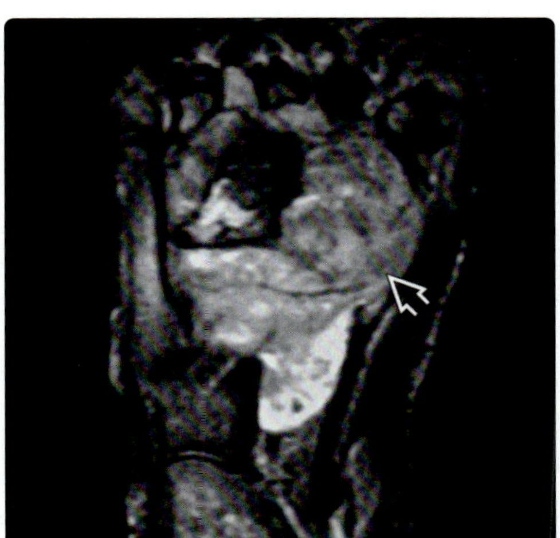

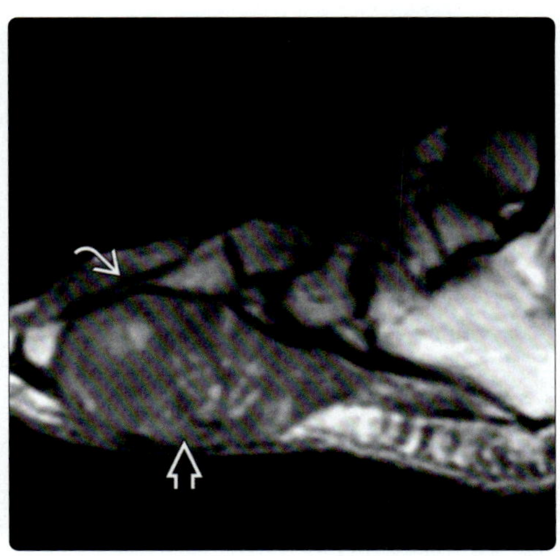

(Left) *Axial T2 FS MR of the same tumor clearly shows the extent of this large tumor on the plantar aspect of the forefoot and midfoot.* **(Right)** *Sagittal T1 MR of the same tumor shows a large plantar mass indenting the 4th metatarsal bone . MR is better than US at showing the extent of large tumors, particularly in complex anatomical areas, such as the foot.*

(Left) *Clinical photograph shows a mass ➔ on the anteromedial aspect of the thigh that had been growing slowly for 10 months. Clinically, the mass was attached to the quadriceps tendon.* **(Right)** *Longitudinal US of the same swelling shows a large soft tissue mass ➔ with prominent acoustic attenuation leading to poor visibility of the deep aspects of the tumor ➔. This finding is suspicious of a dense, fibrous-type tumor.*

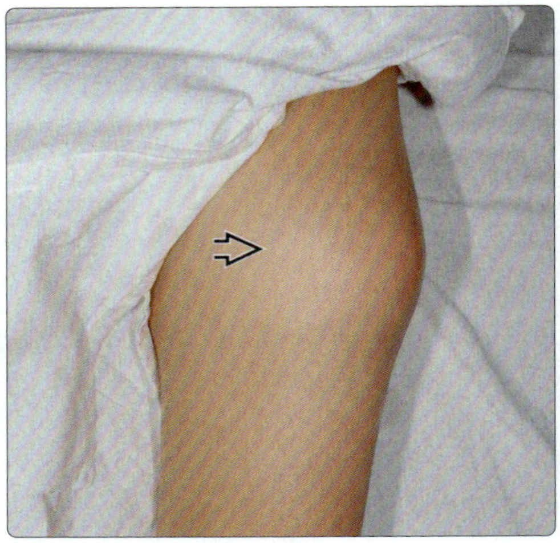

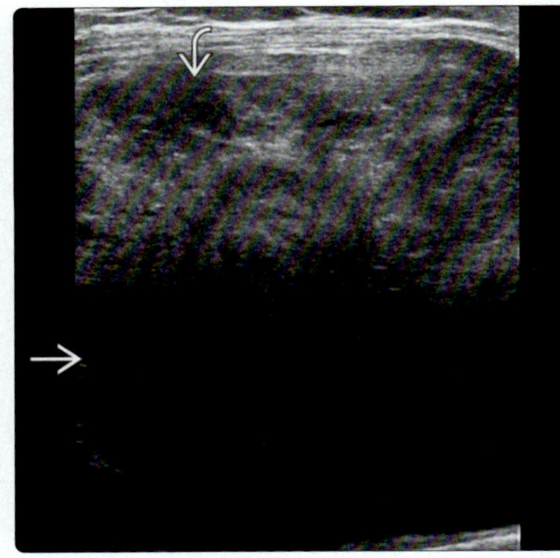

(Left) *Longitudinal color Doppler US of the same tumor shows moderate tumoral hypervascularity.* **(Right)** *Transverse T1 C+ FS MR of the same tumor shows how the moderately enhancing tumor mass ➔ is nearly completely encasing the femoral shaft ➔ without visible cortical erosion. MR is complimentary to US in this respect. There is an enhancing vessel within the femoral medullary canal. Histology revealed pleomorphic undifferentiated sarcoma.*

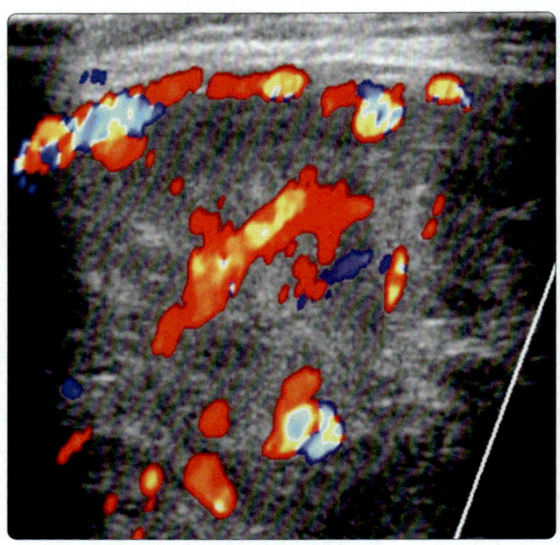

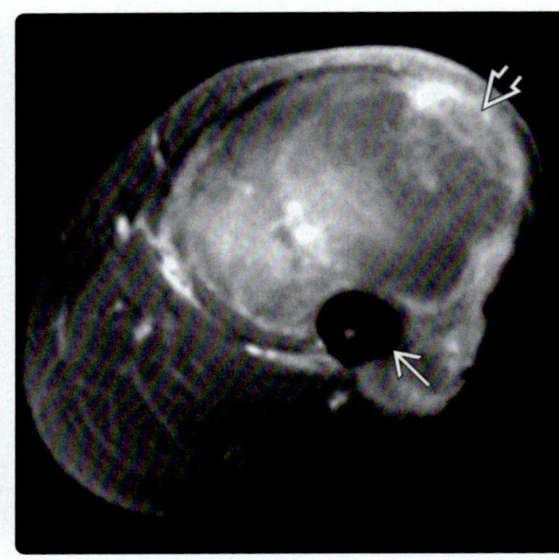

(Left) *Transverse US of a 63-year-old man with a large mass over the anterior aspect of the knee is shown. The lesion is heterogeneous with areas of calcification ➔ and dense, fibrous-type tissue ➔.* **(Right)** *Doppler US shows mild to moderate vascularity. Pathology following percutaneous biopsy showed high-grade spindle cell sarcoma with myofibroblastic differentiation.*

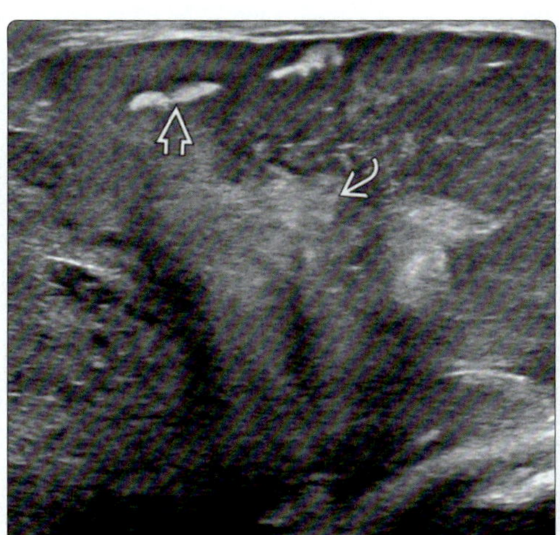

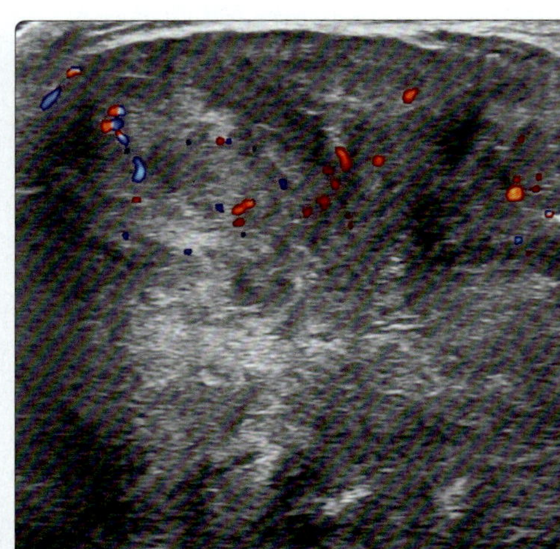

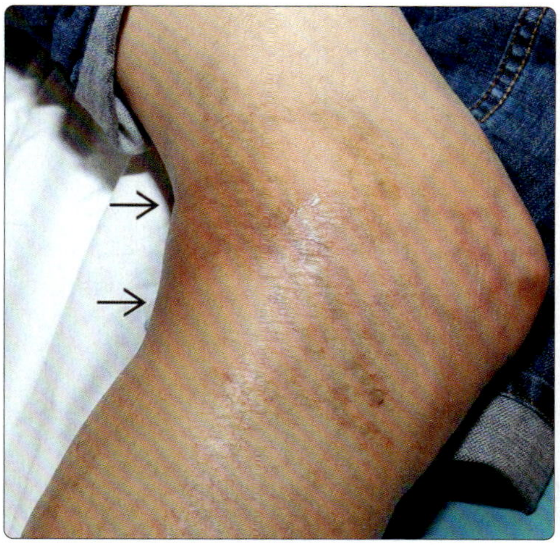
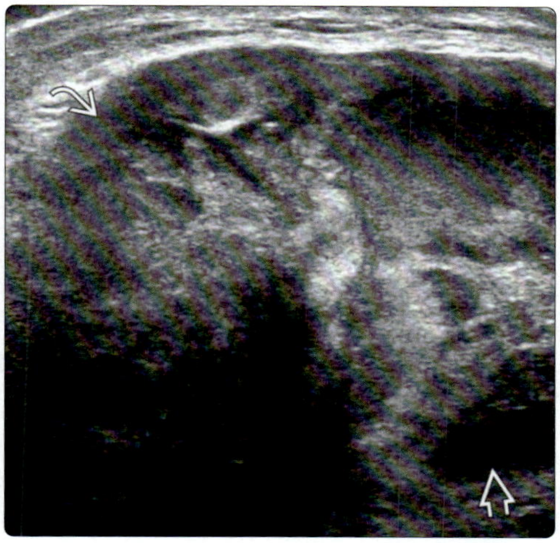

(Left) *Clinical photograph shows a patient with a large popliteal fossa swelling ➡ that had been increasing in size for several months.* **(Right)** *Transverse US of the same swelling shows a large, heterogeneous tumor mass ➡ partially encasing the popliteal artery ➡. The popliteal vein overlying the artery is compressed by the tumor and is not seen. US-guided biopsy for histology revealed a leiomyosarcoma.*

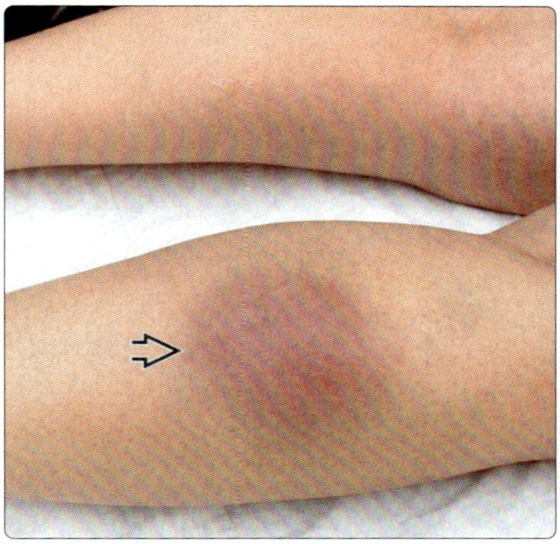
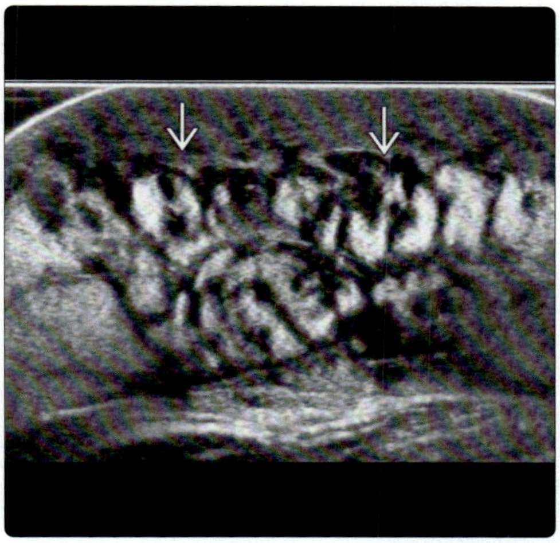

(Left) *Clinical photograph shows a patient with calf swelling ➡, redness, and tenderness that had been present for 1 month. Infection was suspected clinically.* **(Right)** *Transverse US of the same swelling shows a poorly defined area of quite intense subcutaneous edema and swelling ➡. No fluid collection is present, and there is no extension deep to the investing fascia. The US diagnosis was atypical subcutaneous infection.*

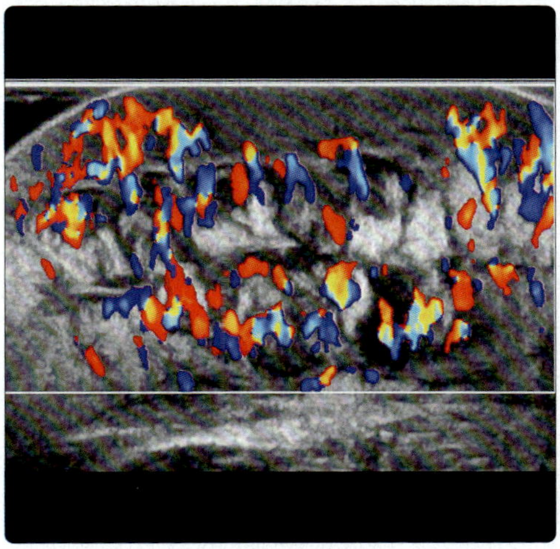
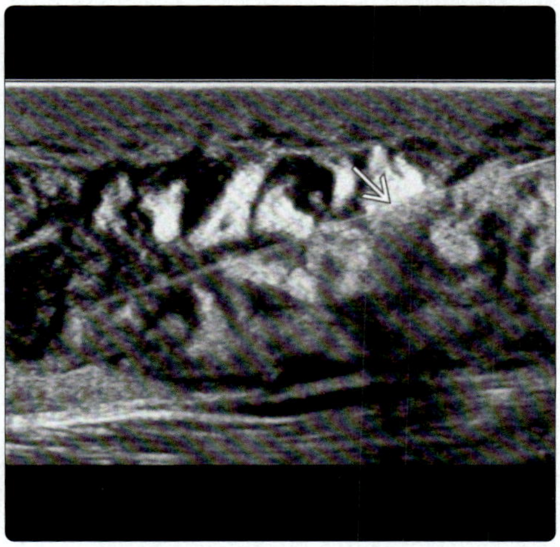

(Left) *Transverse color Doppler US of the same tumor shows intense tumoral vascularity.* **(Right)** *Longitudinal US-guided needle biopsy ➡ was performed in view of atypical findings. Histology revealed a peripheral T-cell malignant lymphoma. Although not classified under soft tissue sarcoma, this is another malignant tumor of soft tissue.*

Bone Tumor

KEY FACTS

TERMINOLOGY

- Bone tumor: Neoplastic growth of tissue in bone
- Benign tumor: Tumor that does not metastasize
- Malignant tumor: Tumor that does metastasize
- Primary bone tumor: Tumor arising de novo in bone
- Secondary tumor: Tumor spreading to bone from malignant primary tumor located elsewhere

IMAGING

- US is not routinely used in standard work-up of bone tumors
  - Helpful as problem solver in particular instances
  - Used to guide biopsy of accessible bone tumors
- US can assess
  - Nature of juxtacortical soft tissue
  - Tumor arising from bone cortex or intramedullary component once cortex is breached
  - Tumor matrix (if cortex is breached)
  - Lymph node metastases (rare)
  - Bone tumor recurrence
- Radiography is most useful investigation
  - Radiographic abnormality is present in nearly all symptomatic bone tumors
  - Used to characterize tumor type and broadly grade aggressiveness of bone tumors
- MR used for
  - Local staging of bone sarcoma &/or assessing chemotherapy response
  - Characterizing and further evaluating benign bone tumors

TOP DIFFERENTIAL DIAGNOSES

- Osteomyelitis
- Healing insufficiency fracture

DIAGNOSTIC CHECKLIST

- Consider tumor arising from bone in any deep-seated soft tissue mass

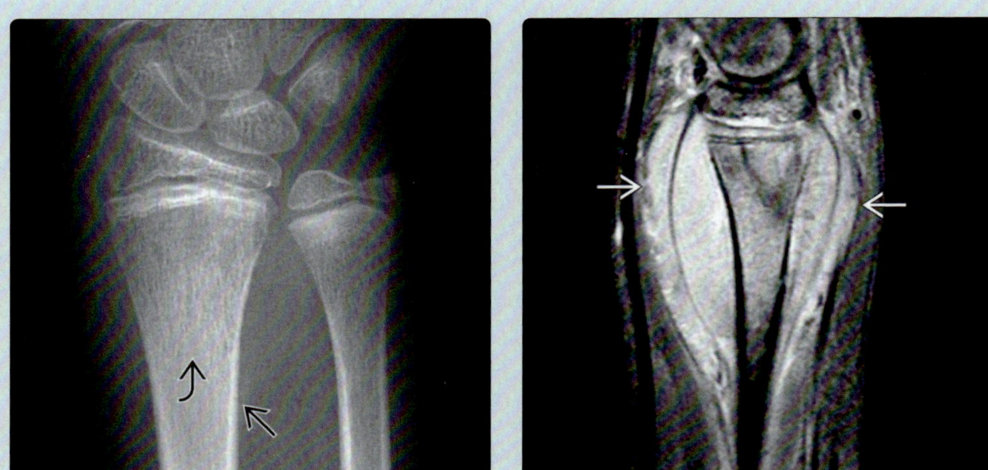

(Left) *Frontal radiograph of a skeletally immature child with a 4-week history of wrist pain and swelling shows a permeative osteolytic pattern → of the distal radial metaphysis indicative of an aggressive bone lesion. There is a faint periosteal reaction →.* **(Right)** *Sagittal T2* GRE MR obtained in the same patient the following day shows an aggressive bone tumor → of the distal radial metaphysis with a surprisingly large extraosseous component. The tumor clearly extends beyond the physis.*

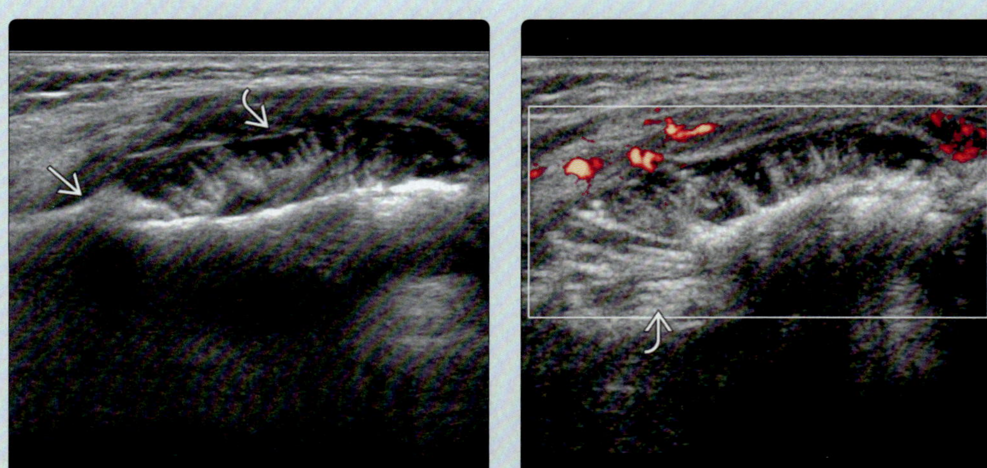

(Left) *Longitudinal US of the same lesion shows hypoechoic tumor matrix with radiating spicules → of new bone suggestive of an osteogenic tumor. There is an ultrasonic Codman triangle → at the proximal end of the tumor.* **(Right)** *Longitudinal color Doppler US shows mild marginal hyperemia. Note the radiating spicules of bone →. This finding is seen in various bone tumors and is not unique for osteosarcoma, though US-guided biopsy did reveal osteosarcoma in this case.*

TERMINOLOGY

Definitions

- Bone tumor: Neoplastic growth of tissue in bone
  - Benign tumor: Tumor that does not metastasize
  - Malignant tumor: Tumor that does metastasize
  - Primary bone tumor: Tumor arising de novo in bone
  - Secondary tumor: Tumor spreading to bone from malignant primary tumor located elsewhere

IMAGING

General Features

- Best diagnostic clue
  - Irregular bony destruction with soft tissue mass originating from bone
- Location
  - Any bone or bone area, though most commonly involves distal ends of long bones

Radiographic Findings

- Variable radiographic appearances
  - Centrally or eccentrically located within bone
  - Expansile or nonexpansile
  - Osteolytic, osteosclerotic, or mixed pattern
    - ± ground-glass opacification of matrix (fibrous tumor)
    - ± matrix mineralization
      - Chondroid-type mineralization (whorls, curves, popcorn-like)
      - Osteoid-type mineralization (hazy, cloud-like, solid)
  - Well-defined margin (narrow zone of transition)
    - Sclerotic or nonsclerotic rim
      - Well-defined margin favors nonaggressive tumor
  - Ill-defined margin (wide zone of transition)
    - Ill-defined margin favors aggressive tumor
  - ± cortical breach
  - ± extraosseous tumor expansion
  - ± periosteal ± tumoral new bone formation
    - Periosteal new bone formation: Periosteum elevated or stimulated by tumor
    - Tumoral new bone formation: Bone deposition by tumor matrix (osteogenic tumor)
  - Many types of periosteal or tumoral new bone formation
    - Lamellated: Single layered or multilayered ("onion skin")
    - Spiculated, sunburst, or hair-on-end patterns (represents tumoral rather than periosteal new bone)
    - Codman triangle (located at edge of extraosseous tumor expansion)
    - Solid (mature)
  - Type of periosteal reaction is not highly sensitive discriminator of tumor aggressiveness
    - Aggressive tumors may have no periosteal reaction or, occasionally, relatively benign type of periosteal reaction
  - Spiculated, "sunburst," and "hair-on-end" tumoral new bone are highly specific for aggressive tumor

CT Findings

- Very specific for some tumors (e.g., intraosseous lipomas, osteoid osteomas)
- Used to guide biopsy for bone tumor with no extraosseous component

MR Findings

- Helps define local extent and often helps characterize both malignant and benign bone tumors
- For bone sarcoma, MR used to specifically define
  - Intramedullary tumor extent
  - Extramedullary tumor extent
  - Physeal involvement by tumor
  - Joint involvement by tumor
  - Neurovascular bundle involvement by tumor
  - Presence of skip metastases

Ultrasonographic Findings

- US is not routinely used in standard work-up of bone tumors
  - Helpful as problem solver in particular instances
- Bone tumor may be unexpected finding in patient undergoing US for nonspecific pain, discomfort, or mass
- US can visualize component arising from bone cortex or intramedullary component once cortex is breached
- If needed, US-guided biopsy is both accurate and safe
- Osteochondroma
  - Pedunculated or sessile and metaphyseal in location
  - ± overlying hypoechoic cartilage cap
  - ± sequelae or complications of osteochondroma such as
    - Adjacent traumatic myositis or fat injury
    - Adjacent bursitis
    - Hemorrhage due to vascular injury
    - Pseudoaneurysm due to arterial injury
    - Nerve injury
    - Fracture of pedunculated osteochondroma
    - Malignant transformation
      - Affects < 1% of osteochondromas
      - Cartilage cap > 2 cm thick has high sensitivity and specificity for secondary chondrosarcoma
      - This diagnostic criterion is established on CT and MR but is not validated for US
- Tumor matrix
  - US appearances generally mirror radiographic appearances
    - Usually hypoechoic matrix with irregular osteolysis
    - ± thin hyperechoic radiating intratumoral spiculated lines representing matrix ossification
      - These are indicative of aggressive bone tumor
    - ± cystic, hemorrhagic areas or fluid-fluid levels
    - ± irregular hyperechoic matrix mineralization
      - Retained osseous fragments can simulate matrix mineralization
    - ± pulsatile tumoral mass on real-time grayscale imaging
    - Variable hyperemia
    - ± Codman triangle and other types of periosteal reaction
  - Bone metastases are most common bone tumor encountered in routine clinical practice
    - Particularly in elderly patients
    - Have aggressive ultrasonic appearance that mimics bone sarcoma

- Primary tumors that commonly metastasize to bone are
  - Breast, lung, prostate, uterus, thyroid, bladder, kidney, pancreas, liver, and bowel
- US is used to guide biopsy when extraosseous soft tissue mass is present or cortex is breached
  - Hyperemic areas on color Doppler imaging reflect areas of viable tumor
    - These areas should be targeted for biopsy
- Assessment of flow in large feeding artery (femoral, popliteal, subclavian, brachial) by spectral Doppler analysis
  - Can be done for all tumors irrespective of whether primary tumor is visible on US
  - Flow velocities can be calculated
  - Normal arterial flow is triphasic with reversal of flow in diastole
  - Metabolically active tumors may show substantial ↑ in antegrade diastolic flow
    - Generally accompanied by ↓ resistive indices
  - Change in flow dynamics can give indication of chemotherapy response
  - Assessment of intratumoral vessels by US is usually not that helpful
    - Too variable, and even some deep-seated bone tumors may not have demonstrable vascularity on color Doppler imaging
- US is very useful for assessing bone tumor recurrence
  - Recurrence of resected bone sarcomas usually occurs in juxtacortical soft tissue and not in bone
    - Recurrence usually seen as discrete mass in soft tissues close to resection site
  - US is very useful when metallic prosthesis or fixation device in situ
    - US assessment of juxtacortical tissues is not as affected by metal as CT or MR
- US is used to assess lymph node metastases
  - Lymph node metastases are rare with primary bone tumor
  - More common with bone tumor recurrence
  - Can guide fine-needle aspiration for cytology of lymph node to confirm metastatic status

Imaging Recommendations

- Best imaging tool
  - Radiography
    - Radiographic abnormality is visible in nearly all symptomatic benign and malignant bone tumors
    - Radiographic abnormality may be absent in asymptomatic benign and malignant bone tumors
    - Radiography is used to characterize tumor type and grade tumor aggressiveness
  - MR imaging used for
    - Local staging of malignant primary bone tumors &/or assessing chemotherapy response
    - Characterizing and assessing extent and complications of benign bone lesions

DIFFERENTIAL DIAGNOSIS

Osteomyelitis

- Acute osteomyelitis and acute exacerbation of chronic osteomyelitis can simulate aggressive bone tumor on US
- Generally, acute osteomyelitis is associated with more pronounced inflammation than sarcoma
- Clinical and hematological parameters also help make distinction between infection and tumor
- Rapidly progressive symptoms, severe local tenderness, fever, other constitutional symptoms, and leucocytosis favor infection over tumor
- Biopsy may be necessary
- Quiescent chronic osteomyelitis may simulate nonaggressive bone tumor

Healing Insufficiency Fracture

- Bone osteolysis and poorly mineralized callus can simulate bone tumor

PATHOLOGY

Staging, Grading, & Classification

- Grading according to degree of tumor spread and level of cellular differentiation

CLINICAL ISSUES

Presentation

- Most common signs/symptoms
  - Progressive increasing pain over weeks to months
- Other signs/symptoms
  - Enlarging palpable mass
  - Pain on weight bearing ± limp
  - Localized mild-to-moderate-severity deep tenderness

Natural History & Prognosis

- Quite rapid, progressive deterioration if aggressive tumor
- Slow deterioration or quiescent symptoms if nonaggressive tumor

Treatment

- Nonaggressive: Treatment is largely related to symptoms and likelihood of sequelae
- Aggressive tumor: Resection ± limb-sparing reconstruction
  - ± neoadjuvant and postoperative chemotherapy for sarcoma
  - ± reconstruction with metallic prosthesis or allograft

DIAGNOSTIC CHECKLIST

Consider

- Tumor arising from bone in any deep-seated soft tissue mass
  - Ensure that deep margin of any soft tissue tumor is seen
- US has limited but occasionally very useful role in management of bone tumors

SELECTED REFERENCES

1. Saifuddin A et al: Pictorial review: ultrasonography of primary bone tumours. Clin Radiol. 53(4):239-46, 1998

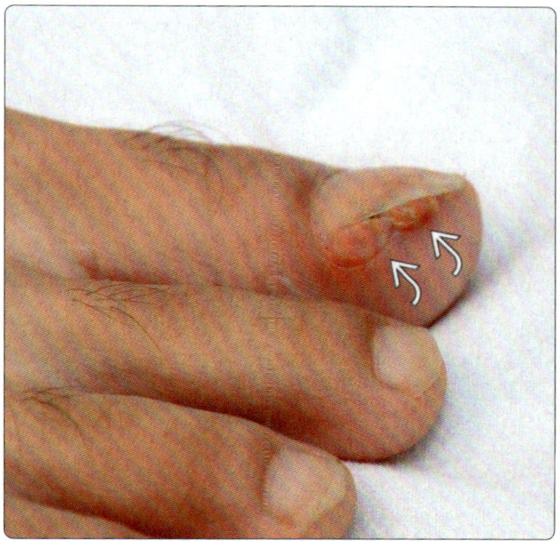

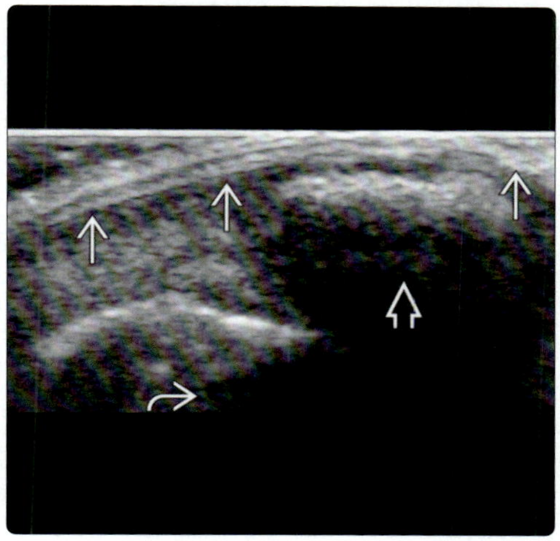

(Left) *Photograph of the medial forefoot shows a seemingly fleshy mass protruding from the undersurface of the big toe nail laterally →. The lateral aspect of the nail is elevated as a result.* **(Right)** *Transverse US shows an echogenic protuberant subungual mass → on the lateral side of the big toe. The mass is lifting the nail → upward. Continuity with the underlying distal phalanx → cannot be fully assessed with US due to acoustic shadowing.*

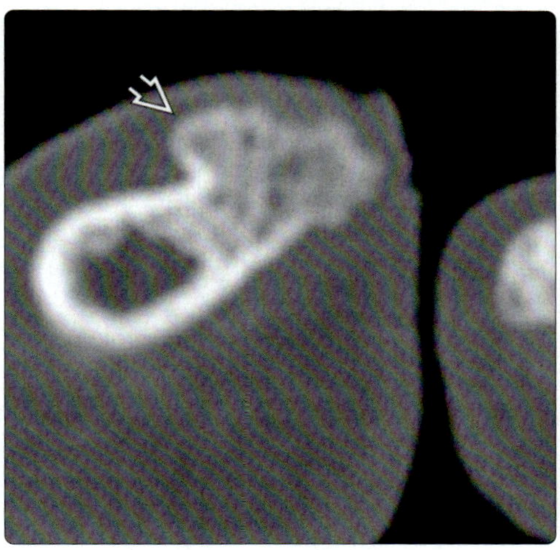

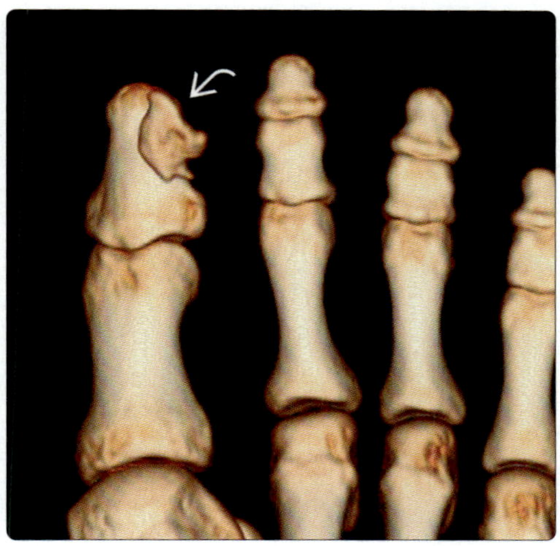

(Left) *Coronal CT with bone window shows an exostosis → arising from the medial aspect of the distal phalanx. Continuity with the medullary canal is not as distinct as that of nonsubungual osteochondromas. This is a feature of subungual exostoses, which are radiologically similar though pathologically distinct from other osteochondromas.* **(Right)** *3D CT reconstruction shows the subungual exostosis →. These nearly always occur on the medial aspect of the big toe and may be induced by trauma.*

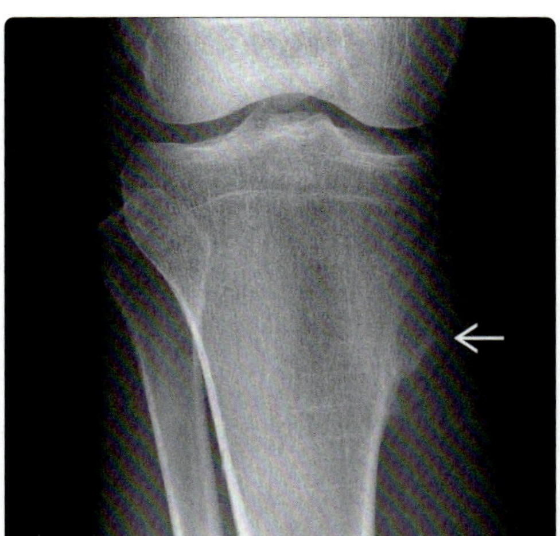

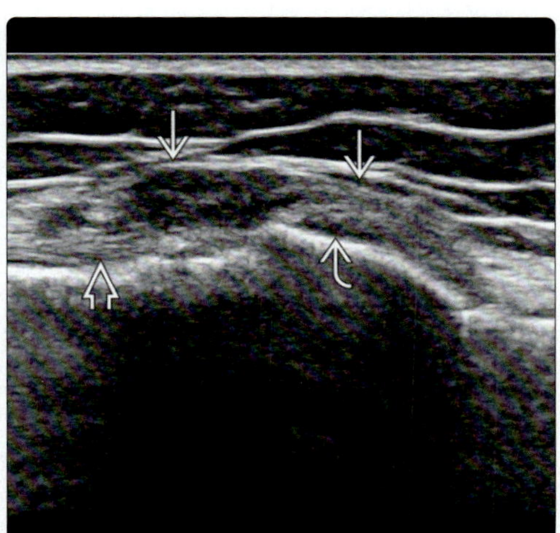

(Left) *Frontal radiograph shows a small sessile osteochondroma located on the medial aspect of the proximal tibial metaphysis →.* **(Right)** *Longitudinal US of the same lesion shows a thick-walled chronic bursitis → overlying this osteochondroma. The medial collateral ligament is seen just proximally →. The very thin hypoechoic layer → is a thin cartilage cap. Adventitious bursa develop in sites where friction occurs. They may become inflamed, infected, or hemorrhagic.*

(Left) *Frontal radiograph in 62-year-old man with medial knee pain shows there is a proximal tibial osteochondroma.* **(Right)** *Transverse US of the medial knee region shows a small pedunculated osteochondroma of the proximal tibial upper metaphysis with an overlying bursa displacing the peroneal muscles. Bursitis and myositis are common causes of symptomatic osteochondroma.*

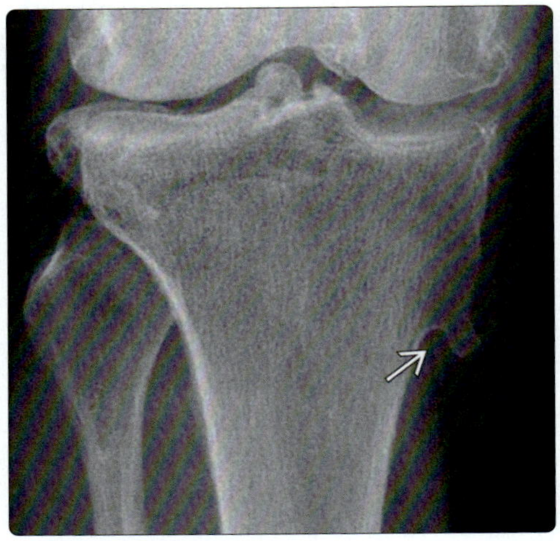

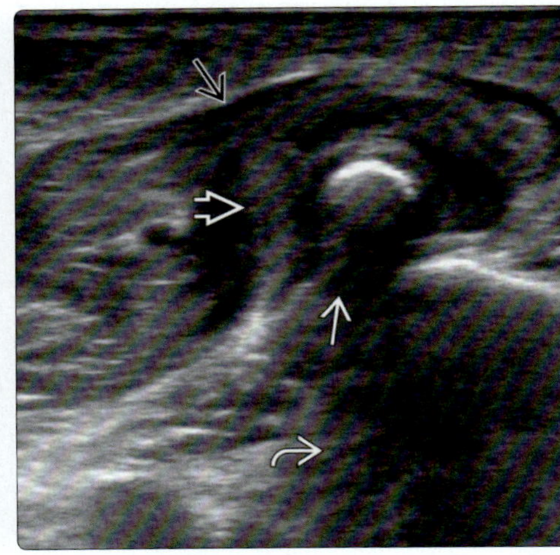

(Left) *Frontal radiograph shows a large pedunculated osteochondroma on the medial side of the distal femoral metaphysis. This lesion was symptomatic and referred for US examination.* **(Right)** *Longitudinal US with extended field of view of the same lesion shows a bursa distended with hemorrhage overlying the femoral osteochondroma. Both chronic bursitis and hemorrhagic bursitis are not uncommon causes of osteochondroma-related pain.*

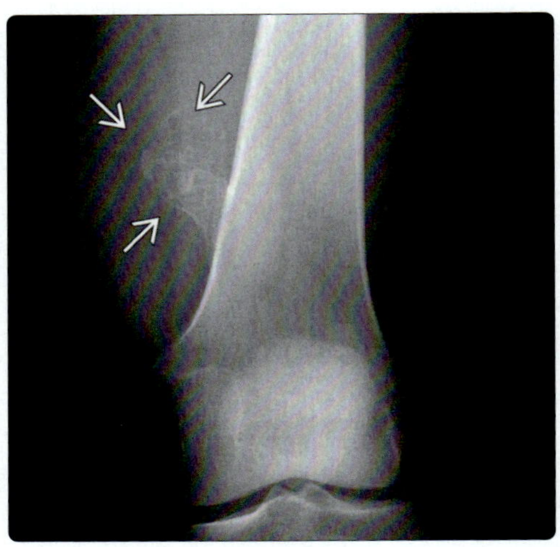

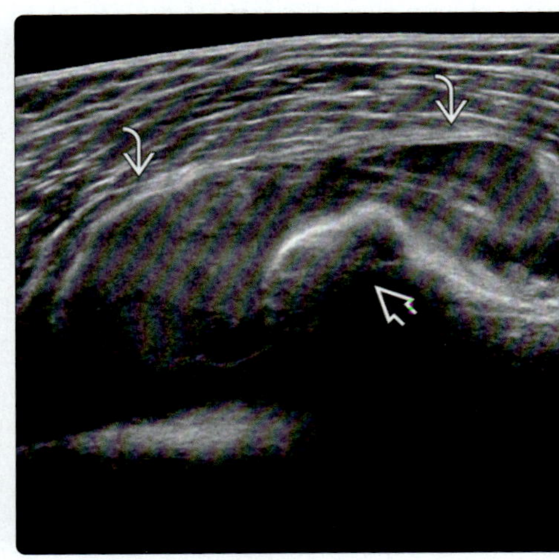

(Left) *More coned-down longitudinal US of the same lesion shows a thin hypoechoic cartilage cap overlying the osteochondroma. There is some gravity-aided sedimentation of hemorrhagic bursal content. The thin bursal wall is visible. No osteochondroma fracture was evident on US.* **(Right)** *Transverse color Doppler US shows mild hyperemia around the hemorrhagic bursitis and helps confirm the nonsolid nature of the bursal content.*

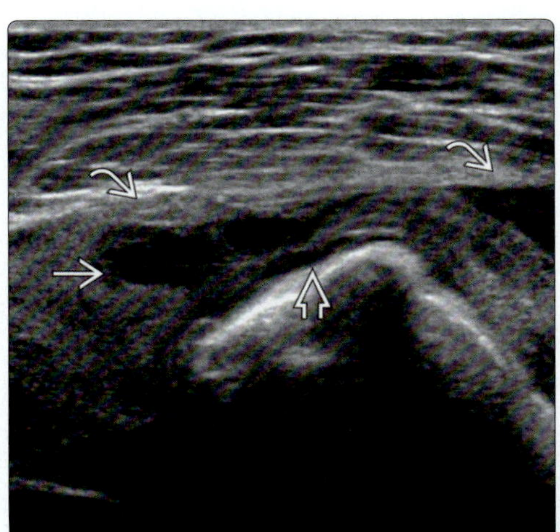

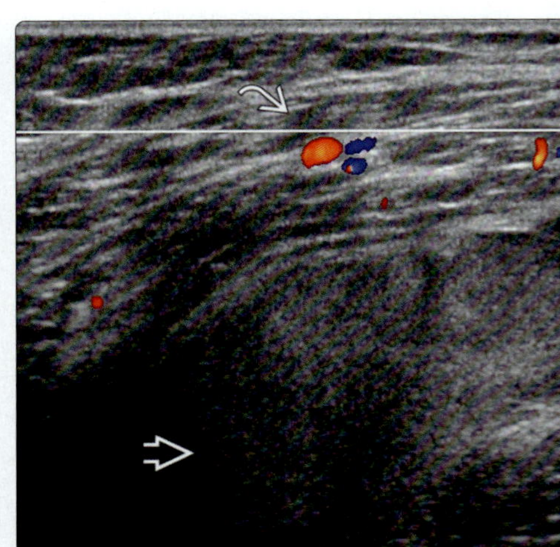

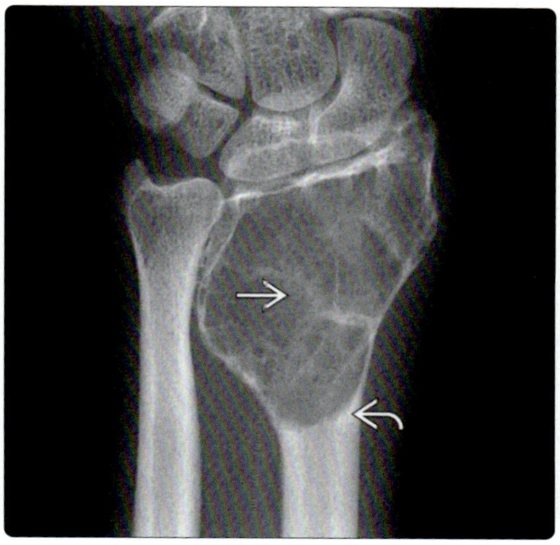

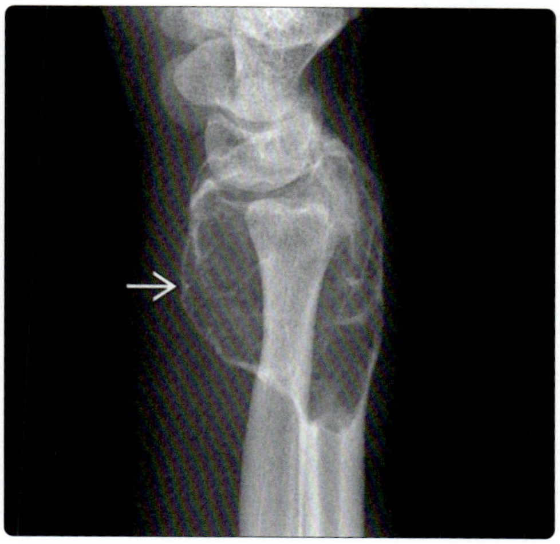

(Left) *Frontal radiograph in a young, skeletally mature patient shows a moderately expansile lytic subarticular tumor with a well-defined, nonsclerotic margin ➡. The tumor has thinned cortices with traversing septa ➡. These appearances are consistent with either a giant cell tumor, a giant cell tumor with secondary aneurysmal bone cyst (ABC) formation, or else a primary ABC.* **(Right)** *Lateral radiograph of the same lesion shows very similar appearances. The cortex ➡ is thinned but not breached.*

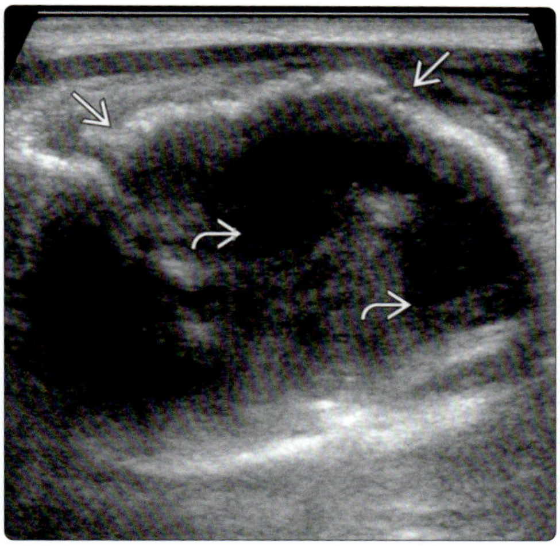

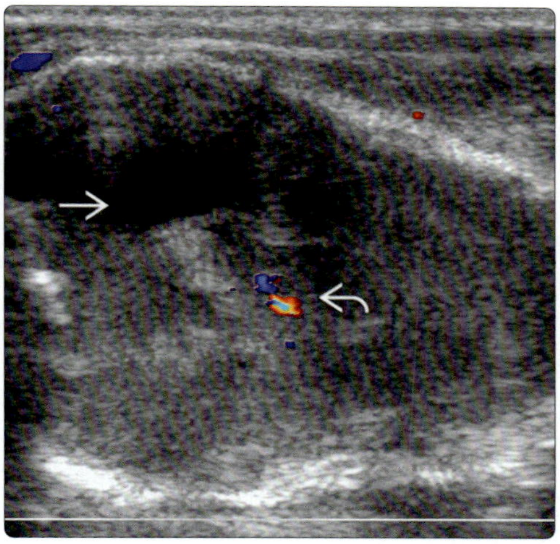

(Left) *Transverse US of the same lesion as previous radiographs shows an expansile distal radius ➡. As the cortex is thinned, one can appreciate the partially cystic nature of the tumor with some fluid levels ➡. No extraosseous tumoral component was present.* **(Right)** *Transverse color Doppler US demonstrates only mild hyperemia of the tumoral matrix ➡. The partially fluid nature of the tumor can be appreciated ➡ even though the cortex was not breached. US-guided biopsy revealed giant cell tumor.*

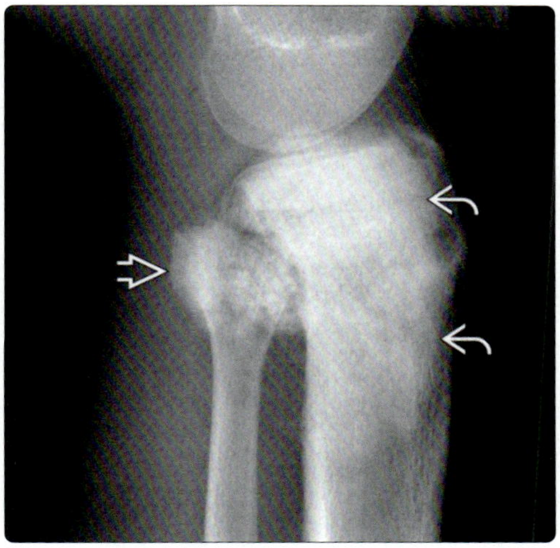

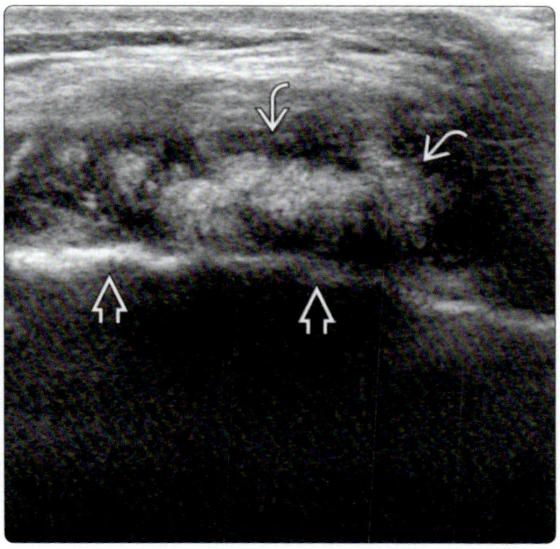

(Left) *Lateral radiograph of the proximal leg in a young patient shows a sclerotic tumor ➡ of the proximal tibia with a large, densely mineralized extraosseous component posteriorly ➡.* **(Right)** *Longitudinal US of the same lesion shows quite densely echogenic extraosseous soft tissue component ➡. This cloud-like echogenicity is suggestive of osteogenic tumor. US-guided percutaneous biopsy (not shown) revealed osteosarcoma. Note the indistinct tibial cortical outline ➡.*

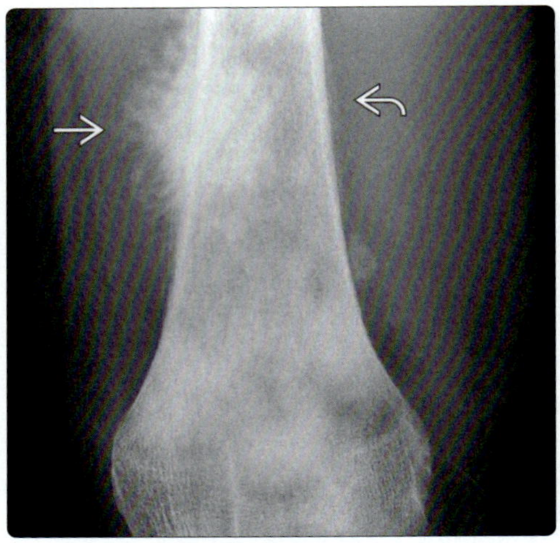

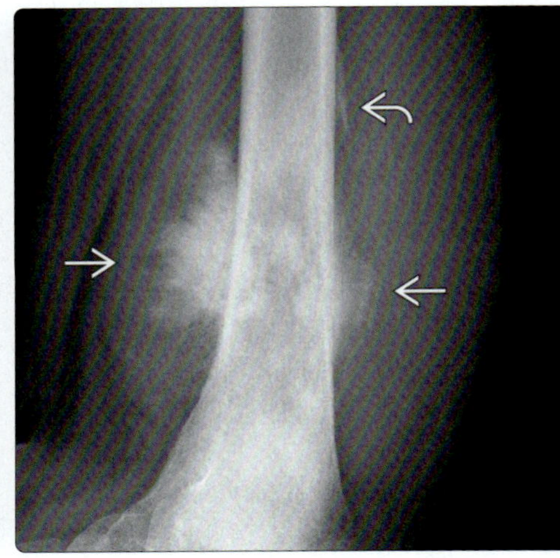

(Left) *Frontal radiograph of the distal femoral diaphysis shows altered bone texture associated with dense, spiculated new bone formation* *medially with lamellar new bone formation laterally* . **(Right)** *Lateral radiograph of the same lesion shows dense, spiculated new bone* *posteriorly and anteriorly with Codman triangle* *proximally. All of these appearances are consistent with osteosarcoma. Most bone tumors are diagnosed and characterized radiographically.*

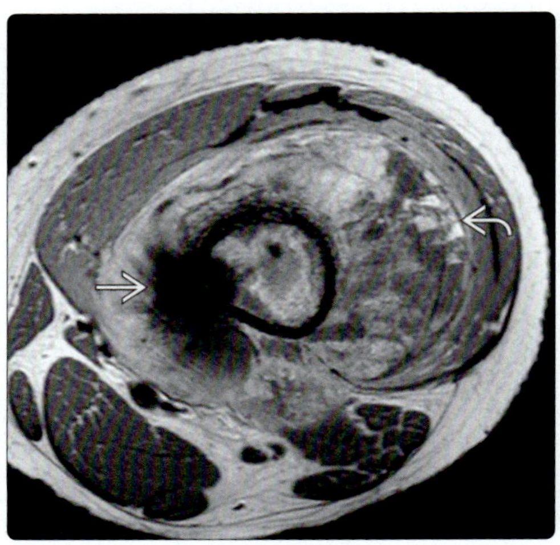

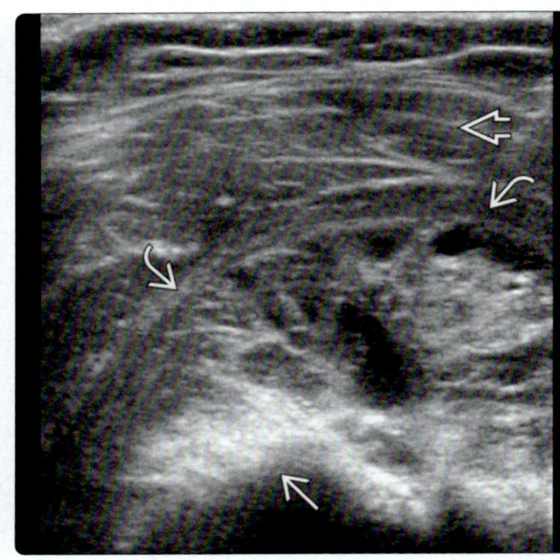

(Left) *Axial PD FSE MR of the same lesion shows a large, extraosseous soft tissue tumor with some fluid-fluid levels anteriorly* *and dense spiculated tumor new bone posteromedially* . **(Right)** *Transverse US of the anteromedial aspect of the same tumor shows dense, spiculated, tumoral new bone formation* *and a large extraosseous soft tissue component* *displacing the vastus medialis muscle* *anteriorly.*

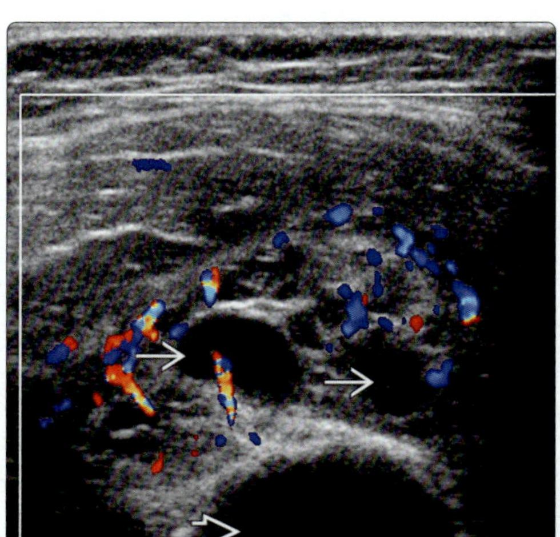

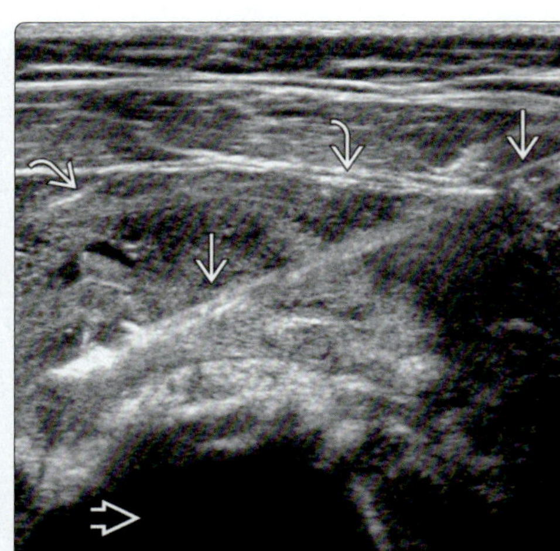

(Left) *Transverse color Doppler US shows moderate hyperemia of the extraosseous soft tissue component interspersed with some cystic areas* . *Hyperemia is a reliable sign of viable tumor that will usually provide a good yield with biopsy. Note the femur* . **(Right)** *Transverse US shows the biopsy needle* *located within the extraosseous tumor mass* *alongside the femur* . *US-guided biopsy is possible in 40% of bone tumors that require biopsy. Histology revealed osteosarcoma.*

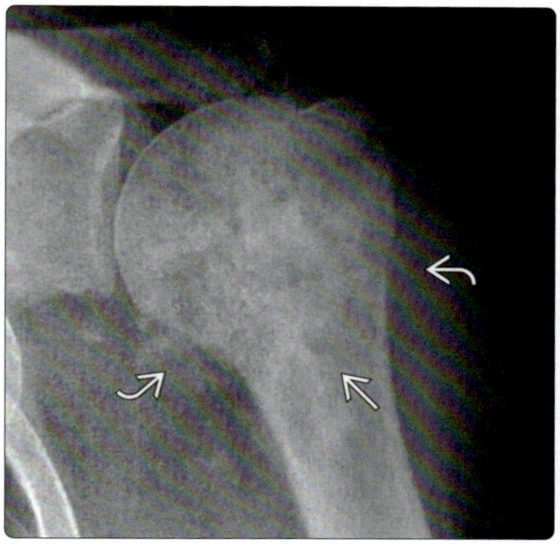

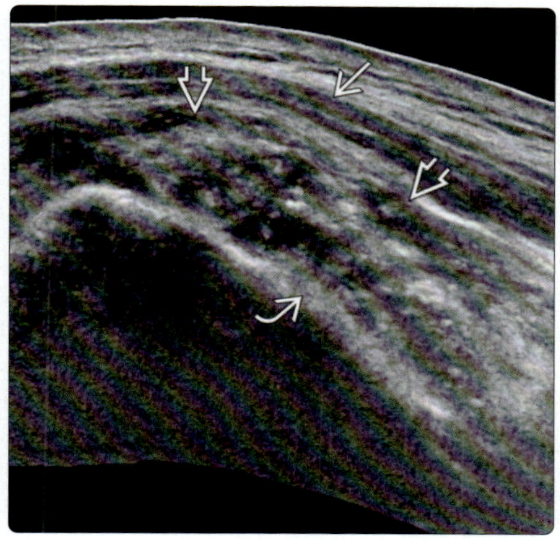

(Left) *Frontal radiograph in a 53-year-old man with shoulder pain shows patchy osteosclerosis of the proximal humeral metaphysis associated with faint spiculated, extraosseous, new bone formation. This is consistent with an aggressive bone lesion.* **(Right)** *Longitudinal extended field-of-view US shows spiculated new bone formation extending from the lateral humeral cortex with a moderate-sized extraosseous soft tissue mass displacing the overlying deltoid muscle.*

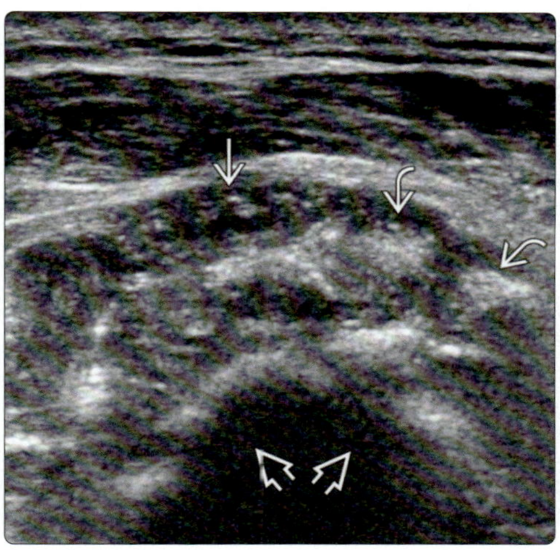

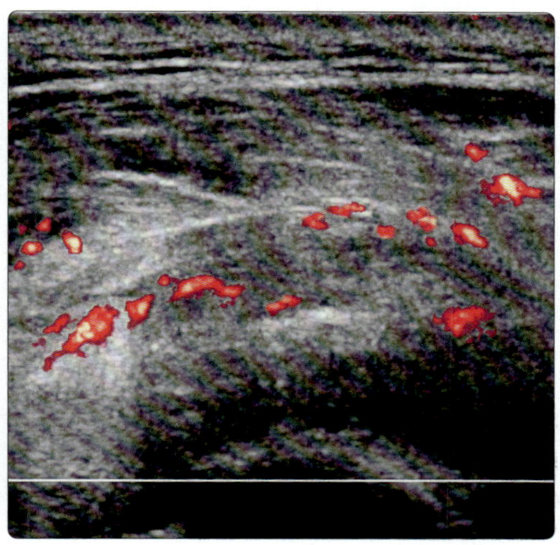

(Left) *Transverse US of the same lesion shows cloud-like echogenic areas within a hypoechoic extraosseous tumor matrix that are suggestive of osteoid deposition together with many smaller echogenic speckles. Note the irregular blurred humeral cortical margin.* **(Right)** *Transverse power Doppler US of the same lesion shows mild hyperemia of the extraosseous component. US-guided percutaneous biopsy (not shown) revealed osteosarcoma.*

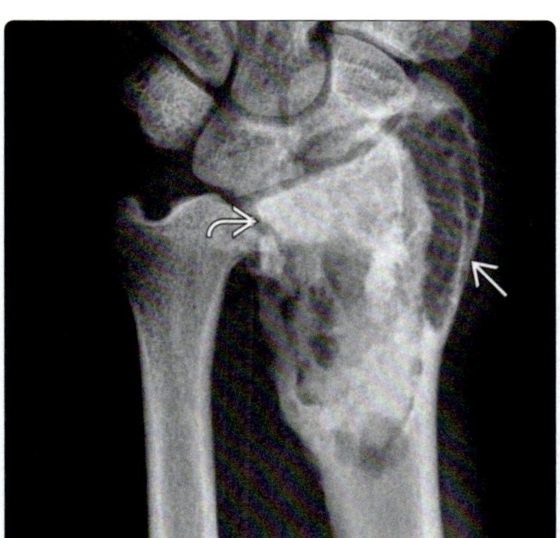

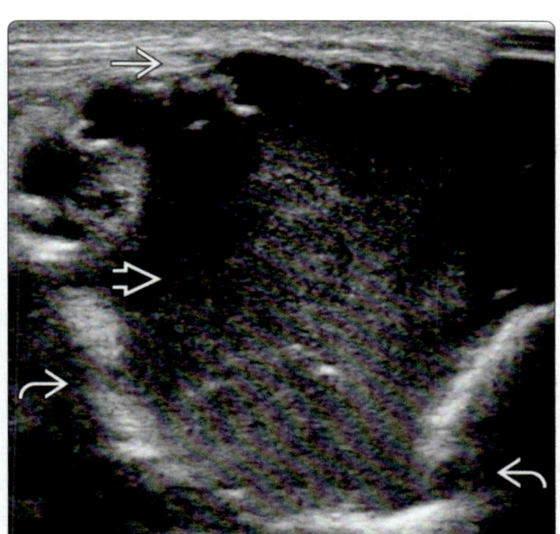

(Left) *Frontal radiograph shows a moderately expansile, partially sclerotic, partially osteolytic, subarticular tumor of the distal radius with a narrow zone of transition. Giant cell tumor was a consideration, but the appearances are atypical.* **(Right)** *Longitudinal US shows the breached dorsal cortex allowing appreciation of the homogeneous almost fluid-like content of the tumor matrix. Note the residual, unsharp, interrupted expanded cortex.*

(Left) *Longitudinal color Doppler US shows moderate marginal hyperemia extending down into the tumor matrix . The presence of tumoral hyperemia is an excellent indicator of viable tumor tissue.* **(Right)** *Longitudinal US shows core biopsy of the bone tumor using a coaxial system with a 15-g coaxial needle transmitting a 16-g Tru-Cut biopsy needle. Multiple biopsies can be taken with minimal disruption of extratumoral tissue. Histology revealed giant cell tumor of bone.*

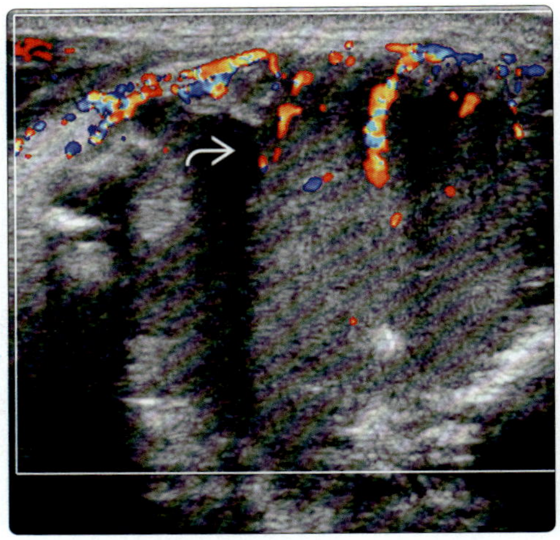

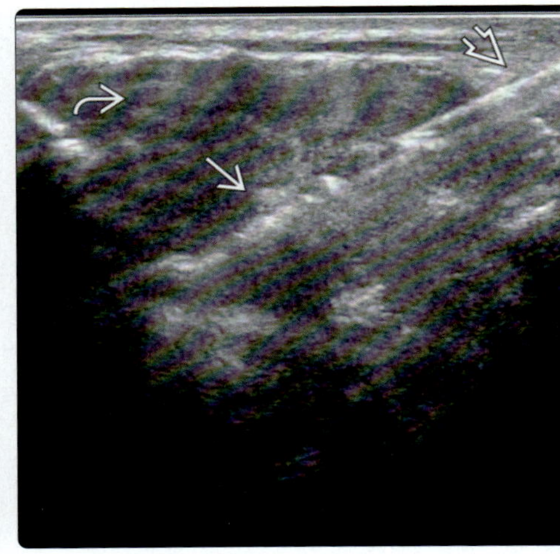

(Left) *Frontal radiograph in an 82-year-old woman shows a mildly expansile tumor of the proximal tibia consistent with fibrous dysplasia. Other than a lucent area in the medial tibial condyle , no particularly suspicious features are present.* **(Right)** *Frontal radiograph of the same lesion 12 months later shows a large osteolytic component with a moderate compression fracture of the proximal tibia. These appearances are compatible with malignant transformation of fibrous dysplasia.*

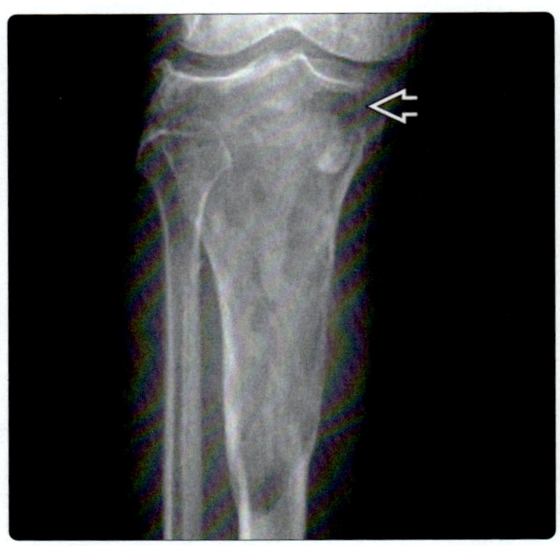

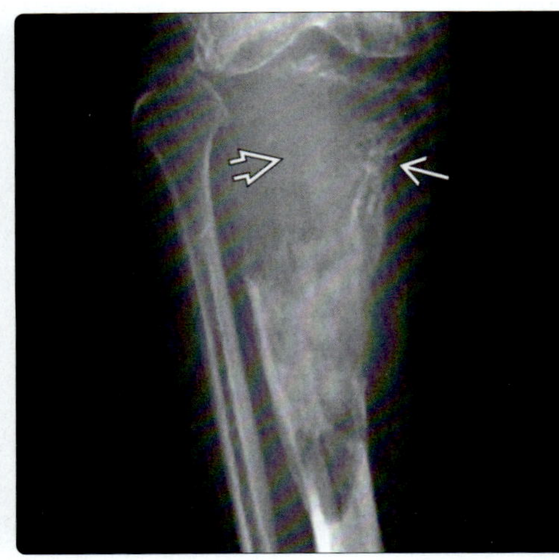

(Left) *Longitudinal US of the same tumor shows osteolysis of the proximal tibial cortex . This allows visibility of the large hypoechoic tumor filling the medullary canal .* **(Right)** *Longitudinal US shows a cystic area within the hypoechoic tumor matrix with a fluid-fluid level . The tumor matrix was moderately hypervascular on color Doppler imaging (not shown). US-guided biopsy revealed high-grade osteosarcoma. An above-knee amputation was performed.*

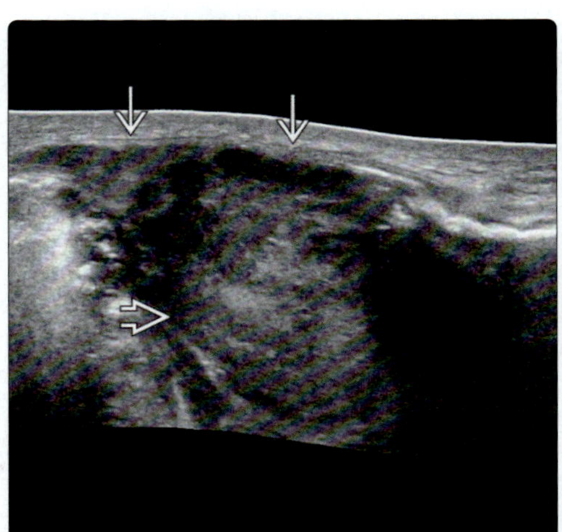

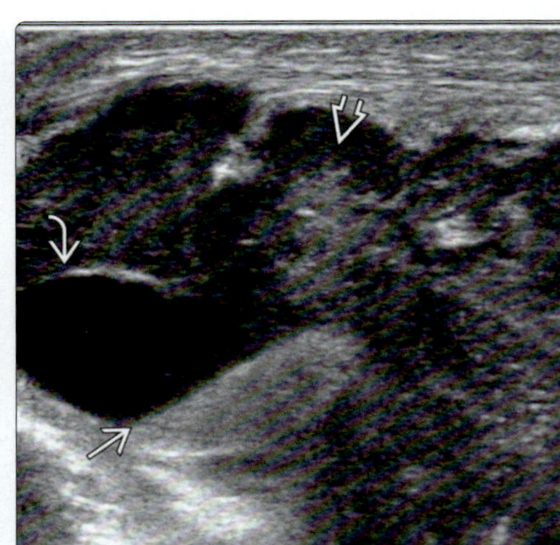

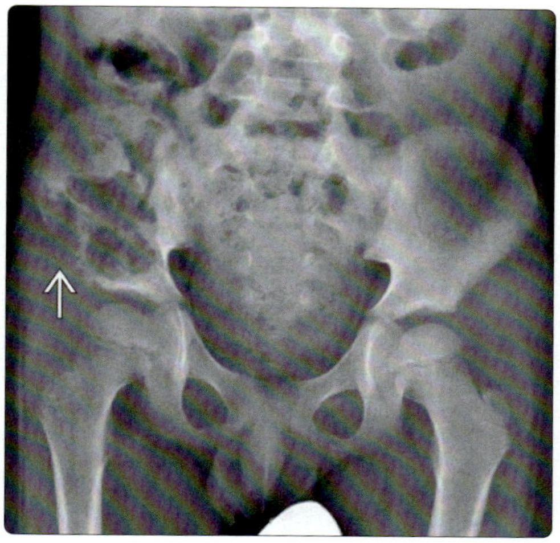

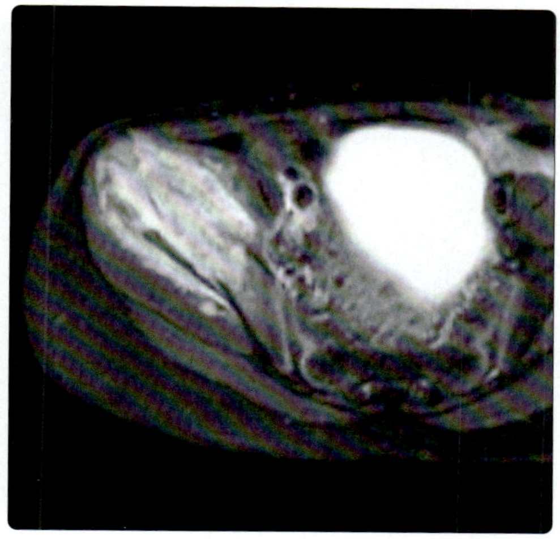

(Left) *Frontal radiograph of the pelvis in a child shows a large, irregular osteolytic area → in the supraacetabular portion of the right hemipelvis, partially obscured by bowel gas.* **(Right)** *Axial T1WI C+ FS MR of the same area shows a large, irregular, mildly expansile mass in the right ilium extending into the iliacus and gluteal muscles. An additional smaller lesion was present in the posterior elements of L5 (not shown).*

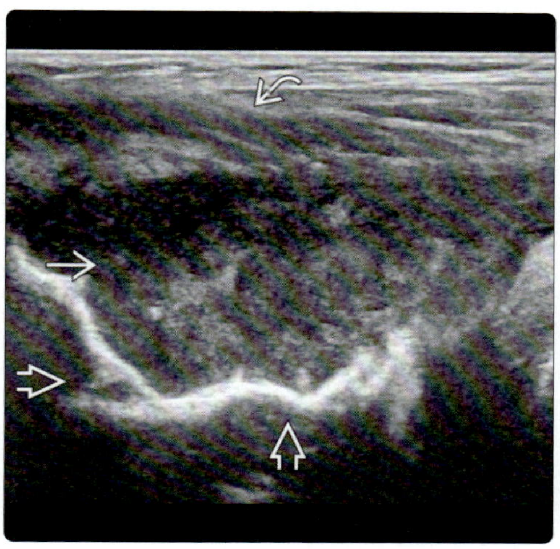

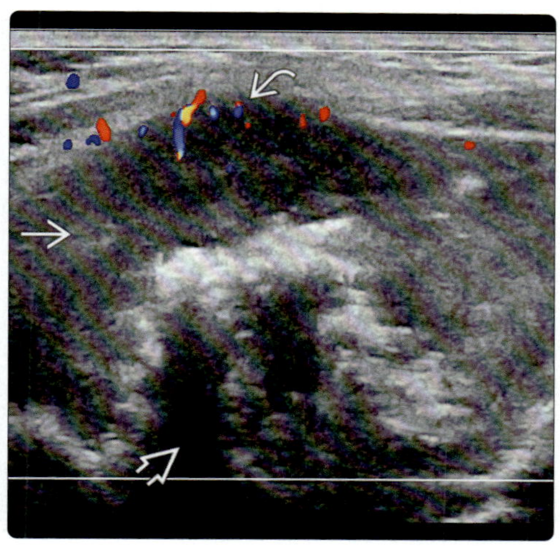

(Left) *Transverse oblique US of the same patient shows irregular osteolysis → of the iliac bone with a large hypoechoic tumor → deep to the gluteal musculature →.* **(Right)** *Longitudinal oblique color Doppler US shows irregular iliac bone osteolysis → surrounded by hypoechoic tumor matrix →. Mild tumor hyperemia → was present on color Doppler imaging. US-guided biopsy (not shown) revealed Langerhans cell histiocytosis of bone.*

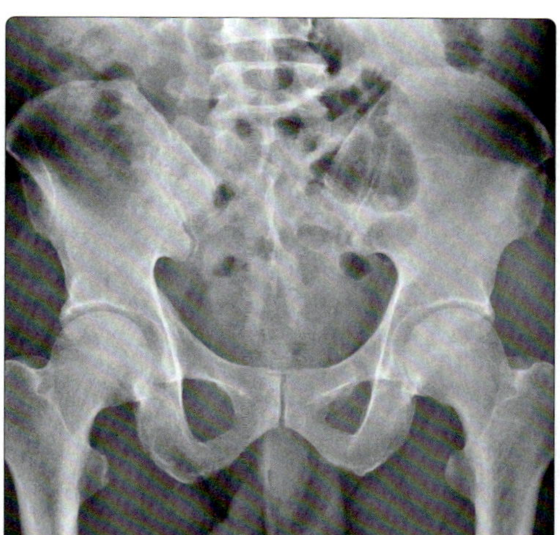

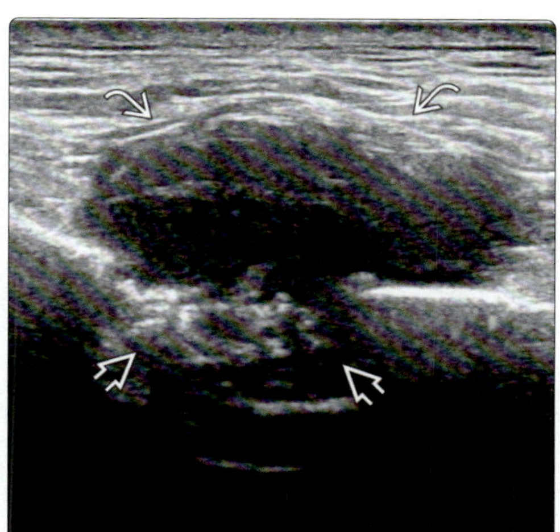

(Left) *Frontal radiograph of the pelvis in a 52-year-old man with myelodysplastic syndrome and recent onset of right hip pain is unremarkable. The patient was referred for US examination.* **(Right)** *Longitudinal oblique US of the gluteal region (following normal examination of the hip region) shows focal irregular osteolysis of the ilium → with a large hypoechoic extraosseous soft tissue component → bulging into the gluteal muscle.*

(Left) *Longitudinal oblique color Doppler US of the same lesion reveals no detectable tumoral or peritumoral hyperemia. This does not preclude the presence of viable tumor tissue. US-guided biopsy revealed blastic transformation of preexisting myelodysplastic syndrome.* **(Right)** *Axial CECT following US and biopsy shows the osteolytic tumor in the medial aspect of the right iliac bone adjacent to the sacroiliac joint.*

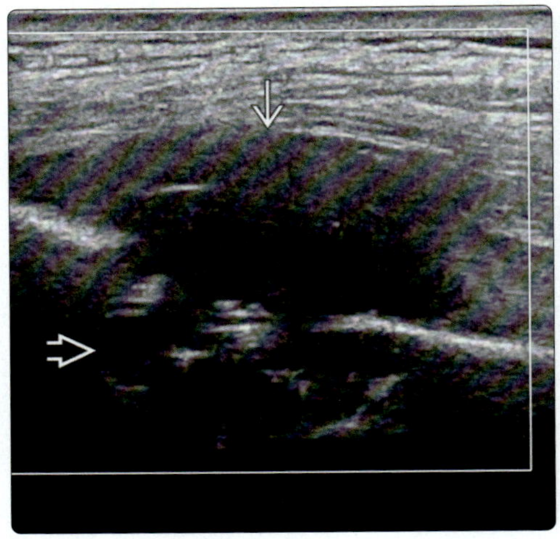

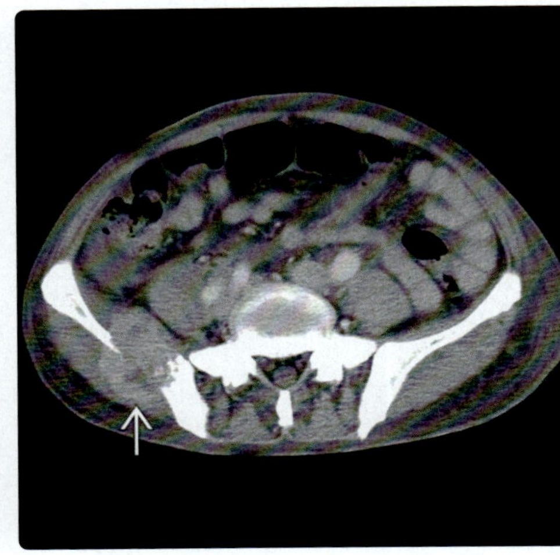

(Left) *Longitudinal US of the hip in a patient with metastatic lung cancer and clinically suspected hip septic arthritis shows irregular supraacetabular osteolysis and adjacent soft tissue swelling. The hip joint is normal.* **(Right)** *Sagittal reformat NECT shows osteolysis of the supraacetabular area. Although US is traditionally not used for investigating bone pathology, it can occasionally provide very helpful information. This painful metastatic deposit was irradiated with good effect.*

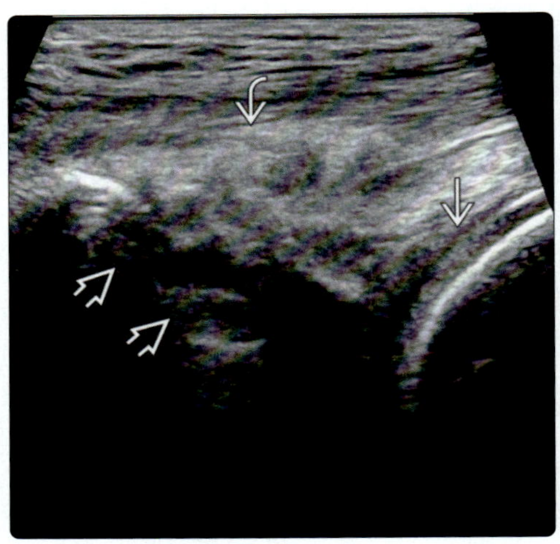

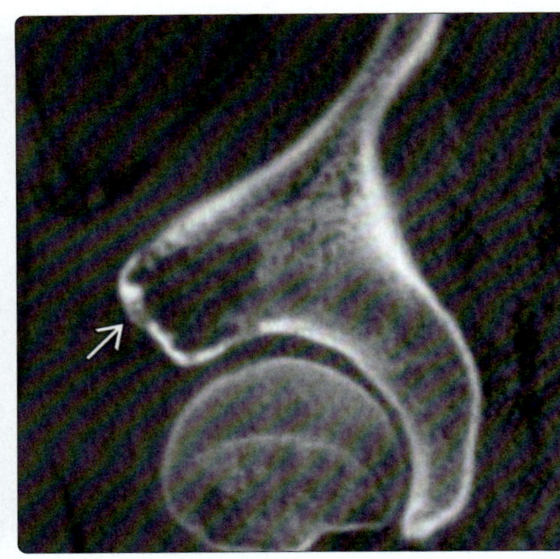

(Left) *Longitudinal US in a patient with previous breast cancer and anterior chest wall pain shows an irregular osteolytic lesion of the sternum suggestive of metastatic disease.* **(Right)** *Subsequent sagittal reconstructed CT 3 weeks later shows a large osteolytic metastatic deposit with extraosseous extension in the mid to lower part of the sternal body. This was confirmed by CT-guided biopsy. The sternum is a common site of breast cancer bony metastases.*

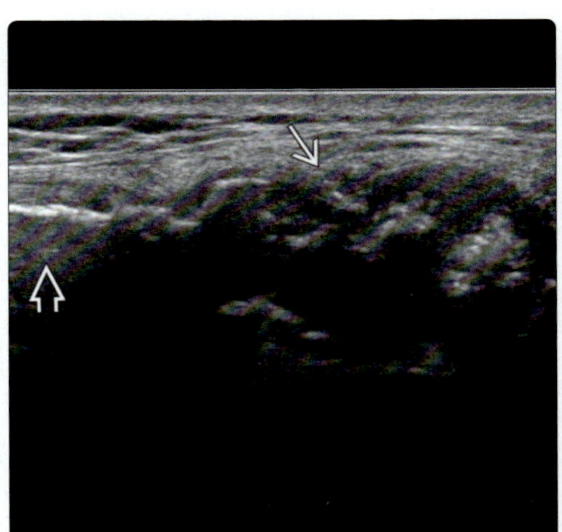

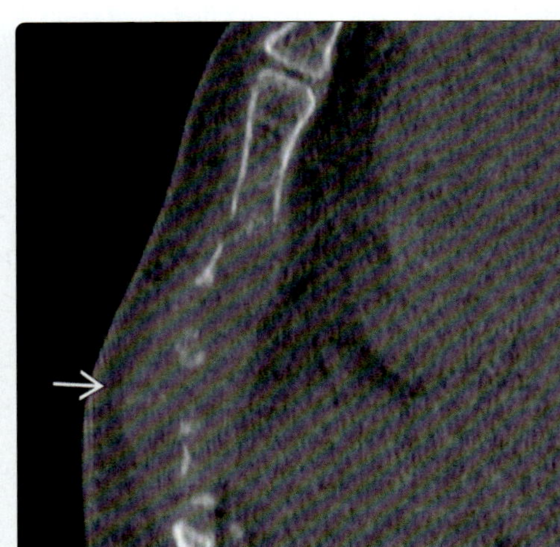

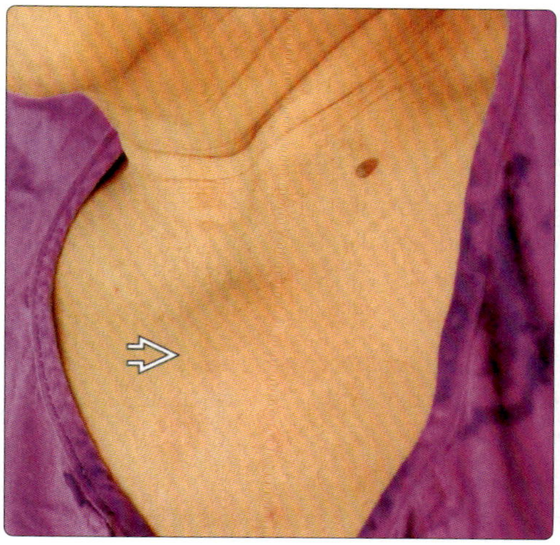
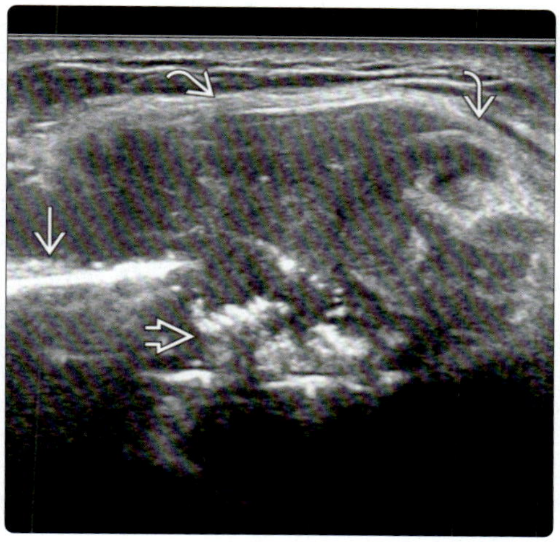

(Left) *Photograph shows a patient with left mastectomy for breast carcinoma 2 years prior and recent onset of sternal swelling.* **(Right)** *Transverse US of the same patient shows irregular osteolysis on the left side of the sternum with a large extraosseous soft tissue component highly suspicious of metastatic deposit. Sternal metastases are common in breast carcinoma and commonly present as a lump due to the superficial location of the sternum.*

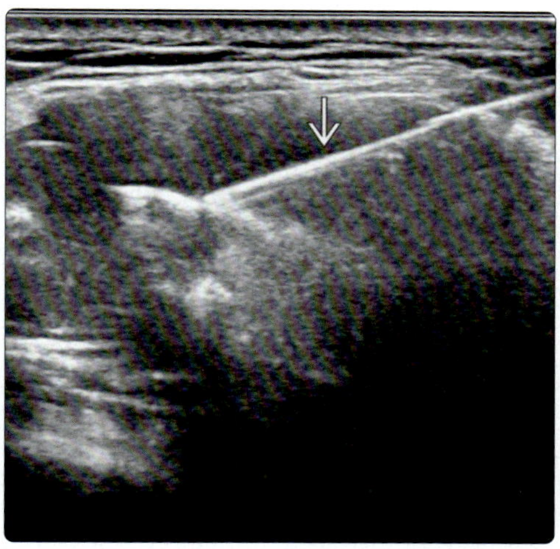
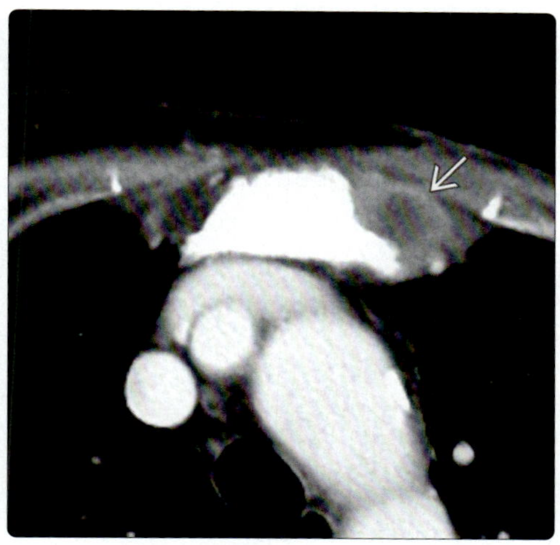

(Left) *Transverse US shows the same patient during percutaneous Tru-Cut needle biopsy of the sternal tumor. Biopsy tissue revealed metastatic breast carcinoma.* **(Right)** *Axial CECT of the same patient in soft window setting shows the metastatic deposit on the left side of the sternum. Subsequent PET/CT revealed that this was an isolated metastatic deposit. The patient received chemoirradiation. Sternotomy is an alternative treatment option for isolated sternal breast metastases.*

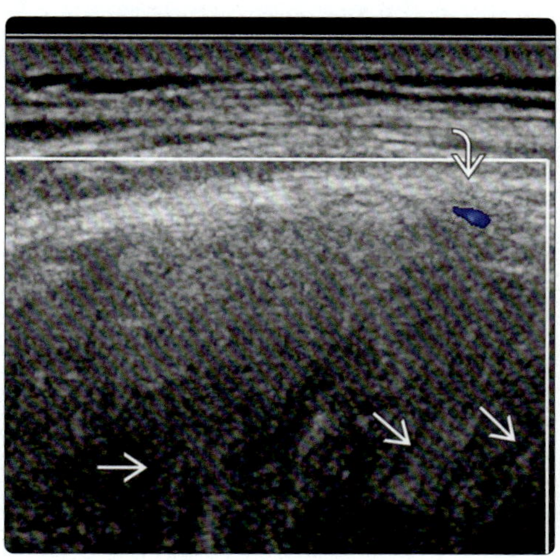
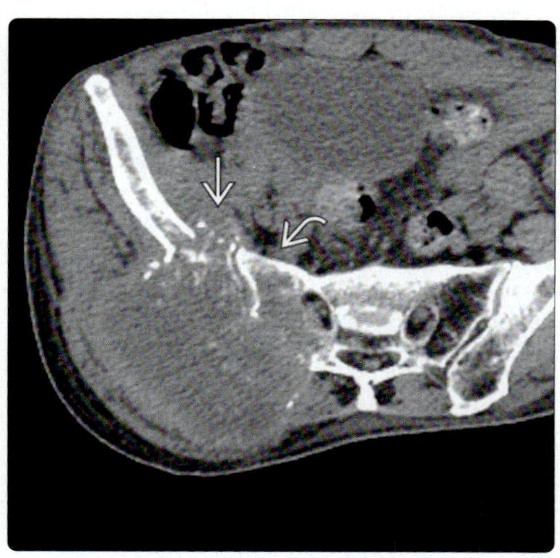

(Left) *Transverse color Doppler US in a patient with clinically suspected soft tissue buttock sarcoma shows a large soft tissue mass with radiating echogenic spicules. These spicules suggest a bony origin, which was confirmed by 7.5-MHz US imaging (not shown). Little demonstrable tumoral hyperemia is present.* **(Right)** *Axial NECT shows a large tumor mass arising from the iliac and sacral bones. Histology revealed primitive neuroectodermal tumor.*

Local Tumor Recurrence

KEY FACTS

TERMINOLOGY

- Local recurrence of tumor following initial planned complete tumor resection
  - If original tumor was not completely resected, this is residual tumor rather than recurrence

IMAGING

- US is as sensitive/specific as MR for detecting bone or soft tissue recurrence
- Examine entire section of limb from which primary tumor was resected
- Presence of discrete soft tissue mass is best diagnostic clue
  - Appearances mirror that of primary tumor
  - Multiple soft tissue tumor recurrences may occur simultaneously
    - If 1 recurrence found, continue careful search for more recurrences
- Consider likelihood of postoperative, nontumoral mass
- For bone tumor, correlate with radiographs if any bone abnormality is detected
- Routinely examine regional lymph nodes
- Arrange either US follow-up or US-guided biopsy of any suspicious mass lesion

TOP DIFFERENTIAL DIAGNOSES

- Postoperative seroma/hematoma
- Postoperative fibrosis/granulation tissue
- Reactive bone change at resection margin

PATHOLOGY

- Several factors may govern risk of recurrence
  - Positive surgical margins and tumor depth, size, and histologic grade

CLINICAL ISSUES

- Most local recurrences occur in first 3 years after diagnosis
- Early detection of local recurrence improves prognosis and morbidity

(Left) *Clinical photograph shows a patient with leiomyosarcoma of the ankle treated with excision several years earlier with skin flap ➡. A nodule ➡ at the posterior aspect of the flap is slowly increasing in size.* **(Right)** *Longitudinal US of the corresponding region shows a heterogeneous, predominantly hypoechoic soft tissue mass at the site of the clinically palpable nodule. The margins are slightly irregular ➡. Features are highly suspicious for local recurrence.*

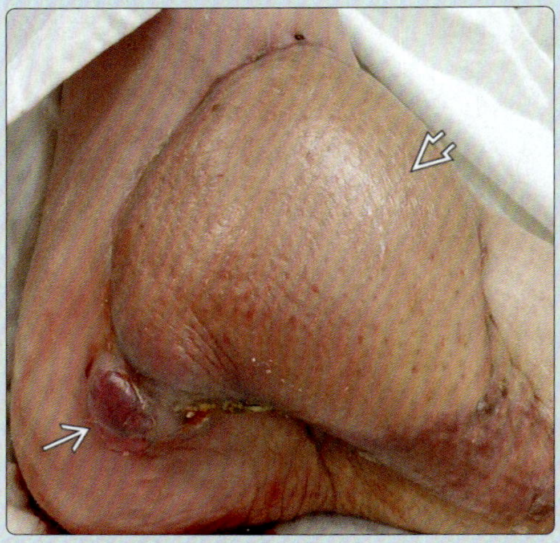

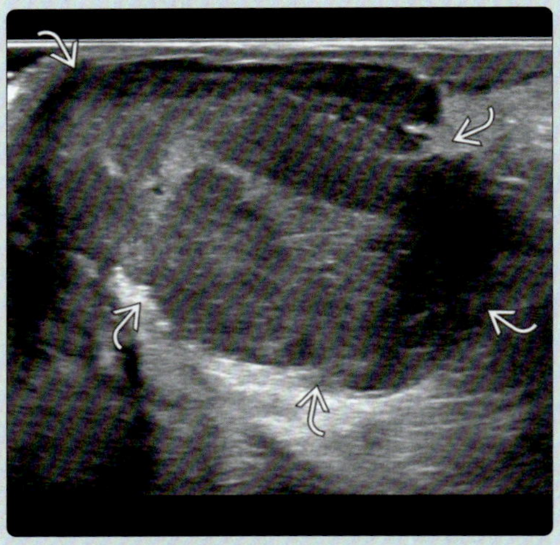

(Left) *Lateral radiograph of a middle-aged patient with a humeral prosthesis for fibrosarcoma resected 3 years earlier shows localized osteolysis ➡ suggestive of tumor recurrence.* **(Right)** *Longitudinal US of corresponding site shows a soft tissue mass ➡ at the anteromedial aspect of prosthetic bone junction ➡ with irregular underlying osteolysis ➡. Percutaneous biopsy confirmed local recurrence. MR and CT would have been limited by metallic artifact while US clearly shows recurrence.*

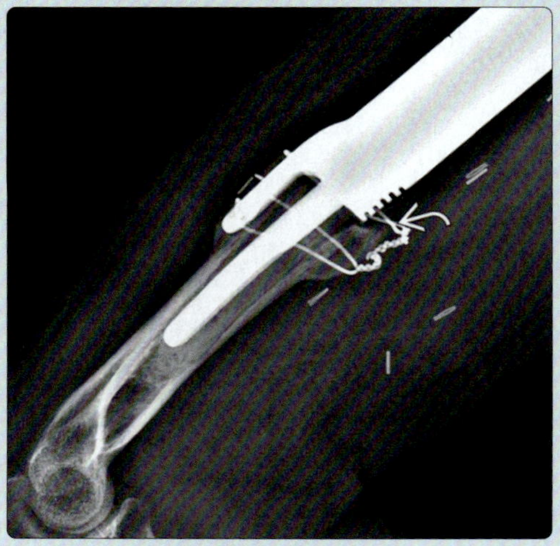

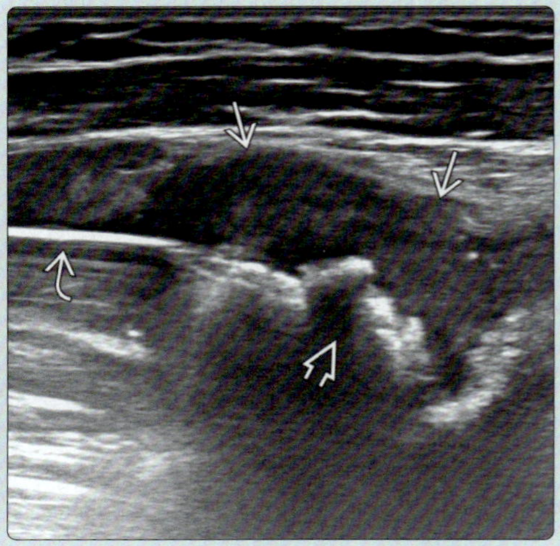

TERMINOLOGY

Definitions

- Local recurrence of tumor following initial planned complete tumor resection
  - If original tumor was not completely resected, this is residual tumor rather than recurrence

IMAGING

General Features

- Best diagnostic clue
  - Presence of discrete soft tissue mass
- Location
  - Most common along tract of tumor excision
    - Anywhere under surgical scar area
  - Local recurrences from superficial tumor commonly occur in muscle layer exposed at time of initial tumor resection
  - Multiple recurrences may simultaneously occur
    - If 1 recurrence found, look carefully for other recurrences
  - Superficial recurrences from deep tumors are generally due to contamination along surgical tract
    - Seeding along tract was much more common with open biopsy
    - Extremely uncommon with closed image-guided percutaneous biopsy
- Size
  - Larger in deeper lesions due to later presentation
- Morphology
  - Mirrors that of primary tumor
    - All malignant bone and soft tissue tumors have propensity for recurrence
    - Some benign bone and soft tissue tumors have propensity for recurrence
      - Fibromatosis
      - Giant cell tumor of bone

Ultrasonographic Findings

- **Soft tissue recurrence**
  - Presence of discrete soft tissue mass at surgical site is best diagnostic clue
    - Hampered by postsurgical or irradiation fibrosis, and disruption of normal anatomical planes
    - Alongside surgical scar and along open biopsy tract are most common sites for recurrence
      - Rare following image-guided percutaneous biopsy
  - Variable echogenicity and vascularity of recurrent tumor but mirrors that of primary tumor
    - Hypoechoic, hyperemic mass with foci of mineralization for osteosarcoma primary
    - Hyperechoic, nonhyperemic tumor without mineralization for liposarcoma primary
    - Myxoid tissue: Well-defined, intratumoral anechoic or hypoechoic areas
    - Calcification: Discrete intratumoral echogenic foci with acoustic shadowing
    - Hemorrhage: Ill-defined intratumoral echogenic areas with acoustic enhancement; fluid levels
    - Usually hypervascular with disorganized vascular pattern on color Doppler imaging
      - However, if primary tumor is hypovascular, such as myxoid chondrosarcoma, recurrence will be hypovascular
  - Associated abnormality
    - Adjacent bone erosion or periosteal reaction
    - ± regional nodal metastasis (not uncommon with recurrent tumor)
- **Bone tumor recurrence**
  - US useful at detecting bone tumor recurrence
  - Most bone tumor recurrences occur in soft tissue and not in bone
  - Most recurrences lie close to prosthetic or allograft implant
    - US is not affected by metallic artifact
  - Discrete soft tissue mass is best diagnostic clue
    - ± association with underlying bony erosion
    - Echogenicity and matrix tumor characteristics mirror those of primary tumor
      - Same characteristics of primary bone tumor (e.g., matrix mineralization in osteosarcoma)
    - Regional metastatic lymphadenopathy is much more common than with primary bone tumor
    - If tumor clearly originates from bone, metachronous malignancy should be considered, especially for osteosarcoma

MR Findings

- T1WI
  - Presence of discrete, enhancing soft tissue mass
  - Signal intensity pattern of recurrence is similar to that of primary tumor
  - Sensitivity 83%; specificity 93%
- T2WI FS
  - Discrete high signal intensity lesion with mass effect
  - Moderate perilesional edema

Imaging Recommendations

- Best imaging tool
  - US, as it is easily available, does not require contrast injection and is not subject to metallic artifact
    - As accurate as MR for detecting bone or soft tissue recurrence
    - Allows ready assessment of regional lymph nodes
    - US-guided biopsy feasible of any suspicious mass
  - For high-grade bone or soft tissue sarcoma (STS), follow-up at 3 months after surgery, then every 6 months for 2 years, then yearly thereafter for 5 years
  - For low-grade bone or STS, follow-up at 6 months after surgery, then yearly thereafter for 3 years
  - However, still not clear whether routine surveillance actually improves survival
- Protocol advice
  - Same protocol is used for both bone and soft tissue recurrence
  - Examine entire section of limb from which primary tumor was resected (e.g., thigh, leg, arm, or forearm)
  - Use parallel series of longitudinal sweeps (with transducer aligned transversely) from proximal to distal along affected limb segment

Local Tumor Recurrence

- Examine patient in both supine and prone positions to ensure that entire circumference of operated segment of limb is examined
- Look for any discrete soft tissue mass, particularly along scar
- Consider likelihood of postoperative, nontumoral masses
  - Chronic collection
  - Organizing hematoma
  - Postoperative fibrosis
  - Localized fat hypertrophy to replace resected muscle
  - Retracted transected muscle belly
  - Reactive changes at bone resection margin
- For bone tumor, correlate with radiographs if any bone abnormality is detected
- Routinely examine regional lymph nodes
  - Fine-needle aspiration for cytology if lymph nodes suspicious
- Either US follow-up or US-guided biopsy of suspicious mass lesion
  - Sample solid component of suspicious tissue
  - 16-g Tru-Cut core biopsy with coaxial system is excellent for tissue typing and histologic grading

DIFFERENTIAL DIAGNOSIS

Postoperative Seroma/Hematoma

- Usually more cystic in appearance, avascular
- Less mass effect
- Tends to remain static or regress on serial imaging

Postoperative Fibrosis/Granulation Tissue

- No mass effect, usually less well defined and more irregular
- Occurs at and tracks along surgical tract
- Does not progress on serial follow-up

Reactive Bone Change at Resection Margin

- Occurs at prosthetic/native bone, allograft/native bone
- New bone formation and localized soft tissue thickening
- Little, if any, mass effect
- Correlate any abnormal US appearances with radiograph
  - Radiography is very helpful in sorting out suspicious bone areas seen on US

PATHOLOGY

General Features

- Etiology
  - Several factors likely to be contributory
    - Unplanned resection
    - Histologic grade of tumor

Gross Pathologic & Surgical Features

- Pathology mirrors that of primary tumor
  - Usually similar grade to primary tumor
  - Infrequently, recurrence may have more aggressive histologic features than primary tumor

CLINICAL ISSUES

Presentation

- Most common signs/symptoms
  - New lump at site of previous surgery
    - Clinical detection of recurrence can be hampered by scar tissue or deep location of recurrence
      - Especially true for irradiated surgical beds as soft tissue may be scarred and indurated
- Other signs/symptoms
  - Pain from bone tumor recurrence
  - Weight loss and other constitutional symptoms probably reflect systemic metastatic disease

Demographics

- Epidemiology
  - 6-14% of patients develop local recurrence after STS treatment
  - Greater risk of recurrence for
    - High-grade > low-grade tumors
    - Larger size of primary tumor
    - Unplanned surgery
    - Positive histologic margins with postoperative radiotherapy associated with risk of recurrence is 5.9x that of tumor with clear resection margins
    - Patients with treated local recurrence have higher likelihood of subsequent local recurrence

Natural History & Prognosis

- Most local recurrences occur in first 3 years after diagnosis
- Local recurrence requires additional surgery, radiotherapy, or amputation
  - Early detection of local recurrence helps avoid extensive surgery or amputation
- Local recurrence of malignant tumors is associated with overall poor prognosis (2-year survival rate of 50-70%)

Treatment

- Well-planned reresection in specialized tumor center
  - ± adjuvant radiotherapy ± chemotherapy
- Radiotherapy is very useful in resected tumor with positive margins
  - Likelihood of radioresistance is high in patients with recurrence and prior radiotherapy treatment
  - Rate of wound complications increases with radiotherapy
- Amputation

DIAGNOSTIC CHECKLIST

Consider

- Regional nodal involvement

Image Interpretation Pearls

- Look for discrete soft tissue mass
- Mirrors primary lesion

SELECTED REFERENCES

1. Loong HH et al: Controversies and consensus of neoadjuvant chemotherapy in soft-tissue sarcomas. ESMO Open. 3(Suppl 1):e000293, 2018
2. Sawamura C et al: How long should we follow patients with soft tissue sarcomas? Clin Orthop Relat Res. 472(3):842-8, 2014

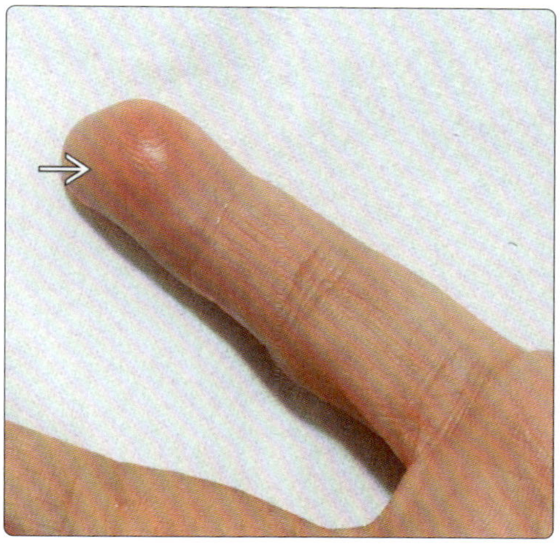

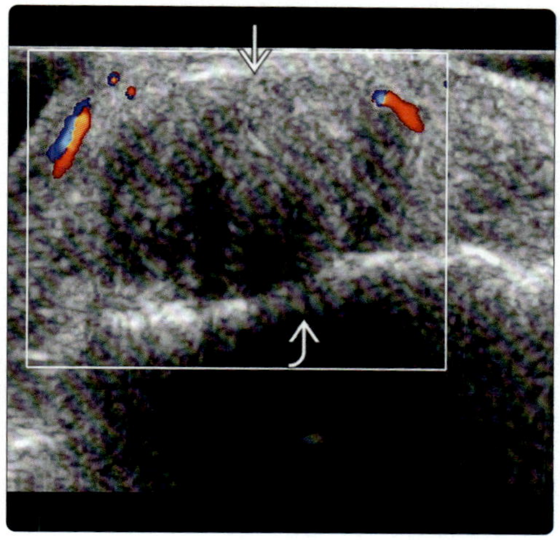

(Left) *Clinical photograph shows a patient with swelling at the tip of the little finger, which had been slowly growing for 1 year.* **(Right)** *Longitudinal color Doppler US of the same patient shows a well-defined, hypovascular mass at the tip of the distal phalanx with mild, posterior bony scalloping. Histology revealed a giant cell tumor. The tumor was excised.*

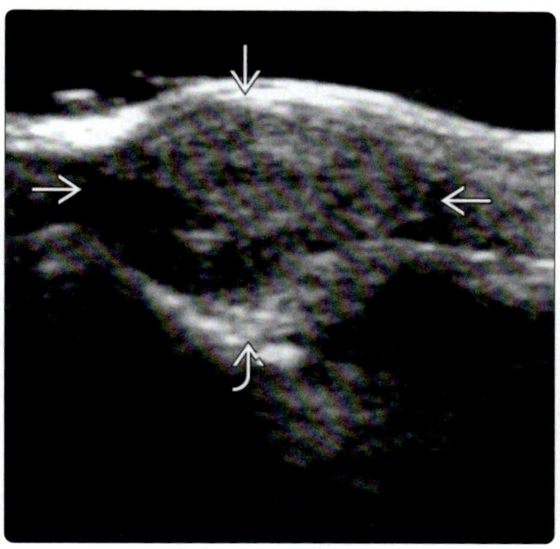

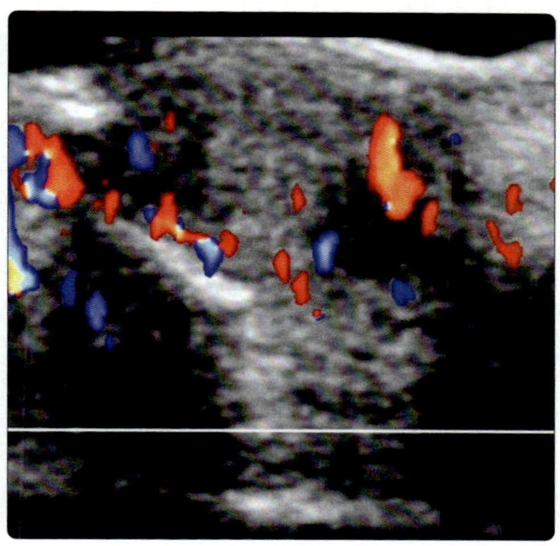

(Left) *Longitudinal US of the same finger 3 years later shows a locally recurrent tumor with more severe bony scalloping. The bony scalloping was likely the result of the previous surgery rather than progressive, tumor-related osteolysis.* **(Right)** *Longitudinal color Doppler US of the same tumor shows moderate hyperemia. In most cases, local recurrence will have similar US characteristics to the original tumor.*

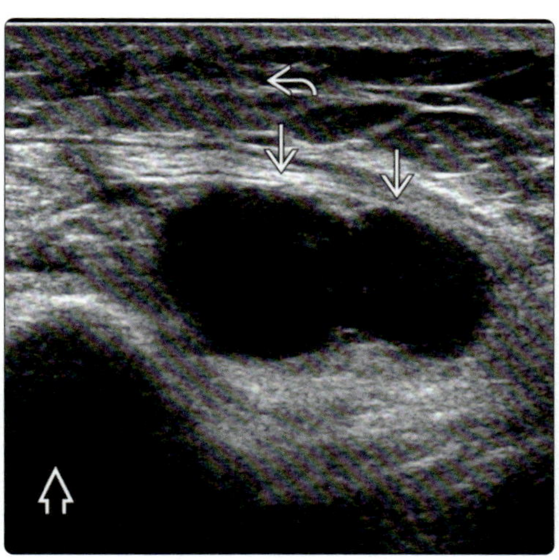

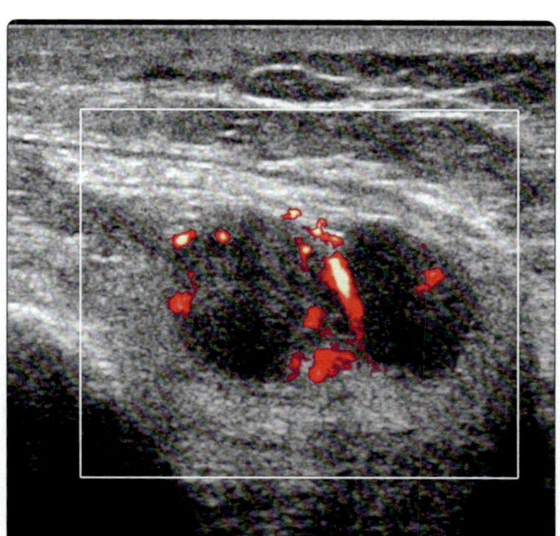

(Left) *Transverse US in a patient with undifferentiated pleomorphic sarcoma of the thigh excised 1 year ago shows a well-defined, hypoechoic mass deep to the surgical scar compatible with local recurrence. The femur is shown. Area of scar is seen as echogenic subcutaneous fat rather than hypoechoic fibrous tissue.* **(Right)** *Transverse power Doppler US of the same tumor shows moderate hyperemia. A discrete mass is the defining characteristic of local recurrence.*

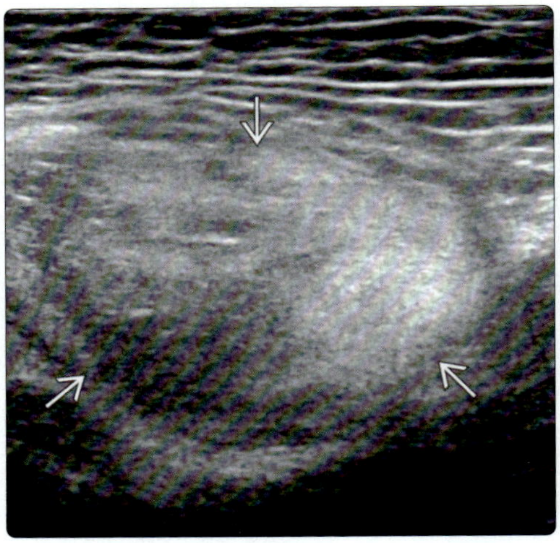

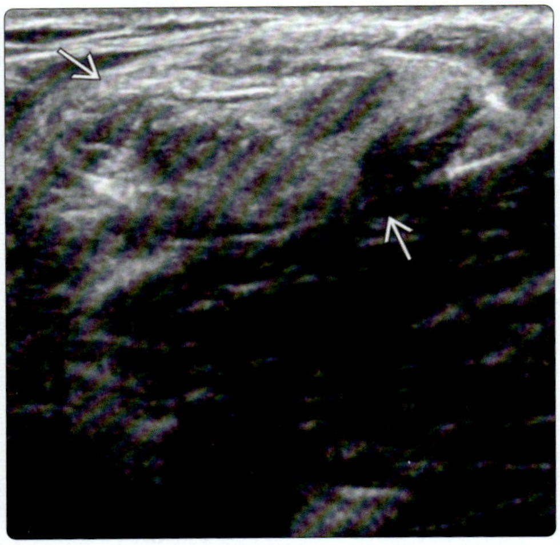

(Left) *Longitudinal US shows quite a large, well-defined, slightly irregular, predominantly echogenic mass* ➡ *in the vastus medialis muscle, which is highly suspicious for an intramuscular liposarcoma. This was confirmed by histology.* **(Right)** *Longitudinal US of the same region in the same patient 5 years later shows a discrete, largely echogenic mass with a well-defined border. This fatty tumor developing at the site of the previous excision was highly suspicious for recurrent liposarcoma* ➡*.*

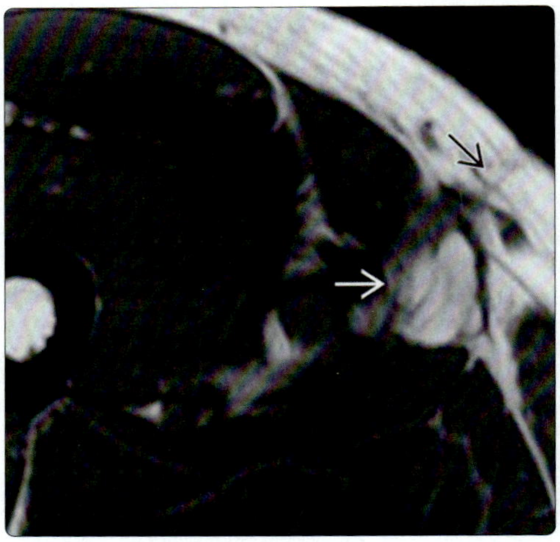

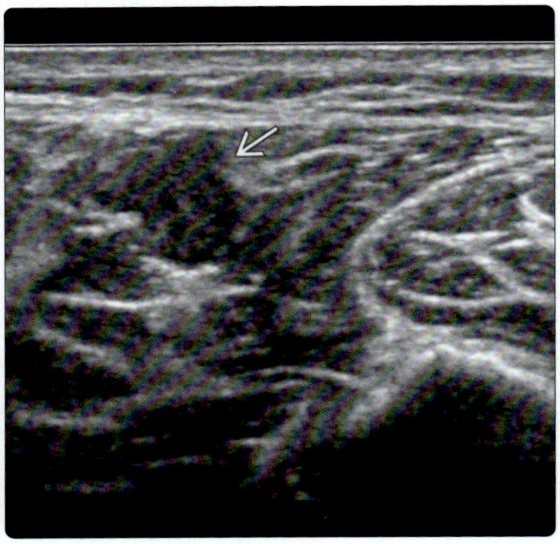

(Left) *Transverse T1 MR of the same patient shows a fatty tumor* ➡ *on the medial aspect of the thigh just deep to the surgical scar* ⇨*. Histology confirmed liposarcoma recurrence.* **(Right)** *Transverse US of a patient with an inflammatory myxohyaline tumor excised years ago shows a well-defined, hypoechoic mass with irregular, lobulated contour* ➡ *in the lateral aspect of soleus, which is suspicious for an intramuscular local recurrence.*

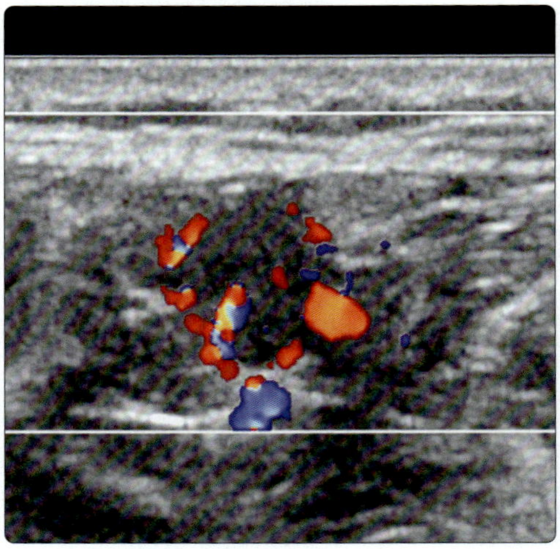

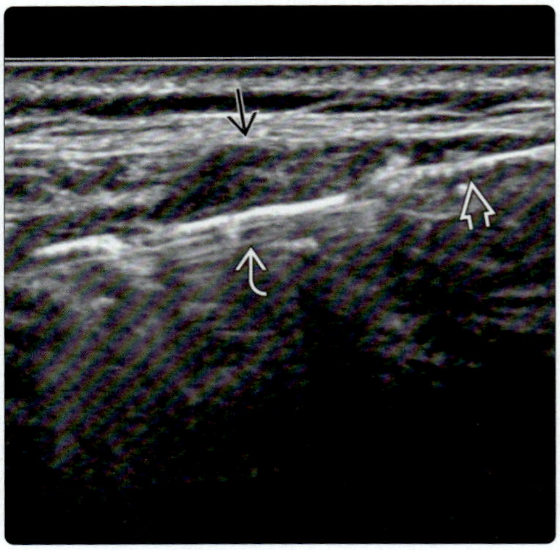

(Left) *Transverse color Doppler US of the same lesion shows severe marginal and intralesional hypervascularity, corresponding to the inflammatory characteristic of the original primary tumor.* **(Right)** *Transverse US of the same patient during needle biopsy is shown. A Tru-cut biopsy needle was inserted through a coaxial needle* ➡*. The specimen slot* ➡ *is present at the center of the lesion* ⇨ *just prior to biopsy. Histology confirmed local recurrence.*

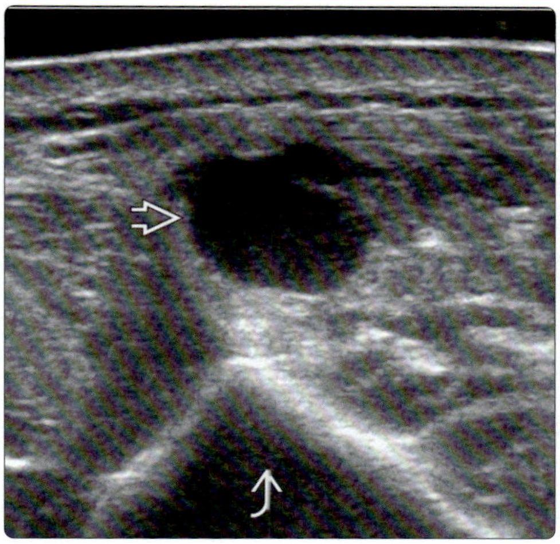
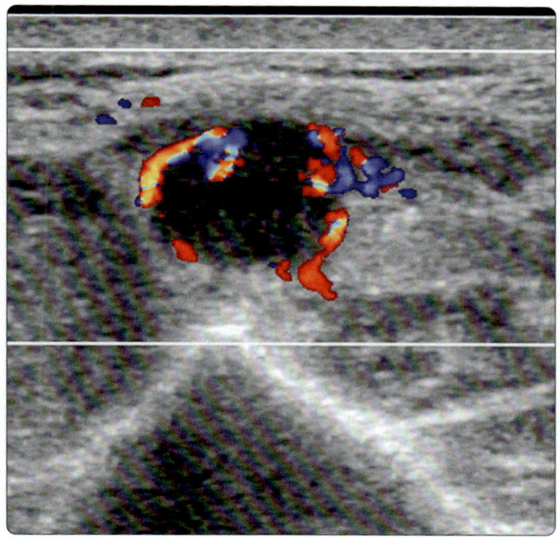

(Left) *Transverse US of the forearm in a patient with subcutaneous dermatofibrosarcoma protuberans resected 3 years earlier shows a discrete hypoechoic nodule ➡ alongside the ulnar shaft ➡ within the extensor carpi ulnaris muscle, consistent with recurrence.* **(Right)** *Transverse color Doppler US of the same tumor shows moderate peripheral hyperemia. Histology confirmed recurrence. Recurrences from subcutaneous tumors may occur within muscle.*

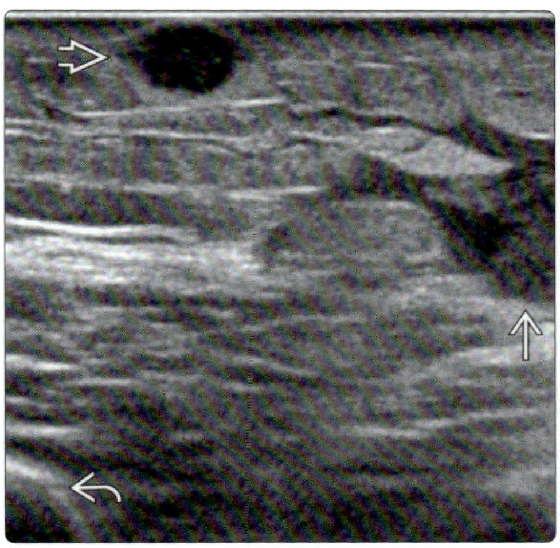
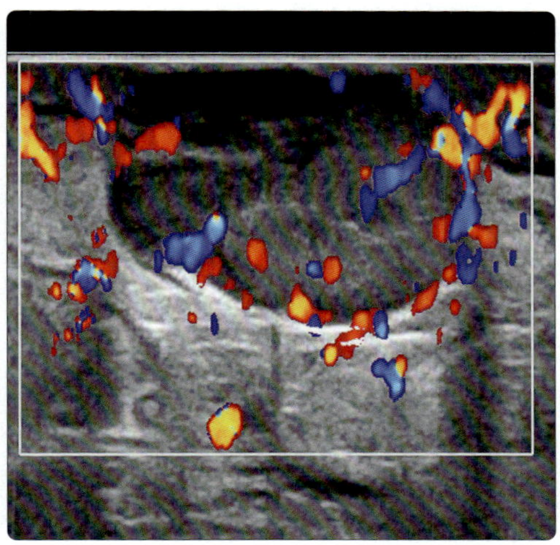

(Left) *Transverse US of the thigh in a patient with heel melanoma and wide local excision performed 2 years earlier shows a discrete hypoechoic nodule ➡ in the superficial subcutaneous layer, suspicious for recurrence. Postsurgical scar ➡ is visible at the more lateral aspect. Part of the femur ➡ is also shown.* **(Right)** *Transverse color Doppler US of a 2nd nodule at the more medial aspect of the thigh shows severe intralesional and peripheral hyperemia, characteristic of melanoma and melanoma recurrence.*

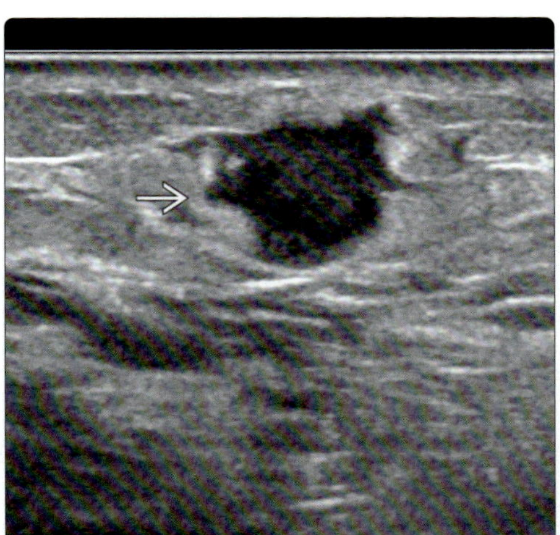
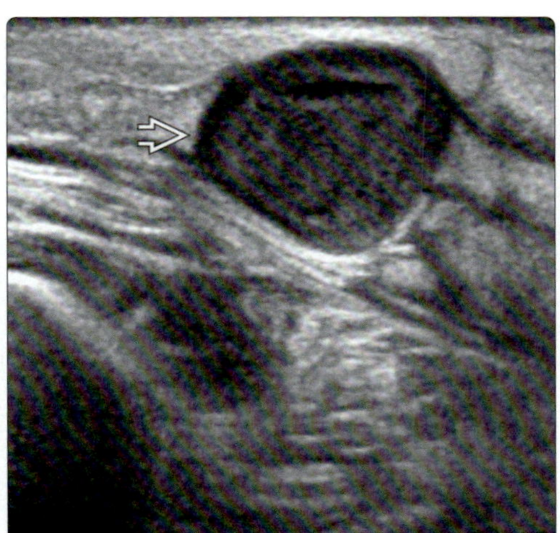

(Left) *Transverse US of the same patient shows a 3rd lesion in the subcutaneous tissue of the thigh ➡, which is again suspicious for recurrence.* **(Right)** *Transverse US shows a 4th lesion ➡ in the superficial subcutaneous layer closer to the femur. All 4 lesions were local recurrent tumors from melanoma. When 1 recurrence is detected, search carefully for other sites of recurrence.*

(Left) *Longitudinal surveillance US in a patient with high-grade femoral osteosarcoma at 6 months post resection shows soft tissue thickening* ➡ *and apparent new bone formation alongside the prosthesis* ➡ *just distal to the bone-prosthesis junction.* **(Right)** *Longitudinal color Doppler US of the same area shows mild hyperemia.*

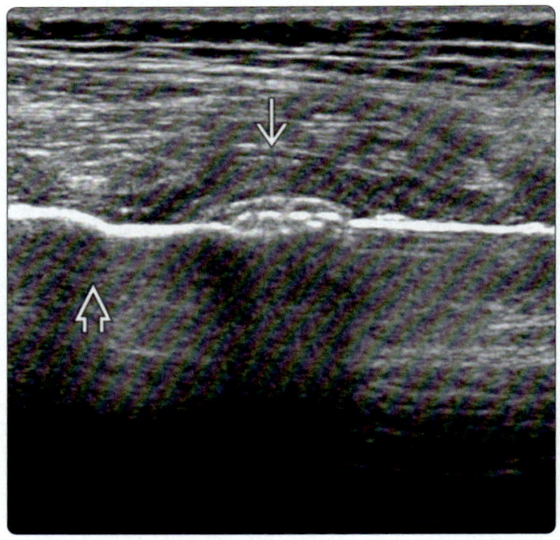

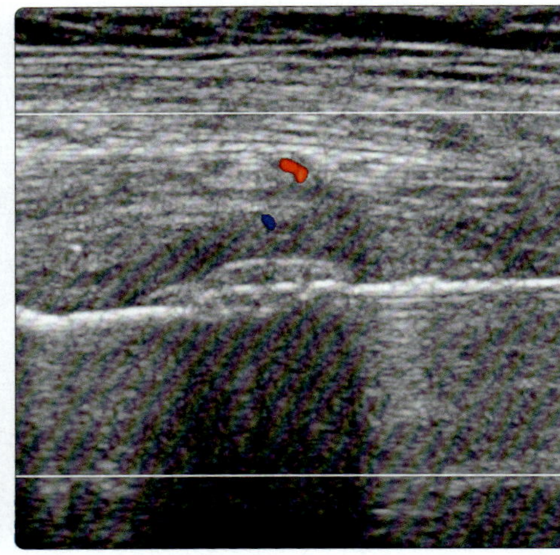

(Left) *Longitudinal US of the same patient also shows some reactive new bone formation with soft tissue thickening* ➡ *at the interface between the prosthesis* ➡ *and bone* ➡*.* **(Right)** *Frontal radiograph of the bone-prosthesis junction in the same patient provides an explanation for the suspicious US appearances due to surgical clips* ➡ *and reactive bone change* ➡ *with no tumor recurrence. Follow-up US 4 weeks later showed no progression.*

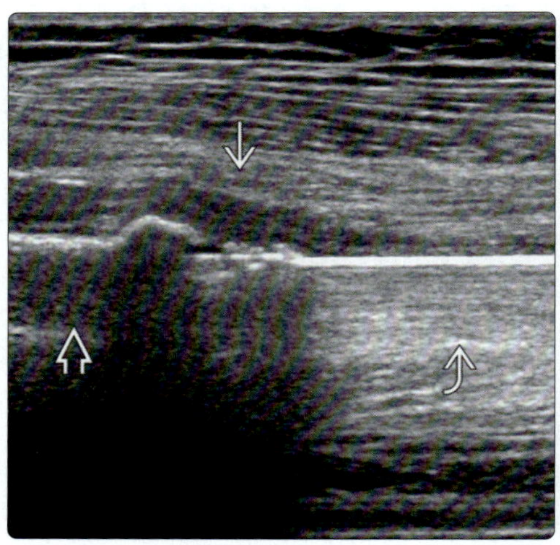

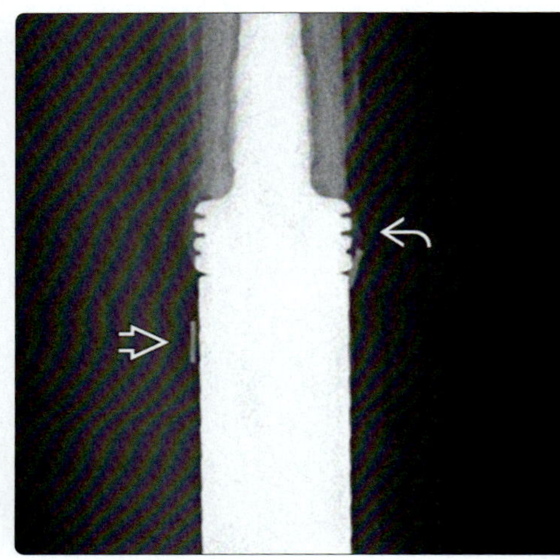

(Left) *Longitudinal surveillance US of the proximal thigh in a young boy with osteosarcoma resected 1 year earlier shows an apparent defect at the interface of the bone* ➡ *and prosthesis* ➡ *with overlying soft tissue thickening* ➡*.* **(Right)** *Frontal radiograph of the same area provides an explanation for US findings, which are exactly as expected due to the design of the prosthesis* ➡*. No recurrence was reported, and a follow-up US 2 months later showed no progression.*

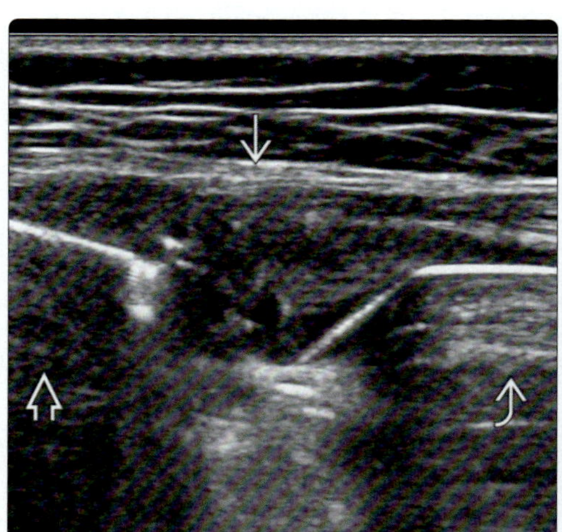

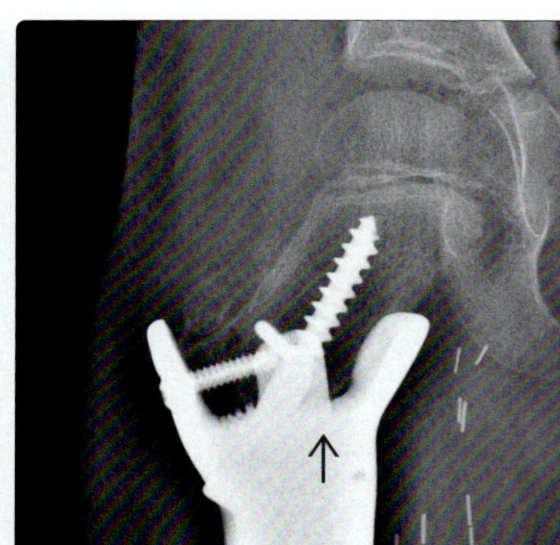

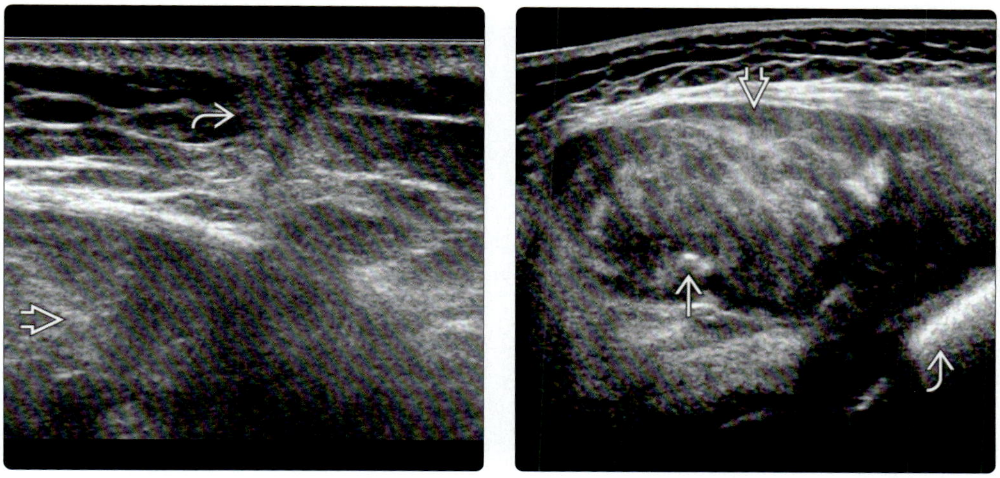

(Left) *Transverse US of the buttock in a patient with proximal femoral chondrosarcoma resected 1 year earlier shows a large, hypoechoic tumor recurrence deep to the surgical scar.* **(Right)** *Transverse extended field-of-view US of the same tumor shows the large, slightly heterogeneous tumor recurrence just above the trochanteric region of the prosthesis. There is some echogenic matrix calcification.*

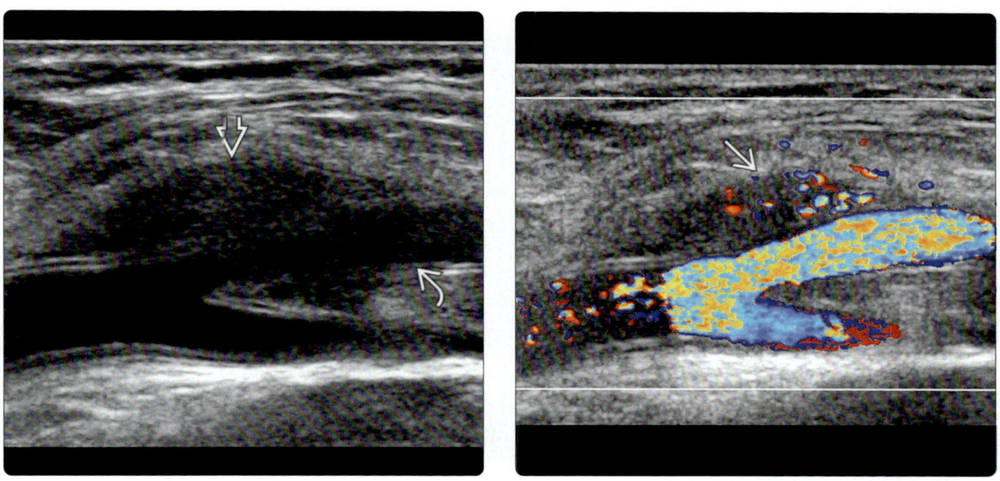

(Left) *Longitudinal US of the anterior groin in the same patient shows regional adenopathy with metastatic infiltration of the superficial femoral artery close to the bifurcation.* **(Right)** *Longitudinal color Doppler US of the same tumor shows mild tumoral hyperemia and arterial patency. Regional lymphadenopathy is quite common in tumor recurrence.*

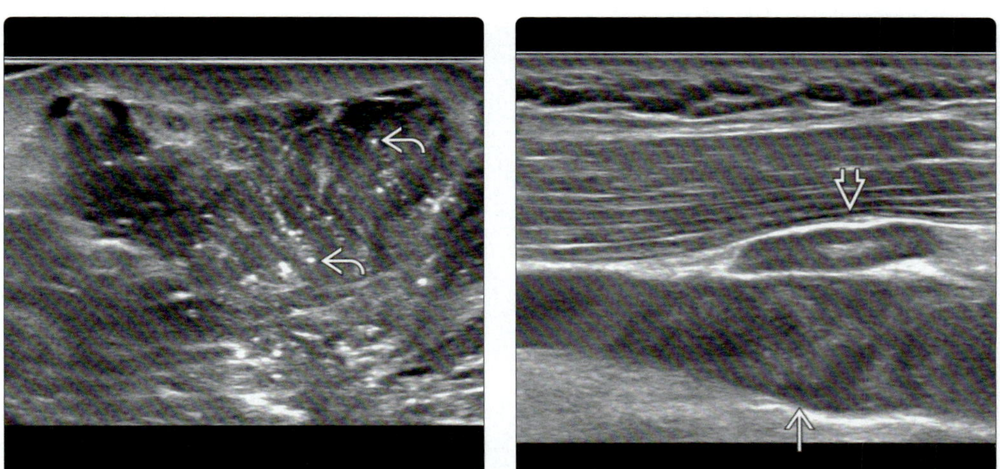

(Left) *Transverse US of the clavicular region in a patient with clavicular chondrosarcoma resected 2 years earlier shows a large, irregular mass at the resection site with foci of calcification, characteristic of chondrosarcoma recurrence, confirmed by histology.* **(Right)** *Longitudinal US of the same patient shows internal jugular vein thrombosis. No hyperemia or calcification was evident in the thrombus, indicating nontumoral thrombosis. A small reactive lymph node is present.*

(Left) *Longitudinal US in a patient with osteosarcoma of the humerus excised 10 years earlier shows a large, osteosarcoma-like mass ➔ with new bone formation ➔ arising from the ipsilateral radial diaphysis. Severe periosteal thickening ➔ extends well beyond the tumor margin.* **(Right)** *Longitudinal US of the same tumor shows sunburst-like new bone formation ➔ within the tumor ➔. This tumor is due to metachronous osteosarcoma rather than local recurrence.*

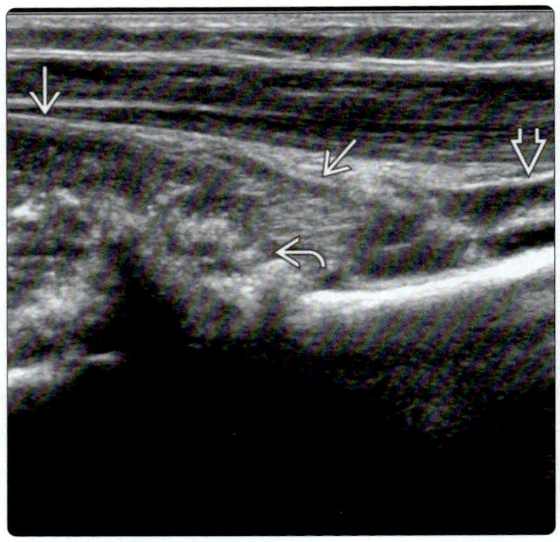

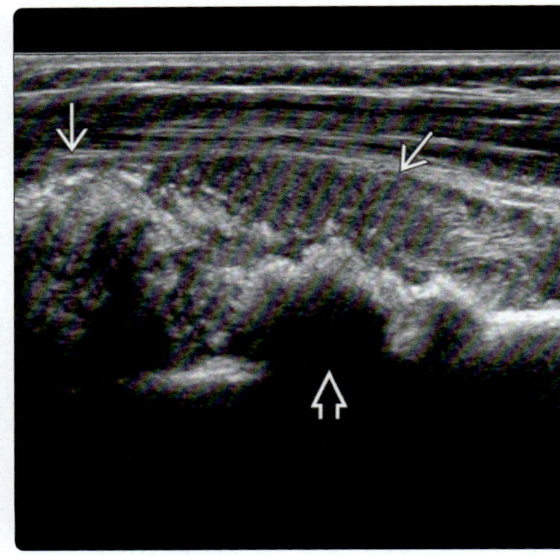

(Left) *Frontal radiograph of a patient with osteosarcoma resected 3 years earlier and increasing knee swelling for 2 months shows permeative osteolysis of both the medial ➔ and lateral ➔ femoral condyles, highly suspicious for tumor recurrence.* **(Right)** *Transverse US in the same patient shows a large tumor recurrence alongside the medial femoral condyle ➔ with matrix mineralization ➔ consistent with osteosarcoma recurrence.*

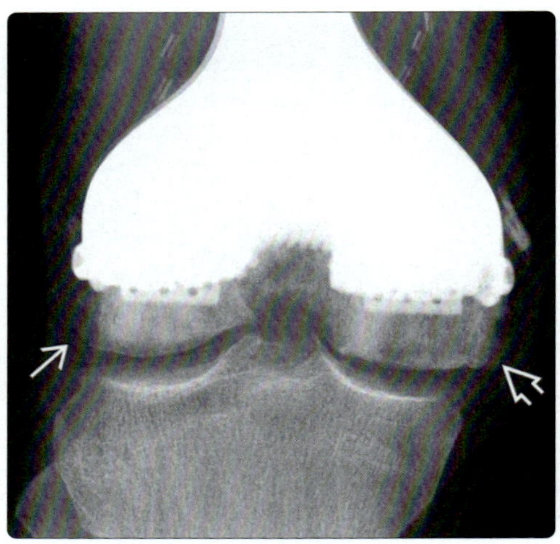

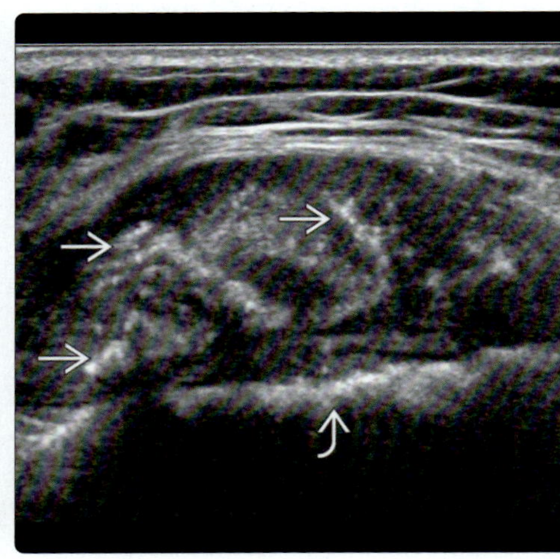

(Left) *Transverse color Doppler US of the same area shows mild peripheral hyperemia of the recurrent tumor ➔.* **(Right)** *Transverse US of the ipsilateral groin in the same patient shows a soft tissue mass with cystic ➔ and calcific ➔ changes compatible with a metastatic node. Mineralization is common in metastatic osteosarcoma. Nodal metastases in bone sarcoma recurrence are not uncommon but rare at initial presentation.*

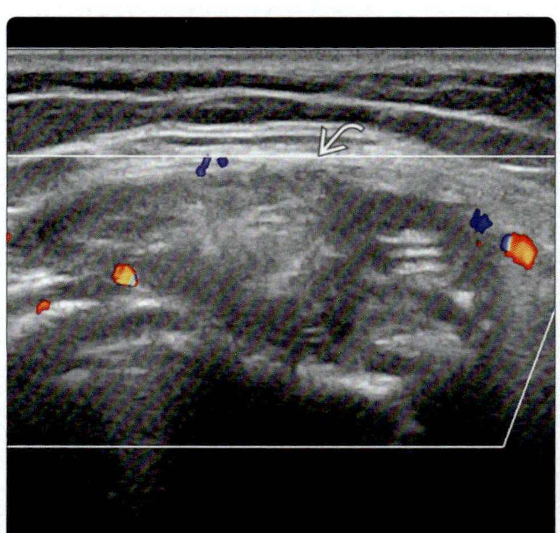

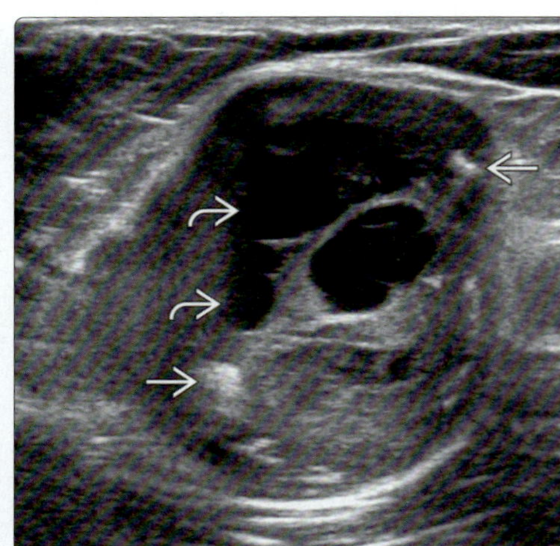

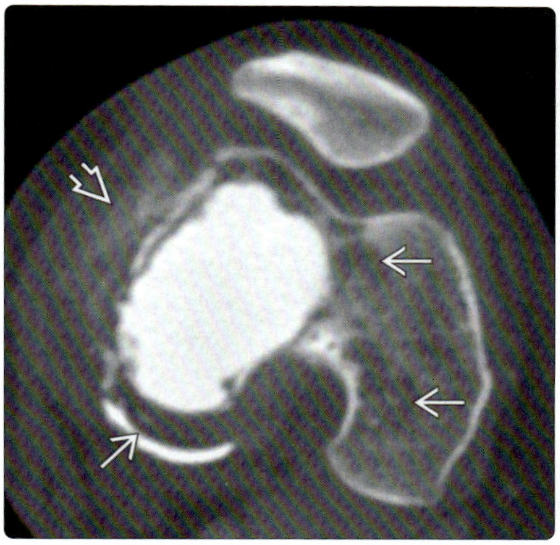

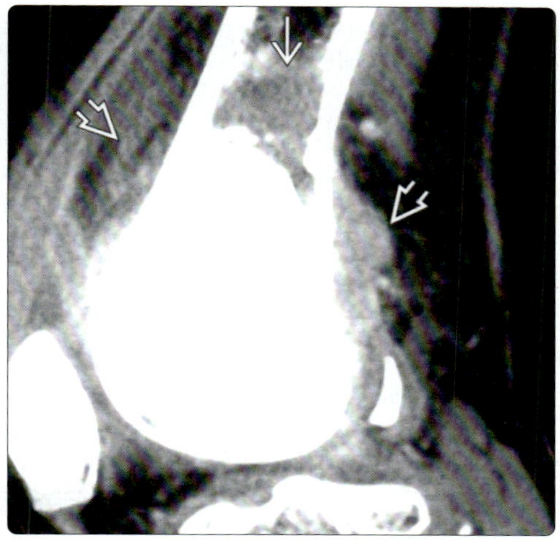

(Left) *Axial NECT in a patient 1 year following resection of initially suspected giant cell tumor shows osteolysis due to tumor recurrence around the bone cement with new bone formation in the soft tissue alongside the lateral femoral condyle.* **(Right)** *Sagittal CT reconstruction in the same patient shows severe intramedullary and juxtacortical tumor recurrence.*

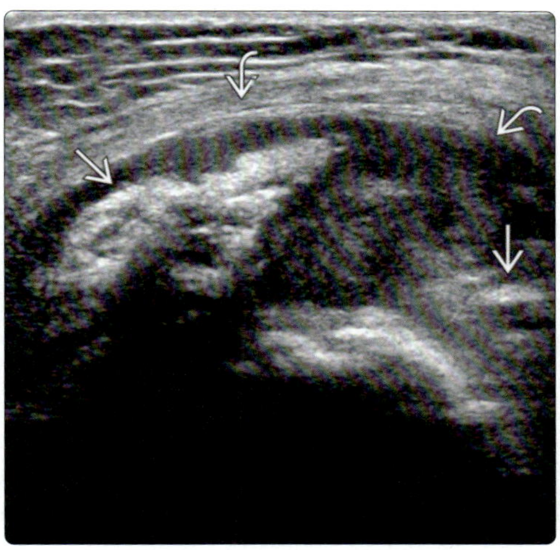

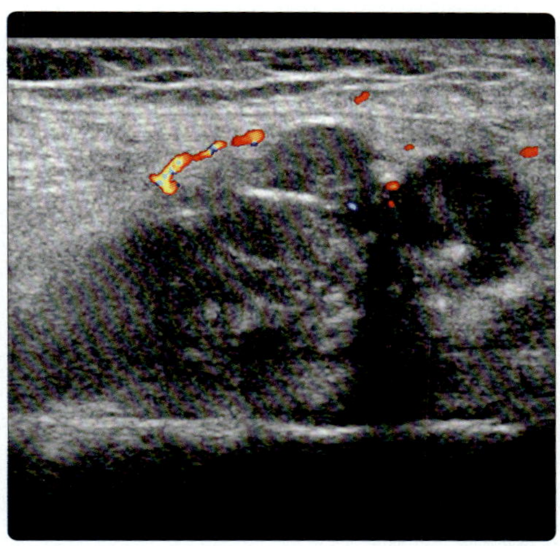

(Left) *Transverse US of the same patient shows prominent new bone formation within the large, hypoechoic area of extraosseous tumor recurrence.* **(Right)** *Longitudinal color Doppler US of the same patient shows mild hyperemia of this tissue with additional small areas of calcification. This aggressive recurrence along with new bone formation prompted review and reclassification of the initial histology from giant cell tumor to giant cell tumor variant of osteosarcoma.*

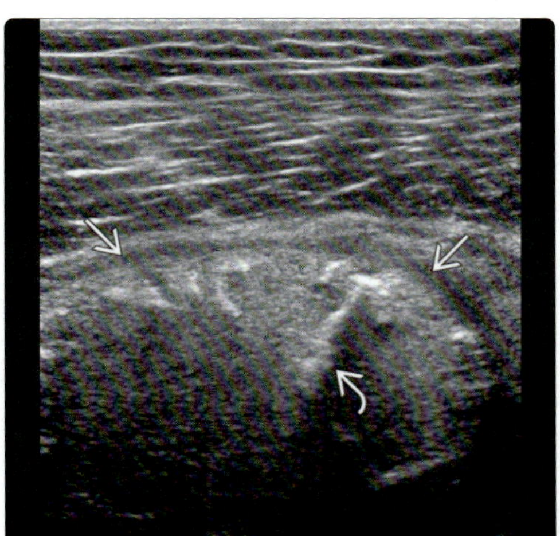

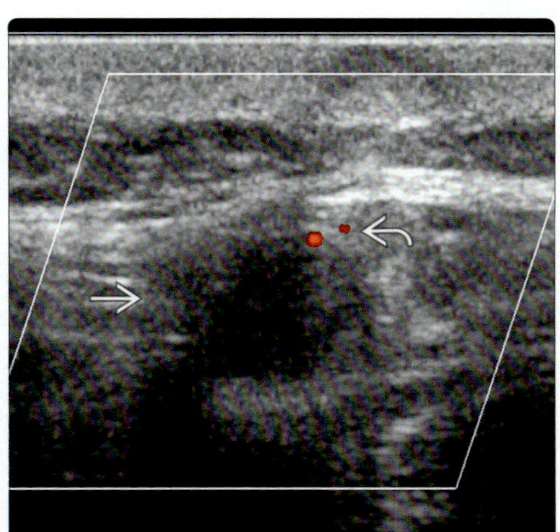

(Left) *Longitudinal US shows a large soft tissue osteosarcoma recurrence with matrix mineralization, which is a characteristic feature of local recurrence of osteosarcoma.* **(Right)** *Transverse color Doppler US of the paraspinal region in a patient with myxoid chondrosarcoma resected 3 years earlier shows an ill-defined, hypoechoic mass at the resection site with minimal vascularity, consistent with a myxoid-type tumor recurrence. This was confirmed histologically.*

Abdominal Wall Hernia

KEY FACTS

CLINICAL ISSUES

- Tissue protruding though defect in abdominal wall
  - Most abdominal wall hernias are point specific
- Midline hernias
  - Epigastric: Usually small
  - Umbilical: In children
  - Paraumbilical: In adults
  - Hypogastric: Rare
- Lateral hernias
  - Spigelian: Defect in transversus abdominis fascia alongside lateral border of rectus abdominis muscle (linea semilunaris) and lateral border of rectus sheath
  - Lumbar: Occur in 2 areas of potential weakness in flank
  - Superior lumbar triangle (Grynfeltt hernia) with erector spinae medially, 12th rib superiorly, and internal oblique muscle laterally
  - Inferior lumbar triangle (Petit hernia) with latissimus dorsi muscle medially, iliac crest inferiorly, and external oblique muscle laterally

DIAGNOSTIC CHECKLIST

- For all abdominal wall masses, exclude presence of hernia
- If hernia confirmed, determine nature of herniated content (extraperitoneal fat, omentum, bowel) and site and size of abdominal wall defect
  - Routinely use Valsalva maneuver ± examination in standing position if symptoms exacerbated by standing
    - Can perform Valsalva by asking patient to blow against back of hand
- Observe for reducibility and consider whether bowel, if present, is obstructed or strangulated
  - Such complications are uncommon in noninguinal abdominal wall hernias
    - Since these hernias have either wide neck or, if neck is narrow, rarely contain bowel

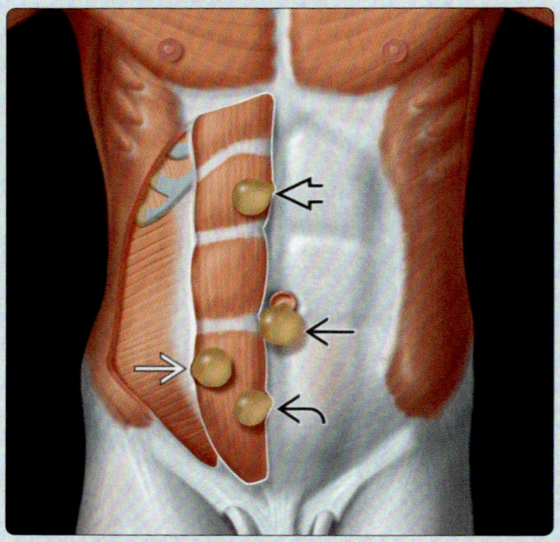

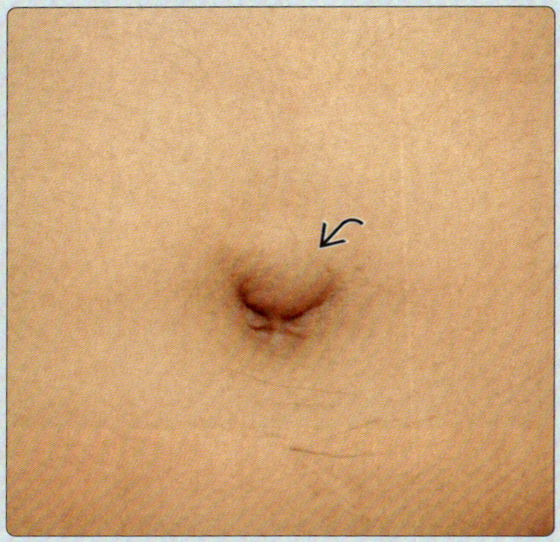

(Left) *Schematic diagram shows a paraumbilical hernia ➙ arising from the umbilicus area. The location of epigastric ➙, spigelian ➙, and hypogastric ➙ hernias is also shown for reference.* **(Right)** *Clinical photograph shows a typical appearance of a small paraumbilical hernia in an adult with swelling ➙ at the superior aspect of the umbilicus.*

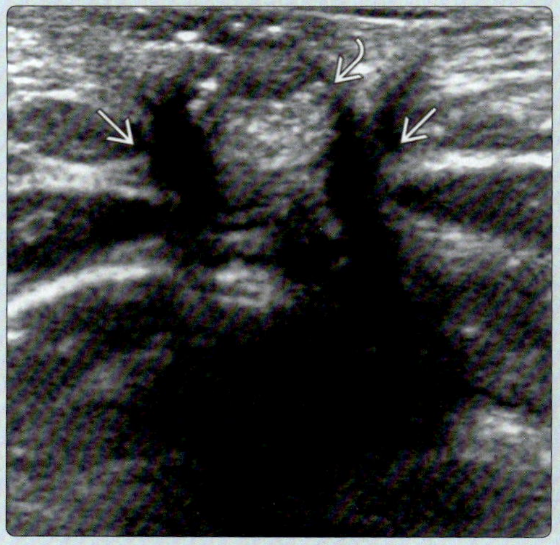

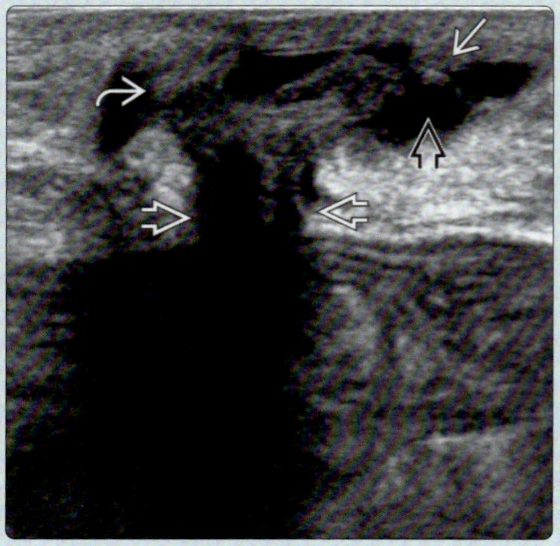

(Left) *Transverse US in the same patient shows a small paraumbilical hernia. The sac contains echogenic fat ➙ and no bowel. The margins of the defect ➙ in the linea alba are clearly seen.* **(Right)** *Transverse US shows a medium-sized paraumbilical hernia ➙ extending through a defect ➙ in the linea alba at the umbilicus. The sac contains both echogenic omental fat ➙ and hypoechoic peritoneal fluid ➙. No bowel content is present.*

TERMINOLOGY

Definitions

- Hernia: Abnormal protrusion of organ or part of organ out of cavity in which it is normally contained
  - Reducible: Hernia that can disappear either spontaneously or with external pressure
  - Obstructed: Hernia leading to mechanical bowel obstruction
  - Strangulated: Hernia causing mechanical constriction of vascular supply to contents of hernia
  - Incarcerated: Hernia that cannot be reduced
- Types of abdominal wall hernia: Midline (epigastric, umbilical, hypogastric), lateral (spigelian, lumbar), and incisional

IMAGING

General Features

- Best diagnostic clue
  - Abdominal wall lump due to tissue protruding though defect in abdominal wall
- Location
  - Midline hernias
    - Epigastric: Between xiphisternum and umbilicus
    - Umbilical: Umbilical or immediate paraumbilical region
    - Hypogastric: Between umbilicus and pubic symphysis
  - Lateral hernias
    - Spigelian
      - Defect along linea semilunaris in transversus abdominis aponeurosis just lateral to rectus sheath
      - Most commonly located near arcuate line
    - Lumbar: Occur in 2 potentially weak areas of flank
      - Superior lumbar triangle (Grynfeltt hernia) bounded by erector spinae medially, 12th rib superiorly, and internal oblique muscle laterally
      - Inferior lumbar triangle (Petit hernia) bounded by latissimus dorsi muscle medially, iliac crest inferiorly, and external oblique muscle laterally
  - Incisional hernia: Located at surgical incisional site
- Morphology
  - Abdominal wall; anatomy
    - Layers: Skin, subcutaneous fat, musculofascial layer, transversalis fascia, and extraperitoneal fat
    - Musculofascial layer
      - Anterior wall: Paired rectus muscles enclosed by rectus sheath (aponeurosis of internal oblique, external oblique, and transversus abdominis muscles), which laterally forms linea semilunaris and centrally fuses to form linea alba
      - Anterolateral-lateral wall: External oblique, internal oblique, and transversus abdominis
      - Posterior wall: Quadratus lumborum and erector spinae muscles
    - Arcuate line: Lower margin of posterior rectus sheath ends at arcuate line, midway between umbilicus and pubic symphysis
    - Distal to arcuate line, rectus muscle is separated from peritoneum only by transversalis fascia, as aponeuroses of internal oblique, external oblique, and transversus abdominis pass anterior to rectus muscle
  - Epigastric hernia
    - Usually small and contains extraperitoneal fat that protrudes through linea alba (fatty hernia of linea alba)
    - May occasionally be large and contains bowel
  - Umbilical hernia
    - Umbilical in children
      - Peritoneal content protrudes through patent umbilical ring
    - Paraumbilical in adults
      - Extraperitoneal fat ± peritoneal contents protrudes at side of umbilical ring
  - Hypogastric hernia
    - Very uncommon
  - Spigelian hernia
    - Defect in transversus abdominis fascia between lateral border of rectus abdominis muscle (linea semilunaris) and lateral border of rectus sheath
      - This section of transversalis fascia is known as spigelian fascia
    - Tends to occur within 6 inches above imaginary line connecting both anterior superior iliac spines
    - Usually extends into subcutaneous tissue, though may pass between transversus abdominis and internal or external oblique fascia or may extend into rectus muscle
      - Can present as flank lump if extends laterally
  - Lumbar hernia
    - Usually painless due to wide hernial neck
  - Incisional hernia
    - Occurs in 0.5-14.0% of postoperative patients
    - May occur years after surgery, though usually occurs within 1st year
    - May go unnoticed by patient and be incidentally detected on imaging
      - Can be very large

CT Findings

- Accurate at assessing presence of hernia, site, and size of abdominal wall defect and identifying sac contents
  - Useful for assessing larger deep-seated hernias

Ultrasonographic Findings

- Grayscale ultrasound
  - Identifies location/size of abdominal wall defect, herniated content (extraperitoneal fat, omentum, bowel), and complications
    - Omentum: Echogenic/hypoechoic tissue without peristalsis
    - Bowel: "Target" echo pattern with strong central echoes due to air in lumen and visible peristalsis
      - May see valvulae conniventes (small bowel) or feculent content (large bowel)
      - Variable appearances due to air-fluid content
  - Reducibility
    - Reduction in hernia size with ↓ in intraabdominal pressure or application of transducer pressure onto hernial sac
    - Irreducible (incarcerated) hernia may proceed to become obstructed or strangulated

- ▫ These complications are uncommon in noninguinal abdominal wall hernias due to wide neck and infrequent bowel content
- o Obstruction
  - – Nonobstructed hernia: Shows active peristalsis and movement of intestinal contents
  - – Obstructed hernia: Irreducible hernia with narrow neck and features of bowel obstruction
    - ▫ Not easy to diagnose with US; better appreciated with radiography or CT
- o Strangulation
  - – No flow on color Doppler imaging within bowel wall or mesentery, swollen bowel wall, aperistalsis, gas within or outside of bowel wall
    - ▫ Not reliable sign, as many nonstrangulated hernia do not show flow on color Doppler imaging
    - ▫ Other signs are also unreliable, making it difficult to assess strangulation on US

Imaging Recommendations

- Best imaging tool
  - o US is accurate and allows dynamic assessment
    - – CT or MR ± performed with Valsalva
- Protocol advice
  - o Identify anatomical layers, localize focal abdominal wall defect, compare with opposite side (if possible), and identify hernial contents
    - – Use light transducer pressure to avoid inhibition of herniation
    - – Valsalva maneuver is used routinely
    - – Examine in standing position if supine position negative or equivocal, particularly if standing clinically accentuates hernia
    - – Measure size of hernial sac as well as size of fascial defect
    - – If bowel content, check for complications (irreducibility/obstruction/strangulation)

DIFFERENTIAL DIAGNOSIS

Abdominal Wall Tumor

- Primary (lipoma, desmoid tumor, endometriosis) or metastatic (scar metastasis, umbilical metastasis)

Abdominal Wall Abscess or Seroma

- No fascial defect present

Rectus Sheath Hematoma

- Post traumatic or spontaneous: Bleeding of epigastric vessels or muscle tear
- No fascial defect present

Divarication of Rectus Abdominis Muscles (Diastasis Recti)

- Bulging of abdominal cavity due to stretching and thinning of linea alba with wide separation of rectus abdominis muscles

PATHOLOGY

General Features

- Etiology
  - o Chronic increased intraabdominal pressure, abdominal distension (ascites) ± muscle laxity (obesity, old age)
    - – Weak abdominal wall musculature
    - – Physical exertion, chronic cough, prostatism, or constipation
  - o Trauma: Blunt force or hyperextension strain
    - – Insufficient to penetrate skin but strong enough to disrupt muscle and fascia
    - – Sudden ↑ in intraabdominal pressure
  - o Postoperative abdominal wall weakness, surgical scar, suture dehiscence
  - o Epigastric hernia may be related to anomalous slips from diaphragm to linea alba region
  - o Umbilical hernia related to periumbilical fascial defect

CLINICAL ISSUES

Presentation

- Most common signs/symptoms
  - o Abdominal lump increasing in size with ↑ in intraabdominal pressure
  - o Reducible swelling with positive cough impulse
  - o Intermittent intestinal obstruction

Demographics

- Age
  - o Umbilical hernia: Young children; paraumbilical hernia: Adults
  - o Epigastric hernia: 20-50 years
  - o Incisional hernia: More frequent in elderly
- Gender
  - o Epigastric hernia is 2x more common in males
- Epidemiology
  - o Most common abdominal wall lesion in US practice

Natural History & Prognosis

- ~ 1/3 of umbilical hernias close within 1 month of birth and rarely persist beyond 3-4 years
- All other hernias persist and frequently enlarge with time

Treatment

- Repair of muscle/fascial defect: Open or laparoscopic technique, meshplasty
- Intestinal obstruction/strangulated hernia; urgent exploratory laparotomy

DIAGNOSTIC CHECKLIST

Consider

- Abdominal wall hernia if posterior margin of any abdominal wall mass cannot be seen on US

Image Interpretation Pearls

- Check for abdominal wall defect, hernial sac contents, peristaltic movement, and vascularity (if bowel)

SELECTED REFERENCES

1. Arend CF: Static and dynamic sonography for diagnosis of abdominal wall hernias. J Ultrasound Med. 32(7):1251-9, 2013
2. Lee RK et al: Ultrasound of the abdominal wall and groin. Can Assoc Radiol J. 64(4):295-305, 2013

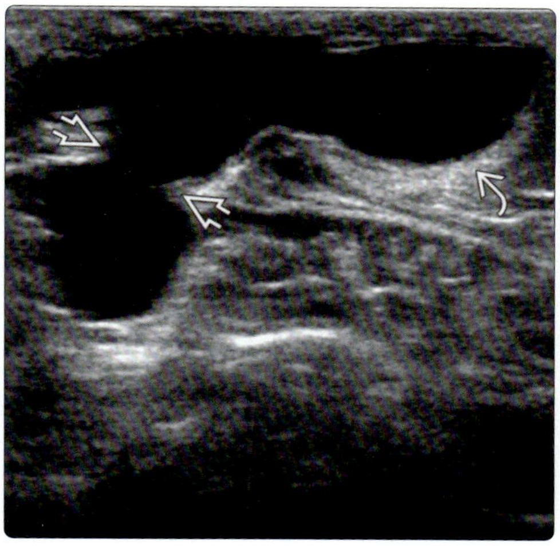

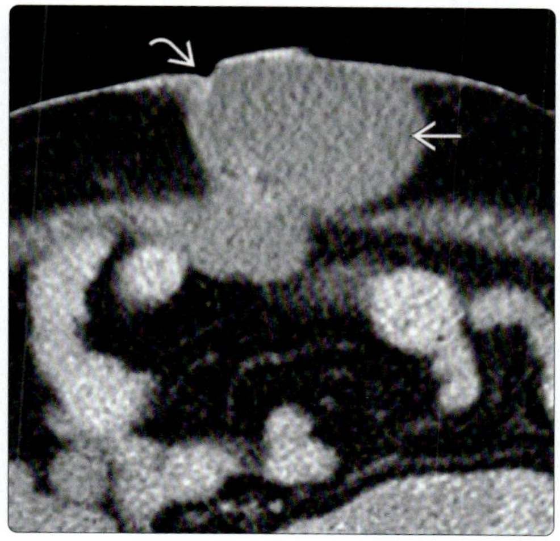

(Left) *Transverse US shows a large umbilical hernia protruding through a defect in the linea alba at the umbilicus. The sac contains peritoneal fluid. This seemed to be encysted, as it was not readily reducible. No bowel content is present.* **(Right)** *Axial NECT of the same patient shows a large paraumbilical hernia. The sac content is made up entirely of encysted peritoneal fluid. No bowel content is present. The hernia bulges to the side of the umbilicus.*

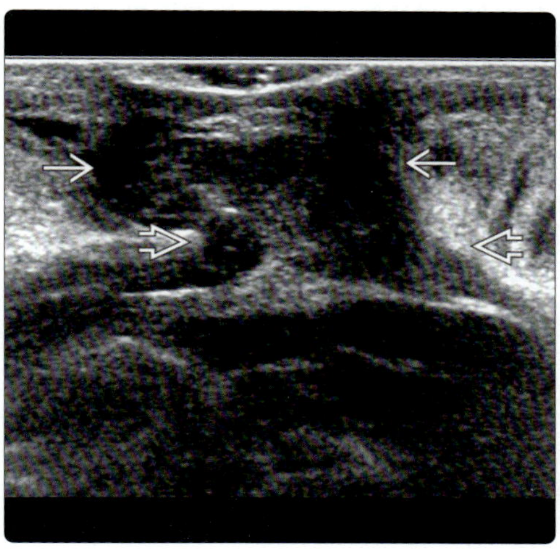

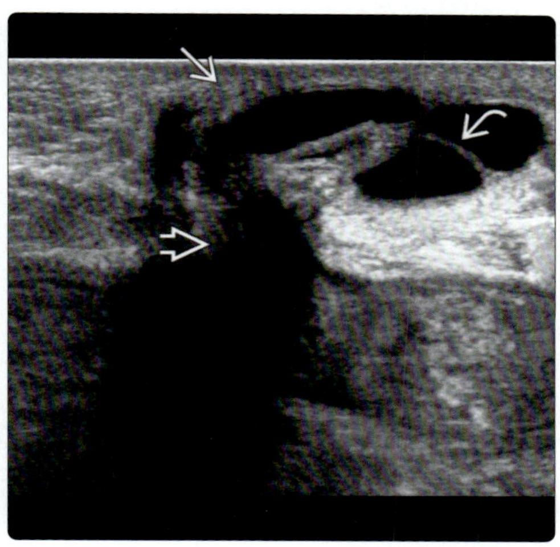

(Left) *Transverse US shows a medium-sized umbilical hernia. The hernial sac is filled with omental fat. The defect in the linea alba at the umbilicus is wide.* **(Right)** *Transverse US shows a medium-sized umbilical hernia. The hernial sac contains strands of omentum as well as encysted peritoneal fluid. No bowel content is present. The hernial neck often shows acoustic shadowing despite the absence of bowel content due to the compactness of tissues in this area.*

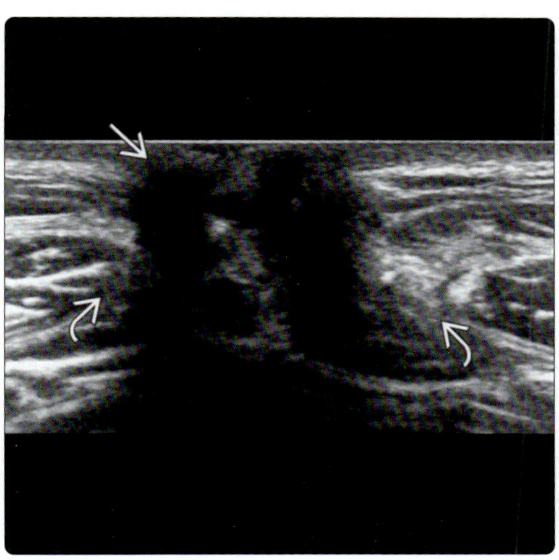

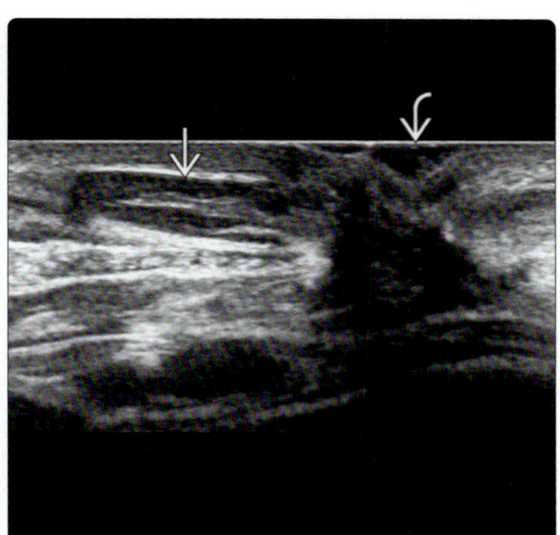

(Left) *Transverse US shows a small umbilical hernia extending though a wide defect in the linea alba.* **(Right)** *Longitudinal US of the same patient shows that the umbilical hernia contains bowel and extends cephalad from the umbilicus within subcutaneous fat. Umbilical hernias in adults are more common in overweight subjects and following pregnancy.*

(Left) *Schematic diagram shows epigastric ⇨ and hypogastric ↩ linea alba hernias arising from the midline. The locations of umbilical → and spigelian ➡ hernias are also shown for reference.* **(Right)** *Transverse US shows a small epigastric hernia ➡ in the supraumbilical region (also called linea alba hernia). This contains hypoechoic extraperitoneal fat. The defect ➡ in the linea alba is clearly seen. Fat may appear as either hypoechoic or hyperechoic on US.*

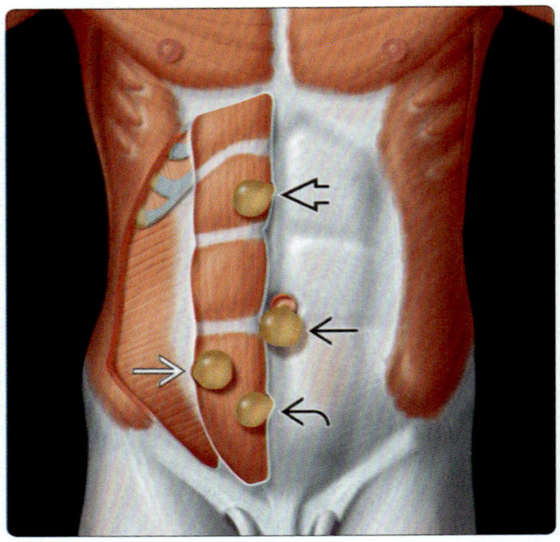

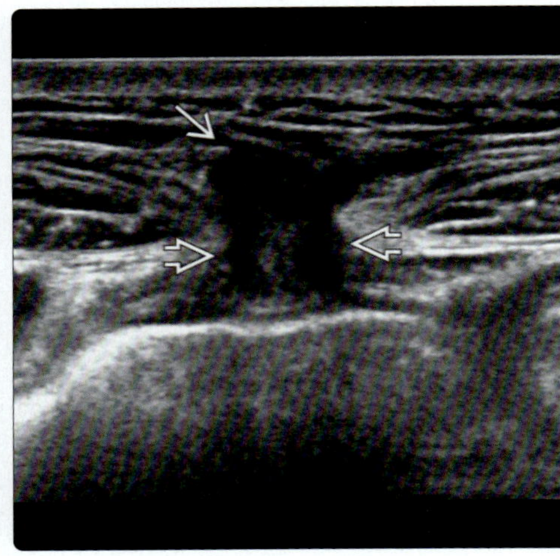

(Left) *Transverse US shows a small epigastric hernia ➡ containing hypoechoic extraperitoneal fat.* **(Right)** *Axial CECT of the same patient obtained several weeks after US examination shows how the wall of the epigastric hernia ➡ is very thin. The hernial sac is larger than on US. This may be due to interim increase in hernial sac size or due to the effect of transducer pressure on sac size.*

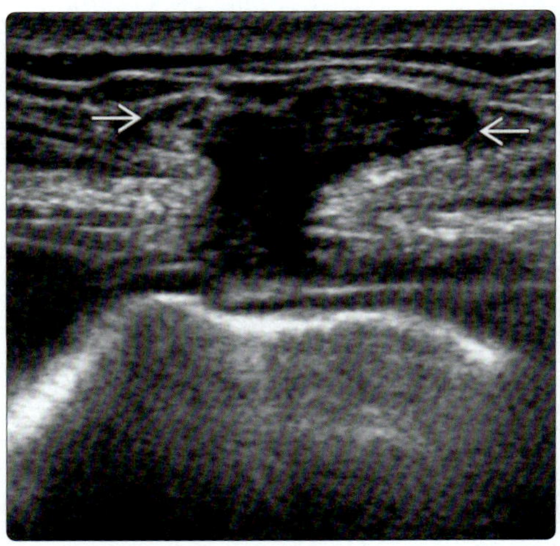

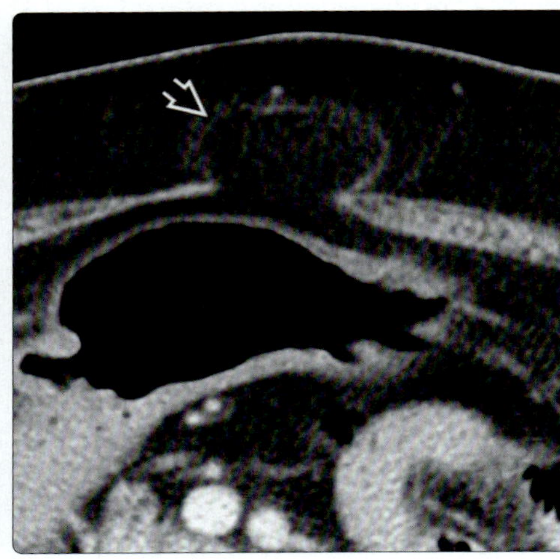

(Left) *Transverse US shows an unusually large epigastric hernia ➡ containing hyperechoic mesenteric fat and a small amount of peritoneal fluid ➡. No bowel is present.* **(Right)** *Transverse US shows another unusually large epigastric hernia ➡ containing mesenteric fat. There is acoustic shadowing present at the hernial neck ➡. This is a common feature of hernias and does not imply bowel content in the hernial sac.*

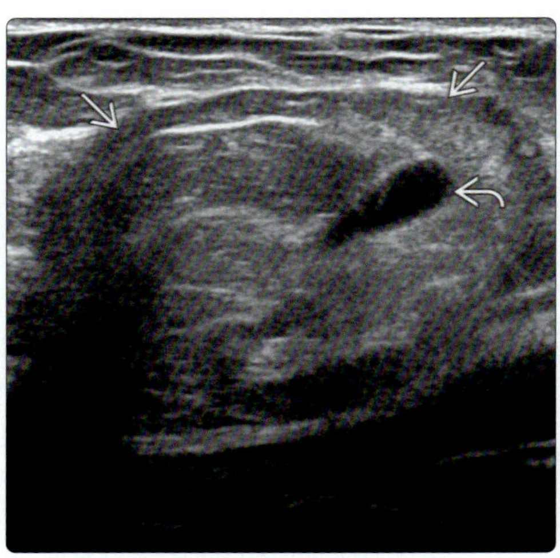

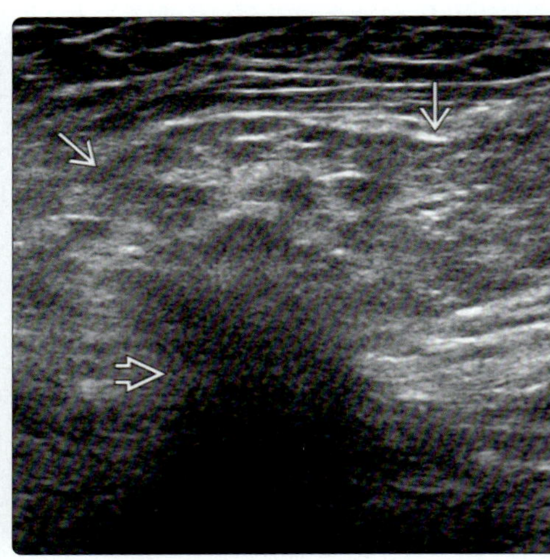

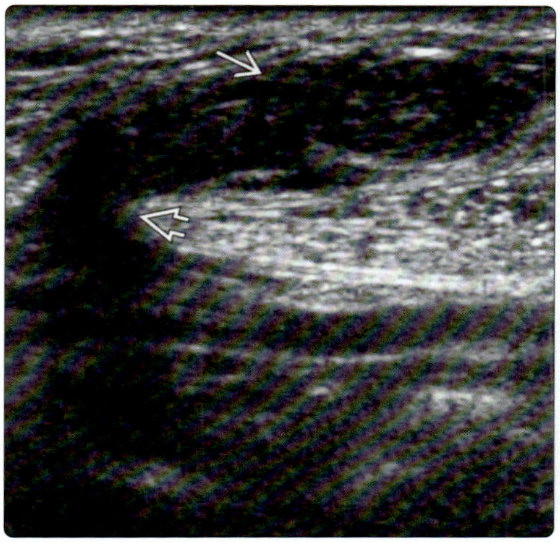

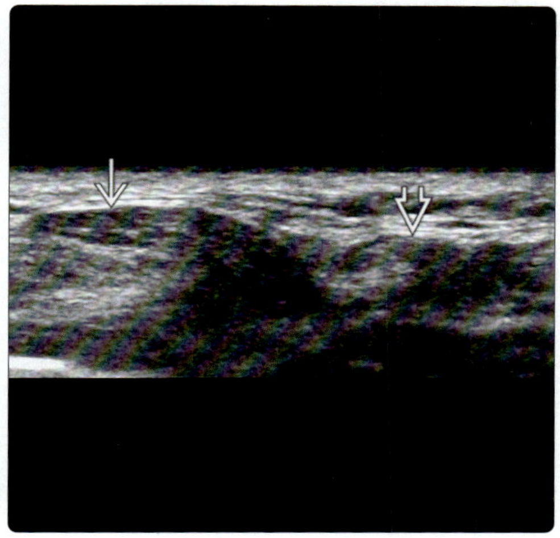

(Left) *Oblique US shows a medium-sized epigastric hernia extending through a small defect in the supraumbilical linea alba. The hernia contains hypoechoic extraperitoneal fat and extends caudally within the subcutaneous tissues.* **(Right)** *Longitudinal US shows 2 separate epigastric hernias. The superior hernia is larger in size than the inferior hernia, and both contain fat. It is uncommon for 2 epigastric hernias to occur in the same patient.*

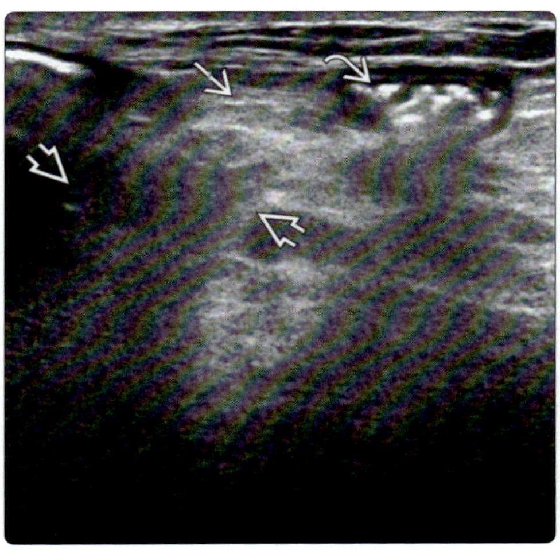

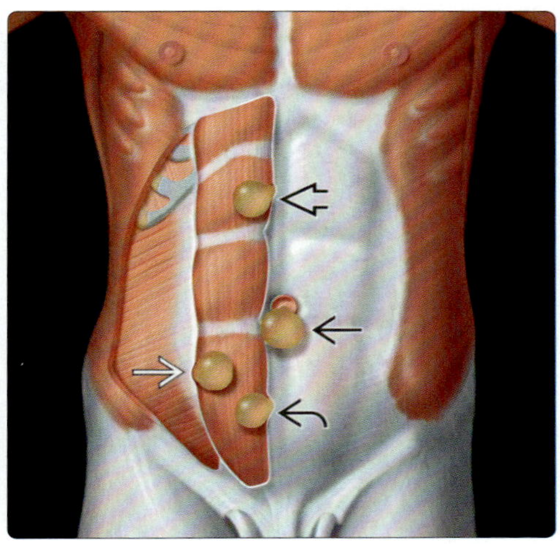

(Left) *Longitudinal US shows a large hypogastric hernia at the infraumbilical region. The sac contains both bowel and mesenteric fat.* **(Right)** *Schematic diagram shows a spigelian hernia extending through the linea semilunaris. The locations of epigastric, umbilical, and hypogastric hernias are also shown for reference.*

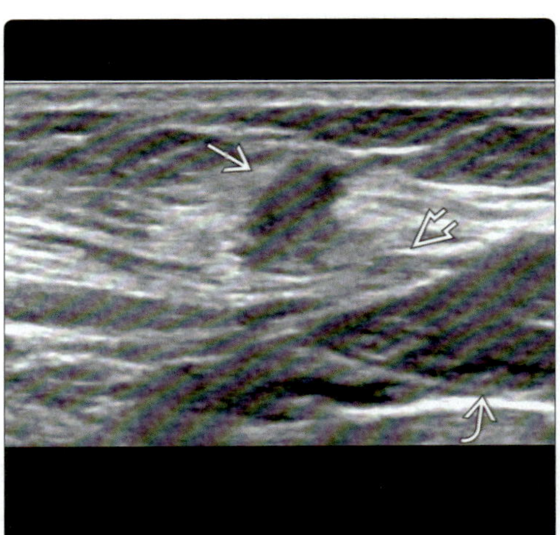

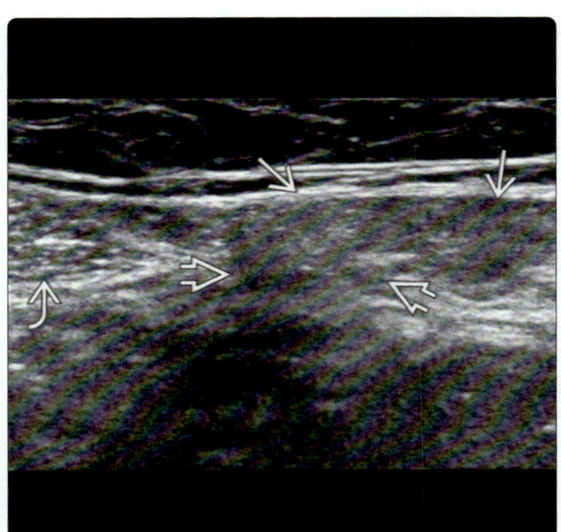

(Left) *Transverse US shows a small spigelian hernia of lower abdominal region extending into the subcutaneous tissues through a small defect at the lateral edge of the rectus abdominis muscle.* **(Right)** *Transverse US shows a medium-sized spigelian hernia extending through a defect in transversus abdominis aponeurosis but staying deep to internal and external oblique aponeuroses. The sac contains mesenteric fat but no bowel. The rectus abdominis muscle is shown.*

(Left) *Schematic diagram shows a superior lumbar hernia (Grynfeltt hernia) arising from the superior lumbar triangle bounded by the erector spinae muscle medially, the 12th rib superiorly, and the internal oblique muscle laterally.* **(Right)** *Schematic diagram shows an inferior lumbar hernia (Petit hernia) arising from the inferior lumbar triangle bounded by the latissimus dorsi muscle medially, the iliac crest inferiorly, and the external oblique muscle laterally.*

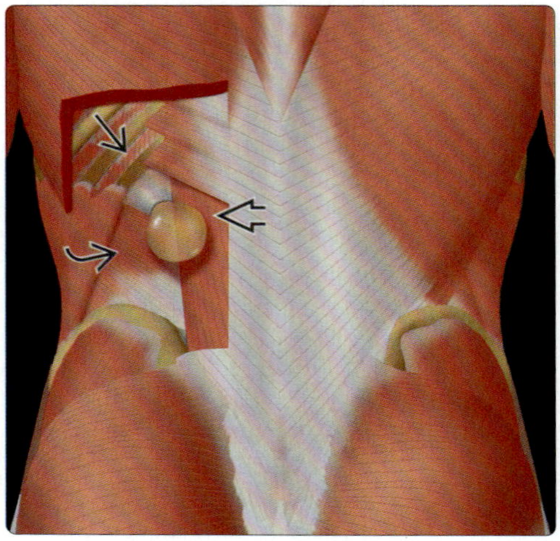

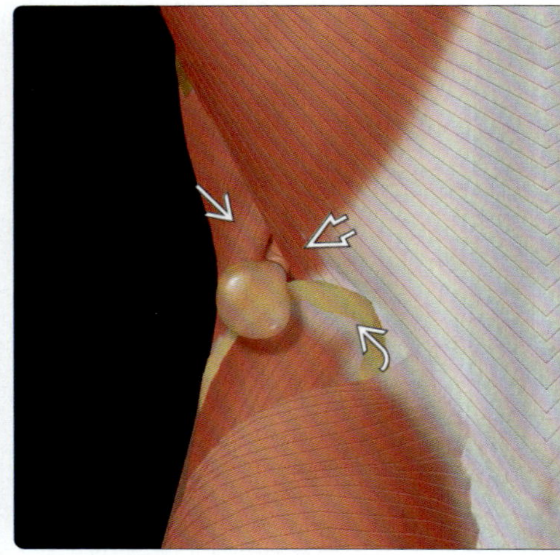

(Left) *Transverse US shows a large lumbar hernia extending through a defect in the superior lumbar triangle. The sac contains both bowel and mesenteric fat.* **(Right)** *Axial NECT of the same patient shows the large lumbar hernia extending through the superior lumbar triangle and containing bowel and intraperitoneal fat. The 12th rib is shown. Lumbar hernias may contain intra- or extraperitoneal content and, therefore, benefit from CT examination.*

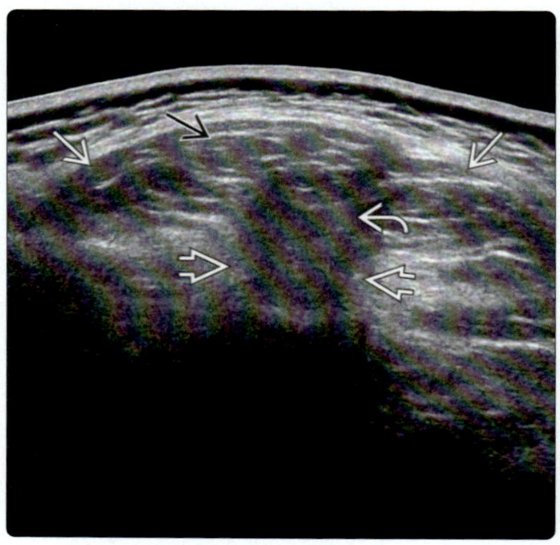

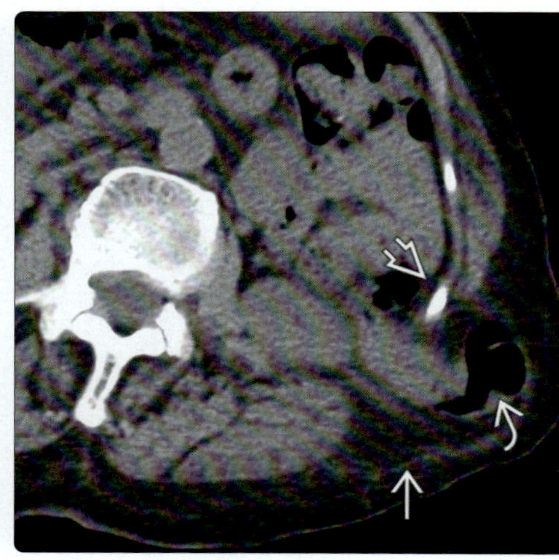

(Left) *Oblique US shows a large lumbar hernia extending through a wide defect in the superior lumbar triangle. The hernia contains hypoechoic fat, but it is not clear on US whether this fat is intraperitoneal or extraperitoneal in origin.* **(Right)** *Axial NECT of the same patient shows the large lumbar hernia containing retroperitoneal fat. The colon, spleen, liver, and kidney may all herniate through either superior or inferior lumbar triangle defects.*

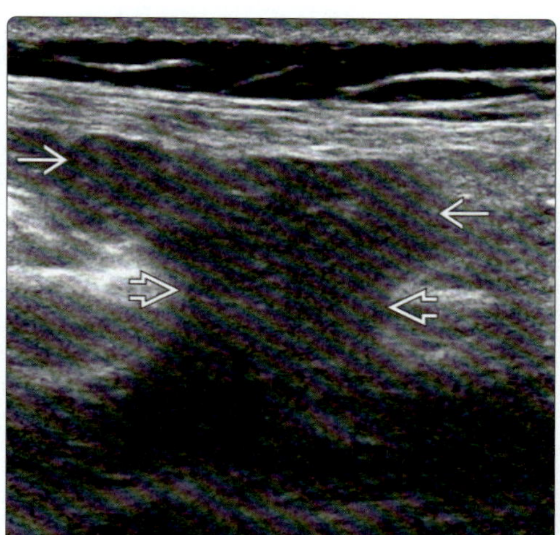

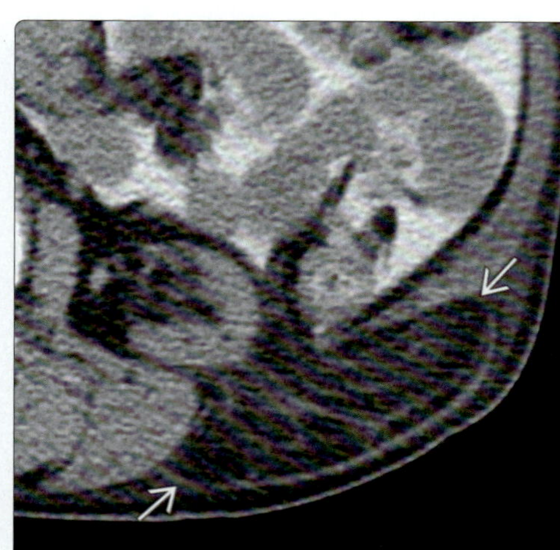

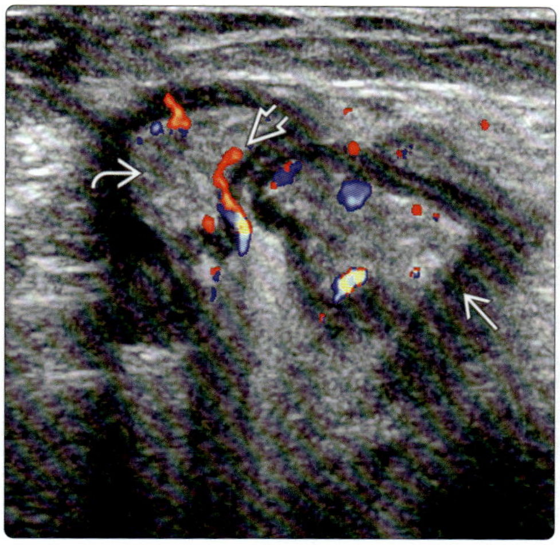

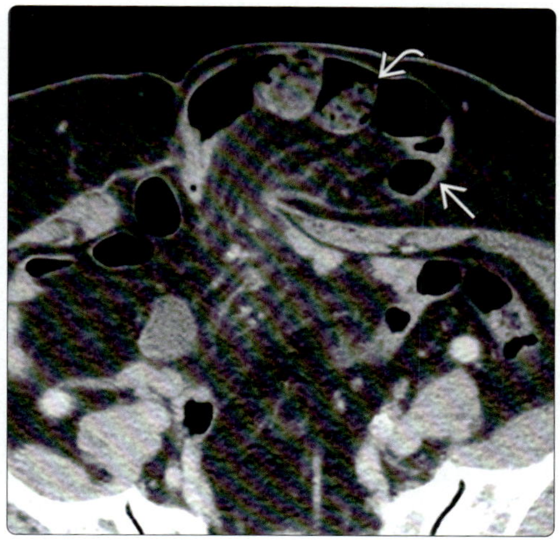

(Left) *Transverse color Doppler US shows a large incisional hernia ➡ extending through a midline abdominal incision. The sac contains mainly bowel ➡. This bowel is well vascularized ➡, but the absence of vascularity on color Doppler imaging does not necessarily imply ischemia.* **(Right)** *Axial NECT of the same patient shows the large incisional hernia ➡ extending through a defect in the anterior abdominal wall. The bowel ➡ within the hernia sac is normal in appearance.*

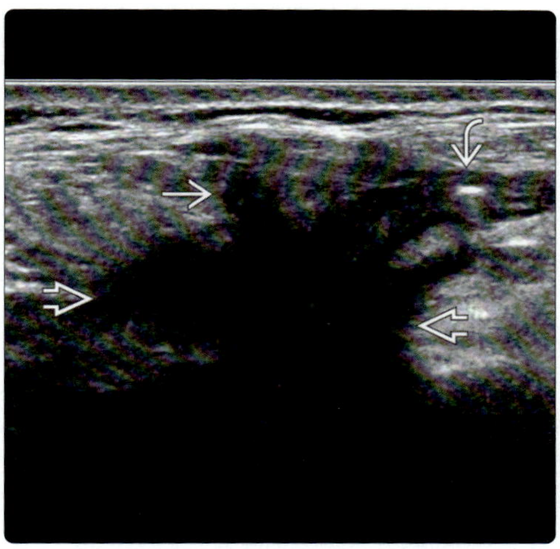

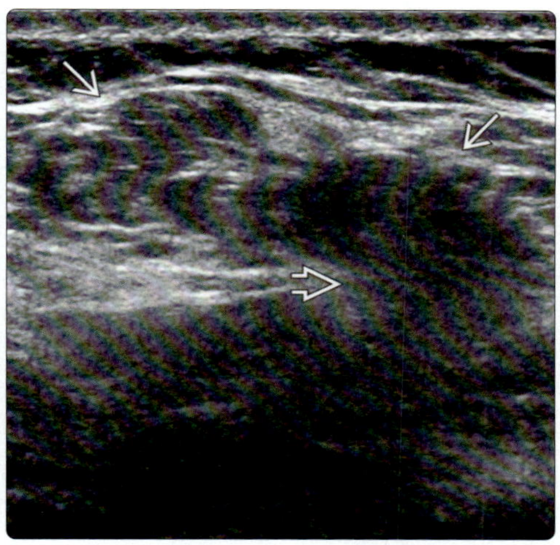

(Left) *Transverse US shows a medium-sized incisional hernia ➡ extending through a defect ➡ at the parastomal site. The sac contains a small amount of gas ➡, which is indicative of some bowel content.* **(Right)** *Transverse US shows a large incisional hernia ➡ extending through a previous surgical incision ➡ in the anterior abdominal wall. Incisional hernias are often large at presentation and tend to enlarge even further with time.*

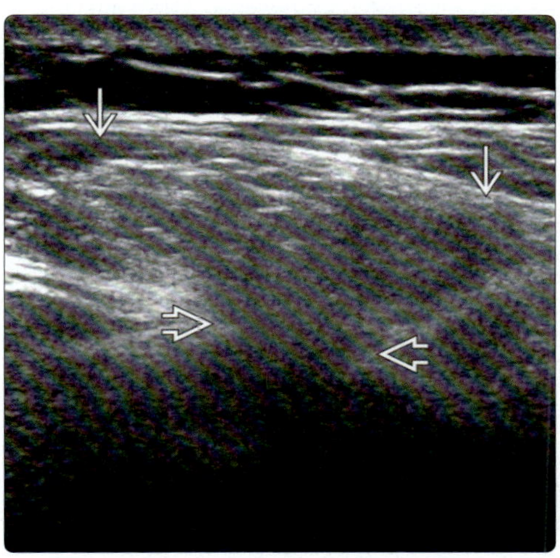

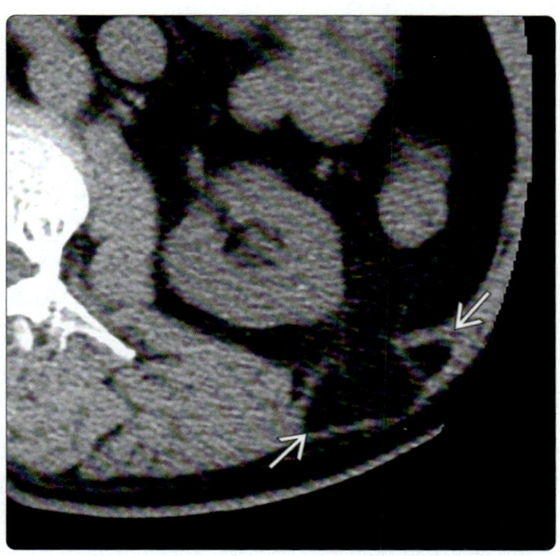

(Left) *Transverse US shows a large incisional hernia ➡ at the site of prior percutaneous nephrostomy ➡. The sac contains the retroperitoneal fat.* **(Right)** *Axial NECT of the same patient shows the large incisional hernia ➡. The sac contains retroperitoneal fat. The site of the hernia is similar to the site of the superior lumbar hernia; however, in this situation, the hernia followed a percutaneous procedure at this site and is therefore deemed an incisional hernia.*

Groin Hernia

KEY FACTS

TERMINOLOGY

- Indirect inguinal hernia, direct inguinal hernia, femoral hernia
- Hernia: Abnormal protrusion of organ or part of organ out of cavity in which it is normally contained

IMAGING

- **Indirect inguinal hernia**: Passes through deep inguinal ring, extends along inguinal canal, and emerges at superficial inguinal ring
- **Direct inguinal hernia**: Passes through transversalis fascia in Hesselbach triangle
- **Femoral hernia**: Passes through femoral canal
- Identify inguinal ligament, inferior epigastric artery, and spermatic cord
- Trace spermatic cord in transverse plane retrogradely from high scrotal area to deep inguinal ring
- Specifically examine deep inguinal ring, Hesselbach triangle, and femoral canal areas
- Examine each area at rest and during Valsalva
    - For Valsalva, ask patient to "blow into back of hand"
- Determine presence of hernia ± laxity of fascia
- Also assess relaxation phase of Valsalva
- ± reexamine with patient standing erect
- US: 97% sensitivity, 86% specificity, and 93% positive predictive value for diagnosis of groin hernia
    - Pure fatty hernia is most difficult to diagnose
    - Slightly lower accuracy for determining specific hernia type
- CT or MR: Useful if US is equivocal

DIAGNOSTIC CHECKLIST

- Indirect inguinal hernia neck: Above inguinal ligament and lateral to inferior epigastric artery
- Direct inguinal hernia neck: Above inguinal ligament and medial to inferior epigastric artery
- Femoral hernia: Below inguinal ligament
- Routinely examine all 3 sites

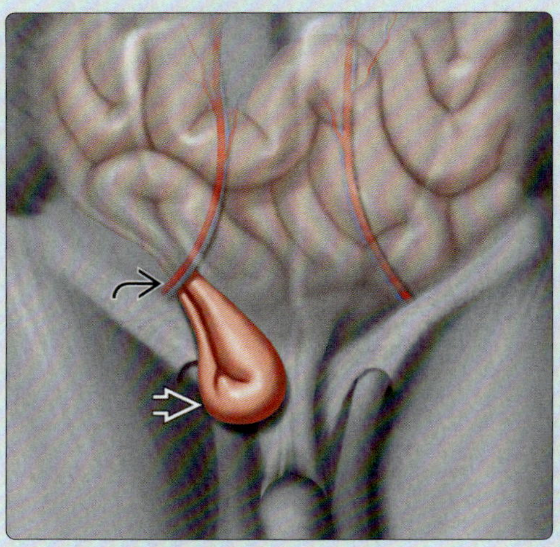

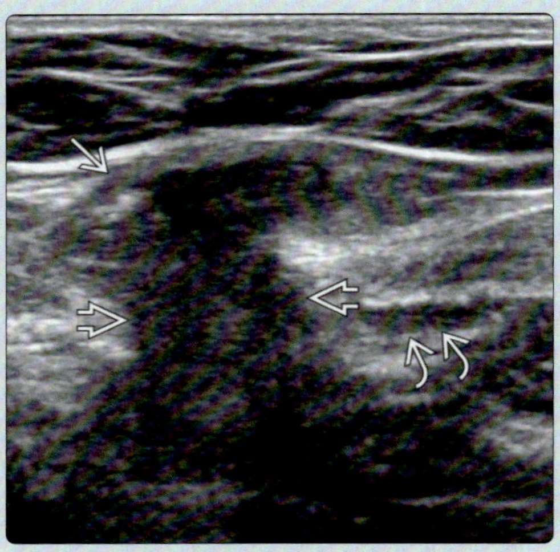

(Left) *Graphic shows a direct inguinal hernia. Note that the neck of a direct inguinal hernia is medial to the inferior epigastric vessels. A direct inguinal hernia passes through the transversalis fascial defect in the Hesselbach triangle.* **(Right)** *Transverse US shows a left direct-type inguinal hernia containing hypoechoic mesenteric fat. The neck of the hernia lies medial to the inferior epigastric vessels.*

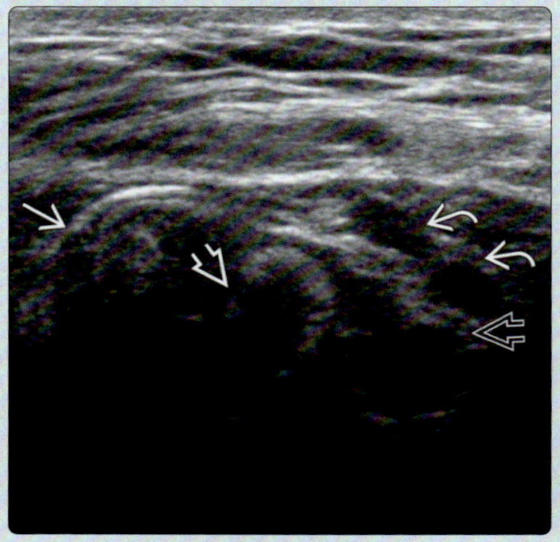

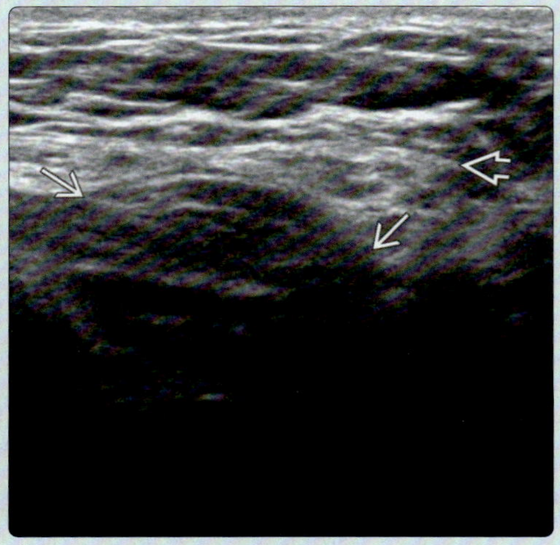

(Left) *Transverse US shows a left direct-type inguinal hernia containing bowel. The neck of the hernia is wide and lies medial to the inferior epigastric vessels. Obstruction is rare in direct inguinal hernias, as they usually have a wide neck.* **(Right)** *Longitudinal US of the same hernia shows that it is a direct-type inguinal hernia that arises cephalad to the inguinal ligament. The inguinal ligament is difficult to appreciate on still images but is more readily appreciable in real time.*

TERMINOLOGY

Synonyms

- Inguinal hernia, indirect inguinal hernia, direct inguinal hernia, femoral hernia

Definitions

- Abnormal protrusion of organ or part of organ out of cavity in which it is normally contained

IMAGING

General Features

- Location
  - **Inguinal canal**
    - ~ 4.0 cm long, ~ 1.2 cm above inguinal ligament
      - Contains spermatic cord and ilioinguinal nerve in males, round ligament of uterus and ilioinguinal nerve in females
    - Floor: Inguinal ligament formed from thickened, free, lowermost border of external oblique aponeurosis
      - Runs from anterior superior iliac spine to pubic tubercle
    - Anterior wall: External oblique fascia
      - Triangular defect medially forms superficial inguinal ring
    - Roof: Internal oblique and transversus abdominus fascia
      - Attached to lateral part of inguinal ligament
      - Medially arch over superficial inguinal ring as conjoint tendon
    - Posterior wall: Transversalis fascia
      - Contains rounded defect midway between anterior superior iliac spine and pubic tubercle, known as deep inguinal ring
    - Inferior epigastric artery from external iliac artery runs just medial to deep inguinal ring on way to rectus sheath
    - Hesselbach triangle located between inferior epigastric artery laterally, rectus sheath medially, and inguinal ligament inferiorly
  - **Femoral canal**
    - ~ 2 cm long, just medial to femoral vein, deep and distal to inguinal ligament
    - Anterosuperior border: Inguinal ligament
    - Medial border: Lacunar ligament (reflection of inguinal ligament to pectineal line)
    - Inferior border: Pectineal ligament (reflection of inguinal ligament over superior pubic ramus)
    - Lateral border: Femoral vein
  - **Indirect inguinal hernia**
    - Passes through deep inguinal ring, extends along inguinal canal, and emerges at superficial inguinal ring
    - Neck is lateral to inferior epigastric artery and above inguinal ligament
    - In males, usually extends along spermatic cord to scrotum
    - In females, may extend along round ligament to labia majora
  - **Direct inguinal hernia**
    - Passes through transversalis fascial defect in Hesselbach triangle
    - Wide neck medial to inferior epigastric artery and above inguinal ligament
    - Does not pass into spermatic cord and generally does not extend to scrotum
  - **Femoral hernia**
    - Passes through femoral canal
    - Narrow neck medial to femoral vein and below inguinal ligament
  - **Double hernia** (a.k.a. saddlebag hernia)
    - Simultaneous occurrence of direct and indirect inguinal hernia in same groin

Radiographic Findings

- Radiography
  - ± small or large bowel obstruction
  - Herniography: Injection of soluble, low osmolar contrast medium into peritoneal cavity

Ultrasonographic Findings

- Grayscale ultrasound
  - Accurate at identifying presence and site of hernia as well as contents of hernial sac (bowel or omentum)
    - Omental fat: Echogenic tissue without peristalsis
    - Bowel: Intestinal loops; target echo pattern with strong central echoes representing air or fluid in lumen ± peristalsis
      - Nonobstructed hernia: Active peristalsis ± movement of intestinal contents
      - Obstructed hernia: No movement of intestinal content in hernia and dilated bowel loops in abdomen, may have peristalsis
  - Reducible hernia
    - Decrease in hernia size with decrease in intraabdominal pressure or application of external pressure to hernial sac with transducer
    - Increase in hernia size during coughing, Valsalva maneuver, or standing
  - Color Doppler
    - Useful to identify inferior epigastric artery
    - Strangulated hernia; absence of vascularity within bowel wall and mesentery but not reliable sign

Imaging Recommendations

- Best imaging tool
  - US is accurate provided that strict systematic approach is used
    - 97% sensitivity, 86% specificity, and 93% positive predictive value for diagnosis of groin hernia
    - Slightly lower accuracy for determining type of hernia (indirect inguinal, direct inguinal, femoral)
  - CT or MR is useful if US is equivocal
    - Use low-dose CT technique if specifically looking for hernia
- Protocol advice
  - 1st identify inferior epigastric artery, spermatic cord (or round ligament), and inguinal ligament
    - Inferior epigastric vessels identified at lateral margin of rectus femoris muscle and traced distally
  - Trace spermatic cord in transverse plane retrogradely from high scrotal area to deep inguinal ring
  - Specifically examine deep inguinal ring, Hesselbach triangle area, and femoral canal

- At each area, examine at rest, and ask patient to perform Valsalva maneuver
  - For Valsalva, ask patient to "blow into back of hand"
  - Look for presence of hernia or laxity (bulging) of fascia (on Valsalva maneuver)
  - Also assess relaxation phase of Valsalva
  - Reexamine specific areas with patient standing if standing accentuates symptoms and supine examination negative
- Routinely examine all 3 hernia areas
  - 2 or more hernias can coexist

DIFFERENTIAL DIAGNOSIS

Lipoma of Spermatic Cord

- No change with Valsalva maneuver, not reducible
- Usually no deep extension into peritoneal cavity

Encysted Hydrocele Canal of Nuck

- Embryologically, processus vaginalis extends from peritoneal cavity to scrotum
  - Normally obliterates in cord, while scrotal component persists as tunica vaginalis
  - Failure of obliteration leads to fluid distension within cord, known as encysted hydrocele canal of Nuck
    - Filled with anechoic fluid, no change with Valsalva maneuver, not reducible
    - No deep extension into peritoneal cavity

PATHOLOGY

General Features

- Etiology
  - Chronic increased intraabdominal pressure, abdominal distension (ascites) ± muscle laxity
    - Weak abdominal musculature, chronic cough, prostatism, constipation, manual labor
  - Indirect inguinal hernia: Congenital due to persistence of processus vaginalis
  - Direct inguinal hernia: Acquired due to weakness of transversalis fascia in Hesselbach triangle
  - Femoral hernia: Weakness of femoral canal

Gross Pathologic & Surgical Features

- Sac contents: Commonly omentum, less commonly small bowel, or uncommonly mobile colon segments (sigmoid, cecum, appendix)
  - Incarcerated hernia: Irreducible hernias
  - Strangulated hernia: Ischemic → infarcted hernia
  - Sliding hernia: Sac contains retroperitoneal organs
    - Urinary bladder, distal ureters, or ascending/descending colon
  - Littre hernia: Meckel diverticulum in sac
  - Richter hernia: Only portion of bowel circumference (antimesenteric portion) in sac

CLINICAL ISSUES

Presentation

- Most common signs/symptoms
  - Groin lump or discomfort (continuous or intermittent)
- Other signs/symptoms
  - Features of intestinal obstruction

Demographics

- Age
  - Indirect inguinal hernia tend to occur in young to middle-aged individuals
  - Prevalence of direct inguinal hernia increases with increasing age
  - Femoral hernia is more common in middle-aged to elderly individuals
- Gender
  - Indirect inguinal hernia is 5-10x more common in males
  - Direct inguinal hernia nearly always occurs in males
  - Femoral hernia is more common in females
    - However, indirect inguinal hernia is most common hernia in females
- Epidemiology
  - 5% of males develop groin hernia
  - 75% of all hernias occur in inguinal region
  - Indirect inguinal hernias are 5x more common than direct inguinal hernias
    - Femoral hernia is much less common
  - Inguinal hernia in children is always result of patent processus vaginalis and is thus indirect hernia extending to scrotal sac
    - Affects 2% of all children
    - Most often seen in premature infants, particularly on right side (75%), bilateral (10%)

Natural History & Prognosis

- Hernia recur after hernia repair in 20% of patients
- Complications: Obstruction, strangulation
  - Indirect inguinal hernia account for 15% of intestinal obstruction cases
  - Obstruction or strangulation is rare with direct inguinal hernia due to wide neck
  - Obstruction or strangulation is more common with femoral hernia due to narrow neck

Treatment

- Laparoscopic or open hernia repair

DIAGNOSTIC CHECKLIST

Consider

- Indirect inguinal hernia neck: Above inguinal ligament and lateral to inferior epigastric artery
- Direct inguinal hernia neck: Above inguinal ligament and medial to inferior epigastric artery
- Femoral hernia: Below inguinal ligament

SELECTED REFERENCES

1. Lee RK et al: High accuracy of ultrasound in diagnosing the presence and type of groin hernia. J Clin Ultrasound. 43(9):538-47, 2015
2. Lee RK et al: Ultrasound of the abdominal wall and groin. Can Assoc Radiol J. 64(4):295-305, 2013
3. Robinson A et al: Meta-analysis of sonography in the diagnosis of inguinal hernias. J Ultrasound Med. 32(2):339-46, 2013

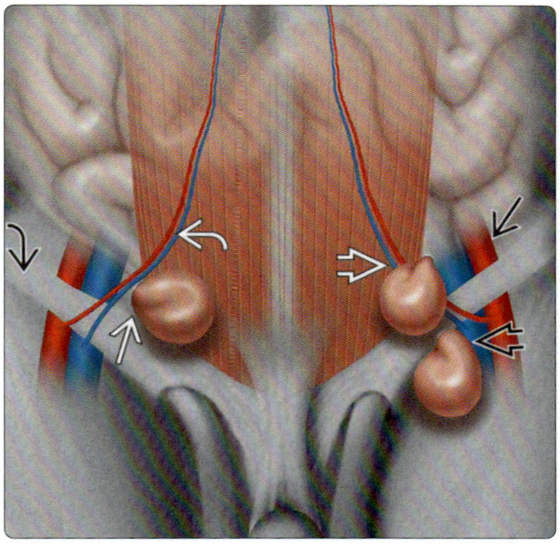

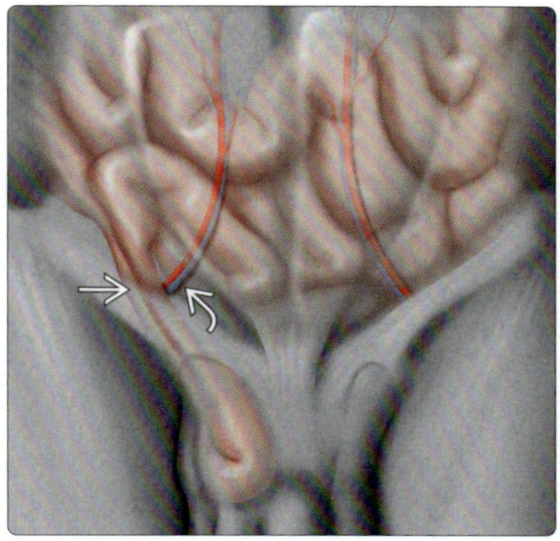

(Left) *Graphic shows 3 types of groin hernia. Direct and indirect inguinal hernias arise above the inguinal ligament and medial and lateral to the inferior epigastric vessels, respectively. Femoral hernias arise below the inguinal ligament and medial to the femoral vessels.* **(Right)** *Graphic shows an indirect inguinal hernia entering the right scrotal sac. Note that the neck of an indirect inguinal hernia lies lateral to the inferior epigastric vessels.*

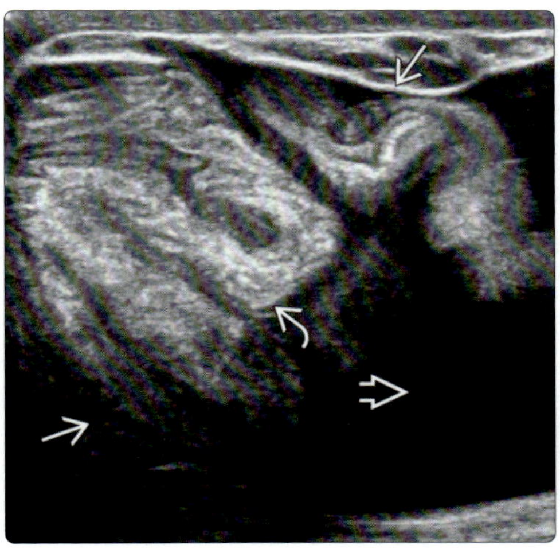

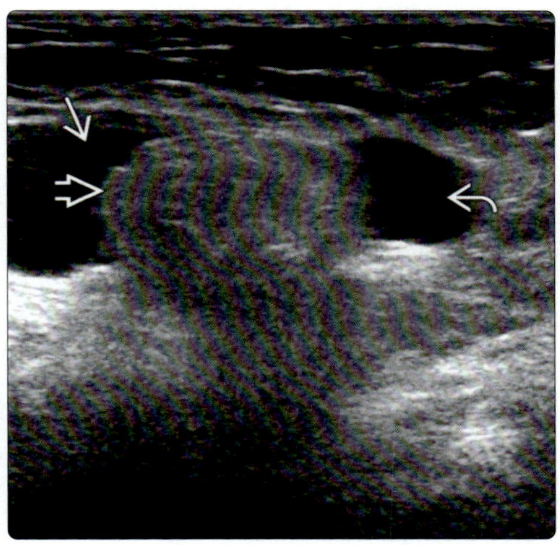

(Left) *Transverse US shows a large indirect-type inguinal hernia extending into the upper scrotal region. There is fluid and bowel within the hernia sac.* **(Right)** *Transverse US shows a small indirect-type inguinal hernia. The sac contains fluid and mesenteric fat.*

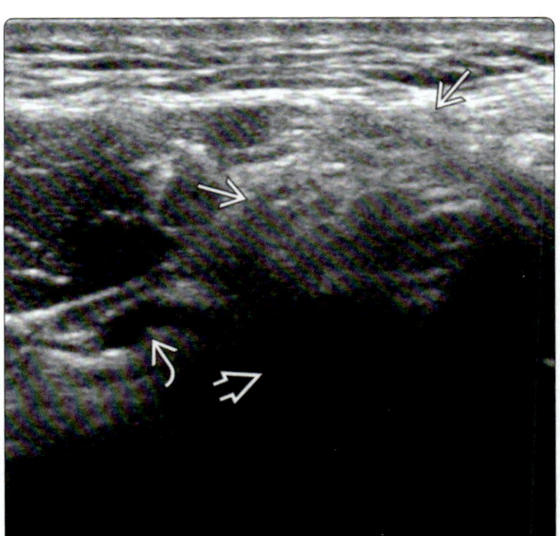

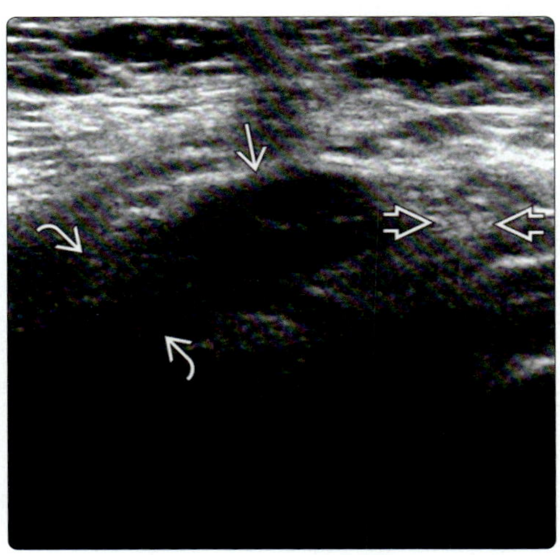

(Left) *Transverse US shows a left indirect-type inguinal hernia containing mesenteric fat and no bowel. The neck of the hernia lies lateral to the inferior epigastric vessels. Place the transducer over the deep inguinal ring before Valsalva maneuver. To-and-fro movement of the hernial sac can usually be observed on real-time imaging.* **(Right)** *Longitudinal US of the same hernia shows that the hernia neck lies cephalad to the inguinal ligament.*

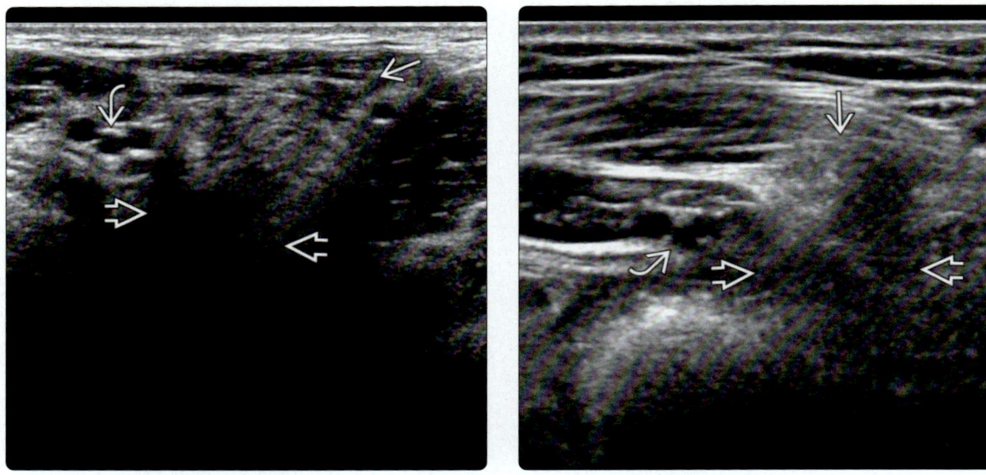

(Left) *Transverse US shows a left indirect-type inguinal hernia* ➡ *containing mesenteric fat. The neck* ➡ *of the hernia lies lateral to the epigastric veins and artery* ➡*.* **(Right)** *Transverse US shows a left indirect-type inguinal hernia* ➡ *containing mesenteric fat. The wide neck* ➡ *of the hernia lies lateral to the epigastric veins and artery* ➡*. Routine identification of pertinent anatomical structures helps determine the type of hernia present.*

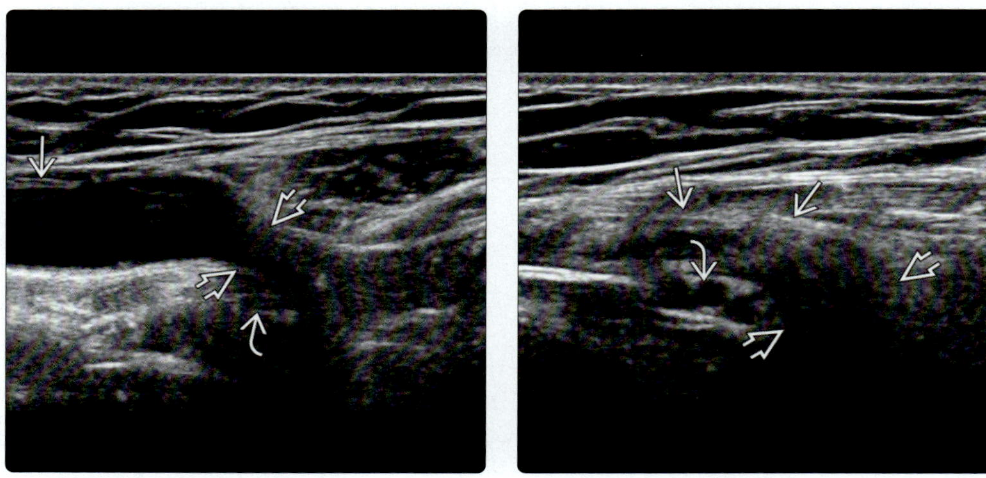

(Left) *Transverse US shows a left indirect-type inguinal hernia* ➡ *containing peritoneal fluid. The narrow hernia neck* ➡ *lies lateral to the epigastric vessels* ➡*, which are not clearly seen.* **(Right)** *Transverse US shows a left indirect-type inguinal hernia* ➡ *containing mesenteric fat. The hernia neck* ➡ *lies lateral to the epigastric veins and artery* ➡*. Indirect inguinal hernias are due to a persistence of the processus vaginalis.*

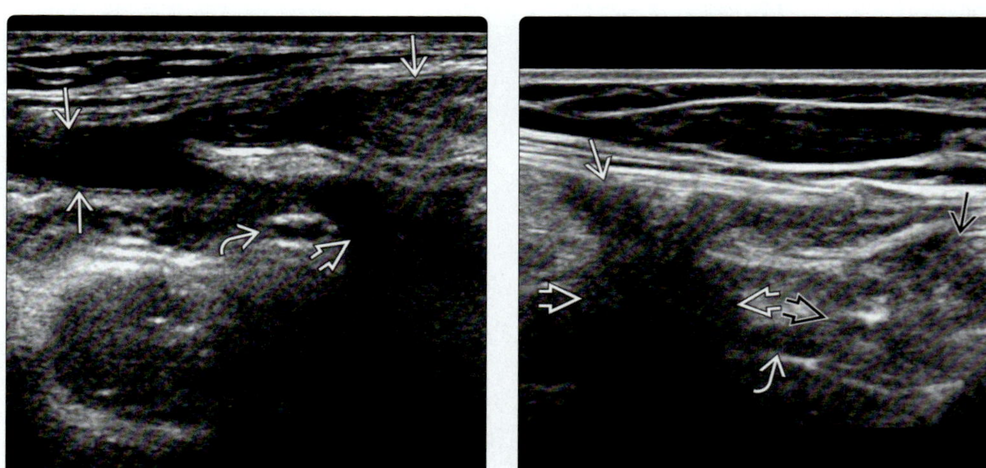

(Left) *Transverse US shows a left indirect-type inguinal hernia* ➡ *containing both fluid and mesenteric fat. The neck* ➡ *of the hernia lies lateral to the epigastric vessels* ➡*.* **(Right)** *Transverse US shows concurrent left direct* ➡ *and indirect* ➡ *inguinal hernias, both containing mesenteric fat. The necks of the indirect* ➡ *and direct* ➡ *hernias lie lateral and medial to the epigastric vessels* ➡*, respectively. This is known as a saddlebag hernia.*

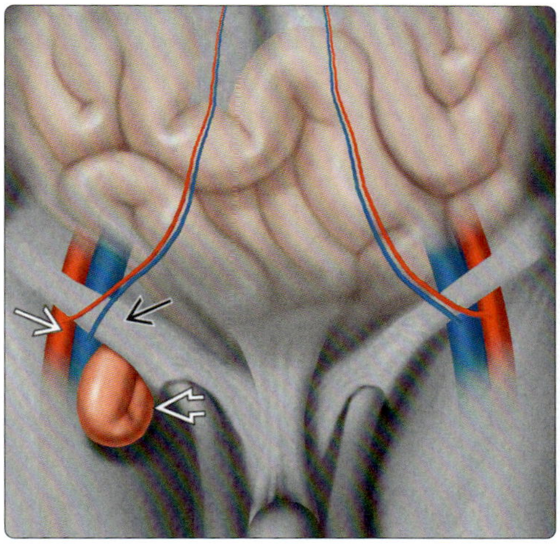

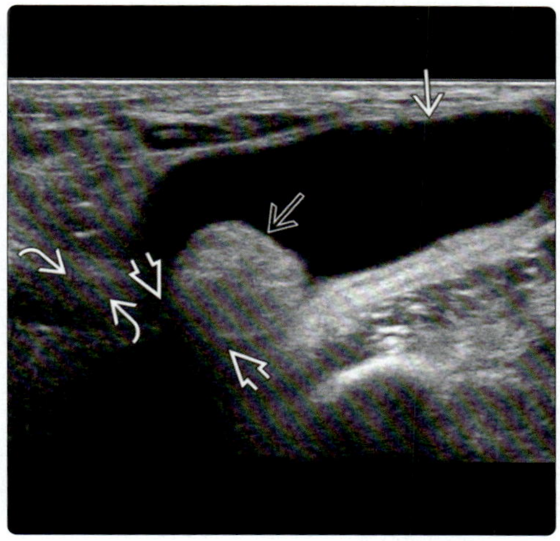

(Left) *Graphic shows a femoral hernia →. Note that the neck of a femoral hernia is medial to the femoral vein → and inferior to the inguinal ligament →.* **(Right)** *Longitudinal US of a femoral hernia → containing fluid and mesenteric fat → shows that the neck → of the hernia lies distal to the inguinal ligament →. The relationship to the inguinal ligament helps mainly to discriminate between direct inguinal and femoral hernias, which are otherwise quite similar.*

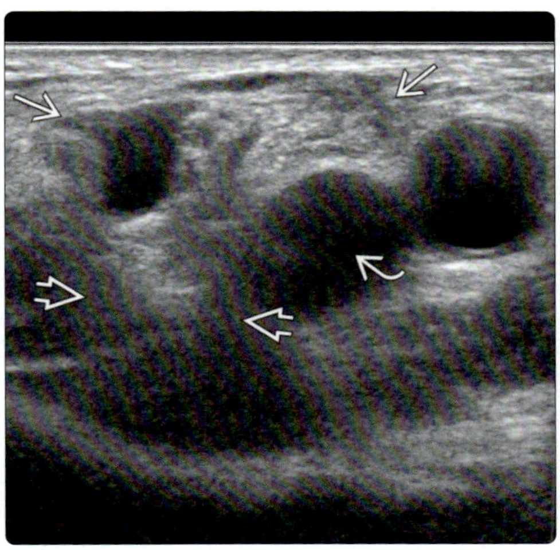

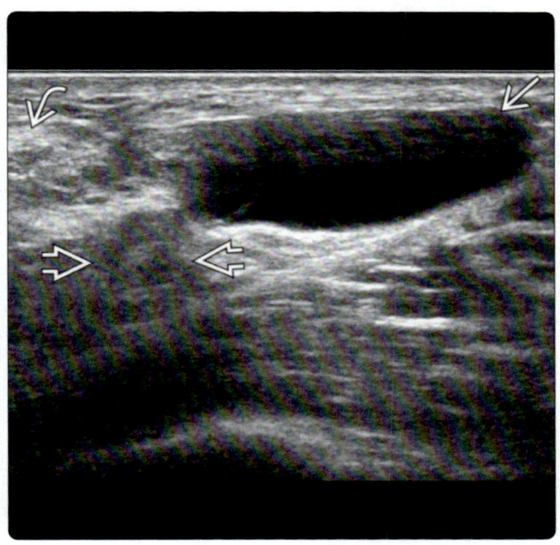

(Left) *Transverse US shows a left femoral hernia → containing mesenteric fat and fluid. The neck → of the hernia lies medial to the femoral vein → and artery.* **(Right)** *Longitudinal US of the same femoral hernia → shows that the hernia neck → lies distal to the inguinal ligament →. The narrow neck of femoral hernias makes them more prone to obstruction than inguinal hernias. If clinically suspicious of hernia, always check the femoral canal medial to the femoral vein as well as the inguinal sites.*

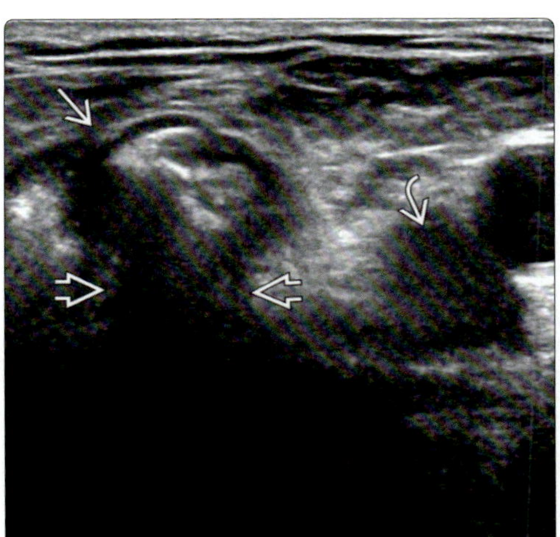

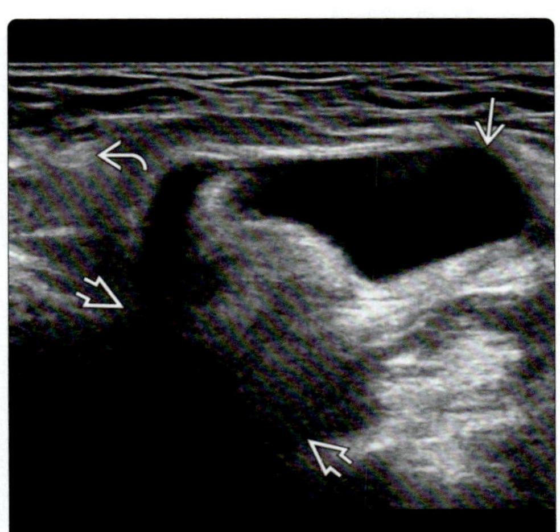

(Left) *Transverse US shows a femoral hernia → containing mesenteric fat. The neck → of the hernia lies medial to the femoral vein → and artery. The femoral nerve (not shown) lies lateral to the artery.* **(Right)** *Longitudinal US of the same femoral hernia → shows the neck → of the hernia lies distal to the inguinal ligament →. Femoral hernias are more common in elderly women; however, the most common hernia in women is still an indirect inguinal hernia.*

SECTION 4

Differential Diagnoses

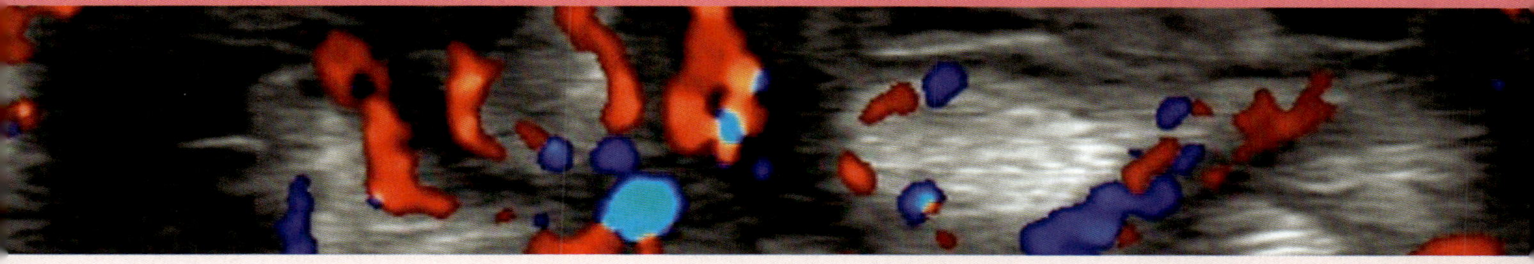

General Lumps and Bumps

Hypoechoic Subcutaneous Mass 812
Hyperechoic Subcutaneous Mass 818
Hypoechoic Muscle Mass 824
Hyperechoic Muscle Mass 830
Cystic Soft Tissue Mass 836
Calcified Soft Tissue Mass 842
Hypervascular Soft Tissue Mass 846

Tendon Abnormalities

Peritendinous Mass 852
Tendon Hypoechogenicity 856
Tendon Hyperechogenicity 860
Tendon Swelling 864

Nerve, Fascia, and Bone

Swollen Nerve 870
Fascial Lesion 876
Bone Surface Lesion 882

Joint Abnormalities

Paraarticular Cystic Mass 888
Synovial Swelling 894
Joint Effusion 900

Chest and Abdominal Wall

Chest Wall Lesion 904
Abdominal Wall Mass 910

DIFFERENTIAL DIAGNOSIS

Common

- Lipoma
- Vascular Anomaly
- Epidermoid Cyst
- Peripheral Nerve Sheath Tumor

Less Common

- Vascular Leiomyoma
- Pilomatricoma
- Fibroma
- Glomus Tumor
- Lymph Node
- Hematoma
- Abscess
- Fat Necrosis
- Rheumatoid Nodule
- Foreign Body Granuloma
- Sweat Gland Tumor
- Metastasis
- Lymphoma
- Sarcoma
  - Dermatofibrosarcoma Protuberans

ESSENTIAL INFORMATION

Key Differential Diagnosis Issues

- Ensure lesion is confined to subcutaneous tissues
  - Lesion location has important implications for diagnosis and treatment
- Use minimal transducer pressure to avoid distorting lesion
  - Copious skin gel will allow stand-off effect for superficial lesions
    - This is better than using gel pad, which will exert pressure on subcutaneous tissues
- Compressibility assessed by transducer pressure
  - Gives indication of firmness of mass, which is useful when distinguishing fatty from nonfatty tissue or inflammatory from noninflammatory tissue

Helpful Clues for Common Diagnoses

- **Lipoma**
  - Variable echogenicity depending on cellularity and fat-water content
    - Usually are mildly hyperechoic or isoechoic to subcutaneous fat
    - Subcutaneous fat usually hypoechoic but this echogenicity can vary for different locations
  - Diagnosis (as always) depends on constellation of findings rather than 1 single feature
    - Well-marginated, encapsulated, compressible mass with multiple fine distinctive echogenic lines parallel to skin surface
    - Usually no vascularity on color Doppler imaging
- **Vascular Anomaly**
  - Comprises both vascular tumor (hemangioma) and vascular malformation
    - Differentiation based on cellular turnover, histologic features, and natural history
  - Variable US appearances
    - Mixed echogenicity depending on relative amount of fatty stroma or dilated vascular channels
    - Dilated vascular or lymphatic spaces may appear as hypoechoic vascular channels or cystic spaces
    - Vascular anomalies with prominent vascular components appear hypoechoic
    - Vascularity varies from no detectable flow to very hypervascular depending on vessel type, size, and flow
      - Slow venous flow best seen on real-time grayscale imaging
    - Moderately compressible ± phleboliths
- **Epidermoid Cyst**
  - Appearances depend on degree of cyst maturation and compactness of keratin ± presence of cyst rupture
  - Well-circumscribed, ovoid or lobulated mass with increased acoustic transmission
  - Most are heterogeneous
    - Layered keratin aggregates ± cholesterol deposition → linear, hyperreflective, disc-like areas
    - Areas lacking keratin → cleft-like or filiform anechoic or hypoechoic areas
    - Less compact keratin → increased hypoechogenicity
  - Most epidermoid cysts are avascular
    - Peripheral and pericystic vascularity is feature of cyst rupture with pericystic inflammation
- **Peripheral Nerve Sheath Tumor**
  - Well-defined homogeneous hypoechoic mass
  - Occasionally heterogeneous
    - Heterogenicity due to areas of myxoid tissue, hemorrhage, fibrosis, or calcification
  - Fusiform or oblong-shaped along course of nerve
    - Entering or exiting nerve seen in majority of cases
    - Entering/exiting nerves may not be seen if arising from small nerve
  - Posterior acoustic enhancement is common
  - Usually mild to moderate vascularity

Helpful Clues for Less Common Diagnoses

- **Vascular Leiomyoma**
  - Benign, solitary smooth muscle neoplasm most commonly occurs in subcutaneous soft tissues of lower limb
  - Usually round or ovoid in shape with long axis parallel to extremity axis
  - Smooth-bordered, homogeneous, hypoechoic mass without perceivable capsule
  - Occurs in close proximity to artery or vein
  - Often hypervascular ± vascular convergence
- **Pilomatricoma**
  - Benign skin tumor arising from cells of hair matrix; more commonly seen in children
  - Well-circumscribed, ovoid, heterogeneous, solid subdermal mass containing matrix calcification, intrinsic vascularity, and peripheral hypoechoic rim (due to connective tissue capsule)
  - Intrinsic vascularity may be peripheral, central, or mixed
- **Fibroma**
  - Localized tumorous collection of collagen with sparse number of fibroblasts
  - Well-defined to irregular hypoechoic mass
  - Minimal to mild intrinsic vascularity

- **Glomus Tumor**
  - Arises from glomus body, which is arteriovenous shunt in dermis that contributes to temperature regulation
  - Most common location: Distal extremities, especially subungual areas
  - Small, well-defined, hypoechoic, hypervascular mass
- **Lymph Node**
  - Enlarged subcutaneous nodes can be due to reactive, infective, metastatic, or lymphomatous nodes
  - Preserved central echogenic fatty hilum and hilar vascularity in reactive, infective, and many lymphoma nodes
  - Obliteration of central echogenic fatty hilum in metastatic → hypoechoic mass
  - Variable peripheral vascularity or mixed hilar and peripheral flow pattern
- **Hematoma**
  - Usually echogenic in acute stages → liquefaction after several days to weeks → hypoechoic
  - Often occurs near bony prominence
- **Abscess**
  - Usually hypoechoic but may be echogenic due to aggregates of inflammatory debris
  - Surrounding edema and hyperemia
  - Moving echoes on real-time imaging
- **Fat Necrosis**
  - Swelling and increased echogenicity of subcutaneous fat in early stages, which becomes more hypoechoic with increasing chronicity
  - ± liquefaction with discrete irregular hypoechoic areas ± calcification ± fat atrophy in later stages
- **Rheumatoid Nodule**
  - Extraarticular manifestation occurring in 2% of rheumatoid arthritis patients
  - Most common in subcutaneous tissue subjected to pressure with deep bony surface
    - Extensor surface forearms, olecranon, and finger knuckles
  - Well-defined, homogeneous, hypoechoic mass with no posterior shadowing
  - Minimal or mild intrinsic vascularity
- **Foreign Body Granuloma**
  - Variable thickness rim of hyperemic and hypoechoic granulation tissue forms around foreign body
    - Makes echogenic foreign body more conspicuous
- **Sweat Gland Tumor**
  - Tumor arises from eccrine or apocrine sweat glands
  - Hypoechoic, heterogeneous nodule of low vascularity ± cystic areas
- **Metastasis**
  - 2.5% of superficial soft tissue are metastases
  - Most common (> 50%) primaries: Breast, bronchus, and melanoma
  - Well-demarcated, irregular or roundish hypoechoic mass
    - Melanoma metastases tend to be distinctly hypoechoic and may be elongated due to spread along lymphatic channels (tail and string signs)
  - Mild to moderate intrinsic vascularity usually present
  - ± features of primary tumor (e.g., myxoid, calcification)
- **Lymphoma**
  - Involvement of superficial tissues uncommon
  - Subcutaneous panniculitis-like T-cell lymphoma is uncommon type of peripheral T-cell lymphoma
- **Sarcoma**
  - 2.5% of superficial soft tissue masses are sarcoma
  - Large (> 5 cm), heterogeneous, hypoechoic mass
  - Usually hypervascular with disorganized vascular pattern on color Doppler imaging
  - Dissimilar to known benign subcutaneous tumors
  - **Dermatofibrosarcoma Protuberans**
    - Low-grade spindle cell mesenchymal sarcoma of dermis and subcutis
    - Most commonly arises in trunk (~ 50%), particularly chest wall, back, and abdominal wall
    - Well-marginated, lobulated, hypoechoic mass
    - Small, echogenic speckles and larger discrete hypoechoic cleft-like areas within mass
    - Posterior enhancement
    - Central and peripheral organized vascularity

Lipoma

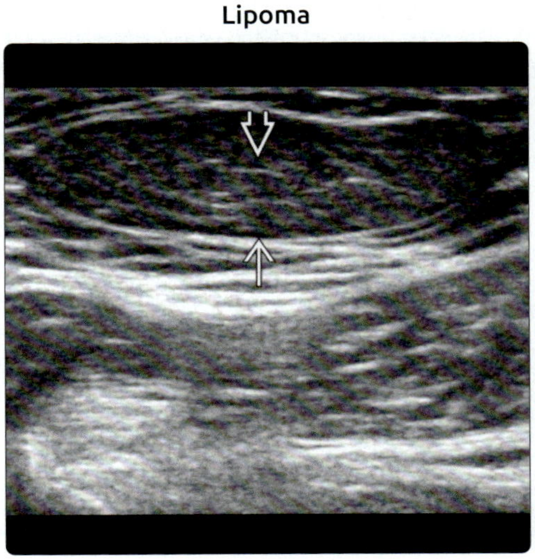

Lipoma

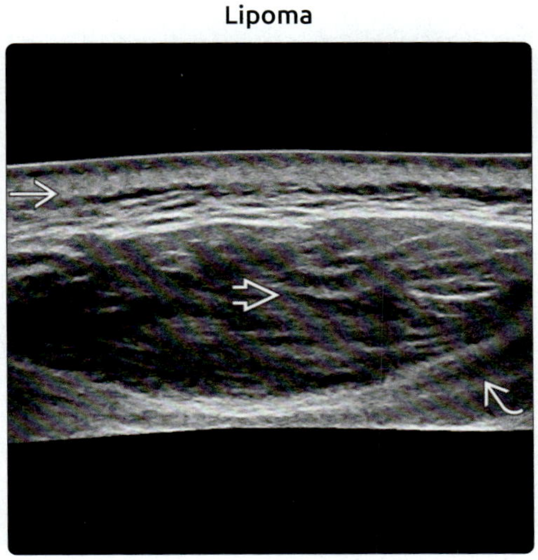

(Left) *Longitudinal US shows an encapsulated, subcutaneous lipoma isoechoic to adjacent subcutaneous fat with fine internal echogenic striations parallel to the long axis of the tumor.* **(Right)** *Extended longitudinal view US of a subcutaneous lipoma shows prominent horizontal striations aligned parallel to the surrounding skin. The echogenicity of the lipoma is similar to the hypoechoic subcutaneous fat and hypoechoic to the dermal and connective tissue.*

(Left) *Longitudinal US shows a well-defined, lobulated, hypoechoic mass lesion ➡ within the subcutaneous tissue. There are tubular hypoechoic areas ➡ connected to prominent central vessel ➡ within the mass.* **(Right)** *Longitudinal color Doppler US of the same lesion shows predominant intralesional venous vascular flow ➡ in keeping with a venous vascular malformation. The supporting stroma ➡ most likely contains many vascular spaces that are too small to be resolved by US.*

Vascular Anomaly

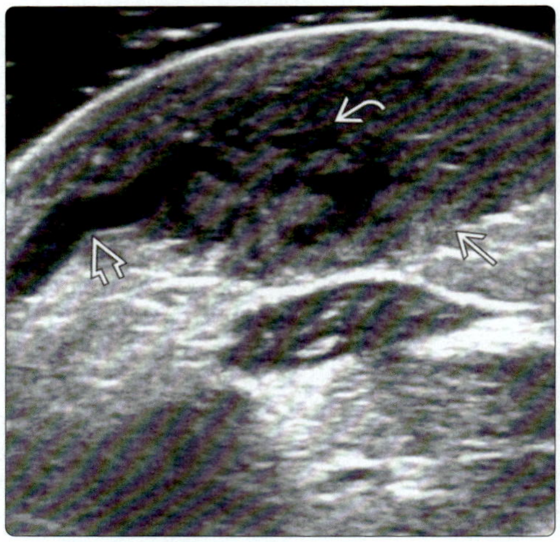

Vascular Anomaly

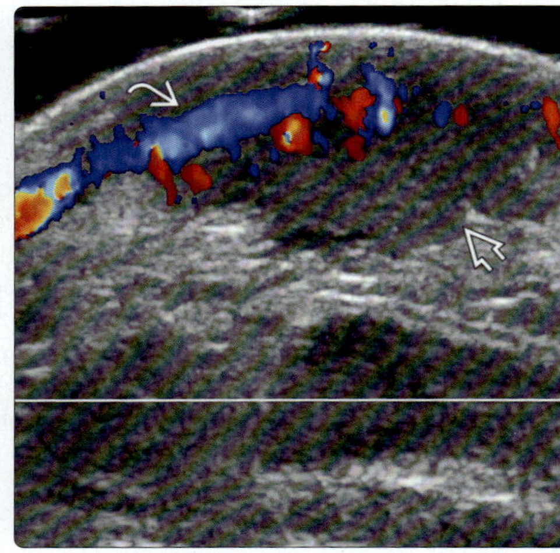

(Left) *Longitudinal US of an epidermoid cyst ➡ shows a fairly homogenous hypoechoic well-capsulated mass with a few small anechoic cystic-like areas ➡ lacking keratin as well as small hyperechoic foci ➡ representing cholesterol deposits. Posterior acoustic enhancement is present.* **(Right)** *Longitudinal color Doppler US of an epidermoid cyst shows a more heterogeneous appearance. There are some small echogenic clefts ➡ due to layered keratin. No intrinsic vascularity is present.*

Epidermoid Cyst

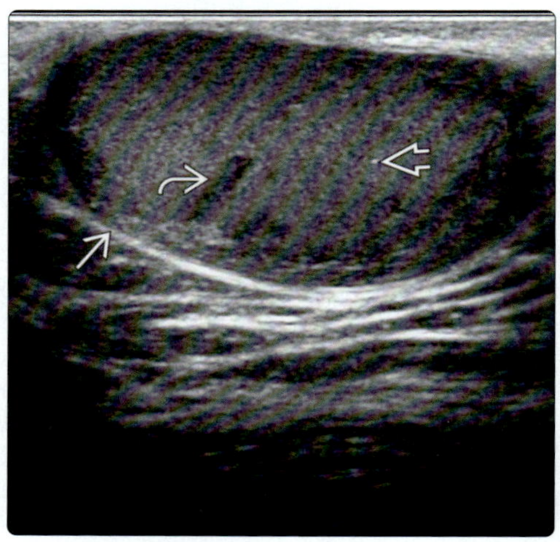

Epidermoid Cyst

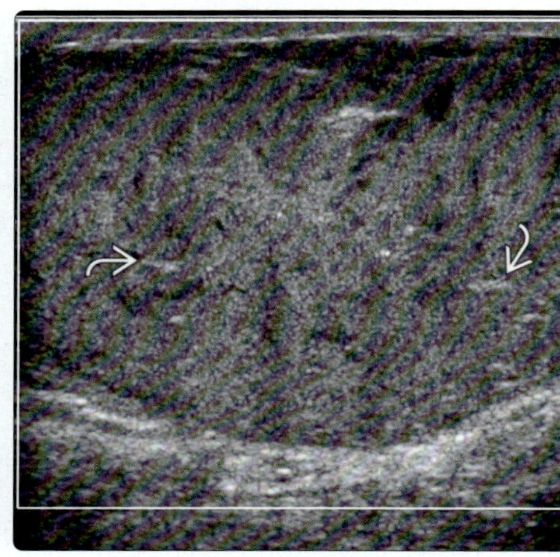

(Left) *Longitudinal US shows a well-defined, subcutaneous hypoechoic mass ➡ with mild posterior acoustic enhancement and a thickened nerve end ➡ in keeping with a nerve sheath tumor arising from a subcutaneous nerve.* **(Right)** *Transverse color Doppler US of a nerve sheath tumor shows a moderate to high degree of disorganized vascularity.*

Peripheral Nerve Sheath Tumor

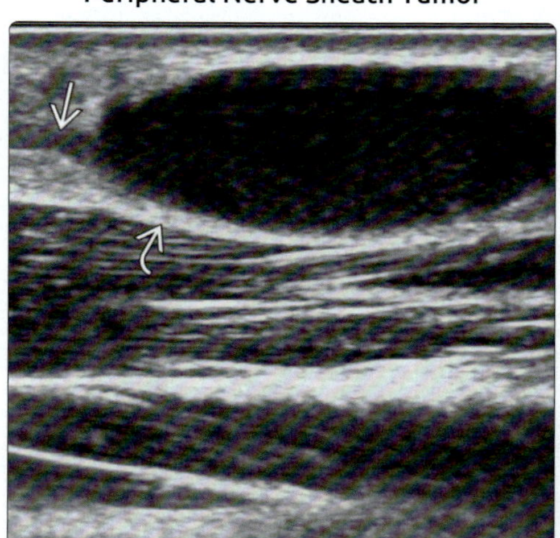

Peripheral Nerve Sheath Tumor

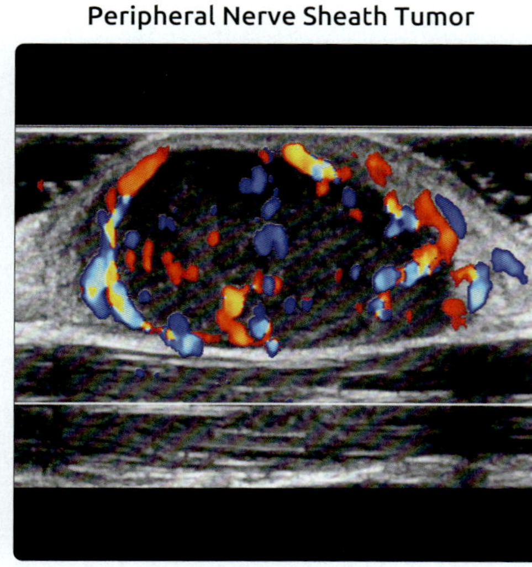

Vascular Leiomyoma

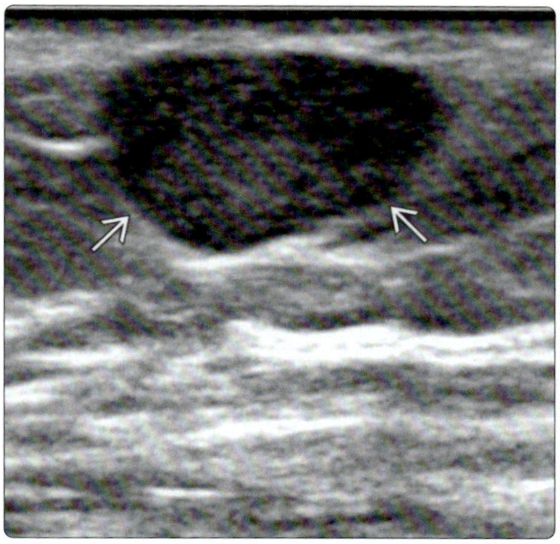

Vascular Leiomyoma

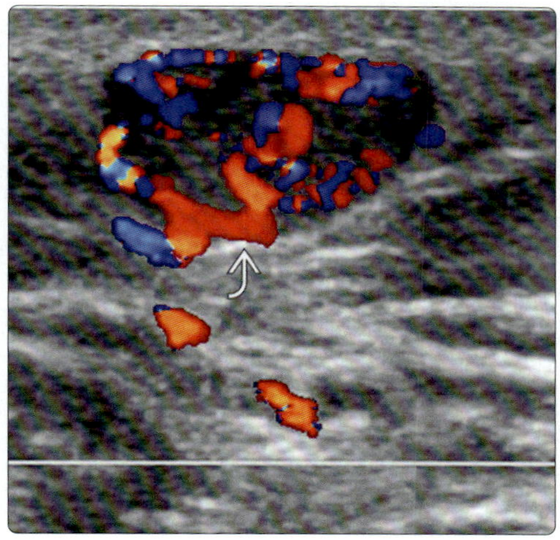

(Left) *Longitudinal US shows a well-defined, smooth bordered, homogeneous, hypoechoic mass* ➡ *in the subcutaneous tissue of the thigh. The long axis of the tumor is parallel to the axis of the lower limb.* **(Right)** *Longitudinal color Doppler US shows a moderate degree of organized hyperemia with peripheral vascular convergence to a point* ➡*. This peripheral vascular convergence pattern is seen in some, but not all, vascular leiomyomas.*

Pilomatricoma

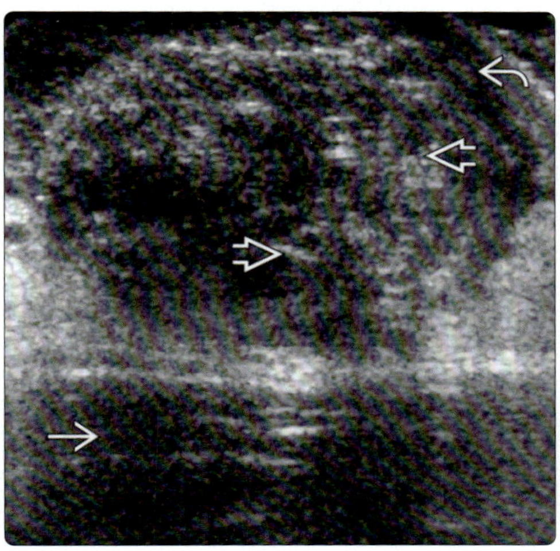

Pilomatricoma

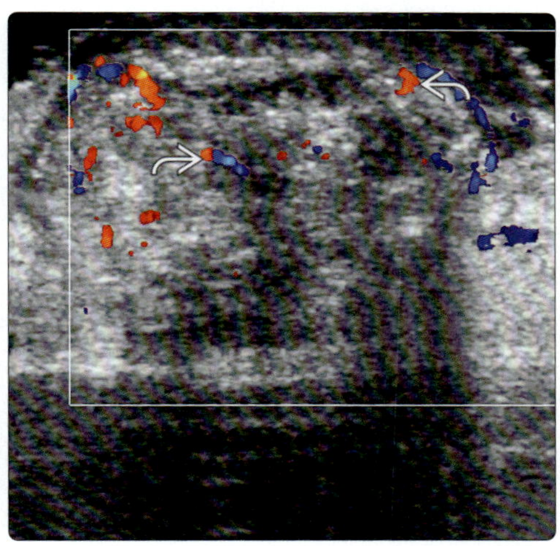

(Left) *Longitudinal US shows a heterogeneous, hypoechoic pilomatricoma containing multiple internal calcifications* ➡ *with mild posterior acoustic shadowing* ➡*. There is a peripheral hypoechoic connective tissue capsule* ➡*.* **(Right)** *Longitudinal color Doppler US of the same pilomatricoma shows a mild to moderate degree of disorganized vascularity* ➡*. Pilomatricoma is a benign tumor arising from cells of the hair matrix and is more commonly seen in children.*

Glomus Tumor

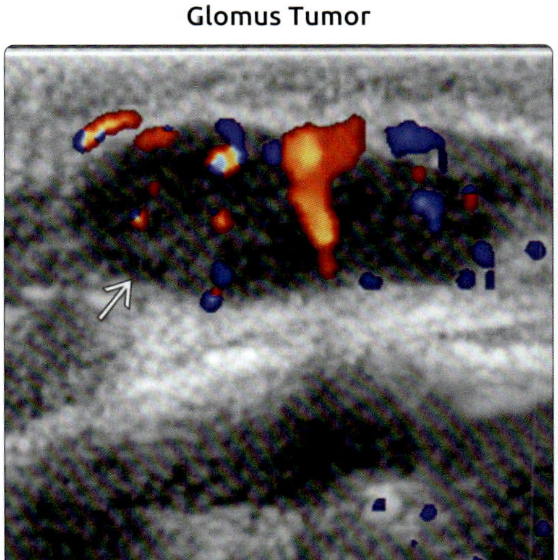

Lymph Node

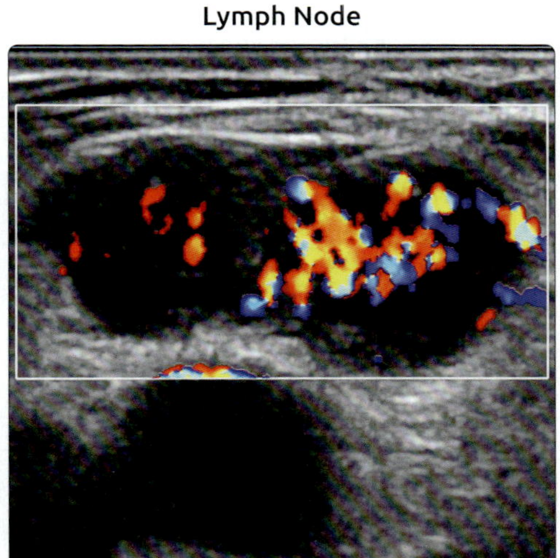

(Left) *Longitudinal color Doppler US of a patient with a painful glomus tumor in the upper limb shows a well-defined, hypoechoic subcutaneous mass* ➡ *with a moderate degree of intratumoral hyperemia.* **(Right)** *Longitudinal US of a patient with known ovarian carcinoma shows an enlarged subcutaneous lymph node on the medial groin region. There is marked cortical hypertrophy with moderate vascularity. Percutaneous biopsy confirmed metastasis from an ovarian primary.*

(Left) *Longitudinal US shows a subcutaneous hematoma with a large, liquefied component and a small peripheral echogenic component. No intrinsic vascularity is evident.* **(Right)** *Transverse US shows a large, irregular subcutaneous abscess containing moderate echogenic internal debris. There is severe edema of the adjacent subcutaneous fat.*

Hematoma

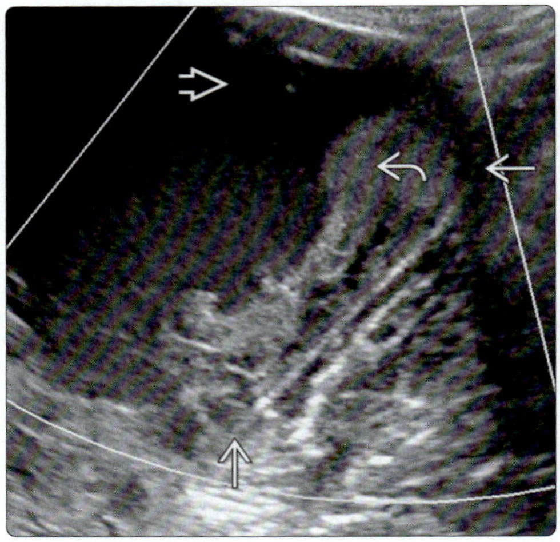

Abscess

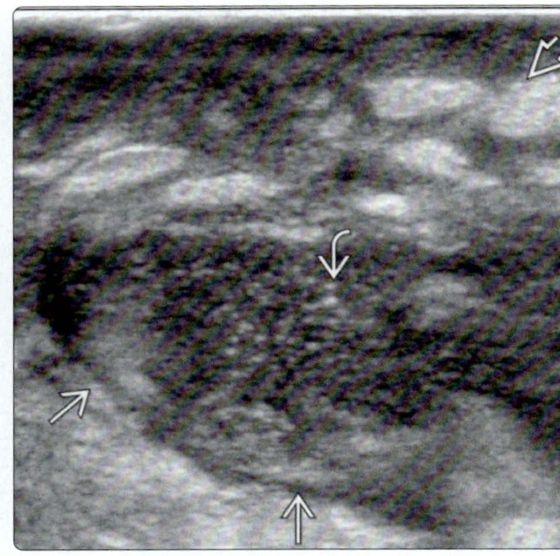

(Left) *Transverse US of posttraumatic fat necrosis shows a heterogeneous, subcutaneous mass composed of an ill-defined, peripheral, hyperechoic rim and a discrete, central hypoechoic component. Areas of established subcutaneous fat necrosis become better defined and more hypoechoic with increasing chronicity.* **(Right)** *Longitudinal US shows an elliptical, hypoechoic rheumatoid nodule superficial to the Achilles tendon in a patient with rheumatoid arthritis.*

Fat Necrosis

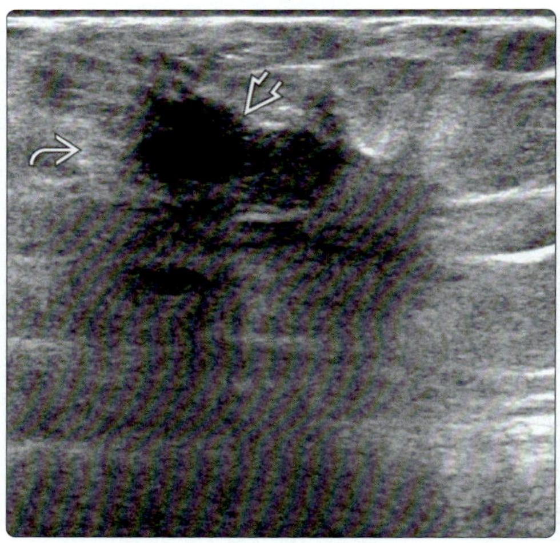

Rheumatoid Nodule

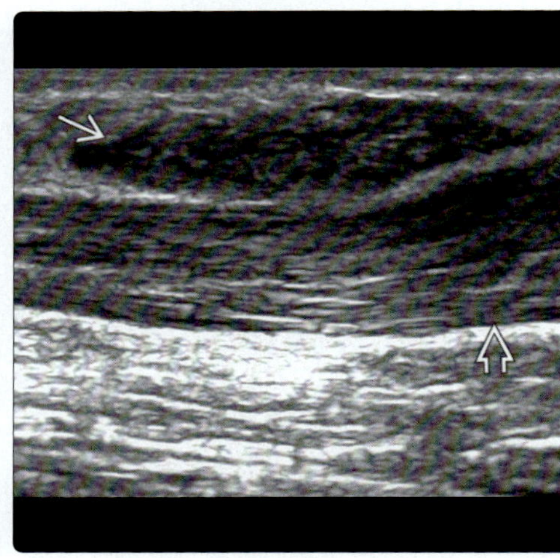

(Left) *Longitudinal US shows a subcutaneous foreign body granuloma of the forearm. The hyperechoic retained wooden fragment is surrounded by a thick, hypoechoic rim of granulation tissue. The mass was moderately vascular on color Doppler imaging (not shown).* **(Right)** *Longitudinal power Doppler US shows a well-defined, ovoid, hypoechoic solid mass with a small cystic area and mild intrinsic vascularity. Excision revealed an eccrine spiradenoma.*

Foreign Body Granuloma

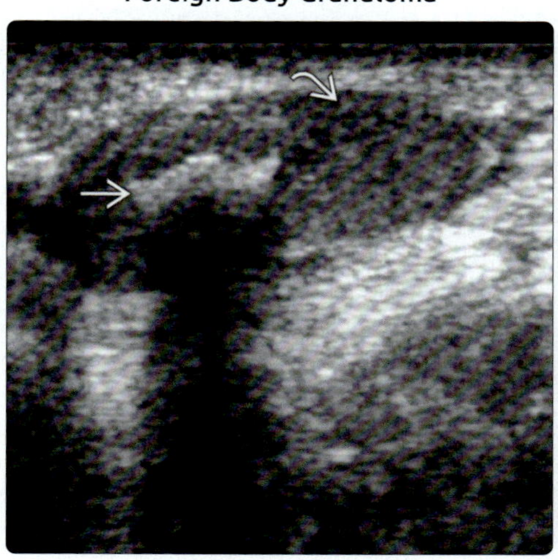

Sweat Gland Tumor

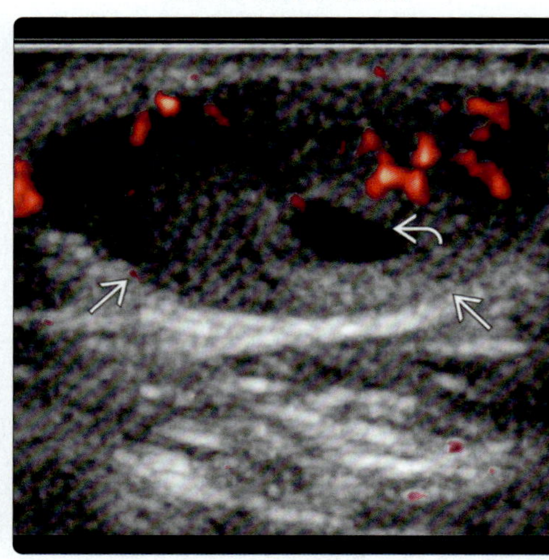

Metastasis

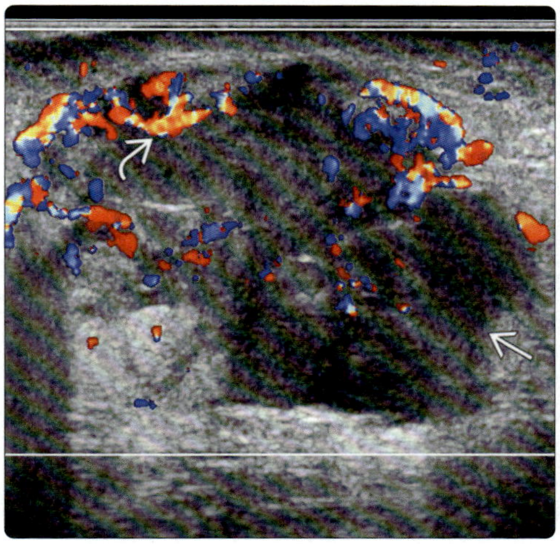

Metastasis

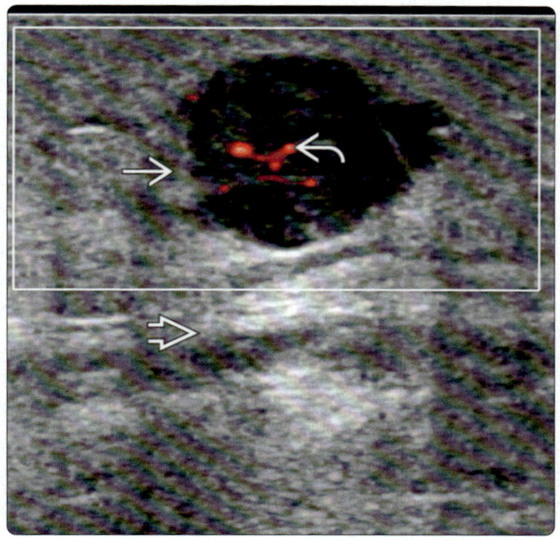

(Left) *Longitudinal power Doppler US shows an irregular subcutaneous melanoma metastasis. Melanoma deposits are typically hypoechoic due to poor beam reflectivity of melanotic tissue. There is moderate tumoral hyperemia.* **(Right)** *Transverse power Doppler US shows a subcutaneous metastasis of the anterior abdominal wall from a malignant nerve sheath tumor in the calf. The mass is irregular and hypoechoic with mild posterior acoustic enhancement and mild hyperemia.*

Sarcoma

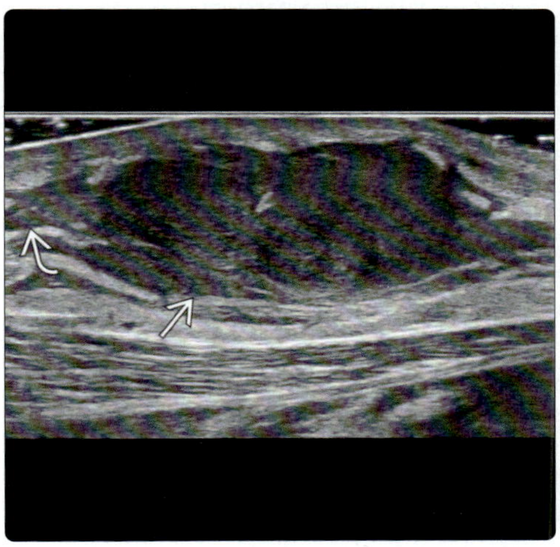

Sarcoma

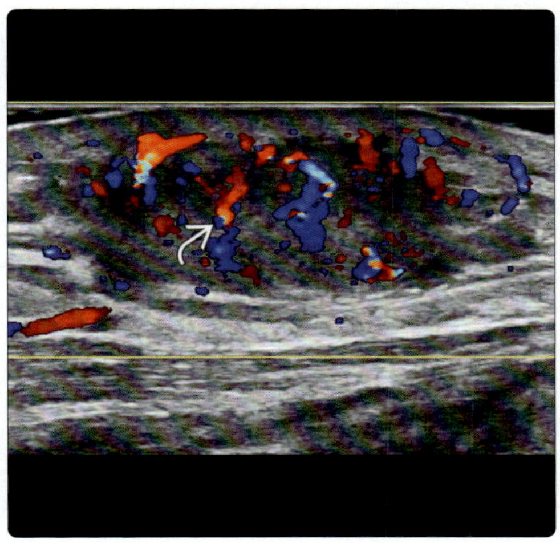

(Left) *Longitudinal US shows a moderate-sized, irregular, hypoechoic subcutaneous mass in the calf. The adjacent subcutaneous tissues are moderately edematous.* **(Right)** *Transverse color Doppler US of the same lesion shows moderate to high intrinsic disorganized vascularity. Percutaneous biopsy for histology revealed a myxoid sarcoma. Superficial sarcoma is rare. The absence of recognizable features of known benign superficial tumors should raise the suspicion of malignancy.*

Dermatofibrosarcoma Protuberans

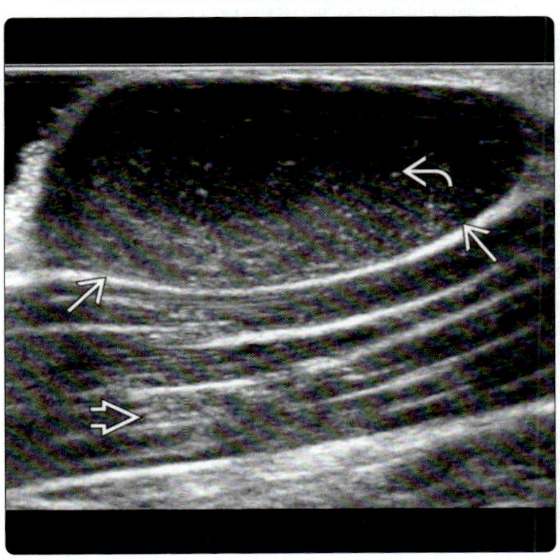

Dermatofibrosarcoma Protuberans

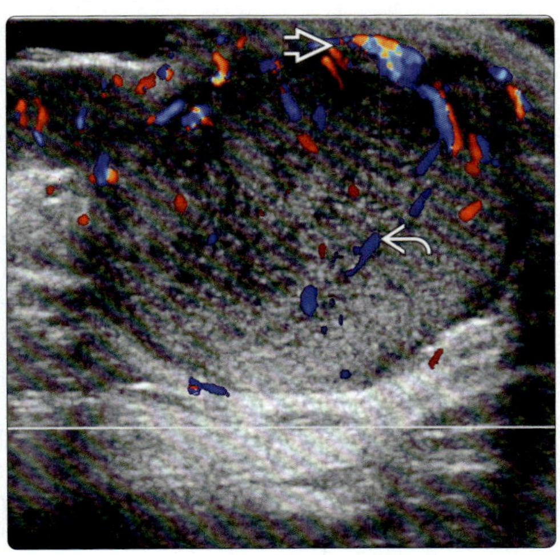

(Left) *Longitudinal US shows a well-defined hypoechoic mass causing focal skin protuberance. It contains multiple fine echogenic foci and shows mild increase in posterior acoustic transmission. These are quite characteristic appearances of dermatofibrosarcoma protuberans.* **(Right)** *Transverse color Doppler US of dermatofibrosarcoma protuberans shows a moderate degree of organized central and peripheral vascularity.*

DIFFERENTIAL DIAGNOSIS

Common

- Lipoma
- Vascular Anomaly
- Epidermoid Cysts

Less Common

- Cellulitis
- Fat Necrosis
- Hematoma
- Panniculitis
- Pilomatricoma
- Abscess
- Gouty Tophus
- Granuloma and Fibroma
- Normal Lymph Node
- Fibrous Hamartoma of Infancy

ESSENTIAL INFORMATION

Key Differential Diagnosis Issues

- Use light transducer pressure
  - Any compression of subcutaneous fat will usually tend to make soft subcutaneous masses less conspicuous
- Subcutaneous fat may vary from hyperechoic to hypoechoic depending on location
- In some regions, particularly gluteal region, subcutaneous fat looks similar to muscle
  - Take care in defining correct layer as this has important diagnostic and surgical implications

Helpful Clues for Common Diagnoses

- **Lipoma**
  - Variable echogenicity depending on cellularity and fat-water content but, in general, isoechoic to fat and hyperechoic to muscle
  - Well-marginated, encapsulated, compressible, fusiform-shaped mass
    - Multiple fine, distinctive echogenic lines parallel to skin surface
    - Usually no detectable vascularity on color Doppler imaging
    - Consider angiolipoma if intrinsic vascularity detected
    - If multiple lipomas present, these tend to be more rounded, more echogenic, with less striation, encapsulation, and compressibility
  - Lipohypertrophy = localized increase in fat accumulation, subcutaneous > > intermuscular
    - Increase in size of fat lobules and depth of subcutaneous fat
    - Usually affects middle-aged women
    - Absence of distinct border or capsule allows distinction of lipohypertrophy from lipoma
- **Vascular Anomaly**
  - Comprises both vascular tumor (hemangioma) and vascular malformation
    - Differentiation based on cellular turnover, histologic features, and natural history
  - Hemangioma
    - Benign vascular tumors of infancy and childhood with cellular proliferation and hyperplasia
    - Grows after birth, proliferates, and then involutes
  - Vascular malformation
    - Due to vascular dysplasia rather than actual tumor with normal endothelial turnover
    - Present at birth and grows commensurately with child without proliferation or involution
    - Classified according to predominant vessel type as either venous, capillary, arterial, lymphatic, or combined type
  - Variable US appearances
    - Mixed echogenicity depending on relative fatty stromal or vascular components
    - Dilated vascular spaces may appear as hypoechoic vascular channels or large hypoechoic cystic spaces (venous lakes)
    - Variable stromal connective tissue component with predominant fatty-type background → appears hyperechoic
    - This echogenic stroma component may contain numerous capillaries too small to be resolved with US
    - Vascularity of vascular anomaly varies from no detectable flow → very hypervascular
    - Moderately compressible ± phleboliths
- **Epidermoid Cysts**
  - Slow-growing subdermal cyst that contains keratin and is lined by stratified squamous epithelium but does not contain dermal elements
  - US appearances depend on cyst maturation and compactness of keratin ± presence of cyst rupture
  - Most are well circumscribed, ovoid or lobulated
  - Increased acoustic transmission
  - Most are heterogeneous and mildly hyperechoic
  - Linear, hyperreflective, disc-like areas represent layered keratin aggregates ± cholesterol deposition
  - Cleft-like or filiform anechoic or hypoechoic areas are due to cyst-like spaces lacking keratin
  - Hypoechoic stalk-like punctum extending from superficial wall of cyst to skin seen in 10% of those referred for US examination
    - Most cysts with clinically apparent puncta are diagnosed clinically and not referred for US examination
  - Most epidermoid cysts are avascular
    - Peripheral and pericystic vascularity is feature of cyst rupture with pericystic inflammation

Helpful Clues for Less Common Diagnoses

- **Cellulitis**
  - Infection of subcutaneous fat
    - Characterized by edema and hyperemia
  - Group A *Streptococcus* and *Staphylococcus aureus* organisms are usually responsible
  - Subcutaneous fat is more echogenic when edematous
  - Subcutaneous edema = nonspecific finding seen with many conditions, e.g., heart failure, venous insufficiency, immobility, and dependency
  - Cellulitis should have combination of edema and hyperemia
    - Thickened interlobular septa
    - ± periseptal fluid or fluid above investing fascia
- **Fat Necrosis**

- Swelling of subcutaneous fat in early stages
  - Hyperechoic edematous fat with loss of normal echogenic striation
  - Majority are due to previous trauma
- Subcutaneous fat becomes more hypoechoic with increasing chronicity
  - ± liquefaction with discrete irregular hypoechoic areas ± calcification ± fat atrophy in later stages
  - Investing fascia and muscle is usually unaffected

- **Hematoma**
  - Echogenic in acute stages
  - ± circular echogenic layering due to sequential deposition of hemorrhage
  - ± linear layering due to separation of cellular and serous components
  - ± moving echoes on real-time imaging
  - ± liquefaction after several days to weeks
  - ± calcification as late feature
- **Panniculitis**
  - Localized inflammation of subcutaneous fat; has many causes
    - Idiopathic, infection, systemic inflammatory disorder, pancreatic disease, drug hypersensitivity, Weber-Christian disease, or myeloproliferative disorder (panniculitis-like T-cell lymphoma)
  - Active stage: Localized area of increased echogenicity and thickness of subcutaneous fat with mild to moderate hyperemia
  - Inactive stage: Atrophy and hypoechogenicity of subcutaneous fat with little or no hyperemia
- **Pilomatricoma**
  - Benign skin tumor arising from cells of hair matrix
  - Mass extending from dermal to subdermal tissues
  - Well-circumscribed, ovoid, heterogeneous, solid subdermal mass with matrix calcification, intrinsic vascularity, and peripheral hypoechoic rim (due to connective tissue capsule)
  - Intrinsic vascularity
  - > children, > head and neck region
- **Abscess**
  - May be echogenic due to aggregates of inflammatory debris
  - Surrounding edema and hyperemia
  - ± moving echoes on real-time imaging
- **Gouty Tophus**
  - Soft or hard tophi depending on concentration of crystals deposited
    - Soft tophi: Echogenic with mild to moderate posterior acoustic shadowing
    - Hard tophi: Echogenic with strong posterior acoustic shadowing
- **Granuloma and Fibroma**
  - Common in gluteal region
    - Frequently follow subcutaneous injection
  - May also occur elsewhere without any recognizable cause
  - Usually hypoechoic, though occasionally hyperechoic
    - Often calcified
- **Normal Lymph Node**
  - May appear echogenic if large central fatty hilum and thin hypoechoic cortex present
    - Particularly in axillary nodes and subcutaneous lymph node on medial aspect proximal thigh
- **Fibrous Hamartoma of Infancy**
  - Uncommon benign superficial fibrous tumor found in early childhood (< 2 years)
  - More common in boys
  - Mainly affects trunk, axilla, and upper extremities
  - Associated with skin pigmentation, hypertrichosis, and eccrine gland hyperplasia
  - Consists of mixture of fibrocollagenous tissue, primitive mesenchymal cells, and mature fat
    - US appearances depend on proportion of these 3 components
  - Ill-defined heterogeneous, predominantly hyperechoic, subcutaneous mass
  - Usually no significant intrinsic vascularity
  - May involve investing fascia, though subfascial extension is rare

Lipoma

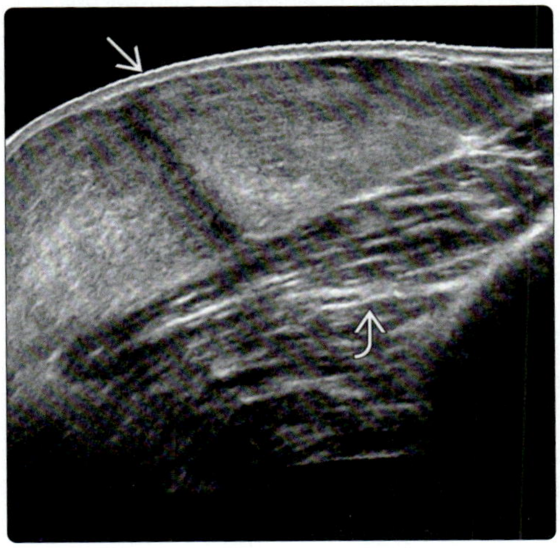

Lipoma

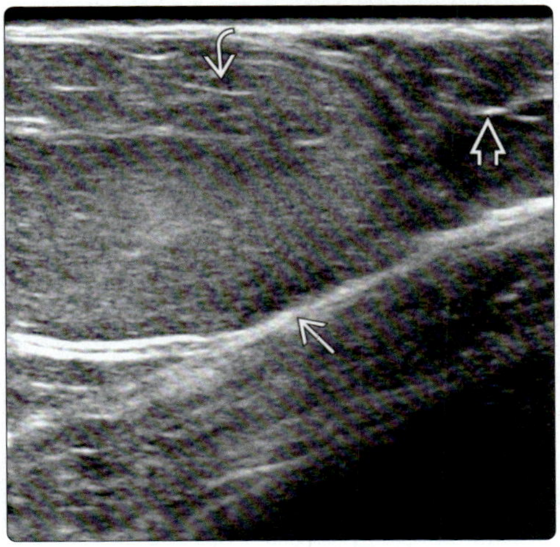

(Left) *Longitudinal US with extended field of view shows a large, hyperechoic subcutaneous lipoma ➡ causing bulging of the skin contour. The underlying deltoid muscle is normal ➡.* **(Right)** *Longitudinal US shows an encapsulated subcutaneous lipoma abutting the investing fascia ➡. The lipoma is hyperechoic to adjacent subcutaneous fat ➡ with fine linear horizontal striations ➡.*

(Left) *Longitudinal US shows a small, rounded, well-defined, hyperechoic subcutaneous lipoma without perceivable linear striation. Such lipomas are often multiple and tend to be located more in the trunk region than in the limbs.* **(Right)** *Transverse color Doppler US of a child with prepatellar port-wine discoloration shows an echogenic subcutaneous mass without detectable intrinsic vascularity, consistent with a slow-flow, capillary-type vascular malformation.*

Lipoma

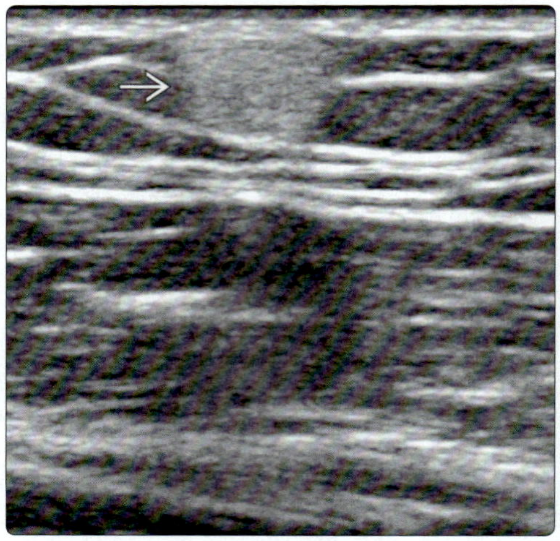

Vascular Anomaly

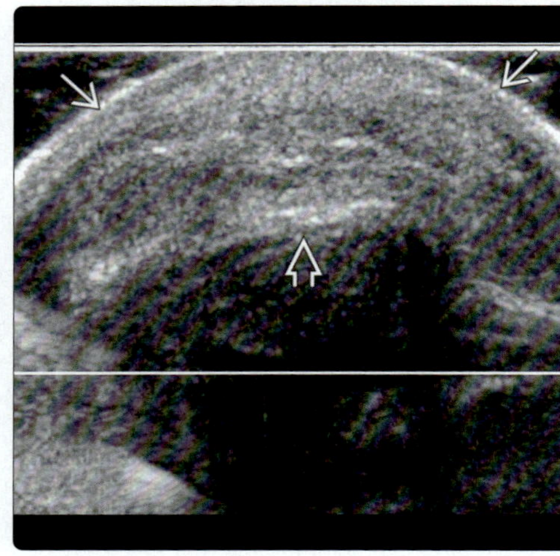

(Left) *Transverse color Doppler US of a 6-month-old infant shows a subcutaneous hemangioma in the chest wall. The hemangioma contains an echogenic fatty stroma and small hypoechoic vascular channels.* **(Right)** *Transverse color Doppler US of the same lesion shows multiple arterial and venous vascular channels embedded within the fatty stroma. This fatty stroma may contain capillaries too small to be resolved on US and only seen on histologic analysis.*

Vascular Anomaly

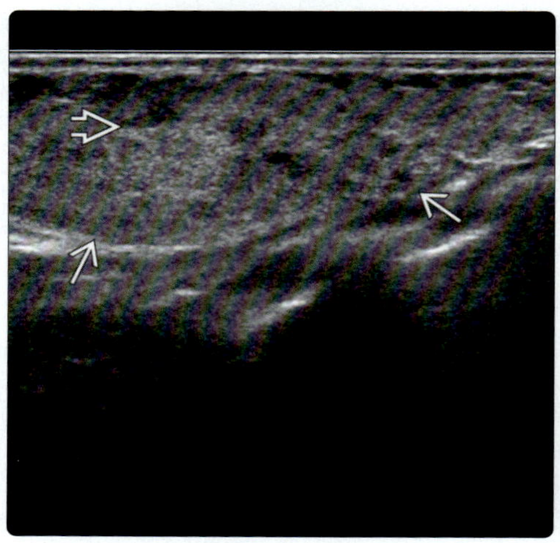

Vascular Anomaly

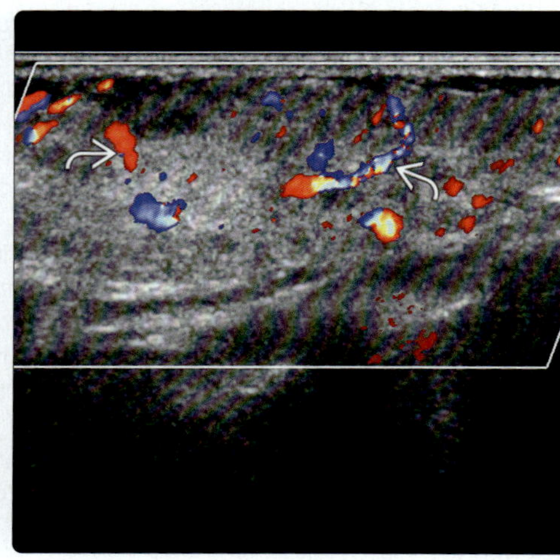

(Left) *Longitudinal US shows typical appearances of an epidermoid cyst with heterogeneous echotexture, dermal attachment, internal floating hyperreflective strands, small hypoechoic areas, and a hypoechoic track extending to a cutaneous punctum.* **(Right)** *Longitudinal US shows an echogenic epidermoid cyst with an almost pseudotestis appearance. The mass is quite homogeneous with scattered internal hypoechoic areas representing cyst-like spaces lacking keratin.*

Epidermoid Cysts

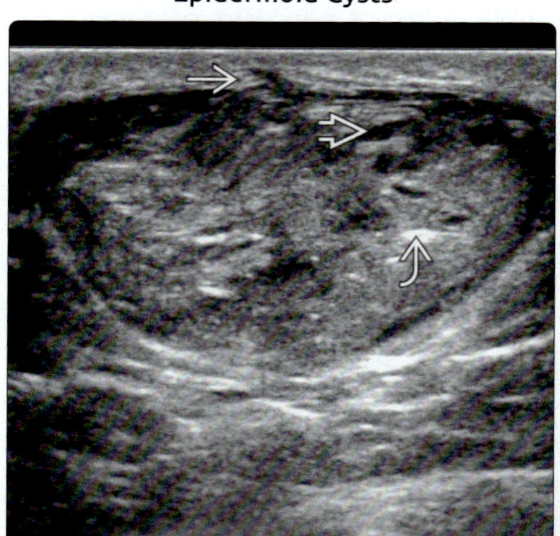

Epidermoid Cysts

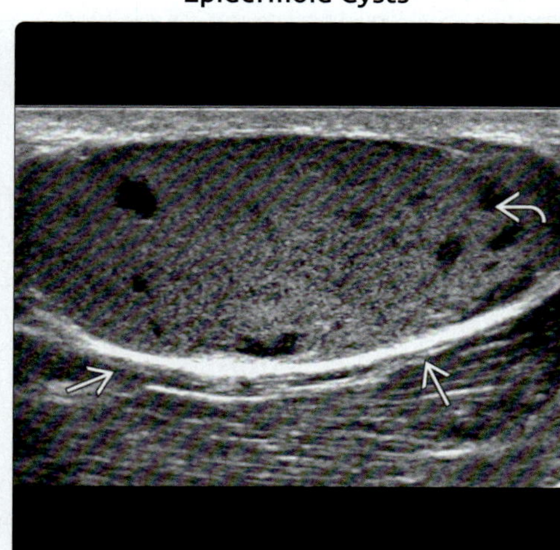

Cellulitis

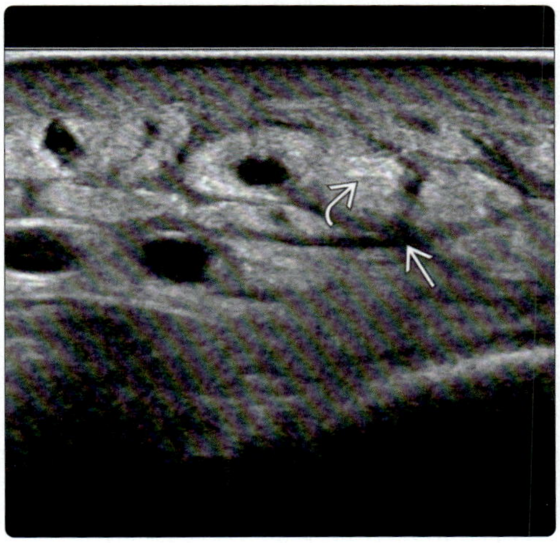

Cellulitis

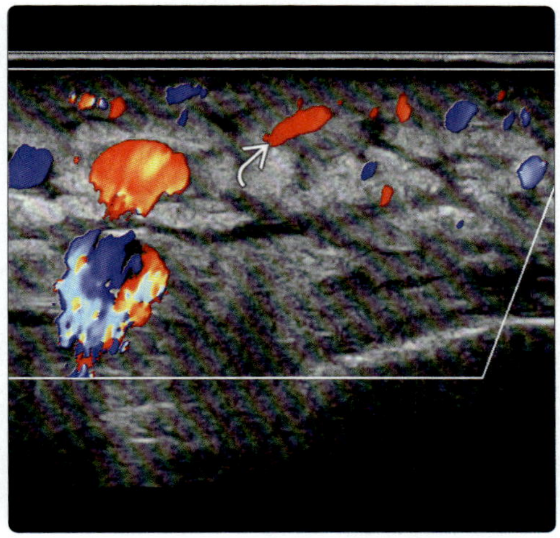

(Left) *Transverse US of the dorsal hand region shows severe subcutaneous edema with diffuse increased echogenicity of the subcutaneous fat lobules ➯, interspersed between thickened hypoechoic septa ➯.* **(Right)** *Transverse color Doppler US of the same patient shows moderate hyperemia ➯ of the subcutaneous tissues overall, consistent with moderate-severity cellulitis. Hyperemia helps differentiate cellulitis from noninfective causes of subcutaneous edema.*

Fat Necrosis

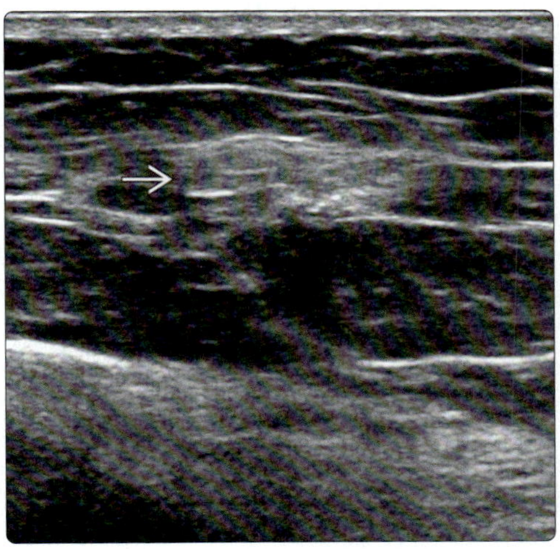

Fat Necrosis

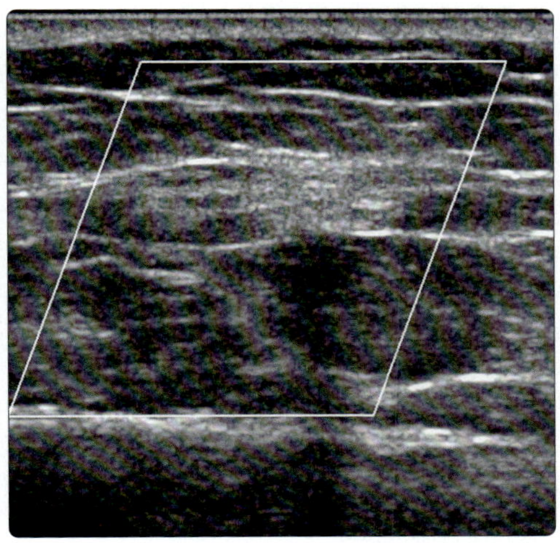

(Left) *Longitudinal US shows an area of hyperechogenicity consistent with posttraumatic subcutaneous edema ➯. There is no mass effect. As the lesion became more chronic, the echogenicity will decrease and the subcutaneous fat may atrophy to a variable degree, all consistent with subcutaneous fat necrosis.* **(Right)** *Longitudinal color Doppler US of the same lesion shows no peripheral or central hyperemia in this early posttraumatic fat edema.*

Hematoma

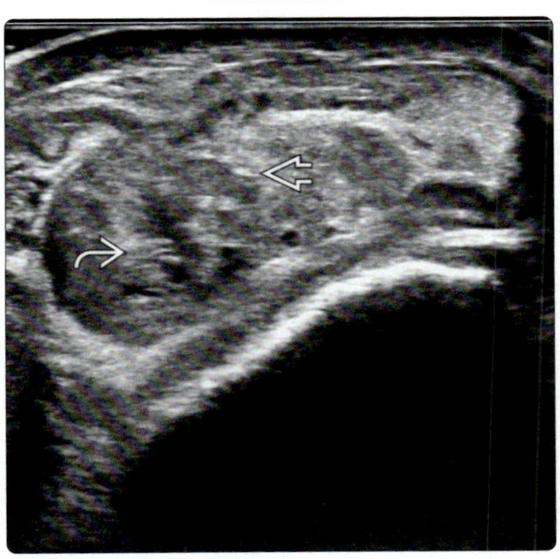

Hematoma

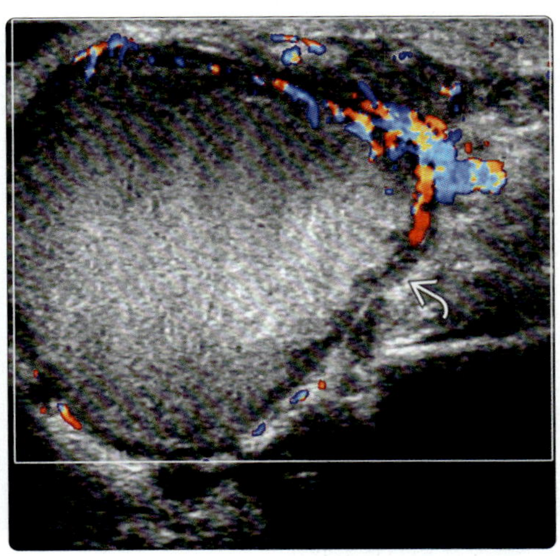

(Left) *Transverse US shows an organizing hematoma in the subcutaneous tissue superficial to the patella. The lesion is marginated and echogenic with intrinsic whorl-like patterns ➯ and echogenic strands ➯. Subcutaneous hematoma tends to occur over bony prominences.* **(Right)** *Transverse color Doppler US shows a subcutaneous organizing hematoma in the forearm. The mass is homogeneous with a peripheral, hypoechoic, moderately vascularized fibrous-type capsule ➯.*

(Left) *Longitudinal US of an immunocompromised patient with fever and leg swelling shows focal subcutaneous edema with multiple prominent echogenic fat lobules ➔ separated by hypoechoic septi ➔.* **(Right)** *Transverse color Doppler US shows moderate hyperemia ➔ of the inflamed subcutaneous fat due to infective panniculitis. The focal nature of panniculitis distinguishes it from more diffuse cellulitis. In another clinical context, panniculitis-like T-cell lymphoma should be considered.*

Panniculitis

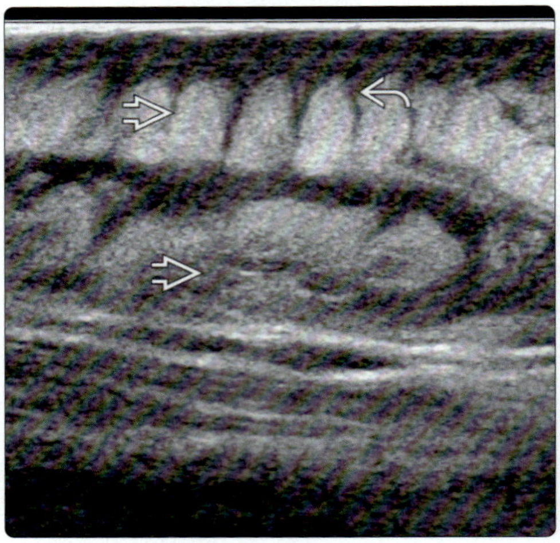

Panniculitis

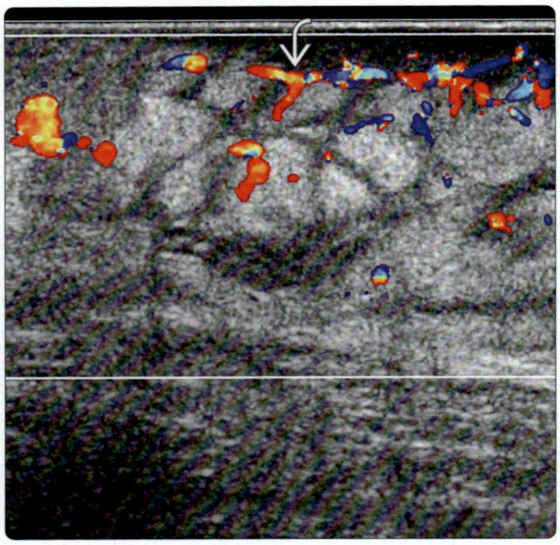

(Left) *Transverse US shows a well-defined, mixed echogenic pilomatricoma ➔ in the subcutaneous tissues. There are multiple scattered echogenic foci with mild posterior shadowing ➔ due to matrix calcification.* **(Right)** *Longitudinal color Doppler US of the same lesion shows mild hyperemia ➔ of the peripheral hypoechoic rim.*

Pilomatricoma

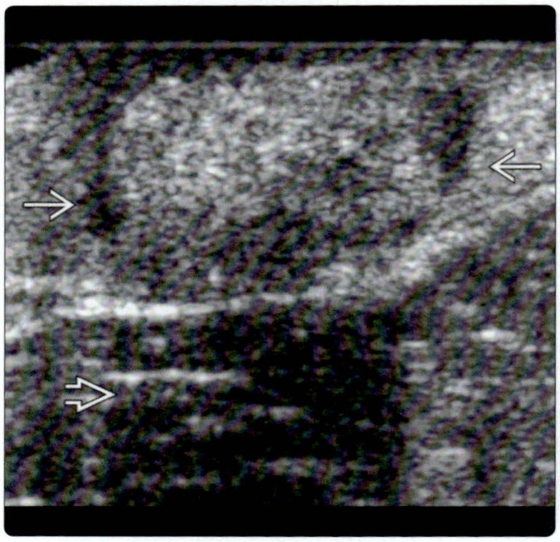

Pilomatricoma

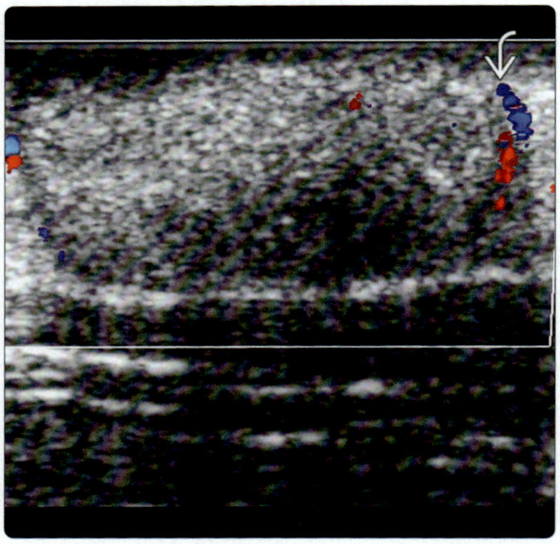

(Left) *Longitudinal US of the finger shows a large heterogeneous tuberculous cold abscess ➔ with internal calcification ➔ and moving echoes (on real-time imaging). This abscess extends from the metacarpophalangeal joint ➔ into the subcutaneous tissues.* **(Right)** *Transverse US of the arm shows an irregular pyogenic abscess ➔ with internal echogenic debris ➔ seen to move on real-time imaging. Microbiology grew Enterococcus and Acinetobacter species.*

Abscess

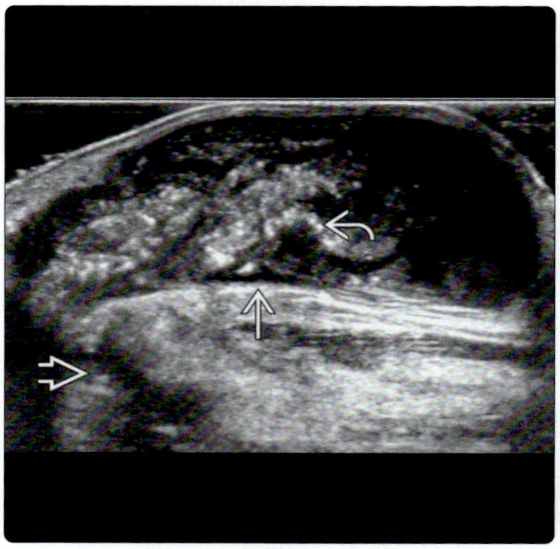

Abscess

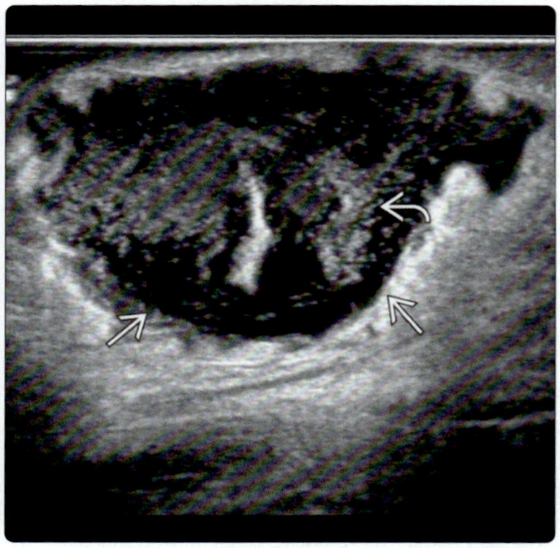

Gouty Tophus

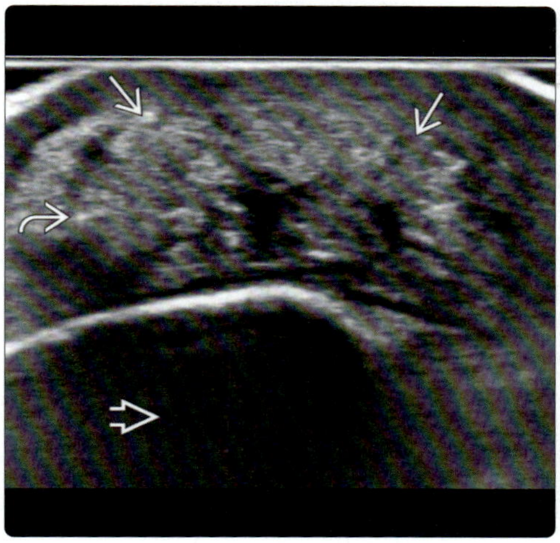

Gouty Tophus

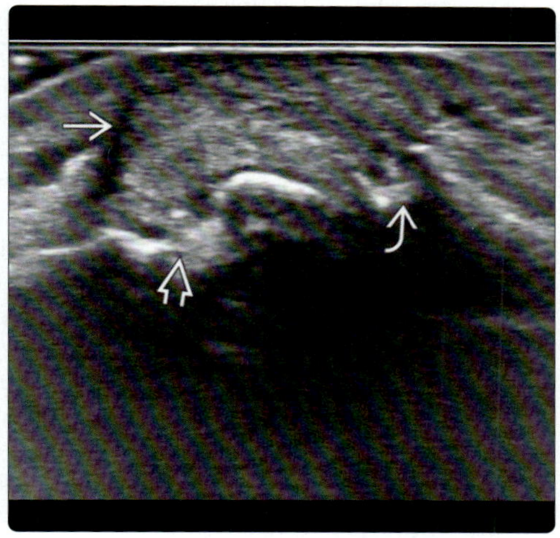

(Left) *Longitudinal US shows a subcutaneous gouty tophus superficial to the lateral malleolus. The gouty tophus contains multiple echogenic gouty crystal aggregates.* **(Right)** *Longitudinal US shows a soft tophus without posterior shadowing within the 1st metatarsophalangeal joint extending into the subcutaneous tissues. There is associated deep marginal erosion of the 1st metatarsal head and base of the proximal phalanx.*

Granuloma and Fibroma

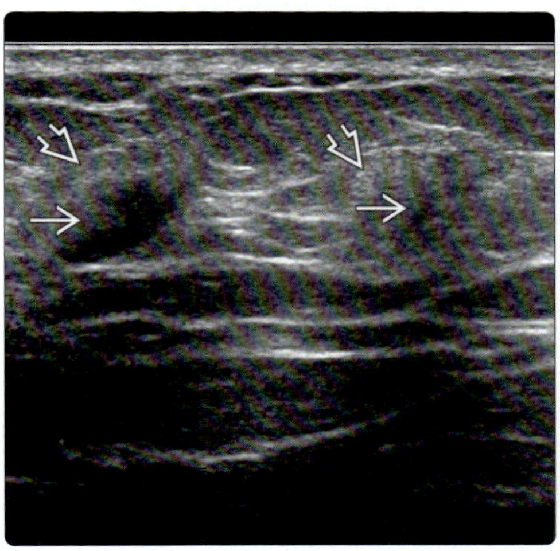

Normal Lymph Node

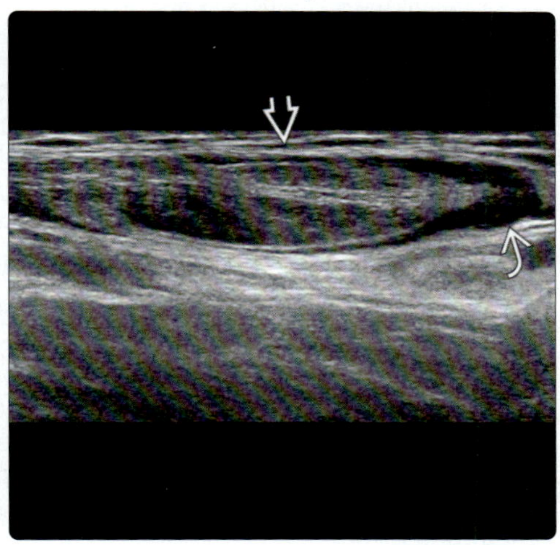

(Left) *Transverse US of the buttock shows 2 subcutaneous injection granulomas, each with a central hypoechoic area and a surrounding hyperechoic area. The different appearances are due to varying amounts of mature (hypoechoic) and immature (hyperechoic) granulation tissue.* **(Right)** *Longitudinal US shows a normal inguinal lymph node with a large central fatty hilum and a thin peripheral hypoechoic cortex. Such subcutaneous nodes are common in the medial proximal thigh.*

Fibrous Hamartoma of Infancy

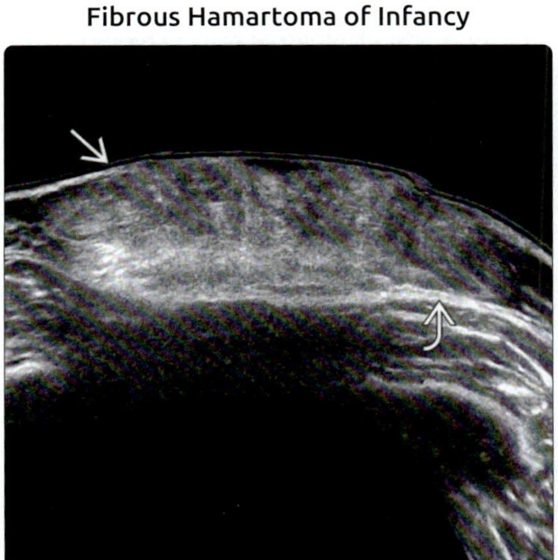

Fibrous Hamartoma of Infancy

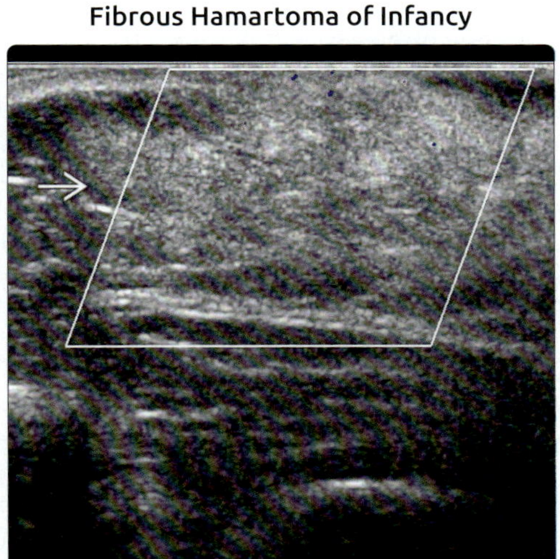

(Left) *Longitudinal US with extended field of view of the anterior thigh mass in an infant shows an ill-defined hyperechoic, heterogeneous subcutaneous mass abutting but not traversing the investing fascia. The varying echogenicity represents mature (hypoechoic) and immature (hyperechoic) fibrous tissue components.* **(Right)** *Transverse localized color Doppler US of the same lesion shows no intrinsic vascularity.*

DIFFERENTIAL DIAGNOSIS

Common

- Hematoma
- Muscle Tear
- Nerve Sheath Tumor

Less Common

- Soft Tissue Sarcoma
- Myxoma
- Fibromatosis
- Abscess
- Metastases
- Lymphangioma
- Intramuscular Extension of Baker Cyst
- Vascular Malformation
- Local Recurrence
- Intramuscular Ganglion Cyst
- Granuloma
- Diabetic Muscle Infarction

Rare but Important

- Parasitic Infection

ESSENTIAL INFORMATION

Key Differential Diagnosis Issues

- Clinical history is important for differentiating any muscle mass
- Acute hematoma and infection are usually readily distinguishable from other muscle masses
- Make every effort to reach diagnosis before resorting to percutaneous biopsy or aspiration
  - Often percutaneous biopsy is not necessary
  - However, if unsure regarding nature of mass, arrange either follow-up or biopsy or request alternative investigation (usually MR)

Helpful Clues for Common Diagnoses

- **Hematoma**
  - Due to trauma, anticoagulation, or vascular malformation
  - Anterior thigh is common location because it is susceptible to compression injury against femoral shaft and muscle contraction injury
  - Initially hyperechoic or isoechoic
    - Becomes more hypoechoic with increasing liquefaction after several days to weeks
    - ± layering or whorled pattern due to sequential episodes of bleeding (expanding hematoma)
  - Mild intramuscular hemorrhage may not be apparent on US, since ill-defined hyperechoic areas of blood blend with hyperechoic muscle
    - No disturbance of muscle architecture with minor bleeding
    - MR is more sensitive than US at detection of tumor hemorrhage
- **Muscle Tear**
  - Discontinuity ± retraction of muscle fibers within muscle bulk or at myofascial junction
    - Myofascial junction most common site of tear
  - ± tear filled with hematoma or fluid
  - ± surrounding muscle and subcutaneous edema
  - Do not confuse with muscle contusion
    - Contusion shows no fiber discontinuity
    - May cross muscle fascial boundaries
    - Ill-defined hyperechogenicity of muscle due to edema ± blood
    - Hyperemia during reparative stage
- **Nerve Sheath Tumor**
  - Arises along course of peripheral nerve
  - Well-defined, fusiform-shaped, hypoechoic mass
    - Anechoic areas due to myxoid accumulation (common)
    - ± additional areas of hyperechogenicity due to hemorrhage, calcification, or fibrosis
  - Posterior acoustic enhancement (mild to moderate)
  - Thickened entering or exiting nerve (very common)
    - Entering or exiting nerve may not be visible if tumor arises from small peripheral nerve
  - Split-fat sign due to separation of muscle fibers with enhanced visibility of fat at edge of tumor
    - Do not confuse with thickened nerve at edge of tumor
  - Mild to moderate hyperemia on color Doppler
    - Very occasionally, minimal demonstrable vascularity
  - Cannot differentiate between schwannoma and neurofibroma based on US findings alone
  - Try, however, to determine whether nerve is centrically or eccentrically placed relative to parent nerve

Helpful Clues for Less Common Diagnoses

- **Soft Tissue Sarcoma**
  - Usually seen as large, well-defined, hypoechoic mass within muscle layer
    - Moderately to highly vascular
    - Occasionally hypovascular or even avascular on color Doppler imaging
    - Hypovascularity does not exclude malignancy
  - Any large (> 5 cm), solid, nonfatty soft tissue tumor should be considered sarcoma unless proven otherwise by biopsy
    - Most large intramuscular masses require biopsy
    - Biopsy not very helpful in distinguishing well-differentiated liposarcoma from lipoma
    - Better to decide likelihood of liposarcoma by clinical and MR criteria as well as histological analysis of full tumor following excision
- **Myxoma**
  - Composed of few spindle-shaped cells within mucoid material supported by loose collagen framework
    - Derived from modified fibroblasts with increased proteoglycan production
  - Hypoechoic, cystic-type, intramuscular mass ± comet-tail artifacts
    - Peripheral rim of increased echogenicity (bright rim sign), corresponding to perilesional muscle edema
    - Triangular hyperechoic area adjacent to mass (bright cap sign), corresponding to muscle atrophy
    - No communication with adjacent joint
    - Main differential diagnosis is myxoid liposarcoma
- **Fibromatosis**
  - Also known as desmoid tumor

- Locally aggressive overgrowth of fibrous tissue with well-defined or infiltrative border
  - Typically absent or minimal vascularity on color Doppler imaging
- **Abscess**
  - More frequent in immunosuppressed patients
  - Many variable appearances
  - Thick-walled, irregular hypervascular outline with hypoechoic, avascular necrotic center (typical)
    - + edema and hyperemia of adjacent inflamed tissues
    - ± intralesional gas locules seen to move on real-time imaging
    - ± enlarged regional lymph nodes
  - US-guided aspiration of liquefied content helps to identify infective organism
- **Metastases**
  - Usually seen in setting of widely disseminated metastatic disease
  - Most common primary tumor is adenocarcinoma of lung
  - Similar in appearance to soft tissue sarcoma though metastases tend generally to be smaller than primary sarcoma
- **Lymphangioma**
  - More common in subcutaneous tissues though often multicompartmental
  - Thin-walled, multiloculated, septated mass
  - Thick walled with debris if complicated by infection or hemorrhage
  - No grayscale flow movement (as in venous vascular malformation)
- **Intramuscular Extension of Baker Cyst**
  - Occasionally extend into calf or distal thigh muscles
    - ± pericystic leakage or intracystic hemorrhage
- **Vascular Malformation**
  - Determine whether low flow (common) or high flow (uncommon)
  - Often multicompartment involvement
  - Thin walled multiseptated with serpiginous cystic spaces ± phleboliths (characteristic)
  - Variable hypo- or hyperechoic stromal component
  - Slow venous flow may be seen only on grayscale, as it is too slow to be detected by color Doppler imaging
- **Local Recurrence**
  - Discrete rounded mass at site of primary tumor excision
  - Need to differentiate from more linear, irregular hypoechoic scar tissue
  - US appearances of recurrence usually mimic those of primary tumor
- **Intramuscular Ganglion Cyst**
  - Extension of ganglion cyst from joint into muscle
  - Similar US appearances to ganglion cyst elsewhere
  - Hypoechoic ± echogenic comet-tail artifacts
  - ± intracystic hemorrhage or pericystic leakage
    - Pericystic leakage usually associated with pericystic inflammation
- **Granuloma**
  - Much more common in subcutaneous tissues, especially gluteal region
  - Usually following injection
    - Occasionally following trauma or foreign body implantation
  - Majority have calcified rim
- **Diabetic Muscle Infarction**
  - Occurs in diabetics with established vasculopathy
  - Ill-defined hypoechoic area becoming better defined after 1 week
    - Muscle architecture remains vaguely detectable within hypoechoic area
    - No necrosis
    - No detectable vascularity at onset → mild to moderate hyperemia after about 1 week

Helpful Clues for Rare Diagnoses

- **Parasitic Infection**
  - Cysticercosis: Most common parasitic infection of soft tissue
    - Well-defined, small, hypoechoic mass with eccentric echogenic nidus representing scolex
    - Diagnosis can be confirmed by therapeutic response

Hematoma

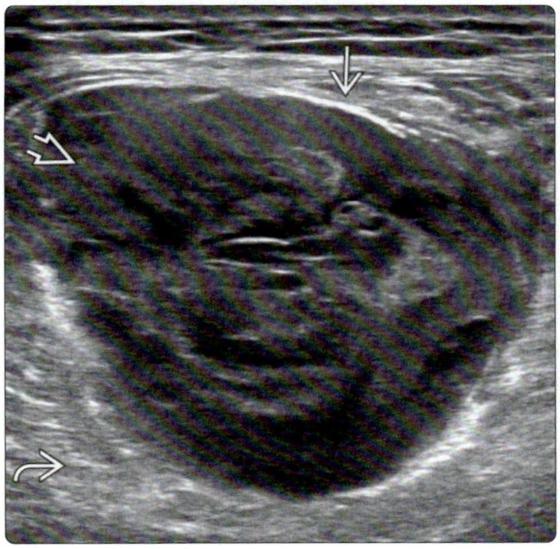

Hematoma

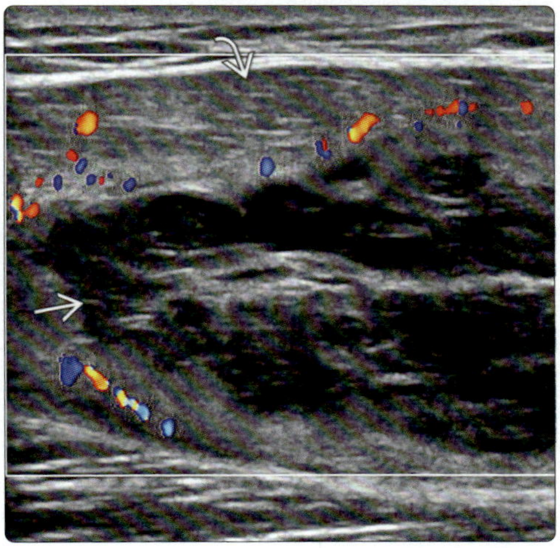

(Left) *Transverse US of 78-year-old man who was overanticoagulated with warfarin shows a large hematoma ➡ within the biceps femoris muscle ⮕. There is early rim calcification ➡. No hyperemia was present.* **(Right)** *Longitudinal US of a 38-year-old man with thigh swelling following a football injury shows a healing hematoma ➡ in the rectus femoris muscle ⮕. There is moderate reparative marginal hyperemia.*

(Left) *Transverse US shows liquefied calf muscle hematoma following trauma 2 months earlier. Note the fine echogenic content suggestive of hemorrhage and the entrapped fluid. Aspiration yielded old blood.* **(Right)** *Transverse US of man with slowly enlarging distal triceps swelling for 4 years shows a well-defined partially cystic mass with a softish-solid component. Mild marginal hyperemia was present. Biopsy revealed hemorrhage and fibrous organization consistent with chronic expanding hematoma.*

Hematoma

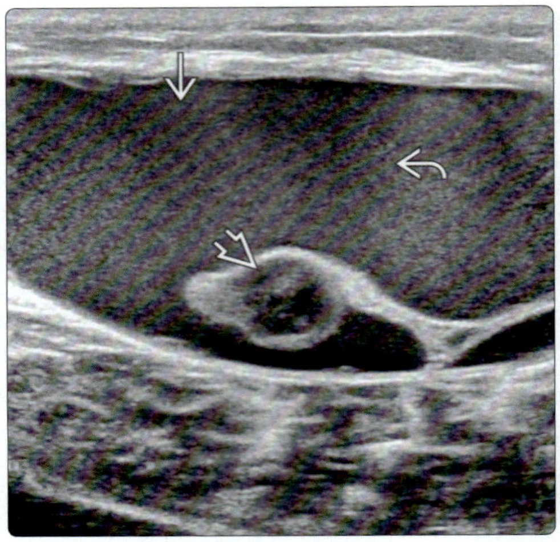

Hematoma

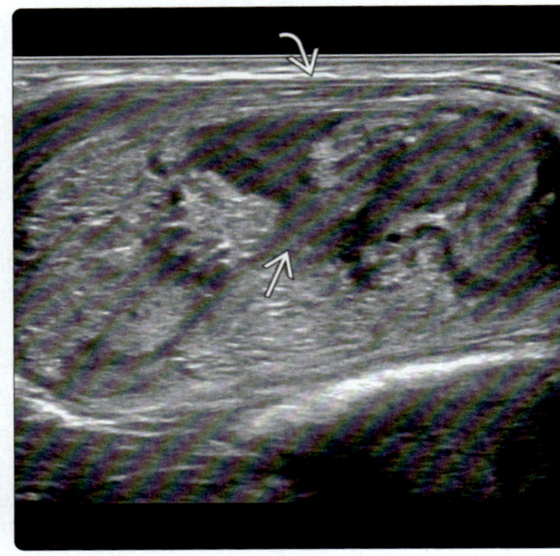

(Left) *Longitudinal US shows a partially liquefied hematoma filling the gap formed by a myofascial tear between the medial belly of the gastrocnemius and the soleus muscles.* **(Right)** *Transverse US of a patient following a kick from a horse shows a large hypoechoic muscle tear within the disorganized, slightly swollen vastus lateralis muscle. The femoral shaft is normal.*

Muscle Tear

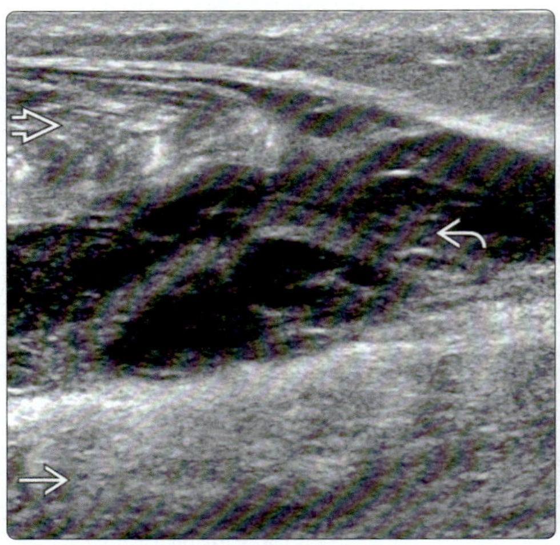

Muscle Tear

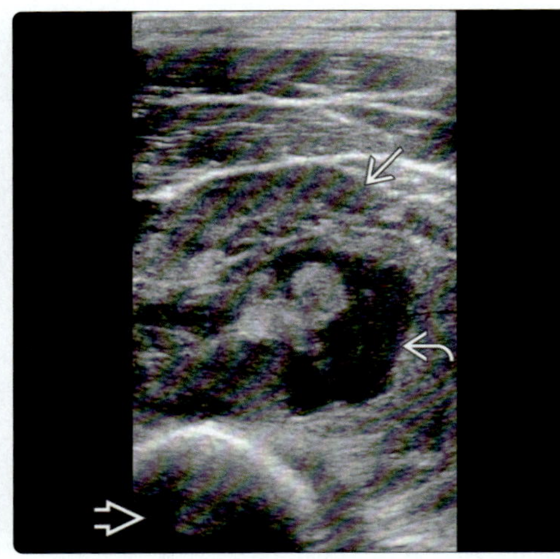

(Left) *Longitudinal US shows a nerve sheath tumor within the sartorius muscle. The adjacent increased echogenicity is due to muscle fiber splaying leading to exaggeration of intramuscular fat (split-fat sign). There is characteristic mild acoustic enhancement.* **(Right)** *Longitudinal US shows a large, partially cystic hypoechoic mass within the medial head of the gastrocnemius. Mild intrinsic vascularity was present (not shown). Histology revealed a monophasic synovial sarcoma.*

Nerve Sheath Tumor

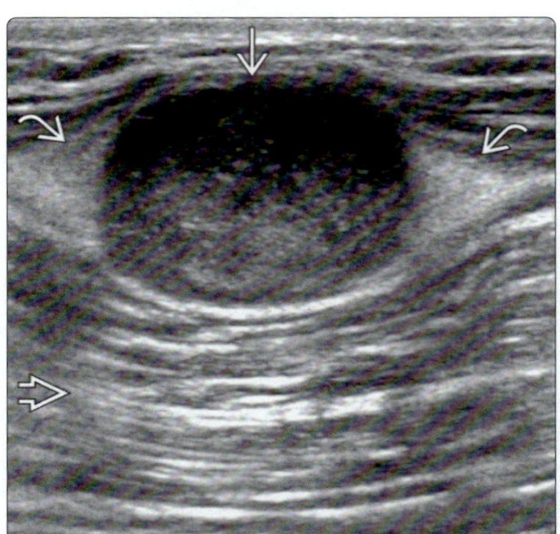

Soft Tissue Sarcoma

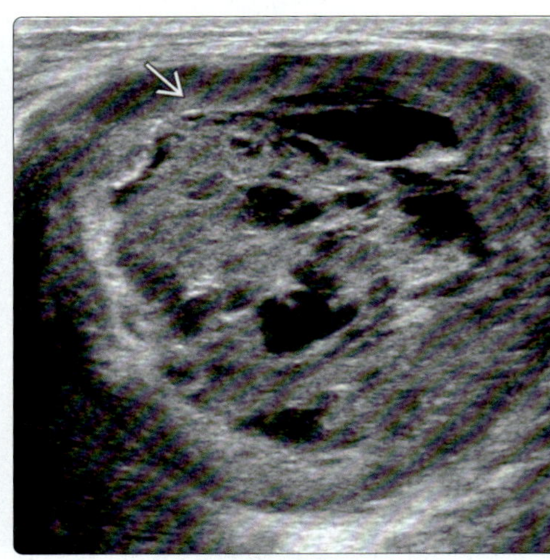

Myxoma

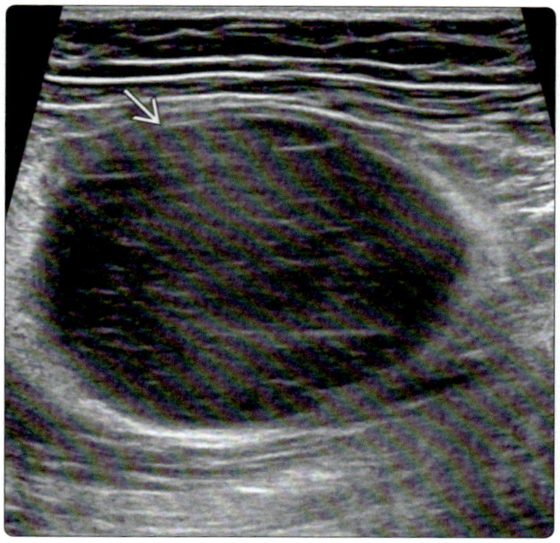

Fibromatosis

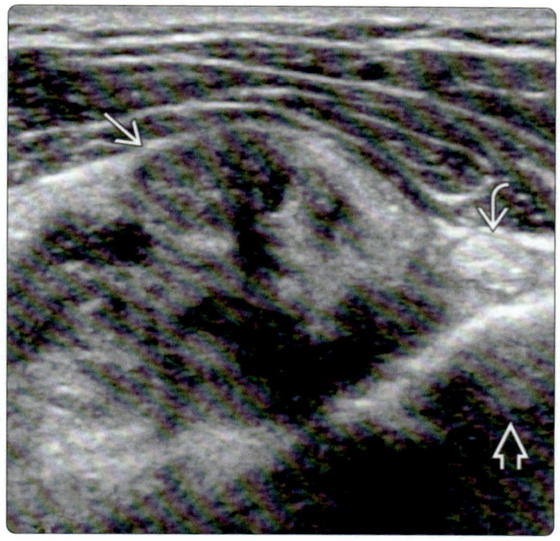

(Left) *Transverse US shows a large intramuscular myxomatous-type mass ➔ in the midthigh region. This may represent either a myxoid-type "ancient" schwannoma or a myxoma. Histology confirmed the latter diagnosis.* **(Right)** *Transverse US shows a well-defined heterogeneous mass ➔ alongside the coracoid process ➔. Biopsy confirmed fibromatosis. The long head of biceps tendon is normal ➔.*

Fibromatosis

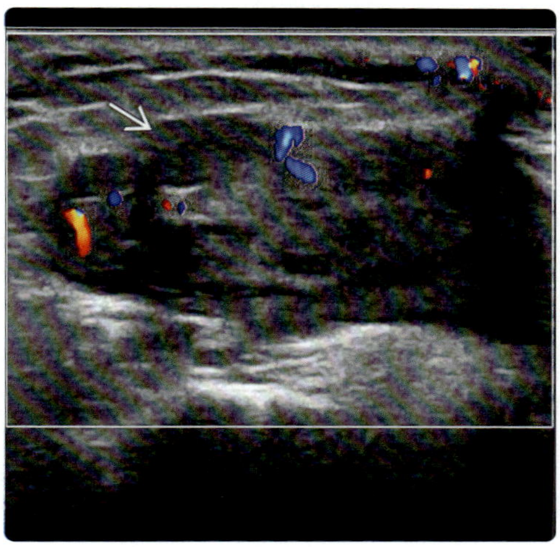

Fibromatosis

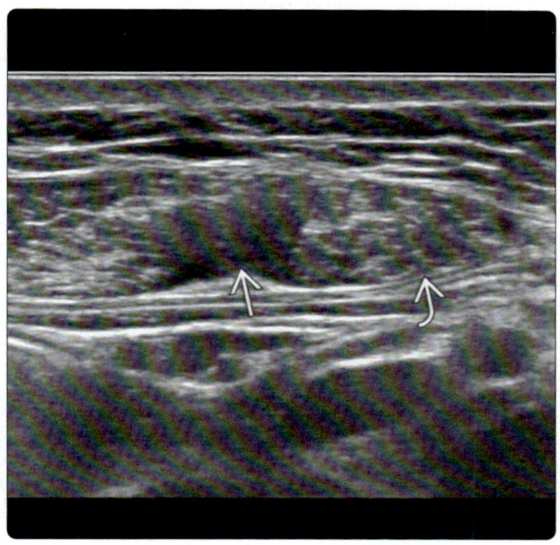

(Left) *Transverse US of the arm shows an area of fibromatosis (desmoid tumor) ➔ in the triceps muscle. This tumor remained unchanged for many years. Mild hyperemia is present.* **(Right)** *Transverse US of 38-year-old woman shows fibromatosis (desmoid tumor) ➔ in the rectus abdominal muscle ➔. The surrounding muscle has moderate fatty atrophy. This fibromatosis reduced slightly in size over a 3-year period. Fibromatosis has a variable US appearance depending on tumor maturity.*

Abscess

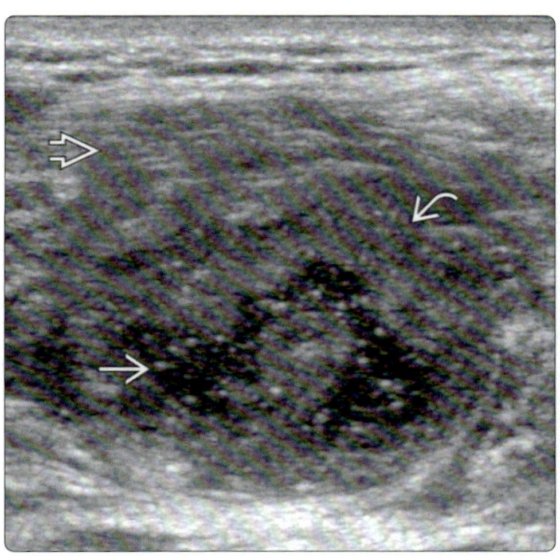

Abscess

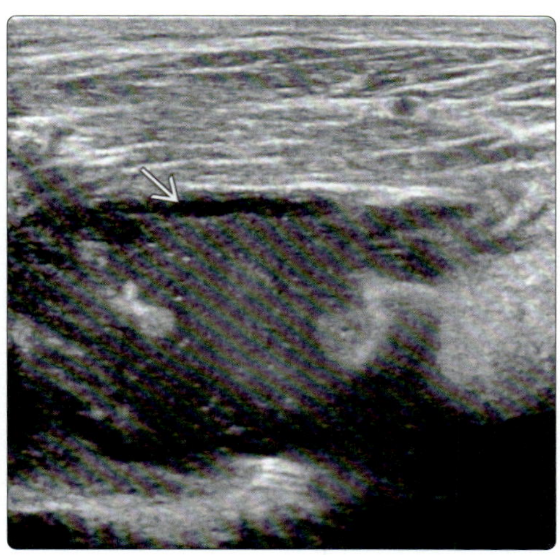

(Left) *Longitudinal US shows an intramuscular abscess ➔. Many small echogenic foci ➔, due to proteinaceous aggregates or gas, are present and seen to move on real-time imaging. The margin is ill-defined with moderate edema of the surrounding muscle ➔.* **(Right)** *Longitudinal US shows a large, irregular fluid collection ➔ present on the posteromedial aspect of the distal femur. Aspiration yielded purulent fluid that grew Staphylococcus aureus.*

(Left) *Transverse US in a patient with acute myeloid leukemia, pancytopenia, and fever shows a medium-sized abscess present within a moderately edematous rectus femoris muscle. Aspiration yielded Pseudomonas aeruginosa.* **(Right)** *Longitudinal US in 68-year-old man with known metastatic nasopharyngeal carcinoma shows a large metastatic deposit within the rectus femoris muscle. This was confirmed by biopsy. Additional imaging of the groin revealed metastatic-type lymph nodes.*

Abscess

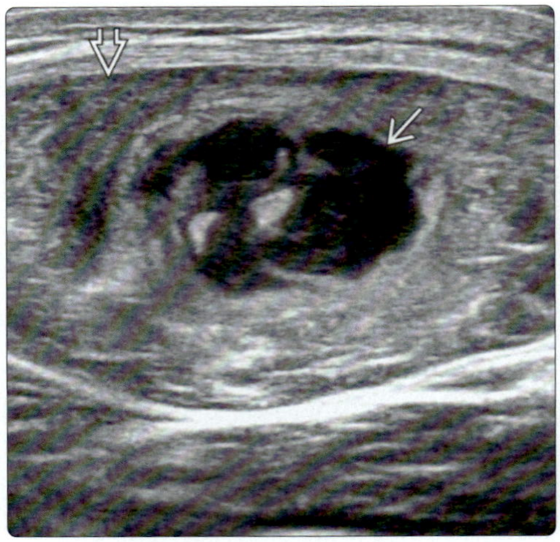

Metastases

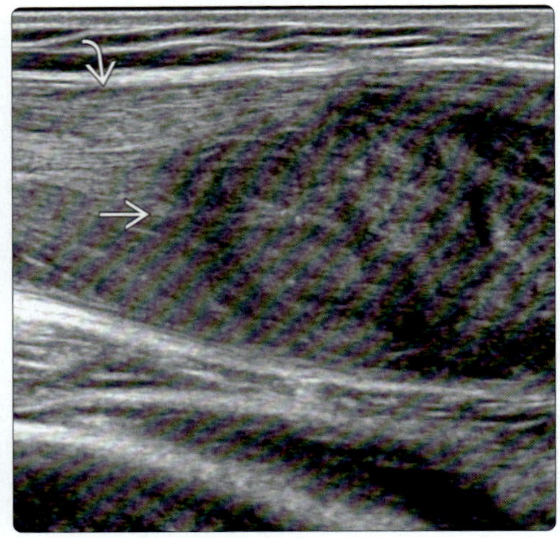

(Left) *Transverse US in an HIV/AIDS patient with known EBV-related liver and adrenal tumors shows a small hypoechoic rectus abdominis muscle tumor with moderate surrounding edema, compatible with an intramuscular metastatic deposit.* **(Right)** *Longitudinal oblique US shows the distal end of a large Baker cyst that has ruptured into the medial belly of the gastrocnemius muscle. This rupture is not acute as shown by the poor wall definition and lack of edema.*

Metastases

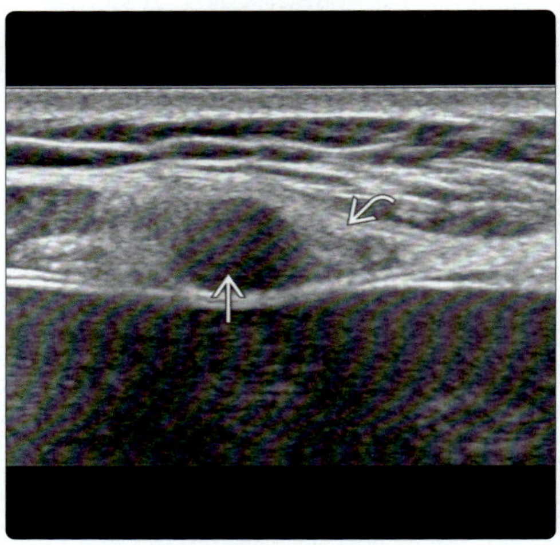

Intramuscular Extension of Baker Cyst

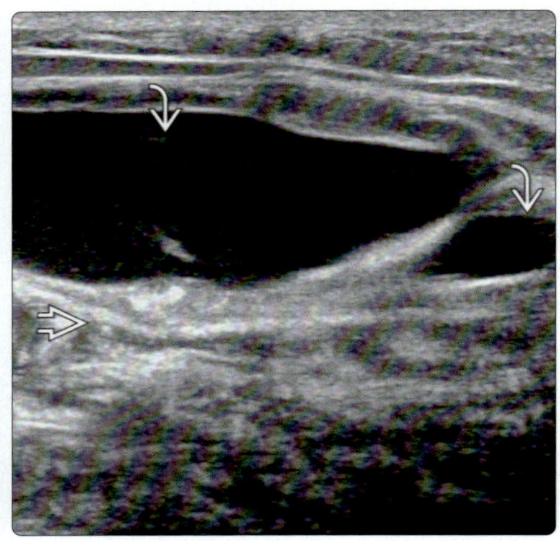

(Left) *Transverse US shows a vascular malformation with large serpiginous vascular channels occupying all of the hypothenar muscle. The 1st metacarpal bone is normal.* **(Right)** *Transverse spectral and color Doppler US of the same lesion shows the vascular malformation to be a moderate- to high-flow lesion with low-resistance, arterial, intralesional flow.*

Vascular Malformation

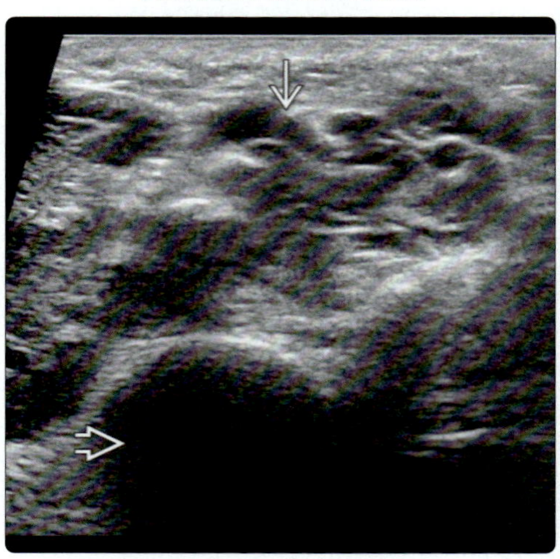

Vascular Malformation

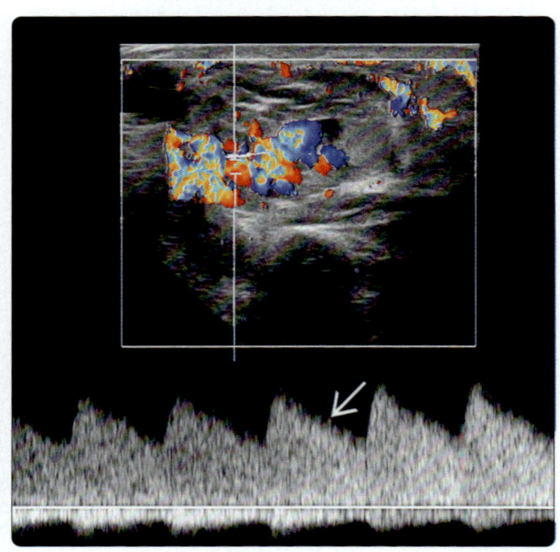

Local Recurrence

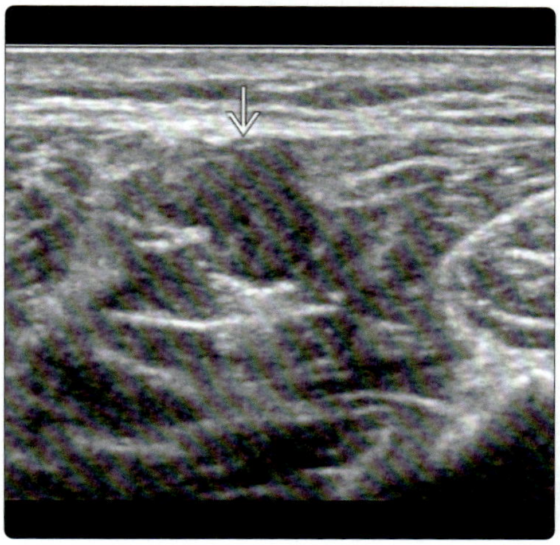

Local Recurrence

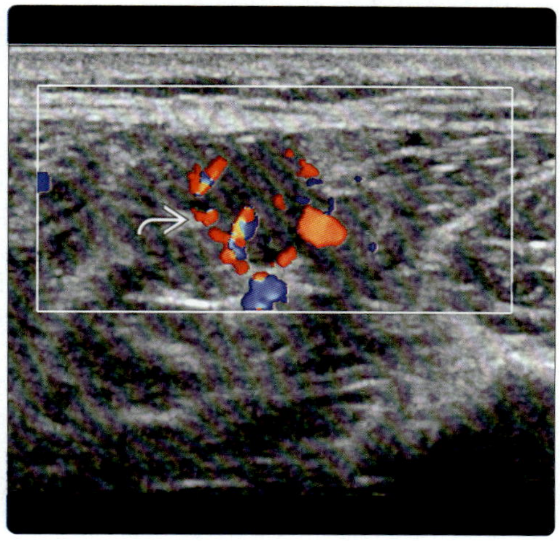

(Left) *Transverse US of a patient with resection of a soleus muscle inflammatory myxohyaline tumor 18 months earlier shows a small hypoechoic intramuscular nodule →. US-guided biopsy confirmed local tumor recurrence.* **(Right)** *Transverse color Doppler of the same tumor shows moderate marginal hyperemia ↪.*

Intramuscular Ganglion Cyst

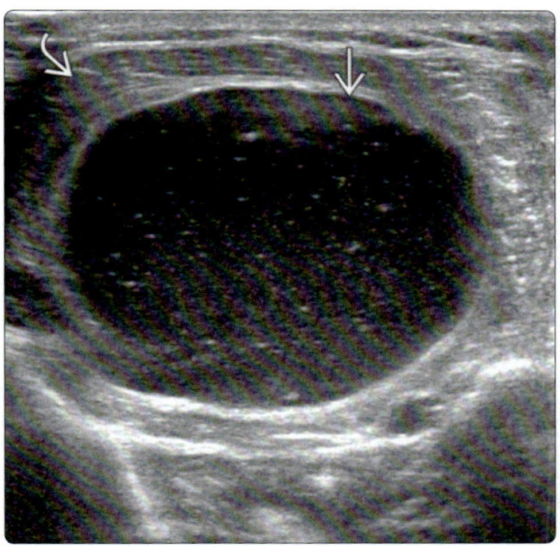

Intramuscular Ganglion Cyst

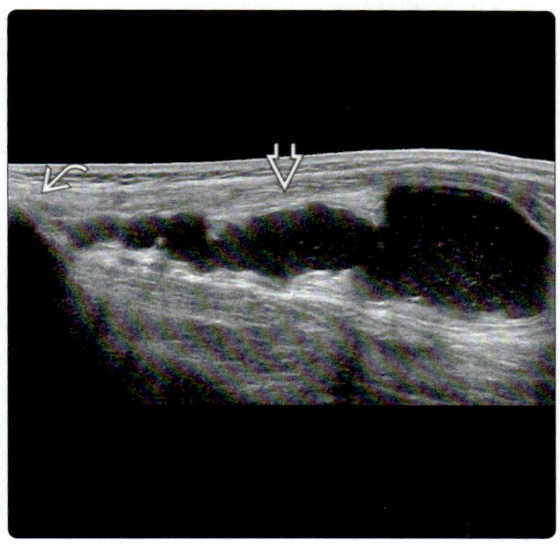

(Left) *Transverse US shows a hypoechoic cystic-type mass → with small echogenic foci within the tibialis anterior muscle ↪.* **(Right)** *Longitudinal US of the same tumor shows a wide hypoechoic tract ⇨ connecting the proximal end of the cyst to the proximal tibiofibular joint ↪. This represents a paraarticular ganglion cyst that has extended into the tibialis anterior muscle.*

Parasitic Infection

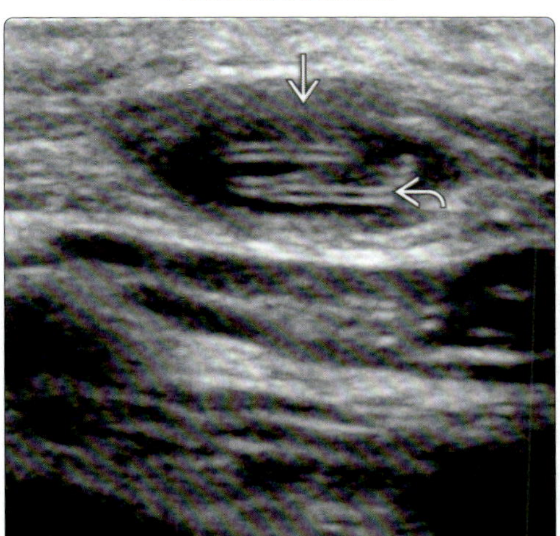

Parasitic Infection

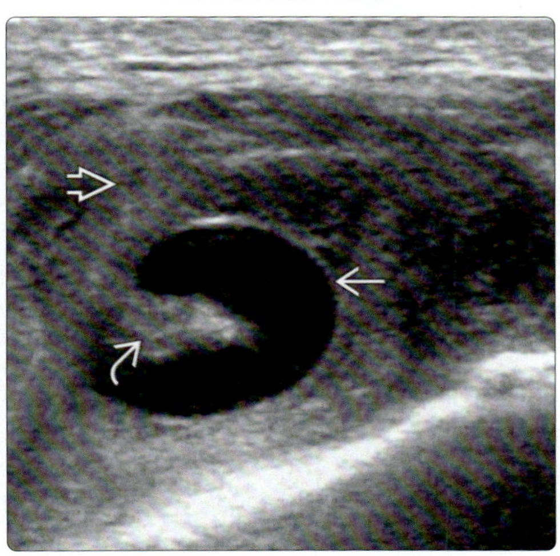

(Left) *Longitudinal US of the hand shows cysticercosis infestation of the thenar muscle with a hypoechoic cyst → containing an echogenic tapeworm ↪. (Courtesy V. Hardas, MD.)* **(Right)** *Transverse US shows intramuscular cysticercosis infestation as a small, hypoechoic intramuscular mass → with an eccentric echogenic nidus ↪ representing the scolex. The surrounding arm musculature is edematous ⇨. (Courtesy V. Hardas, MD.)*

DIFFERENTIAL DIAGNOSIS

Common

- Intermuscular Lipoma
- Intramuscular Lipoma
- Muscle Edema
- Vascular Malformation
- Hematoma

Less Common

- Foreign Body
- Soft Tissue Sarcoma
- Muscle Metastases
- Local Recurrence
- Fibromatosis
- Peripheral Nerve Sheath Tumor
- Myositis
- Rhabdomyolysis
- Venous Dilatation

ESSENTIAL INFORMATION

Key Differential Diagnosis Issues

- Muscle echogenicity varies according to orientation of muscle fiber and transducer and relative amount of lean or fatty tissue
  - Lean muscle tends to be more hypoechoic
  - More fatty muscle (fatty atrophy) tends to be more hyperechoic
- As muscle fibers are aligned in different directions, anisotropy can be used to demarcate apposing muscles

Helpful Clues for Common Diagnoses

- **Inter- or Intramuscular Lipoma**
  - Located deep to investing fascia
  - May be classified as subfascial, intermuscular, intramuscular, or submuscular in location
    - Subfascial: Between investing fascia and muscle
    - Intermuscular: Between muscles (most common)
    - Intramuscular: Within muscle
    - Submuscular: Beneath muscle (i.e., juxtacortical); least common location
  - US appearances of deep lipomas are similar to subcutaneous lipomas but more variable
    - Usually larger than subcutaneous lipomas at presentation
    - Less compressible
    - Occasional vessels detected as opposed to subcutaneous lipomas, which are nearly always avascular
  - Discrete noninfiltrative mass with well-defined convex margins
  - Fusiform or oblong in shape with long axis parallel to skin
  - May be hypoechoic (infrequent)
    - Usually hyperechoic to muscle
  - Fine linear echogenic striations running parallel to long axis of tumor in both transverse and longitudinal planes
  - May show acoustic enhancement as US transmission in fat > muscle
  - No vascularity in or around lipoma on color Doppler
    - Occasionally, several vessels traversing lipoma may be visible
  - Consider malignancy if the following features are present
    - Large mass
    - Thick (> 2 mm) septations
    - Nodular areas of nonlipomatous tissue
    - Cyst-like or necrotic areas within tumor
    - Majority of tumor is nonlipomatous
  - 4 distinct histologic types of liposarcoma
    - Well-differentiated liposarcoma
    - Myxoid liposarcoma
    - Pleomorphic liposarcoma
    - Dedifferentiated liposarcoma
  - MR is helpful for surgical planning and establishing baseline for all deep-seated lipomas
- **Muscle Edema**
  - From inflammation, infection, venous congestion, trauma
  - Muscle becomes hyperechoic; intense edema is hypoechoic
  - Variable hyperemia
  - Usually has associated subcutaneous ± perifascial edema
  - Diffuse intramuscular bleeding can give similar appearance
- **Vascular Malformation**
  - Vascular malformation and hemangioma indistinguishable on US
  - Hemangioma
    - Vascular tumor of childhood: Progressive → regressive stage
    - Regresses usually by puberty
  - Vascular malformation
    - Vascular dysmorphogenesis
    - Enlarges in line with skeletal development
    - May change shape in adulthood due to thrombosis and revascularization
  - 5 different subtypes of vascular malformation depending on predominant vascular component
    - Capillary, venous, arteriovenous, lymphatic, or mixed
  - US is helpful in establishing diagnosis, assessing flow, and determining relative amount of stroma vs. vascular component
- **Hematoma**
  - Intramuscular hematoma may occur due to trauma, anticoagulation, vascular malformation, or sarcoma
    - Anterior thigh is common site, as it is prone to muscle contusion injury
  - Acute hematoma is usually isoechoic or hyperechoic to muscle
    - ± layering due to repeated hemorrhage
    - Becomes more hypoechoic with increasing liquefaction after several days
    - ± sedimentation present after few minutes of rest
  - Mild diffuse intramuscular hemorrhage may be overlooked on US, as ill-defined hyperechoic areas blend with muscle
    - MR is more sensitive than US at detecting mild muscle injury

- Remember that intramuscular sarcomas or vascular malformations may bleed spontaneously or during exercise
  - If no trauma history, always consider likelihood of underlying precipitating condition

Helpful Clues for Less Common Diagnoses

- **Foreign Body**
  - Lodges in subcutaneous tissues much more commonly than in muscle
  - Nearly all foreign bodies are echogenic
  - ± reverberation artifact or acoustic shadowing
  - Bamboo and noncompact wood are difficult to see because they absorb fluid and have echogenicity similar to soft tissue
    - Always repeat US after 1-3 days if foreign body is not detected initially
  - Missed foreign bodies often develop hypoechoic rim
    - Rim is composed of hyperemic granulation and fibrous tissue
    - Makes foreign body more conspicuous
- **Soft Tissue Sarcoma**
  - Occurs most frequently in muscle layer
    - Majority are hypoechoic
    - Occasionally hyperechoic due to fat, fibrosis, or hemorrhage
    - US appearances do not conform to recognized tumor (e.g., nerve sheath tumor, lipoma, vascular malformation, fibromatosis, etc.)
    - If suspected, usually requires percutaneous biopsy
- **Muscle Metastases**
  - Most patients have known primary tumor with disseminated metastases
  - Adenocarcinoma > squamous cell carcinoma
  - Breast, lung, and upper aerodigestive tract are common primary sites
  - Ill-defined, hypoechoic muscle mass aligned along long axis of muscle (typical)
    - US appearances similar to those of primary tumor
  - Often, several muscle metastases occur within same area
  - Percutaneous biopsy is diagnostic and indicated if confirmation required
- **Local Recurrence**
  - Discrete mass at site of primary tumor excision
  - US appearances are similar to those of primary tumor
  - Must be distinguished from reparative postoperative nodular fibrosis
    - Follow-up ± biopsy if in doubt
- **Fibromatosis**
  - May appear echogenic if fibrous tissue component is not fully mature
  - Usually hypoechoic or mixed hypo-/hyperechoic tumor
  - Usually well defined without undue tethering or infiltrative element
  - Minimal or no tumoral hyperemia
  - Evolves slowly on serial imaging
- **Peripheral Nerve Sheath Tumor**
  - Majority are hypoechoic
    - Similar appearance to subcutaneous nerve sheath tumor
    - Tumoral hyperechogenicity indicates internal hemorrhage
- **Myositis**
  - Ill-defined increase in muscle echogenicity and swelling of muscle; decreased definition of muscle architecture ± focal hyperemia
    - Inflammatory myositis is often symmetrical
  - MR is more sensitive than US at detecting mild myositis
- **Rhabdomyolysis**
  - Necrosis of skeletal muscle
    - Contents of injured muscle cells leak into circulation
    - If severe → electrolyte imbalance, acidosis, coagulopathy, hypovolemia, and acute renal failure
  - Many causes, including trauma, strenuous muscle exercise, and prolonged muscle compression
  - Diffusely echogenic muscle or muscles
- **Venous Dilatation**
  - Slow-moving blood → red cell clumping → vascular hyperechogenicity
    - Sedimentation effect may be visible if vein at rest

Intermuscular Lipoma

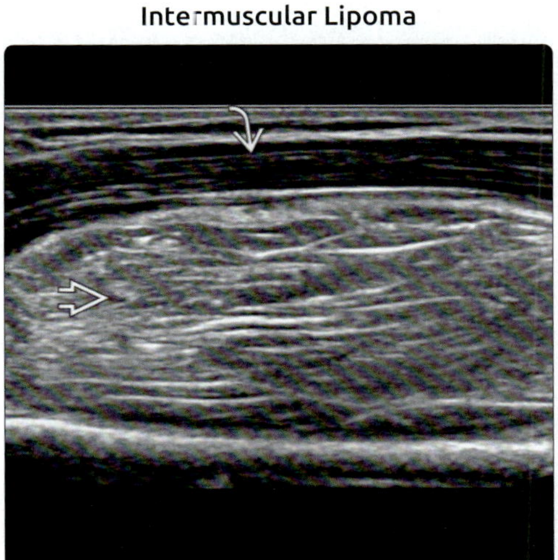

Intermuscular Lipoma

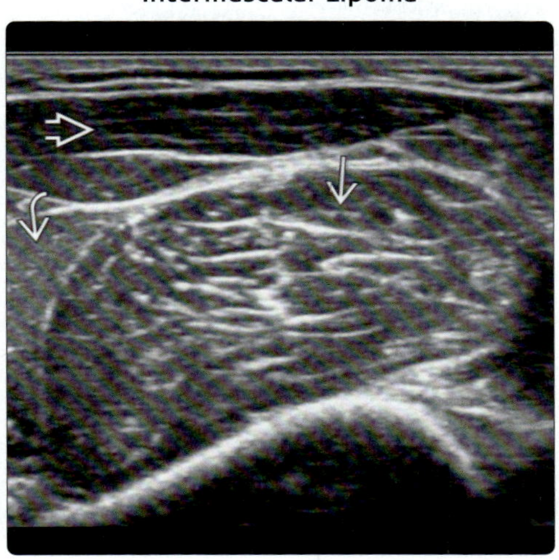

(Left) *Longitudinal US in an 83-year-old man with midarm mass for at least 5 years shows a large intermuscular lipoma ➡ present deep to the biceps muscle ➡ in the midarm. Typical appearances of a lipoma are present with no untoward features.* **(Right)** *Transverse US of the same lesion shows the lipoma ➡ located deep to the biceps ➡ and alongside the brachialis ➡. No biopsy was performed, as it is usually not helpful in this setting. In view of the patient's age, no treatment was undertaken.*

(Left) *Longitudinal US shows a large lipoma on the lateral aspect of the elbow and deep to the brachioradialis muscle. Note the fine linear striations present throughout the tumor, typical of lipoma.* **(Right)** *Transverse US shows a medium-sized intramuscular lipoma within the midportion of the deltoid muscle. The deltoid is a common site for intramuscular lipoma. The humeral head is normal.*

Intermuscular Lipoma

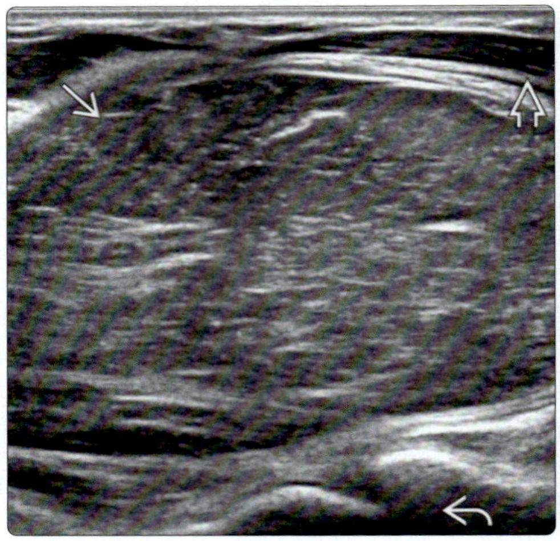

Intramuscular Lipoma

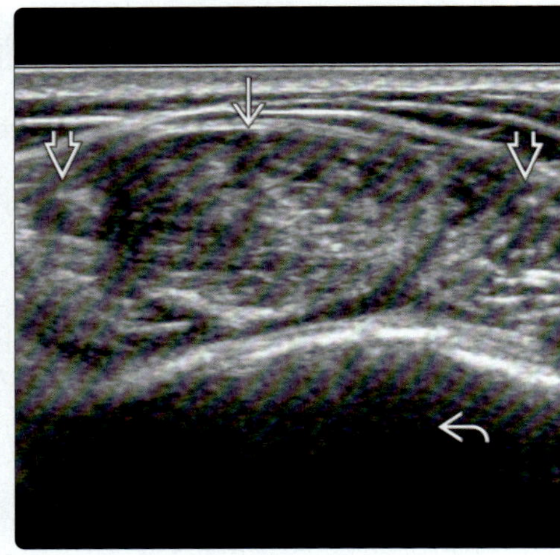

(Left) *Longitudinal US of the same tumor shows the typical mild hyperechogenicity and fine linear striations of a lipoma. There are no untoward features present. There is no need for percutaneous biopsy in this case.* **(Right)** *Longitudinal US shows moderate edema of the subcutaneous tissues as well as the gastrocnemius and soleus muscles of the calf. Edematous muscle is diffusely hyperechoic with preservation of muscle architecture.*

Intramuscular Lipoma

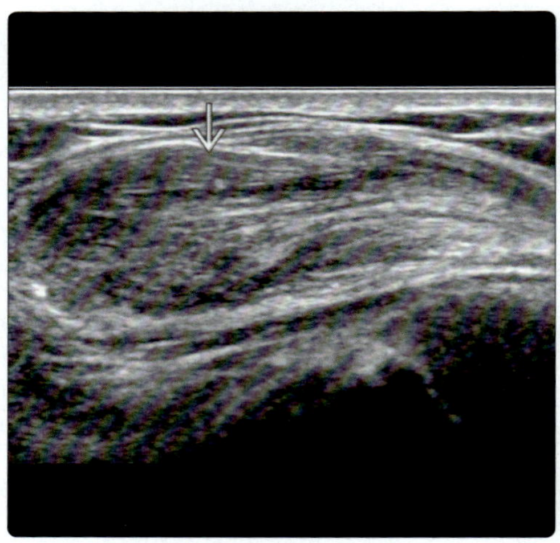

Muscle Edema

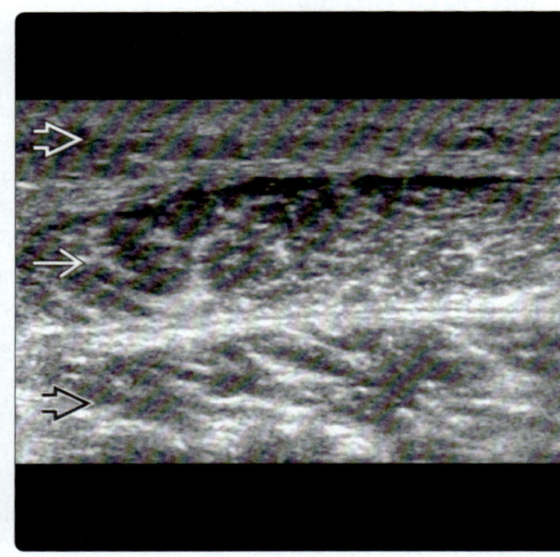

(Left) *Transverse US shows a moderately sized, hyperechoic vascular malformation within the distal part of the triceps alongside the olecranon. The vascular malformation is primarily composed of large vessels. A single phlebolith is present.* **(Right)** *Correlative transverse color Doppler US in the same patient shows slow-flowing vascular channels occupying most of the lesion. The findings are consistent with a venous vascular malformation.*

Vascular Malformation

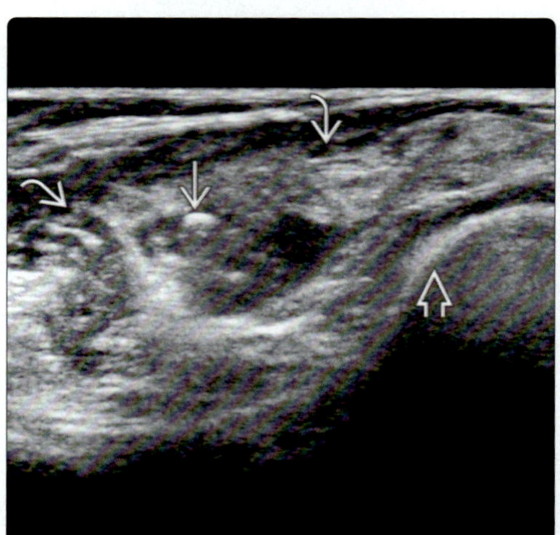

Vascular Malformation

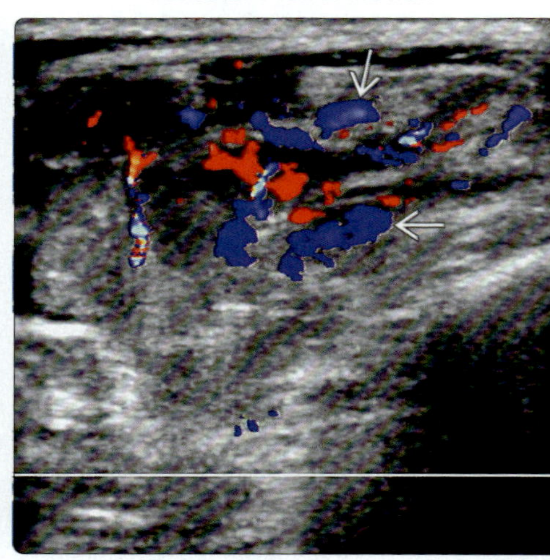

Vascular Malformation

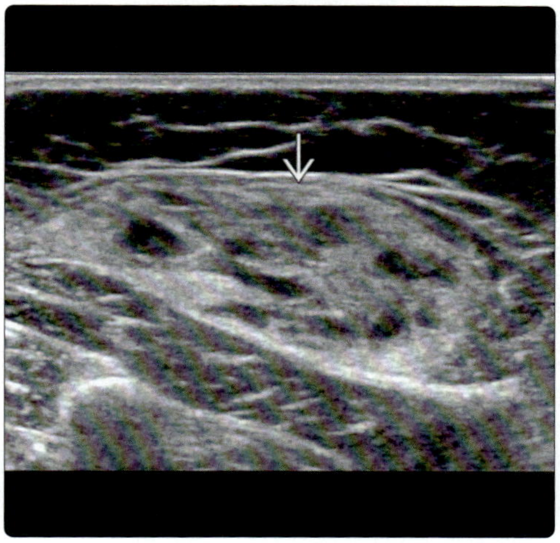

Vascular Malformation

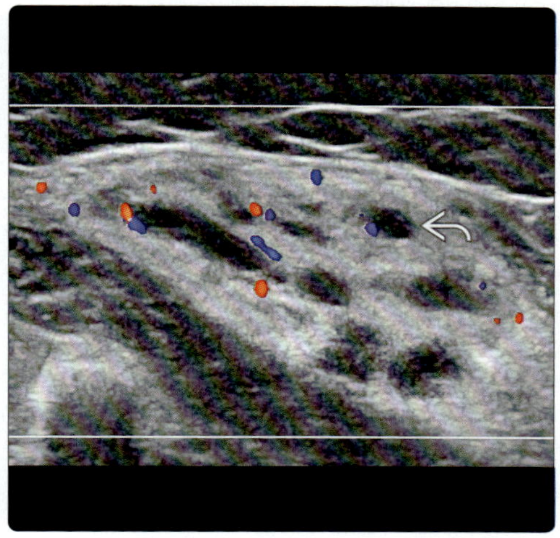

(Left) *Longitudinal US shows a medium-sized, well-defined, intramuscular vascular malformation within the gastrocnemius muscle. The vascular malformation is composed of ~ 25% venous-type channels with ~ 75% echogenic stroma component. No phleboliths are present.* **(Right)** *Longitudinal color Doppler US shows the vascular malformation to be low flow with mild hyperemia. Venous channels often have no demonstrable color flow due to slow flow.*

Hematoma

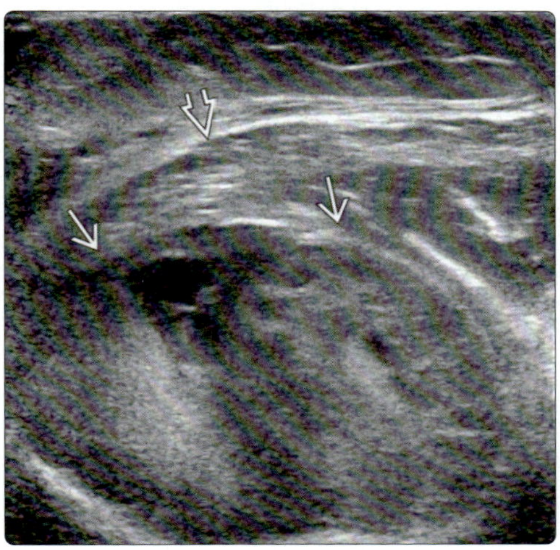

Hematoma

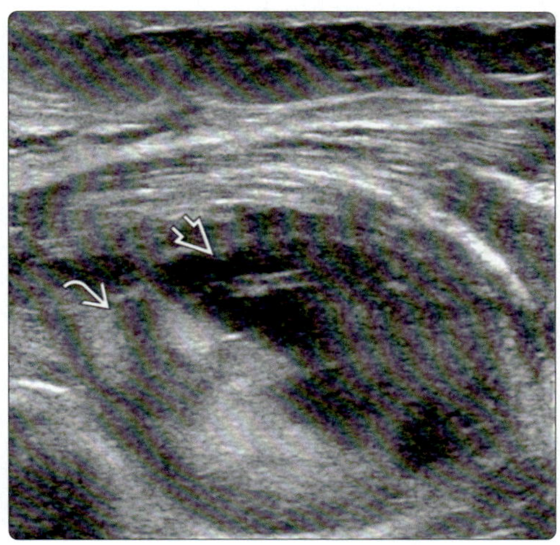

(Left) *Transverse US of an anticoagulated patient shows an acute hyperechoic hematoma within the vastus lateralis muscle.* **(Right)** *Transverse US of the same hematoma after 5 minutes of rest shows sedimentation of content with hypoechogenicity of the more superficial component and hyperechogenicity of the deeper component. Sedimentation is an absolute indicator of fluid content.*

Hematoma

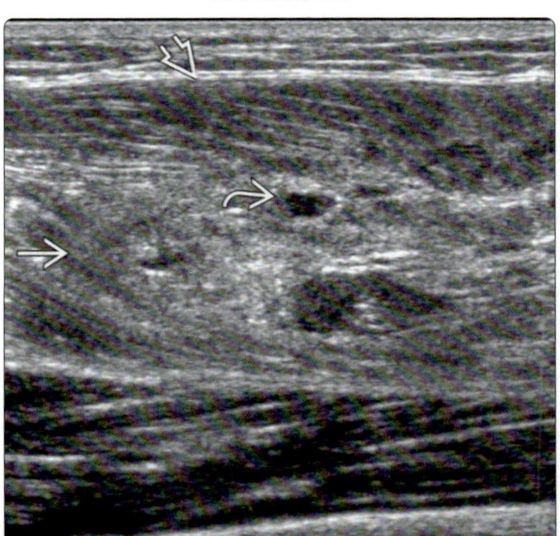

Foreign Body

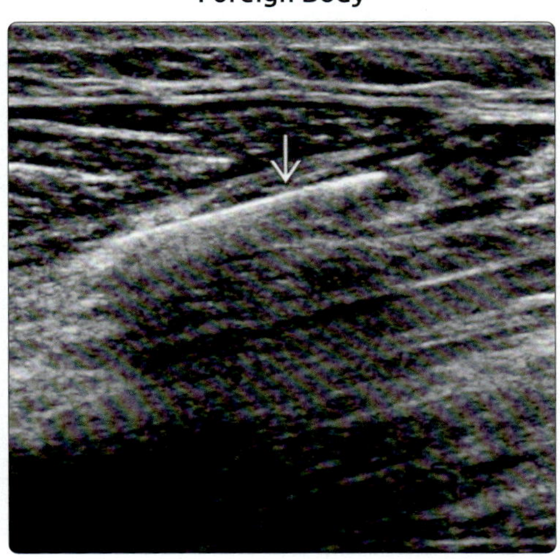

(Left) *Longitudinal US in a 33-year-old man after thigh trauma shows a largely mildly echogenic hematoma in the rectus femoris muscle. Some small liquefied areas are present. There was mild marginal hyperemia.* **(Right)** *Longitudinal US shows a retained needle in the calf muscle, which had been present for many years. During US-guided removal with forceps, the needle was found to be firmly adherent to the adjacent muscle and had to be pried off by rotating the forceps.*

Hyperechoic Muscle Mass

(Left) *Longitudinal US shows a large hyperechoic tumor* ➡ *within the calf musculature. US-guided biopsy for histology revealed a liposarcoma.* **(Right)** *Longitudinal US shows the more distal end of the same tumor* ➡. *Color Doppler US (not shown) showed only minimal tumor vascularity.*

Soft Tissue Sarcoma

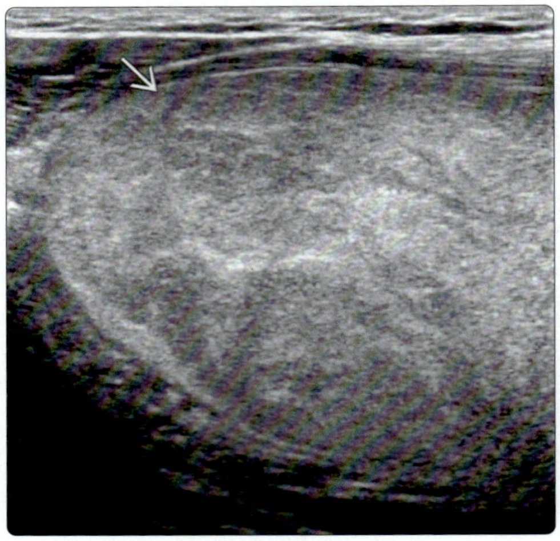

Soft Tissue Sarcoma

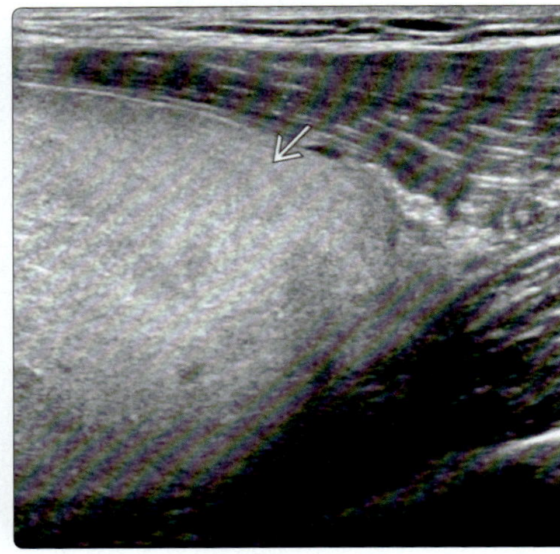

(Left) *Transverse US shows a large echogenic metastasis* ➡ *within the teres major muscle* ➡. *This patient had disseminated shoulder metastases from a nasopharyngeal primary tumor.* **(Right)** *Transverse US of the leg in a 28-year-old woman shows a medium-sized mass* ➡ *at the gastrocnemius myotendinous junction superficially with prominent spiculated muscular interdigitation* ➡. *No hypervascularity was present. Relatively immature fibromatosis may appear slightly echogenic.*

Muscle Metastases

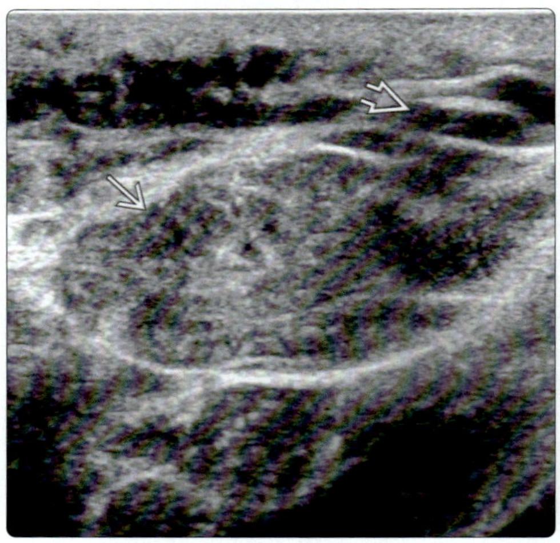

Fibromatosis

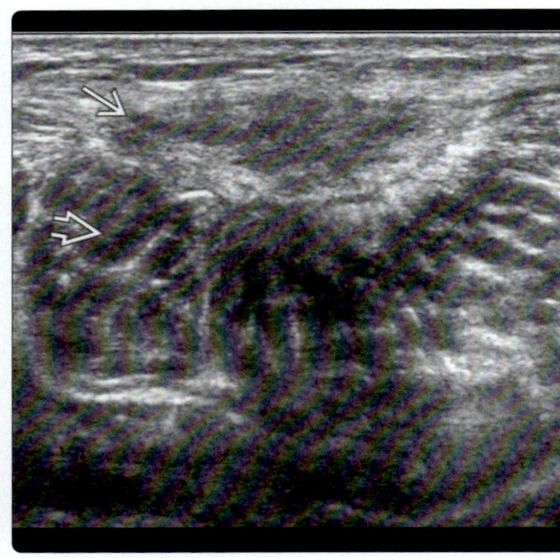

(Left) *Longitudinal US shows a partially hyperechoic nerve sheath tumor* ➡ *of the deep peroneal nerve anterior to the ankle* ➡. *Hyperechogenicity within a nerve sheath tumor usually indicates intratumoral hemorrhage. The deep peroneal nerve is thickened* ➡ *proximally.* **(Right)** *Transverse US shows an echogenic nerve sheath tumor* ➡ *of a small intramuscular nerve in the calf. This histology-confirmed nerve sheath tumor had increased in size, most likely due to intratumoral hemorrhage. This is not a sign of malignancy.*

Peripheral Nerve Sheath Tumor

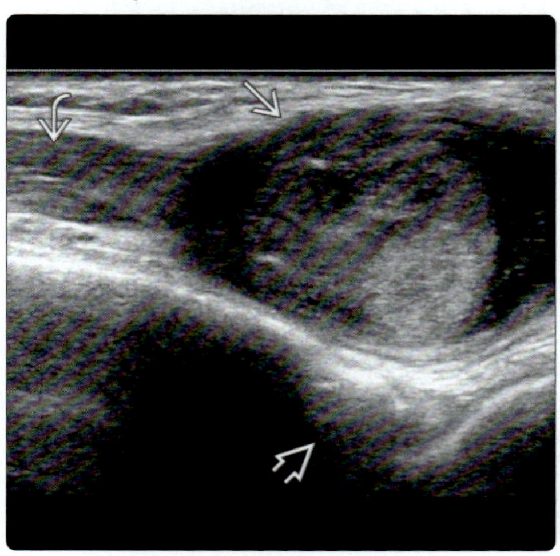

Peripheral Nerve Sheath Tumor

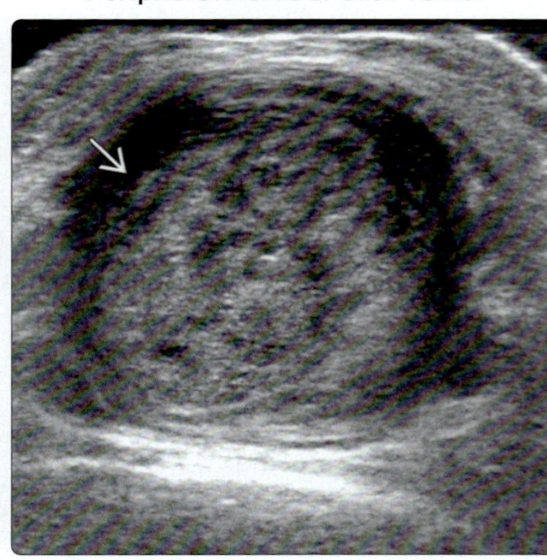

Myositis

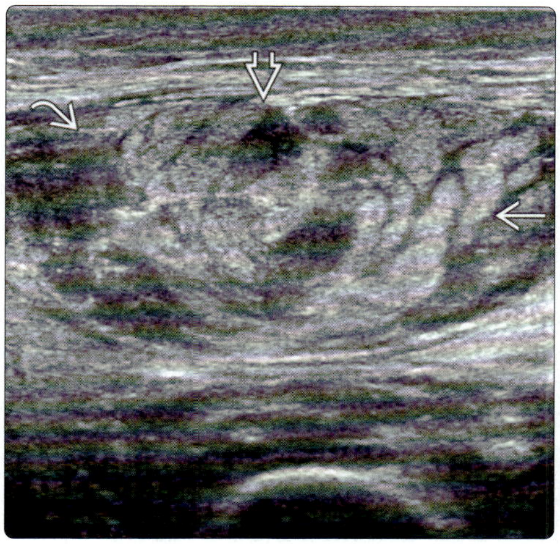

Myositis

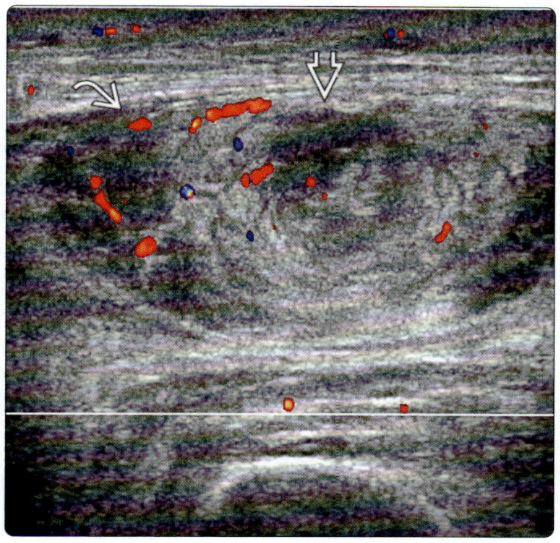

(Left) *Transverse US shows diffuse edema and swelling of the rectus femoris muscle →. The linear muscle striations → are still preserved with some more edematous hypoechoic areas →. Similar changes affected the vastus medialis and lateralis muscles as well as the contralateral thigh.* **(Right)** *Transverse color Doppler US of the same muscle shows moderate hyperemia →. The hypoechoic edematous area is not hyperemic →. The overall appearances indicate an inflammatory myositis.*

Rhabdomyolysis

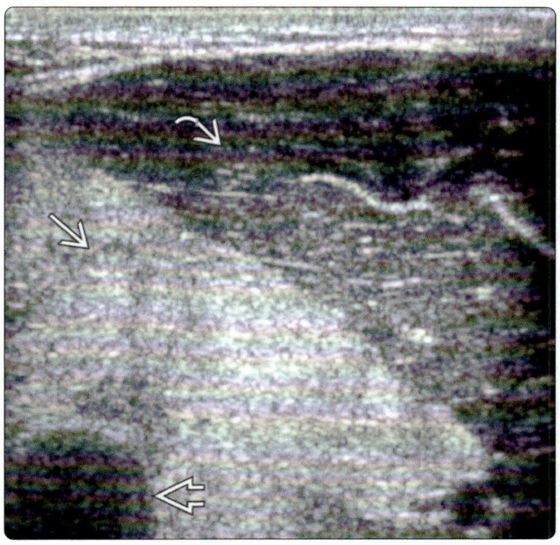

Rhabdomyolysis

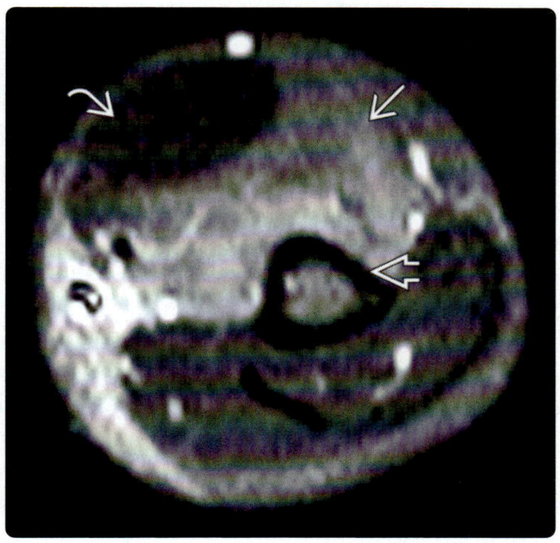

(Left) *Transverse US shows a hyperechoic brachialis muscle → following unaccustomed exertion on a rowing machine, indicative of delayed-onset muscle soreness with rhabdomyolysis. The biceps brachii muscle → is normal. Note the humeral shaft →.* **(Right)** *Correlative axial T2WI MR with fat suppression in the same location shows the edematous brachialis muscle → with a normal biceps brachii muscle →. Note the humeral shaft →.*

Venous Dilatation

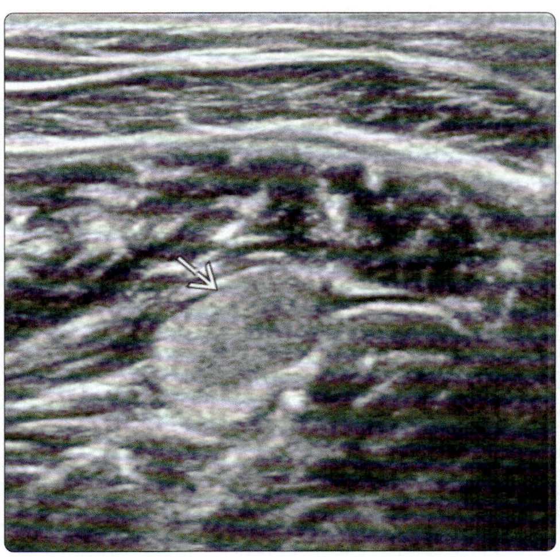

Venous Dilatation

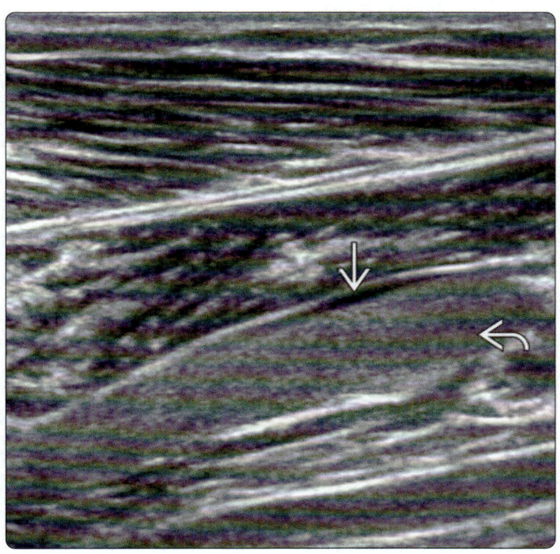

(Left) *Transverse US shows a rounded echogenic structure → within the soleus muscle.* **(Right)** *Longitudinal US of the same patient shows that this echogenic structure → is a dilated deep vein with slow flow. The echogenic content is caused by clumping of slow-moving red blood cells. These intravascular red cell aggregates show early sedimentation superficially →.*

DIFFERENTIAL DIAGNOSIS

Common

- Ganglion Cyst
- Bursal Distension

Less Common

- Abscess
- Hematoma
- Seroma
- Baker Cyst
- Dermoid/Implantation Dermoid
- Parameniscal Cyst
- Metastasis
- Vascular Anomaly
- Lymphatic Malformation
- Fat Necrosis
- Schwannoma
- Myxoma
- Sarcoma
- Lymph Node
- Venous Varix
- Aneurysm
- Pseudoaneurysm

ESSENTIAL INFORMATION

Key Differential Diagnosis Issues

- Adjust gain setting to optimize visualization of internal contents
  - Cystic-type mass is not necessarily anechoic
  - Many cystic lesions contain hyperechoic fluid or echogenic foci
- All cystic-type masses should be evaluated with color Doppler imaging
  - May reveal internal vascularity in hypoechoic lesion initially considered to be cystic
  - May reveal previously unrecognized vascular structures alongside cystic-type mass

Helpful Clues for Common Diagnoses

- **Ganglion Cyst**
  - Most common cystic-type mass
  - Filled with gelatinous material of variable viscosity
  - Occurs alongside joints
  - Synovial fluid squeezed out from joint through capsular defect
  - Usually can see cyst point toward nearby joint
  - Increase in extent through process of cyst enlargement → rupture → consolidation
    - Some can be extensive, serpiginous, multiloculated, and quite removed from joint of origin
    - May extend into muscle and simulate myxoma
  - Fluid-filled mass with stalk extending toward joint
  - Ovoid or irregular configuration with stalk connecting to joint of origin
  - No hyperemia except with recent leakage when surrounding tissues may be mildly hyperemic and edematous due to pericystic inflammation
  - Communication with joint is not always apparent
  - Locate tell-tale stalk of cyst and trace it toward joint of origin
- **Bursal Distension**
  - Inflammation of lining of bursae
  - Synovial bursae: Constant synovial-lined bursae, which occur in defined anatomical locations
  - Adventitial bursae: Nonsynovial-lined bursae that are acquired due to friction between opposing parts
  - Synovial-lined space is not visualized on US unless distended with fluid
  - Most bursae are accessible to US
  - US facilitates image-guided aspiration or injection
  - Bursitis is common, but infective bursitis is uncommon
  - Chronic bursitis is often associated with wall-thickening and internal debris
  - Baker cyst = distended semimembranous, gastrocnemius bursa

Helpful Clues for Less Common Diagnoses

- **Abscess**
  - Usually in immunosuppressed or at-risk patients
  - Majority are hypoechoic; some are hyperechoic if contents are viscous
    - Look for moving echoes within lesion on real-time imaging
    - Surrounding edema or inflammation with peripheral hyperemia
  - Posterior enhancement is characteristic
  - Echo movement on real-time imaging
- **Hematoma**
  - All hematomas start to visibly liquefy after ~ 1 week
    - Some proceed to margination with complete liquefaction
    - Others proceed to gradually reduce in size without margination and liquefaction
    - Hematoma can be complicated by superimposed infection
- **Seroma**
  - Only occurs following tissue disruption due to surgery or trauma
  - No signs or symptoms of infection
  - Do not occur spontaneously
- **Baker Cyst**
  - Fluid-filled sac with neck arising though interspace between medial belly gastrocnemius and semimembranosus tendon
  - Talk-bubble configuration
- **Dermoid/Implantation Dermoid**
  - Implantation of epidermal fragments into subdermal tissues by penetrating injury
  - Epidermis continues to grow and forms cyst lined with stratified squamous epithelium and filled with keratin
- **Parameniscal Cyst**
  - Fluid squeezed from peripheral part of meniscus
  - Usually well-defined and filled with hypoechoic gelatinous-type fluid
  - Normally associated with meniscal tear, nearly always horizontal tear
  - Meniscal tear may not be seen occasionally
- **Metastasis**
  - Metastasis can be cystic or necrotic

- **Vascular Anomaly**
  - Vascular tumor
    - Benign: Hemangioma: Proliferation → stabilization → involution
    - Intermediate: Tufted angioma, kaposiform hemangioendothelioma
    - Malignant: Angiosarcoma
  - Vascular malformation
    - Arterial, venous, capillary, lymphatic, or mixed depending on predominant vascular pattern
    - Best assessed with color and pulsed Doppler US
    - Do not proliferate but change shape due to thrombosis, hemorrhage, or infection
    - Microvasculature looks like supporting stroma because microvessels cannot be resolved
  - Each type of vascular malformation may also
    - Be focal or diffuse, ± ectatic vessels, and ± variable microvascular component
- **Lymphatic Malformation**
  - Unilocular/multilocular
  - Anechoic compressible cysts with thin walls and intervening septa
  - No vascularity demonstrated
- **Fat Necrosis**
  - Fat trauma → fat inflammation → fat necrosis → liquefaction
    - Liquefaction occurs with large areas of fat necrosis
  - ± calcification due to fatty acids combining with calcium to form calcium soap
- **Schwannoma**
  - Majority of peripheral schwannomas have small amount (≤ 10%) of myxoid tissue or fluid component
    - Minority of peripheral schwannomas have larger (> 30%) myxoid tissue or fluid component
    - Cystic component due to necrosis, hemorrhage, or fluid accumulation due to altered tumor barrier
  - Cystic schwannoma tends to be larger and more compressive with shorter duration of neurological symptoms
  - Nerve often shows slight enlargement just proximal and distal to tumor
- **Myxoma**
  - Occurs in muscle
  - Well-defined, roundish hypoechoic mass ± surrounding muscle compression or atrophy ± echogenic colloid aggregates in tumor matrix
- **Sarcoma**
  - Appear cystic if fluid, hemorrhagic, necrotic, or myxoid component
    - Myxoid tumors include liposarcoma, malignant peripheral nerve sheath tumor, and malignant fibrous histiocytoma
- **Lymph Node**
  - Cystic nodes are most commonly due to tumor necrosis or tuberculosis
  - Lymphoma nodes may also appear cystic
- **Venous Varix**
  - Focal dilatation of vein
- **Aneurysm**
  - Popliteal aneurysm is most common appendicular aneurysm
    - Associated with older age, male gender, and atherosclerosis
- **Pseudoaneurysm**
  - Associated with trauma, vascular surgery, or impingement from pedunculated osteochondroma
  - Usually hypoechoic or anechoic structure
  - Thrombus may be seen in lumen
  - Neck connects pseudoaneurysm to artery
  - Characteristic to-and-fro spectral pattern at narrow aneurysm neck
  - Characteristic yin-yang sign on color Doppler imaging
    - Swirling flow seen as different color signals within lesion
  - Lumen may not be fully filled with color due to partial thrombosis

Ganglion Cyst

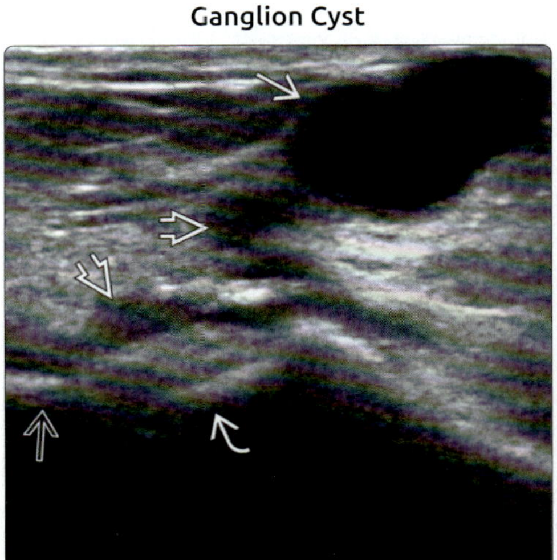

Ganglion Cyst

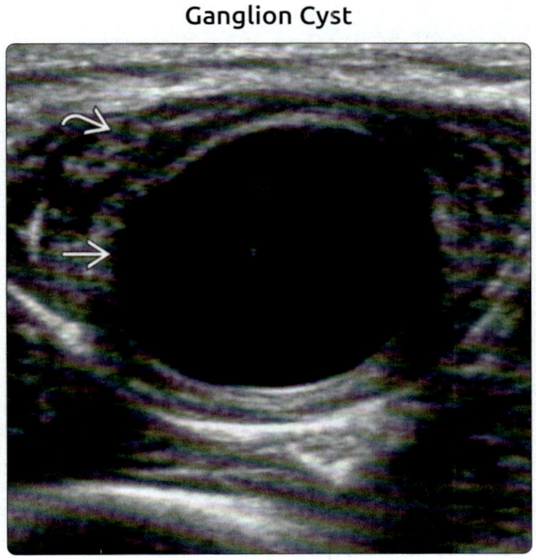

(Left) *Longitudinal US shows a ganglion cyst ➔ on the dorsum of radioscaphoid articulation. The tortuous stalk ➔ extends deeply toward the radioscaphoid articulation. The distal radius ➔ and scaphoid ➔ are normal.* **(Right)** *Transverse US shows an intramuscular ganglion cyst ➔ on the volar aspect of the wrist. The ganglion lies with flexor carpi radialis muscle ➔.*

(Left) *Transverse US of the buttock region shows a large, cystic mass filled with inflammatory material overlying the ischial tuberosity due to a distended ischial tuberosity bursa.* **(Right)** *Transverse US of the elbow region shows a small, cystic mass overlying the olecranon process due to olecranon bursitis. Moderate synovial proliferation is present.*

Bursal Distension

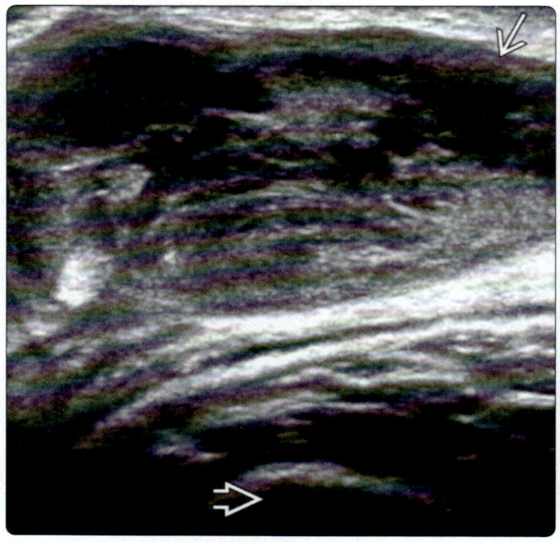

Bursal Distension

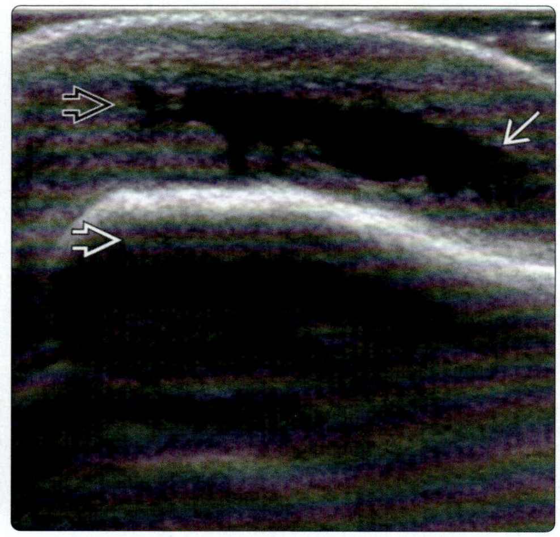

(Left) *Transverse US of the buttock shows a large abscess within the subcutaneous tissues and the gluteal muscle. The sacrum is normal. US-guided aspiration yielded Staphylococcus aureus.* **(Right)** *Transverse US of an intravenous drug user shows a large, irregular abscess cavity with distinctive internal echoes within the gluteus maximus muscle. Percutaneous needle aspiration for culture is being undertaken.*

Abscess

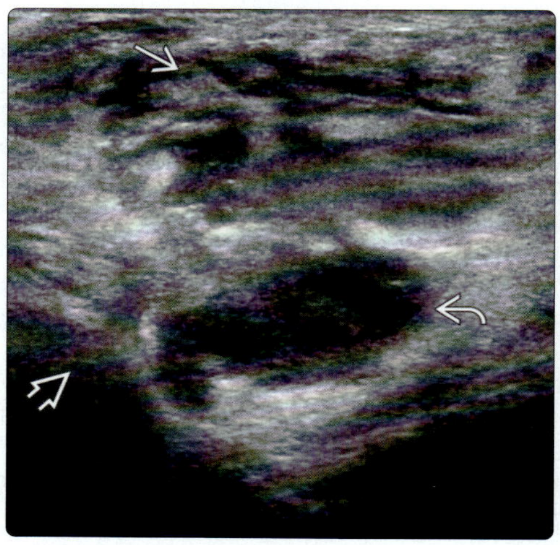

Abscess

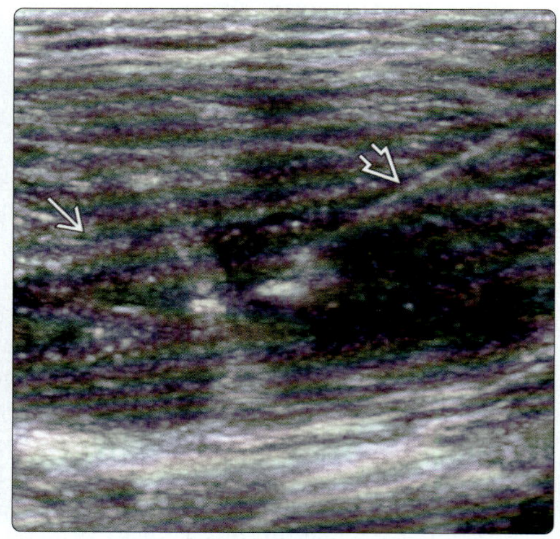

(Left) *Longitudinal US shows a liquefied hematoma with finely speckled fluid within the medial gastrocnemius muscle. There is a small, organized blood clot within the hematoma.* **(Right)** *Longitudinal US of the distal thigh in a patient 2 weeks following fracture fixation of the distal femur shows a medium-sized, cystic collection with internal echogenicity superficial to the distal femur. Aspiration confirmed an infected hematoma.*

Hematoma

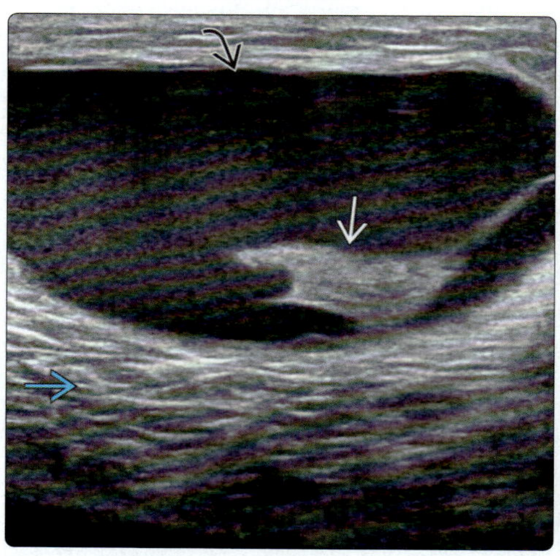

Hematoma

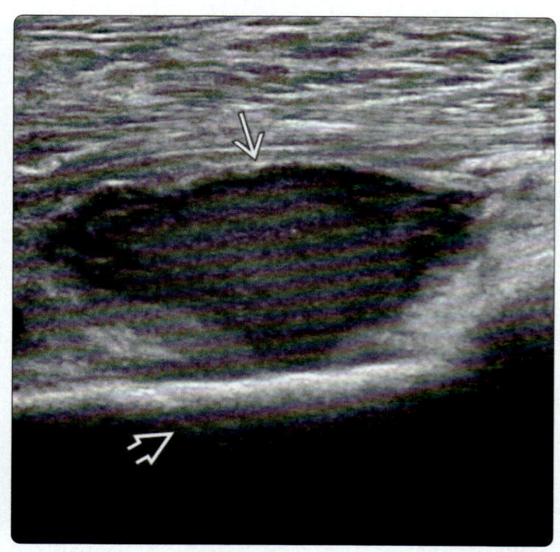

Seroma

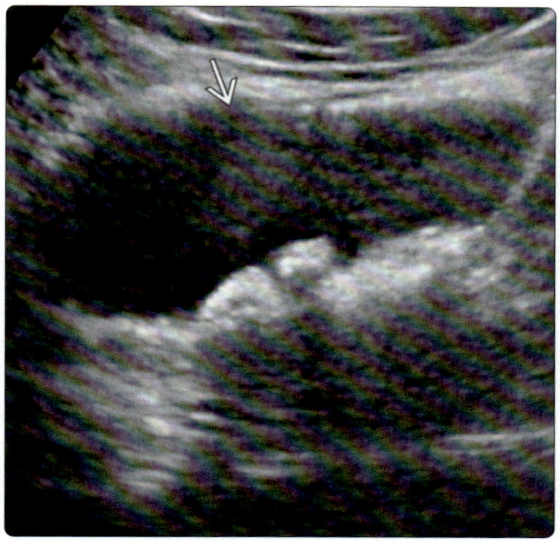

Baker Cyst

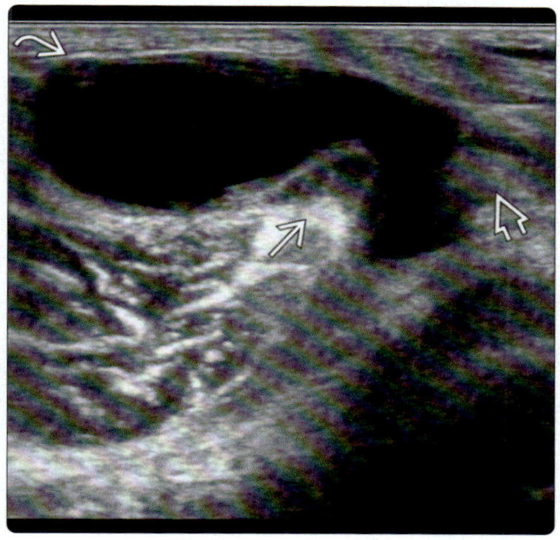

(Left) *Transverse US of the proximal thigh in a patient 1 month following hip replacement shows a large, cystic collection* ➡ *located in the deep subcutaneous region. Aspiration confirmed sterile fluid.* **(Right)** *Transverse US shows a typical Baker cyst* ➡ *with a talk-bubble configuration. The neck of the cyst extends between the semimembranosus tendon* ➡ *and the medial belly of gastrocnemius muscle* ➡.

Dermoid/Implantation Dermoid

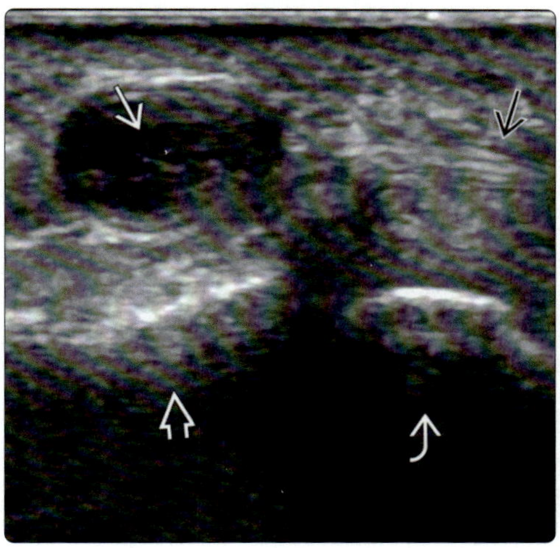

Parameniscal Cyst

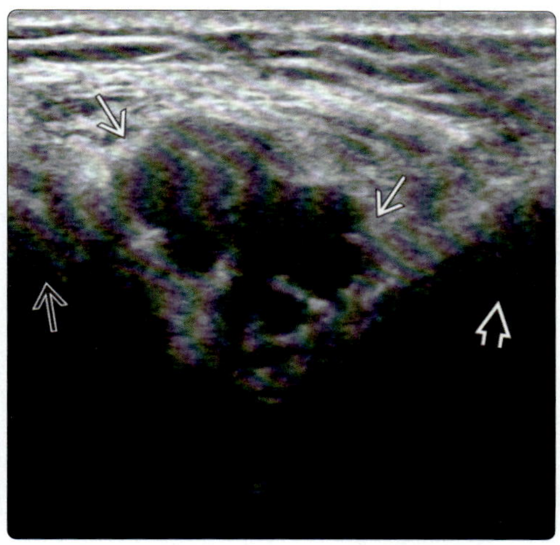

(Left) *Longitudinal US shows a cystic implantation dermoid* ➡ *in the finger pulp. The lesion lies superficial to the flexor digitorum profundus tendon* ⇨. *Note the base of the distal phalanx* ➡ *and the head of the middle phalanx* ➡. **(Right)** *Longitudinal US shows a medium-sized parameniscal cyst* ➡ *between the distal femur* ⇨ *and the proximal tibia* ➡ *at the anterolateral aspect of the knee joint. The communication of the cyst with a meniscal tear is not seen on this US, though it was apparent on other images and on MR.*

Parameniscal Cyst

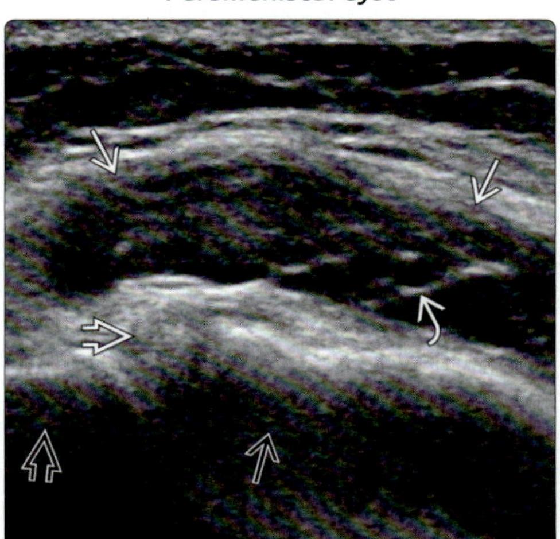

Metastasis

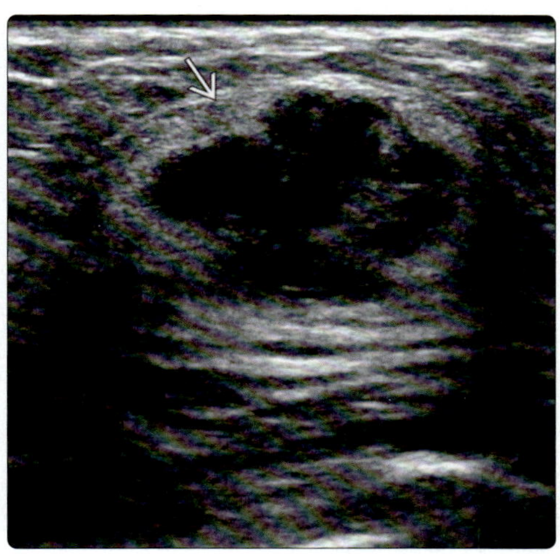

(Left) *Longitudinal US shows a large parameniscal cyst* ➡ *on the medial side of the knee. There is septation* ➡ *within the cyst. Note the periphery of the medial meniscus* ➡ *between the distal femur* ⇨ *and the proximal tibia* ⇨. **(Right)** *Longitudinal US of the chest wall in a patient with high-grade conventional osteosarcoma of the knee shows an irregular, biopsy-proven, subcutaneous, cystic metastatic nodule* ➡.

(Left) *Transverse US of the wrist shows a well-defined, subcutaneous, cystic mass with minimal marginal vascularity (not shown), consistent with slow-flow vascular malformation containing dilated vascular channels. This lies superficial to the distal radius and ulna.* **(Right)** *Longitudinal US of the chest wall shows a poorly defined lymphatic malformation with a cystic component and poorly defined supporting stroma. The underlying ribs are normal.*

Vascular Anomaly

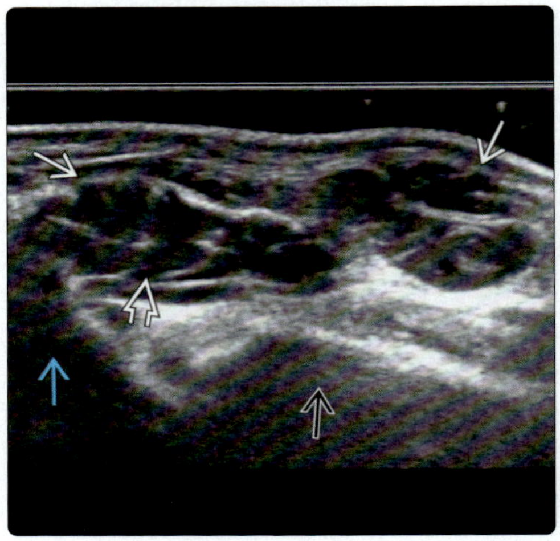

Lymphatic Malformation

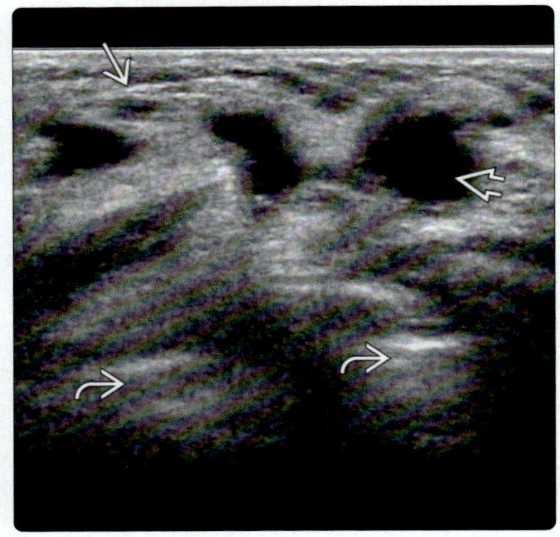

(Left) *Longitudinal US of the buttock 1 month following blunt trauma shows a well-defined, largely cystic, subcutaneous mass with early calcification consistent with liquefied fat necrosis.* **(Right)** *Longitudinal US of the supraclavicular region shows a medium-sized cystic schwannoma of the C7 nerve root. Layered hemorrhage and fibrous septi are present within the tumor. A cystic component is a common feature of nerve sheath tumors. Cystic schwannomas are also known as ancient schwannomas.*

Fat Necrosis

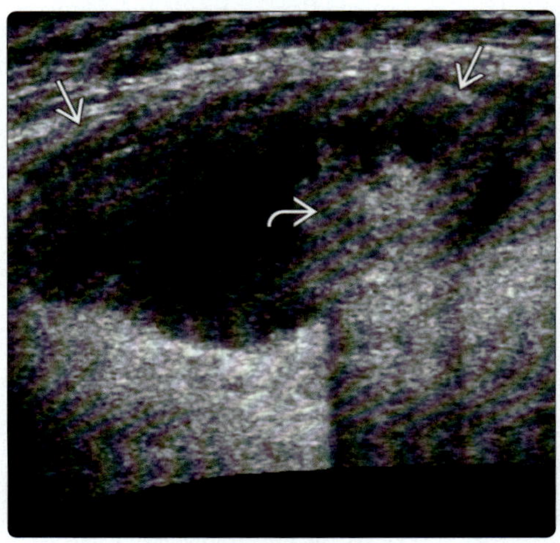

Schwannoma

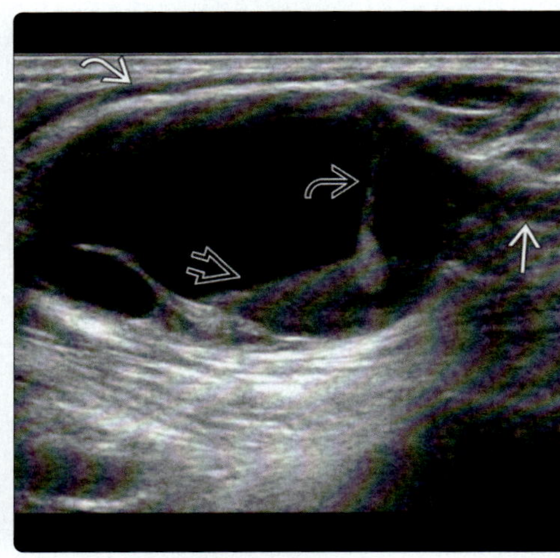

(Left) *Transverse US shows a well-defined myxoid tumor within the thigh muscle. Many small, fine, internal echoes due to colloid aggregates are present. Histology confirmed a myxoma.* **(Right)** *Transverse US shows a well-defined myxoid tumor within the latissimus dorsi muscle. Moderate acoustic enhancement is present. Most of the small, fine, internal echoes in this case represent reverberation artifact from the near wall of the myxoma.*

Myxoma

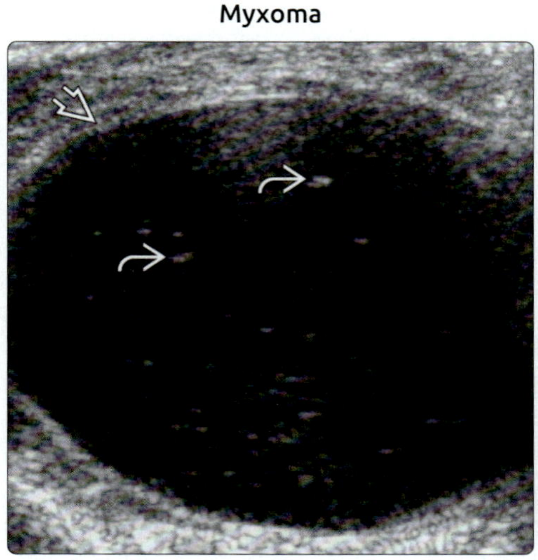

Myxoma

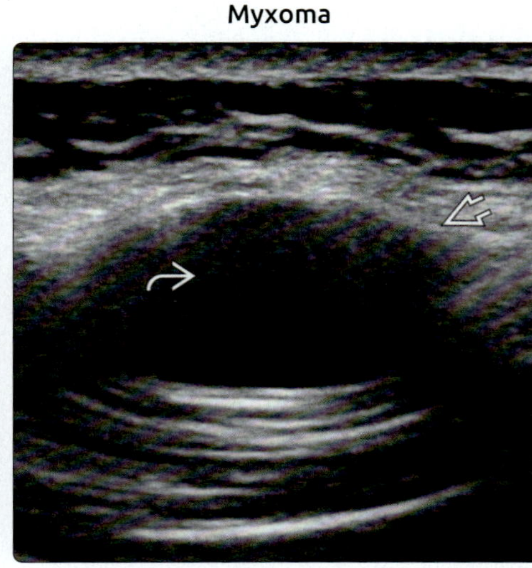

Sarcoma

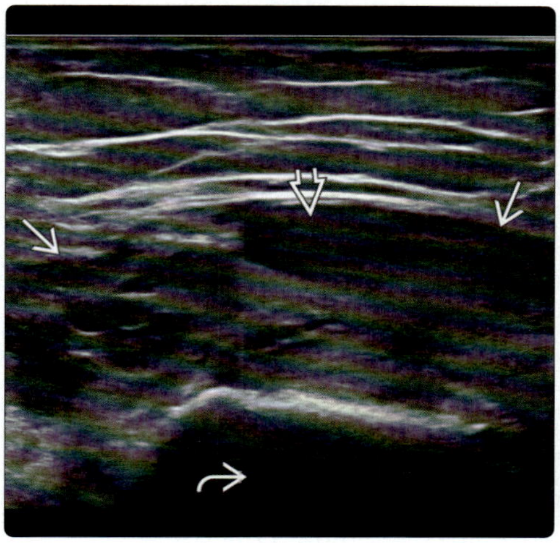

Lymph Node

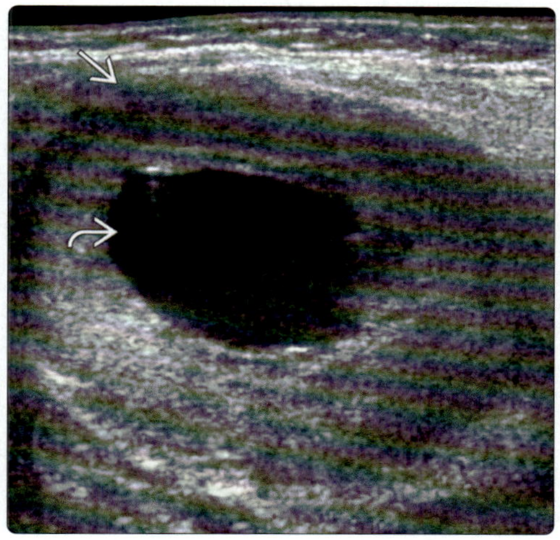

(Left) *Longitudinal US shows a large, subfascial mass ➔ on the medial aspect of the knee alongside the proximal tibia ➔. There are cystic areas ➔ present within the tumor. Percutaneous biopsy revealed an undifferentiated sarcoma.* **(Right)** *Transverse US of the groin region shows a large, metastatic, inguinal node ➔ with a central, cystic necrotic area ➔ from an alveolar soft part sarcoma of the thigh.*

Venous Varix

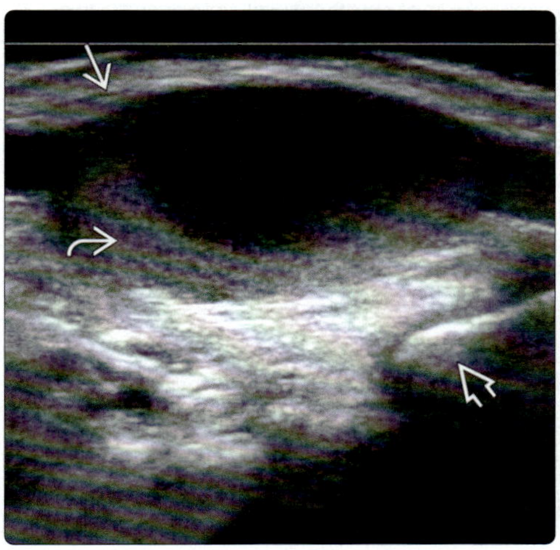

Venous Varix

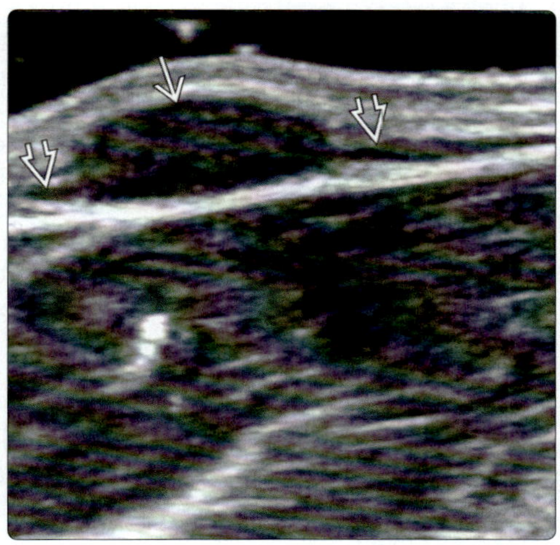

(Left) *Longitudinal US shows a venous varix ➔ on the dorsal aspect of the wrist. There is partially stagnant flow ➔ within the varix, which mimics a hypoechoic mass. Dynamic US with compression/decompression revealed movement within this stagnant flow area. The distal radius is normal ➔.* **(Right)** *Longitudinal US shows a fusiform-shaped venous varix ➔ of a superficial vein ➔ on the medial aspect of the elbow. Internal echoes are due to red cell rouleaux formation.*

Pseudoaneurysm

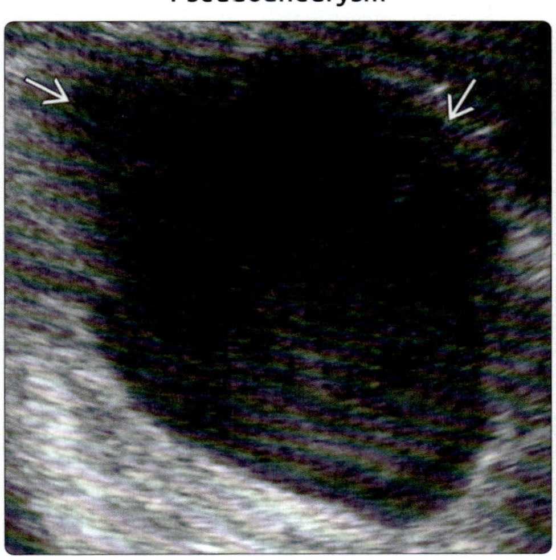

Pseudoaneurysm

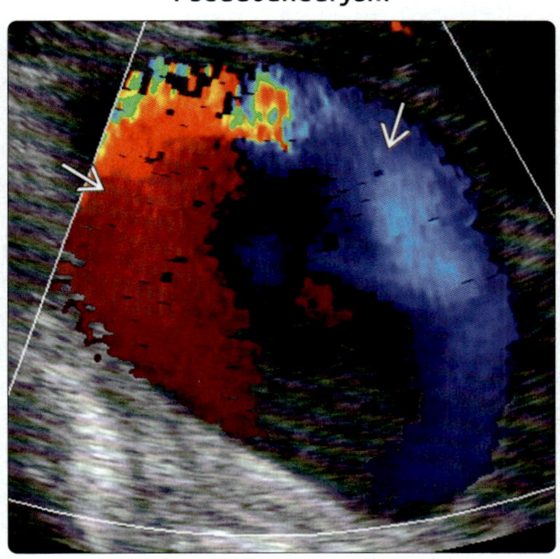

(Left) *Transverse US shows a large cystic lesion ➔ along the path of the superficial femoral artery in the distal thigh medially.* **(Right)** *Transverse US of the same patient shows a typical yin-yang sign ➔. Features are compatible with a pseudoaneurysm of the superficial femoral artery. A to-and-fro spectral pattern was present at the pseudoaneurysm neck (not shown).*

DIFFERENTIAL DIAGNOSIS

Common

- Granuloma
- Venous Vascular Malformation
- Gout or Pseudogout

Less Common

- Fat Necrosis
- Panniculitis
- Pilomatricoma
- Sarcoma
- Soft Tissue Metastases
- Nerve Sheath Tumor
- Fibroma
- Soft Tissue Chondroma
- Hematoma
- Calcifying Aponeurotic Fibroma
- Myositis Ossificans
- Tumoral Calcinosis
- Dermatomyositis
- Calcific Myonecrosis
- Tuberculosis Abscess
- Parasitic Infestation
- Heterotopic Ossification
- Sclerodactyly

ESSENTIAL INFORMATION

Key Differential Diagnosis Issues

- Align transducer obliquely to obtain views not obscured by calcification
- Differential diagnoses
  - Clinical history, e.g., trauma more in favor of fat necrosis, panniculitis, or myositis ossificans (MO)
  - Radiographs ± CT findings are helpful for analyzing type of calcification
  - US features, e.g., size of soft tissue component, hard or soft calcification, vascularity

Helpful Clues for Common Diagnoses

- **Granuloma**
  - Most commonly located in subcutaneous tissues of buttock and frequently multiple
    - Often have previous subcutaneous injection at this site
  - Variable rim calcification → posterior acoustic shadowing
  - Well-defined, rounded, hypoechoic mass
    - Deep margin may not be seen due to dense posterior acoustic shadowing
    - Width dimension is similar to depth dimension, as lesion is rounded in appearance
    - Small to medium sized (typically < 3 cm)
      - Adjacent granulomas may coalesce leading to large, more lobulated appearance
  - Typically no hyperemia on color Doppler imaging
  - ± indentation of adjacent investing fascia or muscle when large
- **Venous Vascular Malformation**
  - Calcified phleboliths common but not invariable feature
    - More common with slow-flow vascular malformations
    - Represent calcified thrombi
    - Small- to medium-sized phleboliths, depending on size of vein
      - Echogenic with acoustic shadowing
      - US more sensitive than radiographs
  - Typical features include vascular channels, compressibility, venous flow ± ill-defined margin
- **Gout or Pseudogout**
  - Located around joints, ligament, and fascial insertions
    - Gout → monosodium urate crystal deposits (tophi)
    - Pseudogout → hydroxyapatite crystal deposits
  - Soft tophi or deposits contain little mineralized component
    - Comet-tail artifacts secondary to crystal aggregates
  - Hard tophi or deposits often heavily mineralized
    - Dense posterior acoustic shadowing limits deep US assessment

Helpful Clues for Less Common Diagnoses

- **Fat Necrosis**
  - Subcutaneous fat is prone to trauma, particularly on anterolateral aspect of thigh, lateral aspect arm, and gluteal region
  - Blunt subcutaneous trauma → injury and necrosis of fat cells
    - → focal swelling and edema of subcutaneous tissues
    - → reparative change with mild localized hyperemia on color Doppler imaging
    - → encapsulated fat necrosis ± heterotopic ossification ± fibrosis
  - More severe forms of fat necrosis tend to undergo calcification and ossification
  - Other typical features: Hyperechoic fat containing hypoechoic spaces (infarcted fat)
- **Panniculitis**
  - Calcification or ossification of fat occurs following panniculitis (panniculitis ossificans)
    - Difficult to distinguish from fat necrosis
- **Pilomatricoma**
  - Benign skin neoplasm with differentiation toward hair matrix
  - Round, well-defined, heterogeneous, hypoechoic mass with partially calcified rim and posterior acoustic shadowing
    - Some lesions may be completely calcified
    - ± peripheral hypoechoic halo ± mild peripheral hyperemia on color Doppler
- **Sarcoma**
  - Benign and malignant soft tissue tumors may calcify
  - Calcification usually intrinsic to tumor, occasionally from bone infiltration
  - Synovial sarcoma is most common malignant soft tissue tumor to calcify; 50% calcify
  - Other sarcomas that contain foci of calcification include
    - Liposarcoma, undifferentiated pleomorphic sarcoma, extraskeletal osteosarcoma, leiomyosarcoma, alveolar soft part sarcoma
- **Soft Tissue Metastases**
  - Soft tissue metastases that contain foci of calcification usually arise from
    - Adenocarcinoma of colon, stomach, or pancreas

- Medullary carcinoma of thyroid and ovarian carcinoma
- **Nerve Sheath Tumor**
  - Calcification uncommon though more frequent in ancient schwannoma
- **Fibroma**
  - Subcutaneous fibrous nodules may calcify or ossify
- **Soft Tissue Chondroma**
  - Often quite heavily calcified at periphery, limiting visibility
    - Well-defined hypoechoic rim of cartilage visible beyond echogenic calcified front
    - Mild or absent hyperemia on color Doppler
- **Hematoma**
  - Minority will show peripheral rim-like calcification
  - Feature of chronic hematomas
- **Calcifying Aponeurotic Fibroma**
  - a.k.a. juvenile aponeurotic fibroma
  - Locally aggressive fibroblastic tumor of children and young adults, usually located in hands and feet
    - Occur alongside dense fibrous connective tissue (fascia or periosteum)
    - May erode bone
  - Nodular or ill-defined, infiltrating mass with fine, stippled calcification
    - ~ 50% recur after resection
- **Myositis Ossificans**
  - Pseudoinflammatory tumor of muscle; 3 types
    - MO progressiva due to hereditary disorder in children
    - Traumatic MO circumscripta seen in young patients following trauma
    - Nontraumatic MO circumscripta also in young patients though no trauma history
      - Sometimes following burns with cerebrospinal disorder or haemophilia
  - Muscle tear with hemorrhage, then cystic transformation
    - 1st week: Heterogeneous hypoechoic irregular focus within muscle
    - 2-6 weeks: Peripheral echogenic rim
    - 2 months: Hyperechoic band adjacent and parallel to cortex or irregular echogenic foci
- **Tumoral Calcinosis**
  - Calcium deposits in periarticular (> shoulder, hip, and elbow) soft tissues
  - Lobulated, well-defined, calcified mass along extensor surfaces of large joints
  - Seen in renal failure ± familial component
- **Dermatomyositis**
  - Idiopathic inflammatory myopathy and dermopathy
  - Especially large muscle groups, upper > lower limbs
  - Associated malignancy, especially elderly men (30% risk)
  - Typical feature is diffuse calcification of symptomatic muscles
- **Calcific Myonecrosis**
  - Posttraumatic condition in which single muscle is replaced by enlarging mass with central liquefaction and peripheral calcification
  - Probably due to compartment syndrome
  - ± decades between trauma and symptom onset
  - Typical feature is muscle calcification at anterior compartment of leg
- **Tuberculosis Abscess**
  - "Cold abscess" with insidious onset
  - Contains calcifying caseating granulomatous material
- **Parasitic Infestation**
  - Cysticercosis
    - Multiple intermuscular "rice grain" calcifications oriented along direction of muscle fibers
  - Dracunculiasis (Guinea worm)
    - Small crescentic calcifications in feet, hands, groin
- **Heterotopic Ossification**
  - More common around hips or elbows, following trauma or joint replacement
- **Sclerodactyly**
  - Subcutaneous and periarticular calcification associated with tightening of skin, subcutaneous atrophy, acroosteolysis
  - Often associated with scleroderma; also associated with graft-vs.-host disease and Raynaud disease

Granuloma

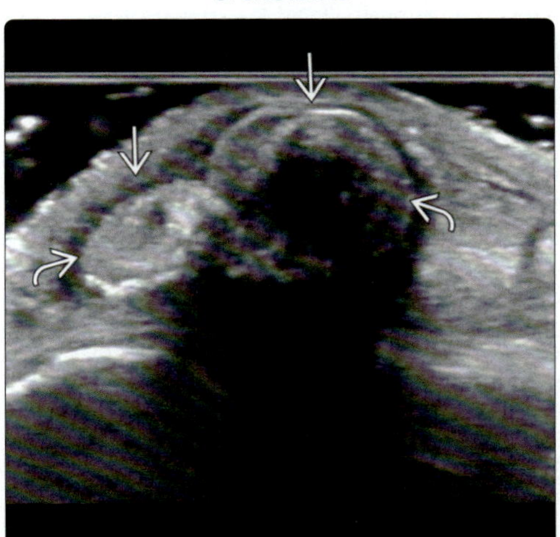

Venous Vascular Malformation

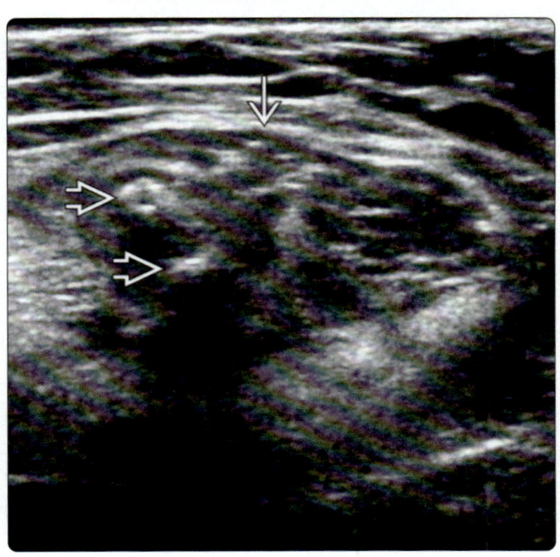

(Left) *Transverse US shows a bilobed, small, subcutaneous injection granuloma ➡ of the buttock. There is mild calcification ➡ present. Other injection granulomas may be much more heavily calcified.* **(Right)** *Transverse US shows a well-defined intramuscular forearm mass ➡. Echogenic phleboliths ➡ with mild posterior acoustic shadowing are present. Mild color Doppler flow (not shown) was present, consistent with slow-flow venous vascular malformation.*

(Left) *Oblique US of the knee shows a well-defined gouty tophus with echogenic crystal aggregates on the lateral aspect of the distal femur . This is a soft tophus. Hard tophi show more pronounced and denser calcification. The echogenic foci are due to crystal aggregates.* **(Right)** *Transverse US of a buttock shows an area of fat necrosis seen as a well-defined, subcutaneous, hypoechoic mass . There is prominent calcification with dense acoustic shadowing .*

Gout or Pseudogout

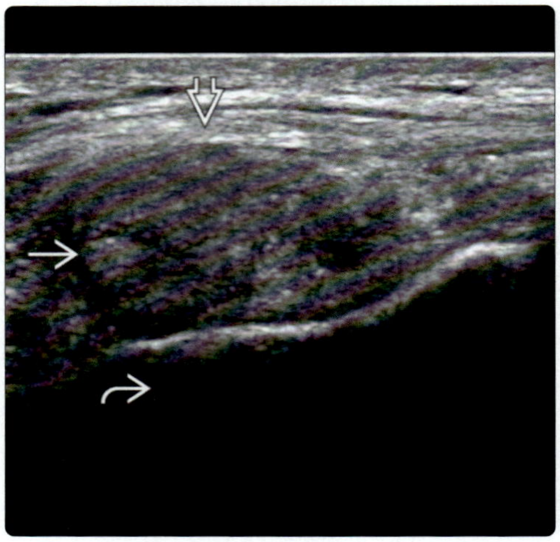

Fat Necrosis

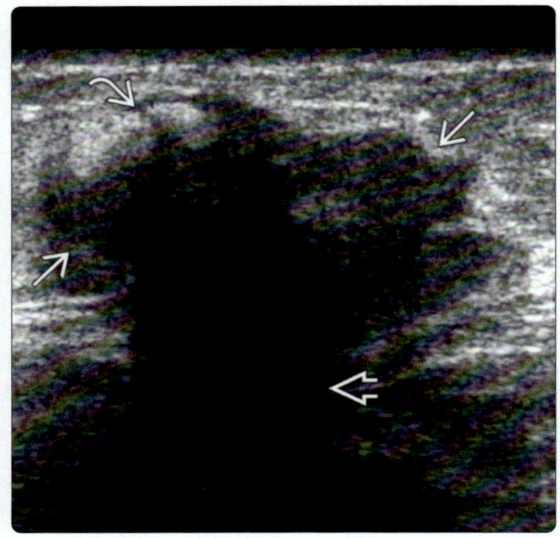

(Left) *Longitudinal US of a nontraumatic tender leg nodule shows a hypoechoic, heavily calcified subcutaneous nodule. Color Doppler imaging (not shown) showed minimal perilesional hyperemia. Histology was compatible with panniculitis ossificans.* **(Right)** *Transverse US of pilomatricoma shows a heterogeneous subcutaneous mass containing multiple lumpy calcifications associated with moderate posterior shadowing. There is a thin, hypoechoic peripheral rim .*

Panniculitis

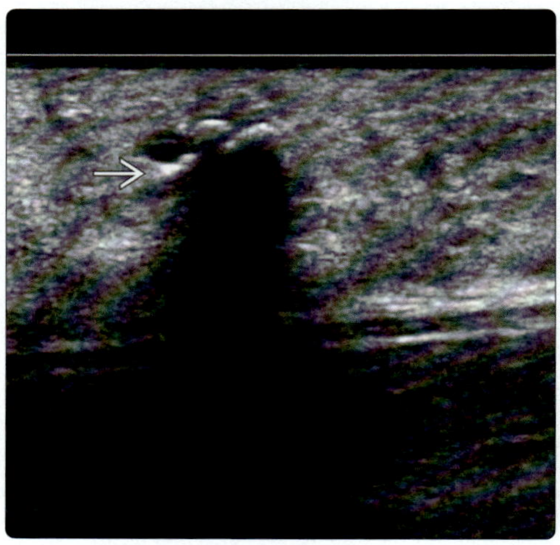

Pilomatricoma

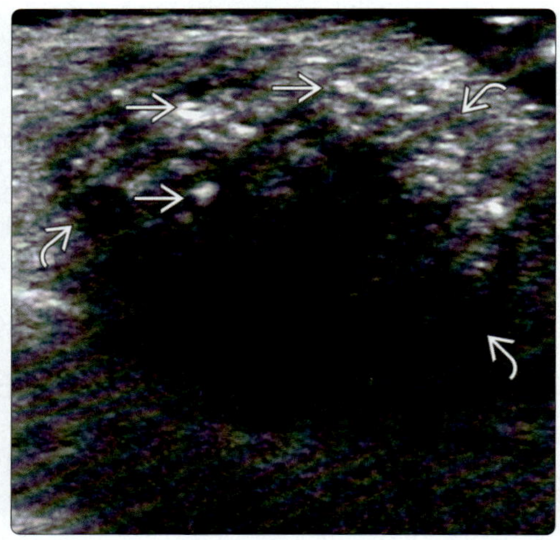

(Left) *Transverse US shows a large tumor mass in the distal thigh adjacent to the femur . Multiple foci of calcification are present within the mass. Biopsy confirmed a spindle cell sarcoma.* **(Right)** *Transverse US of a paraspinal mass shows a discrete hypoechoic mass with several small calcific foci adjacent to the transverse process . Biopsy showed metastatic squamous cell carcinoma from a known lung primary.*

Sarcoma

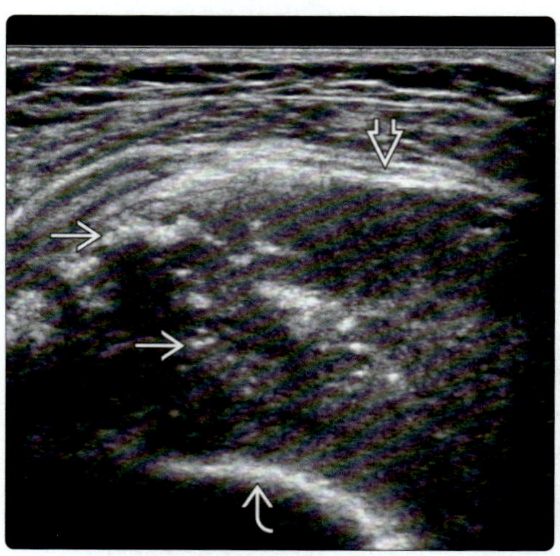

Soft Tissue Metastases

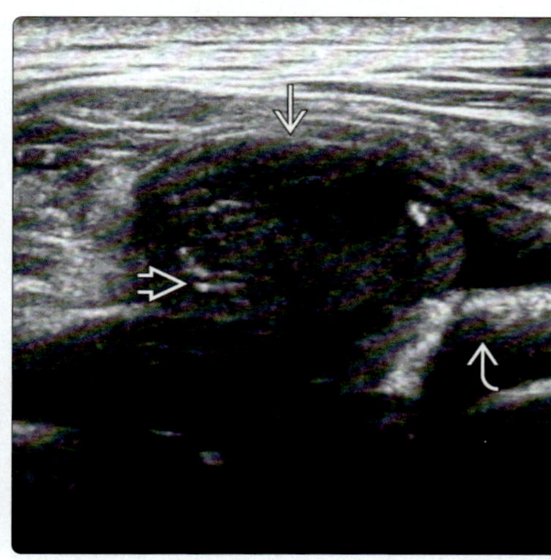

Nerve Sheath Tumor

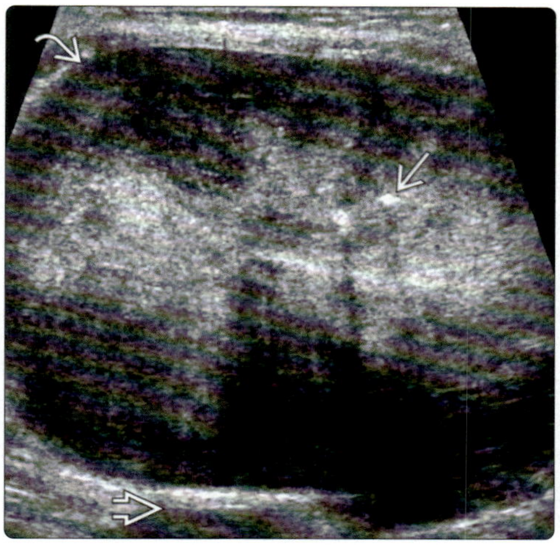

Soft Tissue Chondroma

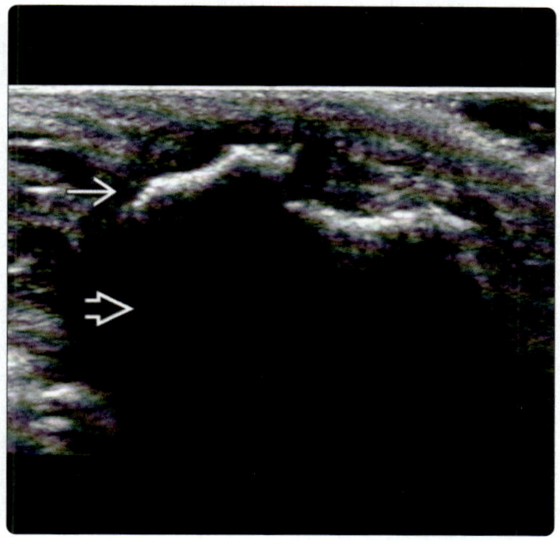

(Left) *Longitudinal US of a mass on the lateral aspect of the leg, which had been slowly growing for 10 years, shows a large peroneal nerve sheath tumor ➡ with small foci of calcification ➡. The fibula ➡ is deep to the tumor.* **(Right)** *Transverse US of the wrist shows a large, heavily calcified mass ➡ with severe posterior acoustic shadow ➡. Biopsy confirmed a chondroma.*

Hematoma

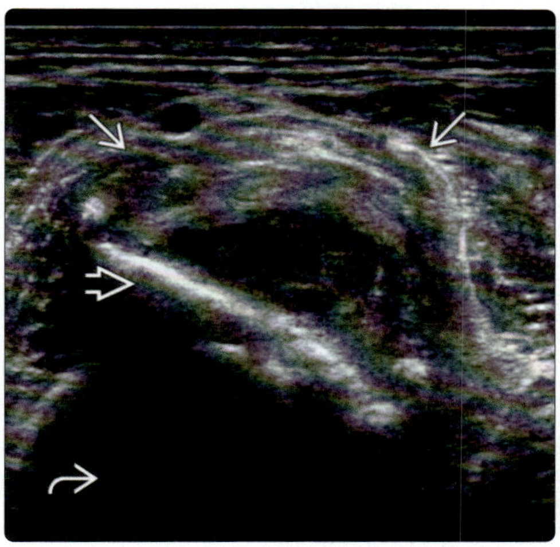

Myositis Ossificans

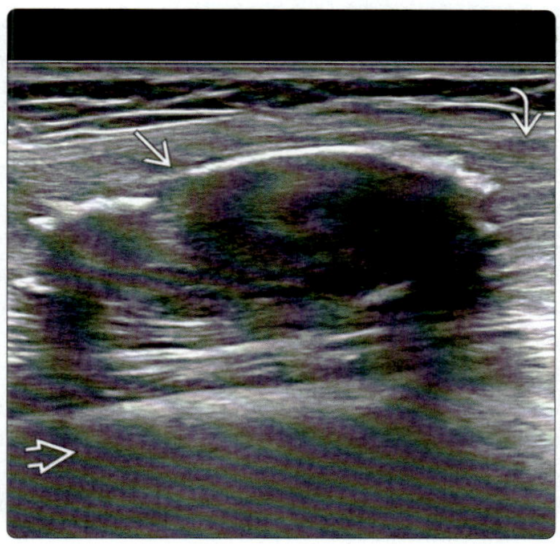

(Left) *Transverse US of the midfemur shows a moderate-sized chronic hematoma ➡ with dystrophic calcification ➡ adjacent to the femur ➡. The center of the hematoma seems partially liquefied.* **(Right)** *Transverse US shows a middle-aged female with mildly painful thigh mass for 4 weeks. There is a well-calcified mass ➡ within mildly edematous vastus lateralis muscle ➡ along the femur ➡. Clinically, the mass was decreasing in size and the pain settling. Features are consistent with myositis ossificans.*

Tumoral Calcinosis

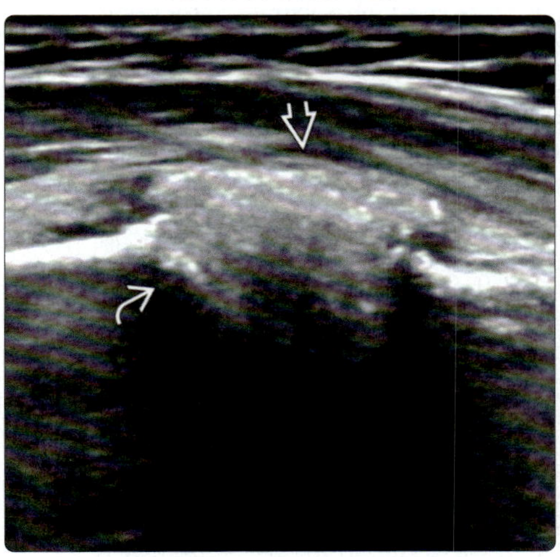

Tuberculosis Abscess

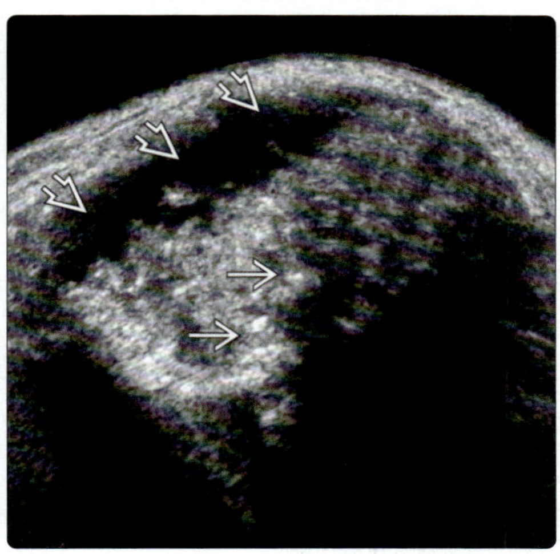

(Left) *Longitudinal US shows the edge of a larger calcified mass in an elderly female with renal osteodystrophy and proximal arm swelling. Note tumoral calcinosis ➡ with cortical erosion ➡. Bone erosion can occur with tumoral calcinosis at nutrient artery sites.* **(Right)** *Transverse US shows a large, partially liquefied ➡ soft tissue mass in the thigh with dependent echogenic debris ➡. US-guided aspiration for culture confirmed a TB abscess.*

DIFFERENTIAL DIAGNOSIS

Common

- Vascular Anomaly
- Nerve Sheath Tumor
- Soft Tissue Sarcoma

Less Common

- Glomus Tumor
- Granulation Tissue or Inflammatory Mass
- Vascular Leiomyoma
- Pilomatricoma
- Metastases
- Fibroma
- Giant Cell Tumor of Tendon Sheath
- Panniculitis
- Focal Myositis
- Lymph Nodes
- Lymphoma
- Diabetic Muscle Infarction

ESSENTIAL INFORMATION

Key Differential Diagnosis Issues

- Vascularity can be assessed on grayscale, color Doppler, and power Doppler imaging
  - Grayscale useful for assessing slow venous flow
  - Color Doppler can assess flow direction but is less sensitive than power Doppler in detection of flow
    - Power Doppler does not assess flow direction
  - Spectral analysis for vascular malformations
- Minimal transducer pressure to avoid effacing vasculature
- Note whether vascularity pattern is predominantly central or peripheral, organized, chaotic, or mixed pattern
- Color Doppler is not quite as sensitive to hyperemia as contrast enhancement on MR or CT
  - Contrast enhancement on MR or CT is also dependent on features, such as permeability and interstitial flow

Helpful Clues for Common Diagnoses

- **Vascular Anomaly**
  - For all vascular anomalies, determine
    - Size and location of lesion
    - Predominant vascular type (arterial/venous/capillary)
    - Multiplicity of lesions
    - Relative amount of vascular vs. stromal tissue
    - Presence of clots or phleboliths
    - Flow characteristics
  - 2 main types of vascular anomaly: Vascular tumor and vascular malformation
  - Vascular tumor
    - Infantile hemangioma (presents after birth)
      - → proliferation → stabilization → involution
    - Congenital hemangioma (present at birth)
      - Rapidly involuting congenital hemangioma
      - Noninvoluting congenital hemangioma
      - Partially involuting congenital hemangioma
    - Tufted angioma and kaposiform hemangioendothelioma
    - Angiosarcoma
  - Vascular malformation (congenital but most are not evident at birth)
    - Commensurate growth
    - More rapid growth precipitated by puberty, pregnancy, trauma
    - 4 main types of vascular malformation with considerable overlap: Capillary, venous, lymphatic, and arteriovenous
- **Nerve Sheath Tumor**
  - Hypoechoic, typically fusiform-shaped
  - ± anechoic areas due to myxoid deposition
  - ± hyperechoic areas due to hemorrhage, calcification, or fibrosis
  - Posterior acoustic enhancement
  - Mild to moderate vascularity, which blanches with transducer pressure
    - Infrequently, no detectable vascularity
- **Soft Tissue Sarcoma**
  - Usually hypervascular and large
    - ± necrosis ± hemorrhage
  - Occasionally hypo- or avascular
    - Particularly fibrosarcoma or liposarcoma
  - Alveolar soft part sarcoma may mimic vascular malformation on imaging

Helpful Clues for Less Common Diagnoses

- **Glomus Tumor**
  - Most common location: Distal extremities, especially subungual areas
    - Hypoechoic, hypervascular nodule between nail and cortex of distal phalanx
  - Arises from glomus body, which is arteriovenous shunt in dermis that contributes to temperature regulation
  - Glomus tumors arising outside of dermis that do not normally contain glomus cells
    - May arise from perivascular cells that can differentiate into glomus cells
    - Small, well-defined, hypoechoic, hypervascular mass
- **Granulation Tissue or Inflammatory Mass**
  - Hypoechoic irregular tissue
  - Seen in low-grade infection, reparative scar tissue, or around foreign body
  - Usually moderately to highly vascularized
- **Vascular Leiomyoma**
  - Benign solitary smooth muscle neoplasm that most commonly occurs in subcutaneous soft tissues of lower extremity
    - 2/3 involve lower leg or ankle in close proximity to artery or vein
  - Well-defined smooth margin
  - Usually round or ovoid in shape with long axis parallel to extremity axis
  - Hypoechoic, homogeneous without perceivable capsule
  - Mild to moderate posterior enhancement
  - Hypervascular with peripheral vascular convergence
- **Pilomatricoma**
  - Benign skin tumor arising from cells of hair matrix
  - Mass extending from dermal to subcutaneous tissues
  - Well-circumscribed, ovoid, heterogeneous, solid subdermal mass containing matrix calcification, intrinsic vascularity, and peripheral hypoechoic rim

- Hypoechoic peripheral rim due to capsule
- Intrinsic vascularity that may be peripheral, central, or mixed
- US appearances sufficiently specific to make diagnosis in most cases

- **Metastases**
  - Manifestation of end-stage metastatic disease
  - US appearances are similar to soft tissue sarcoma, but metastases present at smaller size than sarcomas
    - Metastases more painful than sarcomas
  - Adenocarcinoma is most common cell type
    - Carcinoma of breast and lungs are most common primaries
  - Most common site: Lower limb muscles
    - Biopsy is necessary for diagnosis
- **Fibroma**
  - Localized tumorous collection of collagen with sparse number of fibroblasts
    - Nuchal fibromas in midline nape of neck
    - Tendon sheath fibromas of hands and feet
- **Giant Cell Tumor of Tendon Sheath**
  - Most are eccentrically located to tendon and grow along tendon sheath
  - Hypoechoic, peritendinous in location
  - Moderately to highly vascular
- **Panniculitis**
  - Localized inflammation of subcutaneous fat due to infection, systemic inflammatory disorder, myeloproliferative disorder, pancreatic disease, drug hypersensitivity, or Weber-Christian disease
  - Localized area of altered echogenicity and thickening of subcutaneous fat
    - Early/active stage: Thickening and hyperechogenicity of subcutaneous fat with mild to moderate hyperemia
    - Chronic/inactive stage: Atrophy and hypoechogenicity of subcutaneous fat with little or no hyperemia
- **Focal Myositis**
  - Ill-defined area of affected muscle is hyper- or hypoechoic with hyperemia
    - Underlying muscle architecture visible in early stages
  - Manifestation of generalized inflammatory arthropathy, such as Behçet disease
  - MR is more sensitive than US for diagnosis
- **Lymph Nodes**
  - Lymph node enlargement seen with reactive, infective, metastatic, or lymphomatous nodes
  - Degree and pattern of hypervascularity depends on etiology of nodal enlargement
  - Reactive, infective, and many lymphoma nodes show moderately increased hilar flow pattern
    - Elongated shape, preservation of central fatty hilum, smooth border, with absence of matting or necrosis
    - Tuberculous nodes may show necrosis, calcification, ± capsular vascularity
  - Malignant nodes show rounded shape, obliteration of central fatty hilum, ± extracapsular spread with matting or necrosis
    - Variable capsular vascularity or mixed hilar and capsular flow pattern
- **Lymphoma**
  - Classified as Hodgkin or non-Hodgkin lymphoma
    - Non-Hodgkin lymphoma classified as T-cell or B-cell neoplasms
  - 30% of lymphoma is extranodal with most common musculoskeletal involvement being bone and muscle
  - Involvement of superficial tissues uncommon
    - Subcutaneous panniculitis-like T-cell lymphoma is uncommon type of peripheral T-cell lymphoma
  - Moderate to high vascularity
- **Diabetic Muscle Infarction**
  - Seen in end-stage diabetes with established vasculopathy
  - Hypovascular stage → hypervascular stage after ~ 10 days
    - Ill-defined hypoechoic area evolves into better-defined hypoechoic area
    - Underlying muscle architecture remains visible
    - Hyperemia feature of recovery phase of diabetic muscle infarction

Vascular Anomaly

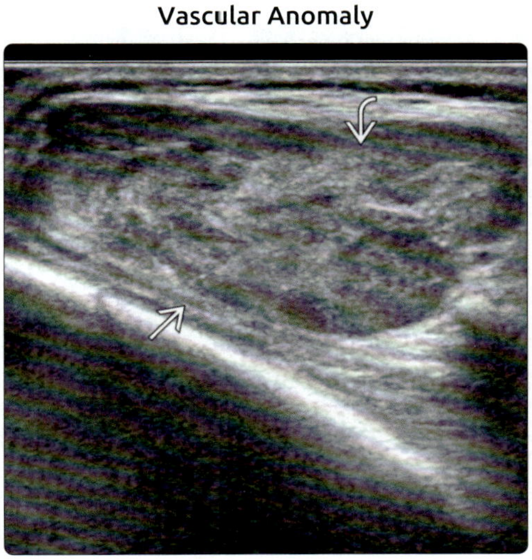

Vascular Anomaly

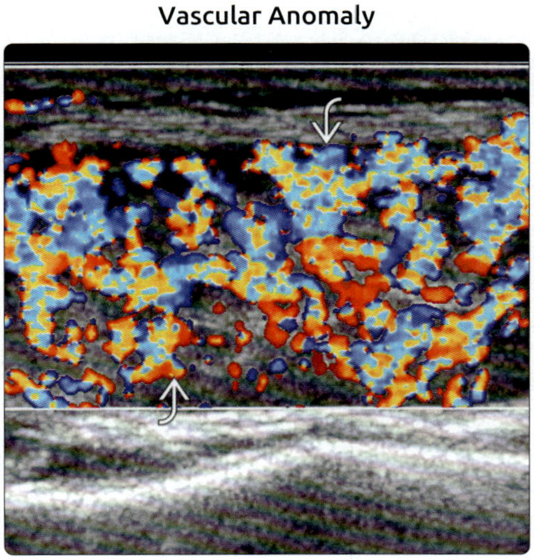

(Left) *Transverse US shows a well-defined, lobulated soft tissue mass ➡ with predominant fatty stroma ➡ in the cheek of an infant. There are only a few visible vascular channels evident.* **(Right)** *Longitudinal color Doppler US of same lesion shows a high-flow vascular anomaly ➡. Clinical course was consistent with a high-flow infantile hemangioma. Most of the arteries are too small to be resolved on grayscale US.*

(Left) *Longitudinal US shows a well-defined, subcutaneous hypoechoic mass* ➡ *with mild posterior acoustic enhancement and thickened nerve end* ➡ *in keeping with a nerve sheath tumor arising from a subcutaneous nerve.* **(Right)** *Transverse color Doppler US of a nerve sheath tumor shows a high degree of disorganized vascularity* ➡*.*

Nerve Sheath Tumor

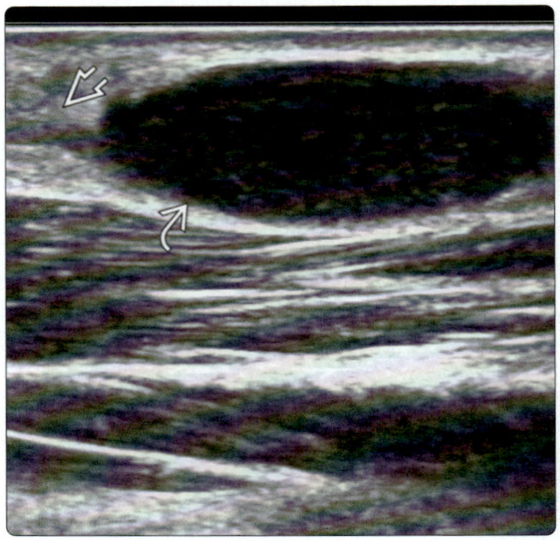

Nerve Sheath Tumor

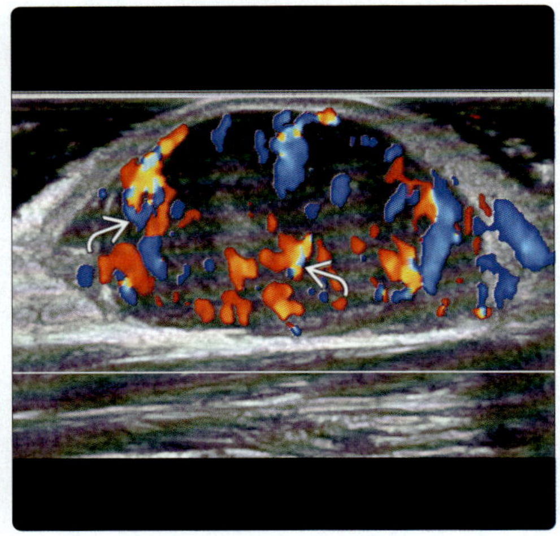

(Left) *Transverse US using a 7.5-MHz transducer shows a large soft tissue mass* ➡ *with multiple cystic spaces* ➡ *in the anterior muscle compartment of the thigh, closely abutting the femur* ➡*.* **(Right)** *Transverse color Doppler US of the same tumor using a high-frequency transducer shows moderate intratumoral hyperemia* ➡*. This solid component was targeted for percutaneous biopsy. Histology revealed a high-grade pleomorphic spindle cell sarcoma.*

Soft Tissue Sarcoma

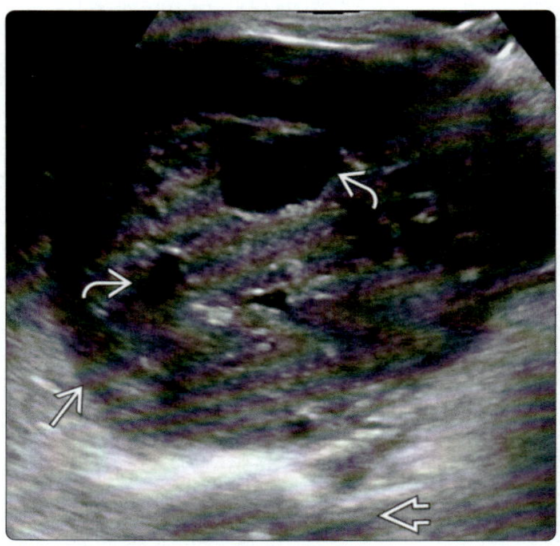

Soft Tissue Sarcoma

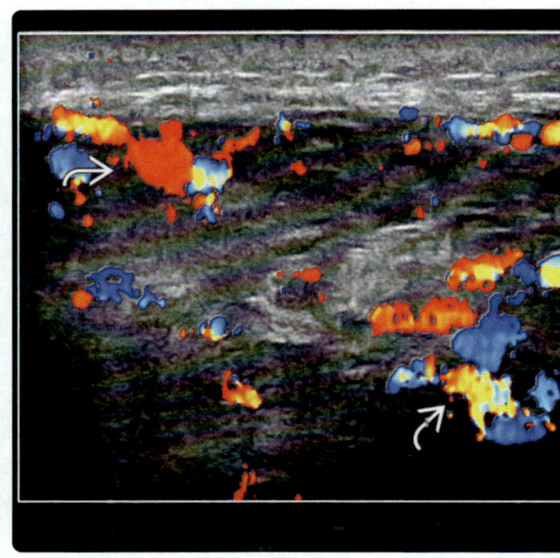

(Left) *Longitudinal US of a patient with a painful nodule in the upper limb shows a well-defined hypoechoic mass* ➡ *in the subcutaneous soft tissue.* **(Right)** *Longitudinal color Doppler US of the same patient shows moderate intratumoral hyperemia* ➡*. The lesion was excised, and histology revealed a glomus tumor.*

Glomus Tumor

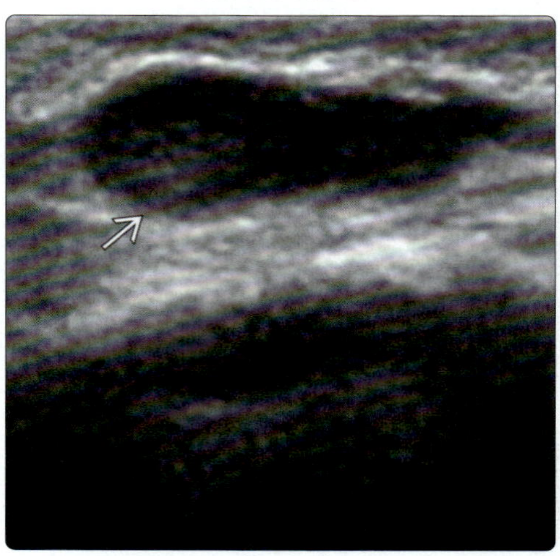

Glomus Tumor

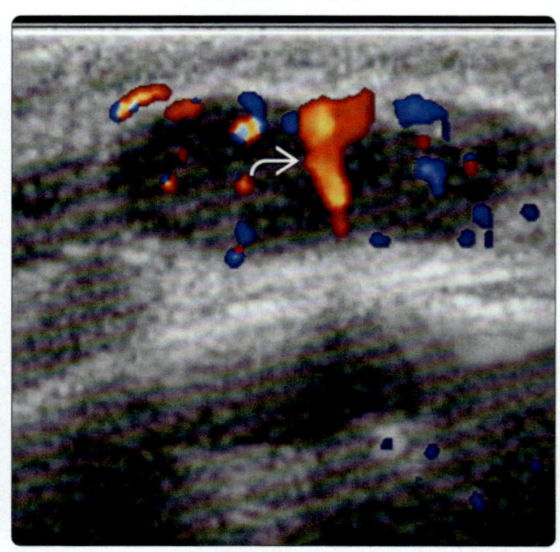

Granulation Tissue or Inflammatory Mass

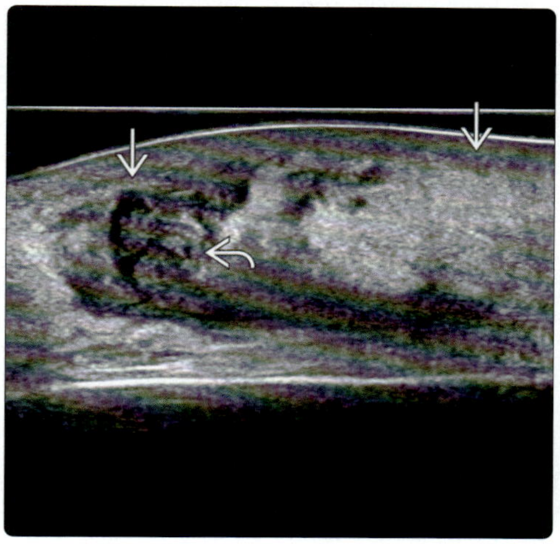

Granulation Tissue or Inflammatory Mass

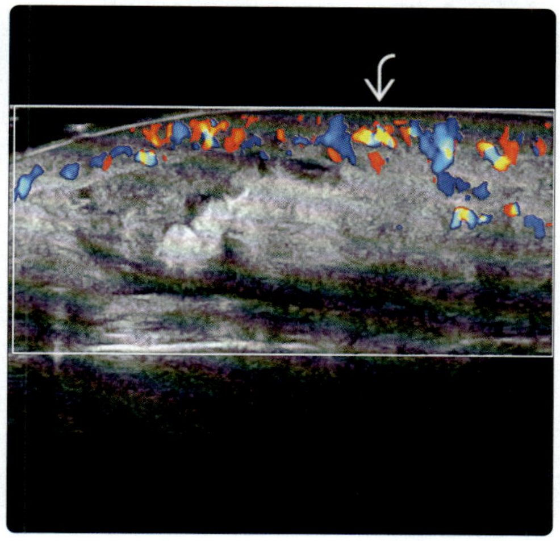

(Left) *Longitudinal US of the leg in a patient with clinically suspected shin infection shows a large subcutaneous mass with marked surrounding edema. There are hypoechoic areas within the affected area, but no discrete liquefaction is present.* **(Right)** *Longitudinal color Doppler US of the same lesion shows a moderate degree of marginal hyperemia due to active soft tissue inflammation. Percutaneous biopsy revealed a chronic inflammatory granulomatous mass, most likely due to tuberculosis.*

Vascular Leiomyoma

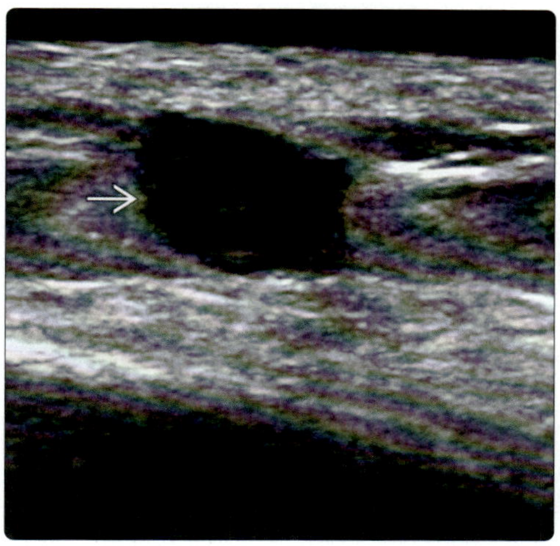

Vascular Leiomyoma

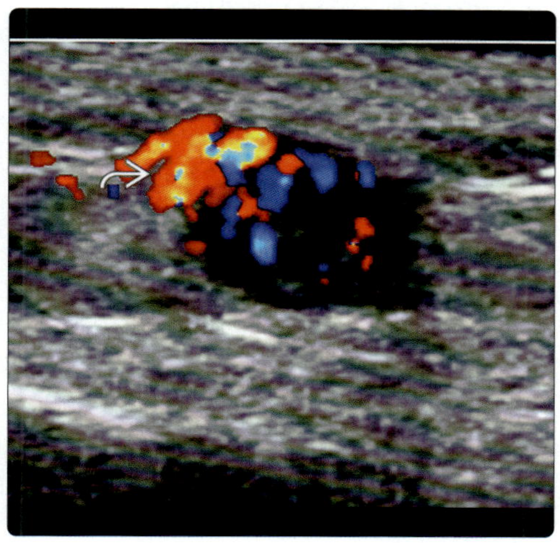

(Left) *Transverse US shows a subcutaneous vascular leiomyoma of the ankle region with a lobulated, mildly angular border. Note the characteristic hypoechoic echotexture of a vascular leiomyoma.* **(Right)** *Transverse color Doppler US of the same tumor reveals moderate peripheral and central vascularity with vascular convergence to the periphery of the nodule. This is a recognized feature of vascular leiomyoma, though it is not seen in all cases.*

Pilomatricoma

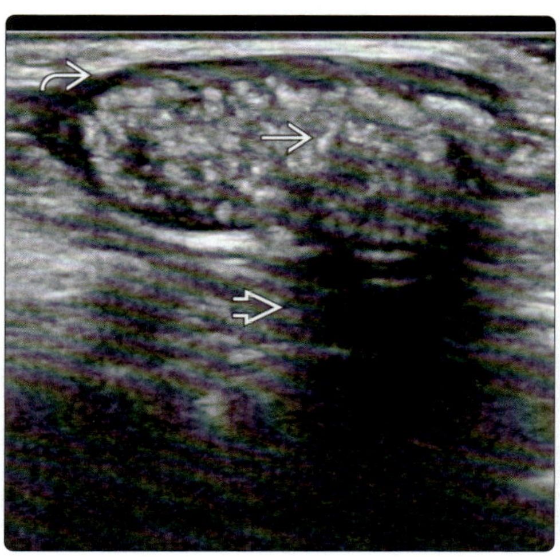

Pilomatricoma

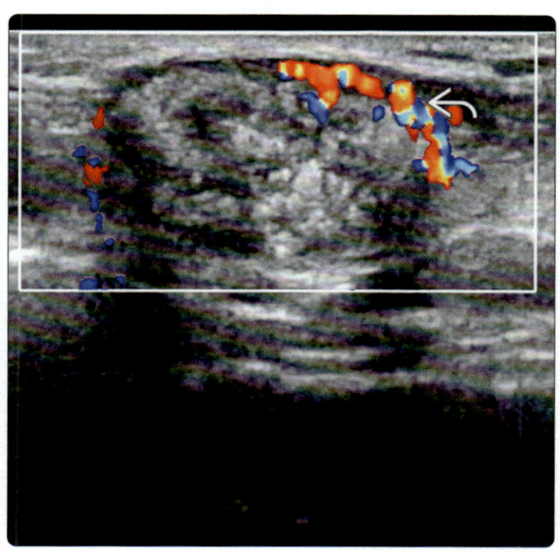

(Left) *Longitudinal US shows a well-defined, hypoechoic pilomatricoma with foci of matrix calcification, mild posterior shadowing, and a hypoechoic peripheral rim. Pilomatricoma are more common in children.* **(Right)** *Longitudinal color Doppler US of the same tumor shows mild to moderate peripheral vascularity.*

Hypervascular Soft Tissue Mass

(Left) *Longitudinal US shows a well-defined, oval, subcutaneous metastasis ➔ in the thigh region in a patient with heel melanoma. Melanoma metastases are typically markedly hypoechoic and associated with moderate posterior acoustic enhancement ➔. Note the small cystic areas within the mass ➔. There is no tail or string sign.* **(Right)** *Longitudinal power Doppler US of the same tumor shows moderate to high intratumoral hyperemia with a chaotic, mainly peripheral vascular pattern ➔.*

Metastases

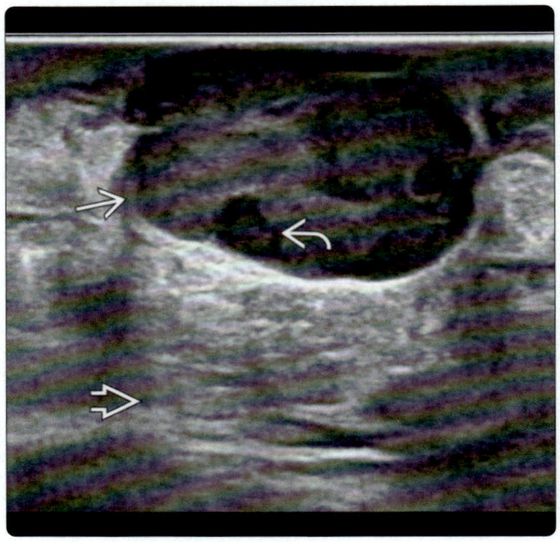

Metastases

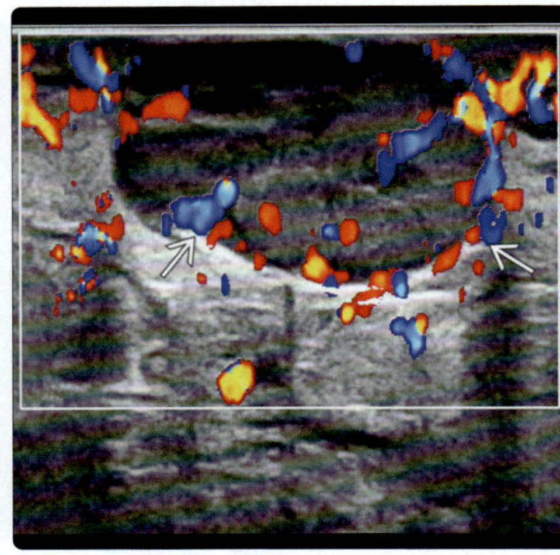

(Left) *Transverse US shows an irregular, hypoechoic, intramuscular soft tissue mass ➔ in the arm region due to fibromatosis. There is a band of tumor extending toward the skin ➔.* **(Right)** *Transverse color Doppler US shows a mild degree of hyperemia ➔ within this area of fibromatosis. Note the characteristic hypoechoic appearance and irregular configuration of fibromatosis. Most fibromatosis tumors show minimal to mild hypervascularity on color Doppler imaging.*

Fibroma

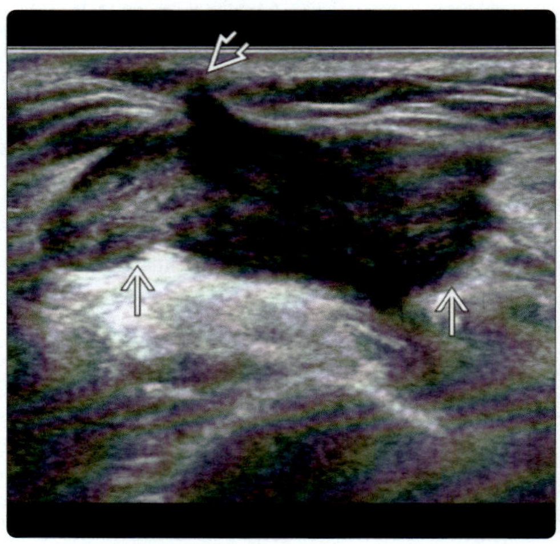

Fibroma

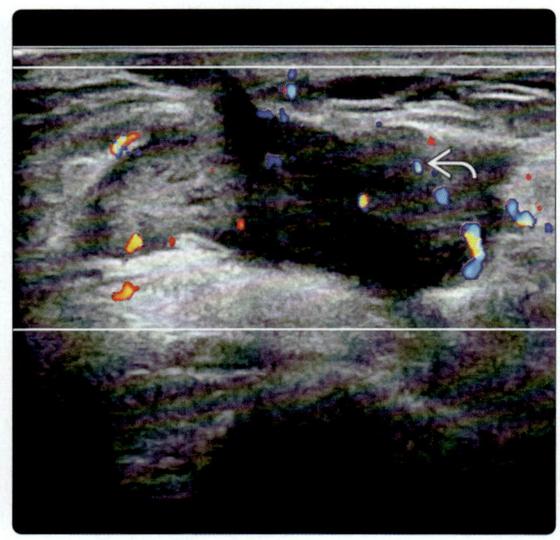

(Left) *Longitudinal US shows a hypoechoic soft tissue mass ➔ closely abutting the flexor tendon of the ring finger ➔. No metacarpal-phalangeal joint ➔ involvement is present.* **(Right)** *Transverse color Doppler US of the same tumor shows moderate intratumoral vascularity ➔. The mass is partially encasing the flexor tendon ➔. Histology revealed a giant cell tumor. Imaging determines whether this arises from the tendon sheath or from the joint.*

Giant Cell Tumor of Tendon Sheath

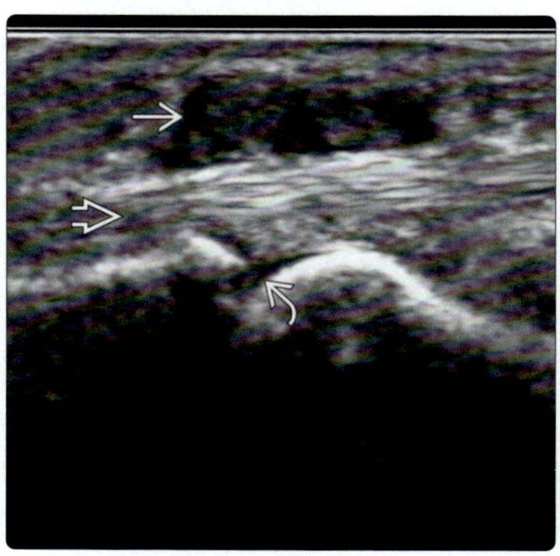

Giant Cell Tumor of Tendon Sheath

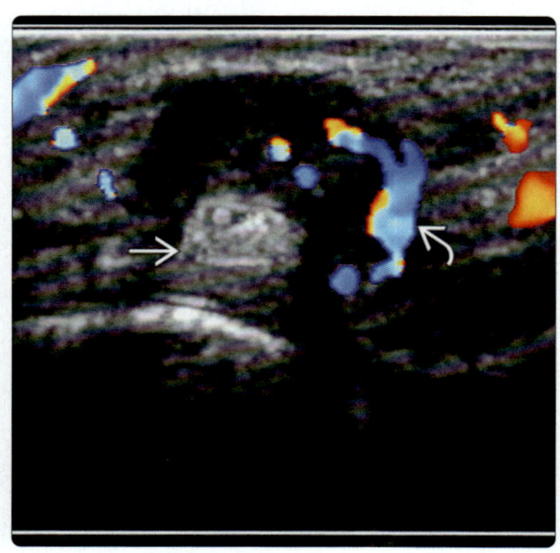

Focal Myositis

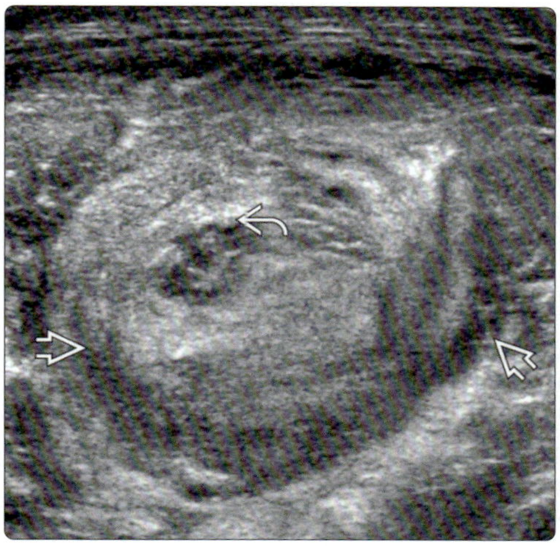

Focal Myositis

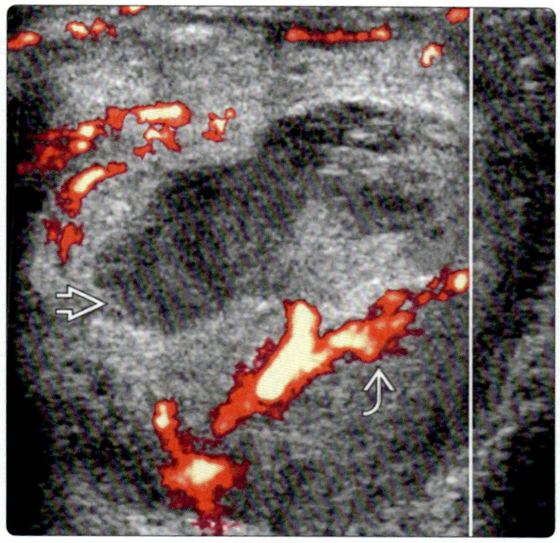

(Left) *Transverse US in an immunocompromised patient shows intense semimembranosus muscle edema ➔ with focal liquefaction ➔ due to infective myositis. There is moderate surrounding muscle and subcutaneous edema.* **(Right)** *Transverse power Doppler US of the same lesion shows moderate hyperemia ➔ of the infected area. There is a hypoechoic, avascular area ➔ centrally due to early abscess formation.*

Lymphoma

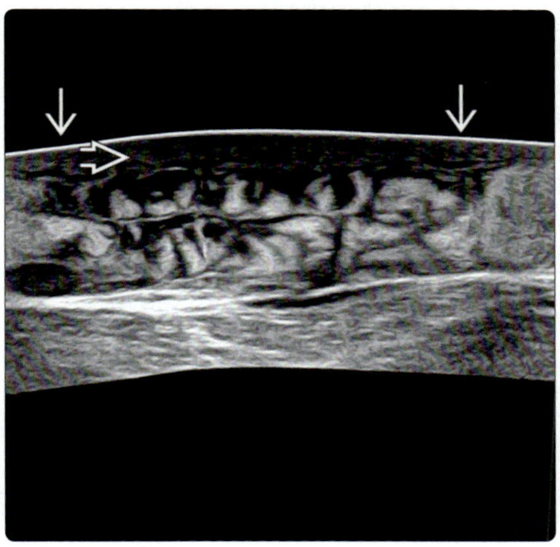

Lymphoma

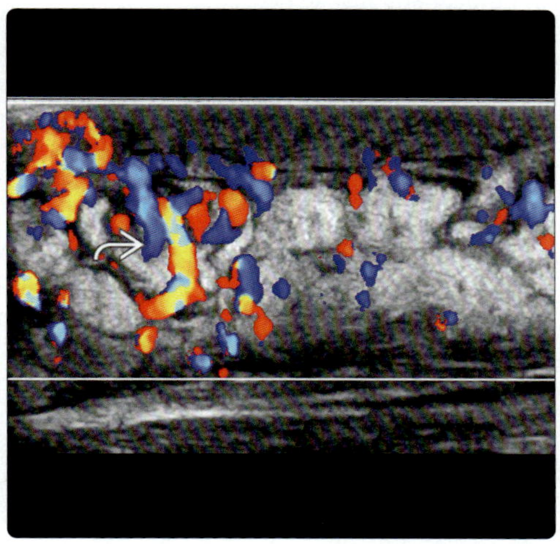

(Left) *Extended view-of-field US shows a poorly defined area of quite intense subcutaneous edema and swelling ➔. No fluid collection is present, and there is no extension deep to the investing fascia. Moderate perilesional subcutaneous edema ➔ is present. The US diagnosis was an atypical subcutaneous infection.* **(Right)** *Transverse color Doppler US shows intense intratumoral hypervascularity ➔. Biopsy revealed peripheral T-cell lymphoma. This can masquerade as an infection.*

Diabetic Muscle Infarction

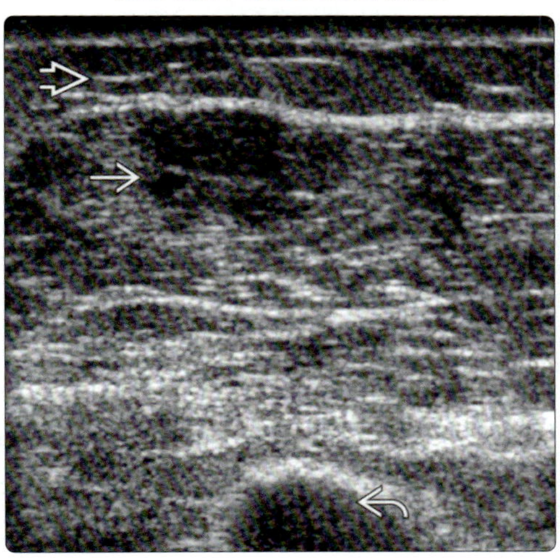

Diabetic Muscle Infarction

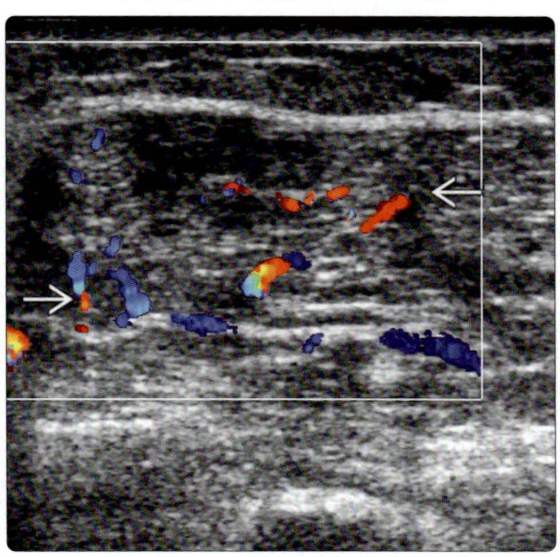

(Left) *Transverse US in a diabetic patient shows a well-defined hypoechoic area ➔ with broadly preserved background muscle architecture due to resolving biceps brachii muscle infarction. There is no subcutaneous edema ➔. The humeral shaft is normal ➔.* **(Right)** *Transverse color Doppler US of the same lesion shows moderate hyperemia around the infarcted hypoechoic area ➔. Hyperemia occurs in the recovery stage of diabetic muscle infarction.*

DIFFERENTIAL DIAGNOSIS

Common

- Ganglion Cyst
- Giant Cell Tumor of Tendon Sheath
- Tenosynovitis
- Annular Pulley Thickening
- Gouty Tophus

Less Common

- Nerve Sheath Tumor
- Extensor Retinaculum Thickening
- Bursitis
- Synovial Sarcoma

ESSENTIAL INFORMATION

Key Differential Diagnosis Issues

- Consider whether affected tendon has tendon sheath
  - Only some tendons, notably those around wrist and ankle, have tendon sheaths
  - Some tendon areas, such as patellar and Achilles tendons, have incomplete tendon sheaths known as paratenon
  - Some tendon areas, such as extensor tendons at fingers and flexor capri ulnaris tendon, do not have tendon sheath
  - Tenosynovitis and giant cell tumor of tendon sheath can only occur in tendon with tendon sheath or paratenon
- Relationship of mass to tendon ± tendon sheath important
  - Dynamic assessment by tendon movement helpful in this respect
- Color Doppler allows differentiation of hypervascular, hypoechoic synovial proliferation from hypoechoic tendon fluid

Helpful Clues for Common Diagnoses

- **Ganglion Cyst**
  - Filled with gelatinous-type material of variable consistency
  - Anechoic, cystic-like structure closely related to joint
    - Common sites are wrists, metacarpophalangeal areas, and tarsus
  - Usually originates from joint and extends to peritendinous location
    - Usually points toward or is closely related to particular joint
    - Usually neck or stalk of ganglion cyst cannot be actually followed into joint
    - Try to identify joint of origin, as this has treatment implications
      - Usually you will see part of ganglion pointing toward particular joint rather than clear joint continuity
  - If originating from tendon sheath, has no extension to adjacent joint
  - ± comet-tail artifacts within cyst due to colloid aggregates
  - ± leakage of or hemorrhage within cyst
- **Giant Cell Tumor of Tendon Sheath**
  - Composed of multinucleated cells and fibroblast-like cells that may have hemosiderin deposits
    - Pathologically, giant cell tumor of tendon sheath identical to pigmented villonodular synovitis
  - Occurs mainly in hand, wrist, and foot
    - Though may occur anywhere tendon sheaths exist
  - All giant cell tumors contact or partially encase tendon
    - Complete encasement less common
    - Tendon movement not affected since tumor expands on outer margin of tendon sheath
  - Hypoechoic tumor, which may be irregular, fusiform, or rounded in outline
    - Tends to extend along long axis of tendon sheath
    - Posterior acoustic enhancement, cystic areas, or calcification (Ca^{++}) are not seen
  - Usually moderately vascular, though, infrequently no demonstrable flow on color Doppler imaging
  - May be multifocal
  - High recurrence rate
  - Can mimic ganglion cyst
    - Increase US gain to review subtle internal echoes present in giant cell tumor
    - Use minimal or no transducer pressure to detect subtle vascularity in giant cell tumor
- **Tenosynovitis**
  - Inflammation or infection of tendon sheath with secondary inflammation of tendon
  - Acute exudative tenosynovitis
    - Usually due to acute infection
    - Fluid with low-level echoes within tendon sheath
    - Echogenic speckles within fluid due to aggregation of purulent debris
    - Compression and release useful to see movement of pus
    - Hyperemia around, rather than within, tendon sheath
    - Tendon sheath often ruptures with spread of infection into peritendinous tissues
    - Tendon usually only mildly thickened ± indistinct margins
  - Acute nonexudative tenosynovitis
    - Due to inflammation or infection
    - Synovial proliferation predominates over tendon sheath effusion
    - Hyperemia around, more than within, sheath
  - Chronic active tenosynovitis
    - Due to inflammation or infection
    - Hyperemia within, rather than around, sheath
  - Chronic inactive tenosynovitis
    - Mild thickening of tendon sheath with clear tendon sheath effusion
    - ± tendon sheath adhesions with entrapped, noncompressible fluid
- **Annular Pulley Thickening**
  - Pulleys are series of fibrous slings that maintain flexor tendons in close proximity to fingers during flexion
  - Pulley thickening usually affects A1 pulley, which is located at metacarpophalangeal joint level
    - Leads to restriction of flexor tendon movement known as "trigger finger"
  - Normal pulley seen as thin, hypoechoic band looping over tendon from bony attachment

- Normally ~ 0.5 mm thick at area most distant from bone
- In "trigger finger," A1 pulley thickens to 1-2 mm
  - ± hyperemia of pulley
  - ± tendinosis or tenosynovitis of flexor tendons
- Dynamic imaging allows depiction of flexor tendon movement restriction
  - Tendon movement may be clearly restricted passive movement or one may see fairly normal passive movement with lifting up of pulley
  - Superficialis tendon restricted more than profundus tendon
    - This distinction not usually possible on US
- **Gouty Tophus**
  - Crystal deposition in proximity of tendon is common
    - Soft or hard tophi depending on level of crystal deposition
    - Crystal aggregates → comet-tail artifacts
  - Usually large, irregular, heterogeneous, hypoechoic mass with calcified foci and hyperemia
  - Calcified foci within mass is characteristic
    - Giant cell tumor has no Ca^{++}
  - ± intratendinous crystal deposition
  - Multiple joint involvement
  - Typical clinical history
  - ± raised serum urate

Helpful Clues for Less Common Diagnoses

- **Nerve Sheath Tumor**
  - Often occur in proximity to tendons particularly at wrist and ankle
  - Arise from any nerve, even small cutaneous nerve
  - Fusiform-shaped, hypoechoic with posterior acoustic enhancement, moderate vascularity, and nerve continuity
    - ± thickening of entering or exiting parent nerve (common)
    - If tumor arises from small peripheral nerve, entering or exiting nerve may not be visible
    - ± areas of myxoid accumulation (common)
    - ± areas of hemorrhage or Ca^{++} (less common)
    - Schwannoma tends to be more eccentrically located relative to main parent nerve than neurofibroma
- **Extensor Retinaculum Thickening**
  - Most common at 1st extensor compartment of wrist (de Quervain disease, "baby wrist")
  - Associated with postpartum status, aromatase inhibitors, tamoxifen ingestion, and diabetes
  - Thickening (1-1.5 mm) of extensor retinaculum; normal < 0.5 mm thick
    - ± variable peritendinitis with hyperemia and edema of adjacent tissues
    - ± tendon sheath effusion
  - Associated with variable tendinosis of abductor pollicis longus and extensor pollicis brevis
    - Leads to restriction of tendon movement; demonstrable on real-time US
- **Bursitis**
  - Chronic frictional trauma or overuse
  - Bursa usually located alongside tendons
    - Distended hypoechoic bursa ± echogenic fluid
    - Thickened wall ± internal septations ± hemorrhage
    - Soft tissue/subcutaneous tissue inflammation with hyperemia
    - Acute: Hyperemia of peribursal soft tissue on color Doppler imaging
    - Chronic: ± Ca^{++} ± intrabursal hyperaemia, which may mimic mass
- **Synovial Sarcoma**
  - May occur in proximity to tendons particularly at wrist and hand
  - Large, heterogeneous, hypoechoic mass with well-defined margin
  - Usually located in paraarticular region, sometimes surrounding tendon
  - Usually hypervascular
  - Posterior acoustic enhancement, cystic areas, or Ca^{++} may be seen

Ganglion Cyst | **Ganglion Cyst**

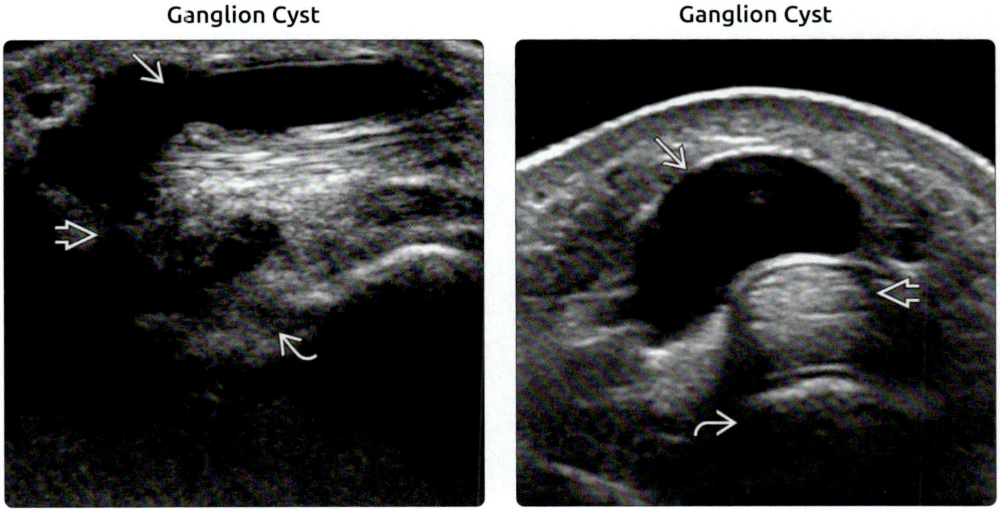

(Left) *Longitudinal US shows an elongated ganglion ➡ on the radiovolar aspect of the wrist. The stalk ➡ of the ganglion, which overlies the flexor carpi radialis tendon, extends deep toward the triscaphe articulation ➡.* **(Right)** *Transverse US of another patient shows a subcutaneous unilocular ganglion ➡ located volar to the flexor digitorum tendon ➡ of the middle finger close to the metacarpophalangeal joint ➡ and A1 pulley. Such "retinacular ganglia" typically do not communicate with a joint.*

(**Left**) *Longitudinal US of the middle finger shows a large, hypoechoic, solid soft tissue mass overlying the proximal ➔ phalanx and encasing the flexor digitorum tendon ➔. Histology confirmed giant cell tumor.* (**Right**) *Transverse US shows a giant cell tumor ➔ occupying the carpal tunnel with encasement (> 180° of circumference) of the flexor digitorum tendons ➔. Hyperemia was present on color Doppler imaging (not shown).*

Giant Cell Tumor of Tendon Sheath

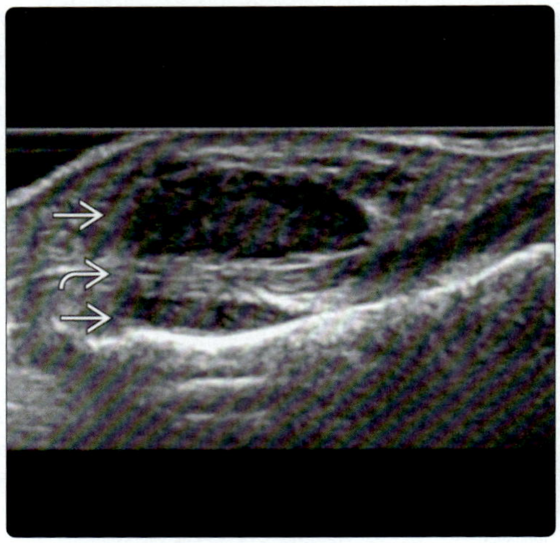

Giant Cell Tumor of Tendon Sheath

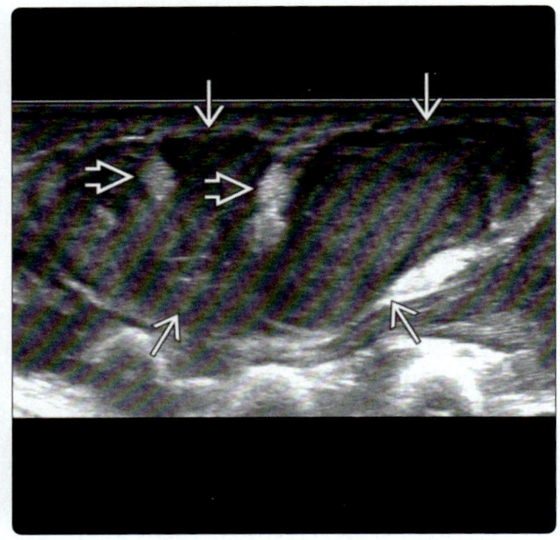

(**Left**) *Transverse US of flexor pollicis longus tendon shows severe synovial thickening ➔ surrounding the mildly swollen tendons, suggestive of chronic tenosynovitis.* (**Right**) *Transverse color Doppler US at the same level as previous image shows moderate peritendinous hyperemia, suggestive of chronic active tenosynovitis. Color Doppler allows one to distinguish between synovial proliferation and echogenic synovial fluid.*

Tenosynovitis

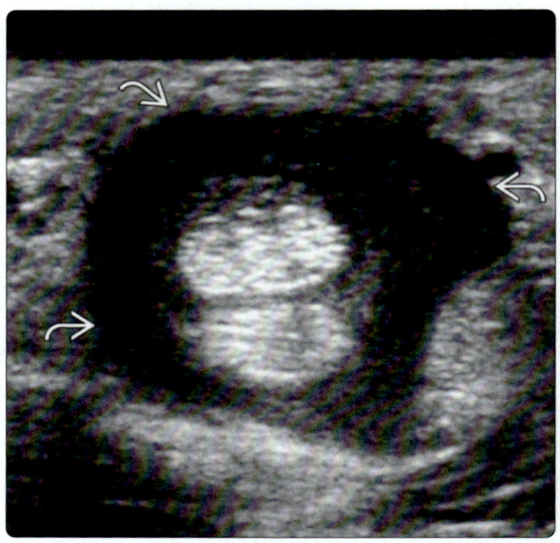

Tenosynovitis

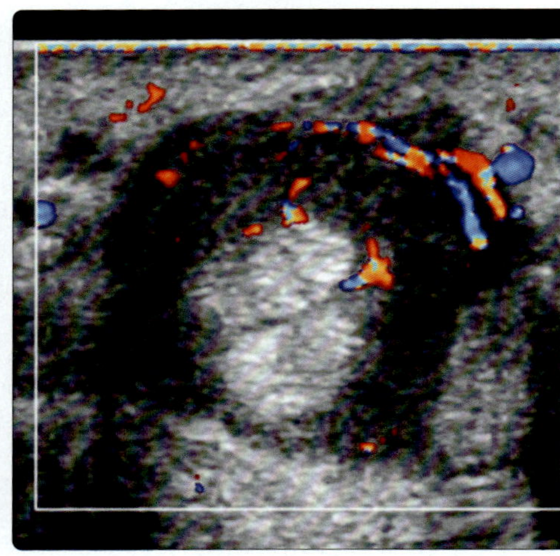

(**Left**) *Transverse US shows diffuse thickening (1.3 mm) of the A1 annular pulley surrounding the flexor digitorum tendon ➔ of the middle finger in a patient with trigger finger. The normal thickness of the A1 pulley, which is measured at 10 or 2 o'clock ➔, is < 0.5 mm.* (**Right**) *Longitudinal US of the same patient shows thickening of A1 pulley ➔. Pulley thickening is less easily appreciated on longitudinal than on transverse US. Passive movement of the tendon ➔ may show movement restriction.*

Annular Pulley Thickening

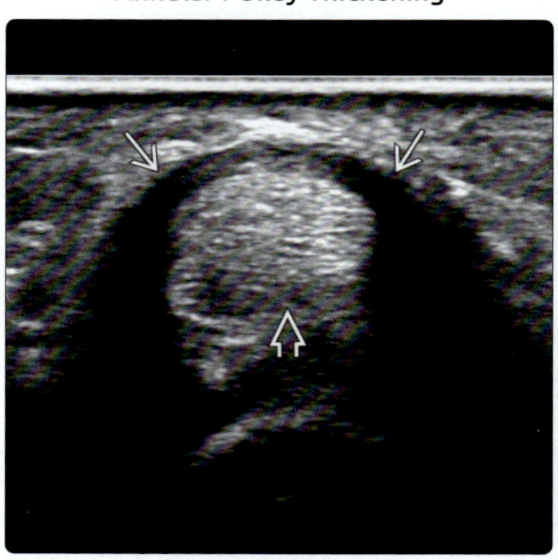

Annular Pulley Thickening

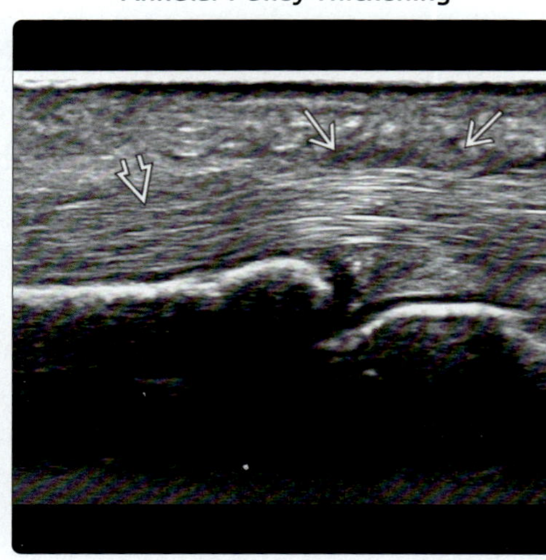

Gouty Tophus

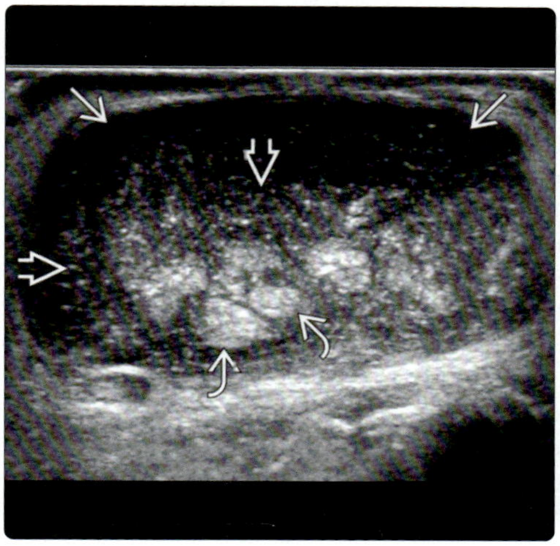

Gouty Tophus

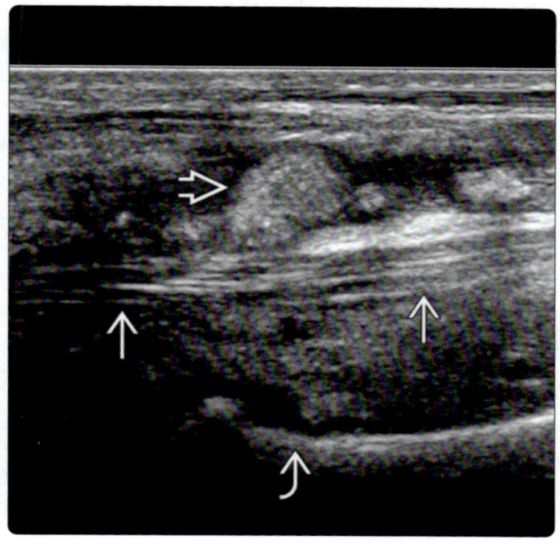

(Left) *Transverse US shows very severe gouty tenosynovitis of the flexor tendons of the hand. The tendon sheath is severely distended ➡ with finely echogenic fluid ("urate sand") ➡, which surrounds the flexor tendons ➡.* **(Right)** *Longitudinal US of the hand shows severe gouty tenosynovitis at the metacarpal ➡ level. The flexor digitorum tendon sheath is distended with fluid and large crystal aggregates ➡, which surround the tendon ➡.*

Nerve Sheath Tumor

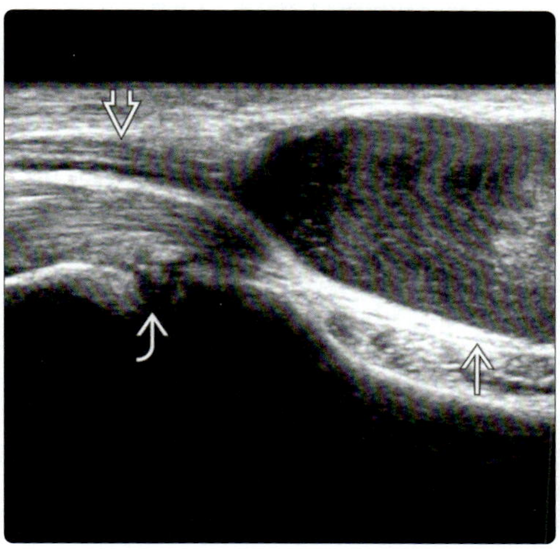

Extensor Retinaculum Thickening

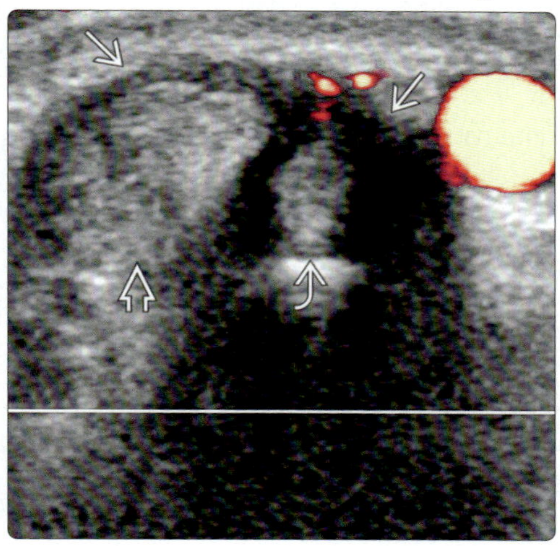

(Left) *Longitudinal US of the wrist shows a mass ➡ arising from the median nerve proximal to the carpal tunnel. The median nerve is slightly thickened ➡. Note the radiocarpal articulation ➡. Histology confirmed a schwannoma.* **(Right)** *Transverse color Doppler US shows mild thickening and minimal hyperemia of 1st extensor compartment retinaculum ➡. There is severe abductor pollicis longus tendinosis ➡ with a normal extensor pollicis brevis tendon ➡.*

Bursitis

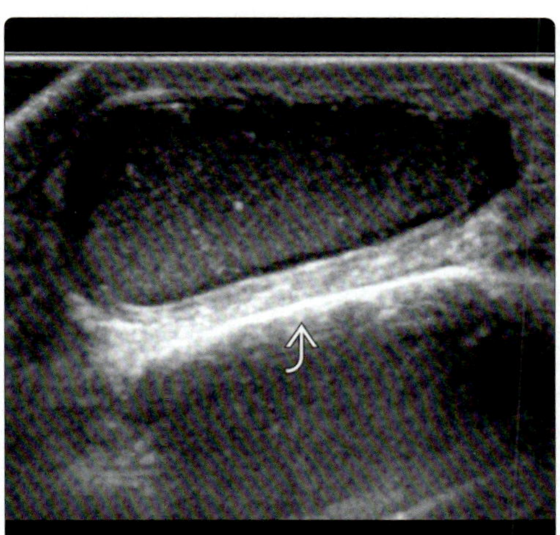

Bursitis

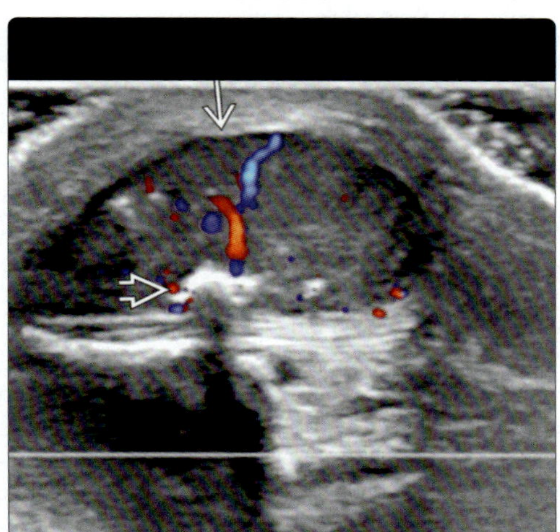

(Left) *Longitudinal oblique US shows a well-defined elbow mass located between the olecranon ➡ and the skin, compatible with olecranon bursitis. The bursa is distended by fluid with low-level echoes, suggesting high proteinaceous or hemorrhagic content.* **(Right)** *Longitudinal US shows a well-defined soft tissue mass located in the prepatellar bursa ➡ with small calcifications ➡ and moderate hypervascularity. Features are compatible with prepatellar bursitis due to gout (confirmed by US-guided biopsy).*

DIFFERENTIAL DIAGNOSIS

Common

- Anisotropy
- Tendinosis
- Tendon Tear

Less Common

- Intratendinous Vessel
- Xanthomatosis
- Amyloidosis

ESSENTIAL INFORMATION

Key Differential Diagnosis Issues

- Be familiar with normal appearances of tendon and insertion before commenting on abnormality
  - Avoid anisotropy
- Changing position of limb can optimize tendon visibility
- Always confirm abnormality on 2 views
- Dynamic assessment to view tendon movement is useful
- Color Doppler is useful to assess tendon vascularity and distinguish between tears and vascular channels
  - Stretching tendon may compress vascularity
  - No pressure on tendon
  - Normal tendons show no internal vascularity on color Doppler imaging
- Some tendons more prone to neovascularization than others
  - Achilles, posterior tibialis, and patellar tendons and tendons originating in common extensor tendon are prone to vascular ingrowth with disease
  - Rotator cuff tendons rarely show vascular ingrowth

Helpful Clues for Common Diagnoses

- **Anisotropy**
  - Reflectivity of normal tendon, nerve, and muscle is dependent on angle of transducer
  - Tendon best seen when incident beam is perpendicular to tendon
  - Any alteration from perpendicular irrespective of direction can lead to loss of reflectivity (anisotropy)
    - Tendinosis and tears also lead to loss of echogenicity, so ensure that US beam is maintained at right angles to tendon
    - Tilt probe at different angles to confirm whether abnormality is real or not
    - Dynamic scan with tendon movement can also help to confirm abnormality
- **Tendinosis**
  - Failed wound healing cascade after microtear sustained either during specific injury or repetitive microtrauma due to overuse
  - Pathologically represents degenerative process
    - Hypercellularity and irregular vascular ingrowth
    - Collagen fiber disorganization and thinning
    - ↑ interfibrillar glycosaminoglycan deposition
    - Inflammatory cell infiltrate not feature of tendinosis
  - Tendons more prone to tendinosis in upper limb
    - Supraspinatus, infraspinatus, subscapularis, and long head of biceps
    - Common tendon extensor origin, extensor carpi ulnaris, abductor pollicis longus, and extensor pollicis brevis
  - Tendons more prone to tendinosis in lower limb
    - Hamstrings at ischial tuberosity attachment; gluteus minimus and medius at greater trochanteric insertion
    - Quadriceps, patellar, Achilles, peroneal, and posterior tibialis tendons
  - Seen on US as tendon thickening, hypoechogenicity, and loss of normal fibrillar pattern ± focal hyperemia
  - May be focal or diffuse
  - Tendon thickening
    - Tendon cross-sectional area larger than expected (compare with asymptomatic side if normal in appearance)
  - Loss of normal fibrillar pattern and hypoechogenicity
    - Due to proteoglycan matrix deposition between tendon fibrils
  - Hyperemia
    - Vessels not present within normal tendons
    - Level of hyperemia is proportional to level of tendinosis disease activity
    - May be related to disease progression or repair; reduction in hyperemia correlates with decrease in symptoms
    - Subjectively graded as mild, moderate, or severe
  - Secondary signs of tendinosis
    - More common in severe tendinosis
    - Tendon sheath effusion (if tendon sheath present)
    - Adjacent bursitis (Achilles, common extensor)
    - Cortical irregularity ± hyperostosis at insertional area
    - Intrinsic tendon calcification
    - Variable degree of peritendinitis
      - Edema and hyperemia of peritendinous tissues
      - Important feature of tendinosis to note, as it will respond to antiinflammatory medication
- **Tendon Tear**
  - US accurately detects macroscopic tendon tears
  - May occur at insertion (avulsive type), tendon surface, or within tendon substance (intrasubstance)
  - Tendon tears are strongly associated with tendinosis
    - Likelihood of tendon tear increases with increasing severity of tendinosis
    - Tendon tears occur in those areas where tendinosis occurs
    - Tears are uncommon in normal tendons
  - Tears tend to happen at specific sites for each tendon
    - Supraspinatus is most commonly torn tendon (avulsive-type tear anterior fibers)
    - Achilles tendon tears occur at junction of proximal and middle 1/3 of tendon
    - Common extensor tendon tears are intrasubstance or deep avulsive-type tears
    - Biceps tendon tears are often longitudinal tears at proximal aspect of bicipital groove
    - Peroneal and posterior tibialis tendons typically located at retromalleolar and inframalleolar regions
  - Partial-thickness tear
    - Focal discontinuity of tendon not involving all tendon fibers

  - Sharp, linear, hypoechoic defect within area of tendinosis
  - Lack of complete tendon retraction is most important feature to distinguish partial from complete tear
  - Tears usually on background of tendinosis
  - Occur with moderate to severe tendinosis
  - Uncommon to see tears in normal healthy tendons
  - Should not be confused with dilated vascular channels within hyperemic tendon; color Doppler imaging is helpful
  - Should not be confused with proteoglycan accumulation; focal hypoechoic area with bulging tendon contour
  - ± dystrophic calcification, adjacent bursitis, reactive bone changes
- Full-thickness tear
  - Tendon discontinuity with anechoic or hypoechoic area
  - Retraction gap may be filled with echogenic blood, making tear less conspicuous
  - Often swollen tendon ends, especially in small finger tendons
  - Collapsed synovial sheath may be mistaken for attenuated tendon (applies to smaller tendons, e.g., finger flexors and biceps tendon)
  - Proximal and distal stumps may show severe tendinosis with dystrophic calcification
  - Measure gap and position of stumps useful for surgery

Helpful Clues for Less Common Diagnoses

- **Intratendinous Vessel**
  - Should not be confused with partial intrasubstance tear
  - Dilated vascular channels within hyperemic tendon due to tendinosis
  - Vessels normally not present within tendons
  - Level of hyperemia is proportional to level of tendinosis disease activity
  - May be related to reparative process
  - Color Doppler imaging is diagnostic
- **Xanthomatosis**
  - Tendon xanthomas are manifestation of familial hyperlipidemia
    - Xanthomas composed of lipid-laden foamy histiocytes, cholesterol ester, and giant cells
  - Appear most commonly as fusiform hypoechoic enlargement of tendon
    - Less commonly discrete hypoechoic areas (due to more concentrated cholesterol deposition)
  - Achilles tendon > > patellar tendon > extensor tendons of hands and feet
    - In Achilles tendon, discrete small calcific foci may also be seen along length of tendon
    - Hyperemia on color Doppler imaging is not feature
  - Xanthoma may regress with lipid-lowering treatment
  - Useful marker given its specific relationship to atherosclerosis
    - Lipid composition of xanthoma is similar to atherosclerotic plaques
    - Some patients with dyslipidemia may present with xanthoma
  - Xanthoma may be difficult to distinguish from tendinosis; helpful, though not very specific, distinguishing US features of xanthomatosis are
    - Symmetrical appearances
    - Calcific foci removed from insertional area
    - More discrete areas of hypoechogenicity
    - No hyperemia on color Doppler imaging
    - Known hypercholesterolemia
- **Amyloidosis**
  - Inherited or secondary (chronic renal failure, multiple myeloma, chronic inflammation)
  - Overproduction of fibrous protein aggregates, which become insoluble and deposit in tissues
  - Heart, lungs, kidneys, spleen, gastrointestinal tract, nerves, and tendons
  - → tendon thickening and hypoechogenicity
    - Supraspinatus > biceps > hand and wrist tendons
    - ± nodular tendon sheath or bursal thickening
    - Predisposes to tendon tears

Anisotropy

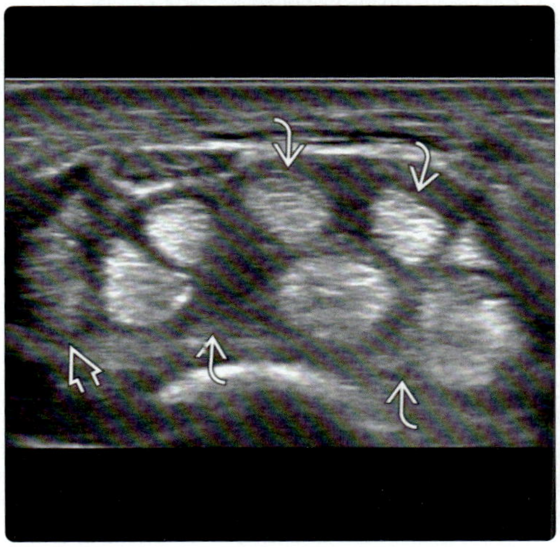

Anisotropy

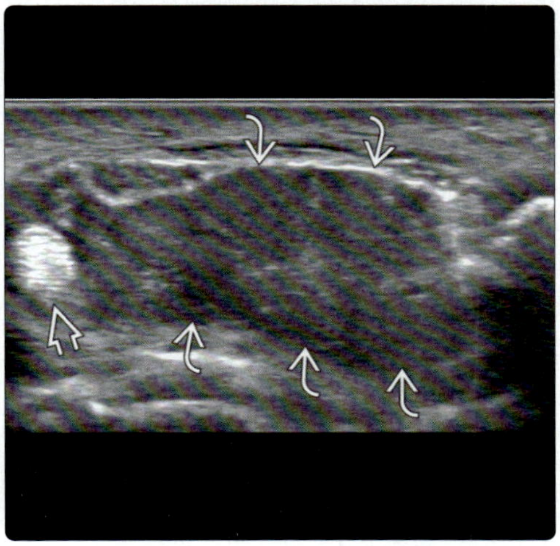

(Left) *Transverse US of carpal tunnel shows normal echogenicity of flexor tendons ➔. The flexor pollicis longus tendon ➔ is less well defined at this insonation angle.* **(Right)** *Transverse US at the same level shows that tendon echogenicity changes from hyperechoic to hypoechoic in almost all the tendons ➔ when the transducer is not aligned at a right angle to the tendon. The flexor pollicis longus tendon ➔ is more well defined at this angle.*

Tendon Hypoechogenicity

(Left) *Longitudinal US of distal patellar tendon shows a focal thickened, hypoechoic tendon ➔ near the superior pole of patella with more severe involvement of superficial fibers and less in the deeper fibers ➔, consistent with severe patellar tendinosis.* **(Right)** *Longitudinal US of the distal Achilles tendon inserting at posterior superior calcaneal tuberosity ➔ shows diffuse severe tendon thickening ➔ with slightly heterogeneous hypoechoic in appearance. No dystrophic calcification is present.*

Tendinosis

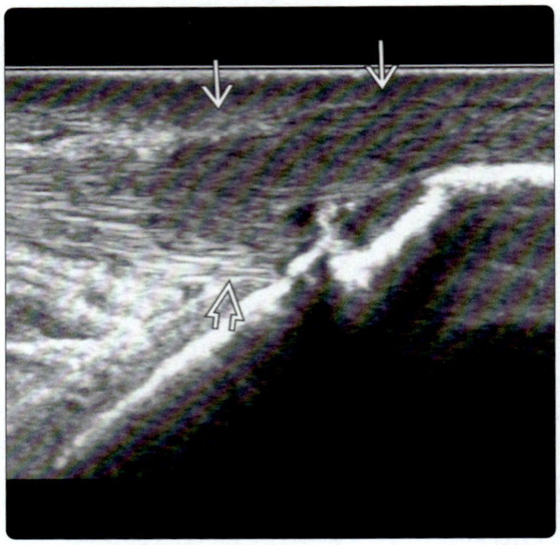

Tendinosis

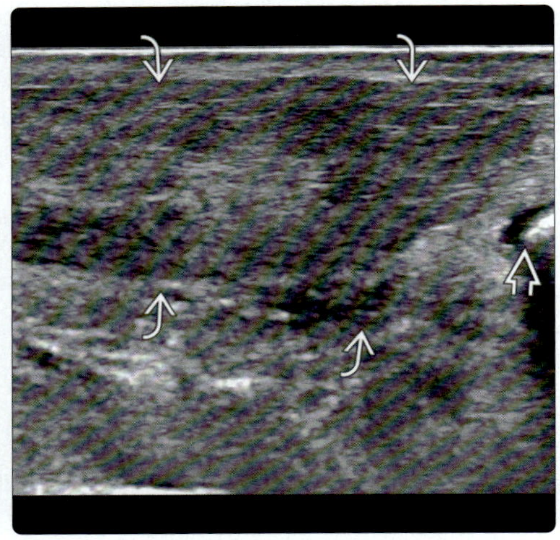

(Left) *Longitudinal US of the adductor longus tendon shows multiple foci of calcification ➔ close to the adductor longus origin at the pubic ramus ➔. The tendon ➔ is moderately swollen and hypoechoic, consistent with tendinosis.* **(Right)** *Transverse US of the retromalleolar tibialis posterior tendon shows severe tendinosis with a thickened heterogeneous tendon ➔. The small hypoechoic area ➔ with an ill-defined margin within the tendon is consistent with localized proteoglycan accumulation.*

Tendinosis

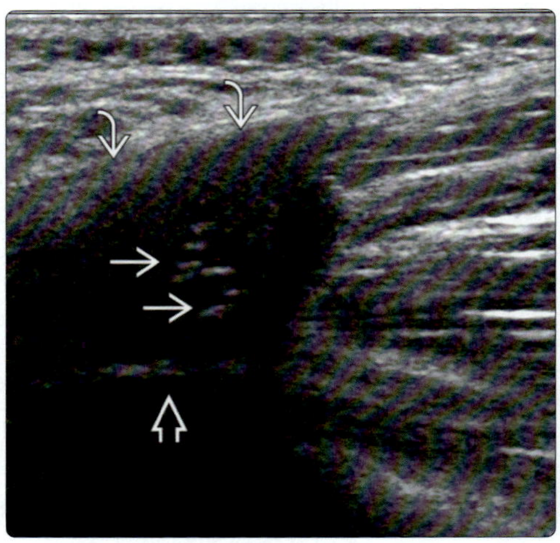

Tendinosis

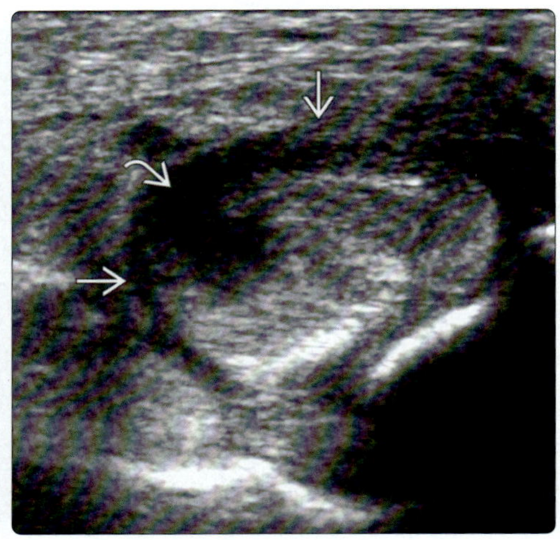

(Left) *Longitudinal US shows tendinosis and fusiform thickening of the proximal Achilles tendon ➔. A small longitudinal intrasubstance partial tear is located within the central portion of the tendon ➔.* **(Right)** *Longitudinal US of the Achilles tendon shows a complete tear at the proximal 1/3 of the tendon ➔. The proximal ➔ and distal ➔ tendon ends are retracted. The soleus muscle deep to the tear ➔ is normal.*

Tendon Tear

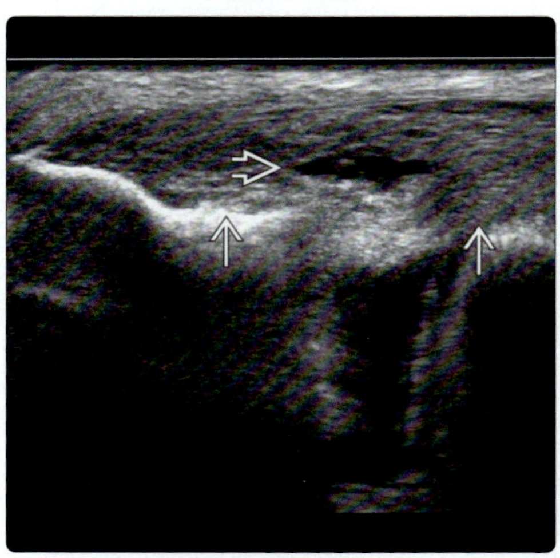

Tendon Tear

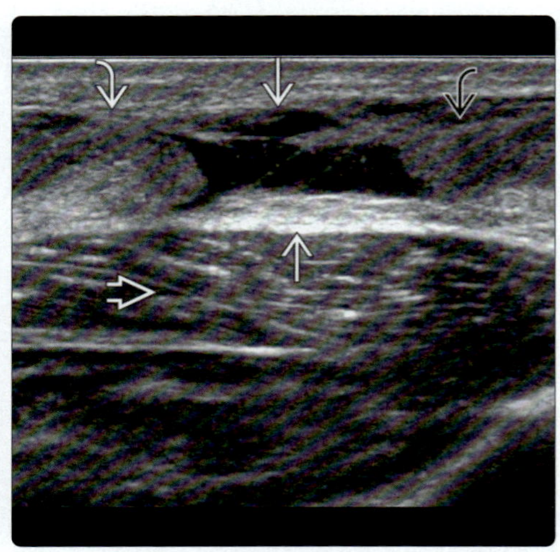

Intratendinous Vessel

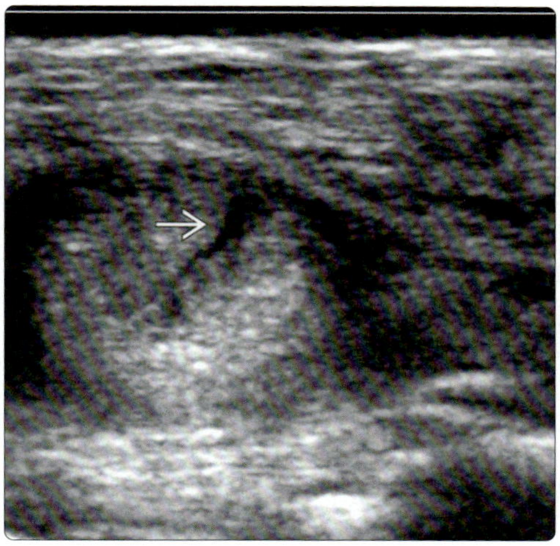

Intratendinous Vessel

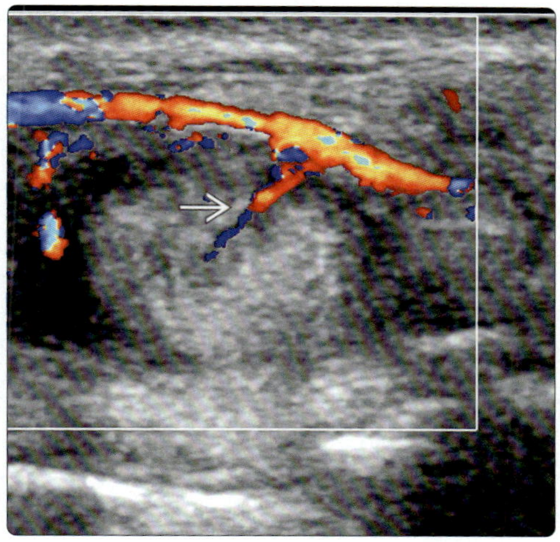

(Left) *Transverse US of the posterior tibialis tendon shows a linear, discrete, hypoechoic line extending from the surface to the center of the tendon →, suspicious of a partial-thickness tear.* **(Right)** *Transverse color Doppler US of the corresponding region shows that the linear, discrete, hypoechoic line represents a vascular channel →. Always check any suspected partial tears with color Doppler US.*

Xanthomatosis

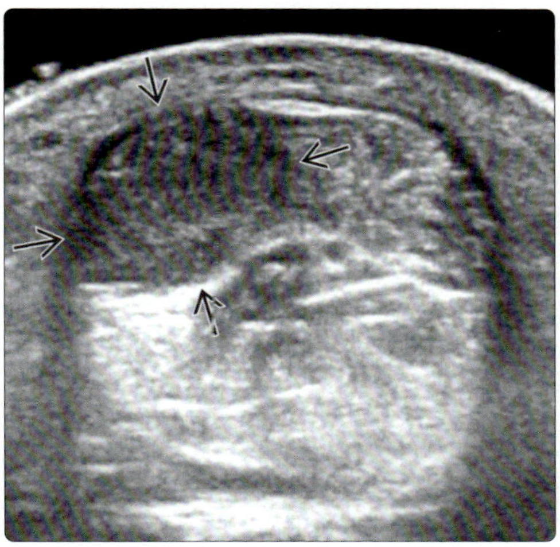

Xanthomatosis

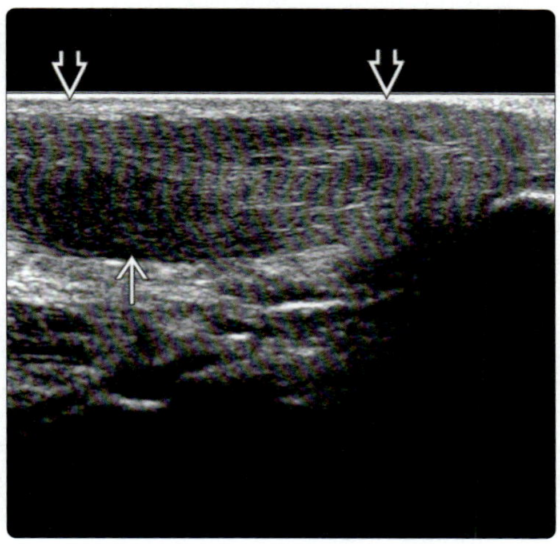

(Left) *Transverse US of the distal 1/3 of the Achilles tendon in a patient with familial hypercholesterolemia shows focal round hypoechoic area → in a normal-appearing tendon suggestive of an xanthoma.* **(Right)** *Longitudinal US of the Achilles tendon in a patient with familial hypercholesterolemia shows that the tendon → is diffusely thickened with a more focal hypoechoic area → on the deeper aspect of the tendon due to a localized cholesterol deposition (xanthoma).*

Xanthomatosis

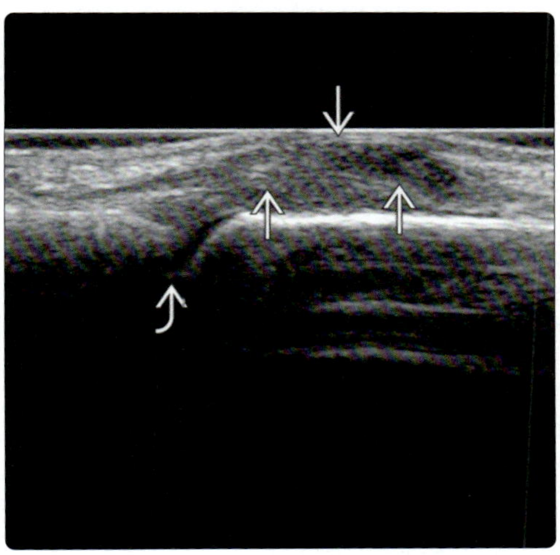

Xanthomatosis

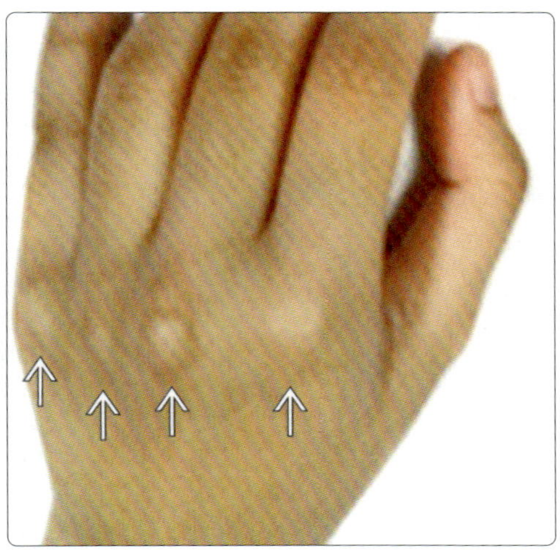

(Left) *Longitudinal US in the same patient shows moderate hypoechoic thickening → of the extensor tendon hood over the metacarpophalangeal joint → due to a tendon xanthoma.* **(Right)** *Clinical photo of the same patient shows xanthomas overlying the metacarpophalangeal joints →.*

Tendon Hyperechogenicity

DIFFERENTIAL DIAGNOSIS

Common

- Calcific Tendinitis
- Calcification Within Tendinosis
- Gout

Less Common

- Tendon Tear
- Avulsion Injury of Tendon
- Accessory Ossicle
- Suture Material Within Tendon
- Foreign Body Within Tendon

ESSENTIAL INFORMATION

Key Differential Diagnosis Issues

- Normal tendons are moderately echogenic with distinct fibrillar pattern
- Areas of tendon hyperechogenicity are usually due to calcification or crystal deposition within tendon substance
  - Alternatively gas, blood, or suture
- US is more sensitive than radiography or MR at detecting calcification or crystal deposition within tendon
- Calcification or crystal deposition varies from small, hyperechoic foci with comet-tail artifacts to large, confluent areas of dense hyperechogenicity with dense, acoustic shadowing
  - Small crystal aggregates not visible radiographically

Helpful Clues for Common Diagnoses

- **Calcific Tendinitis**
  - Due to deposition of hydroxyapatite crystal within tendon substance
  - Leads to severe pain and restriction of movement
    - Can affect any tendon, though some tendons are particularly prone to calcific tendinitis
    - Age: 40-60 years
    - Formative, latent (resting), and resorptive phases
      - Pain particularly during resorptive phase
  - Tendons around shoulder are most commonly affected
    - Supraspinatus > infraspinatus > subscapularis tendon
  - Other less commonly affected tendons are
    - Hamstring tendons at ischial tuberosity
    - Rectus femoris tendon near iliac attachment
    - Adductor tendons near pelvic attachment
    - Gluteus medius and minimus tendons at greater trochanteric insertion
    - Quadriceps tendon
    - Flexor hallucis longus tendon
    - Flexor carpi ulnaris tendon
  - Focal hyperechoic area of varying size and shape
    - Arch-shaped (most common)
    - Punctuate or fragmented
    - Nodular
    - Cystic type (uncommon)
  - ± acoustic shadowing depending on size
  - ± associated tendinosis (mild, moderate, or severe)
  - ± tendon tear
  - ± pericalcific color flow on Doppler imaging
    - Doppler flow suggests resorptive phase and good response to needling
  - ± intraarticular, intrabursal or intraosseous crystal deposition
  - Several tendons may be affected simultaneously
  - Treatment
    - Analgesia and local heat (self-limiting condition)
    - US-guided needling for both hard and soft calcification
    - Aspiration lavage and aspiration especially for soft calcification
    - ± US-guided steroid or long-acting local anesthetic to bursa alongside tendon
    - Extracorporeal shock wave therapy
    - Acetic acid iontophoresis; applying direct current to drive ions through tissues; injection without needle
- **Calcification Within Tendinosis**
  - Dystrophic calcification seen in Achilles tendon > > quadriceps and patellar tendons > posterior tibialis tendon
    - Site specificity allows easy differentiation from calcific tendinitis, which mainly affects rotator cuff tendons
  - Affects tendons close to insertional area
  - Usually associated with background tendinosis
  - Discrete areas of calcification within tendon from few mm to few cm
    - Mature, solid-appearing with well-defined margins and little, if any, associated pain or tenderness
    - Strong propensity to ossify
    - Not possible to aspirate, unlike calcific tendinitis
  - Strong posterior acoustic shadowing
  - Color Doppler flow is feature of tendinosis rather than calcific deposition
  - Usually associated with reactive bony changes at insertional area
    - Cortical irregularity
    - Cortical hyperostosis and spur formation
  - May occur in asymptomatic subjects with no other US features of tendinosis
    - Particularly in Achilles and quadriceps tendons
- **Gout**
  - Crystal deposition within tendon substance or alongside tendon
  - Usually associated with gouty tophi within tendon sheath and tendon sheath effusion
  - Most crystal deposition tends to be around tendons, ligaments, and joints
    - Intratendinous deposition is common feature of chronic gouty arthropathy
    - Rare feature of calcium pyrophosphate crystal deposition
  - Focal tendon thickening
    - Loss of normal fibrillar pattern
  - Intratendinous crystal aggregates seen as comet-tail artifacts within tendon
    - ± posterior acoustic shadowing depending on severity of crystal deposition
    - ± mild peritendinous or intratendinous hyperemia
  - Normally affects tendons of smaller joints, including wrist/hand or ankle/foot
    - Less common in larger tendons

- Typical clinical history ± raised plasma urate
- Usually associated with gouty tophi
  - Extraarticular gouty tophi depending on level of crystal deposition
  - Usually large, irregular, heterogeneous, hypoechoic mass with calcified foci and hyperemia
  - Soft tophi: Hyperechoic with no acoustic shadowing
  - Hard tophi: Hyperechoic with intense acoustic shadowing
  - Quite well-defined, irregular margins with variable hypoechoic halo ("clump of wet sugar")
  - Groups of tophi at single site common
- Usually associated with gouty arthropathy
  - Variable appearance of joint fluid
  - Hyperechoic aggregates in synovial fluid
  - Fine punctuated echogenicity ("urate sand") or large echogenic aggregates (snowstorm appearance)

Helpful Clues for Less Common Diagnoses

- **Tendon Tear**
  - Often associated with hemorrhage
    - Acute hemorrhage may be echogenic on US
    - Makes tear less readily visible on US
  - May occur at insertion (avulsive type), at tendon surface, or within tendon substance (intrasubstance)
  - Look for associated secondary signs of tear
    - Flattening or indentation of tendon surface
    - Loss of continuity of tendon fibrillar pattern
    - Widening of gap on dynamic movement
    - ± compression of hematoma in gap on dynamic scanning
  - Chronic tears may also be filled with echogenic fat, particularly in Achilles tendon
  - Partial-thickness tear
    - Focal discontinuity of tendon not involving all tendon fibers
    - Incomplete tendon retraction
    - Occurs with moderate to severe tendinosis
    - ± dystrophic calcification
    - ± bursitis
    - ± reactive bony changes
  - Full-thickness tear
    - Collapsed synovial sheath may be mistaken for attenuated tendon, particularly in smaller tendons (e.g., finger flexors and biceps tendon)
  - Peritendinitis
    - Edema and hyperemia of peritendinous tissues
- **Avulsion Injury of Tendon**
  - Commonly at extensor digitorum of fingers, Achilles tendon, tibialis posterior tendon
  - Ossified fragment can be seen at distal end of tendon mimicking ossification within tendon
  - Look for loss of normal fibrillation distal to ossified fragment
- **Accessory Ossicle**
  - Most common accessory ossicles within tendon are accessory navicular bone (within posterior tibialis tendon) and os peroneum (within peroneus longus)
  - May be confused with dystrophic ossification
    - Accessory ossicle has smooth convex superficial border, while dystrophic ossification is more irregular
  - Fractures or displacement of os peroneum are clue to peroneus longus tendon tear
- **Suture Material Within Tendon**
  - Nonabsorbable material visible many years after tendon repair
    - Easily recognized as tubular echogenic material within tendon substance
- **Foreign Body Within Tendon**
  - US is excellent at detecting foreign bodies
  - Foreign body echogenicity depends on composition (e.g., wood, glass, metal) and shape
    - Most foreign bodies are hyperreflective with posterior acoustic shadowing or reverberation artifact
  - If US is negative in suspicious acute clinical setting, repeat US 1 week later, as hypoechoic inflammatory response or granulation tissue may make foreign body more conspicuous

Calcific Tendinitis

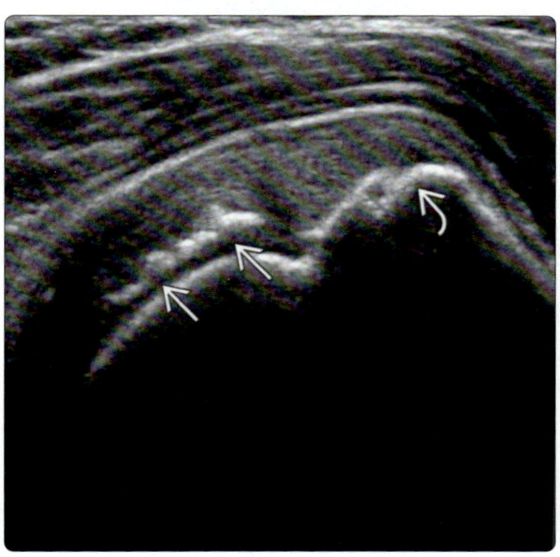

Calcific Tendinitis

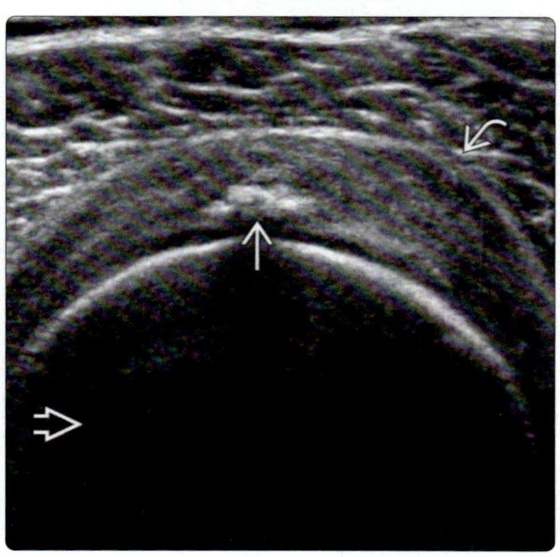

(Left) *Longitudinal US shows a large area of calcific tendinitis ➡ just proximal to the insertional area of the supraspinatus tendon ➡.* **(Right)** *Transverse US shows a smaller focus of calcification ➡ within the supraspinatus tendon ➡ close to the insertional area, compatible with calcific tendinitis. The humerus ➡ is normal.*

(Left) *Longitudinal US shows severe nonconfluent calcific tendinitis within the supraspinatus tendon obscuring the insertional area.* **(Right)** *Longitudinal US shows severe deposition of hydroxyapatite crystals within the biceps tendon at the proximal end of the bicipital groove. This is soft amorphous calcification suited to aspiration. Calcific tendinitis tends to affect those tendon areas most affected by tendinosis. Note resultant acoustic shadowing across humeral head.*

Calcific Tendinitis

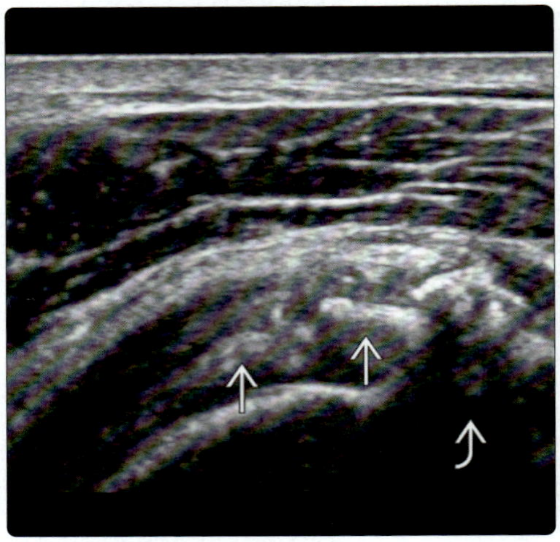

Calcific Tendinitis

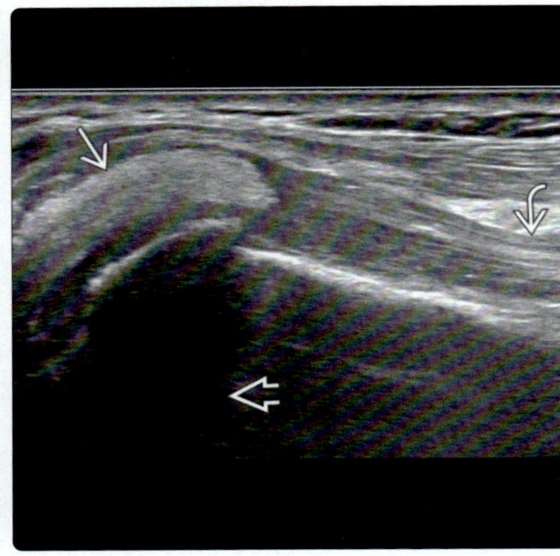

(Left) *Longitudinal US shows a focal area of calcifications within the distal insertion of the quadriceps tendon. Moderate tendinosis is present. Tendinosis is a frequent, but not invariable, accompaniment of dystrophic calcifications.* **(Right)** *Longitudinal US shows a large area of ossification within the origin of patellar tendon.*

Calcification Within Tendinosis

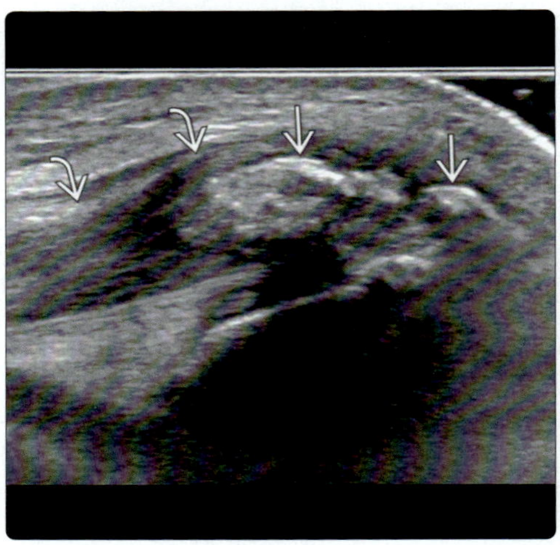

Calcification Within Tendinosis

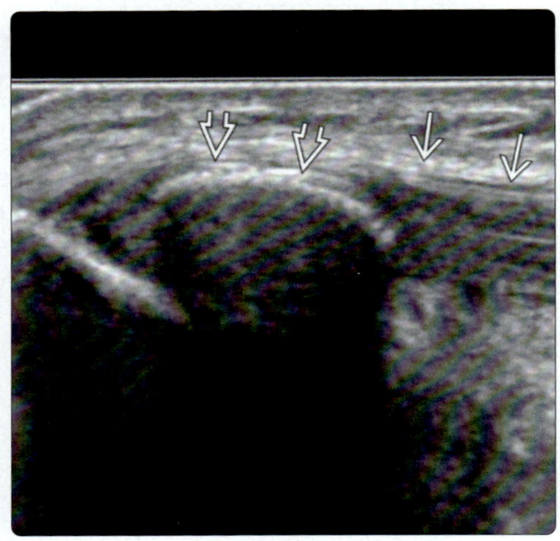

(Left) *Longitudinal US shows echogenic material encasing and infiltrating the extensor digitorum tendon, compatible with severe crystal deposition.* **(Right)** *Transverse US of the extensor carpi ulnaris tendon shows the gouty echogenic material distending the extensor carpi ulnaris tendon sheath. Crystals are also deposited within the tendon leading to tendon hyperechogenicity, though no acoustic shadow is present.*

Gout

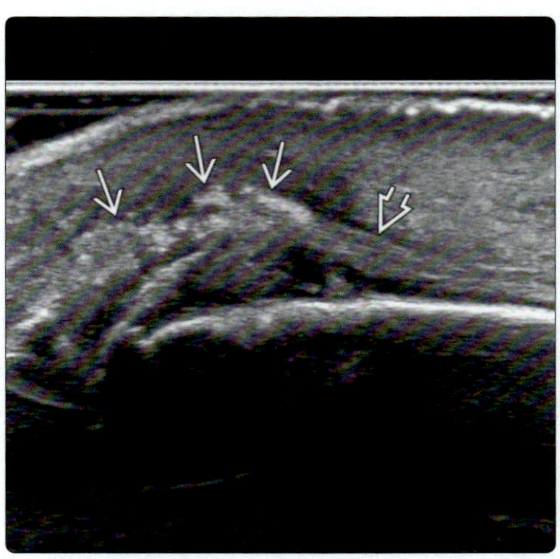

Gout

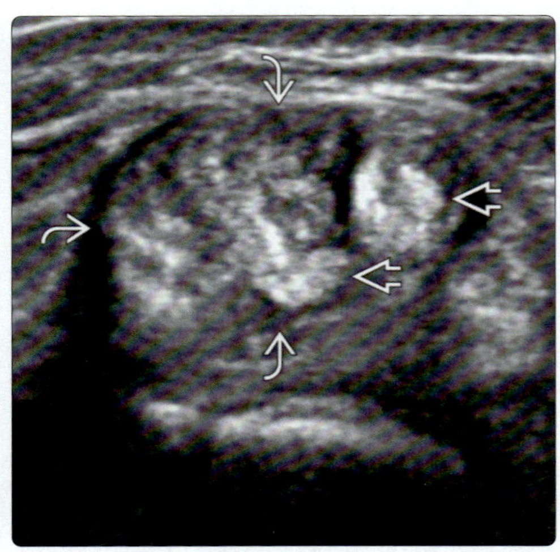

Tendon Tear

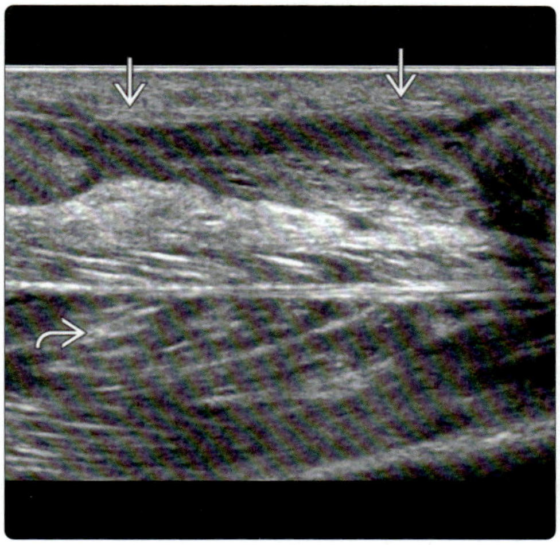

Tendon Tear

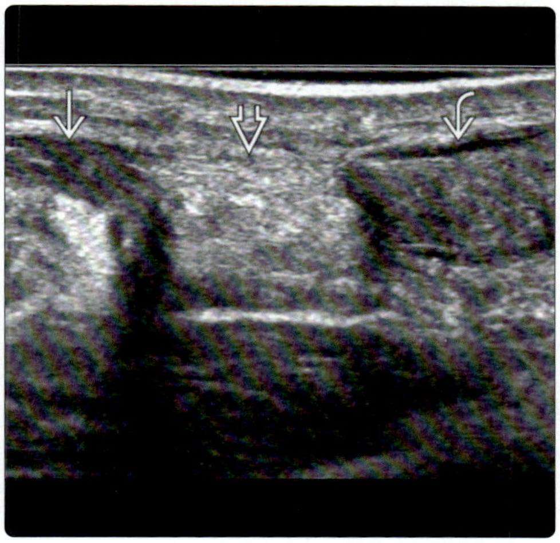

(Left) *Longitudinal US shows a complete chronic rupture of the Achilles tendon. There is a large retraction gap. The chronic tear is filled with slightly echogenic fibrous tissue. The soleus muscle is mildly atrophic.* **(Right)** *Longitudinal US of an acute full-thickness Achilles tendon tear shows a gap filled with echogenic edematous fat. The swollen proximal and distal ends of the tendon are shown.*

Avulsion Injury of Tendon

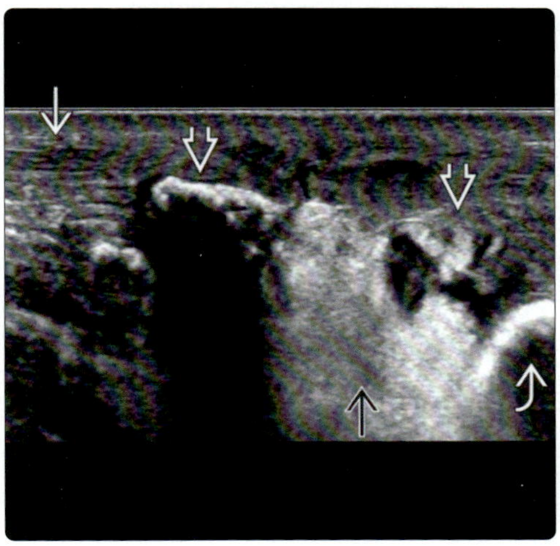

Accessory Ossicle

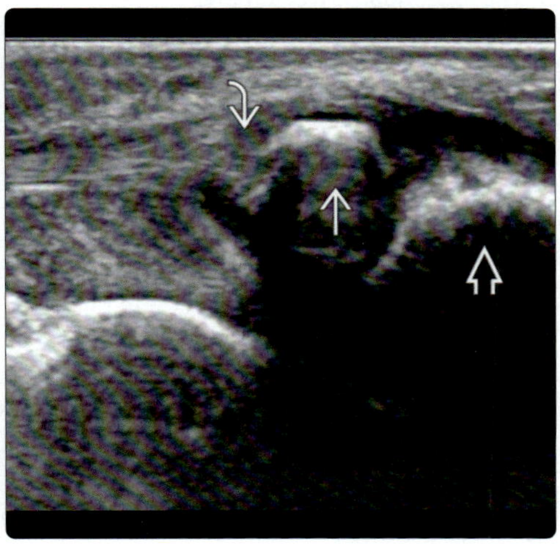

(Left) *Longitudinal US of the Achilles tendon shows a fragmented avulsion fracture from the posterior calcaneal tuberosity insertion with proximal retraction. The gap is filling up with echogenic hematoma.* **(Right)** *Longitudinal US shows a normal accessory bone near the insertion of the posterior tibialis tendon into the navicular bone. The distal end of the posterior tibialis tendon normally expands near to the bony insertion.*

Suture Material Within Tendon

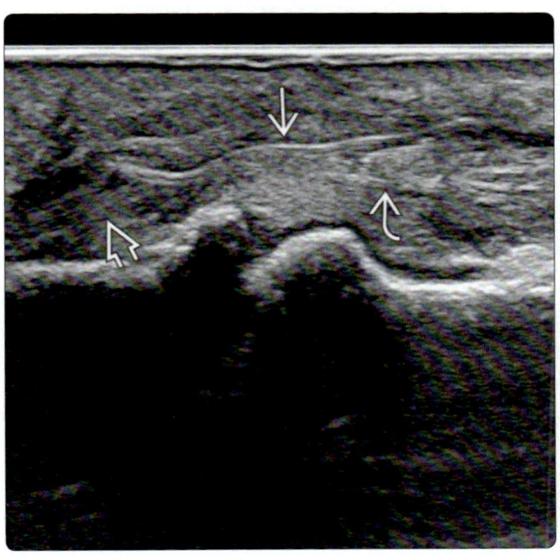

Foreign Body Within Tendon

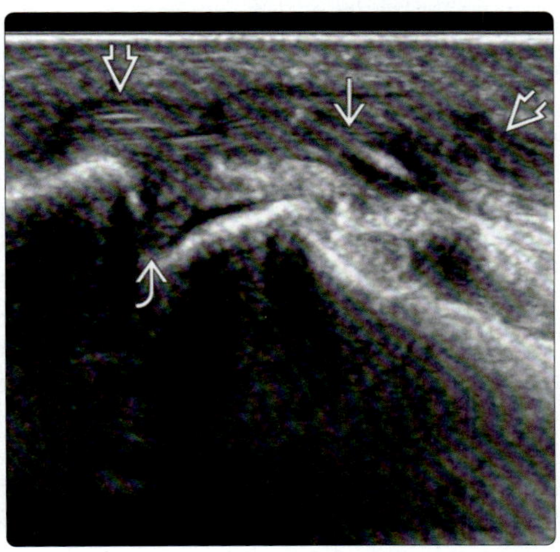

(Left) *Longitudinal US of the flexor digitorum profundus tendon at the finger shows linear hyperechoic material between the tendon ends due to surgical suture. There is a wide gap between the proximal and distal tendon ends, compatible with retear.* **(Right)** *Longitudinal US of the middle finger shows an echogenic glass splinter foreign body lying obliquely within the flexor digitorum tendon just proximal to the metacarpophalangeal joint.*

DIFFERENTIAL DIAGNOSIS

Common

- Tendinosis

Less Common

- Acute Proliferative Tenosynovitis
- Acute Exudative Tenosynovitis
- Chronic Tenosynovitis (Active)
- Chronic Tenosynovitis (Inactive)
- Gout/Pseudogout
- Xanthoma
- Amyloidosis

ESSENTIAL INFORMATION

Key Differential Diagnosis Issues

- Need to determine whether primary pathology affects tendon or tendon sheath
- Distinction between tendinosis and tenosynovitis has important therapeutic and prognostic implications
    - Tendinosis: Tendon disease with secondary involvement of peritendinous structures
        - Tenosynovitis: Disease of tendon sheath with secondary involvement of tendon
    - Tendinosis results from microtears, proteoglycan deposition, and collagen disorganization
        - Tenosynovitis results from tendon sheath inflammation due to inflammatory mediators, repetitive microtrauma, crystals, or bacteria
    - Tendinosis is not primary inflammatory disease
        - Tenosynovitis is primary inflammatory disease
    - Tendinosis affects any tendon
        - Tenosynovitis only affects tendons with tendon sheath
    - Tendinosis often affects only segment of tendon
        - Tenosynovitis usually affects all of tendon contained by tendon sheath
    - Tendinosis is often subclinical
        - Tenosynovitis is rarely subclinical
    - Tendinosis frequently predisposes to tendon tear or rupture
        - Tenosynovitis infrequently predisposes to tendon tear or rupture
    - Tendinosis treated primarily by rest and is likely to recur
        - Tenosynovitis treated primarily by antiinflammatory agents and is less likely to recur
- Note that mild tendon and tendon sheath swelling occurs in presence of cellulitis or edema as feature of generalized soft tissue swelling
    - Do not misdiagnose as tenosynovitis

Helpful Clues for Common Diagnoses

- **Tendinosis**
    - Tendons affected
        - Rotator cuff: Supraspinatus > infraspinatus > subscapularis > teres minor
        - Long head of biceps near top of bicipital groove (genu of biceps tendon)
        - Lower limbs: Hamstrings and adductor, patellar tendon, Achilles tendon, posterior tibialis tendon
        - Upper limbs: Common extensor tendons of elbow, extensor carpi ulnaris
    - US features
        - Focal, nodular, or diffuse tendon enlargement (ranges from mild to severe)
        - Progressive loss of normal fibrillar echotexture
        - Increasing tendon hypoechogenicity
        - Tendon hyperemia
            - ☐ Common extensor and flexor tendon origins, patellar, Achilles, and posterior tibialis tendons
        - ± partial tear
            - ☐ Intrasubstance, avulsive-type, or full-thickness tears
            - ☐ Common in moderate to severe tendinosis
            - ☐ Should not be confused with dilated vascular channels within hyperemic tendon
        - ± tendon sheath effusion
        - ± peritendinitis with inflammation of soft tissues around tendon or paratenon
        - ± reactive change at bony insertional area
            - ☐ Hyperostotic or resorptive bone change
        - ± dystrophic calcification within tendon
        - ± bursitis (Achilles)

Helpful Clues for Less Common Diagnoses

- **Acute Proliferative Tenosynovitis**
    - Due to inflammation or infection
    - Tendons affected
        - Commonly affects common flexor tendon sheath (ulnar bursa) of hand
            - ☐ Ulnar bursa extends from just proximal to just distal to carpal tunnel
            - ☐ Communicates with flexor digitorum tendon sheath in ~ 10% of normal subjects and radial bursa in ~ 5% of normal subjects
            - ☐ Ulnar bursal infection may also spread to space of Parona (potential space superficial to pronator quadratus muscle, which extends to midforearm)
    - Salient features
        - Synovial proliferation within tendon sheath > tendon sheath fluid
        - Hyperemia predominates around, rather than within, tendon sheath
- **Acute Exudative Tenosynovitis**
    - Due to acute infection ± foreign body following penetrating trauma
    - Salient features
        - Fluid accumulation within sheath with little or no synovial proliferation
        - Speckled echogenic fluid
            - ☐ Speckles due to aggregates of purulent material
        - Mild to moderate tendon swelling
            - ☐ ± indistinct tendon margins if infection established and severe
        - Hyperemia around, rather than within, tendon sheath
        - Infection may spread to surrounding tissues
- **Chronic Tenosynovitis (Active)**
    - Due to inflammation or infection
    - Salient features
        - Hyperemia is predominantly within tendon sheath; synovium is visibly hypervascular

- In acute tenosynovitis, synovium is not visibly hypervascular

- **Chronic Tenosynovitis (Inactive)**
  - Salient features
    - Tendon sheath distension with fluid or synovial proliferation, though little or no hypervascularity
    - If adhesions present, tendon sheath may be markedly distended focally and noncompressible
- **Gout/Pseudogout**
  - Tophi are more common with chronic gout
    - Foot > knee > wrist and hand
    - Occur adjacent to bursa/tendon/ligaments and fascia
    - Range from soft to hard tophi
      - Soft tophi: Little mineralized tissue with scattered crystal aggregates and comet-tail artifacts
      - Hard tophi: Heavily mineralized tissue with only superficial margin being seen due to dense posterior acoustic shadowing
    - Acute phase → intercritical phase
  - Salient features
    - Irregular hypovascular mass in and around tendon with echogenic crystalline aggregates
      - Marginal hyperemia during acute phase
- **Xanthoma**
  - Feature of familial hyperlipidemia
    - Accumulation of foamy histiocytes, cholesterol ester, and giant cells within tendons
  - Tendons affected
    - Achilles tendon > patellar tendon > extensor tendons of hands
  - Salient features of xanthoma
    - Generalized tendon swelling, increased tendon hypoechogenicity, and mild hypervascularity
    - The following US signs are more common in xanthoma
      - Nodular intratendinous hypoechoic areas (xanthoma)
      - Less insertional cortical irregularity
      - More diffuse, intratendinous, dystrophic calcification
      - Less peritendinitis
      - Less tendon tears
      - Less hyperemia
    - Other signs to distinguish from tendinosis
      - Known familial hyperlipidemia
      - Features of atherosclerosis (increased intima-media thickness)
      - Nonpainful tendon swelling
      - Xanthoma will reduce with lipid-lowering treatment
- **Amyloidosis**
  - Accumulation of protein fibrils, which become insoluble and are deposited either in specific tissues (localized amyloidosis) or throughout body (generalized amyloidosis)
  - Extracellular deposition of protein fibrils (amyloid)
    - Over 20 different protein fibrils may produce amyloidosis
    - Diagnosis dependent on histology of biopsy specimen (Congo red staining and apple-green birefringence under polarized light)
  - May be primary or secondary
  - **Primary amyloidosis**
    - Genetically inherited disorder of protein metabolism
  - **Secondary amyloidosis**
    - Usually seen in patients with long-term hemodialysis
      - Cannot filter β2-microglobulin that accumulates
    - Also seen in patients with myeloma and chronic inflammatory states
      - Increased production of protein fibrils
  - Salient features
    - Amyloid deposited in synovium, joint capsule, tendons (supraspinatus, biceps, flexor digitorum tendons), ligaments, nerves, and intervertebral disc
      - Hypoechoic tendon thickening
      - ± tendon sheath or bursal effusion and hypervascularity with hypoechoic, inhomogeneous nodular thickening around tendons
    - Other features are amyloid spondyloarthropathy and lytic paraarticular bone lesions

Tendinosis

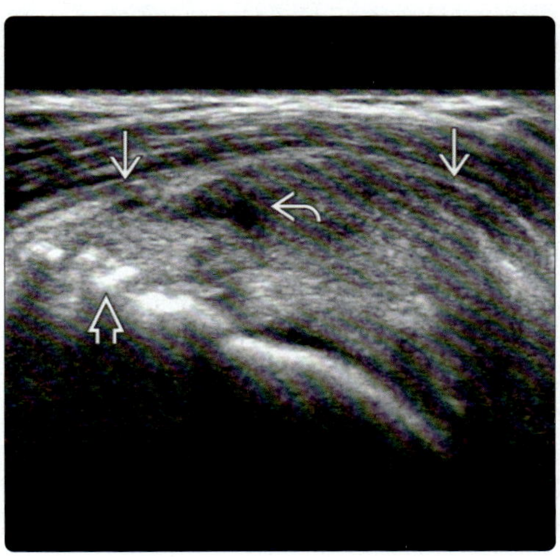

Tendinosis

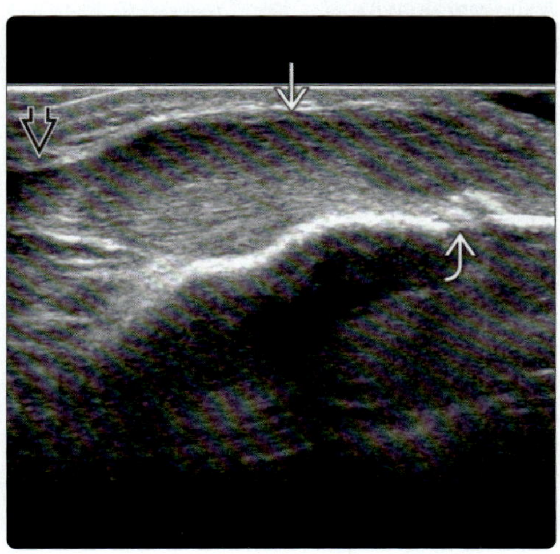

(Left) *Longitudinal US shows a markedly thickened and diffusely hypoechoic supraspinatus tendon ➡ compatible with severe tendinosis. There is insertional cortical irregularity ➡, and there are small tendon tears ➡.* **(Right)** *Longitudinal US of distal patellar tendon shows severe focal thickening with hypoechogenicity and loss of normal fibrillar pattern ➡ due to severe tendinosis. Mild tibial tuberosity cortical irregularity is present ➡. The more normal caliber patellar tendon is seen proximally ➡.*

(Left) *Graphic shows grades of tendinosis. The severity of tendinosis and the degree of surrounding peritendinitis* ➡ *should be independently graded. The likelihood of tendon tear is greater the more severe the tendinosis. The likelihood of clinical response to antiinflammatory medication is greater the more severe the peritendinitis.* **(Right)** *Graphic shows normal tendon structure with tendon sheath* ➡ *shown for illustrative purposes. Most tendons do not have a tendon sheath.*

Tendinosis

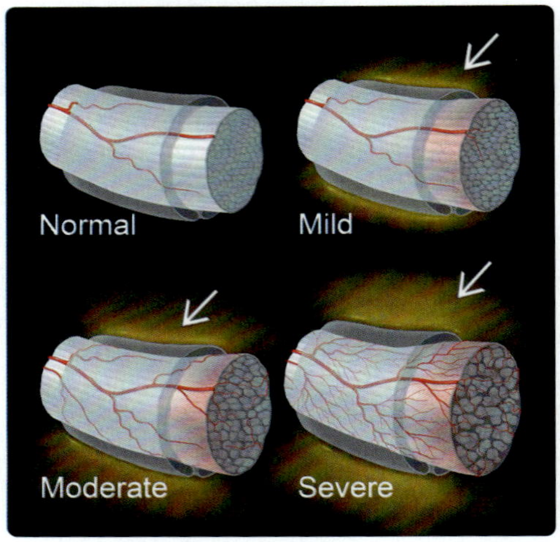

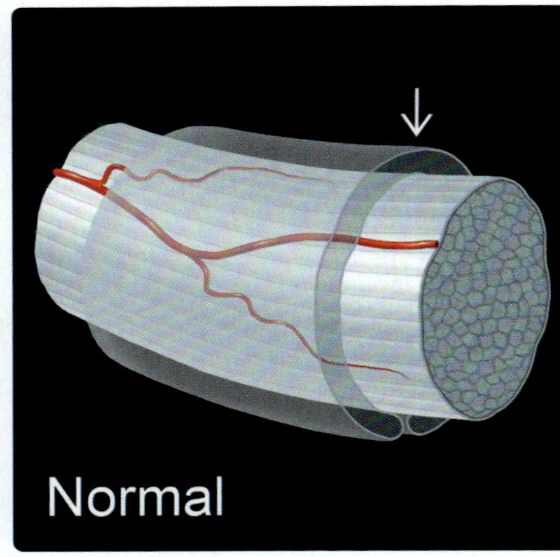

(Left) *Mild tendinosis with mild tendon swelling (due to proteoglycan accumulation) and mild intratendinosis hyperemia is shown. Intrinsic vascularity is not a feature of normal tendons.* **(Right)** *Moderate tendinosis with more severe tendon swelling, proteoglycan deposition, and intratendinosis hyperemia is shown. Although tendinosis and peritendinitis (seen as peritendinous edema and hyperemia)* ➡ *severity are broadly interrelated, both features should be independently graded as mild, moderate, or severe.*

Tendinosis

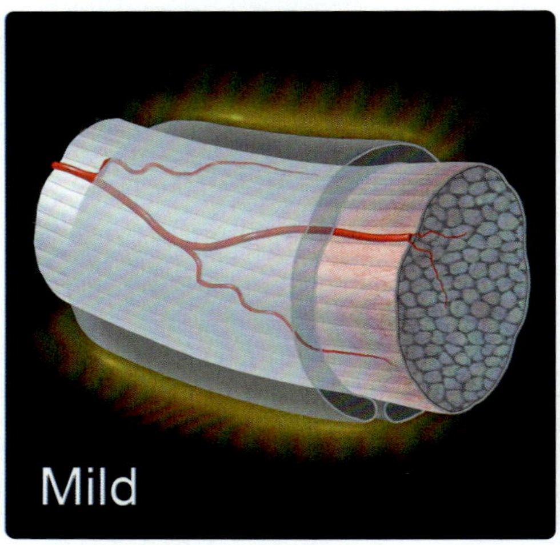

Tendinosis

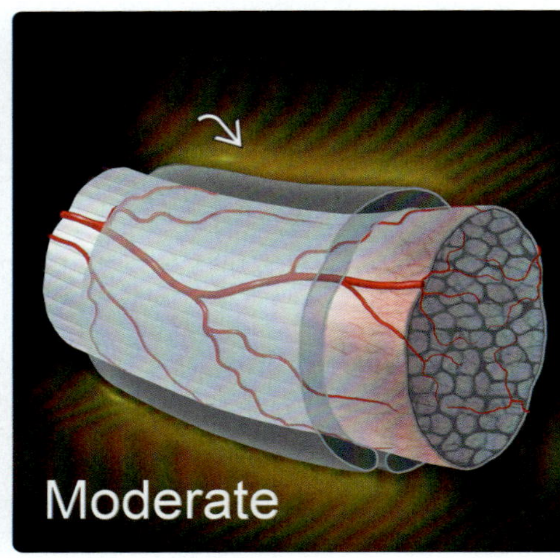

(Left) *Severe tendinosis with more severe tendon swelling and intratendinosis hyperemia* ➡ *(manifest on US as tendon swelling, tendon hypoechogenicity, loss of fibrillar echo pattern, and ↑ tendon hyperemia) is shown. Mild tendon sheath swelling is usually seen, though this is a manifestation of peritendinitis rather than primary tendon sheath inflammation.* **(Right)** *Photograph shows patellar tendinosis. The normal tendon is glistening white* ➡*, and the area with severe tendinosis is yellow* ➡*. (Courtesy C. K. Ming, MD.)*

Tendinosis

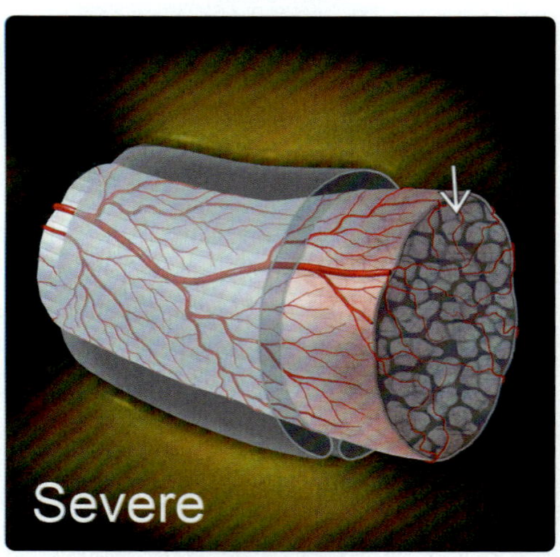

Tendinosis

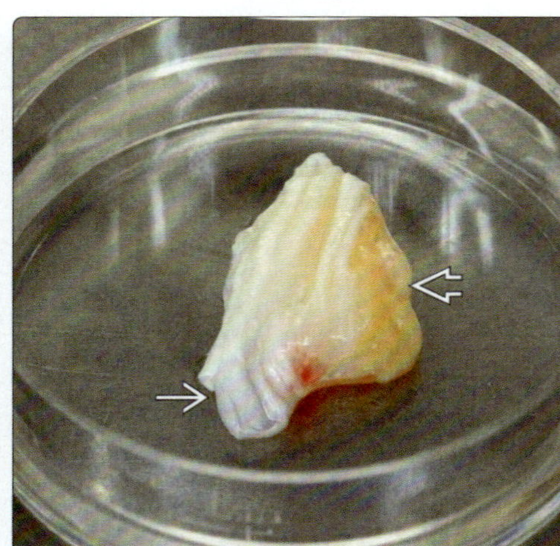

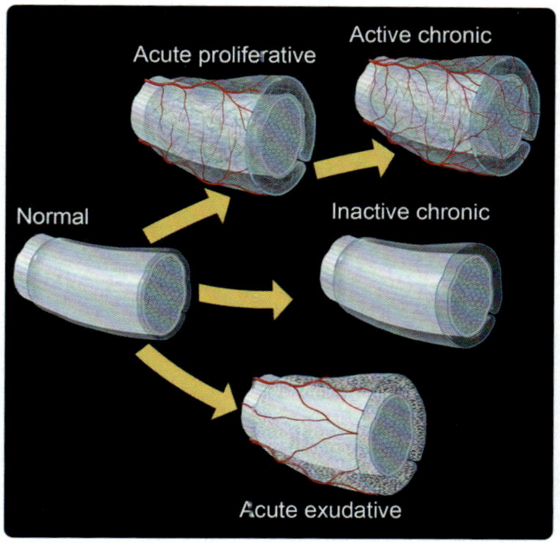

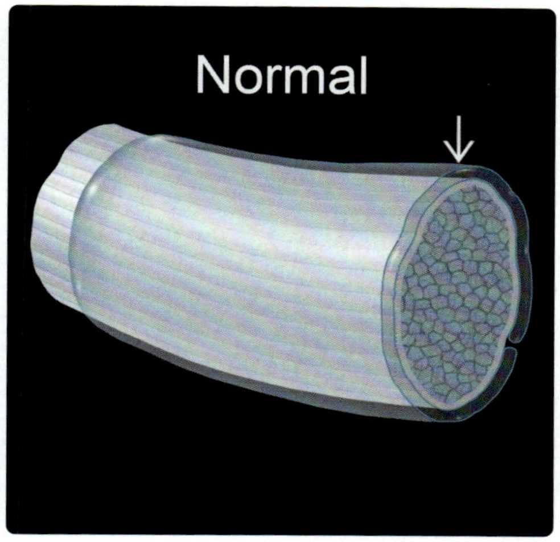

(Left) *Graphic shows different types of tenosynovitis. Tenosynovitis is a disease of the tendon sheath. Unlike tendinosis, which can affect any tendon, tenosynovitis can only affect those tendons with a tendon sheath.* **(Right)** *Graphic shows normal tendon and tendon sheath* ➡ *for comparative purposes. Most, but not all, of the tendons of the wrist/hand and the ankle/foot have a tendon sheath. The long head of biceps has a tendon sheath. The patellar and Achilles tendons have a paratenon.*

Acute Proliferative Tenosynovitis

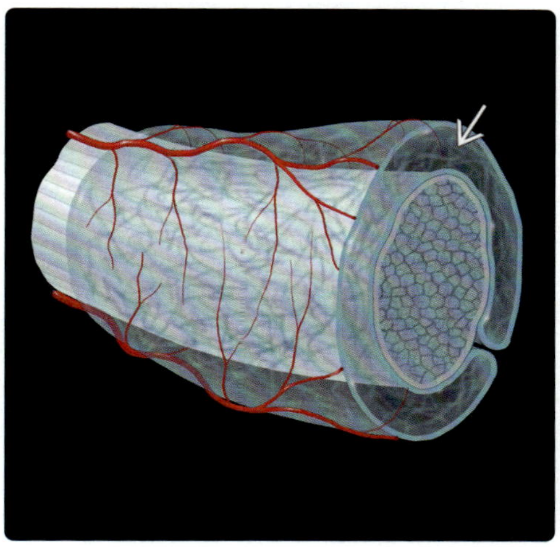

Acute Exudative Tenosynovitis

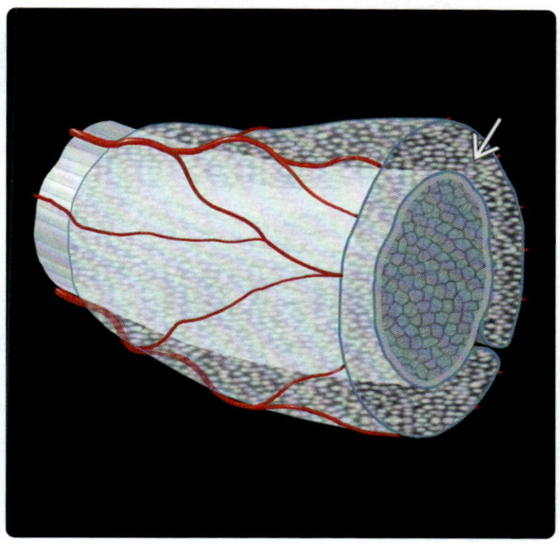

(Left) *Graphic shows acute proliferative tenosynovitis, which is usually due to repetitive trauma or a systemic inflammatory disorder. Tendon swelling occurs due to synovial proliferation* ➡*. Hyperemia is limited to the tendon sheath surface.* **(Right)** *Graphic shows acute exudative tenosynovitis, usually secondary to bacterial infection/penetrating trauma and is rapidly progressive. Purulent exudate* ➡ *can be aspirated. Hyperemia occurs around rather than within the tendon sheath.*

Chronic Tenosynovitis (Active)

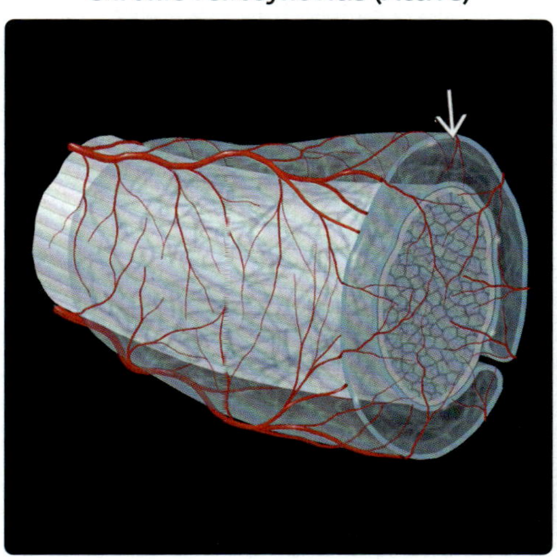

Chronic Tenosynovitis (Inactive)

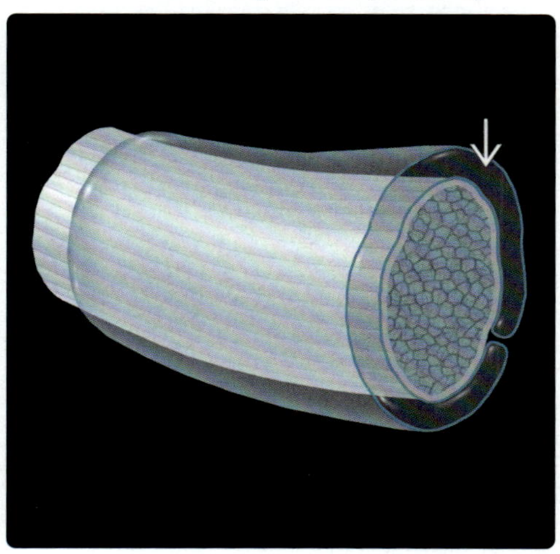

(Left) *Graphic shows chronic proliferative tenosynovitis. Mature hyperemic synovial proliferation* ➡ *fills the distended tendon sheath. Hyperemia may extend into the tendon. The tendon will have normal echogenicity and be only slightly enlarged due to edema.* **(Right)** *Graphic shows chronic inactive tenosynovitis. The tendon sheath is distended with serous fluid* ➡*. This distension may sometimes be severe and localized due to tendon sheath adhesions. Little or no hyperemia is present.*

(Left) *Longitudinal US shows a diffusely thickened and hypoechoic supraspinatus tendon, indicative of moderate to severe tendinosis. Some preservation of the fibrillary pattern is present.* **(Right)** *Longitudinal US of the elbow shows a severely thickened common flexor tendon origin with tendon hypoechogenicity, small foci of dystrophic calcification, a small intrasubstance tear, and moderate insertional cortical irregularity. These are all features of tendinosis.*

Tendinosis

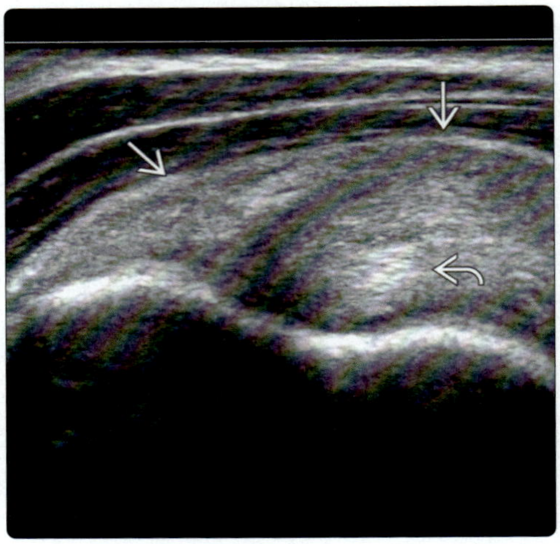

Tendinosis

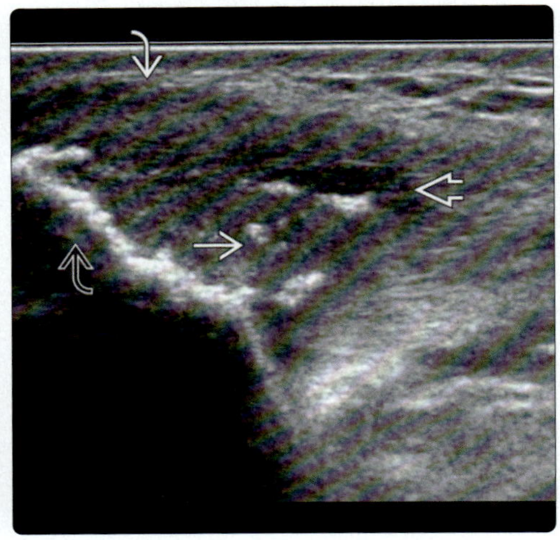

(Left) *Longitudinal US shows moderate-severity supraspinatus tendinosis. Note the hypoechoic anisotropic effect. There is distension of the subacromial-subdeltoid bursa with echogenic fluid, due to proteinaceous fluid content.* **(Right)** *Transverse US shows moderate to severe biceps tendinosis. The tendon is rounded and heterogeneous with an irregular peripheral contour. There is minimal surrounding synovial proliferation.*

Tendinosis

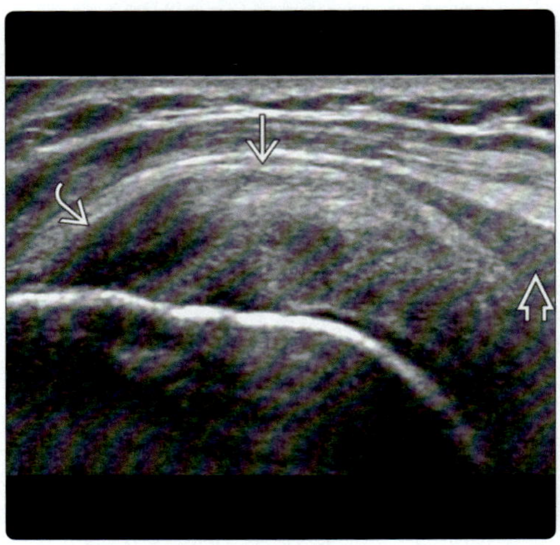

Tendinosis

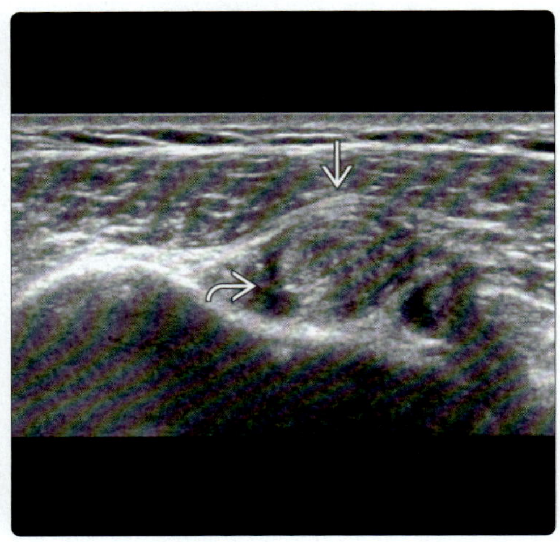

(Left) *Longitudinal US shows acute exudative tenosynovitis of flexor digitorum with distension of the tendon sheath by fluid containing small echogenic aggregates. The flexor tendon is slightly swollen. Purulent fluid was aspirated, which grew Staphylococcus aureus.* **(Right)** *Transverse color Doppler US of the extensor digitorum communis tendon at the wrist shows moderate tenosynovial proliferation as well as hyperemia after a dog bite compatible with active chronic proliferative tenosynovitis.*

Acute Exudative Tenosynovitis

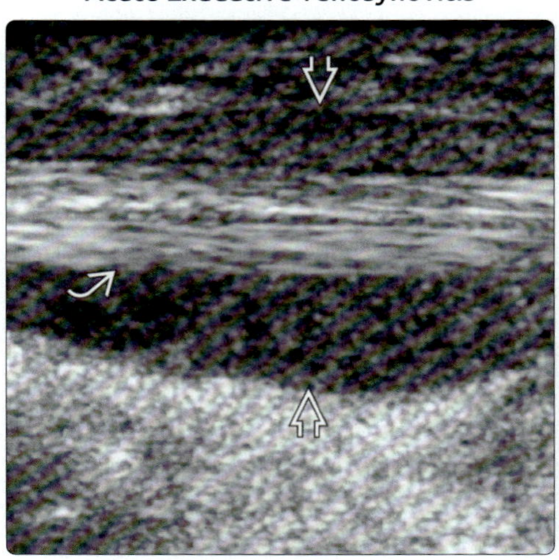

Chronic Tenosynovitis (Active)

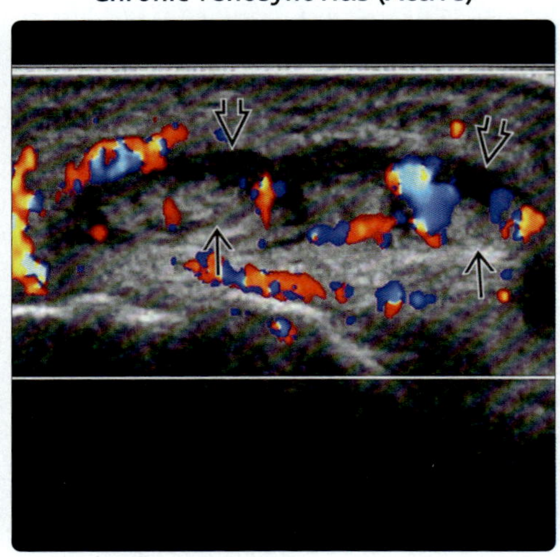

Chronic Tenosynovitis (Active)

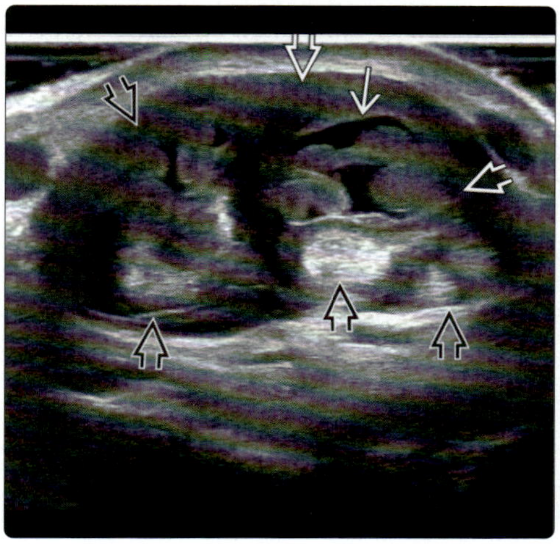

Chronic Tenosynovitis (Active)

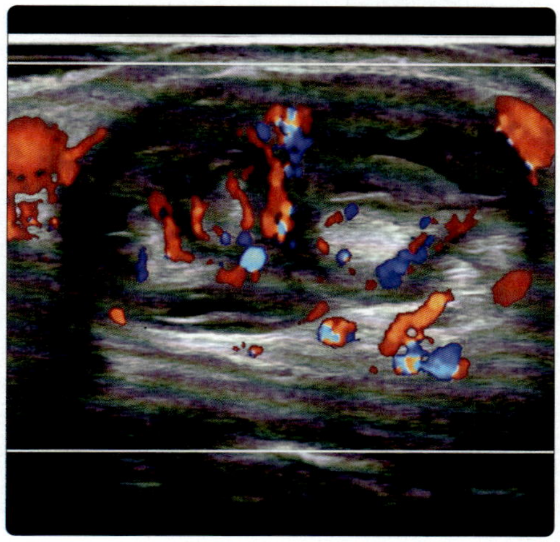

(Left) *Transverse US of a patient with wrist distension for 9 months shows severe tenosynovial proliferation of the extensor digitorum communis tendon sheath associated with a small tendon sheath effusion. The extensor tendons are mildly swollen. This is a secondary feature of more severe tenosynovitis.* **(Right)** *Color Doppler US shows marked tenosynovial hyperemia compatible with active chronic tenosynovitis. Biopsy revealed chronic necrotizing granulomatous inflammation due to TB.*

Chronic Tenosynovitis (Inactive)

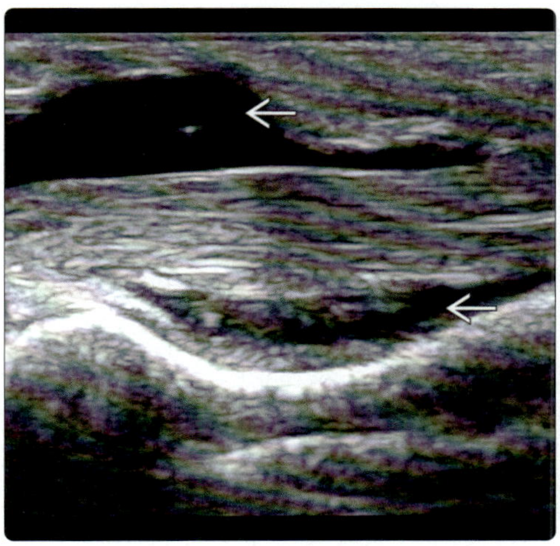

Gout/Pseudogout

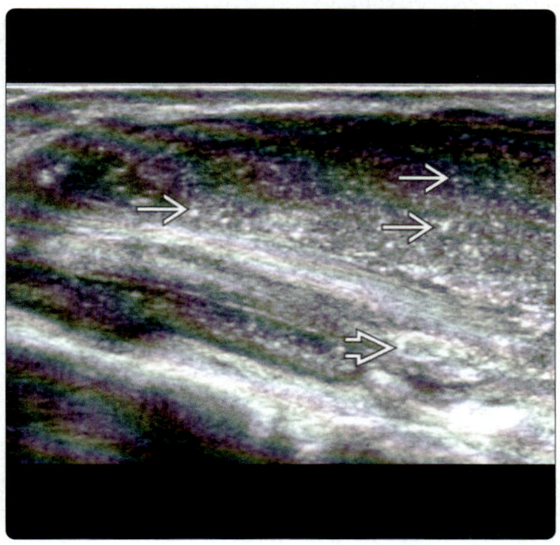

(Left) *Longitudinal US of the ring finger shows a moderate amount of largely anechoic fluid present within the flexor digitorum tendon sheath. US-guided aspiration yielded gelatinous material consistent with inactive chronic tenosynovitis.* **(Right)** *Longitudinal US of the wrist in a patient with known gout shows marked distension of the common extensor tendon sheath with small and large crystal aggregates (snowstorm appearance) due to a gouty tenosynovitis.*

Gout/Pseudogout

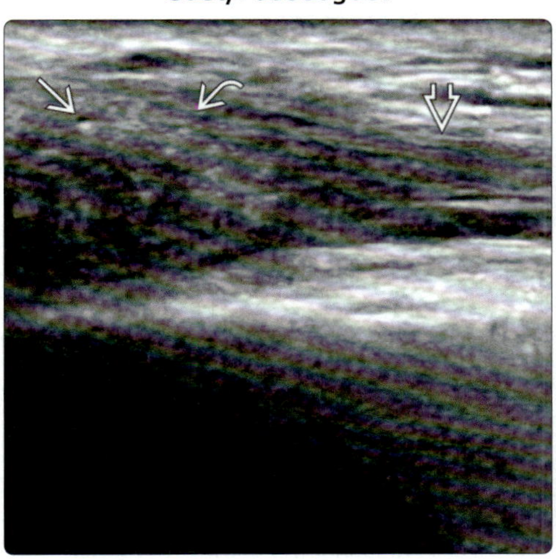

Xanthoma

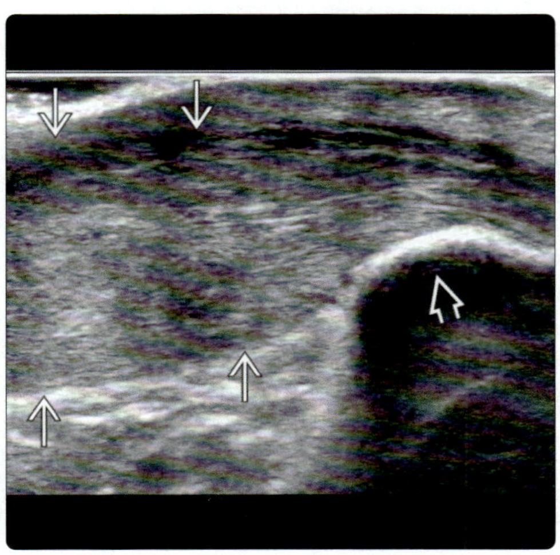

(Left) *Longitudinal US of the distal forearm shows a soft gouty tophus containing a few small echogenic foci with comet-tail artifacts within the flexor digitorum profundus tendon.* **(Right)** *Longitudinal US of the Achilles tendon in a patient with familial hypercholesterolemia shows diffuse tendon swelling close to the calcaneal insertion due to deposition of cholesterol-rich lipid within the tendon.*

DIFFERENTIAL DIAGNOSIS

Common

- Nerve Entrapment
- Nerve Sheath Tumor

Less Common

- Nerve Injury
- Perineural Fibrosis
- Immune-Mediated Peripheral Neuropathy
- Fibrolipomatous Hamartoma
- Intraneural Lipoma
- Intraneural Ganglion
- Metastases
- Acromegaly
- Charcot-Marie-Tooth Disease
- Leprosy

ESSENTIAL INFORMATION

Key Differential Diagnosis Issues

- Best to identify nerve at specific location, then follow it distally or proximally in transverse plane
    - Longitudinal plane useful when assessing for nerve trauma and tumors
- When assessing nerve compression, important to measure cross-sectional area of nerve at specific locations
    - Continuous boundary tracing of nerve provides more accurate area measurement than best-fitting ellipse method
    - Contralateral side is unreliable internal reference as contralateral, subclinical, compressive neuropathy is common
- When assessing noncompressive nerve disorder, comparison with asymptomatic contralateral side is very helpful
- Nerve measurements are ~ 20% smaller on US than MR
    - US measurements do not include epineurium, which is included on MR measurements

Helpful Clues for Common Diagnoses

- **Nerve Entrapment**
    - Site-specific entity
    - Most common sites are in upper limb
        - Median nerve: Carpal tunnel, pronator teres
        - Ulnar nerve: Cubital tunnel, Guyon canal
        - Radial nerve: Supinator inlet
    - Most common sites are in lower limb
        - Sciatic nerve: Piriformis
        - Lateral cutaneous nerve of thigh: Inguinal ligament near anterior superior lilac spine
        - Tibial nerve: Tarsal tunnel
        - Peroneal nerve: Fibular head
    - Swelling of nerve proximal to or at site of compression is overriding sign of nerve compression
    - Infrequently, neural hyperemia may be seen as feature of nerve ischemia
- **Nerve Sheath Tumor**
    - Tumor of peripheral nerve
        - Schwannoma: Tumor of Schwann cells lining axons
        - Neurofibroma: Tumor of connective tissue between axons
    - Hypoechoic, fusiform-shaped tumor of nerve
    - Thickening of entering or exiting nerve
        - Tumors arising from small nerves may not have visible thickening of parent nerve
    - Anechoic areas
        - Due to myxoid or fluid accumulation
        - Some schwannomas may be predominantly anechoic and cyst-like (ancient schwannoma)
    - Posterior acoustic enhancement
    - Moderately hyperemic on Doppler
    - ± hyperechoic areas
        - Due to hemorrhage or calcification
    - Malignant peripheral nerve sheath tumor is more likely if
        - Patient has neurofibromatosis type 1
        - Tumor is large (> 5 cm), ill defined, heterogeneous, &/or rapidly growing
    - FDG PET has quite high (~ 80-90%) sensitivity and specificity for diagnosing malignant peripheral nerve sheath tumor associated with neurofibromatosis

Helpful Clues for Less Common Diagnoses

- **Nerve Injury**
    - Most nerve injuries are clinically obvious
        - Limited need for imaging, particularly in penetrating trauma
        - Imaging indicated in equivocal or chronic cases
    - US of nerve injury is used to assess whether nerve is
        - Not torn (neurapraxia), partially torn (axonotmesis), or completely torn (neurotmesis)
    - Swollen nerve caused by
        - Reparative fibrosis at site of tear
        - Traumatic neuroma: Tangled proliferative mass of axons, Schwann cells, and fibrosis
- **Perineural Fibrosis**
    - Common in brachial plexus following radiotherapy for carcinoma of breast or upper aerodigestive tract
    - Following injury to peripheral nerve or following excision of nerve sheath tumor
    - Localized nerve swelling with hyper- to hypoechoic fibrotic tissue surrounding and obscuring nerve outline
- **Immune-Mediated Peripheral Neuropathy**
    - Chronic inflammatory demyelinating polyneuropathy (CIDP) and multifocal mononeuropathy (MMN) are acquired immune-mediated peripheral neuropathies
        - Aberrant immune response
        - CIDP affects sensory and motor function bilaterally and all 4 limbs
        - MMN is purely motor and usually affects only upper limbs
        - Neural enlargement seen on US, particularly proximally and in nonentrapment sites
        - Examination of vagus nerve in neck (> 2 mm^2), median nerve in midupper arm (> 13.8 mm^2), median nerve in midforearm (> 9.1 mm^2), ulnar nerve in midupper arm (> 12.1 mm^2), and tibial nerve 7 cm above medial malleolus (> 12.2 mm^2) are useful screening points
- **Fibrolipomatous Hamartoma**
    - Benign, fibrofatty malformation of peripheral nerve
    - Most cases present before 30 years of age

- 1/3 associated with finger enlargement or macrodactyly (macrodystrophia lipomatosa)
- Majority involve median nerve
  - Less frequently affects ulnar nerve, radial nerve, brachial plexus, or plantar nerves
- Soft, slow-growing, fusiform swelling of nerve
  - ± increasing pain, tenderness, and diminished sensation
  - ± symptoms of compression neuropathy
- Markedly swollen nerve with coaxial cable-like appearance
- All neural elements are swollen to comparable degree with thickened, hypoechoic fascicles embedded in echogenic stroma background

- **Intraneural Lipoma**
  - Arise from adipose tissue embedded in nerve
  - Most commonly of median or ulnar nerve
  - May cause compressive neuropathy
- **Intraneural Ganglion**
  - May be extraneural or intraneural
  - Accumulation of myxoid fluid encased in fibrous capsule within nerve
    - Due to communication between joint and parent nerve along small articular branch
    - Synovial fluid tracts along small articular branch and accumulates in parent nerve
      - Superficial branch of peroneal nerve is most commonly affected
  - Hypoechoic mass within nerve, which can extend longitudinally along nerve
  - Can cause neural compression → pain, paresthesia, weakness, and muscle atrophy
- **Metastases**
  - Neural metastases usually involve brachial plexus from breast primary
    - Hypoechoic, irregular, hypervascular mass infiltrating brachial plexus
    - Discrete mass, rapid onset and growth, predominant pain symptoms, tumoral hyperemia, and involvement of lower part of plexus all favor metastases over fibrosis
      - Still may need biopsy or fine-needle aspiration for cytology to differentiate from radiotherapy-induced perineural fibrosis
- **Acromegaly**
  - Peripheral nerves ~ 2x as large in acromegalic patients as normal subjects
  - Degree of nerve enlargement related to clinical control, duration, and insulin-like growth factor 1 level
- **Charcot-Marie-Tooth Disease**
  - Heterogeneous group of inherited peripheral nervous system disorders
    - Most common types are autosomal dominant types 1A and 2 and X-linked type
  - Onset usually before 30 years of age
    - Progressive peripheral weakness, ↓ tendon reflexes, peroneal muscle atrophy, pes cavus, and mild sensory loss
  - Median and ulnar nerve hypertrophy
    - Allows differentiation of Charcot-Marie-Tooth type 1A from other types
    - Useful screening tool for next of kin
- **Leprosy**
  - Chronic granulomatous disease due to *Mycobacterium leprae*
  - Affects peripheral nerves, upper respiratory tract mucosa, and skin
  - Wide spectrum of US abnormalities depending on disease duration, activity, ± compression
    - Early disease affects intradermal nerves → no visible US abnormality
    - Active disease → swelling of endoneural space, thickening of individual fascicles, and endoneural hyperemia
    - Longstanding inactive disease → nerves less swollen than in active disease with overall ↑ echogenicity and ↓ visibility of fascicles

Nerve Entrapment

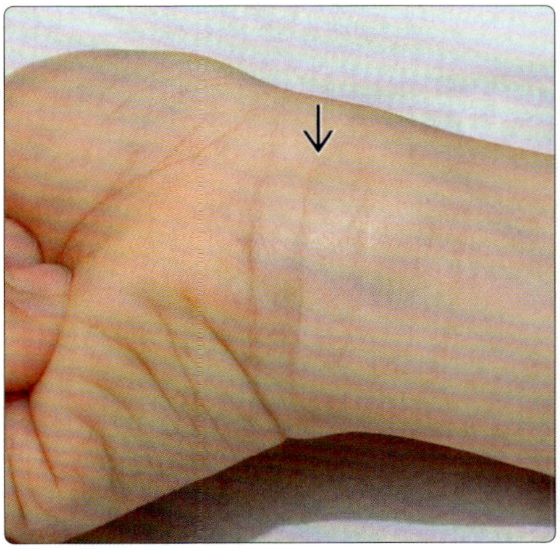

Nerve Entrapment

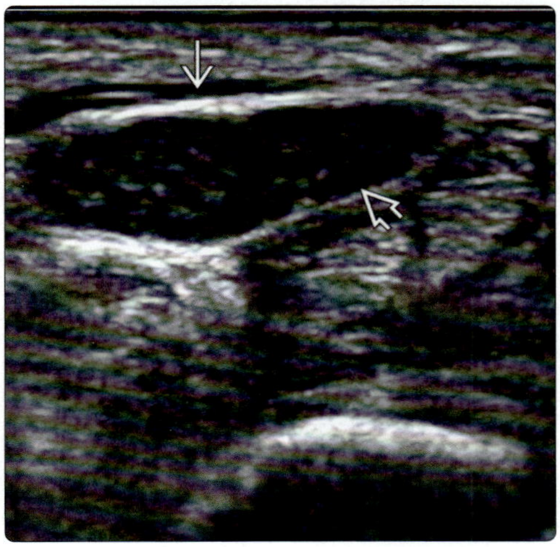

(Left) *Photograph of a patient with symptoms of carpal tunnel syndrome (CTS) shows visible volar wrist swelling ⇨ due to severe median nerve compression proximal to carpal tunnel.* **(Right)** *Transverse US of the same patient shows unusually severe (32 mm²) median nerve swelling ➡ deep to the flexor retinaculum ➡ at the tunnel inlet. In most CTS patients, the median nerve cross-sectional area is ~ 13-16 mm².*

(Left) *Transverse color Doppler US shows moderate hyperemia of the median nerve just proximal to the carpal tunnel. Neural hyperemia is seen in some but not all patients with CTS. It is a sign with low sensitivity but high specificity for CTS.* **(Right)** *Longitudinal US shows moderate swelling of the deep branch of the radial nerve just proximal to the supinator tunnel inlet between the superficial and deep heads of the supinator muscle, compatible with entrapment at the arcade of Frohse.*

Nerve Entrapment

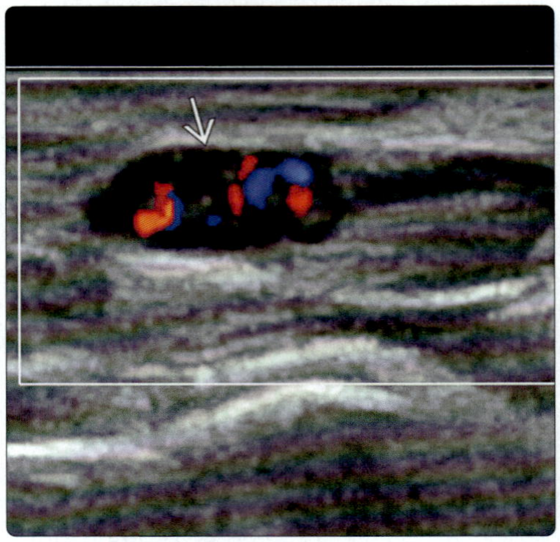

Nerve Entrapment

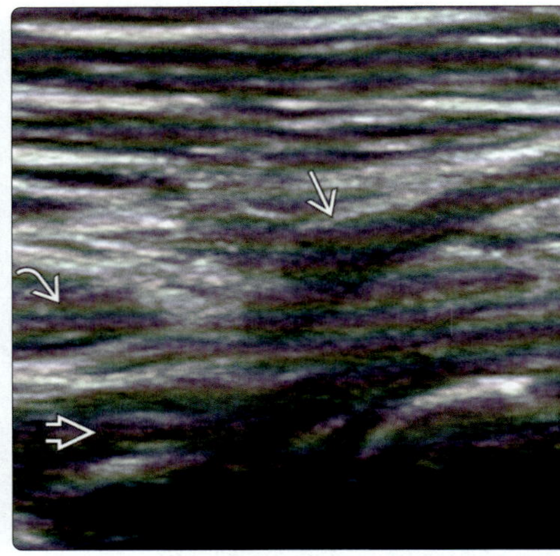

(Left) *Longitudinal US shows a medium-sized, hypoechoic nerve sheath tumor of the medial sural cutaneous nerve between the medial and lateral heads of the gastrocnemius muscle (not shown). The thickened sural nerve alongside the tumor is visible. An underlying vein is moderately compressed.* **(Right)** *Longitudinal color Doppler US in the same patient shows characteristic hyperemia of the nerve sheath tumor.*

Nerve Sheath Tumor

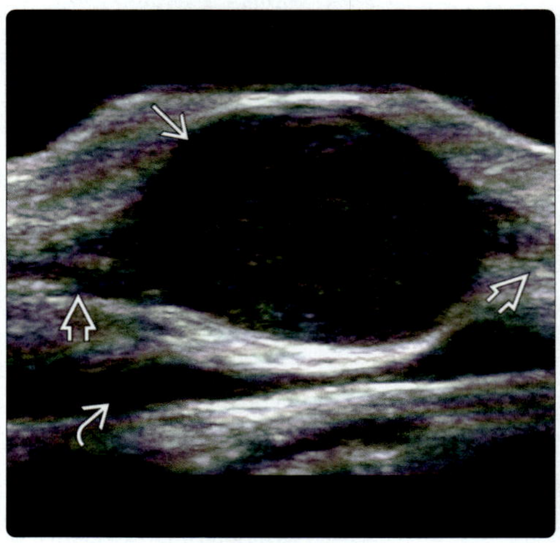

Nerve Sheath Tumor

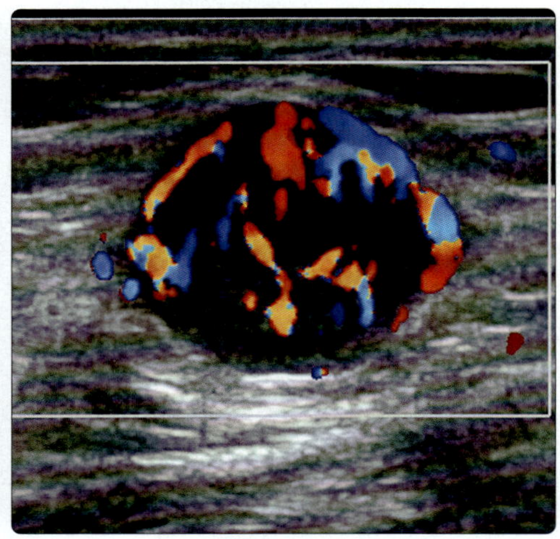

(Left) *Longitudinal US shows a nerve sheath tumor of the deep peroneal nerve anterior to the ankle joint. The proximal part of the deep peroneal nerve is markedly thickened.* **(Right)** *Longitudinal US shows a tumor of the median nerve with solid and myxoid areas. The median nerve just distal to the tumor is slightly thickened.*

Nerve Sheath Tumor

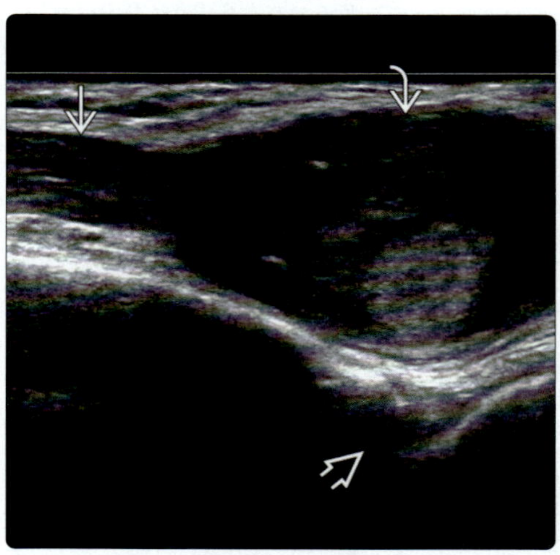

Nerve Sheath Tumor

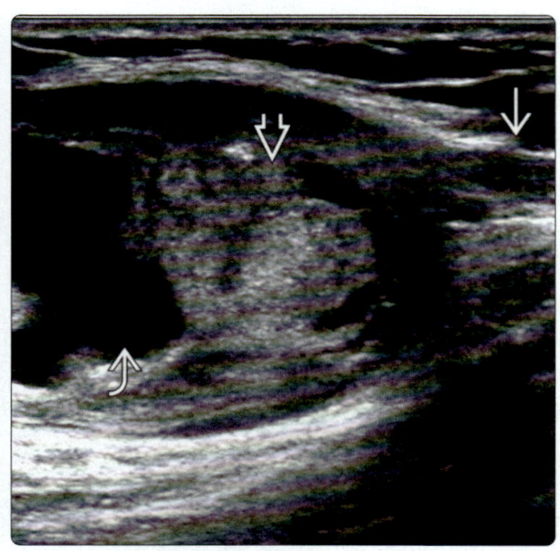

Nerve Sheath Tumor

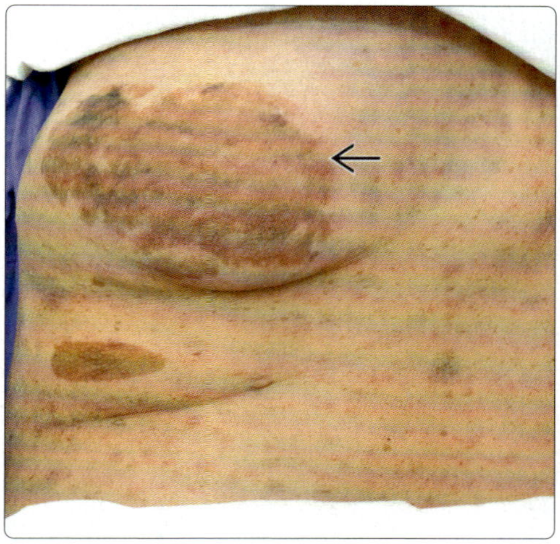

Nerve Sheath Tumor

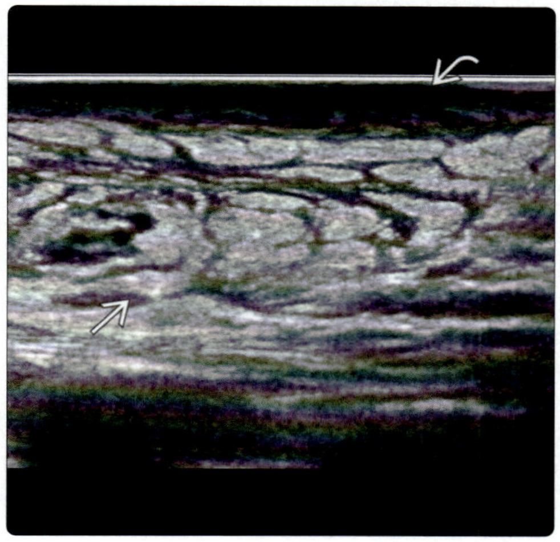

(Left) *Photograph of a patient with known neurofibromatosis shows a mass located deep to a large café au lait mark ⇨ on the lower thoracic region posteriorly. This mass was enlarging clinically.* **(Right)** *Longitudinal US of the same patient shows thickened, hyperechoic subcutaneous tissue ➡ and hypoechoic dermal and subdermal tissue ➡. Biopsy revealed a myxoid spindle cell tumor, consistent with a plexiform neurofibroma. There was no evidence of malignancy.*

Nerve Sheath Tumor

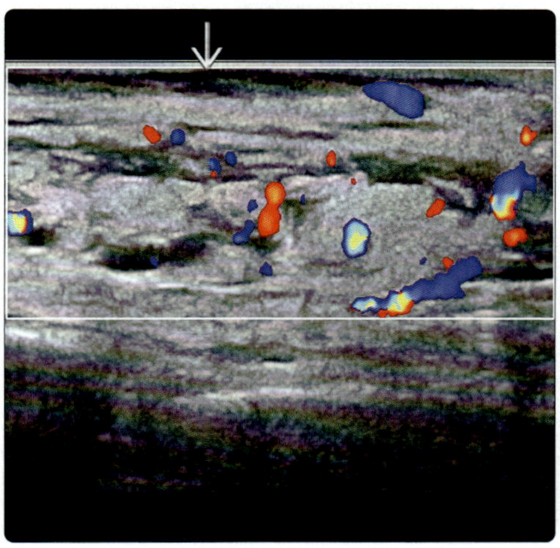

Nerve Sheath Tumor

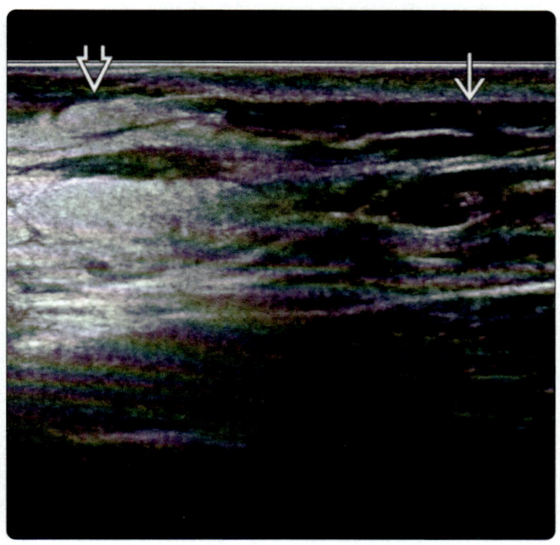

(Left) *Longitudinal color Doppler US of the same patient shows moderate hyperemia of the plexiform neurofibroma ➡.* **(Right)** *Transverse US of the same patient shows normal hypoechoic subcutaneous fat at the end of the plexiform neurofibroma. The juxtaposition between normal hypoechoic subcutaneous fat ➡ and hyperechoic neurofibroma tissue ➡ can be readily appreciated.*

Nerve Sheath Tumor

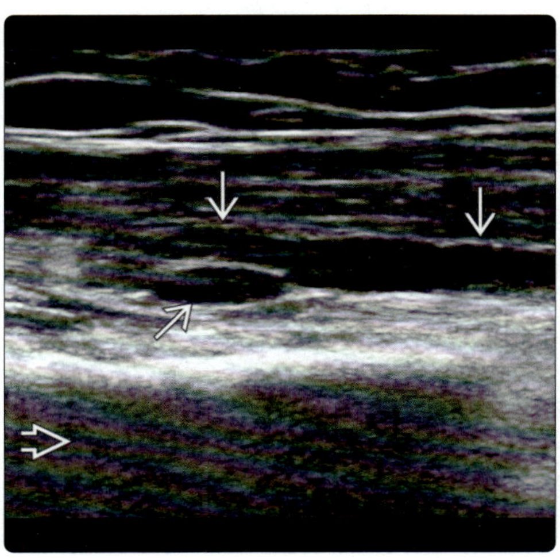

Nerve Sheath Tumor

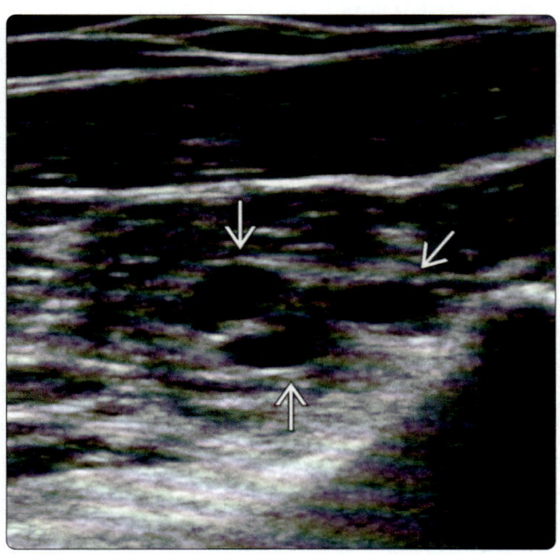

(Left) *Longitudinal US shows elongated plexiform nerve sheath tumor ➡ present in the proximal arm. The tumor was located between the brachialis and biceps muscles. The underlying humeral shaft ➡ is normal.* **(Right)** *Transverse US of the same patient shows the tubular appearance ➡ of the plexiform neurofibroma. Plexiform neurofibromas occur almost invariably in patients with neurofibromatosis.*

(Left) *Transverse US in a patient with a 2-month history of lower limb pain and buttock swelling shows a large mass just external to the sciatic notch.* **(Right)** *Longitudinal US in the same patient shows severe swelling of the sciatic nerve continuous with the more proximally located mass. US-guided biopsy of the mass revealed malignant peripheral nerve sheath tumor. This patient did not have neurofibromatosis.*

Nerve Sheath Tumor

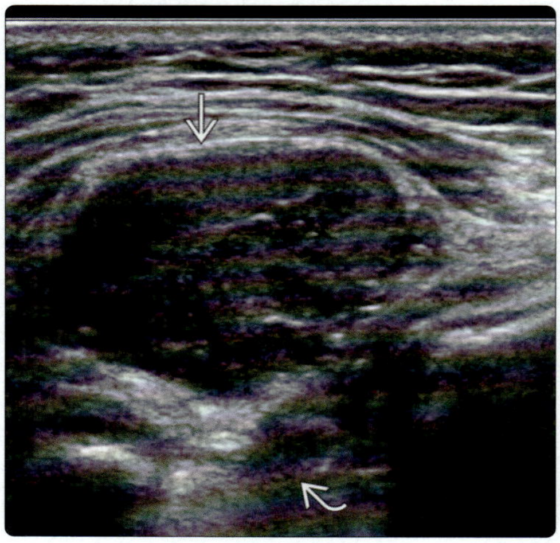

Nerve Sheath Tumor

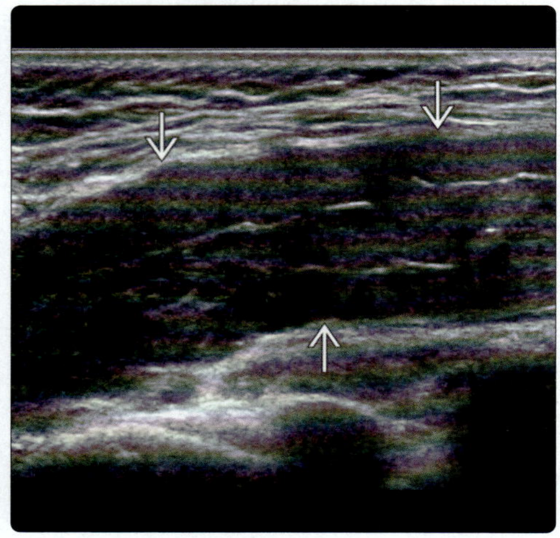

(Left) *Longitudinal US shows the median nerve following a dog bite. The nerve is mildly swollen with injury to the more superficial part. The deep fascicles are intact. The overlying soft tissues are also swollen.* **(Right)** *Longitudinal US in a patient with previous knife injury to the volar wrist shows moderate diffuse swelling of the median nerve due to posttraumatic fibrosis. Injury to the underlying flexor tendons is also apparent.*

Nerve Injury

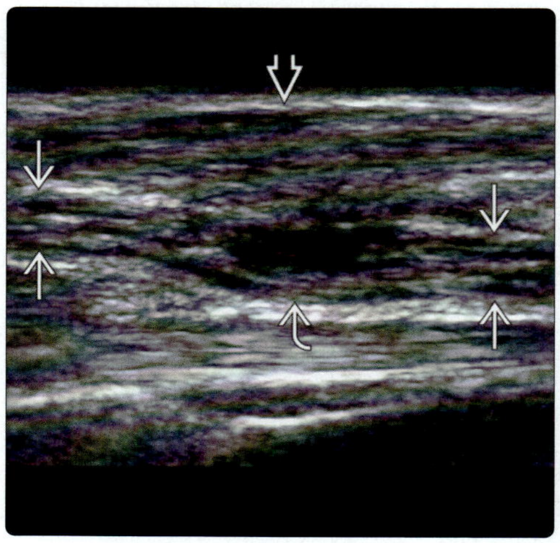

Nerve Injury

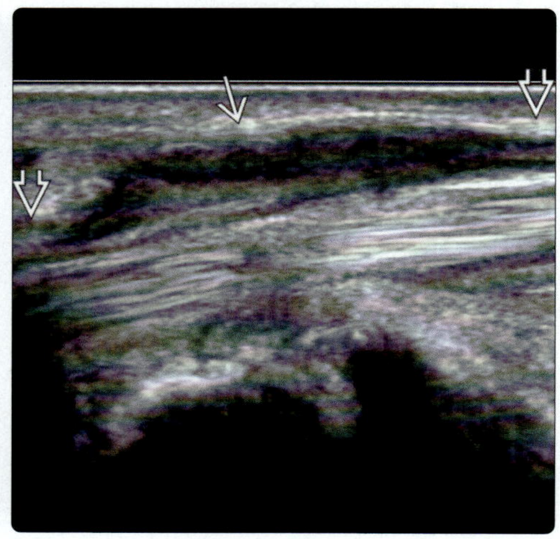

(Left) *Longitudinal US of the brachial plexus shows thickened transected C5 and C6 roots with a large amount of bridging fibrosis between these roots distally.* **(Right)** *Transverse US shows marked swelling of the medial nerve in the midforearm due to postsurgical fibrosis following previous resection of schwannoma.*

Nerve Injury

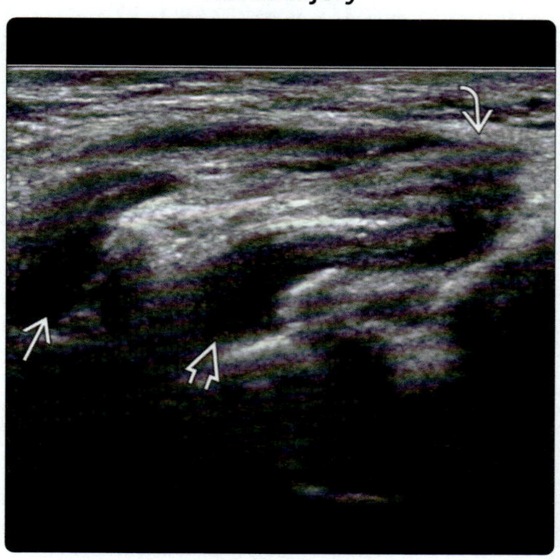

Perineural Fibrosis

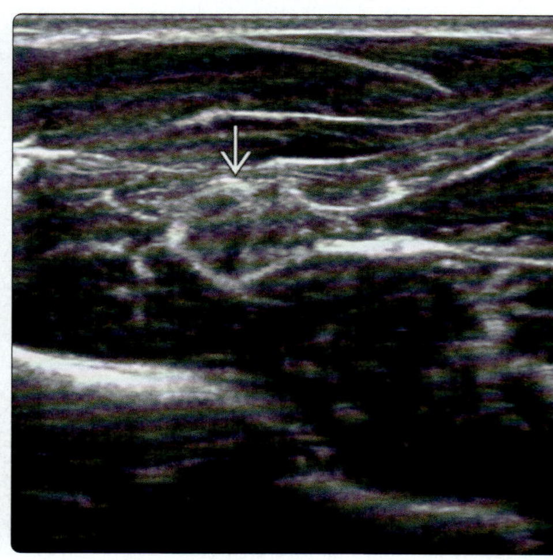

Per neural Fibrosis

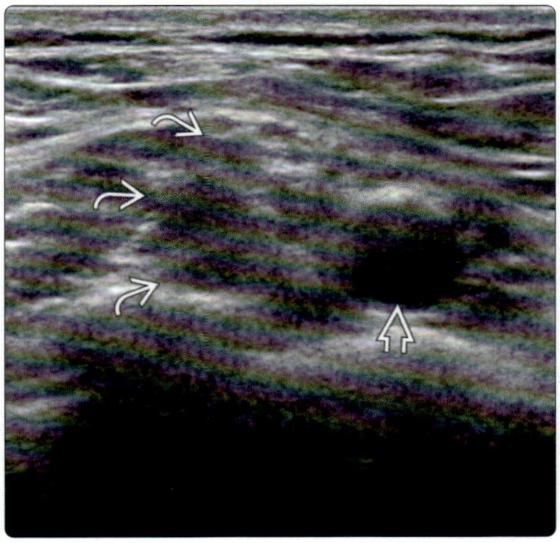

Immune-Mediated Peripheral Neuropathy

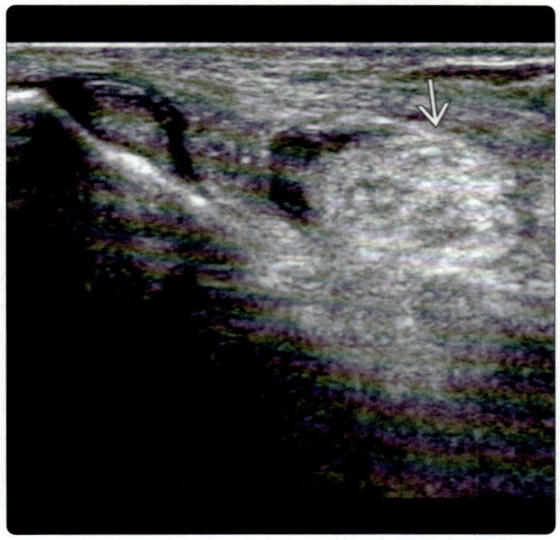

(Left) *Transverse US shows moderate to severe fibrosis of the brachial plexus divisions posterior to the subclavian artery. The divisions are thickened and ill-defined with hypoechogenicity of the perineural fat due to fibrosis.* **(Right)** *Transverse US in a patient with polyneuropathy shows a markedly thickened tibial nerve at the tarsal tunnel. Similar findings were present on the contralateral side. The clinical picture was compatible with chronic inflammatory demyelinating polyneuropathy.*

Fibrolipomatous Hamartoma

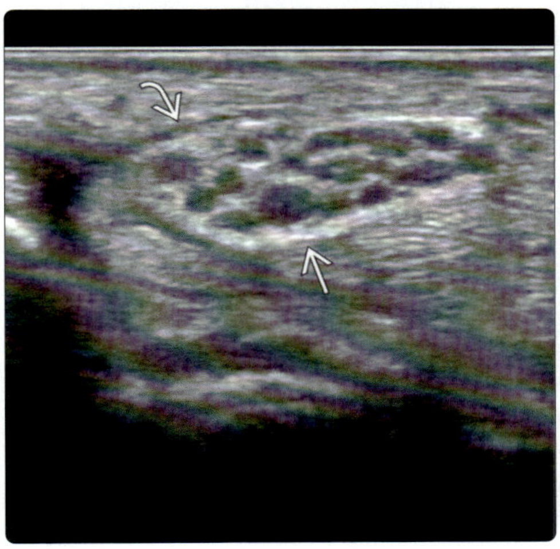

Intraneural Lipoma

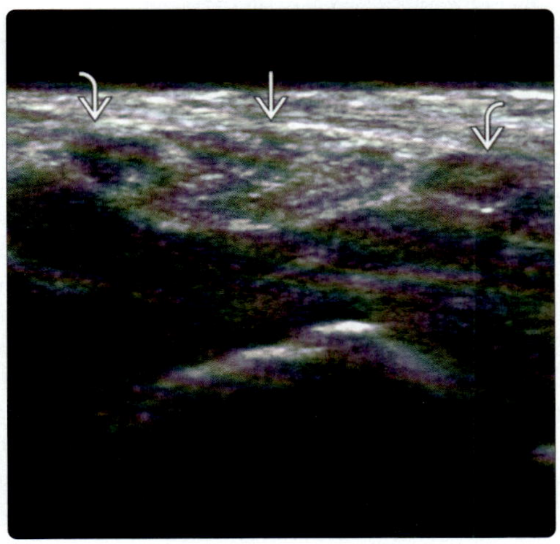

(Left) *Transverse US of carpal tunnel inlet shows a thickened median nerve with thickened, hypoechoic nerve fascicles due to fibrolipomatous hamartoma deep to the flexor retinaculum. The appearance is likened to a coaxial cable.* **(Right)** *Transverse US of the wrist shows a hyperechoic lipoma located between both components of the bifid median nerve.*

Metastases

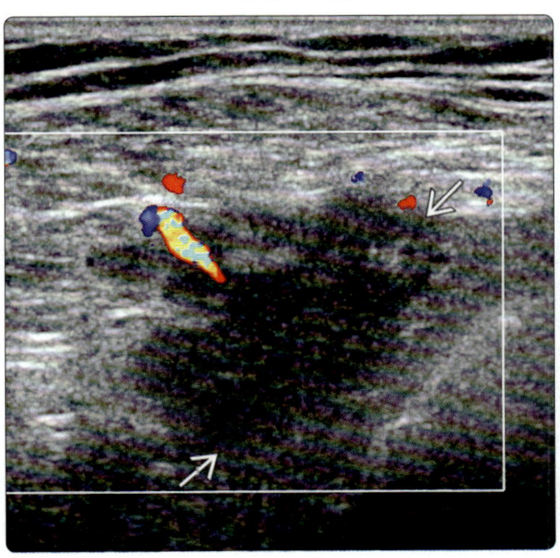

Leprosy

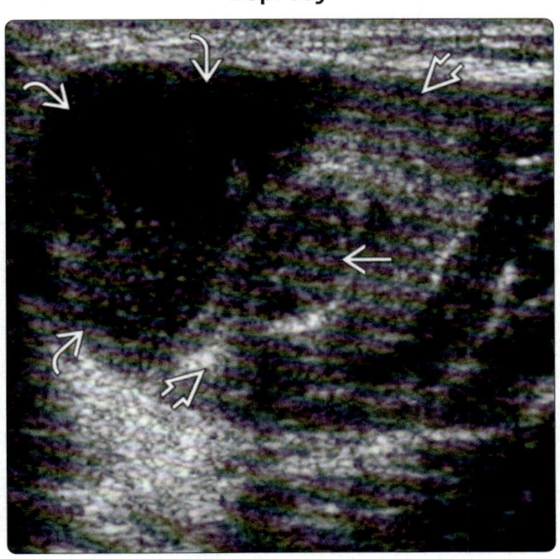

(Left) *Transverse color Doppler US of the supraclavicular region shows metastatic infiltration obscuring the brachial plexus in a patient with known breast carcinoma. Fine-needle aspiration confirmed metastatic infiltration.* **(Right)** *Transverse US proximal to the elbow shows a thickened ulnar nerve with marked surrounding edema, echogenic granulation tissue, and abscess due to leprosy. Perineural abscess is an uncommon feature of leprosy. (Courtesy S. Lodge, MD.)*

DIFFERENTIAL DIAGNOSIS

Common

- Plantar Fasciitis
- Plantar Fibromatosis

Less Common

- Plantar Fascial Tear
- Enthesopathic Fasciitis
- Investing Fascial Tear
- Nodular Fasciitis
- Infective Fasciitis
- Necrotizing Fasciitis
- Eosinophilic Fasciitis
- Dupuytren Disease
- Tumors Occurring Within Fascia

ESSENTIAL INFORMATION

Key Differential Diagnosis Issues

- Plantar fascia is thin, laminated, echogenic aponeurosis stretching between medial calcaneal tuberosity and plantar plates of metatarsophalangeal joints
  - Maintains medial longitudinal arch during weightbearing
  - Comprises 3 bands: Medial, central, and lateral band
    - Central band is strongest and thickest
  - Plantar fascia slightly thinner in patients with pes planus
- Investing fascia is thin, laminated, echogenic tissue enveloping muscles of appendicular skeleton and trunk
- Fascial lesions tend to be site specific
- For most lesions, US is as accurate as MR or CT
- Fascial insertional lesions may be 1st manifestation of inflammatory enthesopathy

Helpful Clues for Common Diagnoses

- **Plantar Fasciitis**
  - Probably caused by repetitive microtrauma ± microvascular injury
  - Common in runners (10%)
    - Other risk factors include faulty shoes, high-arched feet, short Achilles tendon, obesity, and prolonged weightbearing
  - Affects central band attachment to medial calcaneal tuberosity
    - Bilateral in 1/3 of patients
  - Manifested as thickening of plantar fascia at calcaneal insertion
  - > 4.3-mm thickness considered abnormal
    - Hypoechogenicity and thickening of plantar fascia over short segment (~ 10 mm) at calcaneal insertion
    - ± loss of echogenic laminar pattern of plantar fascia ± perifascial edema
    - 4.3 mm = guideline rather than absolute measurement, as range of plantar fascial thickness in patients and normal subjects overlaps
    - Measurements taken at leading edge of calcaneus
    - The greater the plantar fascial thickening, the more likely diagnosis of plantar fasciitis
    - Subclinical fascitis may be present on opposite side; therefore, exercise caution in using contralateral side as normal reference
  - ± plantar calcaneal spur
    - Echogenic bony spur ~ 1-3 mm long at deep surface of plantar fascia
  - Small foci of dystrophic calcification as well as small intrafascial tears can sometimes be seen in thickened fascia at calcaneal insertion
  - ± edema, reduced lobulation, thinning, or early fibrosis of overlying subcutaneous heel fat pad
  - Minimal focal hyperemia in < 10% of patients with plantar fasciitis
  - Associated calcaneal bone marrow edema ± inflammation at plantar fascial insertional area not visible with US
    - Best seen with MR; however, not known to be useful prognostic indicator
  - Many different treatments
    - Stretching
    - Massage (e.g., rolling tennis ball under affected area)
    - Orthosis, night splints
    - US-guided steroid injection
      - Risk of fascial rupture and heel fat pad atrophy
      - Do not inject directly into plantar fascia; inject to perifascial area
    - US-guided, platelet-rich plasma injection
    - Low-level laser therapy
    - Combined cryo-US therapy
    - Extracorporal shock wave therapy
    - Radiofrequency ablation of calcaneal branches of inferior calcaneal nerve
    - US-assisted partial plantar fascial release
  - Decrease in plantar fascial thickness occurs with improvement in symptoms following treatment
- **Plantar Fibromatosis**
  - Focal nodular fibroblastic proliferation of plantar fascia away from calcaneal insertion
    - No specific risk factors identified
  - Most commonly affects central band of plantar fascia in midfoot region
    - Often multiple
    - Bilateral in 1/3 of patients
  - Discrete fusiform-shaded nodule expanding plantar fascia
    - Either hypoechoic (75%) or isoechoic (25%) to plantar fascia
  - Long axis of fibroma usually aligned with long axis of plantar fascia band
  - Posterior acoustic enhancement (20%)
  - Internal vascularity (10%)
  - Calcification or visible tears not feature
  - Does not infiltrate beyond plantar fascia
    - If infiltrative in nature, consider aggressive plantar fibromatosis

Helpful Clues for Less Common Diagnoses

- **Plantar Fascial Tear**
  - Affects localized area of fascia rather than extending across width of fascia
  - Typically affects proximal and middle 1/3 of fascia
  - Acute tears characterized by focal disruption of fascia, perifascial edema, and inflammation

- Chronic tears characterized by focal disruption, tendon thickening, perifascial fibrosis, and hyperemia
- **Enthesopathic Fasciitis**
  - Plantar fascia insertion is not uncommon site of involvement in enthesopathy
    - Central plantar fascial band involvement at calcaneal insertion
    - Lateral plantar fascial band insertion at base of 5th metatarsal bone
  - Plantar fascial thickening, hyperemia with cortical irregularity, osteophytic spurs (enthesophytes), and cortical erosions (entheseal erosions)
- **Investing Fascial Tear**
  - Either complete focal defect in fascia or linear intrasubstance fascial tear
  - Focal defect may result in muscle hernia
    - Accentuated by muscle contraction
    - Often occurs spontaneously in athletic muscular individuals
    - Often occurs near perforating vessel
- **Nodular Fasciitis**
  - Benign proliferation of fibroblasts and myofibroblasts of investing fascia
    - Lesions are generally small and solitary
    - Most commonly involves upper limb
    - Usually no traumatic history and no known cause
- **Infective Fasciitis**
  - Infection centered around investing fascia
    - No fulminant necrotic element present
- **Necrotizing Fasciitis**
  - Advancing soft tissue infection characterized by widespread fascial necrosis
  - May occur after trauma or surgery
  - Either monomicrobial or polymicrobial infection can cause necrotizing fasciitis
    - Group A β-hemolytic *Streptococcus* is common causative organism
  - Thickened disrupted fascia with perifascial fluid
  - Severe subcutaneous and muscle edema
  - ± muscle necrosis
  - ± gas locules due to gas-forming organisms
- **Eosinophilic Fasciitis**
  - Disorder characterized by peripheral eosinophilia and fasciitis
    - Investing fascia and intermuscular fascia of forearm and calf are most commonly affected
  - Thickening and hyperemia of muscle fascia on US
  - US-guided fascial biopsy may be helpful to confirm diagnosis
- **Dupuytren Disease**
  - Fibroproliferative disorder of palms leading to progressive permanent digital contracture
  - Risk factors are old age, male sex, white northern European ancestry, positive family history of Dupuytren disease, and diabetes mellitus
  - Disease deteriorates most rapidly in cases that show earlier age of onset
  - Associated with fibromatosis affecting back of hands, soles of feet, and penis
  - Predominantly affects little, ring, and middle fingers
    - Generally starts as small lump in palm overlying metacarpophalangeal joint area (stage A)
    - Progressive fibrosis of palmar aponeurosis and fingers with fibrous-type cord becoming evident (stage B)
    - Fibrous cord-like extension from palm into fingers with progressive finger contracture (stage C)
  - US generally useful only in stage A disease to ascertain nature of palmar lump
    - In stage B and C disease, diagnosis is clinically obvious, and US is not beneficial
    - Fibrinous digitations between palmar aponeurosis and flexor tendons may be seen
  - Treatment involves either fasciotomy (simple division) of contracted fascia or limited fasciectomy ± under US guidance
- **Tumors Occurring Within Fascia**
  - Fibroma, lipoma, or vascular malformations may arise within or involve fascia
  - Appearances akin to similar tumors arising beyond fascia

Plantar Fasciitis

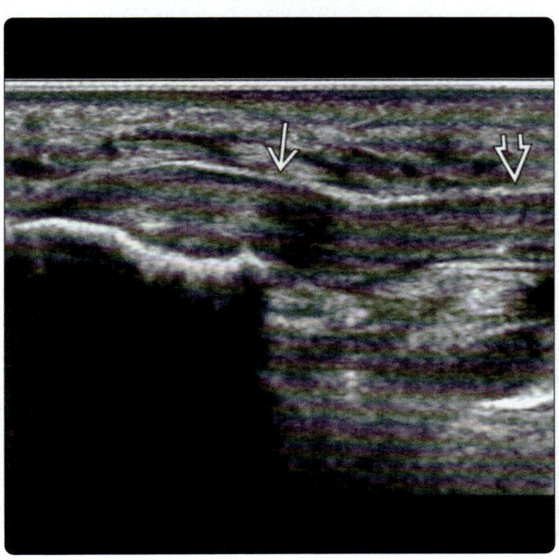

Plantar Fasciitis

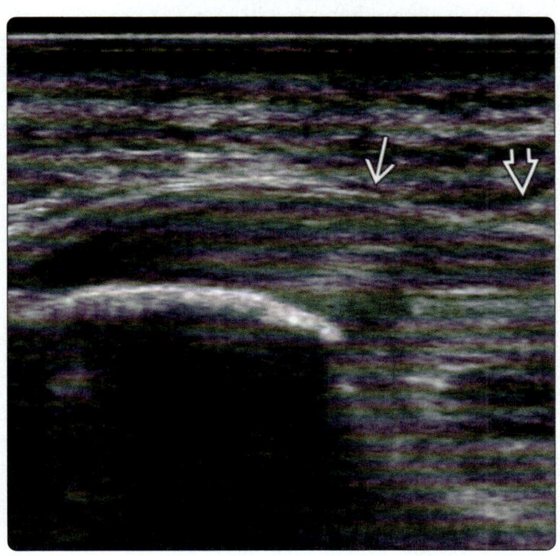

(Left) *Longitudinal US shows mild thickening (4.7 mm)* ➡ *of the plantar fascia* ⇨ *at the anteromedial leading edge of the calcaneum (medial calcaneal tuberosity), consistent with mild plantar fasciitis.* **(Right)** *Longitudinal US shows severe (6.7 mm) thickening* ➡ *of the plantar fascia* ⇨ *at the anteromedial leading edge of the calcaneum, consistent with severe plantar fasciitis.*

(Left) *Photograph shows a solitary plantar fibroma in the midfoot region.* **(Right)** *Longitudinal US shows fusiform-shaped swelling of the plantar fascia in the midfoot region, consistent with a plantar fibroma (plantar fibromatosis). Minimal posterior acoustic enhancement is present.*

Plantar Fibromatosis

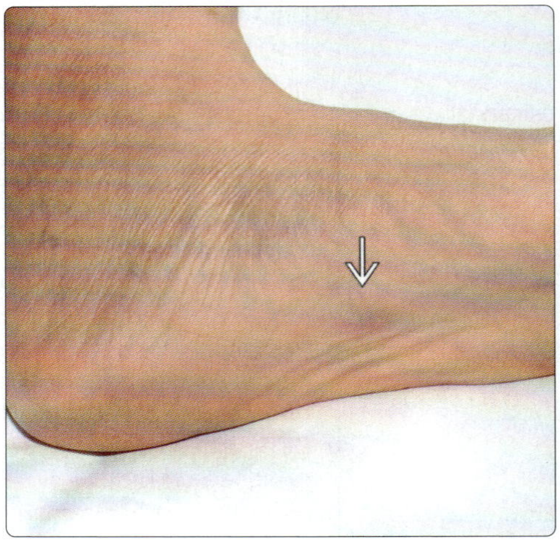

Plantar Fibromatosis

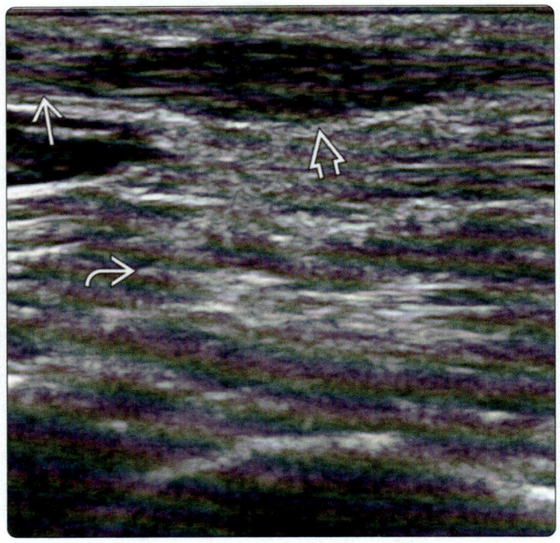

(Left) *Longitudinal US shows localized oblong-shaped thickening of the plantar fascia, consistent with plantar fibroma (plantar fibromatosis). This is an unusually thickened, elongated plantar fibroma. Most plantar fibromas are smaller.* **(Right)** *Longitudinal US shows a fibroma of the more superficial aspect of the plantar fascia. When only part of the plantar fascia is affected, it is usually the more superficial aspect.*

Plantar Fibromatosis

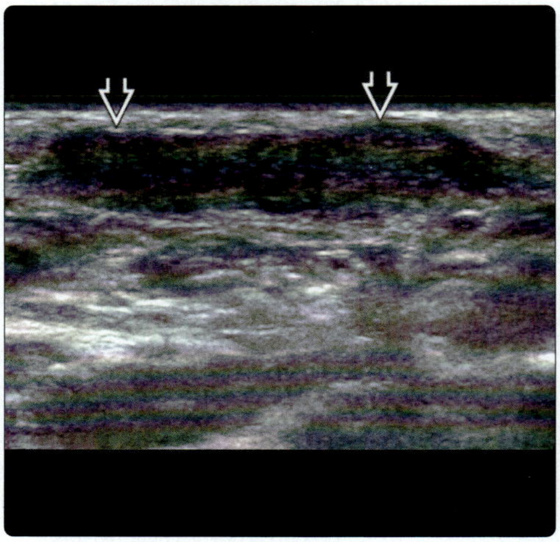

Plantar Fibromatosis

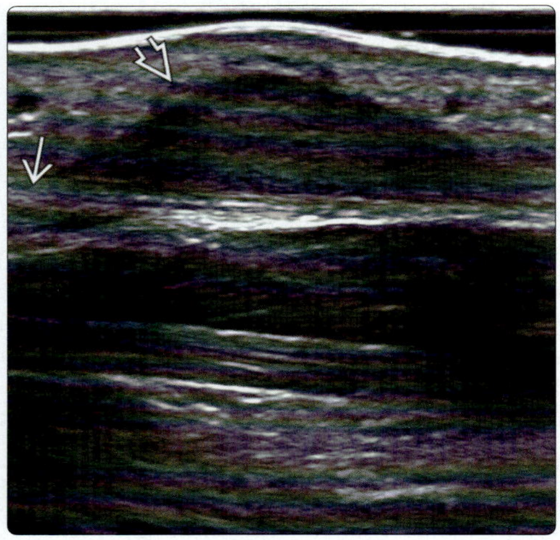

(Left) *Longitudinal US shows localized thinning of the plantar fascia immediately distal to the calcaneum due to a partial-thickness tear. There is moderate perifascial edema. The plantar fascia is not thickened, i.e., there is no plantar fasciitis present.* **(Right)** *Sagittal T1WI MR in the same patient shows a plantar fascial tear just distal to the calcaneal insertion with surrounding fluid and edema.*

Plantar Fascial Tear

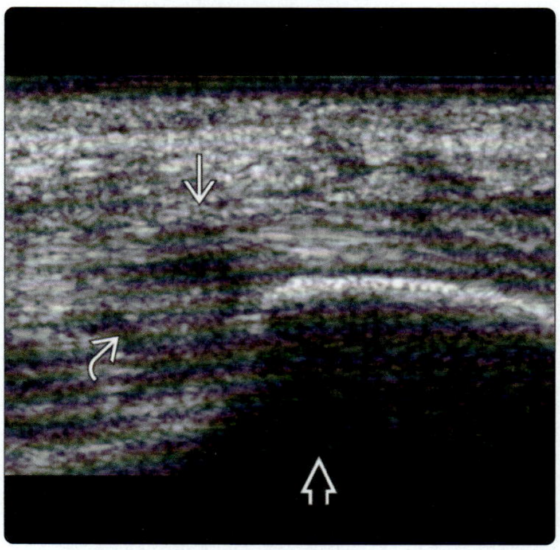

Plantar Fascial Tear

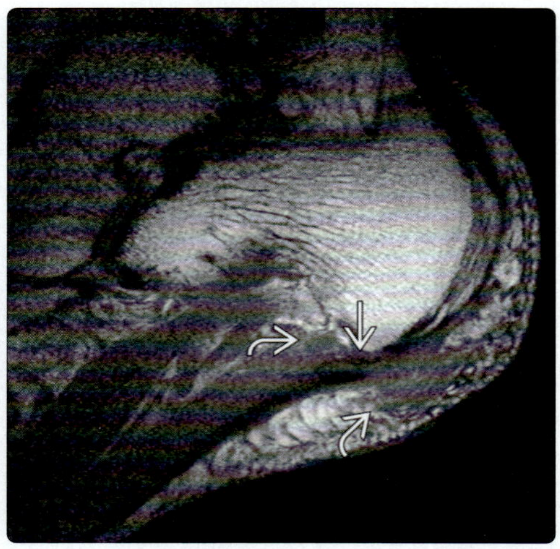

Plantar Fascial Tear

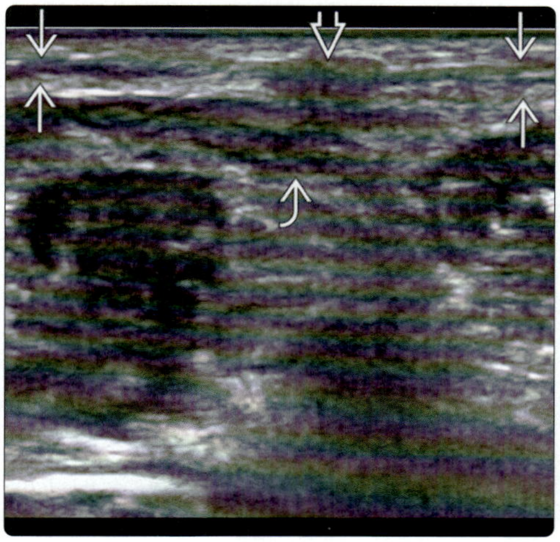

Plantar Fascial Tear

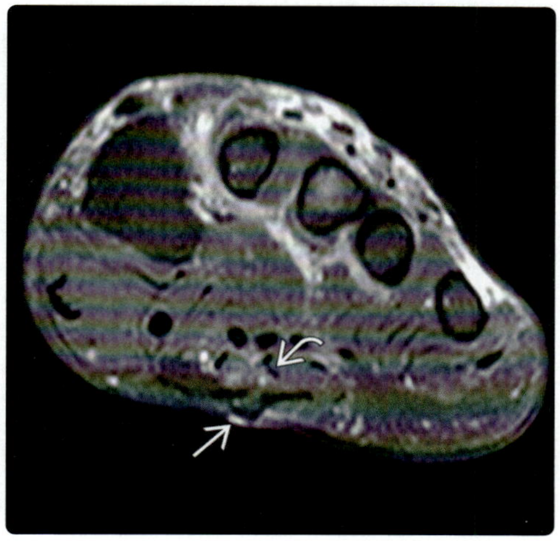

(Left) *Transverse US of the forefoot in a patient with localized trauma to this area 3 months previously shows a slit-like tear ➡ with discontinuity of the plantar fascia ➡. There is herniation of part of the flexor digitorum brevis muscle ➡ through this fascial defect. The remaining plantar fascia is normal.* **(Right)** *Coronal T2WI FS MR of the same patient shows the plantar fascial tear with herniation of part of the flexor digitorum brevis muscle ➡ through this fascial defect ➡.*

Enthesopathic Fasciitis

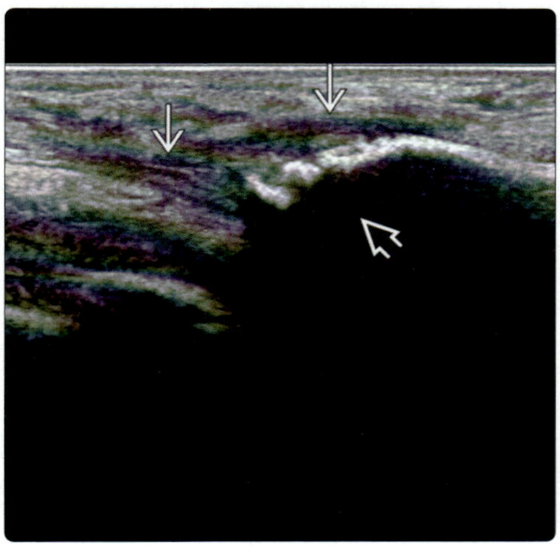

Enthesopathic Fasciitis

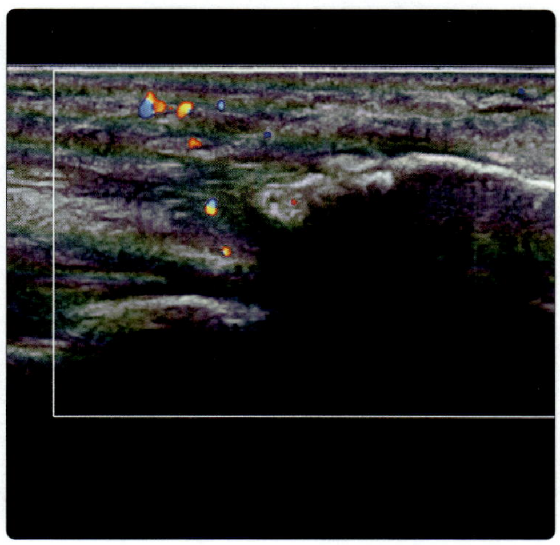

(Left) *Longitudinal US shows moderate cortical irregularity at the medial aspect of the 5th metatarsal base ➡ with thickening of the lateral band of the plantar fascia ➡ just proximal to its 5th metatarsal base insertion, consistent with enthesopathic fasciitis. There were other changes of enthesis elsewhere in the foot. The peroneus brevis insertion more laterally was normal.* **(Right)** *Longitudinal color Doppler US of the same patient shows mild localized hyperemia.*

Investing Fascial Tear

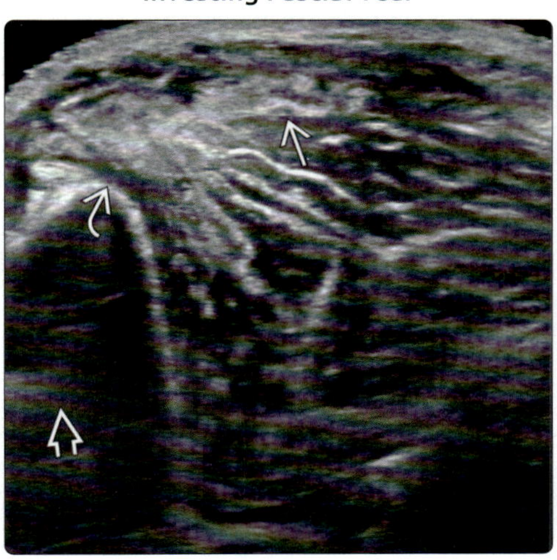

Nodular Fasciitis

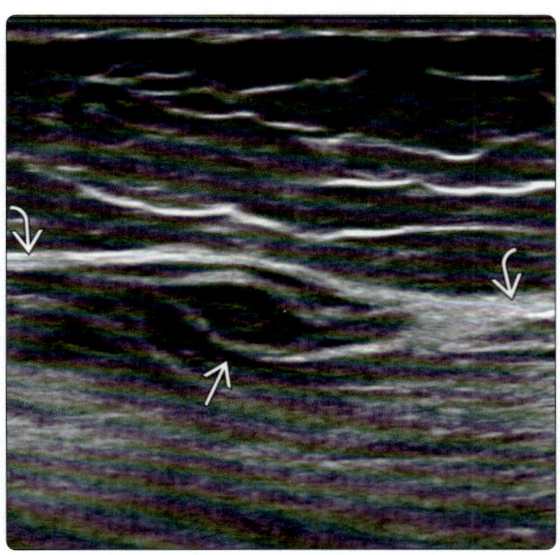

(Left) *Transverse US shows herniation ➡ of the tibialis anterior muscle into the subcutaneous tissues through a defect in the investing fascia ➡ alongside the tibial shaft ➡. This patient had point tenderness at this location, which was aggravated by exercise.* **(Right)** *Longitudinal US in a middle-aged woman with an 8-week history of a mildly tender medial thigh lump shows a hypoechoic swelling ➡ of the investing fascia ➡. Mild hyperemia was present. These appearances favored nodular fasciitis. The mass resolved over 6 months.*

(Left) *Transverse US of the anterior ankle region shows moderate swelling and edema both superficial and deep to the investing fascia of the leg. No liquefaction is present. The tibialis anterior tendon is normal.* **(Right)** *Transverse color Doppler US in the same patient shows moderate surrounding hyperemia. Infective fasciitis is characterized by infection centered around the fascia. Not all infections around a fascia are necrotizing fasciitis.*

Infective Fasciitis

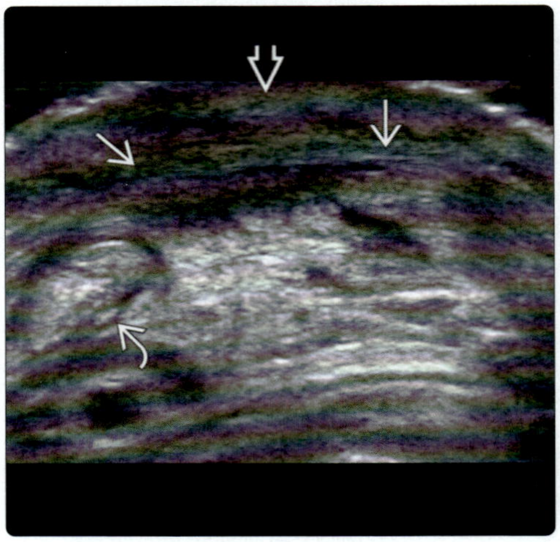

Infective Fasciitis

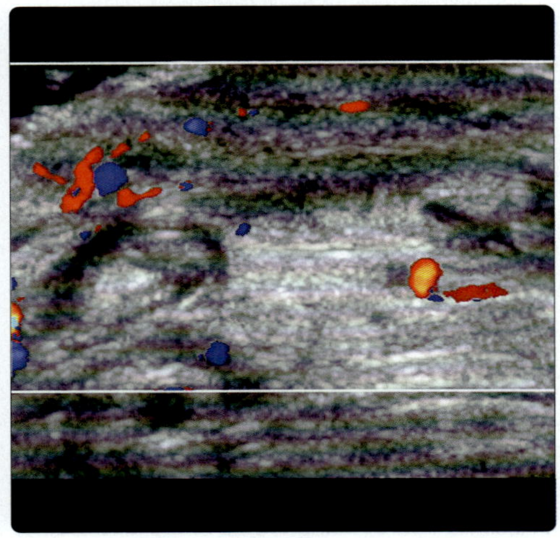

(Left) *Longitudinal US of the thigh shows a lobulated abscess extending through the investing fascia into the vastus lateralis muscle. Surgery confirmed necrotizing fasciitis.* **(Right)** *Transverse US of the forearm shows thickening of the investing fascia and intermuscular fascia in a patient with eosinophilia. Echogenic muscle edema is also present.*

Necrotizing Fasciitis

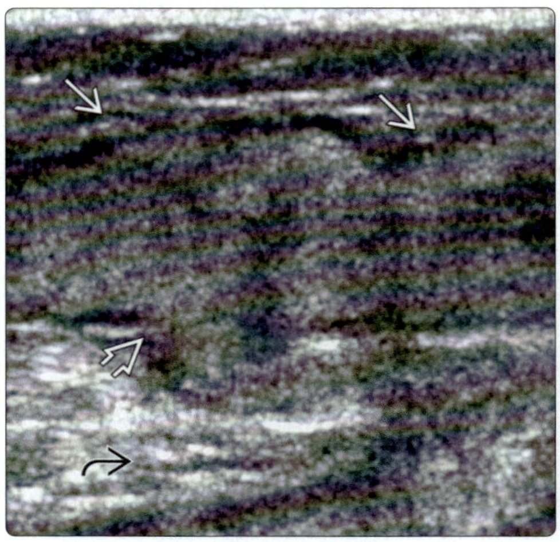

Eosinophilic Fasciitis

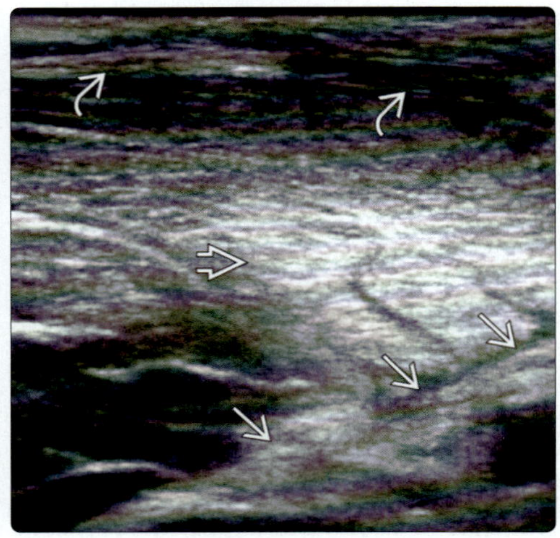

(Left) *Transverse T2WI MR with fat suppression in the same patient shows edema and thickening of the investing fascia and intermuscular fascia. This is characteristic of eosinophilic fasciitis. Biopsy is generally not required.* **(Right)** *Clinical photograph shows an area of nodular thickening at the base of the index and middle fingers with peripheral skin tethering. This is consistent with Dupuytren disease.*

Eosinophilic Fasciitis

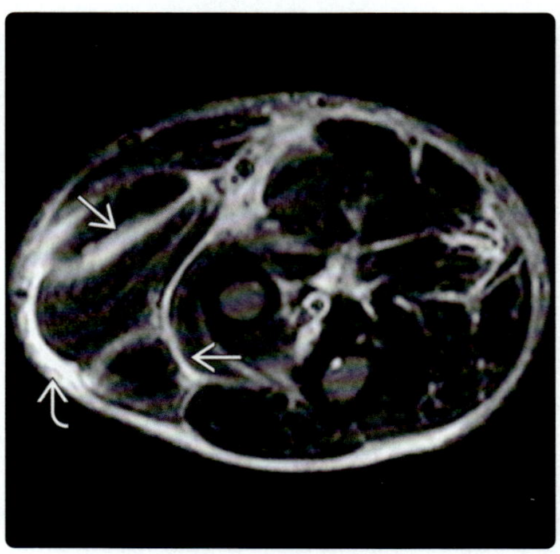

Dupuytren Disease

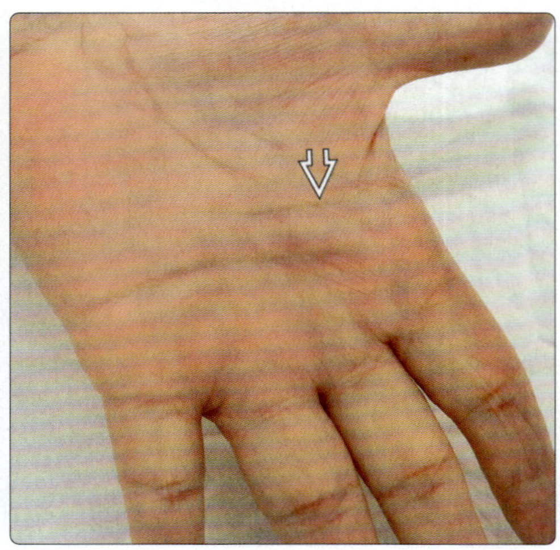

Dupuytren Disease

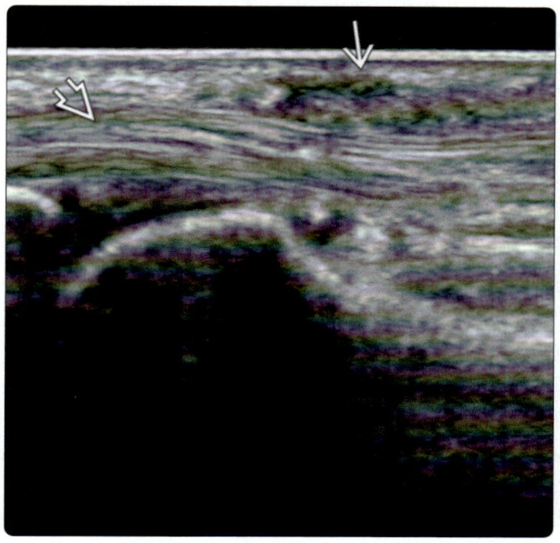

Tumors Occurring Within Fascia

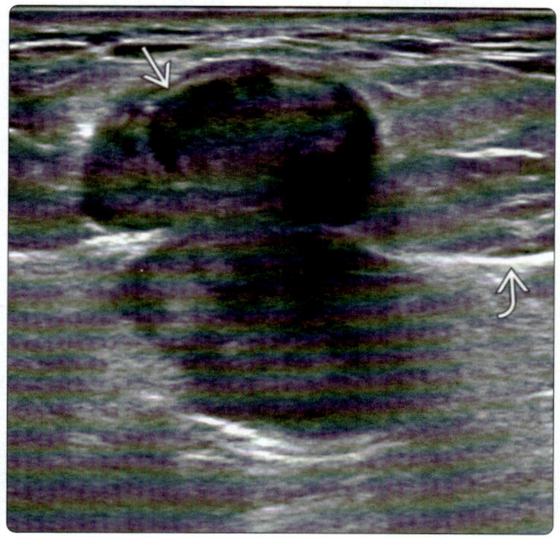

(Left) *Longitudinal US of the palm in the same patient shows localized echogenic nodular thickening of the palmar aponeurosis in contact with the sheath of flexor digitorum tendon. This is consistent with early Dupuytren disease.* **(Right)** *Transverse US in a man with a 9-month history of a thigh mass shows a hypoechoic fibrotic-type mass transgressing and probably arising from the investing fascia. Biopsy revealed a rare primary pigmented villonodular synovitis.*

Tumors Occurring Within Fascia

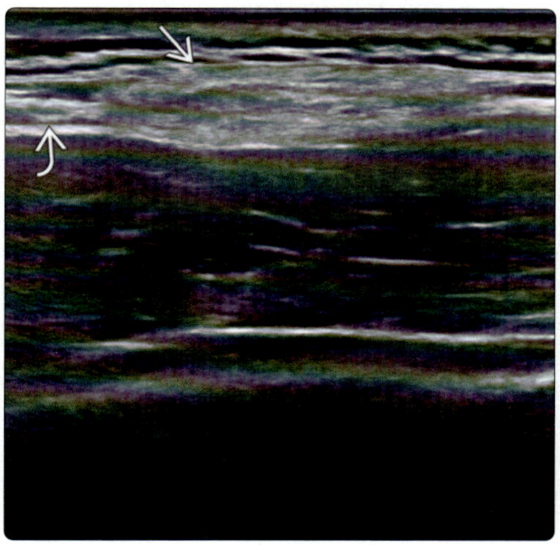

Tumors Occurring Within Fascia

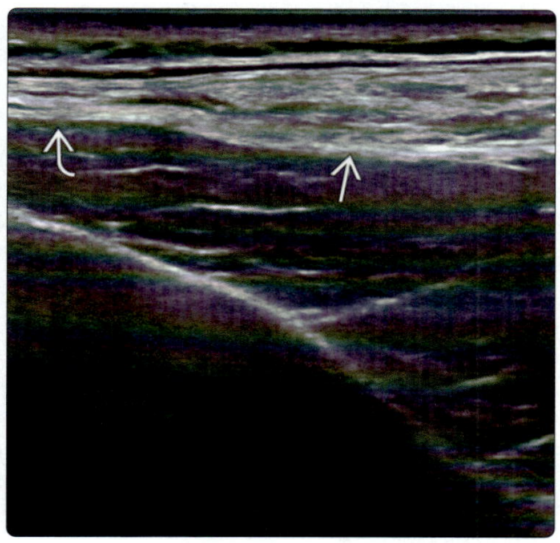

(Left) *Transverse US in a 38-year-old man shows an intrafascial lipoma of the investing fascia overlying the infraspinatus muscle.* **(Right)** *Longitudinal US of the same lesion shows the mass arising from within the investing fascia. CT confirmed an intrafascial lipoma. Lipoma may occasionally occur within the investing fascia.*

Tumors Occurring Within Fascia

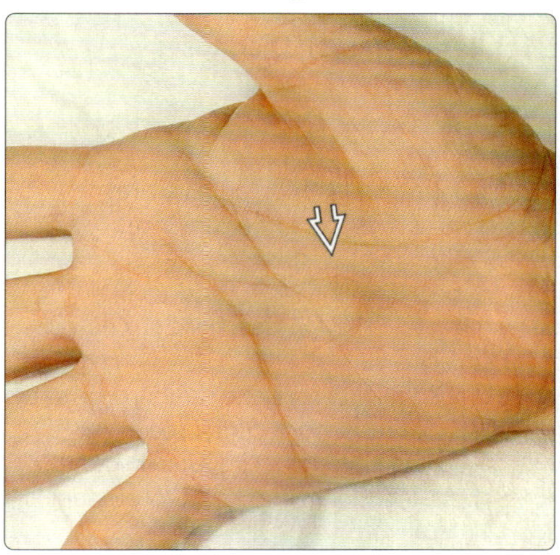

Tumors Occurring Within Fascia

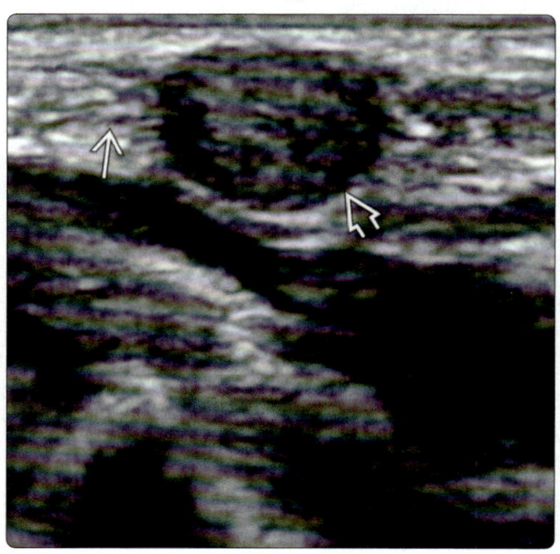

(Left) *Clinical photograph of a young patient with a localized nodule in the midpalm region is shown. There is no skin tethering present.* **(Right)** *Transverse US in the same patient shows a discrete, rounded, nodular, fibrotic-type mass in continuity with the palmar aponeurosis. This is consistent with a palmar fibroma. It is a discrete nodule as opposed to the more diffuse, infiltrative type of nodular fibrosis and tethering seen with Dupuytren disease.*

DIFFERENTIAL DIAGNOSIS

Common

- Osteochondroma
- Other Benign Bone Tumor
- Malignant Bone Tumor

Less Common

- Metastases
- Bone Tumor Recurrence
- Acute Osteomyelitis
- Chronic Osteomyelitis
- Fracture
- Orthopedic Implant
- Cortical Tug Lesion
- Juxtacortical Tumor

ESSENTIAL INFORMATION

Key Differential Diagnosis Issues

- Radiographs or other imaging very helpful when assessing bone surface lesions on US
- US cannot assess intramedullary extent or surrounding bone quality
- Beware of normal bone irregularity at sites of muscle insertion and previous surgery
- Heavy calcification or ossification alongside bone may appear attached to bone on US
- Only portion of bony lesion visible with US
  - Angulation of transducer can help provide more comprehensive evaluation
  - MR or CT is often required for complete evaluation
- Determining etiology of surface bone lesions is often helped by clinical features
  - Age of patient
  - Location of lesion
  - Duration and type of symptoms
- US-guided biopsy of bone tumor is possible if extraosseous mass is present
- US-guided biopsy of intramedullary component of bone tumor or infection is possible via cortical breach if present

Helpful Clues for Common Diagnoses

- **Osteochondroma**
  - Most common bone surface tumor
  - Metaphyseal or metadiaphyseal in location
    - Points away from joint
  - Majority are solitary
    - If multiple and metaphyseal dysplasia indicates multiple hereditary exostoses (diaphyseal aclasis)
  - Variable thickness in hypoechoic cartilage cap
    - Often not visible or barely visible
    - Cap thickness of ≥ 2 cm or progressive thickening of cartilage cap on serial imaging is strongly indicative of secondary chondrosarcoma
  - Most osteochondromas are asymptomatic
  - Symptoms can be due to several causes
  - Reactive myositis secondary to friction between osteochondroma and adjacent muscle
    - Degree of myositis is best evaluated on MR
  - Reactive bursitis
    - Friction between osteochondroma and soft tissue can lead to development of intervening adventitial bursa
  - Pseudoaneurysm
    - Most common vascular complication
    - Arterial rupture → arterial lumen contained by hematoma
    - Characteristic to-and-fro pattern on spectral analysis of narrow neck
    - Characteristic yin-yang bidirectional flow pattern on color Doppler imaging of lumen
  - Bleeding
    - From vascular injury, hemorrhagic bursitis, or myositis
  - Neurological sequelae
    - May cause compressive neuropathy, radiculopathy, or myelopathy depending on location
    - Snapping nerve → reactive neuritis
  - Fracture
    - Affects stalk of pedunculated osteochondroma
  - Malignant transformation more common with
    - Diaphyseal aclasis (likelihood ~ 1%)
    - Increase in age
    - Osteochondromas located in axial skeleton
- **Other Benign Bone Tumor**
  - Most intramedullary benign tumors remain intramedullary
    - May expand bone cortex or fracture → bone surface irregularity
  - Giant cell tumor of bone, not infrequently, has extraosseous soft tissue component
    - Hypervascular ± hemorrhagic cystic areas
  - Other benign juxtacortical tumors
    - Chondroma, fibroma, schwannoma
- **Malignant Bone Tumor**
  - Osteosarcoma is most common malignant bone tumor
    - Usually occurs in childhood or young adulthood
  - Most frequently arises at long bone metaphysis
    - Area of most rapid growth in childhood
  - Hypoechoic, broad-based tumoral outgrowth from bone
  - Radiating seams of tumoral osteoid deposition
  - Moderately hyperemic

Helpful Clues for Less Common Diagnoses

- **Metastases**
  - Can be detected by US if surface based or cortex disrupted
  - Lung and breast are most common primary sites
    - Adenocarcinoma is most common tumor cell type
  - Predilection for diaphyseal location
    - Compromises bone strength as cortical bone is particularly relevant for diaphyseal bone strength
- **Bone Tumor Recurrence**
  - US is very useful at screening for bone tumor recurrence
  - Based on premise that most bone tumor recurrences occur in juxtacortical soft tissues rather than in bone
  - Unlike MR, US assessment of juxtacortical soft tissues is not affected by metallic prosthesis
  - Tumor recurrence usually mirrors imaging characteristics of original tumor

- If recurrence is detected, check regional lymph nodes, as nodal involvement is more frequent with recurrence than with primary bone tumor
- **Acute Osteomyelitis**
  - Most frequently involves metaphyses of long bones in children
  - Fine lamellar periosteal reaction
  - Anechoic fluid accumulation alongside cortex
  - Subperiosteal accumulation of fluid
  - Subperiosteal abscess
  - Color Doppler imaging → hyperemia within and around inflamed periosteum
  - ± concurrent septic arthritis
    - Sympathetic effusion is common
    - Septic arthritis can be excluded by aspiration
  - US-guided aspiration of subperiosteal fluid or abscess for Gram stain and culture
  - MR is best at assessing acute osteomyelitis
    - Unlike MR, acute osteomyelitis is not excluded by normal US examination
- **Chronic Osteomyelitis**
  - Most frequently involves diaphysis of long bones in young to middle-aged adults
  - Coarse or fine periosteal reaction dependent on activity
    - ± inflammatory cortical defect (cloaca)
    - ± sequestrum within extraosseous tissues
    - ± juxtacortical inflammation or abscess
    - ± sinus tract extending toward skin
  - MR best at assessing activity and extent of chronic osteomyelitis
    - US cannot evaluate intramedullary disease
    - US is very helpful if metal implant in situ, which limits assessment by CT or MR
- **Fracture**
  - 4 types of fracture
    - Traumatic, stress, insufficiency, and pathological
  - US is very useful in detecting most radiographically occult traumatic fractures, provided affected bone is assessable to US
    - Also useful in other fracture types though not as sensitive as in traumatic fractures
  - Acute traumatic fracture
    - Focal cortical disruption
    - ± shadowing artifact (quite common) or reverberation artifact ("light house" or "chimney" phenomenon) (uncommon) at fracture site
    - Juxtacortical soft tissue swelling, edema, or hematoma (common)
  - Healing traumatic fracture
    - Hypoechoic callus bridging fracture site with progressive ossification
- **Orthopedic Implant**
  - Plate and screw fixation, bone-allograft anastomosis, etc. can all give rise to suspicious bone surface appearance on US
    - Can mimic localized cortical bone destruction from tumor or infection
  - Correlation with radiography very helpful
  - If still in doubt, follow-up US in 2-4 weeks
- **Cortical Tug Lesion**
  - Cortical hyperostosis at site where strong muscle is attached to bone
  - Most commonly seen at aponeurotic attachments of gluteus maximus into femoral shaft and deltoid into humeral shaft
  - May cause both pain and increased activity on isotope bone scanning
  - Do not confuse with more sinister pathology
- **Juxtacortical Tumor**
  - Soft tissue masses commonly lie against bone, i.e., occur in juxtacortical location
    - Examples include gouty tophus, vascular malformation, nerve sheath tumor, lipoma, etc.
  - Hard tissue masses also less frequently occur in juxtacortical (periosteal) location
    - Examples include chondroma, chondromyxoid fibroma, periosteal osteosarcoma, or periosteal chondrosarcoma

Osteochondroma

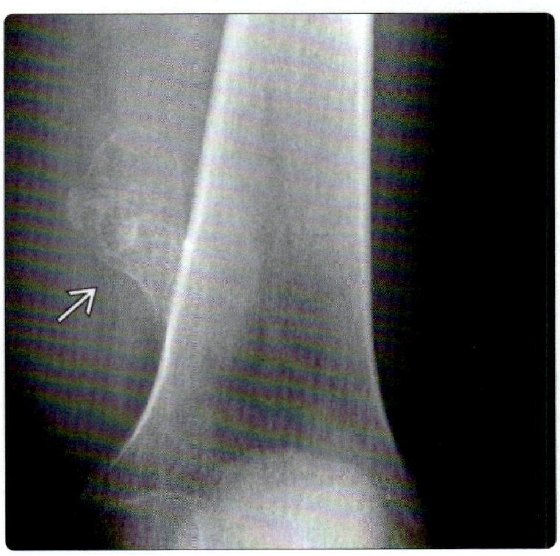

Osteochondroma

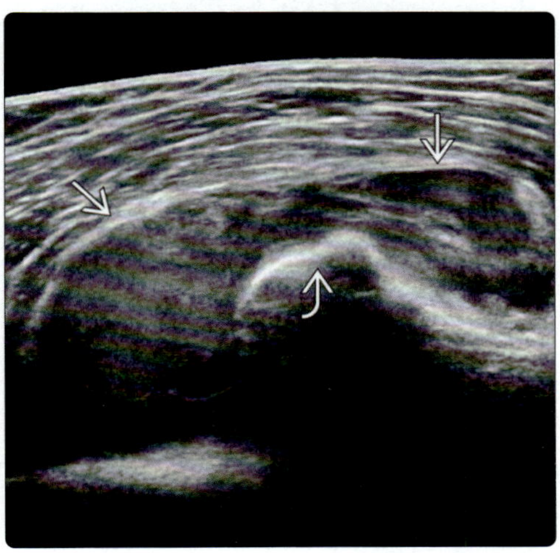

(Left) *Anteroposterior radiograph shows a large, pedunculated osteochondroma ➔ on the medial side of the distal femoral metaphysis. This lesion was symptomatic and referred for US examination.* **(Right)** *Longitudinal US of the same lesion shows bursa distended with hemorrhage ➔ overlying the osteochondroma ➔.*

Osteochondroma

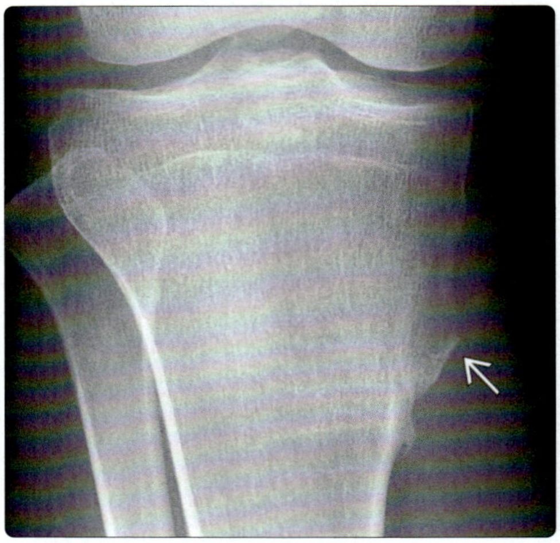

Osteochondroma

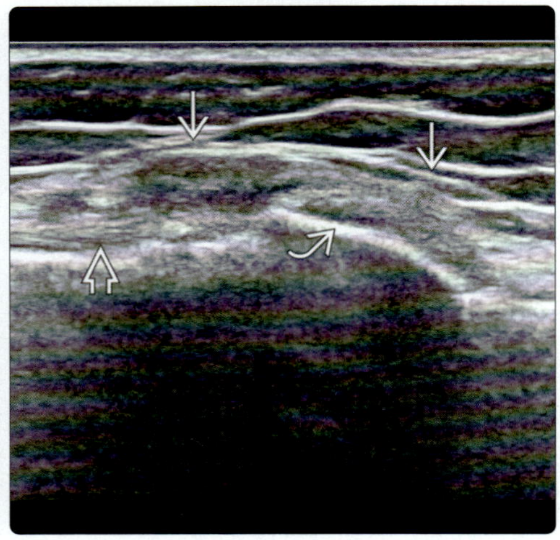

(Left) *Anteroposterior radiograph shows a small, symptomatic sessile osteochondroma on the medial aspect of the proximal tibial metaphysis.* **(Right)** *Longitudinal US of the same lesion shows a thick-walled chronic bursitis overlying this osteochondroma. The medial collateral ligament is seen proximally. The very thin hypoechoic layer is due to a thin cartilage cap. Bursa tend to develop in sites of chronic friction. They are lined by synovium.*

Malignant Bone Tumor

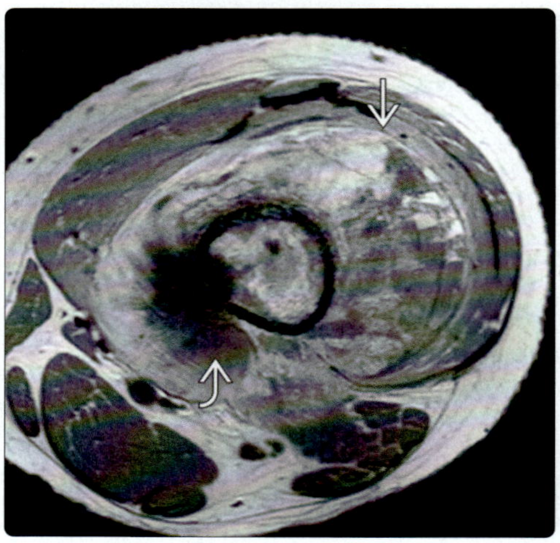

Malignant Bone Tumor

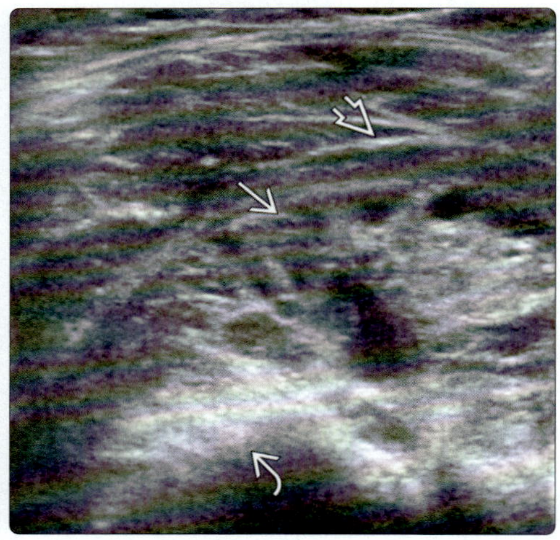

(Left) *Axial PD FSE MR of femur shows a bone tumor with a large, extraosseous soft tissue component tumor anteromedially with fluid-fluid levels and dense, spiculated tumoral new bone formation posterolaterally.* **(Right)** *Transverse US of the anteromedial aspect of the same tumor shows dense, spiculated new bone formation posterolaterally and a large extraosseous soft tissue component displacing the vastus medialis muscle anteriorly. US-guided biopsy confirmed osteosarcoma.*

Malignant Bone Tumor

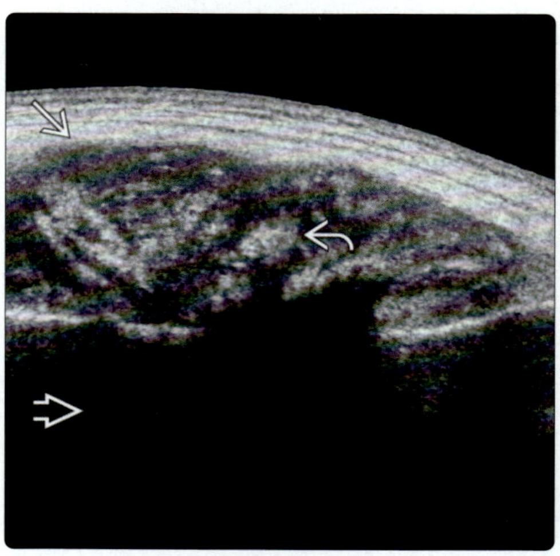

Metastases

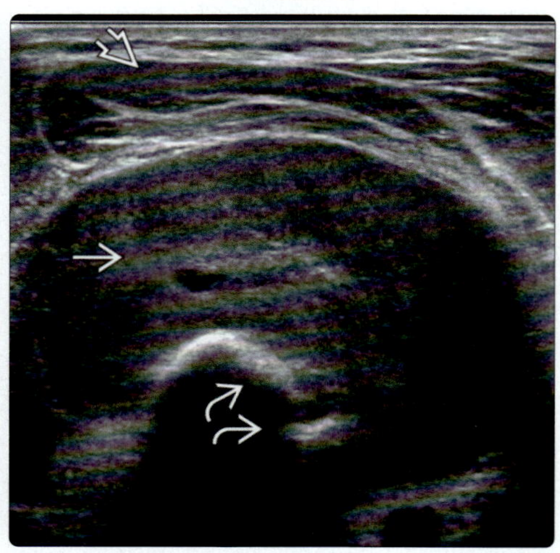

(Left) *Longitudinal US shows a large, hypoechoic osteosarcoma arising from the distal femur. Note the radiating echogenic seams of mineralized osteoid within the extraosseous tumor matrix.* **(Right)** *Transverse US in patient with known metastatic lung cancer and recent onset of elbow pain shows a large soft tissue mass surrounding the proximal radial shaft with early cortical infiltration, consistent with a metastatic deposit. The supinator muscle is displaced.*

Acute Osteomyelitis

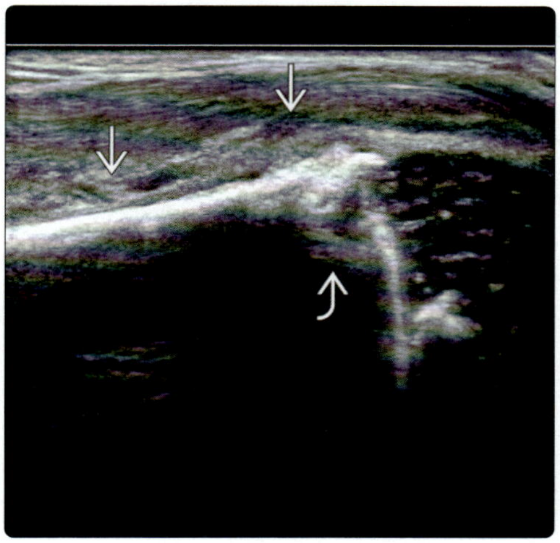

Acute Osteomyelitis

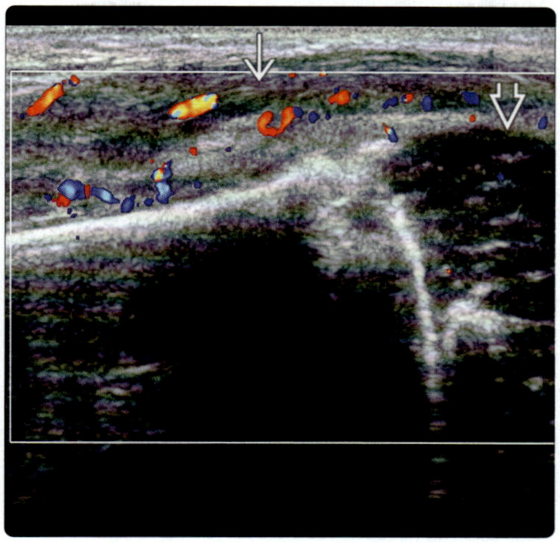

(Left) *Longitudinal US of a neonate with acute osteomyelitis of the femur shows focal cortical irregularity ➔ at the distal femoral metaphysis. This is associated with moderate periosteal soft tissue thickening and a small amount of subperiosteal fluid ➔.* **(Right)** *Longitudinal color Doppler US of the same patient shows moderate hyperemia of the thickened periosteal tissues ➔. The physis and epiphysis ➔ are normal.*

Chronic Osteomyelitis

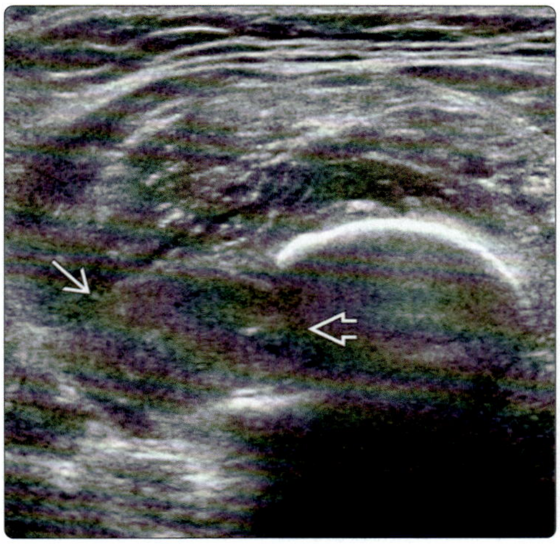

Chronic Osteomyelitis

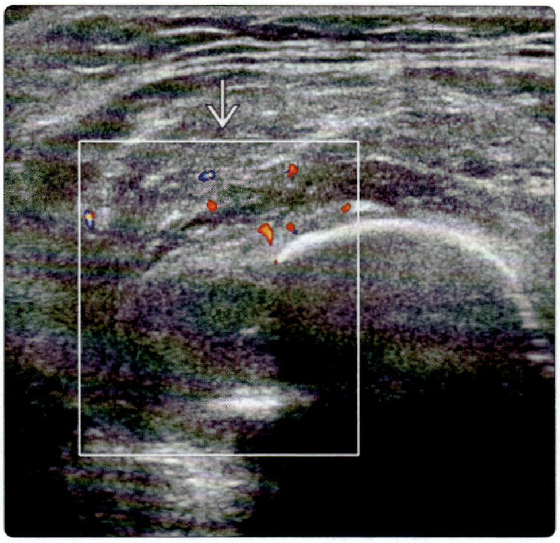

(Left) *Transverse US in a patient with acute exacerbation of chronic osteomyelitis shows an echogenic fluid collection ➔ alongside a large cortical defect ➔ on the medial side of the distal femoral cortex.* **(Right)** *Transverse color Doppler US of the same patient shows mild hyperemia ➔ of the inflamed adjacent soft tissues. US-guided aspiration of the juxtacortical collection yielded Staphylococcus aureus.*

Fracture

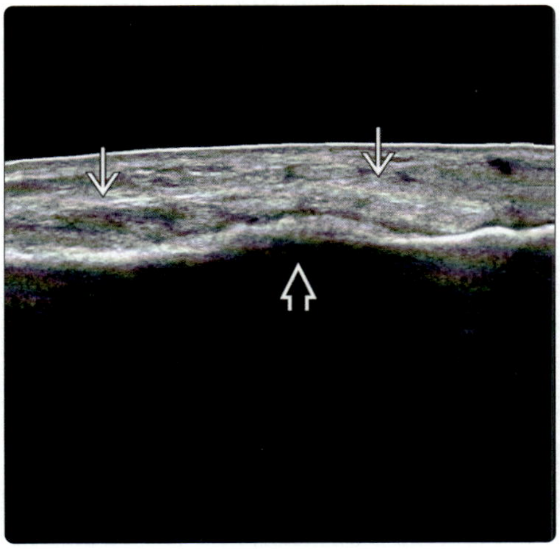

Fracture

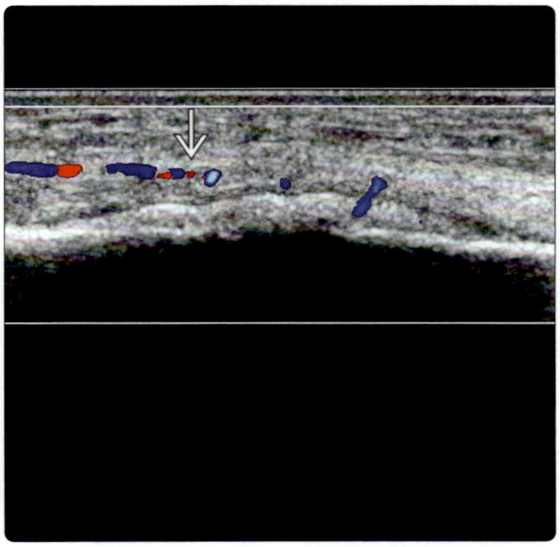

(Left) *Extended field-of-view US in a patient with foot pain shows an active stress fracture of the 2nd metatarsal bone with midshaft cortical hyperostosis ➔ and moderate surrounding soft tissue edema and swelling ➔.* **(Right)** *Longitudinal color Doppler US of the same area shows moderate hyperemia ➔ of the soft tissues around the stress fracture.*

(Left) *Transverse US in a patient with chest wall pain and tenderness shows a healing undisplaced fracture at the anterior aspect of the right 6th rib. The fracture is at the end of the osseous component of the rib close to osteochondral junction. There is early callus formation.* **(Right)** *Transverse US in patient with chest wall pain shows a largely healed nondisplaced rib fracture at the anterolateral aspect of the left 8th rib. No hematoma is present. There is mild adjacent soft tissue inflammation.*

Fracture

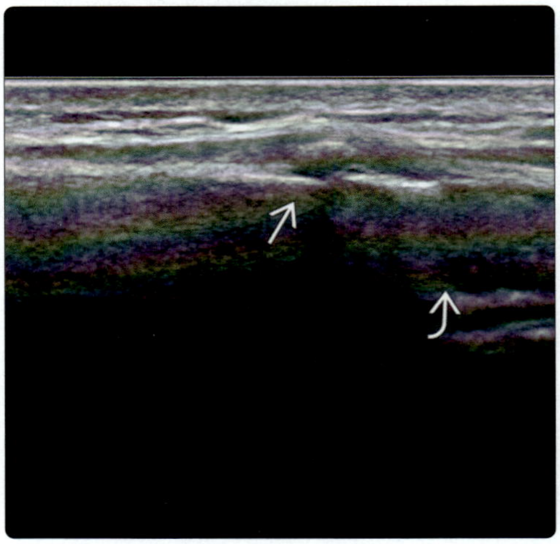

Fracture

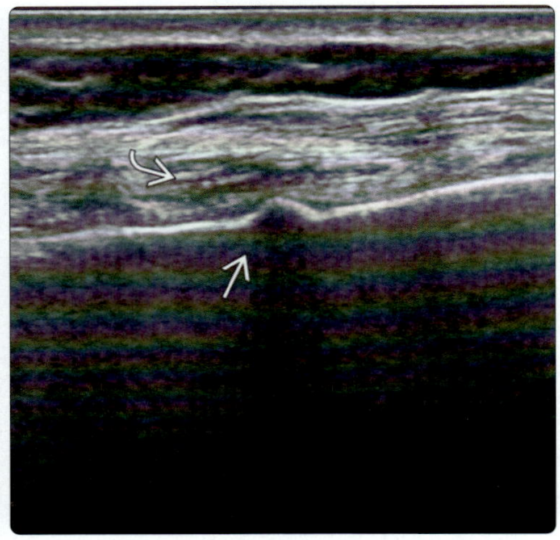

(Left) *Longitudinal US in an elderly female osteoporotic patient with hip and thigh pain shows a small irregular osseous spike on the lateral cortex of the midfemoral shaft. This favors a bisphosphonate-related insufficiency fracture with cortical hyperostosis. There is hypoechoic, poorly mineralized osteoid alongside the area of hyperostosis.* **(Right)** *Subsequent radiography of the same area shows an incomplete cortical insufficiency fracture with cortical hyperostosis in the midfemoral shaft.*

Fracture

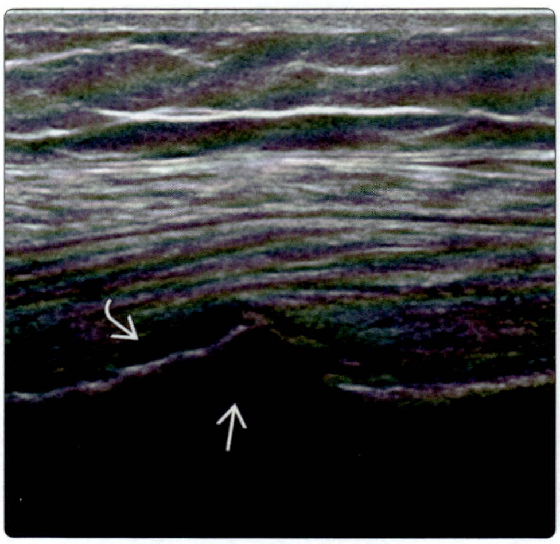

Fracture

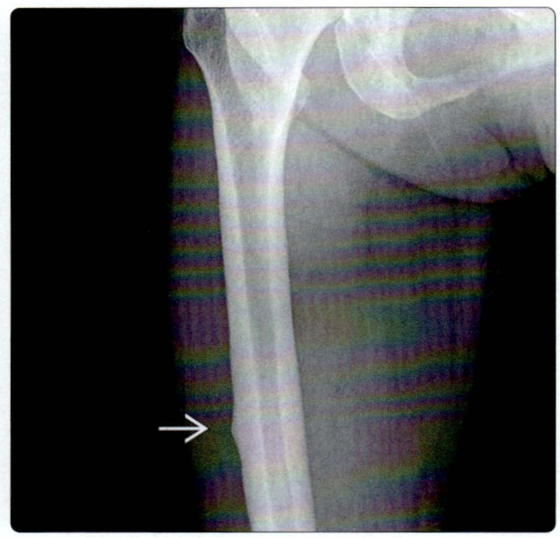

(Left) *Longitudinal US of a patient with an orthopedic implant shows mild reactive new bone formation with soft tissue swelling thickening at the interface between the prosthesis and the bone.* **(Right)** *Frontal radiograph of the same patient provides explanation of US findings, which are due to reactive new bone formation at the bone:prosthesis interface. There is no evidence of tumor recurrence or infection.*

Orthopedic Implant

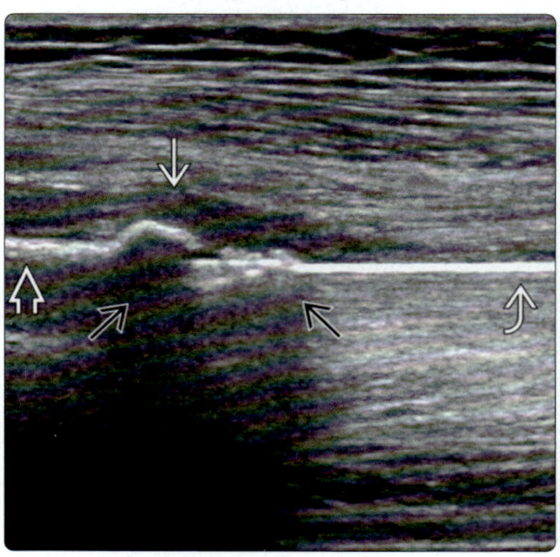

Orthopedic Implant

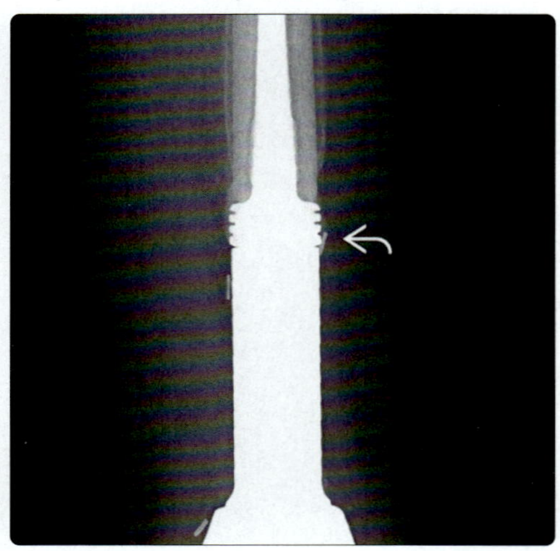

Orthopedic Implant

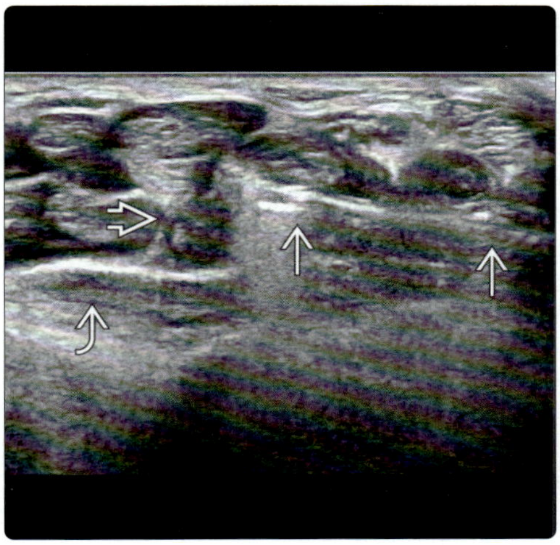

Cortical Tug Lesion

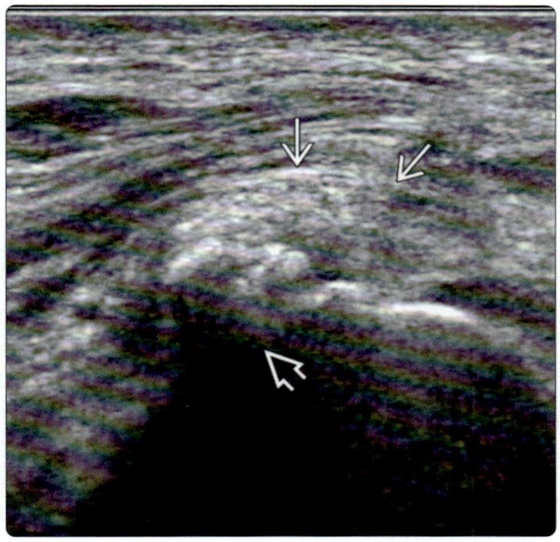

(Left) *Transverse US in a patient shows the medial end of a transverse fixation plate ➡ impinging onto the flexor digitorum profundus myotendinous junction ➡. Repeated friction between the tendon and the fixation plate may well lead to tendon attrition and rupture. The radius ➡ is normal.* **(Right)** *Transverse US shows localized cortical hyperostosis (tug lesion) ➡ on the posterolateral aspect of the femoral shaft due to insertion of the gluteus maximus tendon ➡.*

Juxtacortical Tumor

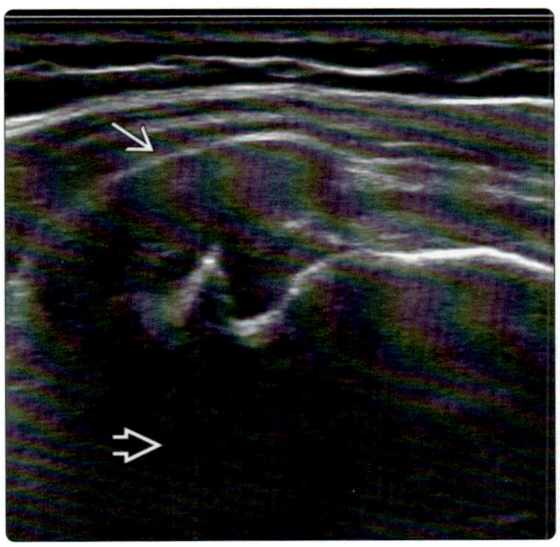

Juxtacortical Tumor

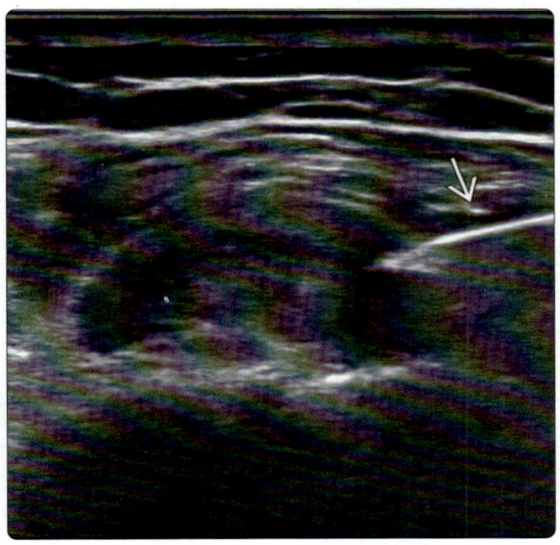

(Left) *Transverse US in patient with shoulder pain shows a mass ➡ arising from the proximal humerus ➡ posteriorly extending into the extraosseous soft tissues just distal to the infraspinatus tendon. In conjunction with a previous MR examination, this favored a fibrous-type tumor.* **(Right)** *Transverse US shows the biopsy needle ➡ in the most peripheral aspect of the mass. Despite anesthetic infiltration, the mass was very tender. Only a single biopsy was taken, which revealed neurofibroma.*

Juxtacortical Tumor

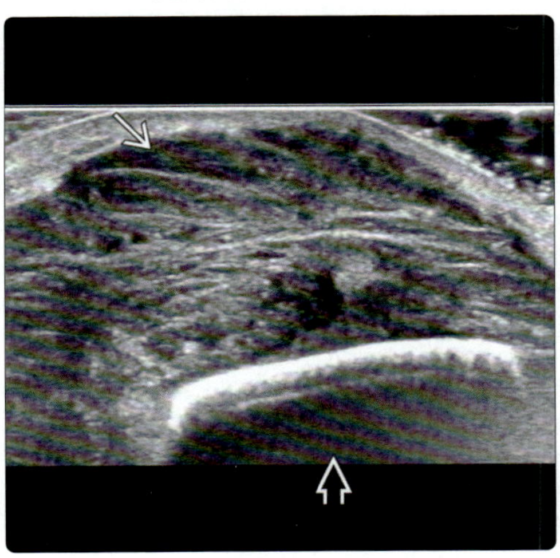

Juxtacortical Tumor

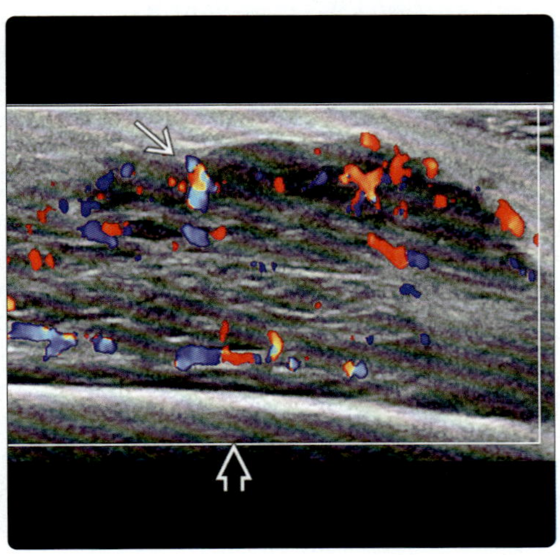

(Left) *Transverse US shows a large pretibial ➡ hypoechoic soft tissue mass ➡. Biopsy revealed non-Hodgkin lymphoma (follicular). Although the tibial cortex seems intact, subsequent MR revealed intramedullary extension.* **(Right)** *Longitudinal color Doppler US of the same mass shows moderate tumoral hyperemia ➡. The tibial cortex seems intact ➡. Most musculoskeletal lymphoma involves either bone or muscle.*

DIFFERENTIAL DIAGNOSIS

Common

- Ganglion Cyst
- Baker Cyst
- Parameniscal Cyst
- Bursal Distension
- Liquefied Hematoma
- Abscess or Collection

Less Common

- Cystic-Looking Soft Tissue Tumors
- Pseudoaneurysm

ESSENTIAL INFORMATION

Key Differential Diagnosis Issues

- Establish whether mass is intra- or extracapsular
  - If extracapsular, does mass communicate with joint?
  - Where does mass communicate with joint?
  - Does it attach to labrum or meniscus, or does it communicate with other part of joint?
  - Sometimes direct continuity with joint is not visible, but mass can be seen pointing toward joint
    - Passive movement of joint may occasionally be helpful to more clearly visualize communication

Helpful Clues for Common Diagnoses

- **Ganglion Cyst**
  - Mucinous fluid contained within fibrous pseudocapsule
    - Not lined by synovial tissue
  - Results from extruded synovial fluid
    - Joint pathology (e.g., synovitis) can occur within ganglia but is uncommon due to closed or limited communication
  - Most common around wrist/hand and foot
    - 80% of soft tissue masses of wrists/hands and feet
    - Dorsal wrist ganglia are more commonly symptomatic, although volar wrist ganglia are overall more common
  - Typical locations reflect weakness in joint capsule
    - Dorsal radiocarpal joint, particularly scapholunate joint
      - Can impinge upon distal end of posterior interosseous nerve
    - Scapho-trapezio-capitate joint
  - Near A1 pulley on middle, index, and ring fingers
    - Does not appear to communicate with metacarpophalangeal joint or flexor tendon sheath
  - Anechoic, well-defined, rounded or irregular mass alongside joint
    - ± comet-tail artifacts due to aggregates of colloid material
    - Infrequently hyperechoic, due to hemorrhage or inflammatory content from joint
  - Typically has neck pointing toward particular joint
    - Usually neck cannot be traced to joint as too small
    - Relevant to identify potential joint of origin as this will aid surgical exploration
  - Most ganglia per se are not painful with pain induced by ganglion leakage
    - Pay particular attention to any inflammation or free fluid around ganglion
  - 50% of ganglia recur after percutaneous aspiration
    - Not possible to determine viscosity of content on grayscale appearances alone
    - Leakage of cyst content can cause pain for 24-36 hours after aspiration
    - Hyaluronidase may facilitate aspiration by making contents less viscous
- **Baker Cyst**
  - Semimembranous-gastrocnemius bursa, which communicates with joint
    - Slit-like opening in capsule close to root of posterior horn medial meniscus
  - Classic talk-bubble configuration on transverse US with beak of bubble emerging between semimembranous tendon and medial head gastrocnemius muscle
    - Slight knee flexion during scanning can accentuate this communication
    - Frequently bilateral
  - Usually extends distally over medial belly gastrocnemius muscle
    - Less commonly spreads proximally and uncommonly spreads into semimembranous or gastrocnemius muscle
  - Has synovial membrane and therefore affected by synovial processes (e.g., PVNS) involving knee joint
    - Any pathological process in knee can be seen in Baker cyst
  - Prone to either hemorrhage or leakage
    - May make Baker cyst symptomatic
  - Hemorrhage
    - Increased speckled echogenicity of cyst fluid
  - Leakage or rupture
    - Free fluid around margin of cyst
    - Angulated arrow-like, rather than rounded, configuration of cyst
    - ± pericystic inflammation
- **Parameniscal Cyst**
  - Cystic mass extending from peripheral aspect of meniscus
  - Usually associated with horizontal meniscal tear
    - Mucinous content (from synovial fluid) expressed as parameniscal cyst
  - Occasionally, no meniscal tear is present
    - Meniscal tear may have healed
    - Cyst may have arisen from severe myxoid degeneration of meniscus in absence of tear
      - Meniscal myxoid content expressed as parameniscal cyst
  - Percutaneous cyst aspiration is helpful, particularly in cases without visible meniscal tear
- **Bursal Distension**
  - 2 types of bursae
  - Synovial bursae: Synovial-lined bursa, which occur at defined anatomical locations
    - Prone to synovial disorders, such as synovitis or synovial osteochondromatosis
  - Adventitial bursae: Non-synovial-lined bursae acquired due to friction between opposing structures

- Some paraarticular bursa more prone to bursitis than others
- Subacromial-subdeltoid bursa
  - Attached to coracoacromial ligament and greater tuberosity distal to tendon insertions
  - Distension strongly associated with rotator cuff tear
  - If bursal wall thickened ± hyperemia, consider inflammatory arthropathy (e.g., SLE, RA) or infection
- Olecranon bursa
  - Between olecranon process and skin
  - Common site of gouty bursitis
  - Commonly inflamed, not commonly infected
- Iliopsoas bursa
  - Anterior to femoral head between hip capsule and iliopsoas muscle/tendon
  - Communicates with hip joint in 15% of patients
  - Consider iliopsoas infection, particularly tuberculous
- Semimembranous bursa
  - Between anterior arm semimembranous tendon complex and medial collateral ligament
  - Located medial to Baker cyst and proximal to pes anserinus bursa
  - Distended bursa has J-shaped appearance
- Pes anserinus bursa
  - Between pes anserinus tendons anteromedial aspect knee and medial collateral ligament
- Prepatellar, superficial, and deep infrapatellar bursae
  - Prepatellar bursa between patella and skin
  - Superficial infrapatellar bursa between tibial tuberosity and skin
  - Deep infrapatellar bursa between distal part patellar tendon and tibia
    - ≤ 3 mm deep distension deep infrapatellar bursa common normal finding

- **Liquefied Hematoma**
  - Usually following surgery
  - Hematoma is solid looking when composed of clotted blood, undergoes progressive liquefaction
  - Collection may be anechoic or of variable echogenicity depending on content
  - Cannot determine infectivity based on echogenicity of content
  - Likelihood of infection increases with greater degrees of surrounding inflammation
    - Aspirate if doubt exists
- **Abscess or Collection**
  - If any abscess or collection around joint, look carefully for septic arthritis or osteomyelitis
  - Clinical signs of infection
  - Echogenic cyst contents with fine or coarse speckles
  - Marginal hyperemia and inflammation
  - If in doubt, aspirate
    - ± biopsy if aspiration not possible or not fruitful

Helpful Clues for Less Common Diagnoses

- **Cystic-Looking Soft Tissue Tumors**
  - Lymphangioma
    - Lobulated, usually anechoic, no color flow
  - Vascular malformation
    - Slow flow or high flow, variable stroma component
  - Cystic schwannoma
    - Visible entering and exiting nerve
  - Myxoma
    - Intramuscular, anechoic ± echogenic, comet-tail artifacts due to colloid aggregates
  - Myxoid sarcoma
    - Myxoid liposarcoma, synovial sarcoma, myxofibrosarcoma, extraskeletal myxoid chondrosarcoma
  - Cystic metastases
    - Feature of disseminated malignancy
    - Usually from either myxoid primary tumor (e.g., gastrointestinal carcinoma) or due to severe tumor necrosis
- **Pseudoaneurysm**
  - Arterial puncture usually either iatrogenic or inadvertent arterial injury in intravenous drug users
  - Yin-yang sign on color Doppler imaging
  - To-and-fro pattern on pulsed Doppler spectral analysis
  - Treat with US-guided compression

Ganglion Cyst

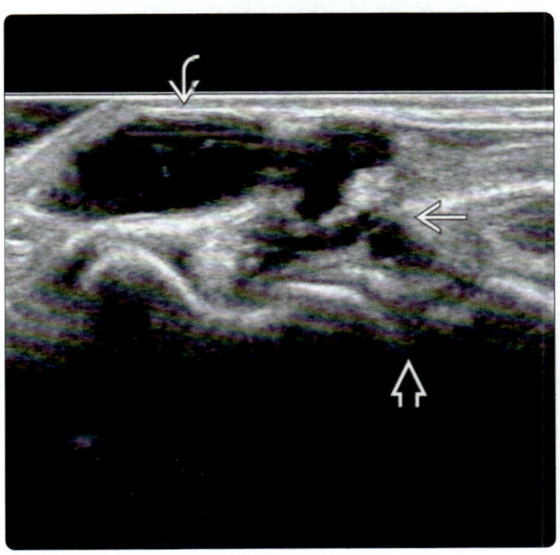

Ganglion Cyst

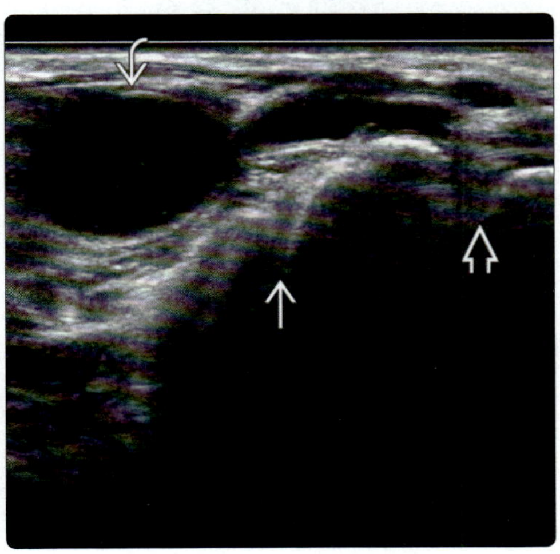

(Left) *Longitudinal US shows a medium-sized ganglion cyst ➡ with an irregular serpiginous tail ➡ extending to the dorsal aspect of the radioscaphoid articulation ➡.* **(Right)** *Longitudinal US shows a medium-sized ganglion cyst ➡ arising from the volar aspect of the scapho-trapezium articulation ➡. The 1st carpo-metacarpal articulation ➡ is normal.*

(Left) *Transverse US of the ankle in a patient with acute traumatic swelling shows an acute ganglion cyst ➡ containing small, echogenic speckles due to colloid aggregates lying alongside the tibia ➡ and pointing toward the ankle joint ➡. This is an acute traumatic ganglion cyst.* **(Right)** *Longitudinal US shows a thick-walled ganglion cyst ➡ on the dorsum of the navicular-cuneiform articulation ➡. Atypical ganglia occur more commonly in the foot and knee regions.*

Ganglion Cyst

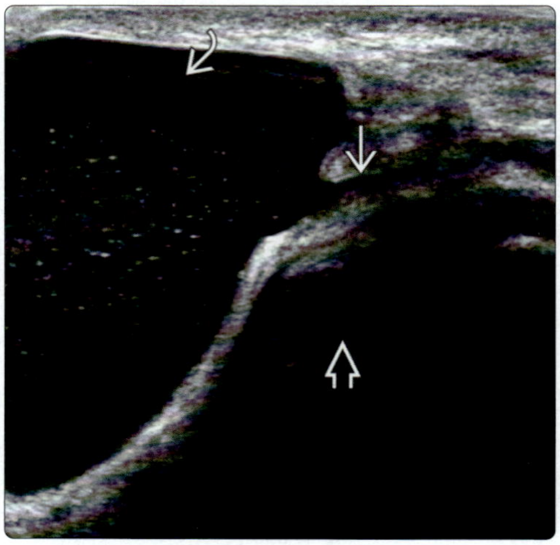

Ganglion Cyst

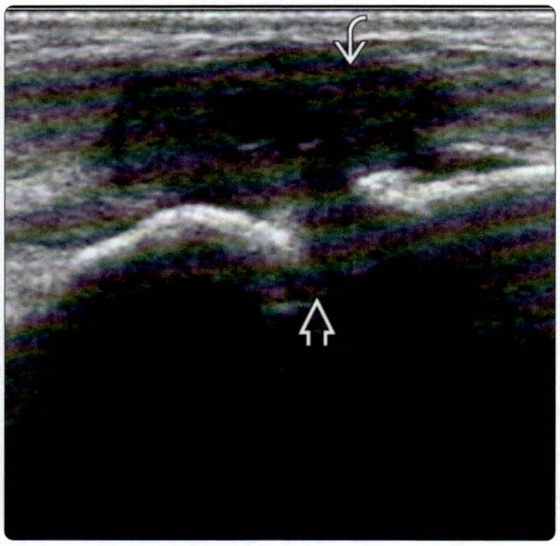

(Left) *Transverse US shows the typical talk-bubble configuration of a Baker cyst ➡. The beak merges between the medial belly of gastrocnemius ➡ and the semimembranous tendon ➡.* **(Right)** *Transverse US shows many small, echogenic foci ➡ within a Baker cyst due to small colloid aggregates. The larger echogenic foci may give rise to comet-tail artifacts.*

Baker Cyst

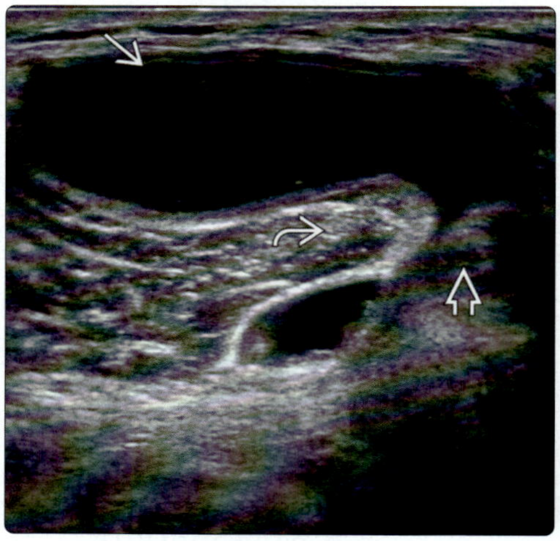

Baker Cyst

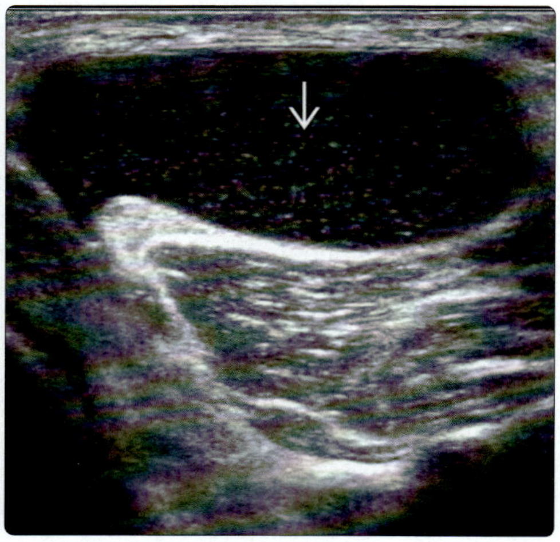

(Left) *Transverse US shows recent hemorrhage within a Baker cyst ➡ with fine, echogenic fluid ➡. The small, hyperechoic ring ➡ may be due to a detached synovial fragment.* **(Right)** *Transverse US shows a Baker cyst containing inflammatory-type synovial proliferation ➡ with embedded, small, calcified foci ➡. This was secondary to an inflammatory knee osteoarthritis with calcified intraarticular debris.*

Baker Cyst

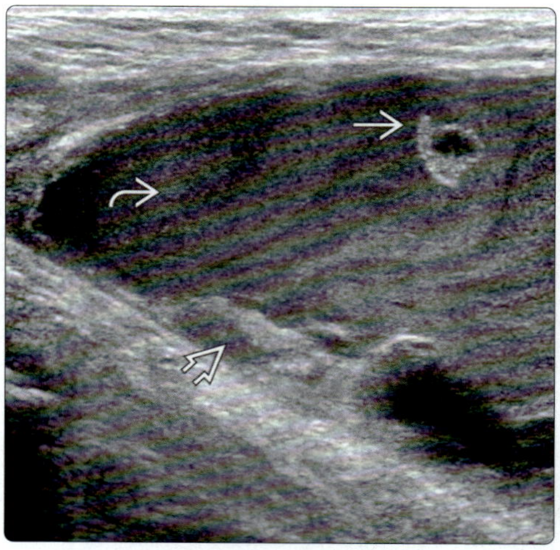

Baker Cyst

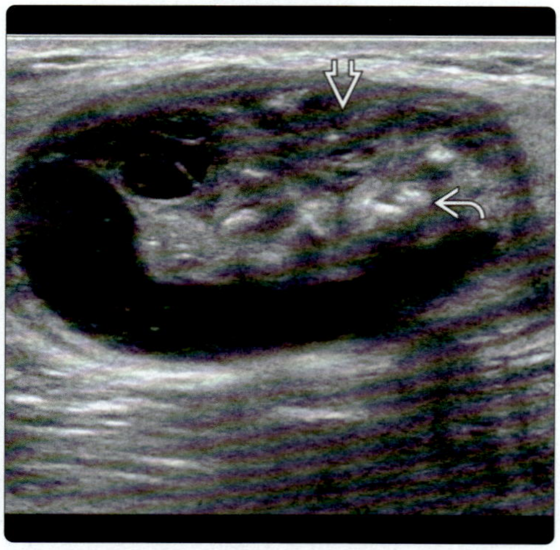

Baker Cyst

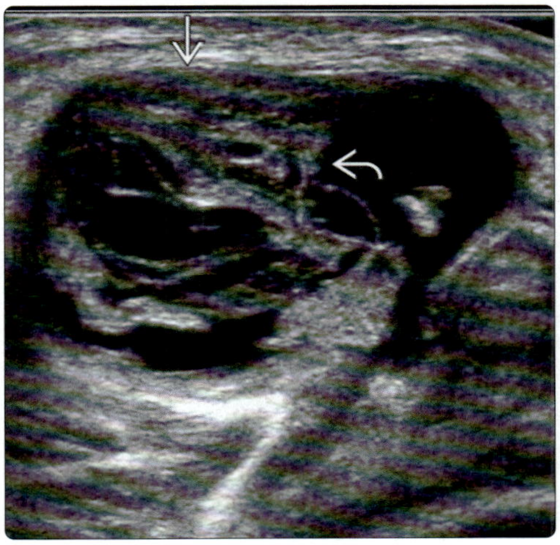

Parameniscal Cyst

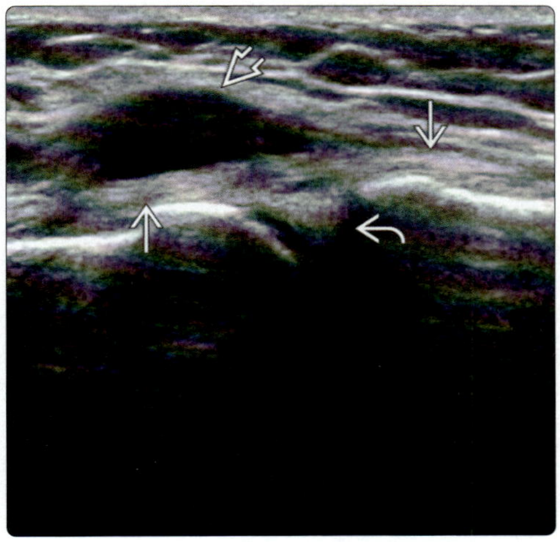

(Left) *Transverse US shows a medium-sized Baker cyst with a large amount of synovial proliferation secondary to an inflammatory-type knee osteoarthritis. The typical talk-bubble configuration is still apparent.* **(Right)** *Longitudinal US of the knee shows a small parameniscal cyst superficial to the medial collateral ligament. This extracapsular component communicated more posteriorly with a deeper intracapsular component extending to the periphery of the medial meniscus.*

Parameniscal Cyst

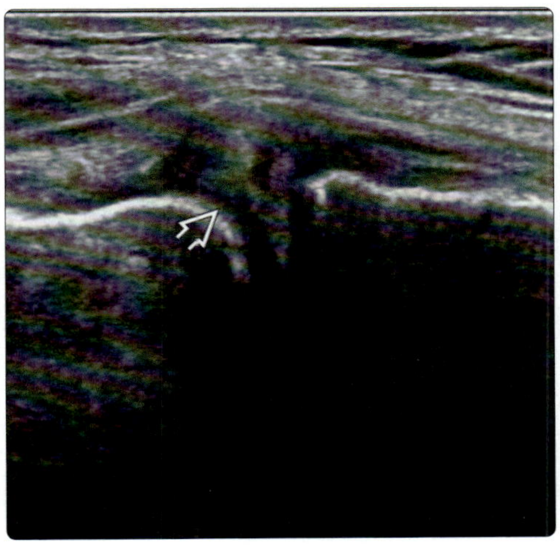

Parameniscal Cyst

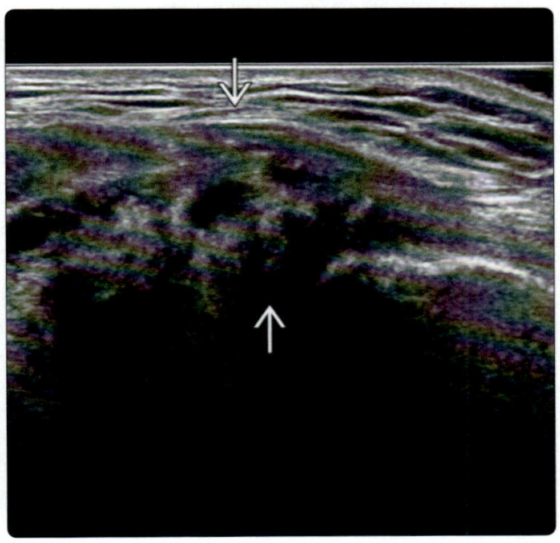

(Left) *Longitudinal US shows a moderately flattened and extruded medial meniscal body with a peripheral horizontal tear.* **(Right)** *Longitudinal oblique US in the same patient at the posteromedial corner of the knee shows a multiloculated, paraarticular cystic mass. The distal end of this cyst contacted the edge of the extruded medial meniscus. The soft tissue component is unusual and is due to localized hemorrhage around the cyst.*

Parameniscal Cyst

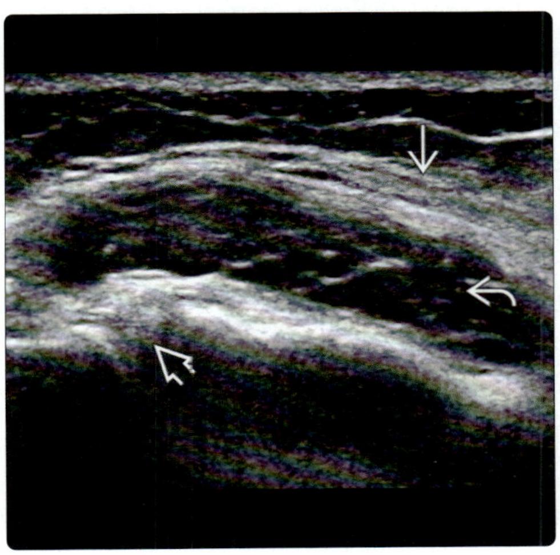

Bursal Distension

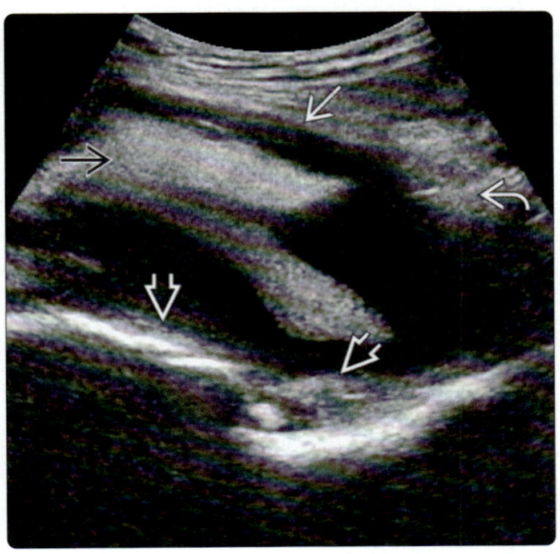

(Left) *Longitudinal US shows a large parameniscal cyst located between the medial collateral ligament and the investing fascia on the medial aspect of the knee.* **(Right)** *Longitudinal US shows a distended iliopsoas bursa anterior to a nondistended hip capsule. The iliopsoas tendon is displaced anteriorly. The bursa contains thick strands of blood clot and fluid. The aspirate grew Staphylococcus aureus.*

(Left) *Transverse US shows a moderately distended olecranon bursa ➔ with frond-like synovial proliferation posterior to the olecranon ➔. The triceps tendon insertion ➔ with its radial extension and the anconeus muscle ➔ are normal.* **(Right)** *Longitudinal color Doppler US shows a subacutely distended adventitial bursa between the iliotibial band ➔ and the lateral aspect of the distal femur ➔ and tibia ➔. Four mL of blood-stained fluid was aspirated, resulting in bursal collapse.*

Bursal Distension

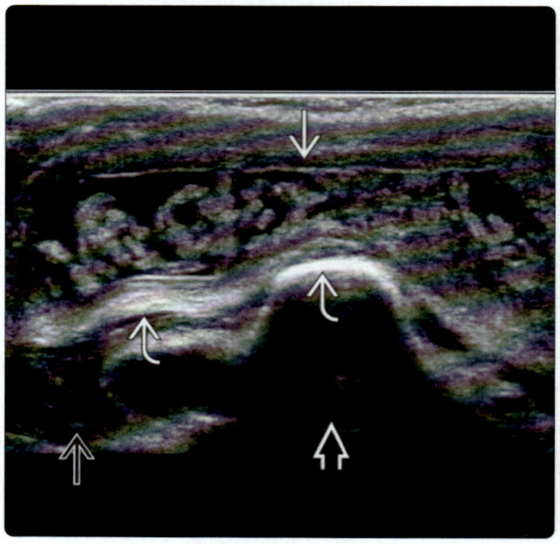

Bursal Distension

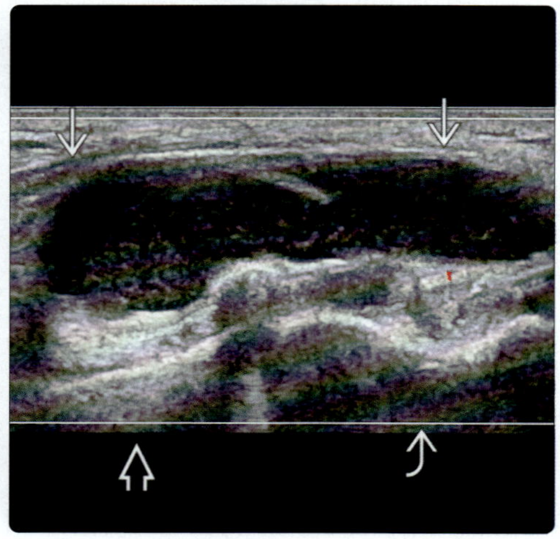

(Left) *Transverse US shows a hemorrhagic pes anserinus bursitis ➔ containing finely echogenic fluid located between the sartorius tendon ➔ and the anteromedial aspect of the proximal tibia ➔.* **(Right)** *Transverse PDWI FS MR in the same patient shows the distended hemorrhagic pes anserinus bursa ➔ deep to the sartorius ➔ and gracilis ➔ tendons. The semitendinous ➔, semimembranous ➔, and medial belly of gastrocnemius ➔ tendons are also shown.*

Bursal Distension

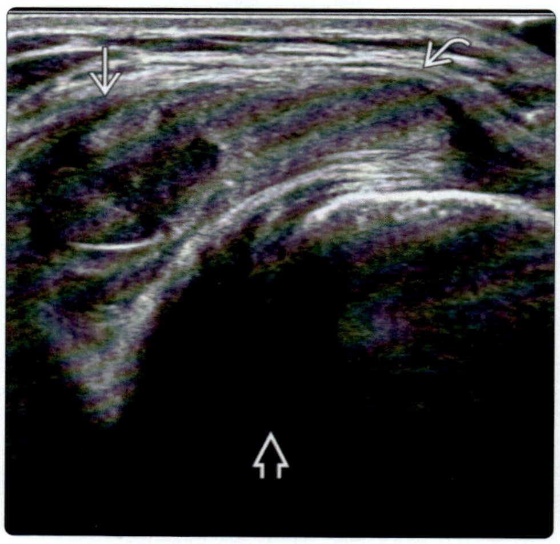

Bursal Distension

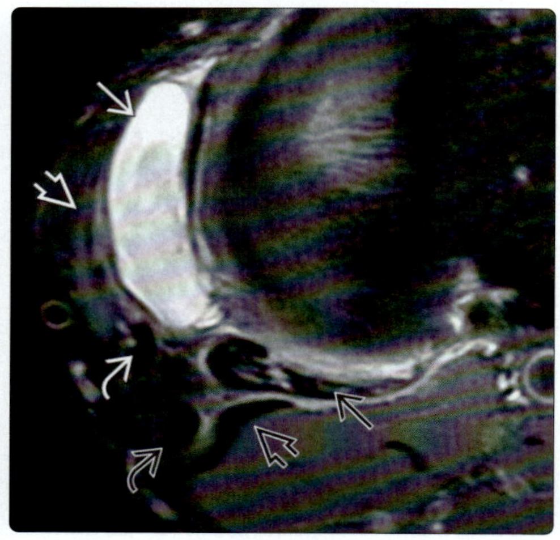

(Left) *Longitudinal US in an elderly man following venepuncture 1 week earlier shows a large hematoma ➔ anterior to the elbow joint ➔, with partial liquefaction, sedimentation, and septation ➔. The brachialis muscle is displaced superficially ➔.* **(Right)** *Transverse US shows a large, partially liquefied hematoma ➔ in the distal thigh. The solid component ➔ is an organizing blood clot. The hematoma is quite chronic with early capsular and internal calcification ➔.*

Liquefied Hematoma

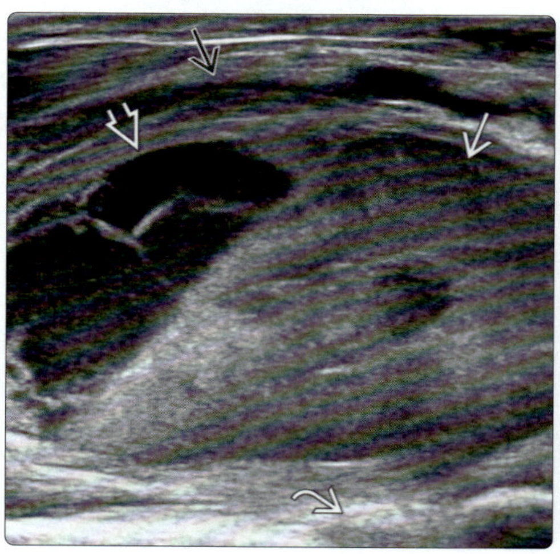

Liquefied Hematoma

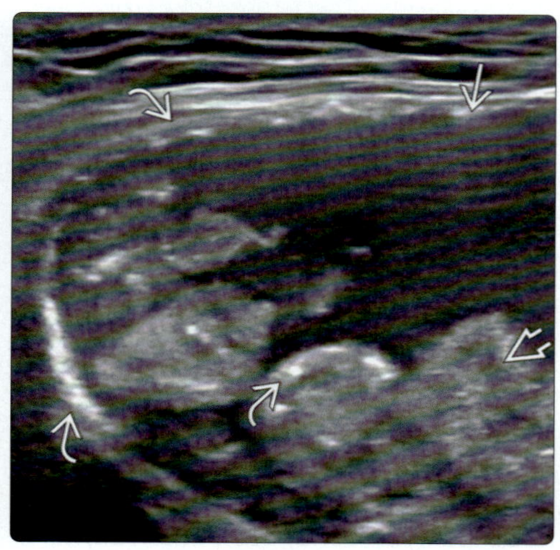

Abscess or Collection

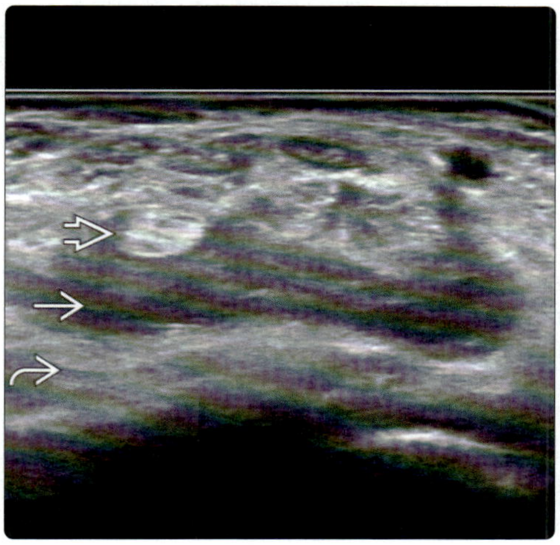

Abscess or Collection

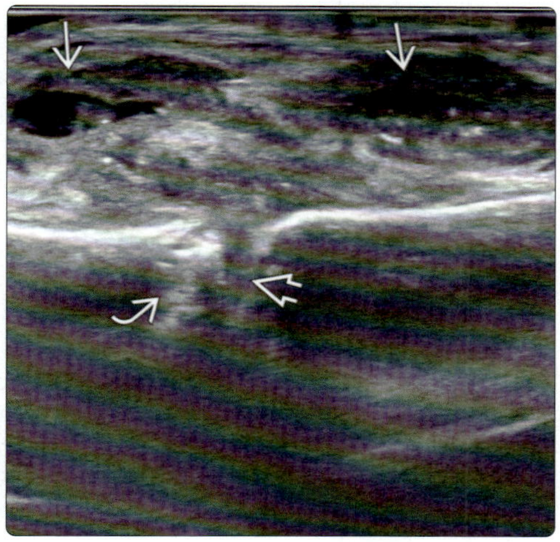

(Left) *Transverse US in patient following penetrating trauma shows a distended space of Parona. This is a potential space between pronator quadratus and flexor digitorum profundus tendon sheath. It communicates with the midpalmar space but not directly with the radial or ulnar wrist bursae* **(Right)** *Transverse US in intravenous drug user shows inflammation and abscess cavities superficial to the infected sternoclavicular joint with osteolysis medial clavicle. Aspirate grew Staphylococcus aureus.*

Abscess or Collection

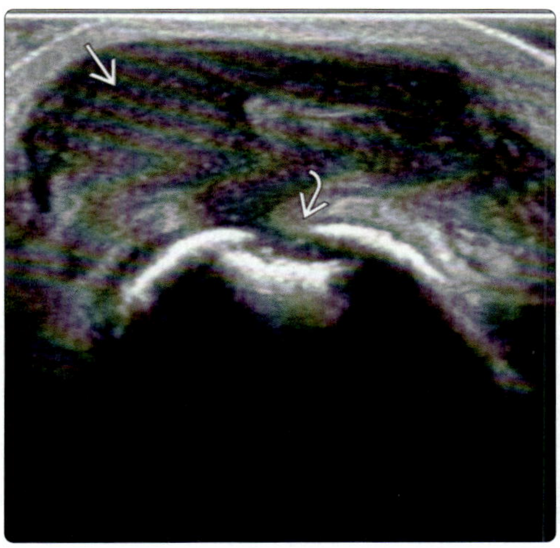

Pseudoaneurysm

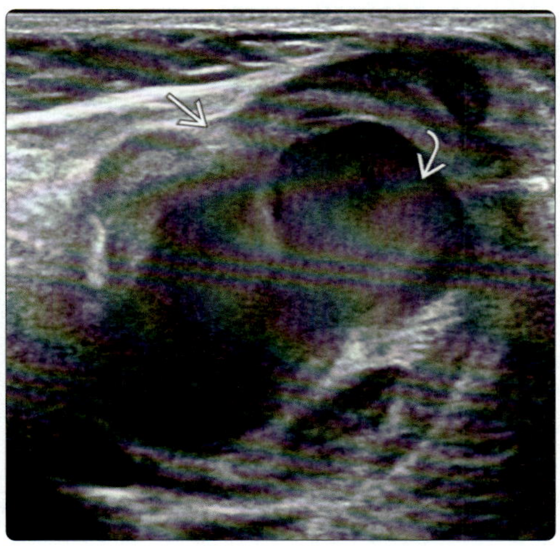

(Left) *Transverse US in a patient with painful elbow swelling shows the abscess cavity of the posterior elbow with cortical osteolysis of olecranon. Subsequent MR confirmed olecranon osteomyelitis.* **(Right)** *Transverse US of the groin region after percutaneous arterial access shows a large femoral artery pseudoaneurysm. The mass was pulsatile with swirling, fine internal echoes on grayscale imaging.*

Pseudoaneurysm

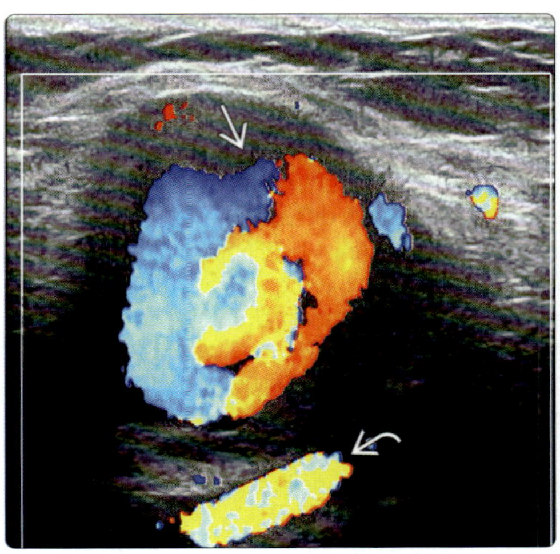

Pseudoaneurysm

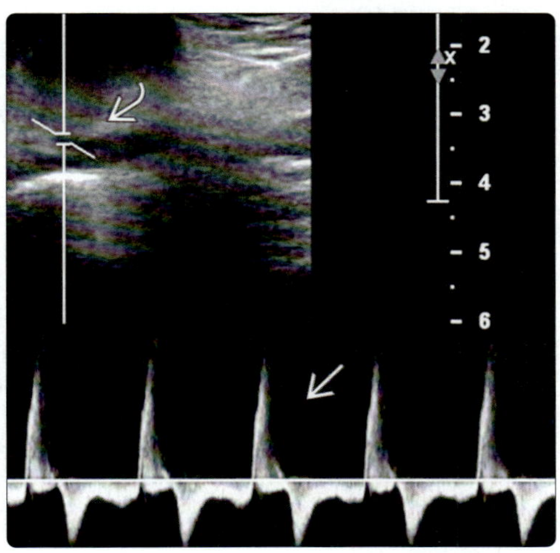

(Left) *Longitudinal US of the same thick-walled pseudoaneurysm as previous image shows yin-yang sign from moving blood in the cavity of pseudoaneurysm, superficial to femoral artery.* **(Right)** *Spectral analysis of pseudoaneurysm neck shows typical to-and-fro spectral pattern. This pattern results from blood flowing inward to pseudoaneurysm cavity during systole and outward to femoral artery during diastole.*

DIFFERENTIAL DIAGNOSIS

Common

- Osteoarthritis
- Inflammatory Arthropathy
- Tophi of Gout or Pseudogout

Less Common

- Pigmented Villonodular Synovitis
- Focal Nodular Synovitis
- Synovial Osteochondromatosis
- Septic Arthritis
- Lipoma Arborescens
- Amyloid
- Synovial Hemangioma
- Synovial Sarcoma
- Synovial Chondrosarcoma

Rare but Important

- Lymphoma

ESSENTIAL INFORMATION

Key Differential Diagnosis Issues

- Diagnosis of synovial mass is based on several features in addition to US characteristics
  - Clinical features
    - Age of patient, disease onset, & clinical course
  - Pattern of disease
    - Which joint or joints are affected & which part of these joints are affected
  - Radiographic features
    - Joint space narrowing, osteophytes, erosions, mineralization, juxtaarticular osteopenia
  - US characteristics
    - Intraarticular or extraarticular mass
    - Degree of synovial thickening
    - Localized or diffuse synovial thickening
    - Degree & character of joint effusion
    - Joint space narrowing, osteophytes, erosions, mineralization
    - Degree of synovial hyperemia
    - Degree of periarticular inflammation
  - Serology: Inflammatory markers, antinuclear factor, rheumatoid factor; antibodies to mutated citrullinated vimentin; antibodies to cyclic citrullinated peptide
  - Histology
    - US-guided biopsy for histology very useful to differentiate many types of synovial overgrowth
    - Cannot distinguish between synovial inflammation secondary to osteoarthritis, inflammatory arthropathy, or nontuberculous infective arthropathy

Helpful Clues for Common Diagnoses

- **Osteoarthritis**
  - Knees, hips, hands, & feet
  - Increasing age, trauma, excessive physical use, genetics, & obesity are all predisposing factors
  - Variable synovial inflammation is more common in osteoarthritis than previously thought
  - Tends to occur close to areas of pathologically damaged bone & cartilage
  - Only modest correlation between severity of osteoarthritis & degree of synovitis
    - Generally easier to quantify degree of periarticular inflammation on MR than US
- **Inflammatory Arthropathy**
  - Synovial proliferation with inflammatory arthropathy
    - Diffuse echogenic or hypoechoic synovial proliferation
  - Osteoarthritis changes (marginal osteophytosis, localized joint space narrowing, meniscal degeneration, articular debris) less pronounced
  - Mild to moderate synovial hyperemia typical
- **Tophi of Gout or Pseudogout**
  - Particularly joints of hands, feet, & knee
    - Gout → urate crystal deposition
    - Pseudogout → calcium pyrophosphate dihydrate deposition
  - Tophi are more common in gout
  - Periarticular location
    - Soft or hard tophi
    - ± synovial or intrameniscal deposition
  - Hyperechoic aggregates in synovial fluid
    - Urate sand or snowstorm appearance
  - Gout: Thin echogenic band on surface of cartilage ("urate icing")
  - Pseudogout: Thin echogenic band in midzone of articular cartilage

Helpful Clues for Less Common Diagnoses

- **Pigmented Villonodular Synovitis (PVNS)**
  - Now known as giant cell tumor of tendon sheath
  - Benign proliferative synovial disorder
    - Affects synovium of joint, bursa, or tendon sheath
  - Knee > hip > ankle > shoulder
  - Joint involvement is diffuse though may predominate in 1 or several areas of joint
  - Usually moderate to high degree of hyperemia
    - May show little or no hyperemia
  - Characteristic MR pattern on gradient echo imaging due to paramagnetic effect of hemosiderin
    - Foci of synovial hypointensity due to hemosiderin
    - Hemosiderin is present in 85% of pigmented villonodular synovitis (PVNS) cases (i.e., lack of hemosiderin does not exclude diagnosis)
- **Focal Nodular Synovitis**
  - Localized to single area of synovium
    - Although histologically identical, focal disease behaves differently from diffuse disease
    - PVNS & focal nodular synovitis (FNS) now considered different entities
    - FNS contains less hemosiderin, does not have frond-like projections & becomes more pedunculated rather than more extensive with growth
- **Synovial Osteochondromatosis**
  - Proliferation & metaplastic transformation of synovium
  - Affects any synovium, including synovium-lined joint, bursa, or tendon
    - Knee > elbow > hip > shoulder
  - Active synovial proliferation & cartilaginous metaplasia → inactive phase
    - Intrasynovial cartilage nodules detach to lie free in joint & ossify

- Nodules may consist of cartilage, cartilage/bone mixture, or mature bone with marrow fat
  - Mineralized nodules are hyperechoic; cartilage nodules are hypoechoic
  - Variable-sized joint effusion
  - ± erosion of adjacent cartilage & bone
    - Especially in less distensible joints (hip)
  - ± secondary degenerative change
- **Septic Arthritis**
  - Impossible to distinguish between tuberculous & pyogenic infection based on imaging alone
    - Phemister radiographic triad of severe periarticular osteopenia, marginal erosions, & progressive ↓ in joint space are not specific for tuberculosis
  - With chronic infection, synovial proliferation predominates with little or no synovial fluid
  - US-guided synovial biopsy to identify infection & causative organism
- **Lipoma Arborescens**
  - Proliferation of subsynovial fat → villous-like synovial swelling & proliferation
    - Probably secondary to inflammatory or traumatic stimuli
  - Usually presents around 50 years of age
    - Monoarticular with predilection for suprapatellar recess of knee
  - Hyperechoic, frond-like, synovial mass that bends & waves during joint movement on real-time scanning
  - MR appearance is more specific with hypertrophied, subsynovial fatty tissue
- **Amyloid**
  - Amyloid → abnormal accumulation of unrelated insoluble protein with tissues
  - Examples: β_2 microglobulin, immunoglobulin light chains, & serum amyloid A protein
    - Chronic hemodialysis → increased β_2 microglobulin
    - Myeloma → increased monoclonal immunoglobulin light chain fragments
    - Chronic inflammatory condition → ↑ acute-phase protein serum amyloid A
    - May also be idiopathic
  - Shoulder, hip, knee, & wrist; usually bilateral
    - Shoulder: Tendon & bursal thickening, periarticular hypoechoic nodules ± bony erosions
- **Synovial Hemangioma**
  - Strictly speaking, most are not hemangiomas
    - Hemangioma = vascular tumor of childhood with proliferative, stable, & involutionary phases
  - Most synovial hemangiomas are synovial vascular malformations
  - More common to see vascular malformation involving synovium than vascular malformation confined to synovium
  - Similar to vascular malformations elsewhere; synovial vascular malformations have predominant vascular or stromal component
  - ± phleboliths (common), ectatic vessels, venous lakes
  - Usually presents first in young adults leading to joint pain, swelling, & repeated hemarthroses
- **Synovial Sarcoma**
  - Named because of similar appearance to synovial tissue under light microscopy
    - However, does not arise from synovium
  - Arises from primitive mesenchymal cells in extraarticular soft tissues
    - < 5% of synovial sarcomas arise from within joint
  - Calcification in up to 50% of synovial sarcomas
    - ± myxoid areas, cystic areas, or hemorrhage
- **Synovial Chondrosarcoma**
  - May occur de novo or in conjunction with synovial chondromatosis
  - Main feature in favor of synovial chondrosarcoma is bony infiltration

Helpful Clues for Rare Diagnoses

- **Lymphoma**
  - Irregular synovial proliferation
  - Focal irregular cortical osteolysis
  - Regional lymphoma-type adenopathy

Osteoarthritis

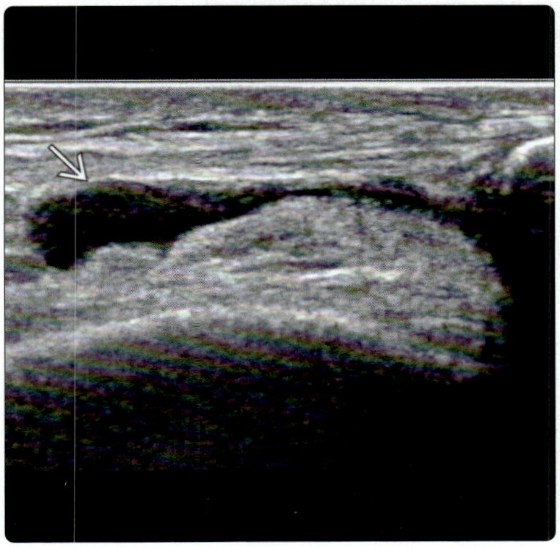

Osteoarthritis

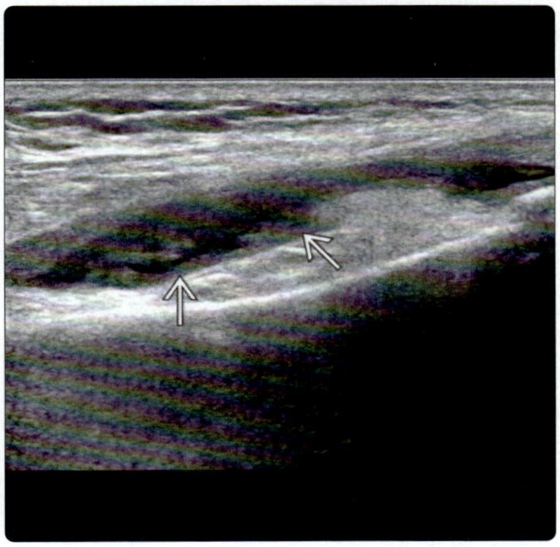

(Left) *Transverse US shows mild synovial thickening ➡ of the medial patellar recess. This, along with a small joint effusion, is a common finding in osteoarthritis.* **(Right)** *Transverse US shows moderate suprapatellar synovitis ➡ with osteoarthritis. Other signs, besides mild to moderate synovitis, are needed to distinguish inflammatory osteoarthritis from inflammatory arthropathy, such as rheumatoid arthritis (RA).*

(Left) *Transverse US shows a small effusion with moderate-severity synovitis ➡ of the lateral patellar recess in a patient with RA. There are some small echogenic foci ➡ present due to crystal aggregates. This crystal deposition, which can be seen in either osteoarthritis or RA, is likely to be aggravating the inflammation.* **(Right)** *Transverse color Doppler US shows moderate synovial hyperemia. The degree of hyperemia is a pointer to the level of synovial inflammatory activity.*

Inflammatory Arthropathy

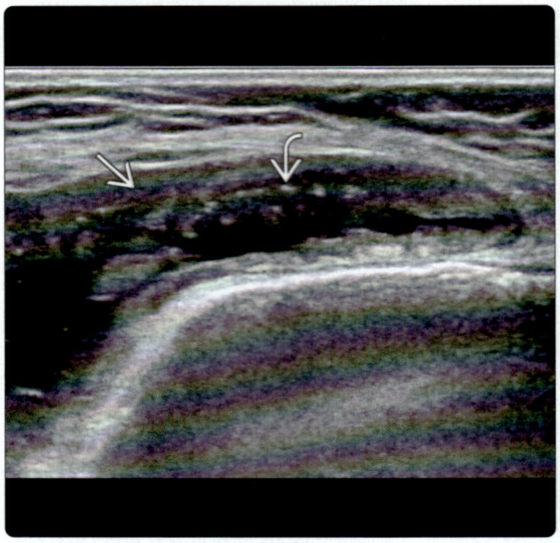

Inflammatory Arthropathy

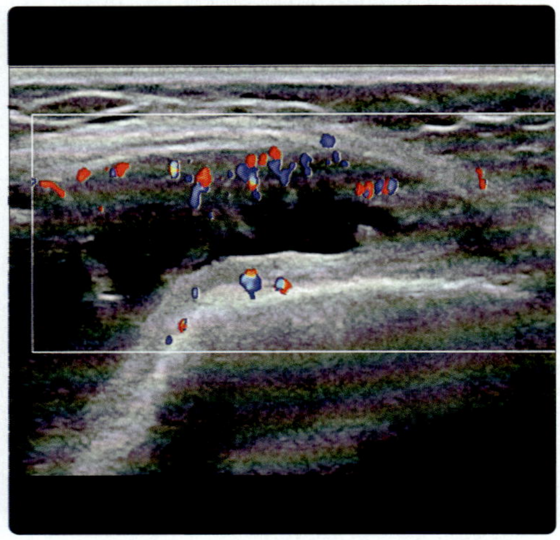

(Left) *Longitudinal US of the medial hindfoot shows a large, hypoechoic mass ➡ centered on the talocalcaneal articulation ➡. Biopsy confirmed pigmented villonodular synovitis (PVNS).* **(Right)** *Longitudinal power Doppler US of the same patient shows the mass ➡ to be highly vascular. This level of hyperemia favors PVNS although not all PVNS tumors are so vascular.*

Pigmented Villonodular Synovitis

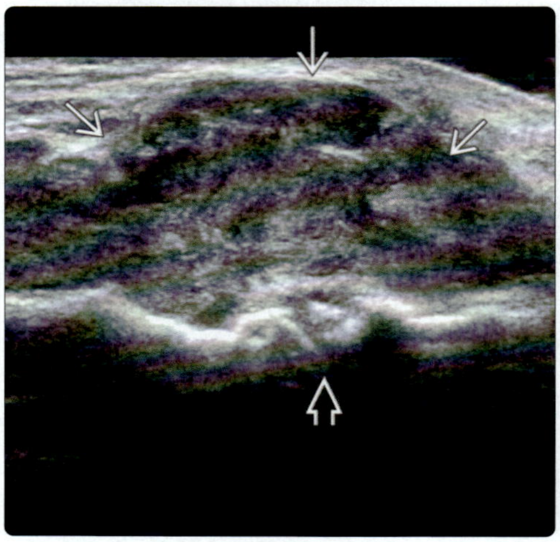

Pigmented Villonodular Synovitis

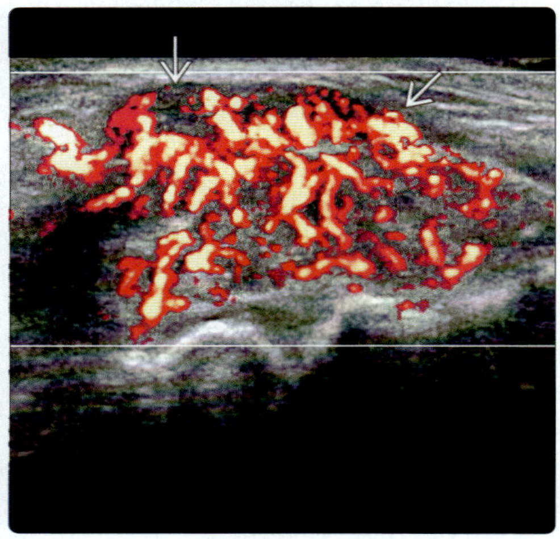

(Left) *Longitudinal US shows a large mass ➡ arising from the lateral aspect of the subtalar joint ➡. This tumor is hyperechoic ➡ in parts due to intratumoral hemorrhage. The mass was moderately hyperemic (not shown).* **(Right)** *Transverse US shows markedly thickened hypoechoic synovial thickening ➡ of the medial patellar recess of the knee. US-guided biopsy revealed PVNS. The hypoechoic nature of the synovium is fairly characteristic of PVNS.*

Pigmented Villonodular Synovitis

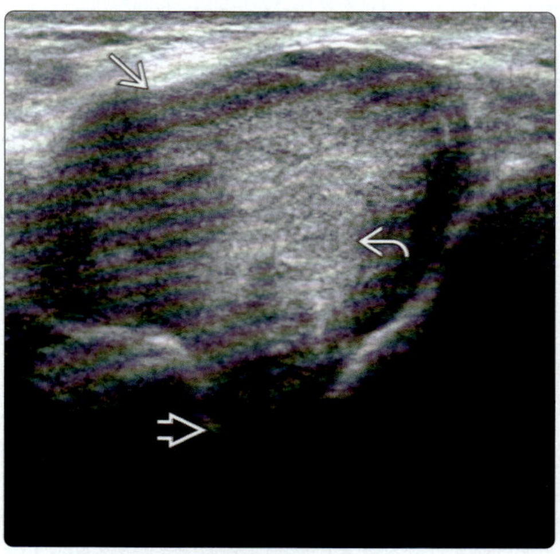

Pigmented Villonodular Synovitis

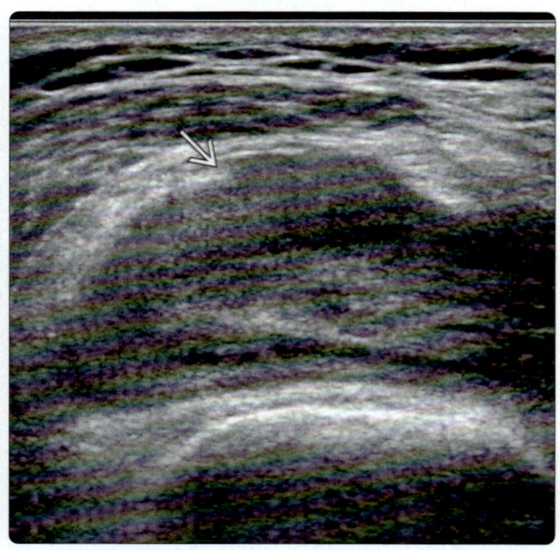

Focal Nodular Synovitis

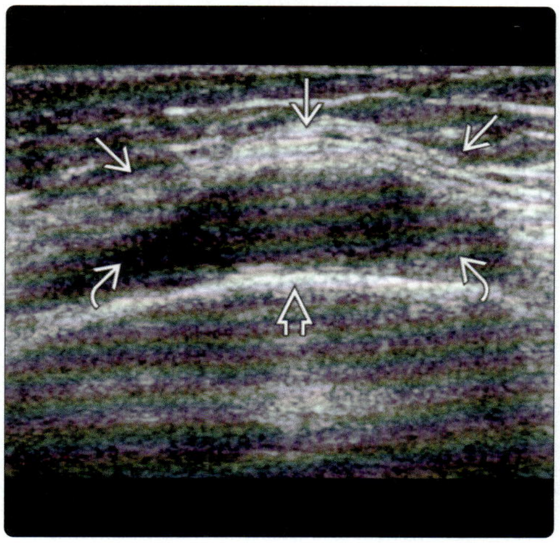

Focal Nodular Synovitis

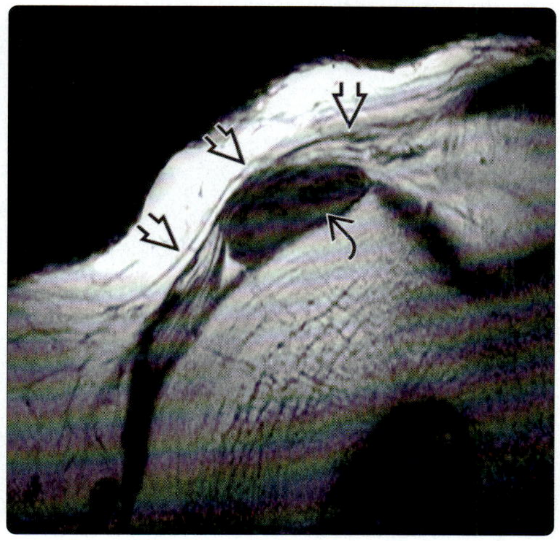

(Left) *Transverse US of the knee shows a fusiform-shaped, hypoechoic mass between the lateral retinaculum and the femoral cortex. The mass was mildly hyperemic on color Doppler imaging. The appearance suggests focal nodular synovitis.* **(Right)** *Axial T2WI MR of the same mass shows a focal, heterogeneous, slightly low signal intensity mass between the lateral retinaculum and the femoral cortex, consistent with nodular synovitis, which was subsequently confirmed histologically.*

Synovial Osteochondromatosis

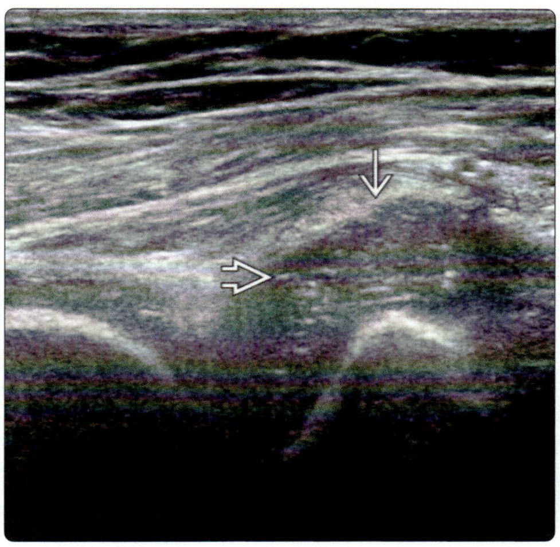

Synovial Osteochondromatosis

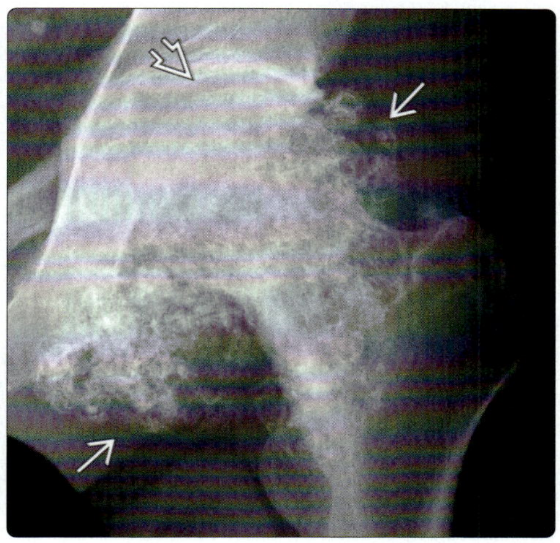

(Left) *Longitudinal US of knee shows thickened synovial-like tissue containing small echogenic calcific-like foci anteromedially, suggestive of synovial osteochondromatosis. Radiograph showed no abnormal calcification. US-guided biopsy for histology revealed synovial chondromatosis.* **(Right)** *Frontal radiograph shows a hip capsule distended with multiple calcific nodules due to synovial osteochondromatosis. Mild to moderate secondary osteoarthritis is present.*

Septic Arthritis

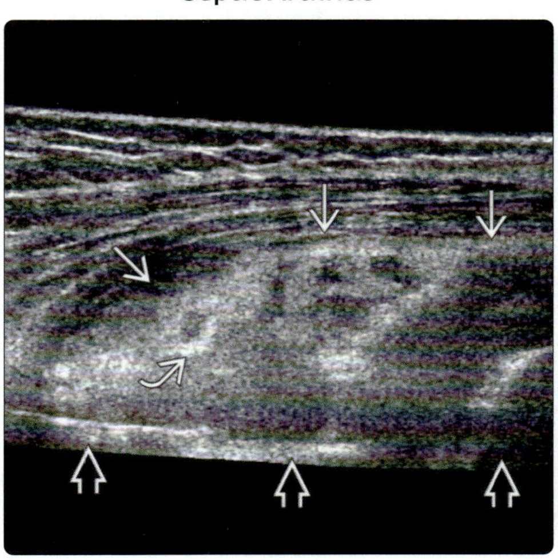

Septic Arthritis

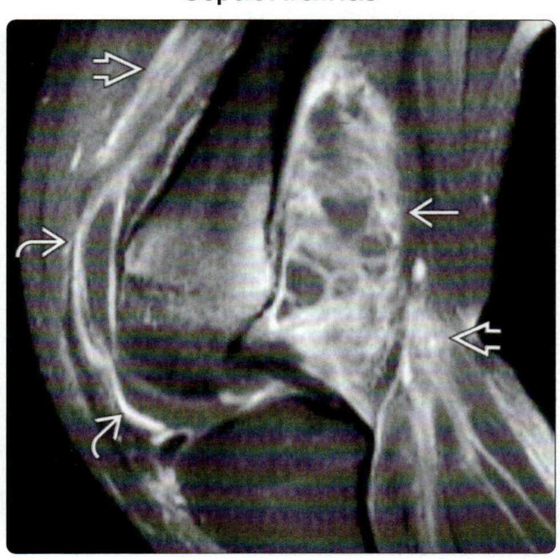

(Left) *Longitudinal US of the knee shows a large, heterogeneous mass with small cystic areas posterior to the distal femur, initially considered to be a soft tissue tumor. Mild hyperemia was present. Biopsy specimen grew Haemophilus influenzae.* **(Right)** *Sagittal T1WI C+ MR in the same patient shows an enhancing mass with synovial enhancement throughout the knee and periarticular inflammation. The inflammatory mass resolved with treatment.*

(Left) *Transverse US shows a severely thickened synovium and capsule of the lateral patellar recess with no effusion. Mild hyperemia was present (not shown). Biopsy revealed tuberculous infection. The femoral condyle is normal.* **(Right)** *Transverse color Doppler US of the knee shows moderate to severe synovial thickening of the medial patellar recess with moderate hyperemia. This patient was known to suffer from disseminated tuberculosis, so no synovial biopsy was performed.*

Septic Arthritis

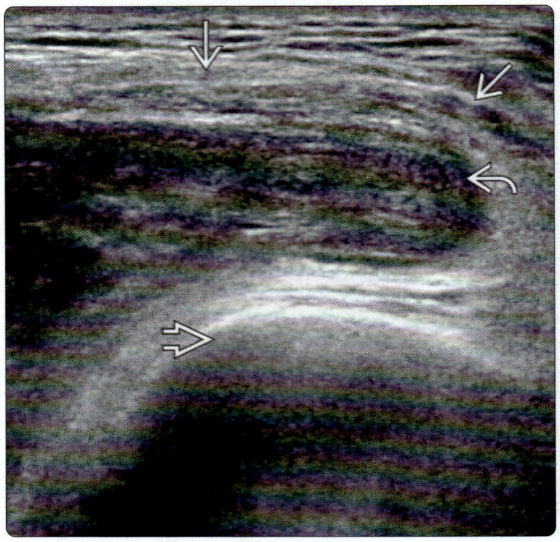

Septic Arthritis

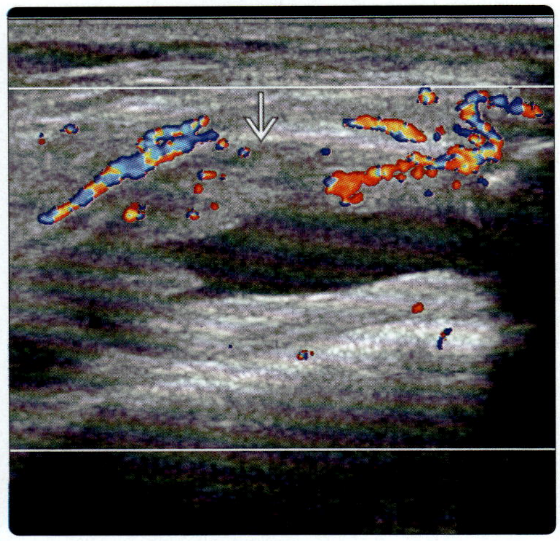

(Left) *Longitudinal US of a patient with chronic knee swelling shows a large effusion with severe frond-like hyperechoic synovial proliferation on the posterior aspect of the suprapatellar pouch. No free nodules are present.* **(Right)** *Axial T1WI MR of the same lesion shows frond-like synovial proliferation involving much of the suprapatellar pouch. These fronds have a similar intensity to subcutaneous fat and are suppressed on fat-suppressed imaging.*

Lipoma Arborescens

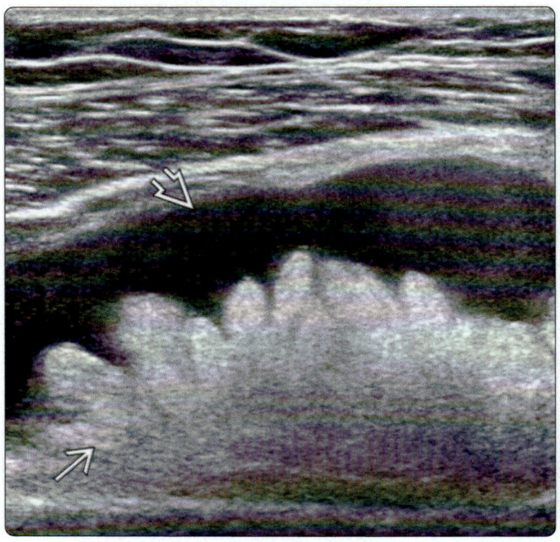

Lipoma Arborescens

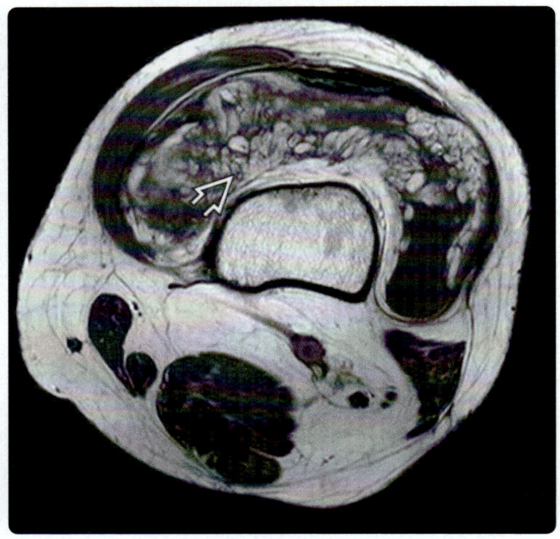

(Left) *Transverse US shows a well-defined, hypoechoic, lobulated, peritendinous mass on the medial side of the ankle in a patient on hemodialysis. Biopsy specimen revealed amyloid.* **(Right)** *Axial T2WI MR with fat suppression shows a venous vascular malformation of the knee, particularly involving the medial retinaculum and extending to the prepatellar region. There is a small joint effusion. This is consistent with synovial hemangioma.*

Amyloid

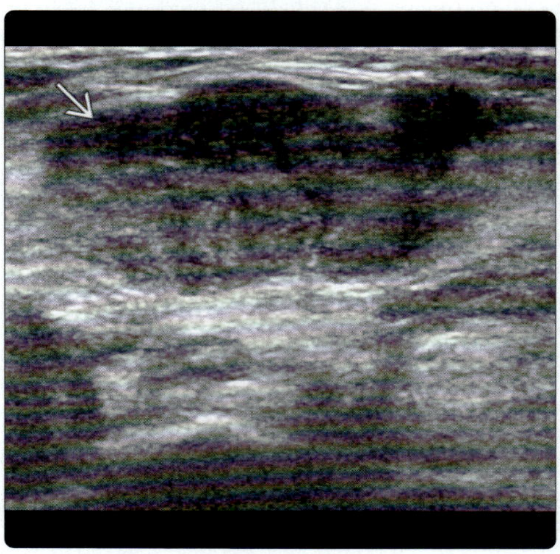

Synovial Hemangioma

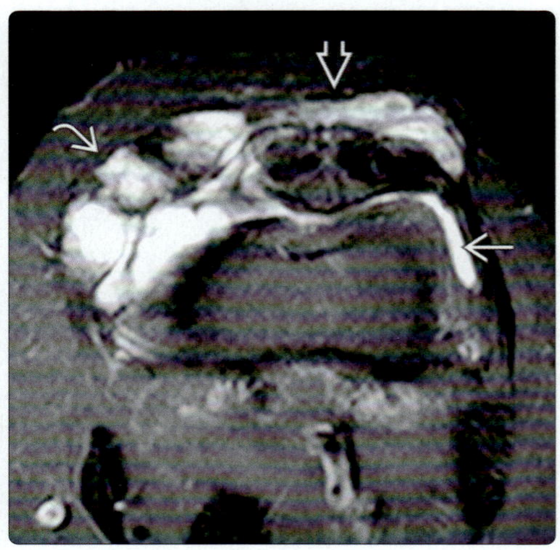

Synovial Sarcoma

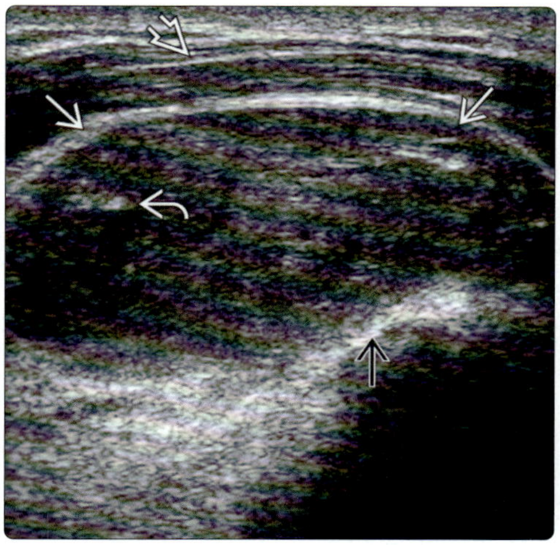

Synovial Sarcoma

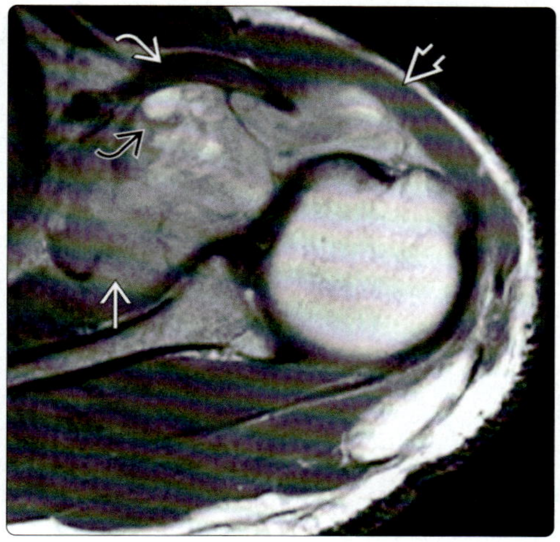

(Left) *Transverse US of the shoulder shows a large, hypoechoic mass with small echogenic areas , which may represent hemorrhage, deep to the deltoid muscle . Note the humerus .* **(Right)** *Axial T2WI MR of the shoulder in the same patient shows a mildly hyperintense mass with small cyst-like areas , deep to the pectoralis and deltoid muscles. Biopsy specimen confirmed synovial sarcoma.*

Synovial Chondrosarcoma

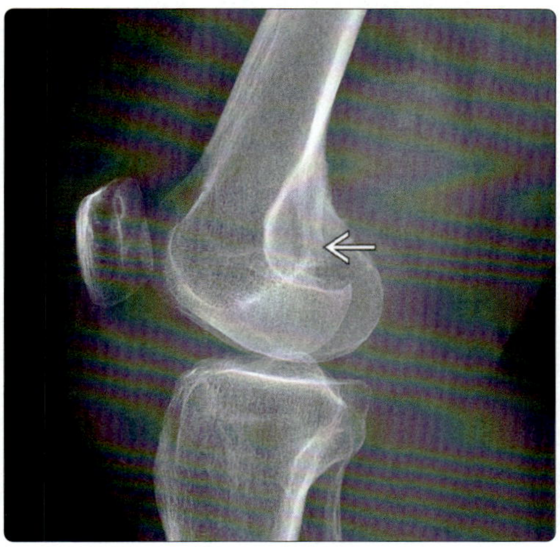

Synovial Chondrosarcoma

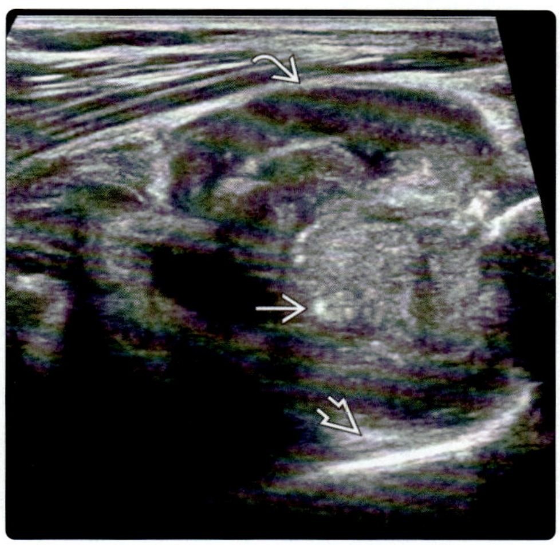

(Left) *Lateral radiograph of the knee in a patient with knee swelling for 20 years shows severe erosion of the posterior aspect of the lateral femoral condyle and supracondylar area .* **(Right)** *Longitudinal US of the same knee shows a large, lobulated mass on the posterior aspect of the knee joint with posterior bone scalloping . Some small calcific-type foci are present . Histology following US-guided biopsy revealed synovial chondrosarcoma.*

Lymphoma

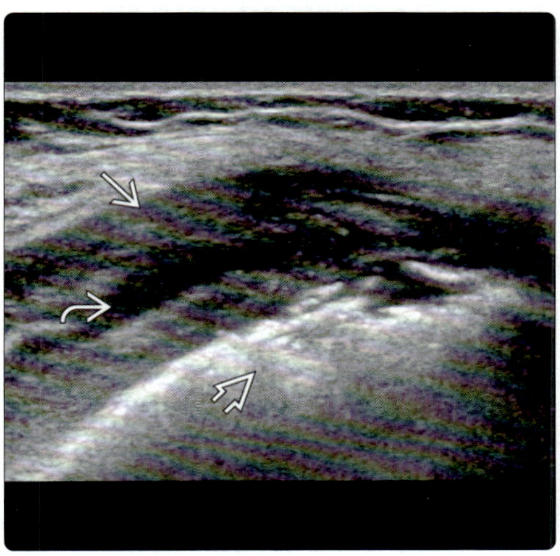

Lymphoma

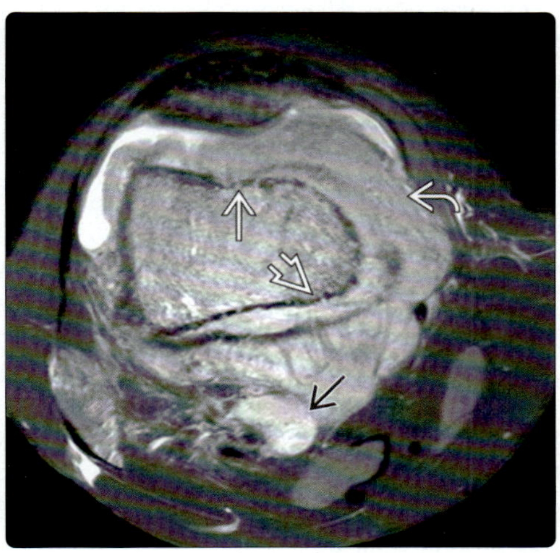

(Left) *Transverse US shows severe synovial thickening with a small knee joint effusion and cortical resorption of the medial femoral condyle. Tuberculosis infection was initially suspected.* **(Right)** *Axial T2WI FS MR of the same patient shows the mass extending from the medullary canal of the distal femur with cortical resorption , periosteal thickening , and synovial thickening . There is an enlarged popliteal lymph node . Synovial biopsy revealed B-cell lymphoma.*

Differential Diagnoses

DIFFERENTIAL DIAGNOSIS

Common

- Osteoarthritis

Less Common

- Inflammatory Arthropathy
- Crystal Arthropathy
- Infective Arthritis
- Hemarthrosis
- Lipohemarthrosis
- Pigmented Villonodular Synovitis
- Synovial Osteochondromatosis

ESSENTIAL INFORMATION

Key Differential Diagnosis Issues

- Characterization of underlying arthropathy relies on disease pattern rather than single US characteristic
  - Speed of onset, joints affected, severity of disease, age of patient, associated features, and radiographic and US appearances
  - Oligoarticular or polyarticular location is feature of both inflammatory or degenerative arthropathy
- US is helpful in assessing disease activity and characterization
- Amount and echogenicity of fluid are useful to note but lack specificity
  - Amount and appearance of joint fluid in osteoarthritis (OA), inflammatory arthropathy, repeated hemarthrosis, and septic arthritis may appear very similar
    - Distinguish among these entities by noting concurrent US features, history, and clinical presentation
    - Check especially for crystal aggregates, which are indicative of crystal arthropathy
    - Crystal arthropathy can coexist with OA, infective arthritis, and other causes of joint effusion
- If joint fluid is hyperechoic, keep joint still for several minutes
  - Facilitates sedimentation of contents, which may help diagnosis
    - If 2 layered, more likely hemorrhagic; infective fluid does not sediment readily
    - If 3 layered, diagnostic of lipohemarthrosis
  - US-guided fluid aspiration is often necessary and very helpful
    - Specimen for crystals needs to be sent fresh since saline or alcohol will dissolve crystals

Helpful Clues for Common Diagnoses

- **Osteoarthritis**
  - Most commonly affects knee, hip, interphalangeal, 1st carpometacarpal, triscaphe, wrist, acromioclavicular, and tarsal joints
  - Joint space narrowing, marginal osteophytosis, capsular swelling, synovitis, and effusion are US signs of OA
  - Variable increase in joint fluid
  - Variable degree of reactive-type inflammatory synovial proliferation
  - ± variable periarticular inflammation
    - Inflammatory component is important to note because this determines response to antiinflammatory medication
  - US is increasingly used to determine inflammatory component of OA and differentiate OA from crystal deposition disease or inflammatory arthropathy
  - Inflammatory element of OA is characterized by
    - Capsular thickening, joint effusion, synovial proliferation, synovial and pericapsular hyperemia, edema of periarticular soft tissues

Helpful Clues for Less Common Diagnoses

- **Inflammatory Arthropathy**
  - Affects any synovial-lined joint
    - Often bilateral and symmetrical
  - Ranges from single (monoarthropathy) to few (oligoarthropathy) to multiple (polyarthropathy) joint involvement
  - Most common finding: Pannus
    - Diffusely thickened hyperechoic or hypoechoic synovium
    - Frond-like or nodular synovial proliferation
    - May resemble echogenic or hypoechoic synovial fluid
  - Effusion may be anechoic or hyperechoic
  - ± echogenic speckles due to precipitated fibrin or inflammatory debris
  - ± marginal erosions
    - Well-defined, small, periarticular defects
    - Coalesce to form larger erosions and more generalized marginal irregularity of joint
    - US is more sensitive than radiography and MR at detecting small joint periarticular erosions
    - US is less sensitive than contrast-enhanced MR
  - Color Doppler imaging
    - Helpful for estimating level of disease activity and distinguishing synovium from joint fluid
    - Hyperemia is more easily appreciated in smaller joints (e.g., hand, wrist, elbow) than larger joints (e.g., shoulder, hip)
    - Active pannus leads to increased color flow
    - Hyperemia is graded semiquantitatively as high, intermediate, or low/absent vascularity
    - Quantitative analysis of hyperemia is feasible, though semiquantitative analysis is adequate for clinical use
  - Other signs include joint space narrowing, subluxation, deformity, and ankylosis
  - Coexistent soft tissue features include tenosynovitis, bursitis, and entrapment neuropathy
- **Crystal Arthropathy**
  - Always consider in any acute arthritis
  - Effusion can be anechoic or hyperechoic
  - Gout results from urate crystal deposition
    - 1st metatarsophalangeal joint is 1st affected joint in 50%; later becomes polyarticular
    - Best diagnostic clue: Identification of echogenic foci with comet-tail artifacts within and around joint
    - ± urate sand (fine punctuate echoes) or sandstorm (larger echogenic aggregates) appearance
    - ± urate deposition on articular cartilage surface as thin echogenic band ("urate icing")

- Soft tophi (no acoustic shadowing) → hard tophi (dense acoustic shadowing)
  - Pseudogout associated with calcium pyrophosphate dihydrate deposition
    - Knee, wrist, scaphotrapezium articulation
    - More often associated with osteoarthritis
    - Crystal deposition in midzone of articular cartilage → more prolific cartilage loss
    - ± chondrocalcinosis of fibrocartilage (menisci, articular disc of triangular fibrocartilaginous complex, glenoid or acetabular labrum)
    - Tophi are less common in pseudogout
- **Infective Arthritis**
  - Gradual-onset joint swelling over 1-2 days
  - Very tender with marked limitation of movement
  - Usually hematogenous spread
  - Joint effusion
    - Cardinal sign of septic arthritis
    - Swirling or moving echoes on real-time imaging
    - Anechoic joint fluid does not exclude septic arthritis
  - For most joints, no effusion equates to no septic arthritis
  - 2 exceptions to rule
    - For nondistensible diarthrodial joints, such as acromioclavicular joint, effusion is minimal
    - If joint capsule is not intact following rupture or recent surgery, no effusion will develop
  - Echogenic fluid due to aggregates of inflammatory debris
  - Findings in more chronic infections
    - Synovial proliferation and capsular thickening
    - Viscosity of joint fluid increases but amount of joint fluid decreases
    - Progressive joint space narrowing (chondrolysis)
    - Marginal erosions (osteolysis), periarticular soft tissue extension
  - US-guided fluid aspiration for Gram stain, culture, and cytology is helpful
  - Ultimate diagnosis rests on isolating pathogens from aspirated joint fluid
- **Hemarthrosis**
  - History of trauma ± associated fracture or known coagulopathy
  - Initially, finely speckled echogenic joint fluid
  - After ~ 5 minutes, shows more or less 2-layered appearance of joint effusion
  - Subacute stage
    - Blood clots → discrete echogenic foci within joint
    - Joint fluid is largely anechoic ± fibrinous adhesions
  - Repeated hemarthroses
    - Variable synovial thickening
      - Common and severe in hemophiliac patients with repeated hemarthroses
      - Synovial hyperemia not usually apparent in hemophiliac arthropathy
    - ± bone overgrowth and secondary osteoarthritis
    - ± progressive joint disorganization
    - ± pseudotumor (expanding hematoma)
- **Lipohemarthrosis**
  - Leakage of marrow fat into joint cavity
    - Presence of lipohemarthrosis implies fracture communicating with joint
  - Initially, finely speckled echogenic joint fluid
  - After 10 minutes → 3-layered appearance of effusion
  - 3 layers that become evident are
    - Fat layer: Least dependent layer (fat)
    - Serum layer: Middle layer (supernatant serum)
    - Cellular layer: Most dependent layer (red blood cell aggregates)
  - ± hypoechoic areas within floating fat due to trapped blood
- **Pigmented Villonodular Synovitis**
  - Diffusely thickened hypoechoic synovium
  - Variable hyperemia ± hemorrhagic joint effusion
- **Synovial Osteochondromatosis**
  - Diffuse synovial proliferation ± small echogenic foci (ossification)
  - Variably echogenic nodules within effusion

Osteoarthritis

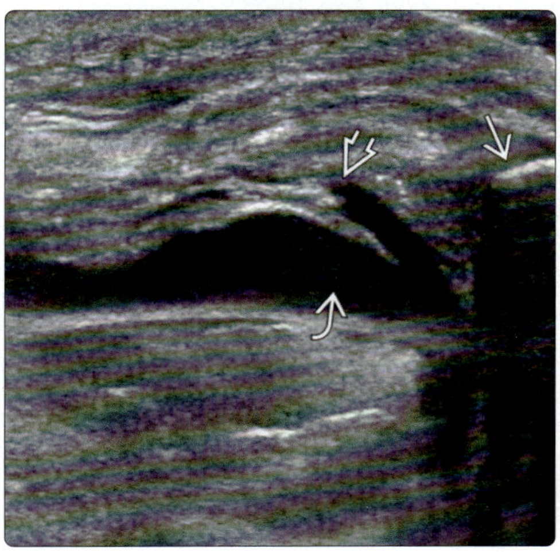

Osteoarthritis

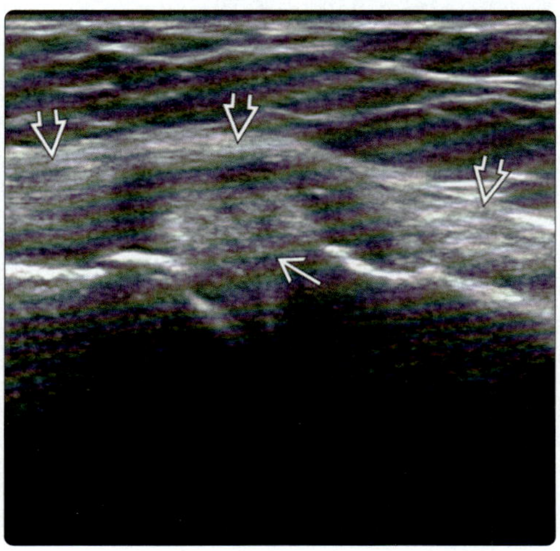

(Left) *Transverse US of the knee shows a mildly distended medial patellar recess ➡ with moderate synovial proliferation extending into a synovial plica ➡. The patella ➡ is normal.* **(Right)** *Longitudinal US shows moderate extrusion of the medial meniscal body ➡, indicative of meniscal degeneration with joint space narrowing due to osteoarthritis. The medial collateral ligament ➡ is displaced.*

(Left) *Transverse US shows both marked echogenic and hypoechoic synovial proliferation distending the medial knee recess in a patient with rheumatoid arthritis. This is not specific for inflammatory arthropathy and may be seen with infection or repeated hemarthrosis. Minimal, if any, synovial fluid is present. Note the distal femur.* **(Right)** *Longitudinal US shows severe synovial proliferation with moderate hyperemia in the suprapatellar pouch of the knee.*

Inflammatory Arthropathy

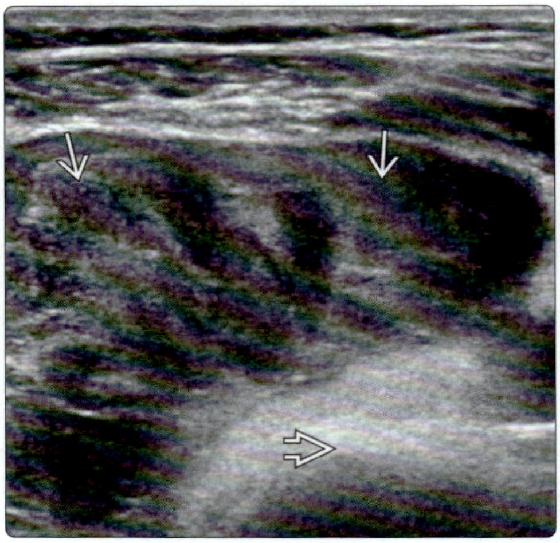

Inflammatory Arthropathy

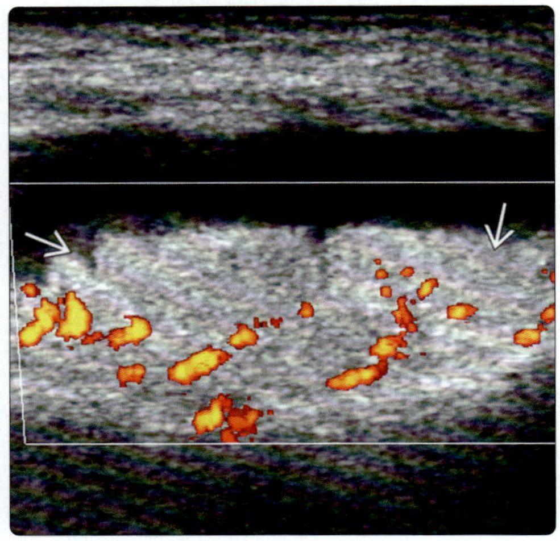

(Left) *Longitudinal US shows an echogenic midzonal band of articular cartilage calcification of the femoral condyle posteriorly, indicative of calcium pyrophosphate deposition. Other images (not shown) showed meniscal chondrocalcinosis.* **(Right)** *Transverse US shows gouty arthritis of the glenohumeral joint. There is a small joint effusion with small echogenic crystal deposits on the humeral articular cartilage surface ("urate icing").*

Crystal Arthropathy

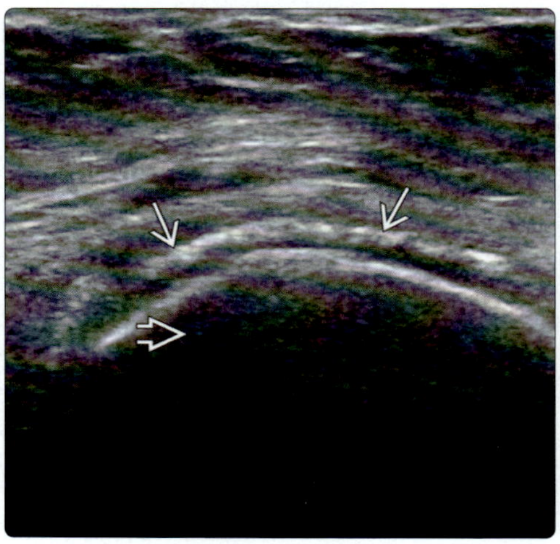

Crystal Arthropathy

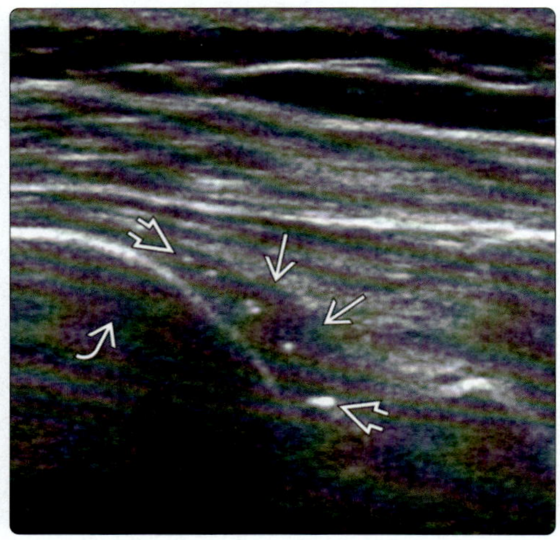

(Left) *Longitudinal US shows an echogenic effusion ("urate sand") of the 1st metatarsophalangeal joint in a patient with chronic gout. There is also a more well-defined intracapsular soft tophus.* **(Right)** *Longitudinal US shows partially hypoechoic, partially hyperechoic fluid distending the ankle joint and displacing intracapsular extrasynovial fat. Aspiration yielded pus, which grew Staphylococcus aureus. The talar dome and distal tibia are normal.*

Crystal Arthropathy

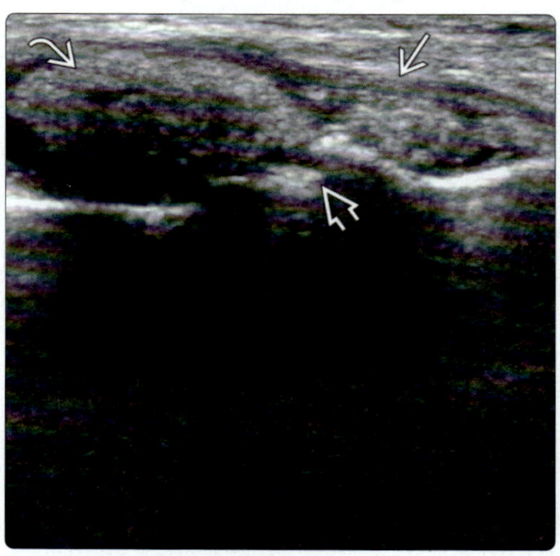

Infective Arthritis

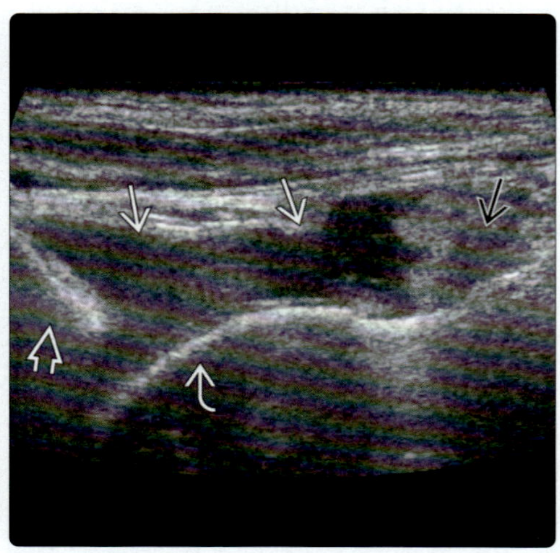

Infective Arthritis

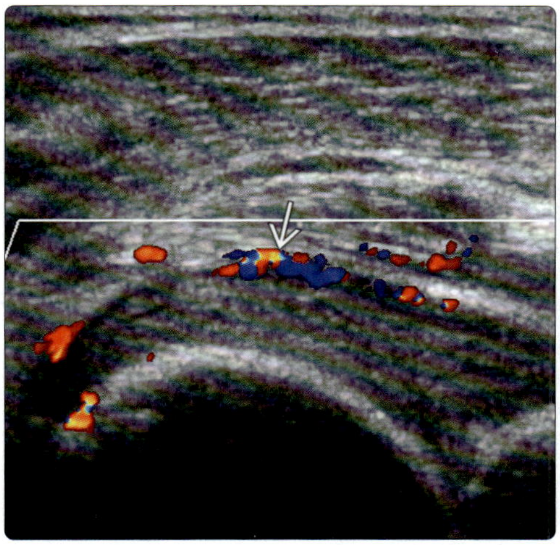

Infective Arthritis

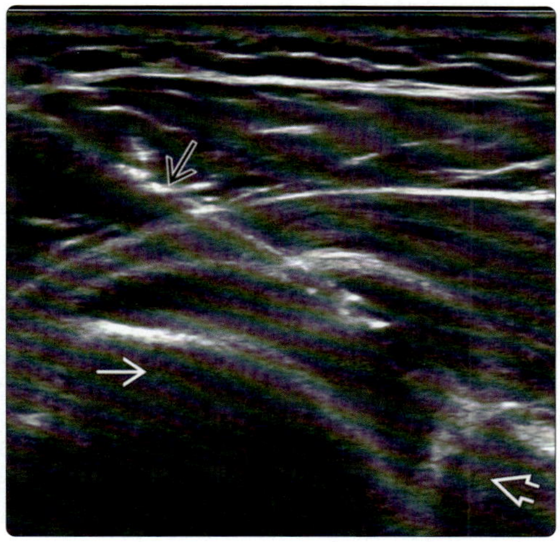

(Left) *Transverse color Doppler US shows moderate peripheral hyperemia* ➡ *of the thickened synovium in patient with chronic septic arthritis.* **(Right)** *Transverse US shows the posterior aspect of the glenohumeral joint, which is distended by synovial proliferation due to acute septic arthritis. The humeral head* ➡ *and glenoid* ➡ *are normal. Needle aspiration* ➡ *was performed, which yielded Staphylococcus aureus.*

Hemarthrosis

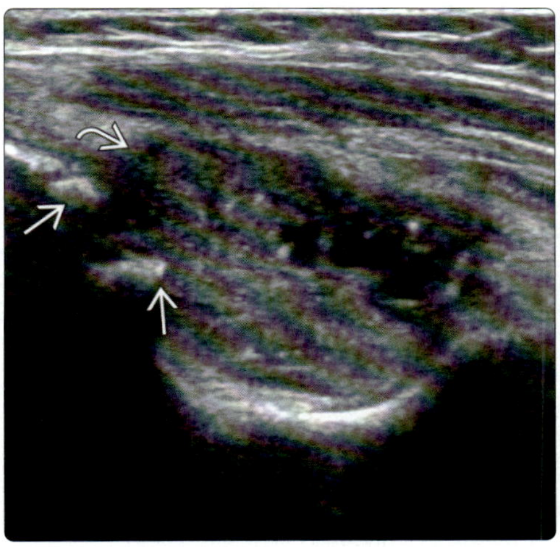

Lipohemarthrosis

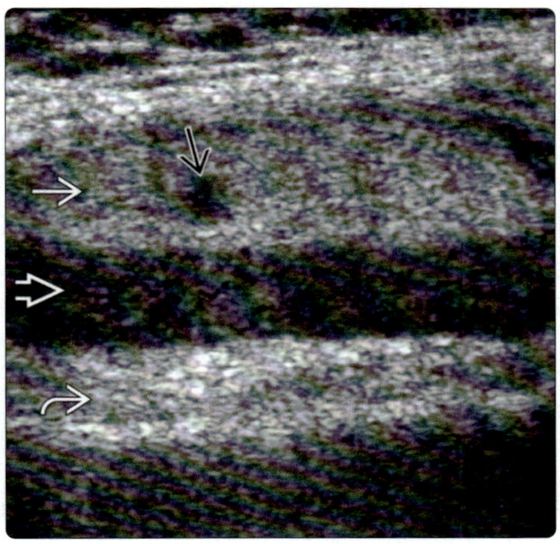

(Left) *Transverse US shows echogenic fluid distending the olecranon fossa of the elbow joint* ➡ *due to acute hemarthrosis in a patient with hemophilia. There is also secondary osteoarthrosis with marginal osteophytosis* ➡*.* **(Right)** *Transverse US of the suprapatellar pouch shows a lipohemarthrosis with fat* ➡*, serum* ➡*, and cellular* ➡ *layers. The hypoechoic area* ➡ *in the fat layer is due to the entrapped blood. This patient had a fracture of the tibial plateau (not shown).*

Lipohemarthrosis

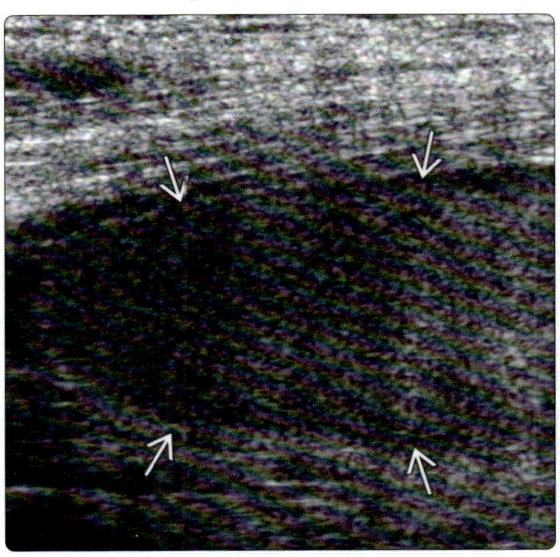

Lipohemarthrosis

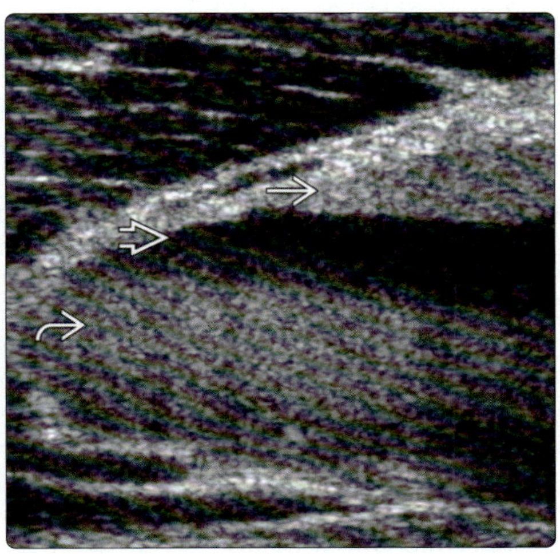

(Left) *Longitudinal US of the knee in a patient with a tibial plateau fracture shows marked distension of the suprapatellar pouch* ➡ *with echogenic fluid.* **(Right)** *Longitudinal US at the same location, after keeping the joint still for 15 minutes, shows separation of the fluid contents into fatty* ➡*, serous* ➡*, and cellular* ➡ *components. Lipohemarthrosis implies communication between joint and marrow cavity and is nearly always due to a bone fracture.*

DIFFERENTIAL DIAGNOSIS

Common

- Acute Rib Fracture
- Healing Rib Fracture
- Rib Metastases

Less Common

- Chest Wall Metastases
- Hematoma
- Lipoma
- Elastofibroma Dorsi
- Fibroma
- Carcinoma, Male Breast
- Gynecomastia
- Accessory Breast
- Prominent Xiphoid Process
- Prominent or Floating Rib
- Costochondritis
- Accessory Muscle
- Lymphatic Malformation
- Vascular Malformation
- Stress Rib Fracture
- Abscess
- Sternoclavicular Joint Arthropathy
- Thyroglossal Duct Cyst

ESSENTIAL INFORMATION

Key Differential Diagnosis Issues

- US is very useful for chest wall lesions as structures are superficial
  - Retroscapular region is only area not accessible
- For rib fractures, examine around painful area only
- Ask patient to place finger over most painful area
  - Caveat: US is less helpful in unconscious, uncooperative patients and young children
- No need to examine all ribs

Helpful Clues for Common Diagnoses

- **Acute Rib Fracture**
  - History of trauma with localized tenderness
  - Sharp break in superficial bone cortex
    - ± displacement or angulation
    - ± adjacent soft tissue swelling or hypoechoic hematoma
    - Wide break in cortex or cortical irregularity suggests pathologic fracture
  - Typically, fractures are vertically aligned to each other
    - Check adjacent ribs alongside rib fracture
  - US will detect 10x more rib fractures than radiography
    - Also able to accurately detect fractures of costal cartilage or costochondral junction
  - US is more accurate than radiography at detecting pneumothorax
    - Pneumothorax seen as loss of normal to-and-fro respiratory motion at pleural surface on real-time scanning over least dependent portion of hemithorax
- **Healing Rib Fracture**
  - Fracture margin is less sharp and more rounded
  - Progressive bridging callus
    - Initially hypoechoic mass alongside fracture
    - Progressive ossification of callus with ↑ echogenicity and ↑ acoustic shadowing
    - After ~ 2 months, only ossified cortical bump remains without visible fracture line
    - Cortical contour will take from months to years to remodel to normal
  - US can accurately discriminate between acute and nonacute rib fracture
- **Rib Metastases**
  - Common site of skeletal metastases
  - Lung, kidney, hepatocellular carcinoma, and bowel are all common primary sites
  - Usually multiple
    - Randomly located as opposed to linear vertical location of rib fractures
  - Can be detected only if cortex is disrupted
    - Metastases usually originate in medullary canal prior to infiltrating cortices
      - US has low sensitivity for detecting early rib metastases
  - If cortex disrupted, US accurate at distinguishing metastatic from traumatic fracture
  - Bone destruction
    - Irregular thinning or loss of echogenic superficial cortex of rib
    - ± soft tissue mass at site of bony destruction

Helpful Clues for Less Common Diagnoses

- **Chest Wall Metastases**
  - Feature of advanced malignancy
  - Muscle metastases relatively uncommon, possibly due to
    - Destruction of tumor cells by muscle motion
    - Muscle pH inhospitable to tumor cells
    - Muscle limits accumulation of tumoral lactic acid that induces tumor neovascularity in other tissues
- **Hematoma**
  - Due to trauma, anticoagulation, strenuous and prolonged coughing, vascular malformation
  - Often within pectoralis or intercostal muscles
  - Mildly echogenic mass ± concentric layering liquefaction ± marginal hyperemia
- **Lipoma**
  - Similar appearances to lipoma elsewhere
  - Hypoechoic, well-defined soft tissue mass with fine echogenic striations
- **Elastofibroma Dorsi**
  - Uncommon; benign, slow-growing, connective tissue tumor
    - Subscapular area (99%); more common in elderly and in females (60%); frequently bilateral
    - More common in persons who perform manual labor
    - Tumor composed of abnormal elastic fibers
    - Generally regarded as fibroblastic reactive process or pseudotumor
  - Large, well-circumscribed tumor
  - Alternating bands of hypoechoic (fat) and hyperechoic (fibrous) tissue
  - Usually only minimal hyperemia
    - Can occasionally be difficult to distinguish from surrounding muscle

- **Fibroma**
  - Hypoechoic subcutaneous nodule with well-defined or ill-defined border
- **Carcinoma, Male Breast**
  - Multilobulated cystic pattern more common than mixed cystic-solid pattern
  - Ill-defined more common than well-defined margin
  - Mildly hypervascular, mainly peripheral
- **Gynecomastia**
  - Growth of glandular tissue in male breast
  - Pseudogynecomastia = ↑ fat in male breast
  - Most cases are idiopathic
  - Identifiable causes include
    - Congenital anorchia, Klinefelter syndrome
    - Hermaphroditism, adrenal carcinoma
    - Liver disorders, drugs (spironolactone, cimetidine, estrogen derivatives)
  - Hypoechoic glandular hyperplasia in retroareolar region
    - Branches out in triangular fashion beyond retroareolar region
    - Progressive glandular proliferation → more diffuse echogenic fibrotic proliferation
- **Accessory Breast**
  - Extra breasts may appear ± nipples or areolae
  - Presence of breast tissue outside normal breast contour
  - Mass anywhere along course of embryologic mammary streak (axilla to inguinal region)
- **Prominent Xiphoid Process**
  - Cartilaginous lower end of sternum, which points more forward than usual
    - If large, may be felt as palpable mass
- **Prominent or Floating Rib**
  - Prominent rib lies more superficial than the rest
  - Patient may feel lump at end of floating rib, mainly 11th or 12th ribs
- **Costochondritis**
  - Swelling of chondral part of rib ± adjacent inflammatory change
  - Transducer probe tenderness
- **Accessory Muscle**
  - Pectoralis major muscle: Many variants
    - Aplasia (Poland syndrome), hypoplasia, pectoralis minimus, pectoralis quartus, pectoralis intermedius, pectoralis anomalous
  - Sternalis: Strap-like accessory muscle along sternocleidomastoid-rectus abdominis line
    - Arises from sternum so overlies medial side of pectoralis major
- **Lymphatic Malformation**
  - Unilocular/multilocular anechoic compressible cysts with thin walls and intervening septa without vascularity
  - Hemorrhage or infection leads to irregular walls, internal debris with fluid-fluid level and thick septa
- **Vascular Malformation**
  - Hypoechoic with heterogeneous echo pattern with dilated channels ± phleboliths
  - Characterized as slow flow or high flow
    - Capillary, venocapillary, venous, or arteriovenous
  - Slow-moving (to-and-fro) blood with red cell aggregates better seen on grayscale US
- **Stress Rib Fracture**
  - US more sensitive than radiography
  - Point specific pain aggravated by exercise
  - Rowers, some track and field athletes, and golfers
- **Abscess**
  - Cystic lesion with posterior acoustic enhancement
  - Adjacent inflammatory change
- **Sternoclavicular Joint Arthropathy**
  - Hypoechoic capsular thickening
  - Check for bone erosion, subluxation, crystals, capsular hyperemia, and pericapsular inflammation or abscess
  - CT usually helpful following initial US assessment
    - Osteoarthritis of sternoclavicular joints is more common than inflammatory, which, in turn, is more common than crystal arthritis or septic arthritis
- **Thyroglossal Duct Cyst**
  - Anywhere along path of thyroglossal duct from tongue base to suprasternal notch
    - Persistence of thyroglossal duct
    - Usually presents in childhood as midline neck mass

Acute Rib Fracture

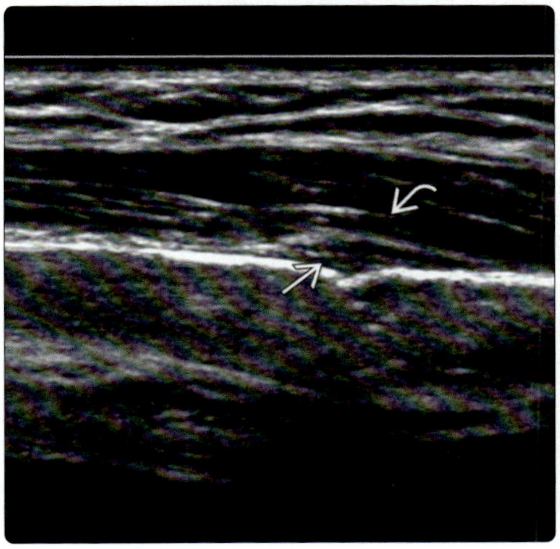

Healing Rib Fracture

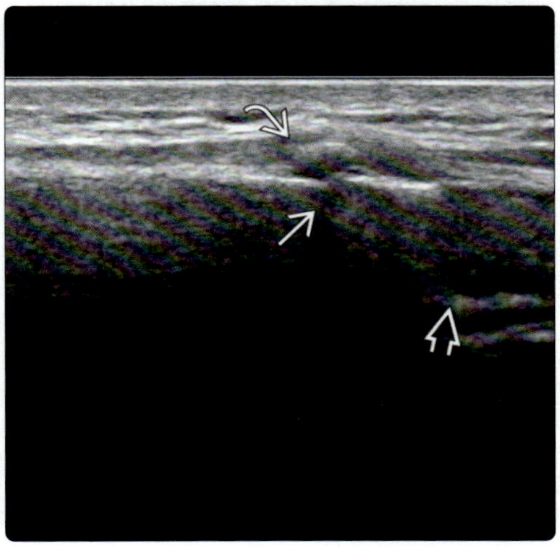

(Left) *Oblique longitudinal US shows an undisplaced acute traumatic fracture ➡ of the echogenic anterior rib cortex with a small adjacent hematoma ➡.* **(Right)** *Longitudinal US shows early healing of anterior undisplaced rib fracture ➡ with echogenic mineralized callus ➡ surrounded by hypoechoic granulation tissue. The fracture is near the costochondral junction ➡.*

Stress Rib Fracture

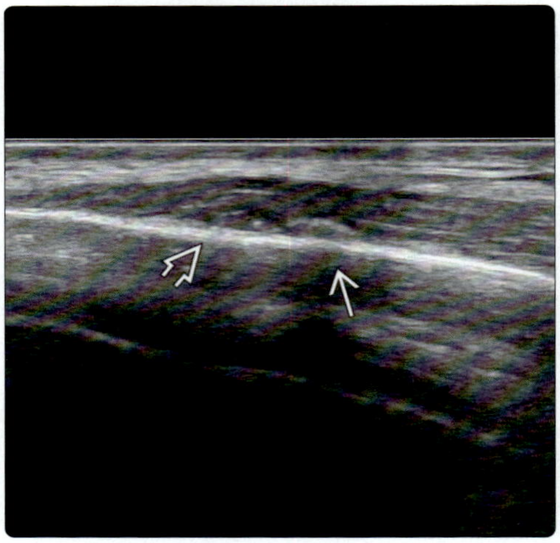

Rib Metastases

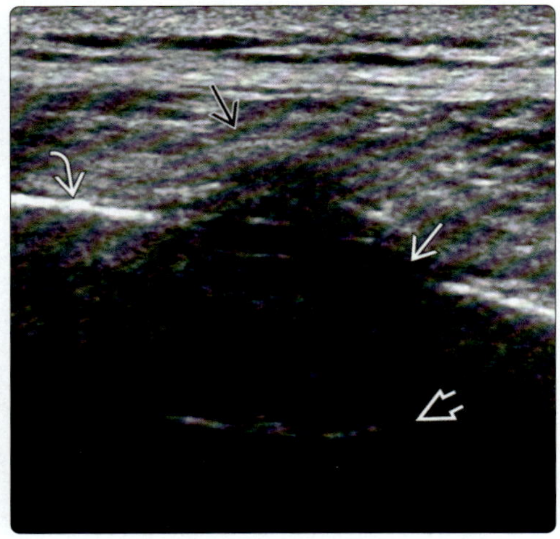

(Left) *Longitudinal US in a surf ski paddler who developed chest wall pain during intensive training shows a stress fracture of right 6th rib in the anterior axillary line. There is localized hyperostosis and soft tissue swelling. No cortical fracture is present.* **(Right)** *Longitudinal US shows a medium-sized, hypoechoic, metastatic deposit destroying the anterior cortex of rib and providing an acoustic window to view the underlying pleura. Mild extraosseous soft tissue extension is present.*

Chest Wall Metastases

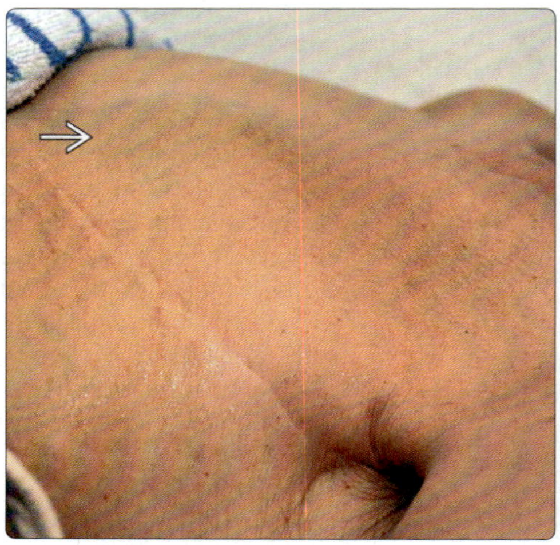

Chest Wall Metastases

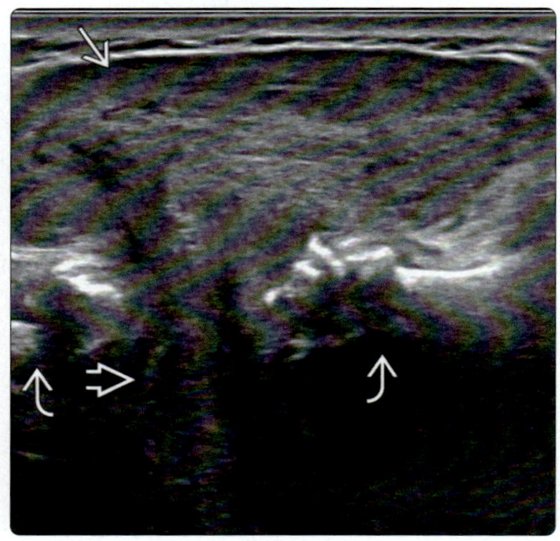

(Left) *Clinical photograph of a patient with prior modified radical mastectomy who presented with a chest wall mass is shown.* **(Right)** *Longitudinal US shows a large tumor emerging from the thorax through an anterior intercostal space and infiltrating 2 opposing ribs. US-guided biopsy revealed metastatic breast carcinoma. There was a large intrathoracic metastatic deposit.*

Chest Wall Metastases

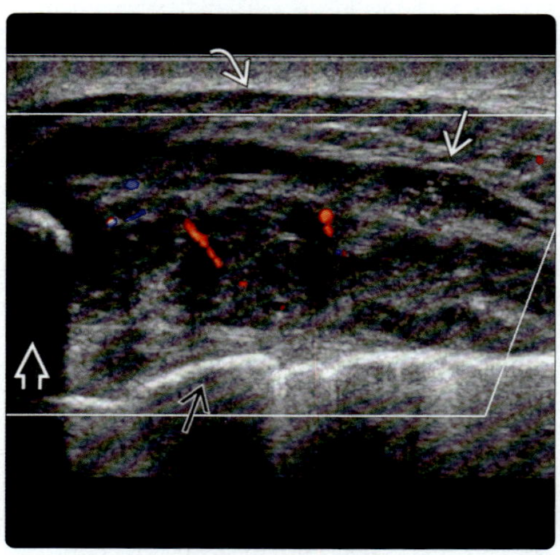

Chest Wall Metastases

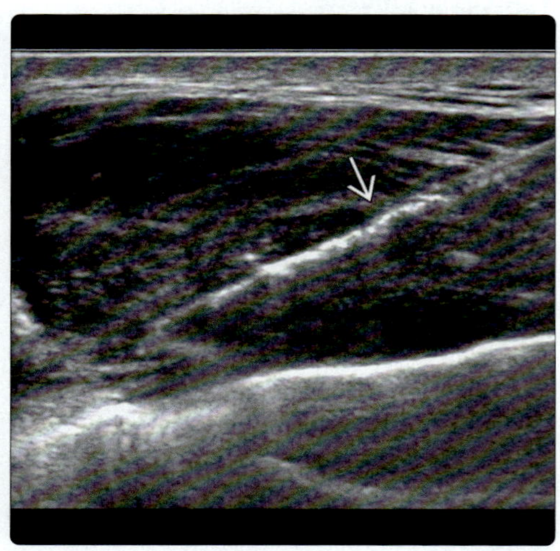

(Left) *Transverse US in a middle-aged man with parasternal mass shows a large soft tissue mass of the anterior chest wall extending to the intercostal space, pleural cavity, and anterior mediastinum. Mild internal vascularity is present. The lesion was encasing the internal mammary artery. Features favored either metastases, lymphoma, or thymoma. The adjacent sternum, pectoralis muscle, and pleura are shown.* **(Right)** *Transverse US during percutaneous core biopsy revealed lymphoma.*

Gynecomastia

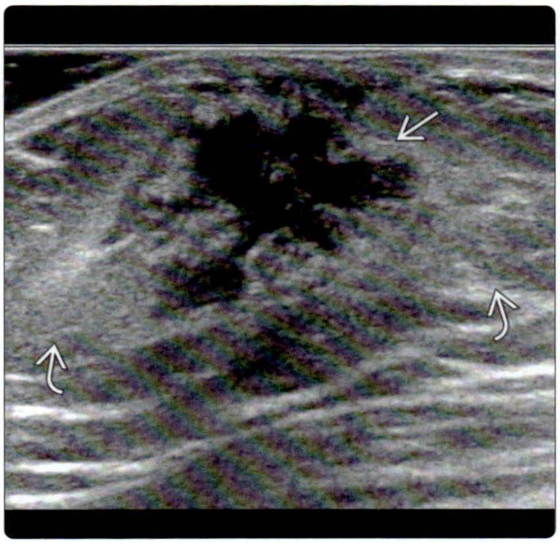

Accessory Breast

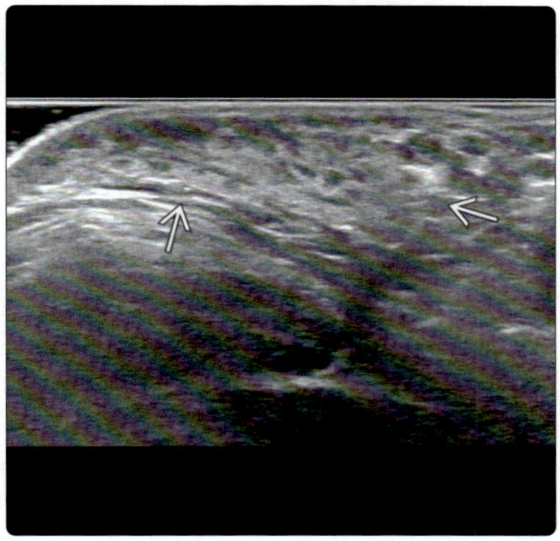

(Left) *Transverse US in a middle-aged man with gynecomastia shows orderly hypoechoic fibroductal tissue* ➔ *embedded in thickened echogenic fat* ➔*, which is very different than breast carcinoma.* **(Right)** *Transverse US in female patient with axillary mass shows a medium-sized area of subcutaneous accessory breast tissue* ➔ *at the inferior aspect of the axilla separate from the normal breast. The mass shows hypoechoic fibroductal tissue interspersed in echogenic fat similar to normal glandular breast tissue.*

Carcinoma, Male Breast

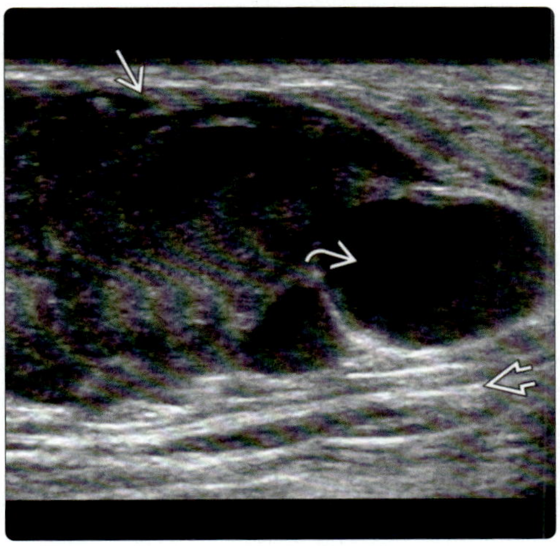

Carcinoma, Male Breast

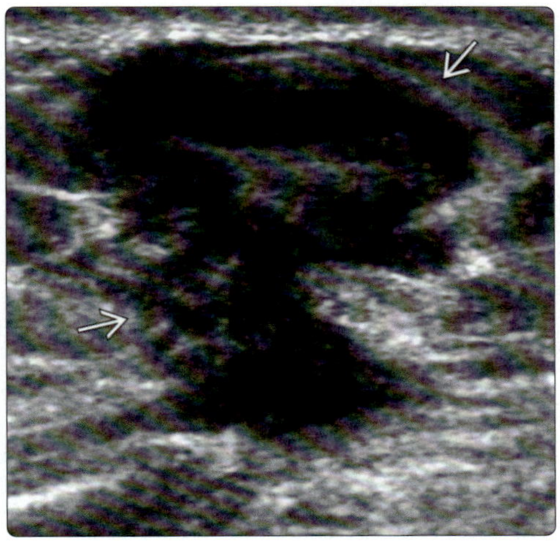

(Left) *Transverse US shows a heterogeneous, mixed solid/cystic mass* ➔ *with distinct margins associated with edema and architectural distortion of the surrounding tissue. There is mild posterior acoustic enhancement* ➔*. Note the cystic component* ➔ *of the mass. Excision confirmed ductal carcinoma.* **(Right)** *Transverse US shows a more solid heterogeneous ductal carcinoma* ➔ *of the male breast with irregular lobulated margins.*

Fibroma

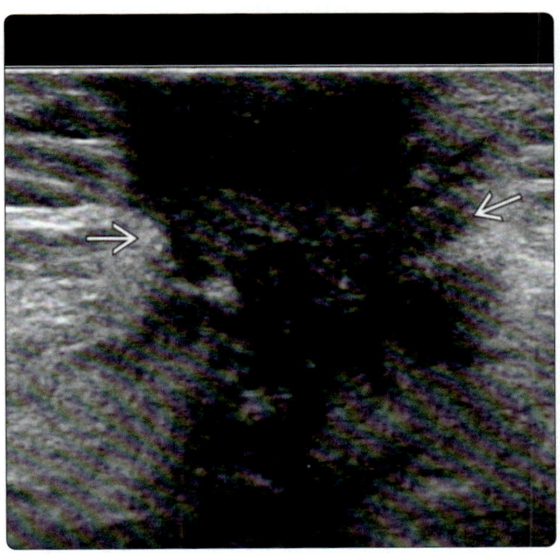

Lipoma

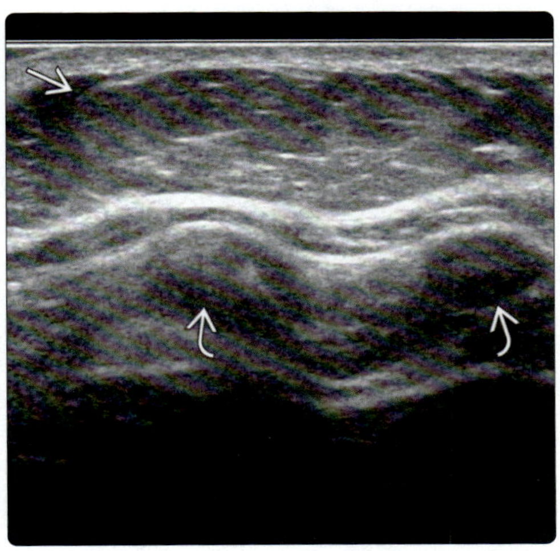

(Left) *Transverse US shows a cutaneous and subcutaneous fibroma* ➔ *in the axillary region. The edges of the typically hypoechoic mass interdigitate with the surrounding fat.* **(Right)** *Longitudinal US shows a well-defined subcutaneous mass* ➔ *with fine linear striations parallel to the long axis, typical of a lipoma. The underlying ribs are seen in transverse section* ➔*.*

(Left) *Transverse US in an elderly woman with a back lump for several months shows a large, typically layered fibrofatty mass arising medial to the scapular tip consistent with elastofibroma dorsi. Appearances are diagnostic without need for biopsy. Ribs and pleura lie deep and the rhomboid muscle is displaced superficially.* **(Right)** *Axial T1WI MR shows a crescent-shaped mass deep to the latissimus dorsi composed of hyperintense (fat) and hypointense (fibrous) strands.*

Elastofibroma Dorsi

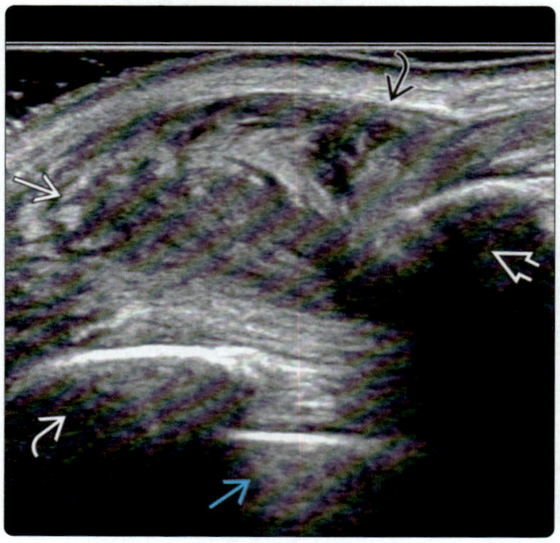

Elastofibroma Dorsi

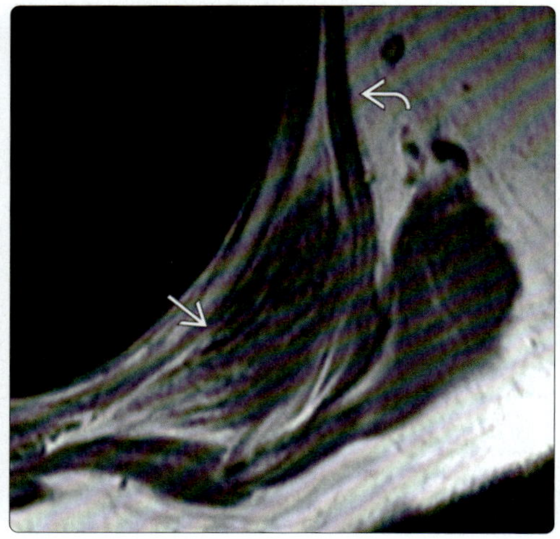

(Left) *Transverse US in a patient with a lower chest wall mass shows a protruding 12th rib, more superficially positioned than the adjacent 11th rib. Both ribs were otherwise normal.* **(Right)** *Longitudinal US in a young adult patient with a mildly tender chronic anterior chest wall lump shows a prominent, anteriorly deviated hypoechoic xiphisternal cartilage at the lower end of the sternum. This anterior deviation of the xiphisternum is a normal variant and corresponds to the symptomatic site.*

Prominent or Floating Rib

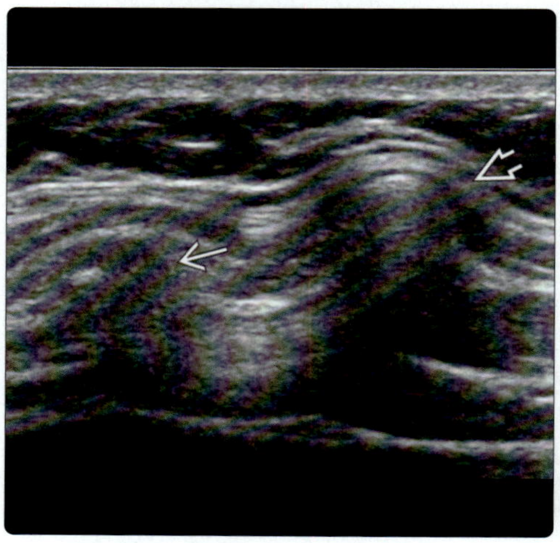

Prominent Xiphoid Process

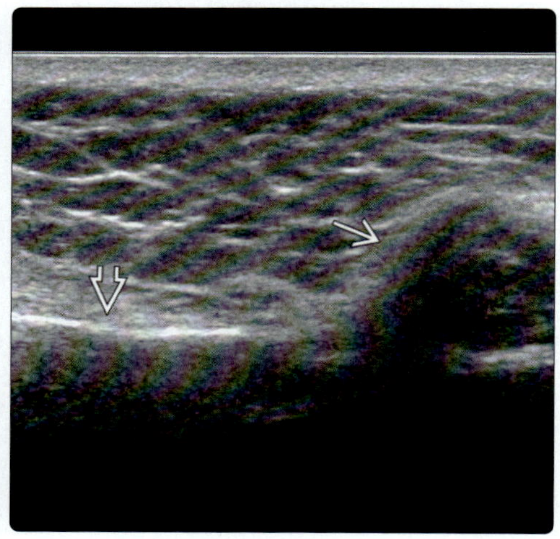

(Left) *Longitudinal US of a man with a left-sided chest wall mass shows a moderately elongated 4th costal cartilage, bulging anteriorly causing clinical prominence. Adjacent costal cartilages were normal.* **(Right)** *Longitudinal US of the right 2nd rib in patient with chest wall pain shows mild swelling of the right 2nd costal cartilage close to the costochondral junction. Although no soft tissue inflammation is evident, there was probe tenderness due to costochondritis. There is an unrelated loculated pleural effusion.*

Prominent or Floating Rib

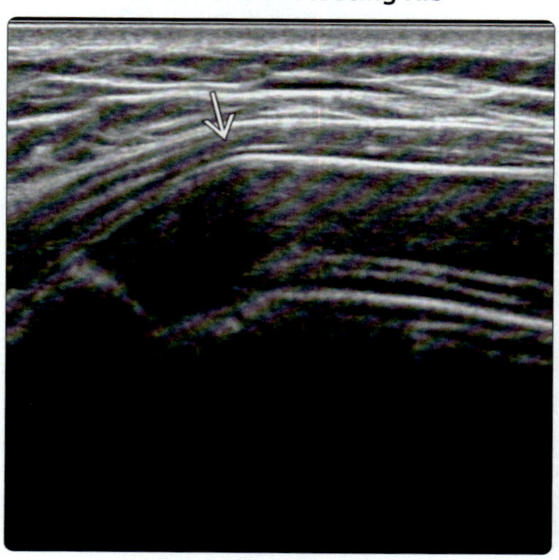

Costochondritis

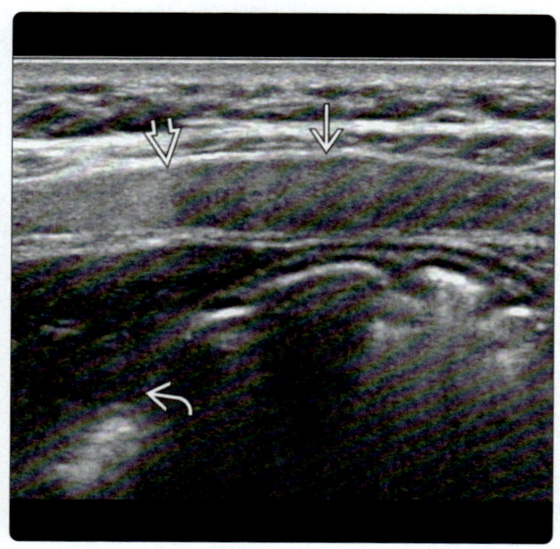

Lymphatic Malformation

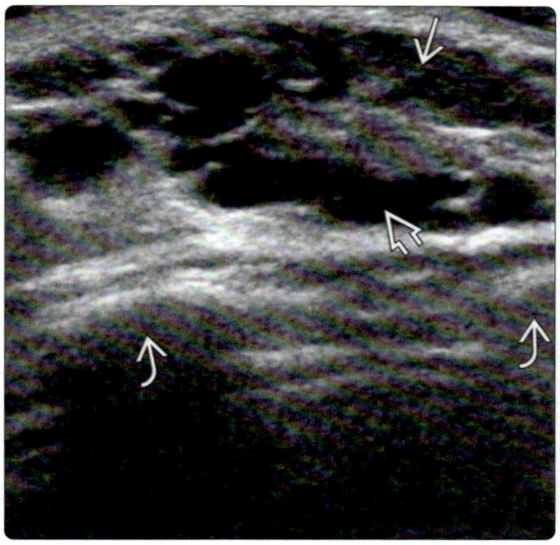

Abscess

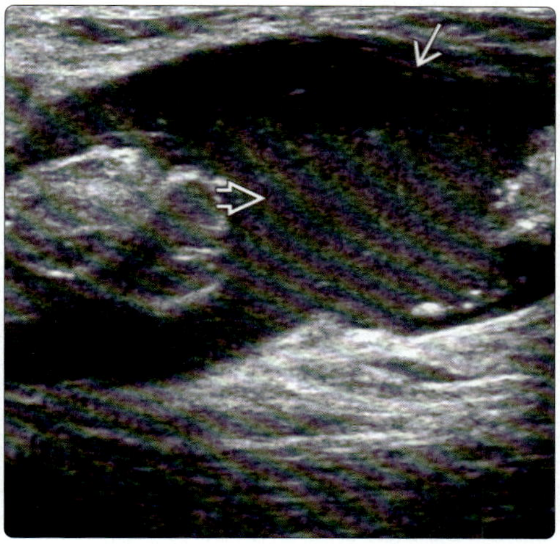

(Left) *Longitudinal US shows a poorly defined lymphatic malformation ➡ of the lower anterior chest wall with a prominent cystic component ➡ and poorly defined swollen supporting stroma. The underlying rib is normal ➡. Color Doppler imaging (not shown) showed no internal vascularity.* **(Right)** *Transverse US shows a recurrent chronic abscess ➡ in the subcutaneous tissues of the lower lateral chest wall. There are fine internal echoes present ➡. US-guided aspirate yielded Staphylococcus aureus.*

Hematoma

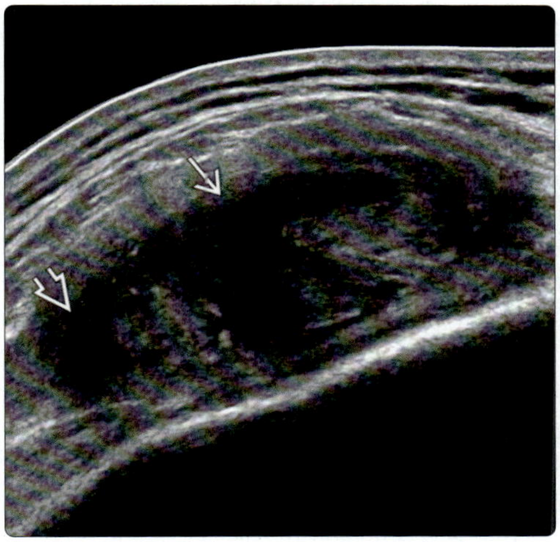

Thyroglossal Duct Cyst

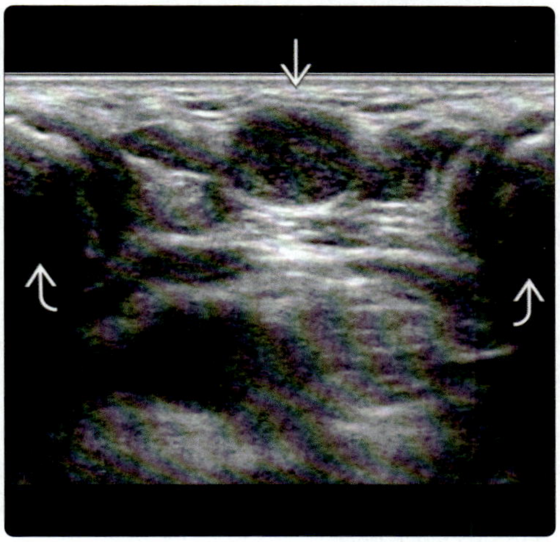

(Left) *Oblique longitudinal US shows a large organizing hematoma ➡ within the latissimus dorsi muscle overlying the posterior aspect of a rib. The hypoechoic component of the hematoma ➡ is due to early liquefaction.* **(Right)** *Transverse US in a child with suprasternal notch lump is shown. This is a hypoechoic nodule ➡ centrally in the sternal notch between the medial ends of the clavicle ➡ consistent with a thyroglossal duct cyst.*

Sternoclavicular Joint Arthropathy

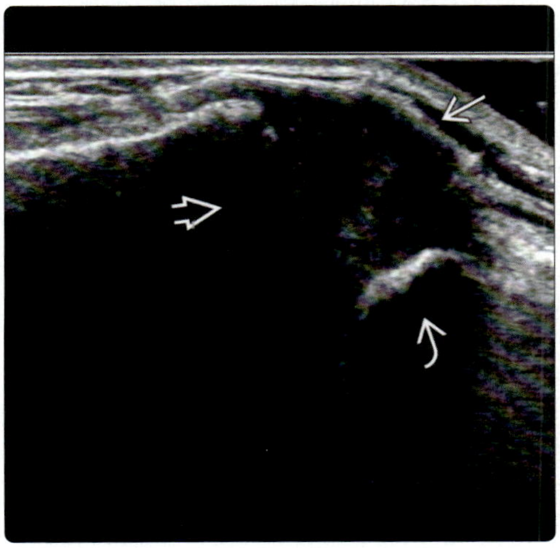

Sternoclavicular Joint Arthropathy

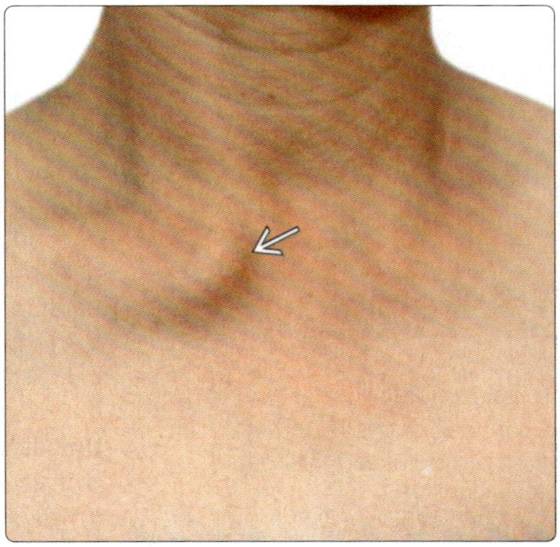

(Left) *Oblique US in a rheumatoid arthritis patient shows moderate inflammatory arthritis of the sternoclavicular joint with prominent capsular thickening ➡ and mild erosion of the medial end of the clavicle ➡. The manubrium ➡ is not eroded. US is good, though CT is better, at discriminating the various forms of sternoclavicular swelling.* **(Right)** *Photograph of the same patient shows how sternoclavicular joint swelling ➡ is easily visible clinically since it lies just beneath the skin.*

DIFFERENTIAL DIAGNOSIS

Common

- Subcutaneous Lipoma
- Incisional Hernia
- Linea Alba Hernia
- Groin Hernia

Less Common

- Spigelian Hernia
- Lumbar Hernia
- Paraumbilical Hernia
- Umbilical Hernia
- Endometrioma
- Undescended Testis
- Enlarged Lymph Node
- Rectus Sheath/Subcutaneous Hematoma
- Fat Necrosis
- Desmoid Tumor
- Muscle Tear
- Divarication of Rectus Abdominis Muscle
- Umbilical Metastasis
- Abscess
- Granuloma
- Umbilical-Urachal Sinus

ESSENTIAL INFORMATION

Key Differential Diagnosis Issues

- Pay close attention to deep margin of any anterior abdominal wall mass because hernias may masquerade as abdominal wall masses
    - Defect may be small relative to size of mass and potentially overlooked
    - Hernia mass may be fixed and not reducible
    - May look like hypoechoic subcutaneous nodule or lipomatous mass
    - Check for continuity with deeper structures
        - Also check for vascular configuration, peristalsis, and perform Valsalva maneuver
        - Valsalva for groin hernia is better with patient standing upright, especially in those situations where supine examination negative and patient symptoms aggravated by weightbearing
    - If deep border of mass not seen, reassess with either CT or MR

Helpful Clues for Common Diagnoses

- **Subcutaneous Lipoma**
    - Often multiple and occur in middle age
    - Fusiform mass with thin, parallel, echogenic internal striations
        - Compressibility similar to adjacent fat
    - Alternatively, well-defined, rounded or ovoid encapsulated echogenic mass without clearly definable linear fine striations
- **Incisional Hernia**
    - Can occur after any abdominal operation
        - Presents months to years after surgery
- **Linea Alba Hernia**
    - Usually occurs in young to middle-aged patients
    - Typically occurs midway between xiphisternum and umbilicus
    - Protrusion of extraperitoneal fat through narrow defect in linea alba
        - Occurs where vessels perforate fascia
    - Hernia though linea alba containing either echo-poor or echogenic fat in epigastric region
        - Can be very small
        - Often no change with Valsalva, as neck of linea alba hernia is tight
    - Fatty contents can strangulate and infarct
- **Groin Hernia**
    - Indirect inguinal hernia is 4x more common than direct inguinal hernia
        - Direct inguinal hernia and femoral hernia have similar prevalence
    - Indirect hernia passes through deep inguinal ring, along inguinal canal, and out of superficial inguinal ring
        - Neck located lateral to inferior epigastric vessels and above inguinal ligament
    - Direct hernia passes through floor of inguinal canal medially (transversalis fascia, Hesselbach triangle)
        - Neck located medial to inferior epigastric vessels and above inguinal ligament
        - Generally smaller than indirect inguinal hernias
    - Femoral hernia passes through femoral canal
        - Neck located medial to femoral vein and below inguinal ligament, usually narrow
        - More common in women
    - US can accurately identify presence, type, and content of groin hernia
        - Content includes fat, omentum, fluid, and bowel

Helpful Clues for Less Common Diagnoses

- **Spigelian Hernia**
    - Defect in aponeurosis at lateral border of rectus abdominis muscle
        - Between rectus abdominis medially and linea semilunaris laterally
    - Most commonly located in lower abdomen ~ 1/3 way between umbilicus and pubic crest
        - Extends deep to external oblique aponeurosis
        - In other words, does not extend into subcutaneous fat but remains within layers of abdominal wall
    - Prone to strangulation as neck narrow
- **Lumbar Hernia**
    - Through defect in posterolateral abdominal wall
    - Spontaneous, postsurgical, or posttraumatic
    - Herniation of retroperitoneal fat ± colon or small bowel
- **Paraumbilical Hernia**
    - In adults
    - Protrudes through side of umbilical ring
    - Extraperitoneal fat ± peritoneal contents
- **Umbilical Hernia**
    - In young children
    - 1/3 of umbilical hernias close within 1 month of birth
    - Rarely persists beyond 3-4 years
    - Peritoneal content protrudes through patent umbilical ring
    - Large neck, therefore unlikely to strangulate
- **Endometrioma**

- Palpable mass often near cesarean section scar or pelvic surgical scar
- Pain ± cyclical pain occurring with menses
- Solid, mainly hypoechoic, heterogeneous nodule
- Infiltrating margins
- Anechoic cystic areas infrequent
- Peripheral hyperechoic ring
- Mild internal vascularity

- **Undescended Testis**
  - US will locate undescended testis in nearly 90% of patients with cryptorchidism
    - 90% of undescended testes located in inguinal canal
    - Remainder located in abdomen and pelvis
- **Enlarged Lymph Node**
  - Features of reactive or malignant node in groin region
- **Rectus Sheath Hematoma**
  - Due to bleeding within rectus sheath from injury to superior or inferior epigastric arteries or direct tear of rectus muscle
  - Inferior epigastric artery arises from external iliac artery
    - Ascends between rectus abdominis muscle and posterior rectus sheath
    - Prone to injury during strong muscle contractions
      - Can occur during bouts of coughing
  - Superior epigastric artery originates from external thoracic artery
    - Descends between rectus abdominis muscle and posterior rectus sheath
- **Subcutaneous Hematoma**
  - Posttraumatic (typical) or spontaneous (uncommon) due to bleeding of subcutaneous vessel
- **Fat Necrosis**
  - Localized, increased echogenicity of subcutaneous fat
  - Irregular, poorly defined margins without discrete mass
  - ± localized swelling ± lobulation of subcutaneous fat due to edema
  - Affected fat area becomes better defined and more hypoechoic with increasing chronicity
  - ± fluid accumulation within affected fat
  - ± calcification of affected fat
- **Desmoid Tumor**
  - Benign fibrous tumor (a.k.a. fibromatosis) with predilection for anterior abdominal wall, especially in postpartum women
    - Hypoechoic tumor often with irregular margin
    - Minimal or no hyperemia
- **Muscle Tear**
  - Affects oblique muscles in lower abdominal quadrants
    - Well-localized mechanical pain
    - Affects athletes and patients with chronic cough
- **Divarication of Rectus Abdominis Muscle**
  - Results from widened linea alba
  - US to exclude presence of true ventral hernia
- **Umbilical Metastasis**
  - Typically spreads from gastrointestinal or genitourinary primary via lymphatic, vascular, or embryologic remnant route
- **Abscess**
  - Depending on maturity & content, echogenicity of fluid may vary from hypoechoic to isoechoic to hyperechoic
    - May mimic solid mass if thick fluid/debris is present
  - Posterior enhancement
  - Echo movement on real-time imaging
  - Wall thickness proportional to chronicity of abscess
  - ± gas locules ± septations ± peripheral hyperemia
- **Granuloma**
  - May arise from retained suture, injection, or chronic infection
  - Hypoechoic mass ± suture material ± mild to moderate vascularity
- **Umbilical-Urachal Sinus**
  - Urachus = ductal remnant from involuted embryologic allantois that extends between bladder dome and umbilicus
    - Failure to involute fully → fusiform fluid-filled structure below umbilicus
      - Sinus tract located between transversalis fascia and peritoneum
    - Usually presents in early childhood as umbilical discharge ± mass ± pain

Subcutaneous Lipoma

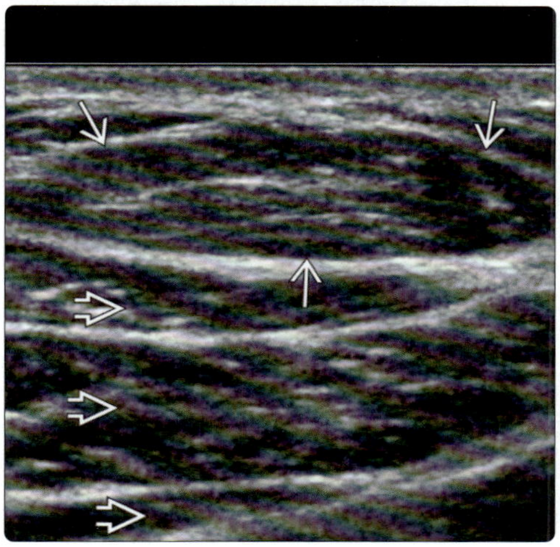

Subcutaneous Lipoma

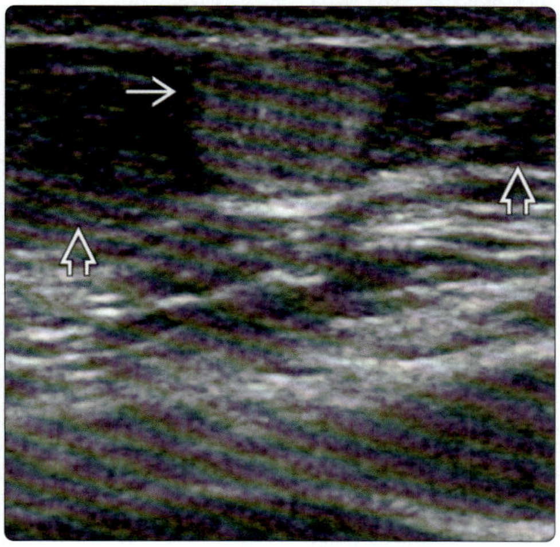

(Left) *Transverse US shows a subcutaneous lipoma ➡ in the lateral abdominal wall. The external oblique, internal oblique, and transversalis muscles ➡ are not involved.* **(Right)** *Transverse US shows a small, rounded, echogenic subcutaneous lipoma ➡ in the anterior upper abdominal wall. The lipoma is distinctly echogenic relative to the adjacent hypoechoic subcutaneous fat ➡.*

(Left) *Transverse US shows loops of small bowel protruding from the peritoneal cavity into the subcutaneous fat through a wide incisional fascial defect in the lower abdomen.* **(Right)** *Transverse US shows a direct inguinal hernia. The neck of the hernia lies above the inguinal ligament and medial to the inferior epigastric vessels. The sac contains fluid and omentum.*

Incisional Hernia

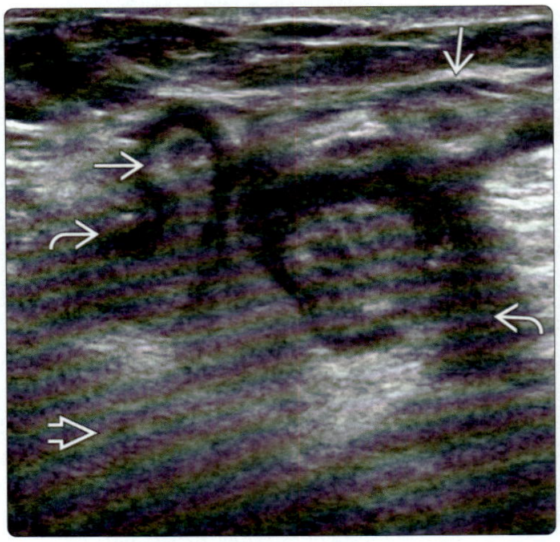

Groin Hernia

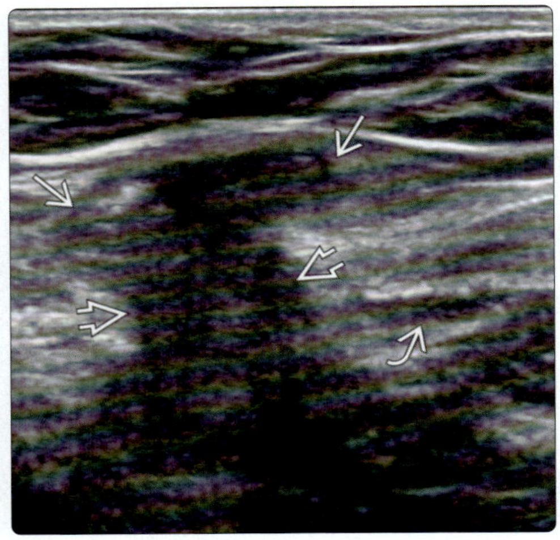

(Left) *Longitudinal US in a patient with point-specific supraumbilical pain shows a very small linea alba hernia containing extraperitoneal fat. Linea alba defect is 1.5 mm wide while sac is 4 mm wide x 5 mm long. This is also called a fatty hernia of linea alba or an epigastric hernia. Bowel is within peritoneal cavity.* **(Right)** *Transverse US shows a larger epigastric hernia containing extraperitoneal fat and mesentery. No bowel herniation is shown. Linea alba defect was 10 mm wide. Note the rectus abdominis muscle.*

Linea Alba Hernia

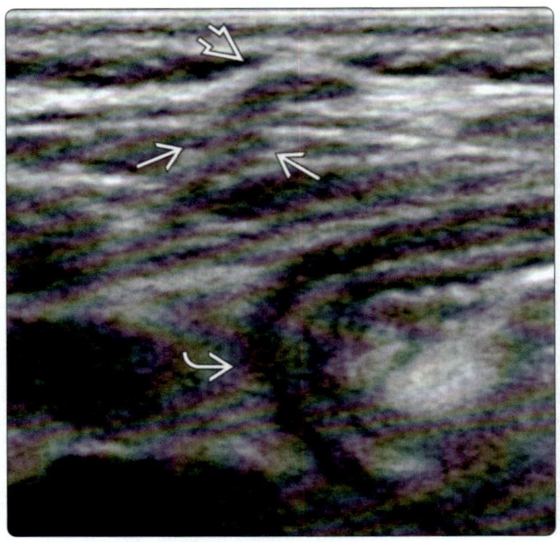

Linea Alba Hernia

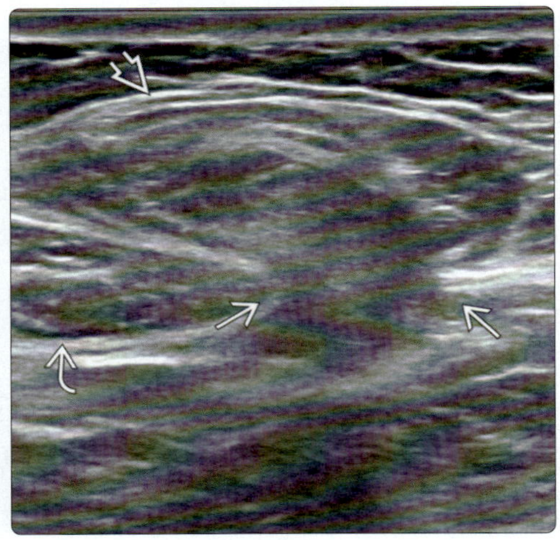

(Left) *Transverse US shows bowel protruding though the umbilicus in this newborn with an umbilical hernia. The neck of this type of hernia is wide, allowing the hernia to be easily reduced and thereby not prone to strangulation.* **(Right)** *Transverse US in female patient with point-specific lower abdominal pain and tenderness, shows a very small Spigelian hernia between the lateral border of rectus abdominis and linea semilunaris. This hernia contains extraperitoneal fat.*

Umbilical Hernia

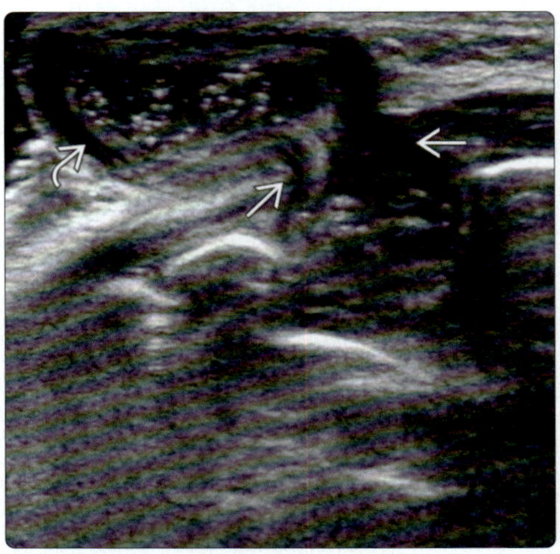

Spigelian Hernia

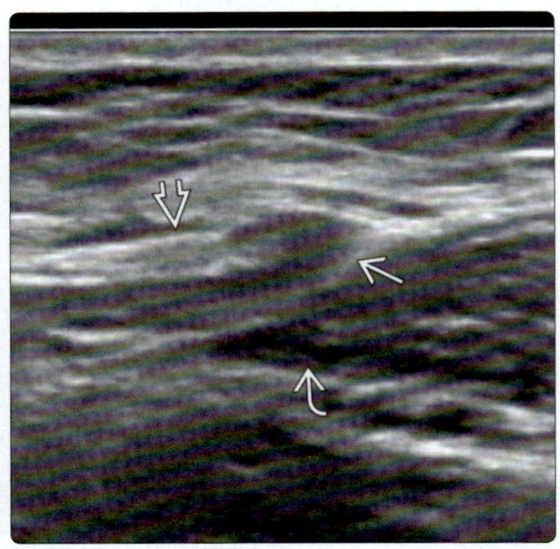

Lumbar Hernia

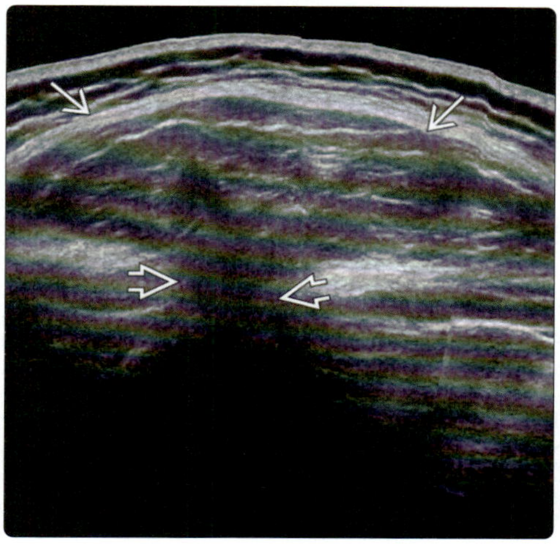

Lumbar Hernia

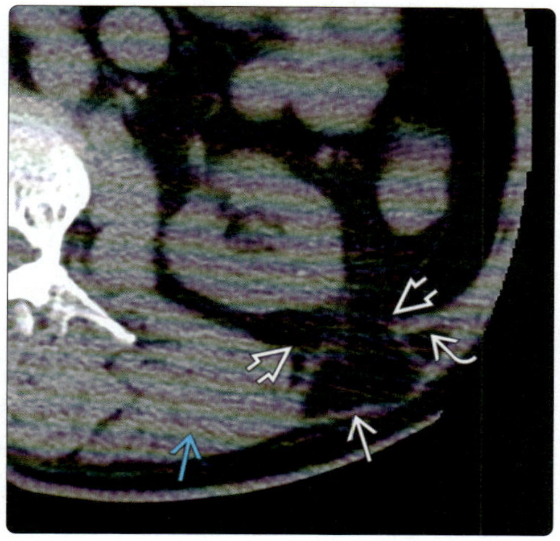

(Left) *Oblique US shows a wide defect in the fascia of the posterior abdominal wall with herniation of retroperitoneal fat through this defect.* **(Right)** *Axial NECT of the same patient shows the large fascial defect posterior to the left kidney. The hernia sac contains retroperitoneal fat. The defect is bounded by the erector spinae muscle medially and the retracted quadratus lumborum muscle laterally.*

Paraumbilical Hernia

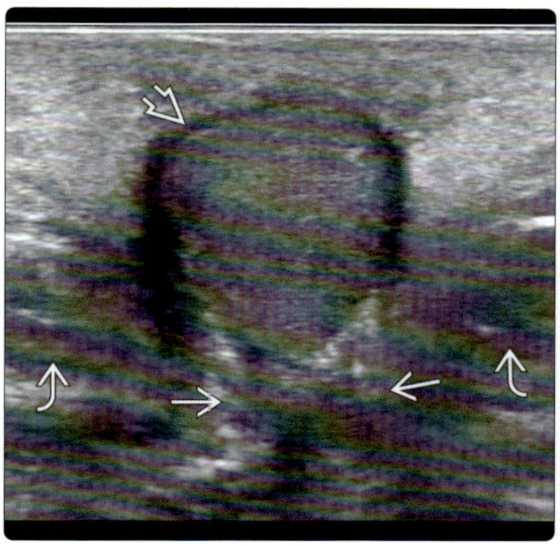

Paraumbilical Hernia

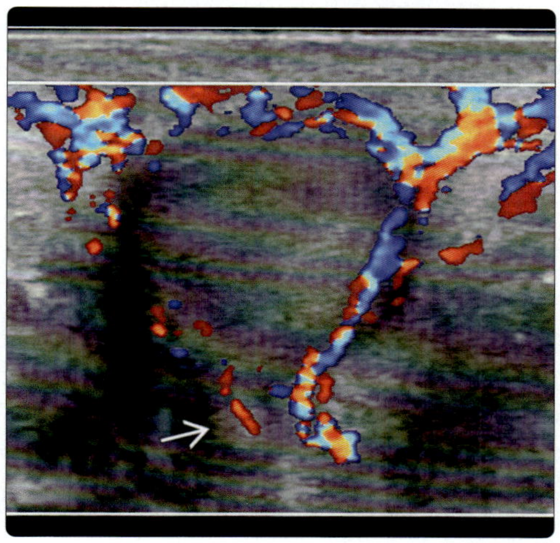

(Left) *Transverse US in a middle-aged woman with painful umbilical nodule shows a medium-sized lipomatous nodule in continuity with the deeper tissues. This nodule located between the rectus abdominis muscles represents herniated extraperitoneal fat.* **(Right)** *Color Doppler imaging pattern of the same nodule helps confirm the herniated nature of this nodule with arteries emerging from deep. This represents herniated infarcted, inflamed extraperitoneal fat. Findings were confirmed surgically.*

Divarication of Rectus Abdominis Muscle

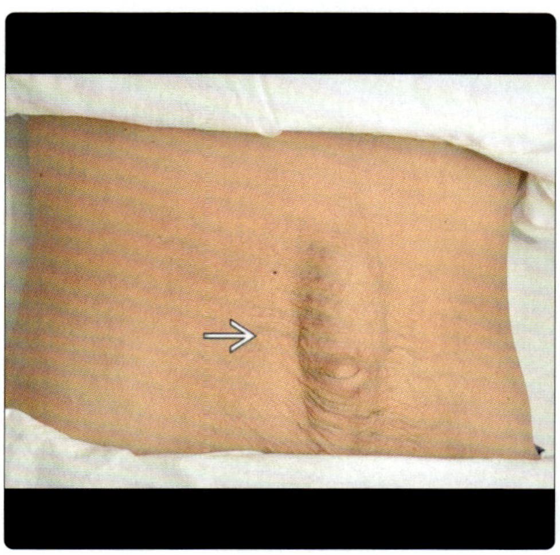

Divarication of Rectus Abdominis Muscle

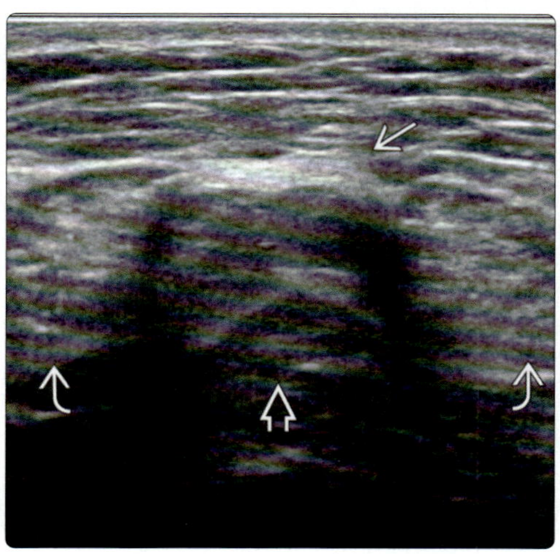

(Left) *Clinical photograph of a 38-year-old woman performed during Valsalva maneuver shows divarication of the rectus abdominis muscles. This condition is also known as diastasis recti.* **(Right)** *Transverse US of the same patient shows a widened (up to 35 mm) linea alba bulging anteriorly between the rectus abdominis muscles. No actual hernia is present as the abdominal content is still contained within its cavity.*

(Left) *Transverse US shows an ill-defined, hypoechoic subcutaneous mass in the lateral abdominal wall. Mild liquefaction is present. There is posterior acoustic shadowing. Biopsy of the mass confirmed fat necrosis. As is commonly the case, there was no recall of any abdominal trauma.* **(Right)** *Transverse US shows a focal superficial tear of the rectus abdominis muscle. There is focal hematoma filling the tear. Note the adjacent subcutaneous edema.*

Fat Necrosis

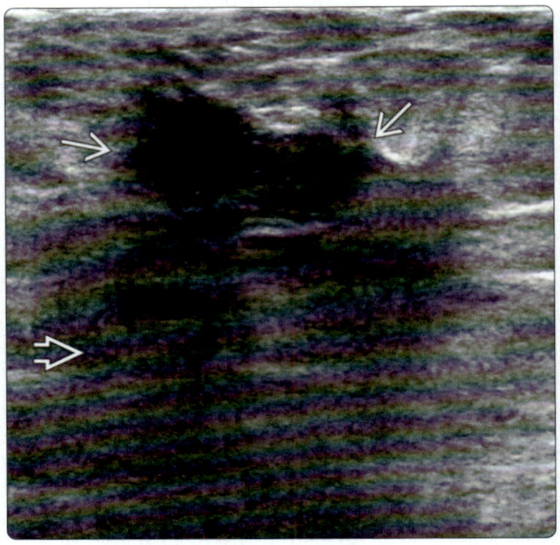

Muscle Tear

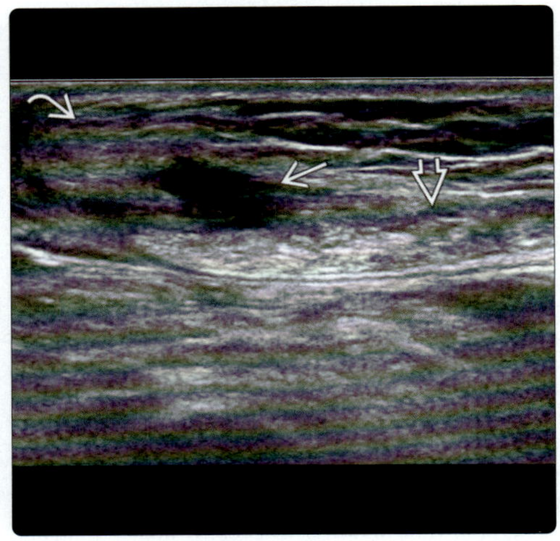

(Left) *Transverse US shows a subcutaneous hematoma in the lateral abdominal wall. There is a large area of liquefaction within the hematoma.* **(Right)** *Transverse US shows a large traumatic hematoma within an edematous rectus abdominis muscle. The hematomas are contained by the rectus sheath. There is a small area of liquefaction.*

Rectus Sheath/Subcutaneous Hematoma

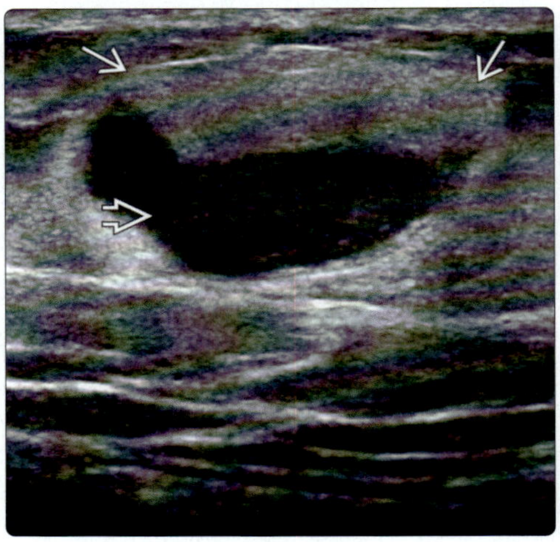

Rectus Sheath/Subcutaneous Hematoma

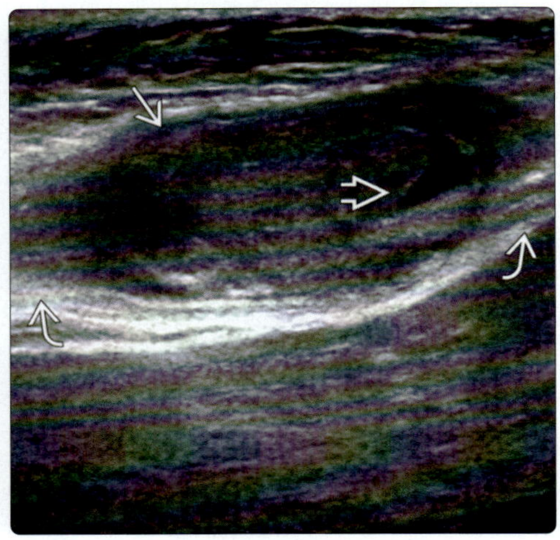

(Left) *Transverse US shows an oval-shaped, hypoechoic testis within the inguinal canal in a patient with cryptorchidism. Most undescended testes are located in the inguinal canal.* **(Right)** *Longitudinal US shows a small stitch granuloma in the deep subcutaneous tissues of the anterior abdominal wall. There is hyperechoic suture material within the hypoechoic granuloma.*

Undescended Testis

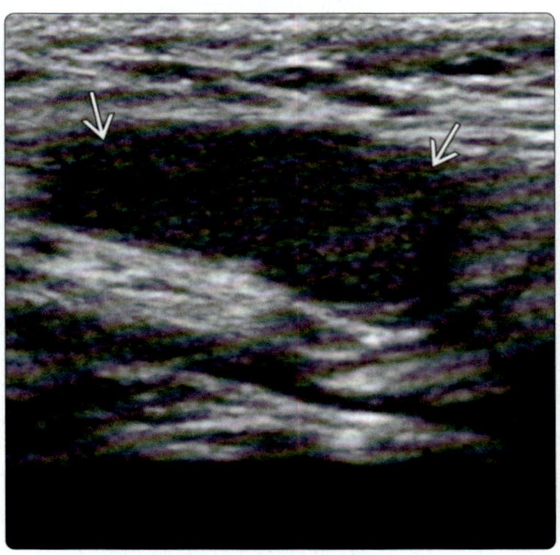

Granuloma

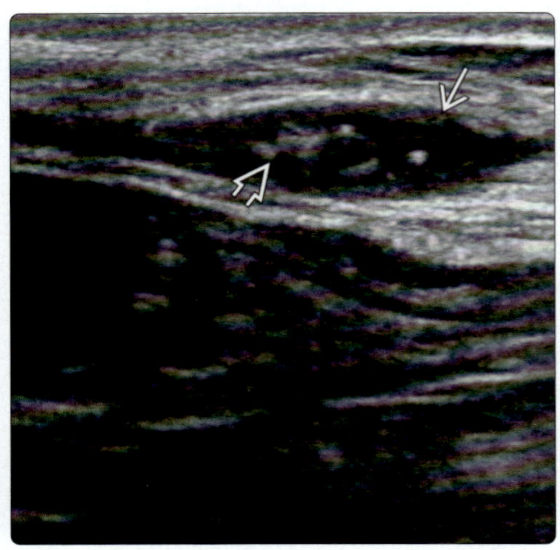

Desmoid Tumor

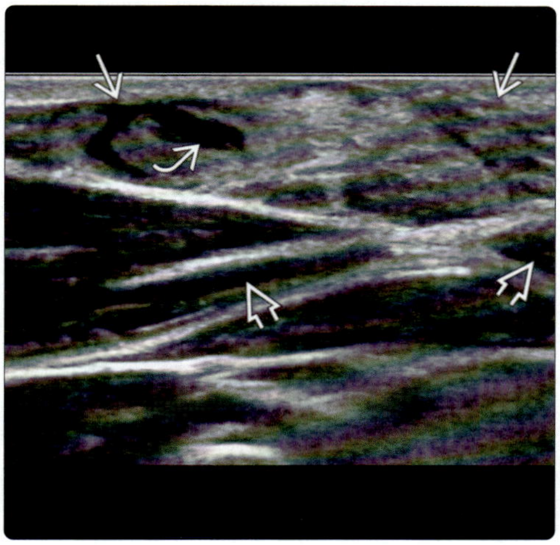

Endometrioma

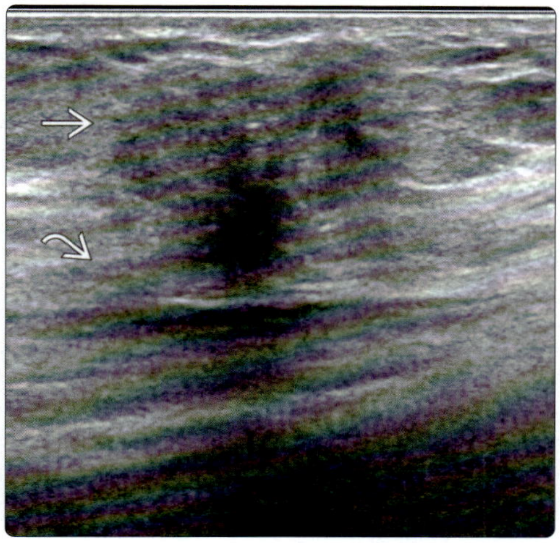

(Left) *Transverse US shows a large subcutaneous desmoid tumor ➡ superficial to the rectus abdominis muscles ➡. The hypoechoic cyst-like component ➡ is due to dense fibrous tissue.* **(Right)** *Transverse US in a middle-aged woman shows a medium-sized, irregular subcutaneous mass ➡ with superficial involvement of rectus abdominis muscle ➡ near the surgical scar (not shown). Overall appearances favored endometriosis, which was confirmed by percutaneous biopsy.*

Abscess

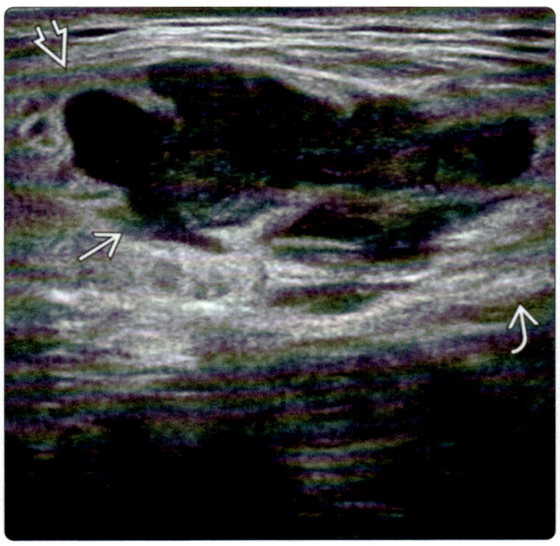

Umbilical-Urachal Sinus

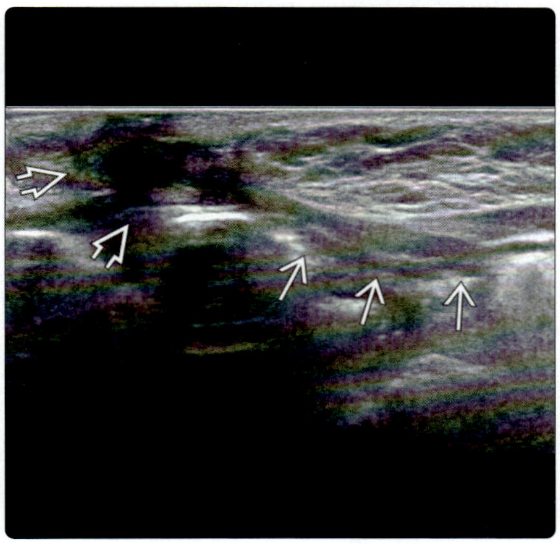

(Left) *Transverse US shows a recurrent chronic abscess ➡ in the lateral abdominal wall. The abscess wall ➡ is thickened with moderate surrounding subcutaneous edema ➡. US-guided aspiration yielded gram-negative bacilli.* **(Right)** *Longitudinal oblique US in a 2-year-old boy with umbilical discharge shows a fibrotic-type umbilical nodule ➡ contiguous with a deeper tract ➡ extending toward the pelvis, consistent with umbilical-urachal sinus. Moderate perilesional hyperemia was present.*

Umbilical Metastasis

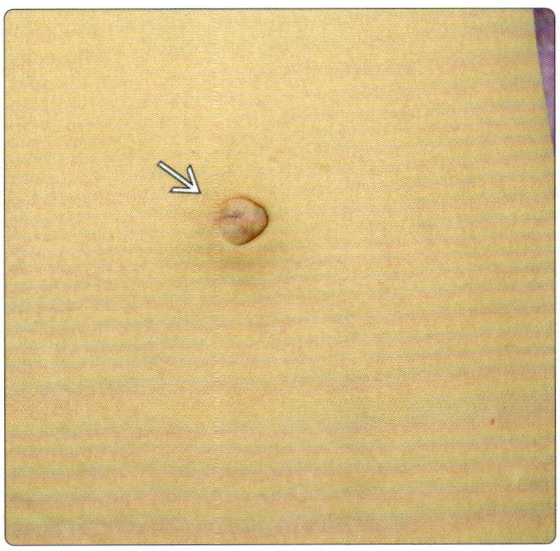

Umbilical Metastasis

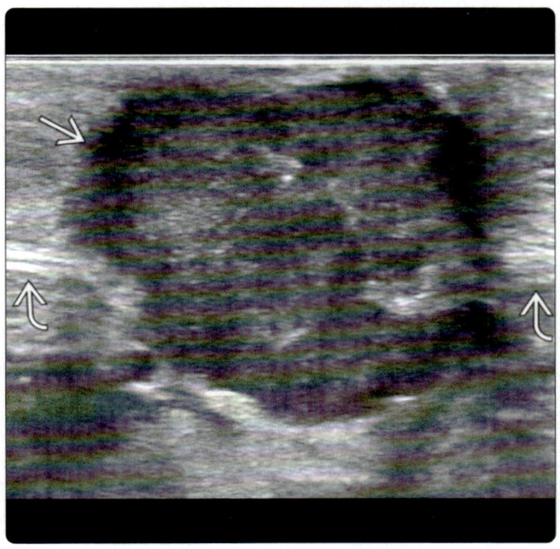

(Left) *Clinical photograph of a 70-year-old woman with ovarian pelvic mass who recently developed an umbilical nodule ➡ is shown.* **(Right)** *Longitudinal US of the same region shows irregular, moderately hyperemic (not shown) umbilical nodule ➡, straddling the linea alba ➡, consistent with a metastasis. US-guided biopsy revealed metastatic tumor of gynecological origin. An umbilical metastasis is known as a Sister Mary Joseph nodule. Discrete noncontiguous lower margin excludes hernia.*

SECTION 5

Interventional Procedures

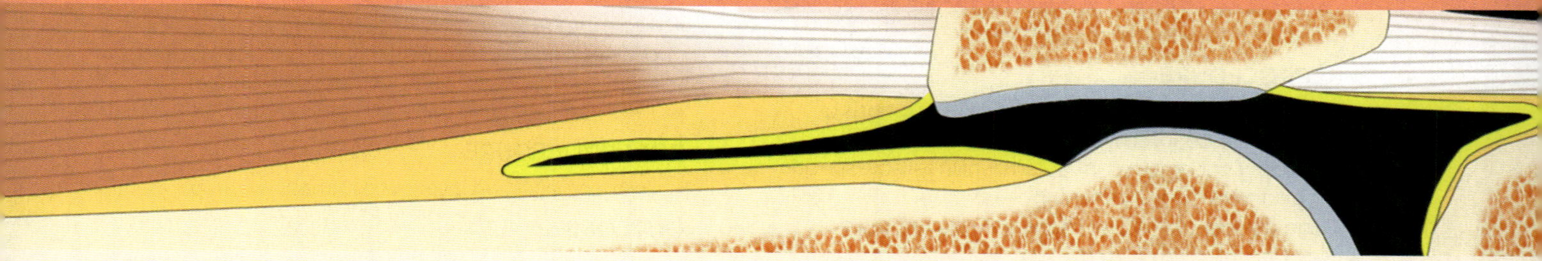

Biopsy

Soft Tissue Tumor Biopsy 918
Bone Tumor Biopsy 924

Joint Procedures

Joint Injection: Upper Limb 932
Joint Injection: Lower Limb 942
Shoulder Procedures 950
Elbow Procedures 958
Hand and Wrist Procedures 964
Hip and Pelvis Procedures 970
Knee Procedures 978
Ankle and Foot Procedures 986

Soft Tissue Tumor Biopsy

KEY FACTS

TERMINOLOGY

- Percutaneous biopsy of soft tissue tumor under US guidance
  - Accurate, safer, and more cost-effective alternative to open biopsy for tissue diagnosis

PREPROCEDURE

- To determine nature and make specific diagnosis of indeterminate soft tissue mass
  - Biopsy not required in cases in which reliable specific diagnosis made by imaging alone
- High-frequency (7-12 MHz) linear array transducer
- Coaxial system preferred
  - 15-g coaxial needle with 16-g Tru-Cut needle

PROCEDURE

- US area of interest to best assess route and select viable areas of lesion before biopsy
  - Target solid area, preferably with evidence of neovascularization on color Doppler
  - Avoid cystic or necrotic areas
  - Obtain 1 biopsy close to edge of tumor
- Avoid important intervening structures, such as uninvolved compartment, neurovascular bundle, tendon, bowel, or other organs, especially in case of pelvic biopsy
- Keep needle path within area of planned surgical resection

POST PROCEDURE

- Local complication is very uncommon

OUTCOMES

- Occasionally, tissue obtained may not be representative of entire tumor
- Theoretical risk of tumor seeding along biopsy pathway
  - Risk much smaller than with open biopsy

Selecting Suitable Area for Biopsy

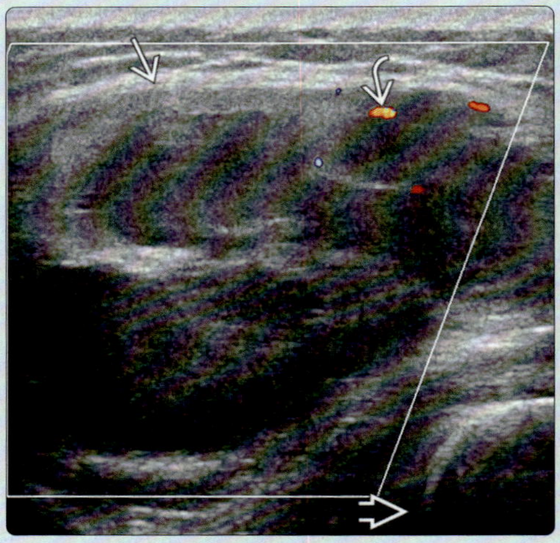

Obtaining Biopsy Close to Tumor Edge

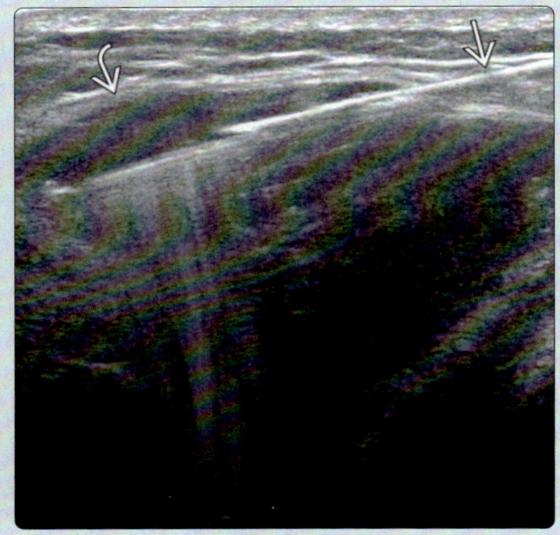

(Left) *Transverse US shows a large (5.1 cm wide x 3.6 cm deep x 8.1 cm long) soft tissue tumor involving the vastus lateralis with partial involvement of the intermedius. Mild hypervascularity is present. Note the femoral shaft.* **(Right)** *Transverse US during percutaneous biopsy with coaxial system is shown. Always obtain at least 1 biopsy close to the tumor edge to optimize yield. US cannot always clearly delineate the area of tumor necrosis centrally.*

Multiple Biopsies Via Single Puncture

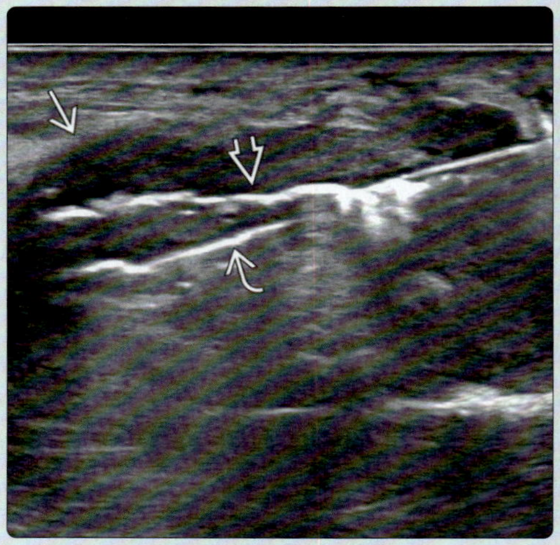

End-Cutting Needle for Small Lesion

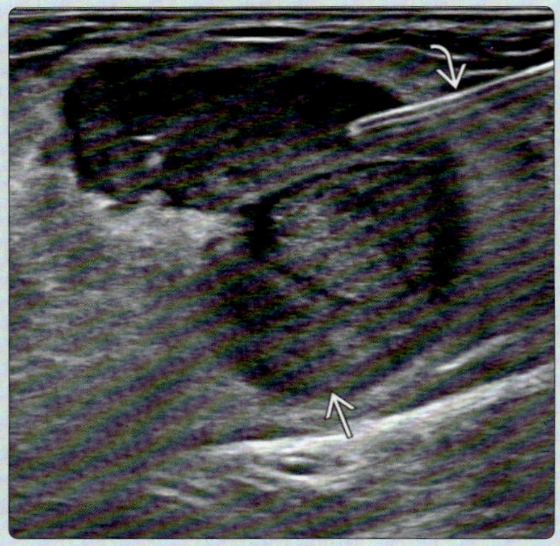

(Left) *Longitudinal US shows a Tru-Cut biopsy of a soft tissue mass in the peroneal muscles. Multiple biopsies are taken along different paths using a single skin puncture and coaxial needle. Note the gas-filled tract from a previous biopsy. Histology revealed pleomorphic undifferentiated sarcoma.* **(Right)** *Longitudinal US shows a 20-g end-cutting needle (Franseen) within a groin lymph node in a patient with melanoma. Histology revealed metastatic melanoma.*

Soft Tissue Tumor Biopsy

TERMINOLOGY

Abbreviations

- Fine-needle aspiration for cytology (FNAC)

Definitions

- Percutaneous biopsy of soft tissue tumor under US guidance

PREPROCEDURE

Indications

- Establish specific histologic diagnosis of soft tissue mass lesion with indeterminate imaging findings
- Confirmation and tumor typing of suspected soft tissue sarcoma
- Confirm or exclude musculoskeletal metastases in patient with known primary tumor
- Confirm or exclude tumor recurrence in patient with prior tumor resection
- Provide tissue for histologic and microbiologic analysis in patient with suspected soft tissue infection
- Provide tissue for histologic analysis in patient with suspected inflammatory myositis

Contraindications

- Uncorrected bleeding diathesis
- Uncooperative patient
- Site inaccessible to US access
- Soft tissue mass too small for percutaneous biopsy
  - Lesions < 1 cm in diameter are generally not amenable to percutaneous biopsy; FNAC more appropriate
- High-flow vascular anomaly or tumor in site inaccessible to postbiopsy tamponade
  - Most highly vascularized tumors can be safely biopsied provided they are located in area where adequate percutaneous tamponade can be applied post biopsy

Getting Started

- Things to check
  - Patient identity, type of procedure, side of procedure, informed consent
  - If patient taking anticoagulants, optimize clotting profile before procedure
    - Optimal level depends on perceived difficulty in applying adequate percutaneous tamponade post biopsy
  - Review prior imaging
    - Accurate specific imaging diagnosis may be made in large number of soft tissue masses, e.g., lipoma, epidermoid cyst, vascular malformation, and nerve sheath tumor
      - Percutaneous biopsy is unnecessary if diagnosis is certain on imaging
      - Biopsy is unnecessary intervention, potentially hazardous, and can be painful (e.g., if nerve sheath tumor is biopsied)
    - In other situations, e.g., distinguishing well-differentiated liposarcoma from lipoma, requires histologic examination of whole tumor mass
    - Plan suitable needle path and decide on area to be sampled, particularly targeting viable tumor area and avoiding neurovascular bundle
- Medications
  - Local anesthetic (2-3 mL 1% lignocaine) to skin
  - Sedation of anxious patient or young children
    - Intravenous midazolam or other benzodiazepine; antidote and resuscitation facilities should be available
- Equipment list
  - High-frequency (7-12 MHz) linear-array transducers should be used routinely
  - Sterile biopsy set
  - Biopsy needle type: Spring-loaded cutting needles are most commonly used
    - Tru-Cut needles employ inner troughed needle, which is advanced into lesion, with outer cylindrical cuff isolating core of tissue
    - End-cutting needles, e.g., Franseen, obtain tissue by isolating core within end of advancing needle cylinder
  - Biopsy needle size
    - Large core (14- to 18-g) biopsy needle
      - Larger biopsy needle provides more tissue for histology, ensuring sufficient tissue is available for immunohistochemical staining and molecular studies
      - As risk of bleeding from easily compressible lesions is low, there is no advantage to using smaller needles for most musculoskeletal biopsies
    - Use finer needle (18- to 22-g) if high risk of postbiopsy bleeding, such as tumor location not being conducive to adequate percutaneous tamponade
  - Coaxial biopsy technique
    - Larger bore cannulated needle, which is placed into tumor under US guidance
      - Biopsy needle inserted through this outer coaxial needle into tumor
    - Coaxial needle 1-g larger than Tru-Cut biopsy needle (e.g., 15-g coaxial needle, 16-g Tru-Cut needle)
    - Ability to obtain multiple cores using only 1 initial puncture
    - Better needle stability and time efficiency when multiple biopsies are taken
    - Reduces trauma to tissues along needle track, reduces likelihood of postbiopsy bleeding, and theoretically reduces risk of tumor seeding along biopsy tract
  - Specimen bottles
    - Specimen can be sent fresh with formalin or alcohol fixation or in sterile bottle

PROCEDURE

Patient Position/Location

- Variable, depending on lesion position
  - Make sure that both patient and radiologist are in comfortable position

Procedure Steps

- Carefully scan region of interest to best assess route for biopsy
- Must ensure that whenever possible compartmental integrity of tumor is maintained
  - Tissue planes can act as natural barriers to spread of tumor or infection

- However, some tissue planes have to be disrupted during biopsy, e.g., need to go through subcutaneous compartment to reach subfascial compartment
  - If tumor is located in one muscle compartment, avoid transgressing another muscle compartment to reach this tumor whenever possible
- Theoretical risk of seeding of tumor along needle tract in particular in case of sarcoma or other malignant tumors
  - In practice, with good technique, this risk is very low and much lower than with open biopsy
  - Biopsy tract has to be removed en bloc with tumor during surgery
  - Surgeon marks site of planned incision on skin with biopsy, whenever possible, being performed at this site
  - Keep needle path within area of planned surgical resection
- Avoid important intervening structures
  - Any neurovascular bundle or tendon as well as peritoneal cavity in case of pelvic mass biopsy
  - Examine adjacent tissue to locate at-risk neurovascular bundle that may not be visible in plane of biopsy
- Select those tumor areas containing more viable representative tissue for sampling
  - Identify solid viable areas using grayscale and color Doppler imaging
    - In general, tumor margins tend to provide more fruitful material than central area of tumor, which may be necrotic
    - If tumor is hypovascular, be certain to sample tumor margin
  - Avoid cystic or necrotic areas as yield will be nondiagnostic
  - Hypervascular areas should be targeted rather than avoided since these most likely contain most metabolically active tumor tissue
    - Risk of bleeding is very low with single puncture and adequate tamponade
    - Take at least 1 biopsy close to capsular edge of tumor
- Sterilize skin over planned area of puncture
- Place probe within sterile probe cover
- Inject local anesthetic
- Insert coaxial needle so that tip just pierces tumor capsule, followed by insertion of core biopsy needle into coaxial needle
  - Sample suitable areas; change direction of coaxial needle within suitable areas
  - In general, 3-4 biopsies are adequate
    - In friable tumors, ≥ 10 biopsies may be required
    - Single intact biopsy core is far better than multiple fragmented cores for pathologic diagnosis
    - Length of biopsy core is more important than width
- Unload tissue cores into specimen bottle
  - Adequate fixation is essential to preserve optimal histologic and cytologic features for proper pathologic examination
  - Minimal tissue handling
    - Tissue autolysis, crushing, drying, and handling artifacts may hinder morphologic interpretation and limit certain immunohistochemical stains
- Rinse biopsy needle tip in normal saline before reinserting into coaxial needle
- Once sufficient biopsies have been obtained, remove coaxial needle and apply local pressure for hemostasis
- Completed within 30 minutes as outpatient procedure
- **Fine-needle aspiration**
  - Local anesthesia is usually not necessary
  - Usually use 21-g intravenous needle
  - Provides cytology rather than histology
  - Usually used in lymph nodes or small lesions
  - Move needle to-and-fro within lesion with simultaneous rotational movement of needle
    - Rotational movement during FNAC is important to optimize cutting action
    - Apply constant gentle suction during FNAC
  - Stop needle suction upon removing needle from lesion
  - Spread aspirate onto specimen slides and fix immediately in 5 mL 95% alcohol solution
  - Alternatively, inject aspirate into specimen bottle containing 5 mL 25% alcohol solution

Alternative Procedures/Therapies

- Surgical
  - Open biopsy or excision

POST PROCEDURE

Expected Outcome

- Uneventful in most circumstances
- Minimal wound discomfort
- Minimal bleeding curtailed by brief period of local pressure for hemostasis

Immediate Aftercare

- Apply simple dressing
  - Dressing can be removed by patient the following day
- Monitor for ~ 1 hour in department for signs of bleeding (swelling, oozing, bruising)

Subsequent Aftercare

- Immediate strenuous exercise involving biopsied area
- Avoid water contact of biopsy site for 24 hours

OUTCOMES

Problems

- Nonrepresentative tissue samples in case of large tumor
  - Close cooperation with pathologist is optimal
- Biopsy may not provide sufficient tissue for cytogenic analysis required to establish treatment protocol
  - Obtain at least 1 biopsy close to tumor capsule

Complications

- Most feared complication(s)
  - Neurovascular bundle injury, peritoneal or thoracic cavity penetration
  - Excessive bleeding
- Delayed complication(s)
  - Infection and bleeding
- Risk of seeding of tumor along needle tract
  - Risk negligible with coaxial system

SELECTED REFERENCES

1. Sperandeo M et al: The role of ultrasound-guided fine needle aspiration biopsy in musculoskeletal diseases. Eur J Radiol. 90:234-244, 2017

Biopsy Set

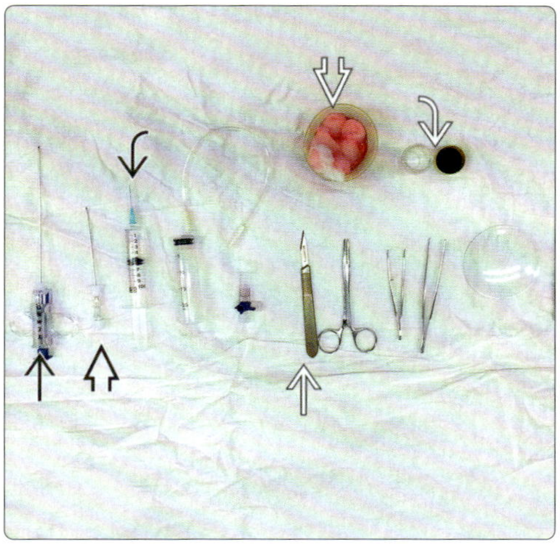

Tru-Cut Biopsy Needle

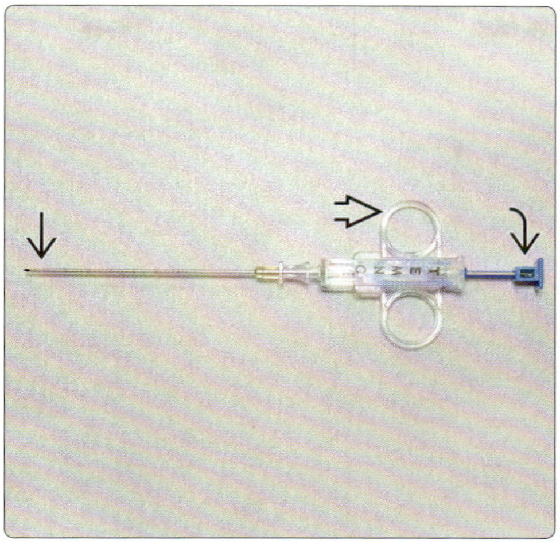

(Left) *Clinical photo of setup for soft tissue tumor biopsy shows the Tru-Cut biopsy needle, coaxial needle, syringe for local anesthetic, scalpel for skin incision, cleansing solution, and specimen bottle.* **(Right)** *Clinical photo of Tru-Cut biopsy needle shows a plunger for controlling the spring-activated mechanism, the handle, and the cutting cannula. The needle is in a loaded status.*

Tru-Cut Biopsy Needle

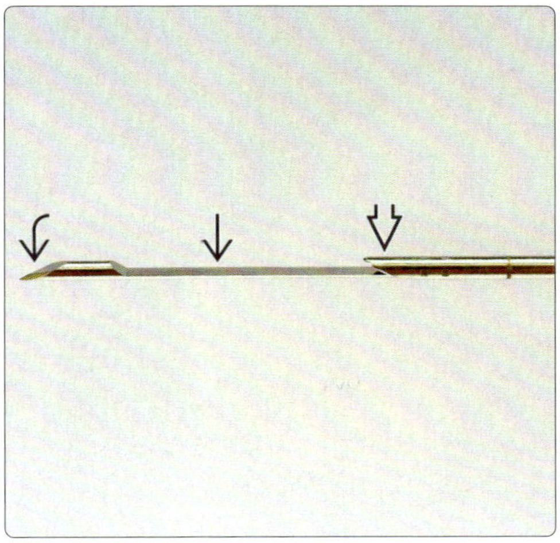

Coaxial Needle

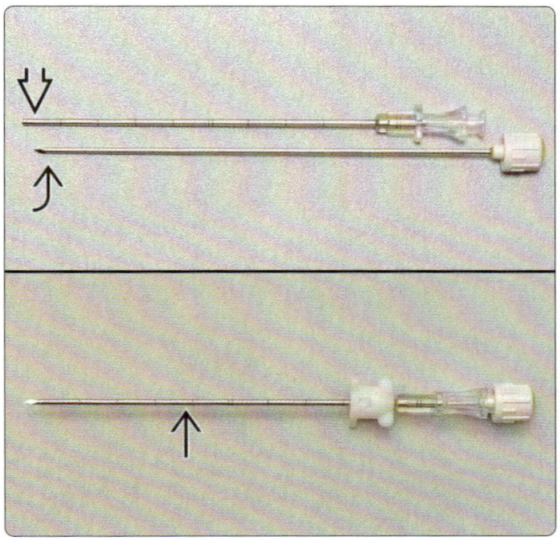

(Left) *Clinical photo shows an open Tru-Cut biopsy needle. Note the specimen slot and the sharp beveled needle tip. Pushing the plunger moves the outer cannula over the specimen slot, isolating a tissue sample.* **(Right)** *Clinical photo shows a coaxial needle. The outer cannula has a blunted end and an inner stylet with a sharp beveled tip. The stylet is placed inside the outer cannula during needle introduction. The coaxial technique allows multiple biopsies to be performed using a single puncture.*

Planning of Biopsy Trajectory

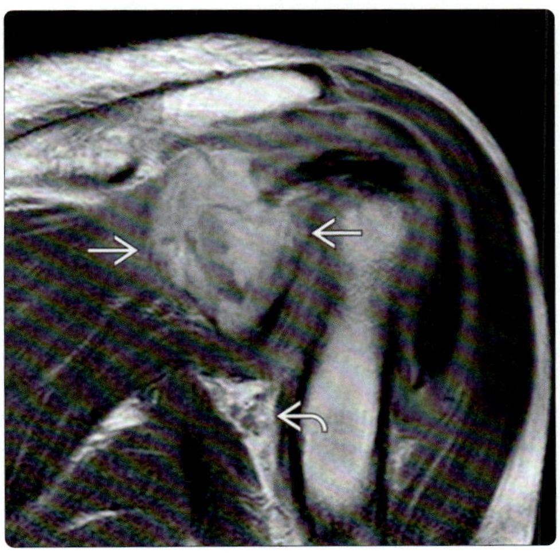

Planning of Biopsy Trajectory

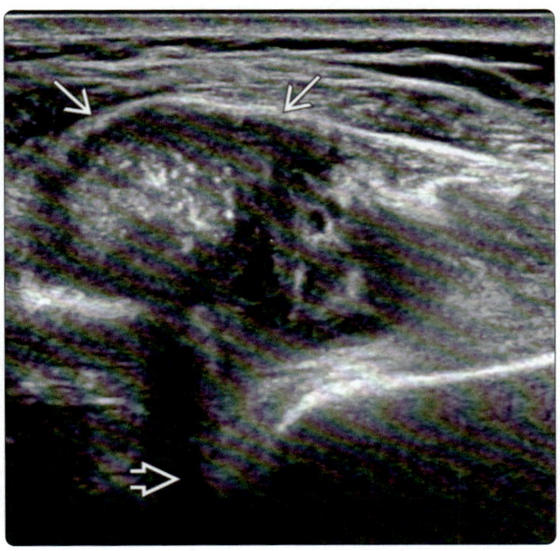

(Left) *Coronal oblique T1WI C+ MR shows an enhancing synovial tumor arising from the glenohumeral joint. The enhancing, soft, solid component should be targeted, avoiding the neurovascular bundle, which is inferior to the tumor.* **(Right)** *Transverse US of the same lesion shows a heterogeneous synovial mass arising from the glenohumeral joint. The biopsy pathway is planned to avoid the axillary neurovascular bundle.*

(Left) *Clinical photo shows the same patient after skin sterilization and application of sterile drapes. Local anesthetic (3 mL of 1% lignocaine) is infiltrated via a 23-g needle to the skin and subcutaneous tissue.* **(Right)** *Transverse US of the same patient shows the 23-g needle for injection of local anesthetic reaching the outer margin of the soft tissue tumor. The pseudocapsule around the tumor can be anesthetized, but the anesthetic needle should not enter the tumor.*

Injection of Local Anesthetic

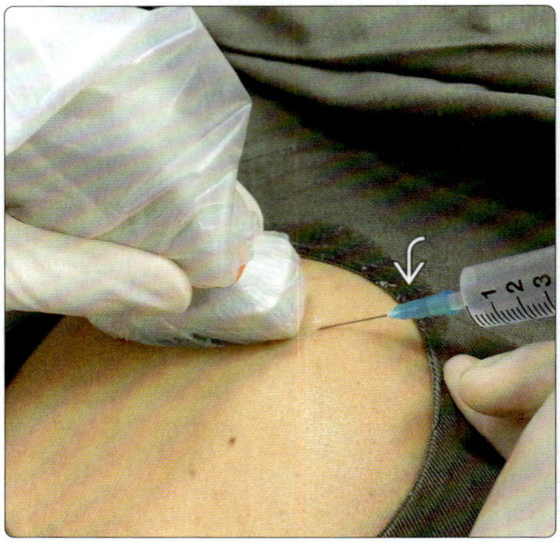

Injection of Local Anesthetic

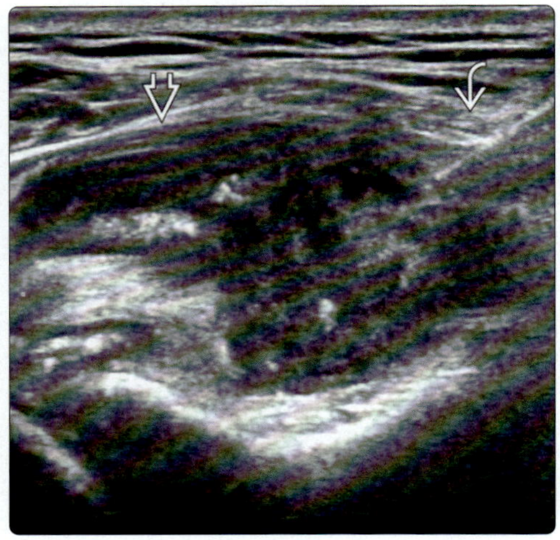

(Left) *Clinical photo shows the same patient after local anesthetic infiltration. A small skin incision is made at the site of the local anesthetic injection.* **(Right)** *Clinical photo shows the same patient with a coaxial needle introduced under US guidance via the small skin incision to the soft tissue tumor.*

Skin Incision for Biopsy

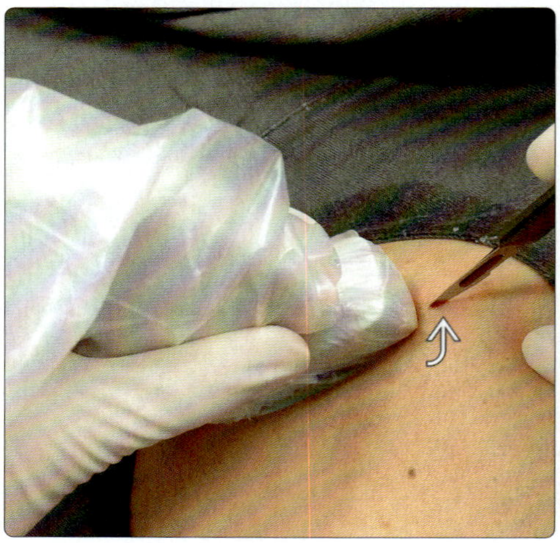

Insertion of Coaxial Needle

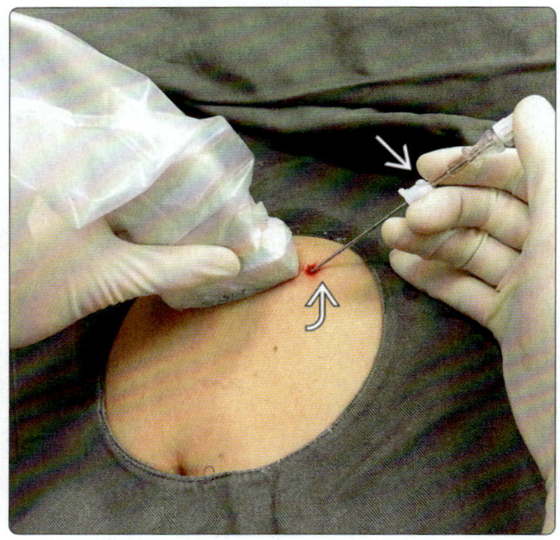

(Left) *Transverse US of the same patient shows the echogenic tip of the coaxial needle just piercing the tumor capsule. The tip of the coaxial system should be left in the outer margin of the tumor to optimize subsequent maneuverability.* **(Right)** *Clinical photo shows the same patient after removal of the inner stylet of the coaxial needle. The Tru-Cut biopsy needle is introduced and locked into the outer cannula of the coaxial needle.*

Insertion of Coaxial Needle

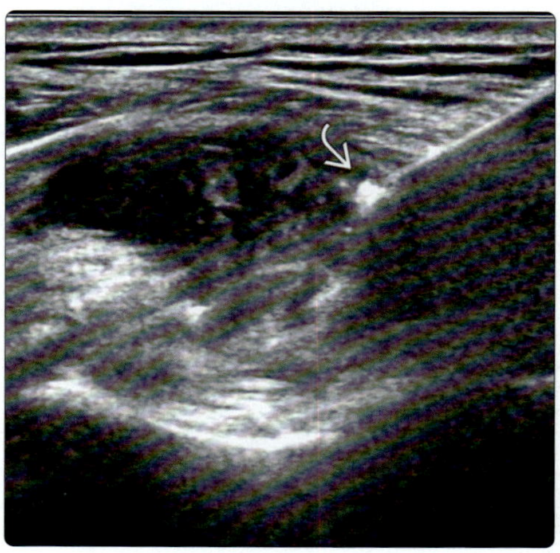

Insertion of Tru-Cut Needle

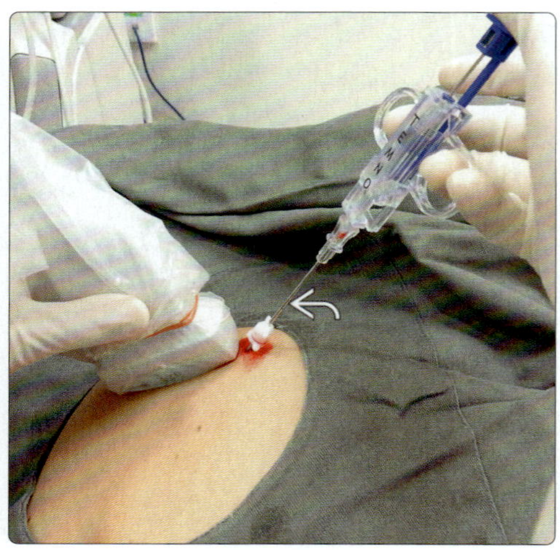

Insertion of Tru-Cut Needle

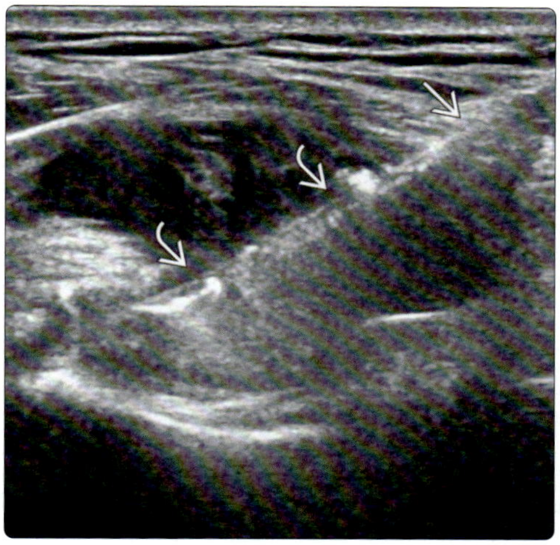

Tissue Core Obtained With Tru-Cut Needle

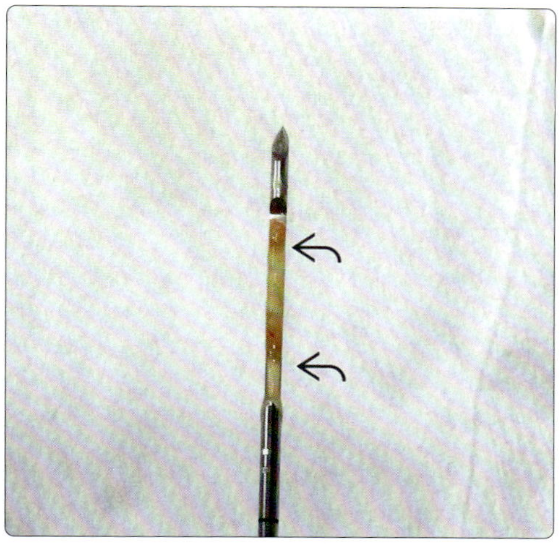

(Left) *Transverse US of the same patient shows advancement of the inner stylet beyond the coaxial needle ➡, exposing the specimen slot ➡ of the needle. Biopsy is performed with further pushing of the plunger of the Tru-Cut biopsy needle.* **(Right)** *Clinical photo shows a tissue core ⇨ within the specimen slot of the Tru-Cut biopsy needle. The specimen is removed from the needle by "washing" it in normal saline. Histology and microbiology revealed tuberculous infection.*

Insertion of Tru-Cut Needle

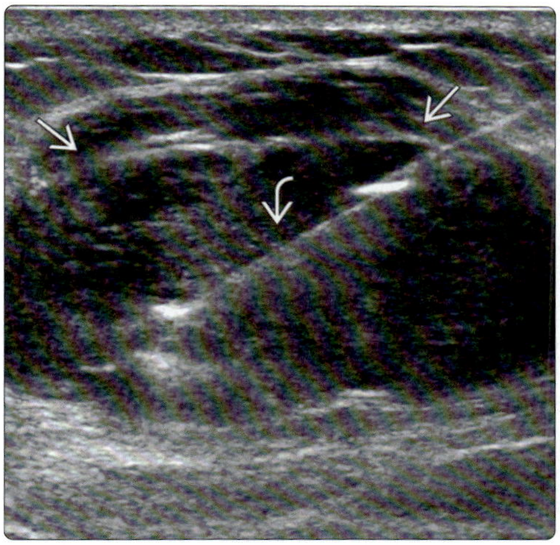

Multiple Biopsies Via Single Puncture

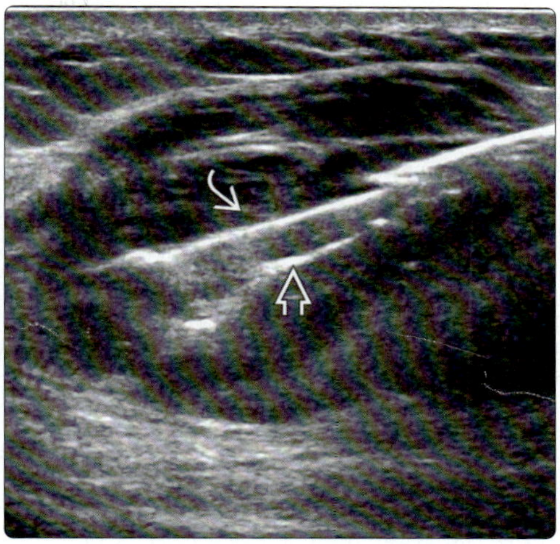

(Left) *Transverse US shows a Tru-Cut biopsy needle ➡ inserted into a soft tissue mass ➡ in the calf. Histology revealed a pleomorphic undifferentiated sarcoma. Note the advancement of the inner stylet of the Tru-Cut biopsy needle.* **(Right)** *Transverse US shows the Tru-Cut biopsy needle ➡ targeting the more superior portion of the tumor. The gas-filled tract ➡ from a previous biopsy is evident. In general, 3-4 nonfriable cores are adequate for histologic analysis.*

Tru-Cut Biopsy of Small Lesion

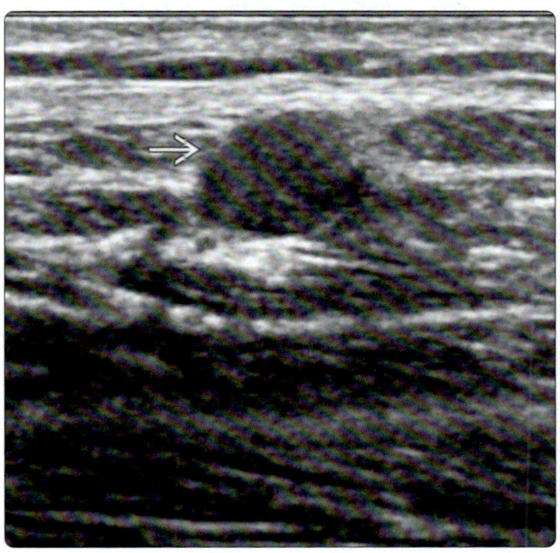

Tru-Cut Biopsy of Small Lesion

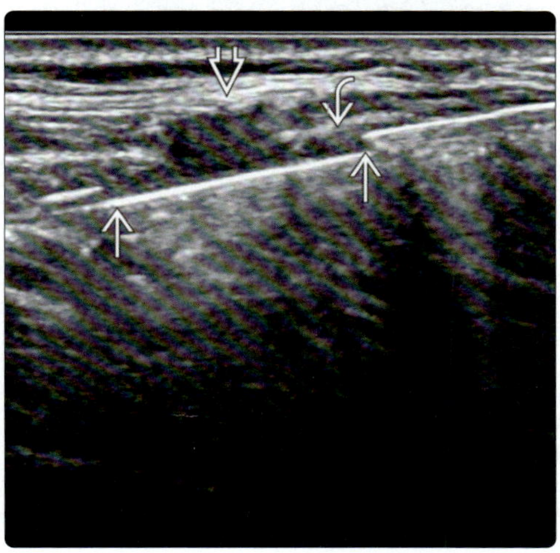

(Left) *Transverse US shows a small (11 mm) hypoechoic tumor ➡ within the soleus muscle in a patient with history of sarcoma. The US appearances may represent tumor recurrence.* **(Right)** *Transverse US shows that the specimen slot ➡ is positioned to include the small tumor ➡ as well as the adjacent normal tissue ➡. During biopsy of small lesions, it may be necessary to include normal tissue on either side of the mass, taking care to avoid any vital structures. Histology revealed tumor recurrence.*

Bone Tumor Biopsy

KEY FACTS

PREPROCEDURE

- 40% of bone tumors that require biopsy are amenable to US-guided percutaneous biopsy
- US-guided needle biopsy of bone tumors has diagnostic accuracy of > 98%

PROCEDURE

- Choose biopsy site bearing in mind site of subsequent surgical excision
  - Can be performed as outpatient procedure
  - If scheduling issues, US-guided core biopsy can be performed prior to MR staging
- Plan approach taking care to avoid any neurovascular bundle or other critical structures
- 3 potential scenarios
  - Extraosseous tumor or no bone cortex overlying lytic tumor: Coaxial needle and Tru-Cut needle technique is used
  - Thin bone cortex overlying lytic tumor: Coaxial needle used to penetrate cortex prior to insertion of Tru-Cut biopsy needle
  - Thinned but intact bone cortex overlying lytic tumor: Trephine biopsy needle used to penetrate cortex prior to insertion of Tru-Cut biopsy needle
- Color Doppler helps identify areas of viable tumor
- Take some cores from just inside edge of tumor
- Usually use 5-cm, 16-g Tru-Cut needle with 10-cm, 15-g coaxial needle
- Properly performed percutaneous bone biopsy is safe procedure with few complications
- **US fusion-guided biopsy**
  - Fuses previously obtained DICOM CT or MR data with real-time US
  - Expands applicability of percutaneous US-guided biopsy to bone lesions not visible on US
  - Diagnostic yield and accuracy of ~ 95%

Malignant Bone Tumor

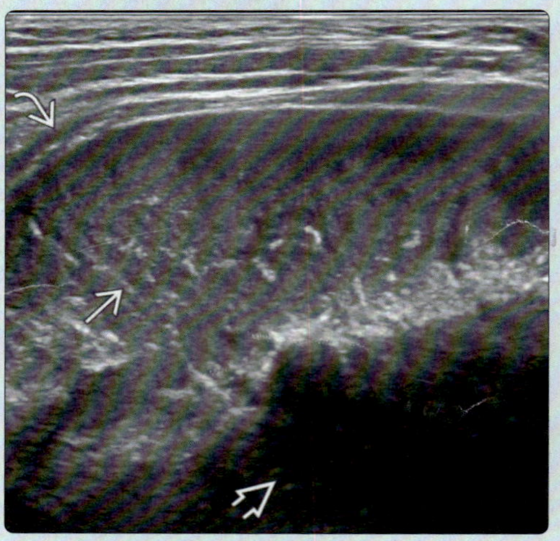

Malignant Bone Tumor

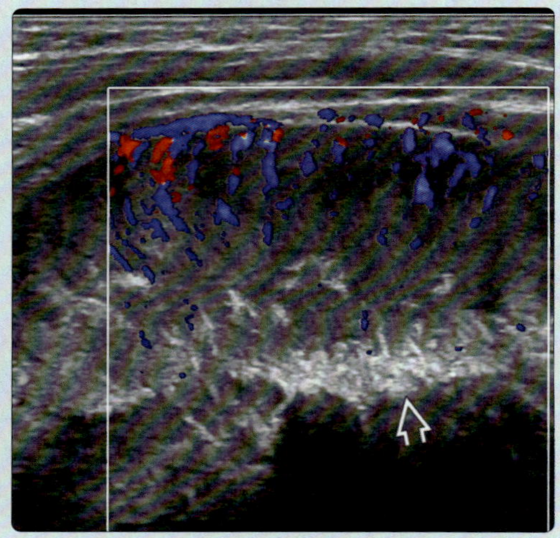

(Left) *Transverse US in 4-year-old boy with large, lytic, iliac bone lesion shows a large, extraosseous mass protruding from the ilium → deep to the iliacus muscle →. There are radiating reams of osteoid tissue → within the mass.* **(Right)** *Color Doppler US shows marked tumoral hyperemia, especially toward the periphery of the tumor. The spiculated tumoral new bone formation → is clearly seen.*

Setup for Bone Biopsy

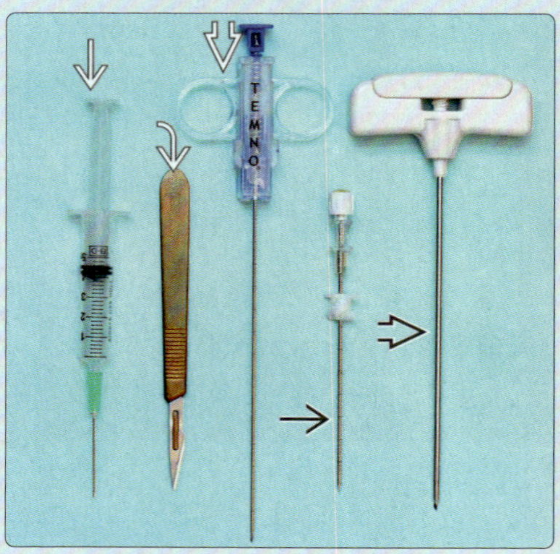

Bone Tumor Biopsy

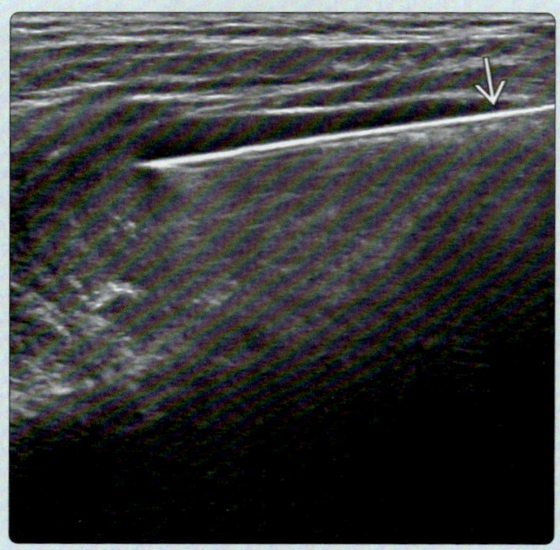

(Left) *Setup for US-guided bone biopsy shows local anesthetic →, scalpel →, Tru-Cut →, coaxial →, and trephine → biopsy needles. Tumor extent and cortical integrity determine type of needle used.* **(Right)** *Longitudinal US shows biopsy needle path →. At least 3 biopsies should be taken targeting mainly hypervascular areas as these are more likely to contain visible tumor. Histology revealed a malignant small round cell tumor, most likely an undifferentiated neuroblastoma.*

TERMINOLOGY

Definitions

- Percutaneous biopsy of tumor originating in bone

PREPROCEDURE

Indications

- Indeterminate or malignant bone lesion, which is either surface based, has extended through cortex, or has thin, intact overlying cortex, is amenable to US-guided biopsy
- 40% of bone tumors requiring biopsy are amenable to US-guided percutaneous biopsy
- US-guided needle biopsy of bone tumors has diagnostic accuracy of > 98%
- Main advantages of US-guided biopsy over CT-guided biopsy
  - Continuous real-time visualization of biopsy needle relative to neurovascular structures is possible
  - Areas of viable tissue can be identified and accurately targeted
  - More expeditious and cost effective, particularly if multiple biopsies are obtained
  - Does not occupy high-end CT technology
- Biopsy is best performed in center where definitive management of patient will take place

Contraindications

- Bone tumor inaccessible to US
- US-guided biopsy tract removed from surgical incision site
  - Ideally, biopsy site and surgical excision should be in same location
    - Only relative contraindication as generally acceptable to proceed if preferred incision does not concur with suitable biopsy tract
    - Risk of biopsy tract contamination with coaxial biopsy system is negligible
    - Nevertheless, ensure that modification is documented

Preprocedure Imaging

- Radiographs
  - Best for bone tumor characterization
- CT
- MR helps identify suitability for US-guided biopsy
  - If scheduling difficulty, US-guided biopsy can be done prior to MR in tumors with large, extraosseous mass
  - Effect of percutaneous biopsy on MR imaging characteristics is minimal
  - Recommendation that MR should always be done before biopsy is carryover from open biopsies in which negative effect on MR imaging was significant

Getting Started

- Things to check
  - Check clotting profile
  - Discuss risk of infection, allergy, and bleeding
  - Obtain informed, signed consent
  - Sedation rarely required
- Medications
  - It is advisable but not necessary to stop antiplatelet therapy, or even anticoagulation, prior to most biopsies
    - Bleeding is rarely problem as adequate tamponade of biopsy site is feasible with most musculoskeletal biopsies
    - Bleeding is much greater concern wherein adequate tamponade is not achievable, e.g., biopsy of inner aspect of iliac bone or deeper areas of spine
      - In these areas, ensure normal coagulation profile
  - Sedation or systemic analgesia usually not required
    - Systemic analgesia required for some CT-guided biopsies, which tend to be more painful due to deep-seated location of bone tumor
    - US-guided biopsy of pathologic fracture, particularly if only small extraosseous component, is often more painful and may require systemic analgesia

PROCEDURE

Patient Position/Location

- Helpful if orthopedic surgeon can mark planned surgery site on skin
  - Allows biopsy tract to be excised at time of surgery
- Biopsy can be performed as outpatient procedure
  - Even young patients (i.e., > 7 years) can undergo percutaneous biopsy under local anesthetic
  - Most very young patients (i.e., < 7 years) will require general anesthesia
  - Do not physically restrain patient to do biopsy
    - Instead, rearrange alternative date for biopsy under general anesthesia
- Perform preliminary US to ensure that tumor is suitable for US-guided needle biopsy and plan best approach to tumor
- Plan approach, taking care to avoid any neurovascular bundle or other critical structures (e.g., spermatic cord)
  - Use color Doppler imaging to identify hypervascular areas within tumor
    - Hypervascular areas contain viable tumor, which should be targeted for biopsy
- Patient is placed in either decubitus, supine, or prone position that will allow best access to planned biopsy path
  - Ensure that patient is comfortable
  - Remember that repositioning patient with pathologic fracture is difficult
    - In patient with pathologic fracture, try to perform biopsy without removing external fixation (backslab) device
- 3 possible US-guided techniques can be considered, depending on integrity of bone cortex
  - Extraosseous tumor or no bone cortex overlying lytic tumor: Coaxial needle and Tru-Cut needle technique is used
  - Very thin bone cortex overlying lytic tumor: Standard coaxial needle used to penetrate very thin cortex, then biopsy tumor with Tru-Cut needle inserted via coaxial needle
  - Thinned (but not very thinned), intact bone cortex overlying lytic tumor: Penetrate cortex using trephine needle
    - Trephine needle stilette is then removed and 16-g Tru-Cut needle passed through trephine cannula into lytic bone tumor

- Check that Tru-Cut biopsy needle is sufficiently long to acquire biopsy once opened within coaxial needle or trephine needle
  - Normally, 15-cm, 16-g Tru-Cut needle is used in conjunction with 10-cm, 15-g coaxial or, less commonly, 11-g trephine needle

Equipment Preparation

- Sterile trolley, drapes, gloves ± gown ± eye protector
- Protective sheath for US transducer
- Sterile cleaning solution and US gel
- 3-4 mL of lignocaine 1% local anesthetic
- 15-g coaxial needle and 16-g Tru-Cut biopsy needle

Procedure Steps

- Cleanse and drape skin
- Local anesthetic to skin ± periosteum
- Small stab incision in skin
- Use dominant hand to guide biopsy needle whenever possible and nondominant hand to hold transducer
- Advance coaxial needle into closest margin of tumor under real-time imaging
  - Only advance needle when needle tip is visible on real-time imaging
  - Use slow-rotating/pushing action of coaxial needle rather than pure pushing action to penetrate investing fascia and pseudocapsule of tumor
  - Maintain finger contact with skin when advancing needle to prevent needle from lunging forward
  - If cortex intact, apply further anesthetic to periosteum prior to cortex penetration
- Position coaxial needle tip in outer periphery of tumor
  - Maximizes ability to angle coaxial needle to biopsy different parts of tumor
- Remove inner stylet of coaxial needle and replace with biopsy needle
- Take biopsy only under real-time imaging
- Avoid cystic and necrotic areas, aiming to target viable tumor tissue for biopsy
  - Take some cores from just inside edge of tumor
- Sampling different parts of tumor ensures representative biopsy sample
- Number of biopsies taken depends on friability of tumor
  - If tumor is not friable, 3 good biopsy cores from different parts of tumor is adequate
  - If tumor is friable, may need to take > 10 biopsies from different parts of tumor, particularly tumor margins
- Apply direct pressure to biopsy site for 5 minutes after procedure or longer (15 minutes) if tumor hypervascular
- Apply simple dressing and observe within department for 1 hour after biopsy
- If no signs of bleeding at 1 hour (swelling, bruising, oozing from wound), patient can be discharged

Alternative Procedures/Therapies

- **US fusion-guided biopsy**
  - Fuses previously obtained DICOM CT or MR data with real-time US
  - Uses navigation system that allows accurate determination of transducer position
- US fusion-guided biopsy is well suited to percutaneous biopsy of bone lesions
  - Suitable for lesions > 1.5 cm with < 2.0 cm transosseous path and without adjacent vulnerable structure
  - Expands applicability for percutaneous US-guided biopsy of bone lesions not visible on US
  - Diagnostic yield (adequacy of biopsy specimen) of 94% and diagnostic accuracy (of specimen relative to final diagnosis) of 95%
- DICOM CT or MR data loaded onto US system
  - 2 electromagnetic sensors clipped onto bracket placed over US transducer
  - US transducer placed in same imaging plane as uploaded CT or MR data
  - 1 or more anatomical landmarks identified on both US and CT/MR
  - Embedded software then fuses US and CT/MR
  - Areas corresponding to those seen on US can be seen on CT or MR
  - Both US and CT/MR data can be viewed side by side or as image overlay where US overlaps CT or MR

OUTCOMES

Complications

- Most feared complication(s)
  - Neurovascular injury can be avoided by
    - Carefully recalling and identifying all nerves and vascular structures within needle path
    - Remembering that large tumors may efface or displace neurovascular bundles, must trace proximally or distally from tumor to identify any neurovascular bundle
    - Check anatomy if unsure regarding presence and location of nerve or vessels in vicinity of tumor
    - Only advancing biopsy needle under real-time imaging
    - Not allowing biopsy needle to pass into normal tissues beyond tumor
- Immediate/periprocedural complication(s)
  - Bent Tru-Cut needle due to forcefully opening biopsy needle against bone
    - Can be avoided by not forcefully opening biopsy needle and planning biopsy tract to avoid any residual tumoral bone
    - Biopsy needle should open freely without pressure
      - Exception being hard, fibrotic tumors
    - If resistance encountered, alter needle direction and continue opening biopsy needle within tumor
    - Alternatively, unlock coaxial Luer lock and allow biopsy needle to slide backward during biopsy
  - Pain is usually minor and, at most, requires mild analgesia for 24 hours
  - Slight bruising at injection site
- Delayed complication(s)
  - Infection and delayed hemorrhage
    - Both complications are interrelated and very rare
  - Properly performed percutaneous bone biopsy is safe procedure with few complications

SELECTED REFERENCES

1. Garnon J et al: Ultrasound-guided biopsies of bone lesions without cortical disruption using fusion imaging and needle tracking: proof of concept. Cardiovasc Intervent Radiol. 40(8):1267-1273, 2017

Metastatic Bone Tumor

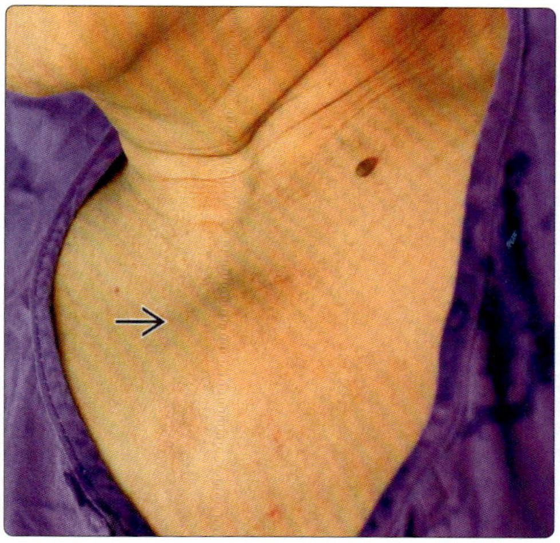

Metastatic Bone Tumor

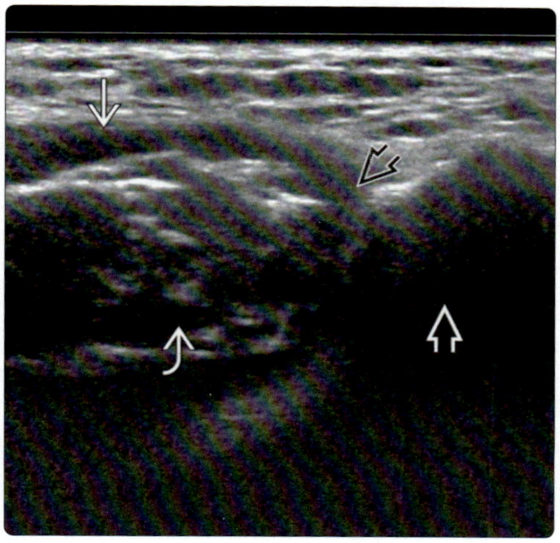

(Left) *Patient with left mastectomy for breast carcinoma 2 years previously shows recent onset of increasing upper chest wall mass →.* **(Right)** *Transverse US in the same patient shows infiltrative tumor with resorption at the medial end of the left clavicle →, abutting the manubrium →. There is a thin, hypoechoic, extraosseous tumor rim →. The manubrium is normal as is the sternomanubrial junction →. In this clinical context, this is highly suspicious of metastatic tumor.*

Metastatic Bone Tumor

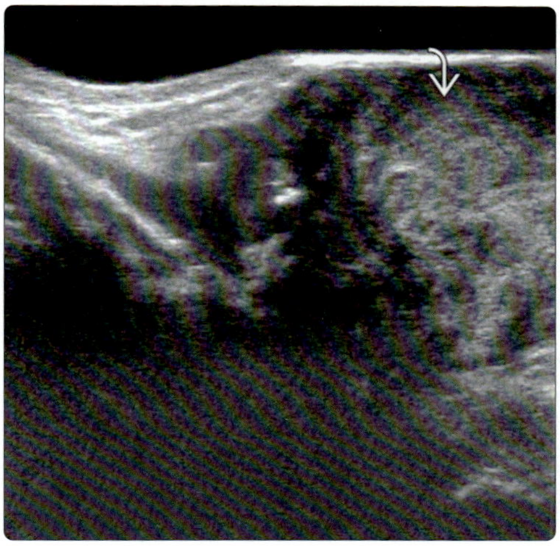

Metastatic Bone Tumor

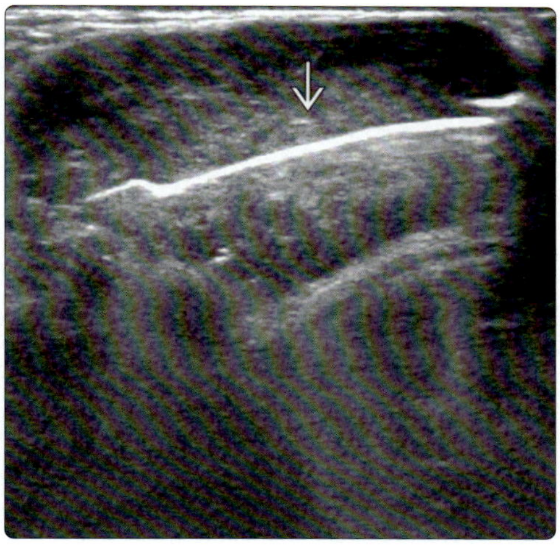

(Left) *Transverse US of the same patient shows a large, extraosseous sternal mass → with complete cortical resorption. This area of the tumor is much more suitable for US-guided biopsy than the medial clavicle.* **(Right)** *Transverse US shows the same sternal tumor during US-guided biopsy →. In chest wall lesions, try to use a horizontal rather than a vertical biopsy approach whenever possible to minimize risk to intrathoracic structures. Histology revealed metastatic breast carcinoma.*

Primary Bone Tumor

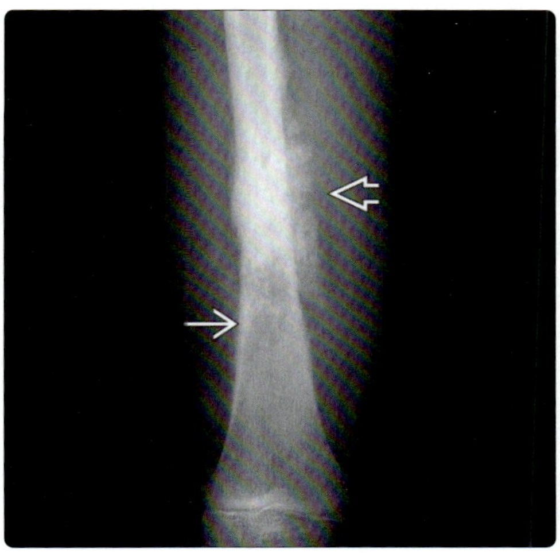

Primary Bone Tumor

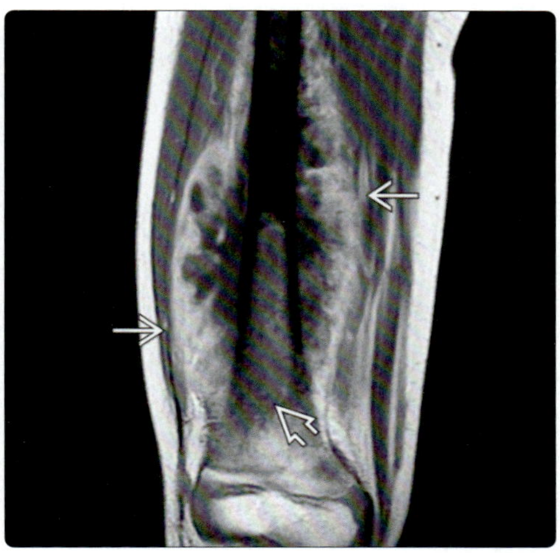

(Left) *Frontal radiograph in a skeletally immature patient shows severe, infiltrative-like osteolysis → with extensive tumoral new bone formation → in the midfemoral diaphysis with soft tissue swelling.* **(Right)** *Coronal T1WI C+ MR of the same patient more clearly shows the extraosseous tumor extent → and a greater degree of intramedullary tumor extent → than seen radiographically. MR imaging is usually advisable prior to any US-guided bone tumor biopsy.*

(Left) *Transverse color Doppler US in the same patient shows a moderately hypervascular, mixed solid/cystic extraosseous tumor infiltrate. The underlying femoral cortex is largely intact.* **(Right)** *Transverse US during US-guided biopsy shows how the more solid component of the extraosseous tumor has been selected for biopsy. The 16-g Tru-Cut needle has been inserted through a 15-g coaxial needle. Histology revealed osteosarcoma.*

Primary Bone Tumor

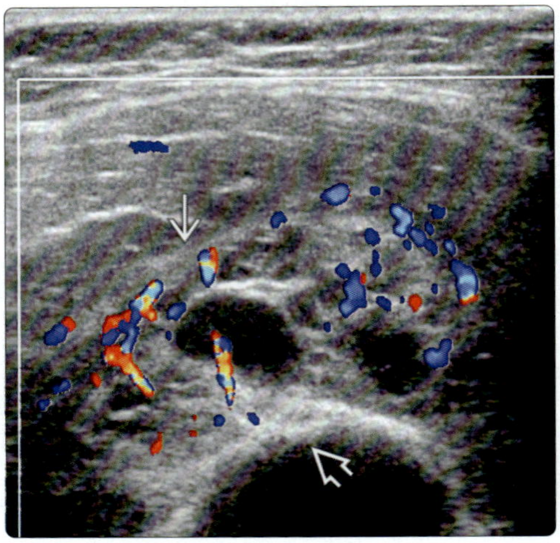

Primary Bone Tumor

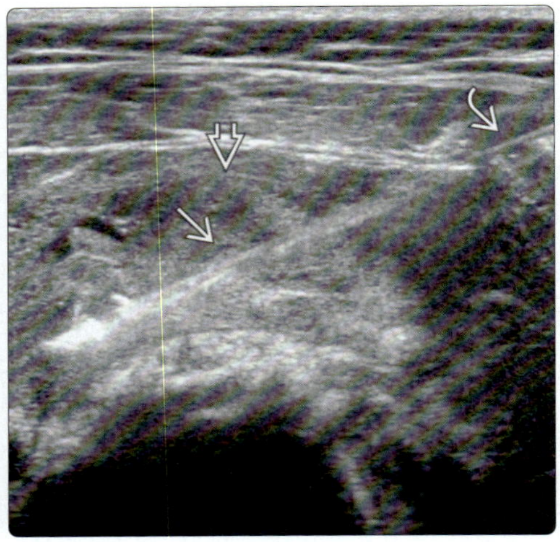

(Left) *Lateral radiograph of the distal humerus in a young patient with clinically suspected infection shows quite marked expansion of the distal humerus, medullary osteolysis, mature-type periosteal new bone formation, and moderate soft tissue swelling, consistent with chronic osteomyelitis.* **(Right)** *Sagittal T1WI C+ FS MR in the same patient shows a medullary sinus tract and abscess extending to the dorsum of the distal humerus, consistent with acute exacerbation of chronic osteomyelitis.*

Chronic Osteomyelitis

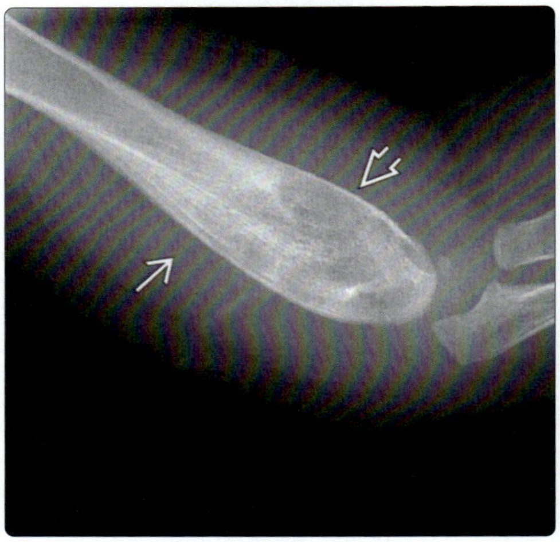

Chronic Osteomyelitis

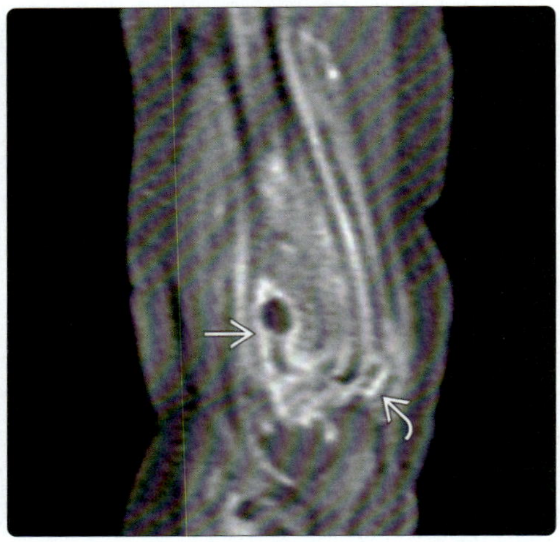

(Left) *Transverse US in the same patient shows thinning and breaching of the distal humeral cortex with focal juxtacortical edematous inflammatory tissue deep to an edematous triceps muscle.* **(Right)** *Transverse US shows the same area during a US-guided biopsy of the juxtacortical inflammatory tissue using a coaxial technique with a 16-g Tru-Cut biopsy needle. Culture of this biopsy material grew Staphylococcus aureus.*

Chronic Osteomyelitis

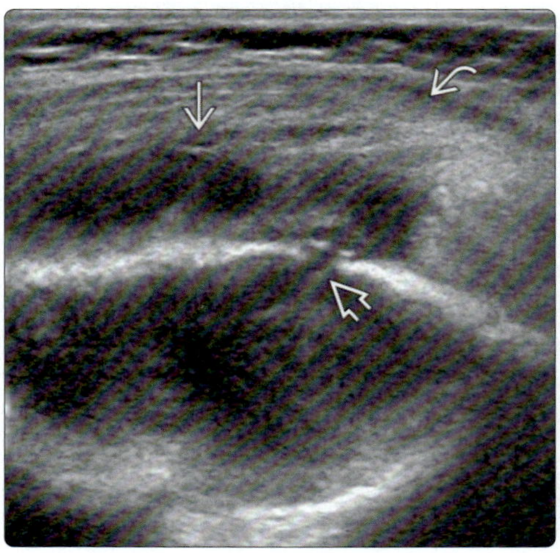

Chronic Osteomyelitis

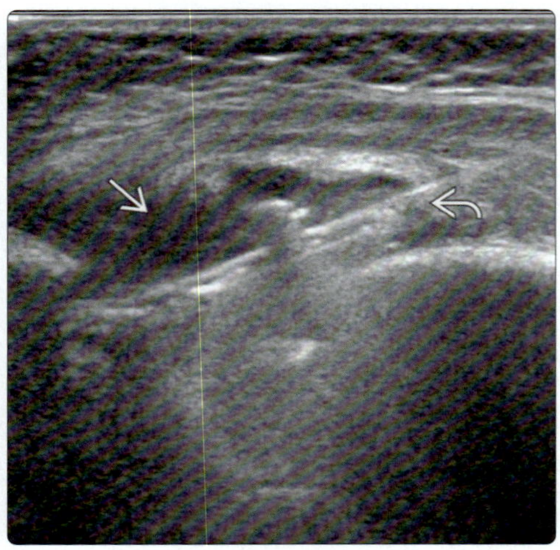

Chronic Osteomyelitis

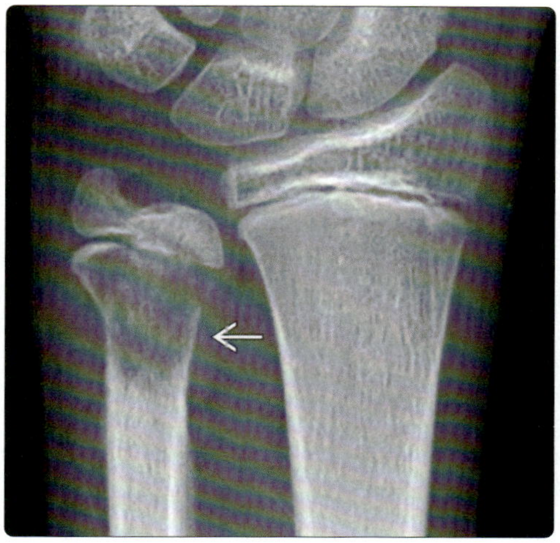

Chronic Osteomyelitis

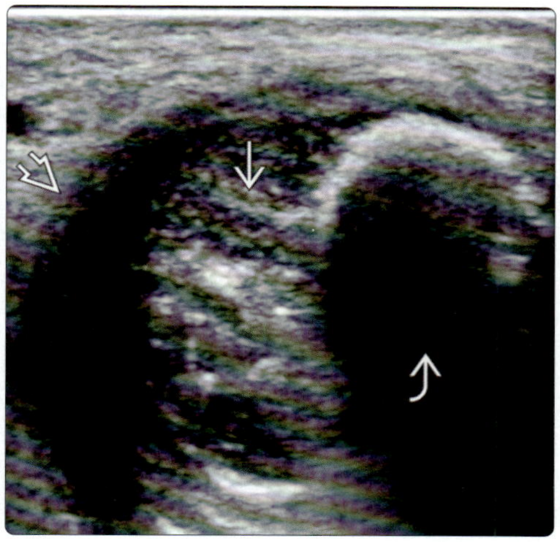

(Left) *Frontal radiograph in a 14 year old with fever and wrist pain shows a lytic lesion with a wide zone of transition in the distal ulnar metaphysis .* **(Right)** *Transverse US of the same patient shows a prominent, solid-appearing, juxtacortical soft tissue mass surrounding the distal ulna with echogenic seams of new bone formation . An outer hypoechoic area representing fluid or severe edema is present . US-guided biopsy revealed chronic osteomyelitis.*

Primary Bone Tumor

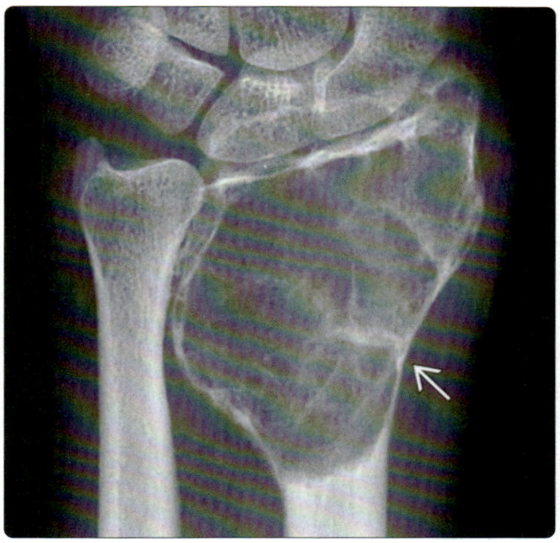

Primary Bone Tumor

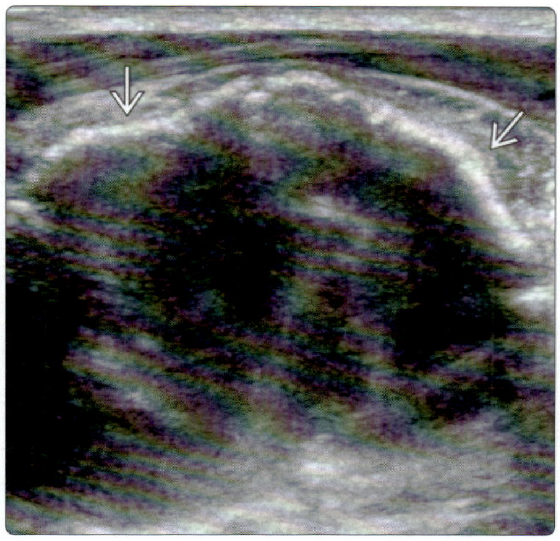

(Left) *Frontal radiograph shows a well-defined, expanded subarticular osteolytic lesion with a soap-bubble appearance in the distal radius, consistent with a giant cell tumor. The cortex is intact but very thinned.* **(Right)** *Transverse US of the same patient shows an expanded, intact cortex of the distal radius. One must rely on other imaging to appreciate that the cortex is thinned and that the tumor is osteolytic. Although the cortex is intact, the tumor is still amenable to US-guided biopsy.*

Primary Bone Tumor

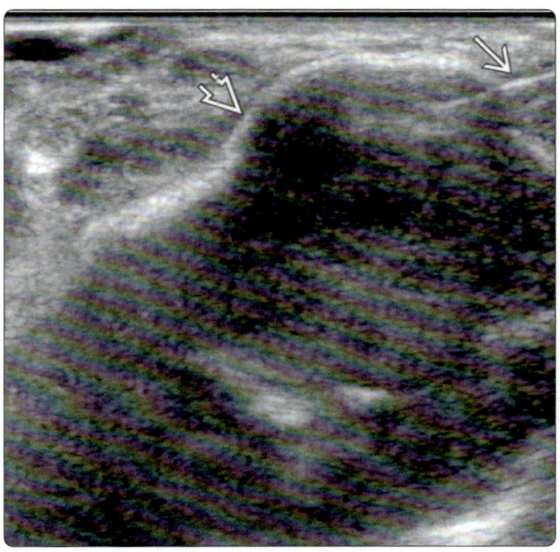

Primary Bone Tumor

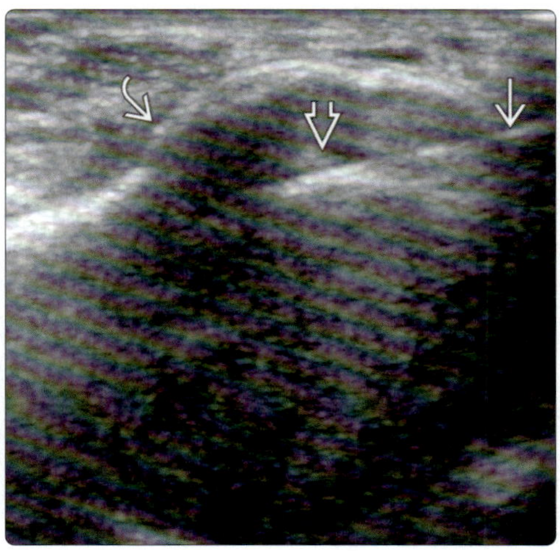

(Left) *Transverse US of the same tumor shows a 15-g coaxial needle just after piercing the thin cortical shell of the distal radial tumor. Gentle sustained pressure/rotation on the coaxial needle allows penetration of the cortex.* **(Right)** *Transverse US of the same tumor shows a 16-g Tru-Cut needle inserted through a 15-g coaxial needle to biopsy the distal radial tumor . Several biopsies from different parts of the tumor revealed giant cell tumor of bone.*

(Left) *Clinical photograph of a 19-year-old man shows swelling* → *on the medial aspect of the forefoot, which had been slowly growing for 3 months.* **(Right)** *Frontal radiograph of the same patient shows a markedly expanded osteolytic tumor with thinned but seemingly intact overlying cortex* → *at the proximal end of the 1st metacarpal bone. Although there is no extraosseous soft tissue component, this tumor is suitable for US-guided biopsy.*

Primary Bone Tumor

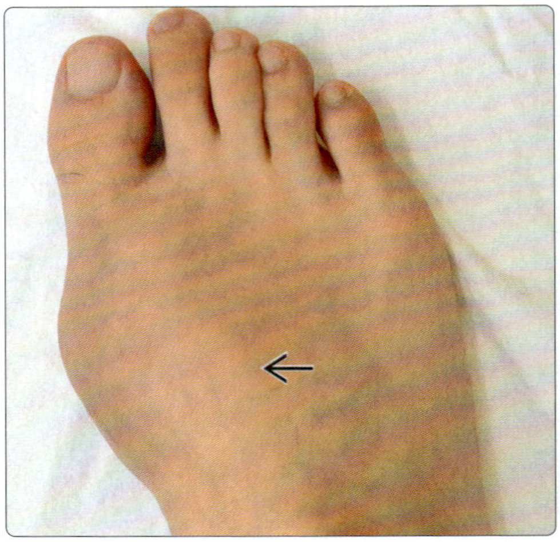

Primary Bone Tumor

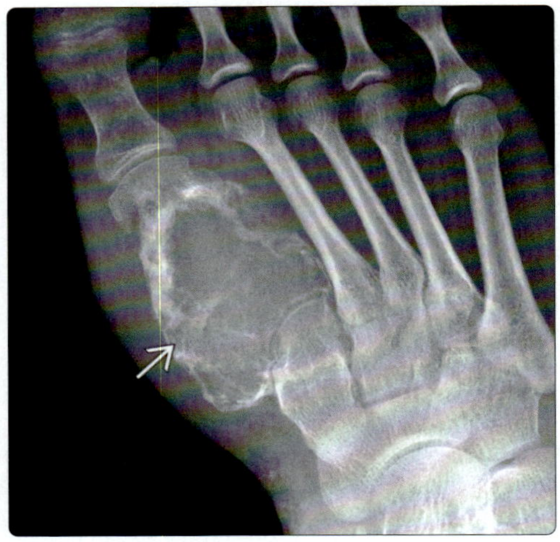

(Left) *Transverse color Doppler US of the same patient shows an osteolytic tumor of the 1st metacarpal bone. The extremely thinned cortex is barely visible on US* →*. This thinned cortex causes so little beam attenuation that the tumor and its vascularity can still be readily appreciated.* **(Right)** *Transverse US of the same patient during bone biopsy shows a 15-g coaxial needle* → *just after piercing the thin cortex* → *overlying the tumor.*

Primary Bone Tumor

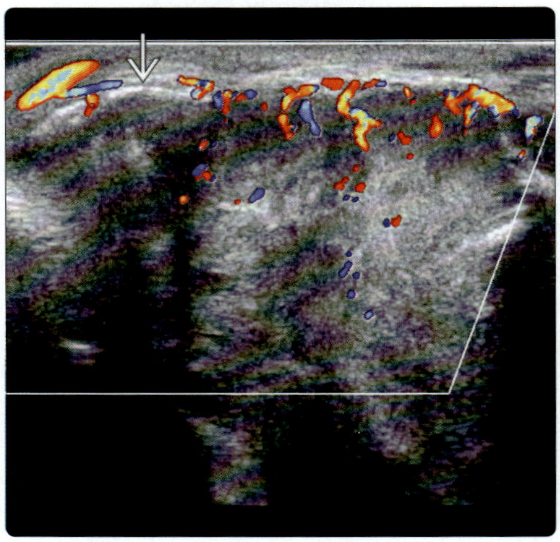

Primary Bone Tumor

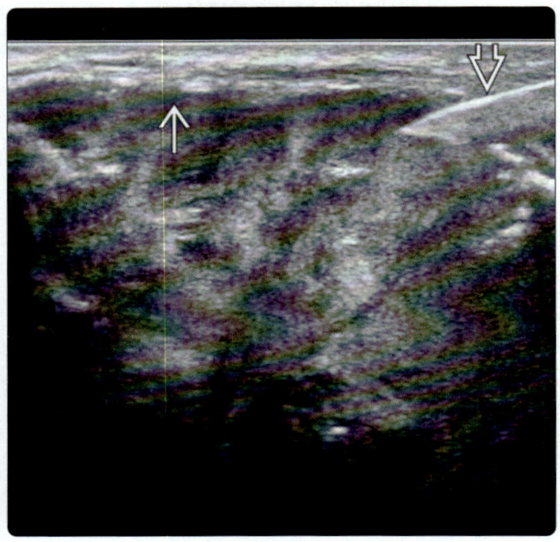

(Left) *Transverse US of the same patient shows a 16-g Tru-Cut needle* → *inserted though the 15-g coaxial needle* → *to biopsy the central part of a metacarpal bone tumor.* **(Right)** *Transverse US of the same tumor shows the 16-g Tru-Cut* → *and coaxial needle* → *repositioned to biopsy a more superficial part of the metacarpal bone tumor. At least 3 biopsies from different parts of the tumor should be obtained. Histology revealed a giant cell tumor.*

Primary Bone Tumor

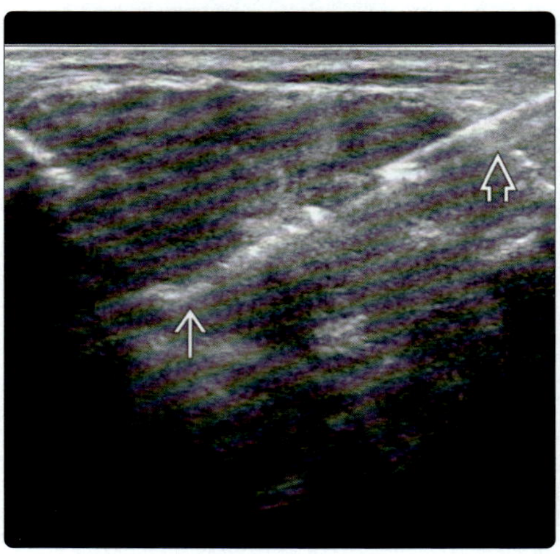

Primary Bone Tumor

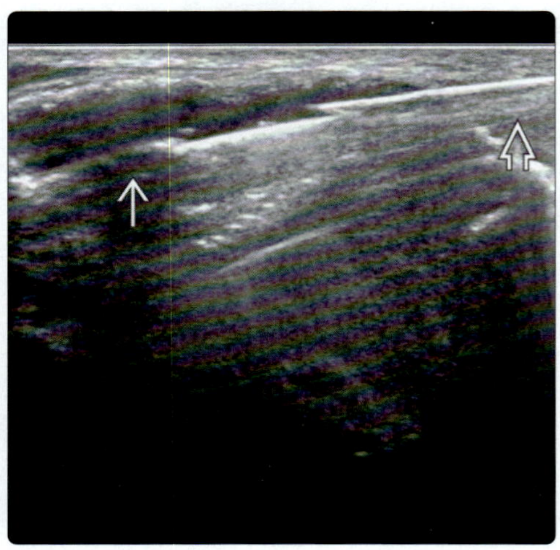

Primary Bone Tumor

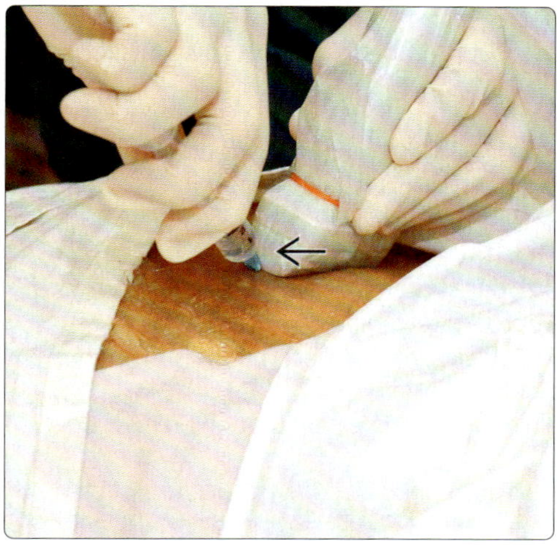

Primary Bone Tumor

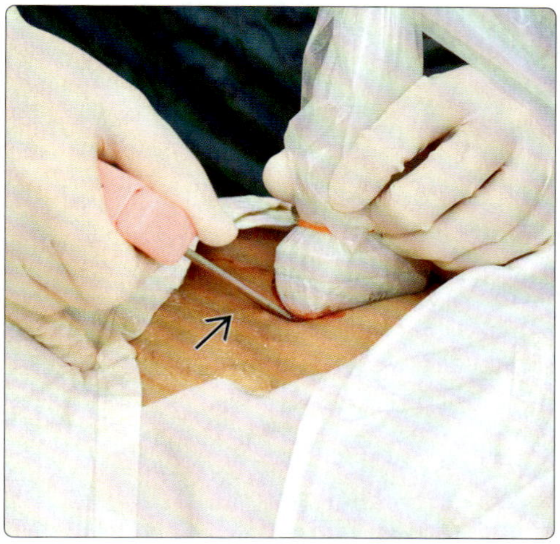

(Left) *Clinical photograph shows local anesthetic infiltration* ⇨ *at the site chosen for percutaneous biopsy of a sacral tumor. The dermal and periosteal tissues should be preferentially infiltrated.* **(Right)** *Clinical photograph shows insertion of an 11-g trephine biopsy needle* ⇨ *following a short stab incision with a scalpel along the chosen biopsy path. Once the cortex of the tumor is penetrated, the inner stylet is removed from the outer cannulated trephine biopsy needle.*

Primary Bone Tumor

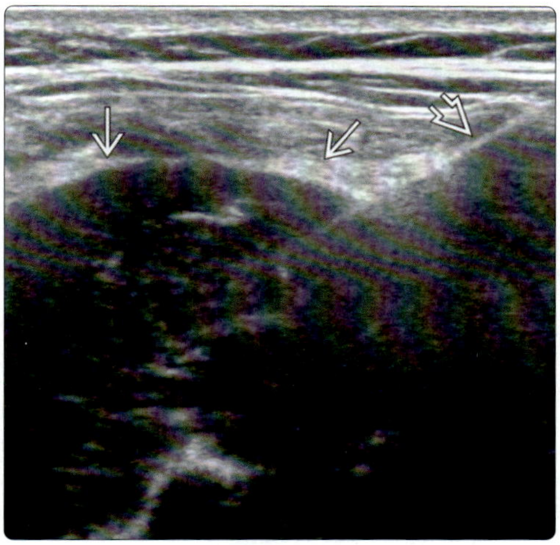

Primary Bone Tumor

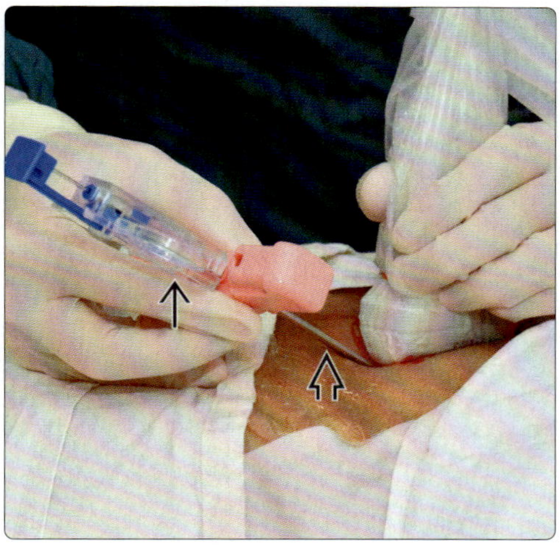

(Left) *Longitudinal US of the same patient shows the trephine needle* ➡ *piercing the thinned, expanded cortex* ➡ *of the sacral tumor. A Tru-Cut needle alone would usually not be able to pierce an intact, thinned bone cortex.* **(Right)** *Clinical photograph shows a 15-g Tru-Cut needle* ⇨ *introduced into the tumor via the cannula of the trephine biopsy needle* ⇨ *following removal of the outer stylet.*

Primary Bone Tumor

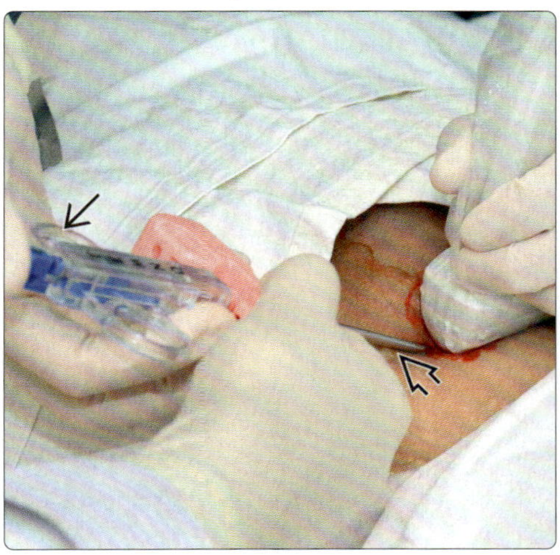

Primary Bone Tumor

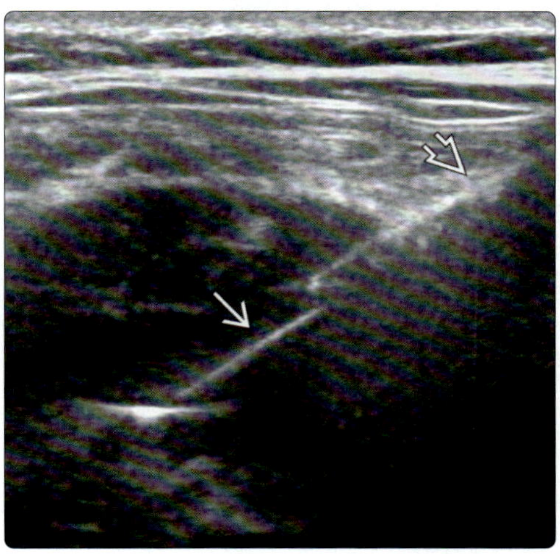

(Left) *Clinical photograph of the same patient during biopsy shows depression of the biopsy needle plunger* ⇨ *under real-time US guidance. An assistant is securing the position of the trephine needle* ⇨ *at this stage.* **(Right)** *Longitudinal US of the same patient shows specimen slot of the Tru-Cut needle* ➡ *extending beyond the tip of the trephine biopsy needle* ➡. *All needle advancement is performed under real-time US guidance. Histology revealed a chordoma.*

Joint Injection: Upper Limb

KEY FACTS

PROCEDURE

- Glenohumeral joint injection
  - Anterior (rotator interval) or posterior approach
    - Anterior approach preferable
      - Less painful & more comfortable position for patient
      - Needle tip positioned on superficial aspect humeral head cartilage in rotator cuff interval
    - Posterior approach
      - Easier to learn & perform
      - Long-axis approach may be either from medial to lateral or from lateral to medial
- Elbow joint injection
  - Posterior or lateral approach
    - Posterior approach preferable as
      - Joint space is larger & easier to perform
      - "Bare area" can be identified for injection
    - Lateral approach is more conventional
      - But joint space is tight laterally
      - "Bare area" can be identified for injection
  - Avoid anterior approach as risk of neurovascular bundle injury
- Wrist joint injection
  - Radiocarpal, distal radioulnar, & midcarpal joints
    - Always dorsal approach
    - Avoid volar approach as risk of neurovascular bundle injury
    - Long axis for radiocarpal & midcarpal joint injections
    - Short axis for distal radioulnar joint injection
    - Try to avoid tendons in radiocarpal & distal radioulnar joint injection
    - Difficult to avoid transversing tendons in midcarpal injection
    - Needle tip positioned on superficial aspect of cartilage
  - Identify & avoid both superficial branch of radial nerve & terminal part posterior interosseous nerve

Needle Placement for Joint Injection

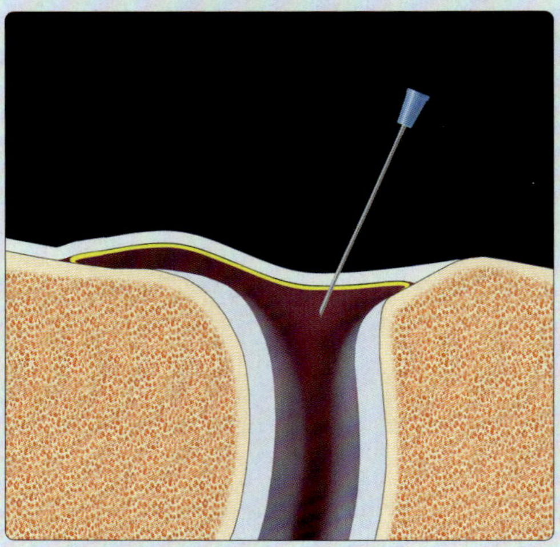

Avoid Cartilage Injury

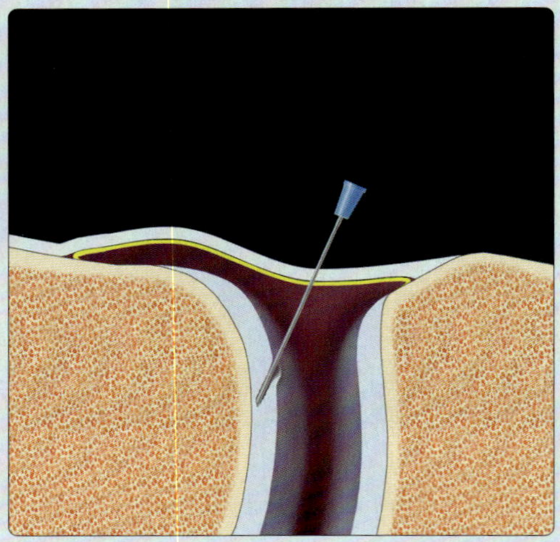

(Left) *The needle tip should just pass through the capsule and synovium into the superficial aspect of the joint. The synovium can be difficult to see clearly if the joint is not distended by fluid. Place the needle tip in the joint cavity between the opposing articular cartilage surfaces.* **(Right)** *The needle tip should not be forced deeply into the joint, particularly if the tip is not well seen as this may cause the injury to the articular cartilage, as shown.*

Other Needle Tip Positions

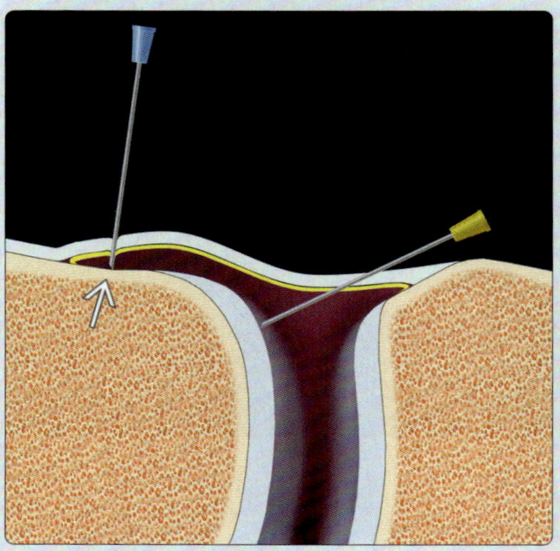

Out of Plane Approach: Practical Tips

(Left) *One can place the needle tip onto the "bare area" ➔ covered by synovium, although it is devoid of articular cartilage (blue needle) as injection can be closely monitored. Not every joint has a "bare area" and alternatively, one can place the needle tip just superficial to the articular cartilage (yellow needle).* **(Right)** *For an out-of-plane approach, to avoid partial leakage of injectant outside the joint (upper diagram), one can use either a needle with a short bevel (left lower) or align the needle obliquely (right lower).*

TERMINOLOGY

Definitions

- Percutaneous joint injection for shoulder, elbow, & wrist joints under US guidance

PREPROCEDURE

Indications

- Diagnostic purpose: Injecting contrast for conventional, CT, or MR arthrography
- Therapeutic purpose: Injecting drugs (steroid, hyaluronic acid, platelet-rich plasma, etc.) to relieve pain & reduce inflammation

Contraindications

- Uncorrected bleeding diathesis
- Uncooperative patient
- Active infection in overlying soft tissues

Getting Started

- Medications
  - Local anesthetic is optional & usually not required
  - Sedation of anxious patient or young children rarely required
    - Intravenous midazolam or other benzodiazepine; antidote should be available
  - Contrast
    - Conventional or CT arthrography: Iodinated contrast, e.g., Omnipaque 300
    - MR arthrography: 0.1 mL of diluted gadolinium in 10 mL normal saline solution
  - Steroid
    - Short (hydrocortisone) or long acting (dexamethasone, methylprednisolone)
  - Hyaluronic acid
  - Platelet-rich plasma
- Equipment list
  - Low- to high-frequency (5-12 MHz) transducers
  - Sterile set
  - Spinal needles
    - 22-g for shoulder & elbow injection
    - 25-g for wrist injection
  - Extension tube (thin, 250 mm long, 8.5-Fr, 0.4 mL dead space, no Luer lock)
  - 3-way connector
  - Syringes for LA, aspiration, & injection

PROCEDURE

Procedure Steps (General)

- Use free-hand technique to advance needle into joint space under real-time imaging
- Can use long-axis (in-plane) or short-axis (out-of-plane) approaches
  - In-plane approach
    - Easier to learn & perform as whole needle track can be well seen but longer needle path from insertion site to target
  - Out-of-plane approach
    - More difficult to learn & perform, needle tip can be seen only close to target
    - Useful when intervening structures on in-plane approach, such as nerves, need to be avoided
- Ensure on real-time imaging that needle tip is within joint space
  - Ideally, needle tip should just pass through synovium & be within superficial part of joint
    - This helps minimize injury to articular cartilage, labrum, or intracapsular ligament
  - Synovium is difficult to separate from subsynovial fat if there is no joint effusion
  - Contacting cartilage with needle tip is not harmful
    - Do not force needle tip into cartilage
  - One can often locate "bare area" between capsule & cartilage where needle tip can be positioned
  - If moderate to large effusion present, aspirate joint fluid prior to injection
- Avoid injection of air by flushing syringe & tubing with injectate
- Free flow of fluid into joint with low-pressure injection & minimal patient discomfort helps affirm intraarticular position of needle tip
- Avoid active exercise for 20 minutes following injection to prevent leakage along needle track

Shoulder (Glenohumeral Joint)

- Posterior approach
  - Various positions can be used
    - Lateral decubitus position with arm mildly adducted
    - Prone position with arm in external rotation, palm facing down, & shoulder resting on pillow
    - Sitting position with patient's back facing you & ipsilateral hand resting on contralateral shoulder (be alert for vasovagal syncope)
  - Scan in transverse plane to identify glenoid, labrum, posterior glenohumeral joint, & humeral head deep to infraspinatus
    - Moving arm passively during scanning will help identify glenohumeral joint
  - Either short-axis or long-axis needle approach can be used
    - For long axis, can approach from medial to lateral or from lateral to medial
    - For medial to lateral approach, needle tip is placed deep to posterior labrum in joint cavity
    - For lateral to medial approach, needle tip is positioned on "bare area" of humeral head posteriorly
      - "Bare area" is covered by synovium but not by articular cartilage
      - Lies deep to infraspinatus muscle & tendon
      - "Bare area" lies superficially, allowing needle tip & injection to be monitored throughout procedure & avoiding injury to labrum & cartilage
- Anterior approach (rotator interval approach): Preferred method
  - Less painful, more comfortable position to patient than posterior approach
    - But technically more difficult than posterior approach, as space is smaller
  - Patient in supine position with arm partially externally rotated

- Avoid extreme external rotation, as it increases tension on anterior shoulder & increases risk of extraarticular leakage
  - Biceps tendon runs in rotator interval between subscapularis & supraspinatus tendons
  - Identify coracoid process, which is medial to rotator interval
  - Needle entry point is just superior to subscapularis tendon, inferior to supraspinatus tendon, & lateral to coracoid process
  - Needle position should be medial to biceps tendon
  - Needle is advanced until tip just superficial to articular cartilage of humeral head
  - Either short-axis or long-axis needle approach can be used

Elbow Joint

- Lateral approach
  - More conventional, classic method & ergonomic in terms of patient positioning, but more difficult & time-consuming to locate injection site
    - Potential risk to radial neurovascular bundle
    - Needle tip placed into smaller space & resistance during injection is usually high
  - Patient sits next to table with flexed elbow resting on table & lateral aspect uppermost
  - Position probe longitudinally along lateral part of elbow
    - Landmarks: Radial head & capitellum
  - Needle tip positioned on "bare area" at lateral aspect of radial head
    - "Bare area" covered by synovium but not by articular cartilage
    - "Bare area" lies just deep to lateral joint capsule, radial collateral ligament, & common extensor tendons
- Posterolateral approach: Preferred method
  - Preferred as no major neurovascular structures posterolaterally
  - Patient positioned with elbow flexed to 90° & forearm fully pronated
    - Access via trochlear-olecranon articulation
  - Position probe sagittally along posterior part of elbow
  - Move transducer laterally until triceps just goes out of view
    - Landmarks: Olecranon fossa of humerus, posterior fat pad, olecranon
  - Needle tip is positioned on "bare area" at olecranon fossa
    - "Bare area" is covered by posterior joint capsule, posterior fat pad, & synovium but not articular cartilage
- Other approaches: Neurovascular injury risk with anterior or medial approaches

Wrist Joint

- Radiocarpal joint
  - Dorsal approach via radioscaphoid joint space
  - Place probe in sagittal plane at level of Lister tubercle distal radius
    - Usually, area just distal to Lister tubercle is free of transversing extensor tendons
    - This area is ideal for needle insertion & avoids injury to extensor tendons
  - Enter 1 cm distal to radioscaphoid joint space
  - Direct needle proximally toward articular cartilage of distal radius
  - Once radial articular cartilage is contacted, do not advance needle tip any further
  - Avoid superficial branch of radial nerve & its terminal medial & lateral branches
    - Lie superficial to 1st extensor compartment (abductor pollicis longus, extensor pollicis brevis) or between 1st & 2nd extensor compartment (extensor carpi radialis longus & brevis) tendons
  - Avoid posterior interosseous nerve
    - Lies deep in interspace between 3rd (extensor pollicis longus) & 4th extensor compartment (extensor digitorum longus) tendons radial to Lister tubercle before it divides into terminal branches ulnarly
- Distal radioulnar joint
  - Dorsal approach
  - Place probe in transverse plane
  - Insert needle from radial side & direct toward ulnar head
  - Insert needle either between 4th (extensor digitorum longus) & 5th (extensor digiti minimi) or between 5th (extensor digiti minimi) & 6th (extensor carpi ulnaris) extensor compartments
    - Try not to transgress tendon with needle, as this is painful & may cause tendon injury
  - Once ulnar articular cartilage is contacted, do not advance needle tip any further
- Midcarpal joint
  - Dorsal approach via lunocapitate joint
  - Place probe in sagittal plane
  - Usually, traversing needle through extensor digitorum tendons cannot be avoided
- Other approaches
  - Neurovascular injury risk with volar approach

OUTCOMES

Complications

- Most feared complication(s)
  - Infection
- Immediate/periprocedural complication(s)
  - Vasovagal reaction
    - Perform procedures with patient lying down whenever possible
  - Bleeding
    - Avoid vascular injury
    - Check & amend clotting profile if abnormal
    - Use thinnest needle possible & minimalist technique in patients with bleeding diathesis
- Delayed complication(s)
  - Pain 1-2 days after steroid injection due to steroid-related microcrystalline synovitis (10%, steroid flare)
    - Will settle with simple oral analgesia
  - Flushing of skin 2-3 days after steroid injection

SELECTED REFERENCES

1. Rastogi AK et al: Fundamentals of joint injection. AJR Am J Roentgenol. 207(3): 484-94, 2016

Shoulder: Posterior Approach

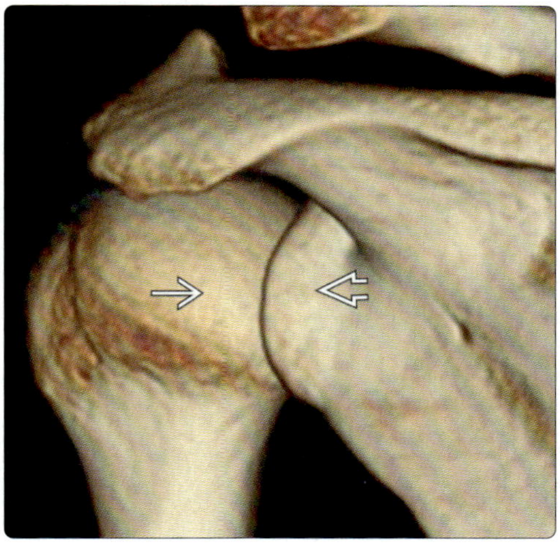

Shoulder: Posterior Approach

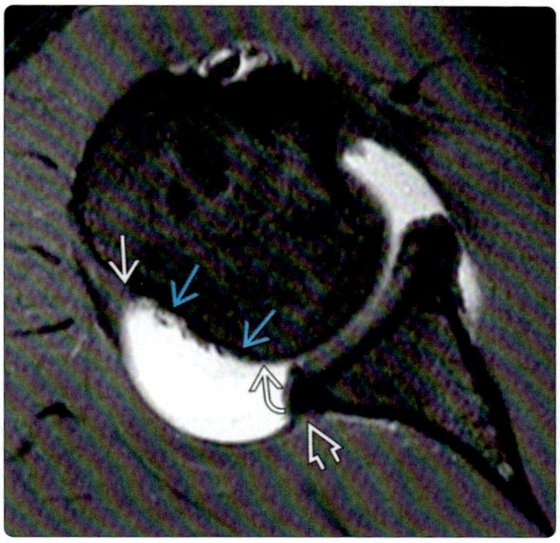

(Left) *Volume-rendered CT shows the bony anatomy of the posterior aspect of the glenohumeral joint between the glenoid ⮕ and humerus ⮕.* **(Right)** *Axial T1 FS MR shows attachment of the posterior capsule to the humeral head ⮕ and glenoid ⮕. There is a quite large "bare area" ⮕ of the humeral head not covered by articular cartilage between the posterior capsular attachment and the posterior margin of humeral head articular cartilage ⮕. The needle tip can be placed against this area.*

Shoulder: Posterior Approach

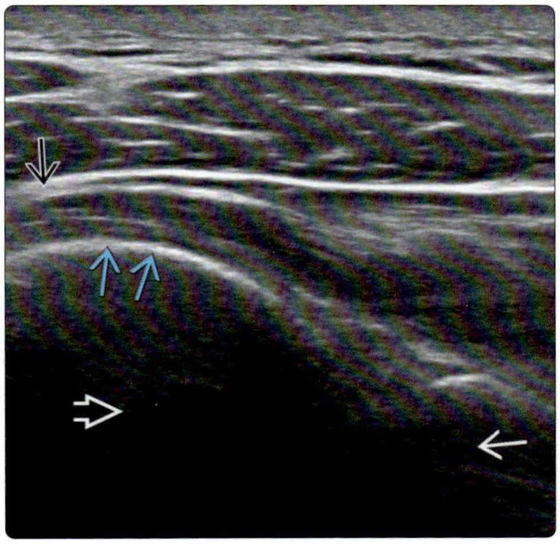

Shoulder: Posterior Approach

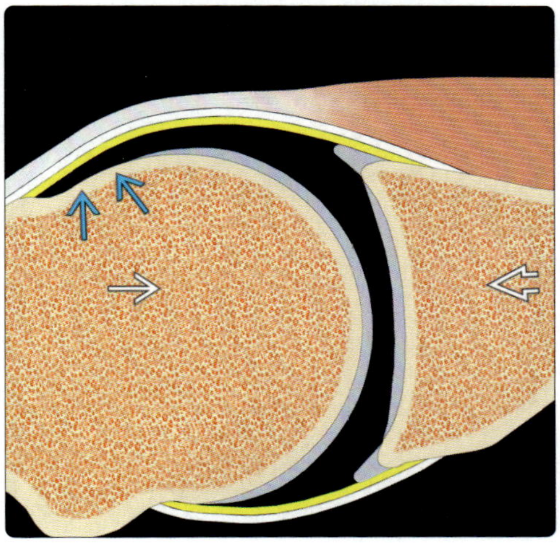

(Left) *Transverse US shows the posterior aspect of the glenohumeral joint between the glenoid ⮕ & humeral head ⮕. The needle tip can be placed against the surface of the "bare area" ⮕ deep to the infraspinatus tendon ⮕ where there is no covering cartilage. This avoids cartilage injury, allows easy monitoring of injection, & reduces procedural pain.* **(Right)** *Schematic diagram shows the posterior aspect of the shoulder joint between the glenoid ⮕ & humeral head ⮕. There is no articular cartilage or labrum at the "bare area" ⮕.*

Shoulder: Posterior Approach (in Plane)

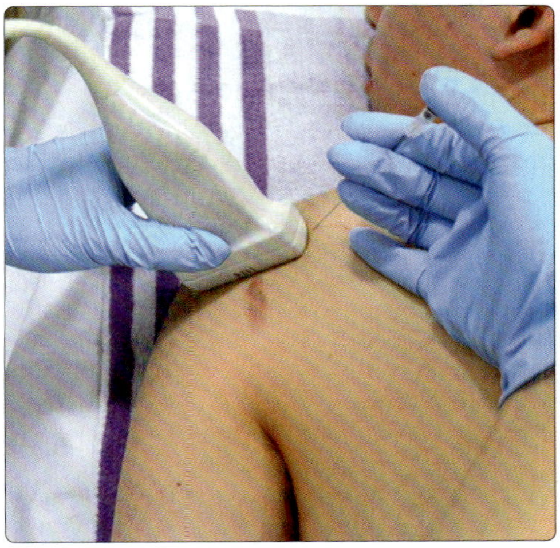

Shoulder: Posterior Approach (in Plane)

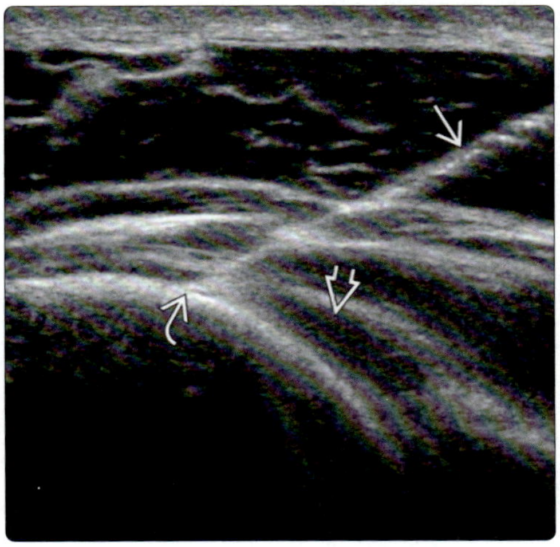

(Left) *Clinical photo shows how to use the in-plane approach to inject the glenohumeral joint via a posterior approach. The needle is inserted at the medial end of the transducer and directed laterally, as shown.* **(Right)** *Transverse US shows the needle tip ⮕ positioned at the "bare area" of the humeral head posteriorly via an in-plane approach. The needle ⮕ can be seen from the insertion site to the target. There is injected fluid ⮕ distending the shoulder joint.*

(Left) *Clinical photo shows how to use the out-of-plane approach to inject the shoulder joint via a posterior approach. The needle is entered at the middle of the US probe, as shown.* **(Right)** *Transverse US shows the needle tip positioned at the "bare area" of the posterior humeral head using an out-of-plane approach. Only the tip of the needle can be seen close to the target. There is injected fluid distending the shoulder joint.*

Shoulder: Posterior Approach (Out of Plane)

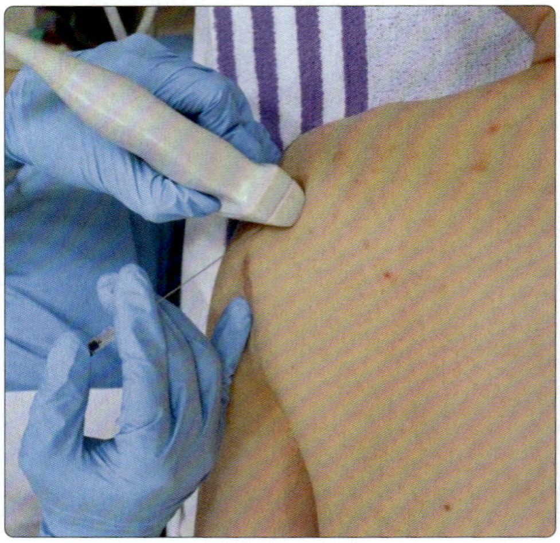

Shoulder: Posterior Approach (Out of Plane)

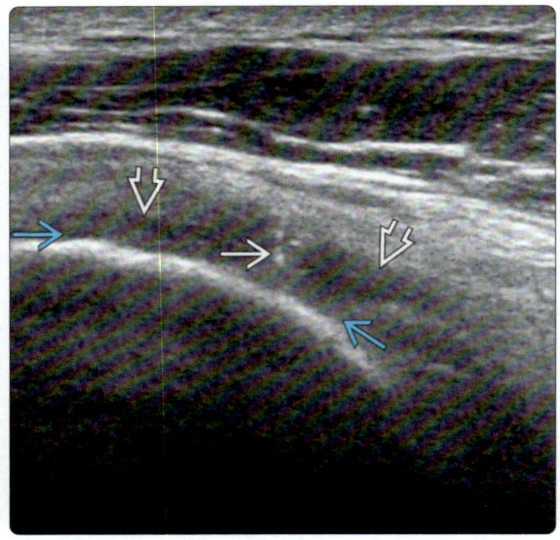

(Left) *Volume-rendered CT shows the bony anatomy of the anterior aspect of the glenohumeral joint between the glenoid and humerus. The coracoid process is also shown.* **(Right)** *Axial T1 FS MR shows the rotator interval at the anterior aspect of the shoulder joint on the medial superior aspect of the humeral head. The biceps tendon lies within the rotator interval. The coracoid process and glenoid are shown.*

Shoulder: Anterior Approach

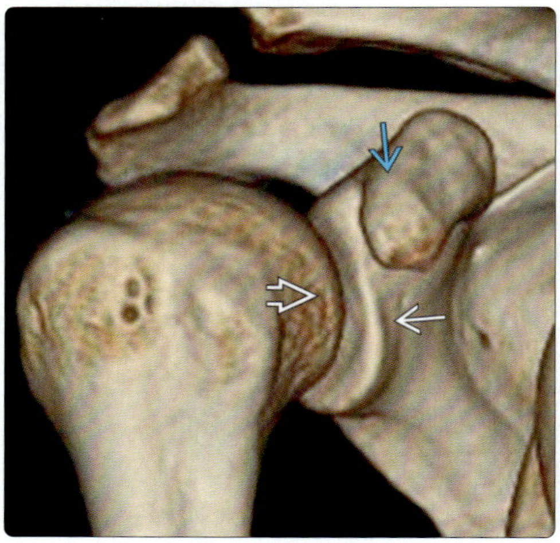

Shoulder: Anterior Approach

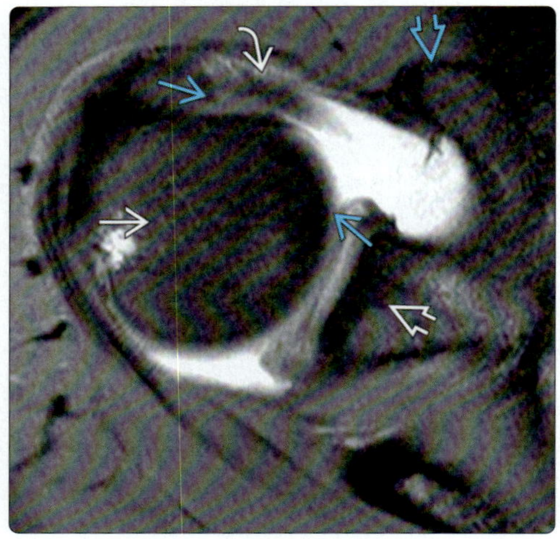

(Left) *Transverse US shows rotator interval area at medial superior aspect of humeral head. The biceps tendon lies laterally and the glenoid medially with gas in the glenohumeral joint. Needle tip should be positioned on articular cartilage at rotator interval.* **(Right)** *Schematic diagram shows the anterior aspect of the shoulder between the glenoid and humeral head. The biceps tendon traverses the rotator interval. The coracoid process and area for needle tip positioning are shown.*

Shoulder: Anterior Approach

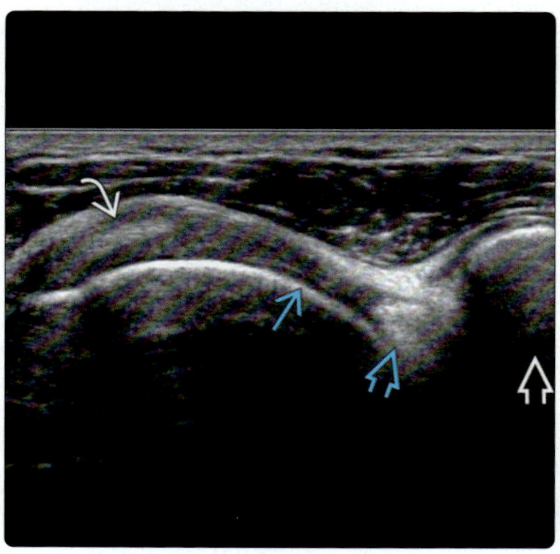

Shoulder: Anterior Approach

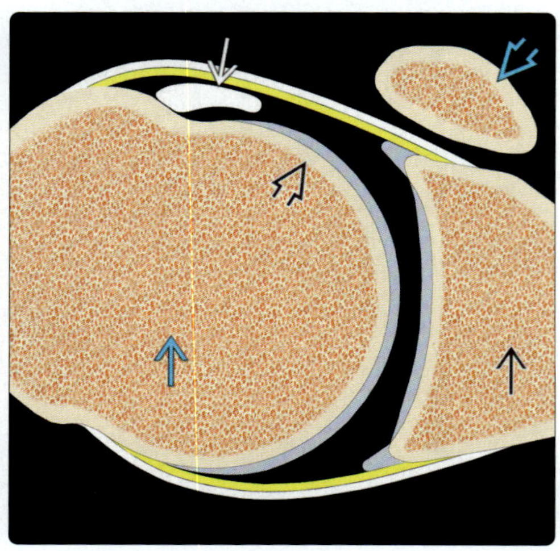

Shoulder: Anterior Approach (in Plane)

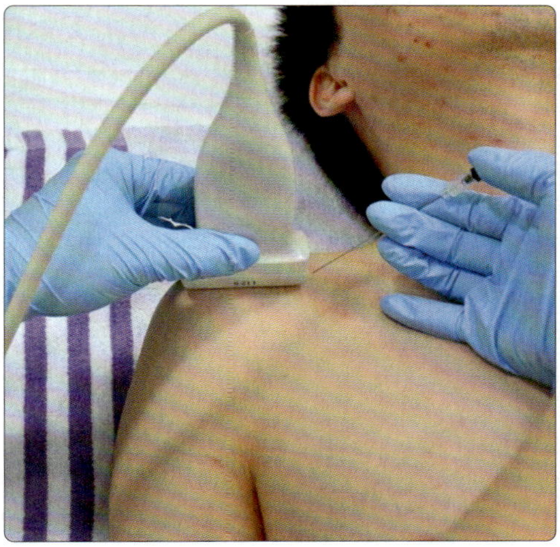

Shoulder: Anterior Approach (in Plane)

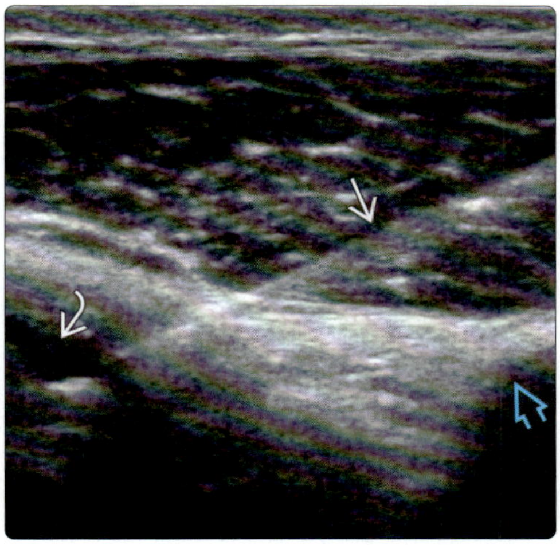

(Left) *Clinical photo shows in-plane approach to glenohumeral joint injection via an anterior approach. The needle is inserted at the medial aspect of the transducer and directed laterally, as shown.* **(Right)** *Transverse US shows the needle with the tip positioned at the rotator interval medial to the biceps tendon (not shown) via an in-plane approach. The needle can be seen from insertion site to target. There is injected fluid distending the shoulder joint. The needle should be lateral to coracoid process.*

Shoulder: Anterior Approach (Out of Plane)

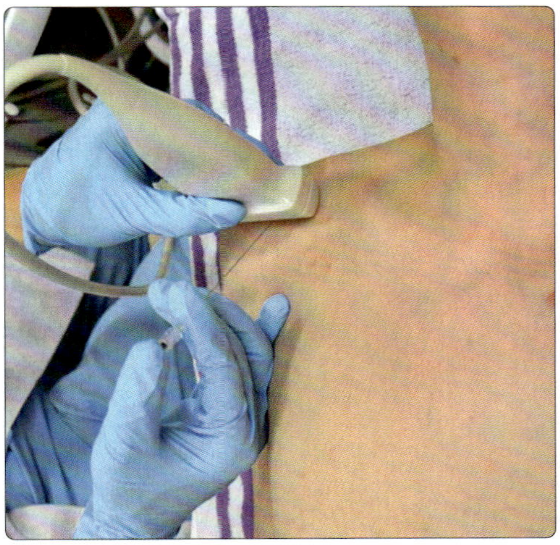

Shoulder: Anterior Approach (Out of Plane)

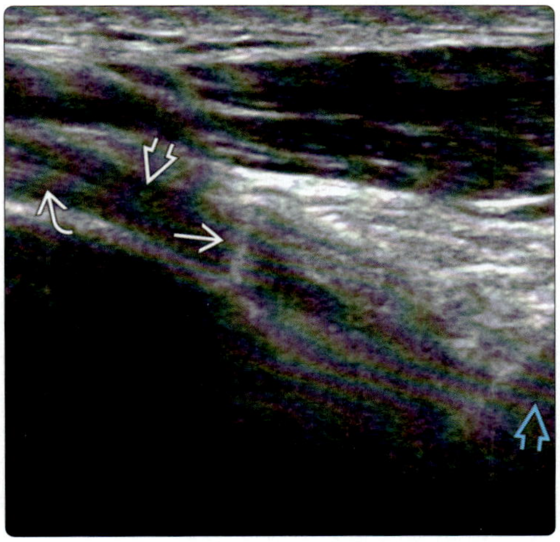

(Left) *Clinical photo shows an out-of-plane anterior approach to the glenohumeral joint via the rotator interval. The needle is inserted at an angle near the middle of the US transducer, as shown.* **(Right)** *Transverse US shows the needle tip positioned at the rotator interval medial to the biceps tendon and lateral to the coracoid process. Only the tip of the needle can be seen within the joint. There is injected fluid distending the shoulder joint.*

Elbow: Lateral Approach

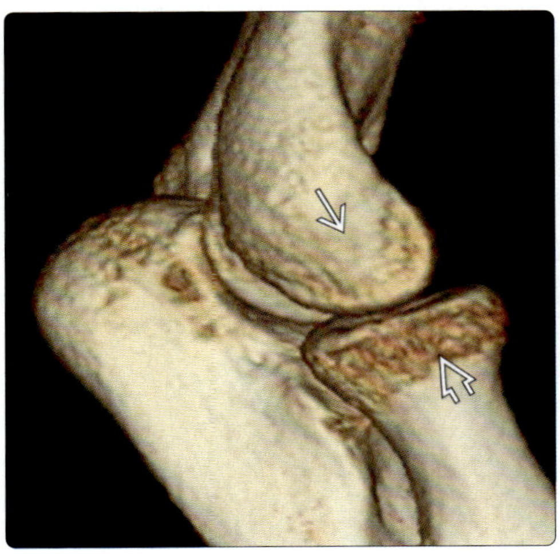

Elbow: Lateral Approach

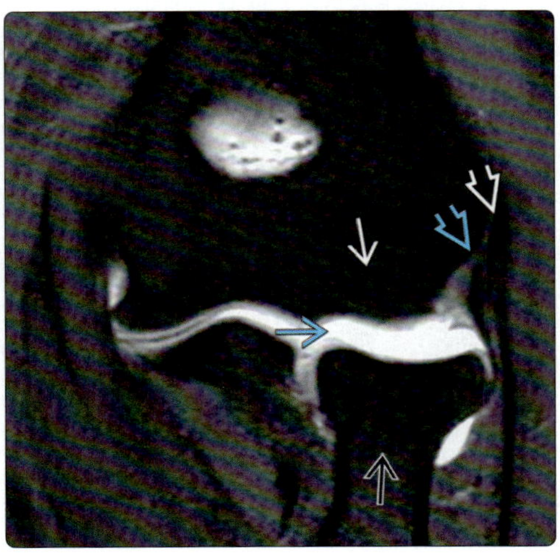

(Left) *Volume-rendered CT shows the bony anatomy at the lateral aspect of the elbow joint (radiocapitellar joint) between the radial head and the capitellum.* **(Right)** *Coronal T1 FS MR shows the radiocapitellar joint at the lateral aspect of the elbow joint between the capitellum and the radial head. The radial collateral ligament and the common extensor tendon origin are shown.*

Elbow: Lateral Approach

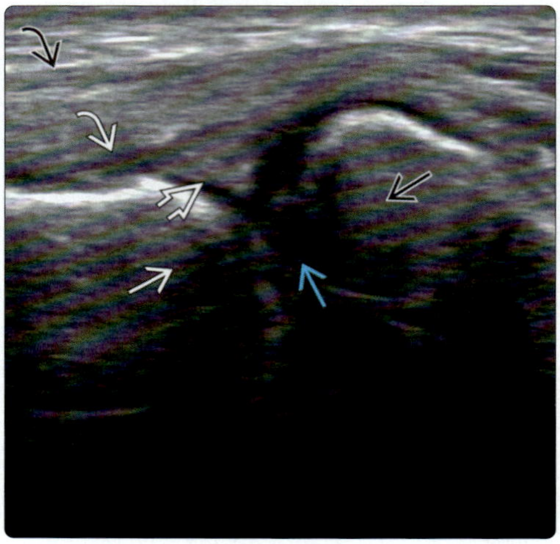

Elbow: Lateral Approach

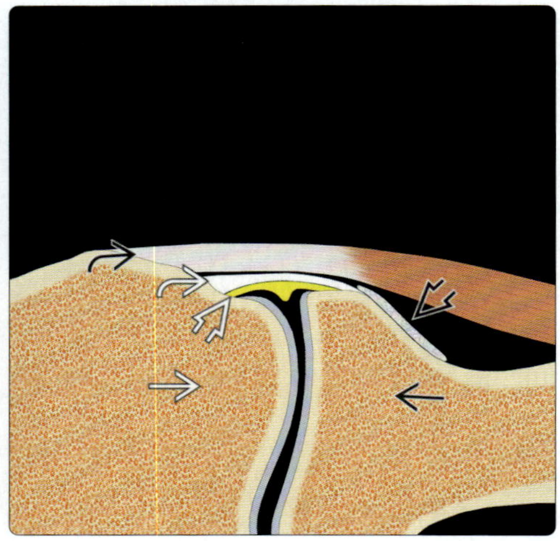

(Left) *Longitudinal US shows the radiocapitellar joint between the capitellum and the radial head. The common extensor tendon and radial collateral ligament lie superficial to the joint. A prominent synovial fold is shown.* **(Right)** *Schematic diagram shows the radiocapitellar joint between the capitellum and radial head. The radial collateral ligament and common extensor tendon origin lie superficial to the joint capsule with a synovial fold more deeply. The annular ligament is also shown.*

Elbow: Lateral Approach

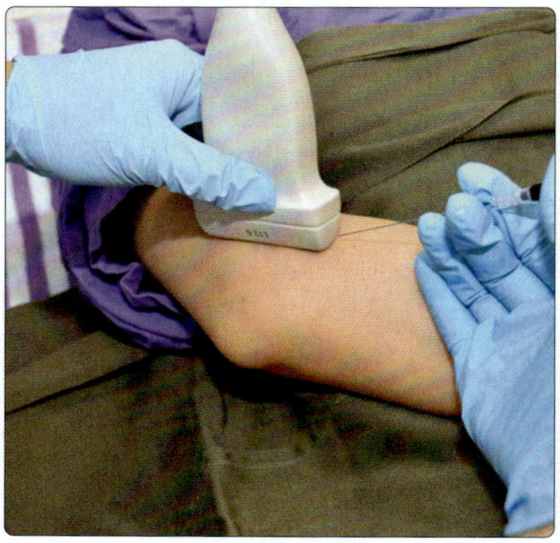

Elbow: Lateral Approach

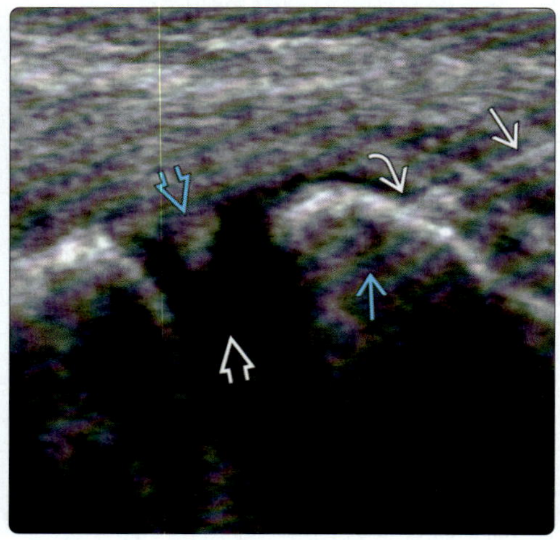

(Left) *Clinical photo shows in-plane lateral approach to inject the elbow joint. The needle should be inserted at the distal aspect of the transducer and directed proximally, as shown.* **(Right)** *Longitudinal US shows the needle tip positioned at the "bare area" on the lateral aspect of the radial head where there is no articular cartilage. The needle is seen from insertion site to the target. There is injected fluid distending the elbow joint and outlining a synovial fold.*

Elbow: Posterior Approach

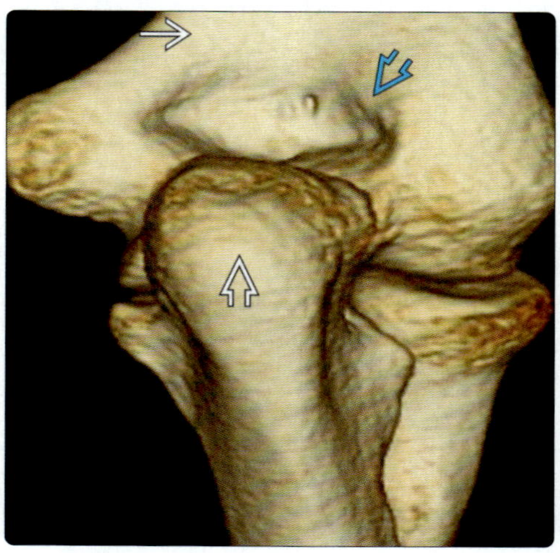

Elbow: Posterior Approach

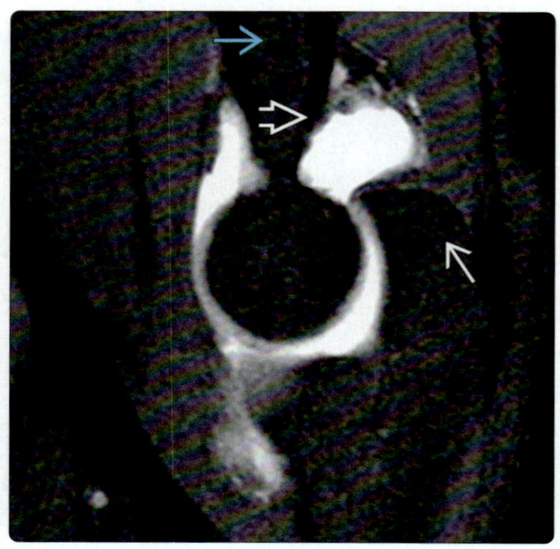

(Left) *Volume-rendered CT shows the bony anatomy of the posterior aspect of the elbow joint (ulnohumeral articulation) between the ulnar olecranon process and the distal humerus. The olecranon fossa of the distal humerus is shown.* **(Right)** *Sagittal T1 FS MR shows the posterior part of the ulnohumeral joint between the olecranon process and the distal humerus. The proximal attachment of the posterior capsule of the elbow joint lies just proximal to the olecranon fossa.*

Elbow: Posterior Approach

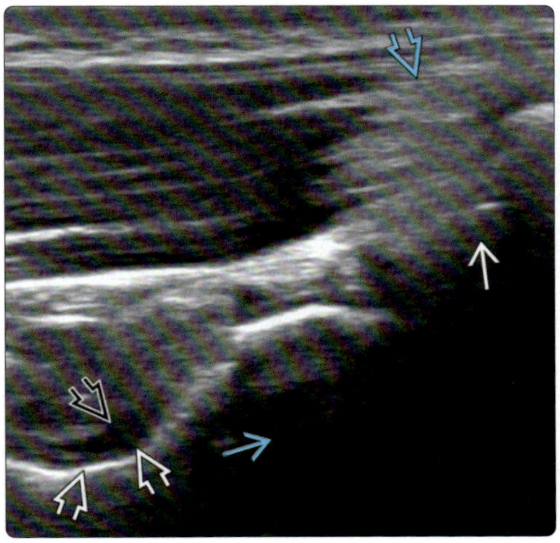

Elbow: Posterior Approach

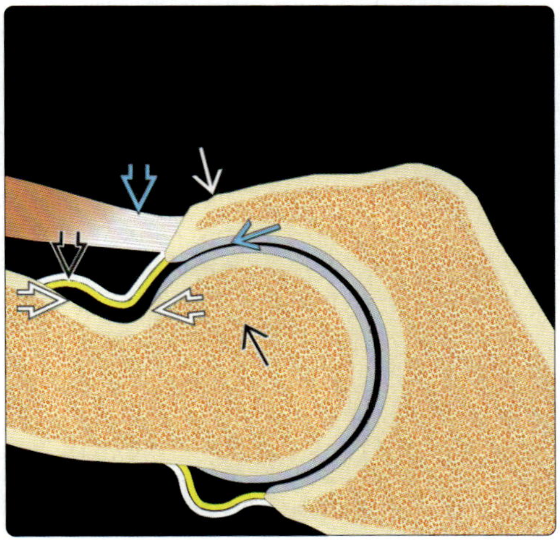

(Left) *Longitudinal US shows the ulnohumeral joint between the olecranon process & distal humerus . The olecranon fossa lies proximal to the joint & is free of cartilage but is covered by synovium. A small amount of physiological fluid is present . Note the triceps tendon .* **(Right)** *Schematic diagram shows the ulnohumeral joint between the olecranon process & distal humerus . The olecranon fossa lies proximal to the joint and is free of cartilage but is covered by synovium . Note the triceps tendon .*

Elbow: Posterior Approach

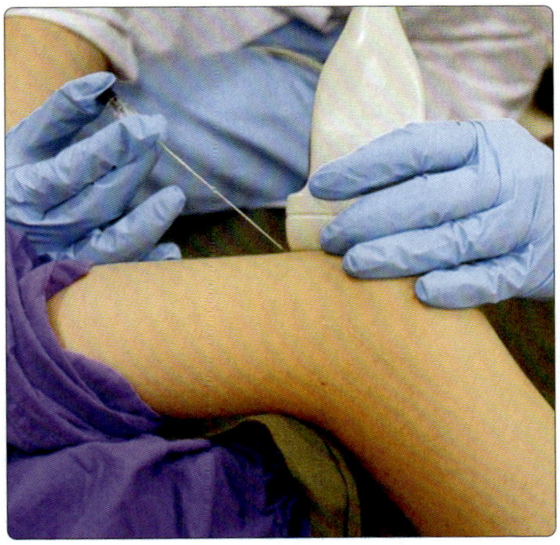

Elbow: Posterior Approach

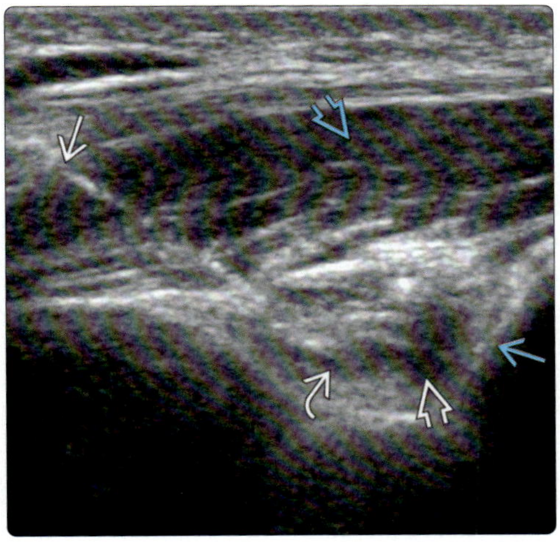

(Left) *Clinical photo shows how to use the in-plane approach to inject the elbow joint via a posterior approach. The needle should be placed at the proximal end of the probe and directed distally.* **(Right)** *Longitudinal US shows the needle tip positioned in the olecranon fossa where there is no overlying cartilage ("bare area"), via an in-plane approach. The whole needle can be seen from insertion site to target. Injected fluid partially distends the elbow joint. The needle has transversed the triceps muscle .*

Wrist Joint: 3 Compartments

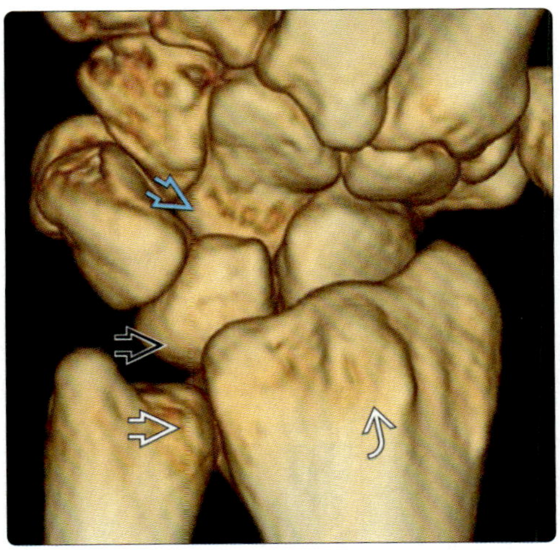

Wrist: Radioscaphoid Joint

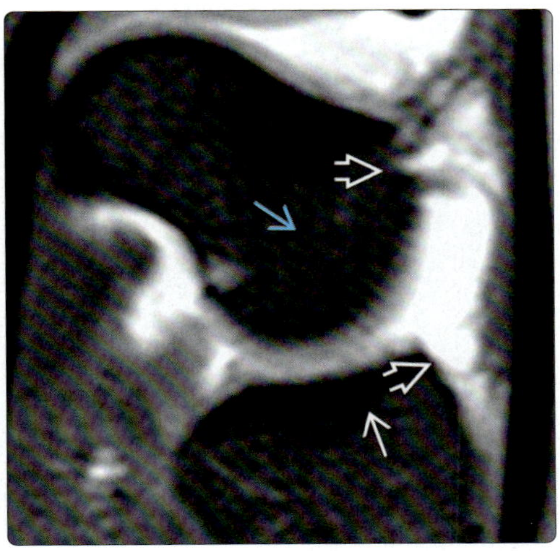

(Left) *Dorsal aspect of the wrist with 3 separated compartments is shown: (1) Radiocarpal joint , (2) distal radioulnar joint , and (3) midcarpal joint . Lister tubercle of the distal radius, which is an important bony landmark for injection, lies just proximal to the scaphoid.* **(Right)** *Sagittal T1 FS MR shows the dorsal aspect of the radioscaphoid joint between the distal radius and the scaphoid . The attachments of the dorsal capsule of the radiocarpal joint are shown .*

(Left) *Axial T1 FS MR shows the dorsal aspect of the wrist just distal to the radioscaphoid joint. There is no tendon transversing that area just distal to Lister tubercle of the distal radius (not shown). Extensor carpi radialis brevis & longus tendons as well as the extensor pollicis longus & extensor digitorum longus tendons are shown.* **(Right)** *Schematic diagram shows Lister tubercle of the distal radius. At this plane, there is usually no tendon between the skin to radiocarpal joint; as such, it represents the best place for needle insertion.*

Wrist: Radiocarpal Joint

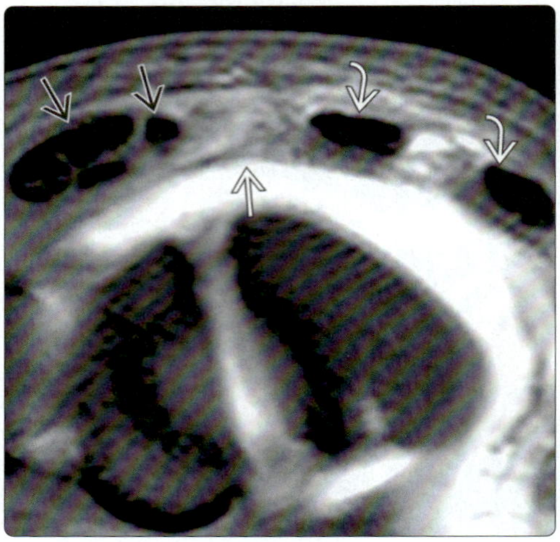

Wrist: Radiocarpal Articulation

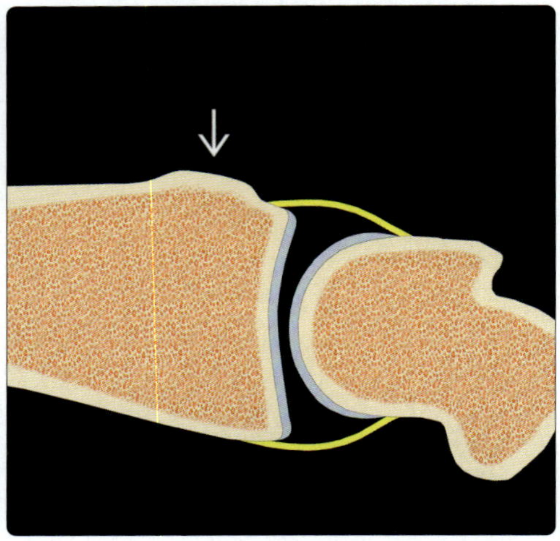

(Left) *Clinical photo shows the in-plane dorsal approach to inject the radioscaphoid joint. The needle is inserted from the distal aspect of the transducer and directed proximally, as shown.* **(Right)** *Longitudinal US along the plane of Lister tubercle shows the needle tip positioned in the dorsal aspect of the radioscaphoid joint just distal to the radial articulating surface. The needle can be well seen from the insertion site to the joint. There is injected fluid distending the radiocarpal joint overlying the scaphoid.*

Wrist: Radioscaphoid Joint

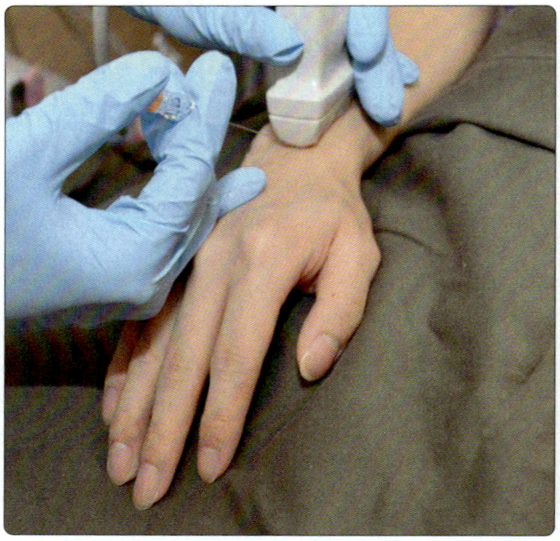

Wrist: Radioscaphoid Joint

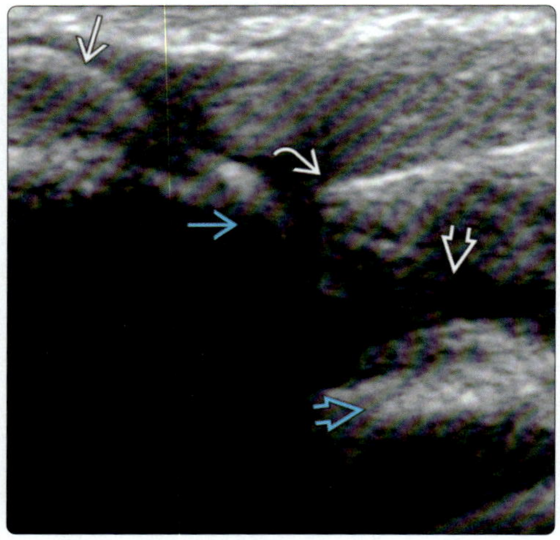

(Left) *T1 FS MR shows the dorsal aspect of the distal radioulnar joint between the distal radius and the ulnar head. The ideal needle path is on either side of the extensor digiti minimi tendon. The extensor digitorum longus and extensor carpi ulnaris tendon are also shown.* **(Right)** *Schematic diagram shows the distal radioulnar joint and preferred paths for needle insertion on either side of the extensor digiti minimi tendon. The extensor digitorum longus and extensor carpi ulnaris tendon are shown.*

Wrist: Distal Radioulnar Joint

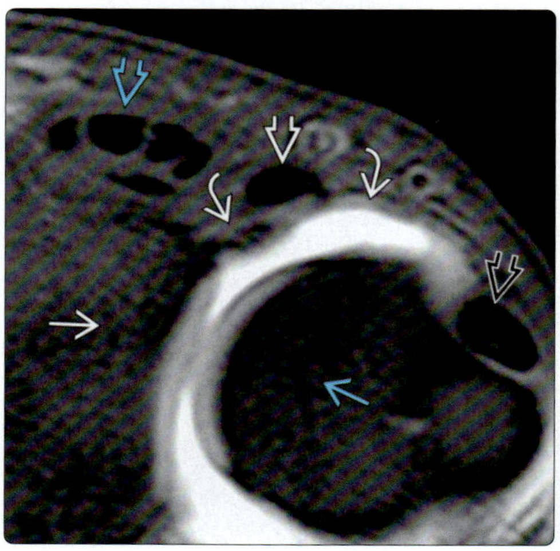

Wrist: Distal Radioulnar Joint

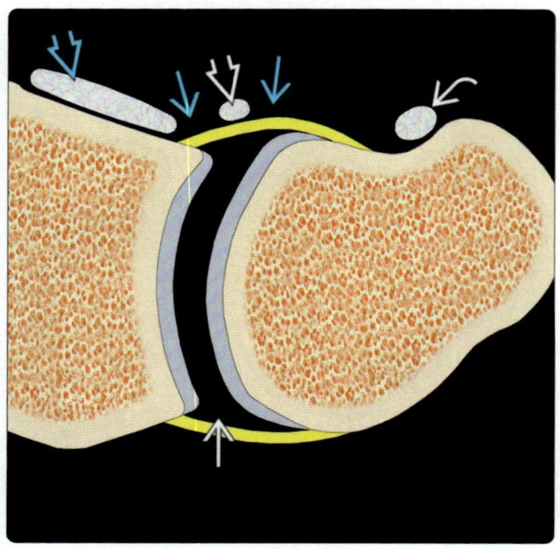

Wrist: Distal Radioulnar Articulation

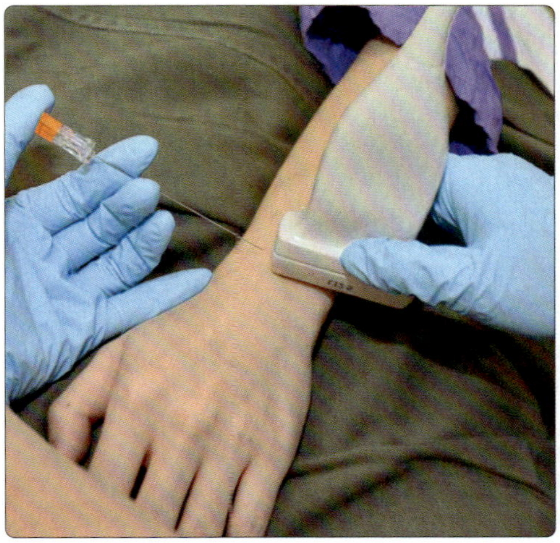

Wrist: Distal Radioulnar Joint

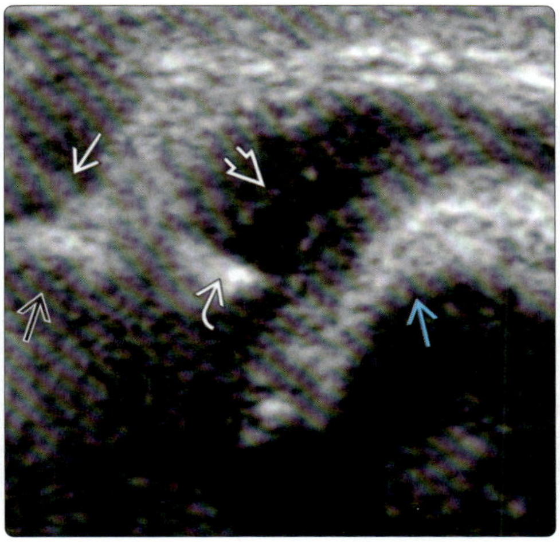

(Left) *Clinical photo shows an in-plane dorsal approach for distal radioulnar joint injection. The needle should be inserted from the radial side of the transducer and directed in an ulnar direction, as shown.* **(Right)** *Transverse US shows the needle tip positioned on the surface of the ulnar head just ulnar to the distal radius. The needle can be seen from the insertion site to the joint. There is injected fluid distending the distal radioulnar joint. All of the extensor tendons are avoided by using this approach.*

Wrist: Midcarpal Joint

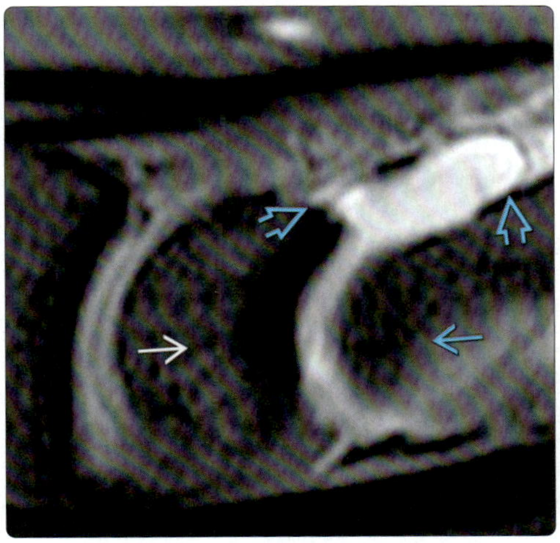

Wrist: Midcarpal Joint

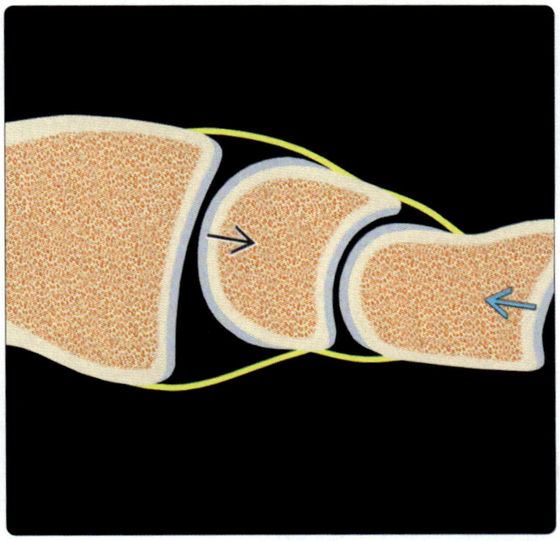

(Left) *Sagittal T1 FS MR shows the dorsal aspect of the lunocapitate joint between the lunate and the capitate. The attachments of the dorsal lunocapitate capsule are shown.* **(Right)** *Schematic diagram shows the lunocapitate joint for midcarpal injection between the lunate and the capitate. The needle should be inserted dorsally.*

Wrist: Midcarpal Joint

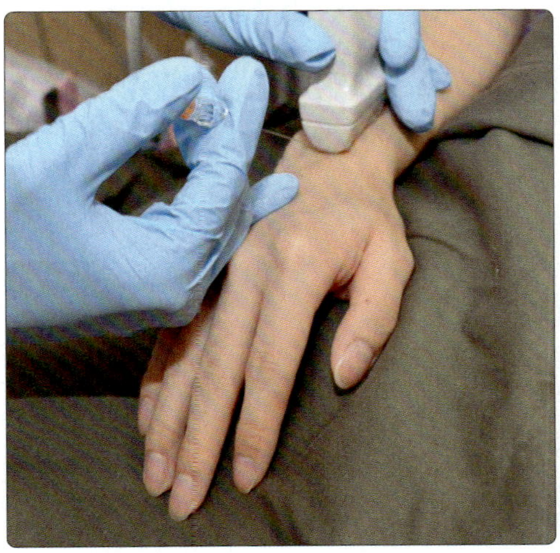

Wrist: Midcarpal Joint

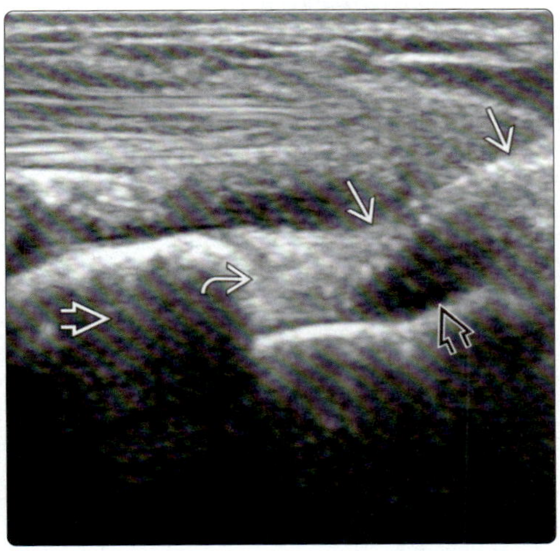

(Left) *Clinical photo shows an in-plane dorsal approach for a midcarpal joint injection. The needle should be inserted at the distal end of the transducer and directed proximally, as shown.* **(Right)** *Longitudinal US shows the needle tip positioned at the lunocapitate joint just distal to the lunate. The needle can be seen from the insertion to the joint. There is injected fluid distending the lunocapitate joint.*

Joint Injection: Lower Limb

KEY FACTS

PROCEDURE

- Hip joint injection
  - Usually anterior approach
    - Longitudinal method: Preferred
      - Needle tip positioned on "bare area" at head-neck junction
    - Transverse approach
      - Needle tip positioned on femoral head cartilage surface
      - Technically more demanding
      - Risk of articular cartilage or labral injury
  - Avoid femoral vessels and nerve, which lie on anteromedial aspect of hip joint
- Knee joint injection
  - Suprapatellar approach: Preferred
    - Needle tip positioned within suprapatellar pouch
      - Can be performed longitudinally or transversely
    - Alternatively, anterolateral or anteromedial parapatellar approaches if lateral or medial patellar recesses contain more fluid than suprapatellar pouch
  - Avoid posterior approach as risk of popliteal vessel and posterior tibial nerve injury
- Ankle joint injection
  - Anterolateral approach: Preferred
    - Needle tip positioned near talar dome cartilage
    - Joint space more capacious in anterolateral than anteromedial aspect of joint
    - Easier to perform as shorter distance from skin to joint than anteromedial approach
  - Anteromedial approach
    - Needle tip positioned near talar dome cartilage
    - Longer distance from skin to joint than anterolateral approach
  - Avoid dorsalis pedis artery and deep peroneal nerve

Hip: Anterior Approach

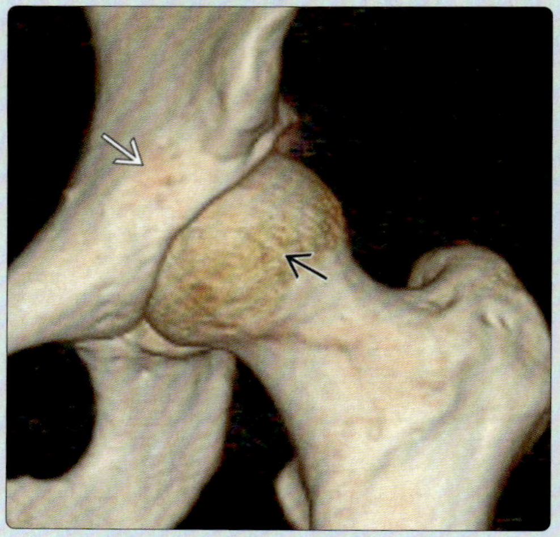

Hip: Anterior Approach

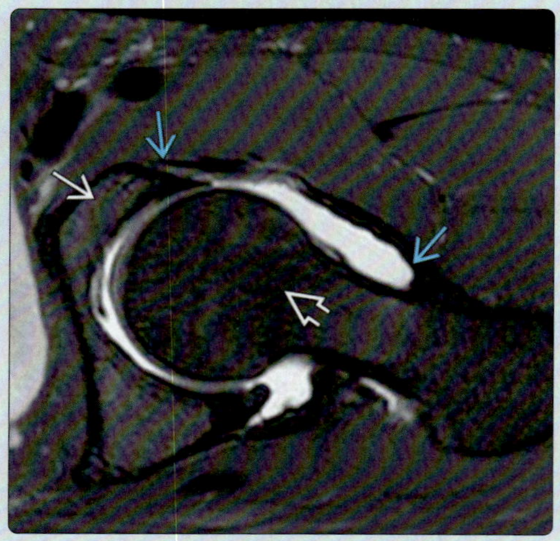

(Left) *CT volume rendering shows bony anatomy on anterior aspect of the hip joint between the acetabulum → and the femoral head →.* **(Right)** *Axial T2 FS MR shows the hip joint between the acetabulum → and femoral head →. The attachments of the anterior capsule to the lip of the acetabulum and to the base of the femoral neck are shown →.*

Hip: Anterior Approach

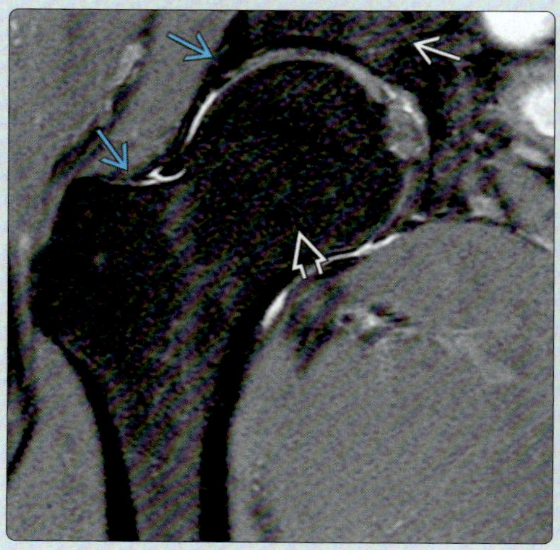

Hip: Anterior Approach

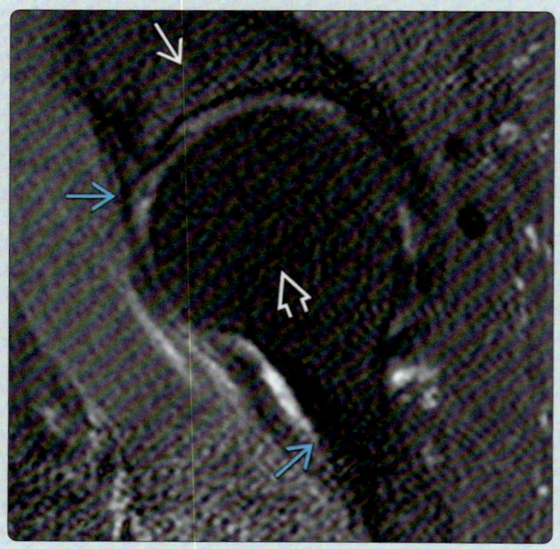

(Left) *Coronal T2 FS MR shows hip joint between acetabulum → and femoral head →. The attachments of the anterior capsule to the lip of the acetabulum and the base of the femoral neck are shown →.* **(Right)** *Sagittal T2 FS MR shows hip joint between acetabulum → and femoral head →. The attachments of the anterior capsule to the lip of the acetabulum and the base of the femoral neck are shown →.*

Joint Injection: Lower Limb

TERMINOLOGY

Definitions

- Percutaneous joint injection for hip, knee, and ankle joints under US guidance

PREPROCEDURE

Indications

- Diagnostic purpose: Injection of contrast for conventional, CT, or MR arthrography
- Therapeutic purpose: Injection of medication (steroid, hyaluronic acid, platelet-rich plasma, etc.) to relieve pain and reduce inflammation

Contraindications

- Uncorrected bleeding diathesis
- Uncooperative patient
- Active infection of overlying skin soft tissues

Getting Started

- Things to check
    - Review any previous relevant imaging
    - Time-out procedure (patient identity, type of procedure, side of procedure)
    - Signed informed consent
        - Discuss risk of infection, allergy, and bleeding
    - If patient taking anticoagulants, optimize clotting profile before procedure
        - How stringently this is applied depends on perceived difficulty in applying adequate tamponade
    - Sedation rarely required
- Medications
    - Local anesthetic is optional and usually not required
    - Sedation of anxious patient of young children
        - Intravenous midazolam or other benzodiazepine; antidote should be available
    - Contrast
        - Conventional or CT arthrography: Iodinated contrast, e.g., Omnipaque 300
        - MR arthrography: 0.1 mL of diluted gadolinium in 10 mL of normal saline solution
    - Steroid
        - Short (hydrocortisone) or long acting (dexamethasone, methylprednisolone)
        - Hyaluronic acid
        - Platelet-rich plasma
- Equipment list
    - Low- to high-frequency (5-12 MHz) transducers
    - Sterile set
    - Spinal needles
        - 22-g for hip, knee, and ankle injection
    - Extension tube (thin, 250 mm long, 8.5-Fr, 0.4 mL dead space, no Luer lock)
    - 3-way connector
    - Syringes for local anesthetic, aspiration, and injection

PROCEDURE

Procedure Steps (General)

- Use free-hand technique to advance needle into joint space under real-time imaging
    - Plan approach carefully, always being mindful of intervening neurovascular structures and tendons
    - Use less preferred approaches when preferred approach is not feasible due to intervening soft tissue infection, bandaging, etc.
- Can use long-axis (in-plane) or short-axis (out-of-plane) approach
    - In-plane approach
        - Easier to learn and perform as whole needle track can be well seen but longer needle path from insertion site to target
    - Out-of-plane approach
        - More difficult to perform as needle tip can be seen only close to target but shorter needle path from insertion site to target
- Ensure on real-time imaging that needle tip is within joint space
    - Ideally, needle tip should just pass through synovium and be within superficial part of joint
    - Synovium may be difficult to separate from subsynovial fat if there is no joint effusion
    - Inserting needle tip too firmly into joint may damage articular cartilage
        - Position needle tip just on surface of articular cartilage
    - One can often find a bare area between cartilage and capsule where needle tip can be positioned
        - Easier to monitor injection at this location
        - However, not every joint has bare area
    - if moderate to large effusion present, aspirate joint fluid prior to injection
- Avoid injection of air by 1st flushing syringe and tubing with injectate
- Free flow of fluid with low-pressure injection and minimal patient discomfort helps affirm intraarticular position of needle tip
- Avoid strenuous exercise to prevent leakage of solution along needle track for 20 minutes
    - Gentle movement of joint helps ensure good dispersion of contrast medium

Hip Joint

- Anterior longitudinal approach: Preferred method
    - Capsule at neck region thickened by zona orbicularis (circular fibers of capsule blending with deeper fibers of iliofemoral ligament)
    - Capsule less thick at femoral head-neck junction
    - Preferred anterior target point is bare area of femoral head-neck junction
        - Place probe longitudinal to hip joint
            - Slightly turn distal end of probe anticlockwise to align along head-neck junction
        - Bare area is covered by synovium but not articular cartilage
        - Covered more superficially by anterior joint capsule and anterior hip muscles
        - Bare area injection can be monitored by real-time US
        - Needle path avoids labrum and therefore minimizes risk of injury
        - Plan approach so as to avoid femoral neurovascular bundle

- Target point for femoral head is superolateral quadrant of femoral head
  - Stop needle tip at surface of articular cartilage to minimize risk of cartilage injury
- Anterior transverse approach
  - Technically more difficult than longitudinal approach
  - Insert needle at lateral aspect of hip joint
    - As entry point is more lateral, this approach carries less risk of femoral neurovascular injury
  - Orientate needle horizontally and insert medially
  - Place needle tip at surface of femoral head or acetabular cartilage
  - Important to avoid injury to labrum, which is close to needle path compared to longitudinal approach
- Other approaches
  - Lateral approach similar to anterior approach targeting either femoral head or head-neck junction
  - Posterior approach is not advisable as much deeper than anterior approach
    - Avoid sciatic nerve

Knee Joint

- Suprapatellar approach: Preferred method
  - Usually, small amount of physiological fluid is present aiding injection
    - Needle tip should be positioned centrally within suprapatellar pouch
  - Can use either longitudinal or transverse approach to suprapatellar pouch
  - Lateral approach to suprapatellar pouch is preferred, as there is no overlying vastus lateralis at distal thigh region
  - Medial approach may be more uncomfortable as overlying vastus medialis muscle needs to be transgressed
- Anteromedial or anterolateral parapatellar approach
  - Preferred when these recesses contain more fluid than suprapatellar recess
  - Place needle into medial or lateral patellar recess
- Other approaches
  - Infrapatellar approach
    - Needle tip positioned on surface of femoral trochlear cartilage
    - Risk of cartilage injury

Ankle Joint

- Anterolateral approach: Preferred method
  - Skin to joint distance is shorter laterally than medially
  - Anterolateral approach is further from dorsalis pedis artery than anteromedial approach
    - Avoids deep peroneal nerve, which lies deep to extensor hallucis longus and lateral to dorsalis pedis artery at ankle joint
      - Tends to move more medially with ankle plantar flexion
    - Does not avoid intermediate and medial dorsal cutaneous nerves (terminal branches superficial peroneal nerve) lying between extensor digitorum tendon and lateral malleolus
  - Technically easier to perform
  - Needle tip positioned just superficial to talar dome articular cartilage
- Anteromedial approach
  - Technically, slightly more difficult as deeper than anterolateral approach
  - Needle tip positioned just superficial to talar dome articular cartilage
  - Saphenous nerve and long saphenous nerve between medial malleolus and tibialis anterior tendon
- Other approaches
  - Posterolateral and posteromedial approaches are technically feasible but deeper than anterior approach, and needle tip monitoring is more challenging
    - Posteromedial approach: Posterior tibial vessel and tibial nerve injury risk
    - Posterolateral approach: Sural nerve injury risk

POST PROCEDURE

Expected Outcome

- Mild discomfort from joint distension

Immediate Aftercare

- Perform arthrographic study as soon as possible
- Contrast will be largely absorbed by 30 minutes

Subsequent Aftercare

- Avoid exercise for 24 hours after corticosteroid injection
  - Maximize therapeutic effect
  - Reduce spread of steroid into surrounding tissues

OUTCOMES

Complications

- Most feared complication(s)
  - Infection
- Immediate/periprocedural complication(s)
  - Vasovagal reaction
    - Perform procedures with patient lying down whenever possible
  - Bleeding
    - Avoid vascular injury
    - Check and amend clotting if abnormal
    - Use thinnest needle possible and minimalist technique in patients with bleeding diathesis
- Delayed complication(s)
  - Pain 1-2 days after steroid injection due to steroid-related microcrystalline synovitis ("steroid flare" in 10% of cases)
    - Settle with simple oral analgesia
  - Flushing of skin 2-3 days after steroid injection
- Other complications
  - Theoretical risk of joint arthropathy with steroid injection, but probably more related to intraarticular lignocaine or excessive use of steroid injection

SELECTED REFERENCES

1. Henning PT: Ultrasound-guided foot and ankle procedures. Phys Med Rehabil Clin N Am. 27(3):649-71, 2016
2. Lueders DR et al: Ultrasound-guided knee procedures. Phys Med Rehabil Clin N Am. 27(3):631-48, 2016
3. Payne JM: Ultrasound-guided hip procedures. Phys Med Rehabil Clin N Am. 27(3):607-29, 2016
4. Rastogi AK et al: Fundamentals of joint injection. AJR Am J Roentgenol. 207(3): 484-94, 2016

Hip: Anterior Transverse Approach

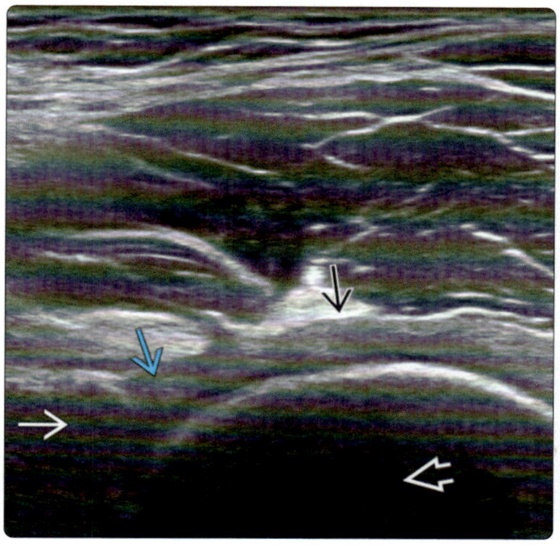

Hip: Anterior Transverse Approach

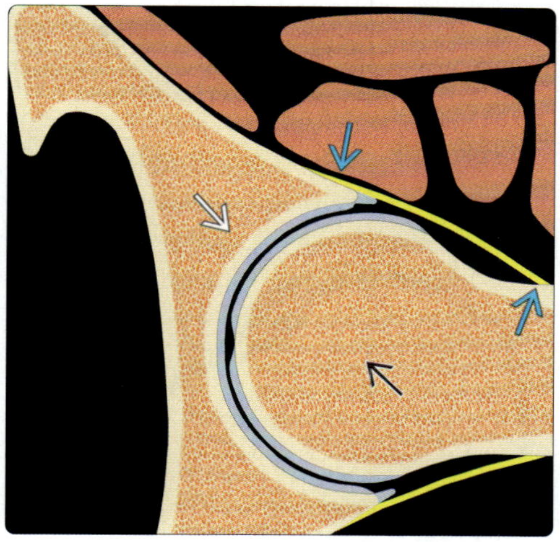

(Left) *Transverse US shows the anterior aspect of the hip joint between the acetabulum and femoral head. The echogenic labrum is shown, which should be avoided during injection. The anterior capsule of the hip joint is shown.* **(Right)** *Schematic diagram shows the anterior aspect of the hip joint between the acetabulum and femoral head. The attachments of the hip capsule to the lip of the acetabulum and the base of the femoral neck are shown.*

Hip: Anterior Transverse Approach

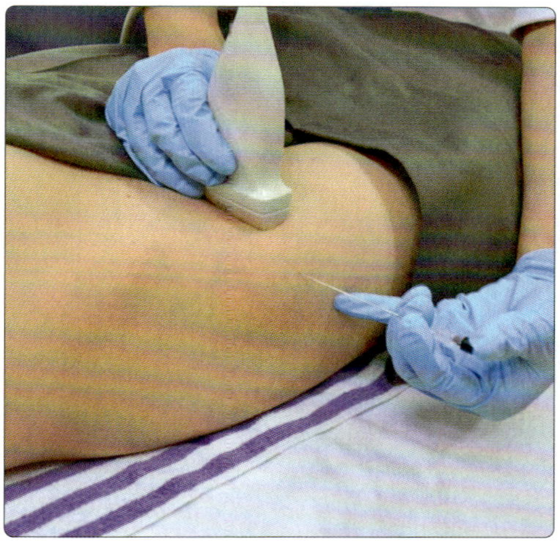

Hip: Anterior Transverse Approach

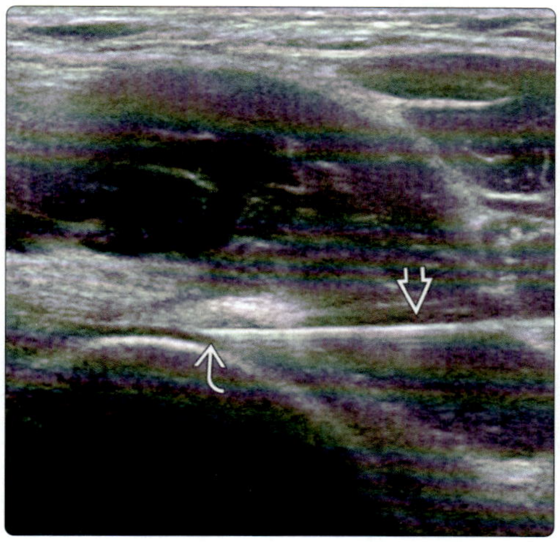

(Left) *Clinical photograph shows needle insertion for transverse lateral approach to the hip joint. The needle is inserted from lateral to medial along the anterior margin of the hip joint.* **(Right)** *Transverse US shows injection of the hip joint via transverse approach. The needle tip is positioned on the surface of the femoral head articular cartilage. The needle is clearly shown in this in-plane approach, as it lies almost parallel to the probe.*

Hip: Anterior Transverse Approach

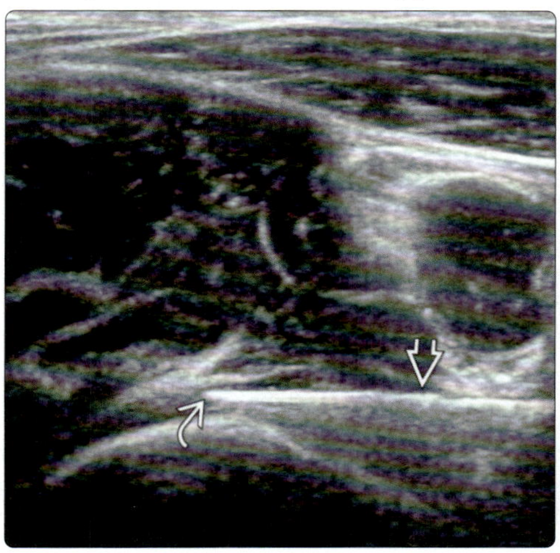

Hip: Anterior Longitudinal Approach

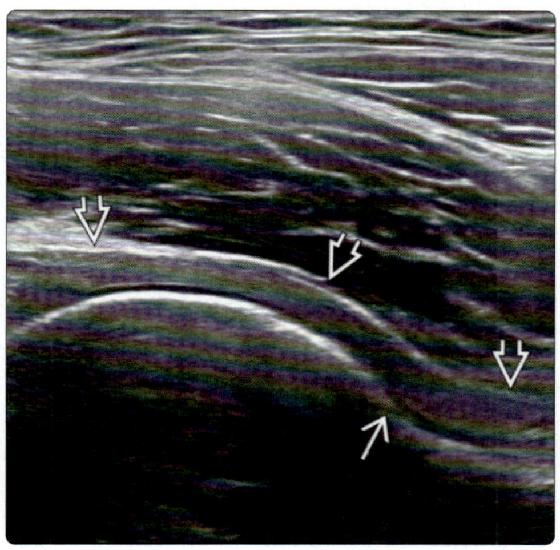

(Left) *Transverse US shows injection of the hip joint via transverse approach. The needle tip is positioned at the surface of the femoral head articular cartilage. The needle is clearly shown in the in-plane approach, as it lies almost lying parallel to the probe.* **(Right)** *Longitudinal US shows the anterior aspect of the femoral head-neck junction. The anterior capsule of the hip joint is shown. The distal end of the anterior capsule is at the base of the femoral neck.*

(Left) *Schematic diagram shows the anterior aspect of hip joint between the acetabulum ➡ and the femoral head ⇨. The attachment of the hip capsule is shown ⇨. Distally, the anterior capsule attaches to the base of the femoral neck.* (Right) *Clinical photograph shows needle position for anterior approach to the hip joint. The needle is inserted from distal to proximal at the anterior aspect of the hip joint.*

Hip: Anterior Longitudinal Approach

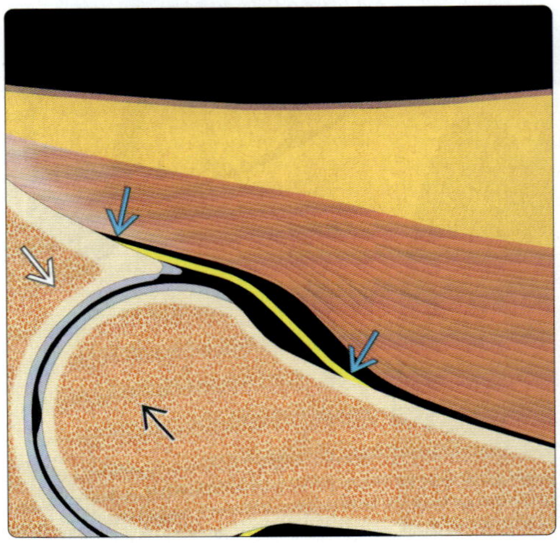

Hip: Anterior Longitudinal Approach

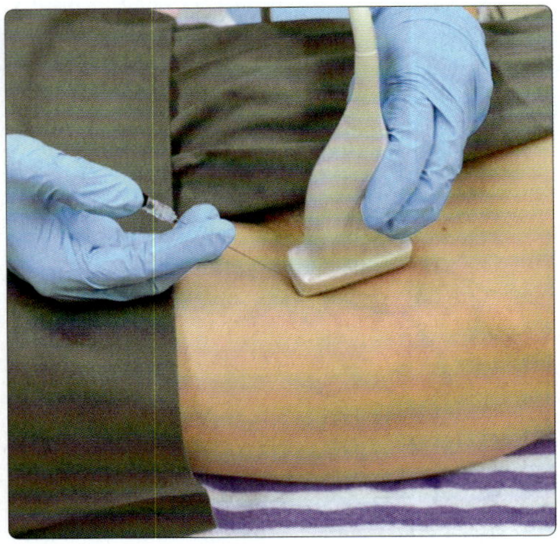

(Left) *Longitudinal US shows injection of the hip joint using an anterior longitudinal approach. Needle tip ➡ is positioned at the femoral head-neck junction ➡. Needle ➡ is less well seen in the anterior approach compared to the transverse approach, as it lies oblique to the US beam.* (Right) *Longitudinal US shows injection of the hip joint using an anterior longitudinal approach. Needle tip ➡ is positioned on the femoral head articular cartilage ➡. The needle ➡ is well shown on this in-plane approach.*

Hip: Anterior Longitudinal Approach

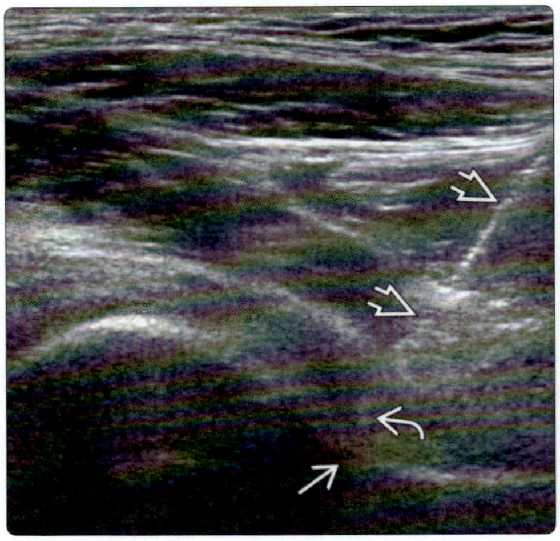

Hip: Anterior Longitudinal Approach

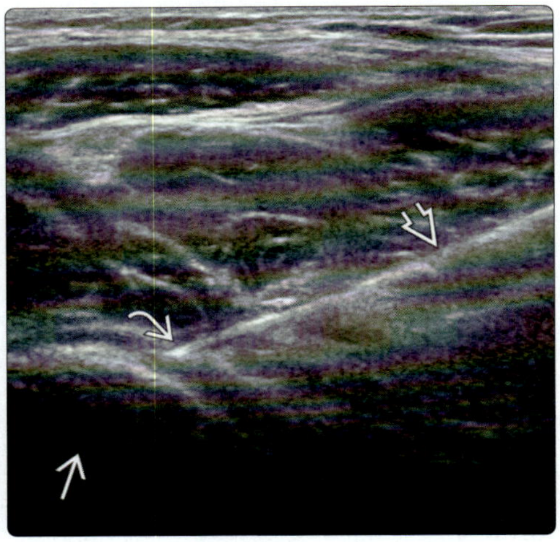

(Left) *Volume-rendered CT shows the bony anatomy of the anterior aspect of the knee joint between the distal femur ➡ and proximal tibia ⇨. The patella is also shown ⇨.* (Right) *Sagittal T2 FS MR shows the knee joint between the distal femur ➡ and the proximal tibia ➡. The suprapatellar pouch of the knee joint is shown ⇨ proximal to the patella ➡.*

Knee: Anterior Suprapatellar Approach

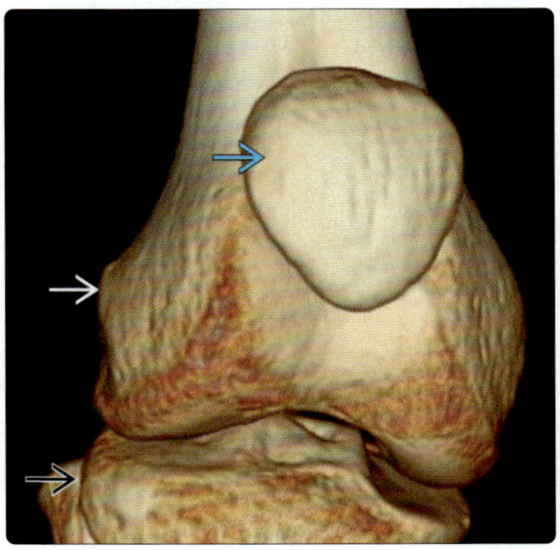

Knee: Anterior Suprapatellar Approach

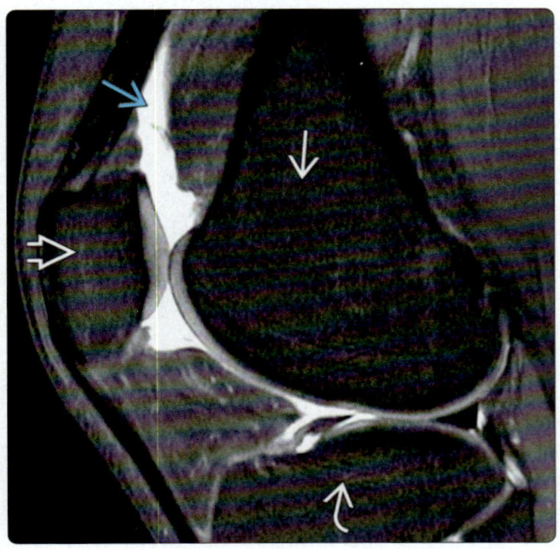

Knee: Anterior Suprapatellar Approach

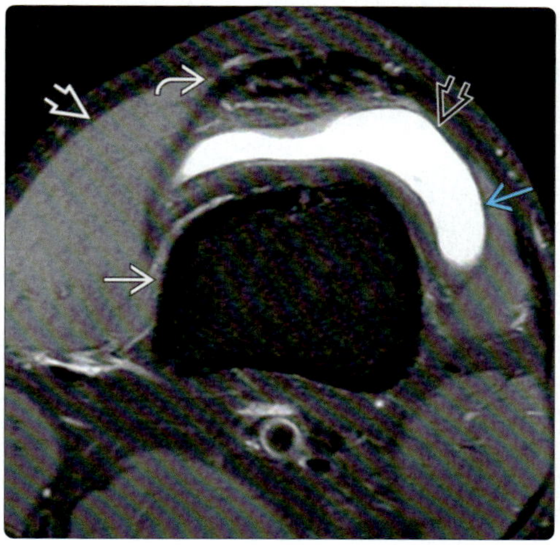

Knee: Anterior Suprapatellar Approach

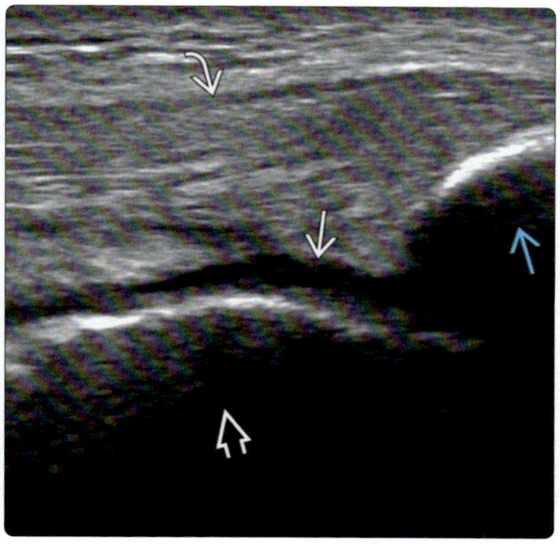

(Left) *Axial T2 FS MR shows the suprapatellar pouch of the knee joint anterior to the distal femur and posterior to the quadriceps tendon. Needle is preferentially inserted anterolaterally, as there is no intervening vastus lateralis muscle. Anteromedially, the vastus medialis muscle covers the suprapatellar pouch.* **(Right)** *Longitudinal US shows a minimally distended suprapatellar pouch posterior to quadriceps tendon, anterior to distal femur, and proximal to the patella.*

Knee: Anterior Suprapatellar Approach

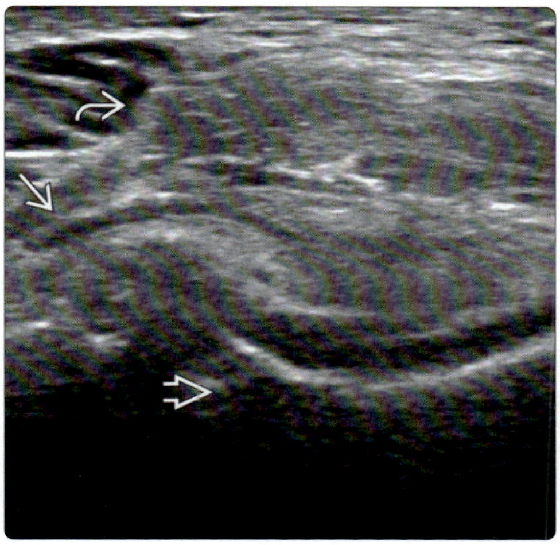

Knee: Anterior Suprapatellar Approach

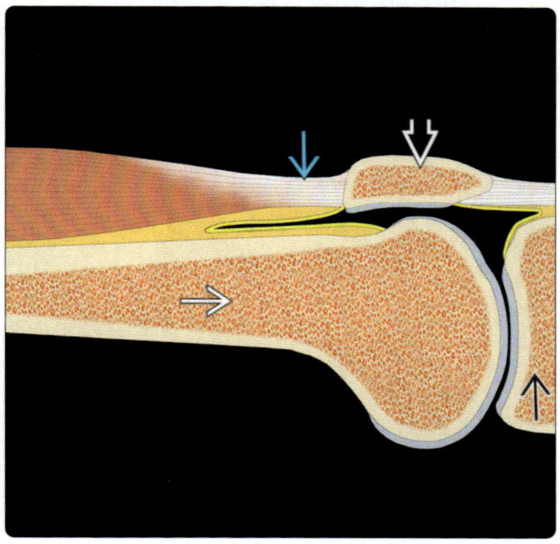

(Left) *Transverse US shows the suprapatellar pouch posterior to the quadriceps tendon and anterior to the distal femur.* **(Right)** *Schematic diagram shows the anterior aspect of the knee joint between the distal femur, patella, and the proximal tibia. The suprapatellar pouch lies proximal to the patella and posterior to the quadriceps tendon.*

Knee: Anterior Suprapatellar (Longitudinal)

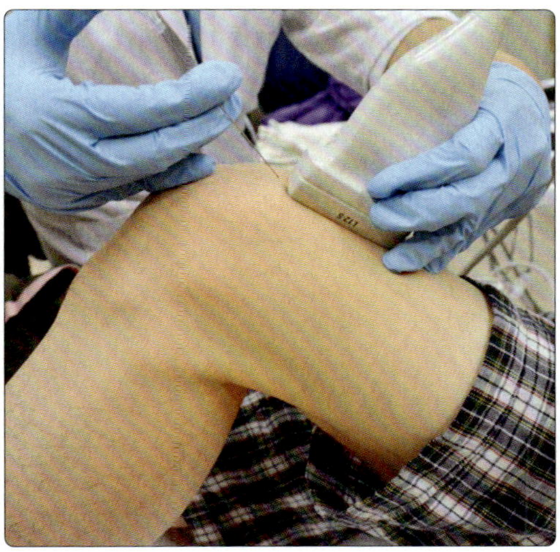

Knee: Anterior Suprapatellar (Transverse)

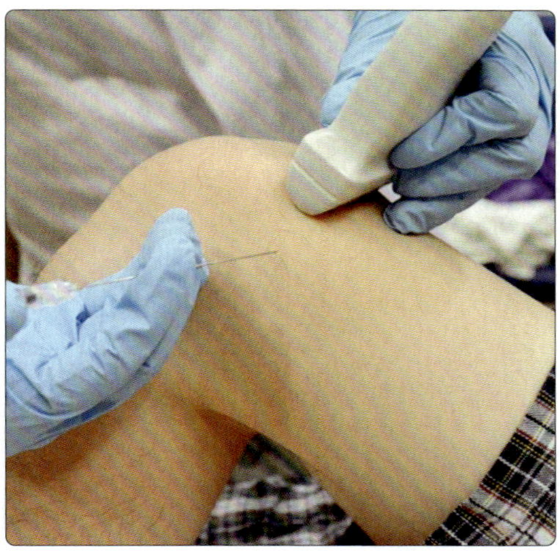

(Left) *Clinical photograph shows approach for anterolateral injection of suprapatellar pouch. The needle is inserted from distal to proximal via the anterolateral aspect of the suprapatellar pouch, as there is no overlying vastus lateralis muscle along this path.* **(Right)** *Clinical photograph shows approach for transverse injection of suprapatellar pouch. Needle is inserted from lateral to medial via the anterolateral aspect of suprapatellar pouch, as there is no overlying vastus lateralis muscle along this path.*

(Left) *Longitudinal US shows injection of knee joint via an anterolateral approach to the suprapatellar pouch. The needle tip is positioned within the distended suprapatellar pouch. The needle path is clearly seen on this in-plane approach.* (Right) *Longitudinal US of same patient after aspiration of joint fluid and during injection of steroid shows the echogenic appearance of steroid injectate.*

Knee: Anterior Suprapatellar Approach

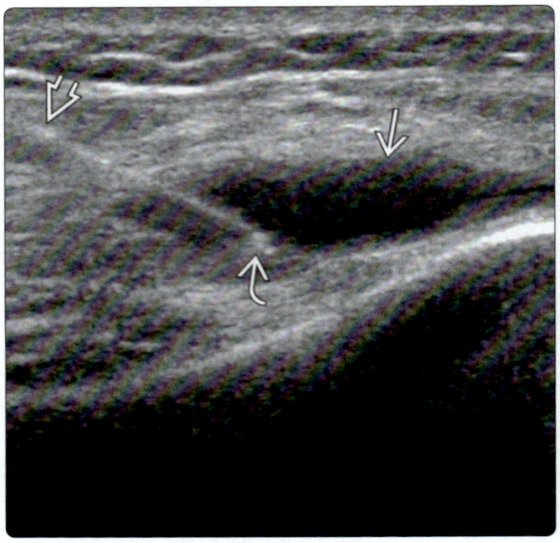

Knee: Anterior Suprapatellar Approach

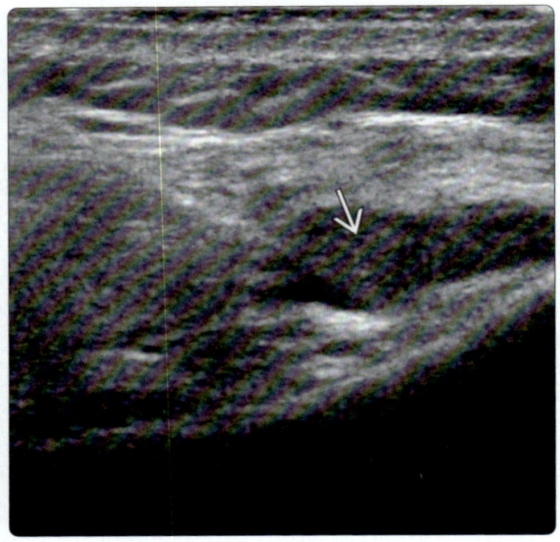

(Left) *Volume-rendered CT shows the bony anatomy of the anterior aspect of the ankle joint between the distal tibia, distal fibula, and talar dome.* (Right) *Sagittal T2 FS MR shows the ankle joint between the distal tibia and talar dome. The attachment of the anterior capsule of the ankle joint is shown.*

Ankle: Anterior Approach

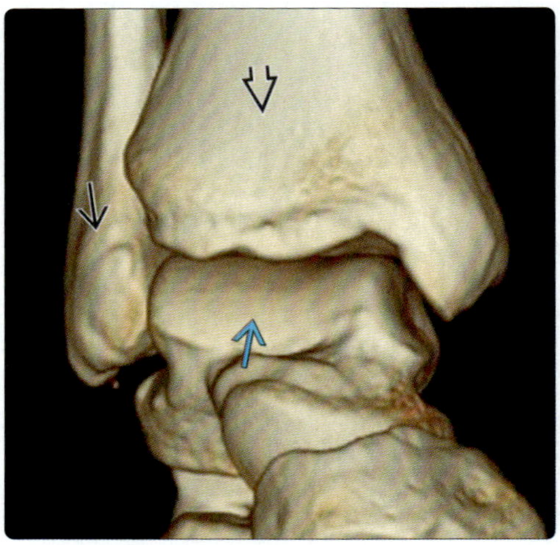

Ankle: Anterior Approach

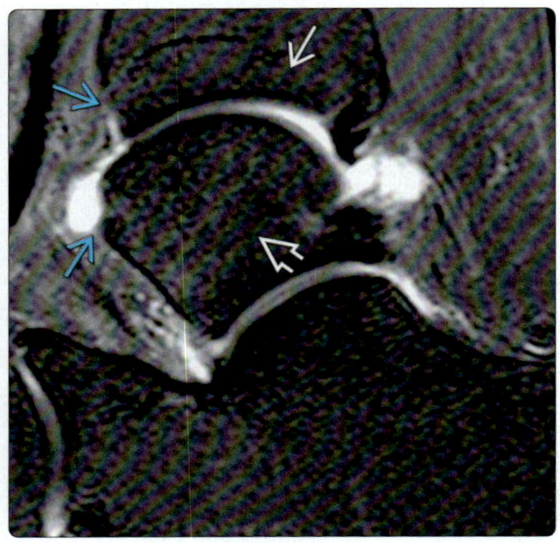

(Left) *Axial T2 FS MR shows the ankle joint just below the joint line. The anterolateral aspect of the ankle joint is closer to the skin than the anteromedial aspect. The lateral and medial malleoli are shown.* (Right) *Schematic diagram shows the anterior aspect of the ankle joint between the distal tibia and talar dome. The attachment of the ankle capsule anteriorly is shown.*

Ankle: Anterior Approach

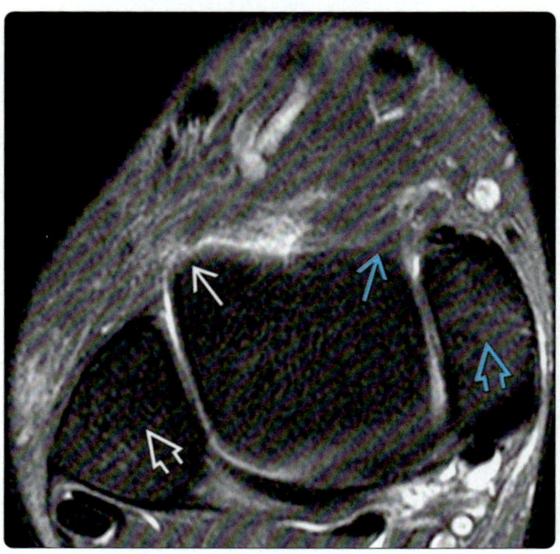

Ankle: Anterior Approach

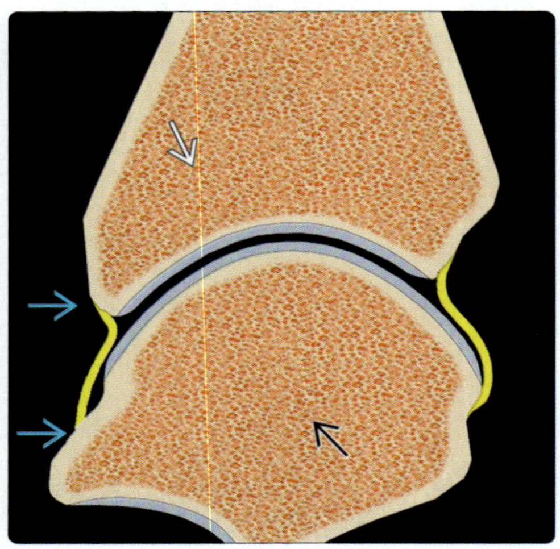

Ankle: Anterior Approach

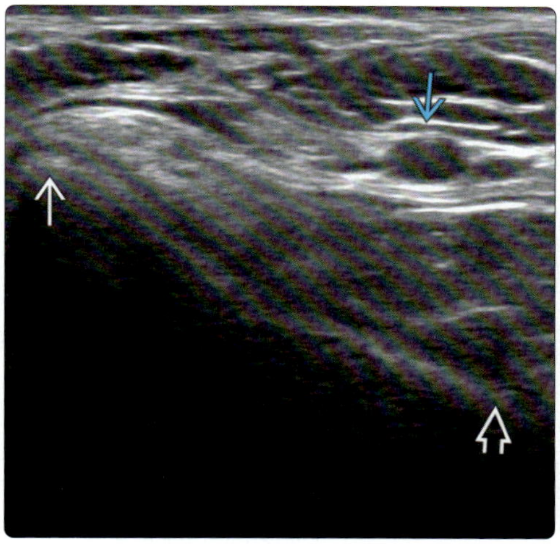

Ankle: Anterior Approach

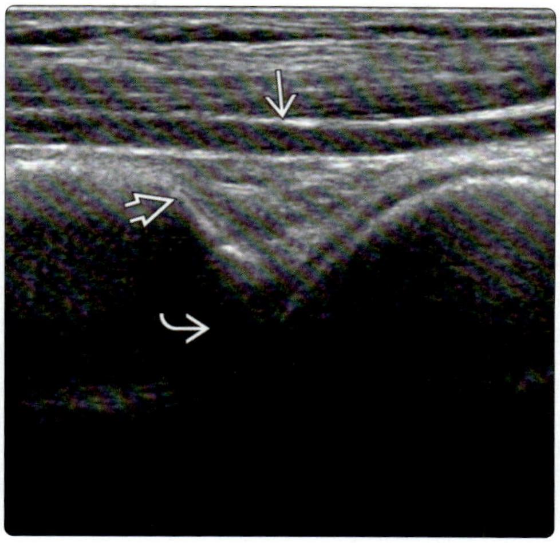

(Left) *Transverse US shows how the anterolateral aspect of the ankle joint/talar dome is closer to the probe than the anteromedial aspect. Anterolateral approach is thus preferred, avoiding injury to the dorsalis pedis artery and the adjacent deep peroneal nerve (not shown).* **(Right)** *Longitudinal US shows the dorsalis pedis artery anterior to the ankle joint. This must be avoided during needle insertion as well as the adjacent deep peroneal nerve (not shown). Note anterior fat pad of the ankle joint.*

Ankle: Anterolateral Approach

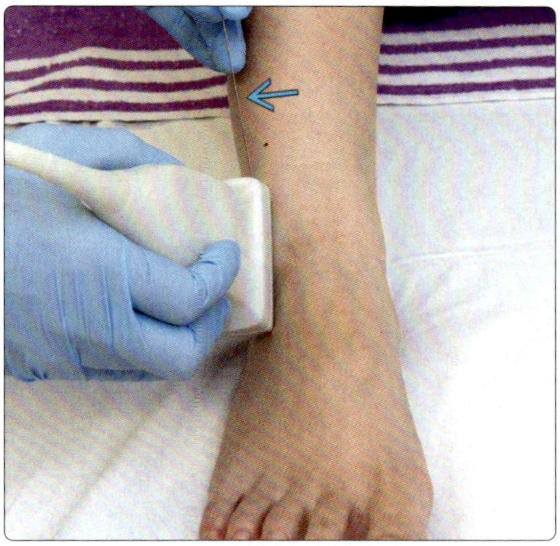

Ankle: Anteromedial Approach

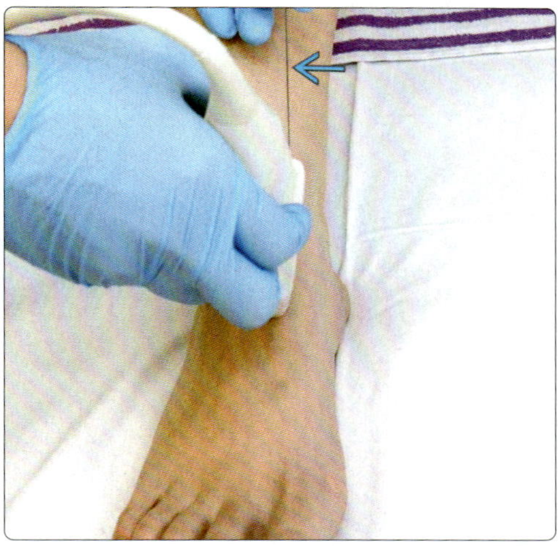

(Left) *The needle is inserted from proximal to distal at the anterolateral aspect of the ankle joint. The anterolateral aspect of the ankle joint is closer to the skin. This approach is lateral to the dorsalis pedis artery and deep peroneal nerve.* **(Right)** *The needle is inserted from proximal to distal at the anteromedial aspect of the ankle joint. This approach is medial to the dorsalis pedis artery and deep peroneal nerve.*

Ankle: Anterolateral Approach

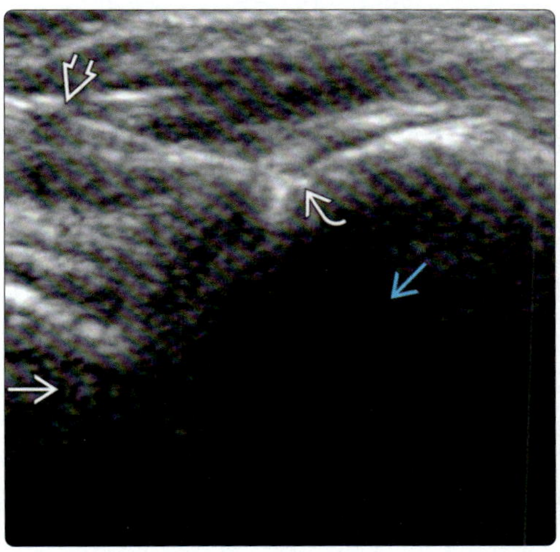

Ankle: Anterolateral Approach

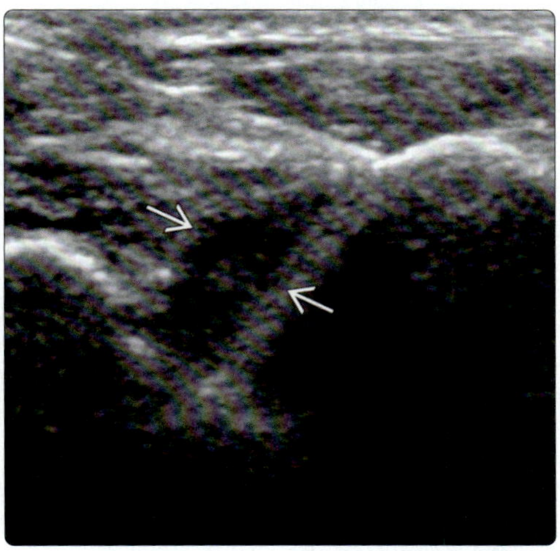

(Left) *Longitudinal US shows injection of the ankle joint via anterolateral approach between the distal tibia and talar dome. The needle tip is positioned on the surface of the talar dome articular cartilage. The needle path can be clearly seen in this in-plane approach.* **(Right)** *Longitudinal US of the same patient shows good distension of the anterior ankle joint by fluid after injection of dilute gadolinium saline mixture.*

KEY FACTS

PREPROCEDURE

- Indication
  - Glenohumeral joint aspiration (suspected infection) or injection (contrast for arthrography; adhesive capsulitis symptom relief)
  - Subacromial-subdeltoid (SA-SD) bursal injection (symptom relief) or aspiration (suspected infection)
  - Calcific tendinitis (symptom relief)
  - Rotator cuff tendinosis or tear (symptom relief and to promote healing)
  - Acromioclavicular joint injection (symptomatic relief)

PROCEDURE

- Glenohumeral joint aspiration or injection
  - Anterior (rotator cuff interval) or posterior approaches
  - Aspiration for infection or crystal analysis
  - Injection for arthrography, osteoarthritis, inflammatory arthropathy, or adhesive capsulitis
- SA-SD bursal injection or aspiration
  - Lateral, anterior, and superior approaches
  - Steroid and long-acting local anesthetic (LA) to SA-SD bursa
- Calcific tendinitis
  - Dry needling and aspiration
  - Percutaneous lavage and aspiration of calcific deposits
  - Single or double-needle techniques
  - Steroid and long-acting LA injected to bursa
- Rotator cuff tendinosis or tear
  - Steroid and LA injection to SA-SD bursa (to reduce inflammation)
  - Localized dry needling of tendinosis or tear (to promote healing)
  - Platelet-rich plasma (PRP) injection of rotator cuff partial tear (to promote healing)
    - No clear proven benefit shown for PRP in tendinosis or rotator cuff tear
- Acromioclavicular joint arthropathy
  - Steroid and LA injection to joint via transverse approach

Glenohumeral Joint Injection via Posterior Approach

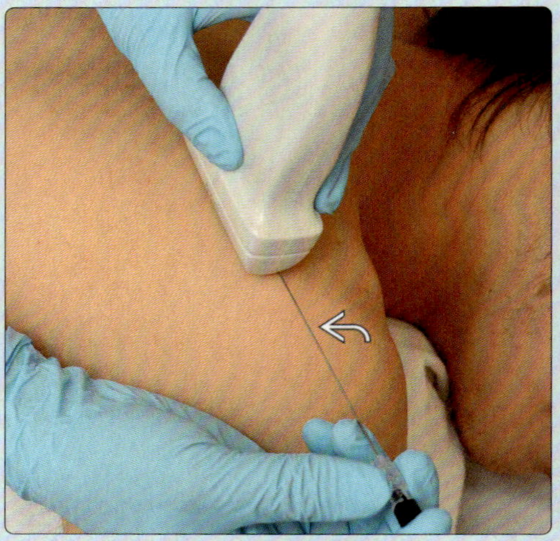

Glenohumeral Joint Injection via Posterior Approach

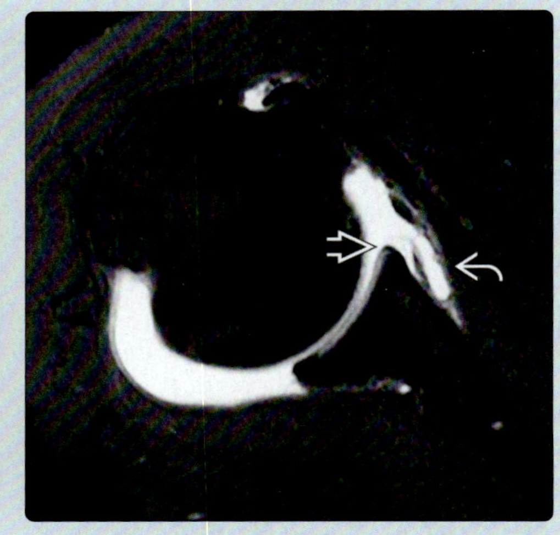

(Left) *Clinical photo of a patient with shoulder dislocation shows needle (22-g spinal) ➔ position for lateral long-axis approach to posterior glenohumeral joint for contrast injection before MR arthrography.* **(Right)** *Subsequent axial MR arthrogram T1 FS (same patient) shows glenohumeral joint distension. Note anterior capsular stripping ➔ and labral avulsion ➔. US-guided MR arthrography is a commonly performed procedure.*

Lavage and Aspiration for Calcific Tendinitis by Single-Needle Technique

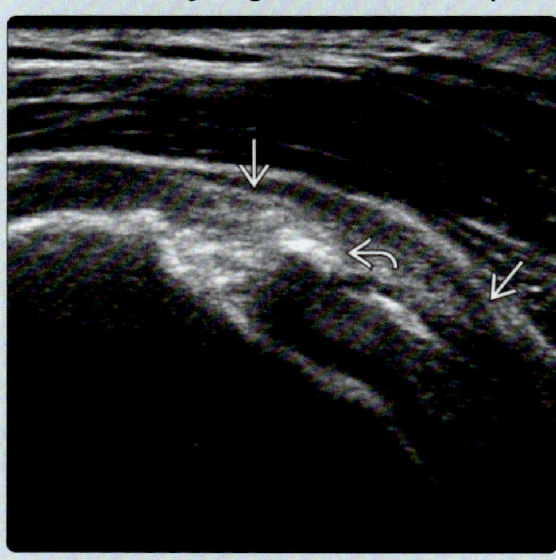

Lavage and Aspiration for Calcific Tendinitis by Single-Needle Technique

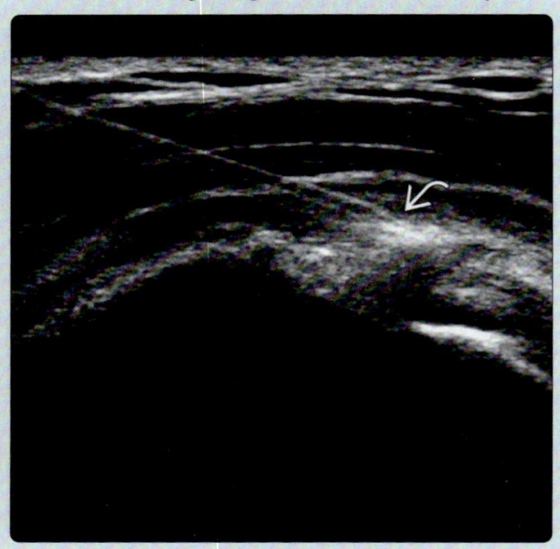

(Left) *Longitudinal US shows moderate-severity calcific tendinitis of the supraspinatus tendon ➔. The largest calcification ➔, calcification with the most focal edema, or calcification that most closely correlates with symptoms are best targeted for percutaneous lavage and aspiration.* **(Right)** *Longitudinal US shows aspiration of calcific content ➔ using an 18-g needle, connected to a syringe containing a saline-lignocaine mixture. Lavage and aspiration are performed in series.*

PREPROCEDURE

Indications

- **Glenohumeral joint injection or aspiration**
  - MR, CT, or conventional arthrography
  - Medication for control of inflammation or pain relief in osteoarthritis, inflammatory arthropathy, or adhesive capsulitis
  - Diagnosis of septic arthritis or crystal arthropathy
- **Subacromial-subdeltoid (SA-SD) bursa injection**
  - For control of inflammation and symptom relief
- **Calcific tendinitis therapy**
  - For control of inflammation, symptom relief, and to promote resorption of calcific deposits
- **Rotator cuff tendinosis or partial tear therapy**
  - For control of symptoms and to promote tissue repair
- **Acromioclavicular (AC) joint injection**
  - For control of inflammation and symptom relief

Contraindications

- Infection of overlying soft tissues
- Uncorrected severe bleeding diathesis

Getting Started

- Things to check
  - Patient identity, type of procedure, side of procedure, informed consent
  - Discuss risk of infection, allergy, bleeding
    - Risk of calcific bursitis for treatment for calcific tendinitis
  - Sedation is rarely required
- Medications
  - Anesthetic
    - Short-acting local anesthetic (LA) for skin: 1-3 mL 1% lignocaine
    - Short- or long-acting anesthetic for intraarticular injection: 1-2% lignocaine or 0.25-0.50% bupivacaine
  - Steroid
    - Short (hydrocortisone) or long acting (methylprednisolone, dexamethasone)
  - Contrast
    - CT arthrography: Iodinated contrast, e.g., Omnipaque 300 ± 0 1 mL of 1:1,000 epinephrine
    - MR arthrography: Diluted 0.1 mL gadolinium in 10 mL normal saline solution
- Equipment list
  - Sterile set
  - Spinal needles
    - 18- to 22-g intravenous or spinal needle
    - Needle size depends on consistency of fluid injected or aspirated
  - Extension tube (thin, 250 mm long, 8.5 Fr, 0.4 mL dead space) and 3-way connector
  - Syringes for LA, aspiration, and injection
  - Sterile specimen bottles

PROCEDURE

Procedure Steps

- General preparation of all percutaneous intervention
  - Sterile technique (operator, probe, patient)
    - LA to skin at puncture site
      - For some quick straightforward procedures (e.g., joint injection), anesthesia optional
- **Glenohumeral joint injection or aspiration**
  - General practical points irrespective of anterior (rotator cuff interval) or posterior approach
    - Use free-hand technique to advance needle into joint space under real-time imaging
      - 18- to 22-g spinal needle for injection or aspiration
      - Larger needle for aspiration of viscous fluid
    - Can use long-axis or short-axis approaches
    - Ensure needle tip is within joint space
      - Aspirate joint fluid prior to injection if moderate to large effusion present
    - Avoid injection of bubbles by first flushing syringe and tubing with injectate
    - Free flow of fluid with low-pressure injection and minimal patient discomfort helps confirm intraarticular position of needle tip
    - Avoid active exercise to prevent leakage of solution along needle track for 20 minutes
  - Posterior approach
    - Various positions can be used
      - Lateral decubitus position; arm mildly adducted
      - Prone position with arm in external rotation, palm facing down, and shoulder resting on pillow
      - Sitting position with back toward you, ipsilateral hand resting on contralateral shoulder (be alert for vasovagal syncope)
    - In longitudinal plane, identify posterior glenohumeral joint, glenoid, labrum, and humeral head deep to infraspinatus musculotendinous junction
      - Slight passive rotation of arm while scanning will help identify glenohumeral joint
    - Either short-axis (preferable) or long-axis needle approach can be used
    - Short-axis approach allows joint to be approached more vertically than with long-axis approach
    - Long axis is preferable as needle path easier to see
      - Can approach from medial or lateral sides
      - Approaching from lateral side allows direction of needle tip deep to posterior larum
      - Lateral approach goes through infraspinatus tendon and therefore may be slightly more painful
    - "Pop" felt when needle transverses joint capsule
  - Rotator interval approach
    - Supine position; arm partially externally rotated
      - Avoid extreme external rotation as it increases tension on anterior shoulder and increases risk of extraarticular leakage
    - Identify coracoid process and adjacent superomedial aspect of humeral head with probe aligned transverse to joint line
    - Needle entry point is just superior to subscapularis tendon and lateral to coracoid process
      - Needle is advanced until tip contacts articular cartilage of humeral head
- **Arthrography**
  - 10-20 mL of contrast solution is injected until joint capsule is satisfactorily distended

- Volume of injection is determined according to patient's comfort level and resistance to injection
- Patients with severe adhesive capsulitis will feel distension after only small volume (5-10 mL) of fluid injected into joint
- Patients following shoulder dislocation can accommodate larger volumes (15-20 mL)

- **Adhesive capsulitis treatment**
  - To reduce inflammation and relieve pain
    - 10 mL bupivacaine 0.25-0.50% mixed with 1-2 mL of 40 mg/mL methylprednisolone
    - Benefits of single injection of bupivacaine outweigh theoretical risk of chondrolysis
    - Other anesthetic agents (ropivacaine 0.5%) is less toxic to human articular chondrocytes in vitro
    - Sodium hyaluronate mixed with anesthetic is alternative if steroids contraindicated
    - Good symptom control after steroid injection, especially in those who have adhesive capsulitis with mainly inflammatory rather than fibrotic component
    - Inflammatory vs. fibrotic component can be assessed with MR prior to intraarticular injection of contrast
      - Benefit from steroid injection likely to be greater during inflammatory phase
  - Hydrodilatation to achieve capsular rupture
    - Potentially more useful in patients with greater fibrotic as opposed to inflammatory component
      - Degree of fibrosis or inflammation can be assessed on MR
    - Larger volume of saline mixed with 5 mL 0.25% or 0.5% bupivacaine and 1-2 mL of 40 mg/mL methylprednisolone/triamcinolone
      - Usually 30-40 mL saline mixture is required for optimal distension/capsular rupture
    - Capsular distension is better seen under fluoroscopy
    - 5 mL of radiopaque contrast into saline mixture
- **SA-SD bursa injection**
  - Examine shoulder in different degrees of rotation with arm by side to optimize bursal fluid accumulation
  - Approaches differ depending on quantity and location of bursal fluid
    - Lateral approach (most frequently used)
      - Probe positioned parallel to long axis of supraspinatus tendon
    - Anterior approach
      - Probe positioned on axial plane in front of shoulder with needle inserted in lateromedial direction
    - Superior approach
      - Used if fluid located in craniomedial aspect of SA-SD bursa between trapezius and upper surface of supraspinatus
      - Probe positioned in supraspinatus fossa along long axis of supraspinatus with lateral edge overhanging acromion
  - Steroid injection
    - Point bevel of needle facing supraspinatus tendon such that needle will glance off surface of tendon and enter bursa
    - 1-2 mL of 40mg/mL methylprednisolone or triamcinolone mixed with 5 mL 0.25% or 0.5% bupivacaine
      - Injection observed in real time with hypoechoic fluid seen to disperse within bursa
      - If needle is appropriately positioned, injectate should flow freely away from needle
      - No injectate should be collected around needle tip
      - Rescan rotator cuff post injection to assess for small bursal surface tears, as injectate will tract into tears and make them more easily visible
  - Good short-term symptom relief after steroid injection
- **Calcific tendinitis therapy**
  - Only painful calcific tendinitis should be treated
  - Preliminary US examination to assess size and distribution of calcific deposits
  - Calcifications best targeted
    - Largest calcification
    - Calcification producing most focal tendon edema
    - Calcification that most closely correlates with symptoms
  - Position patient so that calcification to be treated is well seen and amenable to needle placement
    - Usually supine or lateral decubitus positions
  - Single-needle or double-needle technique
    - 21- to 23-g needle for local anesthesia
    - 18- to 22-g needle for lavage and aspiration of calcium
    - Inject LA (3-5 mL 1% lignocaine) into subcutaneous tissue, SA-SD bursa, and around tendon calcification
    - Avoid introducing air during infiltration of local anesthesia as this will obscure calcification
  - Single-needle technique
    - Introduce lavage needle to center of calcium deposit
      - Needle is connected to syringe containing 5-10 mL sterile saline (70%) and lignocaine 1% (30%)
      - Undertake gentle rotational and to-and-fro movement to grind calcification, followed by injection and aspiration of saline solution
      - Fluid-filled hypoechoic cavity is seen within calcification, expanding when fluid is pushed forward, reducing when pressure is released
      - Calcium is dissolved and cloudy whitish fluid will decompress into syringe when releasing plunger
      - Repeated short series of press and release
      - Syringe exchange is necessary to keep solution clear during procedure
      - Stop lavage when no more calcium can be retrieved
  - Double-needle technique
    - 2 needles are inserted into calcification
    - Tips of these needles are very close to each other; not more than 2-3 mm apart
    - Both needles should be inserted along same coronal plane so that both needles and calcification can be seen simultaneously
      - 1st needle tip is slightly deeper than 2nd
      - Both needles are angulated 25-30°
    - Bevels of needles face each other
      - Bevel of 1st needle faces upward
      - Bevel of 2nd needle faces downward
    - Undertake gentle rotational and to-and-fro movement to grind calcification
    - Inject 5-10 mL of sterile saline (70%) and 1% lignocaine (30%) into calcium via one of needles

- Dissolved calcium content is drained from other (free) needle
- Syringe exchange is necessary to keep solution clear during procedure
- Stop lavage when no more calcium can be retrieved
  - Reposition needle to SA-SD bursa after lavage in both techniques
    - Inject 1-2 mL of 40 mg/mL methylprednisolone or triamcinolone mixed with 5 mL 0.25% or 0.5% bupivacaine
      - This helps to treat and prevent crystal-induced bursitis, which can be very painful
  - Clinical success rate varies from 60-75%
  - Specifically warn patient regarding risk of calcific bursitis after calcific tendinitis therapy, which can be very painful
- **Rotator cuff tendinosis or partial tear therapy**
  - Reduce inflammation
    - Steroid (methylprednisolone 40 mg) and LA (5 mL 0.5% bupivacaine) injection to SA-SD bursa
  - Promote healing
    - Localized dry needling of tendinosis or tear
      - 23-g intravenous needle repeatedly moved to-and-fro within tendinosis area
      - Converts area of quiescent tendinosis or tear into injured, functionally stable area
      - Resultant bleeding leads to deposition of blood cells (stem cells) within tendon or tendon tear
      - Reparative response leads to recruitment of reparative growth factors
    - Platelet-rich plasma (PRP) injection
      - Inject 5 mL PRP concentrate (extracted from patient's own venous blood)
      - PRP is source of circulating mesenchymal stem cells and reparative growth factors
      - Injection to gap between torn tendon ends
      - Although in theory should be helpful, no clear proven benefit in controlled clinical trials of PRP in treatment of tendinosis or tendon tears
      - Also has not been shown to influence retear rate after full-thickness rotator cuff repair
- **AC joint injection**
  - For symptomatic relief in AC joint arthropathy
  - Patient in supine position
  - Probe placed onto skin over superior aspect of AC joint
    - Transverse axis of AC joint
    - Allows excellent visualization of needle tip throughout and less likely to be restricted by large osteophytes
  - Using transverse approach, joint capsule is identified and 23-g needle inserted from anterior into AC joint beneath capsule
    - Identify joint capsule
    - Insert 23-g needle anteriorly into AC joint beneath joint capsule
    - Inject ~ 2 mL 0.25-0.5% bupivacaine and 1 mL of 40 mg/mL methylprednisolone
    - Widening of joint space seen during injection

Findings and Reporting

- Initial US findings
- Approach used, such as anterior or posterior approach for glenohumeral joint injection
- Type and size of needle used
- Dose and type of LA, steroid, contrast agent, or other medication used
- Volume, color, viscosity, nature of joint fluid aspirated
- Any adverse event or immediate complication
- Instruction for aftercare

Alternative Procedures/Therapies

- Medical therapy
- Physiotherapy
- Surgery
- Extracorporeal shock wave therapy (calcific tendinitis)

POST PROCEDURE

Expected Outcome

- Mild distension discomfort after joint injection

Immediate Aftercare

- Perform MR, CT, or conventional arthrography promptly as intraarticular contrast largely absorbed by 45 minutes
- Send aspirate for microbiology, cytology, crystal analysis (need fresh specimen) as appropriate

Subsequent Aftercare

- Avoid shoulder exercise for 24 hours after steroid injection
  - Maximize therapeutic effect
  - Reduce spread of steroid into surrounding tissues

OUTCOMES

Complications

- Most feared complication(s)
  - Infection
- Immediate/periprocedural complication(s)
  - Vasovagal reaction
    - Undertake procedures ideally with patient lying down
  - Bleeding (very uncommon)
- Delayed complication(s)
  - "Steroid flare"
    - Pain 1-2 days after steroid injection (up to 10%) due to steroid-related microcrystalline synovitis
      - Will settle with simple oral analgesia
  - Flushing of skin 2-3 days after steroid injection
    - Settles within 48 hours
  - Tendon rupture
    - Avoid > 3 steroid injections and leave at least 3 weeks between injections
  - SA-SD bursitis due to leakage of crystals into bursa after treatment of calcific tendinitis
    - Can be very painful; patient should be informed of this complication before procedure
    - Routine injection of steroid-lignocaine mixture into bursa after procedure
    - Oral analgesia and antiinflammatory medication

SELECTED REFERENCES

1. Messina C et al: Ultrasound-guided interventional procedures around the shoulder. Br J Radiol. 89(1057):20150372, 2016

(Left) *Clinical photo shows a 22-g spinal needle for intraarticular puncture and injection, 1-mL syringe for drawing MR contrast (which is later added to a 10-mL syringe containing saline) and a 5-mL syringe for local anesthesia.* (Right) *Clinical photo shows positioning for posterior glenohumeral joint injection. Patient is lying in a comfortable semiprone position with the arm externally rotated and abducted. The ipsilateral upper chest and arm are resting on a pillow.*

Needles and Syringes for MR Shoulder Arthrography

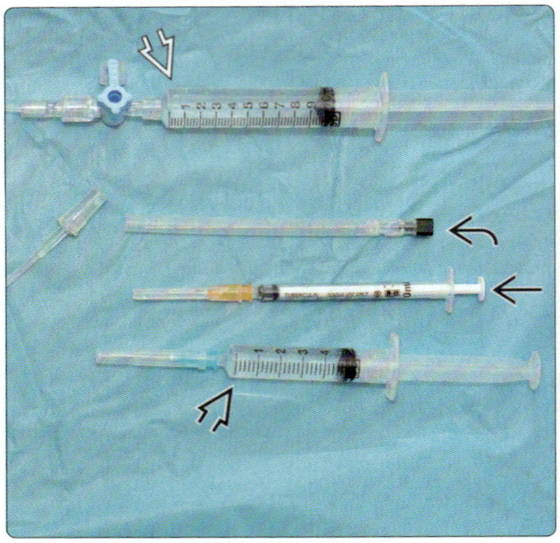

Glenohumeral Joint Injection via Posterior Approach

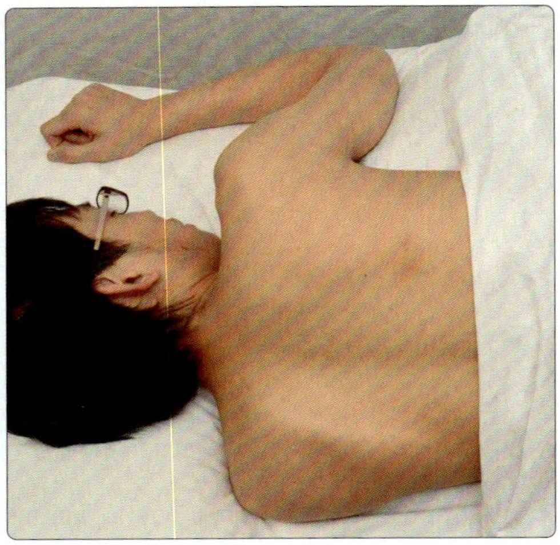

(Left) *Oblique longitudinal US along the musculotendinous junction of the infraspinatus shows the humeral head, glenohumeral joint, glenoid labrum, and glenoid bone.* (Right) *Clinical photo shows the transducer placed transversely across the posterior glenohumeral joint, along the axis of the infraspinatus musculotendinous junction. The injection needle is being introduced from lateral to medial using a long-axis approach.*

Glenohumeral Joint Injection via Posterior Approach

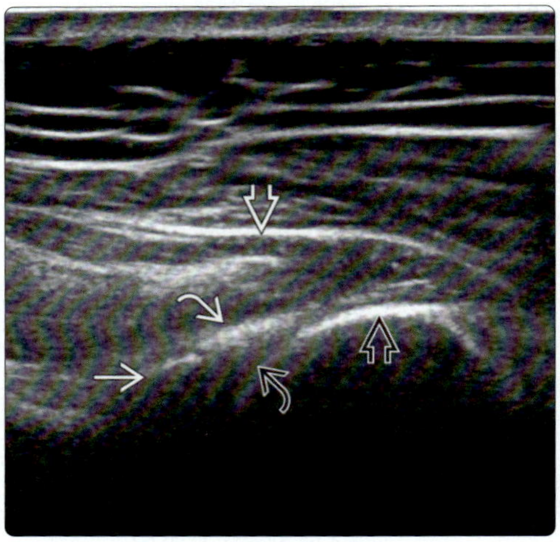

Glenohumeral Joint Injection via Posterior Approach

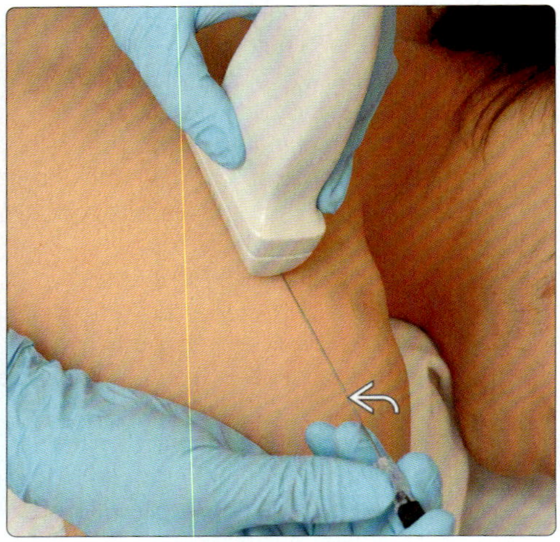

(Left) *US of the same patient shows the tip of the 22-g spinal needle at the posterior glenohumeral joint between the posterior labrum and the humeral head. The needle is introduced in a longitudinal approach. In general, 10-20 mL of contrast is injected to achieve satisfactory joint distension.* (Right) *US of the same patient taken after the injection of 14 mL of contrast shows further joint distension. Patients following shoulder dislocation can accommodate larger volumes.*

Glenohumeral Joint Injection via Posterior Approach

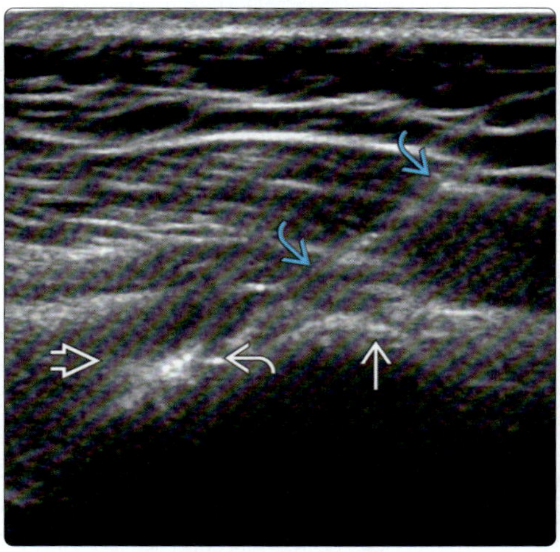

Glenohumeral Joint Injection via Posterior Approach

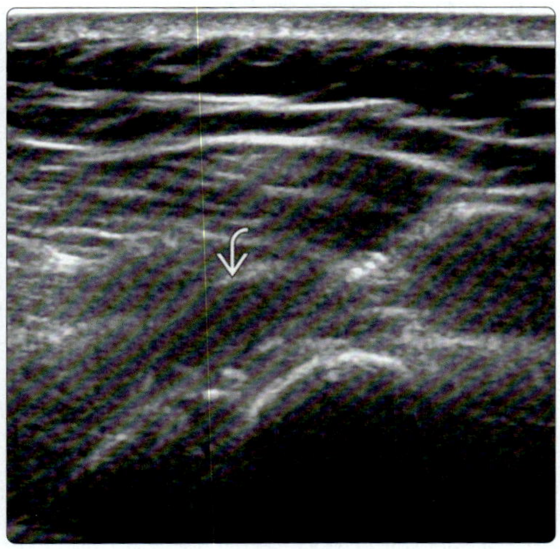

Glenohumeral Joint Injection via Rotator Interval Approach

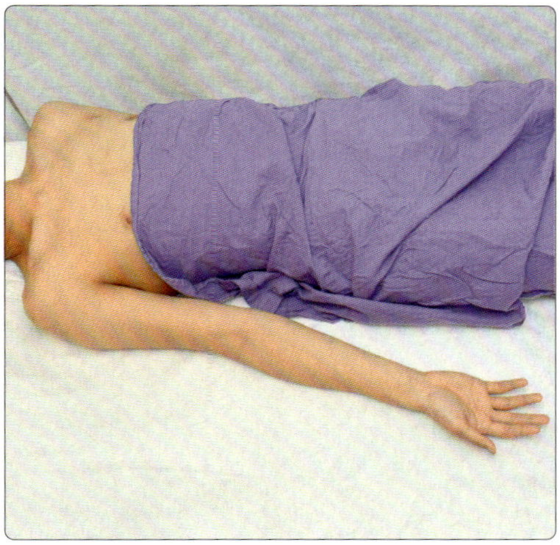

Glenohumeral Joint Injection via Rotator Interval Approach

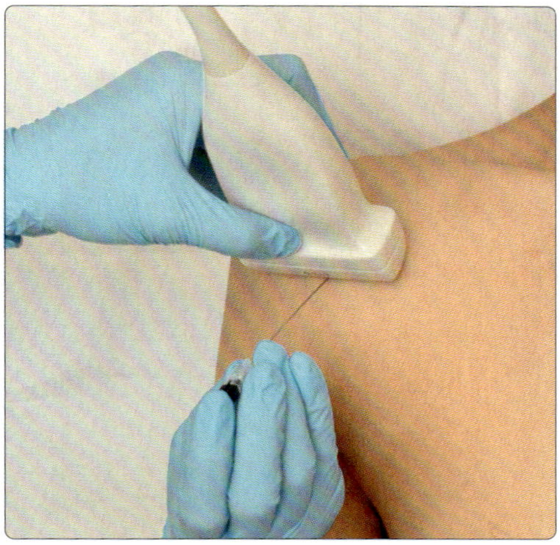

(Left) *Clinical photo shows patient positioning for a rotator cuff interval approach to the glenohumeral joint. Patient lies supine with the arm rotated slightly externally and palm facing upward. Avoid extreme external rotation of the arm.* **(Right)** *Clinical photo shows a transducer placed transversely over the rotator cuff interval. The spinal needle (22-g) is introduced almost perpendicularly into the joint using a short-axis approach.*

Glenohumeral Joint Injection via Rotator Interval Approach

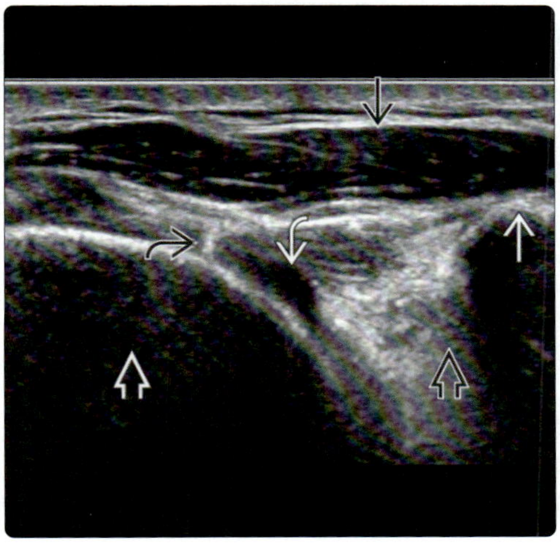

Glenohumeral Joint Injection via Rotator Interval Approach

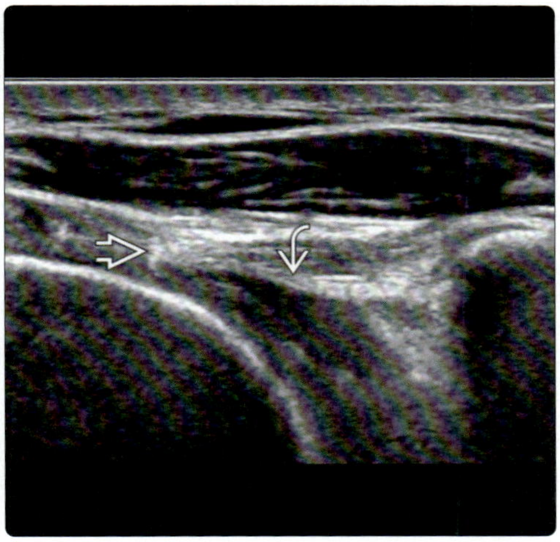

(Left) *Transverse US taken during injection shows the needle tip positioned at the superomedial portion of humeral head at the rotator cuff interval. A small amount of contrast has been injected. Note the coracoid process, glenoid bone, humeral head, and deltoid muscle.* **(Right)** *Transverse US of the same patient shows the spinal needle tip and further distension of the glenohumeral joint.*

Glenohumeral Joint Injection via Rotator Interval Approach

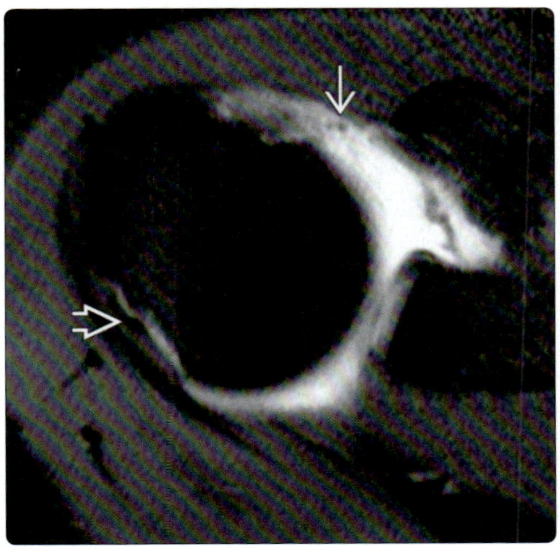

Glenohumeral Joint Injection via Rotator Interval Approach

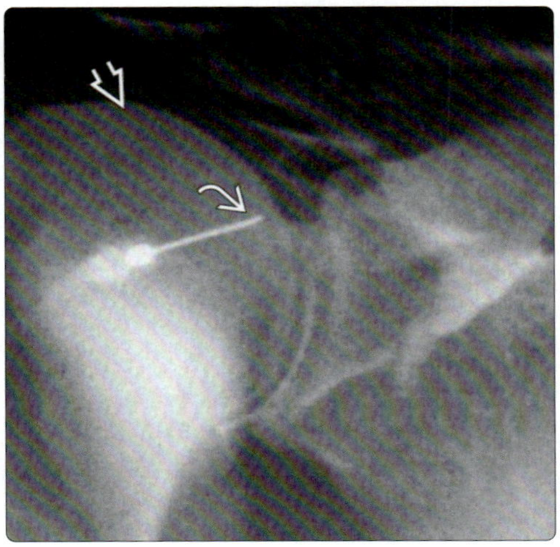

(Left) *Axial T1 FS MR arthrogram of the same patient shows satisfactory glenohumeral joint distension. There is a small Hill-Sachs deformity. Two tiny air bubbles are present in the nondependent portion of the glenohumeral joint.* **(Right)** *Frontal radiograph of the shoulder taken during fluoroscopic-guided contrast injection via rotator interval approach shows a 22-g spinal needle with the needle tip contacting the superomedial portion of the humeral head.*

(Left) *Clinical photo of a patient with septic arthritis shows aspiration of purulent fluid → from the glenohumeral joint via an anterior approach. The approach for joint aspiration is similar to that for joint injection.* **(Right)** *Clinical photo shows positioning for subacromial-subdeltoid bursal injection (lateral approach), which is the most commonly used approach. The probe is positioned along the long axis of the supraspinatus tendon with the needle introduced in a lateral-to-medial direction.*

Glenohumeral Joint Aspiration

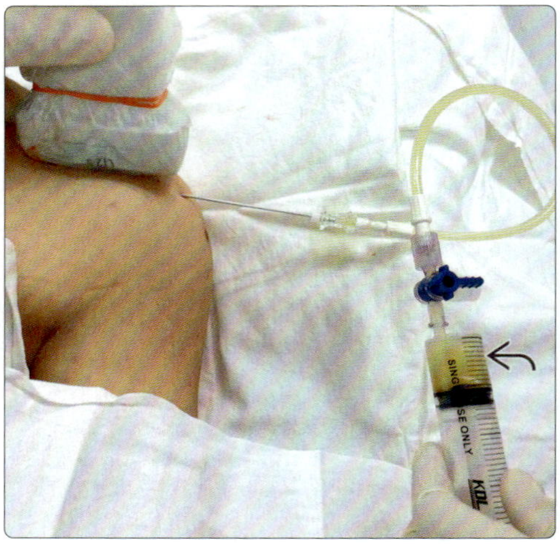

Subacromial Subdeltoid Bursa Injection

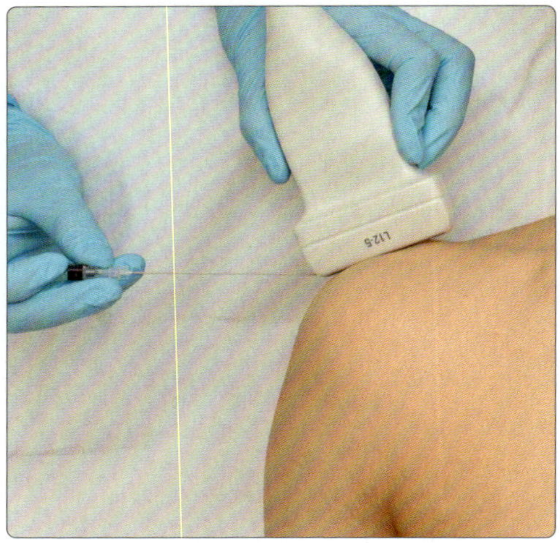

(Left) *Clinical photo shows the positioning for subacromial-subdeltoid bursal injection (superior approach). This approach can be used if fluid has been collected mainly in the cranial part of the bursa.* **(Right)** *Clinical photo shows positioning for subacromial-subdeltoid bursal injection (anterior approach). The transducer is positioned in an axial plane at the front of the shoulder, and the needle is advanced in a lateromedial direction.*

Subacromial Subdeltoid Bursa Injection

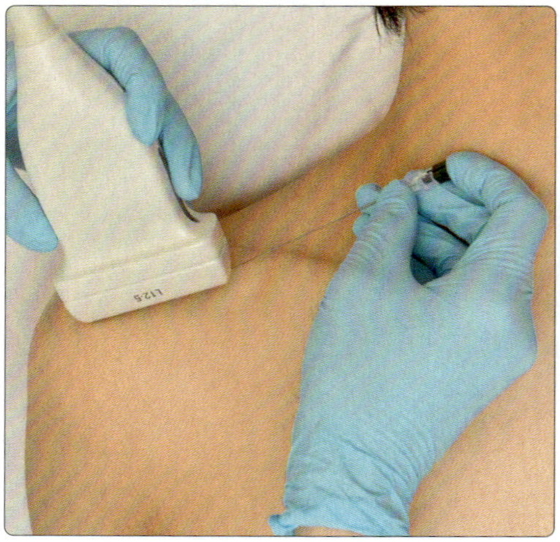

Subacromial Subdeltoid Bursa Injection

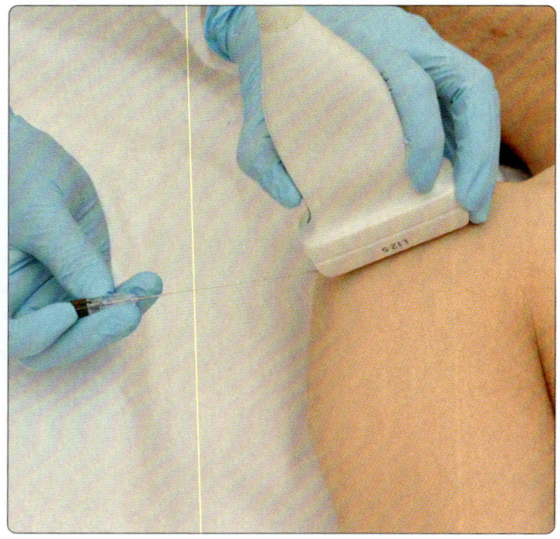

(Left) *Longitudinal US shows the insertion of a 20-g needle → into the subacromial-subdeltoid bursa along a pathway almost parallel to the transducer. Note early distension of the bursa during injection →.* **(Right)** *Longitudinal US shows severe calcific tendinitis of the subscapularis tendon with several large clumps of calcification within the more distal fibers →. Note the surrounding soft tissue edema → due to tendon inflammation.*

Subacromial Subdeltoid Bursa Injection

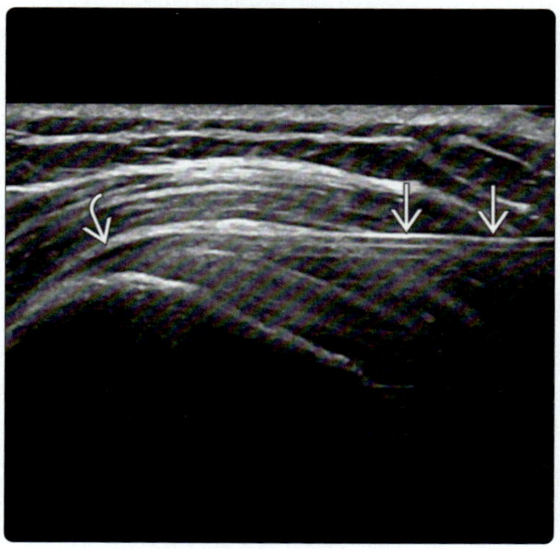

Lavage and Aspiration for Calcific Tendinitis by Single-Needle Technique

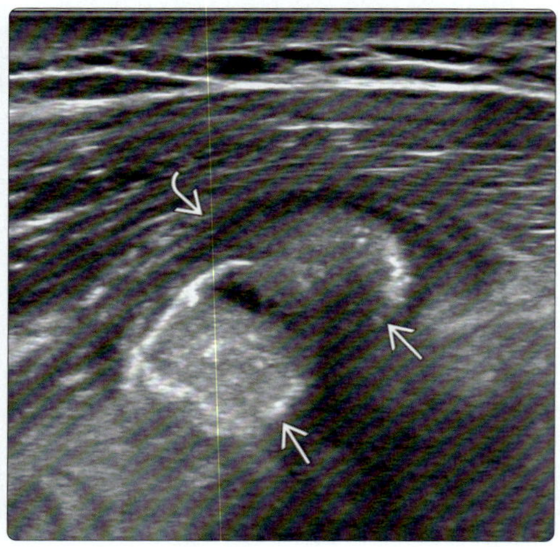

Lavage and Aspiration for Calcific Tendinitis by Single-Needle Technique

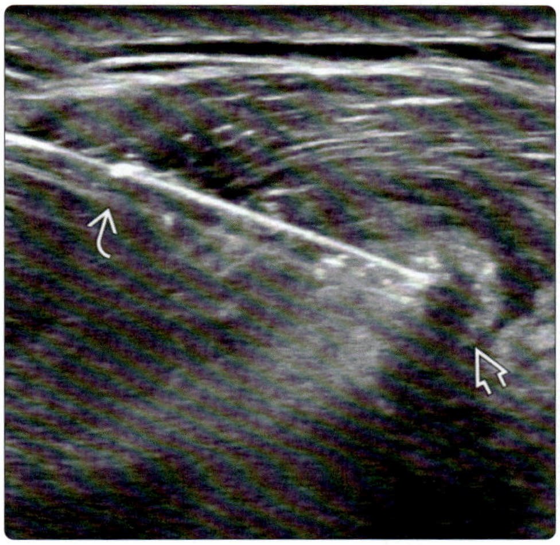

Lavage and Aspiration for Calcific Tendinitis by Single-Needle Technique

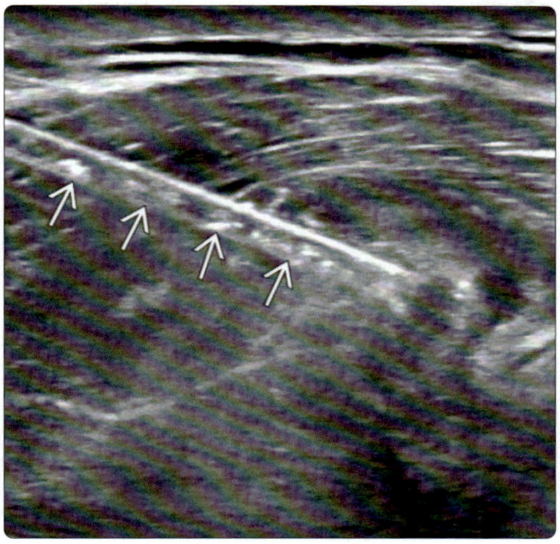

(Left) *Longitudinal US shows the insertion of a 16-g needle into one of the large calcified clumps within the fibers of subscapularis tendon. This was followed by both rotational and to-and-fro movement of the needle to grind the calcification, i.e., dry needling.* **(Right)** *Longitudinal US shows aspiration of the calcific content along the 16-g needle, which is connected to a syringe containing a saline-lignocaine mixture. Lavage followed by aspiration was performed in series.*

Lavage and Aspiration for Calcific Tendinitis by Single-Needle Technique

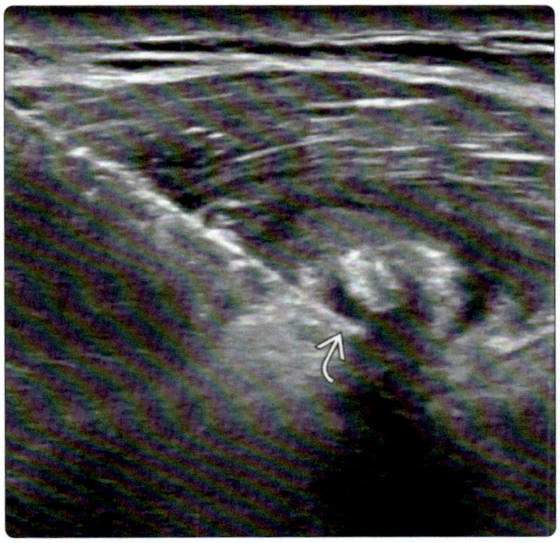

Lavage and Aspiration for Calcific Tendinitis by Single-Needle Technique

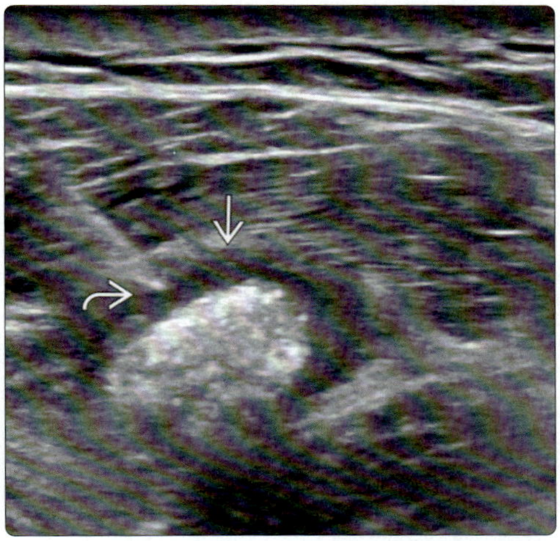

(Left) *Longitudinal US shows the tip of the needle positioned at a different position. Warm saline lavage and aspiration was performed in series.* **(Right)** *Longitudinal US shows injection of steroid into peritendinous soft tissue after a series of lavage and aspiration. The tip of the needle is positioned in the edematous soft tissue surrounding the calcification to reduce inflammation after the procedure. Although calcific material was aspirated, the calcific shell remains similar after the procedure.*

Acromioclavicular Joint Injection

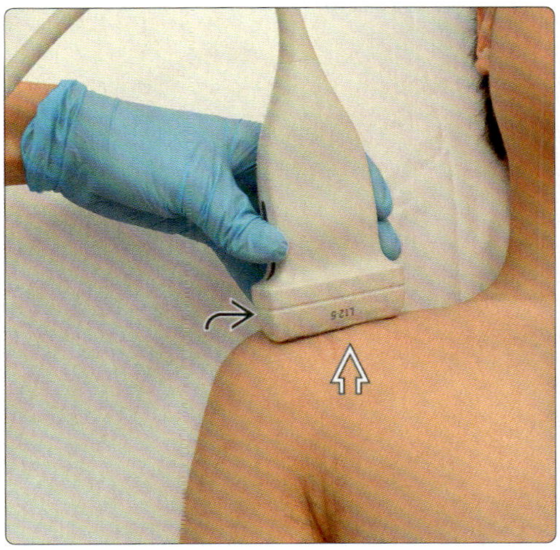

Acromioclavicular Joint Injection

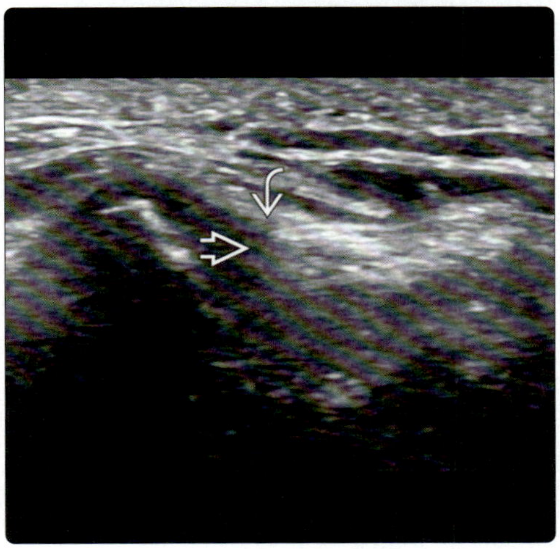

(Left) *Clinical photo shows transverse positioning of transducer over superior aspect of acromioclavicular joint. The point of needle insertion is anterior to the acromioclavicular joint.* **(Right)** *Using a transverse approach, the joint capsule is identified and a 23-g needle inserted anteriorly into the acromioclavicular joint beneath the capsule. 2 mL 0.25-0.5% bupivacaine and 40 mg steroid are injected into the acromioclavicular joint. This is a good treatment for inflammatory-type osteoarthritis.*

Elbow Procedures

KEY FACTS

PREPROCEDURE

- **Indication**
  - Intraarticular aspiration of joint fluid or injection of contrast or medication
  - Tendinosis or ligament injury treatment (dry needling, steroid injection, prolotherapy, autologous blood injection, platelet-rich plasma injection)
  - Paraarticular cyst or bursal aspiration
  - Synovial or intraarticular soft tissue mass biopsy

PROCEDURE

- **Intraarticular aspiration or injection**
  - Posterior (trochlear-olecranon) approach (preferable)
    - Position probe sagittally along posterior part of elbow
  - Anterolateral (radiocapitellar) approach
    - Potential risk to radial neurovascular bundle
    - Needle tip placed into smaller space
- **Tendinosis or ligament injection**
  - Identify abnormal hypoechoic area of tendinosis ± anechoic cleft-like intrasubstance tears within area of tendinosis
  - Treatment is aimed at these areas
- Dry needling: To-and-fro movement of needle within area of tendinosis for 1 minute
- Autologous blood or platelet-rich plasma injection
  - Platelet-rich plasma of no proven benefit over dry needling or autologous blood injection for tendinosis
- Corticosteroid injection
  - Inject peritendinous tissue rather than tendon substance
  - Reduces peritendinitis but has no therapeutic effect on tendinosis per se
    - Ultimately has deleterious effect on tendon
      - Corticosteroids lead to increased fibroblastic senescence, reduced collagen organization, and increased collagen necrosis within tendons
- **Paraarticular cyst aspiration**
  - 18-g or larger needle for more viscous content

US-Guided Aspiration

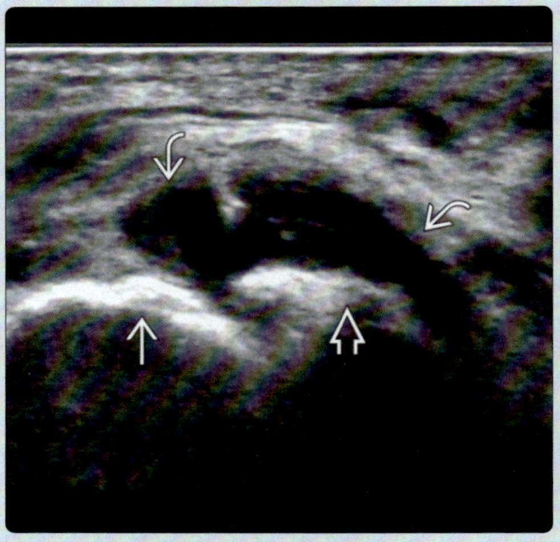

US-Guided Aspiration

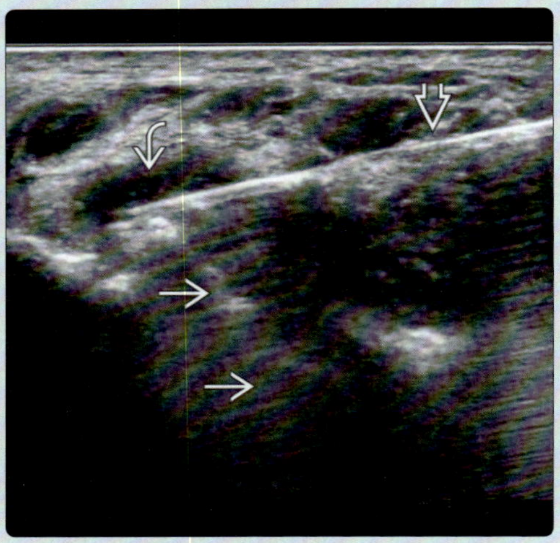

(Left) *Longitudinal US shows the distended posterior elbow recess ➡ superficial to the humerus ➡ and olecranon process ➡. Aspiration was indicated to exclude infection or crystal arthropathy.* **(Right)** *Longitudinal US of the same patient shows an 18-g spinal needle ➡ with reverberation artifacts ➡ within the effusion. The posterior recess is partially collapsed ➡.*

Intraarticular Injection of Contrast

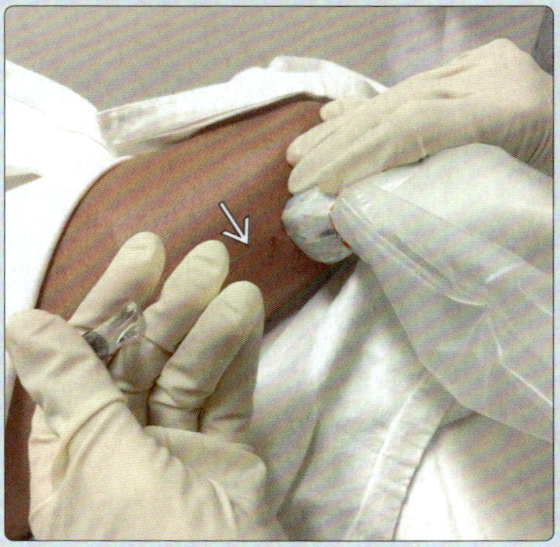

Intraarticular Injection of Contrast

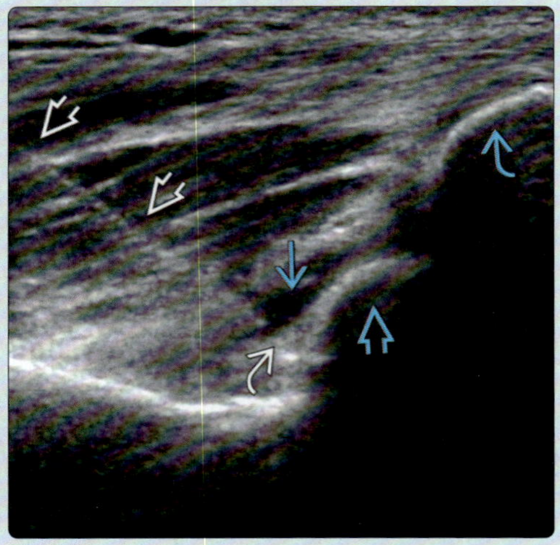

(Left) *Clinical photograph during the US-guided injection of MR contrast into the elbow joint using the posterior approach is shown. The operator is advancing a 25-g spinal needle ➡ under US guidance using a free-hand technique.* **(Right)** *US during the injection of MR contrast into the elbow joint using the posterior approach is shown. The needle ➡ can be seen coursing through the triceps muscle with the tip ➡ at the olecranon fossa where there is fluid ➡ distension. Distal humerus ➡ and olecranon process ➡ are shown.*

PREPROCEDURE

Indications

- Intraarticular aspiration or drainage of fluid
  - Suspected septic arthritis or crystal arthropathy
  - Temporary relief of symptoms due to distension pain
- Intraarticular injection of contrast or medication
  - Arthrography (CT or MR)
  - Medication to relieve pain or inflammation
    - Steroid injection to reduce inflammation in either inflammatory osteoarthritis or inflammatory arthritis, such as rheumatoid arthritis
  - Radiosynovectomy
    - Radioactive isotope (e.g., yttrium-90) injection into joint for radioablation of synovitis in hemophiliac patients or inflammatory joint disease
    - Particularly useful in hemophiliac patients to reduce frequency of joint hemarthroses
      - In hemophilia, synovitis leads to recurrent hemarthroses, which leads to synovial hypertrophy and synovitis; this cycle of progressive deterioration leads to end-stage arthropathy
- Paraarticular aspiration, e.g., ganglion cyst
- Paraarticular soft tissue injection
  - For common extensor tendon origin (CETO) or common flexor tendon origin (CFTO) tendinosis
  - Dry needling, steroid injection, prolotherapy, autologous blood injection, or platelet-rich plasma (PRP) injection
- Biopsy of synovium/intraarticular soft tissue mass

Contraindications

- Local sepsis

Getting Started

- Things to check
  - Time-out procedure (patient identity, type and side of procedure, allergy status)
  - Informed consent form signed and risk of procedure explained
- Medications
  - Local anesthetic for skin (3 mL of 1% lignocaine)
  - Local anesthetic for intraarticular injection
    - 1-2% lignocaine or 0.25-0.50% bupivacaine (Marcaine)
  - Steroid
    - 40 mg/mL of methylprednisolone (Depo-Medrol)
  - Contrast
    - CT arthrogram: 10 mL of iodinated contrast (Omnipaque 300 mg/dL ± 0.1 mL of 1:1,000 epinephrine)
    - MR arthrogram: 10 mL of diluted gadolinium (0.1 mL of gadolinium in 10 mL of normal saline)
- Equipment list
  - Sterile set, 18- to 23-g spinal needle, extension tube, syringes, coaxial needle, and biopsy needle, sterile and specimen bottles

PROCEDURE

Procedure Steps

- **Intraarticular aspiration or injection**
  - General advice irrespective of approach used
    - Flush syringe and injection tubing to avoid injection of bubbles
    - If needle tip is difficult to see during insertion, move needle slightly in and out under real-time imaging to confirm position of needle tip
    - Aspirate as much fluid as possible for decompression
    - If joint fluid is too viscous, consider using larger bore needle, e.g., 16- or 18-g
    - If insufficient aspirate despite good needle tip position, rotate needle 180° to change bevel position or else retract slightly and reposition needle tip
      - If still insufficient aspirate, consider irrigating joint with saline; saline wash is then aspirated and labeled as joint wash
    - Leave needle either free in joint or resting on articular cartilage, but do not push needle into articular cartilage
    - Free flow of fluid with low pressure during injection helps confirm intraarticular position of needle tip
  - Posterolateral approach (trochlear-olecranon articulation)
    - Preferable, as no major neurovascular structures posterolaterally
    - Position patient with elbow flexed to 90° and forearm fully pronated
    - Position probe sagittally along posterior part of elbow
    - Move transducer laterally until triceps is just out of view
  - Anterolateral approach (radiocapitellar articulation)
    - More ergonomic and convenient in terms of patient positioning, but more difficult and time consuming to find injection site
      - Potential risk to radial neurovascular bundle
      - Needle tip placed into smaller space
    - Patient sits next to table with flexed elbow resting on table and lateral aspect uppermost
    - Position probe longitudinally along lateral part of elbow
      - Landmarks: Radial head and capitellum
- **Paraarticular aspiration**
  - Needle advanced into paraarticular cyst or collection under real-time imaging
  - 21-g needle for ganglion cyst aspiration
    - 18-g or larger needle for more viscous content
  - ± hyaluronidase (150 units in 1 mL) prior to aspiration if too viscous
- **Tendinosis treatment**
  - Common procedure for CETO tendinosis, less common for CFTO tendinosis
    - Now widely accepted that tendinosis is not inflammatory condition but fibroblastic and vascular response to failure of normal tendon repair
    - Most patients respond to conservative treatment
    - Intervention indicated in patients refractory to conservative treatment suffering from chronic pain and disability
  - Superficial position of CETO or CFTO is ideally suited to US-guided injection
  - Check particularly whether any tear is present on preliminary US examination as this will have bearing on treatment choice

- Position probe over affected tendon area, including tendinosis or partial tear
- 21- to 25-g needle advanced into tendon under real-time imaging
- **Dry needling: Repeated fenestration of local tendinosis**
  - Fenestration means making holes (fenestra is Latin word for window)
  - Initially infiltrate local anesthetic around tendon
  - Position needle tip centrally within most severe area of tendinosis
  - Move needle tip to and fro within area of tendinosis for 1 minute to create fenestrations, fiber disruption, and internal bleeding
  - Often followed by autologous blood or PRP injection
- **Corticosteroid injection**
  - Historically, steroid is used frequently in treatment of CETO or CFTO tendinosis
  - Local infiltration of 1 mL of triamcinolone acetonide (10 mg/mL) and 1 mL of 2% lignocaine
  - Inject peritendinous tissues rather than tendon itself
  - Leads to significant improvement in pain and grip strength at 6 weeks time, though no significant long-term difference compared with local anesthetic injection or conservative treatment
  - Helps to treat peritendinitis associated with tendinosis, though no known beneficial effect on tendinosis per se
- **Prolotherapy**
  - Treatment of tendinosis, partial tendon tear, or ulnar collateral ligament tear
  - Irritant solution (proliferant, e.g., hyperosmolar dextrose) incites inflammatory response with fibroblast proliferation and collagen repair
  - 0.5 mL of dextrose solution (4 mL of 50% dextrose + 4 mL of normal saline and 2 mL of lignocaine) injected into tendon where there are features of tendinosis or tear
  - Additional 2 mL total of solution injected near bone at supracondylar ridge and annular segment
  - Leads to significant improvement in elbow pain score, grip strength, and range of movement at 8 weeks time
- **Autologous blood**
  - Indicated for tendinosis, partial tendon tear, or ulnar collateral ligament tear
  - First, identify any discrete tear, focal tendinosis, or discrete area of neovascularity
  - Draw 2 mL of autologous blood from contralateral antecubital vein and gently shake to prevent clotting
  - Infiltrate 2 mL of bupivacaine (0.25%) along surface of tendon or ligament
  - Insert needle into most abnormal site of tendon or ligament
    - Usually preceded by dry needling of tendon or ligament
  - Repeat procedure 1x after 4 weeks time
  - Leads to significant reduction in elbow pain score
- **PRP injection**
  - Indicated for tendinosis, partial tendon tear, or ulnar collateral ligament tear
  - PRP contains fibroblast growth factors, which facilitate healing
  - 5 mL of leukocyte-rich PRP injected
  - Apply bandage and moist heat packs every 3 hours for 15 minutes duration for first 24 hours after injection
  - Current evidence indicates that PRP is no better than dry needling or autologous blood injection for elbow tendinosis

- **Intraarticular soft tissue or synovial biopsy**
  - Use coaxial needle to prevent multiple skin punctures
  - Very useful at confirming or excluding septic arthritis if no joint fluid

Findings and Reporting

- Depends on type of interventional procedure performed

POST PROCEDURE

Expected Outcome

- Aspiration of joint fluid
  - Immediate decrease in distension pain and swelling
- Intraarticular injection of medication or contrast
  - May experience synovitis discomfort for 1-2 days
- Paraarticular tendon/ligament injection
  - Slow improvement over 1-2 weeks

Immediate Aftercare

- Perform MR, CT, or conventional arthrography as soon as possible after injection as contrast will be mostly absorbed within 30 minutes
- Send aspirate for microbiology, cytology, crystal analysis (need fresh specimen) as appropriate

Subsequent Aftercare

- Avoid exercise for 24 hours after corticosteroid injection to maximize therapeutic effect and reduce spread of steroid into surrounding tissues
- After treatment of tendinosis or ligament injury, refrain from excessive loading of tendon during week following procedure
  - Avoid taking antiinflammatory medication, which may impede healing process

OUTCOMES

Complications

- Most feared complication(s)
  - Infection
  - Extravasation of contrast/medication
    - Steroid/radiosynovectomy: Fat pad atrophy, skin atrophy, skin depigmentation
- Immediate/periprocedural complication(s)
  - Bleeding, especially for those with bleeding diathesis
  - Neurovascular bundle injury (rare)
    - Ulnar nerve injury from posterior approach
    - Radial nerve injury from anterior radiocapitellar approach
  - Allergy to contrast and medication (rare)

SELECTED REFERENCES

1. Calandruccio JH et al: Autologous blood and platelet-rich plasma injections for treatment of lateral epicondylitis. Orthop Clin North Am. 48(3):351-357, 2017

Posterior Approach

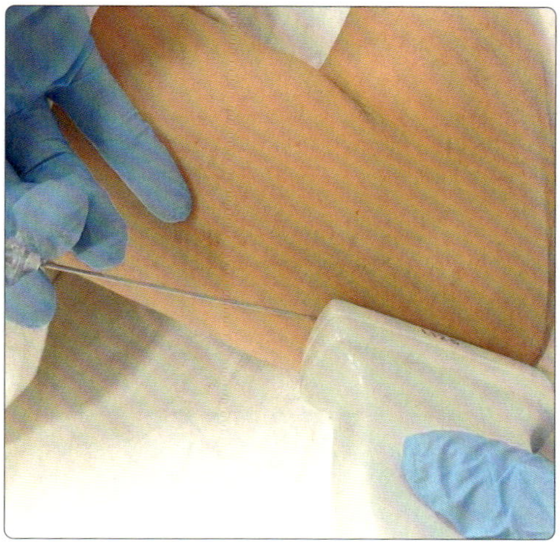

Posterior Approach

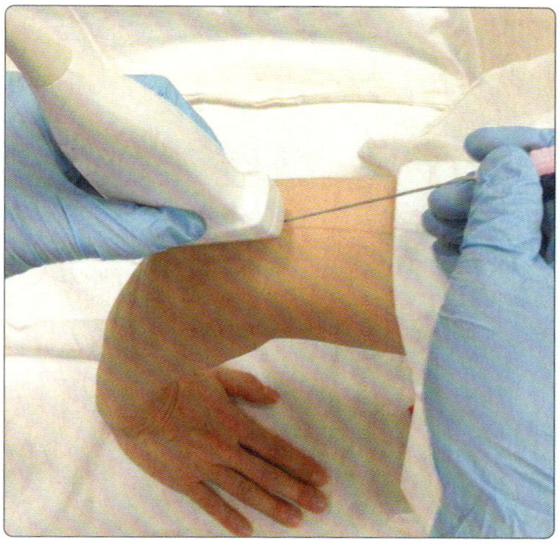

(Left) *Posterior approach shows the flexed elbow either resting on a table or lying on a bed. This approach is common and convenient for the patient. However, it provides less space for redirecting the needle position.* **(Right)** *Posterior approach shows an alternative palm-down, wrist-extended, and elbow-flexed position. This approach allows a wider space for needle puncture and adjustment. This position requires a high level of patient cooperation.*

Anterolateral Approach

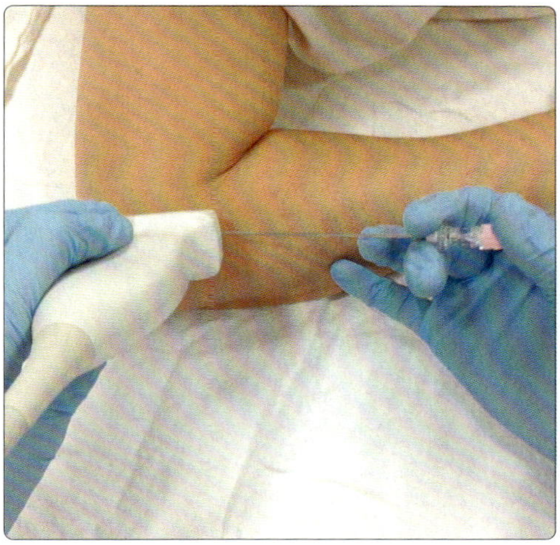

Common Extensor Tendon Origin Injection

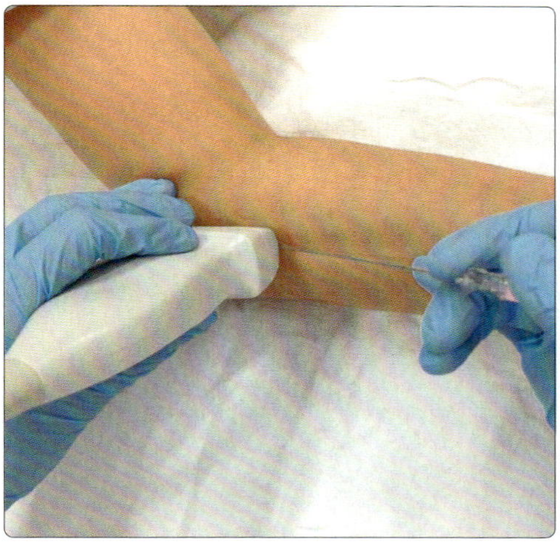

(Left) *Anterolateral approach to radiocapitellar articulation shows how the elbow is flexed and resting on a table. This approach is more difficult and time consuming. The joint space is smaller and the radial nerve and artery are close to the needle path.* **(Right)** *Clinical photograph shows the approach to the common extensor tendon origin (CETO) injection with the elbow slightly flexed and resting on a table under US guidance. The needle is then advanced into the CETO.*

US-Guided Radiosynovectomy

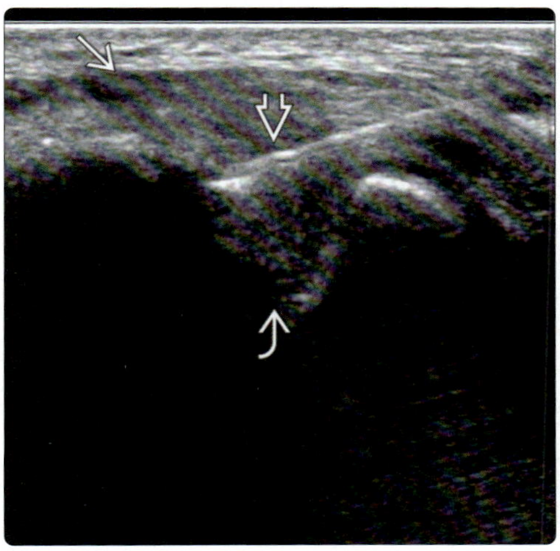

US-Guided Radiosynovectomy

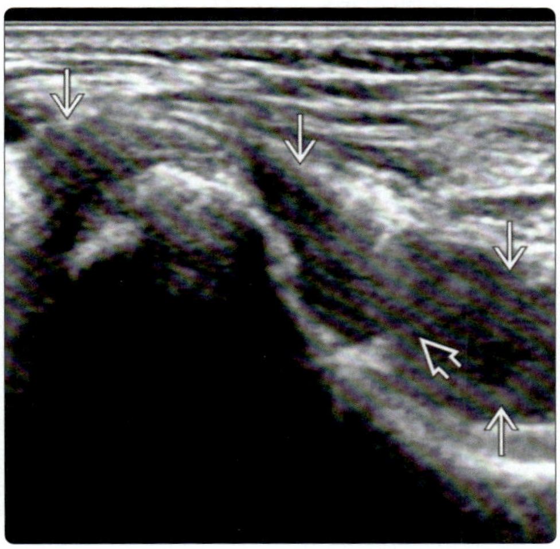

(Left) *US-guided radiosynovectomy of a hemophiliac joint is shown using a 23-g spinal needle ➡ within the moderately eroded lateral radiocapitellar joint ➡ for radioactive yttrium injection. Moderate synovial proliferation ➡ is present.* **(Right)** *US-guided radiosynovectomy of a hemophiliac patient is shown with the tip of the 23-g spinal needle at the posterior aspect of the mildly eroded distal humerus ➡. The posterior recess is distended with thickened hypoechoic synovium ➡.*

(Left) *Clinical photograph of set-up used for injection of contrast or medication shows a syringe with 3-way connector and extension tubing, specimen bottle, syringe with local anesthetic, forceps, sterile solution, and a 22-g spinal needle.* **(Right)** *Clinical photograph shows a commonly used spinal needle with a sharp, beveled edge used to puncture skin and soft tissue. The stylet is removed from the outer sheath only in the joint or cyst.*

US-Guided Aspiration/Injection Sterile Set

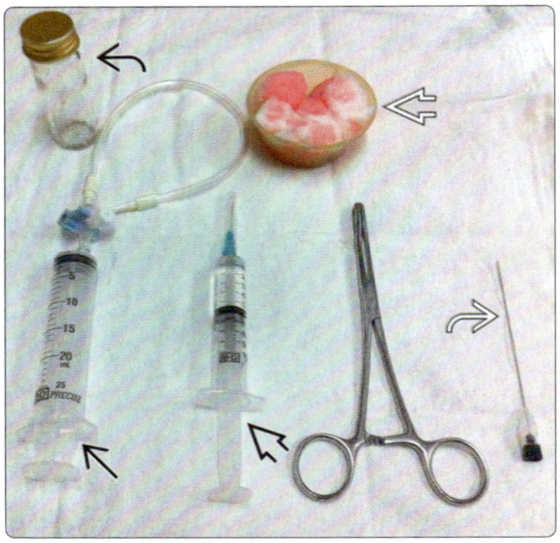

Needle Used for Drainage

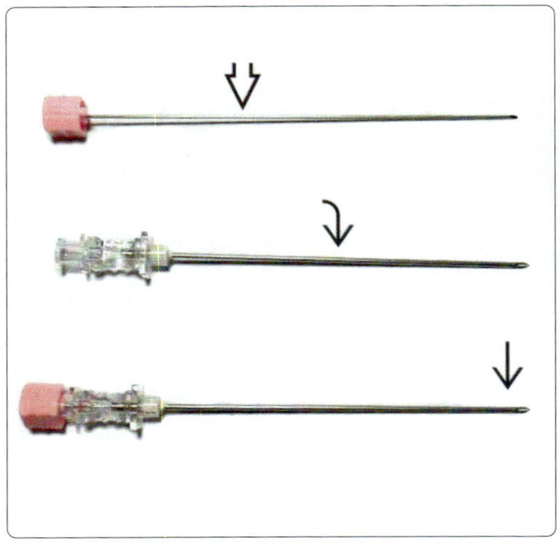

(Left) *Transverse US shows aspiration of a posterior elbow recess in a rheumatoid patient. The recess has been punctured with a 21-g spinal needle under real-time imaging. There is some cortical erosive change at the olecranon tip.* **(Right)** *Transverse US during biopsy of a patient with suspected septic arthritis is shown. The biopsy needle courses through the thickened synovium of the distended posterior recess alongside the olecranon. Culture of the synovium was negative. The final diagnosis was inflammatory arthritis.*

US-Guided Aspiration

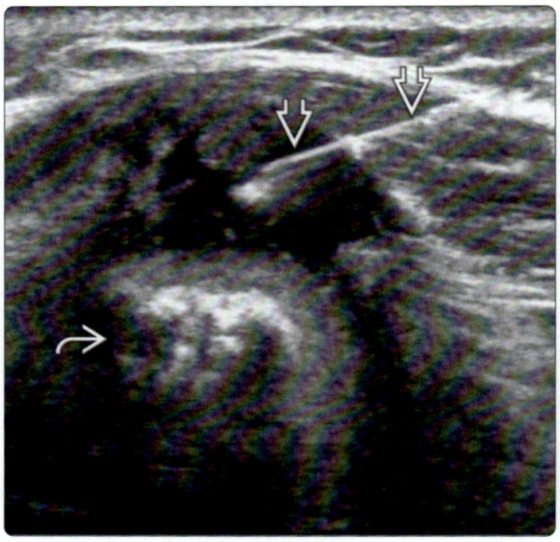

US-Guided Synovial Biopsy

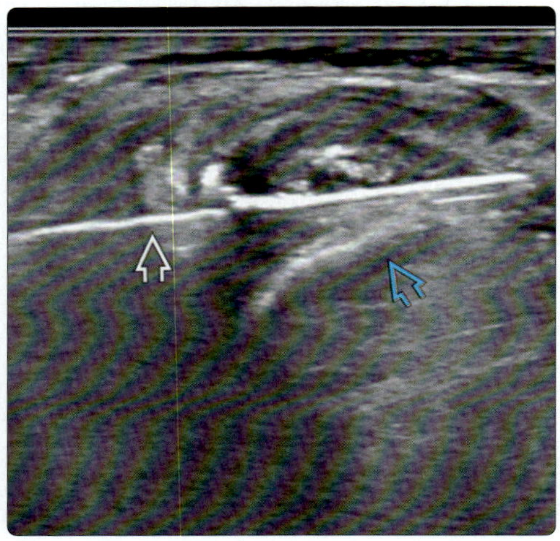

(Left) *US shows a distended anterior elbow recess superficial to the proximal ulna and radius in a rheumatoid patient with suspected superimposed infection.* **(Right)** *US-guided aspiration of the anterolateral aspect of the same elbow with the 21-g spinal needle is shown. Most of the fluid could be aspirated. Microbiology culture showed no bacterial growth.*

US-Guided Aspiration

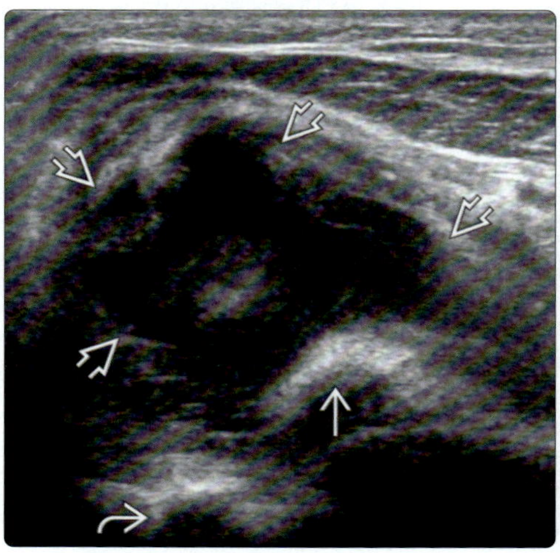

US-Guided Aspiration

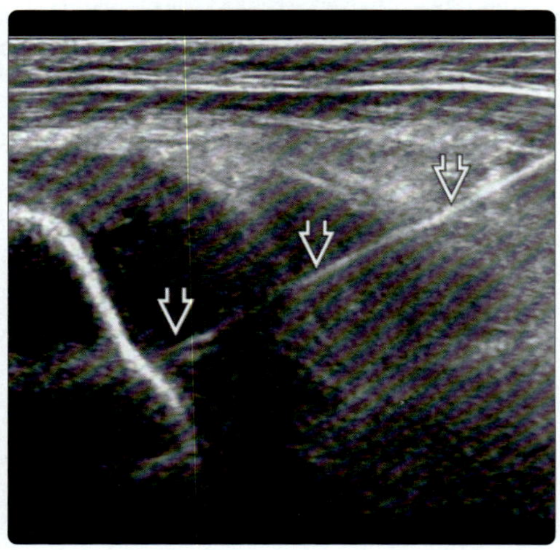

US-Guided Aspiration and Injection

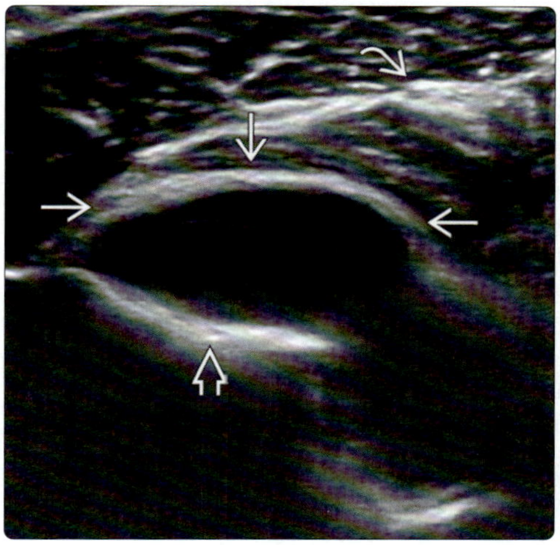

US-Guided Aspiration and Injection

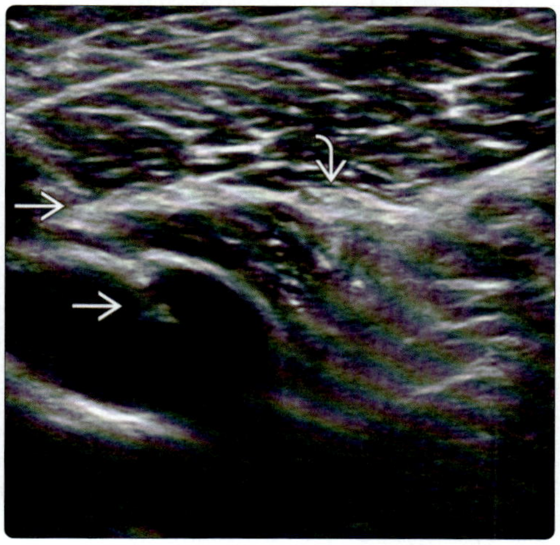

(Left) *US shows a loculated ganglion cyst anterior to the radial neck . The posterior interosseous nerve (PIN) at this level is immediately alongside the supinator muscle lying above the ganglion cyst. (Courtesy S. James, MD.)* **(Right)** *US during injection of steroid using a 22-g spinal needle is shown. The needle path has been selected to avoid injury to the PIN . Steroid injection helps decrease pericystic inflammation and reduces the likelihood of ganglion cyst recurrence. (Courtesy S. James, MD.)*

US-Guided Injection

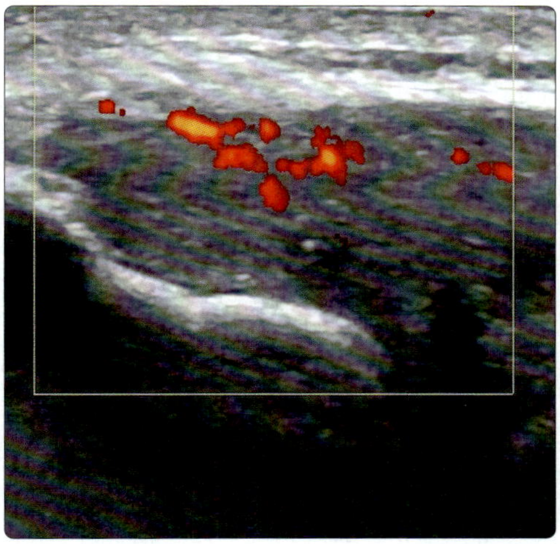

US-Guided Injection

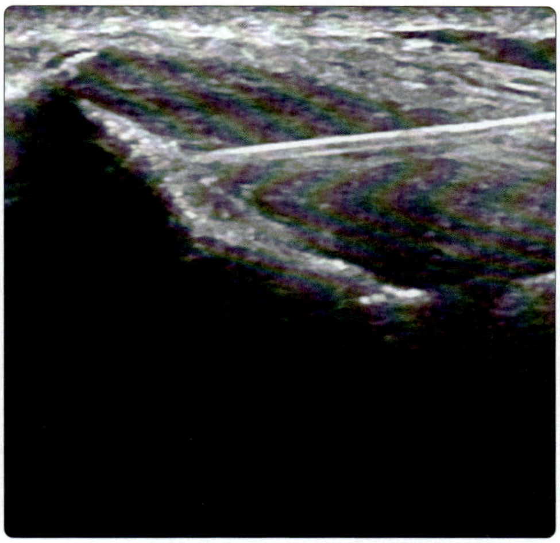

(Left) *Preliminary longitudinal power Doppler US shows severe tendinosis with hyperemia at the CETO.* **(Right)** *Longitudinal US shows US-guided injection of 1 mL (40 mg/mL) methylprednisolone with 0.5 mL (1%) lignocaine deep to the CETO using a 22-g spinal needle after dry needling for 1 minute. Dry needling helps to create fenestrations, disrupts tendon fibrillar structure, and promotes internal bleeding.*

US-Guided Injection

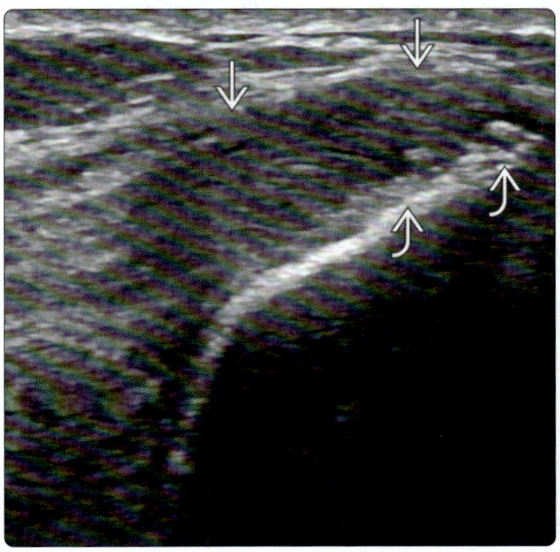

US-Guided Injection

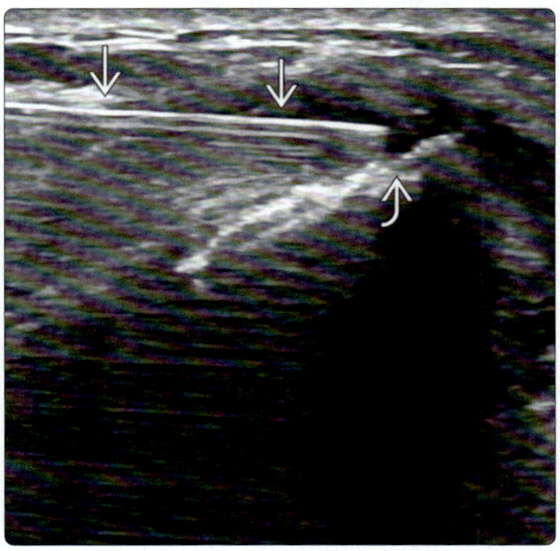

(Left) *Preliminary longitudinal US shows severe common flexor tendon origin (CFTO) tendinosis with severe hypoechoic changes and loss of the normal fibrillation with insertional bony irregularity at the CFTO.* **(Right)** *Longitudinal US shows US-guided injection of 1 mL (40 mg/mL) methylprednisolone with 0.5 mL (1%) lignocaine deep to the CFTO using a 23-g spinal needle after dry needling for 1 minute. Dry needling creates a controlled tendon injury in order to stimulate a healing response.*

Hand and Wrist Procedures

KEY FACTS

PREPROCEDURE

- **Indication**
  - Aspiration of joint fluid
  - Intraarticular injection of medication or contrast agent
  - Percutaneous therapy of ganglion cyst, tendon disease, trigger finger, carpal tunnel syndrome, and Raynaud phenomenon

PROCEDURE

- Joint aspiration or injection
  - Radiocarpal, midcarpal, distal radioulnar joints: Dorsal approach
  - Scaphotrapeziotrapezoid or trapeziometacarpal joints: Dorsoradial approach
- Ganglion cyst aspiration
  - ≥ 19-g needle for more viscous contents
  - No pretest predictor of cyst viscosity
- Tendon disease
  - Approach depends on site of disease
  - Steroid injection: More commonly used
  - Needle tendon pulley release, platelet-rich plasma (PRP) injection: Less commonly used
- de Quervain disease
  - Dorsoradial approach
  - Steroid injection more commonly used
  - Needle tendon retinacular release, PRP injection: Less commonly used
- Trigger finger
  - Palmar approach, short or long axis
  - Steroid injection
  - Percutaneous pulley release
- Carpal tunnel syndrome
  - Palmar approach
  - Steroid injection seems to produce good symptom relief in short (6 weeks) to medium term (6 months)
- Raynaud phenomenon
  - Botox injection to perineurovascular region

(Left) *Transverse US-guided injection of Botox alongside the digital neurovascular bundle ➡ on ulnar side base of the little finger is shown. A 25-g needle ➡ is positioned alongside the neurovascular bundle and Botox is injected ➡. The flexor tendons ⇨ of the little finger are seen alongside the neurovascular bundle.* **(Right)** *Ten target point injections (yellow dots) are performed alongside the radial and ulnar neurovascular bundles at the base of the 5 fingers (7 injection points) and also in the midpalm region (3 injection points).*

Botox Injection for Raynaud Phenomenon

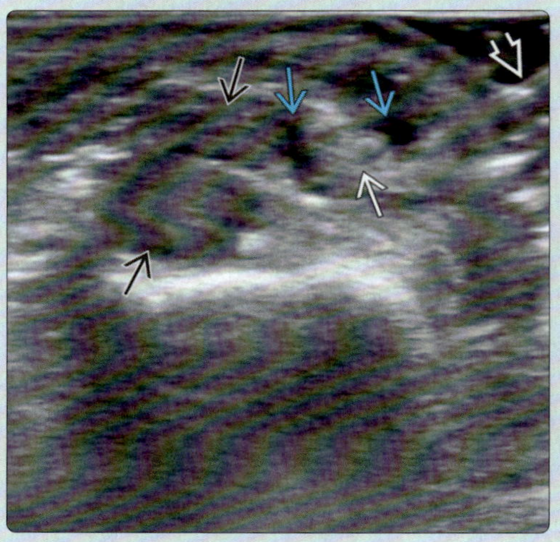

Botox Injection for Raynaud Phenomenon

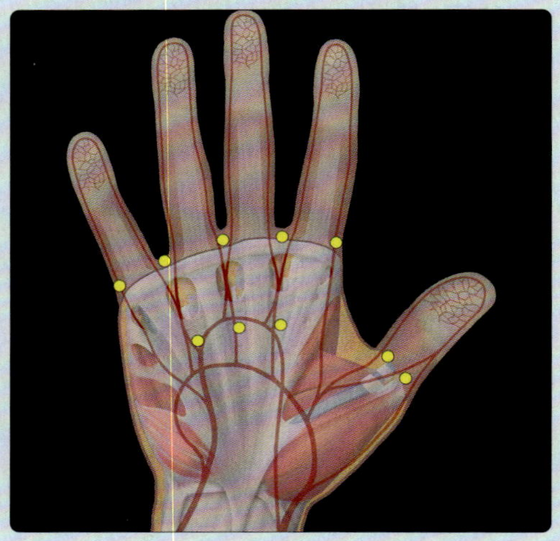

(Left) *Clinical photograph of a patient with Raynaud phenomenon before Botox injection is shown. There is dry gangrene ➡ of the index finger tip and a generalized dusky discoloration of the fingers, particularly the little finger.* **(Right)** *Clinical photograph of the same patient with Raynaud phenomenon several months after Botox injection is shown. The dry gangrene ➡ of the index finger tip has healed. The skin coloration is now normal.*

Botox Injection for Raynaud Phenomenon

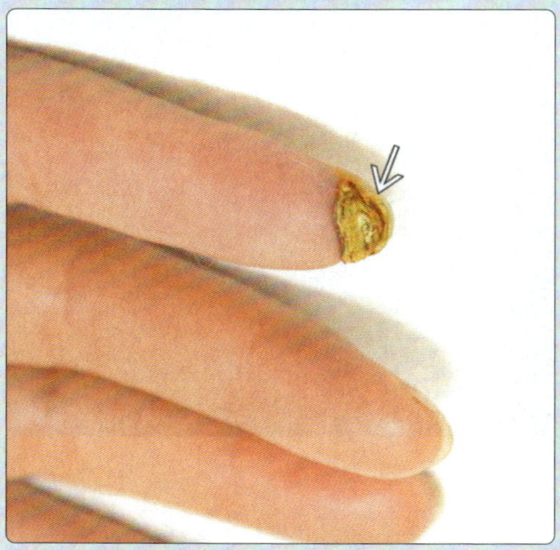

Botox Injection for Raynaud Phenomenon

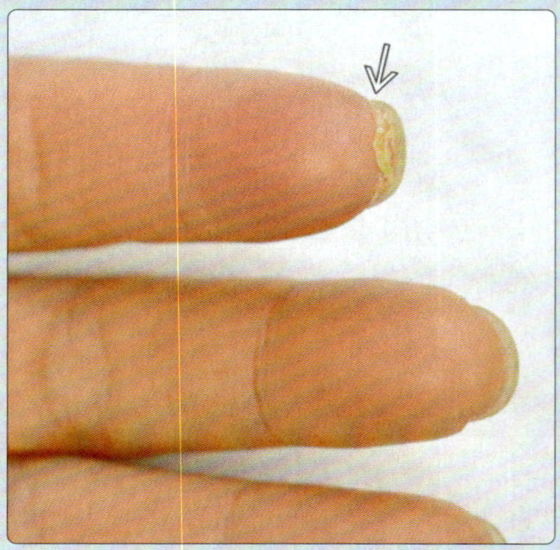

PREPROCEDURE

Indications

- Aspiration or intraarticular injection
  - Suspected septic arthritis or crystal arthropathy
  - Medication for control of inflammation or symptom relief
  - Prior to MR, CT, or conventional arthrography
- Ganglion cyst
  - Aspiration for symptom relief
  - US cannot predict cyst content viscosity
- Tendon disease (tenosynovitis or tendinosis)
  - Aspiration for suspected infective tenosynovitis
  - Injection for symptom relief and functional improvement
- de Quervain disease: Injection or tenotomy for symptom relief and functional improvement
- Trigger finger: Annular pulley injection or percutaneous pulley release for symptom relief and functional improvement
- Carpal tunnel syndrome: Injection for symptom relief

PROCEDURE

Joint Aspiration or Injection

- General clues irrespective of joint aspirated or injected
  - Sterile technique: Skin and underlying soft tissues anesthetized at puncture site
  - Avoid injection of bubbles by 1st flushing syringe and tubing
  - US assessment preprocedure planning best approach that will avoid neurovascular or tendon injury
  - Use free-hand technique to advance 25-g needle into joint space using continuous US guidance
  - Can use long-axis (preferable) or short-axis approach
  - Use smallest needle possible (23- to 27-g) that will comfortably reach joint
    - Larger needle (19-g) used for purulent fluid aspiration
  - Ensure on real-time imaging that needle tip is within joint space and directed toward any joint fluid
  - Free flow of fluid with low-pressure injection helps confirm intraarticular position of needle tip
  - If free-fluid flow is not present, rotate or slightly realign needle
  - Continue monitoring joint on real-time imaging during procedure with realignment of needle if necessary
  - Post injection, apply local pressure and avoid active exercise for 10 minutes to prevent leakage of solution along needle track
- Intraarticular injection of wrist joint
  - 25-g intravenous or spinal needle is used
  - Radiocarpal joint
    - Dorsal approach via radioscaphoid joint space
    - Place probe in sagittal plane at right angles to joint
    - Enter 1 cm distal to radioscaphoid joint space
    - Direct needle proximally toward articular cartilage of distal radius
    - Once cartilage is contacted, tilt needle vertically to enter center of radioscaphoid joint
  - Midcarpal joint
    - Dorsal approach
    - Place probe in sagittal or transverse plane
    - Target either scaphocapitate or triquetrohamate interspaces
  - Distal radioulnar joint
    - Dorsal approach
    - Place probe in transverse plane
    - Position needle alongside radial border of ulnar head a few mm proximal to distal articular surface and directed toward ulnar head
    - When needle contacts ulnar head, angle it more vertically to insert deeper into joint space
  - Scaphotrapeziotrapezoid or trapeziometacarpal joints
    - Dorsoradial approach
    - Place probe in sagittal plane
    - Angle needle 30-45° relative to transducer

Ganglion Cyst Aspiration

- 21-g needle advanced into ganglion under continuous real-time imaging
  - ≥ 19-g needed for more viscous content
- Injecting a little lignocaine into ganglion can loosen cyst contents if initial aspirate is dry due to contents being too viscous
- ± pretreatment with hyaluronidase (150 units in 1-2 mL of normal saline)
  - Hyaluronidase degrades hyaluronic acid, which is macromolecular component of ganglion cyst fluid
  - Inject hyaluronidase to ganglion cyst, leave needle in cyst for 15 minutes, and then aspirate
- ± intracystic injection of steroid (4-10 mg of methylprednisolone) post aspiration
- ± fenestration (puncture) of cyst wall with needle several times during procedure
  - Allows cyst fluid to leak, reducing likelihood of recurrence

Tendon Disease

- Approach is tailored according to affected tendon segment
  - Dorsal or palmar approach
  - Probe usually in sagittal plane, avoiding excessive probe pressure, which may efface tenosynovial fluid
- Needle (21-g) advanced to tendon margin under real-time imaging
  - Alternatively, can enter periphery of tendon substance and slowly withdraw needle, gently aspirating or injecting fluid
- Steroid (4-10 mg of methylprednisolone) or platelet-rich plasma (PRP, 3 mL) injections are potential treatment options for tendinosis

de Quervain Disease

- Wrist in neutral position resting on wedge and slightly ulnar deviated
  - Axial approach preferable
- Steroid injection, PRP injection, or percutaneous needle tendon retinacular release
  - Steroid injection is most commonly used
    - 23-g needle
    - 10 mg of methylprednisolone injected into tendon sheath of 1st extensor compartment deep to retinaculum
- If septum is present (40%), inject tendon sheaths of both subcompartments

- Percutaneous dry needling
  - Dry needling of tendinosis induces bleeding as well as reparative response
  - Converts quiescent nonhealing chronic injury into acute injury with enhanced healing capability and recruitment of reparative stem cells
- PRP injection is less frequently employed than steroid injection
  - 20 mL of patient's whole blood is collected and, following manufacturer's instructions, 3 mL of PRP is prepared
  - 3 mL of PRP is injected into tendon sheath of 1st extensor compartment deep to retinaculum under real-time imaging to ensure even distribution into peritendinous tissues
    - Provides reparative growth factors and stem cells to injured tendon
- No established evidence as to efficacy of 1 treatment over another

Trigger Finger

- Steroid injection or percutaneous release of annular pulley are potential treatment options
  - Hand flat on table, palm upwards
  - Palmar approach, both axial (best for thumb) and sagittal (best for other fingers) planes are used
  - 23-g needle is placed deep to annular pulley
    - 10 mg of methylprednisolone injected deep to annular pulley under real-time imaging to ensure good dispersion
- Percutaneous release of annular pulley
  - Longitudinal approach
  - Insert 19-g intravenous needle deep to annular pulley
    - Needle bent to 15-20° close to hub prior to insertion
    - Rotate bevel sideways to facilitate cutting action
    - Using skin as pivot and downward pressure on attached syringe, cut annular pulley from deep to superficial (i.e., away from tendon)
  - Mean procedure time is < 15 minutes
  - Outcome is comparable to that of surgical release and non-image-guided percutaneous release
  - Much shorter recovery time than surgical release

Carpal Tunnel Syndrome

- Injection is both diagnostic and therapeutic
- Place wrist in slightly dorsiflexed position
- 2 axial volar approaches
  - Ulnar route
  - Radial route, avoiding radial artery and passing deep to flexor tendons
- Using 23-g needle, steroid (10 mg of methylprednisolone) injected around median nerve under continuous US guidance
- 60% of patients get moderate to near-complete symptom relief after single injection
  - Symptom relief sustained for short (6 weeks) to medium term (6 months)
  - 60% of patients require retreatment within 1 year
  - Limited long-term follow-up studies

Raynaud Phenomenon

- Injection of Botox to perineurovascular region of palm and base of digits
- Total of 10 points of injection: Base of 5 fingers and midpalmar region
- Helps to relieve vasospasm

Findings and Reporting

- Depends on interventional procedure performed, but consider reporting these aspects
  - Initial US findings
    - Complete US examination should be performed before any interventional procedure
    - Provide objective (quantitative or semiquantitative) criteria of baseline disease severity, e.g., median nerve caliber or degree of tendinosis/peritendinitis
  - Local anesthetic volume and concentration used
  - Type and size of needle used
  - Volume, color, and viscosity of joint fluid aspirated
  - Volume and type of fluid injected
  - Unexpected adverse occurrence during procedure, e.g., arterial puncture
    - Report even in absence of immediate complications due to potential for later complications
  - Immediate complication, e.g., excessive bleeding

OUTCOMES

Complications

- Most feared complication(s)
  - Infection, e.g., septic arthritis
  - Neurovascular bundle injury
    - de Quervain disease
      - Medial branch of superficial branch of radial nerve typically passes superficial to 1st extensor compartment
      - Lateral branch passes volar to 1st extensor compartment
      - Deep branch of radial artery passes deep to 1st extensor compartment
      - Cephalic vein passes superficial and close to 1st extensor retinaculum
    - Wrist joint injection
      - Terminal part of posterior interosseous nerve between 3rd and 4th extensor compartments
  - Leakage of steroid medication
    - Fat and skin atrophy, skin depigmentation
- Immediate/periprocedural complication(s)
  - Bleeding, especially if bleeding diathesis
    - Apply local pressure
  - Extravasation of contrast or medication
    - Apply local pressure and rest joint for 10 minutes
  - Capsular rupture
  - Allergy to injectate
  - Steroid flare (within 24 hours)

SELECTED REFERENCES

1. Orlandi D et al: Ultrasound-guided procedures around the wrist and hand: how to do. Eur J Radiol. 83(7):1231-1238, 2014
2. Vuillemin V et al: Musculoskeletal interventional ultrasonography: the upper limb. Diagn Interv Imaging. 93(9):665-73, 2012

de Quervain Disease Injection

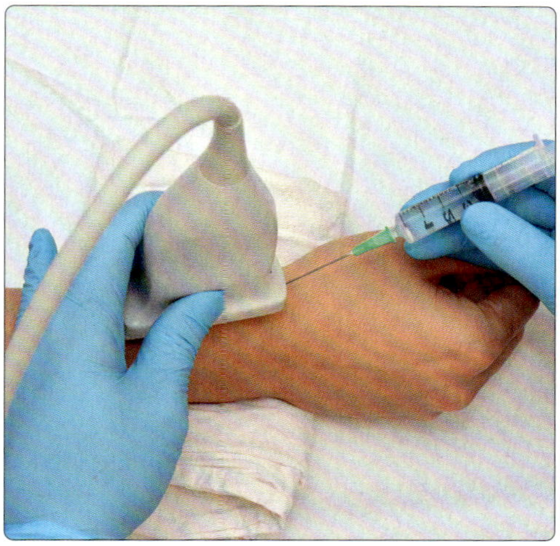

de Quervain Disease Injection

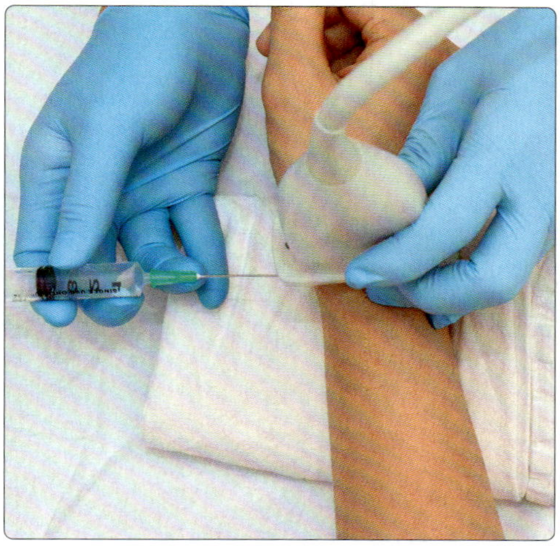

(Left) *Clinical photograph shows the longitudinal approach for injection of de Quervain disease. The wrist is resting on a towel wedge with slight ulnar deviation. The superficial branch of the radial nerve and its branches should be first identified.* **(Right)** *Clinical photograph shows the axial approach for injection of de Quervain disease. If there is a septum present in the 1st extensor compartment (as seen in 40% of subjects), both subcompartments should be injected.*

de Quervain Disease Injection

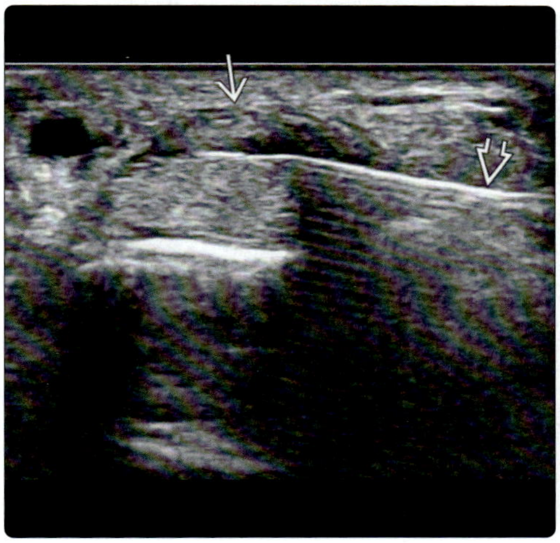

de Quervain Disease Injection

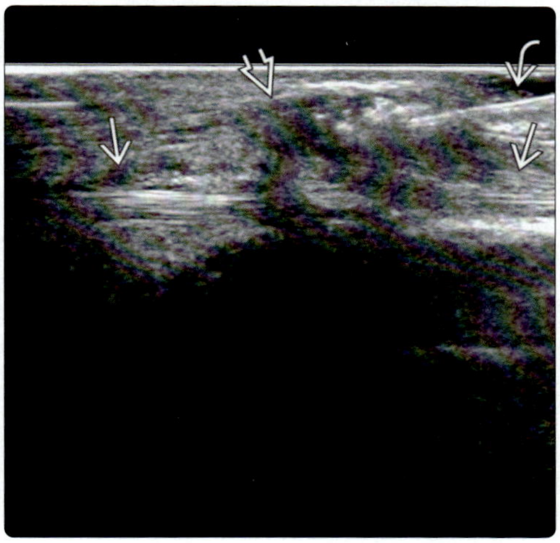

(Left) *Transverse US of a patient with de Quervain disease shows moderate thickening of the extensor retinaculum of the 1st extensor compartment. A 23-g intravenous needle is inserted into the tendon sheath just deep to the retinaculum, and 10 mg of methylprednisolone is injected.* **(Right)** *Longitudinal US shows steroid (10 mg of methylprednisolone) being injected into the tendon sheath of the 1st extensor compartment tendons deep to the retinaculum via a 23-g needle.*

de Quervain Disease Injection

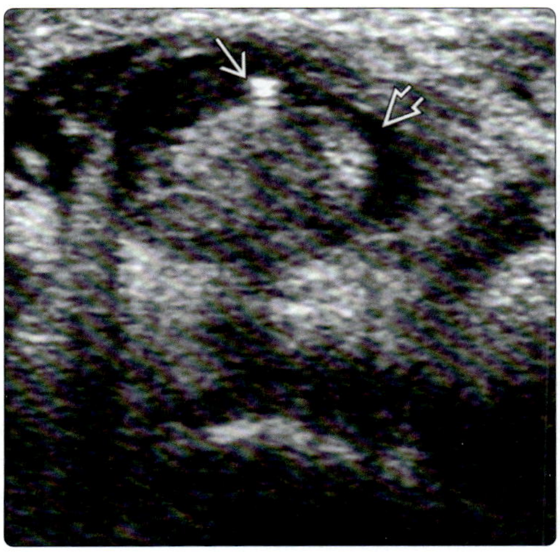

Carpal Tunnel Injection

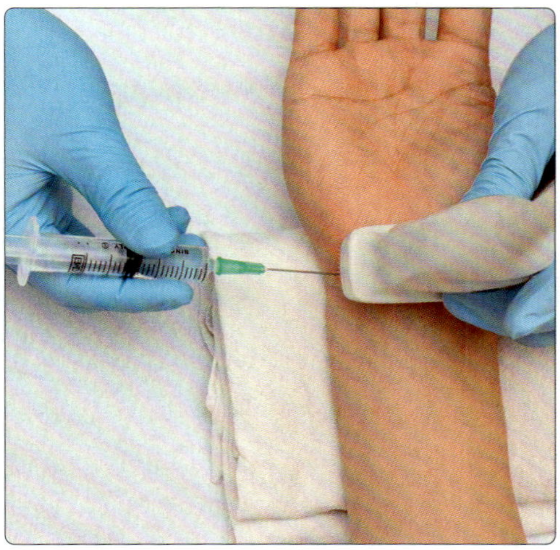

(Left) *Transverse US of a patient with de Quervain disease shows a 23-g intravenous needle advanced into the tendon sheath deep to the thickened retinaculum and steroid (10 mg of methylprednisolone) being injected into the tendon sheath.* **(Right)** *Clinical photo shows steroid (10 mg of methylprednisolone) injection treatment for carpal tunnel syndrome using a transverse approach. The needle (21- or 23-g) is inserted on the ulnar side. An ulnar approach is more commonly performed than a radial approach.*

(Left) *Clinical photograph shows the longitudinal palmar approach for annular pulley injection. This is the most commonly used approach, as access to the area between the pulley and the flexor tendons is easiest. It is also the preferred approach for the pulley-cutting procedure. Placing the wrist on a pad allows the finger to be hyperextended facilitating needle access.* **(Right)** *Clinical photograph shows the transverse oblique palmar approach for annular pulley injection to treat trigger finger.*

Trigger Finger Injection

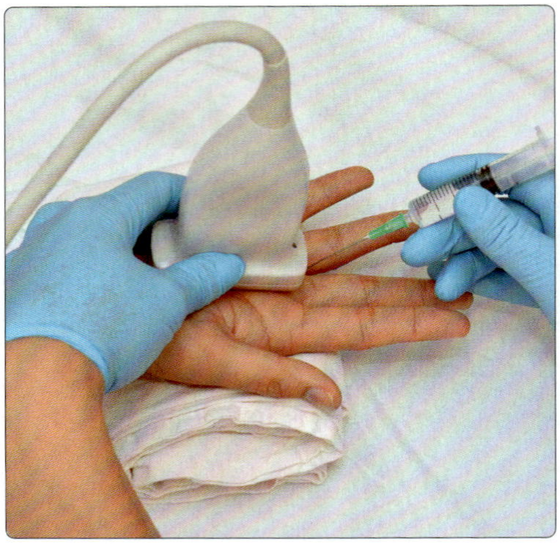

Trigger Finger Injection

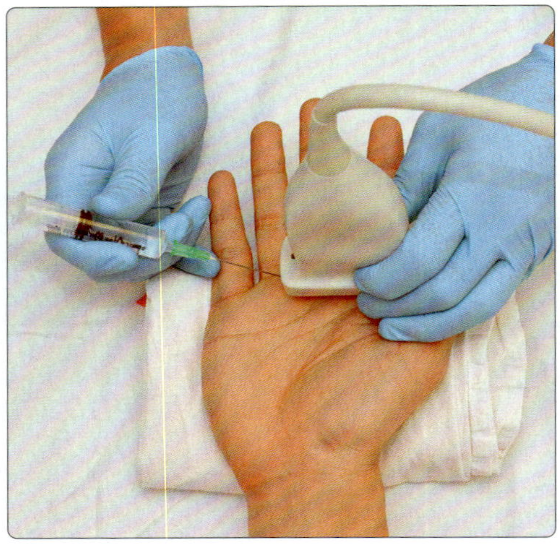

(Left) *Longitudinal US shows steroid (10 mg of methylprednisolone) injection to treat trigger finger. A 23-g needle is placed within the tendon sheath just deep to the mildly thickened A1 pulley.* **(Right)** *Transverse US shows a mild, possibly infective, flexor tenosynovitis of the index finger overlying the proximal phalanx. Fluid from the small tendon sheath effusion was aspirated using a 21-g needle for culture. Fluid culture was sterile.*

Trigger Finger Injection

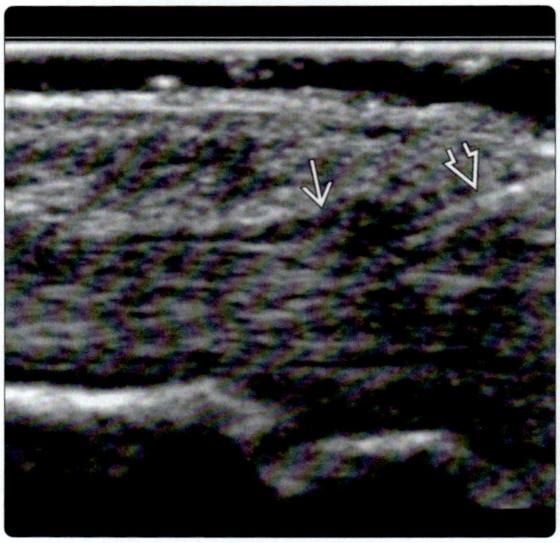

Tenosynovitis Aspiration

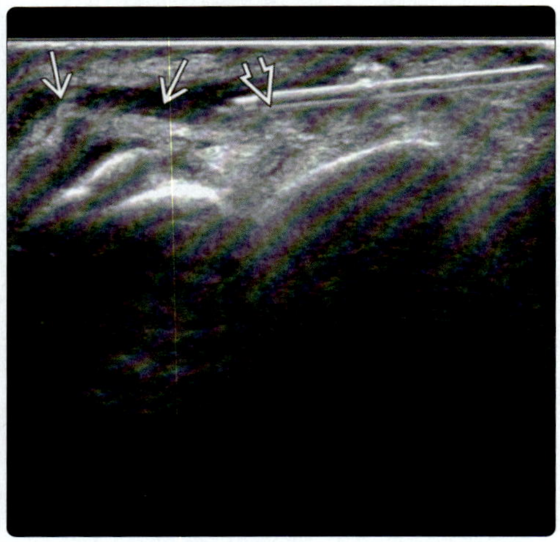

(Left) *Transverse US of the dorsal wrist shows a small ganglion superficial to the scaphoid. A 21-g needle is inserted into this ganglion via a dorsal approach. The needle tip lies within the ganglion.* **(Right)** *Oblique US of the palm shows a large ganglion within the thenar muscle. A 21-g needle is inserted via a palmar approach. The needle tip lies within the ganglion. It is difficult to ascertain the viscosity of ganglion contents with US prior to aspiration.*

Ganglion Aspiration

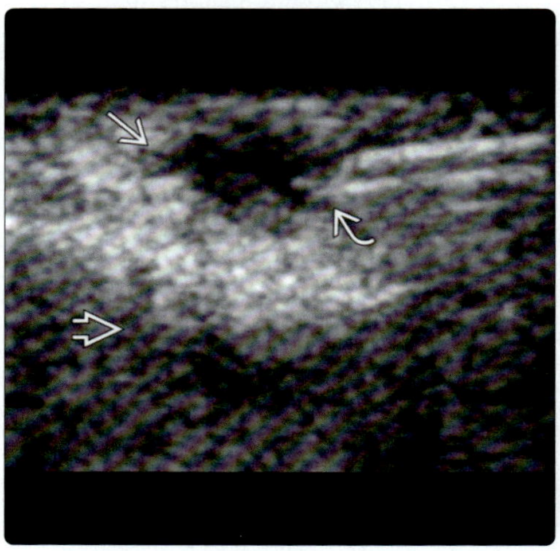

Ganglion Aspiration

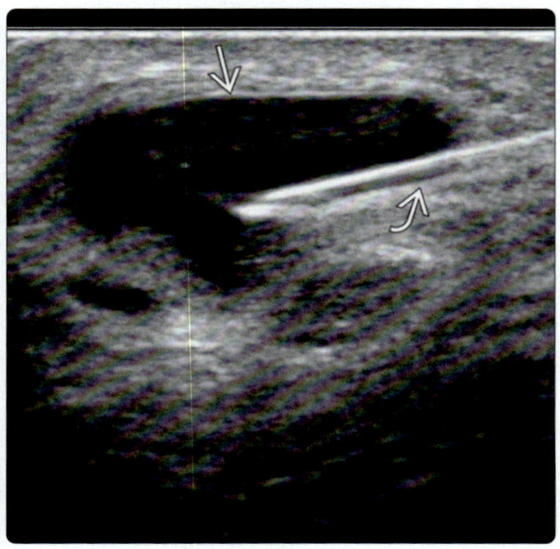

Radiocarpal Joint Injection

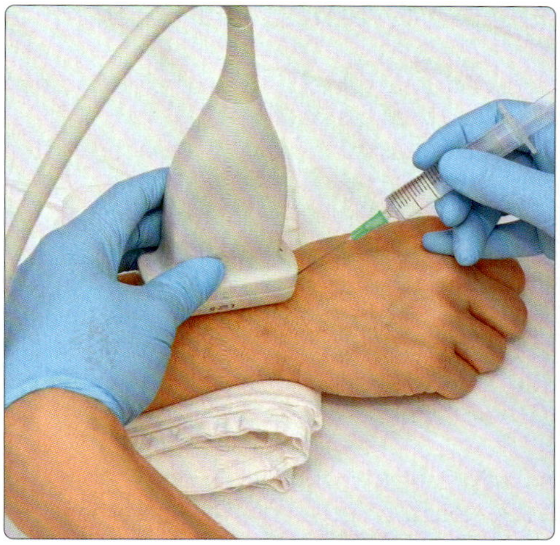

Radiocarpal Joint Injection

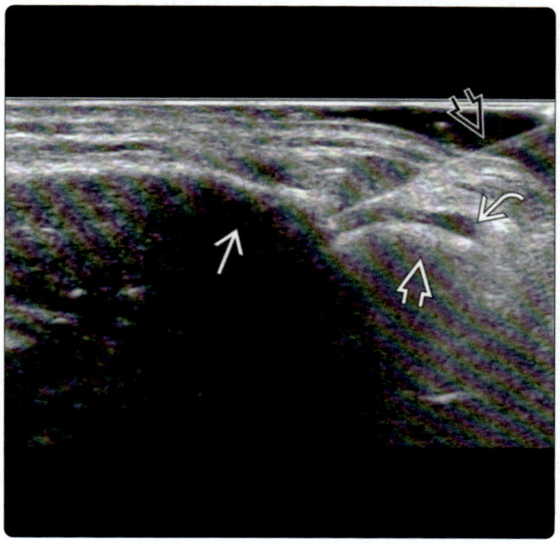

(Left) *Clinical photograph shows a longitudinal dorsal approach for injection of the radiocarpal articulation.* **(Right)** *Longitudinal US shows US-guided contrast injection of the radiocarpal articulation. A 25-g spinal needle lies within the joint space between the distal radius and the proximal pole of the scaphoid. A small amount of contrast has been injected into the joint.*

Radiocarpal Joint Injection

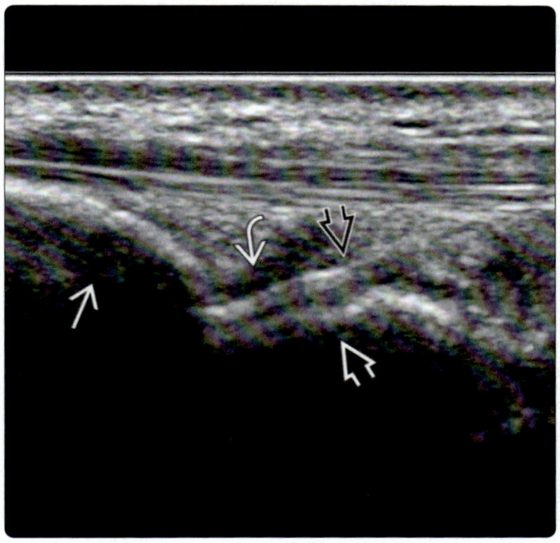

Radiocarpal Joint Injection

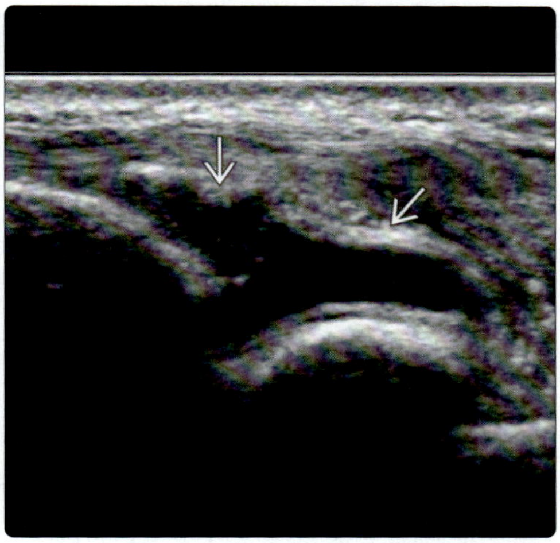

(Left) *Longitudinal US shows US-guided contrast injection into the radiocarpal articulation for MR arthrography. A 25-g spinal needle lies within the joint space between the distal radius and the proximal pole of scaphoid. A small amount of contrast has been injected into the joint.* **(Right)** *Longitudinal US of the same patient shows fluid distension of the radiocarpal joint after the injection has been completed.*

Distal Radioulnar Joint Injection

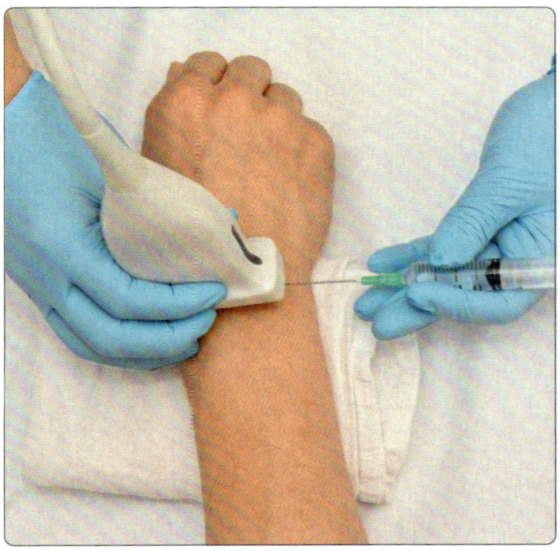

Distal Radioulnar Joint Injection

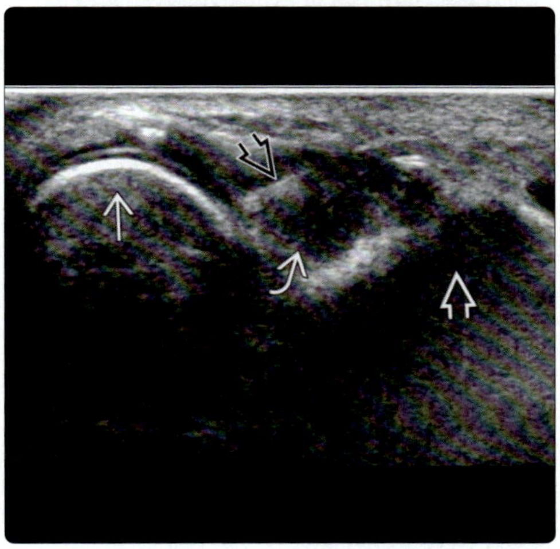

(Left) *Clinical photograph shows the transverse dorsal approach for injection of the distal radioulnar joint.* **(Right)** *Axial US shows US-guided contrast injection for MR arthrography into the distal radioulnar joint. A 25-g spinal needle lies within the joint space between the distal ulna and the distal radius. A small amount of contrast has been injected into the joint.*

KEY FACTS

PROCEDURE

- Hip or sacroiliac (SI) joint injection or aspiration
  - US-guided joint injection more efficient than fluoroscopic-guided joint injection
  - For hip, anterior longitudinal approach usually used, either targeting femoral head or femoral neck
  - For SI joint injection, patient lies prone with joint identified at level of S1 & S2 foramina
- Tendinosis treatment
  - Gluteus medius tendinosis seems most common
  - Also can treat gluteus minimus, hamstring, adductor & iliopsoas tendinosis
  - Fenestration as good as platelet-rich plasma, which may be better than autologous blood injection
  - Steroids treat peritendinitis & are not effective medium to long term
- Paraarticular cyst or bursal aspiration
  - 21-g needle for cyst aspiration or > 19-g needle for more viscous content
  - ± hyaluronidase preaspiration ± steroid post aspiration
- Biopsy of synovial tissue for infection, inflammation, or synovial tumor
  - Synovial biopsy accurate for confirming or excluding joint infection if no fluid to aspirate
  - Histology allows early assessment as to likelihood of TB infection
  - If multinucleated giant cells or caseating granulomas on histology, can treat empirically until culture available in 2-6 weeks
  - Similarly, if no histologic features of TB & gram stain positive, can treat empirically as non-TB infection until culture available in 2-6 days
- Nerve block or neural hydrodissection, diagnostic or therapeutic
  - Lateral femoral cutaneous nerve, ilioinguinal, iliohypogastric, genital branch of genitofemoral nerve, obturator nerve, pudendal & sciatic nerves
- Iliolumbar ligament injection for relief of pain

Hip Aspiration

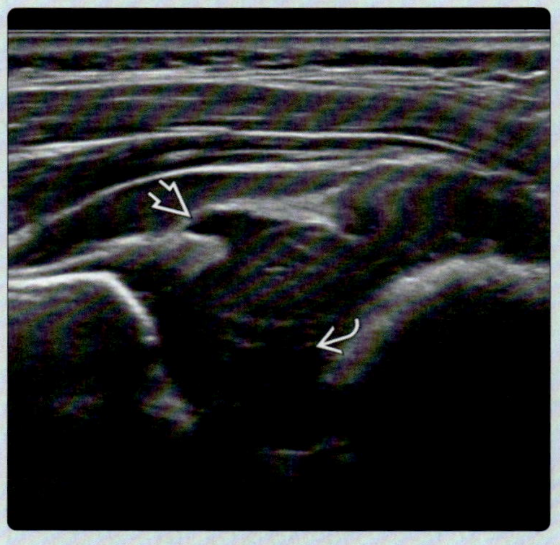

Hip Aspiration

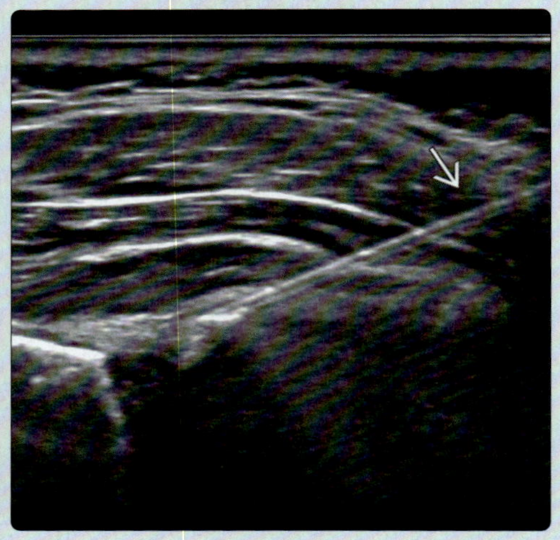

(Left) *Longitudinal US in 3-month-old boy with fever, limited movement, and clinically suspected septic arthritis shows a small hip effusion ➔. The femoral head ➔ is, as yet, not ossified.* **(Right)** *US-guided percutaneous aspiration ➔ yielded a small amount of purulent fluid. This grew Staphylococcus aureus. No sedation was required. US-guided aspiration allows an early and accurate diagnosis of joint infection. Occasionally, septic arthritis is associated with only a small joint effusion.*

Tendinosis Treatment

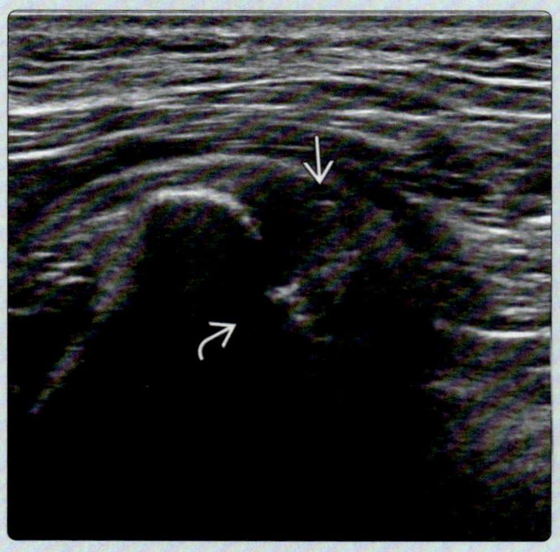

Tendinosis Treatment

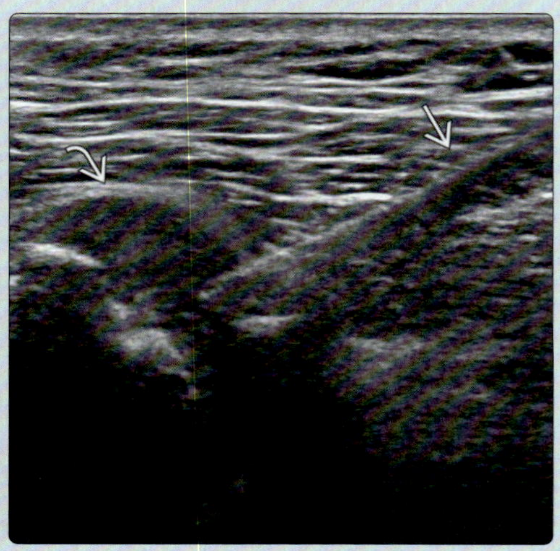

(Left) *Longitudinal US shows moderate to severe tendinosis of the hamstring attachment to the ischial tuberosity. There is thickening of the hamstring tendons ➔ with irregularity of the ischial tuberosity ➔ attachment area.* **(Right)** *Longitudinal US shows 20-g spinal needle ➔ within hamstring tendons ➔ during injection of 8 mL of platelet-rich plasma (PRP). Prior to PRP injection, the hamstring tendons were fenestrated for 1 min by moving the spinal needle to and fro through the tendon substance.*

PREPROCEDURE

Indications

- Aspiration or drainage of hip joint fluid
  - Suspected septic arthritis or crystal arthropathy
- Hip or sacroiliac (SI) injection of contrast medium
- Hip or SI joint injection for relief of pain or inflammation
  - Steroid, hyaluronic acid injection for painful osteoarthritis or inflammatory arthritis
  - Radiosynovectomy (e.g., yttrium-90, half-life 2.7 days) for chronic synovitis, such as hemophiliac arthropathy
- Tendinosis treatment
  - Tendinosis & partial tears of gluteus medius & minimus, adductor, hamstring & psoas tendons
  - Fenestration (dry needling), autologous blood injection (ABI), platelet-rich plasma (PRP) injection to stimulate healing response
    - Prior to tendinosis treatment, patient should refrain from antiinflammatory drugs for 2 weeks so as not to compromise inflammatory phase of early tissue healing
  - Steroid & long-acting local anesthetic injection
- Paraarticular cyst or bursal aspiration
  - Symptomatic or large paralabral cyst or paraarticular ganglion cyst may benefit from aspiration
  - Most clinical "trochanteric bursitis" is due to gluteus medius/minimus tendinosis with peritendinitis rather than bursitis
- Biopsy of synovial tissue for infection, inflammation, or synovial tumor
- Nerve block or neural hydrodissection, diagnostic or therapeutic
  - Lateral femoral cutaneous nerve, ilioinguinal, iliohypogastric, genital branch of genitofemoral nerve, obturator nerve, pudendal & sciatic nerves
- Iliolumbar ligament injection for relief of pain

Getting Started

- Things to check
  - Time-out procedure (patient identity, type, & side of procedure)
  - Discuss risk of infection, allergy, bleeding, neurovascular injury, & (if necessary) tendon rupture
  - Ensure consent form signed
- Medications
  - Local anesthetic for skin: 2-3 mL, 1% lignocaine (optional)
  - Local anesthetic for joint injection: 1% or 2% lignocaine or 0.25-0.50% bupivacaine (Marcaine)
  - Steroid: 40 mg/mL methylprednisolone (Depo-Medrol)
  - Hyaluronic acid (viscosupplementation)
    - Macromolecule occurring naturally in synovial fluid & contributing to its viscoelastic property
    - Concentration & molecular weight of hyaluronic acid is reduced in osteoarthritis
  - Contrast
    - CT arthrogram: 10 mL iodinated contrast (Omnipaque 300 mg/dL ± 0.1 mL of 1:1,000 epinephrine)
    - MR arthrogram: 10 mL of diluted gadolinium solution (0.1 mL of gadolinium in 10 mL of normal saline solution)
  - PRP
    - 20-60 mL of blood withdrawn from arm vein & centrifuged for 15 min to separate platelets
      - □ Anticoagulant added to prevent premature platelet activation
  - ABI
    - 5 mL of blood withdrawn from antecubital vein & injected directly into tendon substance
      - □ Blood & platelets are rich sources of growth factors for tendon healing
- Equipment list
  - Sterile set
  - 18- to 22-g (3.5") spinal needle depending upon type of procedure & viscosity of fluid if aspiration undertaken
    - If viscous aspirate expected, use larger bore needle (e.g., 16 or 18 g)
  - Extension tube, 250 mm 8.5F, 0.4 mL of dead space
  - Syringes for local anesthetic, injection, & aspiration
  - Coaxial needle & biopsy needle
  - Sterile bottles, specimen bottles

PROCEDURE

Procedure Steps

- Aspiration of joint fluid or injection of medication or contrast medium
  - US-guided joint injection is as effective as & quicker than fluoroscopic-guided joint injection
  - **Hip joint injection**
    - Patient lies supine with hip in neutral alignment ± pillow beneath knee to relax anterior part of hip capsule
    - Anterior longitudinal lateral to medial approach usually used, targeting either femoral head or femoral neck
      - □ Target point for femoral head is superolateral quadrant of femoral head
      - □ Capsule at neck region thickened by zona orbicularis
      - □ Target point for femoral neck is femoral head-neck junction
    - Use either 5- to 10-MHz linear-array transducer (smaller patient) or 3- to 5-MHz curved-array transducer (larger patient)
    - Align transducer parallel to femoral neck
    - Use 3.5", 22-g spinal needle & free-hand technique under real-time US guidance
    - Insert needle until bony contour of proximal femur is contacted; may feel distinct pop as capsule is punctured
    - Relax downward pressure on needle & start injection, turning bevel until free flow of injectate occurs from needle tip
    - Needle tip embedded in reflected synovium at femoral neck may impede fluid flow
      - □ May need to withdraw needle tiny fraction into anterior synovial recess
    - For osteoarthritis, inject 1 mg (40 mg) of triamcinolone & 4 mL of 0.5% bupivacaine
  - **SI joint injection**
    - Patient lies prone, & 5.0- to 7.5-MHz linear- or curved-array transducer is used

- Distal end of SI joint is slightly easier to see & access than proximal end of SI joint
  - Nevertheless, both proximal or distal ends can be successfully cannulated in most patients
- Proximal end of SI joint lies at level of S1 foramen
  - With transducer transverse, 1st identify posterior superior iliac spine laterally & transverse process L5 medially
  - Move transducer caudally, depicting medial & lateral sacral crests
  - At level of S1 foramen, hypoechoic cleft laterally represents SI joint
- Distal end of SI joint lies at level of S2 foramen
  - Move transducer caudally in transverse plane until S2 foramen visualized
  - At level of S2 foramen, hypoechoic cleft laterally represents SI joint
- Use 21-g spinal needle (9 cm)
- Needle orientation almost vertical & same orientation as hypoechoic SI joint cleft
- Transverse transducer approach for initial skin puncture
- Then longitudinal position used to guide needle toward SI joint
- Switch back to transverse transducer position when needle tip reaches entrance to SI joint
- Concentrate on contacting hypoechoic cleft of SI joint & avoiding bony spurs
- Once SI joint contacted, needle can be inserted ~ 1 cm further into joint
- Inject 1 mg (40 mg) of triamcinolone & 5 mL of 0.5% bupivacaine

- **Treatment of tendinosis & tears**
  - Most common indication for pelvic/hip US-guided intervention
    - Gluteus medius tendinosis more common (4x) than gluteus minimus tendinosis
    - Hamstring, adductor, & psoas tendinosis
  - Fenestration
    - 20-g spinal needle, 1 mL of lignocaine injected over tendon surface
    - Pass needle 20-40 times through tendon until tendon softened
  - PRP
    - Undertake tendon fenestration
    - Insert needle to deepest, distal part of tendon & inject 3 mL of PRP as needle withdrawn
    - Repeat ~ 10 times until PRP injection complete
  - ABI
    - Undertake tendon fenestration
    - Immediately after fenestration, withdraw 5 mL of blood from antecubital vein & inject into tendon, following similar technique as for PRP
  - Steroid injection
    - 1 mL of 30-40 mg/mL of methylprednisolone or triamcinolone mixed with 8 mL of 0.5% bupivacaine
      - Inject around rather than within tendon, as steroid treats peritendinitis rather than tendinosis per se
      - Only repeat 2x after 3 weeks or more
      - Good short-term but poor long-term clinical effect
  - Overall for chronic gluteal & hamstring tendinosis, fenestration is as effective as PRP
    - PRP may be more effective than ABI, whilst steroids are least effective especially in mid to long term
- **Paraarticular labral/ganglion cyst or bursal aspiration**
  - 21-g needle for cyst aspiration or > 19-g needle for more viscous content
  - ± hyaluronidase (150 U in 2 mL of saline) preaspiration
    - Hyaluronidase is enzyme that degrades hyaluronic acid in paralabral or ganglion cyst
    - Inject hyaluronidase into cyst; leave needle tip within cyst for 15 min & then aspirate
  - ± steroid (1 mL of 30-40 mg/mL of methylprednisolone or triamcinolone) post aspiration
- **Nerve block & neural hydrodissection**
  - Injection 5 mL of short- or long-acting anesthetic around nerve under direct US guidance
  - For neural hydrodissection, 1-2 mL of triamcinolone or dexamethasone, 1 mL of lignocaine, 7 mL of normal saline ± 1 mL of 5% dextrose
    - Low-concentration (5%) dextrose can reduce neuropathic inflammation

POST PROCEDURE

Expected Outcome

- Mild discomfort 1-2 days after contrast or medication injection or cyst aspiration

Immediate Aftercare

- Send any aspirate for microbiology, cytology, crystal analysis (need fresh specimen) as appropriate
- Perform MR, CT, or conventional arthrography within 30 min of contrast injection, as contrast will absorb

Subsequent Aftercare

- After tendinosis treatment, avoid strenuous hip activity for 1 week, gradually building up to normal training activity only after 6 weeks

OUTCOMES

Complications

- Most feared complication(s)
  - Infection, e.g., septic arthritis
    - Avoid traversing infected tissues between skin & joint
  - Extravasation of contrast/medication
    - Steroid or radiosynovectomy agent: Fat pad atrophy, skin atrophy, skin depigmentation
- Immediate/periprocedural complication(s)
  - Neurovascular injury, particularly femoral nerve paraesthesia after hip joint injection (2%)
  - Bleeding, capsular rupture, allergic reaction
- Delayed complication(s)
  - Pain for 1-2 days after steroid injection due to steroid-related microcrystalline synovitis
  - Following cyst aspiration, leakage of cyst contents can induce transient chemical cellulitis

SELECTED REFERENCES

1. Jacobson JA et al: Greater trochanteric pain syndrome: percutaneous tendon fenestration versus platelet-rich plasma injection for treatment of gluteal tendinosis. J Ultrasound Med. 35(11):2413-2420, 2016

Hip Ultrasound

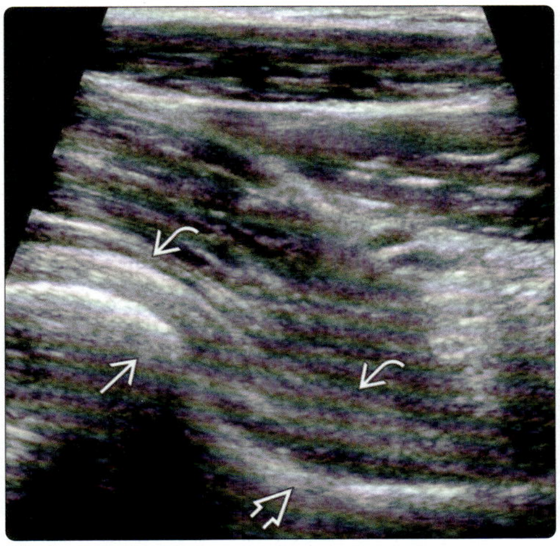

Hip Joint Injection

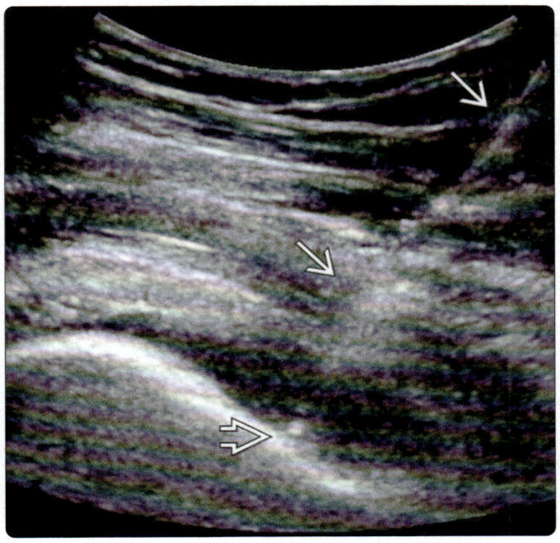

(Left) *Longitudinal oblique US shows a normal femoral head → and neck → with hip capsule →. The capsule is attached to the acetabular rim and the intertrochanteric area. There is a synovial reflection, not seen on US, along the femoral neck.* **(Right)** *Longitudinal oblique US shows 22-g spinal needle → during contrast injection for hip MR arthropathy. The needle tip contacts the head-neck junction →, which is flattened in this patient with femoroacetabular impingement.*

Hip Joint Injection

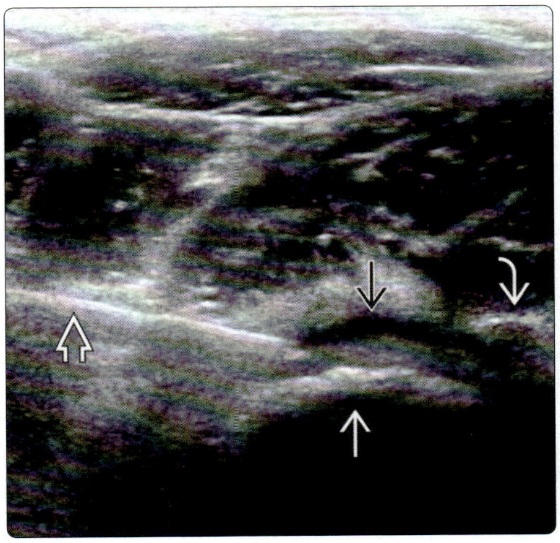

Hip Joint Injection

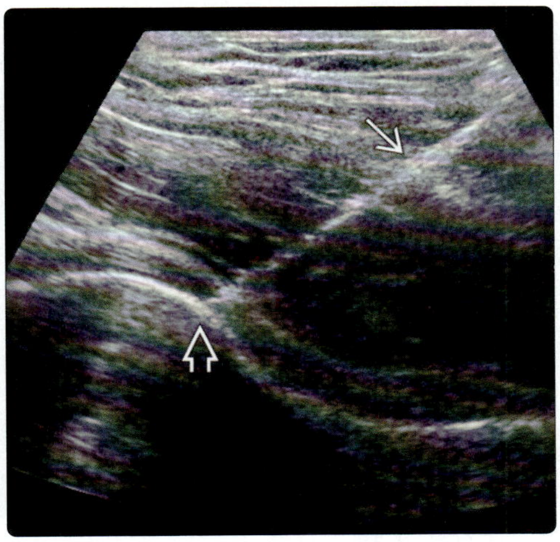

(Left) *Longitudinal oblique US shows an 18-g spinal needle → inserted longitudinally from lateral to medial along the line of the femoral neck. The needle tip lies in the partially distended hip joint →. Part of the femoral neck → and the acetabulum → are seen.* **(Right)** *Longitudinal oblique US shows the position of a 22-g spinal needle → for contrast injection prior to hip MR arthropathy. The needle tip contacts the superolateral aspect of the femoral head →.*

MR Arthrography of Hip

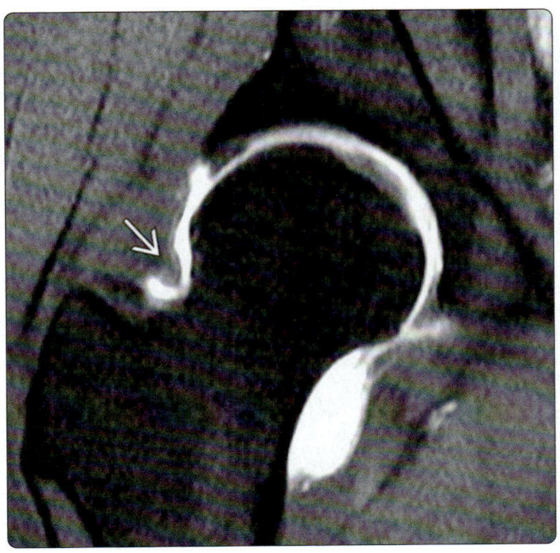

MR Arthrography of Hip

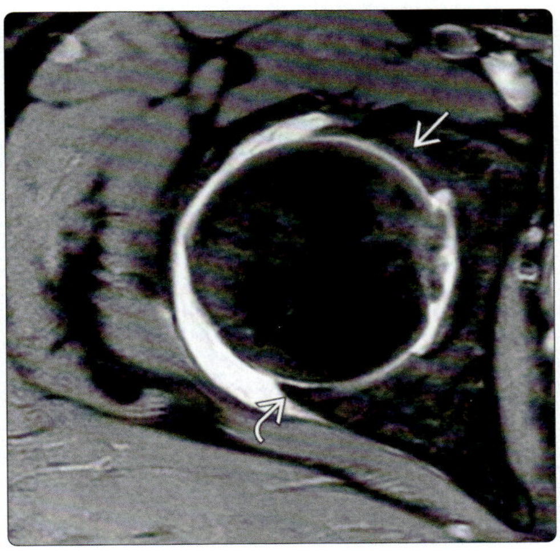

(Left) *Coronal oblique T1WI FS MR shows distended hip capsule following US-guided injection of 10 mL of contrast medium/normal saline mixture. Note the zona orbicularis → (from circular fibers of the hip capsule), which needs to be traversed when the femoral neck is targeted for US-guided hip injection.* **(Right)** *Axial T1WI FS MR arthrography shows the normal acetabular labrum → and the articular cartilage →. MR arthrography is particularly useful for revealing chondral injury.*

(Left) *Photograph shows steroid injection preparation prior to either periarticular or articular injection. Typical injection would be 1 mL (40 mg) of triamcinolone, usually mixed with 2-4 mL of 0.5% bupivacaine (long-acting local anesthetic).* **(Right)** *Photograph shows a typical setup with steroid and long-acting local anesthetic injection. Sterile US gel and a spinal needle (18-22 g, with larger needle for aspiration rather than injection) is used. Local anesthetic is applied to the skin.*

Steroid Injection

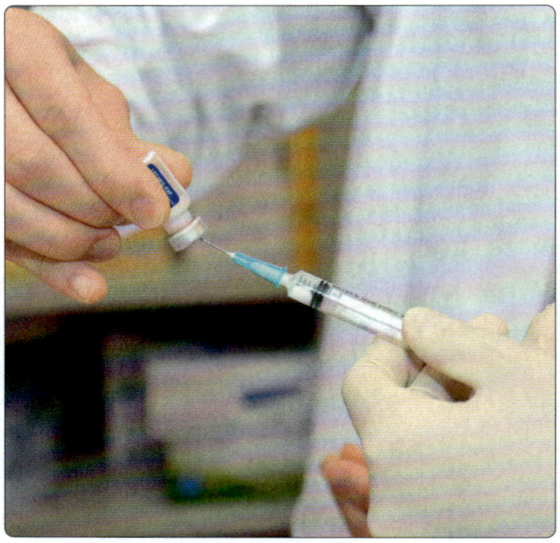

Setup for Hip Joint Injection

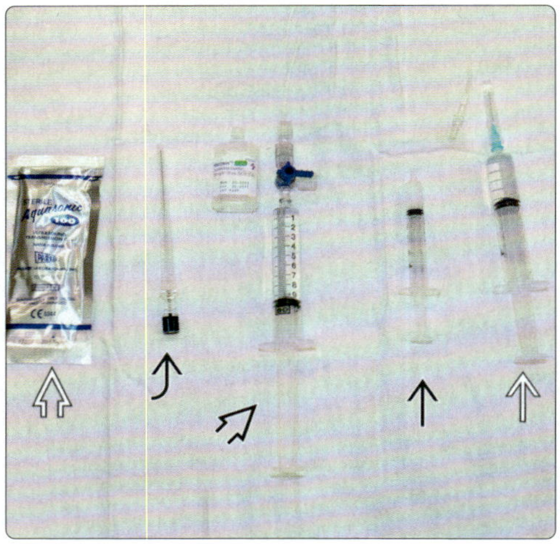

(Left) *Clinical photograph shows skin being cleaned with the injection site draped. Always clean from proposed skin entry site to more peripheral areas rather than vice versa. The US transducer is covered with a sterile probe cover.* **(Right)** *Clinical photograph in the same patient shows infiltration of local anesthetic to skin. One can also infiltrate anesthetic more deeply around the intended site of hip capsule puncture.*

Site Preparation

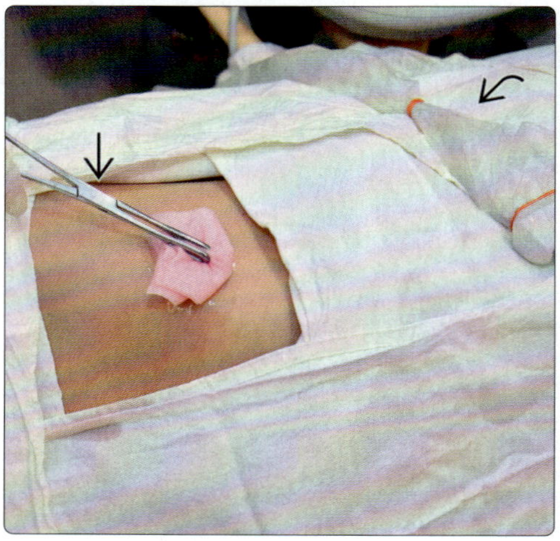

Anesthetic Infiltration

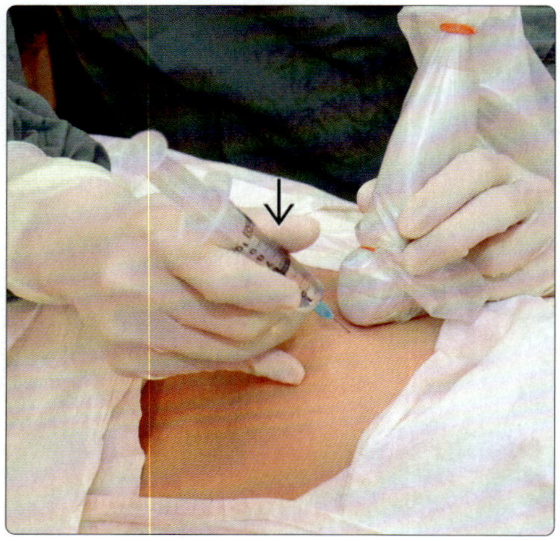

(Left) *Clinical photograph in the same patient shows a 22-g spinal needle being guided toward the hip joint under US guidance. Keep the advancing needle in view on US at all times.* **(Right)** *Longitudinal oblique US shows a faint needle track with the needle tip contacting the midpoint of the femoral neck. About 5 mL of contrast medium has been injected, resulting in mild distension of the hip capsule.*

Hip Joint Injection

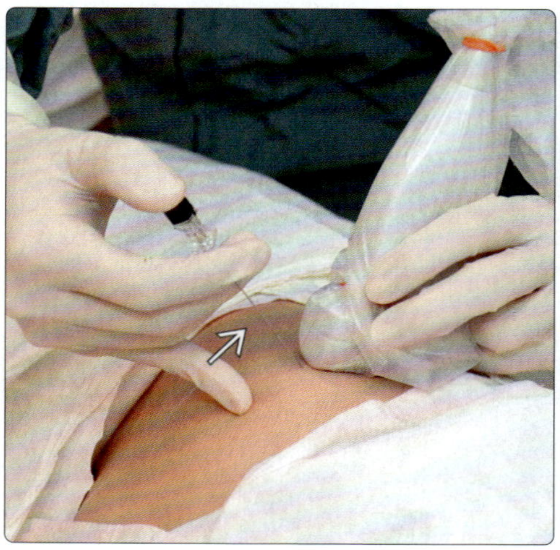

Hip Joint Injection

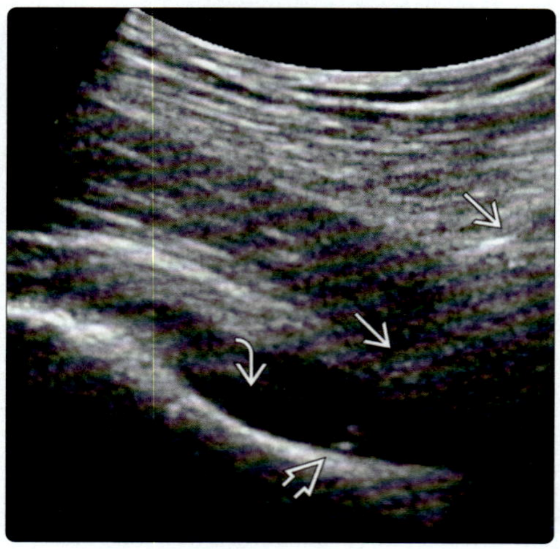

Labral Cyst

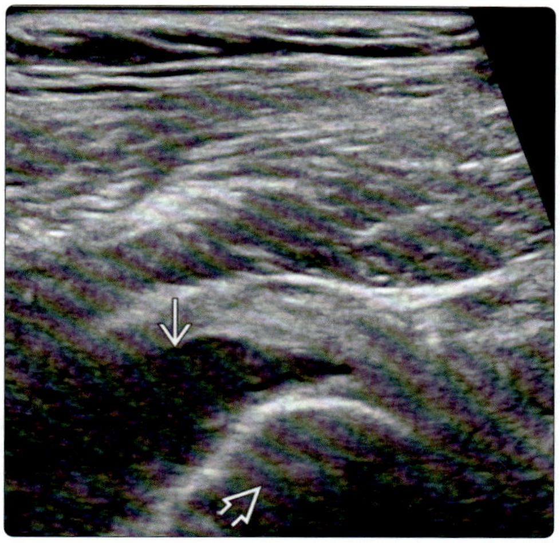

Labral Cyst

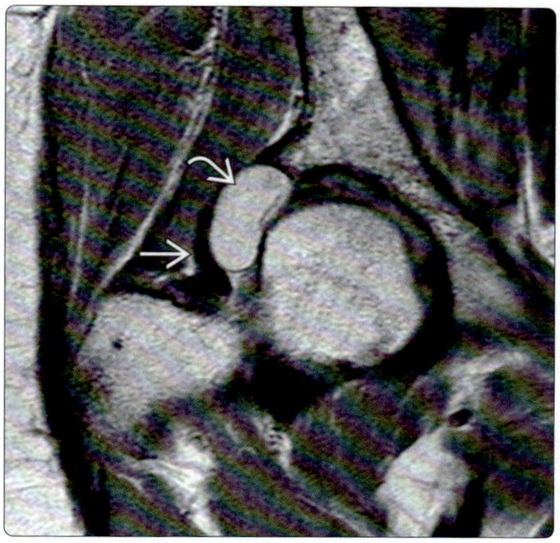

(Left) *Longitudinal US shows a quite large hypoechoic intracapsular labral cyst at the anterosuperior aspect of the hip joint in a patient with hip pain aggravated by abduction for 3 months. The femoral head is normal.* **(Right)** *Coronal oblique intermediate-weighted MR in the same patient shows a large intracapsular labral cyst. The cyst in this case lies deep to the hip capsule. In view of troublesome symptoms, US-guided aspiration was arranged.*

Labral Cyst Aspiration

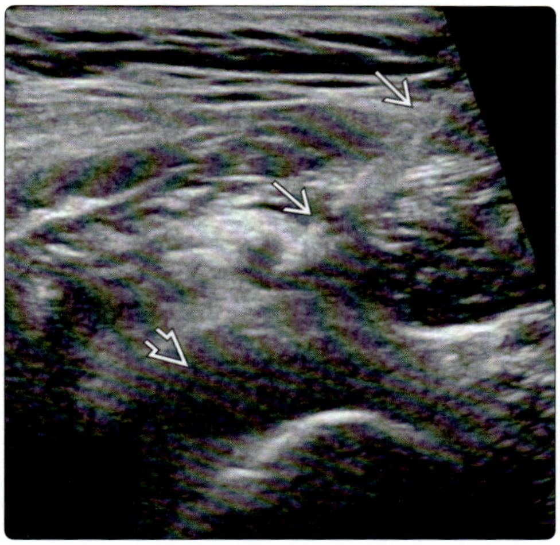

Post Labral Cyst Aspiration

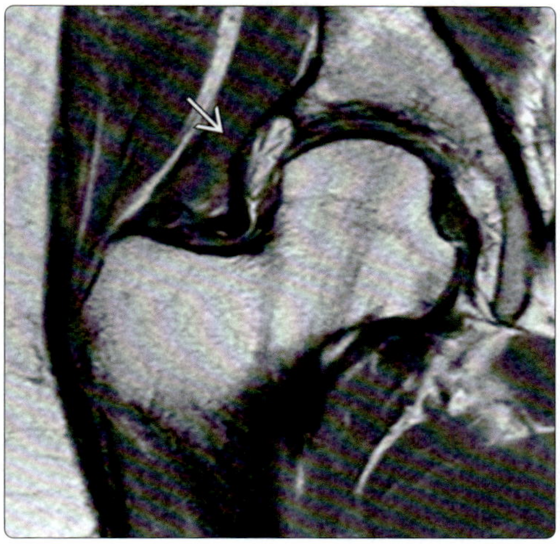

(Left) *Longitudinal US in the same patient shows an 18-g spinal needle approaching the labral cyst prior to cyst aspiration. Only 1 mL of thick, gelatinous fluid could be aspirated from the cyst.* **(Right)** *Coronal oblique PDWI MR in the same patient 2 weeks following US-guided labral cyst aspiration shows that the cyst has markedly reduced in size. Although only 1 mL of fluid could be aspirated from the cyst, it reduced to a greater degree as a result of postprocedure leakage of cyst content.*

Small Labral Cyst

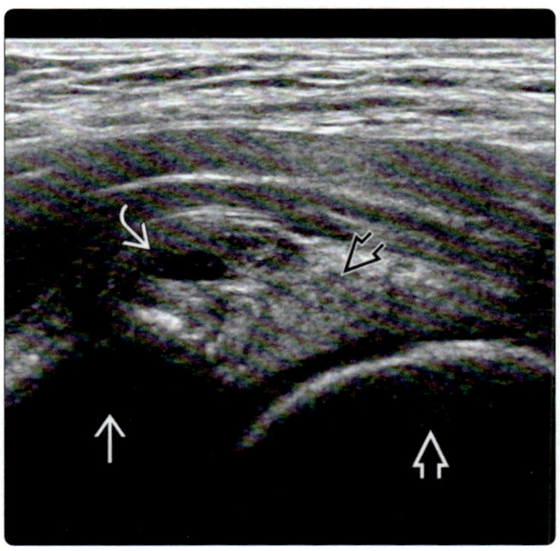

Large Paralabral Cyst

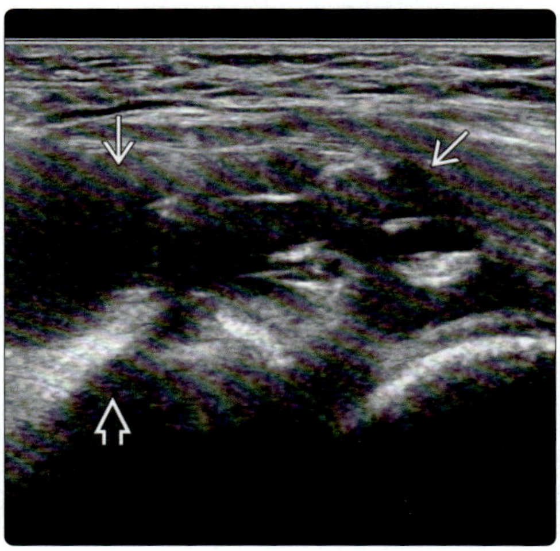

(Left) *Longitudinal oblique US shows a small labral cyst at the superior aspect of the acetabulum. This small cyst is not likely to benefit from aspiration. There is marked thickening of the adjacent capsulolabral tissues along the superior aspect of the hip joint. The femoral head is normal.* **(Right)** *Longitudinal US shows a large, irregular paralabral cyst extending superiorly alongside the acetabulum. This poorly defined extracapsular paralabral cyst has undergone recent leakage.*

(Left) *Axial T2WI FS MR shows mild tendinosis of the gluteus minimus tendon near its insertion into the anterior facet of the greater trochanter. A small avulsive-type tear filled with fluid is also present more laterally.* **(Right)** *Axial T2WI FS MR of the same patient shows moderate peritendinitis with edema surrounding the gluteus minimus tendon just proximal to the insertional area. Peritendinitis is a common accompaniment of active tendinosis.*

Gluteus Minimus Tendinosis and Tear

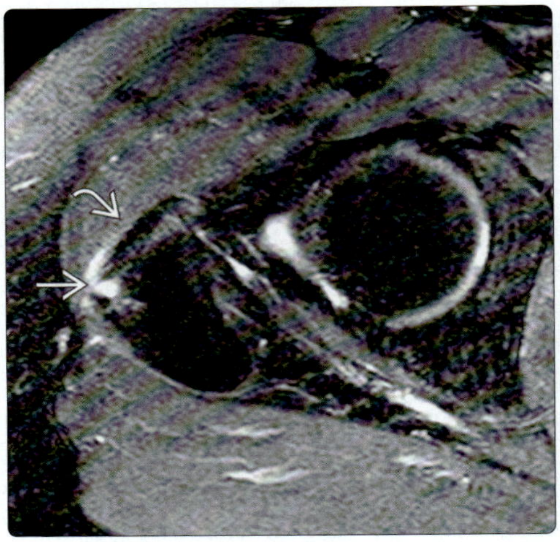

Gluteus Minimus Peritendinitis

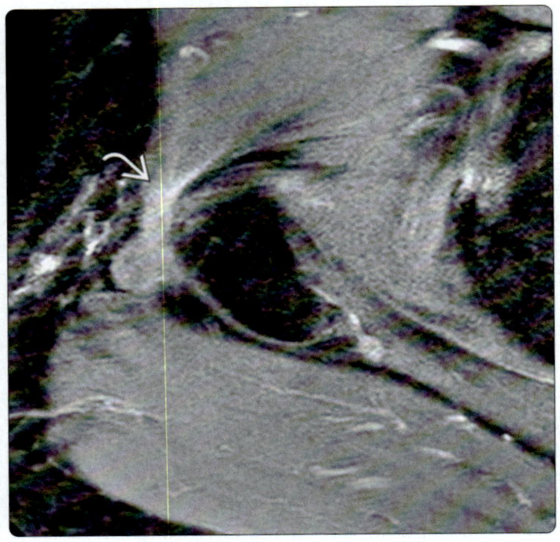

(Left) *Transverse US shows moderate gluteus minimus tendinosis with thickening and hypoechogenicity at the greater trochanteric insertional area. A 20-g spinal needle has been inserted deep to the tendon for steroid (1 mL of triamcinolone 40 mg) and long-acting local anesthetic (1 mL of 0.5% bupivacaine) injection.* **(Right)** *Transverse US in the same patient shows echogenic air loculus at the injection site immediately following needle removal.*

Gluteus Minimus Tendon Injection

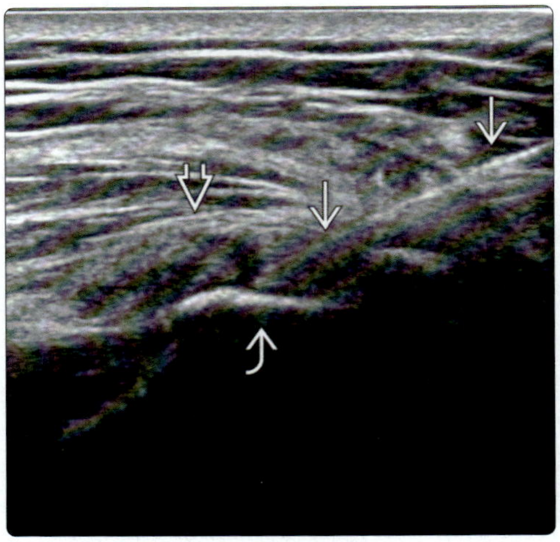

Gluteus Minimus Tendon Injection

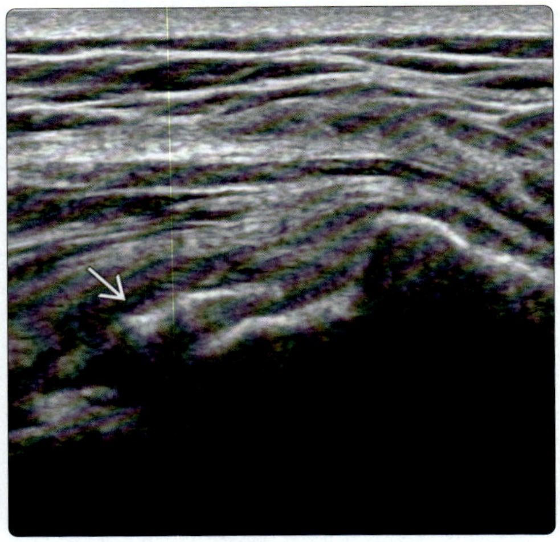

(Left) *Transverse US shows moderate-severity gluteus medius tendinosis with thickening and hypoechogenicity of the tendon near the insertional area into the lateral and superoposterior facets of the greater trochanter. No tendon tear is evident.* **(Right)** *Transverse US in the same patient shows a 20-g spinal needle inserted deep to the peripheral part of the gluteus medius tendon prior to injection of steroid (1 mL of triamcinolone 40 mg) and long-acting local anesthetic (1 mL of 0.5% bupivacaine).*

Gluteus Medius Tendinosis

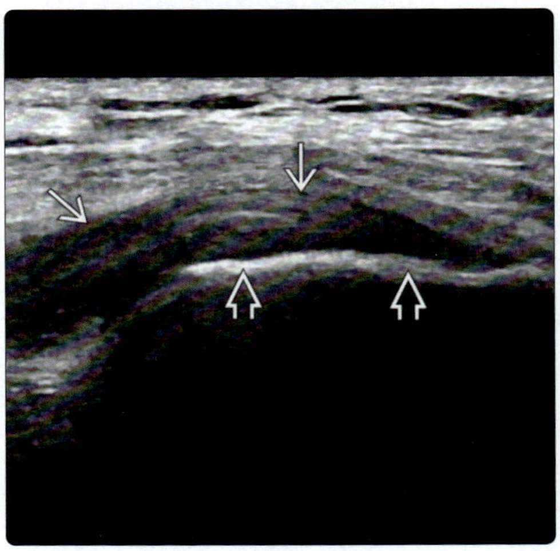

Gluteus Medius Tendon Injection

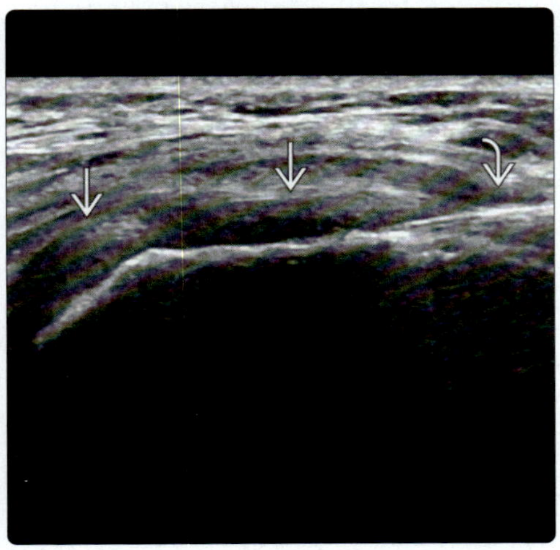

Sacroiliitis

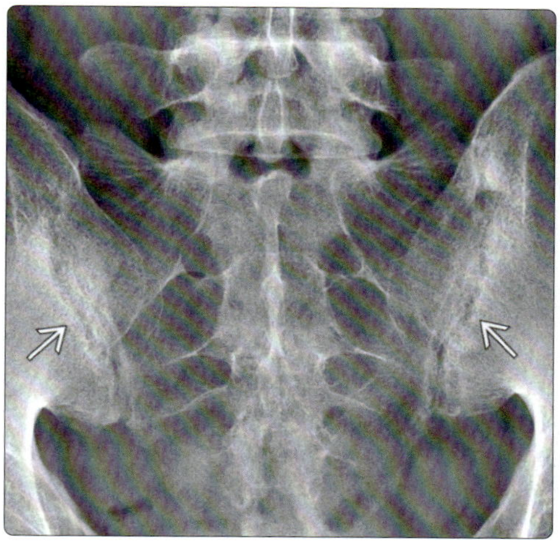

3D CT of Sacrum

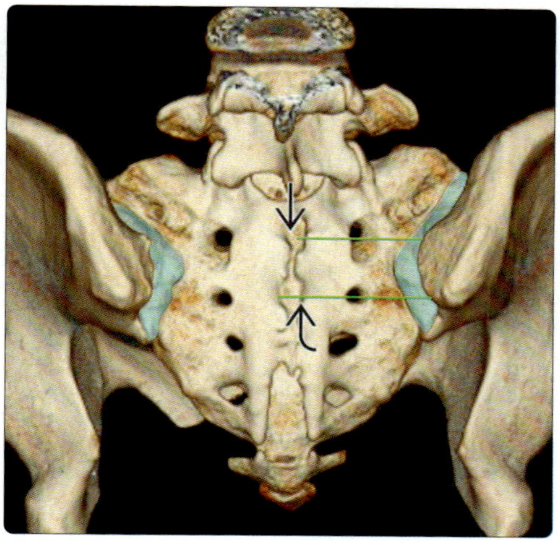

(Left) *Frontal radiograph shows moderate-severity bilateral sacroiliitis ➔. US-guided steroid and long-acting local anesthetic injection is a useful treatment for inflammatory sacroiliitis.* **(Right)** *Coronal 3D CT of the sacrum shows regions of sacroiliac joints (shaded green). The proximal end of the sacroiliac joint is level with the S1 foramen ➔. The distal end of the sacroiliac joint is level with the S2 foramen ➔.*

Upper Aspect of Sacroiliac Joint

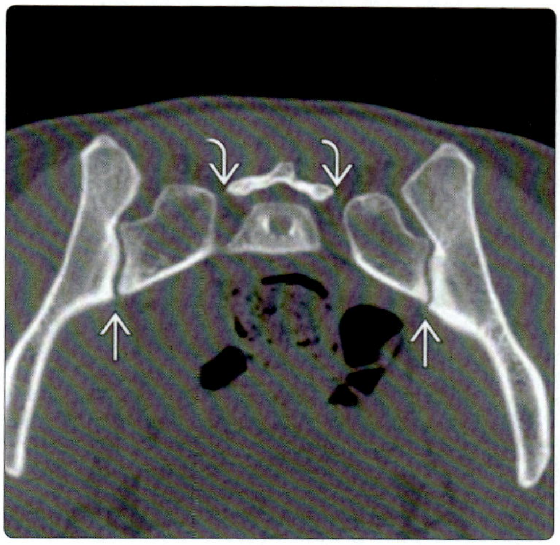

Lower Aspect of Sacroiliac Joint

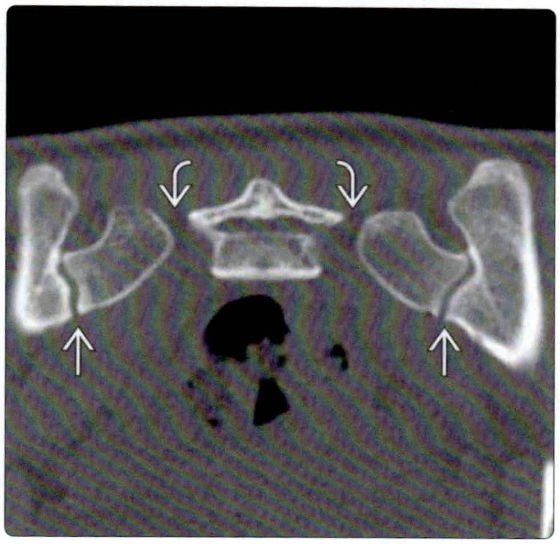

(Left) *Transverse CT shows the proximal end of the sacroiliac joints ➔ at the level of the 1st sacral foramen ➔. Note the lateral needle angulation required to enter the dorsal aspect of the sacroiliac joints.* **(Right)** *Transverse CT shows the distal end of the sacroiliac joints ➔ at the level of the 2nd sacral foramen ➔. Note the lateral needle angulation required to enter the dorsal aspect of the sacroiliac joint. Either CT or US or US-CT/MR fusion imaging can be used for needle insertion to the sacroiliac joint.*

Proximal Part Sacroiliac Joint

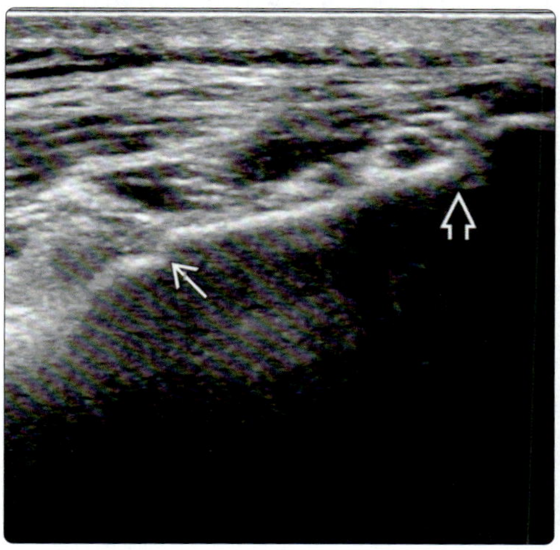

Distal Part Sacroiliac Joint

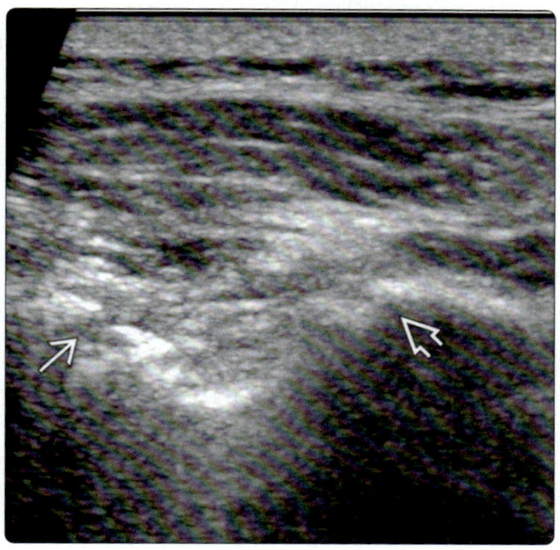

(Left) *Transverse US shows the normal proximal part of the left sacroiliac joint. The S1 neural foramen centrally ➔ and the sacroiliac joint ➔ peripherally are shown.* **(Right)** *Transverse US shows the normal distal part of the left sacroiliac joint. The S2 neural foramen centrally ➔ and the sacroiliac joint ➔ peripherally are shown. The lower part of the sacroiliac joint is generally slightly easier to access with US guidance than the upper part of the sacroiliac joint.*

KEY FACTS

PREPROCEDURE

- **Indications**
  - Intraarticular aspiration of joint fluid or injection of contrast medium or medication
  - Paraarticular tendinosis or ligament tear treatment
    - Dry needling, steroid injection, prolotherapy, autologous blood injection (ABI), platelet-rich plasma (PRP) injection
  - Paraarticular cyst or bursal aspiration
    - Symptomatic Baker cyst or ganglion cyst aspiration
    - Parameniscal cyst: Voluminous and painful cysts may benefit from aspiration ± injection
    - ≥ 19-g needed for more viscous content
  - Biopsy of synovial tissue or intraarticular soft tissue mass

PROCEDURE

- Intraarticular aspiration/drainage of joint fluid or injection
  - Lateral, medial, or suprapatellar recess
  - If no fluid can be aspirated
    - Change position of bevel or reposition needle tip
    - Lower pressure on skin
    - Massage joint fluid toward needle tip
    - Consider changing to larger bore needle
    - Irrigate with saline solution, then aspirate
- Paraarticular tendinosis or ligament treatment
  - Identify most severe area of tendinosis as this will benefit most from percutaneous intervention
  - Dry needling
  - Peritendinous steroid injection: Good short-term but limited long-term benefit
  - Prolotherapy: Reduces pain severity and area of visible tendinosis on US
  - ABI or PRP injection
    - PRP has no proven benefit over dry needling or ABI for tendinosis
- Intraarticular and extraarticular soft tissue biopsy
  - Use coaxial needle to prevent multiple punctures

US-Guided Aspiration and Steroid Injection

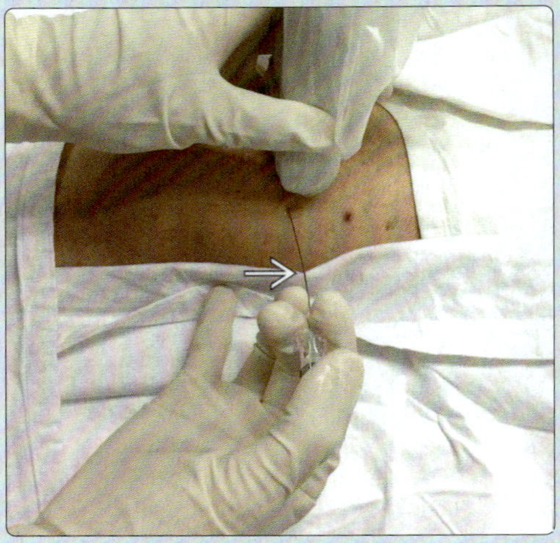

Needle Tip Confirmation

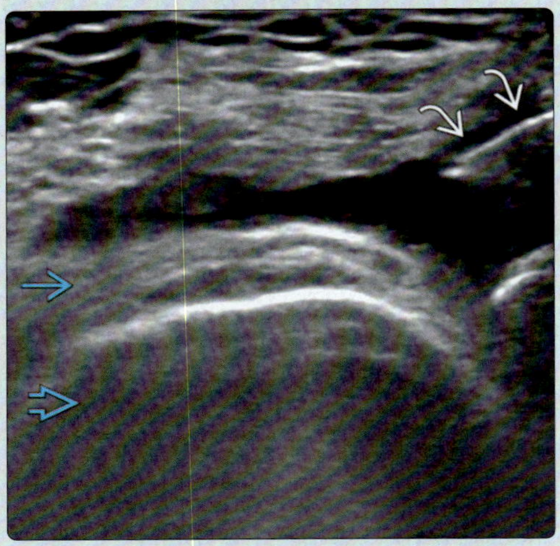

(Left) *US-guided aspiration of joint effusion with chronic swelling and pain due to severe osteoarthritis is shown. A 23-g spinal needle* ➡ *is inserted into the medial patellofemoral recess used under US guidance.* **(Right)** *Transverse US shows the tip of the needle* ➡ *within the effusion distending the medial patellar recess prior to aspiration. The distal femoral shaft* ➡ *and prefemoral subsynovial fat* ➡ *are shown.*

Aspiration of Joint Fluid

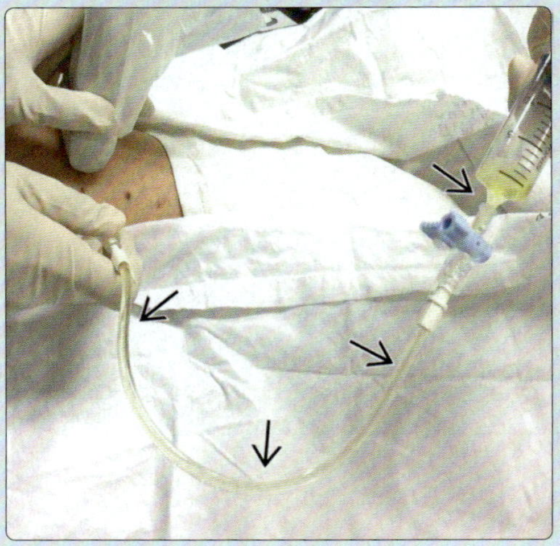

Steroid Injection

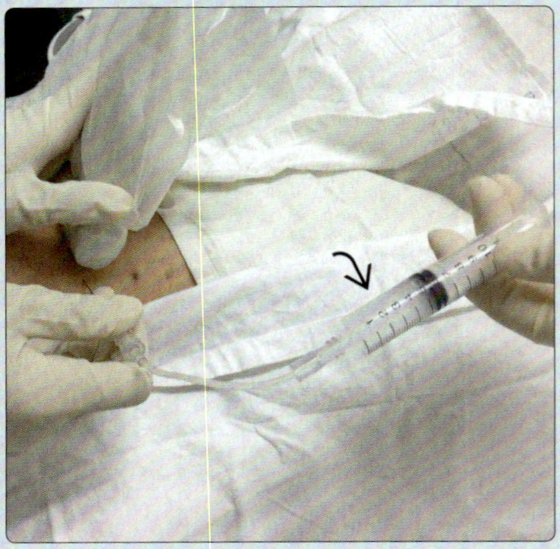

(Left) *Clinical photo during joint aspiration shows straw-colored nonpurulent fluid* ➡ *being aspirated from the joint.* **(Right)** *Clinical photo during injection of steroid* ➡ *and local anesthetic into the knee joint following fluid aspiration is shown. The injectate mixture contains methylprednisolone, bupivacaine, and lignocaine.*

PREPROCEDURE

Indications

- Aspiration or drainage of joint fluid
  - Suspected septic arthritis or crystal arthropathy
  - Temporary relief of symptoms due to distension pain
- Intraarticular injection of contrast medium or medication
  - Medication for relief of pain or inflammation
    - Steroid or hyaluronic acid injection for painful inflammatory osteoarthritis or inflammatory arthritis
  - Radiosynovectomy for chronic synovitis, such as hemophiliac arthropathy and inflammatory arthritis
    - Reduces frequency of recurrent clinical and subclinical hemarthroses in patients with hemophilia
      - Decreases likelihood of developing osteoarthritis
    - Decreases intensity of pain and degree of synovial proliferation
  - Arthrography (conventional, CT, or MR)
    - Stability of osteochondral lesion
      - If rim of contrast insinuates between native bone and osteochondral fragment → unstable → may need surgery
    - Evaluation of postoperative meniscus
- Paraarticular aspiration
  - Symptomatic or large Baker cyst or parameniscal cyst
    - Baker cyst is usually due to knee arthropathy and tense knee effusion
      - Reducing size of effusion by intraarticular aspiration can reduce size of Baker cyst
    - Parameniscal cyst is usually due to horizontal meniscal tear
      - Often treated intraarticularly by partial meniscectomy
    - Both Baker cyst and parameniscal cyst can be treated percutaneously
- Paraarticular tendon injection
  - Patellar/semimembranous/quadriceps/pes anserinus tendinosis injection to stimulate healing response
  - Dry needling, steroid injection, prolotherapy, autologous blood injection (ABI), platelet-rich plasma (PRP) injection
- Biopsy of synovial tissue for suspected infection, inflammation, or synovial tumor

Getting Started

- Things to check
  - Time-out procedure (patient identity, type and side of procedure, allergy status)
  - Risk of infection, allergy, bleeding, vascular and nerve injury ± tendon rupture, unsuccessful outcome discussed
  - Informed consent form is signed
- Medications
  - Local anesthetic for skin: 2-3 mL of 1-2% lignocaine
  - Local anesthetic for intraarticular injection: 1-2% lignocaine or 0.25-0.5% bupivacaine (Marcaine)
  - Steroid: 40 mg/mL of methylprednisolone (Depo-Medrol)
  - Hyaluronic acid ("viscosupplementation")
    - Macromolecule occurring naturally in synovial fluid and contributing to its viscoelastic property
  - Contrast
    - CT arthrogram: 40 mL of iodinated contrast [iohexol (Omnipaque) 300 mg/dL]
    - MR arthrogram: 40 mL of diluted gadolinium solution (0.1 mL gadolinium in 10.0 mL normal saline solution)
- Sterile set; spinal or intravenous needle 18- to 23-g; extension tubing; syringes; ± coaxial needle and biopsy needle; sterile specimen bottles
  - Sterile drapes and probe cover

PROCEDURE

Procedure Steps

- **Intraarticular aspiration of joint fluid or injection of medication or contrast medium**
  - Can aspirate either medial or lateral patellar recess or suprapatellar recess
    - Patient lying supine, knee flexed slightly (small pillow behind knee)
    - For suprapatellar pouch, probe is placed in sagittal plane superior to patella medial or lateral to quadriceps tendon **or** in transverse plane superior to patella with quadriceps tendon seen in center of field
    - Flush needle and tubing with injectate beforehand to avoid bubbles being injected
    - Using free-hand technique, advance needle (18- to 23-g) into joint space under real-time imaging
    - If needle tip not seen, move transducer ± jiggle needle to help confirm needle tip position
    - Should be free flow of fluid during aspiration or injection
  - Aspirate as much as possible for decompression
    - If effusion is too thick, consider changing to larger bore needle, e.g., 15- or 16-g
    - If insufficient aspirate, rotate needle to change position of bevel or reposition needle tip
      - Probe pressure can displace fluid away from needle
      - Use free hand to massage joint fluid toward needle tip under real-time imaging
    - If still insufficient for sampling, irrigate joint with saline solution, then aspirate saline wash and label as joint wash
- **Paraarticular cyst aspiration**
  - Probe placed longitudinally on cyst; larger bore for more viscous content
  - ± hyaluronidase (150 units in 2 mL saline) prior to aspiration
    - Hyaluronidase is human enzyme that degrades hyaluronic acid
    - Inject into cyst, leave needle tip within cyst for 15 minutes, then aspirate
- **Treatment of tendinosis**
  - Probe positioned over area of most severe tendinosis: Most pronounced loss of fibrillation, tendon thickening, and hyperemia
  - Prone position for semimembranous tendon
  - External rotation of knee for pes anserinus tendons
  - **Dry needling**
    - Infiltrate local anesthetic (3-5 mL of 1-2% lignocaine) around tendon
    - Position needle tip within area of tendinosis
    - Move needle tip to and fro through tendinosis area for 1 minute

- May undertake ABI or PRP injection following dry needling
- **Peritendinous steroid injection**
  - 1 mL 30-40 mg/mL methylprednisolone or triamcinolone mixed with 1 mL of 1% lignocaine ± 1 mL of 0.25% bupivacaine
  - Aimed at peritendinous tissue adjacent to tendinosis area
  - Usually can only be repeated 1x or 2x after several weeks
  - Good short-term but poor long-term clinical effect
  - Helps to treat peritendinitis associated with tendinosis but no known effect on tendinosis per se
- **Prolotherapy (or regenerative injection therapy)**
  - 1 mL of 25% dextrose with lignocaine [1 mL 2% lignocaine (20 mg/mL) and 1 mL of 50% dextrose (25 g/50 mL)] injected into tendinosis area
  - 4-6 prolotherapy treatment sessions required on average
  - Prolotherapy reduces pain and tendinosis
- **ABI or PRP injection**
  - For ABI: Blood withdrawn from arm vein and injected directly into tendon after dry needling
    - Amount of blood depends on size of tendon
    - 2-3 mL of blood is injected for patellar tendon
  - For PRP: Larger amount of blood (usually 20 mL) is withdrawn from arm vein
    - Blood is placed into tube and centrifuged for 15 minutes
    - Separates platelets from white cells and red cells
    - Platelet-rich component is used for injection
  - Use Luer lock syringe for injection into tendon
    - Protects against explosive leakage from syringe because of back pressure during injection
    - Gentle controlled pressure and release technique on syringe plunger during injection will also help avoid syringe leakage
  - May require 2 interventions around 4-6 weeks apart
  - No clear benefit of PRP over ABI or dry needling alone for treatment of tendinosis
    - May hasten recovery period of quadriceps tendon tear

- **Intraarticular soft tissue biopsy**
  - Use coaxial needle to prevent multiple joint punctures
  - For suspected malignant tumor, biopsy should be performed at site of subsequent surgery (tumor surgeon marks likely surgical incision site on skin)
  - For mobile intraarticular lesions, can be difficult to obtain core
    - Fix lesion with free hand
    - If lesion in suprapatellar recess, flex knee to fix lesion by tightening surrounding structures

Findings and Reporting

- Depends on interventional procedure performed but consider reporting on these aspects
  - Initial US findings
    - Do not perform any interventional procedure without 1st making full assessment of US findings and reporting clearly on these findings
  - Type and size of needle used
  - Amount and type of local anesthetic used
  - Volume, color, and viscosity of joint fluid aspirated
  - Number of specimen core obtained
    - Record if friable biopsy material only obtained
  - If procedure uneventful and well-tolerated procedure
- Immediate complication, e.g., excessive bleeding
- Specific patient aftercare

POST PROCEDURE

Expected Outcome

- Mild discomfort due to reactive synovitis 1-2 days after contrast or medication injection

Immediate Aftercare

- Perform MR, CT, or conventional arthrography within 30 minutes of contrast injection
- Send aspirate for microbiology, cytology, crystal analysis (need fresh specimen) as appropriate

Subsequent Aftercare

- Avoid knee exercise for 24 hours after corticosteroid injection to maximize therapeutic effect and reduce spread of steroid into surrounding tissues
- After paraarticular soft tissue injection therapy, refrain from tendon-loading activity for 1 week following procedure
  - Avoid taking antiinflammatory medication, which may theoretically impair healing process
- For tendon intervention
  - Rest for 2 weeks with no return to sporting activity/training for 6 weeks

OUTCOMES

Complications

- Most feared complication(s)
  - Infection, e.g., septic arthritis
    - Avoid traversing infected tissues during joint injection
  - Extravasation of contrast or medication
    - Steroid or radiosynovectomy agent: Fat pad atrophy, skin atrophy, skin depigmentation
- Immediate/periprocedural complication(s)
  - Bleeding: Especially in bleeding diathesis, e.g., hemophiliac arthropathy treated with radiosynovectomy
  - Neurovascular injury
- Following ganglion or parameniscal cyst aspiration, leakage of cyst contents through puncture site into surrounding soft tissues can induce chemical-type cellulitis, which lasts ~ 48 hours
- Patellar or quadriceps tendon rarely rupture post-PRP intervention

SELECTED REFERENCES

1. Charousset C et al: Are multiple platelet-rich plasma injections useful for treatment of chronic patellar tendinopathy in athletes? a prospective study. Am J Sports Med. 42(4):906-11, 2014
2. Wong Y et al: Efficacy of yttrium-90 synovectomy across a spectrum of arthropathies in an era of improved disease modifying drugs and treatment protocols. Int J Rheum Dis. 17(1):78-83, 2014
3. Morvan G et al: Interventional musculoskeletal ultrasonography of the lower limb. Diagn Interv Imaging. 93(9):652-64, 2012

Aspiration of Baker Cyst

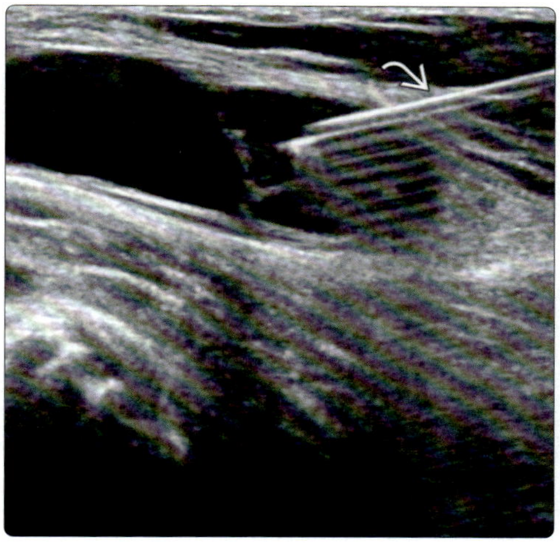

Aspiration of Baker Cyst

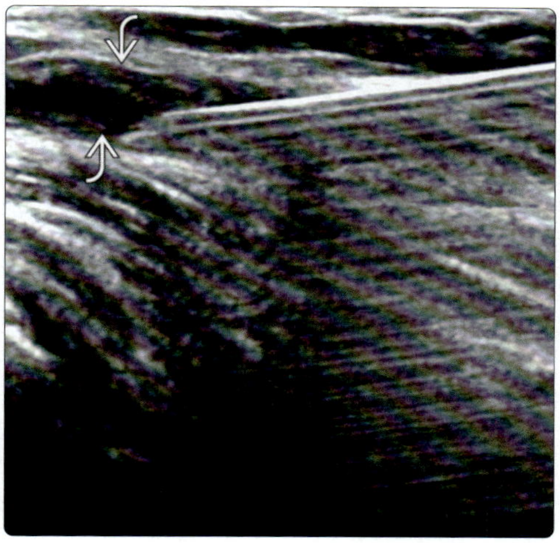

(Left) *Longitudinal US shows a Baker cyst prior to aspiration using a 18-g spinal needle. It is not possible to predict the viscosity of periarticular knee cysts based on US appearances.* **(Right)** *Longitudinal US of the same Baker cyst following aspiration shows almost complete collapse of the cyst. The bursal wall can still be appreciated.*

US-Guided Aspiration

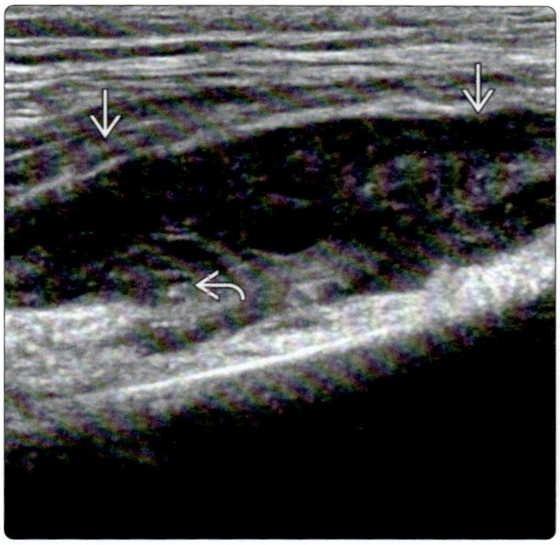

US-Guided Aspiration

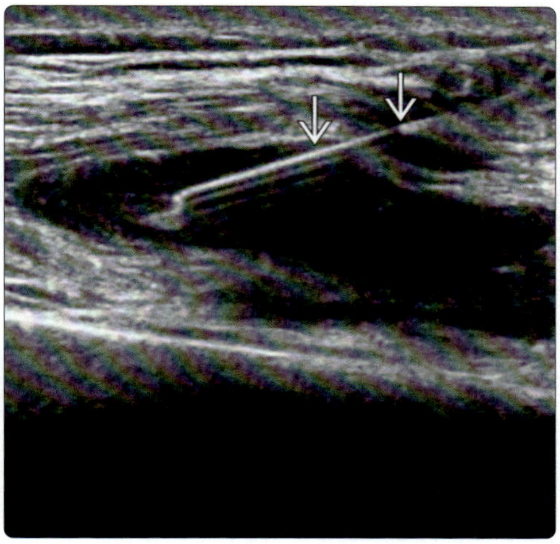

(Left) *Longitudinal US prior to US-guided aspiration of the lateral patellar recess shows severe synovial proliferation with some small, echogenic foci.* **(Right)** *Longitudinal US shows the same recess during aspiration using a 18-g spinal needle for culture and cytology. This revealed monosodium urate crystals consistent with gouty arthritis. No bacterial growth was obtained.*

US-Guided Biopsy

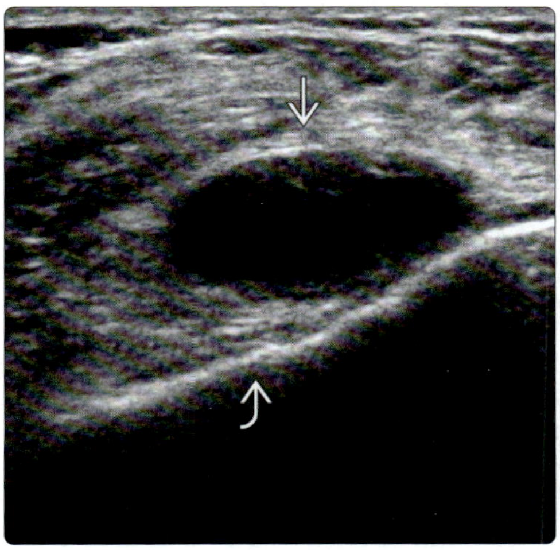

US-Guided Biopsy

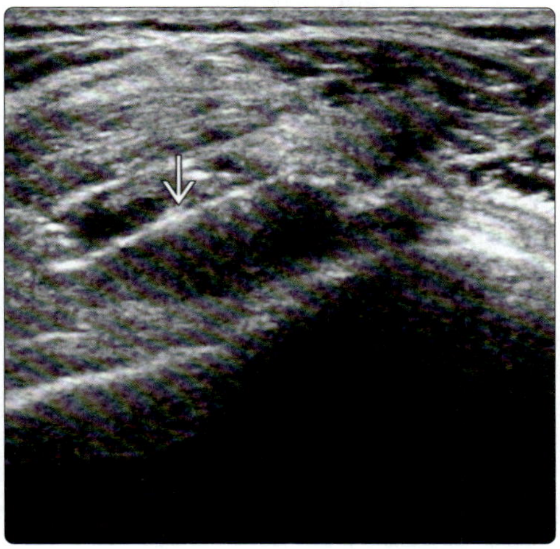

(Left) *Longitudinal US prior to biopsy shows an oval-shaped, hypoechoic, intraarticular mass in the suprapatellar pouch superficial to the distal femur.* **(Right)** *Longitudinal US of the same tumor during US-guided biopsy shows a 16-g biopsy needle that has been inserted through a 15-g coaxial needle. The needle is position in the center of the hypoechoic tumor. Histology confirmed localized pigmented villonodular synovitis (focal nodular synovitis).*

(Left) *Preliminary longitudinal US in a hemophiliac patient shows a hemarthrosis of the suprapatellar pouch with fine echogenicity of the suprapatellar pouch.* **(Right)** *Longitudinal US during aspiration shows a 20-g spinal needle inserted into the hemorrhagic knee effusion. Radioactive yttrium was then injected to reduce synovial proliferation. US-guided injection prevents inadvertent extraarticular injection.*

Radiosynovectomy

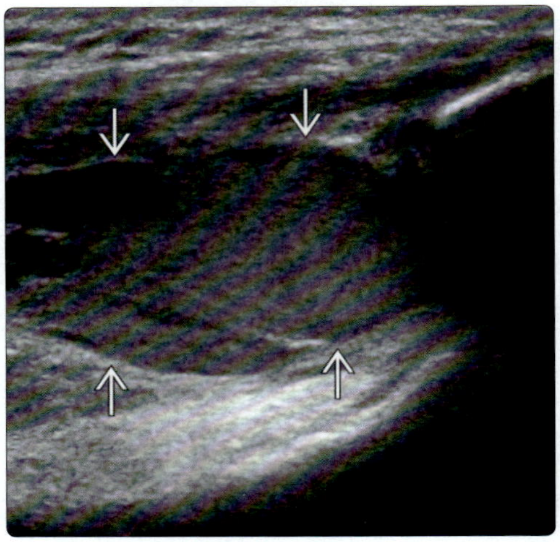

Radiosynovectomy

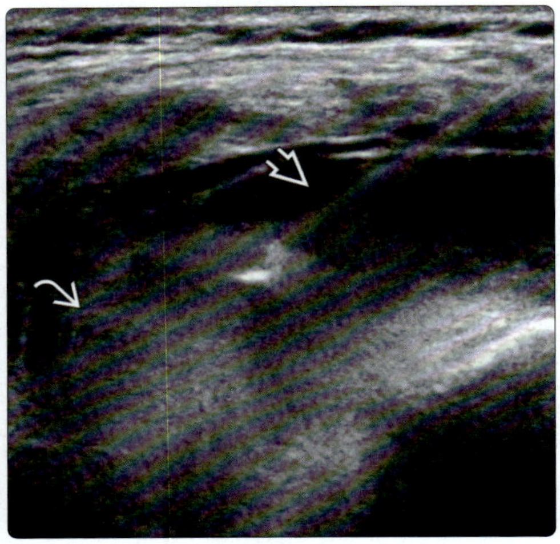

(Left) *Transverse US during dry needling treatment of semimembranous tendinosis shows a 23-g needle present within the hypoechoic semimembranous tendon close to the tibial insertion. (Courtesy S. James, MD.)* **(Right)** *Transverse US shows dry needling treatment of the distal patellar tendon using a 21-g needle inserted into the tendon close to the tibial insertion. The distal patellar tendon is moderately thickened and hypoechoic due to tendinosis. (Courtesy S. James, MD.)*

Dry Needling

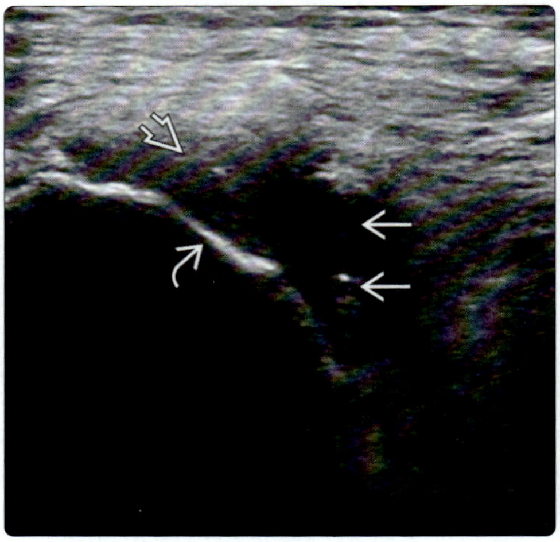

Dry Needling

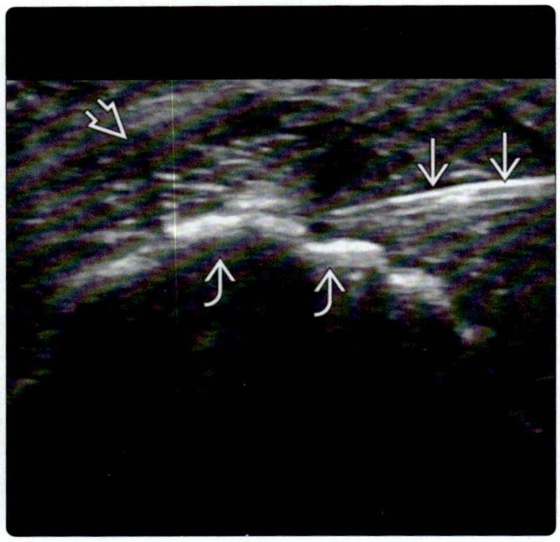

(Left) *Transverse US during dry needling shows patellar tendinosis using a 21-g needle. The fluid present deep to the tendon is within the infrapatellar bursa, which has been distended with lignocaine. Dry needling was followed by autologous blood injection (ABI) (2-3 mL) using the same needle and approach.* **(Right)** *Transverse US of the same patellar tendon after dry needling and ABI shows at least 2 small, hypoechoic fenestration clefts due to injected blood.*

Dry Needling and ABI

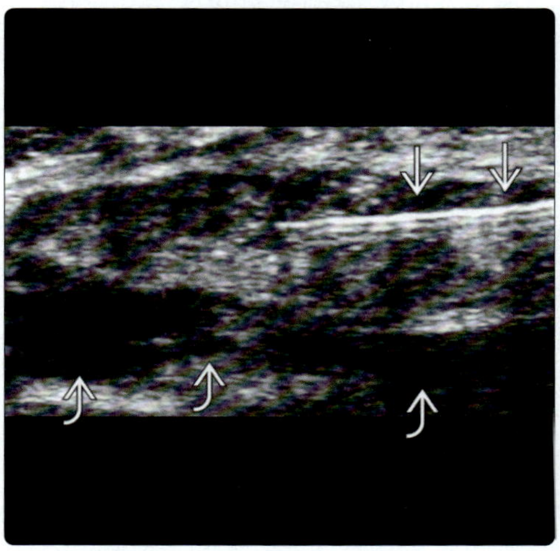

After Dry Needling and ABI

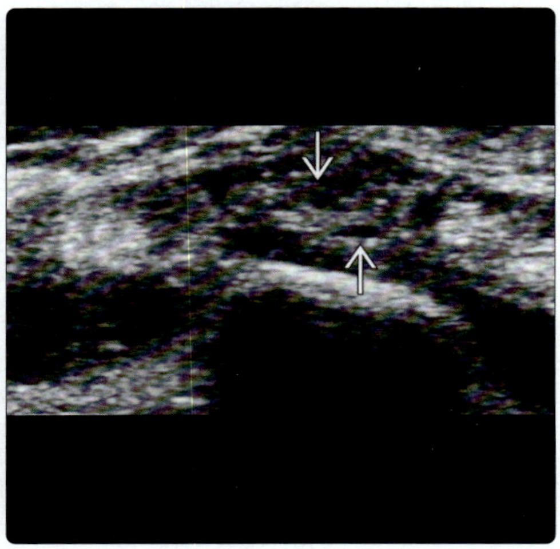

Before Dry Needling and ABI

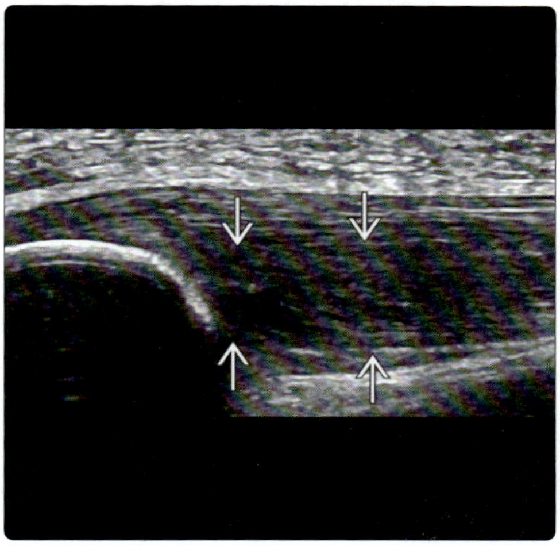

Before Dry Needling and ABI

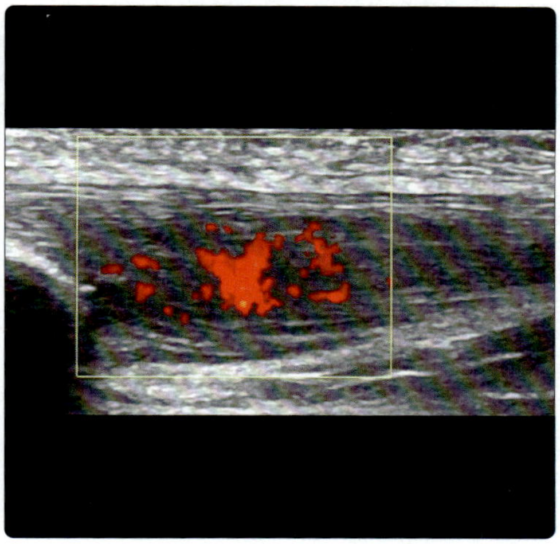

(Left) *Preliminary longitudinal US shows moderate to severe proximal patellar tendinosis, particularly affecting the deep fibers ➡ of the patellar tendon with marked tendon thickening, loss of fibrillar echotexture, and hypoechogenicity.* **(Right)** *Power Doppler US (same patient) of the same area shows moderate to severe hyperemia of the more severe area of tendinosis, indicative of active disease.*

Dry Needling and ABI

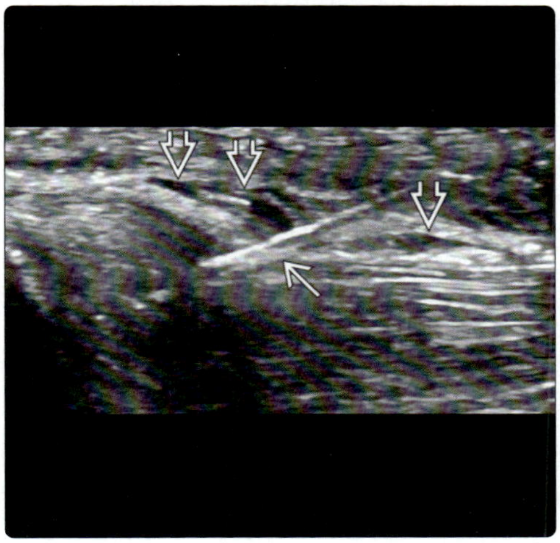

Dry Needling and ABI

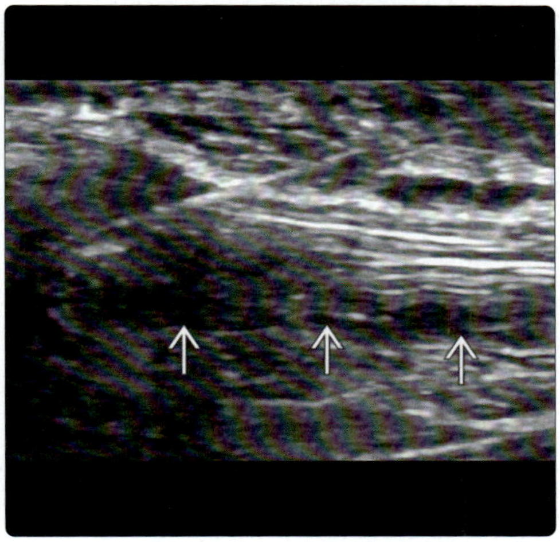

(Left) *Longitudinal US shows the same patient during dry needling using a 21-g needle ➡. Note the thin rim of anesthetic fluid (1% lignocaine) on the tendon surface ➡.* **(Right)** *After dry needling, 2-3 mL of autologous blood was injected into the center of the patellar tendon at the site of most severe tendinosis. Well-defined, linear fenestration clefts are present within the tendon due to this injected fluid ➡.*

PRP Injection

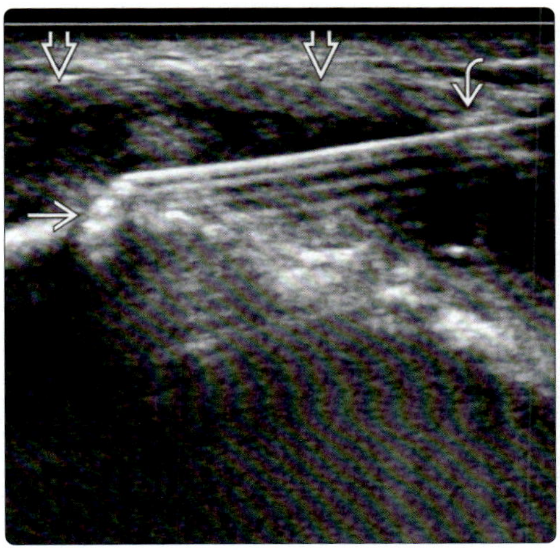

Peritendinous Steroid Injection

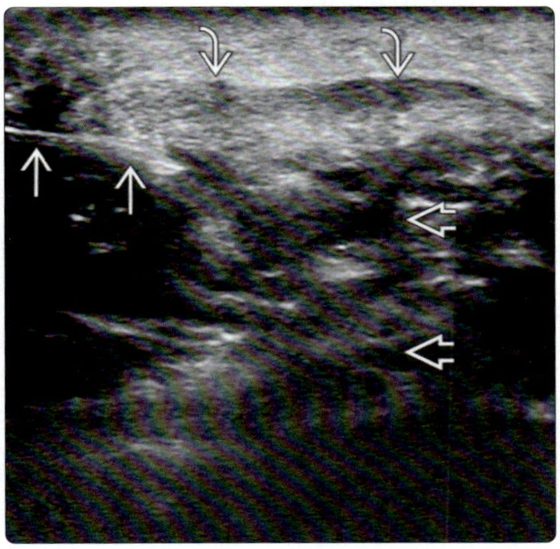

(Left) *Transverse US during injection (18-g intravenous needle ➡) of 6 mL of platelet-rich plasma (PRP) solution into the quadriceps tendon ➡ for treatment of tendinosis is shown. Microbubbles are present at the needle tip ➡.* **(Right)** *Transverse US shows injection of 1 mL (40 mg/mL) of methylprednisolone with 0.5 mL (1%) of lignocaine deep to the patellar tendon ➡ using a 23-g spinal needle ➡ to treat Hoffa fat pad ➡ impingement syndrome. (Courtesy S. James, MD.)*

(Left) *Setup used for knee joint drainage procedure includes drainage bag →, sutures, pigtail catheter ⇒, dilator ⇒, guidewire ➡, syringe for aspiration, diamond needle ➡, scalpel, syringe for local anesthetic, forceps, and cotton buds.* **(Right)** *Preparing the sterile field at the puncture site with chlorhexidine solution minimizes the risk of infection. Sterile drapes are then placed around the puncture site.*

Drainage Set

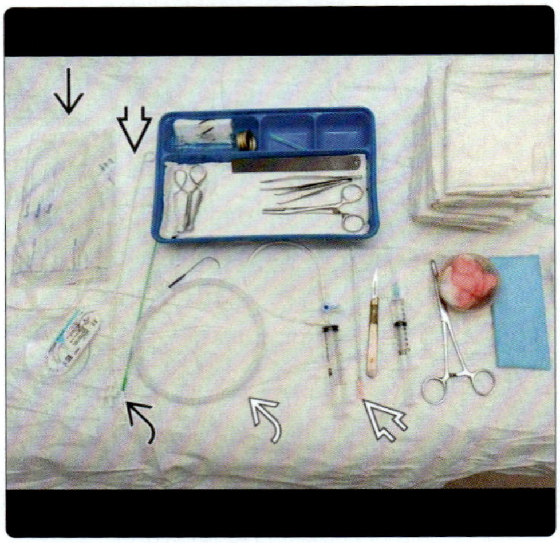

Prepping Site

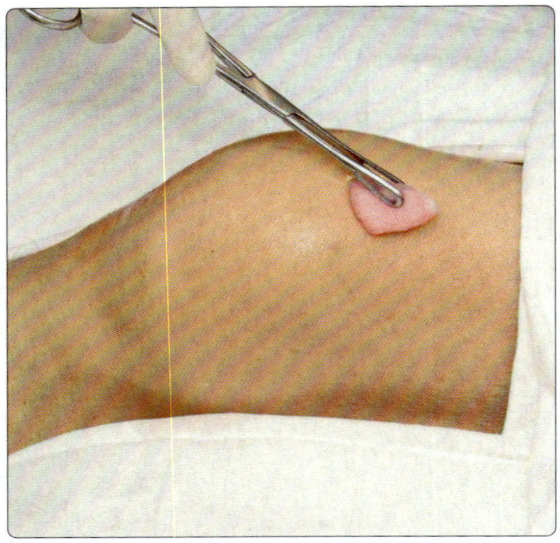

(Left) *After determining a suitable puncture site on preliminary US, the puncture site is anesthetized with local anesthetic (3 mL of 1% lignocaine) using a 23-g hypodermic needle. A sterile probe cover is wrapped around the transducer. Sterile US gel can be used during US scanning.* **(Right)** *A stab incision ➡ of the skin at the anesthetized puncture site is performed. A sharp, small, deep stab incision ensures adequate incision of the subdermal fascia with only a small cutaneous scar.*

Local Anesthetic

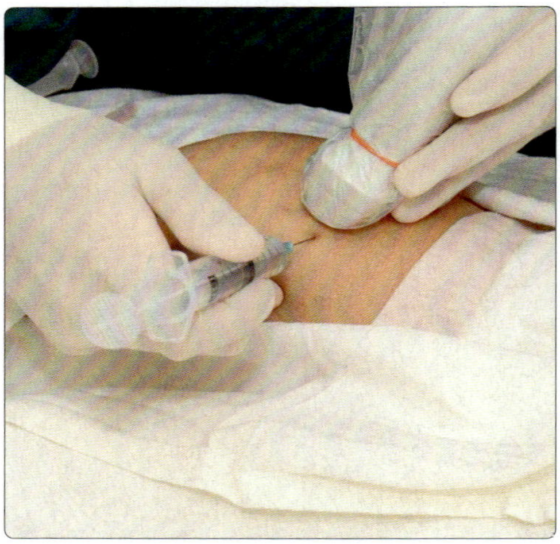

Stab Incision

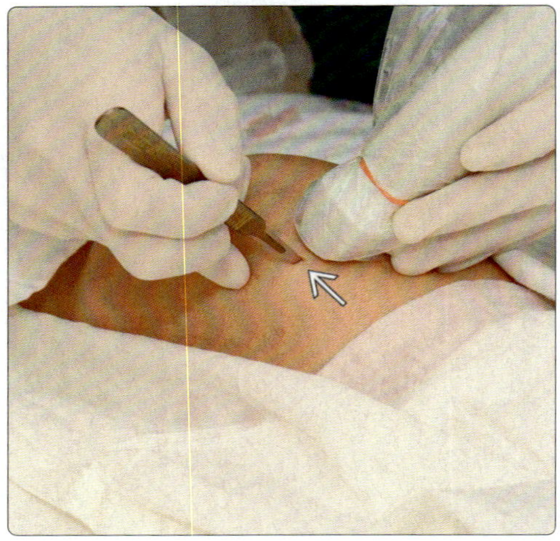

(Left) *Seldinger technique is used to drain the chronic knee effusion due to low-grade infection. An 18-g sheathed needle ➡ is used to puncture the knee capsule under US guidance.* **(Right)** *The needle position can be confirmed by free aspiration of fluid ➡ from the needle. The 1st fluid specimen aspirated is saved and sent for culture and cytology. Try not to aspirate all of the effusion, since this fluid facilitates subsequent guidewire and catheter insertion.*

Seldinger Technique

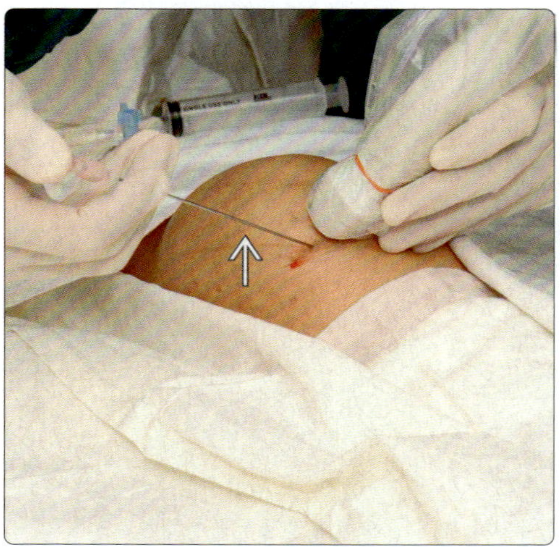

Needle Position Confirmed by Aspiration

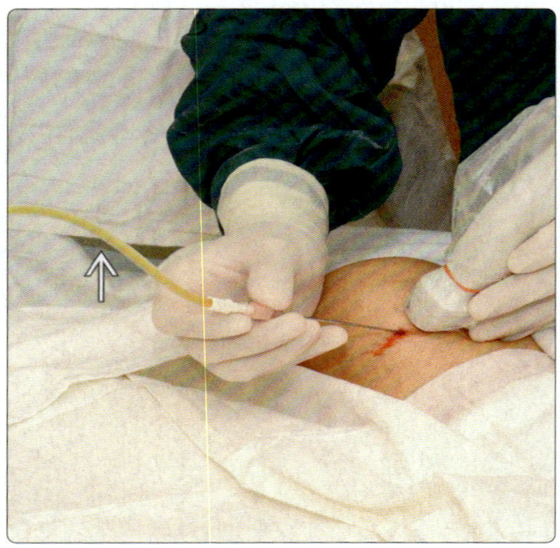

Guidewire Insertion

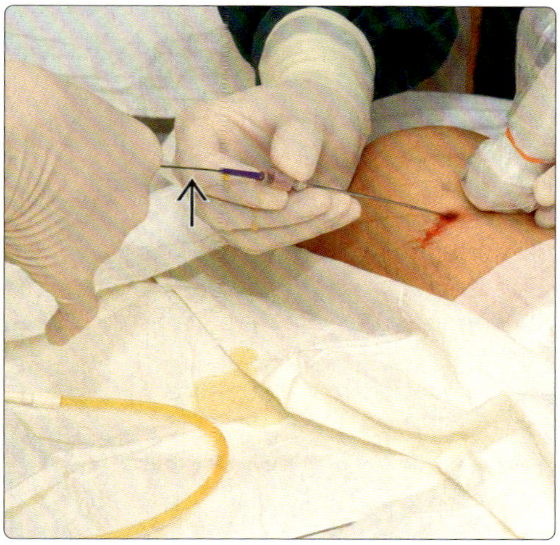

Tract Dilatation

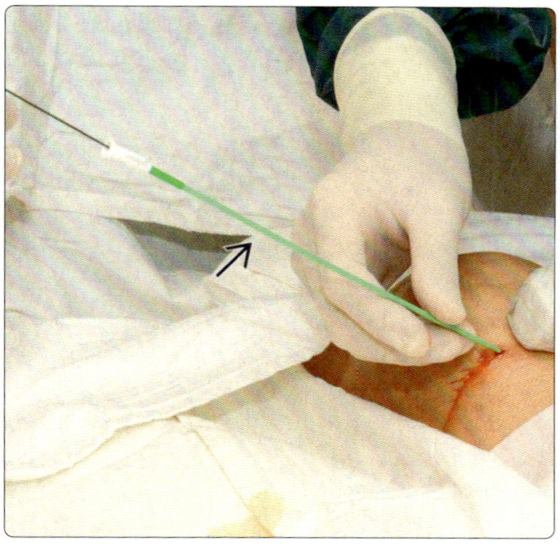

(Left) *After puncturing the knee capsule and confirming the needle tip position with US, a guidewire* ➔ *is inserted through the needle into the knee joint. At this stage, the needle should be kept stable and straight to prevent kinking of the guidewire.* **(Right)** *The needle is removed slowly, leaving the guidewire in situ. The tract is serially dilated with a 6-Fr followed by an 8-Fr dilator* ➔*. The guidewire should be kept stable to prevent it from slipping out.*

Catheter Insertion

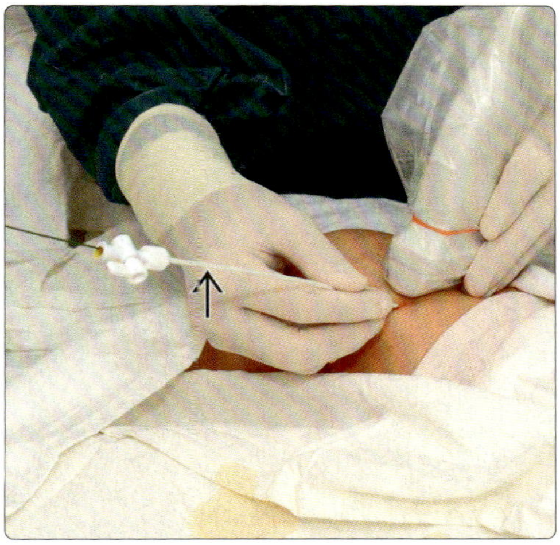

Catheter Fixation

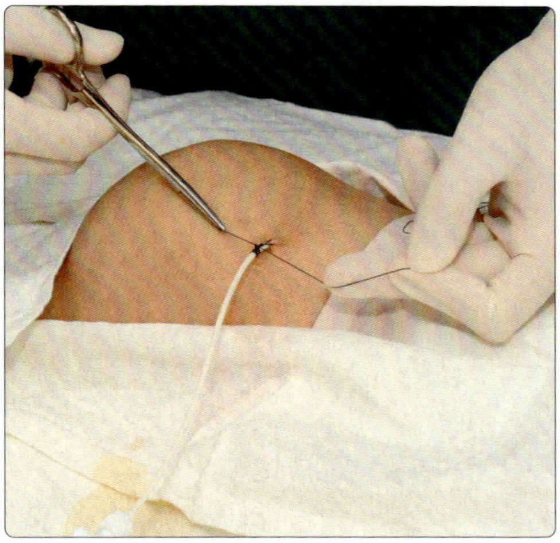

(Left) *After the tract is fully dilated, an 8-Fr side-hole catheter (pigtail catheter)* ➔ *is inserted over the guidewire into the knee joint. US is used to confirm catheter position within the knee joint. Larger catheters are used for collections with more complex, viscous fluid.* **(Right)** *The guidewire is gently pulled out, and the catheter is secured by sutures to the skin. Gauze and plastic tape provide additional catheter protection.*

Confirmation by Aspiration

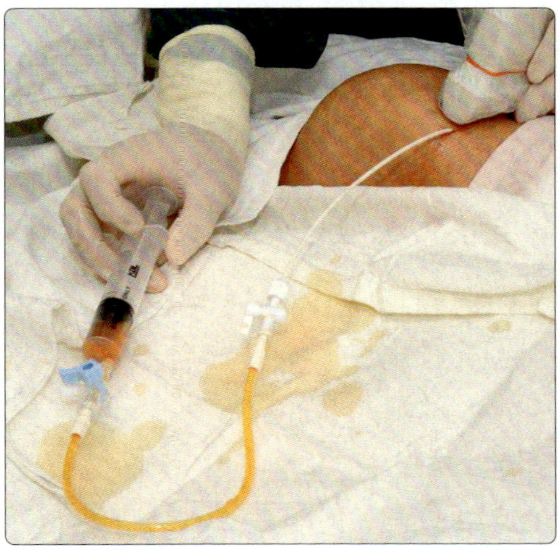

Confirmation by US

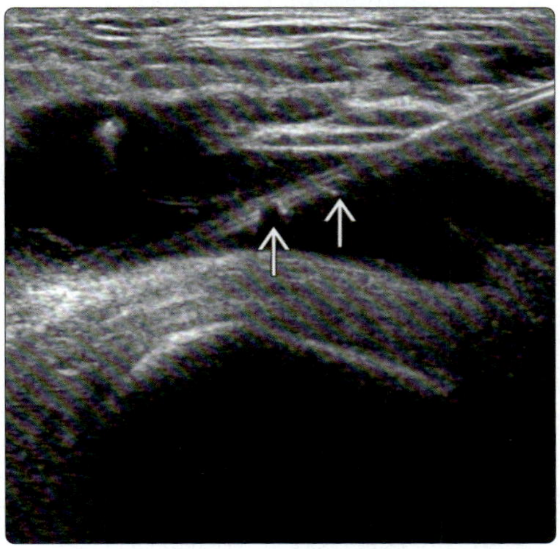

(Left) *A syringe is connected to the catheter through a 3-way stopcock and as much fluid is aspirated as possible. The 8-Fr catheter is then connected to a drainage bag for further gravity-assisted drainage.* **(Right)** *Transverse US after the procedure confirms the position of the catheter within the suprapatellar pouch of the knee joint. The catheter side holes* ➔ *through which knee joint fluid can freely drain are visible.*

KEY FACTS

PREPROCEDURE

- **Indications**
  - Aspiration or injection of joint fluid
  - Intraarticular injection of medication or contrast
  - Paraarticular aspiration or injection
    - Cyst aspiration, peritendon, plantar fasciitis, Morton neuroma, or forefoot bursitis injection

PROCEDURE

- Joint aspiration or injection
  - Diagnostic to determine site of pain or therapeutic to relieve pain
  - Anterolateral approach to ankle preferred
- Ganglion cyst aspiration
  - ≥ 19-g needle for more viscous contents
  - Steroid injection after aspiration is useful even if there is limited content aspirated
- Peritendon injection: Approach depends on site of tendon disease
  - Steroid injection to peritendon region
  - Platelet-rich plasma (PRP) within tendon substance
- For Achilles tendinosis, various treatments and approaches are possible
  - Stimulate healing response: Dry needling, autologous blood, PRP, or prolotherapy
  - Reduce neovascularity: Sclerotherapy, electrocoagulation, or brisement
  - Reduce inflammation: Steroid
- Plantar fasciitis: Medial approach preferable
  - Dry needling, steroid (short acting, preferably)
- Ligament injection
  - PRP within and around ligament substance
- Morton neuroma
  - Steroid ± alcohol ablation (if not responsive)
- Forefoot bursitis: Plantar approach
  - Steroid injection

Plantar Fasciitis Injection

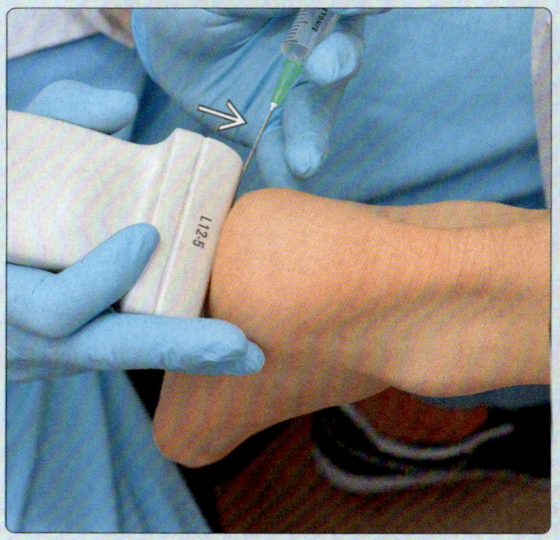

Plantar Fasciitis Injection

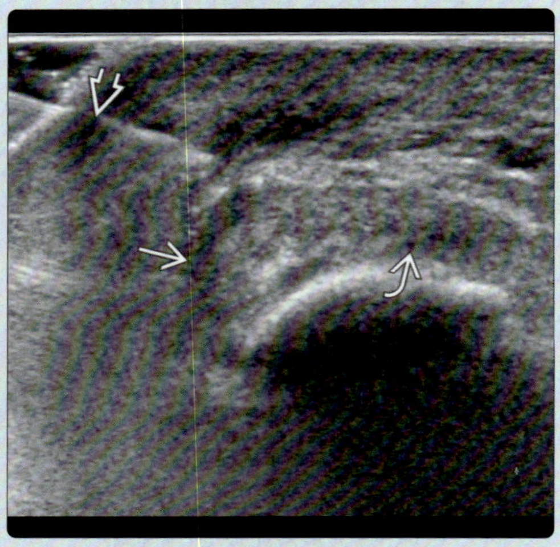

(Left) *Clinical photograph shows transverse approach for injection of plantar fasciitis. The patient lies prone on the examination table. The needle ➡ is inserted medially, almost horizontally.* **(Right)** *Transverse US shows steroid injection for plantar fasciitis. A 23-g intravenous needle ➡ is inserted medially. Short-acting steroid (40 mg of hydrocortisone) ➡ is injected superficial to the plantar fascia ➡.*

US-Guided Foot Ganglion Steroid Injection

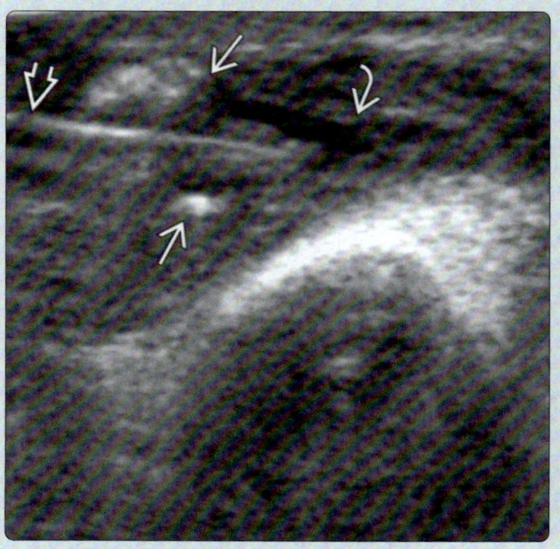

US-Guided Foot Ganglion Steroid Injection

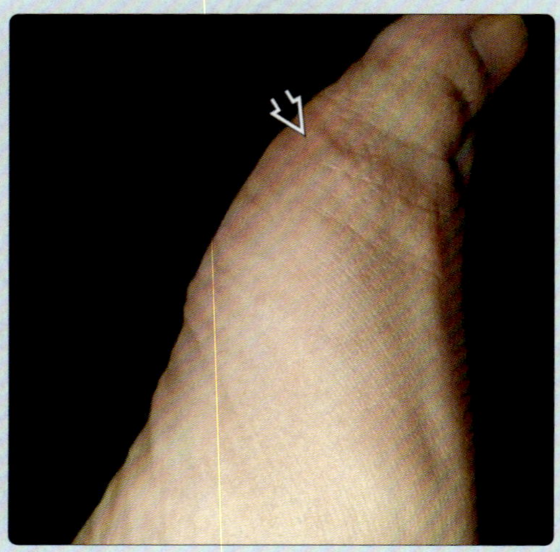

(Left) *Transverse US shows steroid injection of the ganglion ➡. An 18-g intravenous needle ➡ is inserted into the ganglion. Although no fluid could be aspirated, steroid (20 mg of Depo-Medrol) ➡ was injected.* **(Right)** *Clinical photograph of the same patient shows complete resolution of the ganglion after steroid injection, although no fluid could be aspirated before the steroid injection. The site of previous swelling is shown ➡.*

PREPROCEDURE

Indications

- Aspiration or intraarticular injection
  - Suspected septic arthritis or crystal arthropathy
  - Therapeutic for control of inflammation or pain relief
  - Diagnostic to determine origin of pain
  - MR, CT, or conventional arthrography
- Ganglion cyst aspiration ± steroid injection: Pain relief
  - No pretest US predictor of content viscosity
- Peritendon therapy: Pain relief
  - Mainly to Achilles tendon, posterior tibialis, and peroneal tendons
- Plantar fasciitis injection: Pain relief
  - Helpful in cases resistant to standard treatment
- Ligament therapy: Pain relief
- Morton neuroma injection: Pain relief
- Forefoot bursal injection: Pain relief
  - Submetatarsal or intermetatarsal bursitis
- Sinus tarsi injection: Pain relief

Getting Started

- Things to check
  - Check patient identity and type and side of procedure
  - Signed informed consent
    - Achilles tendon intervention: Risk of tendon rupture
    - Plantar fascia injection: Risk of fascial rupture and heel pad atrophy
    - Superficial steroid injection: Skin depigmentation change, particularly in patients who have had multiple injections and those with darker skin
  - Discuss risk of infection, allergy, and bleeding

PROCEDURE

Joint Aspiration or Injection

- Sterile technique
- Tibiotalar joint injection
  - Anterior approach: Patient supine, probe in sagittal plane
    - Avoid anterior tibial neurovascular bundle located just medial to midline
    - Anterolateral side is preferred over anteromedial side as joint is more superficial and capacious and removed from neurovascular bundle
- Subtalar joint injection
  - Lateral approach: Anterior or posterior, needle passing deep to fibula
  - Posterior approach: Patient in prone position
    - Sweep off Achilles tendon
    - Visualize posterior recess of subtalar joint
    - Posteromedial approach preferred to avoid sural nerve
  - Other approach: Medial approach

Ganglion Cyst Aspiration

- 21-g needle advanced into ganglion
  - ≥ 19-g needed for more viscous content
- Injecting small amount of lignocaine into ganglion can loosen cyst content if nothing can be aspirated initially
- ± hyaluronidase injection before treatment
- ± steroid injection post aspiration
- ± fenestration of cyst walls with needle several times during procedure
  - Potentially allows cyst fluid to escape and reduces chance of recurrence

Achilles Tendinosis Treatment

- Patient in prone position with ankle and foot hanging over edge of bed
  - Medial axial approach to avoid sural nerve, which lies in subcutaneous tissues anterior and lateral to Achilles tendon
  - Longitudinal approach
- Stimulate healing response in refractory tendinosis
  - Dry needling: Repeated fenestration of area of localized tendinosis (80% improvement)
    - Often followed by autologous blood injection (ABI) or platelet-rich plasma (PRP) injection
  - ABI or PRP injection: ~ 3 mL injected into area of tendinosis
    - Multiple injections every 2 weeks for 6 weeks seem better than single injection
      - Also, well-vascularized tendons may respond better
  - Followed by complete rest for 3 days → light sports for 3 weeks → normal sports in 3 months
- Reduce inflammation
  - Prolotherapy (or regenerative injection therapy): Local inflammatory response → induces stem cell deposition + vascular sclerosant effect
    - 0.5-1.5 mL of hyperosmolar 15-25% dextrose water injected into tendinopathic area following up to 50 fenestrations
    - 2- to 3-week intervals up to 6x
      - Variable response
    - Practical tips for ABI, PRP, or high-volume stripping
      - Use Luer lock syringes for ABI/PRP/saline injection into tendon to guard against explosive leakage from excessive back pressure
      - Inject blood/PRP/saline using gentle pressure push/release technique to avoid excessive back pressure
  - Peritendinous injection of 20-40 mg of methylprednisolone or triamcinolone mixed with lignocaine ± bupivacaine
  - 21- or 23-g needle is used for injection
    - High frequency of relapse
- Reduce tendon and paratenon neovascularity
  - Sclerotherapy: Polidocanol (2-4 mL at 5 mg/mL) into peritendinous (not tendon) tissue near vessels
    - Good short-term results
    - Multiple treatments often necessary
  - Electrocoagulation (at 20-25 W): Needle placed against vessels (arteries or veins) entering tendon
    - Effect may be short term
  - Brisement: High-volume injection of normal saline (20-40 mL), local anesthetic solution (10 mL of 0.5% bupivacaine), and steroid (25 mg of hydrocortisone) between Achilles tendon and paratenon
    - Aim is to strip paratenon and clear adhesions
    - Mechanical effect ruptures neovasculature
    - Neural ingrowth accompanying neovessels is also potentially damaged

- Reduces pain and improves function in both short and long term

Treatment of Other Tendons and Ligaments

- Approach tailored according to particular tendon and ligament segment injected

Plantar Fasciitis Injection

- Refractory plantar fasciitis
- Patient prone with foot hanging over edge of bed
  - Medial axial approach allows local anesthetic to be injected into skin to reduce overall pain of procedure
    - Medial side selected as plantar fasciitis usually affects central band of plantar fascia insertion into medial calcaneal tuberosity
  - Plantar longitudinal approach more painful (thick plantar skin)
- 23-g needle for injection ± dry needling of plantar fascia initially
- Steroid and local anesthetic mixture then injected deep or superficial to fascia but not into fascia
  - Not possible to inject into fascia
  - Reposition needle during injection to disperse injectate over fascial surface, particularly over medial aspect of fascia close to attachment
- Short-acting (such as hydrocortisone) rather than long-acting steroid preferred in total dose of ~ 40 mg to reduce risk of heel fat pad atrophy

Morton Neuroma Injection

- Main bulk of neuroma mass lies plantar to metatarsal heads
- Dorsal (web space) or plantar approaches
  - Dorsal approach preferable
    - Aim to puncture between toes above skin fold in longitudinal plane
    - Transducer is positioned either dorsal or plantar to web space, depending on which is easier, and aligned parallel to needle (longitudinal)
    - Needle inserted into web space almost parallel to transducer
  - Plantar approach
    - Transverse or longitudinal, more painful than dorsal web space approach
- 23-g needle, steroid injection, typically 40 mg of methylprednisolone along with bupivacaine is used
- Provides several months of pain relief
- After 3 months, if good symptomatic relief from 1st injection, 2nd steroid injection can be given
- If neuroma is large (> 5 mm wide) or does not respond to steroid injection, neuroablative therapy with alcohol injection will help to relieve symptoms

Forefoot Bursal Injection

- Usually plantar approach, rarely dorsal or intermetatarsal approach
- 23-g needle is used for aspiration or steroid injection

Sinus Tarsi Injection

- Probe in coronal oblique plane spanning from anterior process of calcaneus to talar neck → move probe proximally with rotation and tilting to best see sinus tarsi
- Out-of-plane approach to insert needle to depth of subtalar joint

Findings and Reporting

- Consider reporting on these aspects
  - Initial US findings
    - Do not do any interventional procedure without 1st making full assessment of US findings and reporting clearly on these findings
    - Provide objective (quantitative or semiquantitative) criteria of baseline disease severity, e.g., plantar fascia thickness (mm) or degree of tendon/peritendinous vascularity (absent, minimal, mild, moderate, severe)
  - Report on type and amount of local anesthetic used
  - Type and size of needle used
  - Volume, color, and viscosity of joint fluid aspirated
  - Volume and type of fluid injected
  - If only friable material obtained at biopsy
  - Unexpected adverse occurrence during procedure, e.g., arterial puncture
    - Report even though no immediate complications as there is potential for late complication
  - Immediate complication (e.g., excessive bleeding)

POST PROCEDURE

Immediate Aftercare

- MR or CT arthrography within 30 min of arthrography as contrast will absorb
- Joint aspirate should be sent, where appropriate, for
  - Cell count, Gram stain, microscopy and culture (aerobic and anaerobic), and sensitivity
  - Cytology, crystal analysis (fresh specimen for crystals as they are dissolved by water, alcohol, formalin)
- Following tendon injection or fenestration
  - Complete rest for 3 days → light sports for 3 weeks → normal sports in 3 months for Achilles tendon
  - Modify rehabilitation plan according to treatment site, type, and frequency

OUTCOMES

Complications

- Most feared complication(s)
  - Infection, e.g., septic arthritis
    - Do not inject joint if overlying tissue infected
  - Steroid-induced fat pad atrophy, skin atrophy, or depigmentation
  - Tendon or fascial rupture following steroid, PRP, or ABI
- Immediate/periprocedural complication(s)
  - Bleeding, especially for those with bleeding diathesis
  - Extravasation of contrast or medication
    - Rest joint for 20 minutes following injection
  - Capsular rupture
  - Allergy to injectate
  - Injury to neurovascular bundle
- Delayed complication(s)
  - Delayed allergic reaction (uncommon)
  - Tendon or fascia rupture

SELECTED REFERENCES

1. Drakonaki EE et al: Ultrasound-guided intervention in the ankle and foot. Br J Radiol. 89(1057):20150577, 2016

Achilles Tendon Injection

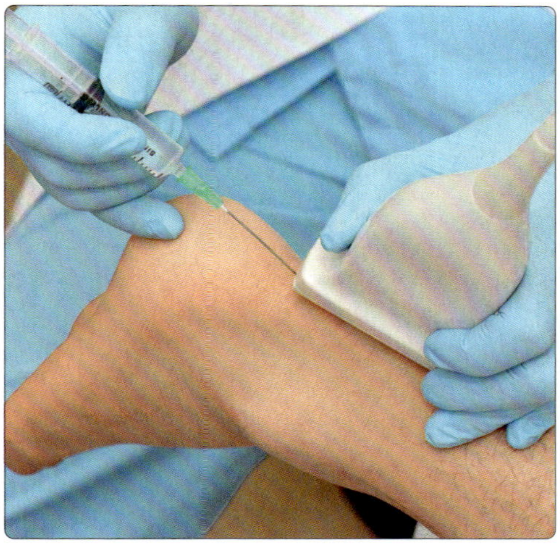

Achilles Tendon Injection

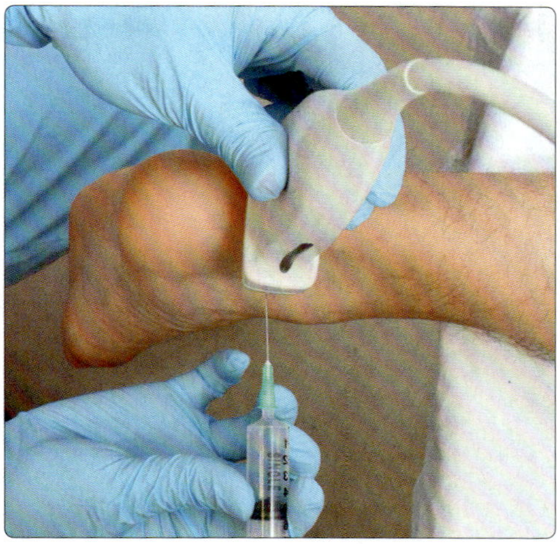

(Left) *Clinical photograph shows longitudinal approach for injection of the Achilles tendon. The patient lies prone on the examination table.* **(Right)** *Clinical photograph shows transverse approach for injection of the Achilles tendon. The needle is inserted medially to avoid the sural nerve, which runs superficial and lateral to the Achilles tendon.*

Achilles Tendon Dry Needling

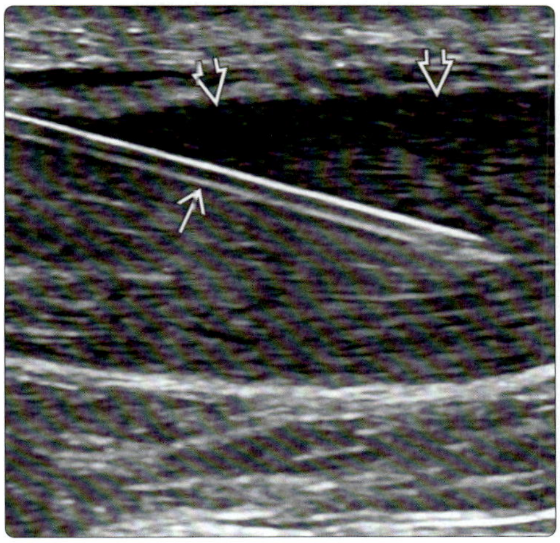

Achilles Tendon PRP Injection After Dry Needling

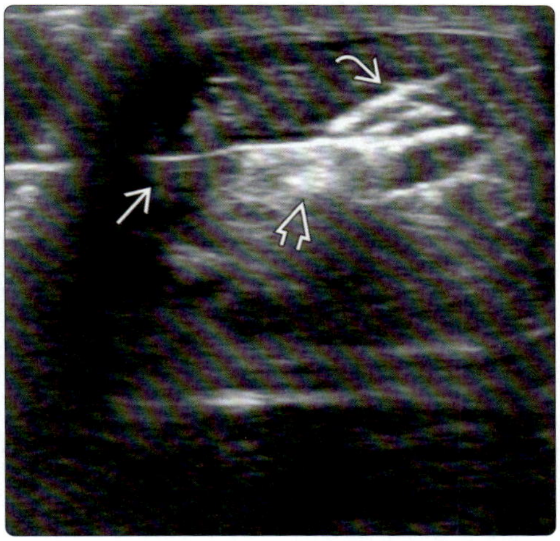

(Left) *Longitudinal US shows dry needling of Achilles tendinosis using a 21-g needle ➡. The hypoechoic rim ➡ superficially is from 1% lignocaine instilled deep to the paratenon prior to dry needling to reduce discomfort during the procedure.* **(Right)** *Transverse US shows platelet-rich plasma (PRP) ➡ injection to Achilles tendinosis using a 21-g needle ➡. The echogenic tracks ➡ are the result of dry needling, which was performed prior to the injection.*

Achilles Tendon High-Volume Injection (Brisement)

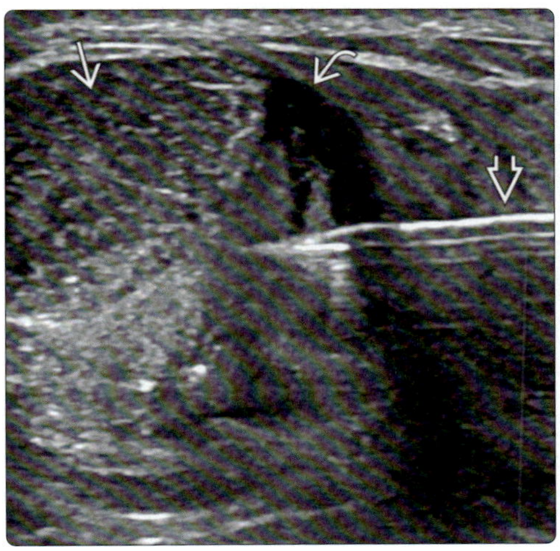

Achilles Tendon High-Volume Injection (Brisement)

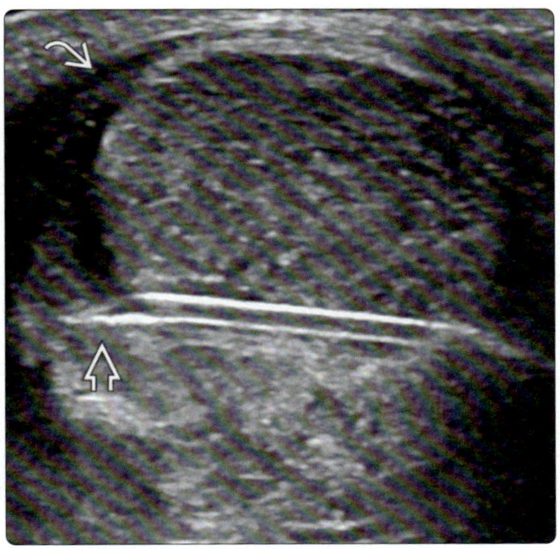

(Left) *Transverse US shows stripping ➡ of the Achilles paratenon ➡ by high-volume injection of a mixture of 10 mL of 0.5% bupivacaine and 40 mL of normal saline using a 21-g intravenous needle ➡.* **(Right)** *Transverse US of the same patient shows advancement of the needle around the tendon to ensure thorough stripping of the paratenon. The bevel ➡ has been turned to face upward to ensure the injection is deep to the paratenon ➡.*

(Left) *Clinical photograph shows dorsal web space approach for injection of Morton neuroma. Patient lies supine on the exam table. The needle is placed between the toes puncturing above the skin fold in a longitudinal plane. The transducer is placed parallel to the needle (longitudinal) either dorsal or plantar to the web space.* **(Right)** *Longitudinal US shows steroid injection of Morton neuroma ➔ via digital web space dorsally. A 23-g needle ➔ is inserted almost parallel to the probe. (Courtesy S. James, MD.)*

Morton Neuroma Injection

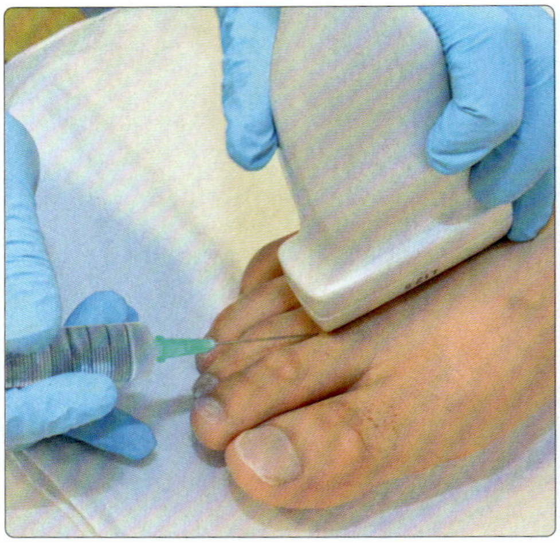

Morton Neuroma Injection

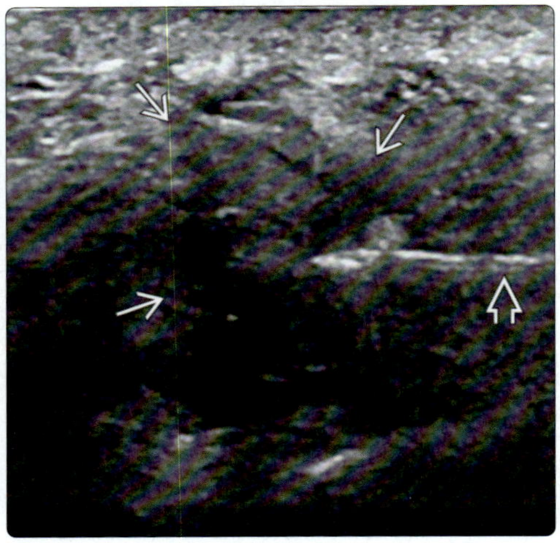

(Left) *Clinical photograph shows plantar longitudinal approach for injection of Morton neuroma. Patient lies prone on the examination table.* **(Right)** *Longitudinal US shows a large Morton neuroma ➔. Steroid (40 mg of methylprednisolone) and local anesthetic (5 mL of 0.5% bupivacaine) injection was performed longitudinally on the plantar side via a 23-g needle ➔. (Courtesy S. James, MD.)*

Morton Neuroma Injection

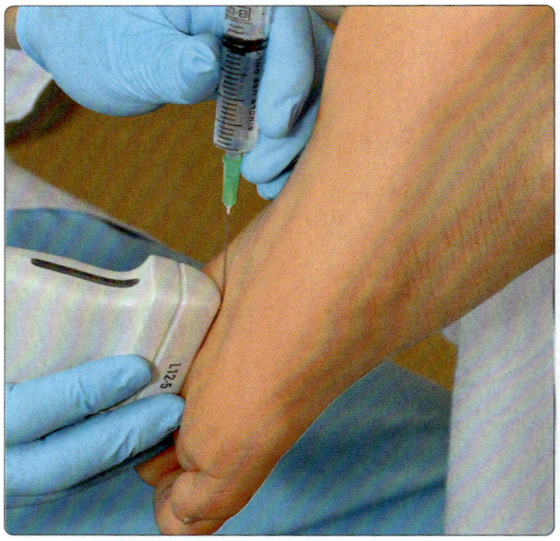

Morton Neuroma Injection

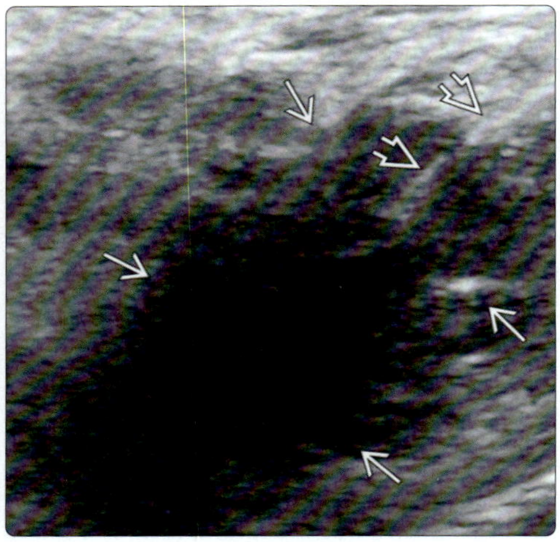

(Left) *Transverse US shows aspiration of posterior tibialis tendon ➔ sheath effusion ➔ via a 23-g intravenous needle ➔. Aspiration was performed before steroid injection. The tendon sheath is distended with synovial proliferation ➔.* **(Right)** *Transverse US of the same patient shows steroid (40 mg of methylprednisolone) injection ➔ into the tendon sheath of the posterior tibialis tendon ➔ via the same 23-g intravenous needle ➔.*

Posterior Tibialis Tendon Sheath Aspiration

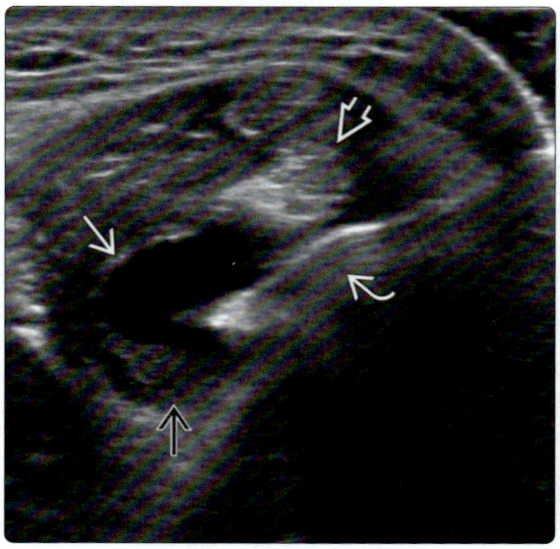

Posterior Tibialis Tendon Sheath Steroid Injection

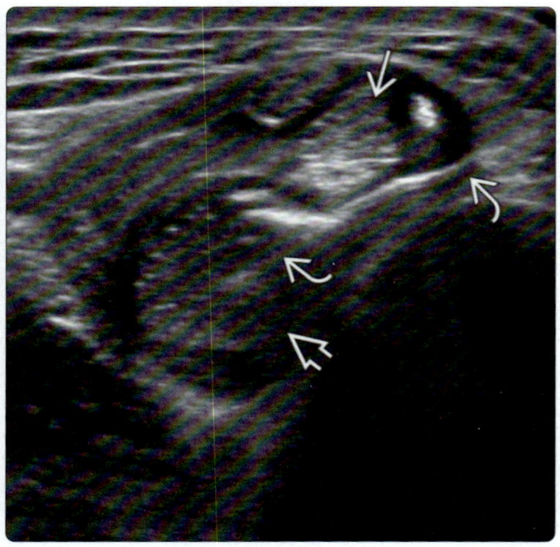

Peroneal Tendon Injection

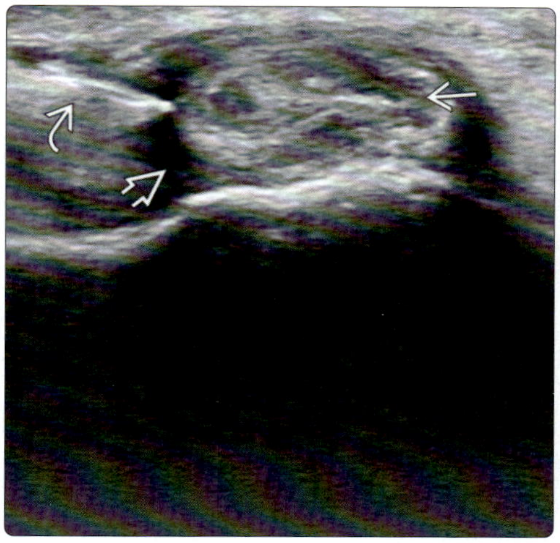

Anterior Talofibular Ligament PRP Injection

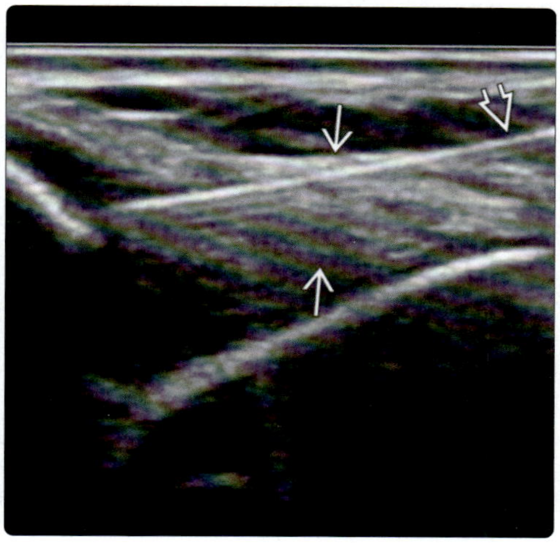

(Left) *Transverse US shows steroid injection (40 mg of methylprednisolone) to the common tendon sheath of the retromalleolar peroneal tendons via a 23-g intravenous needle. The steroid injectate will disperse proximally and distally along the tendon sheath. (Courtesy S. James, MD.)* **(Right)** *Transverse US shows chronic thickening of the anterior talofibular ligament. US-guided PRP injection is performed via a 22-g spinal needle into the ligament.*

Ankle Joint Aspiration and Injection

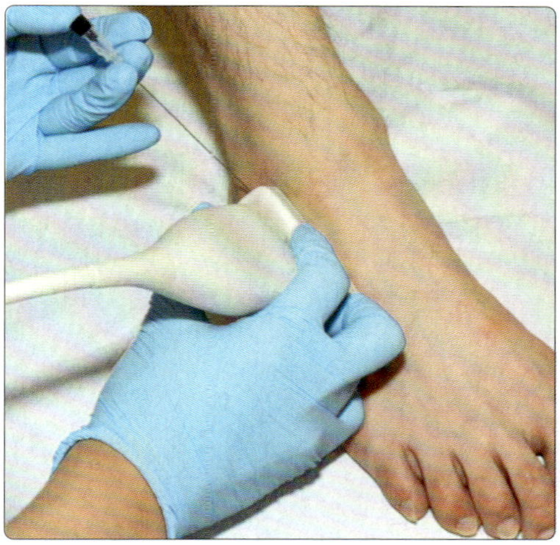

Ankle Joint Aspiration and Injection

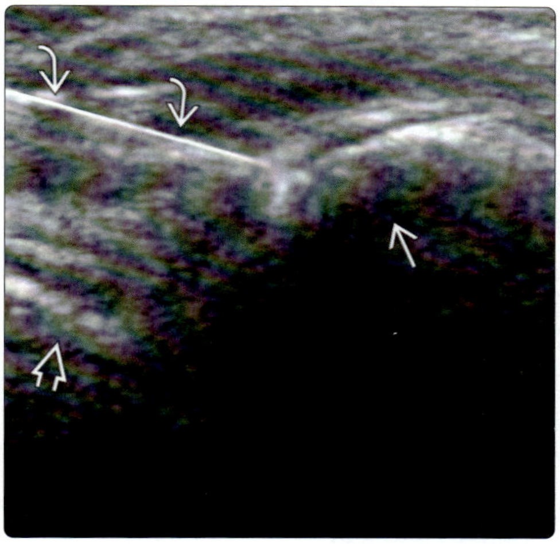

(Left) *Clinical photograph shows longitudinal anterior approach for tibiotalar joint injection or aspiration. Care must be taken to avoid the neurovascular bundle and particularly the deep peroneal nerve, which lies just lateral to the anterior tibial artery at the ankle.* **(Right)** *Longitudinal US shows US-guided injection of tibiotalar articulation. The needle as simulated is inserted between the distal tibia and talus at the talar head-neck junction region.*

Subtalar Joint Injection

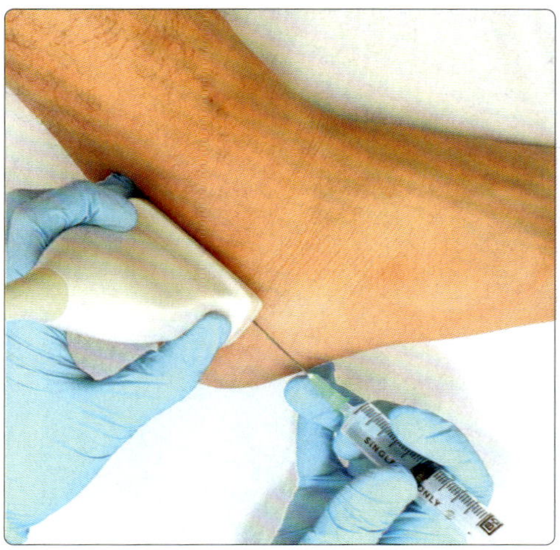

Subtalar Joint Injection

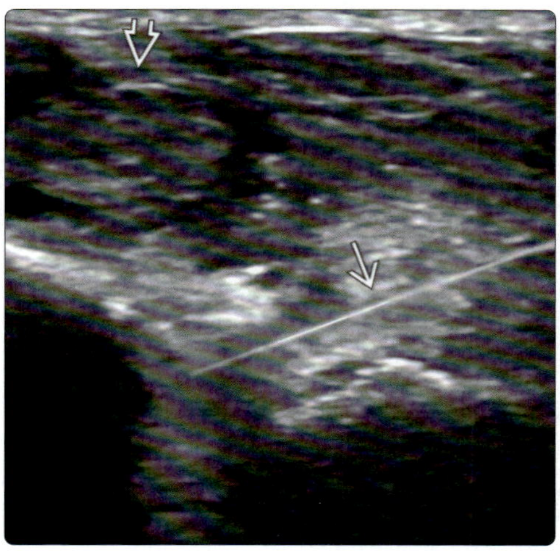

(Left) *Clinical photograph shows longitudinal lateral approach for subtalar joint injection or aspiration. Care must be taken to avoid the peroneal tendons and sural nerve, which lie adjacent to the subtalar articulation laterally. The sural nerve lies deep to the peroneal tendons.* **(Right)** *Longitudinal US shows subtalar joint injection. A simulated 23-g needle is present at the anterior edge of the subtalar joint. The peroneal tendons are seen lying on the left side.*

INDEX

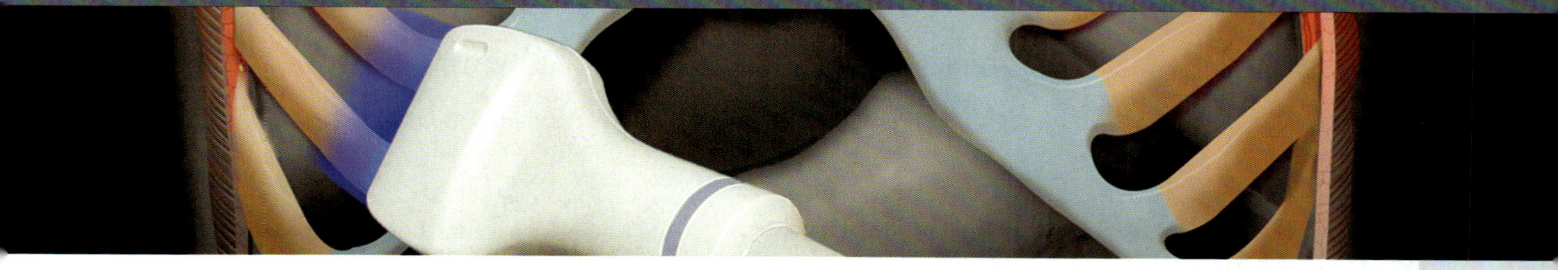

A

A1 pulley, **103**
A2 pulley, **103**
A3 pulley, **103**
A4 pulley, **103**
A5 pulley, **103**
Abdomen, **314, 326**
Abdominal obesity, gout and pseudogout with, **570**
Abdominal wall, **314–325, 326–337**
- anterior, **314, 315, 318, 327, 330**
    - groin, **338**
- anterolateral, **319, 331**
- MR, **325**
- muscles
    - anterior, **326**
    - of back in situ, **317**
    - posterior, **326**
- pathology, **337**
- posterior, **314, 316, 320, 321, 322, 323, 328, 332, 333, 334, 335**

Abdominal wall abscess, abdominal wall hernia vs., **798**
Abdominal wall hernia, **796–803**
- diagnostic checklist, **798**
- differential diagnosis, **798**
- prognosis, **798**

Abdominal wall mass, **910–915**
- abscess, **911**
- desmoid tumor, **911**
- divarication of rectus abdominis muscle, **911**
- endometrioma, **910–911**
- enlarged lymph node, **911**
- fat necrosis, **911**
- granuloma, **911**
- groin hernia, **910**
- incisional hernia, **910**
- inguinal hernia, **910**
- linea alba hernia, **910**
- lumbar hernia, **910**
- muscle tear, **911**
- paraumbilical hernia, **910**
- rectus sheath hematoma, **911**
- Spigelian hernia, **910**
- subcutaneous hematoma, **911**
- subcutaneous lipoma, **910**
- umbilical hernia, **910**
- umbilical metastasis, **911**
- umbilical-urachal sinus, **911**
- undescended testis, **911**

Abdominal wall seroma, abdominal wall hernia vs., **798**
Abdominal wall tumor, abdominal wall hernia vs., **798**
Abductor digiti minimi bursa, tarsus, **278**
Abductor digiti minimi fibers, forming lateral band, median nerve, **159**
Abductor digiti minimi muscle, **96, 103, 110, 111, 168**
- fingers, **132**
- lateral tarsus, **289**
- medial ankle, graphic, **263**
- median nerve, **159**
- sagittal foot overview, **287**
- superficial plantar muscles, **284**
- thumb, **121**

Abductor digiti minimi tendon, **91**
- fingers, **131**

Abductor hallucis insertion, superficial plantar muscles, **284**
Abductor hallucis muscle
- medial ankle, graphic, **263**
- medial tarsus, **288**
- plantar aponeurosis, **286**
- sagittal foot overview, **287**
- superficial plantar muscles, **284**

Abductor pollicis brevis muscle, **96, 98, 103, 106, 108, 109**
- median nerve, **157, 158, 159**
- thumb, **120, 121, 122, 127, 128**

Abductor pollicis longus muscle, **72, 79, 88, 90, 389**
Abductor pollicis longus tendon, **88, 90, 91, 92, 96, 111**
- fingers, **132**
- hand, **117**

Abductor pollicis muscle, **109**
- thumb, **120, 121, 123, 124, 125, 127, 128, 129**

Abductor pollicis tendon, **92**
- hand, **117**

Abscess, **629**
- abdominal wall mass vs., **911**
- chest wall lesion vs., **905**
- cystic soft tissue mass vs., **836**
- hyperechoic subcutaneous mass vs., **819**
- hypoechoic muscle mass vs., **825**
- hypoechoic subcutaneous mass vs., **813**
- low grade, hematoma/seroma vs., **526**
- muscle, muscle infarction vs., **511**
- paraarticular cystic mass vs., **889**
- soft tissue, epidermoid cyst vs., **722**
- superficial metastases, lymphoma, and melanoma vs., **740**

Accessory breast, chest wall lesion vs., **905**
Accessory cephalic vein, **86, 87**
Accessory collateral ligament, fingers, **133**
Accessory lateral collateral ligament, **68**
Accessory lateral gastrocnemius slips, **224**
Accessory medial gastrocnemius slips, **224**
Accessory muscle, chest wall lesion vs., **905**

INDEX

Accessory ossicle, tendon hyperechogenicity vs., **861**
Accessory pollicis longus muscle, **72**
Accessory soleus muscle, **224**
Accompanying artery, **164**
Accompanying vein, **164**
- anterior ankle, **262**
Acetabular labral junction, anterior hip ultrasound, **409**
Acetabular labrum, **173, 174, 178, 179**
- anterior, **174**
Acetabulum, **170, 174**
- anterior column of, **171**
- anterior rim, **171, 173, 174, 176**
- anterior surface, **173**
- medial wall of, **171**
- posterior column of, **171**
- posterior rim of, **171**
- roof, **178, 179**
Achilles tendinosis treatment, procedure, **987–988**
Achilles tendon, **224, 259**
- anterior and posterior leg nerves (graphic), **255**
- axial, **275**
- insertion, posterior ankle panorama, **273**
- lateral ankle, graphic, **269**
- lateral leg (graphic), **232**
- leg arteries
  anterior and ankle (graphic), **238**
  posterior (graphic), **237**
- longitudinal, **274**
- medial ankle, graphic, **263**
- nonrotator cuff tendinosis, **464**
- posterior ankle ultrasound, **426**
- posterior leg
  graphic, **233**
  panoramic, **234**
Acoustic-enhancement artifact (increased through transmission), **365**
Acoustic-shadowing artifact (posterior acoustic shadowing), **365**
Acromegaly, swollen nerve vs., **871**
Acromial branch, thoracoacromial artery, **45**
Acromial plexus, **45**
Acromioclavicular (AC) joint, **4–9, 11, 13, 374**
- injection
  indications, **951**
  procedure, **953**
- ligaments of, **4**
- muscle attachments of, **4**
- transverse US, **8, 9**
- US, **7**
Acromioclavicular ligament, superior and inferior, **25**
Acromion, **7, 8, 9, 17, 18, 25, 26, 27, 37, 42**
- radial nerve, **144**
Acromion process, **12, 13, 25, 28, 30**
Active tenosynovitis, chronic, **485**
Acute exudative tenosynovitis, **485**
- peritendinous mass vs., **852**
- tendon swelling vs., **864**
Acute hematoma, **525**
Acute nonexudative tenosynovitis, peritendinous mass vs., **852**
Acute osteomyelitis, **637**
- bone surface lesion vs., **883**
Acute proliferative tenosynovitis, **485**
- tendon swelling vs., **864**
Acute rib fracture, chest wall lesion vs., **904**
Acute seroma, **525**
Acute traumatic fracture, bone surface lesion vs., **883**
Adductor aponeurosis, thumb, **124, 129**
Adductor brevis muscle, **175, 180, 182–183, 187, 189**
- groin, **340, 341, 342, 343, 346**
Adductor (subsartorial or Hunter) canal, **181, 188**
Adductor digiti minimi muscle
- plantar aponeurosis, **286**
Adductor hallucis muscle, **297**
- deep plantar muscles, **285**
- oblique head, **297**
- transverse head, **297**
Adductor hiatus, **186**
- leg arteries, posterior (graphic), **237**
Adductor longus muscle, **175, 180, 182–183, 186–189, 191, 195, 196, 197, 198, 201, 203**
- groin, **339, 340, 341, 342, 343, 345, 346**
Adductor magnus muscle, **175, 186–189, 191, 197, 204, 205**
- adductor part, **189**
- gluteal muscles, **350, 351, 355**
- groin, **341, 342, 343, 345, 346**
- ischiocondylar portion of, **181, 186, 189**
- leg arteries, posterior (graphic), **237**
Adductor muscles, conjoint of, **343**
Adductor pollicis brevis muscle, **106**
- fingers, **131**
Adductor pollicis muscle, **102, 103, 106, 108, 109**
- fingers, **131, 134, 136**
- median nerve, **158, 159**
- oblique head, **107**
- osseous insertion, **107**
- transverse head, **107**
Adductor tendons and muscles, medial hip ultrasound, **409–410**
Adductor tubercle, **215**
Adenocarcinoma, hypervascular soft tissue mass vs., **847**
Adenopathy. *See* Lymph node abnormality.
Adhesive capsulitis treatment, shoulder, **952**
Adventitious bursae, **669**
Aliasing artifact, **366**
Allograft, bone, bone fracture vs., **542**
Alveolar soft part sarcoma, vascular anomaly vs., **746**
Amyloid, synovial swelling vs., **895**
Amyloidosis
- tendon hypoechogenicity vs., **857**
- tendon swelling vs., **865**
Anconeus epitrochlearis muscle, **162**
Anconeus muscle, **36, 42, 55, 56, 60, 62, 65, 71, 73**
- radial nerve, **144**
Aneurysm
- cystic soft tissue mass vs., **837**
- hematoma/seroma vs., **526**
Angioleiomyoma. *See* Vascular leiomyoma.
Angiolipoma, **713**

INDEX

Angiolymphoid hyperplasia, vascular dilatation or inflammation vs., **622**
Angiomyoma. *See* Vascular leiomyoma.
Angiosarcoma, vascular dilatation or inflammation vs., **622**
Anisotropic artifacts, **364**
- ligament injury vs., **534**
- rotator cuff/biceps tendinosis vs., **443**
Anisotropy, **11**
- tendon hypoechogenicity vs., **856**
Ankle, **258–275**
- Achilles tendon
  - axial, **275**
  - longitudinal, **274**
- anatomy imaging issues, **259**
- anterior ankle, **261**
  - graphic, **260**
- deltoid ligament, **267–268**
- fibular ligaments, **272**
- gross anatomy, **258–259**
  - ligaments, **258**
  - osseous anatomy, **258**
  - retinacula, **258**
  - tendons, **258–259**
- lateral ankle, **270–271**
  - graphic, **269**
- medial and posterior ligaments (graphic), **266**
- medial ankle, **264–265**
  - graphic, **263**
- posterior ankle, panorama, **273**
Ankle and foot procedures, **986–991**
- Achilles tendinosis treatment, procedure, **987–988**
- forefoot bursal injection
  - indications, **987**
  - procedure, **988**
- ganglion cyst aspiration
  - indications, **987**
  - procedure, **987**
- joint aspiration or injection
  - indications, **987**
  - procedure, **987**
- Morton neuroma injection
  - indication, **987**
  - procedure, **988**
- outcomes, **988**
- peritendon therapy, indication, **987**
- plantar fasciitis injection
  - indication, **987**
  - procedure, **988**
- post procedure, **968**
- preprocedure, **987**
- procedure, **987–988**
- sinus tarsi injection
  - indication, **987**
  - procedure, **988**
- tendons and ligaments, treatment
  - indication, **987**
  - procedure, **988**
Ankle (talocrural) joint, **258**
- anterior ankle, **262**
- with fluid, fibular ligaments, **272**
- posterior aspect, medial ankle, **264**
- posterior recess, fibular ligaments, **272**
Ankle ultrasound, **424–431**
- anterior, **425**
  - anterior ankle tendons, **425**
  - anterior recess ankle joint, **425**
  - deep peroneal nerve (DPN), **425**
  - extensor digitorum longus (EDL) tendon, **425**
  - extensor hallucis longus tendon, **425**
  - superior and inferior extensor retinacula, **425**
  - tibialis anterior tendon, **425**
- clinical indications for, **425**
- lateral, **426**
  - anterior talofibular ligament, **426**
  - anterior tibiofibular ligament, **426**
  - calcaneofibular ligament, **426**
  - peroneal tendons, **426**
  - superficial peroneal nerve, **426**
- medial, **425–426**
  - deltoid ligament, **426**
  - flexor digitorum tendon, **426**
  - flexor hallucis longus tendon, **426**
  - tibial nerve and tarsal tunnel, **425**
  - tibialis posterior (TP) tendon, **425–426**
- posterior, **426**
  - Achilles tendon, **426**
  - plantaris tendon, **426**
  - sural nerve, **426**
Annular ligament, **52, 53, 55, 63, 65, 68, 71**
- lateral elbow ultrasound, **382**
Annular pulley thickening, peritendinous mass vs., **852–853**
Antebrachial fascia, **167**
Antecubital fossa, transverse US, **59**
Antecubital vein, **60**
Antecubital vein, proximal forearm, **82**
Anterior abdominal wall, groin, **338**
Anterior and posterior branches, **31**
Anterior and posterior divisions, **31**
Anterior ankle, ultrasound, **425**
Anterior ankle tendons, anterior ankle ultrasound, **425**
Anterior aponeurosis nerve, **77**
Anterior aspect of hip joint, anterior hip ultrasound, **409**
Anterior axillary wall, transverse US, **33**
Anterior band, **68**
Anterior capsule, **8**
Anterior circumflex humeral artery, **37, 45**
Anterior compartment, leg muscles, **225**
Anterior (extensor) compartment, ankle, **258–259**
Anterior coronoid recess, **58**
Anterior cruciate ligament, anterior knee ultrasound, **417**
Anterior elbow ultrasound, **381**
- anterior aspect of elbow joint, **381**
- brachialis, **381**
- distal biceps tendon, **381**
- median nerve, **381**
- radial nerve, **381**
Anterior fat pad, **54, 57**
Anterior femoral cortex, **183, 183–184, 184**
Anterior fibular cortex, peroneal nerve, **257**
Anterior hip ultrasound, **409**

INDEX

Anterior intermuscular septum, panoramic anterolateral leg, **227**
Anterior interosseous artery, **80, 97**
- forearm arteries, **81**
Anterior interosseous artery, hand, **115**
Anterior interosseous nerve, **97, 304**
- median nerve, **151, 152**
Anterior interosseous neurovascular bundle, **74**
- median nerve, **155**
Anterior interosseous vein, **97**
Anterior knee, ultrasound, **417**
Anterior layer of rectus, **327**
Anterior longitudinal ligament, **316**
Anterior longitudinal spinal ligament, **328**
Anterior public cortex, **197**
Anterior recess ankle joint, anterior ankle ultrasound, **425**
Anterior recurrent tibial artery
- anterior and posterior leg nerves (graphic), **255**
- leg arteries, anterior and ankle (graphic), **238**
Anterior recurrent tibial nerve, anterior and posterior leg nerves (graphic), **255**
Anterior superior iliac spine (ASIS), **182, 186, 338, 348**
- groin muscles, **340, 346, 347**
    lateral groin, **346**
- region, anterior hip ultrasound, **409**
Anterior suprapatellar fat pad, **209–210**
Anterior talocalcaneal joint, deltoid ligament, **267**
Anterior talofibular ligament, **258**
- fibular ligaments, **272**
- lateral ankle
    graphic, **269**
    ultrasound, **426**
Anterior tibial artery, **236, 291**
- anterior and posterior leg nerves (graphic), **255**
- anterior ankle, **261**
- anterior compartment of leg, **228, 229**
- anterolateral leg, panoramic, **227**
- lateral compartment of leg, **230**
- leg arteries
    anterior and ankle (graphic), **238**
    posterior (graphic), **237**
- longitudinal anterior tibial vessels, **247**
- spectral Doppler of leg arteries, **248**
- tarsus, **278**
- transverse anterior tibial vessels, **246**
Anterior tibial cortex, anterior ankle, **261**
Anterior tibial vein
- anterior ankle, **261**
- anterior compartment of leg, **228, 229**
- anterolateral leg, panoramic, **227**
- lateral compartment of leg, **230**
- leg veins
    graphic, **239**
    spectral Doppler of, **249**
- longitudinal anterior tibial vessels, **247**
- transverse anterior tibial vessels, **246**
Anterior tibial vessels
- leg vessels
    longitudinal, **247**
    transverse, **246**
- peroneal nerve, **257**
Anterior tibiofibular ligament, **258**
- anterior ankle, **261**
- lateral ankle
    graphic, **269**
    ultrasound, **426**
Anterior tibionavicular ligament, deltoid ligament, **268**
Anterior tibiotalar ligament
- deltoid ligament, **268**
- superficial deltoid, medial and posterior ligaments, **266**
Anterior ulnar recurrent artery, forearm arteries, **81**
Anterolateral calcaneal cortex, dorsal foot, **280**
Anterolateral navicular surface, dorsal foot, **280, 281**
Anteromedial bare area tibia
- anterolateral leg (graphic), **226**
- lateral compartment of leg, **230**
Anteromedial navicular surface, dorsal foot, **280**
Anteromedial talar head, dorsal foot, **280**
- surface, **280, 281**
AP. *See* Abductor pollicis.
APB. *See* Abductor pollicis brevis.
Apex femoral triangle, **195**
- femoral triangle, groin, **339**
APL. *See* Abductor pollicis longus.
Aponeuroses, **315, 327**
Apophysitis, traction, ligament injury vs., **534**
Approach to musculoskeletal ultrasound. *See* Musculoskeletal ultrasound, approach to.
Arcade of Fröhse, radial nerve, **149**
Arches of foot, tarsus, **276**
Arcuate artery, **290, 291**
Arm, **36–43**
- anterior aspect, transverse US, **40**
- anterior compartment of, **36**
- anterolateral aspect, transverse US, **41**
- anteromedial aspect, transverse US, **39**
- axial graphic, **38**
- fascia, **36**
- medial cutaneous nerve of, **31**
- median nerve, **150, 151**
- muscles, **51**
- posterior, **42**
- posterior aspect, transverse US, **43**
- posterior compartment of, **36**
- radial nerve, **142, 145**
    posterior, **144**
- ulnar nerve, **160, 161**
    transverse US, **163**
- upper lateral cutaneous nerve of, **31**
Arm vessels, **44–51**
- arteries, **45**
- axial, right arm arteries and veins, **46**
- brachial artery, color Doppler US, **48**
- brachial vein, longitudinal color Doppler US, **49**
- cephalic vein, transverse and longitudinal US, **51**
- longitudinal US, dynamic, **50**
- proximal arm, transverse US, **47**
Arteries
- arm, **45**
- foot, **290**
- forearm, **81**
- hand, **114**

INDEX

- ribs and intercostal space, **310**
- tarsus, **278**

Arthritis
- infective, joint effusion vs., **900, 901**
- inflammatory, osteoarthritis vs., **552**
- rheumatoid, gout and pseudogout vs., **570**
- septic
  gout and pseudogout vs., **570**
  hemarthrosis and lipohemarthrosis vs., **660**

Arthrography, shoulder, **951–952**
Arthropathies
- crystal, joint effusion vs., **900–901**
- hemophilic, synovial tumor vs., **696**
- inflammatory, joint effusion vs., **900**
- inflammatory arthritis, **558–567**
- osteoarthritis, **550–557**

Articular and paraarticular masses, Baker cyst, paraarticular cystic mass vs., **888**
Articular capsule, knee, **206**
Articular cartilage, **15, 17, 23, 24, 57, 61, 65, 66, 99, 100, 171**
- fingers, **133**

Articular disc, **5, 95, 100**
Ascending branch, deep brachial artery, **45**
Aseptic myonecrosis. *See* Muscle infarction.
ASIS. *See* Anterior superior iliac spine.
Attenuation error artifacts, **365**
- acoustic-enhancement artifact (increased through transmission), **365**
- acoustic-shadowing artifact (posterior acoustic shadowing), **365**

Augmentation, spectral Doppler of leg veins, **249**
Augmented flow, spectral Doppler of leg veins, **249**
AV pulley, **103**
Avulsion, **533**
Avulsion fracture of medial humeral condyle, ligament injury vs., **534**
Avulsion injury of tendon, tendon hyperechogenicity vs., **861**
Axilla, **28–35**
- anatomy relationships
  quadrilateral space, **28**
  triangular space, **28**
- anterior axillary wall, transverse US, **33**
- deeper structures, **30**
- muscles, **29**
- posterior axillary wall, transverse US, **35**
- quadrilateral space, transverse US, **34**
- relationship of neurovascular structures, **31**
- transverse US, **32**

Axillary artery, **28, 31, 32, 33, 45**
Axillary nerve, **12, 28, 42, 169, 302, 303**
- in quadrilateral space, **34**
- relationship of neurovascular structures, **31**
- trunk of, **31**

Axillary nerve, radial nerve, **143, 144**
Axillary neurovascular bundle, **30**
Axillary recess, **13**
Axillary vein, **28, 31, 32, 33**
Axonotmesis, **583**
Azygos vein, **310**

B

Bacteroides, postoperative infection, **654**
Baker cyst, **662–667**
- associated abnormalities, **664**
- cystic soft tissue mass vs., **836**
- diagnostic checklist, **664**
- differential diagnosis, **663–664**
- paraarticular cystic mass vs., **888**
- parameniscal cyst vs., **692**
- prognosis, **664**

Base, axilla, **28**
Basilic vein, **37, 38, 54, 80, 86, 87**
- arm, **44, 46, 47, 48**
- hand, **115**
- proximal forearm, **82**

Beam characteristic artifacts, **364**
- anisotropic artifacts
  ligament injury vs., **534**
  rotator cuff/biceps tendinosis vs., **443**
- anisotropy, **364**
- beam width, **364**
- off-axis lobe, **364**
- slice thickness, **364**

Benign bone tumor, bone surface lesion vs., **882**
Benign peripheral nerve sheath tumor, vascular leiomyoma vs., **735**
Benign soft tissue tumor, soft tissue sarcoma vs., **768**
Bible bump. *See* Ganglion cyst.
Biceps aponeurosis, **55, 67, 77, 78**
- median nerve, **152**

Biceps brachii muscle, **48, 51, 163**
- median nerve, **151, 153**
- radial nerve, **145, 149**

Biceps femoris muscle, **180, 190, 195, 197, 203, 204, 208, 218–219, 221–222**
- gluteal muscles, **350, 351**
- lateral leg (graphic), **232**
- long head, **185, 191, 204, 207, 220, 346**
- posterior leg (graphic), **233**
- short head, **185–186, 191, 207, 220**

Biceps femoris tendon, **221, 357**
- anterior and posterior leg nerves (graphic), **255**
- lateral knee ultrasound, **418**
- lateral leg (graphic), **232**
- leg arteries, posterior (graphic), **237**

Biceps muscle, **38, 47, 51, 54, 61, 62, 73, 77, 304**
- long head, **12, 36, 37, 39, 40, 41**
- median nerve, **152**
- short head, **30, 32, 36, 37, 40, 41**

Biceps tendinosis, rotator cuff/biceps tendon tear vs., **456**
Biceps tendon, **14, 19, 20, 21, 37, 52, 54, 55, 56, 57, 58, 59, 67, 77, 78**
- entering shoulder joint, **20**
- long head, **10, 11, 12, 18, 30, 373**
  ultrasound imaging, **7**
- median nerve, **152**
- radial nerve, **149**

INDEX

- sheath, fluid in, **21**
- short head, **12**
  acromioclavicular joint, **7**
- transverse US, **60**

Biceps tendon tear, **454–461**
- associated abnormalities, **456**
- complete, **455**
- diagnostic checklist, **456**
- differential diagnosis, **456**
- full thickness, **455**
- massive, **455**
- partial, **455**
- partial thickness, **455**
- prognosis, **456**

Bicipital aponeurosis
- median nerve, **151, 152**
- radial nerve, **149**

Bicipital fascia, **36**
Bicipital groove, **19, 25**
- lateral lip of, **19, 21**
- medial lip of, **19, 21**
- upper end of, **19**

Bicipital tendon, **218**
Bicipitoradial bursa, **52**
Bifurcate ligament
- dorsal foot, **280**
  graphic, **279**
- lateral ankle, graphic, **269**
- tarsus, **277**

Biopsy
- bone tumor, **924–931**
  outcomes, **926**
  preprocedure, **925**
  procedure, **925–926**
- soft tissue tumor, **918–923**
  outcomes, **920**
  post procedure, **920**
  preprocedure, **919**
  procedure, **919–920**

Bleeding, bone surface lesion vs., **882**
Blooming artifact ("color bleed"), **366**
Bone fracture, **540–549**
- diagnostic checklist, **542**
- differential diagnosis, **542**
- prognosis, **542**
- staging, grading, & classification, **542**

Bone infection, **636–645**
- diagnostic checklist, **638**
- differential diagnosis, **638**
- prognosis, **638**

Bone sarcoma, bone infection vs., **638**
Bone surface lesion, **882–887**
- differential diagnosis, **882**

Bone tumor, **774–785**
- diagnostic checklist, **776**
- differential diagnosis, **776**
- prognosis, **776**

Bone tumor biopsy, **924–931**
- outcomes, **926**
- preprocedure, **925**
- procedure, **925–926**

Bone tumor recurrence, bone surface lesion vs., **882–883**
Bone-allograft, bone fracture vs., **542**
Bone-prosthetic junction, bone fracture vs., **542**
Bony acetabulum, **174**
Bony anatomy, tarsus, **276–277**
Bowel gas with posterior shadow, **342**
Brachial artery, **32, 37, 38, 39, 54, 55, 56, 58, 66, 77, 80, 81, 163**
- arm, **44, 45, 46, 47, 48**
  branches of, **44**
  color Doppler US, **48**
  Doppler gate in center of, **48**
  spectral trace of, **48**
- deep branch of, **81**
- median nerve, **152, 153, 154, 155**
- proximal forearm, **82**

Brachial fascia, **36**
Brachial plexus, **302–309**
- axilla, **28**
  relations of neurovascular structures, **31**
- axillary area, **302**
- branches, **33**
- cords, **31, 32, 33**
- divisions, **303**
- extraforaminal area, **302**
- infraclavicular area, **302**
- interscalene area, **302**
- lateral cord, **169, 303**
- longitudinal US, **306, 307**
- lower trunk, **303**
- medial cord, **161, 169, 303**
- median nerve
  lateral cord of, **151**
  medial cord of, **151**
- middle trunk, **303**
- MR and US, **308**
- peripheral nerves, **304**
- posterior cord, **169, 303**
- supraclavicular area, **302**
- transverse US, **305, 306, 307, 309**
- trunks, **31, 303**
- upper trunk, **303**

Brachial vein, **32, 37, 38, 39, 40, 59, 163**
- arm, **44, 46, 47, 48, 49, 50**
  Doppler gate in, **50**
  Doppler in, **49**
  longitudinal color Doppler US, **49**
  spectral flow in, **49, 50**
- median nerve, **153**
- radial nerve, **145**

Brachialis muscle, **36, 38, 39, 40, 41, 43, 48, 49, 50, 51, 54, 55, 57, 58, 59, 60, 61, 62, 65, 66, 73, 77, 304**
- anterior elbow ultrasound, **381**
- median nerve, **153, 154**
- radial nerve, **145, 146, 147**

Brachialis tendon, **55, 67**
Brachioradialis muscle, **37, 38, 41, 42, 54, 55, 56, 58, 59, 60, 61, 65, 74, 75, 76, 77**
- proximal forearm, **82**
- radial nerve, **144, 146, 147, 148**

Brachioradialis tendon, **76**

Branchialis tendon, **78**
Brodie abscess, **637**
Bursae, **669**
- cystic collection adjacent, bursitis vs., **670**
- tarsus, **278**

Bursal distension
- cystic soft tissue mass vs., **836**
- paraarticular cystic mass vs., **888–889**

Bursitis, **668–679**
- diagnostic checklist, **670**
- differential diagnosis, **670**
- parameniscal cyst vs., **692**
- peritendinous mass vs., **853**
- prognosis, **670**
- rotator cuff/biceps tendon tear vs., **456**

Bursopathy. *See* Bursitis.

C

C1 pulley, **103**
C2 pulley, **103**
C3 pulley, **103**
C5, **31, 303**
- spinal nerve, **169**

C6, **31, 303**
- spinal nerve, **169**

C7, **31, 303**
- spinal nerve, **169**

C8, **31, 303**
- spinal nerve, **169**
- ventral rami, **161**

Calcaneal fibular ligament, lateral ankle, **270, 271**
Calcaneal spur syndrome. *See* Plantar fasciitis and fibromatosis.
Calcaneal stress, plantar fasciitis and fibromatosis vs., **708**
Calcaneal tuberosity, Achilles tendon
- axial, **275**
- longitudinal, **274**

Calcaneocuboid ligament, dorsal foot ultrasound, **433**
Calcaneofibular ligament, **258**
- fibular ligaments, **272**
- lateral ankle
  - graphic, **269**
  - ultrasound, **426**
- medial and posterior ligaments, **266**

Calcaneus
- anterior tubercle
  - plantar aponeurosis, **286**
- dorsal foot, **282, 283**
- lateral ankle, graphic, **269**
- lateral process
  - plantar aponeurosis, **286**
  - sagittal foot overview, **287**
- lateral tarsus, **289**
- medial and posterior ligaments, **266**
- medial process
  - plantar aponeurosis, **286**
  - sagittal foot overview, **287**
- medial tarsus, **288**

Calcific myonecrosis, calcified soft tissue mass vs., **843**
Calcific tendinitis
- rotator cuff/biceps tendon tear vs., **456**
- tendon hyperechogenicity vs., **860**
- therapy
  - indications, **951**
  - procedure, **952–953**

Calcific tendinosis, rotator cuff/biceps tendinosis vs., **444**
Calcification within tendinosis, tendon hyperechogenicity vs., **860**
Calcified soft tissue mass, **842–845**
- differential diagnosis, **842**

Calcifying aponeurotic fibroma, calcified soft tissue mass vs., **843**
Calcifying epithelioma of Malherbe. *See* Pilomatricoma.
Capitate, **89, 95, 97, 98, 113**
- median nerve, **157, 158**

Capitohamate ligament, **101**
Capsular ligaments, **13**
- sternoclavicular joint, **4**

Capsule
- fingers, **137**
- shoulder, **18**

Carpal bones, wrist compartments, **89**
Carpal tunnel, **88, 390**
- hand, **116**
- median nerve, **150, 157**
- roof, **107**
- tendons outside, **390**

Carpal tunnel syndrome, **150, 602–609**
- diagnostic checklist, **604**
- differential diagnosis, **604**
- indications, **965**
- procedures, **966**
- prognosis, **604**
- staging, grading, & classification, **604**

Carpometacarpal compartment, 1st, **89**
Carpometacarpal joint
- 1st, **89**
- thumb
  - ultrasound imaging, **122**
  - volar hand ultrasound, **397**

Carpometacarpal ligament, **101**
CDH (congenital dysplasia of hip). *See* Developmental hip dysplasia.
Cellulitis, **629**
- hyperechoic subcutaneous mass vs., **818**

Central plantar aponeurosis, plantar aponeurosis, **286**
Central plantar compartment, tarsus, **277**
Central tendon (of diaphragm), **316, 328**
Cephalic vein, **30, 37, 38, 54, 55, 58, 59, 80, 86, 87, 92, 93**
- arm, **44, 46, 51**
  - transverse and longitudinal US, **51**
- proximal forearm, **82**
- transverse US, **83**

Cephalic vein, hand, **115, 117**
Cervical ligament, lateral ankle, graphic, **269**
CETO. *See* Common extensor tendon origin.
CFTO. *See* Common flexor tendon origin.
Charcot-Marie-Tooth disease, swollen nerve vs., **871**
Chest wall anatomy, **310**

INDEX

Chest wall lesion, **904–909**
- abscess, **905**
- accessory breast, **905**
- accessory muscle, **905**
- acute rib fracture, **904**
- costochondritis, **905**
- differential diagnosis, **904**
- elastofibroma dorsi, **904**
- fibroma, **905**
- gynecomastia, **905**
- healing rib fracture, **904**
- hematoma, **904**
- lipoma, **904**
- lymphatic malformation, **905**
- male breast carcinoma, **905**
- muscle metastases, **904**
- prominent or floating rib, **905**
- prominent xiphoid process, **905**
- rib metastases, **904**
- sternoclavicular joint arthropathy, **905**
- stress rib fracture, **905**
- thyroglossal duct cyst, **905**
- vascular malformation, **905**

Chondral injury, focal, osteoarthritis vs., **552**
Chondroid lipoma, **713**
Chondrosarcoma, extraskeletal myxoid, paraarticular cystic mass vs., **889**
Chronic active tenosynovitis, **485**
- peritendinous mass vs., **852**

Chronic hematoma, **525**
Chronic inactive tenosynovitis, **485**
- peritendinous mass vs., **852**

Chronic inflammatory demyelinating polyneuropathy (CIDP), swollen nerve vs., **870**
Chronic joint deformity, nonrotator cuff tendon tears associated with, **474**
Chronic osteomyelitis, **637**
- bone surface lesion vs., **883**

Chronic seroma, **525**
Chronic tenosynovitis (active), tendon swelling vs., **864–865**
Chronic tenosynovitis (inactive), tendon swelling vs., **865**
Circumflex femoral artery, **193**
- descending branch lateral, **193**

Circumflex scapular artery, **37, 45**
Circumflex vessels, **200**
Clavicle, **5, 25**
- acromioclavicular joint, **8, 9**
- distal, acromioclavicular joint, **7**
- lateral end of, acromioclavicular joint, **7**
- medial end of, **5, 6**

Clavicular branch, thoracoacromial artery, **45**
Cloaca, **637**
Collateral branches, **311**
Collateral ligaments
- fingers, **141**
- injury, elbow epicondylitis vs., **496**

Collection, paraarticular cystic mass vs., **889**
Common carpometacarpal compartment, **89**
Common digital tendon sheath, **103**
- fingers, **131**
- thumb, **121**

Common extensor origin, **64**
- longitudinal US, **63**

Common extensor tendon, **52, 53, 55, 63, 73**
- transverse US, **64**

Common extensor tendon origin (CETO), **62, 63, 65**
- lateral elbow ultrasound, **382**

Common femoral artery, **173, 174, 175, 187, 193, 196, 197, 198, 199, 202, 203**
- groin, **339, 341, 344, 345, 347**
  femoral triangle, groin, **341**

Common femoral vein, **173, 174, 175, 184, 187, 189, 194, 197, 200, 201, 202, 203**
- groin, **339, 341, 344, 345, 347**

Common flexor muscle, median nerve, **152**
Common flexor origin
- cubital tunnel, **162**
- longitudinal US, **69**
- transverse US, **70**

Common flexor sheath, **102, 103**
- fingers, **131, 138**

Common flexor tendon, **52, 53, 54, 55, 68, 69, 77, 78, 162**
- median nerve, **152**

Common flexor tendon origin (CFTO), **69, 70**
- medial elbow ultrasound, **381–382**
- tendinosis with peritendinitis, cubital tunnel syndrome vs., **612**

Common iliac artery, **193**
Common iliac vein, **194**
Common interosseous artery, **80**
- forearm arteries, **81**

Common palmar digital artery, hand, **116, 118**
Common peroneal artery, anterior and posterior leg nerves (graphic), **255**
Common peroneal nerve, **185, 192, 195, 219, 254**
- anterior and posterior leg nerves (graphic), **255**
- lateral compartment of leg, **231**
- peroneal nerve, **257**

Common proper digital arteries, hand, **119**
Compartment tendon
- 1st, **92**
- 2nd, **92**

Compartments, tarsus, **277**
Competent perforator, great saphenous veins, **253**
Complete tear
- nonrotator cuff tendon tears, **473**
- rotator cuff/biceps tendon, **455**

Compressed gastrocnemius veins, **251**
Compressed soleal vein, **250**
Condylar cartilage, lateral femoral, **221, 222**
Congenital dysplasia of hip. *See* Developmental hip dysplasia.
Conjoined tendon, **111**
- fingers, **132**

Coracoacromial ligament, **7, 10, 11, 12, 25, 27**
Coracobrachialis muscle, **32, 33, 36, 37, 39, 40, 41, 46, 47, 163**
- radial nerve, **145**
- short head, **30**

Coracoclavicular ligaments, **4, 7**
Coracohumeral ligament, **7, 14, 18, 25**

INDEX

Coracoid process, **11, 12, 21, 27, 37**
- acromioclavicular joint, **7**
Coronoid fossa, **66**
Coronoid process **69**
Cortical irregularity, physiological, bone fracture vs., **542**
Cortical tug lesion, bone surface lesion vs., **883**
Costal cartilage, **311**
- 1st, **5**
Costochondritis, chest wall lesion vs., **905**
Costoclavicular ligaments, **4, 5, 6**
Cruciate ligaments, knee, **207**
- anterior, **208**
Crutch syndrome. *See* Nerve injury.
Crystal arthropathy
- joint effusion vs., **900–901**
- joint infection vs., **648**
Crystal deposition disease
- inflammatory arthritis vs., **560**
- osteoarthritis vs., **552**
Crystalline arthropathy. *See* Gout and pseudogout.
CTS. *See* Carpal tunnel syndrome.
Cubital fossa, median nerve, **152**
Cubital retinaculum, **54, 68, 162**
Cubital tunnel
- medial elbow ultrasound, **382**
- ulnar nerve, **160, 162**
    transverse US, **164**
Cubital tunnel syndrome, **610–613**
- diagnostic checklist, **612**
- differential diagnosis, **612**
- elbow epicondylitis vs., **496**
Cuboid bone, **276**
- dorsal foot, **282, 283**
- lateral ankle, **271**
    graphic, **269**
- lateral tarsus, **289**
- medial tarsus **288**
- sagittal foot overview, **287**
Cuboid tuberosity
- sagittal foot overview, **287**
Cuboideonavicular ligament, dorsal, dorsal foot, **282**
Cuneiform bones, **276**
Cuneiform-cuboid ligament, dorsal foot (graphic), **279**
Cutaneous branches, **31**
CUTS. *See* Cubital tunnel syndrome.
Cystic lesion, paratendinous, tenosynovitis vs., **486**
Cystic metastases, paraarticular cystic mass vs., **889**
Cystic paraarticular tumor, Baker cyst vs., **664**
Cystic schwannoma
- ganglion cyst vs., **682**
- paraarticular cystic mass vs., **889**
Cystic soft tissue mass, **836–841**
- differential diagnosis, **836**
Cystic-looking soft tissue tumors, paraarticular cystic mass vs., **889**

D

DDH. *See* Developmental hip dysplasia.
de Quervain disease injection or tenotomy
- indications, **965**
- procedures, **965–966**
Deep brachial artery, **12, 28, 37, 45, 46**
Deep circumflex artery, **341**
Deep circumflex iliac artery, **193**
Deep circumflex iliac vein, **194**
Deep femoral artery, **192**
Deep femoral vein, **194**
Deep gluteal space, hip ultrasound, **410**
Deep inguinal lymph nodes, **338**
Deep inguinal ring, **401**
- groin, **338**
Deep leg veins, **236**
Deep palmar arch, **109**
- hand, **114, 119**
    radial branch, **115**
    ulnar branch, **115**
Deep palmar arch artery, thumb, **127, 128**
Deep palmar artery, **108**
Deep peroneal nerve, **254, 291**
- anterior and posterior leg nerves (graphic), **255**
- anterior ankle, **261**
    graphic, **260**
    ultrasound, **425**
- anterior compartment of leg, **228**
- anterolateral leg, panoramic, **227**
- lateral compartment of leg, **230**
- peroneal nerve, **257**
- tarsus, **277**
    dorsal foot (graphic), **277**
Deep posterior (flexor) compartment, ankle, **259**
Deep transverse metacarpal ligament, **103**
- fingers, **131**
- thumb, **121**
Deep venous thrombosis
- muscle infarction vs., **512**
- postoperative infection vs., **654**
Defocusing artifact (edge-shadowing artifact), **365**
Delayed-onset muscle soreness. *See* Muscle injury.
Deltoid branch, thoracoacromial artery, **45**
Deltoid ligament
- ankle, **258, 267**
    deep layer, **267, 268**
    medial ankle, **265**
    superficial layer, **267, 268**
- medial ankle ultrasound, **426**
Deltoid muscle, **7, 12, 13, 14, 15, 16, 17, 18, 19, 20, 21, 22, 23, 24, 26, 27, 28, 30, 31, 35, 37, 41, 42**
- anterior belly, **30**
- middle belly, **30**
- posterior belly, **30**
- radial nerve, **144, 145**
Dermatofibrosarcoma protuberans, **730–733**
- differential diagnosis, **731**

INDEX

- prognosis, **731**
Dermatomyositis
- calcified soft tissue mass vs., **843**
- fat injury vs., **504**
Dermoid/implantation dermoid, cystic soft tissue mass vs., **836**
Descending genicular artery, **193**
Desmoid tumor, abdominal wall mass vs., **911**
Developmental hip dysplasia, **576–581**
- diagnostic checklist, **578**
- differential diagnosis, **578**
- prognosis, **578**
DFSP. *See* Dermatofibrosarcoma protuberans.
Diabetes mellitus
- gout and pseudogout with, **570**
- muscle infarction with, **512**
Diabetic muscle infarction. *See also* Muscle infarction.
- hypervascular soft tissue mass vs., **847**
- hypoechoic muscle mass vs., **825**
Diabetic myonecrosis. *See* Muscle infarction.
Diastasis recti, abdominal wall hernia vs., **798**
Digital arteries, **293**
- fingers, **136, 137, 140**
- hand, **118**
  common palmar digital artery, **116, 118**
  common proper digital arteries, **119**
  proper digital arteries, **119**
Digital mucous cyst. *See* Ganglion cyst.
Digital nerve, fingers, **141**
Digital vein
- fingers, **136, 140**
- hand, **118**
Direct inguinal hernia. *See also* Groin hernia.
- groin hernia, **805**
Directional ambiguity/indeterminate flow direction artifact, **366**
Distal biceps tendon, anterior elbow ultrasound, **381**
Distal carpal row, **88**
Distal humerus, **54**
Distal iliotibial band, lateral knee ultrasound, **418**
Distal interphalangeal joint, fingers, **135, 139**
Distal phalanx
- fingers, **139**
- thumb, **125, 126**
Distal radial fracture, nonrotator cuff tendon tears associated with, **474**
Distal radioulnar compartment, **89**
Distal radioulnar joint, **89**
Distal radius, **88**
Distal tibia
- anterior cortex, anterior ankle, **262**
- posteromedial angle, medial ankle, **264**
Distal tibiofibular joint, **258**
Distal ulna, **88**
Distended common femoral vein, **200**
Divarication of rectus abdominis muscles (diastasis recti), abdominal wall hernia vs., **798**
DOMS (delayed-onset muscle soreness). *See* Muscle injury.
Doppler imaging, artifacts in, **365–366**
- aliasing artifact, **366**
- blooming ("color bleed"), **366**
- directional ambiguity/indeterminate flow direction artifact, **366**
- edge, **366**
- flash, **366**
- improper Doppler angle with no-flow artifact, **366**
- mirror (or ghost) image, **366**
- motion artifact, **366**
- noise, **365**
- pseudoflow, **366**
- range ambiguity, **366**
- transducer pressure, **365**
- twinkling, **366**
Dorsal calcaneocuboid ligament, lateral ankle, graphic, **269**
Dorsal carpal arch, hand
- radial branch, **115**
- ulnar branch, **115**
Dorsal compartment, tarsus, **277**
Dorsal cuboideonavicular ligament, dorsal foot, **282**
Dorsal digital nerves, 1st toe, **291**
Dorsal digital vein, **87**
Dorsal digital vessels, thumb, **126**
Dorsal foot
- tarsus, **280–283**
  graphic, **279**
- ultrasound, **433**
  1st metatarsophalangeal joint, **433**
  calcaneocuboid ligament, **433**
  dorsal muscles, **433**
  dorsal talonavicular ligament, **433**
  extensor tendons, **433**
  Lisfranc ligament, **433**
  subtalar joints, **433**
Dorsal hand ultrasound, **397**
- extensor tendon of thumb, **397**
- extensor tendons of fingers, **397**
- UCL of MCP joint of thumb, **397**
Dorsal intercarpal ligament, **101**
Dorsal intermetatarsal ligament, dorsal foot (graphic), **279**
Dorsal interossei muscles, **102, 104, 108, 109, 110, 112, 113**
- 1st, **112**
- 2nd, **112**
- 3rd, **112**
- 4th, **111, 112**
- fingers, **130, 132, 136, 140**
- thumb, **122, 127, 128**
Dorsal Lisfranc ligament, dorsal foot (graphic), **279**
Dorsal metacarpal artery, hand, **115**
- 5th, **115**
Dorsal metacarpal vein, **87**
Dorsal metatarsal arteries, **291**
- 1st, **290**
Dorsal muscles
- dorsal foot ultrasound, **433**
- tarsus, **277**
Dorsal radiocarpal ligament, **101, 113**
Dorsal radioulnar ligament, **100, 101**
Dorsal scaphoquetral ligament, **101**
Dorsal scapular artery, **45**
- anastomoses with intercostal arteries, **45**

INDEX

Dorsal scapular nerve, **31, 169**
Dorsal talonavicular ligament
- dorsal foot, **280**
  graphic, **279**
  ultrasound, **433**
- lateral ankle, graphic, **269**
- medial and posterior ligaments, **266**
Dorsal tarsometatarsal ligament, dorsal foot (graphic), **279**
Dorsal tendons, wrist, **90**
Dorsal ulnar cutaneous nerve, **390**
Dorsal vein, dorsal foot, **280, 283**
Dorsal venous arch, **87**
- dorsal foot, **281**
Dorsal venous plexus tributary, **115**
Dorsal vessels, dorsal foot, **281**
Dorsal wrist joint **389**
Dorsal wrist tendons, **389**
Dorsal wrist ultrasound, **389–390**
- 1st extensor compartment [abductor pollicis longus (APL) and extensor pollicis brevis (EPB)], **389**
- 2nd extensor compartment (extensor carpi radialis longus and extensor carpi radialis brevis), **389**
- 3rd extensor compartment [extensor pollicis longus (EPL)], **389**
- 4th [extensor digitorum (ED), extensor indicis proprius (EIP) and 5th [extensor digitorum minimi (EDM)] extensor compartments, **389**
- 6th extensor compartment [extensor carpi ulnaris (ECU)], **389**
- dorsal ulnar cutaneous nerve, **390**
- dorsal wrist joint, **389**
- dorsal wrist tendons, **389**
- posterior interosseous nerve, **389–390**
Dorsalis pedis artery, **290, 291, 292**
- anterior and posterior leg nerves (graphic), **255**
- anterior ankle, **261, 262**
- dorsal foot, **280, 281, 282**
- leg arteries, anterior and ankle (graphic), **238**
Dorsalis pedis vein
- anterior ankle, **261**
- dorsal foot, **281**
Dorsum hand, transverse US, **113**
Dorsum metatarsal, **298**
Double hernia, groin hernia, **805**
DPN. *See* Deep peroneal nerve.
Drug hypersensitivity, fat injury vs., **504**
Dupuytren disease, fascial lesion vs., **877**

E

ECRB. *See* Extensor carpi radialis brevis muscle.
ECRL. *See* Extensor carpi radialis longus muscle.
ECU. *See* Extensor carpi ulnaris.
ED. *See* Extensor digitorum muscle.
EDC. *See* Extensor digitorum communis.
Edema, subcutaneous, soft tissue infection vs., **630**
Edge artifact, **366**
Edge-shadowing artifact, **365**
EDL. *See* Extensor digitorum longus.
EDM. *See* Extensor digiti minimi.
Effusions associated with intraarticular pathology, Baker cyst, **664**
Elastofibroma dorsi, chest wall lesion vs., **904**
Elastography
- shear wave, **361**
- strain, **361**
Elastography, US, **361** Elbow,
52–71
- anatomy imaging issues, **52**
- anterior
longitudinal US, **57**
  transverse US, **58**
- anterior aspect of, **381**
- anterolateral, transverse and longitudinal US, **61**
- anteromedial, transverse and longitudinal US, **66**
- coronal, **53**
- distal, cross section, **56**
- flexor aspect, **67**
- gross anatomy
  bursae, **52**
  joint capsule attachments, **52**
  ligaments, **52**
  tendons, **52**
- lateral, **62**
  transverse and longitudinal US, **65**
- ligaments around, **68**
- mid, cross section, **55**
- posterior, transverse US, **71**
- proximal, cross section, **54**
- radial nerve, **149**
Elbow epicondylitis, **494–501**
- differential diagnosis, **496**
- prognosis, **496**
- staging, grading, & classification, **496**
Elbow procedures, **958–963**
- intraarticular aspiration or injection
  indications, **959**
  procedure, **959, 960**
- outcomes, **960**
- paraarticular aspiration
  indication, **959**
  procedure, **959**
- post procedure, **960**
- preprocedure, **959**
- procedure, **959–960**
- tendinosis treatment
  autologous blood, **960**
  corticosteroid injection, **960**
  dry needling, **960**
  indications, **959**
  platelet rich plasma, **960**
  procedure, **959–960**
  prolotherapy, **960**
Elbow ultrasound, **380–387**
- anterior, **381**
  anterior aspect of elbow joint, **381**
  brachialis, **381**
  distal biceps tendon, **381**
  median nerve, **381**

radial nerve, **381**
- clinical indications for, **381**
- lateral, **382**
 annular ligament, **382**
 common extensor tendon origin, **382**
 lateral ulnar collateral ligament, **382**
 radial collateral ligament, **382**
 synovial fold, **382**
- medial, **381–382**
 common flexor tendon origin (CFTO), **381–382**
 cubital tunnel, **382**
 medial collateral ligament complex, **382**
- posterior, **382**
 triceps tendon, **382**

Encysted hydrocele, canal of Nück, groin hernia vs., **806**
Endometrioma, abdominal wall mass vs., **910–911**
Endothoracic fascia, **311**
Enterobacter, bone infection associated, **638**
Enterococcus, postoperative infection, **654**
Enthesopathic fasciitis, fascial lesion vs., **877**
Eosinophilic fasciitis, fascial lesion vs., **877**
EPB. *See* Extensor pollicis brevis; Extensor pollicis brevis muscle.
Epicondyle-olecranon groove. *See* Cubital tunnel syndrome.
Epicondylitis
- elbow, **494–501**
 differential diagnosis, **496**
 prognosis, **496**
 staging, grading, & classification, **496**
- ligament injury vs., **534**

Epidermal cyst. *See* Epidermoid cyst.
Epidermal inclusion cyst. *See* Epidermoid cyst.
Epidermoid cyst, **720–725**
- clinical issues, **722**
- diagnostic checklist, **722**
- differential diagnosis, **722**
- hyperechoic subcutaneous mass vs., **818**
- hypoechoic subcutaneous mass vs., **812**
- pilomatricoma vs., **726**
- prognosis, **722**

EPL. *See* Extensor pollicis longus muscle.
Equipment-generated artifact, **365**
Erector spinae muscles, **311, 314, 326, 329**
- gluteal muscles, **351**

Erosive osteoarthritis, inflammatory, inflammatory arthritis vs., **560**
Escherichia coli
- joint infection, **648**
- postoperative infection, **654**

Esophagus, **316, 328**
Extended field-of-view imaging, **361**
Extensor carpi longus muscle, **59**
Extensor carpi radialis accessory muscle, **72**
Extensor carpi radialis brevis muscle, **55, 56, 62, 64, 72, 73, 74, 76, 79, 88, 90, 389**
- proximal forearm, **82**
- radial nerve, **147, 148**

Extensor carpi radialis brevis tendon, **76, 88, 90, 91, 93**
Extensor carpi radialis intermedius muscle, **72**
Extensor carpi radialis longus muscle, **38, 54, 55, 56, 60, 61, 62, 64, 72, 73, 74, 76, 88, 90, 389**
- proximal forearm, **82**
- radial nerve, **147, 148**
- transverse US, **83**

Extensor carpi radialis longus tendon, **88, 90, 91, 93**
- radial nerve, **148**

Extensor carpi radialis tendon, **79**
Extensor carpi ulnaris (ECU), **56, 62, 71, 72, 73, 79, 88, 90, 100, 389**
- transverse US, **95**

Extensor carpi ulnaris tendon, **65, 90, 91, 95, 100, 111**
- fingers, **132**

Extensor compartments, dorsal wrist ultrasound
- 1st, **389**
- 2nd, **389**
- 3rd, **389**
- 4th and 5th, **389**
- 6th, **389**

Extensor digit minimi tendon, **94**
Extensor digiti minimi muscle, **72, 79, 88, 90**
- fingers, **130, 132**
- radial nerve, **148**

Extensor digiti minimi tendon, **88, 90, 91, 95, 111**
- fingers, **130**

Extensor digitorum (ED), **88, 90, 389**
- transverse US, **94**

Extensor digitorum brevis bursa, tarsus, **278**
Extensor digitorum brevis muscle
- ankle
 anterior ankle (graphic), **260**
 fibular ligaments, **272**
- tarsus, dorsal foot, **280, 281, 283**
 graphic, **279**

Extensor digitorum brevis tendon, dorsal foot, **283**
Extensor digitorum communis tendon, **111**
- fingers, **132, 133**

Extensor digitorum longus muscle, **225**
- anterior compartment of leg, **228**
- anterolateral leg
 graphic, **226**
 panoramic, **226**
- lateral compartment of leg, **230**
- lateral leg, **232**
- peroneal nerve, **257**
- posterior leg, panoramic, **234**
- transverse anterior tibial vessels, **246**

Extensor digitorum longus tendon, **258–259**
- anterior ankle, **261**
 graphic, **260**
 ultrasound, **425**
- anterolateral leg (graphic), **226**
- dorsal foot, **280, 282, 283**
- lateral leg (graphic), **232**

Extensor digitorum manus brevis muscle, **72**
Extensor digitorum minimi (EDM), **389**
Extensor digitorum muscle, **55, 56, 62, 65, 72, 73, 76, 79**
- radial nerve, **148**
- slip, **65**

Extensor digitorum tendon, **79, 90, 94, 113, 139, 140**
- insertional area, **139**

- slips, **91, 140**
Extensor expansion
- fingers, **136**
- hand, **102**
- thumb, **120, 129**
Extensor hallucis brevis muscle
- anterior ankle, graphic, **260**
- dorsal foot, **282**
  graphic, **279**
Extensor hallucis longus bursa, tarsus, **278**
Extensor hallucis longus muscle
- anterior ankle, **261, 262**
- anterior compartment of leg, **228**
- anterolateral leg
  graphic, **226**
  panoramic, **226**
- peroneal nerve, **257**
- transverse anterior tibial vessels, **246**
Extensor hallucis longus tendon, **258**
- anterior ankle, **261, 262**
  graphic, **260**
  ultrasound, **425**
- anterolateral leg (graphic), **226**
- dorsal foot, **280, 281, 283**
- lateral leg (graphic), **232**
- medial ankle, graphic, **263**
Extensor hallucis muscle, **225**
Extensor hood, **104, 111**
- fingers, **132, 140**
Extensor indicis muscle, **72, 79, 88, 90, 130**
Extensor indicis proprius (EIP), **389**
Extensor indicis tendon, **90, 91, 111, 132**
Extensor minimi muscle, **62, 73**
Extensor muscle, **77**
Extensor pollicis brevis muscle, **72, 79, 88, 90, 389**
- thumb, **120**
Extensor pollicis brevis tendon, **88, 90, 91, 92, 96, 111**
- fingers, **132**
- hand, **117**
Extensor pollicis longus muscle, **72, 79, 88, 90, 389**
- radial nerve, **148**
- thumb, **120, 122**
Extensor pollicis longus tendon, **76, 88, 91, 93, 94, 111**
- thumb, **126**
Extensor retinaculum, **88, 90, 94, 111, 113**
- anterior and posterior leg nerves (graphic), **255**
- fingers, **132**
- leg arteries, anterior and ankle (graphic), **238**
- thickening, peritendinous mass vs., **853**
Extensor tendons, **113, 141**
- dorsal foot ultrasound, **433**
- of fingers, dorsal hand ultrasound, **397**
- of thumb, dorsal hand ultrasound, **397**
External and internal muscles, **327**
External iliac vessels, groin, **338**
External intercostal muscle, **311**
External oblique muscle, **314, 315, 317, 326, 327, 329**
- groin, **341, 342, 346**
  abdominal wall, groin, **338**
External pudendal artery, **193**
Extraskeletal myxoid chondrosarcoma, paraarticular cystic mass vs., **889**
Extrinsic carpal ligament, **97**
- median nerve, **157**
Extrinsic flexor muscles, **102**
Extrinsic ligaments, wrist, **88**
Exudative tenosynovitis, acute, **485**

F

Fabella, **221**
Facet joint, **311**
False aneurysm. *See* Vascular dilatation or inflammation.
False ribs, **310**
Fascial lesion, **876–881**
- differential diagnosis, **876**
Fat injury, **502–509**
- diagnostic checklist, **504**
- differential diagnosis, **504**
- prognosis, **504**
Fat necrosis
- abdominal wall mass vs., **911**
- calcified soft tissue mass vs., **842**
- cystic soft tissue mass vs., **837**
- hyperechoic subcutaneous mass vs., **818–819**
- hypoechoic subcutaneous mass vs., **813**
- of newborn, subcutaneous, fat injury vs., **504**
Fat pads, **61**
- knee, **206**
Fat trauma, lipoma vs., **714**
Fatty tumor. *See* Lipoma.
FB. *See* Foreign body and injection granuloma.
FCR. *See* Flexor carpi radialis muscle.
FCU. *See* Flexor carpi ulnaris; Flexor carpi ulnaris muscle.
FDP. *See* Flexor digitorum profundus; Flexor digitorum profundus muscle.
FDS. *See* Flexor digitorum superficialis.
Femoral artery, **171**
Femoral canal, **181, 344, 401**
- Cloquet node, femoral triangle, groin, **339**
- groin hernia, **805**
Femoral condyle
- cortex, lateral, **218**
- lateral, **208, 217, 221–222**
  cartilage, **221**
  posterior cortex, **221**
- medial
  articular cartilage, **223**
  cartilage, **215, 220**
  medial surface, **215**
Femoral cortex, **197**
Femoral epicondylar region, posterior surface of, **197**
Femoral head, **170, 174, 178, 196, 352**
- anterior cortex, **173, 174, 344, 347**
- cartilage, **173, 176**
- cortex, **174**
- ossification, **179**
- posterior cortex, **356**

Femoral hernia, **181**. *See also* Groin hernia.
- groin hernia, **805**

Femoral neck, **353, 355**
- cortex, **174**
- groin, **340, 347**
- posterior cortex, **356**

Femoral nerve, **171, 173, 174, 175, 187, 195, 197**
- groin, **339, 341, 344, 345, 347**
  femoral triangle, groin, **339**

Femoral shaft, **187**
- cortex, **177, 179**

Femoral sheath, **181**
- femoral triangle, groin, **339**

Femoral triangle, **181**
- groin, **339**

Femoral trochlear cartilage, anterior knee ultrasound, **417**

Femoral vein, **171, 192**
- compression, **203**

Femoral vessels and nerves, **192–205**

Femur, **210**
- anterior surface, **175**
- distal, **206**
- neck of, **173**
- popliteal surface of, **196, 198**
- trochlear surface, **209**

Fibrocartilage, **100**

Fibrocartilaginous insertion, shoulder, **11**

Fibrohistiocytic synovial tumors, **695**

Fibrolipomatous hamartoma, swollen nerve vs., **870–871**

Fibroma
- calcified soft tissue mass vs., **843**
- chest wall lesion vs., **905**
- hyperechoic subcutaneous mass vs., **819**
- hypervascular soft tissue mass vs., **847**
- hypoechoic subcutaneous mass vs., **812**
- subcutaneous, foreign body and injection granuloma vs., **754**

Fibromatosis
- hyperechoic muscle mass vs., **831**
- hypoechoic muscle mass vs., **824–825**
- soft tissue sarcoma vs., **768**

Fibrous hamartoma of infancy, hyperechoic subcutaneous mass vs., **819**

Fibula
- lateral ankle, **270**
  graphic, **269**
- medial and posterior ligaments, **266**
- medial ankle, graphic, **263**
- popliteal vessel bifurcation, **241**
- proximal, **206**
- soleal veins, **250**
- transverse anterior tibial vessels, **246**
- transverse peroneal vessels, **244**
- transverse posterior tibial vessels, **242**

Fibular collateral ligament, **218, 222**

Fibular cortex, anterior ankle, **261**

Fibular head
- anterior and posterior leg nerves (graphic), **255**
- leg arteries, anterior and ankle (graphic), **238**
- peroneal nerve, **257**

Fibular neck, peroneal nerve, **257**

Field-of-view imaging, extended, **361**

Fine-needle aspiration for cytology, **919, 920**

Fingers, **130–141**
- axial finger and hand dermatomes, graphics, **141**
- extensor aspect
  longitudinal US, **139**
  transverse US, **140**
- ligaments and pulleys, graphic, **132**
- longitudinal US, **134–135**
- metacarpophalangeal joint, longitudinal US, graphic and, **133**
- tendon sheaths, graphic, **131**
- transverse US, **136–137**
  graphic, **138**

Flash artifact, **366**

Flexor carpi radialis muscle, **37, 56, 67, 72, 74, 75, 76, 78, 88**
- median nerve, **155, 156**
- radial nerve, **148**
- transverse US, **83**

Flexor carpi radialis tendon, **88, 91, 96, 97, 99**
- median nerve, **156, 157**

Flexor carpi ulnaris muscle, **42, 55, 56, 67, 70, 72, 75, 78, 88, 91, 161, 162, 165, 166**
- hand, **116**
- humeral head, **164**
- longitudinal US, **84**
- median nerve, **156**
- radial nerve, **144**
- transverse US, **83, 84**
- ulnar head, **70, 164**

Flexor carpi ulnaris tendon, **88, 96, 166, 167**
- ulnar head, **71**

Flexor digital sheaths, **102**

Flexor digiti minimi brevis muscle, **96, 106, 110, 168**
- lateral tarsus, **289**

Flexor digiti minimi muscle, **103, 107**
- deep plantar muscles, **285**
- lateral tarsus, **289**
- median nerve, **159**
- superficial plantar muscles, **284**
- thumb, **121**

Flexor digiti minimi tendon, fingers, **131**

Flexor digitorum accessorius longus muscle, **225**

Flexor digitorum brevis insertion, superficial plantar muscles, **284**

Flexor digitorum brevis muscle
- medial ankle, graphic, **263**
- tarsus
  lateral tarsus, **289**
  medial tarsus, **288**
  plantar aponeurosis, **286**
  superficial plantar muscles, individual slips (graphic), **284**

Flexor digitorum brevis tendon, **297**

Flexor digitorum longus muscle, **224**
- great saphenous veins, **253**
- medial ankle, **264**
- transverse posterior tibial vessels, **242**

Flexor digitorum longus tendon, **259**
- anterior and posterior leg nerves (graphic), **255**

- deltoid ligament, **267**
- leg arteries, posterior (graphic), **237**
- medial ankle, **264, 265**
- medial tarsus, **288**
- plantar aponeurosis, **286**
- sagittal foot overview, **287**
- superficial plantar muscles, **284**
  insertion, **284**

Flexor digitorum profundus muscle, **55, 56, 70, 72, 74, 75, 76, 88, 97, 102, 106, 165, 166**
- fingers, **130, 136**
- longitudinal US, **84**
- median nerve, **155**
- radial nerve, **148**
- transverse US, **83, 84**

Flexor digitorum profundus tendon, **88, 91, 96, 103, 104, 167**
- fingers, **131, 136, 137, 138, 141**
- hand, **118**
- median nerve, **156, 157, 158**
- thumb, **121**

Flexor digitorum superficialis muscle, **55, 56, 67, 70, 72, 74, 75, 76, 78, 88, 97, 102, 106, 165, 166**
- fingers, **130, 136**
- longitudinal US, **84**
- median nerve, **155, 156**
- proximal forearm, **82**
- radial nerve, **148**
- transverse US, **83, 84**

Flexor digitorum superficialis tendon, **88, 91, 96, 103, 166, 167**
- fingers, **131, 136, 137, 141**
- hand, **118**
- median nerve, **157, 158**
- thumb, **121**

Flexor digitorum tendon, **91**
- fingers, **133**
- hand, **116**
- medial ankle ultrasound, **426**

Flexor digitorum tenosynovitis, carpal tunnel syndrome, **603**

Flexor hallucis brevis muscle
- deep plantar muscles
  lateral, **285**
  medial, **285**
- medial and lateral heads, **297**
- medial tarsus, **288**
- sagittal foot overview, **287**
- superficial plantar muscles
  lateral head, **284**
  medial head, **284**

Flexor hallucis longus muscle, **224–225**
- Achilles tendon
  longitudinal, **274**
- anterior compartment of leg, **228**
- anterolateral leg, panoramic, **227**
- fibular ligaments, **272**
- lateral ankle, **270**
- lateral compartment of leg, **228, 231**
- longitudinal anterior tibial vessels, **247**
- longitudinal peroneal vessels, **245**
- longitudinal posterior tibial vessels, **243**
- medial ankle, **264**
- posterior ankle panorama, **273**
- posterior leg
  panoramic, **234**
  transverse, **235**
- spectral Doppler of leg veins, **249**
- transverse peroneal vessels, **244**
- transverse posterior tibial vessels, **242**

Flexor hallucis longus tendon, **259, 297**
- anterior and posterior leg nerves (graphic), **255**
- deep plantar muscles, **285**
- leg arteries, posterior (graphic), **237**
- medial ankle, **264, 265**
  graphic, **263**
  ultrasound, **426**
- medial tarsus, **288**
- sagittal foot overview, **287**
- superficial plantar muscles, **284**
- tibial nerve, **256**

Flexor longus muscle, peroneal nerve, **257**

Flexor pollicis brevis muscle, **96, 98, 106, 107, 108, 109**
- median nerve, **158**
- thumb, **120, 127, 128**

Flexor pollicis brevis tendon, thumb, **125**

Flexor pollicis longus muscle, **72, 74, 75, 76, 88, 96**
- median nerve, **155**
- thumb, **120**
- transverse US, **83**

Flexor pollicis longus tendon, **88, 91, 96, 97, 98, 108, 109**
- median nerve, **157, 158**
- thumb, **123, 125, 127, 128**

Flexor pollicis longus tendon sheath, **102, 103**
- fingers, **131**

Flexor retinaculum, **88, 91, 96, 97, 98, 103, 107, 167, 168, 258**
- fingers, **131**
- hand, **116, 119**
- leg arteries, anterior and ankle (graphic), **238**
- medial ankle, **264**
- median nerve, **157, 158, 159**
- thumb, **121**

Flexor tendon pulleys, volar hand ultrasound, **397**

Flexor tendon sheath
- fingers, **138**
- superficial plantar muscles, **284**
- thumb, **138**

Flexor tendons, **69, 97, 98, 103**
- fingers, **131, 133, 134, 135**
  insertion of, **135**
- hand, **118**
- index finger, **106**
- little finger, **106**
- median nerve, **158**
- middle finger, **106**
- plantar foot ultrasound, **433**
- ring finger, **106**
- thumb, **121**
  with pulley, **129**
- of thumb, volar hand ultrasound, **397**

Flexor-pronator sprain. *See* Elbow epicondylitis.

Floating ribs, **310**
FNAC. *See* Fine-needle aspiration for cytology.
Focal chondral injury, osteoarthritis vs., **552**
Focal myositis, hypervascular soft tissue mass vs., **847**
Focal nodular synovitis, synovial swelling vs., **894**
Foot ultrasound, **432–437**
- clinical indications for, **433**
- dorsal, **433**
  1st metatarsophalangeal joint, **433**
  calcaneocuboid ligament, **433**
  dorsal muscles, **433**
  dorsal talonavicular ligament, **433**
  extensor tendons, **433**
  Lisfranc ligament, **433**
  subtalar joints, **433**
- plantar, **433**
  flexor tendons, **433**
  plantar fascia, **433**
  plantar muscles, **433**
  toe web spaces, **433**
Foot vessels, **290–295**
- anatomy imaging issues, **290**
- dorsal and plantar arteries and nerves, **291**
- dorsalis pedis artery, **292**
- metatarsal and digital arteries, **293**
- plantar arteries, **294–295**
Forearm, **72–79**
- anatomy imaging issues, **72**
- anomalous muscles, **72**
- anterior
  arteries and nerves, **77**
  transverse US, **74**
- anterolateral, transverse US, **76**
- anteromedial, transverse US, **75**
- common extensor tendon, **73**
- common flexor tendon, **78**
- extensors, **72**
- flexors, **72**
- median nerve, **150, 151**
- posterior aspect, transverse US, **79**
- radial nerve, **142, 148**
  posterior, **147**
- ulnar nerve, **160, 161**
  longitudinal US, **166**
  transverse US, **165, 166**
Forearm vessels, **80–87**
- anatomy imaging issues, **80**
- arteries, **81**
- clinical implications, **80**
- distal forearm and hand, arteries, **85**
- gross anatomy, **80**
- proximal, transverse US, **82**
- radial artery, transverse US, **83**
- ulnar artery, transverse and longitudinal US, **84**
- veins, **81**
- venous system
  anterior, **86**
  posterior, **86**
Forefoot, **276**
Forefoot bursal injection
- indications, **987**
- procedure, **988**
Foreign body
- granuloma, hypoechoic subcutaneous mass vs., **813**
- hyperechoic muscle mass vs., **831**
- within tendon, tendon hyperechogenicity vs., **861**
Foreign body and injection granuloma, **752–759**
- clinical issues, **754**
- diagnostic checklist, **754**
- prognosis, **754**
FPB. *See* Flexor pollicis brevis.
FPL. *See* Flexor pollicis longus.
FR. *See* Flexor retinaculum.
Fracture, **541**
- bone surface lesion vs., **882, 883**
- healing insufficiency, bone tumor vs., **776**
- insufficiency, plantar fasciitis and fibromatosis vs., **708**
Full-thickness tear
- nonrotator cuff tendon tears, **473**
- rotator cuff/biceps tendon, **455**
- tendon hyperechogenicity vs., **861**
- tendon hypoechogenicity vs., **857**
Fusion imaging, **361**
Fx. *See* Fracture.

G

Gamekeeper's thumb. *See* Ligament injury.
Ganglion
- carpal tunnel syndrome, **603**
- nerve sheath tumor vs., **592**
Ganglion cyst, **680–689**
- aspiration, ankle and foot
  indications, **965**
  procedures, **965**
- associated abnormalities, **682**
- cystic soft tissue mass vs., **836**
- diagnostic checklist, **682**
- differential diagnosis, **682**
- paraarticular, parameniscal cyst vs., **692**
- paraarticular cystic mass vs., **888**
- peritendinous mass vs., **852**
- prognosis, **682**
Gastrocnemius aponeurosis, **235**
Gastrocnemius muscle, **224**
- gastrocnemius veins, **251**
- great saphenous veins, **253**
- lateral head, **219–222**
- longitudinal peroneal vessels, **245**
- medial head, **215, 220, 222–223**
- popliteal vessels, **240**
  bifurcation, **241**
- small saphenous vein, **252**
- soleal veins, **250**
- transverse peroneal vessels, **244**
- transverse posterior tibial vessels, **242**
Gastrocnemius muscle, lateral, **208, 216, 232**
- anterolateral leg, panoramic, **227**
- posterior leg
  graphic, **233**

panoramic, **234**
transverse, **235**
- tibial nerve, **256**
Gastrocnemius muscle, medial, **208**
- anterolateral leg (graphic), **226**
- posterior leg
graphic, **233**
panoramic, **234**
transverse, **235**
- tibial nerve, **256**
Gastrocnemius muscle tear, muscle injury, **516**
Gastrocnemius tendon, **214, 216**
Gastrocnemius vein, **251**
- popliteal vessel bifurcation, **241**
Gastrocnemius-semimembranosus (Baker cyst), posterior knee ultrasound, **418**
Gemelli muscle, **195**
- gluteal muscles, **351**
Gerdy tubercle, **217**
Ghost-image artifact, **366**
Giant cell tumor
- bone surface lesion vs., **882**
- of tendon sheath
ganglion cyst vs., **682**
hypervascular soft tissue mass vs., **847**
peritendinous mass vs., **852**
Gideon disease. *See* Ganglion cyst.
Glenohumeral joint, **11, 27, 374**
- injection or aspiration
indications, **951**
procedure, **951**
Glenoid, **17, 27, 30**
Glenoid cavity of scapula, **13**
Glenoid fossa, **14**
Glenoid labra, **374**
Glenoid labrum, **10, 13, 27**
Glomus tumor
- hypervascular soft tissue mass vs., **846**
- hypoechoic subcutaneous mass vs., **813**
- nerve sheath tumor vs., **592**
Gluteal fascia, **177**
Gluteal muscles, **348–357**
- bursa, **348**
- graphics, **350**
- greater sciatic foramen, **349**
- ischial region, **355**
- ischial tuberosity, **348, 357**
- lateral hip ultrasound, **410**
- lesser sciatic foramen, **349**
- piriformis, **348**
- posterior sacroiliac ligaments of pelvis, **349**
- region, **352**
- rotator cuff of hip, **348**
- sacroiliac joint, **348–349**
- sagittal, **353**
- sciatic nerve and overview, **351**
- transverse, **354**
- transverse ischial, **356**
Gluteofemoral bursa, **348**
Gluteus maximus muscle, **183, 186, 204, 205, 346**
- gluteal muscles, **348, 350, 351, 352, 353, 354, 355, 356, 357**
Gluteus medius muscle, **177, 178, 179, 186, 346**
- gluteal muscles, **348, 350, 351, 352, 353, 354**
Gluteus medius tendon
- insertion, **353**
- lateral hip ultrasound, **410**
Gluteus minimus muscle, **176, 178, 179, 186, 346**
- gluteal muscles, **348, 352, 353, 354**
Gluteus minimus tendon, **177**
- lateral hip ultrasound, **410**
Golfer's elbow. *See* Elbow epicondylitis.
Gout and pseudogout, **568–575**
- associated abnormalities, **570**
- calcified soft tissue mass vs., **842**
- diagnostic checklist, **570**
- differential diagnosis, **570**
- hemarthrosis and lipohemarthrosis vs., **660**
- prognosis, **570**
- tendon hyperechogenicity vs., **860–861**
- tendon swelling vs., **865**
- tophus
carpal tunnel syndrome, **603**
hyperechoic subcutaneous mass vs., **819**
peritendinous mass vs., **853**
synovial swelling vs., **894**
Gouty arthropathy. *See* Gout and pseudogout.
Gouty tophi. *See* Gout and pseudogout.
Gracilis muscle, **180, 182, 186–187, 189–190, 219, 220**
- gluteal muscles, **350, 351**
- groin, **338, 339, 340, 342, 343, 346**
femoral triangle, groin, **343**
- medial knee ultrasound, **418**
Gracilis tendon, **186, 208, 214–215, 223**
- posterior leg (graphic), **233**
Granulation tissue, hypervascular soft tissue mass vs., **846**
Granuloma
- abdominal wall mass vs., **911**
- calcified soft tissue mass vs., **842**
- hyperechoic subcutaneous mass vs., **819**
- hypoechoic muscle mass vs., **825**
Great saphenous vein, **187, 189, 194, 200, 201, 219, 253, 290**
- groin, **339, 345**
femoral triangle, groin, **339, 345**
- gross anatomy, **236**
- leg veins (graphic), **239**
Greater sciatic foramen, **204, 349, 355**
Greater trochanter, **171, 172, 177, 205, 350, 352, 353, 356**
- medial surface, **355, 356**
- superior surface, **353, 355**
- superior surface of, **177**
- unossified, **178**
Greater tuberosity, **15, 16, 25**
Greater tuberosity of humerus, radial nerve, **144**
Groin, **338–347**
- adductor musculature, **338**
- anterior abdominal wall, **338**
- femoral triangle, central, **344, 345**
- graphics, **339**

- inguinal canal, **338**
- inguinal ligament, **338**
- medial, **343**
- midline, **342**
- muscles, **340**
- nerves, **341**
- nodes, **338**
- osseous anatomy, **338**

Groin hernia, **804–809**
- abdominal wall mass vs., **910**
- diagnostic checklist, **804, 806**
- differential diagnosis, **806**
- prognosis, **806**

Groin hernia ultrasound, **400–407**
- deep inguinal ring, **401**
- femoral canal, **401**
- Hesselbach triangle, **401**
- inferior epigastric vessels, **401**
- spermatic cord (in males), **401**

Guyon canal, **88**
- ulnar nerve, **160**
  - transverse US, **167, 168**

Guyon tunnel, **390**

Gynecomastia, chest wall lesion vs., **905**

H

Haemophilus, bone infection, **638**

Hamartomatous fibrolipoma of median nerve, carpal tunnel syndrome, **603**

Hamate, **89, 95, 110, 113**
- median nerve, **157**

Hamstring origin, **356**

Hamstring tendon origins, hip ultrasound, **410**

Hamstrings, **180**

Hand, **102–113**
- deep dissection of, thumb, **121**
- dermatomes, **141**
- dorsal aspect, **111**
- dorsal interossei, **112**
- dorsum, transverse US, **113**
- extrinsic flexor muscle: digits 2-5, **102**
- hypothenar eminence, transverse US, **110**
- intrinsic muscles
  - hypothenar area, **102**
  - palmar area, **102**
  - thenar area, **102**
- lumbrical and interossei muscles, **104**
- median nerve, **150**
- palm, transverse US, **106**
- palmar aspect, **103**
- palmar interossei, **105**
- radial nerve, **142**
- tendons, **102**
  - extensor expansion (or extensor hood), **102**
  - tendon sheath, **102**
- thenar eminence
  - longitudinal US, **109**
  - transverse US, **108**
- thenar-hypothenar eminence, **107**
- ulnar nerve, **160, 161**

Hand and wrist procedures, **964–969**
- carpal tunnel syndrome
  - indications, **965**
  - procedures, **966**
- de Quervain disease
  - indications, **965**
  - procedures, **965–966**
- ganglion cyst aspiration
  - indications, **965**
  - procedures, **965**
- joint aspiration or injection
  - indications, **965**
  - procedure, **965**
- outcomes, **966**
- preprocedure, **965**
- procedure, **965–966**
- tendon disease
  - indications, **965**
  - procedure, **965**
- trigger finger
  - indications, **965**
  - procedures, **966**

Hand ultrasound, **396–399**
- clinical indications for, **397**
- dorsal, **397**
  - extensor tendon of thumb, **397**
  - extensor tendons of fingers, **397**
  - UCL of MCP joint of thumb, **397**
- volar, **397**
  - carpometacarpal joint of thumb, **397**
  - flexor tendon pulleys, **397**
  - flexor tendons of thumb, **397**
  - metacarpophalangeal and interphalangeal joints, **397**

Hand vessels, **114–119**
- arteries, **114, 118**
- dorsal aspect, transverse US, **117**
- palmar aspect, transverse US, **118**
- vasculature, **115**
- veins, **114**
- wrist and palm, transverse US, **116**

Healing insufficiency fracture, bone tumor vs., **776**

Healing rib fracture, chest wall lesion vs., **904**

Heel fat pad atrophy, plantar fasciitis and fibromatosis vs., **708**

Heel spur syndrome. *See* Plantar fasciitis and fibromatosis.

Hemangioma
- hyperechoic muscle mass vs., **830**
- hyperechoic subcutaneous mass vs., **818**
- pilomatricoma vs., **726**
- synovial, **695**
  - synovial swelling vs., **895**
  - synovial tumor vs., **696**

Hemarthrosis
- joint effusion vs., **900, 901**
- joint infection vs., **648**

Hemarthrosis and lipohemarthrosis, **658–661**
- diagnostic checklist, **660**
- differential diagnosis, **660**
- prognosis, **660**

- staging, grading, & classification, **660**
Hematoma, **524–531**
- acute, **525**
- calcified soft tissue mass vs., **843**
- chest wall lesion vs., **904**
- chronic, **525**
- cystic soft tissue mass vs., **836**
- diagnostic checklist, **526**
- differential diagnosis, **526**
- hyperechoic muscle mass vs., **830–831**
- hyperechoic subcutaneous mass vs., **819**
- hypoechoic muscle mass vs., **824**
- hypoechoic subcutaneous mass vs., **813**
- intravascular organized, **621, 622**
- liquefied, paraarticular cystic mass vs., **889**
- muscle
    soft tissue infection vs., **630**
    soft tissue sarcoma vs., **768**
- organizing, **525**
- postoperative infection vs., **654**
- prognosis, **526**
- rectus sheath, **525**
    abdominal wall hernia vs., **798**
- subacute, **525**
- superficial metastases, lymphoma, and melanoma vs., **740**
Hematoma/seroma, **524–531**
Hemophilia A or B, hemarthrosis and lipohemarthrosis, **660**
Hemophiliac arthropathy, **559**. *See also* Hemarthrosis and lipohemarthrosis.
Hemophiliac pseudotumor, **525**
Hemophilic arthropathy, synovial tumor vs., **696**
Hernia
- abdominal wall, **796–803**
    diagnostic checklist, **798**
    differential diagnosis, **798**
    prognosis, **798**
- groin, **804–809**
    diagnostic checklist, **804, 806**
    differential diagnosis, **806**
    prognosis, **806**
- groin ultrasound, **400–407**
    deep inguinal ring, **401**
    femoral canal, **401**
    Hesselbach triangle, **401**
    inferior epigastric vessels, **401**
    spermatic cord (in males), **401**
Hesselbach triangle, **401**
Heterotopic ossification, calcified soft tissue mass vs., **843**
Hibernoma, **713**
High-grade partial tear, nonrotator cuff tendon tears, **473**
Hindfoot, **276**
Hip, **170–179**
- axial and sagittal sections, **171**
- coronal sections, **172**
- ligaments, **172**
- overview, **173**
- pediatric, **178–179**
Hip and pelvis procedures, **970–977**
- hip or sacroiliac joint injection
    indications, **971**
    procedures, **971–972**
- nerve block & neural hydrodissection
    indications, **971**
    procedures, **972**
- outcomes, **972**
- paraarticular labral/ganglion cyst or bursal
    indications, **971**
    procedures, **972**
- post procedure, **972**
- preprocedure, **971**
- procedure, **971–972**
- tendinosis and tears
    indications, **971**
    procedures, **972**
Hip joint, **176, 347, 356**
- anterior aspect of, anterior hip ultrasound, **409**
Hip joint capsule, **176, 347, 356**
- anterior, **173, 174**
Hip joint injection
- indications, **971**
- procedures, **971**
Hip ultrasound, **408–415**
- anterior, **409**
    acetabular labral junction, **409**
    anterior aspect of hip joint, **409**
    anterior superior iliac spine region, **409**
    dynamic examination for snapping iliopsoas tendon, **409**
    iliopsoas tendon, **409**
    rectus femoris tendon, **409**
- clinical indications for, **409**
- lateral, **410**
    gluteus minimus and medius tendons, **410**
    tensor fascia lata, iliotibial tract, and gluteal muscles, **410**
- medial, **409–410**
    adductor tendons and muscles, **409–410**
- posterior hip and deep gluteal space, **410**
    deep gluteal space proximal to ischial tuberosity, **410**
    ischial tuberosity and hamstring tendon origins, **410**
    posterior thigh musculature, **410**
Honeymoon syndrome. *See* Nerve injury.
Hook of hamate, **98, 110, 116, 168**
- median nerve, **157**
Humeral capitellum, **57, 58, 59, 63, 64, 65**
Humeral head, **7, 15, 16, 17, 18, 19, 20, 21, 22, 23, 24, 26, 27, 30**
Humeral neck, **31**
Humeral shaft, **20, 32, 34, 39, 40, 41, 43, 47, 48, 49, 50, 51, 66, 163**
- radial nerve, **145, 146**
Humeral trochlea, **58, 59, 66, 70**
Humerus, **34, 35, 37, 38, 164, 304**
- capitellum of, **61**
- distal shaft of, **57, 59**
- greater tuberosity of, **42**
- lateral epicondyle, **53, 71**
- lateral supracondylar portion of, **41**

- medial epicondyle, **53, 71**
Hyaline cartilage, **210**
Hyperechoic muscle mass, **830–835**
- differential diagnosis, **830**
Hyperechoic subcutaneous mass, **818–823**
- differential diagnosis, **818**
Hyperemia, tendon hypoechogenicity vs., **856**
Hyperlipidemia, gout and pseudogout with, **570**
Hypertension, gout and pseudogout with, **570**
Hypertrophy of radiohumeral synovial flange, elbow epicondylitis vs., **496**
Hypervascular soft tissue mass, **846–851**
- differential diagnosis, **846**
Hypoechoic muscle mass, **824–829**
- differential diagnosis, **824**
Hypoechoic subcutaneous mass, **812–817**
- differential diagnosis, **812**
Hypothenar eminence, **96, 102**
- transverse US, **110**

I

IG. *See* Foreign body and injection granuloma.
Iliac cortex, **179**
Iliac crest, **186, 317, 329, 351, 352, 355**
Iliac spine
- anterior inferior, **172, 173, 176**
- anterior superior, **172**
Iliacus, **314, 316, 326**
Iliacus muscle, **175, 182, 186, 328**
- groin, **339, 340, 346**
    femoral triangle, groin, **340**
Iliocostalis muscle, **317, 329**
Iliofemoral ligament, **172, 175, 346**
Iliohypogastric nerve, **341**
Ilioinguinal nerve, **341**
Iliopsoas, **181**
Iliopsoas bursa, **171, 669**
Iliopsoas insertion, **175**
Iliopsoas muscle, **173, 174, 176, 196, 197, 198, 200, 201, 202, 203, 344, 346, 347**
- insertion, **316, 328**
Iliopsoas tendon, **173, 174, 175, 356**
- anterior hip ultrasound, **409**
Iliotibial band, **182, 207**
Iliotibial tract, **177, 217, 350, 351, 352**
- anterior and posterior leg nerves (graphic), **255**
- groin muscles, **340**
- lateral hip ultrasound, **410**
- lateral leg (graphic), **232**
- leg arteries, anterior and ankle (graphic), **238**
Ilium, **171, 353, 354**
- lateral cortex of, **354**
Immature hip, physiological, developmental hip dysplasia vs., **578**
Immune-mediated peripheral neuropathy, swollen nerve vs., **870**
Improper Doppler angle with no-flow artifact, **366**
Improper machine setting, artifacts due to, **365**
- equipment-generated, **365**
- range ambiguity, **365**
Inactive tenosynovitis, chronic, **485**
Incisional hernia, abdominal wall mass vs., **910**
Indirect inguinal hernia. *See also* Groin hernia.
- groin hernia, **805**
Infection
- bone, **636–645**
    diagnostic checklist, **638**
    differential diagnosis, **638**
    prognosis, **638**
- joint, **646–651**
    diagnostic checklist, **648**
    differential diagnosis, **648**
    prognosis, **648**
- postoperative, **652–657**
    diagnostic checklist, **654**
    differential diagnosis, **654**
    prognosis, **654**
- soft tissue, **628–635**
    differential diagnosis, **630**
    prognosis, **630**
- soft tissue infection, bone infection vs., **638**
Infective arthritis. *See also* Joint infection.
- joint effusion vs., **900, 901**
Infective fasciitis, **629**
- fascial lesion vs., **877**
Infective myositis, muscle infarction vs., **511**
Infective osteitis, **637**
Inferior acromioclavicular ligament, **7**
Inferior calcaneal nerve, medial ankle, graphic, **263**
Inferior epigastric artery, **193, 347**
Inferior epigastric vein, **194, 347**
Inferior epigastric vessels, **401**
Inferior extensor retinaculum, **258**
- anterior ankle, **261**
    graphic, **260**
    ultrasound, **425**
- dorsal foot, **281**
    graphic, **279**
Inferior fascicle, **216**
Inferior gemellus muscle, **186, 204, 205, 346**
- gluteal muscles, **348, 350, 351, 355, 356**
Inferior glenohumeral ligament complex, **25**
- anterior band, **14**
- axillary pouch, **14**
- posterior band, **14**
Inferior gluteal artery, **193, 205, 349, 356**
Inferior gluteal nerve, **349**
Inferior gluteal vein, **194, 357**
Inferior lateral genicular artery, **193**
Inferior lateral genicular vessels, **218**
Inferior lateral geniculate artery, **217**
- leg arteries, posterior (graphic), **237**
Inferior medial genicular artery, **193**
Inferior medial geniculate artery, **213, 221**
- leg arteries, posterior (graphic), **237**
Inferior patellar tendon, **214**
- anterior and posterior leg nerves (graphic), **255**
- anterolateral leg (graphic), **226**

INDEX

- lateral leg (graphic), **232**
- leg arteries, anterior and ankle (graphic), **238**

Inferior peroneal retinaculum, **258**
- lateral ankle, **271**
    graphic, **269**

Inferior pubic ramus, **338, 343**
Inferior thyroid artery, **45**
Inferior transverse ligament, **258**
- medial and posterior ligaments, **266**

Inferior ulnar collateral artery, forearm arteries, **81**
Inflammatory arthritis, **558–567**
- diagnostic checklist, **560**
- differential diagnosis, **560**
- osteoarthritis vs., **552**
- prognosis, **560**

Inflammatory arthropathy
- joint effusion vs., **900**
- nonrotator cuff tendon tears associated with, **474**
- synovial swelling vs., **894**

Inflammatory erosive osteoarthritis, inflammatory arthritis vs., **560**
Inflammatory mass, hypervascular soft tissue mass vs., **846**
Inflammatory myositis, muscle infarction vs., **511**
Inflammatory synovitis
- carpal tunnel syndrome vs., **604**
- hemarthrosis and lipohemarthrosis vs., **660**

Infrapatellar bursa, **211, 669**
Infrapatellar bursitis (preacher's knee). *See* Bursitis.
Infrapatellar (Hoffa) fat pad, **209, 211**
Infraspinatus muscle, **10, 11, 12, 13, 23, 24, 25, 26, 27, 28, 30, 35, 42**
- radial nerve **144**

Infraspinatus tendon, **11, 12, 13, 14, 18, 23, 28, 30, 373**
- musculotendinous junction, **11**

Infundibular cyst. *See* Epidermoid cyst.
Inguinal canal, **401**
- groin, **338**
- groin hernia, **805**

Inguinal hernia. *See* Groin hernia.
Inguinal ligament, **195, 197, 203, 315, 316, 327, 328, 401**
- groin, **339, 342, 343, 346, 347**
    femoral triangle, groin, **347**

Inguinal lymph nodes, **344**
- deep, **338**
- superficial, **338**

Inguinal ring, deep, **338**
Innermost intercostal muscle, **311**
Insertional area, **15, 17**
- of infraspinatus, **23**
- lateral fibers, **16**
- midfibers, **16**

Insufficiency fracture, **541**
Insulin-dependent diabetes, longstanding, muscle infarction with, **512**
Intercapitular vein, **86, 87**
Interclavicular ligament, **4, 5**
Intercondylar groove, **220**
Intercondylar notch, **223**
Intercostal artery, **311**
Intercostal muscles, **310**
Intercostal nerves, **310, 311**
Intercostal space, **310–313**
- graphics, **311**
- skeletal structures, **310**
- vessels and nerves, **310**

Intercostal vein, **311**
Interligamentous sulcus, **101**
Intermediate antebrachial vein, **86**
Intermediate basilic vein, **86**
Intermediate cephalic vein, **86**
Intermediate cuneiform
- dorsal foot, **281, 283**
- medial tarsus, **288**

Intermetatarsal ligaments, tarsus, **277**
Intermuscular lipoma, hyperechoic muscle mass vs., **830**
Intermuscular septum
- anterior, panoramic anterolateral leg, **227**
- posterior
    lateral compartment of leg, **230, 231**
    panoramic anterolateral leg, **227**

Internal iliac artery, **193**
Internal iliac vein, **194**
Internal intercostal muscle, **311**
Internal mammary artery, **311**
Internal mammary lymph node, **311**
Internal mammary vein, **311**
Internal oblique muscle, **314, 315, 317, 326, 329**
- groin, **341, 342, 346, 347**
    abdominal wall, **338**

Internal pudendal artery, **171**
Internal pudendal vein, **171**
Internal pudendal vessels, **349**
- lesser sciatic foramen, **349**

Internal thoracic artery, **45, 310**
Interossei muscle, **104**
Interosseous artery, **71**
Interosseous compartment, tarsus, **277**
Interosseous ligament, **258**
Interosseous membrane, **74, 75, 79, 97, 224**
- anterior and posterior leg nerves (graphic), **255**
- anterior compartment of leg, **228, 229**
- anterolateral leg, panoramic, **227**
- lateral ankle, graphic, **269**
- lateral compartment of leg, **230**
- leg arteries, anterior and ankle (graphic), **238**
- longitudinal anterior tibial vessels, **247**
- medial and posterior ligaments, **266**
- medial ankle, graphic, **263**
- posterior leg, transverse, **235**
- radial nerve, **148**

Interosseous muscle, **106**
- deep plantar muscles, **285**
- fingers, **141**

Interosseous recurrent artery, forearm arteries, **81**
Interosseous sacroiliac ligament, **349**
Interosseous talocalcaneal ligament
- dorsal foot, **280**
- medial ankle, graphic, **263**

Interosseous vein, **71**
Interphalangeal joints
- thumb, **126**
- volar hand ultrasound, **397**

INDEX

Intersesamoid ligament, **297**
Intertrochanteric line, **172, 175**
Interventional procedures
- ankle and foot procedures, **986–991**
  - Achilles tendinosis treatment, procedure, **987–988**
  - forefoot bursal injection
    - indications, **987**
    - procedure, **988**
  - ganglion cyst aspiration
    - indications, **987**
    - procedure, **987**
  - joint aspiration or injection
    - indications, **987**
    - procedure, **987**
  - Morton neuroma injection
    - indication, **987**
    - procedure, **988**
  - outcomes, **988**
  - peritendon therapy, indication, **987**
  - plantar fasciitis injection
    - indication, **987**
    - procedure, **988**
  - post procedure, **968**
  - preprocedure, **987**
  - procedure, **987–988**
  - sinus tarsi injection
    - indication, **987**
    - procedure, **988**
  - tendons and ligaments, treatment
    - indication, **987**
    - procedure, **988**
- bone tumor biopsy, **924–931**
  - outcomes, **926**
  - preprocedure, **925**
  - procedure, **925–926**
- elbow procedures, **958–963**
  - intraarticular aspiration or injection
    - indications, **959**
    - procedure, **959, 960**
  - outcomes, **960**
  - paraarticular aspiration
    - indication, **959**
    - procedure, **959**
  - post procedure, **960**
  - preprocedure, **959**
  - procedure, **959–960**
  - tendinosis treatment
    - autologous blood, **960**
    - corticosteroid injection, **960**
    - dry needling, **960**
    - indications, **959**
    - platelet rich plasma, **960**
    - procedure, **959–960**
    - prolotherapy, **960**
- hand and wrist procedures, **964–969**
  - carpal tunnel syndrome
    - indications, **965**
    - procedures, **966**
  - de Quervain disease
    - indications, **965**
    - procedures, **965–966**
  - ganglion cyst aspiration
    - indications, **965**
    - procedures, **965**
  - joint aspiration or injection
    - indications, **965**
    - procedure, **965**
  - outcomes, **966**
  - preprocedure, **965**
  - procedure, **965–966**
  - tendon disease
    - indications, **965**
    - procedure, **965**
  - trigger finger
    - indications, **965**
    - procedures, **966**
- hip and pelvis procedures, **970–977**
  - hip or sacroiliac joint injection
    - indications, **971**
    - procedures, **971–972**
  - nerve block & neural hydrodissection
    - indications, **971**
    - procedures, **972**
  - outcomes, **972**
  - paraarticular labral/ganglion cyst or bursal
    - indications, **971**
    - procedures, **972**
  - post procedure, **972**
  - preprocedure, **971**
  - procedure, **971–972**
  - tendinosis and tears
    - indications, **971**
    - procedures, **972**
- joint injection, lower limb, **942–949**
  - outcomes, **944**
  - post procedure, **944**
  - preprocedure, **943**
  - procedure, **943–944**
- joint injection, upper limb, **932–941**
  - outcomes, **934**
  - preprocedure, **933**
  - procedure, **933–934**
- knee procedures, **978–985**
  - intraarticular injection of contrast medium or medication
    - indications, **979**
    - procedure, **979**
  - outcomes, **980**
  - paraarticular aspiration
    - indications, **979**
    - procedure, **979**
  - paraarticular tendinosis treatment, indications, **979**
  - paraarticular tendinosis treatment, procedure, **979–980**
    - ABI or PRP injection, **980**
    - dry needling, **979–980**
    - peritendinous steroid injection, **980**
    - prolotherapy, **980**
  - post procedure, **980**
  - preprocedure, **979**
  - procedure, **979–980**

- shoulder procedures, **950–957**
 acromioclavicular joint injection
 indications, **951**
 procedure, **953**
 calcific tendinitis therapy
 indications, **951**
 procedure, **952–953**
 glenohumeral joint injection or aspiration
 indications, **951**
 procedure, **951**
 outcomes, **953**
 post procedure, **953**
 preprocedure, **951**
 procedure, **951–953**
 rotator cuff tendinosis or partial tear therapy
 indications, **951**
 procedure, **953**
 subacromial-subdeltoid bursa injection
 indications, **951**
 procedure, **952**
- soft tissue tumor biopsy, **918–923**
 outcomes **920**
 post procedure, **920**
 preprocedure, **919**
 procedure, **919–920**
Intraarticular aspiration or injection, elbow
- indications, **959**
- procedure, **959, 960**
Intraarticular disc, **4**
Intramuscular extension of Baker cyst, hypoechoic muscle mass vs., **825**
Intramuscular ganglion cyst, hypoechoic muscle mass vs., **825**
Intramuscular lipoma, hyperechoic muscle mass vs., **830**
Intraneural ganglion, swollen nerve vs., **871**
Intraneural lipoma, swollen nerve vs., **871**
Intrasubstance tendon tears, rotator cuff/biceps tendinosis vs., **443**
Intratendinous vessel, tendon hypoechogenicity vs., **857**
Intravascular angiomatosis. *See* Vascular dilatation or inflammation.
Intravascular organized hematoma, **621, 622**
Intravascular papillary endothelial hyperplasia, **621, 622**. *See also* Vascular dilatation or inflammation.
Intrinsic ligaments, **88**
Investing fascia
- great saphenous veins, **253**
- small saphenous vein, **252**
Investing fascial tear, fascial lesion vs., **877**
Involucrum, **637**
IPEH. *See* Intravascular papillary endothelial hyperplasia.
Ischemic myonecrosis. *See* Muscle infarction.
Ischial spine, **186**
Ischial tuberosity, **171, 205, 348, 351, 355, 356, 357**
- deep gluteal space proximal to, **410**
- hamstring originating from, **355**
- posterior hip ultrasound, **410**
Ischial tuberosity bursa, **669**
Ischial tuberosity (ischiogluteal) bursitis. *See* Bursitis.
Ischiofemoral ligament, **172**
Ischiogluteal bursa, **348**
Ischium, **353**
- posterior cortex, **355**
Ischium posterior cortex, **204**

J

Jogger's heel. *See* Plantar fasciitis and fibromatosis.
Joint, **99, 113**
Joint capsule, **5, 13, 27, 28, 30, 170**
- acromioclavicular joint, **8, 9**
- fingers, **133, 134, 135, 140**
- hip, **173**
Joint deformity, chronic, nonrotator cuff tendon tears associated with, **474**
Joint effusion, **900–903**
- crystal arthropathy, **900**
- differential diagnosis, **900**
- hemarthrosis, **900, 901**
- infective arthritis, **900, 901**
- inflammatory arthropathy, **900**
- lipohemarthrosis, **900, 901**
- osteoarthritis, **900**
- pigmented villonodular synovitis, **900, 901**
- synovial osteochondromatosis, **900, 901**
Joint infection, **646–651**
- diagnostic checklist, **648**
- differential diagnosis, **648**
- prognosis, **648**
Joint injection
- lower limb, **942–949**
 outcomes, **944**
 post procedure, **944**
 preprocedure, **943**
 procedure, **943–944**
- upper limb, **932–941**
 outcomes, **934**
 preprocedure, **933**
 procedure, **933–934**
Joint recess, thumb, **125, 126**
Joint space, thumb, **122**
Junctura tendinum, **111**
- fingers, **132**
Juvenile inflammatory arthropathy. *See* Inflammatory arthritis.
Juxtacortical tumor, bone surface lesion vs., **883**

K

Kaposi syndrome, vascular dilatation or inflammation vs., **622**
Kimura disease, **621, 622**
Klebsiella, postoperative infection, **654**
Knee, **206–223**
- extensor mechanism, **208–213**
- lateral, **216–218**
- medial, **214–215**

INDEX

- posterior, **219–223**
Knee joint, anterior knee ultrasound, **417**
Knee procedures, **978–985**
- intraarticular injection of contrast medium or medication
  indications, **979**
  procedure, **979**
- outcomes, **980**
- paraarticular aspiration
  indications, **979**
  procedure, **979**
- paraarticular tendinosis treatment
  indications, **979**
  procedure, **979–980**
    ABI or PRP injection, **980**
    dry needling, **979–980**
    peritendinous steroid injection, **980**
    prolotherapy, **980**
- post procedure, **980**
- preprocedure, **979**
- procedure, **979–980**
Knee ultrasound, **416–423**
- anterior, **417**
  anterior cruciate ligament, **417**
  femoral trochlear cartilage, **417**
  knee joint, **417**
  patellar bone and patellar tendon, **417**
  patellar retinacula, **417**
  quadriceps tendon, **417**
- clinical indications for, **417**
- lateral, **418**
  biceps femoris tendon, **418**
  distal iliotibial band, **418**
  lateral collateral ligament, **418**
  popliteus tendon, **418**
- medial, **417–418**
  body of medial meniscus, **418**
  medial collateral ligament, **417–418**
  pes anserinus tendons (semitendinosus, gracilis, and sartorius), **418**
  semimembranosus tendon, **418**
- posterior, **418**
  gastrocnemius-semimembranosus (Baker cyst), **418**
  posterior cruciate ligament, **418**
  posterolateral corner structures, **418**
  tibial and peroneal nerves, **418**

L

Labrum, **170, 179**
- 3-o'clock position, **14**
- 6-o'clock position, **14**
- 9-o'clock position, **14**
- 12-o'clock position, **14**
- anterior, **30, 171**
- hip, paraarticular labral/ganglion cyst or bursal aspiration
  indication, **971**
  procedures, **972**
- posterior, **30, 171**
- superior, **172**
Lateral ankle, ultrasound, **426**
Lateral antebrachial cutaneous nerve, **37, 38, 390**
Lateral arcuate ligament, **316, 328**
Lateral bands, **104, 111, 112**
- fingers, **132**
Lateral calcaneal cortex
- fibular ligaments, **272**
- lateral ankle, **270, 271**
Lateral circumflex femoral artery, **193**
Lateral circumflex femoral vein, **194**
Lateral circumflex femoral vessels, **184**
Lateral collateral ligament, lateral knee ultrasound, **418**
Lateral compartment
- ankle, **259**
- leg muscles, **225**
Lateral cuneiform
- dorsal foot, **282**
- lateral tarsus, **289**
- sagittal foot overview, **287**
Lateral cutaneous nerve, forearm, median nerve, **151**
Lateral elbow ultrasound, **382**
- annular ligament, **382**
- common extensor tendon origin, **382**
- lateral ulnar collateral ligament, **382**
- radial collateral ligament, **382**
- synovial fold, **382**
Lateral epicondylalgia. *See* Elbow epicondylitis.
Lateral epicondyle, **54, 61, 63, 64, 65, 71**
Lateral epicondylitis. *See* Elbow epicondylitis.
Lateral femoral condyle, tibial nerve, **256**
Lateral femoral cortex, **185, 189**
Lateral fibular cortex
- anterior compartment of leg, **228**
- anterolateral leg, panoramic, **227**
- lateral compartment of leg, **230, 231**
- peroneal nerve, **257**
Lateral hip ultrasound, **410**
Lateral intermuscular septum, **185, 190**
Lateral knee, ultrasound, **418**
Lateral malleolar artery, **291**
Lateral malleolus
- fibular ligaments, **272**
  posterior cortex, **272**
- lateral ankle, **270**
  posterior aspect, **270**
Lateral patellar retinaculum, **182**
Lateral pectoral nerve, axilla, **28**
Lateral plantar artery, **290, 291**
- lateral tarsus, **289**
- leg arteries, anterior and ankle (graphic), **238**
- medial tarsus, **288**
Lateral plantar compartment, tarsus, **277**
Lateral plantar nerve, **291**
- deep branch, **291**
- medial ankle, **265**
  graphic, **263**
- superficial branch, **291**

- tarsus, **277**
  lateral tarsus, **289**
  medial tarsus, **288**
Lateral plantar neurovascular bundle, lateral tarsus, **289**
Lateral plantar vein
- lateral tarsus, **289**
- medial tarsus, **288**
Lateral retinaculum, **206**
- anterolateral leg (graphic), **226**
- lateral leg (graphic), **232**
Lateral sesamoid, **297**
- deep plantar muscles, **285**
Lateral tarsal artery, **291**
- anterior and posterior leg nerves (graphic), **255**
- dorsal foot, **282, 283**
- leg arteries, anterior and ankle (graphic), **238**
Lateral thoracic artery, **45**
Lateral tibial condyle, **218, 221**
Lateral tibial cortex
- anterior compartment of leg, **228**
- anterolateral leg, panoramic, **227**
- lateral compartment of leg, **230**
- peroneal nerve, **257**
- transverse anterior tibial vessels, **246**
Lateral ulnar collateral ligament (LUCL), **55, 68**
- lateral elbow ultrasound, **382**
Latissimus dorsi muscle, **7, 12, 13, 24, 25, 28, 30, 34, 35, 37, 42, 311, 346**
- radial nerve, **144**
Latissimus dorsi tendon, **12, 37**
Ledderhose disease. *See* Plantar fasciitis and fibromatosis.
Left crus of diaphragm, **316, 328**
Leg muscles, **224–235**
- anatomy imaging issues, **225**
- anterior compartment, **225, 228–229**
- anterolateral leg
  graphics, **226**
  panoramic, **227**
- gross anatomy, **224–225**
- interosseous membrane, **224**
- lateral compartment, **225, 230, 231**
- lateral leg (graphic), **232**
- posterior compartment, **224–225**
- posterior leg
  graphic, **233**
  panoramic, **234**
  transverse, **235**
Leg nerves, **254–257**
- anatomy imaging issues, **254**
- anterior and posterior leg nerves (graphic), **255**
- gross anatomy, **254**
- peroneal nerve, **257**
- tibial nerve, **256**
Leg vessels, **236–253**
- anatomy imaging issues, **236**
- anterior tibial vessels
  longitudinal, **247**
  transverse, **246**
- arteries
  anterior and ankle (graphic), **238**
  gross anatomy, **236**
  posterior (graphic), **237**
  spectral Doppler of, **248**
- gastrocnemius veins, **251**
- great saphenous veins, **253**
- gross anatomy, **236**
- peroneal vessels
  longitudinal, **245**
  transverse, **244**
- popliteal vessels, **240**
  bifurcation, **241**
- posterior tibial vessels
  longitudinal, **243**
  transverse, **242**
- small saphenous vein, **252**
- soleal veins, **250**
- veins
  graphic, **239**
  gross anatomy, **236**
  spectral Doppler, **249**
Leiomyoma, vascular. *See* Vascular leiomyoma.
Leprosy, swollen nerve vs., **871**
Lesser saphenous vein
- anterolateral leg, panoramic, **227**
- fibular ligaments, **272**
- tributaries
  lateral ankle, **270, 271**
Lesser sciatic foramen, **349, 356**
Lesser trochanter, **172, 175, 182, 356**
Lesser tuberosity, **25**
Leukemia
- bone infection vs., **638**
- fat injury vs., **504**
Levator ani muscle, **316, 328**
Ligament, **533**
Ligament injury, **532–539**
- diagnostic checklist, **534**
- differential diagnosis, **534**
- prognosis, **534**
- staging, grading, & classification, **534**
- thumb, **120**
Ligament sprain. *See* Ligament injury.
Ligament strain. *See* Ligament injury.
Ligament tear, **533**
Ligamentum teres, **171, 172**
Linea alba, **314, 315, 326, 327**
Linea alba hernia, abdominal wall mass vs., **910**
Linea aspera, **191**
Linea semilunaris, **314, 315**
Lipoblastoma, **713**
Lipohemarthrosis, joint effusion vs., **900, 901**
Lipohypertrophy, **818**
- lipoma vs., **714**
Lipoma, **712–719**
- carpal tunnel syndrome, **603**
- chest wall lesion vs., **904**
- clinical issues, **714**
- diagnostic checklist, **714**
- differential diagnosis, **714**
- epidermoid cyst vs., **722**
- hyperechoic subcutaneous mass vs., **818**
- hypoechoic subcutaneous mass vs., **812**

- prognosis, **714**
- of spermatic cord, groin hernia vs., **806**
- subcutaneous, fat injury vs., **504**
Lipoma arborescens, **695, 713**
- synovial swelling vs., **895**
Lipomatosis, **713**
- of nerve, **713**
Liposarcoma, myxoid, paraarticular cystic mass vs., **889**
Lisfranc ligament
- dorsal foot ultrasound, **433**
- tarsus, **277**
Lister tubercle, **93, 94**
Little league elbow syndrome, ligament injury vs., **534**
Local tumor recurrence, **786–795**
- clinical issues, **788**
- diagnostic checklist, **788**
- differential diagnosis, **788**
- hyperechoic muscle mass vs., **831**
- hypoechoic muscle mass vs., **825**
- prognosis, **788**
Long dorsal sacroiliac ligament, **349**
Long plantar ligament
- deep plantar muscles, **285**
- lateral ankle, graphic, **269**
- medial and posterior ligaments, **266**
- tarsus, **277**
Long radiolunate ligament, **101**
Long saphenous vein, **192**
Long thoracic nerve, **31, 169**
Longissimus thoracis muscle, **317, 329**
Longitudinal arch of foot, tarsus, **276**
Longitudinal capsular ligament, **171**
Loose areolar tissue, **113**
Lower limb, joint injection, **942–949**
- outcomes, **944**
- post procedure, **944**
- preprocedure, **943**
- procedure, **943–944**
Lower subscapular nerve, **302**
Low-grade abscess, hematoma/seroma vs., **526**
Lumbar hernia, abdominal wall mass vs., **910**
Lumbrical muscle
- fingers, **130, 132, 136, 141**
- hand, digital branch, **118**
- sagittal foot overview, **287**
- superficial plantar muscles, **284**
Lumbricals, **102, 104, 106, 110**
- 1st, **104**
- 4th, **104**
- 2nd, **104**
- 3rd, **104, 111**
- lateral band, **104**
Lunate, **89, 94, 97, 113, 167**
- median nerve, **157, 158**
Lung, **311**
Lunotriquetral ligament, volar portion, **101**
Lymph node abnormality, **760–765**
- differential diagnosis, **762**
Lymph node enlargement. *See* Lymph node abnormality.
Lymph nodes, **28, 32**
- cystic soft tissue mass vs., **837**
- groin, **345, 347**
  femoral triangle, groin, **388**
  inguinal, superficial, **338**
- hypervascular soft tissue mass vs., **847**
- hypoechoic subcutaneous mass vs., **813**
- malignant, nerve sheath tumor vs., **592**
Lymphadenitis. *See* Lymph node abnormality.
Lymphadenopathy. *See* Lymph node abnormality.
Lymphangioma
- hypoechoic muscle mass vs., **825**
- paraarticular cystic mass vs., **889**
Lymphatic malformation
- chest wall lesion vs., **905**
- cystic soft tissue mass vs., **837**
Lymphatics, **339**
Lymphocele, **525**
Lymphoma
- fat injury vs., **504**
- hypervascular soft tissue mass vs., **847**
- hypoechoic subcutaneous mass vs., **813**
- synovial swelling vs., **895**

M

Main collateral ligament, fingers, **133**
Major ligaments, tarsus, **277**
Malaligned fracture, nonrotator cuff tendon tears associated with, **474**
Male breast carcinoma, chest wall lesion vs., **905**
Malignant bone tumor, bone surface lesion vs., **882**
Malignant lymph node, nerve sheath tumor vs., **592**
Malignant skin lesions, dermatofibrosarcoma protuberans, **731**
Malignant transformation, bone surface lesion vs., **882**
Mammary glands and ducts, **311**
Manubrium sternum, **5**
Massive tear, rotator cuff/biceps tendon, **455**
Masson hemangioma. *See* Vascular dilatation or inflammation.
Masson tumor. *See* Vascular dilatation or inflammation.
Medial ankle, ultrasound, **425–426**
Medial antebrachial cutaneous nerve, **38, 39**
Medial antebrachial nerve, **37**
Medial antecubital vein, **59**
Medial arcuate ligament, **316, 328**
Medial brachial cutaneous nerve, **37, 38**
Medial calcaneal nerve, medial ankle, graphic, **263**
Medial circumflex femoral artery, **193**
Medial circumflex femoral vein, **194, 345**
Medial collateral artery, **46**
Medial collateral bursa, **214**
Medial collateral ligament, **258**
- complex, **382**
- medial knee ultrasound, **417–418**
- superficial fibers, **214, 215**
Medial cuneiform
- dorsal foot, **281, 283**
- medial tarsus, **288**

INDEX

Medial cutaneous nerve, axilla, **28**
Medial elbow ultrasound, **381–382**
- common flexor tendon origin (CFTO), **381–382**
- cubital tunnel, **382**
- medial collateral ligament complex, **382**
Medial epicondylalgia. *See* Elbow epicondylitis.
Medial epicondyle, **54, 66, 69, 70, 71, 164**
Medial epicondylitis. *See* Elbow epicondylitis.
Medial femoral circumflex artery, **193**
Medial femoral condylar cartilage, **213**
Medial femoral condyle, **214**
Medial femoral cortex, **187, 189**
Medial hip ultrasound, **409–410**
Medial humeral condyle, avulsion fracture of, ligament injury vs., **534**
Medial intermuscular septum, **42**
- radial nerve, **144**
Medial knee, ultrasound, **417–418**
Medial malleolus
- deltoid ligament, **267, 268**
- medial ankle, **265**
- posterior surface, medial ankle, **264**
- posteromedial surface, medial ankle, **264**
Medial meniscus
- body of, medial knee ultrasound, **418**
- posterior horn, **215**
Medial navicular cortex, medial ankle, **265**
Medial nerve, **163**
- axilla, relationship of neurovascular structures, **31**
Medial patellar retinaculum, **182, 186**
Medial patellofemoral ligament (MPFL), **206, 214**
Medial patellotibial ligament, **214**
Medial patellotibial muscle, **214**
Medial pectoral nerve, axilla, **28**
Medial plantar artery, **290, 291**
- leg arteries, anterior and ankle (graphic), **238**
- tarsus, medial tarsus, **288**
Medial plantar compartment, tarsus, **277**
Medial plantar nerve, **291**
- medial ankle, **265**
  graphic, **263**
- tarsus, **277**
  medial tarsus, **288**
Medial plantar vein, tarsus, medial tarsus, **288**
Medial retinaculum, **206**
- anterolateral leg (graphic), **226**
Medial sesamoid, **297**
- deep plantar muscles, **285**
- superficial plantar muscles, **284**
Medial talar complex, deltoid ligament, **267, 268**
Medial talar tubercle, tibial nerve, **256**
Medial tarsal artery, dorsal foot, **283**
Medial tarsal vein, dorsal foot, **283**
Medial tennis elbow. *See* Elbow epicondylitis.
Medial tibial condyle, **213, 215**
Median arcuate ligament arches, **316, 328**
Median artery, hand, **114, 115**
Median crest (crista) of metatarsal, **297**
Median nerve, **31, 32, 37, 38, 39, 46, 54, 55, 56, 58, 66, 74, 75, 76, 77, 91, 97, 98, 106, 150–159, 169, 302, 304, 390**
- anterior elbow ultrasound, **381**
- branch, **98**
- carpal tunnel, transverse US, **157**
- compression, potential sites, **150**
- course and distribution of median nerve (graphic), **151**
- cubital fossa, **152**
- digital branches, **151**
- hand, **119**
  digital branch, **118**
  muscular branches, **119**
  palmar cutaneous branch, **119**
- longitudinal US, **158**
- nerves, **158**
- palmar nerves, **159**
- proximal forearm, **82**
- radial nerve, **145**
- recurrent branches of, **159**
- superficial branches, **106**
- transverse US, **83, 153–155, 158**
- wrist, transverse US, **156**
Median nerve edema, carpal tunnel syndrome, **603**
Meniscal cyst. *See* Parameniscal cyst.
Menisci, knee, **207**
Meniscofemoral ligament, **214–215**
Meniscotibial ligament, **214, 215**
Metabolic arthritis. *See* Gout and pseudogout.
Metabolic syndrome, gout and pseudogout with, **570**
Metacarpal
- 1st, **89**
- 3rd, **98**
- 3rd finger, **98**
- index finger, **98**
- ring finger, **98**
Metacarpal head, fingers, **133, 134, 136, 140**
Metacarpal shaft
- fingers, **133, 134**
- hand, **118**
Metacarpals
- 1st, **108, 109**
- 2nd, **106, 108, 109, 113**
- 3rd, **106, 113**
- 4th, **106, 113**
- 5th, **106, 110**
Metacarpophalangeal (MCP) joint
- fingers, **133**
- thumb, **123, 125**
  ulnar sesamoid of, **123, 124**
  ultrasound imaging, **123**
- thumb, UCL of, dorsal hand ultrasound, **397**
- volar hand ultrasound, **397**
Metastases
- bone surface lesion vs., **882**
- cystic soft tissue mass vs., **836**
- hypervascular soft tissue mass vs., **847**
- hypoechoic muscle mass vs., **825**
- hypoechoic subcutaneous mass vs., **813**
- muscle
  chest wall lesion vs., **904**
  hyperechoic muscle mass vs., **831**
  soft tissue sarcoma vs., **768**
- rib metastases, chest wall lesion vs., **904**
- swollen nerve vs., **871**

Metastatic infiltration, bone infection vs., **638**
Metastatic tumor. *See* Superficial metastases, lymphoma, and melanoma.
Metatarsal arteries, **293**
Metatarsals, **276–277, 296, 296–301**
- 1st, **296**
  head section and toes, **297**
  metatarsal head section and toes, **297**
- 2nd-4th, **296**
- 5th, **296**
- dorsum, **298**
- lateral ankle, **271**
  graphic, **269**
- lateral tarsus, **289**
- plantar, **299**
- plantar aponeurosis, **286**
- sagittal foot overview, **287**
- transverse, **300**
Metatarsophalangeal joint,1st, **296**
- dorsal foot ultrasound, **433**
Microbubble-enhanced US, **361**
Midcarpal compartment, **89**
Middle collateral artery, **38**
- radial nerve, **144**
Middle cuneiform
- plantar aponeurosis, **286**
- sagittal foot overview, **287**
Middle glenohumeral ligament, **14, 25**
Middle phalanx
- fingers, **135, 138, 139**
  proximal end, **137**
  shaft, **135**
- hand, **118**
Midfoot, **276**
Mirror-image artifacts, **364–365, 366**
Morel-Lavallée lesion, **503**
Morphea, fat injury vs., **504**
Morton foot, **296**
Motion artifact, **366**
Multifocal mononeuropathy (MMN), swollen nerve vs., **870**
Multiple echoes, artifacts due to, **364–365**
- mirror image, **364–365**
- reverberation, **364**
- ring down (resonance), **364**
Muscle abscess, muscle infarction vs., **511**
Muscle contusion. *See* Muscle injury.
Muscle edema, hyperechoic muscle mass vs., **830**
Muscle hematoma. *See also* Muscle injury.
- soft tissue infection vs., **630**
- soft tissue sarcoma vs., **768**
Muscle hernia. *See* Muscle injury.
Muscle infarction, **510–513**
- associated abnormalities, **512**
- diagnostic checklist, **512**
- differential diagnosis, **511–512**
- prognosis, **512**
- soft tissue infection vs., **630**
Muscle injury, **514–523**
- diagnostic checklist, **516**
- differential diagnosis, **516**
- prognosis, **516**
- staging, grading, & classification, **516**
Muscle metastases
- chest wall lesion vs., **904**
- hyperechoic muscle mass vs., **831**
- soft tissue sarcoma vs., **768**
Muscle strain. *See* Muscle injury.
Muscle tear. *See also* Muscle injury.
- abdominal wall mass vs., **911**
- hypoechoic muscle mass vs., **824**
Muscles of back in situ, **317, 329**
Muscular branches, **193**
Musculocutaneous nerve, **37, 38, 39, 40, 47, 77, 169, 302, 303, 304**
- axilla, **28**
- median nerve, **151**
Musculoskeletal artifacts, **364–371**
- attenuation error artifacts, **365**
  acoustic-enhancement artifact (increased through transmission), **365**
  acoustic-shadowing artifact (posterior acoustic shadowing), **365**
- beam characteristic artifacts, **364**
  anisotropic artifacts
    ligament injury vs., **534**
    rotator cuff/biceps tendinosis vs., **443**
  anisotropy, **364**
  beam width, **364**
  off-axis lobe, **364**
  slice thickness, **364**
- Doppler imaging, artifacts in, **365–366**
  aliasing, **366**
  blooming ("color bleed"), **366**
  directional ambiguity/indeterminate flow direction artifact, **366**
  edge, **366**
  flash, **366**
  improper Doppler angle with no-flow artifact, **366**
  mirror (or ghost)-image, **366**
  motion, **366**
  noise, **365**
  pseudoflow, **366**
  range ambiguity, **366**
  transducer pressure, **365**
  twinkling, **366**
- improper machine setting artifacts, **365**
  equipment generated, **365**
  range ambiguity, **365**
- multiple echo artifact, **364–365**
  mirror image, **364–365**
  reverberation, **364**
  ring down (resonance), **364**
- technological aspects, **364**
- velocity error artifacts, **365**
  defocusing (edge shadowing), **365**
  refraction, **365**
  velocity error, **365**
Musculoskeletal ultrasound, approach to, **360–363**
- clinical implications, **361**
- getting started, **360**
- machine setup and scanning, **360**
- newer ultrasound techniques, **361**

- recognizing pathology, **361**
- supplementary imaging techniques, **360–361**
- ultrasound machine, **360**

Musculotendinous junction, **17**
- in infraspinatus muscle, **23**

Mycobacterium, bone infection, **638**
Myeloproliferative disorder, fat injury vs., **504**
Myolipoma, **713**
Myonecrosis, aseptic. *See* Muscle infarction.
Myositis
- hyperechoic muscle mass vs., **831**
- infective or inflammatory, muscle infarction vs., **511**

Myositis ossificans, calcified soft tissue mass vs., **843**
Myxofibrosarcoma, paraarticular cystic mass vs., **889**
Myxoid liposarcoma, paraarticular cystic mass vs., **889**
Myxoid sarcoma, paraarticular cystic mass vs., **889**
Myxoma
- cystic soft tissue mass vs., **837**
- hypoechoic muscle mass vs., **824**
- paraarticular cystic mass vs., **889**

N

Nail, fingers, **139**
Nail base, thumb, **126**
Nail bed, fingers, **139**
Nasal bone fracture, **542**
Navicular bone, **276**
- anterior ankle, **262**
- deltoid ligament, **268**
- dorsal foot, **283**
    - surface, **282**
- lateral ankle, graphic, **269**
- medial and posterior ligaments, **266**
- medial tarsus, **288**
- plantar aponeurosis, **286**
- sagittal foot overview, **287**

Necrotizing fasciitis, **629**
- fascial lesion vs., **877**

Necrotizing vasculitis, fat injury vs., **504**
Neisseria gonorrhoea, joint infection, **648**
Nephropathy, muscle infarction with, **512**
Nerve entrapment
- nerve injury vs., **584**
- radial nerve, **142**
- swollen nerve vs., **870**

Nerve injury, **582–589**
- diagnostic checklist, **584**
- differential diagnosis, **584**
- prognosis, **584**
- staging, grading, & classification, **584**
- swollen nerve vs., **870**

Nerve sheath tumor, **590–601**
- benign peripheral, vascular leiomyoma vs., **735**
- calcified soft tissue mass vs., **843**
- diagnostic checklist, **592**
- differential diagnosis, **592**
- epidermoid cyst vs., **722**
- hypervascular soft tissue mass vs., **846**
- hypoechoic muscle mass vs., **824**
- lymph node abnormality vs., **762**
- nerve injury vs., **584**
- peritendinous mass vs., **853**
- staging, grading, & classification, **592**
- superficial metastases, lymphoma, and melanoma vs., **740**
- swollen nerve vs., **870**

Nerves
- ribs and intercostal space, **310**
- tarsus, **277–228**

Neurapraxia, **583**. *See also* Nerve injury.
Neurofibroma, swollen nerve vs., **870**
Neurofibromatosis type 1, nerve sheath tumor and, **591**
Neurological sequelae, bone surface lesion vs., **882**
Neuropathy, muscle infarction with, **512**
Neurotmesis, **583**
Neurovascular structures, relationship of, **31**
Newer ultrasound techniques, **361**
Nipple, **311**
Nodular fasciitis, fascial lesion vs., **877**
Noise artifact, **365**
Noninfectious fasciitis, soft tissue infection vs., **630**
Nonrotator cuff tendinosis, **462–471**
- diagnostic checklist, **464**
- differential diagnosis, **464**
- prognosis, **464**

Nonrotator cuff tendon tears, **472–483**
- complete, **473**
- diagnostic checklist, **474**
- differential diagnosis, **474**
- full thickness, **473**
- high-grade partial, **473**
- partial thickness, **473**
- prognosis, **474**

Normal longitudinal arch, of foot, tarsus, **276**
Normal lymph node, hyperechoic subcutaneous mass vs., **819**
Normal venous pattern, **50**

O

OA. *See* Osteoarthritis.
Obesity, abdominal, gout and pseudogout with, **570**
Oblique abdominal muscle, gluteal muscles, **352**
Oblique and transverse muscles, **316, 328**
Oblique cord, **68**
Oblique pulley, **103**
Obturator artery, **193**
Obturator externus muscle, **175, 180, 342, 346**
- gluteal muscles, **348**

Obturator internus muscle, **186, 195**
- gluteal muscles, **348, 350, 351**
    - nerve to, **349**

Obturator internus tendon, **204, 355, 356**
Obturator membrane, **186**
Obturator nerve
- anterior branch, **341**

- posterior branch, **341**
Obturator vein, **194**
Off-axis lobe artifact, **364**
Olecranon, **68, 71, 162**
Olecranon bursa, **52, 669**
Olecranon process, **54, 55**
OP. *See* Opponens pollicis.
Opponens brevis muscle, **106**
Opponens digiti minimi brevis muscle, **110**
Opponens digiti minimi muscle, **96, 103, 106, 107, 168**
- fingers, **131**
- median nerve, **158**
Opponens pollicis muscle, **98, 107, 108, 109**
- hand, **116**
- median nerve, **157, 158**
- thumb, **120, 121, 122, 123, 127, 128**
Organizing hematoma, **525**
Orthopedic implant, bone surface lesion vs., **883**
Osteoarthritis, **550–557**
- diagnostic checklist, **552**
- differential diagnosis, **552**
- inflammatory arthritis vs., **560**
- joint effusion vs., **900**
- prognosis, **552**
- staging, grading, & classification, **552**
- synovial swelling vs., **894**
Osteochondroma, bone surface lesion vs., **882**
Osteochondromatosis, synovial, joint effusion vs., **900, 901**
Osteomyelitis, **637**
- acute, **637**
- bone tumor vs., **776**
- chronic, **637**
Osteosarcoma, bone surface lesion vs., **882**

P

Palm, transverse US, **106**
Palmar aponeurosis, **106**
- median nerve, **158**
Palmar arch, hand
- deep, **114, 119**
  radial branch, **115**
  ulnar branch, **115**
- superficial, **114, 119**
  radial branch, **115**
  ulnar branch, **115**
Palmar arch artery, thumb, deep, **127, 128**
Palmar carpal arch
- radial branch, **115**
- ulnar branch, **115**
Palmar digital vein, **86**
Palmar interossei muscle, **102, 105, 110**
- 1st, **105**
- 2nd, **105**
- 3rd, **105**
- fingers, **130, 136**
- hand, **118**
Palmar nerves, median nerve, **159**
Palmar ulnar cutaneous nerve, **390**
Palmaris brevis muscle, **106, 110, 168**
- slip of, **168**
Palmaris longus muscle, **56, 67, 72, 74, 75, 78**
- median nerve, **155**
Palmaris longus tendon, **70, 88, 91, 97**
- median nerve, **156, 157**
Pancreatic carcinoma, fat injury vs., **504**
Pancreatic disease, fat injury vs., **504**
Pancreatitis, fat injury vs., **504**
Panniculitis
- calcified soft tissue mass vs., **842**
- hyperechoic subcutaneous mass vs., **819**
- hypervascular soft tissue mass vs., **847**
- subcutaneous, fat injury vs., **504**
Paraarticular aspiration, elbow
- indication, **959**
- procedure, **959**
Paraarticular cystic mass, differential diagnosis, **888–893**
- abscess or collection, **889**
- Baker cyst, **888**
- bursal distension, **888–889**
- cystic-looking soft tissue tumors, **889**
- ganglion cyst, **888**
- liquefied hematoma, **889**
- parameniscal cyst, **888**
- pseudoaneurysm, **889**
Paraarticular ganglion cyst, parameniscal cyst vs., **692**
Paraarticular tumor, cystic, Baker cyst vs., **664**
Parameniscal cyst, **690–693**
- Baker cyst vs., **663–664**
- cystic soft tissue mass vs., **836**
- diagnostic checklist, **692**
- differential diagnosis, **692**
- paraarticular cystic mass vs., **888**
Parasitic infection, hypoechoic muscle mass vs., **825**
Parasitic infestation, calcified soft tissue mass vs., **843**
Paraspinal muscles, **314, 324, 326, 336**
Paraspinal structures, **326–337**
- pathology, **337**
Paratendinous cystic lesion, tenosynovitis vs., **486**
Paraumbilical hernia, abdominal wall mass vs., **910**
Pariarticular ganglion, Baker cyst vs., **664**
Parietal pleura, **311**
Partial tear, rotator cuff/biceps tendon, **455**
Partial-thickness tear
- nonrotator cuff tendon tears, **473**
- rotator cuff/biceps tendon, **455**
- tendon hyperechogenicity vs., **861**
- tendon hypoechogenicity vs., **856–857**
Patella, **206, 208, 210–211**
- anterior knee ultrasound, **417**
- anterior surface, **209, 213**
- dorsal surface, **212**
- lateral articular facet, **213**
- medial articular facet, **213**
- upper pole, **210**
Patellar retinaculum, **213**
- anterior knee ultrasound, **417**
- lateral, **213, 217**
- medial, **212**

Patellar tendon, **206, 208–209, 211–212, 217**
- anterior knee ultrasound, **417**
- nonrotator cuff tendinosis, **464**
Pathologic fracture, **541**
Pecten, pubis, **338**
Pectineus muscle, **173, 174, 175, 180, 182, 184, 195, 197, 200**
- groin muscles, **339, 340, 341, 343, 344, 346, 347**
Pectoral branch, thoracoacromial artery, **45**
Pectoral nerve, **169**
Pectoralis major muscle, **6, 30, 32, 33, 37, 310, 311**
Pectoralis major tendon, **37**
Pectoralis minor muscle, **30, 31, 32, 33, 310, 311**
Penetrating injury, nonrotator cuff tendon tears associated with, **474**
Perforating artery, **193**
Perforating branch peroneal artery
- anterior and posterior leg nerves (graphic), **255**
- leg arteries, anterior and ankle (graphic), **238**
Perforating vein, **86, 185, 194**
Perforating vessels, **193**
Perforator, great saphenous veins, **253**
Perineural fibrosis, swollen nerve vs., **870**
Peripheral nerve sheath tumor
- hyperechoic muscle mass vs., **831**
- hypoechoic subcutaneous mass vs., **812**
Peritendinitis
- nonrotator cuff tendinosis, **463**
- tendinosis with, nonrotator cuff tendon tears vs., **474**
- tendon hyperechogenicity vs., **861**
Peritendinous mass, **852–855**
- differential diagnosis, **852**
Peroneal artery, **236, 290**
- anterior and posterior leg nerves (graphic), **255**
- anterior compartment of leg, **228, 229**
- anterolateral leg, panoramic, **227**
- lateral ankle, **270**
- lateral compartment of leg, **230, 231**
- leg arteries, posterior (graphic), **237**
- longitudinal anterior tibial vessels, **247**
- longitudinal peroneal vessels, **245**
- popliteal vessel bifurcation, **241**
- posterior leg, transverse, **235**
- soleal veins, **250**
- spectral Doppler of leg arteries, **248**
- tarsus, **278**
- transverse anterior tibial vessels, **246**
- transverse peroneal vessels, **244**
- transverse posterior tibial vessels, **242**
Peroneal communicating nerve, **219**
Peroneal nerve, **257**
- posterior knee ultrasound, **418**
- tibial nerve, **256**
Peroneal tendons, lateral ankle ultrasound, **426**
Peroneal trochlea, lateral ankle, **271**
Peroneal vein
- anterior compartment of leg, **229, 230**
- anterolateral leg, panoramic, **227**
- lateral ankle, **270**
- lateral compartment of leg, **230, 231**
- leg veins (graphic), **239**
- longitudinal anterior tibial vessels, **247**
- longitudinal peroneal vessels, **245**
- popliteal vessel bifurcation, **241**
- soleal veins, **250**
- spectral Doppler of leg arteries, **248**
- spectral Doppler of leg veins, **249**
- transverse anterior tibial vessels, **246**
- transverse peroneal vessels, **244**
- transverse posterior tibial vessels, **242**
Peroneal vessels
- anterolateral leg, panoramic, **227**
- leg vessels
  longitudinal, **245**
  transverse, **244**
- posterior leg, transverse, **235**
Peroneocalcaneus internus muscle, **225**
Peroneus brevis muscle, **225**
- anterolateral leg, panoramic, **227**
- lateral ankle, **270**
- lateral compartment of leg, **230, 231**
- lateral leg (graphic), **232**
Peroneus brevis tendon, **259**
- fibular ligaments, **272**
- lateral ankle, **270, 271**
  graphic, **269**
- medial ankle, graphic, **263**
Peroneus digiti minimi muscle, **225**
Peroneus longus muscle, **225**
- anterior and posterior leg nerves (graphic), **255**
- anterior compartment of leg, **228**
- anterolateral leg
  graphic, **226**
  panoramic, **227**
- lateral compartment of leg, **230, 231**
- lateral leg, **232**
- leg arteries, anterior and ankle (graphic), **238**
- posterior leg, panoramic, **234**
- transverse peroneal vessels, **244**
Peroneus longus tendon, **259**
- anterior and posterior leg nerves (graphic), **255**
- fibular ligaments, **272**
- groove, lateral ankle, **271**
- lateral ankle, **270, 271**
  graphic, **269**
- lateral compartment of leg, **230, 231**
- lateral leg, **232**
- lateral tarsus, **289**
- leg arteries, anterior and ankle (graphic), **238**
- medial ankle, graphic, **263**
- sagittal foot overview, **287**
Peroneus quartus muscle, **225**
Peroneus tertius muscle, **225**
- anterior ankle, **261**
Peroneus tertius tendon, **259**
- anterior ankle, **261**
- dorsal foot, **282**
Peroneus tunnel
- anterior and posterior leg nerves (graphic), **255**
- leg arteries, anterior and ankle (graphic), **238**
Pes anserine, **180–181, 186, 206–207**
Pes anserine bursa, **669**

INDEX

Pes anserine tendon, **182**
- medial knee ultrasound, **418**
Phalangeal apparatus, **296**
Phalanges, **276–277, 296**
Phalanx, fingers, **141**
Phlegmon, **629**
Physiological immature hip, developmental hip dysplasia vs., **578**
Physis, bone fracture vs., **542**
Pigmented villonodular synovitis (PVNS)
- joint effusion vs., **900, 901**
- synovial swelling vs., **894**
Pilomatricoma, **726–729**
- calcified soft tissue mass vs., **842**
- differential diagnosis, **727**
- genetics, **727**
- hyperechoic subcutaneous mass vs., **819**
- hypervascular soft tissue mass vs., **846–847**
- hypoechoic subcutaneous mass vs., **812**
- prognosis, **727**
Pilomatrixoma. *See* Pilomatricoma.
Piriformis muscle, **186, 195, 204, 316, 328, 346**
- gluteal muscles, **350, 351, 353, 355**
Pisiform, **89, 97, 167**
- hand, **116**
- median nerve, **157**
Pisohamate ligament, median nerve, **159**
Pisotriquetral compartment, **89**
PL. *See* Palmaris longus muscle.
Plantar aponeurosis, **277, 286**
- central band
  lateral tarsus, **289**
  sagittal foot overview, **287**
- lateral band
  lateral tarsus, **289**
  sagittal foot overview, **287**
- medial band
  medial tarsus, **288**
  sagittal foot overview, **287**
Plantar arch, **290, 291**
Plantar arteries, **291, 294–295**
Plantar bursitis, plantar fasciitis and fibromatosis vs., **708**
Plantar calcaneocuboid (spring) ligament, tarsus, **277**
Plantar calcaneonavicular ligament, **258**
Plantar compartment, tarsus, **277**
Plantar digital nerves, **291**
Plantar fascia, plantar foot ultrasound, **433**
Plantar fascial tear
- fascial lesion vs., **876–877**
- plantar fasciitis and fibromatosis vs., **708**
Plantar fasciitis and fibromatosis, **706–711**
- clinical issues, **708**
- diagnostic checklist, **708**
- differential diagnosis, **708**
- fascial lesion vs., **876**
- prognosis, **708**
- tarsal tunnel syndrome vs., **616**
Plantar fasciitis injection
- indication, **987**
- procedure, **988**
Plantar fasciopathy. *See* Plantar fasciitis and fibromatosis.
Plantar fat pad, plantar aponeurosis, **286**
Plantar fibroma. *See* Plantar fasciitis and fibromatosis.
Plantar fibromatosis. *See* Plantar fasciitis and fibromatosis.
Plantar foot ultrasound, **433**
- flexor tendons, **433**
- plantar fascia, **433**
- plantar muscles, **433**
- toe web spaces, **433**
Plantar metatarsal, **299**
Plantar muscles, **277**
- deep layers (graphic), **285**
- plantar foot ultrasound, **433**
- superficial layers (graphic), **284**
Plantar plate, **296, 297**
Plantar venous arch, **290**
Plantaris muscle, **220–222, 224**
- posterior leg (graphic), **233**
Plantaris tendon, **259**
- posterior ankle ultrasound, **426**
- posterior leg (graphic), **233**
Podagra. *See* Gout and pseudogout.
Polyarteritis nodosa, fat injury vs., **504**
Popliteal artery, **185, 191, 193, 196, 197, 198, 199, 202, 203, 220, 222–223, 236**
- anterior and posterior leg nerves (graphic), **255**
- anterior compartment of leg, **228**
- leg arteries, posterior (graphic), **237**
- longitudinal anterior tibial vessels, **247**
- popliteal vessels, **240**
- tibial nerve, **256**
Popliteal artery aneurysm, Baker cyst vs., **664**
Popliteal cyst. *See* Baker cyst.
Popliteal hiatus, **216**
Popliteal muscle, **222**
- anterior and posterior leg nerves (graphic), **255**
Popliteal synovial cyst. *See* Baker cyst.
Popliteal tendon, **218, 222**
Popliteal vein, **185, 191, 194, 196, 197, 198, 202, 203, 220, 222, 236**
- anterior compartment of leg, **228**
- gastrocnemius veins, **251**
- leg veins (graphic), **239**
- popliteal vessels, **240**
- small saphenous vein, **252**
Popliteal vessels, **208**
- posterior leg, panoramic, **234**
Popliteofibular ligament, **216**
Popliteomeniscal ligament, **216**
Popliteus muscle, **224**
- leg arteries, posterior (graphic), **237**
- popliteal vessels, **240**
- posterior leg, transverse, **235**
Popliteus tendon, **216**
- lateral knee ultrasound, **418**
Posterior ankle, ultrasound, **426**
Posterior antebrachial cutaneous nerve, **38, 42**
- radial nerve, **144**
Posterior axillary fold, artifact due to, **35**
Posterior axillary wall, transverse US, **35**
Posterior band, **68**
Posterior calcaneus, posterior ankle panorama, **273**

Posterior capsule, **208, 214, 216**
Posterior circumflex humeral artery, **12, 28, 34, 42, 45**
- radial nerve, **144**
Posterior compartment, leg muscles
- deep muscles, **224–225**
- superficial muscles, **224**
Posterior cord, radial nerve, **143**
Posterior cortex
- of medial tibial condyle, posterior leg, panoramic, **234**
- of tibia, posterior leg, panoramic, **234**
Posterior cruciate ligament, **208, 214, 216**
- posterior knee ultrasound, **418**
- synovium, **208**
Posterior cruciate tendon, **216**
Posterior cutaneous nerve, radial nerve, **149**
Posterior elbow ultrasound, **382**
- triceps tendon, **382**
Posterior fat pad, **54, 71**
Posterior femoral cortex, **190, 191, 204**
- linea aspera, **185**
Posterior fibular cortex
- posterior leg
  panoramic, **234**
  transverse, **235**
- tibial nerve, **256**
Posterior glenoid labrum, **27**
Posterior hip, ultrasound, **410**
Posterior horn medial meniscus, **214**
Posterior inferior iliac spine, **355**
Posterior intermuscular septum
- lateral compartment of leg, **230, 231**
- panoramic anterolateral leg, **227**
Posterior interosseous artery, **80**
- forearm arteries, **81**
- hand, **115**
Posterior interosseous nerve, **56, 79, 143, 389–390**
Posterior knee, ultrasound, **418**
Posterior oblique ligament, **214–215**
Posterior sacral foramen, **354**
Posterior sacroiliac ligament, **350**
- of pelvis, **349**
Posterior spinous process, **311**
Posterior subtalar joint, posterior ankle panorama, **273**
Posterior suprapatellar fat pad, **209–210**
Posterior talofibular ligament, **258**
- fibular ligaments, **272**
- lateral ankle, graphic, **269**
- medial and posterior ligaments, **266**
Posterior talus, medial tubercle, medial ankle, **264**
Posterior thigh musculature, hip ultrasound, **410**
Posterior tibial artery, **236, 290**
- anterior and posterior leg nerves (graphic), **255**
- anterior compartment of leg, **228**
- great saphenous veins, **253**
- leg arteries
  anterior and ankle (graphic), **238**
  posterior (graphic), **237**
- longitudinal posterior tibial vessels, **243**
- medial ankle, **264, 265**
- popliteal vessel bifurcation, **241**
- posterior leg, transverse, **235**
- soleal veins, **250**
- spectral Doppler of leg arteries, **248**
- spectral Doppler of leg veins, **249**
- tarsus, **278**
- tibial nerve, **256**
- transverse anterior tibial vessels, **246**
- transverse peroneal vessels, **244**
- transverse posterior tibial vessels, **242**
Posterior tibial cortex
- lateral compartment of leg, **230**
- posterior leg
  panoramic, **234**
  transverse, **235**
- tibial nerve, **256**
Posterior tibial disorders, tarsal tunnel syndrome vs., **616**
Posterior tibial nerve, medial ankle, graphic, **263**
Posterior tibial vein
- anterior compartment of leg, **228**
- great saphenous veins, **253**
- leg veins (graphic), **239**
- longitudinal posterior tibial vessels, **243**
- medial ankle, **264, 265**
- popliteal vessel bifurcation, **241**
- posterior leg
  panoramic, **234**
  transverse, **235**
- soleal veins, **250**
- spectral Doppler of leg arteries, **248**
- spectral Doppler of leg veins, **249**
- tibial nerve, **256**
- transverse anterior tibial vessels, **246**
- transverse peroneal vessels, **244**
- transverse posterior tibial vessels, **242**
Posterior tibial vessels
- longitudinal, **243**
- longitudinal anterior tibial vessels, **247**
- medial ankle, **265**
- posterior leg, transverse, **235**
- transverse, **242**
Posterior tibialis, nonrotator cuff tendinosis, **464**
Posterior tibiofibular ligament, **258**
- lateral ankle, graphic, **269**
- medial and posterior ligaments, **266**
Posterior tibiotalar
- deep deltoid, medial and posterior ligaments, **266**
- deltoid ligament, **267**
- superficial deltoid, medial and posterior ligaments, **266**
Posterior tibiotalar ligament, medial ankle, graphic, **263**
Posterior tubercle of calcaneus, posterior leg, panoramic, **234**
Posterior ulnar recurrent artery, **55, 68, 81, 162**
Posterior wall acetabulum, **356**
Posterolateral corner structures, posterior knee ultrasound, **418**
Posterolateral cortex of tibia, lateral compartment of leg, **231**
Posteromedial intermuscular septum, **191**
Posterosuperior internal impingement, rotator cuff/biceps tendon tear associated with, **456**
Postoperative fibrosis/granulation tissue, local tumor recurrence vs., **788**

Postoperative infection, **652–657**
- diagnostic checklist, **654**
- differential diagnosis, **654**
- prognosis, **654**
Postoperative inflammation, postoperative infection vs., **654**
Postoperative seroma/hematoma, local tumor recurrence vs., **788**
Posttraumatic hemarthrosis or lipohemarthrosis. *See* Hemarthrosis and lipohemarthrosis.
PQ. *See* Pronator quadratus.
Pre-Achilles fat pad, lateral leg (graphic), **232**
Prepatellar bursa, **669**
Prepatellar bursitis (housemaid's knee). *See* Bursitis.
Primary amyloidosis, tendon swelling vs., **865**
Princeps pollicis artery, **108, 109**
- hand, **119**
- thumb, **123, 125, 127, 128**
Princeps pollicis vein, thumb, **125**
Profunda brachii artery, **43**
- radial nerve, **145, 146**
Profunda brachii vein, radial nerve, **145**
Profunda brachii vessels, radial nerve, **146**
Profunda femoral artery, **175, 183, 188, 193, 196, 198, 199, 345**
Profunda femoral vein, **183, 188, 202, 345**
Profunda femoris, perforating vessel, **185**
Proliferative tenosynovitis, acute, **485**
Prominent xiphoid process, chest wall lesion vs., **905**
Prominent/floating rib, chest wall lesion vs., **905**
Promontory, **179**
Pronator quadratus muscle, **72, 74, 75, 96, 97, 166**
- longitudinal US, **84**
- median nerve, **156**
- transverse US, **83, 84**
Pronator teres muscle, **37, 54, 55, 56, 60, 61, 65, 66, 67, 70, 75, 76, 77, 78, 164, 165**
- median nerve, **151, 154**
  humeral head, **152**
- proximal forearm, **82**
- radial nerve, **147**
- transverse US, **83**
Proper digital arteries, hand, **119**
Proteus, postoperative infection, **654**
Proximal carpal row, **88**
Proximal femur, **353**
Proximal interphalangeal joint, **135, 139**
- capsule, **111, 132**
Proximal phalanx
- fingers, **133, 134, 135, 136, 139, 140**
  extent of articular cartilage, **133**
  head, **137**
  shaft, **137**
- thumb, **123, 124, 126, 129**
  base, **123**
Proximal radioulnar joint, **53**
Pseudoaneurysm, **621, 622**. *See also* Vascular dilatation or inflammation.
- bone surface lesion vs., **882**
- cystic soft tissue mass vs., **837**
- paraarticular cystic mass vs., **889**
Pseudoflow artifact, **366**
Pseudoganglion, **31**
Pseudomonas
- bone infection, **638**
- postoperative infection, **654**
Pseudomonas aeruginosa, joint infection, **648**
Psoas, **314, 326**
Psoas major muscle, **316, 328**
- groin muscles, **340**
Psoas minor muscle, **316, 328**
Psoas muscle, **175, 182, 186, 339**
Psoriatic arthropathy. *See* Inflammatory arthritis.
Pubic bone, **338, 344**
- body of, **342, 343**
- superior surface of, **342**
Pubic crest, **338**
Pubic ligament, superior, **338**
Pubic symphysis, **338, 342**
Pubic tubercle, **338, 339, 347**
Pubofemoral ligament, **172**
Pudendal nerve, **349**
- lesser sciatic foramen, **349**
Pudendal vessels, internal, **349**
"Pulled muscle." *See* Muscle injury.
Pulp space, tendon, **135**
Pulvinar, **172**
Pyomyositis, **629**

Q

Quadrate tubercle of posterior cortex of femur, **356**
Quadratus femoris muscle, **186, 195, 204, 346**
- gluteal muscles, **348, 350, 351, 353, 355, 356, 357**
Quadratus lumborum, **314, 316, 326, 328**
Quadratus plantae muscle
- lateral tarsus, **289**
- medial ankle, graphic, **263**
- medial tarsus, **288**
- plantar aponeurosis, **286**
- sagittal foot overview, **287**
- superficial plantar muscles, **284**
Quadriceps, **206**
Quadriceps femoris, **180**
Quadriceps tendon, **209–210**
- anterior knee ultrasound, **417**
- contributing fibers to, **184**
- leg
  anterolateral leg (graphic), **226**
  lateral leg, **232**
Quadrilateral space, **28**
- axilla, **28**
  transverse US, **34**
Quadrilateral space syndrome, **28**

R

Radial artery, **56, 60, 74, 76, 77, 80, 97, 99**
- distal forearm and hand, **85**
- Doppler gate in, **85**
- forearm arteries, **81**
- hand, **114, 115, 117, 119**
  contribution to superficial palmar arch, **119**
  palmar cutaneous branch, **119**
  superficial palmar branch, **116**
- proximal forearm, **82**
- radial nerve, **147**
- spectral trace, **85**
- superficial and deep branch, **92**
- transverse US, **83**

Radial bursa, **103**
- fingers, **131**
- thumb, **121**

Radial collateral artery, **38, 42, 46**
- radial nerve, **144**

Radial collateral ligament, **52, 53, 55, 63, 68, 101**
- lateral elbow ultrasound, **382**

Radial head, **53, 55, 57, 59, 61, 63, 65, 71**
- radial nerve, **147**

Radial nerve, **12, 37, 38, 39, 42, 43, 46, 47, 54, 55, 56, 58, 59, 60, 61, 74, 76, 77, 142–149, 163, 169, 302, 303, 304**
- anterior elbow ultrasound, **381**
- arm, transverse US, **145**
- axilla, **28, 31, 32**
  relationship of neurovascular structures, **31**
- course and distribution of radial nerve (graphic), **143**
- deep branch, **77, 79, 147, 148, 149, 304**
- dorsal digital branches, **149**
- elbow and wrist, **149**
- forearm, transverse US, **148**
- median nerve, **152, 153**
  superficial branch, **155**
- posterior arm, **144**
- posterior forearm, transverse US, **147**
- posterior interosseus branch, **149**
- in spiral groove, **43**
- spiral groove, transverse US, **146**
- subcutaneous branch, **76**
  proximal forearm, **82**
- superficial branch, **77, 147, 148, 149, 304**
  proximal forearm, **82**
  transverse US, **83**
- superficial terminal branch, **143**

Radial recurrent artery, **59, 60**
- forearm arteries, **81**
- radial nerve, **147, 148**

Radial sesamoid bone, thumb, **123, 125**

Radial shaft, **57**
- radial nerve, **148**

Radial styloid process, **89**

Radial tuberosity, **60**

Radial vein, **60, 74, 86, 92**
- hand, **117**
- proximal forearm, **82**
- transverse US, **83**

Radialis indicis artery, hand, **119**

Radiocarpal compartment, **89**

Radiohumeral joint, **63**

Radiohumeral synovial flange, hypertrophy of, elbow epicondylitis vs., **496**

Radioscaphocapitate ligament, **101**

Radioscapholunate ligament, **101**

Radiosynovectomy, indications, **959**

Radioulnar bursa, **52**

Radius, **56, 57, 74, 75, 76, 79, 92, 93, 94, 95, 97, 99, 304**
- distal end, **92**
- median nerve, **155, 156**
- proximal forearm, **82**
- radial nerve, **147, 148**
- transverse US, **83**

Range ambiguity artifact, **365, 366**

Reactive bone change at resection margin, local tumor recurrence vs., **788**

Reactive bursitis, bone surface lesion vs., **882**

Reactive lymph node, superficial metastases, lymphoma, and melanoma vs., **740**

Reactive myositis, bone surface lesion vs., **882**

Rectum, **316, 328**

Rectus abdominis aponeurosis, **342**

Rectus abdominis muscle, **314, 326**
- divarication of, abdominal wall mass vs., **911**
- groin, **338, 342**

Rectus femoris artery, **175**

Rectus femoris muscle, **173, 175, 180, 182, 183, 184, 186, 187, 208–209**
- groin muscles, **339, 340, 346, 347**
- leg
  anterolateral leg (graphic), **226**
  lateral leg (graphic), **232**
- long head, **173, 176**

Rectus femoris tendon, **184, 208, 211**
- anterior hip ultrasound, **409**

Rectus muscle, **315, 327**

Rectus sheath, **315, 327, 342**

Rectus sheath hematoma, **525**
- abdominal wall hernia vs., **798**
- abdominal wall mass vs., **911**

Recurrent articular branch, **254**

Refraction artifact, **365**

Renal stones, gout and pseudogout with, **570**

Retinopathy, muscle infarction with, **512**

Retrocalcaneal bursa, **669**
- Achilles tendon
  axial, **275**
  longitudinal, **274**

Retrocalcaneal fat, Achilles tendon
- axial, **275**
- longitudinal, **274**

Retrocondylar bursa, **214**

"Retrocondylar groove." *See* Cubital tunnel syndrome.

Reverberation artifact, **364**

Rhabdomyolysis, hyperechoic muscle mass vs., **831**

Rheumatic fever, fat injury vs., **504**

Rheumatoid arthritis. *See also* Inflammatory arthritis.
- gout and pseudogout vs., **570**
- nonrotator cuff tendon tears associated with, **474**
Rheumatoid nodule
- gout and pseudogout vs., **570**
- hypoechoic subcutaneous mass vs., **813**
Rhomboid muscle, **311**
Rib fracture, **542**
- healing, chest wall lesion vs., **904**
Rib metastases, chest wall lesion vs., **904**
Ribs, **310–313**
- 1st, **5, 6**
- general anatomy and function, **310**
- graphics, **311**
- muscles, **310**
- skeletal structures, **310**
- vessels and nerves, **310**
Right crus of diaphragm, **316, 328**
Ring-down (resonance) artifact, **364**
Rotator cable
- rotator cuff/biceps tendinosis vs., **444**
- rotator cuff/biceps tendon tear vs., **456**
Rotator cuff, **10**
- cable, imaging pitfalls, **11**
- interval, **10, 18**
  tendinous interspace, **11**
Rotator cuff arthropathy, rotator cuff/biceps tendon tear associated with, **456**
Rotator cuff musculature, **374**
Rotator cuff tendinopathy. *See* Rotator cuff/biceps tendinosis.
Rotator cuff tendinosis or partial tear therapy
- indications, **951**
- procedure, **953**
Rotator cuff tendon, blood supply, **10**
Rotator cuff/biceps tendinosis, **442–453**
- calcific tendinosis, **443**
- diagnostic checklist, **444**
- differential diagnosis, **443–444**
- genetics, **444**
- rotator cuff/biceps tendon tear vs., **456**
- staging, grading, & classification, **444**
Rotator cuff/biceps tendon tear, **454–461**
- associated abnormalities, **456**
- complete, **455**
- diagnostic checklist, **456**
- differential diagnosis, **456**
- full thickness, **455**
- massive, **455**
- partial, **455**
- partial thickness, **455**
- prognosis, **456**
Rotator interval, **373**

S

Sacral median crest, **354**
Sacroiliac joint, **348–349, 354**
Sacroiliac joint injection
- indications, **971**
- procedures, **971–972**
Sacroiliac ligaments, **349**
- long dorsal, **349**
- posterior, **349**
  pelvis, **349**
- short dorsal, **349**
Sacrospinous ligament, **186, 349**
Sacrotuberous ligament, **186, 349, 350, 355**
Saddlebag hernia, **805**
Sagittal band, **111**
- fingers, **132, 133**
Saphenofemoral junction, **200, 201**
Saphenopopliteal junction, small saphenous vein, **252**
Saphenous muscle, **195, 219**
Saphenous nerve, **192, 219, 254**
Saphenous opening, **339**
Saphenous vein
- Doppler, **201**
Sarcoidosis, fat injury vs., **504**
Sarcoma
- alveolar soft part, vascular anomaly vs., **746**
- angiosarcoma, vascular dilatation or inflammation vs., **622**
- bone, bone infection vs., **638**
- calcified soft tissue mass vs., **842**
- cystic soft tissue mass vs., **837**
- hypoechoic subcutaneous mass vs., **813**
- myxoid, paraarticular cystic mass vs., **889**
- soft tissue
  hematoma/seroma vs., **526**
  superficial metastases, lymphoma, and melanoma vs., **740**
- synovial chondrosarcoma, synovial swelling vs., **895**
- synovial sarcoma, synovial swelling vs., **895**
Sartorius muscle, **173, 175, 176, 180, 182, 183, 186, 187, 188, 189, 190, 195, 196, 197, 198, 201, 203, 208, 214, 223**
- groin muscles, **339, 340, 345, 346, 347**
- medial knee ultrasound, **418**
Sartorius tendon, **219**
- leg
  anterolateral leg (graphic), **226**
  posterior leg (graphic), **233**
SA-SD. *See* Subacromial-subdeltoid bursa injection.
Saturday-night palsy. *See* Nerve injury.
Scaphoid, **89, 92, 93, 94, 97, 113**
- distal part, **99**
- hand, **117**
- median nerve, **157**
- proximal part, **99**
- US, **99**
- waist, **99**
Scapholunate ligament, **94, 113**
Scaphotrapezium-trapezoid ligament, **101**
Scapula, **24, 25, 26, 35, 311**
Scapular spine, **25, 28**
Scarred tissue, nonrotator cuff tendon tears vs., **474**
Schwannoma
- cystic, paraarticular cystic mass vs., **889**

INDEX

- cystic soft tissue mass vs., **837**
- swollen nerve vs., **870**

Sciatic nerve, **185, 190, 192, 195, 204, 205, 222, 254, 349, 351, 355, 356, 357**
- transverse, **205**

Sclerodactyly, calcified soft tissue mass vs., **843**
Sebaceous cyst. *See* Epidermoid cyst.
Secondary amyloidosis, tendon swelling vs., **865**
Secondary deposits. *See* Superficial metastases, lymphoma, and melanoma.
Semimembranosus bursae, **214**
Semimembranosus muscle, **181, 186, 187, 189, 190, 196, 197, 198, 205, 207, 214, 219, 220, 346**
- gluteal muscles, **350, 351**
- posterior leg (graphic), **233**

Semimembranosus tendon, **191, 205, 208, 214–215, 220, 223, 357**
- medial knee ultrasound, **418**
- posterior leg (graphic), **233**

Semimembranous bursa, **669**
Semitendinosus muscle, **181, 186, 190, 191, 219, 346**
- gluteal muscles, **351**
- medial knee ultrasound, **418**
- posterior leg (graphic), **233**

Semitendinosus tendon, **208, 214, 223, 357**
Semitendinous muscle, gluteal muscles, **350**
Septic arthritis. *See also* Joint infection.
- bone infection vs., **638**
- gout and pseudogout vs., **570**
- hemarthrosis and lipohemarthrosis vs., **660**
- inflammatory arthritis vs., **560**
- synovial swelling vs., **895**

Sequestrum, **637**
Seroma, **524–531**
- acute, **525**
- chronic, **525**
- cystic soft tissue mass vs., **836**
- diagnostic checklist, **526**
- differential diagnosis, **526**
- prognosis, **526**
- subacute, **525**

Serratus anterior, **310, 311**
Serratus posterior, **351**
Serratus posterior inferior muscle, **317, 329**
Sesamoids, **296**
- sagittal foot overview, **287**

Sesamophalangeal apparatus, **296**
Sharpey fibers at insertional site, **15**
Shear wave elastography, **361**
Short dorsal sacroiliac ligament, **349**
Short plantar ligament
- medial and posterior ligaments, **266**
- tarsus, **277**

Short radiolunate ligament, **101**
Short saphenous vein
- lateral compartment of leg, **230**
- popliteal vessel bifurcation, **241**

Shoulder, **10–27**
- biceps tendon
  - longitudinal US, **20**
  - transverse US, **19**
- deep structures, **13**
- glenohumeral joint, longitudinal and transverse US, **27**
- infraspinatus muscle, longitudinal and transverse US, **23**
- longitudinal US, **15**
- muscles and ligaments (graphic), **12**
- normal labrum (sagittal graphic), **13**
- posterior aspect
  - graphic, **25**
  - longitudinal US, **26**
- posterior glenoid labrum, longitudinal and transverse US, **27**
- rotator cuff interval, graphic and transverse US, **18**
- shoulder joint and posterior aspect (graphic), **25**
- subscapularis muscle, longitudinal US, **21**
- subscapularis tendon, transverse US, **22**
- supraspinatus insertional area, transverse US, **16**
- supraspinatus muscle, transverse and longitudinal US, **17**
- teres minor muscle, ultrasound imaging, **24**

Shoulder procedures, **950–957**
- acromioclavicular joint injection
  - indications, **951**
  - procedure, **953**
- calcific tendinitis therapy
  - indications, **951**
  - procedure, **952–953**
- glenohumeral joint injection or aspiration
  - indications, **951**
  - procedure, **951**
- outcomes, **953**
- post procedure, **953**
- preprocedure, **951**
- procedure, **951–953**
- rotator cuff tendinosis or partial tear therapy
  - indications, **951**
  - procedure, **953**
- subacromial-subdeltoid bursa injection
  - indications, **951**
  - procedure, **952**

Shoulder ultrasound, **372–379**
- acromioclavicular joint, **374**
- biceps tendon, long head of, **373**
- clinical indications, **373**
- glenohumeral joint, **374**
- glenoid labra, **374**
- infraspinatus tendon, **373**
- rotator cuff musculature, **374**
- rotator interval, **373**
- subacromial impingement, dynamic assessment of, **374**
- subacromial-subdeltoid and subcoracoid bursae, **374**
- subscapularis tendon, **374**
- suprascapular/spinoglenoid notches, **374**
- supraspinatus tendon, **373**
- symptomatic area, **374**
- teres minor tendon, **373**

Sinus tarsi
- injection
  - indication, **987**
  - procedure, **988**
- medial ankle, graphic, **263**

Skier's thumb. *See* Ligament injury.

Skin, **311**
Skin metastasis. *See* Superficial metastases, lymphoma, and melanoma.
Slice-thickness artifact, **364**
Slow flow vascular anomaly, epidermoid cyst vs., **722**
Slow spontaneous flow, spectral Doppler of leg veins, **249**
Small saphenous vein, **219, 236, 252, 290**
- great saphenous veins, **253**
- leg veins (graphic), **239**
Snapping iliopsoas tendon, dynamic examination for, anterior hip ultrasound, **409**
Soft tissue abscess, epidermoid cyst vs., **722**
Soft tissue chondroma, calcified soft tissue mass vs., **843**
Soft tissue ganglia, ganglion cyst, **682**
Soft tissue infection, **628–635**
- bone infection vs., **638**
- differential diagnosis, **630**
- prognosis, **630**
Soft tissue injuries
- fat injury, **502–509**
- hematoma/seroma, **524–531**
- ligament injury, **532–539**
- muscle infarction, **510–513**
  soft tissue infection vs., **630**
- muscle injury, **514–523**
Soft tissue mass, hematoma/seroma vs., **526**
Soft tissue metastases, calcified soft tissue mass vs., **842–843**
Soft tissue sarcoma, **766–773**
- differential diagnosis, **768**
- hematoma/seroma vs., **526**
- hyperechoic muscle mass vs., **831**
- hypervascular soft tissue mass vs., **846**
- hypoechoic muscle mass vs., **824**
- prognosis, **768**
- staging, grading, & classification, **768**
- superficial metastases, lymphoma, and melanoma vs., **740**
Soft tissue tumor
- biopsy, **918–923**
  outcomes, **920**
  post procedure, **920**
  preprocedure, **919**
  procedure, **919–920**
- muscle injury vs., **516**
Soleal musculotendinous junction, posterior ankle panorama, **273**
Soleal veins, **250**
- popliteal vessel bifurcation, **241**
- posterior leg, transverse, **235**
- transverse peroneal vessels, **244**
Soleus muscle, **224**
- Achilles tendon
  axial, **275**
  longitudinal, **274**
- anterior and posterior leg nerves (graphic), **255**
- anterior compartment of leg, **229**
- gastrocnemius veins, **251**
- great saphenous veins, **253**
- lateral compartment of leg, **230**
- lateral leg (graphic), **232**
- leg arteries
  anterior and ankle (graphic), **238**
  posterior (graphic), **237**
- longitudinal peroneal vessels, **245**
- longitudinal posterior tibial vessels, **243**
- peroneal nerve, **257**
- popliteal vessel, **240**
  bifurcation, **241**
- posterior ankle panorama, **273**
- posterior leg
  panoramic, **234**
  transverse, **235**
- small saphenous vein, **252**
- soleal veins, **250**
- spectral Doppler of leg arteries, **248**
- spectral Doppler of leg veins, **249**
- tibial nerve, **256**
- transverse peroneal vessels, **244**
- transverse posterior tibial vessels, **242**
Spatial compound US, **360**
Spermatic cord (in males), groin hernia, **401**
Spigelian hernia, abdominal wall mass vs., **910**
Spinalis thoracis muscle, **317, 329**
Spindle cell lipoma, **713**
Spine of scapula, **26**
Spinoglenoid notch, **28, 374**
- imaging approaches, **11**
Spinous process, **317, 329**
Spiral groove, radial nerve, **144, 146**
Spondyloarthropathy (SpA). *See* Inflammatory arthritis.
Spontaneous hemarthrosis, **559**. *See also* Hemarthrosis and lipohemarthrosis.
Spring ligament, **258**
- deep plantar muscles, **285**
- deltoid ligament, **267**
- medial ankle, **265**
- superomedial, medial and posterior ligaments, **266**
Staphylococcus, bone infection, **638**
Staphylococcus aureus
- joint infection, **648**
- postoperative infection, **654**
Stener lesion. *See also* Ligament injury.
- thumb, **120**
  graphics, **129**
Stenosing tenosynovitis, **485**
Sternalis, **310**
Sternoclavicular (SC) joint, **4–9**
- arthropathy, chest wall lesion vs., **905**
- ligaments of, **4**
  anterior, **4**
- longitudinal US, **6**
- muscle attachments, **4**
- transverse US, **5**
Sternocleidomastoid muscle, **6**
- clavicular insertion, **6**
- sternal end, **6**
- sternal insertion, **6**
Sternothyroid, tendon, **6**
Sternum, **6, 310, 311**
Strain elastography, **361**
Streptococcus infection, fat injury vs., **504**

Streptococcus pyogenes, joint infection, **648**
Stress fracture, **541**
Stress reaction of bone, **541**
Stress rib fracture, chest wall lesion vs., **905**
STS. *See* Soft tissue sarcoma.
Student's elbow (olecranon bursitis). *See* Bursitis.
Styloid process of radius, hand, **117**
Subacromial impingement, dynamic assessment of, **374**
Subacromial-subdeltoid bursa, **374, 669**
- imaging approaches, **11**
Subacromial-subdeltoid bursa injection
- indications, **951**
- procedure, **952**
Subacromial-subdeltoid fat plane, **10, 11**
Subacute hematoma, **525**
Subacute seroma, **525**
Subclavian artery, **6, 45**
Subclavian vein, **6**
Subcoracoid bursa, **374**
Subcutaneous edema, soft tissue infection vs., **630**
Subcutaneous fat, **20, 196, 198, 210–212, 311**
Subcutaneous fat necrosis
- foreign body and injection granuloma vs., **754**
- of newborn, fat injury vs., **504**
Subcutaneous fibroma, foreign body and injection granuloma vs., **754**
Subcutaneous hematoma, abdominal wall mass vs., **911**
Subcutaneous lipoma
- abdominal wall mass vs., **910**
- fat injury vs., **504**
Subcutaneous metastasis. *See also* Superficial metastases, lymphoma, and melanoma.
- foreign body and injection granuloma vs., **754**
Subcutaneous olecranon bursa, **52**
Subcutaneous panniculitis, fat injury vs., **504**
Subcutaneous tissue, **39, 40, 41, 57, 163**
- great saphenous veins, **253**
Subdeltoid bursa, **13**
Subdeltoid peribursal fat, **15, 16, 17, 18, 19, 22, 23, 26**
Subgastrocnemius bursa, **214**
Subgluteus medius bursa, **172, 669**
Subgluteus minimus bursa, **669**
Subinguinal space, groin, **338**
Subscapular nerve, axilla, **28**
Subscapularis muscle, **7, 10, 12, 14, 19, 21, 30, 33, 37, 311**
Subscapularis tendon, **11, 14, 21, 22, 27, 374**
- lateral fibers, **22**
- midfibers, **22**
- musculotendinous junction, **11**
Subtalar joint
- dorsal foot ultrasound, **433**
- injection, **987**
Subtendinous areolar tissue, fingers, **136, 138**
Subtendinous olecranon bursa, **52**
Superficial calcaneal bursitis, **669**
Superficial circumflex artery, **197**
Superficial circumflex iliac artery, **193**
Superficial deltoid ligament, medial ankle, graphic, **263**
Superficial epigastric artery, **193, 200, 344, 345**
Superficial femoral artery, **175, 188, 189, 192, 193, 196, 197, 198, 199, 200, 202, 203, 345**
Superficial femoral vein, **188, 189, 194, 196, 197, 198, 199, 202, 203, 345**
Superficial gluteal vein, **194**
Superficial inguinal ring, groin, **338**
Superficial leg veins, **236**
Superficial metastases, lymphoma, and melanoma, **738–743**
- diagnostic checklist, **740**
- differential diagnosis, **740**
- prognosis, **740**
Superficial palmar arch, **110**
- hand, **114, 119**
  - radial branch, **115**
  - ulnar branch, **115**
Superficial peroneal nerve, **254**
- anterior and posterior leg nerves (graphic), **255**
- anterolateral leg, panoramic, **227**
- lateral ankle ultrasound, **426**
- lateral branch, **291**
- lateral compartment of leg, **230**
- medial branch, **291**
- tarsus, **278**
Superficial posterior compartment, ankle, **259**
Superficial radial nerve, **390**
Superficial thrombophlebitis, **621, 622**
- fat injury vs., **504**
Superficial vein, forearm, **87**
Superficial vessels
- dorsal foot, **281**
- thigh, **183**
Superior acromioclavicular ligament, **4, 7, 8, 9**
Superior extensor retinaculum, **258**
- anterior ankle, **261**
  - graphic, **260**
  - ultrasound, **425**
- dorsal foot, graphic, **279**
- leg arteries, anterior and ankle (graphic), **238**
Superior fascicle, popliteomeniscal portion, **216**
Superior gemellus muscle, **186, 204, 346**
- gluteal muscles, **348, 351, 355**
Superior glenohumeral ligament, **14, 25**
Superior gluteal artery, **193, 349**
Superior gluteal nerve, **349**
Superior labrum anterior to posterior tears, rotator cuff/biceps tendon tear associated with, **456**
Superior lateral genicular artery, **193**
Superior medial genicular artery, **193**
- leg arteries, posterior (graphic), **237**
Superior peroneal retinaculum, **258**
- lateral ankle, **270**
  - graphic, **269**
Superior pubic ligament, **342**
Superior pubic ramus, **173, 338**
- anterior cortex of, **347**
Superior scapular transverse ligament, **12**
Superior thoracic artery, **45**
Superior transverse scapular ligament, **12, 13, 25, 28, 30**
Superior ulnar collateral artery, **38, 46**
- forearm arteries, **81**
Supinator muscle, **56, 57, 60, 63, 74, 76, 79**
- proximal forearm, **82**

INDEX

- radial nerve, **143, 147, 149**
Supplementary imaging techniques, **360–361**
Suprapatellar bursa, **208–209**
Suprapatellar fat pad, **210**
Suprascapular artery, **28, 45**
- infraspinatus branch, **28, 45**
Suprascapular nerve, **28, 169, 302**
- branch and vessels, **30**
- infraspinatus branch in spinoglenoid notch, **13, 30**
- in suprascapular notch, **13, 30**
Suprascapular notch, **28, 374**
Supraspinatus muscle, **7, 10, 12, 13, 25, 26, 28, 30, 42**
- central tendon of, **26**
- insertional area
    longitudinal US, **15**
    transverse US, **17**
- musculotendinous junction of, **27**
- radial nerve, **144**
- transverse and longitudinal US, **17**
Supraspinatus tendon, **11, 12, 13, 14, 15, 16, 17, 18, 25, 26, 28, 30, 373**
Supraspinatus-infraspinatus junction, focal, **11**
Sural nerve, **219, 254, 291**
- lateral compartment of leg, **230**
- posterior ankle ultrasound, **426**
- tarsus, **278**
Sustentaculum talus
- deltoid ligament, **267**
- medial ankle, **265**
    calcaneus, **265**
Suture material within tendon, tendon hyperechogenicity vs., **861**
Sweat gland tumor, hypoechoic subcutaneous mass vs., **813**
Swollen nerve, **870–875**
- differential diagnosis, **870**
Symmetrical polyarthropathy, gout and pseudogout vs., **570**
Symphysis pubis, **339**
Symptomatic area, **374**
Synovial bursae, **669**
Synovial chondrosarcoma, **695**
- synovial swelling vs., **895**
Synovial fibroma, **695**
Synovial fluid, medial ankle, **264**
Synovial fold, lateral elbow ultrasound, **382**
Synovial hemangioma, **695**
- synovial swelling vs., **895**
- synovial tumor vs., **696**
Synovial lipoma, **695**
Synovial membrane, **13**
Synovial osteochondromatosis, **695**
- joint effusion vs., **900, 901**
- synovial swelling vs., **894–895**
Synovial sarcoma, **695**
- paraarticular cystic mass vs., **889**
- peritendinous mass vs., **853**
- synovial swelling vs., **895**
Synovial sheath, lateral ankle, **270**
Synovial swelling, **894–899**
- amyloid, **895**
- differential diagnosis, **894**
- focal nodular synovitis, **894**
- inflammatory arthropathy, **894**
- lipoma arborescens, **895**
- lymphoma, **895**
- osteoarthritis, **894**
- pigmented villonodular synovitis, **894**
- septic arthritis, **895**
- synovial chondrosarcoma, **895**
- synovial hemangioma, **895**
- synovial osteochondromatosis, **894–895**
- synovial sarcoma, **895**
- tophi of gout or pseudogout, **894**
Synovial tumor, **694–705**
- clinical issues, **696**
- diagnostic checklist, **696**
- differential diagnosis, **696**
- pigmented villonodular synovitis, synovial swelling vs., **894**
- prognosis, **696**
Synovitis
- hip, transient, joint infection vs., **648**
- pigmented villonodular, joint effusion vs., **900, 901**
- synovial tumor vs., **696**
Synovium, **18**
Systemic inflammatory disorder, fat injury vs., **504**
Systemic lupus erythematosus. *See also* Inflammatory arthritis.
- fat injury vs., **504**

T

T1, **31, 303**
- spinal nerve, **169**
- ventral rami, **161**
Takayasu arteritis, fat injury vs., **504**
Talar dome
- anterior ankle, **262**
- anterior lateral corner, anterior ankle, **261**
- posterior margin, posterior ankle panorama, **273**
Talocalcaneal interosseous ligament, lateral ankle, graphic, **269**
Talofibular ligament, medial ankle, graphic, **263**
Talonavicular joint
- anterior ankle, **262**
- deltoid ligament, **268**
- medial ankle, **265**
Talus
- deltoid ligament, **267, 268**
- dorsal foot, **283**
    lateral surface, **282**
- head
    anterior ankle, **262**
    medial ankle, **265**
- lateral tubercle
    fibular ligaments, **272**
    posterior ankle panorama, **273**
- medial and posterior ligaments, **266**

INDEX

- medial ankle, **265**
  - graphic, **263**
- neck
  - anterior ankle, **262**
  - deltoid ligament, **268**
  - fibular ligaments, **272**
  - medial ankle, **265**

Tarsal coalition, tarsal tunnel syndrome vs., **616**
Tarsal tunnel, medial ankle ultrasound, **425**
Tarsal tunnel syndrome, **614–619**
- differential diagnosis, **616**

Tarsus, **276–289**
- anatomy imaging issues, **278**
- arches of foot **276**
- arteries, **278**
- bony anatomy, **276–277**
- bursae, **278**
- clinical implications, **278**
  - alignment, **278**
  - foot motion, **278**
  - malalignment, **278**
- compartments, **277**
- distribution of weight bearing, **276**
- dorsal foot, **280–283**
  - graphic, **279**
- lateral tarsus, **289**
- major ligaments, **277**
- medial tarsus, **288**
- musculature, **277**
- nerves, **277–228**
- plantar aponeurosis, **286**
- plantar muscles
  - deep layers (graphic), **285**
  - superficial layers (graphic), **284**
- sagittal foot overview, **287**

TCL. *See* Transverse carpal ligament.
Tendinitis. *See* Tenosynovitis.
Tendinopathy. *See* Tendinosis.
Tendinosis
- bursitis vs., **670**
- of common extensor tendon origin, **495, 496**
- of common flexor tendon origin, **495, 496**
- nonrotator cuff, **462–471**
  - diagnostic checklist, **464**
  - differential diagnosis, **464**
  - prognosis, **464**
- with peritendinitis, nonrotator cuff tendon tears vs., **474**
- rotator cuff/biceps, **442–453**
  - calcific tendinosis, **443**
  - diagnostic checklist, **444**
  - differential diagnosis, **443–444**
  - genetics, **444**
  - rotator cuff/biceps tendon tear vs., **456**
  - staging, grading, & classification, **444**
- tendon hypoechogenicity vs., **856**
- tendon swelling vs., **864**
- tenosynovitis vs., **486**

Tendinous inscription, **315, 327**
Tendon disorders
- elbow epicondylitis, **494–501**
- nonrotator cuff tendinosis, **462–471**
- nonrotator cuff tendon tears, **472–483**
- rotator cuff/biceps tendinosis, **442–453**
  - calcific tendinosis, **443**
  - diagnostic checklist, **444**
  - differential diagnosis, **443–444**
  - genetics, **444**
  - staging, grading, & classification, **444**
- rotator cuff/biceps tendon tear, **454–461**
- tenosynovitis, **484–493**

Tendon edges, **11**
Tendon hyperechogenicity, **860–863**
- differential diagnosis, **860**

Tendon hypoechogenicity, **856–859**
- differential diagnosis, **856**

Tendon injury, nerve injury vs., **584**
Tendon pulley system
- fingers, **130, 131, 133, 135, 136, 137, 138**
  - annular pulley, **130**
  - extensor expansion, **130**
- thumb, **120, 121, 125**

Tendon rupture. *See* Nonrotator cuff tendon tears.
Tendon sheaths
- fingers, **130, 131**
- hand, **102**

Tendon sprain. *See* Nonrotator cuff tendon tears.
Tendon strain. *See* Nonrotator cuff tendon tears.
Tendon swelling, **864–869**
- differential diagnosis, **864**

Tendon tear
- bursitis vs., **670**
- intrasubstance, rotator cuff/biceps tendinosis vs., **443**
- muscle injury vs., **516**
- nonrotator cuff, **472–483**
  - complete, **473**
  - diagnostic checklist, **474**
  - differential diagnosis, **474**
  - full thickness, **473**
  - high-grade partial, **473**
  - partial thickness, **473**
  - prognosis, **474**
- rotator cuff/biceps tendon. *See* rotator cuff/biceps tendon tear.
- tendon hyperechogenicity vs., **861**
- tendon hypoechogenicity vs., **856–857**

Tendon thickening, tendon hypoechogenicity vs., **856**
Tendonitis. *See* Tenosynovitis.
Tennis elbow. *See* Elbow epicondylitis.
Tennis heel. *See* Plantar fasciitis and fibromatosis.
"Tennis leg." *See* Muscle injury.
Tenosynovial giant cell tumor, **695**
Tenosynovitis, **484–493**
- acute exudative, **485**
- acute proliferative, **485**
- carpal tunnel syndrome vs., **604**
- chronic active, **485**
- chronic inactive, **485**
- diagnostic checklist, **486**
- differential diagnosis, **486**
- nonrotator cuff tendinosis vs., **464**
- peritendinous mass vs., **852**

- prognosis, **486**
- stenosing, **485**
Tensor fascia lata muscle, **173, 176, 180, 182, 183**
- gluteal muscles, **348, 350, 351**
- groin muscles, **339, 340, 346**
- lateral hip ultrasound, **410**
Teres major muscle, **7, 12, 13, 25, 28, 30, 32, 33, 34, 35, 37, 42, 46, 311**
- radial nerve, **144**
Teres minor muscle, **10, 12, 13, 14, 24, 25, 28, 30, 34, 35, 42**
- insertional area of, **24**
- nerve to, **31**
- radial nerve, **144**
- ultrasound imaging, **24**
Teres minor musculotendinous junction, **24**
Teres minor tendon, **11, 12, 14, 24, 28, 373**
Teres muscle, long head, **37**
Terminal phalanx, tendon, **135, 138**
Terminal tendon, **111**
- fingers, **132**
Thenar eminence, **96, 102**
- longitudinal US, **109**
- thumb
    longitudinal US, **128**
    transverse US, **127**
- transverse US, **108**
Thenar muscle atrophy, carpal tunnel syndrome, **603**
Thenar muscles
- median nerve, **151**
- thumb, **120**
Thenar-hypothenar eminence, **107**
Thigh
- arteries, **193**
    color Doppler, **198**
    spectral Doppler, **199**
- lateral cutaneous nerve of, **192**
- nerves, **195**
- veins, **194, 202**
- vessels
    longitudinal, **196**
    transverse, **197**
Thigh muscle, **180–191**
- anterior, **182**
    overview, **183**
- anterior compartment, **180**
    central, **184**
    lateral, **185**
- medial, **186, 189**
    adductor canal, **188**
    compartment, **180–181**
    overview, **187**
- posterior, **186, 191**
    compartment, **180–181**
    overview, **190**
Thoracic spinal nerves, **310**
Thoracoacromial artery, **45**
- acromial branch, **45**
Thoracodorsal artery, **45**
Thoracodorsal nerve, **302**
- axilla, **28**
3D ultrasound, **361**
Thrombophlebitis, superficial, fat injury vs., **504**
Thrombosed persistent median artery, carpal tunnel syndrome, **603**
Thumb, **120–129**
- carpometacarpal joint, ultrasound imaging, **122**
- deep dissection of hand, **121**
- dorsal hand ultrasound
    extensor tendon of, **397**
    flexor tendons of, **397**
    UCL of MCP joint of, **397**
- extensor aspect, ultrasound imaging, **126**
- metacarpophalangeal joint, ultrasound imaging, **123**
- palmar aspect, ultrasound imaging, **125**
- Stener lesion, graphics, **129**
- thenar eminence
    longitudinal US, **128**
    transverse US, **127**
- ulnar collateral ligament, ultrasound imaging, **124**
Thyrocervical trunk, **45**
Thyroglossal duct cyst, chest wall lesion vs., **905**
Tibia
- lateral ankle, **270**
    graphic, **269**
- medial and posterior ligaments, **266**
- medial ankle, graphic, **263**
- popliteal vessel bifurcation, **241**
- popliteal vessels, **240**
- posterior ankle panorama, **273**
- proximal, **206**
- transverse peroneal vessels, **244**
- transverse posterior tibial vessels, **242**
Tibial artery, posterior leg, transverse, **235**
Tibial condylar cortex
- lateral, **218**
Tibial condyle, **209, 211**
- cortex, medial, **215**
- lateral, **217**
Tibial cortex, **223**
Tibial nerve, **185, 192, 195, 196, 197, 198, 219, 220, 254, 256**
- anterior and posterior leg nerves (graphic), **255**
- anterior compartment of leg, **228**
- lateral compartment of leg, **230**
- longitudinal posterior tibial vessels, **243**
- medial ankle, **264, 265**
- medial ankle ultrasound, **425**
- posterior knee ultrasound, **418**
- posterior leg, transverse, **235**
- tarsus, **277**
Tibial tuberosity, **209, 211**
Tibialis anterior muscle, **225**
- anterior compartment of leg, **228, 229**
- anterolateral leg
    graphic, **226**
    panoramic, **227**
- lateral compartment of leg, **230**
- lateral leg, **232**
- longitudinal anterior tibial vessels, **247**
- peroneal nerve, **257**
- spectral Doppler of leg arteries, **248**

- spectral Doppler of leg veins, **249**
- transverse anterior tibial vessels, **246**

Tibialis anterior tendon
- ankle, **258**
  - anterior ankle, **261, 262**
    - graphic, **260**
  - medial ankle, **264**
    - graphic, **263**
- anterior ankle ultrasound, **425**
- dorsal foot, **280, 281, 283**
- leg
  - anterior compartment of leg, **229**
  - anterolateral leg
    - graphic, **226**
    - panoramic, **227**
  - lateral leg, **232**

Tibialis posterior muscle, **224**
- anterior compartment of leg, **228, 229**
- lateral compartment of leg, **230, 231**
- leg arteries, posterior (graphic), **237**
- longitudinal anterior tibial vessels, **247**
- longitudinal peroneal vessels, **245**
- longitudinal posterior tibial vessels, **243**
- posterior leg, transverse, **235**
- soleal veins, **250**
- spectral Doppler of leg arteries, **248**
- spectral Doppler of leg veins, **249**
- tibial nerve, **256**
- transverse anterior tibial vessels, **246**
- transverse peroneal vessels, **244**
- transverse posterior tibial vessels, **242**

Tibialis posterior tendon, **259**
- anterior and posterior leg nerves (graphic), **255**
- deep plantar muscles, **285**
- deltoid ligament, **267, 268**
- leg arteries, posterior (graphic), **237**
- medial ankle **265**
  - graphic, **263**
  - ultrasound, **425–426**
- medial tarsus, **288**

Tibiocalcaneal band, superficial deltoid
- medial and posterior ligaments, **266**

Tibiocalcaneus internus muscle, **224**
Tibiofibular joint, **206**
Tibionavicular band, superficial deltoid
- medial and posterior ligaments, **266**

Tibioperoneal trunk, **236**
Tibiospring, superficial deltoid
- medial and posterior ligaments, **266**

Tibiotalar joint injection, **987**
Tissue harmonic imaging, **360–361**
Toe web spaces, plantar foot ultrasound, **433**
Toes, **296–301**
- plantar and dorsum, **301**

Tophaceous gout. *See* Gout and pseudogout.
Trabecular microfracture, **541**
Traction apophysitis, ligament injury vs., **534**
Transducer pressure artifact, **365**
Transient synovitis of hip, joint infection vs., **648**
Transversalis fascia, **314, 326**
Transverse abdominis, groin, **338, 341, 342**
Transverse arch of foot, tarsus, **276**
Transverse band, **68**
Transverse carpal ligament, **603**
Transverse cervical artery, **45**
Transverse intermuscular septum, transverse posterior tibial vessels, **242**
Transverse ligament, **12, 19, 37, 172**
Transverse metatarsal, **300**
Transverse metatarsal ligaments, tarsus, **277**
- dorsal foot (graphic), **279**

Transverse processes, **351**
Transverse spinous process, **311**
Transverse thoracic muscle, **311**
Transversus abdominis muscle, **314, 317, 326, 327, 329**
Trapeziocapitate ligament, **101**
Trapeziotrapezoid ligament, **101**
Trapezium, **89, 98, 99**
- hand, **117**
- median nerve, **157**
- thumb, **122**

Trapezius muscle, **26, 311**
Trapezoid, **89**
- median nerve, **157**

Trauma, fat, lipoma vs., **714**
Traumatic fat necrosis, **503**
Triangular fibrocartilage, graphics and US, **100**
Triangular ligament, **111**
- fingers, **132**

Triangular space, **28**
Triceps muscle, **12, 38, 39, 43, 47, 48, 54, 55, 68, 71, 145, 146, 153, 162, 163, 164**
- lateral head, **13, 25, 28, 30, 31, 36, 37, 38, 42, 43**
- long head, **25, 28, 30, 31, 34, 35, 36, 38, 42, 144**
- medial head, **36, 38, 43**

Triceps tendon, **12, 38, 42, 52, 54, 55, 68, 71, 162**
- lateral head, **13, 25, 28, 30**
- long head, **25, 28, 30**
- posterior elbow ultrasound, **382**

Trigger finger injection or percutaneous release
- indications, **965**
- procedures, **966**

Trigger thumb, **120**
Triphasic Doppler waveform, spectral Doppler of leg arteries, **248**
Triquetrocapitate ligament, **101**
Triquetrohamate ligament, **101**
Triquetrum, **89, 97, 100, 113, 157, 167**
Triradiate cartilage, **178**
Trochanteric bursa, **172, 669**
Trochanteric (hip) bursitis. *See* Bursitis.
Trochlea, **69**
Trochlear groove, **210**
Trochlear surface, articular cartilage, **213**
True ribs, **310**
Trunk, groin, **338**
TTS. *See* Tarsal tunnel syndrome.
Tuberculosis, fat injury vs., **504**
Tuberculosis abscess, calcified soft tissue mass vs., **843**
Tumor
- bone, **774–785**
  - diagnostic checklist, **776**

differential diagnosis, **776**
prognosis, **776**
- occurring within fascia, fascial lesion vs., **877**
- soft tissue, muscle injury vs., **516**
- synovial, **694–705**
clinical issues, **696**
diagnostic checklist, **696**
differential diagnosis, **696**
prognosis, **696**
Tumoral calcinosis, calcified soft tissue mass vs., **843**
Twinkling artifact, **366**

U

Ulna, **55, 56, 60, 65, 70, 74, 75, 79, 89, 94, 95, 97, 165, 166, 304**
- Doppler gate in, **84**
- median nerve, **156**
- olecranon process of, **164**
- radial nerve, **148**
- radial notch, **55**
- transverse and longitudinal US, **84**
Ulnar artery, **56, 60, 75, 77, 80, 97, 98, 106, 110, 165, 166, 167, 168**
- deep palmar arch, distal forearm and hand, **85**
- distal forearm and hand, **85**
- forearm arteries, **81**
- hand, **114, 115, 116, 119**
contribution to deep palmar arch, **119**
superficial branch, **116**
- longitudinal US, **84**
- median nerve, **157, 158**
- proximal forearm, **82**
- superficial palmar arch, distal forearm and hand, **85**
- transverse US, **83, 84**
Ulnar bursa, **103**
- fingers, **131, 138**
- thumb, **121**
Ulnar collateral ligament (UCL), **52, 53, 55, 69, 100**
- anterior band, **69**
- of MCP joint of thumb, dorsal hand ultrasound, **397**
- posterior band, **67**
- thumb, **120, 124, 129**
- of thumb
injury. *See* Ligament injury.
proximal retraction. *See* Ligament injury.
- transverse band, **67, 78**
Ulnar coronoid, **53, 66**
Ulnar head, **100**
Ulnar nerve, **31, 32, 37, 38, 39, 46, 47, 54, 55, 56, 68, 70, 71, 91, 97, 98, 160–169, 163, 169, 302, 303, 304, 390**
- arm, **160, 161**
transverse US, **163**
- axilla, **28**
relationship of neurovascular structures, **31**
- cubital tunnel, **160, 162**
transverse US, **164**
- deep branch, **110, 161, 168**
- digital branches, **161**
palmar, **169**
- forearm, **160, 161**
longitudinal US, **166**
transverse US, **165, 166**
- Guyon canal, transverse US, **167, 168**
- hand, **160, 161**
deep branch, **116, 119**
superficial branch, **116, 119**
- longitudinal US, **84**
- median nerve, **157, 159**
- palmar branch, **161**
cutaneous, **169**
- proximal and distal, **169**
- radial nerve, **145, 149**
dorsal branch, **149**
dorsal digital branches, **149**
- superficial branch, **106, 110, 168, 169**
- transverse US, **83, 84**
- wrist, transverse and longitudinal US, **166**
"Ulnar neural groove." *See* Cubital tunnel syndrome.
Ulnar neuropathy at elbow. *See* Cubital tunnel syndrome.
Ulnar recurrent artery, **54**
Ulnar sesamoid, thumb, **125**
Ulnar styloid process, **100**
Ulnar vein, **97, 98, 167**
- hand, **116**
- longitudinal US, **84**
- proximal forearm, **82**
- transverse US, **83, 84**
Ulnocapitate ligament, **101**
Ulnocarpal ligament, **100**
Ulnolunate ligament, **101**
Ulnotriquetral ligament, **101**
Ultrasound elastography, **361**
Ultrasound machine, **360**
Ultrasound report, writing, **438–439**
- differential diagnosis, **438**
- grading severity of disease, **438**
- measurements, **438**
Umbilical hernia, abdominal wall mass vs., **910**
Umbilical metastasis, abdominal wall mass vs., **911**
Umbilical-urachal sinus, abdominal wall mass vs., **911**
Umbilicus, **315, 327**
Undescended testis, abdominal wall mass vs., **910**
Upper limb
- arm, **36–43**
- arm vessels, **44–51**
- axilla, **28–35**
- joint injection, **932–941**
outcomes, **934**
preprocedure, **933**
procedure, **933–934**
- shoulder, **10–27**
- sternoclavicular and acromioclavicular joints, **4–9**
Upper subscapular nerve, **302**
Urethra, **316, 328**

V

Vascular anomaly, **744–751**
- clinical issues, **746**
- cystic soft tissue mass vs., **837**
- diagnostic checklist, **746**
- differential diagnosis, **746**
- ganglion cyst vs., **682**
- hyperechoic subcutaneous mass vs., **818**
- hypervascular soft tissue mass vs., **846**
- hypoechoic subcutaneous mass vs., **812**
- lymph node abnormality vs., **762**
- prognosis, **746**

Vascular dilatation or inflammation, **620–627**
- diagnostic checklist, **622**
- differential diagnosis, **622**

Vascular leiomyoma, **734–737**
- diagnostic checklist, **735**
- differential diagnosis, **735**
- genetics, **735**
- hypervascular soft tissue mass vs., **846**
- hypoechoic subcutaneous mass vs., **812**
- lymph node abnormality vs., **762**
- nerve sheath tumor vs., **592**
- prognosis, **735**

Vascular malformation
- chest wall lesion vs., **905**
- hyperechoic muscle mass vs., **830**
- hyperechoic subcutaneous mass vs., **818**
- hypervascular soft tissue mass vs., **846**
- hypoechoic muscle mass vs., **825**
- lipoma vs., **714**
- paraarticular cystic mass vs., **889**

Vascular tumor, hypervascular soft tissue mass vs., **846**

Vastus intermedius muscle, **180, 183, 184, 185, 187, 190, 210**
- groin muscles, **340**

Vastus intermedius tendon, **208**

Vastus lateralis muscle, **177, 183, 184, 185, 186, 190, 191**
- groin muscles, **340**
- leg
    - anterolateral leg (graphic), **226**
    - lateral leg (graphic), **232**

Vastus lateralis tendon, **208**

Vastus medial tendon, **208**

Vastus medialis muscle, **180, 182, 183, 184, 186, 187, 188, 189, 190, 191, 197, 203, 209, 210, 214**
- groin muscles, **340, 345**
- leg, anterolateral leg (graphic), **226**

Vastus medialis obliquus muscle, **214**

Vegetant intravascular hemangioendothelioma. *See* Vascular dilatation or inflammation.

Veins
- arm, **44**
- foot, **290**
- forearm, **81**
- hand, **114**
- ribs and intercostal space, **310**

Velocity error artifact, **365**
- defocusing (edge shadowing), **365**
- refraction, **365**
- velocity error, **365**

Venous dilatation, hyperechoic muscle mass vs., **831**

Venous system, forearm
- anterior, **86**
- posterior, **87**

Venous valve leaflets, **196**

Venous varix, **621, 622**
- cystic soft tissue mass vs., **837**

Venous vascular malformation, calcified soft tissue mass vs., **842**

Vertebral artery, **45**

Vertebral body, **311**

Visceral pleura, **311**

Volar carpal ligament, **91**
- hand, **119**

Volar hand ultrasound, **397**
- carpometacarpal joint of thumb, **397**
- flexor tendon pulleys, **397**
- flexor tendons of thumb, **397**
- metacarpophalangeal and interphalangeal joints, **397**

Volar ligaments, **101**

Volar plate
- fingers, **133, 134, 135**
    - membranous portion, **133**
    - thick portion, **133**
- thumb, **123**

Volar radioulnar ligament, **100, 101**

Volar wrist, **96, 390**
- tendons, **91**
- transverse US, **97**
- US, **98**

Volar wrist ultrasound, **390**
- lateral antebrachial cutaneous nerve, **390**
- median nerve (MN) and carpal tunnel, **390**
- palmar ulnar cutaneous nerve, **390**
- superficial radial nerve, **390**
- tendons outside carpal tunnel, **390**
- UN and Guyon tunnel, **390**
- volar wrist joint, **390**

von Recklinghausen disease, nerve sheath tumor and, **591**

W

Weber-Christian disease, fat injury vs., **504**

Weight bearing, distribution of, tarsus, **276**

Wheelchair syndrome. *See* Nerve injury.

Wrist, **88–101**
- anatomic spaces, **88**
- carpal bones and wrist compartments, **89**
- dorsal
    - ligaments, **101**
    - transverse US, **93**
- dorsal tendons and tendon sheath, **90**
- extensor carpi ulnaris, transverse US, **95**
- extensor digitorum, transverse US, **94**
- ligaments, **88**

- median nerve, **150, 156**
- muscles and tendons, **88**
- osseous structures, **88**
- radial aspect, transverse US, **92**
- radial nerve, **149**
- retinacula, **88**
- scaphoid, US, **99**
- tendons: relations to dorsal and volar, **91**
- triangular fibrocartilage, graphics and US, **100**
- ulnar nerve, **166**
- volar, **96**
  ligaments, **101**
  transverse US, **97**
  US, **98**

Wrist ultrasound, **388–395**
- clinical indications for, **389**
- dorsal, **389–390**
  1st extensor compartment [abductor pollicis longus (APL) and extensor pollicis brevis (EPB)], **389**
  2nd extensor compartment (extensor carpi radialis longus and extensor carpi radialis brevis), **389**
  3rd extensor compartment [extensor pollicis longus (EPL)], **389**
  4th [extensor digitorum (ED), extensor indicis proprius (EIP) and 5th [extensor digitorum minimi (EDM)] extensor compartments, **389**
  6th extensor compartment [extensor carpi ulnaris (ECU)], **389**
  dorsal ulnar cutaneous nerve, **390**
  dorsal wrist joint, **389**
  dorsal wrist tendons, **389**
  posterior interosseous nerve, **389–390**
- volar, **390**
  lateral antebrachial cutaneous nerve, **390**
  median nerve (MN) and carpal tunnel, **390**
  palmar ulnar cutaneous nerve, **390**
  superficial radial nerve, **390**
  tendons outside carpal tunnel, **390**
  UN and Guyon tunnel, **390**
  volar wrist joint, **390**

X

Xanthoma
- nonrotator cuff tendinosis vs., **464**
- tendon swelling vs., **865**

Xanthomatosis, tendon hypoechogenicity vs., **857**
Xiphoid process, prominent, chest wall lesion vs., **905**

Z

Zona orbicularis, **171, 172, 174**